U0485527

朱作言文集

（上卷）

汪亚平　胡　炜　孙永华　崔宗斌　主编

科 学 出 版 社

北　京

内容简介

朱作言先生研究团队于20世纪80年代初期首创农业动物基因工程研究，在1985年发表了世界第一篇鱼类基因转移研究文章，享有世界声誉。本书从该团队发表的200多篇文章中，选取了100篇编辑成册，分上卷和下卷出版。本书的主线是基于转基因鱼理论模型的建立、转基因鲤研制及品系选育、转植基因功能，以及转基因鱼食用和生态安全研究。本书内容共包括7个部分。第一部分和第二部分建立了转基因鱼研究的完整理论模型，并以此为基础成功研制转"全鱼"生长激素基因黄河鲤家系，系统揭示了转植基因的整合、表达和遗传规律。第三部分从摄食、代谢、生长、免疫和游泳行为等方面，详细描述了转基因黄河鲤与野生黄河鲤的生物学特性。第四部分系统评估了转基因鱼在实验室及模拟自然环境条件下的生态学效应。第五部分系统研究了鱼类生长激素对性腺发育的调控作用、育性控制技术及研制不育转基因鱼的策略。第六部分包括草鱼全基因组草图绘制、草鱼病原病毒分子生物学和抗病草鱼基因工程育种的基础研究。第七部分主要以斑马鱼为模型，对鱼类基因工程的生物学原理、技术拓展和应用进行了全方位的探讨。为了便于广大读者阅读，英文文章后面均附有中文摘要。

本书内容是迄今为止关于转基因鱼，乃至整个转基因动物研究领域最系统、最全面和最深入的研究总结，可供鱼类学、鱼类生理学、鱼类内分泌学、农业动物基因工程育种等领域的科技工作者和师生们学习和参考。

图书在版编目(CIP)数据

朱作言文集：全2册/汪亚平等主编. —北京：科学出版社，2019. 3

ISBN 978-7-03-060461-3

Ⅰ. ①朱…　Ⅱ. ①汪…　Ⅲ. ①鱼类–基因工程–文集　Ⅳ. ①Q959. 403-53

中国版本图书馆CIP数据核字(2019)第014010号

责任编辑：王　静　罗　静　王　好 / 责任校对：郑金红

责任印制：肖　兴 / 封面设计：无极书装

科学出版社 出版

北京东黄城根北街16号

邮政编码：100717

http://www.sciencep.com

北京通州皇家印刷厂 印刷

科学出版社发行　各地新华书店经销

*

2019年3月第　一　版　开本：787×1092 1/16

2019年3月第一次印刷　印张：90 3/4

字数：2 110 000

定价：880.00元（全2册）

(如有印装质量问题，我社负责调换)

序　言

《朱作言文集》收录了朱作言教授及其同事们在中国科学院水生生物研究所围绕鱼类基因转移所开展的研究工作，包括转基因鱼的研制、转基因鱼的基础生物学研究、转基因鱼的遗传和生态安全研究及对未来鱼类基因工程育种的展望等。

朱作言教授是全球鱼类转基因研究的开创者。20 世纪 80 年代中期，他带领团队率先成功研制快速生长的转基因鱼，他的这一研究开创了鱼类转基因育种的新领域。他建立了转基因鱼研究的完整理论模型，系统揭示了转基因的整合、表达和遗传规律。接下来，他带领团队克隆了草鱼的 β-肌动蛋白基因启动子和鲤生长激素基因，构建了全部由鲤科鱼类基因组元件构成的“全鱼”转基因载体，并成功研制出转“全鱼”生长激素基因黄河鲤家系，系统开展了外源转基因的生物学效应和转基因鲤的生态安全研究。朱作言团队独立完成了从转基因鱼的理论模型建立到生态安全评估的全链条研究。从这一角度来说，朱教授团队无疑是在全球范围内同领域中工作最为完整的一个研究团队。这些系统研究代表这一团队对转基因鱼生物学领域的巨大贡献。

朱作言教授还带领团队详细研究了转基因鱼的生态安全对策。他的团队一方面开发了具有普适效应的转基因鱼生殖控制策略，另一方面应用三倍体策略获得了完全不育的转基因黄河鲤。利用这些策略，将能够有效解决转基因鱼应用的遗传和生态安全问题，对于转基因鱼的商品化有着重要意义。

该文集共包含朱作言教授团队在过去 30 年内所产出的 100 篇同行评审论文，由 7 个非常具有逻辑性的部分组成，每一部分的主要内容简介如下。

第一部分：鱼类基因转移模型建立。朱作言团队在全球首次成功研制转基因鱼，并在多种重要经济鱼类中成功研制转生长激素基因鱼。他们延续了中国在鱼类细胞核移植方面的独创性工作，是利用这一工具开展早期发育过程中基因表达等基础生物学问题的世界权威。

第二部分：外源基因的整合与表达。接下来，他的团队通过一系列工作发展和优化了鱼类转基因技术。值得一提的是，这一团队深入研究了转基因在受体基因组中传代特征。他的团队永远保持技术先进性，将很多最为先进的基因操作技术引入到多种鱼类中，如转座子、Cre/loxP 重组系统、Gal4/UAS 转录激活系统、TALEN 和 CRISPR/Cas9 基因组编辑工具、RNA 干扰等(同时见第二、五、七部分的部分论文)。

第三部分和第四部分：外源基因表达的生理功能和转基因黄河鲤的生态安全研究。他的团队利用整体手段从摄食、代谢、生长到发育、免疫和游泳行为等方面详细描述了生长激素基因转基因黄河鲤。特别值得一提的是，他的团队花费了大量精力，详尽地评估了转基因鱼在不同条件下的生态学效应。这些研究为未来转基因鱼的产业化在食品安全和生态安全方面提供了大有裨益的信息。

第五部分：鱼类生殖调控与转基因鱼的育性控制研究。此外，他的团队还系统研究了生长激素对性腺发育的作用及开发不育转基因鱼的技术。这将有助于为转基因鱼的产业化提供有效的生态安全保障。

第六部分：抗病草鱼基因工程技术育种研究。他的团队进行了多种抗病的基因工程鲤和草鱼基础研究，这些工作为利用遗传改良手段开发抗病鱼新品系奠定了科学基础。他的团队完成了草鱼的全基因组草图绘制，这是对鱼类生物学领域的巨大贡献。这不仅为草鱼的遗传选育提供了基因组信息，而且将为鲤科鱼类的生物学研究提供全方位的帮助。

第七部分：鱼类基因转移育种技术展望。他们进一步利用斑马鱼模型开展了一些在水产育种中非常有价值的转基因研究，如激活生长激素信号通路、耐低氧、高 omega-3 多不饱和脂肪酸含量等。

总之，朱作言教授团队在多个方面都做出了重要的科学贡献，正如他们通过以上 7 个部分所展示的。他们的研究是将基础科研和水产应用科学相结合的典范——从技术研发到研制快速生长和抗病的转基因鱼，这是现代水产业亟须解决的两个重大问题。他们同时对鱼类生物学的很多基础生物学问题做出了多方面的贡献，如细胞核移植、草鱼全基因组测序、鱼类生物技术的研发和基因工程技术的优化等。该文集的出版，无疑是转基因鱼研究领域的一个重要里程碑。我们热切地期待着这本书的及时出版。

博士

新加坡国立大学荣誉教授

博士

新加坡国立大学教授

2018 年 12 月

Preface

The present monograph represents a major contribution to the transgenic fish study by Professor Zuoyan Zhu and his scientific colleagues at the Institute of Hydrobiology of the Chinese Academy of Sciences, Wuhan, China. The major topics include the development and generation of transgenic fish, the basic biological study of transgenic fish, the genetic and ecological studies of transgenic fish, and the perspective on the future development of fish genetic engineering.

Prof. Zhu is a pioneer in the field of the transgenic fish study. In the mid-1980s, his group succeeded in the generation of the first-case of fast-growing transgenic fish in the world, which initiated a new research field of transgenic fish biology. His team has established a complete research model of transgenic fish, including the mechanisms of transgene integration, expression, and inheritance. Subsequently, his team cloned grass carp *β-actin* promoter and common carp *growth hormone* (*gh*) gene, and generated an "all-fish" transgene construct which is composed of *Cyprinidae*-derived genomic elements. Based on the construct, his team successfully generated an "all-fish" *gh*-transgenic line of Yellow River common carp and conducted a series of studies concerning the biological effects of *gh*-transgene and the ecological evaluation of *gh*-transgenic Yellow River common carp. Zhu's team has accomplished the full-chain study from the theoretical model to the ecological evaluation of transgenic fish, and it should be the team that has done the complete study of transgenic fish worldwide. These studies are comprehensive and represent a major contribution to the field of transgenic fish biology by the Zhu team.

The Zhu team has thoroughly studied the biosafety containment of transgenic fish. On the one hand, his team developed some general approaches for reproductive control, and on the other hand, his team explored the triploidy to manipulate sterility in common transgenic. By using these strategies, it is possible to effectively solve the genetic and ecological problems in the application of transgenic fish, which is of great importance for the commercialization of these new species.

The monograph, compiling of 100 peer-reviewed papers by the Zhu team in the past 30 years, is well organized into 7 logical sections. The contents of each section are briefly highlighted below.

The 1st section: the establishment of a transgenic fish model. Zhu's group was the first one in the world to generate transgenic fish and has established *gh*-transgenic fish in many species, including aquaculture important species. They continued the strength of fish nuclear transplantation work initiated from China and are a world authority in this technology to study fundamental biological questions to investigate early gene expression in development.

The 2nd section: the transgene's integration and expression. His group has conducted a series of works to develop and optimize the transgenic technology. In particular, his groups carried out extensive studies to characterize the fate of transgenes in the host for multiple generations. His group always followed the most updated technological development in genetic engineering and has introduced many state-of-the-art gene modification technologies in various fish models, such as use of transposons, Cre/loxP recombination, Gal4/UAS binary system, genome editing by TALEN and CRISPR/Cas9, RNA interference, etc. (also Sections 2, 5, 7).

The 3rd and 4th sections: the physiological function of the expressed transgene and the ecological assessment of the transgenic Yellow River common carp. His group uses a holistic approach to characterize *gh*-transgenic fish from feeding, metabolism, growth to development, immunity, swimming behavior, etc. In particular, his group has conducted tremendous studies in the assessment of ecological consequence of *gh*-transgenic fish under various conditions. These studies have provided valuable information on food safety and ecological impact for future application of transgenic fish in aquaculture.

The 5th section: fish reproduction regulations and the control of transgenic fish fertility. Also, his group has conducted a series of experiments to investigate the hormonal effect on gonad development and the possibility for development of sterility in transgenic fish; this may help to develop an effective transgene terminator when transgenic fish is used in aquaculture.

The 6th section: the genetic engineering breeding of disease-resistant grass carp. His group also researched various genetically modified carp for disease resistance. These works provide the scientific basis for using the genetic modification approaches to develop disease-resistant fish strains in aquaculture. His group completed the draft genome sequence of the grass carp and this is an important contribution in fish biology, which is not only important genomic resources for selection of better carp strains in aquaculture, but also facilitates future biological studies in carp and other fish in all aspects.

The 7th section: the prospect of transgenic fish breeding. They also made a few interesting transgenic studies using the zebrafish model fish to test some potentially important features in aquacultures, such as the constitutively activated growth hormone pathways, hypoxia resistance and production of fish flesh containing high omega-3 fatty acids.

Overall, Zhu's group has made important scientific contributions in several areas as they classified in the content. Their research is an excellent combination of fundamental research and translational potential in aquaculture, as evident from the technology development to generate growth-accelerated and disease-resistant transgenic fish, the two most important aspects in modern aquaculture. They also made numerous contributions in fundamental fish biology, such as nuclear transplantation, carp genome sequencing, technical development and optimization of various genetic engineering tools for fish, etc. The monograph represents a major milestone in this field, and we look forward to the timely publication of this monograph.

PhD

Emeritus Professor, National University of Singapore

and

PhD

Professor, National University of Singapore

Dec. 2018

前　言

——从鱼类细胞核移植到基因转移

20 世纪五六十年代，科学家们发展了用于低等脊椎动物胚胎显微操作的细胞核移植技术，研究细胞核的全能性，以及胚胎发育中细胞核与细胞质的相互作用。其中，代表性科学家有美国的 R. Briggs 和 T. J. King，英国的 J. B. Gurdon (2012 年诺贝尔生理学或医学奖获得者)和我国的 T. C. Tung (童第周)。美英学者的实验对象主要是两栖类，而童第周先生的实验对象主要是鱼类，而且是在不同种的鱼类之间，探讨受体卵母细胞质与外来异种细胞核在促进胚胎发育中的作用。20 世纪 70 年代，我们在童第周先生指导下完成的重要研究成果之一，是发现了鱼类的卵细胞质可以在一定程度上配合外来的异种细胞核，共同促进胚胎早期发育；甚至在个别情况下，可以获得性成熟的个体，即“核质杂种鱼”，由此尝试开启新一代养殖鱼类育种的细胞核移植(克隆)技术。

也是在 20 世纪五六十年代，生物学研究出现了巨大的飞跃。以解析 DNA 双螺旋结构为先导的分子生物学兴起，引发了生命科学一日千里的蓬勃发展。新理论、新技术不断涌现，大大加深了人们对生命过程和生命本质的认识。半个多世纪以来，分子生物学研究的成就，就像 19 世纪达尔文物种演化论的创立一样，为生物学和现代生命科学竖起了光芒万丈的灯塔，为相关学科的深入发展指明了航向，也为以分子生物学为基础的现代生物技术及产业的发展夯实了宽广和坚实的基础。在这一大背景下，鱼类基因工程育种研究应运而生。

纵观人类文明和科学技术进步的历史，农业育种实践由引种驯化、品种选育、杂交育种、植物组织培养或动物人工授精和胚胎分割扩繁、染色体组的倍性操作，到细胞核移植(克隆)技术，从群体和个体的层面向细胞水平逐步深入。20 世纪 80 年代初期，以基因克隆为代表的 DNA 分子操作技术逐渐成熟，为品种改良从遗传物质的源头，即 DNA 和基因水平的精准操作提供了可能性。我们很幸运地在中国科学院水生生物研究所率先进行了第一次鱼类基因转移的成功尝试。中国科学院水生生物研究所的前身，是 20 世纪 30 年代初期建立的中央研究院所辖的自然历史博物馆，后来逐渐演替为中央研究院的动植物研究所、动物研究所，直至中国科学院成立后组建为水生生物研究所，并由上海迁至武汉。中国科学院水生生物研究所积淀了数十年的鱼类生物学研究成果，在养殖鱼类资源发掘、人工繁殖、品种驯化、杂交选育和细胞遗传育种方面有着独特的优势，鱼类基因工程育种由此发源乃属水到渠成。早在 20 世纪 70 年代中后期，我们在非常困难的研究条件下，利用显微操作技术，把内蒙古瓦氏雅罗鱼(*Leuciscus waleckii*，冷水性鱼类)的总 DNA 注射到广东鲮鱼(*Cirrhinus molitorella*，温水性鱼类)的受精卵，探索了从遗传物质载体——DNA 分子层面入手，提高鱼类耐寒性的可行性。此时，我虽然萌发了基因转移育种的概念，但苦于没有克隆的基因而无法付诸行动。1981~1983 年，我有机会去欧美学习，先到了英国南安普顿大学细胞遗传学家 Norman Maclean 博士实验室考察，向他介绍了鱼类细胞核移植实验的成果，就计划开展基因转移研究的设想做了广泛交流；接着去英国帝国癌症研究基金会(ICRF)转录实验室主任 Robert Kamen 博士的实验室学习分子克隆技术，他后来又推荐我到美国波士顿遗传研究所，与 Edward F. Fritsch (*Molecular Cloning* 作者之一)合作，从事人胰岛素和促红细胞生成素(erythropoietin，EPO)基因克隆研究，以全面掌握分子克隆技术。可是，当我于 1983 年按期回国的时候，还是没有获得所需的鱼类克隆基因，世界上也没有任何一个实验室完成了鱼类基因的克隆。可喜的是，我得到了美国国家卫生研究院 D. H. Hamer 博士赠予的人生长激素重组 DNA 载体。有了它，鱼类基因转移研究终于可以扬帆起航。从 1984 年起，我们把生长激素基因片段与载体质粒 DNA 分离，显微注射到鲫(*Carassius auratus*)等多种鱼类的受精卵，获得了具有明显快速生长效应的转基因鱼；通过随后几年的研究，建立了完整的转基因鱼研究模型，并得到了国际学术界的认可。后来，我带领研究团队克隆了鲤科(Cyprinidae)鱼类的 β-肌动蛋白基因(与 P. Hackett 教授合作)和生长激素基因(与 T. T. Chen 教授合作)，构建了“全鱼”基因表达载体，培育了生长快、饵料利用率高、生态和食用安全的转基因鲤，我们后来命名它为“冠鲤”，以彰显其卓越的养殖特性。本文集选编了我们研究团队的研究论文 100 篇，分上下两册，作为我们研究工作的一份总结。

这是在中国科学院水生生物研究所及淡水生态与生物技术国家重点实验室开创性的一项系统性研究工作，时任所长刘建康院士是启动这项研究最重要的支持者和推动者。早期有李国华、何玲和稍后加入的谢岳峰三位年轻助手跟我一起白手起家，因陋就简，克服重重困难，创建了第一个鱼类基因转移研究实验室，它也是我国最早的几个分子生物学实验室之一。在随后的 30 多年里，先后有几十位同事和同学们参与这些研究，并与本项研究共同成长。我衷心感谢汪亚平、胡炜、孙永华、崔宗斌 4 位研究员，他们不仅是我们研究团队的主力，而且共同主编了本文集。我衷心感谢张甫英、黄容等副研究员，许克圣、陈尚萍、廖兰杰等高级工程师，黎明、李勇明、祁德运、甄建设等实验助理的出色工作和辛勤付出。我衷心感谢河南省农业厅水产科学研究院张西瑞、冯建新和王宇锋高级工程师在提供黄河鲤底栖品种和养殖上的鼎力帮助。我衷心感谢武汉多福科技农庄股份有限公司和大湖水殖股份有限公司高瞻远瞩地按国家《农业转基因生物安全管理条例》与《农业转基因生物安全评价管理办法》的规定建立了合格的鱼类基因工程育种养殖试验基地，包括构建面积为 6.7 hm^2 的人工模拟试验湖泊；配合我们开展了转“全鱼 *gh*”基因黄河鲤的食品安全与生态安全评估试验，包括中间试验、环境释放试验和生产性试验。我还要衷心感谢中国科学院、科学技术部、国家自然科学基金及湖北省等不同渠道的研究经费资助。本文集出版得到了淡水生态与生物技术国家重点实验室的资助。

最后我想强调的是，从本文集所呈现的 30 余年的研究轨迹不难看出，转基因育种技术实质上是杂交育种的逻辑延伸和技术跃迁。转基因育种技术可理解为一个优良性状目的基因分子与一个生物个体整套基因组的杂交，是建立在现代分子生物学基础上的定向、精准、可预测、可追踪的分子杂交育种技术。由于种种原因，这一在我国开创并具有完全自主知识产权的鱼类基因工程育种新技术，在推广应用上却遇到了重重困难。“莫道浮云终蔽日，总有云开雾散时”，我坚信，科学精神终将驱散困扰公众的形形色色的迷雾。历史终将证明，转基因技术一定会在现代农业发展中发挥无可比拟和不可替代的重要作用。

中国科学院水生生物研究所　研究员
北京大学　博雅讲席教授
《中国科学》《科学通报》　总主编
中国科学院　发展中国家科学院　院士
2018 年 12 月

Introduction

——From nuclear transplantation to gene transfer of fish

The 1950s and 1960s, scientists developed a technique of nuclear transplantation for micro-manipulating lower vertebrate embryos to study the totipotency of nucleus and the interaction between nucleus and cytoplasm in early embryonic development. The representative scientists were R. Briggs and T. J. King of USA, J. B. Gurdon of UK (Nobel laureat in Physiology or Medicine in 2012) and T. C. Tung of China. The subjects of these experiments by American and British scholars are mainly amphibians. However, Professor T. C. Tung mainly utilized fish to explore the role of the enucleated oocyte with interspecies donor nucleus in promoting embryonic development. In the 1970s, one of the major achievements of our team under the guidance of Professor T. C. Tung was the discovery that fish egg cytoplasm with interspecies donor nucleus was able to go through the early development and, in some cases, the whole ontogenies, resulting in "nucleocytoplasmic hybrid fish". Thus, we attempted to develop a new generation of the technology of nuclear transplantation (cloning) for fish breeding in aquaculture.

During the same period, there was an enormous breakthrough in biological sciences. The rise of molecular biology evoked by the discovery of the structure of DNA double helix triggered the explosive development of life sciences. The emergence of new theories and technologies has greatly promoted people's understanding of the nature of life and the process of adapting to the surrounding environment. For more than half a century, the achievements in molecular biology, just like Darwin's theory of evolution in the 19th century, have shone as a beacon light for biology and modern life science, pointed out the direction of in-depth development of related disciplines, and laid a broad and solid foundation for modern biotechnology and industry. In this context, the genetic engineering of fish breeding came into being.

Throughout the history of human civilization and scientific and technological advances, the practice of agricultural breeding has evolved from introduction and domestication, variety breeding, crossbreeding, plant tissue culture, artificial animal fertilization and embryonic bisection for propagation, chromosomal ploidy operation to nuclear transplantation (cloning), gradually deepening from the level of population and individuals to the level of cells. In the early 1980s, techniques of DNA manipulation represented by gene cloning matured gradually, providing the possibility for variety improvement by way of accurately manipulating the genetic material itself. We were fortunate to have taken the lead in the first successful attempt of fish gene transfer at the Institute of Hydrobiology, Chinese Academy of Sciences. The predecessor of the Institute of Hydrobiology, Chinese Academy of Sciences was the Natural History Museum of Academia Sinica established in 1930. Later, it gradually became the Institute of Zoology and Botany in 1934 and the Institute of Zoology in 1944 of Academia Sinica, and was reorganized to form the Institute of Hydrobiology after the establishment of Chinese Academy of Sciences in 1950, and eventually moved from Shanghai to its current location in Wuhan in 1954. The Institute of Hydrobiology has accumulated decades of research achievements in fish biology and has unique advantages in the exploitation of cultured fish resources, artificial reproduction, variety domestication, crossbreeding and cytogenetic breeding, from which fish breeding by genetic engineering just happened that way. In the middle of the late 1970's, under very difficult research conditions, we used the micro-manipulation technique to explore the possibility of DNA transfer in fish: injecting the total DNA from *Leuciscus waleckii* (cold water fish) into the fertilized eggs of *Cirrhinus molitorella* (warm water fish) to improve cold tolerance of fish. At this time, the concept of transgenic breeding came into my mind, but I was unable to put it into practice because there was no cloned gene available for us. From 1981 to 1983, I had the opportunity to study in Europe and in USA. I first visited Dr. Norman Maclean, a professor of cytogenetics at the University of Southampton, UK and introduced him the findings of nuclear transplantation of fish. We made extensive exchanges on the idea of planning to carry out gene transfer in fish. I then studied molecular biology and gene cloning from Dr. Robert Kamen, director of Gene Transcription Laboratory of Imperial Cancer Research Fund laboratories in London. Dr. Robert Kamen then recommended me to work in the Genetics Institute of Boston, USA with Dr. Edward F. Fritsch (one of the authors of *Molecular Cloning*) in cloning human insulin and erythropoietin (EPO) genes to master the molecular cloning technology. However, when I returned to my homeland on schedule in 1983, I still did not obtain the necessary fish genes, and in fact no laboratory in the world had completed the cloning of any fish gene. Fortunately, I received a gift of recombinant human growth hormone DNA vector from Dr. D. H. Hamer of National Institutes of Health, USA. With it, fish gene transfer research finally started. From 1984 onwards, we isolated the growth hormone gene fragment from the vector plasmid DNA and injected it into the fertilized eggs of crucian carp and other kinds of fish to obtain transgenic fish with obvious rapid growth performance. We established a transgenic fish research model in the subsequent years, which was recognized by the international academic community. Later, we cloned β-actin gene (with Professor Perry Hackett) and growth hormone gene (with Professor T. T. Chen) of Cyprinidae, made a series of corresponding "all-fish" gene expression vector and generated the transgenic carp with rapid growth performance, high rate of feed utilization, ecological safety for cultivation and food safety for consumption. We now named it "crown carp" to highlight its outstanding characteristics for breeding and cultivation. This book has compiled 100 research papers from our research team, divided into two volumes, as a summary of our research work.

The systematic research has been mainly carried out in the Institute of Hydrobiology of the Chinese Academy of Sciences and the State Key Laboratory of Freshwater Ecology and Biotechnology. The then director, academician Dr. Jiankang Liu, was the most important supporter and promoter for the launch of the initiatives. In the early stage, Guohua Li, Ling He, and Yuefeng Xie, three young assistants who joined together with me, started from scratch and overcame countless difficulties to establish the first fish gene transfer research laboratory, which is also one of the earliest molecular biology laboratories in China. Over the next 30 years, dozens of colleagues and over ×× graduate students participated in these studies and grew up together with this program. I would like to thank Professors Yaping Wang, Wei Hu, Yonghua Sun and Zongbin Cui, who are not only the main force of our research team but also co-editors of this monograph. I sincerely thank Associate Professors Fuying Zhang, Rong Huang and others, the senior engineers Kesheng Xu, Shangping Chen, Lanjie Liao, and research assistants Ming Li, Yongming Li, Deyun Qi, Jianshe Zhen for their excellent work. I sincerely thank senior engineers Xirui Zhang, Jianxin Feng and Yufeng Wang of Henan Academy of Fishery Sciences for their great help in providing Yellow River common carp for stating varieties and assistance of fish breeding and cultivation. I sincerely thank Wuhan Duofu Technological Farm Co., Ltd. and the Dahu Aquaculture Co., Ltd. for their foresight in establishing a qualified experimental genetic engineering fish farm, including the construction of 6.7 hm^2 artificially simulated lake, in accordance with the provisions of the national regulations on safety of agricultural GMOs and the administrative measures on safety assessment of agricultural GMOs. In cooperation with them, we carried out the food safety and environmental assessment test of the "all-fish *gh*" transgenic Yellow River common carp, including restricted field testing, enlarged field testing, and productive testing. I would also like to thank the funding supports from the Chinese Academy of Sciences, the Ministry of Science and Technology, the National Natural Science Foundation of China and Hubei province. This book is published with the support of the State Key Laboratory of Freshwater Ecology and Biotechnology.

Last but not least, I would like to point out that transgenic breeding technology is essentially a logical extension and technological transition of the traditional crossbreeding technology, which is not difficult to be figured out from the 30-year's research presented in this book. Transgenic breeding technology can be understood as the hybridization between the target gene coding for the desired trait and a whole genome of an organism that needs to be improved. It is a type of molecular crossbreeding technology based on modern molecular biology, but more directional, accurate, predictable and traceable when compared with the traditional crossbreeding. For various reasons, this new technology, which was initiated in our country with all independent intellectual property rights, has encountered many difficulties in its popularization and application. “Do not say that the clouds always blot out the sun. There’s always a time when the mist clears away. ” I firmly believe that the spirit of science will eventually dispel all kinds of mist that troubles the public. History will eventually prove that transgenic technology will surely play an incomparable and irreplaceable role in the development of modern agriculture.

Professor of IHB, CAS

Peking University Boya Chair Professor

Editor General, *Science China* and *Science Bulletin*

Academician of CAS & TWOS

Dec. 2018

目　　录

上　卷

第一部分　鱼类基因转移模型建立

第二部分　外源基因的整合与表达

第三部分　外源基因表达的生理功能

第四部分　转基因黄河鲤的生态安全研究

第五部分　鱼类生殖调控与转基因鱼的育性控制研究

下 卷

第六部分 抗病草鱼基因工程技术育种研究

第七部分 鱼类基因转移育种技术展望

第一部分 鱼类基因转移模型建立

Novel Gene Transfer into the Fertilized Eggs of Goldfish (*Carassius auratus* L. 1758)

Z. ZHU G. LI L. HE S. CHEN

Institute of Hydrobiology, Academia Sinica, Luojiashan, Wuhan 430072

Summary Novel gene which was microinjected into fertilized eggs of the goldfish replicated during the embryogenesis. A proportion of the novel gene has been integrated into the host DNS of the 50-day-old transgenic goldfish.

Zusammenfassung

Übertragung von Fremdgenen in befrucbtete Eier des Goldfisches (Carassius auratus L. 1758)

Durch Mikroinjektion in befruchtete Eier des Goldfisches eingebrachte Fremdgene wurden während der Embryogenese repliziert. Ein Teil der Fremd-DNS ist in die DNS der 50 Tage alten Goldfische integriert worden.

Résumé

Transmission des gènes aliens dans des ceufs fertilisés du Cyprin doré (Carassius auratus L. 1758)

De gènes nouveaux sont microinjectés dans des ceufs fertilisés des embryons de *Carassius auratus*. Ces gènes sont répliqués. Un part de la DNS nouveau était intégré dans la DNS des *Carassius auratus* originaux à ce moment agés 50 journées.

In recent years, purified genes have been microinjected into fertilized eggs or embryos of mammals, amphibians, and insects species in order to assess the fate and expression of the novel gene during the development in whole animal system [for review, see (3)]. Results of experiments performed on *Xenopus laevis* demonstrated that the microinjected, cloned DNA sequences could be replicated and faithfully transcribed during embryogenesis (2,7). PALMITER *et al*. (1982) (5) introduced into the pronuclei of fertilized eggs of mice a hybrid gene containing the promoter of the mouse metallothionein-1 gene fused to the structural gene for rat growth hormone. Some transgenic mice grew significantly larger than their littermates. Thus it is hoped to use the gene transfer method to develop rapidly growing strains of domestic

文章发表于 *Journal of Applied Ichthyology*, 1985, 1(1): 31-34

animals. Compared with mammals and amphibians, fish, being at a lower level in evolution among the vertebrates, are much more suitable for micromanipulation during embryogenesis. Our previous experiences in nucleo-transplantation in fish (1) indicated that the affinity of the nucleus and cytoplasm among different species or even different genera of fish and the adaptability of fish embryonic development to its environment are greater than among the other vertebrates. It is naturally imaginable that a superfish could be bred by the microinjection of a cloned and properly modified fish growth hormone gene. To this end, as a first step, we performed an experiment of novel gene transfer into the fertilized eggs of gold fish to assay its stability and existence during the host embryogenesis and in the adult fish as well.

Novel Gene Preparation: The cloned DNA sequence used here for microinjection came from a recombinate plasmid pBPVMG-6 (6) (Fig. 1), a gift of Dr. D. HAMER at NHI, U.S.A. A 9.4 kb linear DNA fragment of BPVMG was isolated from pBPVMG-6 with a procedure of restriction enzyme *Bam*H1 digestion, 0.7% agarose gel electrophoresis separation, phenol and chloroform extraction, and alcohol precipitation. It was finally dissolved in 88 mM NaCl and 10 mM Tris HCl (pH 7.5) at an injection concentration of 27 ng/μl.

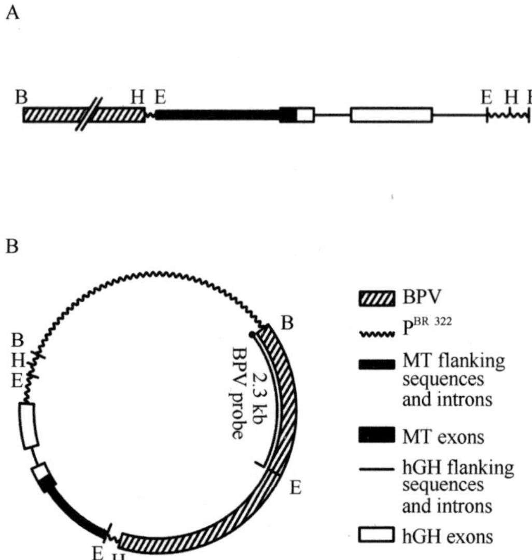

Fig. 1 The structure of the novel gene: (A) 9.4 kilobase pair (kb) BPVMG linear sequence as a novel gene to be microinjected into gold fish eggs consists of 69% BPV (bovine papilloma virus) transforming sequence followed by an MG hybrid gene. The MG contains an EcoR1/Bgl2 fragment of the promoter of the mouse MT-1 (metallothionein-1) gene fused to the BamH1 site of a hGH (human growth hormone) mini-gene that includes all the coding sequence and the first of its four intervening sequences. (B) Recombinate plasmid pBPVMG-6 is an MG hybrid gene inserted into a pBRBPV vector (modified and reconstructed according to PAVLAKIS and HAMER1983; 6)

The Preparation of Fertilized Eggs of Gold Fish: In April, 1-2 year old gold fish, males and females, were induced for reproduction by injection of a homogenate of carp pituitary gland (1 mg pituitary gland/1 kg body weight). About 10 hours later, the eggs were squeezed out and inseminated artificially. The chorion of the eggs was removed after 10 minute treatment in 0.25% trypsine solution, and the naked eggs were then transferred into Holtfreter's solution (4) in an agar layered Petri dish.

The Novel Gene Introduction: A glass micro-needle with a tip inner diameter of 3 μm for microinjection was drawn with a needle drawing apparatus (Leitz). A plastic tube connected the micro-needle with a 50 ml syringe. After loading the DNA solution into the microneedle, the micromanipulater was carefully adjusted to insert the tip of the micro-needle into the central position of the germinal disc of the fertilized egg. A 1-2 nl dose of DNA solution (about 7×10^6 copies of BPVMG molecules) was then released into each egg before the first cleavage occurred by applying a proper pressure to the syringe. The microinjected eggs were allowed to develop in Holtfreter's solution up to blastula stage and then transferred into "aged" boiling water (boiled and then left at room temperature for more than three days).

Fate of the Novel Gene in Fish Development: More than 3000 gold fish eggs received the microinjection during the spawnning season. Random samples of 20-30 embryos were collected immediately after injection and at the specified stages of embryogenesis. The total DNA in the embryos was isolated by the procedure of RUSCONI and SCHAFFNER (1981) (7). DNA equivalents of one embryo at progressive stages of development were dropped onto a nitrocellulose membrane filter for dot hybridization against a ^{32}P-labeled nicktranslated probe of a 2.3 kb fragment of BPV sequence. Neither this nor its homologous sequence, so far as we are aware, exists in fish DNA. Similarly, DNA equivalents of four embryos were separated by 0.8% agarose gel electrophoresis and Southern blotting on nitrocellulose filters, and hybridized against the same hot probe (8, 9). The replication of the introduced novel gene began at the early stages of cleavage, and the strongest signal of replication appeared from the late blastula to early neurula stages. This was followed by its selective degradation. The physical form of the injected linear fragment of DNA sequence changes dramatically within the host eggs. Linear DNA molecules rapidly ligate and replicate in the form of a circular or concatenating polymer linears in the fertilized eggs. After the neurula stage, the beginning of organogenesis in the fish embryo, most of the hybridizable sequences co-migrated with the high molecular weight DNA rather than as 9.4 kb linear or circular DNA.

The Existence of the Novel Gene in 50-day-old Fish: Six of the 50-day-old fish developed from the microinjected eggs were taken randomly. The total DNA in the fish was extracted as described above except that the gut was discarded prior to homogenization. The result of dothybridization indicated that in 3 of 6 fish examined, DNA hybridized to the hot probe as a 2.3 kb BPV sequence, i. e. the ratio of positive transgenic fish at the age of 50 days is about 50%. DNA patterns analyzed by restriction enzyme digestion and Southern blotting indicate that (1) a novel gene hybridized to the probe migrates with the host chromosome DNA in 0.6% agarose gel electrophoresis; (2) DNA digests with restriction enzyme BamH1 giving only a 9.4 kb hybridizable band, suggesting that the novel gene maintains its original size and the same sticking ends of BamH1; (3) restriction enzyme Xho1 does not cut the novel gene at all but definitely digests the host chromosome DNA generating a hybridizable band longer than 9.4 kb. This seems to suggest that the remaining novel gene, having undergone replication and selective degredation in embryogenesis, has integrated into the host chromosome DNA of the 50-day-old transgenic fish.

In conclusion, our result reported here demonstrated that the novel gene, microinjected into the fertilized eggs of gold fish, replicates in different physical forms and selectively degrades during the host embryogenesis. Yet, migration of some novel gene in the form of macromolecules could actually be detected in the 50-day-old transgenic gold fish, which seems to represent the proportion of the novel gene that had been integreted into the host chromosome DNA. In conclusion our experiment has revealed a possible method for investigating the expression of a novel gene in fish development and its stability in the offspring.

Acknowledgement The project was supported by the Science Fund of the Chinese Academy of Sciences.

References

1. [*], 1980. Nuclear transplantation in teleosts. Scientia Sinica 23 (4), 517-523.
2. BENDIG, M., 1981: Persistence and expression of histone genes injected into *Xenopus* eggs in early development. Nature (London) 292, 65-67.
3. ETKIN, L. D.; BERARDION, A.D., 1983: Eukaryotic Genes, N. Maclean *et al* Eds. (Butterworths, London) pp. 148-150.
4. GRAND, C. G.; GORDON, M.; CAMERON, G., 1941: Neoplasm studies VII. Cell types in tissue culture of fish melanotic tumors compared with mammalian melanomas. Cancer Res. 1, 660-666.
5. PALMITER, R. D.; BRINSTER, R. L.; HAMMER, R. E.; TRUMSAUER, M. E.; ROSENFELD, M. G.; BIRNBERG, N. C.; EVANS, R. M., 1982: Dramatic growth of mice that develop from eggs microinjected with metallothionein-growth hormone fusion genes. Nature (London) 300, 611-615.
6. PAVLAKIS, G. N.; HAMER, D. H., 1983: Regulation of a metallothionein-growth hormone hybrid gene in bovine papilloma virus. Proc. Natl. Acad. Sci. U.S.A. 80, 397-401.
7. RUSCONI, S.; SCHAFFNER, W., 1981: Transformation of frog embryos with a rabbit-globin gene. Proc. Natl. Acad. Sci. U.S.A. 78, 5010-5055.
8. SOUTHERN, E. M., 1975: Detection of specific sequences among DNA fragments separated by gel electrophoresis. J. Mol. Biol. 98, 503-517.
9. WAHL, G. M.; STERN, M.; STARK, G. R., 1979: Efficient transfer of large DNA fragments from agarose gels to

* Research Group of Cytogenetics, Institute of Zoology, Academia Sinica; Research Group of Somatic Cell Genetics, Institute of Hydrobiology, Academic Sinica; and Research Group of Nuclear Transplantation, Chang Jiang Fisheries Research Institute, State Fisheries General Board.

diazobenzyl-oxymethyl-paper and rapid hybridization by using dextran sulfate. Proc. Natl. Acad. Sci. U.S.A. 76, 3683-3687.

外源基因向金鱼受精卵中的转移

朱作言 李国华 何 玲 陈尚萍

中国科学院水生生物研究所，武汉 430072

摘 要 本文首次报道了采用显微注射方法将外源基因导入金鱼受精卵的实验结果。导入的外源基因随着胚胎的发育进行了复制。进一步研究发现，在 50 日龄的转基因鱼中，部分外源基因已被整合到受体鱼的染色体 DNA 中。

人生长激素基因在泥鳅受精卵显微注射转移后的生物学效应

朱作言　许克圣　李国华　谢岳峰　何　玲

中国科学院水生生物研究所，武汉 430072

1982 年底，美国 Palmiter 和 Brinster 等人[1]报道了给小鼠受精卵的雄性原核注射大鼠生长激素基因从而培育成功“超级鼠”的实验结果，揭开了用外源基因转移方法培育动物新品系的序幕。两栖类的受精卵具有接受、整合和表达外源基因的潜力[2-5]，我们推测，作为低等脊椎动物的鱼类，这种潜力可能比高等脊椎动物的更大一些。而且鱼的怀卵量一般都很大，并在体外受精，因而基因操作手术比较容易。近两年来，我们开始了外源基因在鱼类受精卵内转移的研究，追踪了外源基因在鲫鱼胚胎发育过程中的行为，初步证明了外源基因在 50 日龄受体鱼基因组内的整合作用[6]。近来，英国 Maclean 博士实验室开始了对虹鳟(*S. trutta*)早期胚胎转移金属鳌合蛋白-I (metallothionein-I，简写为 MT-I)基因的实验。加拿大、美国和日本等国家的几个实验室正在着手开始或计划用生长激素基因对鲑鳟鱼类的受精卵进行外源基因的转移研究。可见，不同国家的研究者们竞相用基因操作的手段，试图开拓培育鱼类优良品系的新途径，但迄今尚未见有实质性生物学效应的报道。

材料和方法

催产用的泥鳅(*Misgurnus anguillicaudatus* (Cantor))从武汉市大东门和关山集贸市场购得。经注射鲤鱼脑下垂体催青、挤卵和人工授精后，用 0.25%的胰蛋白酶溶液除去受精卵卵膜，裸卵在 Holtfreter 氏溶液中接受外源基因注射的显微手术。注射的重组 D N A 顺序 pMhGH，系由 Hamer 博士赠送的 pBPVMG-6 剪切后重新克隆而成。它带有一个删除了后 3 个内含子的人生长激素基因，其 5′ 端冠以大鼠 MT-I 基因的启动基因顺序，然后克隆在质粒 pBR322 的 EcoRI 位点上，总长度为 7.88 千碱基对(kb)，其中人生长激素基因的顺序为 1.59 kb，MT-I 基因的启动基因顺序为 1.90 kb。重组质粒 pMhGH DNA 经氯化铯密度梯度离心纯化，在 pBR322 顺序上的 BamHI 位点开环而线型化；酒精沉淀后溶于注射缓冲液(NaCl 88mM；Tris HCl 10 mM，pH 7.5)备用，其浓度为 30 μg/ml。简易显微注射装置系由我所自制的一个三维可调支架、塑料导管、注射器和一根玻璃毛细管针组成。玻璃毛细管针尖的内径为 3 μm。显微注射手术在受精卵第一次卵裂发生之前进行，进针位置在胚盘的中央，每卵接受 1-2 nl 的外源基因溶液。在受体卵发育的不同

文章发表于《科学通报》, 1986, 31(5): 387-387

阶段(从 1 细胞期至摄食开始分 10 个阶段)随机取样制备总 D N A[5]，经 0.7%琼脂糖胶电泳后转移到硝酸纤维素滤膜上[7]，并与用 ^{32}P 标记的 MT-I 基因的启动基因顺序探针杂交[8]，检测外源基因在受体泥鳅胚胎发育过程中的行为。鱼苗开始摄食后饲养在玻璃鱼缸内，定期换水和投喂鱼虫。本实验的对照组除了不接受显微注射外源基因的手术之外，其他的饲养管理、样品处理与检测程序均和实验组相同。

结果和讨论

DNA 分子杂交的结果表明，在受体泥鳅卵发育过程中，外源基因的行为基本上与它在受体鲫鱼卵中的行为一致[6]。即外源基因一旦被导入受精卵之后，其形态立即发生深刻的变化：除一部分保持原来的线型单体构象之外，其他的转换为紧密的超环、松驰的大环、线型二聚体乃至多聚体等多样化构象，并开始缓慢的复制。从囊胚中期开始，外源基因在受体胚胎内复制的速度急剧加快，至胚孔封闭和神经胚形成期达到高峰，然后伴随着选择性的降解。但是，以大分子形式出现的外源基因信号显然依旧存在于摄食开始期的受体鱼苗的 DNA 样品之中。在对照样品中，即使是在胚孔封闭期，也检测不出任何外源基因信号。

观察摄食3-5天的鱼苗，发现实验组外源基因转移的个体一般均比对照组的小，估计这一现象是显微手术本身的机械损伤造成的。摄食两周左右，对照组群体生长比较整齐，用肉眼鉴别不出个体之间的大小差异；但实验组群体内的大小表现出明显分化，10%以上的外源基因转移个体体长为对照组的1.5倍，估计其体重为对照组的3倍(此时无法称重)。在一批注射后43天的实验里，12尾对照组泥鳅平均体重35.1 mg(大的46 mg，小的25 mg)，实验组内有2尾分别重160 mg和120 mg，为对照组平均值的4.6和3.4倍。在另外一批实验里，实验后135天取样称重，9尾对照泥鳅平均体重0.77 g(大的0.9g，小的0.7g)，实验组内有3尾分别重2.78 g、2.20 g和2.10 g，为对照组平均值的3.8、3.2和3.1倍。由于材料鱼系从市场购得，虽然形态学检查属于同一物种，但远非纯系实验动物，以致“同胎”的对照组群体内亦存在着个体大小上的差别。但是和实验组内快速生长的外源基因转移个体相比较，对照组内的这种差别是有限的。因此，我们认为在泥鳅个体生长的一定时期(至少在135天之内)，实验组内部分个体3-4.6倍于对照组的生长速度，很可能是由于人生长激素基因在受体内整合和表达的一种生物学效应。在分子水平上为这种生物学效应提供证据的实验正在进行之中。

最近，Gill 等人[9]报道了给幼龄的太平洋鲑鱼(*Oncorhynchus kisutch*)定期注射用重组 DNA 方法生产的鸡或牛生长激素(5 μg 激素/g 体重/周)的实验，42 天之后，实验组体重的增加为对照组增加的两倍，其效果与注射天然的牛生长激素的一致。说明某些高等脊椎动物的生长激素有加快鱼类生长的生物学效应。可以预期，用显微注射等手段把外源生长激素基因导入鱼的受精卵内，一旦这些遗传信息分子在受体染色体组内得到了稳定的整合和忠实的表达，便可以培育出快速生长的鱼类新品系。推而广之，随着基因克隆技术的不断发展，各种优良性状的基因终将陆续被分离出来。转移这些优良性状的基因，定将为鱼类的抗性(如抗病、抗寒等)育种开辟新途径。

参 考 文 献

[1] Palmiter, R., Brinster, R., Hammer, R., frumbauer, M., Rosenfeld, M., Birnberg, N. and Evans, R., Dramatic growth of mice that develop from eggs microinjected with metallothionein-growth hormone fusion. genes, *Nature* (London), 300(1982), 611-615.

[2] Bendig, M., Persistence and expression of histone genes injected into *Xenopus* eggs in early development, *Nature* (London), 292(1981), 65-67.

[3] Bendig, M. & Williams, J., Replication and expression of *Xenopus laevis* tadpole and adult-globin genes injected into fertilized *Xenopus* eggs, *Proc. Natl. Acad. Sci. USA*, 80(1983), 6197-6201.

[4] Etkin, L., Pearman, B., Roberts, M., & Bektesh, S., Replication integration and expression of exogenous DNA injected into fertilized eggs of *Xenops laevis*, *Differentiation*, 26(1984), 194-202.

[5] Rusconi, S., & Schaffner, W., Transformation of frog embryos with a rabbit-globin gene, *Proc. Natl. Acad. Sci. USA*, 80(1981), 397-401.

[6] Zhu, Z., Li, G., He, L., & Chen, S., Novel gene transfer into the fertilized eggs of gold fish (*Carassius auratus* L. 1758). *J. Applied Ichthyology*, 1(1985), 31-34.

[7] Southern, E. M., Detection of specific sequences among DNA fragments separated by gelelectrophoresis, *J. Mol. Biol.*, 98 (1975), 503-517.

[8] Maniatis, T., Fritsch, E. & Sambrook, J., *Molecular Cloning*, (*a Laboratory Manual*), Cold Spring Harbor Laboratory, 1982, 387-389.

[9] Gill, J., Sumpter, J., Donaldson, E., Dye, H., Souza, L., Berg, T., Wypych, J., & Langley, K., Recombinant chicken and bovine growth hormones accelerate growth in aquacultured juvenile Pacific salmon *Oncorhynchus kisutch, Biotechnology*, 3 (1985), 643-646.

Biological Effects of Human Growth Hormone Gene Microinjected into the Fertilized Eggs of Loach

Zhu Zuoyan Xu Kesheng Li Guohua Xie Yuefeng He Ling

Institute of Hydrobiology, Academia Sinica, Wuhan 430072

Abstract At the end of 1982, Palmiter, Brinster *et al.* reported that mice of superb growth had been created through microinjection of a hybrid gene coding for rat growth hormone into the male pronuclei of mouse fertilized eggs. This work ushered in a new approach to develop new breeds of domestic animals with a method of foreign gene transformation. We infer that fish, on a lower level of evolution in vertebrata, would probably possess the potency of accepting, integrating, and expressing foreign genes as amphibia does. And this kind of potency in fish would probably be even higher than in the more advanced vertebrates. In addition, fish egg is favorable for micromanipulation of gene transformation since the egg is easy to collect in quantities and its fertilization is external. During the past two years wo have initiated a project of foreign gene transfer into fertilized eggs of gold fish and carp, investigated the behavior of the configuration of injected foreign gene sequence in the development of the host embryo, and detected, preliminarily, its integration into host genome in 50-day-old fish. Recently N. Maclean at the Southampton University of U. K. started to transfer a gene coding for mouse metallothonein-1 into rainbow trout embryo. It is said that a few laboratories in Canada, United States, and Japan are starting or going to start to do an almost same project of transferring growth gene into fertilized eggs of fish. We can clearly see from all information that investigators from different countries are eager to pave the way of applying genetic engineering to the creating of new aquacultured fish strain of high qualities.

Introduction of Novel Genes into Fish

N. Maclean D. Penman Z. Zhu*

Department of Biology, Southampton University, Hampshire, England, U. K.
*Institute of Hydrobiology, Wuhan, Hubei Province, P. R. China

The experimental introduction of novel gene sequences into various fish species is discussed and the longer term outlook for such work appraised. Current methods are critically discussed, together with the various possible fates of the inserted novel sequences. Some positive results from our two laboratories are presented, together with an indication of parallel work being pursued in other laboratories.

Novel genes have been successfully introduced, expressed, and transmitted via germ line cells to progeny in the mouse[1,2], the amphibian *Xenopus laevis*[3], and in the fruit fly *Drosophila melanogaster*[4]. In this review we discuss the introduction of novel vertebrate genes into various fish species, together with some preliminary evidence for expression of these genes. We also consider the various problems associated with such experiments, together with their likely, value both to scientific understanding and to commercial fish production.

Fish as Transgenic Animals

Fish lend themselves to experimental introduction of novel genes. Fertilisation of eggs is external and is easily carried out by artificial stripping of cock and hen fish and mixing of eggs and milt immediately or after some delay. Eggs are numerous and in many species quite large, that is more than 1 mm in diameter, rendering injection of material by micromanipulation relatively straightforward. Eggs are easily maintained after fertilisation and in many warm water species development is very rapid although in the rainbow trout (*Salmo gairdneri*) development in water at 10℃ may take about 24 days to hatching[5]. Work with fish thus avoids many of the difficulties of the mammalian egg, such as difficult procurement, a brief period of possible *in vitro* culture, and the necessity for reintroduction into the reproductive tract of a receptive female. But the mammal does offer one advantage denied to experiments on fish eggs (or indeed insect or amphibian eggs also), namely that it is possible to microinject directly into the egg nucleus. The zygote nucleus of fish eggs is very small and not readily visible, and so injection is only possible into the area of the nucleus in

文章发表于 *Bio-Technology*, 1987, 5(3): 257-261

almost all cases actually into the perinuclear cytoplasm[6]. This implies that injected DNA molecules must first find their way into the nucleus before incorporation into the nuclear genome is possible. An interesting way round the problem is to attempt injection directly into the large 4C nucleus of the primary oocyte (also know as the germinal vesicle). This approach has been successfully adopted by Ozata *et al*[7]. In which copies of cloned chicken crystallin genes were injected into nuclei of oocytes of medaka (*Oryzias latipes*). Injected oocytes were cultured, being fertilised *in vitro*, and embryos 7 days post-fertilization were analyzed and shown to include a proportion positive for copies of the injected sequence.

A further problem with microinjection is passage of the needle through the egg coat, the chorion. In some fish species this is very tough and impermeable[8] and remarkably resistant to digestion. Some chorions are readily removed by brief digestion followed by mechanical removal, and, in species such as the goldfish *Carassius auratus*[9] and loach *Misgurnus anguillicaudtus*[10] in which development to hatching is very rapid, eggs may be dechorionated prior to injection without any deleterious effects on future development[8]. However, it has not yet proved feasible to dechorionate salmonid eggs and ways must be found to inject through the chorion. One laboratory has developed a procedure for boring a hole in the chorion of the trout egg prior to insertion of the injection needle[11], while we (N.M. and D.P.) have depended on making needles with strong points and injecting soon after fertilisation, when the chorion is less hard than it finally becomes.

The Production of Transgenic Fish

The genes to be used

The possibility of producing transgenic animals results entirely from the technology of gene cloning. Eukaryotic and prokaryotic genes have now been cloned into bacterial or phage vectors in great variety, and the first question to be addressed is to ask what gene sequence is desirable and in what form. Genes may be cloned either as genomic or cDNA copies, the former including, for the most part, all intron sequences, promotor sequences and some flanking sequence both up and down stream from the coding sequence. cDNA (complementary DNA) results from reverse transcription of message, and includes therefore only the sequences there represented, that is, it lacks introns, promotors, and flanking sequences. There is therefore always some doubt whether such sequences will be expressed if integrated even if they are spliced to another promoter. If the aim of the exercise is to achieve strong expression of the sequence it is undoubtedly best to use a genomic copy, together with maximal amounts of flanking sequence. The latter results from the known importance in expression of enhancer sequences[12] that often lie considerable distances up and down stream from a coding sequence and seem to be important in tissue specific expression. Some work has been done on fish using cDNA sequences[11], and is has been demonstrated that the human growth hormone construct employed is indeed expressed in monkey tissue culture cells[13].

The question of the probable expression of the injected sequence and its likely utilisation by the transgenic organism also arises. There is little point in the introduction of a gene sequence coding for a plant chloroplast protein into a fish, since almost certainly the protein product would have no function in the fish. But many genes have been remarkably conserved in the course of evolution and indeed mammalian growth hormones are known to work effectively when injected into fish[14]. There is then a broad and ever increasing range of cloned genes from which to choose. The sequences with which satisfactory expression is most likely are those whose effects in the whole animal are not dependent on tissue specific expression. Thus genes coding for polypeptide hormones are likely to give a satisfactory expression since their products are secreted from the cellular side of synthesis, while those coding for globin or muscle myosin, for example, require selective expression in a particular tissue. Although there is now good evidence that tissue specific expression does not require integration at a particular chromosomal locus[15], it is likely that some sites of integration are incompatible with satisfactory expression.

Table 1 Some laboratories active in work on transgenic fish

Laboratory	Area of Interest
G. Dixon, Department of Biochemistry University of Calgary, Calgary, Canada	Rainbow trout gene library
L. Gedamu, Department of Biology University of Calgary, Calgary, Canada	Rainbow trout gene library, and isolation of trout metallothionein gene
D. A. Powers, T. T. Chen, R. Sonstegard Department of Biology, McMaster University Hamilton, Ontario, Canada	Isolation of rainbow trout growth hormone gene
S. Sekine, Tokyo Research Laboratory Tokyo Research Laboratory, Kyowa Hakko Kogyo Co. Ltd Tokyo, Japan	Cloning and expression of chum salmon cDNA in *E. coli*
D. Chourrout, Laboratory of Fish Genetics I. N. R. A., Paris, France	Gene transfer into rainbow trout
S. Valla, University of Trondheim Institute for Biotechnology, Trondheim, Norway	Atlantic salmon gene library
K. M. Gautvik, E. Rokkones, Institute of Medical Biochemistry Oslo, Norway	Transfer of human growth hormone genes into rainbow trout and Atlantic salmon
K. Ozato, Biological Laboratory, Yoshida College Kyoto, University, Kyoto, Japan	Transfer of chicken crystallin genes into oocytes of medaka

The use of genes derived from fish is an interesting possibility and is dependent on the use of gene banks constructed from fish species. Such banks (libraries) have been made for genomic DNA from rainbow trout[16], atlantic salmon (Valla, personal communication), and carp and grass carp by one of us (Z.Z.). In addition a cDNA Iibrary has been constructed for chum salmon (*Oncorhynchus keta*) by Sekine *et al.*[17]

Whether the genes are of piscine or other origin it may be desirable to increase the chances of good expression by splicing the coding sequence to a strong promoter from another gene. The promotor of the mouse metallothionein I gene has been widely used to this end, especially since it is effective in a wide variety of tissues in the normal animal, represents a protein product with a wide distribution in nature, and may be specifically induced by heavy metals (cadmium, zinc or copper) or corticosteriod, if necessary. The promoter of the animal virus SV40 is also attractive as a universal promoter.

Another stratagem which has been employed to increase the chances of integration is to splice the gene to be introduced between sequences from an animal viral genome, especially the terminal repeat sequences found in some viruses that effect the efficient integration of the viral genome into host DNA. Bovine papilloma virus sequences have been so used, but it now appears that fairly efficient integration is achieved without them.

Introducing the genes into the fish

Assuming that a sequence has been selected, it is necessary to determine how it is to be introduced into the fish. The form in which the gene is to be introduced is clearly important. Should it be linear of circularized, should it include the cloning vector DNA or not, and how many copies should be introduced? In the opinion of most of those working in the production of transgenics at the moment, the DNA should be linear, should not include vector DNA, and should consist of about 10^6 copies per injected cell (actual numbers vary from 10^5 to 10^9 in different laboratories). There is the theoretical choice of injecting into oocytes or eggs, but, as far as fish are concerned, injection into the presumed nuclear area of a newly fertilised egg is the favoured approach, and there is real doubt whether oocytes can be satisfactorily brought to hatching in many species (but note ref. 7, for an interesting exception).

Possible fates of injected sequences

The possible outcome of experiments in which novel genes are injected into fertilised fish eggs is complex, and we will here briefly review the possibilities.

1. *Most of the injected DNA is degraded prior to possible integration.* Since cell cytoplasm is richly endowed with both exo and endonucleases, there is no doubt that most injected DNA is digested. But there is good evidence that such digestion is not immediate.

2. *The DNA may persist, or some copies may persist, for long periods, but still without integration.* In the laboratory of one of us (Z.Z.) work has been undertaken to investigate the fate of DNA injected into eggs of the loach (*Misgurnus* sp.). It is clear that injected DNA, at least in some circumstances, is replicated and undergoes quite rapid conformational changes

in the cell: only after many days do the small molecules of injected DNA disappear entirely from the embryos (Fig. 1), leaving in some cases, copies associated with high molecular weight DNA, It is known from work on other transgenic systems (Etkin, personal communication), that single or multiple copies of injected genes may persist for long periods and be distributed to many tissues. In particular there is a tendency for such genes to form concatamers. This clearly implies that evidence for true integration must include restriction digestions and Southern transfers that show the flanking sequences of this introduced gene are contiguous with chromosomal DNA.

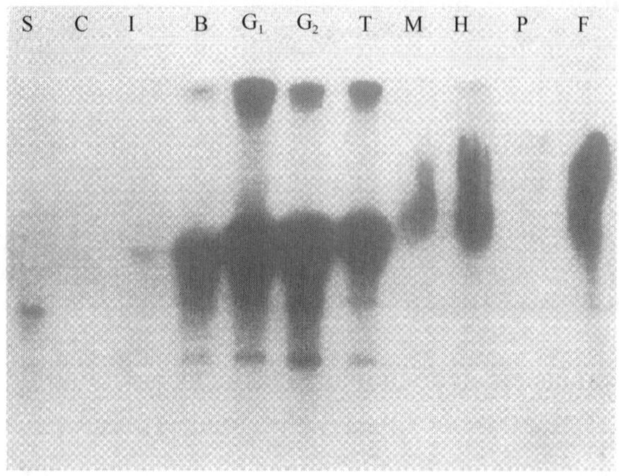

Figure 1 Conformational changes in the DNA of a foreign gene (hGH gene) during the embryogenesis of host egg of loach. DNA was extracted and after transfer hybridized with a ^{32}P-labeled 400 bp fragment from hGH gene sequence. S, hGH gene fragment, as injected; 1, soon after microinjection (30-40 minutes, one cell stage) the foreign gene sequence is recovered in three different forms (presumably linear, covalently circular, and supercoiled). B, blastula stage; G_1 and G_2, early and late gastrula stage respectively. The replication of the foreign gene sequence in host egg is accelerated from B to G_2 stages. T, tail-bar stage; M, muscle reaction stage, H, heart beating stage; P, pigment forming stage and F, feeding stage. From M stage onward, some copies of the foreign gene associated with high molecular weight DNA and that of low molecular weight is gradually degraded. C, control sample from non-injected G_2 stage

3. *Integration after partial degradation.* Since degradative enzymes will attack injected DNA, it is clearly possible that a sequence representing only part of the injected DNA may be integrated. Again, the clear necessity is to use restriction enzyme analysis and a DNA sequence whose ends are defined by specific endonuclease sensitive sites. Such nucleases should then yield a fragment of molecular weight precisely equal to that of the injected sequence.

4. *Integration of one complete copy.* The least complex situation is where integration of one complete copy of the injected sequence has occurred. This is the likely result in the experimental situation analysed in Fig. 2. The precise site of integration will not usually be known, only its association at either terminus with high

molecular weight genomic DNA. A recent publication reveals the interesting possibility that injected sequences can persist in a sort of pseudochromosome, in which the novel sequences are associated with some centromeric DNA sequences of the host organism but not integrated into a true chromosome. These authors injected chimaeric plasmids into fertilised mouse eggs and found a complex pattern of restriction enzyme fragments when the transgenic mice were analyzed after birth. This outcome was thought to follow from transposition of genomic DNA under the influence of viral genes included in the insert. Whether such an outcome is indeed permanent and is likely to involve expression remains to be demonstrated.

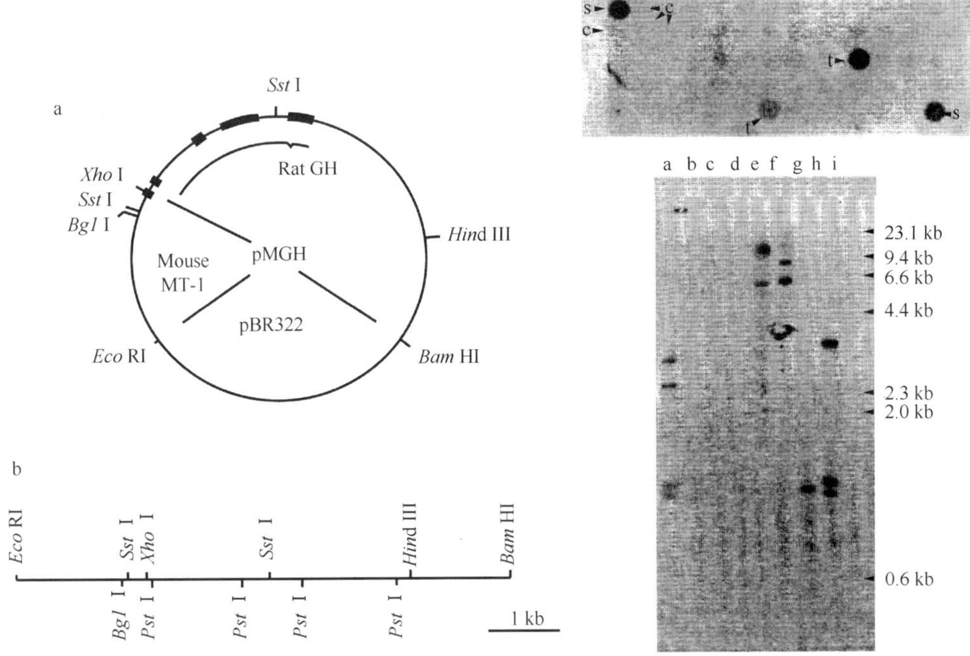

Figure 2 Data demonstrating incorporation of a rat growth hormone/mouse metallothionein gene construct into the genome of the rainbow trout (reference 19, by permission). (a) The cloning vector pMrGH[1] (8.9kb). (b) The 6.6kb *Eco* RI-*Bam* HI MTrGH fragment used for microinjection into trout eggs. (c) Dot blots (10 μg DNA from late yolk-sac alevins or feeding fry) probed with ^{32}P-labelled MTrGH DNA. c=control, s=hybridisation standard, t=putative transgenic. (d) Autoradiograph of Southern transfer of DNA (10 μg per lane) from putative transgenic rainbow trout, probed with ^{32}P-labelled MTrGH DNA. (a) (Control trout DNA + MTrGH) × *Pst* I, (b) TI147 DNA × *Bam* HI + *Eco* RI, (c) TI147 DNA × *Xho* I, (d) TI147 DNA × *Hinf* I, (e) TI157 DNA × *Bam* HI + *Eco* RI, (f) TI157 DNA × Xho I, (g) TI157 DNA × *Hinf* I, (h) TI157 DNA × *Pst* I, (i) Control trout DNA × *Hind* III

5. *Integration of many complete copies*. Since multiple copies are injected, it will not be surprising to find that multiple copies may be integrated; these may be integrated at many different sites, or as tandem repeats at one site. In either case a complex pattern of bands will be obtained with restriction enzyme analysis.

6. *Integration after some rounds of cell division.* Such a situation will lead to the production of transgenic mosaics, either because the novel genes are integrated into different sites in different tissues, or because they are integrated in some tissues and not in others. The most serious outcome of mosaicism is likely to be lack of integration into germ line cells and so lack of germ line transmission to progeny of the novel gene.

7. *Integration without expression.* One of the early fears of would-be genetic engineers was that integrated novel genes would never be expressed. Happily, this is not the case. But lack of expression of integrated novel sequences may occur, either because of inactivation through DNA methylation, or lack of a satisfactory promoter, or integration close to chromosomal heterochromatin.

8. *Adequate expression.* The outcome of integration may be satisfactory expression of the injected sequence in terms of good quantities of the precise protein coded by the novel gene. In some cases this must be tissue specific; in others, such as growth hormone gene, it need not be. The protein may require post-translational modification and such accurate processing is necessary for the recovery of an active product. Evidence for expression may be sought in many ways and at different levels. RNA may be recovered from specific tissues and assayed for the messenger RNA appropriate to the injected sequence by Northern blot hybridisation with a suitable probe. Alter-natively, assays for the specific protein may be used. These will almost always have to be immunologically based, assuming always that discrimination between the product of the novel gene and any native protein is possible. It is possible to rely on phenotypic expression such as weight gain following growth hormone gene injection (Fig. 3), but this is not adequate in itself.

9. *Inappropriate expression.* Some DNA sequences are only likely to be effective if expressed in a tissue specific way, and transgenic experiments afford examples where such genes are expressed appropriately[15] or inappropriately[19]. But other sequences may be more or less ubiquitously expressed and the gene product secreted into plasma.

10. *Genetic damage due to recombinational integration.* It is at least theoretically likely that a percentage of transgenics may carry genetic impairment due to loss of DNA through recombinational exchange or integration into a crucial sequence. The very large genomes of vertebrates with a substantial proportion of DNA being redundant, may explain the apparent unimportance of this problem.

11. *Germ line transmission.* The final goal of most transgenic work is the germ line transmission of the injected gene. Although now established in mouse, *Xenopus* and fruit fly, this goal remains to be confidently realised with fish, but there seems no ground for doubting that such transmission will follow from the present work and one of us (Z.Z.) has recent unpublished evidence for germ line transmission in goldfish. Early fears that integrated sequences might be rapidly lost seem not realised.

Genes in use, or for possible future use

Clearly as more organisms are used to provide gene libraries, an ever increasing number of genes become available. Those that either have been used, or offer promise in fish transgenic work, are: somatotrophin genes (growth hormone), metallothionein genes, somatotrophin releasing factor genes, globin genes, 'antifreeze' genes, 'disease resistance' genes, 'digestive enzyme' genes, (possibly relevant to the use of agricultural biproducts as fish food).

Methods of assaying integration and expression

Evidence for integration must rely on restriction enzyme digestion of DNA from individual post-larval fish, and Southern blot hybridisation with radio labelled copies of the injected sequence. Such hybridization can be delayed until a percentage of the injected fish are negative on dot blot analysis, indicating that non-integrated sequences have been degraded. Dot-blot positive samples can then be analyzed on Southerns and evidence obtained that intact sequences are now integrated into high molecular weight genomic DNA.

Methods for assaying expression of injected sequences are more complex. If the gene has an obvious phenotypic effect, such as that of a growth hormone, then some indication of expression can be sought from studies on growth rate and size of transgenics as compared to controls. One of us (Z.Z.) has obtained such evidence for the loach. (Fig. 3). But we would stress that growth rate in fish is quite variable and rigorous proof of expression must depend on other criteria. Possible approaches are detection of specific messenger RNA on Northern blots, or by *in situ* hybridization and immunoassays for the protein product, assuming that the

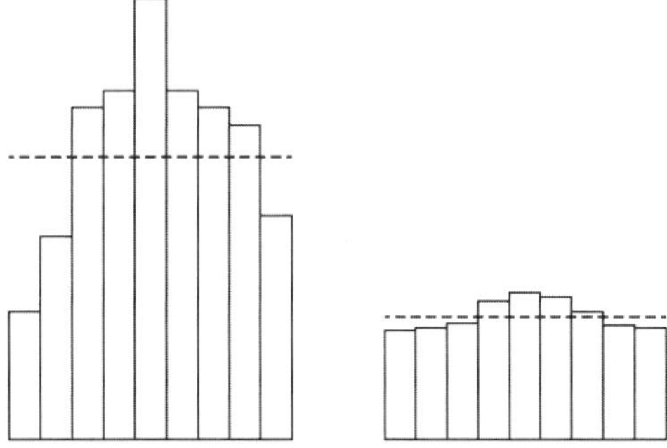

Figure 3 The body-weights of loach injected with human growth hormone (hGH) gene (left) and their control siblings (right) at the age of 135 days old. The hGH gene was microinjected into the decapsulated fertilized eggs before the first cleavage stage. Each egg received 10^6 copies of hGH gene in a linear form, and the eggs were allowed to develop into adults. The cultural conditions for both the injected and none injected groups were exactly the same. Each column represents the body weight of one individual loach: the dashed line shows the average weight of each group of loach

protein is readily distinguished from any homologous molecule-important in experimental design. It is worth noting that if grown fish are assayed by biopsy for transgenic expression. the product sought must be found in blood, either in cells or plasma. More intricate tissue biopsy on large fish may be possible but might risk survival of the transgenic individual.

Success to date

Both of our laboratories have been successful in demonstrating the production of transgenic fish. In the first instance, N.M. and D.P. working with the rainbow trout, detected transgenics as about 5% of fish hatching from injected eggs[20]. Figure 2 shows a Southern blot restriction analysis of DNA from one positive fish demonstrating the integration of an entire injected sequence in high molecular weight genomic DNA. Larger injected fish are now being analyzed for transgenic expression by blood biopsy.

The group of Z.Z. in China has produced transgenics in a number of different fish species, including goldfish, common carp, mirror carp, mud carp, silver crucian, loach, and Wuchang fish. The novel gene, associated with high molecular weight DNA, has been clearly detected by Southern blotting in adult samples such as 2-year-old goldfish and 135-day-loach. In one batch of 7-month goldfish, five of the six fish tested were positive transgenics. The second generation from a transgenic goldfish has recently been shown to include positive individuals.

Other research groups working on transgenic fish projects include those of Chourrout *et al.*[8] who have demonstrated persistence of injected psV507 (human growth hormone cDNA + SV40 promoter and plasmid) in high molecular weight DNA of rainbow trout embryos. These authors have not carried out a restriction analysis so detailed evidence of integration is lacking, but the injected DNA was finally detected in association with a high molecular weight fraction.

Another interesting approach has been taken by Ozata *et al.*[7] in experiments involving the injection of cloned chicken crystallin genes into oocyte nuclei of the small teleost, medaka (*Oryzias latipes*). They introduced a genomic crystallin gene inserted and circularized in a cloning plasmid into oocyte nuclei, and screened 7 day old embryos both for the gene and its crystallin products. Although the Southern blot analyses used do not readily distinguish integrated from non-integrated copies, the evidence for expression of the multiple injected copies is itself most interesting. The expression observed was not tissue specific but included the lens as a site of synthesis.

Outlook

Initially, it may be possible to improve growth rate in commercially important fish species. In addition, other genes likely to improve fish as exploited domesticated species will undoubtedly be used. It also seems that fish may constitute good expression vectors, and so may be used by those interested in the mechanisms of eukaryotic gene regulation. We should stress that work on transgenic fish should be conducted distant from rivers and lakes and with secure containment facilities for the fish, to ensure that genetically modified animals do not find their way into natural ecosystems.

References

1. Palmiter, R. D., Brinster, R. L, Hammer, R. E., Trumbauer, M. E., Rosenfeld, M. G., Brinberg, N. C. m, and Evans, R. M. 1982. Dramatic growth of mice that develop from eggs microinjected with metal-lothionein growth hormone fusion genes. Nature 300: 611-615.
2. Brinster, R. L., Chen, H. Y., Trumbauer, M. E., Yagle, M. K. and Palmiter, R. D. 1985. Factors affecting the efficiency of introducing foreign DNA into mice by microinjecting eggs. Proc. Nat. Acad. Sci. USA. 82: 4438-4442.
3. Etkin, L. D. and Balcells, S. 1985. Transformed *Xenopus* embryos as a transient expression system to analyze gene expression at the midblastula transition. Developmental Biology 108: 173-178.
4. Spradling, A. C. and Rubin, G. M. 1983. The effect of chromosomal position on the expression of the *Drosophila* xanthine dehydrogenase gene. Cell 34: 47-57.
5. Ballard, W. W. 1973. Normal embryonic stages for salmonid fishes, based on *Salmo gairdneri* (Richardson) and *Salvelinus fontanalis* (Mitchell), J. Exp. Zool. 184: 7-26.
6. Maclean, N. and Talwar, S. 1984. Injection of cloned genes into rainbow trout eggs. J. Emb. Exp. Morph: 82: 187.
7. Ozato, K., Kondoh, H., Inohara, H., Iwamatsu, T., Wakamatsu, Y., and Okada, T. S. 1986. Production of transgenic fish: introduction and expression of chicken crystalline gene in medaka embryos. Cell Differ.
8. Krogh, A. and Ussing, H. H. 1937. A note on the permeability of trout eggs to D_2O and H_2O. J. Exp. Biol. 14: 35-37.
9. Zhu, Z., Li, G., He, L., and Chen, S. 1985. Novel gene transfer into the fertilized eggs of gold fish (*Carassiuss auratus*). A. angew. Ichthyol 1: 31-34.
10. Zhu, Z., Xu, K., Li, G., Xie, Y., and He, L. 1986. Biological effects of human growth hormone gene microinjected into the fertilized eggs of loach. *Misgurnus anguillicaudatus*. Kexue Tongbao Academia Sinica. 31: 988-990.
11. Chourrout, D., Guyomard, R., and Houdebine, L. 1986. High efficiency gene transfer in rainbow trout (*Salmo gairdneri rich*) by microinjection into egg cytoplasm. Aquaculture 51: 143-150.
12. Khoury, G. and Gruss, P. (1983). Enhancer elements. Cell 33: 313-314.
13. Lupker, J. H., Roskam, W. G., Miloux, B., Liauzun, P., Yaniv, M., and Jouannau, J. 1983. Abundant excretion of human growth hormone by recombinant-plasmid-transformed monkey kidney cells. Gene 24: 281-287.
14. Gill, J. A., Sumpter, J. P., Donaldson, E. M., Dye, H. M., Souza, L., 1985 Berg, T., Wypych, J. and Langley, K. 1985. Recombinant chicken and bovine growth hormones accelerate growth in aquacultured juvenile Pacific Salmon *Oncarhynchus kisutch*. Bio/Technology 3: 643-646.
15. Scholnick, S. B., Morgan, B. A., and Hirsch, J. 1983. The cloned dopa decarboxylase gene is developmentally regulated when reintegrated into the *Drosophila* genome. Cell 34: 37-45.
16. States, J. C., Conor, W., Wosnick, M. A., Aiken, J. M., Gedamu, L., and Dixon, G. H. 1982. Nucleotide sequence of a protamine component C11 gene of *Salmo gairdneri*. Nucleic Acids Res. 10: 4551-4553.
17. Sekine, S., Mizukami, T., Nishi, T., Kuana, Y., Saito, A., Sato, M., Itoh, S., and Kawauhi, H. 1985. Cloning and expression of cDNA for salmon growth hormone in *Escherichia coli*. Proc. Nat. Acad. Sci. USA 82: 4306-4310.
18. Rassoulzadegan, M., Leopold, P., Vailly, J., and Cuzin, F. 1986. Germ line transmission of automous genetic elements in transgenic mouse strains. Cell 46: 513-9.
19. Lacy, E., Roberts, S., Evans, E. P., Burtenshaw, M. D., and Constantini, F. D. 1983. A foreign β-globin gene in transgenic mice: integration at abnormal chromosomal positions and expression in inappropriate tissues. Cell 34: 343-358.
20. Maclean, N., Penman, D., and Talwar, S. 1986. Introduction of novel genes into the rainbow trout. EIFAC/FAO Symposium on Selection, Hybridization and Genetic Engineering in Fish, K. Tiews (ed).

鱼类的外源基因导入

N. Maclean　D. Penman　朱作言*

Department of Biology, Southampton University, Hampshire, England, U. K.
*中国科学院水生生物研究所，武汉，湖北，中国

摘　要　本文报道了在多种鱼类导入外源基因的实验结果，评估了此项研究的应用前景，深入探讨了外源基因整合的各种类型及其可能的生物学效应。文中详细介绍了我们两个实验室获得的重要实验结果，对同行实验室取得的类似研究进展进行了比较分析。

转基因鱼模型的建立

朱作言　许克圣　谢岳峰　李国华　何　玲

中国科学院水生生物研究所，武昌

摘　要　本研究运用显微注射方法，把带有小鼠重金属螯合蛋白(MT-1)基因启动顺序与人生长激素基因顺序的重组 DNA 片段，注入鲤鱼、鲫鱼和泥鳅等的受精卵内，在由此发育的受体中，50%以上整合了外源基因，成为转基因鱼。其中约有一半左右的转基因个体具有表达外源基因、合成人生长激素的能力，并显示出不同程度的快速生长效应。转基因鱼已通过有性生殖将外源基因传递给子代，后者亦具有表达人生长激素的能力。至此，本研究首次建立了一个高效、简洁而且完整的转基因鱼模型，它为鱼类基因工程定向育种新技术奠定了实验基础。同时，本文还就鱼类基因转移和育种的一系列理论问题进行了深入讨论。

关键词　鱼；生长激素；基因转移；整合表达；性腺传递

通过显微注射等手段，把克隆的基因片段导入哺乳类的受精卵原核，使之与受体基因组整合，培育出了携带外源遗传信息的“转基因动物”，如转基因小鼠[1]、兔、猪、羊等。这类研究除了理论上的意义之外，在实践上有可能开辟家养动物定向育种的新途径，也有可能建立微生物、真核细胞之外的第三大克隆基因表达系统，即以大动物为“加工厂”，大量合成和分泌贵重而且安全的基因工程产品，如人胰岛素、生长激素等。正因为如此，转基因动物研究已引起了从事遗传育种、生物工程的研究工作者和生物技术企业家们的浓厚兴趣。

鱼类属低等脊椎动物，是研制转基因动物的极好材料。这不仅因为我们曾阐述过的鱼类怀卵量大、受精卵体外发育、易于操作等优点[2.3]，更在于鱼类育种的单性发育技术日趋成熟，因而为培育遗传性状稳定的转基因动物新品系提供了哺乳类、鸟类等实验动物所缺乏的前提条件。为此，我们对鱼类基因转移进行了系统研究，以常见的几种淡水鱼为材料，建立了一个完整的转基因鱼模型。

一、受精卵的基因转移

本模型研究使用了鲤鱼(*Cyprinus carpio haematopterus* Tem. et Schl.)包括镜鲤和红鲤、鲫鱼(*Carassius auratus auratus* Linnaeus)、银鲫(*Carassius auratus gibelio* Bloch)和泥鳅(*Misgurnus anguillicaudatus* Cantor)。它们多在春末夏初繁殖。经人工催青产卵、授精，0.25%的胰蛋白酶(Difco)消化去除卵壳后，将裸卵移至 Holtfreter 氏培养液内，在第一次

文章发表于《中国科学》(B 辑), 1989, 19(2): 147-155

卵裂前施行导入外源基因的显微注射手术(图 1)。每卵接受 1-2 nl 外源基因溶液，约含 10^6 个冠以小鼠重金属螯合蛋白(MT-1)基因启动顺序的人生长激素 (hGH)基因拷贝。除了使用 9.4 kb 碱基对的 BPVMG[4]和 7.88 kb 的 pMhGH[5]两个含 hGH 基因的重组 DNA 顺序之外，还使用了 5.5 kb 的 MhGI 顺序 (人生长激素与人胰岛素基因重组质粒，朱作言、谢岳峰，未发表资料)作为转移用的外源基因，即目的基因(图 2)。

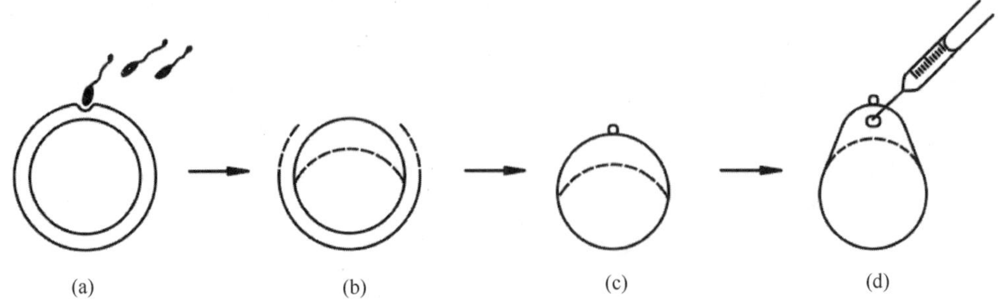

图 1 鱼受精卵显微注射模式图

((a)鱼卵人工授精，(b)用胰蛋白酶去除卵壳，(c)裸卵移至 Holtfreter 氏溶液中，(d) 显微注射法将外源基因引入鱼卵胚盘中)

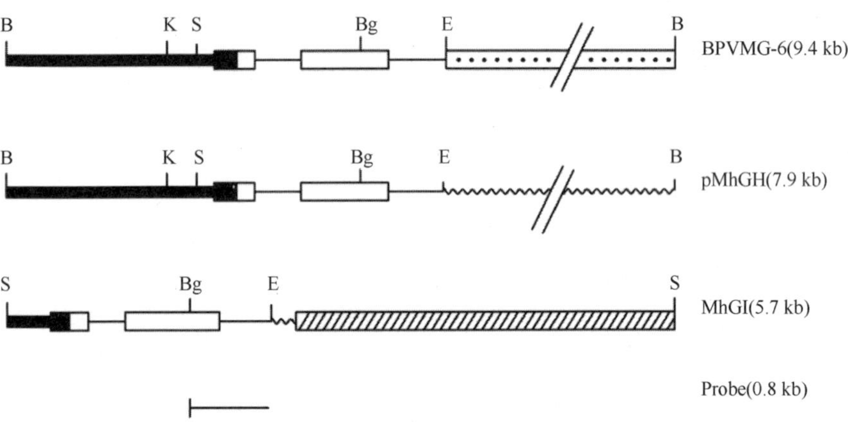

图 2 转移所使用的外源基因

(BPVMG-6(9.4kb)为带小鼠重金属螯合蛋白(MT-1)基因启动顺序的人生长激素(hGH)基因，并与牛乳头状瘤病毒(BPV)的转化顺序融合。pMhGH(7.9kb)为克隆在质粒 pBR322 EcoRI 位点上的 MT-hGH 融合基因顺序。MhGI(5.7kb)为上述 MT-hGH 与人前胰岛素原基因(I)融合基因顺序。取 hGH 基因 3′端 0.8kb 顺序为杂交用探针。——小鼠 MT-1 基因 5′端侧翼顺序，—— 小鼠 MT-1 基因启动顺序，——人 GH 基因外显子，—— 人 GH 基因内含子及 3′端侧翼顺序，——BPV 转化顺序，—— pBR322，——人前胰岛素原基因；B——BamHI；Bg——Bg*l*I；E——EcoRI；K——KpnI；S-SacI)

显微注射后，受体卵在 Holtfreter 氏液中培养发育。胚胎发育至原肠期后，将培养液逐渐用暴气的冷开水稀释；发育至心跳期后，将胚胎转移到暴气的冷开水中直至形成鱼苗。鱼苗摄食 5 天后，一部分移入水族箱饲养观察，另一部分则养殖在小池塘备用。来自同一对亲本的对照组，除了不接受外源基因注射之外，其他处理程序与实验组相同。

二、外源基因扩增和整合作用

1. 原肠胚细胞的原位杂交

受精卵发育为中期原肠胚时，用 0.7%的 NaCl 盐液清洗 5 次，用玻璃丝割下胚盘，移入无 Ca^{2+}的 Holtfreter 氏液中使细胞分散开来。用微吸管收集细胞并转移至硝酸纤维素滤膜上。经 0.5 mol/L 的 NaOH 浸润变性两次，每次 5 min，再用 1.5 mol/L 的 NaCl(pH7.4)中性处理两次，每次 5 min，接着用 2 × SSC 浸润固定一次，晾干，80℃烘烤 2 h，然后与 ^{32}P 缺口转译标记的 hGH 基因片段探针杂交[6](^{32}P-α-dCTP 购自英国 Amersham 公司，DNA 多聚酶 1 和本实验所用的其他限制性核酸内切酶大部分购自美国 New England Biolabs 公司，部分购自北京华美生物工程公司和北京友谊医学科技公司)。实验组 24 个原肠胚均呈杂交阳性，注入的 hGH 基因拷贝数约扩增了 10^3 倍；对照组 6 个原肠胚均呈杂交阴性(图 3)。原肠胚细胞原位分子杂交结果表明，通过显微注射可以有效地把外源基因导入鱼类受精卵内，而后者又为前者的复制和扩增提供了必需的条件。

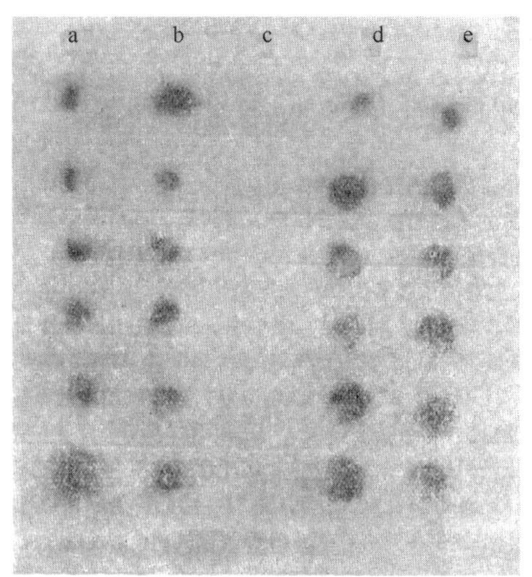

图 3 原肠胚细胞原位杂交放射自显影图谱

(a、b、d、e 为行为注射外源基因组样品，c 为行为对照组样品)

2. 外源基因在受体胚胎发育中的行为

我们曾初步报道过受体胚胎发育过程中，外源基因复制、降解、聚合和整合的复杂行为[2]。在受体卵发育的各个不同阶段，随机取样，制备总 DNA。经与 hGH 基因片段探针进行斑点杂交或 Southern 氏吸印转移杂交[7]，表明注入的线性 DNA 顺序以多种物理构型(线型、环状或寡聚体等形式)进行复制和扩增，至胚孔封闭期达到峰值。尾芽期之后，杂交检测出的外源基因拷贝数递减，但分子量却越来越大。血液循环期之后，胚体内的外源基因随总 DNA 大分子一同迁移，此时已见不到或很少见到小分子形式外源基因的踪迹(图 4)。

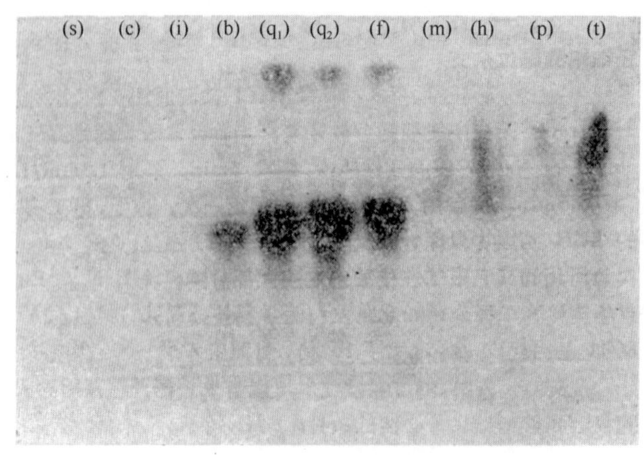

图 4　pMhGH 重组质粒导入受精卵后的变化

(在胚胎发育的各个不同阶段随机取 3 个样品，制备总 DNA，进行 Southern 氏吸印转移杂交。(s) 注射的 pMhGH/Bam HI DNA 样品(注射 3 个卵的量)，(c) 对照组原肠胚期样品，(b) 囊胚期，(q_1) 原肠早期，(q_2) 原肠末期，(f) 尾芽期，(m) 肌肉效应期，(h) 心跳期，(p)色素出现期，(t)鱼苗摄食期)

3. 外源基因与受体基因组的整合作用

从受体幼鱼至成鱼的不同时期，制备肌肉、红细胞、肝脏和精巢等组织的总 DNA，与 hGH 基因片段探针进行斑点杂交，约 50%的个体呈杂交阳性(表 1)。此结果与导入卵母细胞核的相似[8]，高于其他有关鱼类基因转移的报道结果[9]。Southern 氏吸印转移杂交结果表明，杂交带与总 DNA 一同迁移。将注射 MhGI/SacI 重组 DNA 片段的受体红鲫鱼肌肉总 DNA 样品，分别用 SacI 和 KpnI 消化(SacI 识别注入的 MhGI 重组 DNA 片段的两个黏性末端顺序，而 KpnI 在 MhGI 顺序内无识别位点)，经琼脂糖凝胶电泳和 Southern 氏吸印转移杂交后，在 SacI 消化产物泳道上显示一条与注射样品分子量相同的杂交带，而在 KnpI 消化产物泳道上则显示出一片连续的杂交带，其迁移率介乎于未经酶切样品(杂交带与总 DNA 泳道位置相同)与 SacI 消化样品之间(图 5)。由此可见，随总 DNA 同时迁移的外源基因顺序已与受体基因组发生了整合作用。外源基因的整合作用发生在胚胎发生的晚期，整合后的外源基因随着胚胎发育而分布到身体的各个组织和器官之中。若外源基因整合到性腺细胞染色体组内，则可通过有性过程传递给子代。我们称整合了外源基因的个体为转基因鱼。

表 1　受体鱼外源基因检测率

	检测数(尾)	阳性数(尾)	阳性率(%)
鲫鱼	64	33	51.6
镜鲤	89	46	51.7
泥鳅	10	4	40
红鲫	66	28	42.4
银鲫	23	16	69.6
红鲤	8	7	87.5

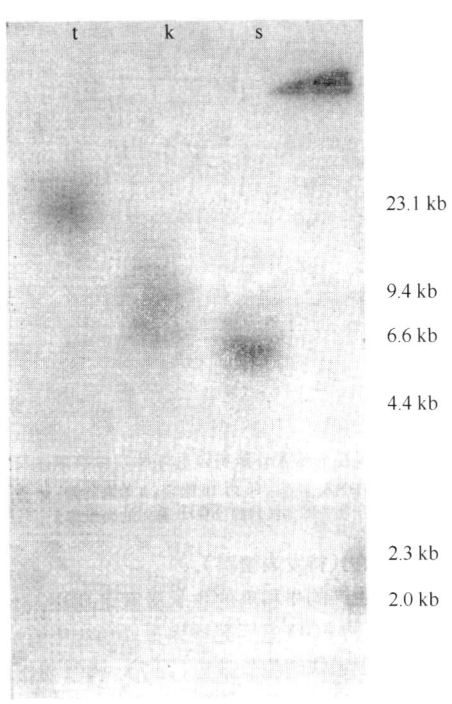

图 5 转 MhGI DNA 片段的红鲫鱼 DNA Southern 氏吸印转移杂交图谱

(t 为总 DNA，未经酶消化；k 为总 DNA，经 Kpn I 消化；s 为总 DNA，经 Sac I 消化)

三、外源基因的表达作用

从不同发育阶段提取总核酸样品，经琼脂糖凝胶电泳，显示出胚胎诸发育阶段均含有丰富的 28S，18S，5SRNA 和信使 RNA。但是，分子杂交表明，只有当胚胎发育至原肠末期以后，即从肌肉效应期开始，才能检测出 hGH 基因的转录本(图 6)。如上所述，在胚胎发育的这一阶段，外源 hGH 基因大多尚未与受体基因组整合，其丰富的转录本似为胚胎细胞核外“游离”外源基因转录的结果。此外，随着原肠化开始器官分化，外源基因起始表达与胚胎器官分化同时出现是巧合还是有内在必然的联系，是一个十分令人寻味的问题。当然，摄食阶段的鱼苗样品中已不存在“游离”外源基因，此时的 hGH 基因转录本则是被整合了的 hGH 基因的转录产物。

对注射 hGH 基因的受精卵发育而成的鲤鱼进行放射免疫测定，hGH 放射免疫药盒由北京市神经外科研究所提供。在 20 尾 35 日龄的幼鱼肌肉组织匀浆上清液样品和 8 尾 913 日龄成鱼的血清样品中，分别检测出了 10 尾和 3 尾含人生长激素，阳性检出率接近 50%。同时，幼鱼体内人生长激素含量明显高于成鱼(表 2)。实验结果同样表明，转基因鲤鱼和鲫鱼的子一代亦具有合成人生长激素的能力。

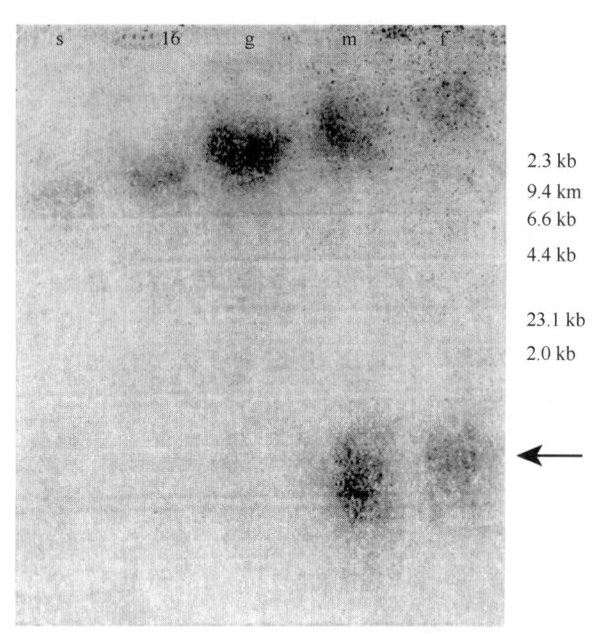

图 6 鱼受精卵导入外源基因后胚胎期总核酸杂交图

(s 为注射的 pMhGH/BamH I DNA 样品，16 为 16 胞期，g 为原肠期，m 为肌肉效应期，f 为摄食期。箭头所指处即为转移的 hGH 基因的转录本)

有实验证明，基因工程技术生产的牛和鸡的生长激素定期注入幼鲑体内之后，有加速受体鱼生长的作用[10]。我们曾报道了转 hGH 基因泥鳅体重剧增的实验结果[4]。在本模型研究中，转 hGH 基因银鲫。镜鲤同样表现出快速生长效应(图 7)。转基因银鲫的快速生长效应比转基因鲤鱼的明显；同一种转基因鱼在幼鱼阶段的快速生长效应又比其成鱼阶段明显。转 hGH 基因鲫鱼不仅体长增加，背部肌肉增厚，而且体高与体长之比由对照组的 38.5%增加为 53.3%，很容易从形态上加以鉴别(图 8)。

表 2　转基因鲤鱼体内人生长激素(hGH)放免测定结果

检测鱼龄(天)	检测鱼数(尾)	表达鱼数(尾)	表达量范围
35	20	10	26.1-50 ng/0.2 g 肌肉组织
913	8	3	1.1-4.0 ng/ml 血清

四、外源基因向子一代的传递

转 hGH 基因的鲫鱼和鲤鱼性腺成熟后，经过自然产卵或人工催青繁殖了子一代。由转 hGH 基因鲫鱼(6♀×1♂)自然产卵繁殖的子一代群体中，分别检测了 30 尾幼鱼的总 DNA 样品，分子杂交证明均含有 hGH 基因，但各样品的外源基因拷贝数不一。由于这个子代群体来源于 6 尾转基因雌鱼和 1 尾转基因雄鱼，无法对这种拷贝数差异进行分析。后来用 1 尾转 hGH 基因雌性鲤鱼与 1 尾普通的雄性鲤鱼进行人工催青繁殖，并对 18 尾子代幼鱼 DNA 做分子杂交检查，结果只有 9 尾(50%)为携带 hGH 基因阳性个

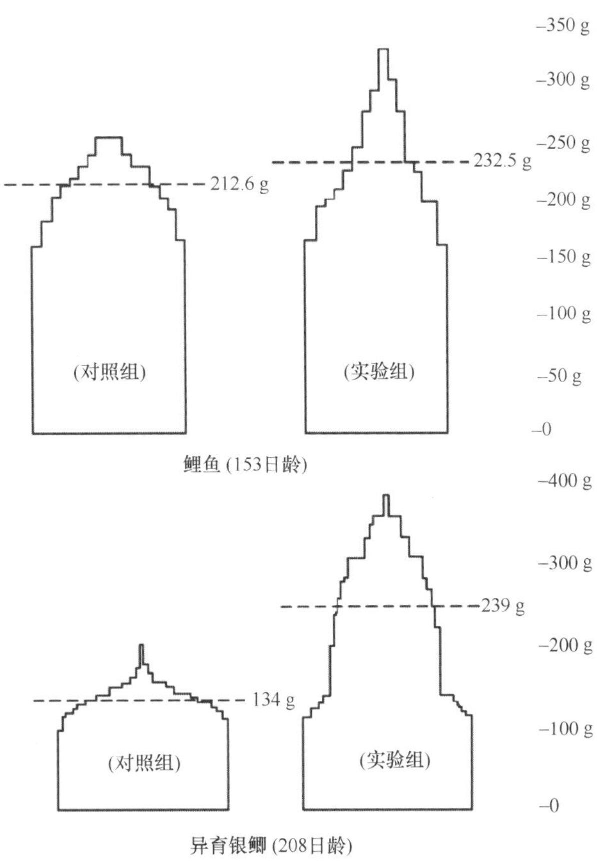

图 7 转人生长激素基因(hGH)银鲫、镜鲤的快速生长效应

(153 日龄镜鲤实验组(16 条)平均体重为 232.5 g，对照组(17 条)平均体重为 212.6 g，增长幅度为 9.4%；208 日龄异育银鲫实验组(41 条)平均体重为 239 g，对照组(41 条)平均体重为 134 g，增长幅度为 78%)

图 8 转人生长激素(hGH)基因鲫鱼外部形态的变化

(上为实验组鲫鱼，下为对照组鲫鱼)

体。令人惊奇的是，有 1 尾子代幼鲤携带的 hGH 基因拷贝数不仅大大高于同窝“兄妹”，而且也明显超出亲本的检测值。在 7 月龄的密养的这一子代群体中，有 3 尾巨型个体，最大的体重为同窝“兄妹”平均体重的 7.7 倍。

五、总结与讨论

带有小鼠 MT-1 基因启动顺序的 hGH 基因重组 DNA 片段，通过显微注射导入几种淡水鱼受精卵细胞质之后，随受体胚胎发育以多种物理构型进行复制扩增，至胚孔封闭期达到高峰。此后，扩增的大部分外源基因逐渐降解，只有小部分在胚胎发育晚期与受体基因组整合，由此形成转基因鱼个体。转基因鱼个体占存活的受体总数的 50%以上。受体胚胎原肠化和器官分化开始以后，无论是“游离”状态 (肌肉效应期)，还是整合状态(摄食期) 的 hGH 基因，均表达其 mRNA 转录本；在转基因鱼苗和成鱼体内合成了人生长激素，它对泥鳅、银鲫和鲤鱼有促进生长的作用。更重要的是，外源基因可以通过性腺系传递给子代，由此建立了鱼类基因转移的模型。

哺乳动物基因转移研究发现，几乎只有当外源基因导入受精卵原核内才能与受体基因组整合，形成转基因哺乳动物，而导入卵细胞质的外源基因转移是无效的[11]。由于不可能同时获得大量的哺乳动物受精卵和大量同步发情，可供“借腹怀胎”的假孕母体，加上把基因导入受精卵原核和后续的体外培养，在假孕母体着床等程序复杂，因而不易对外源基因在哺乳动物胚胎发育诸阶段的行为进行跟踪，以致无法判断注入卵细胞质的外源基因何时被降解，也无法了解注入卵原核内的外源基因何时以什么形式与受体基因组整合。鱼类基因转移的情况则不相同，人们可以同时获得大量的受精卵，在第一次卵裂前大量地进行细胞质内的显微注射，受体卵的管理方便易行。本模型研究结果指出，外源基因在原肠末期之前的受体卵内大量复制扩增，如果存在着内源核酸酶的降解作用，被降解的部分与扩增的拷贝数相比是微不足道的。原肠末期以后，可能是内源核酸酶大量表达，也可能是因为器官分化而失去了“游离”基因自主复制的条件，大量积累的外源 DNA 片段总量逐渐减少，但最终仍有少量与受体基因组整合。和哺乳动物胚胎发育相比，鱼类胚胎发育速度要快几倍、十几倍乃至几十倍。鲫鱼受精卵第一次卵裂需 45-50 min (20℃)，后来的细胞分裂时程更加短暂。鱼类胚胎细胞核在胚胎发育过程中如此频繁地“解体”与重建，外源 DNA 片段在此期间持续复制和扩增，为后者掺入受体基因组提供了良好的机会。这也许是对注入鱼类卵细胞质的外源基因不但可以被整合，而且整合率高于哺乳动物基因转移的一种合乎逻辑的解释。

诚然，我们尚不清楚外源基因与受体基因组整合作用产生的确切机制，但本模型研究证实了这种整合是发生在胚胎主要器官分化以后的一个渐进过程。可以推测，由此形成的转基因鱼是携带外源基因的嵌合体，即不同组织或器官之间，同一器官的不同细胞之间所整合的外源基因在数量和质量(整合位点)是不均一的。可以设想只有那些不挠乱受体“管家基因”结构与功能的整合的受体，才有可能建成胚体；进而，只有那些不挠乱在发育阶段表达的“旺势基因”结构与功能的整合的胚体，才有可能持续发育为转基因鱼苗；在转基因鱼苗中，只有那些适宜位点的整合方可表达。我们称满足上述三种条件的整合为“有效整合”，无表达作用的整合为“沉默整合”，挠乱受体“管家基因”和主要“旺势基因”的整合为“毒性整合”。沉默整合的个体不具外源基因赋予的表现型，而毒性整合的个体则生长发育受阻、畸形甚至夭折。在我们的实验中，转基因鱼(包括有效整合和沉默整合)占受体存活总数的 50%以上，其中具表达作用的有效整合约占转基因个体总数的 50%左右，而且高拷贝整合的不一定高表达，反之，高表达的个体也不一定携带大量的外源基因拷贝。

从已经繁殖的转基因鲫鱼和鲤鱼子一代，证实了外源基因可以通过性腺系细胞传递给后代。但是，由于转基因鱼亲本是携带外源基因的嵌合体，外源基因在各个性腺细胞基因组内的整合的数量与位点不尽相同，由此导致子代群体在携带外源基因(包括数量和整合位点)上出现分离。从定向育种的角度考虑，从转基因鱼子代中挑选有效整合的雌性个体是至关重要的一步。因为从理论上说，通过雌核发育及生理转性技术制种，可以大大限制转基因鱼后代的分离程度，缩短建立纯系的世代周期，将外源基因有效整合和最适表达的个体稳定成为经济性状优良的养殖新品系。所以，转基因鱼模型的建立，为鱼类定向育种高技术投射了一道希望的曙光。

本模型研究还初步证明，hGH 基因一旦表达，可以在转基因泥鳅、银卿和鲤鱼体内行使促进生长的生物学功能。这三种鱼在自然条件下的最终大小不同，银鲫大于泥鳅，鲤鱼大于银鲫；而转 hGH 基因个体的快速生长效应却相反，转基因泥鳅的快速生长效应大于银鲫，银鲫的又大于鲤鱼。联系到转基因哺乳动物亦有类似情况，已培育出快速生长的巨型转基因小鼠，而未见体大如牛的转基因猪问世，我们推测，受体动物体内表达的外源生长激素的促生长效应，似与受体物种固有的最终大小呈反变关系。当然，这一推断是否成立，尚有待更多的实验加以证实。最后必须指出，本模型研究选择 hGH 基因为目的外源基因，只是考虑到取材和检测的方便，事先并不知道也未期待过人生长激素在鱼体内的促生长作用。为了培育快速生长的养殖鱼类新品系，hGH 基因并不是理想的目的基因。因为我们不明确人生长激素对鱼体代谢究竟会产生什么影响，它到底能在多大程度上干预鱼类的个体发育；转 hGH 基因的鱼类一旦被释放到自然水域时，它会在多大程度上影响水体生态平衡？更重要的是人们的心理承受能力问题，经常食用人生长激素含量高的鱼是否安全？即使安全的话又能否被人们所接受？所以，本模型研究还有另外一部分内容，即克隆鱼类的生长因子基因和高效表达启动顺序等元件，有关内容将另作报道。我们相信，转移鱼类自身的生长因子基因，并求适量表达，对于培育快速生长的养殖鱼类新品系将有实际意义。

参考文献

[1] Palmiter, R. *et al.*, *Nature*, 300(1982), 611-615.

[2] Zhu, Z. *et al.*, *J. Applied Ichthyology*, 1(1985), 31-34.

[3] Maclean, N., Penman. D., Zhu, Z., *Bio/Technology*, 5(1987), 257-261.

[4] Pavlakis, G. N. & Hamer, D. H., *Proc. Natl. Acad. Sci. U.S.A.*, 80(1983), 397-401.

[5] Zhu, Z. *et al.*, *Kexue Tongbao*, 31(1986), 988-990.

[6] Maniatis, T. *et al.*, *Molecular Cloning, a Laboratory Manual*, Cold Spring Harbor Laboratory, 1982, 387-389.

[7] Southern, E, M., *J. Mol. Biol.*, 98(1975), 503-517.
[8] Ozato, K., *et al.*, *Cell Differ.*, 19(1986), 237-244.
[9] Dunham, R. A. *et al.*, *Transactions of the American Fisheries Society*, 116(1987), 87-91.
[10] Gill, J. A. *et al.*, *Bio/Technology*, 3(1985), 643-646.
[11] Hammer, R. E. *et al.*, *J. Anim. Sci.*, 63(1986), 269-278.

The Establishment of a Transgenic Fish Model

Zhu Zuoyan Xu Kesheng Xie Yuefeng Li Guohua He Ling

Institute of Hydrobiology, Academia Sinica, Wuchang

Abstract In this study, the recombinant DNA fragment with the promoter sequence of MT-1 gene and human growth hormone gene was injected into the fertilized eggs of common carp, crucian carp and loach. More than 50% of the developed receptors integrate foreign genes and become transgenic fish. About half of the transgenic individuals have the ability to express foreign genes and synthesize human growth hormone, and show different degrees of rapid growth effect. Transgenic fish have transmitted exogenous genes to offspring through sexual reproduction, and the latter also have the ability to express human growth hormone. At this point, a highly efficient, simple and complete model of transgenic fish has been established for the first time. It lays an experimental foundation for the new technology of directional breeding of fish gene engineering. At the same time, a series of theoretical issues on fish gene transfer and breeding were discussed in depth.

第二部分
外源基因的整合与表达

The β-actin gene of carp (*Ctenopharyngodon idella*)

Z. Liu Z. Zhu K. Roberg A. J. Faras K. S. Guise
A. R. Kapuscinski P. B. Hackett

Department of Genetics and Cell Biology and Institute of Human Genetics, University of Minnesota, St Paul. MN 55108-1095, USA

The nucleotide sequence is very similar to that of the β-actin gene of *Cyprinus carpio* (Liu *et al*., GENE submitted). The proximal promoter elements (nt 121-188) are capitalized as are the exons (beginning at 211, ending at the protein termination site at 3323). There are two poly (A) sites at 3860 and 3950. The AUG initiation site is at 1606 (exon 2). The a.a. sequence of the two carp genes is identical; there are 20 silent nt changes in the coding region. GenBank number M25013.

草鱼 β-肌动蛋白基因克隆

刘占江 朱作言 K. Roberg A. J. Faras K. S. Guise
A. R. Kapuscinski P. B. Hackett

Department of Genetics and Cell Biology and Institute of Human Genetics, University of Minnesota, St Paul. MN 55108-1095, USA

摘 要 草鱼 β-肌动蛋白基因序列与鲤的相似性较高。本研究克隆了长度为 4030 bp 的草鱼 β-actin 基因序列，最近的启动子元件(位于 121-188 nt)和外显子(始于 211 nt，终止于 3323 nt)均用大写字母表示。在第 3860 和 3950 nt 处各有一个 poly (A)信号，1606 nt 处有一个 AUG 起始位点。编码的氨基酸序列与来源于鲤鱼的序列一致性较高，但在编码区与鲤序列相比存在 20 个同义突变。GenBank 序列号为 M25013。

文章发表于 *Nucleic Acids Research*, 2015, 17(14): 5850

tttgatcaaaatcgcctaggccttg ttcttcagctagtctagcttcccct tctttcactctcgagttgcaagaaa gcaagtgtagcaatgtgcacgcgac 100
agccgggtgtgtgacgctggaCCAA Tcagagggcagagctccgaaagttt aCCTTTTATGGctagagccggcata tgccgtcaTATAAaagagctcgccc 200
agcttttcaaCCTCACTTTGAGCTC CTCCACACGCAGCTAGTGCGGAATA TCATCAGCTTGTAACCCATTCTCTT AAGTCGACAAACCCCCCCAAACCTA 300
AGgtgagttgatttttcagctttta ttgcatttttagttcattaatatta atttaaacctgtaattatgatgtat aattaaaactggataaagaaattag 400
actaagttaccggtcttttcgcttt taagttttaactcctgcttcaaaac ggtagttatttattatgtagttatt tttatgtaaatgtttgctgtattat 500
cataactttatgactgtactggaca tgtcaggtggaaacgacggtatccg ttgtaggcacgacattaaatgggcc ggtgtgaaataagtgttcgcctttc 600
ttaacatttaagatgtgctctgatt aaccgtgctttaacagctataaact tgacttgacagttttaagggtattt atttgtgaggcatgttacacacttg 700
atggatagccggcatgggaagttct ttgagcaggcaagtgctgcagaagg gtgtgacctagtttagctagccggc taaccagcgttcaccttctgtaaac 800
ctggtgaatctaattcacttatgcg atcactaaattatcaaagattggtt attagttcactttcaacacacccaa catttttagtgaatactggcgcttt 900
ataatgtagtatgcagcatgcatcc cttcagtctttccttaaaactgaag atagtcaagtatttcattcacgtaa gcgtttcgaaatataaggtgtggtc 1000
aattttaataactggggtaaagtgt tccggttgcatttaagcttttataa ttgtgtcgaacagtaatttttctta aaactaattgataaagactagaata 1100
taaactgaatgggctgatggtgatc tctcaagtgcttcggcgtttagtct gctcttatggagtcatttgaagtga ctgcagatctgtgacgcagaatgtt 1200
gggcagacacccgtcgaaattcggt tgtgtaattgataccaggcgaggat gaagaggatgtaaaacttcattcgt gtagaatttagggagtggccctggc 1300
gtgatgaatgttcgaaatctgttcc tttttactgaaccatacgacactgg ctgagtgccacaccgccggcagccg caaacgcgtcaatccattgCCTTTT 1400
ATGGtaataatgagagaatgcagag ggacttcctttgtctggcatatctg aggcgcgcattgtcactctagcacc cactagcggtcagactgcagaacac 1500
atgaaacaggaagttgactccacat ggtcacatgctcactgacactttct tacatggcagcagtgcacttcaaaa acactttcctctctcttttacagtt 1600
cagCCATGGATGATGAAATTGCCGC ACTGGTTGTTGACAACGGATCCGGT ATGTGCAAAGCCGGATTCGCTGGAG ATGATGCTCCCCGTGCTGTCTTCCC 1700
ATCCATCGTGGGTCGTCCCAGACAT CAGgtgagaaacggaggatagtttc gggctgaccaataataagaattttc atgctattttctcattcctaaacat 1800
tttacaaaaattaacatgctttcct tcattacagGGCGTCATGGTCGGTA TGGGACAGAAGGACAGCTATGTTGG TGACGAGGCTCAGAGCAAGAGAGGT 1900
ATCCTGACCCTGAAGTACCCCATCG AGCACGGTATTGTCACCAACTGGGA CGATATGGAGAAGATCTGGCATCAC ACCTTCTACAACGAGCTGCGTGTTG 2000
CCCCAGAGGAGCACCCCGTCCTGCT CACAGAGGCCCCCCTGAACCCCAAA GCCAACAGGGAAAAGATGACACAGg ttggtttttggctagtaaatggtgc 2100
tttgaagtctcttgtctgtcctgtt acctcacttaagttctccttttcat tcgttcacttcctccaggctttgtt tcctctgagctcctgagtttctcat 2200
cttttgctggaacgcaggttatcta tacttttgcctgcctgttttgcagt ctcctccgcactctgattctttatg cacttttgtttctttactctagatt 2300
tccaactaacccctgcatgggtgtg gatgtgctataactttttgagcatc cgttaacttgtcactctcattacag ATCATGTTCGAGACCTTCAACACCC 2400
CCGCCATGTACGTTGCCATCCAGGC TGTGCTGTCCCTGTATGCCTCTGGT CGTACCACTGGTATCGTGATGGACT CTGGTGATGGTGTCACCCACACTGT 2500
GCCCATCTACGAGGGTTATGCCCTG CCCCATGCCATCCTCCGTCTGGACT TGGCTGGCCGTGACCTGACTGACTA CCTCATGAAGATCCTCACCGAGAGA 2600
GGCTACAGCTTCACCACCACAGCTG AGAGGGAAATTGTCCGTGACATCAA GGAGAAGCTCTGCTATGTGGCTCTT GACTTCGAGCAGGAGATGGCCACTG 2700
CTGCTTCCTCCTCCTCCCTGGAGAA GAGCTATGAGCTGCCTGACGGACAG GTCATCACCATTGGCAATGAGAGGT TCAGGTGCCCAGAGGCTCTGTTCCA 2800
GCCATCCTTCTTGGgtaggtttcct ggacaaacgttacctggtgtgtatg tactctagaattgaagaactaaggt taaccttctctctcgctctgcagGT 2900
ATGGAGTCTTGCGGTATCCATGAGA CCACCTTCAACTCCATCATGAAGTG TGACGTCGATATCCGTAAAGACCTG TATGCCAACACTGTATTGTCTGGTG 3000
GTACCACAATGTACCCTGGCATTGC TGACAGGATGCAGAAGGAGATCACA TCCCTGGCCCCTAGCACAATGAAAA TCAAGgtgagcatgtgatctgaact 3100
ttgacccttacctctcacatcagtt tgtaacttaatgcatatggcaactc ctgcattgtgctaatcatttgtttc tccacagATCATCGCCCCACCTGAG 3200
CGTAAATACTCTGTCTGGATCGGAG GTTCCATCCTGGCCTCCCTGTCCAC CTTCCAGCAGATGTGGATTAGCAAG CAGGAGTATGACGAGTCTGGACCAT 3300
CCATCGTCCACCGCAAATGCTTCta aaccggactgttaccacttcacgcc gactcaaactgcgcagagaaaaact tcaaacgacaacattggcatggctt 3400
ttgttattttggcgcttgactcagg atctaaaaactggaacggtgaaggt gacggcaatgtttttttggcaaata agcatccccgaagttctacaatgca 3500
tctgaggactcaatggtttttgttt tgtttctttagtcattccaaatgtt tgttaaatgcattgttccgaaactt atttgcctctatgaaggctgcccag 3600
taattgggagcatacttaacattgt agtattgtatgtaaattatgtaaca aaacaatgtctgggtttttgtactt tcagccttaaaatcttgggtttttt 3700
tttttttttaatttttttttctttg ttccaaaaaactaagctttaccatt caagatgtaaaggtatccattctcc ccctggcatattgtaaaagctgtgt 3800
ggaacgtggcggtgcagacatttgg tggggccaacctgtacactgactaa ttcaaatccAATAAAagtgcacatg tgtaagacatcatattcctgtgtga 3900
cttcctgtgttggtgctgaatgaac ttgagtagaaggatattaggtttAA TAAAactagtcttgggttttcttag tgttgtattcacattttattgagta 4000
tgccctttggttccagagcactttg gtgta 4030

Isolation and Characterization of β-actin Gene of Carp (*Cyprinus carpio*)

Zhanjiang Liu[1] Zuoyan Zhu[1,2] Kevin Roberg[1] Anthony Faras[3,6]
Kevin Guise[4,6] Anne R. Kapuscinski[5] Perry B. Hackett[1,6]

1 Department of Genetics and Cell Biology, University of Minnesota, St Paul, MN 55108, USA
2 Institute of Hydrobiology, Academia Sinica, Wuhan, Hubei, People's Republic of China
3 Department of Microbiology, University of Minnesota, Minneapolis, MN 55455, USA
4 Department of Animal Science, University of Minnesota, St Paul, MN 55108, USA
5 Department of Fisheries and Wildlife, University of Minnesota, St Paul, MN, USA
6 Institute of Human Genetics, University of Minnesota, Minneapolis, MN 55455, USA

Abstract A β-actin gene of carp (*Cyprinus carpio*) was isolated from a genomic EMBL3 library. The nucleotide sequence of the gene indicates six exons spanning 3.6 kb. Southern blot hybridization of restriction endonuclease digests of carp genomic DNA indicate that there are two copies of the β-actin isotype and several other species of actin genes. The transcriptional start site is 85 bp and 24 bp downstream respectively from consensus CCAAT and TATA promoter elements. The organization of the carp β-actin gene is identical to that of chicken, human, and rat genes in terms of size, exon/intron locations and junctions and in having a translationally silent first exon. The fish gene is 90% and 99% conserved at the nucleotide and amino acid levels, respectively, with land vertebrate β-actin genes. Northern blot analysis of β-actin gene expression indicated that the gene is highly expressed in brain, less so in muscle, and much less so in liver cells. The putative β-actin proximal promoter of carp, identified by the conservation of known actin regulatory sequences, is transcriptionally active in both mammalian and piscine cells.

Keywords genomic library; promoter; mapping; tissue-specific expression; evolutionary conservation

Actin is one of the most highly conserved proteins in evolution (Pollard and Cooper, 1986) due to its central role as the basic component of cellular microfilaments. At least six isoforms of actin have been identified in vertebrates. These include two more divergent cytoplasmic actins, β and γ, and four muscle-type α-actins (skeletal, cardiac, aorta-type smooth muscle, and stomach-type smooth muscle actins) (Vandekerckhove and Weber, 1979). Muscle-type actins are tissue specific and participate in muscle contractions. Cytoplasmic, β-and γ-actins are expressed in many, if not all, tissue types, participate in a variety of cellular functions (Clarke and Spudich, 1977), and are more closely related to the actins found in lower eukaryotes than are the vertebrate muscle actins.

文章发表于 *DNA Sequence the Journal of DNA Sequencing & Mapping*, 1990, 1(2):125-136

Owing to its role in establishing cell structure, actin proteins and their mRNAs are relatively abundant. This abundance suggests a strong transcriptional promoter (Gunning *et al.*, 1987; Liu *et al.*, 1990). We are interested in using the wide tissue specificity of the *β*-actin gene promoter in the construction of expression vectors for gene transfer into fish, a very useful model system for studies of gene regulation (Powers, 1989).Transparent fish eggs and embryos which normally develop outside of the female have several advantages for studying the cascade of gene expression at early stages of growth. Since the gene is apparently well conserved, and the promoter elements for several *β*-actin genes have been partially characterized (Frederickson *et al.*, 1989; Ng *et al.*, 1989) we initiated a search for the *β*-actin genes in two species of carp for which we had genomic libraries in order to obtain functional enhancer/promoter complexes.

Here we characterize the putative *β*-actin gene from the common carp (*Cyprinus carpio*) and compare its regulatory components and expression to that of other vertebrate actin genes. Our results indicate a surprising conservation of the regulatory sequences in the 5′-flanking and first-intron sequences to the homologous sequences of land vertebrates and, as expected, the promoter elements are active in both fish and mammalian tissue culture cell lines. The organization of the gene into exons and the positioning of the protein coding sequences is nearly the same as in land vertebrates and the amino acid sequence of the carp *β*-actin gene 99% conserved between carp, mouse, humans and birds.

Results and Discussion

Identification of a carp *β*-actin gene

A single recombinant clone, CA16, containing a 13.5 kb insert was isolated from a carp liver genomic DNA library by screening with a chicken cDNA probe. Restriction enzyme mapping and Southern-blot hybridization analyses were used to characterize the clone (fig. 1). The isolated clone contains sequences corresponding to the entire *β*-actin coding region plus 4.5 kb upstream sequence and about 5 kb of downstream sequence. Exons 2-6 were determined by homology comparison with the corresponding exons of the chicken *β*-actin gene using the Intelligenetics GENALIGN program. Exon 1 and intron 1 were determined by homology comparison of carp sequences with grass carp sequences (Liu *et al.*, 1989) and by S1 nuclease mapping and primer extension (see p. 132).

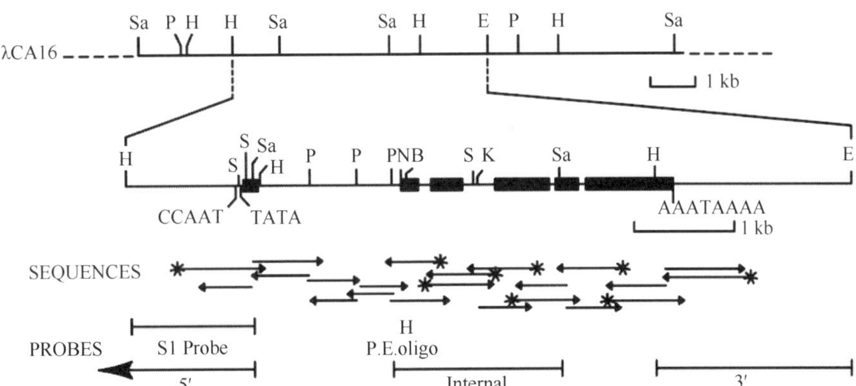

Figure 1 Map of the lambda (λ) CA16 clone of the carp β-actin gene. The top two lines show the restriction enzyme map of the β-actin gene. E, *Eco*RI: H, *Hin*dIII: K, *Kpn*I; P, *Pst*I, S, *Sst*I, Sa, *Sal*I. Positions of the CCAAT, TATA and AATAAAA transcription regulatory regions are indicated on the second line; note the change in scale. The sequencing strategy is indicated in the third set of lines; sequences beginning with *s were derived from clones made by the reverse cloning procedure (Liu and Hackett, 1989). The nuclease S1 (S1) and primer extension (PE oligo) probes used for mapping the transcriptional initiation site (see fig. 4) and the 5′-specific, internal, and 3′-specific probes used to count the number of genomic sequences homologous to the CA16 gene (see fig. 2) are shown at the bottom

Hybridization analysis of carp genomic DNA suggests the presence of multiple sequences related to the carp β-actin gene (fig. 2). *Eco*RI, *Pst*I, *Hin*dIII, *Bgl*II, and *Sal*I generates at least eight bands that hybridize with probes specific for the 5′-flanking and protein-coding portions of the gene (see fig. 1). In contrast, the 3′-specific probe hybridizes to only two DNA bands with the same restriction endonucleases. Of the bands shown in fig. 2, those corresponding to the CA16 clone are the 1.1 kb *Hin*dIII, 4.06 kb *Hin*dIII and 3.0 kb *Pst*I fragments identified with the 5′-probe; the 4.06 kb *Hin*dIII fragment identified by the internal probe; and the 3.3 kb *Hin*dIII, 6.7 kb *Sal*I and the triplet 4.7/5.6/6.6 kb *Pst*I fragments identified by the 3′-probe. Identification of multiple non-β-actin genes with 5′ and internal probes, but not with 3′-specific probes, is common for other vertebrate actin genes identified with, β-actin probes (Nakajima-lijima *et al.*, 1985).

The 6.7 kb *Sal*I fragment identified by the internal probe is very weak, presumably due to substantial methylation at this site. The triplet *Pst*I fragments are due to incomplete digestion at the *Pst*I site which forms the left-hand border of the internal probe. With the exception of the 6.7 kb *Sal*I band, the genomic fragments homologous to the CA16 gene are relatively heavy in the blot shown in fig. 2. Bands which are more intensely labelled than these known bands presumably are composed of more than one hybridizing fragment. We suppose that the less intense bands come from fragments with partial length homology or with greater sequence divergence. Multiple copies of actin genes in vertebrates have been noted previously (Minty *et al.*, 1983; Hightower and Meagher, 1986). The 5′ and internal, β-actin probes often hybridize to other actin genes such as those encoding α-actin and γ-actin (Erba *et al.*, 1988) whereas 3′-actin probes display a greater specificity (Ponte *et al.*, 1983; Yaffe *et al.*, 1985; Kim *et al.*, 1989).

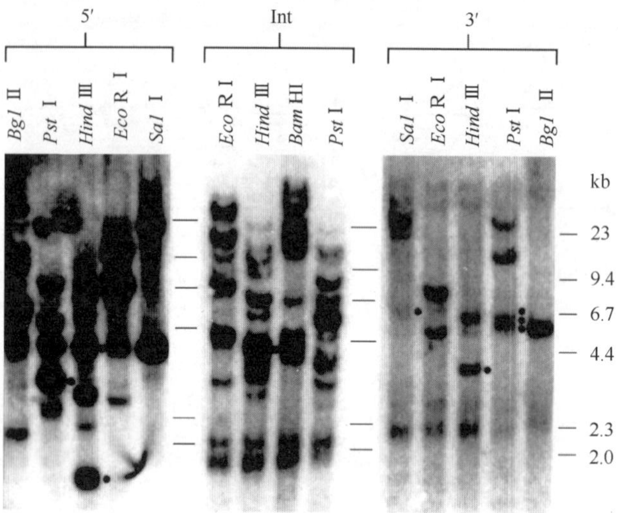

Figure 2 Southern hybridizations of carp β-actin gene sequences to carp genomic DNA cleaved with several restriction enzymes indicated at the top of each lane of the agarose gels. The left-hand gel was probed with the 5′-specific probe, the middle gel with the internal probe and the right-hand gel with the 3′-specific probe (probes are shown in fig. 1). The same gel was used for hybridization with the 5′ and 3′ probes; the middle gel was run in parallel. Lambda (λ) DNA cleaved with *Hind*III was used for size markers (right margin) which were visualized by ethidium bromide staining. The dots to the right of selected bands correspond to fragments derived from the CA16 β-actin gene clone

Sequence of a putative carp β-actin gene

The nucleotide sequence of the presumed carp β-actin gene is given in fig. 3. The gene consists of 6 exons with the protein-coding sequence beginning close to the 5′ end of exon 2, as is the case with the human, chick, and rat β-actin genes (Kost *et al.*,1983; Nudel *et al.*, 1983; Nakajima-lijima *et al.*,1985; Ng *et al.*, 1985). The boundaries of exons 2-6 were defined by comparison with the chicken, β-actin sequence. Since the first exon for all vertebrate β-actin genes currently known are non-coding, the sequence conservation of exon 1 is relatively low. As shown in Table 1, all intron-exon boundaries are identical to those of chicken, β-actin including those which diverge from the usual AG/GT motif in the donor site. At the intron-1/exon-2 junction, the carp sequence is like that of chicken (Kost *et al.*, 1983) and not similar to that in the rat (Nudel *et al.*, 1983). A 1.3kb first intron divides the 5′-untranslated region and is located 90-91 nt downstream from the transcriptional initiation site, and 2 nt upstream of translational initiation codon AUG. In the chicken, human, and rat, β-actin genes, the exon-1/intron-1 boundaries are located 91, 78, 74 nucleotides, respectively, downstream from their transcriptional initiation sites. The first intron of the β-actin gene is large in all species, 0.8 kbp in human, 0.9 kbp in chicken, and 0.8 kpb in rat. Introns 2-5 are all smaller in the carp, β-actin gene than in the chicken gene.

Table 1 Sequence of carp β-actin gene organization

Exon	Intron	Size	Intron bondaries		
			5′		3′
1		90 nt			
	1	1264 nt	AG/GT	...	AG/CC
2		125 nt			
	2	106 nt	AG/GT	...	AG/GG
3		240 nt			
	3	303 nt	AG/GT	...	AG/AT
4		439 nt			
	4	93 nt	GG/GT	...	AG/GT
5		182 nt			
	5	101 nt	AG/GT	...	AG/AT
6		698 nt			
		778 nt*			

* depending on which AAUAAA polyadenylation signal is used

Identification of exon-1 and the transcriptional initiation site

The transcriptional initiation site was determined as described in Materials and Methods (p. 134). A 23 oligonucleotide primer complementary to nts 75-97 of exon 2 (see fig. 3) was hybridized to total RNA from carp brain and muscle and extended with reverse transcriptase (fig. 4A). A 187/188 nt primer extension product was evident in RNA from brain cells (fig. 4A, lane 13) and two products of 187/188 nt and 156 nt were obtained with RNA from muscle cells (fig. 4A, lane 6). Since the 5′ end of exon 2 was 74 nt upstream from the primer, the consistent 187/188 nt primer extension product suggests an additional 5′-untranslated exon of 90/91 nt is upstream of exon 2. A search of the 5′-upstream sequences revealed potential CCAAT and TATA promoter proximal elements plus a candidate 5′-exon with a 3′-splice site that corresponded to that seen in other vertebrate β-actin genes. Primer extention for the 187/188 nt product indicated that the transcriptional initiation site was at the CC residues (nts 210/211, fig. 3). The identification of that initiation site was confirmed by S1 nuclease analysis. The 1.1 kbp *Hin*dIII/*Sal*I fragment, which overlaps the putative start point, was hybridized to carp brain total RNA, digested with nuclease S1, and the products were displayed on a sequencing gel (fig. 4B). The resulting 73 nt product indicated that the 5′ end of exon-1 was at the C at position 211 in the sequence shown in fig. 3. A prominent S1 product of 47 nt was not observed with brain RNA, suggesting that the 156 nt primer extension product from muscle RNA was either an artifact, possibly by inhibition of elongation due to secondary structure of the 5′ end of the RNA

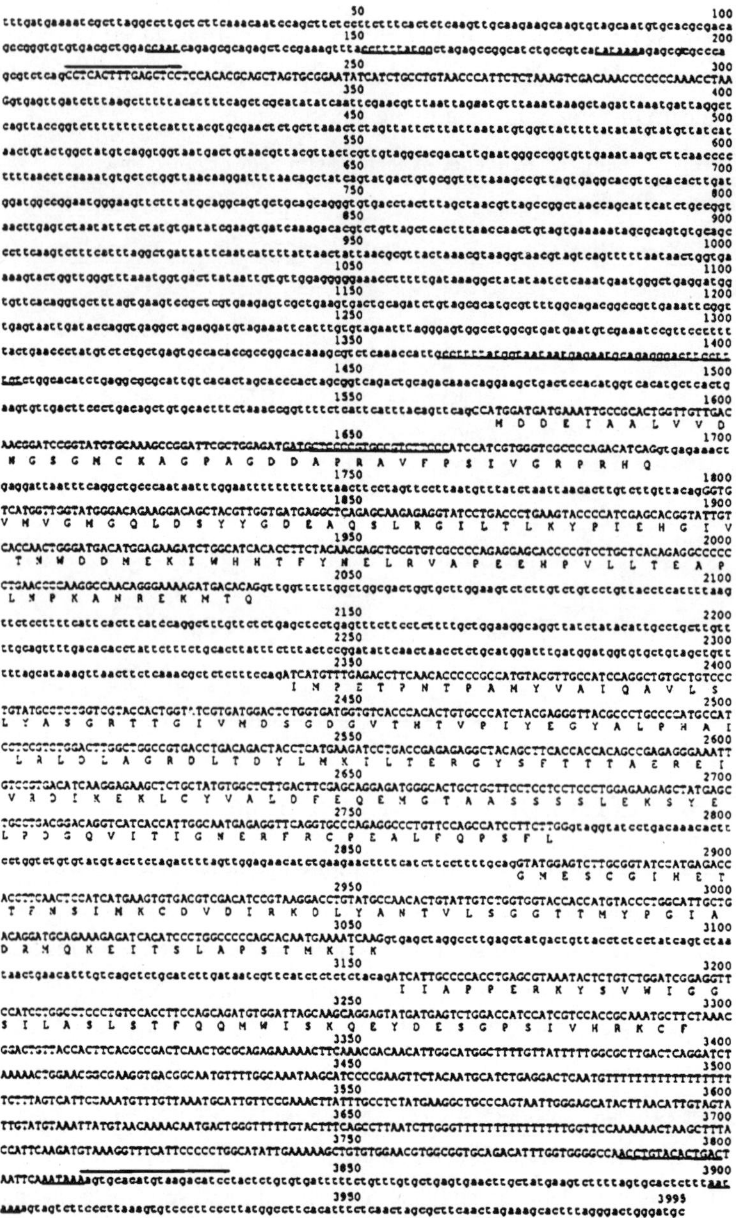

Figure 3 The 3995 nt sequence of the carp *β*-actin gene and its flanking sequences from the CA16 clone. For the nucleic acid sequence, lower case letters designate non-transcribed flanking sequences, non-translated leader and trailer RNA sequences and introns. Upper case letters represent exon sequences to the first poly(A) signal. The predicted *β*-actin translational product from the gene is indicated by the sequence broken into codons with the appropriate amino acid indicated above the sequence. The following sequences are designated by underlining: the CCAAT box (121-125), CCTTTTATGG (CArG) box (151-160) and TATAAA box (183-188) proximal promoter elements; conserved, intron-1 enhancer sequences (1362-1404) the 23 nt primer extension oligonucleotide (complementary to 1640-1662); CCTGTACACTGAC *β*-actin down-regulation sequence (3787-3799); and two potential AAUAAA polyadenylation signals (3807-3812 and 3897-3903). The two PCR primers, 210-228 and 3811-3832, are overlined

(McKnight *et al.*, 1982) in this reaction, or heterogeneity in the muscle RNAs (see below, fig. 8). Characteristic promoter elements are present in the 5′-flanking region of the gene (Wasylyk, 1988): a consensus TATAAA sequence is located at 22-27 nt, upstream from the transcriptional initiation site, and a CCAATC sequence was found at 84-89 nt upstream from the start site (see fig. 4). The results of primer extension and S1 mapping, taken together with phylogenetic comparisons of other β-actin genes, strongly indicate that transcription of the carp β-actin gene begins at position 211 on the sequence shown in fig. 3. The gene would direct the synthesis of an hnRNA precursor of 3.6-3.7 kb and an mRNA of about 1.8 or 1.9 kb, depending on which poly(A) site was used and assuming a poly(A) sequence of about 100 nt.

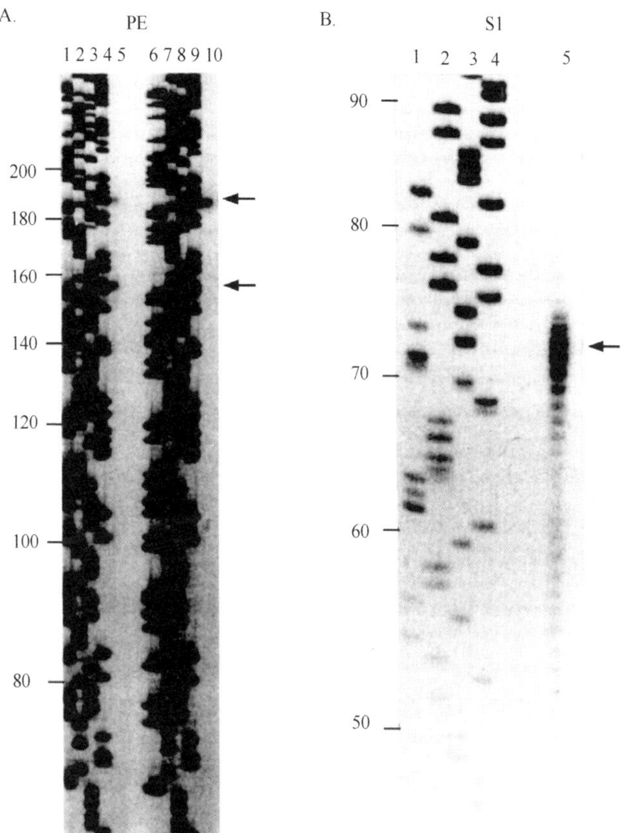

Figure 4 Transcriptional start-site mapping. The PE and S1 probe locations are shown in figs 1 and 3. (A) Primer extension of the 23 nt primer with total RNA from either muscle (lane 5) or brain (lane 10) was done as described in Materials and Methods. Two products were found with muscle RNA of 156 and 186/187 nt, whereas brain RNA gave just a single 186/187 nt product (arrows). Lanes 1-4 and 6-9 are Sanger sequence ladders used for size determination as indicated by the scale. (B) Nuclease S1 mapping with a uniformly- labelled DNA Probe (see Methods and Materials) hybridized to muscle RNA. Lanes 1-4, sequence ladder size markers

Comparison of carp, chicken, human, and rat *β*-actin genes

The nucleotide sequences of *β*-actin genes are very conserved through evolution. The evidence that the carp gene reported above is a true *β*-actin gene rather than some other actin isoform rests on four conserved homology regions:

(1) The coding exons of the carp gene are very similar to that of other, *β*-actin genes and less so with other non-*β*-actin genes (comparisons not shown); the strongest conservation in the coding region was found between carp and chicken; and slightly less conservation between carp and either the human or the rat genes (Table 2).

Table 2 Conservation of selected vertebrate *β*-actin genes

Homology of with	chicken (%)	human (%)	rat (%)
Carp nt sequences	91.0	88.5	87.9
Carp aa sequences	99.5	99.0	99.0

(2) The conserved nucleotide sequence of the promoter region is very similar to that for chicken, human, and rat *β*-actin genes (fig. 5A) (Kost *et al*., 1983; Nudel *et al*., 1983; Nakajima-lijima *et al*., 1985; Ng *et al*., 1985). In particular the CC $(A/T)_6$GG, or CArG sequence (Minty and Kedes, 1986), is found in all actin sequences including the carp gene promoter.

(3) The 40 nt sequence immediately upstream of the AATAAA poly(A) signal, reportedly involved in the down-regulation of expression of the *β*-actin gene in mouse during myogenesis (Deponti-Zilli, Seiler-Tuyns and Paterson, 1988), is conserved (fig. 5C).

(4) Two blocks of sequences of 20 nt and 16 nt are completely conserved in the introns of carp and other *β*-actin genes (Kawamoto *et al*., 1988; Ng *et al*.,1989) with 13 nt of divergent sequence in between (fig. 5B). In carp the conserved intron sequences are located 1055 nt downstream of the 5′-splice junction and 161 nt upstream of the 3′-splice junction of intron-1. We have compared all the actin genes in the GENBANK and EMBL data bases and found that the conserved intron sequence only appears in *β*-actin genes. The considerable conservations of sequences in the 5′-flanking region, first intron, and 3′-untranslated region of the carp and other *β*-actin genes suggested that these sequences may be involved in regulation, perhaps tissue-specific or developmental stage-specific expression of the *β*-actins.

The carp *β*-actin gene proximal sequences direct transcription

The homology of the putative carp *β*-actin sequence with that of land vertebrates, the completeness of the protein coding region for the gene giving a protein 99% similar to those of other vertebrates, the conservation of all splicing junctions, and the identity of the presumed transcriptional control signals in the carp sequence with other, previously identified transcriptional control signals for *β*-actin genes was persuasive evidence that we had cloned a functional β-actin gene. To further verify our conclusion, we (1) used the derived carp sequence to design two primers for PCR amplification of carp cell mRNA; and (2) used the identified transcriptional control sequences to drive transcription of a reporter gene in transfected mouse and fish tissue culture cells.

```
                 CONSERVATION OF CARP β-ACTIN GENE REGULATORY SEQUENCES

CONSERVED PROXIMAL PROMOTER SEQUENCES:

Human: CCAATCAG cGTgcgccGTTCCGAAAGTT GCCTTTTATGGCtcGAGcGgCcGCggCgGC GCCcTATAAAA
       :::::::: ::  ::::::::::::: :: ::::::::::::::::: :::::: : ::: : :::::::::
Rat  : CCAATCAG cGCccgccGTTCCGAAA TT GCCTTTTATGGCtcGAGtGgCcGCtgTgGC GYCcTATAAAA
       ::::::::  ::    :: ::::::::::   ::::::::::: : : ::: :: : ::: : :::::::::
Chick: CCAATCAGa GCggcgcGCTCCGAAAGTTt CCTTTTATGGCaaGGGcGgCgGCggCgGCgGCCcTATAAAA
       ::::::::: ::    : ::::::::::::: :::::::::: : :   : : :::: : :: : : :::::::
Carp : CCAATCAGa GC cagaGCTCCGAAAGTTtACCTTTTATGGCta  GaGcCgGCatCtGCcGCCaTATAAAA
       ::::::::: :: : : ::::::::::::: ::::::::: ::   : : :::: : :: ::: :::::::
CONS : CCAATCAGacGYgc—cGYTCCGAAAGTTtRCCTTTTATGGCt-gRG—GgC-GC-gYgGC-GYCCTATAAAA
       ========         =========================                      =======
                                  **********

CONSERVED FIRST INTRON SEQUENCES:

Human: TgtTTGCCTTTTATGGTAATaa  CGcGGccGgcccG G CTTCCTTTGTC ccC aATC T
        :::::::::::::::::::::   ::::: : :: : : ::::::::::: :::  : : :
Rat  : CgtTTGCCTTTTATGGTAATaaT  GcGGctGtcctGcG CTTCCTTTGTC ccCtgAgCtT
       :  ::::::::::::::::: :  : :. : : : : ::::::::::: ::  : : :
Chick: CcaTTGCCTTTTATGGTAATcgTgCGaGAggGcgcaGgGaCTTCCTTTGTC  cCaaATC T
       :::::::::::::::::::: :  :::: :    ::::::::::::::::  :: ::: :
Carp : CcaTTGCCTTTTATGGTAATaaT  GaGAatGcagaGgGaCTTCCTTTGTCtcgCacATC T
       :  :::::::::::::::::: : :: :    :::::::::::::: : :: ::: :
CONS : Y--TTGCCTTTTATGGTAATaaT cG-GR—G—c-GgGaCTTCCTTTGTC ccCa-ATC T
       =  =================               =============
          **********

CONSERVED 3' UNTRANSLATED REGION SEQUENCES:

Human: GGcTtACCTGTACACTGACTTGAgaCCagtt GAATAAAA
       ::::  ::::::::::::::::::::::       :::::::
Mouse: GGcTgGCCTGTACACTGACTTGAgaCC       AATAAAA
       ::    :::::::::::::: :  ::       :::::::
Chick: GGgCtACCTGTACACTGACTTGAgaCCagttcAAATAAAA
       :: : :::::::::::::: : :   :     ::::::::
Carp : GGcCaACCTGTACACTGAC TAAttC      AAATAAAA
       :: : :::::::::::::: ::: :      ::::::::
CONS : GG-Y-ACCTGTACACTGACTTRA—CCagtt RAATAAAA
          =================        ========
```

Figure 5 Phylogenetically conserved sequences in the carp β-actin gene. All comparisons were done using the Intelligenetics GENALIGN algorithm and β-actin sequences from the rat, chicken, mouse, human from the NIH GENBANK and EMBL data bank (releases 58 and 17, respectively) with the carp sequence given in fig. 3. In all comparisons, completely conserved residues are capitalized. CONS, consensus sequence; Y, pyrimidine; R, purine; consensus bases that are not completely conserved are in lower case (in the consensus sequences), any base; space, spacing for maximal alignment. (A) Comparison of proximal promoter sequences. The conserved CCAAT, $CC(A/T)_6GG$, and TATA motifs are all highly conserved as shown by capital letters in the consensus sequence. The $CC(A/T)_6$.GG sequence found in both the proximal promoter and the first intron is indicated with asterisks. (B) Comparison of conserved intron sequences implicated in enhancer activity. (C) Comparison of 3′-untranslated sequences implicated in down-regulation of β-actin gene expression

As described in Materials and Methods (p. 134), we used a sequence complementary to the 3′ end of the carp gene close to the first poly(A) signal and a second extending into the gene from the 5′ end indentified by S1 mapping and primer extension (see fig. 3) for PCR amplification (Frohman, Dush and Martin, 1988). The results using both 5 and 50 μg of total RNA are shown in fig. 6. We were not able to collect RNA from the same common carp (Wuhan, China) as the source of the library; so we isolated RNA from common carp cells obtained from Dr Boaz Moav (Tel Aviv, Israel). The 5 μg sample provided a clean, single band of about 1800 bp in length, the expected sum of the lengths of the exons between the two primers. The 50 μg RNA samples reproducibly gave a poorer signal of the same size. The cDNA was cloned, sequenced and, to our surprise, showed several third base changes in the coding sequence (data not shown) but in all other ways confirmed our identification of the carp β-actin genes. We presume that the few changes in sequence were due to slight evolutionary drift in sequences between separate isolations of common carp fish in different localities. The PCR amplified sequence was cleaved with six restriction enzymes; *Pst*I and *Eco*RI did not cut the DNA, *Bam*HI, *Bgl*II, and *Hin*dIII cleaved once and *Sal*I cleaved twice. All fragments were of the expected size (see fig. 6).

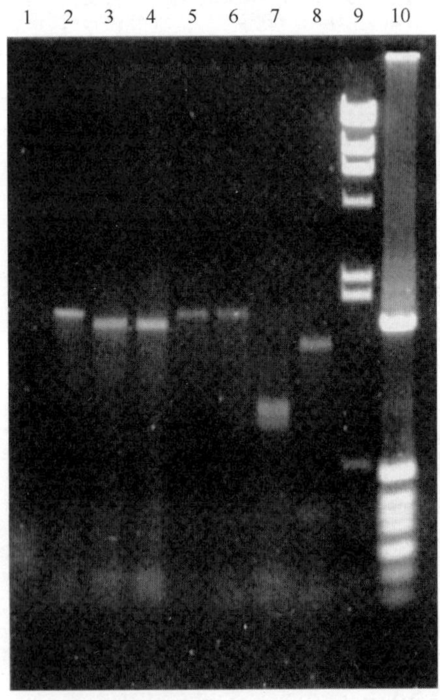

Figure 6 PCR amplification of total carp RNA with carp β-actin specific primers. Aproximately 100 ng of isolated amplified DNA was examined by electrophoresis on an agarose gel with pBR322 cleaved with *Hin*fI (lane 10) and lambda (λ) cleaved with *Hin*dIII (lane 9) size markers. Lane 1, DNA from 50 μg cytoplamic RNA; lane 2, DNA from 5 μg RNA; lanes 3-8, DNA from the 5 μg RNA sample cleaved respectively with *Bam*HI, *Hin*dIII, *Pst*I, *Eco*RI, *Sal*I and *Bgl*II

To further validate the identity of the putative carp β-actin gene, we fused the 5′-flanking sequences of the isolated gene to the CAT reporter gene. The fusions shown in fig. 7 indicate that the first 68 base pairs of exon 1 plus either 1100 or 193 nt of 5′-flanking sequence could drive CAT gene transcription in either mouse L-cells (see fig. 7) or carp cells (liu, unpublished) in tissue culture. Since the first exon does not have an AUG initiation codon, translation begins at the normal initiation codon for the chloramphenicol acetyltransferase. Cells were harvested 60 h after transfection and assayed for CAT activity (see Materials and Methods (p. 134). An equal amount of extract, determined by the Bradford (1976) procedure, was used in each assay. The assays were repeated five times and the averages for each construct are given. The results were the same in fish and mouse cells. The variation between experiments was less than 20% for each construct. The data indicate the following: (1) The putative carp β-actin gene promoter (constructs 1 and 2) is very active in tissue cultured cells, enhancing transcription 1000-fold above background (construct 3). (2) The first 193 base pairs of 5′-flanking sequence, containing the CCAAT, CArG, and TATA motifs conserved in all β-actin promoters (see fig. 5) are sufficient for dramatic initiation of transcription and are not affected by an additional 1000 bp of distal flanking sequence.

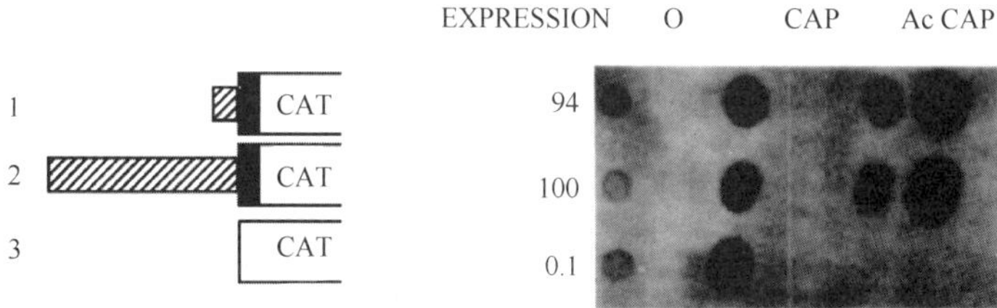

Figure 7 Transcriptional activity of carp β-actin proximal promoter sequences. Either just the CAT gene (construct 3) or two constructs (1 and 2) containing 5′-flanking sequences of the carp β-actin gene were fused at the *Sal*I site in exon-1 to the bacterial CAT gene, cloned into the polycloning site of pUC119, transfected into mouse L-cells, and assayed for expression. Construct 1, 193 bp of flanking sequence plus 68 bp of exon-1 fused to CAT; construct 2, 1100 bp of flanking sequence plus 68 bp of exon-1 fused to CAT; construct 3 lacks all carp sequences. O, chromatographic origin; CAP, $I^{14}Cl$ chloramphenicol; AcCAP acetylation at carbon-1 (left spot); and carbon-3 (right spot). The relative expression values are the averages of at least five independent experiments, normalized to the average of construct 2

Tissue-preference expression of the carp β-actin gene

To determine relative levels of expression of actin mRNA, we first used the coding region of the carp β-actin gene as a probe (see fig. 1) to determine the expression of this gene. We found a 1.8 kb actin RNA band with liver RNA, one major 1.8 kb band and a minor 1.6 kb band with brain RNA, and a diffuse set of RNAs ranging between 1.3 and 1.8 kb from muscle RNA (fig. 8, left). For greater resolution of β-actin RNA expression we used the 3′-untranslated region (3′-UTR) of the β-actin gene which should be isotype-specific (see fig. 2; Ponte *et al.*, 1983; Yaffe *et al.*, 1985; Erba *et al.*, 1988; Kim *et al.*, 1989) and which should show the steady-state amounts of *β*-actin mRNA in the three different tissues. Total carp cellular RNA, 10 μg, from each tissue was electrophoresed through a 1% agarose-formaldehyde gel, transferred to a nitrocellulose filter, and hybridized to the *β*-actin-specific, 3′-UTR probe. An autoradiograph of the filter is shown in fig. 8, right panel. The results clearly revealed the presence of a single 1.8 kb, *β*-actin mRNA in all tissues, both muscle and non-muscle (fig. 8, right), which was the upper band detected with the coding region probe (fig. 8, left). The level of *β*-actin varied between carp tissues, it was expressed highest in brain, intermediate in muscle, and lowest in liver. The autoradiogram was scanned with an optical densitometer. The expression in brain is 2.9-fold higher than in muscle, and 7.3-fold higher than in liver. These results are consistent with the report that the highest expression of *β*-actin in mouse is in the brain (Erba *et al.*, 1988).

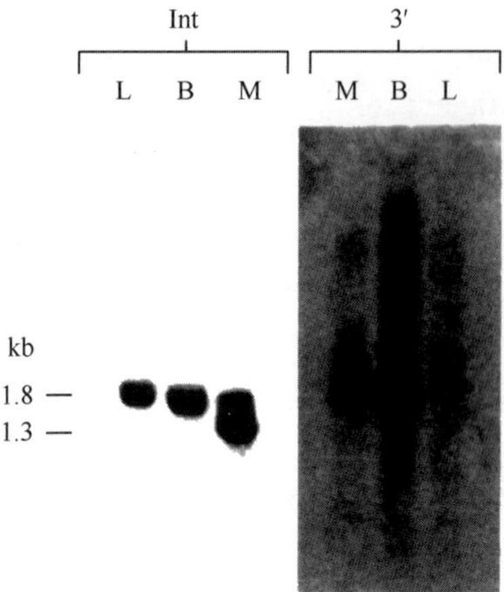

Figure 8 Northern hybridization of carp *β*-actin probes to total carp RNA from carp liver (L), brain (B) and muscle (M) tissues. The RNAs were probed with the internal (INT) and 3′-specific (3′) probes shown in fig. 1, 5-fold more liver RNA was loaded when the internal probe was used in order to have a detectable signal. Sizing of the RNAs was determined using rRNA markers

Conclusions

(1) A carp *β*-actin gene spanning 3.6 kb and containing 6 exons, as in all other vertebrate *β*-actin genes, has been isolated and sequenced. (2) There appears to be at least eight actin-like gene sequences in carp. (3) The carp *β*-actin gene shares strong sequence similarity with all other characterized vertebrate *β*-actin genes, not only in the coding region, but also in the promoter region, a portion of the first intron, and in the 3′ non-translated region of exon 6. The 40 nt conserved region in the 3′ untranslated sequence may mediate down-regulation of *β*-actin during myogenesis. The conserved sequences in the promoter region, as well as the

first intron (Liu *et al*, 1990), have transcriptional promoter activity. (4) The carp β-actin gene is preferentially expressed in brain tissues while the expression in muscle and liver tissues is much lower. We are currently testing the use of the identified, conserved sequences in expression vectors for use in studies of gene expression in transgenic fish.

Materials and Methods

Screening of the carp genomic library

A partial *Mbo*I library of carp liver genomic DNA fragments cloned in lambda (λ) EMBL3 phage (constructed at the Institute of Hydrobiology, Wuhan, China) was screened by filter hybridization to a chicken β-actin cDNA (Kost *et al.*, 1983) probe. A single recombinant clone, CA16, was identified and isolated by repeated plaque purification (Kaiser and Murray, 1984).

Nucleotide sequencing

The restriction endonuclease sites of the 13.5 kb carp DNA insert in the CA16 clone were mapped. Nucleotide sequences were determined by the dideoxyribonucleotide chain-termination method (Sanger, Nicklen and Coulson, 1977) from overlapping restriction fragments as well as from overlapping fragments using the reverse cloning procedure (Liu and Hackett, 1989). Both strands of the 3995 bp sequence containing the gene were completely sequenced. Junctions between overlapping fragments were sequenced using bridge clones generated by reverse cloning (Liu and Hackett, 1989). Only one strand of the cDNA was sequenced since we already had both strands of the genomic DNA.

Determination of the transcriptional initiation site

A 23 nt synthetic oligonucleotide probe, complementary to a phylogenetically conserved sequence at the end of exon 2, beginning 70 nt downstream from the presumptive AUG start codon, was used to prime reverse transcription (Kingston, 1987; Williams and Mason, 1985) by AMV reverse transcriptase (Lifesciences Inc.) of total RNA from carp brains and muscles. The labelled cDNA product was subjected to electrophoresis through a 8% polyacrylamide in 8M urea gel. The size of the extended primer was determined by comparison with a known sequencing ladder. S1 nuclease analysis was performed (Greene, 1987) to confirm the transcriptional initiation sites. A 1.1 kb fragment which overlaps the putative initiation site was continuously labelled (Greene, 1987) and hybridized to 20 μg total brain RNA, digested with S1 nuclease (BRL) at room temperature or 30℃ as specified.

Southern-blot analysis

Carp muscle genomic DNA was prepared using standard procedures (Maniatis, Fritsch and Sambrook, 1982), digested with various restriction enzymes according to manufacturer's instructions, and subjected to electrophoresis on a 0.8% agarose gel. The DNA was transferred to a nitrocellulose filter (Southern, 1975) and hybridized to radiolabelled probes of 3′ non-translated region, 5′ non-translated region and coding region made by nick translation (Maniatis, Fritsch and Sambrook, 1982).

Northern-blot analysis

Carp RNA was prepared from muscle, brain, and liver by the method of Chomczynski and Sacchi (1987). Total RNA was electrophoresed through a 1% agarose formaldehyde gel and transferred to nitrocellulose filters for hybridization with labelled DNA probes in the presence of 50% formamide (Williams and Mason, 1985).

cDNA cloning and amplification by polymerase chain reaction

Total cellular RNA was isolated from tissue-cultured common carp cells obtained from Dr. Boaz Moav (Tel Aviv, Israel); RNA from the same common carp fish cells as the library was made was not available from Wuhan, China. The carp RNA was prepared as described above, and copied by the RNA-dependent DNA polymerase from avian myloblastosis virus (Lifesciences, Inc.). Full-length cDNAs were amplified by polymerase chain reaction (PCR) using the 5′-specific (5′-CCTCACTTTGAGCTCCTCC-3′) and the 3′-specific (5′-GGAT-GTCTTACATGTGCACTT-3′) oligonucleotides according to the procedure of Frohman, Dush and Martin (1988). The PCR amplified DNAs were cloned into the pUC118 (Vieira and Messing, 1987), and transformed into *E. coli* JM101. The PCR-amplified clone was sequenced as described above.

CAT Analysis

An expression vector consisting of the pUC119 vector, the putative proximal promoter from the isolated β-actin gene, and the bacterial chloramphenicol acetyltransferase (CAT) gene from the pRSV-CAT construct of Gorman *et al.* (1982) was transfected (Lopata, Cleveland and Sollner-Webb, 1984) into both carp cells and mouse L-cells; 60h after transfection, cell extracts were prepared for thin layer chromatography analysis of CAT enzymatic activity (Gorman *et al.*, 1982). The radioactivity in the acetylated chloramphenicol spots was determined by cutting the spots off the thin layer plate and counting in a Beckman L1201 scintillation counter.

Acknowledgements We thank Ling He, Yuefeng Xie, Guohua Li and Kesheng Xu (Institute of Hydrobiology, Wuhan, People's Republic of China) for help in the preparation of the carp genomic library; Steve Hughes for providing his chicken β-actin gene clone; Aris Moustakas and Boaz Moav for discussions and reading the manuscript; Peter Saurugger and Darrin Johnson (University of Minnesota Molecular Biology Computing Center) for help with the Intelligenetics Programs; Kris Kirkeby and Toots McTavish for preparing the figures; and Carol Heiser for typing the manuscript. This work was supported by grants from NIH and the Minnesota Sea Grant College Program, supported by the NOAA Office of Sea Grant, DOC to P.H.Z.L was supported by a stipend from Minnesota Sea Grant; Z.Z. was supported by a University of Minnesota Hill Visiting Professorship. The sequence has been deposited in the GenBank database under the accession number M241134.

References

Bradford, M.M. (1976). A rapid and sensitive method for quantitation of microgram quantities of protein utilizing the principle of protein-dye binding. *Anal. Biochem*. 72, 248-254.

Chomczynski, P. and Sacchi, N. (1987). Single-step method of RNA isolation by acid guanidinium thiocyanate-phenol-chloroform extraction. *Anal. Biochem*. 162, 156-159.

Clarke, M. and Spudich, J. (1977). Nonmuscle contractile proteins: the role of actin and myosin in cell motility and shape determination. *Ann. Rev. Biochem*. 46, 797-822.

DePonti-Zilli, L., Seiler-Tuyns, A. and Paterson, B.M. (1988). A 40-base-pair sequence in the 3′ end of the β-actin gene regulates β-actin mRNA transcription during myogenesis. *Proc. Natl Acad. Sci. USA* 85, 1389-1393.

Erba, H.P., Eddy, R., Shows, T., Kedes, L and Gunning, P. (1988). Structure, chromosome location, and expression of the human γ-actin gene: differential evolution, location, and expression of the cytoskeletal β-actin and γ-actin genes. *Mol. Cell. Biol*. 8, 1775-1789.

Frederickson, R.M., Micheau, M.R., Iwamoto, A. and Miyamoto, N.G. (1989). 5′ Flanking and first intron sequences of the human β-actin gene are required for efficient promoter activity. *Nucl. Acids Res*. 17, 253-271.

Frohman, M.A., Dush, M.K. and Martin, G.R. (1988). Rapid production of full-length cDNAs from rare transcripts: amplification using a single gene-specific oligonucleotide primer. *Proc. Natl. Acad. Sci. USA* 85, 8998-9002.

Gorman, G.M., Merlino, G.T., Willingham, M.C., Pastan, I. and Howard, B.H. (1982). The Rous sarcoma virus long terminal repeat is a strong promoter when introduced into a variety of eukaryotic cells by DNA-mediated transfection. *Proc. Natl. Acad. Sci. USA* 79, 6777-6781.

Greene, J.M. (1987). S1 analysis of mRNA using single-stranded DNA probes. In *Current Protocols in Molecular Biology* (F.M, Ausubel et al., eds) pp, 4.6.1-4.6.9. Harvard Medical School, Cambridge, MA.

Gunning, P., Leavitt, J., Muscat, G., Ng, S.-Y. and Kedes, L. (1987). A human β-actin expression vector system directs high-level accumulation of antisene transcripts. *Proc. Natl. Acad. Sci. USA* 84, 8431-4835.

Hightower, R.C. and Meagher, R.B. (1986). The molecular evolution of actin. *Genetics* 114, 315-332.

Kaiser, K, and Murray, N.E. (1984). The use of phage lambda replacement vectors in the construction of representative genomic DNA libraries. In *DNA Cloning: A Practical Approach* (D.M. Glover, ed.) pp. 1-47 IRL Press, Oxford.

Kawamoto, T., Makino, K., Niwa, H., Sugiyama, H., Kimura, S., Amemura, M., Nakata, A. and Kakunaga, T. (1988). Identification of the human β-actin enhancer and its binding factor. *Mol. Cell. Biol*. 8, 267-272.

Kim, E., Waters, S.H., Hake, L.E. and Hecht, N.B, (1989). Identification and developmental expression of a smooth-muscle γ-actin in postmeiotic male germ cells of mice. *Mol. Cell, Biol*. 9, 1875-1881.

Kingston, R.E. (1987). Primer extension. In *Current Protocols in Molecular Biology* (F.M. Ausubel *et al*., eds) pp. 4.8.1-4.8.3. Harvard Medical School, Cambridge, MA.

Kost, T.A., Theodorakis, N. and Hughes, S.H. (1983). The nucleotide sequence of the chick cytoplasmic β-actin gene. *Nucl. Acids Res.* 11, 8287-8301.

Liu Z., Moav, B., Faras, A., Guise, K., Kapuscinski, A. and Hackett, P. B. (1990). Functional analysis of the control elements that affect expression of the β-actin gene of carp. *Mol. Cell. Biol.* 63, (July, in press).

Liu, Z., Zhu, Z., Roberg, K., Faras, A.J., Guise, K., Kapuscinski, A.R. and Hackett, P.B. (1989). The β-actin gene of carp (*Ctenopharyngodon idella*) *Nucl. Acids Res.* 14, 5850.

Liu, Z. and Hackett, P.B, (1989). Rapid generation of subclones for DNA sequencing using the reverse cloning procedure. *Bio/Techniques* 7, 722-728.

Lopata, M.A., Cleveland, D.W. and Sollner-Webb, B. (1984). High level transient expression of a chloramphenicol acetyltransferase gene by DEAE-dextran mediated DNA transfection coupled with a dimethyl sulfoxide or glycerol shock treatment. *Nucl. Acids Res.* 12, 5707-5717.

Maniatis, T., Fritsch, E.F. and Sambrook, J. (1982). Molecular Cloning: A Laboratory Manual. Cold Spring Harbor Laboratory, Cold Spring Harbor, NY.

McKnight, S.L (1982). Functional relationships between transcriptional control signals of the thymidine control signals of the thymidine kinase gene of herpes simplex virus. *Cell* 31, 355-365.

Minty, A.J., Alonzo, S., Guenet, J-L. and Buckingham, M.E. (1983). Number and organization of actin-related sequences in the mouse genome. *J. Mol. Biol.* 167, 77-101.

Minty, A.J. and Kedes, L. (1986). Upstream regions of the human cardiac actin gene that modulate its transcription in muscle cells: presence of an evolutionary conserved repeated motif. *Mol. Cell. Biol.* 6, 2125-2136.

Nakajima-lijima, S., Hamada, H., Reddy, P. and Kakunaga, T. (1985). Molecular structure of the human cytoplasmic β-actin gene: interspecies homology of sequences in the introns. *Proc. Natl. Acad. Sci. USA* 82, 6133-6137.

Ng, S-Y., Gunning, P., Eddy, R., Ponte, P., Leavitt, J., Shows, T. and Kedes, L. (1985). Evolution of the functional human β-actin gene and its multi-pseudogene family: conservation of non-coding regions and chromosome dispersion of pseudogenes. *Mol. Cell. Biol.* 5, 2720-2732.

Ng, S-Y., Gunning, P., Liu, S-H., Leavitt, J. and Kedes, L. (1989). Regulation of the human β-actin promoter by upstream and intron domains. *Nucl. Acids Res.* 17, 601-615.

Nudel, U., Zakut, R., Shani, M., Neuman, S. Levy, Z. and Yaffe, D. (1983). The nucleotide sequence of the rat cytoplasmic β-actin gene. *Nucl. Acids Res.* 11, 1759-1771.

Pollard, T.D. and Cooper, J.A. (1986). Actin and actin-binding proteins. A critical evaluation of mechanisms and functions. *Ann. Rev. Biochem* 55, 987-1035.

Ponte, P., Gunning, P., Blau, H. and Kedes, L. (1983). Human actin genes are single copy for α-skeletal and α-cardiac actin but multicopy for β- and γ-cytoskeletal genes: 3′ untranslated regions are isotype specific but are conserved in evolution. *Mol. Cell. Biol.* 3, 1783-1791.

Powers, D. (1989). Fish as model systems-*Science* 246, 352-358.

Sanger, F., Nicklen, S. and Coulson, A.R. (1977). DNA sequencing with chain terminating inhibitors. *Proc. Natl. Acad. Sci. USA* 74, 5463-5467.

Southern, EM. (1975). Detection of specific sequences among DNA fragments separated by gel electrophoresis. *J. Mol. Biol.* 98, 503-517.

Vandekerckhove, J. and Weber, K. (1979). The complete amino acid sequence of actins from bovine aorta, bovine heart, bovine fast skeletal muscle, and rabbit slow skeletal muscle. *Differentiation* 14, 123-133.

Vieira, J. and Messing, J. (1988). Production of single-stranded plasmid DNA. *Meth. Enzymol.* 153, 3-11.

Wasylyk, B. (1988). Transcription elements and factors of RNA polymerase B promoters of higher eukaryotes. *CRC Crit. Rev. Biochem.* 23, 77-120.

Williams, J.G. and Mason, P.J. (1985). Hybridisation in the analysis of RNA. In *Nucleic Acid Hybridisation: A Practical Approach.* (B.D. Hames and S.J. Higgins, eds) pp. 139-160. IRL Press, Oxford.

Yaffe, D., Nudel, U., Mayer, Y. and Neuman, S. (1985). Highly conserved sequences in the 3′ untranslated region of mRNAs coding for homologous protein in distantly related species. *Nucl. Acids Res.* 13, 3723-3737.

鲤β-肌动蛋白基因的分离和特征分析

刘占江[1] 朱作言[1,2] Kevin Roberg[1] Anthony Faras[3,6]
Kevin Guise[4,6] Anne R. Kapuscinski[5] Perry B. Hackett[1,6]

1 Department of Genetics and Cell Biology, University of Minnesota, St Paul, MN 55108, USA
2 Institute of Hydrobiology, Academia Sinica, Wuhan, Hubei, People's Republic of China
3 Department of Microbiology, University of Minnesota, Minneapolis, MN 55455, USA
4 Department of Animal Science, University of Minnesota, St Paul, MN 55108, USA
5 Department of Fisheries and Wildlife, University of Minnesota, St Paul, MN, USA
6 Institute of Human Genetics, University of Minnesota, Minneapolis, MN 55455, USA

摘　要　本研究从基因组 EMBL3 文库中分离得到鲤(*Cyprinus carpio*)β-肌动蛋白基因，其核苷酸序列长 3.6 kb，包含 6 个外显子。利用限制性内切酶消化鲤基因组 DNA，Southern 印迹杂交表明，存在两个同型的 β-肌动蛋白基因拷贝和其他几种类型的肌动蛋白基因。该基因的启动子元件 CCAAT 框和 TATA 框分别位于转录起始点上游-85bp 和-24bp。在基因长度、外显子/内含子位置、连接位点和第一外显子的翻译沉默方面，鲤 β-肌动蛋白基因与鸡、人和大鼠是一致的。与脊椎动物比较，鱼类 β-肌动蛋白基因在核苷酸和氨基酸水平上的保守性分别具有 90%和 99%。Northern 印迹分析表明，β-肌动蛋白基因高表达于脑，较少表达于肌肉，在肝细胞极少量表达。分析已知的肌动蛋白调控序列的保守性，鲤 β-肌动蛋白近端启动元件在哺乳动物和鱼细胞中都具有转录活性。

Primary-Structural and Evolutionary Analyses of the Growth-Hormone Gene from Grass Carp (*Ctenopharyngodon idellus*)

Zuoyan ZHU[1,2] Ling HE[2] Thomas T. CHEN[1]

1 Center of Marine Biotechnology, University of Maryland, Baltimore, USA
2 Institute of Hydrobiology, Chinese Academy of Sciences, Hubei, Peoples Republic of China

Abstract The growth-hormone (GH) gene of grass carp, one of the fastest-growing species of farmed fish, was isolated and the DNA sequenced. Only one GH gene is found in this species. This gene, which is 2.5 kb in length, has five exons and four introns, in common with all of the mammalian and the recently published common-carp GH genes. In the course of vertebrate evolution, the total lengths of the intron and the non-coding region of exon 5 of the GH gene have been shortened by 40-70%, whereas the encoding exons of the gene have been slightly increased. The more closely related species exhibit the closest sequence similarity in their GH genes. For example, the similarity of the exons is 84.1-93.2% between grass carp and common carp (within the same family of Syprinedae), 43.5-82.1% between grass carp and rainbow trout (in different orders of Teleostei) and 45.8-58.6% between grass carp and rat (in different grades of Vertebrata). In addition, similar DNA domains, such as thyroid-hormone-receptor-complex-binding site and cell-type-specific *cis* elements involved in regulation of expression of rat and human GH genes, have been localized in the corresponding regions of the grass-carp GH gene.

Keywords grass carp; head kidney; cDNA; EST; immune-related gene

Growth hormone (GH) is a polypeptide with the physiological function of stimulating the growth rate of vertebrates, especially in the early stages of individual development. Since 1981, the nucleotide sequences of the genes encoding GH have been identified in several mammals, including rat (ra) [1], human (h) [2], cattle (c) [3], pig (p) [4] and sheep (s) [5]. The overall length of the GH gene in these mammals is 1.6-2.0 kb and they consist of five exons and four introns. Among lower vertebrates, the GH gene has been isolated and the DNA sequence determined in rainbow trout (rt; *Salmo gairdneri)* [6], Atlantic salmon (as; *Salmo salar)* [7] and recently common carp (cc; *Cyprinus carpio)* [8]. Interestingly, the GH genes of the first two fish species studied are about twice the size of their mammalian counterparts. This difference in size is due to the first, third and, especially, the fourth introns being greatly lengthened, and, in addition, one more intron has been introduced in the last exon. This

文章发表于 *Febs Journal*, 1992, 207(2): 643-648

finding made a notable impact on the study of GH-gene evolution and encouraged scientists to check more GH genes in other fish species, since fish are the most numerous and divergent group of vertebrates. Consequently, grass carp (gc), a fresh-water species, which is also one of the fastest-growing species of farmed fish, was the subject of our study. Investigation of GH-gene structure is fundamental to our understanding of the regulation of gene expression and of the hormone's physiological function during the development in such an important species. In this paper, we report the results of *gc*GH-gene cloning, sequence characterization and evolutionary comparison.

Materials and Methods

Construction of a grass-carp gene library

The methods employed were mainly from Maniatis *et al.* [9], and the enzymes used were purchased from New England BioLabs. High-molecular-mass genomic DNA was extracted from the liver of grass carp from East Lake, Wuhan, China. After partial digestion with restriction enzyme *Mbo*I, the total DNA was fractionated, the 12-20 kb fragments were recovered and purified by sucrose-density-gradient centrifugation, followed by phenol/ chloroform extraction. The DNA fragments were ligated with the λEMBL3A *Bam*HI cloning vector (insert/vector 5 ∶ 1) by T4 DNA ligase and packaged *in vitro* with two lysogen extracts. The packaged recombinant bacteriophages were infected and amplified in *Escherichia coli* LE392. The packaging and amplification were performed at the Genetics Institute, Cambridge, MA, USA. This resulted in a grass-carp genomic library which covered 99.999% of its genomic information [10].

Screening for the *gc*GH gene

The molecular-biology kits and enzymes were purchased from Promega, BRL, USB and Bio-Rad and used according to manufacturers' instructions. [γ-^{32}P] ATP, [α-^{32}P] dATP and ^{35}S-labelled deoxyadenosine 5′[α-thio] triphosphate were purchased from Amersham and NEN. Based upon a similarity comparison of GH cDNA from five fish species [11-16], the conserved domains of nucleotide sequences were found in both 5′ and 3′ regions. Four nucleotide oligomers were synthesized: 5′-AACACCTGCACCAGCTGGCCGCA-3′ and 5′-CCCTGGATTCCAACTGCACCCTG-3′ matching with each corresponding 5′ and 3′ conserved region in ccGH cDNA; 5′-ACCTGTTGCCTGAGGAACGCAGACAGCTGA-3′ and 5′-TGTCTCGACCTTGT^{G}/$_{C}$CATGTCCTTCTTGAAGCA-3′ matching with each corresponding 5′ and 3′ conserved region in *rt*GH cDNA. The 5′ terminus was labelled with [γ-^{32}P] ATP by T4 polynucleotide kinase. The procedure for identification of a *gc*GH-gene clone was modified from that of Benton and Davis [17]. About 4 × 10^5 recombinant clones were screened by plaque hybridization *in situ* against the four ^{32}P-labelled probes simultaneously, and 152 putative plaques were identified. The λ-phage particles around the 152 positive areas identified were pooled as an enriched library of the *gc*GH gene. After two rounds of small-scale repeated screening of the enriched library, two candidates of the *gc*GH gene clone, λGCMZ1 and λGCMZ4, were isolated and identified by hybridization against each of the four ^{32}P-labelled oligomer probes separately.

Sequencing of the *gc*GH gene

By analyses of restriction digestion and Southern hybridization, the DNA fragment containing the *gc*GH gene in GCMZ1 was narrowed down to a 3.8-kb *Hin*dIII-*Pst*I fragment, within which *Acc*I and *Sac*I restriction sites were found (Fig. 1 A). The *Hin*dIII-*Sac*I 5′ fragment (2.2 kb) and the *Sac*I-*Pst*I 3′ fragment (1.6 kb) were then subcloned into the multiple cloning site of pUC118/119, named *pgc*GH9 and *pgc*GH 19, respectively. A deletion series was constructed corresponding to every 200-300 bp by digestion from each end of the two fragments with exonuclease *Bal*31 (removing 70 bp/min at 30℃) followed by mung- bean-nuclease treatment. This generated a whole set of overlapping fragments of *gc*GH-gene deletions. These fragments were then cloned in pUC118/119, in the multiple cloning site of the fragments *Hin*dIII-*Hin*dII, *Sac*I-*Hin*dII or *Pst*I-*Sma*I, for nucleotide sequencing. For each of the pUC118/119 overlapping subclones, 2-4 μg denatured mini-prep DNA was annealed with forward or reverse primer, for the sequencing reaction catalyzed by T7 DNA polymerase or sequenase™ based on the chain-termination DNA-sequencing method [18]. Both strands of the inserts of *pgc*GH9 and *pgc*GH19 were sequenced in this way.

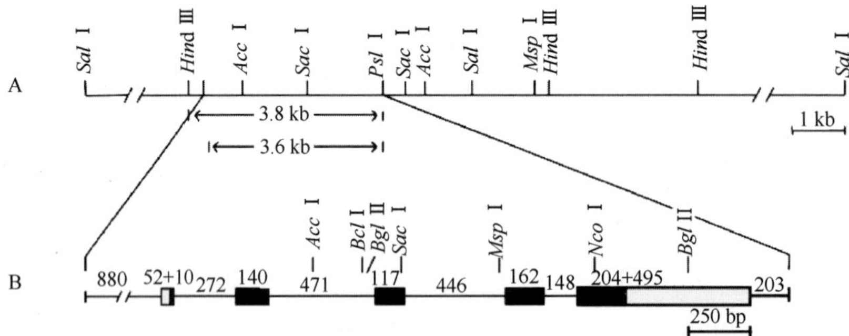

Fig. 1 Restriction map of a 12-kb *Sal*I fragment of λGCMZ1 or λGCMZ4 (A) and the transcription unit of the *gc*GH gene and the flanking fragments of sequence determined in Fig. 2 (B). Boxes are exons (open boxes are untranslated regions) and lines are introns and flanking sequences. Numbers are base pairs

Comparative analysis of the sequence data of GH genes

Comparative analysis of the sequence data of GH genes was mediated by the Wisconsin GCG Sequence Analysis Software Package, Version 6.2. The global alignment program was used for comparing exons and the introns; the local alignment program was used for comparing the regional domains. Exons 5 and 6 of the *rt*GH gene were treated as one exon (exon 5) for comparison with other GH genes.

Results and Discussions

Primary structure of the *gc*GH gene

λGCMZ1 and λGCMZ4 are two different clones, but both contain the same 12-kb *Sal*I fragment in which the *gc*GH gene is located. Restriction mapping of the gene and the flanking sequences is shown in Fig. 1 B. When the five exons and four introns in the *gc*GH gene had tentatively been determined by alignment of the sequence data of *pgc*GH9 and *pgc*GH19 with *cc*GH *c*DNA, the nucleotide-sequence data of *gc*GH *c*DNA [19] and the *cc*GH gene [8] were published one after another, which facilitated analysis of the organization of the *gc*GH gene. The primary structure and organization of the *gc*GH gene is shown in Fig. 2. By sequence alignment, the *gc*GH and *cc*GH genes made a perfect match in the putative promoter region and the first exon; thus, the putative transcription start point of *gc*GH gene could be assigned by analogy with the *cc*GH gene (Fig. 3). In the 5′ direction, a typical TATA box (TATAAA) is located 24 bp upstream of the transcription start point, but another consensus sequence of a CAT box is not found even up to –880 bp from the putative transcription start point. In the 3′ direction, the last 699 bp of the *gc*GH *c*DNA were found to match 100% with exon 5 of the *gc*GH gene, thus the transcription termination site was determined as being 12 bp downstream of the corresponding mRNA polyadenylation signal, ATTAAA. The transcription unit of the *gc*GH gene consists, therefore, of 2517 bp with five exons interrupted by four introns. The sizes of exons 1-5 are 62, 140, 117, 162 and 699 bp, and of introns 1-4, 272, 471, 446 and 148 bp. There are 52 bp in exon 1 and 498 bp in exon 5 as untranslated regions. All four introns start as gt and end as ag. The coding region of the *gc*GH gene encodes 210 amino acid residues of premature GH.

The grass-carp genome contains one GH gene

There are two mismatches between the *gc*GH gene reported here and the previously reported *gc*GH *c*DNA sequence [19]: TCT, 1012-1014 in the *gc*GH gene and TGT in the corresponding *gc*GH *c*DNA; AGCC, 1581-1583 in the *gc*GH gene and CCAG in the corresponding *gc*GH *c*DNA. These probably arise from sequencing errors, because codon 73 (TCT) which codes for serine in the *gc*GH gene is conserved in 10 other fish GH amino acid sequences. In order to check for the possible existence of polymorphism of the *gc*GH gene in the grass carp, its genomic DNA was digested with *Hin*dIII plus one of six other restriction

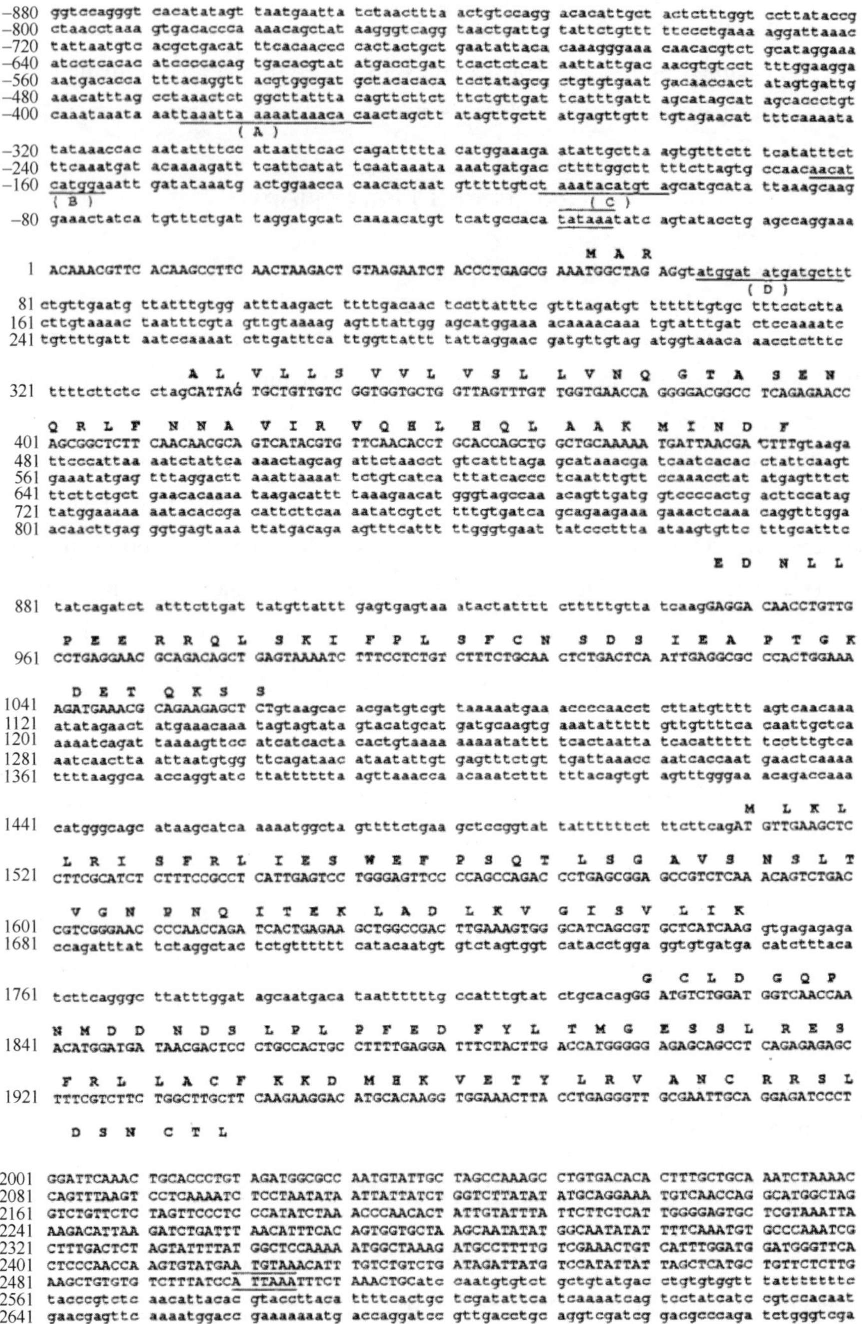

Fig. 2 The nucleotide sequence of the *gc*GH gene and the flanking regions. Exons are shown as capital letters; introns and flanking regions are shown as lower-case letters. The TATA box and the polyadenylation signal, ATTAAA, are shown between lines. By aligning the *gc*GH and *cc*GH genes, the putative *gc*GH cap site (+1) was determined 24 bp downstream of the TATA box. The underlined sequences A-D are possible regulatory regions (see Fig. 3). The encoded amino acid residues of premature *gc*GH are presented in bold capital letters above the corresponding codons

```
gcGH -64   tgattaggatgcatcaaaacatgttcatgccacatataaa
             :::: ::::::: :::::::::::::::::::::::::
ccGH -62   gcattaagatgcattaaaacatgttcatgccacatataaa
                         *
tatcagtatacctgagccaggaaaACAAACGTTCACAAGCCTTCAACTAA
::::::  :::::::::: : ::::::::: :::::::::  : ::::::
tatcag..tacctgagcctgaaaaACAAACATTCACAAGCTCTTAACTAA

GACTGTAAGA....ATCTACCCTGAGCGAAATGGCTAGAGgtatggatat
: ::: ::::     ::::::::::::::::::::::::::::::  : :
GCCTGCAAGAGTTTGTCTACCCTGAGCGAAATGGCTAGAGgtatgtctgt

gatgctttctgttgaatgttatttgt.ggatttaagacttttgac 117
:::::::::::::: :::::::::::  : : ::: ::: :
gatgctttctgttgcatgttgtttgtagcacttaggacatggattt 122
```

Fig. 3 The 5′-region alignment of the *gc*GH (positions-64-117) and *cc*GH genes (positions-62-122 [19]). The 5′-flanking region of the first intron is shown as lower-case letters, and the first exon is shown in capital letters (the coding region is in bold capitals). The TATA box is underlined. The cap site is marked by an asterisk

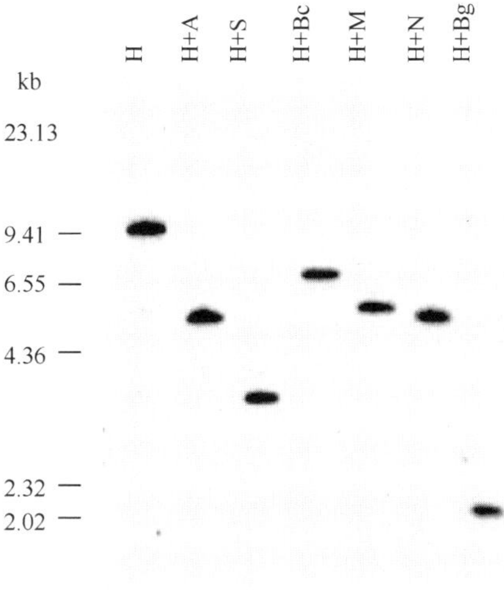

Fig. 4 Southern-blot hybridization of grass-carp liver DNA against a 357-bp *Nco*I-*Bgl*II probe of exon 5 of the *gc*GH gene, which consists of the most conserved region of the GH genes. Each lane represents 10 μg grass-carp DNA digested completely by restriction enzymes. All seven sets of digests resulted in a unique hybridization band indicating that no polymorphism of the GH gene is observed in this species. A, *Acc*I; Bc, *Bcl*I; Bg, *Bgl*II; H. *Hin*dIII; M, *Msp*I; N, *Nco*I; S, *Sac*I. The numbers shown (kbp) reprcsent the sizes of markers

Table 1 Nucleotide-sequence similarity among exons of different GH genes

Gene compared with the *gc*GH gene	Similarity within exons/%				
	1	2	3	4	5
*cc*GH	88.7	92.8	93.2	92.0	84.1
*rt*GH	43.5	70.7	72.3	82.1	70.5
*ra*GH	45.8	51.9	51.2	58.6	47.5

enzymes which have one or two recognition sites in the gene. Fig. 4 shows the Southern blot of these digestions following separation on 0.8% agarose-gel electrophoresis. The hybridization probe used was the ^{32}P-random-labelled 357-bp *Nco*I-*Bgl*II fragment of exon 5 of the *gc*GH gene, which includes the most-conserved region among all the GH genes. Each of the six sets of restriction digestions generated an unique hybridization band. From such a comprehensive restriction-digestion assay, it can be concluded that no polymorphism of the *gc*GH gene exists in this species. In addition, the two clones of the *gc*GH gene, λGCMZ1 and λGCMZ4, differ from each other in size of the genomic DNA inserts but share a common 12-kb *Sal*I restriction fragment in which the *gc*GH gene is located. Previous work has indicated that there are two GH genes in chum salmon [11], rainbow trout [20,21] and common carp [8,15,16], but we found only one GH gene in the grass-carp genome. The obvious difference between these fish species is that common carp, salmon and rainbow trout are tetraploid [22,23] and grass carp is a diploid [24]. It is significant that polymorphism of the GH gene is observed only in tetraploid species.

Comparative analysis of GH genes of grass carp and other species

To our knowledge, there are altogether nine GH genes from the Vertebrata that have been isolated and sequenced (Fig. 5). Five of them are from mammals, including rat, human, cattle, pig and sheep. The other four are from fish, including rainbow trout, Atlantic salmon, common carp and grass carp. The size and exon/intron arrangement are similar in both the *gc*GH gene (and the *cc*GH gene) and mammalian GH genes, but are quite different from two other fish (Atlantic salmon and rainbow trout) GH genes. It is clear that, in general, the ratio of exon/intron arrangement of the GH gene in the Vertebrata is 5∶4, with the exception of 6∶5 in Atlantic salmon and rainbow trout. It is very difficult to explain why the *as*GH and *rt*GH genes have such an unique exon/intron arrangement and are double the size of introns 2 and 4. These two species, however, migrate between fresh water and salt water, and their GH are involved in osmoregulation [25] in addition to growth enhancement. It will be very interesting to examine other fresh-water/salt-water-living species for this property of one more intron interrupting exon 5 of the GH genes. During the course of vertebrate evolution, the length of introns and the 3′ untranslated exon of the GH gene have changed such that in mammals they are shortened by about one-third. On the other hand, the length of the exons is much more stable (e.g. exons 1, 3, 4 and 5) or even slightly enlarged (e.g. exon 2). The similarity of nucleotide sequences of the exons between the *gc*GH gene and other GH genes are listed in Table 1. The similarity of exons between grass carp and common carp (within the same family of Syprinedae) is 84.1-93.2%, between grass carp and rainbow trout (in different orders of Teleostei) 43.5-82.1%, and between grass carp and rat (in different grades of Vertebrata) 45.8-58.6%, The similarity of the 5′ putative regulatory region between the *gc*GH and *cc*GH genes is 73.2%, but between the *gc*GH and *rt*GH genes and between the *gc*GH and *ra*GH genes the similarities are much reduced (less than 40%). Thus, GH genes are highly conserved in the coding regions among the closely related species, as well as in the putative promoter region and in some regions that are adjacent to the exons (data not shown). On the other hand, more distantly related species have less similarity between the genes. It is likely that the primary structure of GH genes in the Vertebrata could serve as a model for the study of molecular evolution.

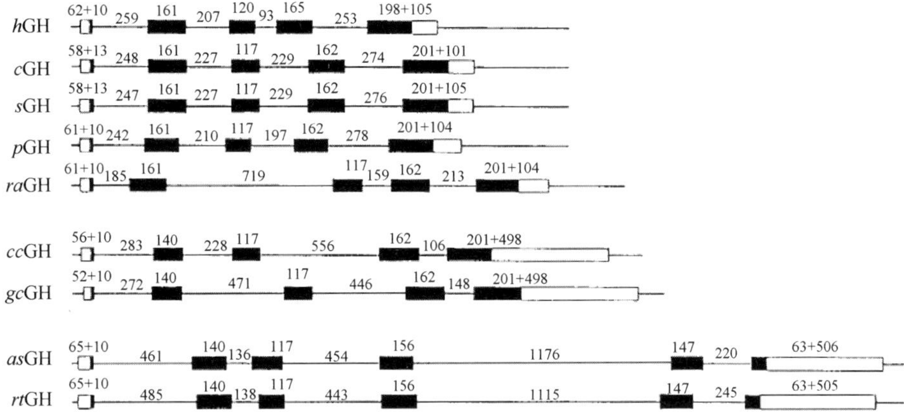

Fig. 5 Primary structure comparison of GH genes of the Vertebrata. Exons are shown as boxes (open boxes are untranslated regions) and introns and flanking regions are shown as lines. Numbers are base pairs. Atlantic salmon [7]; cattle [3]; common carp [19]; grass carp; human [2]; pig [4]; rat [1]; rainbow trout [6]; sheep [5]

The putative regulation domains in *gc*GH

Extensive study of the regulation of mammalian GH gene expression has been carried out. The glucocorticoid and thyroid hormones play important roles in regulating transcription of *ra*GH mRNA [26]. The binding sites for both hormones on either the upstream control region or the first intron of *h*GH or *ra*GH genes have been determined [27,28]. The tissuespecific expression domains of *ra*GH *cis*-acting elements were also precisely located in the *ra*GH gene [29]. In the first intron and the upstream region of the *gc*GH gene, four regions (A, B, C and F), similar to the corresponding regulatory domains or the adjacent sequences in the mammalian GH gene, could be found. In addition, the conserved TATA box (D) and the coding sequence in the first exon (E) were also found (Fig. 6). Region A is a typical sequence for the thyroid-hormone-receptor-complex-binding site in the *ra*GH gene. Regions B and C are similar to those of tissue-specific *cis*-acting elements of the *ra*GH gene. Sequences similar to the glucocorticoid-hormone-binding site have not been identified in the fish GH genes; however, region F in the *gc*GH gene is very similar to the *ra*GH gene in the first intron, which follows immediately after the glucocorticoid-*trans*-acting-element-binding site. This can be explained by the fact that glucocorticoid regulation of the GH gene has developed after the divergence of fish and tetrapods. In addition, domain C and region F are also present in the corresponding *cc*GH gene. Although direct evidence is currently lacking about whether any hormonal control of fish-GH-gene expression occurs, the results presented in this paper suggest that such regulation may occur.

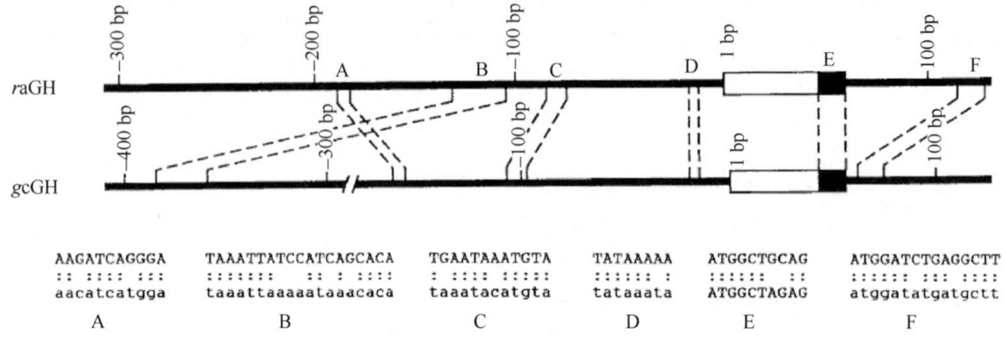

Fig. 6 Alignment of the first introns and the upstream part of the *gc*GH and *ra*GH genes. (A) Thyroid-hormone-receptor-complex-binding site in the *ra*GH gene; (B, C) cell-type-specific *cis*-acting elements in the *ra*GH gene; (D) TATA box; (E) the coding regions of the first exons; (F) sequence immediately following the glucocorticoid-hormone-receptor-complex-binding site in the *ra*GH gene. The *ra*GH-gene sequence data are from [1]

Acknowledgements Z. Z. is grateful to Dr Perry Hackett, Dept. of Genetics and Cell Biology, University of Minnesota, USA, for valuable advice and discussion when part of this project was being carried out in his laboratory supported by the University Hill Visiting Professorship. We thank Dr C.-M. Lin for synthesizing oligonucleotides and P. Z. Mao for technical assistance in screening λGCMZ1 and λGCMZ4. This project was financially supported by the Chinese Research Fund of National Development Program (the 7th five-year plan) to Z. Z. and the Maryland Sea Grant to T. T. C and Z. Z. This article is contribution no. 192 from the Center of Marine Biotechnology, University of Maryland, and the Institute of Hydrobiology.

References

1. Barta, A., Richards, R., Baxter, J. D. & Shine J. (1981) *Proc. Natl. Acad . Sci. USA* 78, 4867-4871.
2. DeNoto, F. M., Moore, D. D. & Goodman, H. M. (1981) *Nucleic Acids Res.* 9, 3719-3730.
3. Woychik, R. P., Camper, S. A., Lyons, R. H., Horowitz, S., Goodwin, E. C. & Rottman, E. M. (1982) *Nucleic Acids Res.* 10, 7197-7210.
4. Vize, P. D. & Wells, J. R. E. (1987) *Gene* 55, 339-344.
5. Orian, J. M., O'Mahoney, J. V. & Brandan, M. R. (1988) *Nucleic Acids Res.* 16, 9046.
6. Agellon, L. B., Davies, S. L., Chen, T. T. & Powers, D. A. (1988) *Proc. Natl. Acad. Sci. USA* 85, 5136-5140.
7. Johansen, B., Christian, O. & Valla, S. (1989) *Gene (Amst.)* 77, 317-324.
8. Chiou. C.-S., Chen, H.-T. & Chang, W.-C. (1990) *Biochim. Bio-Phys. Acta* 1087, 91-94.
9. Maniatis, T., Fritsch, E. F. & Sambrook, J. (1982) *Molecular cloning: a laboratory manual,* Cold Spring Harbor Laboratory Press, Cold Spring Harbor, New York.

10. Zhu, Z., He, L., Xie, Y. & Li, G. (1990) *Acta Hydrobiol. Sin.* 14, 176-178.
11. Sekine, S., Mizukami, T., Nishi, T., Kuwana, Y., Saito, A., Sato,M., Itoh, S. & Kawauchi, H. (1985) *Proc. Natl. Acad. Sci. USA* 82, 4306-4310.
12. Agellon, L. B. & Chen, T. T. (1986) *DNA (NY)* 5, 463-471.
13. Nicoll, C. S., Steiny, S. S., King, D. S., Nishioka, R. S., Mayer, G. L., Eberhardt, N. L., Boxter, J. D., Yamanaka, M. K., Miller, J. A., Sckilkamer, J. J., Schilling, J. W. & Johnson, L. K. (1987) *Gen. Comp. Endocrinol.* 68, 387-399.
14. Hew, C. L., Trinh, K. Y., Du, S. J. & Song, S. (1989) *Fish Physiol. Biochem.* 7, 375-380.
15. Chao, S.-C., Pan, F.-M. & Chang, W.-C. (1989) *Biochim. Biophys. Acta* 1007, 233-236.
16. Koren, Y., Sarid, S., Ber, R. & Daniel, V. (1989) *Gene (Amst.)* 77, 317-324.
17. Benton, W. D. & Davis, R. W. (1977) *Science* 196, 180-182.
18. Singer, F., Nicklen, S. & Colson, A. R. (1977) *Proc. Nut1. Acad. Sci. USA* 74, 5463-5467.
19. Ho, W. K. K., Tsang, W. H. & Dias, N. P. (1989) *Biochem. Biophys. Res. Commun.* 161, 1239-1243.
20. Agellon, L. B., Davies, S. L., Lin, C.-M., Chen, T. T. & Powers, D. A. (1989) *Mol. Reprod. Dev.* 1, 11-17.
21. Rentier-Delrue, F., Swennen, D., Mercier, L., Lion, M., Benrubi, O. & Martial, J. A. (1989) *DNA (NY)* 8, 109-117.
22. Ohno, S. (1974) in *Animal cytogenetics* (John, B., ed.) vol. 4, pp.1-91, Bortraeger, Berlin.
23. Uyeno, T. & Smith, G. R. (1972) *Science* 175, 644-646.
24. Zan, R. G. & Son, Z. (1979) *Acta Genet. Sinin.* 6, 205-210.
25. Clarke, W. C., Farmer, S. W. & Hartwell, K. M. (1977) *Gen. Comp. Endocrinol.* 33, 174-178.
26. Martial, J. A., Seeburg, P. H., Guenzi, D., Goodman, H. M. & Baxter, J. D. (1977) *Proc. Natl. Acad. Sci. USA* 74, 4293-4295.
27. Slater, E. P., Rabenau, O., Karin, M., Baxter, J. D. & Beato, M. (1985) *Mol. Cell. Biol.* 5, 2984-2992.
28. Brent, G. A., Larsen, P. R., Harney, J. W., Keonig, R. J. & Moore, D. D. (1989) *J . Biol. Chem.* 264, 178-182.
29. Nelson, C., Albert, V. R., Elsholtz, H. P., Lu, L. I.-W. & Rosenfeld, G. (1988) *Science* 239, 1400-1405.

草鱼生长激素基因的初级结构和演化分析

朱作言[1,2] 何 玲[2] Thomas T. CHEN [1]

1 Center of Marine Biotechnology, University of Maryland, Baltimore, USA
2 中国科学院水生生物研究所，湖北，中国

摘　要 草鱼作为水产养殖行业生长速度最快的鱼种之一，其生长激素基因已获得分离和测序。在草鱼中仅发现一个生长激素基因。与所有哺乳动物和最近发现的鲤生长激素基因相似，这个基因长约 2.5 kb，拥有 5 个外显子和 4 个内含子。在脊椎动物演化过程中，生长激素基因的内含子和第五外显子的非编码序列长度缩减了 40-70%，而外显子编码序列长度则有所增加。亲缘关系相近的物种中生长激素基因的序列也更为接近。例如，草鱼和鲤 GH 的外显子相似度为 84.1-93.2%(同属 Cyprinedae 科)，草鱼和虹鳟 GH 的外显子相似度为 43.5-82.1%(不同目的真骨鱼类)，而草鱼和大鼠的外显子相似度仅为 45.8-58.6%(不同等级的脊椎动物)。此外，与大鼠和人类生长激素基因的表达调控相关的甲状腺激素受体复合物结合位点和细胞类型特异性顺式元件，也定位到了草鱼生长激素基因的相应位置。

Time Course of Foreign Gene Integration and Expression in Transgenic Fish Embryos

Zhao Haobin Chen Shangping Sun Yonghua Wang Yaping Zhu Zuoyan

State Key Laboratory of Freshwater Ecology and Biotechnology, Institute of Hydrobiology, Chinese Academy of Sciences, Wuhan 430072

Abstract Using a nuclear transplantation approach, the integration and expression of the green fluorescent protein (*GFP*) gene in the embryogenesis of transgenic loach (*Misgurnus anguillicaudatus* Cantor) have been studied. The *GFP* gene expression is first observed at the gastrula stage, which is consistent with the initiation of cell differentiation of fish embryos. The time course of the foreign gene expression is correlated with the regulatory sequences. The expression efficiency also depends on the gene configuration: the expression of pre-integrating circular plasmid at early embryos is higher than that of the linear plasmid. The integration of the *GFP* gene is first detected at the blastula stage and lasts for quite a long period. When two types of different plasmids are co-injected into fertilized eggs, the behavior of their integration and expression is not identical.

Keywords gene integration and expression; time course; nuclear transplantation; fish; *GFP*

In 1982, Palmiter *et al.*[1] generated the first batch of transgenic super-mice, which shows the great potential of the transgenic animal technology and points out the new path for animal breeding. In 1984, Zhu et a1. [2] generated the human growth hormone (*hGH*)-transgenic fish. These fish showed "faster-growing and less-eating" effect and had the potential to become a new farming strain. However, some fundamentally theoretic questions concerning transgenic animals need to be answered, for example, the time course of the foreign gene integration and expression, the stability of the inheritance and so on.

When does the foreign gene start to integrate into host genome? The model of transgenic fish reveals that the integration is a gradual process: the integration is taken place at mid- and late-embryogenesis and the integrated genes are randomly distributed at host genome[3]. Nonetheless, little is known when exactly the gene integration occurs and whether such integration is stable. These questions have been the main attention of the present study concerned.

Fish embryonic nuclear transplantation technique, established by Tung *et al.* in the 1960s[4], was used in this experiment. Tung and his collaborators had made huge progress on

文章发表于《中国科学通报》(英文版), 2000, 45(8): 734-739

setting up the interactions between fish nucleus and cytoplasm and the breeding principle taking advantage of the "nuclear-cytoplasmic hybridization". In the early 1980s, the first batch of cloned fish was derived from kidney cells by nuclear transplantation[5]. If the foreign gene could stably integrate into the host genome at some stages of embryogenesis and then the embryonic cells were used as donor to be transferred into oocytes, it would be possible to generate homogenous transgenic fish. The cloned fish would have the same genotype in all cells of the body. This would overcome the mosaicism of the transgenes and may also help to determine the timing of the transgene integration.

The *GFP* was first found on deep-sea jellyfish. This protein emits green fluorescent light when it is activated under the blue UV light. The *GFP* gene has been cloned and is now widely used as a reporter gene in many experimental systems, including bacteria, yeast, plants, *C. elegance, Drosophila*, zebrafish, *Xenopus* and mammals[6]. The *GFP* gene is used as reporter in this study to detect the existence and expression of foreign gene in fish embryogenesis.

1. Materials and Methods

(i) Fish

Loach (*Misgurnus anguillicaudatus* Cantor) used in the experiment was purchased from Dadongmen Market, Wuchang. And the artificial spawning was induced by intraperitoneal injection of common carp pituitory homogenates.

(ii) Constructs

Five different gene constructs were used in the experiment. They are XIG[7], ccMTeGFP, pCAeGFP, pCMVeGFP and pMhGH. ccMTeGFP was made by subcloning a 660-bp common carp *MT* gene promoter into *Xho*I and *Bam*HI cut pEGFP-1 (GenBank #: U55761). pCAeGFP was generated by subcloning a 1.2-kb common carp actin gene promoter into *Hin*dIII and *Sal*I cut pEGFP-1. The pCMVeGFP was constructed by inserting a CMV promoter into *Hin*dIII and *Bam*HI cut pEGFP-1. The ccMTeGFP was a kind gift from K. M. Chan, Hong Kong Chinese University. Other constructs were made in this laboratory.

(iii) Manipulation of eggs and embryos

The inseminated eggs were obtained using the dry method. The transgenic embryos were generated by micro-injection according to the method of Zhu[2]. The cells of transgenic embryos were dissociated and the nucleus were then micro-injected into the mature egg according to the method of Tung *et al.*[4]

(iv) Detection of *GFP* gene expression

Live *GFP* transgenic fish embryos were examined using an inverted fluorescent microscope under blue UV light (480 nm).

(v) DNA preparation

The fry genomic DNA was extracted using phenol/chloroform method. The embryonic DNA was prepared using the same method with modifications. 100 μL DNA extraction buffer[3] (including 10 μg/mL proteinase K) was added to each embryo in a micro-centrifugal tube at 50℃ for 3 h. Then phenol/chloroform was used and the embryonic DNA was finally purified using ethanol precipitation.

(vi) PCR reactions

The detection of *hGH* gene by PCR was carried out using the method described by Li *et al.*[8]. The reaction mixture (25 μL in total volume) for detecting *GFP* gene contained 0.5 ng template DNA, 50 μmol/L dNTPs, 20 pmol primers, 1 unit of Taq enzyme. The reaction conditions were as follows: hot start (94℃ for 2 min), denaturing at 94℃ for 30 s, annealing at 58℃ for 30 s and extending at 72℃ for 1 min. This cycle was repeated for 30 times and finally the reaction was incubated at 72℃ for 5 min. The primers for detecting *GFP* gene were derived from its structural DNA. The 5′ primer is 5′-AGCAAGGGCGAGGAGCT-GTT-3′ and the 3′ primer is 5′-TCCATGCCGAGAGTGATCCC-3′. The length of the expected PCR fragment is 698 bp. 0.8% agarose gel was used to separate PCR products and the results were visualized using a UVP GDS8000 gel image system.

(vii) Statistic analysis

The results were statistically analyzed using the student *t*-test.

2. Results

(i) Expression of the *GFP* gene at different stages of fish embryogenesis

Circular plasmid XIG was injected into eggs. The results of the *GFP* expression at the embryonic stages were summarized in table 1. *GFP* gene expression was first detected at mid-gastrula stage and continued until fry stage. The expression efficiency of *Bgl*I-linearized XIG plasmid in fish embryos was slightly lower than that of circular XIG: there were about 15.0%, 27.5% and 52.5% embryos expressing linearized XIG at mid-gastrula, tail bud stage and muscular reaction stage, respectively. The *GFP* gene was expressed at late-gastrula stage of fish embryos that were injected with linearized pCAeGFP and pCMVeGFP. However, the expression of linearized ccMTeGFP started at myotome stage. These results demonstrate that gene expression depends on the gene configurations and their regulatory sequences. The expression efficiency of circular plamid was higher than that of the linearized plasmid. When the promoters are different, the timing of the expression is also different among the plasmids. The expression of XIG was initiated earlier than that of pCAeGFP and pCMVeGFP, and ccMTeGFP was the construct being expressed at very late stage. The fluorescent intensity of the *GFP* expression is also different: pCMVeGFP > pCAeGFP > XIG > ccMTeGFP.

Table 1 Expression of circular XIG in embryogenesis of transgenic loach

Developmental stage	Gastrula	Tail bud	Heart beating	Melanoid eye	Fry
No. of observed	80	51	40	40	75
No. of *GFP* expressed	54	43	38	39	74
(%)	67.5	84.3	95.0	97.5	98.7

The expression of the *GFP* protein was not evenly distributed (fig. 1(a)). Patches of green fluorescence were randomly observed at different parts of transgenic loach embryos and frys, for example, in yolk sack, head, heart, belly, back, tail and eyeballs.

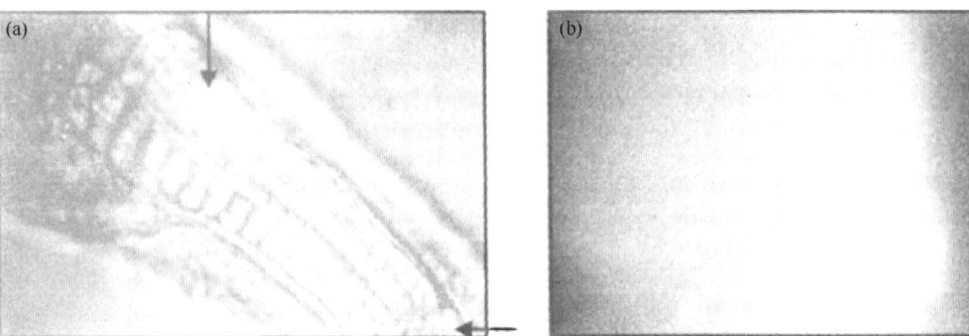

Fig. 1 Expression of *GFP* gene in transgenic embryo and nuclear transplant embryo. (a) Transgenic embryo. Arrows point to the green fluorescent spots of *GFP*; (b) nuclear transplant embryo

(ii) PCR detection of *GFP* gene in transgenic embryos

Plasmid ccMTeGFP was linearized by *Xho*I cutting and then injected into fertilized loach eggs. The *GFP* gene existence at different embryonic stages was examined by PCR, and the results were shown in table 2 and fig. 2(a). All transgenic embryos at blastula stage have *GFP* gene existence. The percentage of *GFP* positive embryos goes down as embryonic development proceeding further: 90% at muscular reaction stage and 80% at fry stage. This result shows that the foreign gene injected into eggs was gradually degraded during the process of embryonic development, only a small portion of gene could be integrated into host genome.

Table 2 PCR results of ccMTeGFP in transgenic embryos

Developmental stage	Blastula	Muscular reaction	Heart beating	Fry
No. of samples	10	10	10	10
Positive No.	10	9	8	8
(%)	100	90	80	80

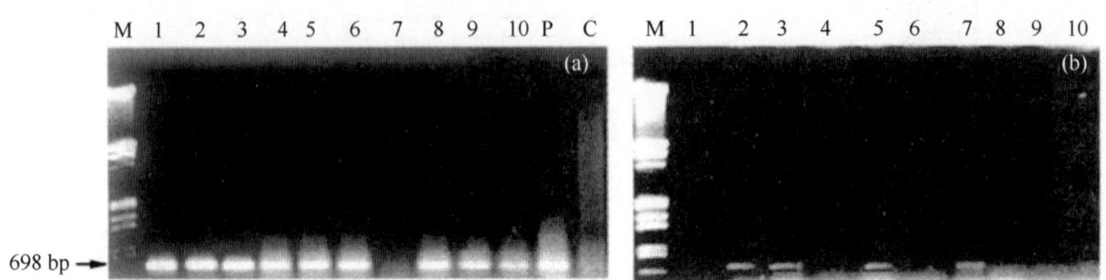

Fig. 2 PCR results of transgenic embryos and nuclear transplant embryos. (a) PCR results of *GFP* gene transferred embryos at muscular reaction stage; (b) PCR results of nuclear transplanted embryos at blastula stage. M, λDNA (*Hin*dIII/*Eco*RI); P, plasmid of ccMTeGFP; C, control; 1-10, samples

(iii) Co-injection of XIG and pMhGH

If there was correlation between two genes co-injected into the host, a reporter gene might be used to monitor/represent the other gene's behavior in fish embryogenesis. *Bgl*I-linearized XIG and *Bam*HI-linearized pMhGH were mixed (1∶1, *w/w*), and micro-injected into fertilized eggs. The resultant embryos with or without green fluorescence at late embryogenesis were separated. The genomic DNA was then isolated at fry stage. The presence of the *hGH* gene was detected by PCR and the result was summarized in table 3 (fig. 3). Among 31 *GFP* positive fry, only 25 individuals presented *hGH* gene. In contrast, in 35 *GFP*-free individuals there were 22 fry presented *hGH* gene. There was no significant difference between two groups in terms of the *hGH* gene presence (t=1.59, P>0.05, student t-test). It is concluded that the expression of the *GFP* protein (XIG) cannot be used to indicate the co-transferred *hGH* gene integration at fry stage, which is inconsistent with the transgenic mice result by Beermann *et al.*[9].

Table 3 PCR results of *hGH* gene in coinjected fishes

Fluorescence	Total	Positive No. for *hGH* gene	Positive ratio for *hGH* gene (%)
With	31	25	80.6 a)
Without	35	22	62.8 a)

a) P>0.05

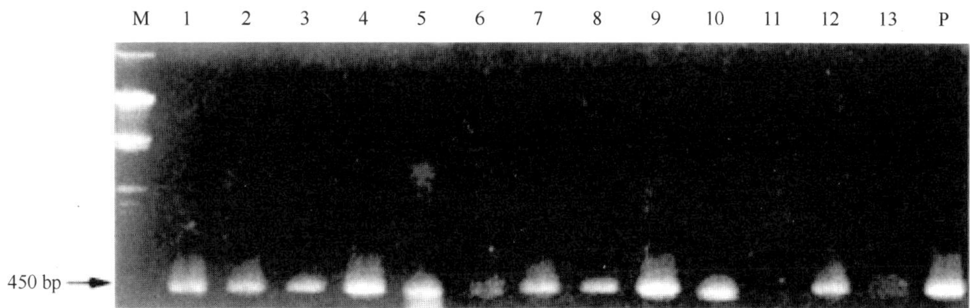

Fig. 3 *PCR* results of *hGH* gene in loach coinjected with XIG and pMhGH. M, λDNA (*Hin*dIII/*Eco*RI); P, plasmid of pMhGH; 1-7, fishes with fluorescence; 8-13, fishes without fluorescence

(iv) Expression of the *GFP* gene in nuclear transplant loach embryos

GFP gene was micro-injected into loach fertilized eggs. At different stages of embryogenesis, *GFP* expression embryos were selected and the embryonic cells were dissociated. The nucleus from the dissociated embryonic cells was transplanted into loach eggs and the results are shown in table 4.

Table 4 Development of nuclear transplanted embryos

Plasmids	Donor stages	No. of transplanted	No. at blastula (%)	No. at gastrula (%)	No. at neural plate (%)	No. at tail bud (%)	No. at muscular reaction (%)	No. at fry (%)
XIG	blastula	389	152 (39.1)	11 (2.8)	8 (2.1)	4 (1.0)	2 (0.5)	2 (0.5)
ccMTeGFP	gastrula	90	30 (33.3)	1 (1.1)	1 (1.1)	1 (1.1)	0	0
	myotome	91	45 (49.5)	2 (2.2)	1 (1.1)	1 (1.1)	1 (1.1)	0
	muscular reaction	113	20 (17.7)	2 (1.8)	0	0	0	0
pCMVeGFP	gastrula	75	17 (22.7)	2 (2.7)	2 (2.7)	0	0	0
	myotome	110	28 (25.5)	2 (1.8)	1 (0.9)	1 (0.9)	0	0

In the case of XIG-transgenic blastula cells served as donors for nuclear transplantation, one resultant embryo at tail-bud stage expressed *GFP*. The green fluorescence distributed evenly in the whole embryo (fig. 1(b)). In the case of ccMTeGFP-transgenic myotome cells served as donors for nuclear transplantation, one resultant embryo at neural plate stage expressed *GFP* but the green fluorenscent spots patched unevenly in the embryo (data not shown). In the case of pCMVeGFP-transgenic myotome cells served as donors for nuclear transplantation, one resultant embryo at myotome stage also showed unevenly patches (data not shown). These results demonstrate that there is gene integration at gastrula stage in one case, but in other cases foreign genes are not yet integrated even at myotome stage.

The nuclei of ccMTeGFP-transgenic gastrula cells were transplanted into unfertilized eggs. The resultant blastulas were examined by PCR for the *GFP* gene presence. About 40%

blastulas were *GFP* positive (fig. 2(b)). This reveals that there are about 40% cells in the gene-transferred blastula with the foreign gene; fish developed from this blastula must be transgenic mosaicism.

3. Discussion

(i) Time course of the foreign gene expression

The expression of foreign genes in fertilized eggs starts from gastrula stage. The starting point of the foreign gene expression depends on the promoter used; the expression of a gene regulated by early promoter is initiated earlier than that governed by late promoter. For example, plasmid DNA XIG in loach embryo started to be expressed from mid-gastrula stage; pCAeGFP and pCMVeGFP was expressed from late gastrula; ccMTeGFP DNA was the one being expressed latest, starting from myotome stage. In the XIG plasmid, *GFP* cDNA is driven by a promoter from the elongation factor *EF1α* gene[7]. At gastrula stage, cells have already started to differentiate and transcription of corresponding genes has taken place. The *EF1α* promoter must be an "early promoter" for the elongation factor is essential for the process of protein synthesis. It is thus not surprising that the elongation factor promoter drives *GFP* expression earlier than the other plasmid DNAs in this study.

Gene expression also depends on its configuration. The efficiency of circular plasmid expression is higher than the linearized counterpart. This might be due to the fact that circular plasmid DNA is not easily degraded, and there are plenty of gene copies existing in cells, thus may enhance the opportunity for the association of transcription factors. But the exact mechanism remains to be explored.

(ii) Timing of foreign gene integration

It is known that foreign genes have undergone amplification, degradation, polymerization and integration into host genome during the embryogenesis. Foreign genes are normally integrated into the host genome in head-tail or randomly polymerized forms. In addition, some are extra-chromosomal existing as macromolecules. Linearized DNA injected into fertilized eggs can replicate and amplify in multiple configurations, and this mode of action reaches a plateau at the completion of gastrulation. After tail-bud stage, molecular weight of the foreign gene increases as its copy number decreases. Southern blotting experiments show that the foreign genes co-migrate with genomic DNA[3,10,11], suggesting that gene integration takes place at late embryogenesis. In this study, *GFP* gene was detected by PCR in all embryos at blastula stage and then the percentage of *GFP* positive embryos went down. This suggests that the foreign gene disappeared through degradation after blastula stage in some embryos. Even in an individual embryo, the existence of the foreign gene is not homogeneous, resulting a mosaic embryo. For example, there were only 40% embryonic cells at gastrula stage being *GFP* gene detectable. The other cells in the same embryo either did not have foreign gene distributed, or if there had been the existence of the foreign gene, but it had already degraded. When the nucleus from a transgenic XIG embryo at blastula stage was used for nuclear transplant, an embryo was produced with homogeneous *GFP* expression, thus suggesting that the foreign gene has already integrated into donor genome and the integration starts from blastula stage. As far as we know, this is the first report of the gene integration at blastula stage. The *GFP* expression was not homogeneous in a nuclear transplant fish whose nuclear came from a transgenic *GFP* embryonic cell at myotome stage. This may be due to two possibilities, either foreign gene did not integrate into donor genome or such integration was not stable enough to survive. In conclusion, gene integration may start from blastula stage and continues for a quite long period. This is a very sophisticated process and the exact mechanism remains to be explored.

(iii) Relation between two co-transferred genes' integration

In 1991, Beermann *et al.*[9] reported that two genes inserted into mouse genome at the same site when the genes were co-transferred, the expression of the reporter melanoid gene could represent the integration of the other co-transferred one. In this report, equal amounts of *Bam*HI-linearized *hGH* gene and *Bgl*I-linearized *GFP* gene were co-transferred. The integration ration of *hGH* gene in fish with fluorescence was higher than that without fluorescence, but this was not statistically significant ($P>0.05$). In other words, there is no correlation in gene integration between co-transferred genes in fish. One gene's integration in genome does not represent the other gene's integration. In fish gene transfer, the foreign gene is normally injected into cytoplasma, whilst in mammals, foreign gene is introduced directly into nucleus. Other reason is that fish embryonic cells divide much faster than their mammal counterparts and thus fish embryos develop more rapidly. In Beermann *et al.*'s experiment, co-transferred genes have the same *Sal*I-cohesive ends. These two genes may ligate to each other well before gene integration, resulting in the integration into the same site in host genome. When the co-transferred genes do not have the same adhesive ends as in this study, the genes may integrate at more than one site. Even when there is a single integration site, multiple copy gene transfer may cause the inconsistency of gene integration or expression[12].

References

1. Palmiter, R. D., Brinster, R. L., Hammer, R. E. *et al.*, Dramatic growth of mice that develop from eggs microinjected with metallothionein-growth hormone fusion genes, Nature, 1982, 300: 611.
2. Zhu, Z., Li, G., He, L. *et al.*, Novel gene transfer into the fertilized eggs of goldfish (*Carassius auratus* L. 1758), Z Angew Ichthyol, 1985, 1: 31.
3. Zhu, Z., Xu. K., Xie, Y. *et al.*, A model of transgenic fish, Science in China, Ser. B (in Chinese), 1989, (2): 147.

4. Tung, T. C., Wu, S. C., Tung, Y. F. *et al.*, Nuclear transplantation in fish, Kexue Tongbao (in Chinese), 1963(7): 60.
5. Chen, H., Yi, Y., Chen, M. *et al.*, Studies on the developmental potentiality of cultured cell nuclei of fish, Acta Hydrobiol. Sinica (in Chinese), 1986, 10(1): 1.
6. Misteli, T., Spector, D. L., Applications of the green fluorescent protein in cell biology and biotechnology, Nature Biotech., 1997, 15: 961.
7. Amsterdam, A., Lin, S., Hopkins, N., The *aequorea victoria* green fluorescent protein can be used as a reporter in live zebrafish embryos, Dev. Biol., 1995, 171: 123.
8. Li, G., Cui, Z., Zhu, Z. *et al.*, Introduction of foreign gene carried by sperms, Acta Hydrobiol. Sinica (in Chinese), 1996, 20(3): 242.
9. Beermann, F., Ruppert. S., Hummler, E. *et al.*, Tyrosinase as a marker for transgenic mice, Nucleic Acids Res., 1991, 19(4): 958.
10. Xie, Y., Liu, D., Zou, J. *et al.*, Gene transfer in the fertilized eggs of loach via electroporation in fish, Acta Hydrobiol. Sinica (in Chinese), 1989, 13(4): 387.
11. Zou, J., Xie, Y., Liu. D. *et al.*, Foreign gene expression during embryogenesis of crucian carp, Acta Hydrobiol. Sinica (in Chinese), 1991, 15: 372.
12. Devlin, R. H., Yesaki, T. Y., Donaldson, E. M. *et al.*, Production of germline transgenic pacific salmonids with dramatically increased growth performance, Can. J. Fish Aquat. Sci., 1995, 52: 1376.

转植基因在鱼胚胎中整合和表达的时程研究

赵浩斌　陈尚萍　孙永华　汪亚平　朱作言

中国科学院水生生物研究所，淡水生态与生物技术国家重点实验室，武汉 430072

摘　要　借助细胞核移植显微注射技术，以绿色荧光蛋白基因为报告基因，泥鳅为实验材料，研究了外源基因在鱼类胚胎发育过程中的表达与整合。结果表明，外源基因在鱼类胚胎中，从原肠胚期开始表达，与鱼类胚胎分化时期相吻合。转植外源基因的表达时间与启动调控顺序有关；表达效率与基因构型有关，受体胚胎早期未整合的环形质粒的表达效率高于线性质粒。外源基因的整合最早从囊胚期开始，并持续相当长的时间。两种不同结构的基因共转移时，其整合与表达不存在一致性。

Transgenes in F_4 pMThGH-Transgenic Common Carp (*Cyprinus carpio* L.) are Highly Polymorphic

ZENG Zhiqiang ZHU Zuoyan

State Key Laboratory of Freshwater Ecology and Biotechnology, Institute of Hydrobiology, Chinese Academy of Sciences, Wuhan 430072

Abstract To gain information on the integration pattern of pMThGH-transgene, 50 transgenes were recovered from F_4 generation of pMThGH transgenic common carp (*Cyprinus carpio* L.) and 33 recovered genes were analyzed. The restriction maps of these recovered genes were constructed by digestion with five kinds of enzymes. These transgenes can be classified into 4 types according to their restriction maps. Only one type of transgenes maintains its original molecular form, whereas the other three types are very different from the original one and vary each other on both molecular weight and restriction maps. This implies that the sequences of most transgenes have been deleted and/or rearranged during integration and inheritance. The results of PCR amplification and Southern blot hybridization indicate that *MThGH* in Type I transgene keeps intact but most of its sequence has been deleted in other three types. All these results suggest that transgenes in F_4 generation of transgenic carp are highly polymorphic. Two DNA fragments concerning integration site of transgenes were cloned from recovered transgenes, and found to be homologous to the 5′UTR of *β*-actin gene of common carp and mouse mRNA for receptor tyrosine kinase (*RTK*), respectively.

Keywords transgenic; common carp; integration pattern; transgene recovery

The introduction of recombinant DNA into the blastocyst cavity of animal to produce transgenic animals becomes a more and more important technique in biological study. Transgenic technology offers exciting possibilities to generate precise animal models for human genetic diseases, to revolutionize animal breeding and to produce large quantities of economically important proteins by means of farming transgenic animals. On the other hand, the molecular mechanisms of transgene integration and inheritance are still not fully understood, so it is difficult for researchers to introduce foreign gene into a specific site on the host genome and to control its proper expression in founder transgenic animal produced by microinjection. Southern blot hybridization[1], genomic screening[2,3], and *in situ* hybridization[4,5] have been used to study the integration pattern of transgenes in host animals. Moreover, the integration copies and level of transgene expression in transgenic offsprings have been intensively studied

文章发表于《科学通报》(英文版), 2001, 46(2): 143-148

recently[6]. The major conclusions are summarized as follows: (i) Transgenes always integrate as Concatemers in a head-to-tail manner. (ii) There is no regularity on the number and location of integration sites, except that a few homologous recombination. (iii) Nucleotide sequence of transgene may be deleted, rearranged and disturbed with the host sequence. Nonetheless, little is known about how exactly the gene integration occurs, whether there is any inevitability on the integration events of transgene, and what the inheritance of transgene behaves like via reproduction. These questions await further study.

Transgenic fish have a direct application in fish genetics and breeding, meanwhile it can serve as an animal model for studying the integration and inheritance behavior of transgene. The integration and germ line transmission of transgene[7] has been proved. In a recent report, integration of foreign gene was observed at as early as the blastula stage[8]. The F_4 generation of pMThGH-transgenic common carp was obtained by self-breeding from the first generation to the next one. As far as we know, whether transgene existence is stable and whether its sequence has been changed in the host genome through several generations have not been reported. In this note we try to gain insight into the detailed mechanisms of transgene integration and inheritance by recovering transgenes from F_4 generation pMThGH-transgenic fish and analyzing them with a modified plasmid rescue technique.

1. Materials and Methods

(i) Production of F_4 pMThGH-transgenic fish

Recombinant plasmid of pMThGH was constructed by capping a mouse metallothionein-1 gene promoter to human growth hormone gene sequences, and then inserted into pBR322 at the *Eco*RI site (fig. 1). DNA of pMThGH was linearized by digestion with *Bam*HI and microinjected into the fertilized eggs of common carp to produce the founder transgenics[7]. After being confirmed by dot blotting or PCR, transgenic fishes were naturally mated and gave birth to their offspring and then led to the F_4 generation. At the age of two months old, the average body weight for 17 F_4 transgenic individuals was 7.59 g. The largest one (26.35 g) was selected for analysis.

(ii) Restriction digestion with *Bam*H I and determination of the enriched region of pMThGH-transgene

Genomic DNA was extracted from the liver of transgenic fish. 10 μg of DNA was completely digested with restriction endonuclease *Bam*H I, and separated by 0.8% agarose gel electrophoresis. Southern blotting against α-^{32}P-dCTP labeled pMThGH probe was carried out as described by Maniatis *et al*[9]. The molecular weight of transgene in single copy was figured out by the location of hybridization band.

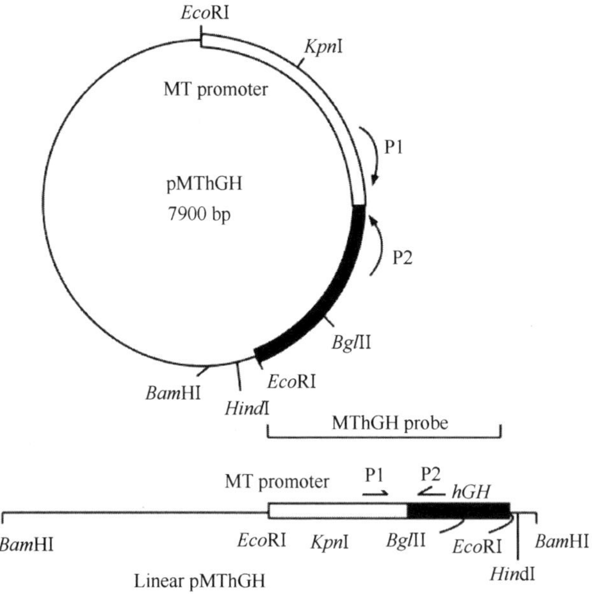

Fig. 1 Structure of transgene pMThGH and linearized DNA for microinjection. Location of PCR primer P1, P2 mid *MThGH* probe used for this study are also shown in the diagram

(iii) Recovering pMThGH-transgene

(1) Partial digestion One of our aims in this note is to clone the host sequences flanking transgene or the sequences of integration sites. By partial digestion with *Bam*HI, host sequence joining the linear transgene at its end with *Bam*HI site may be cloned. 40 μg of genomic DNA were digested with *Bam*HI for the varied period of time. The procedure can be briefly described as follows: 1/5 volume of total reaction solution were taken out at intervals of 2nd, 5th, 10th, 15th, 30th min, then the reaction was stopped by adding EDTA to a final concentration of 0.05 mol/L.

(2) Recovery of DNA fragments The result of Southern blot hybridization shows that the molecular weight of single pMThGH-transgene released from concatemer is 7.5-8.0 kb (data not shown). Considering part of transgenes may have changed in sequence length after several generations, DNA fragments ranging from 5 to 10 kb were recovered and purified from low-melting agarose gel. Recovered DNA was diluted in 100 μL TE buffer (10mmol/L Tris·HCl, 1 mmol/L EDTA, pH 8.0) at a concentration of 40 μg/mL.

(3) Circularization of recovered DNA fragments A modified method[10] for DNA circularization was employed here. 200 μL DNA fragments at a final concentration of 2 μg/mL were self-ligated with 1 U T_4 ligase at 16℃ for 8 h. The circularized DNA was extracted with phenol-chloroform, precipitated by ethanol and re-suspended in 10 μL TE buffer (10 mmol/L Tris·HCl, 1 mmol/L EDTA, pH 8.0).

(4) Transformation into *E. coli* 10 μL of circularized DNA was used to transform 100 μL

of DH5α competent cell using $CaCl_2$ treatment procedure. Transformed cells were spread on LB plates with 50 μg/mL ampicillin. pMThGH DNA was transformed as the positive control.

(iv) Classification of recovered transgenes

50 clones were obtained and 33 clones were randomly picked up for further analysis. To classify these clones, plasmid DNA were mini-prepared, and doubly digested with *Eco*RI and *Bam*HI. Classification was performed according to the electrophoresis patterns of restriction fragments resulting from double digestion.

(v) Mapping

Recovered plasmid DNA were firstly digested with *Bam*HI and *Hin*dIII, respectively, 1/3 of each linear DNA continued to be digested with *Eco*RI, *Bgl*II or *Kpn*I. Digestion DNA fragments were separated by 1.0% agarose gel electrophoresis, stained with ethidium bromide and visualized under UVP GDS8000 system (UVP Ltd., UK). Restriction maps for five kinds of restriction endonucleases were constructed on the basis of digestion fragment sizes.

(vi) Detection for *MThGH* fragment in recovered transgenes by PCR amplification and Southern blot hybridization

The detection of *MThGH* fragment by PCR was performed using the method described by Zhao *et al*[8]. Sense primer P_1 and anti-sense primer P_2 were located at the *MT-1* promoter and *hGH* sequence, respectively (fig. 1). The expected size of PCR product was 450 bp. For Southern blotting, various recovered transgenes digested with *Bam*HI were separated by 0.8% agarose gel electrophoresis, and transferred onto nylon membrane. DNA hybridization against DIG-labeled *MTh*GH and detection were carried out according to the user's manual (DIG DNA Labeling and Detection Kit, Boehringer Mannheim).

(vii) DNA sequencing of the regions flanking the recovered transgenes

To confirm that recovered transgenes had trapped the host DNA sequences at their ends, two DNA fragments, B/H (1 kb) on the left side of Type II transgene and B/B (0.5 kb) on the right side of Type III transgene, were subcloned into the appropriate sites on pUC18. Recombinants were named B5(B-H) and B6(B-B'), respectively. DNA sequences of these two subfragments were determined using the dideoxy sequencing method. The data collection was automatically performed on the ABI 310 Genetic Analyzer (PE Applied Biosystems). Searching for homologous sequences to the determined sequences in nucleotide sequence database (GenBank + EMBL + DDBJ) were carried out using database search program "BLAST 2.0".

2. Results

(i) Recovery of transgene

50 ampicillin-resistant colonies were recovered from genomic DNA of a F_4 transgenics. 33 colonies were randomly selected for analysis. The 33 transgenes can be classified into four types according to the results of double digestion (fig. 2). Detailed classifying results are shown as follows. Type I: Nos. 2, 15, 16, 19, 21, 26 (6 in total, 18.0%), their electrophoresis patterns were the same as that of pMThGH; Type II: Nos. 1, 3-5, 8, 10, 12-14, 17, 18, 20, 23-25, 27, 28, 30-33, (21 in total, 63.6%); Type III: Nos. 6, 7; Type IV: Nos. 9, 11, 22, 29.

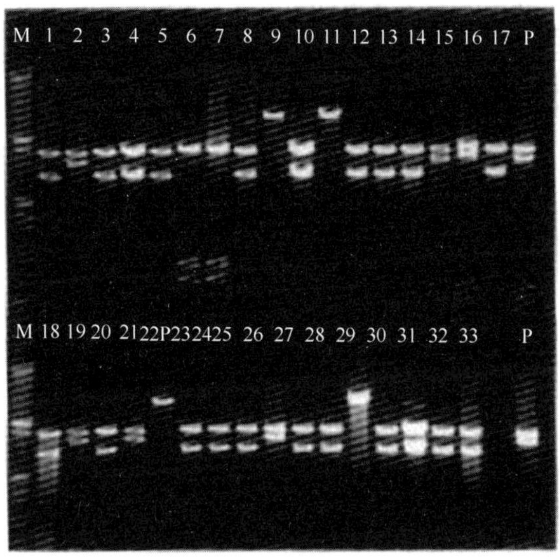

Fig. 2 The electrophoresis pattern of plasmid DNA of recovered colonies after double digestion with *Eco*R I and *Bam*H I. P, pMThGH; 1-33, recovered colonies

(ii) The restriction maps of recovered transgenes

The restriction maps for four types of recovered transgenes were constructed by digestion with *Bam*HI, *Hin*dIII, *Eco*RI , *Bgl*II and *Kpn*I (fig. 3). The map for Type I is the same as that for pMThGH, maps for the other three are very different from that for Type I.

(iii) Results of PCR amplification and Southern blot analysis for *MThGH* fragment

PCR products with 450 bp in length did not appear in all types of transgene except for Type I as revealed on electrophoresis gel (fig. 4), while all recovered transgenes could hybridize against *MThGH* probe (fig. 5). These results indicated that the sequences of most transgenes were quite different from their original form. The fact is that the analyzed 450 bp fragment changed significantly while those regions responsible for plasmid replication and *E. coli* ampicillin resistance still kept intact.

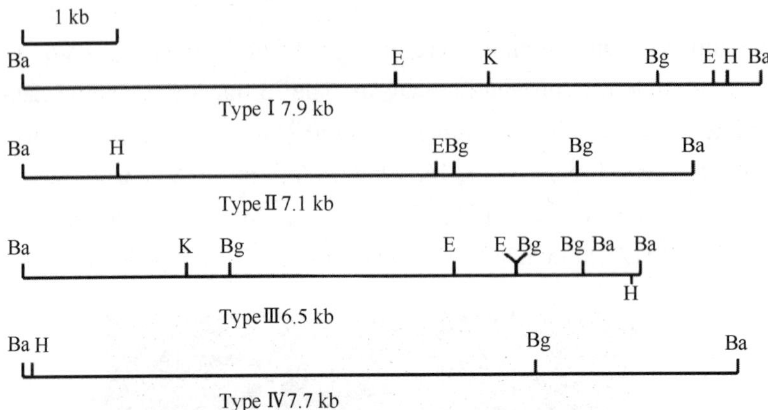

Fig. 3 Restriction maps of recovered transgenes. Type I, Original form; Types II-IV deficient form. Ba, *Bam*HI; E, *Eco*RI; K, *Kpn*I ; Bg, *Bgl*II; H, *Hin*dIII

(iv) Sequencing results of DNA fragments at the flanks of two transgenes

470 bp of DNA sequence from the *Bam*HI to *Hin*dIII sites on Type II transgene was determined. The homology between its 438 bp and the promoter or 5′ untranslation region (5′UTR) of common carp β-actin gene on the piscine *GFP* expression vector FRMwg (accession number in GeneBank: AFI709151) was pronounced (98% identity, fig. 6). The full length of B/B fragment in Type III transgene was sequenced and 509 bp was determined. The

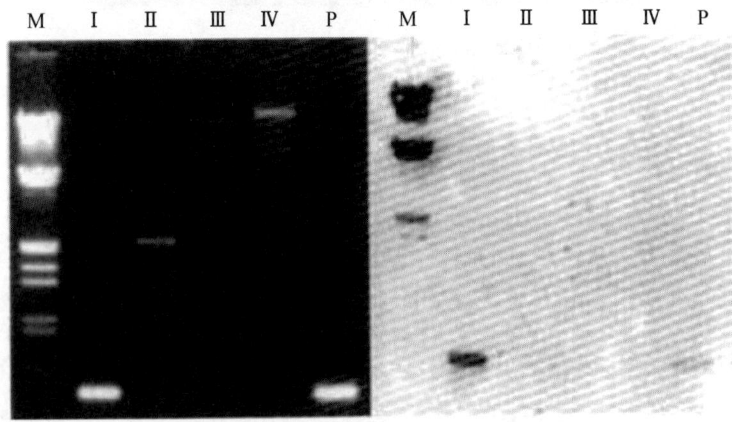

Fig. 4 Detection far *MThGH* in recovered transgenes by PCR and confirnnation by Southern blotting. DIG-labeled *MThGH* and λDNA were used as probes. M, λDNA (*Hin*dIII/*Eco*RI); P, plasmid of pMThGH; Type I, original form; Types II-IV, deficient forms

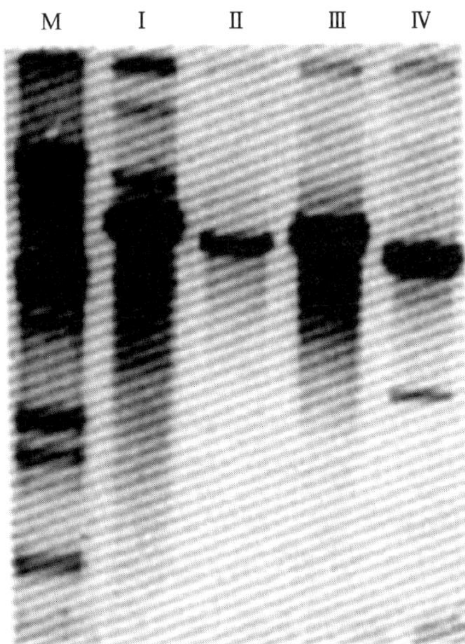

Fig. 5 Southern hybridization analysis of recovered transgene. DIG-labeled *MThGH* and λDNA were used as probes. M, λDNA (*Hin*dIII/*Eco*RI); R, plasmid of pMThGH; Type L, Original form; Types II-IV, deficient forms

```
                          BamH I
B5(B→H)(plus)     1   GGATCCCTAA GCGATTTTCA TCAAAATCGC TGTTTTTTGT TTGCGAAGTT CAATACGTTG
FRMwg(minus)          954   CCTAA GCGATTTTCA TCAAAATCGC TGTTTTTTGT TTGCGAAGTT CAATACGTTG

B5(B→H)          61   TTTTCCGTAT TTGCGTATTT TGTTGTGATA ACGCCAATAA GGCTCTTTTC AGCCTTTCAA
FRMwg           899   TTTTCCGTAT TTGCGTATTT TGTTGTGATA ACGCCAATAA GGCTCTTTTC AGCCTTTCAA

B5(B→H)         121   AGAGCCTGTG CAAAGTGCTA GTATTGGTCA TAGTGATGGA CACCTCCTTG ATCCTGTGCT
FRMwg           839   AGAGCCTGTG CAAAGTGCTA GTATTGGTCA TAGTGATGGA CACCTCCTTG ATCCTGTGCT

B5(B→H)         181   GTGTTTACAA TCTAACACAA CAGCAGCAGC AGCTCTGTAG GTCATTTACC TCTTATTGTA
FRMwg           779   GTGTTTACAA TCTAACACAA CAGCAGCAGC AGCTCTGTAG GTCATTTACC TCTTATTGTA
                                                                   *          *
B5(B→H)         141   TTCACAATAA TGGCATAATA CTGCTCTGTG GTTATCTACC AAGTTACCAA GTACAATAGG
FRMwg           719   TTCACAATAA TGGCATAATA CTGCTCTGTG GTTATCTACC AAG-TACCAA GTACAATACG
                                     *          *
B5(B→H)         301   GGTATTTTTC ATTGGAAATG AGGATTAGTG ACATGATTGT ATACTTAAGG AGCAACTAGC
FRMwg           660   GGTATTTTTC ATT-GAAATG AGGATTACTG ACATGATTGT ATACTTAAGG AGCAACTAGC
                                                                        *
B5(B→H)         361   TGGTCTGAGT TCAGTAGGTG ATGTTGTCAG AGAAAGTGTA GTGAAACTTG TTAATGAATC
FRMwg           601   TGGTCTGAGT TCAGTAGGTG ATGTTGTCAG AGAAAGTGTA GTGAAACTTG -TAATGAATC
                        *          *
B5(B→H)         421   AG-CTTGCATT CCTTGTAATT GTTAAGCTGA TTACAATTAA AATCACCAGG AGTGATATTA 47
FRMwg           542   AGGCTTGCATT CC-TGTAATT GT 521
```

Fig. 6 Nucleotide sequence comparison between B/H fragment in the flank of Type II transgene and the promoter or 5′ UTR of common carp *β*-actin gene on the piscine *GFP* expression vector FRMwg. Restriction site at the end of fragment is underlined. Base difference and gaps are indicated by asterisks

```
                               BamH I                Hind III
B6(B→B')(plus)          1   GGATCCGATT CTAGAGCGGC CGCAAGCTTA CTAGCTTTCA ACAACTCACA ACTTTGCGAC
M mRNA for RTK(minus)  298                                      AGCTTTCA ACAACTCACA ACTTTGCGAC

B6(B→B')               61   TTCCCGCTCG CATGGTCCAC TCGCTCTTGT TTACAAGTTG GCGGCAAGGA GAAACACCAC
M mRNA for RTK        270   TTCCCGCTCG CATGGTCCAC TCGCTCTTGT TTACAAGTTG GCGGCAAGGA GAAACACCAC

B6(B→B')              121   AGAAGCAGGC GGTAACAGTC TCATTTCTGT CTGAGCACAG GGAGGGTTAA GTTCCTTTTT
M mRNA for RTK        210   AGAAGCAGGC GGTAACAGTC TCATTTCTGT CTGAGCACAG GGAGGGTTAA GTTCCTTTTT

B6(B→B')              181   CCTGTTTCCT TTGCAGATTA GGATGGGAAA GGCTGTATCT TAAAGGCACT TGGTATCAGC
M mRNA for RTK        150   CCTGTTTCCT TTGCAGATTA GGATGGGAAA GGCTGTATCT TAAAGGCACT TGGTATCAGC

                                          *
B6(B→B')              241   AGGGCTTGGG GCATAGCGAG CCCTATCCAT CTTGCCCTTC ATCCAAGGCT TATCTTCTGC
M mRNA for RTK         90   AGGGCTTGGG GCAGAGCGAG CCCTATCCAT CTTGCCCTTC ATCCAAGGCT TATCTTCTGC

B6(B→B')              301   TCCTGCTCCG GCTCCTGCTC CTGCCTTAAC TGGATTGTGG GGCAGAGGGA TCTTTGTTAC
M mRNA for RTK         30   TCCTGCTCCG GCTCCTGCTC                                          11

B6(B→B')              361   AAGTAAGGTC CTGGTCAGCA TTTTCAGGAA CAATAGGGGT ATCCTCTCAT AGGCCAGGAA
                      421   TTGAATAACA GCCCTCCACC TATGTATGGT ATGCGATGAG GACCAGCCCT TGCAGGCTAA
                      481   GCTGTTCTCT GGGGTGCATT TGGGATCC                                509
                                                    BamH I
```

Fig. 7 Nucleotide sequence of B/B fragment on the flank of Type III transgene and alignment of its 288 bp with the sequence of mouse mRNA for receptor tyrosine kinase. Restriction sites are underlined. The single difference base is indicated by an asterisk

blast research result showed that the homology between DNA sequence of B/B fragment and mouse mRNA for receptor tyrosine kinase (accession number in DBJ: D13738.1) was significant. 288 bp of the former sequence (base positions 33-320) matched the latter with base identity of 99% (fig. 7).

3. Discussion

(i) The polymorphism of transgene in transgenic fish

The restriction maps for these four types of transgenes showed that only a few transgenes maintained their original construction for the founder transgenics, most transgenes are totally different from their original form in product sizes and endonucleases recognition sites in sort, number and location. This may imply that part of the sequences in most transgenes have been deleted and/or rearranged during the course of integration and inheritance, and appear to be very polymorphic.

Sequencing results confirm that some sequences in recovered transgenes Type II and Type III not only were deleted, but also were interrupted by host sequences. Similar evidence was also found in the early reported case. In the case of early postimplantation embryo lethality due to DNA rearrangements in a transgenic mouse strain (*HUGH*/3), approximately five copies of transgenes were arrayed in tandem but interrupted at least twice by mouse cellular sequences [2]. These results lead us to propose that multiple copies of transgenes may not be simply arranged in tandem and directly inserted into the chromosome at integration site, the course of integration may involve homologous and/or illegitimate recombination between transgene and transgene, and between transgene and genome. Some transgenes sequences will be deleted, arranged and/or interrupted by host sequences. Whether matters involving the polymorphism of transgenes take place in founder transgenics, or during the inheritance to the following generations, or in both of case, needs further studies.

Due to the complex composition of fragments in digested genomic DNA, it is very inefficient to self-ligate these fragments and then transform circularized transgenes into *E. coli* cell despite target fragments being enriched. The recovered transgenes are only a fraction of transgenes with molecular weight ranging from 5 to 10 kb, and those regions responsible for ampicillin selection and plasmid replication are intact. These results strongly suggest that transgenes in F_4 transgenics are highly polymorphic.

(ii) Function of transgenes in F_4 transgenics

PCR results showed that only in Type I transgene can the structure of *MThGH* fragment remain intact. Part of *MThGH* sequence in other three types of transgene is confirmed to be lost by Southern blot. Of course, transgene will be expressed and results in proper biological consequences if only there are a few intact transgenes integrated where are suitable for expression. In the cases of transgenic mouse[11], drosophila[12] and fish[7], it has been demonstrated that the expression level of transgene is significantly affected by the location of transgene. Body weight of the F_4 transgenic fish selected for this project was 3.5 times as large as the average of 17 individuals. This fact argues that the dramatic fast growing effect of this transgenics was due to a few intact transgenes and their proper location on chromosomes. We believe that some transgene integrations in this individual are genetically functional as described in our previous paper[7].

(iii) The method of plasmid rescue useful to analysing the structural feature of transgene integration site

To gain information on the sequences of integration sites, probes from transgene were used to screen genomic DNA library[2,3]. It is well known that constructing and screening genomic DNA library are complex and expensive. It sounds impracticable to construct genomic DNA libraries for every transgenic individual in case that we need to analyze the transgene integration and inheritance for several transgenics.

It is well believed that transgenes, arraying in a head to-tail manner, are integrated into host genome at random sites. Single copy of transgene can be released from concatemer by digestion with proper restriction enzyme. Released transgene at both ends of concatemer always traps the host sequences or sequences of integration site at its ends. Based on this feature, transgenes can be directly recovered from genome of transgenic animal. This method can be applied further in gene transfer studies because the whole vector sequence may be reserved in most cases for purposes of properties screen and functional sequence protection from being deleted. In this note, two DNA fragments concerning integration sites were cloned using this method, and we have demonstrated its usefulness.

Acknowledgements We are grateful to Wang Yaping for his constructive suggestion and discussion, and to Hu Wei, Sun Yonghua and Wang Wei for their kindly assistance throughout the whole project. This work was supported by the National Natural Science Foundation of China (Grant No. 39730290).

References

1. Brinster, R. L., Chen, H. Y., Trumbauer, M. E., Somatic expression of Herpes Thymidine Kinase in mice following injection of a fusion gene into eggs, Cell, 1981, 27: 223.
2. Covarrubias, L., Nishida, Y., Mintz, B., Early postimplantation embryo lethality due to DNA rearrangements in a transgenic mouse strain, Proc. Natl. Aced. Sci. USA, 1986, 83: 6020.
3. Hamada, T., Sasaki, H., Seki, R., *et al.*, Mechanism of chromosomal integration of transgenes in microinjected mouse eggs: sequence analysis of genome-transgene and transgene-transgene junctions at two loci, Gene, 1993, 128: 197.
4. He, X., Liu, G., Chen, Q., *et al.*, Exogenous gene localization on chromosomes of transgenic pig, Acta Zoologica Sinica (in Chinese), 1999, 26: 241.
5. Miao, C., Lu, G., Wang, Y., *el al.*, Mapping BNLF-1 transgene on transgenic mouse progeny chromosomes hy fluorescence *in situ* hybridization, Acta Genetic Sinica (in Chinese), 1998, 25: 422.
6. Zhang, J., Lao, W., Zhang, X., *et al.*, Genetics and expression stability of exogenous gene construct in transgenic mice, Acta Genetic Sinica (in Chinese), 1999, 26: 135.
7. Zhu, Z., Xu, K., Xie, Y., *et al.*, A model of transgenic fish, Scientia Sinica, Ser. B (in Chinese), 1989(2): 147.
8. Zhao, H., Chen, S., Sun, Y., *et al.*, Time course of foreign gene integration and expression in transgenic fish embryos, Chinese Science Bulletin, 2000, 45(8): 734.
9. Maniatis, T., Fritsch, E. F., Sambrook, J., Molecular cloning, A Laboratory Manual, 2nd ed. New York: Cold Spring Harbor Laboratory Press, 1982, 9.31-9.37.
10. Collins, F. S., Weissman, S. M., Directional cloning of DNA fragments at a large distance from an initial probe: A circularization method, Proc. Natl. Acad. Sci. USA, 1984, 81: 6812.
11. Palimiter, R. D., Brinster, R. L., Transgenic mice (Minireview), Cell, 1985, 41: 343.
12. Howard-Flander, P., West, S. C., Stsiak, A., Role of recA protein spiral filament in genic recombination. Nature, 1984, 309: 215.

转植基因在转基因鲤 F_4 群体中的多态性整合

曾志强　朱作言

中国科学院水生生物研究所，淡水生态与生物技术国家重点实验室，武汉 430072

摘　要　为了探索 pMThGH 转基因鲤 F_4 基因组中转植基因的整合模式，我们获取了 50 个转植基因位点克隆，并分析了其中的 33 个克隆。根据限制性酶切图谱，这些整合的转植基因可分为 4 种类型。其中只有一类保留了原有的分子结构，而其他三类在分子量和限制性酶切图谱上已改变。这表明大部分转植基因序列在整合和遗传时都会丢失或重排。PCR 扩增和 Southern blot 的结果表明第一类克隆的序列未变，而其他三类的序列出现了大段丢失。上述结果说明转基因鲤 F_4 的转植基因序列呈现高度多态型。我们分析了转植基因整合位点的旁侧 DNA 顺序，发现一个与鲤 β-肌动蛋白基因 5′ UTR 同源，另一个与小鼠 *RTK* mRNA 同源。

Sequences of Transgene Insertion Sites in Transgenic F_4 Common Carp

Bo Wu　Yong-Hua Sun　Ya-Ping Wang　Yan-Wu Wang　Zuo-Yan Zhu

State Key Laboratory of Freshwater Ecology and Biotechnology, Institute of Hydrobiology, Chinese Academy of Sciences, Wuhan 430072

In our previous study, fast-growing transgenic common carp (*Cyprinus carpio*) were produced by microinjection of *Bam*HI-digested plasmid pMThGH into common carp fertilized eggs (Zhu *et al*., 1989). To keep the genetic diversity of transgenic group, we produced the transgenic F_1 offspring by crossing one transgenic female with four transgenic males. The transgenic F_4 common carp were raised by hybridization between the transgenic male and transgenic female generation by generation. In F_4 generation, 100% of the offspring carried pMThGH-transgene and the transgenic offspring had improved growth rate and feed utilization efficiency compared with the controls (Fu *et al*., 1998). However, up to the present, integration information of the transgene in transgenic F_4 fish has been lacking.

Since firstly described by Perucho *et al*. (1980), the technique of plasmid rescue has been utilized in the study of various transgenic species, but it has not, to our knowledge, been employed previously in the study of transgenic fish. In this study, by use of the technique of plasmid rescue, the sequences of transgene insertion sites in transgenic fish were characterized in two heaviest individuals of 1-year-old transgenic F_4 fish. Total DNAs were partially digested with *Bam*HI, gel recovered, and treated with T4 DNA ligase. The self-ligated DNA was transformed into *E. coli*, and ampicillin-resistant clones were selected for plasmid preparation and analysis.

Among the recovered plasmids, in spite of those having the same configuration as the original pMThGH, the other five types have aberrant configurations, which were quite different from the original configuration. These aberrant plasmids were considered harboring common carp genomic sequences at insertion sites. Based on restriction analysis and Southern hybridization, host DNA fragments from three aberrant types (type I, II, III) of plasmid were subcloned into pUC19 and sequenced.

We found that these three types of host DNA fragments all have high homology to the so-called ‘house-keeping’ genes of mammals or fish. In typeI, a 3.2 kb fragment next to the transgene was sequenced from both ends. The sequence of nt. 116-481 from one end was homologous to 5′ regulatory sequence of the *mouse phosphoglycerate kinase-1* gene (gi|200323|gb|M18735.1|) with 99% identity (to see Genebank, accession number AF353996).

文章发表于 *Transgenic Research*, 2004, 13(2): 95-96

The sequences of nt. 116-481 and nt. 334-458 from the other end were homologous to *mouse phosphoglycerate kinase-1b 3′* downstream regulation region including the polyA signal (gi|53670|emb|X15340.1|) with 99 and 97% identity, respectively (to see Genebank, accession number AF353995). This implied that the host DNA sequence of type I was the *phosphoglycerate kinase-1* homologue of common carp.

In type II, a 609 bp DNA fragment adjacent to the transgene was sequenced and found 98% homologous to the upstream of *human epidermal keratin 14 (KRT14)* gene promoter region (gi|533529|gb|U11076.1|HSU11076) (to see Genebank, accession number AF353994). This suggested that the host DNA sequence of type II was common carp *KRT14* like sequence.

In type III, two DNA fragments adjacent to the transgene were analyzed. For one fragment, nt. 1-230 and nt. 241-686 were homologous to the promoter and 5′ UTR region of *common carp β-actin* gene (gi|5881101|gb|AF170915.1|AF170915) with 100 and 99% identity, respectively. For the other fragment, a 694 bp DNA sequence was found 99% homologous to *common carp β-actin* gene intron A (gi|213041|gb|M24113.1|). It is obvious that the sequence adjacent to the transgene of A-6 was *common carp β-actin* gene.

In view of the results, it seems that the integration events tend to occur at the regulatory and coding sequences in common carp genome. This interpretation is fairly reasonable, since these regions on chromosome are regularly loosened, and the loosened DNA regions facilitate the integration of foreign gene, while those non-coding sequences often emerge as compact regions on chromosome. In our study, however, the transgenic fish appeared not to suffer from any insertion mutagenesis. This may be due to that each integrated transgene was a hemizygous one, since hybridization between the transgenic F_4 fish and the non-transgenic ones did not generate 100% transgenic offspring.

Acknowledgements This work was supported by the Projects of Development Plan of the State Key Fundamental Research (G2000016109) from the Ministry of Science and Technology, China and the National Natural Science Foundation of China (NSFC 90208024).

References

1. Fu C, Cui Y, Hung SSO and Zhu Z (1998) Growth and feed utilization by F_4 human growth hormone transgenic carp fed diets with different protein levels. *J Fish Biol* 53: 115-129.
2. Perucho M, Hanahan D, Lipsich L and Wigler M (1980) Isolation of the chicken thymidine kinase gene by plasmid rescue. *Nature* 285: 207-210.
3. Zhu Z, Xu K, Xie Y, Li G and He L (1989) A model of transgenic fish. *Sci Sin* B: 147-155.

转植基因在转基因鲤 F_4 中的插入位点序列

吴 波 孙永华 汪亚平 王燕舞 朱作言

中国科学院水生生物研究所，淡水生态与生物技术国家重点实验室，武汉 430072

在先前的工作中，我们通过显微注射获得了快速生长的转基因鲤。在保持转基因群体的遗传多样性，我们利用 1 尾转基因雌鱼和 4 尾转基因雄鱼杂交，获得了 F1 转基因群体。随后，我们通过转基因鱼的自交，获得了 F_4 转基因鱼群体。在 F_4 群体中，全部后代携带转植基因 pMThGH，其后代生长效率及饲料利用效率都超过了对照鱼。但是，在 F_4 转基因鱼中转植基因的整合信息仍不清楚。

Characterization of Transgene Integration Pattern in F_4 hGH-Transgenic Common Carp (*Cyprinus carpio* L.)

Bo WU[1] Yong Hua SUN[1] Yan Wu WANG[1,2]
Ya Ping WANG[1] Zuo Yan ZHU[1]

1 State Key Laboratory of Freshwater Ecology and Biotechnology, Institute of Hydrobiology, Chinese Academy of Sciences, Wuhan 430072
2 College of Life Science, Wuhan University, Wuhan 430072

Abstract The integration pattern and adjacent host sequences of the inserted pMThGH-transgene in the F_4 *hGH*-transgenic common carp were extensively studied. Here we show that each F_4 transgenic fish contained about 200 copies of the pMThGH-transgene and the transgenes were integrated into the host genome generally with concatemers in a head-to-tail arrangement at 4-5 insertion sites. By using a method of plasmid rescue, four hundred copies of transgenes from two individuals of F_4 transgenic fish, A and B, were recovered and clarified into 6 classes. All classes of recovered transgenes contained either complete or partial pMThGH sequences. The class I, which comprised 83% and 84.5% respectively of the recovered transgene copies from fish A and B, had maintained the original configuration, indicating that most transgenes were faithfully inherited during the four generations of reproduction. The other five classes were different from the original configuration in both molecular weight and restriction map, indicating that a few transgenes had undergone mutation, rearrangement or deletion during integration and germline transmission. In the five types of aberrant transgenes, three flanking sequences of the host genome were analyzed. These sequences were common carp *β-actin* gene, common carp DNA sequences homologous to *mouse phosphoglycerate kinase-1* and *human epidermal keratin 14*, respectively.
Keywords transgenic common carp; plasmid rescue; germline transmission, integration pattern

Introduction

After the development of the transgenic "supermouse" by introduction of a novel growth hormone gene construct into mice fertilized eggs [1], gene transfer has been extensively studied in numerous species [2]. The integration, expression and germline transmission of transgenes is the prerequisite of animal transgenesis [3]. In fish embryos, since the novel gene is commonly injected into the egg cytoplasm and the early embryonic cell cleavage is

文章发表于《细胞研究》(英文版), 2005, 15(6): 447-454

fairly rapid, the integration of foreign gene occurs during embryogenesis, resulting in transgenic mosaics [4,5]. The germline transmission of transgenes in fish has been described by different authors [6-9]. However, little is known about the transgene's integration sites and stability or integrity in the host fish following reproduction.

In a tested model, transgenic common carp were produced by microinjecting a recombinant plasmid pMThGH into common carp fertilized eggs. To keep the genetic diversity of transgenic offspring and to combine different valuable traits of transgenic founders, we produced the transgenic F_1 offspring by crossing one transgenic male with four transgenic females. The F_4 pMThGH-transgenic common carp was raised through hybridization between transgenic males and transgenic females generation by generation. In F_4 generation, 100% of the offspring contained pMThGH-transgene and displayed normal expression of transgene [10]. Compared with the controls, the F_4 transgenic common carp have improved growth rate and feed utilization efficiency, just like the transgenic founders [11]. In addition, we obtained seven cross-genus cloned fish derived from F_4 transgenic common carp nuclei and goldfish eggs [12]. However, up to the present, comprehensive information about the status of the transgene copies in F_4 transgenic fish has been lacking. Since the technique of plasmid rescue was firstly described by Perucho *et al* [13], it has been utilized in the study of transgenic mice [14-17], transgenic tomato [18] and transgenic *Drosophila* [19], but it has rarely, to our knowledge, been employed previously in the study of transgenic fish. Recently, we briefly reported three integration site sequences in F_4 transgenic fish [20]; however, the detailed integration pattern of transgene in F_4 transgenic fish needs to be clarified. Here we adopted the method of plasmid rescue to study the copy number, manner and sites of transgene integration in the F_4 pMThGH-transgenic fish. This study will provide a comprehensive understanding of the integration pattern and the inheritance stability of transgenes after four-generation transmission in transgenic fish.

Materials and Methods

Production of F_4 pMThGH-transgenic fish

A recombinant DNA fragment of 3.5 kb, containing a mouse metallothionein-1 (*MT*) promoter and regulation region and human growth hormone "mini-gene" (*hGH* with introns 2, 3 and 4 deleted), was cloned into pBR322 at the *Eco*RI site. The resulting plasmid, pMThGH (Fig. 1A), was digested by *Bam*HI and microinjected into the fertilized eggs of common carp to produce the transgenic founders [4]. After being confirmed by dot blotting and PCR, a female transgenic founder with significant growth enhancement was mated with 4 male transgenic founders and the pool of F_1 offspring was produced. The transgenic ones of F_1 offspring were mated each other to produce the F_2 generation, and the F_3 generation was subsequently produced by random mating among the transgenic ones of F_2 offspring. A pair of transgenic F_3 individuals was mated to produce the F_4 offspring (Fig. 2). Two heaviest individuals (named as fish A and B) of the F_4 fish in one-year age were sampled and used in the present study.

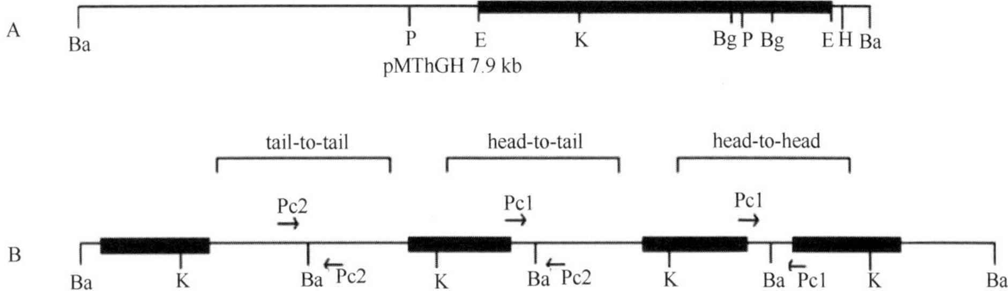

Fig. 1 The plasmid pMThGH and sketch map of various joint transgenes. (A) Physical map of pMThGH. Thick line refers to *MThGH* gene. Fine line is the vector sequence of pBR322. Ba, *Bam*HI; E, *Eco*RI; H, *Hin*dIII; P, *Pst*I; K, *Kpn*I; Bg, *Bgl*II. (B) Different types of joint transgenes. Pc1 and Pc2 refer to the primers used for PCR detection of the joint transgenes

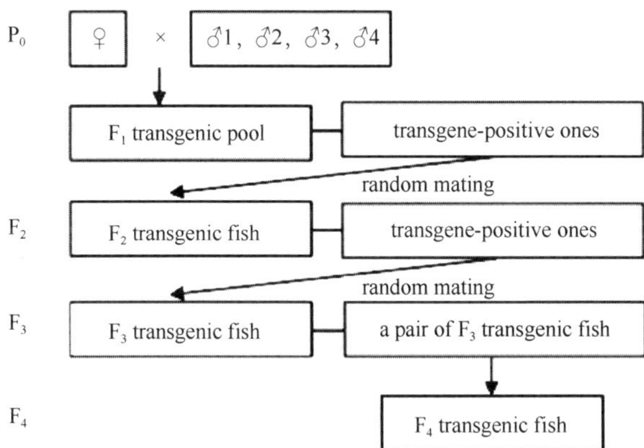

Fig. 2 Diagram of the breeding history of F_4 transgenic fish

Dot blotting and PCR analysis of transgene in host fish

High-molecular-weight genomic DNA was isolated from tail fin of the F_4 fish as described [4]. In order to identify the copy number of the transgene in F_4 fish, dot blotting was carried out with 10 μg genomic DNA of each fish against DIG-labeled pMThGH probes (DIG DNA Labeling and Detection Kit, Roche). Quantity control was equally carried out using ten-fold serially diluted pMThGH DNA, with the initiative quantity of 50 ng. Primer labeling, dot blotting and detection were performed according to the user's manual. PCR primers were designed to determine concatenate status of transgene in host fish. The sequences were Pc1 (5′-GTTAAATTGCTAACGCAGTCAGGC-3′) and Pc2 (5′-TGATGTCGGCGATATAGGCG-3′), respectively. Three combinations of primers were used for PCR amplification: Pc1 alone, Pc2 alone and Pc1+Pc2. If the concatenates were formed in a head-to-head manner, Pc1 alone would give specific PCR amplification; if in a tail-to-tail manner, Pc2 alone would give specific amplification; if in a head-to-tail manner, Pc1+Pc2 would give specific amplification. The expected amplification length of Pc1 alone is 643 bp, Pc2 alone is 159 bp and Pc1+Pc2 is 401 bp (Fig. 1B). PCR reaction parameters were as follows: 1 U of Taq DNA polymerase (MBI), 400 μM of each primer, and 100 ng of genomic DNA in a total volume of 25 μl; 94℃, 4 min, 30 cycles of 94℃, 30 sec; 60℃, 30 sec; and 72℃, 1 min.

Genetic analysis of the F_4 transgenic fish

In the two series, fish A and B were each artificially fertilized with the sperm from non-transgenic common carp. For each series, 100 hatched fry were randomly selected for extraction of total DNA. Total DNA was extracted and PCR assay was carried out to detect the presence of the transgene in hatched fry. Sense primer P_+ (5′-GGTAAGCGCCCCTAAAATCC-3′) was located across the end of exon 1 and the beginning of intron 1 and the anti-sense primer P_- (5′-TTGAAGATCTGCCCAGTCCG-3′) was at the exon 2 of the *hGH* mini-gene. The length between the two primers on the *hGH* mini-gene was 747 bp. PCR reactions consisted of 30 cycles: 30 sec at 94℃, 30 sec at 58℃, 1 min at 72℃, with a 2 min initial 94℃ denaturation and a 5 min final 72℃ elongation. Based on the transgene positive ratios in two series, the number of integration sites in each F_4 transgenic fish was deduced according to the Mendel's law. Statistical analysis was carried out as goodness-of-fit test for discrete random variables.

Recovery of the pMThGH-transgene

Transgenes were recovered by a modified plasmid rescue technique [21]. 10 μg of the genomic DNA was partially digested with 4U *Bam*HI (Promega) in 100 μl total reaction volume. At time interval of 2, 5, 10, 15, 30 min, every 1/5 volume of the reaction was collected and the reaction was stopped by addition of EDTA to a final concentration of 0.05 M. Digested DNA fragments ranging from 4 to 12 kb were recovered and purified with glassmilk (Biostar). The recovered DNA fragment in 1.8 μg/ml was self-circularized with 3 U of T_4 DNA ligase (Promega) at 16℃ overnight. The circularized DNA was transformed into competent cells of *E. coli* Top10F strain. Transformed cells were spread onto LB plates containing 50 μg/ml ampicillin. The circularized pUC19 DNA was transformed as a positive control. In addition, *Kpn*I digested genomic DNA was circularized and transformed into *E. coli* as well.

Classification, mapping, Southern hybridization and PCR of recovered transgenes

Plasmid DNA of each clone of the recovered transgene was prepared and doubly digested with *Eco*RI and *Bam*HI for classification. The classified clones are named as fish A- and fish B- series. Physical mapping of the clones was carried out with two groups of double digestion of the plasmid DNA: (1) *Bam*HI plus one of *Eco*RI, *Bgl*II, *Pst*I, *Kpn*I or *Hin*dIII, and (2) *Hin*dIII plus one of *Eco*RI, *Bgl*II, *Pst*I or *Kpn*I. Based on the restriction maps, each class of the transgene DNA was digested with appropriate enzymes for Southern hybridization: A1, A2, B1 and B2 were digested with *Bam*HI and *Eco*RI; A3 was digested with *Pst*I; A4 and B3 were digested with *Bam*HI; A5 and B4 was digested with *Hin*dIII and *Bgl*II; A6 was digested with *Hin*dIII and *Eco*RI. Southern hybridization of the digests was performed against DIG-labeled pBR322-absent *MThGH* probe. PCR analysis of the recovered transgenes was carried out as described previously.

Characterization of host genome DNA adjacent to the recovered transgenes

Based on the restriction map and Southern DNA hybridization, host DNA fragments adjacent to the transgene were deduced and subcloned into pUC19. DNA sequences of these fragments were determined using the dideoxy sequencing method with M13 universal primers. The data collection was automatically performed on the ABI 310 Genetic Analyzer (PE Applied Biosystems). According to the sequence data of these fragments, three sets of primers were designed for amplification of the supposed adjacent DNA fragments among F_4 transgenic fish and non-transgenic fish. The primer sequences are listed as follows: for fragment 1, 5′-GAATTCTACCGGGTAGGGGA-3′ and 5′-TATCTAATCCCACCCCACCC-3′; for fragment 2, 5′-GGATGGATACCCGGCTGGAA-3′ and 5′-TTGGGGCTAAGCCTGGG-CTA-3′; for fragment 3, 5′-GCCACTAAATCACACTGTCCTTGG-3′ and 5′-CTGCAGTCA-CTTCAGCGACTCTT-3′. The PCR amplification reactions were conducted with parameters similar to those for transgene detection, but with PCR cycles of annealing temperature of 55℃ and elongation time of 2 min.

Results

Copy number and concatenation of transgene in host fish

The result of dot blotting showed that the signal density of two individuals of F_4 fish was indicative of 5 ng pMThGH (Fig. 3A). Since the genome size of common carp haploid genome is about 1.7 pg [22], the number of diploid genome equivalent in 10 μg of common carp genomic DNA can be calculated as 10 μg/ 1.7 pg/2 = 3×10^6. Since the size of the plasmid pMThGH is 7.88 kb and 1 copy of 1 kb DNA molecular weights about 10^{-6} pg, it can be calculated that 5 ng pMThGH contains 6.3×10^8 (5 ng/7.88×10^{-6} pg) copies of the plasmid molecules. It can there fore be estimated that each genome of the F_4 fish contains approximately 200 ($6.3\times10^8/3\times10^6$) copies of transgene.

In the three combinations of PCR analysis with the primers Pc1 and Pc2, Pc1+Pc2 produced the specific amplification band of 401 bp (Fig. 3B), while the other two combinations (Pc1 or Pc2 alone) did not produce any specific amplification. Since the injected plasmid was linearized with *Bam*HI, those plasmids linked in a head-to-tail manner could be amplified merely with the primers Pc1 and Pc2 (see Fig. 1B). If the injected plasmids were linked in a head-to-head or tail-to-tail manner, Pc1 or Pc2 alone will give specific amplification. The results revealed that pMThGH transgenes were tandemly arrayed with a head-to-tail manner in host fish, resulting in transgene concatemers in the F_4 genome.

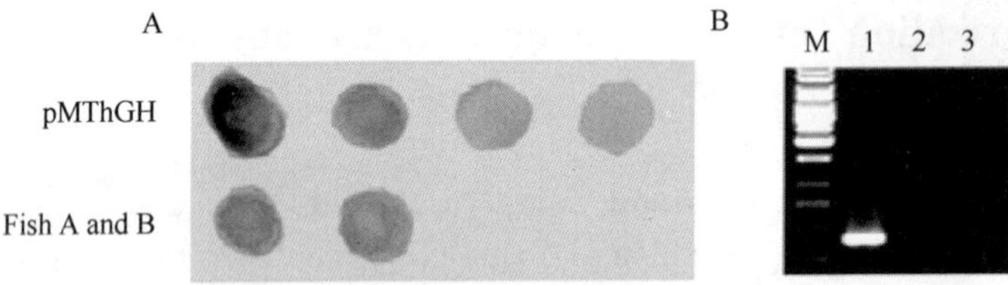

Fig. 3 Copy number and concatenation of transgene in host fish. (A) The upper series refer to dot blotting against serially diluted pMThGH DNA of ten-fold with an initiative quantity of 50 ng. The lower series refer to dot blotting against 10 μg of genomic DNA of fish A and B. (B) PCR results of transgenic fish with the primers Pc1 and Pc2. M: molecular marker 1kb DNA ladder (MBI co.); 1, Pc1+Pc2; 2, Pc1 alone; 3, Pc2 alone

Insertion numbers of transgenes in F_4 transgenic fish

In two series of mating experiments from fish A and B with non-transgenic fish, the positive ratios of pMThGH in the hybrid fry were each 96% and 94%. According to the Mendel's law, if there is only one transgene insertion in each fish, the positive ratio will be 50%; if there are one or more homozygous integration sites, the positive ratio will be 100%; if there are two or more separate hemizygous integration sites, the positive ratio will be $1-(1/2)^n$ (n indicates the number of integration sites). Since the positive ratios were 96% and 94%, the number of integration sites could be deduce as 4-5, which is accordant to the goodness-of-fit test ($P<0.05$).

Plasmid rescue

Several hundred colonies containing DNA from fish A and B were picked up from *Bam*HI digested genomic DNA, among which two hundred clones for each fish were analyzed by *Eco*RI and *Bam*HI digestion. The transgenes from fish A were classified into six classes while those from fish B into four classes according to the electrophoretic patterns. The details were shown in Tab. 1.

Tab. 1 Types of transgene colonies recovered from fish A and B

Fish A	Fish B
A-1, 7.9 kb (166 in total, 83%), the electrophoretic pattern was the same as that of pMThGH	7.9 kb (169 in total, 84.5%), the electrophoretic pattern was the same as that of pMThGH
13 kb (1 in total, 0.5%)	13kb(3 in total, 1.5%)
5.6 kb (3 in total, 1.5%)	6.5 kb (27 in total, 13.5%)
6.5 kb (26 in total, 13%)	7.1 kb (1 in total, 0.5%)
7.1 kb (3 in total, 1.5%)	
8.4 kb (1 in total, 0.5%)	

On the other hand, when *Kpn*I was used for digestion, several hundreds of colonies from two individuals were grown on LB plates containing ampicillin. 30 clones were randomly picked up from the plates and analyzed by multiple-restriction-enzyme digestion, which revealed that 73% (22/30) of the recovered transgenes were the same as the plasmid pMThGH on molecular configuration. This result gave another evidence that the transgenes were arranged as head-to-tail arrays in the host genome, since only tandemly arrayed transgenes could be recovered by *Kpn*I digestion (see Fig. 1B).

Restriction maps of recovered transgenes

The restriction maps for the ten classes of the recovered transgenes were made as described in Fig. 4. According to the restriction maps, classes A1 and B1 were the same as pMThGH, which was used for producing the transgenic founders, while the other types were different from pMThGH on both molecular weight and composition. Between the two series, the classes B2, B3 and B4 are the same as A2, A4 and A5, respectively. The variation of transgenes in host fish suggested that parts of the sequences in pMThGH transgene had been rearranged or deleted during the course of integration and four-generation transmission.

MThGH fragment detected by PCR analysis and Southern blot

All the classes of the recovered transgenes could be hybridized against *MThGH* probe (Fig. 5A, B), indicating that all the recovered transgenes maintained complete or partial *MThGH* sequence. PCR analysis showed that A1, A2, A3, A4, A5, B1, B3 and B4 gave specific amplification band identical to pMThGH. However, A6 and B2 did not produce any specific amplification (Fig. 5C). This suggested that deletion or mutation must have occurred in the PCR primer binding sequences in those transgenes. We noticed that although the restriction map of B2 was the same as A2, PCR analysis showed different results, indicating that primer binding sequences in B2 having changed a lot in base composition but not in length.

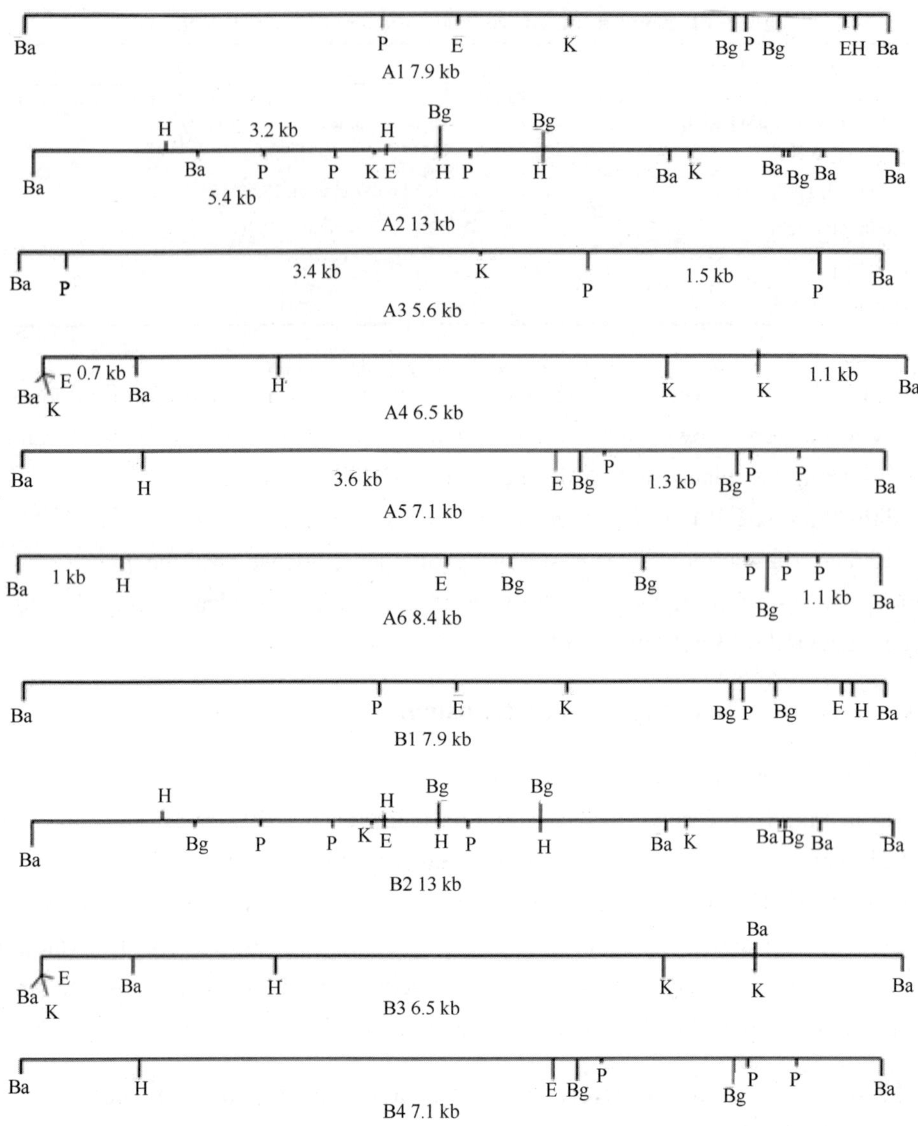

Fig. 4 Restriction analysis of six types of recovered plasmids from fish A and four types of recovered plasmids from fish B. Ba, *Bam*HI; E, *Eco*RI; H, *Hin*dIII; P, *Pst*I; K, *Kpn*I; Bg, *Bgl*II

Host DNA fragment adjacent to the transgenes

Three types of host DNA fragments adjacent to the transgenes in both fish A and B were subcloned into pUC19 and analyzed. Firstly, transgene fragments in the recovered clones were examined by Southern hybridization using DIG-labeled *MThGH* probe (Fig. 5). In A1 and B1, there was a 3.5 kb *Eco*RI hybridization fragment, i.e. the transgene *MThGH* retains its integrity. In the aberrant classes, hybridization signals appeared at a 5.4 kb *Bam*HI-*Eco*RI fragment in A2 and B2, a 3.4 kb *Pst*I fragment in A3, a 4.7 kb *Bam*HI fragment in A4 and B3, and a 3.6 kb *Hin*dIII-*Bgl*II fragment in A5 and B4, respectively. In A6, both 5.2 kb and 3.2 kb *Hin*dIII-*Eco*RI fragments were detected using the *MThGH* probe. We refer to these hybridized fragments as "hot fragments".

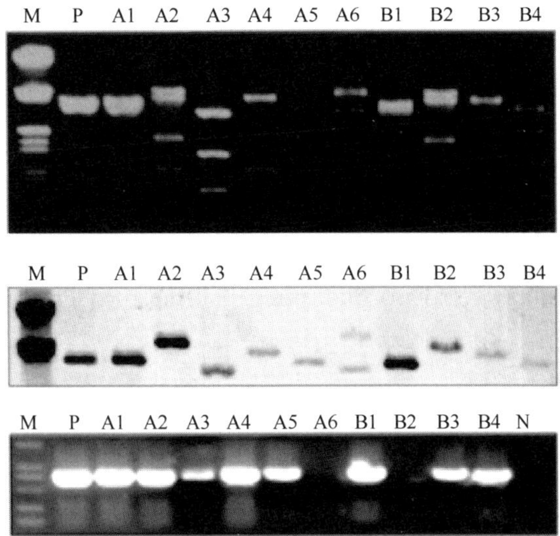

Fig. 5 Location and existence of the *MThGH*-fragment in recovered transgene clones of F_4 transgenic fish. (A) Analysis of enzyme digestion of transgene clones. M, λDNA/*Eco*RI+*Hin*dIII marker; P, pMThGH was digested with *Bam*HI and *Eco*RI; A1, A2, B1 and B2 were digested with *Bam*HI and *Eco*RI, A3 was digested with *Pst*I, A4 and B3were digested with *Bam*HI, A5 and B4 was digested with *Hin*dIII and *Bgl*II, A6 was digested with *Hin*dIII and *Eco*RI. (B) Southern blot analysis after enzyme-digestion. Probes are *MThGH* and 1 DNA marker labeled with DIG. M, λDNA/*Eco*RI+*Hin*dIII marker; pMThGH and transgene clones were digested as described above. (C) PCR analysis of recovered clones with P_+ and P_-. M, DL2000 marker; P, positive control; N, negative control

In A2, a 3.2 kb *Hin*dIII fragment next to the "hot fragment" was subcloned and sequenced from both ends. A 567 bp fragment from one end was sequenced in which NT 116-481 was homologous to 5′ regulatory sequence of the *mouse phosphoglycerate kinase-1* gene (gi|200323-|gb|M18735.1|) with 99% identity (Genebank, accession number AF353996). A 547 bp fragment from the other end was sequenced, whose NT 145-313 and NT 334-458 were homologous to the 3′ downstream regulation region of *mouse phosphoglycerate kinase-1b*, including the polyA signal (gi|53670|emb|X15340.1|) with 99% and 97% identity, respectively (Genebank, accession number AF353995). This implied that the 3.2 kb *Hin*dIII fragment of A2 was the phosphoglycerate kinase-1 homologue of common carp.

In A3, a 1.5 kb *Pst*I fragment next to the 3.4 kb *Pst*I "hot fragment" was analyzed. A 609 bp DNA fragment from one end was found 98% homologous to the upstream of *human epidermal keratin 14* (*KRT14*) gene promoter region (gi|533529|gb|U11076.1|HSU11076) (Genebank, accession number AF353994). The sequence from the other end could not be identified because of GC rich clusters.

In A6, a 1.0 kb *Hin*dIII-*Bam*HI fragment at one end of "hot fragment" was analyzed. A 698 bp DNA sequence from *Hin*dIII→*Bam*HI direction was homologous to the promoter and 5′UTR region of *common carp β-actin* gene (gi|5881101|gb|AF170915.1 |AF170915) with 100% and 99% identity at NT 1-230 and NT 241-686, respectively. A 1.1 kb *Bgl*II-*Bam*HI fragment at the other end of "hot fragment" was also analyzed. It was found that a 694 bp DNA sequence at *Bgl*II-*Bam*HI direction was 99% homologous to *common carp β-actin* gene intron A (gi|213041|gb|M24113.1|). It is obvious that the sequence adjacent to the transgene of A6 was *common carp β-actin* gene.

Furthermore, with each set of primers designed according to the fragment sequences, PCR analysis gave unique and distinct amplification among the F_4 transgenic fish and the non-transgenic fish (data not shown), indicating that the recovered adjacent sequences did belong to the genome of common carp.

Discussion

The present investigation provides new information about transgenes in transgenic common carp through four generations of transmission.

Germline transmission of transgene to F_4 offspring

We previously found that the integration of foreign genes occurred during the embryogenesis, resulting in random multiple-site integration of transgene and transgenic mosaicism [4]. Since the F_4 transgenic fish were raised from one transgenic male and four transgenic females, F_4 transgenics would inevitably contain various integration patterns. In this study, we found that each F_4 transgenic genome contained about 200 copies of transgene, which were concatameric arrays at 4-5 insertion sites. It was also found that the transgene concatemers existed as head-to-tail arrays, which is quite consistent to the order of repetitive DNA in normal genome. This implies that the transgene concatemers were formed with a mechanism similar to the formation of repetitive DNA in animal genome. Since the transgene concatemers were presumably formed in the egg or early embryo of injected founders [4], it is reasonable to assume that the transgene concatemers in F_4 genome were transmitted from the transgenic founders *via* germline.

Among all the recovered transgenes, although they were characterized into 6 classes, more than 83% maintained their original configuration. Since the transgenes were recovered through *Bam*HI digestion, self-ligation, transformation and ampicillin selection, it could be concluded that these transgenes in F_4 genome maintained the *Bam*HI cohesive ends, the plasmid replicon and Amp^r sequences. In addition, most of the recovered transgenes were the same as the injected plasmid in both molecular weight and multiple-restriction-enzyme recognition. These results revealed that most transgenes were inherited faithfully through four generation transmission. As a result of the faithful germline transmission of pMThGH transgene, F_4 transgenic common carp would as expected have inherited the genetic traits such as growth enhancement [10].

Polymorphism of transgene in F_4 genome

According to restriction maps, except A1 and B1 (with original configuration), other classes (less than 17% in proportion) were distinctly different from the original configuration of pMThGH in molecular weight and restriction enzyme recognition, suggesting that parts of the sequences in aberrant transgenes had been rearranged or deleted during the course of integration and inheritance. Although B2 was the same as A2 based on restriction analysis, different results were obtained from PCR analysis, suggesting that mutations or rearrangements occurred and there were some changes which could not be detected by restriction analysis. In addition, all the four classes of transgenes recovered from fish B could be found in fish A and the proportion of each among the four classes was similar between the two F_4 individuals. This could be explained by that the polymorphic integration patterns of transgenes in these two F_4 individuals were derived from the transgenic parents.

According to the present study, the possible reasons for polymorphism of transgene integration in F_4 genome could be speculated as following: (1) Independent assortment of chromorsomes during four-generation transmission. Since the transgenic founders received randomly and mosaically integrated transgenes as described by our previous studies [4], the transgenic offspring derived from hybridization between transgenic individuals will contain various integration patterns. In previous studies, e.g., it was found that two or three integration sites existed in the genome of P_0 transgenic common carp [23] and widespread mosaicism existed in transgenic founders of rainbow trout (*Oncorhynchus mykiss*) [24]. (2) Recombination during the course of integration and germline transmission. Recombination is thought to be very important for the integration of transgene [25,26]. Massive rearrangement of genomic DNA including deletion or translocation was observed at the integration site and the flanking region of the transgene in transgenic rice and transgenic rat [25,27]. The present study showed that 5 classes among 6 classes of the transgenes underwent great changes in restriction maps, implying that recombination occurs between transgene and transgene, or between transgene and host genome. Moreover, since there were several kinds of transgene concatemers in the transgenic founders, this might allow the generation of alleles with different repeat numbers due to misalignment during genetic recombination. (3) Mutation or rearrangement of transgene during germline transmission. In our study, although it was found that the class B2 was the same as A2 on both molecular weight and restriction map, mutations or rearrangement occurred and resulted in different results of PCR products.

Analysis of integration sites

Multiple integration sites of transgene existed in the F_4 genome, suggesting that the integration of transgene occurred randomly in the transgenic founders. For example, transgene of class A2 was integrated into the common carp sequence homologous to the *mouse phosphoglycerate kinase-1* gene, that of class A3 was flanked by common carp sequence homologous to *human epidermal keratin 14* (*KRT14*) gene and that of A6 was integrated into common carp *β-actin* gene. It is surprising that although these transgenes were integrated into coding or regulatory sequences of some housekeeping genes, the resulted fish did not suffer from any insertional mutagenesis. It may be due to that those transgenic fish carry inserted transgenes in a hemizygous status. In previous studies, however, the existence of integration hotspots had been proposed in some cases. DNA topoisomerase I or II binding sites were found to cluster around the junctions; short, purinerich tracts, some short forward and reverse overlapping sequences or short, direct repeats consisting of 4-6 bp, AT-rich S/MAR were present, either at the junction site or in the immediate flanking regions [15,26-30]. In the present study, similar results were found. Among the identified three sequences adjacent to the transgenes of F_4 transgenic fish, a 188 bp tract in A3 (NT 313-501) was found 94% homologous to the "MIR" repeat family (gi|10280853:c51526-51336), and it may be involved in the course of transgene integration.

Due to the limitation of the nature of plasmid rescue, we could not recover those transgenes which had lost the plasmid replicon or ampcilin resistance region. This resulted in the underestimation of transgene classes. However, the present study demonstrated a relatively comprehensive situation of the transgene integration pattern in F_4 transgenic fish.

Acknowledgements We express our appreciation to Prof. Norman Maclean for his valuable comments. This work was supported by the Major State Basic Research Development Program of China (No. 2004CB117406 and G2000016109) and the National Natural Science Foundation of China (No. 90208024 and 39823003).

References

1. Palimiter RD. Dramatic growth of mice that developed from eggs microinjected with metalothionein-growth hormone fusion genes. Nature 1982; 300: 680-3.
2. Boyd AL, Samid D. Molecular biology of transgenic animals. J Anim Sci 1993; 71 Suppl 3: 1-9.
3. Gordon JW. Studies of foreign genes transmitted through the germ lines of transgenic mice. J Exp Zoology 1983; 228: 313-24.
4. Zhu Z, Xu K, Xie Y, Li G, He L. A model of transgenic fish. Scientia Sinica (Series B) 1989; 2: 147-55.
5. Zhu ZY, Sun YH. Embryonic and genetic manipulation in fish. Cell Res 2000; 10: 17-27.
6. Patricia C, Christiane N, Nancy H. High-frequency germ-line transmission of plasmid DNA sequences injected into fertilized zebrafish eggs. Proc Natl Acad Sci USA 1991; 88: 7953-7.
7. Hew CL, Davies PL, Fletcher G. Antifreeze protein gene transfer in Atlantic salmon. Mol Mar Biol Biotechnol 1992;

1: 309-17.

8. Chen TT, Kight K, Lin CM, *et al.* Expression and inheritance of RSVLTR-rtGH1 complementary DNA in the transgenic common carp, *Cyprinus carpio*. Mol Mar Biol Biotechnol 1993; 2: 88-95.
9. Wei Y, Xu K, Xie Y, *et al.* Inheritance of human growth hormone gene in carp (*Cyprinus carpio* Linnaes). Aquaculture 1993; 111: 312.
10. Sun YH, Chen SP, Wang YP, Zhu ZY. The onset of foreign gene transcription in nuclear-transferred embryos of fish. Science in China (Ser. C) 2000; 43: 597-605.
11. Fu C, Cui Y, Hung SSO, Zhu Z. Growth and feed utilization by F_4 human growth hormone transgenic carp fed diets with different protein levels. J Fish Biology 1998; 53: 115-29.
12. Wu B, Sun YH, Wang YP, Wang YW, Zhu ZY. Sequences of transgene insertion sites in transgenic F_4 common carp. Transgenic Res 2004; 13: 95-6.
13. Perucho M, Hanahan D, Lipsich L, Wigler M. Isolation of the chicken thymidine kinase gene by plasmid rescue. Nature 1980; 285: 207-10.
14. Kiessling U, Becker K, Strauss M, Schoeneich J, Geissler E. Rescue of a tk-plasmid from transgenic mice reveals its episomal transmission. Mol Gen Genet 1986; 204: 328-33.
15. Rassoulzadegan M, Leopold P, Vailly J, Cuzin F. Germ line transmission of autonomous genetic elements in transgenic mouse strains. Cell 1986; 46: 513-9.
16. Grant SG, Jessee J, Bloom FR, Hanahan D. Differential plasmid rescue from transgenic mouse DNAs into *Escherichia coli* methylation-restriction mutants. Proc Natl Acad Sci USA 1990; 87: 4645-9.
17. Zoraqi G, Spadafora C. Integration of foreign DNA sequences into mouse sperm genome. DNA Cell Biol 1997; 16: 291-300.
18. Rommens CM, Rudenko GN, Dijkwel PP, *et al.* Characterization of the Ac/Ds behaviour in transgenic tomato plants using plasmid rescue. Plant Mol Biol 1992; 20: 61-70.
19. Hersberger M, Kirby K, Phillips JP, *et al.* A plasmid rescue to investigate mutagenesis in transgenic *D. melanogaster*. Mutat Res 1996; 361: 165-72.
20. Sun YH, Chen SP, Wang YP, Hu W, Zhu ZY. Cytoplasmic impact on cross-genus cloned fish derived from transgenic common carp (*Cyprinus carpio*) nuclei and goldfish (*Carassius auratus*) enucleated eggs. Biol Reprod 2005; 72: 510-5.
21. Collins FS, Weissman SM. Directional cloning of DNA fragments at a large distance from an initial probe: A circularization method. Proc Natl Acad Sci USA 1984; 81: 6812-6.
22. Gold JR, Ragland CJ, Schliesing LJ. Genome size variation and evolution in North American *cyprinid* fishes. Genetics, Selection, Evolution 1990; 22: 11-29.
23. Wang Y, Hu W, Wu G, *et al.* Genetic analysis of "all-fish" growth hormone gene transferred carp (*Cyprinus carpio* L.) and its F_1 generation. Chinese Science Bulletin 2001; 46: 143-7.
24. Penman DJ, Iyengar A, Beeching AJ, *et al.* Patterns of transgene inheritance in rainbow trout (*Oncorhynchus mykiss*). Mol Reprod Dev 1991; 30: 201-6.
25. Covarrubias L, Nishida Y, Mintz B. Early postimplantation embryo lethality due to DNA rearrangements in a transgenic mouse strain. Proc Natl Acad Sci USA 1986; 83: 6020-4.
26. Makarova IV, Tarantul VZ, Gazarian KG. Structural features of integration site of foreign DNA in the transgenic mouse genome. Molecular Biology (Mosk) 1988; 22: 1553-61.
27. Takano M, Egawa H, Ikeda JE, Wakasa K. The structures of integration sites in transgenic rice. Plant J 1997; 11: 353-61.
28. Hamada T, Sasaki H, Seki R, Sasaki Y. Mechanism of chromosomal integration of transgenes in microinjected mouse eggs: sequence analysis of genome-transgene and transgene-transgene junctions at two loci. Gene 1993; 128: 197-202.
29. Sawasaki T, Takahashi M, Goshima N, Morikawa H. Structures of transgene loci in transgenic Arabidopsis plants obtained by particle bombardment: junction regions can bind to nuclear matrices. Gene 1998; 218: 27-35.

30. Kohli A, Griffiths S, Palacios N, *et al*. Molecular characterization of transforming plasmid rearrangements in transgenic rice reveals a recombination hotspot in the CaMV 35S promoter and confirms the predominance of microhomology mediated recombination. Plant J 1999; 17: 591-601.

转植基因在转基因鲤 F_4 中的整合位点特征

吴 波[1] 孙永华[1] 王燕舞[1,2] 汪亚平[1] 朱作言[1]

1 中国科学院水生生物研究所，武汉 430072

2 武汉大学，武汉 430072

摘 要 本文对转 pMThGH 基因鲤 F_4 的基因整合图式及其邻近序列进行了深入研究。我们的研究发现，每尾转基因 F_4 个体都包含了约 200 个 pMThGH 基因拷贝，外源基因多以头尾相接的串连体形式整合。利用质粒拯救的方法，我们从两尾 F_4 转基因鱼 A、B 中回收到 400 个基因拷贝，并且将其分为 6 种类型。所有回收到的转基因拷贝都包含完整或者部分的 pMThGH 基因序列。第一类具备 pMThGH 质粒最初的特征，这一类分别占了从 A、B 两尾鱼中回收的 83%、84.5%的克隆序列，这说明经过了四代繁殖，大多数的转植基因都被忠实地传递。另外五类在分子量和酶切位点的种类、数量、位置等特征上均与原始基因不同，说明少量的转植基因在整合或生殖传递过程中发生了突变、重排或缺失。在这五类变化的类型中我们对从受体基因组中获得的转植基因的侧翼序列进行了分析，这些序列分别对应鲤的 β-肌动蛋白基因，以及与小鼠 *phosphoglycerate kinase-1* 基因同源和与人表皮角蛋白 14 同源的鲤鱼基因组序列。

Integration of Double-Fluorescence Expression Vectors into Zebrafish Genome for the Selection of Site-Directed Knockout/Knockin

Yuping Wu[1] Guangxian Zhang[1] Qian Xiong[1] Fang Luo[1] Caimei Cui[1]
Wei Hu[2] Yanhong Yu[1] Jin Su[1] Anlong Xu[1] Zuoyan Zhu[2]

1 State Key Laboratory of Biocontrol, The Open Laboratory of Marine Functional Genomics of State High-Tech Development, College of Life Sciences, Sun Yet-Sen (Zhong shan) University, Guangzhou 510275

2 State Key Laboratory of Freshwater Ecology and Biotechnology, Institute of Hydrobiology, Chinese Academy of Sciences, Wuhan 430072

Abstract Production of zebrafish by modifying endogenous growth hormone (GH) gene through homologous recombination is described here. We first constructed the targeting vectors pGHT1.7k and pGHT2.8k, which were used for the knockout/knockin of the endogenous GH gene of zebrafish, and injected these two vectors into the embryos of zebrafish. Overall, the rate of targeted integration with the characteristic of germ line transmission in zebrafish was 1.7×10^{-6}. In one experimental patch, the integrating efficiency of pGHT2.8k was higher than that of pGHT1.7k, but the lethal effect of pGHT2.8k was stronger than that of pGHT1.7k. The clones with the correct integration of target genes were identified by a simple screening procedure based on green fluorescent protein (GFP) and RFP dual selection, which corresponded to homologous recombination and random insertion, respectively. The potential homologous recombination zebrafish was further bred to produce a heterozygous F_1 generation, selected based on the presence of GFP. The potential targeted integration of exogenous GH genes into a zebrafish genome at the P_0 generation was further verified by polymerase chain reaction and Southern blot analysis. Approximately 2.5% of potential founder knockout and knockin zebrafish had the characteristic of germ line transmission. In this study, we developed an efficient method for producing the targeted gene modification in zebrafish for future studies on genetic modifications and gene functions using this model organism.

Keywords GH gene; integration; knockin; knockout; zebrafish

Introduction

Zebrafish offers many advantages as an experimental organism and has become an important model for the study of vertebrate development and other functional analyses because of

文章发表于 *Marine Biotechnology*, 2006, 8(3): 304-311

its ease of use in forward genetics, embryonic manipulation, and transgenic analysis (Amsterdam *et al.*, 1999; Traver *et al.*, 2003). In addition, transparent zebrafish embryos are well suited to manipulations involving DNA or mRNA injection, cell labeling, and transplantation (Gibbs and Schmale, 2000; Ju *et al.*, 2004). Efficient and rapid approaches to inactivating genes of known sequence have been developed and successfully applied in zebrafish (Wienholds *et al.*, 2003). Antisense morpholino-modified oligonucleotides designed to block translation or splicing of specific mRNAs can be used to knock down the function of genes identified on the basis of their sequence (Nasevicius and Ekker, 2000; Boonanuntanasarn *et al.*, 2002; Heasman, 2002). Injection of such oligonucleotides into newly fertilized eggs has highly specific effects on a wide variety of genes. Large-scale mutagenesis screening methods are available for zebrafish, allowing the identification of important genes in development and diseases (Lin *et al.*, 1994; Gaiano *et al.*, 1996; Cui *et al.*, 2003). However, in comparison with the mouse, for which the gene knockout method was developed a long time ago (Bradley *et al.*, 1984), for zebrafish the method to produce a line of targeted gene modification through homologous recombination between the introduced DNA and the corresponding chromosomal segments is lacking. After the scheduled completion of the zebrafish genome project, targeted genetic manipulation in zebrafish will become even more desirable in the biological research community.

It is generally believed that one can develop the targeted gene modification through homologous recombination in embryonic stem cells of zebrafish. However, the culture of embryonic stem cells capable of producing germ line transmission has been a technical bottleneck for a long time in targeted gene modification in zebrafish, although some cultured embryonic cells of zebrafish with embryonic characteristics have been described (Fan *et al.*, 2004). It remains to be determined if these cells will be able to produce germ line transmission after long-term culture, which is required for genetic manipulations involved in homologous recombination and selection.

Site-specific gene knockout or knockin in mice has been an important tool for the analysis of gene function *in vivo* (Koller and Smithies, 1992; Soriano, 1995; Yang and Seed, 2003; Austin *et al.*, 2004). However, the process is resource intensive and the frequencies of successful targeting are typically low. Although some of the bottlenecks in generating genetically engineered mice have been addressed by recent advances, such as techniques to speed the generation of targeting vectors (Angrand *et al.*, 1999; Testa *et al.*, 2003), major limitations remained until now. Gene targeting by homologous recombination allows precise genetic manipulation. However, the complexity and expense of the process often discourage investigators from applying it routinely. To circumvent this problem, we attempted to develop an efficient method for the production of a new line of zebrafish by targeting the endogenous gene and allowing deliberate modification of the gene of interest.

In this study, we developed an efficient integration technology that can be utilized to manipulate genetic characteristics of zebrafish, to potentially produce the gene-modified zebrafish line by gene knockout or knockin.

Materials and Methods

Zebrafish strain

Zebrafish of the AB strain were obtained from the Institute of Hydrobiology, Chinese Academy of Science.

DNA manipulations

The 15-kb growth hormone gene (BX005440.3 22150 to 37150) of zebrafish, containing the GH gene, its promoter, and the 5′ flanking sequence was amplified by genomic polymerase chain reaction (PCR) with specific primers. A 2.5-kb DNA fragment from the GH gene (BX005440.3 44660 to 47160) and 1.7-kb DNA fragment from the GH gene coding sequence (BX005440.3 35250 to 37150) were amplified separately. The promoter of the β-actin gene (AL929031.3 146351 to 147092) was isolated similarly. To be used as reporter genes, green fluorescent protein (GFP) and red fluorescent protein (RFP) coding sequences were amplified by PCR from the pEGFP-N1 vector and pDsRed-N1 vector (Clontech) with primers that are unique sequences at the vector end.

Construction of the double-fluorescence selection cassette

To make the final conditional targeting, PCR products of two reporter genes were constructed with different ends. The GFP gene was flanked by a pair of *lox*P sequences (ATAACTTCGTATAATGTATGCTATACGAAGTTAT). All PCR products were cloned to pGEM-T Easy vector (Promega). The double-fluorescent selection cassette contained the following sequences: the GFP gene flanked with *lox*P sites (ATAACTT CGTATAATGTATG-CTATACGAAGTTAT), the promoter of β-actin gene of zebrafish, and the RFP gene. The reporter and selection cassette were ligated together by different restriction endonucleases sites at the primer ends, and cloned to the pBluescript KS vector (Stratagene).

Construction of targeting vectors

The targeting vector pGHT1.7k (pGHT2.8k) was constructed by ligating the 2.5-kb GH gene fragment to the 5′ end of the GFP gene in the double-fluorescent selection cassette and 1.7-kb (or 2.8-kb) DNA fragment to the site between the 3′ end of the β-actin promoter and 5′ end of the RFP gene in the cassette. The vectors were linearized by digestion with *Bss*HII, purified by extraction with (25 ∶ 24 ∶ 1) phenol-chloroform-isoamyl alcohol (Sigma) and ethanol precipitation, dissolved in 10 mM Tris-HCl (pH 7.5), 1 mM EDTA to a concentration of 50 ng/μl, and stored at 4℃.

Restriction sites of targeting vectors were located in the following positions: *Apa*I flanked with RFP; *Spe*I and *Sca*II flanked with the 1.7-kb or 2.8-kb fragment; *Eco*RI and *Bam*HI flanked with GFP; *Cla*I and *Hin*dIII flanked with the 2.5-kb homologous arm; *Hin*dIII and *Bam*HI flanked with the β-actin promoter.

Preparation of fertilized eggs and microinjection

On the night before microinjection, females were placed in mating cages with males. After initiation of the light cycle, the females and males were removed from the mating cage separately as soon as breeding activity commenced. Before the microinjection, the sperms were squeezed from the males to a clean and dry plate, and mixed slightly with the eggs squeezed from the females. The eggs were immediately placed in Holtfreter's solution.

The linearized targeting vectors pGHT1.7k and pGHT2.8k (about 50 pg, 8×10^{-6}copies) were microinjected into zebrafish embryos at the one-cell stage of development, respectively (Zhu *et al.*, 1985). An individual egg expressing GFP was identified and selected using a fluorescein isothiocyanate filter on a stereo fluorescent microscope. Positive integration embryos were cultured in Holtfreter's solution at 28℃ until hatching. Hatched embryos were then placed in water and reared to the adult stage.

Fluorescence screening and imaging

Life embryos were observed and photographed with a Leica MZFL III stereo fluorescence dissection microscope equipped with a camera. The Leica microscope is equipped with a standard 4′,6-diamidino-2-phenylindole (DAPI) (Leica-DAPI) or a modified fluorescein isothiocyanate (FITC) filter set (Leica-GFP) with excitation from 395 to 455 nm and emission above 480 nm. Embryos were observed and screened at about 56 h post-fertilization (hpf). To immobilize embryos for photography or detailed observation, the embryos were placed in water with a final concentration of 2×10^{-5} eugenol in a 60-mm Petri dish. Anesthetized embryos, with or without an intact chorion, became immobile in about 1 min and remained viable in this solution for at least 30 min. After observation, the embryos were moved to fresh water and regained their mobility in several minutes. Photography of adults anesthetized in a 1×anesthetic solution was accomplished with excitation from a slide projector equipped with a 480-nm band-pass filter and a camera with a 558-nm long-pass filter.

Potential endogenous gene-modified zebrafish

Surviving embryos were raised to sexual maturity and crossed with noninjected mates. The P_0 germ-line chimeras were identified by the production of F_1 embryos that expressed GFP, and the resultant F_1 offspring were assessed for germ-line transmission of the targeting by PCR and Southern blot analysis. F_1 fish that hatched from the microinjected eggs were raised to sexual maturity. F_2 progeny was produced from the inbreeding of F_1 fish, by checking the GFP expression on the stereo fluorescent microscope and by testing genomic DNA extracted from pooled offspring by PCR and Southern blot analysis.

PCR and Southern blot analysis

Genomic DNA was extracted from the fish fin. PCR was carried out using oligonucleotide primers that were positioned in the flanking region of the 5′ homologous arm and the positive marker gene (Figure 1). The forward primer P1 is 5′-TCACAGATTTGGTG-TACATTAGGTAACT-3′. The reverse primer P2 is 5′-ATGATATAGACGTTGTGGCTG-TTGTAG-3′. The reaction mixture contained 200 ng of template, 1.5 mM $MgCl_2$, 10 nmol of dNTPs, 25 pmol of each primer, and 1.2 U of *Taq* polymerase in a total volume of 20 μl. The thermocycler was programmed as follows: 95℃ for 2 min; 10 cycles of 94℃ for 30 s, 60℃ for 1 min, 72℃ for 3 min; and 72℃ for 6 min; 25 cycles of 94℃ for 30 s, 55℃ for 1 min, 72℃ for 3 min; and finally 72℃ for 10 min. PCR products were separated by electrophoresis on 1% agarose gel, stained with ethidium bromide, and transferred to a Hybond-N+ nylon membrane (Amersham Life Science, Arlington Heights, IL) with 0.4 N NaOH. Hybridization with ^{23}P-labeled probe was performed at 65℃ overnight in

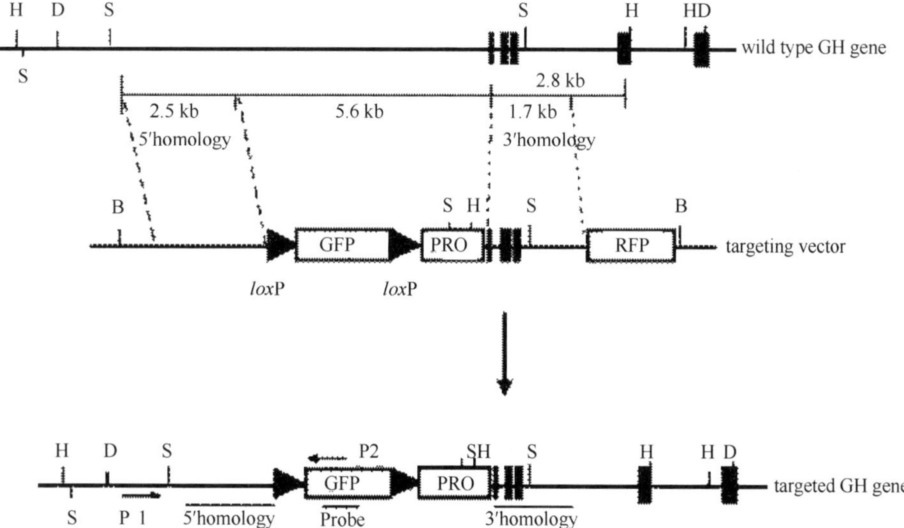

Fig. 1 Strategy of knockout GH promoter and knockin *β*-actin promoter. Boxes represent exons; transcription is from left to right. D, *Dra*III; GH, growth hormone; H, *Hin*dIII; PRO, *β*-actin promoter; S, *Sca*I. First line: structure of the growth hormone gene. Second line: the targeting vector, one contains the 2.5-kb (5′ homologous arm) and 1.7-kb (3′ homologous arm) fragments of the GH gene, while another contains the 2.5-kb (5′ homologous arm) and 2.8-kb (3′ homologous arm) fragments of the GH gene. The GFP gene flanked by *lox*P sites (filled triangles) is used as a positive selection marker, and the RFP gene is used as a negative selection marker. The pBluescript II SK (+) plasmid (Invitrogen) is used as the backbone vector. The third line shows the putative genomic structure following homologous recombination at the GH locus. P1: Forward primer; P2: reverse primer; probe: for Southern blot analysis

hybridization solution (5×saline sodium citrate [SSC], 0.02% sodium dodecyl sulfate [SDS], 1% blocking reagent [Boehringer Mannheim, Germany], 0.1% *N*-lauroylsarcosine), followed by several washes at a maximum stringency of 0.1×SSC/ 0.1% SDS at 65℃, and exposed to X-ray film at –70℃ for 48 h. The probe was labeled using the Random Primer DNA Labeling Kit (TaKaRa, cat. no. 6045), and a part of GFP-ORF from pEGFP-N1 (nt 670-1140) was used as the template, according to the manufacturer's specifications.

Results

Construction of targeting vectors

To develop a strategy for GH gene knockout/knockin in zebrafish (Thomas *et al.*, 1986; Mansour *et al.*, 1988), we designed two targeting vectors for the subsequent genetic manipulation (Figure 1). Based on the analysis of the genomic organization of the GH gene of zebrafish, a 2.5-kb DNA fragment, approximately 8 kb upstream of the GH promoter in the 5′-flanking region, was amplified and cloned into the pGEM-T Easy vector. The 1.7-kb and 2.8-kb DNA fragments in the 3′-flanking region of the GH promoter in the GH gene were amplified and cloned to the pGEM-T Easy vector, respectively. The 1.7-kb DNA fragment was comprised of exons I, II, and III and introns I and II and partial of intron III of the GH gene, while the 2.8-kb fragment included the entire exons I, II, III and partial exon IV and introns I, II, III of the GH gene. We then constructed the two targeting vectors. pGHT1.7k and pGHT2.8k that could be used in modifying the endogenous gene. The recombination- promoting targeting vectors in our study contained the *β*-actin promoter GFP as a positive selection reporter gene on the location between the 2.5-kb (5′ homologous arm) and 1.7-kb (3′ homologous arm) fragments, and between the 2.5-kb (5′ homologous arm) and 2.8-kb (3′ homologous arm) fragments, respectively. In addition, the RFP gene was used as a negative selection reporter gene for the two targeting vectors (Figure 2A, B).

Targeted insertion of vector DNA into genome in zebrafish embryos and isolation of individual recombinant colonies

To determine if the embryos could incorporate linearized plasmid DNA, the two targeting vectors (pGHT 1.7k and pGHT 2.8k) were introduced into the one-cell stage embryos by microinjection. After 56 h, the gene-integrated fish were detected by screening GFP, and fry expressing only GFP but not RFP were selected. There are three outcomes for recombination between the vector and the genome: (1) homologous recombination, (2) random insertion, and (3) partial homologous recombination at one end but not the other. Only the first is the desired, site-specific recombination. Among the fry expressing GFP, a small number of the positive fry expressed GFP in a cluster of cells, which suggested integration. However, a larger portion expressed GFP in the mosaic pattern, which could be due to expression from unintegrated

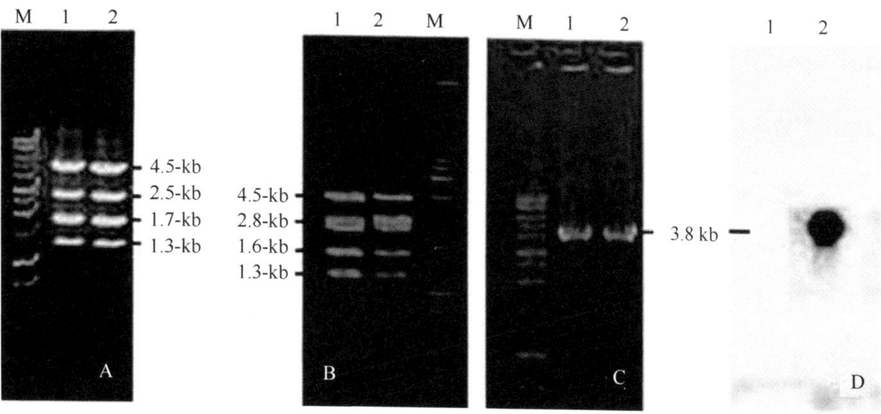

Fig. 2 Electrophoresis of targeting vectors with *Eco*RI digestion and verification of the potential homologous recombination. (A) Electrophoresis of targeting vector pGHT1.7k with *Eco*RI digestion. Lane M: 1-kb marker; lanes 1, 2: pGHT1.7k; 4.54-kb fragment: RFP 1.54-kb and pBSK 3-kb; 2.5-kb fragment: homologous arm; 1.7-kb fragment: homologous arm; 1.64-kb GFP and 68-bp *lox*P (not labeled separately because of space limitations); 1.3-kb fragment: β-actin promoter. (B) Electrophoresis of targeting vector pGHT2.8k with *Eco*RI digestion. Lane M: 1-kb marker; lanes 1, 2: pGHT2.8K; 4.5-kb fragment: RFP 1.54-kb and pBSK 3-kb; 2.8-kb fragment: homologous arm, 2.5-kb fragment: homologous arm (not labeled separately because of space limitations); 1.6-kb fragment: GFP and *lox*P; 1.3-kb fragment: β-actin promoter. (C) PCR products for identifying targeting event. Lane M: 1-kb marker; lanes 1, 2: 3.8-kb specific product produced by P1/P2 primer from potential gene-modified zebrafish genome using pGHT1.7k. (D) Southern blot analysis of DNA from putative gene-modified zebrafish. Lane 1 indicates no expected band. Lane 2 indicates the expected Southern blot result with the GFP486probe. PCR products were separated by electrophoresis on a 1% agarose gel, stained with ethidium bromide, and transferred to a Hybond-N$^+$ nylon membrane (Amersham Life Science, Arlington Heights, IL), hybridized with ^{23}P-labeled probe GFP486, which was labeled using the Random Primer DNA Labeling Kit (TaKaRa, cat. no. 6045), and a part of GFP-ORF from pEGFP-N1 (nt670-1140) was used as the template, according to the manufacturer's specifications

plasmids and/or random integration. The frequency of initial integration screening in zebrafish embryos was approximately 3×10 (130 positive fries out of 41,918 microinjections, Table 1) in our study and was observed in each patch of different microinjections.

Table1 Result of individual trials in each patch of microinjections

Injection no.	Injected eggs	Green embryos
1	1730	4
2	750	4
3	3000	11
4	5521	3
5	5451	3
6	1136	6
7	5206	40
8	19,124	59
Total	41,918	130

Verification of the potential homologous recombination using PCR assay and Southern blot analysis

Potential homologous recombination (orthotopic integration) can be detected by PCR using primers that are positioned in the flanking region of the 5′ homologous arm and the positive marker gene. The representative results based on this knockout/knockin strategy are shown in Figure 2C, D. We introduced the GFP/RFP dual-selection cassette into a targeting vector containing the zebrafish *β*-actin gene promoter. To identify the correct targeting events, we designed a simple approach based on PCR, by using primers P1/P2 for screening positive targeting embryos for the correct targeting events at the desired site. Green fluorescence colonies were first screened for the presence of recombination, while any colonies with both GFP and RFP or only RFP, which usually resulted from random insertion, were discarded. For pGHT 1.7k and pGHT 2.8k gene-targeting vectors, a specific 3.8-kb DNA fragment resulted from the PCR assay (Figure 2C). Furthermore, a total of 46 fries were analyzed by Southern blot analysis and 1 of 46 (2%) gave rise to the expected result (Figure 2D). These results suggested that the *β*-actin promoter could be knocked in the fish genome, by modifying the endogenous GH gene through homologous recombination.

The characteristic of germ line transmission

Zebrafish with the desired knockout/knockin gene were obtained by homologous recombination in this study. By using pGHT 1.7k targeting vector, a 3×10^{-3} (130/41,918) initial screening rate was achieved, resulting in 130 gene-modified zebrafish fries (56 h post-fertilization) expressing GFP only, which indicated the targeting events of potential homologous recombination occurred. The site-directed integration fries with one cluster of cells expressing GFP, not RFP, came from the same ancestor of potential homologous recombination (Figure 3A-C). The incidence of random integration (expressing both GFP and RFP in the same pattern; Figure 3D-F) was about 60%. Fries with GFP distribution scattered throughout the whole body were observed; among these some parts of body expressed both GFP and RFP, and some parts of the body expressed only GFP. For the GFP-positive zebrafish, 120 of the identified embryos were raised to sexual maturity and three were confirmed to be germ-line chimeras by the production of F_1 embryos expressing GFP only. Of these, one female produced 3142 eggs, of which 231 fries fully developed to express GFP throughout the whole body (Figure 3G-I), but not in a scattered distribution pattern. The survival rate of F_1 offspring was 26% (61/231), which was lower than that of control (about 50%). This phenomenon suggested that the GFP expressed in the living cells had only a small effect on the growth of zebrafish.

Overall, the rate of knockin with the characteristic of germ line transmission was achieved as follows: One fish out of 41,918 gave progeny that were all green (2.4×10^{-5}); in this study, only 230 out of 3142 offspring (7.3%) were green. This meant that the overall rate was about $(2.4\times10^{-5})\times(7.3\times10^{-2})=1.7\times10^{-6}$, about two fish per million.

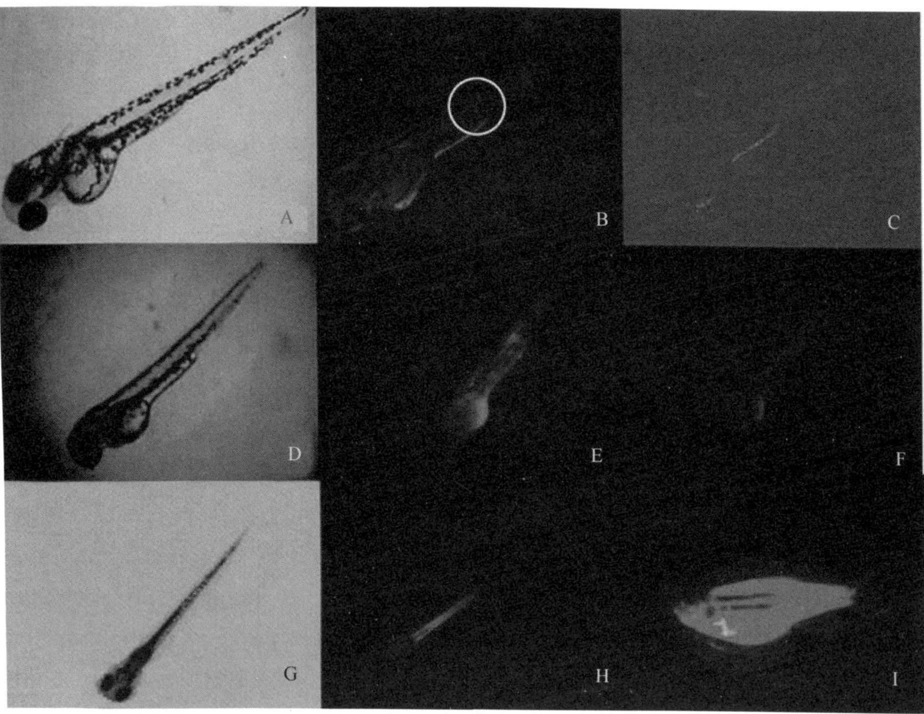

Fig. 3 Potential gene-targeted zebrafish selected by the double-fluorescence selection method. In the fry expressing only GFP, targeting events of homologous recombination occurred. (A) View in white light. (B) View at 480 nm; the green spot is circled in yellow. (C) View at 558 nm. (D, E, F) GFP and RFP expression after microinjection by random integration. (E, F) The same fry expressing GFP and RFP in the same locus at 480 nm and 558 nm, respectively, and (D) observed in white light. (G, H) F_1 fry that express GFP through the whole body. (G) View in white light; (H) view at 480 nm; (I) F_1 adult viewed at 480 nm

Efficiency comparison of two targeting vectors

Based on our study, the efficiency of two targeting vectors was different. We compared the efficiency of the two vectors pGHT1.7k and pGHT2.8k using the same experimental procedure for microinjecting them into the fertilized eggs, respectively (Table 2). In the patch of microinjections, the targeting efficiency of pGHT2.8k was higher than that of pGHT1.7k. The former reached 2.5×10^{-3}(1 out of 400), and the latter was 2×10^{-3} (1 out of 500). However, there was a stronger lethal effect for pGHT2.8k than pGHT1.7k.

Therefore, we have established an efficient method that can be used to introduce genetic alterations into the fish embryos to create lines of fish that possess the desired characteristics.

Table 2 Comparison of zebrafish embryo survival rate between two targeting vectors pGHT 1.7k and pGHT 2.8k (300 eggs operated at one time)

Constructs	75%Epiboly stage 8hpf	Bud stage 10hpf	14 Somite stage 16hpf	20 Somite stage 19hpf	Protruding mouth stage 72hpf
pGHT1.7k	270/280 96%	250/270 93%	248/250 99%	200/248 81%	150/200 75%
pGHT2.8k	246/280 88%	214/246 87%	154/214 72%	130/154 84%	78/130 60%

hpf, Hours post-fertilization

Discussion

Zebrafish that possess the desired genetic alteration can be selected based on fluorescence and propagated for further genetic and functional analyses. When the selected zebrafish are mated, they may transfer the genetic alteration to the eggs or sperm as they become sexually mature. The gene targeting approach by homologous recombination described here could produce fertile, diploid offspring. The F_1 progeny continued to express GFP through the whole body in a pattern different from that of the founder zebrafish. Most F_2 progeny expressed GFP in a pattern identical to F_1 zebrafish, whereas some F_2 could not express GFP. A stable potential knockout/knockin line could be generated and bred to homozygosity until the F_3 generation. The rate of homologous recombination with the characteristic of germ line transmission was 1.7×10^{-6} in this study, which was close to the results previously described (Cui *et al.*, 2003). Even so, the method established here can still be regarded as efficient and practical for the manipulation of genes in zebrafish without embryonic stem (ES) cells.

One factor that is likely to strongly influence gene targeting efficiency is the length of homologous arms. An extended stretch of both homologous arms may transform transient contacts between extrachromosomal DNA molecules and chromosomal sequences into relatively long-lived associations, and hence may promote recombination (Rong and Golic, 2000). Although the GH targeting vector in our study possessed relatively short (a pair of 1.7-kb and 2.5-kb fragments and another pair of 2.5-kb and 2.8-kb fragments, respectively) homologous arms, the embryos underwent potential homologous recombination at a frequency that was comparable to that reported for mouse ES cells. The limited data available from zebrafish lead us to suggest that homologous arms of 1.7-kb to 2.8-kb fragments of vectors may be sufficient for efficient targeting. These results indicate that the zebrafish line may provide the basic tools needed to develop a gene-knockout/knockin approach by modifying endogenous genes.

The conventional gene targeting strategy developed in fish was intended to enable construction of a positive-negative selection vector containing a positive marker gene (*neo*) and a negative marker gene (*tk*) and examination of its expression in fish cells. The premise of this strategy is the existence of the ES cell lines for fish and well-developed cell transfer technologies in fish, but it has not been achieved until now. To circumvent this difficulty, we designed this study by constructing double-fluorescent selection gene targeting vectors. In these vectors, the GFP was used as positive marker gene and the RFP as a negative marker gene. The recombination between the vector and the genome *in vivo* was reflected by the presence of green fluorescence excitation. Since the potential presence of homologous recombination at the early developmental stages of zebrafish embryos had been demonstrated in our study, we microinjected the targeting vector into the zebrafish embryos directly after fertilization in the one-cell stage. In this way, we could obtain the gene-modified zebrafish by two steps, faster than that via the conventional way used in the mouse. Gene targeting by homologous recombination allows precise genetic manipulation. However, the complexity and expense of the process often discourage investigators from applying it routinely. We have simplified the approach of homologous recombination (orthotopic integration) by using the GFP and RFP dual system as the positive and negative markers at the primary screening, further confirmed by PCR and Southern blot analysis.

Our approach may provide the obvious advantage of enabling researchers to produce lines of fish that overexpress a particular gene or knock out a specific gene. Such an approach can be employed to permanently manipulate genetic characteristics that are important for biological functions such as fertility, growth, immunity, and diseases. The technology is also valuable for basic studies of gene function during fish development.

The system can be used to modify endogenous genes for the needs of special lines. Although more data must be collected to accurately assess the efficiency of homologous recombination, the evidence here suggested that our procedure may provide substantially higher targeting frequencies. The method also provides a simpler approach to more sophisticated genomic manipulations, such as conditional knockouts, knockins, and large-scale chromosomal engineering.

The results from the current study may establish the foundation for the employment of knockout/knockin approach to modify endogenous genes in zebrafish. Our study provides an efficient method for altering specific genetic characteristics of fish in order to make a particular line more suitable for further genetic and functional analyses in biological community. Having established the gene modification method, one can expect that a true gene knockout/knockin approach may be established for the cost-effective large-scale manipulation of zebrafish.

Acknowledgements The authors wish to thank Professor Yaping Wang, Professor Shangping Chen, and Mrs Ming Li for their expert technical assistance. This work was supported by the National "863" Project (No. 2004AA626080), and a project from the National Natural Science Foundation of China (No. 30371113). Additional support was received from the Commission of Science and Technology of Guangdong Province (No. 2003C20303) and Key Project of Commission of Science and Technology of Guangdong Province (04205408).

References

Amsterdam A, Burgess S, Golling G, Chen W, Sun Z, Townsend K, Farrington S, Haldi M, Hopkins N (1999) A large-scale insertional mutagenesis screen in zebrafish. Genes Dev 13, 2713-2724

Angrand PO, Daigle N, van der Hoeven F, Scholer HR, Stewart AF (1999) Simplified generation of targeting constructs using ET recombination. Nucl Acids Res 27, 16

Austin CP, Battey JF, Bradley A, Bucan M, Capecchi M, Collins FS, Dove WF, Duyk G, Dymecki S, Eppig JT, Grieder FB, Heintz N, Hicks G, Insel TR, Joyner A, Koller BH, Lloyd KC, Magnuson T, Moore MW, Nagy A, Pollock JD, Roses AD, Sands AT, Seed B, Skarnes WC, Snoddy J, Soriano P, Stewart DJ, Stewart F, Stillman B, Varmus H, Varticovski L, Verma IM, Vogt TF, von Melchner H, Witkowski J, Woychik RP, Wurst W, Yancopoulos GD, Young SG, Zambrowicz B (2004).The knockout mouse project. Nat Genet 36, 921-924

Boonanuntanasarn S, Yoshizaki G, Takeuchi Y, Morita T, Takeuchi T (2002) Gene knock-down in rainbow trout embryos using antisense morpholino phosphorodiamidate oligonucleotides. Mar Biotechnol 4, 256-266

Bradley A, Evans MJ, Kaufman MH, Robertson E (1984) Formation of germ-line chimaeras from embryoderived teratocarcinoma cell lines. Nature 309, 255-256

Cui Z, Yang Y, Kaufman CD, Agalliu D, Hackett PB (2003) RecA-mediated, targeted mutagenesis in zebrafish. Mar Biotechnol 5, 174-184

Fan L, Crodian J, Collodi P (2004) Production of zebrafish germline chimeras by using cultured embryonic stem (ES) cells. Methods Cell Biol 77, 113-119

Gaiano N, Amsterdam A, Kawakami K, Allende M, Becker T, Hopkins N (1996) Insertional mutagenesis and rapid cloning of essential genes in zebrafish. Nature 383, 829-832

Gibbs PDL, Schmale MC (2000) GFP as a genetic marker scorable throughout the life cycle of transgenic zebra fish. Mar Biotechnol 2, 107-125

Heasman J (2002) Morpholino oligos: making sense of antisense? Dev Biol 243, 209-214

Ju B, Huang H, Lee KY (2004) Cloning zebrafish by nuclear transfer. Methods Cell Biol 77, 403-411

Koller BH, Smithies O (1992) Altering genes in animals by gene targeting. Annu Rev Immunol 10, 705-730

Lin S, Gaiano N, Culp P, Burns JC, Friedmann T, Yee JK, Hopkins N (1994) Integration and germ-line transmission of a pseudotyped retroviral vector in zebrafish. Science 265, 666-669

Mansour SL, Thomas KR, Capecchi MR (1988) Disruption of the proto-oncogene *int*-2 in mouse embryo-derived stem cells: a general strategy for targeting mutations to non-selectable genes. Nature 336, 348-352

Nasevicius A, Ekker SC (2000) Effective targeted gene ''knockdown'' in zebrafish. Nat Genet 26, 216-219

Rong YS, Golic KG (2000) Gene targeting by homologous recombination in *Drosophila*. Science 288, 2013-2018.

Soriano P (1995) Gene targeting in ES cells. Annu Rev Neurosci 18, 1-18

Testa G, Zhang Y, Vintersten K, Benes V, Pijnappel WW, Chambers I, Smith AJ, Smith AG, Stewart AF (2003) Engineering the mouse genome with bacterial artificial chromosomes to create multipurpose alleles. Nat Biotechnol 21, 443-447

Thomas KR, Folger KR, Capecchi MR (1986) High frequency targeting of genes to specific sites in the mammalian genome. Cell 44, 419-428

Traver D, Paw BH, Poss KD, Penberthy WT, Lin S, Zon LI (2003) Transplantation and *in vivo* imaging of multilineage engraftment in zebrafish bloodless mutants. Nat Immunol 4, 1238-1246

Wienholds E, van Eeden F, Kosters M, Mudde J, Plasterk RH, Cuppen E (2003) Efficient target-selected mutagenesis in zebrafish. Genome Res 13, 2700-2707

Yang Y, Seed B (2003) Site-specific gene targeting in mouse embryonic stem cells with intact bacterial artificial chromosomes. Nat Biotechnol 21, 447-451

Zhu Z, Li G, He L, Chen S (1985) Novel gene transfer into the fertilized eggs of goldfish (Carassius auratus L. 1758). Z Angew Ichthyol 1, 31-34

双荧光素表达载体用于斑马鱼基因组中的定点敲除或敲入筛选

Yuping Wu[1] Guangxian Zhang[1] Qian Xiong[1] Fang Luo[1]
Caimei Cui[1] 胡 炜[2] Yanhong Yu[1] Jin Su[1] Anlong Xu[1] 朱作言[2]

1 中山大学生命科学学院，有害生物控制与资源利用国家重点实验室，国家高技术研究发展计划海洋生物功能基因组开放实验室，广州，510275

2 中国科学院水生生物研究所，淡水生态与生物技术国家重点实验室，武汉，430072

摘 要 本文通过同源重组获得了内源生长激素(GH)基因修饰的斑马鱼。我们首先构建了用于斑马鱼内源 GH 基因的敲除和敲入的打靶载体 pGHT1.7k 和 pGHT2.8k，然后将这两种载体显微注射到斑马鱼的胚胎。打靶载体在斑马鱼的生殖细胞系中的定点整合效率为 1.7×10^{-6}。研究发现 pGHT2.8k 载体的定点整合效率高于 pGHT1.7k，但是 pGHT2.8k 载体整合导致的致死效应比 pGHT1.7k 更强。基于绿色荧光(GFP)和红色荧光(RFP)双色筛选，可以简捷地鉴定打靶载体发生整合的克隆，存在同源重组和随机插入两种方式。将发生同源重组的候选斑马鱼繁殖传代，通过 GFP 标记筛选 F_1 杂合子。通过 PCR 和 Southern blot 分析，确定基因组可能整合外源 GH 基因的 P_0 斑马鱼。大约有 2.5% 的斑马鱼亲本在生殖腺中发生打靶载体的敲除或敲入。本研究研发的斑马鱼靶基因定点修饰的方法，可望用于模式生物斑马鱼的遗传修饰和基因功能研究。

Site-Directed Gene Integration in Transgenic Zebrafish Mediated by Cre Recombinase Using A Combination of Mutant *Lox* Sites

Wei-yi Liu Yun Wang Yao Qin Ya-ping Wang Zuo-yan Zhu

State Key Laboratory of Freshwater Ecology and Biotechnology, Institute of Hydrobiology, Chinese Academy of Sciences, Graduate School of the Chinese Academy of Sciences, Wuhan, 430072

Abstract With current gene-transfer techniques in fish, insertion of DNA into the genome occurs randomly and in many instances at multiple sites. Associated position effects, copy number differences, and multiple gene interactions make gene expression experiments difficult to interpret and fish phenotype less predictable. To meet different fish engineering needs, we describe here a gene targeting model in zebrafish. At first, four target zebrafish lines, each harboring a single genomic *lox71* target site, were generated by zebrafish transgenesis. The zygotes of transgenic zebrafish lines were coinjected with capped Cre mRNA and a knockin vector *pZklox66RFP*. Site-specific integration event happened from one target zebrafish line. In this line two integrant zebrafish were obtained from more than 80,000 targeted embryos (integrating efficiency about 10^{-4} to 10^{-5}) and confirmed to have a sole copy of the integrating DNA at the target genome site. Genomic polymerase chain reaction analysis and DNA sequencing verified the correct gene target events where *lox71* and *lox66* have accurately recombined into double mutant *lox72* and wild-type *loxP*. Eachintegrant zebrafish chosen for analysis harbored the transgene *rfp* at the designated *egfp* concatenates. Although the Cre-mediated recombination is site specific, it is dependent on a randomly placed target site. That is, a genomic target cannot be preselected for integration based solely on its sequence. Conclusively, an *rfp* reporter gene was successfully inserted into the *egfp* target locus of zebrafish genome by Cre-*lox*-mediated recombination. This site-directed knockin system using the *lox71/lox66* combination should be a promising gene-targeting platform serving various purposes in fish genetic engineering.

Keywords cre-*lox*; gene integration; site-specific recombination; transgenic zebrafish

Introduction

Gene targeting, defined as the introduction of site-specific modifications into the genome by homologous recombination, has revolutionarized the field of genetic engineering and

文章发表于 *Marine Biotechnology*, 2007, 9(4): 420-428

allowed the analysis of diverse aspects of gene function *in vivo* (Muller 1999). Homologous recombination in mammalian cells between an artificial targeting vector and an endogenous gene was achieved at a very low frequency (Smithies *et al.* 1985). Thereafter, gene targeting based on homologous recombination in embryonic stem (ES) cells of mammalian developed rapidly (Thomas and Capecchi 1987; Skchwartzberg *et al.* 1989).

Recently the most versatile and widely applied strategy to gene targeting was based on the Cre-*loxP* recombination system (Kilby *et al.* 1993; Torres and Kuhn 1997). Cre recombinase catalyzes reciprocal site-specific recombination between two *loxP* sites (Sauer and Herdenson 1990). The *loxP* sequence is composed of an asymmetric 8-bp spacer flanked by 13-bp inverted repeats, and Cre protein binds to the 13-bp repeats mediating the recombination within the 8-bp spacer (Austin *et al.* 1981; Lee and Saito 1998). Cre recombinase can excise any intervening sequence flanked by *loxP* sites at high efficiency, which makes the Cre-*loxP* system a widely used method to achieve conditional targeted deletion, inversion, translocations, and other modifications in chromosomal or episomal DNA (Sauer and Henderson 1988; Gu *et al.* 1994; Buchholz *et al.* 2000). Also, the Cre-*lox* system has been used for site-specific integration or replacement of transfected DNA into a chromosomally positioned *lox* site (Sauer and Hendenson 1990; Feng *et al.* 1992). The integration reaction is inefficient with wild type *loxP* sites due to the reexcision of the recombined product, and therefore mutant *lox* sites have been developed to increase the efficiency of Cre mediated insertion or replacement (Albert *et al.* 1995; Araki *et al.* 1997). Targeted integration using the left element right element (LE/RE) mutant *lox* sites proved to be a useful strategy in embryonic stem cells (Araki *et al.* 2002; Liu *et al.* 2004).

In fish engineering, we transferred the human growth hormone (GH) gene into the gold fish and got the first transgenic gold fish which grows faster than the wild control (Zhu *et al.* 1985). Thereafter, we analyzed the integration patterns and mechanism of foreign transgenes in transgenic fish and found that insertion of DNA into the genome occurs randomly and in many instances at multiple sites (Zhu and Sun 2000). With current gene-transfer techniques in fish, the random integration of foreign transgenes was still a limit in fish genetic engineering. Associated position effects, copy number differences, and multiple gene interactions make gene expression experiments difficult to interpret and fish phenotype less predictable (MacLean *et al.* 1987; Lin 2000). To explore new gene transfer tools and strategies, we have tried the homologous recombination combined with positive and negative selection in common carp spleen cells to produce gene-targeting fish. However, the low recombination efficiency (10^{-6}) and the limited target cells led to the failure of producing gene-targeting fish (unpublished data). We also developed double fluorescence expression vector to mediate knockout/knockin in zebrafish although the efficiency was also low (10^{-6}) (Wu *et al.* 2006). Gene transfer based on *Sleeping Beauty* transposon was introduced to produce transgenic fish at a higher integrating efficiency (Davidson *et al.* 2003). RecA recombinase also showed efficient homologous recombination for targeted mutagenesis in zebrafish (Cui *et al.* 2003). Cre-mediated gene deletion in zebrafish suggested Cre-*loxP* system could efficiently work in zebrafish (Dong and Stuart 2004; Pan *et al.* 2005), which may hold potential uses for a knockin strategy in fish. Recently homologous recombination in embryonic stem cells of zebrafish also showed the potential for cell-mediated gene targeting in spite of many limitations need to be concurred (Fan *et al.* 2006).

Despite of some progress in homologous recombination and Cre-*lox* recombination in fish, more efficient transgenesis methods were desired both for gene function research and aquaculture application. We tried to produce some founder transgenic fish for the goal of site-specific integration where *lox*-flanked reporter gene was driven by tissue specific promoters. Site-specific insertion of any desired genes into that predefined sites in fish genome may circumvent the unwanted position effects caused by randomly integration of transgenes. Site-specific gene integration can be applied to improve important aquaculture traits and production by replacing the desired genes to known sites of founder transgenic fish. Also, site-specific gene integration can be explored to study gene function by accurate gain-of-function research which is much better than normal transgenic methods. All these applications can be used in other aquaculture species just as other genetic tools such as morpholino, TILLING, and so forth (Dahm and Geisler 2006).

This study was designed mainly to test two hypotheses. First, when a gene without a promoter is used to replace a preexisting gene in the genome, site-specific recombination will result in expression of the replacement transgene at the same rate as the replaced (recombined out) gene. Second, a highly efficient gene-targeting platform can be constructed to precisely regulate the desired gene expression in fish. Here we utilized zebrafish as a model to investigate the possibility of Cre-*lox*-mediated integration of foreign gene into the specific locus of fish genome.

Materials and Methods

Experimental fish

Zebrafish (AB strain) were reared at approximately 28.5℃, under a 14-h light/10-h dark photoperiod. After fertilization, eggs were collected and cultured in an aquarium according to *the zebrafish book* (Westerfield, M 1989). Fluorescence was observed with an Olympus SZX-12 microscope.

DNA constructs

Plasmid of *pCMVlox71eGFP* was constructed to produce transgenic zebrafish. Using *pCMVeGFP* as the template, CMV promoter was amplified with a pair of primers of $P_{CMV\text{-}G}$ and $P_{CMV\text{-}M}$ (Table 1), digested with *Bam*HI and *Bgl*II, and cloned to the *Bgl*II site of *pZeroAmp* (accession no. AY569776) to form *pZA-CMV*. Using *pCMVeGFP* as the template, a specific polymerase chain reaction (PCR) product *lox71-egfp-polyA* was amplified with primers $P_{loxP71+}$ and P_{71G-}. *Lox71-egfp-polyA* was inserted into *Bam*HI site of *pZeroKan* (accession no. AY569777) to form *pZK-lox71eGFP*, which was digested with *Bam*HI and *Xho*I and the *lox71-egfp* fragment was inserted into *pZA-CMV* to form *pCMVlox71e*GFP. Plasmid of *pZKlox66RFP* was constructed as the targeting vector. Using *pXeXRFP* as the template, a specific PCR product containing *rfp* flanked by *lox66* was amplified with primers $P_{lox66R+}$ and P_{66R-}. The PCR fragment was digested with *Bgl*II and inserted into *Bgl*II site of *pZeroKan* to form the plasmid *pZKlox66RFP*. Capped Cre mRNA was used to mediate gene deletion efficiently in cultured cell (Plas *et al.* 2003). So plasmid *pZKCre* was constructed as the template to transcribe and cap *cre* mRNA *in vitro*. The small *Sac*I-*Xho*I fragment of plasmid *pCCre* was inserted into *pZeroKan* to form *pZKCre*. The sequences of all the *lox* sites contained in these plasmids were confirmed by DNA sequencing.

Table 1 Primers used in the experiment

$P_{lox71G+}$	ATAGGATCCTACCGTTCGTATAGCATACATTATACGAAGTTATACCGGTCGCCACCATG
$P_{lox71G-}$	TTTAGATCTACGCCTTAAGATACATTGATG
$P_{CMV\text{-}G}$	TTTTAGATCTACGCGTGGAGCTAGTTATTA
$P_{CMV\text{-}M}$	TTTTGGATCCGGAGGCTGGATCGGT
$P_{lox66R+}$	TTTAGATCTATAACTTCGTATAGCATACATTATACGAACGGTACCATGGTGCGCTCCTC
P_{66R-}	TTTAGATCTGATCTGATCTAGAGGATC
P_{gfp+}	TCCAGGAGCGCACCATCTT
P_{gfp-}	TGCTCAGGTAGTGGTTGTCGG
P_{CMV+}	GCAGTACATCAATGGGCGTGGAT
P_{R-}	TCGCATGAACTCCTTGATGACGTT
P_{rfp+}	TCCAAGAACGTCATCAAGGAGTT
P_{rfp-}	AGCTTCAGGGCCTTGTGGAT

Generation of stable transgenic *pCMVlox71eGFP* zebrafish lines that hold the *cmv-lox71-egfp* integration site

100ng/μl of plasmid *pCMVlox71eGFP* solutions were microinjected into zebrafish embryos at the one-cell stage via drawn glass microcapillary pipettes attached to a micromanipulator. Injection was driven by compressed air. Microinjection volumes were estimated at 1 nl per embryo, based on calibrations using known quantities of solution. Zebrafish showing strong green fluorescence during the segmentation stage were selected as the P_0 generation. Mating the P_0 generation with wild-type zebrafish produced F_1 progeny, and mating F_1 generation with wild-type zebrafish produced F_2 progeny. Maternal effects of green fluorescence might confuse the selection of positive transgenic zebrafish, so genomic PCR was utilized to facilitate and confirm the selection process. Genomic PCR was manipulated to verify the transgenic zebrafish with primers P_{gfp+} and P_{gfp-}. P_{gfp+} was designed at 281 bp of *egfp* and P_{gfp-} was designed at 589 bp of *egfp*. The expected PCR bands were 329 bp.

Production of *in vitro* transcribed cre mRNA

The *pZKCre* vector was linearized with *Apa*I (MBI Fermentas, Lithuania), purified using a PCR purification kit (V-gene, China), and used as DNA templates for *in vitro* transcription reaction. Transcription and Purification of mRNA were carried out with the SP6 Message Machine Kit (Ambion, Austin, TX) according to the manufacturer's instructions. RNA was stored at –70℃ in small aliquots.

Coinjection of cre mRNA and knockin vector *pZKlox66RFP*

One-cell embryos of the F_3 *CMVlox7-egfp* zebrafish were coinjected with Cre mRNA and knockin vector *pZKlox66RFP*. In preliminary experiments, the injection dose of *pZKlox66RFP* was 100 pg per embryo, and the injection dose of Cre mRNA was 1.0, 5.0, 10.0, 20.0, 50.0, and 100.0 pg per embryo. For coinjection experiment, a nominal concentration of 10 pg Cre mRNA and 100 pg *pZKlox66RFP* per embryo yielded less toxicity to zebrafish embryos, whereas an equivalent amount of control *pZKlox66RFP* injection group yielded embryos indistinguishable from wild-type embryos.

PCR analysis and DNA sequencing

Total genomic DNA of two red-fluorescent embryos at 4 dpf (days postfertilization) were extracted with the lysis buffer [pH 8.0 10 mM Tris.Cl, pH 8.0 0.1 M EDTA, 2% sodium dodecyl sulfate (SDS), 20 μg/ml of RNase]. The genomic DNA was dissolved in 20 μl of ddH_2O. The PCR was performed in a 25-μl solution containing 5 μl of genome DNA, 200 μM dNTPs, 10 pmol of primers, 1.0 U of *Taq* polymerase (Fermentas, Lithuania), 50 mM KCl, 1.5 mM $MgCl_2$, and 0.001% gelatin in 10 mM Tris-HCl, pH 8.3. The primers for *egfp* are P_{gfp+} and P_{gfp-}, the primers for *rfp* are P_{rfp+} and P_{rfp-}, and the primers for the recombination fragment are P_{cmv+} and P_{R-}. A GenAmp PCR System 9700 (Perkin Elmer, Waltham, MA) was used with the following program: a predenaturation at 94℃ for 5 min, 30 cycles of amplification (94℃ for 30 s, 57℃ for 30 s, 72℃ for 30 s) and a final extension at 72℃ for 7 min. The PCR products were separated using 1.0% agarose gel electrophoresis. The PCR products were purified, cloned into the T-vector (Takara, Japan), and sequenced.

Results

Description of the gene integration strategy

Our site-directed, gene-targeting system in zebrafish was mainly based on two vectors:

one was the target vector *pCMVlox71eGFP*, the other was the knockin vector *pZKlox66RFP*. The experimental strategy was designed to remove many of the variables that have been implicated to cause gene silencing in other systems, namely, the generation of the inverted repeat, multiple transgene copies, and variable chromosomal locations. The gene-targeting strategy is depicted in Figure 1 and as follows:

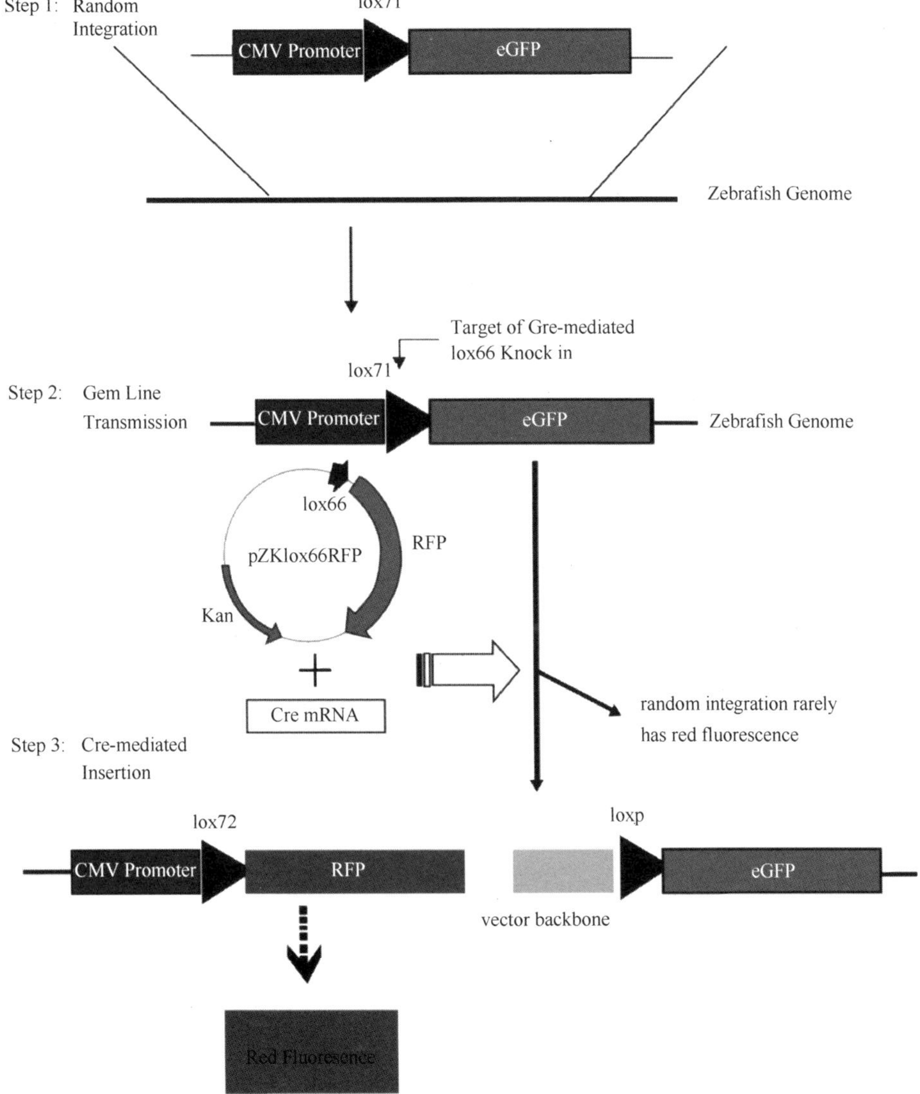

Figure 1 Three-step strategy to construct a site-specific gene integrating platform in zebrafish. Step1: structure of the target vector and its random integration into zebrafish genome; step 2: germline transmission of the transgene; step 3: Cre recombinase-mediated integration of knockin vector into the targeted site previously introduced into the zebrafish genome. The right integrant zebrafish could show red fluorescence

Step 1: Random integration. By microinjection of the target vector pCMV*lox71* eGFP, the target vector was randomly integrated into the zebrafish genome. Following the effective integration of the transgene, green fluorescence zebrafish were selected as the positive parental generation to produce the transgenic lines.

Step 2: Germ line transmission. By mating the P_0 green fluorescence zebrafish with the wild-type zebrafish, the F_1 generation embryos of green fluorescence were selected. The positive fish were mated with wild-type zebrafish until zebrafish with a stably integrated reporter gene were identified, which became founders for further studies.

Step 3: Cre-mediated gene insertion. The reversibility of the recombination reaction catalyzed by Cre recombinase can lower the targeting efficiency. The mutant *lox*, *lox71*, has 5 bp mutated in the left 13-bp repeat, and the mutant *lox*, *lox66*, has 5 bp mutated in the right 13-bp repeat (Siegel *et al.* 2001). Recombination between a chromosomally located *lox71* site and a *lox66* site on a targeting plasmid results in site-specific integration of the plasmid, producing a double mutant *lox* site with mutated sequences in both the 13-bp repeats and a wild-type *loxP* site. Since the binding affinity of the *lox* mutant *lox72* with Cre recombinase was reduced, the integrated plasmid was stably retained (Araki *et al.* 1997).

The zygotes of transgenic zebrafish were coinjected with a capped mRNA coding Cre recombinase and the knockin vector *pZKlox66RFP*. The latter consisted of a *lox66* flanked RFP coding sequence without eukaryotic promoter. The expression of Cre mRNA can induce a site-specific recombination event between the *lox71* site in the zebrafish genome locus and the *lox66* site of the knockin vector, allowing the integration of the whole vector. Red fluorescence was used to select the targeted embryos in which Cre-mediated insertion occurred. Genomic PCR analysis and the PCR product sequencing were used to verify the right targeting event but not foreign gene's random insertion.

Stable transgenic *pCMVlox71eGFP* zebrafish lines holding *lox71* flanked *egfp* target transgene

Utilizing fluorescence proteins as reporters is a convenient and effective way in zebrafish to various goals (Wan *et al.* 2002; Linney and Udvadia 2004). We use GFP and RFP as the report markers for the convenient selection of transgenic lines. A total of 112 P_0 positive transgenic zebrafish, which showed green fluorescence under 480-nm excitation light, were obtained and matured. Mating these zebraifish with wild-type zebrafish, we obtained 6 P_0 female transgenic zebrafish in which transgene was integrated into germ cells. By mating F_1 generation with wild-type zebrafish, we obtained the F_2 generation. By analyzing F_2 generation with PCR detection, we obtained 4 F_1 female transgenic zebrafish in which the progeny was near 50% positive. By mating the heterozygous *lox71egfp+/lox71egfp-* F_2 generation with wild-type zebrafish, we obtained F_3 generation served as the targeted embryos.

Owing to the carryover of GFP protein from the mother (Kane and Kimmel 1993), all of the embryos showed the green fluorescence during the early embryo development (Figure 2).

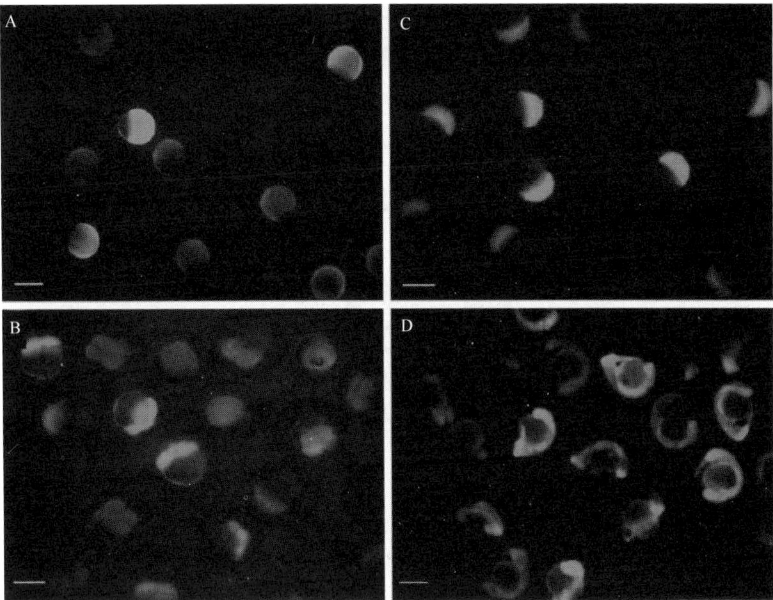

Figure 2 F_3 heterozygous transgenic zebrafish embryos derived from the mating of F_2 female *pCMVlox71eGFP* with wild male fish. Owing to the carryover of GFP protein from the mother, green fluorescence can be visualized at the very early stage and seemingly 100% positive. (A-D) The green fluorescence was observed at different developmental stages. (A) Zygotes; (B) four-cell stages; (C) blastula stage; (D) 24-somite stages

Site-directed integration of *rfp* reporter gene

We took the heterozygous *lox71gfp+/lox71gfp* F_3 generation as the targeted zebrafish zygotes. By coinjecting 10 ng/μl of capped mRNA coding Cre recombinase and 100 ng/μl of knockin vector *pZKlox66RFP*, totally we injected about 80,000 embryos and two red fluorescence zebrafish were finally obtained from one of the four transgenic lines (Figure 3A and D). Surprisingly, the two zebrafish also showed green fluorescence exactly at the sameposition of red fluorescence (Figure 3C and F), which suggested that the two fluorescent proteins were expressed at the same region. These phenotypes for overlapped expression of RFP and GFP indicated the genotypes of the two zebrafish should be *lox71gfp-lox72rfp* concatermer, which implied the incomplete gene replacement and site-directed *lox* recombination. Both *egfp* and *rfp* could be detected by PCR in the two red embryos also confirmed the point (data not shown).

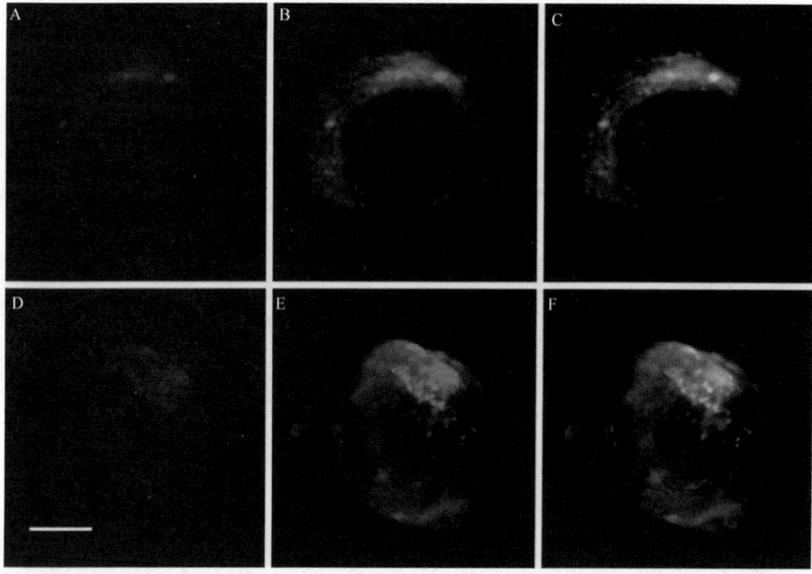

Figure 3 The right gene targeted zebrafish showed both green fluorescence and red fluorescence at the same position. (A-C) The same integrant zebrafish at 10-somite stage. (D-F) The other integrated zebrafish at 10-somite stage. (C) Merged picture of A and B. (F) Merged picture of C and D. As shown, the fluorescence overlapped and the green fluorescence was still present, which implied incomplete gene integration in the genome locus. Scale bar=250 μm

Molecular analysis of integrant zebrafish

In some cases, a promoterless foreign gene integrates into a genome behind an endogenous promoter and is expressed accordingly, which is referred to as promoter trapping (Wadman *et al.* 2005). To exclude the possibility of promoter trapping in red fluorescent zebrafish, the genomic DNA from RFP-positive embryos was isolated and PCR analysis was taken to verify the gene-targeting event. The forward primer was P_{cmv+} designed at the position of CMV promoter 586bp, and the reverse primer was P_{rfp-} designed at the position of RFP 17bp (Figure 4A). Both primers were specific to CMV promoter and *rfp* gene, respectively. The expected 359-bp PCR products were amplified with P_{cmv+} and P_{rfp-} in red fluorescence zebrafish but not in the control (Figure 4B). The identities of PCR amplified gene specific products was confirmed by DNA sequencing (Figure 4C). As was expected, recombination reaction between *lox71* and *lox66* led to double mutant *lox72* and wild-type *loxP*.

Discussion

We produced a stable transgenic *pCMVlox71eGFP* zebrafish line holding a *lox71* and enhanced *gfp* driven by CMV promoter. The capped Cre mRNA and the knockin vector *pZKlox66RFP* were coinjected into F_3 generation heterozygous zebrafish zygotes and we successfully obtained the targeted zebrafish embryos that showed red fluorescence at the position of green fluorescence located. Genomic PCR analysis and DNA sequencing verified that the *lox66* flanked *rfp* gene precisely integrated into the *lox71* site of zebrafish genome.

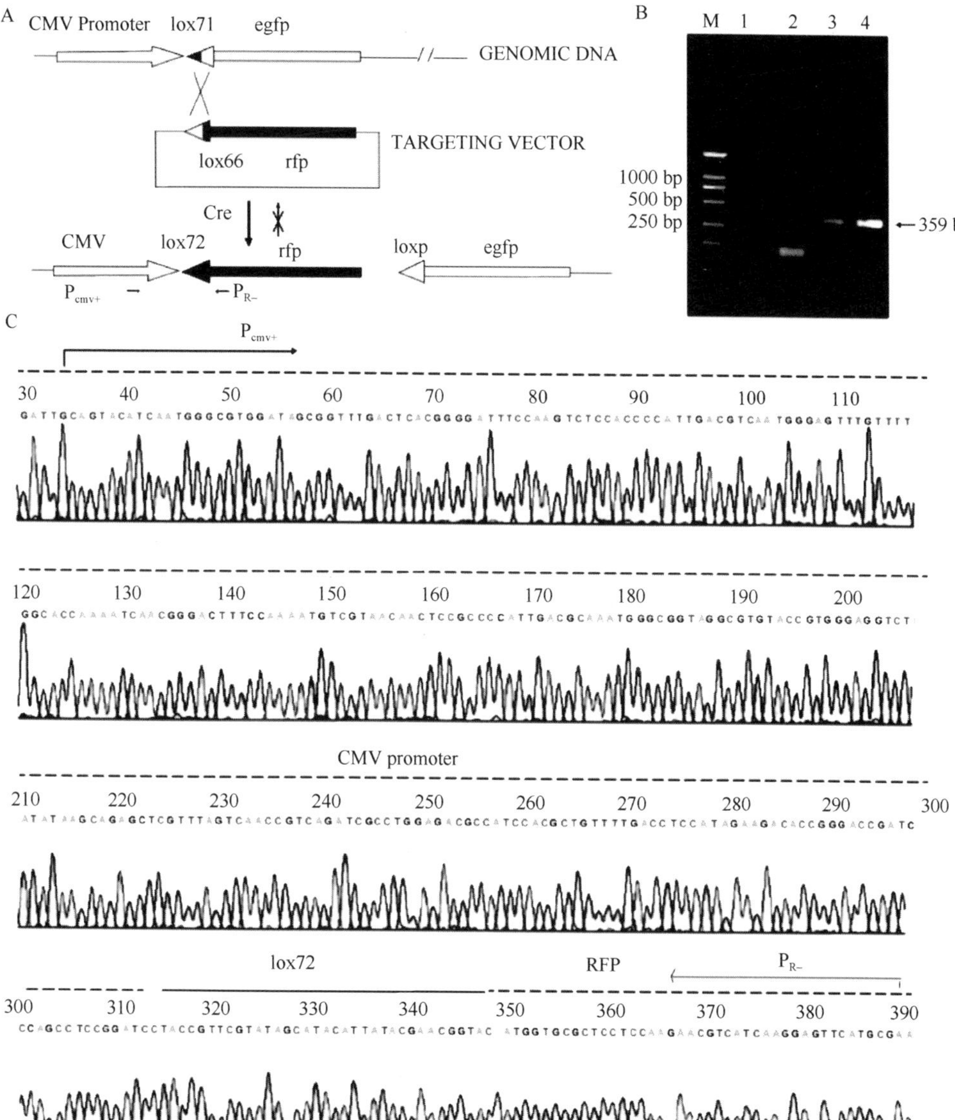

Figure 4 PCR analyses of genomic DNA from embryos of Cre mRNA-mediated gene targeting. (A) Expected DNA structure for site-specific integration. The small arrows indicate the positions and directions of the PCR primers. The expected sizes of PCR products are indicated. (B) Agarose electrophoresis of genomic PCR. *Lanes 1-4* represent genomic DNA templates from different embryo samples. 1, Wild-type embryos; 2, non-red embryos; 3 and 4, red fluorescence embryos. (C) DNA sequencing of the PCR products confirmed the right gene-targeting event. P_{cmv+}: 33-55 bp, *lox72*: 314-347 bp, P_{R-}: 368-391 bp

We have shown that a mutated *lox* site, *lox71*, can be targeted for site-specific recombination by a second mutated *lox* sequence, *lox66*, to produce a stable double mutant. When *lox71* and *lox66* recombined into double mutant *lox72* and wild-type *loxP*, the resulting integration is predicted to be stable due to the likely poor function of the double mutant *lox72* site formed in the process (Araki *et al.* 2002). However, the integrate efficiencies varied with different *egfp* target sites, the same as in mouse (Vooijs *et al.* 2001). The data that both targeted embryos were from one transgenic line indicated the chromosome environment of different targeting sites might influence the gene-targeting efficiency. Also from the consistent fluorescence observation we evaluated that RFP and GFP were expressed nearly at the same levels. These data supported the hypothesis that this target site could express different desired genes and obtain basically same expression levels, which should be very useful to identify a new gene by gain-of-function.

However, we also found the strategy is not very efficient. In total, about 80,000 eggs from four transgenic lines were injected and roughly 20,000 eggs per line. It turned out that the overall integration efficiency was about 10^{-4} to 10^{-5}, which was much less than gene deletion based on *loxP* in zebrafish (Dong and Stuart 2004; Pan *et al.* 2005). During the process of RFP screening, we also find some RFP-positive embryos now and then, but nearly all of those RFP are expressed in a very small area of body, which were precluded from our view because the correct targeted embryos are supposed to be CMV-*rfp* genotype where red fluorescence should be expressed in a large part of body or whole body. There are several limited steps that can be improved and optimized. First, the system as described in the work requires co-injection of Cre mRNA and the knockin vector, which may lead to the low efficiency of the successful gene-targeting event. Injection of *lox*-containing plasmids with mRNA encoding Cre recombinase can lead to a large number of recombination products that can reduce the efficiency of the site-specific recombination of the sequence in the plasmid with the genomic target site. Moreover, if there were multiple genomic target sites, recombination between them can occur. Toxic effects preclude increasing the level of Cre in the cell (Loonstra *et al.* 2001). Hence, a single-copy transgenic zebrafish line with the *lox66/lox71* flanking sequence may be more efficient (Bronson *et al.* 1996; Rahman *et al.* 2000). It is also highly advisable to construct some Cre-expressing transgenic zebrafish lines. Second, the excision of target transgenes among *lox71*s on the chromosome also showed that recombination reaction between two *lox71* is much easier than that of *lox71* and *lox66* according to our observation.

To conclude, four *cmv-lox71-egfp* transgenic zebrafish lines, in which *cmv* was the promoter and *lox71* flanks *egfp*, have already been produced. Site-specific integration into these lines will enable insertion of any desired exogenous gene fragments at predetermined locations in zebrafish genome and may thereby allow avoidance of the so-called “position effects” by the selection of specific effective insertion sites. Our results also showed that a transgene can be delivered into a specific chromosome position and be expressed at a predicted level. It seemed likely that target sites can be found such that the chromosomal context will yield a higher and more consistent level of transgene expression. Although it was very low in this system, the integration efficiency can be improved greatly by producing single-copy target strain. Taken together, Cre-*lox* recombination offers a method for precise insertion of single-copy DNA into genomic targets, which holds great potential for gene-transfer studies in fish.

Acknowledgements We express our appreciation to Gang Wu for plasmids of pZeroKan and pZeroAmp, Gary Stuart for his valuable comments, and reviewers for critical suggestions. This study was supported by the National 973 Project of China (Grant 2004CB117406), the National Natural Science Foundation of China (Grant 30000090), and the Frontier Science Projects Program from the Institute of Hydrobiology, Chinese Academy of Sciences (Grant No. 220311).

References

Albert H, Dala EC, Lee E, Ow DW (1995) Site-specific integration of DNA into wild type and mutant lox sites placed in the plant genome. Plant J 7, 649-659

Araki K, Araki M, Yamamura K (1997) Targeted integration of DNA using mutant lox sites in embryonic stem cells. Nucleic Acids Res 25, 868-872

Araki K, Araki M, Yamamura K (2002) Site-directed integration of the cre gene mediated by Cre recombinase using a combination of mutant lox sites. Nucleic Acid Res 30, e103

Austin S, Ziese M, Sternberg N (1981) A novel role for site-specific recombination in maintenance of bacterial replicons. Cell 25, 729-736

Bronson S, Plaehn E, Kluckman K, Hagaman J, Maeda N, Smithies O (1996) Single-copy transgenic mice with chosen-site integration. Proc Natl Acad Sci USA 93, 9067-9072

Buchholz F, Refaeli Y, Trumpp A, Bishop J (2000) Inducible chromosomal translocation of *AML1* and *ETO* genes through Cre/*loxP*-mediated recombination in the mouse. EMBO Rep 1, 133-139

Cui Z, Yang Y, Kaufman C, Agalliu D, Hackett P (2003) Rec-A mediated, targeted mutagenesis in zebrafish. Mar Biotechnol 5, 174-184

Dahm R, Geisler R (2006) Learning from small fry: the zebrafish as a genetic model organism for aquaculture fish species. Mar Biotechnol 8, 329-345

Davidson A, Balciunas D, Mohn D, Shaffer J, Hermanson S, Sivasubbu S, Cliff M, Hackett P, Ekker S (2003) Efficient gene delivery and gene expression in zebrafish using the Sleeping Beauty transposon. Dev Biol 263, 191-202

Dong J, Stuart G (2004) Transgene manipulation in zebrafish by using recombinases. Methods Cell Biol 77, 363-379

Fan L, Moon J, Crodian J, Collodi P (2006) Homologous recombination in zebrafish ES cells. Transgenic Res 15, 21-30

Feng Y, Seibler J, Alami R, Eisen A, westerman KA, Leboulch P, Fiering S, Bouhassira E (1992) Site-specific chromosomal integration in mammalian cells: highly efficient Cre recombinase-mediated cassette exchange. J Mol Biol 292, 779-785

Gu H, Marth J, Orban P, Mossomann H, Rajewsky K (1994) Deletion of a DNA polymerase β gene segment in T cells using cell type-specific gene targeting. Science 265, 103-106

Kane D, Kimmel C (1993) The zebrafish midblastula transition. Development 119, 447-456

Kilby N, Snaith M, Murray A (1993) Site-specific recombinases: tools for genome engineering. Trends Genet 9, 413-421

Lee G, Saito I (1998) Role of nucleotide sequences of loxp spacer region in Cre-mediated recombination. Gene 216, 55-65

Lin S (2000) Transgenic zebrafish. Meth Mol Biol 136, 375-383

Linney E, Udvadia A (2004) Construction and detection of fluorescent germline transgenic zebrafish. Methods Mol Biol

254, 271-288

Liu WY, Wang YP, Zhu ZY (2004) Screening and application of loxp mutants. High Tech Lett 167, 102-105

Loonstra A, Vooijs M, Beverioo H, Allak B, Drunen E, Kanaar R, Berns A, Jonkers J (2001) Growth inhibition and DNA damage induced by Cre recombinase in mammalian cell. Proc Natl Acad USA 98, 9209-9214

Maclean N, Penman D, Zhu ZY (1987) Introduction of novel genes into fish. Bio/technology 5, 257-261

Muller U (1999) Ten years of gene targeting: targeted mouse mutants, from vector design to phenotype analysis. Mech Dev 82, 3-21

Pan X, Wan H, Chia W, Tong Y, Gong ZY (2005) Demonstration of a site directed recombination in transgenic zebrafish using the Cre/*loxP* system. Trans Res 14, 217-223

Plas D, Ponsaerts P, Tendeloo V, Bockstaele D, Berneman Z, Merregaerta J (2003) Efficient removal of LoxP-flanked genes by electroporation of Cre-recombinase mRNA. Biochem Biophys Res Commun 305, 10-15

Rahman M, Hwang G, Razak S, Sohm F, Maclean N (2000) Copy number related transgene expression and mosaic somatic expression in hemizygous and homozygous transgenic tilapia. Transgenic Res 9, 417-427

Sauer B, Henderson N (1988) Site-specific DNA recombination in mammalian cell by the Cre recombinase of bacteriophage P1. Proc Natl Acad USA 85, 5166-5170

Sauer B, Hendenson N (1990) Targeted insertion of exogenous DNA into the eukaryotic genome by the Cre recombinase. New Biol 441-449

Siegel R, Jain R, Bradbury A (2001) Using an *in vivo* phagemid system to identify non-compatible loxp sequences. FEBS Lett 505, 467-473

Smithies O, Gregg R, Boggs S, Koralewski M, Kucherlapati R (1985) Insertion of DNA sequences into the human chromosomal beta-globin locus by homologous recombination. Nature 317, 230-234

Thomas K, Capecchi M (1987) Site-directed mutagenesis by gene targeting in mouse embryo-derived stem cells. Cell 51, 503-512

Torres R, Kuhn R (1997) *Laboratory Protocols for Conditional Gene Targeting*. (Oxford: Oxford University Press)

Vooijs M, Jonkers J, Berns A (2001) A highly efficient ligand-regulated Cre recombinase mouse line shows that *loxp* recombination is position dependent. EMBO Report 2, 292-297

Wadman S, Klark K, Hackett P (2005) Fishing for answers with transposons. Mar Biotechnol 7(3), 135-141

Wan H, He J, Ju B, Yan T, Lam T, Gong ZY (2002) Generation of two-color transgenic zebrafish using the green and red fluorescent protein reporter genes gfp and rfp. Mar Biotechnol 4(2), 146-154

Westerfield M (1989) *The Zebrafish Book; A Guide for the Laboratory Use of zebrafish* (Brachydanio rerio). (Eugene, OR: University of Oregon Press)

Wu Y, Zhang G, Xiong Q, Luo F, Cui C, Hu W, Yu Y, Su J, Xu A, Zhu ZY (2006) Integration of double fluorescence expression vectors into zebrafish genome for the selection of site-directed knockout/knockin. Mar Biotechnol 8(3), 304-311

Zhu ZY, Sun YH (2000) Embryonic and genetic manipulation in fish. Cell Res 10(1), 17-27

Zhu ZY, Li GH, He L, Xu KS (1985) Novel gene transfer into the fertilized eggs of goldfish. Z Angew Ichthyol1, 31-34

斑马鱼中 cre/*lox* 系统介导的基因定点整合

刘维一　王　蕴　秦　瑶　汪亚平　朱作言

中国科学院大学，中国科学院水生生物研究所淡水生态与生物技术国家重点实验室，武汉，430072

摘　要　随着鱼类基因转植技术的发展，转植 DNA 片段在受体基因组中的随机性整合和多位点整合的情况逐渐显现。关联位置效应、拷贝数目的不同、以及多基因相互作用导致基因表达实验结果难以解释，鱼类表型也难以预测。为了满足不同鱼类基因工程的需求，我们在本文中介绍了一种斑马鱼的基因打靶模型。首先，通过转基因技术建立四个靶标斑马鱼家系，每个家系具有单个基因组 *lox71* 靶位点。将 Cre mRNA 和敲入载体 *pZklox66RFP* 共注射到转基因斑马鱼家系的受精卵。每个靶标斑马鱼家系都发生了特异性位点的整合。在这个家系中，从 20,000 个靶标胚胎中获得了 2 尾整合的斑马鱼(整合效率为 10^{-4} 到 10^{-5})，并且证实在靶向基因组位点处为单拷贝整合。用基因组聚合酶链反应分析和 DNA 测序证实了基因打靶的准确性，即 *lox71* 和 *lox66* 准确地重组插入到双突变的 *lox72* 和野生型 *loxP* 位点。选择整合的斑马鱼在特定的 *egfp* 区域转入 *rfp* 来进行分析。虽然 Cre 介导的重组位点特异，但是它也依赖于随机分配的靶位点。因此，基因组打靶不能只基于它的序列进行整合位点的预选。最后，通过 Cre-*lox* 介导的重组系统，*rfp* 报告基因成功插入到斑马鱼基因组的 *egfp* 靶位点。这种通过 *lox71*/*lox66* 组合进行定向位点敲入的系统将会为鱼类基因工程中的各种用途提供一个有发展前景的基因打靶平台。

Inhibition of No Tail (*ntl*) Gene Expression in Zebrafish by External Guide Sequence (EGS) Technique

De-Sheng Pei Yong-Hua Sun Yong Long Zuo-Yan Zhu

State Key Laboratory of Freshwater Ecology and Biotechnology, Institute of Hydrobiology, Chinese Academy of Sciences, Wuhan 430072

Group of Environmental Genomics, Institute of Hydrobiology, Chinese Academy of Sciences, Wuhan 430072

Abstract External guide sequence (EGS) technique, a branch of ribozyme strategy, can be enticed to cleave the target mRNA by forming a tRNA-like structure. In the present study, no tail gene (*ntl*), a key gene participating in the formation of normal tail, was used as a target for ribonuclease (RNase) P-mediated gene disruption in zebrafish *in vivo*. Transient expression of pH1-m3/4 ntl-EGS or pH1-3/4 ntl-EGS produced the full no tail phenotype at long-pec stage in proportion as 24 or 35%, respectively. As is expected that the fulllength *ntl* mRNA of embryos at 50% epiboly stage decreased relative to control when injected the embryos with 3/4 EGS or m3/4 EGS RNA. Interestingly, *ntl* RNA transcripts, including the cleaved by EGS and the untouched, increased. Taken together, these results indicate that EGS strategy can work in zebrafish *in vivo* and becomes a potential tool for degradation of targeted mRNAs.

Keywords external guide sequence; no tail; RNase P; zebrafish

Introduction

In recent years of post-genome era, to elegantly elucidate the function of many genes becomes the focus in genetics and molecular biology. Gene knockdown methodologies play an important role in both gene targeting applications and gene therapy. External guide sequence (EGS) technique, a branch of ribozyme strategy, can be enticed to cleave any target mRNA that forms a tRNA-like structure by endogenous ribonuclease (RNase) P. Such oligonucleotides that simulated three-quarters of a tRNA was called 3/4 EGS. This method has been successfully used to down-regulate the expression of genes in bacteria, human and plant cells in tissue culture [1-3]. Further studies indicate that 3/4 EGS without anticode loop had the same function to cleave the target mRNA, which was termed minimized 3/4 EGS or m3/4 EGS [4]. Encouraged by these results, we tried to determinate the function of genes in zebrafish using EGS strategy. To date, there is no report of application of EGS technique in zebrafish *in vivo* or *in vitro*. We choose to use the zebrafish system to develop this knockdown

---

文章发表于 *Molecular Biology Reports*, 2008, 35(2): 139-143

strategy not only because this model system offers several significant advantages over other laboratorial model, such as its short sex-maturity cycle, high reproductive capacity, transparent eggs, etc. [5-7], but also this model system helps to determine the functions of many mammalian and human genes for strong conservation of nucleotide sequence.

To initially test this knockdown strategy, the zebrafish *ntl* gene [8] was chosen as a target to knockdown its expression because this phenotype is easily observable. In the present study, we have extended our earlier studies by successfully demonstrating the efficacy of EGS for disrupting the expression of *ntl* gene in zebrafish.

Materials and Methods

Selection of targeted sites on the *ntl* mRNA

To eliminate synthesis of the encoded protein, the EGS-mediated cleavage in the mRNA should be positioned close to the start codon to ensure complete non-translatability. RNADRAW program [9] was used to assist the design of EGSs against the *ntl* mRNA of zebrafish (NM_131162). Site was chosen in looped regions to provide accessible regions in the target RNA. In addition, the acceptor and D-stem equivalent in 3/4 EGS or m3/4 EGS were designed to contain 7 and 4 bp complementary to the target RNA as described by Gopalan *et al.* [4]. The targeting sites of *ntl* mRNA is shown in Fig. 1B.

Construction of plasmids and preparation of RNA oligonucleotide

The primer pairs of SP6-3/4 EGS and SP6-m3/4 EGS (Table 1) were annealed and filled-in with Klenow DNA polymerase. After purification with DNA purification kit (Promega), the products were used as templates to produce RNA transcripts of 3/4 EGS and m3/4 EGS *in vitro* with SP6 RNA polymerase (Ambion). To construct the plasmids of pH1-3/4 EGS and pH1-m3/4 EGS, the primer pairs of H1-3/4 EGS and H1-m3/4 EGS (Table 1) were also annealed, filled-in using Klenow DNA polymerase, then digested with *Pst*I and *Hin*dIII, and cloned into pUC18-CMVeGFP downstream of the H1 promoter. They were used to analyze the phenotype of zebrafish at the late development stage *in vivo*.

Microinjection and fluorescent microscopic observations

The concentration of RNA oligonucleotide (3/4 EGS or m3/4 EGS) and plasmid (pH1-3/4 EGS or pH1-m3/4 EGS) was normalized to 100 and 50 ng/μl in sodium–Tris buffer (44 mM NaCl, 5 mM Tris-HCl, pH 7.0) with spectrophotometer, respectively. A mutant pH1-3/4 EGS harboring seven bases mutation of acceptor D-stem (GGCAGAC to CCGTCTG) was used as control for microinjection. The microinjection was performed as previously

Table 1 Primers used in the present study

Names	Sequences
SP6-3/4EGS	5-ATTTAGGTGACACTATAGAAGTACTTGAGCGGGTTGTGGTCCCGCTTCTA-3 5-TGGTGTCTGCCCTAGGATTCGAACCTAGAAGCGGGACCACAACCCGCTCA-3
SP6-m3/4EGS	5-ATTTAGGTGACACTATAGAAGTACTTGCTAGGTTCGAATCCTAGGGCAGACACCA-3 5-TGGTGTCTGCCCTAGGATTCGAACCTAGCAAGTACTTCTATAGTGTCACCTAAAT-3
H1-3/4EGS	5-AACTGCAGTACTTGAGCGGGTTGTGGTCCCGCTTCTAGGTTCGAATCCTAGG-3 5-GGGAAGCTTAAAAAATGGTGTCTGCCCTAGGATTCGAACCTAGAAGCGGGAC-3
H1-m3/4EGS	5-GGTACTTGCTAGGTTCGAATCCTAGGGCAGACACCATTTTTA-3 5-AGCTTAAAAATGGTGTCTGCCCTAGGATTCGAACCTAGCAAGTACCTGCA-3
ntl1	5-CTCGGTCCTGCTGGATTTTG-3 5-TGTGGTCTGGGACTTCCTTGT-3
ntl2	5-TGTCAAAGCAACAGTATCCAACG-3 5-GCTTTGGGGTTCGGGTTT-3
GAPDH	5-GTGTAGGCGTGGACTGTGGT-3 5-TGGGAGTCAACCAGGACAAATA-3

described by Zhu and Sun [10]. Each fertilized egg at the single cell stage received 2 nl of microinjection solution after the chorion was removed by 0.25% trypsin solution. The injected embryos were maintained in Holtfreter's solution (0.35% NaCl, 0.01% KCl, and 0.01% $CaCl_2$) at 28.5℃. Fluorescent images were captured by using an Olympus SZX12 microscope and a digital camera with a green fluorescent protein (GFP) (Chroma 41015) filter cassette with 488 nm excitation.

RNA preparation and RT-PCR analyses

Total RNAs from 50 injected zebrafish embryos at 50% epiboly and at 96 h post-fertilization [11] were isolated by using the SV total RNA kit (Promega). cDNAs were synthesized with poly (dT)18 primers using avian myeloblastoma virus transcriptase (Promega). All cDNA concentrations were adjusted to 80 ng/μl based on spectrophotometric measurements at 260 nm. RT-PCR was performed with ntl1 or ntl2 primers (Table 1) in a final volume of 25 μl containing 2 μl of template with the program: 94℃ 5 min; then 94℃ 15 s, 53℃ 30 s, 72℃ 50 s, 24 cycles (before the plateau); 72℃ 5 min, 4℃ hold. GAPDH was used as the endogenous reference. The agarose gel was scanned by GeneGenius and analyzed by genesnap software (Syngene).

Results

Design and construction of the EGS

With the help of RNADRAW program, we identified a single-stranded region containing the start codon (from 1 to 15 nt) as an ideal target for EGS binding and subsequent cleavage by RNase P (Fig. 1B). The three-terminal CCA sequence was added to 3/4 or m3/4 EGS for substrate recognition [12]. The 3/4 EGS is about 55 nt in length and is complementary to 11 nt in the target mRNA. The acceptor and D-stem contain 7 and 4 bp in the target RNA respectively (Fig. 1C), which is the essence of substrate specificity of eukaryotic RNase P. The m3/4 EGS is similar with 3/4 EGS, only without anticodon loop and variable loop (Fig. 1D). In pH1-m3/4 ntl-EGS and pH1-3/4 ntl-EGS plasmids, GFP cassette, driven by CMV promoter, is located in the downstream of EGS sequence (Fig. 2A). The sequencing result shows that pH1-m3/4 ntl-EGS and pH1-3/4 ntl-EGS plasmids were constructed successfully (Fig. 2B, C).

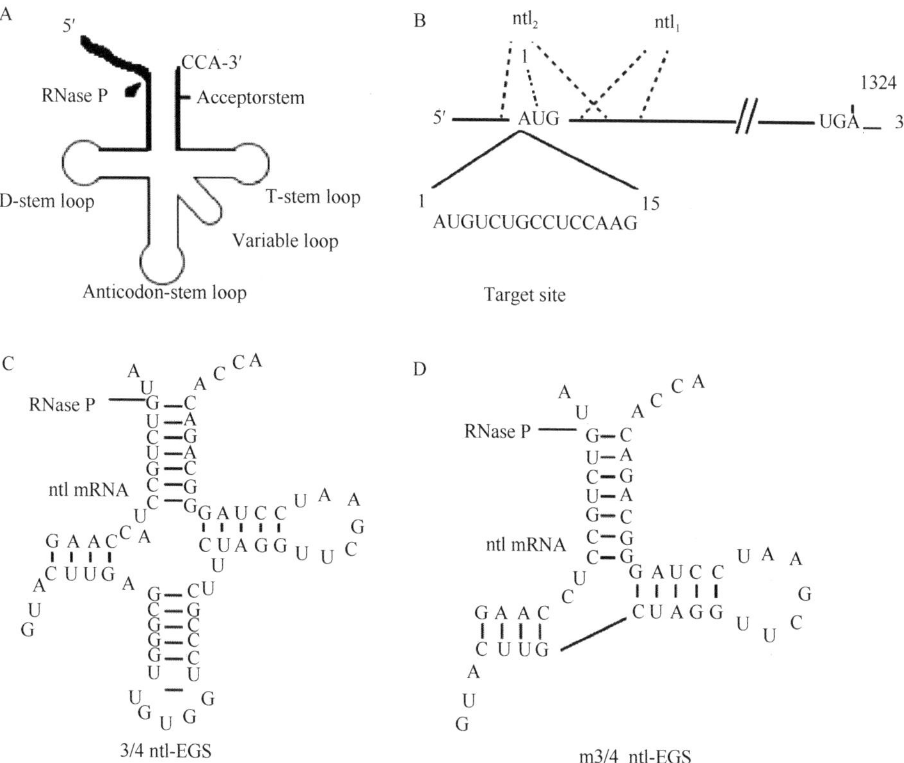

Fig. 1 RNase P-mediated cleavage of *ntl* mRNA. (A) The structure of a pre-tRNA, a typical substrate for RNase P. The *arrow* indicated the site of cleavage by RNase P. (B) Map of the *ntl* mRNA showing the EGS target site (not drawn to scale). ntl1 and ntl2 represented the PCR fragments amplified by ntl1 and ntl2 primer pairs (Table 1), respectively. (C) A pre-tRNA-like structure formed by the *ntl* mRNA and 3/4 *ntl*-EGS. (D) A minimized pre-tRNA-like structure formed by the *ntl* mRNA and m3/4 *ntl*-EGS. The cleavage sites of ntl mRNA by RNase P were indicated in (C) and (D)

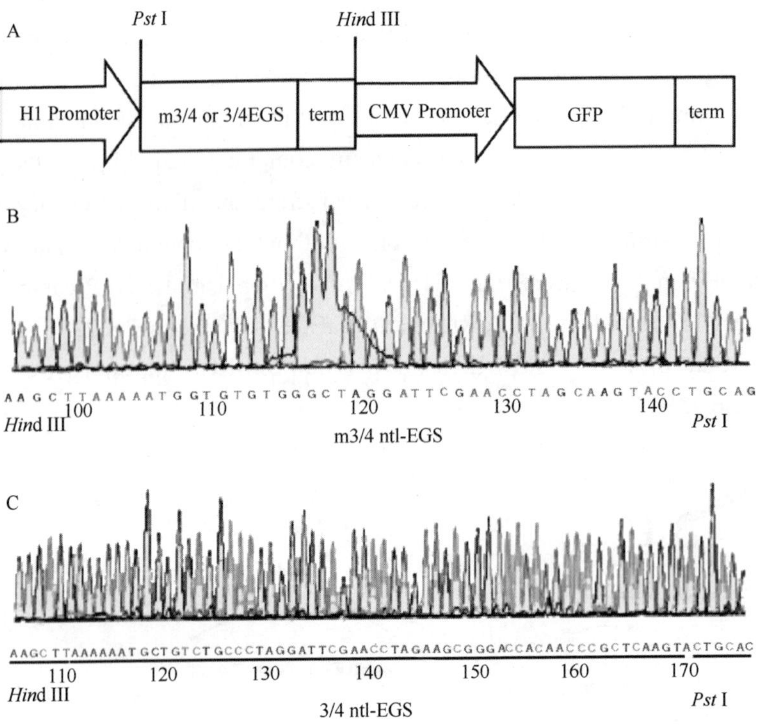

Fig. 2 Maps of the pH1-m3/4 ntl-EGS and pH1-3/4 ntl-EGS constructs. (A) The main structure of pH1-m3/4 ntl-EGS and pH1-3/4 ntl-EGS. The primer pairs of H1-3/4 EGS and H1-m3/4 EGS were annealed, filled-in using Klenow DNA polymerase, digested with *Pst*I and *Hin*dIII, and cloned into pUC18-CMVeGFP downstream of the H1 promoter. The terminator (term) of EGS was five thymines and the terminator (term) of GFP was SV40 polyA site. (B) and (C) were the sequencing results of m3/4 EGS and 3/4 EGS. The sites of *Pst*I and *Hin*dIII were marked

Phenotype of zebrafish eggs injected with pH1-3/4 EGS and pH1-m3/4 EGS

The embryos injected with pH1-m3/4 ntl-EGS, pH1-3/4 ntl-EGS and the mutant pH1-3/4 EGS all produced GFP. Though injection of the mutant pH1-3/4 EGS plasmid showed the no tail phenotype to a very limited extent (5%), twenty-four or thirty-five percent of fry developing from zebrafish fertilized eggs injected with pH1-m3/4 ntl-EGS or pH1-3/4 ntl-EGS showed the full no tail phenotype, respectively. This indicates that pH1-3/4 ntl-EGS is more effective than pH1-3/4 ntl-EGS in knocking down the *ntl* gene (Fig. 3). Our result in present study shows that EGS strategy does work in zebrafish *in vivo*, but the knockdown effect is not powerful to some extent.

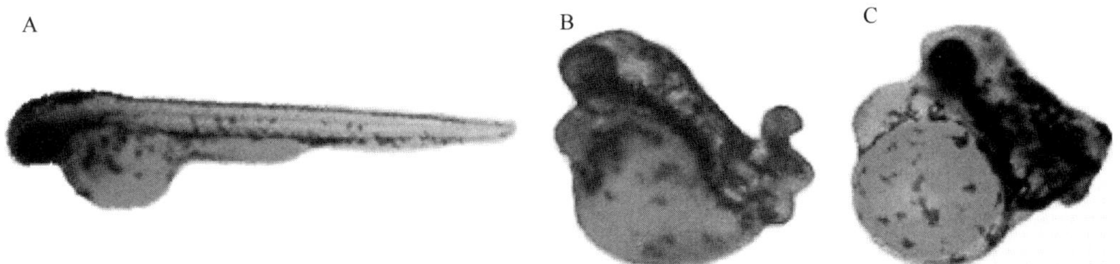

Fig. 3 The no tail phenotype caused by pH1-m3/4 ntl-EGS and pH1-3/4 ntl-EGS. (A) Normal zebrafish injected with the mutant pH1-3/4 EGS at 48 h post-fertilization. (B) and (C) No tail phenotypic changes at 48 h post-fertilization injected with pH1-m3/4 ntl-EGS or pH1-3/4 ntl-EGS vector, respectively. Green signals represented the GFP driven by CMV promoter

Effects of EGSs on *ntl* mRNA levels

To quantificationally analyze the knockdown effect of *ntl* gene, RT-PCRs were performed by ntl1 primers or ntl2 primer pairs, respectively. The ntl1 primer pairs, designed from the regions after the cut sites, were used to detect all *ntl* transcripts, including the cutting part and no cutting part by RNase P. While ntl2 primer pairs, spanning the cut site, were used to detect only the *ntl* transcripts without cleavage by RNase P. In the embryos injected with 3/4 EGS or m3/4 EGS, all *ntl* transcripts detected by ntl1 primer pairs were dramatically more than the control at 50% epiboly stage. Whereas, *ntl* transcripts detected by ntl2 primer pairs showed the adverse results (Fig. 4B). Those results indicates that EGS strategy can knock down the *ntl* mRNA levels but such degradation function may stimulate the over-expression of *ntl* mRNA from the feedback pathway. Some no tail phenotypic fries injected with m3/4 EGS RNA were also assayed. Interestingly, they showed no difference compared to controls (Fig. 4A). The reason may come from the limited effect time of m3/4 EGS RNA.

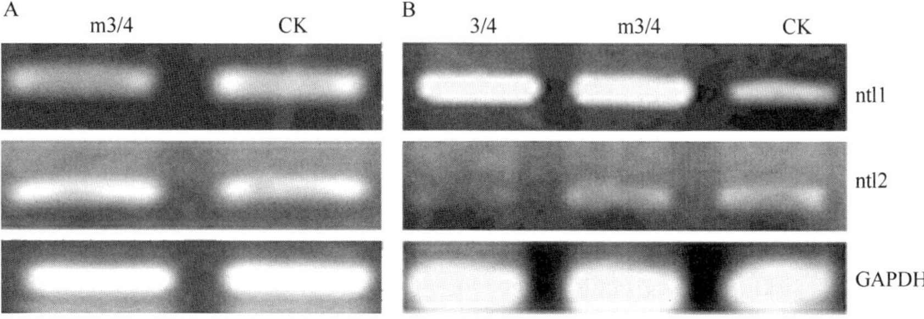

Fig. 4 RT-PCR analysis of *ntl* transcripts. (A) RT-PCR analysis of the *ntl* transcripts injected with m3/4 EGS RNA at 96 h of embryogenesis. (B) RT-PCR analysis of the *ntl* transcripts injected with m3/4, 3/4 EGS RNA at 50% epiboly stage. ntl1 primer pairs were the primers after the cut sites, while ntl2 primer pairs were the primers spanning the cut sites. *CK* denoted the control and *GAPDH* was used as endogenous reference

Discussion

External guide sequence technique, a novel approach, can specifically cleave target molecular by RNase P. Previous studies have demonstrated that RNA inactivation of targeted mRNA by EGS *in vivo* can be more effective than gene inactivation by conventional antisense DNA oligonucleotides [13,14], which have a different gene inactivation mechanism by endogenous RNase H [15,16]. Reduced off-site cleavage by EGS is also an advantage of EGS relative to RNAi and antisense technologies [17,18]. In RNAi technique, expression of the RISC must be induced, may not be induced in all cell types, and may be saturated by multiple mRNA targets [19]. These limitations do not appear for EGS. Moreover, previous research reported that the effect of EGSs was apparent in 24 h after transfection but the effect of siRNAs took at least 48-96 h to become evident in cell culture [19,20]. Morpholinos have been shown to block the translation of mRNA and is now used routinely in zebrafish for reverse genetic screens during embryonic development [21]. However, the cost of morpholinos is too higher for many laboratories.

In present study, we firstly applied the EGS technique for gene inactivation in zebrafish *in vivo*. Though the effect of *ntl* inactivation is not powerful to some extent, we are sure that EGS technique can work in fish *in vivo* and has potential application for studying of gene function. Data in this study indicate that EGS RNA works well in zebrafish embryos at 50% epiboly stage, but its inactivation function gradually decays for RNA degradation. As a result, *ntl* transcripts reach to normal level in zebrafish fry at 96 h of embryogenesis. Since EGS RNA may make the compensation of *ntl* mRNA from the feedback pathway, all *ntl* transcripts, including the cutting part and no cutting part by RNase P, increased in zebrafish embryos at 50% epiboly stage. But the complete *ntl* mRNA of no cutting part decreased relative to the control, indicating the successful inactivation effect caused by EGS. Whole *in situ* hybridization assay, a popular approach for detection of knockdown study, must cause a fake result in present study. The better assay method is to detect the protein output using *ntl* antibody. However, mercantile antibody of *ntl* protein is absent. Once we get the antibody, such assay will become our future study. In our on-going study, multiple well-known genes such as cyp26 and foxi1 are selected and multiple cleavage sites are designed with EGS strategy to validate further the effects of disrupting gene function.

Acknowledgements All authors are grateful to Ms Ming Li for supplying experimental materials and Dr Zong-Bin Cui for fruitful discussions and critical reading of the manuscript. This work was supported by the State Key Fundamental Research of China (Grant No. 2004CB117406) and the National Natural Science Foundation of China (Grant No. 90208024).

References

1. Liu F, Altman S (1996) Requirements for cleavage by a modified RNase P of a small model substrate. Nucleic Acids Res 24: 2690-2696
2. Piron M, Beguiristain N, Nadal A, *et al* (2005) Characterizing the function and structural organization of the 5′ tRNA-like motif within the hepatitis C virus quasispecies. Nucleic Acids Res 33: 1487-1502
3. Rangarajan S, Raj ML, Hernandez JM, *et al* (2004) RNase P as a tool for disruption of gene expression in maize cells. Biochem J 380: 611-616
4. Gopalan V, Vioque A, Altman S (2002) RNase P: variations and uses. J Biol Chem 277: 6759-6762
5. Grunwald DJ, Eisen JS (2002) Headwaters of the zebrafish—emergence of a new model vertebrate. Nat Rev Genet 3: 717-724
6. Haffter P, Granato M, Brand M, *et al* (1996) The identification of genes with unique and essential functions in the development of the zebrafish, *Danio rerio*. Development 123: 1-36
7. Key B, Devine CA (2003) Zebrafish as an experimental model: strategies for developmental and molecular neurobiology studies. Methods Cell Sci 25: 1-6
8. Schulte-Merker S, Ho RK, Herrmann BG, *et al* (1992) The protein product of the zebrafish homologue of the mouse T gene is expressed in nuclei of the germ ring and the notochord of the early embryo. Development 116: 1021-1032
9. Matzura O, Wennborg A (1996) RNAdraw: an integrated program for RNA secondary structure calculation and analysis under 32-bit Microsoft Windows. Comput Appl Biosci 12: 247-249
10. Zhu ZY, Sun YH (2000) Embryonic and genetic manipulation in fish. Cell Res 10: 17-27
11. Kimmel CB, Ballard WW, Kimmel SR, *et al* (1995) Stages of embryonic development of the zebrafish. Dev Dyn 203: 253-310
12. Altman S (1990) Nobel lecture. Enzymatic cleavage of RNA by RNA. Biosci Rep 10: 317-337
13. Plehn-Dujowich D, Altman S (1998) Effective inhibition of influenza virus production in cultured cells by external guide sequences and ribonuclease P. Proc Natl Acad Sci USA 95: 7327-7332
14. Guerrier-Takada C, Altman S (2000) Inactivation of gene expression using ribonuclease P and external guide sequences. Methods Enzymol 313: 442-456
15. Nyce JW, Metzger WJ (1997) DNA antisense therapy for asthma in an animal model. Nature 385: 721-725
16. Sandrasagra A, Leonard SA, Tang L, *et al* (2002) Discovery and development of respirable antisense therapeutics for asthma. Antisense Nucleic Acid Drug Dev 12: 177-181
17. Jackson AL, Linsley PS (2004) Noise amidst the silence: off-target effects of siRNAs? Trends Genet 20: 521-524
18. Snove O Jr, Holen T (2004) Many commonly used siRNAs risk off-target activity. Biochem Biophys Res Commun 319: 256-263
19. Dreyfus DH, Matczuk A, Fuleihan R (2004) An RNA external guide sequence ribozyme targeting human interleukin-4 receptor alpha mRNA. Int Immunopharmacol 4: 1015-1027
20. Zhang H, Altman S (2004) Inhibition of the expression of the human RNase P protein subunits Rpp21, Rpp25, Rpp29 by external guide sequences (EGSs) and siRNA. J Mol Biol 342: 1077-1083
21. Nasevicius A, Ekker SC (2000) Effective targeted gene ‘knockdown’ in zebrafish. Nat Genet 26: 216-220

EGS 介导抑制斑马鱼 *ntl* 基因的表达

裴得胜　孙永华　龙　勇　朱作言

中国科学院水生生物研究所淡水生态与生物技术国家重点实验室，武汉 430072
中国科学院水生生物研究所环境基因组，武汉 430072

摘　要　外源性指导序列技术(EGS)属核酸酶应用策略的一个分支，它可以通过形成一种 tRNA 样的结构从而诱导剪切目标 mRNA。在本研究中，*no tail(ntl)* 这个参与正常尾部形成的关键基因作为核糖核酸酶 P 介导的剪切靶标。在斑马鱼胚胎 long-pec 期，瞬时表达的 pH1-m3/4 ntl-EGS 和 pH1-3/4 ntl-EGS 分别产生了 24%和 35%的完全无尾表型。正如预期的那样，胚胎注射了 3/4 EGS 或者 m3/4 EGS RNA 后，比较于对照组而言，在 50%外包期，胚胎中的全长 *ntl* mRNA 减少了。有趣的是，包括被 EGS 剪切的和未被剪切的 *ntl* RNA 转录本都增加了。总的看来，这些结果显示 EGS 策略能够在斑马鱼体内工作，并且成为一种降解目标 mRNAs 的潜在工具。

Non-Homologous End Joining Plays A Key Role in Transgene Concatemer Formation in Transgenic Zebrafish Embryos

Jun Dai[1,2] Xiaojuan Cui[1,2] Zuoyan Zhu[1] Wei Hu[1]

1. State Key Laboratory of Freshwater Ecology and Biotechnology, Institute of Hydrobiology, Chinese Academy of Sciences, Wuhan 430072

2. Graduate School of the Chinese Academy of Sciences, Beijing 100049

Abstract This study focused on concatemer formation and integration pattern of transgenes in zebrafish embryos. A reporter plasmid based on enhanced green fluorescent protein (eGFP) driven by Cytomegalovirus (CMV) promoter, pCMV-pax6in-eGFP, was constructed to reflect transgene behavior in the host environment. After removal of the insertion fragment by double digestion with various combinations of restriction enzymes, linearized pCMV-pax6in-eGFP vectors were generated with different combinations of 5′-protruding, 3′-protruding, and blunt ends that were microinjected into zebrafish embryos. Repair of double-strand breaks (DSBs) was monitored by GFP expression following religation of the reporter gene. One-hundred-and-ninety-seven DNA fragments were amplified from GFP-positive embryos and sequenced to analyze the repair characteristics of different DSB end combinations. DSBs involving blunt and asymmetric protruding ends were repaired efficiently by direct ligation of blunt ends, ligation after blunting and fill-in, or removed by cutting. Repair of DSBs with symmetric 3′-3′ protrusions was less efficient and utilized template-directed repair. The results suggest that non-homologous end joining (NHEJ) was the principal mechanism of exogenous gene concatemer formation and integration of transgenes into the genome of transgenic zebrafish.

Key words transgene; concatemer; DSB (double strand breaks); NHEJ (non-homologous end joining); zebrafish

Introduction

The mechanism by which exogenous genes are integrated into the host genome has been a major concern of transgene research. In an effort to develop effective methods for site-directed integration and enhance the efficiency of stable transgene integration, researchers have analyzed transgene flanking sequences to identify potential integration hotspots (1,2). However, studies on transgene integration in mam mals have shown that

文章发表于 *International Journal of Biological Sciences*, 2010, 6(7): 756-768 integration appears to be a random process and, although sequences in integration sites have some common structural features, no so-called integration hotspots exist. Furthermore, transgenes are prone to forming concatemers prior to integration (3,4,5,6). The integrated foreign DNA appears mainly as random end-to-end concatemers (4,7,8,9,10,11,12). There are 3 types of concatemers: head-to-tail, head-to-head, and tail-to-tail. Head-to-tail concatemers can be stably integrated into the chromosomes (9). Head-to-head and tail-to-tail concatemers are observed when the concentration of linear foreign molecules in the host nucleus is high; these types of concatemers are unstable and can change into head-to-tail concatemers (9). In transgenic mammals, repair mechanism associated with double strand breaks (DSBs) may affect transgene behavior, as well (3,5). Studies of transgenic fish have also shown that exogenous genes had consistently integrated as head-to-tail concatemers into the hostfish genome and the integration process exhibits characteristics of non-homologous recombination (1,2,12,13). Taken together, those studies suggest that DSB repair is related to the molecular mechanism of concatemer formation during exogenous gene integration in transgenic fish.

Cellular genomes are usually sensitive to ionizing radiation (IR), DNA mutagens, and other physical, chemical and biological factors that result in DNA damage in the form of DSBs (14,15). For example, V(D)J recombination during lymphocyte maturation in the immune system of higher organisms generates a large number of endogenous DSBs (16,17,18). As DSBs are considered to be the most severe form of DNA damage, they must be repaired immediately to prevent cell death (19,20,21).

Cells repair DSBs by two mechanisms: homologous recombination (HR) and non-homologous end joining (NHEJ) (22,23,24). HR is the principle mechanism for DSB repair in bacteria and lower eukaryotes, such as yeast (25,26), while NHEJ is the main mechanism used in higher eukaryotes (27). Although NHEJ is the predominant mechanism for DSB repair in mammals (28,29), it is an error-prone repair process (30,31,32,33,34). Since the NHEJ machinery directly ligates the ends of DSBs, and there is little or no homologous sequence, it frequently leads to the loss or insertion of base pairs (22,26).

The majority of research on NHEJ in zebrafish embryos has concentrated on gene expression patterns and functions of the NHEJ repair complex core components. It has been shown that injection of the Ku70 morpholino does not affect zebrafish embryogenesis, but exposure of Ku70 morpholino-injected embryos to low doses of ionizing radiation results in marked cell death throughout the developing brain, spinal cord and tail (35). In addition, Li *et al* found that V(D)J recombination in the lymphocytes of zebrafish was different from that in mammals in that the signal ends were lost and the coding ends did not form circle-hairpin structures (36). Zhong found that early immunoglobulin gene rearrangement occurred during the maturation process of zebrafish eggs; however, there was no ligation of immunoglobulin gene coding ends, presumably because of the instability of NHEJ-associated components in zebrafish lymphocytes (37). These findings suggest that zebrafish have an NHEJ repair complex that is similar to that in mammals. The studies described above, however, did not systematically investigate the characteristics of NHEJ or the molecular mechanism of concatemer formation during exogenous gene integration.

In this report, we characterized the end ligation process and DSB repair via NHEJ in zebrafish embryos. To monitor DSB repair, we constructed a reporter plasmid with a CMV promoter and an enhanced green fluorescent protein (eGFP) coding sequence, having a linker sequence between them. The linker contained pax6in and SV40 poly(A) to prevent the CMV promoter from driving eGFP expression. The reporter plasmid was digested with different pairs of restriction endo nucleases to remove the linker and generate linearized vectors with non-homologous ends. Linearized vectors were then microinjected into zebrafish embryos, and the ability to ligate and repair different fragment ends was monitored by restoration of GFP expression. In addition, the sequences of 197 junction regions were determined in order to analyze their ability to repair different fragment ends. The results indicated that the NHEJ repair mechanism in zebrafish is similar to that in mammals in that repair occurs through direct ligation of DNA ends, rat her than depending on multiple homologous sequences. Based on these results, we propose a two-step mechanism for the integration of exogenous genes via NHEJ repair in zebrafish. Specifically, after microinjection the NHEJ complex would first process DSBs between exogenous gene molecules and form concatemers. As the concentration of exogenous gene DSBs decreases, the NHEJ repair mechanism would begin to skew towards endogenous DSBs generated in the host genome during successive generations of cellular division. Since endogenous DSBs are distributed on various sites in the chromosomes, ligation between transgene DSBs and endogenous DSBs would result in random integration of exogenous genes. We presume that ligases, polymerases and endonucleases related to DNA repair are involved in this process.

Materials and Methods

Zebrafish and microinjection

Wild type zebrafish (AB) were maintained according to the established guide for zebrafish care (38) in a glass aquarium with a circulating water system maintained at a constant temperature of 28℃. The ratio of illumination time to darkness time was 12hrs:12hrs. Fish handling and embryo generation were performed in accordance with IACUC regulations. Fifteen minutes after artificial fertilization, zebrafish embryos were microinjected with DNA at a concentration of 100 ng/μl.

Construction of pCMV-pax6in-eGFP plasmid

With consideration to the study of NHEJ repair mechanism in mammalian cells (28), our zebrafish NHEJ reporter plasmid, pCMV-pax6in-eGFP, was derived from the pEGFP-N1 vector (Clontech, Mountain View, CA, USA). The pEGFP-N1 vector backbone was PCR amplified using reverse primer P1 (5′-GAGCTCGAGAATTCACTA AACCAGCTCT-3′) with *Sac*I, *Xho*I and *Eco*RI restriction sites and forward primer P2 (5′-AAGCTTGGTA-CCCACGTGGGATCCACCGGT CGCCACC-3′) with *Hin*dIII, *Kpn*I, *Pma*CI and *Bam*HI restriction sites (Figure 1). Underlined text and strikethroughs correspond to the various restriction endonuclease target sequences. The amplified vector backbone was digested with *Sac*I and *Hin*dIII (Fermentas, MBI, Shenzhen, China). The PCR amplified 1.2 kb pax6 fragment was digested with *Sac*I and *Eco*RI, and an amplified SV40 poly(A) fragment was digested with *Eco*RI and *Hin*dIII. The three digested amplification products, vector, pax6 fragment, and SV40 poly(A) fragment were ligated and transformed into DH5α competent bacteria to obtain the NHEJ reporter plasmid pCMV-pax6 in-eGFP (Figure 1).

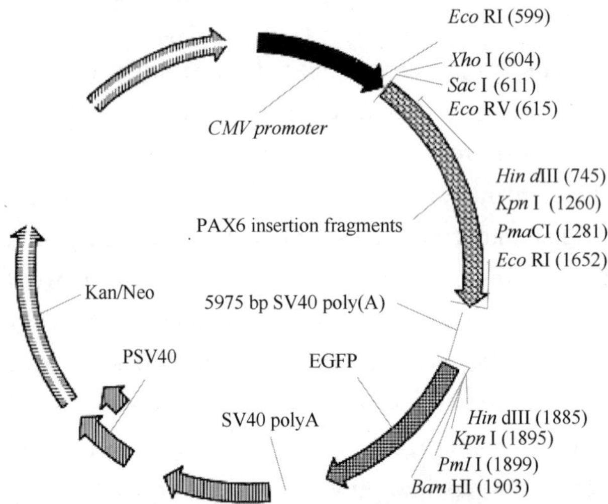

Figure 1 Features of the plasmid pCMV-pax6in-eGFP. The plasmid contains a CMV promoter and the enhanced green fluorescent protein (eGFP) coding sequence, separated by a linker sequence. The linker contains pax6in and SV40 poly(A)

NHEJ reporter assay

The pCMV-pax6in-eGFP reporter plasmid was digested with seven different pairs of restriction endonucleases to produce linearized vectors with different DNA ends. The corresponding enzyme combinations were used to generate DSBs: 5′-5′ (*Xho*I+*Hin*dIII and *Hin*dIII+*Eco*RI), 3′-3′ (*Kpn*I+*Sac*I), 5′-3′ (*Eco*RI+*Kpn*I), 5′-B (*Hin*dIII+*Eco*RV), 3′-B (*Kpn*I+*Eco*RV) and B-B (*Pma*CI+*Eco*RV), where 5′, 3′, and B denote digestions that result in

5′-protruding, 3′-protruding and blunt ends, respectively. After removal of the insertion fragment by restriction endonuclease digestion, linearized vectors were recovered and injected into artificially fertilized zebrafish embryos, and the fluorescence was monitored 10 hours after sample injection. Ligation of the linearized fragment ends between the CMV promoter and eGFP coding sequence was expected to result in GFP expression and a readout of a functional NHEJ repair mechanism (Figure 2).

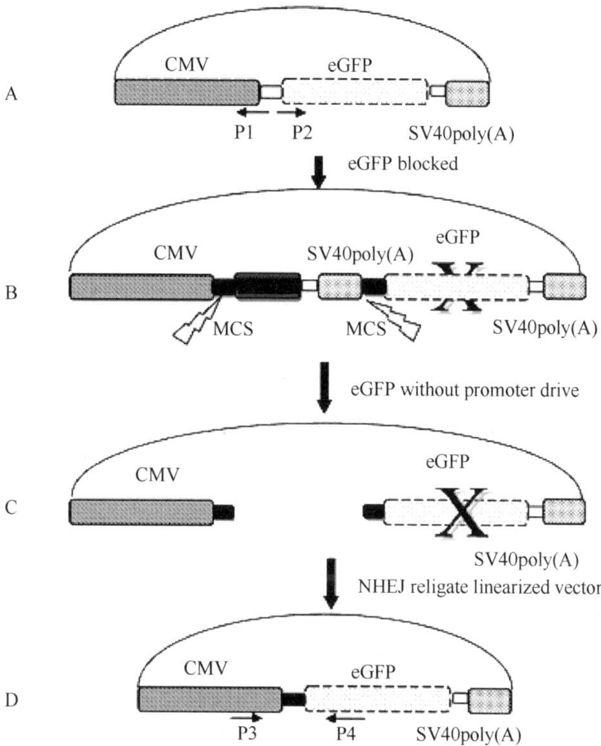

Figure 2 Schema of the NHEJ reporter assay. A: The pEGFP-N1 vector backbone was PCR amplified using reverse primer P1 and forward primer P2. B: pCMV-pax6in-eGFP plasmid in which the pax6 fragment and SV40 poly(A) was ligated with the pEGFP-N1 vector backbone, in order to prevent the CMV promoter from driving eGFP expression. C: The insertion fragment was removed from the pCMV-pax6in-eGFP plasmid through double digestion, and the linearized plasmid was injected into zebrafish embryos or eggs. D: The linearized vector was rejoined by NHEJ in zebrafish embryos or eggs

Each linearized vector was injected into zebrafish embryos separately; 10 hours later, twenty GFP-emitting embryos were collected for DNA extraction and sequencing. Each group of embryos were ground separately with a plastic rod, and 500 μL of DNA extraction solution [10 mM Tris·HCl (pH 8.0), 300 mM NaCl, 10 mM EDTA, 2.0% (*w/v*) SDS] was added. Samples were incubated for 3 hours at 65℃ to extract the genomic DNA. The fragment between the religated CMV promoter and eGFP was amplified by using the forward primer P3 (5′-AGAGCTGGTTTAGTGAA-3′) targeting the CMV promoter and the reverse primer

P4 (5′-TGCCGTTCTTCTGCTTGTC-3′) targeting the eGFP coding sequence; the predicted amplified fragment length was 531 bp. The amplified PCR fragment was ligated into pMD18-T vector (Takara, Shiga, Japan), and 30 clones from each group were selected for sequencing (Invitrogen, Carlsbad, CA, USA).

The mechanism by which exogenous transgenes are processed in zebrafish embryos may require expression of the zygotic gene that processes DSBs; alternatively, residual maternal substances in eggs may be utilized to repair exogenously introduced DSBs. To analyze the mechanism for processing exogenous genes in zebrafish embryos, unfertilized eggs were microinjected with linearized fragments digested with 5′-5′ (*Xho*I+*Hin*dIII), 3′-3′ (*Kpn*I+*Sac*I) and 5′-5′ (*Hin*dIII+*Eco*RI). Samples were collected 15 minutes after microinjection, and DNA extraction solution was added. DNA extracted from 20 embryos was amplified with primers P3 and P4 to detect NHEJ reaction products. In addition, 20 unfertilized zebrafish eggs were placed into 1 × PBS and triturated with a plastic grind rod. Homogenized samples were incubated with 100 ng of linearized vector digested with 5′-5′ (*Xho*I+*Hin*dIII), 3′-3′ (*Kpn*I+*Sac*I) or 5′-5′ (*Hin*dIII+*Eco*RI). After incubation for 15 minutes, DNA extraction solution was added to stop the reaction and extract DNA. PCR amplification with primers P3 and P4 was used to determine if ligation occurred between the CMV promoter and eGFP.

Results

NHEJ reporter plasmid

To monitor NHEJ DSB repair in zebrafish embryos, we constructed pCMV-pax6in-eGFP, in which pax6 cDNA and SV40 poly(A) terminator fragments were inserted between the CMV promoter and eGFP coding sequence to block eGFP expression. The use of SV40 poly(A) terminator was necessary for eliminating background fluorescence from the plasmid; use of a fragment without the SV40 poly(A) terminator did not completely eliminate the green fluorescence expression (data not shown). The pax6 fragment contained a number of restriction sites on each end to facilitate generation of restriction fragment ends with various combinations of 5′-protruding, 3′-protruding or blunt ends. After removal of the insertion fragment through double digestion, ligation of non-homologous ends via NHEJ DSB repair restored GFP expression from the reporter plasmid.

Microinjection of NHEJ reporter plasmid

Zebrafish embryos that were microinjected with the intact pCMV-pax6in-eGFP plasmid, including pax6 fragment and SV40 poly(A) terminator, did not express GFP. However, embryos injected with pCMV-pax6in-eGFP after removal of the insertion fragment exhibited increased GFP fluorescence (Figure 3). DNA extracted from GFP expressing embryos was amplified with primers P3 and P4 and yielded the expected ~531 bp fragment which resulted from religation of the CMV promoter and eGFP (Figure 4). This result suggested that direct ligation occurred between DSBs in linearized fragments after injection into the zebrafish embryo. In addition, the amplified fragment length demonstrated that the ligation direction between transgenes occurred in a head-to-tail manner.

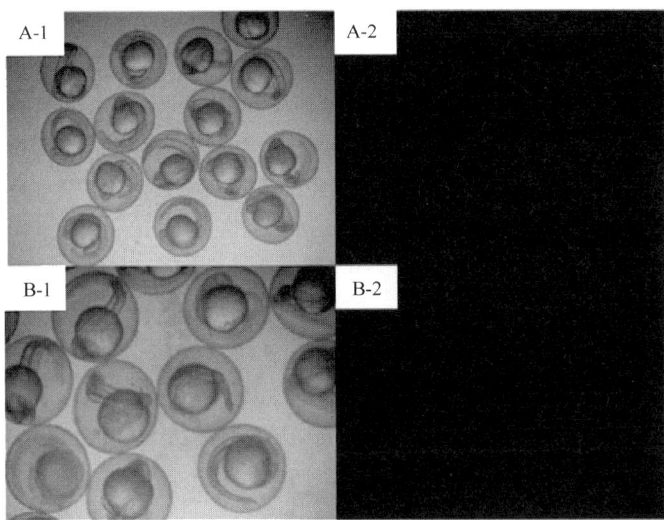

Figure 3 Zebrafish embryos were microinjected with circular type plasmid pCMV-pax6in-eGFP and linearized plasmid which had the insertion fragment removed from between the CMV promoter and the eGFP reporter gene. A: Zebrafish embryos microinjected with circular type plasmid pCMV-pax6in-eGFP (no fluorescence was observed). B: Zebrafish embryos microinjected with linearized plasmid which had the insertion fragment removed from between the CMV promoter and the eGFP reporter gene in the pCMV-pax6in-eGFP plasmid. Re-expression of eGFP was observed as the embryo emitted fluorescence

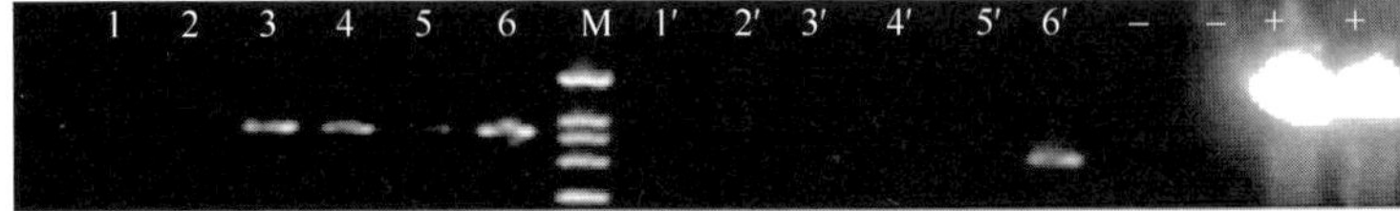

Figure 4 Gel electrophoresis of the PCR products amplified from the genomes of six fluorescent zebrafish embryos which were microinjected with linearized pCMV-pax6in-eGFP plasmids. A band of 531 bp length was amplified from all the fluorescent zebrafish embryo genomes, which represented the expected product from a religation event between the CMV promoter and eGFP. Lanes 1-6: PCR products of the amplified beta-actin gene, used for genome quality control. Lanes 1′-6′: PCR products of the junction sites between the CMV promoter and eGFP. –: negative control in which the template was the wild type zebrafish genome DNA. +: positive control in which the template was circular type pCMV-pax6in-eGFP plasmid. M: DNA molecular weight marker

Characteristics of DSB repair in zebrafish embryos

The pCMV-pax6in-eGFP reporter plasmid was digested with seven different pairs of restriction enzymes to generate seven groups of linearized vectors with different forms of DSBs. Following microinjection into zebrafish embryos, DNA was extracted and sequenced, yielding 197 sequences that crossed the junction site of the CMV promoter and eGFP. We found that the majority of linearized vectors were directly ligated and formed head-to-tail junctions; only a minority of the ends resulted in recombination through as few as one or even no base pair homology.

From the 197 PCR products, we were able to deduce the repair mechanism for different pairs of blunt and/or protruding ends. The characteristics of junction region formation were typical of the NHEJ repair mechanism. According to the ends formed, linearized vectors could be divided into groups with blunt ends, asymmetrical protruding ends and symmetrical protruding ends (Figures 5 and 6). For linearized vectors with blunt ends and asymmetrical ends, the primary repair mechanisms were direct blunt-end ligation, ligation after blunting of protruding ends or fill-in and removal post-ligation (shown with green numbers) (Figure 5). In most cases, ligation was homology-independent and the terminal sequences were kept intact, i.e. no additional bases were added or deleted on the basis of terminal sequences. For DSBs with two blunt ends, B-B (*Pma*CI+*Eco*RV), 89.5% of the ends were found to have been

A

*Pma*C I +EcoR V										*Xho* I											
				*Eco*R V			*Sac* I								*Eco*R I						
				A	T	C	G	A	G	C	T	C	G	A	G	A	A	T	T	C	3′
			2	T	A	G	C	T	C	G	A	G	C	T	C	T	T	A	A	G	5′
	*Pma*C I	C	G	26																	
		A	T																		
BamH I		C	G																		
		C	G																		
		T	A		1																
		A	T																		
		G	C																		
		G	C																		
		5′	3′																		

B

Hind III+*Eco*R V										*Xho* I											
				*Eco*R V			*Sac* I								*Eco*R I						
				A	T	C	G	A	G	C	T	C	G	A	G	A	A	T	T	C	3′
	Hind III		A1	T	A	G	C	T	C	G	A	G	C	T	C	T	T	A	A	G	5′
			A	25																	
			G																		
			C																		
			T																		
		A	T			C1															
	Kpn I	C	G																		
		C	G																		
		A	T	1																	1
		T	A																		
		G	C																		
*Pma*C I		G	C																		
		T	A																		
		G	C																		
		C	G																		
		A	T																		
	*Bam*H I	C	G																		
		C	G																		
		T	A																		
		A	T																		
		G	C																		
		G	C																		
		5′	3′																		

C

Kpn I +*Eco*R V										*Xho* I											
				*Eco*R V			*Sac* I								*Eco*R I						
				A	T	C	G	A	G	C	T	C	G	A	G	A	A	T	T	C	3′
	Kpn I			T	A	G	C	T	C	G	A	G	C	T	C	T	T	A	A	G	5′
		C		18																	
		A		4																	
		T		2																	
		G		2																	
*Pma*C I		G	C	2																	
		T	A																		
		G	C																		
		C	G																		
		A	T																		
	*Bam*H I	C	G																		
		C	G																		
		T	A																		
		A	T																		
		G	C																		
		G	C																		
		5′	3′																		2

D

Kpn Ⅰ +EcoR I			*Eco*R I						CMV promoter											
				A	A	T	T	C	A	C	T	A	A	A	C	C	A	G	C	3′
								G	T	G	A	T	T	T	G	G	T	C	G	5′
		C		16																
	Kpn Ⅰ	A		2																
		T			1															
		G																		
		G	C	4		1														
		T	A	C1																
*Pma*C Ⅰ		G	C		1			1												
		C	G																	
		A	T	1																
		C	G																	
		C	G																	
	*Bam*H Ⅰ	T	A																	
		A	T																	
		G	C																	
		G	C																	
		5′	3′																	

Figure 5 Sequence analysis of the junction site between CMV and eGFP isolated from transgenic zebrafish embryos microinjected with blunt-end (*Pma*CI+*Eco*RV) and asymmetrical end (*Hin*dIII+*Eco*RV, *Kpn*I+*Eco*RV, *Kpn*I+*Eco*RI) linearized vectors. Nucleotide sequences of the ends of the indicated substrates are shown along the axes. DNA strands are illustrated in the 5′-3′ direction (from bottom to top and from left to right). Arabic numerals indicate the position of the junction, occurrence frequency of specific ligation types. Ligation types are shown with different colours: direct ligation of blunt-end and ligation after blunting of protruding ends, or fill-in and removal by excision post-ligation are shown with green numbers; correct base-pairings following repair are shown with orange numbers; incorrect base-pairings following repair attempt are shown with red numbers; blue numbers represent the repair of bases where more than 15 bps were lost following base repair. The nucleotide sequence of these junctions starts with the last protruding or extension nucleotide of the left enzyme and the first protruding or extension nucleotide of the upper enzyme. For example, the green number 18 shown in boldface in *Kpn*I+*Eco*RV indicates the type of this junction site is direct ligation, and it appeared 18 times in a total of 30 sequenced DNA ends; the sequence of this junction site is 5′-GTACATG-3′. This denotation scheme is preserved in Figure 6

directly ligated, while linearized vectors with one blunt and one protruding 5′ end also exhibited 86% direct ligation. The amount of direct ligation for 3′-B (*Kpn*I+*Eco*RV) and 5′-3′(*Eco*RI+*Kpn*I) combinations was lower for both (60% and 53.3%), and in this two linearized vectors contained *Kpn*I 3′-protruding ends; the proportions followed incorrect pairing between corresponding base pairs (shown with red numbers) were quite high, reaching 30% and 43.3%, respectively.

A

Hind III +*Xho* I								*Eco*R I						
			Xho I											
				T	C	G	A	G	A	A	T	T	C	3′
	Hind III							C	T	T	A	A	G	5′
			A	11			2							
			G	3	1									
			C		1									
			T	1										
		A	T	4										
	Kpn I	C	G											
		C	G											
		A	T											
		T	A											
		G	C											
*Pma*C I		G	C											
		T	A											
		G	C											
		C	G											
		A	T											
	*Bam*H I	C	G											
		C	G											
		T	A											
		A	T											
		G	C											
		G	C											
		5′	3′											

B

Kpn Ⅰ +*Sac* Ⅰ						*Xho* Ⅰ											
			Sac Ⅰ								*Eco*R Ⅰ						
								C	G	A	G	A	A	T	T	C	3′
				T	C	G	A	G	C	T	C	T	T	A	A	G	5′
	Kpn Ⅰ	C					17										
		A				3											
		T															
		G					1	2									
*Pma*C Ⅰ		G	C														
		T	A														
		G	C				1										
		C	G														
		A	T														
	*Bam*H Ⅰ	C	G														
		C	G														
		T	A														
		A	T					2									
		G	C														
		G	C														
		5′	3′														

C

Hind III +*EcoR* I			*Eco*R I						CMV promoter											
				A	A	T	T	C	A	C	T	A	A	A	C	C	A	G	C	3′
								G	T	G	A	T	T	T	G	G	T	C	G	5′
			A	9	2			6					1							
	*Hin*d III		G																	
			C			2														
			T			2														
		A	T	7																
		C	G																	
		C	G																	
	Kpn I	A	T																	
		T	A																	
		G	C																	
		G	C																	
		T	A																	
*Pma*C I		G	C																	
		C	G																	
		A	T																	
		C	G																	
		C	G																	
	*Bam*H I	T	A																	
		A	T																	
		G	C																	
		G	C																	
		5′	3′														1			

Figure 6 Sequence analysis of the junction site between CMV and eGFP isolated from transgenic zebrafish embryos microinjected with symmetrical end (*Hin*dIII+*Xho*I, *Hin*dIII+*Eco*RI, *Kpn*I+*Sac*I) linearized vectors

For linearized vectors generated with symmetrical protruding ends, 5′-5′ (*Xho*I+*Hin*dIII) and 5′-5′ (*Eco*RI+*Hin*dIII) (Figure 6), 47.8% (11/23) and 30% (9/30) of the ends were repaired, respectively, by direct ligation of blunt ends following fill-in. In addition, repair after ligation involved base deletion in 17.4% (4/23) and 43.3% (13/30) of the cases, respectively. On the contrary, ligation after end blunting and fill-in or removal by excision repair post-ligation was not observed for linearized vectors with symmetrical 3′-protruding ends, 3′-3′ (*Kpn*I+*Sac*I). The majority of ligation types for this type of vector was template-directed repair and depended on microhomology of terminal bases. Following pairing of complementary bases, extra bases were excised, and the ends were finally religated; 84.6% (22/26) were ligated with the correct base pairing (shown with orange numbers) compared to 15.4% (4/26) with incorrect base pairing. The above results indicated that 3′-protruding ends were quite active in ligation reactions and the repair proportion of correct base pairing in this process was high, suggesting that the invasion of free radical -OH at the 3′ end required highly homologous base pairs.

Based on the junction sites of head-to-tail-ligated junctions formed by the repair of seven different kinds of DSBs, we divided repair types into twoclasses: (I) terminal base homology-independent repair, including direct ligation of blunt ends, ligation after blunting of protruding ends and fill-in or removal by excision post-ligation of protruding ends; and (II) terminal base homology-dependent repair, which included ligation following template-directed repair. The first repair type acted on all DSB combinations except for the 3′-3′ type, which were primarily repaired by the second mechanism. In addition, template-directed repair could be further divided into two groups, depending on whether bases were correctly or incorrectly paired. The repair mechanism for each type of DSB combination is summarized in Table 1.

Table 1 Ligation types of the seven groups of different DNA end joining

DSBs type	**NHEJ repair type and proportion**		
DSBs types generated by pairs of restriction endonucleases	Direct ligation of blunt ends,ligation after blunting of protruding ends and fill-in or cut-off after ligation	Template-directed repair(correct base pairing)	Others (incorrect base pairing and base loss of more than 15bp)
*Xho*I+*Hin*dIII(5′-5′)	11/23 (ligation after fill-in)+ 4/23 (excision after ligation){65.0%}	4/23(**17.5%**)	4/23(**17.5%**)
*Eco*RI+*Hin*dIII(5′-5′)	9/30 (ligation after fill-in)+(6/30+7/30) (excision after ligation){73.3%}	2/30(**6.7%**)	6/30(**20.0%**)
*Hin*dIII+*Eco*RV(5′-B′)	25/29 (ligation after fill-in){86.0%}	0	4/29(**14.0%**)
*Eco*RV+*Pma*C I(B′-B′)	26/29 (direct ligation){89.5%}	0	3/29(**10.5%**)
*Eco*RV+*Kpn*I(B′-3′)	18/30 (fill-in after ligation){60.0%}	2/30(**6.7%**)	10/30(**33.3%**)
*Kpn*I+*Sac*I(3′-3′)	0	22/26(**84.6%**)	4/26(**15.4%**)
*Kpn*I+*Eco*RI(3′-5′)	16/30 (fill-in after ligation){53.3%}	0	14/30(**46.7%**)

NHEJ complex pre-exist in zebrafish unfertilized eggs

To test if the machinery for NHEJ was present in unfertilizd eggs, zebrafish eggs were injected with linearized vectors, or homogenates of unfertilized eggs were incubated with linearized vectors. The DNA from both cases was extracted and amplified with primers P3 and P4. In both cases, the amplified PCR product corresponded to the religation between the CMV promoter and eGFP (Figure 7). This result suggested that the NHEJ machinery for repairing DSBs was present in the mature eggs and was not a result of embryogenesis.

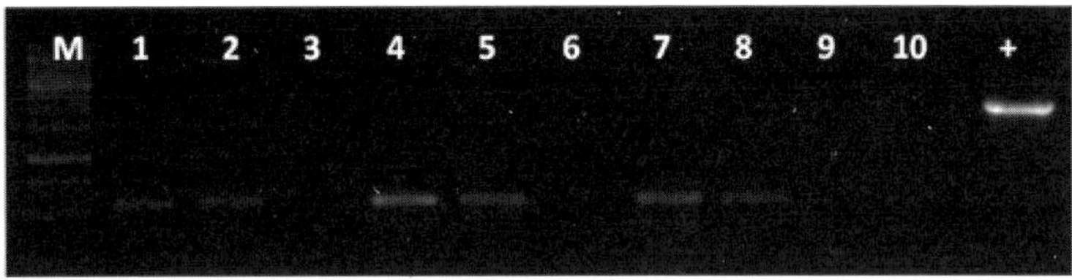

Figure 7 NHEJ effects in unfertilized zebrafish eggs. 1, 4, 7 junction site PCR product amplified from the genomes of fluorescent embryos microinjected with linearized DNA. 2, 5, 8 junction site PCR product amplified from homogenates of unfertilized eggs incubated with linearized DNA. 3, 6, 9 linearized vector used as the PCR template. 1-3: *Xho*I+*Hin*dIII linearized DNA; 4-6: *Kpn*I+*Sac*I linearized DNA; 7-9: *Hin*dIII+*Eco*RI linearized DNA. M: DNA molecular weight marker. DNA, 10: template was the wild type zebrafish genome. + : template was the pCMV-pax6in-eGFP plasmid. A band of 531 bp length was amplified from all the fluorescent zebrafish embryo genomes, which was the expected product for the event of religation between the CMV promoter and eGFP

Discussion

In our previously published studies on the F_4 hGH-transgenic common carp, the integration pattern and host sequences adjacent to the inserted pMThGH-transgene were extensively studied by using methods of plasmid rescue and Southern analysis. We found that the transgenes had consistently integrated into the host genome with concatemers (1-2). Many other researchers have previously reported that linearized exogenous genes form concatemers prior to integration into the host genome when used in microinjection (3-8). Many studies have explained the mechanism of transgene integration, but the mechanisms of concatemer formation have not yet been clearly elucidated. Studies of transgenic fish have shown that the integration process of exogenous genes shows characteristics of non-homologous recombination (1,2,12,13).

This study analyzed the sequence characteristics of 197 PCR products isolated from transgenic zebrafish embryos, all of the sequence in the junction region examined was the product of direct ligation, a typical characteristic of NHEJ repair. This indicated that NHEJ plays an important role in concatemer formation of DSB repair in zebrafish. In contrast to repair mechanisms in mammals, the efficiency of single-strand ligation was quite high. These results suggest that although the NHEJ repair mechanism in zebrafish appears to be similar to that in higher eukaryotes, it also retains some distinct characteristics.

Linearized vectors with double blunt ends showed the highest frequency of ligation without base pair insertions or deletions, and appeared to be the optimal substrate for NHEJ repair.

The repair ratio of substrates containing 3′-protruding ends, especially those with 3′-3′, was sufficiently less than the corresponding ligation of B-B and 5′-B. The results suggested that the presence of a 3′ end may reduce the efficiency of the direct ligation reaction.

It is technically challenging to treat 3′-protruding ends with polymerase, but 5′-protruding ends are receptive to polymerase activity to fill-in overhangs. This may be the reason why 3′- and 5′-protruding ends have different effects on the DSB repair efficiency (39). On the other hand, the free 5′-PO_4 group from the protruding end base appeared to be more prone to attack and subsequent direct ligation with the vicinal 3′-OH group; but, when the 5′-PO_4 group was absent, the 3′-OH group preferentially served as the primer for polymerase. Since direct ligation of DSBs is the main target of the NHEJ repair mechanism, our results support the hypothesis that NHEJ is the principal mechanism for exogenous gene concatemer formation and integration of transgenes into host genome in zebrafish.

The correction rate for NHEJ repair of asymmetric protruding ends in zebrafish is quite high, and the proportions of base loss and insertion are quite small. This feature is consistent with the results for NHEJ repair in *Kluyveromyces lactis*, as reported by Kegel *et al*. (22). However, the repair efficiencies without nucleotide deletion and insertion in the three combinations we examined, 5′-5′ (*Hin*dIII+*Eco*RI and *Xho*I+*Hin*dIII) and 3′-3′ (*Kpn*I+*Sac*I), was significantly less when compared with that of asymmetric protruding end and blunt-end combinations, suggesting that when polarity of protruding ends is the same, the ligation reaction may be hindered and result in base loss or insertion through other NHEJ processes.

Although 5′-5′ protruding ends displayed no ligation effects after end fill-in or cleavage in mammalian cells (3), our study found that the efficiencies of ligation after protruding end cleavage were higher than 30%. In addition, the proportion of fill-in and cleavage events following joining of 5′-5′ protruding ends was significantly smaller in zebrafish, as compared to that in mammals.

Researchers have hypothesized that significant differences exist between the NHEJ complexes of zebrafish and mammals (35,36,37). Our results indicated that there was, at least, a small divergence among the NHEJ repair pathways of zebrafish and mammals.

The proper repair events we observed for linearized vectors injected into unfertilized eggs or incubated with homogenates of unfertilized eggs indicated that the NHEJ complex is present and functional in zebrafish eggs, and it can mediate the ligation of exogenous DSB ends. In addition, we predict that this NHEJ complex also plays a principal role in repairing the large quantity of endogenous DSBs generated by the rapid cleavage rate that occurs in the early stage of zebrafish embryogenesis.

Similar conclusions have been obtained from concatemer type and transgene integration patterns in three-year-old P_0 common carp tissues expressing an antisense sGnRH transgene (40). In those studies, we found that concatemers were the primary form of transgenes in the genome, and the ligation type mediating concatemer formation was similar to what we found in zebrafish here (data not shown).

Taken together, the above data demonstrated that fish utilize a DSB repair mechanism that is similar to that used by mammals. Since DSB-directed ligation appears to be the main form of NHEJ repair mechanism, our results indicate that NHEJ is the principal mechanism for exogenous gene concatemer formation and integration of transgenes into the host genome of zebrafish. We propose the following two-step model to explain these mechanisms.

The NHEJ repair components functioning in zebrafish embryos could repair DSBs with high efficiency, even *in vitro*. The extremely high concentration of exogenous genes microinjected into zebrafish embryos would result in large accumulations of DSBs. Following microinjection, the NHEJ complex would first process DSBs between exogenous gene moleculesand form concatemers. As the concentration of exogenous gene DSBs decreases, the NHEJ repair mechanism would begin to skew towards endogenous DSBs generated in the host genome during successive generations of cellular division. Since endogenous DSBs are distributed on various sites in the chromosomes, ligation between transgene DSBs and endogenous DSBs would result in random integration of exogenous genes.

Foreign genes introduced by microinjection inevitably form concatemers and randomly integrate into the host chromosomes. Many studies have shown that the transgenes are efficiently expressed, only with fewer integrated copy numbers of the transgene and allowing for their stable transmission in the germ-line of fish. Our results suggest that it is necessary to modify the end of a transgene before microinjection, in order to prevent concatemer formation and increase the integration efficiency and expression level of the transgene.

Acknowledgements We thank Mrs. Shangpin Chen for assistant in microinjection during the study. This work was supported by the Development Plan of the state Key Fundamental Research of China (grant number 2007CB109206), National Natural Science Foundation (grant numbers 30930069, 30430540) and the '863' High Technology Project of China (grant number 2006AA10Z141).

Conflict of Interests The authors have declared that no conflict of interest exists.

References

1. Zeng ZQ and Zhu ZY. The molecular polymorphism of transgenes in F_4 generation red carp transfected with

pMThGH gene. *Chinese Sci Bulletin*. 2000; 45: 1957-1962

2. Wu B, Sun YH, Wang YW, Wang YP and Zhu ZY. Characterization of transgene integration pattern in F_4 hGH-transgenic common carp (*Cyprinus carpio*). *Cell Res*. 2005; 15: 447-454.
3. Roth DB and Wilson JH. Non-homologous recombination in mammalian cells: role for short sequence homologies in the joining reaction. *Mol Cell Biol*. 1986; 6: 4295-4304.
4. Hamada T, Sasaki H, Seki R, Sakaki Y. Mechanism of chromosomal integration of transgenes in microinjected mouse eggs: sequence analysis of genome-transgene and transgene-transgene junctions at two loci. *Gene*. 1993; 128: 197-202.
5. Chan AWS, Homan EJ, Ballou LU, Burns JC and Bremel R. Transgenic cattle produced by reverse-transcribed gene transfer in oocyte. *Proc Nat Acad Sci*. 1998; 95: 14028-14033.
6. Bill CA and Summers J. Genomic DNA double-strand breaks are targets for hepadnaviral DNA integration. *Proc Nat Acad Sci*. 2004; 101: 11135-11140.
7. Houdebine LM and Chourrout D. Transgenesis in fish. *Experientia*.1991; 47(9): 891-897
8. Bishop JO and Smith P. Mechanism of chromosomal integration of microinjected DNA. *Mol Biol Med.* 1989; 6(4): 283-298.
9. Folger KR, Wong EA, Wahl G and Capecchi MR. Patterns of integration of DNA microinjected into cultured mammalian cells: evidence for homologous recombination between injected plasmid DNA molecules. *Molecular and cellular biology*. 1982; 2(11): 1372-1387.
10. Dellaire G, Yan J, Little KC, Drouin R and Chartrand P. Evidence that extrachromosomal double-strand break repair can be coupled to the repair of chromosomal double-strand breaks in mammalian cells. *Chromosoma* 2002; 111(5): 304-312.
11. Roth DB, Wilson JH. Relative rates of homologous and nonhomologous recombination in transfected DNA. *PNAS*, 1985; 82(10): 3355-3359.
12. Penman DJ, Iyengar A, Beeching AJ, Rahman A, Sulaiman Z and Maclean N. Patterns of transgene inheritance in rainbow trout (*Oncorhynchus mykiss*). *Mol Reprod and Dev*. 1991; 30: 201-206.
13. Hsiao CD, Hsieh FJ and Tsai HJ. Enhanced expression and stable transmission of transgenes flanked by inverted terminal repeats from adeno-associated virus in zebrafish. *Dev Dyn*. 2001; 220: 323-36.
14. Jillian LY, O'neil NJ and Rose AM. Homologous recombination is required for genome stability in the absence of DOG-1in Caenorhabditis elegans. *Genetics*. 2006; 173: 697-708.
15. Siim P, Burkert JS, Martin J, *et al*. Alternative Induction of Meiotic Recombination From Single-Base Lesions of DNA deaminases. *Genetics*. 2009; 182: 41-54.
16. Lieber RM. Immunoglobulin diversity: rearranging by cutting and repairing. *Curr Biol*. 1996; 6: 134-136.
17. Vangent DC, Ramsden DA and Gellert M. The RAG1 and RAG2 proteins establish the 12/23 rule in V(D)J recombination. *Cell*. 1996; 85: 107-113.
18. Ramsden DA, Van Gent DC and Gellert M. Specificity in V(D)J recombination: new lessons from biochemistry and genetics. *Curr Opin Immunol*. 1997; 8: 114-120.
19. Yoshida K, Kondoh G, Matsuda Y, Habu T, Nishimune Y and Morita T. The mouse RecA-like gene Dmc1 is required for homologous chromosome synapsis during meiosis. *Mol Cell*. 1998; 1: 707-718.
20. Couteau F, Belzile F, Horlow C, Grandjean O, Vezon D and Doutriaux MP. Random chromosome segregation without meiotic arrest in both male and female meiocytes of a Dmc1 mutant of Arabidopsis. *Plant Cell*.1999; 11: 1623-1634.
21. Kenichiro M, Shinohara A and Shinohara M. Forkhead-associated domain of yeast Xrs2, a homolog of human Nbs1, promotes non-homologous end joining through interaction with a ligase IV partner protein, Lif1. *Genetics*. 2008; 179: 213-225.
22. Kegel A, Martinez P, Carter SD and Åström SU. Genome wide distribution of illegitimate recombination events in Kluyveromyces lactis. *Nucleic Acids Res*. 2006; 34: 1633-1645.
23. Jin L, Wen TJ and Schnable PS. Role of RAD51 in the repair of MuDR-induced double-strand breaks in Maize (*Zea*

mays L.). *Genetics*. 2007; 178: 57-66.

24. Mansour WY, Schumacher S, Rosskopf R *et al.* Hierarchy of non-homologous end joining, single-strand annealing and gene converse at site-directed DNA double-strand breaks. *Nucleic Acids Res*. 2008; 36: 4088-4098.
25. Pastwa E and Błasiak J. Non-homologous DNA end joining. *Acta Biochimica Polonica*. 2003; 50: 891-908.
26. Xin Y and Abram G. Ku-dependent and Ku-independent end joining pathways lead to chromosomal rearrangements during double-strand break repair in Saccharomyces cerevisiae. *Genetics*. 2003; 163: 843-856.
27. Kanaar R, Hoeijmakers J and Van Gent DC. Molecular mechanisms of DNA double-strand break repair. *Trends in Cell Biol.* 1998; 8: 483-489.
28. Liang F, Han M, Romanienko MP and Jasin M. Homology-directed repair is major double-strand break repair pathway in mammalian cells. *Proc Nat Acad Sci*. 1998; 95: 5172-5177.
29. Mia RL, Meier B, Lee TW, Hall J and Ahmed S. End joining at Caenorhabditis elegans telomeres. *Genetics*. 2008; 180: 741-754.
30. Lees-Miller SP, Godbout R, Chan D *et al.* Absence of p350 subunit of DNA-activated protein kinase from a radiosensitive human cell line. *Science*. 1995; 5201: 1183-1185.
31. Haber JE, Leung WY. Lack of chromosome territoriality in yeast: promiscuous rejoining of broken chromosome ends. *Proc Nat Acad Sci*.1996; 93: 13949-13954.
32. Lieber MR, Ma Y, Pannicke U and Schwarz K. Mechanism and regulation of human non-homologous DNA end-joining. *Nat Rev Mol Cell Biol*. 2003; 4: 712-720.
33. Jason AS, Waldman BC and Waldman AS. A Role for DNA mismatch repair protein Msh2 in error-prone double-strand-break repair in mammalian chromosomes. *Genetics*. 2005; 170: 355-363.
34. Wyman C, Kanaar R. DNA double-strand break repair: all's well that ends well. *Annu Rev. of Genet*. 2006; 40: 363-383.
35. Catherine L, Bladen WKL, Dynan WS and Kozlowski DJ. DNA damage response and Ku80 function in the vertebrate embryo. *Nucleic Acids Res*. 2005; 9: 3002-3010.
36. Li Z and Chang Y. V(D)J recombination in zebrafish: Normal joining products with accumulation of unresolved coding ends and deleted signal ends. *Mol Immunol*. 2007; 44: 1793-1802.
37. Zhong H, Li Z, Lin S and Chang Y. Initiation of V(D)J recombination in zebrafish (*Danio rerio*) ovaries. *Mol Immunol.* 2007; 44: 1784-1792.
38. Westerfield M. The zebrafish book: a guide for the laboratory use of zebrafish danio (*brachydanio*) rerio. Eugene: University of Oregon Press, 2000.
39. Kornberg A. DNA replication. San Francisco: WH Freeman & Co, 1980.
40. Hu W, Li SF, Tang B, Wang YP, Lin HR, Liu XC, Zou J and Zhu ZY. Antisense for gonadotropin-releasing hormone reduces gonadotropin synthesis and gonadal development in transgenic common carp (*Cyprinus carpio*). *Aquaculture*. 2007; 271: 498-506.

非同源末端连接在转植基因串联体形成中起关键作用

戴　军[1,2]　崔小娟[1,2]　朱作言[1]　胡　炜[1]

1 中国科学院水生生物研究所，武汉 430072
2 中国科学院大学，北京 100049

摘　要　本研究主要关注转基因斑马鱼中转植基因串联体的形成及其在受体基因组中的整合模式。我们构建了 CMV 启动子驱动 EGFP 的报告质粒 pCMV-pax6in-eGFP，研究转植基因在斑马鱼基因组中的整合行为。当用不同组合的两种限制性内切酶消化质粒后，我们得到的线性化 pCMV-pax6in-eGFP 报告质粒末端具有不同的特性，包括：5′端突出、3′端突出及平末端，将这些不同末端的线性化报告质粒注射至斑马鱼胚胎。通过观察 GFP 的表达与否，检测双链断裂末端(DSBs)是否发生修复。我们分离得到了 197 个发出绿色荧光的斑马鱼胚胎中的 CMV 与 eGFP 的连接处序列，通过分析这些序列特征，研究了不同 DSBs 组合间修复的特点。根据实验结果我们认为，非同源末端连接(NHEJ)是转基因斑马鱼中外源基因形成串联体及转植基因整合进宿主基因组的主要作用机制，斑马鱼胚胎 NHEJ 复合体主要是通过平末端直接连接，突出末端平端化后连接，或以连接后补平切平方式修复 3′-3′末端组合之外的其他组合的 DSBs，而修复 3′-3′组合的突出末端的 DSBs 则主要采用依赖末端碱基微同源性进行的模板引导型修复后连接方式。

Integration Mechanisms of Transgenes and Population Fitness of GH Transgenic Fish

HU Wei　ZHU ZuoYan

State Key Laboratory of Freshwater Ecology and Biotechnology, Institute of Hydrobiology, Chinese Academy of Sciences, Wuhan 430072

Abstract It has been more than 20 years since the first batch of transgenic fish was produced. Five stable germ-line transmitted growth hormone (GH) transgenic fish lines have been generated. This paper reviews the mechanisms of integration and gene targeting of the transgene, as well as the viability, reproduction and transgenic approaches for the reproductive containment of GH-transgenic fish. Further, we propose that it should be necessary to do the following studies, in particularly, of the breeding of transgenic fish: to assess the fitness of transgenic fish in an aqueous environment with a large space and a complex structure; and to develop a controllable on-off strategy of reproduction in transgenic fish.

Keywords transgene; integration mechanism; GH transgenic fish; population fitness

Over 35 different fish species have been used for gene transfer studies since the first batch of transgenic fish [1], and several stable germ-line transmitted growth hormone (GH) transgenic fish lines have been generated, including growth-enhanced transgenic common carp (*Cyprinus carpio*) [2], Atlantic salmon (*Salmo salar*), Coho salmon (*Oncorhynchus kisutch*) [3,4], mud loach (*Misgurrus mizolepis*) [5], tilapia (*Oreochromis* sp.) and tilapia (*Oreochromis nilotius*) [6,7]. In cooperation with Hunan Normal University, we produced a transgenic triploid fish with a growth rate that is 15% faster than that of the control carp, with a forage utilization rate increased by 11.1%. The transgenic triploid fish is completely sterile [8]. Such completely sterile transgenic triploid carp such as the all-fish GH transgenic carp may be suitable for industrialized conditions. The US Food and Drug Administration (FDA) is evaluating the ecological effects of current and potential uses of the field release of GH transgenic Atlantic salmon, which would promote the commercial farming of transgenic fish. No transgenic animals have been released into a natural environment for commercialized cultivation as food. Transgenic fish would be the breakthrough model as the first commercial farming transgenic animal. We review the mechanisms of integration and controllable regulatory expression of transgenes and the population fitness assessments of GH transgenic fish. We will conclude by providing suggestions for the thorough study of genetic engineering in breeding fish for potential commercial uses.

文章发表于《中国科学: 生命科学》, 2010, 53(4): 401-408

1 Studies of the Mechanisms of Integration and Gene Targeting in Transgenic Fish

1.1 Mechanisms of transgene integration

The mechanism by which exogenous genes are integrated into the host genome has been a major topic of transgenic research. Previous studies have indicated that the transgene randomly integrates into fish and at diverse integration sites [1]. The exogenous genes tend to form long-fragment concatamers. Southern blot analysis indicated that the transgenes were organized in head-to-tail, head-to-head or tail-to-tail concatemers in the genome of the host fish [3,9,10]. The high copy numbers of transgene concatemers might exist as heterochromatin in the host genome, which would be subject to gene silencing [11,12]. The multi-copy integration of transgene concatemers are potentially toxic to the host fish [7]. The number of transgene copies within an array exerts a repressive influence on expression, with several studies reporting a decrease in the level of expression per copy as the copy number increases, while a reduction in the copy number results in a marked increase in expression of the transgene [13]. Our research results in all-fish GH transgenic common carp showed that the transgene was randomly integrated into a single locus of the host chromosome. Three transgene copies exist in both the F3 and F5 generations of "all fish" GH transgenic common carp, suggesting that the transgene in "all-fish" GH gene transgenic common carp was faithfully passed down from one generation to the next. Additionally, PCR and sequencing analysis indicated that the foreign genes integrated in transgenic carp were head-to-head and head-to-tail (unpublished data). We presume that the transgenes are expressed in a copy-number dependent manner. The fewer the copy numbers of the integrated transgene in the host chromosome, the higher the efficiency of transgene expression. Lower copy numbers would also result in stable transmission of the transgenes in germ-lines through different generations. However, the sizes of transgenic animals did not correlate well with the copy numbers of the transgenes present [3].

During our investigation of the mechanism of the concatamer formation and integration pattern of transgenes, we found that foreign genes were integrated into the host chromosome in the 2-step recombination process of nonhomologous end joining (NHEJ). The NHEJ repair components functioning in fish eggs repair double strand breaks (DSBs) with high efficiency. The extremely high concentration of exogenous genes microinjected into fish embryos result in large accumulations of DSBs. Following microinjection, the NHEJ complex first processed DSBs between exogenous gene molecules. As the concentration of exogenous gene DSBs decreased, the NHEJ repair mechanism began to skew towards endogenous DSBs generated in the host genome during successive generations of cellular division. Because endogenous

DSBs are distributed at various sites in the chromosomes, ligation between transgene DSBs and endogenous DSBs resulted in random integration of exogenous genes. We presume that ligase, polymerase and endonuclease related to DNA repair are involved in this process (unpublished data). Hence, foreign genes inevitably formed concatamers and randomly integrated into the host chromosomes by microinjection. The transgenes were efficiently expressed with fewer integrated copy numbers of the transgene, allowing for their stable transmission into the germ-line of fish. Additionally, there is a lower probability of functional integration of the transgene in the fish in each generation, therefore, it becomes necessary to perform gene transfers with a large sample to produce stable and functionally integrated transgenes in fish.

1.2 Gene targeting in transgenic fish

Due to the random integration of transgenes, the site of insertion cannot be precisely defined or confirmed. The copy number at the site of insertion also cannot be controled, which leads to lower ratios of functional integration and lower efficiencies of transgene expression. The highly random process of integration is toxic for the host fish. The expression of transgenes cannot be pre-configured when transgenic fish are produced by conventional methods, making generation of a stable transgenic fish line with desirable traits for aquaculture a difficult task. Thus, targeted and controllable transgene expression in fish has been a hot research focus in recent years.

Many studies have also revealed that the efficiency of homologous recombination in transgenic fish is extremely low. Fan *et al.* [14] studied the targeted insertion of a plasmid by homologous recombination in zebrafish embryonic stem (ES) cell cultures. Two colonies of cells were confirmed to have undergone homologous recombination out of a total of approximately 1×10^6 cells [14]. Chen *et al.* [15] found that the probability of homologous recombination was only 10^{-5} in a carp epithelioma papulosum cell line. Wu *et al.* [16] described the production of zebrafish by modifying an endogenous GH gene through Cre-lox-mediated homologous recombination with two integrants, and the rate of targeted integration with germline transmission in zebrafish was 1.7×10^{-6}. Another study in zebrafish screened more than 80000 targeted embryos and confirmed that they were able to obtain lines with a sole copy of the integrating DNA at the target genome site [17]. However, their integration efficiency was approximately 10^{-4} to 10^{-5}. The underlying reasons for the extremely low rates of homologous recombination of exogenous genes remain unknown.

Our recent study suggests that the existing NHEJ complex in fish embryos causes concatamer formation in targeting vectors and interferes with homologous recombination (unpublished data), which may account for the extremely low efficiency of homologous recombination in transgenic fish. Thus, new techniques have been developed to achieve directional gene interference in zebrafish. Recent approaches used in the zebrafish model take advantage of the insertion or deletion of bases associated with NHEJ to introduce mutations into target sites with high efficiency. Utilization of this mechanism has resulted in directional interference of gene expression and a gain in phenotype coincident with gene knock-out [18,19].

With conventional gene-transfer techniques in fish, insertion of DNA into the genome randomly occurs and in many instances at multiple sites. Associated position effects, copy number differences, and multiple gene interactions all result in lower efficiency of functional integration and limit the application of fish genetic engineering. Therefore, it is essential for transgenic fish breeding to improve the integration efficiency of the transgene.

The medaka fish Tol2 transposable element belongs to the hAT family of transposons, and is active in a variety of vertebrates. As the zebrafish genome does not contain this element, transgenic fish were created with fertilized eggs co-injected with mRNA transcribed *in vitro*, using cDNA of the Tol2 transcript as a template and a DNA plasmid harboring a Tol2 mutant, which contains a deletion in the putative transposase gene but retains necessary *cis* sequences. This method was demonstrated to be highly efficient in creating transgenic fish expressing fluorescent reporter genes or cloned genes under the control of specific promoters [20,21]. Urasaki *et al.* [22] also developed an *in vivo* transposition system using Tol2 and created genome-wide insertions from a single-copy donor. The Tols transposable element is a practical genetic tool which reduces the time for the production of transgenic fish lines.

2 Population Fitness of GH Transgenic Fish

Fitness means that any animal whose form or behavior facilitates the avoidance of predators or escape when attacked by predators will have a greater probability of surviving to breed and therefore greater probability of producing offspring [23]. Fitness provides relevant parameters to assess the potential environmental risk of transgenic fish, and most fitness research has been performed with GH transgenic fish.

2.1 Viability of transgenic fish

Dunham *et al.* studied the predator avoidance of GH transgenic channel catfish (*Ictalurus punctatus*) and found that overall predator avoidance was also better for non-transgenic individuals [24]. They suggested that any ecological effect would be unlikely because the increased susceptibility of transgenic channel catfish to predators would most likely decrease or eliminate the transgenic genotype if they were to be released into nature [24]. Similar results were also found in a study by Abrahams and Sutterlin [25]. Sundström *et al.* found that larger sized GH transgenic Coho salmon fry remained closer to the surface than normal Coho salmon both during feeding and predatory attacks [26]. In nature, where predators may attack from above (birds) or below(fish), this kind of behavior may translate into higher risk of predation, which could increase mortality and lower the fitness of the transgenic fish [26]. We conducted a series of predation trials with a paired-contrast design to test the differences in anti-predator ability between "all-fish" growth hormone transgenic and non-transgenic common carp. We showed that the young transgenic fish suffered higher mortality due to predation than the controls in both small and large size-matched trials, and thus possess lower antipredator defenses (unpublished data). All of the results suggested that GH transgenic fish would experience lower survival fitness in any environment where there is higher risk of predation.

The higher predation mortality of transgenic fish is related to the swimming characteristics of transgenic fish. In the natural environment, fish usually escape attack from predators by swimming [27]. The highest swimming speed of any fish species determines their ability to survive in a natural environment [28]. Some work has indicated that GH-transgenic common carp and salmon juveniles are inferior swimmers with significantly decreased critical swimming speeds compared with control fish of the same size [29,30]. This may affect the ability of the transgenics to escape under the threat of predators. Similar results were also found in a study by Lee *et al.* [31] which showed that growth-enhanced GH-transgenic salmon juveniles had critical swimming speeds half those of the size-matched control fish.

Variations in morphological characteristics affect fish swimming performance. Previous studies have reported significant differences in morphology between the GH-transgenic salmonids and the controls [32-35]. Ostenfeld *et al.* [36] and Lee *et al.* [31] suggested that changes in body shape were responsible for reduced critical swimming speeds (Ucrit) of GH-transgenic salmonids. We found that transgenic fish have significantly deeper heads, longer caudal lengths of the dorsal region, longer standard lengths (LS) and shallower body and caudal regions, and shorter caudal lengths of the ventral region. Swimming speeds are related to the combination of deeper body and caudal regions, longer caudal lengths of the ventral region, shallower head depths, shorter caudal lengths of dorsal region and LS. These findings suggest that morphological variations which are poorly suited to produce maximum thrust and minimum drag in GH-transgenic *C. carpio* may be responsible for their decreased swimming abilities in comparison with nontransgenic controls [37]. Both the morphological characteristics and the swimming performance of fish are related to maintenance of position in a water current and migration to spawning habitats [38]. We presumed that variations in morphology and the inferior swimming characteristics of transgenic fish would seriously affect their fitness in a natural environment.

Appetite and feeding behavior may be additional factors related to the higher predation mortality of transgenic fish. In salmonids, growth hormone (GH) stimulates growth, appetite and the ability to compete for food. Increased GH levels in GH-transgenic Coho salmon

Oncorhynchus kisutch (Walbaum) increases their competitive abilities through higher feeding motivation [32,39,40]. We recently measured food consumption, movement levels and feeding hierarchies of juvenile transgenic common carp and their size-matched non-transgenic conspecifics with a limited food supply. The transgenic fish had a 73.3% higher movement level as well as a higher feeding order, and consumed 1.86 times as much food pellets as their nontransgenic counterparts. The "all fish" GH transgenic common carp have a number of altered feeding related behaviors, e.g. increased food consumption and feeding order, time spent in the feeding areas and the movement level [41]. GH-transgenic salmon are more active [42] and appear more risk-prone than non-transgenic fish, which may potentially increase their mortality from predation in the wild [43-45]. The appetite and feeding behavior of transgenic fish increase their competitive ability to obtain food resources, thus enhancing population viability. However, the much higher predation mortality of transgenic fish relative to non-transgenic fish remains a higher threat to the transgenic population.

Respiratory metabolism, including metabolic rate, aerobic scope, the capacity for sustained aerobic exercise and tolerance of hypoxia, is a fundamental component of fish physiology and ecology, and is a major fitness trait related to the viability of fish.

GH transgenic fish including Atlantic salmon (*Salmo salar*), tilapia (*Oreochromis* sp.) and Coho salmon (*Oncorhynchus kisutch*), were found to have a greater oxygen uptake rate than control fish during routine feeding conditions. Additionally, transgenic fish have a greater daily feed intake in comparison to control fish [42,46-48]. We found that the mean daily feed intake of GH transgenic common carp was 2.12 times greater than control fish during 4 days of feeding. The average oxygen uptake of GH transgenic fish was 1.32 times greater than control fish within 96 h of starvation, but was not significantly different from controls between 96 and 144 h of starvation. At the same time, GH transgenic fish did not deplete energy reserves at a faster rate than did the controls, as the carcass energy contents of the two groups following a 60-d starvation period were not significantly different [49]. We investigated the point of asphyxiation of transgenic fish and control fish under routine conditions and during starvation periods. The average asphyxiation point of transgenic fish was the same as control fish under starvation, but was significantly higher in controls after feeding (unpublished data), which suggested that transgenic fish would have lower viability than control fish under hypoxia stress conditions.

The GH transgene does not modify the standard metabolic rate (SMR) of common carp, a fact of substantial ecological significance. During times of low food abundance, GH transgenic common carp would be at no greater risk of suffering from food deprivation than non-transgenic carp. Additionally, when food is in abundance, GH transgenic common carp increase nutrient intake and specific dynamical action (SDA) to facilitate rapid growth. Therefore, GH transgenic common carp have increased energy storage, which facilitates survival during periods of low food availability. However, both the greater oxygen uptake and higher asphyxiation point of GH transgenic fish relative to control fish suggested that the transgenic fish would experience a lower overall population viability.

There are a number of factors which affect the viability of fish and the interaction between the phenotype and the environment is complex. Hence, it is problematic to estimate the complex interaction and viability of transgenic fish in the natural ecosystem on the basis of the simple phenotypic differences observed in the laboratory between the transgenic and wild-type fish. Therefore, the evaluation of the potential ecological risks posed by transgenic fish is typically conducted in an aqueous environment with a large space and a complex structure which mimics as closely as possible the natural or wild environment [8,50].

2.2 Reproductive fitness traits of GH transgenic fish

The gonad is an essential organ for the generation of sperm and ovum in vertebrates. Gonad development is related to the growth and reproductive performance of fish. The age of sexual maturity is a key reproductive fitness parameters for GH transgenic fish [51]. However, there is conflicting evidence concerning gonad development and the age at sexual maturity in fast-growing GH transgenic fish, with few recent studies.

Reproduction-growth trade-off is highly significant in fish [52]. General life history theory predicted that rapid growth should lead to early maturation [53]. However, Tsikliras *et al.* found that the cost of round sardinella growing faster is reduced fecundity [54]. GH transgenic Coho salmon sexually mature 1 or 2 years earlier than non-transgenic wild-strain Coho salmon in the laboratory environment [3,55]. The female gonado-somatic index (I_G) was found to be lower in GH transgenic Nile tilapia than in non-transgenic siblings in both mixed and separate culture conditions. Transgenic male I_G values were found to be higher in mixed culture and lower in separate culture than that of their non-transgenic siblings [7].

It is unclear as to why there are differences in ages of sexual maturity between GH transgenic fish and control fish. No studies concerning the underlying mechanism have been found. It has been suggested that the mechanism lies in the energy allocation between reproduction and growth. Normal salmon with high growth rates and sufficient energy reserves sexually mature 1 year earlier than the majority of fish in the population. It would seem that GH transgenic Coho salmon meet the growth and physiology cues necessary to allow early maturity, but their enhanced growth rate allows a normal body size to be reached prior to maturation [3]. Rahman *et al.* [7] thought that growth and the prevailing competition for food amongst transgenic individuals limits the energy available for gonad development when transgenics and their non-transgenic siblings were separately cultured. The reduced gonadal development observed in non-transgenic male tilapia in mixed culture might have occurred due to lack of adequate food energy to maintain reproductive development [7].

It has become increasingly clear that somatic growth is tightly interconnected with nutrition and reproductive status. The biological action of GH is not only to control growth but also gonadal development [56]. In female common carp, the application of various concentrations of different steroids on a primary culture of pituitary cells led to release of GH [57]. As somatotrophs and gonadotrophs are co-localized in the proximal pars distalis in telleosts, GnRH peptides released from neurosecretory terminals are located in the proximity of somatotrophs [56]. Blaise demonstrated *in vitro* evidence of GnRH being a factor in GH release in rainbow trout, which is related to variable permissive effects of IGF-I with the sexual cycle [58]. GH has a surge release at ovulation in goldfish and is a factor in egg maturation in *Heteropneustes fossilis* [59,60]. Male goldfish also have surge release of LH and GH at the time of spermeation and spawning [59]. All of these results suggested that GH release was stimulated by GnRH and had been identified as co-gonadotropin regulation of fish reproduction.

GH is an important factor in the sexual development of various species. Phosphoenol-pyruvate carboxykinase (PEPCK)-bGH (bovine GH) transgenic female mice, which express very high GH levels, are fertile. The infertility in female mice transgenic for the PEPCK.bGH gene construct is due to luteal failure, which in turn appears to be caused by inadequate PRL secretion after mating [61]. In studying the GH receptor gene knockout (GHR-KO) mice, Chandrashekar *et al.* found that the circulating LH response to GnRH treatment was significantly attenuated, and basal and LH- stimulated testosterone release from the isolated testes of GHR-KO mice were decreased [62]. In GH receptor (GHR) and GH-binding protein (GHBP) knockout (KO) mice, the number of follicles per ovary was markedly reduced, and both the reproductive function and ovulation rate are measurably decreased compared with wild-type animals. IGF-I treatment is not able to rescue either fertility or ovarian responsiveness to exogenous gonadotropins [63]. The weights of testes and epididymii were significantly reduced in GHR-KO mice. The intratesticular testosterone levels and the testosterone response to LH treatment were attenuated in GHR gene-disrupted mice. Elongated spermatids appeared later in the testes of GHR-KO mice than in the testes of normal mice [64].

IGFs are recognized as mediators of the growth-promoting actions of GH. Several lines of evidence showed that insulin-like growth factor I (IGF-I) is a pivotal factor in the interactions between the somatotropic and gonadotropic axes. IGF-I is involved in the direct regulation of GTH subunit genes during sexual maturation. In particular, IGF-I differentially modulates sGnRH-induced GTH subunit gene expression, depending on the reproductive stage [65]. A gonad-specific IGF subtype in teleosts has been identified, which highlights the significance of IGF in teleost gonadal development and reproduction [66]. Growth- and reproduction-related genes from the hypothalamus-pituitary-gonadal (HPG) axis or the GH/IGF axis were expressed during the very early development of embryos and the early larval stages [67], which suggested that the interactions between HPG and GH/IGF begin in the very early development of fish embryos.

In essence, studying the reproductive and gonadal developmental traits of GH transgenic fish provides population fitness parameters for evaluating ecosafety. GH transgenic fish provide a useful model system for studying the interaction mechanism between the somatotropic and gonadotropic axes of fish.

2.3 Reproductive containment strategies of transgenic fish

The potential ecological risks posed by transgenic fish are closely related to their reproduction characteristics. The optimal safeguard against potential ecological risks is to render transgenic fish completely sterile. Ecological risk evaluation and prevention strategies are fundamentally significant components of breeding research [8]. Transgenic fish breeding research sought to suppress the expression of various critical genes that are related to gonad development and sex maturation, the goal being to inhibit gonad development and to produce sterile transgenic fish [68].

Gonadotropin-releasing hormone (GnRH), a member of the conservative neuro-decapeptide family linking the brain and the reproductive system, is a primary factor in controlling and maintaining gonad development and the reproduction function by stimulating the synthesis and release of gonadotropin (GtH) in the pituitary. *GnRH* is usually selected as the target gene to produce sterile transgenic fish. Uzbekova *et al.* [69] reported that they prepared a recombinant vector containing anti-sense DNA complementary to Atlantic salmon (*Salmo salar*) *GnRH*3 cDNA driven by the specific promoter Pab derived from a corresponding *GnRH3* gene. They tried to produce sterile transgenic fish by inhibiting the endogenous *GnRH3* expression. However, they were unable to inhibit gonad development [69]. The authors suggested that a stronger promoter and 100% matched complementary *GnRH* RNA may be required to induce sterility [69]. Maclean *et al.* reported that partially or completely sterile male tilapia were produced when they suppressed the *GnRH3* expression via anti-sense sequences of *GnRH3* driven by the strong, constitutively expressing β-actin promoter [70]. Four forms of GnRH are found in common carp [71,72]. According to the expression pattern of *GnRH* genes in dissected brain areas, pituitary glands and gonads of juveniles, mature and regressed common carp, we chose GnRH3 as the target gene, and prepared a recombinant construct for microinjection, which contains anti-sense a DNA fragment targeting *GnRH3* mRNAs driven by a carp beta actin (β-actin) gene promoter. We were able to produce completely sterile transgenic fish [73]. It is noteworthy that GnRH precursors are usually encoded by multiple genes, and different GnRH serve as different factors for stimulating the synthesis and release of GtH. Different GnRH may have different effects on reproduction between male and female fish. Therefore, before disrupting GnRH for sterility induction, it is necessary to establish which GnRH type predominantly controls GtH release in the species of interest to target the correct GnRH type for inactivation [74].

GtH is a signal regulated by GnRH in the HPG axis. In teleost fish, as in other vertebrates, there are two forms of GtH, generally referred to as FSH and LH. GtHs are glycoprotein hormones composed of a common alpha subunit and a hormone-specific β subunit, which confers its biological specificity. The hormonal activity is expressed after a noncovalent association between these two subunits. GtH regulates gonadal development in vertebrates. Therefore, GtH is another target gene which is used to develop sterile transgenic fish. Maclean *et al.* prepared a tilapia LH beta subunit gene anti-sense construct, but they could not produce sterile transgenic fish [70]. However, targeted disruption of the LH beta subunit in mice led to hypogonadism, defects in gonadal steroidogenesis and infertility [75].

Huang *et al.* [76] reported using RNA interference (RNAi) to knockdown *GtHα*, *FSHβ* and *LHβ* genes. The RNA or protein expression levels of the GtH genes were monitored by RT-PCR, Southern blotting, and GFP analyses. The expression of *GtH* mRNA was more efficiently suppressed by *GtHα* RNAi expression compared with the other two subunits. They suggested that *GtHα* RNAi may be an essential factor in the further development of sterility technology of transgenic fish for biosafety purposes [76]. However, we discovered a new alternative splicing transcript of the GtH alpha subunit from the pituitary glands and ovaries of common carp. The novel transcript encodes a special protein which interacts with both FSHβ and LHβ and might be a key factor in regulating the different levels of FSH and LH during the different reproductive cycles [77]. Therefore, it may be difficult or problematic to select *GtH* as a target gene for producing sterile transgenic fish.

Primordial germ cells (PGCs) are the founder cells of both female and male gametes, from which all sexually reproducing organisms arise. During the early embryonic development of fish, as in other vertebrates, PGCs initially arise outside the gonadal region and migrate to the genital ridges, eventually coalescing with their somatic counterpart, which gives rise to ovaries in females and testes in males. Therefore, the initial PGCs form the basis for gonad development and maturation. Dead end (*dnd*) is one of the necessary genes for migration and survival of PGCs, and adult fish would be sterile without the Dnd protein [78]. Slanchev *et al.* [79] sectioned fish derived from embryos in which the germ line had been ablated by using *dnd* morpholino oligonucleotide (MO), and they found that in the experimental fish the gonadal structure appeared smaller in size and histologically uniform. At 90 days post-fertilization (*dpf*), no gonad-like structures were observed in *dnd* MO-injected fish, suggesting that the gonads degenerated in the absence of PGCs. The results implied that in zebrafish germ cells are not required for the formation of the gonads but, rather, are essential for its differentiation and survival [79]. Therefore, those genes related to the migration and survival of PGCs are primary targets for the production of sterile transgenic fish.

3 Summary and Prospects

It has been more than 20 years since the first transgenic fish was produced. GH transgenic fish with enhanced growth traits might be the breakthrough model for the industrialization of agricultural transgenic animals. However, compared with agriculture crops, there are significantly less functional fish genes related to benefit. *GH* is one of the few genes currently used in transgenic fish breeding. It may be essential for further investigations transgenic fish breeding to identify and select additional genes or to obtain bioregulatory information related to fish characteristics that would be economically beneficial.

Highly efficient integration and controllable expression of transgenes would ensure the success of global transgenic fish breeding. As mechanisms of transgene integration are further understood and enhanced gene targeting technologies are developed, it may become possible to establish technologies for the stable, highly efficient integration and controllable expression of transgenes.

The ecological risk evaluation of transgenic fish is generally employed with the analysis of a single phenotype. The analysis of a single phenotype of transgenic fish is helpful for the recognition of their fitness in the natural ecosystem and consequently sheds light on the evaluation of the potential ecological risks posed by transgenic fish. However, the viability, competitive ability and reproductive fitness of transgenic fish in the natural ecosystem are the result of interactions between many complicated factors. The evaluation of the potential ecological risks posed by transgenic fish would be optimally conducted in an aqueous environment with a large space and a complex structure [8,50]. We designed and constructed an artificially simulated lake with an area of 6.7 hm^2 to study the survival, growth, reproduction fitness of GH transgenic common carp in a natural ecosystem. This environment has allowed us to obtain to conduct a number of studies concerning the fitness and potential ecological risks of transgenic fish [8].

Producing sterile transgenic fish radically resolves the potential ecological risks of transgenic fish. There have been reports concerning the production of founder generations of sterile transgenic fish. However, questions remain as to how to restore the fertility of sterile transgenic fish or how to stably inherit the sterile trait in different generations of transgenic fish.

This work was supported by the Development Plan of the State Key Fundamental Research of China (Grant Nos. 2007CB109205, 2007CB109206) and the National Natural Science Foundation of China (Grant No. 30930069).

References

1 Zhu Z, Li G, He L, *et al*. Novel gene transfer into the fertilized eggs of gold fish (*Carassius auratus*). J Appl Ichthyol, 1985, 1: 31-34

2 Wang Y, Hu W, Wu G, *et al*. Genetic analysis of 'all-fish' growth hormone gene transferred carp (*Cyprinus carpio* L.) and its F_1 generation. Chinese Sci Bull, 2001, 46: 1174-1177

3 Devlin R H, Biagi C A, Yesaki T Y. Growth, viability and genetic characteristics of GH transgenic Coho salmon strains. Aquaculture, 2004, 236: 607-632

4 Fletcher G L, Shears M A, Yaskowiak E S, *et al.* Gene transfer: potential to enhance the genome of Atlantic salmon for aquaculture. Aust J Exp Agric, 2004, 44: 1095-1100

5 Nam Y K, Cho Y S, Cho H J, *et al.* Accelerated growth performance and stable germ-line transmission in androgenetically derived homozygous transgenic mud loach, *Misgurnus mizolepis*. Aquaculture, 2002, 209: 257-270

6 Martínez R, Arenal A, Estrada M P, *et al.* Mendelian transmission, transgene dosage and growth phenotype in transgenic tilapia (*Oreochromis hornorum*) showing ectopic expression of homologous growth hormone. Aquaculture, 1999, 173: 271-283

7 Rahman M A, Ronyai A, Engidaw B Z, *et al.* Growth and nutritional trials on transgenic Nile tilapia containing an exogenous fish growth hormone gene. J Fish Biol, 2001, 59: 62-78

8 Hu W, Wang Y, Zhu Z. Progress in the evaluation of transgenic fish for possible ecological risk and its containment strategies. Sci China C-Life Sci, 2007, 50: 573-579

9 Zeng Z Q, Zhu Z Y. The molecular polymorphism of transgenes in F_4 generation red carp transfected with pMThGH gene. Chinese Sci Bull, 2000, 45: 1957-1962

10 Wu B, Sun Y H, Wang Y W, *et al.* Characterization of transgene integration pattern in F_4 hGH-transgenic common carp (*Cyprinus carpio*). Cell Res, 2005, 15: 447-454

11 Dorer D R. Do transgene arrays form heterochromatin in vertebrates? Transgenic Res, 1997, 6: 3-10

12 Martin D I K, Whitelaw E. The vagaries of variegating transgenes. Bio Essays, 1996, 18: 919-923

13 Garrick D, Fiering S, Martin D I, *et al.* Repeat-induced gene silencing in mammals. Nat Genet, 1998, 18: 56-59

14 Fan L, Moon J, Crodian J, *et al.* Homologous recombination in zebrafish ES cells. Transgenic Res, 2006, 15: 21-30

15 Chen S, Hong Y, Schartl M. Development of a positive-negative selection procedure for gene targeting in fish cells. Aquaculture, 2002, 214: 67-79

16 Wu Y, Zhang G, Xiong Q, *et al.* Integration of double-fluorescence expression vectors into zebrafish genome for the selection of site-directed knockout/knockin. Mar Biotechnol, 2006, 8: 304-311

17 Liu W Y, Wang Y, Qin Y, *et al.* Site-directed gene integration in transgenic zebrafish mediated by Cre recombinase using a combination of mutant Lox sites. Mar Biotechnol, 2007, 9: 420-428

18 Doyon Y, McCammon J M, Miller J C, *et al.* Heritable targeted gene disruption in zebrafish using designed zinc-finger nucleases. Nat Biotech, 2008, 26: 702-708

19 Meng X, Noyes M B, Zhu L J, *et al.* Targeted gene inactivation in zebrafish using engineered zinc-finger nucleases. Nat Biotech, 2008, 26: 695-701

20 Kawakami K, Shima A, Kawakami N. Identification of a functional transposase of the Tol2 element, an Ac-like element from the Japanese medaka fish, and its transposition in the zebrafish germ lineage. Proc Natl Acad Sci USA, 2000, 97: 11403-11408

21 Kawakami K, Takeda H, Kawakami N, *et al.* A transposon-mediated gene trap approach identifies developmentally regulated genes in zebrafish. Dev Cell, 2004, 7: 133-144

22 Urasaki A, Asakawa K, Kawakami K. Efficient transposition of the Tol2 transposable element from a single-copy donor in zebrafish. Proc Natl Acad Sci USA, 2008, 105: 19827-19832

23 Lind J, Cresswell W. Determining the fitness consequences of antipredation behavior. Behav Ecol, 2005, 16: 945-956

24 Dunham R A, Chitmanat C, Nichols A, *et al.* Predator avoidance of transgenic channel catfish containing salmonid growth hormone genes. Mar Biotechnol, 1999, 1: 545-551

25 Abrahams M V, Sutterlin A. The foraging and antipredator behavior of growth-enhanced transgenic Atlantic salmon. Anim Behav, 1999, 58: 933-942

26 Sundström L F, Devlin R H, Johnsson J I, *et al.* Vertical position reflects Increased feeding motivation in growth hormone transgenic Coho salmon (*Oncorhynchus kisutch*). Ethology, 2003, 109: 701-712

27 Videler J J. Fish Swimming. London: Chapman and Hall, 1993

28 Swanson C, Young P S, Cech Jr J C. Swimming performance of delta smelt: maximum performance, and behavioral and kinematic limitations on swimming at submaximal velocities. J Exp Biol, 1998, 201: 333-345

29 Li D, Fu C, Hu W, *et al.* Rapid growth cost in "all-fish" growth hormone gene transgenic carp: Reduced critical swimming speed. Chinese Sci Bull, 2007, 52: 1501-1506

30 Anthony P, Farrell W B, Devlin R H. Growth-enhanced transgenic salmon can be inferior swimmers. Can J Zool, 1997, 75: 335-337

31 Lee C G, Devlin R H, Farrell A P. Swimming performance, oxygen consumption and excess post-exercise oxygen consumption in adult transgenic and ocean-ranched Coho salmon. J Fish Biol, 2003, 62: 753-766

32 Devlin R H, Johnsson J I, Smailus D E, *et al.* Increased ability to compete for food by growth hormone-transgenic coho salmon *Oncorhynchus kisutch*. Aquac Res, 1999, 30: 479-482

33 Stevens E D, Devlin R H. Gut size in GH transgenic coho salmon is enhanced by both the GH transgene and increased food intake. J Fish Biol, 2005, 66: 1633-1648

34 Don Stevens E, Sutterlin A. Gill Morphometry in growth hormone transgenic Atlantic salmon. Environ Bio Fishes, 1999, 54: 405-411

35 Seiler S M, Keeley E R. Morphological and swimming stamina differences between Yellowstone cutthroat trout (*Oncorhynchus clarkii bouvieri*), rainbow trout (*Oncorhynchus mykiss*), and their hybrids. Can J Fish Aquat Sci, 2007, 64: 127-135

36 Ostenfeld T H, McLean E, Devlin R H. Transgenesis changes body and head shape in Pacific salmon. J Fish Biol, 1998, 52: 850-854

37 Li D, Hu W, Wang Y, *et al.* Reduced swimming abilities in fast-growing transgenic common carp *Cyprinus carpio* associated with their morphological variations. J Fish Biol, 2009, 74: 186-197

38 Boily P, Magnan P. Relationship between individual variation in morphological characters and swimming costs in brook charr (*Salvelinus fontinalis*) and yellow perch (*Perca flavescens*). J Exp Biol, 2002, 205: 1031-1036

39 Johnsson J I, Björnsson B T. Growth hormone increases growth rate, appetite and dominance in juvenile rainbow trout, *Oncorhynchus mykiss*. Anim Behav, 1994, 48: 177-186

40 Jonsson E, Johnsson J I, Bjornsson B T. Growth hormone increases predation exposure of rainbow trout. Proc R Soc B, 1996, 263: 647-651

41 Duan M, Zhang T, Hu W, *et al.* Elevated ability to compete for limited food resources by all-fish growth hormone transgenic common carp *Cyprinus carpio*. J Fish Biol, 2009, 75: 1459-1472

42 Don Stevens E, Sutterlin A, Cook T. Respiratory metabolism and swimming performance in growth hormone transgenic Atlantic salmon. Can J Fish Aquat Sci, 1998, 55: 2028-2035

43 Sundstrom L F, Lohmus M, Johnsson J I, *et al.* Growth hormone transgenic salmon pay for growth potential with increased predation mortality. Proc Biol Sci, 2004, 271 (Suppl 5): S350-352

44 Vandersteen Tymchuk W E, Abrahams M V, *et al.* Competitive ability and mortality of growth-enhanced transgenic coho salmon fry and parr when foraging for food. T Am Fish Soc, 2005, 134: 381-389

45 Biro P A, Abrahams M V, Post J R. Direct manipulation of behaviour reveals a mechanism for variation in growth and mortality among prey populations. Anim Behav, 2007, 73: 891-896

46 Cook J T, McNiven M A, Richardson G F, *et al.* Growth rate, body composition and feed digestibility/conversion of growth-enhanced transgenic Atlantic salmon (*Salmo salar*). Aquaculture, 2000, 188: 15-32

47 McKenzie D J, Martínez R, Morales A, *et al.* Effects of growth hormone transgenesis on metabolic rate, exercise performance and hypoxia tolerance in tilapia hybrids. J Fish Biol, 2003, 63: 398-409

48 Leggatt R A, Devlin R H, Farrell A P, *et al.* Oxygen uptake of growth hormone transgenic coho salmon during starvation and feeding. J Fish Biol, 2003, 62: 1053-1066

49 Guan B, Hu W, Zhang T, *et al.* Metabolism traits of 'all-fish' growth hormone transgenic common carp (*Cyprinus carpio* L.). Aquaculture, 2008, 284: 217-223

50 Devlin R H, Sundström L F, Muir W M. Interface of biotechnology and ecology for environmental risk assessments of transgenic fish. Trends Biotechnol, 2006, 24: 89-97

51 Muir W M, Howard R D. Possible ecological risks of transgenic organism release when transgenes affect mating success: Sexual selection and the Trojan gene hypothesis. Proc Natl Acad Sci USA, 1999, 96: 13853-13856

52 Roff D A, Heibo E, Vøllestad L A. The importance of growth and mortality costs in the evolution of the optimal life history. J Evol Biol, 2006, 19: 1920-1930

53 Stearns S C. The Evolution of Life Histories. New York: Oxford University Press, 1992

54 Tsikliras A, Antonopoulou E, Stergiou K. A phenotypic trade-off between previous growth and present fecundity in round sardinella Sardinella aurita. Popul Ecol, 2007, 49: 221-227

55 Bessey C, Devlin R H, Liley N R, *et al.* Reproductive performance of growth-enhanced transgenic coho salmon. T Am Fish Soc, 2004, 133: 1205-1220

56 Canosa L F, Chang J P, Peter R E. Neuroendocrine control of growth hormone in fish. Gen Compa Endocr, 2007, 151: 1-26

57 Degani G, Boker R, Jackson K. Growth hormone, gonad development, and steroid levels in female carp. Comp Biochem Physiol C Pharmacol Toxicol Endocrinol, 1996, 115: 133-140

58 Blaise O, Le Bail P Y, Weil C. Permissive effect of insulin-like growth factor I (IGF-I) on gonadotropin releasing-hormone action on *in vitro* growth hormone release, in rainbow trout (*Oncorhynchus mykiss*). Comp Biochem Physiol A: Physiology, 1997, 116: 75-81

59 Yu K L, Peter R E. Changes in brain levels of gonadotropin-releasing hormone and serum levels of gonadotropin and growth hormone in goldfish during spawning. Can J Zool, 1991, 69: 182-188

60 Sarang M, Lal B. Effect of piscine GH/IGF-I on final oocyte maturation *in vitro* in *Heteropneustes fossilis*. Fish Physiol Biochem, 2005, 31: 231-233

61 Cecim M, Kerr J, Bartke A. Infertility in transgenic mice overexpressing the bovine growth hormone gene: luteal failure secondary to prolactin deficiency. Biol Reprod, 1995, 52: 1162-1166

62 Chandrashekar V, Bartke A, Coschigano K T, *et al.* Pituitary and testicular function in growth hormone receptor gene knockout mice. Endocrinology, 1999, 140: 1082-1088

63 Bachelot A, Monget P, Imbert-Bollore P, *et al.* Growth hormone is required for ovarian follicular growth. Endocrinology, 2002, 143: 4104-4112

64 Keene D E, Suescun M O, Bostwick M G, *et al.* Puberty is delayed in male growth hormone receptor gene-disrupted mice. J Androl, 2002, 23: 661-668

65 Ando H, Luo Q, Koide N, *et al.* Effects of insulin-like growth factor I on GnRH-induced gonadotropin subunit gene expressions in masu salmon pituitary cells at different stages of sexual maturation. Gen Compa Endocr, 2006, 149: 21-29

66 Wang D-S, Jiao B, Hu C, *et al.* Discovery of a gonad-specific IGF subtype in teleost. Biochem Biophys Res Co, 2008, 367: 336-341

67 Lu Y, Hu W, Zhu Z. Gene expression profiles of growth and reproduction related genes during the early development of common carp (*cyprinus carpio* L.). Acta Hydrobiol Sin, 2009, 33: 1126-1131

68 Hu W, Wang Y, Zhu Z. A perspective on fish gonad manipulation for biotechnical applications. Chinese Sci Bull, 2006, 51: 1-6

69 Uzbekova S, Chyb J, Ferriere F, *et al.* Transgenic rainbow trout expressed sGnRH-antisense RNA under the control of sGnRH promoter of Atlantic salmon. J Mol Endocrinol, 2000, 25: 337-350

70 Maclean N, Molina G H A, Ashton T, *et al.* Reversibly-sterile fish via transgenesis. ISB News Rep, 2003, 3-5

71 Li S, Hu W, Wang Y, *et al.* Cloning and experssion analysis in mature individuals of salmon gonadotropin-releasing hormone (sGnRH) gene in common carp. Acta Genet Sin, 2004, 31: 1072-1081

72 Li S, Hu W, Wang Y, *et al.* Cloning and expression analysis in mature individuals of two chicken type-II GnRH (cGnRH-II) genes in common carp (*Cyprinus carpio*). Sci China C Life Sci, 2004, 47: 349-358

73 Hu W, Li S, Tang B, *et al.* Antisense for gonadotropin-releasing hormone reduces gonadotropin synthesis and gonadal development in transgenic common carp (*Cyprinus carpio*). Aquaculture, 2007, 271: 498-506

74 Wong A C, Van Eenennaam A L. Transgenic approaches for the reproductive containment of genetically engineered fish. Aquaculture, 2008, 275: 1-12

75 Ma X, Dong Y, Matzuk M M, *et al.* Targeted disruption of luteinizing hormone β-subunit leads to hypogonadism,

defects in gonadal steroidogenesis, and infertility. Proc Natl Acad Sci USA, 2004, 101: 17294-17299
76 Huang W T, Hsieh J C, Chiou M J, *et al*. Application of RNAi technology to the inhibition of zebrafish GtHalpha, FSHbeta, and LHbeta expression and to functional analyses. Zoolog Sci, 2008, 25: 614-621
77 Wang Y, Hu W, Liu W Y, *et al*. Identification and characterization of a novel splice variant of gonadotropin alpha subunit in the common carp *Cyprinus carpio*. J Fish Biol, 2007, 71: 1082-1094
78 Ciruna B, Weidinger G, Knaut H, *et al*. Production of maternalzygotic mutant zebrafish by germ-line replacement. Proc Natl Acad Sci USA, 2002, 99: 14919-14924
79 Slanchev K, Stebler J, de la Cueva-Méndez G, *et al*. Development without germ cells: The role of the germ line in zebrafish sex differentiation. Proc Natl Acad Sci USA, 2005, 102: 4074-4079

转基因鲤中的转植基因整合机制及群体适应性

胡　炜　朱作言

中国科学院水生生物研究所淡水生态与生物技术国家重点实验室，武汉 430072

摘　要　转基因鱼诞生至今已有 20 多年，目前已有 5 种快速生长的转 GH 基因鱼建立了稳定遗传家系，转 GH 基因鱼很可能成为转基因农业养殖动物产业化的突破口。本文简要综述了转植基因的整合机制和定点整合研究，以及转 GH 基因鱼的生存力、性腺发育特性及育性控制对策等研究进展。在此基础上，提出大规模开展与经济性状相关的鱼类功能基因筛选，建立高效稳定整合与可控表达的鱼类转基因操作技术，在大而结构复杂的水环境中研究转基因鱼的种群适合度，建立转基因鱼可控生殖开关等是鱼类基因工程育种研究深入发展的关键。

Gene Transfer and Mutagenesis Mediated by *Sleeping Beauty* Transposon in Nile Tilapia (*Oreochromis niloticus*)

Xiaozhen He[1,2] Jie Li[1,2] Yong Long[1] Guili Song[1] Peiyong Zhou[3] Qiuxiang Liu[3] Zuoyan Zhu[1] Zongbin Cui[1]

1 The Key Laboratory of Aquatic Biodiversity and Conservation, Institute of Hydrobiology, Chinese Academy of Sciences, Wuhan 430072

2 Graduate University of the Chinese Academy of Sciences, Beijing 100049

3 Qingdao Tilapia seed multiplication farm, Jiaozhou 266317

Abstract The success of gene transfer has been demonstrated in many of vertebrate species, whereas the efficiency of producing transgenic animals remains pretty low due to the random integration of foreign genes into a recipient genome. The *Sleeping Beauty* (SB) transposon is able to improve the efficiency of gene transfer in zebrafish and mouse, but its activity in tilapia (*Oreochromis niloticus*) has yet to be characterized. Herein, we demonstrate the potential of using the SB transposon system as an effective tool for gene transfer and insertional mutagenesis in tilapia. A transgenic construct pT2/tiHsp70-SB11 was generated by subcloning the promoter of tilapia heat shock protein 70 (tiHsp70) gene, the SB11 transposase gene and the carp β-actin gene polyadenylation signal into the second generation of SB transposon. Transgenic tilapia was produced by microinjection of this construct with *in vitro* synthesized capped SB11 mRNA. SB11 transposon was detected in 28.89% of founders, 12.9% of F1 and 43.75% of F2. Analysis of genomic sequences flanking integrated transposons indicates that this transgenic tilapia line carries two copies of SB transposon, which landed into two different endogenous genes. Induced expression of SB11 gene after heat shock was detected using reverse transcription PCR in F2 transgenic individuals. In addition, the Cre/loxP system was introduced to delete the SB11 cassette for stabilization of gene interruption and biosafety. These findings suggest that the SB transposon system is active and can be used for efficient gene transfer and insertional mutagenesis in tilapia.

Keywords *Sleeping Beauty*; transposon; nile tilapia; gene transfer; mutagenesis

Introduction

The tilapia (*Oreochromis niloticus*) has become one of the most important farming fish around the world due to their large size, tender meat and delicate flavor, rapid growth, tolerance to high stocking density and relative resistance to various diseases (Amal and

文章发表于 *Transgenic Research*, 2013, 22(5): 913-924

Zamri-Saad 2011; Gupta and Acosta 2004). During the past decades, tilapia has been widely used in transgenic researches for growth enhancement (Brem *et al.* 1988; Martinez *et al.* 2000), germ cell transplantation (Farlora *et al.* 2009), islet cell xenotransplantation and humanized insulin production (Pohajdak *et al.* 2004), and potential biofactories in the production of valuable pharmaceutical products (Maclean *et al.* 2002). However, most of tilapiine strains originated from the tropics are less resistant to the low-temperature (Cnaani *et al.* 2000). Generation of cold-resistant transgenic tilapia lines is under way in different laboratories around the world, although the success of these efforts remains to be published. In addition, the tilapia has been widely accepted as a laboratory animal for dissection of molecular mechanisms underlying various biological processes such as salinity stress response (Fiol *et al.* 2011; Sandra *et al.* 2000). Currently, the whole genome sequences of tilapia are successfully assembled and thousands of genes identified are waiting for further functional assessments (http://cichlid.umd.edu/cic hlidlabs/kocherlab/genomebrowsers.html) (Soler *et al.* 2010).

Since the generation of transgenic goldfish through growth hormone gene transfer has been succeeded (Zhu *et al.* 1985), many of other transgenic fish are successfully constructed, including rainbow trout (Chourrout *et al.* 1986), medaka (Ozato *et al.* 1986), tilapia (Martinez *et al.* 1996), salmon (Fletcher *et al.* 1988), zebrafish (Stuart *et al.* 1988), channel catfish (Dunham *et al.* 1987), and common carp (Zhang *et al.* 1990). Meanwhile, a number of gene transfer approaches including direct microinjection (Zhong *et al.* 2002), retrovirus infection (Linney *et al.* 1999), electroporation (Xie *et al.* 1993), sperm-mediation (Khoo *et al.* 1992) and particle-gun bombardment (Yamauchi *et al.* 2000) were successfully used to deliver foreign DNA into different fish species, but direct microinjection of DNA into the fertilized eggs at one-cell stage has been the prevalent method. However, the microinjection method is very laborious and time-consuming, and usually leads to the integration of foreign DNA in only about 5% of the surviving injected embryos (Davidson *et al.* 2003). Electroporation appears to be an effective mean of foreign gene transfer into fish embryos, but this approach remains not widely used by researchers because the rate of foreign gene integration in transgenic fish produced by this method varies widely. Recently, a few of vertebrate transposon systems including the *Sleeping Beauty* (Ivics *et al.* 1997), Tol2 (Kawakami 2007; Koga *et al.* 1996) and *PiggyBac* (Cary *et al.* 1989), are shown to markedly improve the integration rate of foreign genes in transgenic animals. Furthermore, *Tol2* transposon system was recently used to improve the transgenesis efficiency in tilapia (Fujimura and Kocher 2011). These transposable elements can carry a large DNA fragment that contains all components required for mutagenesis of a target genome. Therefore, transposable systems have attracted extensive attention due to their potential applications of gene transfer, mutagenesis and gene therapy.

The SB transposon belongs to the Tc1/mariner superfamily of DNA transposable elements and is reconstructed from an inactive transposable sequence in salmonids (Ivics *et al.* 2004, 1997). The SB transposon can remobilize from one DNA location to another through a cut-and-paste mechanism and specifically recognizes a TA dinucleotide. It has been shown that the SB system can mediate a long-term expression in a wide range of vertebrate species including mouse (Dupuy *et al.* 2002; Su *et al.* 2008), zebrafish (Davidson *et al.* 2003), medaka (Grabher *et al.* 2003), Xenopus laevis (Sinzelle *et al.* 2006), and pig (Jakobsen *et al.* 2010). In addition, the integrated SB transposons can serve as a molecular tag for efficient identification of disrupted DNA sequences in a target genome. However, it remains unknown whether the SB transposon system functions in tilapia.

In this study, we investigated the potential of using the SB system in tilapia. Given that tilapia is of great importance in world fisheries and widely used as a model for the assessment of gene functions, findings from this study would provide strong evidence that the SB system is suitable for efficient gene delivery and transmission as well as large-scale mutagenesis in tilapia.

Materials and Methods

Maintenance of tilapia

The NEW GIFT Nile tilapia (*Oreochromis niloticus*) is a new, nationally certificated strain in china that was selected for 14 years and 9 generations from the base strain of GIFT Nile tilapia (Li *et al.* 2010). The parental GIFT strain of Nile tilapia and their transgenic offsprings were maintained at the Qingdao tilapia seed multiplication farm (Qingdao, China) in 80 m^3 round concrete ponds supplied with circulating warm water (27-30℃) and continuous aeration, and fed with commercial feeds twice a day.

Generation of transgenic construct and capped SB11 mRNA

To construct a transgenic vector pT2/tiHsp70-SB11 (Fig. 2A), the LoxP-tiHsp70-SB11-poly(A)-LoxP cassette flanked by two splicing acceptor signals (SA and SA′ in different orientations) from a previous study in our lab (Song *et al.* 2012a) was subcloned into the pT2/HB vector (Cui *et al.* 2002). The tilapia Hsp70 promoter was obtained from the tilapia genome (Molina *et al.* 2001) using PCR primers pTi-f and pTi-r and inserted at the *Nhe* I/*Age* I sites to drive the expression of SB11 transposase gene. The SA signal containing three stop codes (TGAATTAGTGA) for different reading frames in the exon sequence of *Cyprinus carpio* β-actin gene was amplified using primers pSA-f and pSA-r. The SA′ signal was obtained from the mouse hypoxanthine phosphoribosyltransferase gene (HPRT) by annealing two oligonucleo-tides pSA′-f and pSA′-r, which contains two restriction enzyme sites *Sac* II and *Bgl* II and a transcriptional terminator element (TT). Once a transposon has integrated into an intron or exon, the SA and SA′ signals can interrupt the proper splicing of endogenous transcripts from genes in two orientations. Capped SB11 mRNA was in vitro transcribed from the linearized pSB11RNAX (http://www.cbs.umn.edu/labs/perry/) using the mMESSAGE mMACHINE kit from Ambion (USA) according to the manufacturer's instructions. All primers used are listed in Table S1.

Artificial insemination and microinjection

A number of mature brood fish were collected for artificial insemination. Females were selected by gently stripping their abdomen for ripened eggs with a uniform size. Males were selected according to the pink color on their fins. Before artificial insemination, the abdomens of females with ripened eggs were dried with cotton towel and squeezed gently, starting behind the pectoral fins and moving toward the tail. The eggs were collected in a dry plastic basin and immediately mixed with fresh semen by stirring with a piece of feather for 2 min. Fertilized eggs were washed three times with culture water to remove excess semen and debris, and then incubated in petri dishes. Twenty minutes later, the eggs were checked under a stereomicroscope and the success of fertilization was determined by the blastodisc formation. Eggs showing a high fertilization rate of more than 80% were used for microinjection. Microinjection was performed following a previous protocol (Rahman and Maclean 1992). Each tilapia embryo at one cell stage was microinjected with about 3 nl of solution containing 100 ng/μl of capped SB11 mRNA and 20 ng/μl of circular pT2/tiHsp70-SB11 plasmids at the blasto-derm-yolk interface as described previously in zebrafish (Hyatt and Ekker 1999).

Development of an efficient system for hatching tilapia eggs

The Nile tilapia is a mouth brooder. Immediately after fertilization, female fish pick up the eggs in their mouths and hold them throughout embryogenesis and a few days after hatching. The eggs are kept suspended by constant water flow through the breathing of brood fish. Therefore, the most important requirement for artificial hatching system is to keep the eggs in gentle motion. According to this notion, we have developed a convenient system for artificial hatching of tilapia eggs. After microinjection, 500-1,000 injected eggs were transferred to a flask containing incubation water (Li *et al.* 2003) at 28-30℃ and half of incubation water was replaced every 24 h. Developing embryos were carefully moved into Petri dishes to get ride of dead eggs at a 12 h-interval. It takes 4-5 days for the eggs to hatch out under such conditions. After hatching, larvae were further hold in the flask for 2 days until they developed their swimming ability.

Isolation of genomic DNA and detection of transgene

Total genomic DNA was isolated from the caudal fin of 6-month-old fish. The fish were marked by cutting and numbering the rays in their dorsal fins. Caudal fins were cut into small pieces with sterile scissors and incubated at 55℃ overnight in a protein lysis buffer containing 10 mM Tris (pH 8.0), 100 mM EDTA (pH 8.0), 0.5% SDS, and 200 μg/ml Proteinase K. Isolation of high-molecular-weight DNA was performed following the previous protocol (Rahman and Maclean 1992). The DNA quality was examined by running on a 1% *w/v* agarose gel containing 0.5 μg/ml ethidium bromide and DNA concentration was determined by spectrophotometry.

The polymerase chain reaction (PCR) was performed for screening positive tilapia carrying the SB transposon. Several PCR primers were designed according to the coding sequence of SB11 transposase gene (Table S1). The sensitivity and specificity of different forward and reverse primer combinations were tested by addition of transgenic plasmids to the wild-type genomic tilapia DNA. Previous study has shown that the haploid genome size of Nile tilapia is 1.2 pg or about 10^9 bp (Hinegard and Rosen 1972). Each reaction was performed in a 25μl volume containing 100 ng genomic DNA as template, 1 μl of transgenic plasmids (0, 1, 5, 10, 20, 50 or 100 copies), 0.2 mM dNTP, 0.2 μM of each primer and 1 unit of DNA polymerase from Biostar. The PCR was performed under the conditions: 1 cycle at 94℃ for 5 min; 30 cycles at 94℃ for 30 s, 60℃ for 30 s and 72℃ for 90 s; 1 cycle at 72℃ for 10 min. PCR products were checked by agarose electrophoresis and ethidium bromide staining. Based on these data, three sets of primers were selected for the screening of positive fish. The PCR program was 40 cycles of 95℃ for 30 s, 55℃ for 30 s, 72℃ for 1 min, and a final extension at 72℃ for 10 min. The PCR products were run on a 1% agarose gel containing 0.5 μg/ml ethidium bromide.

Transcriptional expression of *SB11 gene* in F2 tilapia

Two-month old F2 individuals were treated at 37℃ for 1 h and caudal fins were cut for the isolation of total RNA using the TRIZOL reagent from Invitrogen. RNA samples were digested with RNase-free DNase I from Promega. The RNA integrity and quality were determined by electrophoresis and spectrophotometer. The cDNAs were transcribed from 2 μg of total RNA using the RevertAidTM First Strand cDNA Synthesis Kit from Fermentas in a reaction volume of 20 μl.

A PCR program (30 cycles of 94℃ for 30 s, 58℃ for 30 s, 72℃ for 1 min, and a final extension at 72℃ for 10 min) was used to detect the transcription of SB11 gene. The reaction was conducted in a total reaction volume of 25 μl containing 0.5 μl of synthesized cDNAs as template and 50 pg of the primers p104f/p496r. The PCR products were run on a 1% agarose gel containing 0.5μg/ml ethidium bromide and sequenced. The primer set pTi-f1/p80r was used to determine whether the genomic DNA was completely digested by RNase-free DNase I. The primer set p-actin-f/p-actin-r was used as control primers to amplify the cDNA of β-actin gene in tilapia. All primers used are listed in Table S1.

Integration analysis of SB transposons

A splinkerette PCR approach (Uren *et al.* 2009) with some modifications was used for identification of SB transposon integration sites in the tilapia genome. Genomic DNAs isolated from caudal fins of transgenic offspring individuals were digested with *Sau3*AI or *Nla*III at a concentration of 50 ng/μl. A splinkerette oligonucleotides were denatured and equimolar amounts of Long-strand and Short-strand adaptors were annealed by heating to 80℃ for 5 min and allowing them to cool to room temperature. A DNA ligation was done in a 20μl volume containing 50 μM adaptors and 300 ng digested genomic DNA. The primary PCR was performed by using the ligation mixture as a template and the primer set Splink 1 and IR/DR (L2/R2) under the following conditions: 1 cycle at 95℃ for 3 min; 10 cycles at 95℃ for 15 s and 70℃ (–0.5℃/cycle) for 2 min; 20 cycles at 95℃ for 15 s and 65℃ for 2 min; 1 cycle at 72℃ for 10 min. The secondary and nested PCR was performed by using the primary PCR product as a template and primers Splink 2 and IR/DR (L3/R3) under the conditions: 1 cycle at 95℃ for 10 min; 30 cycles at 95℃ for 15 s, 61℃ for 30 s and 72℃ for 2 min; 1 cycle at 72℃ for 8 min. PCR products were run on a 1.5% agarose gel and the main DNA bands were cut out of agarose gels, purified with a gel extraction kit, and subcloned into the pZero2.0-TOPO vector for sequencing. All primers used are listed in Table S1.

Once the transposition events were identified by splinkerette PCR, the integrity of integrated transposons can be determined using PCR primers on the IR/DR (L), IR/DR (R) and genome DNA adjacent to the inserted transposon. Two transposition events were detected in the transgenic line. All primers used for characterization of integrated transposons are listed in Table S1.

Results

Strategies for transgenesis and mutagenesis

The overall strategy for SB-mediated transgenesis and mutagenesis in tilapia is presented in Fig. 1. A pT2/tiHsp70-SB11 plasmid was constructed for the gene transfer (Fig. 2A), which contains the tiHsp70 promoter that can markedly derive the expression of the SB11 transposase gene at 37℃, but not at 28℃ in cultured cells and zebrafish embryos. The inducible expression of SB11 is expected to result in the remobilization of integrated insertions and cause an efficient and large-scale mutation in tilapia genome. Fertilized eggs were obtained by artificial insemination. The circular plasmid of pT2/tiHsp70-SB11 was co-injected with capped SB11 mRNA into the embryos at one-cell stage. Injected embryos

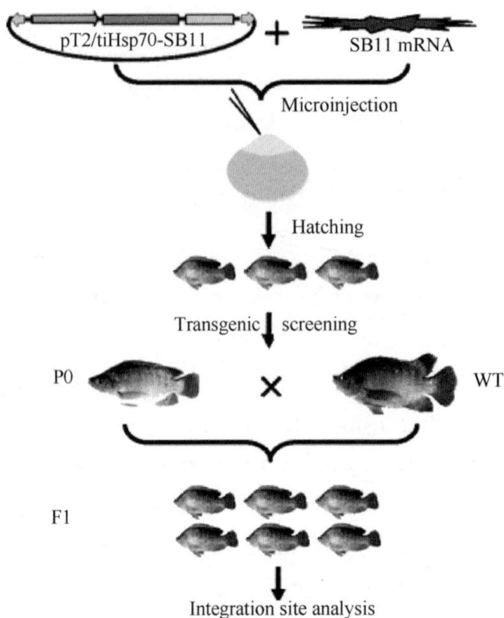

Fig. 1 The strategy for generation of transgenic tilapia by SB transposition. The transgenic vector pT2/tiHsp70-SB11 was microinjected with SB capped mRNA into one-cell stage fertilized eggs. Injected eggs were hatched out and reared to adult fish for transgenic screening. Positive fish (P0) were crossed with wild type fish to produce F1 offspring and integration site in transgenic F2 was examined using splinkette PCR. The F2 offspring were produced by the intercross of positive F1 fish

were then collected and cultured in a hatching system designed according to the hatching habit of tilapia under natural circumstance. This system includes a plastic aquarium containing 50 l of embryo medium, four round bottom flasks placed on a steel scaffold, nylon silk nets tied on the neck of the flasks to prevent eggs and larvae from overflowing, a water pump to supply water flow at 0.1 l/min, air stones connected to an air pump to ensure the concentration of dissolved oxygen and a aquarium heater to control water temperature (Fig. 2B). This simple hatching system works very well and gives a hatching rate of more than 80% for wild type (Table 1). When the fish fries were all hatched and started feeding, they were moved to a standard culture system that is suitable for tilapia culture in the farm.

Table 1 Efficiency of the hatching system

Source of eggs	Numbers of eggs obtained by artificial fertilization	Numbers of fertilized eggs	Numbers of hatched larvae	Fertilization rate (%)	Hatching rate (%)
Wild type-1[a]	636	512	420	80.50	82.03
Wild type-2	527	456	422	86.53	92.54
Wild type-3	544	487	412	89.52	84.60
Microinjected[b]	421	348	129	82.66	37.07

[a] Fertilized eggs were obtained by artificial fertilization

[b] Fertilized eggs were microinjected with the mixture of transgenic construct and capped SB11 mRNA

SB-mediated transgenesis and germline transmission

Genomic DNA samples were extracted from caudal fin clips when the transgenic tilapias (G0) were reared to mature for about 6 months. The quality of DNA was examined by electrophoresis on a 1% agarose gel containing ethidium bromide and the DNA concentration was determined by spectrophotometry. A PCR strategy was used to screen SB transposons incorporated into the genome of transgenic tilapia. Six PCR primers were designed against the coding sequence of SB11 gene (Fig. S1A). The specificity of six primer sets was tested by amplification of wild type genomic DNA containing a certain copies of transgenic plasmids (Fig. S1B) and three primer sets (pXf/pYr, p88f/pYr and p88f/p89r) were used for screening positives of SB-transgenic tilapia. As shown in Fig. S2A and Table 2, the SB transposon was detected in 28.89% of P0 transgenic fish, a transgenic rate that is much higher than that (less than 5% in general) in plasmid-injected transgenic fish. P0 were crossed with wild type fish to produce F1 offspring and 12.79% of F1 fish carried the SB gene in the genome (Fig. S2B and Table 2). The F2 offspring were produced by the intercross of positive F1 fish and 43.75% of F2 fish carried the SB gene in the genome (Fig. S2C and Table 2). These findings indicate we have successfully generated a transgenic tilapia line with stable transmission of the tiHSP70-SB transposon.

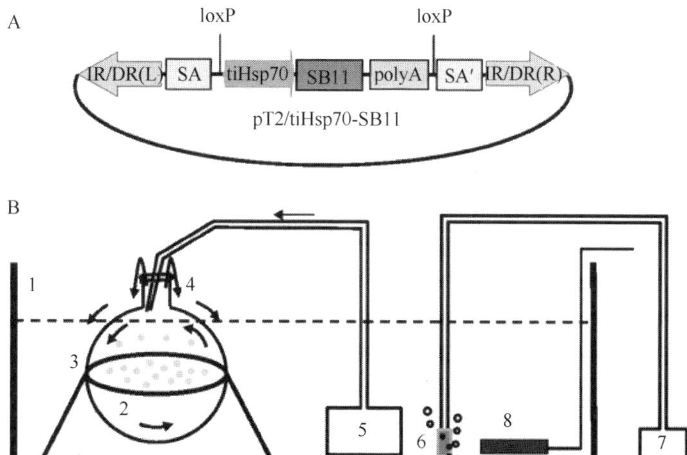

Fig. 2 Schematic diagrams of transgenic vector and hatching apparatus. (**A**) Transgenic vector pT2/tiHsp70-SB11 was generated by subcloning two splicing acceptors (SA and SA′ from different species) and SB expression cassette driven by the tilapia Hsp70 promoter into the pT2/HB vector. SA: a modified splicing acceptor from the carp (*Cyprinus carpio*) β-actin intron1/exon2; SA′: a splicing acceptor from mouse HPRT gene; loxP: the binding site of Cre recombinase; tiHSP70: tilapia HSP70 promoter; SB11: coding sequence of SB11; poly A: the carp β-actin poly A; IR/DR(R) and IR/DR(L): left and right arms of the SB transposon. (**B**) Apparatus used for hatching tilapia embryos. *1*: plastic aquarium (60×40×40 cm^3), *2*: round bottom flask, *3*:scaffold made of steel wire, *4*: nylon silk net, *5*: aquarium pump, *6*: air stone, *7*: air pump, *8*: aquarium heater. *Arrows* indicate the direction of water flow. *Filled little circles* in the flask indicate developing tilapia embryos

Table 2 Screening of transgenic tilapia by PCR

Generations of transgenic fish	Total numbers of mature fish	Numbers of positive fish	Transgenesis frequency (%)
P0	45	13	28.89
F1	219	28	12.79
F2	91	42	43.75

Transcriptional expression of integrated *SB11* gene

To ascertain whether the *SB11* gene integrated in the tilapia genome can be expressed, partial caudal fins of two-month-old F2 fish were cut with sterilized scissors after heat shock at 37℃ for 1 h. Several primer sets were designed according to the sequence of pT2/tiHsp70-SB11 (Fig. 3A) and RT-PCR assays were conducted to detect the transcription of *SB11* gene. The SB transposon was detected in the genomic DNA using the primer set pTi-f/p80r (Fig. 3B1), but

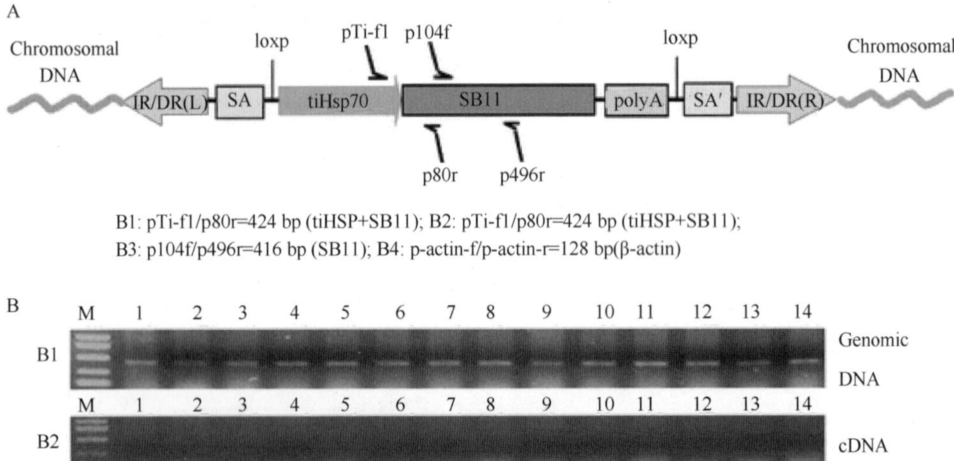

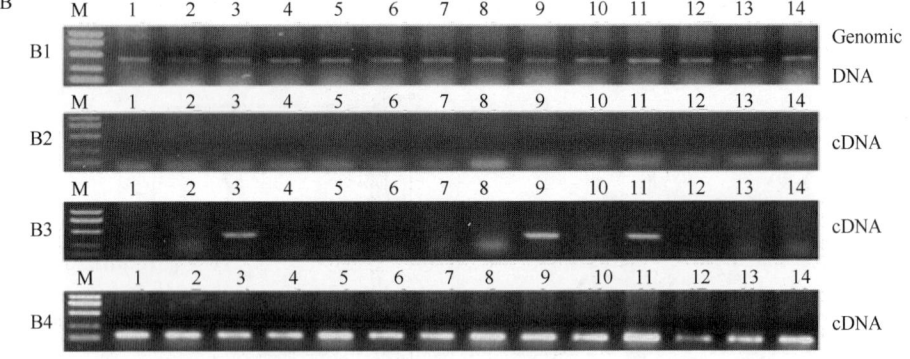

Fig. 3 RT-PCR detection of SB11 transcription in F2 fish. (A) Primer design on the vector. pTi-f1: (A) primer on tilapia HSP70 promoter from vector; p80r, p104f, p596r were primers on SB11 CDS; p-actin-f and p-actin-r were primers on tilapia β-actin CDS. (B) Detection of SB11 transcription in transgene-positive F2 tilapia after heat-shock. *B1*: PCR conducted by using the template of genomic DNA and the primer of pTi-f/p80r to confirm the transgene cassette in these tilapias; *B2*: PCR conducted by using the template of cDNA and the primer of pTi-f/p80r to confirm the fully digestion of DNA from genomic; *B3*: PCR conducted by using the template of cDNA and the primer of p104f/p496r to determine whether the SB11 were transcript or not; *B4*: PCR conducted by using the template of cDNA and the primer of p-actin-f/p-actin-r to detect the quality of cDNA; 1-14 represent 14 transgene-postive fish picked by random

not in cDNA templates synthesized from total RNAs (Fig. 3B2), indicating that genomic DNA was completely digested with RNase-free DNase I. SB11 expression was detected in three of fourteen individuals after heat shock using the primer set p104f/p496r (Fig. 3B3), but not in other individuals under normal rearing temperature (data not shown). In addition, the expression of endogenous β-actin gene was detected in all of samples using the primer set p-actin-f/p-actin-r (Fig. 3B4). These findings indicate the integrated SB11 gene can be normally transcribed in F2 transgenic tilapia. Thus, we have obtained a transgenic tilapia line with stable integration, transmission and inducible expression of SB11 gene, which is suitable for insertional mutagenesis and large-scale screening of functional genes in tilapia.

Characterization of transposon insertion sites

Although an increased transgenic frequency from P0 is noted in SB-transgenic tilapia, the copy number of SB transposons in transgenic genomes and the insertion sites remain to be characterized. The F2 positive individuals were selected for analysis of insertion sites by using the splinkerette-mediated PCR. The genomic DNAs were digested with *Nla*III, ligated to linkers and amplified with PCR to obtain the junction fragments containing partial transposon and its right flanking genomic DNA (Fig. 4A). Two junction fragments from two positive individuals were obtained and sequenced (Fig. 4B). Analysis by blasting the tilapia nucleotide sequence database (http://cichlid.umd.edu/cichlidlabs/kocherlab/bouillabase.html) indicates that these two junction sequences contain one TA dinucleotide, partial transposon and tilapia genomic DNA sequence, which are typical indicators of SB-mediated transposition events.

The integrity of SB transposon at the insertion site in the transgenic line was further determined by PCR amplification of the left junction sequences containing partial IR/DR(L) and flanking genomic DNA with primer sets p111#-f/L3 or p48#-f/L3, respectively (Fig. 5A, B). DNA sequencing of these two DNA fragments indicate that the SB transposons have stably integrated at a TA site in the genome of the tilapia line (Fig. 5C, D). These findings provide the direct evidence that SB transposon is able to mediate the germline integration of foreign genes in tilapia.

Discussion

The completion of tilapia sequencing project will soon uncover thousands of genes without known functions, so the main task of tilapia post-genome era is to decipher the function of genes. The SB transposon system has been successfully utilized for transgenesis and insertional mutagenesis of several vertebrates (Clark *et al.* 2004; Collier *et al.* 2005; Sivasubbu *et al.* 2006), but its activity remains unclear in tilapia. In the study, we have generated transgenic tilapia lines with stable integration and transmission of SB transposase gene under the control of tilapia Heat shock 70 gene promoter. The SB transposon system is highly active in mediation of DNA transposition in tilapia and the transposon-mediated gene transfer can markedly improve the efficiency of transgene integration in comparison with the approach of plasmid microinjection. Since a few copies of SB transposon are usually found in a target genome of transgenic animals (Kawakami *et al.* 2004) and coding exons make up 1-2% of most vertebrate genome (Venter *et al.* 2001), there is less opportunity to directly disrupt the expression of an endogenous gene by a few transposon insertions. The inducible expression of SB gene upon heat stress will allow the controlled remobilization of integrated SB transposons from their original insertion sites and thus increase the possibility of mutating a gene across the tilapia genome. Findings of this study suggest that the SB transposon system is an ideal tool for mediating efficient gene transfer in tilapia. In addition, the pT2/tiHsp70-SB11 transgenic lines can be used as a platform for large-scale screening of genes with various functions in tilapia.

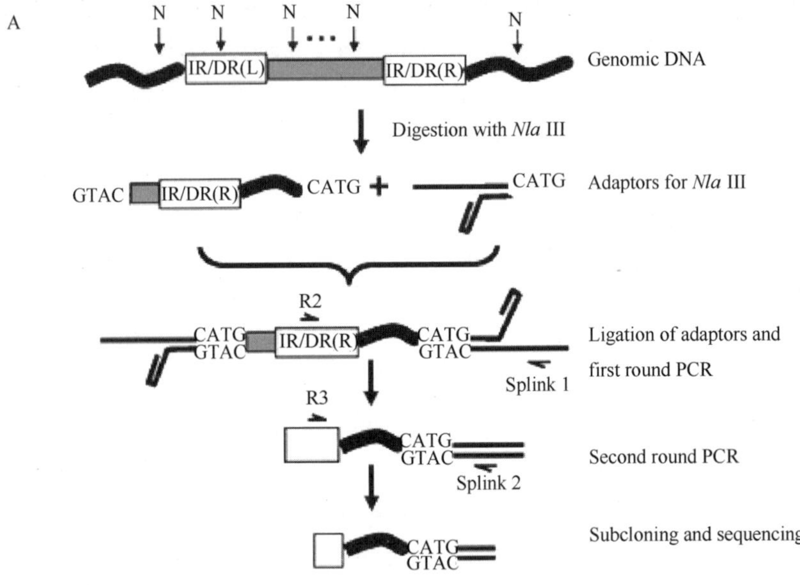

B

Sample	Junction Sequence	Insertion site	454 embryonic transcriptome GMAP
111#	Acttcaactg TA taggcactga	Sacffold_445: 147070	pCONTIG08602_c_sf (exon, scaffold_445:146786..147100, -strand)
48#	Acttcaactg TA tatgcctcga	Sacffold_1: 2400590	pCONTIG26801_c_sf (exon, scaffold_1:2400516..2400912, +strand)

Fig. 4 Analysis of transposon integration site in F1 transgenic tilapia. (A) The schematic diagram of splinkerette PCR assays. Genomic DNA carrying the transgene was digested with *Nla*III and ligated to adapters, followed by two rounds of PCR. (B) Two transposition events obtained from F1 individuals

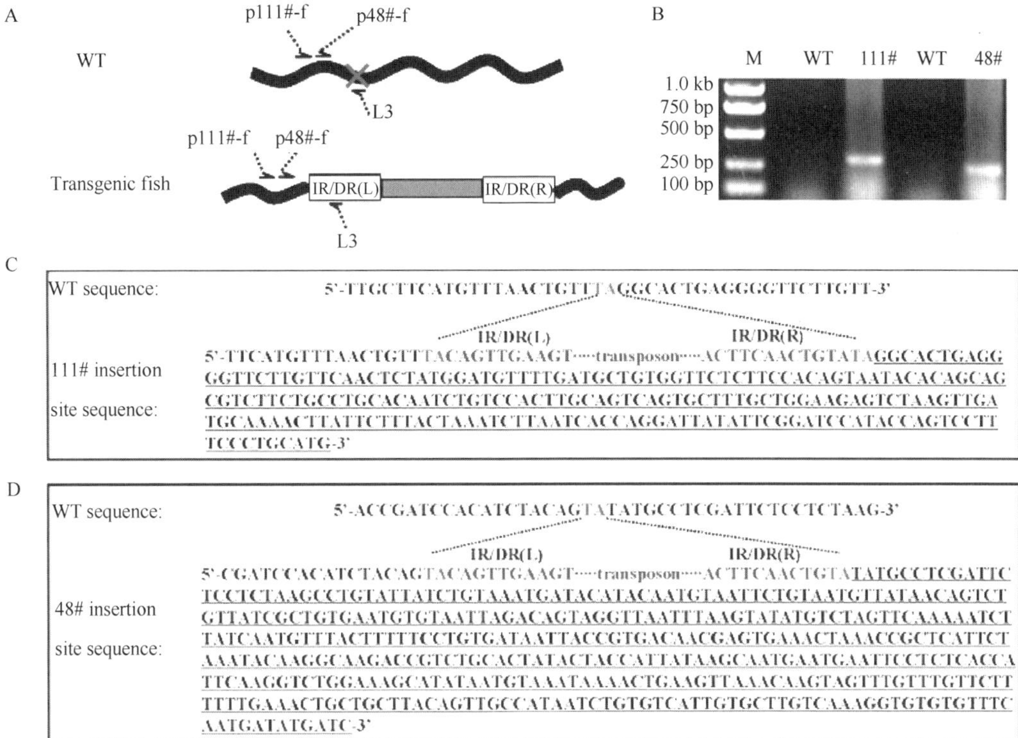

Fig. 5 Integrity of integrated SB transposons in the genome of transgenic tilapia. (A) Locations of PCR primers. Primers p111#-f and p48#-f were designed according to the sequences of genomic DNA near the insertion sites in two transgenic lines. L3 is a primer on the left IR/DR of the SB transposon. (B) PCR assays. M, DNA Marker; WT, PCR assays of wild type fish with primer sets p111#-f/L3 or p48#-f/L3; 111#-1, PCR assays of 111# transgenic fish with primer set p111#-f/L3; 148#-1, PCR assays of 48# transgenic fish with primer set p48#-f/L3. PCR products were purified, inserted into a TA cloning vector and sequenced. (C) Insertion site sequences in wild type (WT) and 111# transgenic fish. TA dinucleotides are in red. SB transposon sequences are in blue. *Underlined* sequence represents the junction genomic DNA sequence from 111# transgenic fish. (D) Insertion site sequences in wild type (WT) and 48# transgenic fish. TA dinucleotides are in red. SB transposon sequences are in blue. *Underlined* sequence represents the junction genomic DNA sequence from 48# transgenic fish

Many species of fish were utilized for transgenesis through approaches including microinjection, electroporation, sperm-or transopson-mediated gene transfer (Chen 1995). Although microinjection of cloned DNA into fertilized fish eggs has been the most widely and successfully used method for generating transgenic fish, injection of linear DNA into embryos usually results in concatomerization of the DNA and random integration into the genome. These concatomer integration events frequently lead to rearrangements, deletions, duplications, or translocations of the host DNA at the integration site (Collier and Largaespada 2007; Toneguzzo *et al.* 1988), which can cause undesired effects of germline transgenesis. In addition, the expression levels of concatomerized transgenes are often very low due to partial methylation, inactivation and other mechanisms of silencing and only a multicopy array can achieve expression levels rivaling the endogenous gene (Henikoff 1998; Collas 1998; Manuelidis 1991; Dobie *et al.* 1997). Moreover, the analysis of the insertion site is difficult due to the multiple copy inserts and the host sequence arrangements.

In contrast, transposon-mediated gene transfer exhibits several advantageous properties. First, gene transfer by transposition is able to generate offspring with multiple independent insertions that can be segregated by breeding, thus increasing the number of useful transgenic events. Second, transposons with a target site choice for integration can, in most cases, deliver a specific DNA fragment into a target site without alteration of endogenous genomic sequences. Third, the use of transposons can overcome the most important barrier to gene transfer into somatic cells of animals. Forth, a few of transposon systems have been widely used for transgenesis and insertional mutagenesis in many species including Caenorhabditis elegans, Drosophila, zebrafish, medaka fish, Xenopus, chickens, mice and rats (Niemann *et al.* 2012). Since transposon-mediated mutations are molecularly tagged, the identification of mutated genes is greatly facilitated and the remobilization of integrated transposons allows the generation of new gene insertions simply by breeding of transgenic animals. However, the use of transposons can improve the transgenesis in commercially important species is yet at an early stage, although transgenic pigs produced using SB transposon systems have been reported (Jakobsen *et al.* 2010).

In the study, we have demonstrated the activity of SB transposon in transgenesis and mutagenesis of tilapia. The production of transgenic tilapia is laborious, time consuming, inefficient and cost intensive due to the difficulty in collecting fertilized eggs and the non-transparency of embryos. The success of generating transgenic tilapia in this study is attributable to the use of artificial insemination and the design of an effective system for hatching fertilized eggs. The tilapia Hsp70 promoter (Molina *et al.* 2001) was used to drive the expression of SB11 transposase after heat shock at 37℃, which allows the conditional remobilization of integrated traps from non-coding sites to new locations and thus increases the opportunity of trapping and mutating endogenous genes. However, the transcriptional expression of integrated transgene may be affected by many of endogenous factors such as the different insertion sites, the modification of transgene, and even the physiological status of transgenic individuals, so the induced transcription of SB11 gene was only detected in three out of fourteen individuals. In addition, the changeable water in fishing ponds may lead to the leaky expression of SB11 gene in a stable transgenic line and thus the undesired excision of integrated SB transposons. To circumvent this problem, the Cre/loxP system (Van Duyne 2001) was introduced into the SB transposon to conditionally delete the SB11 gene for stabilization of disrupted genes. These elements have been found to work very well in zebrafish in our previous studies (Song *et al.* 2012a; Song *et al.* 2012b). Thus, the application of this SB transposon-mediated transgenic system appears to bring significant improvements with regard to efficiency, precision and safety of gene transfer.

The debate on the safety of genetically modified organism (GMO) has never stopped since the generation of transgenic mice in 1985 (Overbeek *et al.* 1985). Transgenic cattle, goats, sheep and pigs have been successfully developed for commercial production of highly valuable human therapeutics (Robl *et al.* 2007). The Food and Drug Administration (FDA) of America has recently approved a transgenic Atlantic salmon that grows twice as fast as wild Atlantics (Marris 2010), indicating a positive outlook for implementing transgenic technology on the area of aquaculture; however, three aspects of public concerns with regard to food safety, ecological impacts and genetics threats should be carefully evaluated before the release of a GMO. To avoid these problems, transgenic tilapia from this study were reared in ponds with very secure facilities to prevent the escape of transgenic tilapia. The transgenic tilapia will be utilized only for insertional mutagenesis and evaluation of the SB activity in tilapia and all of non-positive fish were killed after PCR detection.

Acknowledgements We are grateful to Drs. Qinjin Xu and I-Farn Lei for their careful reading and suggestions. We thank all other members in Qingdao tilapia seed multiplication farm (Qingdao, China) for their helps with the maintenance of experimental tilapia and all other members in Dr. Cui's laboratory for helpful suggestions and technical assistance. This work was funded by grants from the National High-tech R&D (863) Program (#2007AA10Z164 to Z. Cui) and the National Natural Science Foundation of China (#31101892 to Y. Long).

References

Amal M, Zamri-Saad M (2011) Streptococcosis in Tilapia (*Oreochromis niloticus*): a review. Pertanika J Trop Agric Sci 34(2): 195-206

Brem G, Brenig B, Horstgen-Schwark G, Winnacker EL (1988) Gene transfer in tilapia (*Oreochromis niloticus*). Aquaculture 68(3): 209-219

Cary LC, Goebel M, Corsaro BG, Wang HG, Rosen E, Fraser MJ (1989) Transposon mutagenesis of baculoviruses: analysis of Trichoplusia ni transposon IFP2 insertions within the FP-locus of nuclear polyhedrosis viruses. Virology 172(1): 156-169

Chen TT (1995) Transgenic fish and aquaculture. In: Bagarinao TU, Flores EEC (eds) Towards sustainable aquaculture in Southeast Asia and Japan. SEAFDEC Aquaculture Department, Illoilo, Philippines, pp 81-89

Chourrout D, Guyomard R, Houdebine LM (1986) High-efficiency gene-transfer in rainbow-trout (Salmo-Gairdneri Rich) by microinjection into egg cytoplasm. Aquaculture 51(2): 143-150

Clark KJ, Geurts AM, Bell JB, Hackett PB (2004) Transposon vectors for gene-trap insertional mutagenesis in vertebrates. Genesis 39(4): 225-233

Cnaani A, Gall GAE, Hulata G (2000) Cold tolerance of tilapia species and hybrids. Aquacult Int 8(4): 289-298

Collas P (1998) Modulation of plasmid DNA methylation and expression in zebrafish embryos. Nucleic Acids Res 26(19): 4454-4461

Collier LS, Largaespada DA (2007) Transposons for cancer gene discovery: Sleeping Beauty and beyond. Genome biology 8

Collier LS, Carlson CM, Ravimohan S, Dupuy AJ, Largaespada DA (2005) Cancer gene discovery in solid tumours using transposon-based somatic mutagenesis in the mouse. Nature 436(7048): 272-276

Cui Z, Geurts AM, Liu G, Kaufman CD, Hackett PB (2002) Structure-function analysis of the inverted terminal repeats of the Sleeping Beauty transposon. J Mol Biol 318(5): 1221-1235

Davidson AE, Balciunas D, Mohn D, Shaffer J, Hermanson S, Sivasubbu S, Cliff MP, Hackett PB, Ekker SC (2003) Efficient gene delivery and gene expression in zebrafish using the Sleeping Beauty transposon. Dev Biol 263(2): 191-202

Dobie K, Mehtali M, McClenaghan M, Lathe R (1997) Variegated gene expression in mice Trends in genetics. TIG13(4): 127-130

Dunham RA, Eash J, Askins J, Townes TM (1987) Transfer of the metallothionein human growth-hormone fusion gene into channel catfish. T Am Fish Soc 116(1): 87-91

Dupuy AJ, Clark K, Carlson CM, Fritz S, Davidson AE, Markley KM, Finley K, Fletcher CF, Ekker SC, Hackett PB, Horn S, Largaespada DA (2002) Mammalian germline transgenesis by transposition. Proc Natl Acad Sci USA 99(7): 4495-4499

Farlora R, Kobayashi S, Franca L, Batlouni S, Lacerda S, Yoshizaki G (2009) Expression of GFP in transgenic tilapia under the control of the medaka β-actin promoter: establishment of a model system for germ cell transplantation. Anim Reprod 6(3): 450-459

FiolDF, Sanmarti E, LimAH, KultzD (2011) A novel GRAIL E3 ubiquitin ligase promotes environmental salinity tolerance in euryhaline tilapia. Biochim Biophys Acta 4: 439-445

Fletcher GL, Shears MA, King MJ, Davies PL, Hew CL (1988) Evidence for antifreeze protein gene-transfer in Atlantic Salmon (Salmo-Salar). Can J Fish Aquat Sci 45(2): 352-357

Fujimura K, Kocher TD (2011) Tol2-mediated transgenesis in tilapia (*Oreochromis niloticus*). Aquaculture 319(3-4): 342-346

Grabher C, Henrich T, Sasado T, Arenz A, Wittbrodt J, Furutani-Seiki M (2003) Transposon-mediated enhancer trapping in medaka. Gene 322: 57-66

GuptaMV, Acosta BO (2004) A review of global tilapia farming practices. Aquac Asia 9: 7-12

Henikoff S (1998) Conspiracy of silence among repeated transgenes. Bioessays 20(7): 532-535

Hinegard R, Rosen DE (1972) Cellular DNA content and evolution of teleostean fish. Am Nat 106(951): 621

Hyatt TM, Ekker SC (1999) Vectors and techniques for ectopic gene expression in zebrafish. Methods Cell Biol 59: 117-126

Ivics Z, Hackett PB, Plasterk RH, Izsvak Z (1997) Molecular reconstruction of Sleeping Beauty, a Tc1-like transposon from fish, and its transposition in human cells. Cell 91(4): 501-510

Ivics Z, Kaufman CD, Zayed H, Miskey C, Walisko O, Izsvak Z (2004) The Sleeping Beauty transposable element: evolution, regulation and genetic applications. Curr Issues Mol Biol 6: 43-55

Jakobsen JE, Li J, Kragh PM, Moldt B, Lin L, Liu Y, Schmidt M, Winther KD, Schyth BD, Holm IE, Vajta G, Bolund L, Callesen H, Jorgensen AL, Nielsen AL, Mikkelsen JG (2010) Pig transgenesis by Sleeping Beauty DNA transposition. Transgenic Res 20(3): 533-545

Kawakami K (2007) Tol2: a versatile gene transfer vector in vertebrates. Genome Biol 8(Suppl 1): S7

Kawakami K, Takeda H, Kawakami N, Kobayashi M, Matsuda N, Mishina M (2004) A transposon-mediated gene trap approach identifies developmentally regulated genes in zebrafish. Dev Cell 7(1): 133-144

Khoo HW, Ang LH, Lim HB, Wong KY (1992) Sperm cells as vectors for introducing foreign DNA into zebrafish. Aquaculture 107(1): 1-19

Koga A, Suzuki M, Inagaki H, Bessho Y, Hori H (1996) Transposable element in fish. Nature 383(6595): 30

Li YH, Bai JJ, Jian Q, Ye X, Lao HH, Li XH, Luo JR, Liang XF (2003) Expression of common carp growth hormone in the yeast *Pichia pastoris* and growth stimulation of juvenile tilapia (*Oreochromis niloticus*). Aquaculture 216(1-4): 329-341

Li SF, Tang SJ, Cai WQ (2010) RAPD-SCAR Markers for genetically improved NEW GIFT Nile Tilapia (*Oreochromis niloticus niloticus* L.) and their application in strain identification. Zoologic res 31(2): 147-153

Linney E, Hardison NL, Lonze BE, Lyons S, DiNapoli L (1999) Transgene expression in zebrafish: a comparison of retroviral-vector and DNA-injection approaches. Dev Biol 213(1): 207-216

Maclean N, Rahman MA, Sohm F, Hwang G, Iyengar A, Ayad H, Smith A, Farahmand H (2002) Transgenic tilapia and the tilapia genome. Gene 295(2): 265-277

Manuelidis L (1991) Heterochromatic features of an 11-megabase transgene in brain cells. Proc Natl Acad Sci USA 88(3): 1049-1053

Marris E (2010) Transgenic fish go large. Nature 467(7313): 259

Martinez R, Estrada MP, Berlanga J, Guillen I, Hernandez O, Cabrera E, Pimentel R, Morales R, Herrera F, Morales A, Pina TC, Abad Z, Sanchez V, Melamed P, Lleonart R, delaFuente J (1996) Growth enhancement in transgenic tilapia by ectopic expression of tilapia growth hormone. Mol Mar Biol Biotech 5(1): 62-70

Martinez R, Juncal J, Zaldivar C, Arenal A, Guillen I, Morera V, Carrillo O, Estrada M, Morales A, Estrada M (2000) Growth efficiency in transgenic tilapia (*Oreochromis* sp.) carrying a single copy of an homologous cDNA growth hormone. Biochem Biophys Res Commun 267(1): 466-472

MolinaA, DiMartino E, Martial JA, Muller M (2001) Heat shock stimulation of a tilapia heat shock protein 70 promoter is mediated by a distal element. Biochem J 356(Pt 2): 353-359

Niemann H, Petersen B, Kues W, Carnwath JW (2012) Recent progress in the production of transgenic pigs. Xenotransplantation 19(1): 13

Overbeek PA, Chepelinsky AB, Khillan JS, Piatigorsky J, Westphal H (1985) Lens-specific expression and developmental regulation of the bacterial chloramphenicol acetyltransferase gene driven by the murine alpha A-crystallin promoter in transgenic mice. Proc Natl Acad Sci USA 82(23): 7815-7819

Ozato K, Kondoh H, Inohara H, Iwamatsu T, Wakamatsu Y, Okada TS (1986) Production of transgenic fish—introduction and expression of chicken delta-crystallin gene in medaka embryos. Cell Differ Dev 19(4): 237-244

Pohajdak B, Mansour M, Hrytsenko O, Conlon JM, Dymond LC, Wright JR Jr (2004) Production of transgenic tilapia with Brockmann bodies secreting [desThrB30] human insulin. Transgenic Res 13(4): 313-323

Rahman MA, Maclean N (1992) Production of Transgenic Tilapia (*Oreochromis niloticus*) by one-cell-stage microinjection. Aquaculture 105(3-4): 219-232

Robl JM, Wang Z, Kasinathan P, Kuroiwa Y (2007) Transgenic animal production and animal biotechnology. Theriogenology 67(1): 127-133

Sandra O, Le Rouzic P, Cauty C, Edery M, Prunet P (2000) Expression of the prolactin receptor (tiPRL-R) gene in tilapia *Oreochromis niloticus*: tissue distribution and cellular localization in osmoregulatory organs. J Mol Endocrinol 24(2): 215-224

Sinzelle L, Vallin J, Coen L, Chesneau A, Du Pasquier D, Pollet N, Demeneix B, Mazabraud A (2006) Generation of trangenic *Xenopus laevis* using the Sleeping Beauty transposon system. Transgenic Res 15(6): 751-760

Sivasubbu S, Balciunas D, Davidson AE, Pickart MA, Hermanson SB, Wangensteen KJ, Wolbrink DC, Ekker SC (2006) Genebreaking transposonmutagenesis reveals an essential role for histone H2afza in zebrafish larval development. Mech Develop 123(7): 513-529

Soler L, Conte MA, Katagiri T, Howe AE, Lee BY, Amemiya C, Stuart A, Dossat C, Poulain J, Johnson J, Di Palma F, Lind-blad-Toh K, Baroiller JF, D'Cotta H, Ozouf-Costaz C, Kocher TD (2010) Comparative physicalmaps derived from BAC end sequences of tilapia (*Oreochromis niloticus*). BMC genomics 11: 636

Song G, Li Q, Long Y, Gu Q, Hackett PB, Cui Z (2012a) Effective gene trapping mediated by sleeping beauty transposon. PLoS ONE 7(8): e44123

Song G, Li Q, Long Y, Hackett PB, Cui Z (2012b) Effective expression-independent gene trapping and mutagenesis mediated by sleeping beauty transposon. Journal of genetics and genomics = Yi chuan xue bao 39(9): 503-520

Stuart GW, Mcmurray JV, Westerfield M (1988) Replication, integration and stable germ-line transmission of foreign sequences injected into early zebrafish embryos. Development 103(2): 403-412

Su Q, Prosser HM, Campos LS, Ortiz M, Nakamura T, Warren M, Dupuy AJ, Jenkins NA, Copeland NG, Bradley A, Liu P (2008) A DNA transposon-based approach to validate oncogenic mutations in themouse. Proc Natl Acad Sci USA 105(50): 19904-19909

Toneguzzo F, Keating A, Glynn S, Mcdonald K (1988) Electric field-mediated gene-transfer—characterization of DNA transfer and patterns of integration in lymphoid-cells. Nucleic Acids Res 16(12): 5515-5532

Uren AG, Mikkers H, Kool J, van derWeyden L, Lund AH, Wilson CH, Rance R, Jonkers J, van Lohuizen M, Berns A, Adams DJ (2009) A high-throughput splinkerette-PCR method for the isolation and sequencing of retroviral insertion sites. Nat Protoc 4(5): 789-798

Van Duyne GD (2001) A structural view of cre-loxp site-specific recombination. Annu Rev Biophys Biomol Struct 30: 87-104

Venter JC, Adams MD, Myers EW*et al* (2001) The sequence of the human genome. Science 291(5507): 1304-1351

Xie YF, Liu D, Zou J, Li GH, Zhu ZY (1993) Gene-transfer via electroporation in fish. Aquaculture 111(1-4): 207-213

Yamauchi M, Kinoshita M, Sasanuma M, Tsuji S, Terada M, Morimyo M, Ishikawa Y (2000) Introduction of a foreign gene into medakafish using the particle gun method. J Exp Zool 287(4): 285-293

Zhang PJ, Hayat M, Joyce C, Gonzalezvillasenor LI, Lin CM, Dunham RA, Chen TT, Powers DA (1990) Gene-transfer, expression and inheritance of prsv-rainbow trout-Gh Cdna in the common carp, *Cyprinus-carpio* (Linnaeus). Mol Reprod Dev 25(1): 3-13

Zhong JY, Wang YP, Zhu ZY (2002) Introduction of the human lactoferrin gene into grass carp (*Ctenopharyngodon idellus*) to increase resistance against GCH virus. Aquaculture 214(1-4): 93-101

Zhu Z, He L, Chen S (1985) Novel gene transfer into the fertilized eggs of gold fish (*Carassius auratus* L. 1758). J Appl Ichthyol 1(1): 31-34

尼罗罗非鱼"睡美人"转座子介导的基因转移和基因突变

何小镇[1,2]　历　洁[1,2]　龙　勇[1]　宋桂丽[1]　周培勇[3]
刘秋香[3]　朱作言[1]　崔宗斌[1]

1 中国科学院水生生物研究所，水生生物多样性与保护重点实验室，武汉　430072
2 中国科学院大学，北京　100049
3 青岛罗非鱼良种场，胶州　266317

摘　要　转基因技术已经成功应用于许多脊椎动物中。然而，由于外源基因插入基因组的随机性，生产转基因动物的效率仍然低下。睡美人转座子能够改进斑马鱼和小鼠中的基因转植效率，但是其活性在罗非鱼中仍然未知。在本研究中，我们证明了睡美人转座子系统在罗非鱼中可以作为基因转植和插入突变的有效工具。将罗非鱼热休克蛋白基因的启动子、SB11 转座酶基因及鲤肌动蛋白基因的多聚腺苷酸信号插入到第二代转座子系统中，构建了一个转基因构建体 pT2 /tiHsp70-SB11。利用人工合成加帽的 SB11 mRNA 在体外显微注射罗非鱼胚胎，获得了转基因罗非鱼，并在 28.89%的亲本、12.9%的 F1 个体及 43.75%的 F2 个体中检测到了 SB11 转座子。通过对整合转座子的侧翼基因组序列分析，发现转基因罗非鱼品系拥有两个拷贝的转座子，分别插入在两个不同的内源基因中。在 F2 转基因个体中，通过反转录 PCR 检测到了 SB11 基因在热激条件下能够被诱导表达。此外，为提高基因插入的稳定性和生物安全性，引入了 Cre/loxP 系统，以删除 SB 转座子的外源基因。以上结果表明，SB 转座子系统在罗非鱼中具有生物活性，可以作为基因转植和插入突变的有效工具。

Targeted Expression in Zebrafish Primordial Germ Cells by Cre/LoxP and Gal4/UAS Systems

Feng Xiong Zhi-Qiang Wei Zuo-Yan Zhu Yong-Hua Sun

State Key Laboratory of Freshwater Ecology and Biotechnology, Institute of Hydrobiology, Chinese Academy of Sciences, Wuhan 430072

University of Chinese Academy of Sciences, Beijing 100049

Abstract In zebrafish and other vertebrates, primordial germ cells (PGCs) are a population of embryonic cells that give rise to sperm and eggs in adults. Any type of genetically manipulated lines have to be originated from the germ cells of the manipulated founders, thus it is of great importance to establish an effective technology for highly specific PGC-targeted gene manipulation in vertebrates. In the present study, we used the Cre/loxP recombinase system and Gal4/UAS transcription system for induction and regulation of mRFP (monomer red fluorescent protein) gene expression to achieve highly efficient PGC-targeted gene expression in zebrafish. First, we established two transgenic activator lines, *Tg*(*kop:cre*) and *Tg*(*kop:KalTA4*), to express the Cre recombinases and the Gal4 activator proteins in PGCs. Second, we generated two transgenic effector lines, *Tg*(*kop:loxP-SV40-loxP-mRFP*) and *Tg*(*UAS:mRFP*), which intrinsically showed transcriptional silence of mRFP. When *Tg*(*kop:cre*) females were crossed with *Tg(kop:loxP-SV40-loxP-mRFP)* males, the loxP flanked SV40 transcriptional stop sequence was 100% removed from the germ cells of the transgenic hybrids. This led to massive production of PGC-specific mRFP transgenic line, *Tg*(*kop:loxP-mRFP*), from an mRFP silent transgenic line, *Tg*(*kop: loxP-SV40-loxP-mRFP*). When *Tg*(*kop:KalTA4*) females were crossed with *Tg*(*UAS:mRFP*) males, the hybrid embryos showed PGC specifically expressed mRFP from shield stage till 25 days post-fertilization (pf), indicating the high sensitivity, high efficiency, and long-lasting effect of the Gal4/UAS system. Real-time PCR analysis showed that the transcriptional amplification efficiency of the Gal4/UAS system in PGCs can be about 300 times higher than in 1-day-pf embryos. More importantly, when the *UAS:mRFP- nos1* construct was directly injected into the *Tg(kop:KalTA4)* embryos, it was possible to specifically label the PGCs with high sensitivity, efficiency, and persistence. Therefore, we have established two targeted gene expression platforms in zebrafish PGCs, which allows us to further manipulate the PGCs of zebrafish at different levels.

Keywords Gal4/UAS; Cre/loxP; primordial germ cells; zebrafish

---

文章发表于 *Marine Biotechnology*, 2013, 15(5): 526-539

Introduction

In vertebrates, the primordial germ cells (PGCs) are the common ancestors of spermatozoa and oocytes. PGCs are distinct from somatic cells during early embryonic development, and they undergo dynamic migration, proliferation, and differentiation, resulting in final formation of mature oocyte and sperm in adults which could transmit genetic information between generations through reproduction. By using the zebrafish as an animal model, previous studies have shown that a batch of factors contribute to the early specification and development of PGCs. For instance, *vasa* was the first PGC-specific factor that was found in zebrafish (Braat *et al.* 2000; Yoon *et al.* 1997), *nanos*-related gene is required for the proper migration and survival of zebrafish PGCs (Koprunner *et al.* 2001), and *dead end* is necessary for zebrafish PGCs migration and survival through mainly counteracting miRNA mediated silencing (Ketting 2007; Kedde *et al.* 2007; Weidinger *et al.* 2003; Liu and Collodi 2010). Besides, the PGC-originated cells contribute to and are integrated in later gonad development (Slanchev *et al.* 2005; Youngren *et al.* 2005). PGCs are ideal target cells for genetic manipulation since they could be specifically labeled, removed and isolated during early development (Koprunner *et al.* 2001; Saito *et al.* 2008; Ciruna *et al.* 2002; Weidinger *et al.* 2003). Most importantly, they will give rise to adult gametes which provide genetic materials necessary to form a whole organism. However, as an excellent model for studies of vertebrate development and genetics, zebrafish still lacks a highly efficient and highly specific PGC-targeted gene expression platform under conditional control.

A group of DNA recombination induced activation systems, i.e., Cre/loxP, Flp/FRT and PhiC31 att/int systems, utilizes tissue-specific expressed recombinases to direct recombination between pairs of recognition DNA targets and to activate a gene-of-interest in specific tissues (Branda and Dymecki 2004). Among those, the Cre/loxP system is the one that was discovered the earliest and applied the most widely (Abremski *et al.* 1983; Shaikh and Sadowski 1997; Hoess and Abremski 1985). Cre/loxP system has been applied to genetic modification and expression induction in various animals and plants, including mice (Lakso *et al.* 1992; Gu *et al.* 1994), *Drosophila* (Siegal and Hartl 1996) and *Arabidopsis* (Vergunst *et al.* 1998). In 2005, functionality of the Cre/loxP system in zebrafish was demonstrated by injection of *Cre* mRNA in early embryos (Langenau *et al.* 2005; Pan *et al.* 2005). Since then, several *Cre* transgenic zebrafish lines have been established using heat shock promoter (Yoshikawa *et al.* 2008; Thummel *et al.* 2005; Le *et al.* 2007) and promoters specific for a few cell types or tissues, such as hematopoietic progenitors (Wang *et al.* 2008), neural progenitors (Seok *et al.* 2010; Kroehne *et al.* 2011), and oocytes (Liu *et al.* 2008). Those studies demonstrate that the Cre/loxP system works in zebrafish either after heat shock or in certain tissues. Our previous studies also demonstrated that a pair of mutated loxP sites could be used for site-directed gene integration in zebrafish (Liu *et al.* 2007). Nevertheless, there has been no application of the Cre/loxP system in zebrafish PGCs, an ideal cell type for genetic manipulation. By using monomer red fluorescent protein (*mRFP*) as a reporter and *askopos* (*kop*) promoter as an effective germ cell-targeted promoter (Blaser *et al.* 2005), we report our use of the Cre/loxP system for PGC-targeted manipulation of mRFP expression in zebrafish.

Gain-of-function approaches are commonly used for genetic analysis in animal models. Zebrafish is often manipulated at embryonic stage through injection the mRNA encoding proteins of interest. However, it is difficult to confine the mRNA expression in specific tissue and later developmental stages of zebrafish. Thus, mRNA overexpression method is inappropriate for gene function analysis in a tissue-specific and long lasting manner in zebrafish. Fortunately, the use of the Gal4/UAS system in zebrafish can overcome these two shortcomings, as tissue-specific expression of Gal4 transcriptional activator could restrict the expression of target gene in specific tissues, and lasting activation of UAS sequence by Gal4 protein could drive durable gene expression in target tissues. The Gal4/UAS system usually utilizes two transgenic animal lines, an activator line and an effector line. In the activator line, tissue-specific promoter/enhancer is used to drive expression of Gal4 transcriptional activator in given types of cells. Whereas, in the effector line, upstream activation sequence (UAS) is used to drive the gene of interest, and the UAS sequence could be merely recognized and activated by the Gal4 transcriptional activator. Generally, the gene of interest is transcriptionally silent in the effector line, and it will be tissue specifically activated in the transgenic offspring once the effector line is crossed with the activator line (Guarente *et al.* 1982; Traven *et al.* 2006; Carey *et al.* 1989; Giniger *et al.* 1985; Phelps and Brand 1998). The Gal4/UAS system was firstly utilized in the genetic studies of *Drosophila* and it has been popularly used in this animal model. Until recently, zebrafish has taken advantage of the Gal4/UAS system for promoter trapping and for targeted gene expression in a few tissues (Davison *et al.* 2007; Asakawa *et al.* 2008; Scott *et al.* 2007). In consideration of the activation efficiency and possible toxic effect of the heterogeneously expressed Gal4 protein, the Gal4/UAS system was optimized in zebrafish (Distel *et al.* 2009). In Distel *et al.*'s study, a fusion construct containing Kozak translational upstream sequence and a Gal4 protein with attenuated repeats of the VP16 transactivator (TA) core sequence, KalTA4, proved to be highly potent but low in toxicity. In the present study, we utilized the optimized Gal4/UAS system for targeted expression in zebrafish PGCs. We will also compare and discuss the application range of the Gal4/UAS and Cre/loxP systems in zebrafish PGCs.

Materials and Methods

Zebrafish

Embryos were obtained from the natural mating of zebrafish of the *AB* genetic background (from the China Zebrafish Resource Center, Wuhan, China) and maintained, raised, and staged as previously described (Kimmel *et al.* 1995; Westerfield 1995). The experiments involving zebrafish were performed under the approval of the Institutional Animal Care and Use Committee of the Institute of Hydrobiology, Chinese Academy of Sciences.

Preparation of DNA constructs

The PGC-specific Cre expression construct, pTol2(kop-CreUTRnos1, CMV-EGFP-SV40), is a Tol2 transposon-based, bipartite construct consisting of zebrafish *askopos* (*kop*) promoter and *nanos1* (*nos1*) 3′ untranslated region (UTR) regulated Cre expression cassettes as well as a cis-linked *CMV* promoter and *SV40* poly(A)-regulated EGFP reporter. We amplified *nos1* 3′UTR with primers of nos1-3′UTR-F and nos1-3′UTR-R from the *kop*-EGFP-F-*nos1*-3′UTR vector (Blaser *et al.* 2005) with introducing three enzyme sites *Bsi*WI, *Mlu*I, and *Bsp*EI at the 5′-end of *nos1* 3′ UTR, as well as three enzyme sites *Xho*I, *Apa*I, and *Bgl*II at the 3′-end for subsequent insertion of other fragments, and subcloned it into the pMD18-T vector (TaKaRa). On the base of this construct, firstly, we inserted the left arm of Tol2 amplified with primers of TL-F and TL-R from pDestTol2pA2 (Urasaki *et al.* 2006; Kwan *et al.* 2007) at the *Bsi*WI and *Mlu*I sites. Secondly, we added *kop* promoter with one more enzyme site *Asc*I at the 3′-end amplified with primers of kop-F and kop-R from the *kop*-EGFP-F-*nos*1-3′UTR vector (Blaser *et al.* 2005) at the *Mlu*I and *Bsp*EI sites. Thirdly, we fused *Cre* open reading frame (ORF) amplified with primers of Cre-F and Cre-R from pBS185 CMV-Cre vector (Addgene) at the *Asc*I and *Bsp*EI sites. Then the reporter cassette (*CMV:EGFP:SV40*) amplified with primers of CES-F and CES-R from pCMVEGFP (Liu *et al.* 2005) was inserted at *Xho*I and *Apa*I sites. Finally, the right arm of Tol2 amplified with primers of TR-F and TR-R from pDestTol2pA2 (Urasaki *et al.* 2006) was added at the *Apa*I and *Bgl*II sites. The resulted DNA construct was herein named as pkop: Cre.

The loxP-containing construct, pTol2(kop-loxP-SV40loxP-mRFP-UTRnos1, CMV-EGFP-SV40), was constructed by replacing the *Cre* ORF of pkop: Cre with two loxP sites flanked *SV40* poly(A) sequence and one *mRFP* ORF. The loxP-flanked *SV40* poly(A) fragment was obtained by PCR amplification using the two primers, loxpSV40-F and loxpSV40-R, containing synthetic loxP sequences, and was subcloned into the pMD18-T vector (TaKaRa), resulting in the pMD18-T-loxP-SV40-loxP construct. The loxP-flanked *SV40* poly(A) fragment and *mRFP* ORF were fused into a strand of DNA with *Asc*I and *Bsp*EI at the two ends using four primers, lsl-mRFP-F, lsl-mRFP-m-R, lsl-mRFP-m-F and lslmRFP-R, by overlap extension PCR (Horton *et al.* 1989). The PCR product was cut with *Asc*I and *Bsp*EI, and was inserted into the pkop: Cre plasmid from which the *Cre* ORF was removed, resulting in the pkop: loxP-SV40-loxP-mRFP construct.

The PGC-specific Gal4 expression construct, pTol2(kop-KalTA4-UTRnos1, CMV-EGFP-SV40), was also constructed on the base of the pkop: Cre plasmid. We amplified *KalTA4* ORF with primers of KalTA4-F and KalTA4-R from the construct TK5xC (Distel *et al.* 2009) with *Asc*I and *Bsp*EI at the ends, and replaced the *Cre* ORF of pkop: Cre with *KalTA4* ORF at the *Asc*I and *Bsp*EI sites, resulting in the pkop: KalTA4 construct. The UAS effector construct, pTol2(UAS-mRFP-UTRnos1, CMV-EGFPSV40), was constructed using a similar construction as the pkop: loxP-SV40-loxP-mRFP construct. The 4×UAS sequences and *mRFP* ORF were fused into a strand of DNA with the addition of *Mlu*I and *Bsp*EI at the ends using four primers, UAS-mRFP-F, UAS-mRFP-m-R, UAS-mRFP-m-F and 4UAS-mRFP-R, by overlap extension PCR, and the *kop-Cre-nos1* expression cassette of pkop: Cre was replaced by the *UAS-mRFP-nos1* fusion fragment at the *Mlu*I and *Bsp*EI sites, resulting in the pUAS:mRFP construct.

For transient expression assay, the construct pUAS: mRFP-SV40 was constructed by replacing the *CMV* promoter of PCS2+ vector (Addgene) with UAS–mRFP fusion fragment amplified with primers of UASmRFP-F and UASmRFP-R from pUAS:mRFP construct at the *Sal*I and *Xho*I sites. The construct pCMV:loxP-SV40loxP-mRFP-SV40 was constructed by amplifying loxpSV40-loxp-mRFP fusion fragment with adding the *Bam*HI and *Xho*I sites at the ends with primers of loxpmRFP-F and loxpmRFP-R from pkop: loxP-SV40loxP-mRFP, and the PCR product was inserted into PCS2+ vector (Addgene) at the *Bam*HI and *Xho*I sites.

Microinjection and generation of transgenic zebrafish lines

To generate transgenic lines, zebrafish embryos were injected with approximately 1 nL of a DNA/RNA solution containing 25 ng/μL Tol2 transposase mRNA and 25 ng/μL Tol2-based transgenic construct at one-cell stage (Urasaki *et al.* 2006). The embryos showing EGFP expression were screened and raised to adulthood and the 3-month-old fish (F0) were crossed to wildtype fish for analyzing germline transmission by screening the offspring under a fluorescence microscope (Olympus MVX10). F1 embryos expressing EGFP were raised to adulthood and were then screened in the same way to establish stable transgenic lines. For transient expression assay, DNA construct (25 ng/μL) was directly injected into one-cell stage embryos, and the injected embryos were allowed to develop at 28℃ for analysis of fluorescent protein expression.

PCR analysis and DNA sequencing

To verify Cre-mediated recombination events in *Tg(kop: loxP-SV40-loxP-mRFP)* after crossing with female *Tg(kop:cre)*, PCR was performed with primers of kopmRFP-F1 (F1) from zebrafish *kop* promoter and kopmRFP-R1 (R1) from the *mRFP* ORF. To test the efficiency of the recombination in different tissues of the hybrid founder fish, a pair of primers were designed and used for PCR analysis, primer kop-SV40-F2 (F2) locating in zebrafish *kop* promoter and primer kop-SV40-R2 (R2) locating in the loxp-flanked *SV40* poly(A) sequence. In addition, to confirm the presence of Cre transgene fragment, genomic DNA from different tissues was examined by PCR with Cre-specific primers, Cre-F3 and Cre-R3. The following cycling conditions were used: denaturation of the DNA template at 94℃ for 5 min; 30 cycles of 94℃ for 30 s, 57℃ for 30 s, 72℃ for 1 min, and a final 72 ℃ extension for 10 min. All the PCR products were analyzed by agarose gel electrophoresis. The PCR products resulting from primers F1 and R2 were gel-extracted, subcloned into the pMD18-T vector (TaKaRa), and sequenced.

RNA whole-mount *in situ* hybridization

To generate *Kal4TA4*, *Cre*, and *mRFP* antisense RNA probes for whole-mount *in situ* hybridization, we used three forward primers: Kal4TA4-specific primer (Probe-KalTA4F), Cre-specific primer (Probe-Cre-F), mRFP-specific primer (Probe-mRFP-F), and the same reverse primer of T3-nos1 3′UTR-R which introduced T3 promoter sequences in the 5′-end from *nos1* 3′UTR sequences, for PCR amplification of partial sequences of Kal4TA4, Cre, and mRFP from pkop*:KalTA4*, pkop*:Cre*, and pkop*:loxP-SV40-loxP-mRFP*, respectively. The PCR products served as templates to synthesize antisense RNA probes using the digoxigenin RNA labeling kit (Roche) and T3 RNA polymerase (Promega), and purified using mini-Spin Columns (Sigma). One-color whole-mount *in situ* hybridization was carried out as previously described (Thisse *et al.* 2004) and NBT/BCIP substrate solution (Roche) was used in the color reaction to detect the probes.

Quantitative real-time PCR

Fifty embryos at 1 day post-fertilization (pf) from testing lines were collected and homogenized in TRIZOL reagent (Invitrogen) to isolate total RNA. For the quantitative RT-PCR analysis, 1 μg of total RNA was used as templates for the first strand cDNA synthesis with oligo(dT) and Superscript II (Invitrogen). Then quantitative real-time PCR was carried out with SYBR green dye and TaqMan analyses were performed using Applied Biosystems 7900 Real-Time PCR System according to the manufacturer's directions. Relative expression levels of *EGFP* or *mRFP* were calculated using $2^{\wedge}$(-delta delta C(T)) method with β-actin as the internal control. Variance analysis (one-way ANOVA) was employed for comparing the mean differences between the experimental groups. A P value <0.05 was considered statistically significant (Livak and Schmittgen 2001). The primer sequences used to perform the PCR amplication are listed in Table S1.

Results

Generation of transgenic lines for Cre/loxP system

In order to test the Cre/loxP system in zebrafish PGCs, we first designed and constructed a PGC-specific Cre expression vector, pkop: Cre (Fig. 1a). The construct contains two bicistronic expression cassettes, *kop:Cre:nos1* in which the Cre recombinase gene is driven by germ cell-specific *kop* promoter (Blaser *et al.* 2005) and *nanos1* 3′UTR (Koprunner *et al.* 2001), and *CMV:EGFP:SV40* in which enhanced green fluorescent protein (EGFP) gene is driven by *CMV* promoter and *SV40* poly(A) signal. The *CMV:EGFP:SV40* expression cassette was used as a fluorescent reporter to facilitate transgenic screening. Among 40 randomly selected transgenic founders, four produced EGFP-positive F1 embryos (Table 1). We further raised one batch of transgenic F1 embryos to establish a homozygous transgenic line, *Tg(kop:cre)*^{ihb7}.

To assay the expression of Cre mRNA in *Tg(kop:cre)* embryos, we fixed the transgenic embryos at different stages and performed whole mount *in situ* hybridization using *cre* antisense probe. As shown in Fig. 1c, prominent signals were visible in the germplasm at the cleavage planes in the four-cell stage embryos. Later in development, *cre*-expressed cells distributed in four separate groups near the margin of shield stage embryos (Fig. 1d), migrated to the dorsal side to form two groups on both sides of the notochord during somitogenesis (Fig. 1e), and located in future genital ridge of 1-day-pf embryos (Fig. 1f). These revealed that the *Tg(kop:cre)* embryos could specifically express Cre recombinase in the PGCs from early stage during development.

We then designed and constructed a loxP transgene construct, *pkop:loxP-SV40-loxP-mRFP*, which also contains two bicistronic expression cassettes (Fig. 1b). The first expression cassette contains a loxP flanked *SV40* poly(A) signal between the *kop* promoter and the *mRFP* gene, while the second one is a *CMV:EGFP:SV40* fluorescent reporter. Among 48 transgenic founders, we identified three that produced green fluorescent F1 embryos (Table 1). A homozygous transgenic line, *Tg(kop:loxP-SV40-loxP:mRFP)*^{ihb10}, which carries the above construct was established. The transgenic line showed strong EGFP expression in the whole body but no mRFP expression in any tissue (data not shown).

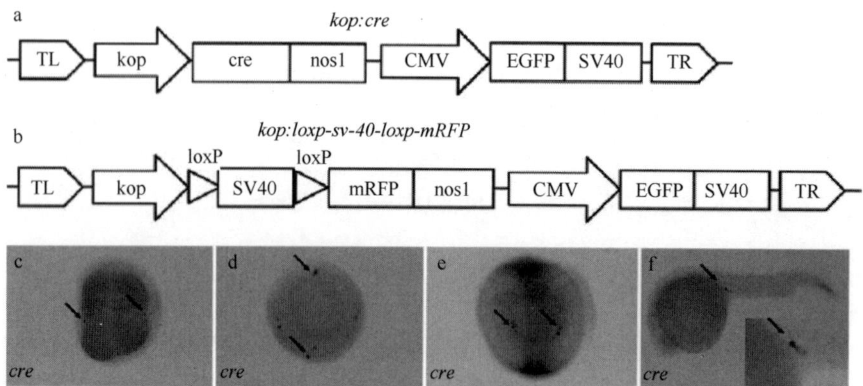

Fig. 1 Generation of transgenic zebrafish lines of the Cre/loxP system. Schema of the *kop:cre* construct (**a**). Schema of the *kop:loxP-SV40loxP-mRFP* construct (**b**). Spatio-temporal expression pattern of *cre* mRNA in four-cell stage (**c**), shield stage (**d**), early somite stage (**e**), and 1 day pf (**f**) embryos of the *Tg(kop:cre)* line. *Arrows* in **c-f** point to the PGC specifically expressed *cre* mRNA signals

Table 1 Generation of four stable transgenic lines

	No. of injected eggs	No. of hatchlings	No. of P0 fish analyzed[a]	No. of germline transmitters[b]
Tg(*kop:Cre*)	510	102(20%)	40	4(10%)
Tg(*kop:loxp-SV40-Loxp-mRFP*)	490	123(25%)	48	3(6%)
Tg(*kop:KalTA4*)	550	88(16%)	45	4(9%)
Tg(*UAS:mRFP*)	520	94(18%)	36	4(11%)

[a]The number of P0 fish that crossed with wildtype zebrafish

[b]The number of P0 fish that transmitted the transgene insertions to the F1 generation

High-efficient PGC-targeted excision of *SV40* stop sequence in Cre/loxP hybrid fish

To determine whether Cre mediated loxP recombination in the germ cells of zebrafish, we conducted cross-hybridization between homozygous *Tg(kop:cre)* females and homozygous *Tg(kop:loxP-SV40-loxP-mRFP)* males for the maternal effect of the *kop* promoter (Blaser *et al.* 2005). The hybrid embryos were examined for EGFP and mRFP expression. Although all the embryos were shown to be EGFP positive in the whole body, no red fluorescence could be observed in those embryos, which should be due to the strict maternal activity of the *kop* promoter that drives mRFP expression (Blaser *et al.* 2005). These hybrid embryos were raised to adulthood and they were designated as the hybrid founders. The genomic DNA from various tissues of the hybrid founders was prepared, and PCR was carried out to examine the excision of *SV40* stop sequences between two loxP sites with primer pair F1 and R1, which locate in the kop promoter and *mRFP* coding region, respectively. The expected amplification bands before and after Cre-mediated recombination

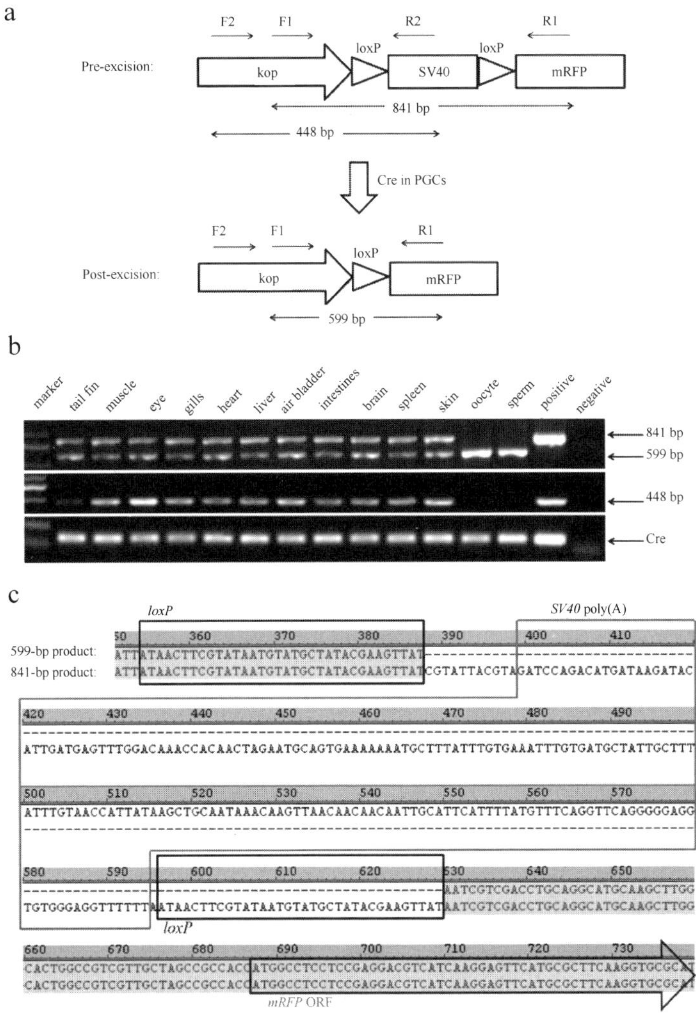

Fig. 2 Cre recombinase mediates efficient excision of *SV40* stop signal in zebrafish germline. **a** Schema of the polymerase chain reaction primer positions and product size. **b** Cre-mediated recombination in transgenic fish was detected by PCR analysis. PCR was performed for different tissues of the hybrid founders, *Tg*(*kop:cre, kop:lox-PSV40-loxP-mRFP*), using primers F1 and R1 with expected sizes of PCR products before and after Cre-mediated excision. Primers F2 and R2 amplified 488-bp unexcised *SV40* poly(A) fragment. Presence of Cre DNA was also amplified by PCR using a pair of Cre-specific primers. The 599-bp recombined fragment and 841-bp full-length transgene fragment of *kop:loxP-SV40-loxP-mRFP* existed in all types of somatic cells in *Tg*(*kop:cre, kop:loxP-SV40-loxP-mRFP*). However, in the oocyte and sperm samples, only the 599-bp recombined fragment can be detected but not 841-bp full-length transgene fragment and 488-bp unexcised *SV40* poly(A) fragment. PCR was also performed to check the presence of the *kop:cre* transgene fragment, using Crespecific primers, Cre-F3 and Cre-R3. Samples: DL2000 DNA marker (Fermentas); tail fin; muscle; eye; gills; heart; liver; air bladder; intestines; brain; spleen; skin; oocyte; sperm; positive (PCR control with pkop:loxP-SV40-loxP-mRFP or pkop:cre plasmids); negative (wildtype embryos). **c** Sequence traces from genotyping. After excision by Cre recombinase, the stop signal and one *loxP* site was precisely excised from the genomic DNA in the hybrid embryos

are 841 bp and 599 bp, respectively (Fig. 2a). In all of the somatic tissues examined, both the 841-bp pre-excision fragment and the 599-bp post-excision fragment could be detected, suggesting Cre-mediated recombination in the hybrid founders in various cell types (Fig. 2b). In the sperm or oocyte samples, intriguingly, only the post-excision fragment of 599 bp could be detected. Sequencing of the 841-bp and 599-bp fragments revealed that the loxP-flanked *SV40* transcriptional stop sequences were precisely excised from the transgenic loci in the germ cells (Fig. 2c). To confirm the excision efficiency of the Cre-mediated recombination in the hybrid founders, additional PCR was performed with primer pair F2 and R2, which locate in the *kop* promoter and *SV40* poly(A) sequence, respectively (Fig. 2a). As expected, a 488-bp unexcised *SV40* poly(A) fragment existed in all types of somatic tissues, whereas no specific amplicon could be obtained from the genomic DNA of the sperm or oocyte (Fig. 2b). This confirmed that the *loxP*-flanked *SV40* transcriptional stop sequence was completely removed in the germ cells, and thus Cre-mediated recombination achieved at an efficiency of 100% in the germline of the hybrid founders. Therefore, the hybrid founders reproduced from *Tg*(*kop:cre*) females and *Tg*(*kop:loxP- SV40-loxP-mRFP*) males could be considered as a special transgenic fish, which contains *Tg*(*kop:cre, kop:loxP-SV40-loxP-mRFP*) somatic cells and *Tg*(*kop:cre, kop:loxP-mRFP*) germ cells (Fig. 3a). In contrast, Cre-mediated recombination was not detected in the hybrid founders resulted from the cross between male *Tg*(*kop: Cre*) fish and female *Tg*(*kop:loxP- SV40-loxP-mRFP*) fish (data not shown).

To verify the germline-specific excision of loxP-flanked *SV40* transcriptional stop sequence in *Tg*(*kop:loxP-SV40-loxP-mRFP*) offspring, the mRFP silent hybrid founders were incrossed, and the resulting embryos were checked for expression of mRFP in germ cells (Fig. 3a). Interestingly, although the hybrid founders did not show PGC-specific expression of mRFP by fluorescent observation (Fig. 3b, b') and RNA in situhybridization (Fig. 3c), the hybrid offspring showed PGC-specific expression of mRFP at both protein and mRNA levels (Fig. 3d, d', e). Therefore, through a germline targeted Cre/loxP system, we were able to achieve massive production of transgenic embryos, *Tg*(*kop:loxP-mRFP*), from a transgenic silent line, *Tg*(*kop:loxP-SV40-loxP-mRFP*).

Generation of transgenic lines for Gal4/UAS system

In order to realize the PGC-targeted expression in zebrafish embryos by Gal4/UAS system, we first designed and constructed a PGC-specific Gal4 expression vector, pkop: KalTA4 (Fig. 4a). The construct contains two bicistronic expression cassettes, *kop:KalTA4-nos1*, in which the *KalTA4* activator protein gene is driven by germ cell-targeted *kop* promoter and *nanos1* 3'UTR, and the fluorescent reporter *CMV:EGFP-SV40*. Among 45 randomly selected transgenic founders, four produced EGFP-positive transgenic F1 embryos (Table 1). We further raised one batch of transgenic F1 embryos and set up a homozygous line of

*Tg(kop:KalTA4)^ihb8^*. The expression of *KalTA4* mRNA in *Tg(kop:KalTA4)* embryos was exactly similar to the expression of *cre* mRNA in *Tg(kop:cre)* embryos as shown in Fig. 1c, d (data not shown).

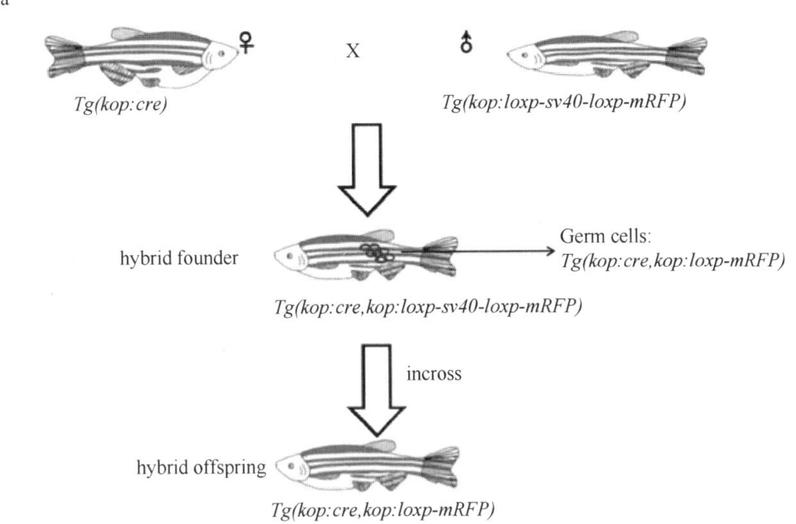

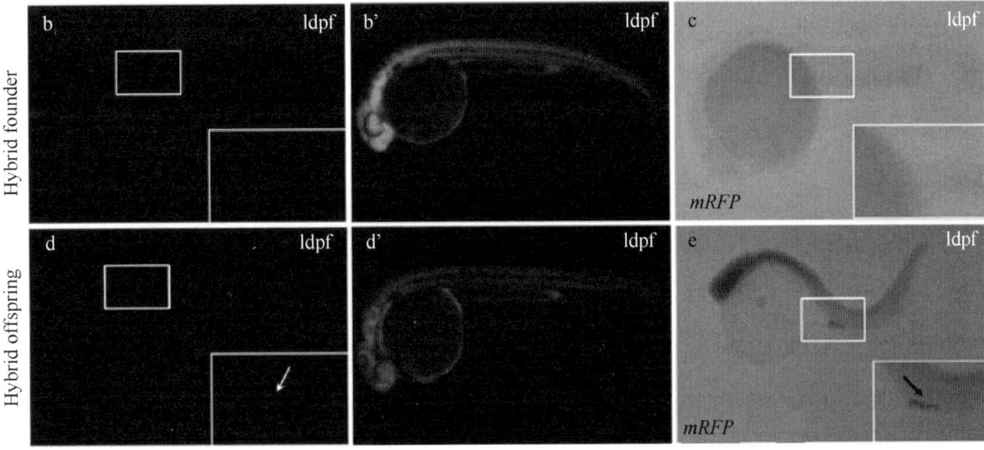

Fig. 3 Conditional activation of PGC-specific mRFP expression in the transgenic zebrafish by the cre/loxP system. **a** The schema for the genotype of the cre/loxP system. The genotype is *Tg*(*kop:cre, kop:loxP-mRFP*) in the germ cells of progeny from a cross between female *Tg*(*kop:cre*) and *Tg*(*kop:lo-xPSV40-loxP-mRFP*) males. **b-e** mRFP expression in the hybrid founders and the hybrid offspring resulted from the Cre/loxP system. Neither PGC-specific mRFP protein (**b**, **b'**) nor *mRFP* mRNA (**c**) was detected in the hybrid founder at 1 day pf. Weak PGC-specific expression of mRFP proteins (**d**, **d'**) and *mRFP* mRNA (**e**) was observed in the hybrid offspring at 1 day pf. *Arrows* point to the PGC specially expressed mRFP proteins (**d**) or mRNA signals (**e**)

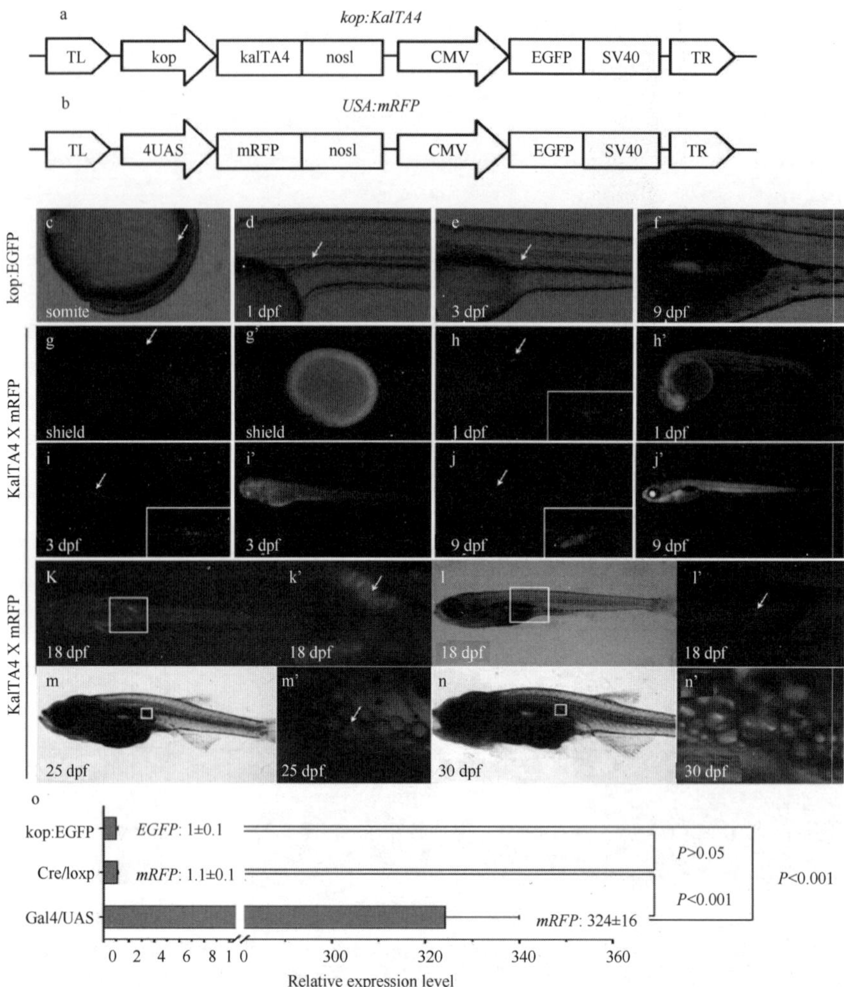

Fig. 4 PGC-specific *mRFP* expression in the embryos induced by the Gal4/UAS system. **a** Schema of the *kop:KalTA4* construct. **b** Schema of the *UAS:mRFP* construct. **c-f** Weak expression of EGFP was observed in progeny of *Tg*(*kop:EGFP*) females from early somite stage (**c**), at 1 day pf (**d**), 3 days pf (**e**), but not at 9-day-pf embryos (**f**). **g-n** mRFP expression in the embryos from a cross between *Tg*(*kop:KalTA4*) female and *Tg*(*UAS:mRFP*) male. Strong expression of mRFP was detectable under a fluorescence microscope, from shield stage in PGCs (**g**, **g'**). Strong mRFP expression could be visualized at 1 day pf (**h**, **h'**), 3 days pf (**i**, **i'**), 9 days pf (**j**, **j'**). At 18 days pf, mRFP expression continued strongly in most gonad cells of some larvae (**k**, **k'**), and in a few gonad cells of other larvae (**l**, **l'**). mRFP expression was still detectable by 25 days pf in a few larvae (**m**, **m'**), and was undetectable by 30 days pf in all of the embryos (**n**, **n'**). o Comparison of *mRFP* and *EGFP* mRNA expression levels at 1 day pf among *Tg*(*kop:EGFP*) embryos, *Tg*(*kop:cre, kop:loxP-SV40-loxP-mRFP*) embryos, and a cross between *Tg*(*kop:KalTA4*) female and *Tg*(*UAS: mRFP*) male by real-time PCR, using β-actin mRNA as internal control. The *mRFP* mRNA expression levels of embryos from incross of *Tg*(*kop:cre, kop:loxP-SV40-loxP-mRFP*) and *Tg*(*kop:EGFP*) were similar ($P>0.05$) to each other. Significant difference ($P<0.001$) of *mRFP* mRNA expression levels was found between embryos from a cross between female *Tg*(*kop:KalTA4*) and male *Tg*(*UAS:mRFP*) with others. *Arrows* point to PGC specifically expressed EGFP or mRFP signals

We then designed and constructed a UAS-dependent transgene construct, pUAS:mRFP, which also contains two bicistronic expression cassettes (Fig. 4b). The first expression cassette *UAS:mRFP-nos1* contains a four Gal4-binding-site (4×UAS)-driven *mRFP* and *nanos1* 3'UTR, while the second one is a *CMV:EGFP-SV40* fluorescent reporter. Among 36 transgenic founders, we screen out four which could produce EGFP-positive F1 embryos (Table 1). A homozygous transgenic line, *Tg(UAS:mRFP)*^{ihb9}, that carries the above construct, was established. The transgenic line showed strong EGFP expression but no mRFP expression (data not shown).

High-efficient PGC-targeted expression in hybrid embryos of *Tg(UAS-mRFP)* and *Tg(kop:KalTA4)*

Due to the maternal effect of the *kop* promoter (Blaser *et al.* 2005), the hybrid embryos resulted from *Tg*(*kop:cre*) females, and *Tg*(*kop:loxP-SV40-loxP-mRFP*) males did not show PGC-specific mRFP expressions, although the *lox*P-flanked *SV40* stop signal was removed from the transgene in the PGCs with high efficiency. Thus, we were curious about the hybrid embryos which resulted from *Tg*(*kop:KalT4*) and *Tg*(*UAS:mRFP*). Prior to testing this, we first generated several *Tg*(*kop:EGFP*)^{ihb8} lines that carried the *kop:EGFP-nos1* expression construct (a gift of Prof. Raz) (Blaser *et al.* 2005). In all of the *Tg*(*kop:EGFP*) lines we tested, the PGC-specific *EGFP* expression was only detectable from early somite stage and lasted to 3 days pf under the fluorescent microscope in our lab (Fig. 4c-f). Given the activation potential of the Gal4/UAS system to amplify the expression of target genes downstream which has been used in many organisms, we were interested whether the Gal4/UAS system could drive and even amplify mRFP expression in zebrafish PGCs. In the embryos from a cross between *Tg*(*kop:KalTA4*) females and *Tg*(*UAS:mRFP*) males, intriguingly, PGC-specific mRFP expression was detectable under the same fluorescent microscope at as early as shield stage (Fig. 4g, g'), and the expression tended to be stronger and stronger along development (Fig. 4h-j, h'-j'). Strong mRFP expression could be visualized specifically in PGCs of the hybrid embryos from 1 day pf and lasted till 15 days pf. Later, the intensity of mRFP varied among different larvae from a few fluorescent cells in the gonad to almost whole fluorescing gonad at 18 days pf. mRFP was still detectable by 25 days pf in most embryos, and is undetectable by 30 days pf (Fig. 4m, m', n, n').

In order to compare the transcriptional levels among different activation systems, i.e., the Cre/loxP and Gal4/UAS systems, the mRFP transcriptional levels at 1 day pf among embryos from *Tg*(*kop:Cre*, *kop:loxP-SV40-loxP-mRFP*) female and a cross between female *Tg*(*kop:KalTA4*) and male *Tg*(*UAS:mRFP*) were compared with the embryos of *Tg*(*kop:EGFP*) female by real-time PCR. When β-actin was used as an internal control, as expected, mRFP mRNA expression levels of the embryos from incross of *Tg*(*kop:cre, kop:loxP-SV40-*

loxP-mRFP) was similar to the embryos of *Tg*(*kop: EGFP*) (*P*>0.05), since both types of embryo were driven by the *kop* promoter and the *nanos1* 3′UTR. Interestingly, we found that the mRFP RNA transactivation level in the embryos from a cross between *Tg*(*kop:KalTA4*) females and *Tg*(*UAS:mRFP*) males were about 300-fold stronger than that in the *Tg*(*kop:EGFP*) embryos or the embryos from incross of *Tg*(*kop:cre, kop:loxP-SV40-loxP-mRFP*) (*P*<0.001) (Fig. 4o). These demonstrate that the Gal4/UAS system could not only activate the gene expression but also dramatically amplify the expression level in zebrafish PGCs.

mRFP mRNA spatio-temporal expression pattern in Cre/loxP and Gal4/UAS hybrid fish

The detection of mRFP in PGCs under a fluorescence microscope is not indicative of how early the PGC-specific Gal4 protein can drive transcription of mRFP under UAS, since accumulation of sufficient mRFP protein for visual detection is delayed with respect to initial mRFP transcription. Besides, it is possible that sites of mRFP expression are missed because *mRFP* expression level is undetectable by the eyes under a fluorescence microscope. To determine more accurately the spatio-temporal distribution of mRFP in the Cre/loxP- and Gal4/UAS-activated embryos, we observed the *mRFP* transcripts by whole-mount *in situ* hybridization on embryos from incross of *Tg*(*kop:cre, kop:loxP-SV40-loxP-mRFP*), and a cross between female *Tg*(*kop: KalTA4*) and male *Tg*(*UAS:mRFP*), and utilized the *vasa* expression as a control to mark the location of germplasm and PGCs (Fig. 5a-e). In the embryos from the incross of *Tg*(*kop:cre, kop:loxP-SV40-loxP-mRFP*), the *mRFP* transcripts were provided maternally and enriched in the distal end of the first two cleavage furrows of the four-cell stage embryos (Fig. 5f), indicating that the *kop* promoter was already activated during oogenesis of the hybrid founders and the loxP flanked *SV40* stop sequences were completely excised in the germline genome of the hybrid founders. The PGC-specific expression of *mRFP* was obvious in the following developmental stages (Fig. 5g-j). Whereas, in the embryos from a cross between *Tg*(*kop:KalTA4*) females and *Tg*(*UAS:mRFP*) males, the germplasm-specific *mRFP* mRNA was undetectable at the four-cell stage (Fig. 5k), and PGC-specific *mRFP* expression was only detectable starting at the sphere stage in four separate groups in the hybrid embryos (Fig. 5l). The *mRFP*-positive cells in all the testing embryos migrated and formed to four separate clusters, spaced around the embryo, and generally near the margin at shield stage (Fig. 5h, m), and then clustered into two groups on either side of notochord at somite stage (Fig. 5i, n). At 1 day pf, the *mRFP* transcripts located bilaterally where the yolk ball meets the yolk tube (Fig. 5j, o). The expression pattern of *mRFP* transcripts in both Cre/loxP and Gal4/UAS hybrid fish resembles that described for the zebrafish *vasa* mRNA, which is a specific marker for germ cells. These strongly supported that *mRFP* was expressed in PGCs of the hybrid embryos resulted from the two systems.

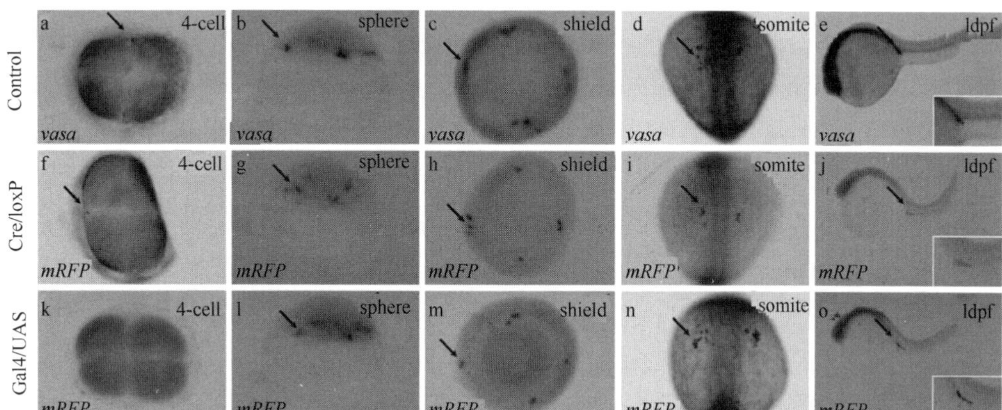

Fig. 5 Spatio-temporal expression pattern of *mRFP* mRNA in the embryos resulted from Cre/loxP and Gal4/UAS systems. **a** Wholemount *in situ* analysis was performed to determine the distribution of vasa (**a-e**) mRNA and *mRFP* mRNA in embryos from incross of *Tg(kop:Cre, kop:loxP-SV40-loxP-mRFP)* (**f-j**) and a cross between *Tg(kop:KalTA4)* female and *Tg(UAS:mRFP)* male (**k-o**) from four-cell stage to 1 day pf stage. The developmental stages are: four-cell stage (**a**, **f**, **k**), sphere stage (**b**, **g**, **l**), shield stage (**c**, **h**, **m**), early somite stage (**d**, **i**, **n**), and 1 day pf (**e**, **j**, **o**), respectively. The *arrows* point to the transcripts enriched to the cleavage furrows and the PGCs, respectively

Transient PGC-targeted expression of mRFP by direct Injection of different expression constructs

Since both the Cre/loxP and Gal4/UAS systems could specifically activate targeted gene expression in PGCs of the hybrid embryos, we then asked whether it is possible to activate *mRFP* expression in the activator embryos, i.e., *Tg(kop:KalTA4)* embryos or *Tg(kop:cre)* embryos, by direct injection of effector constructs, such as *UAS:mRFP-nos1* and *UAS:mRFP-SV40*, or *CMV:loxP-SV40-loxP-mRFP-SV40*. As expected, by injecting *kop:EGFP-nos1* construct directly into wild type embryos, it was impossible to observe PGC-specific expression of EGFP due to the maternal effect of the *kop* promoter (Fig. 6a). However, it was very efficient to label the PGCs of the host embryos with mRFP expression by injecting *UAS:mRFP-nos1* or *UAS:mRFP-SV40* constructs into *Tg*(*kop:KalTA4*) embryos, with the former one of PGC-specific expression of mRFP (Fig. 6b, red arrowhead) and the latter one of some ectopic expression of mRFP other than PGC-specific expression (Fig. 6c, white and red arrowhead). These indicate that the KalTA4 activator proteins were maternally expressed and thus could possibly activate the mRFP expression in somatic cells when *UAS:mRFP*-SV40 construct was injected. When *CMV:loxP-SV40-loxP-mRFP-SV40* was injected into *Tg*(*kop:cre*) embryos, PGC-specific expression of mRFP could be detected in all of the injected embryos, but most embryos showed ectopic expression of mRFP in somatic cells other than PGCs (Fig. 6d, white and red arrowhead). This result was consistent to the fact that *cre* mRNA was expressed maternally in the *Tg*(*kop:cre*) embryos and thus it was only restricted to the PGCs during later developmental stages (Fig. 1c-f). The PGC-specific expression of mRFP in the *Tg*(*kop:KalTA4*) embryos injected with *UAS:mRFP-nos1* was then analyzed in detail. The PGC-specific red fluorescence was visualized strongly and exclusively in PGCs from early somite stage in the injected embryos (Fig. 6e). With the embryonic development, mRFP was still specifically and strongly expressed in the PGCs of the injected embryos from 1 day pf to 9 days pf (Fig. 6f-h), indicating that by using the Gal4/UAS system, it is possible to visualize the zebrafish PGCs with high sensitivity and persistence by direct injection of a certain DNA construct.

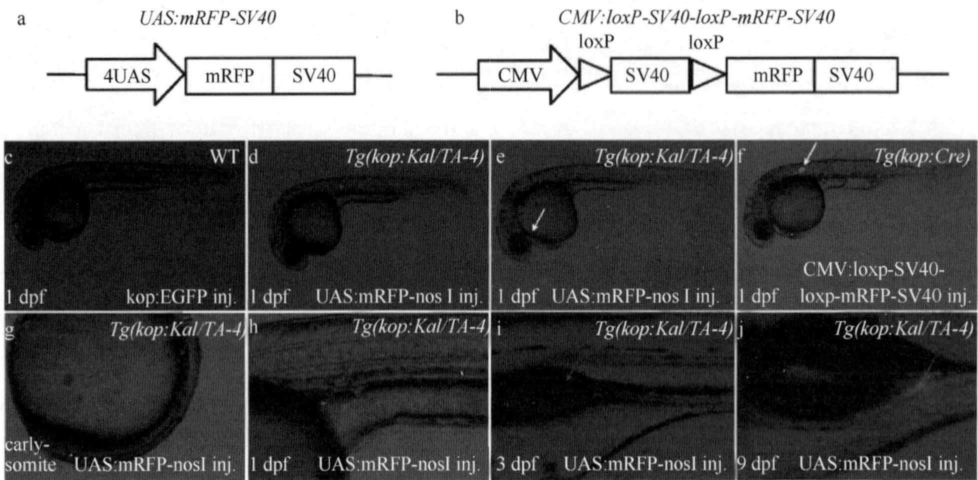

Fig. 6 Transient expression of fluorescent proteins in zebrafish embryos by injection of different expression constructs. **a** Schema of construct *UAS:mRFP*-SV40. **b** Schema of construct *CMV:loxP-SV40-loxP-mRFP-SV40*. **c** EGFP could not be detectable in wildtype embryos by injecting *kop:EGFP-nos1* construct. **d**, **g-j** mRFP was visualized strongly and specifically in the PGCs of *Tg*(*kop:KalTA4*) embryos after direct injection of *pUAS:mRFP* vector from early somite stage (**g**) to 9 days pf (**j**). e mRFP was not only detected in the PGCs but also in other somatic cells after injection of *pUAS:mRFP-SV40* vector into *Tg*(*kop:KalTA4*) embryos. **f** mRFP was observed in PGCs and other types of cells in *Tg*(*kop:cre*) embryos after injection of pCMV:loxPSV40-loxP-mRFP-SV40 construct. The *red arrows* mark the PGC- specific mRFP expression, and the *white arrows* show the ectopic expression of mRFP in somatic cells

Discussion

Germ cells are unique and important for sexually reproducing eukaryotes, since they are the only cell population which could inherit genetic materials between generations. To study the development of germ cells has long been one of the focuses of cell and developmental biologists. Besides this, any type of genetically manipulated lines originated from the germ cells of the manipulated founders, thus it is of great importance to establish an effective technology for highly specific PGC-targeted gene manipulation in vertebrates. Cell type-specific genetic modification using the induction system is an invaluable tool for studies of distinct cell lineages. In the present study, we utilized two induction systems, the Cre/loxP recombination system and Gal4/UAS transcription system, and *mRFP* gene as a reporter, to realize high-efficient PGC-targeted gene expression, in the notable animal model, zebrafish.

In vertebrates, only a few germ cell-specific Cre transgenic lines have been established in the mouse model, which allow germ cell-specific genetic modifications to harbor visual markers to detect germ cells *in vivo* and to conduct conditional knockout in germ cells (Gallardo *et al.* 2007; Hammond and Matin 2009; Lomeli *et al.* 2000; Ohinata *et al.* 2005; Suzuki *et al.* 2008). However, the recombinant efficiencies induced by germ cell-expressed Cre in mice were varied between 25% (for *Nanos*3-*Cre*) and 60% (for *TNAP-Cre*). In zebrafish, although Cre-mediated global excision has been demonstrated by injection of *cre* mRNA into the zygotes of some transgenic lines, such as *Tg*(*mylz2:loxP-EGFPloxP*)^{gz3} (Pan *et al.* 2005) and *Tg*(*rag2-loxP-dsRED2-loxP-EGFP-mMyc*) (Langenau *et al.* 2005), there is no report of PGC-specific Cre lines and PGC-specific Cre-mediated recombination. The only germ cell type Cre transgenic zebrafish line is *Tg*(*zp3:cre,krt8:rfp*), which was constructed using oocyte specific *zp3* promoter and not suitable for studying the development of zebrafish PGCs and conducting PGCs manipulation (Liu *et al.* 2008). In the present study, we utilized an upstream *kop* promoter and a downstream 3′UTR of the nanos1 gene, which confer specific EGFP expression in PGCs in the *Tg*(*kop:EGFP*) transgenic embryos (Blaser *et al.* 2005), to drive PGC-specific expression of Cre in *Tg*(*kop:cre*). As a result, we observed the exquisite specificity of the *cre* mRNA expression pattern in the PGCs of the transgenic embryos. This PGC-specific Cre transgenic line of zebrafish will provide a platform to perform PGC conditional gene function analysis and PGC-targeted gene modifications. In addition, to verify the function of PGC specifically expressed Cre recombinase in *Tg*(*kop:cre*), we established a loxP transgenic line, *Tg*(*kop:loxP-SV40-loxP-mRFP*). By crossing the *Tg*(*kop:cre*) females and *Tg*(*kop: loxP-SV40-loxP-mRFP*) males, we demonstrated that the PGCs specifically expressed Cre provided by *Tg*(*kop:cre*) is capable of conditional activation of loxP-blocked genes in the hybrid embryos. In our study, the efficiency of Cre-mediated germline recombination achieved 100%, which was strongly supported by two types of PCR analysis. In all of the previously described PGC-specific cre transgenic mouse lines, the Cre-mediated recombination happened in other tissues besides the germ cells (Suzuki *et al.* 2008; Gallardo *et al.* 2007; Lomeli *et al.* 2000; Ohinata *et al.* 2005). In our study, Cre-meditated recombination was also detected in other somatic tissues, which was due to the maternal expression of *Cre* mRNA, just similar to the study of *Tg*(*zp3:cre,krt8:rfp*) (Liu *et al.* 2008). Nevertheless, we have generated a PGC-specific Cre transgenic line in zebrafish and demonstrated a PGC-specific recombination with an effeciency of 100%. In future studies, it is possible to use a drug-inducible Cre^{ERT2} (John *et al.* 2008) to eliminate the maternal effect of Cre recombinase in *Tg*(*kop:cre*).

Although many stable transgenic lines were established that express Gal4 in a tissue-specific manner with the development of an enhancer trapping system, which enabled investigators to regulate gene expression in specific cell types and tissues (Davison *et al.* 2007), no PGC-specific Gal4 transgenic line has been reported in zebrafish. In *Drosophila*, using the promoter and 3′UTR of the *nanos* RNA, the *nos-Gal4-VP16* transgene was generated for the activation of a germline-specific expression of targets (Van Doren *et al.* 1998). In the present study, we generated the *Tg*(*kop:KalTA4*) transgenic line, in which the KalTA4 activator was driven by the *kop* promoter and *nanos1* 3′UTR, and *KalTA4* mRNA was expressed specifically in PGCs. To date, researchers have generated transgenic lines to label the PGCs in live embryos of zebrafish (Blaser *et al.* 2005; Fan *et al.* 2008; Knaut *et al.* 2002; Krovel and Olsen 2002). Nevertheless, these transgenic constructs and/or approaches still did not provide ideal tools for labeling, isolation, and genetic manipulation of zebrafish PGCs, because (1) direct injection of these constructs into zygotes could not distinctly label the PGCs, (2) the germ cell targeted promoter used in those studies is rather large in size and it is not convenient to use them for genetic manipulations, (3) the expression levels of fluorescent proteins driven by these constructs were not elevated enough for genetic manipulation (from our own experiments). In the present study, by utilizing the transcriptional amplification speciality of the Gal4/UAS system, we demonstrated the hybrid embryos resulted from *Tg*(*kop:KalT4*) females and *Tg*(*UAS:mRFP*) males showed germ cells specifically expressed mRFP from shield stage till 25 days pf. Interestingly, the level of PGC-specific mRFP expression in 1-day pf embryos is as high as 300 times in the Gal4/UAS hybrid embryos, so that strong mRFP expression could be visualized specifically in PGCs of the hybrid embryos resulted from the Gal4/UAS system. Although the expression delay of transgene induced by the Gal4/UAS system in zebrafish was demonstrated in a previous study (Zhan and Gong 2010), here we observed that *mRFP* mRNA could be detected in PGCs at sphere stage, close to the time when PGCs segregated (2.75 h pf), which suggests that it did not required a long recovery time for transcriptional activation when the Gal4/UAS system was used in zebrafish PGCs. Moreover, compared with that, the PGC-specific mRFP expression cannot be detectable at 9 days pf in the embryos of transgenic lines *Tg*(*kop:EGFP*), the hybrid embryos showed germ cell-specific mRFP expression from shieldstage till 25 days pf, indicating the high sensitivity, high efficiency, and long-lasting effect of this system. More importantly, by direct injection of the *UAS:mRFP-nos1* construct into the *Tg*(*kop:KalTA4*) zygotes, it was possible to label the PGCs of host embryos with mRFP from early somite stage till 9 days pf. To our knowledge, it is the first time that zebrafish PGCs could be exclusively and distinctly labeled by direct injection of an expression construct. Moreover, the small size of the UAS regulatory region (<100 bp) allows the expression cassette to be conveniently used as a genetic marker or a selection tool for further PGC manipulations.

In conclusion, we have established two induction systems, Cre/loxP recombination system and Gal4/UAS transcription system, to realize high-efficient PGC-targeted gene expression in zebrafish. These transgenic zebrafish lines will provide powerful tools to conduct conditional gene activation and inactivation systems for studies of gene function in PGCs. In addition, they have some other obvious applications. First, *Tg*(*kop:cre*) would be useful to develop some conditional transgenic lines that may display early defects or mortality because of the global overexpression of certain functional genes. Second, as the Cre-mediated excision achieved an efficiency of 100% in germ cells of the Cre/loxP hybrid founders, *Tg*(*kop:cre*) can serve as a transgenic fish model to investigate the feasibility of self excision of transgenes from transgenic fish to address the proprietary issue as well as ecological concerns on transgene contamination. As demonstrated in transgenic plants (Luo *et al.* 2007), it is possible to engineer transgenic fish to retain a beneficial phenotype from the transgene while the transgene is programmed to be excised from the gerMline by the gerMline-specific Cre/loxP system; thus the released transgenic fish will not be able to produce transgenic offspring even if they accidently escaped to a natural body of water. Third, owing to the strong and specific mRFP expression in PGCs of the hybrid embryos which resulted from *Tg*(*kop: KalT4*) and *Tg*(*UAS:mRFP*), it offers a useful tool for tracking and manipulating germ cells during zebrafish development. Finally, the expression cassette, *UAS:mRFP-nos1*, would be useful to facilitate screening of the candidates for germ line-transmitting founders if *Tg*(*kop:KalTA4*) zygotes are used for microinjection.

Acknowledgements The authors thank Prof. Erez Raz for providing the kop-EGFP-F-nos1-3′UTR construct and Prof. Chi-Bin Chien for sharing the pDestTol2pA2 vector. This work was supported by the China 973 Project (2010CB126306 and 2012CB944504) and the National Science Fund for Excellent Young Scholars of NSFC (31222052).

References

Abremski K, Hoess R, Sternberg N (1983) Studies on the properties of P1 site-specific recombination: evidence for topologically unlinked products following recombination. Cell 32(4): 1301-1311

Asakawa K, Suster ML, Mizusawa K, Nagayoshi S, Kotani T, Urasaki A, Kishimoto Y, Hibi M, Kawakami K (2008) Genetic dissection of neural circuits by Tol2 transposon-mediated Gal4 gene and enhancer trapping in zebrafish. Proc Natl Acad Sci USA 105(4): 1255-1260

Blaser H, Eisenbeiss S, Neumann M, Reichman-Fried M, Thisse B, Thisse C, Raz E (2005) Transition from non-motile behaviour to directed migration during early PGC development in zebrafish. J Cell Sci 118(Pt 17): 4027-4038

Braat AK, van de Water S, Goos H, Bogerd J, Zivkovic D (2000) Vasa protein expression and localization in the zebrafish. Mech Dev 95(1-2): 271-274

Branda CS, Dymecki SM (2004) Talking about a revolution: the impact of site-specific recombinases on genetic analyses in mice. Dev Cell 6(1): 7-28

Carey M, Kakidani H, Leatherwood J, Mostashari F, Ptashne M (1989) An amino-terminal fragment of GAL4 binds DNA as a dimer. J Mol Biol 209(3): 423-432

Ciruna B, Weidinger G, Knaut H, Thisse B, Thisse C, Raz E, Schier AF (2002) Production of maternal-zygotic mutant zebrafish by germ-line replacement. Proc Natl Acad Sci USA 99(23): 14919- 14924

Davison JM, Akitake CM, Goll MG, Rhee JM, Gosse N, Baier H, Halpern ME, Leach SD, Parsons MJ (2007) Transactivation from Gal4-VP16 transgenic insertions for tissue-specific cell labeling and ablation in zebrafish. Dev Biol 304(2): 811-824

Distel M, Wullimann MF, Koster RW (2009) Optimized Gal4 genetics for permanent gene expression mapping in zebrafish. Proc Natl Acad Sci USA 106(32): 13365-13370

Fan L, Moon J, Wong TT, Crodian J, Collodi P (2008) Zebrafish primordial germ cell cultures derived from vasa: : RFP transgenic embryos. Stem Cells Dev 17(3): 585-597

Gallardo T, Shirley L, John GB, Castrillon DH (2007) Generation of a germ cell-specific mouse transgenic Cre line, Vasa-Cre. Genesis 45(6): 413-417

Giniger E, Varnum SM, Ptashne M (1985) Specific DNA binding of GAL4, a positive regulatory protein ofyeast. Cell 40(4): 767-774

Gu H, Marth JD, Orban PC, Mossmann H, Rajewsky K (1994) Deletion of a DNA polymerase beta gene segment in T cells using cell type-specific gene targeting. Science 265(5168): 103-106

Guarente L, Yocum RR, Gifford P (1982) A GAL10-CYC1 hybrid yeast promoter identifies the GAL4 regulatory region as an upstream site. Proc Natl Acad Sci USA 79(23): 7410-7414

Hammond SS, Matin A (2009) Tools for the genetic analysis of germ cells. Genesis 47(9): 617-627

Hoess RH, Abremski K (1985) Mechanism of strand cleavage and exchange in the Cre-lox site-specific recombination system. J Mol Biol 181(3): 351-362

Horton RM, Hunt HD, Ho SN, Pullen JK, Pease LR (1989) Engineering hybrid genes without the use of restriction enzymes: gene splicing by overlap extension. Gene 77(1): 61-68

John GB, Gallardo TD, Shirley LJ, Castrillon DH (2008) Foxo3 is a PI3K-dependent molecular switch controlling the initiation of oocyte growth. Dev Biol 321(1): 197-204

Kedde M, Strasser MJ, Boldajipour B, Oude Vrielink JA, Slanchev K, le Sage C, Nagel R, Voorhoeve PM, van Duijse J, Orom UA, Lund AH, Perrakis A, Raz E, Agami R (2007) RNA-binding protein Dnd1 inhibits microRNA access to target mRNA. Cell 131(7): 1273-1286

Ketting RF (2007) A dead end for microRNAs. Cell 131(7): 1226-1227

Kimmel CB, Ballard WW, Kimmel SR, Ullmann B, Schilling TF (1995) Stages of embryonic development of the zebrafish. Dev Dyn 203(3): 253-310

Knaut H, Steinbeisser H, Schwarz H, Nusslein-Volhard C (2002) An evolutionary conserved region in the vasa 3'UTR targets RNA translation to the germ cells in the zebrafish. Curr Biol 12(6): 454-466

Koprunner M, Thisse C, Thisse B, Raz E (2001) A zebrafish nanosrelated gene is essential for the development of primordial germ cells. Genes Dev 15(21): 2877-2885

Kroehne V, Freudenreich D, Hans S, Kaslin J, Brand M (2011) Regeneration of the adult zebrafish brain from neurogenic radial glia-type progenitors. Development 138(22): 4831-4841

Krovel AV, Olsen LC (2002) Expression of a vas: EGFP transgene in primordial germ cells of the zebrafish. Mech Dev 116(1-2): 141-150

Kwan KM, Fujimoto E, Grabher C, Mangum BD, Hardy ME, Campbell DS, Parant JM, Yost HJ, Kanki JP, Chien CB (2007) The Tol2kit: a multisite gateway-based construction kit for Tol2 transposon transgenesis constructs. Dev Dyn 236(11): 3088-3099

Lakso M, Sauer B, Mosinger B Jr, Lee EJ, Manning RW, Yu SH, Mulder KL, Westphal H (1992) Targeted oncogene activation by site-specific recombination in transgenic mice. Proc Natl Acad Sci USA 89(14): 6232-6236

Langenau DM, Feng H, Berghmans S, Kanki JP, Kutok JL, Look AT (2005) Cre/lox-regulated transgenic zebrafish model with conditional myc-induced T cell acute lymphoblastic leukemia. Proc Natl Acad Sci USA 102(17): 6068-6073

Le X, Langenau DM, Keefe MD, Kutok JL, Neuberg DS, Zon LI (2007) Heat shock-inducible Cre/Lox approaches to induce diverse types of tumors and hyperplasia in transgenic zebrafish. Proc Natl Acad Sci USA 104(22): 9410-9415

Liu W, Collodi P (2010) Zebrafish dead end possesses ATPase activity that is required for primordial germ cell development. FASEB J 24(8): 2641-2650

Liu WY, Wang Y, Sun YH, Wang YP, Chen SP, Zhu ZY (2005) Efficient RNA interference in zebrafish embryos using siRNA synthesized with SP6 RNA polymerase. Dev Growth Differ47(5): 323-331

Liu WY, Wang Y, Qin Y, Wang YP, Zhu ZY (2007) Site-directed gene integration in transgenic zebrafish mediated by cre recombinase using a combination of mutant lox sites. Mar Biotechnol (NY) 9(4): 420-428

Liu X, Li Z, Emelyanov A, Parinov S, Gong Z (2008) Generation of oocyte-specifically expressed cre transgenic zebrafish for female gerMline excision of loxP-flanked transgene. Dev Dyn 237(10): 2955-2962

Livak KJ, Schmittgen TD (2001) Analysis of relative gene expression data using real-time quantitative PCR and the 2(-Delta Delta C(T)) Method. Methods 25(4): 402-408

Lomeli H, Ramos-Mejia V, Gertsenstein M, Lobe CG, Nagy A (2000) Targeted insertion of Cre recombinase into the TNAP gene: excision in primordial germ cells. Genesis 26(2): 116-117

Luo K, Duan H, Zhao D, Zheng X, Deng W, Chen Y, Stewart CN Jr, McAvoy R, Jiang X, Wu Y, He A, Pei Y, Li Y (2007) 'GM-genedeletor': fused loxP-FRT recognition sequences dramatically improve the efficiency of FLP or CRE recombinase on transgene excision from pollen and seed of tobacco plants. Plant Biotechnol J 5(2): 263-274

Ohinata Y, Payer B, O'Carroll D, Ancelin K, Ono Y, Sano M, Barton SC, Obukhanych T, Nussenzweig M, Tarakhovsky A, Saitou M, Surani MA (2005) Blimp1 is a critical determinant of the germ cell lineage in mice. Nature 436(7048): 207-213

Pan X, Wan H, Chia W, Tong Y, Gong Z (2005) Demonstration of site directed recombination in transgenic zebrafish using the Cre/loxP system. Transgenic Res 14(2): 217-223

Phelps CB, Brand AH (1998) Ectopic gene expression in *Drosophila* using GAL4 system. Methods 14(4): 367-379

Saito T, Goto-Kazeto R, Arai K, Yamaha E (2008) Xenogenesis in teleost fish through generation of germ-line chimeras by single primordial germ cell transplantation. Biol Reprod 78(1): 159-166

Scott EK, Mason L, Arrenberg AB, Ziv L, Gosse NJ, Xiao T, Chi NC, Asakawa K, Kawakami K, Baier H (2007) Targeting neural circuitry in zebrafish using GAL4 enhancer trapping. Nat Methods 4(4): 323-326

Seok SH, Na YR, Han JH, Kim TH, Jung H, Lee BH, Emelyanov A, Parinov S, Park JH (2010) Cre/loxP-regulated transgenic zebrafish model for neural progenitor-specific oncogenic Kras expression. Cancer Sci 101(1): 149-154

Shaikh AC, Sadowski PD (1997) The Cre recombinase cleaves the lox site in trans. J Biol Chem 272(9): 5695-5702

Siegal ML, Hartl DL (1996) Transgene coplacement and high efficiency site-specific recombination with the Cre/loxP system in *Drosophila*. Genetics 144(2): 715-726

Slanchev K, Stebler J, de la Cueva-Mendez G, Raz E (2005) Development without germ cells: the role of the germ line in zebrafish sex differentiation. Proc Natl Acad Sci USA 102(11): 4074-4079

Suzuki H, Tsuda M, Kiso M, Saga Y (2008) Nanos3 maintains the germ cell lineage in the mouse by suppressing both Bax-dependent and-independent apoptotic pathways. Dev Biol 318(1): 133-142

Thisse B, Heyer V, Lux A, Alunni V, Degrave A, Seiliez I, Kirchner J, Parkhill JP, Thisse C (2004) Spatial and temporal expression of the zebrafish genome by large-scale *in situ* hybridization screening. Methods Cell Biol 77: 505-519

Thummel R, Burket CT, Brewer JL, Sarras MP Jr, Li L, Perry M, McDermott JP, Sauer B, Hyde DR, Godwin AR (2005) Cre-mediated site-specific recombination in zebrafish embryos. Dev Dyn 233(4): 1366-1377

Traven A, Jelicic B, Sopta M (2006) Yeast Gal4: a transcriptional paradigm revisited. EMBO Rep 7(5): 496-499

Urasaki A, Morvan G, Kawakami K (2006) Functional dissection of the Tol2 transposable element identified the minimal cis-sequence and a highly repetitive sequence in the subterminal region essential for transposition. Genetics 174(2): 639-649

Van Doren M, Williamson AL, Lehmann R (1998) Regulation of zygotic gene expression in *DrosophilaP* primordial germ cells. Curr Biol 8(4): 243-246

Vergunst AC, Jansen LE, Hooykaas PJ (1998) Site-specific integration of Agrobacterium T-DNA in Arabidopsis thaliana mediated by Cre recombinase. Nucleic Acids Res 26(11): 2729-2734

Wang L, Zhang Y, Zhou T, Fu YF, Du TT, Jin Y, Chen Y, Ren CG, Peng XL, Deng M, Liu TX (2008) Functional characterization of Lmo2-Cre transgenic zebrafish. Dev Dyn 237(8): 2139-2146

Weidinger G, Stebler J, Slanchev K, Dumstrei K, Wise C, Lovell-Badge R, Thisse C, Thisse B, Raz E (2003) dead end, a novel vertebrate germ plasm component, is required for zebrafish primordial germ cell migration and survival. Curr Biol 13(16): 1429-1434

Westerfield M (1995) The zebrafish book: a guide for the laboratory use of zebrafish (*Danio rerio*). University of Oregon, Eugene

Yoon C, Kawakami K, Hopkins N (1997) Zebrafish vasa homologue RNA is localized to the cleavage planes of 2- and 4-cell-stage embryos and is expressed in the primordial germ cells. Development 124(16): 3157-3165

Yoshikawa S, Kawakami K, Zhao XC (2008) G2R Cre reporter transgenic zebrafish. Dev Dyn 237(9): 2460-2465

Youngren KK, Coveney D, Peng X, Bhattacharya C, Schmidt LS, Nickerson ML, Lamb BT, Deng JM, Behringer RR, Capel B, Rubin EM, Nadeau JH, Matin A (2005) The Ter mutation in the dead end gene causes germ cell loss and testicular germ cell tumours. Nature 435(7040): 360-364

Zhan H, Gong Z (2010) Delayed and restricted expression of UAS regulated GFP gene in early transgenic zebrafish embryos by using the GAL4/UAS system. Mar Biotechnol (NY) 12(1): 1-7

Cre/LoxP 和 Gal4/UAS 系统介导斑马鱼胚胎原始生殖细胞的定位表达

熊　凤　魏志强　朱作言　孙永华

中国科学院水生生物研究所，武汉 430072

中国科学院大学，北京 100049

摘　要　在斑马鱼和其他脊椎动物中，原始生殖细胞(PGCs)是将要发育为成熟精子和卵子的一群胚胎细胞。任何类型的基因操作品系必须源于原始生殖细胞的有效操作，因此，建立一种具有高度特异性的脊椎动物 PGC 靶基因操纵技术至关重要。在本研究中，我们采用 Cre/loxP 重组酶系统和 Gal4/UAS 转录诱导系统，在斑马鱼 PGC 中诱导单体红色荧光蛋白 (mRFP)基因的特异性表达。首先，我们建立了两个转基因激活品系：*Tg*(*kop:cre*)和 *Tg*(*kop:KalTA4*)，用于 PGC 中 Cre 重组酶和 Gal4 激活蛋白的特异表达。同时，我们构成了两个转基因效应品系：*Tg*(*kop:loxP-SV40-loxP-mRFP*)和 *Tg*(*UAS:mRFP*)，这两个品系中的 mRFP 均表现为转录沉默。当 *Tg*(*kop:cre*)雌性与 *Tg*(*kop:loxP-SV40-loxP-mRFP*)雄性杂交，SV40 转录终止序列将被其两侧的 loxP 位点定点切除，激活效应品系 *Tg*(*kop:loxP-SV40-loxP-mRFP*)中 mRFP 基因的表达，从而获得大量 PGC 特异性表达的转基因品系 *Tg*(*kop:loxP-mRFP*)。当 *Tg*(*kop:KalTA4*)雌性与 *Tg*(*UAS:mRFP*)雄性杂交，杂交胚胎从胚盾到受精后 25 天的 PGC 中都能高效且特异地表达 mRFP，表明 Gal4/UAS 系统具有灵敏度高、效率高及持久性等特点。实时定量 PCR 分析表明，在受精后 1 天的胚胎中，Gal4/UAS 系统在 PGC 中的转录放大效率达到 300 倍，可以更灵敏、更高效和更持久地标记原始生殖细胞。由此，我们在斑马鱼 PGC 中建立了两套基因靶向表达平台，这使得我们能够进一步在不同品系对斑马鱼 PGC 进行遗传操作。

Inheritable and Precise Large Genomic Deletions of Non-Coding RNA Genes in Zebrafish Using TALENs

Yun Liu[1,4] Daji Luo[2,3] Hui Zhao[1,4] Zuoyan Zhu[2] Wei Hu[2]
Christopher H. K. Cheng[1,4]

1 School of Biomedical Sciences, The Chinese University of Hong Kong, Hong Kong
2 State Key Laboratory of Freshwater Ecology and Biotechnology, Institute of Hydrobiology, Chinese Academy of Sciences, Wuhan
3 Department of Genetics, School of Basic Medical Sciences, Wuhan University, Wuhan
4 School of Biomedical Sciences Core Laboratory, The Chinese University of Hong Kong Shenzhen Research Institute, Shenzhen

Abstract Transcription activator-like effector nucleases (TALENs) have so far been applied to disrupt protein-coding genes which constitute only 2-3% of the genome in animals. The majority (70-90%) of the animal genome is actually transcribed as non-coding RNAs (ncRNAs), yet the lack of efficient tools to knockout ncRNA genes hinders studies on their *in vivo* functions. Here we have developed novel strategies using TALENs to achieve precise and inheritable large genomic deletions and knockout of ncRNA genes in zebrafish. We have demonstrated that individual miRNA genes could be disrupted using one pair of TALENs, whereas large microRNA (miRNA) gene clusters and long non-coding RNA (lncRNA) genes could be precisely deleted using two pairs of TALENs. We have generated large genomic deletions of two miRNA clusters (the 1.2 kb *miR-17-92* cluster and the 79.8 kb *miR-430* cluster) and one long non-coding RNA (lncRNA) gene (the 9.0 kb *malat1*), and the deletions are transmitted through the germline. Taken together, our results establish TALENs as a robust tool to engineer large genomic deletions and knockout of ncRNA genes, thus opening up new avenues in the application of TALENs to study the genome *in vivo*.

Introduction

TALENs are artificial nucleases that consist of the DNA binding domain from transcription activator like effectors (TALE) and a catalytic domain from FokI nuclease [1-2]. TALEs bind to DNA through the repeat domain, with one TALE repeat recognizing one DNA base. The base specificity is determined by the 12th and 13th amino acids called repeat-variable di-residues (RVDs) in each TALE repeat, with RVDs NI, NG, HD and NN recognizing adenine (A), thymine (T), cytosine (C) and guanine (G), respectively [3-4]. These RVDs-DNA pairings bring the FokI nuclease to a predetermined genomic locus to create DNA double-strand breaks (DSB).

文章发表于 *PloS One*, 2013, 8(10): e76387

Repair of the DSB through the error-prone non-homologous end-joining pathway leads to small indels at the break site, thus enabling targeted gene disruption. So far, TALENs have been employed to disrupt specific genomic loci in yeast [5], worms[1,6-7], plants [8], zebrafish[9-13], medaka[14], rat[15], *Xenopus*[16], pig [17] as well as in cell lines[18-20] and human stem cells [21].

The indel-mutations generated by TALENs are often small (<30 bp) [22]. So far, these small indel-mutations have been used to disrupt the open reading frames (ORF) of the protein coding genes which constitute only 2-3% of the animal genome. The majority (70-90%) of the animal genome is actually transcribed, producing thousands of ncRNAs such as miRNAs, lncRNA, small interfering RNAs and PIWI-interacting RNAs [23-24]. Different from the protein coding genes, the ncRNA genes contain no ORFs [25]. The functional regions of some ncRNAs such as lncRNAs are often unknown [25]. Other ncRNAs such as miRNAs often occur as gene clusters. Therefore, small sequence alterations may not be sufficient to disrupt the functions of these ncRNAs. Increasing evidence have demonstrated the functional roles of ncRNAs in a wide range of biological processes [26-28], yet the lack of efficient tools to knockout ncRNA genes in most animal species hinders studies on their *in vivo* functions.

Zebrafish is an important animal model to investigate gene functions. Besides the protein-coding genes, many ncRNA genes in the zebrafish genome are also transcribed [29-33]. In this study, we have developed strategies using TALENs to achieve precise and inheritable large genomic deletions and knockout of ncRNA genes in zebrafish.

Materials and Methods

Zebrafish husbandry

AB zebrafish used in this study were maintained at 28℃ in the zebrafish facility of the Chinese University of Hong Kong and the Institute of Hydrobiology. All animal experiments were conducted in accordance with the guidelines and approval of the respective Animal Research and Ethics Committees of the Chinese University of Hong Kong and the Institute of Hydrobiology.

Construction of customized TALENs

The pCS2-TALEN-ELD/KKR plasmids were constructed as described [16]. Using the modified TALENs vectors, highly effective customized TALENs recognizing 12-31 bp half-sites could be assembled in five days. The whole procedure involves two digestion-ligation steps. The protocol to assemble TALENs was modified from a previous study [34]. The modular plasmids (60 ng each) were digested and ligated in a 10 μl volume containing 1 μl BsaI buffer (NEB buffer 4), 0.6 μl BsaI (6 U, NEB), 0.6 μl T4 ligase (1200 U, NEB) and

0.4 μl of 25 mM ATP. The reaction was performed on a PCR machine for 6 cycles of 20 min at 37℃ and 10 min at 16℃, followed by heating to 50℃ for 5 min and then to 80℃ for 5 min. Thereafter, 1 μl Plasmid Safe DNase (10 U, Epicentre) was added and digested for 30 min. Five μl of the final products were used to transform competent cells. Five white clones were analyzed. The assembled array plasmids were isolated from the correct clones. The second digestion and ligation step was performed in 10 μl volumes containing 60 ng of each array plasmid, pCS2-TALEN-KKR or pCS2-TALEN-ELD, the last repeat plasmid, 1 μl of Esp3I buffer (NEB buffer 3), 0.6 μl of Esp3I (6 U, NEB), 0.4 μl of T4 ligase (800 U, NEB), and 0.4 μl of 25 mM ATP. The reaction was performed on a PCR machine for 6 cycles of 20 min at 37℃ and 10 min at 16℃, followed by heating to 50℃ for 5 min and then to 80℃ for 5 min. Five μl of the final products were used to transform competent cells. Plasmids were isolated from the correct clones for DNA sequencing. The zebrafish nanos-3′UTR was cloned into the 3′ end of the pCS2-TALEN-ELD/KKR coding sequence between *Xba*I and *Not*I. These vectors would be provided upon request.

TALEN mRNA preparation, microinjection and mutation detection

To prepare capped TALEN mRNA, the TALEN expression vectors were linearized by *Not*I and transcribed using the Sp6 mMESSAGE mMACHINE Kit (Ambion). TALEN mRNAs (100-500 pg) were microinjected into one-cell stage zebrafish embryos. The number of normal and deformed embryos were recorded at 24 hours post fertilization (day 1) and 48 hours post fertilization (day 2) (Table S1 in File S1). Two days after injection, genomic DNA was isolated from 8-10 pooled larvae with normal morphology. The target genomic region was amplified by limited cycles of PCR and subcloned to into pMD18-T (Takara) [35]. The mutation was analyzed by PCR or by sequencing. The primers used in this study are listed in Table S2 in File S1.

Mutation and deletion frequency analysis

To determine the mutation frequency of each targeted locus and the fragment deletion frequency induced by the TALEN pair, genomic DNA was isolated from three replicates of 8-10 pooled zebrafish embryos at 48 hours post fertilization. For mutation frequency analysis, each target locus was amplified and subcloned into pMD18-T. Thirty-two single colonies were analyzed for each sample by PCR or subsequent sequencing. For genomic fragment deletion frequency analysis, real-time PCR was performed on an ABI PRISM 7900 Sequence Detection System (Applied Biosystems) using the SYBR Green I Kit. Standard curves were generated by serial dilution of the plasmid DNA. The genomic fragment deletion frequency was calculated as the copy number of genomic DNA with deletions divided by that of the total genomic DNA.

Suppression of target GFP expression by wild-type or mutated miR-1-2

The GFP sensor plasmid containing three imperfect complementary sites to miR-1-2 and the RFP indicator plasmid expressing the wild-type miR-1-2 precursor were from a previous study [35]. The miR-1-2 precursor mutants were generated by mutation PCR and cloned into the 3′ end of the pSP64T+dsRed coding sequence between *Eco*RI and *Xho*I. The 3′UTR sequence of *cnn2* was cloned into pCS2-GFP-F vector between the *Xba*I and *Xho*I. GFP sensor (100 pg) was coinjected with the wild-type or mutated miR-1-2 (300 pg) into one-cell stage embryos. The embryos were photographed at 24 hours post fertilization.

Screening of founders

The TALEN injected embryos were raised to adulthood and outcrossed with wild-type fish. For *miR-1-1*, genomic DNA from 32 pooled F1 embryos of each founder was amplified and subcloned. Thirty-two single clones were analyzed by PCR and sequencing. For genomic fragment deletions, 24 or 32 F1 embryos were collected from each founder and genomic PCR was performed to detect fragment deletion from each single embryo. The fragment deletions were subsequently confirmed by sequencing.

Results

Targeted disruption of individual miRNA genes using a single talen approach

MiRNAs are small non-coding RNAs regulating their targets by post-transcriptional mechanisms [26]. Loss-of-function of genes involved in the miRNA biogenesis pathway has revealed wide range biological functions of these tiny RNAs [26,37]. Here we investigate whether TALENs could be used to knockout miRNAs in zebrafish.

Using our optimized TALEN platform [16], we have assembled TALENs for two zebrafish miRNAs (*miR-1-1* and *miR-1-2*). We placed the miRNA seed (a critical region for miRNA-mRNA pairing) at the spacer region where indels often occur. The assembled TALENs induced somatic mutations with high frequencies of up to 97% (Fig. 1A and Fig. S1). Nearly all these mutations altered the miRNA seed sequences, thus leading to loss-of-function of the miRNAs. Moreover, the indels also altered the hairpin structures of the pre-miRNAs (Fig. 1B), conceivably leading to aberrant miRNA biogenesis. Consistent with these predictions, functional studies indicated that the wild-type pre-miR-1-2 but not the pre-miR-1-2 mutants could effectively suppress the target GFP reporters containing either three imperfect complementary sites to *miR-1-2* or the 3′UTR sequence of a reported *miR-1-2* target gene [36] (Fig. 1C and Fig. 1D).

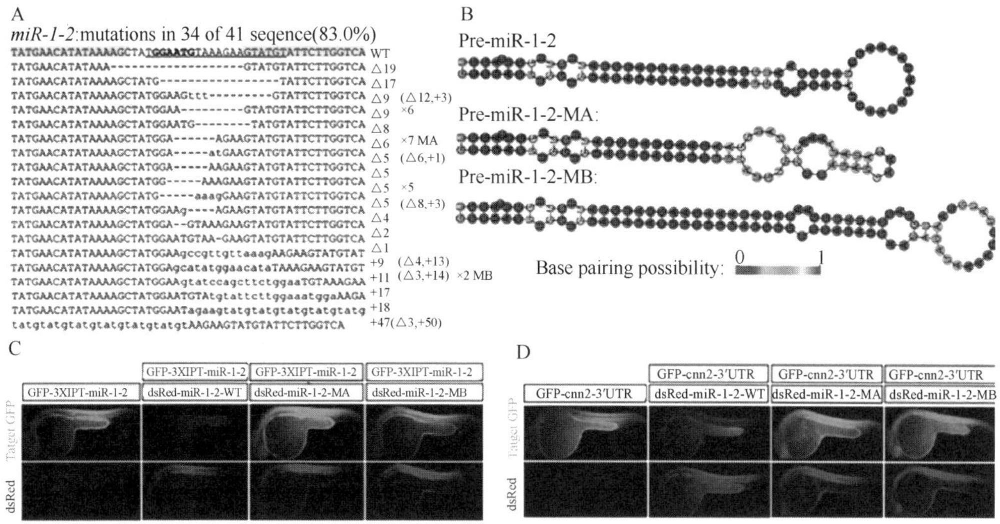

Figure 1 Targeted disruption of zebrafish *miR-1-2*. (A) Frequency and spectrum of TALEN induced *miR-1-2* mutations. The TALEN binding sites are shown in yellow background. DNA sequence encoding the mature miR-1-2 is underlined, with the seed sequence in bold. Deletions are indicated by dash lines and insertions are indicated by lowercase letters. The sizes of the insertions (+) or deletions (Δ) and the number of times each mutant allele appearing are shown on the right side of the mutant allele. (B) The hairpin structure of the wild-type pre-miR-1-2 and two pre-miR-1-2 mutants (MA and MB) in Panel A. (C-D) Functional suppression of target GFP expression by the wild-type or two mutated pre-miR-1-2. Messager RNA of GFP sensor (GFP-3XIPT-miR-1-2 or GFP-*cnn2*-3′UTR) and RFP indicator expressing wild-type or mutant pre-miR-1-2 was co-injected into one-cell stage zebrafish embryos. Pictures were taken at 24 hours after injection. GFP-3XIPT-miR-1-2, GFP sensor containing three imperfect complementary sites to miR-1-2; GFP-*cnn2*-3′UTR, GFP sensor containing the 3′UTR sequence of *cnn2* (a miR-1-2 target gene); dsRed-miR-1-2, RFP indicator expressing mature miR-1-2

Precise large genomic deletions of miRNA clusters using a dual talen approach

More than half of the miRNA genes occur as gene clusters in many vertebrates genome [38]. We thereafter asked whether TALENs could be used to knockout miRNA gene clusters. To knockout gene clusters of large genomic regions, we have devised a strategy that creates two DSBs simultaneously on each side of the targeted genomic fragment using two pairs of TALENs. Repair of the DSB by ligation of the broken ends would lead to disruption of each targeted locus and deletion of the flanked genomic fragment (Fig. 2A). To test this dual TALEN strategy we designed two pairs of TALENs to delete the zebrafish *miR-17-92* cluster (1.2 kb) and the *miR-430* cluster (79.8 kb) (Fig. 2B and Fig. 2F). While the *miR-17-92* cluster encodes 6 miRNAs, the *miR-430* cluster is the largest *miRNA* cluster

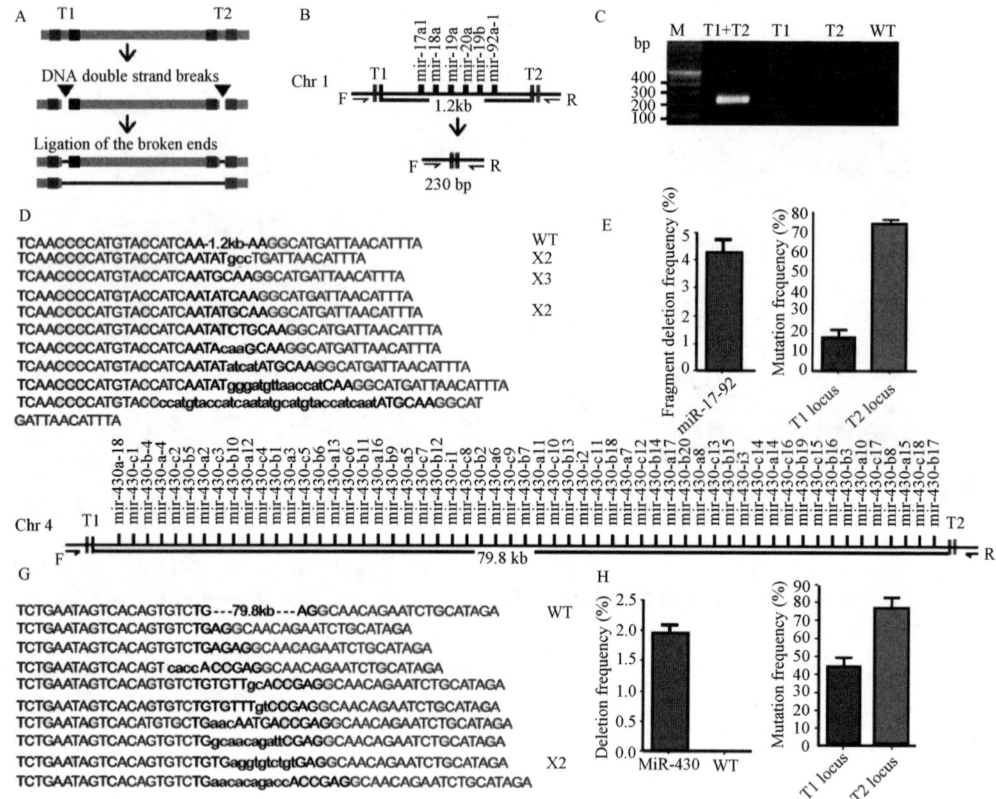

Figure 2 Precise large genomic deletions of miRNA gene clusters. (A) Schematic diagram of the large genomic deletion strategy using dual TALEN. The TALEN binding sites are shown in color. T1, TALEN pair 1; T2, TALEN pair 2. (B) Schematic representation of the zebrafish *miR-17-92* cluster and the designed TALENs for its deletion. After genomic deletions, a band of about 230 bp is expected to be amplified by PCR using the designed primers. (C) Genomic PCR showing deletion of the *miR-17-92* cluster. Gel picture showing PCR amplification of genomic DNA isolated from the pooled zebrafish embryos microinjected with two pairs of TALENs, one pair of TALENs or the wild-type control. (D) Sequencing results confirmed deletion of the *miR-17-92* cluster. Alien nucleotides inserted are indicated by lowercase letters. The number of times each mutant allele appearing are shown on the right side of the mutant allele. (E) Deletion frequency of the *miR-17-92* cluster and indel-mutation frequency of the T1 and T2 loci. Genomic DNA was isolated from pooled zebrafish embryos injected with the two pairs of TALENs. Data shown are mean values ± S.E.M from three replicates. (F) Schematic representation of the zebrafish *miR-430* cluster and the designed TALENs. (G) Sequencing results confirmed deletion of the *miR-430* cluster. Alien nucleotides inserted are indicated by lowercase letters. (H) Deletion frequency of the *miR-430* cluster and indel-mutation frequency of the T1 and T2 loci. Genomic DNA was isolated from pooled zebrafish embryos injected with the two pairs of TALENs. Data shown are mean values ± S.E.M from three replicates

in vertebrates encoding 57 miRNAs. To create two concurrent DSBs on each targeted genomic region, mRNA of two pairs of TALENs were co-injected into one-cell stage zebrafish embryos. Two days after injection, genomic DNA was isolated from 8-10 pooled embryos. Primers were designed to detect deletion of the targeted genomic fragment (Table

S2, Fig. S2 and Fig. S3 in File S1). PCR amplification of genomic DNA isolated from the pooled embryos indicated successful deletion of these miRNA clusters (Fig. 2C and Fig. S3 in File S1). Sequencing of the PCR products confirmed such successful deletions (Fig. 2D and Fig. 2G). All these deletions occurred accurately between the two targeted sites and there were none or few alien nucleotides retained after the large genomic deletions. Taken together, these results indicate that this dual TALEN strategy provides a powerful approach to generate precise large deletions in the genome.

The fragment deletion frequency of each targeted miRNA gene cluster and the indel-mutation frequency of each targeted locus produced by the dual TALEN approach were systematically analyzed. To evaluate the fragment deletion frequency, two pairs of primers were used to amplify genomic DNA with fragment deletions and a nearby undisrupted genomic region (Table S2 in File S1). Quantitative real-time PCR was performed to calculate the copy number of genomic DNA with fragment deletions with respect to that of the undisrupted genomic DNA. The mean deletion frequency was 4.2% for the *miR-17-92* cluster and 1.9% for the *miR-430* cluster in zebrafish embryonic cells respectively (Fig. 2E and Fig. 2H). The dual TALEN treatment produced indel mutations on each targeted locus with high frequency (Fig. 2E, Fig. H and Table S3 in File S1). For both gene clusters, the fragment deletion frequency is lower than that of the indel mutations.

Precise large genomic deletions of lncRNA gene using the dual talen approach

Apart from small RNAs, the animal genome also produces large RNA transcripts called lncRNAs longer than 200 nucleotides that do not code for proteins [27, 39]. In zebrafish, two recent studies annotated more than 1000 lncRNA genes that are expressed during early development [32-33]. To determine whether this dual TALEN strategy is also applicable to delete lncRNA genes, we have designed two pairs of TALENs to delete the 9.0 kb genomic region encoding the lncRNA *malat1* (Fig. 3A and Fig. 3B). Using this approach we obtained deletion of *malat1* in zebrafish somatic cells with a mean deletion frequency of 2.0% (Fig. 3D and Fig. S4 in File S1). The genomic sequence between the two TALEN targeting sites was accurately deleted (Fig. 3C). These results indicate that the dual TALEN strategy is a powerful approach to knockout lncRNA genes.

Germline transmission of knockout genotypes

To test whether these knockout genotypes were inheritable, we have outcrossed the P0 fish raised from the TALEN injected embryos. For *miR-1-1*, 5 out of the 6 P0 fish transmitted the *miR-1-1* indel mutations with high frequencies (Fig. 4A). For the *miR-17-92* cluster, we have genotyped 24 individual F1 embryos from each of the 6 outcrossed P0 fish. Three out of the 6

P0 fish transmitted the 1.2 kb *miR-17-92* deletions to 1/24, 3/24 and 6/24 F1 embryos respectively (Fig. 4B). For *malat1*, 4 out of the 32 P0 fish transmitted the 9.0 kb *malat1* deletions to 1/32, 2/32, 4/32 and 5/32 F1 embryos respectively (Fig. 4D).

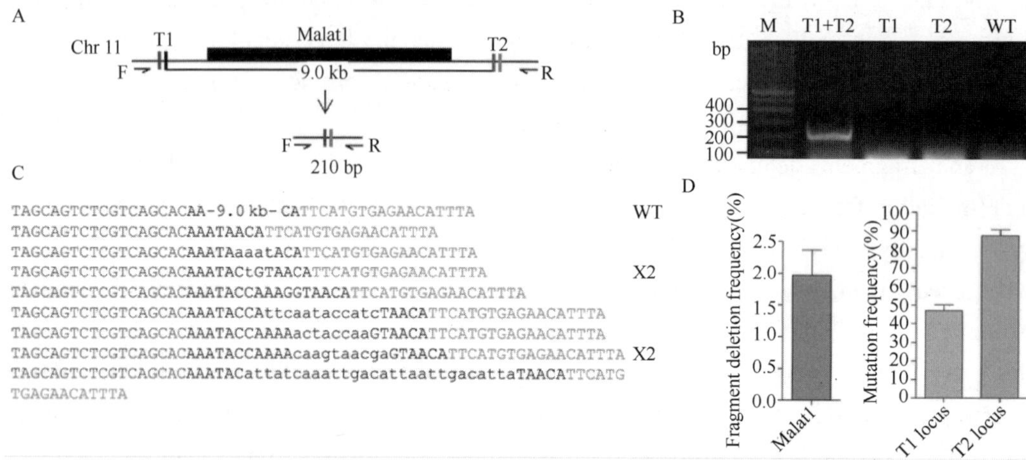

Figure 3 Precise large genomic deletion of a lncRNA gene. (A) Schematic representation of the zebrafish lncRNA *malat1* and the TALENs designed for its deletion. After the genomic deletion, a band of about 210 bp is expected to be amplified by PCR using the designed primers. (B) Genomic PCR showing the deletion of *malat1*. Gel picture showing PCR amplification of genomic DNA isolated from the pooled zebrafish embryos microinjected with two pairs of TALENs, one pair of TALENs or the wild-type control. (C) Sequencing results confirming *malat1* deletions. Alien nucleotides inserted are indicated by lowercase letters. The number of times each mutant allele appearing are shown on the right side of the mutant allele. (D) Deletion frequency of the *malat1* and mutation frequency of the two targeted loci. Genomic DNA was isolated from pooled zebrafish embryos injected with the two pairs of TALENs. Data shown are mean values ± S.E.M from three replicates

For the larger *miR-430* cluster deletions, we have outcrossed 32 P0 fish but failed to obtain germline deletions, suggesting that large genomic deletions are not efficiently transmitted. To overcome this limitation, we have incorporated the zebrafish *nanos*-3′UTR into the TALEN construct (Fig. S5 in File S1). By screening 23 P0 fish raised from the TALEN-*nanos*-3′UTR injected embryos, we have obtained 1 founder transmitted the *miR-430* deletion to 3/32 embryos (Fig. 4C). These data suggested that the *nanos*-3′UTR may increase the germline transmission frequency of large genomic deletions.

Discussion

Recently the ENCODE project estimated more than 70% of the human genome is transcribed into ncRNAs [40-41], indicating that ncRNAs represent a substantial portion of the transcriptome. However, because of the lack of an efficient ncRNA gene knockout platform in vertebrates, only a few of the ncRNA genes have been inactivated in mouse. The

A Germline transmission of *miR-1-1* mutations [F0:5/6]

```
       WT: TATGAACAAGAGCAGCTATGGAATGTAAAGAAGTATGTATCCCAGGTGA
Founder 1: TATGAACAAGAGCAGCTATG---------GAAGTATGTATCCCAGGTGA   (4/32)
Founder 2: TATGAACAAGAGCAGCTATGGA---------AGTATGTATCCCAGGTGA  (11/31)
Founder 3: TATGAACAAGAGCAGCTATGG------AAGAAGTATGTATCCCAGGTGA   (15/32)
Founder 4: TATGAACAAGAGCAGCTATGGAAT--AAAGAAGTATGTATCCCAGGTGA   (10/32)
Founder 5: TATGAACAAGAGCAGCTATGGA---------AGTATGTATCCCAGGTGA  (11/31)
Founder 5: TATGAACAAGAGCAGCTATGGAagtatgagaagaaGTATGTATCCCAGG   (2/31)
```

B Germline transmission of *miR-17-92* cluster deletions [F0:3/6]

```
       WT: TCAACCCCATGTACCATCAA-1.2kb-AAGGCATGATTAACATTTA
Founder 1: TCAACCCCATGTACCATCAAGGCATGATTAACATTTA              (1/24)
Founder 2: TCAACCCCATGTACCATCAATACATGCAAGGCATGATTAACATTTA     (3/24)
Founder 3: TCAACCCCATGTACCATCAATATCaagagGCAAGGCATGATTAACATTTA (6/24)
```

C Germline transmission of *miR-430* cluster deletions [F0:1/23]

```
       WT: TCTGAATAGTCACAGTGTCTG-79.8kb-AGGCAACAGAATCTGCATAGA
Founder 1: TCTGAATAGTCAgaaggcaaccaccgagCAGAATCTGCATAGA        (3/32)
```

D Germline transmission of *malat1* deletions [F0:4/32]

```
       WT: TAGCAGTCTCGTCAGCACAA-9.0kb-CATTCATGTGAGAACATTTA
Founder 1: TAGCAGTCTCGTCAGCACAAATACCATTCATGTGAGAACATTTA              (1/32)
Founder 2: TAGCAGTCTCGTCAGCACAAATACATGTGAGAACATTTA                   (2/32)
Founder 3: TAGCAGTCTCGTCAGCACAAATACtGTAACATTCATGTGAGAACATTTA         (4/32)
Founder 4: TAGCAGTCTCGTCAGCACAAATACCAAAAGGAGACGTAACATTCATGTGAGAACATTTA (5/32)
```

Figure 4 Germline transmission of the knockout genotypes. (A) Germline transmission of the *miR-1-1* mutations. The TALEN binding sites are shown in yellow background. DNA sequence encoding the mature miR-1-1 is underlined, with the seed sequence in bold. Deletions are indicated by dash lines and insertions are indicated by lowercase letters. The F0 germline deletion ratio is shown in square brackets. The ratios of mutated/total analyzed sequences from the pooled F1 embryos of each founder are shown in brackets. (B-D) Germline transmission of the *miR-17-92* deletions (B), *miR-430* deletions (C) and malat1 deletions (D). The F0 germline deletion ratios are shown in square brackets and the F1 inheritance ratios are shown in brackets

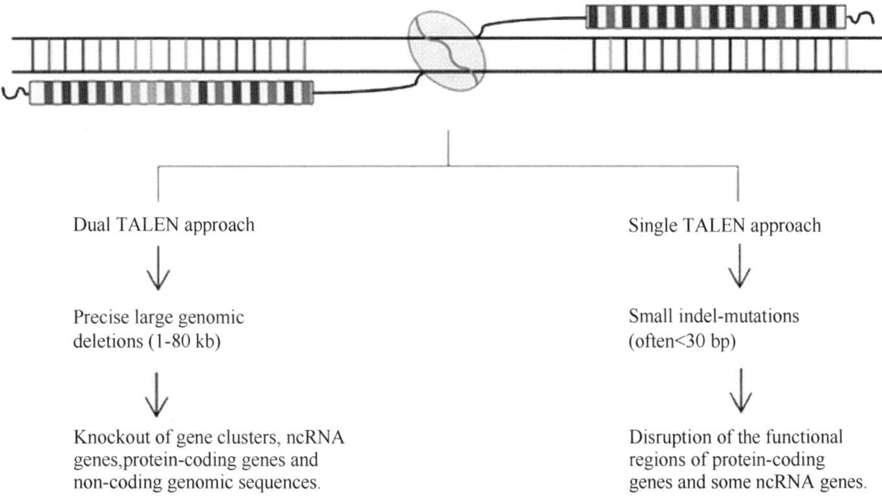

Figure 5 Schematic diagram illustrating knockout strategies using dual or single TALEN approaches. The dual TALEN approach could generate precise large genomic deletions (1-80 kb), allowing knockout of large gene clusters, ncRNA genes, protein-coding genes and non-coding regulatory sequences. The single TALEN approach could produce small indel-mutations of the targeted genomic loci (often <30 bp), allowing disruption of individual protein-coding genes or some individual ncRNA genes

ability of engineered nucleases to target specific genomic loci across species makes them an attractive option to address this issue. However, the engineered nucleases have been mainly used to induce small indel mutations. Because the functional regions of ncRNA gene are often unknown, knockout of an entire ncRNA gene is preferred to eliminate gene function. Here we test whether TALENs could be used to knockout of ncRNA genes in zebrafish.

Using zinc finger nucleases and TALENs, targeted fragment deletions have been recently reported in cell lines [17, 42-44]. More recently, genomic deletions of an 800 bp fragment using TALENs have been reported in silkworm [7]. During the time this manuscript was reviewed, others have also reported that large genomic deletions could be efficiently engineered in zebrafish using a dual TALEN strategy [45-47]. In this study, we have demonstrated that the dual TALEN approach is a robust means to generate inheritable large genomic deletions in zebrafish. First, large genomic deletions (1-80 kb) could be efficiently generated in a few days and the deletions could be readily detected by genomic PCR. Second, all deletions have accurately occurred and none or few alien nucleotides are retained after the genomic deletions (Fig. S6 in File S1). Third, although the fragment deletion frequency is lower than that of the indel-mutation, these large deletions of about 9 kb could be efficiently transmitted through the germline.

Analysis of the transcriptome revealed that many small regulatory RNA are transcribed from the eukaryotic genome [40,48]. miRNA are small RNA molecules regulating the decay and translation of their target mRNAs. Mature miRNAs are generated from cleavage of the hairpin structure of pre-miRNAs by the RNAse III enzymes dicer [49]. In zebrafish, loss-of-function mutations of dicer in the maternal and zygotic embryos lead to multiple developmental defects followed by death on day 5 post-fertilization [50], indicating that miRNAs as a whole play essential roles in embryogenesis. But the exact functional roles of individual miRNAs or miRNAs clusters remain elusive. Moreover, most miRNAs expressed in zebrafish adult tissues may also regulate other physiological processes. In this study, we have demonstrated that large miRNA gene clusters could be precisely deleted using two pairs of TALENs, whereas individual miRNA genes could be inactivated by destruction of the miRNA hairpin structure and miRNA seed using one pair of TALENs, establishing TALEN as a robust method to knockout miRNA genes. Further studies will be performed to analyze the phenotypic changes resulting from deleting of individual miRNA and miRNA clusters.

Apart from small RNAs, the animal genome also produces many lncRNAs [27,39]. An increasing number of reports revealed that lncRNAs regulate many cellular processes ranging from maintaining embryonic stem cell pluripotency to epigenetic silencing of large genomic regions [28,39]. In zebrafish, more than 1000 lncRNA genes are expressed during early development [32-33]. Moreover, morpholino knockdown of lncRNAs lead to abnormal brain morphogenesis [33]. However, most lncRNAs are not well conserved and their functional regions are mostly unknown, thus limiting the applicability of the morpholino knockdown approach to study these lncRNAs. The dual TALEN strategy described in this study provides a powerful tool to knockout these lncRNA genes.

Germline transmission of the knockout genotypes is critical to obtain the homozygous gene knockout animals. In this study, we have demonstrated that deletions as long as 9.0 kb could be efficiently transmitted through the germline. To improve the germline integration efficiency of larger genomic deletions, we have incorporated the zebrafish *nanos*-3′UTR into the TALEN construct. The *nanos* 3′UTR was reported to protect mRNA from degradation in the primordial germ cells [51,52], thus may improve the germ cell target efficiency. Using this TALEN-*nanos*-3′UTR construct, we have demonstrated that large genomic deletions of the miR-430 cluster could be transmitted through the germline. This strategy might also be applicable to engineer large genomic deletions in other species.

In summary, we have generated high frequency indel mutations of individual miRNA genes and precise large genomic deletions of non-coding RNA gene clusters in zebrafish using TALENs. Moreover, these deletions are inheritable. We propose that the same TALEN approach could be employed as a molecular tool to study the *in vivo* functions of ncRNA genes in other species. Moreover, the two TALEN strategies described in this study also allow one to knockout other ncRNA genes, entire protein coding genes, genomic regulatory sequences and to disrupt a small functional genomic region depending on the needs (Fig. 5).

Acknowledgements We thank Professor Antonio J. Giraldez of Yale University for providing us with the miRNA related vectors and Professor Daniel Voytas of University of Minnesota for the TALENs golden gate toolkit. We thank Professor Igor Dawid of the NIH for critical reading of the manuscript. We also thank Josie Lai, Kathy Sham, Peng Wu and Chengrong Zhong for expert technical assistance.

Funding Statement This work was supported by the Research Grant Council of Hong Kong (462312 to CHKC), the National Natural Science Foundation of China (31172394 to CHKC, 30900853 to DL and 30930069 to WH) and the Chinese Academy of Sciences (KSCX2-EW-N-004 to WH). The funders had no role in study design, data collection and analysis, decision to publish, or preparation of the manuscript.

References

1. Christian M, Cermak T, Doyle EL, Schmidt C, Zhang F, *et al.* (2010) Targeting DNA double-strand breaks with TAL effector nucleases. Genetics 186: 757-761
2. Li T, Huang S, Jiang WZ, Wright D, Spalding MH, *et al.* (2011) TAL nucleases (TALNs): hybrid proteins composed of TAL effectors and FokI DNA-cleavage domain. Nucleic Acids Res 39: 359-372
3. Boch J, Scholze H, Schornack S, Landgraf A, Hahn S, *et al.* (2009) Breaking the code of DNA binding specificity of TAL-type III effectors. Science 326: 1509-1512
4. Moscou MJ, Bogdanove AJ (2009) A simple cipher governs DNA recognition by TAL effectors. Science 326: 1501
5. Li T, Huang S, Zhao X, Wright DA, Carpenter S, *et al.* (2011) Modularly assembled designer TAL effector nucleases

for targeted gene knockout and gene replacement in eukaryotes. Nucleic Acids Res 39: 6315-6325

6. Wood AJ, Lo TW, Zeitler B, Pickle CS, Ralston EJ, *et al.* (2011) Targeted genome editing across species using ZFNs and TALENs. Science 333: 307
7. Ma S, Zhang S, Wang F, Liu Y, Liu Y, *et al.* (2012) Highly Efficient and Specific Genome Editing in Silkworm Using Custom TALENs. PLoS One 7: e45035
8. Li T, Liu B, Spalding MH, Weeks DP, Yang B (2012) High-efficiency TALEN-based gene editing produces disease-resistant rice. Nat Biotechnol 30: 390-392
9. Huang P, Xiao A, Zhou M, Zhu Z, Lin S, *et al.* (2011) Heritable gene targeting in zebrafish using customized TALENs. Nat Biotechnol 29: 699-700
10. Sander JD, Cade L, Khayter C, Reyon D, Peterson RT, *et al.* (2011) Targeted gene disruption in somatic zebrafish cells using engineered TALENs. Nat Biotechnol 29: 697-698
11. Dahlem TJ, Hoshijima K, Jurynec MJ, Gunther D, Starker CG, *et al.* (2012) Simple methods for generating and detecting locus-specific mutations induced with TALENs in the zebrafish genome. PLoS Genet 8: e1002861
12. Cade L, Reyon D, Hwang WY, Tsai SQ, Patel S, *et al.* (2012) Highly efficient generation of heritable zebrafish gene mutations using homo- and heterodimeric TALENs. Nucleic Acids Res 40: 8001-8010
13. Moore FE, Reyon D, Sander JD, Martinez SA, Blackburn JS, *et al.* (2012) Improved somatic mutagenesis in zebrafish using transcription activator-like effector nucleases (TALENs). PLoS One 7: e37877
14. Ansai S, Sakuma T, Yamamoto T, Ariga H, Uemura N, *et al.* (2013) Efficient Targeted Mutagenesis in Medaka Using Custom-Designed Transcription Activator-Like Effector Nucleases (TALENs). Genetics.
15. Tesson L, Usal C, Menoret S, Leung E, Niles BJ, *et al.* (2011) Knockout rats generated by embryo microinjection of TALENs. Nat Biotechnol 29: 695-696
16. Lei Y, Guo X, Liu Y, Cao Y, Deng Y, *et al.* (2012) Efficient targeted gene disruption in Xenopus embryos using engineered transcription activator-like effector nucleases (TALENs). Proc Natl Acad Sci U S A 109: 17484-17489
17. Carlson DF, Tan W, Lillico SG, Stverakova D, Proudfoot C, *et al.* (2012) Efficient TALEN-mediated gene knockout in livestock. Proc Natl Acad Sci USA 109: 17382-17387
18. Mussolino C, Morbitzer R, Lutge F, Dannemann N, Lahaye T, *et al.* (2011) A novel TALE nuclease scaffold enables high genome editing activity in combination with low toxicity. Nucleic Acids Res 39: 9283-9293
19. Miller JC, Tan S, Qiao G, Barlow KA, Wang J, *et al.* (2011) A TALE nuclease architecture for efficient genome editing. Nat Biotechnol 29: 143-148
20. Reyon D, Tsai SQ, Khayter C, Foden JA, Sander JD, *et al.* (2012) FLASH assembly of TALENs for high-throughput genome editing. Nat Biotechnol 30: 460-465
21. Hockemeyer D, Wang H, Kiani S, Lai CS, Gao Q, *et al.* (2011) Genetic engineering of human pluripotent cells using TALE nucleases. Nat Biotechnol 29: 731-734
22. Hwang WY, Fu Y, Reyon D, Maeder ML, Tsai SQ, *et al.* (2013) Efficient genome editing in zebrafish using a CRISPR-Cas system. Nat Biotechnol 31: 227-229
23. Kapranov P, Willingham AT, Gingeras TR (2007) Genome-wide transcription and the implications for genomic organization. Nat Rev Genet 8: 413-423
24. Amaral PP, Dinger ME, Mercer TR, Mattick JS (2008) The eukaryotic genome as an RNA machine. Science 319: 1787-1789
25. Gutschner T, Baas M, Diederichs S (2011) Noncoding RNA gene silencing through genomic integration of RNA destabilizing elements using zinc finger nucleases. Genome Res 21: 1944-1954
26. Bartel DP (2009) MicroRNAs: target recognition and regulatory functions. Cell 136: 215-233
27. Mercer TR, Dinger ME, Mattick JS (2009) Long non-coding RNAs: insights into functions. Nat Rev Genet 10: 155-159
28. Huarte M (2012) LncRNAs have a say in protein translation. Cell Res. Doi: 10.1038/cr.2012.169
29. Wienholds E, Kloosterman WP, Miska E, Alvarez-Saavedra E, Berezikov E, *et al.* (2005) MicroRNA expression in zebrafish embryonic development. Science 309: 310-311

30. Chen PY, Manninga H, Slanchev K, Chien M, Russo JJ, *et al.* (2005) The developmental miRNA profiles of zebrafish as determined by small RNA cloning. Genes Dev 19: 1288-1293

31. Wei C, Salichos L, Wittgrove CM, Rokas A, Patton JG (2012) Transcriptome-wide analysis of small RNA expression in early zebrafish development. RNA 18: 915-929

32. Pauli A, Valen E, Lin MF, Garber M, Vastenhouw NL, *et al.* (2012) Systematic identification of long noncoding RNAs expressed during zebrafish embryogenesis. Genome Res 22: 577-591

33. Ulitsky I, Shkumatava A, Jan CH, Sive H, Bartel DP (2011) Conserved function of lincRNAs in vertebrate embryonic development despite rapid sequence evolution. Cell 147: 1537-1550

34. Cermak T, Doyle EL, Christian M, Wang L, Zhang Y, *et al.* (2011) Efficient design and assembly of custom TALEN and other TAL effector-based constructs for DNA targeting. Nucleic Acids Res 39: e82

35. Sanjana NE, Cong L, Zhou Y, Cunniff MM, Feng G, *et al.* (2012) A transcription activator-like effector toolbox for genome engineering. Nat Protoc 7: 171-192

36. Mishima Y, Abreu-Goodger C, Staton AA, Stahlhut C, Shou C, *et al.* (2009) Zebrafish miR-1 and miR-133 shape muscle gene expression and regulate sarcomeric actin organization. Genes Dev 23: 619-632

37. Giraldez AJ, Cinalli RM, Glasner ME, Enright AJ, Thomson JM, *et al.* (2005) MicroRNAs regulate brain morphogenesis in zebrafish. Science 308: 833-838

38. Thatcher EJ, Bond J, Paydar I, Patton JG (2008) Genomic organization of zebrafish microRNAs. BMC Genomics 9: 253

39. Rinn JL, Chang HY (2012) Genome regulation by long noncoding RNAs. Annu Rev Biochem 81: 145-166

40. Djebali S, Davis CA, Merkel A, Dobin A, Lassmann T, *et al.* (2012) Landscape of transcription in human cells. Nature 489: 101-108

41. Birney E, Stamatoyannopoulos JA, Dutta A, Guigo R, Gingeras TR, *et al.* (2007) Identification and analysis of functional elements in 1% of the human genome by the ENCODE pilot project. Nature 447: 799-816

42. Sollu C, Pars K, Cornu TI, Thibodeau-Beganny S, Maeder ML, *et al.* (2010) Autonomous zinc-finger nuclease pairs for targeted chromosomal deletion. Nucleic Acids Res 38: 8269-8276

43. Lee HJ, Kim E, Kim JS (2010) Targeted chromosomal deletions in human cells using zinc finger nucleases. Genome Res 20: 81-89

44. Chen F, Pruett-Miller SM, Huang Y, Gjoka M, Duda K, *et al.* (2011) High-frequency genome editing using ssDNA oligonucleotides with zinc-finger nucleases. Nat Methods 8: 753-755

45. Gupta A, Hall VL, Kok FO, Shin M, McNulty JC, *et al.* (2013) Targeted chromosomal deletions and inversions in zebrafish. Genome Res 23: 1008-1017

46. Xiao A, Wang Z, Hu Y, Wu Y, Luo Z, *et al.* (2013) Chromosomal deletions and inversions mediated by TALENs and CRISPR/Cas in zebrafish. Nucleic Acids Res 41: e141

47. Lim S, Wang Y, Yu X, Huang Y, Featherstone MS, *et al.* (2013) A simple strategy for heritable chromosomal deletions in zebrafish via the combinatorial action of targeting nucleases. Genome Biol 14: R69

48. Landgraf P, Rusu M, Sheridan R, Sewer A, Iovino N, *et al.* (2007) A mammalian microRNA expression atlas based on small RNA library sequencing. Cell 129: 1401-1414

49. Kim VN, Han J, Siomi MC (2009) Biogenesis of small RNAs in animals. Nat Rev Mol Cell Biol 10: 126-139

50. Giraldez AJ, Cinalli RM, Glasner ME, Enright AJ, Thomson JM, *et al.* (2005) MicroRNAs regulate brain morphogenesis in zebrafish. Science 308: 833-838

51. Mishima Y, Giraldez AJ, Takeda Y, Fujiwara T, Sakamoto H, *et al.* (2006) Differential regulation of germline mRNAs in soma and germ cells by zebrafish miR-430. Curr Biol 16: 2135-2142

52. Koprunner M, Thisse C, Thisse B, Raz E (2001) A zebrafish nanos-related gene is essential for the development of primordial germ cells. Genes Dev 15: 2877-2885

TALEN 介导非编码 RNA 大基因片段的精准缺失及其可遗传性

刘 云[1,4] 罗大极[2,3] Hui Zhao[1,4] 朱作言[2] 胡 炜[2] Christopher H. K. Cheng[1,4]

1 香港中文大学，生物医学院，香港

2 中国科学院水生生物研究所，淡水生态与生物技术国家重点实验室，武汉

3 武汉大学，基础医学院，武汉

4 香港中文大学深圳研究所，深圳

摘 要 转录激活因子样效应物核酸酶(TALENs)现已广泛用于突变编码基因，但编码蛋白的基因仅占动物基因组的 2-3%，而绝大多数的基因(70-90%)实际上是非编码RNA(ncRNAs)，由于缺乏有效的敲除ncRNA基因的手段，ncRNA基因在生物体内功能的研究受到了极大的阻碍。本文研制出了一种新的方法，通过TALENs在斑马鱼中实现了定向、精确且可遗传的大段基因缺失，从而敲除ncRNA基因。本研究证实，可用一对TALENs破坏单个miRNA，使用两对TALENs可以对microRNA基因簇和长非编码RNA基因(lncRNA)实现精确敲除。我们在成功实施了两个microRNA基因簇(1.2 kb *miR-17-92* cluster 和 79.8 kb *miR-430* cluster)和一个长非编码RNA(9.0 kb *malat1*)中的大片段缺失，而且这些缺失可通过生殖细胞系传代。综上所述,本研究建立了一种新的TALENs技术，可以高效特异地完成基因大片段缺失和ncRNA基因的敲除，开辟了利用TALENs开展基因组在体研究的新方向。

Efficient Ligase 3-Dependent Microhomology-Mediated End Joining Repair of DNA Double-Strand Breaks in Zebrafish Embryos

Mu-Dan He[1,2] Feng-Hua Zhang[1,2] Hua-Lin Wang[1,2] Hou-Peng Wang[1]
Zuo-Yan Zhu[1] Yong-Hua Sun[1]

1 State Key Laboratory of Freshwater Ecology and Biotechnology, Institute of Hydrobiology, Chinese Academy of Sciences, Wuhan 430072
2 University of the Chinese Academy of Sciences, Beijing 100049

Abstract DNA double-strand break (DSB) repair is of considerable importance for genomic integrity. Homologous recombination (HR) and non-homologous end joining (NHEJ) are considered as two major mechanistically distinct pathways involved in repairing DSBs. In recent years, another DSB repair pathway, namely, microhomology-mediated end joining (MMEJ), has received increasing attention. MMEJ is generally believed to utilize an alternative mechanism to repair DSBs when NHEJ and other mechanisms fail. In this study, we utilized zebrafish as an *in vivo* model to study DSB repair and demonstrated that efficient MMEJ repair occurred in the zebrafish genome when DSBs were induced using TALEN (transcription activator-like effector nuclease) or CRISPR (clustered regularly interspaced short palindromic repeats)/Cas9 technologies. The wide existence of MMEJ repair events in zebrafish embryos was further demonstrated via the injection of several *in vitro*-designed exogenous MMEJ reporters. Interestingly, the inhibition of endogenous ligase 4 activity significantly increased MMEJ frequency, and the inhibition of ligase 3 activity severely decreased MMEJ activity. These results suggest that MMEJ in zebrafish is dependent on ligase 3 but independent of ligase 4. This study will enhance our understanding of the mechanisms of MMEJ *in vivo* and facilitate inducing desirable mutations via DSB-induced repair.
Keywords double-strand breaks repair; transcription activator-like effector nuclease (TALEN); CRISPR (clustered regularly interspaced short palindromic repeats)/Cas9; microhomology-mediated end joining (MMEJ); ligase 3; ligase 4

1. Introduction

In eukaryotic cells, DNA double-strand breaks (DSBs) are common genomic events, and the proper repair of DSBs is of considerable importance for the preservation of genomic

文章发表于 *Mutation Research*, 2015, 780: 86-96

integrity. Homologous recombination (HR) and non-homologous end joining (NHEJ) are considered as two major pathways through which DSBs are repaired. Generally, HR ensures high-fidelity DNA repair by using the undamaged sister chromatid or homologous DNA as a template to accurately repair the DNA DSBs and is most active in the late S and G2 phases of the cell cycle. Proteins involved in HR include Rad51, RPA, XRCC2, BRCA1, BRCA2, *etc.* [1-3]. By contrast, NHEJ is the most straightforward way to repair a DSB, as the broken ends are simply rejoined regardless of the genetic sequences at the break. Efficient NHEJ is most active during G1 and early S phase, and it requires Ku70/Ku80 heterodimers, DNA-PK catalytic subunits (PKcs), DNA ligase 4 (Lig4), and X-ray repair cross-complementing (XRCC4) protein [4,5].

Microhomology-mediated end joining (MMEJ), a less well-characterized repair mechanism also known as alternative NHEJ (a-NHEJ), is believed to be a Ku-independent end joining process that is mediated by base pairing between microhomologous sequences of approximately 2-25 nucleotides [5-8]. MMEJ is an error-prone process that often results in deletions or translocations. The existence of MMEJ was first demonstrated in mammalian cells [9], and MMEJ was long considered a "back-up" mechanism to repair DSBs when NHEJ and other mechanisms fail. However, recent studies indicate that MMEJ occurs even when canonical NHEJ is functional [10,11]. Several *in vitro* studies have shown that MMEJ utilizes Ku-independent repair machinery [12-14] and still exhibits robust activity in DNA-PKc-deficient cells [15,16]. MMEJ was once recognized as a type of single-strand annealing (SSA), an error-prone variant of HR, which requires Rad52 and occurs between sequence repeats within the same or between two heterologous chromosomes [17,18]. However, despite its similarity to SSA, MMEJ requires shorter stretches of homology (2-25 bp versus ≥ 30 bp for SSA) [19] and occurs in the genetic absence of RAD52 [20]. Studies in mammalian cells have shown that PARP-1, ligase 3, CtIP, and Mre11 can promote MMEJ [7, 11,21-24]. Nevertheless, how often MMEJ occurs during vertebrate development and which type of ligase is required for MMEJ repair *in vivo* remain largely unknown.

The generation and repair of DNA DSBs are vital for normal development, and studies in vertebrate systems have made substantial contributions to our understanding of DNA repair. For instance, homologous DNA recombination is essential for the viability of vertebrate cells [25,26], and NHEJ components mediate DNA repair to promote the survival of irradiated cells during zebrafish embryogenesis [27]. Due to its external development and embryo transparency, zebrafish is not only a notable model for studying development and human disease, it is also helpful for *in vivo* analysis of DNA DSB repair [28-30]. The NHEJ, HR, and SSA machineries are also thought to be conserved from zebrafish to mammals.

Several site-directed mutagenesis technologies, such as zinc-finger nucleases (ZFNs), transcription activator-like effector nucleases (TALENs), and clustered regulatory interspaced short palindromic repeat (CRISPR)/Cas-based RNA-guided DNA endonucleases

(CRISPR/Cas), are emerging as powerful tools for genetic modifications in zebrafish and other animal models [31,32]. These technologies commonly induce DNA DSBs at specific genomic locations, which stimulate the error-prone NHEJ repair pathway and therefore introduce target-specific mutagenesis. However, due to the error-prone DNA repair process induced by these technologies, the exact post-repair sequences at target sites are generally unpredictable. It remains unclear whether MMEJ repair can be actively utilized in these site-specific nuclease technologies. Here, we used zebrafish embryos as an *in vivo* model and designed several exogenous and endogenous assays to analyze MMEJ repair at certain target sites. Our study revealed that not only do MMEJ events occur widely throughout early zebrafish development but that ligase 3-dependent MMEJ can also be efficiently utilized to repair DSBs in zebrafish embryos.

2. Materials and Methods

2.1 Fish husbandry

All the zebrafish used in this study were maintained and raised at the China Zebrafish Resource Center (CZRC, http://zfish.cn), Wuhan, China. A wild-type strain (*AB* strain) was used for most of the studies, and the *lig4*$^{fh302/+}$ mutant was kindly provided by Dr. Cecilia Moens (Fred Hutchinson Cancer Research Center, Seattle, USA) and deposited at CZRC. Zebrafish experiments were performed under the approval of the Institutional Animal Care and Use Committee of the Institute of Hydrobiology, Chinese Academy of Sciences.

2.2 Design of TALENs and CRISPR/Cas9

TALEN modules were kindly provided by Dr. Bo Zhang (Peking University, China) and assembled according to the "Unit Assembly" method [33]. Two endogenous gene loci, *nanog* (Ensembl ID: ENSDARG00000075113) and *flt4* (Ensembl ID: ENSDARG00000015717), were chosen for TALEN- or CRISPR/Cas9- induced DSB analysis. The left and right recognition sequences of *nanog* TALENs are 5′-TAACCCATCTTAT-3′ and 5′-ACATGAGCCCGT-3′, respectively, and the spacer sequence is 5′-CATGCATATGCAT-3′, which contains an *Nde*I enzyme site and direct repeats of 5′-ATGCAT-3′. The TALEN left and right arms of *flt4* are 5′-CCGTTCAATGTGT-3′ and 5′-GTGAGAAGCCATCGT-3′, respectively, and the spacer sequence is 5′-CTCAGAGCTCAGAGG-3′, which contains a *Sac*I restriction enzyme site and direct repeats of 5′-CTCAGAG-3′ (Fig. 1A).

The Cas9 and gRNA backbone plasmids were kindly provided by Dr. Jing-Wei Xiong (Peking University, China), deposited at the CZRC [34]. The gRNA target site of *flt4* is 5′-AATGTGTCTCAGAGCTCAG<u>AGG</u>-3′, in which the PAM sequence is underlined.

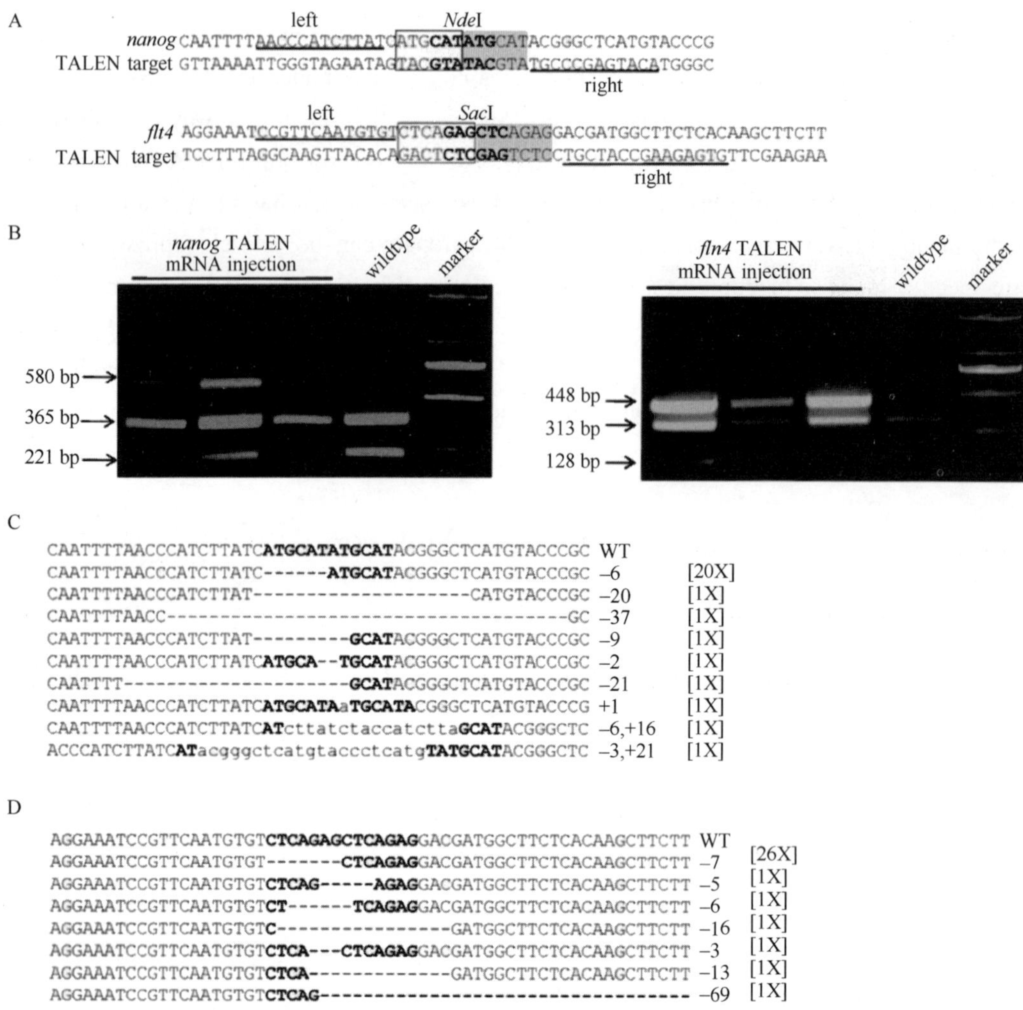

Fig. 1 Detection of MMEJ repair of endogenous DSBs induced by TALEN in zebrafish embryos. (A) The TALEN targets, spacer sequences, and flanking sequences of zebrafish *nanog* and *flt*4. The left and right binding sequences of TALEN are underlined; the direct nucleotide repeats in the spacer region are marked by boxes and shadows; and the *Nde*I and *Sac*I cutting sites used for the mutation test are marked in bold. (B) Detection of mutations in founder embryos *via* PCR and enzyme digestion. The appearance of the upper bands indicates the potential mutation at the TALEN target sites, which are uncleaved by the enzymes. (C) and (D) Sequencing results of the uncleaved bands of the *nanog* (C) and *flt*4 (D) TALEN samples. The wild-type (WT) target sequence is shown at the top; deletions are indicated by dashes, and insertions are indicated by lowercase letters. The sizes of the insertions (+) or deletions (−) are indicated to the right of each mutant allele, and the number of times that each mutant allele was isolated is shown in brackets. Note that 71.4% (20/28) of the *nanog* TALEN clones were shown to be microhomologous sequence deletion mutants (the 5′-ATGCAT-3′ deletion), and 81.2% (26/32) of the *flt4* TALEN clones were shown to be microhomologous sequence deletion mutants (the 5′-CTCAGAG-3′ deletion)

2.3 Construction of plasmids

Two types of exogenous MMEJ plasmids, pSmaI-EGFP and pEcoRV-EGFP, were designed. First, the EGFP coding sequence was sub-cloned into the pCS2+vector (Addgene), resulting in the pCS2+EGFP construct. Second, inverse PCR was used to amplify pCS2+EGFP with the primers SmaI-F and SmaI-R as described previously [35], and the PCR products were digested by *Sma*I and ligated to generate the pSmaI-EGFP vector. In the pSmaI-EGFP vector, an additional repeat sequence, "5′-GGGCGACACCC-3′", was inserted into the wild-type EGFP, with a *Sma*I cleavage site between the repeat sequences. The same method was used to construct the pEcoRV-EGFP using the primers EcoRV-F and EcoRV-R and the sequence of repeats is 5′-ATCAAGGTGAACTTCAAGAT-3′.

pSmaI-EGFP was further amplified with the SmaI-EGFP-ClaI-6- F/SmaI-EGFP- ClaI-6-R, SmaI-EGFP-ClaI-12-F/SmaI-EGFP-ClaI-12-R, and SmaI-EGFP-ClaI-20-F/SmaI- EGFP-ClaI-20-R primer pairs through inverse PCR, and several MMEJ assay constructs with different spacer lengths (pSmaI-EGFP-6 bp, pSmaI-EGFP-12 bp, and pSmaI-EGFP-20 bp) were constructed. All of the inserted spacer sequences contained a *Cla*I restriction site. The pClaI-EGFP-9bp construct was also constructed through inverse PCR by amplifying pSmaI-EGFP using the primers ClaI-EGFP-F1 and ClaI-EGFP-R1. All the primers used for PCR are listed in Table 1.

Table 1 Primer sequences used in this study

Primer name	Sequence(5′-3′)
SmaI-F	tccccсGGGCGACACCCTGGTGA
SmaI-R	tccccсGGGTGTCGCCCTCGAAC
EcoRV-F	ccggatATCAAGGTGAACTTCA
EcoRV-R	ccggatATCTTGAAGTTCACCT
SmaI-EGFP-ClaI-6-F	cgatgggcgacaccctggtga
SmaI-EGFP-ClaI-6-R	atgggtgtcgccctcgaact
SmaI-EGFP-ClaI-12-F	gatatagggcgacaccctggtgaac
SmaI-EGFP-ClaI-12-R	gatttcgggtgtcgccctcgaact
SmaI-EGFP-ClaI-20-F	gatatagctagggcgacaccctggtga
SmaI-EGFP-ClaI-20-R	gatttcgtacgggtgtcgccctcgaact
ClaI-EGFP-F1	gtacgacaccctggtgaccgc
ClaI-EGFP-R1	tcgatgggtgtcgccctcga
lig3-F	ccatcgatatcggctaacaatgaagact
lig3-R	tgctctagaccattactaagcacgaggg
part-lig3-F1	ccatcgattatgcccagtctttgg
part-lig3-R1	cggaattcggttaggaacttcca
part-lig4-F1	cgggatccacagtttcttccgtgtcttct
part-lig4-R1	ccatcgataccataagccattcgctccct
EGFP-F1	cggaattctggtgagcaagggcgaggag
EGFP-R1	tgctctagattacttgtacagctcgtcc

Continued

Primer name	Sequence(5′-3′)
DN-lig4-F	ccatcgatatggccaaagagacagatatgttt
DN-lig4-R	ccgctcgagtgtttcagtgttttgattttc
DN-lig3-F	ccatcgattatgcccagtctttggcttg
DN-lig3-R	tgctctagatcaattcttcactacgcctg
nanog-TALEN-F	caggtgatgtaaatgggtcggta
nanog-TALEN-R	tatcgcgtcgagtgtacgcatg
flt4-TALEN-F	ccctgaaatgtgcaaccaa
flt4-TALEN-R	gctgagaaggctagaagtaa
Normalizing forward primer (Norm-F)	atcatggccgacaagcagaagaacg
Normalizing reverse primer (Norm-R)	cggcggcggtcacgaactcc
HR and SSA-Rep-F	tgaccaccctgacctacg
Repair reverse primer (Rep-R)	caccttgatgccgttcttctgc
nanog-3730-F	taattaggtgtttagtacgt
nanog-3730-R	ctctgacacttgcgggtaca
flt4-3730-F	tactattacattcctaatggc
flt4-3730-R	atgttgctgtaccgtgttgg

To clone zebrafish *l*ig3 cDNA and the dominant-negative forms of zebrafish ligase 3 (DN-lig3) and ligase 4 (DN-lig4), reverse transcriptase polymerase chain reaction (RT-PCR) was performed using total RNA isolated from wild-type embryos at 24 h post-fertilization (hpf). The primers lig3-F and lig3-R were used to amplify the full length cDNA of *lig3*, DN-lig3-F and DN-lig3-R were used to amplify the cDNA for the N-terminal of Lig3, and DN-lig4-F and DN-lig4-R were used to amplify the cDNA for the C-terminal of Lig4 [36]. All the PCR products were cloned into the pCS2+vector, resulting in pCS2+lig3, pCS2+DN-lig3 and pCS2+DN-lig4. To validate the efficiency of *lig3* morpholino (MO) and *lig4* MO, cDNA encoding the upstream sequence of *lig3* or *lig4* was partially amplified using the primers part-lig3-F1 and part-lig3-R1 or part-lig4-F1 and part-lig4-R1, and the PCR product was ligated to the pCS2+EGFP construct, resulting in pCS2+lig3-EGFP and pCS2+lig4-EGFP. All the constructs were confirmed by DNA sequencing.

The constructs pCS2+lig4, pCS2+I-SceI, an NHEJ reporter pNHEJ, and an SSA assay construct, pSSA were kindly provided by Dr. Jun Chen (Zhejiang University, China) [30].

2.4 mRNA, gRNA, MO, and plasmid injection

TALEN-targeting plasmids for *nanog* and *flt4*, as well as pCS2+DN-lig4, pCS2+DN-lig3, and pCS2+I-SceI, were linearized with *Not*I (Takara, Japan) and transcribed *in vitro* using the mMESSAGE mMACHINE SP6 kit (Ambion) following the manufacturer's protocol. The *cas9* mRNA and gRNA were synthesized as described previously [34]. *lig4* MO

5′-TTGCAGAAGACACGGAAGAAACTGT-3′ and *lig3* MO 5′-CCCTGCCAAGCCAAAGA-CTGGGCAT-3′ were synthesized from Gene Tools (Oregon, USA). *In vitro*-synthesized mRNAs, gRNA, MOs, and certain DNA constructs were microinjected into one-cell-stage embryos at doses between 100 and 500 pg per embryo for mRNA injection, 50 pg for gRNA injection, 1500 pg per embryo for MO injection, and between 10 and 50 pg for DNA injection, as previously described [37]. The injected zebrafish embryos were raised at 28.5℃ in 0.3 × Danieau's solution [38]. The fluorescent embryos were observed through an Olympus MVX10 fluorescence microscope. The F1 embryo populations potentially targeted by *nanog* TALEN were screened by PCR and verified by Sanger sequencing.

2.5 Western blot analysis

For western blotting, total protein was extracted from the corresponding injected embryos at 10 hpf using sodium dodecyl sulfate (SDS) sample buffer. The anti-GFP antibody was purchased from ABclone (China), and the anti-β-actin antibody (Bioss, China) was used as a loading control.

2.6 Mutation detection by enzyme digestion and sequencing

Genomic DNA was extracted and pooled from 30 embryos injected with different TALEN mRNA or *flt4* gRNA/*cas9* mRNA, and 3 pools were analyzed. A 586-bp DNA fragment containing the *nanog* target site was amplified using the primers nanog-TALEN-F and nanog-TALEN-R. A 448-bp fragment containing the *flt4* target site was amplified using the primers flt4-TALEN-F and flt4-TALEN-R. The PCR products were digested with the restriction enzymes *Nde*I or *Sac*I, and the uncleaved bands were gel-purified, subcloned, and sequenced. The genomic DNA of each of the 30 embryos injected with MMEJ assay constructs was amplified using the SP6 and T3 primers. The PCR products were digested with *Sma*I or *Eco*RV, and the undigested bands were gel-purified, subcloned, and sequenced.

2.7 Detection of fluorescent cells by flow cytometry

A total of 20 embryos injected with each reagent at 10 hpf were collected. The embryonic cells were dissociated through digestion with trypsin solution (0.5% trypsin, 0.22 mM EDTA) at room temperature for 20 min, and the embryonic cells were collected via centrifugation at 300×*g* for 4 min. The collected cells were resuspended and detected *via* flow cytometry (FACSAria III, BD, USA).The flow cytometry assays were repeated 3 times for each sample, and $P < 0.05$ was considered statistically significant.

2.8 Real-time PCR analysis

The relative repair frequency of SSA was assayed using real-time PCR as previously described [30]. Genomic DNA was extracted from a total of 30 embryos at 24 hpf, which were co-injected with uncut SSA/I-SceI mRNA and DN-lig4 mRNA. Quantitative real-time PCR was carried out using a Bio-Rad CFX96/C1000 Real-time PCR machine. The amount of injected DNA was normalized using normalizing primers, Norm-F and Norm-R, and the SSA repair frequency was quantified using the repair primers HR, SSA-Rep-F, and Rep-R. The relative frequencies of specific PCR products were calculated using the $2^{-\Delta\Delta C(T)}$ method using the normalized products the internal control [39]. The real-time quantitative PCR assays were repeated 3 times for each sample, and $P < 0.05$ was considered statistically significant.

3. Results

3.1 Detection of MMEJ in zebrafish embryos *in vivo*

We have focused on the study of fish nuclear transfer and nuclear reprogramming [40-42], and we are interested in studying totipotency-related factors in zebrafish. Nanog is one of the major totipotency-related factors, playing an important role in the maintenance of cell pluripotency in mammals [43,44]. However, the developmental role of *nanog* during embryogenesis is largely unknown. Therefore, the TALEN method was used to generate zebrafish *nanog* knockout mutants. We designed a target site located at exon1 of the *nanog* gene, immediately downstream of the start codon. The target sequence harbors an *Nde*I site, and the spacer sequence contains a direct repeat of 5′-ATGCAT-3′ (Fig. 1A). The mutations were detected through PCR and *Nde*I digestion, and approximately 30% of the PCR products could not be digested (Fig. 1B). The uncleaved bands were recovered for sequencing. Although 28 clones were analyzed and 9 types of mutations were found, surprisingly, 20 clones (71.4%) were found to be 5′-ATGCAT-3′ deletion mutants (Fig. 1C). This type of DNA repair is identical to MMEJ repair, which uses microhomologous sequences during the alignment of broken ends before joining, resulting in deletions flanking the original break [5,6]. To identify mutations transmitted throughout the germline, we raised the injected founder embryos to adulthood and outcrossed them with wild-type zebrafish. Seventy *nanog* TALEN founders were screened, 21 of which transmitted the target mutations to their F1 offspring, demonstrating successful and efficient germline transmission. Among the 21 germline transmitters, 15 were 5′-ATGCAT-3′ deletion mutants (Table 2), suggesting that MMEJ induced by TALEN can be inherited through germline transmission in zebrafish.

Table 2　Percentage of different mutation types in TALEN-induced nanog F1 mutant fish

NO.	Mutation type	Fish number(s)	Percentage(%)
1	"ATGCAT" deletion	15	71.4
2	"GCA" mutated to "ACGGGCTCATGTACCCTCATG"	1	4.8
3	"A" insertion	1	4.8
4	"AT" deletion	2	9.5
5	"CA" deletion	1	4.8
6	"GCATAT" mutated to "CTTATCTACCATCTTA"	1	4.8

To confirm that the MMEJ-mediated repair of endogenous genes is not unique to the *nanog* gene, TALEN targeting of another gene, *flt4*, was performed, with the repetitive sequence 5′-CTCAGAG-3′ in the spacer region of the TALEN target sites (Fig. 1A). The resulting mutations were similarly detected through PCR and *Sac*I digestion, and approximately 40% of the PCR products could not be cleaved (Fig. 1B). The uncleaved bands were recovered for sequencing. Of the 32 sequenced clones, 81.2% (26) were found to be 5′-CTCAGAG-3′ deletion mutants (Fig. 1D), thus resembling the repair by MMEJ. This result indicated that MMEJ was similarly used to repair the TALEN-induced DSBs in the *flt*4 gene at a relatively high frequency. These results suggest that MMEJ may be efficiently used to repair DSBs during embryonic development in zebrafish.

3.2　MMEJ repair of exogenous reporters in zebrafish embryos

To verify the prevalence and study the mechanism of MMEJ repair in zebrafish, we constructed two MMEJ assay plasmids: pSmaI-EGFP and pEcoRV-EGFP. pSmaI-EGFP was constructed by inserting an 11-bp microhomology sequence, 5′-GGGCGACACCC-3′, downstream of nucleotide 347 of wild-type EGFP, with an *Sma*I enzyme site formed between the repeat sequences (Fig. 2A). As expected, embryos injected with circularized pSmaI-EGFP did not show any green fluorescence on a fluorescence microscope because the EGFP coding sequence was frame-shifted by the inserted microhomologous fragment (Fig. 2B). However, zebrafish embryos injected with the linearized plasmid pSmaI-EGFP/*Sma*I showed strong EGFP expression throughout the entire embryo (Fig. 2B), suggesting that the coding sequence of EGFP was repaired to the correct frame by the machinery of the host embryos. To verify the sequence modification after linearized construct injection, the EGFP fragment was PCR-amplified and digested by *Sma*I. In the control group, the pSmaI-EGFP-injected products could be fully digested by *Sma*I, whereas the pSmaI-EGFP/SmaI-injection group showed a strong band corresponding to undigested products (Fig. 2C). The undigested products were further subcloned and sequenced, and the sequencing results demonstrated that 30.8% (4/13) of the undigested clones resulted from a deletion of the repeat sequence, 5′-GGGCGACACCC-3′, which resembles typical MMEJ repair. This led to the restoration of the wild-type EGFP coding sequence in pSmaI-EGFP (Fig. 2D). These results suggest that zebrafish embryos harbor MMEJ machinery to repair the injected SmaI-digested pSmaI-EGFP.

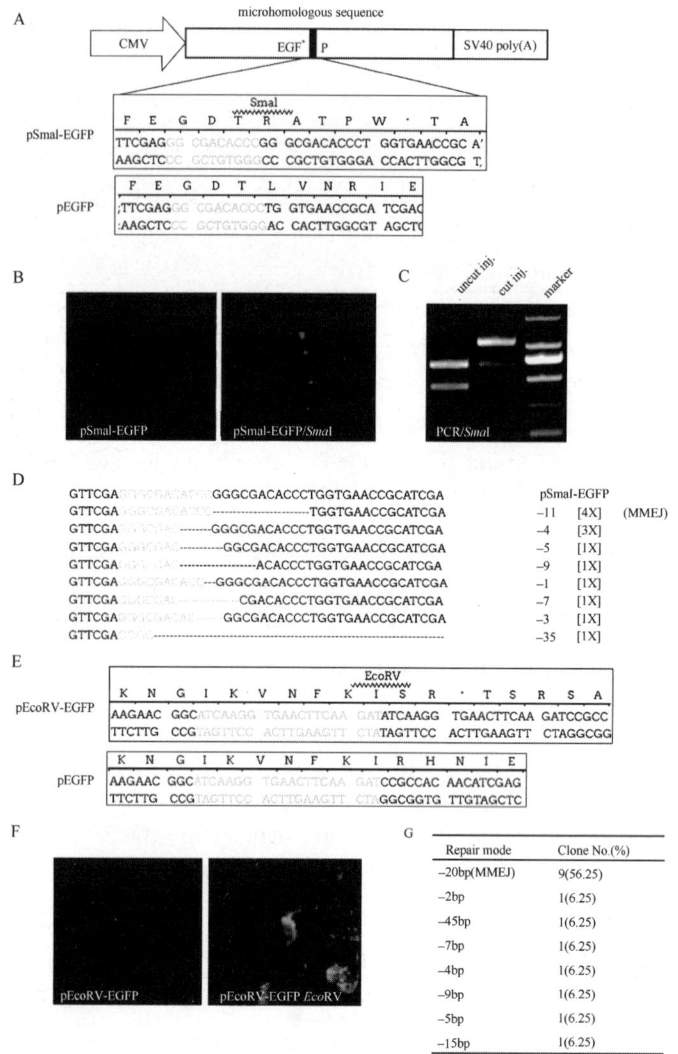

Fig. 2 Detection of MMEJ repair by injecting exogenous MMEJ assay reporter constructs. (A) and (E) present schematic diagrams of the MMEJ assay reporter constructs pSmaI-EGFP (A) and pEcoRV-EGFP (E). The EGFP reading frame is destroyed by the insertion of a microhomologous sequence, and its detailed sequence is shown below. The red letters indicate the inserted repeat nucleotides, and the light green letters indicate the original sequence. (B) and (F) show that no fluorescence was detected in embryos injected with the circularized plasmids, as the EGFP was frame-shifted by the insertion of the repeat sequences. By contrast, fluorescence was detected in embryos injected with the linearized plasmids, indicating that MMEJ repair was used to restore the EGFP coding sequences. (C) The result of *Sma*I digestion of PCR products. The products from uncut plasmid-injected embryos were completely digested, and most of the PCR products resulting from the linearized plasmid-injected embryos were uncleaved. (D) Sequencing results for the pSmaI-EGFP/*Sma*I-injected PCR products. Uncleaved bands were recovered, and 30.8% (4/13) of the linearized pSmaI-EGFP plasmids were shown to be repaired by MMEJ. (G) Statistics of the repair modes detected in the pEcoRV-EGFP/*Eco*RV-injected embryos. As shown in the table, 56.2% (9/16) of the DSBs were repaired through the deletion of the 20-bp repeat sequence, which corresponds to MMEJ repair

To further clarify the existence of MMEJ repair, we constructed another MMEJ assay construct, pEcoRV-EGFP, by inserting the microhomology sequence 5′-ATCAAGGT-GAACTTCAAGAT-3′ downstream of the original sequence, which resulted in a novel *Eco*RV restriction site (Fig. 2E). As with the pSmaI-EGFP experiment, injection of the circularized pEcoRV-EGFP did not cause expression of any EGFP protein, whereas injection of the digested pEcoRVEGFP led to strong EGFP expression (Fig. 2F). The results from the digestion of PCR products and EGFP fragment sequencing further confirmed the occurrence of MMEJ repair following the injection of EcoRV-digested pEcoRV-EGFP (Fig. 2G). These results revealed that MMEJ is frequently used to repair exogenous DNA sequences in zebrafish embryos.

3.3 Zebrafish MMEJ is dependent on ligase 3 but independent of ligase 4

To quantitatively assay the repair efficiency, we injected 50 pg of each of the following constructs, pCMV:EGFP, a previously described NHEJ reporter [30], pSmaI-EGFP/*Sma*I, and pEcoRV-EGFP/*Eco*RV, into one-cell-stage embryos (Fig. 3A), and EGFP-expressing cells were detected and analyzed by flow cytometry. As shown in Fig. 3B, although 27.6% of the cells expressed EGFP in the pCMV:EGFP-injected embryos, only 5.8%, 4.1%, and 2.3% of the cells were EGFP-positive in the NHEJ reporter, pEcoRVEGFP/*Eco*RV, and pSmaI-EGFP/*Sma*I groups, respectively (Fig. 3B). Thus, we have established a method to quantify the relative repair efficiencies of the corresponding MMEJ and NHEJ reporters.

Ligase 3 and ligase 4 are two major ligases that mediate DNA-end ligation in DSB repair [45]. In yeast and human cells, LIG4 mediates double-strand ligation during NHEJ [46,47], and Lig3 is considered a necessary factor for homologous end ligation [7,21,22]. We next determined which ligases are required for MMEJ activity in zebrafish embryos. To block the activities of Lig3 and Lig4 in zebrafish, translational MOs against *lig4* [30] and *lig3* were designed and synthesized. The *lig3* MO targeted the "ATG" of the *lig3* coding sequence, and the *lig4* MO targeted the 5′-UTR of zebrafish lig4. As expected, the *lig*3 MO efficiently blocked the expression of a lig3-EGFP fusion protein (Fig. S1A and B), and the *lig4* MO efficiently blocked the expression of a lig4-EGFP fusion protein (Fig. S1C and D). Moreover, the *lig3* and *lig4* MOs did not interfere with each other's expression, highlighting the effectiveness and specificity of these morpholinos. To investigate the roles of Lig3 and Lig4 in MMEJ, we co-injected the *lig4* or *lig3* MO with exogenous MMEJ reporters into one-cell-stage embryos. Although Lig4 is maternally expressed [30], *lig4* MO should be able to block the translation of both maternal and zygotic transcripts, thereby leading to decreased Lig4 protein levels. As a result, fluorescent intensity was clearly decreased in *lig3* MO-coinjected embryos and significantly increased in *lig4* MO-coinjected embryos (Fig. 3C and D). Fluorescent cells in each group were detected by flow cytometry and statistically analyzed. Interestingly, the relative ratios of the EGFP-positive cells in the pSmaI-EGFP/SmaI and pEcoRV-EGFP/EcoRV groups were reduced to 40% and 28% upon *lig3* knockdown and increased by 2.4 and 2.8-fold upon *lig4* knockdown, respectively (Fig. 3E and F). In addition, to test the effect of *lig3* and *lig4* overexpression on the efficiency of MMEJ, we coinjected the *lig3* or *lig4* mRNA with exogenous MMEJ reporters and detected the relative ratio of fluorescent cells by flow cytometry. Statistical results show that the relative ratios of the EGFP-positive cells in the pSmaI-EGFP/*Sma*I and pEcoRV-EGFP/*Eco*RV groups were increased to 1.6 and 3.2-fold through overexpression of *lig3* and reduced to 74% and 80% through overexpression of *lig4*, respectively (Fig. 3E and F). This result suggests that Lig3 is required for and promotes MMEJ repair of exogenous DSBs in zebrafish embryos, whereas Lig4 is unnecessary for and suppresses the MMEJ activity.

To test whether Lig3 and Lig4 are required for MMEJ repair of the endogenous genome, *lig3* and *lig4* MOs were co-injected with *nanog* TALEN mRNA into one-cell-stage embryos. The ratio of MMEJ repair (5′-ATGCAT-3′ deletion) among all of the repair products was calculated and compared for each MO. As expected, attenuating Lig3 activity significantly decreased the frequency of MMEJ repair, and inhibiting Lig4 activity led to a significantly higher MMEJ frequency in *nanog* TALEN samples (Fig. 3G). Similarly, the frequency of MMEJ repair was significantly decreased by co-injection of *lig3* MO but significantly increased by co-injection of *lig4* MO in the *flt*4 TALEN experiment (Fig. 3H). These data further support the requirement for Lig3 and the dispensability of Lig4 in MMEJ repair of endogenous DSBs in zebrafish. We also noticed that knockdown of *lig*4 led to early developmental defects, whereas knockdown of *lig*3 did not cause visible developmental defects (Fig. 3C and D).

The relative roles of Lig3 and Lig4 in MMEJ and NHEJ in zebrafish were further verified using dominant-negative forms of Lig3 or Lig4, as well as with *lig*4 mutants. We designed and utilized two dominant-negative forms of zebrafish Lig4 (DN-Lig4) and Lig3 (DNLig3) (Fig. S1E). The XRCC4 and BRCT binding domains of Lig4 were cloned, and the overexpression of this mRNA induces a dominant negative effect, as reported in cultured human cells [48]. Similar to DN-*lig4*, the N-terminal PARP binding domain of Lig3 was cloned into pCS2+, resulting in DN-*lig3*. The mRNA of DN-*lig4* or DN-*lig3* was co-injected with the pEcoRV-EGFP/*Eco*RV construct. As shown in supplementary Fig. S1F, co-injection of DN-*lig4* mRNA increased the frequency of fluorescent embryos, and co-injection of DN-*lig3* mRNA dramatically reduced the fluorescence intensity. Flow cytometry analysis of the injected embryos further supported these results (Fig. S1G). The mRNA of DN-*lig4* or DN-*lig3* was also co-injected with *nanog* TALEN and *flt*4 TALEN mRNA, respectively. As shown in Fig. S1H and I, the relative frequency of MMEJ repair was increased by inhibiting Lig4 activity and reduced by attenuating Lig3 activity, which is consistent with the result of exogenous MMEJ reporter. Moreover, wild-type or lig4 mutant

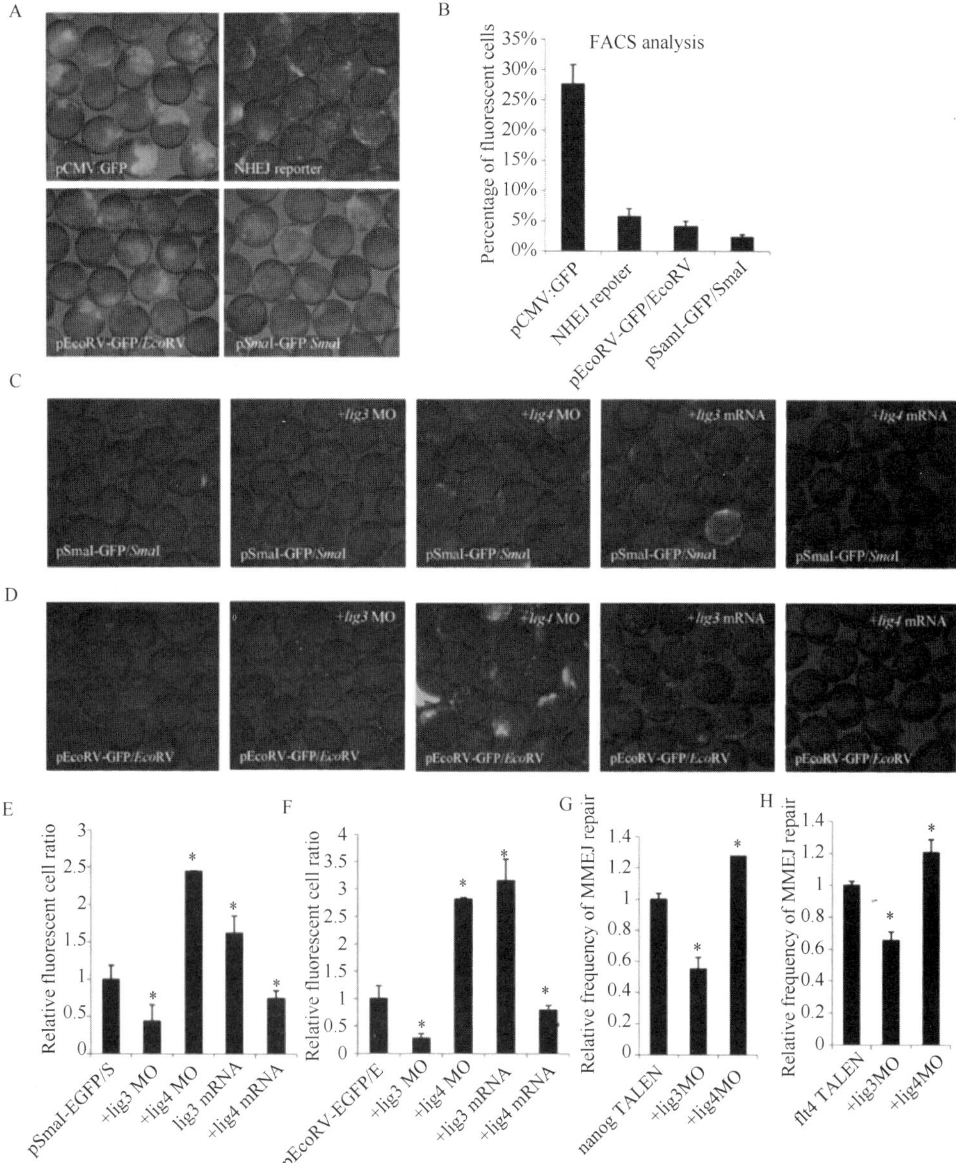

Fig. 3 Lig3 but not Lig4 is required for MMEJ repair. (A) EGFP expression in embryos injected with pCMV:EGFP, NHEJ reporter, pSmaI-EGFP/*Sma*I, or pEcoRV-EGFP/*Eco*RV. (B) The ratio of fluorescent cells from embryos shown in (A) detected by flow cytometry. (C) and (D) Fluorescent observation of MMEJ repair from the pSmaI-EGFP/*Sma*I (C) or pEcoRV-EGFP/*Eco*RV (D) injected embryos, as well as the embryos coinjected with *lig3* MO, *lig4* MO, *lig3* mRNA, or *lig4* mRNA. (E) and (F) The statistical results of the ratio of fluorescent cells in (C) and (D) assessed by flow cytometry. (G) *nanog* TALEN mRNA was co-injected with *lig3* MO or *lig4* MO, and the relative frequency of MMEJ repair in each group was analyzed by Sanger sequencing. (H) *flt*4 TALEN mRNA was co-injected with *lig3* MO or *lig4* MO, and the relative frequency of MMEJ repair in each group was analyzed by Sanger sequencing. The relative frequency of MMEJ repair in the control group was normalized to 1 in (E)-(H). The bars in (E)-(H) represent mean values of 3 replicates, and the asterisks above the bars indicate $P < 0.05$ compared to control group

embryos were injected with the linearized pSmaI-EGFP and pEcoRV-EGFP constructs, and the ratios of fluorescent cells were detected by flow cytometry. Not only was the fluorescent intensity stronger in the *lig4*-mutant embryos than in the wild-type embryos (Fig. S2A and B), but the ratios of fluorescent cells were also increased by 127% and 76% in the mutant embryos (Fig. S2C and D).

As SSA is also a DSB repair pathway that proceeds via the annealing of homologous sequences [6,19], we further tested whether the repair events described above occurred through SSA repair. We co-injected embryos with a mixture containing DN-*lig4* mRNA, an uncut SSA assay construct and I-*Sce*I mRNA [30], wherein *in vivo* expressed I-*Sce*I can cut the SSA plasmid and produce a DSB with microhomologous ends. As shown in Fig. S2E and F, there was a clear decrease in the intensity of green fluorescence and SSA repair frequency in the embryos co-injected with DN-*lig4* mRNA compared with those without DN-*lig4* mRNA injection. This result further supports the previous conclusion that Lig4 is required for SSA repair in zebrafish embryos [30], and this mechanism was distinctly different from that observed in the MMEJ assay (Fig. S1F and G). Therefore, zebrafish embryos utilize a Lig3-dependent but Lig4 independent mechanism to repair DSBs via MMEJ, which is distinct from Lig4-dependent SSA repair.

3.4 Factors affecting the efficiency of MMEJ

First, we intended to analyze the impact of spacer length on MMEJ activity, since most micro-homologous break ends are not directly adjacent to and there is often a spacer between two repeats in *in vivo* system. We inserted spacer sequences of different sizes, i.e., 6, 12, and 20 bp, between the two repeat sequences of pSmaI-EGFP as described previously, resulting in the 3 plasmids, pSmaI-EGFP-6 bp, pSmaI-EGFP-12 bp, and pSmaI-EGFP-20 bp, which contain a novel *Cla*I site in the spacer sequences. After injection of the *Cla*I linearized plasmids, green fluorescence appeared to decrease with increasing spacer length, i.e., in the pSmaI-EGFP-6 bp/*Cla*I-, pSmaI-EGFP-12 bp/*Cla*I-, and pSmaI-EGFP-20 bp/*Cla*I-injected embryos (Fig. 4A). This result suggests that MMEJ activity is strongly influenced by the spacing between the two microhomologous sequences. Genomic DNA was collected from the pSmaI-EGFP-6 bp/*Cla*I-injected embryos, and the mutation status at the cut site was determined as described above. Compared with a MMEJ repair frequency of 30.8% in the pSmaIEGFP/*Sma*I-injection experiment (Fig. 2D), only 21.1% (4/19) of the clones resulted from MMEJ repair in the pSmaI-EGFP-6 bp/*Cla*I injection group (Fig. 4B). These results demonstrate that the efficiency of MMEJ decreases along with increasing spacer length between the repeat sequences.

Second, we wondered whether the efficiency of MMEJ would be affected when the cutting site is biased to one side repeat, since in the aforementioned constructs, the cutting site

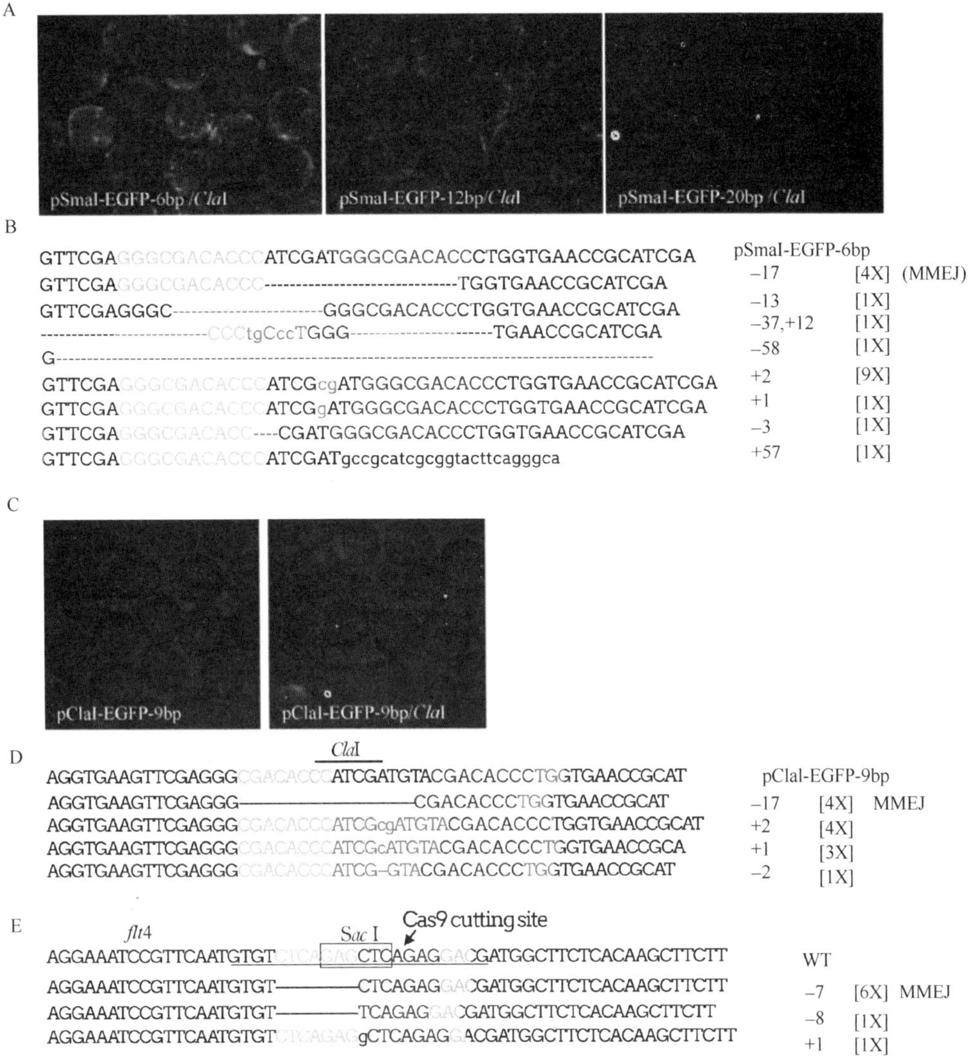

Fig. 4 The efficiency of MMEJ is affected by the spacer length between two repeats but not by the accuracy of the location of the cutting site. (A) Fluorescence observation of embryos injected with the MMEJ assay reporters with different spacer lengths, 6 bp, 12 bp or 20 bp. (B) Sequencing results of the PCR products from the pSmaI-EGFP-6 bp/*Sma*I-injected embryos. In total, 21.1% (4/19) of the linearized plasmids were repaired *via* MMEJ. The direct repeats of the two microhomologs are marked in blue or red. (C) Fluorescence observation of the uncut or linearized pClaI-EGFP-9bp-injected embryos. (D) Sequencing results of the PCR products from the pClaI-EGFP-9 bp/ClaI-injected embryos. Note that although the *Cla*I cutting site is closer to the first repeat sequence of 5′-CGACACCC-3′, a total of 33.3% (4/12) of the pClaIEGFP-9 bp/ClaI plasmids were still repaired through MMEJ. (E) Sequencing results of CRISPR/Cas9 targeted mutations on *flt*4 gene. The underlined sequence represents the gRNA target site, which contains direct repeat sequences, and the boxed sequence shows the *Sac*I enzyme site used for mutation testing. The presumptive Cas9 cutting site is biased to the second repeat sequence (arrow). 75.0% (6/8) of the *flt*4 CRISPR/Cas9-targeted clones were shown to be the MMEJ mutants (the 5′-CTCAGAG-3′ deletions). (For interpretation of the references to color in this figure legend, the reader is referred to the web version of this article.)

is just at the junction of two repeats or in the middle of the spacer. We constructed a reporter pClaI-EGFP-9 bp, which was generated through inverse PCR by amplifying pSmaI-EGFP and a ClaI-containing 9 bp spacer formed between the repeats, and the cutting site is biased to the left repeat. Fluorescence can be still observed in the linearized pClaI-EGFP-9 bp injected embryos (Fig. 4C), and sequencing results show that 33.3% (4/12) of junction sites were repaired in the way of MMEJ (Fig. 4D). While considering the endogenous MMEJ events, the aforementioned TALEN likely cuts at the junction of the direct repeats, we designed a CRISPR/Cas9 target site at *flt*4 gene and the cutting site of Cas9 was located to the sequence on the right repeat. The mutation sequences were detected using the method same to TALEN, and there are still 75% (6/8) clones resulted from MMEJ repair (Fig. 4E). In addition, our laboratory has attempted to knock out dozens of genes using CRISPR/Cas9 technology, with the target sites chosen randomly, and we found that 3-6-bp-mediated MMEJ was frequently used to repair the DSBs in 8 genes (Table S1). Therefore, the bias of cutting site does not affect the occurrence and efficiency of MMEJ.

4. Discussion

In contrast to well-characterized NHEJ, the genetic requirements of MMEJ have not been fully identified in eukaryotes. MMEJ is thought to be most evident when HR and/or NHEJ are not available for repair, e.g., due to mutations in mouse Ku86 or the deletion of the yeast *KU70* and *RAD52* genes [49-51]. In *Drosophila melanogaster*, *P*-element-excised DNA DSBs are repaired through a non-conservative pathway involving the annealing of microhomologies within the 17-nt overhangs in *spn-A* mutants [52]. The end-to-end fusion of short telomeres in an Arabidopsis *ku tert* mutant occurs by MMEJ [53]. In mammalian cells, however, MMEJ was observed in the presence of normal NHEJ function when using truncation mutants of the V(D)J recombination proteins Rag1 and Rag2 during recombination [54,55]. Several recent studies suggested that MMEJ may be utilized to repair TALEN or CRISPR/Cas9-induced DSBs in different animal models [56,57]. While this manuscript was under review, it was reported that precise in-frame knock-in could be achieved in zebrafish by using short homologous sequences (10-40 bp) [58], although it is unknown whether MMEJ, SSA or both were involved in that process.

In the present study, we utilized several designed endogenous and exogenous DSB reporters to clearly demonstrate the occurrence of MMEJ repair in zebrafish embryos even in the presence of NHEJ and HR machineries and regardless of whether the DSBs contained blunt ends (exogenous reporters) or cohesive ends (endogenous DSBs generated by TALEN or CRISPR/Cas9). MMEJ repair occurs in early zebrafish embryos at a relatively high frequency due to the rapid and robust cell division at early developmental stages. As with MMEJ repair in other eukaryotes, the zebrafish MMEJ repair pathway utilizes microhomologous sequences during the alignment of broken ends before joining, thereby resulting in one microhomologous sequence deletion at the original break. The spacing between the two homologs strongly affects the MMEJ repair efficiency, and MMEJ activity drops dramatically with increasing spacer length. Whereas, the bias of cutting site does not affect the efficiency of MMEJ.

Because we can use zebrafish to easily knock down or overexpress certain factors, as well as to conveniently produce DNA DSBs at the target genome sites using site-directed mutagenesis technologies, such as TALEN and CRISPR/Cas9 [32], zebrafish may serve as a model with which to study the mechanisms of DNA repair *in vivo*. Previous studies concerning which ligase is required for MMEJ repair produced quite controversial conclusions in different systems. For instance, MMEJ has been shown to be partially dependent on Lig4 in yeast [12,59]. However, *in vitro* studies in human cells indicate that DNA Lig3 and Lig1 are required for MMEJ [5,7,60]. MMEJ was also considered to be a type of SSA, called micro-SSA, because both repair DNA homologous ends result in a repeat sequence deletion [61,62]. In our study, we utilized different tools (morpholinos, dominant-negative proteins, mutants) and different approaches (endogenous DSBs and exogenous vectors) to study the relative roles of Lig4 and Lig3 in MMEJ repair in zebrafish embryos. Our study shows that *lig3* but not *lig4* is necessary for MMEJ repair in zebrafish embryos. Zebrafish *lig4* is maternally expressed and is used for repairing the DSBs in the way of NHEJ, which could competitively occur with other repair modes, such as MMEJ. Therefore, attenuation of Lig4 activity would decrease the frequency of NHEJ and enhance the MMEJ repair activity, and *vice versa*. This leads us to conclude that MMEJ utilizes a Lig3 dependent but Lig4-independent repair mechanism in zebrafish embryos, which is distinctly different from the mechanism used in either SSA or NHEJ repair. Although *lig3*-dependent MMEJ repair frequently occurs in zebrafish early development, knockdown of *lig4* but not *lig3* led to strong developmental defects. This suggests that *lig4*-dependent NHEJ is much more vital for the global DSBs repair events in the rapid development of early zebrafish embryos.

MMEJ has also been suggested to be highly mutagenic, likely contributing to genome instability in certain cancers, e.g., bladder cancer and leukemia [15,63,64]. According to a recent study using deep sequencing analysis, MMEJ is also believed to play important roles in the mechanism that generates novel chimeric RNAs [65]. Zebrafish have proven to be an important disease model for biomedical research over the last few decades. Due to the obvious advantages of zebrafish, such as large spawning, transparent embryos, ease of use, and *ex utero* development, zebrafish may serve as an ideal model with which to study human diseases caused by MMEJ. Additionally, given the high efficiency of MMEJ repair in CRISPR/Cas9-induced mutagenesis in our study, we suggest that these targeted nuclease technologies can be combined with the features of MMEJ repair to generate zebrafish mutants with desirable sequences.

Acknowledgements The authors thank Drs. Bo Zhang, Jing-Wei Xiong, Cecilia Moens and Jun Chen for providing various reagents. This work was supported by the National Science Fund for Excellent Young Scholars of NSFC (grant number 31222052), the China 863 High-Tech Program Grant 2011AA100404 and the FEBL grant 2011FBZ23. The funding agencies have no role in study design; in the collection, analysis, and interpretation of data; in the writing of the report; and in the decision to submit the paper for publication.

References

[1] K. Hiom, Coping with DNA double strand breaks, DNA Repair 9 (2010) 1256-1263.

[2] A.C. Magwood, M.J. Malysewich, I. Cealic, M.M. Mundia, J. Knapp, M.D. Baker, Endogenous levels of Rad51 and Brca2 are required for homologous recombination and regulated by homeostatic re-balancing, DNA Repair (Amst.) 12 (2013) 1122-1133.

[3] H. Yan, T. Toczylowski, J. McCane, C. Chen, S. Liao, Replication protein A promotes 5′→3′ end processing during homology-dependent DNA double-strand break repair, J. Cell. Biol. 192 (2011) 251-261.

[4] E. Weterings, D.J. Chen, The endless tale of non-homologous end-joining, CellRes. 18 (2008) 114-124.

[5] L. Deriano, D.B. Roth, Modernizing the nonhomologous end-joining repertoire: alternative and classical NHEJ share the stage, Annu. Rev. Genet. 47 (2013) 433-455.

[6] M. McVey, S.E. Lee, MMEJ repair of double-strand breaks (director's cut): deleted sequences and alternative endings, Trends Genet. 24 (2008) 529-538.

[7] L. Liang, L. Deng, S.C. Nguyen, X. Zhao, C.D. Maulion, C. Shao, J.A. Tischfield, Human DNA ligases I and III, but not ligase IV, are required for microhomology-mediated end joining of DNA double-strand breaks, Nucleic Acids Res. 36 (2008) 3297-3310.

[8] L.S. Symington, J. Gautier, Double-strand break end resection and repair pathway choice, Annu. Rev. Genet. 45 (2011) 247-271.

[9] D.B. Roth, J.H. Wilson, Nonhomologous recombination in mammalian cells: role for short sequence homologies in the joining reaction, Mol. Cell. Biol. 6 (1986) 4295-4304.

[10] Y. Zhang, M. Jasin, An essential role for CtIP in chromosomal translocation formation through an alternative end-joining pathway, Nat. Struct. Mol. Biol.18 (2011) 80.

[11] M. Lee-Theilen, A.J. Matthews, D. Kelly, S. Zheng, J. Chaudhuri, CtIP promotes microhomology-mediated alternative end joining during class-switch recombination, Nat. Struct. Mol. Biol. 18 (2011) 75-79.

[12] J.L. Ma, E.M. Kim, J.E. Haber, S.E. Lee, Yeast Mre11 and Rad1 proteins define a Ku-independent mechanism to repair double-strand breaks lacking overlapping end sequences, Mol. Cell. Biol. 23 (2003) 8820-8828.

[13] F. Fattah, E.H. Lee, N. Weisensel, Y. Wang, N. Lichter, E.A. Hendrickson, Ku regulates the non-homologous end joining pathway choice of DNA double-strand break repair in human somatic cells, PLoS Genet. 6 (2010) e1000855.

[14] E. Feldmann, V. Schmiemann, W. Goedecke, S. Reichenberger, P. Pfeiffer, DNA double-strand break repair in cell-free extracts from Ku80-deficient cells: implications for Ku serving as an alignment factor in non-homologous DNA end joining, Nucleic Acids Res. 28 (2000) 2585-2596.

[15] J. Bentley, C.P. Diggle, P. Harnden, M.A. Knowles, A.E. Kiltie, DNA double strand break repair in human bladder cancer is error prone and involves microhomology-associated end-joining, Nucleic Acids Res. 32 (2004) 5249-5259.

[16] B. Corneo, R.L. Wendland, L. Deriano, X. Cui, I.A. Klein, S.Y. Wong, S. Arnal, A.J. Holub, G.R. Weller, B.A. Pancake, S. Shah, V.L. Brandt, K. Meek, D.B. Roth, Rag mutations reveal robust alternative end joining, Nature 449 (2007) 483-486.

[17] Y. Ma, H. Lu, K. Schwarz, M.R. Lieber, Repair of double-strand DNA breaks by the human nonhomologous DNA end joining pathway: the iterative processing model, Cell Cycle 4 (2005) 1193-1200.

[18] P. Pfeiffer, W. Goedecke, S. Kuhfittig-Kulle, G. Obe, Pathways of DNA double-strand break repair and their impact on the prevention and formation of chromosomal aberrations, Cytogenet. Genome Res. 104 (2004) 7-13.

[19] N. Sugawara, G. Ira, J.E. Haber, DNA length dependence of the single-strand annealing pathway and the role of Saccharomyces cerevisiae RAD59 in double-strand break repair, Mol. Cell. Biol. 20 (2000) 5300-5309.

[20] H. Wang, Z.C. Zeng, T.A. Bui, E. Sonoda, M. Takata, S. Takeda, G. Iliakis, Efficient rejoining of radiation-induced DNA double-strand breaks in vertebrate cells deficient in genes of the RAD52 epistasis group, Oncogene 20 (2001) 2212-2224.

[21] H.C. Wang, B. Rosidi, R. Perrault, M.L. Wang, L.H. Zhang, F. Windhofer, G. Iliakis, DNA ligase III as a candidate component of backup pathways of nonhomologous end joining, Cancer Res. 65 (2005) 4020-4030.

[22] S. Oh, A. Harvey, J. Zimbric, Y. Wang, T. Nguyen, P.J. Jackson, E.A. Hendrickson, DNA ligase III and DNA ligase IV carry out genetically distinct forms of end joining in human somatic cells, DNA Repair 21 (2014) 97-110.

[23] L.N. Truong, Y. Li, L.Z. Shi, P.Y. Hwang, J. He, H. Wang, N. Razavian, M.W. Berns, X. Wu, Microhomology-mediated End Joining and Homologous Recombination share the initial end resection step to repair DNA double-strand breaks in mammalian cells, Proc. Natl. Acad. Sci. U. S. A. 110 (2013) 7720-7725.

[24] M. Audebert, B. Salles, P. Calsou, Involvement of poly(ADP-ribose) polymerase-1 and XRCC1/DNA ligase III in an alternative route for DNA double-strand breaks rejoining, J. Biol. Chem. 279 (2004) 55117-55126.

[25] E. Sonoda, M. Takata, Y.M. Yamashita, C. Morrison, S. Takeda, Homologous DNA recombination in vertebrate cells, Proc. Natl. Acad. Sci. U. S. A. 98 (2001) 8388-8394.

[26] H. Tauchi, J. Kobayashi, K. Morishima, D.C. van Gent, T. Shiraishi, N.S. Verkaik, D. vanHeems, E. Ito, A. Nakamura, E. Sonodo, M. Takata, S. Takeda, S. Matsuura, K. Komatsu, Nbs1 is essential for DNA repair by homologous recombination in higher vertebrate cells, Nature 420 (2002) 93-98.

[27] C.L. Bladen, W.K. Lam, W.S. Dynan, D.J. Kozlowski, DNA damage response and Ku80 function in the vertebrate embryo, Nucleic Acids Res. 33 (2005) 3002-3010.

[28] J.Z. Sandrini, G.S. Trindade, L.E. Nery, L.F. Marins, Time-course expression of DNA repair-related genes in hepatocytes of zebrafish (*Danio rerio*) after UV-B exposure, Photochem. Photobiol. 85 (2009) 220-226.

[29] D.S. Pei, P.R. Strauss, Zebrafish as a model system to study DNA damage and repair, Mutat. Res. Fund. Mol. 743 (2013) 151-159.

[30] J. Liu, L. Gong, C. Chang, C. Liu, J. Peng, J. Chen, Development of novel visual-plus quantitative analysis systems for studying DNA double-strand break repairs in zebrafish, J. Genet. Genomics 39 (2012) 489-502.

[31] D. Ye, Z. Zhu, Y. Sun, Fish genome manipulation and directional breeding, Sci. China Life Sci. 58 (2015) 170-177.

[32] T.O. Auer, F. Del Bene, CRISPR/Cas9 and TALEN-mediated knock-in approaches in zebrafish, Methods 69 (2014) 142-150.

[33] P. Huang, A. Xiao, M.G. Zhou, Z.Y. Zhu, S. Lin, B. Zhang, Heritable gene targeting in zebrafish using customized TALENs, Nat. Biotechnol. 29 (2011) 699-700.

[34] N. Chang, C. Sun, L. Gao, D. Zhu, X. Xu, X. Zhu, J.W. Xiong, J.J. Xi, Genome editing with RNA-guided Cas9 nuclease in zebrafish embryos, Cell Res. 23 (2013) 465-472.

[35] F. Xiong, Z.Q. Wei, Z.Y. Zhu, Y.H. Sun, Targeted expression in zebrafish primordial germ cells by Cre/loxP and Gal4/UAS systems, Mar Biotechnol. (N.Y.) 15 (2013) 526-539.

[36] H. Ochiai, N. Sakamoto, K. Fujita, M. Nishikawa, K. Suzuki, S. Matsuura, T.Miyamoto, T. Sakuma, T. Shibata, T. Yamamoto, Zinc-fingernuclease-mediated targeted insertion of reporter genes for quantitative imaging of gene expression in sea urchin embryos, Proc. Natl. Acad. Sci. U. S. A. 109 (2012) 10915-10920.

[37] C.-Y. Wei, H.-P. Wang, Z.-Y. Zhu, Y.-H. Sun, Transcriptional factors Smad1 and Smad9 act redundantly to mediate zebrafish ventral specification downstream of Smad5, J. Biol. Chem. (2014).

[38] J. Shih, S.E. Fraser, Characterizing the zebrafish organizer: microsurgical analysis at the early-shield stage, Development 122 (1996) 1313-1322.

[39] K.J. Livak, T.D. Schmittgen, Analysis of relative gene expression data using real-time quantitative PCR and the $2^{-\Delta\Delta C(T)}$ method, Methods 25 (2001) 402-408.

[40] Z.Y. Zhu, Y.H. Sun, Embryonic and genetic manipulation in fish, Cell Res. 10 (2000) 17-27.

[41] Y.H. Sun, S.P. Chen, Y.P. Wang, W. Hu, Z.Y. Zhu, Cytoplasmic impact on cross-genus cloned fish derived from transgenic common carp (*Cyprinus carpio*) nuclei and goldfish (*Carassius auratus*) enucleated eggs, Biol. Reprod. 72 (2005) 510-515.

[42] Y. Sun, S. Chen, Y. Wang, Z. Zhu, The onset of foreign gene transcription in nuclear-transferred embryos of fish, Sci. China Life Sci. 43 (2000) 597-605.

[43] J. Wang, D.N. Levasseur, S.H. Orkin, Requirement of Nanog dimerization for stem cell self-renewal and pluripotency, Proc. Natl. Acad. Sci. U. S. A. 105 (2008) 6326-6331.

[44] K. Mitsui, Y. Tokuzawa, H. Itoh, K. Segawa, M. Murakami, K. Takahashi, M. Maruyama, M. Maeda, S. Yamanaka, The homeoprotein Nanog is required for maintenance of pluripotency in mouse epiblast and ES cells, Cell 113 (2003) 631-642.

[45] V. Kukshal, I.K. Kim, G.L. Hura, A.E. Tomkinson, J.A. Tainer, T. Ellenberger, Human DNA ligase III bridges two DNA ends to promote specific intermolecular DNA end joining, Nucleic Acids Res. (2015).

[46] P. Ahnesorg, P. Smith, S.P. Jackson, XLF interacts with the XRCC4-DNA ligaseIV complex to promote DNA nonhomologous end-joining, Cell 124 (2006) 301-313.

[47] T.E. Wilson, U. Grawunder, M.R. Lieber, Yeast DNA ligase IV mediates non-homologous DNA end joining, Nature 388 (1997) 495-498.

[48] P.Y. Wu, P. Frit, S. Meesala, S. Dauvillier, M. Modesti, S.N. Andres, Y. Huang, J. Sekiguchi, P. Calsou, B. Salles, M.S. Junop, Structural and functional interaction between the human DNA repair proteins DNA ligase IV and XRCC4, Mol. Cell. Biol. 29 (2009) 3163-3172.

[49] X. Yu, A. Gabriel, Ku-dependent and Ku-independent end-joining pathways lead to chromosomal rearrangements during double-strand break repair in Saccharomyces cerevisiae, Genetics 163 (2003) 843-856.

[50] S.J. Boulton, S.P. Jackson, Components of the Ku-dependent non-homologous end-joining pathway are involved in telomeric length maintenance and telomeric silencing, EMBO. J. 17 (1998) 1819-1828.

[51] F. Liang, M. Jasin, Ku80-deficient cells exhibit excess degradation of extrachromosomal DNA, J. Biol. Chem. 271 (1996) 14405-14411.

[52] M. McVey, D. Radut, J.J. Sekelsky, End-joining repair of double-strand breaks in Drosophila melanogaster is largely DNA ligase IV independent, Genetics 168 (2004) 2067-2076.

[53] M. Heacock, E. Spangler, K. Riha, J. Puizina, D.E. Shippen, Molecular analysis of telomere fusions in Arabidopsis: multiple pathways for chromosomeend-joining, EMBO J. 23 (2004) 2304-2313.

[54] R. Rai, H. Zheng, H. He, Y. Luo, A. Multani, P.B. Carpenter, S. Chang, The function of classical and alternative non-homologous end-joining pathways in the fusion of dysfunctional telomeres, EMBO J. 29 (2010) 2598-2610.

[55] C.T. Yan, C. Boboila, E.K. Souza, S. Franco, T.R. Hickernell, M. Murphy, S.Gumaste, M. Geyer, A.A. Zarrin, J.P. Manis, K. Rajewsky, F.W. Alt, IgH class switching and translocations use a robust non-classical end-joining pathway, Nature 449 (2007) 478-482.

[56] S.J. Gratz, A.M. Cummings, J.N. Nguyen, D.C. Hamm, L.K. Donohue, M.M. Harrison, J. Wildonger, K.M. O'Connor-Giles, Genome engineering of Drosophila with the CRISPR RNA-guided Cas9 nuclease, Genetics 194 (2013) 1029-1035.

[57] A. Yasue, S.N. Mitsui, T. Watanabe, T. Sakuma, S. Oyadomari, T. Yamamoto, S.Noji, T. Mito, E. Tanaka, Highly efficient targeted mutagenesis in one-cell mouse embryos mediated by the TALEN and CRISPR/Cas systems, Sci. Rep. 4 (2014) 5705.

[58] Y. Hisano, T. Sakuma, S. Nakade, R. Ohga, S. Ota, H. Okamoto, T. Yamamoto, A.Kawahara, Precise in-frame integration of exogenous DNA mediated byCRISPR/Cas9 system in zebrafish, Sci. Rep. 5 (2015) 8841.

[59] K. Lee, S.E. Lee, Saccharomyces cerevisiae Sae2 and Tel1-dependent single-strand DNA formation at DNA break promotes microhomology-mediated end joining, Genetics 176 (2007) 2003-2014.

[60] D. Simsek, E. Brunet, S.Y. Wong, S. Katyal, Y. Gao, P.J. McKinnon, J. Lou, L.Zhang, J. Li, E.J. Rebar, P.D. Gregory, M.C. Holmes, M. Jasin, DNA ligase III promotes alternative nonhomologous end-joining during

chromosomal translocation formation, PLoS Genet. 7 (2011) e1002080.

[61] A. Decottignies, Microhomology-mediated end joining in fission yeast is repressed by pku70 and relies on genes involved in homologous recombination, Genetics 176 (2007) 1403-1415.

[62] L. Glover, J. Jun, D. Horn, Microhomology-mediated deletion and gene conversion in African trypanosomes, Nucleic Acids Res. 39 (2011) 1372-1380.

[63] E. Mattarucchi, V. Guerini, A. Rambaldi, L. Campiotti, A. Venco, F. Pasquali, F. Lo Curto, G. Porta, Microhomologies and interspersed repeat elements at genomic breakpoints in chronic myeloid leukemia, Genes Chromosomes Cancer 47 (2008) 625-632.

[64] Y. Zhang, J.D. Rowley, Chromatin structural elements and chromosomal translocations in leukemia, DNA Repair (Amst.) 5 (2006) 1282-1297.

[65] K. Ritz, B.D.C. van Schaik, M.E. Jakobs, E. Aronica, M.A. Tijssen, A.H.C. van Kampen, F. Baas, Looking ultra deep: Short identical sequences and transcriptional slippage, Genomics 98 (2011) 90-95.

斑马鱼胚胎中微同源末端介导断裂 DNA 双链的连接修复

何牡丹[1,2]　张峰华[1,2]　王华林[1,2]　王厚鹏[1]　朱作言[1]　孙永华[1]

1 中国科学院水生生物研究所，淡水生态与生物技术国家重点实验室，武汉 430072

2 中国科学院大学，北京 100049

摘　要　DNA 双链断裂修复机制在保持基因组的稳定性中扮演者重要角色。在 DNA 双链断裂的修复机制中，同源重组(HR)和非同源末端连接(NHEJ)被认为是两种主要的修复机制。近年来，另外一种 DNA 修复机制，微同源介导的末端连接(MMEJ)已经得到越来越多的关注。MMEJ 通常被认为是一种当 NHEJ 或其他修复机制失败时机体所采取的另一种修复 DNA 双链断裂的机制。本研究中，我们利用斑马鱼作为体内模型研究 DNA 双链断裂的修复机制，证实了高效的 MMEJ 修复机制存在于经 TALEN 或 CRISPR/Cas9 技术诱导的断裂 DNA 双链的修复。通过注射几个体外设计的外源 MMEJ 报告基因，进一步证明了 MMEJ 修复事件广泛存在于斑马鱼胚胎中。更有趣的是，抑制内源 DNA 连接酶 4 的活性可以显著增加 MMEJ 的活性；而干扰 DNA 连接酶 3 的活性则严重降低 MMEJ 的活性。这些结果表明，斑马鱼中 MMEJ 依赖连接酶 3 而非连接酶 4。本研究将增加我们对体内 MMEJ 修复机制的理解，并将有利于我们有效利用 DNA 双链断裂诱导的修复机制产生人为设计的突变。

第三部分
外源基因表达的生理功能

Hormonal Replacement Therapy in Fish: Human Growth Hormone Gene Function in Hypophysectomized Carp

Zongbin Cui[1] Zuoyan Zhu[1,2]

1 Institute of Hydrobiology, Chinese Academy of Sciences, Wuhan, 430072, China
2 Departments of Molecular and Cell Biology and Zoology, University of Aberdeen, Aberdeen AB9 2TN, Scotland

Abstract Transgenic common carp, *Cyprinus carpio,* produced by the microinjection of fertilized eggs with a linearized chimeric plasmid pMThGH, a human growth hormone (hGH) gene with a mouse metallothionein-1 (MT) gene promoter in pBR322, were used to produce F_1 and F_2 transgenics. Following hypophysectomy of the transgenic F_2 common carp, non-transgenic common carp and non-transgenic crucian carp, growth was monitored for up to 110 days. In addition, recombinant hGH was injected subcutaenously into a group of the non-transgenic crucian carp. Growth rate analyses indicated that (1) hypophysectomy of non-transgenic common carp and crucian carp results in the cessation of growth, (2) hGH administration can stimulate the growth of hypophysectomized crucian carp and (3) hypophysectomized hGH-transgenic common carp continue to grow in the absence of their own growth hormone, suggesting that the hGH-transgene is being expressed in tissues other than the pituitary.

Keywords common carp; crucian carp; growth hormone; growth hormone gene; transgenic fish; hormonal replacement; hypophysectomy

Introduction

Growth hormone (GH) is a single chain polypeptide of about 22 kDa, synthesized in, and secreted from, the anterior lobe of the pituitary gland of vertebrates. In mammals, GH is required for somatic growth and development; GH triggers target cells to synthesize somatomedin (insulin like growth factor, IGF) which stimulates growth of muscle, bone and connective tissue. A number of studies have shown that administration of mammalian GH enhances the growth rate of fish including killifish *Fundulus heteroclitus* (Pickford and Thompson 1948), Atlantic salmon *Salmo salar* (Komourdjian *et al.* 1976) and common carp (Adelman 1977). Recombinant GHs have similar effects in promoting growth in Pacific salmon, *Oncorhynchus kisutch* (Gill *et al.* 1985) and crucian carp *Carassius auratus gibelio* (Xu *et al.* 1991). Moreover, growth in hypophysectomized black bullhead, *Ictalurus melas*

文章发表于 *Fish Physiology & Biochemistry*, 1993, 12(2):161-169

was stimulated by the injection of bovine GH (bGH) at a dose of 1 μg/g body weight (BW)/2 days (Kayes 1977). It is evident that fish respond with growth enhancement to treatment of most, if not all, kinds of GH. For this reason, GH genes or their cDNAs have been attracting attention for use in studies on fish transgenesis, both as models and for the future application to aquaculture. The hGH gene with a mouse MT gene promoter (pMThGH) was used to produce the first batch of transgenic goldfish, *Carassius auratus auratus* (Zhu *et al.* 1985), and since then to produce transgenic loach *Misgurnus anguillicaudatus*, common carp (Zhu *et al.* 1986, 1989) and salmonids (Rokkones *et al.* 1989). More recently, rainbow trout *Oncorhynchus mykiss* GH cDNA sequence coding for the mature hormone mounted with the Rous sarcoma virus-long terminal repeat promoter (pRSV-rtGHcDNA) was used to make transgenic common carp (Zhang *et al.* 1990). Chinook salmon, *Oncorhynchus tshawytscha* GH cDNA linked to ocean pout antifreeze protein gene promoter (opAFP-GHc) was used to produce transgenic Atlantic salmon, *Salmo salar* (Du *et al.* 1992). Bovine GH cDNA with RSV promoter (pRSV/bGH) was employed to produce transgenic northern pike, *Esox lucius* (Gross *et al.* 1992). In these studies, heterogeneous GHs were detected and fast growth rates were noted in some of the founder fish (P_0) as well as in the transgenic F_1 descendant. For example, hGH was quantified in the pMThGH-transgenic crucian carp embryos (2.0-5.8 ng/embryo), the transgenic P_0 common carp (1.1-4.0 ng/ml serum) and F_1 common carp (2.1-6.4 ng/ml serum) with monoclonal hGH-antibody by radio-immunoprecipitation assay. Some of the P_0 and F_1 transgenic individuals showed 1.4-4 times growth enhancement over the control (Zhu *et al.* 1989; Wei *et al.* 1992; Zhu 1992). Nevertheless, GH-transgene expression in fish has been a disputed area. GH genes or cDNAs were also transferred into fish eggs in other laboratories (Chourrout *et al.* 1986; Maclean *et al.* 1987; Dunham *et al.* 1987; Brem *et al.* 1988; Inoue *et al.* 1990). Though the authors unambiguously demonstrated the transgenes' integration into the host's genome, no transgene expression or growth enhancement of the founder fish were subsequently reported. It is necessary for us to verify whether the hGH is playing a role in the transgenics' growth performance and how, or to what extent, this effect is exerted. To this end, we carried out physiological studies on the transgenic and control fishes, i.e. removing the endogenous source of GH by hypophysectomy and then checking the effect of hGH replacement. This study would in turn prove additional evidence for the pMThGH-transgene expression in the transgenic fish.

Materials and Methods

Experimental design

Two sets of experiment were carried out. In the first, hGH-transgenic common carp and control non-transgenic common carp were hypophysectomized. Expression of the hGH gene in the hypophysectomized transgenics, if any, and its function in replacing endogenous fish GH could thus be analyzed. In the second, the pituitary was removed or a sham-operation performed on crucian carp. Some of the hypophysectomized crucian carp were injected subcutaneously with recombinant hGH (provided by Professor L. H. Guo, Shanghai Cell Biology Institute, Chinese Academy of Sciences), and the effects of the recombinant hGH on growth rate was subsequently monitored.

Experimental fish

The pMThGH-transgenic common carp (F_2) and crucian carp used in the experiments were produced in our laboratory and raised at the fish farm of the Institute of Hydrobiology, Chinese Academy of Sciences, Wuhan, China (Wei *et al.* 1992). The fish were 5 to 6-months old at the start of the experiment. The plasmid pMThGH is a pBR322 derived, chimeric clone of a human GH "mini-gene" with its introns 2, 3 and 4 deleted, linked to a mouse MT-1 gene promoter and regulation region (Fig. 1A). The construct was linearized by *Bam*HI digestion and microinjected into the fertilized eggs of common carp. Subsequently transgenic common carp (P_0) were raised in 1985. The descendant F_1 (a non-transgenic male × transgenic female) and F_2 (a transgenic F_1 male x transgenic F_1 female) generations were raised in 1988 and 1990. DNA dot blotting and hybridization indicated that 22.7% of 180-day F_1 carp contained the hGH gene in the total DNA of red blood cells with 5-200 copies per cell. In the F_2, 66.6% of 20-day fry contained the hGH gene with 2-80 copies per cell. DNA-Southern blotting and hybridization established recovery of the 3488 bp MThGH/*Eco*Rl band from kidney and spleen DNA samples of an F_2 individual (Fig. 1B). The F_2 individuals used in this study were determined to be transgenic descendants by both tail-fin and red blood cell DNA hybridization against a probe of 0.8 kb hGH *Bgl*II/*Eco*RI fragment.

Hypophysectomy

Hypophysectomy was accomplished using a modification of Chavin's (1956) method (Fig. 2). Fish was placed on a home-made operating table, and the trunk and tail covered with a pad of wet gauze to keep the fish in position and moist. One side of the branchiostegal membrane was cut open and the operculum was lifted with a pair of medical forceps to enlarge the gill cavity. A longitudinal 3-5 mm cut was made posterior to the edge of the orbit in the mid-dorsal epithelium. The underlying tissues were parted until the parasphenoid bone and the preotic bone were laid bare. The front edge of the preotic bone and the adjacent part of the parasphenoid bone were pared away and removed. A constant flow of saline solution was introduced to keep the operating area free of blood. The inferior and posterior rectus muscles were carefully parted with a glass needle to expose the pituitary gland. A glass pipette with gentle suction was carefully directed towards this position and the pituitary was then sucked away. The sham-operation fish received the same surgical treatment except that the pituitary was not removed. The surgically treated fish received antibiotic treatment before being transferred into a fish tank containing 30 ppm of benzylpenicillium. At the end of experiment, all the hypophysectomized fish were killed and dissected to confirm that the hypophysectomy was complete.

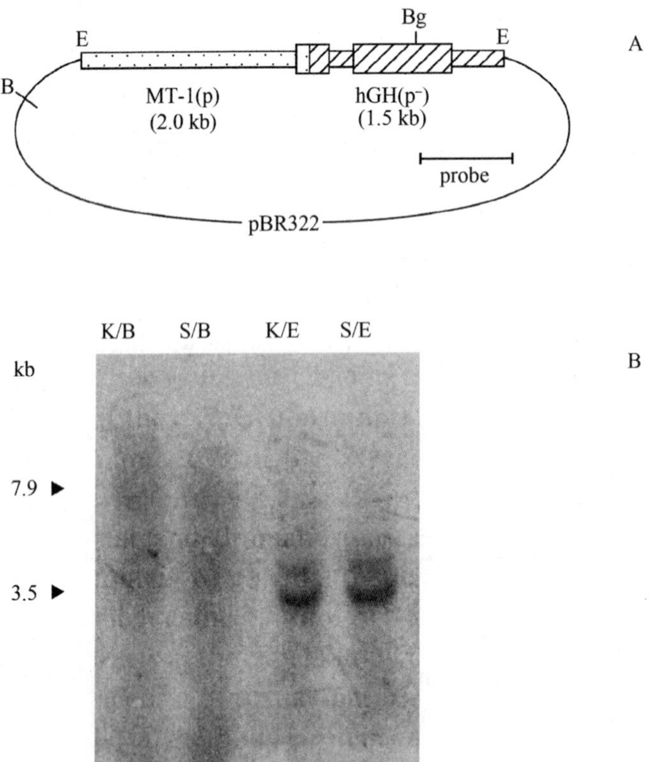

Fig. 1 The schematic diagram of pMThGH (modified from Zhu *et al.* 1985) (A) and its existence in a transgenic common carp F_2 (B). The 7.9 kb pMThGH/*Bam*HI fragment was used for producing the founders (P_0) of transgenic common carp. Total DNA from kidney and spleen of a transgenic F_2 common carp was digested with *Bam*HI or *Eco*RI. After 0.8% agarose gel electrophoresis, the samples were blotted onto nitrocellulose filter and hybridized against ^{32}P-labelled probe of hGH gene (a 0.8 kb *Bgl* II/*Eco*RI fragment). The 7.9 kb fragment of transgene was not clearly recovered from the *Bam*HI digestion which suggested that most of the *Bam*HI sites at the ends of the transgene were modified when the transgenes were integrated into the host genome or inherited by the offspring. However, a 3.5 kb of *Eco*RI band, containing the MThGH fragment, was recovered from kidney and spleen DNA samples. A minor band of *Eco*RI of about 4.2 kb was also hybridized against the hGH probe. The reason for this anomaly is not clear. The data indicate that the modification of foreign DNA in the descendants of transgenic fish occurred mainly at both ends and the inside integrity of the transgene remained. B: *Bam*HI; Bg: *Bgl*II; E: *Eco*RI; K/B: kidney DNA (6 μg) digested with *Bam*HI; K/E: kidney DNA (6 μg) digested with *Eco*RI; S/B: spleen DNA (6 μg) digested with*Bam*HI; S/E: spleen DNA (6 μg) digested with *Eco*RI; kb: kilo base pairs

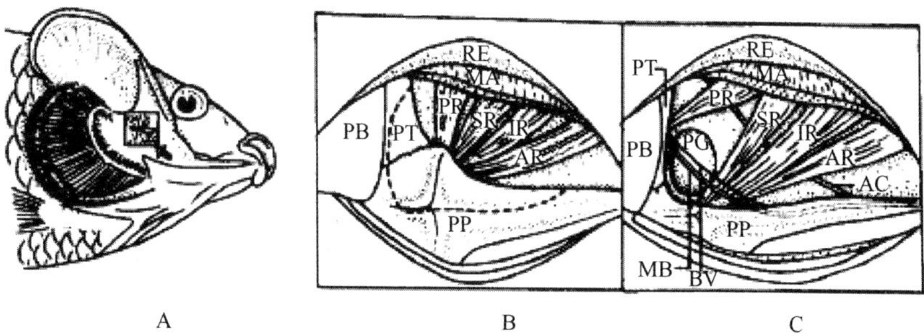

Fig. 2 Hypophysectomy of common carp and crucian carp. (A) One side of the operculum of the fish was lifted. The square on the epithelium indicates operation field which is enlarged in part (B) and (C). (B) The preotic and parasphenoid bones were laid bare. Part of each bone (dotted lines) was removed. (C) The inferior and posterior rectus muscles were parted and the pituitary gland aspirated. AC-artery carotid interna, AR-anterior rectus muscle, BV-branch of vein rostralis, IR-inferior rectus muscle, MA-muscle adductor arcus palatini, MB-maxillary branch of the trigeminal nerve, PB-pseudobranch, PG-pituitary gland, PP-parasphenoid bone, PT-preotic bone, PR-posterior rectus muscle, RE-reflected epithelium, SR-superior rectus muscle

Statistical analysis

Both body weight and fork length of each fish were measured every 10-days; specific growth rate (SGR) and condition factor (CF) were employed to analyze growth performance. The SGR was determined for both body weight and fork length by the formula $[(\ln V_1 - \ln V_0)/(t_1 - t_0)] \times 1000$. Where $\ln V_0$ and $\ln V_1$ are the natural logarithms of body weight or fork length at the start (t_0) and the end (t_1) of the time interval. A factor of 1000 instead of 100 was used (Higgs *et al.* 1975). Statistical analysis to test for differences among different groups of crucian carp was made using one-way ANOVA followed by Duncan's multiple comparison test. Student's t-test was used to analyze growth between different groups of common carp. The CF was calculated by the formula $(\text{weight}/\text{length}^{2.81}) \times 100$. The exponent 2.81 was determined by regressing log weight on log length for all data points. The relationship was linear ($r^2 = 0.86$ in the crucian carp and $r^2 = 0.97$ in the transgenic F_2 common carp with a slope of 2.81). One-way ANOVA was used to compare CF between all groups at each sampling time and to compare CF within each group across all sampling times.

Results

Survival rates of hypophysectomized fish

Survival rates among hypophysectomized fish are summarized in Table 1. The operating wound was not a key factor affecting fish survival rate since 100% of the sham-operated fish survived for 45 days after the operation. Water temperature, however, appeared to play an important role in survival, with higher rearing temperatures resulting in higher mortalities of hypophysectomized fish. In addition, the survival rate was species specific. At the same

Table 1　Survival rates of hypophysectomized fish

Fish/operation	Rearing temp(℃)	Number operated on	Survivall rates (%)		
			2 days	25 days	45 days
Crucian carp/Sm	15	15	100	100	100
Crucian carp/Hx	15	15	100	93.3	93.3
Crucian carp/Sm	25	15	100	100	100
Crucian carp/Hx	25	15	100	67.7	53.5
Common carp/Sm	15	10	100	100	100
Common carp/Hx	15	10	100	70	60
Common carp/Sm	25	10	100	100	100
Common carp/Hx	25	20	100	40	15
Common carp,TF_2/Hx	25	10	100	50	40

Hx: Hypophysectomy; Sm: Sham-operation;TF_2: Transgenic F_2

rearing temperature (15℃), survival rate of hypophysectomized crucian carp (93.3% at 45days) was higher than that of the hypophysectomized common carp (60% at 45 days). A survival rate of up to 87 days at 25℃ was recorded for hypophysectomized hGH-transgenic F_2 common carp (40%) when compared to that of the hypophysectomized non-transgenic carp (15%).

Growth in F_2 transgenic common carp

Altogether, 10 hGH-transgenic F_2 common carp and 20 non-transgenic common carp were successfully hypophysectomized. At 87 days post-hypophysectomy, 4 of the transgenic F_2 carp and 3 of the non-transgenic controls had survived. Fish growth performance in body weight and fork length was measured every 10 days (Table 2). Growth of transgenic F_2 carp continued after hypophysectomy ($r^2 = 0.97$, $n = 4$) while growth in control fish ceased ($r^2 = 0.4$, $n = 3$). The trans genic F_2 carp increased in both body weight and fork length, whereas the fork length of the control fish had changed little, and the body weight had decreased. Body weight and fork length-SGRs (SGR_W and SGR_L) in transgenic F_2 carp were significantly greater than that in the control fish (p <0.01 for SGR_W and $p < 0.05$ for SGR_L).

The CF of the transgenic F_2 carp was retarded for a short time (20 days), possibly due to the effect of surgical trauma. During this period, the fish ate little and the body weight decreased, but the fork length did not change. After a recovery period, however, the CF increased; both body weight and fork length increased with time but the body weight increased prior to the fork length. About 50 days after the operation, the CF had recovered completely and no significant difference was found with the CFs measured before the operation ($p > 0.05$). In contrast, the CF of the control group had not recovered ($p < 0.01$) (Fig. 3).

The effect of hGH therapy on hypophysectomized crucian carp

Three groups of crucian carp (12 individuals/group) were sham-operated (group A, Sham) or hypophysectomized (group B and C). Thirty days after the operation, group C was subcuntaneously injected with recombinant hGH at 2 μg/g BW/10 days for 80 days (Hx + GH). At the termination of day 110, the number of survivors in groups A, B (Hx) and C were 10, 8 and 8, respectively.

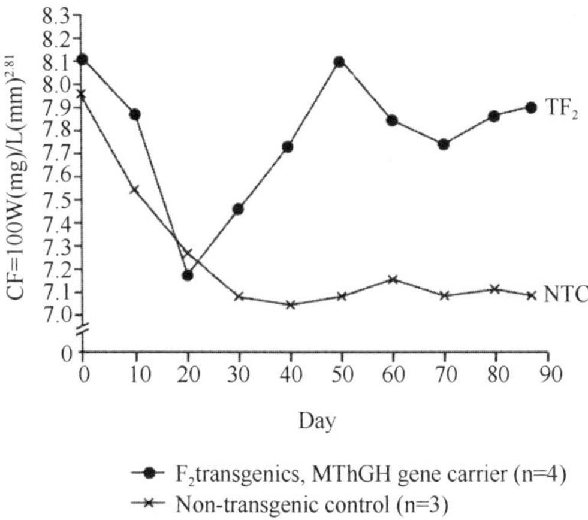

Fig. 3 The effect of hypophysectomy on the condition factor (CF) of the pMThGH-transgenic F_2 common carp (TF_2) and non-transgenic control (NTC). The mean values are plotted

The growth performance of crucian carp is summarized in Table 3. Growth in body weight and fork length was measured every 10 days, from day 30 to 110. In group A, fish growth continued (r^2 =0.91, n = 10) while in group B it ceased (r^2 = 0.54, n = 8). In group C, fish growth ceased during the first 30 days but subsequently recovered following subcutaneous injection of hGH (r^2 = 0.85, n = 8). The body weight in group B decreased, with little changing fork length. However, in comparison, the body weight and fork length of groups A and C continued to increase. The SCR_W and SGR_L in group B was significantly lower than that in group A ($p < 0.01$ for SGR_W and $p < 0.05$ for SGR_L) and in grou C ($p < 0.01$ for both SGR_W and SGR_L). There was no significant difference for SGR_w and SGR_L between groups A and C (p >0.05).

An effect of surgical trauma was also observed in crucian carp. During the first 30 days, the CF of each group declined sharply with the decrease only in body weight. Subsequently, the CF of each group changed in different ways. In group A, the CF stopped declining and stabilized at a constant level; both the body weight and the fork length increased simultaneously. In group B, the CF continued to decline for 70 days. In group C, the CF was increased following two injections of hGH, while the subsequent six injections brought the CF to the highest level among the three groups; the body weight increased prior to fork length (Fig. 4).

Discussion

The pituitary produces a number of hormones which regulate metabolism, growth and reproduction in fish, and hypophysectomy will affect the animal's viability. Moreover, any damage caused to the adjacent tissues, the branch of vein rostralis and the maxillary branch of trigeminal nerve in particular, results in high mortality. The survival rate of the hypophysectomized common carp at 87 days was 15% while that for the hypophysectomized crucian carp at 110 days was 66.7%. Other experiments demonstrated that common carp were much more susceptible than crucian carp to the absence of pituitary, resulting in higher mortality (data not shown). Parwez *et al.* (1984) proposed that teleost fishes could be grouped under two broad categories: those in which the presence of pituitary is essential for survival in freshwater because hypophysectomy impairs their osmoregulatory function; and those where ability to survive in freshwater is not impaired by hypophysectomy. The results from the present study suggest that common carp is a member of the former group while crucian carp may be closer to the latter.

Rearing temperature is another factor that affects survival of hypophysectomized fish. When the rearing temperature was raised from 15℃ to 25℃, the role of the pituitary in fish survival becomes more apparent. This implies that the faster metabolic rate of fish at high temperatures requires more precise regulation by pituitary hormones. On the other hand, even at 25℃ the hypophysectomized F_2 transgenic fish exhibited a much higher survival rate (40%) than that of the control fish (15%), suggesting that the hGH-transgene expressed in the F_2 transgenics and the hGH played a role in maintaining the survival of the hypophysectomized fish.

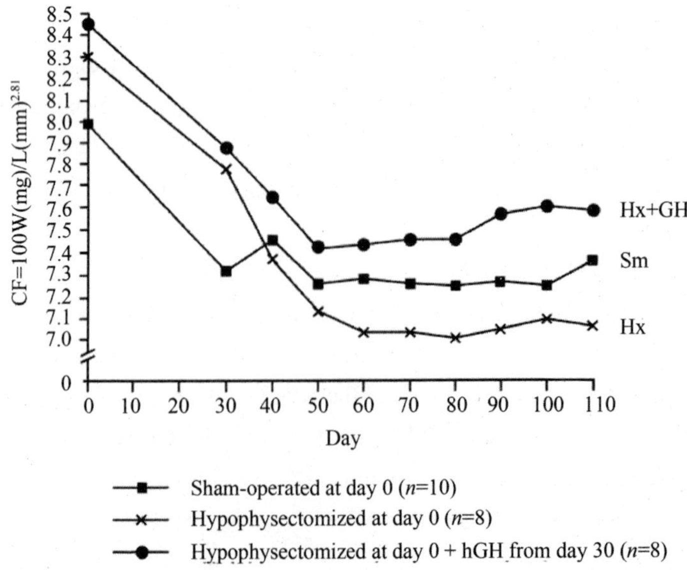

Fig. 4 The effect of hypophysectomy and hGH-injection on the condition factor (CF) of the crucian carp. The mean values are plotted. Sm: sham-operated; Hx: hypophysectomized at day 0; Hx + GH: hypophysectomized at day 0 and 8 injections of hGH of 2 μg/g BW/10 days since day 30

Table 2 Effect of hypophysectomy on growth of transgenic F_2 (TF_2) and non-transgenic common carp (NTC)

	Days 0	Days 10	Days 20	Days 30	Days 40	Days 50	Days 60	Days 70	Days 80	Days 87	SGR by Day 87
						Body weight (g)					
TF_2	88.0±2.5	84.3±2.1	78.5±2.8	82.3±3.1	86.2±3.2	90.6±3.4	91.2±3.5	92.4±3.7	95.3±3.1	97.0±3.1	1.12±0.12
NTC	83.5±1.7	80.2±1.4	79.0±1.7	78.5±1.6	77.4±1.3	77.8±1.6	78.1±1.5	78.0±1.6	77.8±1.4	78.3±1.6	−0.74±0.14**
						Fork length (mm)					
TF_2	138.5±2.9	138.2±3.0	138.4±3.1	138.8±1.6	139.0±2.9	139.5±3.2	140.3±3.0	141.5±3.2	142.7±3.3	143.5±4.1	0.40±0.08
NTC	138.2±2.4	137.8±2.0	138.0±2.3	138.4±2.4	138.2±2.5	138.5±2.3	138.1±2.6	138.2±2.6	138.6±2.4	138.5±2.9	0.02±0.04*

Data are shown as mean ± SEM(n=4 for TF_2 and 3 for NTC); SGR: Specific growth rate in body weight or fork length (‰ day); Day 0 = day of hypophysectomy; **compared with the TF_2 group, t-test: $p<0.01$; * compared with the TF_2 group, t-test: $p<0.05$

Table 3　Growth of hypophysectomized crucian carp (Hx), hypophysectomized carp injected with hGH (Hx+GH) and sham-operated carp (Sm)

	Days 0	Days 30	Days 40	Days 50	Days 60	Days 70	Days 80	Days 90	Days 100	Days 110	SGR by Day 110
Body weight (g)											
Sm											
(*n*=10)	68.6±1.6	66.8±1.5	67.1±1.6	68.4±1.6	69.0±1.4	69.5±1.6	70.1±1.7	70.8±1.8	71.2±1.9	71.4±2.0	0.72±0.10**
Hx											
(*n*=8)	72.4±0.9	71.0±1.0	71.8±0.9	70.4±0.9	69.5±0.9	69.1±0.7	68.7±0.8	68.3±0.9	68.6±1.0	68.5±1.1	−0.45±0.04
Hx+GH											
(*n*=8)	71.4±1.5	69.8±1.4	68.4±1.4	66.0±1.2	68.5±1.4	69.5±1.4	70.4±1.5	71.2±1.5	72.4±1.5	73.5±1.6	0.64±0.05**,†
Fork length (mm)											
Sm											
(*n*=10)	129.0±1.5	134.8±1.5	134.7±2.0	135.2±2.1	135.0±2.2	135.4±2.5	135.2±2.2	135.8±1.7	136.2±1.8	137.2±1.6	0.23±0.02*
Hx											
(*n*=8)	130.0±2.2	131.8±2.3	132.1±2.1	132.0±2.0	132.6±2.1	131.5±2.1	132.0±1.9	132.4±2.0	132.7±1.7	132.5±2.0	0.09±0.11
Hx+GH											
(*n*=8)	129.0±1.9	130.5±2.3	130.7±2.4	131.2±4.1	132.2±2.5	132.8±2.5	133.5±2.5	133.8±2.5	134.0±2.4	135.0±2.4	0.42±0.01**,†

Data are shown as mean ± SEM; hupophysectomy was carried out on day 0. The Hx + GH group was given 8 injections of hGH (2 μg/g BW) every 10 days after day 30; SGR: Specific growth rate; **comprated with the Hx group, Duncan-test: $p<0.01$; * compared with the Hx group, Duncan-test: $p<0.05$; † compared with the Sm group, Duncan-test: $p>0.05$

Hypophysectomy resulted in a cessation of growth in both common carp and crucian carp. This is in agreement with previous studies (Pickford 1957; Ball 1969). Subcutaneous injection of recombinant hGH into hypophysectomized crucian carp restored growth performance. Hypophysectomized transgenic F_2 common carp continued to grow in the absence of their own GH suggesting that the hGH-transgene is expressed in the F_2 transgenics. Although we have not yet examined the level of hGH in transgenic F_2 common carp, the amount of hGH in the founder (P_0) and the transgenic F_1 offsprings were measured as 1.1-6.4 ng/ml serum (see Introduction). It is reasonable to assume that restoration of growth in the hypophysectomized transgenic F_2 common carp was due to expression of the hGH transgene. To our knowledge, this is the first time that the GH-transgene function has been verified in transgenic animals. The hGH-stimulated growth restoration, however, was less apparent than that exhibited in sham-operated crucian carp. This might indicate that hGH is not as efficient as fish GH in stimulating fish growth. Hypophysectomy also removes a number of hormones. The hGH could at most replace the endogenous GH but could not compensate for the loss of other kinds of pituitary hormones in maintaining fish homeostasis.

It is believed that GH synthesis and secretion is under developmental, seasonal and even daily regulation. In GH-transgenic fish, the mechanisms of heterogeneous GH synthesis are more complicated and little understood, because the transgenes have a mosaic distribution in different organs. The copy numbers and integration sites in each type of host cell are not identical (Zhu *et al.* 1989). Transgene expression is regulated by the fusion promoter (Bearzotti *et al.* 1992), and is also restricted by many other unknown factors. These factors may include the flanking sequences of the host chromosome where the transgene integration occurred randomly and the nuclear proteins of the different cell types in which the transgene mosaic is distributed. It is also supposed that DNA methylation of the rat GH-transgene resulted in transcriptional inactivity and absence of phenotypic expression (Maclean*et al.* 1992). In addition, only a few monoclonal GH-antibodies have been available to increase the accuracy of GH measurements. The growth performance of the GH-transgenic fish is restricted by rearing conditions, as the Minnesota transgenic fish group evaluated quantitatively under carefully controlled rearing conditions (Gross *et al.* 1992). At present, it is hard to establish the relationship between the transgene's copy number, the detectable, if any, heterogeneous GH level and the growth rate in GH-transgenic fish.

It has been reported that the administration of GH to salmonid fish stimulates an increase in body length prior to an increase in body weight, resulting in a short-term decline in the CF (Komourdjian *et al.* 1976; Clarke *et al.* 1977; Gill *et al.* 1985). The early decline of the CF in the hypophysectomized transgenic F_2 common carp probably resulted from the stress of the operation. The hGH, expressed in the hypophysectomized transgenic F_2 common carp, preferentially stimulated growth in body weight compared to fork length. This result matches well with that reported by Adelman (1977) and Teskeredzic *et al.* (1991) where CF increased as a result of injection of bGH or porcine GH into common or leather carp. It has been suggested that the effect of GH treatment on the CF in fish is species specific. Cyprinids may express different responses to GH treatment than salmonids. The CF in hypophysectomized transgenic F_2 common carp recovered completely within 50 days of the operation (Fig. 3). This is also consistent with our observation in which the pMThGH-transgenic common carp doubled in body weight with a little increase of the body length (Zhu 1992). In hypophysectomized crucian carp (Fig. 4), administration of recombinant hGH stimulated increases in body weight and fork length simultaneously so that the CF was restored to a stable level following operation.

Acknowledgements We gratefully acknowledge K. Xu and Y. Wei for provision of the transgenic F_2 common carp, Professor L. H. Guo for the hGH and Dr. D. Hamer (NIH, U.S.A.) for recombinant plasmid of pBPVMG-6 from which the pMThGH recloned. This project was financially supported by the Chinese Research Fund of National Development Programme (7th five-year plan) and the Chinese High-Tech Development Programme (Biotechnology) to Z. Z.

References

Adelman, I.A. 1977. Effect of bovine growth hormone on growth of carp (*Cyprinus carpio*) and the influences of temperature and photoperiod. J. Fish. Res. Bd. Can. 34: 509-515.

Ball, J.N. 1969. Prolactin (fish prolactin or paralactin) and growth hormone. *In* Fish Physiology. Vol. 2, pp. 207-240. Edited by W.S. Hoar and D.J. Randall. Academic Press, New York.

Bearzotti, M., Perrot, E., Michard-Vanhee, C., Jolivet, G., Attal, J., Theron, M-C., Puissant, C., Dreano, M., Kopchick, J.J., Powell, R., Gannon, F., Houdebine, L-M. and Chourrout, D. 1992. Gene expression following transfection of fish cells. J. Biotechnol. 26: 315-325.

Brem, G., Brenig, B., Horstgen-Schwark, G. and Winnacker, E.L. 1988. Gene transfer in tilapia (*Oreochromis niloticus*).Aquaculture 68: 209-219.

Chavin, W. 1956. Pituitary-adrenal control of melanization in xanthic goldfish, *Carassius auratus* L. J. Exp. Zool. 133: 1-45.

Chourrout, D., Guyomard, R. and Houdebine L-M. 1986. High efficiency gene transfer in rainbow trout (*Salmo gairdneri* Rich.) by microinjection into egg cytoplasm. Aquaculture 51: 143-150.

Clark, W.C., Farmer, S.W. and Hartwell, K.M. 1977. Effect of teleost pituitary growth hormone on growth of *Tilapia mossambica* and on growth and seawater adaptation of sockeye salmon (*Oncorhynchus nerka*) Gen. Comp. Endocrinol. 33: 174-178.

Du, S.J., Gong, Z., Fletcher, G.L., Shears, M.A., King, M.J., Idler, D.R. and Hew, C.L. 1992. Growth enhancement in transgenic Atlantic salmon by the use of an "all fish" chimeric growth hormone gene construct. Bio/Technology 10: 176-180.

Dunham, R.A., Eash, J., Askins, J. and Townes T.M. 1987. Transfer of the metallothionein-human growth hormone fusion gene into channel catfish. Trans. Am. Fish. Soc. 116:87-91.

Gill, J.A., Sumpter, J.P., Donaldson, E.M., Dye, H.M., Souza, L., Berg, T., Wypych, J. and Langley, K. 1985. Recombinant chicken and bovine growth hormone accelerate growth in aquacultured juvenile Pacific salmon (*Oncorhynchus kisutch*). Bio/Technology 3: 643-646.

Gross, M.L., Schneider, J.F., Moav, N., Moav, B., Alvarez,C., Myster, S.H., Liu, Z., Hallerman, E.M., Hackett, P.B., Guise, K.S., Faras, A.J. and Kapuscinski, A.R. 1992. Molecular analysis and growth evaluation of northern pike (*Esox lucius*) microinjected with growth hormone genes. Aquaculture 103: 353-273.

Higgs, D.A., Donaldson, E.M., Dye, H.M. and McBride, J.R. 1975. A preliminary investigation of the effect of bovine growth hormone on growth and muscle composition of coho salmon (*Oncorhynchys kisutch*). Gen. Comp. Endocrinol. 27: 240-253.

Inoue, K., Yamashita, S., Hata, J.-i., Kabeno, S., Asada, S., Nagahisa, E. and Fujita, T. 1990. Electroporation as a new technique for producing transgenic fish. Cell Diff. Dev. 29: 123-128.

Kayes, T. 1977. Effect of hypophysectomy, beef growth hormone replacement therapy, pituitary autotransplantation, and environmental salinity on growth in the black bullhead (*Ictalurus melas*). Gen. Comp. Endocrinol. 33: 371-381.

Komourdjian, M.P., Saunders, R.L. and Fenwick, J.C. 1976.The effect of porcine somatotropin on growth and survival in seawater of Atlantic salmon (*Salmo salar*). Can. J. Zool. 54: 531-535.

Maclean, N., Penman, D. and Zhu, Z. 1987. Introduction of novel genes into fish. Bio/Technology 5: 257-261.

Maclean, N., Iyengar, A., Rahman, A., Sulaiman, Z. and Penman, D. 1992. Transgene transmission and expression in rainbow trout and tilapia. Mol. Mar. Biol. Biotehnol. 1: 355-365.

Parwez, I., Goswami, S.V. and Sundararaj, B.I. 1984. Effects of hypophysectomy on some osmoregulatory parameters of the catfish, *Heteropneustes fossilis* (Bloch). J. Exp. Zool. 229: 375-381.

Pickford, G.E. and Thompson, E.F. 1948. The effects of purified mammalian growth hormone on the killifish *Fundulus heteroclitus* (Linn). J. Exp. Zool. 109: 367-383.

Pickford, G.E. 1957. The chromatophore hormones of the pituitary. *In* The Physiology of the Pituitary Gland of Fishes. pp. 32-59. Edited by G.E. Pickford and J.M. Atz. New York Zoology Society, New York.

Rokkones, E., Alestrom, P., Skjervold, H. and Gautvik, K.M. 1989. Microinjection and expression of a mouse metallothionein human growth hormone fusion gene in fertilized salmonid eggs. J. Comp. Physiol. 158B: 751-758.

Teskeredzic E., Tomec M., Hacmanjek M., McLean E., Teskeredzic Z. and Donaldson, E.M. 1991. The effect of porcine somatotropin therapy upon growth and body composition of leather carp (*Cyprinus carpio* L.) maintained at suboptimal temperatures. Riv. Ital. Acquacol. 26: 131-135.

Xu, K., Wei, Y., Guo L. and Zhu, Z. 1991. The effects of growth enhancement of human growth hormone gene transfer and human growth hormone administration on crucian carp (*Carassius auratus gibelio* Bloch). Acta Hydrobiol. Sin.15: 31-34.

Wei, Y., Xie, Y., Xu, K., Li, G., Liu, D., Zou, J. and Zhu, Z. 1992. Heredity of human growth hormone gene in transgenic carp (*Cyprinus carpio* L). Chin. J. Biotechnol. 8: 140-144.

Zhang, P., Hayat, M., Joyce, C., Gonzalez-villasenor, L.I., Lin, C.M., Dunham, R.A., Chen, T.T. and Powers, D.A.1990. Gene transfer, expression and inheritance of pRSV-rainbowtrout-GH cDNA in the common carp, *Cyprinus carpio* (Linnaeus). Mol. Reprod. Develop. 25: 3-13.

Zhu, Z., Li, G., He, L. and Chen, S, 1985. Novel gene transfer into the fertilized eggs of goldfish (*Carassius auratus* L. 1758). Z. angew. Ichthyol. 1: 31-34.

Zhu, Z., Xu, K., Li, G., Xie, Y. and He, L. 1986. Biological effects of human growth hormone gene microinjected into the fertilized eggs of loach (*Misgurnus anguillicaudatus*). Kexue Tongbao 31: 988-990.

Zhu, Z., Xu, K., Xie, Y., Li, G. and He, L. 1989. A model of transgenic fish. Scientia Sinica (Series B) 2: 147-155.

Zhu, Z. 1992. Growth hormone gene and the transgenic fish. *In* Agricultural Biotechnology. pp. 106-116. Edited by C.B. You and Z.L. Chen, China Scientific and Technological Press, Beijing.

鱼类的激素替代疗法：人生长激素基因在垂体切除的鲤中的功能

崔宗斌[1] 朱作言[1,2]

1 中国科学院水生生物研究所，中国武汉 430072

2 阿伯丁大学分子细胞生物学与动物学研究中心，苏格兰阿伯丁 AB9 2TN

摘　要　将人类生长激素基因 hGH 与鼠金属硫蛋白-1(MT-1)基因启动子进行连接重组获得嵌合质粒 pMThGH。对鲤受精卵显微注射该线性化嵌合质粒，培养受精卵使其生长并繁殖获得 F_2 转 hGH 基因鲤鱼品系。接着对转 hGH 基因鲤 F_2、非转基因鲤及非转基因鲫分别进行垂体切除，实时监测三组实验鱼的生长状态至切除垂体 110 天后，并且对非转植基因鲫组进行皮下注射重组蛋白 hGH。对实验鱼的生长速率进行分析，结果显示：(1)垂体切除后，非转基因鲤和非转基因鲫的生长停止；(2)注射 hGH 可以刺激垂体切除后的非转基因鲫的生长；(3)切除转 hGH 基因鲤的垂体后，在缺乏内源 GH 条件下其仍可继续保持生长，表明转植的 hGH 基因在组织中而不是在垂体器官中得到了表达并具有促进鱼生长的作用。

Food Consumption and Energy Budget in MThGH-Transgenic F_2 Red Carp (*Cyprinus carpio* L. red var.)

CUI Zongbin ZHU Zuoyan CUI Yibo LI Guohua XU Kesheng

Institute of Hydrobiology, Chinese Academy of Sciences, Wuhan 430072

Abstract Growth hormone (GH) is a multi-function hormone secreted from meso-adenohypophysis. The effects of GH on promoting animal growth and development are unsuspicious. Recombinant plasmid pMThGH, a human GH (hGH) gene capped with a mouse MT-1 gene promoter, has been used to produce the first batch of transgenic goldfish *Carassius auratus auratus* and loach *Misgumus anguillicaudatus*[1,2]. The transgene's integration, expression and transmission were confirmed and faster growing rates were noted in some of the founder fish (P_0) and the F_1 descendants[3]. However, no study has so far been reported on the physiological mechanism of growth enhancement in the GH-transgenic fish. The present study aims to give a bioenergetic analysis of the trait of "faster-growing and less-eating" shown by the MThGH-transgenic red carp. This study formed, in part, theoretical foundation of genetic-breeding in fish.

Keywords transgenic fish; food consumption; energy budget

1 Materials and Methods

Two groups of red carp (*Cyprinus carpio* L. red var.) aged 2 months were used in the experiment. The experimental group was transgenic F_2 red carp being produced in our laboratory and the control group was non-transgenics of the same species being produced in Guan Qiao Experimental Fish Farm of the Institute. Most chemical reagents were purchased from the Sigma Chemical Company.

Prior to the experiment of 7d, each group of 30 individuals was randomly picked up and transferred into a temperature-controlled room. The ambient temperature was gradually adjusted to (27.3 ± 0.15)℃ (mean ± SE.) and the photoperiod was kept 12h per day. The fish were fed with live tubificid worms (mainly *Limnodrilus hoffmeisteri*), and were starved for two days before the commencement of the experiment. On the first day of the experiment, the fish were blotted out of the surface water on a piece of filter paper and weighed. Individuals of 15 in each group were separately transferred into plexiglas fish tanks containing 12L of aged tap water. Another 15 individuals of each group were sacrificed in order to determine the

文章发表于 *Chinese Science Bulletin*, 1996, 41(7):591-596

initial levels of the fish dry matter, chemical components and energy content. At the end of the experiment, fin DNA samples of all individuals in both groups were prepared and the DNA polymerase chain reaction (PCR) technique with MTl-hGH primers was employed to determine the transgenic positives. Fourteen F_2 individuals in the 15 sacrificed and 12 F_2 individuals in the 15 tank-rearing group were confirmed to be the MThGH-transgenics, i.e. transgenic ratio in the F_2 was 86.7%. And only the MThGH-positives were referred to the transgenic F_2 while statistical analysis of the experimental group was carried out. All of the control groups were the PCR-negatives. The fish were fed once a day for 21 d. A small excess of net weight of the live tubificid worms, of which the surface water was blotted out with a piece of filter paper, was used for each feeding. The uneaten worms were reweighed and removed the next day. Correction of weight loss of the worms in 24 h, 7.39% ± 1.54 (mean ± S.E.), was determined from a control tank containing no fish and a subsample of worms was dried for determining dry matter and energy content. Faeces produced by each fish were collected twice a day and dried at 70℃.

Water in each tank was replenished once a week. Energy loss in fish excreta was estimated from the amounts of ammonia and urea measured prior to and post each replenishment[4]. The ammonia and urea excretion by the worms, being determined from the control tank, was deducted from the fish rearing tanks. The energy value for ammonianitrogen was 24.83 J/mg and for urea-nitrogen was 23.03 J/mg[5]. At the end of the experiment, the fishes were sacrificed, weighed and then dried at 70℃. The dried fish carcass, tubificid worms and faeces were homogenized separately and stored at –20℃ for chemical analysis. The energy contents of fish, tubificid worms and faeces in each tank were determined using a Phillipson micro-bomb calorimeter. Protein content of the carcass was measured with Folin-phenol reagent. Lipid content was determined by the method of chloroform-methanol extraction.

Kruskal-Wallis test was used to analyze the differences of food consumption, growth rate, biochemical composition and energy budget between the F_2 transgenics and the control. The specific growth rate in wet weight (SGRw, %/d) was calculated as SGRw = $100 \times (\ln W_t - \ln W_0)/t$ and the feeding rate was expressed as $C/[(W_t - W_0) \times t]$, where W_t and W_0 were the final and initial wet weight of fish, respectively, t was the days of the experimental duration and C was the total energy of the eaten food. Specific growth rates in energy, protein and dry matter were calculated similarly, replacing W_t and W_0 with the corresponding final and initial factors. Conversion efficiency was expressed as percentage of gain in wet weight, dry weight, energy and protein in the total eaten food. The content of dry matter was expressed as percentage of fish wet weight. The content of chemical components was expressed as the percentage of fish dry weight. The content of energy was expressed as the value released from the Phillipson micro-bomb calorimeter while every gram of dry matter was completely fired.

2 Results and Discussions

2.1 Comparison of food consumption and growth between the transgenic F_2 and the control red carp

Feeding rates of the transgenic F_2 were significantly lower than that of the non-transgenic control red carp ($p<0.05$). However, the specific growth rates of the transgenic F_2 in wet weight ($p<0.01$), dry weight ($p<0.01$), energy ($p<0.01$) and protein ($p<0.001$) and the conversion efficiencies of the transgenic F_2 in wet weight ($p<0.05$), dry weight ($p<0.05$), energy ($p<0.05$) and protein ($p<0.001$) were higher than those of the non-transgenic control (table 1). In other words, the MThGH-transgenic carp is physiologically characterized as "faster-growing and less-eating" It was reported that GH-treatment can not only stimulate the growth but also improve the food conversion efficiency in coho salmon *Oncorhynhus kisutch*. We have successfully quantified hGH in the transgenic F_2 red carp (data not shown here) so that the lower feeding rate and the growth enhancement of the transgenics should result from MThGH-transgene expression in the fish body.

Table 1 Comparison of feeding rate (FR, $kJ \cdot g^{-1} \cdot d^{-1}$), specific growth rate (SGR, %/d) and conversion efficiency (CE, %) between the transgenic F_2 and control red carp

	F_2 transgenics	Control red carp	Kruskal-Wallis test
Fish No.	n=12	n =15	
FR	0.603±0.014	0.649±0.012	p <0.05
SGRw	6.82±0.16	6.04 ±0.14	p <0.01
SGRd	7.34±0.18	6.63±0.13	p <0.01
SGRe	8.14±0.20	7.17±0.15	p <0.01
SGRp	7.59±0.16	6.48±0.16	p <0.001
CEw	30.96±0.70	28.75 ±0.54	p <0.05
CEd	43.83±1.24	40.43 ±0.65	p <0.05
CEe	41.44 ±0.89	38.13±0.85	p <0.05
CEp	57.38±0.96	47.41 ±1.08	p <0.001

Values are expressed as mean±S.E. SGRw, SGR in wet weight; SGRd, SGR in dry weight; SGRe, SGR in energy; SGRp, SGR in protein; CEw, CE in wet weight; CEd, CE in dry weight; CEe, CE in energy; CEp, CE in protein

2.2 Comparison of energy budget between the transgenic F_2 and the control red carp

The energy budget equation for fish is written as $G= C–F–U–R$, where G is the energy channelled to growth, C is the energy from food, F is the energy lost in faeces, U is the energy lost in nitrogenous excretion and R is the energy channelled to metabolism[6]. The causes for growth enhancement of fish can be; i) increase in food consumption, ii) decrease of loss in faecal production, iii) decrease of loss in nitrogenous excretion, iv) reduction of the energy channelled to metabolism, and v) a combination of the above causes. In this study, food consumption (C), faecal production (F), energy lost in nitrogenous excretion (U) and energy channelled to growth (G) were measured and determined directly. Energy channelled to metabolism (R) was calculated as the difference between C and other components of the energy budget. The energy budget equation was 100 C=8.9F + 0.63U + 49.03R + 41.44G for the transgenic F_2 and 100 C=7.37F + 1.14U + 53.36R + 38.13G for control. Compared with the control group, the transgenic F_2 had a higher proportion of food energy channelled to F ($p<0.05$) and a significantly lower proportion of food energy channelled to U ($p<0.001$). The proportion of food energy channelled to G was higher ($p<0.05$) and that channelled to R was significantly lower ($p<0.01$) (table 2). Analysis of energy budget indicated that both faecal production and nitrogenous excretion represented a small proportion of the eaten food energy. The energy channelled to metabolism, however, was a main budget which held about 50% of the total eaten food energy. The significantly lower proportion of food energy channelled to metabolism in the transgenic F_2 carp was consequently a main cause of the improved growth performance. To sum up, in comparison with the control carp, the transgenic F_2 has saved 6.62% of the total eaten food energy, part of which was used for growth improvement. The trait of "faster-growing and less-eating" implies that the GH-transgenic fish are applicable to aquaculture.

Table 2 Comparison of energy budget between the transgenic F_2 and control red carp

	F_2 transgenics	Control red carp	Kruskal-Wallis test
Fish No.	n=12	n = 15	
G/C (%)	41.44±0.89	38.13 ±0.85	$p<0.05$
R/C (%)	49.03±0.91	53.36 ±0.98	$p<0.01$
F/C (%)	8.90±0.53	7.37±0.30	$p<0.05$
U/C (%)	0.63±0.04	1.14±0.08	$p<0.001$

Values are expressed as mean±S.E. G, Growth; R, metabolism; F, faecal production; U, nitrogenous excretion; C, food consumption

2.3 Comparison of chemical composition and energy content between the transgenic F_2 and the control red carp

At the start of experiment, there was no significant difference in chemical composition and energy content of dry matter between the transgenic F_2 and the control red carp. This may result from a crowded rearing condition where both groups of fish had been subjected to oxygen and food shortage before the experiment. At the end of the experiment, however, the protein content in the transgenic F_2 red carp was significantly higher than that in the control.

Other factors such as dry matter, lipid and energy content showed no differences between these two groups (table 3), GH stimulates protein synthesis. Bovine GH was responsible for increasing the content of muscle protein in coho salmon. In this study, the transgenic F_2 showed a significantly higher conversion efficiency in protein (table 1), a significantly lower proportion of food energy channelled to *U* (table 2) and a significantly higher protein content of dry matter (table 3). These data demonstrated that the expressed human GH in the transgenic F_2 played a role of promoting protein synthesis in red carp. On the other hand, we have found no difference in lipid content of dry matter between the two groups. There were some inconsistent reports on GH function on lipid synthesis in fish. In several teleost species, GH was found to be lipolytic. For example, GH increased the plasma concentration of free fatty acid in goldfish *Carassius auratus* and injection of bovine GH significantly reduced lipid content of muscle in coho salmon. These findings were in contrast to results that bovine GH treatment cannot alter the concentrations of serum lipid and cholesterol and the content of hepatic lipid' in tilapia *Oreochromis mossambicus*. Moreover, bovine GH given daily to rainbow trout *Salmo gairdneri* significantly increased plasma free fatty acid levels without apparent effects on muscle lipid corftent. The different results about the effect of GH on lipid metabolism may result from the differences of the GH sources, the fish species and the experimental duration.

Table 3 Comparison of body composition (%) and energy content ($kJ·g^{-1}$) between the transgenic F_2 and control red carp

	F_2 transgenics	Control red carp	Kruskal-Wallis test
Start of experiment			
Fish No.	$n = 14$	$n = 15$	
Dry matter	19.65 ±0.26	19.53±0.24	p >0.05
Protein	59.93±0.57	58.40±0.51	p >0.05
Lipid	12.05±0.55	11.90±0.29	p >0.05
Energy	18.71 ±0.25	19.27±0.13	p >0.05
End of experiment			
Fish No.	$n = 12$	$n = 15$	
Dry matter	21.95±0.36	22.11 ±0.23	p >0.05
Protein	63.19±1.24	56.61 ±0.53	p <0.001
Lipid	23.32±0.51	23.97 ±0.57	p >0.05
Energy	22.16±0.29	21.60±0.22	p >0.05

Values are expressed as mean ± S.E.

Acknowledgement The study was conducted in the State Key Laboratory of Freshwater Ecology and Biotechnology.

References

1 Zhu, Z., Li, G., He, L. *et al.*, Novel gene transfer into the fertilized eggs of goldfish (*Carassius auratus L.* 1758), *Z. angew. lchthyol.*, 1985, 1: 31

2 Zhu, Z., Xu, K., Li, G. *et al.*, Biological effects of human growth hormone gene microinjected into the fertilized eggs of loach *Misgumus anguillicaudatus* (Cantor), *Kexue Tongbao* (*Chinese Science Bulletin*), 1986, 31(14): 988.

3 Zhu, Z., Xu, K., Xie, Y. *et al.* A model of transgenic fish, *Scientia* Sinica, Ser. B, 1989, 2: 147.

4 Chaney, A. L., Marbach, E. P., Modified reagents for determination of urea and ammonia, *Cli. Chem.*, 1962, 8: 130

5 Elliott, J. M., Energy losses in the waste products of brown trout (*Salmo trutta L.*), *J. Anim. Ecol.*, 1976, 45: 561

6 Brett, J. R., Groves, T. D. D., Physiological energetics, in *Fish Physiology* (eds. Hoar, W. S., Randall, D. J., Brett, J. R.), Vol. VIII, London: Academic Press, 1979, 279-352

转基因红鲤 F_2 的食物消耗与能量收支

崔宗斌 朱作言 崔奕波 李国华 许克圣

中国科学院水生生物研究所，武汉 430072

摘 要 生长激素是一种由脑垂体分泌的多功能激素，具有促进动物生长发育的功能。有研究表明，小鼠 MT-1 基因启动子驱动人生长激素 hGH 表达的重组质粒 pMThGH，已用于研制转基因金鱼和泥鳅，检测到转植基因的整合、表达和性腺传递，并在 P_0 和 F_1 的一些个体中观察到快速生长效应。然而，迄今为止尚未见文献报道转 GH 基因鱼快速生长的生理机制。本研究旨在为转 MThGH 基因红鲤“吃得少而长得快”特征进行生物能量学分析，研究结果将为鱼类遗传育种理论提供一定的理论基础。

Whole-Body Amino Acid Pattern of F_4 Human Growth Hormone Gene-Transgenic Red Common Carp (*Cyprinus carpio*) Fed Diets with Different Protein Levels

Cuizhang Fu[1] Yibo Cui[1] Silas S. O. Hung[2] Zuoyan Zhu[1]

1 State Key Laboratory of Freshwater Ecology and Biotechnology, Institute of Hydrobiology, The Chinese Academy of Sciences, Wuhan, Hubei 430072, People's Republic of China

2 Department of Animal Science, University of California, One Shield Avenue, Davis, CA 95616-8521, USA

Abstract F_4 generation of human growth hormone (*hGH*) gene-transgenic red common carp, and the non-transgenic controls were fed for 8 weeks on purified diets with 20%, 30% or 40% protein. Analysis of whole-body amino acids showed that the proportions of lysine, leucine, phenylalanine, valine and alanine, as percentages of body protein, increased significantly, while those of arginine, glutamic acid and tyrosine decreased, with increases in dietary protein level in at least one strain of fish. Proportions of the other amino acids were unaffected by the diets. The proportions of lysine and arginine were significantly higher, while those of leucine and alanine were lower in the transgenics than in the controls in at least one diet group. Proportions of the other amino acids were unaffected by strain. The results suggest that the whole-body amino acid profile of transgenic carp, when expressed as proportions of body protein, was in general, similar to that of the non-transgenic controls.

Keywords transgenic fish; amino acid composition; growth hormone; *Cyprinus carpio*

1. Introduction

Transgenic fish have been produced in a variety of species using different foreign gene constructs (Zhu, 1992; Hackett, 1993; Devlin *et al.*, 1994). Many investigators have reported that the growth hormone (GH)-transgenic fish exhibited higher growth rates than the controls (Zhu *et al.*, 1986, 1989; Chourrout *et al.*, 1986; Dunham *et al.*, 1987; Rokkones *et al.*, 1989; Zhang *et al.*, 1990; Du *et al.*, 1992; Devlin *et al.*, 1994). However, information on body composition of transgenic fish is limited (Chatakondi *et al.*, 1995; Cui *et al.*, 1996b; Fu *et al.*, 1998), and there has been only one report on the body amino acid composition of transgenic fish, which suggested that the contents of 14 of the 18 muscle amino acids were higher in the transgenic carp containing rainbow trout GH gene than in the controls (Chatakondi *et al.*,

文章发表于 *Aquaculture*, 2000, 189(3): 287-292

1995). In a previous paper, we reported the growth, feed utilization and proximate body composition of F_4 generation of human growth hormone (hGH) transgenic carp fed diets with different protein levels in an 8-week growth trial (Fu *et al.*, 1998). The study showed that transgenic carp had higher growth rates, protein retention, and contents of body dry matter and protein than the controls. The purpose of the present study was to compare whole-body amino acid pattern between the F_4 *hGH*-transgenic and the non-transgenic red common carp.

2. Materials and Method

Details of fish, experimental systems and procedures were described in Fu *et al.* (1998). P_0 transgenic fish were produced by microinjection of the *hGH* gene capped with a mouse metallothionein promoter into the fertilized eggs of common carp (*Cyprinus carpio*-red variety). The F_4 transgenic carp were produced in 1996 bybreeding the F_3 generation following spawning induction. A batch of non-transgenic control carp (red variety) was also produced on the same day. The polymerase chain reaction (PCR) method based on the procedure of Li *et al.* (1996) was used to detect the transgene-positive individuals. Results of PCR on 90 transgenic carp revealed that 100% of the tested fish were positive. At the age of 4 weeks, 300 transgenic and 300 control carps were transferred to the laboratory and fed an equal mixture of the three experimental diets. The growth experiment was initiated when the fish were approximately 8 weeks old and weighed 1.75 g.

Three isoenergetic purified diets, with crude protein levels of 20%, 30% and 40%, and carbohydrate levels of 50%, 40% and 30%, respectively, were formulated (see Table 1), and proximate composition and gross energy of the diets were presented in Fu *et al.* (1998). For amino acid analysis, triplicate samples of each diet were hydrolyzed with 6 N HCl for 24 h at 100℃ in tubes sealed after nitrogen flushing. After dilution and filtration, a small aliquot was derivitized with phenyisothiocyanate. The derivitized amino acids were analyzed by reverse phase liquid chromatography using a Pico-tag Amino Acid Analysis System (including Waters 510 pump, 717 autosampler, 996 photodiode array detector, 470 scanning fluorescence detector and 2010 work station; Millipore, Milford, MA, USA). Tryptophan was not analyzed in this study. The amino acid compositions of the diets are presented in Table 1.

The growth experiment was conducted in 18 circular fiberglass tanks (diameter 70cm, water volume 90 l), which were part of a recirculation system. Each tank had a flat bottom with a central drain. The water temperature varied from 26℃ to 29℃ (mean 27.5℃, SE 0.9℃). Photoperiod was controlled by artificial lighting with the light. period between 07:00 and 19:00.

Thirty individuals of either transgenic or control carp were weighed and put into each tank, and three tanks were assigned randomly to each combination of strain and diet. The growth experiment lasted for 8 weeks and the fish were fed to satiation twice a day at 08:00

and 16:00. At the end of the experiment, the fish were not fed for 1 day and weighed. Twenty fish from each tank were killed and used for the analysis of body composition. Amino acid analysis of fish body was as described for the diets. All analyses were carried out in triplicates.

The effects of diet and fish strain were analyzed by two-way analysis of variance (ANOVA). Multiple comparisons were made based on least-square means (Cui *et al.*, 1996a). Differences were regarded as significant when $P<0.05$.

Table 1 Formulation of the experimental diets

Ingredients(%)	Dietary protein (%)		
	20	30	40
Casein	16.0	24.0	32.0
Gelatin	4.0	6.0	8.0
Cod liver oil	3.0	3.0	3.0
Soybean oil	3.0	3.0	3.0
Dextrin	50.0	40.0	30.0
Others[a]	24	24	24
Amino acid composition (% dry weight)			
Essential			
Lys	1.5	2.2	3.0
Met	0.7	1.1	1.5
Thr	0.8	1.2	1.6
Leu	1.8	2.6	3.6
Ile	0.7	1.1	1.3
His	0.6	0.9	1.0
Arg	0.9	1.4	1.9
Phe	0.6	0.8	1.2
Val	1.2	1.8	2.4
Non-Essential			
Cys	0.1	0.1	0.1
Ser	1.2	1.7	2.3
Gly	1.5	2.1	2.9
Pro	2.7	4.0	5.5
Ala	1.0	1.4	1.9
Glu	3.7	5.5	7.3
Asp	0.9	1.4	1.9
Tyr	0.8	1.2	1.7

a. Others: 1% vitamin premix and 4% mineral premix (Fu *et al.*, 1998), 16% cellulose, 2% carboxymethyl-cellulose and 1% chromic oxide

3. Results and Discussion

In general, the amino acid profiles of body protein in both the transgenic carp and the controls (Tables 2 and 3) were similar to those reported for common carp or other cyprinids (Schwarz and Kirchgessner, 1988; Gatlin, 1987). Of the essential amino acids, regardless of strain, the proportions of lysine, leucine, phenylalanine and valine as percentages of body protein, increased significantly, while that of arginine decreased, with increases in dietary protein level. Proportions of other essential amino acids were unaffected by the diets (Table 2). Of the non-essential amino acids, the proportion of glutamic acid decreased with increases in dietary protein level in both the transgenics and controls. The proportion of alanine in the transgenics increased with increases in dietary protein level, and the proportion of tyrosine in the controls decreased with increases in dietary protein level. Proportions of the other non-essential amino acids were unaffected by the diets (Table 3).

Few papers have reported the effect of dietary protein levels on the body amino acid composition in fish. Schwarz and Kirchgessner (1988) reported that dietary protein level had no obvious influence on the body amino acid pattern in common carp. The results were somewhat different from those of the present study. Several differences in experimental conditions may partly account for the different results between the two studies. In Schwarz and Kirchgessner's (1988) study, fish were housed at a mean water temperature of 24℃ and

Table 2 Whole body essential amino acid pattern (% protein) of control and transgenic carp fed diets with different protein levels (mean±SE, n=3)[a]

	Dietary protein(%)						ANOVA(P-value)effect			
	20		30		40					
	Control	Transgenics	Control	Transgenics	Control	Transgenics	Overall	Diet	Strain	Diet×strain
Lys	7.4±0.0^{Xa}	7.7±0.0^{Ya}	7.5±0.1^{Xa}	7.9±0.1^{Yb}	8.0±0.1^{Xb}	8.4±0.1^{Yc}	0.0000	0.0000	0.0002	0.6655
Met	2.9±0.1	3.1±0.1	2.9±0.0	3.1±0.2	3.2±0.1	3.3±0.1	0.3325	0.2278	0.1382	0.9948
Thr	4.3±0.2	4.4±0.2	4.5±0.5	4.4±0.2	4.8±0.1	4.5±0.1	0.8706	0.5784	0.7885	0.6864
Leu	7.5±0.1^{Xa}	6.7±0.1^{Ya}	7.7±0.3^{a}	7.5±0.2^{b}	8.9±0.5^{Xb}	8.0±0.3^{Yb}	0.0039	0.0013	0.0106	0.4395
Ile	4.3±0.3	4.4±0.1	4.6±0.2	4.2±0.1	5.1±0.3	4.4±0.1	0.1699	0.1513	0.0767	0.2823
His	2.8±0.1	2.9±0.0	2.6±0.1	2.9±0.1	2.7±0.0	2.8±0.0	0.2477	0.1789	0.0609	0.0848
Arg	7.9±0.2^{a}	7.9±0.0	7.5±0.2^{a}	7.8±0.2	6.9±0.1^{Xb}	7.5±0.1^{Y}	0.0100	0.0048	0.0238	0.0719
Phe	4.0±0.2^{a}	4.1±0.2^{a}	4.6±0.5^{a}	4.9±0.2^{b}	3.2±0.0^{b}	3.6±0.1^{a}	0.0055	0.0008	0.2482	0.8781
Val	4.8±0.1^{a}	4.9±0.0^{ab}	5.1±0.1^{b}	4.7±0.1^{a}	5.2±0.1^{b}	5.1±0.1^{b}	0.0256	0.0266	0.1587	0.0434
Total essential amino acids										
	45.8±0.2^{a}	46.0±0.3^{a}	47.1±0.8^{b}	47.4±0.2^{b}	47.2±0.7^{b}	47.4±0.1^{b}	0.0431	0.0083	0.4790	0.9737

a. Letters after each value indicate results of pair-wise comparisons. Different upper case letters (XY) indicate significant differences between fish strain within diet; different lower case letters (abc) indicate significant differences between diets within strain as ranked by least square means

Table 3 Whole body non-essential amino acid pattern (% protein) of control and transgenic carp fed diets with different protein levels (mean±SE, n=3)[a]

	Dietary protein(%)						ANOVA(P-value)effect			
	20		30		40					
	Control	Transgenics	Control	Transgenics	Control	Transgenics	Overall	Diet	Strain	Diet×strain
Cys	0.4±0.0	0.4±0.0	0.4±0.0	0.4±0.0	0.5±0.0	0.4±0.0	0.1469	0.1569	0.1097	0.5119
Ser	4.6±0.1	4.5±0.0	4.4±0.1	4.3±0.0	4.5±0.2	4.6±0.1	0.1446	0.0606	0.4787	0.5104
Asp	6.4±0.1	7.2±0.1	6.8±0.1	6.8±0.1	7.1±0.2	6.8±0.2	0.0874	0.7431	0.3249	0.0242
Glu	17.1±0.1[a]	17.2±0.1[a]	16.2±0.1[b]	16.5±0.3[a]	15.2±0.2[b]	15.8±0.3[b]	0.0022	0.0002	0.2142	0.8918
Pro	5.5±0.2	5.2±0.1	5.6±0.2	5.5±0.2	5.4±0.3	5.7±0.1	0.5550	0.5291	0.7762	0.3343
Gly	8.9±0.1	9.0±0.1	8.7±0.4	8.6±0.1	8.8±0.1	8.4±0.1	0.1815	0.1295	0.3013	0.3595
Ala	7.3±0.1[X]	6.7±0.1[Ya]	7.6±0.2[X]	6.9±0.2[Ya]	7.6±0.1	7.5±0.0[b]	0.0022	0.0061	0.0016	0.0942
Tyr	4.0±0.2[a]	4.0±0.1	3.4±0.2[b]	3.6±0.1	3.6±0.1[ab]	3.6±0.1	0.1188	0.0270	0.7421	0.4933
Total non-essential amino acids	54.2±0.1[a]	54.0±0.4[a]	53.1±0.9[ab]	52.6±0.2[b]	52.8±0.7[b]	52.6±0.1[b]	0.0581	0.0129	0.4052	0.9039

a. Letters after each value indicate results of pair-wise comparisons. Different upper case letters (XY) indicate significant differences between fish strain within diet; different lower case letters (ab) indicate significant differences between diets within strain as ranked by least square means

received a fixed feeding rate. The initial weight of the fish was 170 g. In the present study, fish were fed to satiation twice a day at a mean water temperature of 27.5℃. The initial weight of the fish was only 1.75 g. Formulations of the experimental diets were also different.

Of the essential amino acids, the proportion of lysine was significantly higher in transgenics than in controls regardless of diets. The proportion of leucine was significantly lower in the transgenics than in the controls fed with 20% and 40% protein diets, but was unaffected by strain in fish fed with the 30% protein diet. The proportion of arginine was significantly higher in the transgenics than in the controls fed with 40% protein diet, but was unaffected by strain in fish fed with diets that have lower protein levels. Proportions of the other essential amino acids and total essential amino acids were unaffected by strain (Table 2).

Of the non-essential amino acids, the proportion of alanine was significantly lower in the transgenics than in the controls fed with 20% and 30% protein diets. Proportions of the other non-essential amino acids and total non-essential amino acids were unaffected by strain (Table 3) .

Chatakondi *et al.* (1995) reported that the contents of 14 of the 18 muscle amino acids were higher in the transgenic carp containing rainbow trout GH gene than in the controls. In the present study, only 4 of the 17 body amino acids analyzed were significantly different between the transgenics and the controls. Amino acid composition was expressed as grams per 100 g of muscle in Chatakondi *et al.* (1995), and as gramsper 100 g of protein in the present study. As the body protein content was significantly higher in the transgenics than in the controls in the present study, the contents of most amino acids, when expressed as grams per unit body weight, would be expected to be higher in the transgenics than in the controls.

In conclusion, results of the present study suggested that the whole-body amino acid profile of the transgenic carp, when expressed as proportions of body protein, was in general, similar to that of the non-transgenic controls.

Acknowledgements We would like to thank T. Storebakken for providing the vitamin and mineral premix, and S. Xie, Z. Cui, W. Lei, Y. Yang and Y. Wang for their help during the experiment. This work was supported by The National Natural Science Foundation of China through grants to Y. Cui (project no. 39625006) and Z. Zhu (39823003), and the '863 High Technology Project' of China through a grant to Z. Zhu.

References

Chatakondi, N., Lovell, R.T., Duncan, P.L., Hayat, M., Chen, T.T., Powers, D.A., Weete, J.D., Cummins, K., Dunham, R.A., 1995. Body composition of transgenic common carp, *Cyprinus carpio*, containing rainbow trout growth hormone gene. Aquaculture 138, 99-109.

Chourrout, D., Guyomard, R., Houdebine, L.-M., 1986. High efficiency gene transfer in rainbow trout (*Salmo gairdneri* Rich.) by microinjection into egg cytoplasm. Aquaculture 51, 143-150.

Cui, Y., Hung, S.S.O., Zhu, X., 1996a. Effect of ration and body size on the energy budget of juvenile white sturgeon. J. Fish Biol. 49, 863-876.

Cui, Z., Zhu, Z., Cui, Y., Li, G., Xu, K., 1996b. Food consumption and energy budget in *MThGH*-transgenic F_2 red carp (*Cyprinus carpio* L. red var.). Chin. Sci. Bull. 41, 591-596.

Devlin, R.H., Yesaki, T.Y., Biagy, C.A., Donaldson, E.M., Swanson, P., Chan, W.K., 1994. Extraordinary salmon growth salmon growth. Nature 371, 209-210.

Du, S.J., Gong, Z., Fletcher, G.L., Shears, M.A., King, M.J., Idler, D.R., Hew, C.L., 1992. Growth enhancement in transgenic Atlantic salmon by the use of an 'all fish' chimeric growth hormone gene construct. Biotechnology 10, 176-181.

Dunham, R.A., Eash, J., Askins, J., Townes, T.M., 1987. Transfer of metallothionein-human growth hormone fusion gene into channel catfish. Trans. Am. Fish. Soc. 116, 87-91.

Fu, C., Cui, Y., Hung, S.S.O., Zhu, Z., 1998. Growth and feed utilisation by F_4 human growth hormone transgenic carp fed diets with different protein levels. J. Fish Biol. 53, 115-129.

Gatlin, D.M. III, 1987. Whole-body amino acid composition and comparative aspects of amino acid nutrition of the goldfish, golden shiner and fathead minnow. Aquaculture 60, 223-229.

Hackett, P.B., 1993. The molecular biology of transgenic fish. In: Hochachka, P., Mommsen, R. (Eds.), Biochemistry and Molecular Biology of Fishes vol. 2 Elsevier, Amsterdam, pp. 207-240.

Rokkones, E., Alestrom, P., Skjervold, H., Gautvik, K.M., 1989. Microinjection and expression of a mouse metallothionein human growth hormone fusion gene in fertilized salmonid eggs. J. Comp. Physiol. 158, 751-758.

Schwarz, F.J., Kirchgessner, M., 1988. Amino acid composition of carp (*Cyprinus carpio* L.) with varying. protein and energy supplies. Aquaculture 72, 307-317.

Zhang, P., Hayat, M., Joyce, C., Gonzalez-Villasenor, L.I., Lin, R.A., Dunham, R.A., Chen, T.T., Powers, D.A., 1990. Gene transfer, expression and inheritance of pRCV-rainbow trout-GHcDNA in the common carp *Cyprinus carpio*. Mol. Reprod. Dev. 25, 13-25.

Zhu, Z., 1992. Generation of fast growing transgenic fish: methods and mechanisms. In: Hew, C.L., Fletcher, G.L. (Eds.), Transgenic Fish. World Scientific Press, Singapore, pp. 92-119.

Zhu, Z., Xu, K., Li, G., Xie, Y., He, L., 1986. Biological effects of human growth hormone gene microinjected into the fertilized eggs of loach *Misgurnus anguillicaudatus*. KeXue TongBao 31, 988-990, (Science Bulletin, Academia Sinica).

Zhu, Z., Xu, K., Xie, Y., Li, G., He, L., 1989. A model of transgenic fish. Sci. China, Ser. B2, 147-155.

不同蛋白水平饲料对转基因红鲤 F_4 全鱼氨基酸分布模式的影响

傅萃长[1] 崔奕波[1] Silas S. O. Hung[2] 朱作言[1]

1 淡水生态与生物技术国家重点实验室，中国科学院水生生物研究所，武汉，430072，中国
2 Department of Animal Science, University of California, One Shield Avenue, Davis, CA 95616-8521, USA

摘　要　对转人生长激素基因红鲤 F_4 和非转基因对照组分别连续投喂 8 周含 20%、30%和 40%蛋白的饲料后，测定每条鱼的氨基酸含量。检测结果显示，随着饲料蛋白含量的增加，赖氨酸、亮氨酸、苯丙氨酸、缬氨酸和丙氨酸占鱼体全身蛋白比例有明显增加，但精氨酸、谷氨酸和络氨酸比例有所下降，其他氨基酸含量不受摄食蛋白水平影响。相比对照组，转基因鱼赖氨酸和精氨酸显著增高，亮氨酸和丙氨酸则较低，其他氨基酸水平并不受影响。就全鱼蛋白比例而言，转基因鱼全鱼氨基酸水平与非转基因鱼无显著差异。

Transgene for Growth Hormone in Common Carp (*Cyprinus carpio* L.) Promotes Thymus Development

GUO Qionglin WANG Yaping JIA Weizhang ZHU Zuoyan

State Key Laboratory of Freshwater Ecology and Biotechnology, Institute of Hydrobiology, Chinese Academy of Sciences, Wuhan 430072

Abstract The transgenic carp were produced by microinjection of CAgcGHc into the fertilized eggs. Observation of the thymus development between the transgenics and nontransgenic controls was carried out. The thymus of one-year-old transgenics F1 showed a great increase in both size and weight. The unilateral thymus of the transgenics weighed from190 to 295 mg with average 218.6 mg, whereas the unilateral thymus of the controls weighed 20-81 mg with average 42.5 mg; i.e. the thymus weight in the transgenics was 5.14 fold over that in the controls. The index of thymus/body weight in the transgenics was 2.97 fold over the controls. Light microscopy observation indicated that the thymus of the transgenics well developed with the thickened outer region and compactly arranged thymocytes, while the thymus in the controls were degenerating with the thinned outer region, scattered thymocytes and groups of fatty cells. Further analysis with the electron microscopy revealed that proliferous cells in the transgenics were mainly small lymphocytes and no pathological changes were found. The results confirmed that the "All-fish" GH-transgene promotes thymus development and thymocyte proliferation, and retards thymus degeneration. The study has laid a foundation for further analysis of the immunobiological function in GH-transgenic carp.

Keywords "All-fish" growth hormone gene; growth hormone (GH); transgenic carp; thymus; immunity

Since the first batch of transgenic fish was born in China[1,2], many laboratories all over the world have turned to the study of growth hormone (GH), antifreeze protein and lactoferrin transgenic fish[3-7], in order to gain new farming strains with the traits of "fast-growing", "anti-adversity" and "anti-disease". In addition to promoting growth and metabolism, as an immunostimulant or immunomodulating therapeutic agent, GH has received extensive attention[8,9]. Murphy *et al.* pointed out that GH is a cytokine, and plays an important role in lymphopoiesis and haematopoiesis as a thymopoietic factor, T-cell-stimulating factor and haematopoietic growth factor[9]. In mammals, GH has been shown to stimulate thymocyte maturation and differentiation, activate phagocytes and counter glucocorticoid-induced apoptosis of T cell progenitors[10,11]. Similarly, in fish, GH promotes haematopoiesis, enhances

文章发表于《科学通报》(英文版), 2003, 48(16):1764-1770

phagocytosis, natural killer cell activity , antibody production[10] and the resistance of fish body to bacterial pathogen[12].

Studies on transgenic animals often include the observation of the physiological and pathological changes caused by some important genes of higher expression *in vivo*, in order to confirm the relations between these genes and the growth, immunity, anti-disease or disease occurrence of animals. In mammals, the studies on transgenicmice have mainly focused on the gene transfer of GH, insulin-like growth factor I (IGF-I) and some important cytokines[13-16]. Dialynas *et al.* reported that bGH transgene in mice increased the absolute thymic weight and the mitogenic responses of splenocyte to PHA, LPS and ConA, but the relative thymic weight was not altered[13]. Up to now, the effects of GH transgene on thymus structure and thymocyte development are still unclear. There is no unanimous opinion on immune action. In this study, we examined the effects of "All-fish" GH-tranegene on thymus development in carp (*Cyprinus carpio* L.), in order to understand the immunobiological function on "All-fish" GH transgenic carp and search for the relations between GH and the structure and function of immune system in common carp.

1. Materials and Methods

(i) Gene construct pCAgcGHc. "All-fish" gene construct pCAgcGHc was there combinant construct of grass carp (*Ctenopharyngodon idellus*) growth hormone cDNA (*gcGHc*) driven by β-actin gene promoter of common carp (*Cyprinus carpio* L.), which was modified from the construct pCAgcGH[17,18]. pCAgcGHc was digested with *Eco*RI and *Hin*dIII for gene transfer.

(ii) Gene transfer. Yellow River carp (*Cyprinus carpio* L.) were supplied by the Henan Institute of Aquaculture, Henan Province. Around the late April, fertilized eggs of Yellow River carp were obtained by artificial spawning and insemination. Microinjection was carried out before the first cell division according to the method of Zhu *et al*[19].

(iii) Detection of transgene and breeding of transgenic fish. P_0 generation of transgenics was produced by microinjection of *CAgcGHc* into the fertilized eggs of Yellow River carp. Sperm was obtained from the transgenic fish of P_0 generation and DNAs were extracted with the routine method. PCR was performed to examine the existence of *CAgcGHc*-transgene. Two primers for PCR were PII (5′-TGGCGTGATGAATGTCG-3′) and Pc (5′-AACACGTATGACTGC-3′). *CAgcGHc* positive fishes in sperm were selected for artificial spawning and inseminaction[18]. F_1 generation of transgenics was produced as follows: P_0 transgenic ♂ (sperm revealed to be *CAgcGHc* positive) × non-transgenic ♀. F_1 transgenics and the age matched controls were farmed under the same sitution of Wang *et al*[18].

(iv) Measurement of body length, body weight, thymus weight and count of the index of thymus/body weight. Around the late April of the next year, 5 one-year-old transgenics and

5 age-matched controls (the controls) were both randomly obtained and the body length, body weight, thymus weight (in unilateral side) were measured, and index of thymus/body weight [thymusweight (mg)/body (g)×10^5] was counted.

(v) Sample preparations for light microscopy. Samples forhistological examination were collected from the thymus of F_1 transgenics and the controls. All samples were preserved in Bouin's fixative, routinely sectioned (6-7 μm), stained with haematoxylin and eosin (H&E), and observed under a microscope.

(vi) Sample preparations for electron microscopy. For transmission electron microscopy, the thymus tissues (about 1 mm^3) of F_1 transgenics and the controls were fixed with 2.5% glutaraldehyde in 0.2 mol/L sodium cacodylate, post-fixed in 1% buffered osmium tetroxide, dehydrated in a series of graded ethanols and embedded in 812 medium. Ultrathin sections were stained with uranyl acetate and lead citrate, and examined with a H-600 electron microscope.

2. Results

(i) Body length and body weight of F_1 transgenic fish. The body length and body weight of F_1 transgenics and the controls were both randomly obtained, and measured (Table 1). Results showed that the body weight of the transgenics was from 987 to 1400 g with average 1173 g, while that of the controls was from 600 to 750 g with average 676.2 g. The body weight of the transgenics significantly exceeded that of the controls. The body length was not obviously altered in the two groups.

Table 1 Body length and body weight of F_1 transgenics and the controls

		1	2	3	4	5	Average
Transgenic fish	Body length/cm	38.5	40.6	36.8	34.1	37.6	37.52
	Body weight/g	1050	987	1400	1250	1180	1173
		1	2	3	4	5	Average
Control fish	Body length/cm	37.3	38.9	34.9	33.1	30.4	34.92
	Body weight/g	685	669	677	750	600	676.2

(ii) Thymus weight and index of thymus/body weight in F_1 transgenic fish. The thymus weight and thymus weight index of F_1 transgenics and the controls were measured and counted (Table 2). Results showed that the unilateral thymus of the transgenics weighed from 190 to 295 mg with average 218.6 mg, whereas the unilateral thymus of the controls weighed from 20 to 81 mg with average 42.5 mg; i.e. the thymus weight in the transgenics was 5.14 fold over that in the controls. The index of thymus/body weight in the transgenics was 2.97 fold over that in the controls. Both thymus weight and the index of thymus/body weight of the transgenics significantly exceeded that of the controls.

Table 2　Thymus weight and index of thymus /body weight of F_1 transgenics and the controls

		1	2	3	4	5	Average
Transgenic fish	Thymus weight/mg	295/196 245.5	195/214 204.5	195/259 227	205/190 197.5	237/200 218.5	218.6
	Body weight/g	1050	987	1400	1250	1180	1173
	Weight index	23381	20719	16214	15800	18517	18926
		1	2	3	4	5	Average
Control fish	Thymus weight/mg	40/50 45	23/34 28.5	20/42 31	42/53 47.5	40/81 60.5	42.5
	Body weight/g	685	669	677	750	600	676.2
	Weight index	6569	4260	4579	6333	10083	6365

Thymus weight(mg) in unilateral side and index of unilateral thymus/bady weight (mg/g$\times10^5$)

(iii) Anatological characters of the thymus in F_1 transgenic fish. The thymus of the controls was situated as a paired organ on either side in the dorsal part of the opercular cavity beneath superficial epithelium, and it remained visible in one of the dorsal upper corners of the opercular cavity. The appearance and weight in bilateral thymus were inconsistent. Generally, the thymus of the controls appeared to be smaller, irregular conical or oval in shape in light-grey colours (Fig. 1); while the thymus of the transgenics appeared larger (Fig. 1), irregular oval or quadrilateral in shape in cream colours, and it was easily observed in exposed opercular cavity.

Fig. 1　The thymus of F_1 transgenic and control fish

(iv) Histological observation of the thymus in F_1 transgenic fish. The thymus superficial of common carp was a typical epithelium (containing mucous goblet cells) under which there were granulocytes. Both of them composed the external epithelial surface (EES) of thymus. Thymus parenchyma is divided into the outer region and the inner region (similar to the cortex and medulla of thymus described in mammals). In the thymus of the controls, the EES was thinner with few mucous goblet cell and granulocyte (Fig. 2(a)). Thymic lobe was indistinct. The bilateral thymus was degenerating in different degree and in peripheral part of thymus. The greater part of thymus degenerated with the thinned outer region, decreased thymocytes and groups of myoid cells and fatty cells (Fig. 2(b)-(d)). In the inner region, post-capillary venule became empty and thymus corpusles were unclear.

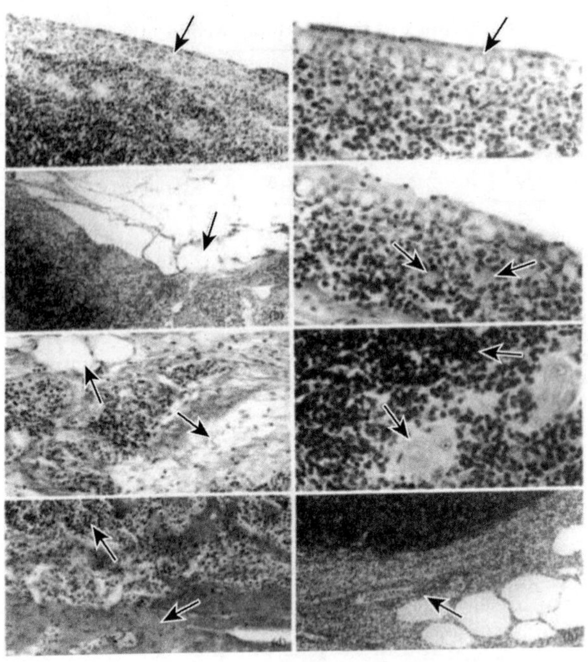

Fig. 2 (a)-(d) The thymus sections of the controls, stained with H&E. (a) Thinner external epithelial surface, with few mucous cell and granulocyte (×200); (b) degeneration in the peripheral part of thymus (×100); (c) scattered thymocytes and groups of fatty cells (×200); (d) groups of myoid cells in thymus (×200); (e)-(h) the thymus sections of the transgenics, stained with H&E; (e) mucous cells (×400); (f) granulocytes (×400); (g) compactly arranged thymocytes and the thymus corpuscle in the juncture of outer and inner region (×400); (h) a great amount of thymocytes in the thymus capsule and the connective tissues of outside thymus (×400)

In F_1 transgenic fish, the thymus well developed, EES thickened with a layer of continuous and stable mucous cells (Fig. 2(e)) and increased granuocytes. The granuocytes were large in size with an oval, dark-stained nucleus in the center or accentre of cells and the eosinophage granules in cytoplasm (Fig. 2(f)). The outer region of thymus also thickened with compactly arranged thymocytes. For this reason, epithelial-reticular cells were relatively indistinct. Thymus corpuscles were found (Fig. 2(g)). Thymocytes were more numerous in the inner region, so were those in the connective tissue of outside thymus, in the thymus capsule passed by blood vessels and in the interlobular septum (Fig. 2(h)). The proliferative thymocytes were mainly small lymphocytes filled with heterochromatin. Medium and large lymphocytes with euchromatin were also seen. The postcapillary venule was clear in the inner region, while the capillaries were abundant in the outer region.

(v) Ultrastructural observation of the thymus in F_1 transgenic fish. The thymocytes of the controls scattered in both outer and inner regions (Fig. 3(a)) with normal character. Groups of fatty cells were found, containing fattydrops in varied numbers and sizes in cytoplasm. Epithelial-reticular cells were relatively morenumerous. The lystic-reticular cells were often detected with a large amount of lystic structures (internal cysts) containing degenerative organelles and cellular debris (Fig. 3(b)), invaried size and electron density in cytoplasm. Besides this, the granuocytes were also observed, and contained two types of cytoplasmic granules: round shape, with the contains in higher and uniform electron density, similar to lysosome; long-rod shape, with crystal body-like substances (Fig. 3(c)).

The thymocytes of F_1 transgenic fish proliferated strongly and arranged compactly. They were divided into three types: (1) smallly mphocytes, the most numerouscells, with a higher nuclear/cytoplasmic ratio and abundant heterochromatin; (2) large lymphocytes, with a lower nuclear/cytoplasmic ratio and abundant euchromatin; and (3) medium lymphocytes. No pathological microstructural changes were detected in proliferous thymocytes (Fig. 3(d)). Lymphoblasts in mitosiswere obviously observed. The epithelial-reticular cells were extended in branching processes with enchromatin-enriched nucleus and lysosome-filled cytoplasm. The special phagocytes with an irrugular, enchromatin-enriched nucleus were often found. In the cytoplasm of this phagocytes, besides mitochondria and rough endoplasmic reticulum, the different sized lysosomes, engulfed small lymphocytes and damaged cellular debris were also seen (Fig. 3(e)). They were often located around blood vessels. Shown by the electron m icroscopy (TEM), the thymus corpusles were clear, and characterized by a concentric arrangement of epithelial-reticular cells surrounding an obviously hypertrophied central cell (Fig. 3(f)). The cystic cells were also observed in corpucles. Structurally, the corpuscles should be the degenerated ones.

3. Discussion

The observation of "All-fish" GH transgene effect on thymus development in carp indicated that the thymus of F_1 transgenic fish showed a great increase in both size and weight. The thymus weight in the transgenics was 5.14 fold over that in the controls. The index of thymus/body weight in the transgenics was 2.97 fold over that in the controls. Light microscopy observation revealed that the thymus of GH-transgenic fish well developed with obviously thickened outer region and compactly arranged thymocytes. Further analysis with electron microscopy revealed that proliferative cells were mainly small lymphocytes, no pathological changes were found and the thymic structure was normal. By sexual reproduction, the exogenous "All-fish" GH gene has been passed from P_0 transgenic fish into F_1 generation successfully. Up to now, it is the first report that GH affects the thymus and thymocytes in fish, and GH transgene promotes the thymocyte development in animals and leads to the immunobiological function of fish.

After puberty, the thymus in mammals degenerated with age, whereas in fish the thymus degenerated at different ages. For example, Tatner and Lu reported respectively that the thymus in rainbow trout (*Salmo garidneri*) and grass carp (*Ctenopharyngodon idellus*) started to degenerate over 15 and 12 months[20,21]. In addition, seasonal degeneration also occurred in thymus of fish[22]. Therefore, besides the age factor, the thymus degeneration of the controls should also be associated with the season factor. Because GH gene was expressed highly *in vivo*, the thymus of the transgenics did not only avoid degeneration, but also develop well with the thymocytes proliferating significantly. With regard to the mechanism of GH effect on the immune cells in mammals, Sprang and Bazan pointed out that GH is structurally similar to a number of cytokines, including interleukin (IL)-2, IL-4, IL-5, granulocyte-colony stimulating factor (G-CSF), granulo-cyte-macrophage-colony stimulating factor (GM-CSF) and the interferons (ITF)[23]. GH receptors are producedby many cells in the body including haematopoietic and lymphoid cells[9,13]. When bound to the GH receptors (GHR) of hepatic/extrahapatic tissues, GH induces to produce IGF-Ⅰ, which promotes the tissue development by endocrine/paracrine, autocrine; GH, bound to the GHR, also acts directly on extrahapatic tissues promoting tissue development[24]. What's important is GHR belongs to Ⅰ type of cytokine/haematopoietin receptor family[9,25]. Ⅰ type of cytokines has not only common structural features, but also the same effects. The functional similarity has been explained in their receptors, because to a great extent they depend on the common chains of type Ⅰ of cytokine receptor and the common pathways of signal transducer induced by cytokine receptors. For example, GH, erythropoietin (EPO), thrombopoietin (TPO), Prolactin (PRL), IL-2, IL-3, IL-5, IL-6, IL-7, IL-9, G-CSF and GM-CSF were reported to participate in the signal transcription of T cell receptors and to transmit the signal of T cell proliferation by activating different Janus kinase (Jak) to further activate the signal tranducer and activator oftranscription 5 (Stat5)[16,25]. On one hand, GH could exert its effects when it was bound to the GHR, IL-2R and other Ⅰ type of cytokine receptors of thymocytes; on the other hand, an Ⅰ type of cytokine could also exert its effects when boundto their receptors and other Ⅰ type of cytokine receptors (including GHR). This is one of the mechanisms that GH functions as a thymopoietic factor, T-cell-stimulating factor and haemotopoietic growth factor and this is also an important qualification that GH is regarded as a cytokine. In fact, GH, PRL and thyroxine *in vivo* and *in vitro* have been shown to increase the secretion of thymic hormone in epithelial-reticular cells (such as thymosin, thymopoietin, etc.), associated with T cell proliferation[26]. It is notable that the thymocyte itself could produce IL-1, IL-2, IL-4, IL-6, and express IL-2 receptor[26]. Thus it can be seen that GH, as both hormone and cytokine, shows significant multi-effectness and net workness in promoting the proliferation of T lymphocytes. Of course, due to the late started studies on the cytokine in fish, up to now, further investigation has still been lacking for the action mechanism of GH on the immune system and the relation between the GH and other cytokine in fish. But, it has been confirmed that the sequence analyses for GH gene show high homologous between fish and other mammals and the molecular structure of GH in fish also contains four counter-parallel arranged spiral fragments[27]. According to the recent reports about GH acting *in vitro* as a phagocyte activating factor[28], enhancing leukocyte mitogenesis and phagocytosis[29,30], and in this study on "All-fish" GH-transgene promoting thymocyte proliferation, it is suggested that there might be a certain relation (similar to mammals) between the GH/GHR and other cytokine/cytokine receptors in fish. Based on our results and the positive cross reaction on blood lymphocyte membrane in grass carp with anti-human IL-2R monoclonal antibody[31], we realize that there might exist GHR and/or other cytokine receptors on thymocyte membranein fish.

Our observation revealed that, in degenerative thymus of the controls, the thymocytes scattered, the granulocytes with granules and the epithelial-reticular cells with internal cysts can be seen. According to the microstructure of two types of the granules, this granulocyte should be the neutrophilic granulocyte, whose action was to engulf and digest the dead thymocytes. The epithelial-reticular cells, with a great amount of internal cysts containing the degenerative organelles and cellular debris, were more numerous and arranged in groups. Analysed structurally, its action could also be similar to that of neutrophilic granulocyte. Though the typical thymus corpusle and the phagocyte which engulfed and damaged small lymphocytes were found in transgenic thymus, this phagocyte was not distinguished from the epithelial-reticular cell, macrophage or atypical nurse cell. In addition, it is difficult to observe the epithelial-reticular cell with internal cysts and the neutrophilic granulocytes in the transgenic thymus. Although this precise mechanism of the function and phenomena on fish remains unknown, it might be correlated with thymocyte survival and selection in mammals. Murphy *et al.* realized that GH can induce both IL-7 and stem cell factor (SCF), produced by thymic epithelial cells. IL-7 is a key factor for γδTCR rearrangement. T-cell progenitors undergo T-cell receptor (TCR) rearrangement, acquire CD_4 and CD_8, and then under gothymic selection based on affinity for the major histocompatibility complex (MHC). After the thymics selection, mature CD_4/CD_8 single-positive cells leave the thymus. GH also appears to promote early thymocyte (pro-T_1 and pro-T_2) survival[9]. This might also be one of the important reasons why GH-transgene promotes thymocyte development and retards thymus degeneration.

Undoubtedly, the "All fish" GH transgene effectively promotes thymus developmemt and thymocyte proliferation, and retards thymus degeneration. The observational results suggest that the advantage of exogenous GH transgene might also lie in enhancing the immune function and the resistance of fish body to pathogen. In addition to enhancing physiological function, in fact, every mapulated gene transfer (specially immune-related genes) might bring about some pathological changes. Because immune reaction shows its dual effects, further research and investigation are necessary to overall understanding of the immunobiological function of GH transgene. Due to the infiltration of a great amount of neutrophilic granuocytes and macrophages into tissues, IL-5, TNF and GM-CSF transgenic mice were accomplished by acidophilic granuemia, chronic inflammatory arthritis or other pathological symptoms[15,16]. It is reported that the pneumonia was shown in GH-transgenic pigs[18]. As a colony stimulating factor[9], GH might also increase phagocyte infiltration and phagocytosis, which could sometimes result in inflammation.

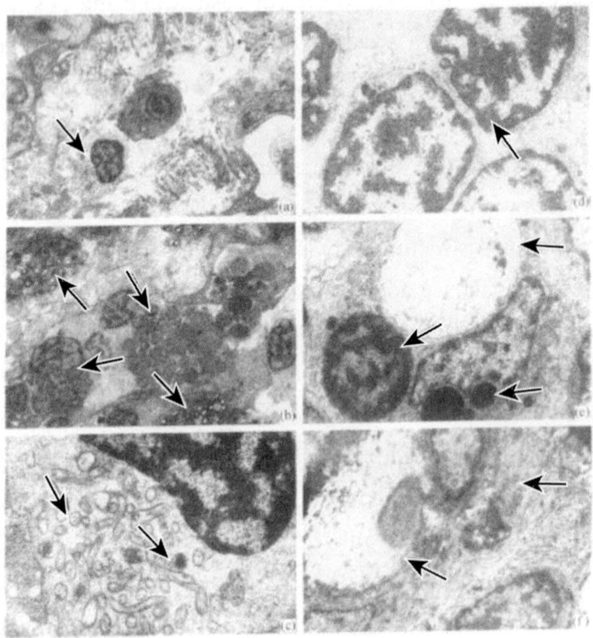

Fig. 3 (a)-(c) The thymic ultrastructure of the controls, shown by TEM. (a) Scattered lymphocytes and increased connective tissues (×4000); (b) fatty cells and cystic reticular cells (×3500); (c) two types of granules in granulocyte (×20 000); (d)-(f) the thymic ultrastructure of the transgenics, shown by TEM; (d) the different shapes of lymphocytes with normal structure (×15 000); (e) the phagocyte and the engulfed, digested cellular debris (×12 000); (f) the degenerative cell in the centre of thymus corpuscle (×10 000)

Acknowledgements This work was supported by the State High Technology Program and the Special Funds of the Major State Basic Research of China (Grant Nos. 2001AA212281 and 2001CB109006), and the National Natural Science Foundation of China (Grant Nos. 30130050 and 30070588).

References

1. Zhu, Z., Li, G., He, L. *et al.*, Nover gene transfer into the fertilized eggs for goldfish (*Carassium auratus* L), Z. Angew. Chthyol., 1985, 1: 31-34.
2. Zhu, Z., Xu, K., Li, G. *et al.*, Biological effects of human growth hormone gene microinjected into the fertilized gene of loach *Misgurnus anguillicaudatus* (Cantor), Kexue Tongbao, 1986, 31(14): 988.
3. Dunham, R. A., Eash, J., Adkins, J. *et al.*, Transfer of the metallothione in human Growth hormone gene into channel catfish,Trans. Amer. Fish. Soc., 1987, 116: 87-91.
4. Penman, D. J., Beeching, A. J., Penn, S. *et al.*, Factors affecting survial and integration following microinjection of novel DNA into rainbow trout eggs, Aquaculture, 1990, 85: 35-50.

5. Fletcher, G. L., Shears. M. A., King, M. J. *et al.*, Evidence for antifreeze protein gene transfer in Atlantic salmon (*Salmo salar*), Can. J. Fish. Aqaut. Sci., 1988, 45: 352-357.

6. Shear, M. A., Flecher, G. L., Hew, C. L. *et al.*, Transfer, expression and stable inheritance of antifreeze protein gene in Atlantic salmon (*Salmo salar*), Mol. Mar. Biol. Biotechnol., 1991, 1(1): 58-63.

7. Zhong, J., Zhu, Z., Resistance to GCHV of hLFc-transgenic grass carp, Acta Hydrobiol. Sin. (in Chinese), 2001, 25(5): 528-530.

8. Sakai, M., Current research status of fish immunostimulant, Aquaculture, 1999, 172: 63-92.

9. Murphy, W. J., Longo, D. L., Growth hormone as an immunomodulating therapeutic agent, Immunol., Today, 2000, 21(5): 211-213.

10. Harris, J., Bird, D. J., Mini Review. Modulation of the fish immune system by hormones, Vet. Immunol. Immunopathol., 2000, 77: 163-176.

11. Warwick-Davies, J., Lowrie, D. B., Cole, P. J., Growth hormone is a human macrophage activating factor-priming of human monocytes for enhanced release of H_2O_2, J. Immunol., 1995, 154: 1909-1918.

12. Sakai, M., Kajita, Y., Kobayashi, M. *et al.*, Immunostimulating effcct of growth hormone: *in vivo* administration of growth hormone in rainbow trout enhances resistance to Vibrio anguillarum infection, Vet. Immunol. Immunopathol., 1997, 57:1-6.

13. Dialynas, E., Brown-Borg, H., Bartke, A., Immune function in transgenic mice overexpressing growth hormone (GH) releasing hormone, GH or GH antagonist, Proc. Society for Exp. Biol. Med., 1999, 221: 178-183.

14. Mathew, L. S., Hammer, R. E., Behringer, R. R. *et al.*, Growth enhancement of transgenic mice expressing human insulin-like growth factor Ⅰ, Endocrinolgy, 1988, 123: 2827-2833.

15. Sun, W., Wang, H., Methods of Cytokine Research (in Chinese), Beijing: China People's Health Press, 1999, 436-690.

16. Jin, B., Cellular and Molecular Immunology (in Chinese), Beijing: China Science Press, 2001, 133-202.

17. Zhu, Z., Growth hormone gene and the transgenic fish, Agricultural biotechnology (eds. You, C. B., Chen, Z. L.,), Beijing: China Science and Technology Press, 1992, 106-116.

18. Wang, Y., Hu, W., Wu, G. *et al.*, Genetic analysis of "All-fish" growth hormone gene transferred carp (*Cyprinus carpio* L.) and its F_1 generation, Chinese Science Bulletin, 2001, 46(14): 1174-1177.

19. Zhu, Z., Xu, K., Xie, Y. *et al.*, A model of transgenic fish, Science in China, Ser. B (in Chinese), 1989(2): 147-155.

20. Tatner, M. F., Manning, M. J., Growth of the lymphoid organs in rainbow trout, *Salmo garidneri* from one to fifteen months of age, J. Zool., 1983, 199(4): 503-520.

21. Lu, Q., Histological studies of the thymus in Grass carp, *Ctenopharyngodon idellus*, Acta Hydrobiol. Sin. (Abstract in English), 1991, 15(4): 327-331.

22. Manning, M. J., Fish, Immunology: A Comparative Approach (ed. Turner, R. J.), England: John Wiley & Sons Ltd, 1994, 69-100.

23. Sprang, S. R., Bazan, J. F., Cytokine structural taxonomy and mechanisms of receptor engagement, Curr. Opin. Struc. Biol.,1993, 3: 815-827.

24. Lupu, F., Terwilliger, J. D., Lee, K. *et al.*, Roles of growth hormone and insulin-like growth factor 1 in mouse postnatal growth, Dev. Biol., 2001, 229(1): 141-162.

25. Moutoussamy, S., Kelly, P. A., Finidori, J., Growth-hormone-receptor and cytokine-receptor-family signaling, Eur. J. Biochem.,1998, 225: 1-11.

26. Kendall, M. D., Function anatomy of the thymic microenvironment., J. Anat., 1991, 177: 1-29.

27. Wang, W., Wang, Y., Zhu, Z., Studies on growth hormone gene engineering in fish, Proc. in Natural Sci. (in Chinese), 2002, 12(5): 456-460.

28. Calduch-Giner, J. A., sitja-bobadilla, A., A lvarz-Pellitero, P . *et al.*, Growth hormone as an *in vitro* phagocyte-activating factor in the gilthead sea bream (*Sparus aurata*), J. Endocrinol., 1997, 146:459-467.

29. Sakai, M., Kobayashi, M., Kawauchi, H., Mitogenic effect of growth hormone and prolactin on chum salmon *Oncorhynchus keta* leukocytes *in vitro*, Vet. Immunol. Immunopathol., 1996, 53:185-189.

30. Sakai, M., Kobayashi, M., Kawauchi, H., Enhancement of chemiluminescent responses of phagocytic cells from rainbow trout, *Oncorhynchus mykiss* by injection of growth hormone, Fish. Shellfish Immunol., 1995, 5: 375-379.

31. Guo, Q., Identification of an interleukin-2 substance in splenocyte culture supernatant of Grass carp, *Ctenopharyngodon idellus* and Chinese soft-shelled turtle, *Trionyx sinensis*, Acta Hydrobiol. Sin. (Abstract in English), 2001, 25(1): 21-27.

生长激素转植基因促进鲤的胸腺发育

郭琼林　汪亚平　贾伟章　朱作言

中国科学院水生生物研究所，武汉　430072

摘　要　采用显微注射方法获得转 CAgcGHc 基因鲤群体。转基因鱼与野性型对照鱼胸腺的比较发现，一龄 F_1 转基因鱼的胸腺在大小和重量上均显著高于对照鱼。转基因鱼单侧胸腺重 190 至 295mg(平均 218.6mg)，对照为 20 至 81mg (平均 42.5mg)，转基因鱼的胸腺重量是对照鱼的 5.14 倍，胸腺/体重指数是对照鱼的 2.97 倍。显微镜观察发现转基因鱼的胸腺拥有更厚实的外围区域及致密的细胞，对照鱼的胸腺则比较稀疏并分散较多的脂肪细胞，电镜显示转基因鱼的增殖细胞主要是小淋巴细胞且无病理性的变化。本研究证实转“全鱼”GH 基因促进了胸腺的发育和细胞增殖，并迟滞胸腺细胞的退化。本研究为进一步研究转 GH 基因鲤的免疫功能奠定了基础。

Effects of the "all-fish" Growth Hormone Transgene Expression on Non-specific Immune Functions of Common Carp, *Cyprinus carpio* L.

Wen-Bo Wang[1,2,3] Ya-Ping Wang[1] Wei Hu[1] Ai-Hua Li[1] Tao-Zhen Cai[1] Zuo-Yan Zhu[1] Jian-Guo Wang[1]

1 State Key Laboratory of Freshwater Ecology and Biotechnology, Institute of Hydrobiology, Chinese Academy of Sciences, Wuhan 430072

2 Institute of Hydrobiology, Jinan University, Guangzhou 510632

3 Graduate School of the Chinese Academy of Sciences, Beijing 100049

Abstract This study investigated non-specific immune functions of the F_2 generation of "all-fish" growth hormone transgenic carp, *Cyprinus carpio* L. Lysozyme activity was 145.0 (±30.7) U ml^{-1} in the transgenic fish serum and 105.0 (±38.7) U ml^{-1} in age-matched non-transgenic control fish serum, a significant difference ($P<0.01$). The serum bactericidal activity in the transgenics was significantly higher than that in the controls ($P<0.05$), with the percentage serum killing of 59.5% (±6.83%) and 50.8% (±8.67%), respectively. Values for leukocrit and phagocytic percent of macrophages in head kidney were higher in transgenics than controls ($P<0.05$). However, the phagocytic indices in the transgenics and the controls were not different. In addition, the mean body weight of the transgenics was 63.4 (±6.65) g, much higher than that of the controls [39.2 (±3.30) g, $P<0.01$]. The absolute weight of spleen of the transgenics [0.13 (±0.03) g] was higher than that of the controls [0.08 (±0.02) g, $P<0.01$]. However, there was no difference in the relative weight of spleen between the transgenics and the controls, with the spleen mass index being 0.21% (±0.02%) and 0.20% (±0.03%), respectively. This study suggests that the "all-fish" growth hormone transgene expression could stimulate not only the growth but also the non-specific immune functions of carp.

Keywords "all-fish" growth hormone gene; transgenic fish; growth hormone; non-specific immune functions

1. Introduction

In 1984, the first batch of growth hormone (GH) transgenic fish was produced in China (Zhu *et al.*, 1985). Since then, many laboratories all over the world have turned to the study of transgenic fish. In the early studies, the recombinant gene was *MThGH*, i.e., the human GH

文章发表于 *Aquaculture*, 2006, 259(1-4):81-87

gene (*hGH*) under the transcriptional control of the mouse metallothionein-1 (*MT-1*) promoter (Pavlakis and Hammer, 1983). Due to biosafety and bioethical issues, the *MT-1* promoter that needs heavy metal ion induction and the *hGH* that encodes a human protein are not appropriate for the purpose of human food production. For this reason, Zhu *et al.* (1989) proposed the idea of constructing an "all-fish" transgene which contains only piscine sequences. Transgenic fish have been studied now for 20 years and the research results are very extensive, including growth effects, mechanism of transgene integration, and energy, body composition, cultivation, and ecological safety of transgenic fish (Zhu and Zeng, 2000). However, the immunology of the GH transgenic common carp remains unknown. What pleiotropic effects have been induced by the exogenous GH gene expression on the immune functions of carp? Are these effects beneficial or harmful to host fish? Is resistance to disease of the host fish altered by GH transgene expression? Investigation of these questions will enrich basic research on GH transgenics and supply information relating to the prevention and control of fish disease.

GH, produced by the hypothalamus-pituitary-adrenal (HPA) axis, is a pleiotropic hormone regulating many aspects of fish physiology, including growth (Donaldson *et al.*, 1979), osmoregulation (Sakamoto *et al.*, 1997), and reproduction (Trudeau, 1997). Recently, the relationship between the HPA axis and the immune system in fish has been given considerable attention by investigators. There is increasing evidence that GH also exerts an immunostimulatory function in fish (Calduch-Giner *et al.*, 1997; Harris and Bird, 2000; Yada and Nakanishi, 2002). For example, in some teleost fishes, GH can promote lymphopoiesis, leukocyte mitogenesis, natural killer cell activity, antibody production, serum haemolytic activity and resistance to disease (Sakai *et al.*, 1996; Harris and Bird, 2000; Yada and Nakanishi, 2002; Johansson *et al.*, 2004). However, Jhingan *et al.* (2003) found that GH suppresses immune function in GH transgenic coho salmon.

Teleost fishes possess a variety of specific and non-specific defence mechanisms against invading organisms. When a pathogen penetrates the physical barriers of the animal, the first lines of defence are those of the non-specific immune system. Chemical defences, such as serum lysozyme, complement components, lectins and C-reactive protein, attack the pathogen, or may opsonise it for further destruction by the cellular components of the non-specific immune system (Secombes, 1996; Yano, 1996). The present work studies the effects of "all-fish" GH transgene expression on the non-specific immune functions of common carp.

2. Materials and Methods

2.1 Production of transgenic fish

The "all-fish" gene construct *pCAgcGHc* included the grass carp (*Ctenopharyngodon idellus*) GH cDNA (*gcGHc*) driven by the β-actin gene promoter of common carp (*Cyprinus *carpio* L.), which was modified from the construct *pCAgcGH* (Zhu, 1992; Wang *et al.*, 2001). The production of P_0 and F_1 "all-fish" GH transgenics was described in detail by Wang *et al.* (2001). In brief, P_0 transgenics were produced by microinjection of *pCAgcGHc* into the fertilized eggs of Yellow River carp (*C. carpio* L.) (for details of gene transfer, see Zhu *et al.*, 1989). (1) F_1 transgenics were produced as follows: P_0 transgenic ♂ (sperm revealed to be *pCAgcGHc* positive) × non-transgenic ♀. (2) F_2 transgenics were produced by a similar method in 2004: F_1 transgenic ♂ (sperm revealed to be *pCAgcGHc* positive) × non-transgenic ♀. On the same day, a batch of non-transgenic control carp (Yellow River carp) was also produced. The fry of F_2 transgenics and controls were reared in separate ponds at the Institute of Hydrobiology, Chinese Academy of Sciences (Wuhan, Hubei province, China). Polymerese chain reaction (PCR) was performed on F_2 transgenics according to Wang *et al.* (2001) to detect the *pCAgcGHc*-transgene positive fish. PCR-positive fish and controls were randomly taken to use as the experimental fish in this study (n=10 for each group).

2.2 Rearing of the experimental fish

After PCR detection, the 10 transgenics and 10 controls were transferred to two identical tanks (1800 l each) in a recirculation system, supplied with a constant flow of water (flow rate 2.5 l min^{-1}). The temperature range of the water during the experimental period was 22-26℃. They were fed twice daily with a carp feed without growth additive until use. The feed was produced by the Institute of Hydrobiology, Chinese Academy of Sciences.

2.3 Bacteria

Micrococcus lysodeikticus, purchased from Sigma Chemical Co. (St Louis, MO, USA), was stored at –20℃ and used for determining the lysozyme activity. *Vibrio fluvialis* (XS 91-24-3) and *Aeromonas hydrophila* (XS 91-4-1) were originally isolated from a virulent outbreak of bacterial hemorrhagic septicemia in a carp farm in China and stored in the laboratory at –20℃, with the former used for the bactericidal assay and the latter for the phagocytic ssay.

2.4 Sampling

The 10 transgenics and 10 controls were sampled after 3.5 months of rearing. The fish were rapidly placed in water containing a concentration of anaesthetic (0.2 g l^{-1}; MS 222; Sigma) for about 10 s, then weighed and blood was sampled from the caudal vessel. An aliquot of each blood sample was stored on ice for 30 min, centrifuged at 400×*g* for 10 min at 4℃, and the serum collected was stored at –80℃ for analyses. The remainder of the blood sample was used to determine leukocrit values. The head kidney and spleen of each fish were aseptically dissected after blood sampling, with the former placed in Leibovitz-15 medium (Sigma) and stored at 4℃ for the phagocytic assay, and the latter examined for spleen mass index directly.

2.5 Lysozyme assay

Lysozyme activity was measured according to Parry *et al.* (1965), using a turbidity assay in which 0.2 mg ml^{-1} lyophilized *M. lysodeikticus* in 0.04 M sodium phosphate buffer (pH 5.75) was used as substrate. Five microlitres of fish serum was added to 3 ml of the bacterial suspension, and the reduction in absorbance at 540 nm determined after 0.5 and 4.5 min incubations at 22℃. One unit of lysozyme activity was defined as a reduction in absorbance of 0.001 per min.

2.6 Serum bactericidal activity

Serum bactericidal activity was assessed as described by Ainsworth *et al.* (1995) with some modifications. *V. fluvialis* (XS 91-24-3) was grown in 100 ml of tryptone soy broth (TSB) in a rotary shaker (200 rpm) at 28℃ for 24 h. The bacterium was washed with 0.5% saline and the bacterial density adjusted to 10^6 bacteria ml^{-1} in saline by spectrophotometer (OD 620), then 50 μl of serum was mixed with 50 μl of suspension of *V. fluvialis* (XS 91-24-3). After 5 h of incubation at 22℃, 10-fold and 100-fold dilutions of the mixture (10 μl per dilution) were plated on tryptone soy agar (TSA) plates. After 24 h of incubation at 28℃, colony-forming units (CFUs) on the TSA plates were counted. The results were expressed as the percentage serum killing using the formula: percentage serum killing = 100 – (CFUnumber×100) × (CFU number in controls) $^{-1}$.

2.7 Leukocrit value

Blood was collected in heparinized microleukocrit tubes and centrifuged in a swinging-bucket rotor at 400×*g* for 20 min. The packed leukocyte layer height was measured using a vernier caliper, and the leukocrit value [white blood cell (WBC) volume] was defined as the percentage of the leukocyte layer height to total blood height inside the tube.

2.8 Phagocytic assay

Macrophage cells and suspensions of *A. hydrophila* (XS 91-4-1) were prepared according to the methods of Secombes (1990) and Thompson *et al.* (1996), respectively. The phagocytic assay was performed using the method of Sarder *et al.* (2001). In brief, the head kidney tissue was dissected from the fish and pushed through a 100 μl nylon mesh into L-15 medium containing 20 I.U. ml^{-1} of heparin. Cell suspensions were layered onto a 34%/51% Percoll (Sino-American Biotechnology Co., Shanghai, China) gradient and centrifuged at 400×*g* for 25 min at 4℃. The macrophage-enriched band was collected and washed with L-15 medium and the cell numbers were adjusted to 5×10^6 cells ml^{-1} L-15 medium containing 1% penicillin/streptomycin. Two hundred microlitre aliquots of the suspension were added to circles of clean glass slides (6 wells per slide, 15 mm diameter per well) and incubated for 3 h at 22℃, after which the glass slides were washed with L-15 medium to remove the nonadherent cells. Each of the resultant macrophage monolayer was incubated together with 200 μl of *A. hydrophila* (XS 91-4-1) bacterium suspension (5×10^7 cells ml^{-1} sterile saline) for 1 h at 22℃. The macrophage monolayers then were washed with L-15 medium, fixed with formaldehyde and stained with Giemsa. Finally, 200 macrophages were examined microscopically and the number of macrophages containing phagocytosed bacteria, as well as the number of phagocytosed bacteria, were counted. The results were expressed as phagocytic percent and phagocytic index, where: phagocytic percent = the number of the macrophages containing phagocytosed bacteria $\times$ 200^{-1} $\times$ 100 and phagocytic index = the number of phagocytosed bacteria $\times$ 200^{-1}.

2.9 Spleen mass index

Spleens of fish were excised, washed with saline, dried using filter paper and then weighed. The spleen mass index was calculated using the following formula: spleen mass index = (spleen weight) (body weight)$^{-1}$ $\times$ 100.

2.10 Statistical analysis

All results are expressed as means±standard errors (S.E.), and the effects of the "all-fish" GH transgene expression were analysed using a Student's *t*-test (P<0.05). All analyses were carried out on 10 fish per group.

3. Results

3.1 Lysozyme activity

The range of lysozyme activity in the transgenics was between 100 and 200 U ml^{-1}, with a mean of 145.0 (±30.7) U ml^{-1}; while the range in the controls was between 50 and 175 U ml^{-1}, with a mean of 105.0 (±38.7) U ml^{-1}. Lysozyme activity in the transgenics was significantly higher than that in the controls (P<0.01).

3.2 Bactericidal activity

The average number of CFUs on the TSA plates was 9.50 (±1.82) in the transgenics and 11.5 (±2.16) in the controls (P<0.01). The percentage serum killing in the transgenics was 59.5 (±6.83), higher than that in the controls [50.8 (±8.67); P<0.05].

3.3 Leukocrit

The leukocrit value in the transgenics was significantly higher than that in the controls (P<0.05), with values of 4.22% (±0.37%) and 3.87% (±0.39%), respectively.

3.4 Phagocytic activity

The phagocytic percent of macrophages in head kidney was 45.73 (±5.25) in the transgenics and 39.20 (±7.21) in the controls (Table 1; P<0.05). With respect to the phagocytic index, there was no statistical difference between the transgenics and the controls.

Table 1 Comparison of phagocytic activity of macrophages in head kidney between the F_2 "all fish" GH transgenic carp (*C. carpio* L.) and control carp

		No.										Means±S.E.	*t*-test
		1	2	3	4	5	6	7	8	9	10		
Phagocytic percent (%)	Transgenic carps	47.25	51.75	41.50	45.00	54.50	47.50	41.00	49.50	39.50	39.75	45.73±5.25	P<0.05
	Control carps	43.75	39.75	38.50	29.75	47.00	33.75	49.75	44.00	38.00	27.75	39.20±7.21	
Phagocytic index	Transgenic carps	2.80	2.53	3.27	2.78	2.74	3.78	3.36	2.84	2.78	2.15	2.90±0.46	P>0.05
	Control carps	3.13	1.94	2.33	3.57	2.78	2.57	3.53	2.26	3.61	3.03	2.87±0.60	

3.5 Spleen mass index

As shown in Table 2, both the spleen weight and the body weight of the transgenics were higher than those of the controls (P<0.01), but the spleen mass indices were not different between the transgenics and the controls, with values of 0.21% (±0.02%) and 0.20% (±0.03%), respectively.

Table 2 Comparison of spleen mass index between the F_2 "all fish" GH transgenic carp (*C. carpio* L.) and control carp

		No.										Means±S.E.	*t*-test
		1	2	3	4	5	6	7	8	9	10		
Spleen weight (g)	Transgenic carps	0.12	0.15	0.11	0.12	0.19	0.15	0.13	0.12	0.09	0.13	0.13±0.03	P<0.01
	Control carps	0.07	0.09	0.08	0.09	0.11	0.09	0.07	0.06	0.08	0.06	0.08±0.02	
Body weight(g)	Transgenic carps	56.9	65.7	58.4	58.3	75.7	70.2	60.5	63.2	55.8	69.4	63.4±6.65	P<0.01
	Control carps	36.9	41.2	37.5	39.8	44.5	41.5	38.6	39.5	40.4	32.1	39.2±3.30	
Spleen mass index (%)	Transgenic carps	0.21	0.23	0.19	0.21	0.25	0.21	0.21	0.19	0.16	0.19	0.21±0.02	P>0.05
	Control carps	0.19	0.22	0.21	0.23	0.25	0.22	0.18	0.15	0.20	0.19	0.20±0.03	

4. Discussion

Lysozyme is an important innate humoral factor of the fish immune system, having an antibacterial effect both by attacking the bacterial cell wall, thereby causing lysis, and by stimulating phagocytosis of bacteria. It is produced by neutrophils and macrophages, of which binding sites for GH have been found on both (Calduch-Giner *et al.*, 1995). In the present study, we observed higher lysozyme activity in transgenic fish than in age-matched control fish (Fig. 1). Was this the result of stimulation of neutrophils and macrophages induced by transgene expression? We presume that in the F_2 transgenic fish, the expressive production of the foreign GH gene may possess the same biological functions as that of the endogenous GH gene. A similar conclusion can be found in other articles (Cui, 1998; Dialynas *et al.*, 1999; Wang *et al.*, 2001). In aquaculture, GH usually acts as an immunostimulant to enhance the non-specific immunity of fish. When flounder (*Paralichthys olivaceus*) was fed with recombinant yeast (*Saccharomyces cerevisiae*) containing salmon (*Oncorhynchus keta*) GH, the serum lysozyme activity in fish was greatly improved, and this effect was dose-dependent (Wang and Zhang, 2000). In addition, endogenous GH can also regulate lysozyme activity. After brown trout (*Salmo trutta*) was transferred from freshwater to seawater, the GH level in fish blood rose, accompanied with an increase in serum lysozyme activity (Marc *et al.*, 1995). Although both endogenous GH and exogenous GH transgene expression can stimulate lysozyme activity, whether their mechanisms are the same requires further study.

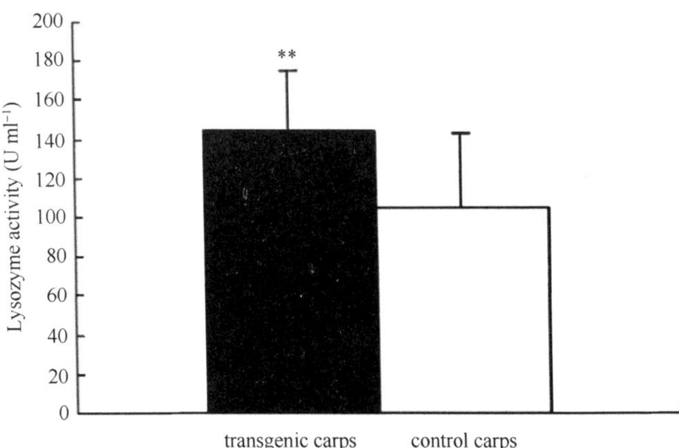

Fig. 1 Comparison of serum lysozyme activity between the F_2 "all fish" GH transgenic carp (*C. carpio* L.) and control carp. Bar graphs indicate means ± S.E. (n = 10). $^{**}P < 0.01$

There are many aspects exerting stimulative or suppressive effects on the immune system in fish, one of which is growth rate or size. It can influence not only innate immune functions in fish, but also their disease susceptibility (Magnadóttir, 2006). Owing to lack of size-matched controls in our experimental design, the increased non-specific immune response presented in this study may have been caused directly by transgene expression or indirectly through the effect of transgene expression on size of individual fish.

It is well known that lysozyme performs a bacteriolytic function only on Gram-positive bacteria, whereas the *V. fluvialis* (XS91-24-3) used in the bactericidal assay was a Gram-negative bacterium, which would not be dissolved by lysozyme. Therefore, the elevation of the serum bactericidal activity in the transgenics (Fig. 2) was unlikely to have been caused by the increased lysozyme activity. Then, what was responsible for this? Fish size may be taken into account and, on the other hand, besides lysozyme, the complement components in fish serum also play an important role in killing pathogens. It was reported by Doong (2000) that after injection with black porgy (*Acanthopagrus schlegeli*) GH, both the lysozyme activity and the alternative complement activity in black porgy were improved. But in the present study, whether GH transgene expression has caused the activation of the complement system and thereby promoted the bactericidal activity in the transgenics requires further investigation.

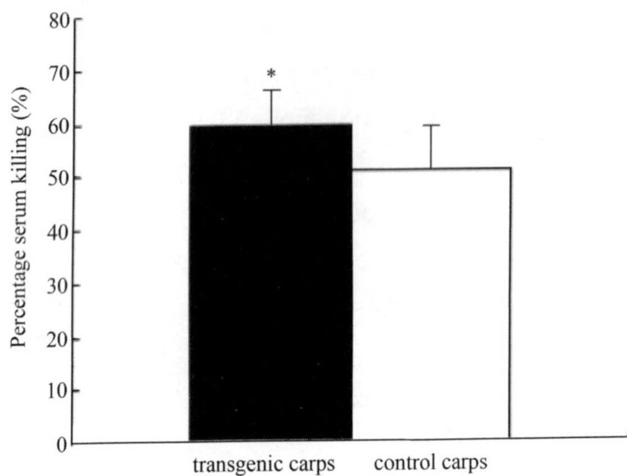

Fig. 2 Comparison of percentage serum killing between the F_2 "all fish" GH transgenic carp (*C. carpio* L.) and the control carp. Bar graphs indicate means ± S.E. (n = 10). $^*P < 0.05$

There are two main issues when it comes to the cellular immunity: one is the number of immune cells involved in the immune response, and the other is the activity of these cells. Leukocrit value and phagocytic activity of macrophages in head kidney are two typical indices to assess the amount and activity of the immune cells. The leukocrit value in the transgenics was higher than that in the controls (Fig. 3; P<0.05), indicating that GH transgene expression accelerated the leukocyte proliferation. Among the leukocytes, lymphocytes and monocytes are important immune cells mediating specific and non-specific immune response in fish, respectively. Thus, the enhanced leukocytosis in the transgenics supports the finding of Yada *et al.* (2004), who suggested that GH showed stimulatory effects not only on lymphopoiesis, but also on proliferation of other leukocytes in trout (*O. mykiss*) *in vitro*, resulting in an increase in the total blood cell number. Under common status, the amount of

GH in a fish is not fully saturated, so that the expressive GH production by the transgene has a chance to exert its biological effects on the fish (Donaldson *et al.*, 1979). Perhaps this is a reason why GH transgene expression could stimulate the leukocyte proliferation. It is reported that as an *in vitro* phagocyte-activating factor, GH can enhance the phagocytic activity of macrophages (Calduch-Giner *et al.*, 1997). Increased phagocytic activity and phagocytic index have been observed after rainbow trout (*O. mykiss*) was treated with chum salmon (*O. keta*) GH *in vitro* (Sakai *et al.*, 1996). Similarly, GH transgene expression also stimulated the phagocytosis of macrophages in head kidney (Table 1). As was mentioned previously, macrophages can produce lysozyme to kill pathogens, including the *A. hydrophila* used in the phagocytic assay (Grinde, 1989). Therefore, it can be inferred that the phagocytic activity promoted in the transgenics may correlate with their increased lysozyme activity.

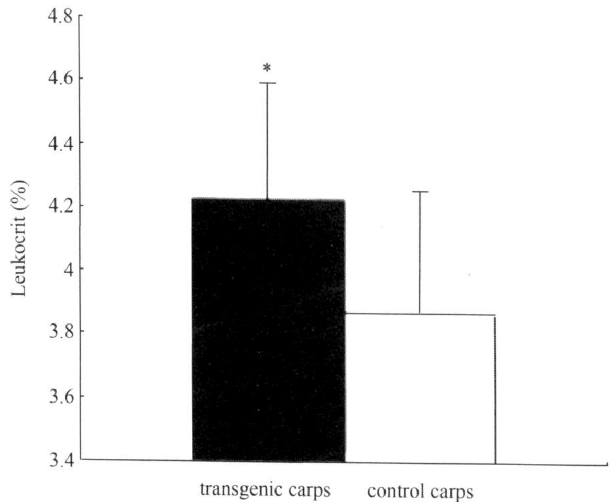

Fig. 3 Comparison of leukocrit value between the F_2 "all fish" GH transgenic carp (*C. carpio* L.) and the control carp. Bar graphs indicate means ± S.E. ($n = 10$). $^*P < 0.05$

In mammals, it has been documented that bovine GH transgenic mice showed significant increases in spleen weight and mitogenesis of splenic cells after a mitogenic stimulation (Dialynas *et al.*, 1999). In the present study, GH transgene expression increased the absolute weight of fish body and spleen, but did not increase the relative weight of spleen, as shown by spleen mass index (Table 2), which closely match the findings derived from the human GH transgenic carp (*C. carpio* L.) (Cui, 1998). As one of the important immune organs in fish, the spleen contains a variety of immune cells, such as lymphocytes, macrophages and granulocytes. Hence, the accelerated development of spleen may be beneficial to the immunity of fish.

In conclusion, the non-specific immune functions of common carp were significantly enhanced after "all-fish" GH transgene expression, but the underlying molecular mechanism requires more study. With growth and immunity enhancement, "all-fish" GH transgenic carp is a good breed of fish for aquaculture, and is expected to become the first transgenic animal to come into the market in China (Zhu and Zeng, 2000; Wu *et al.*, 2003).

Acknowledgements The authors would like to thank Liu Weiyi and Cao Ting for their help in collecting the experimental fish. This work was supported by the National Natural Science Foundation of China (Grant No. 30130050), the National Basic Research Program of China (Grant No. 2001CB109006) and the National High Technology Research and Development Program of China (Grant No. 2004AA213120).

References

Ainsworth, A.J., Rice, C.D., Xue, L., 1995. Immune responses of channel catfish, *Ictalurus punctatus* (Rafinesque), after oral or intraperitoneal vaccination with particulate or soluble *Edwardsiella ictaluri* antigen. J. Fish Dis. 18, 397-409.

Calduch-Giner, J.A., Sitja-Bobadilla, A., Alvarez-Pellitero, P., Perez-Sanchez, J., 1995. Evidence for a direct action of GH on haemopoietic cells of a marine fish, the gilthead sea bream (*Sparus aurata*). J. Endocrinol. 146, 459-467.

Calduch-Giner, J.A., Sitja-Bobadilla, A., Alvarez-Pellitero, P., Perez- Sanchez, J., 1997. Growth hormone as an *in vitro* phagocyte-activating factor in the gilthead sea bream (*Sparus aurata*). Cell Tissue Res. 287, 535-540.

Cui, Z., 1998. Biosafety assessment of GH-transgenic common carp (*Cyprinus carpio* L.). PhD dissertation. Institute of Hydrobiology, Chinese Academy of Sciences, Wuhan, China. 54 pp.

Dialynas, E., Brown-Borg, H., Bartke, A., 1999. Immune function in transgenic mice overexpressing growth hormone (GH) releasing hormone, GH or GH antagonist. Proc. Soc. Exp. Biol. Med. 221, 178-183.

Donaldson, E.M., Fagerlund, U.H.M., Higgs, D.A., McBride, J.R., 1979. Hormonal enhancement of growth. In: Hoar, W.S., Randall, D.J., Brett, J.R. (Eds.), Fish Physiology, vol. VIII. Academic Press, New York, pp. 455-597.

Doong, J., 2000. Effects of recombinant growth hormone on dietary protein assimilation and immunity in the black porgy (*Acanthopagrus schlegeli*). MSc thesis. Institute of Marine Biology, National Sun Yat-sen University, Kaohsiung, Taiwan, ROC. 4 pp.

Grinde, B., 1989. Lysozyme from rainbow trout, *Salmo gairdneri* Richardson, as an antibacterial agent against fish pathogens. J. Fish Dis. 12, 95-104.

Harris, J., Bird, D.J., 2000. Modulation of the fish immune system by hormones. Vet. Immunol. Immunopathol. 77, 163-176.

Jhingan, E., Devlin, R.H., Iwama, G.K., 2003. Disease resistance, stress response and effects of triploidy in growth hormone transgenic coho salmon. J. Fish Biol. 63, 806-823.

Johansson, V., Winberg, S., Jönsson, E., Hall, D., Björnsson, B.T., 2004. Peripherally administered growth hormone increases brain dopaminergic activity and swimming in rainbow trout. Horm. Behav. 46, 436-443.

Magnadóttir, B., 2006. Innate immunity of fish (overview). Fish Shellfish Immunol. 20, 137-151.

Marc, A.M., Quentel, C., Severe, A., Le Bail, P.Y., Boeuf, G., 1995. Changes in some endocrinological and non-specific immunological parameters during seawater exposure in the brown trout. J. Fish Biol. 46, 1065-1081.

Parry, R.M., Chandan, R.C., Shahani, K.M., 1965. A rapid and sensitive assay of muramidase. Proc. Soc. Exp. Biol. Med. 119, 383-386.

Pavlakis, G.N., Hammer, D.H., 1983. Regulation of a metallothionein-growth hormone hybrid gene in bovine papilloma virus. Proc. Natl. Acad. Sci. U. S. A. 80, 397-401.

Sakai, M., Kobayashi, M., Kawauchi, H., 1996. *In vitro* activation of fish phagocytic cells by GH, prolactin and somatolactin. J. Endocrinol. 151, 113-118.

Sakamoto, T., Shepherd, B.S., Madsen, S.S., Nishioka, R.S., Siharath, K., Richman III, N.H., Bern, H.A., Grau, E.G., 1997. Osmoregulatory actions of growth hormone and prolactin in an advanced teleost. Gen. Comp. Endocrinol. 106, 95-101.

Sarder, M.R.I., Thompson, K.D., Penman, D.J., McAndrew, B.J., 2001. Immune responses of Nile tilapia (*Oreochromis niloticus* L.) clones: I. Non-specific responses. Dev. Comp. Immunol. 25, 37-46.

Secombes, C.J., 1990. Isolation of salmonid macrophages and analysis of their killing activity. In: Stolen, J.S., Fletcher, T.C., Anderson, D.P., Kaattari, S.L., Rowley, A.F. (Eds.), Techniques in Fish Immunology. SOS Publications, Fair

Haven, NJ, pp. 137-154.

Secombes, C.J., 1996. The non-specific immune system: cellular defense. In: Iwama, G., Nakanishi, T. (Eds.), The Fish Immune System: Organism, Pathogen and Environment. Academic Press, London, pp. 63-103.

Thompson, K.D., Tatner, M.F., Henderson, R.J., 1996.. Effects of dietary (n-3) and (n-6) polyunsaturated fatty acid ratio on the immune response of Atlantic salmon, *Salmo salar* L. Aquac. Nutr. 2, 21-31.

Trudeau, V.L., 1997. Neuroendocrine regulation of gonadotrophin II release and gonadal growth in the goldfish, *Carassius auratus*. Rev. Reprod. 2, 55-68.

Wang, H., Zhang, P., 2000. Effects of recombinant yeast on the non-specific immune activities of *Paralichthys olivaceus*. Oceanol. Limnol. Sin. 31, 631-635 (in Chinese, with English abstract).

Wang, Y., Hu, W., Wu, G., Sun, Y., Chen, S., Zhang, F., Zhu, Z., Feng, J., Zhang, X., 2001. Genetic analysis of "all-fish" growth hormone gene transferred carp (*Cyprinus carpio* L.) and its F_1 generation. Chin. Sci. Bull. 46, 1174-1177.

Wu, G., Sun, Y., Zhu, Z., 2003. Growth hormone gene transfer in common carp. Aquat. Living Resour. 16, 416-420.

Yada, T., Nakanishi, T., 2002. Interaction between endocrine and immune systems in fish. Int. Rev. Cyt. 220, 35-92.

Yada, T., Misumi, I., Muto, K., Azuma, T., Schreck, C.B., 2004. Effects of prolactin and growth hormone on proliferation and survival of cultured trout leucocytes. Gen. Comp. Endocrinol. 136, 298-306.

Yano, T., 1996. The non-specific immune system: humoral defense. In: Iwama, G., Nakanishi, T. (Eds.), The Fish Immune System: Organism, Pathogen and Environment. Academic Press, London, pp. 106-157.

Zhu, Z., 1992. Growth hormone gene and the transgenic fish. In: You, C.B., Chen, Z.L. (Eds.), Agricultural Biotechnology. China Science and Technology Press, Beijing, pp. 106-116.

Zhu, Z., Zeng, Z., 2000. Open a door for transgenic fish to market. Biotechnol. Inf. 1, 1-6 (in Chinese, with English abstract).

Zhu, Z., Li, G., He, L., Chen, S., 1985. Novel gene transfer into the fertilized eggs of goldfish (*Carassius auratus* L. 1758). J. Appl. Ichthyol. 1, 31-34.

Zhu, Z., Xu, K., Xie, Y., Li, G., He, L., 1989. A model of transgenic fish. Sci. Sin., B 2, 147-155.

"全鱼"生长激素转植基因的表达对鲤非特异性免疫功能的影响

Wen-BoWang[1,2,3]　汪亚平[1]　胡　炜[1]　李爱华[1]　蔡桃珍[1]
朱作言[1]　汪建国[1]

1 中国科学院水生生物研究所，武汉　430072
2 暨南大学，广州　510632
3 中国科学院大学，北京　100049

摘　要　本文研究了转"全鱼"生长激素基因鲤 F_2 的非特异性免疫功能。转基因鱼的溶酶体活性与非转基因鱼相比具有显著差异(P<0.01)，分别为 145.0 (±30.7) U ml^{-1} 和 105.0 (±38.7) U ml^{-1}。转基因鱼血清的杀菌能力高于对照(P<0.05)，血清杀菌百分比分别为 59.5% (±6.83%)和 50.8% (±8.67%)。转基因鱼头肾组织中的白细胞百分比和吞噬性巨噬细胞的百分比显著高于对照鱼(P<0.05)，而吞噬细胞百分比在两种鱼种没有显著差异。另外，转基因鱼的平均体重为 63.4 (±6.65) g，显著高于对照鱼[39.2 (±3.30) g, P<0.01]。转基因鱼脾脏的绝对重量[0.13 (±0.03) g]也显著高于对照鱼[0.08 (±0.02) g, P<0.01]，然而转基因鱼和对照鱼的相对脾脏重量却没有显著差异，脾脏指数分别为 0.21% (±0.02%)和 0.20% (±0.03%)。本研究结果表明，"全鱼"生长激素基因的表达不仅能促进鲤的生长，而且还能增强其非特异性免疫功能。

Rapid Growth Cost in "all-fish" Growth Hormone Gene Transgenic Carp: Reduced Critical Swimming Speed

LI DeLiang[1,2] FU CuiZhang[1] HU Wei[1] ZHONG Shan[1,3]
WANG YaPing[1] ZHU ZuoYan[1]

1 State Key Laboratory of Freshwater Ecology and Biotechnology, Institute of Hydrobiology, Chinese Academy of Sciences, Wuhan 430072
2 Graduate University of Chinese Academy of Sciences, Beijing 100049
3 Department of Genetics, Basic Medical School of Wuhan University, Wuhan 430071

Abstract Evidence has accumulated that there is a trade-off between benefits and costs associated with rapid growth. A trade-off between growth rates and critical swimming speed (U_{crit}) had been also reported to be common in teleost fish. We hypothesize that growth acceleration in the F_3 generation of "all-fish" growth hormone gene (*GH*) transgenic common carp (*Cyprinus carpio* L.) would reduce the swimming abilities. Growth and swimming performance between transgenic fish and non-transgenic controls were compared. The results showed that transgenic fish had a mean body weight 1.4-1.9-fold heavier, and a mean specific growth rate (SGR) value 6%-10% higher than the controls. Transgenic fish, however, had a mean absolute U_{crit} (cm/s) value 22% or mean relative U_{crit} (BL/s) value 24% lower than the controls. It suggested that fast-growing "all-fish" *GH*-transgenic carp were inferior swimmers. It is also supported that there was a trade-off between growth rates and swimming performance, i.e. faster-growing individuals had lower critical swimming speed.

Keywords transgenic fish; growth; swimming performance; trade-off; growth hormone; critical swimming speed.

Since the first successful case of transgenic fish was achieved[1], transgenic technology has made great advance in fish, and fish may be considered the best candidate for the first marketable transgenic animal for human consumption[2,3]. Before such a product can enter the marketplace, many aspects of the bio-safety of transgenic fish have to be carefully evaluated. In China, researchers have generated stable lines of fast-growing "all-fish" growth hormone gene (*GH*) transgenic common carp[3,4]. The assessment of the environmental impacts and food safety of fast-growing transgenic fish is under way. Locomotion ability is an important component of fitness[5]. The knowledge of fast-growing *GH*-transgenic fish swimming abilities is important to assessing the eco-safety of this strain.

There is a growing list of ectotherms in which individual growth rates are optimized to

文章发表于《科学通报》(英文版), 2007, 52(11):1501-1506

suit local conditions rather than maximized as fast as possible[6]. Evidence has accumulated that there is a trade-off between benefits and costs associated with rapid growth[5,7,8]. Recent studies reported that a trade-off between growth rates and swimming performance may be common in teleost fish. Kolok and Oris[9] first reported that faster-growing individuals in a minnow (*Pimephales promelas*) turned out to have slower critical swimming speeds (a measure of aerobic swimming). The same results were later demonstrated in studies on rainbow trout (*Oncorhynchus mykiss*)[10,11], the Atlantic Silverside (*Menidia menidia*)[12,13], threespine sticklebacks (*Gasterosteus aculeatus*)[14], and fast-growing *GH*-transgenic coho salmon (*Oncorhynchus kisutch*)[15,16].

Fast-growing transgenic juvenile coho salmon grow more than twice as fast by length as the controls, but have critical swimming speeds half those of the controls of the same length[15]. Further study showed that fastgrowing transgenic adult coho salmon displayed significantly lower critical swimming speeds than the adult controls, and transgenic fish are less economical swimmers[16]. However, other studies did not find any significant differences in swimming abilities between *GH*-transgenic fish and controls[17,18]. Such differences may be attributed in part to the degree of GH enhancement that exists between different *GH*-transgenic strains and species. Further tests with other fast-growing *GH*-transgenic fish may serve to ultimately clarify this.

In this study, we tested a hypothesis that growth acceleration in the F_3 generation of "all-fish" *GH*-transgenic common carp would reduce the swimming abilities. The critical swimming speed (U_{crit}) method was used to measure the maximum swimming speed[19], which is an easy mean of measuring swimming performance and involves swimming fish at incrementally increasing speeds until exhaustion[20].

1. Materials and Methods

1.1 Growth experiment

P_0 "all-fish" *GH*-transgenic carp were produced by micro-injection of pCAgcGHc into the fertilized eggs of common carp (Yellow River carp variety). "All-fish" gene construct pCAgcGHc was the recombinant construct of grass carp, *Ctenopharyngodon idellus* (Cuvier & Valenciennes) growth hormone cDNA (gcGHc) driven by β-actin gene promoter of common carp (pCA)[4]. The F_3 generation of "all-fish" *GH*-transgenic fish and non-transgenic controls used in the growth experiment were the siblings produced from crosses between a wild type female and an F_2 transgenic male of a fast-growing transgenic germ-line on April 14, 2005. Frequencies of transgene transmission to F_3 progeny from this line were about 50% (unpublished). At the age of 106 d, 500 transgenic fish and 500 control fish from the same pond were randomly selected, and then distributed into two equal size rectangular ponds with an area of about 660 m^2 (length×width: 33 m×20 m) and a water depth of 1.0 m for each pond in Duofu Technology Farm, Wuhan, China. The experimental fish were fed to satiation with the same diet throughout the experiment. Data of body weights (BW) of transgenic and control fish were collected by 4-time samplings when the fish was 160, 190, 221 and 251 d old. Each individual was weighted nearest to 0.01 g. In order to compare the growth rates of transgenic fish with that of the controls, specific growth rate was calculated for each fish by the equation 100×(Ln Final BW – Ln Intial BW)/number of growing days.

1.2 Swimming test

The major difficulty in swimming test lied in the way to achieve size-matched transgenic and non-transgenic fish because of higher growth rates of transgenic fish[16,17]. To this end, the F_3 generation of "all-fish" *GH*-transgenic carp was transferred to indoor recycling-water fish-reared system in the laboratory on Jun 15, 2005, and wild-type carp with the same age from Institute of Aquaculture, Henan Province were transferred to laboratory three months later. In this study, the critical swimming speed (U_{crit}) tests were conducted using a Brett-type swim tunnel[19] at the Fish Gene-Engineering Laboratory of Institute of Hydrobiology, the Chinese Academy of Sciences. Regular calibration of the water velocity within the swim chamber (125 cm in length; 14 cm in inside diameter) was performed by ultrasonic flow-meter with a digital read-out unit (cm/s). A fine-meshed grid was placed in the front of the swimming chamber in order to produce a rectilinear uniform water-flow. At the end of the chamber, there is a stainless steel electrode (0-12 V) to drive the fish to swim against the water flow if necessary. On January 4-23, 2006, U_{crit} of 16 size-matched F_3 transgenic fish and non-transgenic controls (for fish biometrics, see Appendix) were tested with the water temperature varying between 8.3℃ and 11.2℃ (mean 9.58, S.E. 0.26℃). Prior to the start of the experiment, each swimming-tested individual was starved for 12 h (08:30-20:30), and then was placed in the swim chamber to habituate a water velocity of 10 cm/s overnight. After a 12-14-h habituation period, the fish was forced to swim against a water velocity of 20 cm/s for 10-min, and then to swim against a water velocity that increased by 10 cm/s every 20-min. A fish was judged to be fatigued when it could no longer be persuaded to swim off the grid at the rear of the swim chamber. After the swimming challenge, each fish was anaesthetized using Eugenol, and then the live weight was weighed nearest to 0.01 g and the body length, maximum body depth, and maximum body width were measured nearest to 0.01 cm. U_{crit} was expressed as absolute U_{crit} (cm/s) and relative U_{crit} (BL/s). Absolute U_{crit} was calculated with the formula given by Brett[19]. U_{crit} (cm/s)=$u_i+(t_i/t_{ii}\times u_{ii})$, where u_i is the highest velocity maintained for the prescribed period (cm/s); u_{ii} is the velocity increment (10 cm/s); t_i is the time (min) fish swim at the "fatigue" velocity; t_{ii} is the prescribed period of swimming (20-min). Relative U_{crit} was calculated using the formula: U_{crit} (BL/s)= U_{crit} (cm/s)/body length. In order to determine whether fast-growing transgenic fish were morphologically different from controls, the condition factor was determined for each fish by the equation $100\times BW/BL^3$, where BW and BL are wet weight and body length, respectively.

1.3 Statistical analysis

Mann-Whitney *U* test was made to test the differences between strains in growth performance. One-way analysis of variance (ANOVA) was done to test the differences between strains in temperature and physical parameters. Analysis of covariance (ANCOVA) was done to examine the differences in critical swimming speeds. We used temperature and body length as covariates for absolute U_{crit}, and used only temperature as covariate for relative U_{crit}. The differences were regarded as significant when $P<0.05$.

2. Results

2.1 Growth performance

Body weight and specific growth rate (SGR) were significantly higher in transgenic fish than in non-transgenic controls (Table 1). Transgenic fish had a mean body weight 1.4-1.9-fold heavier than the control fish, and a mean SGR value 6%-10% higher than the controls for different growth phrases.

Table 1 Body weight (BW, g) and specific growth rate (SGR) for F_3 "all-fish" *GH*-transgenic carp and non-transgenic controls at different phases of growth in the growth experiment (mean ± S.E.)

Growth phase(d)	Variables	Fish strain				Mann-Whitney *U* test for the strain effect	
		Transgenics		Controls		*Z*-value	*P*-value
160	BW	316.82±10.22	*n*=33	223.05±3.58	*n*=43	6.58	<0.001
	SGR	4.02±0.02		3.81±0.01		6.58	<0.001
190	BW	396.64±9.53	*n*=53	247.07±9.60	*n*=27	6.50	<0.001
	SGR	3.50±0.01		3.26±0.02		6.50	<0.001
221	BW	525.64±13.18	*n*=45	282.85±9.54	*n*=47	7.87	<0.001
	SGR	3.14±0.01		2.85±0.01		7.87	<0.001
251	BW	531.15±0.63	*n*=46	307.24±10.46	*n*=46	8.07	<0.001
	SGR	2.77±0.01		2.55±0.01		8.07	<0.001

2.2 Swimming test

Body weight of experimental fish used in this experiment was size-matched ($P>0.68$), but transgenic fish had significantly larger body length and total length, and lower condition factor than controls ($P<0.05$) (Table 2). Although there were slightly differences in the test temperature for individual fish (see Appendix), the mean was not different ($P>0.31$) (Table 2).

The absolute and relative critical swimming speeds of transgenic fish were significantly lower than controls ($P<0.01$) (Appendix; Table 3). Transgenic fish had a mean U_{crit} (cm/s) value 22% or mean U_{crit} (BL/s) value 24% lower than the controls.

Table 2 Temperature and physical parameters for F_3 "all-fish" *GH*-transgenic carp and non-transgenic controls (mean ± S.E.) used in the swimming test

Fish strain	Temperature(℃)	Body weight(g)	Body length (cm)	Total length (cm)	Condition factor(g/cm^3)
Transgenics(n=8)	9.85±0.37	73.19±1.93	15.11±0.09	18.75±0.10	2.12±0.04
Controls(n=8)	9.30±0.37	73.77±0.82	14.84±0.06	18.10±0.12	2.26±0.02
ANOVA for the strain effect					
F-value	1.093	0.180	4.880	8.270	9.297
d.f.	1,14	1,14	1,14	1,14	1,14
P-value	0.314	0.680	0.044	0.012	0.009

Table 3 Critical swimming speeds (U_{crit}) for F_3 "all-fish" *GH*-transgenic carp and non-transgenic controls (mean ± S.E.) in the swimming test

Fish strain	U_{crit}(cm/s)	U_{crit}(BL/s)
Transgenics(n=8)	43.58±1.79	2.89±0.13
Controls(n=8)	53.09±1.28	3.58±0.09
ANCOVA for the strain effect		
F-value	9.575	21.974
d.f.	1,12	1,13
P-value	0.009	<0.001

3. Discussion

Previous studies showed that the F_1 generation of "all-fish" *GH*-transgenic carp had a mean body weight 1.6-fold heavier than controls[4], and the F_2 generation had a mean body weight 1.8-2.5-fold heavier and a mean SGR value 10%-13% higher than controls (unpublished). In this study, the F_3 generation had a mean body weight 1.4-1.9-fold heavier and a mean SGR value 6%-10% greater than controls. It suggested that the growth- promoting effect of "all-fish" *GH*-gene could persist in the F_3 generation. The same growth-promoting effect of foreign *GH*-gene persisting in the F_4 generation had also been demonstrated in our previous study with human *GH*-transgenic carp[21]. Growth-promoting effects of exogenous growth hormone gene had been observed in a number of *GH*-transgenic fishes, including *GH*-transgenic cyprinid fish[22-24], *GH*-transgenic salmonid fish[25-27], *GH*-transgenic channel catfish (*Ictalurus punctatus*)[28], *GH*-transgenic tilapia (*Oreochromis niloticus*)[29], *GH*-transgenic Ayu (*Plecoglossus altivelis*)[30] and *GH*-transgenic rohu (*Labeo rohita*)[31].

In this study, "all-fish" *GH*-transgenic fish had a mean U_{crit} (cm/s) value 22% or a mean U_{crit} (BL/s) value 24% smaller than controls. This is in agreement with the fact that fast-growing *GH*-transgenic fish was inferior swimmers, as demonstrated in previous studies on "all-fish" *GH*-transgenic salmon[15,16]. It further supported the conclusion that there was a trade-off between growth rates and swimming performance, i.e. faster-growing individuals had slower critical swimming speeds, as had been reported in teleost fish[9-13] and anurans[5,32,33]. The differences in morphology between *GH*-transgenic fish and controls may be responsible for lower swimming performance in transgenic fish. Deep-bodied fish with large median fins had been reported to be superior burst swimmers whereas streamlined fish could sustain swimming for long periods of time[34]. The dorsal caudal peduncle and abdominal regions were distinctly enhanced in transgenic salmon as compared to controls[35]. Fast-growing "all-fish" *GH*-transgenic carp showed much thicker muscles on backside and obvious hunch behind the head[4]. In this study, transgenic fish had a significantly larger body length and total length, and lower condition factor than the controls ($P<0.05$). It suggested that fast-growing *GH*-transgenic fish might be an excellent model to disentangle the functional link between morphology and swimming performance.

Somatic development may compromise rapid growth because most cells lose the ability to divide and contribute further towards growth once they differentiate and take up their mature function[36]. In fishes, a fast growing strain of pumpkinseed sunfish (*Lepomis gibbosus*) had a delayed onset of mineralization in their cranial bones relative to a slow growing strain[37]. Further study found a negative correlation between scale strength (in terms of ability to resist being pierced) and growth rate that was consistent both within and among populations using pumpkinseed from six populations known to differ in their intrinsic growth rates[38]. This trade-off between growth rate and scale strength may have fitness consequences in terms of likelihood of surviving predation attempts or swimming efficiency. This may in part explain the trade-offs between growth rates and critical swimming speeds found in fast-growing coho salmon[15,16] and in the F_3 generation of "all-fish" *GH*-transgenic carp of this study.

The difference in swimming ability has important implications for ecological risk assessments of fast-growing *GH*-transgenic fish, since the swimming performance characteristic influences many life-history characteristics, including catching prey, escaping predators, and migrating up spawning habitats[20]. Previous studies provided evidence that physiological trade-offs between growth rates and swimming performance translate into fitness trade-offs by affecting the escape performance of silversides under threat of predation, i.e. the difference in swimming performance is correlated directly with vulnerability to predators[12,39]. Fast-growing *GH*-transgenic channel catfish displayed poorer anti-predatory behavior than non-transgenic controls[40]. The same effect was also found in behavior studies of fast-growing *GH*-transgenic Atlantic salmon (*Salmo salar*)[41] and fast-growing *GH*-transgenic coho salmon[42-44]. These results show that fast-growing *GH*-transgenic fish display poorer anti-predatory behavior, and that they will suffer high mortality rates under natural conditions. Further studies are needed to test the hypothesis that the difference in predation vulnerability for fast-growing *GH*-transgenic fish and controls lies in their swimming performance, but not their attraction to predators.

To sum up, this study showed that rapid growth of the F_3 generation of "all-fish" *GH*-transgenic carp is at the cost of reduced critical swimming speed.

Acknowledgement The authors would like to thank XIE SouQi, LEI Wu, YANG YunXia, ZHU XiaoMing, and NIE GuangHan for their help in designing and assembling the Brett-type swim tunnel.

References

1 Zhu Z, Li G, He L, *et al.* Novel gene transfer into the fertilized eggs of goldfish (*Carassius auratus* L. 1758). J Appl Ichthyol, 1985, 1: 31-34

2 Zbikowska H. Fish can be first-advances in fish transgenesis for commercial applications. Transgenic Res, 2003, 12(4): 379-389

3 Fu C, Hu W, Wang Y, *et al.* Developments in transgenic fish in the People's Republic of China. Rev Sci Tech Off Int Epiz, 2005, 24(1): 299-307

4 Wang Y, Hu W, Wu G, *et al.* Genetic analysis of 'all-fish' growth hormone gene transferred carp (*Cyprinus Carpio* L.) and its F_1 generation. Chin Sci Bull, 2001, 46(14): 1174-1177

5 Arendt J D. Reduced burst speed is a cost of rapid growth in anuran tadpoles: Problems of autocorrelation and inferences about growth rates. Funct Ecol, 2003, 17(3): 328-334

6 Arendt J D. Adaptive intrinsic growth rates: An intergration across taxa. Q Rev Biol, 1997, 72(2): 149-177

7 Morgan I J, Metcalfe N B. Deferred costs of compensatory growth after autumnal food shortage in juvenile salmon. Proc R Soc Lond B, 2001, 268(1464): 295-301

8 Mangel M, Stamps J. Trade-offs between growth and mortality and the maintenance of individual variation in growth. Evol Ecol Res, 2001, 3(5): 583-593

9 Kolok A S, Oris J T. The relationship between specific growth rate and swimming performance in male fathead minnows (*Pimephales promelas*). Can J Zool, 1995, 73(11): 2165-2167

10 Gregory T R, Wood C M. Individual variation and interrelationships between swimming performance, growth rate, and feeding in juvenile rainbow trout (*Oncorhynchus mykiss*). Can J Fish Aquat Sci, 1998, 55(7): 1583-1590

11 Gregory T R. Wood C M. Interactions between individual feeding behaviour, growth, and swimming performance in juvenile rainbow trout (*Oncorhynchus mykiss*) fed different rations. Can J Fish Aquat Sci, 1999, 56(3): 479-486

12 Billerbeck J M, Lankford T E, Conover D O. Evolution of intrinsic growth and energy acquisition rates. I. Tradeoffs with swimming performance in *Menidia menidia*. Evolution, 2001, 55(9): 1863-1872

13 Munch S B, Conover D O. Nonlinear growth cost in Menidia menidia: Theory and empirical evidence. Evolution, 2004, 58(3): 661-664

14 Álvarez D, Metcalfe N B. Catch-up growth and swimming performance in threespine sticklebacks (*Gasterosteus aculeatus*): Seasonal changes in the cost of compensation. Can J Fish Aquat Sci, 2005, 62(9): 2169-2176

15 Farrell A P, Bennett W, Devlin R H. Growth-enhanced transgenic salmon can be inferior swimmers. Can J Zool, 1997, 75(2): 335-337

16 Lee C G, Devlin R H, Farrell A P. Swimming performance, oxygen consumption and excess post-exercise oxygen consumption in adult transgenic and ocean-ranched coho salmon. J Fish Biol, 2003, 62(4): 753-766

17 Stevens E D, Sutterlin A, Cook T. Respiratory metabolism and swimming performance in growth hormone transgenic Atlantic salmon. Can J Fish Aquat Sci, 1998, 55(9): 2028-2035

18 McKenzie D J, Martínez R, Morales A, *et al*. Effects of growth hormone transgensis on metablic rate, exercise performance and hypoxia tolerance in tilapia hybrids. J Fish Biol, 2003, 63(2): 398-409

19 Brett J R. The respiratory metabolism and swimming performance of young sockeye salmon. J Fish Res Bd Can, 1964, 21(5): 1183-1226

20 Plaut I. Critical swimming speed: Its ecological relevance. Comp Biochem Physiol, 2001, 131(1): 41-50

21 Fu C, Cui Y, Hung S S O, *et al*. Growth and feed utilization by F_4 human growth hormone transgenic carp fed diets with different protein levels. J Fish Biol, 1998, 53(1): 115-129

22 Zhu Z Y, Xu K S, Li G H, *et al*. Biological effects of human growth hormone gene microinjected into the fertilized eggs of loach, *Misgurus anguillicaudatus* (Cantor). Chin Sci Bull, 1986, 31(14): 988-990

23 Zhu Z Y, Xu K S, Xie Y F, *et al*. A model of transgenic fish. Sci China Ser B (in Chinese), 1989, (2): 147-155

24 Cui Z B, Zhu Z Y, Cui Y B, *et al*. Food consumption and energy budget in MThGH transgenic F_2 red carp (*Cyprinus carpio* L. red var.). Chin Sci Bull, 1996, 41(7): 591-596

25 Chourrout D, Guyomard R, Houdebine L M. High efficiency gene transfer in rainbow trout (*Salmo gairdneri Rich.*) by microinjection into egg cytoplasm. Aquaculture, 1986, 51(1): 143-150

26 Devlin R H, Biagi C A, Yesaki T Y. Growth, viability and genetic characteristics of GH transgenic coho salmon strains. Aquaculture, 2004, 236(1-4): 607-632

27 Cook J T, McNiven M A, Richardson GF, *et al*. Growth rate, body composition/feed digestibility conversion of growth-enhanced transgenic Atlantic salmon (*Salmo salar*). Aquaculture, 2000, 188(1-2): 15-32

28 Dunham R A, Eash J, Askins J, *et al*. Transfer of metallothionein-human growth hormone fusion gene into channel catfish. Trans Am-Fish Soc, 1987, 116(1): 87-91

29 Rahman M, Ronyai A, Engidaw B Z, *et al*. Growth and nutritional trials on transgenic nile tilapia containing an exogenous piscine growth hormone gene. J Fish Biol, 2001, 59(1): 62-78

30 Cheng C, Liu K, Lau E, *et al*. Growth promotion in Ayu (*Plecoglossus altivelis*) by gene transfer of the rainbow trout growth hormone gene. Zool Stud, 2002, 41(3): 303-310

31 Venugopal T, Anathy V, Kirankkumar S, *et al*. Growth enhancement and food conversion efficiency of transgenic fish *Labeo rohita*. J Exp Zool, 2004, 301A(6): 477-490

32 Arendt J, Hoang L. Effect of food level and rearing temperature on burst speed and muscle composition of western spadefoot toad (*Spea hammondii*). Funct Ecol, 2005, 19(6): 982-987

33 Dayton G H, Saenz D, Baum K A, *et al*. Body shape, burst speed and escape behavior of larval anurans. Oikos, 2005, 111(3): 582-591

34 Swain D P. The functional basis of natural-selection for vertebral traits of larvae in the stickleback *Gasterosteus aculeatus*. Evolution, 1992, 46(4): 987-997

35 Ostenfeld T H, McLean E, Devlin R H. Transgenesis changes body and head shape in Pacific salmon. J Fish Biol, 1998, 52(4): 850-854

36 Lankford T E, Billerbeck J M, Conover D O. Evolution of intrinsic growth and energy acquisition rates. II. Tradeoffs with vulnerability to predation in Menidia menidia. Evolution, 2001, 55(9): 1873-1881

37 Arendt J D, Wilson D S. Population differences in the onset of cranial ossification in pumpkinseed (*Lepomis gibbosus*), a potential cost of rapid growth. Can J Fish Aquat Sci, 2000, 57(2): 351-356

38 Arendt J D, Wilson D S, Stark E. Scale strength as a cost of rapid growth in sunfish. Oikos, 2001, 93(1): 95-100

39 Lankford T E, Billerbeck J M, Conover D O. Evolution of intrinsicgrowth and energy acquisition rates. II. Tradeoffs with vulnerability topredation in *Menidia menidia*. Evolution, 2001, 55(9): 1873-1881

40 Dunham R A, Chitmanat C, Nichols A. Predator avoidance of transgenicchannel catfish containing salmonid growth hormone genes. Mar Biotech, 1999, 1(6): 545-551

41 Abrahams M V, Sutterlin A. The foraging and anti-predator behavior of growth-enhanced transgenic Atlantic salmon. Anim Behav, 1999, 58(5): 933-942

42 Sundström L F, Devlin R H, Johnsson J I, et al. Vertical position reflects increased feeding motivation in growth hormone transgenic coho salmon (*Oncorhynchus kisutch*). Ethology, 2003, 109(8): 701-712

43 Sundström L F, Lõhmus F, Johnsson J I, et al. Growth hormone transgenic coho salmon pay for growth potential with increased predation mortality. Proc R Soc Lond B, 2004, 271: S350-352

44 Sundström L F, Lõhmus F, Devlin R H. Selection on increased intrinsic growth rates in coho salmon (*Oncorhynchus kisutch*). Evolution, 2005, 59(7): 1560-1569

转“全鱼”生长激素基因鲤快速生长与临界游泳速度降低的关联性

李德亮[1,2] 傅萃长[1] 胡 炜[1] 钟 山[1,3] 汪亚平[1] 朱作言[1]

1 中国科学院水生生物研究所淡水生态与生物技术国家重点实验室，武汉 430072

2 中国科学院大学，北京 100049

3 武汉大学基础医学院遗传学系，武汉 430071

摘 要 研究表明快速生长在赋予生物的优势与劣势之间存在权衡。硬骨鱼类的生长率和临界游泳速度(U_{crit})之间也存在这种权衡。我们认为 F_3 转“全鱼”生长激素基因鲤的快速生长可能会降低其游泳能力。本文测定了转基因鱼及其对照鱼的生长率与游泳速度。结果发现转基因鱼的平均体重是对照鱼的 1.4-1.9 倍，平均特定生长率比对照鱼高 6-10%，而其绝对临界游泳速度和相对临界游泳速度的平均值分别比对照鱼低 22%和 24%。本研究提示具有快速生长效应的转基因黄河鲤的游泳能力低于对照鲤，这也证实了生长率和游泳能力之间存在权衡，即同一物种的个体生长率越高，临界游泳速度则越低。

Growth and Energy Budget of F_2 'all-fish' Growth Hormone Gene Transgenic Common Carp

C. Fu D. Li W. Hu Y. Wang Z. Zhu

State Key Laboratory of Freshwater Ecology and Biotechnology, Institute of Hydrobiology, Chinese Academy of Sciences, Wuhan 430072

Abstract The growth and energy budget for F_2 'all-fish' growth hormone gene transgenic common carp *Cyprinus carpio* of two body sizes were investigated at 29.2℃ for 21 days. Specific growth rate, feed intake, feed efficiency, digestibility coefficients of dry matter and protein, gross energy intake (I_E), and the proportion of I_E utilized for heat production (H_E) were significantly higher in the transgenics than in the controls. The proportion of I_E directed to waste products [faecal energy (F_E) and excretory energy loss (Z_E+U_E) where Z_E is through the gills and U_E through the kidney], and the proportion of metabolizable energy (M_E) for recovered energy (R_E) were significantly lower in the transgenics than in the controls. The average energy budget equation of transgenic fish was as follows: 100 I_E = 19.3 F_E + 6.0(Z_E+ U_E)+ 45.2 H_E + 29.5 R_E or 100 M_E =60.5 H_E + 39.5 R_E. The average energy budget equation of the controls was: 100 I_E = 25.2 F_E + 7.4(Z_E+ U_E)+35.5 H_E + 31.9R_E or 100 M_E = 52.7 H_E + 47.3 R_E. These findings indicate that the high growth rate of 'all-fish' transgenic common carp relative to their non-transgenic counterparts was due to their increased feed intake, reduced lose of waste productions and improved feed efficiency. The benefit of the increased energy intake by transgenic fish, however, was diminished by their increased metabolism.

Keywords common carp; energy budget; fast-growth; growth hormone gene; transgenic fish

Introduction

With the expansion of the global population and over-fishing, increased aquaculture output is needed to meet man's increasing demands for high quality fish protein. Genetically modified fishes (transgenic fishes) offer the opportunity to improve both the production and characteristics of conventional fish strains currently exploited in aquaculture. Since the first batch of transgenic fishes was achieved by Zhu *et al.* (1985), transgenic fishes have been produced in a variety of species using different foreign gene constructs (Zhu *et al.*, 1992; Devlin *et al.*, 1994; Rahman & Maclean, 1999; Zhong *et al.*, 2002; Mao *et al.*, 2004). At

文章发表于 *Journal of Fish Biology*, 2007, 70(2):347-361

present, the biotechnology has sufficiently advanced with respect to producing transgenic animals, and rapidly growing fishes [growth hormone (GH) gene transgenic fishes] may be considered the best candidates for the first marketable transgenic animals for human consumption (Zbikowska, 2003).

Fast-growing GH-transgenic fishes have been achieved in many laboratories throughout the world (Zhu *et al.*, 1986, 1989; Devlin *et al.*, 1994; Fu *et al.*, 1998; Rahman & Maclean, 1999; Nam *et al.*, 2001; Wang *et al.*, 2001; Cheng *et al.*, 2002; Devlin *et al.*, 2004; Venugopal *et al.*, 2004). Increased feed intake and improved feed conversion efficiency have been suggested as the mechanisms for the fast-growth of the GH-gene transgenic fishes (Cui *et al.*, 1996a; Fu *et al.*, 1998; Cook *et al.*, 2000*a*; Nam *et al.*, 2001; Rahman *et al.*, 2001; Venugopal *et al.*, 2004). Energy digestibility in fast-growing transgenic fishes *v.* non-transgenic controls has been found to be either similar (Rahman *et al.*, 2001) or greater in the transgenic fishes (Cui *et al.*, 1996*a*; Fu *et al.*, 1998; Cook *et al.*, 2000a). Studies on energy intake and utilization in transgenic fishes are needed to make further generalizations on growth-promoting mechanisms that result from a foreign GH-gene.

The energy budget of an animal is an input-output model that balances energy intake (I_E) with energy expenditures for metabolism, growth and excretion. Energy input can be approximated by I_E derived from foods, and the energy output is related to metabolism in terms of respiratory demand and excretory losses by way of faecal and non-faecal pathways (Jobling, 1994). Reports on the energy budget of transgenic fishes are limited to one preliminary study (Cui *et al.*, 1996a), in which a comparison was made of the energy budget of the F_2 generation of human GH-transgenic common carp *Cyprinus carpio* L. and control fish fed live tubificid worms in a 3 week experiment. In this regard, the preceding study found that the proportion of metabolizable energy allocated for recovered energy [(growth) was significantly higher in the human GH-transgenic fish than in the controls (Cui *et al.*, 1996a)].

The purpose of this study was to compare the growth and energy budgets of F_2 'all-fish' GH-gene transgenic common carp of two body sizes to the growth and energy budgets of non-transgenic counterparts. Such knowledge would be useful for further understanding of growth-promoting mechanisms that result from foreign GH-genes. Also the findings would be helpful for bio-safety assessments and for future aquaculture management of fast-growing transgenic fishes.

Material and Methods

Source of fish

P_0 'all-fish' growth hormone gene transgenic carp were produced by micro-injection of pCAgcGHc into the fertilized eggs of common carp (Yellow River carp variety). 'All-fish' gene construct pCAgcGHc was the recombinant construct of grass carp, *Ctenopharyngodon idellus* (Cuvier & Valenciennes) growth hormone cDNA (gcGHc) driven by β-actin gene promoter of common carp (pCA) (Wang *et al.*, 2001). F_2 transgenic fish used in this experiment were produced from crosses between a wild type female and a F_1 transgenic male of a fast-growing transgenic germ-line which was produced previously in the laboratory on 25 April 2004. At the age of 4 weeks, transgenic carp were transferred to the laboratory. The tail fin DNA of these fish was extracted and the polymerase chain reaction (PCR) method was used to detect the transgene positive individuals based on the procedure of Wang *et al.* (2001). A batch of non-transgenic control common carp was also produced on 15 April 2004. At the age of 10 weeks, the controls were also transferred to the laboratory. During the acclimation period, the fish were fed the experimental diet. The diet was formulated to contain 38% crude protein, and Cr_2O_3 was added as an indigestible indicator for the digestibility determinations. The ingredient and chemical composition of the diet are shown in Table I. The diets were made into 4-6 mm pellets using a pelleting machine (Haiguang Electrical Machinery Company, Shanghai, China) and oven-dried at 60℃. Triplicate samples of each diet were used for chemical analysis.

Fish-rearing facilities

The growth experiment was conducted in 20 circular fibreglass tanks (diameter 150 cm and water volume 1000 l). Water was aerated in a common reservoir, and a gentle water flow (*c.* 5 l min^{-1}) was introduced to each tank. During the experimental period, dissolved oxygen was >7 mg l^{-1}, ammonia-nitrogen was <0.15 mg l^{-1}, and pH ranged from 7.0 to 7.3. The water temperature varied between 28 and 32℃ (mean ± s.e. 29.2 ± 0.2℃). Photoperiod was controlled by artificial lighting and the light period extended between 0700 and 1900 hours.

Growth experiment

After acclimation, two body sizes of transgenic fish and the non-transgenic controls (size × fish strain: 2 × 2) were tested, with a mean initial mass of 138 g (n = 14) and 208 g (n = 14). There were no significant differences in mass between the transgenic fish and the controls for the 138 g size-class and the 208 g size-class (Table II). Ten transgenic fish and 10 control fish were selected, and then distributed randomly into 20 tanks. Prior to the start of the experiment, the fish were starved for 1 day to evacuate their gut. The live mass of each fish was measured to the nearest to 0.01 g after removal of excess water. Two fish from each group (size × fish strain: 2 × 2) were killed to determine the initial whole body proximate composition and energy content. Fish were hand-fed to satiation twice a day at 0800 and 1600 hours throughout the experiment. During each feeding, water flow was stopped, and a quantity of diet in excess of what the fish would consume in 2 h was provided. Two hours after feeding, the uneaten feed was collected by pipetting, dried and weighed. A sub-sample of diet was weighed each day and dried to constant mass at 70℃ to determine the dry matter content. Leaching rate of feed in water was determined by placing a weighed quantity of feed in a tank without fish for 2 h, collecting the feed, drying to constant mass and reweighing. The leaching rate was used to adjust the amount of the feed intake. Fish faeces were collected by siphoning from the tank bottom twice a day at 1000 and 1800 hours and dried at 70℃. Only faeces that remained intact were kept for chemical analysis. Ammonia-N and urea-N excretion from each fish in each tank were measured during the second week of the experiment. During the measurement, water flow was stopped, and the fish were fed once to satiation and then starved for 3 days to allow ammonia and urea to accumulate to measurable levels. Aeration continued during this period, ensuring a high level of dissolved oxygen. Water was sampled before and after this period. The ammonia-N and urea-N were determined by the method of Chaney & Marbach (1962). The growth experiment lasted for 21 days in August 2004.

At the end of the experiment, all fish were killed following weighing. Sacrificed fish were autoclaved at 120℃ for 1 h, dried at 70℃ for *c*. 48 h and homogenized. Per cent nitrogen was determined using the Kjeldahl method (AOAC, 1984), Lipid concentration was determined by Lambert & Dehnel's (1974) modification of the chloroform–methanol extraction technique, ash by combustion at 550℃ for 12 h, and energy by bomb calorimetry (Gentry Instruments Inc., Aiken, SC, U. S. A.). Protein content was calculated from % nitrogen content multiplied by 6.25. Contents of protein, energy and Cr_2O_3 were determined for the faecal sample from each tank.

Calculations and Statistical Analysis

The following parameters of growth performance were calculated: specific growth rate (G_W) (% day^{-1})=$100(\ln M_t - \ln M_0)t^{-1}$=, feed intake ($I_F$) (% g day^{-1}) = $100 I_{Td}(\sqrt{M_t M_0})^{-1} t^{-1}$ (Richardson *et al.*, 1985), feed efficiency (E_F) = $(M_t - M_0)$ $(I_{Td})^{-1}$ and protein retention efficiency (E_P) = $100(M_{Pt} - M_{P0})$ $(I_{TP})^{-1}$where M_t is the final and M_0 the initial wet masses of the fish, M_{Pt} is final and M_{P0} the initial masses of the body protein, t is the experimental period (21 days), I_{Td} is total dry feed consumption during the experimental period and I_{TP} is total mass of protein consumed.

Apparent digestibility coefficients (D_A) of dry matter, protein and energy were calculated using Cr_2O_3 as an inert indicator: D_A of dry matter (%) = $100(1-C_1 C_2^{-1})$ and D_A of protein or energy (%) =$100[1-(C_1 N_2)(C_2^{-1} N_1^{-1})]$, where C_1 is % Cr_2O_3 in feed, C_2 is % Cr_2O_3 in faeces, N_1 is % protein or J mg^{-1} energy in feed, and N_2 is % protein or J mg^{-1} energy in faeces.

Energy loss in nitrogenous excretion was calculated from ammonia and urea excretion using the equivalents of 24.83 J mg^{-1} for ammonia-nitrogen and 23.03 J mg^{-1}

for urea-nitrogen (Elliott, 1976a). Energy budget was calculated using the approach of Elliott (1976b). The excretory loss of each fish was expressed as a percentage of energy intake during the period when nitrogen excretion was measured. The excretory loss was then extrapolated to the whole experimental period based on the mean daily energy intake. Terminology and symbols for the energy budget followed those proposed by NRC (1981), with slight modifications: I_E, gross energy intake; F_E, faecal energy; Z_E+U_E, excretory (non-faecal) energy loss Z_E through the gills and U_E through the kidneys; H_E, heat production (metabolism); R_E, recovered energy (growth); M_E, metabolizable energy. I_E, F_E, Z_E+U_E and R_E were determined directly, and metabolism was calculated by difference: $H_E=I_E-F_E-(Z_E+U_E)-R_E$. Metabolizable energy was calculated as: $M_E=H_E+R_E$. The following parameters of energy budgets were calculated: I_E (J g^{-1} day^{-1}) $=I_{TE}(\sqrt{M_t M_0})^{-1}t^{-1}$, R_E (J g^{-1} day^{-1}) $=(E_t-E_0)(\sqrt{M_t M_0})^{-1}t^{-1}$, F_E (J g^{-1} day^{-1}) $=I_E$ (1-D_A of energy) and Z_E+U_E (J g^{-1} day^{-1}) $=(24.83\ N_A) + (23.03\ N_U)$, where I_{TE} is total energy intake during the experimental period, E_t is the final and E_0 the initial body energy content, N_A is ammonia and N_U urea nitrogenous excretion of the fish for each day.

The effects of fish size and fish strain on the parameters of growth performance and energy budgets were analysed by two-way ANOVA. Possible differences between groups were examined using a multiple range test (Newman-Keuls) after ANOVA. ANCOVA was used to remove the effect of size to test the effects of fish size and fish strain on whole body proximate compositions, using final mass as the covariate (Shearer, 1994). Differences were regarded as significant when $P<0.05$. Log_{10} transformations were used to achieve homogeneity variance of all mass measurements, food consumption and gross energy intake. Arcsine square root transformations were applied to percentage data. A percentage was converted to a proportion, this was square rooted and then the arcsine was taken.

Results

Growth performance

Both body size and fish strain affected significantly final mass, G_W, total feed consumption and feed efficiency, but protein retention efficiency was not affected by size and fish strain (Tables II and III). Initial mass was significantly affected by body size, but was not affected by fish strain. Feed intake was significantly higher in the transgenic fish than in the controls, but was not affected by body size (Table II). Values for final mass, G_W, total feed consumption, feed intake and feed efficiency were significantly higher in the transgenic carp than in the controls for each size-class (Tables II and III).

Table I Ingredient and proximate composition of the experimental diet

Ingredients	%
White fishmeal (from Russia)*	27
Soybean meal*	33.5
Corn starch*	25.84
Soybean oil*	5
Mineral premix†	5
Vitamin premix‡	0.55
Vitamin C§	0.01
Choline chloride¶	0.1
Cellulose	2
Cr_2O_3	1
Proximate analysis (% dry mass)	
Dry matter (%)	95.9
Crude protein (%)	38.0
Lipid (%)	8.1
Ash (%)	12.4
Gross energy (kJ g^{-1})	17.7

*Fish meal, soybean meal, corn starch and soybean oil were commercially available

†Mineral premix (mg kg^{-1} diet): NaCl, 500; $MgSO_4 \cdot 7H_2O$, 7500; $NaH_2PO_4 \cdot H_2O$, 12 500; KH_2PO_4, 16 000; $Ca(H_2PO_4)_2 \cdot H_2O$, 100 000; $FeSO_4 \cdot 7H_2O$, 1250; $C_6H_{10}CaO_6 \cdot 5H_2O$, 1750; $ZnSO_4 \cdot 7H_2O$, 176.5; $MnSO_4 \cdot H_2O$, 81; $CuSO_4 \cdot 5H_2O$, 15.5; $CoSO_4 \cdot 7H_2O$, 0.5; KI, 1.5

‡Vitamin premix (mg kg^{-1} diet): thiamin, 20; riboflavin, 20; pyridoxine, 20; cyanocobalamine, 2; folic acid, 5; calcium pantothenate, 50; inositol, 100; niacin, 100; biotin, 5; ascorbic acid, 111; vitamin A, 110; vitamin D_3, 20; vitamin E (DL-a-tocopherolacetate), 100; vitamin K_3 (menadione sodium bisulphite), 10

§Coated vitamin C from the Sunhy Biology Company, Wuhan, China

¶Choline chloride of the diet ingredient was on a dry basis

Table II Mean ± S.E. body mass, specific growth rate (G_W), total dry feed consumption (I_{Td}) and feed intake (I_F) for transgenic carp and the controls of different size-classes

Size (g)	Fish strain	*n*	Initial mass (g)	Final mass (g)	G_w (% day^{-1})	I_{Td} (g $fish^{-1}$)	I_F (% g day^{-1})
138	Controls	5	135.6 ± 1.2^{a}	262.0 ± 5.4^{Xa}	3.1 ± 0.1^{X}	180.8 ± 12.2^{Xa}	4.6 ± 0.3^{X}
	Transgenics	5	139.6 ± 2.4^{a}	440.8 ± 14.8^{Ya}	5.5 ± 0.2^{Ya}	334.7 ± 14.5^{Ya}	6.4 ± 0.2^{Y}
208	Controls	5	211.5 ± 2.6^{b}	396.3 ± 17.3^{Xb}	3.0 ± 0.2^{X}	280.1 ± 10.8^{Xb}	4.6 ± 0.1^{X}
	Transgenics	5	205.0 ± 4.6^{b}	582.3 ± 27.2^{Yb}	5.0 ± 0.2^{Yb}	457.4 ± 10.2^{Yb}	6.3 ± 0.2^{Y}
ANOVA (*P*-value)							
Effect							
Size			0.0000	0.0000	0.0299	0.0000	0.9424
Fish strain			0.9174	0.0000	0.0000	0.0000	0.0000
Size×fish strain			0.0458	0.0548	0.3194	0.1696	0.6625

Letters after each value indicate results of pair-wise comparisons. Different upper case letters (X Y) indicate significant differences between fish strain for each size; different lower case letters (a b) indicate significant differences between sizes for each strain ($P<0.05$)

Table III Feed efficiency (E_F), protein retention efficiency (E_P) and apparent digestibility coefficients for dry matter, protein and energy for transgenic carp and the controls of different size-classes (values are means ±S.E.)*

					Apparent digestibility coefficient (%)		
Size (g)	Fish strain	*n*	E_F	E_P	Dry matter	Protein	Energy
138	Controls	5	$0.70{\pm}0.03^{X}$	29.2±0.6	$65.1{\pm}0.6^{X}$	$76.6{\pm}0.6^{X}$	$72.7{\pm}1.5^{Xa}$
	Transgenics	5	$0.90{\pm}0.01^{Ya}$	30.2±0.3	$71.8{\pm}1.1^{Y}$	$84.8{\pm}0.8^{Y}$	$80.4{\pm}1.0^{Y}$
208	Controls	5	$0.66{\pm}0.04^{X}$	28.1±1.4	$68.5{\pm}1.9^{X}$	$78.7{\pm}1.4^{X}$	$77.4{\pm}1.2^{b}$
	Transgenics	5	$0.82{\pm}0.04^{Yb}$	28.4±1.2	$73.7{\pm}1.6^{Y}$	$83.2{\pm}1.1^{Y}$	81.2±1.4
ANOVA (*P*-value)							
Effect							
Size			0.0286	0.1480	0.0366	0.8415	0.0343
Fish strain			0.0000	0.5277	0.0001	0.0000	0.0002
Size × fish strain			0.4171	0.7036	0.5569	0.0459	0.1342

* Letters after each value indicate results of pair-wise comparisons. Different upper case letters (X Y) indicate significant differences between fish strain for each size; different lower case letters (a b) indicate significant differences between sizes for each strain ($P<0.05$)

Apparent digestibility coefficient

Both body size and fish strain significantly affected apparent digestibility coefficients for dry matter and energy. Protein digestibility coefficients were significantly higher in the transgenic carp than in the controls, but were not affected by body size (Table III). Digestibility coefficients of dry matter and protein were significantly higher in the transgenic carp than in the controls for each size-class. Energy digestibility coefficient was significantly higher in the transgenic fish than in the controls for the 138 g size-class, but there was only marginal significant difference for the 208 g size-class (P= 0.0563).

Energy budget

The energy budgets for transgenic fish and the controls are given in Table IV. Gross energy intake (I_E), the proportion of I_E used for heat production (H_E), the proportion of I_E for metabolism (M_E) and the proportion of M_E for H_E were significantly higher in the transgenic fish than in the controls for each of the size-classes. The proportion of I_E used for faecal energy (F_E) was also significantly lower in the transgenic carp than in the controls for the 138 g size-class, but there was only a marginal significant difference for the 208 g size-class (P= 0.0563). The proportion of I_E used for the excretory energy loss (Z_E+U_E) was significantly lower in the transgenic carp than in the controls for the 138 g size-class, but there was no significant difference for the 208 g size-class. The proportion of I_E for recovered energy (R_E) was significantly lower in the transgenic carp than in the controls for the 208 g size-class, but there was no significant difference for the 138 g size-class. The proportions of I_E for waste products (F_E+Z_E+U_E) and the proportions of M_E for R_E were significantly lower in the transgenic fish than in the controls for each size-class. The average energy budget equation of transgenic fish was as follows: 100 I_E = 19.3 F_E + 6.0(Z_E+ U_E)+ 45.2 H_E + 29.5 R_E or 100 M_E = 60.5 H_E + 39.5 R_E. The average energy budget equation of the controls was: 100 I_E = 25.2 F_E + 7.4(Z_E+ U_E)+ 35.5 H_E + 31.9R_E or 100 M_E = 52.7 H_E + 47.3 R_E.

Table IV Energy budgets for transgenic carp and the controls of different size-classes (values are means ± S.E.)*†

Size (g)	Fish strain	n	I_E (J g^{-1} day^{-1})	F_E	Z_E+U_E	H_E	R_E	$F_E+Z_E+U_E$	M_E	H_E	R_E
				% I_E						%M_E	
138	Controls	5	807.5±46.5^{X}	27.3±1.5^{X}	6.6±0.4^{Xa}	33.8±0.5^{X}	32.4±1.3	33.8±1.3^{X}	66.2±1.3^{X}	51.1±1.2^{X}	48.9±1.2^{X}
	Transgenics	5	1136.8±32.7^{Y}	19.6±1.0^{Y}	4.6±0.1^{Ya}	44.5±1.6^{Y}	31.3±1.0^{a}	24.2±1.0^{Y}	75.8±1.0^{Y}	58.6±1.6^{Y}	41.4±1.6^{Ya}
208	Controls	5	817.0±20.2^{X}	22.6±1.2	8.4±0.7^{b}	37.7±1.1^{X}	31.3±0.6^{X}	31.0±1.1^{X}	69.0±1.1^{X}	54.6±1.0^{X}	45.4±1.0^{X}
	Transgenics	5	1118.2±29.7^{Y}	18.8±1.4	7.8±0.1^{b}	46.1±2.2^{Y}	27.3±0.9^{Yb}	26.6±1.5^{Y}	73.4±1.5^{Y}	62.7±1.9^{Y}	37.3±1.9^{Yb}
			ANOVA (P-value)								
Effect											
Size			0.8877	0.0343	0.0000	0.0435	0.0153	0.8891	0.8891	0.0453	0.0101
Fish strain			0.0000	0.0002	0.0014	0.0000	0.0149	0.0000	0.0000	0.0000	0.0000
Size × fish strain			0.6659	0.1342	0.0259	0.3490	0.1285	0.0271	0.0271	0.3490	0.8129

* I_E, gross energy intake; F_E, faecal energy; Z_E+ U_E, excretory energy loss (Z_E through the gills and U_E through the kidneys); H_E, heat production; R_E, recovered energy; M_E, metabolizable energy

† Letters after each value indicate results of pair-wise comparisons. Different upper case letters (X Y) indicate significant differences between fish strain for each size; different lower case letters (a b) indicate significant differences between sizes for each strain (P<0.05)

Body composition and energy content

The whole body proximate compositions and energy contents for transgenic fish and the controls are presented in Table V. Body contents of dry matter and protein were significantly lower in the transgenic carp than in the controls, but were not affected by body size. Lipid content was significantly affected by body size, but not affected by fish strain. Both body size and fish strain did not significantly affect ash content and energy content.

Table V Whole body proximate compositions and energy contents per unit wet mass for transgenic carp and the controls of different size-classes (values are means ± S.E.)*

Size (g)	Fish strain	*n*	Dry matter (%)	Protein (%)	Lipid (%)	Ash (%)	Energy (kJ g^{-1})
Initral							
138	Controls	2	27.4±0.5	16.3±0.3	8.1±0.5	2.6±0.1	6.2±0.2
	Transgenics	2	28.9±0.0	14.6±0.2	10.8±1.4	2.0±0.2	7.7±0.7
208	Controls	2	27.9±1.0	16.1±0.2	8.6±0.4	2.7±0.3	6.3±0.1
	Transgenics	2	27.2±0.5	13.9±0.1	10.5±0.3	2.3±0.2	6.5±0.1
Final							
138	Controls	5	$29.0±0.1^{X}$	$16.1±0.1^{X}$	$9.8±0.2^{a}$	2.6±0.1	7.1±0.1
	Transgenics	5	$26.3±0.5^{Y}$	$13.3±0.1^{Y}$	9.8±0.1	2.4±0.1	6.3±0.1
208	Controls	5	$30.5±0.3^{X}$	$16.1±0.2^{X}$	$11.5±0.4^{b}$	2.6±0.1	7.3±0.1
	Transgenics	5	$26.7±0.4^{Y}$	$13.4±0.1^{Y}$	10.3±0.5	2.5±0.1	6.1±0.1
ANCOVA (*P*-value)							
Effect							
Size			0.2384	0.3321	0.0018	0.4259	0.4381
Fish strain			0.0280	0.0000	0.1649	0.1275	0.1045
Size × fish strain			0.1535	0.8809	0.0337	0.8772	0.0643

*Letters after each value indicate results of pair-wise comparisons. Different upper case letters (X Y) indicate significant differences between fish strain for each size; different lower case letters (a b) indicate significant differences between sizes for each strain ($P< 0.05$)

Discussion

In the present study, on average, F_2 'all-fish' GH-transgenic carp had specific growth rates for wet mass that were 67-77% higher than noted for control fish. The growth-promoting effects of exogenous growth hormone genes have also been observed in a number of other GH-transgenic fishes, including GH-transgenic cyprinids (Zhu *et al.*, 1986, 1989; Cui *et al.*, 1996a; Fu *et al.*, 1998; Nam *et al.*, 2001; Wang *et al.*, 2001; Dunham *et al.*, 2002), GH-transgenic salmonids (Chourrout *et al.*, 1986; Devlin *et al.*, 1994; Cook *et al.*, 2000a), GH-transgenic channel catfish *Ictalurus punctatus* (Rafinesque) (Dunham *et al.*, 1987), GH-transgenic Nile tilapia *Oreochromis niloticus* L. (Rahman *et al.*, 2001), GH-transgenic ayu *Plecoglossus altivelis* Temminck & Schlegel (Cheng *et al.*, 2002) and GH-transgenic rohu *Labeo rohita* Hamilton (Venugopal *et al.*, 2004).

This study showed that higher feed intakes and feed efficiencies were mainly responsible for the fast-growth of the F_2 'all-fish' GH-transgenic carp. Increased feed intake and improved feed efficiency have also been reported in other fast-growing transgenic fishes (Fu *et al.*, 1998; Cook *et al.*, 2000a; Nam *et al.*, 2001; Rahman *et al.*, 2001; Venugopal *et al.*, 2004).

Previous studies have shown that human GH-transgenic carp had higher protein retention efficiency (E_P) than the controls (Fu *et al.*, 1998). By contrast, this study found that there was no difference in protein utilization between 'all-fish' GH-transgenic carp and the controls. The dissimilar findings were most likely due to two factors. First, the human GH-gene was used in the previous study, whereas the grass carp GH-gene was used in the present study. Second, < 2 g juvenile fish were used in the previous study, while >138 g fish were used in the present study. With respect to protein digestibility coefficients that have been observed between fast-growing transgenic fishes and non-transgenic controls, studies have reported either no difference (Rahman *et al.*, 2001) or higher protein digestibility coefficients in transgenic fishes (Fu *et al.*, 1998; Cook *et al.*, 2000a). In this study, protein digestibility coefficients were significantly higher in transgenic fish than in the controls. More studies are needed on this topic so that generalizations can be made on how fast-growing transgenic fishes digest and utilize dietary protein for growth. The problem of nutrient leaching from faeces may have resulted in higher digestibility coefficient values in the present results. Faeces collected from water may result in higher digestibility values than those determined using faeces obtained directly from the fishes (Inaba *et al.*, 1963; Spyridakis *et al.*, 1989; Hajen *et al.*, 1993). Cho & Slinger (1979), however, found similar digestibility coefficients when faeces were obtained by dissection and from faecal collection columns. Hajen *et al.* (1993) suggested that faecal collection by the settling column appeared to be reliable for estimating digestibility. Although faecal collection by stripping avoids the problem of leaching, errors may arise from contamination of faeces with urine and body mucus, and a reduction of the transit time of intestinal contents, resulting in underestimates of digestibility coefficients (Austreng, 1978; Cho *et al.*, 1982). Hence, there are errors associated with all faecal collection methods and these may have been greater in this study than in studies where the faeces were collected by intestinal dissection and faecal collection columns.

Studies that have examined routine oxygen consumption (Mo_2), in fast-growing transgenic salmonids have found that transgenic fishes have elevated oxygen consumption rates relative to non-transgenic controls (Stevens *et al.*, 1998; Cook *et al.*, 2000b; Herbert *et al.*, 2001; Lee *et al.*, 2003; McKenzie *et al.*, 2003). Leggatt *et al.* (2003) also found that GH-transgenic fishes had an oxygen uptake that was 1.4 times greater than control fishes, when fishes that had eaten the same percent body mass were compared. The results of the present study were in accord with those of previous studies, once it was noted that the proportion of I_E that was used for heat production (H_E) was significantly higher in transgenic carp than in the controls. This suggests that fast-growing transgenic carp also had increased routine oxygen consumption in contrast to the non-transgenic controls, although oxygen consumption was not directly determined in this study. In the present study, metabolism was calculated indirectly by difference from other components of the energy budget. The accuracy of metabolism estimates depends on the accuracy of the directly estimated components. Estimates of gross intake energy and recovered energy for growth should be accurate. Faecal production and nitrogenous excretion accounted for only small proportions of feed energy, and errors in the estimates of these components are unlikely to affect the estimate of metabolism greatly (Cui *et al.*, 1996b).

In the present study, the proportion of metabolizable energy (M_E) that was allocated for recovered energy (R_E) was significantly lower in the transgenic fish than in the controls. These results suggest that 'all-fish' GH-transgenic carp were not efficient in converting metabolizable energy to growth compared with the controls. Previous studies, however, have found that the proportion of M_E that was allocated for R_E was significantly higher in the human GH-transgenics than in the controls (Cui *et al.*, 1996a). The differences in findings can most likely be attributed to the magnitude of feed intake and growth acceleration observed for the transgenic carp in the present study. The feeding rate values of human GH-transgenic carp were 7.6% higher than those of the controls in the experiment of Cui *et al.* (1996a), but were 26.4% higher than those of the controls in this study. The specific growth rate (G_W) values of human GH-transgenic carp were 19% higher than those of the controls in Cui *et al.* (1996a), whereas they were 67-77% higher than those of the controls in this study.

Cui & Liu (1990) calculated an average energy budget for several teleosts based on 14 published budgets for fishes fed maximum rations which was: $100\ (I_E\text{–}F_E\text{–}Z_E\text{–}U_E) = 60\ H_E + 40\ R_E$. In this study, the average budget found for transgenic fish was: $100\ (I_E\text{–}F_E\text{–}Z_E\text{–}U_E) = 60.5\ H_E+39.5\ R_E$. Thus, despite their high growth rates, transgenic carp were not found to be especially efficient in converting M_E for growth compared with other fishes.

Cui *et al.* (1996b) compared the growth rates and energy budget of white sturgeon *Acipenser transmontanus* Richardson with those of two teleosts; the European minnow *Phoxinus phoxinus* (L.), a small-sized fish, and the southern catfish *Silurus meridionalis* Chen, a medium-sized fish. In this regard, they found that the white sturgeon had the highest growth rate and the lowest waste production, but the allocation of food energy to metabolism was highest in the European minnow and lowest in the southern catfish, while allocation to growth followed the reverse order. This study also found that 'all-fish' GH-transgenic carp had the higher growth rate, the lower waste production and the higher allocation of food energy to metabolism. The control fish, however, had the higher allocation of food energy to growth. These findings indicate that 'all-fish' transgenic carp adopted a strategy of a high feeding rate and low loss in waste production in order to enhance their growth rate compared with the controls.

In the present study, the body contents of dry matter and protein were significantly lower in 'all-fish' GH-gene transgenic fish than in the controls. There were no significant differences, however, in ash, lipid and energy contents between transgenic fish and controls. These results differ from those of a previous experiment on human GH-gene transgenic carp. In the latter case, the transgenic fish showed significantly higher contents of dry matter and protein, but lower contents of lipid than did the controls (Fu *et al.*, 1998). The discrepancy in findings can most likely be attributed to the magnitude of growth acceleration in the common carp of this study. The G_W values of GH-transgenic carp were 19-25% higher than those of the controls in the previous experiment, whereas they were 67-77% higher than those of the controls in this study. Significantly lower body contents of dry matter, protein, lipid and energy have also been observed in fast-growth transgenic Atlantic salmon *Salmo salar* L., where the G_W values of the transgenic fish were 180% higher than those of the controls (Cook *et al.*, 2000a) and also in transgenic Nile tilapia, where the mean mass gains of the transgenic fish were over four times greater than those of the controls (Rahman *et al.*, 2001).

In conclusion, the findings of this study indicate that 'all-fish' GH-transgenic carp grew rapidly in response to foreign GH-gene administration by increasing feed intake and reducing loss in waste production (increased feed efficiency). The benefit of the increased energy intake by transgenic fish was diminished, however, by their increased metabolism.

Acknowledgements The authors would like to thank two anonymous reviewers for their constructive comments on the manuscript, and also thank S. Xie, W. Lei, Y. Yang and X. Zhu for their help during the experiment. We are grateful to Y. Cui, who first introduced fish bioenergetics in China and trained the first author. This work was supported by the Projects of the National Natural Science Foundation of China (NSFC 30130050 and 30400056), the State '863' High-Tech Project (2004AA213120), and Development Plan of the State Key Fundamental Research (2001CB109006) from the Ministry of Science and Technology, China. C. Fu and D. Li contributed equally to the study.

References

AOAC (1984). Official Methods of Analysis of the Association of Official Analytical Chemists, 14th edn. Arlington, VA: Association of official Analytical Chemists.

Austreng, E. (1978). Digestibility determination in fish using chromic oxide marking and analysis of contents from different segments of the gastrointestinal tract. Aquaculture 13, 265-272.

Chaney, A. L. & Marbach, E. P. (1962). Modified reagents for determination of urea and ammonia. Clinical Chemistry 8, 30-132.

Cheng, C., Liu, K., Lau, E., Yang, T., Lee, C., Wu, J. & Chang, C. (2002). Growth promotion in Ayu (*Plecoglossus altivelis*) by gene transfer of the rainbow trout growth hormone gene. Zoological Studies 41, 303-310.

Cho, C. Y. & Slinger, S. J. (1979). Apparent digestibility measurement in feedstuffs for rainbow trout. In Finfish Nutrition & Fish Feed Technology, Vol. II (Halver, J. E. & Tiews, K., eds), pp. 239-247. Berlin: Heenemann.

Cho, C. Y., Slinger, S. J. & Bayley, H. S. (1982). Bioenergetics of salmonid fishes: energy intake, expenditure and productivity. Comparative Biochemistry and Physiology 73B, 25-41.

Chourrout, D., Guyomard, R. & Houdebine, L. M. (1986). High efficiency gene transfer in rainbow trout (*Salmo gairdneri* Rich.) by microinjection into egg cytoplasm. Aquaculture 51, 143-150.

Cook, J. T., McNiven, M. A., Richardson, G. F. & Sutterlin, A. M. (2000a). Growth rate, body composition/feed digestibility conversion of growth-enhanced transgenic Atlantic salmon (*Salmo salar*). Aquaculture 188, 15-32.

Cook, J. T., Sutterlin, A. M. & McNiven, M. A. (2000b). Effect of food deprivation on oxygen consumption/body composition of growth-enhanced transgenic Atlantic salmon (*Salmo salar*). Aquaculture 188, 47-63.

Cui, Y. & Liu, J. (1990). Comparison of energy budget among six of teleosts. III. Growth rate and energy budget. Comparative Biochemistry and Physiology 97A, 381-384.

Cui, Z., Zhu, Z., Cui, Y., Li, G. & Xu, K. (1996a). Food consumption and energy budget in ThGH transgenic F_2 red carp (*Cyprinus carpio* L. red var.). Chinese Science Bulletin 41, 591-596.

Cui, Y., Hung, S. S. O. & Zhu, X. (1996b). Effect of ration and body size on the energy budget of juvenile white sturgeon. Journal of Fish Biology 49, 863-876.

Devlin, R. H., Yesaki, T., Biagy, C. A., Donaldson, E. M., Swanson, P. & Chan, W. K. (1994). Extraordinary salmon growth. Nature 371, 209-210.

Devlin, R. H., Biagi, C. A. & Yesaki, T. Y. (2004). Growth, viability and genetic characteristics of GH transgenic coho salmon strains. Aquaculture 236, 607-632.

Dunham, R. A., Eash, J., Askins, J. & Townes, T. M. (1987). Transfer of metallothionein-human growth hormone fusion gene into channel catfish. Transactions of the American Fisheries Society 116, 87-91.

Dunham, R. A., Chatakondi, N., Nichols, A. J., Kucuktas, H., Chen, T. T., Powers, D. A., Weete, J. D., Cummins, K. & Lovell, R. T. (2002). Effect of rainbow trout growth hormone complementary DNA on body shape, carcass yield, and carcass composition of F_1 and F_2 transgenic common carp (*Cyprinus carpio*). Marine Biotechnology 4, 604-611.

Elliott, J. M. (1976a). Energy losses in the waste products of brown trout (*Salmo trutta* L.). Journal of Animal Ecology 45, 561-580.

Elliott, J. M. (1976b). The energetics of feeding, metabolism and growth of brown trout (*Salmo trutta* L.). Journal of Animal Ecology 45, 923-948.

Fu, C., Cui, Y., Hung, S. S. O. & Zhu, Z. (1998). Growth and feed utilization by F_4 human growth hormone transgenic carp fed diets with different protein levels. Journal of Fish Biology 53, 115-129.

Hajen, W. E., Beames, R. M., Higgs, D. A. & Dosanjh, B. S. (1993). Digestibility of various feedstuffs by post-juvenile chinook salmon (*Oncorhynchus tshawytscha*) in sea water. 1. Validation of technique. Aquaculture 112, 321-332.

Herbert, N. A., Armstrong, J. D. & Björnsson, B. Th. (2001). Evidence that growth hormone induced elevation in routine metabolism of juvenile Atlantic salmon is a result of increased spontaneous activity. Journal of Fish Biology 59, 754-757.

Inaba, D., Ogino, C., Takamastu, T., Ueda, T. & Kurokawa, K. (1963). Digestibility of dietary components in fishes-II. Digestibility of dietary protein and starch in rainbow trout. Bulletin of the Japanese Society of Scientific Fisheries 29, 242-244.

Jobling, M. (1994). Fish Bioenergetics. London: Chapman & Hall.

Lambert, P. & Dehnel, P. A. (1974). Seasonal variations in biochemical composition during the reproductive cycle of the intertidal gastropod Thais lamellosa Gmelin (*Gastropoda, Prosobranchia*). Canadian Journal of Zoology 52, 305-318.

Lee, C. G., Devlin, R. H. & Farrell, A. P. (2003). Swimming performance, oxygen consumption and excess post-exercise oxygen consumption in adult transgenic and ocean-ranched coho salmon. Journal of Fish Biology 62, 753-766.

Leggatt, R. A., Devlin, R. H., Farrell, A. P. & Randall, D. J. (2003). Oxygen uptake of growth hormone transgenic coho salmon during starvation and feeding. Journal of Fish Biology 62, 1053-1066.

Mao, W., Wang, Y., Wang, W., Wu, B., Feng, J. & Zhu, Z. (2004). Enhanced resistance to Aeromonas hydrophila infection and enhanced phagocytic activities in human lactoferrin-transgenic grass carp (*Ctenopharyngodon idellus*). Aquaculture 242, 93-103.

McKenzie, D. J., Martinez, R., Morales, A., Acosta, K., Taylor, E. W., Steffensen, J. F. & Estrada, M. P. (2003). Effects of growth hormone transgenesis on metabolic rate, exercise performance and hypoxia tolerance in tilapia hybrids. Journal of Fish Biology 63, 398-409.

Nam, Y. K., Noh, J. K., Cho, Y. S., Cho, H. J., Cho, K. N., Kim, C. G. & Kim, D. S. (2001). Dramatically accelerated growth and extraordinary gigantism of transgenic mud loach *Misgurnus mizolepis*. Transgenic Research 10, 353-362.

NRC (National Research Council) (1981). Nutritional Energetics of Domestic Animals and a Glossary of Energy Terms, 2nd edn. Washington, DC: National Academy Press.

Rahman, M. & Maclean, N. (1999). Growth performance of transgenic tilapia containing an exogenous piscine growth hormone gene. Aquaculture 173, 333-346.

Rahman, M., Ronyai, A., Engidaw, B. Z., Jauncey, K., Hwang, G. L., Smith, A., Roderick, E., Penman, D., Varadi, L. & Maclean, N. (2001). Growth and nutritional trials on transgenic Nile tilapia containing an exogenous piscine growth hormone gene. Journal of Fish Biology 59, 62-78.

Richardson, N. L., Higgs, D. A., Beames, R. M. & McBride, J. R. (1985). Influence of dietary calcium, phosphorous, zinc and sodium phytate level on cataract incidence, growth, and histopathothology in juvenile chinook salmon (*Oncorhynchus tshawytscha*). Journal of Nutrition 115, 553-567.

Shearer, K. D. (1994). Factors affecting the proximate composition of cultured fishes with emphasis on salmonids. Aquaculture 119, 63-88.

Spyridakis, P., Metailler, R., Gabaudan, J. & Riaza, A. (1989). Studies on nutrient digestibility in European sea bass (*Dicentrarchus labrax*). 1. Methodological aspects concerning faeces collection. Aquaculture 77, 61-70.

Stevens, E. D., Sutterlin, A. & Cook, T. (1998). Respiratory metabolism and swimming performance in growth hormone transgenic Atlantic salmon. Canadian Journal of Fisheries and Aquatic Sciences 55, 2028-2035.

Venugopal, T., Anathy, V., Kirankkumar, S. & Pandian, T. J. (2004). Growth enhancement and food conversion efficiency of transgenic fish *Labeo rohita*. Journal of Experimental Zoology 301A, 477-490.

Wang, Y., Hu, W., Wu, G., Sun, Y., Chen, S., Zhang, F., Zhu, Z., Feng, J. & Zhang, X. (2001). Genetic analysis of "all-fish" growth hormone gene transferred carp (*Cyprinus carpio* L.) and its F_1 generation. Chinese Science Bulletin 46, 1174-1177.

Zbikowska, H. M. (2003). Fish can be first-advances in fish transgenesis for commercial applications. Transgenic Research 12, 379-389.

Zhong, J., Wang, Y. & Zhu, Z. (2002). Introduction of the human lactoferrin gene into grass carp (*Ctenopharyngodon idellus*) to increase resistance against GCH virus. Aquaculture 214, 93-101.

Zhu, Z., Li, G., He, L. & Chen, S. (1985). Novel gene transfer into the fertilized eggs of goldfish (*Carassius auratus* L. 1758). Journal of Applied Ichthyology 1, 31-34.

Zhu, Z., Xu, K., Li, G., Xie, Y. & He, L. (1986). Biological effects of human growth hormone gene microinjected into the fertilized eggs of loach, *Misgurnus anguillicaudatus* (Cantor) Chinese Science Bulletin 31, 988-990.

Zhu, Z., Xu, K., Xie, Y., Li, G. & He, L. (1989). A model of transgenic fish. Scientia Sinica B 2, 147-155.

Zhu, Z., He, L. & Chen, T. T. (1992). Primary-structural and evolutionary analyses of growth-hormone gene from grass carp (*Ctenopharyngodon idellus*). European Journal of Biochemistry 207, 643-648.

转“全鱼”生长激素基因鲤 F_2 的生长和能量收支

傅萃长　李德亮　胡　炜　汪亚平　朱作言

中国科学院水生生物研究所，淡水生态与生物技术国家重点实验室，武汉　430072

摘　要　本研究对两种大小规格的转“全鱼”生长激素基因鲤的生长和能量收支进行了研究。实验在 29.2℃条件下进行，为期 21 天。转基因鲤的特定生长率、摄食量、饲料转化效率、干物质和蛋白质的消化效率、摄取的总能量(I_E)及其用于产生热量(H_E)的比例均显著高于对照鱼。鱼类的代谢废物能包括排粪能(F_E)和排泄能($Z_E + U_E$)，其中 Z_E 是指通过鳃排出的能量，U_E 是指通过肾排出的能量。转基因鱼总能量中转化为代谢废物能的比例，以及恢复能量(R_E)占代谢能(M_E)的比例均显著低于对照鱼。转基因鱼平均能量收支方程为：$100\ I_E = 19.3\ F_E + 6.0(Z_E+U_E)+45.2\ H_E+29.5\ R_E$，或者是 $100\ M_E = 60.5\ H_E+39.5\ R_E$。对照鱼的平均能量收支方程为：$100\ I_E = 25.2\ F_E+7.4(Z_E+U_E)+35.5\ H_E+31.9R_E$，或者是 $100\ M_E = 52.7\ H_E+47.3\ R_E$。研究结果表明，增加摄食量，减少废物代谢以提高食物转化效率是转“全鱼”生长激素基因鲤快速生长的主要机制。通过加速新陈代谢，转基因鱼减弱其增加的能量摄取优势。

Fast-Growing Transgenic Common Carp Mounting Compensatory Growth

C. Fu　D. Li　W. Hu　Y. Wang　Z. Zhu

State Key Laboratory of Freshwater Ecology and Biotechnology, Institute of Hydrobiology, Chinese Academy of Sciences, Wuhan 430072

Abstract Compensatory growth is a phase of accelerated growth apparent when favourable conditions are restored after a period of growth depression. To investigate if F_2 common 'all-fish' growth hormone gene transgenic common carp (*Cyprinus carpio*) could mount compensatory growth, a 9 week study at 29℃ was performed. The control group was fed to satiation twice a day throughout the experiment. The other two groups were deprived of feed for 1 or 2 weeks, respectively, and then fed to satiation during the re-feeding period. At the end of the experiment, the live masses of fish in the deprived groups were still significantly lower than those of the controls. During the re-feeding period, size-adjusted mean specific growth rates and mean feed intakes were significantly higher in the deprived fish than in the controls, indicating a partial compensatory growth response in these fish. No significant differences were found in food conversion efficiency between the deprived and control fish during re-feeding, suggesting that hyperphagia was the mechanism responsible for increased growth rates. The proximate composition of the deprived fish at the end of the experiment was similar to that of the control fish. This study is, to our knowledge, the first to report that fast-growing transgenic fish can achieve partial compensation of growth following starvation.

Keywords common carp; compensatory growth; fast-growing; growth hormone gene; starvation; transgenic fish

Introduction

Compensatory (or catch-up) growth is a well-documented form of compensation, being reported across a wide range of taxa (Wilson & Osbourn, 1960; Sibly & Calow, 1986; Arendt, 1997; Ali *et al.*, 2003). The general pattern is for fishes that have experienced a period of retarded growth to then enter a phase of growth acceleration when conditions improve (Wilson & Osbourn, 1960; Jobling, 1994; Metcalfe & Monaghan, 2001; Ali *et al.*, 2003). Depending on the degree of recovery, compensatory growth can be classified into three types, i.e. over compensation, full compensation and partial compensation. Over compensation

文章发表于 *Journal of Fish Biology*, 2007, 71(Supplement sb):174-185 occurs when the fishes that had experienced a restricted ration achieve a greater size at the same age than non-restricted fishes. Full compensation means that the deprived fish regain the mass trajectory shown by control fish. In partial compensation, the deprived fishes fail to achieve the same size at the same age as non-restricted contemporaries, but do show relatively rapid growth rates, and may have better food conversion efficiencies during the realimentation period (Jobling, 1994; Ali *et al.*, 2003).

The occurrence of compensatory growth has been reported for a restricted number of fish taxa: most studies of compensatory growth in fish have been carried out on cold-water species, and reports on warm-water species are few (Ali *et al.*, 2003). Studies on compensatory growth in fish have yielded inconsistent results. Full compensation and partial compensation have been observed in most studies, such as those on species of Cyprinidae (Russell & Wootton, 1992; Wieser *et al.*, 1992; Qian *et al.*, 2000; Xie *et al.*, 2001; Zhu *et al.*, 2001, 2004, 2005; van Dijk *et al.*, 2005), species of Salmonidae (Miglavs & Jobling, 1989a, b; Jobling *et al.*, 1993; Nicieza & Metcalfe, 1997; Johansen *et al.*, 2001; Maclean & Metcalfe, 2001; Nikki *et al.*, 2004), species of Ictaluridae (Kim & Lovell, 1995; Gaylord & Gatlin, 2001; Zhu *et al.*, 2004, 2005), species of Cichlidae (Wang *et al.*, 2000) and species of Sparidae (Eroldoğanv *et al.*, 2006). Over compensation was observed in *Lepomis* hybrids by Hayward *et al.* (1997), but it seems to be a rare outcome. No growth compensation was observed in common carp, *Cyprinus carpio* L. (Schwarz *et al.*, 1985) and Atlantic cod, *Gadus morhua* Linnaeus (Jobling *et al.*, 1994).

Since the first batches of transgenic fish were achieved by Maclean & Talwar (1984) and Zhu *et al.* (1985), transgenic fish have been produced in a variety of species using different foreign gene constructs (Zhu *et al.*, 1992; Devlin *et al.*, 1994; Rahman & Maclean, 1999; Zhong *et al.*, 2002; Mao *et al.*, 2004). Growth hormone gene (GH) transgenic fish have been reported to exhibit higher growth rates than the controls (Zhu *et al.*, 1986, 1989; Devlin *et al.*, 1994; Fu *et al.*, 1998; Rahman & Maclean, 1999; Nam *et al.*, 2001; Wang *et al.*, 2001; Devlin *et al.*, 2004; Venugopal *et al.*, 2004; Raven *et al.*, 2006). Information on growth, feed utilization, proximate body composition and metabolic rate of fast-growth transgenic fish has also been reported in many studies (Cui *et al.*, 1996; Fu *et al.*, 1998, 2000; Cook *et al.*, 2000a, b; Rahman *et al.*, 2001, Leggatt *et al.*, 2003; McKenzie *et al.*, 2003; Venugopal *et al.*, 2004). However, until now, no information has been published to examine whether compensatory growth occurs in fast-growing GH-transgenic fish.

The purpose of the present study was to examine compensatory growth responses in feeding and growth in individually held F_2 common carp (*C. carpio*), transgenic for an 'all-fish' GH. Such knowledge would be useful for the understanding of feeding biology, bio-safety assessment and future aquaculture management of fast-growing GH-transgenic fish.

Material and Methods

P_0 'all-fish' GH-transgenic carp were produced by microinjection of pCAgcGHc into the fertilized eggs of common carp (Yellow River carp variety). 'All-fish' gene construct pCAgcGHc was the recombinant construct of grass carp (*Ctenopharyngodon idellus* Valenciennes) growth hormone complementary DNA (gcGHc) driven by β-actin gene promoter of common carp (pCA) (Wang *et al.*, 2001). F_2 'all-fish' GH-transgenic fish used in this experiment were produced from crosses between a wild type female and a F_1 'all-fish' GH-transgenic male of a fast-growing transgenic strain which was produced previously in our laboratory. Frequencies of transgene transmission to F_2 progeny from this line were *c.* 50%, and body mass of F_2 transgenic fish was *c.* 2.5 times that of controls at age 200 days (unpubl. data). At the age of 4 weeks, 100 F_2 transgenic carp were transferred to the laboratory from Duofu Technology Farm, China, and fed with the experimental diet. DNA was extracted from tail fin clips of these fish, and the polymerase chain reaction method was used to detect the transgene-positive individuals based on the procedure of Wang *et al.* (2001). The growth experiment was initiated when the fish were *c.* 8 weeks old and weighed between 87 and 939 g (Table I).

Table I Body mass (g) of transgenic carp at different times of the experiment*

Group	Initial mass	Mass at week 3(start of re-feeding)	Final mass	*n*
C	89.15±0.13	304.67±1.36	891.67±7.62	6
S1	89.93±0.05	213.00±0.49	779.14±2.42	7
S2	90.14±0.06	151.86±0.25	630.00±2.92	7
ANCOVA	using initial length as covariate			
F	0.18	93.11	7.06	6
d.f.	2,16	2,16	2,16	7
P	0.8343	0.0000	0.0063	7

* C, control fish that were fed throughout the experiment; S1, fish that were starved for 1 week during week 3 and fed for the rest of the experiment; S2, fish that were starved for 2 weeks during weeks 2 and 3 and fed for the rest of the experiment (mean ±s.e.). *F*, fisher's *F*-ratio; d.f., degrees of freedom; *P*, probability value

In this study, eight circular fibreglass tanks (diameter 150 cm and water volume 1000 l) were used, with each tank divided into four equal aquaria by perforated fibreglass partitions. Fish in adjacent aquaria were visible to one another. The growth experiment was conducted in 30 aquaria. Water was aerated in a common reservoir, and a gentle water flow (*c.* 5 l min^{-1}) was introduced to each aquarium. During the experimental period, dissolved oxygen was >7 mg l^{-1}, ammonia-nitrogen was <0.15 mg l^{-1} and pH varied from 7.0 to 7.3. The water temperature varied from 27 to 32℃ (mean 29, s.e. 0.18℃). The photoperiod was controlled by artificial lighting with the light period between 0700 and 1900 hours.

After acclimation for 1 week, 30 fish (mean mass 89.9, s.e. 0.38 g) were assigned at random to three feeding groups. Each group included 10 aquaria with one fish per aquarium. Prior to the start of the experiment, the fish were starved for 1 day to evacuate their gut. The live mass and standard length of each fish was measured, blotted of excess water, weighed to 0.01 g and measured to 0.01 cm. The control fish (C) were hand-fed to satiation twice a day at 0800 and 1600 hours throughout the experiment with formulated dry pellets manufactured by Shunde Fish Feed Limited Company, Guangdong, China. The proximate composition of the experiment diet was 36.7% for protein, 5.6% for lipid, 5.1% for ash and 21 J mg^{-1} for energy. The S1 group experienced 1 week of starvation during the third week of the experiment and the S2 group experienced 2 weeks of starvation during the second and the third weeks of the experiment prior to being fed to satiation twice a day in week 4. The growth experiment lasted for 9 weeks from 25 June to 28 August 2004. During each feeding, a quantity of diets in excess of what the fish consumed in 2 h were provided. Two hours after feeding, the uneaten feed was collected, dried and weighed. Water flow was stopped during feeding.

The chemical composition and energy content of fish were separately determined in week 3 and week 9. At the end of week 3 (before re-feeding), three fish from each group were randomly taken and killed. At the end of week 9 (at the end of the experiment), all fish were killed following weighing. The fish killed were dried to constant mass at 70℃ and homogenized. Nitrogen was determined using the Kjeldahl method (AOAC, 1984), lipid concentration was determined by Lambert and Dehnel's (1974) modification of the chloroform–methanol extraction technique, ash by combustion at 550℃ and energy by bomb calorimetry (Gentry Instruments Inc., Aiken, SC, U.S.A.). Protein content was calculated from nitrogen content by multiplying with 6.25.

The following parameters of growth performance were calculated: specific growth rate (G_w) (% day^{-1}) = 100× ($\ln W_t - W_0$)/t; feed intake (% body mass day^{-1}) = $100 \times C/[(W_0+W_t)/2]t$; feed conversion efficiency (%) = $100 \times (W_t - W_0)/C$, where W_t and W_0 are final and initial masses (g), t is the feeding duration (day) and C is dry feed consumption (g).

In fish, G_w and the rate of food consumption are both functions of body mass (Jobling, 1994; Wootton, 1998). Consequently, the comparisons of control and deprived fish had to be based on adjusted mean masses by the appropriate ANCOVA in which body mass acts as the covariate (Huitema, 1980; Ruohonen *et al.*, 2001; Zhu *et al.*, 2004).

ANCOVA was used to test for differences between treatments in body mass at the start of the experiment, start of re-feeding and at the end of the experiment, using initial length as covariate. The trajectories of body mass, G_w and feed intake over the course of the experiment were analysed by repeated measures ANCOVA, after the appropriate log transformations. Body mass at the start of each re-feeding week was used as a covariate to assess the treatment effect on G_w and feed intake. The trajectory in feed conversion efficiency was analysed by repeated measures ANOVA. The proximate composition of the fish at the end of the experiment was analysed by single factor ANOVA. Pre-planned comparisons tested for the overall effect of deprivation (control *v*. mean of S1 and S2) and for the effect of the mildest deprivation (control *v*. S1). Differences were regarded as significant when $P< 0.05$.

Results

One fish in the control group died during the experiment. There was no significant difference in the initial body mass of the fish among all feeding groups (Table I). At the start of re-feeding, differences in fish mass were significant among groups, and body mass in S1 and S2 groups was 70 and 50% of the control body mass, respectively. After re-feeding for 6 weeks, the final body masses among all feeding groups were still significantly different (Table I and Fig. 1). Thus, any compensatory growth response in live mass was, at best, partial during the experiment.

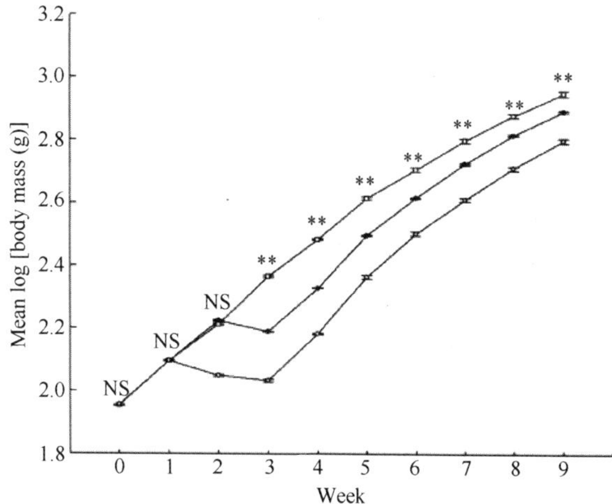

Fig. 1 Mean log (body mass) at weekly intervals adjusted for initial length in three groups of transgenic carp subjected to different periods of starvation. C, control group (n = 6) without starvation; S1, the group starved for 1 week during the third week of the experiment (n = 7); S2, the group starved for 2 weeks during the second and third weeks of the experiment (n = 7). Results of pre-planned statistical contrasts: NS, no significant difference between C and mean of treatments S1 and S2; **, significant difference between C and mean of S1 and S2 and between C and S1. Differences were regarded as significant when $P < 0.05$. Error bars represent s.e. —▫—, C; —•—, S1; —▫—, S2

Body mass was significantly affected by feeding regime, extent of re-feeding with time and their interaction, and G_W and feed intake were significantly affected by feeding regime and time but not by their interaction (Table II). Food conversion efficiency was significantly affected by time but not by feeding regime and the interaction between feeding regime and time (Table II).

Table II Results of repeated measures ANCOVA or ANOVA of weekly changes in body mass, specific growth rate (G_w), feed intake and food conversion efficiency during the re-feeding period in transgenic carp subjected to different feeding regimes. Initial body length was used as covariate for the analysis on body mass, and mass at start of each re-feeding week was used as covariate for the analysis on G_w and feed intake*

	Treatment effect			Week effect			Treatment × week interaction		
Variables	*F*	d.f.	*P*	*F*	d.f.	*P*	*F*	d.f.	*P*
Body mass	41.27	2,16	0.0000	21.76	5,102	0.0000	27.00	5,102	0.0000
G_W	34.16	2,16	0.0000	63.98	5,102	0.0000	0.62	5,102	0.7911
Feed intake	32.55	2,16	0.0000	29.72	5,102	0.0000	0.66	5,102	0.7547
Food conversion efficiency	1.94	2,16	0.1486	17.80	5,102	0.0000	0.50	5,102	0.9033

**F*, fisher's *F*-ratio; d.f., degrees of freedom; *P*, probabolity value

Mean G_w was significantly higher in the deprived groups than in the control during each week after re-feeding (Fig. 2). The G_w in S1 fish was significantly higher than those in the control from week 4 to week 8 ($P < 0.001$), but there was no significant difference between S1 fish and the control in the last week of re-feeding. The G_w in S2 fish was significantly higher than those in the control during all weeks of re-feeding ($P < 0.001$). Mean feed intake was significantly higher in the deprived groups than in the control during each week after re-feeding (Fig. 3). Food conversion efficiency was not significantly different among all feeding groups (Fig. 4). Since the G_w lines for the S1 and S2 groups were approaching that of the C group (Fig. 2), the compensation response for group S1 may have been near completion, but this had not yet occurred for group S2.

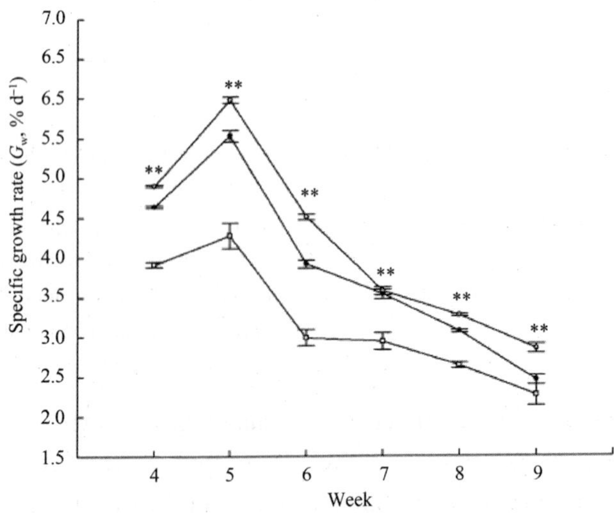

Fig. 2 Adjusted mean specific growth rate in three groups of transgenic carp during re-feeding weeks. ANCOVA used mass at start of each re-feeding week as covariate for adjustment of mean. Treatment abbreviation and symbols as in Fig 1. —▫—, C; —•—, S1; —▫—, S2

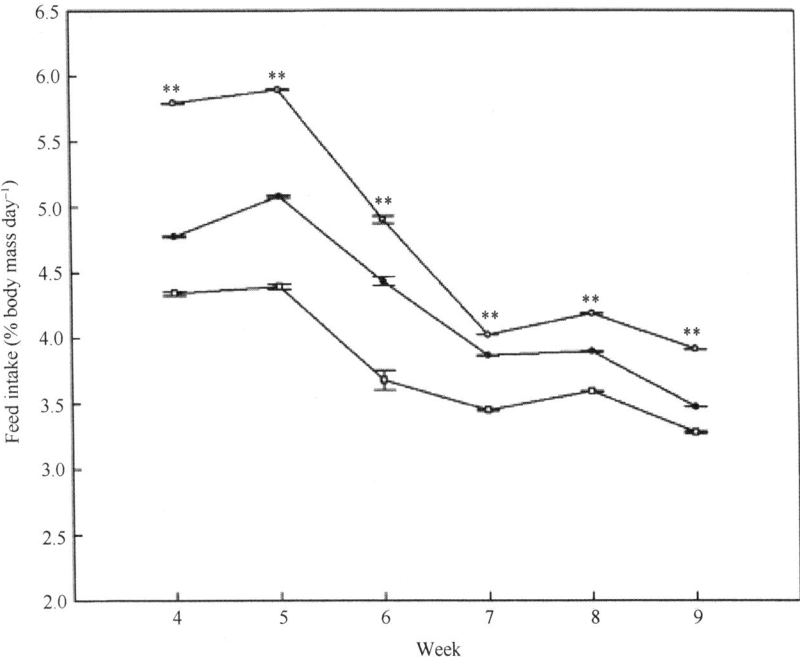

Fig. 3 Adjusted mean feed intake in three groups of transgenic carp during re-feeding weeks. ANCOVA used mass at start of each re-feeding week as covariate for adjustment of mean. Treatment abbreviation and symbols as in Fig. 1. ─□─, C; ─■─, S1; ─○─, S2

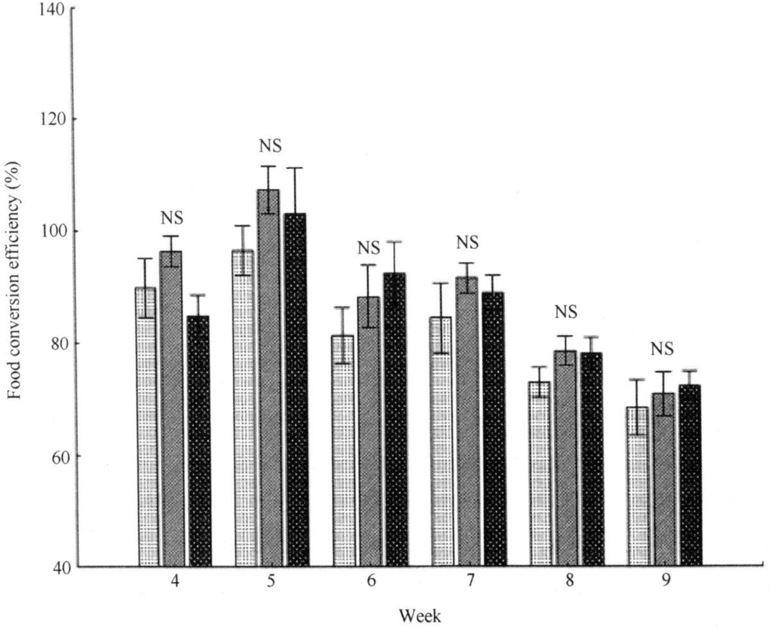

Fig. 4 Food conversion efficiency in three groups of transgenic carp during re-feeding weeks. Treatment abbreviation and symbols as in Fig. 1. ▦, C; ▨, S1; ■, S2

At the start of re-feeding period, there were no significant differences in the concentrations of protein and ash among all feeding groups. Concentrations of dry matter, lipid, energy and the ratio of lipid to lean body mass in S2 fish were significantly lower than those in the control, while no significant differences in these variables were found between S1 fish and the control. At the end of the re-feeding period, there were no significant differences in all variables of body composition among all feeding groups (Table III).

Table III Body composition (mean ±S.E.) per unit wet mass of transgenic carp at the end of week 3 and at the end of week 9 in three feeding groups: (1) fish fed continuously throughout the experiment (C); (2) fish deprived from food for 1 week then fed to satiation (S1); (3) fish deprived from food for 2 weeks then fed to satiation (S2)*

Periods	Variables	C	S1	S2	*P*
Week 3(n=3)	Dry matter (%)	26.16±0.19[a]	25.62±0.25[a]	24.17±0.02[b]	0.0473
	Protein (%)	14.20±0.09	13.91±0.15	13.65±0.03	0.3865
	Lipid (%)	9.73±0.07[a]	9.02±0.17[a]	7.88±0.03[b]	0.0056
	Ash (%)	2.23±0.04	2.37±0.01	2.64±0.04	0.0586
	Energy (J mg^{-1})	6.51±0.07[a]	6.03±0.03[ab]	5.75±0.03[b]	0.0143
	Lipid /LBM	0.59±0.00[a]	0.55±0.01[a]	0.48±0.00[b]	0.0073
Week 9(C:n=6; S1 or S2:n=7)	Dry matter (%)	28.70±0.03	29.05±0.06	29.51±0.03	0.2866
	Protein (%)	14.04±0.01	14.44±0.02	14.38±0.01	0.1934
	Lipid (%)	12.85±0.03	12.97±0.08	13.36±0.03	0.6849
	Ash (%)	1.82±0.01	1.64±0.02	1.77±0.00	0.2357
	Energy (J mg^{-1})	7.15±0.01	6.96±0.02	710±0.01	0.5221
	Lipid /LBM	0.81±0.00	0.81±0.01	0.83±0.00	0.8968

* Lipid/LBM (lean body mass) is the ratio of lipid to sum of protein and ash. Means with different letters in the same row were significantly different (P< 0.05). *P*, probability value

Discussion

Compensatory growth is identified by being significantly faster than the growth rate of control fishes that have not experienced growth depression, held under comparable conditions (Ali *et al.*, 2003). As the body mass of food-restricted fish is usually smaller than that of the controls at the start of the re-feeding period, higher growth rates would be expected among restricted fish because of the body size dependency of growth rate (Jobling, 1983). Thus, care must be taken in interpreting the data on partial compensatory growth. In this study, G_W and feed intake data were adjusted for size effects, and the analysis revealed that there was a partial compensatory response in the deprived fish. A limited capacity for compensatory growth has also been observed in other fish (Jobling *et al.*, 1993; Paul *et al.*, 1995; Wang *et al.*, 2000; Tian & Qin, 2003; Eroldoğanv *et al.*, 2006). For non-transgenic carp, previous study has found that no growth compensation was observed (Schwarz *et al.*, 1985). Schwarz *et al.* (1985) studied the effects of protein or energy restriction on growth performance of carp. Both restricted groups grew more slowly than the control, but both showed positive growth rates. The duration of the restriction period was determined by the time taken to reach a target mass set in the experiment so that treatment groups and controls started the re-alimentation period at approximately the same mass. During re-alimentation, previously restricted groups had the same growth rate as the control group, with no evidence of a compensatory growth response. The lack of a compensatory growth response may have reflected inadequate rates of feeding in the re-alimentation period, or the effects of the deprivations may not have been sufficient to trigger a compensatory response (Ali *et al.*, 2003). In this study, transgenic carp in the deprived group were still showing hyperphagia and higher G_w in the last week of re-feeding, indicating that the compensatory response had not completely abated. The mass trajectories actually did appear to be converging in Fig. 1, suggesting that full compensation may have been achieved eventually. Again, various lengths of re-feeding would have been necessary to determine this. These relationships need to be empirically evaluated by examining different starvation and re-feeding periods in the future.

In this study, there was also a tendency for both G_w and feed intake during re-feeding to increase with the length of deprivation. In three cyprinids, growth rate during re-feeding increased in proportion to the length of starvation (Wieser *et al.*, 1992). In the minnow, compensatory growth was detected in fish that were starved for 16 days, but not in fish that were starved for 4 days (Russell & Wootton, 1992). Similar pattern was also observed in other fish (Wang *et al.*, 2000; Zhu *et al.*, 2005). Thus, the magnitude of compensatory growth tends to be dependent on the severity of undernutrition.

Hyperphagia is the main mechanism involved in the compensatory growth response, although increased food conversion efficiencies or behavioural adjustments might play a role (Ali *et al.*, 2003). Hyperphagia has been found in fish species during compensatory growth (Miglavs & Jobling, 1989a, b; Russell & Wootton, 1992; Hayward *et al.*, 1997; Wang *et al.*, 2000; Xie *et al.*, 2001; Nikki *et al.*, 2004), whereas improved food conversion efficiency has not been widely observed in fish showing compensatory growth (Dobson & Holmes, 1984; Russell & Wootton, 1992; Qian *et al.*, 2000; Eroldoğanv *et al.*, 2006). In fact, many studies detected no significant differences in food conversion efficiency between the control group and the deprived groups during re-feeding periods (Kim & Lovell, 1995; Haywood *et al.*, 1997; Speare & Arsenault, 1997; Xie *et al.*, 2001). In this study, hyperphagia was observed in the deprived fish after re-feeding, but there were no significant differences in food conversion efficiency. Thus, enhanced food intake during the re-feeding periods was the major cause for compensatory growth in transgenic carp.

In this study, the proximate composition of the deprived fish at the end of the experiment was similar to that of the control fish, although the fresh masses diverged from controls.

Similar patterns were also observed in other fish (Zhu *et al.*, 2004). It suggests that defence of body composition has priority over defence of the growth trajectory in fresh mass (Ali *et al.*, 2003).

In summary, this study confirmed that the fast-growing transgenic carp can mount a compensatory growth response. Hyperphagia was suggested to be the mechanism responsible for compensatory growth response. In the future, further studies to make generalizations on compensatory growth in transgenic fish should include non-transgenic fish for comparison and also include a more complex experimental design, which included different periods of re-feeding.

Acknowledgements The authors would like to thank two anonymous reviewers for their constructive comments on the manuscript and also thank Shouqi Xie, Wu Lei, Yunxia Yang and Xiaoming Zhu for their help during the experiment. We are grateful to Yibo Cui, who first introduced Fish Bioenergetics in China and trained the first author. This work was supported by the Projects of the National Natural Science Foundation of China (NSFC 30130050 and 30400056), the State '863' High-Tech Project (2004AA213120) and Development Plan of the State Key Fundamental Research (2001CB109006) from the Ministry of Science and Technology, China.

References

Ali, M., Nicieza, A. & Wootton, R. J. (2003). Compensatory growth in fishes: a response to growth depression. Fish and Fisheries 4, 47-190.

AOAC (Association of Official Analytical Chemists) (1984). Official Methods of Analysis of the Association of Official Analytical Chemists, 14th edn, pp. 1018. Virginia: AOAC.

Arendt, J. D. (1997). Adaptive intrinsic growth rates: an integration across taxa. Quarterly Review of Biology 72, 149-177.

Cook, J. T., McNiven, M. A., Richardson, G. F. & Sutterlin, A. M. (2000a). Growth rate, body composition/feed digestibility reconversion of growth-enhanced transgenic Atlantic salmon (*Salmo salar*). Aquaculture 188, 15-32.

Cook, J. T., Sutterlin, A. M. & McNiven, M. A. (2000b). Effect of food deprivation on oxygen consumption/body composition of growth-enhanced transgenic Atlantic salmon (*Salmo salar*). Aquaculture 188, 47-63.

Cui, Z., Zhu, Z., Cui, Y., Li, G. & Xu, K. (1996). Food consumption and energy budget in ThGH transgenic F_2 red carp (*Cyprinus carpio* L. red var.). Chinese Science Bulletin 41, 591-596.

Devlin, R. H., Yesaki, T., Biagy, C. A., Donaldson, E. M., Swanson, P. & Chan, W. K. (1994). Extraordinary salmon growth. Nature 371, 209-210.

Devlin, R. H., Biagi, C. A. & Yesaki, T. Y. (2004). Growth, viability and genetic characteristics of GH transgenic coho salmon strains. Aquaculture 236, 607-632.

Van Dijk, P. L. M., Hardewig, I. & Hölker, F. (2005). Energy reserves during food deprivation and compensatory growth in juvenile roach: the importance of season and temperature. Journal of Fish Biology 66, 167-181.

Dobson, S. H. & Holmes, R. M. (1984). Compensatory growth in the rainbow trout, Salmo gairdneri Richardson. Journal of Fish Biology 25, 649-656.

Eroldoğanv, O. T., Kumlu, M., Kiris, G. A. & Sezer, B. (2006). Compensatory growth response of *Sparus aurata* following different starvation and refeeding protocols. Aquaculture Nutrition 12, 203-210.

Fu, C., Cui, Y., Hung, S. S. O. & Zhu, Z. (1998). Growth and feed utilization by F_4 human growth hormone transgenic carp fed diets with different protein levels. Journal of Fish Biology 53, 115-129.

Fu, C., Cui, Y., Hung, S. S. O. & Zhu, Z. (2000). Whole-body amino acid pattern of F_4 human growth hormone gene-transgenic red common carp (*Cyprinus carpio*) fed diets with different protein levels. Aquaculture 189, 287-292.

Gaylord, T. G. & Gatlin, D. M. (2001). Dietary protein and energy modifications to maximize compensatory growth of channel catfish (*Ictalurus punctatus*). Aquaculture 194, 337-348.

Hayward, R. S., Noltie, D. B. & Wang, N. (1997). Use of compensatory growth to double hybrid sunfish growth rates. Transactions of the American Fisheries Society 126, 316-322.

Huitema, B. E. (1980). The Analysis of Covariance and Alternatives. New York: J. Wiley & Sons.

Jobling, M. (1983). Growth studies with fish overcoming the problem of size variation. Journal of Fish Biology 22, 153-157.

Jobling, M. (1994). Fish Bioenergetics. London: Chapman & Hall.

Jobling, M., Jorgensen, E. H. & Siikavuopio, S. I. (1993). The influence of previous feeding regime on the compensatory growth response of maturing and immature Arctic charr, *Salvelinus alpinus*. Journal of Fish Biology 43, 409-419.

Jobling, M., Meloy, O. H., Dos Santos, J. & Christiansen, B. (1994). The compensatory growth response of the Atlantic cod: effects of nutritional history. Aquaculture International 2, 75-90.

Johansen, S. J. S., Ekli, M., Stanges, B. & Jobling, M. (2001). Weight gain and lipid deposition in Atlantic salmon, *Salmo salar*, during compensatory growth: evidence for lipostatic regulation? Aquaculture Research 32, 963-974.

Kim, M. K. & Lovell, R. T. (1995). Effect of feeding regimes on compensatory weight gain and body tissue changes in channel catfish, *Ictalurus punctatus* in ponds. Aquaculture 135, 285-293.

Lambert, P. & Dehnel, P. A. (1974). Seasonal variations in biochemical composition during the reproductive cycle of the intertidal gastropod *Thais lamellosa* Gmelin (Gastropoda, Prosobranchia). Canadian Journal of Zoology 52, 305-318.

Leggatt, R. A., Devlin, R. H., Farrell, A. P. & Randall, D. J. (2003). Oxygen uptake of growth hormone transgenic coho salmon during starvation and feeding. Journal of Fish Biology 62, 1053-1066.

Maclean, A. & Metcalfe, N. B. (2001). Social status, access to food and compensatory growth in juvenile Atlantic salmon. Journal of Fish Biology 58, 1331-1346.

Maclean, N. & Talwar, S. (1984). Injection of cloned genes into rainbow trout eggs. Journal of Embryology and Experimental Morphology (later retitled Development) 82, 187.

Mao, W., Wang, Y., Wang, W., Wu, B., Feng, J. & Zhu, Z. (2004). Enhanced resistance to *Aeromonas hydrophila* infection and enhanced phagocytic activities in human lactoferrin-transgenic grass carp (*Ctenopharyngodon idellus*). Aquaculture 242, 93-103.

Mckenzie, D. J., Martinez, R., Morales, A., Acosta, J., Morales, R., Taylor, E. W., Steffensen, J. F. & Estrada, M. P. (2003). Effects of growth hormone transgenesis on metabolic rate, exercise performance and hypoxia tolerance in tilapia hybrids. Journal of Fish Biology 63, 398-409.

Metcalfe, N. B. & Monaghan, P. (2001). Compensation for a bad start: grow now, pay later. Trends in Ecology and Evolution 16, 255-260.

Miglavs, I. & Jobling, M. (1989a). The effect of feeding regime on proximate body composition and patterns of energy deposition in juvenile Arctic charr, *Salvelinus alpinus*. Journal of Fish Biology 35, 1-11.

Miglavs, I. & Jobling, M. (1989b). Effects of feeding regime on food consumption, growth rates and tissue nucleic acids in juvenile Arctic charr, *Salvelinus alpinus*, with particular respect to compensatory growth. Journal of Fish Biology 34, 947-957.

Nam, Y. K., Noh, J. K., Cho, Y. S., Cho, H. J., Cho, K. N., Kim, C. G. & Kim, D. S. (2001). Dramatically accelerated growth and extraordinary gigantism of transgenic mud loach *Misgurnus mizolepis*. Transgenic Research 10, 353-362.

Nicieza, A. G. & Metcalfe, N. B. (1997). Growth compensation in juvenile Atlantic salmon: responses to depressed temperature and food availability. Ecology 78, 2385-2400.

Nikki, J., Pirhonen, J., Jobling, M. & Karjalainen, J. (2004). Compensatory growth in juvenile rainbow trout, *Oncorhynchus mykiss* (Walbaum), held individually. Aquaculture 235, 285-296.

Paul, A. J., Paul, J. M. & Smith, R. L. (1995). Compensatory growth in Alaska yellowfin sole, *Pleuronectes asper*, following food deprivation. Journal of Fish Biology 46, 442-448.

Qian, X., Cui, Y., Xiong, B. & Yang, Y. (2000). Compensatory growth, feed utilization and activity in gibel carp, following feed deprivation. Journal of Fish Biology 56, 228-223.

Rahman, M. & Maclean, N. (1999). Growth performance of transgenic tilapia containing an exogenous piscine growth hormone gene. Aquaculture 173, 333-346.

Rahman, M., Ronyai, A., Engidaw, B. Z., Jauncey, K., Hwang, G. L., Smith, A., Roderick, E., Penman, D., Varadi, L. & Maclean, N. (2001). Growth and nutritional trials on transgenic Nile tilapia containing an exogenous piscine growth hormone gene. Journal of Fish Biology 59, 62-78.

Raven, P. A., Devlin, R. H. & Higgs, D. A. (2006). Influence of dietary digestible energy content on growth, protein and energy utilization and body composition of growth hormone transgenic and non-transgenic coho salmon (*Oncorhynchus kisutch*). Aquaculture 254, 730-747.

Ruohonen, K., Kettunen, J. & King, J. (2001). Experimental design in feeding experiments. In Food Intake in Fish (Houlihan, D., Boujard, T. & Jobling, M., eds), pp. 88-107. Oxford: Blackwell Scientific Publications.

Russell, N. R. & Wootton, R. J. (1992). Appetite and growth compensation in the European minnow, *Phoxinus phoxinus* (Cyprinidae) following short term of food restriction. Environmental Biology of Fishes 34, 277-285.

Schwarz, F. J., Plank, J. & Kirchgessner, M. (1985). Effects of protein or energy restriction with subsequent realimentation on performance of carp (*Cyprinus carpio* L.). Aquaculture 48, 23-33.

Sibly, R. M. & Calow, P. (1986). Physiological Ecology of Animals. Oxford: Blackwell Scientific Publications.

Speare, D. J. & Arsenault, G. J. (1997). Effects of intermittent hydrogen peroxide exposure on growth and columnaris disease prevention of juvenile rainbow trout (*Oncorhynchus mykiss*). Canadian Journal of Fisheries and Aquatic Sciences 54, 2653-2658.

Tian, X. & Qin, J. (2003). A single phase of food deprivation provoked compensatory growth in barramundi *Lates calcarifer*. Aquaculture 224, 169-179.

Venugopal, T., Anathy, V., Kirankkumar, S. & Pandian, T. J. (2004). Growth enhancement and food conversion efficiency of transgenic fish *Labeo rohita*. Journal of Experimental Zoology 301A, 477-490.

Wang, Y., Cui, Y., Yang, Y. X. & Cai, F. S. (2000). Compensatory growth in hybrid tilapia, *Oreochromis mossambicus* × *O. niloticus*, reared in seawater. Aquaculture 189, 101-108.

Wang, Y., Hu, W., Wu, G., Sun, Y., Chen, S., Zhang, F., Zhu, Z., Feng, J. & Zhang, X. (2001). Genetic analysis of "all-fish" growth hormone gene transferred carp (*Cyprinus Carpio* L.) and its F_1 generation. Chinese Science Bulletin 46, 1174-1177.

Wieser, W., Krumschnabel, G. & Ojwang-Okwor, J. P. (1992). The energetics of starvation and growth after refeeding in juveniles of three cyprinid species. Environmental Biology of Fishes 33, 63-71.

Wilson, P. N. & Osbourn, D. F. (1960). Compensatory growth after undernutrition in mammals and birds. Biological Review 35, 324-363.

Wootton, R. J. (1998). Ecology of Teleost Fishes. Dordrecht: Kluwer.

Xie, S., Zhu, X., Cui, Y., Lei, W., Yang, Y. & Wootton, R. J. (2001). Compensatory growth in the gibel carp following feed deprivation: temporal patterns in growth, nutrient deposition, feed intake and body composition. Journal of Fish Biology 58, 999-1009.

Zhong, J., Wang, Y. & Zhu, Z. (2002). Introduction of the human lactoferrin gene into grass carp (*Ctenopharyngodon idellus*) to increase resistance against GCH virus. Aquaculture 214, 93-101.

Zhu, X., Cui, Y., Ali, M. & Wootton, R. J. (2001). Comparison of compensatory growth responses of juvenile threespined stickleback and minnow following similar food deprivation protocols. Journal of Fish Biology 58, 1149-1165.

Zhu, X., Xie, S., Zou, Z., Lei, W., Cui, Y., Yang, Y. & Wootton, R. J. (2004). Compensatory growth and food consumption in gibel carp, *Carassius auratus gibelio*, and Chinese longsnout catfish, *Leiocassis longirostris*, experiencing cycles of feed deprivation and re-feeding. Aquaculture 241, 235-247.

Zhu, X., Xie, S., Lei, W., Cui, Y., Yang, Y. & Wootton, R. J. (2005). Compensatory growth in the Chinese longsnout catfish, *Leiocassis longirostris* following feed deprivation: temporatory patterns in growth, nutrient deposition, feed intake and body composition. Aquaculture 248, 307-314.

Zhu, Z., Li, G., He, L. & Chen, S. (1985). Novel gene transfer into the fertilized eggs of goldfish (*Carassius auratus* L. 1758). Journal of Applied Ichthyology 1, 31-34.

Zhu, Z., Xu, K., Li, G., Xie, Y. & He, L. (1986). Biological effects of human growth hormone gene microinjected into the fertilized eggs of loach, *Misgurnus anguillicaudatus* (Cantor). Chinese Science Bulletin 31, 988-990.

Zhu, Z., Xu, K., Xie, Y., Li, G. & He, L. (1989). A model of transgenic fish. Scientia Sinica Series B 2, 147-155.

Zhu, Z., He, L. & Chen, T. T. (1992). Primary-structural and evolutionary analyses of growth-hormone gene from grass carp (*Ctenopharyngodon idellus*). European Journal of Biochemistry 207, 643-648.

快速生长转基因鲤的补偿生长效应

傅萃长　李德亮　胡　炜　汪亚平　朱作言

中国科学院水生生物研究所，淡水生态与生物技术国家重点实验室，武汉　430072

摘　要　补偿生长是指当生长受到一段时间抑制后，再恢复至有利条件后表现出快速生长效应的现象。为了研究转“全鱼”生长激素基因鲤 F_2 是否具有补偿生长能力，在 29℃水温条件下设计了一个为期 9 周的实验。在实验过程中，对照组进行一天两次的饱食投喂，两个实验组分别饥饿 1 周和 2 周，然后在再投喂阶段进行饱食投喂。实验结束时，两个实验组的个体体重显著低于对照组。在再投喂阶段，实验组的平均特定生长率和摄食量显著高于对照组，表明实验组表现出部分的补偿生长效应。而实验组和对照组的食物转化效率并无显著差异，说明食欲增强是生长率增加的主要原因。实验组的鱼体成分与对照组类似。本研究第一次报道了快速生长转基因鱼具有部分补偿生长能力。

Metabolism Traits of ‘all-fish’ Growth Hormone Transgenic Common Carp (*Cyprinus carpio* L.)

Bo Guan[1,2] Wei Hu[1] Tanglin Zhang[1] Yaping Wang[1] Zuoyan Zhu[1]

1 State Key Laboratory of Freshwater Ecology and Biotechnology, Institute of Hydrobiology, Chinese Academy of Sciences, Wuhan 430072

2 Graduate School of the Chinese Academy of Sciences, Beijing 100049

Abstract Transgenic animals with improved qualities have the potential to upset the ecological balance of a natural environment. We investigated metabolic rates of ‘all-fish’ growth hormone (GH) transgenic common carp under routine conditions and during starvation periods to determine whether energy stores in transgenic fish would deplete faster than controls during natural periods of starvation. Before the oxygen uptake was measured, the mean daily feed intake of transgenic carp was 2.12 times greater than control fish during 4 days of feeding. The average oxygen uptake of GH transgenic fish was 1.32 times greater than control fish within 96 h of starvation, but was not significantly different from controls between 96 and 144 h of starvation. At the same time, GH transgenic fish did not deplete energy reserves at a faster rate than did the controls, as the carcass energy contents of the two groups following a 60-d starvation period were not significantly different. Consequently, we suggest that increased routine oxygen uptake in GH transgenic common carp over that of control fish may be mainly due to the effects of feeding, and not to an increase in basal metabolism. GH transgenic fish are similar to controls in the regulation of metabolism to normally distribute energy reserves during starvation.

Keywords transgenic; growth hormone (GH); common carp (*Cyprinus carpio* L.); oxygen uptake

1. Introduction

Stable lines of several species of transgenic fish have been generated, including growth-enhanced ‘all-fish’ GH transgenic common carp (*Cyprinus carpio* L.) in China (Wang *et al*., 2001), growth-enhanced transgenic Atlantic salmon (*Salmo salar*) in Canada (Du *et al*., 1992; Cook *et al*., 2000a), growth-enhanced transgenic coho salmon (*Oncorhynchus kisutch*) in Canada (Devlin *et al*., 1995), and growth-enhanced transgenic tilapia (*Oreochromis sp*.) in Britain and Cuba (Rahman *et al*. 1998; Martinez *et al*., 1996). However, biosafety issues have hindered the marketability of transgenic fish and they remain unavailable in the marketplace (Maclean and Laight, 2000; Wu *et al*., 2003; Zbikowska, 2003; Hu *et al*., 2007).

文章发表于 *Aquaculture*, 2008, 284(1-4):217-223

Several pleiotropic effects of transgenic GH genes have been reported in fish, such as changing the non-specific immune functions (Wang *et al.*, 2006b), impairing swimming ability (Farrell *et al.*, 1997; Li *et al.*, 2007), skeletal abnormalities (Ostenfeld *et al.*, 1998), and regulating sea water adaptability (Seddiki *et al.*, 1996). Accordingly, the fitness and adaptability of transgenic fish must be carefully predicted before introduction to a natural ecosystem in order to conserve ecological balance and minimize potential ecological risk.

Respiratory metabolism, including metabolic rate, aerobic scope, the capacity for sustained aerobic exercise and tolerance of hypoxia, is an important part of fish physiology and ecology (Fry, 1971; Mckenzie *et al.*, 2003). Some researchers have attempted to investigate the effects of GH transgene on respiratory metabolism traits, which are believed to contribute to whether energy stores in transgenic fish would deplete faster than controls during natural periods of starvation (Cook *et al.*, 2000c; Mckenzie *et al.*, 2003). Other types of GH transgenic fish including Atlantic salmon (*S. salar*), tilapia (*Oreochromis* sp.) and coho salmon (*O. kisutch*), were found to have a greater oxygen uptake rate than control fish during routine feeding conditions (Stevens *et al.*, 1998; Cook *et al.*, 2000a, c; Mckenzie *et al.*, 2003; Leggatt *et al.*, 2003). Additionally, transgenic fish have a greater daily feed intake in comparison to control fish (Fu *et al.*, 1998; Cook *et al.*, 2000b; Rahman *et al.*, 2001; Leggatt *et al.*, 2003). Furthermore, there is contrary evidence as to whether GH transgene modifies the metabolic rate during starvation (Cook *et al.* 2000a; Leggatt *et al.*, 2003). It is not clear whether the increased oxygen uptake in transgenic fish is due to a modification of the standard metabolic rate (SMR) by the GH transgene. Consequently, the respiratory metabolism of the GH transgenic common carp may be one of the urgent scientific problems to be researched, as it would be helpful in the evaluation of the fitness of transgenic fish.

Fu *et al.* (2007) suggested that fast-growing transgenic carp may have increased routine oxygen consumption in contrast to the nontransgenic controls, which was calculated indirectly by difference from other components of the energy budget. However the respiratory metabolism traits of 'all-fish' GH transgenic common carp remain unclear.

Fish have evolved the capability to endure long-term food shortage by reducing metabolic rate and energy expenditures (Mayzaud, 1976; O'Connor *et al.*, 2000). Due to seasonal and spatial differences in food availability, it is very important to predict how 'all-fish' GH transgenic common carp might adjust respiratory metabolism to meet conditions found in natural waters. The probability and extent of ecological impact of a transgenic animal is in part dependent upon the fitness of transgenic individuals and on the ability to cope with changes in food abundance (Cook *et al.*, 2000c).

Consequently, the present work investigated differences in respiratory metabolism traits between 'all-fish' GH transgenic common carp and non-transgenic controls in order to clarify the fitness of transgenic fish and to assess potential ecological risk.

2. Methods

2.1 Experimental fish

The 'all-fish' gene construct *pCAgcGHc* included the grass carp (*Ctenopharyngodon idellus*) GH cDNA (*gcGHc*) driven by the β-actin gene promoter of common carp (*C. carpio* L.) (Wang *et al.*, 2001). P_0 'all-fish' growth hormone transgenic carp were produced by microinjection of *pCAgcGHc* into the fertilized eggs of common carp (Yellow River variety) (Wang *et al.*, 2001). F_1 transgenic carp were produced as follows: P_0 transgenic ♂ (sperm revealed to be *pCAgcGHc* positive)×non-transgenic ♀. F_3 transgenic carp used in this experiment were produced by a similar method: F_2 transgenic ♂ (sperm revealed to be *pCAgcGHc* positive)×non-transgenic ♀ on 14 April 2005. On the same day, the same non-transgenic female carp and one control male carp were used to produce the control fish offspring. F_3 transgenic carp and control carp were reared in field ponds. About 600 3-month-old F_3 transgenic carp were transferred to an indoor water-recycling fish-rearing system (WRFRS) on 15 July 2005, while about 300 control carp were also transferred to the WRFRS on 5 August 2005. Abundant living plankton food in field ponds could help to enhance the growth of control carp during the 20 days. So size-matching of transgenic and control fish was undertaken. The *pCAgcGHc* transgene positive fish were identified by PCR according to Wang *et al.* (2001). The transgenic and control fish were acclimated to the indoor WRFRS system for at least 1 month, fed twice daily with an approximately equal ration, before the following experimental study.

2.2 Facilities

An improved flow-through respirometer system (FRS) in an airconditioned room was used to measure oxygen uptake, as described previously (Liu *et al.*, 2000). The FRS system was composed of several separate 20-liter chambers, each of which had a water inlet on the top, a water outlet on the side, a feces outlet in the bottom, and easilyopened lid with a removable air vent on the top. The easily-opened lid with a removable air vent was designed to be convenient for opening and sealing each chamber. Water was circulated through each chamber by a central pump and was decontaminated by a sponge filter and recharged with air before distribution to each chamber. Dissolved oxygen (DO) concentrations in the inflow and outflow of each chamber were measured with an YSI550A Dissolved Oxygen Analyzer (YSI USA).

The long-term starvation experiments were performed in 8 circular fiberglass tanks (diameter 150 cm and volume 1000 L) of WRFRS. Each tank was divided into four equal enclosures by perforated fiberglass partitions.

2.3 Respiratory metabolism experiment

2.3.1 Routine metabolic rate (RMR)

The RMR is defined as the mean rate of oxygen uptake observed in fish under experimental conditions allowing only random activity and protection from outside stimuli (Fry, 1971).

After 1 day of starvation in WRFRS, experimental fish were weighed, and size-matched transgenic and control fish were selected. They were converted from WRFRS tanks into each respiration chamber of FRS with minimal stress. Each group had 3 replicates (n=3). The number of fish in each respiration chamber was 5, and the mean mass of the control and transgenic fish was 32.45±1.45 g and 31.86±1.58 g ('±' is defined as S.E.). The mean water temperature throughout the experiment was 28.6±0.2℃. To reduce outside stimuli during measurement of oxygen uptake, each chamber was shaded with a black shield at one side. During the first 4 d, the fish were fed to satiation twice daily at 9:00 and 16:00 h with a diet comprised of 36.7% protein, 5.6% lipid, 5.1% ash, and 21 J/mg energy. The surplus food was collected 1 h after feeding, and was dried and weighed. After completion of the last feeding period, respiration chambers were sealed to measure oxygen uptake during a 144-h starvation period. Water flow rates to each chamber and dissolved oxygen concentrations (measured to the nearest 0.01 mg/L) in the water at the inflow and the outflow were measured every 2 h over the starvation period. Fish were taken out after a 144-h starvation period, and then biological oxygen demand (BOD) of each respiration chamber were obtained through measuring the water flow rates and dissolved oxygen concentrations in the water at the inflow and the outflow.

2.3.2 The effect of body weight on SMR

The minimum observable metabolic rate when at rest is conventionally defined as the SMR (Fry, 1971). SMR was defined as the oxygen uptake of starved, resting fish at a constant temperature.

SMR was estimated on a group of transgenic and control fish within a weight range 32 to 132 g. The seven transgenic groups (n=7), and the eight control groups (n=8 were subjected to the same protocol (4-d feeding period followed by 4-d starvation period) as the 2.3.1. section described above, however, the oxygen uptake was only measured between 96 and 120 h of starvation, as an estimate of SMR. The number of fish in each FRS chamber was based upon the volume of the chamber and the weight of experimental fish (the total weight of fish in each chamber was not more than 150 g). Mean water temperature was 28.6℃±0.2℃.

2.4 Effect of starvation on bio-energetics

The initial wet weight of all fish was measured to the nearest 0.01 g before they were distributed in 8 tanks. Both transgenic carp and control carp were divided into 3 groups for collecting experimental fish at the 1st, 30th, and 60th day, and each group was divided into 5 sub groups according to mean weights of 30 g, 40 g, 50 g, 60 g, and 80 g. These 30 sub groups were distributed into 30 separate enclosures. Experimental fish were subjected to a 60-d starvation period in the indoor WRFRS. During the 60-d starvation experiment in the indoor WRFRS, dissolved oxygen concentration was above 7 mg/L, pH was between 7.0 and 7.6, and water temperature varied between 16℃ and 24℃. The initial wet weight of all fish was measured to the nearest 0.01 g. One transgenic group and one control group was removed, weighed, and frozen at −20℃ for bio-energetics analysis, at the 1st, 30th, and 60th day during the starvation period. At the end of the experiment, the carcasses of the fish were steamed in an autoclave at 120℃ for 1 h, dried at 70℃ for approximately 48 h, and homogenized. Energy content of the carcasses of each fish was determined by bomb calorimetry (Gentry Instruments Inc., Aiken, SC, U.S.A.).

2.5 Data analysis

The oxygen uptake of a fish of known weight was calculated according to the following equation:

$$VO_2 = \frac{([O_2]_{inflow} - [O_2]_{outflow}) \times V - BOD \times V}{N}$$

$$MO_2 = \frac{([O_2]_{inflow} - [O_2]_{outflow}) \times V - BOD \times V}{W}$$

Where VO_2 is the oxygen uptake per fish (mg O_2/h), MO_2 is the weight-specific oxygen uptake of the fish (mg O_2/kg/h), $[O_2]_{inflow}$ is DO of the inflow water, $[O_2]_{outflow}$ is DO of the outflow water, V is the water flow rate of each chamber, BOD is the biological oxygen demand of each chamber without experimental fish, N is the number of fish in each chamber, and W is the total fish biomass of each chamber.

The oxygen uptake per fish was related to body weight according to the power relationship using the following equation:

$$VO_2=aW^b$$

Where W denotes the body weight (g) of the fish, b is the weight exponent, and a is the weight coefficient.

One-way ANOVA was used to analyze the RMR between transgenic fish and control fish. ANCOVA was used to account for the effect of weight on the SMR of transgenic and control fish, using wet weight as the covariate. ANCOVA GML homogeneity-of-slopes model and

ANCOVA were used to test the effect long starvation on dry matter and energy between transgenic fish and control fish, using starvation time and initial wet weight as covariates. Differences were regarded as significant when $P<0.05$. Comparisons between transgenic fish and control fish were performed with STATISTICA 6.0 and illustrations were performed with SigmaPlot 8.0.

3. Results

3.1 Routine metabolic rate

After 4 d of feeding to satiation, the mean daily feed intake averaged 4.62±0.21% of body mass, and 2.11±0.21% of body mass for size-matched transgenic and control common carp, respectively ($P=0.0012$). Oxygen uptake by transgenic and control fish gradually decreased and reached an apparent steady state after 96 h of starvation (Fig. 1). Within the first 96 h of starvation, transgenic fish had a significantly higher mean oxygen uptake than control fish (Fig. 1). Between 96 and 144 h of starvation, the difference in oxygen uptake of transgenic and control fish was not significant (Fig. 1). According to these research results, RMR was estimated as the average oxygen uptake within the first 24 h of starvation; SMR was estimated as the mean oxygen uptake between 96 and 120 h of starvation. As shown in Fig. 2, transgenic fish had 1.50 times significantly greater ($P<0.0001$) RMR (590±31 mg/kg/h) when compared with size-matched control fish (390±33 mg/kg/h). There was not a significant difference ($P=0.41$) between the SMR of the two size-matched groups (250±7 mg/kg/h for transgenic fish and 240±8 mg/kg/h for control fish).

3.2 The effect of body weight on SMR

The oxygen uptake per fish of transgenic and control common carp was related to fish body weight according to the power relationship (Fig. 3). After oxygen uptake per fish and body weight were logarithmically transformed, GML homogeneity-of-slopes and ANCOVA were used to compare the slope of the regression line and the mean oxygen uptake per fish. The slope of the regression line was not significantly steeper ($P=0.67$) for the transgenic fish compared to the control, indicating that the transgenic fish had the same increase in oxygen uptake per fish with increase in body weight (Fig. 3).

In the same way, using one-way ANCOVA with weight as a covariate, body size did not significantly ($P=0.17$) effect weight-specific oxygen uptake, and there was no significant difference ($P=0.17$) between adjusted mean standard oxygen uptake of transgenic and control fish (250±10 mg/kg/h and 240±19 mg/kg/h, respectively).

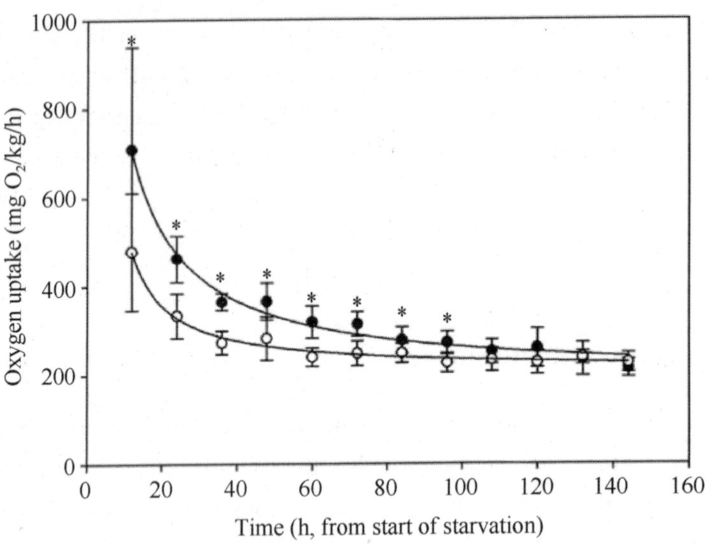

Fig. 1 Oxygen uptake of transgenic (solid circles) (n=3) and control (open circles) common carp (n=3) every 12 h during the 144-h starvation period. Regression equations: transgenic fish, $y=189+7464x^{-1}-15,327x^{-2}$ (r^2=0.99); control fish, $y=209+2496x^{-1}+8879x^{-2}$ (r^2=0.98). * indicates a significant difference between transgenic and control fish (P<0.05)

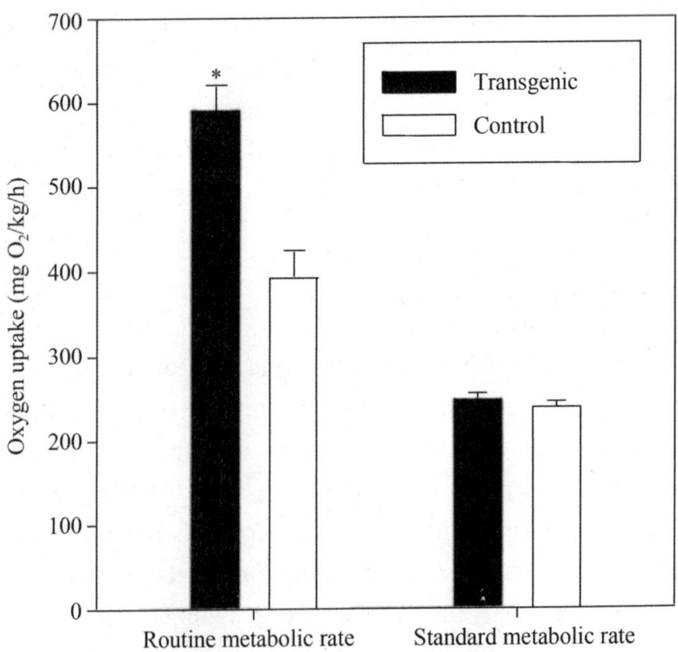

Fig. 2 Measurement of oxygen uptake during initial 24 h of starvation and between 96 and 120 h of starvation, giving the routine metabolic rate and standard metabolic rate, respectively, of transgenic (n=3) and control common carp (n=3). * indicates a significant difference between transgenic and control fish (P<0.05)

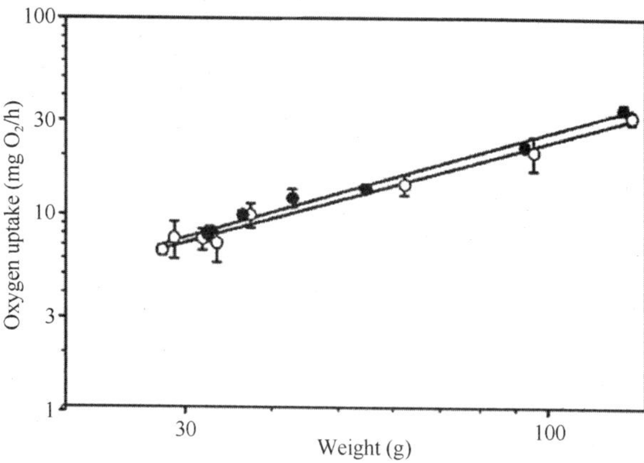

Fig. 3 The effect of body weight on standard oxygen uptake of transgenic (solid circles) (n=7) and control (open circles) fish (n=8). Regression equations: transgenic fish, y=0.23$x^{1.02}$ (r^2 =0.99); control fish, y=0.26$x^{0.97}$ (r^2 =0.99)

3.3 Effect of starvation on bio-energetics

Following the 60-d starvation period, the dry matter weight and energy content of transgenic and control fish carcasses were measured in relation to starvation time and initial wet body weight (Figs. 4 and 5), and were best described by multiple regression with starvation time and initial wet body weight as independent variables (Table 1). ANCOVA GML homogeneity-of-slopes mode and ANCOVA was used to test the effects of these variables on carcass dry matter weight and energy content following the 60-d starvation (Tables 2 and 3). We found that the decline slopes in dry matter content or energy content between two groups did not significantly vary with starvation time and initial wet body weight, and that the results were demonstrated by values for interactions to test homogeneity of the slopes in the parallelism test (P>0.05, Tables 2 and 3). Although adjusted means of dry matter content of transgenic fish were significantly lower than that of control fish (9.42±0.12 g/fish for transgenic fish and 9.78±0.12 g/fish for control fish, P=0.001, ANCOVA, Table 2), adjusted means of energy content between groups was not significantly (207.95±11.49 KJ/fish for transgenic fish and 198.42±10.38 KJ/fish for control fish, P=0.067, ANCOVA, Table 3) different because transgenic fish had a higher adjusted mean energy value per unit dry matter content than that of Measurements of body composition of control fish (21.64±0.19 KJ/g for transgenic fish and 20.91±0.18 KJ/g for control fish, P=0.023, ANCOVA).

4. Discussion

4.1 Routine metabolic rate

In our studies, the RMR of transgenic common carp was significantly higher than that of controls, and the mean daily feeding rate of transgenic common carp was significantly higher

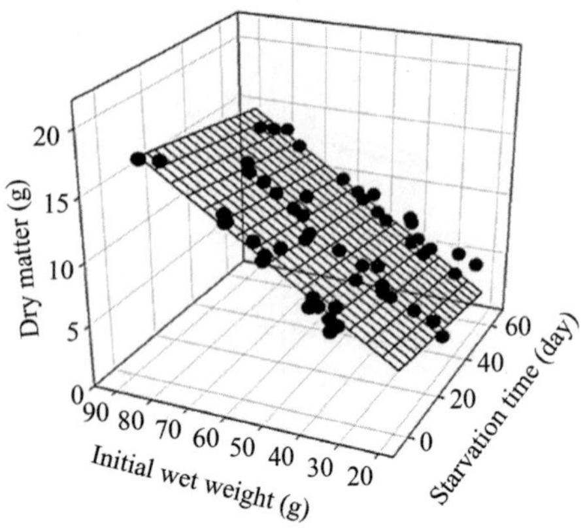

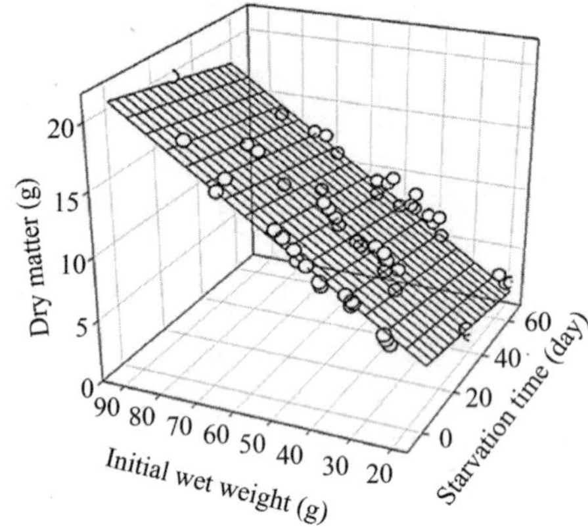

Fig. 4 Dry matter content in relation to starvation time and initial wet weight of transgenic (solid circles) (n=55) and control common carp (open ci rcles) (n=57)

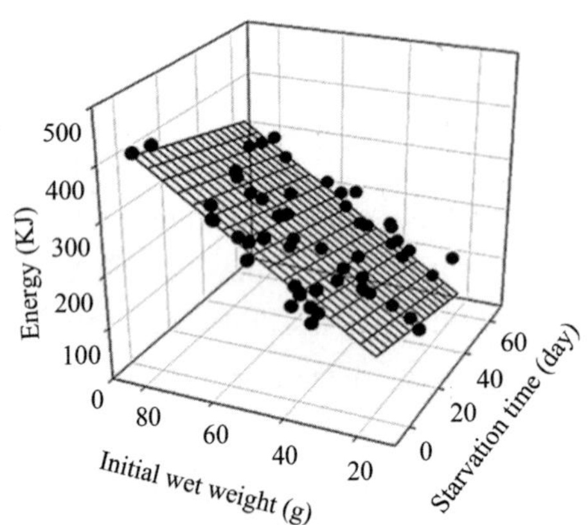

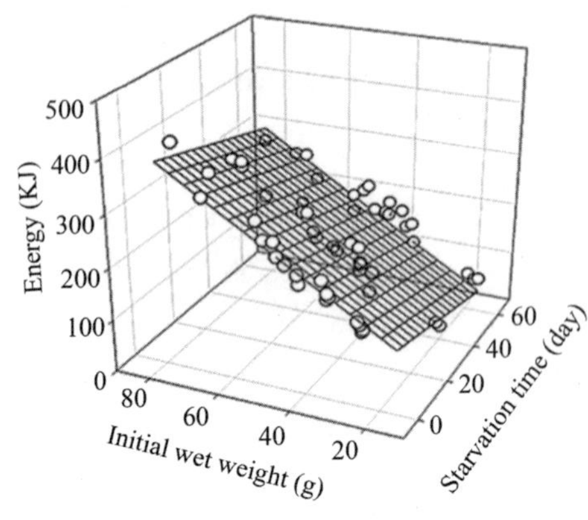

Fig. 5 Energy content in relation to starvation time and initial wet weight of transgenic (solid circles) (n=55) and control common carp (opencircles) (n=57)

Table 1 Measurements of body composition of transgenic and control fish in relation to initial wet body weight and time of starvation

Z (g or kJ)/fish	Experiment group	b_0	b_1	b_2	R^2
Dry matter	Transgenic	1.30	–0.037	0.19	0.94
	Control	2.04	–0.043	0.19	0.93
Energy	Transgenic	33.98	–1.20	4.22	0.91
	Control	53.32	–1.30	4.02	0.81

Regression equation: $Z=b_0+b_1X+b_2Y$ where X is starvation time, Y is initial wet weight, Z is dry weight or energy, and b_0, b_1, and b_2 are regression coefficients

Table 2 Results from an ANCOVA with dry matter content as the response variable, starvation time and initial wet weight as covariates, and group (transgenic and control) as fixed factors

	df	*F*	*P*
Slope homogeneity			
Group×Starvation time	1	0.17	0.68
Group×Initial wet weight	1	1.21	0.27
Group×Starvation time×Initial wet weight	1	1.01	0.32
Error	100		
ANCOVA			
Group	1	22.10	<0.001
Error	104		

Values for interactions to test homogeneity of the slopes are provided (r^2= 0.94 for the model)

Table 3 Results from an ANCOVA with energy content as the response variable, starvation time and initial wet weight as covariates, and group (transgenic and control) as fixed factors

	df	*F*	*P*
Slope homogeneity			
Group×Starvation time	1	0.87	0.36
Group×Initial wet weight	1	0.36	0.55
Group×Starvation time×Initial wet weight	1	1.67	0.20
Error	100		
ANCOVA			
Group	1	3.41	0.067
Error	104		

Values for interactions to test homogeneity of the slopes are provided (r^2= 0.89 for the model)

than that of control fish. Studies of RMR in other transgenic fish species, including coho salmon and Atlantic salmon, found similar results (Stevens *et al.*, 1998; Cook *et al.*, 2000a; Leggatt *et al.*, 2003). Therefore, transgenic fish have demonstrated an elevated routine oxygen uptake in order to meet their metabolic needs. Reasons for this may have to do with an increased feeding rate and food conversion efficiency in the transgenic fish (Devlin *et al.*, 1995; Cook *et al.*, 2000c; Cui *et al.*, 1996; Venugopal *et al.*, 2004; Fu *et al.*, 2007). While the mechanisms for this are not completely clear, the ability to digest food, and to absorb and transport nutrition may be better in transgenic fish (Stevens *et al.*, 1999; Stevens and Devlin, 2000, 2005; Fu *et al.*, 1998; Cook *et al.*, 2000c). GH transgenic coho salmon were found to have more intestinal surface area supports and more numerous and longer pyloric caecae than controls, which could increase intestinal absorption and support more rapid growth (Stevens *et al.*, 1999; Stevens and Devlin, 2000, 2005). These structural changes may modify the physiological digestive function in these fish, by increasing intestinal amino acid absorption (Farmanfarmaian and Sun, 1999), or protein digestion ability (Fu *et al.*, 1998). Moreover, extrinsic GH can influence lipolysis and gluconeogenesis (O'Connor *et al.*, 1993), and protein synthesis and lipid turnover (Oommen and Johnson, 1998; Fauconneau *et al.*,1996). Consequently, GH transgenic fish have a greater ability to digest food and absorb nutrients, are able to convert larger proportions of food to body composition, resulting in and overall increased food conversion efficiency. In conclusion, it appears that GH transgenic common carp have increased oxygen demand to adjust to increased feed intake and increased food efficiency.

Previous experiments suggested that the increased routine oxygen uptake in transgenic fish may be due in part to increased swimming activity for higher feeding motivation (Leggatt *et al.*, 2003; Hallerman *et al.*, 2006). So Leggatt *et al.* (2003) suggested that the increased routine oxygen uptake in transgenic fish appeared attributable to the effects of feeding, acclimation ability, and activity level.

Deitch *et al.* (2006) found that cardiac function of GH transgenic Atlantic salmon is enhanced, possibly compensating for increased metabolic demands. However, there was no increase in gill surface area of the transgenic salmon, which appears to be a limiting factor in oxygen uptake and metabolic capacity. Although the cardiorespiratory traits of GH transgenic common carp are not well understood, culture density and oxygen supply should be well monitored during commercial aquaculture due to higher routine oxygen demand in these fish.

4.2 Standard metabolic rate

The similarities in SMR and energy change during starvation between the GH transgenic and control common carp suggests that the growth hormone transgene does not change basal metabolic rates of common carp. SMR is important because it is the minimum metabolic rate required to sustain life and therefore gives an indication of the overall physiological condition of the fish. Although SMR is difficult to measure directly, measurements of oxygen consumption of resting fish are thought to provide a close approximation. We found that both transgenic and control carp lowed down their activity under starvation, staying on the bottom of each respirometer chamber and seldom proceeding activity. Because these carp could not be completely resting in the respiration chamber, SMR may be marginally overestimated due to faint activity level.

There are differing opinions in the literature as to whether the GH transgene can modify the metabolic rate of fish during starvation. Cook *et al.* (2000b) found that the metabolic rate of GH transgenic Atlantic salmon was higher than that of controls after 8-wk starvation periods. In contrast to this study, rates of oxygen uptake in transgenic and size-matched control coho salmon were found to be similar after a 4-d starvation (Leggatt *et al.*, 2003). Leggatt *et al.* (2003) suggested that differences in starvation oxygen uptake among transgenic and control coho salmon juvenile appeared attributable to the effects of feeding, acclimation ability, and activity level, and not to a difference in basal metabolism.

4.3 Effect of starvation on bio-energetics

By investigating the effect of long-term starvation on energy and dry matter content, we hoped to reveal the fitness of transgenic individuals faced with food shortage, and to indirectly reveal their metabolism physiology traits during starvation. Although adjusted means of dry matter content of transgenic fish were significantly lower than those of control fish, the adjusted means of energy content between them was not significantly different because transgenic fish had a higher adjusted mean energy value per unit dry matter content than that of control fish. However, throughout most of the 60-d food deprivation, the measured decline slopes in dry matter content and energy content were not significantly different between transgenic and control common carp. Much research has revealed that most energy expenditure during long starvation periods goes toward basal metabolic rate (Warren and Davis, 1967; Brett and Groves, 1979; Cui and Wootton, 1988). Therefore, since the energy expenditure of GH transgenic and control common carp declined similarly during starvation, this study indirectly indicates that the GH transgene does not modify the basal metabolic traits of common carp.

When deprived of food, animals employ various responses at the behavioral, physiological, biochemical, and molecular levels, prolonging the period in which energy reserves can support basal metabolism (Wang *et al.*, 2006a). Fish can regulate metabolism through the endocrine system when deprived of food (Brett and Groves, 1979). A central step in this endocrine pathway is the growth hormone (GH) and insulin-like growth factor-I (IGF-I) axis (Duan, 1998). It is not clear how GH transgenic fish regulate metabolism through the endocrine system to adapt to food shortage. Further study is needed to investigate how GH transgenic common carp could regulate metabolism through GH-IGF axis when faced with food deprivation or nutritional depletion, regardless of their high level secretion of GH (unpublished data). Fish undergo natural periods of food deprivation throughout a normal life cycle and have consequently evolved the capability to endure prolonged food shortages. An important survival strategy during such times is to reduce energy expenditures by decreasing activity and lowering metabolic rate (Beamish, 1964; Hogendoorn,1983; Mehner and Wieser,1994; Land and Bernier,1995; Cook *et al.*, 2000b; O'Connor *et al.*, 2000). Consequently, transgenic common carp released into a natural ecosystem may adjust to naturally occurring food shortages by reducing energy expenditures by decreasing activity and lowering metabolic rate as control carp do. Although the risk of extrapolating laboratory data to ecological risks should be carefully considered (Devlin *et al.*, 2004; Sundström *et al.*, 2007), these physiological traits have important ecological significance for transgenic carp in natural waters.

5. Conclusion

The increased RMR of growth hormone transgenic common carp appears to be a function of greater feeding levels and increased ability to process and absorb ingested food, a reflection of fast growth traits of transgenic carp. It appears that there is no additional basal metabolic cost of the GH transgene. During periods of starvation, GH transgenic common carp seem to have the same rate of energy depletion, and may not be at an advantage or disadvantage during seasonal or spatial food shortages occurring in natural waters, at least in terms of energy expenditure.

Acknowledgements This work was financially supported by the Development Plan of the state Key Fundamental Research of China (Grant No. 2007CB109205, No. 2007 CB109206), National Natural Science Foundation (Grant No. 30430540) and the '863' High Technology Project (Grant No. 2006AA10Z141).

References

Beamish, F.W.H., 1964. Influence of starvation on standard and routine oxygen consumption. Transactions of the American Fisheries Society 93, 103-107.

Brett, J.R., Groves, T.D.D., 1979. Physiological energetics. In: Hoar, W.S., Randall, D.J., Brett, J.R. (Eds.), Fish Physiology. Academic Press, New York, pp. 279-352.

Cook, J.T., McNiven, M.A., Richardson, G.F., Sutterlin, A.M., 2000a. Growth rate, body composition and feed digestibility/conversion of growth-enhanced transgenic Atlantic salmon (*Salmo salar*). Aquaculture 188, 15-32.

Cook, J.T., McNiven, M.A., Sutterlin, A.M., 2000b. Metabolic rate of pre-smolt growth-enhanced transgenic Atlantic salmon (*Salmo salar*). Aquaculture 188, 33-45.

Cook, J.T., Sutterlin, A.M., McNiven, M.A., 2000c. Effect of food deprivation on oxygen consumption and body composition of growth-enhanced transgenic Atlantic salmon (*Salmo salar*). Aquaculture 188, 47-63.

Cui, Y., Wootton, R.J., 1988. The metabolic rate of the minnow, *Phoxinus phoxinus* (L.) (Pisces:Cyprinidate), in relation to ration, body size and temperature. Functional Ecology 2, 157-161.

Cui, Z., Zhu, Z., Cui, Y., Li, G., Xu, K., 1996. Food consumption and energy budget in MThGH-transgenic F_2 red carp (*Cyprinus carpio* L. red var.). Chinese Science Bulletin 41, 591-596.

Deitch, E.J., Fletcher, G.L., Petersen, L.H., Costa, I.A.S.F., Shears, M.A., Driedzic, W.R., Gamperl, A.K., 2006. Cardiorespiratory modifications, and limitations, in post-smolt growth hormone transgenic Atlantic salmon (*Salmo salar*). Journal of Experimental Biology 209, 1310-1325.

Devlin, R.H., Yesaki, T.Y., Donaldson, E.M., Hew, C.L., 1995. Transmission and phenotypic effects of an antifreeze/GH gene construct in coho salmon (*Oncorhynchus kisutch*). Aquaculture 137, 161-169.

Devlin, R.H., D'Andrade, M., Mitchell, U., Carlo, A.B., 2004. Population effects of growth hormone transgenic coho salmon depend on food availability and genotype by environment interactions. PNAS 101 (25), 9303-9308.

Du, S.J., Gong, Z.Y., Fletcher, G.L., Shears, M.A., King, M.J., Idler, D.R., Hew, C.L., 1992. Growth enhancement in transgenic Atlantic salmon by the use of an all-fish chimeric growth hormone gene construct. Bio/Technology 10, 176-181.

Duan, C., 1998. Nutritional and developmental regulation of insulin-like growth factors infish. Journal of Nutrition 128,

306-314.

Farmanfarmaian, A., Sun, L.Z., 1999. Growth hormone effects on essential amino acid absorption, muscle amino acid profile, N-retention and nutritional requirements of striped bass hybrids. Genetic Analysis: Biomolecular Engineering 15, 107-113.

Farrell, A.P., Bennett, W., Devlin, R.H., 1997. Growth-enhanced transgenic salmon can be inferior swimmers. Canadian Journal of Zoology 75, 335-337.

Fauconneau, B., Mady, M.P., LeBail, P.Y., 1996. Effect of growth hormone on muscle protein synthesis in rainbow trout (*Oncorhynchus mykiss*) and Atlantic salmon (*Salmo salar*). Fish Physiology and Biochemistry. 15, 49-56.

Fry, F.E.J.,1971. The effect of environmental factors on the physiology of fish. In: Hoar, W.S., Randall, D.J. (Eds.), Fish Physiology VI. Academic Press, New York, pp. 1-98.

Fu, C., Cui, Y., Hung, S.S.O., Zhu, Z., 1998. Growth and feed utilization by F_4 human growth hormone transgenic carp fed diets with different protein levels. Journal Fish Biology 53, 115-129.

Fu, C., Li, D., Hu, W., Wang, Y., Zhu, Z., 2007. Growth and energy budget of F_2 'all-fish' growth hormone gene transgenic common carp. Journal Fish Biology 70, 347-361.

Hallerman, E.M., McLean, E., Fleming, I.A., 2006. Effects of growth hormone transgenes on the behavior and welfare of aquaculturedfishes: a review identifying research needs. Applied Animal Behaviour Science 104 (3-4), 265-294.

Hogendoorn, H., 1983. Growth and production of the African catfish, *Clarias lazera* (C. and V.): III. Bioenergetic relations of body weight and feeding level. Aquaculture 35, 1-17.

Hu, W., Wang, Y.P., Zhu, Z.Y., 2007. Progress in the evaluation of transgenic fish for possible ecological risk and its containment strategies. Science in China Series C: life Sciences 50 (5), 573-579.

Land, S.C., Bernier, N.J., 1995. Estivation: mechanisms and control of metabolic suppression. In: Hochachka, P.W., Mommsen, T.P. (Eds.), Biochemistry and Molecular Biology of Fishes, vol. 5. Elsevier Science, Amsterdam, pp. 381-412.

Leggatt, R.A., Devlin, R.H., Farrell, A.P., Randall, D.J., 2003. Oxygen uptake of growth hormone transgenic coho salmon during starvation and feeding. Journal Fish Biology 62, 1053-1066.

Li, D., Fu, C., Hu, W., Zhong, S., Wang, Y., Zhu, Z., 2007. Rapid growth cost in 'all-fish' growth hormone gene transgenic carp: reduced critical swimming speed. Chinese Science Bulletin 52 (11), 1501-1506.

Liu, J., Cui, Y., Liu, J., 2000. Resting metabolism and heat increment of feeding in mandarin fish (*Siniperca chuatsi*) and Chinese snakehead (*Channa argus*). Comparative Biochemistry and Physiology, A Comparative Physiology 127,131-138.

Maclean, N., Laight, R.J., 2000. Transgenic fish: an evaluation of benefits and risks. Fish and Fisheries 1 (2), 146-172.

Martinez, R., Estrada, M.P., Berlanga, J.I., Hernandez, O., Pimentel, R., Morales, R., Herrera, F., Fuente, J., *et al.*,1996. Growth enhancement in transgenic tilapia by ectopic expression of tilapia growth hormone. Molecular Marine Biology and Biotechnology 5, 62-70.

Mayzaud, P., 1976. Respiration and nitrogen excretion of zooplankton. IV. The influence of starvation on the metabolism and the biochemical composition of some species. Marine Biology 37, 47-58.

Mckenzie, D.J., Martinez, R., Morales, A., Acosta, J., Morales, R., Taylor, E.W., Steffensen, J.F., Estrada, M.P., 2003. Effects of growth hormone transgenesis on metabolic rate, exercise performance and hypoxia tolerance in tilapia hybrids. Journal of Fish Biology 63, 398-409.

Mehner, T., Wieser, W., 1994. Energetics and metabolic correlates of starvation in juvenile perch (*Perca fluviatilis*). Journal of Fish Biology 45, 325-333.

O'Connor, K.I., Taylor, A.C., Metcalfe, N.B., 2000. The stability of standard metabolic rate during a period of food deprivation in juvenile Atlantic salmon. Journal of Fish Biology 57 (1), 41-51.

O'Connor, P.K., Reich, B., Sheridan, M.A., 1993. Growth hormone stimulates hepatic lipid mobilizationin rainbow trout, *Oncorhynchus mykiss*. Journal of Comparative Physiology B 163, 427-431.

Oommen, O.V., Johnson, B., 1998. Metabolic effects of ovine growth hormone in a teleost, *Anabas testudineus*. Annals of the New York Academy of Sciences 839, 380-381.

Ostenfeld, T., Mclean, H.E., Devlin, R.H., 1998. Transgenesis changes body and head shape in Pacific salmon. Journal of Fish Biology 52, 850-854.

Rahman, M.A., Mak, R., Ayad, H., Smith, A., Maclean, N., 1998. Expression of a novel piscine growth hormone gene results in growth enhancement in transgenic tilapia (*Oreochromis niloticus*). Transgenic Research 357-369.

Rahman, M., Ronyai, A., Engidaw, B.Z., Jauncey, K., Hwang, G.L., Smith, A., Roderick, E., Penman, D., Varadi, L., Maclean, N., 2001. Growth and nutritional trials on transgenic Nile tilapia containing an exogenous piscine growth hormone gene. Journal of Fish Biology 59, 62-78.

Seddiki, H., Boeuf, G., Maxime, V., Peyraud, C., 1996. Effects of growth hormone treatment on oxygen consumption and seawater adaptability in Atlantic salmon parr and pre-smolts. Aquaculture 148, 49-62.

Stevens, E.D., Devlin, R.H., 2000. Intestinal morphology in growth hormone transgenic coho salmon. Journal of Fish Biology 56, 191-195.

Stevens, E.D., Devlin, R.H., 2005. Gut size in GH-transgenic coho salmon is enhanced by both the GH transgene and increased food intake. Journal of Fish Biology 66, 1633-1648.

Stevens, E.D., Sutterlin, A., Cook, T., 1998. Respiratory metabolism and swimming performance in growth hormone transgenic Atlantic salmon. Canadian Journal of Fisheries and Aquatic Sciences 55, 2028-2035.

Stevens, E.D., Wagner, G.N., Sutterlin, A., 1999. Gut morphology in growth hormone transgenic Atlantic salmon. Journal of Fish Biology 55, 517-526.

Sundström, L.F., Lõhmus, M., Tymchuk, W.E., Devlin, R.H., 2007. Gene-environment interactions influence ecological consequences of transgenic animals. PNAS 104, 3889-3894.

Venugopal, T., Anathy, V., Kirankumar, S., Pandian, T.J., 2004. Growth enhancement and food conversion efficiency of transgenic fish, (*Labeo rohita*). Journal of Experimental Biology 301A, 477-490.

Wang, T., Hung, C.C.Y., Randall, D.J., 2006a. The comparative physiology of food deprivation: from feast to famine. Annual review of physiology 68 (1), 223-251.

Wang, W., Wang, Y., Hu, W., Li, A., Cai, T., Zhu, Z., Wang, J., 2006b. Effects of the 'all-fish' growth hormone transgene expression on non-specific immune functions of common carp, *Cyprinus carpio* L. Aquaculture 259 (1-4), 81-87.

Wang, Y., Hu, W., Wu, G., Sun, Y., Chen, S., Zhang, F., Zhu, Z., Feng, J., Zhang, X., 2001. Genetic analysis of 'all-fish' growth hormone gene transferred carp (*Cyprinus carpio* L.) and its F_1 generation. Chinese Science Bulletin 46, 1174-1177.

Warren, C.E., Davis, G.E., 1967. Laboratory studies on the feeding bioenergetics and growth in fish. In: Gerking, S.D. (Ed.), The Biological Basis of Freshwater Fish Production. Blackwell, Oxford, pp. 175-214.

Wu, G., Sun, Y., Zhu, Z., 2003. Growth hormone gene transfer in common carp. Aquatic Living Resources 16, 416-420.

Zbikowska, H.M., 2003. ish can be first advances in fish transgenesis for commercial applications. Transgenic Research 12, 379-389.

转"全鱼"生长激素基因鲤的代谢特征

管　波[1,2]　胡　炜[1]　张堂林[1]　汪亚平[1]　朱作言[1]

1 中国科学院水生生物研究所，淡水生态与生物技术国家重点实验室，武汉　430072

2 中国科学院大学，北京　100049

摘　要　采用循环流水呼吸仪连续测定方法，比较研究了转生长激素基因黄河鲤和对照黄河鲤的呼吸代谢。研究发现，摄食后 0-96h 转基因鱼和对照鱼的平均耗氧率分别是 0.385±0.144 mg/g/h 和 0.291±0.082 mg/g/h，前者的平均耗氧率是后者的 1.32 倍，存在显著性差异(P<0.05)；饥饿 96h 以后，96h 至 144h 转基因鱼和对照鱼的平均耗氧率分别为 0.234±0.037 mg/g/h 和 0.237±0.015 mg/g/h，两者无显著性差异(P>0.05)。进一步研究转基因鱼与对照鱼经 60 天饥饿后的干重和能量变化，发现两者之间干重变化及能量消耗无显著性差异(P>0.05)。上述研究表明，外源生长激素基因没有改变黄河鲤的标准代谢特征，转基因鱼在摄食条件下耗氧率的明显增加，可能主要是由于摄食率提高、摄食后转基因鱼对食物的消化和吸收能力增强，以及营养物质转化能力提高等综合因素引起。

Enzyme-Linked Immunosorbent Assay of Changes in Serum Levels of Growth Hormone (cGH) in Common Carps (*Cyprinus carpio*)

WU Gang[1] CHEN LiHua[2] ZHONG Shan[1,3] LI Qi[2] SONG ChaoJun[2] JIN BoQuan[2] ZHU ZuoYan[1]

1 State Key Laboratory of Freshwater Ecology and Biotechnology, Institute of Hydrobiology, Chinese Academy of Sciences, Wuhan 430072

2 Department of Immunology, Fourth Military Medical University, Xi'an 710032

3 College of Life Science, Wuhan University, Wuhan 430072

Abstract The aim of the present study was to purify the common native carp growth hormone (ncGH), produce monoclonal antibodies (mAbs) to common native carp growth hormone (ncGH), and further enhance the sensitivity of enzyme-linked immunosorbent assays (ELISA) for ncGH. Additionally, we investigated changes in serum ncGH levels in carps raised in different environmental conditions. The recombinant grass carp (*Ctenopharyngodon idella*) growth hormone was purified and used as antigen to immunize the rabbit. The natural ncGH was isolated from the pituitaries of common carp. SDS-PAGE and Western blot utilizing the polyclonal anti-rgcGH antibody confirmed the purification of ncGH from pituitaries. Purified ncGH was then used as an immunogen in the B lymphocyte hybridoma technique. A total of 14 hybridoma cell lines (FMU-cGH 1-14) were established that were able to stably secrete mAbs against ncGH. Among them, eight clones (FMU-cGH 1-6, 12 and 13) were successfully used for Western blot while nine clones (FMU-cGH 1-7, 9 and 10) were used in fluorescent staining and immunohistochemistry. Epitope mapping by competitive ELISA demonstrated that these mAbs recognized five different epitopes. A sensitive sandwich ELISA for detection of ncGH was developed using FMU-cGH12 as the coating mAb and FMU-cGH6 as the enzyme labeled mAb. This detection system was found to be highly stable and sensitive, with detection levels of 70 pg/mL. Additionally, we found that serum ncGH levels in restricted food group and in the net cage group increased 6.9-and 5.8-fold, respectively, when compared to controls, demonstrating differences in the GH stress response in common carp under different living conditions.

Keywords common carp; growth hormone; monoclonal antibodies; ELISA; starvation; net cage

Growth hormone (GH) plays a crucial role in stimulating and controlling the growth, metabolism and differentiation of many cell types, by modulating mRNA and protein

文章发表于《中国科学:生命科学》, 2008, 51(2):157-163

synthesis[1]. In fish, GH participates in almost all the major physiological processes in the body, including(1) the regulation of ionic and osmotic balance, (2) lipid, protein, and carbohydrate metabolism, (3) skeletal and soft tissue growth, (4) reproduction, and (5) immune function[2]. Some of these effects are brought about indirectly via insulin-like growth factors (IGF), which are produced in the liver and other peripheral tissues; other effects result from direct action on target cells[3,4]. GH receptors (GHRs) are expressed in many tissues, such as the liver, muscle, adipose tissue, bone, cartilage, and brain, with the highest levels of expression generally found in the liver[5-7].

Various assay methods, such as radioimmunoassay (RIA) and enzyme-linked immunosorbent assay (ELISA), have been established to detect GH serum levels[8-11]. Although RIA is sensitive and specific, its use is limited by safety concerns of radioactive isotope contamination, as well as concerns regarding radioactivity attenuation. ELISA is a simple method, with no radioactivity concerns, however its use for GH quantitation has been limited by the lack of two or more monoclonal antibodies (mAbs) able to recognize different epitopes of GH molecule. To date, most ELISA systems have been based on the use of one monoclonal or polyclonal Ab. To our knowledge, no ELISA mAb system has been reported which utilizes two Abs for GH detection[12,13].

In the current study, 14 hybridoma cell lines, able to stably secrete mAbs against cGH, were established. Among them, two mAbs were selected to further develop a cGH-specific ELISA sandwich to detect native GH in carp. Our data suggested that serum cGH levels of common carps varied, depending upon living conditions of carp.

1 Materials and Methods

1.1 Purification of recombinant grass carp GH

Medium (3 L) containing recombinant grass carp growth hormone (rgcGH) was prepared as previously described[14]. The medium was centrifuged (10 min, 4000×*g*) and the supernatant was ultrafiltrated utilizing a 50 kDa membrane (Millipore) and concentrated to 200mL with a 3-kDa ultrafiltration membrane (Millipore). The protein solution was adjusted to pH 7.0 with Tris, and then loaded onto a DEAE Sepharose Fast Flow column (Amersham Biosciences), which was preequilibrated in 20 mmol/L pH 7.0 phosphate buffer. The column was washed with the same buffer and the binding proteins were eluted with a linear gradient of 0 to 1mol/L NaCl in the presence of 20 mmol/L pH 7.0 phosphate buffer. The eluate was analyzed on a 12% SDS-PAGE gel. Fractions containing rgcGH were desalted with a Sephadex G-25 column (Amersham Biosciences), pre-equilibrated with 10 mmol/L NH_4HCO_3 and the nlyophilized. The purified rgcGH was used as the antigen to immunize rabbits.

1.2 Preparation of polyclonal antibody in rabbits

Purified rgcGH was prepared in complete Freund's adjuvant (FCA), and then injected subcutaneously at 20 sites on the thigh of two New Zealand white rabbits. The first booster injection was given with rgcGH prepared in incomplete Freund's adjuvant (FIA) 15 days later, followed by booster injections every 15 days there after. Bleedings were performed 45 days after the initial immunization and continued every 15 days. The serum was separated and the titer was measured by ELISA using purified rgcGH as the detection antigen, with serum from non-injected rabbits used as control. High-titer antibodies containing serum were stored at −20℃.

1.3 Preparation and purification of the native common carp growth hormone (ncGH)

Frozen pituitaries (approximately 5 g) were homogenized in extraction buffer (50 mL) containing 50 mmol/L Tris·HCl pH 9.0, 500 mmol/L NaCl, 1 mmol/L PMSF, 1% streptomycin sulphate, followed by the addition of 1mol/L $CaCl_2$ (50 mL) and 10% Dextran sulphate (2 mL). The crude solution was centrifuged at low speed (15 min, 4℃, 200 r/min) followed by high speed centrifugation (10 min, 4℃,10 000×*g*) to obtain the final supernatant, which was then loaded onto a Sephadex G-25 column (Amersham Biosciences), pre-equilibrated with a solution of 20 mmol/L Tris·HCl pH 7.5, 250 mmol/L NaCl, 1 mmol/L $MgCl_2$, 1 mmol/L $CaCl_2$ and 1 mmol/L $MnCl_2$ and eluted by the same buffer. The eluate was loaded onto the ConA Sepharose column (Amersham Biosciences) to rescue additional hormones, and the fraction containing GH was eluted with a buffer containing 20mmol/L Tris·HCl pH 7.5, 500 mmol/L NaCl, 1 mmol/L $MgCl_2$, 1 mmol/L $CaCl_2$ and 1 mmol/L $MnCl_2$. The solution was concentrated utilizing a 5 kDa ultrafiltration membrane (Millipore) and then loaded onto a Sephadex G-25 column pre-equilibrated with a solution containing 20 mmol/L Tris·HCl pH 8.0. A desalting solution was then loaded onto a DEAE Sepharose Fast Flow column (Amersham Biosciences) pre-equilibrated with 20mmol/L Tris·HCl pH 8.0. Binding proteins were eluted with a linear gradient, (0 to 1 mol/L NaCl), in the presence of 20 mmol/L Tris·HCl pH 8.0. Fractions were then analyzed by 12% SDS-PAGE and those fractions containing ncGH were mixed with equal volumes of 3 mol/L $(NH_4)_2SO_4$ and loaded onto a Phenyl Sepharose 6 Fast Flow column (Amersham Biosciences) pre-equilibrated with 1.5 mol/L $(NH_4)_2SO_4$. Protein fractions were eluted with a linear gradient (1.5 to 0 mol/L $(NH_4)_2SO_4$) in the presence of 20 mmol/L Tris·HCl pH 8.0 and were once again analyzed by 12% SDS-PAGE. The ncGH fraction was concentrated to a volume of about 1 mL. The crude ncGH was purified utilizing a Sephacryl S-100 High Resolution column, pre-equilibrated with 10 mmol/L NH_4HCO_3. Purified ncGH was then subjected to 12% SDS-PAGE and Western blot analysis with the polyclonal anti-rgcGH antibody.

1.4 Production of hybridomas

Female BALB/c mice (8 weeks old) were immunized with 20 μg ncGH in FCA by subcutaneous (s.c.) injection. Subsequent s.c. immunizations were carried out twice with 20 μg ncGH in FIA and intra-peritoneal (i.p.) injections, without adjuvant respectively, at 3-week-intervals. Mice were bled from caudal vein 10 days after the third immunization and anti-serum titers were determined by indirect ELISA. Immunized mice were boosted with 20 μg of antigens by i.p. injection. Splenocytes from immunized mice and SP2/0 myeloma cells, cultured in RPMI 1640 containing 20% fetal calfserum (FCS), were fused, three days later, in the presence of PEG (MW4000, Merk, Germany). Positive hybrids were screened by ELISA and subcloned four times by limiting dilution. Monoclonal antibodies were produced either from supernatants of the hybridoma culture or from ascites of BALB/c mice in which hybridomas had been injected intraperitoneally. Ig isotypes were identified using an Isotype kit (Sigma, M-5907). Epitope specificities of mAbs to cGH were analyzed by competitive ELISA.

1.5 Indirect ELISA

In order to screen Abs reacting with cGH, 96-well plates (NUNC, Nagel Inc., Roskilde, Denmark) were coated with 5 μg/mL of cGH in coating buffer (0.05 mol/L carbonate/bica-rbonate buffer, pH 9.5) and incubated (4℃, overnight). Plates were washed (3×) with 0.01 mol/L PBS / 0.05% Tween-20 and blocked with a 0.1% BSA solution. Various dilutions of immunized rabbit or murinesera or supernatants from hybridoma cultures were then added to the wells. Plates were incubated (1 h, 37℃) and washed (3×), then incubated with 100 μL of horseradish peroxidase (HRP)-conjugated goat anti-rabbit or anti-mouse antibody (Vector, USA), and diluted 1/5000 in PBS (1 h, 37℃). Plates were then washed (3×) and 100 μL of 2,2′-azobis-3-ethylbenzthiazoline-6-sulfonic(ABTS, Vector, USA) was added to each well prior to reading (410 nm) with an automatic plate reader (Bio-Rad).

1.6 Western blot

Purified ncGH was denatured, subjected to SDS-PAGE then transferred onto PVDF membranes (BioRad), and blocked (4℃, overnight) in TBS (20 mmol/L Trizma, 0.5 mol/L NaCl, pH 7.5) containing 5% non-fat milk. Membranes were incubated (room temperature, 2 h) with mAbs against ncGH. After washing, membranes were incubated with HRP-labeled goat anti-mouse Ig diluted in antibody buffer (1∶1000, 2 h). An enhanced Chemilumin-escence (ECL, Roche, USA) kit was used to detect bands.

1.7 Immunohistochemistry

Sections from carp pituitary (20 μm) were used in immunohistochemical analysis of ncGH expression. Several Ab dilutions were tested to find the optimal staining concentration. The staining procedure, without protease treatment, has been previously described. Briefly, sections were washed (20 min) in 0.01 mol/L PBS, pretreated with methanol-H_2O_2 to remove endogenous peroxidase, rinsed in 0.1 mol/L PBS and incubated with goat serum (30 mL/L, 40 min). Samples were then incubated with primary mouse anti-ncGH antibody (1∶200, 25 μg/mL, 4℃, 24 h) followed by secondary biotin labeled sheep anti-mouse IgG (1∶200, room temperature,3 h, Vector, USA). Sections were subsequently incubated with avidin-biotin peroxidase complex (ABC Vector, USA, 1∶100 dilution, room temperature, 1 h), then washed (3×, 10 min). A colorimetric reaction utilizing diaminobezidin (DAB) was used to develop the sections. Negative control tissue sections utilized anti-staphylococcus enterotoxin D (SED) mAb as the primary Ab.

1.8 Purification, HRP-conjugating of mAbs and sandwich ELISA

Selected mAbs were purified from mouse ascites fluids. IgG was purified from ascites fluid utilizing a HiTrap protein G column (Amersham Biosciences). Purified mAbs were conjugated with HRP using peroxidase, as previously described[15]. In order to detect ncGH, a sandwich ELISA was developed using FMU-cGH12 as the coating mAb and FMU-cGH6 as the enzyme-labeled mAb. In brief, 100 μL of anti-cGH mAb (5 μg/mL in 0.05 mol/L carbonate/bicarbonate buffer, pH 9.5) was added to each ELISA plate well (Nunc. Maxisorp), incubated overnight (4℃) and washed (3×) with 0.01 mol/L PBS containing 0.1% (*v/v*) Tween-20 (PBS/Tween). Plates were then blocked with a 0.1% BSAs olution. Standard ncGH serial dilutions with PBS containing 0.1% BSA and 0.1% (*v/v*) Tween-20 or serum samples of carp or grass carp were added to the wells and incubated (1 h, 37℃). After extensive washing with PBS containing 0.1% (*v/v*) Tween-20 (PBS/Tween), wells were incubated (1 h, 37℃) with FMU-cGH6 conjugated with HRP and diluted in PBS containing 3% PEG. Wells were washed once again (3×) and 100 μL of ABTS substrate was added to each well. Plates were read at 410 nm on a microplate reader. The coefficient variation of inter-assay and intra-assay was measured respectively. Additionally, levels of prolactin, thyroidstimulating hormone, follicle stimulating hormone, and luteinizing hormone were also measured.

1.9 Fish and experimental design

Farm-raised common carps (one year old, n=45) were divided randomly into three groups: group 1 (control, n=15) was transferred into a 10×10×0.8 m^3 pool; group 2 (Net cage, n=15) was kept in a 31.5×0.8 m^3 net cage, within the same pool as group 1. Both groups 1 and

2 received normal amounts of food. Group 3 (Food restriction, n=15) was placed in a pool similar to that of group 1, but was not fed artificially. Three months later, blood samples of each group were collected from the caudal vasculature, kept on ice and later centrifuged (3000×g, 4℃, 5 min) to obtain serum samples which were then examined by sandwich ELISA system.

1.10 Statistical analysis

Serum GH levels were analyzed using a Student-Newman-Keuls-q test. Average serum GH levels between different groups were expressed as $x \pm sd$; $P<0.05$ was considered significant.

2 Results

2.1 Identification of purified ncGH

A total of 18 mg of the target protein was purified from pituitaries. Western blot analysis utilizing a polyclonal Ab against rgcGH confirmed that the 23 kDa protein identified was ncGH (Figure 1). The ncGH purification rate was 87%.

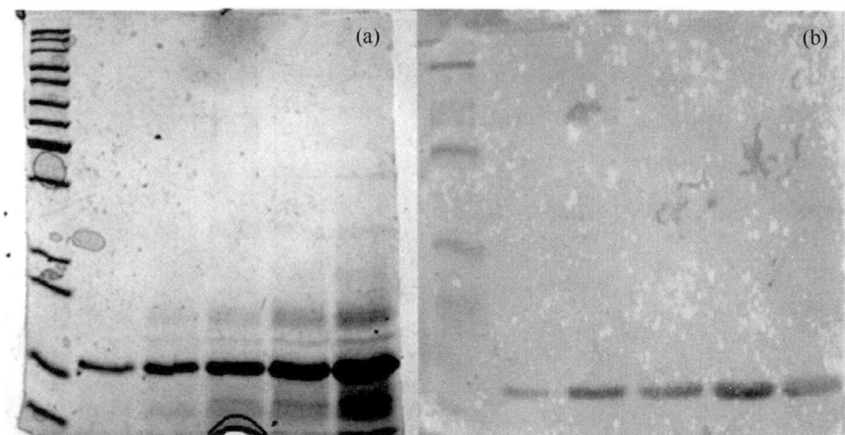

Figure 1 Identification of ncGH purified from common carp pituitaries (87% purification rate) under denaturing conditions. (a) SDS-PAGE of purified ncGH; (b) Western blot analysis of ncGH, utilizing a polyclonal Ab against rgcGH. Lane 1, molecular marker; lanes 2-6, representative samples at different concentrations

2.2 Production of mAbs specific for ncGH

A total of 14 hybridoma cell lines were established which could stably secrete mAbs against ncGH (FMU-cGH 1-14). In almost all cages, the Ig subclass of mAbs secreted by the hybridomas was IgG1 (κ), with the exception of FMU-cGH 4 and FMU-cGH 14, which secreted IgG2b (κ).

Titers of ascites fluid containing mAbs ranged between 10^{-7} and 10^{-5}, as measured by indirect ELISA. Secreted mAbs from FMU-cGH 1-6, 12 and 13 were successfully used for Western blot, and FMU-cGH 1-7, 9 and 10 were used in fluorescent staining and immunohistochemistry. Western blot analysis using specific mAbs further confirmed that the molecular weight of ncGH was 23 kDa (Figure 2). Immunostaining data localized ncGH staining to the cytoplasm, as membrane and nuclei remained immuno negative (Figure 3).

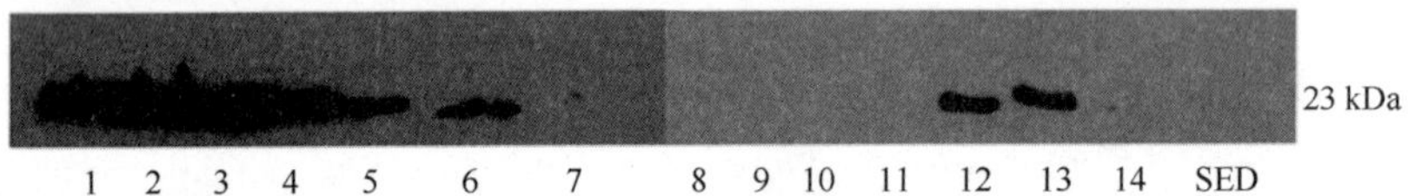

Figure 2 Identification of mAbs able to react with ncGH under denaturing conditions. Lanes 1-14, FMU-cGH 1-14, respectively

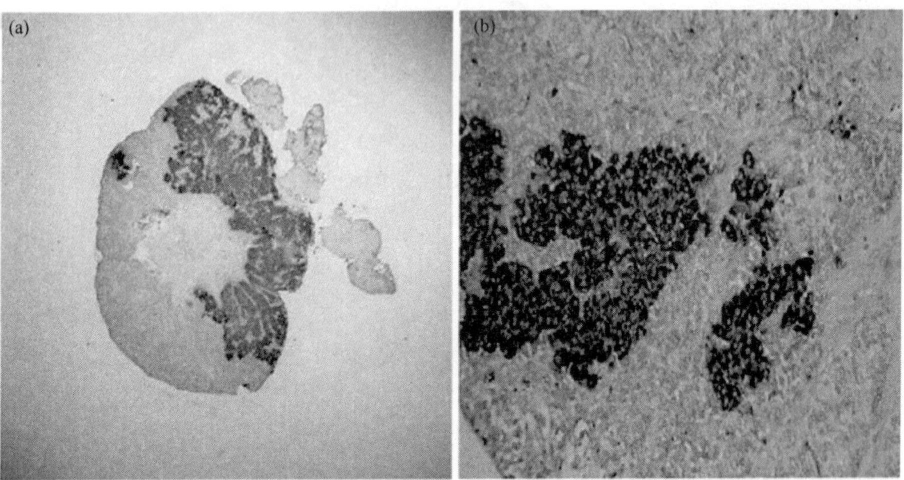

Figure 3 Distribution of ncGH in carp pituitary. A positive signal was seen mainly in the cytoplasm. (a) Whole pituitary sagittal section (40×); (b) pituitary sagittal cross-section (400×)

2.3 Development of ELISA for ncGH quantitation

Epitope mapping by competitive ELISA suggested that mAbs recognized five different epitopes. The first was recognized by FMU-cGH 1, 3, 8, 12 and 13; the second by FMU-cGH 6 and 7 and the third by FMU-cGH 9 and 11. FMU-cGH 4 and 10 recognized unique epitopes (Figure 4). A sandwich ELISA for detecting ncGH was developed using FMU-cGH12 as the coating mAb and FMU-cGH6 as the enzyme-labeled mAb. As seen in Figure 5, the ELISA system was very sensitive, with a limitation of 70 pg/mL. The inter-assay and intra-assay coefficient variation was 7% and 5%, respectively. The ELISA system was also able to detect GH concentrations in grass carp. However it was not able to detect other substances existing in the pituitary, such as prolactin, thyroid stimulating hormone, follicle stimulating hormone, and luteinizing hormone (n=3 independent experiments).

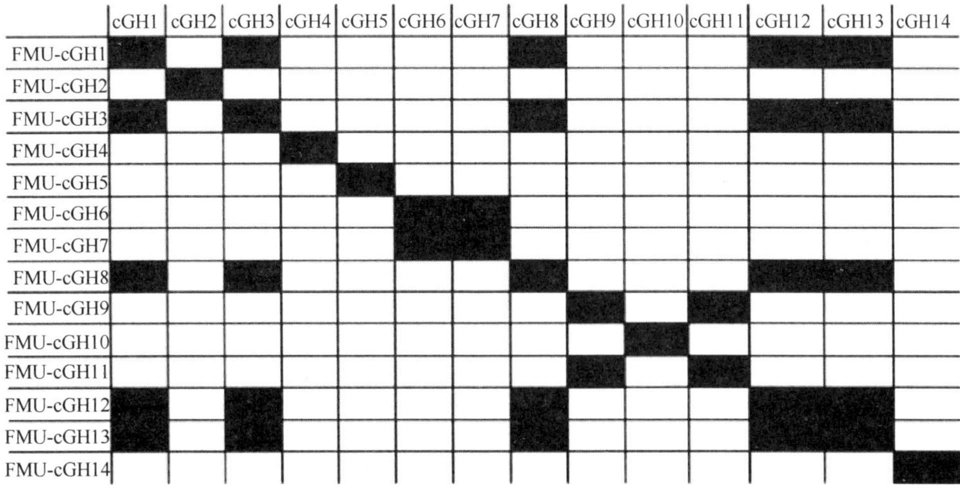

Figure 4 Epitope mapping by competitive ELISA. Both mAbs recognized the same epitope

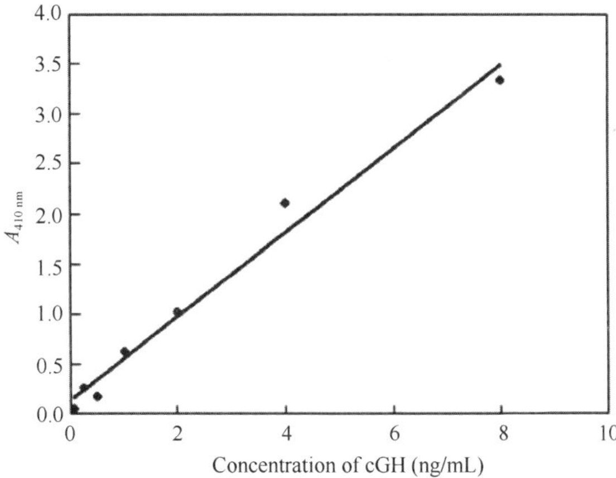

Figure 5 Sandwich ELISA standard curve for detecting ncGH. The detection limit was approximately 70 pg/mL (R^2=0.9827)

2.4 Elevated serum GH levels in carps subjected to food restriction or to net cage living conditions

The average serum ncGH levels of group 1 (control), group 2 (net cage), and group 3 (food restriction) were 0.73±0.74, 4.21±4.41 and 5.04±5.04 ng/mL, respectively. Compared with the control group, serum ncGH levels of common carp living in a net cage or subjected to food restriction increased 5.8 fold (P<0.05) and 6.9 fold (P<0.05), respectively. As shown in Figure 6, of the 15 fish in the control group, only four had serum ncGH levels >1.5 ng/mL of ncGH, with 2.2 ng/mL recorded as the highest serum ncGH level. Contrarily, fish in the net cage (n=11) and in the food restriction group (n=9) had serum cGH levels≥1.5 ng/mL, with the highest levels of serum cGH measured at 16.0 and 14.0 ng/mL respectively (Figure 6).

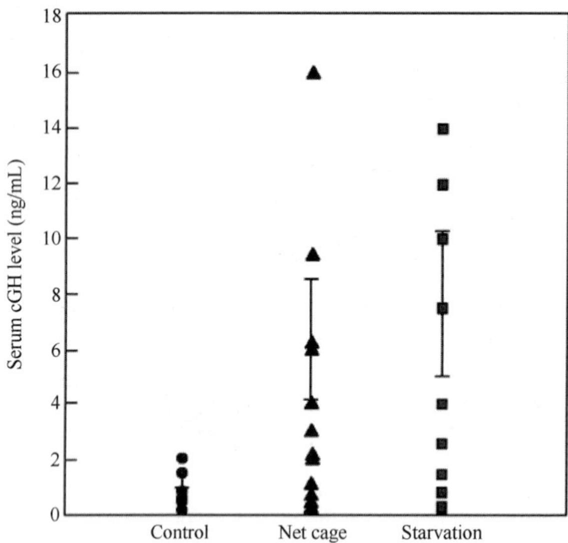

Figure 6 Serum growth hormone levels of different groups ($x\pm sd$). Of the 15 fish in the control group, only 4 displayed ncGH serum levels ⩾ 1.5 ng/mL, while in the net cage and restricted food groups, serum ncGH levels greater or equal to 1.5 ng/mL were noted in 11 and 9 fish, respectively

3 Discussion

In teleosts, GH, one of the pituitary hormones, is essential for the maintenance of growth. In our study, we produced a set of mAbs against ncGH, and confirmed that eight of the cell clones could be used in Western blot while nine clones could be used in both fluorescent staining and immunohistochemistry. Furthermore, we found that these mAbs recognized five different epitopes on the ncGH molecule. Based on the results we got above, a sensitive sandwich ELISA system with the limitation of 70 pg/mL against ncGH was developed successfully.

The development of GH isolation and purification, and the preparation of an antiserum to GH from immunized rabbits, a radioimmuno assay (RIA) was successfully developed, in 1983, to detect pituitary or serum GH of carp and goldfish (*Carassius auratus*)[16]. The detection limitation was 5 ng/mL. Soon afterwards, a RIA assay for detecting GH of teleosts, such as salmon (*Oncorhynchus keta*), chinook salmon (*Oncorhynchus tshawytscha*), European eel (*Anguilla anguilla*), and African catfish (*Arius africanus* Günther) was also developed[8-11]. However, this system had obvious limitations, such as radioisotope contamination and the short half life of labeled tracers.

One of the advantages of the ELISA method is that it avoids potential radioactive contamination. The first sandwich ELISA kit for measuring oncorhynchid GH using mAbs was reported in 1991[12]. While this ELISA recognized GH in pituitary extracts or plasma of rainbow trout (*Oncorhynchus mykiss*), coho salmon (*Oncorhynchus kisutch*), and Chinook salmon, it did not recognize GH in carp, goldfish or other important teleosts. Additionally, it was less than 1.56 ng/mL. A new competitive ELISA to measure plasma GH levels in channel catfish was reported in 2003[17]. In this system, plates were coated with 25 μg of channel catfish (*Ictalurus punctatus*) GH in carbonate buffer. The standard hormone or plasma samples were then mixed with rabbit anti-cfGH, followed by the addition of HRP-labeled goat anti-rabbit antibody.

Specificity and stability are well-established advantages to using an ELISA two-mAb system that recognizes different epitopes on the same molecule. For we know, the ELISA system we developed for detecting ncGH was the first one basing on two monoclonal antibodies which recognize different epitopes on the ncGH molecule. Besides specificity and stability, the ELISA displayed a very high sensitivity with the detection limitation of 70 pg/mL, which enabled us to further investigate the changes of serum ncGH levels in common carps living in different conditions.

Previous research has suggested that the stress response in teleost fish is, in many ways, similar to that of terrestrial vertebrates[18-20]. Growth hormone is one of the indicators of the teleost stress response. In the current study, we found a 6.9-fold increase in serum ncGH levels in the restricted food group and a 5.8-fold increase in the net cage group, compared to control. These data are consistent with previous reports in other fish living under similar conditions, such as salmon, rainbow trout, (*Oncorhynchus mykiss*), European eel, and black seabream (*Acanthopagrus schlegeli*)[9,21-23].

In conclusion, the ELISA system we developed for detecting ncGH was very sensitive, stable and easy to perform since we selected monoclonal antibodies with high affinity and recognizing different epitopes of cGH from a number of monoclonal antibodies against ncGH we raised. This ELISA enabled us to compare serum ncGH levels in carps lived in different stress conditions. Compared with the control, serum ncGH level significantly increased in carps living either in starvation condition or in net cage. These novel mAbs to ncGH may be useful in future studies of ncGH, to further elucidate the relationship between GH and other hormones and the dynamic expression of cGH in the potential development of GH gene-transgenic carps.

References

1 Carter-Su C, Schwartz J, Smit L S. Molecular mechanism of growth hormone action. Annu Rev Phys, 1996, 58: 187-207

2 Reinecke M, Bjornsson B T, Dickhoff W W, *et al*. Growth hormone and insulin-like growth factors in fish: Where we are and where to go. Gen Comp Endocrinol, 2005, 142: 20-24

3 Wabitsch M, Hauner H, Heinze E, *et al*. The role of growth hormone/insulin-like growth factor in adipocyte differentiation. Metabolism,1995, 44: 45-49

4 Waters M J, Rowlinson S W, Clarkson R W, *et al*. Signal transduction by the growth hormone receptor. Proc Soc Exp Biol Med, 1994, 206: 216-220

5 Mertani H C, Waters M J, Jambou R, *et al*. Growth hormone receptor binding protein in rat anterior pituitary. Neuroendocrinology, 1994, 59: 483-494

6 Perez Sanchez J, Smal J, Le Bail P Y. Location and characterization of growth hormone binding sites in the central nervous system of a teleost fish (*Oncorhynchus mykiss*). Growth Regul, 1992, 1: 145-152

7 Sun X, Zhu S, Chan S S, *et al*. Identification and characterization of growth hormone receptors in snakehead fish (*Ophiocephalus argus* Cantor) Liver. Gen Comp Endocrinol, 1997, 108: 374-385

8 Lescroart O, Roelants I, Mikolajczyk T, *et al*. A Radioimmuno assay for African catfish growth hormone: Validation and effects of substances modulating the release of growth hormone. Gen Comp Endocrinol, 1996, 104: 147-155

9 Marchelidon J, Schmitz M, Houdebine L M, *et al*. Development of a radioimmunoassay for European eel growth hormone and application to the study of silvering and experimental fasting. Gen Comp Endocrinol,1996, 102: 360-369

10 Le Bail P Y, Sumpter J P, Carragher J F, *et al*. Development and validation of a highly sensitive radioimmunoassay for chinook salmon (*Oncorhynchus tshawytscha*) growth hormone. Gen Comp Endocrinol,1991, 83: 75-85

11 Bolton J P, Takahashi A, Kawauchi H, *et al*. Development and validation of a salmon growth hormone radio-immunoassay. Gen Comp Endocrinol, 1986, 62: 230-238

12 Farbridge K J, Leatherland J F. The development of a noncompetitive enzyme-linked immunosorbent assay for oncorhynchid growth hormone using monoclonal antibodies. Gen Comp Endocrinol, 1991, 83: 7-17

13 Chen S, Yang F, He L, *et al*. Generation and characterization of monoclonal antibodies against grass carp. Asian Fish Sci, 1996, 9: 183-189

14 Wang W, Sun Y H, Wang Y P, *et al*. Expression of grass carp growth hormone in the yeast *Pichia pastoris*. Acta Genet Sin, 2003, 30: 301-306

15 Tijssen P, Kurstak E. Highly efficient and simple methods for the preparation of peroxidase and active peroxidase-antibody conjugates for enzyme immunoassays. Anal Biochem, 1984, 136: 451-457

16 Cook A F, Wilson S W, Peter R E. Development and validation of a carp growth hormone radioimmunoassay. Gen Comp Endocrinol,1983, 50: 335-347

17 Drennon K, Moriyama S, Kawauchi H, *et al*. Development of an enzyme-linked immunosorbent assay for the measurement of plasma growth hormone (GH) levels in channel catfish (*Ictalurus punctatus*): Assessment of environmental salinity and GH secretogogues on plasma GH levels. Gen Comp Endocrinol, 2003, 133: 314-322

18 Wendelaar Bonga S E. The stress response in fish. Physiol Rev, 1997, 77: 591-625

19 Rotllant J, Balm P H, Perez-Sanchez J, *et al*. Pituitary and interregnal function in gilthead sea bream (*Sparus aurata* L., Teleostei) after handling and confinement stress. Gen Comp Endocrinol, 2001, 121: 333-342

20 Auperin B, Baroiller J F, Ricordel M J, Fostier, *et al*. Effect of confinement stress on circulating levels of growth hormone and two prolactins in freshwater-adapted tilapia (*Oreochromis niloticus*). Gen Comp Endocrinol, 1997, 108: 35-44

21 Wagner G F, McKeown B A. Development of salmon growth hormone radioimmunoassay. Gen Comp Endocrinol, 1986, 62: 452-458

22 Sumpter J P, Le Bail P Y, Pickering A D, *et al*. The effect of starvation on growth and plasma growth hormone concentration of rainbow trout, *Oncorhynchus mykiss*. Gen Comp Endocrinol, 1991, 83: 94-102

23 Deng L, Zhang W M, Lin H R, *et al*. Effects of food deprivation on expression of growth hormone receptor and proximate composition in liver of black seabream *Acanthopagrus schlegeli*. Comp Biochem Physiol B Biochem Mol Biol, 2004, 137: 421-432

ELISA 法检测鲤血清 GH 水平的变化

吴　刚[1]　CHEN LiHua[2]　钟　山[1,3]　LI Qi[2]　SONG ChaoJun[2]
JIN BoQuan[2]　朱作言[1]

1 中国科学院水生生物研究所淡水生态与生物技术重点实验室，武汉　430072
2 第四军医大学免疫学，西安　710032
3 武汉大学生命科学院，武汉　430072

摘　要　本研究的目的是纯化鲤的生长激素(ncGH)蛋白，制备 ncGH 的单克隆抗体，建立灵敏的 ncGH 酶联免疫检测方法。同时，采用该方法，检测在不同环境条件下，鲤血清中 ncGH 的含量变化。首先，利用草鱼生长激素酵母表达载体，从表达产物中分离纯化草鱼生长激素蛋白，将其作为抗原免疫兔子，获得草鱼 GH 多克隆抗体(anti-rgcGH)；ncGH 蛋白从鲤的脑垂体中分离获得。在此基础上，利用 anti-rgcGH，通过蛋白质印迹实验和电泳分离方法，分离纯化鲤脑垂体中的 ncGH。然后，以 ncGH 作为抗原，免疫小鼠，用于筛选单克隆抗体。从杂交瘤细胞中筛选获得 14 株单克隆抗体，分别识别 5 个不同的抗原表位。其中，8 个克隆体成功应用于蛋白质印迹实验，9 个克隆体成功应用于荧光染色实验和免疫组织化学实验。经配对分析，从 3 对组合中筛选到一对具有最高灵敏度的单克隆抗体，用于 ELISA 分析，其检测灵敏度达到 70 pg/mL。在饥饿组和网箱投喂组与大塘投喂对照组的对比实验中，鲤血清中 ncGH 水平分别增加了 6.9 倍和 5.8 倍，此结果表明鲤生长激素的应激反应随生长条件不同而不同。

The Hematological Response to Exhaustive Exercise in 'All-Fish' Growth Hormone Transgenic Common Carp (*Cyprinus carpio* L.)

Deliang Li[a,b] Cuizhang Fu[a,c] Yaping Wang[a] Zuoyan Zhu[a] Wei Hu[a]

1 State Key Laboratory of Freshwater Ecology and Biotechnology, Institute of Hydrobiology, Chinese Academy of Sciences, Wuhan 430072

2 College of Animal Science and Technology, Hunan Agricultural University, Changsha 410128

3 Ministry of Education Key Laboratory for Biodiversity Science and Ecological Engineering, Institute of Biodiversity Science, Fudan University, Shanghai 200433

Abstract This study examined the hematological profiles and responses to exhaustive exercise in 'all-fish' growth hormone (GH) transgenic common carp (*Cyprinus carpio* L.) compared to non-transgenic controls. Transgenic fish had significantly smaller erythrocytes (length, width, surface area and volume), with a larger surface to volume ratios than non-transgenics. There were no significant differences in hematocrit, blood hemoglobin concentrations, plasma glucose, lactate, cholesterol and protein concentrations between the two strains at rest, but the transgenic fish had significantly lower resting mean corpuscular hemoglobin concentrations (MCHC) or higher plasma triglyceride concentrations than controls. Exhaustive exercise did not induce significant changes in hematocrit, hemoglobin concentrations and MCHC in either strain. The patterns of elevation and recovery in plasma glucose and lactate were the same for transgenic and control fish, although lactate levels were lower in the transgnic fish than controls at 1 h and 2 h post-exercise. There was a significant elevation in plasma cholesterol and triglyceride concentrations for 1 h post-exercise in transgenics but not in non-transgenic controls. Plasma protein concentrations were not altered by exhaustive exercise in transgenics but were lower from 2 h post-exercise in non-transgenic controls. These findings suggest that the 'all-fish' GH transgene significantly affects the erythrocyte morphology and the patterns in hematological response to exhaustive exercise of common carp, especially for lipid metabolism. The greater reliance on triglycerides and a greater erythrocyte surface to volume ratio may help transgenic carp to recover from exhaustive exercise, but further research on muscle metabolism and aerobic metabolic rate are needed in order to confirm this conclusion.

Keywords common carp; transgenic; exhaustive exercise; hrowth hormone (GH); hematology

文章发表于 *Aquaculture*, 2011, 311(1-4):263-268

1. Introduction

Since the first transgenic fish were produced by Zhu *et al.* (1985), gene transfer studies have been conducted in over 35 different species of fish (Zbikowska, 2003; Hallerman *et al.*, 2007). The objective of applying transgenic techniques to fish that has received the most attention is growth enhancement (Maclean and Laight, 2000). Many countries are discussing or developing safety assessment strategies for fast-growing transgenic fish, and growth hormone (GH) transgenic fish are near the point of widespread commercial use (Niiler, 2000; Fu *et al.*, 2005; Hu *et al.*, 2007). Many studies have shown that in addition to enhancing growth rates, transgenic fish overexpressing GH can show a broad range of pleiotropic effects on morphology, physiology, metabolism, immunology and behavior (Fu *et al.*, 1998, 2007a, b; Devlin *et al.*, 2004; 2006; Wang *et al.*, 2006; Li *et al.*, 2007, 2009; Guan *et al.*, 2008). Transgenic fish can be a useful model for examining the effects of GH on many physiological processes (Devlin *et al.*, 2004).

Fast-growing transgenic Atlantic salmon (*Salmo salar* L.) overexpressing GH were shown to have slight variations in erythrocyte morphology, but the effects on hematological parameters, including erythrocyte concentration, hematocrit, total blood hemoglobin and some secondary indices, were not significantly different from non-transgenic controls (Cogswell *et al.*, 2002; Wang *et al.*, 2005; Deitch *et al.*, 2006). In Arctic charr (*Salvelinus alpinus* L.), GH transgenic individuals have decreased plasma triglycerides concentration but similar plasma cholesterol concentrations compared with controls (Krasnov *et al.*, 1999). There were no significant differences in hematocrit and mean corpuscular hemoglobin concentrations (MCHC) between transgenic Atlantic salmon and controls in response to stress with 45 s of air exposure, but post-stress hemoglobin levels were 14% higher in the transgenic salmon (Deitch *et al.*, 2006). There are no reports on physiological recovery to exhaustive exercise in fast-growing GH transgenic fish.

For common carp, the level of plasma GH in 'all-fish' GH transgenic individuals was about 50 to 250-fold that of non-transgenics dependent on water temperature (Zhong *et al.*, 2009a, b), and these fish differed from non-transgenics in their metabolism, physiology and behavior (Wang *et al.*, 2006; Fu *et al.*, 2007a, b; Li *et al.*, 2007, 2009; Guan *et al.*, 2008; Duan *et al.*, 2009), so it is speculated that there may be differences between two strains in manners in response to exhaustive exercise. Therefore, the aims of the present study were: 1) to provide baseline hematological information in 'all-fish' growth hormone transgenic carp and 2) to assess if and how the hematological patterns for transgenic carp in response to exhaustive exercise are modified compared to controls. Information on this variability would provide important insights into the ecological differences/ requirements between transgenic carp and controls with respect to exercise performance

(Kieffer, 2000).

2. Materials and Methods

2.1 Experimental animals

P_0 'all-fish' GH gene transgenic carp were produced by micro-injection of the pCAgcGH construct into fertilized eggs of common carp (Yellow River carp variety). This construct contains the recombinant sequence of the grass carp *Ctenopharyngodon idellus* (Cuvier & Valenciennes) growth hormone gene driven by a common carp β-actin gene promoter (Wang *et al.*, 2001). F_5 'all-fish' GH transgenic carp used in the present study were produced on April 25, 2007, from crosses between a wild-type female and a F_4 'all-fish' GH transgenic male of a fast-growing transgenic strain as reported in previous studies (Fu *et al.*, 2007a; Li *et al.*, 2007). A batch of non-transgenic controls was also produced on the same day. At the age of 16 weeks, the transgenic fish were transferred to our lab and held under flow through conditions in 1000 l tanks. Transgene positive individuals were identified by the PCR procedure of Wang *et al.* (2001), using DNA extracted from tail fins. Non-transgenic controls were size-matched fish transported to the laboratory at approximate 20 weeks of age. During the acclimation period, all fish were fed to satiation twice daily with commercial dry pellets for carp. Well aerated water was supplied to rearing tanks. Dissolved oxygen was > 7 mg l^{-1}, ammonia-nitrogen was < 0.15 mg l^{-1} and pH varied from 7.0 to 7.3. Photolight was controlled by artificial lighting, with the light period between 0700 and 1900 h. All experimental procedures were approved by the Animal Care Committee of the Institute of Hydrobiology, Chinese Academy of Sciences.

2.2 Experiment I – erythrocyte measurements

For the present study, 20 transgenic and 23 control carp were used. Fish were anesthetized with 0.25 ml l^{-1} Eugenol (purity ≥ 99.99%), and peripheral blood samples (0.2 ml/fish) were collected from the caudal vessels using heparinized syringes (1.25×10^5 U heparin sodium in 100 ml 0.85% NaCl). The blood samples were stabilized with 12.5 U of heparin sodium per 1 ml blood. The fish were then blotted to remove excess water, and their body weights and standard lengths were measured (to nearest 0.01 g and 0.1 cm, respectively). The blood was diluted 200-fold with 0.85% NaCl and blood cells were counted with a Neubauer hemocytometer under a microscope. Blood smears were air-dried and stained with Wright-Giemsa solution. The lengths (a) and widths (b) of 30 erythrocytes per smear were measured under a light microscope (Cart Zeiss Shanghai Co. Ltd.) with a video camera linked to computer with image-analysis software (ACT-2U, version 1.1, Nikon). The surface areas (S) and volumes (V) of the cells were calculated using the formulae: $S = (\pi \cdot a \cdot b) / 4$

and $V = (\pi \cdot a \cdot b^2) / 6$ (Cal *et al.*, 2005).

2.3 Experiment II – physiological response to exhaustive exercise

Two weeks prior to experimentation, pre-selected fish were evenly distributed to six circular fiberglass tanks (diameter 150 cm and water volume 1000 l), with 6 transgenics and 6 controls per tank. Transgenic fish were distinguished from controls by different tail fin clips. The day before sampling or exercise, all fish from a selected tank were randomly placed into small, individual flow-through aquaria (Fu *et al.*, 2007b) that allowed access to each fish without disturbing the others. The fish were not fed while in these aquaria. Aerated and dechlorinated water was supplied to each individual aquarium at a flow rate of 5 l min^{-1}. These flow rates did not induce swimming in the fish. The water temperature varied from 16.2 to 17.2℃ (mean 16.7, S.E. 0.15℃).

After at least 24 h acclimation, 48 fish were subjected to exhaustive exercise. Eight transgenics and 10 controls from the 24 remaining pre-selected fish were sampled at rest for baseline measurements. The exercise protocols were as follows: the fish were netted individually from the acclimation aquaria, transferred to a circular fiberglass tank (diameter 150 cm and water depth 50 cm) and were immediately forced to swim by manual chasing to exhaustion. This is a standard technique used to stress fish in physiological studies (Wood, 1991; Milligan, 1996; Kieffer *et al.*, 2001; Iwama *et al.*, 2004). Fish were considered to be physically exhausted when they lost equilibrium and were no longer responsive to manual chasing (Baker *et al.*, 2005). The duration of exercise for each fish was not determined, but ranged between about 7 and 10 min. After exhaustive exercise, the fish were quickly transferred from the exercise tank back to their respective acclimation aquaria and assigned at random to recover for 0, 1, 2 or 4 h. Six fish were included in each recovery group, except for the 2 h recovery group of transgenic fish, because one blood sample was hemolytic.

Blood samples were quickly taken when the designated recovery periods were completed. Sampled fish were quickly anesthetized using Eugenol (0.25 ml l^{-1}). When gill irrigation had ceased, fish were removed from the anesthetic and were bled (about 1.5 ml) from the caudal vessels using 2 ml heparinized syringes. Then the fish were blotted of excess water, and their body weights and standard lengths measured. The blood samples were immediately treated as described below. In order to reduce the potential effect of diurnal rhythm, all samplings were carried out between 1400 and 1600 h.

An aliquot of each blood sample was used to determine hematocrit and hemoglobin. The remainder was centrifuged at 2600 × *g* for 5 min at 4℃ to collect plasma, which was stored at −80℃ for later measurement of glucose, lactate, triglyceride, cholesterol and total protein content.

2.4 Analytical techniques

Hematocrit were measured in duplicate by collecting blood in 5 μl capillary tubes, and centrifugation for 10 min. Total blood hemoglobin concentrations were estimated spectrophotometrically (at 540 nm) using the cyanomethahemoglobin method (Maker Technology Co. Ltd. Chengdu, China). MCHC was calculated by dividing hemoglobin concentration by hematocrit. The following measurements were performed with commercial kits purchased from Nanjing Jiancheng Bioengineering Institute, Nanjing, China: plasma glucose (glucoseoxidase-peroxidase method); plasma lactate and cholesterol levels; plasma triglyceride (GPO-PAP method), and total protein concentration (Bradford method).

2.5 Statistical analysis

All values are presented as means ± S.E. and all data were tested for normality using Shapiro-Wilk's *W* test and for homogeneity of variances using Levene's test. Where necessary, log10-transformations were performed. When normality or equality of variance assumptions were not met, a non-parametric test were used. In Experiment I, differences between strains in fish weight and length were analyzed using the Mann-Whitney *U* test and all comparisons of optical erythrocyte parameters between strains were performed using *t*-tests. In Experiment II, differences of each physiological parameter between strains for each sample time were compared using *t*-tests or Mann-Whitney *U* tests. Two-way ANOVAs were carried out to test for the overall effects of strain and time on all parameters measured. For each strain, a one-way ANOVA with post-hoc Dunnett's test was used to compare physiological parameters at each sample time to values for resting fish. All analyses were carried out using the Statistica 6.0 software, with $P < 0.05$ considered statistically significant.

3. Results

3.1 Erythrocyte measurements

Average body weight and standard length did not differ between transgenics ($n = 20$ and 24 weeks old) and non-transgenics ($n = 23$ and 24 weeks old) (152.80 ± 5.32 g and 19.48 ± 0.25 cm vs. 148.36 ± 2.59 g and 18.77 ± 0.12 cm in controls, respectively; $P > 0.05$).

There were no significant differences between transgenic common carp and non-transgenic controls in erythrocyte count (Table 1). The erythrocytes of transgenics were shorter in both length (by 3%) and width (by 6%) than those of controls, but the decrease in cell length and width were proportionate; that is, the erythrocyte length to width ratios in transgenics were similar to those in non-transgenic controls (Table 1). However, compared with non-transgenic controls, erythrocytes in transgenic carp had significantly smaller surface area (9%) and volume (14%), and larger surface to volume ratios (6%).

Table 1 Erythrocyte count and dimensions for 'all-fish' growth hormone transgenic carp and non-transgenic controls

Variable (units)	Controls* (n = 23)	Transgenics* (n = 20)	Trans/con ratio	P value
Erythrocyte count (million mm^{-3})	1.25 ± 0.04	1.32 ± 0.03	1.06	0.14
Cell length (μm)	13.37 ± 0.10	12.91 ± 0.12	0.97	< 0.01
Cell width (μm)	8.61 ± 0.04	8.12 ± 0.06	0.94	< 0.01
Cell length–width ratio	1.56 ± 0.01	1.60 ± 0.02	1.03	0.10
Cell surface area ($μm^2$)	90.38 ± 0.83	82.25 ± 0.96	0.91	< 0.01
Cell volume ($μm^3$)	520.99 ± 6.64	446.77 ± 7.59	0.86	< 0.01
Cell surface–volume ratio	0.18 ± 0.00	0.19 ± 0.00	1.06	< 0.01

*Each value for cell dimensions represents the mean of those means taken of 30 erythrocyte dimensions for each fish

3.2 Physiological response to exhaustive exercise

At sampling, there were no significant differences ($P > 0.05$) in weight between transgenic carp (n = 32 and 26 weeks old) and non-transgenic controls (n = 34 and 26 weeks old) (149.26 ± 3.17 g vs. 141.85 ± 3.02 g, respectively) but transgenic fish were significantly longer than controls (19.31 ± 0.15 cm vs. 18.55 ± 0.14 cm, respectively, $P < 0.05$).

Mean resting hematocrit values and hemoglobin concentrations were similar between the two strains and neither parameter was affected by sampling time or strain following exhaustive exercise (Fig. 1A, B). At rest, MCHC of transgenics was significantly lower than in non-transgenic controls, but there were no differences in MCHC among sampling times or between strains during recovery (Fig. 1C).

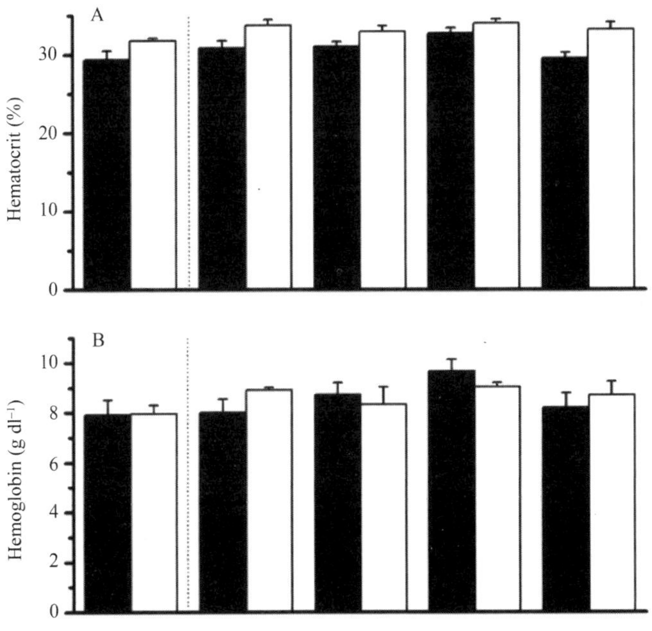

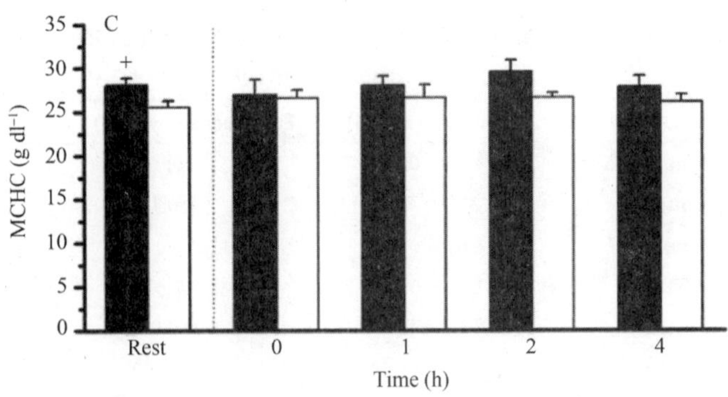

Fig. 1 Changes in hematocrit (A), hemoglobin (B) and MCHC (C) of growth hormone transgenic carp (open bars) and non-transgenic controls (black bars) before and after exhaustive exercise (dotted line). Values are means ± S.E. ($n = 6$ for each recovery group, except for the 2 h recovery group of transgenics, $n = 5$). The plus sign (+) denotes a significant difference between strains within the time interval; $P < 0.05$

There were no significant differences between strains in plasma glucose or lactate at rest (Fig. 2A, B). The exhaustive exercise evoked a significant increase in plasma glucose compared to resting levels, with a return to resting levels with 4 h of recovery, that did not differ between strains (Fig. 2A, two-way ANOVA, interaction terms $P > 0.05$). A similar pattern of increase and recovery was seen for plasma lactate (Fig. 2B, two-way ANOVA, interaction terms $P > 0.05$), except that lactate concentrations were significantly lower in transgenics than in non-transgenic controls at 1 and 2 h post-exercise.

Mean plasma cholesterol concentrations in resting transgenics were not significantly different from those in controls. Post-exercise plasma cholesterol concentrations increased significantly to 3.53 mM, then gradually declined and returned to resting levels after 2 h of recovery in transgenics. Exhaustive exercise did not induce significant changes of plasma cholesterol concentrations in non-transgenic controls (Fig. 3A).

Mean resting triglyceride concentrations were significantly greater in transgenics than in non-transgenic controls (Fig. 3B). Plasma triglyceride concentrations in transgenics were significantly evaluated immediately following exhaustive exercise (0 and 1 h), but had returned to resting levels at 2 h post-exercise; no such effect was observed in the controls. At 0 and 1 h post-exercise, the plasma triglyceride concentrations were greater in transgenics than in non-transgenic controls, although the differences were only significant at 1 h post-exercise (Fig. 3B).

Mean plasma protein concentrations in transgenics and in non-transgenic controls did not differ from each other at rest. Following exercise, there were no significant changes in plasma protein concentrations for transgenics, but the values were significantly lower than resting levels in non-transgenic controls at 2 h and 4 h post-exercise (Fig. 3C).

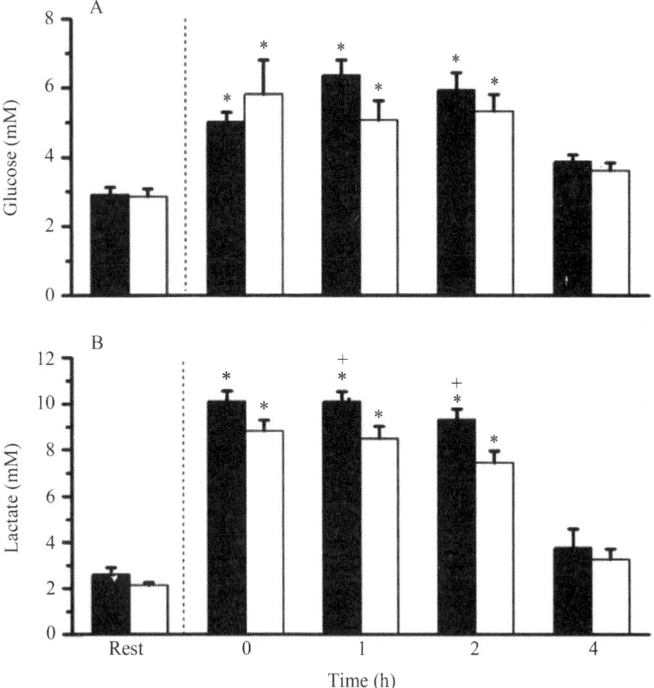

Fig. 2 Changes in plasma glucose (A) and lactate (B) of 'all-fish' growth hormone transgenic carp (open bars) and non-transgenic controls (black bars) before and after exhaustive exercise (dotted line). Values are means ± S.E. ($n = 6$ for each recovery group, except for the 2 h recovery group of transgenics, $n = 5$). An asterisk (*) denotes a significant difference from fish at rest, and the plus sign (+) denotes a significant difference between strains within the time interval; $P < 0.05$

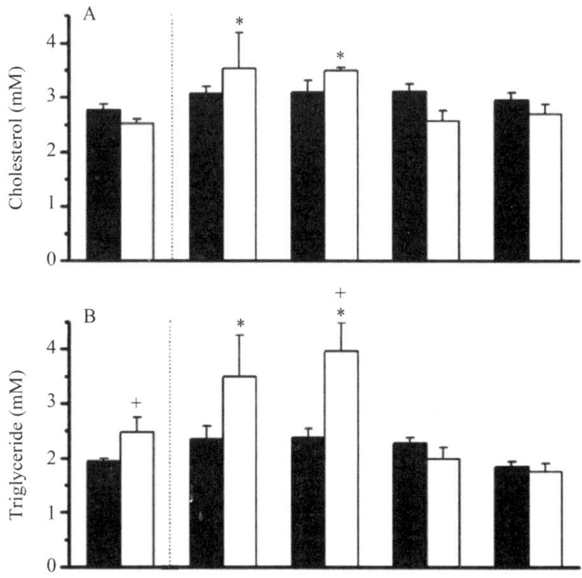

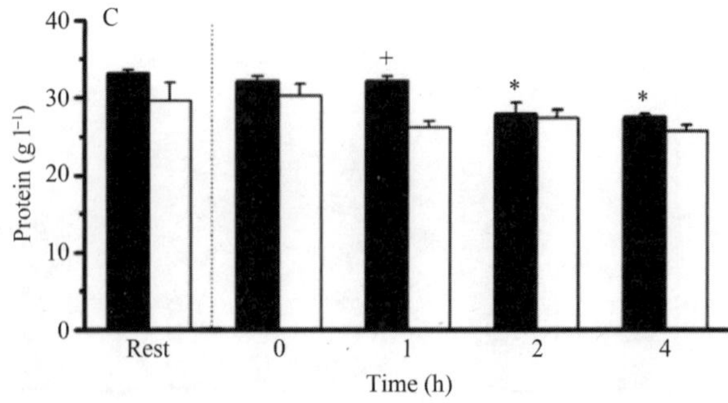

Fig. 3 Changes in plasma cholesterol (A), triglyceride (B) and protein (C) of 'all-fish' growth hormone transgenic carp (open bars) and non-transgenic controls (black bars) before and after exhaustive exercise (dotted line). Values are means ± S. E. ($n = 6$ for each recovery group, except for the 2 h recovery group of transgenics, $n = 5$). An asterisk (*) denotes a significant difference from fish at rest, and the plus sign (+) denotes a significant difference between strains within the time interval; $P < 0.05$

4. Discussion

4.1 Erythrocyte measurements

In the present study, 'all-fish' GH transgenic carp had significant decreases in all erythrocyte variables including cell length, width, surface area and volume; however, the cell surface to volume ratio was greater. Similar results were reported in fast-growing transgenic Atlantic salmon by Cogswell *et al.* (2002), who speculated that the alterations in transgenic Atlantic salmon erythrocyte morphology were likely an adaptive mechanism to meet increased metabolic requirements because the increase in the cell surface area to volume ratio would facilitate more rapid oxygen diffusion into and out of the cells (Benfey, 1999, Lay and Baldwin, 1999, Cogswell *et al.*, 2002;). This could apply to 'all-fish' GH transgenic carp as well, which have also been shown to have elevated routine metabolic rates (Guan *et al.*, 2008).

4.2 Physiological response to exhaustive exercise

At rest, 'all-fish' GH transgenic carp had similar hematocrit and hemoglobin, but lower MCHC, compared to non-transgenic controls, with all values within normal ranges for carp (Schwaiger *et al.*, 2000; Wojtaszek *et al.*, 2002; Harikrishnan *et al.*, 2003; Walencik and Witeska, 2007). However, Wang *et al.* (2005) found that 'all-fish' GH transgenic carp had higher hematocirts than age-matched siblings, but transgenics were also significantly heavier and body weight has been shown to affect hematocrit (Houston, 1990, 1997). None of these hematological patterns changed immediately following exhaustive exercise or during recovery in either strain in our study. Since there is a lack of reports on GH transgenic fish responding to exhaustive exercise, we could not compare our results with other studies. However, hematological parameters of GH transgenic Atlantic salmon showed little response to air exposure stress, except that hemoglobin increased significantly (Deitch *et al.*, 2006). Therefore, our results suggest that blood hematology parameters do not change drastically in GH transgenic fish similar to that noted in non-transgenic controls in response to exhaustive exercise.

Resting plasma glucose and lactate concentrations did not differ between transgenic and non-transgenic carp, and values were comparable to those from other studies on carp (Pottinger, 1998; Ruane *et al.*, 2001, 2002). Both groups exhibited a typical secondary metabolic stress response to exhaustive exercise, as indicated by elevated plasma glucose and lactate concentrations (Barton, 2002; Iwama *et al.*, 2004; Portz *et al.*, 2006), but transgenic fish appeared to clear lactate more quickly than controls. This may reflect enhanced aerobic respiration facilitated by the greater relative surface area to volume ratio of their erythrocytes (Milligan *et al.*, 2000; Kieffer *et al.*, 2001).

Compared with non-transgenic controls, transgenic carp showed similar resting plasma cholesterol concentrations and significantly higher resting plasma triglyceride concentrations. In Arctic charr, similar plasma cholesterol concentrations were found between GH transgenics and controls, but the transgenic fish had decreased plasma triglyceride concentrations (Krasnov *et al.*, 1999). Previous studies in carp (Ruane *et al.*, 2001) and other fish species (Lidman *et al.*, 1979; Ellsaesser and Clem, 1987) as well as our current results in control carp show that plasma cholesterol and triglyceride levels are not responsive to acute stress. A very interesting result of the present study was that the plasma cholesterol and triglyceride concentrations in transgenic fish both increased significantly immediately following exercise (0 h) and at 1 h post-exercise, then returned to resting levels by 2 h of recovery. Previous studies indicated that circulating lipids come from three sources: (1) newly absorbed from food; (2) recently processed in the liver and transported in the blood; and (3) or mobilized from storage sites (Sheridan, 1988, 1990). Many fishes rely on oxidation of lipid released from muscle adipose tissue to fuel recovery (Wang *et al.*, 1994; Keins and Richter, 1998; Richards *et al.*, 2002a, b). Therefore, the elevated plasma triglyceride and cholesterol concentrations in transgenic carp may indicate that these fish mobilize more lipids to meet the increased energy demand from the exhaustive exercise disturbance and that more esterification of nonesterified fatty acids into triglyceride occurred in the blood. Taken together, the findings may also suggested that transgenic carp showed greater reliance on lipid (especially triglyceride) to fuel basal and post-exercise metabolism than non-transgenic controls. However, further work on muscle metabolism (e.g. glycogen and triglycerides) and/or aerobic metabolic rate should be undertaken to determine the exact roles of lipid and erythrocyte surface area to volume ratio on muscle recovery in fast-growing transgenic fish. If they both facilitate faster recovery from exhaustive exercise, it would give these fish an advantage on predator-prey interactions compared with non-transgenic carp.

In this study, plasma protein concentrations did not change following exhaustive exercise in transgenics, but were significantly lower than resting levels from 2 h post-exercise in non-transgenic carp. Several studies have reported that crowding (Yin *et al.*, 1995; Ruane *et al.*, 2002) or infection by pathogens (Harikrishnan *et al.*, 2003) reduce total plasma protein concentrations in carp. Significant changes in the amounts and activities of the serum proteins in fish during stress suggest that stressed fish may have decreased disease resistance and become more susceptible to infection (Yin *et al.*, 1995). Therefore, if the difference in patterns of plasma protein displayed in response to exhaustive exercise reflected the difference in post-stress disease resistance between two strains, further study is required.

5. Conclusion

'All-fish' GH transgenic carp had significantly smaller erythrocyte dimensions and larger erythrocyte surface to volume ratios compared with non-trangenics. The patterns of measured hematological characteristics in transgenic carp responding to exhaustive exercise were not different from those in non-transgenic controls, with the exception that transgenic fish exhibited expedited recovery in plasma lactate and greater reliance on lipid (especially triglyceride) to fuel basal and post-exercise metabolism. However, further research is required to clarify the relationship between erythrocyte morphology and lipid mobilization to response to exhaustive exercise and muscle recovery.

Acknowledgement This work was financially supported by the Development Plan of the State Key Fundamental Research of China (Grant No. 2007CB109205, 2009CB118804) and National Natural Science Foundation (Grant No. 30930069). We also thank three anonymous referees for their helpful comments on early versions of this manuscript.

References

Baker, D.W., Wood, A.M., Litvak, M.K., Kieffer, J.D., 2005. Haematology of juvenile *Acipenser oxyrinchus* and *Acipenser brevirostrum* at rest and following forced activity. J. Fish Biol. 66, 208-221.

Barton, B.A., 2002. Stress in fishes: a diversity of responses with particular reference to changes in circulating corticosteroids. Integr. Comp. Biol. 42, 517-525.

Benfey, T.J., 1999. The physiology and behavior of triploid fishes. Rev. Fish. Sci. 7, 36-67.

Cal, R.M., Vidal, S., Camacho, T., Piferrer, F., Guitian, F.J., 2005. Effect of triploidy on turbot haematology. Comp. Biochem. Physiol. 141A, 35-41.

Cogswell, A.T., Benfey, T.J., Sutterlin, A.M., 2002. The hematology of diploid and triploid transgenic Atlantic salmon (*Salmo salar*). Fish Physiol. Biochem. 24, 271-277.

Deitch, E.J., Fletcher, G.L., Petesen, L.H., Costa, I.A.S.F., Shears, M.A., Driedzic, W.R., Gamperl, A.K., 2006. Cardiorespiratory modifications, and limitations, in post-smolt growth hormone transgenic Atlantic salmon *Salmo salar*. J. Exp. Biol. 209, 1310-1325.

Devlin, R.H., Biagi, C.A., Yesaki, T.Y., 2004. Growth, viability and genetic characteristics of GH transgenic coho salmon strains. Aquaculture 236, 607-632.

Devlin, R.H., Sundstrom, L.F., Muir, W.M., 2006. Interface of biotechnology and ecology for environmental risk assessments of transgenic fish. Trends Biotechnol. 24, 89-97.

Duan, M., Zhang, T., Hu, W., Sundstroms, L.F., Wang, Y., Li, Z., Zhu, Z., 2009. Elevated ability to compete for limited food resources by 'all-fish' growth hormone transgenic common carp *Cyprinus carpio*. J. Fish Biol. 75, 1459-1472.

Ellsaesser, C.F., Clem, L.W., 1987. Blood serum chemistry measurements of normal and acutely stressed channel catfish. Comp. Biochem. Physiol. 88A, 589-594.

Fu, C., Cui, Y., Hung, S.S.O., Zhu, Z., 1998. Growth and feed utilization by F_4 human growth hormone transgenic carp fed diets with different protein levels. J. Fish Biol. 53, 115-129.

Fu, C., Hu, W., Wang, Y., Zhu, Z., 2005. Developments in transgenic fish in the People's Republic of China. Rev. Sci. Tech. 24, 299-307.

Fu, C., Li, D., Hu, W., Wang, Y., Zhu, Z., 2007a. Growth and energy budget of F_2 'all-fish' growth hormone gene transgenic common carp. J. Fish Biol. 70, 347-361.

Fu, C., Li, D., Hu, W., Wang, Y., Zhu, Z., 2007b. Fast-growing transgenic common carp mounting compensatory growth. J. Fish Biol. 71 (Suppl B), 174-185.

Guan, B., Hu, W., Zhang, T.L., Wang, Y.P., Zhu, Z.Y., 2008. Metabolism traits of 'all-fish' growth hormone transgenic common carp (*Cyprinus carpio* L.). Aquacultrue 284, 217-223.

Hallerman, E.M., Mclean, E., Fleming, I.A., 2007. Effects of growth hormone transgenes on the behavior and welfare of aquacultured fishes: a review identifying research needs. Appl. Anim. Behav. Sci. 104, 265-294.

Harikrishnan, R., Nisha Rani, M., Balasundaram, C., 2003. Hematological and biochemical parameters in common carp, *Cyprinus carpio*, following herbal treatment for *Aeromonas hydrophila* infection. Aquaculture 221, 41-50.

Houston, A.H., 1990. Blood and circulation. In: Schreck, C.B., Moyle, P.B. (Eds.), Methods for Fish Biology. American Fisheries Society, Bethesda, Maryland, pp. 273-334.

Houston, A.H., 1997. Review: are the classical hematological variables acceptable indicators of fish health? Trans. Am. Fish. Soc. 126, 879-894.

Hu, W., Wang, Y.P., Zhu, Z.Y., 2007. Progress in the evaluation of transgenic fish for possible ecological risk and its containment strategies. Sci. China B 50, 573-579.

Iwama, G.K., Afonso, L.O.B., Todgham, A., Ackerman, P., Nakano, K., 2004. Are hsps suitable for indicating stressed states in fish? J. Exp. Biol. 207, 15-19.

Keins, B., Richter, E.A., 1998. Utilization of skeletal muscle triacylglycerol during postexercise recovery in humans. Am. J. Physiol. Endocrinol. Metab. 275, 332-337.

Kieffer, J.D., 2000. Limits to exhaustive exercise in fish. Comp. Biochem. Physiol. 126A, 161-179.

Kieffer, J.D., Wakefield, A.M., Litvak, M.K., 2001. Juvenile sturgeon exhibit reduced physiological responses to exercise. J. Exp. Biol. 204, 4281-4289.

Krasnov, A., Ågrenb, J.J., Pitkänena, T.I., Mölsä, H., 1999. Transfer of growth hormone (GH) transgenes onto Arctic charr (*Salvelinus alpinus* L.) II. Nutrient partitioning in rapidly growing fish. Genet. Anal. Biomol. Eng. 15, 99-105.

Lay, P.A., Baldwin, J., 1999. What determines the size of teleost erythrocyte? Correlations with oxygen transport and nuclear volume. Fish Physiol. Biochem. 20, 31-35.

Li, D., Fu, C., Hu, W., Wang, Y., Zhu, Z., 2007. Rapid growth cost in 'all-fish' growth hormone gene transgenic carp: reduced critical swimming speed. Chin. Sci. Bull. 52, 923-926.

Li, D., Hu, W., Wang, Y., Zhu, Z., Fu, Z., 2009. Reduced swimming abilities in fast-growing transgenic common carp *Cyprinus carpio* associated with their morphological variations. J. Fish Biol. 74, 186-197.

Lidman, U., Dave, G., Johansson-Sjobeck, M.L., Larsson, A., Lewander, K., 1979. Metabolic effects of cortisol in the European eel, *Anguilla anguilla* (LeSueur). Comp. Biochem. Physiol. 63B, 339-344.

Maclean, N., Laight, J., 2000. Transgenic fish: an evaluation of benefits and risks. Fish Fish. 1, 146-172.

Milligan, C.L., 1996. Metabolic recovery from exhaustive exercise in rainbow trout. Comp. Biochem. Physiol. 113A, 51-60.

Milligan, L.C., Hooke, G.B., Johnson, C., 2000. Sustained swimming at low velocity following a bout of exhaustive exercises enhances metabolic recovery in rainbow trout. J. Exp. Biol. 203, 921-926.

Niiler, E., 2000. FDA, researchers consider first transgenic fish. Nat. Biotechnol. 18, 143.

Portz, D.E., Woodley, C.M., Cech, J.J., 2006. Stress-associated impacts of short holding on fishes. Rev. Fish Biol. Fish. 16, 125-170.

Pottinger, G.T., 1998. Changes in blood cortisol, glucose and lactate in carp retained in angler's keepnets. J. Fish Biol. 53, 728-742.

Richards, J.G., George, J.F., Heigenhauser, G.J.F., Wood, C.M., 2002a. Lipid oxidation fuels recovery from exhaustive exercise in white muscle of rainbow trout. Am. J. Physiol. Regul. Integr. Comp. Physiol. 282, 89-99.

Richards, J.G., Mercad, A.J., Clayton, C.A., Heigenhauser, G.J.F., Wood, C.M., 2002b. Substrate utilization during graded aerobic exercise in rainbow trout. J. Exp. Biol. 205, 2067-2077.

Ruane, N.M., Huisman, E.A., Komen, J., 2001. Plasma cortisol and metabolite level profiles in two isogenic strains of common carp during confinement. J. Fish Biol. 59, 1-12.

Ruane, N.M., Carballo, E.C., Komen, J., 2002. Increased stocking density influences the acute physiological stress response of common carp *Cyprinus carpio* (L.). Aquac. Res. 33, 777-784.

Schwaiger, J., Spieser, O.H., Bauer, C., Ferling, H., Mallow, U., Kalbfus, W., Negele, R.D., 2000. Chronic toxicity of nonylphenol and ethinylestradiol: haematological and histopath-ological effects in juvenile Common carp (*Cyprinus carpio*). Aquat. Toxicol. 51, 69-78.

Sheridan, M.A., 1988. Lipid dynamics in fish: aspects of absorption, transportation, deposition and mobilization. Comp. Biochem. Physiol. 90B, 679-690.

Sheridan, M.A., 1990. Regulation of lipid metabolism in poikilothermic vertebrates. Comp. Biochem. Physiol. 107B, 495-508.

Walencik, J., Witeska, M., 2007. The effects of anticoagulants on hematological indices and blood cell morphology of common carp (*Cyprinus carpio* L.). Comp. Biochem. Physiol. 146C, 331-335.

Wang, Y., Heigenhauser, G.J.F., Wood, C.M., 1994. Integrated responses to exhaustive exercise and recovery in rainbow trout white muscle: acid–base, phosphogen, carbohydrate, lipid, ammonia, fluid volume and electrolyte metabolism. J. Exp. Biol. 195, 227-258.

Wang, Y., Hu, W., Wu, G., Sun, Y., Chen, S., Zhang, F., Zhu, Z., Feng, J., Zhang, X., 2001. Genetic analysis of 'all-fish' growth hormone gene transferred carp (*Cyprinus carpio* L.) and its F_1 generation. Chin. Sci. Bull. 46, 1174-1177.

Wang, W., Wang, Y., Hu, W., Li, A., Cai, T., Zhu, Z., Wang, J., 2005. Investigation on several haematological indexes of growth hormone transgenic carp. High Technol. Lett. 15, 89-93.

Wang, W.B., Wang, Y.P., Hu, W., Li, A.H., Cai, T.Z., Zhu, Z.Y., Wang, J.G., 2006. Effects of the 'all-fish' growth hormone transgene expression on non-specific immune functions of common carp, *Cyprinus carpio* L. Aquaculture 259, 81-87.

Wojtaszek, J., Dziewulska-Szwajkowska, D., Łoziska-Gabska, M., Adamowicz, A., Dugaj, A., 2002. Hematological effects of high dose of cortisol on the carp (*Cyprinus carpio* L.): cortisol effect on the carp blood. Gen. Comp. Endocrinol. 125, 176-183.

Wood, C.M., 1991. Acid-base and ion balance, metabolism and their interactions after exhaustive exercise in fish. J. Exp. Biol. 160, 285-308.

Yin, Z., Lam, T.J., Sin, Y.M., 1995. The effect of crowding stress on the non-specific immune response in fancy carp (*Cyprinus carpio* L.). Fish Shellfish Immunol. 5, 519-528.

Zbikowska, H.M., 2003. Fish can be first-advances in fish transgenesis for commercial applications. Transgenic Res. 12, 379-389.

Zhong, S., Wang, Y.P., Pei, D.S., Luo, D.J., Liao, L.J., Zhu, Z.Y., 2009a. A one-year investigation of the relationship between serum GH levels and the growth of F_4 transgenic and non-transgenic common carp *Cyprinus carpio*. J. Fish

Biol. 75, 1092-1100.

Zhong, S., Luo, D.J., Wang, Y.P., Chen, Z., Guan, B., Liao, L.J., Zhu, Z.Y., 2009b. Study on serum GH expression of transgenic and control common carp under hungry condition. Acta Hydrob. Sin. 6, 1046-1050.

Zhu, Z., Li, G., He, L., Chen, S., 1985. Novel gene transfer into the fertilized eggs of goldfish (*Carassius auratus* L. 1758). J. Appl. Ichthyol. 1, 31-34.

转“全鱼”生长激素基因鲤对高强运动的血液学反应

李德亮[1,2]　傅萃长[1,3]　汪亚平[1]　朱作言[1]　胡　炜[1]

1 中国科学院水生生物研究所，淡水生态与生物技术国家重点实验室，武汉　430072

2 湖南农业大学，动物科技学院，长沙　410128

3 复旦大学，生物多样性科学研究所，生物多样性和生态工程教育部重点实验室，上海　200433

摘　要　本研究比较了转“全鱼”生长激素基因鲤和非转基因鲤疲劳运动的血液学反应。转基因鱼的红细胞(长度、宽度、表面积和体积)比非转基因鱼小。正常状态下，转基因鱼和非转基因鱼的血球密度，血液的血红蛋白浓度，血浆的葡萄糖、乳糖、胆固醇和蛋白质浓度都没有显著差异，但其红细胞平均血红蛋白浓度(MCHC)显著低于对照鱼，血浆甘油三酯浓度显著高于对照鱼。在两个品系中，疲劳运动都不会改变血球密度、血红蛋白浓度和MCHC。虽然运动后1 h和2 h转基因鲤血浆的葡萄糖和乳糖浓度低于对照鱼，但其恢复过程与对照鱼相同。运动后1 h转基因鲤血浆胆固醇和甘油三酯含量显著升高，但对照鱼并没有升高。血浆蛋白质含量在转基因鱼中不受疲劳性运动的影响，但其在非转基因鱼中从运动后2 h开始降低。结果表明，转植“全鱼”生长激素基因显著影响了鲤的红细胞形态，血液特别是脂肪代谢对疲劳运动的响应模式也发生了改变。转基因鲤更倾向于依赖甘油三酯和较大的红细胞表面积/体积比从疲劳运动中恢复，该结论还需要进一步开展肌肉代谢及有氧代谢研究来验证。

Acute and Chronic Un-Ionized Ammonia Toxicity to 'All-Fish' Growth Hormone Transgenic Common Carp (*Cyprinus carpio* L.)

GUAN Bo[1,2] HU Wei[1] ZHANG TangLin[1] DUAN Ming[1,2] LI DeLiang[1,2] WANG YaPing[1] ZHU ZuoYan[1]

1 State Key Laboratory of Freshwater Ecology and Biotechnology, Institute of Hydrobiology, Chinese Academy of Sciences, Wuhan 430072

2 Graduate School of Chinese Academy of Sciences, Beijing 100049

Abstract Ammonia is toxic to fish in natural and artificial waters. We evaluated the acute (96 h) and chronic (21 d) toxicity of un-ionized ammonia to GH transgenic common carp (*Cyprinus carpio* L.) and non-transgenic common carp using a static-renewal bioassay. The 24, 48, 72 and 96 h median lethal concentrations (LC_{50}) of un-ionized ammonia were slightly lower in transgenic carp (2.64, 2.44, 2.28 and 2.16 mg N/L, respectively) than in non-transgenic carp (2.70, 2.64, 2.52 and 2.33 mg N/L, respectively). Similarly, the median lethal time (LT_{50}) was significantly shorter for transgenic carp (1.41, 7.91 and 117.42 h) than for non-transgenic common carp (2.53, 14.06 and 150.44 h) following exposure to 3.86, 3.29, or 2.09 mg N/L, respectively. Moreover, the mortality of transgenic carp was significantly higher than that of non-transgenic carp at all un-ionized ammonia concentrations ((0.91 ± 0.12), (0.48 ± 0.06) and (0.12 ± 0.01) mg N/L) during the 21 d chronic toxicity test. Our results suggest that GH transgenic carp are less tolerant of un-ionized ammonia than non-transgenic carp. Our data are useful for evaluating potential environmental risk, optimizing stocking density in intensive aquaculture and establishing water quality criteria for ammonia in aquaculture.

Keywords transgenic common carp; growth hormone; un-ionized ammonia; toxicity; mortality

Ammonia is widespread in the aquatic environment due to agricultural run-off, industrial run-off and decomposition of biological waste. High levels of ammonia nitrogen may occur during intensive fish culture as a result of excretion and decomposition of feces and residual feed. Ammonia exists primarily in two forms in aqueous solutions, un-ionized ammonia (NH_3) and the ammonium ion (NH_4^+) [1]. Un-ionized ammonia, a neutral molecule, is the more toxic form because it is able to diffuse across the epithelial membranes of aquatic organisms much more readily than the charged ammonium ion [1]. Numerous studies have investigated acute and chronic ammonia toxicity in different fish species. High levels of ammonia are acutely

文章发表于 *Chinese Science Bulletin*, 2010, 55(35):4032-4036

toxic to fish, causing convulsions, coma and death [2]. Low levels of ammonia are also toxic to fish, causing chronic sublethal effects, including impairment of swimming performance [3], histopathologic alteration of the gills, liver and kidney [4], inhibition of the immune response and increased susceptibility to pathogens [5,6] and negative effects on growth [7]. Furthermore, several studies have documented a relationship between stocking density, growth and survival and water quality (including ammonia) during intensive fish culture in recirculating systems or ponds [8,9]. Taken together, these studies suggest that ambient ammonia represents an environmental stressor in both natural and artificial waters.

Although over 30 kinds of transgenic fish have been generated to date, none have been cultivated commercially [10,11]. Biosafety issues have hindered the marketability of transgenic fish and they remain unavailable in the market-place [12,13]. Growth hormone (GH) transgenic fish have enhanced growth and feed conversion efficiency, traits that are advantageous during intensive culture. However, GH acts on many processes in addition to growth. For example, transgenic GH genes have a range of pleiotropic effects in fish, including changes in non-specific immune function [14], swimming ability [15], skeletal structure [16] and sea water adaptability [17]. It is unclear whether these changes affect the survival of transgenic fish that are released into the wild, intentionally or otherwise. The survivability of transgenic fish is important part for objectively evaluating the potential ecological risk [12,13]. Moreover, the fish culture model and technical standards of fast growing transgenic fish compatible with good quality traits, is very important to ensure healthy development of their future industrialization. Ambient ammonia is a well-known environmental stressor. However, little is known about the ammonia tolerance of GH transgenic fish. We evaluated the acute and chronic toxicity of un-ionized ammonia to 'all-fish' GH transgenic common carp. Our results provide insight into the potential environmental risk, as well as providing guidance for optimizing stocking densities and establishing water quality criteria for ammonia in aquaculture.

1. Materials and Methods

1.1 Experimental fish

The 'all-fish' gene construct *pCAgcGH* contains the grass carp (*Ctenopharyngodon idellus*) GH gene (*gcGH*) driven by the *β-actin* gene promoter in common carp. P_0, F_1, F_2, F_3, F_4 and F_5 'all-fish' GH transgenic carp were produced as reported by Guan *et al.* [18]. The F_4 transgenic carp and sibling carp used in this experiment were produced by cross-fertilization between F_3 transgenic males and normal non-transgenic females on April 14, 2006. The offspring were reared in field ponds after initial feeding. Approximately 2000 60-d-old F_4 transgenic juveniles and sibling juveniles were transferred to an indoor water-recycling fish-rearing system on June 15, 2006. The fish were acclimated to the indoor water-recycling fish-rearing system for at least 1 month before the study began. During the acclimation period, the fish were fed twice daily with an approximately equal ration. Dissolved oxygen was >7 mg O_2/L, total ammonia nitrogen was <0.4 mg/L, pH ranged between 7.0 and 7.6, water temperature ranged between 26 and 29.5℃ and the fish were held under a 12 h light/12 h dark photoperiod, controlled by artificial lighting.

We prepared total DNA from each fish using PCR templates based on the Chelex 100 method [19]. The F_4 and F_5 *pCAgcGH* transgene positive carp were identified by PCR using the primers GHF1 (5′-CGGTTTTCTCATTCATTTACAGT-3′) and GHR1 (5′-TACCATCTACAACATCGTTCCTA-3′) under the following conditions: 5 min pre-denaturation at 95℃ followed by 35 cycles of a 10 s denaturation at 95℃, 30 s annealing at 56℃ and 15 s extension at 72℃. This was followed by a final extension for 5 min at 72℃. The PCR products were analyzed on an agarose electrophoresis gel to isolate the DNA band of interest (283 bp).

We used F_4 transgenic and sibling carp for the acute ammonia toxicity trials in 2006 and F_5 transgenic and sibling carp for the chronic ammonia toxicity trials in 2007. The F_5 transgenic carp were produced and acclimated to the experimental conditions following the methods outlined above for the F_4 transgenic carp.

1.2 Water quality analysis

Dissolved oxygen (DO) and temperature were measured using a YSI 550A Dissolved Oxygen Analyzer (YSI: Yellow Springs, OH, USA). We measured water pH using a Delta 320 pH meter (Mettler-Toledo, Shanghai, China). Total ammonia nitrogen concentrations were determined using Nessler's spectrophotometric method [20]. Concentrations of un-ionized ammonia and total ammonia are given in terms of nitrogen (mg N/L) [1].

The un-ionized ammonia fraction of the total ammonia concentration was calculated using the general formula for bases [21]:

$$[NH_3]=\frac{[NH_3+NH_4^+]}{1+10^{(pKa-PH)}}$$

where p*Ka* is the ammonia-total ammonia equilibrium constant and calculated as: p*Ka* = 0.09018 + 2729.92/*T* (*T* = temperature in Kelvin).

1.3 Acute ammonia toxicity trials

We assessed the effect of short-term ammonia exposure using static and renewal bioassays to estimate the median lethal concentrations (LC_{50}) and median lethal times (LT_{50}) for transgenic and non-transgenic juvenile carp. Based on a preliminary test, we prepared 6 stock test solutions (9.69, 13.67, 19.21, 27.48, 41.43 and 54.16 mg N/L, NH_3 + NH_4^+) by serial dilution of reagent grade NH_4Cl with fresh dechlorinated water. Each stock solution (~879.2 L) was prepared and stored in a 1500-L plastic tank. The actual total ammonia concentration was measured spectrophotometrically using Nessler's method. The pH of each stock solution (including the control solution) was maintained at ~8.00 by addition of 0.47 mol/L Na_2CO_3 and 1 mol/L HCl prior to the toxicity trials. The concentration of un-ionized ammonia depended on the measured total ammonia concentrations, pH and temperature. The control or transgenic groups were marked by shearing pterygiophores at 7th d prior to the experiment. The fish were starved for 2 d prior to the experiment. After 1 d of starvation, experimental fish were weighed; the mean mass of the control was (1.57 ± 0.05) g and that of transgenic fish was (1.68 ± 0.05) g (ANOVA, P >0.05). About 80 L of stock solution at each concentration were transferred to 120 L plastic tanks prior to toxicity trials. Three replicate tanks were used for each group exposed to one of 6 ammonia concentrations or control. On July 14, 2006, we randomly selected the marked fish. These fish were gently transferred the test tanks from the stock tanks. The 20 non-transgenic juveniles and 20 transgenic juveniles were transferred to the same test tank. All the tanks were aerated (DO >7.00 mg/L). We measured pH, temperature and dissolved oxygen in the test tanks at 4 h intervals. If the pH of the test solution fluctuated, it was immediately adjusted with 0.47 mol/L Na_2CO_3 and 1 mol/L HCl. We recorded mortalities every hour. Individuals were assumed to be dead when they were immobile and showed no response when touched with a glass rod. The test solutions were renewed every 24 h.

1.4 Chronic ammonia toxicity trials

We tagged ~140 F_5 transgenic carp with wire tags (Northwest Marine Technology, Inc., USA) 2 weeks before the experiment. The ~140 sibling non-transgenic carp were untagged. Both groups were acclimated to the indoor water-recycling fish-rearing system. On August 18, 2007, the chronic ammonia toxicity test experiment was conducted in 4 circular plastic tanks (diameter = 150 cm; volume = 1500 L). Each tank contained ~927 L fresh dechlorinated water. We added one of 4 concentrations of reagent grade NH_4Cl (0, 7.06, 35.28 or 70.57 g) to each tank. The actual total ammonia concentration in each tank was measured spectrophotometrically using Nessler's method. Each test solution (including the control solution) was adjusted to pH 8.00 with 0.47 mol/L Na_2CO_3 or 1 mol/L HCl. The 32 control juveniles and 32 transgenic juveniles (the mean mass of the control = (5.36 ± 1.00) g; and that of transgenic fish = (5.53 ± 1.13) g; ANOVA, P >0.05), were randomly sampled and gently transferred to the same aerated test tank. All the tanks were aerated (DO > 7.00 mg/L). We measured the pH, temperature and dissolved oxygen at 24 h intervals and the test solutions were renewed every 24 h during the 21 d trial period. These fish were fed to satiation twice daily at 8:30 and 14:30 with commercial bloodworm (midge larvae).

1.5 Data analysis

The LC_{50} or LT_{50} values and their 95% confidence intervals (CI) were calculated using the PROBIT procedure (STATISTICA 6.0). The difference between two LC_{50} (or LT_{50}) values was deemed significant if there was no overlap of the 95% CI [22]. Results are presented as the mean ± standard error. We used the chi-square test to compare the mortality (%) of transgenic and non-transgenic carp under chronic ammonia toxicity. A P value < 0.05 was regarded as significant for all statistical analyses. All analyses were conducted using STATISTICA 6.0. The data were graphed using SigmaPlot 8.0.

2. Results

2.1 Acute ammonia toxicity trials

During the 96 h acute toxicity experiment, the pH varied between 7.95 and 8.06 and the temperature varied between 28.5 and 29.5℃. Dissolved oxygen levels were relatively constant and remained close to saturation (>7.00 mg/L). The measured total ammonia concentrations were 0.21 (control), 9.69, 13.67, 19.21, 27.48, 41.43 and 54.17 mg N/L. The un-ionized ammonia test concentrations were 0.015 (control), 0.75, 0.10, 1.46, 2.09, 3.29 and 3.86 mg N/L. There was no difference (ANOVA, P = 0.11) in the mean starting weight of the transgenic (1.68 ± 0.05 g) and non-transgenic (1.57 ± 0.05 g) groups. We observed no mortality at the three lowest concentrations (≤1.46 mg NH_3-N /L) or in the non-transgenic. The 24, 48, 72 and 96 h LC_{50} for un-ionized ammonia were slightly lower in transgenic carp (2.64, 2.44, 2.28 and 2.16 mg/L) than in the non-transgenic carp (2.70, 2.64, 2.52 and 2.33 mg/L), but the differences were not significant (Table 1). The median lethal time decreased as the concentration of ammonia increased. The LT_{50} for transgenic carp (117.42, 7.91 and 1.41 h) was significantly shorter than for their sibling common carp (150.44, 14.06 and 2.53 h) when exposed to 2.09, 3.29, or 3.86 mg N/L un-ionized ammonia, respectively (Table 2).

2.2 Chronic ammonia toxicity trials

During the 21-d test period, dissolved oxygen was > 7 mg O_2/L, pH ranged between 7.9 and 8.1 and the water temperature varied between 24℃ and 25℃. There was no difference (ANOVA, P = 0.24) in the mean starting weight of the non-transgenic (5.36 ± 1.00 g) and transgenic (5.53 ± 1.13 g) fish. The measured concentrations of un-ionized ammonia were (0.010 ± 0.002) (control), (0.91 ± 0.12), (0.48 ± 0.06) and (0.12 ± 0.01) mg/L. The mortality of transgenic carp was significantly higher than that of non-transgenic carp at the same concentration of un-ionized ammonia (Figure 1).

Table 1 The un-ionized ammonia LC_{50} for transgenic and non-transgenic common carp[a)]

Experiment group	24 h LC_{50} (95% CI)*	48 h LC_{50} (95% CI)*	72 h LC_{50} (95% CI)*	96 h LC_{50} (95% CI)*
Transgenic	2.64(2.40-2.89)	2.44(2.20-2.70)	2.28(2.10-2.47)	2.16(2.02-2.32)
Control	2.70(2.48-2.93)	2.64(2.39-2.94)	2.52(2.33-2.75)	2.33(2.20-2.47)

a) *, Concentrations of un-ionized ammonia are given in mg N/L; CI, confidence interval

Table 2 The LT_{50} value for transgenic and non-transgenic carp held in different concentrations of un-ionized ammonia

Un-ionized ammonia (mg N/L)	LT_{50} value (h) of transgenic carp(95% CI) [a)]	LT_{50} value (h) of control carp(95% CI)
2.09	117.42 (110.58-126.16)*	150.44(136.80-169.77)
3.29	7.91 (6.85-8.94)*	14.06 (12.51-15.74)
3.86	1.41 (1.09-1.72)*	2.53 (2.00-3.03)

a) *, The difference between two LC_{50} values was deemed significant when the 95% confidence intervals did not overlap (CI)

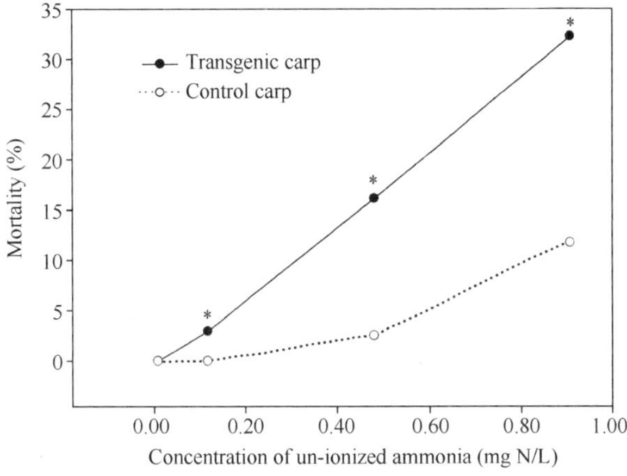

Figure 1 Mortality (%) of transgenic and non-transgenic carp under chronic ammonia toxicity. *, Significant difference between transgenic fish and non-transgenic fish (chi-square test, $P<0.05$)

3. Discussion

The median lethal times (LT_{50}) for transgenic carp were significantly shorter than for the non-transgenic common carp during the 96 h acute un-ionized ammonia toxicity test. Compared to their non-transgenic siblings, the GH transgenic carp also displayed higher mortality during the 21 d chronic un-ionized ammonia toxicity test. Furthermore, the mean body weight of the transgenic carp did not increase, whereas the non-transgenic carp gained weight throughout the study (unpublished data). Thus, our results suggest that GH transgenic common carp are more sensitive to un-ionized ammonia than normal carp.

The survival of transgenic fish is an important factor when objectively evaluating their potential ecological risk [12,13]. Ammonia is widespread in the aquatic environment and has been one of the important environmental stress factors for the survivability [2]. For example, ammonia exposure impairs the escape response in brown trout (*Salmo trutta* L.) and may also affect predation rate, social interactions and predator-prey relationships [23]. Despite the large body of research examining the effects of ammonia, little is known about the ammonia tolerance of GH transgenic fish. Our results suggest that GH transgenic common carp may be even more susceptible to ammonia than genetically unmodified carp. Thus, higher ammonia concentrations are likely to have a negative effect on the fitness of transgenic fish in natural waters.

Ammonia is one of the major limiting factors for the growth and survival of fish during intensive fish culture [24]. Therefore, it is important to understand the effects of stocking density on fish growth and the relationship between stocking density and water quality (including ammonia) in recirculating systems or ponds [8,9]. Biswas *et al.* [9] proposed that density dependent ambient ammonium is a key factor when optimizing the stocking density of common carp in small holding tanks. The current study revealed that the ammonia toxicity tolerance of GH transgenic carp is lower compared with control carp. We have also observed that GH transgenic common carp have a greater daily feed intake and a higher average oxygen uptake after feeding in comparison with non-transgenic fish [18]. Consequently, we recommend that the stocking density of GH transgenic fish be reduced during intensive fish culture. Therefore, further studies should be conducted to reveal the relationship between stocking density on growth and survival of GH transgenic fish and water quality (including ammonia).

It is intriguing that GH transgenic carp are significantly more sensitive to un-ionized ammonia than non-transgenic carp. Ammonia is toxic to all vertebrates, typically because elevated NH_4^+ displaces K^+ and depolarizes neurons, causing activation of *N*-methyl-*D*-aspartate (NMDA) type glutamate receptor (NMDA-R), which leads to an influx of excessive Ca^{2+} and subsequent cell death in the central nervous system [2,25]. Therefore, we might ask if there is an association between the over-expression of GH and NMDA-R in the brain of transgenic carp. Bhat *et al.* [26] showed that the NMDA sub (1) receptor subunit is co-localized in many cell types including LH, FSH, GH, TSH and PRL cells in the anterior pituitary of rats. Le Grevès *et al.* [27,28] found that GH elicits an increase in the hippocampal gene transcript of the GH receptor (GHR) and the NMDA receptor subunits NR1, NR2A and NR2B in rats. Furthermore, the authors noted a significant positive correlation between the levels of the GHR and the NR2B gene transcripts. Together, these observations are consistent with a hypothesis that the hormone-induced effect on the NMDA receptor subunits is mediated by GHRs [27]. Moreover, there are other lines of evidence to suggest that the NMDA receptor in the anterior pituitary can regulate GH secretion in the rhesus monkey

(*Macaca mulatta*) [29], rats [30] and rainbow trout (*Oncorhynchus mykiss*) [31]. Therefore, we hypothesize that over-expression of GH mediated directly by the GHR or indirectly by IGF-1 receptors in the brain causes an increase in the expression of the NMDA receptor. Excessive activation of NMDA receptors may be responsible for the increase in mortality in GH transgenic carp, relative to non-transgenic animals, following exposure to un-ionized ammonia.

In summary, GH transgenic common carp are less tolerant to un-ionized ammonia than non-transgenic carp. The molecular mechanism underlying the reduced ammonia toxicity tolerance traits of GH transgenic carp and optimization of stocking density of transgenic carp depended on ammonia toxicity will need further investigation.

Acknowledgements This work was supported by the National Basic Research Program of China (2007CB109205, 2009CB118804) and the National Natural Science Foundation of China (30623001).

References

1. USEPA. Update of ambient water quality criteria for ammonia–technical version-1999. EPA-823-F-99-024. USEPA, Washington DC, USA, 1999
2. Randall D J, Tsui T K N. Ammonia toxicity in fish. Mar Pollut Bull, 2002, 45: 17-23
3. Shingles A, McKenzie D J, Taylor E W, *et al.* Effects of sublethal ammonia exposure on swimming performance in rainbow trout (*Oncorhynchus mykiss*). J Exp Biol, 2001, 204: 2691-2698
4. Benli A K, Köksal G, Özkul A. Sublethal ammonia exposure of Nile tilapia (*Oreochromis niloticus* L.): Effects on gill, liver and kidney histology. Chemosphere, 2008, 72: 1355-1358
5. Das P C, Ayyappan S, Jena J K, *et al.* Acute toxicity of ammonia and its sub-lethal effects on selected haematological and enzymatic parameters of mrigal, *Cirrhinus mrigala* (Hamilton). Aquac Res, 2004, 35: 134-143
6. Liu C H, Chen J C. Effect of ammonia on the immune response of white shrimp *Litopenaeus vannamei* and its susceptibility to *Vibrio alginolyticus*. Fish Shellfish Immunol, 2004, 16: 321-334
7. El-Shafai S A, El-Gohary F A, Nasr F A, *et al.* Chronic ammonia toxicity to duckweed-fed tilapia (*Oreochromis niloticus*). Aquaculture, 2004, 232: 117-127
8. Sumagaysay-Chavoso N S, San Diego-McGlone M L. Water quality and holding capacity of intensive and semi-intensive milkfish (*Chanos chanos*) ponds. Aquaculture, 2003, 219: 413-429
9. Biswas J K, Sarkar D, Chakraborty P, *et al.* Density dependent ambient ammonium as the key factor for optimization of stocking density of common carp in small holding tanks. Aquaculture, 2006, 261: 952-959
10. Zhu Z, Li G, He L, *et al.* Novel gene transfer into the fertilized eggs of goldfish (*Carassius auratus*L. 1758). Z Angewlchthyol, 1985, 1: 31-34
11. Hallerman E M, McLean E, Fleming I A. Effects of growth hormone transgenes on the behavior and welfare of aquacultured fishes: A review identifying research needs. Appl Anim Behav Sci, 2007, 104: 265-294
12. Devlin R H, Sundstrom L F, Muir W M. Interface of biotechnology and ecology for environmental risk assessments of transgenic fish. Trends Biotechnol, 2006, 24: 89-97
13. Hu W, Wang Y P, Zhu Z Y. Progress in the evaluation of transgenic fish for possible ecological risk and its containment strategies. Sci China Ser C-Life Sci, 2007, 50: 1-7
14. Wang W, Wang Y, Hu W, *et al.* Effects of the 'all-fish' growth hormone transgene expression on non-specific

immune functions of common carp, *Cyprinus carpio* L. Aquaculture, 2006, 259: 81-87
15. Li D L, Fu C Z, Hu W, *et al.* Rapid growth cost in 'all-fish' growth hormone gene transgenic carp, Reduced critical swimming speed. Chinese Sci Bull, 2007, 52: 1501-1506
16. Ostenfeld T, Mclean H E, Devlin R H. Transgenesis changes body and head shape in Pacific salmon. J Fish Biol, 1998, 52: 850-854
17. Seddiki H, Boeuf G, Maxime V, *et al.* Effects of growth hormone treatment on oxygen consumption and seawater adaptability in Atlantic salmon parr and pre-smolts. Aquaculture, 1996, 148: 49-62
18. Guan B, Hu W, Zhang T L, *et al.* Metabolism traits of 'all-fish' growth hormone transgenic common carp (*Cyprinus carpio* L.). Aquaculture, 2008, 284: 217-223
19. Walsh P S, Metzger D A, Higuchi R. Chelex 100 as a medium for simple extraction of DNA for PCR-based typing from forensic material. Biotechniques, 1991, 10: 506-513
20. APHA (American Public Health Association, American Water Works Association and Water Pollution Control Federation). Standard Methods for the Examination of Water and Wastewater, 19th ed. New York: American Public Health Association, 1998. 1038
21. Emerson K, Russo R C, Lund R E, *et al.* Aqueous ammonia equilibrium calculations: Effect of pH and temperature. J Fish Res Board Can, 1975, 32: 2379-2383
22. Mason G A, Johnson M W, Tabashnik B E. Susceptibility of *Liriomyza sativae* and *L. trifolii* (Diptera: Agromyzidae) to permethrin and fenvalerate. J Econ Entomol, 1987, 80: 1262-1266
23. Tudorache C, Blust R, De Boeck G. Social interactions, predation behaviour and fast start performance are affected by ammonia exposure in brown trout (*Salmo trutta* L.). Aquat Toxicol, 2008, 90: 145-153
24. Colt J. Water quality requirements for reuse systems. Aquac Eng, 2006, 34: 43-156
25. Monfort P, Kosenko E, Erceg S, *et al.* Molecular mechanism of acute ammonia toxicity: Role of NMDA receptors. Neurochem Int, 2002, 41: 95-102
26. Bhat G K, Mahesh V B, Chu Z W, *et al.* Localization of the *N*-methyl-*D*-aspartate R sub(1) receptor subunit in specific anterior pituitary hormone cell types of the female rat. Neuroendocrinology, 1995, 62: 178-186
27. Le Grevès M, Steensland P, Le Grevès P, *et al.* Growth hormone induces age-dependent alteration in the expression of hippocampal growth hormone receptor and *N*-methyl-*D*-aspartate receptor subunits gene transcripts in male rats. Proc Natl Acad Sci USA, 2002, 99: 7119-7123
28. Le Grevès M, Le Grevès P, Nyberg F. Age-related effects of IGF-1 on the NMDA-, GH-and IGF-1-receptor mRNA transcripts in the rat hippocampus. Brain Res Bull, 2005, 65: 369-374
29. Rizvi S S R, Altaf S. Differential effects of *N*-methyl-*D*-aspartate receptor stimulation on growth hormone secretion at specific stages of postnatal development of the male rhesus monkey. Life Sci, 2000, 67: 783-797
30. Tena-Sempere M, Pinilla L, Gonzalez L C, *et al.* Regulation of Growth Hormone (GH) secretion by different glutamate receptor subtypes in the rat. Amino Acids, 2000, 18: 1-16
31. Holloway A C, Leatherland J F. The effects of *N*-methyl-*D*, *L*-aspartate and gonadotropin-releasing hormone on *in vitro* growth hormone release in steroid-primed immature rainbow trout, *Oncorhynchus mykiss*. Gen Comp Endocrinol, 1997, 107: 32-43

急性和慢性氨中毒对转“全鱼”生长激素基因鲤的影响

管 波[1,2] 胡 炜[1] 张堂林[1] 段 明[1,2] 李德亮[1,2]
汪亚平[1] 朱作言[1]

1 中国科学院水生生物研究所，淡水生态与生物技术国家重点实验室，武汉 430072
2 中国科学院大学，北京 100049

摘 要 自然水体和人工水体中的氨氮对鱼类是有毒的。我们采用静水更新式生物测试方法，研究了非离子态氨对转基因鲤和对照鲤的96 h急性毒性实验和21 d慢性毒性实验。非离子态氨急性毒性实验发现，在不同的胁迫时间下(24、48、72和96 h)，转基因鲤与对照鲤的非离子态氨氮半数致死浓度没有显著差异，分别为2.64、2.44、2.28、2.16 mg N/L和2.70、2.64、2.52、2.33 mg N/L；在不同的胁迫浓度下(3.86、3.29和2.09 mg N/L)，转基因鲤的半数致死时间显著短于对照鲤，分别为1.41、7.91、117.42 h和2.53、14.06、150.44 h。21 d的慢性毒性实验发现，在不同非离子态氨氮浓度胁迫下((0.91±0.12)、(0.48±0.06)和(0.12±0.01) mg N/L)，转基因鲤的死亡率均显著性高于对照鲤。上述研究表明，转基因鲤对氨氮胁迫的耐受能力比对照鲤差，此研究结果将为评估转基因鲤潜在生态风险、确定转基因鲤集约化养殖密度和制定养殖水体氨氮安全指标提供了重要科学参数。

Effects of Growth Hormone (GH) Transgene and Nutrition on Growth and Bone Development in Common Carp

Tingbing Zhu[1,2] Tanglin Zhang[1] Yaping Wang[1] Yushun Chen[3]
Wei Hu[1] Zuoyan Zhu[1]

1 State Key Laboratory of Freshwater Ecology and Biotechnology, Institute of Hydrobiology, Chinese Academy of Sciences, Wuhan
2 The University of Chinese Academy of Sciences, Beijing
3 Aquaculture and Fisheries Center, University of Arkansas, Pine Bluff, Arkansas

Abstract Limited information is available on effects of growth hormone transgene and nutrition on growth and development of aquatic animals. Here, we present a study to test these effects with growth-enhanced transgenic common carp under two nutritional conditions or feeding rations (i.e., 5% and 10% of fish body weight per day). Compared with the nontransgenic fish, the growth rates of the transgenic fish increased significantly in both feeding rations. The shape of the pharyngeal bone was similar among treatments, but the transgenic fish had relatively smaller and lighter pharyngeal bone compared with the nontransgenic fish. Calcium content of the pharyngeal bone of the transgenic fish was significantly lower than that of the nontransgenic fish. Feeding ration also affected growth rate but less of an effect on bone development. By manipulating intrinsic growth and controlling for both environment (e.g., feeding ration) and genetic background or genotype (e.g., transgenic or not), this study provides empirical evidence that the genotype has a stronger effect than the environment on pharyngeal bone development. The pharyngeal bone strength could be reduced by decreased calcium content and calcification in the transgenic carp.

Natural selection favors submaximal rather than maximal growth. The importance of body size and growth rate in ecological interactions are widely recognized, and both of them are frequently used as surrogates for fitness (Biro *et al.*, 2006). Empirical evidence for increased advantages (e.g., survival rate, fecundity, and competitive ability) in fast growing individuals and/or populations was well documented (e.g., Roff, 1992; Stearns, 1992; Urban, 2007). It is therefore difficult to explain the maintenance of genetic variation in growth rates (Mangel and Stamps, 2001). One explanation is that there is a trade - off related to the fast growth. This cost may pose a selective pressure that makes fast growth animals evolve to the submaximization in growth rate (Arendt, 1997).

Costs of fast growth can be roughly grouped as: behavioral, physiological, or

文章发表于 *Journal of Experimental Zoology Part A Ecological Genetics & Physiology*, 2013, 319(8):451-460 developmental (Arendt *et al.*, 2001). Behavioral costs include increased risk taking (Nicieza and Metcalfe, 1999; Munch and Conover, 2003), higher activity level (Sundt-Hansen *et al.*, 2009), etc. Physiological costs include increased energy investment on growth (Lankford *et al.*, 2001), weakened swimming ability (Billerbeck *et al.*, 2001), etc. Developmental costs include reduced acquisition of mature function in skeletal muscle (Ricklefs *et al.*, 1994). Costs of fast growth were also found in growth enhanced transgenic common carp *Cyprinus carpio* (L. 1758). For example, the transgenic carp show poorer swimming ability (Li *et al.*, 2007) and show increased routine metabolic rates (Guan *et al.*, 2008), but they are more vulnerable to predators (Duan *et al.*, 2010) and more susceptible to un-ionized ammonia stress (Guan *et al.*, 2010). However, little information is available on developmental cost of fast growth in transgenic carp limiting our ability to evaluate the effect of growth rate on fitness.

Skeletal development as a cost of rapid growth has received limited attention. Compared with a slow growing strain, a fast growing strain of pumpkinseed sunfish *Lepomis gibbosus* (L. 1758) had a delayed onset of mineralization in their cranial bones of approximately 2 days (Arendt and Wilson, 2000). Subsequent research on pumpkinseed sunfish showed weaker scale strength in the fast growing strain (Arendt *et al.*, 2001). In addition, variations were found in body and head shape of growth enhanced transgenic coho salmon *Oncorhynchus kisutch* (Walbaum, 1792) compared with the control (Ostenfeld *et al.*, 1998). The fastest growth transgenic coho salmon showed overgrowth of cartilage, which resulted in abnormal morphologic changes of head, fin, jaw, and opercula (Devlin *et al.*, 1995).

Bone development can also be affected by environmental factors such as nutrient condition. For example, supplementation of dietary minerals can increase vertebral strength and reduce the prevalence of vertebral deformities in fast growing Atlantic salmon *Salmo salar* (L. 1758) smolt (Fjelldal *et al.*, 2009). Negative effects on bone development can occur under extremely high nutrient condition. Fast growth induced by high nutrient intake alone can cause severe lameness, bone defects, and deformity in animals (e.g., poultry) that have not been selected for fast growth (Julian, 1998). However, other studies have found no effects of nutrition on bone development in some other animals (Walstra, 1980; Ott and Asquith, 1986).

The present study aims to determine the effects of both GH (growth hormone) transgene and nutrition on growth and pharyngeal bone development of common carp. Carp feed primarily on gastropod molluscs that they crush using their pharyngeal jaws. This makes the pharyngeal bone one of the most important bones for feeding in carp. A hypothesis was raised that the pharyngeal bone would be underdeveloped as a cost of fast growth in GH transgenic common carp. Related research questions include: (1) what are the effects of GH transgene and nutrition on growth of common carp? (2) What are the effects of GH transgene and nutrition on pharyngeal bone development in common carp? (3) What is the relationship between growth and pharyngeal bone development in common carp?

Materials and Methods

Fast-growing genotypes of P_0 GH-transgenic *C. carpio* were initially produced by microinjection of the *pCAgcGH* into fertilized eggs of *C. carpio* (Yellow River variety) in 2000 (Wang *et al.*, 2001). The all-fish gene construct *pCAgcGH* was a recombinant construct of grass carp *Ctenopharyngodon idella* (Valenciennes 1844) growth hormone gene (*gcGH*), whose expression is driven by the β-actin gene promoter of *C. carpio* (*pCA*) (Wang *et al.*, 2001). Ttransgenic carp were produced as follows: P_0 transgenic♂ (sperm revealed to be *pCAgcGH* positive) × wild type♀, then the F1, F2, F3, and F4 generation were produced (Zhong *et al.*, 2012). The body weight of the F1, F2, and F3 transgenic carp were as 1.6 times (Wang *et al.*, 2001), 1.8-2.5 times (Li, 2008), and 1.4-1.9 times (Li *et al.*, 2007) as that of the nontransgenic carp respectively, under hatchery-reared conditions. The transgenic carp used in the present study are the F5 generation which were produced by cross-breeding between wild type female and F4 homozygote transgenic male of a fast-growing transgenic strain on April 2009, in Institute of Hydrobiology, Chinese Academy of Sciences. The nontransgenic carp were wild type carp and hatched at the same time under normal conditions. After the first feeding, fry were transported to Duofu Technology Farm (Wuhan, Hubei, China) and stocked in 10 rectangular cement pools (4×2×1m) with 1,500 individuals each pool. The nontransgenics were fed to satiation with frozen chironomid *Chironomus* sp. larvae twice a day, whereas the transgenics were fed roughly half of the amount every day to allow size-matching with the nontransgenics. This feeding regime was maintained until their gape size matching the size (1.61 ± 0.12 mm, mean ± 1 SE) of the commercial pellets.

Growth trial

The experimental fish (transgenics and nontransgenics) were fed in low and high rations, respectively. The low ration was 5% of body weight per day and the high ration was 10% of body weight per day. Therefore, there were four groups of experimental fish based on genotype and feeding ration: 5% ration transgenics, 5% ration nontransgenics, 10% ration transgenics, and 10% ration nontransgenics. A total of 20-22 experimental fish were selected for each treatment, and the body sizes of them were similar among different treatments (Table 1). In each treatment, fish were stocked in 90-L net cages (45 × 45 × 65 cm^3) separately, and the net cages were set in the same rectangular cement pool (10 × 5× 1.5 m). Thus, four pools were used in total. The net cages were adjacent to each other, so visual and olfactory contacts among fish were allowed within the same treatment but not physical contact. The current experimental design was based on a number of considerations.

Table 1 The total length (L), body weight (W), and condition factor (CF) of the transgenic and nontransgenic common carp with 5% and 10% feeding rations, at the beginning of the experiment

	5% Ration		10% Ration	
	Transgenic	Nontransgenic	Transgenic	Nontransgenic
n	20	21	22	21
L(cm)	8.34 ± 0.12^{ab}	8.24 ± 0.08^{b}	8.56 ± 0.21^{ab}	8.69 ± 0.08^{a}
W(g)	9.30 ± 0.39^{ab}	8.01 ± 0.24^{b}	10.69 ± 0.83^{a}	10.16 ± 0.33^{a}
CF(g cm^{-3})	1.57 ± 0.02^{b}	1.42 ± 0.01^{c}	1.65 ± 0.03^{a}	1.54 ± 0.02^{b}

Data are presented as mean ±1 SE

Means followed by different superscripts in the same row are significantly different ($P < 0.05$)

For instance, fish growth is often influenced by social factors. So the intrinsic growth rate of fish is best measured in isolated cultural condition (Arendt *et al.*, 2001). However, carp are gregarious species that do not behave normally in isolation. Following the design of Arendt *et al.* (2001), the present design made the best compromise between complete isolation and complete contact. The experimental units were kept in natural temperature and light conditions (30°07′N, 114°22′E, Wuhan, Hubei). The experiment was carried out for 53 days starting from October 7, 2009. All fish were fed twice (09:00 and 15:00) daily with manufactured pellets. Total length (L, cm) and body weight (W, g) were measured every other week, in order to monitor the growth dynamics and for timely adjustment of the food amount. At the end of the experiment, all the experimental fish were measured for the L and W under an anesthesia with overdosed eugenol, and then were frozen stored until later measurements of the pharyngeal bone development. This procedure was approved by the Institute of Hydrobiology Institutional Animal Care and Use Committee (Approval ID: Keshuizhuan 08529).

Pharyngeal bone measurement

After the above proceedings, the whole fish of all experimental fish were boiled for 90 sec, and the pharyngeal bone was then easily removed (Hendrixson *et al.*, 2007). Length (PL, mm) and width (PW, mm) of the right pharyngeal bone were measured to the nearest 0.01 mm using caliper. Both left and right pharyngeal bone were dried at 60° until gaining a constant weight (about 72 h), and the dry weight (W_p, mg) was measured to the nearest 0.1 mg.

Calcium content in the pharyngeal bone was analyzed using an inductively coupled plasma atomic emission spectrometer (ICP-AES) (Hendrixson, 2002). First, dried samples of the pharyngeal bone were nitrified by the concentrated nitric acid. Then the nitrified solutions were diluted 1,000 times with double-distilled water, and were sent to the Centre of Analysis and Test of Wuhan University (Wuhan, China) for analysis of calcium using the ICP-AES (IRIS Intrepid II XSP, Thermo Electron Corporation).

Data calculation

Fish specific growth rates in total length (SGR_L) and body weight (SGR_W), condition factor (CF), and the pharyngeal bone calcium content (Ca content) were calculated as:

$$SGR_L\ (\%d^{-1})=100\times(\ln L_2-\ln L_1)d^{-1} \quad (1)$$

$$SGR_W(\%d^{-1})=100\times(\ln W_2-\ln W_1)d^{-1} \quad (2)$$

$$CF(g\cdot cm^{-3})=100\times WL^{-3} \quad (3)$$

$$\text{Ca content}(\%)=100\%\times W_{Ca}\times W_p^{-1} \quad (4)$$

where, L is the total length (cm), and L_2 and L_1 were the total final and initial total length (cm), respectively; d is the experimental period in days (d); W is the body weight (g), W_2 and W_1 were the final and initial body weight (g), respectively; W_{Ca} is the weight of calcium in the pharyngeal bone; W_p is the dry weight of the pharyngeal bone.

The ratio of the length to width of the pharyngeal bone (PL/PW) was also calculated.

Data analysis

The above-mentioned parameters L, W, CF, SGR_L, SGR_W, PL, PW, PL/PW, W_p, W_{Ca} and Ca content in the pharyngeal bone were measured and calculated individually for group comparisons. As calcium was measured as a percentage, an arcsine square root transformation was done prior to statistical analysis. Duncan's multiple-range tests were used to compare the differences among mean values. Analysis on PL and PW were also compared with analysis of covariance (ANCOVA) using final L as the covariate. Similar analysis was conducted on W_p except using final W as the covariate. Two-way ANOVAs were used to compare the effects of GH transgene (i.e., transgenic or nontransgenic) and feeding ration (5% or 10%) on fish growth and the pharyngeal bone development. The relationships between PL and L, PW and L, and W_p and W were explored by liner regression analysis. A test for common slope was used to compare coefficients in regression equations for transgenic and nontransgenic fish. Pearson correlation was used to test the relationship between calcium content of pharyngeal bone and growth rate. All the data were analyzed with SPSS 13.0 (Systat Inc., Chicago, IL). Accepted levelof statistical significance was $P< 0.05$ for all the tests.

Results

Growth

The dynamics of L, W, CF, SGR_L and SGR_W of the transgenics and nontransgenics with 5% and 10% feeding ration are showed in Figures 1 and 2. The transgenics showed significant superiority over the nontransgenics in growth (Table 2). Both the transgenics and nontransgenics grew consistently faster in the 10% ration than in 5% ration (Table 2). As the temperature declined, The SGR_L and SGR_W of all groups decreased gradually (Fig. 2).

No interaction effect was found between genotype and ration on SGR_L($F_{1,80}$=0.697, P = 0.406) and SGR_W ($F_{1,80}$ = 0.542, P = 0.464). Genotype significantly affected SGR_L ($F_{1,82}$ = 100.655, P< 0.001) and SGR_W ($F_{1,82}$ = 208.057, P< 0.001). At the 5% ration SGR_L was 1.64 times and SGR_W 1.73 times greater in transgenics than in controls. At the 10% ration SGR_L was 1.36 times and SGR_W 1.65 times greater in transgenics than in controls. There was also an effect of ration on SGR_L ($F_{1,82}$= 40.873, P< 0.001) and SGR_W ($F_{1,82}$ = 28.310, P< 0.001). The transgenics with 10% ration showed an increase of 18.5% in SGR_L and 18.6% in SGR_W than the transgenics with 5% ration, while the nontransgenics with 10% ration showed an increase of 42.4% in SGR_L and 24.4% in SGR_W than the nontransgenics with 5% ration.

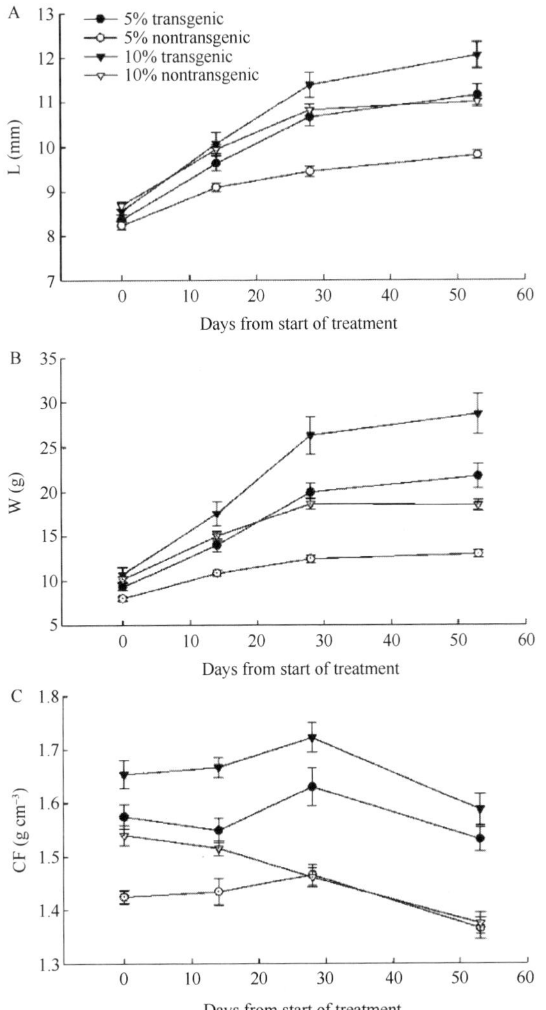

Figure 1 The dynamics of mean (±1 SE) (A) total length (L), (B)body weight (W), and (C) condition factor (CF) of the transgenic and nontransgenic common carp with 5% and 10% feeding rations. Both feeding rations were applied over 53 days

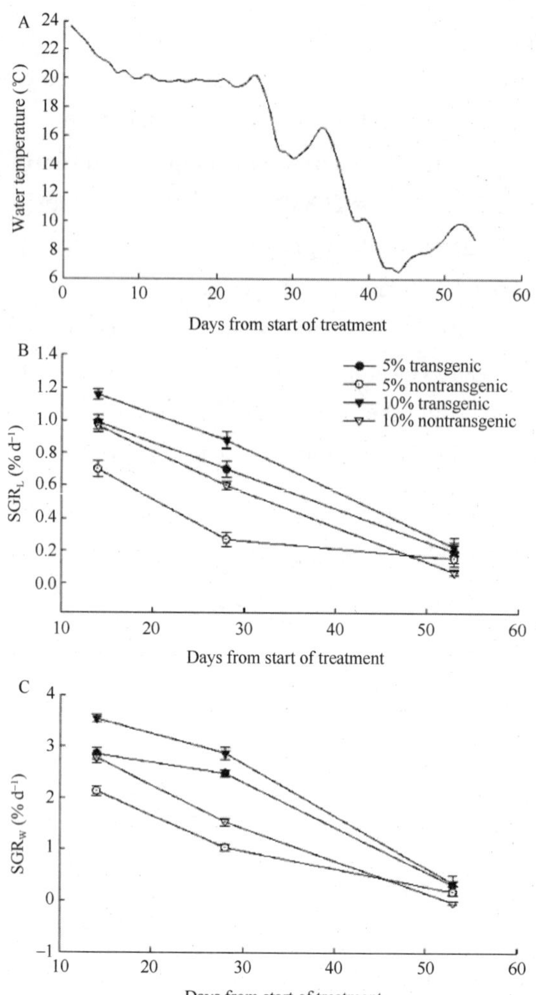

Figure 2 (A) Water temperature during the 53 days experimental period. The dynamics of the mean (±1 SE) specific growth rates in(B) total length (SGR_L) and (C) body weight (SGR_W) of the transgenic and nontransgenic common carp with 5% and 10% feeding rations. Both feeding rations were applied over 53 days

Table 2 The total length (*L*), body weight (*W*), condition factor (CF), and specific growth rates in total length (SGR_L) and body weight (SGR_W)of the transgenic and nontransgenic common carp with 5% and 10% feeding rations, at the end of the experiment.

	5% Ration		10% Ration	
	Transgenic	Nontransgenic	Transgenic	Nontransgenic
n	20	21	22	21
L(cm)	11.16 ± 0.23^b	9.80 ± 0.10^c	12.05 ± 0.29^a	11.01 ± 0.12^b
W(g)	21.71 ± 1.38^b	12.94 ± 0.42^c	28.66 ± 2.25^a	18.46 ± 0.61^b
CF(g cm^{-3})	1.53 ± 0.02^a	1.37 ± 0.02^b	1.59 ± 0.03^a	1.38 ± 0.02^b
SGR_L(% d^{-1})	0.54 ± 0.02^b	0.33 ± 0.01^d	0.64 ± 0.02^a	0.47 ± 0.02^c
SGR_W(% d^{-1})	1.56 ± 0.06^b	0.90 ± 0.03^d	1.85 ± 0.06^a	1.12 ± 0.05^c

Data are presented as mean±1 SE

Means followed by different superscripts in the same row are significantly different ($P < 0.05$)

Morphology of pharyngeal bone

Data on the PL, PW, and PL/PW of pharyngeal bone of the transgenics and nontransgenics are showed in Table 3. No interaction effect was found between genotype and ration on PL ($F_{1,80}$=0.540, P=0.464). Both genotype and ration affected PL (genotype, $F_{1,82}$=9.988, P=0.002; ration, $F_{1,82}$=13.181, P<0.001).

Positive relationships were found between PL and L, as well as PW and L (Fig. 3). Difference in the regression slopes between PL and L among genotypes was significant in 5% ration ($F_{1,39}$ = 8.095, P = 0.007) but not significant in 10% ration ($F_{1,41}$= 0.085, P = 0.773). When using fish final total length as the covariate, the ANCOVA results showed that the adjusted PL of the transgenics was significantly shorter than that of the nontransgenics (Table 3). The adjusted PW of the transgenics was also shorter although the difference was not significant at the 5% ration (Table 3). There was no significant difference in PL/PW among the treatments (Table 3).

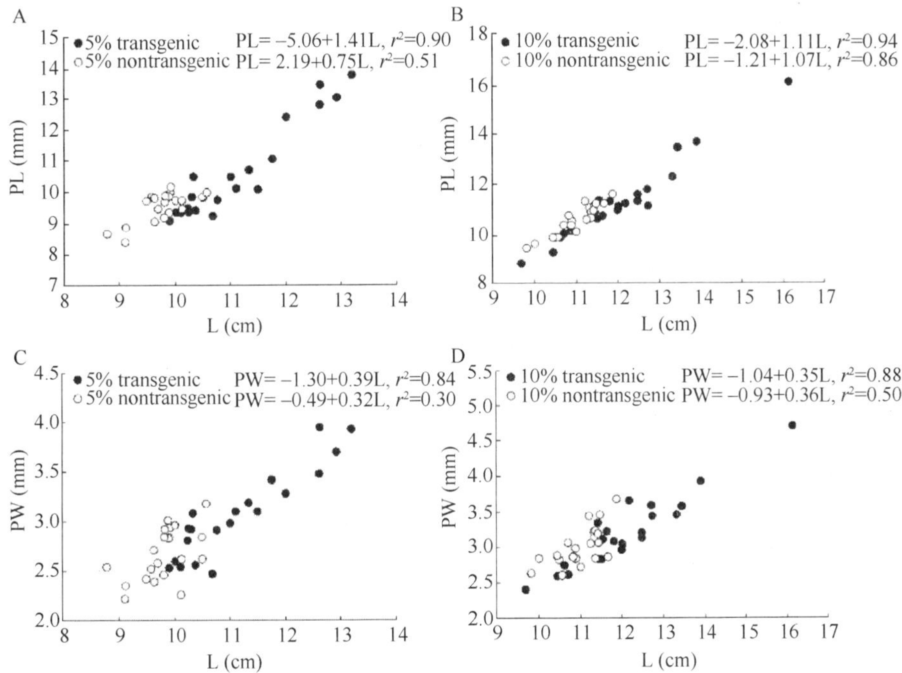

Figure 3 Relationships between the pharyngeal bone length (PL) and fish total length (L) of the transgenic and nontransgenic common carp in (A) 5% and (B) 10% feeding rations, and relationships between the pharyngeal bone width (PW) and fish total length (L) of the transgenic and nontransgenic common carp with (C) 5% and (D) 10% feeding rations. Both feeding rations were applied 53 days

Table 3 The adjusted means of length (PL) and width (PW), and length/width (PL/PW) of right pharyngeal bone of the transgenic and nontransgenic common carp with 5% and 10% feeding rations, at the end of the experiment

	5% Ration		10% Ration	
	Transgenic	Nontransgenic	Transgenic	Nontransgenic
n	20	21	22	21
Adjusted PL (mm)	10.51[b]	10.98[a]	10.08[c]	10.58[b]
Adjusted PW (mm)	3.02[a]	3.11[a]	2.82[b]	3.02[a]
PL/PW	3.48±0.04	3.60±0.06	3.55±0.04	3.51±0.04

Data are presented as mean ±1 SE

Adjusted means of PL and PW were estimated using ANCOVA with final total length as the covariate; means followed by different superscripts in the same row are significantly different ($P < 0.05$)

Weight and calcium content of pharyngeal bone

The absolute W_p of the transgenics was significantly heavier than that of the nontransgenics under both 5% and 10% rations (Table 4). But when using fish final body weight as the covariate, the ANCOVA results showed that the adjusted W_p of the transgenics was significantly lower than that of the nontransgenics (Table 4). No interaction effect was found between genotype and ration on W_p ($F_{1,80}$ =0.075, P = 0.785). Both genotype and ration affected W_p (genotype, $F_{1,82}$ = 16.911, P < 0.001; ration, $F_{1,82}$= 8.179, P = 0.005). Positive relationships were found between W_p and W (Fig. 4). There was no significant difference in regression slopes between W_p and W among genotypes (5% ration: $F_{1,39}$= 0.800, P = 0.007; 10% ration: $F_{1,41}$= 0.063, P =0.804).

Table 4 The absolute and adjusted means of weight (W_p), and Ca content of both sides pharyngeal bone of the transgenic and nontransgenic common carp with 5% and 10% feeding rations, at the end of the experiment

	5% Ration		10% Ration	
	Transgenic	Nontransgenic	Transgenic	Nontransgenic
n	20	21	22	21
Absolute W_p (mg)	31.46 ± 2.21[ab]	22.83 ± 0.68[c]	36.53 ± 2.98[a]	28.99 ± 0.95[b]
Adjusted W_p (mg)	29.88[b]	32.97[a]	25.66[c]	31.74[a]
Ca content (%)	18.00 ± 0.21[b]	19.11 ± 0.28[a]	18.34 ± 0.25[b]	19.59 ± 0.17[a]

Data are presented as mean ± 1 SE

Adjusted W_p was estimated using ANCOVA with final body weight as the covariate; means of Ca content of were not adjusted as final body weight has no significant effects (P = 0.284) on Ca content; means followed by different superscripts in the same row are significantly different (P < 0.05)

The calcium content of pharyngeal bone of the transgenics was significantly lower than that of the nontransgenics for both 5% and 10% rations (Table 4). There was no significant effect of ration on pharyngeal bone calcium content ($F_{1,82}$ = 3.140, P = 0.080). No interaction effect was found between genotype and ration ($F_{1,80}$= 0.095, P = 0.759) on pharyngeal bone calcium content. Under the same feeding ration, calcium content of pharyngeal bone of the experimental fish was negatively related to growth rate significantly (5% ration: $r = -0.506$, $n = 41$, $P = 0.001$; 10% ration: $r = -0.346$, $n = 43$, $P = 0.023$) (Fig. 5A,B). However, there was no significant correlation between calcium content and growth rate within genotype (transgenic: $r = -0.023$, $n = 42$, $P = 0.886$; nontransgenic: $r = 0.245$, $n = 42$, $P = 0.118$) (Fig. 5C, D).

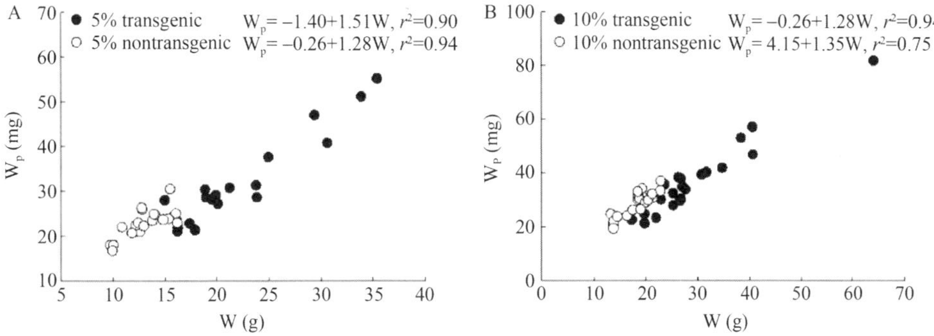

Figure 4 Relationships between the final pharyngeal bone weight (W_p) and body weight (W) of the transgenic and nontransgenic common carp with (A) 5% and (B) 10% feeding rations. Both feeding rations were applied 53 days

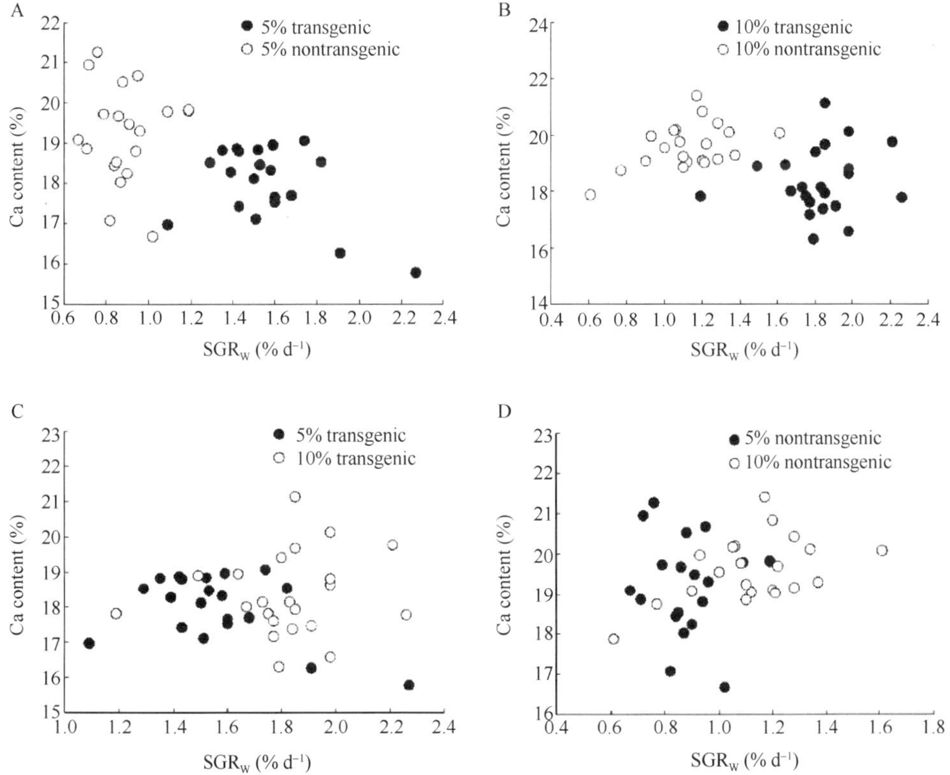

Figure 5 Relationships between the calcium content of pharyngeal bone (Ca content) and specific growth rates of body weight (SGR_W) of experimental fish belonging to the same feeding ration (A-5% ration; B-10% ration) or same genotype (C-transgenic; D-nontransgenic).Both feeding rations were applied 53 days

Discussion

Somatic growth

The transgenic carp showed faster growth than the nontransgenic carp even in the low feeding ration. This is similar to another growth study on transgenic (with GH gene from chum salmon *Oncorhynchus keta* Walbaum, 1792) common carp (Liu *et al.*, 2010). In the later period of the current experiment, fish feeding activities were reduced and stopped due to the low temperature (<10°), which may have caused the reduced and even negative growth.

Pharyngeal bone development

Pharyngeal bone growth rate did not match with somatic growth rate in the transgenic carp. The adjusted length and weight of the pharyngeal bone of the transgenic carp were significantly shorter and lower than those of the nontransgenic carp, respectively. The lengths of the tibia and femur in rabbit legs also decreased with increasing growth rate (Gondret *et al.*, 2005). There is an unbalanced relationship between somatic growth rate and bone growth (abnormal head shape, reduced growth rate in skull or operculum, etc.) in transgenic fish as a result of over expression of the growth hormone gene (Ostenfeld *et al.*, 1998; Rahman *et al.*, 1998; Dunham and Chatakondi, 2002). In a previous study of our laboratory, we had found that transgenic carp had 6.36 times higher serum GH level than the nontransgenic fish (Duan *et al.*, 2011), which may have resulted in the reduced growth rate of the pharyngeal bone in the transgenic carp. It could be due to a de-coupling between somatic growth (strongly stimulated by GH) and pharyngeal bone growth (less strongly stimulated by GH) similar to that found for coho salmon eye and brain growth (Devlin *et al.*, 2012; Kotrschal *et al.*, 2012).

Pharyngeal bone calcium content is reduced by GH transgene. Previous studies had found remarkable effects of GH transgene on bone development. For example, early adult "all-fish" GH transgenic Nile tilapia *Oreochromis niloticus* (L. 1758) showed incomplete bone mineralization, and late adult transgenic tilapia were observed to have skeletal abnormalities (Lu *et al.*, 2009). There are four possible explanations for decreased calcification in the transgenic carp. The first one is lower calcification rate in the transgenic carp. Fast growth must be coupled with fast cellular proliferation, while cell differentiation would be delayed (Starck, 1994). The calcification of pharyngeal bone in the transgenic carp may give way to fast cellular proliferation. The second one is different allocations of calcium to metabolic demands and skeletal development between the transgenic and nontransgenic carp. Flik *et al.* (1993) found that growth hormone treated Mozambique tilapia *Oreochromis mossambicus* (Peters 1852) was slightly although not significantly lower in calcium contents of both scales and bones than the control fish (scales: 4.80 ± 0.62 vs. 5.15 ± 0.52 mmol/g, bones: 5.04 ± 0.36 vs. 5.20 ± 0.45 mmol/g). However, plasma calcium concentrations were higher in growth hormone treated rainbow trout *Oncorhynchus mykiss* (Walbaum, 1792) than in the control fish (Takagi *et al.*, 1992). These studies indicate that the allocation of calcium may be changed by the growth hormone treatment. The third one is delayed in onset of bone calcification in the transgenic carp. Assuming growth of the collagen component of the bone is not delayed in its onset, then a delay in the start of mineralization means more collagen is present at any one time. This will result in a lower content of calcium. The fourth one is different rate of collagen growth and mineralization. Palmer (1981) argued that mineralization occurs at a nearly constant rate in all organisms regardless of growth rate. If mineralization happens at a fixed rate but the collagen component grows faster in fast growing transgenic carp (i.e., the bone keeps pace with the overall size of the fish), then it will also result in a lower percentage of calcium.

Interestingly a high feeding ration showed no additional cost to bone development. The present study also tested whether high feeding ration affects bone development. Food and growth hormone were both found to be influential on calcium metabolism. For example, when feeding ration was 5% of body weight, the mean branchial calcium influx (11.71 ±7.14 μmol h^{-1}100 g^{-1}) of male Mozambique tilapia was significantly higher than that (5.80 ± 2.64 μmol h^{-1}100 g^{-1}) of those receiving 2% feeding ration (Flik *et al.*, 1993). Significant difference in plasma calcium concentrations was found between GH treated rainbow trout and control fish, but no difference was found among the treated rainbow trout that received different GH doses (Takagi *et al.*, 1992). These studies indicate that calcium supply will increase with the increase of feeding ration, but bone calcium content may not increase significantly. Both environmental and genetic factors affect bone development, but they do not contribute equally (e.g., Vicente-Rodriguez *et al.*, 2008). The present research found no significant difference in pharyngeal bone calcium content among rations. A possible explanation is that the onset and rate of calcium deposition were mainly genetically driven, and food ration may have a weaker influence on calcium content. In the present study, the scale of ration used might be too small to make such a weak influence significant. Combined with comparisons between genotypes and rations, it is reasonable to infer that the difference of pharyngeal bone calcium content between genotypes was due to different allocations of calcium to metabolic demands and skeletal calcification, but allocation of calcium is relatively stable within a genotype. The possible reason is that the allocation pattern of calcium is adapted to fast growth and as part of the genetic information in the transgenic carp. Therefore, the calcium differences between transgenic and nontransgenic carp may be a direct effect of GH level but not directly linked to differences in growth rate.

Conclusions

In the current study, rapid growth of transgenic carp due to GH transgene resulted in relatively smaller pharyngeal bone with relatively lower calcium content. Feeding ration did not show significant effect on pharyngeal bone development. Reduced Ca content (i.e., reducing bone strength) in pharyngeal bone may affect the transgenic carp's ability in molluscs predation, but it needs more future confirmatory studies.

Acknowledgements The two journal reviewers provided constructive suggestions in improving the early draft of the manuscript.

References

Arendt JD. 1997. Adaptive intrinsic growth rates: an integration across taxa. Q Rev Biol 72:149-177.

Arendt JD, Wilson DS. 2000. Population differences in the onset of cranial ossification in pumpkinseed (*Lepomis gibbosus*), a potential cost of rapid growth. Can J Fish Aquat Sci 57:351-356.

Arendt JD, Wilson DS, Stark E. 2001. Scale strength as a cost of rapid growth in sunfish. Oikos 93:95-100.

Billerbeck JM, Lankford TE, Conover DO. 2001. Evolution of intrinsic growth and energy acquisition rates. I. Trade - offs with swimming performance in *Menidia menidia*. Evolution 55:1863-1872.

Biro PA, Abrahams MV, Post JR, Parkinson EA. 2006. Behavioural trade offs between growth and mortality explain evolution of submaximal growth rates. J Anim Ecol 75:1165-1171.

Devlin RH, Yesaki TY, Donaldson EM, Du SJ, Hew CL. 1995. Production of germline transgenic pacific salmonids with dramatically increased growth performance. Can J Fish Aquat Sci 52:1376-1384.

Devlin RH, Vandersteen WE, Uh M, Stevens ED. 2012. Genetically modified growth affects allometry of eye and brain in salmonids. Can J Zool 90:193-202.

Duan M, Zhang T, Hu W, *et al.* 2010. Increased mortality of growth enhanced transgenic common carp (*Cyprinus carpio* L.) under short term predation risk. J Appl Ichthyol 26:908-912.

Duan M, Zhang T, Hu W, *et al.* 2011. Behavioral alterations in GH transgenic common carp may explain enhanced competitive feeding ability. Aquaculture 317:175-181.

Dunham RA, Chatakondi N. 2002. Effect of rainbow trout growth hormone complementary DNA on body shape, carcass yield and carcass composition of F1 and F2 transgenic common carp (*Cyprinus carpio*). Mar Biotechnol 4:604-611.

Fjelldal PG, Hansen T, Breck O, *et al.* 2009. Supplementation of dietary minerals during the early seawater phase increase vertebral strength and reduce the prevalence of vertebral deformities in fast-growing under-yearling atlantic salmon (*Salmo salar* L.) smolt.Aquacul Nutr 15:366-378.

Flik G, Atsma W, Fenwick JC, *et al.* 1993. Homologous recombinant growth hormone and calcium metabolism in the tilapia, *Oreochromis mossambicus*, adapted to fresh water. J Exp Biol 185:107-119.

Gondret F, Larzul C, Combes S, Rochambeau Hd. 2005. Carcass composition, bone mechanical properties, and meat quality traits in relation to growth rate in rabbits. J Anim Sci 83:1526-1535.

Guan B, Hu W, Zhang T, Wang Y, Zhu Z. 2008. Metabolism traits of 'all-fish' growth hormone transgenic common carp (*Cyprinus carpio* L.). Aquaculture 284:217-223.

Guan B, Hu W, Zhang T, *et al.* 2010. Acute and chronic un-ionized ammonia toxicity to "all-fish" growth hormone transgenic common carp (*Cyprinus carpio* L.). Chin Sci Bull 55:4032-4036.

Hendrixson HA. 2002. Stoichiometry of freshwater fish in relation to allometry and phylogeny, and the role of skeleton in fish stoichiometry [MS Thesis]. St Paul, MN: University of Minnesota.

Hendrixson HA, Sterner RW, Kay AD. 2007. Elemental stoichiometry of freshwater fishes in relation to phylogeny, allometry and ecology. J Fish Biol 70:121-140.

Julian RJ. 1998. Rapid growth problems: ascites and skeletal deformities in broilers. Poultry Sci 77:1773-1780.

Kotrschal A, Sundstrom LF, Brelin D, Devlin RH, Kolm N. 2012. Inside the heads of David and Goliath: environmental effects on brain morphology among wild and growth-enhanced coho salmon *Oncorhynchus kisutch*. J Fish Biol 81:987-1002.

Lankford TE, Billerbeck JM, Conover DO. 2001. Evolution of intrinsic growth and energy acquisition rates. II. Trade-offs with vulnerability to predation in *Menidia menidia*. Evolution 55:1873-1881.

Li D. 2008. Primary studies on the biology of 'all - fish' growth hormone transgenic common carp (*Cyprinus carpio* L.) [PhD Thesis]. Institute of Hydrobioligy, Wuhan, China: Chinese Academy of Science (in Chinese).

Li D, Fu C, Hu W, *et al.* 2007. Rapid growth cost in "all - fish" growth hormone gene transgenic carp: Reduced critical swimming speed.Chin Sci Bull 52:1501-1506.

Liu C, Chang Y, Liang L, *et al.* 2010. Effects of feed shortage on growth competition and gonad development in transgenic carp and wild carp. Acta Ecol Sin 30:5975-5982 (in Chinese).

Lu J, Li J, Furuya Y, *et al.* 2009. Efficient productivity and lowered nitrogen and phosphorus discharge load from gh - transgenic tilapia (*Oreochromis niloticus*) under visual satiation feeding. Aquaculture 293:241-247.

Mangel M, Stamps J. 2001. Trade - offs between growth and mortality and the maintenance of individual variation in growth. Evol Ecol Res 3:583-593.

Munch SB, Conover DO. 2003. Rapid growth results in increased susceptibility to predation in *Menidia menidia*. Evolution 57:2119-2127.

Nicieza AG, Metcalfe NB. 1999. Costs of rapid growth: the risk of aggression is higher for fast - growing salmon. Funct Ecol 13:793-800.

Ostenfeld TH, McLean E, Devlin RH. 1998. Transgenesis changes body and head shape in pacifc salmon. J Fish Biol 52:850-854.

Ott EA, Asquith RL. 1986. Influence of level of feeding and nutrient content of the concentrate on growth and development of yearling horses. J Anim Sci 62:290-299.

Palmer AR. 1981. Do carbonate skeletons limit the rate of body growth.Nature 292:150-152.

Rahman MA, Mak R, Ayad H, Smith A, Maclean N. 1998. Expression of a novel growth hormone gene results in growth enhancement in transgenic tilapia. Transgenic Res 7:357-369.

Ricklefs RE, Shea RE, Choi IH. 1994. Inverse relationship between functional maturity and exponential growth rate of avian skeletal muscle: a constraint on evolutionary response. Evolution 48:1080-1088.

Roff DA. 1992. The evolution of life histories: theory and analysis. New York, USA: Chapman and Hall.

Starck JM. 1994. Quantitative design of the skeleton in bird hatchlings: does tissue compartmentalization limit posthatching growth rates? J Morphol 222:113-131.

Stearns SC. 1992. The evolution of life histories. Oxford, UK: Oxford University Press.

Sundt-Hansen L, Neregård L, Einum S, *et al.* 2009. Growth enhanced brown trout show increased movement activity in the wild. Funct Ecol 23:551-558.

Takagi Y, Moriyama S, Hirano T, Yamada J. 1992. Effects of growth hormones on bone formation and resorption in rainbow trout (*Oncorhynchus mykiss*), as examined by histomorphometry of the pharyngeal bone. Gen Comp Endocr 86:90-95.

Urban MC. 2007. The growth-predation risk trade-off under a growing gape-limited predation threat. Ecology 88:2587-2597.

Vicente-Rodríguez G, Ezquerra J, Mesana MI, *et al.* 2008. Independent and combined effect of nutrition and exercise on bone mass development. J Bone Miner Metab 26:416-424.

Walstra P. 1980. Growth and carcass composition from birth to maturity in relation to feeding level and sex in dutch landrace pigs[PhD thesis]. Communications Agricultural University, Wageningen,The Netherlands.

Wang Y, Hu W, Wu G, *et al.* 2001. Genetic analysis of "all - fish" growth hormone gene transferred carp (*Cyprinus carpio* L.) and its F1 generation. Chin Sci Bull 46:1174-1177.

Zhong C, Song Y, Wang Y, *et al.* 2012. Growth hormone transgene effects on growth performance are inconsistent among offspring derived from different homozygous transgenic common carp (*Cyprinus carpio* L.). Aquaculture 356-357:404-411.

营养状况对转基因鲤生长及骨骼发育的影响

朱挺兵[1,2]　张堂林[1]　汪亚平[1]　陈宇顺[3]　胡　炜[1]　朱作言[1]

1 淡水生态与生物技术国家重点实验室，中国科学院水生生物研究所，武汉

2 中国科学院大学，北京

3 Aquaculture and Fisheries Center, University of Arkansas, Pine Bluff, Arkansas

摘　要　营养状况对转基因鲤的生长和发育的影响尚不清楚。本文研究了两种营养条件，对转生长激素基因鲤生长、咽骨形态及钙质含量的影响。在两种营养条件下(每天投喂量分别为体重的 5%和 10%)，转基因鲤的生长率都显著高于非转基因鱼。虽然两种鱼咽骨的形状相似，但转基因鲤的咽骨比非转基因鱼的咽骨小而轻。转基因鲤咽骨的钙质含量低于非转基因鱼。虽然投饲率也能影响生长率，但其对骨骼发育的作用很小。由于转基因鲤咽骨的钙质含量和钙化程度低，其强度也相应降低。本研究通过调控环境(投饲率)和遗传背景或基因型(转基因和非转基因)，初步证实在对咽骨发育的影响方面，基因型的作用比环境的作用更强。

Increased Food Intake in Growth Hormone-Transgenic Common Carp (*Cyprinus carpio* L.) May Be Mediated by Upregulating Agouti-Related Protein (AgRP)

Chengrong Zhong[1,2] Yanlong Song[1,2] Yaping Wang[1] Tanglin Zhang[1]
Ming Duan[1] Yongming Li[1] Lanjie Liao[1] Zuoyan Zhu[1] Wei Hu[1]

1 State Key Laboratory of Freshwater Ecology and Biotechnology, Institute of Hydrobiology, Chinese Academy of Sciences, Wuhan 430072
2 University of Chinese Academy of Sciences, Beijing 100049

Abstract In fish, food intake and feeding behavior are crucial for survival, competition, growth and reproduction. Growth hormone (GH)-transgenic common carp exhibit an enhanced growth rate, increased food intake and higher feed conversion rate. However, the underlying molecular mechanisms of feeding regulation in GH-transgenic (TG) fish are not clear. In this study, we observed feeding behavior of TG and non-transgenic (NT) common carp, and analyzed the mRNA expression levels of *NPY*, *AgRP I*, orexin, *POMC*, *CCK*, and *CART I* in the hypothalamus and telencephalon after behavioral observation. We detected similar gene expression levels in the hypothalamus of TG and NT common carp, which had been cultured in the field at the same age. Furthermore, we tested the effects of GH on hypothalamus fragments *in vitro* to confirm our findings. We demonstrated that TG common carp displayed increased food intake and reduced food consumption time, which were associated with a marked increase in hypothalamic *AgRP I* mRNA expression. Our results suggest that elevated GH levels may influence food intake and feeding behavior by upregulating the hypothalamic orexigenic factor *AgRP I* in GH-transgenic common carp.

Keywords growth hormone; *AgRP*; *NPY*; good intake; transgenic; common carp

1. Introduction

In aquaculture, enhancing growth is a key objective to improve fish production (Raven *et al*., 2008). Gene transfer has been employed as a rapid and effective approach to realize this objective. Since the first batch of transgenic fish was produced (Zhu *et al*., 1985), transgenesis has been used in many fish species (Hu and Zhu, 2010). To date, several stable lines of growth-enhanced transgenic fishes have been generated, including Atlantic salmon (*Salmo salar*) (Fletcher *et al*., 2004), coho salmon (*Oncorhynchus kisutch*) (Devlin *et al*., 2004), mud loach (*Misgurnus mizolepis*) (Nam *et al*., 2002), tilapia (*Oreochromis* sp.) (Martinez *et al*.,

文章发表于 *General & Comparative Endocrinology*, 2013, 192(9):81-88

1999; Rahman *et al.*, 1998) and common carp (*Cyprinus carpio* L.) (Wang *et al.*, 2001; Zhong *et al.*, 2012). During recent years, it has become clear that elevated growth hormone (GH) levels in transgenic fish induce a wide range of effects apart from growth promotion (Devlin *et al.*, 1994; Du *et al.*, 1992; Martinez *et al.*, 1996; Nam *et al.*, 2001; Rahman *et al.*, 1998; Wang *et al.*, 2001; Zhong *et al.*, 2012), including altered metabolism (Guan *et al.*, 2008; McKenzie *et al.*, 2003; Stevens *et al.*, 1998), swimming performance (Farrell *et al.*, 1997; Lee *et al.*, 2003; Stevens *et al.*, 1998), anti-predator behavior (Abrahams and Sutterlin, 1999; Duan *et al.*, 2010; Dunham *et al.*, 1999) and growth-related neuroendocrine regulation (Raven *et al.*, 2008). In addition, GH treatment resulted in increased food intake and feeding behavior in normal fish (Johnsson and Bjornsson, 1994). Similarly, GH-transgenic (TG) fish, with constantly high levels of GH, exhibit accelerated feeding motivation, higher social status, and increased feeding competitive ability (Abrahams and Sutterlin, 1999; Devlin *et al.*, 1999; Duan *et al.*, 2009, 2011; Sundstrom *et al.*, 2003).

Feeding behavior in fish is regulated by numerous environmental factors as well as by complex homeostatic mechanisms, which are mediated by interactions between central and peripheral signals. Peripheral endocrine and metabolic factors convey information about nutritional status to the brain, and then the brain processes this information and induces neurons to secrete neuropeptides to control feeding behavior (Volkoff and Peter, 2006). In the brain, the hypothalamus is the primary center of feeding regulation. A large number of neuropeptides regulate food intake and energy homeostasis, including orexigenic factors, such as neuropeptide Y (*NPY*), Agouti-related protein (*AgRP*) and orexins; and anorexigenic factors, such as pro-opiomelanocortin (*POMC*) and cocaine-and amphetamine-regulated transcript (*CART*) (Volkoff *et al.*, 2009). In goldfish (*Carassius auratus*) and catfish (*Ictalurus punctatus*), intracerebroventricular (ICV) injections of *NPY* stimulated food intake (Lopez-Patino *et al.*, 1999; Silverstein and Plisetskaya, 2000). Furthermore, the hypothalamic expression of *NPY* mRNA increased in response to fasting in goldfish and salmon (Narnaware and Peter, 2001; Silverstein *et al.*, 1998). *AgRP*, an antagonistic gene of melanocortin receptor 3 and 4 (*MC3R* and *MC4R*), participates in the regulation of energy homeostasis and feeding by blocking melanocortin signaling (Cerda-Reverter and Agulleiro, 2011; Stutz *et al.*, 2005). Moreover, fasting led to upregulation of *AgRP* mRNA in the hypothalamus of goldfish and zebrafish (*Danio rerio*) (Cerda-Reverter and Peter, 2003; Song *et al.*, 2003). Two *AgRP* genes, *AgRP I* and *AgRP II*, have been identified in common carp (Wan *et al.*, 2012). Orexins consist of two peptides, orexin A and orexin B encoded by a same precursor, prepro-orexin. Orexins play a role in the feeding physiology of fish, as demonstrated by the fact that ICV injections of orexin A and orexin B stimulate feeding behavior and food consumption in goldfish (Volkoff *et al.*, 1999). *POMC*, the precursor of melanocortin, is mainly processed to alpha-melanocyte-stimulating hormone (*α-MSH*) and β-endorphin (Cerda-Reverter and Agulleiro, 2011). *POMC I* and *POMC II* have already been identified in common carp

(Arends *et al.*, 1998). The melanocortin system has been illustrated to play a critical role in the regulation of food intake and body weight in fish (Cerda-Reverter and Agulleiro, 2011). *MC4R*, a G protein-coupled receptor of *α-MSH*, possesses seven transmembrane-domain regions and mediates *α-MSH* to tonic restraint on feeding (Cerda-Reverter *et al.*, 2003; Ringholm *et al.*, 2002). In common carp, two forms of pre-pro-*CART* derived from separate genes have been identified (Wan *et al.*, 2012). As a well-known appetite inhibiting factor (Volkoff and Peter, 2000, 2001), *CART* mRNA expression undergoes postprandial increases in common carp (Wan *et al.*, 2012). Cholecystokinin (*CCK*), a satiety signal mainly produced in the gut and brain, decreases food intake and feeding behavior (Himick and Peter, 1994; Volkoff, 2006). Corticotrophin releasing hormone (*CRH*) is one of the hypothalamic hormone which controls stress axis and also acts as an anorexigenic factor for satiety regulation. *CRH* and *CRH* receptor 1 (CRH-R1) have been cloned and characterized in common carp (Huising *et al.*, 2004).

GH directly or indirectly regulates appetite and food intake through the interaction with hypothalamic neuropeptides. GH and GH receptor (*GHR*) are present in regions known to participate in the regulation of feeding behavior, energy balance and motivation (Gossard *et al.*, 1987; Hojvat *et al.*, 1982; Lobie *et al.*, 1993). Besides, GH positively regulates *AgRP* and *NPY* (Bohlooly *et al.*, 2005; Chan *et al.*, 1996) which are appetite-stimulating neuropeptides. On the other hand, appetite-regulation peptides can influence GH secretion. For instance, *NPY* negatively regulates GH production through the stimulation of somatostatin (*SS*)-producing cells (Minami *et al.*, 1998; Peng *et al.*, 1993). In zebrafish, suppression of *AgRP I* decreased the expression of GH mRNA in pituitary (Zhang *et al.*, 2012). CCK may have direct effects on the pituitary gland and also acts on GH production by reducing *SS* mRNA (Canosa and Peter, 2004; Himick *et al.*, 1993). Furthermore, changes in GH levels are tightly coupled to nutritional status (Canosa *et al.*, 2007). For example, reduced food intake or starvation causes GH to rise, while GH returns to baseline after resumed feeding (Gabillard *et al.*, 2006; Pierce *et al.*, 2005; Shimizu *et al.*, 2009).

Feeding behavior influences the ability of fish to compete for food resources which is crucial for survival, competition and growth. Therefore, it is an important fitness parameter for evaluating the ecological effects of TG fish. Besides, the mechanism of GH in feeding regulation is still unknown. In our study, we compared the feeding behavior of TG common carp and non-transgenic (NT) fish, and analyzed the mRNA expression levels of *NPY*, *AgRP*, orexin, *POMC*, *CCK*, *CART*, *MC4R*, *CRH* and *CRH-R1*. In addition, we tested the effects of GH on hypothalamus fragments *in vitro* to confirm our findings. We demonstrated that TG common carp exhibited increased food intake and reduced food consumption time, which were associated with marked increases in hypothalamic and telencephalic *AgRP I* mRNA expression. We speculate that elevated GH levels may influence food intake and feeding behavior by upregulating the hypothalamic orexigenic factor *AgRP I*.

2. Materials and Methods

2.1 Experimental fish

P_0 'all-fish' GH-transgenic common carp were produced by the microinjection of fertilized eggs with the gene construct pCAgcGH, which contained the grass carp (*Ctenopharyngodon idellus*) *GH* gene (gcGH) driven by the β-actin gene promoter of common carp (*C. carpio* L.). The transmission rates from F_1 to F_3 followed the expected Mendelian ratio (unpublished data). The GH-transgenic fish used in this study were from the TG2 family as previously described (Zhong *et al.*, 2012).

2.2 Behavioral observations

Experimental fish were maintained indoor in a re-circulating system at the Institute of Hydrobiology, China. Incoming nonchlorinated water was automatically filtered, sterilized and constantly aerated to ensure high water quality. Throughout the experiment, the water temperature, dissolved oxygen and pH were 25-29℃, 70-85% and 7.4-7.8, respectively. Artificial lighting with a photoperiod set between 07:00 and 19:00 h were provided. Fish were fed with a commercial diet (2 mm diameter pellets; diet containing approximately 42% protein and 3.6% lipids). An approximate 3% of body weight ration of pellets per fish were fed twice a day.

Seven individuals of each genotype, five months post fertilization (mpf), were separately cultured in 14 tempered glass aquaria measuring 50 × 25 × 50 cm with a water depth of 35 cm. The average weight of TG fish was 20.9 ± 0.55 g, ranging from 20.1 g to 23.6 g, and that of NT fish was 21.0 ± 0.56 g, ranging from 20.3 g to 23.3 g. The facility was the same as described in our previous study (Duan *et al.*, 2011).

The experiment consisted of two main parts: (1) a treatment period (days 1-7), and (2) an observation period (days 8-11). Observations began 2 min before the fish were offered pellets in the tank. Each fish was fed with 36 pellets one by one in a 30 s interval time during an 18 min feeding session, twice a day (07:30-12:00 h and 14:00-18:30 h). Two days later, after fasting for a night, all fish were sacrificed for the subsequent experiment.

All experimental procedures employed were approved by the Institute of Hydrobiology Institutional Animal Care and Use Committee (Approval ID: keshuizhuan 08529).

2.3 *In vitro* experiment

NT common carp (n = 12) were anesthetized with 0.25 ml l^{-1} Eugenol (purity≥99.99%, Sinopharm Chemical Reagent Co. Ltd., Shanghai, China). The hypothalamus was dissected from each fish; 1 ml cell culture medium was added to each 1.5 ml Eppendorf tube. The tissue samples were aseptically cut, evenly mixed and separated to 24-well plates with common carp GH at final concentrations of 20, 80 and 200 ng ml^{-1}, respectively. The 24-well plates were incubated at 28.6℃ for 2 and 4 h, respectively. After that, the fragments from each well were collected and centrifuged at 4℃, 4000 rpm for 10 min, the supernatant was removed, 1 ml TRIzol Reagent (Invitrogen, Carlsbad, CA, USA) was added to each tube, the samples were mixed and stored at –70℃ for subsequent analysis.

2.4 Measurement of serum GH levels by ELISA

TG (n = 12, 5 mpf) and NT common carp (n = 12, 5 mpf) were cultured in the indoor re-circulating system for the measurement of serum GH. Six individuals of each genotype were fasted for two weeks and another six fish of each genotype were fed normally. All 24 fish were anesthetized as described above. Sera from these fish were collected and processed for the measurement of serum GH levels. An enzyme-linked immunosorbent assay (ELISA) was performed as previously described (Wu *et al.*, 2008).

2.5 Expression studies

Total RNA samples were isolated from the hypothalamus and telencephalon of each fish from the behavioral observation experiment. Furthermore, TG (n = 30, 5 mpf) and NT common carp (n = 30, 5 mpf) were divided into four tanks in the indoor re-circulating system. Following the acclimation period, two tanks were food deprived for two weeks and two tanks were maintained on the regular feeding schedule. The hypothalamus of all 60 fish were also dissected for the gene expression analysis. Another 20 individuals of each genotype, 5 mpf, cultured outdoors and fed regularly, were also sacrificed after fasting for overnight.

RNA extractions were performed using TRIzol Reagent (Invitrogen) according to the manufacturers' protocol. The RNA amount and purity were determined by spectrophotometry and electrophoresis. The quality of RNA samples was assessed by measuring the ratio of sample absorbance at 260 and 280 nm. Only RNA samples with a ratio between 1.8 and 2.1 were used.

After DNase I (Promega, Madison, WI, USA) treatment, 1 μg total RNA was used for the first cDNA transcription. The reverse transcription system contained 1 μg RNA, 4 μl 5× RT buffer, 1 μl ReverTra-Ace (100 U), 0.5 μl RNase inhibitor (20 U), 1 μl dNTPs (10 mM) and 2 μl random primers 9 (25 pm/μl; TaKaRa, Japan), to a total volume of 20 μl. PCR was conducted under the following conditions: 30℃ for 10 min, 42℃ for 60 min and 95℃ for 5 min. Real-time PCR was performed as previously described (Guan *et al.*, 2011). The primers used in this study are listed in Table 1. The POMC primers used for amplification are specific to both *POMC I* and *POMC II*. The primer sequences of *CRH*, *CRH-R1*, *AgRP II*, *CART II*,

MC4R were previously published (Huising *et al.*, 2004; Wan *et al.*, 2012).

Table 1 Primer sequences used in this study

Primer	F	R
GH	5′-TGCTATTGTCGGTGGT-3′	5′-CTGTCTGCGTTCCTCA-3′
GHR	5′-CCACAACACGCAAGTCT-3′	5′-CCAGTCCGTTTCCACA-3′
NPY	5′-TGCTTGGGAACTCTAACGGAA-3′	5′-GACCTTTTGCCATACCTCTGC-3′
AgRP I	5′-CCGTGCATCCCTCATCAGC-3′	5′-GCTACGGCAGTAGCAGAAGGC-3′
Orexin	5′-AATCCTGACGATGGGAAAGAG-3′	5′-TCGTGGTTTTAGCGACAAGTG-3′
CART I	5′-AGGGTGCCGAAATGGACT-3′	5′-CAACATCGCATGTAGGAACG-3′
CCK	5′-CAGAATCATCTCCACCAAAGG-3′	5′-TCCATCCCAAGTAATCTCTGTC-3′
POMC	5′-AACCCCTTCTCACGCTCTTC-3′	5′-AACACCACCCACCCTCTTTT-3′
β-actin	5′-GATGATGAAATTGCCGCACTG-3′	5′-ACCAACCATGACACCCTGATGT-3′

Relative quantification experiments were conducted on 96-well plates under the following conditions: 95℃ for 5 min, followed by 40 cycles at 95℃ for 15 s, 58℃ for 15 s and 72℃ for 45 s. Samples were analyzed in duplicates and experiments were repeated at least twice. In all cases, a no template-negative control (in which cDNA was replaced by water) was included.

Expression levels were compared using the relative Ct ($\Delta\Delta$CT) method. Briefly, the average CT of the reference gene (*β-actin*) was subtracted from the average CT of the gene of interest to determine the ΔCT for each sample. The ΔCT of the calibrator (control fish or non-treated group) was then subtracted from the ΔCT of each of the samples to determine the $\Delta\Delta$CT. This number was then used to determine the amount of mRNA relative to the calibrator and normalized by *β-actin*. Data were provided as fold changes in expression relative to the reference gene, and compared to a calibrator sample from the control group. The average fold of all control samples were taken and set at 1. The experimental groups were then normalized relative to the control group. These values (1 for the control and other values for the experimental groups) were then statistically compared.

2.6 Data analysis

Student's *t*-tests with $P< 0.05$ were used to detect significant differences in food intake. Gene expression levels between GH-transgenic and NT common carp, in the *in vitro* experiment, and gene expression levels between GH-treated and non-treated groups were also analyzed by the Student's *t*-test. Statistical comparisons between fasted and fed TG and NT carp were carried out with one-way analysis of variance (ANOVA) using GraphPad Prism 5.0 (GraphPad Software, San Diego CA, USA). In the expression studies, real-time PCR data were acquired and analyzed using ABI Prism 7000 SDS 1.1 software (Life Technology, CA, USA). Data were considered reliable when the standard deviation (SD) of three replicate reactions was <15%. All samples are expressed as ratios of the specific target gene to *β-actin* and normalized as mRNA levels of NT fish or the non-treated group.

3. Results

3.1 Food intake

During the experimental period, we measured the average food intake of TG and NT common carp, which were 0.71 ± 0.024 g and 0.59 ± 0.03 g, respectively (Fig. 1A). TG common carp consumed more food during the experiment. The average maximum food intake in 0.5 h was 0.98 ± 0.05 g and 0.69 ± 0.043 g in TG and NT common carp, respectively (Fig. 1B). The time to consume 0.5 g food was also measured, in TG fish it was 8.65 ± 0.72 min and 14.76 ± 1.8 min in the control group (Fig. 1C). TG common carp exhibited higher and faster food intake.

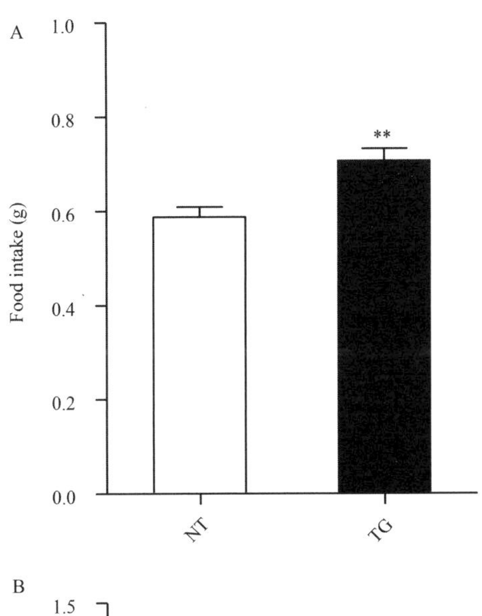

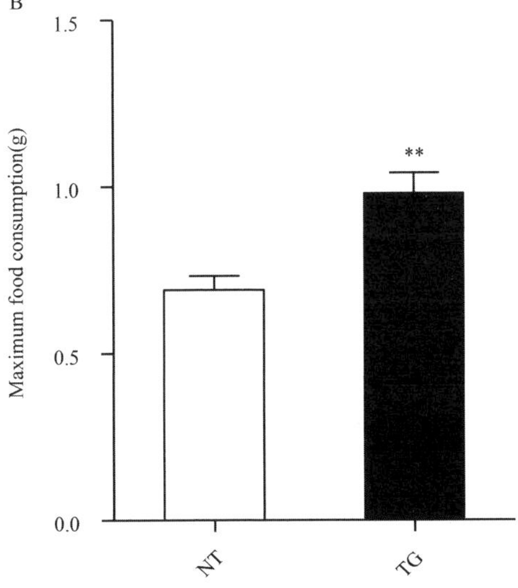

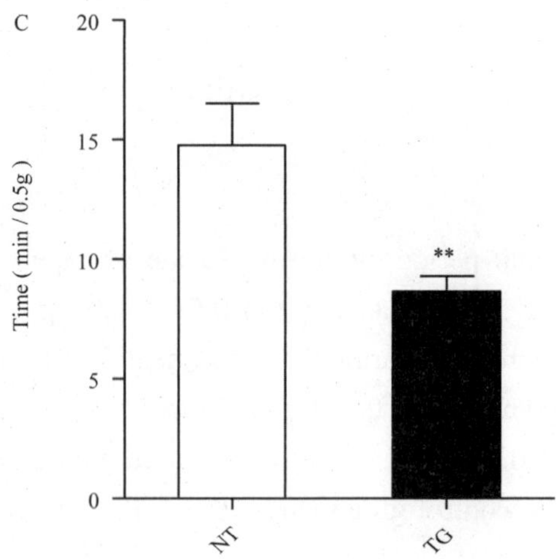

Fig.1 Food intake, maximum food consumption, and time taken to consume an equal amount of food by GH-transgenic (TG) and non-transgenic (NT) common carp. (A) Average food intake during the experimental period for TG and NT fish. For TG fish, the average food intake was 0.71 ± 0.024 g, and 0.59 ± 0.03 g for NT fish. (B) Average maximum food consumption in 0.5 h by TG and NT common carp. For TG fish, the average maximum food intake was 0.98 ± 0.05 g, compared to 0.69 ± 0.043 g for NT fish. (C) Average time taken to consume 0.5 g food by TG and NT common carp. TG fish took 8.65 ± 0.72 min to consume 0.5 g food, and NT fish took 14.76 ± 1.8 min. Asterisks represent significant differences (Student's *t*-test, $P< 0.001$)

3.2 Serum GH levels

For control fish, serum growth hormone levels during fasting were 3.10 ± 0.89 ng ml^{-1}. In satiety status when fed normally, the serum growth hormone levels were 0.2 ± 0.12 ng ml^{-1}. In TG fish, serum growth hormone levels were 7.74 ± 2.90 ng ml^{-1} and 9.81 ± 2.3 ng ml^{-1}, respectively (Fig. 2A).

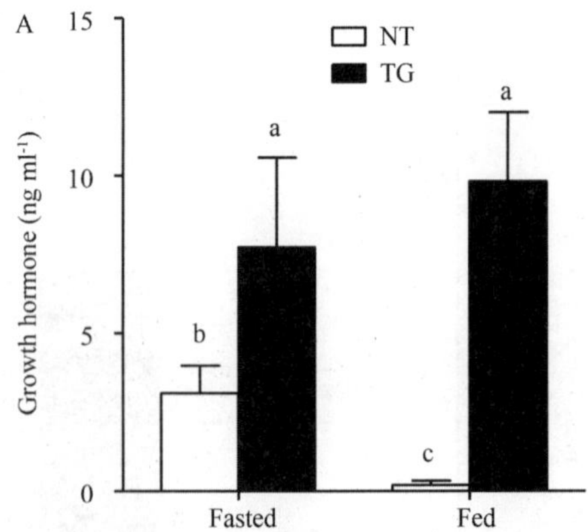

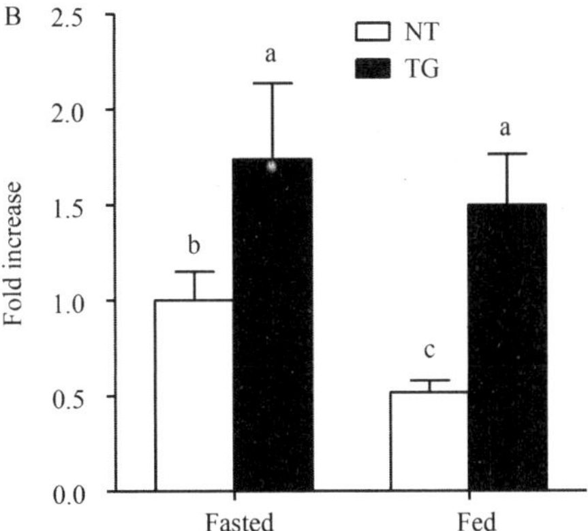

Fig. 2 Serum GH levels and hypothalamic *AgRP I* mRNA expression levels in fasted and fed common carp. (A) In NT common carp, serum GH level during fasting was 3.10 ± 0.89 ng ml^{-1}, the serum growth hormone level in satiety status when fed normally was 0.2 ± 0.12 ng ml^{-1}. In TG common carp, serum GH levels were 7.74 ± 2.90 ng ml^{-1} and 9.81 ± 2.3 ng ml^{-1}, respectively. (B) Expression levels in fasted NT common carp were normalized to 1. The mRNA expression level of hypothalamic *AgRP I* in the fed NT common carp is significantly lower (0.52) compared to the fasted NT fish (1). The mRNA expression level of hypothalamic *AgRP I* in fasted and fed TG common carp is 1.74-and 2.90-fold higher when compared to their NT counterparts respectively. Significant differences among groups are represented by different character (P< 0.05)

3.3 Hypothalamic *AgRP I* mRNA expression in fasted and fed fish

The mRNA expression level of hypothalamic *AgRP I* in the fed NT common carp is significantly lower (0.52) compared to the fasted NT fish (1). The mRNA expression level of hypothalamic *AgRP I* in fasted and fed TG common carp is 1.74-and 2.90-fold higher when compared to their NT counterparts respectively (Fig. 2B).

3.4 Hypothalamic GH and GHR mRNA expression

TG common carp displayed significantly higher hypothalamic *GH* mRNA expression, which was about 137.7-fold higher than in control fish. There was no significant difference in hypothalamic *GHR* mRNA expression levels between TG and NT common carp (Fig. 3).

3.5 Gene expression of behavior-observed fish

We analyzed gene expression levels in TG (n = 7) and NT (n = 7) common carp after observing their feeding behavior. The relative gene expression levels of *NPY*, *AgRP I*, orexin, *CCK*, *CART I* and *POMC* in the telencephalon and hypothalamus were detected. Significantly higher expression levels of the orexigenic gene *AgRP I* were observed in both the telencephalon and hypothalamus in TG common carp, which were 2.97 ± 0.72 and 2.62 ± 0.43-fold higher than in NT control fish, respectively. Other appetite regulation genes (*NPY*, orexin, *CCK*, *CART I* and *POMC*) showed no significant differences between the two groups (Fig. 4A and B).

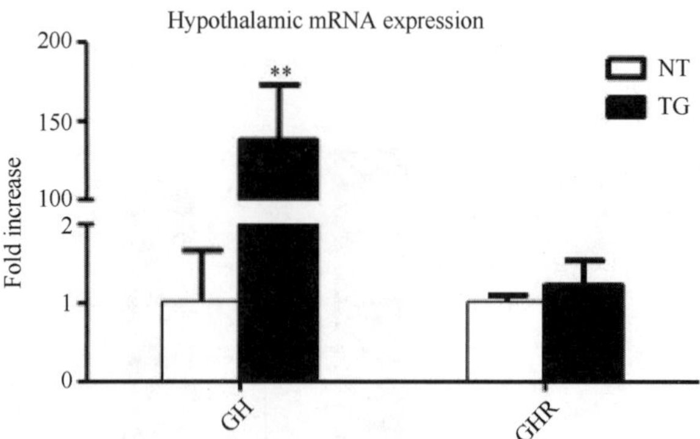

Fig. 3 Hypothalamic *GH* and *GH* receptor (*GHR*) mRNA expression levels in TG and NT common carp. Hypothalamic *GH* mRNA expression was 137.7-fold higher in TG fish than in NT fish, while there was no significant difference in hypothalamic *GHR* mRNA expression between TG and NT common carp. Expression levels in NT fish were normalized to 1. Data are presented as mean ± SEM. Asterisks represent significant differences (Student's *t*-test, $P< 0.001$)

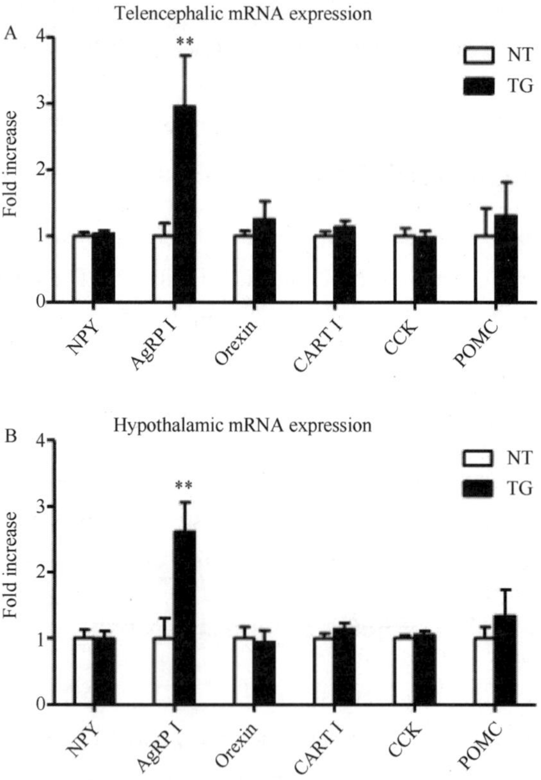

Fig. 4 Telencephalic and hypothalamic gene expression levels in TG and NT common carp after feeding behavior observation. (A) Telencephalic mRNA expression levels of orexigenic factors *NPY*, orexin, *AgRP I* and anorexigenic factors *CART I*, *CCK* and *POMC*. *AgRP I* was 2.97 ± 0.72-fold higher in TG fish (n = 7) than in NT fish (n = 7). (B) Hypothalamic mRNA expression levels of orexigenic factors *NPY*, orexin, *AgRP I* and anorexigenic factors *CART I*, *CCK* and *POMC*. *AgRP I* was 2.62 ± 0.43-fold higher in TG fish than in NT fish. Expression levels in NT fish were normalized to 1. Data are presented as mean ± SEM. Asterisks represent significant differences (Student's *t*-test, $P< 0.001$)

3.6 Gene expression of fish cultured outdoors

We analyzed hypothalamic gene expression levels in TG (n = 20) and NT (n = 20) common carp, which had been cultured outdoors. A significantly higher expression level of *AgRP I* was observed in TG fish, which was increased 2.14 ± 0.39-fold compared to NT fish. The relative gene expression levels of *NPY*, orexin, *CCK*, *CART I*, *CART II*, *POMC*, *MC4R*, *CRH* and *CRH-R1* showed no significant differences between the two groups (Fig. 5).

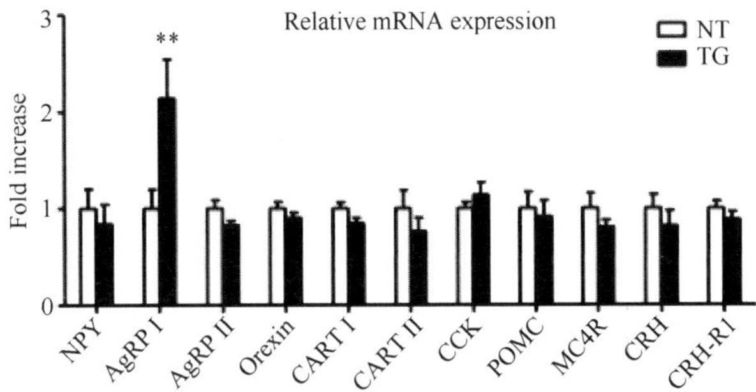

Fig. 5 Hypothalamic gene expression levels in TG (n = 20, 5 mpf) and NT (n = 20, 5 mpf) common carp fed normally when cultured outdoors. The relative gene expression levels of *NPY, AgRP I, AgRP II*, orexin, *CCK, CART I, CART II, POMC, MC4R, CRH* and *CRH-R1* in the hypothalamus. A higher expression level of *AgRP I* was observed in the hypothalamus of TG fish, which was 2.14 ± 0.39-fold higher than in NT fish. Expression levels in NT fish were normalized to 1. Data are presented as mean ± SEM. Asterisks represent significant differences (Student's t-test, P< 0.001)

3.7 *In vitro* experiment

NT common carp hypothalamic (n = 12) fragments were treated with GH at 20 ng ml^{-1}, 80 ng ml^{-1}, 200 ng ml^{-1}, and cultured for 2 h and 4 h, respectively. Changes in gene expression levels were detected in GH-treated fish at every concentration at each culture time, and also in the non-treated groups. When treated with 20 ng ml^{-1} of GH for 2 h, the mRNA expression level of *NPY* increased 2.10 ± 0.31-fold compared to non-treated group. There were no significant changes for the other tested concentrations of GH (80 ng ml^{-1} and 200 ng ml^{-1} for 2 h; 20 ng ml^{-1}, 80 ng ml^{-1} and 200 ng ml^{-1} for 4 h; Fig. 6A). For *AgRP I*, when treated with GH at 80 ng ml^{-1} and 200 ng ml^{-1} for 2 h, the expression levels of *AgRP I* were increased 1.85 ± 0.30 and 2.93 ± 0.34-fold compared to the control, respectively. When treated with 20 ng ml^{-1} GH for 4 h, the expression levels of *AgRP I* were increased 2.63 ± 0.23-fold compared to the control (Fig. 6B). There was no significant difference in *AgRP II* mRNA expression levels between the GH-treated and non-treated groups (Fig. 6C). The relative gene expression levels of orexin, *CCK*, *CART I*, *CART II*, *POMC* and *MC4R* showed no significant differences between the GH-treated and non-treated groups (Supplemental Fig. 1).

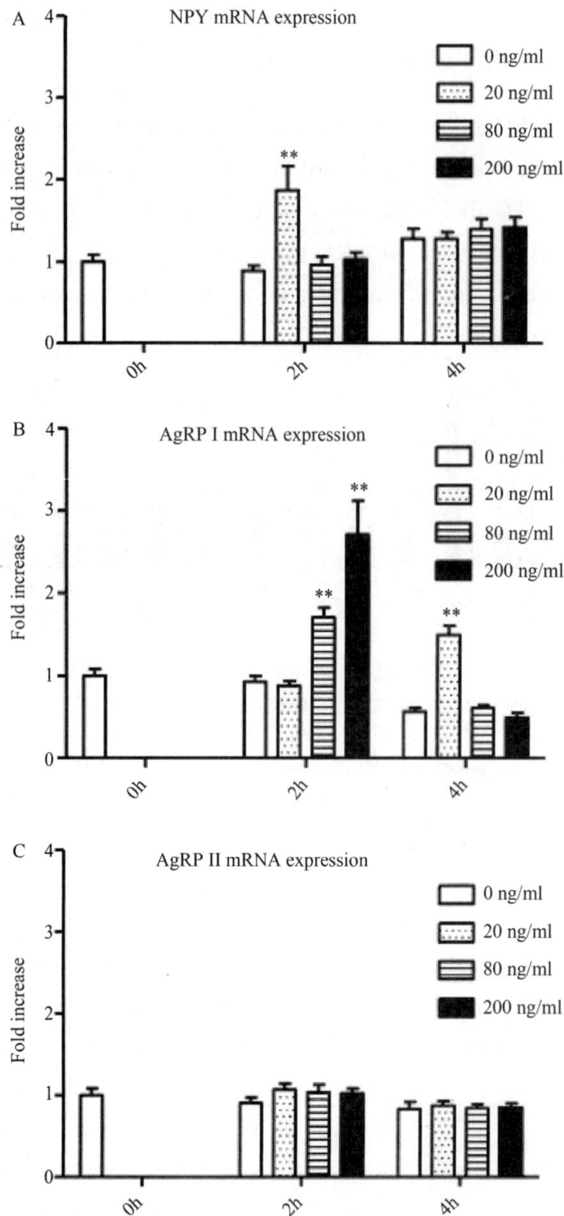

Fig. 6 Relative expression levels of *NPY, AgRP I* and *AgRP II* in GH-treated hypothalamus fragments. (A) *NPY* mRNA expression in cultured hypothalamus fragments treated with GH at final concentration of 20 ng ml^{-1}, 80 ng ml^{-1}, 200 ng ml^{-1}, for 2 h and 4 h, respectively. When treated with 20 ng ml^{-1} of GH for 2 h, the mRNA expression levels increased 2.10 ± 0.31-fold, compared to the non-treated group. (B) *AgRP I* mRNA expression in cultured hypothalamus fragments treated with GH at final concentration of 20 ng ml^{-1}, 80 ng ml^{-1}, 200 ng ml^{-1}, for 2 h and 4 h, respectively. When treated with GH at 80 ng ml^{-1} and 200 ng ml^{-1} for 2 h, the expression level of *AgRP I* increased 1.85 ± 0.30 and 2.93 ± 0.34-fold, compared to control levels, respectively. When treated with 20 ng ml^{-1} GH for 4 h, the expression levels of *AgRP I* increased to 2.63 ± 0.23-fold, compared to the control. (C) *AgRP II* mRNA expression in cultured hypothalamus fragments treated with GH at final concentration of 20 ng ml^{-1}, 80 ng ml^{-1}, 200 ng ml^{-1}, for 2 h and 4 h, respectively. Expression levels in control for 0 h were normalized to 1. Data are presented as mean ± SEM. Asterisks represent significant differences (Student's *t*-test, $P< 0.001$)

4. Discussion

4.1 GH increased food intake

In fish, it is well established that administration of exogenous GH increases food intake and feeding behavior (Johnsson and Bjornsson, 1994; Markert *et al.*, 1977). Similarly, elevated GH in transgenic salmon induced heightened feeding motivation, increased food intake and feeding behavior (Abrahams and Sutterlin, 1999; Devlin *et al.*, 1999; Sundstrom *et al.*, 2003). In our previous study, we also found that TG common carp consumed more food during competition with normal common carp of the same size, and that this was associated with higher social status and enhanced feeding motivation (Duan *et al.*, 2011). In this study, we demonstrated that under separate culture conditions, the maximum food intake in 0.5 h was significantly higher in TG individuals, and the food consumption time was also significantly shorter than in control fish. These results suggest that overexpressed GH may inherently enhance appetite and feeding motivation in common carp, which result in increased food intake and improve the ability to compete for limited food resources when cultured with control fish.

4.2 GH and hypothalamic feeding regulation

Circulating GH levels are influenced by the nutritional status of fish (Canosa *et al.*, 2007). Our study demonstrated that two weeks fasting elevated serum GH levels in NT common carp, while serum GH levels were very low in satiated fish. These results are in concordance with the studies in goldfish, flounder, rainbow trout and salmons (Canosa *et al.*, 2005; Fuentes *et al.*, 2012; Gabillard *et al.*, 2006; Pierce *et al.*, 2005; Shimizu *et al.*, 2009). Next, we observed high expression levels of serum GH in both fasted and fed TG common carp, while there is no significant difference between them (Fig. 2A). On the other hand, we observed that the mRNA expression of *AgRP I* was decreased in fed NT common carp, while there is no significant difference between fasted and fed TG common carp although both are significantly higher than their NT counterparts (Fig. 2B). Interestingly, this mRNA expression pattern is similar to the changes in serum GH levels in NT and TG common carp. In our study, the hypothalamic *GH* mRNA expression was significantly higher in TG common carp compared with NT individuals. As the *GH* transgene is driven by the β-actin promoter, the *GH* mRNA expression was also significantly higher in other tissues, such as muscle, liver, gonad (unpublished data) compared to NT fish. This may provide an explanation for the constant high level of serum GH in TG common carp which is not influenced by feeding condition. Our results suggested that the altered GH secretion pattern may change the mRNA expression levels of hypothalamic orexigenic factor *AgRP I* in TG common carp. The constant overexpression of GH transgene can result in increased food intake and feeding behavior (Abrahams and Sutterlin, 1999; Devlin *et al.*, 1999; Duan *et al.*, 2009, 2011; Sundstrom *et al.*, 2003), and GH can also regulate food intake by acting on the feeding control centre of the central nervous system (CNS) (Bohlooly *et al.*, 2005). Therefore, we speculate that the constant high levels of hypothalamic and serum GH may have a direct or indirect impact on the feeding regulation centre in the hypothalamus, thereby enhancing appetite and feeding motivation.

4.3 GH upregulated *AgRP I*

We did not detect any differences in the mRNA expression of *CART*, orexin and *CCK* in the hypothalamus and telecephelon of TG and NT common carp. We also found no significant changes in the expression of these genes when hypothalamus fragments were treated with GH *in vitro*. Similarly, there were no significant differences in the expression of *NPY* and *CCK* in TG coho salmon (Raven *et al.*, 2008). However, hypothalamic *AgRP I* mRNA expression in TG common carp was drastically upregulated in comparison with the control group, independent of whether these fish were bred indoors or outdoors, fasted or fed. Examination of the GH-treated hypothalamus fragments revealed that incubation with GH (80 ng ml^{-1} or 200 ng ml^{-1} for 2 h; 20 ng ml^{-1} for 4 h) could promote the expression of *AgRP I*. For *NPY*, no significant difference was found between the transgenic and control fish. *In vitro*, the mRNA expression levels of *NPY* only increased when incubated with 20 ng ml^{-1} of GH for 2 h, while there were no significant changes with other GH concentrations (80 ng ml^{-1} and 200 ng ml^{-1} for 2 h; 20 ng ml^{-1}, 80 ng ml^{-1} and 200 ng ml^{-1} for 4 h). These results show that *AgRP I*, instead of *NPY*, was responsive to the action of GH overexpression in transgenic common carp. A previous study revealed that *AgRP* and *NPY* are synthesized in the same neurons, and both are orexigenic neuropeptides in the CNS (Hahn *et al.*, 1998; Volkoff and Peter, 2006), indicating that *AGRP* and *NPY* are regulated in parallel. However, in TG common carp, the expression levels of *NPY* and *AgRP* were different, which may be the result of *NPY* and *AgRP* being involved in two different neuropeptide regulation pathways (Cerda-Reverter and Agulleiro, 2011; Murashita *et al.*, 2009). Some studies also demonstrated that GH could promote the expression of hypothalamic *AgRP*. Kamegai *et al.* found that *NPY*/*AgRP* neurons expressed *GHR* mRNA (Kamegai *et al.*, 1996), which was activated following systemic GH administration (Minami *et al.*, 1992). Furthermore, when the *GH* transgene was overexpressed in the CNS of mice, it could stimulate food intake, which was associated with a marked increase in *AGRP* (Bohlooly *et al.*, 2005). Moreover, the orexigenic effect of ghrelin, an endogenous GH secretagogue, was partly mediated by *AgRP*. After injection of ghrelin in *GHR*-deficient mice, neither the mRNA expression levels of *AgRP* or food intake changed, while both *AgRP* and food intake in the control group vastly increased (Egecioglu *et al.*,

2006). It has also been shown that *AgRP* can compensate for the loss of *NPY*. GH-releasing peptide-2 (*GHRP*-2) treated *NPY*-deficient mice still retained a strong appetite, and there was an increased level of *AgRP* expression (Tschop *et al*., 2002). Collectively, elevated GH levels could play a central role in regulating food intake via the hypothalamic *AgRP* neuropeptide pathway.

It has been revealed that *AgRP* is an important orexigenic neuropeptide in goldfish and zebrafish (Cerda-Reverter and Peter, 2003; Song *et al*., 2003). Two *AgRP* genes, *AgRP I* and *AgRP II* have been identified in common carp (Wan *et al*., 2012). In this study, both *in vitro* and *in vivo* data showed that only *AgRP I* was upregulated by GH, while *AgRP II* was not affected. In phylogenetic analysis, fish *AgRP I* and mammalian *AgRP* were grouped in the same cluster, indicating a conserved function of *AgRP I*. *AgRP II* did not share syntenies with *AgRP* in mammal, which might be a result of whole genome tetraplodization events (Västermark *et al*., 2012). In zebrafish, *AgRP I* morpholino oligonucleotides (MOs) injection reduced somatic growth rate which were associated with a marked decrease in pituitary *GH* mRNA expression. In contrast, suppression of *AgRP II* expression did not affect linear growth (Zhang *et al*., 2012). Therefore, we postulate that GH may regulate feeding behavior in common carp through upregulating *AgRP I* in the hypothalamus. *AgRP* increases food intake by functioning as an antagonist to the melanocortin 3 and 4 receptors. However, we found no differences in the expression of *POMC* and *MC4R* mRNA between TG and NT fish. In GH-treated hypothalamus fragments, we were unable to find any significant changes in the expression levels of these two genes. Some other studies have also demonstrated the orexigenic role of *AgRP* may act through other pathway besides inhibiting *MC4R* (Hagan *et al*., 2000; Kim *et al*., 2002).

In conclusion, we demonstrated that the expression of hypothalamic *AgRP I*, but not *NPY*, was higher in GH-transgenic common carp, suggesting that the constant high levels of GH may influence the *AgRP I* neuropeptide pathway. Therefore, we speculate that GH-induced hyperphagia in transgenic common carp may be partially due to direct action upon the appetite control centers in the hypothalamus. The mechanisms of GH in regulating appetite are a complex network. Besides its action on the CNS, GH may affect feeding behavior indirectly through a peripheral pathway. It is possible that changes in the feeding behavior of GH-transgenic carp may be the results of both CNS and peripheral regulation. Further research is required to clarify the underlying molecular mechanisms of appetite regulation in GH-transgenic fish.

Acknowledgements This work was supported financially by National Natural Science Foundation (Grant No. 30930069), the Development Plan of the State Key Fundamental Research of China (Grant No. 2010CB126302), the Key Research Program of the Chinese Academy of Sciences (Grant No. KSCX2-EW-N-004) and the '863' High Technology Project

(Grant No. 2011AA100404).

Appendix A. Supplementary data Supplementary data associated with this article can be found, in the onling version, at http://dx.doi.org/10.1016/j.ygcen.2013.03.024.

References

Abrahams, M.V., Sutterlin, A., 1999. The foraging and antipredator behaviour of growth-enhanced transgenic Atlantic salmon. Anim. Behav. 58, 933-942.

Arends, R.J., Vermeer, H., Martens, G.J., Leunissen, J.A., Wendelaar Bonga, S.E., Flik, G., 1998. Cloning and expression of two proopiomelanocortin mRNAs in the common carp (*Cyprinus carpio* L.). Mol. Cell. Endocrinol. 143, 23-31.

Bohlooly, Y.M., Olsson, B., Bruder, C.E., Linden, D., Sjogren, K., Bjursell, M., *et al.*, 2005. Growth hormone overexpression in the central nervous system results in hyperphagia-induced obesity associated with insulin resistance and dyslipidemia. Diabetes 54, 51-62.

Canosa, L.F., Chang, J.P., Peter, R.E., 2007. Neuroendocrine control of growth hormone in fish. Gen. Comp. Endocrinol. 151, 1-26.

Canosa, L.F., Peter, R.E., 2004. Effects of cholecystokinin and bombesin on the expression of preprosomatostatin-encoding genes in goldfish forebrain. Regul. Pept. 121, 99-105.

Canosa, L.F., Unniappan, S., Peter, R.E., 2005. Periprandial changes in growth hormone release in goldfish: role of somatostatin, ghrelin, and gastrin-releasing peptide. Am. J. Physiol. Regul. Integr. Comp. Physiol. 289, R125-R133.

Cerda-Reverter, J.M., Agulleiro, M.J., Sanchez, R.G.R.E., Ceinos, R., Rotllant, J., 2011. Fish melanocortin system. Eur. J. Pharmacol. 660, 53-60.

Cerda-Reverter, J.M., Cerda-Reverter, J.M., Ringholm, A., Schioth, H.B., Peter, R.E., 2003. Molecular cloning, pharmacological characterization, and brain mapping of the melanocortin 4 receptor in the goldfish: involvement in the control of food intake. Endocrinology 144, 2336-2349.

Cerda-Reverter, J.M., Peter, R.E., 2003. Endogenous melanocortin antagonist in fish: structure, brain mapping, and regulation by fasting of the goldfish agoutirelated protein gene. Endocrinology 144, 4552-4561.

Chan, Y.Y., Steiner, R.A., Clifton, D.K., 1996. Regulation of hypothalamic neuropeptide-Y neurons by growth hormone in the rat. Endocrinology 137, 1319-1325.

Devlin, R.H., Biagi, C.A., Yesaki, T.Y., 2004. Growth, viability and genetic characteristics of GH-transgenic coho salmon strains. Aquaculture 236, 607-632.

Devlin, R.H., Johnsson, J.I., Smailus, D.E., Biagi, C.A., Jonsson, E., Bjornsson, B.T., 1999. Increased ability to compete for food by growth hormone-transgenic coho salmon *Oncorhynchus kisutch* (Walbaum). Aquaculture Res. 30, 479-482.

Devlin, R.H., Yesaki, T.Y., Biagi, C.A., Donaldson, E.M., Swanson, P., Chan, W.K., 1994. Extraordinary salmon growth. Nature 371, 209-210.

Du, S.J., Gong, Z.Y., Fletcher, G.L., Shears, M.A., King, M.J., Idler, D.R., *et al.*, 1992. Growth enhancement in transgenic Atlantic salmon by the use of an all fish chimeric growth-hormone gene construct. Biotechnology 10, 176-181.

Duan, M., Zhang, T.L., Hu, W., Guan, B., Wang, Y.P., Li, Z.J., *et al.*, 2010. Increased mortality of growth-enhanced transgenic common carp (*Cyprinus carpio* L.) under short-term predation risk. J. Appl. Ichthyol. 26, 908-912.

Duan, M., Zhang, T.L., Hu, W., Sundstrom, L.F., Wang, Y.P., Li, Z.J., *et al.*, 2009. Elevated ability to compete for limited food resources by 'all-fish' growth hormone transgenic common carp *Cyprinus carpio*. J. Fish Biol. 75, 1459-1472.

Duan, M., Zhang, T.L., Hu, W., Li, Z.J., Sundstrom, L.F., Zhu, T.B., *et al.*, 2011. Behavioral alterations in GH transgenic common carp may explain enhanced competitive feeding ability. Aquaculture 317, 175-181.

Dunham, R.A., Chitmanat, C., Nichols, A., Argue, B., Powers, D.A., Chen, T.T., 1999. Predator avoidance of transgenic channel catfish containing salmonid growth hormone genes. Mar. Biotechnol. 1, 545-551.

Egecioglu, E., Bjursell, M., Ljungberg, A., Dickson, S.L., Kopchick, J.J., Bergstrom, G., *et al.*, 2006. Growth hormone receptor deficiency results in blunted ghrelin feeding response, obesity, and hypolipidemia in mice. Am. J. Physiol. Endocrinol. Metab. 290, E317-E325.

Farrell, A.P., Bennett, W., Devlin, R.H., 1997. Growth-enhanced transgenic salmon can be inferior swimmers. Can. J. Zool.: Rev. Can. Zool. 75, 335-337.

Fletcher, G.L., Shears, M.A., Yaskowiak, E.S., King, M.J., Goddard, S.V., 2004. Gene transfer: potential to enhance the genome of Atlantic salmon for aquaculture. Aust. J. Exp. Agric. 44, 1095-1100.

Fuentes, E.N., Kling, P., Einarsdottir, I.E., Alvarez, M., Valdes, J.A., Molina, A., *et al*., 2012. Plasma leptin and growth hormone levels in the fine flounder (*Paralichthys adspersus*) increase gradually during fasting and decline rapidly after refeeding. Gen. Comp. Endocrinol. 177, 120-127.

Gabillard, J.C., Kamangar, B.B., Montserrat, N., 2006. Coordinated regulation of the GH/IGF system genes during refeeding in rainbow trout (*Oncorhynchus mykiss*). J. Endocrinol. 191, 15-24.

Gossard, F., Dihl, F., Pelletier, G., Dubois, P.M., Morel, G., 1987. *In situ* hybridization to rat-brain and pituitary-gland of growth-hormone cDNA. Neurosci. Lett. 79, 251-256.

Guan, B., Hu, W., Zhang, T., Wang, Y., Zhu, Z., 2008. Metabolism traits of 'all-fish' growth hormone transgenic common carp (*Cyprinus carpio* L.). Aquaculture 284, 217-223.

Guan, B., Ma, H., Wang, Y.P., Hu, Y.L., Lin, Z.P., Zhu, Z.Y., *et al*., 2011. Vitreoscilla hemoglobin (*VHb*) overexpression increases hypoxia tolerance in zebrafish (*Danio rerio*). Mar. Biotechnol. 13, 336-344.

Hagan, M.M., Rushing, P.A., Pritchard, L.M., Schwartz, M.W., Strack, A.M., Van Der Ploeg, L.H., *et al*., 2000. Long-term orexigenic effects of AgRP (83-132) involve mechanisms other than melanocortin receptor blockade. Am. J. Physiol. Regul. Integr. Comp. Physiol. 279, R47-R52.

Hahn, T.M., Breininger, J.F., Baskin, D.G., Schwartz, M.W., 1998. Coexpression of Agrp and NPY in fasting-activated hypothalamic neurons. Nat. Neurosci. 1, 271-272.

Himick, B.A., Golosinski, A.A., Jonsson, A.C., Peter, R.E., 1993. CCK/gastrin-like immunoreactivity in the goldfish pituitary: regulation of pituitary hormone secretion by CCK-like peptides *in vitro*. Gen. Comp. Endocrinol. 92, 88-103.

Himick, B.A., Peter, R.E., 1994. CCK/gastrin-like immunoreactivity in brain and gut, and CCK suppression of feeding in goldfish. Am. J. Physiol. 267, R841-R851.

Hojvat, S., Baker, G., Kirsteins, L., Lawrence, A.M., 1982. Growth-hormone (GH) immunoreactivity in the rodent and primate CNS-distribution, characterization and presence post-hypophysectomy. Brain Res. 239, 543-557.

Hu, W., Zhu, Z.Y., 2010. Integration mechanisms of transgenes and population fitness of GH transgenic fish. Sci. China-Life Sci. 53, 401-408.

Huising, M.O., Metz, J.R., van Schooten, C., Taverne-Thiele, A.J., Hermsen, T., Verburg-van Kemenade, B.M., *et al*., 2004. Structural characterisation of a cyprinid (*Cyprinus carpio* L.) CRH, CRH-BP and CRH-R1, and the role of these proteins in the acute stress response. J. Mol. Endocrinol. 32, 627-648.

Johnsson, J.I., Bjornsson, B.T., 1994. Growth-hormone increases growth-rate, appetite and dominance in juvenile rainbow-trout, *Oncorhynchus mykiss*. Anim. Behav. 48, 177-186.

Kamegai, J., Minami, S., Sugihara, H., Hasegawa, O., Higuchi, H., Wakabayashi, I., 1996. Growth hormone receptor gene is expressed in neuropeptide Y neurons in hypothalamic arcuate nucleus of rats. Endocrinology 137, 2109-2112.

Kim, M.S., Rossi, M., Abbott, C.R., AlAhmed, S.H., Smith, D.M., Bloom, S.R., 2002. Sustained orexigenic effect of Agouti related protein may be not mediated by the melanocortin 4 receptor. Peptides 23, 1069-1076.

Lee, C.G., Devlin, R.H., Farrell, A.P., 2003. Swimming performance, oxygen consumption and excess post-exercise oxygen consumption in adult transgenic and ocean-ranched coho salmon. J. Fish Biol. 62, 753-766.

Lobie, P.E., Garciaaragon, J., Lincoln, D.T., Barnard, R., Wilcox, J.N., Waters, M.J., 1993. Localization and ontogeny of growth-hormone receptor gene-expression in the central-nervous-system. Dev. Brain Res. 74, 225-233.

Lopez-Patino, M.A., Guijarro, A.I., Isorna, E., Delgado, M.J., Alonso-Bedate, M., de Pedro, N., 1999. Neuropeptide Y has a stimulatory action on feeding behavior in goldfish (*Carassius auratus*). Eur. J. Pharmacol. 377, 147-153.

Markert, J.R., Higgs, D.A., Dye, H.M., Macquarrie, D.W., 1977. Influence of bovine growth-hormone on growth-rate, appetite, and food conversion of yearling coho salmon (*Oncorhynchus kisutch*) fed 2 diets of different composition. Can. J. Zool.: Rev. Can. Zool. 55, 74-83.

Martinez, R., Arenal, A., Estrada, M.P., Herrera, F., Huerta, V., Vazquez, J., *et al.*, 1999. Mendelian transmission, transgene dosage and growth phenotype in transgenic tilapia (*Oreochromis hornorum*) showing ectopic expression of homologous growth hormone. Aquaculture 173, 271-283.

Martinez, R., Estrada, M.P., Berlanga, J., Guillen, I., Hernandez, O., Cabrera, E., *et al.*, 1996. Growth enhancement in transgenic tilapia by ectopic expression of tilapia growth hormone. Mol. Mar. Biol. Biotechnol. 5, 62-70.

McKenzie, D.J., Martinez, R., Morales, A., Acosta, J., Morales, R., Taylor, E.W., *et al.*, 2003. Effects of growth hormone transgenesis on metabolic rate, exercise performance and hypoxia tolerance in tilapia hybrids. J. Fish Biol. 63, 398-409.

Minami, S., Kamegai, J., Sugihara, H., Hasegawa, O., Wakabayashi, I., 1992. Systemic administration of recombinant human growth-hormone induces expression of the c-fos gene in the hypothalamic arcuate and periventricular nuclei in hypophysectomized rats. Endocrinology 131, 247-253.

Minami, S., Kamegai, J., Sugihara, H., Suzuki, N., Wakabayashi, I., 1998. Growth hormone inhibits its own secretion by acting on the hypothalamus through its receptors on neuropeptide Y neurons in the arcuate nucleus and somatostatin neurons in the periventricular nucleus. Endocr. J. 45, S19-S26.

Murashita, K., Kurokawa, T., Ebbesson, L.O.E., Stefansson, S.O., Ronnestad, I., 2009. Characterization, tissue distribution, and regulation of agouti-related protein (AgRP), cocaine-and amphetamine-regulated transcript (CART) and neuropeptide Y (NPY) in Atlantic salmon (*Salmo salar*). Gen. Comp. Endocrinol. 162, 160-171.

Nam, Y.K., Cho, Y.S., Cho, H.J., Kim, D.S., 2002. Accelerated growth performance and stable germ-line transmission in androgenetically derived homozygous transgenic mud loach, *Misgurnus mizolepis*. Aquaculture 209, 257-270.

Nam, Y.K., Noh, J.K., Cho, Y.S., Cho, H.J., Cho, K.N., Kim, C.G., *et al.*, 2001. Dramatically accelerated growth and extraordinary gigantism of transgenic mud loach *Misgurnus mizolepis*. Transgenic Res. 10, 353-362.

Narnaware, Y.K., Peter, R.E., 2001. Effects of food deprivation and refeeding on neuropeptide Y (NPY) mRNA levels in goldfish. Comp. Biochem. Physiol. B Biochem. Mol. Biol. 129, 633-637.

Peng, C., Chang, J.P., Yu, K.L., Wong, A.O.L., Vangoor, F., Peter, R.E., *et al.*, 1993. Neuropeptide-Y stimulates growth-hormone and gonadotropin-ii secretion in the goldfish pituitary–involvement of both presynaptic and pituitary cell actions. Endocrinology 132, 1820-1829.

Pierce, A.L., Shimizu, M., Beckman, B.R., Baker, D.M., Dickhoff, W.W., 2005. Time course of the GH/IGF axis response to fasting and increased ration in Chinook salmon (*Oncorhynchus tshawytscha*). Gen. Comp. Endocrinol. 140, 192-202.

Rahman, M.A., Mak, R., Ayad, H., Smith, A., Maclean, N., 1998. Expression of a novel piscine growth hormone gene results in growth enhancement in transgenic tilapia (*Oreochromis niloticus*). Transgenic Res. 7, 357-369.

Raven, P.A., Uh, M., Sakhrani, D., Beckman, B.R., Cooper, K., Pinter, J., *et al.*, 2008. Endocrine effects of growth hormone overexpression in transgenic coho salmon. Gen. Comp. Endocrinol. 159, 26-37.

Ringholm, A., Fredriksson, R., Poliakova, N., Yan, Y.L., Postlethwait, J.H., Larhammar, D., *et al.*, 2002. One melanocortin 4 and two melanocortin 5 receptors from zebrafish show remarkable conservation in structure and pharmacology. J. Neurochem. 82, 6-18.

Shimizu, M., Cooper, K.A., Dickhoff, W.W., Beckman, B.R., 2009. Postprandial changes in plasma growth hormone, insulin, insulin-like growth factor (IGF)-I, and IGF-binding proteins in coho salmon fasted for varying periods. Am. J. Physiol. Regul. Integr. Comp. Physiol. 297, R352-R361.

Silverstein, J.T., Breininger, J., Baskin, D.G., Plisetskaya, E.M., 1998. Neuropeptide Ylike gene expression in the salmon brain increases with fasting. Gen. Comp. Endocrinol. 110, 157-165.

Silverstein, J.T., Plisetskaya, E.M., 2000. The effects of NPY and insulin on food intake regulation in fish. Am. Zool. 40, 296-308.

Song, Y., Golling, G., Thacker, T.L., Cone, R.D., 2003. Agouti-related protein (AGRP) is conserved and regulated by metabolic state in the zebrafish, *Danio rerio*. Endocrine 22, 257-265.

Stevens, E.D., Sutterlin, A., Cook, T., 1998. Respiratory metabolism and swimming performance in growth hormone transgenic Atlantic salmon. Can. J. Fish. Aquat. Sci. 55, 2028-2035.

Stutz, A.M., Morrison, C.D., Argyropoulos, G., 2005. The agouti-related protein and its role in energy homeostasis. Peptides 26, 1771-1781.

Sundstrom, L.F., Devlin, R.H., Johnsson, J.I., Biagi, C.A., 2003. Vertical position reflects increased feeding motivation in growth hormone transgenic coho salmon (*Oncorhynchus kisutch*). Ethology 109, 701-712.

Tschop, M., Statnick, M.A., Suter, T.M., Heiman, M.L., 2002. GH-releasing peptide-2 increases fat mass in mice lacking NPY: indication for a crucial mediating role of hypothalamic agouti-related protein. Endocrinology 143, 558-568.

Västermark, A., Krishnan, A., Houle, M.E., Fredriksson, R., Cerdá-Reverter, J.M., Schiöth, H.B., 2012. Identification of distant Agouti-like sequences and reevaluation of the evolutionary history of the Agouti-related peptide (AgRP). PloS One 7, e40982.

Volkoff, H., 2006. The role of neuropeptide Y, orexins, cocaine and amphetaminerelated transcript, cholecystokinin, amylin and leptin in the regulation of feeding in fish. Comp. Biochem. Physiol. Part A: Mol. Integr. Physiol. 144, 325-331.

Volkoff, H., Bjorklund, J.M., Peter, R.E., 1999. Stimulation of feeding behavior and food consumption in the goldfish, *Carassius auratus*, by orexin-A and orexin-B. Brain Res. 846, 204-209.

Volkoff, H., Peter, R.E., 2000. Effects of CART peptides on food consumption, feeding and associated behaviors in the goldfish, *Carassius auratus*: actions on neuropeptide Y-and orexin A induced feeding. Brain Res. 887, 125-133.

Volkoff, H., Peter, R.E., 2001. Characterization of two forms of cocaine–and amphetamine-regulated transcript (CART) peptide precursors in goldfish: molecular cloning and distribution, modulation of expression by nutritional status, and interactions with leptin. Endocrinology 142, 5076-5088.

Volkoff, H., Peter, R.E., 2006. Feeding behavior of fish and its control. Zebrafish 3, 131-140.

Volkoff, H., Xu, M., MacDonald, E., Hoskins, L., 2009. Aspects of the hormonal regulation of appetite in fish with emphasis on goldfish, Atlantic cod and winter flounder: notes on actions and responses to nutritional, environmental and reproductive changes. Comp. Biochem. Physiol. Part A: Mol. Integr. Physiol. 153, 8-12.

Wan, Y., Zhang, Y., Ji, P., Li, Y., Xu, P., Sun, X., 2012. Molecular characterization of CART, AgRP, and MC4R genes and their expression with fasting and re-feeding in common carp (*Cyprinus carpio*). Mol. Biol. Rep. 39, 2215-2223.

Wang, Y.P., Hu, W., Wu, G., Sun, Y.H., Chen, S.P., Zhang, F.Y., *et al.*, 2001. Genetic analysis of "all-fish" growth hormone gene transferred carp (*Cyprinus carpio* L.) and its F_1 generation. Chin. Sci. Bull. 46, 1174-1178.

Wu, G., Chen, L., Zhong, S., Li, Q., Song, C., Jin, B., *et al.*, 2008. Enzyme-linked immunosorbent assay of changes in serum levels of growth hormone (cGH) in common carps (*Cyprinus carpio*). Sci. China Ser. C Life Sci. 51, 157-163.

Zhang, C., Forlano, P.M., Cone, R.D., 2012. AgRP and POMC neurons are hypophysiotropic and coordinately regulate multiple endocrine axes in a larval teleost. Cell Metab. 15, 256-264.

Zhong, C.R., Song, Y.L., Wang, Y.P., Li, Y.M., Liao, L.J., Xie, S.Q., *et al.*, 2012. Growth hormone transgene effects on growth performance are inconsistent among offspring derived from different homozygous transgenic common carp (*Cyprinus carpio* L.). Aquaculture 356, 404-411.

Zhu, Z., Li, G., He, L., Chen, S., 1985. Novel gene transfer into the fertilized eggs of goldfish (*Carassius auratus* L. 1758). J. Appl. Ichthyol. 1, 31-34.

转基因鲤摄食量的增加可能由 AgRP 的上调介导

钟成容[1,2] 宋焱龙[1,2] 汪亚平[1] 张堂林[1] 段 明[1] 李勇明[1] 廖兰杰[1]
朱作言[1] 胡 炜[1]

1 中国科学院水生生物研究所，武汉 430072
2 中国科学院大学，北京 100049

摘 要 鱼类的食物摄取和摄食行为对其生存、竞争、生长和繁殖至关重要。转生长激素基因鲤表现出快速生长、饵料摄入增加和食物转化效率高的特征。然而，调控转基因鲤摄食的分子机制尚不清楚。本研究中，我们分析了转基因鲤和对照鱼摄食行为的差异，检测了下丘脑和端脑中 *NPY*、*AgRP I*、orexin、*POMC*、*CCK* 和 *CART I* 等基因 mRNA 的表达量。结果显示，养殖于池塘中同龄的转基因鲤与对照鱼下丘脑中食欲相关基因的表达水平接近。生长激素体外孵育下丘脑碎片证实了上述结果。进一步的研究发现，转基因鲤表现出摄食量增加和食物消化时间缩短的特征，转基因鱼的这种食欲特征与其下丘脑中 *AgRP I* mRNA 表达量的显著升高密切相关。我们的研究结果表明，转基因鲤中生长激素通过上调下丘脑中促食欲因子 *AgRP I* 基因的表达，进而影响其摄食行为。

第四部分
转基因黄河鲤的生态安全研究

Progress in the Evaluation of Transgenic Fish for Possible Ecological Risk and Its Containment Strategies

HU Wei WANG YaPing ZHU ZuoYan

State Key Laboratory of Freshwater Ecology and Biotechnology, Institute of Hydrobiology, Chinese Academy of Sciences, Wuhan 430072

Abstract Genetically improved transgenic fish possess many beneficial economic traits; however, the commercial aquaculture of transgenic fish has not been performed till date. One of the major reasons for this is the possible ecological risk associated with the escape or release of the transgenic fish. Using a growth hormone transgenic fish with rapid growth characteristics as a subject, this paper analyzes the following: the essence of the potential ecological risks posed by transgenic fish; ecological risk in the current situation due to transgenic fish via one-factor phenotypic and fitness analysis, and mathematical model deduction. Then, it expounds new ideas and the latest findings using an artificially simulated ecosystem for the evaluation of the ecological risks posed by transgenic fish. Further, the study comments on the strategies and principles of controlling these ecological risks by using a triploid approach. Based on these results, we propose that ecological risk evaluation and prevention strategies are indispensable important components and should be accompanied with breeding research in order to provide enlightments for transgenic fish breeding, evaluation of the ecological risks posed by transgenic fish, and development of containment strategies against the risks.

Keywords transgenic fish; ecological risks; countermeasures

Over 30 kinds of transgenic fish that provide high-quality food proteins have been successfully developed till date since the first batch of transgenic fish came out[1]; these include many important breeds such as carps, tilapia, catfish, and salmonids. The genetically improved transgenic fish possess desirable characteristics such as fast growth, high efficiency in forage transformation, cold resistance, and strong disease resistance[2]. However, till date, no transgenic fish have been released into a natural water body for commercialized cultivation as food; one of the reasons for this is the possible ecological risk posed by these fish. From a scientific perspective, these risks include the following. On one hand, the transferred gene may alter the population fitness of transgenic fish, lending them stronger viability and competitiveness to be the dominant species in the natural ecosystem. They may then occupy the ecological niche of other wild fish, leading to an alteration in the fish population structure

文章发表于《中国科学:生命科学》, 2007, 50(5):573-579

that could influence species diversity. On the other hand, transgenic fish may hybridize with species of close kinship in the natural ecosystem; consequently, the transferred gene may drift and pollute the wild-type gene library, thus affecting genetic diversity. Therefore, how do we evaluate the potential ecological risks of transgenic fish in an objective and all-inclusive manner? How do we take relevant containment strategies to control the possible ecological risks posed by transgenic fish in order to protect the ecosystem? This not only has become an urgent and challenging problem in the research on transgenic fish and a bottleneck to be cleared for their commercialization, but also becomes the basis for persistent research on transgenic fish breeding in the future. For these purposes, this paper is based on the subject of growth hormone (*GH*) transgenic fish with rapid growth characteristics. First we briefly review the *status quo* in the evaluation of ecological risks posed by transgenic fish by analyzing a single phenotype and performing fitness analysis, and carrying on mathematical model deduction. Then we expound some new ideas, and evaluate the latest findings from the ecological risk posed by transgenic fish by using an artificially simulated ecosystem of a lake in the field. At the same time, we also explain the strategies and principles involved in employing the triploid pathway to control the ecological risks posed by transgenic fish. Therefore, this paper proposes some enlightenment for the evaluation and countermeasure study of the ecological risks caused by the transgenic fishes.

1 Evaluation of the Ecological Risks Posed by Transgenic Fish

1.1 Research strategy in the analysis of single phenotype and mathematical model development

Although "case-by-case analysis" has become a common approach in the evaluation of the ecological risks posed by genetically modified organisms (GMO), no comprehensive theory or reliable method has been developed till date for the evaluation of the possible effect and the potential threat of transgenic fish on the ecosystem and biological diversity. Many studies related to ecological risk evaluation were based on the *GH*-transferred transgenic fish and employed the analysis of single phenotype; in other words, following the analysis of the transferred *GH* gene, the differences and similarities between the transgenic and wild-type fish were analyzed with regard to ingestion, growth, behavior, and other phenotypes in order to investigate the alteration of the fitness of the transgenic fish if they escaped or were released. Consequently, potential ecological risks posed by the transgenic fish were evaluated.

Food resources and the ability to gain food are the basis for fish to survive, grow, and reproduce. *Onchorhynchus kisutch* transferred with the *GH* gene appears to possess a better appetite and better ability to compete for food than the control fish[3]. Under conditions of abundant food, the transgenic fish with transferred *GH* grow faster than the control fish and can inhibit the growth of the latter through competition when bred together. Under conditions of food scarcity, the population of the *GH*-transferred fish was close to eradication or was extinct; meanwhile, in a mixed-breeding environment, the control fish population was close to eradication. However, when the control fish population was bred in the absence of transgenic fish, 72% of the fish could survive with a continuous increase in the population biomass[4]. An energetics study of the fast-growing "all-fish" transgenic carp with *GH* revealed that the food intake rate of the transgenic carp was apparently higher than that of the control carp[5]. The population of any species is in a dynamic balance in the natural ecosystem and is affected by food, space, and other ecosystem resources; in other words, in the natural ecosystem with abundant food resources, transgenic fish with *GH* may pose a larger threat to the survival and growth of the wild-type species due to its superior ability to compete for food. However, in the natural ecosystem with food scarcity, the wild-type species have better adaptability to the residential ecosystem due to the long-term natural selection of the wild-type species; consequently, the survival ability of the *GH*-transgenic fish may be weaker than that of the wild-type fish.

Swimming performance is one of the most important phenotypes of fish[6,7], and the highest swimming speed of the fish determines their ability to obtain food, search for mates, escape predation, and avoid an unfavorable environment; thus it is a factor that affects its survival and propagation in the natural environment[8-10]. The adolescent fish of *O. kisutch* transferred with *GH* possess a borderline swimming speed half that of the control fish with a similar size; the adult transgenic fish also has a significantly smaller borderline swimming speed[11,12]. Further, the death rates of both the fast-growing *Ictalurus punctatus* and *O. kisutch* transferred with the *GH* gene are markedly higher than that of the control fish[13,14]. The reason for the higher death rate of the transgenic fish is probably their lower swimming performance than the control fish, thus leading to their diminished ability to escape predation compared with the control fish. Our study revealed that the average body weight of the F_3 generation of the "all-fish" *GH* transgenic fish was 1.4-1.9 times that of the control fish with a specific growth rate that was 6%-10% higher than the control fish. However, the average values of the absolute and relative borderline swimming speed of the transgenic fish were 22% and 24% lower than the control fish, respectively[15]. In addition, the rates of the death of adolescent and adult transgenic fish caused by predation were markedly higher than those of the control fish[16]. Study of the evolutionary biology demonstrated that there might exist trade-offs between growth and swimming performance and that the higher the individual growth rate within the same species, the lower the borderline swimming speed or stroke speed[17-21]. This suggested that the fast-growth phenotype of the *GH*-transferred fish might lead to decreases in the swimming performance and the ability to escape predation risks, thereby reducing the fitness of the population and decreasing the risk to the ecosystem to a certain extent.

In addition to the above-mentioned studies on feeding, growth, swimming, and the ability to escape predation, the ecological risk evaluation of transgenic fish also involves a number of phenotype and fitness studies, such as those on morphology, physiology, metabolism, and anti-adversity[22]. It should be noted that the potential ecological risk posed by transgenic fish is the result of synthetic action between the complete fitness of the transgenic fish and the selection of the natural environment. The analysis of a single phenotype of transgenic fish is helpful for the recognition of their fitness in the natural ecosystem and consequently sheds light on the evaluation of the potential ecological risks posed by transgenic fish. However, all the studies reported till date were conducted indoors or in the small artificial outdoor facilities, and a great difference exists between the experimental conditions and a natural water body. Further, the analysis of a single phenotype could not indicate all the phenotypic parameters that might affect the fitness of transgenic fish in the natural ecosystem. Therefore, it may be difficult to extrapolate the result of the experimental analysis of a single fitness parameter to the conditions that transgenic fish have to survive the natural environmental factors in a complex ecosystem.

Viability and reproduction are very important parameters that affect the survival and multiplication of fish populations. Interbreeding is one of the ways for a transferred gene to drift away, and the survival rate, the size and age at sexual maturation, absolute puberty, and breeding selection ability are key factors that determine the drifting degree of the transferred gene. Therefore, the viability and reproduction of the population in the natural ecosystem constitute a very important fitness parameter in evaluating the ecological risks posed by transgenic fish. Muir and Howard[23,24] selected 6 fitness parameters closely related to viability and reproduction, namely, the viability of young fish, the viability of adult fish, the age of sexual maturity, and at each mature age, the fecundity of female fish, the inseminating ability and the copulation advantages of male fish. They used *Oryzias latipes* as a fish model to raise the "Trojan gene hypothesis" in evaluating the ecological risks posed by transgenic fish by establishing a mathematical model. This is currently a unique mathematical model for evaluating such risks. This model hypothesized that transgenic fish had characteristics similar to those of the wild-type fish, such as age of sexual maturity, fecundity, inseminating ability, rate of capture by predators, and life-span. However, the *GH* transgene might lead to a lower survival rate of *GH*-transgenic fish than the wild-type fish, and male transgenic fish might have a higher successful mating rate due to larger body size. Analysis of the model revealed that release of transgenic fish would result in the extinction of wild-type fish[23]. The introduction of this model attracted much attention and led to debates within the academic circle regarding the safety problem posed by transgenic fish[25]. Factors such as complex ecosystem, evolution, and random factors had a great effect on the transgene fate. However, in this model, the 6 parameters were measured without taking these important factors into consideration; hence, this model had an apparent defect and required revision[26]. Therefore, some scholars believed that the deduction of the ecological risks posed by transgenic fish from this model was unsuitable for the evaluation of the ecological risk posed by the rapidly growing transgenic fish in a population of cultivated fish[2].

Recently, Devlin and other scholars[22] summarized the status quo of research on the evaluation of the ecological risks posed by transgenic fish. They believed that the interaction between phenotype and environment had a very complex effect. Hence, it would be extremely difficult to estimate the complex interaction in the natural ecosystem following the escape or release of the transgenic fish on the basis of the simple phenotypic differences observed in the laboratory between the transgenic and wild-type fish. Even if it was possible to estimate the results in a natural ecosystem, they would be unreliable because in an indoor artificial facility, many limitations exist in the experimental water-body space, the structure of the aqueous organisms was simple, and the food was provided artificially rather than naturally. Therefore, the evaluation of the ecological risks posed by transgenic fish should be conducted in an aqueous environment with a large space and a complex structure.

The most natural and reliable aqueous environment for the evaluation of the ecological risks posed by transgenic fish is the natural ecosystem. However, when these ecological risks are unknown, it is obviously erroneous to release transgenic fish into the natural ecosystem for the evaluation of the ecological risks. Devlin *et al.*[22] proposed that the fish that had certain phenotypes similar to those of the transgenic fish could be released into the natural ecosystem for a substitution test and the ecological risks can then be evaluated. A study in which the transgenic fish are substituted with other fish such as sterile triploid transgenic fish, fish with embedded GH, or domesticated fish might help recognize a few of the potential ecological risks of the transgenic fish. However, the phenotype of the fish for the substitution test cannot completely resemble that of the transgenic fish; therefore, certain differences in fitness exist between the two types in the natural ecosystem, and the corresponding results of the research could not determine the ecological risks to the natural ecosystem. Therefore, there are certain limitations in the substitution test of fish that need to be overcome for the evaluation of the transgenic fish in the natural ecosystem.

1.2 Construction of artificial lakes

On the basis of an overall analysis of the biological characteristics and the key factors that are closely related to the ecological risks of the fast-growing transgenic fish, the Institute of Hydrobiology, Chinese Academy of Sciences, proposed a new concept of conducting safety research on the ecosystem of transgenic fish in an artificially simulated ecosystem. In May 2002, an artificially simulated lake with an area of 6.7 hectares was designed and constructed for the test, and all safety equipments were used. This lake was originally a piece of lowland in the water system of Liangzihu Lake. The researchers eliminated the accumulated water, deepened the lake, and increased the embankment in order to make the test zone a completely isolated area. In order to increase the space heterogeneity of the test lake and to investigate the utility and environmental choice of the transgenic fish, the lake bottoms with different depths were designed, and different water environments were created. This was done by transplanting submerged plants in certain areas, to ensure that the lake had abyssal regions, superficial regions, and regions with or without grass. There were no aquatic creatures in the test lake prior to water storage except some weeds and a few snails. When water began to collect in the lake, a net with mesh size of 380 μm was used to screen the water source to prevent the entry of bigger aquatic macroinvertebrates and fish into the lake. In order to rapidly enhance the multiplication of organisms within the water body of the test lake, rye grass was planted in the bottom of the lake as an organic fertilizer and fish forage. From a neighboring lake (belonging to the water system of Liangzihu lake), a few submerged plants, molluscs, shrimps, and fish of different species were artificially transplanted to the test lake in order to simulate the ecosystem structure of a natural lake.

Following the construction of the environment and the biotic community of the lake for the test, a follow-up monitoring of the physical and chemical environment and the biotic community of the test lake was performed from 2003 till date; data related to the structure and function of this ecosystem was collected in order to obtain basic data for the study of the effect of transgenic carp on the biotic community and the ecosystem of the lake. This study had revealed that, after the transitive adaptation for 3 years (2002-2004), the artificially transplanted submerged plants were thriving well, and the molluscs, shrimps and fish were alive and had multiplied to form populations. This indicated that these artificially released and transplanted organisms had been stable in the system of the test lake. Further, the fish fauna in the artificially simulated lake was composed of 12 families and 23 genera with carp fish forming the majority and accounting for approximately 65.2% of the total species; this is similar to the composition of the fish fauna found in the lakes located middle-and downstream of the Yangtze River in China[27,28]. Therefore, the artificially constructed lake had a species composition representative of the lakes located middle-and downstream of the Yangtze River.

In 2005, researchers released the all-fish *GH* transgenic common carps and the control carps into the artificially simulated test lake for a long-term follow-up analysis of the survival, growth, reproduction, and the drift of the transferred gene of the all-fish *GH* transgenic carps and the control carps in the artificially simulated natural ecosystem. They investigated the effect of transgenic fish on the structure of the biotic community and the ecosystem of the lake, and then evaluated the ecological risks posed by fast-growing carps wholly transferred with the *GH* gene in an overall, systematic and objective manner. Currently, relevant studies are in progress.

2 Study of the Containment Strategies Against the Ecological Risks of Transgenic Fish

The above analysis indicates that the potential ecological risks posed by the transgenic fish are closely related to the reproduction characteristics. Therefore, if sterile transgenic fish were developed, the genetic drift of the transferred gene caused by copulation could be controlled; moreover, transgenic fish that escape into the natural water system would be prevented from forming a viable population completely countering the threat posed to the ecosystem by transgenic fish.

2.1 Strategy for the artificial induction of triploid fish

An odd chromosome complement may lead to the failure of meiosis and impaired development of the reproductive organs. Therefore, triploid fish are sterile in general. The traditional techniques of artificially creating sterile fish such as triploid fish by physical and chemical measures have become one of the strategies to control the ecological risks posed by transgenic fish. However, the artificial rate of creating triploid fish could not reach 100%; moreover, among the artificially created triploid fish, the reproductive organs of some individuals might develop normally and produce mature sperms or ova with normal function[29-31]. Therefore, a few individuals with the ability to breed exist among the artificially created triploid fish and can pose potential ecological risks. Further, the method of artificial induction cannot satisfy the need for a large-scale commercial production of transgenic fish. Consequently, difficulties exist in the use of this technique and cannot be overcome by the method of artificial induction in order to control such potential risks.

2.2 Strategy to cultivate sterile triploid fish by hybridization between different ploids and the sterile mechanism

In a study conducted in the Hunan Normal University, *Carassius auratus* red var ($2n = 100$) was treated as the female parent and *Cyprinus carpio* L ($2n = 100$) as the male parent for hybridization. In the first generation (F_1) of hybridization, it was found that some fish could foster offspring. The selfing of the F_1 generation fish produced the F_2 generation of diploid fish; some fish could produce diplontic ova and sperms that could produce an amphiprotic allotetraploid after fertilization in the F_3 generation to foster offspring. This is the first artificially cultivated amphiprotic fertilizable allotetraploid organism in the fish or even the vertebrate family[32], and 15 continuous generations have been obtained till date.

Aiming at reducing the shortcomings of artificial induction in triploid fish, the Institute of Hydrobiology, Chinese Academy of Sciences, cooperated with the Hunan Normal

University in order to control the potential ecological risks posed by transgenic fish by conducting hybridization between the diplontic carp and the allotetraploid carp *Cyprinus carpio*; a transgenic triploid fish was developed by hybridization between the different ploids. The growth rate of the transgenic triploid fish was 15% faster than that of the control carp, and the utilization rate of the forage increased by 11.1%, so the transgenic triploid fish possessed some beneficial characteristics. In particular, the developmental characteristics of the sexual glands of the triploid fish included 3 states of abortion, namely, sterile ovary, sterile spermary, and sterile fat[33]. A direct slice-making method of the spermary cells was employed to study the mating of spermatocyte chromosome of the triploid fish during the first meiosis metaphase, and it was found that the spermatocyte of the triploid fish formed 50 bivalents and 50 univalents. During the meiotic process, the bivalents and the univalents coexisted and caused disorder in the mating and separation of the homologous chromosomes, finally leading to the production of aneuploid generative cells and the infertility of the triploid fish[34]. Therefore, hybridization between the carp and *Cyprinu scarpio* can produce sterile transgenic triploid fish, thus overcoming the limitations of the artificial induction of triploid pathway and completely countering the ecological threats posed by the transgenic carp with the *GH* gene.

3 Summary and Prospect

The potential ecological risk is a bottleneck in the persistent development of transgenic techniques for improvement of fish species, and the establishment of acomprehensive theory and methodology for evaluation of the ecological risk is the key to clear this bottleneck in transgenic fish breeding research. The evaluation of the ecological risk in the artificially simulated natural ecosystem presents a completely novel method of studying transgenic species. The successful construction of artificially simulated test lake implemented this concept, thus providing an important test platform for the study of the ecological risk posed by all-fish *GH*-transgenic fish in the natural ecosystem in an overall, systematic and objective manner. It also established the basis for obtaining an ecosystem model for predicting the risk to the ecosystem due to the fast-growing *GH*-transgenic fish.

After analyzing the history of breeding studies on transgenic fish and the industrial bottleneck from both realistic and long-term perspectives, we believe that the evaluation of this risk and the strategy against it are a necessary and important part of such studies and must be performed along with the breeding of transgenic fish. From the perspective of science, completely sterile transgenic triploid fish of carp and *Cyprinus carpio* have been bred through hybridization between ploids; this indicates that the industrialization conditions are suitable for all-fish *GH* transgenic carp. However, it should be noted that tetraplont fish are very difficult to obtain, and the successful induction of the tetraplont fish as actual species for cultivation has rarely been reported. Therefore, the development of a new strategy (such as the study of antisense gonadotropin releasing hormone (GnRH)-transgenic fish in which fertility can be controlled) for controlling the ecological risk posed by transgenic fish is very important and urgent.

Acknowledgements We thank Dr. Zhang Tanglin for assistance in preparing this manuscript.

References

1 Zhu Z, Li G, He L, *et al.* Novel gene transfer into the fertilized eggs of goldfish (*Carassius auratus* L. 1758). Z Angew Ichthyol, 1985, 1: 31-34

2 Maclean N, Laight R J. Transgenic fish: An evaluation of benefits and risks. Fish Fish, 2000, 1: 146-172

3 Devlin R H, Johnsson J I, Smailus D E, *et al.* Increased ability to compete for food by growth hormone-transgenic coho salmon *Oncorhynchus kisutch* (Walbaum). Aquac Res, 1999, 30: 479-482

4 Devlin R H, D'Andrade M, Uh M, *et al.* Population effects of growth hormone transgenic coho salmon depend on food availability and genotype by environment interactions. Proc Natl Acad Sci USA, 2004, 101: 9303-9308

5 Fu C, Li D, Hu W, *et al.* Growth and energy budget of F_2 "all-fish" growth hormone gene transgenic common carp. J Fish Biol, 2007, 70: 347-361

6 Plaut I. Critical swimming speed: Its ecological relevance. Comp Biochem Phys A, 2001, 131: 41-50

7 Reidy S P, Kerr S R, Nelson J A. Aerobic and anaerobic swimming performance of individual Atlantic Cod. J Exp Biol, 2000, 203: 347-357

8 Young P S, Cech J J. Improved growth, swimming performance, and muscular development in exercise-conditioned young-of-the-year striped bass *Morone saxatilis*. Can J Fish Aquat Sci, 1993, 50: 703-707

9 Drucker E G. The use of gait transition speed in comparative studies of fish locomotion. Am Zool, 1996, 36: 555-566

10 Swanson C, Young P S, Cech J J. Swimming performance of delta smelt: Maximum performance, and behavioral and kinematic limitations on swimming at submaximal velocities. J Exp Biol, 1998, 201: 333-345

11 Farrell A P, Bennett W, Devlin R H. Growth-enhanced transgenic salmon can be inferior swimmers. Can J Zool, 1997, 75: 335-337

12 Lee C G, Devlin R H, Farrell A P. Swimming performance, oxygen consumption and excess post-exercise oxygen consumption in adult transgenic and ocean-ranched coho salmon. J Fish Biol, 2003, 62: 753-766

13 Dunham R A, Chitmanat C, Nichols A, *et al.* Predator avoidance of transgenic channel catfish containing salmonid growth hormone genes. Mar Biotechnol, 1999, 1: 545-551

14 Sundstrom L F, Lohmus M, Johnsson J, *et al.* Growth hormone transgenic salmon pay for growth potential with increased predation mortality. Proc R Soc Lond B (Suppl) 2004, 271: 350-352

15 Li D, Fu C, Hu W, *et al.* Rapid growth cost in "all-fish" growth hormone gene transgenic carp: Reduced critical swimming speed. Chin Sci Bull, 2007, 52(11): 1501-1506

16 Zhang T, Hu W, Duan M, *et al.* The *gh*-transgenic common carp (*cyprinus carpio*): Growth performance, viability and predator avoidance.In: the Seventh International Congress on the Biology of Fish, St John's, Newfoundland, Canada, July 18-22, 2006

17 Kolok A S, Oris J T. The relationship between specie growth rate and swimming performance in male fathead minnows (*Pimephales promelas*). Can J Zool, 1995, 73: 2165-2167

18 Gregory T R, Wood C M. Individual variation and interrelationships between swimming performance, growth rate, and feeding in juvenile rainbow trout (*Oncorhynchus mykiss*). Can J Fish Aquat Sci, 1998, 55: 1583-1590

19 Gregory T R, Wood C M. Interactions between individual feeding behavior, growth, and swimming performance in juvenile rainbow trout (*Oncorhynchus mykiss*) fed different rations. Can J Fish Aquat Sci, 1999, 56: 479-486

20 Billerbeck J M, Lankford T E, Conover D O. Evolution of intrinsic growth and energy acquisition rates. I. Trade-offs with swimming performance in *Menidia menidia*. Evolution, 2001, 55: 1863-1872

21 Arendt J D. Reduced burst speed is a cost of rapid growth in anuran tadpoles: Problems of autocorrelation and inferences about growth rates. Funct Ecol, 2003, 17: 328-334

22 Devlin R H, Sundstrom L F, Muir W M. Interface of biotechnology and ecology for environmental risk assessments of transgenic fish. Trends Biotechnol, 2006, 24: 89-97

23 Muir W M, Howard R D. Possible ecological risks of transgenic organism release when transgenes affect mating success: Sexual selection and the Trojan gene hypothesis. Proc Natl Acad Sci USA, 1999, 96(24): 13853-13856

24 Muir W M，Howard R D. Fitness components and ecological risk of transgenic release: A model using Japanese Medaka (*Oryzias latipes*). Am Natur, 2001, 158: 1-16

25 Stokstad E. Engineered fish: Friend or foe of the environment? Science, 2002, 297: 1797-1798

26 Kapuscinski A R. Current scientific understanding of the environmental biosafety of transgenic fish and shellfish. Rev Sci Tech, 2005,24: 309-322

27 Zhang T, Li Z, Guo Q. Investigations on fishes and fishery of four lakes along the middle and lower basins of the Changjiang River. Acta Hydrobiol Sin (in Chinese), 2007, 31: in press

28 Zhang T L, Fang R L, Cui Y B. Comparisons of fish community diversity in five lake areas under different levels of fishery development. Acta Hydrobiol Sin (in Chinese), 1996, 20 (Suppl): 191-199

29 Wu Q J, Ye Y Z. Fish Genetics-breeding (in Chinese). Shanghai: Shanghai Science & Technology Press, 1998. 235-243

30 Wu C, Ye Y, Chen R. Genome manipulation in carp (*Cyprinus carpio*). Aquaculture, 1986, 54: 57-61

31 Wu C, Ye Y, Chen R. An artificial multiple triploid carp and its biological characteristics. Aquaculture, 1993, 111: 255-262

32 Liu S J, Liu Y, Zhou G, *et al*. The formation of tetraploid stocks of red crucian carp × common carp hybrids as an effect of interspecific hybridization. Aquaculture, 2001, 192(3-4): 171-186

33 Liu S J, Hu F, Zhou G J, *et al*. Gonadal structure of triploid crucian carp produced by crossing allotetraploid hybrids of *Carassium auratus* red var. (female)×*Cyprinus carpio* (male) with Japanese crucian carp (*Carassius auratus* Cavieri T. et S). Acta Hydrobiol Sin (in Chinese), 2000, 24(4): 301-306

34 Zhang C, He X X, Liu S J, *et al*. Chromosome pairing in meiosis I in allotetraploid hybrids and allotriploid crucian carp. Acta Zool Sin (in Chinese), 2005, 51(1): 89-94

转基因鱼的生态风险评价及其防范策略研究进展

胡　炜　汪亚平　朱作言

中国科学院水生生物研究所，淡水生态与生物技术国家重点实验室，武汉　430072

摘　要　遗传改良的转基因鱼具有许多优良经济性状，但转基因鱼迄今尚未进行商业化养殖，主要原因之一在于对转基因鱼逃逸或释放到自然水体中可能产生的生态风险的担忧。本文以具有快速生长特性的转生长激素基因鱼为对象，分析了转基因鱼潜在生态风险的实质，简要综述了单因子表型与适合度分析和数学模型推演等研究转基因鱼生态风险的现状，阐述了利用人工模拟生态系统开展转基因鱼生态风险评价的新思路及最新研究成果，同时评述了采用三倍体途径控制转基因鱼生态风险的策略及原理。在此基础上，提出生态风险评价与生态风险防范策略是转基因鱼育种研究体系中不可或缺的重要组成，必须与育种研究同步进行的观点，以期为转基因鱼育种及生态风险评价和对策研究提供启示。

Reduced Swimming Abilities in Fast-Growing Transgenic Common Carp *Cyprinus carpio* Associated with Their Morphological Variations

D. Li[1,2] W. Hu[1] Y. Wang[1] Z. Zhu[1] C. Fu[2]

1 State Key Laboratory of Freshwater Ecology and Biotechnology, Institute of Hydrobiology, Chinese Academy of Sciences, Wuhan 430072

2 Graduate University of Chinese Academy of Sciences, Beijing 100049

Abstract Critical swimming speeds (U_{crit}) and morphological characters were compared between the F_4 generation of GH-transgenic common carp *Cyprinus carpio* and the non-transgenic controls. Transgenic fish displayed a mean absolute U_{crit} value 22.3% lower than the controls. Principal component analysis identified variations in body shape, with transgenic fish having significantly deeper head, longer caudal length of the dorsal region, longer standard length (L_S) and shallower body and caudal region, and shorter caudal length of the ventral region. Swimming speeds were related to the combination of deeper body and caudal region, longer caudal length of the ventral region, shallower head depth, shorter caudal length of dorsal region and L_S. These findings suggest that morphological variations which are poorly suited to produce maximum thrust and minimum drag in GH-transgenic *C. carpio* may be responsible for their lower swimming abilities in comparison with non-transgenic controls.

Keywords cyprinidae; growth; morphology; swimming performance; trade-off; transgene

Introduction

Transgenic fish technology offers the opportunity to enhance global fish production. Since the first batch of growth hormone gene (GH) transgenic fishes was produced by Zhu *et al.* (1985), GH-transgenic fishes with fast-growing phenotypic traits have been produced in a variety of species throughout the world, such as common carp *Cyprinus carpio* L. (Zhu, 1992; Wang *et al.*, 2001), tilapia *Oreochromis* sp. and *Oreochromis nilotius* (L.) (Martinez *et al.*, 1996; Rahman *et al.*, 1998), mud loach *Misgurrus mizolepis* Günther (Nam *et al.*, 2001), Atlantic salmon *Salmo salar* L. (Du *et al.*, 1992) and coho salmon *Oncorhynchus kisutch* (Walbaum) (Devlin *et al.*, 1994, 1995). With the advance of transgenic fish technology, fishes may be considered the best candidate for the first marketable transgenic animal for human consumption (Zbikowska, 2003). Before such a product can enter the marketplace,

文章发表于 *Journal of Fish Biology*, 2009, 74(1):186-197

environmental risk assessments of transgenic fishes have to be carefully performed (Hu *et al.*, 2007; Sundström *et al.*, 2007). Knowledge of the physiology, morphology and behaviour in fast-growing GH-transgenic fishes is important for predicting the potential environmental risks posed by transgenics (Devlin *et al.*, 2006).

The information on swimming ability has important implications for ecological risk assessments of fast-growing GH-transgenic fishes, since the swimming performance is an important adaptation in fishes, which influences many life-history characteristics including prey capture, escape from predators, maintenance of position in a water current and migration to spawning habitats (Beamish, 1978; Plaut, 2001). GH-transgenic *O. kisutch* displayed significantly lower critical swimming speeds (U_{crit}) than non-transgenic controls (Farrell *et al.*, 1997; Lee *et al.*, 2003). Studies on swimming performance of GH-transgenic *S. salar* showed that transgenic fish were inferior (Deitch *et al.*, 2006) or equivalent swimmers (Stevens *et al.*, 1998). McKenzie *et al.* (2003), however, found no significant difference in swimming abilities between the GH-transgenic *Oreochromis* sp. hybrids and the controls. Such differences may be attributed in part to the degree of GH enhancement that exists between different GH-transgenic fish strains and species.

Variations in morphological characters can affect fish swimming performance (Webb, 1982, 1984; Taylor & Foote, 1991; Hawkins & Quinn, 1996; Blake, 2004; Seiler & Keeley, 2007). Previous studies have reported significant differences in morphology between the GH-transgenic salmonids and the controls (Ostenfeld *et al.*, 1998; Stevens & Devlin, 1999, 2005; Stevens *et al.*, 1999). Ostenfeld *et al.* (1998) and Lee *et al.* (2003) suggested that changes in body shape would be responsible for reduced U_{crit} of GH-transgenic salmonids. No studies, however, have explored the relationships between swimming performance and associated morphology in GH-transgenic fishes and non-transgenic controls.

'All-fish' GH-transgenic *C. carpio* were produced by micro-injection of pCAgcGHc into their fertilized eggs (Yellow River variety). All-fish gene construct pCAgcGHc was the recombinant construct of grass carp *Ctenopharyngodon idella* (Valenciennes) growth hormone cDNA (gcGHc) driven by *β*-actin gene promoter of *C. carpio* (pCA) (Wang *et al.*, 2001). A stable line of the GH-transgenic *C. carpio* has been obtained in the laboratory (Fu *et al.*, 2005). All-fish GH-transgenic *C. carpio* have a higher growth rate than non-transgenic controls (Wang *et al.*, 2001). Compensatory growth and energy budget of the GH-transgenic fish have also been evaluated (Fu *et al.*, 2007a, b). Recently, a preliminary study found that all-fish F_3 GH-transgenic *C. carpio* displayed reduced swimming ability compared with non-transgenic controls (Li *et al.*, 2007).

In this study, the hypothesis that variations in morphological characters may play an important role in differences in swimming abilities between fast-growing all-fish growth hormone gene (GH) transgenic *C. carpio* and non-transgenic wild types was tested.

Materials and Methods

Source of fish

F_4 all-fish GH-transgenic *C. carpio* used in this experiment were produced on 21 April 2006, from crosses between a wild-type female and a F_3 all-fish GH-transgenic male of a fast-growing transgenic strain, of which each generation transmits transgene DNA as a stable Mendelian trait. Frequencies of transgene transmission to F_4 progeny from this line were *c*. 50%, and body mass of F_4 transgenic fish were about twice that of the controls at age 69 days (unpubl. data). A batch of non-transgenic controls was also produced on the same day. At the age of 20 weeks, transgenic fish were transferred to indoor recycled water fish rearing systems in the laboratory. The tail-fin DNA of these fish was extracted and the polymerase chain reaction (PCR) method was used to detect the transgene positive individuals based on the procedure of Wang *et al*. (2001). Size-matched fish were obtained by transferring non-transgenic control fish aged *c*. 24 weeks to the laboratory. Trials of swimming performance were initiated when the fish were *c*. 28 weeks old.

Swimming test

The U_{crit} tests were conducted using a Brett-type swim tunnel (Brett, 1964), the swimming chamber was 1.25 m in length and 140 mm in inside diameter. The U_{crit} of 20 size-matched transgenic fish and non-transgenic controls were tested. Given that temperature is an important determinant of U_{crit}, a constant experimental temperature (23.1℃) was maintained and was essentially the same for both transgenic fish and controls (Table I). Prior to the start of the experiment, each test fish was starved for 24 h. After being placed in the chamber, the experimental fish was acclimated to a water velocity of 20 cm s^{-1} for 1 h. Thereafter, the fish was forced to swim against a water velocity that increased by 10 cm s^{-1} every 20 min until fatigue velocity (when fish ceased swimming against a downstream electrified grid). Once the swimming trials were completed, fish were anaesthetized, weighed (M), measured for maximum body depth (D_{Bmax}), maximum body width (W_{Bmax}) and for 11 additional morphological measurements. In this study, U_{crit} (cm s^{-1}) was calculated according to the formula described by Brett (1964) as follows: $U_{crit} = u_i + (t_i t_{ii}^{-1} u_{ii})$, where u_i is the highest velocity maintained for the prescribed period (cm s^{-1}), u_{ii} is the velocity increment (10 cm s^{-1}), t_i is the time (min) fish swam at the 'fatigue' velocity and t_{ii} is the prescribed period of swimming (20 min). Cross-sectional area (a) of the fish was calculated based on the equation: $a = \pi \times (0.5\ D_{Bmax}) \times (0.5\ W_{Bmax})$. Since the cross-sectional area of 16 transgenic fish and 16 controls was >10% of that of the inner tube, their U_{crit} were corrected for the blocking effect (Bell & Terhune, 1970).

Table I Water temperature, body mass (M), standard length (L_S) and critical swimming speeds (U_{crit}) of F_4 all-fish GH-transgenic *Cyprinus carpio* and non-transgenic controls used in the swimming tests (values are mean±S.E.)

Fish strain	n	Temperature (℃)	M (g)	L_S (mm)	U_{crit} (mm s^{-1})
Controls	20	23.11±0.12	202.37±5.41	206.9±2.2	942.7±22.8
Transgenics	20	23.14±0.17	207.35±4.48	217.8±1.8	770.0±16.6
ANOVA					
F		0.02	0.50	149	37.45
d.f.		1.38	1.38	1.38	1.38
P		0.88	0.48	$<$0.01	$<$0.01

Morphological data collection

For each individual fish, 11 morphological measurements were collected. Ten landmarks followed the truss method (Strauss & Bookstein, 1982) modified by Taylor & Foote (1991). They consisted of two head depths and eight distances between six truss network landmark points (Fig. 1). The 10 landmarks used by Hawkins & Quinn (1996) were adopted. Standard length (L_S) was also measured. All distances were measured by electronic digital calliper to the nearest 0.1 mm.

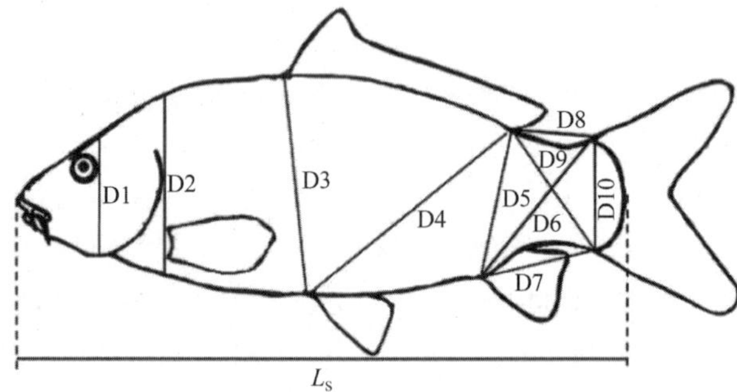

Figure 1 Morphological characters taken on the swimming GH-transgenic *Cyprinus carpio* and non-transgenic controls tested in swimming trials. D1, head depth at posterior margin of eye; D2, head depth at posterior margin of opercular membrane; D3, distance between dorsal fin's anterior insertion and pelvic fin's insertion; D4, distance between dorsal fin's posterior insertion and pelvic fin's insertion; D5, distance between dorsal fin's posterior insertion and anal fin's insertion; D6, distance between anal fin's insertion and caudal fin's dorsal origin; D7, distance between anal fin's insertion and caudal fin's ventral origin; D8, distance between dorsal fin's posterior insertion and caudal fin's dorsal origin; D9, distance between dorsal fin's posterior insertion and caudal fin's ventral origin; D10, distance between caudal fin's dorsal origin and caudal fin's the ventral origin; L_S, distance between snout tip and posterior end of caudal peduncle (standard length)

Statistical analysis

ANOVA was used to analyse the difference between transgenic fish and the controls for temperature, M, L_S and U_{crit}, which were confirmed to meet the assumption of normality and equality of variance by Kolmogorov-Smirnov test and Levene's test, respectively.

To remove size-dependant variation, all morphological variables were $\log_{10}$ transformed and were corrected for size effects using Burnaby's method (Burnaby, 1966; Rolf & Bookstein, 1987). This correction method removes the isometric growth related variation by performing a principal component analysis (PCA) of data, setting the first principal component (PC) scores to zero and then reserving the PCA processes, back-transforming the principal component scores to a size-free data set so that patterns in allometric growth variation (i.e. shape) among the groups could be detected (Sheehan *et al.*, 2005). Multivariate analysis of variance (MANOVA) was performed to assess body shape differences between the transgenic fish and the controls using the size-adjusted morphological characters. A PCA on the covariance matrix of size-adjusted morphological characters was carried out to summarize body shape differences between strains. With this ordination technique, the correlated multivariate data set was reduced into a smaller set of composite variables (PC) with a limited loss of information (McGarigal *et al.*, 2000). For PCA, scores of factors with eigenvalue greater than the mean eigenvalue of all factors were used in further analysis. As there is no absolute rule for factor loading cut offs, the cut off value 0.5 was used to determine if a component-loading coefficient was strong. The differences in mean factor scores between strains were examined by ANOVA. A general linear model (GLM) with U_{crit} as dependent variable, strain as a categorical factor and the PC axes which the differences were detected in mean factor scores between strains as independent variables, was used to reveal possible effects of morphology on swimming performance. The relationship between morphological characters and swimming abilities was explored by regressing U_{crit} from each fish on its corresponding factor score from the principal axes.

All analyses were performed with SPSS 13.0 (SPSS Inc., Chicago, IL, U.S.A.), Origin 6.1 (Origin Lab Corp., Northampton, MA, U.S.A.) and NTSYS 2.0 (Applied Biostatistics Inc., Port Jefferson, NY, U.S.A.). Differences were regarded as significant when $P < 0.05$.

Results

Swimming performance

The M of experimental fish were size-matched ($P > 0.05$), but transgenic fish had significantly longer L_S than the controls ($P < 0.01$) (Table I). There were not any significantly relationships ($P > 0.05$), however, between L_S and U_{crit} for transgenics ($r^2 = 0.01$), for controls ($r^2 = 0.13$) or for both ($r^2 = 0.07$), so one-way ANOVA was carried out to compare U_{crit} between transgenic fish and controls. The U_{crit} of transgenic fish were significantly lower than those of the controls ($P < 0.01$) (Table I). Transgenic fish had a mean U_{crit}

value 22.3% lower than non-transgenic controls.

Morphological comparison

The analysis based on 11 size-adjusted morphological characters indicated there was significant difference in body shape between transgenic fish and non-transgenic controls (MANOVA, Wilk's $\lambda = 0.33, F_{10,29} = 5.83, P < 0.01$). Using PCA, 11 morphological characters were condensed into two PCs, which were extracted accounting for 68.72% of the total variance. There was no significant relation between any principal axis and L_S (for all correlations, $r^2 \leqslant 0.06, P \geqslant 0.05$), which showed that the effects of size were removed effectively by Burnaby's method. PC1 accounted for 48.68% of variance and summarized variation in head depth (D1), body depth (D3), caudal region (D5, D6, D7, D8 and D10) and L_S (Table II). Transgenic fish were essentially separated from non-transgenic controls on PC1, because they have significantly smaller body depth (D3), caudal depth (D5 and D10), caudal length of ventral region (D6 and D7) and larger head depth (D1), caudal length of dorsal region (D8) and L_S (ANOVA, $F_{1,38} = 51.38, P < 0.001$; Fig. 2). PC2 accounted for 20.04% of variance and summarized variation in head depth (D1 and D2), body depth (D3), caudal region (D7, D8 and D9) and L_S (Table II). But it was not effective to identify difference between strains (ANOVA, $F_{1,38} = 1.17, P > 0.05$; Fig. 2).

Table II Loading coefficients and percentage of total variance from a principal component analysis with 11 size-adjusted morphological characters (See Fig. 1) of F_4 all-fish GH-transgenic *Cyprinus carpio* and non-transgenic controls used in the swimming tests (strong loadings indicated in bold)

Morphological characters	PC1	PC2
D1	**−0.619**	**0.721**
D2	−0.122	**0.625**
D3	**0.624**	0.464
D4	0.254	0.295
D5	**0.932**	−0.046
D6	**0.621**	0.163
D7	**0.567**	−0.334
D8	**−0.716**	**−0.643**
D9	0.133	**−0.676**
D10	**0.641**	0.257
SL	**−0.503**	**0.543**
Variation (%)	48.68	20.04

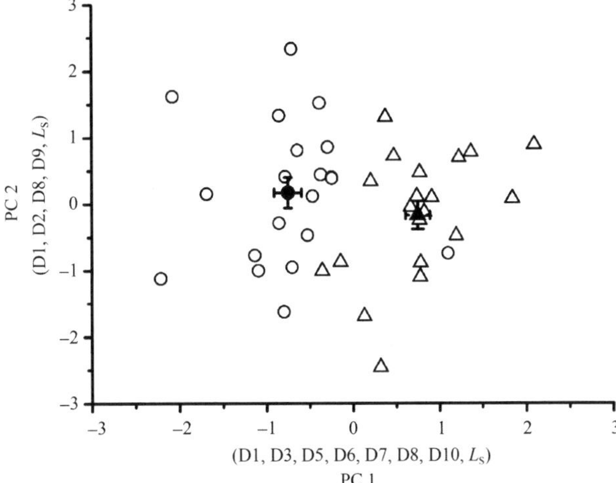

Figure 2 Principal component (PC) scores from a principal component analysis with 11 size-adjusted morphological characters (see Fig. 1) of F_4 all-fish GH-transgenic *Cyprinus carpio* (○) and non-transgenic controls (△) used in the swimming tests for PC1 *v*. PC2. The mean PC score for each strain is indicated by larger solid symbol (S.E. of the mean, horizontal and vertical bars)

Relationship between swimming performance and morphology

GLM analysis revealed that morphology defined by PC1 had significant effects on U_{crit} ($F_{1,37}$= 5.01, P < 0.05). There was a significant positive relationship between U_{crit} and PC1 (Fig. 3). Higher U_{crit} was related to the combination of deeper body and caudal region, longer caudal length of ventral region, shallower head depth, shorter caudal length of dorsal region and L_S found in control fish.

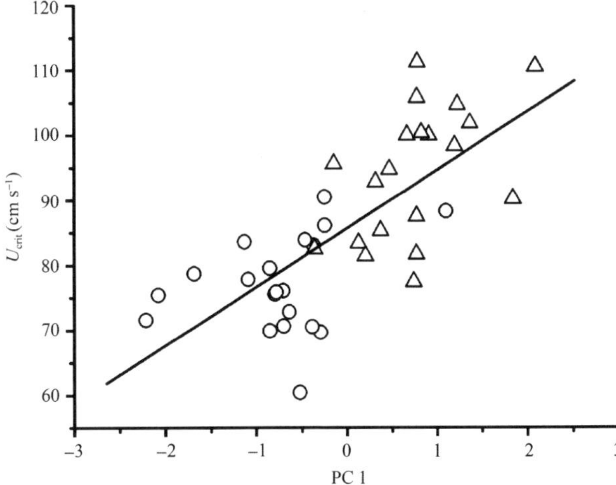

Figure 3 The relationship between critical swimming speed (U_{crit}) and body shape variations as defined by PC1 from the principal component analysis with 11 size-adjusted morphological characters (see Fig. 1) of F_4 all-fish GH-transgenic *Cyprinus carpio* (○) and non-transgenic controls (△) used in the swimming tests. The curve was fitted (both groups) by y = 85.63 + 8.99 x (r^2= 0.52, P < 0.01). The regressions for individual treatments tested separately were not significant (control: r^2= 0.16, P > 0.05; transgenic: r^2= 0.10, P > 0.05)

Discussion

The mean U_{crit} for non-transgenic controls (94 cm s^{-1}) recorded in the present study were within the range reported for *C. carpio*. In a previous study, where experimental conditions were similar to the present ones, the mean U_{crit} for *C. carpio* were between 70 and 100 cm s^{-1} depending on the flume length (Tudorache *et al.*, 2007). Fast-growing F_4 all-fish GH-transgenic *C. carpio* have a mean U_{crit} value 77 cm s^{-1}, displaying significantly reduced swimming ability. This result is in agreement with previous reports that GH-transgenic fishes were inferior swimmers (Farrell *et al.*, 1997; Lee *et al.*, 2003; Deitch *et al.*, 2006; Li *et al.*, 2007). It is generally assumed that U_{crit} reflect maximum oxygen consumption capability (Farrell & Steffensen, 1987) and they could provide a relative index by which the physical status of the fishes can be quantified and compared (Brauner *et al.*, 1994; Plaut, 2001). So the reduced U_{crit} may be reflection of GH-transgenic *C. carpio*'s poor physical status. In addition, there is a growing list of ectotherms for which there is evidence that individual growth rates are optimized to suit local conditions rather than maximized to be as fast as possible (Arendt, 1997; Mangel & Munch, 2005; Stamps, 2007). Evidence has accumulated that there is a trade-off between benefits and costs associated with rapid growth in teleosts and anurans (Kolok & Oris, 1995; Gregory & Wood, 1998, 1999; Morgan & Metcalfe, 2001; Arendt, 2003; Arendt & Hoang, 2005; Dayton *et al.*, 2005; Biro *et al.*, 2004, 2006). The present result which showed fast-growing GH-transgenic *C. carpio* having significantly reduced swimming ability may further support that trade-offs between growth rates and swimming performance, *i.e.* faster-growing individuals had slower U_{crit}, may be common in teleosts (Kolok & Oris, 1995; Gregory & Wood, 1998, 1999; Billerbeck *et al.*, 2001; Munch & Conover, 2004; Álvarez & Metcalfe, 2005; Arnott *et al.*, 2006).

This study found that GH-transgenic *C. carpio* had significant morphological variations in head depth, body depth, caudal section and L_S compared with their non-transgenic controls. Previous studies have shown that GH-transgenic salmonids and *C. carpio* displaying morphological variations in body and head shape, muscle, gill and gut morphology (Ostenfeld *et al.*, 1998; Stevens & Devlin, 1999, 2005; Stevens *et al.*, 1999; Hill *et al.*, 2000; Wang *et al.*, 2001; Dunham *et al.*, 2002). Changes in body shape may greatly alter fish hydrodynamics (Webb, 1984; Sagnes *et al.*, 1997). Ostenfeld *et al.* (1998) and Lee *et al.* (2003) suggested that changes in body shape would affect fish hydrodynamics, hence drag coefficients and U_{crit} for GH-transgenic salmonids. The present study provided the first evidence that fast-growing GH-transgenic *C. carpio* possessing a relatively smaller body depth, caudal depth, caudal length of ventral region and larger head depth, caudal length of dorsal region and L_S displayed reduced swimming ability in comparison to the controls. These results were consistent with the previous results which all found intraspecific differences in U_{crit} and found the superior swimmer had deeper body and larger caudal regions (Taylor & Foote, 1991; Hawkins & Quinn, 1996; Seiler & Keeley, 2007). U_{crit} represent the maximum sustainable speeds of fishes, and at these speeds fishes predominantly utilize a body–caudal-fin swimming mode (Fisher & Hogan, 2007). Deeper body sections move more water and thus produce greater thrust (Law & Blake, 1996). So it is easy to understand the non-transgenic *C. carpio* with deeper body and caudal region may generate stronger propulsion. Moreover, their greater body and caudal depth could function to minimize recoil energy losses (Webb, 1984; Blake, 2004). For transgenic *C. carpio*, deeper head may have increased their resistance to water and swimming drag, given fusiform body may reduce swimming drag (Webb, 1975). So the present results indicated that the changes of morphological characters in GH-transgenic *C. carpio* may be responsible for their inferior prolonged swimming ability. Otherwise, areas and ratios of fish body and fins play an important role in the generation of thrust (Blake *et al.*, 2005; Fisher & Hogan, 2007; Nanami, 2007), so further studies should be performed to compare these morphological characters between fast-growing transgenic *C. carpio* and controls, and to explore the linkage between the these morphological characters and swimming performance. In addition, changes in body shape have been related to habitat use in freshwater fish populations (Robinson & Wilson, 1994; Snorrason *et al.*, 1994). Relationships between morphology and habitat use have been correlated with intraspecific variations in swimming performance associated with habitat use (McLaughlin & Grant, 1994; Bourke *et al.*, 1997; Dynes *et al.*, 1999; Boily & Magnan, 2002). Morphological variations linked with reduced swimming ability may make GH-transgenic *C. carpio* have a disadvantage for overcoming natural selection pressures when competing with wild-type individuals in their natural habitat. Further studies are needed to clarify this hypothesis.

In conclusion, this study showed that reduced U_{crit} were the cost of rapid growth of F_4 all-fish GH-transgenic *C. carpio*. Relatively shallower body and caudal region, shorter caudal length of the ventral region and deeper head, longer caudal length of the dorsal region and L_S in GH-transgenic *C. carpio* correlated with reduced swimming ability in contrast with the controls. The findings of this study suggest that morphological variations which are poorly suited to produce maximum thrust and minimum drag in GH-transgenic *C. carpio* may be responsible for their lower swimming abilities in comparison with non-transgenic controls and fast-growing GH-transgenic fish might be an excellent model to disentangle the functional link between morphology and swimming performance.

Acknowledgements This work was financially supported by National Natural Science Foundation (Grant No. 30400056 and 30430540) the Development Plan of the state Key Fundamental Research of China (Grant No. 2007CB109205, 2007CB109206), and the '863' High Technology Project (Grant No. 2006AA10Z141).

References

Álvarez, D. & Metcalfe, N. B. (2005). Catch-up growth and swimming performance in threespine sticklebacks (*Gasterosteus aculeatus*): seasonal changes in the cost of compensation. *Canadian Journal of Fish Aquatic Sciences* 62, 2169-2176.

Arendt, J. D. (1997). Adaptive intrinsic growth rates: an integration across taxa. *The Quarterly Review of Biology* 72, 149-177.

Arendt, J. D. (2003). Reduced burst speed is a cost of rapid growth in anuran tadpoles: problems of autocorrelation and inferences about growth rates. *Functional Ecology* 17, 328-334.

Arendt, J. & Hoang, L. (2005). Effect of food level and rearing temperature on burst speed and muscle composition of western spadefoot toad (*Spea hammondii*). *Functional Ecology* 19, 982-987.

Arnott, S. A., Chiba, S. & Conover, D. O. (2006). Evolution of intrinsic growth rate: metabolic costs drive trade-offs between growth and swimming performance in *Menidia menidia. Evolution* 60, 1269-1278.

Beamish, F. W. H. (1978). Swimming capacity. In Fish Physiology, Vol. 7 (Hoar, W. H. & Randall, D. J., eds), pp. 101-187. New York, NY: Academic Press.

Bell, W. H. & Terhune, L. D. B. (1970). Water tunnel design for fisheries research. *Fisheries Research Board of Canada Technical Reports* 195, 1-69.

Billerbeck, J. M., Lankford, T. E. & Conover, D. O. (2001). Evolution of intrinsic growth and energy acquisition rates. I. Tradeoffs with swimming performance in *Menidia menidia. Evolution* 55, 1863-1872.

Biro, P. A., Abrahams, M. V., Post, J. R. & Parkinson, E. A. (2004). Predators select against high growth rates and risk-taking behaviour in domestic trout populations. *Proceedings of the Royal Society B* 271, 2233-2237.

Biro, P. A., Abrahams, M. V., Post, J. R. & Parkinson, E. A. (2006). Behavioural trade-offs between growth and mortality explain evolution of submaximal growth rates. *Journal of Animal Ecology* 75, 1165-1171.

Blake, R. W. (2004). Fish functional design and swimming performance. *Journal of Fish Biology* 65, 1193-1222.

Blake, R. W., Law, T. C., Chan, K. H. S. & Li, J. F. Z. (2005). Comparison of the prolonged swimming performances of closely related, morphologically distinct three-spined sticklebacks *Gasterosteus* spp. *Journal of Fish Biology* 67, 834-848.

Boily, P. & Magnan, P. (2002). Relationship between individual variation in morphological characters and swimming costs in brook charr (*Salvelinus fontinalis*) and yellow perch (*Perca flavescens*). *Journal of Experimental Biology* 205, 1031-1036.

Bourke, P., Magnan, P. & Rodrigues, M. A. (1997). Individual variations in habitat use and morphology in brook charr. *Journal of Fish Biology* 51, 783-794.

Brauner, C. J., Iwama, G. K. & Randalld, J. (1994). The effect of short-duration seawater exposure on the swimming performance of wild and hatchery-reared juvenile coho salmon (*Oncorhynchus kisutch*) during smoltification. *Canadian Journal of Fisheries and Aquatic Sciences* 51, 2188-2194.

Brett, J. R. (1964). The respiratory metabolism and swimming performance of young sockeye salmon. *Canadian Journal of Fisheries and Aquatic Sciences* 21, 1183-1226.

Burnaby, T. P. (1966). Growth-invariant discriminant functions and generalized distances. *Biometrics* 22, 96-110.

Dayton, G. H., Saenz, D., Baum, K. A., Langerhans, R. B. & Dewitt, T. J. (2005). Body shape, burst speed and escape behavior of larval anurans. *Oikos* 111, 582-591.

Deitch, E. J., Fletcher, G. L., Petersen, L. H., Costa, I. A. S. F., Shears, M. A., Driedzic, W. R. & Gamperl, A. K. (2006). Cardiorespiratory modifications, and limitations, in post-smolt growth hormone transgenic Atlantic salmon Salmo salar. *Journal of Experimental Biology* 209, 1310-1325.

Devlin, R. H., Yesaki, T., Biagy, C. A., Donaldson, E. M., Swanson, P. & Chan, W. K. (1994). Extraordinary salmon growth. *Nature* 371, 209-210.

Devlin, R. H., Yesaki, T. Y., Donaldson, E. M., Du, S. J. & Hew, C. L. (1995). Production of germline transgenic Pacific salmonids with dramatically increased growth performance. *Canadian Journal of Fisheries and Aquatic Sciences* 52, 1376-1384.

Devlin, R. H., Sundstrom, L. F. & Muir, W. M. (2006). Interface of biotechnology and ecology for environmental risk assessments of transgenic fish. *Trends in Biotechnology* 24, 89-97.

Du, S. J., Gong, Z. Y., Fletcher, G. L., Shears, M. A., King, M. J., Idler, D. R. & Hew, C. L. (1992). Growth enhancement in transgenic Atlantic salmon by the use of an 'all-fish' chimeric growth hormone gene construct. *Biotechnology* 10, 176-181.

Dunham, R. A., Chatakondi, N., Nichols, A. J., Kucuktas, H., Chen, T. T., Powers, D. A., Weete, J. D., Cummins, K. & Lovell, R. T. (2002). Effect of rainbow trout growth hormone complementary DNA on body shape, carcass yield, and carcass composition of F_1 and F_2 transgenic common carp (*Cyprinus carpio*). *Marine Biotechnology* 4, 604-611.

Dynes, J., Magnan, P., Bernatchez, L. & Rodriguez, M. A. (1999). Genetic and morphological variations between two forms of lacustrine brook charr. *Journal of Fish Biology* 47, 775-787.

Farrell, A. P. & Steffensen, J. F. (1987). An analysis of the energetic cost of the branchial and cardiac pumps during sustained swimming in trout. *Fish Physiology and Biochemistry* 4, 73-79.

Farrell, A. P., Bennett, W. & Devlin, R. H. (1997). Growth-enhanced transgenic salmon can be inferior swimmers. *Canadian Journal of Zoology* 75, 335-337.

Fisher, R. & Hogan, J. D. (2007). Morphological predictors of swimming speed: a case study of pre-settlement juvenile coral reef fishes. *Journal of Experimental Biology* 210, 2436-2443.

Fu, C., Hu, W., Wang, Y. & Zhu, Z. (2005). Developments in transgenic fish in the people's Republic of China. *Revue Scientifique et Technique-office International des Epizooties* 24, 299-307.

Fu, C., Li, D., Hu, W., Wang, Y. & Zhu, Z. (2007a). Fast-growing transgenic common carp mounting compensatory responses. *Journal of Fish Biology* 71 (Suppl. B), 174-185.

Fu, C., Li, D., Hu, W., Wang, Y. & Zhu, Z. (2007b). Growth and energy budget of F_2 'all-fish' growth hormone gene transgenic common carp. *Journal of Fish Biology* 70, 347-361.

Gregory, T. R. & Wood, C. M. (1998). Individual variation and interrelationships between swimming performance, growth rate, and feeding in juvenile rainbow trout (*Oncorhynchus mykiss*). *Canadian Journal of Fisheries and Aquatic Sciences* 55, 1583-1590.

Gregory, T. R. & Wood, C. M. (1999). Interactions between individual feeding behaviour, growth, and swimming performance in juvenile rainbow trout (*Oncorhynchus mykiss*) fed different rations. *Canadian Journal of Fisheries and Aquatic Sciences* 56, 479-486.

Hawkins, D. K. & Quinn, T. P. (1996). Critical swimming velocity and associated morphology of juvenile coastal cutthroat trout (*Oncorhynchus clarki clarki*), steelhead trout (*Oncorhynchus mykiss*), and their hybrids. *Canadian Journal of Fisheries and Aquatic Sciences* 53, 1487-1496.

Hill, J. A., Kiessling, A. & Devlin, R. H. (2000). Coho salmon (*Oncorhynchus kisutch*) transgenic for a growth hormone gene construct exhibit increased rates of muscle hyperplasia and detectable levels of differential gene expression. *Canadian Journal of Fisheries and Aquatic Sciences* 57, 939-950.

Hu, W., Wang, Y. & Zhu, Z. (2007). Progress in the evaluation of transgenic fish for possible ecological risk and its containment strategies. *Science in China Series C: Life Sciences* 50, 573-579.

Kolok, A. S. & Oris, J. T. (1995). The relationship between specific growth rate and swimming performance in male fathead minnows (*Pimephales promelas*). *Canadian Journal of Zoology* 73, 2165-2167.

Law, T. C. & Blake, R. W. (1996). Comparison of the fast-start performance of closely related, morphologically distinct three spine sticklebacks (*Gasterosteus* spp.). *The Journal of Experimental Biology* 199, 2595-2604.

Lee, C. G., Devlin, R. H. & Farrell, A. P. (2003). Swimming performance, oxygen consumption and excess post-exercise oxygen consumption in adult transgenic and ocean-ranched coho salmon. *Journal of Fish Biology* 62, 753-766.

Li, D., Fu, C., Hu, W., Wang, Y. & Zhu, Z. (2007). Rapid growth cost in 'all-fish' growth hormone gene transgenic carp: reduced critical swimming speed. *Chinese Science Bulletin* 52, 923-926.

Mangel, M. & Munch, S. B. (2005). A life-history perspective on short-and long-term consequences of compensatory growth. *American Naturalist* 166, E155-E176.

Martinez, R., Estrada, M. P., Berlanga, J., Guillen, I., Hernandez, O., Cabrera, E., Pimentel, R., Morales, R., Herrera, F., Morales, A., Pina, J. C., Abad, Z., Sanchez, V., Melamed, P., Lleonart, R. & De La Fuente, J. (1996). Growth enhancement in transgenic tilapia by ectopic expression of tilapia growth hormone. *Marine Molecular Biology and Biotechnology* 5, 62-70.

McGarigal, K., Cushman, S. & Stafford, S. (2000). *Multivariate Statistics for Wildlife and Ecology Research*. New York, NY: Springer-Verlag.

McKenzie, D. J., Martinez, R., Morales, A., Acosta, K., Taylor, E. W., Steffensen, J. F. & Estrada, M. P. (2003). Effects of growth hormone transgenesis on metabolic rate, exercise performance and hypoxia tolerance in tilapia hybrids. *Journal of Fish Biology* 63, 398-409.

McLaughlin, R. L. & Grant, J. W. A. (1994). Morphological and behavioural differences among recently-emerged brook charr, *Salvelinus fontinalis*, foraging in slow vs. fast-running water. *Environmental Biology of Fishes* 39, 289-300.

Morgan, I. J. & Metcalfe, N. B. (2001). Deferred costs of compensatory growth after autumnal food shortage in juvenile salmon. *Proceedings of the Royal Society B* 268, 295-301.

Munch, S. B. & Conover, D. O. (2004). Nonlinear growth cost in *Menidia menidia*: theory and empirical evidence. *Evolution* 58, 661-664.

Nam, Y. K., Noh, J. K., Cho, Y. S., Cho, H. J., Cho, K. N., Kim, C. G. & Kim, D. S. (2001). Dramatically accelerated growth and extraordinary gigantism of transgenic mud loach *Misgurnus mizolepis*. *Transgenic Research* 10, 353-362.

Nanami, A. (2007). Juvenile swimming performance of three fish species on an exposed sandy beach in Japan. *Journal of Experimental Marine Biology and Ecology* 348, 1-10.

Ostenfeld, T. H., McLean, E. & Devlin, R. H. (1998). Transgenesis changes body and head shape in Pacific salmon. *Journal of Fish Biology* 52, 850-854.

Plaut, I. (2001). Critical swimming speed: its ecological relevance. *Comparative Biochemistry and Physiology A* 131, 41-50.

Rahman, M. A., Mak, R., Ayad, H., Smith, A. & Maclean, N. (1998). Expression of a novel piscine growth hormone gene results in growth enhancement in transgenic tilapia (*Oreochromis niloticus*). *Transgenic Research* 7, 357-369.

Robinson, B. W. & Wilson, D. S. (1994). Character release and displacement in fishes: a neglected literature. *American Naturalist* 144, 596-627.

Rolf, F. J. & Bookstein, F. L. (1987). A comment on shearing as a method for 'size correlation'. *Systematic Zoology* 36, 356-367.

Sagnes, P., Gaudin, P. & Statzner, B. (1997). Shifts in morphometrics and their relation to hydrodynamic potential and habitat use during grayling ontogenesis. *Journal of Fish Biology* 50, 846-858.

Seiler, S. M. & Keeley, E. R. (2007). Morphological and swimming stamina differences between Yellow cutthroat trout (*Oncorhynchus clarkia bouvieri*), rainbow trout (*Oncorhynchus mykiss*) and their hybrids. *Canadian Journal of Fisheries and Aquatic Sciences* 64, 127-135.

Sheehan, T. F., Kocik, J. F., Cadrin, S. X. & Legault, C. M. (2005). Marine growth and morphometrics for three populations of Atlantic salmon from eastern Maine, USA. *Transactions of the American Fisheries Society* 134, 775-788.

Snorrason, S. S., Skulason, S., Jonsson, B., Malmquist, H. J., Sandlund, O. T. & Lindem, T. (1994). Trophic specialization in Arctic charr *Salvelinus alpinus* (Pisces: Salmonidae): morphological divergence and ontogenic niche shifts. *Biological Journal of the Linnean Society* 52, 1-18.

Stamps, J. A. (2007). Growth-mortality tradeoffs and 'personality traits' in animals. *Ecology Letters* 10, 355-363.

Stevens, E. D. & Devlin, R. H. (1999). Intestinal morphology in growth hormone transgenic coho salmon. *Journal of Fish Biology* 56, 191-195.

Stevens, E. D. & Devlin, R. H. (2005). Gut size in GH-transgenic coho salmon is enhanced by both the GH transgene and increased food intake. *Journal of Fish Biology* 66, 1633-1648.

Stevens, E. D., Sutterlin, A. & Cook, T. (1998). Respiratory metabolism and swimming performance in growth hormone transgenic Atlantic salmon. *Canadian Journal of Fisheries and Aquatic Sciences* 55, 2028-2035.

Stevens, E. D., Wagner, G. N. & Sutterlin, A. (1999). Gut morphology in growth hormone transgenic Atlantic salmon. *Journal of Fish Biology* 55, 517-526.

Strauss, R. E. & Bookstein, F. L. (1982). The truss: body form reconstructions in morphometrics. *Systematic Zoology* 31, 113-135.

Sundström, L. F., Lõhmus, F., Tymchuk, W. E. & Devlin, R. H. (2007). Gene environment interactions influence ecological consequences of transgenic animals. *Proceedings of the Royal Society* B 104, 3889-3894.

Taylor, E. B. & Foote, C. J. (1991). Critical swimming velocities of juvenile sockeye salmon and kokanee, the anadromous and non-anadromous forms of *Oncorhynchus nerka* (Walbaum). *Journal of Fish Biology* 38, 407-419.

Tudorache, C., Viaenen, P., Blust, R. & Boech, G. D. E. (2007). Longer flumes increase critical swimming speeds by increasing burst-glide swimming duration in carp *Cyprinus carpio*, L. *Journal of Fish Biology* 71, 1630-1638.

Wang, Y., Hu, W., Wu, G., Sun, Y., Chen, S., Zhang, F., Zhu, Z., Feng, J. & Zhang, X. (2001). Genetic analysis of 'all-fish' growth hormone gene transferred carp (*Cyprinus carpio* L.) and its F_1 generation. *Chinese Science Bulletin* 46, 1174-1177.

Webb, P. W. (1975). Hydromechanics and energetics of fish propulsion. *Bulletin of the Fisheries Research Board of Canada* 190, 109-119.

Webb, P. W. (1982). Locomotor patterns in the evolution of Actinopterygians fishes. *American Zoologist* 22, 329-342.

Webb, P. W. (1984). Body form, locomotion and foraging in aquatic vertebrates. *American Zoologist* 24, 107-120.

Zbikowska, H. M. (2003). Fish can be first-advances in fish transgenesis for commercial applications. *Transgenic Research* 12, 379-389.

Zhu, Z. Y. (1992). Growth hormone gene and the transgenic fish. In *Agricultural Biotechnology* (You, C. B. & Chen, Z. Z., eds), pp. 106-116. Beijing: China Science and Technology Press.

Zhu, Z., Li, G., He, L. & Chen, S. (1985). Novel gene transfer into the fertilized eggs of goldfish (*Carassius auratus* L. 1758). *Journal of Applied Ichthyology* 1, 31-34.

快速生长转基因鲤的游泳能力降低与其形体变化的相关性

李德亮[1,2]　胡　炜[1]　汪亚平[1]　朱作言[1]　傅萃长[2]

1 中国科学院水生生物研究所，淡水生态与生物技术国家重点实验室，武汉　430072

2 中国科学院大学，北京　100049

摘　要　本研究比较了转生长激素基因鲤 F_4 和非转基因鱼的临界游泳速度及形态指标。转基因鱼的临界游泳速度比对照鱼低 22.3%。主成分分析发现转基因鱼的体形发生了变化。与对照鱼相比，其头部变高，上尾鳍变长，躯干和尾部变窄，下尾鳍变短。游泳速度与体形指标密切相关，躯干和尾柄越高，下尾鳍越长，头高越小，上尾鳍及体长越短，游泳速度越快。上述研究结果表明，转基因鲤的形态改变后，不利于产生最大推力和最小阻力，是其游泳能力降低的原因。

Elevated Ability to Compete for Limited Food Resources by 'All-Fish' Growth Hormone Transgenic Common Carp *Cyprinus carpio*

Ming Duan[1,2] Tanglin Zhang[1] Wei Hu[1] L. F. Sundström[3] Yaping Wang[1] Zhongjie Li[1] Zuoyan Zhu[1]

1 State Key Laboratory of Freshwater Ecology and Biotechnology, Institute of Hydrobiology, Chinese Academy of Sciences, Wuhan 430072, China

2 Graduate University of Chinese Academy of Sciences, Beijing 100049, China

3 Department of Fisheries and Oceans, University of British Columbia, Centre for Aquaculture and Environmental Research, 4160 Marine Drive, West Vancouver, British Colombia V7V 1N6, Canada

Abstract Food consumption, number of movements and feeding hierarchy of juvenile transgenic common carp *Cyprinus carpio* and their size-matched non-transgenic conspecifics were measured under conditions of limited food supply. Transgenic fish exhibited 73.3% more movements as well as a higher feeding order, and consumed 1.86 times as many food pellets as their non-transgenic counterparts. After the 10 day experiment, transgenic *C. carpio* had still not realized their higher growth potential, which may be partly explained by the higher frequency of movements of transgenics and the 'sneaky' feeding strategy used by the non-transgenics. The results indicate that these transgenic fish possess an elevated ability to compete for limited food resources, which could be advantageous after an escape into the wild. It may be that other factors in the natural environment (*i.e.* predation risk and food distribution), however, would offset this advantage. Thus, these results need to be assessed with caution.

Keywords competition; foraging; growth hormone (GH); transgene

Introduction

The use of growth hormone (GH) treatment (McLean & Donaldson, 1993) and altered GH gene expression in fishes by transgenesis (Zhu *et al.*, 1985; Gross *et al.*, 1992; Devlin *et al.*, 1994) have been explored as a way to increase aquaculture production and efficiency. Fast-growing fishes also have been considered as the best candidates for the marketing of transgenic animals for human consumption (Zbikowska, 2003). Released or escaped fast-growing transgenic fishes, however, pose an ecological risk that should be of concern (Devlin *et al.*, 2006). Assessment of the environmental effects of fast-growing transgenic

文章发表于 *Journal of Fish Biology*, 2009, 75(6):1459-1472

fishes is therefore urgently needed (Kapuscinski *et al.*, 2007).

Under natural conditions, acquiring large size through rapid growth can provide fitness advantages (Arendt, 1997). Rapid growth reduces vulnerability to gape-limited predators, allows earlier exploitation of novel resources (Werner & Gilliam, 1984) and can result in sexual maturity at a younger age (Alm, 1959). The increase in growth rate observed after GH treatment or transgenesis comes primarily from an elevated food intake acting through increased appetite (Johnsson & Björnsson, 1994) and foraging activity (Jönsson *et al.*, 1996), leading to higher competitive ability, at least in salmonids (Devlin *et al.*, 1999). GH-transgenic fishes also appear to be better at using low-quality food and have higher feed-conversion efficiency (Fu *et al.*, 1998; Venugopal *et al.*, 2004; Raven *et al.*, 2006). GH-transgenic fishes, however, do not always benefit from elevated GH production. In coho salmon *Oncorhynchus kisutch* (Walbaum), transgenic individuals were prone to spending more time and to investing more energy on unprofitable prey due to reduced discrimination (Sundström *et al.*, 2004a). They experienced lower survival in the presence of predators (Sundström *et al.*, 2004b, 2005), and may experience reduced disease resistance (Jhingan *et al.*, 2003) and impaired swimming ability (Farrell *et al.*, 1997; Li *et al.*, 2007). Distinct effects observed among species and strains under different conditions reveal the need for risk assessments to be performed on a case-by-case basis (Devlin *et al.*, 2001, 2006).

Many natural water bodies are diverse and complex, in which the resources are often unevenly distributed. Individuals of the same species require similar resources and therefore compete when these resources are in limited supply (Begon *et al.*, 1996). The prevalence of dominant-subordinate relationships in animal societies suggests that there is an adaptive significance to achieving a high rank when in a group (Huntingford & Turner, 1987). Feeding rank within a group has been used as an indirect measure of social rank in a number of studies involving salmonids flatfishes and tilapias (McCarthy *et al.*, 1992, 1999; Carter *et al.*, 1994; Jobling & Baardvik, 1994; Shelverton & Carter, 1998). Foraging theory predicts that individuals attempt to maximize their net energy gain by foraging in patches with high densities of preferred prey (Stephens & Krebs, 1996). This energy maximization premise assumes that the forager adopts a strategy, largely depending on the ambient foraging conditions, which maximizes its long-term fitness (Townsend & Winfield, 1985). Motivation to feed increases with hunger, which also affects short-term behavioural changes such as increased search rate (Colgan, 1973; Dill, 1983; Höjesjö *et al.*, 1999; Vehanen, 2003).

There are numerous publications on the feeding behaviour of GH-transgenic salmonids (Abrahams & Sutterlin, 1999; Devlin *et al.*, 1999; Sundström *et al.*, 2004a, 2007a; Tymchuk *et al.*, 2005). As far as is known, however, there have not been any reports on the feeding behaviour of the 'all-fish' GH-transgenic common carp *Cyprinus carpio* L. Here, in a short-term experiment, the ability of hatchery-reared juvenile GH-transgenic *C. carpio* to compete for limited food resources was compared with size-matched non-transgenic counterparts. Such information would be useful for understanding the survival component of fitness of these transgenic fish under competitive conditions and also for assessing potential ecological risk of the transgenic individuals should they escape or be released into the wild.

Materials and Methods

Source of fish

Fast-growing genotypes of P_0 'all-fish' GH-transgenic *C. carpio* were initially produced by microinjection of the *pCAgcGHc* into the fertilized eggs of *C. carpio* (Yellow River variety). The all-fish gene construct *pCAgcGHc* was a recombinant construct of grass carp *Ctenopharyngodon idella* (Valenciennes) growth hormone cDNA (*gcGHc*), whose expression is driven by the *β*-actin gene promoter of *C. carpio* (*pCA*) (Wang *et al*., 2001). The F_1, F_2 and F_3 generation were, respectively, 1.6 times (Wang *et al*., 2001), 1.8-2.5 times (W. Hu, unpubl. data) and 1.4-1.9 times (Li *et al*., 2007) the body mass of non-transgenic counterparts under hatchery-reared conditions, showing how the growth enhancement remains relatively stable across generations. The F_5 generation transgenic and non-transgenic fish were produced from crosses between a wild-type female and an F_4 hemizygous transgenic male of a fast-growing transgenic strain on 25 April 2007. Frequencies of transgene transmission to F_5 progeny from this line were c. 50% (W. Hu, unpubl. data). Siblings were used to minimize effects of genetic differences and maternal and paternal effects. After their emergence in Duofu Technology Farm, Wuhan, China, the first-feeding fry, containing a mix of the two genotypes, were transferred to four concrete rectangular pools (8 m^3) with a rearing density of 100 individuals m^{-3} each. A month later, the mixed populations were transferred to an indoor re-circulating system and reared in four circular fibreglass tanks (diameter 150 cm, volume 1000 l) at the Institute of Hydrobiology, Chinese Academy of Sciences. Thereafter, the *pCAgcGHc* transgene-positive fish were identified by the polymerase chain reaction (PCR) method following Wang *et al.* (2001). The two genotypes then were reared separately in two of the fibreglass tanks as described above. Non-transgenics were fed to satiation with frozen chironomid *Chironomus sp.* larvae twice daily, whereas the transgenics were fed roughly the same amount every other day to allow size-matching and avoid the possible effects of size on competitive ability (Huntingford *et al*., 1990). This feeding regime was maintained until the fish gape sizes matched the size of the artificial food pellets (mean ± S.E. 4.6 ± 0.3 mg; comprised of 33.2% protein, 9.1% lipid, 11.4% ash and 18.1 J mg^{-1} energy).

Experimental procedures

The non-commercial research facility at the Fish Behavioral Ecology Laboratory of the Institute of Hydrobiology had multiple containment screen systems and was specially designed to prevent transgenic fish from escaping to nature. The experiment consisted of two main parts: (1) an acclimation period (days 1-7) and (2) an observation period (days 8-10). The water inlet to the research facilities was at 10 m depth in the Liangzi Lake near the laboratory. Incoming water was aerated and filtered with an adjustable internal filter. During the experiment, the water temperature was 26.1 ± 0.3℃, the air temperature was 25.2 ± 0.9℃, the dissolved oxygen was 6.6 ± 0.1 mg l^{-1} and the pH was 6.9 ± 0.01 (mean ± S.E.).

Treatment period

One hundred transgenic and 100 non-transgenic individuals were placed separately into two rectangular aquaria (210 × 35 × 35 cm in size) on 1 September 2007, and were fed 2% of total body mass with the artificial food pellets twice daily as described above. On day 1 (5 September), 12 individuals of each genotype (size-matched within pairs) were anesthetized with eugenol. Body mass (M, to the nearest 0.01 g) and total length (L_T nearest 0.01 cm) were measured. Individual fish in each pair were identified visually through randomly marking them with white (a saturated solution of white titanium oxide and water was used) (Southwood & Henderson, 2000) or blue (alcian blue) dye (Nicieza & Metcalfe, 1999; Westerberg *et al.*, 2004) by subcutaneous injection. Next, the 12 pairs were transferred to 12 tempered glass aquaria measuring 35 × 35 × 35 cm with a water depth of 20 cm to acclimate. Two fluorescent lights were fixed 1 m above each aquarium [a 12 L : 12 D (on 0700 and off 1900 hours) photoperiod was used]. A black plastic screen was used to cover all sides of each aquarium to minimize disturbance and to maintain similar levels of illumination. In addition, each aquarium had a white gravel substratum and a white-bottomed Petri dish (diameter 10 cm) located at the centre of the aquarium bottom. The dish marked the feeding area and clumped the food (*i.e.* prevented food pellets from drifting away). A tube from the outside of the screen was fixed above the feeding area 2 cm under the water surface allowing sinking food pellets to be administered in the exact location each time without any disturbance to the fish (Abrahams, 1989; Tymchuk *et al.*, 2005).

During the acclimation period, a pilot study was undertaken to ascertain the maximum number of food pellets consumed by a single fish at each feeding trial. An individual was estimated to be satiated when it refused to grab, or did not swallow five consecutive pellets which then sank gently down to the bottom of the aquarium [the method was modified from Johnsson & Björnsson (1994)]. Neither of the two genotypes would consume > 30 food pellets during a 10 min interval. In the light of these results, each pair was fed with 10 pellets at one time, but twice daily (0800-1100 and 1400-1700 hours) during the 6 day acclimation. Uneaten food items and faeces were removed by a glass siphon after 10 min. The fish were deprived of food for 1 day (day 7) before the observations commenced on day 8.

Observations

The same feeding procedures described above were repeated during the 3 day behavioural observations (from 13 September to 15 September).

Behaviours of the fish were recorded by a digital videocassette recorder (Sony, HDR-HC1E Handy Camera; www.sony.net) placed 50 cm above the aquarium. The order of observations was randomly allocated among the aquaria so as to remove any bias and order effects. For each aquarium, the video camera was started, and then 10 food pellets were poured into the Petri dish through the feeding tube at once (no fish fed during this period). Behaviours of the two individuals then were recorded for 10 min. Observations were made twice daily, at 0800-1100 and 1400-1700 hours during the feeding periods. Consequently, this procedure yielded six repeated measures for competitive ability for each of the 12 pairs during the 3 day observations. Four main behavioural variables of the two genotypes were noted from the video: feeding order, number of food pellets consumed, number of movements (when a fish moved more than its L_T following by a less active period of at least 1 s) and the time spent in the feeding area (when the entire fish remained over the Petri dish). After the last observation on day 10 (15 September), L_T and M of the 24 individual fish were again measured under light anaesthesia.

Data analysis

The specific growth rate for L_T was calculated as $G_L = 100\ (\ln L_{T2} - \ln L_{T1})\ t^{-1}$ (Ricker, 1979), where L_{T2} and L_{T1} were the final and initial L_T (cm) and t is the experimental period (10 days, consisting of the 7 day treatment period and the 3 day observations). The condition factor (K) was calculated as $K = 100ML_T^{-3}$.

Most data were normally distributed and had homogeneous variance, so paired t-tests and a general linear model (GLM) were used. There was some variance heterogeneity for the number of movements, so these data were square-root transformed prior to statistical analysis. Rank data on feeding order were tested with the Wilcoxon signed ranks test. For evaluating correlations between two behavioural variables, Spearman's rank correlation (r_s) and partial correlation analysis were used. Differences were regarded as significant when $P < 0.05$. All the data were analysed with SPSS 15.0 (www.spss.com) and described as mean ± S.E.

Results

Growth

At the beginning of the experiment, there was no significant difference in L_T between the two genotypes (paired t-test, d.f. = 11, $P > 0.05$), but transgenic fish were heavier (paired t-test, d.f. = 11, $P < 0.001$), and consequently had a higher K (paired t-test, d.f. = 11, $P < 0.001$) (Table I). During the 10 day experiment, there was not a statistically significant difference (ANCOVA, with initial M as a covariate, d.f. = 1, 21, $P > 0.05$. adjusted means:

transgenics, 0.52 ± 0.09% day^{-1}, non-transgenics, 0.49 ± 0.09% day^{-1}; Fig. 1) although the G_L of transgenic *C. carpio* (0.53 ± 0.07% day^{-1}) was higher than that of non-transgenic carp (0.48 ± 0.08% day^{-1}). Initial *M* had no significant effect on the G_L (ANCOVA, d.f. = 1, 21, $P > 0.05$). Compared with the growth potential of these two genotypes under rich (natural and artificial foods) hatchery conditions (Li *et al.*, 2007), transgenic fish attained on average 9.0% of their growth potential and non-transgenic fish attained 29.4% of their growth potential under these competitive conditions.

Table I Total length (L_T), body mass (*M*) and condition factor (*K*) of transgenic and non-transgenic *Cyprinus carpio* before the experiment (mean ± S.E.)

Genotype	*n*	L_T (cm)	*M* (g)	*K*
Transgenic	12	6.75 ± 0.04	4.39 ± 0.15*	1.42 ± 0.03*
Non-transgenic	12	6.76 ± 0.03	3.71 ± 0.08	1.20 ± 0.02

n, number of fish

*, A significant difference between the two genotypes, $P< 0.05$ (paired *t*-test)

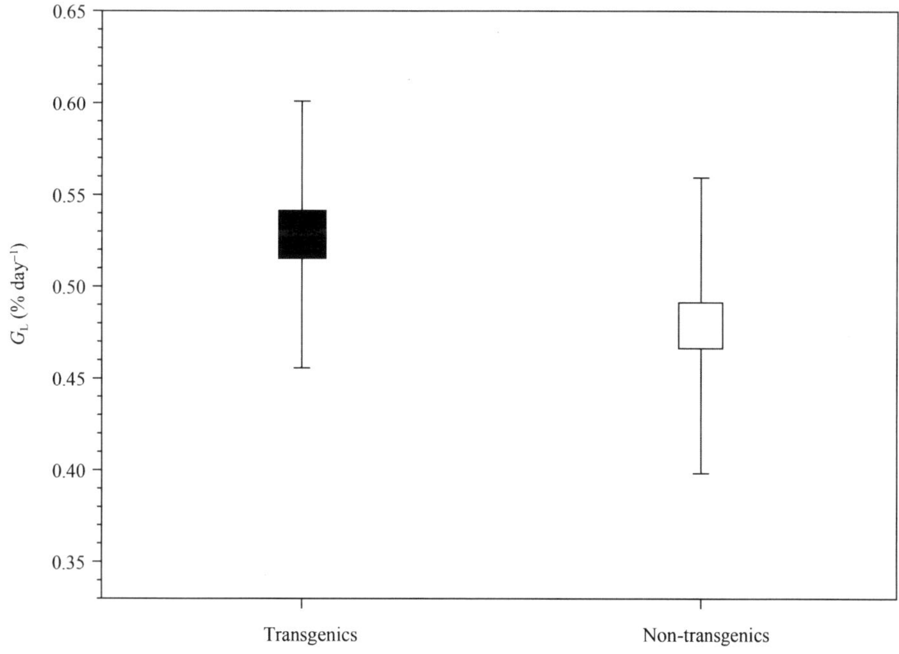

Fig. 1 Mean ± S.E. specific growth rate for total length (G_L) of transgenic (n = 12) (■) and non-transgenic (n = 12) (□) *Cyprinus carpio* during the 10 day experiment

Food consumption

Overall, out of 60 pellets provided, the number of food pellets consumed by transgenic *C. arpio* was 38.5 ± 2.9 during the 3 day observation period, which was significantly more (repeated measures ANOVA, between-subject genotype; d.f. = 1, 22, $P < 0.001$) than that consumed by non-transgenic fish (20.7 ± 3.1). This higher food intake was consistent over time (within-subject trial; d.f. = 5, 110, $P > 0.05$), with no significant interaction between trial and genotype (d.f. = 5, 110, $P > 0.05$), indicating that the transgenic fish maintained a higher consumption rate during the six consecutive observations (Fig. 2).

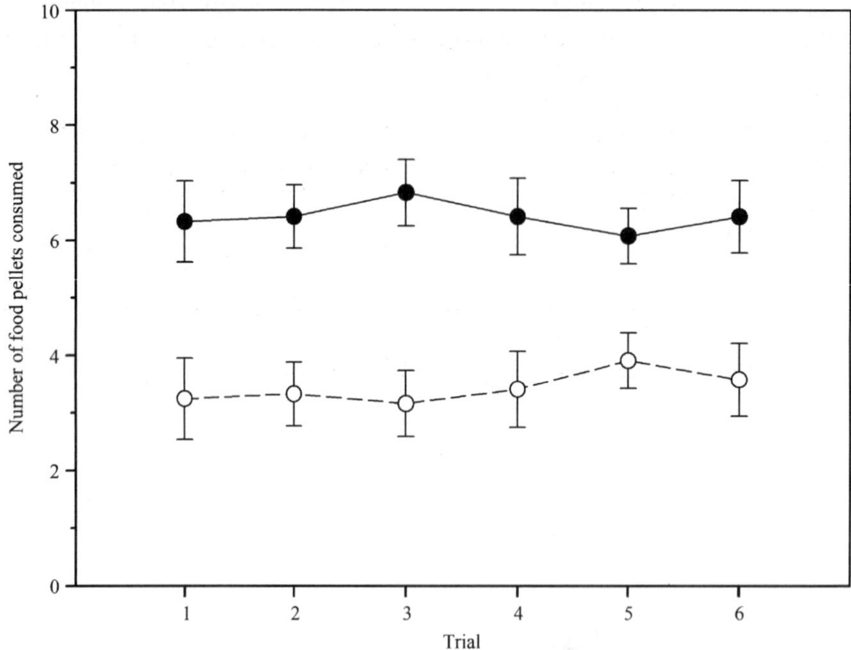

Fig. 2 Mean ± S.E. amount of food consumed by transgenic (n = 12) (●) and non-transgenic (n = 12) (○) *Cyprinus carpio* during each of six consecutive trials. All of the differences between the two genotypes in each trial were significant ($P < 0.05$)

Feeding order

The individuals were ranked according to their scores obtained for feeding order. The first-feeding fish obtained a score of 2, the last (second-feeding) one was given a score of 1 and if the fish did not feed it was given a score of 0. Overall, the scores obtained for transgenics (1.6 ± 0.1) were significantly higher (Wilcoxon signed ranks test, $z = -2.149$, $P < 0.05$) than those for the non-transgenics (1.1 ± 0.2), indicating that transgenics were faster to grasp the food pellets.

Time spent in the feeding area

Transgenic *C. carpio* spent > 50% of their time in the feeding area (306.5 ± 42.6 s), which was significantly more (repeated measures ANOVA, between-subject genotype; d.f. = 1, 22, $P < 0.01$) than that for the non-transgenic *C. carpio* (143.7 ± 25.4 s).

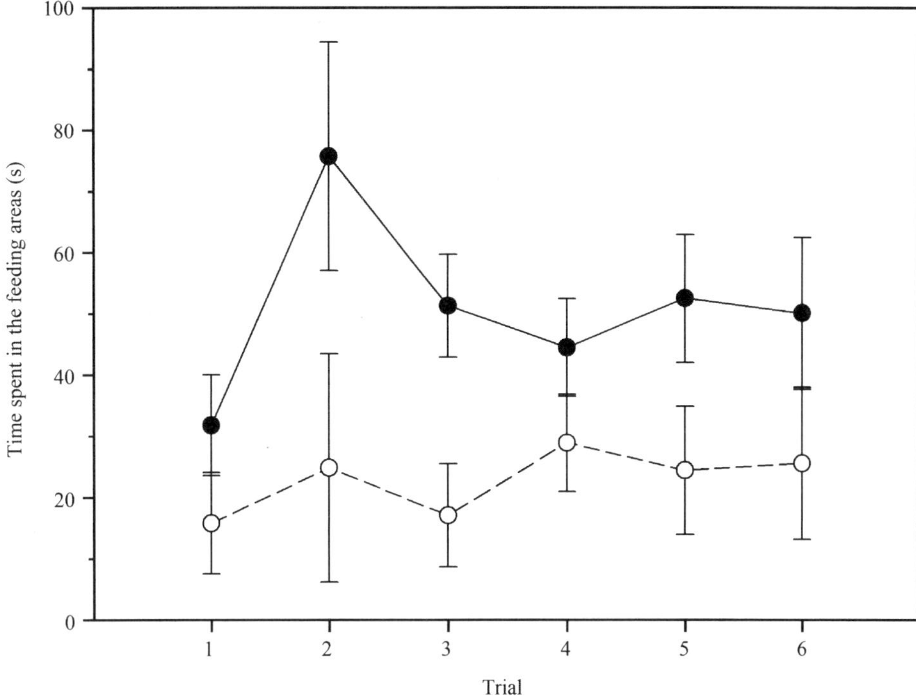

Fig. 3 Mean ± S.E. amount of time spent in the feeding areas for transgenic ($n = 12$) (●) and non-transgenic ($n = 12$) (○) *Cyprinus carpio* during each of six consecutive trials. All of the differences between the two genotypes in each trial were significant ($P < 0.05$)

Trial had no significant effect on the time spent in the feeding area (within-subject trial, d.f. = 5, 110, $P > 0.05$), nor was the interaction between trial and genotype significant (d.f. = 5, 110, $P > 0.05$; Fig. 3). These data indicate that transgenic *C. carpio* often monopolized the feeding areas, and thus they were dominant, whereas the non-transgenic fish were subordinate. In addition, dominant *C. carpio* were often observed to swim freely in aquaria, whereas the subordinate individuals were residing in the corners.

Movement level

Overall, the number of movements by transgenic fish averaged 353.2 ± 32.1, which was significantly higher (repeated measures ANOVA, between-subject genotype, d.f. = 1, 22, $P < 0.01$) than that of non-transgenic fish, 203.8 ± 29.5.

Trial had a significant effect (within-subject trial, d.f. = 5, 110, $P \leqslant 0.01$) on the number of movements, but the pattern was similar for the two genotypes (d.f. = 5, 110, $P > 0.05$; Fig. 4).

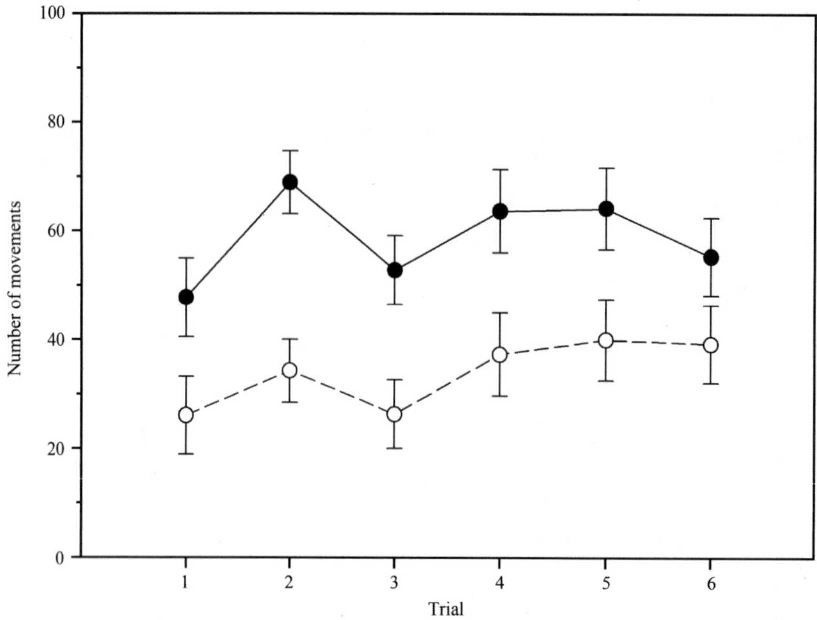

Fig. 4 Mean ± S.E. number of movements of transgenic (n = 12) (●) and non-transgenic (n= 12) (○) *Cyprinus carpio* during each of six consecutive trials. All of the differences between the two genotypes in each trial were significant ($P < 0.05$)

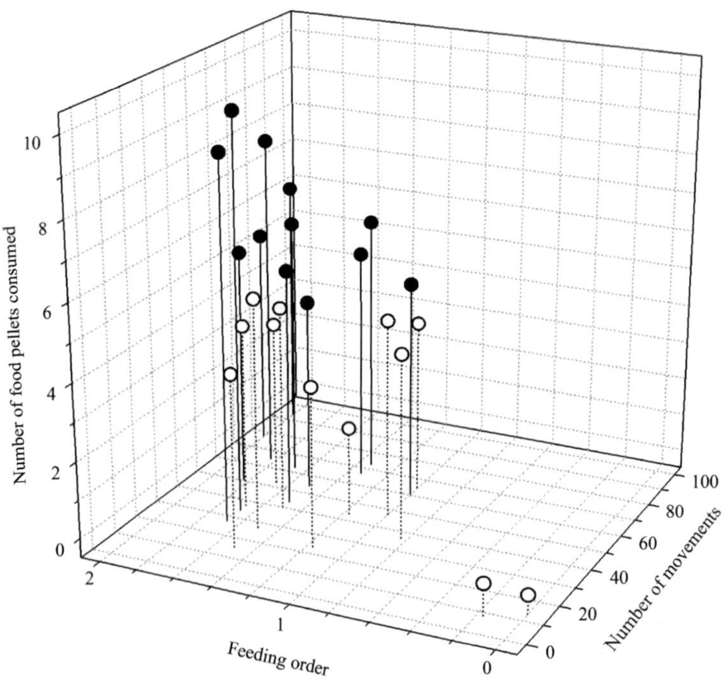

Fig. 5 Correlations between mean number of movements, social status (feeding order) and food consumption (number of food pellets consumed) for transgenic (n = 12) (●) and non-transgenic (n = 12) (○) *Cyprinus carpio* across six consecutive trials

Relationships between food consumption, feeding order and movement level

There was a positive correlation between the food consumption and the frequency of movement of the two genotypes ($r_s = 0.437$, $n = 24$, $P < 0.05$), indicating that the higher the number of movements, the more food was consumed. Moreover, there was a significant positive correlation between feeding order and food consumption ($r_s = 0.547$, $n = 24$, $P < 0.01$), indicating that the earlier the food acquisition, the more food was consumed. In order to determine which factor among frequency of movement and feeding order more strongly affected the quantity of food consumed, partial correlation was used. After controlling for feeding order, no clear correlation was found between frequency of movement and the number of food pellets consumed ($r_s = -0.017$, $n = 21$, $P > 0.05$). On the other hand, when movement level was controlled for, there was a significant positive correlation between feeding order and the number of the food pellets consumed by the two genotypes ($r_s = 0.664$, $n = 21$, $P \leqslant 0.001$). These results indicate that feeding order had a much stronger effect on the food consumption than did the frequency of movement (Fig. 5).

Discussion

In the present study, expression of the all-fish GH transgene altered a number of feeding-related behaviours in *C. carpio*, mainly increasing food consumption and feeding order, time spent in the feeding areas and the frequency of movement. These results are highly consistent with previous studies on growth hormone function in salmonids, both from GH injection (Johnsson & Björnsson, 1994; Jönsson *et al.*, 1996, 1998) and GH transgenesis (Abrahams & Sutterlin, 1999; Devlin *et al.*, 1999; Tymchuk *et al.*, 2005). The present study suggests that GH influences competition for food through increased feeding motivation in *C. carpio* in a similar way.

The GH transgene induces over-expression of GH, which may elevate metabolism and appetite in GH-transgenic *C. carpio* (Fu *et al.*, 2007), which in turn enhances feeding motivation and induces short-term behavioural changes such as increased search rate (Colgan, 1973; Dill, 1983; Höjesjö *et al.*, 1999; Vehanen, 2003). Quite similarly, GH injection in rainbow trout *Oncorhynchus mykiss* (Walbaum) was found to increase foraging due to hunger (Jönsson *et al.*, 1998), it may be that the higher hunger levels may translate to transgenic fishes being less choosey when it comes to for aging. Constant searching for prey may increase activity and consequently result in higher energy costs for the individual. Fu *et al.* (2007) observed an increased food intake for the F_2 GH-transgenic *C. carpio* of the same strain as employed in the present study, and suggested that their elevated food intake partly compensate for a higher metabolism originating from greater activity. In Fu *et al.* (2007) study, they cultured the two genotypes separately, and sufficient food resources were supplied, and thus the growth of the two genotypes did not interfere competitively, and the transgenic individuals may not significantly express their agonistic behaviour potential towards cohorts (Devlin *et al.*, 2004). Besides, the *M* of the fish used in the study of Fu *et al.* (2007) was > 138 g, whereas in the present study, the *M* of juvenile transgenic fish was <5 g, so the behaviours of these two different animals were probably different. In addition, the transgenic fish were fed with a restricted ration before the acclimation period (*i.e.* feeding every other day), which may also be a factor that could affect the transgene effects on the feeding motivation due to some physiological alterations. In addition, the short duration of the present experiment may affect this result. The transgenic *C. carpio*, however, did not realize their growth potential under restricted food availability conditions during the 10 day experiment, which parallels results of studies on transgenic *O. kisutch* where transgenics did not reach their highest growth potential under semi-natural environments (Sundström *et al.*, 2004a, 2007b).

According to Metcalfe (1986), less competitive individuals may use a feeding strategy other than that used by stronger competitors. Instead of attempting to maximize food intake, less competitive individuals may try to minimize energetic expenditure by, for example, decreasing their activity. The energetic cost to obtain one food pellet may therefore be significantly different between the two *C. carpio* genotypes. Under conditions of limited food supply, the benefits of the increased energy intake by transgenic fish, however, were diminished by their increased movements and metabolism, which may partly explain why the juvenile GH-transgenic *C. carpio* did not express fast growth potential during the short-term (10 days) experiment. The results further suggest that the initial size differences between the two genotypes are not the main cause of the higher competitive ability found in transgenic *C. carpio*.

Feeding rank within a group has been used as an indirect measure of social rank in a number of studies involving salmonids, flatfishes and tilapiines (McCarthy *et al.*, 1992, 1999; Carter *et al.*, 1994; Jobling & Baardvik, 1994; Shelverton & Carter, 1998). Once individuals recognize a conspecific's rank (Utne-Palm & Hart, 2000) or when the ranks are settled (Tindo & Dejean, 2000; Forkman & Haskell, 2004), costly overt aggression may not be necessary, thereby decreasing the risk of fatal injury or stress that may lead to increased susceptibility to pathogens (Huntingford & Turner, 1987). In the present study, the treatment period was 6 days, and there was adequate time to establish a feeding hierarchy. Therefore, the feeding order and the time spent in the feeding area were used as metrics to evaluate the feeding hierarchy (McCarthy *et al.*, 1999; Bailey *et al.*, 2000). Transgenic *C. carpio* were much faster to occupy the feeding areas and to obtain the food pellets. Furthermore, they spent more time in the feeding areas, also indicating that the juvenile transgenic *C. carpio* were dominant over the non-transgenic individuals. Similarly, dominant three-spined sticklebacks *Gasterosteus aculeatus* L. consumed more food and grew faster, but subordinate individuals could also gain body mass by sneaking access to food (Sneddon *et al.*, 2006). Similarly, non-transgenic fish often were observed to carefully approach the feeding area and sneakily fed upon the remaining food while the transgenic individuals were approaching fast and swallowing several food pellets while staying in the area for a considerable time.

Environmental conditions are likely to determine whether it would pay for a fish to be highly aggressive and monopolize the food resource or to avoid aggressive encounters and obtain food in a sneaky manner (Sneddon *et al.*, 2006). In the present study, fish were kept in small and simple tank environments where it presumably was easy for dominant transgenic fish to dominate the food source, but in nature conditions would be more complex (Devlin *et al.*, 2004). Höjesjö *et al.* (2005) observed that subordinate Atlantic salmon *Salmo salar* L. fed by briefly invading the space occupied for most of the time by the dominant brown trout *Salmo trutta* L. in a controlled stream channel environment, and the *S. salar* using such sneaky strategy fed no less than did the dominants. Under natural conditions, food availability would be less predictable and predation risk may influence feeding decisions so that the advantage of the transgenic fish could be lessened or increased depending on their responses under such complex environments. Arendt (1997) argued that there were many functions that may be compromised when intrinsic growth rates were accelerated to meet evolutionary challenges. Conversely, if selection acts to improve any of these functions, intrinsic growth rates might be reduced. Thus, further work on the behavioural and developmental alteration of the transgenic fish in complex environments is needed.

The present study is a first approximation of answering whether transgenic *C. carpio* compete effectively for food. *Cyprinus carpio*, however, are different from salmonids in many important aspects of foraging behaviour, often spending much time foraging together for food items such as aquatic invertebrates. Hence, further research should focus on how much time the fish spend foraging, how likely they are to deplete food resources and how likely they are to engage in agonistic behaviour if food is limiting.

In the present study, the all-fish GH-transgenic *C. carpio* exhibited elevated feeding motivation, higher frequency of movement, earlier food capture tendency and higher feeding rank, and thus increased the ability to compete for limited food resources in comparison with non-transgenic conspecifics. These results suggest that such altered feeding behaviour would probably benefit the transgenic individuals in natural ecosystems. Environmental complexity such as food abundance and predation risk levels, however, generally reduces hormonal effects on behaviour (Sloman & Armstrong, 2002) and can greatly alter the phenotypic effects of the GH transgene (Sundström *et al.*, 2007b). Muir & Howard (2001), however, presented a comprehensive method by using the wild-type and transgenic Japanese medaka *Oryzias latipes* (Temminck & Schlegel) to determine environmental risk of transgenic fish that they called the 'net fitness approach'. They grouped various aspects of an organism's life cycle into six net fitness components: juvenile viability, adult viability, age at sexual maturity, female fecundity, male fertility and mating advantage. The present study only examined a single fitness-related trait, the ability to engage in interference competition. The entire life cycle, however, needs to be studied to determine overall fitness of transgenic *C. carpio* because once released into the natural environment, the impact may be permanent and irreversible (Muir & Howard, 1999, 2001). Furthermore, genotype-by-environment interactions will play a crucial role determining risk levels posed by transgenic fishes to natural ecosystems (Sundström *et al.*, 2005; Devlin *et al.*, 2006). Thus, the results of this study should be used with caution when assessing the risk of genetically modified *C. carpio*.

Acknowledgements The authors wish to thank K. V. Radhakrishnan and S. Ye for their constructive comments on the manuscript and also thank W. Li, B. Guan and D. Li for their help during the experiment. This work was financially supported by the Development Plan of the State Key Fundamental Research of China (Grant No. 2007CB109205) and the National Natural Science Foundation (Grant Nos. 30770377 and 30830025).

References

Abrahams, M. V. (1989). Foraging guppies and the ideal free distribution: the influence of information on patch choice. *Ethology* 82, 116-126.

Abrahams, M. V. & Sutterlin, A. (1999). The foraging and antipredator behaviour of growth-enhanced transgenic Atlantic salmon. *Animal Behaviour* 58, 933-942.

Alm, G. (1959). Relation between maturity, size and age in fishes. *Report of the Institute of Freshwater Research, Drottningholm* 40, 4-145.

Arendt, J. D. (1997). Adaptive intrinsic growth rates: an integration across taxa. *The Quarterly Review of Biology* 72, 149-177.

Bailey, J., Alanärä, A. & Brannas, E. (2000). Methods for assessing social status in Arctic charr. *Journal of Fish Biology* 57, 258-261.

Begon, M., Harper, J. L. & Townsend, C. R. (1996). *Ecology*, 3rd edn. Oxford: Blackwell Scientific.

Carter, C. G., McCarthy, I. D., Houlihan, D. F., Johnstone, R., Walsingham, M. V. & Mitchell, A. I. (1994). Food consumption, feeding behaviour and growth of triploid and diploid Atlantic salmon, *Salmo salar* L., parr. *Canadian Journal of Zoology* 72, 609-617.

Colgan, P. W. (1973). Motivational analysis of fish feeding. *Behaviour* 45, 38-66.

Devlin, R. H., Yesaki, T. Y., Biagi, C. A. & Donaldson, E. M. (1994). Extraordinary salmon growth. *Nature* 371, 209-210.

Devlin, R. H., Johnsson, J. I., Smailus, D. E., Biagi, C. A., Jönsson, E. & Björnsson, B. T. (1999). Increased ability to compete for food by growth hormone-transgenic coho salmon *Oncorhynchus kisutch* (Walbaum). *Aquaculture Research* 30, 479-482.

Devlin, R. H., Biagi, C. A., Yesaky, T. Y., Smailus, D. E. & Byatt, J. C. (2001). Growth of domesticated transgenic fish: a growth-hormone transgene boosts the size of wild but not domesticated trout. *Nature* 409, 781-782.

Devlin, R. H., D'Andrade, M., Uh, M. & Biagi, C. A. (2004). Population effects of growth hormone transgenic coho salmon depend on food availability and genotype by environment interactions. *Proceedings of the National Academy of Sciences of the United States of America* 101, 9303-9308.

Devlin, R. H., Sundström, L. F. & Muir, W. M. (2006). Interface of biotechnology and ecology for environmental risk assessments of transgenic fish. *Trends in Biotechnology* 24, 89-97.

Dill, L. M. (1983). Adaptive flexibility in the foraging behaviour of fishes. *Canadian Journal of Fisheries and Aquatic Sciences* 40, 398-408.

Farrell, A. P., Bennett, W. & Devlin, R. H. (1997). Growth-enhanced transgenic salmon can be inferior swimmers. *Canadian Journal of Zoology* 75, 335-337.

Forkman, B. & Haskell, M. J. (2004). The maintenance of stable dominance hierarchies and the pattern of aggression: support for the suppression hypothesis.*Ethology* 110, 737-744.

Fu, C., Cui, Y., Hung, S. S. O. & Zhu, Z. (1998). Growth and feed utilization by F_4 human growth hormone transgenic carp fed diets with different protein levels. *Journal of Fish Biology* 53, 115-129.

Fu, C., Li, D., Hu, W., Wang, Y. & Zhu, Z. (2007). Growth and energy budget of F_2"all-fish" growth hormone gene transgenic common carp. *Journal of Fish Biology* 70, 347-361.

Gross, M. L., Schneider, J. F., Moav, N., Moav, B., Alvarez, C., Myster, S. H., Liu, Z., Hallerman, E. M., Hackett, P. B., Guise, K. S., Faras, A. J. & Kapuscinski, A. R.(1992). Molecular analysis and growth evaluation of northern pike (*Esox lucius*) microinjected with growth hormone genes. *Aquaculture* 103, 253-273.

Höjesjö, J., Johnsson, J. & Axelsson, M. (1999). Behavioural and heart rate responses to food limitation and predation risk: an experimental study on rainbow trout. *Journal of Fish Biology* 55, 1009-1019.

Höjesjö, J., Armstrong, J. D. & Griffiths, S. W. (2005). Sneaky feeding by salmon in sympatry with dominant brown trout. *Animal Behaviour* 69, 1037-1041.

Huntingford, F. A. & Turner, A. K. (1987). *Animal Conflict*. London: Chapman & Hall.

Huntingford, F. A., Metcalfe, N. B., Thorpe, J. E., Graham, W. D. & Adams, C. E. (1990). Social dominance and body size in Atlantic salmon parr, *Salmo salar* L. *Journal of Fish Biology* 36, 877-881.

Jhingan, E., Devlin, R. H. & Iwama, G. K. (2003). Disease resistance, stress response and effects of triploidy in growth hormone transgenic coho salmon. *Journal of Fish Biology* 63, 806-823.

Jobling, M. & Baardvik, B. M. (1994). The influence of environmental manipulations on inter-and intra-individual variation in food acquisition and growth performance of Arctic charr, *Salvelinus alpinus*. *Journal of Fish Biology* 44, 1069-1087.

Johnsson, J. I. & Björnsson, B. Th. (1994). Growth hormone increases growth rate, appetite, and dominance in juvenile rainbow trout, *Oncorhynchus mykiss*. *Animal Behaviour* 48, 177-186.

Jönsson, E., Johnsson, J. I. & Björnsson, B. Th. (1996). Growth hormone increases predation exposure of rainbow trout. *Proceedings of the Royal Society* B 263, 647-651.

Jönsson, E., Johnsson, J. I. & Björnsson, B. T. (1998). Growth hormone increases aggressive behavior in juvenile rainbow trout. *Hormones and Behavior* 33, 9-15.

Kapuscinski, A. R., Hayes, K. R., Li, S. & Dana, G. (2007). *Environmental Risk Assessment of Genetically Modified Organisms*, Vol. III. Wallingford: CABI Publishers.

Li, D., Fu, C., Hu, W., Zhong, S., Wang, Y. & Zhu, Z. (2007). Rapid growth cost in "all-fish" growth hormone gene transgenic carp: reduced critical swimming speed. *Chinese Science Bulletin* 52, 1501-1506.

McCarthy, I. D., Carter, C. G. & Houlihan, D. F. (1992). The effect of feeding hierarchy on individual variability in daily feeding of rainbow trout, *Oncorhynchus mykiss* (Walbaum). *Journal of Fish Biology* 41, 257-263.

McCarthy, I. D., Gair, D. J. & Houlihan, D. F. (1999). Feeding rank and dominance in *Tilapia rendalli* under defensible and indefensible patterns of food distribution. *Journal of Fish Biology* 55, 854-867.

McLean, E. & Donaldson, E. M. (1993). The role of growth hormone in the growth of poikilotherms. In *The Endocrinology of Growth, Development, and Metabolism in Vertebrates* (Schreibman, M. P., Scanes, C. G. & Pang, P. K. T., eds), pp. 43-71. San Diego, CA: Academic Press.

Metcalfe, N. B. (1986). Intraspecific variation in competitive ability and food intake in salmonids: consequences for energy budgets and growth rates. *Journal of Fish Biology* 28, 525-531.

Muir, W. M. & Howard, R. D. (1999). Possible ecological risks of transgenic organism release when transgenes affect mating success: sexual selection and the Trojan gene hypothesis. *Proceedings of the National Academy of Sciences of the United States of America* 96, 13853-13856.

Muir, W. M. & Howard, R. D. (2001). Fitness components and ecological risks of transgenic release: a model using Japanese medaka (*Oryzias latipes*). *The American Naturalist* 158, 1-16.

Nicieza, A. G. & Metcalfe, N. B. (1999). Costs of rapid growth: the risk of aggression is higher for fast-growing salmon. *Functional Ecology* 13, 793-800.

Raven, P. A., Devlin, R. H. & Higgs, D. A. (2006). Influence of dietary digestible energy content on growth, protein and energy utilization and body composition of growth hormone transgenic and non-transgenic coho salmon (*Oncorhynchus kisutch*). *Aquaculture* 254, 730-747.

Ricker, W. E. (1979). Growth rates and models. In *Fish Physiology*, Vol. VIII (Hoar, W. S., Randall, D. J. & Brett, J. R., eds), pp. 677-743. New York, NY: Academic Press.

Shelverton, P. A. & Carter, C. G. (1998). The effect of ration on behaviour, food consumption and growth in juvenile greenback flounder (*Rhombosolea tapirina*: Teleostei). *Journal of the Marine Biological Association of the United Kingdom* 78, 1307-1320.

Sloman, K. A. & Armstrong, J. D. (2002). Physiological effects of dominance hierarchies: laboratory artefacts or natural phenomena. *Journal of Fish Biology* 61, 1-23.

Sneddon, L. U., Hawkesworth, S., Braithwaite, V. A. & Yerbury, J. (2006). Impact of environmental disturbance on the stability and benefits of individual status within dominance hierarchies. *Ethology* 112, 437-447.

Southwood, T. R. E. & Henderson, P. A. (2000). *Ecological Methods*. Oxford: Blackwell Science.

Stephens, D. W. & Krebs, J. R. (1996). *Foraging Theory*. Princeton, NJ: Princeton University Press.

Sundström, L. F., Lõhmus, M., Devlin, R. H., Johnsson, J. I., Biagi, C. A. & Bohlin, T. (2004a). Feeding on profitable and unprofitable prey: comparing behaviour of growth-enhanced transgenic and normal coho salmon (*Oncorhynchus kisutch*). *Ethology* 110, 381-396.

Sundström, L. F., Lõhmus, M., Johnsson, J. I. & Devlin, R. H. (2004b). Growth hormone transgenic salmon pay for growth potential with increased predation mortality. *Proceedings of the Royal Society* B 271, S350-S352.

Sundström, L. F., Lõhmus, M. & Devlin, R. H. (2005). Selection on increased intrinsic growth rates in coho salmon *Oncorhynchus kisutch*. *Evolution* 59, 1560-1569.

Sundström, L. F., Lõhmus, M., Johnsson, J. I. & Devlin, R. H. (2007a). Dispersal potential is affected by growth-hormone transgenesis in coho salmon (*Oncorhynchus kisutch*). *Ethology* 113,403-410.

Sundström, L. F., Lõhmus, M., Tymchuk, W. E. & Devlin, R. H. (2007b). Gene-environment interactions influence ecological consequences of transgenic animals. *Proceedings of the National Academy of Sciences of the United States of America* 104, 3889-3894.

Tindo, M. & Dejean, A. (2000). Dominance hierarchy in colonies of *Belonogaster juncea juncea*. *Insectes Sociaux* 47, 158-163.

Townsend, C. R. & Winfield, I. J. (1985). The application of optimal foraging theory to feeding behaviour in fish. In *Fish Energetics. New Perspectives* (Tytler, P. &Calow, P., eds), pp. 67-98. Sydney: Croom Helm.

Tymchuk, W. E., Abrahams, M. V. & Devlin, R. H. (2005). Competitive ability and mortality of growth-enhanced transgenic coho salmon fry and parr when foraging for food. *Transactions of the American Fisheries Society* 134, 381-389.

Utne-Palm, A. C. & Hart, P. J. B. (2000). The effects of familiarity on competitive interactions between threespined sticklebacks. *Oikos* 91, 225-232.

Vehanen, T. (2003). Adaptive flexibility in the behaviour of juvenile Atlantic salmon: short-term responses to food availability and threat from predation. *Journal of Fish Biology* 63, 1034-1045.

Venugopal, T., Anathy, V., Kirankumar, S. & Pandian, T. J. (2004). Growth enhancement and food conversion efficiency of transgenic fish *Labeo rohita*. *Journal of Experimental Zoology* 301A, 477-490.

Wang, Y., Hu, W., Wu, G., Sun, Y., Chen, S., Zhang, F., Zhu, Z., Feng, J. & Zhang, X. (2001). Genetic analysis of "all-fish" growth hormone gene transferred carp (*Cyprinus carpio* L.) and its F_1 generation. *Chinese Science Bulletin* 46, 1174-1177.

Werner, E. E. & Gilliam, J. F. (1984). The ontogenetic niche and species interactions in size-structured populations. *Annual Review of Ecology and Systematics* 15, 393-425.

Westerberg, M., Staffan, F. & Magnhagen, C. (2004). Influence of predation risk on individual competitive ability and growth in Eurasian perch, *Perca fluviatilis*. *Animal Behaviour* 67, 273-279.

Zbikowska, H. M. (2003). Fish can be first–advances in fish transgenesis for commercial applications. *Transgenic Research* 12, 379-389.

Zhu, Z., Li, G., He, L. & Chen, S. (1985). Novel gene transfer into the fertilized eggs of goldfish (*Carassius auratus* L. 1758). *Zeitschrift für Angewandte Ichthyologie* 1, 31-34.

转"全鱼"生长激素基因鲤增强了对有限食物资源的竞争能力

段 明[1,2] 张堂林[1] 胡 炜[1] Sundström L. Fredrik[3] 汪亚平[1] 李钟杰[1] 朱作言[1]

1 中国科学院水生生物研究所，淡水生态与生物技术国家重点实验室，武汉 430072，中国
2 中国科学院大学，北京 100049，中国
3 Department of Fisheries and Oceans, University of British Columbia, Centre for Aquaculture and Environmental Research, 4160 Marine Drive, West Vancouver, British Colombia V7V 1N6, Canada

摘 要 在食物资源有限的条件下，我们比较分析了转生长激素基因鲤和非转基因鱼的摄食量、活动水平和摄食等级。与非转基因鱼相比，转基因鱼具有更高的摄食等级，活动水平提高了73.3%，摄食量提高了86%。经过10天的室内实验，转基因鲤未表现出更高的生长潜能，这可能与转基因鲤的高活动水平，以及非转基因鱼所采取的"偷食策略"有关。上述结果显示，在有限食物资源供给下，转基因鱼提高了摄食竞争力，这可能有利于释放或逃逸到野外的转基因鱼的存活力。然而，自然环境中的其他因素(如捕食风险和食物分布)，也可能会抵消这种优势，因此本研究结果需要谨慎利用。

Increased Mortality of Growth-Enhanced Transgenic Common Carp (*Cyprinus carpio* L.) Under Short-Term Predation Risk

M. Duan[1,2] T. Zhang[1] W. Hu[1] B. Guan[1,2] Y. Wang[1] Z. Li[1] Z. Zhu[1]

1 State Key Laboratory of Freshwater Ecology and Biotechnology, Institute of Hydrobiology, Chinese Academy of Sciences, Wuhan 430072

2 Graduate University of Chinese Academy of Sciences, Beijing 100049

Abstract There is strong evidence that genetic capacity for growth evolves toward an optimum rather than an absolute maximum. This implies that fast growth has a cost and that tradeoffs occur between growth and other life-history traits. In this study, we conducted a series of predation trials with a paired-contrast design to test the differences in anti-predator ability between growth-enhanced transgenic and non-transgenic common carp (*Cyprinus carpio* L.). We showed that young transgenic fish suffered higher predation mortality than control carp in both small-bodied and large-bodied size-matched trials, and thus possessed lower anti-predator ability. Our results suggest that a trade-off exists between growth rate and survival such that rapid growth entails a cost in terms of mortality.

Introduction

In the early 1980s, the gene construct *MThGH* human growth hormone (GH) gene driven by mouse metallothionein-1 (*MT-1*) promoter was microinjected into the fertilized eggs of goldfish (*Carassius auratus* L.), producing a first batch of fastgrowing transgenic fish (Zhu *et al*., 1985). Under the consideration of bio-safety and bio-ethics, *MThGH* construct is not encouraged for use in the breeding of fish. Thus the recombinant transgene construct, *pCAgcGH*, made up of all fish (Cyprinidae) genomic elements (e.g. the grass carp GH structural gene spliced under a common carp *β*-actin gene promoter) was transferred into common carp (*Cyprinus carpio* L.), which led to the birth of all fish GH-transgenic common carp (Zhu *et al*., 1992). In order to differentiate the gene construct *MThGH*, the recombinant construct *pCAgcGH* is often called the 'all fish' GH gene. The GH-transgenic common carp grew 1.4-1.9 times that of non-transgenic counterparts under hatchery conditions (Wang *et al*., 2001; Li *et al*., 2007). GH-transgenic fish are of great commercial interest in aquaculture, but also raise concerns with regard to the potential impacts of escaped or released GH-transgenic

文章发表于 *Journal of Applied Ichthyology Zeitschrift Für Angewandte Ichthyology*, 2010, 26(6): 908-912 fish on aquatic ecosystems (Maclean and Laight, 2000). To predict these risks, it is crucial to obtain empirical data on the relative fitness of GH-transgenic and non-transgenic fish under simulated natural conditions.

Rapid growth is often perceived as being beneficial, especially in juvenile stages (Sogard, 1997). In many fish species, fast growth should increase fitness since larger fish at a given age can achieve higher dominance status, reduce their susceptibility to predators (Juanes *et al.*, 2002) and increase their reproductive potential (Fleming, 1996). Evidence has accumulated indicating a trade-off between benefits and costs associated with rapid growth (Arendt, 1997; Mangel and Stamps, 2001; Morgan and Metcalfe, 2001; Munch and Conover, 2003; Biro *et al.*, 2005), suggesting that limitations exist with the concept that faster growth is always better. Recent studies with fast-growing fish have shown decreased escape performance (Lankford *et al.*, 2001; Munch and Conover, 2003), less vigilance and more willingness to predation risk when foraging (Johnsson *et al.*, 1996; Abrahams and Pratt, 2000), and greater predation mortality (Biro *et al.*, 2004) with increases in growth rate. Similar results were found in GH-transgenic fish relative to their non-transgenic counterparts (Farrell *et al.*, 1997; Abrahams and Sutterlin, 1999; Sundström *et al.*, 2004). These findings indicate that the fitness advantage of rapid growth may be balanced by a variety of associated costs.

Predation is an important factor structuring fish populations in nature (Whoriskey and Fitzgerald, 1985; Persson, 1991). By killing prey, predators also can control prey populations, drive some types of prey to extinction, and alter the relative and absolute abundances of prey (Sih *et al.*, 1985). Small fish are more vulnerable to predators, and few individuals will survive to maturation in the presence of various predators (Sogard, 1997). For juvenile fishes, an innate ability to recognize and avoid natural predators can have fitness advantages (Sogard, 1997). It is therefore essential to obtain data on the survival fitness of GH-transgenic and non-transgenic fish exposed to predators during their earlier life-history stages.

In this study, we used survival after short-term risk of predation to compare the anti-predator ability of GH-transgenic and non-transgenic common carp at the juvenile stage. Such information would be useful for further examination of the trade-off between rapid growth and mortality. Also, this would be essential for evaluating potential impacts that transgenic fish may have on aquatic ecosystems.

Materials and Methods

Source of fish

P_0 GH-transgenic common carp were initially produced by micro-injection of *pCAgcGH* into fertilized eggs of common carp (Yellow River variety). Gene construct *pCAgcGH* was a recombinant construct of the grass carp *Ctenopharyngodon idella* (Valenciennes) growth hormone gene (*gcGH*), whose expression is driven by the *β*-actin gene promoter of common carp (*pCA*). The transgene was integrated, expressed, and inherited in the subsequent F_1 and F_2 generations in a Mendelian fashion. Experimental GH-transgenic fish (T) and their non-transgenic all-siblings (NT) used in this study were F_3 generation progeny, produced by crossing a wild type female with an F_2 transgenic heterozygous male of a fastgrowing transgenic germ line. Fertilized embryos from this cross were incubated on 14 April 2005, which theoretically yields 50% T and NT offspring each. Hatched first-feeding fry were reared in eight outdoor rectangle concrete tanks (each 10 m^2) at different stocking rates. Fish first fed on plankton and then consumed commercial diets when over 1.8 cm TL. At 8 weeks of age they were transported to the laboratory aquaria for identification. The left or right ventral fin was partially clipped and used to identify the *pCAgcGH* transgene-positive fish by the PCR method following Wang *et al.* (2001). All experimental T (positive) and NT (negative) fish were selected from those individuals that completely regenerated the clipped fins after more than 4 weeks.

Mandarin fish, *Siniperca chuatsi* (Basilewsky), was chosen as a predator in this study. It is a demersal large-sized piscivorous predator, widely distributed in many rivers and lakes of China. The fish have peculiar feeding habits and eat only live larvae of fish and shrimps when first feeding and refuse to consume dead prey or artificial diets during all life-history stages (Liang *et al.*, 1998). Age 1^+ mandarin fish in Liangzi Lake were captured with trap-nets and 15 individuals (mean mass ± SD: 482.3 ± 15.6 g) were selected for use in this study. Prior to the experiment, the mandarin fish were given at least 1 week to acclimate to confined concrete-tank conditions and fed with live crucian carp (*Carassius auratus* L.) juveniles once a day during acclimation.

Experimental procedures

The non-commercial research facility at the Fish Behavioral Ecology Laboratory of the Institute of Hydrobiology in China had multiple containment screen systems and was specially designed to prevent transgenic fish from escaping into nature.

A paired-contrast design tested differences in anti-predator ability of T and NT fish: Mixed schools containing an equal number of size-matched T and NT fish were exposed simultaneously to a predator. Two predation trials on T and NT fish were conducted from 18 August to 9 September and different trial durations were determined based upon the previous observations. The first trial (I) lasted 5 days, with a small-bodied size-matched population (T = 60, NT = 60) in one tank (see below). The second trial (II) lasted 3 days, with a large-bodied size-matched school (T = 30, NT = 30) in another tank. Prior to each trial, the left ventral fin was completely clipped for each T fish whereas the right ventral fin was clipped for each NT fish so that the two genotypes could be differentiated visually. To verify that the difference in predation mortality between T and NT fish was not derived from the fin-marking position (left vs right) and other non-treatment factors such as size, predator, and rearing facilities, an additional two predation trials (III and IV) were conducted using only NT fish separated into two groups and marked by an alternating ventral fin clip (left or right).

In all trials, experimental fish were reared in large, concrete tanks (3.5 m × 2.3 m × 1.5 m) to minimize confinement of prey and to permit them to express their natural flight response. A net-cover was put on each tank to keep aerial predators such as birds and cats away. Fifteen earthen flowerpots (three in each corner and three in the middle) were placed in each tank to provide refuga for prey. Experimental fish were maintained under ambient temperature (mean ± SD: 23.6 ± 2.6℃) and photoperiod conditions, and fed a commercial diet twice (09.00 h and 14.00 h) daily. Tanks were supplied with flow-through filtered natural water drawn from Liangzi Lake, with pH 7.9 and dissolved oxygen > 6.9 mg/l. In each trial, three mandarin fish were chosen randomly as predators and placed into the experimental tank. During each trial, predation was the only source of mortality for both T and NT fish.

Prior to each trial, experimental fish were lightly anaesthetized with Eugenol, weighed (±g) and total length measured (± cm); the same procedure was carried out for the survivors at the end of each trial. Condition factor (CF) was calculated as: CF (g/cm^{-3}) = 100 × W/L^3, Where W (g) is body mass and L (cm) is total length.

Data analysis

The differences between two mean values in total length, body mass and condition factor were examined by Students, t-test. The chi-square test was used to evaluate differences in mortality rate between two subpopulations in each predation trial. Differences were regarded as significant when $P < 0.05$. All statistical analyses were performed using the STATISTICA program by StatSoft, Inc. (1995).

Results

Data on initial total length, body mass and condition factor of experimental fish in four predation trials are shown in Table 1. In these morphological measures, there were no significant differences between two size-matched subpopulations at the start of each predation trial (Students' t-test, all $P > 0.05$; Table 1).

In trial I with small-bodied size-matched population, mortality rate of the T fish was 88.3%, significantly higher than that (61.7%) of the NT fish ($x^2 = 11.38$, $P = 0.001$, Fig. 1). Similar results were found in trial II with the large-bodied sizematched population; mortality rate of the T fish was about two times that of the T fish (53.3% vs 26.7%, $x^2 = 4.44$, $P = 0.035$, Fig. 1).

In contrast to trial I, mortality rate of the trial III NT fish with the left ventral fin clipped was same as NT fish with their right ventral fin clipped, as occurred in trial IV contrasted to trial II (Fig. 1), which showed that the size-matched individuals of the same genotype possessed the same anti-predator ability. This also demonstrated that the fin-clip positions (i.e. left vs right) and other non-treatment factors (approximate size match, predator selection and experimental facilities, etc.) did not yield differences in predation mortality, and that differences in mortality between T and NT fish can be attributed to genotype in trials I and II.

Table 1 Data (mean ± SD) on initial total length (*L*), body mass (*W*) and condition factor (CF) of GH-transgenic (T) and non-transgenic (NT) common carp *Cyprinus carpio* in four predation trials

Trial	Genotype	Ventral fin clipped	Number of fish	*L* (cm)	*W* (g)	CF (g cm^{-3})
I	T	Left	60	6.63 ± 0.34	4.0 ± 0.6	1.365 ± 0.005
	NT	Right	60	6.63 ± 0.23	4.0 ± 0.4	1.365 ± 0.003
II	T	Left	30	8.03 ± 0.32	7.8 ± 0.9	1.504 ± 0.129
	NT	Right	30	8.12 ± 0.23	7.2 ± 0.8	1.351 ± 0.085
III	NT	Left	60	6.41 ± 0.18	3.7 ± 0.4	1.386 ± 0.100
	NT	Right	60	6.37 ± 0.18	3.6 ± 0.5	1.374 ± 0.098
IV	NT	Left	30	8.36 ± 0.27	8.2 ± 1.1	1.397 ± 0.102
	NT	Right	30	8.39 ± 0.24	8.3 ± 1.0	1.399 ± 0.079

Notes: I, small-bodied size-matched population with T and NT subpopulations; II, large-bodied size-matched population with T and NT subpopulations; III, small-bodied size-matched population with only NT fish; and IV, large-bodied size-matched population with only NT fish.For all parameters there were no significant differences between T and NT fish examined by Students *t*-test at the level of $P = 0.05$

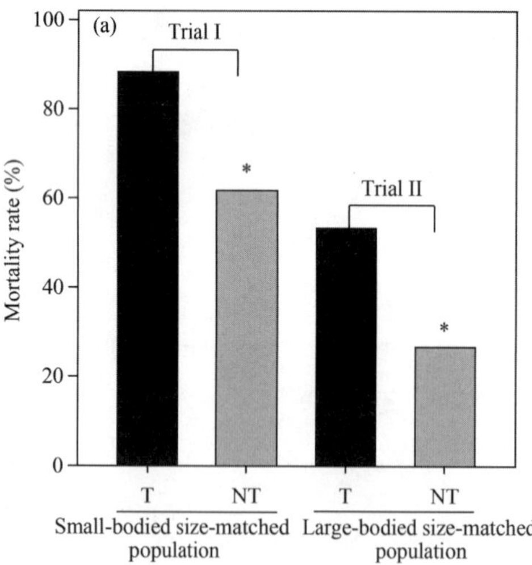

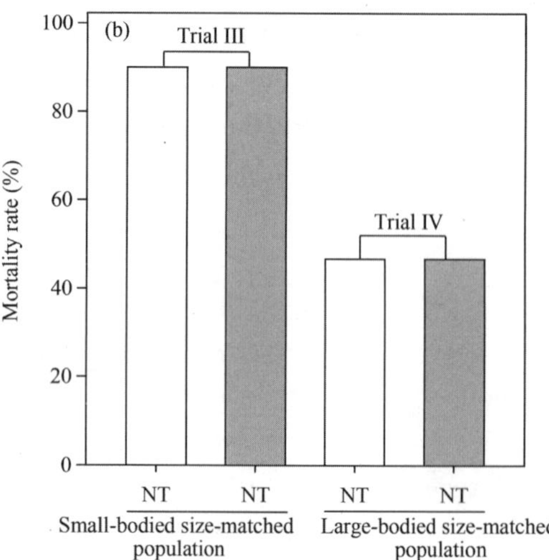

Fig. 1 Mortality rates of GH-transgenic (T) and non-transgenic (NT) common carp *Cyprinus carpio* at end of (a) trial I (5 days, small-bodied size-matched population) and trial II (3 days, large-bodied size-matched population); (b) trial III (5 days; small-bodied size-matched population) and trial IV (3 days; large-bodied size-matched population) with three predator mandarin fish. Black and open bars = left ventral fin clipped; gray bars = right ventral fin clipped. Asterisks = significant differences in mortality rate between the two genotypes in the same predation trials (Chi-square test, $P < 0.05$)

Discussion

In the present study, replicated treatments in trials I and II were not designed because of the scarcity, difficulty in size match, and high cost of GH-transgenic common carp. We took advantage of a limited supply of the GH-transgenic fish. Our results demonstrate that GH-transgenic carp juveniles suffer increased predation mortality and possess lower anti-predator ability. These findings are consistent with the study of Sundström *et al.* (2004), who showed that predation mortality of first-feeding fry of GH-transgenic coho salmon, *Oncorhynchus kisutch* (Walbaum), is much higher than that of non-transgenic control fry under near-natural conditions. Similar results were obtained in another study showing increased susceptibility to predators for GH-transgenic channel catfish, *Ictalurus punctatus* (Rafinesque), at an early life stage (Dunham *et al.*, 1999).

In contrast to the present study, Tymchuk *et al.* (2005) found that GH-transgenic coho salmon at the parr stage did not suffer higher rates of mortality than non-transgenic counterparts and that they maintained their considerably enhanced growth when feeding in the presence of a predator. Johnsson and colleagues also found that GH-implanted wild brown trout (*Salmo trutta* L.) grew faster than control trout in the wild without suffering increased mortality (Johnsson *et al.*, 1999; Johnsson and Björnsson, 2001), although the effect of GH-treatment was stronger in the hatchery where food was more abundant (Johnsson *et al.*, 2000). Those studies differ from the present study in that the experiments were conducted over a relatively long time period (over 3 weeks), in which fast-growing GH-implanted or GH-transgenic fish may be allowed to attain the expected benefits of large size. However, in those studies, there actually was an overall tendency for higher mortality in GH-transgenic or GH-implanted fish than in control fish. In addition, experiments on Atlantic salmon (*Salmo salar* L.) suggest that maintenance of energy reserves over winter is critical for fitness (Bull *et al.*, 1996); previous work on brown trout indicated that GH-implanted fish increase growth rate at the cost of reduced energy reserves (Johnsson *et al.*, 1999, 2000). Consequently, high GH expression in GH-transgenic or GH-implanted fish may eventually increase mortality by reducing energy reserves necessary for survival under harsh conditions.

Our results suggest that trade-offs exist between growth rate and fitness in the transgenic common carp and rapid growth entails a cost in terms of mortality. There are several possible explanations for these findings. In parallel experiments in earthen ponds where supplemental feed was applied, the same groups of transgenic common carp used in this study grew 40% to 90% faster than non-transgenic siblings (Li *et al.*, 2007). To maintain increased rates of growth, fish must make trade-offs between foraging effort and risk of predation (Abrahams and Dill, 1989). Previous studies have indicated that GH-transgenic fish have a greater overall appetite and a much higher consumption rate (Abrahams and Sutterlin, 1999; Devlin *et al.*,

1999; Fu *et al.*, 2007; Duan *et al.*, 2009) and thereby appear more active and risk-prone to forage (Abrahams and Sutterlin, 1999; Sundström *et al.*, 2003) than non-transgenic fish. Moreover, increased foraging activity level (Abrahams and Sutterlin, 1999) may reduce vigilance and / or increase the probability of encountering or being detected by predators (Werner and Anholt, 1993; Anholt and Werner, 1998). Finally, some studies have indicated that GH-transgenic common carp and salmon juveniles can be inferior swimmers, with significantly decreased critical swimming speeds compared with control fish of the same size (Farrell *et al.*, 1997; Li *et al.*, 2007), which may affect their escape performance. Similar results were also found in the study of Farrell *et al.* (1997), which showed that growth-enhanced GH-transgenic salmon juveniles had critical swimming speeds half those of size-matched control fish.

The present study indicated that GH-transgenic common carp at the juvenile stage showed lower survival under a shortterm risk of predation. However, it is difficult to extrapolate our limited results into the natural environment because environmental complexity can greatly alter the phenotypic effects of the GH transgene (Sundström *et al.*, 2007). Thus, we should use the results with caution when assessing ecological risk of GH-transgenic common carp. Future research is needed to understand key variables (Devlin *et al.*, 2006) influencing survival fitness for risk assessment of transgenic fish because disadvantages in one attribute can be offset by advantages in another (Muir and Howard, 1999).

Acknowledgements We wish to give our deep thanks to two anonymous reviewers for their helpful comments in the critical reviews of this manuscript. This work was financially supported by the Development Plan of the State Key Fundamental Research of China (Grant No. 2007CB109205) and National Natural Science Foundation (Grant Nos. 30770377 and 30830025).

References

Abrahams, M. V.; Dill, L. M., 1989: A determination of the energetic equivalence of the risk of the predation. Ecology 70, 999-1007.

Abrahams, M. V.; Pratt, T. C., 2000: Hormonal manipulations of growth rate and its influence on predator avoidance-foraging trade-offs. Can. J. Zool. 78, 121-127.

Abrahams, M. V.; Sutterlin, A., 1999: The foraging and antipredator behaviour of growth-enhanced transgenic Atlantic salmon. Anim. Behav. 58, 933-942.

Anholt, B. R.; Werner, E. E., 1998: Predictable changes in predation mortality as a consequence of changes in food availability and predation risk. Evol. Ecol. 12, 729-738.

Arendt, J. D., 1997: Adaptive intrinsic growth rates: an integration across taxa. Q. Rev. Biol. 72, 149-177.

Biro, P. A.; Abrahams, M. V.; Post, J. R.; Parkinson, E. A., 2004: Predators select against high growth rates and risk-taking behaviour in domestic trout populations. Proc. R. Soc. Lond. B 271, 2233-2237.

Biro, P. A.; Post, J. R.; Abrahams, M. V., 2005: Ontogeny of energy allocation reveals selective pressure promoting risk-taking behaviour in young fish cohorts. Proc. R. Soc. Lond. B 272, 1443-1448.

Bull, C. D.; Metcalfe, N. B.; Mangel, M., 1996: Seasonal matching of foraging effort to anticipated energy requirements in anorexic juvenile salmon. Proc. R. Soc. Lond. B 263, 13-18.

Devlin, R. H.; Johnsson, J. I.; Smailus, D. E.; Biagi, C. A.; Jönsson, E.; Björnsson, B. T., 1999: Increased ability to compete for food by growth hormone-transgenic coho salmon *Oncorhynchus kisutch* (Walbaum). Aquac. Res. 30, 479-482.

Devlin, R. H.; Sundström, L. F.; Muir, W. M., 2006: Interface of biotechnology and ecology for environmental risk assessments of transgenic fish. Trends Biotechnol. 24, 89-97.

Duan, M.; Zhang, T.; Hu, W.; Sundström, L. F.; Wang, Y.; Li, Z.; Zhu, Z., 2009: Elevated ability to compete for limited food resources by 'all-fish' growth hormone transgenic common carp *Cyprinus carpio*. J. Fish Biol. 75, 1459-1472.

Dunham, R. A.; Chitmanat, C.; Nichols, A.; Argue, B.; Powers, D. A.; Chen, T. T., 1999: Predator avoidance of transgenic channel catfish containing salmonid growth hormone genes. Mar. Biotechnol. 1, 545-551.

Farrell, A. P.; Bennett, W.; Devlin, R. H., 1997: Growth-enhanced transgenic salmon can be inferior swimmers. Can. J. Zool. 75, 335-337.

Fleming, I. A., 1996: Reproductive strategies of Atlantic salmon: ecology and evolution. Rev. Fish Biol. Fish. 6, 379-416.

Fu, C.; Li, D.; Hu, W.; Wang, Y.; Zhu, Z., 2007: Growth and energy budget of F_2 "all-fish" growth hormone gene transgenic common carp. J. Fish Biol. 70, 347-361.

Johnsson, J. I.; Björnsson, B. T., 2001: Growth-enhanced fish can be competitive in the wild. Funct. Ecol. 15, 654-659.

Johnsson, J. I.; Petersson, E.; Jönsson, E.; Björnsson, B. T.; Järvi, T., 1996: Domestication and growth hormone alter antipredator behaviour and growth patterns in juvenile brown trout, *Salmo trutta*. Can. J. Fish. Aquat. Sci. 53, 1546-1554.

Johnsson, J. I.; Petersson, E.; Jönsson, E.; Järvi, T.; Björnsson, B. T., 1999: Growth hormone-induced effects on mortality, energy status and growth a field study on brown trout (*Salmo trutta*). Funct. Ecol. 13, 514-522.

Johnsson, J. I.; Jönsson, E.; Petersson, E.; Järvi, T.; Björnsson, B. T., 2000: Fitness-related effects of growth investment in brown trout under field and hatchery conditions. J. Fish Biol. 57, 326-336.

Juanes, F.; Buckel, J. A.; Scharf, F. S., 2002: Feeding ecology of piscivorous fishes. In: Handbook of fish biology and fisheries, Vol. 1. P. J. B. Hart, J. D. Reynolds (Eds), Blackwell Science, Oxford, pp. 267-283.

Lankford, T. E.; Billerbeck, J. M.; Conover, D. O., 2001: Evolution of intrinsic growth and energy acquisition rates. II. Tradeoffs with vulnerability to predation in *Menidia menidia*. Evolution 55, 1873-1881.

Li, D.; Fu, C.; Hu, W.; Zhong, S.; Wang, Y.; Zhu, Z., 2007: Rapid growth cost in ''all-fish'' growth hormone gene transgenic carp: reduced critical swimming speed. Chin. Sci. Bull. 52, 1501-1506.

Liang, X.; Liu, J.; Huang, B., 1998: The role of sense organs in the feeding behaviour of Chinese perch. J. Fish Biol. 52, 1058-1067.

Maclean, N.; Laight, R. J., 2000: Transgenic fish: an evaluation of benefits and risks. Fish Fish. 1, 146-172.

Mangel, M.; Stamps, J., 2001: Trade-offs between growth and mortality and the maintenance of individual variation in growth. Evol. Ecol. Res. 3, 583-593.

Morgan, I. J.; Metcalfe, N. B., 2001: Deferred costs of compensatory growth after autumnal food shortage in juvenile salmon. Proc. R. Soc. Lond. B 268, 295-301.

Muir, W. M.; Howard, R. D., 1999: Possible ecological risks of transgenic organism release when transgenes affect mating success: Sexual selection and the Trojan gene hypothesis. Proc. Natl. Acad. Sci. USA 96, 13853-13856.

Munch, S. B.; Conover, D. O., 2003: Rapid growth results in increased susceptibility to predation in *Menidia menidia*. Evolution 57, 2119-2127.

Persson, L., 1991: Behavioral response to predators reverses the outcome of competition between prey species. Behav. Ecol. Sociobiol. 28, 101-105.

Sih, A.; Crowley, P.; McPeek, M.; Petranka, J.; Strohmeier, K., 1985: Predation, competition, and prey communities: a review of field experiments. Annu. Rev. Ecol. Evol. Syst. 16, 269-312.

Sogard, S. M., 1997: Size-selective mortality in the juvenile stage of teleost fishes: A review. Bull. Mar. Sci. 60, 1129-1157.

Sundström, L. F.; Devlin, R. H.; Johnsson, J. I.; Biagi, C. A., 2003: Vertical position reflects increased feeding motivation in growth hormone transgenic coho salmon (*Oncorhynchus kisutch*). Ethology 109, 701-712.

Sundström, L. F.; Lõhmus, M.; Johnsson, J. I.; Devlin, R. H., 2004: Growth hormone transgenic salmon pay for growth potential with increased predation mortality. Proc. R. Soc. Lond. B 271, S350-S352.

Sundström, L. F.; Lõhmus, M.; Tymchuk, W. E.; Devlin, R. H., 2007: Gene-environment interactions influence ecological consequences of transgenic animals. Proc. Natl. Acad. Sci. USA 104, 3889-3894.

Tymchuk, W. E.; Abrahams, M. V.; Devlin, R. H., 2005: Competitive ability and mortality of growth-enhanced transgenic coho salmon fry and parr when foraging for food. Trans. Am. Fish. Soc. 134, 381-389.

Wang, Y.; Hu, W.; Wu, G.; Sun, Y.; Chen, S.; Zhang, F.; Zhu, Z.; Feng, J.; Zhang, X., 2001: Genetic analysis of "all-fish" growth hormone gene transferred carp (*Cyprinus carpio* L.) and its F_1 generation. Chin. Sci. Bull. 46, 1174-1177.

Werner, E. E.; Anholt, B. R., 1993: Ecological consequences of the trade-off between growth and mortality rates mediated by foraging activity. Am. Nat. 142, 242-272.

Whoriskey, F. G.; Fitzgerald, G. J., 1985: The effects of bird predation on an estuarine stickleback (Pisces: Gasterosteidae) community. Can. J. Zool. 63, 301-307.

Zhu, Z.; Li, G.; He, L.; Chen, S., 1985: Novel gene transfer into the fertilized eggs of goldfish (*Carassius auratus* L. 1758). J. Appl. Ichthyol. 1, 31-34.

Zhu, Z.; He, L.; Chen, T. T., 1992: Primary-structural and evolutionary analyses of growth-hormone gene from grass carp (*Ctenopharyngodon idellus*). Eur. J. Biochem. 207, 643-648.

快速生长的转基因鲤的反捕食死亡率增加

段　明[1,2]　张堂林[1]　胡　炜[1]　管　波[1,2]　汪亚平[1]
李钟杰[1]　朱作言[1]

1 中国科学院水生生物研究所，淡水生态与生物技术国家重点实验室，武汉　430072
2 中国科学院大学，北京　100049

摘　要　大量研究表明，动物的内禀生长率并不会趋向于最大绝对值而是趋向于一个最佳值，即动物的快速生长可能是有代价的，其生长率和其他生活史特征之间存在着某种权衡。本研究设计了一系列的配对反捕食试验，以此来检验快速生长转基因鲤和非转基因鲤之间的反捕食能力。结果发现，无论是大个体组或小个体组，转基因幼鲤的反捕食死亡率均比非转基因鲤显著提高，证实了转基因鲤具有较低的反捕食能力。本研究验证了动物生长和存活间存在权衡的演化理论，也就是说，转基因鲤的快速生长会导致反捕食能力的降低。

Behavioral Alterations in GH Transgenic Common Carp May Explain Enhanced Competitive Feeding Ability

Ming Duan[1] Tanglin Zhang[1] Wei Hu[1] Zhongjie Li[1] L. Fredrik Sundström[3] Tingbing Zhu[1,2] Chengrong Zhong[1,2] Zuoyan Zhu[1]

1 State Key Laboratory of Freshwater Ecology and Biotechnology, Institute of Hydrobiology, Chinese Academy of Sciences, Wuhan 430072, China
2 Graduate University of Chinese Academy of Sciences, Beijing 100049, China
3 Department of Animal Ecology/Evolutionary Biology Centre, Uppsala University, Norbyvägen 18D, SE-752 72 Uppsala, Sweden

Abstract The aim of the present study was to clarify the role of GH transgenesis and its production in social interactions in juvenile common carp *Cyprinus carpio*. With food pellets provided sequentially, the serum GH levels and behavioral effects were measured in 14 pairs of size-matched 'all-fish' GH-transgenic and non-transgenic common carp. In six consecutive observations during 3 days, transgenic fish had a higher movement level as well as a higher social status, being 2.69 times as aggressive, two minutes before and after the 10-min feeding session compared to non-transgenic fish. Transgenic fish also were more than 1.74 times as likely to consume each pellet. During the 8-day experiment, transgenic fish had 4.09 times higher specific growth rate in body weight as well as 6.36 times higher serum GH level than the non-transgenic fish. These results show that GH transgenesis promotes over-expression of GH and alters behaviors in juvenile common carp, thereby increasing their ability to compete and gain food resources, presumably to meet a higher intrinsic growth rate, which gives direct evidence for the GH-induced elevation in feeding competitive ability of GH-transgenic common carp. Understanding these relationships would not only help evaluating potential ecological effects of the escaped/released transgenic fishes, but also help using potential aquaculture of this growth-enhanced strain.

Keywords transgene; growth hormone (GH); foraging; competition; growth; common carp *Cyprinus carpio*

1. Introduction

Both the use of growth hormone (GH) treatment (McLean and Donaldson, 1993) and altered GH gene expression by transgenesis (Devlin *et al.*, 1994; Gross *et al.*, 1992; Zhu *et al.*, 1985) can increase the growth rate in fishes, which provides a key biotechnological opportunity to increase global aquaculture production efficiency and yield (Devlin *et al.*,

---

文章发表于 *Aquaculture*, 2011, 317(1-4):175-181

2006). However, commercial transgenic fish derived from non-fish gene constructs and 'all-fish' constructs raises public concerns about bio-safety and bio-ethics (Kapuscinski *et al.*, 2007; Nam *et al.*, 2008). Recently, 'all-fish' gene transfer has been presented and recommended to aquaculture (Nam *et al.*, 2001). In 'all-fish' GH-transgenic fish, all constructs are derived from fish including the promoter, enhancer and coding sequence. GH-transgenic strains would not only be adopted for use in commercial aquaculture in near future, but also provide useful model systems for studying the consequences of growth enhancement from genetic, physiological and ecological standpoints (Devlin *et al.*, 2006; Nam *et al.*, 2008).

GH is a major promoter of post-natal growth and has a key role in the metabolism and regulating the use of nutrients in tissue synthesis of vertebrates (Jönsson *et al.*, 2003; Steele and Evock-Clover, 1993). In teleosts, the growth-promoting effects of GH are well documented (Devlin *et al.*, 2006; Jönsson and Björnsson, 2002; McLean and Donaldson, 1993). Behavioral effects of GH administration in normal fish include increased competitive feeding ability, higher physical activity, increased aggression and decreased risk sensitivity (Jönsson and Björnsson, 2002). This supports the hypothesis that GH-treatment increases metabolic demands and feeding motivation, which in turn increases appetite and competitive ability. Similarly, GH-transgenic salmons can possess elevated chronic levels of GH (Devlin *et al.*, 2000), which results in increased ability to compete for contested food resources (i.e., point food resources, Devlin *et al.*, 1999), higher locomotion and more risk-taking (Abrahams and Sutterlin, 1999; Sundström *et al.*, 2003), in addition to a feeding motivation (Sundström *et al.*, 2004a). Similar results were also found in GH-transgenic common carp *Cyprinus carpio* reared in ponds and aquaria, which can possess higher and constant serum GH levels (Zhong *et al.*, 2009), which results in elevated ability to compete for limited food resources (i.e., clumped food resources, Duan *et al.*, 2009), besides a higher food conversion efficiency (Fu *et al.*, 2007) and increased feeding motivation (Duan *et al.*, 2009). GH-transgenic fishes also appear to be better at using low-quality food and have higher feed-conversion efficiency (Fu *et al.*, 1998; Raven *et al.*, 2006; Venugopal *et al.*, 2004). GH-transgenic fishes, however, do not always benefit from elevated GH production. In coho salmon *Oncorhynchus kisutch*, GH-transgenic individuals were prone to spending more time and to investing more energy on unprofitable prey due to reduced discrimination (Sundström *et al.*, 2004a). They experienced lower survival in the presence of predators (Sundström *et al.*, 2004b; Sundström *et al.*, 2005), and may experience reduced disease resistance (Jhingan *et al.*, 2003) and impaired swimming ability (Farrell *et al.*, 1997). Similarly, GH-transgenic common carp spent more time defending territory and taking risks (M. Duan, unpublished), and lower swimming ability (Li *et al.*, 2007) increased anti-predation mortality and decreased survival fitness (Duan *et al.*, 2010). Also in mammals, GH has been shown to modulate behavior. Somewhat inconsistent, GH-treatment decreases motor activity in rats (Alvarez and Cacabelos, 1993; Kelly, 1983; Stern *et al.*, 1975), while GH-transgenic mice displayed higher locomotor activity in a novel environment than non-transgenic controls (Söderpalm *et al.*, 1999). Moreover, GH increases aggression in wild male mice (Matte, 1981). Therefore, these discrepancies in GH effects may not only be due to the different routes of administration, but may also depend on the protocol used. In addition, although GH has similar stimulatory effects on growth in most vertebrates examined so far, the underlying causes may vary among species. Factors such as increased feed intake, elevated feed conversion efficiency and improved foraging efficiency may, to a varying degree, contribute to the growth increase. Thus, the relationship between GH, growth, and behavior is also likely to differ among species.

The prevalence of dominant-subordinate relationships in animal societies suggests that there is an adaptive significance to achieve a high rank when in a group (Huntingford and Turner, 1987). An evolutionary game theory model by Enquist and Leimar (1987) predicts that the costs of fighting will increase when resource value increases and, the probability of victory for an animal will increase when resource value is increased only to that animal. Aggressiveness is often shown by dominant individuals in interference competition (Metcalfe, 1986). For instance, in trout, the dominants can aggressively defend a food source and thereby largely reduce feeding and consequently growth of the subordinates (Metcalfe, 1986).

However, it remains unclear as to whether aggressiveness, fighting ability and or social status of GH-transgenic carp is modified by GH expression in a similar way. In the present study, with food pellets provided sequentially (point food resource), serum GH level, food consumption, aggressive level, fighting ability and social status were measured in size-matched pairs consisting of one ‘all fish’ GH-transgenic and one non-transgenic common carp. The results provide behavioral and endocrine evidence on the competitive feeding ability of the GH-transgenic fishes, which may be of immense importance in appraising potential effects of possible future releases/escapes of transgenic fishes into nature.

2. Materials and Methods

Experimental animals were maintained at a specially constructed facility at the Institute of Hydrobiology. This facility was designed to eliminate the possibility of fish escape with the emplacement of multiple containment screen systems. All experimental procedures employed were approved by the Institute of Hydrobiology Institutional Animal Care and Use Committee (Approval ID: keshuizhuan 08529).

2.1 Source of fish

P_0 'all-fish' GH-transgenic carp were initially produced by microinjection of the *pCAgcGH* into the fertilized eggs of common carp *Cyprinus carpio* (Yellow River carp variety). The 'all-fish' gene construct *pCAgcGH* was a recombinant construct of grass carp *Ctenopharyngodon idellus* (Valenciennes) growth hormone gene (*gcGH*), whose expression is driven by the *β*-actin gene promoter of common carp (*pCA*) (Wang *et al.*, 2001). F_1 transgenic carp were produced as follows: P_0 transgenic ♂ (sperm revealed to be *pCAgcGH* positive) × non-transgenic ♀. The F_1, F_2 and F_3 generation were, respectively, 1.6 times (Wang *et al.*, 2001), 1.8-2.5 times (W. Hu, unpublished) and 1.4-1.9 times (Li *et al.*, 2007) the body mass of non-transgenic counterparts under hatchery-reared conditions, showing how the growth enhancement remains relatively stable across generations. The F_4 generation transgenic and non-transgenic fish were produced from crosses between a wild-type female and an F_3 hemizygous transgenic male of a fast-growing transgenic strain on 21 April 2008. Frequencies of transgene transmission to F_4 progeny from this line were *c.* 50% (W. Hu, unpublished). Siblings were used to minimize effects of genetic differences and maternal/paternal effects (Metcalfe *et al.*, 1989). After emergence (28 April) in Duofu Technology Farm, Wuhan, China, the first-feeding fry containing a mix of the two genotypes, were transferred to an earthen pond (area 667 m^2, water depth 1 m) with a rearing density of 100 ind. m^{-3} each. A month later, 2000 individuals of the mixed populations were transferred to an indoor re-circulating system and reared in two circular fiberglass tanks (diameter 150 cm, volume 1000 l) at the Institute of Hydrobiology, Chinese Academy of Sciences. Thereafter, the *pCAgcGH* transgene-positive fish were identified by the polymerase chain reaction (PCR) method following Guan *et al.* (2010). The two genotypes then were reared separately in two of the fiberglass tanks as described above. The two genotypes were fed to satiation with frozen chironomid *Chironomus* sp. larvae twice daily, and this feed regime was maintained until the fish gape sizes matched the size of the artificial food pellets (weight 5.52 ± 0.16 mg, length 2.89 ± 0.05 mm, diameter 1.54 ± 0.02 mm, mean ± SE; comprised of 36.0% protein, 12.1% lipid, 11.3% ash and 18.2 J mg^{-1} energy). The water temperature varied from 25.0 to 27.4℃ during this period.

2.2 Experimental procedures

To ascertain the maximum number of food pellets consumed and the time spent in swallowing a single pellet by a fish, a pilot study was undertaken in 10 aquaria (20 × 20 × 20 cm, water depth 15 cm) prior to the start of the experiment. An individual was estimated to be satiated when it refused to grasp, or did not swallow five consecutive pellets which then sank gently down to the bottom of the aquarium, the method was modified from

Johnsson and Björnsson (1994). After six feeding observations in three consecutive days, transgenic carp (n = 5) were fed to satiation with 25.3 ± 1.3 (mean ± SE) food pellets during a 10-min interval, while 16.0 ± 2.6 pellets satiated the non-transgenic individuals (n = 5). The number of pellets consumed by one transgenic and one non-transgenic individual was over 40. Neither of the two genotypes spent more than 30 s in handling (swallowing) a single pellet, so, 20 pellets were supplied one by one at a 30-s interval for the subsequent experiment.

The experiment consisted of two main parts: (1) a treatment period (days 1-5), and (2) an observation period (days 6-8). Another indoor re-circulating system was used for the experiment. Incoming non-chlorinated water was automatically filtrated, sterilized and constantly aerated to ensure high water quality. Throughout the experiment, the water temperature, dissolved oxygen and pH were maintained at 25.3 ± 0.1℃, 8.5 ± 0.1 mg $l^{-1}$and 7.7 ± 0.1 (mean ± SE), respectively.

2.2.1 Treatment period

On day 1 (25 July) at 0900 h, 14 individuals of each genotype (size-matched within pairs, see Table 1) were anesthetized with 0.25 ml l^{-1} Eugenol (purity ⩾ 99.99%, Sinopharm Chemical Reagent Co. Ltd, Shanghai, China). Body weight (nearest 0.01 g) and total length (nearest 0.01 cm) were measured. Individual fish in each pair were identified visually through randomly tagging them with colored beads (diameter 2.33 ± 0.04 mm, weight 14.28 ± 0.33 mg, mean ± SE) on nylon thread (diameter 0.25 mm) inserted through the dorsal musculature, anterior to the dorsal fin and then tied in a surgical knot (Bailey *et al.*, 2000; Johnsson and Björnsson, 1994; Metcalfe *et al.*, 1989). Next, the 14 pairs were transferred to 14 tempered glass aquaria measuring 50 × 25 × 50 cm with a water depth of 35 cm to acclimate. Two fluorescent lights (8 W) were fixed 15 cm above the water level on each aquarium (12 L ∶ 12 D, 0700 h: 1900 h). All aquaria had their three walls covered with an opaque, black plastic screen to prevent visual contact among neighboring fish and also to minimize disturbance and maintain similar levels of illumination; the frontal wall was not covered, to facilitate video recording. A Petri dish (diameter 12 cm) was placed at the nearest short end of each aquarium bottom. The water column above the dish was defined as the feeding area. A plastic funnel from the outside of one of the lateral walls was fixed above the feeding area, 5 cm above the water surface, allowing food pellets to be administered to the feeding area each time, without any disturbance to the fish.

Each pair was fed with 20 pellets one by one at a 30-s interval time during a 10-min feeding session, twice daily (0800 h-1100 h and 1400 h-1700 h), for a period of 2-4 days. The fish were deprived of food for 1 day (day 5) before the observations commenced on day 6.

Table 1 Total length (L), body weight (W), and condition factor (CF) of transgenic and non-transgenic common carp before and after the 8-day experiment, and specific growth rate for body weight (SGR_W) and total length (SGR_L) of the two genotypes during the 8-day experiment (mean ± SE)

	Before		After	
	Transgenics (n = 14)	Non-transgenics (n = 14)	Transgenics (n = 14)	Non-transgenics (n = 14)
L (cm)	6.95 ± 0.10	6.97 ± 0.08	7.37 ± 0.08*	7.10 ± 0.07
W (g)	4.17 ± 0.11	4.11 ± 0.13	4.87 ± 0.15*	4.26 ± 0.14
CF (g cm^{-3})	1.25 ± 0.03	1.21 ± 0.02	1.22 ± 0.03	1.19 ± 0.01
SGR_W (% day^{-1})	–	–	1.92 ± 0.11*	0.47 ± 0.16
SGR_L (% day^{-1})	–	–	0.74 ± 0.07*	0.24 ± 0.07

*Denotes a significant difference between the two genotypes, $P < 0.05$ (paired t-test)

2.2.2 Observations

The same feeding procedures described above were repeated during the 3-day behavioral observations (days 6-8, from 30 July to 1 August).

Behaviors of the fish were recorded by a digital videocassette recorder (Sony, HDR-HC1E Handy Camera; Sony Corp., Tokyo, Japan) placed in front of the aquarium. The order of observations was randomly allocated among the aquaria so as to remove any bias and order effects. For each aquarium, the video camera was started 2 min prior to the 10-min feeding session as described above. The video recording was stopped 2 min after the feeding session. Thus, behaviors of each pair were recorded for 14 min. Observations were made twice daily, at 0800-1100 and 1400-1700 h during the feeding periods. Consequently, this procedure yielded six repeated measures for competitive ability for each of the 14 pairs during the 3-day observations. Three main behavioral parameters of the two genotypes were noted from the video: Food consumption (number of food pellets consumed of a maximum of 20), number of movements (when a fish moved more than its total length followed by a less inactive period of at least 1 s), and the number and type of overt aggressive actions (i.e. bites/attacks and chases/retreats) that each fish initiated and received and whether it was the winner or loser. A fish was classified as a loser of an agonistic interaction when it adopted a submissive posture or moved away, either by retreating or escaping from the opponent that was then classified as the winner (Oliveira *et al*., 1996). On the 9th morning (2 August), total length and body weight of the 28 individual fish were again measured under a light anesthesia with 0.25 ml l^{-1} Eugenol (purity ≥ 99.99%).

2.2.3 Detection of serum GH

After the morphological parameters measurement, 0.5 ml fresh blood samples were collected from the caudal veins by syringes. They were maintained at 4℃ for 4 h and then centrifuged at 6000 rpm for 5 min. The supernatant (blood serum samples) was then stored at −70℃ for Elisa analysis, which was performed as follows: the coating antibody and enzyme-labeled antibody were specially made from different rat *anti-cGH* monoclonal antibodies (*mAbs*). This sensitive sandwich Elisa for detection of *ncGH* was developed using *FMU-cGH12* as the coating *mAb* and *FMU-cGH6* as the enzyme-labeled *mAb*, which was found to be the most sensitive to date with detection levels of 70 pg ml^{-1} (Wu *et al.*, 2008). A 96 well plate was coated overnight at 4℃ with 100 μl of a 1 ∶ 500 solution of coating antibodies (rat *anti-cGH mAbs*). Afterwards, the wells were washed with phosphate buffered saline (PBS) with 0.05% Tween 20. Each well was filled with 300 μl of blocking buffer [PBS with 1% bovine serum albumin (BSA)] and the plate was incubated for 2 h at 37℃. After washing, 60 μl blood serum samples of both standard preparation and negative control were added to the wells. After incubating the plate for 1 h at 37℃, the wells were washed and 100 μl horseradish peroxidase-labeled rat *anti-cGH mAbs* solution was added to each well with final 1 ∶ 800 dilution. After incubation for 45 min at 37℃, the solution was removed and all wells were washed six times. Then, 100 μl chromogenic solution (ortho-phenylenediamine) was added to each well. After incubation for 15-30 min at 37℃, the enzymatic reaction was stopped by 50 μl of 2 M sulphuric acid. Finally, the absorbance (490 nm) was measured using BioTek ELx 800 reader (BioTek Instruments, Inc., USA).

2.3 Statistical analyses

The specific growth rate for total length (SGR_L) was calculated following Ricker (1979) as:

$$SGR_L\,(\%\ day^{-1}) = 100 \times (\ln L_2 - \ln L_1)\ d^{-1}$$

where L_2 and L_1 were the final and initial total length (cm), d is the experimental period (8 day, consisting of the 5-day treatment period and the 3-day observations).

Condition factor (CF) was calculated as:

$$CF\ (g\ cm^{-3}) = 100 \times WL^{-3}$$

where L and W were the total length (cm) and body weight (g) of the two genotypes.

Dominance index (I_D) was calculated from the aggression data using the following equation:

$$I_D = A_V \times A_{V+D}^{-1}$$

where A_{V+D} is the total number of aggressive interactions (victory + defeat) in which a fish was involved and A_V is the number of aggressive interactions in which that fish was the winner (Oliveira and Almada, 1996a, 1996b). Individual values for I_D could range between 0 (lost all encounters) and 1 (won all encounters).

All mean values are presented with the accompanying standard errors (SE). Proportional data (I_D) were arc-sine transformed prior to statistical analysis. Most data were normally distributed and had homogeneous variance; so paired *t*-test and GLM were used. Movement data were not normally distributed, but this was resolved by a square-root transformation before analysis. For evaluating correlations between two behavioral variables, Spearman's rank correlation and partial correlation analysis were used. Differences were considered as significant when $P < 0.05$. All the data were analyzed with SPSS 15.0 (Systat Inc., Chicago, IL, USA).

3. Results

3.1 Serum growth hormone level

The mean serum GH level of transgenic carp (16.91 ± 2.05 ng ml^{-1}) was 6.36 times higher (paired t-test, $t_{13} = 7.251$, $P < 0.001$) than that of non-transgenics (2.66 ± 1.58 ng ml^{-1}), and the GH level ranged 8.02-31.13 ng ml^{-1} and 0.93-5.58 ng ml^{-1} in GH-transgenic and non-transgenic common carp, respectively.

3.2 Movement level in feeding and non-feeding sessions

The number of movements of transgenic fish during the feeding session averaged 174.3 ± 6.6, which was 2.09 times higher (one-way repeated measures ANOVA, genotype as between-subjects, $F_{1,26} = 94.432$, $P < 0.001$) relative to the non-transgenic fish of 83.5 ± 6.6. Trial had no significant effect ($F_{5,130} = 0.704$, $P = 0.662$) on the number of movements but the pattern differed between the two genotypes (trial × genotype, $F_{5,130} = 4.664$, $P = 0.001$; Fig. 1).

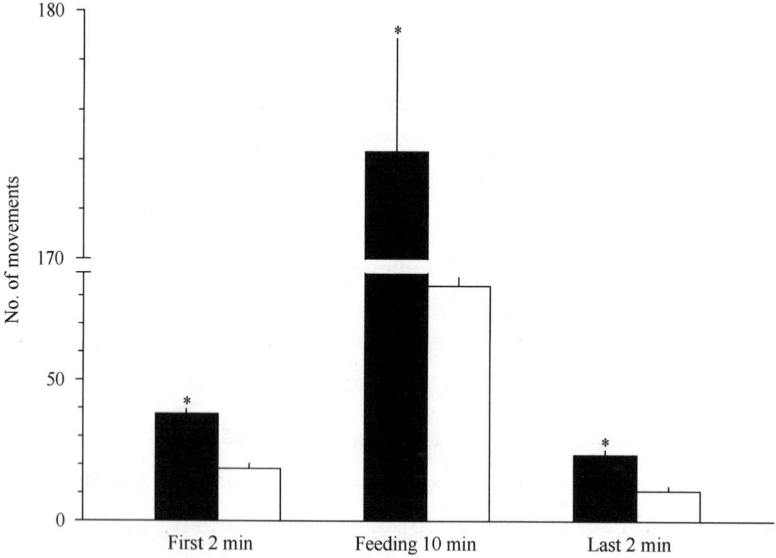

Fig. 1 The mean ± SE amount of movements of transgenic (■, $n = 14$) and non-transgenic common carp (□, $n = 14$) during 10-min feeding periods and first and last 2 min observations. Differences between the first and last 2 min observations were significant, * denotes a significant difference between the two genotypes, $P < 0.05$

The number of movements by transgenic fish was higher than for the non-transgenic fish (square-root transformed, one-way repeated measures ANOVA, genotype as between-subjects, $F_{1,26} = 63.409$, $P < 0.001$) during both first 2 min session and last 2 min session, however, the movement level was higher before the feeding session than after the feeding session (within-trial as within-subjects, $F_{1,26} = 68.187$, $P < 0.001$), but the pattern was similar for the two genotypes (genotype × within-trial, $F_{1,26} = 1.476$, $P = 0.235$; Fig. 1).

3.3 Aggression level and social status in feeding and non-feeding sessions

Overall, during six 14-min observations for 14 pairs, 2217 overt aggressions were observed. The number of aggressive actions initiated by transgenic fish (110.4 ± 10.8) was 2.69 times higher (paired t-test, $t_{13} = 11.564$, $P < 0.001$) than that of the non-transgenic fish (41.1 ± 5.6).

The mean dominance index obtained for transgenic carp (0.74 ± 0.01) was 2.85 times higher (arc-sine transformed, paired t-test, $t_{13} = 17.965$, $P < 0.001$) than that of non-transgenics (0.26 ± 0.01; Fig. 2).

The dominance index obtained for transgenic fish was higher than that of non-transgenic fish (one-way repeated measures ANOVA, genotype as between-subjects, $F_{1,26} = 412.607$, $P < 0.001$) during both first 2 min session and last 2 min session, and the index was consistent before and after the 10-min feeding session (within-trial as within-subjects, $F_{1,26} = 0.052$, $P = 0.822$). However, the pattern was not similar for the two genotypes (genotype × within-trial, $F_{1,26} = 5.197$, $P = 0.031$; Fig. 2).

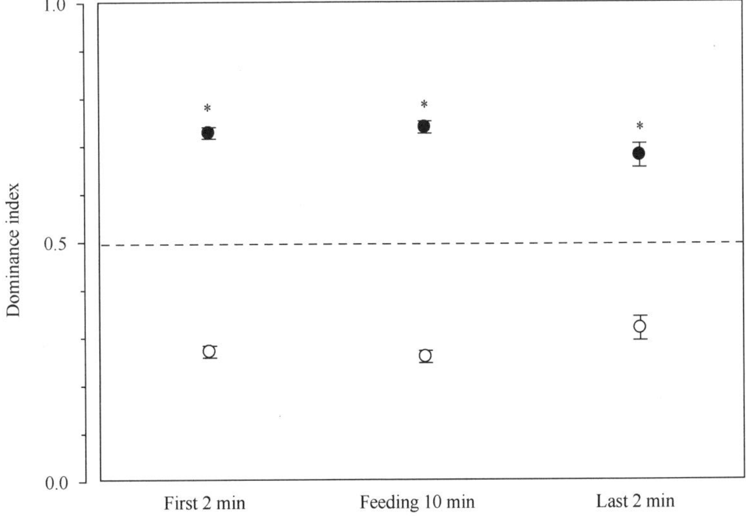

Fig. 2 The mean ± SE of dominance index (I_D) of transgenic (●, $n = 14$) and non-transgenic common carp (○, $n = 14$) during 10-min feeding periods and first and last 2 min observations. * denotes a significant difference between the two genotypes, $P < 0.05$

3.4 Food consumption

For each fish in the 14 pairs, food consumption during the consecutive feeding trials was first summarized for each of the 20 sequentially supplied pellets. As there were six feeding trials, a fish in a pair could consume a maximum of six pellets for any of the sequential pellets given (Devlin *et al.*, 1999). These data were then averaged for the 14 pairs and are presented in Fig. 3.

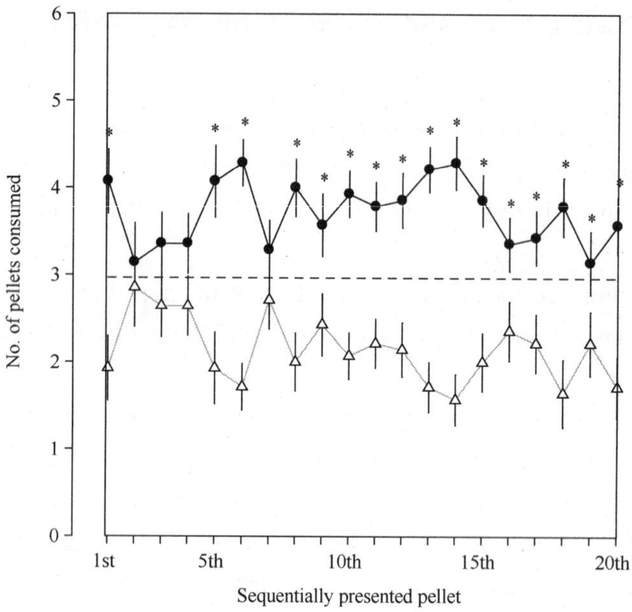

Fig. 3 The mean ± SE number of pellets consumed by transgenic (●, n = 14) and non-transgenic common carp (△, n = 14) over 6 repeated feeding trials, in relation to feeding sequence. Up to 20 pellets were given sequentially to each pair (x-axis), and it was noted whether the transgenics or non-transgenics took each pellet. The maximum number of pellets consumed at any time (y-axis) was 6 (= the number of feeding trials). * denotes a significant difference between the two genotypes, $P < 0.05$

Analysis of the food consumption at the initiation of feeding (calculated for the first pellet) showed that the transgenic fish took them more often than the non-transgenic fish (67.9% and 32.1% of the presented pellets, respectively, paired t-test, $t_{13} = 2.895$, $P = 0.013$; Fig. 3). Thus, the initial consumption of the transgenic fish was 2.11 times higher than that of the non-transgenic fish.

Overall, out of 120 pellets provided to each pair, the number of food pellets consumed by transgenic fish averaged 74.4 ± 2.4 during the 3-day observation which was higher ($t_{13} = 5.780$, $P < 0.001$) than that of non-transgenic fish (42.7 ± 3.1). The food consumption by transgenic common carp was consistently higher over the 10-min feeding session (Fig. 3).

3.5 Relationships between social status, food intake and movement level

There was a remarkable positive correlation between the number of pellets consumed and the number of movements of the fish (Spearman rank correlation, $r = 0.797$, $P < 0.001$, $n = 28$). Moreover, there was a significant positive correlation between social status and food consumption ($r = 0.835$, $P < 0.001$, $n = 28$) indicating that, the higher the social status possessed, the more food was obtained. In order to determine which major factor among movement level and social status affected the quantity of food consumed, we used partial correlation. After controlling for dominance index, no clear correlation was found between the number of movements and the number of food pellets consumed ($r = 0.151$, $P = 0.425$, $n = 25$). On the other hand, when number of movements was controlled, there was a significant positive correlation between dominance index and the number of pellets consumed by the two genotypes ($r = 0.588$, $P = 0.001$, $n = 25$). These results indicate that social status had a much stronger impact on the food consumption than did the movement level (Fig. 4).

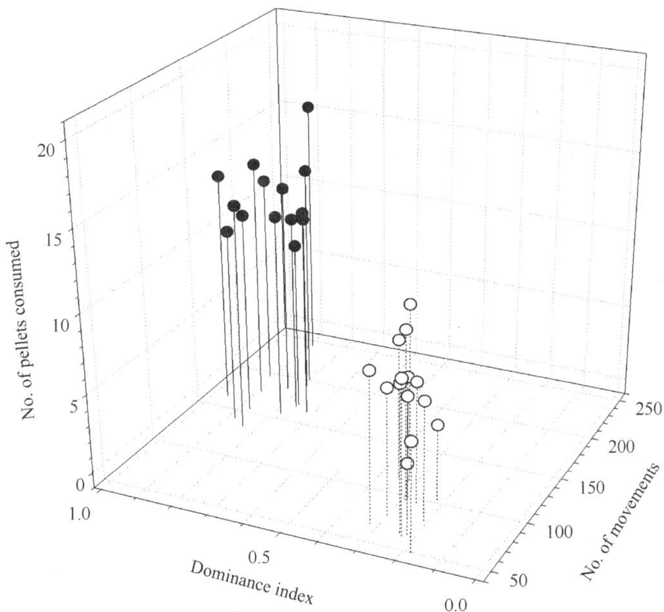

Fig. 4 Correlations between social status (dominance index, x-axis), food intake (number of pellets consumed, y-axis) and movement level (number of movements, z-axis) of the mean amount of the transgenic (•, $n = 14$) and non-transgenic (○, $n = 14$) common carp with all of 6 consecutive trials combined

3.6 Growth

At the beginning of the experiment, there was no significant difference in total length (paired t-test, $t_{13} = -0.884$, $P = 0.393$), body weight ($t_{13} = 2.104$, $P = 0.06$) and condition factor (t_{13} = 1.870, P = 0.084) between the two genotypes, respectively (Table 1). After the 8-day experiment, the SGR_W of transgenic common carp (1.92 ± 0.11% day^{-1}) was 4.09 times higher than that of non-transgenic carp (0.47 ± 0.16 % day^{-1}, t_{13} = 7.326, P < 0.001; Table 1). The transgenic fish (0.74 ± 0.07% day^{-1}) also grew faster in length than their non-transgenic ounterparts (0.24 ± 0.07% day^{-1}, t_{13} = 13.199, P < 0.001; Table 1).

4. Discussion

In the present study, juvenile GH-transgenic common carp had higher serum GH level, elevated movement activity, higher social status, enlarged food consumption rate and increased growth rate compared with size-matched non-transgenic counterparts. The GH-transgenic common carp possessed elevated levels of GH, which probably stimulated higher feeding motivation and appetite, thereby leading to altered behaviors, which may be a main cause of feeding priority and growth advantage. These results are highly consistent with previous studies on GH function in salmonid fishes both from GH implants (Johnsson and Björnsson, 1994, Jönsson *et al.*, 1996, 1998) and GH transgenesis (Abrahams and Sutterlin, 1999; Devlin *et al.*, 1999; Tymchuk *et al.*, 2005). The present study suggests that GH influences competition for point food by increasing motivation in common carp in a similar way, showing that the better competitors had the better growth. In addition, the behavioral alterations of the GH transgenesis in common carp are clear.

This study shows that F_4 'all-fish' GH-transgenic common carp have much higher serum GH levels than their non-transgenic siblings, which is in agreement with other previous reports (Devlin *et al.*, 2000; Mori and Devlin, 1999; Zhong *et al.*, 2009). However, the results are different from those of a study on the behavioral effects of the GH-injection on the juvenile rainbow trout *Oncorhynchus mykiss* (Jönsson *et al.*, 1998), in which the authors found that although the aggression level was elevated, the fighting ability and the social status were not increased by the GH. In the present study, it became evident that GH transgenesis could increase the serum GH level and fighting ability of the common carp, thereby increasing their dominance status to gain more contested food pellets, which in turn improved their specific growth rate. However, previous studies showed that the effect of GH administration might be depended on level of dietary protein and food provided. Under optimal conditions, with sufficient amounts of food and protein provided, GH had little or no further effect on growth rate, but where amounts of food or protein were limited, a remarkable effect of GH was observed in juvenile common carp (Fine *et al.*, 1996), consistently, a similar GH effect was observed in juvenile rainbow trout (Sakata *et al.*, 1993). Furthermore, Raven *et al.* (2008) argued that ration-unrestricted GH-transgenic coho salmon displayed approximately 2-fold higher GH concentration than that in non-transgenic individuals, whereas ration-restricted GH-transgenic salmon (held to control growth rates) had GH levels further elevated to approximately 5-fold higher than in non-transgenic salmon and revealed that strong nutritional modulation of insulin-like growth factor-I (IGF-I) production remained even in the presence of constitutive ectopic GH expression in these GH-transgenic salmon. Under natural conditions, food availability would be less predictable and predation risk may influence feeding decisions so that the advantage of the transgenic fish could be lessened or increased depending on their responses under such complex environments. Arendt (1997) argued that there were many functions that may be compromised when intrinsic growth rates were accelerated to meet evolutionary challenges. Conversely, if selection acts to improve any of these functions, intrinsic growth rates might be reduced. Thus, further work on the behavioral and developmental alteration of the transgenic fish in complex environments is needed.

In the present study, GH-transgenic common carp exhibit a much higher movement level. It was documented that GH administered directly into the third ventricle increased swimming activity in juvenile rainbow trout, indicating that GH may act directly on the central nervous system to influence behavior (Jönsson *et al.*, 2003). Generally, pituitary GH expression can be regulated by negative feedback controls (Ágústsson and Björnsson, 2000; Szabo *et al.*, 1995). But in our previous study, it was reported that the regulatory mechanism of serum growth hormone of transgenic common carp was different with that of the nontransgenics, because its expression level was not impacted by pituitary feedback inhibition mechanism but related to the control mode of β-actin gene promoter in the transgene (Zhong *et al.*, 2009). Therefore, expression of the GH may be having something to do with the transgenic construct. In other words, GH implants may not completely mimic the effect of the GH transgenesis. It seems that there may be some differences in the triggering mechanism of the behavior between the GH transgenesis and GH implants. In addition, however, carp are different from salmon in many important aspects of foraging behavior, often spending much time foraging together for food items such as aquatic invertebrates, flowers, etc. Hence, further research should focus on how much time the fish spend foraging, and how likely they are to deplete food resources if food is limiting.

Aggression can play a major role in determining individual survival and reproductive success (Komnicki, 1988). In cases where phenotypes differ from the norm for the species, the transformed fish might not be constrained by the limitations that normally keep the species in check but might compete in an entirely different niche and be able to access previously unavailable resources (Devlin *et al.*, 2006). The most aggressive individuals are also often the most successful resource exploiters (Alanärä *et al.*, 2001; Metcalfe, 1986). Such a shift may affect resource allocation within an ecosystem significantly (Devlin *et al.*, 2006). In our previous study, we found that the F_5 transgenic common carp exhibited a higher feeding order, however, we did not observe aggression during the study period due to the longer acclimation as well as the clumped food resources supply (Duan *et al.*, 2009). In contrast, the intensive aggressions between the two genotypes were observed throughout the present experiment (before, during and after the feeding session), and the transgenic common carp initiated most aggressive actions and won more conflicts, indicating that the transgenic fish possessed higher social status whatever the food distributions were.

The juvenile GH-transgenic common carp were more active before and after the feeding period as well as during the point food given sequentially, which may elevate the encounter rate between the two genotypes in pair besides enhancing the chance to acquire food resources (Arendt, 1997). The transgenic juveniles won most conflicts in the present study, which resulted in a higher dominance in aggressive interactions and in turn increased the competitive feeding ability. However, aggressive behavior entails the energetic costs of swimming when initiating an attack or escape (Metcalfe, 1986; Noakes and Grant, 1992; Puckett and Dill, 1985). The benefits for the aggressor can include the general strengthening of dominant-subordinate relationships as well as priority of access to food items. On the other hand, aggression involves additional costs for the recipients, which are often displaced from the most profitable territories, where food availability or quality are high or energetic costs low (Metcalfe, 1986; Wootton, 1985). Moreover, the recipients are often prone to suffer physical damage that can lead to disease infection (Turnbull *et al.*, 1996) or physiological effects (Abbott and Dill, 1989), which in turn can cause reduced appetite or food conversion efficiency (Jobling and Wandsvik, 1983). Therefore, to some extent, fish should balance the benefits from continuous growth with the potential costs of acquiring energy (Nicieza and Metcalfe, 1999). For instance, there is a trade-off between swimming performance and specific growth rate in some GH-transgenic fishes (Farrell *et al.*, 1997; Li *et al.*, 2007) as well as in some non-transgenic fish (Kolok and Oris, 1995).

5. Conclusion

Overall, the results clearly showed that the juvenile 'all fish' GH-transgenic common carp were superior in fighting ability and activity with much higher serum GH level, which gives direct evidence for the growth hormone-induced elevation in feeding competitive ability of GH-transgenic common carp. It is possible that transgenic fish possess novel phenotypes and a genetic capacity not previously available to the species, which could facilitate adaptation to a more fit phenotypic state (Devlin *et al.*, 2006). Nevertheless, evolutionary theory would suggest that wild fish should have achieved a local fitness maximum through selection, and that anthropogenically induced genetic changes from transgenesis should not often result in enhanced fitness. This has been interpreted to mean that the risk of harm from escaped transgenic fish is low (Knibb, 1997).

Acknowledgements The study was supported by the Development Plan of the State Key Fundamental Research of China (Grant No. 2007CB109205), the National Natural Science Foundation of China (Grant Nos. 30770377 and 30830025), and the Shanghai Postdoctoral Scientific Program(Grant No. 10R21420500).

References

Abbott, J.C., Dill, L.M., 1989. The relative growth of dominant and subordinate juvenile steelhead trout (*Salmo gairdneri*) fed equal rations. Behaviour 108, 104-113.

Abrahams, M.V., Sutterlin, A., 1999. The foraging and antipredator behaviour of growth-enhanced transgenic Atlantic salmon. Anim. Behav. 58, 933-942.

Ágústsson, T., Björnsson, B.T., 2000. Growth hormone inhibits growth hormone secretion from the rainbow trout pituitary *in vitro*. Comp. Biochem. Physiol. C Pharmacol. Toxicol. Endocrinol. 126, 299-303.

Alanärä, A., Burns, M.D., Metcalfe, N.B., 2001. Intraspecific resource partitioning in brown trout: the temporal distribution of foraging is determined by social rank. J. Anim. Ecol. 70, 980-986.

Alvarez, X.A., Cacabelos, R., 1993. Influence of growth hormone (GH) and GH-releasing factor on locomotor activity in rats. Peptides 14, 707-712.

Arendt, J.D., 1997. Adaptive intrinsic growth rates: an integration across taxa. Q. Rev. Biol. 72, 149-177.

Bailey, J., Alanara, A., Brannas, E., 2000. Methods for assessing social status in Arctic charr. J. Fish Biol. 57, 258-261.

Devlin, R.H., Yesaki, T.Y., Biagi, C.A., Donaldson, E.M., 1994. Extraordinary salmon growth. Nature 371, 209-210.

Devlin, R.H., Johnsson, J.I., Smailus, D.E., Biagi, C.A., Jönsson, E., Björnsson, B.T., 1999. Increased ability to compete for food by growth hormone-transgenic coho salmon *Oncorhynchus kisutch* (Walbaum). Aquac. Res. 30, 479-482.

Devlin, R.H., Swanson, P., Clarke,W.C., Plisetskaya, E., Dickhoff,W.,Moriyama, S., Yesaki, T.Y., Hew, C.-L., 2000. Seawater adaptability and hormone levels in growth-enhanced transgenic coho salmon, *Oncorhynchus kisutch*. Aquaculture. 191, 367-385.

Devlin, R.H., Sundström, L.F., Muir, W.M., 2006. Interface of biotechnology and ecology for environmental risk assessments of transgenic fish. Trends Biotechnol. 24, 89-97.

Duan, M., Zhang, T., Hu, W., Sundström, L.F., Wang, Y., Li, Z., Zhu, Z., 2009. Elevated ability to compete for limited food resources by 'all-fish' growth hormone transgenic common carp *Cyprinus carpio*. J. Fish Biol. 75, 1459-1472.

Duan, M., Zhang, T., Hu,W., Guan, B.,Wang, Y., Li, Z., Zhu, Z., 2010. Increasedmortality of growth-enhanced transgenic common carp (*Cyprinus carpio* L.) under short-term predation risk. J. Appl. Ichthyol. 26, 908-912.

Enquist, M., Leimar, O., 1987. Evolution of fighting behavior: the effect of variation in resource value. J. Theor. Biol. 127, 187-205.

Farrell, A.P., Bennett, W., Devlin, R.H., 1997. Growth-enhanced transgenic salmon can be inferior swimmers. Can. J. Zool. 75, 335-337.

Fine,M., Zilberg, D., Cohen, Z., Degani, G.,Moav, B., Gertler, A., 1996. The effect of dietary protein level, water temperature and growth hormone administration on growth and metabolism in the common carp (*Cyprinus carpio*). Comp. Biochem. Physiol. A Mol. Integr. Physiol. 114, 35-42.

Fu, C., Cui, Y., Hung, S.S.O., Zhu, Z., 1998. Growth and feed utilization by F_4 human growth hormone transgenic carp fed diets with different protein levels. J. Fish Biol. 53, 115-129.

Fu, C., Li, D., Hu, W., Wang, Y., Zhu, Z., 2007. Growth and energy budget of F_2 'all-fish' growth hormone gene transgenic common carp. J. Fish Biol. 70, 347-361.

Gross, M.L., Schneider, J.F., Moav, N., Moav, B., Alvarez, C., Myster, S.H., Liu, Z., Hallerman, E.M., Hackett, P.B., Guise, K.S., Faras, A.J., Kapuscinski, A.R., 1992. Molecular analysis and growth evaluation of northern pike (*Esox lucius*) microinjected with growth hormone genes. Aquaculture 103, 253-273.

Guan, B., Hu, W., Zhang, T., Duan, M., Li, D., Wang, Y., Zhu, Z., 2010. Acute and chronicun-ionized ammonia toxicity to 'all-fish' growth hormone transgenic common carp (*Cyprinus carpio* L.). Chinese Sci. Bull. 55, 4032-4036.

Huntingford, F.A., Turner, A.K., 1987. Animal Conflict. Chapman and Hall, London. Jhingan, E., Devlin, R.H., Iwama, G.K., 2003. Disease resistance, stress response and effects of triploidy in growth hormone transgenic coho salmon. J. Fish Biol. 63, 806-823.

Jobling, M., Wandsvik, A., 1983. Effect of social interactions on growth rates and conversion efficiency of Arctic charr, *Salvelinus alpinus* L. J. Fish Biol. 22, 577-584.

Johnsson, J.I., Björnsson, B.T., 1994. Growth hormone increases growth rate, appetite, and dominance in juvenile rainbow trout, *Oncorhynchus mykiss*. Anim. Behav. 48,177-186.

Jönsson, E., Björnsson, B.T., 2002. Physiological functions of growth hormone in fish with special reference to its influence on behaviour. Fisher. Sci. 68, 742-748.

Jönsson, E., Johnsson, J.I., Björnsson, B.T., 1996. Growth hormone increases predation exposure of rainbow trout. Proc. R. Soc. Lond. B 263, 647-651.

Jönsson, E., Johnsson, J.I., Björnsson, B.T., 1998. Growth hormone increases aggressive behavior in juvenile rainbow trout. Horm. Behav. 33, 9-15.

Jönsson, E., Johansson, V., Björnsson, B.T., Winberg, S., 2003. Central nervous system actions of growth hormone on brain monoamine levels and behavior of juvenile rainbow trout. Horm. Behav. 43, 367-374.

Kapuscinski, A.R., Hayes, K.R., Li, S., Dana, G. (Eds.), 2007. Environmental Risk Assessment of Genetically Modified Organisms, Vol. III. CABI Publishers, Wallingford, UK.

Kelly, P.H., 1983. Inhibition of voluntary activity by growth hormone. Horm. Behav. 17, 163-168.

Knibb,W., 1997. Risk from genetically engineered and modified marine fish. Transgenic Res. 6, 59-67.

Kolok, A.S., Oris, J.T., 1995. The relationship between specific growth rate and swimming performance in male fathead minnows (*Pimephales promelas*). Can. J. Zool. 73, 2165-2167.

Komnicki, A. (Ed.), 1988. Population Ecology of Individuals. Princeton University Press, Princeton, New Jersey.

Li, D., Fu, C., Hu, W., Zhong, S., Wang, Y., Zhu, Z., 2007. Rapid growth cost in "all-fish" growth hormone gene transgenic carp: reduced critical swimming speed. Chinese Sci. Bull. 52, 1501-1506.

Matte, A.C., 1981. Growth hormone and isolation-induced aggression in wild male mice. Pharmacol. Biochem. Behav. 14, 78-85.

McLean, E., Donaldson, E.M., 1993. The role of growth hormone in the growth of poikilotherms. In: Schreibman, M.P., Scanes, C.G., Pang, P.K.T. (Eds.), The Endocrinology of Growth, Development, and Metabolism in Vertebrates. Academic Press, San Diego, pp. 43-71.

Metcalfe, N.B., 1986. Intraspecific variation in competitive ability and food intake in salmonids: consequences for energy budgets and growth rates. J. Fish Biol. 28, 525-531.

Metcalfe, N.B., Huntingford, F.A., Graham, W.D., Thorpe, J.E., 1989. Early social status and the development of life-history strategies in Atlantic salmon. Proc. R. Soc. Lond. B 236, 7-19.

Mori, T., Devlin, R.H., 1999. Transgene and host growth hormone gene expression in pituitary and nonpituitary tissues of normal and growth hormone transgenic salmon. Mol. Cell. Endocrinol. 149, 129-139.

Nam, Y.K., Noh, J.K., Cho, Y.S., Cho, H.J., Cho, K.N., Kim, C.G., Kim, D.S., 2001. Dramatically accelerated growth and extraordinary gigantism of transgenic mud loach *Misgurnus mizolepis*. Transgenic Res. 10, 353-362.

Nam, Y.K., Maclean, N., Hwang, G., Kim, D.S., 2008. Autotrasngenic and allotransgenic manipulation of growth traits in fish for aquaculture: a review. J. Fish Biol. 72, 1-26.

Nicieza, A.G., Metcalfe, N.B., 1999. Costs of rapid growth: the risk of aggression is higher for fast-growing salmon. Funct. Ecol. 13, 793-800.

Noakes, D.L., Grant, J.W.A., 1992. Feeding and social behaviour of brook and lake Charr. In: Thorpe, J.E., Huntingford,

F.A. (Eds.), The Importance of Feeding Behavior for the Efficient Culture of Salmonid Fishes. World Aquaculture Society, Baton Rouge, LA, pp. 13-20.

Oliveira, R.F., Almada, V.C., 1996a. On the (in) stability of dominance hierarchies in the cichlid fish *Oreochromis mossambicus*. Aggress. Behav. 22, 37-45.

Oliveira, R.F., Almada, V.C., 1996b. Dominance hierarchies and social structure in captive groups of the Mozambique tilapia *Oreochromis mossambicus* (Teleostei ichlidae). Ethol. Ecol. Evol. 8, 39-55.

Oliveira, R.F., Almada, V.C., Canario, A.V.M., 1996. Social modulation of sex steroid concentrations in the urine of male cichlid fish *Oreochromis mossambicus*. Horm. Behav. 30, 2-12.

Puckett, K.J., Dill, L.M., 1985. The energetics of feeding territoriality in juvenile coho salmon (*Oncorhynchus isutch*). Behaviour 92, 97-111.

Raven, P.A., Devlin, R.H., Higgs, D.A., 2006. Influence of dietary digestible energy content on growth, protein and energy utilization and body composition of growth hormone transgenic and non-transgenic coho salmon (*Oncorhynchus kisutch*). Aquaculture 254, 730-747.

Raven, P.A., Uh, M., Sakhrani, D., Beckman, B.R., Cooper, K., Pinter, J., Leder, E.H., Silverstein, J., Devlin, R.H., 2008. Endocrine effects of growth hormone over-expression in transgenic coho salmon. Gen. Comp. Endocrinol. 159, 26-37.

Ricker, W.E., 1979. Growth rates and models. In: Hoar, W.S., Randall, D.J., Brett, J.R. (Eds.), Fish Physiology, Vol. VIII. Academic Press, New York, pp. 677-743.

Sakata, S., Noso, T., Moriyama, S., Hirano, T., Kawauchi, H., 1993. Flounder growth hormone: isolation, characterization, and effects on growth of juvenile rainbow trout. Gen. Comp. Endocrinol. 89, 396-404.

Söderpalm, B., Ericsson, M., Bohlooly-Y, M., Engel, J.A., Törnell, J., 1999. Bovine growth hormone transgenic mice display alterations in locomotor activity and brain monoamine neurochemistry. Endocrinology 140, 5619-5625.

Steele, N.C., Evock-Clover, C.M., 1993. Role of growth hormone in growth of homeotherms. In: Schreibman, M.P., Scanes, C.G., Pang, P.K.T. (Eds.), The Endocrinology of Growth, Development and Metabolism in Vertebrates. Academic Press, San Diego, pp. 73-90.

Stern, W.C., Miller, M., Jalowiec, J.E., Forbes, W.B., Morgane, P.J., 1975. Effects of growth hormone on brain biogenic amine levels. Pharmacol. Biochem. Behav. 3, 1115-1118.

Sundström, L.F., Devlin, R.H., Johnsson, J.I., Biagi, C.A., 2003. Vertical position reflects increased feeding motivation in growth hormone transgenic coho salmon (*Oncorhynchus kisutch*). Ethology 109, 701-712.

Sundström, L.F., Lõhmus, M.,Devlin, R.H., Johnsson, J.I., Biagi, C.A., Bohlin, T., 2004a. Feeding on profitable and unprofitable prey: comparing behaviour of growth-enhanced transgenic and normal coho salmon (*Oncorhynchus kisutch*). Ethology 110, 381-396.

Sundström, L.F., Lõhmus, M., Johnsson, J.I., Devlin, R.H., 2004b. Growth hormone transgenic salmon pay for growth potential with increased predation mortality. Proc. R. Soc. Lond. B 271, S350-S352.

Sundström, L.F., Lõhmus, M., Devlin, R.H., 2005. Selection on increased intrinsic growth rates in coho salmon *Oncorhynchus kisutch*. Evolution 59, 1560-1569.

Szabo, M., Butz, M.R., Banerjee, S.A., Chikaraishi, D.M., Frohman, L.A., 1995. Autofeed-back suppression of growth hormone (GH) secretion in transgenic mice expressing a human GH reporter targeted by tyrosine hydroxylase 5'-flanking sequences to the hypothalamus. Endocrinology 136, 4044-4048.

Turnbull, J.F., Richards, R.H., Robertson, D.A., 1996. Gross, histological and scanning electron microscopic appearance of dorsal fin rot in farmed Atlantic salmon, *Salmo salar* L., parr. J. Fish Dis. 19, 415-427.

Tymchuk,W.E., Abrahams,M.V., Devlin, R.H., 2005. Competitive ability and mortality of growth-enhanced transgenic coho salmon fry and parr when foraging for food. Trans. Am. Fish. Soc. 134, 381-389.

Venugopal, T., Anathy, V., Kirankumar, S., Pandian, T.J., 2004. Growth enhancement and food conversion efficiency of transgenic fish *Labeo rohita*. J. Exp. Zool. 301A, 477-490.

Wang, Y., Hu, W., Wu, G., Sun, Y., Chen, S., Zhang, F., Zhu, Z., Feng, J., Zhang, X., 2001. Genetic analysis of "all-fish" growth hormone gene transferred carp (*Cyprinus carpio* L.) and its F_1 generation. Chinese Sci. Bull. 46, 1174-1177.

Wootton, R.J., 1985. Energetics of reproduction. In: Tytler, P., Calow, P. (Eds.), Fish Energetics: New Perspectives. Croom

Helm, London, pp. 231-254.

Wu, G., Chen, L., Zhong, S., Li, Q., Song, C., Jin, B., Zhu, Z., 2008. Enzyme-linked immunosorbent assay of changes in serum levels of growth hormone (*c*GH) in common carps (*Cyprinus carpio*). Sci. China Ser C 51, 157-163.

Zhong, S., Luo, D.,Wang, Y., Chen, Z., Guan, B., Liao, L., Zhu, Z., 2009. Study on serum GH expression of transgenic and control common carp under hungry condition. Acta Hydrobiol. Sin. 33, 1046-1050 (In Chinese with English abstract).

Zhu, Z., Li, G., He, L., Chen, S., 1985. Novel gene transfer into the fertilized eggs of goldfish (*Carassius auratus* L. 1758). Z. Angew. Ichthyol. 1, 31-34.

转基因鲤摄食竞争力增强的行为学机制

段　明[1]　张堂林[1]　胡　炜[1]　李钟杰[1]　Sundström L. Fredrik[3]　朱挺兵[1,2]
钟成容[1,2]　朱作言[1]

1 中国科学院水生生物研究所，淡水生态与生物技术国家重点实验室，武汉　430072，中国
2 中国科学院大学，北京　100049，中国
3 Department of Animal Ecology/Evolutionary Biology Centre, Uppsala University, Norbyvägen 18D, SE-752 72 Uppsala, Sweden

摘　要　为阐明转植基因在转基因鲤社群行为中的作用，在连续投喂颗粒饵料的条件下，比较了 14 对转基因鲤幼鱼和同样大小的非转基因鲤(对照鱼)的血清生长激素水平及相关行为特征。在 3 天共 6 次的持续性观察中，转基因鱼具有更高的活动水平和社群地位，侵犯频次是对照鱼的 2.69 倍，在 10 分钟的投喂期间和投喂前后 2 分钟内，转基因鲤的侵犯行为均显著高于对照鱼。转基因鱼的摄食频率是对照鱼的 1.74 倍。在 8 天实验结束后，转基因鱼体重的特定生长速率是对照鱼的 4.09 倍，血清生长激素水平是对照鱼的 6.36 倍。实验结果表明，转植的生长激素基因促进了生长激素的过量表达，这会改变转基因幼鲤的各种行为特征，从而促使转基因鱼摄食竞争力的增强，进而提高了特定生长率。本研究揭示了快速生长转基因鲤摄食竞争力增强的行为学机制。研究结果不仅有助于科学评估逃逸或释放到野外的转基因鱼潜在的生态效应，还有助于上市转基因鱼的规模化养殖。

Risk-Taking Behaviour May Explain High Predation Mortality of GH-transgenic Common Carp *Cyprinus carpio*

Ming Duan[1,2] Tanglin Zhang[1] Wei Hu[1] Songguang Xie[2] L. F. Sundström[3]
Zhongjie Li[1] Zuoyan Zhu[1]

1 State Key Laboratory of Freshwater Ecology and Biotechnology, Institute of Hydrobiology, Chinese Academy of Sciences, Wuhan 430072, P. R. China

2 Key Laboratory of Biodiversity and Conservation of Aquatic Organisms, Institute of Hydrobiology, Chinese Academy of Sciences, Wuhan 430072, P. R. China

3 Department of Animal Ecology, Evolutionary Biology Centre, Uppsala University, Norbyvägen 18D, SE-752 72 Uppsala, Sweden

Abstract The competitive ability and habitat selection of juvenile all-fish GH-transgenic common carp *Cyprinus carpio* and their size-matched non-transgenic conspecifics, in the absence and presence of predation risk, under different food distributions, were compared. Unequal-competitor ideal-free-distribution analysis showed that a larger proportion of transgenic *C. carpio* fed within the system, although they were not overrepresented at a higher-quantity food source. Moreover, the analysis showed that transgenic *C. carpio* maintained a faster growth rate, and were more willing to risk exposure to a predator when foraging, thereby supporting the hypothesis that predation selects against maximal growth rates by removing individuals that display increased foraging effort. Without compensatory behaviours that could mitigate the effects of predation risk, the escaped or released transgenic *C. carpio* with high-gain and high-risk performance would grow well but probably suffer high predation mortality in nature.

Keywords food distribution; foraging; growth hormone; predation risk; transgene

Introduction

Numerous varieties of genetically engineered fishes are being developed for aquaculture production around the world (Nam *et al.*, 2008). Growth hormone (GH)-transgenic fishes would not only be adopted for use in commercial aquaculture in the near future (FDA, 2012), but also provide useful model systems for studying the consequences of enhancement from genetic, behavioural, physiological, evolutionary and ecological standpoints (Devlin *et al.*, 2006; Nam *et al.*, 2008). Escaped or released GH-transgenic fishes, however, raise public concerns with regard to the potential effects on aquatic ecosystems (Kapuscinski *et al.*, 2007; Nam *et al.*, 2008; Hu & Zhu, 2010).

文章发表于 *Journal of Fish Biology*, 2013, 83(5):1183-1196

Growth is of great evolutionary interest (Sibly *et al*., 1985; Sogard, 1997). In many fish species, rapid growth should increase fitness as faster growing fishes are more dominant, stronger competitors, outgrow gape-limited predators faster and can consume a greater range of potential prey (Werner & Gilliam, 1984; Arendt, 1997; Juanes *et al*., 2002). They may also have higher reproductive potential (Fleming, 1996) and reach sexual maturity earlier (Alm, 1959). Substantial genetic variation in growth rates within and among species, however, has been documented, demonstrating that organisms rarely grow at their physiological maximum and theory suggests that maximization of growth is unlikely (Arendt, 1997). An increasing amount of research suggests that faster growth does not always increase fitness, mainly from various trade-offs that limit the benefit of rapid growth (Arendt, 1997; Mangel & Stamps, 2001; Morgan & Metcalfe, 2001; Munch & Conover, 2003; Biro *et al*., 2005; Devlin *et al*., 2006; Duan *et al*., 2010; Dmitriew, 2011). One major trade off is between rapid growth and survival, acting through heightened activity during foraging (Abrahams & Dill, 1989; Lima & Dill, 1990).

Fishes are capable of expressing flexible growth in response to food availability (Sebens, 1987; Jobling, 1995). As fast-growing transgenic fishes possess higher feeding motivation compared to non-transgenic fishes (Devlin *et al*., 1999; Duan *et al*., 2009), they may have different phenotypic and behavioural responses to food availability. When subjected to low food availability in a microcosm, populations of coho salmon *Oncorhynchus kisutch* (Walbaum 1792) experienced population crashes or extinctions when fast-growing transgenic individuals were present (Devlin *et al*., 2004), indicating that food availability could have population-wide consequences. In rainbow trout *Oncorhynchus mykiss* (Walbaum 1792), fast-growing juveniles had to shift the optimal trade-offs between growth and survival to a more 'high-gain and high-risk' phenotype (Johnsson, 1993). This strategy has been interpreted as boldness, a personality trait, where bold individuals are more risk prone compared with shy individuals (Wilson *et al*., 1993; Sneddon, 2003; Stamps, 2007). For instance, variation in personality traits (*e.g*. boldness) influences risk-taking, diet and consequently growth and fitness in pumpkinseed *Lepomis gibbosus* (L. 1758) (Wilson *et al*., 1993), as well as in the poeciliid *Brachyrhaphis episcope* (Steindachner 1878) (Brown & Braithwaite, 2004; Brown *et al*., 2007). Similar results have been found in GH-treated and GH-transgenic salmonids, which were more willing to take risks that incurred higher predation (Johnsson *et al*., 1996; Jönsson *et al*., 1996; Abrahams & Sutterlin, 1999; Sundström *et al*., 2004; Tymchuk *et al*., 2005).

All-fish GH-transgenic common carp *Cyprinus carpio* L. 1758 were produced two decades ago (Zhu *et al*., 1992). Previous studies have shown that such *C. carpio* are more active and aggressive, show higher feeding motivation, achieve higher social status and elevated individual ability to compete for limited food resources (point and clumped food) compared with non-transgenic conspecifics (Duan *et al*., 2009, 2011). At the same time, they are inferior swimmers (Li *et al.*, 2007, 2009) and suffer higher predation mortality (Duan *et al.*, 2010) compared with non-transgenics. These effects are probably mediated by behaviours influencing habitat selection under different food distributions and the decision to take risks when the reward is food. There have not been any reports on the competitive ability for food of the transgenic *C. carpio* at the population level. Generally, if transgenic *C. carpio* are to retain their growth advantage under natural conditions, it is assumed that they would need to be more effective at competing for food than non-transgenic *C. carpio* to meet their increased metabolic requirements (Hu & Zhu, 2010). Ideal free distribution theory can be used to test the relative competitive abilities of fishes (Grand, 1997; Grand & Dill, 1997), which has been used to assess variation in competitive abilities among non-transgenic and transgenic *O. kisutch* populations (Tymchuk *et al.*, 2005). In an ideal free distribution, individuals are distributed so that none would benefit by switching sites (Fretwell & Lucas, 1970; Fretwell, 1972), and the proportion of individuals at each site matches the proportion of resources in each site (input matching; Parker, 1974). The ideal-free-distribution theory has been further developed by considering the situation where not all individuals are equal but instead differ in competitive ability (Sutherland & Parker, 1985; Parker & Sutherland, 1986).

Here, the competitive ability and habitat selection between the transgenic *C. carpio* and their non-transgenic conspecifics, in the absence and presence of predation risk, under different food distributions were compared. It is helpful to explain the behavioural costs and trade-offs that cause fishes to grow below their physiological maximum, as well as to assess the potential effects on the natural environment of released or escaped GH-transgenic fishes.

Materials and Methods

Experimental animals were maintained at a specially constructed facility at the Fish Behavioural Ecology Laboratory of the Institute of Hydrobiology (IHB), Chinese Academy of Sciences (CAS). This facility was designed to minimize the risk of *C. carpio* escaping by the use of multiple containment screen systems. All experimental procedures employed were approved by the Institute of Hydrobiology Institutional Animal Care and Use Committee (Approval ID: keshuizhuan 08529).

Source of fishes

P_0 all-fish GH-transgenic Yellow River *C. carpio* were initially produced by microinjection of the *pcagcgh* into the fertilized eggs of *C. carpio* (Yellow River variety). The all-fish gene construct *pcagcgh* was a recombinant construct of grass carp *Ctenopharyngodon idella* (Valenciennes 1844) GH gene (*gcgh*), the expression of which is driven by the β-actin gene promoter of *C. carpio* (pca) (Wang *et al.*, 2001). F_1 *C. carpio* were produced as follows:

P_0 transgenic male (sperm revealed to be *pcagcgh* positive) × non-transgenic female. The F_1, F_2 and F_3 generation were, respectively, 1.6 times (Wang *et al.*, 2001), 1.8-2.5 times (W. Hu, unpubl. data) and 1.4-1.9 times (Li *et al.*, 2007) the body mass of non-transgenic counterparts under hatchery-reared conditions, showing how the growth enhancement remains relatively stable across generations. The F_4 generation transgenic and non-transgenic *C. carpio* were produced on 21 April 2008 from crosses between a wild-type female and an F_3 hemizygous transgenic male of a fast-growing transgenic strain. Frequencies of transgene transmission to F_4 progeny from this line were *c.* 50% (W. Hu, unpubl. data). Siblings were used to minimize effects of genetic differences and maternal or paternal effects (Metcalfe *et al.*, 1989). After emergence (28 April) in Duofu Technology Farm, Wuhan, China, the first-feeding fry containing a mix of the two genotypes were transferred to an earthen pond (area 667 m^2, water depth 1 m) with a rearing density of 100 individuals m^{-3} each. A month later, 2000 individuals of the mixed populations were transferred to an indoor recirculating system and reared in two circular fibreglass tanks (diameter 150 cm, volume 1000 l) at the IHB, CAS. Thereafter, the *pcagcgh* transgene-positive *C. carpio* were identified by PCR following Guan *et al.* (2010). The two genotypes were then reared separately in the two fibreglass tanks as described above. Non-transgenics were fed to satiation with frozen chironomid *Chironomus* sp. larvae twice daily, whereas the transgenics were fed roughly the same amount every other day to allow size-matching and avoid the possible effects of size on competitive ability (Huntingford *et al.*, 1990). This feed regime was maintained until the *C. carpio* gape sizes matched the size of the artificial food pellets (mean±s.e. mass 5.52±0.16 mg, length 2.89±0.05 mm, diameter 1.54±0.02 mm; comprising 36.0% protein, 12.1% lipid, 11.3% ash and 18.2 J mg^{-1} energy). In a feeding trial, Duan *et al.* (2011) showed that the transgenic *C. carpio* used in the present experiments consumed on average 58% more pellets than did non-transgenics (mean±S.E. 25.3±1.3 *v.* 16.0±2.6). The water temperature varied from 25.0 to 27.4℃ during this period.

The mandarin fish *Siniperca chuatsi* (Basilewsky 1855) was chosen as the predator. It is a demersal large-sized piscivorous predator, widely distributed in many rivers and lakes of China. *Siniperca chuatsi* consume only live larvae of fishes and shrimp when first-feeding and refuse to consume dead prey or artificial diets during their entire life-history stages (Liang *et al.*, 1998). Age 1 year *S. chuatsi* in Liangzi Lake were captured with trap nets and three individuals were selected for use in this study (mean±S.E. total length, L_T, 26.3±1.1 cm, mass 297.4±25.7 g). Prior to the experiment, the predators were given at least 1 week to acclimate to confined concrete-tank conditions.

Experimental procedures

To test the responses of transgenic and non-transgenic *C. carpio* to food resources and predation risk, the experiment was conducted in four tempered glass aquaria (175×35×35 cm in size; Fig. 1) with a water depth of 20 cm. Eight fluorescent lights (8 W) were fixed 15 cm above the water level on each aquarium (a 12L : 12D on at 0700 hours and off at 1900 hours). All aquaria had their four walls covered with an opaque, black plastic screen to minimize disturbance and maintain similar levels of illumination. Each aquarium was divided into five sections: two non-predator or predator sections in the two ends (50 cm length each), two safe or risky sections (25 cm length each) near the two ends and one buffer section (25 cm length each) in the middle of the aquarium (Fig. 1). There were several holes (2 cm diameter) in the two screens between non-predator or predator and safe or risky sections through which fishes could both smell and see each other, but neither the prey nor the predator were able to go through the holes. In addition, each aquarium had a white gravel substratum and two white-bottomed Petri dishes (15 cm) located at the centre of each of the two feeding sections considered as feeding areas (Fig. 1). Two plastic funnels from the outside of one of the lateral walls were fixed above each feeding area, 5 cm above the water surface, allowing food pellets to be administered to the feeding area each time, without any disturbance to the fishes (Fig. 1).

The experiment consisted of two main parts: (1) a treatment period (days 1-5) and (2) an observation period (days 6-7) and was carried out in an indoor recirculating system. Incoming non-chlorinated water was automatically filtered, sterilized and constantly aerated to ensure high water quality. Throughout the experiment, the water temperature, dissolved oxygen and pH were maintained at 26.4±0.3℃, 7.1±0.1 mg l^{-1} and 7.6±0.1 (mean±S.E.), respectively.

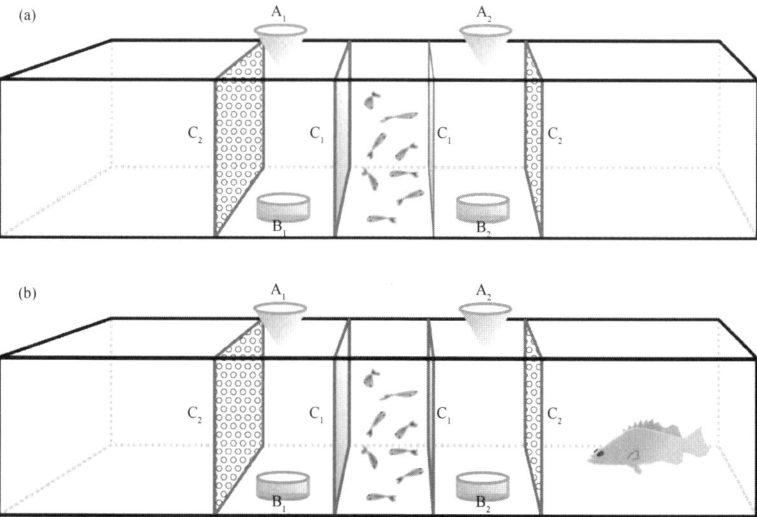

Fig. 1 Schematic representation of the apparatus used for the experiment. A_1 and A_2 are the feeding funnels (food sources); food abundance ratio in $B_1 : B_2 = 1 : 2$ (0.6 g, 100 pellets:1.2 g, 200 pellets); C_1 is the non-porous transparent glass screen and C_2 is the porous transparent glass screen (mesh size, 2 cm); (a) in the absence of a predator, which could test the feeding competition of the two genotypes and (b) in the presence of a predator, smell and visual contacts between the prey *Cyprinus carpio* and predator exist, but none of the *C. carpio* could pass through the porous screens (C_2), which could test whether the transgenic *C. carpio* would be more risk-taking when foraging

Treatment period

On day 1 (20 August) at 0800 hours, four size-matched groups (eight *C. carpio* in each group), with four individuals of each genotype (size-matched as two and two; see Tables 1 and 2), were anaesthetized with $0.25\,\text{ml}\,\text{l}^{-1}$ eugenol (purity ≥ 99.99%, Sinopharm Chemical Reagent Co. Ltd; http://en.reagent.com.cn/). Body mass (*M*, to the nearest 0.01 g) and L_T (to the nearest 0.01 cm) were measured. Individual *C. carpio* from each genotype were identified visually by tagging with different coloured beads (mean ± S.E. diameter 2.3 ± 0.2 mm, mass 14.3 ± 0.3 mg) on nylon thread (diameter 0.25 mm) inserted through the dorsal musculature, anterior to the dorsal fin and then tied in a surgical knot (Duan *et al.*, 2011). Thereafter, the eight *C. carpio* were transferred to the buffer section of each aquarium to acclimate overnight.

The following day (day 2) at 0800 hours, two screens of each buffer section were gently removed and after 3 min the *C. carpio* were fed pellets as described above. In each aquarium, one feeder provided 100 food pellets (total of 0.6 g), one pellet every 5 s during a 20 min feeding session, while the other feeder provided twice that amount by providing two pellets each time (200 food pellets, 1.2 g). Thereafter, the *C. carpio* were gently pushed back into the buffer sections and the screens were put back in place. Uneaten food items and faeces were removed by a glass siphon. This feeding regime was repeated twice daily (0800-1100 and 1400-1700 hours), for a period of 3 days (days 2-4). In the late afternoon of day 5, a *S. chuatsi* was introduced into the predator sections next to the high-food abundance habitat in any two of the four systems. The prey *C. carpio* were deprived of food for 1 day (day 5) before the observations commenced on day 6.

Observations

The same feeding procedures described above were repeated during the two-day behavioural observations. Behaviours of the prey *C. carpio* were recorded by a digital videocassette recorder (Sony, HDR-HC1E Handy Camera; www.sony.net) placed 1 m above the aquarium.

The order of observations was randomly allocated among the four aquaria so as to remove any bias and order effects among tanks. For each aquarium, feeding and video recording started 10 min after the buffer screens were removed. Observations were made twice daily, at 0800-1100 and 1400-1700 hours (*i.e.* two trials for each group) during the feeding periods. Upon completion of the second trial on the first day, the location of the predator and the food supply was reversed, and another two trials were conducted the following day. On the eighth day morning (27 August), L_T and *M* of the 32 *C. carpio* were again measured under a light anaesthesia with $0.25\,\text{ml}\,\text{l}^{-1}$ eugenol (purity ≥ 99.99%). Thereafter, another four size-matched groups and two predators were randomly selected and the procedures repeated as described above from 28 August to 4 September and from 4 to 11 September. Consequently, for each treatment, two size-matched groups were filmed for each trial set, which yielded six repeated trials for each of the two treatments during the two-day observations, and six different size-matched groups with the two genotypes were used in total (Tables 1 and 2).

Table 1 Total length (L_T), body mass (M), condition factor (K) and tag colour (T_C) of transgenic and non-transgenic *Cyprinus carpio* used in the absence of a predator (mean±S.E.)

Group	Genotype	L_T (cm)	M (g)	K (g·cm^{-3})	T_C
1	Transgenic	6.62±0.98	3.40±0.18	1.17±0.04	Blue
1	Non-transgenic	6.42±0.13	3.35±0.27	1.26±0.01	Yellow
2	Transgenic	6.49±0.14	3.17±0.13	1.16±0.03	Blue
2	Non-transgenic	6.32±0.11	3.00±0.08	1.19±0.05	Yellow
3	Transgenic	6.61±0.13	3.32±0.19	1.15±0.01	Yellow
3	Non-transgenic	6.60±0.11	3.27±0.18	1.35±0.03	Red
4	Transgenic	6.61±0.14	3.32±0.22	1.15±0.03	Yellow
4	Non-transgenic	6.49±0.14	3.25±0.24	1.18±0.02	Red
5	Transgenic	6.36±0.11	3.11±0.16	1.20±0.01	Yellow
5	Non-transgenic	6.23±0.13	2.99±0.12	1.23±0.04	Blue
6	Transgenic	6.72±0.26	3.69±0.33	1.21±0.05	Yellow
6	Non-transgenic	6.72±0.26	3.59±0.43	1.16±0.02	Blue

Table 2 Total length (L_T), body mass (M), condition factor (K) and tag colour (T_C) of transgenic and non-transgenic *Cyprinus carpio* used in the presence of a predator (mean±S.E.)

Group	Genotype	L_T (cm)	M (g)	K (g·cm^{-3})	T_C
1	Transgenic	6.54±0.14	3.26±0.26	1.16±0.02	Blue
1	Non-transgenic	6.40±0.17	3.14±0.24	1.20±0.03	Yellow
2	Transgenic	6.49±0.11	3.22±0.12	1.18±0.02	Blue
2	Non-transgenic	6.32±0.08	3.20±0.13	1·27±0.01	Yellow
3	Transgenic	6.78±0.05	3.78±0.09	1.22±0.02	Yellow
3	Non-transgenic	6.58±0.09	3.52±0.13	1.24±0.02	Red
4	Transgenic	6.96±0.11	3.85±0.11	1.14±0.02	Yellow
4	Non-transgenic	6.89±0.86	3.87±0.13	1.19±0.01	Red
5	Transgenic	6.45±0.19	3.24±0.30	1.20±0. 03	Yellow
5	Non-transgenic	6.31±0.18	3.10±0.35	1.21±0.02	Blue
6	Transgenic	6.28±0.14	3.10±0.26	1.24±0.04	Yellow
6	Non-transgenic	6.14±0.07	2.78±0.08	1.20±0.01	Blue

Spatial analysis of the two genotypes was completed from the video. For each trial, location of each individual in the aquarium was captured every 30 s during the 20 min, which was used to determine the spatial distribution of competitors at the risky high and safe low-food sources. Only *C. carpio* that were feeding at the safe or risky sections were counted in the spatial distribution. Individuals that were feeding could be identified by their movement, which included fast darts across the tank to capture a food item or a stationary position under the food supply with short darts to capture the food. Non-feeding *C. carpio* generally maintained a stationary position in the tank unless chased by another individual.

Data analysis

The specific growth rate was calculated as $G_M = 100\ (\ln M_2 - \ln M_1)\ t^{-1}$ (Ricker, 1979), where M_2 and M_1 were the final and initial M (g) and t is the experimental period (7 days, consisting of the 5 day treatment period and the 2 day observations). The condition factor (K) was calculated as $K = 100 M L_T^{-3}$.

To test the effects of genotype and predation risk on proportion of individuals feeding, habitat choice, and the specific growth rate of the transgenic and non-transgenic *C. carpio*, two-way ANCOVA were used with M_1 as a covariate. In the absence of a predator, the difference between the proportion of *C. carpio* feeding at the high-food source that were transgenic and the proportion of transgenic *C. carpio* in the entire feeding population was obtained for each group, and then compared to an expected difference of zero with a one-sample *t*-test. If the transgene had no effect on feeding behaviour, it was expected that the difference between the proportion of *C. carpio* feeding at the high-food source that were transgenic and proportion of transgenic *C. carpio* in the feeding population would be zero (Sutherland & Parker, 1985; Parker & Sutherland, 1986; Tymchuk *et al.*, 2005). The differences between the proportion of transgenic and non-transgenic *C. carpio* feeding were tested by one-way ANCOVA with M_1 as a covariate.

All the data were normally distributed and had homogeneous variance. Differences were regarded as significant when $P<0.05$. All the data were analysed with SPSS 15.0 (http://www-01.ibm.com/software/analytics/spss/) and described as mean ± S.E.

Results

Population competitive ability and predation avoidance

Overall, there was a significant difference between the proportion of the two genotypes that was feeding (ANCOVA, $F_{1,19}=5.561$, $P<0.05$), however, a smaller proportion of both genotypes fed in the presence of the predator ($F_{1,19}=11.843$, $P<0.01$; Fig. 2). Initial M had no effect on the proportional feeding of the two genotypes ($F_{1,19}=2.021$, $P>0.05$), nor was the interaction between genotype and predation risk significant ($F_{1,19}$=0.009, P>0.05). In the absence of the predator, the percent feeding in the transgenic *C. carpio* averaged 79.0±5.7%, while the per cent feeding in the non-transgenic *C. carpio* was 68.7±7.9%, and these values were not significantly different (ANCOVA, $F_{1,9}$=0.119, P>0.05). In the presence of the predator, the percent feeding in the transgenic *C. carpio* averaged 59.0±6.2%, which was higher than that in the non-transgenic *C. carpio* (42.2±6.6%, $F_{1,9}$=7.545, P<0.05; Fig. 2).

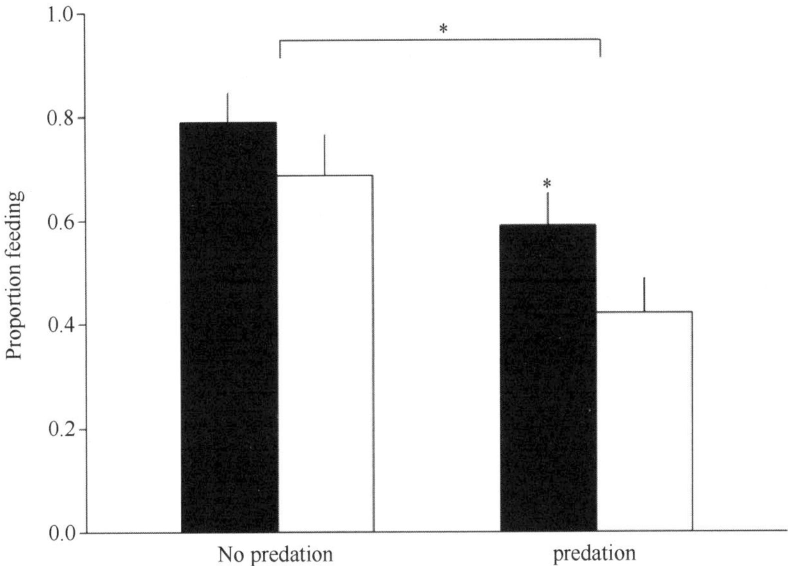

Fig. 2 Proportion of transgenic (n=4) (■) and non-transgenic (n=4) (□) *Cyprinus carpio* feeding in the absence or presence of a predator during the experiment. *, a significant difference between the two genotypes or between the treatments (P<0.05). The whisker on each bar indicates S.E.

In the absence of the predator, there was 79.9±2.8% *C. carpio* feeding at the high-food supply habitat [Fig. 3(a)]. If the GH transgene had no effect on feeding behaviour that is dependent on food supply level, it would be expected that the difference between the proportion of *C. carpio* feeding at the high-food source that were transgenic and the proportion of transgenic *C. carpio* in the entire feeding population would be zero. The per cent of *C. carpio* feeding at the high-food source that were transgenic (53.7±3.5%) was not significantly higher than the percent of transgenic *C. carpio* in the feeding population (54.3±4.4%, t-test, t_5= −1.178, P>0.05), indicating no strong influence of the transgene on access to the higher-quantity food source [Fig. 3(a)].

In the presence of the predator, fewer *C. carpio* of both genotypes fed compared to when the predator was absent [$F_{1,19}$=119.004, P<0.001; Fig. 3(b)]. The number of transgenic *C. carpio* feeding either at low-food and no-risk or high-food and high-risk habitat was much higher than that of the non-transgenic *C. carpio* feeding in the same habitat [ANCOVA, both of $P<0.05$; Fig. 3(b)]. In the high-food and high-risk habitat, the percent feeding in the transgenic *C. carpio* (5.4±0.7%) was much higher than that in the non-transgenic individuals [1.1±0.4%; Fig. 3(b)], indicating that transgenic *C. carpio* were less risk-sensitive at the high-food resource. Initial body *M* had no significant effects on the habitat choice of the two genotypes ($F_{1,19}=0.148$, $P>0.05$), nor was interaction between genotype and predation risk significant ($F_{1,19}=1.778$, $P>0.05$), indicating that transgenic *C. carpio* were not more likely to spend time in the risky zone.

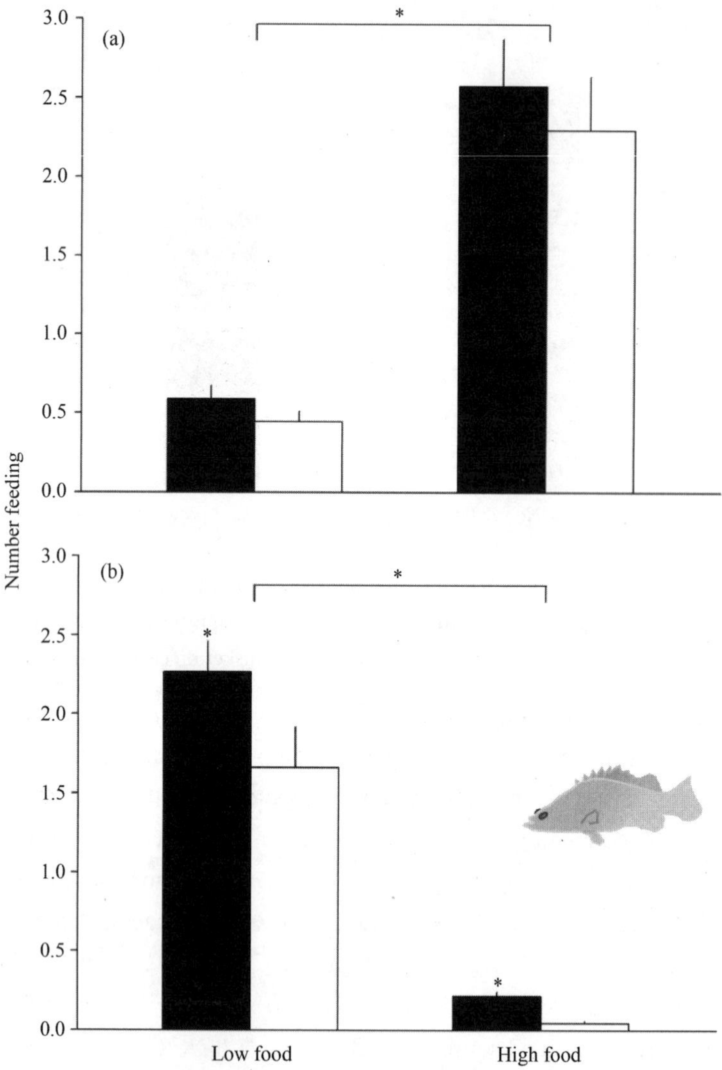

Fig. 3 Mean+S.E. number of transgenic ($n=4$) (■) and non-transgenic ($n=4$) (□) *Cyprinus carpio* feeding at both the low and high-food abundance location (a) in the absence and (b) in the presence of a predator. *, a significant difference between the two genotypes or between the two feeding areas in the same treatment ($P<0.05$)

Growth

At the beginning of the experiment, there was no significant difference in L_T, M and K between the two genotypes in six different size-matched groups (ANOVA, all $P>0.05$; Tables 1 and 2).

Genotype had significant effects on the growth of the experimental *C. carpio* (ANCOVA, $F_{1,19}=46.122$, $P<0.001$; Fig. 4). During the seven-day experiment, the G_W of transgenic *C. carpio* (3.24±0.09% day^{-1}) was 1.62 times that of the non-transgenic *C. carpio* (2.00±0.18% day^{-1}) in the absence of the predator. In the presence of the predator, the G_M of transgenic *C. carpio* (3.27±0.15% day^{-1}) was 3.41 times higher than that of non-transgenic *C. carpio* (0.96±0.45% day^{-1}). Initial M and predation risk had no effects on G_M of the two genotypes, nor was the interaction between genotype and predation risk significant (all $P>0.05$; Fig. 4).

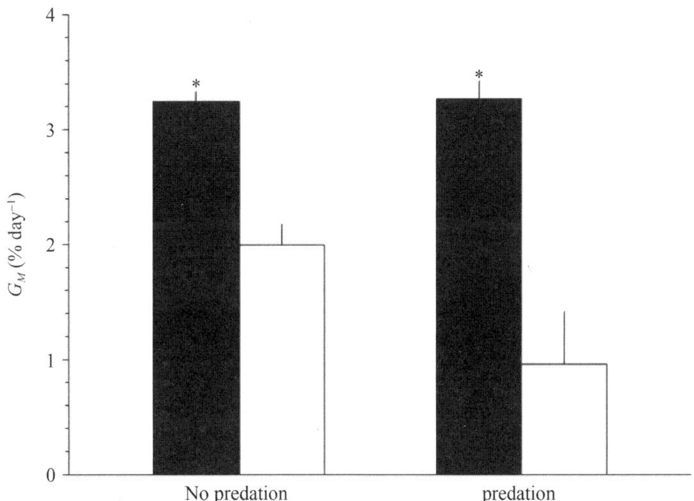

Fig. 4 Mean+S.E. specific growth rates of body mass (G_M) for the transgenic (■) and non-transgenic (□) *Cyprinus carpio* in the absence and in the presence of a predator. *, a significant difference between the two genotypes ($P<0.05$)

Discussion

In this study, transgenic *C. carpio* were competitively equal to their non-transgenic conspecifics as there was no difference in the proportion of transgenic *C. carpio* feeding at the high-food source compared to the proportion of transgenic *C. carpio* within the group. Furthermore, there was no difference between the proportions of the two genotypes that fed at the lower or higher quantity food resources without predation. The results were similar to the previous study on *O. kisutch* at the fry stage in that they were not overrepresented at a higher-quantity food resource and thus exhibited equal competitive ability to controls (Tymchuk *et al.*, 2005). Moreover, in stream microcosms, first-feeding GH-transgenic and non-transgenic Atlantic salmon *Salmo salar* L. 1758 fry were equally likely to be dominant (Moreau *et al.*, 2011). In contrast, juvenile GH-transgenic *C. carpio* and GH-transgenic *O. kisutch* both display elevated ability to compete for pointed and clumped food resources (Devlin *et al.*, 1999; Duan *et al.*, 2009, 2011). These previous results were obtained under limited food supply in single paired tests (one transgenic v. one non-transgenic fish), which did not consider variable food distribution and population effects. In this study, conditions were more complex with four transgenic and four non-transgenic *C. carpio* in each group, which probably mimicked the foraging-associated behaviors such as schooling behaviour (Huntingford *et al.*, 2010) and visual interference (Grosenick *et al.*, 2007) seen in nature. Besides, in this feeding system, food resources in the two habitats were supplied simultaneously over a 20 min time period, which formed a temporal clumping of resources that may reduce dominant defensibility. According to resource defence theory 'synchrony hypothesis' (Grant & Kramer, 1992), when a large number of prey items are encountered simultaneously, less competitive ability group members are able to exploit the remaining resources for the period of time the dominant individual is incapacitated whilst handling prey, and in this way benefit from group membership more than when resources are encountered sequentially and are therefore more defensible. In previous work, GH-transgenic *C. carpio* were better at using low-quality food and had higher feed-conversion efficiency (Fu *et al.*, 1998), which could partly explain why transgenic *C. carpio* grew much faster than their non-transgenic equal competitors during the study period.

Transgenic *C. carpio* were significantly more willing to risk exposure to a predator when foraging for food, and realized their rapid growth potential as a consequence in this study. This may provide direct behavioural evidence for a trade-off between rapid growth and high predation mortality in transgenic *C. carpio*, mediated by their anti-predator behaviour when foraging, indicating that the genetic capacity for growth evolves towards an optimum rather than an absolute maximum. Previous studies on the same fast-growing transgenic strain showed that they increased movements (Duan *et al.*, 2009, 2011), reduced swimming speed (Li *et al.*, 2009) and possessed lower anti-predator ability (higher predation mortality) when directly facing predators (Duan *et al.*, 2010). Similar results were found for fast-growing domestic *O. mykiss* that were more willing to forage in productive and risky habitats, which resulted in higher predation mortality compared to wild individuals, suggesting a trade-off between growth rate and survival such that rapid growth entailed a cost in terms of mortality (Biro *et al.*, 2006). One possible explanation was that *O. mykiss* had to be more active in order to gain access to more food to meet their higher metabolic requirement which increased risk of predation. Indeed, although predation risk resulted in decreased feeding movements in the two genotypes, there were more transgenic *C. carpio* feeding than non-transgenics at the high-food and high-risk habitat, which was consistent with theoretical expectations about how animals should integrate the risk of predation into their foraging decisions (Werner & Gilliam, 1984). The results, however, parallel studies on GH-treated and GH-transgenic salmonids that were bolder towards a predator when foraging for food (Johnsson *et al.*, 1996; Jönsson *et al.*, 1996; Abrahams & Sutterlin, 1999), indicating that levels of GH represent a balance between the positive effects of growth and the increased mortality associated with changes in anti-predator behaviour. Based on a field study, Biro *et al.* (2006) suggested that one of the personalities, boldness, may be the major determinant of predation mortality in *O. mykiss*, and therefore the major mechanism for selection against maximum growth rates. In this study, however, the two genotypes did not have direct contact with the predator, indicating that these individuals should be less willing to forfeit profitable activities in order to reduce the probability of death from predation. As a consequence, the GH-transgenic *C. carpio* should be less sensitive to the indirect effects of predator intimidation.

It seems that escaped or released transgenic *C. carpio* with high-gain and high-risk performance would grow well but suffer high predation mortality in nature. It is worth noting that hungry animals are more risk-prone and willing to pay the costs of predation with greater motivation to take the benefits (*e.g.* food) of risk-taking behaviour (Lima & Dill, 1990). In this study, transgenic *C. carpio* were fed with a restricted ration before the acclimation period (*i.e.* feeding every other day), which may also be a factor that could affect the transgene effects on the risk-taking behaviour due to hunger. Under natural conditions, however, food availability would be less predictable and predation risk may influence feeding decisions so that the advantage of transgenic fishes could be lessened or increased depending on their response under such complex environments. Several recent field studies suggest that the relation between boldness, growth and mortality can be quite variable in natural populations (Adriaenssens & Johnsson, 2013). The effects found on domesticated *O. mykiss* may be specifically influenced by a history of directional selection for growth in captivity, dragging boldness and aggression along, whereas natural selection is often stabilizing with fluctuating selection pressures. Arendt (1997) argued that there were many functions that may be compromised when intrinsic growth rates were accelerated to meet evolutionary challenges. Conversely, if selection acts to improve any of these functions, intrinsic growth rates might be reduced.

Overall, the results clearly showed that the juvenile all-fish GH-transgenic *C. carpio* maintained a faster growth rate, and were more willing to risk exposure to a predator when foraging, as well as were competitively equal to their non-transgenic conspecifics. Short-term, small-scale experiments, however, are likely to overestimate the importance of behavioural variation (Sutherland, 1996; Lima, 1998; Schmitz, 2001). Because these experiments did not have sufficient realism to allow results to be extrapolated to natural situations (Sutherland,

1996; Lima, 1998), or allow expression of compensatory behaviours that could mitigate effects of predation risk (Lind & Cresswell, 2005), the present results could not make convincing connections between behaviour, growth and survival at a population or community level in nature. Further work on the behavioural and developmental alterations of transgenic fishes in complex environments is urgently needed.

Acknowledgements The study was financially supported by the National Natural Science Foundation of China (Grant Nos. 31200423 and 30970553) and the Development Plan of the State Key Fundamental Research of China (Grant No. 2009CB119205).

References

Abrahams, M. V. & Dill, L. M. (1989). A determination of the energetic equivalence of the risk of predation. *Ecology* 70, 999-1007.

Abrahams, M. V. & Sutterlin, A. (1999). The foraging and antipredator behaviour of growth-enhanced transgenic Atlantic salmon. *Animal Behaviour* 58, 933-942.

Adriaenssens, B. & Johnsson, J. I. (2013). Natural selection, plasticity and the emergence of a behavioural syndrome in the wild. *Ecology Letters* 16, 47-55.

Alm, G. (1959). Connection between maturity, size and age in fishes. *Report of the Institute of Freshwater Research Drottningholm* 40, 4-145.

Arendt, J. D. (1997). Adaptive intrinsic growth rates: an integration across taxa. *Quarterly Review of Biology* 72, 149-177.

Biro, P. A., Post, J. R. & Abrahams, M. V. (2005). Ontogeny of energy allocation reveals selective pressure promoting risk-taking behaviour in young fish cohorts. *Proceedings of the Royal Society* B272, 1443-1448.

Biro, P. A., Abrahams, M. V., Post, J. R. & Parkinson, E. A. (2006). Behavioural trade-offs between growth and mortality explain evolution of submaximal growth rates. *Journal of Ecology* 75, 1165-1171.

Brown, C. & Braithwaite, V. A. (2004). Size matters: a test of boldness in eight populations of the poeciliid *Brachyrhaphis episcopi*. *Animal Behaviour* 68, 1325-1329.

Brown, C., Jones, F. & Braithwaite, V. A. (2007). Correlation between boldness and body mass in natural populations of the poeciliid *Brachyrhaphis episcopi*. *Journal of Fish Biology* 71, 1590-1601.

Devlin, R. H., Johnsson, J. I., Smailus, D. E., Biagi, C. A., Jönsson, E. & Björnsson, B. T. (1999). Increased ability to compete for food by growth hormone-transgenic coho salmon *Oncorhynchus kisutch* (Walbaum). *Aquaculture Research* 30, 479-482.

Devlin, R. H., D'Andrade, M., Uh, M. & Biagi, C. A. (2004). Population effects of growth hormone transgenic coho salmon depend on food availability and genotype by environment interactions. *Proceedings of the National Academy of Sciences of the United States of America* 101, 9303-9308.

Devlin, R. H., Sundström, L. F. & Muir, W. M. (2006). Interface of biotechnology and ecology for environmental risk assessments of transgenic fish. *Trends in Biotechnology* 24, 89-97.

Dmitriew, C. M. (2011). The evolution of growth trajectories: what limits growth rate? *Biological Reviews* 86, 97-116.

Duan, M., Zhang, T., Hu, W., Sundström, L. F., Wang, Y., Li, Z. & Zhu, Z. (2009). Elevated ability to compete for limited food resources by 'all-fish' growth hormone transgenic common carp *Cyprinus carpio*. *Journal of Fish Biology* 75, 1459-1472.

Duan, M., Zhang, T., Hu, W., Guan, B., Wang, Y., Li, Z. & Zhu, Z. (2010). Increased mortality of growth-enhanced transgenic common carp (*Cyprinus carpio* L.) under short-term predation risk. *Journal of Applied Ichthyology* 26, 908-912.

Duan, M., Zhang, T., Hu, W., Li, Z., Sundström, L. F., Zhu, T., Zhong, C. & Zhu, Z. (2011). Behavioral alterations in GH transgenic common carp may explain enhanced competitive feeding ability. *Aquaculture* 317, 175-181.

Fleming, I. A. (1996). Reproductive strategies of Atlantic salmon: ecology and evolution. *Reviews in Fish Biology and Fisheries* 6, 379-416.

Fretwell, S. D. (1972). *Populations in a Seasonal Environment*. Princeton, NJ: Princeton University Press.

Fretwell, S. D. & Lucas, H. L. Jr. (1970). On territorial behavior and other factors influencing habitat distribution in birds. I. Theoretical development. *Acta Biotheoretica* 19, 16-36.

Fu, C., Cui, Y., Hung, S. S. O. & Zhu, Z. (1998). Growth and feed utilization by F_4 human growth hormone transgenic carp fed diets with different protein levels. *Journal of Fish Biology* 53, 115-129.

Grand, T. C. (1997). Foraging site selection by juvenile coho salmon: ideal free distributions of unequal competitors. *Animal Behaviour* 53, 185-196.

Grand, T. C. & Dill, L. M. (1997). The energetic equivalence of cover to juvenile coho salmon (*Oncorhynchus kisutch*): ideal free distribution theory applied. *Behavioral Ecology* 8, 437-447.

Grant, J. W. A. & Kramer, D. L. (1992). Temporal clumping of food arrival reduces its monopolization and defence by zebrafish, *Brachydanio rerio*. *Animal Behaviour* 44, 101-110.

Grosenick, L., Clement, T. S. & Fernald, R. D. (2007). Fish can infer social rank by observation alone. *Nature* 445, 429-432.

Guan, B., Hu, W., Zhang, T., Duan, M., Li, D., Wang, Y. & Zhu, Z. (2010). Acute and chronic un-ionized ammonia toxicity to 'all-fish' growth hormone transgenic common carp (*Cyprinus carpio* L.). *Chinese Science Bulletin* 55, 4032-4036.

Hu, W. & Zhu, Z. (2010). Integration mechanisms of transgenes and population fitness of GH transgenic fish. *Science China Life Sciences* 53, 401-408.

Huntingford, F. A., Metcalfe, N. B., Thorpe, J. E., Graham, W. D. & Adams, C. E. (1990). Social dominance and body size in Atlantic salmon parr, *Salmo salar* L. *Journal of Fish Biology* 36, 877-881.

Huntingford, F. A., Andrew, G., Machenzie, S., Morera, D., Coyle, S. M., Pilarczyk, M. & Kadri, S. (2010). Coping strategies in a strongly schooling fish, the common carp *Cyprinus carpio*. *Journal of Fish Biology* 76, 1576-1591.

Jobling, M. (1995). *Environmental Biology of Fishes*. London: Chapman & Hall.

Johnsson, J. I. (1993). Big and brave: size selection affects foraging under risk of predation in juvenile rainbow trout, *Oncorhynchus mykiss*. *Animal Behaviour* 45, 1219-1225.

Johnsson, J. I., Petersson, E., Jönsson, E., Björnsson, B. T. & Järvi, T. (1996). Domestication and growth hormone alter antipredator behaviour and growth patterns in juvenile brown trout, *Salmo trutta*. *Canadian Journal of Fisheries and Aquatic Sciences* 53, 1546-1554.

Jönsson, E., Johnsson, J. I. & Björnsson, B. T. (1996). Growth hormone increases predation exposure of rainbow trout. *Proceedings of the Royal Society* B263, 647-651.

Juanes, F., Buckel, J. A. & Scharf, F. S. (2002). Feeding ecology of piscivorous fishes. In *Handbook of Fish Biology and Fisheries*, Vol. I (Hart, P. J. B. &Reynolds, J. D., eds), pp. 267-283. Oxford: Blackwell Science.

Kapuscinski, A. R., Hayes, K. R., Li, S. & Dana, G. (2007). *Environmental Risk Assessment of Genetically Modified Organisms*, Vol. III. Wallingford: CABI Publishers.

Li, D., Fu, C., Hu, W., Wang, Y. & Zhu, Z. (2007). Rapid growth cost in 'all-fish' growth hormone gene transgenic carp: reduced critical swimming speed. *Chinese Science Bulletin* 52, 1501-1506.

Li, D., Hu, W., Wang, Y., Zhu, Z. & Fu, C. (2009). Reduced swimming abilities in fast-growing transgenic common carp *Cyprinus carpio* associated with their morphological variations. *Journal of Fish Biology* 74, 186-197.

Liang, X., Liu, J. & Huang, B. (1998). The role of sense organs in the feeding behaviour of Chinese perch. *Journal of Fish Biology* 52, 1058-1067.

Lima, S. L. (1998). Stress and decision making under the risk of predation: recent developments from behavioral, reproductive, and ecological perspectives. *Advances in the Study of Behavior* 27, 215-290.

Lima, S. L. & Dill, L. M. (1990). Behavioral decisions made under the risk of predation: a review and prospectus. *Canadian Journal of Zoology* 68, 619-640.

Lind, J. & Cresswell, W. (2005). Determining the fitness consequences of antipredation behavior. *Behavioral Ecology* 16, 945-956.

Mangel, M. & Stamps, J. (2001). Trade-offs between growth and mortality and the maintenance of individual variation in growth. *Evolutionary Ecology Research* 3, 583-593.

Metcalfe, N. B., Huntingford, F. A., Graham, W. D. & Thorpe, J. E. (1989). Early social status and the development of life-history strategies in Atlantic salmon. *Proceedings of the Royal Society* B236, 7-19.

Moreau, D. T. R., Fleming, I. A., Fletcher, G. L. & Brown, J. A. (2011). Growth hormone transgenesis does not influence territorial dominance or growth and survival of first-feeding Atlantic salmon *Salmo salar* in food-limited stream microcosms. *Journal of Fish Biology* 78, 726-740.

Morgan, I. J. & Metcalfe, N. B. (2001). Deferred costs of compensatory growth after autumnal food shortage in juvenile salmon. *Proceedings of the Royal Society* B268, 295-301.

Munch, S. B. & Conover, D. O. (2003). Rapid growth results in increased susceptibility to predation in *Menidia menidia*. *Evolution* 57, 2119-2127.

Nam, Y. K., Maclean, N., Hwang, G. & Kim, D. S. (2008). Autotransgenic and allotransgenic manipulation of growth traits in fish for aquaculture: a review. *Journal of Fish Biology* 72, 1-26.

Parker, G. A. (1974). The reproductive behaviour and the nature of sexual selection in *Scatophaga stercoraria* L. (Diptera: Scatophagidae). IX. Spatial distribution of fertilisation rates and evolution of male search strategy within the reproductive area. *Evolution* 28, 93-108.

Parker, G. A. & Sutherland, W. J. (1986). Ideal free distributions when individuals differ in competitive ability: phenotype-limited ideal free model. *Animal Behaviour* 34, 1222-1242.

Ricker, W. E. (1979). Growth rates and models. In *Fish Physiology*, Vol. VIII (Hoar, W. S., Randall, D. J. & Brett, J. R., eds), pp. 677-743. New York, NY: Academic Press.

Schmitz, O. J. (2001). From interesting details to dynamical relevance: toward more effective use of empirical insights in theory construction. *Oikos* 94, 39-50.

Sebens, K. (1987). The ecology of indeterminate growth in animals. *Annual Review of Ecology and Systematics* 18, 371-407.

Sibly, R., Calow, P. & Nichols, N. (1985). Are patterns of growth adaptive? *Journal of Theoretical Biology* 112, 553-574.

Sneddon, L. U. (2003). The bold and the shy: individual differences in rainbow trout (*Oncorhynchus mykiss*). *Journal of Fish Biology* 62, 971-975.

Sogard, S. M. (1997). Size-selective mortality in the juvenile stage of teleost fishes: a review. *Bulletin of Marine Science* 60, 1129-1157.

Stamps, J. A. (2007). Growth-mortality tradeoffs and 'personality traits' in animals. *Ecology Letters* 10, 355-363.

Sundström, L. F., Lõhmus, M., Johnsson, J. I. & Devlin, R. H. (2004). Growth hormone transgenic salmon pay for growth potential with increased predation mortality. *Proceedings of the Royal Society* B271, S350-S352.

Sutherland, W. J. (1996). *From Individual Behaviour to Population Ecology*. Oxford: Oxford University Press.

Sutherland, W. J. & Parker, G. A. (1985). Distribution of unequal competitors. In *Behavioural Ecology: Ecological Consequences of Adaptive Behaviour* (Sibly, R. M. & Smith, R. H., eds), pp. 255-275. Oxford: Blackwell Scientific Publications.

Tymchuk, W. E., Abrahams, M. V. & Devlin, R. H. (2005). Competitive ability and mortality of growth-enhanced transgenic coho salmon fry and parr when foraging for food. *Transactions of the American Fisheries Society* 134, 381-389.

Wang, Y., Hu, W., Wu, G., Sun, Y., Chen, S., Zhang, F., Zhu, Z., Feng, J. & Zhang, X. (2001). Genetic analysis of 'all-fish' growth hormone gene transferred carp (*Cyprinus carpio* L.) and its F_1 generation. *Chinese Science Bulletin* 46, 1174-1177.

Werner, E. E. & Gilliam, J. F. (1984). The ontogenetic niche and species interactions in size-structured populations. *Annual Review of Ecology and Systematics* 15, 393-425.

Wilson, D. S., Coleman, K., Clark, A. B. & Biederman, L. (1993). Shy-bold continuum in pumpkinseed sunfish (*Lepomis gibbosus*): an ecological study of a psychological trait. *Journal of Comparative Psychology* 107, 205-260.

Zhu, Z., He, L. & Chen, T. T. (1992). Primary-structural and evolutionary analyses of growth-hormone gene from grass carp (*Ctenopharyngodon idellus*). *European Journal of Biochemistry* 207, 643-648.

冒险行为可能导致转基因鲤的反捕食死亡率增加

段　明[1,2]　张堂林[1]　胡　炜[1]　谢松光[2]　L. Fredrik Sundström[3]
李钟杰[1]　朱作言[1]

1 中国科学院水生生物研究所，淡水生态与生物技术国家重点实验室，武汉　430072，中国
2 中国科学院水生生物研究所，水生生物多样性与保护重点实验室，武汉　430072，中国
3 Department of Animal Ecology, Evolutionary Biology Centre, Uppsala University, Norbyvägen 18D, SE-752 72 Uppsala, Sweden

摘　要　在不同食物资源分布与捕食风险存在的情况下，比较了转生长激素基因鲤和同样大小的非转基因鲤间的群体摄食竞争力和栖息地选择行为。不平等竞争者自由分布模型分析表明，转基因鱼参与摄食的比例较高，但它们并没有在高质食物资源占有数量上的优势。转基因鲤发挥出了生长潜能，生长更快速，具有显著频繁的冒险摄食活动。本研究支持了捕食风险总是通过去除摄食努力较强的个体，从而对动物最大生长率具有反选择作用的进化生物学假说。进一步地，若转基因鱼没有演化具有降低被捕食风险的一系列补偿行为，那么转基因鲤如果逃逸或释放到野外生境，虽然会快速生长，但是它们这种“高风险-高收益”的大胆行为个性会更多地遭遇捕食者，可能付出被大量捕食从而种群消亡的代价。

Transgenic Common Carp Do Not Have the Ability to Expand Populations

Hao Lian[1,2] Wei Hu[1] Rong Huang[1] Fukuan Du[1]
Lanjie Liao[1] Zuoyan Zhu[1] Yaping Wang[1]

1 State Key Laboratory of Freshwater Ecology and Biotechnology, Institute of Hydrobiology, The Chinese Academy of Sciences, Wuhan 430012
2 University of Chinese Academy of Sciences, Beijing 100049

Abstract The ecological safety of transgenic organisms is an important issue of international public and political concern. The assessment of ecological risks is also crucial for realizing the beneficial industrial application of transgenic organisms. In this study, reproduction of common carp (*Cyprinus carpio*, CC) in isolated natural aquatic environments was analyzed. Using the method of paternity testing, a comparative analysis was conducted on the structure of an offspring population of "all-fish" growth hormone gene-transgenic common carp (af*gh*-CC) and of wild CC to evaluate their fertility and juvenile viability. Experimental results showed that in a natural aquatic environment, the ratio of comparative advantage in mating ability of af*gh*-CC over wild CC was 1∶1, showing nearly identical mating competitiveness. Juvenile viability of af*gh*-CC was low, and the average daily survival rate was less than 98.00%. After a possible accidental escape or release of transgenic CC into natural aquatic environments they are unable to monopolize resources from eggs of natural CC populations, leading to the extinction of transgenic CC. Transgenic CC are also unlikely to form dominant populations in natural aquatic environments due to their low juvenile viability. Thus, it is expected that the proportion of af*gh*-CC in the natural environment would remain low or gradually decline, and ultimately disappear.

Introduction

Transgenic fish technology was first introduced in the 1980s and has been used for the genetic improvement of farmed fish [1,2]. In the past 20 years, important progress has been made in the research of transgenic fish breeding for the purposes of increasing growth rates. Several fast-growing transgenic fish strains have been bred, showing attractive commercial prospects [3-13]. However, there are currently no transgenic fish strains used in commercial production for human consumption.

Food security and the ecological safety of transgenic organisms are of great public

---

文章发表于 *PloS One*, 2013, 8(6): e65506

concern and represent the last obstacles preventing transgenic fish from entering the market. Compared with food safety assessment, ecological security assessment poses a more difficult task. It is generally believed that fertility and viability are key fitness parameters for evaluating the ecological safety of transgenic fish[14,15]. In recent years, a number of transgenic fish viability-related traits have been reported, including characteristics of the appetite and feeding behavior of transgenic salmon and common carp[16-19], swimming ability [20-22], respiratory metabolism characteristics[23], viability under dissolved oxygen and ammonia nitrogen stress [24,25] and juvenile prey mortality [15,26,27]. Studies on transgenic fish fertility mainly involve the transgenic fish gonad index [5,7], the first sexual maturation time [28], sperm ejaculation volume, sperm motility and mating behavior [29]. Results of comparative studies of single factors of viability and fertility showed that transgenic fish exhibited lower viability and fertility [15,20,22,27-29].

However, Muir *et al.* [30] obtained opposing results in a study on transgenic medaka (*Oryzias latipes*). By assessing a population genetics model based on the findings of fertility and viability of transgenic medaka, they proposed the "Trojan gene hypothesis". This hypothesis predicts that because of the significant fertility advantage of their transgenic medaka and the low viability of its juveniles, once transgenic medaka were released into a natural aquatic environment, transgenic and wild medaka populations would be extinct within 50 generations [30,31]. This hypothesis has resulted in a strong public response, but its universality has been questioned by the academic community [32].

We examined and analyzed the fertility and juvenile viability of af*gh*-CC and CC in an isolated pond. The study aimed to obtain fitness parameters of fertility and viability of the two populations in a natural state, thus further assessing the ecological risk for the environment into which af*gh*-CC could be released.

Materials and Methods

Experimental fish

In this study, male parents were used from the same family population of two-year old transgenic fish. Transgenic males were heterozygotes and wild-type males were their full-sib controls, while female parents were three-year wild-type females, provided by the Zhan Dian Breeding Farm of Fisheries Institute of Sciences, Henan Province, China. Ethical approval for the work was obtained from Expert Committee of Biomedical Ethics, Institute of Hydrobiology of the Chinese Academy of Sciences. The Reference number obtained was 091110-1-303.

Artificial reproduction and reproduction experiment in a natural ecosystem

Artificial insemination was conducted using the dry fertilization method. Fertilized eggs were hatched in culture dishes at an average water temperature of 18.5℃. The fertilization rate was calculated when fertilized eggs developed into the segmentation phase and myocomma appeared. The hatching rate was calculated when juveniles emerged from the membrane (0 days).

The reproduction experiment was conducted in an outdoor isolated-pond as a natural ecosystem. The pond had an area of approximately 1400 m^2 and a depth of 1.5-2.0 m. Six transgenic males (T1-T6), six wild-type males (N1-N6) and six wild-type females (W1-W6) constituted a natural reproductive population. The reproductive population was put into the experimental pond one month before the breeding season to perform natural spawning. Parents were removed after mating which was continued for eight days. Fertilized eggs and juveniles hatched and grew under the natural conditions of the pond. During the experiment, no human intervention, such as enriching oxygen, changing water or feeding were conducted to fully simulate the natural process of breeding. Forty-day-old juvenile offspring were used for paternity testing and PCR detection of transgenes.

DNA sample preparation and transgene detection

Genomic DNA of fin rays or fry were prepared using the phenol-chloroform extraction method. The PCR reaction (10 μL) contained 1 μL genomic DNA (50 ng/μL), 1 μL 10 × buffer (containing Mg^{2+}), 0.5 μL Taq DNA polymerase (1 U/μL), 0.4 μL dNTPs (2.5 mmol/L each), 0.4 μL each of forward and reverse primers (10 μmol/L) and 6.3 μL ddH_2O. The thermocycling program consisted of pre-denaturation at 94℃ for 3 min, 30 cycles of amplification (denaturation at 94℃ for 30 s, annealing at 58℃ for 30 s and extension at 72℃ for 40 s), and a final extension at 72℃ for 5 min. PCR products were detected on a 1% agarose gel by electrophoresis.

Microsatellite marker screening and genotyping

Thirty candidate marker loci were selected from carp microsatellite MFW series [40], KOI series [41] and HLJ series [42]. The 18 parent fish described above were taken as the detection group for microsatellite marker screening. The PCR reaction system (25 μL) contained 1 μL genomic DNA (50 ng/μL), 2.5 μL 10 × buffer (including Mg^{2+}), 0.8 μL dNTPs (2.5 mmol/L each), 0.8 μL each of forward and reverse primers (10 μmol/L), 1 μL Taq DNA polymerase (1 U/μL) and 18.1 μL ddH_2O. The PCR reaction program consisted of pre-denaturation at 94℃ for 3 min, 30 cycles of amplification (denaturation at 94℃ for 30 s, annealing at 58℃ for 30 s and extension at 72℃ for 40 s), and a final extension at 72℃ for 5 min. Genotyping of PCR products was detected using the LI-COR 4300 DNA gel electrophoresis system. The polymorphic information content (PIC) and heterozygosity (H) at each locus were analyzed using POPGEN (Version 1.32) software. Nine appropriate SSR loci were screened for paternity testing (Table 1). Marker genotyping of paternity testing samples was conducted using the same method.

Table 1 Primers used for PCR amplification

Locus	Primer⁺	Primer⁻	H	PIC
MFW1	mGTCCAGACTGTCATCAGGAG	GAGGTGTACACTGAGTCACGC	0.880	0.867
MFW9	mGATCTGCAAGCATATCTGTCG	ATCTGAACCTGCAGCTCCTC	0.788	0.763
MFW11	mGCATTTGCCTTGATGGTTGTG	TCGTCTGGTTTAGAGTGCTGC	0.799	0.777
MFW15	mCTCCTGTTTTGTTTTGTGAAA	GTTCACAAGGTCATTTCAGC	0.853	0.836
HLJ38	mCACAGAACGCATCAGTAA	TGTAAACCTTCAACCTCC	0.860	0.844
MFW19	mGAATCCTCCATCATGCAAAC	GCACAAACTCCACATTGTGCC	0.838	0.817
MFW18	mGTCCCTGGTAGTGAGTGAGT	GCGTTGACTTGTTTTATACTAG	0.708	0.665
MFW26	mCCCTGAGATAGAAACCACTG	CACCATGCTTGGATGCAAAAG	0.855	0.837
MFW29	mGTTGACCAAGAAACCAACATGC	GAAGCTTTGCTCTAATCCACG	0.726	0.728
Pll-Pc	CATTTACAGTTCAGCCATGGCTAGA	AGCACCACCGACAACAGCACTAATG	—	—

NOTE:+ “m” represents the M13 sequence (CACGACGTTGTAAAACGAC)

Paternity testing and statistical analysis

Paternity testing was conducted using Cervus 2.0 software. Allele frequencies (P), exclusion probabilities (PE), cumulative chance of exclusion (CCE), natural logarithm of likelihood ratio (LOD) values and delta values at each locus were calculated. The likelihood ratio of each parental candidate [43] was counted. Delta values of assumed parents were calculated by a simulation program to ensure assessment of parentage with high statistical confidence [44]. Significant differences between populations were detected by *t*-test or chi-square test.

Results

Genetics and reproductive biology characteristics of transgenic CC

Transgenic CC carried recombinant af*gh*, which is the grass carp growth hormone gene driven by the CC β-actin promoter (pCAgcGH) (Figure 1). P0 transgenic males obtained by microinjection were hybridized with wild-type females to obtain F1 transgenic fish heterozygous groups [10]. Fast-growing F1 transgenic heterozygous males were selected to hybridize with wild-type females to screen for a fast-growing transgenic fish F2 family whose unit points were integrated. In this family group, transgenes showed Mendelian segregation (1∶1) (transgenic fish:non-transgenic fish), where transgenic fish were transgenic heterozygotes. We conducted selfing of the F2 transgenic fish for the screening of the homozygous transgenic family; F2 transgenic males hybridized with wild-type females to establish F3 transgenic fish-segregated populations. Similarly, transgenic fish were transgenic heterozygotes. Transgenic fish and non-transgenic fish showed 1∶1 segregation. Transgenes were passed between generations in a Mendelian way.

Experimental results of artificial fertility-hatching showed that reproductive biology characteristics of F3 transgenic fish were similar to those of the controls. Three F3 transgenic males and three full-sib non-transgenic males were used to fertilise eggs of the same wild-type female, respectively. Average fertilization rates of transgenic fish and non-transgenic fish were 90.37 ± 3.48% and 92.82 ± 5.24%, respectively (t = 0.68; P = 0.54; df = 4). Their average hatching rates were 85.60 ± 8.05% and 85.20 ± 10.36%, respectively (t = –0.05; P = 0.96; df = 4). Under laboratory conditions, no significant differences were found in the fertility between transgenic and non-transgenic fish (Table 2). Transgene detection by PCR was conducted on the offspring population of transgenic males. In offspring populations of three transgenic males, ratios of transgenic fish were 48% (48/100) (χ^2 = 0.08; P = 0.78), 47% (47/100) (χ^2 = 0.18; P = 0.67) and 51% (51/100) (χ^2 = 0.02; P = 0.89). The segregation of transgenes in the offspring population was consistent with a Mendelian segregation ratio of 1∶1 (Table 2). These results indicate that transgenes passed steadily between generations of transgenic fish, that sperm carrying transgenes and their controls had the same level of fertilization ability, and they did not impact on the early development of embryos.

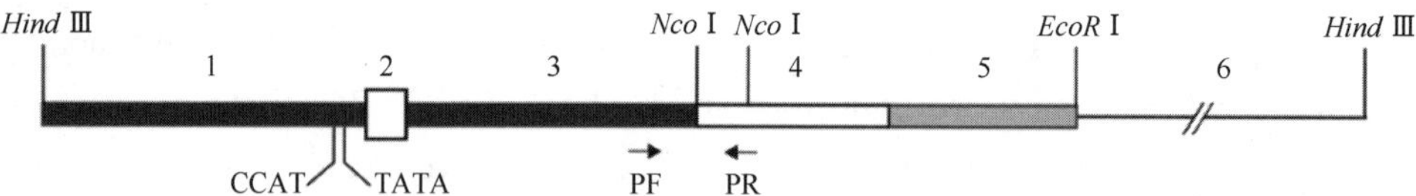

Figure 1 pCAgcGH structure diagram. 1: carp β-actin gene 5′-flanking sequence; 2: carp β-actin gene first exon; 3: carp β-actin gene first intron; 4: grass carp *GH* gene sequence; 5: grass carp *GH* gene 3′-flanking sequence; 6: plasmid pUC118. PF and PR indicate PCR primers of transgenes

Table 2 Fertilization, hatching rates and transgene segregation

	Non-transgenic males			Transgenic males			P
Fertility (%)	87.05	94.15	97.27	94.15	89.66	87.31	>0.05
Hatchability (%)	73.38	92.68	89.55	94.15	84.48	78.17	>0.05
Segregation (%)	—	—	—	48	47	51	>0.05

Composition of the natural fertility population and their offspring population

From the segregated populations of two-year old F3 transgenic fish described above, six transgenic males (T1-T6) and six full-sib non-transgenic males (N1-N6), plus six wild-type females aged three years (W1-W6) were selected to constitute a fertility population including six females and 12 males. The weight distribution of transgenic males was 1.31-2.63 kg, and average weights of three small individuals (T1-T3) and three large individuals (T4-T6) were 1.49 ± 0.16 kg and 2.55 ± 0.09 kg, respectively. Non-transgenic males showed a smaller weight distribution (0.80-1.17 kg), and the average weight was 0.98 ± 0.15 kg; the weight distribution of wild-type females was 3.50-4.84 kg, and the average weight was 4.22 ± 0.53 kg (Table 3).

Six females and 12 males described above were released into an isolated pond before the reproductive season. In late March, experimental fish started spawning, and spawning was continued for eight days. After spawning, reproductive parents were removed from the pond, fertilized eggs and fry were naturally hatched and grown in the pond without human intervention. After 40 days, 1200 juveniles with an average weight of 0.28 ± 0.03 g were randomly sampled from the offspring population for paternity testing.

Marking and typing were conducted on nine SSR loci. Complete SSR typing data were obtained from 1138 juvenile samples. The paternity testing results showed that these 1138 offspring were from 72 parental combinations (12 × 6), that is, six females successfully mated with 12 males and produced offspring populations ranging from 5 to 44 offspring (Table 3).

Table 3 Number of offspring of all parental combinations

Wild-type females (body weight,kg)	Non-transgenic males						Transgenic males					
	N1 (0.87)	N2 (0.87)	N3 (0.80)	N4 (1.05)	N5 (1.17)	N6 (1.10)	T1 (1.62)	T2 (1.31)	T3 (1.54)	T4 (2.46)	T5 (2.56)	T6 (2.63)
W1 (4.08)	9	15	16	20	44	19	12	12	9	5	16	8
W2 (3.89)	21	10	14	12	18	21	11	8	22	18	22	5
W3 (3.50)	9	20	8	22	33	17	19	6	13	10	9	9
W4 (4.14)	15	7	18	32	37	33	10	15	5	13	17	11
W5 (4.84)	35	27	18	10	34	13	8	19	23	31	8	13
W6 (4.84)	7	10	14	19	22	15	8	8	15	9	12	5

Juvenile viability of af*gh*-CC descendants

According to the results of paternity testing, six transgenic males in the testing samples produced 444 offspring (Table 3). Transgene PCR results showed 54 transgenic fish and 390 non-transgenic fish. In offspring populations of transgenic males, the transgenic fish ratio was 12.16%, strongly deviating from the theoretical value of 50% (Table 4). The juvenile viability of descendants of af*gh*-CC was significantly lower than that of their non-transgenic full-sib controls (χ^2 = 148.37; P = 0.00). During the 40-48 day experiment, the relative viability of transgenic juveniles was 13.85%, and the average daily relative viability was 0.98.

In the offspring population of transgenic males at different sizes, ratios of transgenic fish showed no significant difference. Among 223 offspring of small transgenic males (T1-T3), there were 29 transgenic fish, and the ratio of transgenic fish was 13.00%. Among 221 offspring of large transgenic males (T4-T6), there were 25 transgenic fish, and the ratio of transgenic fish was 11.31% (Table 4). The size of transgenic males did not impact on their offspring viability (χ^2 = 0.30; P = 0.59).

Table 4 Number of offspring form male parents

	N1	N2	N3	N4	N5	N6	T1	T2	T3	T4	T5	T6
Transgenic offspring	—	—	—	—	—		9	8	12	9	13	3
Non-transgenic offspring	96	89	88	115	188	118	59	60	75	77	71	48
Sum	96	89	88	115	188	118	68	68	87	86	84	51

Effect of body size on reproductive success in af*gh*-CC males

Offspring populations of transgenic males (T1-T6) at different sizes were compared. The numbers of offspring from small males (T1-T3) and large males (T4-T6) were 223 and 221, respectively (Table 4), and ratios in offspring populations of transgenic males were 50.23% and 49.77%, respectively. A chi-square test showed that the number of offspring of T1-T3 and T4-T6 were not significantly different ($\chi^2 = 0.01$; $P = 0.95$) (Figure 2a).

Since no significant differences were found in the offspring viability, population size and composition of T1-T3 and T4-T6, the offspring population from T1-T6 was compared with that from non-transgenic males (N1-N6). The number of offspring from T1-T6 was 444, and that from N1-N6 was 694 (Table 4), while their ratios of offspring population were 39.01% and 60.09%, respectively. The number of offspring from T1-T6 was significantly lower than that from N1-N6 ($\chi^2 = 27.80$; $P = 0.00$) (Figure 2b).

Taking into account the high mortality of transgenic juveniles, the number of surviving offspring of transgenic males could not accurately reflect the number of fertilized eggs, and a direct comparison of the number of surviving offspring would underestimate the fertility of transgenic males. Therefore, twice the number of non-transgenic offspring of transgenic males

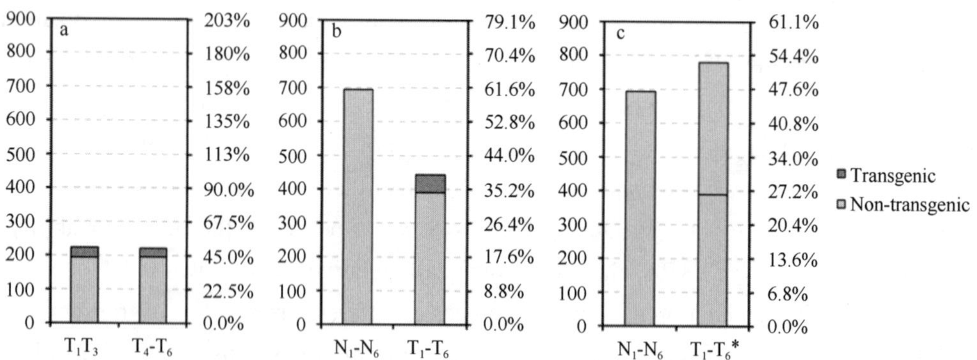

Figure 2 Number and ratio of each offspring population and the significance test between offspring populations. a: T1-T3 are the numbers of offspring of small transgenic males, T4-T6 are the numbers of offspring of large transgenic males; b: N1-N6 are the numbers of offspring of wild-type males, T1-T6 are the numbers of offspring of transgenic males; c: N1-N6 are the numbers of offspring of wild-type males, T1-T6* are the numbers of offspring of transgenic males after correction. Left vertical axis shows the number of individuals; right vertical axis shows the ratio of offspring populations

was taken as the corrected value of the number of offspring from transgenic males; that is 780 (2×390) offspring. Results of the comparison after this correction showed that the ratio of offspring from T1-T6 in the population was 52.92%, and that from N1-N6 was 47.08% (Figure 2c). Based on this finding, the comparative advantage in mating ability of transgenic carp to wild-type carp was 1∶1, with no significant difference between the two groups (χ^2 = 0.18; P = 0.67).

Discussion

The "Trojan gene hypothesis" proposed by Muir *et al.* [30] assumes that transgenic males would be significantly larger than control males, thus holding a dominant position when competing for a mate. In contrast, the hypothesis states the viability of offspring from transgenic fish is significantly lower than that of wild-type fish. The combination of these two factors would lead to the extinction of the entire population, and transgenes show the so-called "Trojan effects" [30,31]. In this study of af*gh*-CC and wild CC, paternity testing of 1138 offspring showed that transgenic and wild-type males, regardless of their individual sizes, were successfully involved in breeding, including small males. This result differs substantially from the situation observed in medaka [31]. Carp follow the reproductive strategy of "group spawning" and have the characteristic of batch spawning (batch spawning and promiscuity of carp account for this similarity) [33-35]. The different reproductive strategies of carp and medaka are probably the main reason for the different results described above. These results suggest that the role of male size in competitive mating arenas and its effects on the structure of their offspring populations is different in fish employing different reproductive strategies. Large individuals do not result in an advantage in fertility. In a study of af*gh*-modified Atlantic salmon (*Salmo salar*), even the opposite phenomenon was found, with the fertility of large transgenic fish significantly decreased [36]. Based on findings from different transgenic fish species, it is hard to draw a general conclusion on the ecological risk assessment of transgenic fish. Targeted case analyses are still required for the ecological risk assessment of transgenic fish.

Male fertility is an embodiment of the competitive ability of mating and sperm. It is essentially the ability of males to fertilize eggs. It is difficult to accurately observe competitive breeding behavior under natural conditions. In this experiment, the number of surviving offspring of males was used to calculate the number of fertilized eggs, further calculating the relative fertility advantages of transgenic and wild-type males. As described in the results section of this article, because of the significant difference in the viability between transgenic and wild-type juveniles, the number of surviving transgenic juveniles could not truly reflect the number of fertilized eggs of transgenic males. A direct comparison of the number of surviving offspring between transgenic and wild-type males would underestimate the actual fertility of transgenic males. Artificial insemination-incubation experiments have shown that heterozygous transgenic males produce an equal number of transgenic and wild-type sperms with the same fertilization capacity. Therefore, it would be more accurate to estimate the total number of fertilized eggs by using the number of wild-type surviving offspring of transgenic males. The fertility comparison indicated that the fertility ratio of transgenic males to wild-type males was 1∶1, showing no significant differences. A larger body size of transgenic carp did not confer any advantage in breeding competitiveness.

Fast-growing transgenic fish generally exhibit lower juvenile viability [8,5,37,38]. The increased metabolism of fast-growing transgenic fish may be the main cause of high juvenile mortality [9]. Juvenile viability differs between different transgenic fish species. The daily viability of 3-day-old transgenic medaka reached 91.5-93.0% [37], while all juveniles of transgenic coho salmon died within a few weeks due to food shortage [38]. In the present study, the viability of 40-48-day-old transgenic carp was 13.8%, and the average daily viability was 98.0%. Because most transgenic juveniles die at an early stage of initial feeding, the daily viability in the early stage will be lower. According to the ecological risk assessment model of transgenic fish established by Muir *et al.* [30,39], if the relative fertility advantage was 1∶1 and the daily survival rate was 98.00% or less, transgenic fish populations would survive. Transgenic carp differ from transgenic medaka in that they do not present “Trojan effects” leading to the extinction of the population.

Conclusions

Our results show that in a natural aquatic environment, fast-growing transgenic common carp have no advantages in fertility, and that their juvenile viability is low. We suggest that transgenic carp escaped or released into the natural aquatic environment would be unable to monopolize the eggs of natural CC populations, thus leading to population extinction. Moreover, transgenic CC are incapable of forming dominant populations in natural aquatic environments due to their low juvenile viability. Therefore, we predict that the ratio of transgenic CC in natural populations would remain at low levels or gradually decline, and ultimately disappear.

Acknowledgements This work was financially supported by grants from the Development Plan of the State Key Fundamental Research of China (Grant No. 2009CB118701), and the National Natural Science Foundation of China (Grant No. 30771664).

References

1. Zhu Z, Li G, He L, Chen S (1985) Novel gene transfer into the fertilized eggs of goldfish (*Carassius auratus* L. 1758).

Z Angew Ichthyol 1: 31-34.
2. Zhu Z, Xu KS, Li GH, Xie YF (1986) The biological effect of human growth hormone when microinjected into the fertilized eggs of loach. Chin Sci Bull 31: 387-389.
3. Du SJ, Gong Z, Fletcher GL, Shears MA, King MJ, *et al.* (1992) Growth enhancement in transgenic Atlantic salmon by the use of an "all-fish" chimeric growth hormone gene constructs. Biotechnology 10: 176-181.
4. Tsai HJ, Tseng FS, Liao IC (1995) Electroporation of sperm to introduce foreign DNA into the genome of loach *(Misgurnus anguillicaudatus*). Can J Fish Aquat Sci 52: 776-787.
5. Rahman MA, Mak R, Ayad H, Smith A, Maclean N (1998) Expression of a novel piscine growth hormone gene results in growth enhancement in transgenic tilapia (*Oreochromis niloticus*). Transgenic Res 7: 357-370.
6. Rahman MA, Maclean N (1999) Growth performance of transgenic tilapia containing an exogenous piscine growth hormone gene. Aquaculture 173: 333-346.
7. Rahman MA, Ronyai A, Engidaw BZ, Jauncey K, Hwang GL, *et al.* (2001) Growth and nutritional trials of transgenic Nile tilapia containing an exogenous fish growth hormone gene. J Fish Biol 59: 62-78.
8. Cook JT, McNiven MA, Richardson GF, Sutterlin AM (2000a) Growth rate, body composition and feed digestibility/conversion of growth enhanced Atlantic salmon (*Salmo salar*). Aquaculture 188: 15-32.
9. Cook JT, McNiven MA, Sutterlin AM (2000b) Metabolic rate of presmolt growth enhanced transgenic Atlantic salmon (*Salmo salar*). Aquaculture 188: 33-45.
10. Wang YP, Hu W, Wu G, Sun Y, Chen S, *et al.* (2001) Genetic analysis of "all-fish" growth hormone gene transferred carp (*Cyprinus carpio* L.) and its F1 generation. Chin Sci Bull 46: 226-229.
11. Nam YK, Noh JK, Cho YS, Cho HJ, Cho KN, *et al.* (2001) Dramatically accelerated growth and extraordinary gigantism of transgenic mud loach (*Misgurnus mizolepis*). Transgenic Res 10: 353-362.
12. Kobayashi S, Morita T, Miwa M, Lu J, Eudo M, *et al.* (2007) Transgenic Nile tilapia (*Oreochromis niloticus*) over-expressing growth hormone show reduced ammonia excretion. Aquaculture 270: 427-435.
13. Hallerman EM, McLean E, Fleming IA (2007) Effects of growth hormone transgenes on behavior and welfare of aquacaltured fishes: a review identifying research needs. Appl Anim Behav Sci 104: 265-294.
14. Devlin RH, Donaldson EM (1992) Containment of genetically altered fish with emphasis on salmonids. In: Hew CL, Fletcher GL, editors. Singapore: World Scientific. pp. 229-265.
15. Sundström LF, Lôhmus M, Johnsson J, Devlin RH (2004) Growth hormone transgenic salmon pay for growth potential with increased predation mortality. Proc R Soc Lond B 271: S350-S352.
16. Devlin RH, Johnsson JI, Smailus DE, Biagi CA, Jönsson E, *et al.* (1999) Increased ability to compete for food by growth hormone-transgenic coho salmon *Oncorhynchus kisutch* (Walbaum). Aquaculture Research 30: 479-482.
17. Fu C, Li D, Hu W, Wang Y, Zhu Z (2007) Growth and energy budget of F2 "all-fish" growth hormone gene transgenic common carp. J Fish Biol 70: 347-361.
18. Duan M, Zhang T, Hu W, Sundström LF, Wang Y, *et al.* (2009) Elevated ability to compete for limited food resources by 'all-fish' growth hormone transgenic common carp *Cyprinus carpio*. J Fish Biol 75: 1459-1472.
19. Duan M, Zhang T, Hu W, Li Z, Sundström F, *et al.* (2011) Behavioral alterations in GH transgenic common carp may explain enhanced competitive feeding ability. Aquaculture 317: 175-181.
20. Farrell AP, Bennett W, Devlin RH (1997) Growth-enhanced transgenic salmon can be inferior swimmers. Can J Zool 75: 335-337.
21. Lee CG, Devlin RH, Farrell AP (2003) Swimming performance, oxygen consumption and excess post-exercise oxygen consumption in adult transgenic and ocean-ranched coho salmon. J Fish Biol 62: 753-766.
22. Li D, Hu W, Wang Y, Zhu Z, Fu C (2009) Reduced swimming abilities in fast-growing transgenic common carp *Cyprinus carpio* associated with their morphological variations. J Fish Biol 74:186-197.
23. Guan B, Hu W, Zhang T, Wang Y, Zhu Z (2008) Metabolism traits of 'all-fish' growth hormone transgenic common carp (*Cyprinus carpio* L.). Aquaculture 284: 217-223.
24. McKenzie DJ, Martínez R, Morales A, Acosta J, Morales R, *et al.* (2003) Effects of growth hormone transgenesis on metabolic rate, exercise performance and hypoxia tolerance in tilapia hybrids. J Fish Biol 63: 398-409.

25. Guan B, Hu W, Zhang TL, Duan M, Li DL, *et al.* (2010) Acute and chronic un-ionized ammonia toxicity to 'all-fish' growth hormone transgenic common carp (*Cyprinus carpio* L.). Chin Sci Bull 55: 4032-4036.
26. Zhang T, Hu W, Duan M (2006) The *gh*-transgenic common carp (*Cyprinus carpio*): growth performance, viability and predator avoidance. The Seventh International Congress on the Biology of Fish, St John's, Newfoundland, Canada, 18-22.
27. Duan M, Zhang T, Hu W, Guan B, Wang Y, *et al.* (2010) Increased mortality of "all-fish" growth hormone transgenic common carp (*Cyprinus carpio* L.) under a short-term risk of predation. J Appl Ichthyol 26: 908-912.
28. Bessey C, Devlin RH, Liley NR, Biagi CA (2004) Reproductive Performance of Growth-Enhanced Transgenic Coho Salmon. Trans Am Fish Soc 133: 1205-1220.
29. Fitzpatrick JL, Akbarashandiz H, Sakhrani D, Biagi CA, Pitcher TE, *et al.* (2011) Cultured growth hormone transgenic salmon are reproductively out-competed by wild-reared salmon in semi-natural mating arenas. Aquaculture 312: 185-191.
30. Muir WM, Howard RD (1999) Possible ecological risks of transgenic organism release when transgenes affect mating success: Sexual selection and the Trojan gene hypothesis. Proc Natl Acad Sci USA 96: 13853-13856.
31. Howard RD, Andrew DeWoody J, Muir WM (2004) Transgenic male mating advantage provides opportunity for Trojan gene effect in a fish. Proc Natl Acad Sci USA 10: 2934-2938.
32. Maclean N, Laight RJ (2000) Transgenic fish: An evaluation of benefits and risks. Fish and Fisheries 1: 146-172.
33. Li DS (1961) Inland fish farming. In: Xin Hailian fishery technical college (Agriculture), editor. pp. 11.
34. Yang ZJ (1981) The artificial propagation of carps in three seasons. Freshwater fishery 1: 31-32.
35. Li MD (1989) Fish Ecology. Tianjin Technology Translated Press Company Press, 266 p.
36. Gage MJG, Stockly P, Parker GA (1995) Effects of alternative male mating strategies on characteristics of sperm production in Atlantic salmon (*Salmo salar*). Philos Trans R Soc London Ser B 350: 391-399.
37. Muir WM, Howard RD (2002) Assessment of possible ecological risks and hazards of transgenic fish with implications for other sexually reproducing organisms. Transgenic Res 11: 101-114.
38. Devlin RH, D'Andrade M, Uh M, Biagi CA (2004) Population effects of growth hormone transgenic coho salmon depend on food availability and genotype by environment interactions. Proc Natl Acad Sci USA 101: 9303-9308.
39. Howard RD, Andrew DeWoody J, Muir WM (2004) Transgenic male mating advantage provides opportunity for Trojan gene effect in a fish. Proc Natl Acad Sci USA 10: 2934-2938.
40. Crooijmans RPMA, Bierbooms VAF, Komen J, Van der Poel JJ, Groenen MAM (1997) Microsatellite markers in common carp (*Cyprinus carpio* L.). Anim Genet 28: 129-134.
41. David L, Jinggui F, Palanisamy R, Hillel J, Lavi U (2001) Polymorphism in ornamental and common carp strains (*Cyprinus carpio* L.) as revealed by AFLP analysis and a new set of microsatellite markers. Mol Gen Genomics 266: 353-362.
42. Quan YC, Sun XW, Liang LQ (2005) Genetic diversity of four breeding populations of common carps revealed by SSR. Zoological research/Dongwuxue Yanjiu 26: 595-602.
43. Bekkevold D, Hansen MM, Loeschcke V (2002) Male reproductive competition in spawning aggregations of cod (*Gadus morhua*, L.). Mol Ecol 11: 91-102.
44. Slate J, Marshall T, Pemberton, J (2000) A retrospective assessment of the accuracy of the paternity inference program CERVUS. Mol Ecol 9: 801-808.

转基因鲤不具备种群扩张能力

连 灏[1,2] 胡 炜[1] 黄 容[1] 杜富宽[1] 廖兰杰[1] 朱作言[1] 汪亚平[1]

1 中国科学院水生生物研究所，淡水生态与生物技术国家重点实验室，武汉 430072
2 中国科学院大学，北京 100049

摘 要 繁殖力和生存力是影响鱼类种群演化的两个重要因素，也是评价转基因鱼生态安全的关键适合度参数。本文研究了在隔离的自然水体中鲤自然繁殖的结果，采用亲子鉴定的方法，对转"全鱼"生长激素基因鲤和野生对照鲤的子代群体结构进行了比较分析，评价两者的繁殖力及其子代幼鱼的存活力。实验结果显示，在自然水体中，转"全鱼"生长激素基因鲤相对于野生型鲤具有相同的繁殖竞争力，但其子代幼鱼存活力低下。因此，转"全鱼"生长激素基因鲤逃逸或释放到自然水体中，不可能对鲤自然种群的卵资源产生毁灭性掠夺，进而导致种群的灭绝；另一方面，由于转"全鱼"生长激素基因鲤的子代存活力低下，不可能形成优势种群，其在鲤自然种群中占据的比率将维持在一个较低水平，或逐渐减少最终消亡。

Rapid Growth Increases Intrinsic Predation Risk in Genetically Modified *Cyprinus carpio*: Implications for Environmental Risk

L. ZHANG[1,2]   R. E. GOZLAN[3]   Z. LI[1]   J. LIU[1]
T. ZHANG[1]   W. HU[1]   Z. ZHU[1]

1 State Key Laboratory of Freshwater Ecology and Biotechnology, Institute of Hydrobiology, Chinese Academy of Sciences, Wuhan, 430072, China
2 Graduate School University of Chinese Academy of Sciences, Beijing, 100049, China
3 Unité Mixte de Recherche Biologie des Organismes et Ecosystèmes Aquatiques (IRD207, CNRS 7208, MNHN, UPMC), Muséum National d'Histoire Naturelle, 75231, Paris Cedex, France

Abstract The intrinsic effect of feeding regime on survival and predation-induced mortality was experimentally tested in genetically modified (GM) *Cyprinus carpio* and wild specimens. The results clearly indicate a knock-on effect of the GH gene (*gcGH*) introduction into the *C. carpio* genome on their vulnerability to predation. The experiments unequivocally showed that it is the genetic nature of the *C. carpio* rather than its size that affects the risk of predation. In addition, fed *C. carpio* were more susceptible to predation risk. Thus, the study characterizes the existence of a trade-off between somatic growth and predator avoidance performance. Current research inEurope suggests that high uncertainty surrounding the potential environmental effects of escapee transgenic fishes into the wild is largely due to uncertainty in how the modified gene will be expressed. Understanding variables such as the cost of rapid growth on antipredator success would prove to be pivotal in setting up sound risk assessments for GM fishes and in fully assessing the environmental risk associated with GM fish escapees.

Keywords carp; *Ctenopharyngodon idellus*; GM; growth hormone; invasive; risk assessment; transgenic

Introduction

At a time where genetically modified (GM) fishes will inevitably find their way to the dinner table (Britton & Gozlan, 2013; Ledford, 2013), the environmental implications of farming GM fishes remain. From a commercial viewpoint alone, faster growth characteristics of transgenic fishes have been a central consideration as they allow faster attainment of market size and optimization of production cost (Zhu, 1992; Devlin *et al*., 1995). From an ecological perspective, it is currently challenging to understand the effects of such GM fishes

文章发表于 *Journal of Fish Biology*, 2014, 84(5): 1527-1538

on wild fish populations and overall on aquatic ecosystem function. There have been many studies on a range of species that show the potential transfer of genes from engineered plants and animals to their wild counterparts but there is still a lack of information about the ecological interactions of GM organisms released into native aquatic communities.

Most GM fish aquaculture research has focused on salmonid species owing to the extent of this aquaculture production in countries that possess the transgenic technology (Breton & Uzbekova, 2000; Takeuchi *et al.*, 2001; Devlin *et al.*, 2004; Mori *et al.*, 2007). With the emergence of other economic partners such as China into the worldstage, a whole newmarket for GM fishes has opened up with the use of cyprinid species (Zhu *et al.*, 1985). With a current annual freshwater fish production of *c.* 54 Mt (Zhao *et al.*, 2011), China is becoming the most significant contributor to world fisheries (*i.e.* 17% of the global capture) and aquaculture (*i.e.* 61.4% of global aquaculture production). Recent work has focused on GM common carp *Cyprinus carpio* L. 1758 as it constitutes one of the most farmed species in Asia and eastern Europe and has shown the growth potential of GM *C. carpio* over wild specimens (Guan *et al.*, 2008; Duan *et al.*, 2009), with, under hatchery conditions, *c.* 1.9 times the growth of their wild counterparts (Wang *et al.*, 2001; Li *et al.*, 2007).

At the same time recent studies have shown that a large number of invasive non-native fish species are cyprinid species such as topmouth gudgeon *Pseudorasbora parva* (Temminck & Schlegel 1846), nasus *Chondrostoma nasus* (L. 1758) or *C. carpio* itself (Gozlan, 2008; Gozlan *et al.*, 2009, 2010). The reason for this is likely to befound in the manner that cyprinid and in particular *C. carpio* farming is performed. Because of the relatively low market value of *C. carpio* production when compared, for example, with salmonid species, facilities are far less biosecure often consisting of earthen ponds on floodplains or net cages in lakes, thus limiting infrastructurecosts. These types of farming infrastructures are not considered as closed systems and the introduction of GM *C. carpio* into these environments would inevitably lead to escapees into natural open systems such as lakes or rivers (Maclean & Laight, 2000).

As most GM fish research focuses on enhancing individual growth, it is vital to characterize the existence of behavioural and ecological trade-offs associated with such rapid growth (Lima & Dill, 1990). Several studies have revealed a direct increase drisk of predation in fishes showing naturally faster growth rate in wild populations (Gotthard, 2000; Lankford *et al.*, 2001; Biro *et al.*, 2006). Given that rapid growth in fishes is linked to higher growth-hormone levels that stimulate appetite, a difference in boldness from fast-growing individuals when compared with slow-growing ones has been measured as a behavioural syndrome driving individuals to maintain feeding activity even under increasing predation risk (Lima & Dill, 1990; Biro *et al.*, 2006; Ioannou *et al.*, 2008; Nyqvist *et al.*, 2013). Thus, there is a natural trade-off against selection of fast-growing individuals (Arendt, 1997; Sundström *et al.*, 2004a, b), showing that rapid growth cannot be used as a surrogate for fitness as is often suggested (Stearns & Koella, 1986; Perrin & Rubin, 1990; Stearns, 1992).

Optimal for aging and game theories, which are thought to provide a reliable basis for prediction, are based on the assumption that individuals within animal populations always behave to maximize their own chances of survival and reproduction (*i.e.* maximizing their fitness), nomatter how much the environment changes (Sutherland, 1996). Individuals that grow faster should reap the benefits of large size earlier in life and experience higher survival and thus fitness (Roff, 1992; Stearns, 1992; Sogard, 1997). In fishes, there are numerous well-known benefits of rapid growth, including enhanced resistance to starvation during the winter, lower susceptibility to predators and higher competitive ability (Arendt, 1997; Devlin *et al.*, 1999). If faster somatic growth was truly better, natural selection would be expected to have driven growth rates to the maximum possible allowed by physiological or phylogenetic constraints (Stephen & Arnott, 2005), which is not the case as many animals, including fishes (Ali *et al.*, 2003), display growth rates in nature that are kept below the physiological maximum (Arendt & Wilson, 1997, 1999; Gotthard, 2000). This indicates that the fitness advantage of rapid growth is naturally balanced against other life-history trait costs (Arendt, 1997) such as decrease descape performance (Sibly *et al.*, 1985; Lankford *et al.*, 2001; Munch & Conover, 2003), less vigilance and more willingness to risk predation exposure when foraging (Johnsson *et al.*, 1996; Abrahams & Pratt, 2000; Tymchuk & Devlin, 2005a, b; Nyqvist *et al.*, 2012) and greater predation mortality (Biro *et al.*, 2004).

Until now, the oreticians have overlooked the intrinsic underlying physiological cost of rapid growth (Stephen & Arnott, 2005) as most studies have only looked at the increased risk of predation incurred by individuals when foraging (Lima & Dill, 1990). Here, the intrinsic effect of feeding status on survival and predation-induced mortality was experimentally tested in GM *C. carpio* and wild specimens. Understanding variables such as the cost of rapid growth on antipredator success would prove to be pivotal in setting up sound risk assessments (RA) for GM fishes (Devlin *et al.*, 2006) and in fully assessing the environmental risk associated with GM fish escapees.

Materials and Methods

Source of *C. carpio*

The all-fish gene of growth hormone transgenic carp construct *pCAgcGH* contains the grass carp *Ctenopharyngodon idellus* (Valenciennes 1844) GH gene (*gcGH*) driven by the *β*-actin gene promoter in *C. carpio*. P_0, F_1, F_2, F_3, F_4 and F_5 all-fish GH-transgenic *C. carpio* were produced as reported by Guan *et al.* (2008). In May 2010 the F_5 GH-transgenic *C. carpio* (T) were produced by cross-fertilization between F_4 transgenic males and normal non-transgenic *C. carpio* (NT) wild-type females captured in the flood plain of the Yangtze River. Non-transgenic *C. carpio* were produced by cross-fertilization of males and females in F_4 sibling fishes to minimize the effects of genetic differences and maternal and paternal effects. Both

NT and T groups of *C. carpio* were reared in the contained ponds in Wuhan Duofu High-Tech Company Ltd. At the start of July, at least 1 month before the start of the experimental trials, both groups of *C. carpio* were transferred to an indoor water-recycling fish-rearing facility and reared in circular fibreglass tanks (diameter 150 cm and volume 1000 l). *Cyprinus carpio* were acclimated under a 12L︰12D photoperiod regime and were fed twice daily at 0900 and 1600 hours with frozen *Chironomus* sp. larvae. Water temperature ranged between 27.9 and 31.7℃, dissolved oxygen was 6.75-7.35 mg l^{-1} and pH ranged between 8.27 and 8.45.

In addition, the Mandarin fish *Siniperca chuatsi* (Basilewsky 1855) species was selected as anatural predator of *C. carpio*. It is a demersal piscivorous predator, feeding exclusively on live fishes and widely distributed in many rivers and lakes in China (Liang *et al.*, 1998). All predators were captured from the wild in the Yangtze floodplain to maximize natural predatory behaviorand taken back to the indoor facility at the same time in July and reared in the same fiberglass test tanks as the *C. carpio*. *Siniperca chuatsi* were given at least 1 month before the start of the experiment to acclimate to the experimental environment and were fed once a day during that period with live juvenile *C. carpio*.

Experimental protocol

A paired-contrast design was performed to test the effect on population of genotype (T *v.* NT), and food consumption (fed *v.* unfed) on vulnerability to predation. Thus, 24 h before any treatment all test individuals (predator and prey) were deprived of food. Thirty minutes before some trials, groups of *C. carpio* were allowed an unrestricted meal and thus labelled as fed *C. carpio* in these respective trials. Each trial consisted of a set of three predators of similar sizesper tank [mean±S.E. mass=283±33.57 g; mean±S.E. total length (L_T)=26.41±0.90 cm]and one of the four groups of prey. The four groups of tested *C. carpio* were set as follows: (1) NT fed (n=30) and T fed (n=30), (2) NT unfed (n=30) and T unfed (n=30), (3) NT unfed (n=30) and NT fed (n=30) and (4) T unfed (n=30) and T fed (n=30). Each trial was replicated at least four times and a maximum of six times. All *C. carpio* used in the trials were selected to be of similar mass and L_T (NT *C. carpio* mean±S.E. L_T =4.99±0.35 cm; mean±S.E. mass 2.02±0.35 g and T *C. carpio* mean±S.E. L_T =4.82±0.42 cm; mean±S.E.mass 1.81±0.43 g). Schools of *C. carpio* were introduced into the net cage within each trial arena and allowed 15 min acclimation. Net cages were then removed and *S. chuatsi* were allowed to prey.

All fish were observed and recorded visually during each trial and the experiment was stopped when *c.* 50% of *C. carpio* were captured. The time of each trial was ranged between 1.5 and 2 h. For each trial, each specific group was marked by fin clipping of either right or left ventral fins. Ventral fin side for each marked group was randomly selected between replicated trials to avoid confusion. All *C. carpio* were anaesthetized with Eugenol at the concentration of 20 mg l^{-1} before marking (Velisek *et al.*, 2005), weighing and measuring. Fin clipping was performed as it generates extremely limited pain when compared with external tagging in the fish muscle and it is particularly suitable to the small size of *C. carpio*.

In addition, clipped fins re-grow fairly rapidly and the effect of clipping the ventral fin on swimming ability is limited as ventral fins in *C. carpio* are not used for propulsion and escape behaviour. No other source of mortality apart from predation was noticed during the whole period of experimentation. All ethical considerations for the welfare of *C. carpio* during the regulated procedure were taken into account according to the institutional ethical regulations in place. In each experiment, a cut-off point of 50% mortality was used to minimize *C. carpio* suffering; a lower cut-off point would have represented an arbitrary end point rather than true predator preference. GM *C. carpio* have a strong desire to feed so they experience an increased risk of predation during foraging. A bare environment was used in each experiment to represent the natural situation in most Chinese lakes that, as a result of extensive fisheries exploitation, are fairly shallow, flat and characterized by the absence of macrophytes or overhanging vegetation on the banks. Thus, the lack of shelter was environmentally realistic and ensured that the experiments tested susceptibility to predation of *C. carpio*, not the antipredator ability (Munch & Conover, 2003).

Statistical analyses

The relative vulnerability of *C. carpio* from contrasting treatments was estimated using frequencies of captured and non-captured individuals during the various trials. The frequency of captured individuals into mortality estimates was then translated. Mortalities from contrasting treatments were then analysed as 2×2 contingency tables (χ^2-test in cross table) to test treatment effects. A significance level of $P \leqslant 0.05$ was used for all tests. Statistical analyses were performed using SPSS 16.0 (www-01.ibm.com/software/analytics/spss/downloads/) and graphics were made with Excel and origin75 software. The number of *C. carpio* in each trial was based on the natural density of *C. carpio* in Chinese lakes. The number of replicates (between four and six) was the minimum number of replicates to statistically characterize significant differences based on the intrinsic variability (*i.e.* noise) in the data whilst minimizing the ethical impact of the study (*i.e.* reduction).

Differential vulnerability or predator preference?

As prey selection from predators could have resulted more from a preference for fed *C. carpio* that are visibly fatter than the unfed ones, rather than from the ability of prey to avoid predation, in a separate no-choice predation trial experiment, *C. carpio* evasiveness was measured independently from predator selectivity. In this evasiveness experiment the predators wereallowed to forage exclusively on either NT or T *C. carpio* (NT mean±S.E. L_T =6.89±0.35 cm; mass=4.78±0.92 g and T mean±S.E. L_T =6.86±0.33 cm; mass=4.85±0.80 g). Trials were initiated after the paired-contrast experiment by stocking each replicate tank with three new *S. chuatsi* that were deprived of food for 24 h. A school of each type of *C. carpio* (*n*=20) was introduced into a net cage within each tank and allowed 15min to acclimate. Net cages were then removed and *S. chuatsi* were allowed to prey for 100 min. In order to minimize interference with the running of the experiment, every 5-10 min, each tank was photographed to count the number of test *C. carpio* left in each trial. Thus, survival curves were established to characterize the mortality trade-off for each *C. carpio* type.

Kaplan-Meier (K-M) survival probabilities were used to construct survivorship functions for each *C. carpio* treatment. Statistical analyses were performed with the SURVIVAL programme (Steinberg & Colla, 1988; Li *et al.*, 2008). Non-parametric Mantel-Cox log-ranktests were then used to test whether survivorship curves obtained by the K-M method differed between the four groups of carp (Li *et al.*, 2007). Median lethal time (t_{L50}) and maximum-likelihood estimates of survival probability were calculated for each treatment using the non-parametric K-M product-limit method (Kaplan & Meier, 1958). *Z*-tests were used to test whether significant differences existed in t_{L50} of each treatment *C. carpio*: $z=[S_1(t)-S_2(t)]\left(\sqrt{\{x^2[S_1(t)]+x^2[S_2(t)]\}}\right)^{-1}$, where x=S.E. and z is the value that could define whether there is a significant difference as the function of *P*-value. There is a significant difference when $z < 0.05$, whereas there is no significant difference when $z \geqslant 0.05$; $S_i(t)$ is the estimated time of the t_{L50} of the test *C. carpio* in group i through the SURVIVAL analysis. Statistical analyses were performed again using SPSS 16.0 and graphics were made with Exceland origin 75 software.

Results

Predation mortalities

Size-matched groups of T and NT *C. carpio* genotypes differed significantly in their ability to avoid predators under both fed and unfed conditions, with fast-growing T genotypes being highly vulnerable to predation relative to slow-growing NT genotypes.In the trial with the unfed *C. carpio*, mortality rate of T *C. carpio* was significantly higher than that of the NT *C. carpio* [χ^2 =5.842, d.f. = 1, P<0.05; Fig. 1(a)]. Similar results were found in trials with fed T and NT *C. carpio*, where T *C. carpio* were more likely to be captured by predators than the NT *C. carpio* [χ^2 =7.065, d.f. = 1, P<0.01; Fig. 1(b)]. T and NT *C. carpio* that consumed meals suffered significantly higher predation mortality than their unfed counterparts. In T *C. carpio*, fed individuals were more likely to be captured by the predators than their unfed T *C. carpio* [χ^2 =6.559, d.f. = 1, P<0.05; Fig. 1(c)]; feeding also significantly reduced escape performance in NT *C. carpio* attacked by predators [χ^2 =5.376, d.f. = 1, P<0.05; Fig. 1(d)].

Differential vulnerability or predator preference?

At the population level, unfed T *C. carpio* were captured faster by predators than unfed NT *C. carpio* [χ^2 =5.281, d.f. = 1, P<0.05; Fig. 2(b)]. Similar results were found when *C. carpio* were allowed to feed before the trial [χ^2 =6.237, d.f. = 1, P<0.05; Fig. 2(a)], suggesting that T *C. carpio* are less able to escape predation than NT individuals (Fig. 2). Thus,

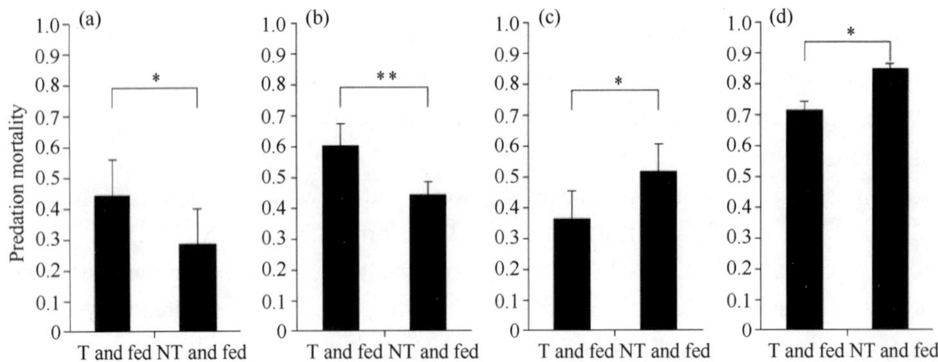

Fig. 1 Effect of *Cyprinus carpio* genotype (*i.e.* transgenic, T or natural, NT) on predation vulnerability when either unfed for 24 h before the experiment or with unlimited feeding 30 min before the experiment. Experimental set-up included three *Siniperca chuatsi* of similar sizes, and either (a), (b) mixed schools (n =60) of transgenic and non-transgenic individuals with different feeding status [(a) n = 4; (b): n = 5] or (c), (d) single genotyped schools (n = 60) also of different feeding status [(c) n = 5; (d) n = 4]. Mortalities from contrasting treatments were then analysed as 2×2 contingency tables (χ^2-test in cross table) to test treatment effects. Significant differences in mortalities between treatments were tested using a χ^2-test (*P < 0.05 and **P < 0.01). Values are mean±S.E.

survival times for fed T individuals were significantly lower than those for the unfed individuals [χ^2 =7.225, d.f. = 1, P<0.01; Fig. 2(d)] and fed NT *C. carpio* [χ^2 =19.546, d.f. = 1, P<0. 001; Fig. 2(c)]. Survival times offed and unfed *C. carpio* indicated that consumption of meals reduced evasiveness in both genotypes (Fig. 2) and unfed T *C. carpio* were captured faster by the predators than fed NT *C. carpio* even though the difference between them was not significant [χ^2 =1.821, d.f. = 1, P>0.05; Fig. 2(e)]. The t_{L50} of the four treatment groups NT fed, T fed, NT unfed and T unfed was 20, 10, 50 and 30 min, respectively (Table I), with the t_{L50} of NT fed individuals being significantly lower than that of the unfed individuals (P<0.001); the same for the T fed *C. carpio* (P<0.001). Finally, the t_{L50} of T fed *C. carpio* was also significantly lower than that of the NT fed (P<0.001) and the same for T unfed *v.* NT unfed *C. carpio* (P<0.01). Analysis of failure times indicates that selection against *C. carpio* that display genetic potential for rapid growth could beat least partially explained by lower escape potential.

Discussion

The results clearly indicate a knock-on effect of the GH gene (*gcGH*) introduction into the *C. carpio* genome on their vulnerability to predation. The experiment sunequivocally show that it is the genetic nature of the *C. carpio* rather than its size that affects the risk of predation. In addition, the satiety level of the *C. carpio* also increased the predation risk regardless of any specific genetic makeup. Thus, the study characterizes the existence of a trade-off between somatic growth and predator avoidance performance (Lankford *et al.*, 2001; Li *et al.*, 2007). Similar results were obtained in brown trout *Salmo trutta* L. 1758, in which simple injection of growth hormone influenced antipredator behaviour (Johnsson *et al.*, 1996; Jönsson *et al.*, 1998), making fast-growing *S. trutta* more risk-prone, significantly decreasing their critical swimming speeds and thus making them more vulnerable to predation (Farrell *et al.*, 1997). Here, the fed-unfed trials suggested that large meals represented a hidden additional cost of growth for the *C. carpio*. Previous research has argued that high rates of food

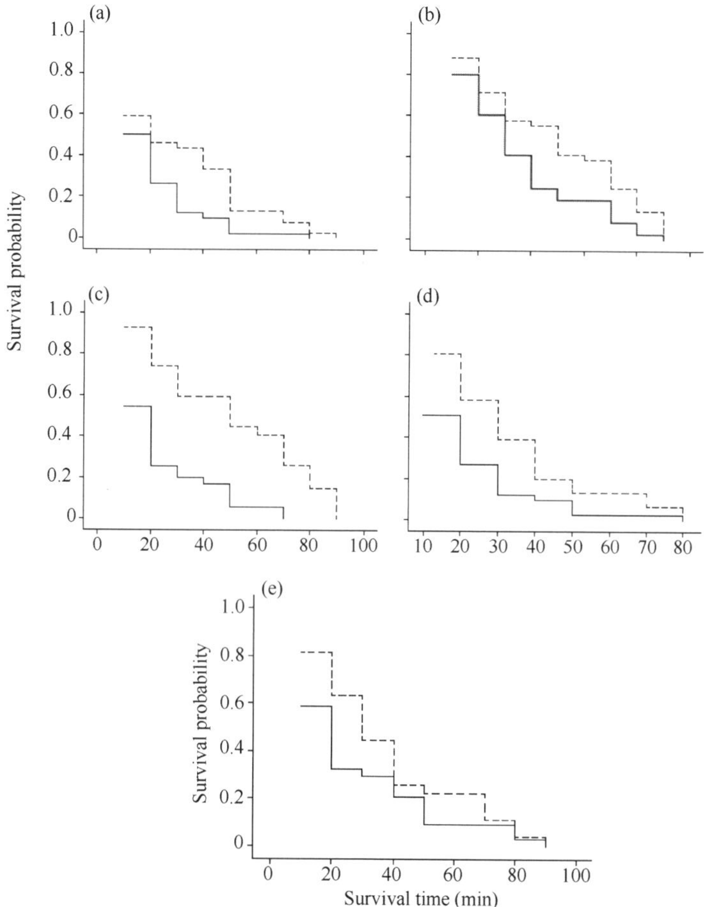

Fig. 2 Survival curves of *Cyprinus carpio* genotype (*i.e.* transgenic, T or natural, *i.e.* non-transgenic, NT) on predation vulnerability when either unfed for 24 h before the experiment or with unlimited feeding 30 min before the experiment. In each of the four experiments, three *Siniperca chuatsi* of similar sizes were allowed to prey on a monotyped school (n = 20) of either transgenic fed (n = 4 replicates) or transgenic unfed (n = 4) or non-transgenic fed (n = 6) or non-transgenic unfed (n = 5) *C. carpio*. Comparison of the survival times between (a) T and fed (____) and NT and fed (_ _ _); (b) T and unfed (____) and NT and unfed (_ _ _); (c) NT and fed (____) and NT and unfed (_ _ _); (d) T and fed (____) and T and unfed (_ _ _) and (e) T and unfed (____) and NT and fed (_ _ _)

consumption and rapid growth in fishes drive metabolism to its aerobic maximum, thereby producing a conflict with allocation of oxygen to other functions such as swimming (Priede, 1985; Wieser & Medgyesy, 1990; Wieser, 1991). Other studies on *Salmo salar* L. 1758 and *Ictalurus punctatus* (Rafinesque 1818) show that the overexpression in growth hormone-stimulated fishes has enhanced forage motivation, lowered resolution of the food selection, increased activity level and reduced clustering trends (Abrahams& Sutterlin, 1999; Dunham *et al.*, 1999; Sundström *et al.*, 2003, 2004a, b; Tymchuk & Devlin, 2005a, b).

It is key to differentiate between farmed fishes and wild populations of *C. carpio* as both environments have separate selection pressures. Most research on growth-enhanced fishes, either through the use of growth hormones or the use of a modified genome, has an economic applied endpoint, which is to produce marketable fishes at a faster pace, thus lowering production cost. In such farm environments, fishes are not subjected to predation and thus, as seen in rainbow trout, *Oncorhynchus mykiss* (Walbaum 1792), for example, growth rates in domestic strains can reach three times those observed in wild strains (Johnsson *et al.*, 1996). This differential in growth is achieved by a greater motivation to feed with elevated foraging rates and also increased aggression (Fleming *et al.*, 2002). On the contrary, in natural environments, foraging competition is an important evolutionary pressure, so is predation. Thus, in the wild, the selection for rapid growth is counter selected by individuals with better predator avoidance capabilities (Arendt, 1997; Post & Parkinson, 2001; Munch & Conover, 2003; Stoks *et al.*, 2005) and this trade-off explains the absence of individuals with maximized growth rates in nature (Werner & Gilliam, 1984; Werner & Anholt, 1993; Anholt & Werner, 1995; Gotthard, 2000; Biro *et al.*, 2003a, b, 2004, 2006; Stoks *et al.*, 2005).

Table I Mean survival time and median lethal time (t_{L50}) expressed as the survival time of 50% of a given school of *Cyprinus carpio* genotype (*i.e.* transgenic, T or natural, *i.e.* non-transgenic, NT) either unfed for 24 h before the experiment or with unlimited feeding 30 min before the experiment. The experimental set-up included three *Siniperca chuatsi* of similar size. The superscript lowercase letters in the median estimate column represent significant differences between the two values

Identification	Type	Mean			Median (t_{L50})				
		Estimate	S.E.	95% C.L.	Estimate	S.E.	95% C.L.	*Z*-value	Pair
A	NT and fed	27.40	2.98	21.56-33.25	20.0^{a}	2.92	14.26-25.73	<0.0001	A-B
B	NT and unfed	49.44	4.67	40.27-58.61	50.0^{b}	5.91	38.40-61.59	0.0059	B-C
C	T and unfed	35.83	3.85	28.28-43.37	30.0^{c}	4.22	21.71-38.28	<0.0001	C-D
D	T and fed	20.47	2.33	15.89-35.88	10.0^{d}	0.00	0.00-0.00	0.0006	D-A

In light of the results of this study, the key environmental risk lies with the potential for transgenic fishes to escape aquaculture farms and contaminate wild natural stocks. Because of the relatively low economic value of *C. carpio* compared with salmonid species, their culture facilities tend to be less biosecure and often consist of earthen ponds located on the floodplain or fish cages in reservoirs, as these limit infrastructure costs. Such farming infrastructures are not closed systems and the use of transgenic *C. carpio* in them would inevitably lead to their escape into the wild. *De facto*, it is crucial to fully understand the costs of rapid growth in a natural context where predation is a strong evolutionary pressure. Current research in Europe suggests that high uncertainty surrounding the potential environmental effects of escapee transgenic fishes into the wild is largely due to uncertainty in how the modified gene will be expressed (Britton& Gozlan, 2013). The development of regulatory guidelines in Europe that are underpinned by GM fish RAs has resulted in transgenic fishes being assessed as per any other alien species being proposed for aquaculture production. The risk posed by transgenic escapees is not solely reflected in the escapees themselves and their competition ability with wild strains but first and foremost in their ability to transfer the GH gene to wild *C. carpio*. If the transgenic *C. carpio* is considered as a non-native species only, as currently suggested by the European Food Safety Authority (Britton & Gozlan, 2013), then based on their significant increased susceptibility to predation, it could be imagined that their chances of survival in the wild are limited compared with natural strains and therefore that the associated risk would be considered moderate (Copp *et al.*, 2009). In the meantime, if transgenic *C. carpio* can hybridize with wild individuals, then this represents the GH gene pollution element that should be scrutinized as part of the environmental RA rather than only the GH gene host. Although the effect of rapid growth on predation could be life-history stage dependent (Fleming *et al.*, 2000) and the long-term effects of transgenic fishes could not only be determined by performance during their early life history but also dependent on their fitness throughout their whole life (Muir & Howard, 1999; Sundström *et al.*, 2004b), such genetic pollution in wild *C. carpio* populations would probably result in an imbalance in the natural prey-predator relationship at least during the juvenile stage. Thus, it would induce an increase of the environmental risk illustrated by a predicted decline in the stock of wild *C. carpio*.

Acknowledgements This work was financially supported by the National Natural Science Foundation of China (Grant No. 30970553) and the Development Plan of the State Key Fundamental Research of China (Grant No. 2007CB109205). We would like to thank S. Lek and S. Ye for constructive suggestions in data analysis.

References

Abrahams, M. V. & Pratt, T. C. (2000). Hormonal manipulations of growth rate and its influence on predator avoidance-foraging trade-offs. Canadian Journal of Zoology 78, 121-127.

Abrahams, M. V. & Sutterlin, A. (1999). The foraging and antipredator behaviour of growth-enhanced transgenic Atlantic salmon. Animal Behaviour 58, 933-942.

Ali, M., Nicieza, A. G. & Wootton, R. J. (2003). Compensatory growth in fishes: a response to growth depression. Fish and Fisheries 4, 147-190.

Anholt, B. R. & Werner, E. E. (1995). Interaction between food availability and predation mortality mediated by adaptive behavior. Ecology 76, 2230-2234.

Arendt, J. D. (1997). Adaptive intrinsic growth rates: an integration across taxa. Quarterly Review of Biology 72, 149-177.

Arendt, J. D. & Wilson, D. S. (1997). Optimistic growth: competition and an ontogenetic niche-shift select for rapid growth in pumpkinseed sunfish (*Lepomis gibbosus*). Evolution 51, 1946-1954.

Arendt, J.D. & Wilson, D. S. (1999). Countergradient selection for rapid growth in pumpkinseed sunfish: disentangling ecological and evolutionary effects. Ecology 80, 2793-2798.

Biro, P. A., Post, J. R. & Parkinson, E. A. (2003a). From individuals to populations: risk-taking by prey fish mediates mortality in whole-system experiments. Ecology 84, 2419-2431.

Biro, P. A., Post, J. R. & Parkinson, E. A. (2003b). Population consequences of a predator-induced habitat shift by trout in whole-lake experiments. Ecology 84, 691-700.

Biro, P. A., Abrahams, M. V., Post, J. R. & Parkinson, E. A. (2004). Predators select agains thigh growth rates and risk-taking behaviour in domestic trout populations. Proceedings of the Royal Society B 271, 233-237.

Biro, P. A., Abrahams, M. V. & Post, J. R. (2006). Behavioural trade-offs between growth and mortality explain evolution of submaximal growth rates. Journal of Animal Ecology 75, 1165-1171.

Breton, B. & Uzbekova, S. (2000). Evaluation des risques biologiques lies a la disseminationde poissons genetiquement modifies dans les milieux naturels. Comptes Rendus del'Academie d'Agriculture de France 86, 67-76.

Britton, R. E. & Gozlan, R. E. (2013). Transgenic fish: European concerns over GM salmon. Nature 498, 171.

Copp, G. H., Vilizzi, L., Mumford, J., Godard, M. J., Fenwick, G. & Gozlan, R. E. (2009). Calibration of FISK, an invasiveness screening tool for non-native freshwater fishes. Risk Analysis 29, 457-467.

Devlin, R. H., Yesaki, T. Y., Donaldson, E. M., Du, S. J. & Hew, C. L. (1995). Production of germline transgenic Pacific salmonids with dramatically increased growth-performance. Canadian Journal of Fisheries and Aquatic Sciences 52, 1376-1384.

Devlin, R. H., Johnsson, J. I., Smailus, D. E., Biagi, C. A., Jonsson, E. & Björnsson, B. T. (1999). Increased ability to compete for food by growth hormone-transgenic coho salmon *Oncorhynchus kisutch* (Walbaum). Aquaculture Research 30, 479-482.

Devlin, R. H., Biagi, C. A. & Yesaki, T. Y. (2004). Growth, viability and genetic characteristics of GH transgenic coho salmon strains. Aquaculture 236, 607-632.

Devlin, R. H., Sundström, L. F. & Muir, W. M. (2006). Interface of biotechnology and ecology for environmental risk assessments of transgenic fish. Trends in Biotechnology 24, 89-97.

Duan, M., Zhang, T., Hu, W., Sundström, L. F., Wang, Y., Li, Z. & Zhu, Z. (2009). Elevatedability to compete for limited food resources by "all-fish" growth hormone transgenic common carp *Cyprinus carpio*. Journal of Fish Biology 75, 1459-1472.

Dunham, R. A., Chitmanat, C., Nichols, A., Argue, B., Powers, D. A. & Chen, T. T. (1999). Predator avoidance of transgenic channel catfish containing salmonid growth hormone genes. Marine Biotechnology 1, 545-551.

Farrell, A. P., Bennett, W. & Devlin, R. H. (1997). Growth-enhanced transgenic salmon can be inferior swimmers. Canadian Journal of Zoology 75, 335-337.

Fleming, I. A., Hindar, K., Mjoelneroed, I. B., Jonsson, B., Balstad, T. & Lamberg, A. (2000). Lifetime success and interactions of farm salmon invading à native population. Proceedings of the Royal Society B 267, 1517-1523.

Fleming, I. A., Agustsson, T., Finstad, B., Johnsson, J. I. & Björnsson, B. T. (2002). Effects of domestication on growth physiology and endocrinology of Atlantic salmon (*Salmo salar*). Canadian Journal of Fisheries and Aquatic Sciences 59, 1323-1330.

Gotthard, K. (2000). Increased risk of predation as a cost of high growth rate: an experimental test in the speckled wood butterfly, Pararge aegeria. Journal of Animal Ecology 69, 896-902.

Gozlan, R. E. (2008). Introduction of non-native freshwater fish: is it all bad? Fish and Fisheries 9, 106-115.

Gozlan, R. E., Britton, J. R., Cowx, I. & Copp, G. H. (2009). Current knowledge on non-native freshwater introductions. Journal of Fish Biology 76, 751-786.

Gozlan, R. E., Andreou, D., Asaeda, T., Beyer, K., Bouhadad, R., Burnard, D., Caiola, N., Cakic, P., Djikanovic, V., Esmaeili, H. R., Falka, I., Golicher, D., Harka, A., Jeney, G., Kováč,V., Musil, J., Nocita, A., Povz, M., Poulet, N., Virbickas, T., Wolter, C., Tarkan, A. S., Tricarico, E., Trichkova, T., Verreycken, H., Witkowski, A., Zhang, C.-G., Zweimueller, I. & Britton, J. R. (2010). Pan-continental invasion of Pseudorasbora parva: towards a better understanding of freshwater fish invasions. Fish and Fisheries 11, 315-340.

Guan, B., Hu, W. & Zhang, T. L. (2008). Metabolism traits of "all-fish" growth hormone transgenic common carp (*Cyprinus carpio* L.). Aquaculture 284, 217-223.

Ioannou, C. C., Payne, M. & Krause, J. (2008). Ecological consequences of the bold-shy continuum: the effect of predator boldness on prey risk. Oecologia 157, 177-182.

Johnsson, J. I., Petersson, E., Jönesson, E., Björnsson, B. T. & Jàrvi, T. (1996). Domestication and growth hormone alter antipredator behaviour and growth patterns in juvenile brown trout, *Salmo trutta*. Canadian Journal of Fisheries and Aquatic Sciences 53, 1546-1554.

Jönsson, E., Johnsson, J. I. & Björnsson, B. T. (1998). Growth hormone increases aggressive behavior in juvenile rainbow trout. Hormones and Behaviour 33, 9-15.

Kaplan, E. L. & Meier, P. (1958). Nonparametric estimation from incomplete observations. Journal of the American Statistical Association 53, 457-481.

Lankford, T. E., Billerbeck, J. M. & Conover, D. O. (2001). Evolution of intrinsic growth andenergy acquisition rates. II. Tradeoffs with vulnerability to predation in Menidia menidia. Evolution 55, 1873-1881.

Ledford, H. (2013). Transgenic salmon nears approval. Nature 497, 17-18.

Li, D., Fu, C., Hu, W., Zhong, S., Wang, Y. & Zhu, Z. (2007). Rapid growth cost in "all-fish" growth hormone gene transgenic carp: reduced critical swimming speed. Chinese Science Bulletin 52, 1501-1506.

Li, C. X., Jiang, L. N., Shao, Y. & Wang, W. L. (2008). Biostatistics, 4th edn. Henan: Science Press.

Liang, X., Liu, J. & Huang, B. (1998). The role of sense organs in the feeding behaviour of Chinese perch. Journal of Fish Biology 52, 1058-1067.

Lima, S. & Dill, L. M. (1990). Behavioral decisions made under the risk of predation: a review and prospectus. Canadian Journal of Zoology 68, 619-640.

Maclean, N. & Laight, R. J. (2000). Transgenic fish: an evaluation of benefits and risk. Fish and Fisheries 1, 46-72.

Mori, T., Hiraka, I., Kurata, Y., Kawachi, H., Mano, N., Devlin, R. H., Nagoya, H. & Araki, K.(2007). Changes in hepatic gene expression related to innate immunity, growth and iron metabolism in GH-transgenic amago salmon (*Oncorhynchus masou*) by cDNA subtraction and microarray analysis, and serum lysozyme activity. General and Comparative Endocrinology 151, 42-54.

Muir, W. M. & Howard, R. D. (1999). Possible ecological risks of transgenic organism release when transgenes affect mating success: sexual selection and the Trojan gene hypothesis. Proceedings of the National Academy of Sciences 96, 13853-13856.

Munch, S. B. & Conover, D. O. (2003). Rapid growth results in increased susceptibility to predation in Menidia menidia. Evolution 57, 2119-2127.

Nyqvist, M. J., Gozlan, R. E., Cucherousset, J. & Britton, J. R. (2012). Boldness syndrome in a solitary predator is independent of body size and growth rate. PLoS One 7, e31619.

Nyqvist, M. J., Gozlan, R. E., Cucherousset, J. & Britton, J. R. (2013).Absence of acontext-general behavioural syndrome in a solitary predator. Ethology 119, 156-166.

Perrin, N. & Rubin, J. F. (1990). On dome-shaped norms of reaction for size-to-age at maturity in fish. Functional Ecology 4, 53-57.

Post, J. R. & Parkinson, E. A. (2001). Energy allocation in young fish: allometry and survival. Ecology 82, 1040-1051.

Priede, I. G. (1985). Metabolic scope in fishes. In Fish Energetics: New Perspectives (Tytler, P. & Calow, P., eds), pp. 33-64. London: Croom Helm.

Roff, D. A. (1992). The Evolution of Life Histories: Theory and Analysis. NewYork, NY: Chapman& Hall.

Sibly, R. M., Calow, P. & Nichols, N. (1985). Are patterns of growth adaptive? Journal of Theoretical Biology 112, 553-574.

Sogard, S. M. (1997). Size-selective mortality in the juvenile stage of teleost fishes: a review. Bulletin of Marine Science 60, 1129-1157.

Stearns, S. C. (1992). The Evolution of Life Histories. New York, NY: Oxford University Press.

Stearns, S. C. & Koella, J. C. (1986). The evolution of phenotypic plasticity in life-history traits: predictions of reaction norms for age and size at maturity. Evolution 40, 893-913.

Steinberg, D. & Colla, P. (1988). SURVIVAL: A Supplementary Module for SYSTAT. Evanston, IL: SYSTAT, Inc.

Stephen, D. W. & Arnott, R. (2005). Evolution of intrinsic growth rate: metabolic costs drivetrade-offs between growth and swimming performance in Menidia menidia. Evolution 60, 1269-1278.

Stoks, R., De Block, M., Van De Meutter, F. & Johansson, F. (2005). Predation cost of rapidgrowth: behavioural coupling and physiological decoupling. Journal of Animal Ecology 74, 708-715.

Sundström, L. F., Devlin, R. H., Johnsson, J. I. & Biagi, C. A. (2003). Vertical position reflects increased feeding motivation in growth hormone transgenic coho salmon (*Oncorhynchus kisutch*). Ethology 109, 701-712.

Sundström, L. F., Lõhmus, M., Johnsson, J. I. & Devlin, R. H. (2004a). Growth hormone transgenic salmon pay for growth potential with increased predation mortality. Proceedings of the Royal Society B 271, 350-352.

Sundström, L. F., Lõhmus, M., Johnsson, J. I. & Devlin, R. H. (2004b). Feeding on profitable and unprofitable prey: comparing behaviour of growth-enhanced transgenic and normal coho salmon (*Oncorhynchus kisutch*). Ethology 110, 381-396.

Sutherland, W. J. (1996). From Individual Behaviour to Population Ecology. Oxford: Oxford University Press.

Takeuchi, Y., Yoshizaki, G. & Takeuchi, T. (2001). Production of germ-line chimeras in rainbow trout by blastomere transplantation. Molecular Reproduction and Development 59, 380-389.

Tymchuk, W. E. & Devlin, R. H. (2005a). Growth differences among first and second generation hybrids of domestical and wild rainbow trout (*Oncorhynchus mykiss*). Aquaculture 245, 295-300.

Tymchuk, W. E. & Devlin, R. H. (2005b). Competitive ability and mortality of growth-enhanced transgenic Coho salmon fry and parr when foraging for food. Transactions of the American Fisheries Society 134, 381-389.

Velisek, J., Svobodova, Z., Piackova, V., Groch, L. & Nepejchalova, L. (2005). Effects of clove oil anaesthesia on common carp (*Cyprinus carpio L.*). Veterinární Medicína 50, 269-275.

Wang, Y., Hu, W., Wu, G., Sun, Y., Chen, S., Zhang, F., Zhu, Z., Feng, J. & Zhang, X. (2001). Genetic analysis of all-fish growth hormone gene transferred carp (*Cyprinus carpio* L.) and its F_1 generation. Chinese Science Bulletin 46, 1174-1177.

Werner, E. E. & Anholt, B. R. (1993). Ecological consequences of the trade-off between growth and mortality rates mediated by foraging activity. The American Naturalist 142, 242-272.

Werner, E. E. & Gilliam, J. F. (1984). The ontogenetic niche and species interactions in size structured populations. Annual Review of Ecology and Systematics 15, 393-425.

Wieser, W. (1991). Limitations of energy acquisition and energy use in small poikilotherms: evolutionary implications. Functional Ecology 5, 234-240.

Wieser, W. & Medgyesy, N. (1990). Aerobic maximum for growth in the larvae and juveniles of a cyprinid fish, Rutilus rutilus (L.): implications for energy budgeting in small poikilotherms. Functional Ecology 4, 233-242.

Zhao, Y., Gozlan, R. E. & Zhang, C. (2011). Out of sight out of mind: Current knowledge of Chinese cave fishes. Journal of Fish Biology 79, 1545-1562.

Zhu, Z. (1992). Generation of fast growing transgenic fish: methods and mechanisms. In Transgenic Fish (Hew, C. L. & Fletcher, G. L., eds), pp. 92-119. Singapore: World Scientific Publishing Co. Pte. Ltd..

Zhu, Z., Li, G., He, L. & Chen, S. (1985). Novel gene transfer into the fertilized eggs of goldfish (*Carassius auratus* L. 1758). Zeitschrift für Angewandte Ichthyologie 1, 31-34.

转基因鲤的快速生长增强了其自身被捕食的风险：环境风险的关联性

张丽红[1,2] R. E. Gozlan[3] 李钟杰[1] 刘家寿[1] 张堂林[1] 胡 炜[1] 朱作言[1]

1中国科学院水生生物研究所，淡水生态与生物技术国家重点实验室，武汉 430072，中国
2中国科学院大学，北京 100049，中国
3 Unité Mixte de Recherche Biologie des Organismes et écosystèmes Aquatiques (IRD207, CNRS 7208, MNHN, UPMC), Muséum National d'Histoire Naturelle, 75231, Paris Cedex, France

摘 要 本研究测试了喂养方式对转基因鲤和对照鲤在生存，以及捕食诱发死亡风险的影响。结果表明，转植生长激素基因导致转基因鲤具有更高的被捕食风险，影响转基因鱼被捕食的风险是其遗传特性而不是它的体型大小，并且喂养的转基因鱼更易被捕食，从而揭示了在生长和逃避捕食间存在着利弊关系。欧洲目前的研究表明，逃逸到自然环境中的转基因鱼对环境的潜在影响具有不确定性，这种不确定性是由转植基因表达的不确定性引起的。揭示快速生长和逃避捕食之间的利弊关系等因素，对于全面客观评估转基因鱼类逃逸后与环境风险的关系至关重要。

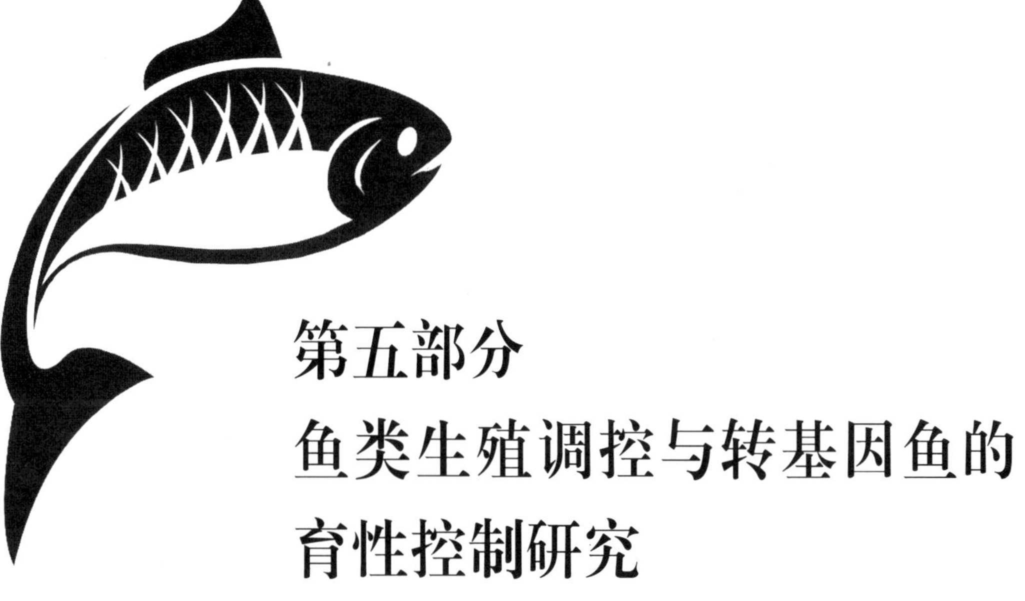

第五部分
鱼类生殖调控与转基因鱼的育性控制研究

A Perspective on Fish Gonad Manipulation for Biotechnical Applications

HU Wei WANG Yaping ZHU Zuoyan

State Key Laboratory of Freshwater Ecology and Biotechnology, Institute of Hydrobiology, the Chinese Academy of Sciences, Wuhan 430072

Abstract The gonad is an essential organ for generating sperm and ova in vertebrates. This review describes several pilot studies on gonad gene manipulation and development in fish. With antisense RNA techniques, we suppressed the gonad development, and thus the fertility, of an antisense gonadotropin-releasing hormone (sGnRH) transgenic common carp. Then, using a tissue-specific exogenous gene excision strategy with sexual compensation, we knocked out the gonad-specific transgene. Under the control of the rainbow trout protamine promoter, the transgenic fish expressed the reporter gene eGFP specifically in the spermary. These results indicate that the fish gonad is a new model organ that can improve contemporary biotechnology experiments. Herein we discuss the potential of fish gonad manipulation for resolving important biosafety problems regarding transgenic fish generation and producing the new transgenic animal bioreactor.

Keywords fish; transgene; gonad manipulation

Fish make attractive candidates for conducting studies in gene expression, regulation, function, and genetic modification, since they reach sexual maturity relatively quickly, are highly fecund, utilize an external fertilization strategy, develop externally and are simple to modify genetically. Since 1984, when the first transgenic fishes were produced[1], many researchers have modified fish with genetically desirable traits, such as growth enhancement, high feed conversion, disease and resistance, and cold tolerance[2-6]. However, none has become a successful commercial transgenic product because of inherent biosafety and societal concerns. Ecosafety assessment and the related gene manipulation technology itself are the most important issues to be resolved for the industrialization potential of transgenic fish.

In addition to producing nutrient proteins for food consumption, transgenic fish may constitute a novel bioreactor for pharmaceutical drug processing. Several research groups are studying this potential[3,7-9], but are encountering some technical challenges in their initial studies. They are working toward resolving these to develop a novel strategy for bioreactor development.

Manipulation of genes expressed in the gonad could be used for regulating gamete

文章发表于《科学通报》(英文版), 2006, 51(1): 1-6

production, as well as establishing a new type of bioreactor model for generating pharmaceutical proteins. Towards this end, our research group is working on gene regulation via gonad manipulation in fish to solve the biosafety problems attributed to transgenic fish, and to create new bioreactor species option.

1. Growth-enhanced transgenic fish and biosafety concerns

In 1984, a research group at the Institute of Hydrobiology of the Chinese Academy of Sciences started studying fish gene engineering breeding. They have thus gained an integrated view on the theory and technology underlying transgenic fish production[1-10]. Consequently, great strides have been made on fast-growing fish carrying all-fish growth hormone gene. According to a middle-scale trial of all-fish CagcGH-transgenic common carp in China, the average weight and feed conversion of the transgenic common carp offspring were respectively 42% and 18.5% greater than those in the control group. Transgenic fish cultivation could produce significant economic benefits over non-transgenic wild-type breeds in aquaculture[11]. Meanwhile, current research focuses on producing fast-growing transgenic rainbow trout, Atlantic trout, and tilapia in other countries. Although the field of fish transgenics has been applied successfully to aquaculture, none of these fish has been developed into a commercial product[3]. As a genetically modified organism, the transgenic fish is considered a biohazard, which has garnered great negative attention worldwide. The specific biosafety issues include food and environmental safety (ecosafety). Some people worry that biological changes induced as a by-product of the genetic engineering process might enable transgenic fish to absorb and concentrate toxins that wild fish cannot absorb. Environmental concerns include the possibility that transgenic species could escape their holding pens and spread novel traits into the ecosystem by breeding with wild conspecifics. Currently, there is research aimed at trying to minimize the food safety concerns regarding transgenic fish production by using all-fish gene elements. The food safety test results thus far have revealed no harmful effects on growth or development of transgenic fish consumption in mice, regardless of mouse breed[12]. In fact, genetic engineering could potentially reduce certain fish-specific food safety dangers. This leaves ecosafety issues at the forefront of industrialization problems with which research must contend.

Conventional aquaculture practices have raised a number of environmental issues including pollution and the impact of transgenics on wild fish populations. Currently, the ecosafety concerns regarding the potential escape of transgenic fish include the following issues: (i) escaped transgenic fish could interbreed and compete in the wild with native fish, that in turn may lead to genetic introgression of the transgene into the wild gene pool, and (ii) dominance of the transgenic species, due to superior traits. This may lead to changes in the structuring of fish communities, biological diversity and ecosystem balance.

Thus far there is no standard criteria for evaluating the potential influence or risk of transgenic fishes to the ecosystem. Some researchers have attempted to evaluate the potential biosafety problems by establishing various mathematical models. The Trojan gene hypothesis, raised by Muir and Howard is among the most favored[13-15]. They stated that male transgenic fish are substantially larger than wild type individuals, due to the expression of an exogenous growth hormone gene. If these transgenic fish escaped, or were released, and wild females selected the male transgenic fish preferentially on the basis of size, then the transgene would be sexually selected for and propagated as well. Thus introduction of transgenic fish into wild populations might lead to extinction of wild fish groups[14,16]. However, there are assumptions intrinsic to this hypothesis that may be seriously flawed. Namely, large male fishes do not always have a mating advantage[17]. Moreover, research on the transgenic carp indicates that transgenic males were not the preferred sperm donors (unpublished data in our laboratory). Maclean and coworkers challenged Muir's hypothesis, stating that the computer simulation, based on preliminary data from medaka, should not be used to make blanket predictions about the effects of escape by growth-enhanced transgenic fish of more commercially important species[3].

The transgenic fish biosafety issues are closely related to their breeding characteristics. Producing sterile transgenic fishes can prevent introgression of the transgene into the wild gene pool, thus minimizing the potential risk to the wild fish groups and the ecosystem. There are two general ways that gene technology can induce sterility in fish. One method is to produce sterile transgenic fish with the polyploidy manipulation, and the other method is to produce transgenic fish with reversible fertility by the antisense-gonadotropin-releasing hormone (GnRH) strategy.

Triploid fish are frequently sterile. Thus, generating triploid fish is generally considered the preferred method by aquaculturists for achieving transgenic fish containment. Triploid fish are produced most commonly by heat-or pressure-shocking the eggs after fertilization, which prevents the expulsion of the second polar body in the fertilized egg. However, not all triploid fish are completely sterile. The success rates vary between 10 and 95% triploidy, depending upon the species, shock conditions and egg batch quality. Furthermore, the tripoloid sterile transgenic fish usually have a stunted growth rate[3,18,19].

In collaboration with Prof. LIU and coworkers at the Hunan Normal University, our research group successfully produced sterile triploid transgenic fishes through diploid-tetraploid crosses. These sterile fishes maintain the beneficial traits of fast growth and high feed conversion. However, we'll have to point out that tetraploid fishes are very difficult to get; and the Liu *et al.*'s research group is the first one to report the fertile tetraploid crucian-carp in China and even the world-wide[20-21]. The triploid strategy is not suitable for all species of transgenic fish. New triploid-producing techniques will be required to make this process reproducible.

2. Antisense transgenic strategy to control the gonad development

The antisense strategy for producing transgenic fish by manipulating gonad development consists of two parts: (i) to suppress the expression of some critical genes that are related to gonad development and sexual maturity, and inhibiting gonad development, and (ii) reversing the sterility of transgenic fish by administering exogenous hormone to produce brood stock. Thus, the antisense transgenic strategy could be used to generate reversible sterile transgenic fish.

The GnRH, a member of the conservative neuro-decapeptide family linking brain and reproduction system, plays an important role in controlling and maintaining gonad development and reproductive function by stimulating the synthesis and release of gonadotropin (GtH) in the pituitary gland[22]. In mammals, inhibiting or depleting GnRH leads to a loss of reproductive function or a failure of gonad development. A typical example is the hypogonadal (*hpg*) mouse, in which deleting the GnRH gene abolishes its ability to synthesize gonadotropins. The *hpg* mouse has an infantile reproductive system, that causes complete sterility. However, when hypothalamic tissue from the fetal mice is transplanted, its reproductive functions are restored[23-25].

Antisense transgenic technology is based on antisense RNA technology. Briefly, the antisense DNA segment, which is inserted downstream of the promoter, transcribes into antisense RNA, then hybridizes to the target sequence and inhibits the target gene expression after being digested by the endoribonuclease. The antisense transgenic technology has achieved commercial success in transgenic plants[26]. Therefore, this strategy may inhibit gonadal development of transgenic fishes and interrupt their breeding capability by suppressing the expression of GnRH.

In 2001, Uzbekova and coworkers described a recombinant vector that they constructed which contained antisense DNA complementary to Atlantic salmon (*Salmo salar*) sGnRH cDNA, driven by the sGnRH Pab promoter. Subsequently, they tried to manipulate the gonad development of the transgenic fish by inhibiting the endogenous sGnRH expression, but were unsuccessfulin their efforts[27].

We found that the common carp GnRH precursors are encoded by multiple genes and cloned two type genes of GnRH (named salmon GnRH, sGnRH and chicken type-II GnRH, cGnRH-II, respectively) as well as their cDNA[28,29]. Then, according to the expression pattern of GnRH genes in dissected brain areas, pituitary and gonad of juvenile, mature and regressed common carp, we hypothesized that the sGnRH gene plays a crucial role in the hypothalamus-pituitary-gonad axis for regulating gonad development and sexual maturity[28-29]. Using sGnRH as the target gene, we prepared a recombinant construct (pCAsGnRHpc-antisense) for microinjection that contains the antisense sGnRH DNA fragment targeting the sGnRH mRNAs and the carp β-actin gene promoter. We transferred antisense-GnRH mRNA into the common carp fertilized eggs, and determined that roughly 30% of the founders did not develop gonads. This study indicates that inhibiting GnRH expression in specific brain areas can cause sterility by inhibiting gonad development. Moreover, we determined that the fertility could be restored by exogenous hormone administration. Ours represents a novel transgenic fish line that is fertile physiologically but sterile genetically. This combination addresses directly, and could resolve, the biosafety issues surrounding transgenic fish generation.

3. Gonad-specific exogenous gene excision via sexual compensation

Gonad-specific exogenous gene excision via sexual compensation is a technology to delete the foreign gene from the germ cell line of the transgenic fishes, to ensure that their offspring do not inherit the transgene. This technology consists of a recombinant construction containing two core elements (a T7 RNA Polymerase/Promoter system and Cre/*loxP* site-specific recombinant system) and a spermary-specific promoter (protamine promoter).

The T7 RNA Polymerase is a monochain enzyme encoded by the T7 phage gene1 that recognizes the T7 promoter to transcribe mRNA in a high efficiency[30]. The T7 phage possesses 17 promoters, among which Φ10 is the most efficient[31]. In 1985 and 1986, Tabor and coworkers created an efficient protokaryotic T7 RNA polymerase/promoter system[32]. This technology was later applied to eukaryotes as an inducible promoter system[33,34].

Mediation of site-specific DNA recombination by the Cre/*loxP* system was originally described in the bacteriophage P1. Two components are involved: (i) a 34-bp DNA sequence containing two 13-bp inverted repeats and an asymmetric 8-bp spacer region (5′-ATAAC-TTCGTATAATGTATGCTATACGAAGTTAT-3′) termed *loxP* (locus of X-over in P1) that targets recombination, and (ii) a 343 amino acid monomeric protein termed Cre recombinase that mediates the recombination event. Any DNA sequence flanked by two *loxP* sites will either be excised (*loxP* sites in same orientation) or inverted (*loxP* sites in opposite orientation) in the presence of the Cre recombinase[35].

By inducing Cre recombinase expression, or by injecting Cre recombinase directly, the transgene flanked by two *loxP* sites in the same orientation is excised. The Cre/*loxP* system was firstly untilized to excise transgenes from plants. Briefly, the Cre DNA sequence, driven by a tissue-specific inducible promoter, and the other foreign genes are flanked by two *loxP* sites in the same orientation, so that gene deletion in specific tissues will occur if chemically induced[36-38]. Foreign gene excision eliminates the biosafety issue of propagating transgenic genes, but renders the production of a transgenic line impossible.

To overcome the shortcomings of the above technology using one line, we introduced the idea of gonad-specific exogenous gene deletion via sexual compensation, and successfully used it in the transgenic zebrafish. The principle is to produce two transgenic zebrafish using the T7 RNA polymerase and the Cre/*loxP* site-specific recombinant systems. One transgenic zebrafish line specifically expresses the T7 RNA polymerase mRNA under the control of protamine promoter in the spermary, and the other one expresses Cre recombinase mRNA driven by the T7 promoter. In this study, we used genes encoding red and green fluorescent proteins (RFP and GFP, respectively), flanked by two *loxP* sites in the same orientation, as our recombinant constructs.

When crossing the above two transgenic zebrafish lines, the T7 RNA polymerase is expressed specifically in the gonads and self-activates the T7 promoter to transcribe and express the Cre recombinase, which, in turn, leads to the excision of the entire foreign gene between the loxP sites in the gonad of the offspring. The transgenic animals produced by this system retain the beneficial traits of the transgene, and exhibit no biosafety problems, because the transgene is specifically excised from the gonads of the offspring.

4. Secretion of transgenic spermary-specific fish bioreactor

Researchers have been using transgenic animals, also called bioreactors, to generate economical, abundant, safe, and efficacious supplies of therapeutic proteins (human wild-type and genetically engineered) since the late 1980s. Research on animal bioreactors has enjoyed great progress and success over the past decade, and represents a sophisticated use of recombinant biology procedures. The goal is to obtain the desired amount or type of protein without harming the animal. This has been accomplished most readily by targeting the mammary glands to express the recombinant proteins in the milk. For example, the antitrypsin protein was expressed in the transgenic sheep mammary latex, at a maximal concentration of 37.5 g/L, which is over 50% of the total latex protein concentration[39]. The concentrations of the whey acidic protein (WAP) and the human protein C (HPC) in transgenic pig latex were 1 g/L[40,41]. Although the use of bioreactors for protein production has a high societal appeal, there are only a few cases that have gone to Phase II or III clinical trials. The difficulties facing mammary bioreactor research include a limited number of zygotes, internal *in vivo* fertilization and development in mammals, difficult gene transfer techniques, low transfer efficiency, and high cost. Moreover, only a few transgenic individuals actually express the transgene successfully, because the foreign genes were randomly inserted into the genome. This makes the inserted gene vulnerable to the adjacent genomic elements, which is known as a "positional effect". In addition, a requisite long research timeframe also limits the success of this procedure[42].

Fish represent an ideal experimental model for conducting transgenic studies because of their accessible *in vitro* fertilization and developmental characteristics. Compared to the mammalian bioreactor, the transgenic fish bioreactor has easily obtainable zygotes, a convenient and established transgenic technology, a short research period, a high gene ration-expanding speed, and is relatively inexpensive to maintain. Some researchers have tried to study the transgenic fish bioreactor to produce human proteins in the ovum, embryo and muscle for medical applications[8,9,43]. However, these researches are still hampered similar to the limitations of the transgenic mammary bioreactor, except for the utilization of the fish per se. Even if one kind of transgenic fish line has been produced to be a bioreactor, a new stable transgenic fish line must be produced each time a different foreign protein is needed. In short, this technology is disadvantaged by a longer research period, and the production of the expected protein is limited by the seasonal zygotes of the transgenic fishes.

Our research group has developed a new idea for secreting recombinant proteins from a bioreactor fish gonad (spermary). Two components for this strategy include: (i) producing a spermary-specific transgenic fish line to specifically express the reporter gene eGFP for large-scale screening, and (ii) site-specific recombination for gene targeting in fish. Accordingly, we can express a different foreign protein easily by exchanging the target DNA sequence with the reporter gene eGFP flanked by *loxP* sites via Cre-*loxP* gene targeting strategy. In this manner, we will make a spermary-specific bioreactor to produce various foreign proteins in order to meet the requirement for market.

Common carp, one of the main aquaculture species in China, grows rapidly, matures within one year in most regions, and produces 10-20 g of sperm per matured male fish at each time. Overall, 750 kg sperm can be obtained from the common carp cultured in a hectare. Moreover, male fish can produce sperm over several seasons during a year under artificial conditions. For these reasons, we selected common carp as our research material to explore the potential of the spermary specific transgenic fish bioreactor.

For our initial experiments, we injected a plasmid construct containing a *lox*71-P-flanked eGFP into the embryos of common carp. After the transgenic carp matured, we verified mRNA and protein expression using RT-PCR, Western blots and fluorescence microscopy, respectively. This study indicated that the foreign gene could be expressed specifically in fish spermary via the protamine promoter, and demonstrated the feasibility of the secretion transgenic spermary-specific fish bioreactor. In addition, we used a Cre-*loxP*-mediated recombination construct to insert an eGFP reporter gene into the known eGFP target *lox*71 in the zebrafish genome. We then co-injected the target zebrafish lines with capped Cre mRNA and a knock-in vector designed to integrate the *lox*66-flanked RFP gene into the genomic *lox*71 target site. The results demonstrated that the *lox*71/*lox*66 mutant was able to direct irreversible recombination for site-specific gene targeting technology in fish, providing a solid basis for using a secretion transgenic spermary-specific fish bioreactor.

5. Perspective

With antisense RNA knock-out and suppression techniques, we and others have demonstrated that the expression of certain genes is essential to proper gonad evelopment and sexual maturity. By the strategy of gonad-specific exogenous gene excision via sexual compensation, the entire foreign gene in the gonad of the offspring will be specifically excised. All sorts could safeguard against current and future biosafety problems. The application of genetic engineering to fish to produce transgenic bioreactors for scarce pharmaceutical resources is a sophisticated use of genetic engineering and a powerful tool for aiding society.

With the rapid development of the human function genome project, the demand for various valuable and/or medical albumen products is increasing rapidly. The protein products from fish bioreactors would be safer, as there are no reports of viruses or prions infecting both humans and fish. Thus transmission of disease is highly unlikely. Compared with the mammalian bioreactors, protein expression in fish is less likely to be saturated, because of its distant genetic relationship with humans. Cold water fish species, such as salmonids develop at around 10℃, and are advantageous incubators for certain heat-unstable proteins. Mammals do not have this temperature tolerance range. In addition, transgenic fish bioreactors are not culturally religious taboos, while pig and ox bioreactors were prohibited in some countries. Therefore, the transgenic fish bioreactor will be highly useful to biomedical and pharmaceutical research and tothe aquaculture industry. However, some difficulties remain that must be addressed. For example, fish gonad development can be interrupted by the antisense transgenic technique in the brain, but a more efficient approach to recovering the lost fertility in the sterile transgenic fish is needed. Moreover, we urgently need to identify a better tissue-specific promoter that can function in early embryogenesis and is expressed in both male and female fishes. This promoter would be excised ideally at 100% efficiency. Meanwhile, if the spermary-specific promoter is more efficient, it can be highly applicable to generating a spermary-specific transgenic fish bioreactor. In addition, research on exchange gene targeting in fish, which is more promising and practical than the current knock-in gene targeting, is still on-going. Nevertheless, we believe that the fish gonad will be a new model with great potential for contributing directly to contemporary biotechnology.

Acknowledgements This work was financially supported by the National Natural Science Foundation of China (Grant No. 30430540 & 30130050), the "863" High Technology Project (Grant No. 2004AA213121) and the Development Plan of the State Key Fundamental Research of China (Grant No. 2001 CB109006).

References

1. Zhu, Z., Li, G., He, L. *et al.*, Novel gene transfer into the fertilized eggs of goldfish (*Carassius auratus* L. 1758). Z. Angew. Ichthyol.,1985, 1: 31-34.

2. Wang, R., Zhang, P., Gong, Z. *et al.*, Expression of the antifreeze protein gene in transgenic goldfish (*Carassius*

auratus) and its implication in cold adaptation, Mol. Mar. Biol. Biotechnol., 1995, 4(1): 20-26.

3. Maclean, N., Laight, R. J., Transgenic fish: an evaluation of benefits and risks, Fish and Fisheries, 2000, 1: 146-172.
4. Nam, Y. K., Noh, J. K., Cho, Y. S. *et al.*, Dramatically accelerated growth and extraordinary gigantism of transgenic mud loach, *Misgurnus mizolepis*, Transgenic Res., 2001, 10(4): 353-362.
5. Zhong, J., Wang, Y., Zhu, Z., Introduction of the human lactoferrin gene into grass carp (*Ctenopharyngodon idellus*) to increase resistance against GCH virus, Aquaculture, 2002, 214: 93-101.
6. Mao, W., Wang, Y., Wang, W. *et al.*, Enhanced resistance to Aeromonas hydrophila infection and enhanced phagocytic activities in human lactoferrin-transgenic grass carp (*Ctenopharyngodon idellus*), Aquaculture, 2004, 242: 93-103.
7. Rudolph, N. S., Biopharmaccutical production in transgenic livestock, Trends Biotechnol., 1999, 17, 367-374.
8. Morita, T., Yoshizaki, G., Kobayashi, M. *et al.*, Fish eggs as bioreactors: The production of bioactive luteinizing hormone in transgenic trout embryos, Transgenic Res., 2004, 13: 551-557.
9. Hwang, G., Müller, F., Rahman, M. *et al.*, Fish as Bioreactors: Transgene expression of human coagulation factor VII in fish embryos, Mar. Biotechnol., 2004, 6: 485-492.
10. Zhu, Z., Xu, K., Xie, Y. *et al.*, A model of transgenic fish, Scientia Sinica B (in Chinese), 1989, 2: 147-155.
11. Wang, Y., Hu, W., Wu, G. *et al.*, Genetic analysis of "All-fish" growth hormone gene transferred Yellow River carp (*Cyprinus carpio*. L) and its F1 generation, Chinese Science Bulletin, 2001, 46: 1174-1177.
12. Zhang, F., Wang, Y., Hu, W. *et al.*, Physiological and pathological analysis of mice fed with "all-fish" gene transferred Yellow River carp, High Technol. Lett. (in Chinese), 2000, 7: 12-17.
13. Knibb, W., Risk from genetically engineered and modified marine fish, Transgenic Res., 1997, 6: 59-67.
14. Muir, W. M., Howard, R. D., Possible ecological risks of transgenic organism release when transgenes affect mating success: Sexual selection and the Trojan gene hypothesis, Proc. Natl. Acad. Sci. USA., 1999, 96(24): 13853-13856.
15. Muir, W. M., Howard, R. D., Assessment of possible ecological risks and hazards of transgenic fish with implications for other sexually reproducing organisms, Transgenic Res., 2002, 1(2): 101-114.
16. Stokstad, E., Engineered fish: Friend or foe of the environment? Science, 2002, 297: 1797-1798.
17. Gage, M. J. G., Stockly, P., Parker, G. A., Effects of alternative male mating strategies on characteristics of sperm production in the Atlantic salmon (*Salmo salar*): Theoretical and empirical investigations. Philos. Trans. R. Soc. Lond. Ser. B, 1996, 350: 391-399.
18. Suresh, A. V., Sreenan, R. J., Biochemical and morphological correlates of growth in diploid and triploid rainbow trout, J. Fish Biol.,1998, 52: 588-599.
19. Razak, S. A., Hwang, G., Rahman, M. A. *et al.*, Growth performance and gonadal development of growth enhanced transgenic tilapia *Oreochromis niloticus* following heat-shock-induced triploidy, Mar. Biotechnol., 1999, 1: 533-544.
20. Liu, S. J., Liu, Y., Zhou, G. *et al.*, The formation of tetraploid stocks of red crucian carp × common carp hybrids as an effect of interspecific hybridization, Aquaculture, 2001, 192(3-4): 171-186.
21. Sun, Y., Liu, S., Zhang, C. *et al.*, The chromosome number and gonadal structure of f9-f11 allotetraploid crucian carp, Acta Genetica Sinica (in Chinese), 2003, 30(5): 414-418.
22. Fernald, R. D., White, R. B., Gonadotropin-releasing hormone genes: Phylogeny, structure and functions, Front. Neuroendocrinol.,1999, 20: 224-240.
23. Mason, A. J., Hayflick, J. S., Zoeller, R. T. *et al.*, A deletion truncating the gonadotropin-releasing hormone gene is responsible for hypogonadism in the hpg mouse, Science, 1986, 234: 1366-1371.
24. Mason, A. J., Pitts, S. L., Nikolics, K. *et al.*, The hypogonadal mouse: Reproductive functions restored by gene therapy, Science, 1986, 234: 1372-1378.
25. Singh, J., O'Neill, C., Handelsman, D. J., Induction of spermatiogenesis by androgens in gonadotropin-deficient (*hpg*) mice, Endocrinology, 1995, 136: 5311-5321.
26. Oeller, P. W., Lu, M. W., Tanlor, P. L. *et al.*, Reversible inhibition of tomato fruit senescence by antisense RNA.

Science, 1991, 254: 437-439.

27. Uzbekova, S., Chyb, J., Ferriere, F. *et al.*, Transgenic rainbow trout expressed sGnRH-antisense RNA under the control of sGnRH promoter of Atlantic salmon, J. Mol. Endocrinol., 2000, 25: 337-350.

28. Li, S., Hu, W., Wang, Y. *et al.*, Cloning and expression analysis in mature individuals of two chicken type-II GnRH (cGnRH-II) genes in common carp (*Cyprinus carpio*), Science in China, Ser. C, 2004, 47(4): 349-358.

29. Li, S., Hu, W., Wang, Y. *et al.*, Cloning and expression analysis in mature individuals of salmon gonadotropin-releasing hormone (sGnRH) gene in common carp, Acta Genetica Sinica (in Chinese), 2004, 31(10): 107-1081.

30. Moss, B., Elroy-Stein, O., Mizukami, T. *et al.*, New mammalian expression vectors, Nature, 1990, 348: 91-92.

31. Rosenberg, A. H., Studier, F. W., T7 RNA polymerase can direct expression of influenza virus cap-binding protein (PB2) in *Escherichia coli*, Gene, 1987, 59(2-3): 191-200.

32. Studier, F. W., Moffatt, B. A., Use of bacteriophage T7 RNA polymerase to direct selective high-level expression of cloned genes, J. Mol. B., 1986, 189(1): 113-130.

33. McBride, K. E., Schaaf, D. J., Daley, M. *et al.*, Controlled expression of plastid transgenes in plants based on a nuclear DNA-encoded and plastid-targeted T7 RNA polymerase. Proc.Natl. Acad. Sci. USA, 1994, 91(15): 7301-7305.

34. Tornaletti, S., Patrick, S. M., Turchi, J. J. *et al.*, Behavior of T7 RNA polymerase and mammalian RNA polymerase II at site-specific cisplatin adducts in the template DNA, J. Biol. Chem., 2003, 278(37): 35791-35797.

35. Hoess, R. H., Wierzbicki, A., Abremski, K., The role of the loxP spacer region in P1 site-specific recombination, Nucleic Acids Res., 1986, 14(5): 2287-2300.

36. Zuo, J., Niu, Q. W., Moller, S. G. *et al.*, Chemical-regulated, site-specific DNA excision in transgenic plants, Nat. Biotechnol., 2001, 19(2): 157-161.

37. Hoa, T. T., Bong, B. B., Huq, E. *et al.*, Cre/lox site-specific recombination controls the excision of a transgene from the rice genome,Theor. Appl. Genet., 2002, 104: 518-525.

38. Mlynárová, L., Nap, J. P., A self-excising Cre recombinase allows efficient recombination of multiple ectopic heterospecific lox sites in transgenic tobacco, Transgenic Res., 2003, 12(1): 45-57.

39. Wright, G., Carver, A., Cottom, D. *et al.*, High level expression of active human alpha-1-antitrypsin in the milk of transgenic sheep, Biotechnology (NY), 1991, 9(9): 830-834.

40. Wall, R. J., Pursel, V. G., Shamay, A. *et al.*, High-level synthesis of a heterologous milk protein in the mammary glands of transgenic swine, Proc. Natl. Acad. Sci. USA, 1991, 88(5): 1696-1700.

41. Velander, W. H., Johnson, J. L., Page, R. L. *et al.*, High-level expression of a heterologous protein in the milk of transgenic swine using the cDNA encoding human protein C, Proc. Natl. Acad. Sci. USA, 1992, 89(24): 12003-12007.

42. Li, N., Li, Q., Liu, J. *et al.*, The current status of research and development of animal transgenic technology in pharmaceutical industry, High Technol. Lett. (in Chinese), 2000, 11: 106-110.

43. Gong, Z., Wan, H., Tay, T. L. *et al.*, Development of transgenic fish for onamental and bioreactor by strong expression of fluorescent proteins in the skeletal muscle, Biochem. Biophys. Res. Commun., 2003, 308: 58-63.

鱼类性腺操作生物技术的应用前景

胡　炜　汪亚平　朱作言

中国科学院水生生物研究所, 淡水生态与生物技术国家重点实验室, 武汉 430072

摘　要　脊椎动物性腺是精卵发生的场所，是生命繁衍的关键器官。本文评述了通过调控特定基因表达与鱼类性腺发育，培育性腺发育被完全抑制的转反义GnRH基因鲤、特异剔除鱼类性腺中的转植基因，以及在鱼类性腺中特异表达外源基因的研究进展，认为鱼类性腺有望发展成为一种实施生物技术操纵的新型载体，并探讨了鱼类性腺发育调控在转基因鱼的生态安全对策研究及新型生物反应器的研制等方面的应用前景。

Cloning and Expression Analysis in Mature Individuals of Two Chicken Type-II GnRH (cGnRH-II) Genes in Common Carp (*Cyprinus carpio*)

LI Shuangfei HU Wei WANG Yaping ZHU Zuoyan

State Key Laboratory of Freshwater Ecology and Biotechnology, Institute of Hydrobiology, Chinese Academy of Sciences, Wuhan 430072

Abstract Gonadotropin-releasing hormone (GnRH) is a conservative neurodecapeptide family, which plays a crucial role in regulating the gonad development and in controlling the final sexual maturation in vertebrate. Two differing cGnRH-II cDNAs of common carp, namely cGnRH-II cDNA1 and cDNA2, were firstly cloned from the brain by rapid amplification of cDNA end (RACE) and reverse transcription-polymerase chain reaction (RT-PCR). The length of cGnRH-II cDNA1 and cDNA2 was 622 and 578 base pairs (bp), respectively. The cGnRH-II precursors encoded by two cDNAs consisted of 86 amino acids, including a signal peptide, cGnRH-II decapeptide and a GnRH-associated peptide (GAP) linked by a Gly-Lys-Arg proteolytic site. The results of intron trapping and Southern blot showed that two differing cGnRH-II genes in common carp genome were further identified, and that two genes might exist as a single copy. The multi-gene coding of common carp cGnRH-II gene offered novel evidence for gene duplication hypothesis. Using semi-quantitative RT-PCR, expression and relative expression levels of cGnRH-II genes were detected in five dissected brain regions, pituitary and gonad of common carp. With the exception of no mRNA2 in ovary, two cGnRH-II genes could be expressed in all the detected tissues. However, expression levels showed an apparent difference in different brain regions, pituitary and gonad. According to the expression characterization of cGnRH-II genes in brain areas, it was presumed that cGnRH-II might mainly work as the neurotransmitter and neuromodulator and also operate in the regulation for the GnRH releasing. Then, the expression of cGnRH-II genes in pituitary and gonad suggested that cGnRH-II might act as the autocrine or paracrine regulator.

Keywords gonadotropin-releasing hormone (GnRH); gene cloning; expression analysis; common carp (*Cyprinus carpio*)

Gonadotropin-releasing hormone (GnRH), the pivotal signal molecular of hypothalamic-pituitary-gonad (HPG) axis, plays a crucial role in gonadal development and maintenance of reproduction function of vertebrates by stimulating the anterior pituitary releasing gonadotropin (GtH). Mammal GnRH (mGn-RH) was firstly identified from porcine and ovine in the 1970s[1,2], which was originally named luteinizing hormone releasing hormone, viz LHRH. To date, 16 GnRH variants

文章发表于《中国科学: 生命科学》, 2004, 47(4): 349-358

have been characterized from the protochordate and various vertebrate nervous tissues[3], which are commonly named according to the species in which they are first characterized with the exception of mGnRH[3,4]. Owing to having full-conserved positions 1, 4, 9 and 10, all the GnRHs can be grouped into a conserved neurodecapeptide family. With the exception of the amino acid of position 3 of lamprey GnRH-I (lGnRH-I) and of position 2 of guinea pig GnRH (gpGnRH) being tyrosine, all the others have also full-conserved positions 2 and 3. Compared with mGnRH, the subsequently identified GnRHs show amino acid substitution in one or multi-positions, in which the substitution frequency of position 8 is the highest[3-6]. With the increase of GnRH molecule variants and the isolation of GnRH genes from different species, the distributions of GnRH molecular variants show two distinct characteristics: one is that manifold GnRHs are coexpressed in nearly all the brain tissues of organisms detected; the other is that cGnRH-II coexists always with one or two forms of GnRH in a species[3-11]. So, cGnRH-II gene is highly conservative during the vertebrate evolution.

Like other vertebrates, both the gonad development and reproduction of teleost are also controlled by the HPG axis, and GnRH is the first and crucial signal molecule of the HPG axis[12]. cGnRH-II exists in the brain tissues of all the fishes, in which cDNA sequences of GnRH have been characterized, and are distributed mainly in the midbrain. Both the function of cGnRH-II and the cycle variations of expression levels during gonad development are still controversial. However, the species-specific GnRH, for salmon GnRH (sGnRH) in Masu salmon (*Onchorhynchus masou*)[8], sea bream GnRH (sbGnRH) in striped bass (*Morone saxatilis*)[11], catfish GnRH (cfGnRH) in African catfish (*Clarias gariepinus*)[13], is distributed mainly in hypothalamic and pre-optic areas and operates as the central regulator of GtH synthesis and release. In addition, the third form of GnRH exists in the nervous center of some fish in the form of sGnRH[7,9]. The functions of multiform GnRHs existing in teleost nerve centre are still unclear, however, the conclusions from immuno-histochemistry assay of various GnRH variants and stimulating pituitary releasing GtH *in vitro*[14-16] show that species-specific GnRH is much more important than cGnRH-II in activating the anterior pituitary releasing GtH.

Common carp (*Cyprinus carpio*), a familiar economical freshwater fish in China, is used as the model of endocrine regulation of freshwater fish[16-17]. sGnRH stimulating the carp GtH and growth hormone (GH) secretion *in vitro*[16] and the sGnRH content in discrete brain areas in different development stages have been assayed by radioimmunoassay (RIA) method[17], but the molecule mechanism of regulating gonad development and reproduction of common carp has not been reported. In our laboratory, cGnRH-II and sGnRH genes have been cloned from common carp brain tissue for the first time, and all of them are encoded by two different gene loci. In this study, the isolation and identification of two differing cGnRH-II cDNAs and genes in the common carp are reported, and at the same time, expression and relative expression levels of cGnRH-II genes are assayed in five dissected brain regions, pituitary and gonad by RT-PCR. The research results offer novel evidence for tetraplo-idization research in teleost and the theory foundation for understanding further the function and regulation mechanism of cGnRH-II genes in HPG axis.

1. Material and Methods

1.1 Fish and sampling

Common carp from Ningxia Fisheries Research Institute were maintained in the Guanqiao Experimental Station of the Institute of Hydrobiology, the Chinese Academy of Sciences. Fishes were transferred to the laboratory before experiment. After fishes were bled, the whole brain and pituitary were firstly removed, and then a littlc gonad was carefully incised. All of the samples were placed on ice, and then each brain area was dissected immediately. Each sample from five brain regions, pituitary and gonad, weighing about 30 mg, was used to extract total RNA. The Gonadosomatic index (GSI) of common carp sampling from the recruit population was 11.5% and 7.3% for female and male, respectively.

1.2 Isolation of total RNA and genomic DNA

Total RNA was extracted using the SV total RNA isolation systems (*Promega*) referring to the protocols. Using two-steps isolation methods, high quality genomic DNA was isolated from liver tissue according to ref. [18]. After the quality and concentrations of total RNA and genomic DNA were assayed by agarose gel electrophoresis and optical density reading at 260 and 280 nm, the RNA and DNA were loaded in batches and frozen at –40℃.

1.3 3′ RACE and RT-PCR

Due to the cGnRH-II peptide and proteolytic site (Gly-Lys-Arg) being very conservative, degenerated primers (DPs) (cDP1 5′-CA(A/G)CA(C/T)TGGTCICA(C/T)GGITG-3′ based on the 1-7 amino acids and cDP2 5′-TGGTA(C/T)CCIGGIGG(A/G)AA(A/G)AG-3′ based on the 7-10 amino acids of cGnRH-II peptide and proteolytic site (Gly-Lys-Arg)) were designed. The Oligo(dT) adaptor primer (AP) used for the first strain cDNA synthesis was GACCACGCGTA TCGATGTCGAC$(T)_{16}$V, and the specific AP was 5′-GACCACGCGTATCGATGTCG-3′. The first strain cDNA synthesis and first round PCR amplification of 3′ RACE were performed with the mRNA selective PCR kit (Takara) referring to the protocols. The PCR conditions were as follows: 90℃ denaturation for 2 min, running 35 cycles of 90℃ 30 s, 55℃ 30 s, 72℃ 1 min, and 72℃ elongation for 5 min. The second round PCR with the template of 0.5 μL reaction mix of the first PCR was processed with cDP2 and AP.

To obtain the cGnRH-II cDNAs with complete coding region, two sense DPs (cDP3

5′-ACT(G/C)AACC(G/A)(T/C)(T/C)(G/C)ACTT(G/C)AGG-3′ and cDP4 5′-GG(G/A)TTACCAA(G/C)ACCAGGACT(T/G)C-3′) were designed according to the complete cGnRH-II cDNA sequences from goldfish (*Carassius auratus*) and Roach (*Rutilus rutilus*). At the same time, two reverse gene specific primers (GSP) (cGSP1 5′-TGGGCTGATGCATCATCTC-3′ and cGSP2 5′-GTTGGACACAGGCTGTTGCG-3′) were designed according to two part cGnRH-II cDNA sequences of common carp obtained by 3′ RACE. Four primers for RT-PCR were used to amplify 5′ coding region and part 5′ un-coding region, which corresponded with two cGnRH-II cDNAs fragments in common carp.

1.4 Intron trapping

Genomic sequences of two cGnRH-II genes were separately confirmed by PCR amplification with two sets of primers (cDP3 and cGSP1) and (cDP4 and cGSP2). The PCR reaction volume was 50 μL, containing 1 μg genomic DNA, 1×buffer (with Mg^{2+} added), 200 μmol/L of each dNTP, 0.5 μmol/L of each primers and 2.0 unit *Pyrobest*™ DNA Polymerase (Takara). PCR reactions were performed in a Perkin-Elmer DNA GeneAmp PCR System 9700, with a denaturation step of 94℃ for 2 min; followed by 35 cycle of 94℃ 30 s, 55℃ 30 s, and 72℃ for 2 min; and a final elongation at 72℃ for 5 min.

1.5 Subcloning and sequencing

PCR products were separated by agarose gel electrophoresis, and the incised gels were purified using the DNA gel extraction kit (MBI). The desired DNA fragments were subcloned into pMD18-T vector (Takara). The recombinant positive colonies were screened using M13(±) primers. DNA sequences of these fragments were determined using the dideoxy sequencing method with M13 universal primers. The data was automatically collected on the ABI PRISM 377-96 Genetic Analyzer (PE Applied Bio-systems), and analyzed and aligned using the DNAstar software.

1.6 Southern blot analysis

According to the cDNA sequences of cGnRH-II genes, the sense specific primers (cGSP3 5′-TCACAGTAGAGGAACTACTAC-3′ and cGSP4 5′-GCACAGTAGAGGAGAATATC-3′), and reverse specific primer (cGSP5 5′-TGGAAATCCCGTATGAGGGC-3′) were designed. Two sets of primer (cGSP3 and cGSP5, cGSP4 and cGSP5) were used to amplify the part fragments of cGnRH-II gene1 and gene2 in common carp, respectively. The desired fragments were recruited, quantified and used for the template of probe DIG-labeling, which both spanned intron1 and intron2 of cGnRH-II genes. Genomic DNA extracted from liver tissue of common carp was digested by restriction endonucleases *Eco*RI, *Hin*dIII and *Bam*HI, respectively, and electrophoresed on a 0.8% agarose gel without ethidium bromide (EB), each lane with about 10 μg digested DNA. After denaturation and neutralization, DNA was transferred to nylon membranes by vacuum blotter (Bio-RAD, model 785), and fixed by baking at 120℃ for 30 min. Two pieces of nylon membrane were separately hybridized using two differential labeled probes. The concrete procedures of hybridization and signal detection referred to the user's manual of DIG High Primer DNA Labeling and Detection starter Kit I (Boehringer Mannheim).

1.7 Expression analysis of two cGnRH-II genes in mature common carp individuals

3 μg total RNA isolated from dissected brain areas, pituitary and gonad in female and male individuals was used to synthesize the first strain cDNA. 10 μL of the first stand cDNA mixture was used as PCR amplification template. Two sets of primer (cGSP1 and cGSP3, cGSP2 and cGSP4) could specifically amplify part fragments of cGnRH-II cDNA1 and cDNA2. The reverse transcript products without RNA were used as the template of negative control. To analyze the relative expression level of cGnRH-II genes in different tissues, expression of beta actin (β-actin) gene in midbrain was as the reference at the same time. The primers for β-actin cDNA amplification (β(+) 5′-CACTGTGCCCATCTACGAG-3′ and β(−) 5′-CTGCATCCTGTCAGCAATGC-3′) were positioned in exon3 and exon4 of β-actin gene, respectively. The expectant fragment length of β-actin cDNA amplification was 460 bp. PCR reactions with a total volume of 50 μL consisted of 35 cycles of 90℃ for 20 s, 55℃ for 30 s, 72℃ for 50 s, with a 1 min initial 90℃ denaturation and a 5 min final 72℃ elongation. PCR products were separated by 1.3% agarose gel electrophoresis.

2. Results

2.1 Cloning of two differing cGnRH-II cDNAs in common carp

By nested-PCR, the agarose gel electrophoresis showed a specific band, about 400 bp. The specific fragment was incised, reclaimed and subcloned into pMD18-T vector. Then, five positive colonies were sequenced. Sequence analysis showed that every cDNA fragment contained the open reading frame of cGnRH-II gene 3′ end, but the part nucleotide sequences of four colonies were dissimilar to the other. Two kinds of cDNA fragment, the length being 398 and 403 bp barring the Poly(A) sequences, shared 90.0% similitude of nucleotide sequences. The same result was obtained by sequencing multiple single clones and sequencing repeatedly, so the correctness was credible.

Both RT-PCR products with two sets of primers (cDP3 and cGSP1, cDP4 and cGSP2 for amplifying the 5′ coding region of cGnRH-II cDNA1 and cDNA2, respectively) showed an exclusive fragment, with the length of about 400 bp. The specific fragments were reclaimed, subcloned into pMD18-T vector and sequenced. The sequence results bore out that the fragments for cGnRH-II cDNA1 and cDNA2 were 404 and 357 bp in length. Two cDNA fragments contained the 5′ coding region and part 5′ non-coding sequences of cGnRH-II gene. Therein, putting the 398 bp cDNA fragment obtained by 3′ RACE and the 404 bp cDNA fragment obtained by RT-PCR together and omitting the identical nucleotide sequences of the overlapping, the cDNA including complete coding sequences, 622 bp in length, was gotten, which was called cGnRH-II cDNA1 (GenBank accession NO. AF147400). The corresponding mRNA and genomic DNA sequences were called mRNA1 and gene1, respectively. In a similar way, the other cDNA, 578 bp in length, was gotten according to the 403 bp cDNA fragment obtained by 3′ RACE and the 357 bp cDNA fragment obtained by RT-PCR, which was named cGnRH-II cDNA2 (GenBank accession No: AY189961). The corresponding mRNA and genomic DNA sequences were called mRNA2 and gene2, respectively. Both of them had an incomplete noncoding region (the length being 119 bp for cDNA1 and 73 bp for cDNA2) at 5′ ends, an open reading frame coding 86 amino acids, a TGA termination codon, and a 3′ non-coding region containing the poly(A) tailing signal (fig. 1).

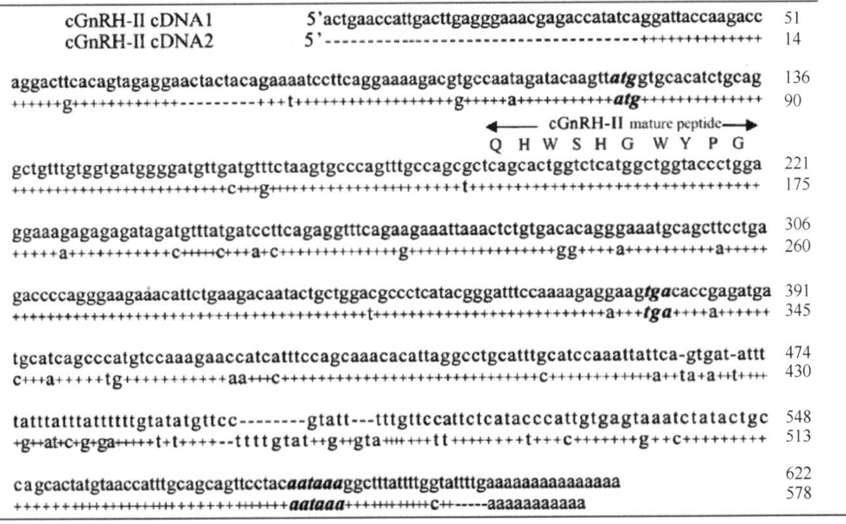

Fig. 1 Sequences alignment of cGnRH-II cDNA1 and cDNA2 in common carp. '+' signs the uniform sequences of two cDNAs; '–' signs the cDNA lacking the corresponding sequences. *atg, tga* and *aataaa* represent the initiation codon, termination codon and poly(A) tailing signal, respectively

2.2 Intron trapping for two differential cGnRH-II genes

Agarose gel electrophoresis analysis of PCR products gained by two sets of primers showed specific fragments. The sequencing results confirmed that the length of fragment amplified with cDP4 and cGSP1 was 1338 bp, namely cGnRH-II gene1 (GenBank accession

No. AY148223), and that the length of the other fragment amplified with cDP4 and cGSP2 was 1593 bp, namely cGnRH-II gene2 (GenBank accession No. AY246698). The exon sequences of the cGnRH-II gene1 and gene2 were completely similar with the corresponding cGnRH-II cDNA1 and cDNA2, respectively.

The nucleotide organization of cGnRH-II gene1 and gene2 revealed three obvious characteristics in common carp. (i) both genes contained the complete coding sequences, and spanned four exons. This characterization testified further that all the GnRH variants in different vertebrate groups shared a common organization, viz. 4 exons and 3 introns. (ii) cGnRH-II gene1 and gene2 shared the same 5′ and 3′ splice site, which was the 5′ GT and 3′ AG intron splice sites. (iii) The length and sequence identity of three introns in both genes were differing. However, the junction site of exon and intron was highly conservative, and the length and encoding characterization of exon 2 and exon 3 were nearly identical.

2.3 Southern blot analysis

In order to testify the intron trapping results and try to find out the copy number of cGnRH-II gene1 and gene2 in common carp genome, Southern blot analysis was done. Genomic DNA from liver tissue was separately digested by restriction endonucleases *Eco*RI, *Hin*dIII and *Bam*HI. After transferring the digested genomic DNA, the nylon films were separately hybridized with different labeled probes, which were synthesized according to the templates of partial cGnRH-II gene1 and gene2. The Southern blot results showed that each lane loading the digested genomic DNA produced a single hybridizing band, but the molecular size of hybridizing band in different electrophoresis lane was differential (fig. 2). Since the partial cGnRH-II gene fragments used as

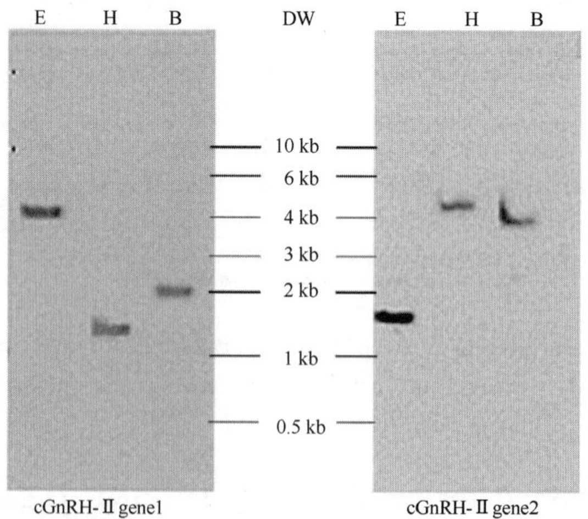

Fig. 2 Southern blot analysis of two cGnRH-II genes in common carp. The capital letters E, H and B represent genomic DNA separately digested with *Eco*RI, *Hin*dIII and *Bam*HI. The corresponding lane with 10 μg digested DNA show the specific hybridizing band. Molecular weight marker (mw) is 1 kb ladder

the probe templates had not the appropriate sequences which could be recognized by *Eco*RI, *Hin*dIII and *Bam*HI, Southern blot result could illuminate the specificity of the hybridizing signal. The conclusion of Southern blot further identified the correctness that there were two differing cGnRH-II in common carp genome, suggesting that each gene existence might be as a single copy.

2.4 Expression of two cGnRH-II genes in brain regions, pituitary and gonad of mature individuals

Total RNA was isolated from dissected brain regions, pituitary, ovary and testis in mature common carp, the reverse transcript products of which could be amplified by two sets of primer (cGSP3 and cGSP1, cGSP4 and cGSP2 for cGnRH-II cDNA1 and cDNA2, respectively) (fig. 3).

The results of RT-PCR analysis showed that cGnRH-II gene1 and gene2 were co-expressed in the dissected brain regions of mature common carp. However, expression levels of two cGnRH-II genes in telencephalon and midbrain were much higher than that of other brain regions.

In the pituitary of male and female common carp individuals, two cGnRH-II mRNAs were assayed, but the expression levels of cGnRH-II gene1 were higher than that of cGnRH-II gene2. At the same time, the expression levels of two cGnRH-II genes in male individuals were higher than that in female individuals. In the testis, two cGnRH-II genes were co-expressed, but expression levels of cGnRH-II gene1 were higher than that of gene2. In the ovary, cGnRH-II mRNA1 was detected solely, not mRNA2, and expression level of cGnRH-II gene1 was lower than that in testis.

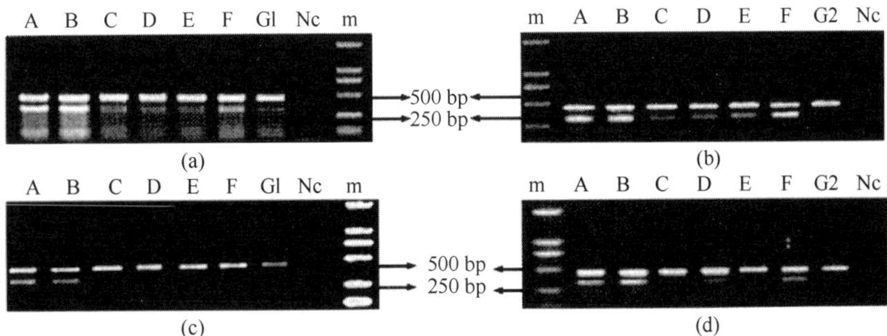

Fig. 3 Expression of two cGnRH-II genes in different brain regions, pituitary and gonad of mature common carp. (a) and (b) Expression of cGnRH-II gene1 in male and female common carp, respectively; (c) and (d) expression of cGnRH-II gene2 in male and female common carp, respectively. A, Telencephalon; B, midbrain; C, hypothalamus; D, cerebellum; E, medulla-spinal; F, pituitary; G1, testis; G2, ovary; Nc: negative control without template. Molecular weight marker (m) was DL 2000. The length of PCR products for sGnRH mRNA1, mRNA2 and β-actin mRNA were 346, 337 and 460 bp, respectively. The loading volume of agarose gel electrophoresis was 8, 5 μL for cGnRH-II mRNAs and β-actin mRNA

3. Discussion

This paper reports the clone of two differing cDNAs encoding the cGnRH-II precursor from brain tissues of common carp for the first time. The nucleotide sequence identity of two cGnRH-II cDNAs was 81.2%, but the nucleotide sequences identity of the complete encoding region was 95.0%. The cGnRH-II precursor encoded by two cDNAs contained 86 amino acid residues, and theamino acids identity of two cGnRH-II precursors was 91.9%. The cGnRH-II precursor was composed of a 24 amino acids signal peptide, cGnRH-II decapeptide, and a 49 amino acids GAP linked by the processing site (Gly-Lys-Arg). The structure characteristics of cGnRH-II precursors in common carp are similar with the other GnRH variants[8,10,19,20]. The results showed that cGnRH-II gene and other GnRH genes might evolve from a common ancestral molecule.

The amino acid sequences of common carp cGnRH-II precursors encoded by cDNA1 and cDNA2 were compared with that of some identified cGnRH-II precursors (table 1), such as the precursors of teleost fishes roach (*Rutilus rutilus*)[19], goldfish (*Carassius auratus*)[20], rainbow trout (*Oncorhynchus mykiss*)[21], medaka (*Oryzias latipes*)[22], amphibian bullfrog (*Rana catesbeiana*)[23], mammal tree shrew (*Tupaia glis belangeri*)[24] and human (*Homo sapiens*)[25]. The result showed that the amino acid homology of cGnRH-II precursors between cyprinoids was 81.4%. However, when comparing with some other teleosts cGnRH-II precursors, the amino acid homology of cGnRH-II precursors in teleosts was only 45.3%. The comparison results of amino acid sequences of cGnRH-II precursors from different vertebrates showed that the cGnRH-II decapeptide and adjacent processing site (Gly-Lys-Arg) were very conservative, and that two amino acids were invariable in the signal peptides, which were the initiative amino acid (methionine, Met) and fourteenth amino acid (leucine, Leu) and that the amino acid sequences of GAP were entirely altered. So, both the encoding characteristic of cGnRH-II cDNA and the sequences of cGnRH-II decapeptide and processing site were entirely conservative in vertebrate evolution. However, the amino acid divergence of cGnRH-II signal peptide and GAP between vertebrates from different evolution lineages was much higher than that between neighboring species. Then, it was presumed that the function of cGnRH-II peptide might change in different evolution lineages for adapting the natural selection during evolution.

Two cGnRH-II genes in common carp obtained by intron trapping included four exons and three introns, which was consistent with all the reported GnRH genes of different variants in vertebrates. The first exon contained only the 5′-UTR; the second exon encoded the single peptide, the cGnRH-II decapeptide followed by the processing site and the N-terminus of GAP; the third exon encoded the middle part of GAP; and the last exon included C-terminus of GAP and the 3′-UTR region[10,11]. The fragment length of intron1 and intron3 in cGnRH-II

Table 1 Amino acid sequences alignments of some representative cGnRH-II precursors

	Signal peptide	nRH-II decapeptide	GAP
Common carp precursor 1	MVHICRLFVVMGMLMFLSAQF—ASA	QHWSHGWYPG GKR	EIDVYDPSE ---------- ----
Common carp precursor 2	MVHICRLFVVMGMLLCLSAQF—ASS	QHWSHGWYPG GKR	EIDVYDTSE ---------- ----
Goldfish precursor 1	MVHICRLFVVMGMLLCLSAQF—ASS	QHWSHGWYPG GKR	EIDVYDSSE ---------- ----
Goldfish precursor 2	MVHICRLFVVMGMLMFLSVQF—ASS	QHWSHGWYPG GKR	EIDVYDPSE ---------- ----
Roach	MVHICRLLVLMGMLLCLSAQF—ASS	QHWSHGWYPG GKR	EIDIYDTSE ---------- ----
Rainbow trout	MVSVARLVFMLGLLLCLGAQL—SSS	QHWSHGWYPG GKR	ELDSFTTSE ---------- ----
Nile tilapia	M-CVSRLALLLGLLLCVGAQL—SFA	QHWSHGWYPG GKR	ELDSFGTSE ---------- ----
Gilthead seabream	M-CVSRLVLLLGLLLCVGAQL—SNG	QHWSHGWYPG GKR	ELDSFGTSE ---------- ----
Medaka	M---SRLVLLLGVLLYVGAQL—SQA	QHWSHGWYPG GKR	ELDSF- - - E - ---------- ----
Bullfrog	MACQRHLLFLLLVLFAVSTQL—SHG	QHWSHGWYPG GKR	ELDMPASPE ---------- ----
Tree shrew	MASSMLGFLLLLLLL-MAAHPGPSEA	QHWSHGWYPG GKR	ASNSPQDPQSALRPPAP ------
Human	MASSRRG--LLLLLL-LTAHLGPSEA	QHWSHGWYPG GKR	ALSSAQDPQNALRPPGRA LDTA
Consensus seqeuences	* *	* * * * * * * * * * * * *	
	GAP		
Common carp precursor 1	VSEEIKLCDTGKCSFLRP ------------------ QGRNILKTILLDALIRDFQ ---- KRK		
Common carp precursor 2	VSEEIKLCEAGKCSYLRP ------------------ QGRNILKTILLDALIRDFQ ---- KRK		
Goldfish precursor 1	VSGEIKLCEAGKCSYLRP ------------------ QGRNILKTILLDAIIRDSQ ---- KRK		
Goldfish precursor 2	VSEEIKLCNAGKCSFLIP ------------------QGRNILKTILLDALTRDFQ ---- KRK		
Roach	VSGEIKLCEAGKCSYLRP ------------------ QGRNILKTILLDALIRDFQ ---- KRK		

Continued

	Signal peptide	nRH-II decapeptide	GAP
Rainbow trout	ISEEIKLCEAGECSYLRP - - - - - - - - - - - - - - - - - - QRRNILKNVILDALAREFQ - - - - KRK		
Nile tilapia	ISEEIKLCEAGECSYLRP - - - - - - - - - - - - - - - - - -QRRSILRNILLDALARELQ - - - - KRK		
Gilthead seabream	ISEEIKLCEAGECSYLTP - - - - - - - - - - - - - - - - - - QRRSVLRNILLDALARELQ - - - - KRK		
Medaka	VSEEMKLCETGECSYMRP - - - - - - - - - - - - - - - - - QRRSFLRNIVLDALARELQ - - - - KRK		
Bullfrog	VSEEIKLCEGEECAYLRN - - - - - - - - - - - - - - - - - - PRKNLLKNILADVLARQLQ - - - K - K		
Tree shrew	- - SAAQTAHSFRSAALASPEDSVPWEGRTTAGWSLRRKQHLMRTLLSAAGAPRPAAVPI-KP		
Human	AGSPVQTAHGLPSDALAPLDDSMPWEGRTTAQWSLHRKRHLARTLLTAAREPRPAPPSSNKV		
Consensus seqeuences			

The GenBank accession numbers of corresponding cDNA encoding goldfish cGnRH-II precursor 1, precursor 2, Roach, Rainbow trout, Nile tilapia, Gilthead seabream, Medaka, Bullfrog, Tree shrew and Human cGnRH-II precursor were U30386, U40567, U60668, AFl25973. ABl01666, U30325, AB041330, AFl86096, U63327 and NM_001501. "*" means the consensus amino acids sequences of all the enumerative cGnRH-II precursors, and "-" means that the corresponding cGnRH-II precursor had not the corresponding amino acid.

gene1 and gene2 was similar, and the nucleotide sequence identity of intron1 and intron3 was 70.7 and 70.3%, respectively. However, the divergence of fragment length of intron2 was obvious, and the nucleotide sequence identity was reduced to 20.9%. Thus it can be seen that the basic structure of two cGnRH-II genes was similar, but the fragment length and sequence identity of introns were different.

According to geneticist Ohno's evolution notion, gnathostome vertebrates underwent two rounds of genome duplication. The first genome duplication separated in vertebrate from early agnathan vertebrates, and the second genome duplication resulted in the diversity of vertebrates (the genome duplication being the one-to-four rule)[26,27]. After that, Ohno added a third round of genome duplication to his original two (the genome duplication being the one-to-four-to-eight rule in fish)[27], which occurred just after the separation of the lobed-fin fishes that were the ancestor of all the land-based vertebrates. Due to having double chromosomes compared with most other cyprinoids, common carp was thought as the typical model of tetraploidization. sGnRH and cGnRH-II genes were firstly cloned in Goldfish[28], whereafter, the second sGnRH and cGnRH-II genes encoding the same sGnRH ans cGnRH-II precursor were confirmed in Goldfish[20,29]. At the same time, the prolactin encoded by two different cDNAs was identified in goldfish[30]. Furthermore, the same GnRH precursor encoded by two different genes was found in some salmonids[31,32]. But in other vertebrates barring teleost, the phenomenon of multi-gene encoding the same GnRH precursor was not detected[23-25]. So, the isolation of two cDNAs encoding cGnRH-II precursor and cGnRH-II genes offered novel evidence for the gene duplication theory and genome tetraploidization in common carp.

Both the research on the distribution of cGnRH-II peptide and the expression of cGnRH-II gene in brain region of teleost showed that the expression pattern of cGnRH-II gene in some teleost fishes, cGnRH-II peptide and mRNA existing mainly in midbrain, was similar to that of other vertebrates[33-35]. So, it was proposed that the function of cGnRH-II might mainly act as the neurotransmitter and/or neuromodulator[5]. It was found by assaying the expression of two goldfish cGnRH-II genes in all dissected brain areas and gonad that the expression pattern of cGnRH-II genes was exceptional[20,36]. cGnRH-II mRNAs could be detected in all the brain regions, and cGnRH-II mRNA1, no cGnRH-II mRNA2, was expressed in the the ovary and testis of goldfish[20,36]. In common carp, the expression pattern of two cGnRH-II genes was that they could be expressed in all the dissected brain regions, pituitary and gonad with the exception of having no cGnRH-II mRNA2 in ovary, which was similar to that of goldfish. However, the expression levels of two cGnRH-II genes were apparently different, being the highest in telencephalon and midbrain and the lowest in ovary and testis. Due to the co-existence of two cGnRH-II genes in all the brain regions, expressed predominately in telencephalon and midbrain, it was presumed that the regulatory function of cGnRH-II genes in HPG axis in common carp did not work as that of species specific GnRH. At the same time, the sGnRH gene expressed predominantly in telencephalon and hypothalamus of common carp was detected in our laboratory (unpublished data), which also supported the above-mentioned hypothesis. So, according to the expression characteristic of cGnRH-II genes in brain areas, it was presumed that the cGnRH-II might mainly work as the neurotransmitter and neuromodulator and, therewith, might operate in the regulation for the GnRH releasing. Then, the expression of cGnRH-II genes in pituitary and gonad suggested that cGnRH-II might relate with stimulating the releasing of other hormones, such as growth hormone, and act as the autocrine or paracrine regulator.

Acknowledgements This work was supported by the '863' High Technology Project of China (Grant No. 2001AA213101, 2001AA212281), the National Natural Science Foundation of China (Grant No. 200102006) and the '973' Project of the Ministry of Science and Technology (Grant No. 2001CB109006).

References

1. Matsuo, H., Baba, Y., Nair, R. M. G. *et al.*, Structure of the porcine LH- and FSH-releasing hormone (I)- The proposed amino acid sequence, Biochem. Biophys. Res. Commun., 1971, 43: 1334-1339.
2. Burgus, R., Butcher, M., Amoss, M. *et al.*, Primary structure of ovine hypothalamic lutenizing hormone-releasing factor (LRF), Proc. Natl. Acad. Sci. USA, 1972, 69: 278-282.
3. Somoza, G. M., Miranda, L. A., Strobl-Mazzulla, P. *et al.*, Gonadotropin-releasing hormone (GnRH): From fish to mammalian brains, Cellular and Molecular Neurobiology, 2002, 22(5/6): 589-609.
4. Fernald, R. D., White, R. B., Gonadotropin-releasing hormone genes: Phylogeny, structure and functions, Frontiers in Neuroendocrinology, 1999, 20: 224-240.
5. Lin, X. W., Otto, C. J., Peter, R. E., Evolution of neuroendocrine peptide systems: Gonadotropin-releasing hormone and somatostatin, Comparative Biochemistry and Physiology C, 1998, 119: 375-388.

6. Lin, H. R., The structural variants and functional diversity of gonadotropin-releasing hormone in vertebrate, Acta Zoological Sinica (in Chinese), 1998, 44(2): 226-234.
7. Adams, B. A., Vickers, E. D., Warby, C. *et al.*, Three forms of gonadotropin-releasing hormone, including a novel form, in a basal salmonid, *Coregonus clupeaformis*, Biol. Reprod., 2002, 67: 232-239.
8. Amano, M., Oka, Y., Aida, K. *et al.*, Immunocytochemical demonstration of salmon GnRH and chicken GnRH-II in the brain of Masu salmon (*Onchorhynchus masou*), J. Comp. Neurol., 1991, 314: 587-597.
9. Dubois, E. A., Zandbergen, M. A., Peute, J. *et al.*, Evolutionary development of three gonadotropin-releasing hormone (GnRH) systems in vertebrates, Brain Research Bulletin, 2002, 57: 413-418.
10. White, S. A., Bond, C. T., Francis, R. C. *et al.*, A second gene for gonadotropin-releasing hormone: cDNA and expression pattern in the brain, Proc. Natl. Acad. Sci. USA, 1994, 91: 1423-1427.
11. Chow, M. M., Kight, K. E., Gothilf, Y. *et al.*, Multiple GnRHs presents in a teleost species are encoded by separate genes: Analysis of the sbGnRH and cGnRH-II genes from the striped bass, Morone saxatilis, J. Mol. Endocrinol., 1998, 21: 277-289.
12. Peter, R. E., Chang, J. P., Nahoriuk, C. S. *et al.*, Interaction of catecholamines and GnRH in regulation of gonadotropin secretion in teleost fish, Recent Prog. Horm. Res., 1986, 42: 513-548.
13. Schulz, R. W., Bosma, P. T., Zandbergen, M. A. *et al.*, Two gonadotropin releasing hormones in the African catfish, *Clarias gariepinus*: Localization, pituitary receptor binding, and gonadotropin release activity, Endocrinology, 2000, 133: 1569-1577.
14. Kobayashi, M., Amano, M., Kim, M. H. *et al.*, Gonadotropin-releasing hormone and gonadotropin in goldfish and masu salmon, Fish Physiology and Biochemistry, 1997, 17: 1-8.
15. Rosenblum, P. M., Goos, H. J. T., Peter, R. E., Regional distribution and *in vitro* secretion of salmon and chicken-II GnRH gonadotropin-releasing hormones from the brains and pituitary of juvenile and adult goldfish, *Carassius auratus*, Gen. Comp. Endocrinol., 1994, 93: 369-379.
16. Lin, X. W., Lin, H. R., *In vitro* studies of the effect of salmon GnRH on the growth hormone secretion by the pituitary of common carp (*Cyprinus carpio* L.), Acta Zoological Sinica (in Chinese), 1994, 40(1): 30-38.
17. Wang, L., Lin, H. R., Distribution and variations of sGnRH in discrete brain areas from common carp (*Cyprinus carpio* L.) of different ages and gonad conditions, Zoological Research (in Chinese), 1998, 19(3): 197-202.
18. Lu, S. D., Current Protocols for Molecular Biology (in Chinese), 2nd ed., Beijing: Peking Union Medical College Press, 1999: 102-104.
19. Penlington, M. C., Williams, M. A., Sumpter, J. P. *et al.*, Isolation and characterisation of mRNA encoding the salmon- and chicken-II type gonadotropin-releasing hormones in teleost fish *Rutilus rutilus* (Cyprinidae), J. Mol. Endocrinol., 1997, 19: 337-346.
20. Lin, X. W., Peter, R. E., Cloning and expression pattern of a second [His5Trp7Tyr8] gonadotropin-releasing hormone (chicken GnRH-II) mRNA in goldfish: Evidence for two distinct genes, Gen. Comp. Endocrinol., 1997, 170: 262-272.
21. von Schalburg, K. R., Harrower, W. L., Sherwood, N. M., Regulation and expression of gonadotropin-releasing hormone in salmon embryo and gonad, Mol. Cell. Endocrinol., 1999, 157(1-2): 41-54.
22. Okubo, K., Amano, M., Yoshiura, Y. *et al.*, A novel form of gonadotropin-releasing hormone in the medaka, *Oryzias latipes*. Biochem. Biophys. Res. Commun, 2000, 276(1): 298-303.
23. Wang, L., Yoo, M. S., Kang, H. M. *et al.*, Cloning and characterization of cDNAs encoding the GnRH1 and GnRH2 precursors from bullfrog (*Rana catesbeiana*), J. Exp. Zool., 2001, 289(3): 190-201.
24. Kasten, T. L., White, S. A., Norton, T. T. *et al.*, Characterization of two new preproGnRH mRNAs in the tree shrew: First direct evidence for mesencephalic GnRH gene expression in a placental mammal, Gen. Comp. Endocrinol., 1996, 104(1): 7-19.
25. White, R. B., Eisen, J. A., Kasten, T. L. *et al.*, Second gene for gonadotropin-releasing hormone in humans, Proc. Natl. Acad. Sci. USA, 1998, 95(1): 305-309.
26. Ohno, S., Evolution by gene duplication, Berlin-Heidelberg-New York: Springer-Verlag, 1970.
27. Meyer, A., Schart, M., Gene and genome duplication in vertebrates: The one-to four (-to eight in fish) rule and the evolution of novel gene functions, Current Opinion in Cell Biology, 1999, 11: 699-704.

28. Lin, X. W., Peter, R. E., Expression of salmon gonadotropin-releasing hormone (GnRH) and chicken GnRH-II precursor messenger ribonucleic acids in the brain and ovary of goldfish, Gen. Comp. Endocrinol., 1996, 101: 282-296.
29. Suetake, H., Yoshiura, Y., Kikuchi, K. *et al.*, Two salmon gonadotropin-releasing hormone genes and their differential expression in the goldfish *Carassius auratus*, Fisheries Science, 2000, 66: 49-57.
30. Chan, Y. H., Cheng, K. W., Yu, K. L. *et al.*, Identification of two prolacti n cDNA sequences from a goldfish pituitary cDNA library, Biochimica et Biophysica Acta, 1996, 1307: 8-12.
31. Ashihara, M., Suzuki, M., Kubokawa, K. *et al.*, Two differing precursor genes for the salmon-type gonadotropin-releasing hormone exist in salmonids, J. Mol. Endocrinol., 1995, 15(1): 1-9.
32. Higa, M., Kitahashi, T., Okada, H. *et al.*, Distinct promoter sequences of two precursor genes for salmon gonadotropin-releasing hormone in masu salmon, J. Mol. Endocrinol., 1997, 19: 149-161.
33. Gothilf, Y., Munoz-Cueto, J. A., Sagrillo, C. A. *et al.*, Three forms of gonadotropin-releasing hormone on a perciform fish (*Sparus aurata*): Complementary deoxyribonucleic acid characterization and brain localization, Biol. Reprod., 1996, 55: 636-645.
34. Zandbergen, M. A., Kah, O., Bogerd, J. *et al.*, Expression and distribution of two gonadotropin-releasing hormones in the catfish brain, Neuroendocrinology, 1995, 62: 571-578.
35. Ferriere, F., Uzbekova, S., Breton, B. *et al.*, Two different messenger RNAs for salmon gonadotropin-releasing hormone are expressed in rainbow trout (*Oncorhynchus mykiss*) brain, Gen. Comp. Endocrinol., 2001, 124(3): 321-332.
36. Yu, K. L., He, M. L., Chik, C. C. *et al.*, mRNA expression of gonadotropin-releasing hormones (GnRHs) and GnRH receptor in goldfish, Gen. Comp. Endocrinol., 1998, 112: 303-311.

鲤两个 cGnRH-II 基因的克隆及表达分析

黎双飞　胡　炜　汪亚平　朱作言

中国科学院水生生物研究所，武汉 430072

摘　要　促性腺释放激素包含一类保守的神经十肽家族，在调控脊椎动物性腺发育和性成熟过程中发挥重要作用。本研究首次通过 RACE 和 RT-PCR 从鲤脑中克隆获得两种不同的 cGnRH-II 的 cDNA，分别命名为 cGnRH-II cDNA1和 cGnRH-II cDNA2。cGnRH-II cDNA1和 cGnRH-II cDNA2长度分别为622和578个碱基对。cGnRH-II 前体是由两种 cDNA 编码的86个氨基酸，包括一个信号肽，cGnRH-Ⅱ十肽和促性腺激素释放激素相关肽(GAP)由一个甘氨酸-赖氨酸-精氨酸蛋白水解位点相连。通过内含子捕获和 Southern blot 进一步鉴定了鲤基因组中两种不同的 cGnRH-Ⅱ基因，并且两个基因可能以单拷贝的形式存在。鲤 cGnRH-II 基因的多基因编码为“基因复制”理论提供了新的证据。采用半定量 RT-PCR 检测了两个 cGnRH-Ⅱ基因在成体鲤的脑区、垂体和性腺中的表达和相对表达水平。脑区、垂体和性腺组织中均检测到两种 cGnRH-II 基因的表达，但表达水平在不同的脑区、垂体和性腺存在明显的差异，而在卵巢中没有检测到 cGnRH-II 基因的表达。根据两个 cGnRH-Ⅱ基因的广泛表达特点，推测 cGnRH-II 可能主要作为神经递质或神经调节因子发挥作用，同时促进 GnRH 的释放，而垂体和性腺合成的 cGnRH-Ⅱ基因可能起自分泌或旁分泌调节因子的作用。

Identification and Characterization of A Novel Splice Variant of Gonadotropin α Subunit in the Common Carp *Cyprinus carpio*

Y. WANG　W. HU　W.-y. LIU　Y.-p. WANG　Z.-y. ZHU

State Key Laboratory of Freshwater Ecology and Biotechnology, Institute of Hydrobiology, Chinese Academy of Sciences, Graduate School of the Chinese Academy of Sciences, Wuhan 430072

Abstract In this study, an alternative splicing transcript *GtH-α291* was identified by RT-PCR, which is 291 nt and exists not only in the pituitary but also in the ovary in common carp *Cyprinus carpio*. The analysis of GtH-α291 amino acid sequence by the SignalP server predicted that the 'missing segment' might characterize as a signal peptide. In the secretion experiment, GtH-α357 subunit could be secreted out of HeLa cells while GtH-α291 could not, which confirmed the prediction. Co-immunoprecipitation assay proved that GtH-α291 subunit is able to interact with both FSH-β and LH-β as GtH-α357 does. This is the first report concerning an alternative splicing transcript of a GtH α subunit. Further studies are necessary to elucidate the specific role of this variant in the regulation of gonadal development and sexual maturation.

Keywords alternative splicing; common carp; gonadotropin; signal peptide

Introduction

Alternative splicing of pre-mRNAs is a powerful and versatile regulatory mechanism that can exert quantitative control over gene expression and influence the functional diversification of proteins (Black, 2003). Because of this, alternative splicing contributes to major developmental decisions and also to the fine tuning of gene function (Lopez, 1998). Alternatively spliced mRNA is also found from the gene related to the reproduction process. For example, a gonadotropin-releasing hormone (GnRH) RNA splicing product has been identified in cultured GnRH neurons and mouse hypothalamus, as well as in the mutant *hpg* mouse (Zhen *et al.*, 1997). Son *et al.* (2003) also observed that the precise excision of intron A and the joining of exons of GnRH serves as a key regulatory step in the synthesis of the GnRH prohormone. Alternatively spliced variants of the follicles stimulating hormone (FSH) receptor gene are also present in the human testis (Song *et al.*, 2002). Recent evidence indicates that there exists a sexual dimorphic expression pattern of a splice variant of zebrafish *Danio rerio*

文章发表于 *Journal of Fish Biology*, 2010, 71(4): 1082-1094

(Hamilton) vasa during gonadal development (Krovel & Olsen, 2004). To date, however, no alternatively spliced variant of gonadotropin (GtH) α mRNA has been identified.

GtH is a pituitary glycoprotein hormone that regulates gonadal development in vertebrates. In mammals, FSH and luteinizing hormone (LH) from the pituitary gland, as well as chorionic gonadotropin (CG) from the placenta are categorized as GtHs (Kamei *et al.*, 2003). In teleosts, as in other vertebrates, there are two forms of GtH, traditionally referred to as FSH and LH (Van Der Kraak *et al.*, 1998). GtHs are glycoprotein hormones composed of a common α subunit and a hormone-specific β subunit, which confers its biological specificity. The hormonal activity is expressed only after a non-covalent association between these two subunits (Pierce & Parsons, 1981).

As previously reported, only a single α-subunit gene has been identified in the human and bovine genomes (Fiddes & Goodman, 1981; Godine *et al.*, 1982; Goodwin *et al.*, 1983; Burnside *et al.*, 1988), whereas a novel human glycoprotein hormone α subunit-related gene was identified as glycoprotein-α2 (GPA2) based on unique sequence similarity to the α subunit of glycoprotein hormones (Hsu *et al.*, 2002). Moreover, two α subunits have been reported in some species including salmonids and goldfish *Carassius auratus* (L.) (Itoh *et al.*, 1990; Swanson *et al.*, 1991; Gen *et al.*, 1993; Kobayashi *et al.*, 1997). In salmonid pituitary glands, there are two different active α subunits that share 72% identity in their amino acid sequence. Both salmonid α subunits, upon association with corresponding β subunits, give rise to functionally active GtH (Suzuki *et al.*, 1988; Itoh *et al.*, 1990). In common carp, two highly similar 357 bp α subunit cDNAs (α1 and α2, which share 96% identity) are composed of three exons (2, 3 and 4) and encoding 118 amino acids (Chang *et al.*, 1988; Huang *et al.*, 1992). Despite the homology, these two cDNAs are believed to be derived from different genes and encode proteins that differ in seven amino acid residues, three in the signal peptide and four in the mature polypeptide (Huang *et al.*, 1991).

In this study, a new GtH α subunit transcript was discovered in common carp and its general physiological properties were determined.

Materials and Methods

Fish

Common carp used in the experiments were captured from a fish pond at the Institute of Hydrobiology, Chinese Academy of Sciences in Wuhan, Hubei Province, during the spring of 2004.

RNA extraction, cDNA synthesis and reverse transcription PCR

Pituitaries and ovaries of the common carp were collected prior to reproduction. Total RNA was extracted using the SV Total RNA Isolation System Kit (Promega, Madison, WI, U.S.A.). The first cDNA chain was obtained by reverse transcription using a random primer. GtH α subunit cDNA was obtained by PCR using the following primers: P1: tttaagcttatgttttggacaagatatgc, P2: tttgaattcttaagacttatgatagtagcag. A GenAmp PCR System 9700 (Perkin Elmer, Waltham, MA, U.S.A.) was used with the following programme: a pre-denaturation at 94℃ for 5 min, 30 cycles of amplification (94℃ for 30 s, 62℃ for 30 s, 72℃ for 40 s) and a final extension at 72℃ for 5 min. The PCR products were separated with 1.5% agarose gel electrophoresis, purified with a glassmilk kit (MBI, Vilnius, Lithuania) and cloned into the pMD-18T vector (Takara, Otsu, Shiga, Japan). After transformation, four clones [4,14,16 and 25; Fig. 1(a)] were sequenced (Sangon, Shanghai, China).

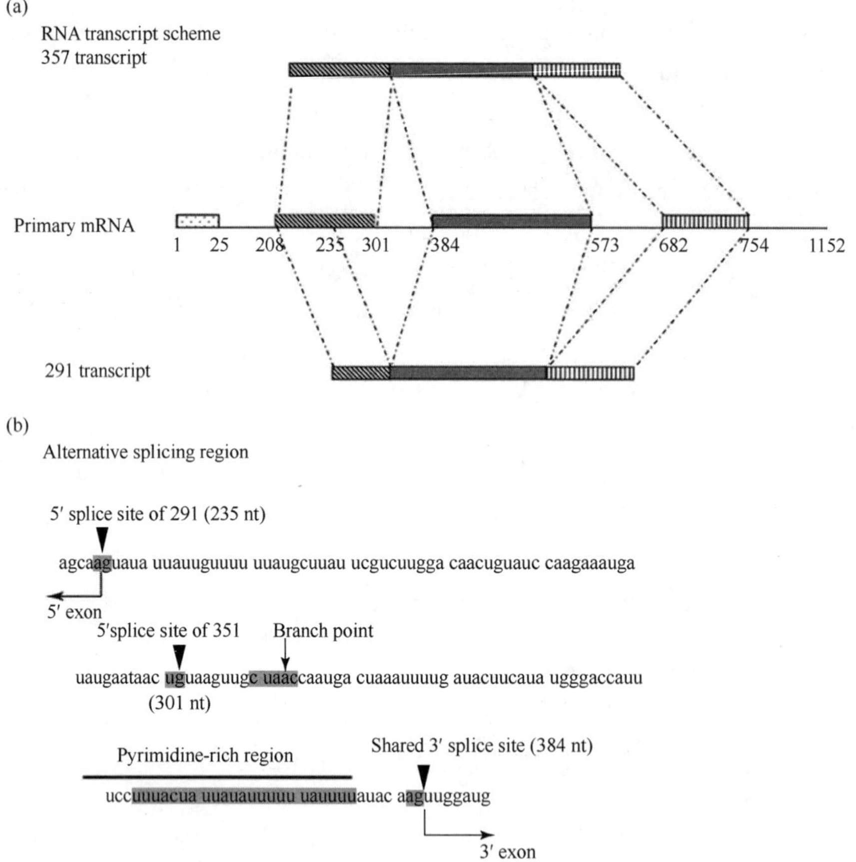

Fig. 1 Schematic summary of *GtH-α291* and the analysis of alternative splicing region compared with *GtH-α357*. (a) The schematic structure of *GtH-α357* and *GtH-α291*. The published *GtH-α357* sequence was composed of exon 2, 3 and 4 and was 357 bp in size. In addition, the novel transcript *GtH-α291* was composed of exon 3, 4 and partial exon 2. Partial sequence (235…301) of the second exon of *GtH-α357* was spliced out during the maturation of *GtH-α291*. (b) The alternative splicing region of *GtH-α291* began at 236 bp as GU and ended at the usual 384 AG, shares a common 3′ intron/exon conjunction site with *GtH-α357*. The novel intron begins at GU (236 nt) and ends at the end of AG (383 nt), has a pyrimidine-rich region upstream 3′ splice site and a conserved CUAAC branch point (310 nt)

Confirmation by reverse transcription PCR

Another primer P1′ gctggagcaattggatgtga was designed, with which only the novel splice variant was obtained. Reverse transcription PCR (RT-PCR) was repeated with the pituitary RNA using the different combinations of primers: P1 and P2, P1′ and P2, and the mixture primers of P1, P1′, P2 in different proportion. P2 was as the same reverse primer, while P1 and P1′ were both forward one. Further RT-PCR confirmation was made using the ovary RNA with the similar combination of primers. The PCR programme was as follows: a pre-denaturation at 94℃ for 5 min, 40 cycles of amplification (94℃ for 30 s, 62℃ for 30 s, 72℃ for 1 min) and a final extension at 72℃ for 5 min. The PCR products were analysed with 2.0% agarose gel electrophoresis. The segment obtained in these experiments was confirmed by sequencing.

Signal peptide sequence prediction

Firstly, amino acid sequences of the both α subunit protein were analysed by protein-protein blast (http://www.ncbi.nlm.nih.gov/BLAST/). The 'missing segment' of the alternatively spliced version normally functions as part of the signal sequence and the N-terminal part of the 'normal' mature α subunit protein. In order to identify this important piece of information and check if this deletion destroys the signal sequence property, the two amino acid sequences were checked at the SignalP server (http://www.cbs.dtu.dk/services/SignalP/).

Expression of GtH α splice variant in HeLa cells and western blot analysis

GtH-α291 cDNA and GtH-α357 cDNA were digested from *pT-α291* and *pT-α357* and cloned into the expression vector pHM6 to form *pHM6-α291* and *pHM6-α357*. HeLa cells were maintained in DMEM supplemented with 10% foetal bovine serum at 37℃ in a humidified atmosphere of 5% CO_2. Each of cDNA expression constructs were transfected into HeLa cells using Lipofectamine 2000 Reagent (Invitrogen, Carlsbad, CA, U.S.A.). After 48 h of transfection, the medium and the cell lysate were collected for western blot analysis. The protein samples were separated by 15% SDS-polyacrylamide gel electrophoresis. The separated proteins were transferred to NC membranes (Millipore, Billerica, MA, U.S.A.). The membranes were incubated with monoclonal anti-HA antibody (Santa Cruz Technology, Santa Cruz, CA, U.S.A.) for 2 h at room temperature. After reaction with peroxidase-conjugated immunopure goat anti-mouse IgG secondary antibodies (Pierce Biotechnology, Rockford, IL, U.S.A.), proteins were visualized with DAB kit (Zhongshan, Beijing, China).

Co-immunoprecipitation

Co-immunoprecipitation (Co-IP) was carried out in separate samples. Four eukaryotic expression vectors were constructed in the assay. The vectors *pHM6-α291* and *pHM6-α357* expressed a HA epitope tag N-terminally, and the *pCMV-FSHβ* and *pCMV-LHβ* vectors expressed a Flag-tag N-terminally. HeLa cells were maintained in DMEM supplemented with 10% foetal bovine serum at 37℃ in a CO_2 incubator. *PHM6-α291/pCMV-FSHβ*, *pHM6-α291/pCMV-LHβ*, *pHM6-α357/pCMV-FSHβ* and *pHM6-α357/pCMV-LHβ* plasmids were separately co-transfected into HeLa cells using Lipofectamine 2000 Reagent (Invitrogen). The cultured supernatant was subsequently collected. Protein G beads coupled with monoclonal anti-HA (Santa Cruz Technology) was used to co-immunoprecipitate the complex. Protein G beads without anti-HA was used as a negative control. All the co-immunoprecipitation procedures were performed according to the manipulation manual of ProFound™ Mammalin Co-Immunoprecipitation Kit (Pierce). Monoclonal anti-HA antibody (mAb)-protein G-linked beads were used to immunoprecipitate HA-tagged proteins from the extracts of transfected cells. The precipitation profound was detected by western blot analysis using the monoclonal anti-Flag antibody (Stratagene, La Jolla, CA, U.S.A.).

Results

Identification of an alternative splicing transcript of GtH α subunit

As reported, the primary transcript of common carp GtH α subunit was 1152 nt, comprising four exons (denoted as 1, 2, 3, and 4; 1…25, 203…301, 384…573, 682…1152) and three introns (26…202, 302…383, 574…681) (Huang *et al.*, 1991). The coding sequence was 357 nt in length, composed of three exons (2, 3, and 4; 208…301, 384…573, 682…754) and encoding 118aa. The intron/exon scheme is shown in Fig. 1(a). In the present study, four GtH α clones were obtained by RT-PCR and sequenced. *GtH-(α1)-16* was identical in size to the reported *GtH-α1* (NCBI number: X56497), while *GtH-(α1)-25* was only 291 bp, and was thus named *GtH-α291* [Fig. 2(a)]. To make sure the existence of the new transcript, the RT-PCR experiment was repeated and the *GtH-α291* transcript was obtained again. After sequence alignment analysis with *GtH-α357*, *GtH-α291* was found to lack 66 nt (236…301), therefore, it should be a novel alternative splicing transcript of *GtH-α357* [Fig. 2(b)]. This splicing transcript *GtH-α291* contains a partial exon 2 (208…235) and shares a common 3′ intron/exon conjunction site with *GtH-α357* [Fig. 1(a)]. The novel intron begins at GU (236 nt) and ends at the end of intron 2 AG (383 nt) [Fig. 1(b)], it has a pyrimidine-rich region upstream 3′ splice site and a conserved CUAAC branch point (310 nt), which completely follows the typical mRNA alternative splicing model in a eukaryotic (Horowitz & Krainer, 1994).

The novel transcript GtH-α291 is expressed both in pituitary and ovary cells

To compare the expression level of *GtH-α291* with *GtH-α357*, a *GtH-α291* specific forward primer P1' was designed. The 5′ ten nucleotides of P1′ were the same as the 19 to 28 nt

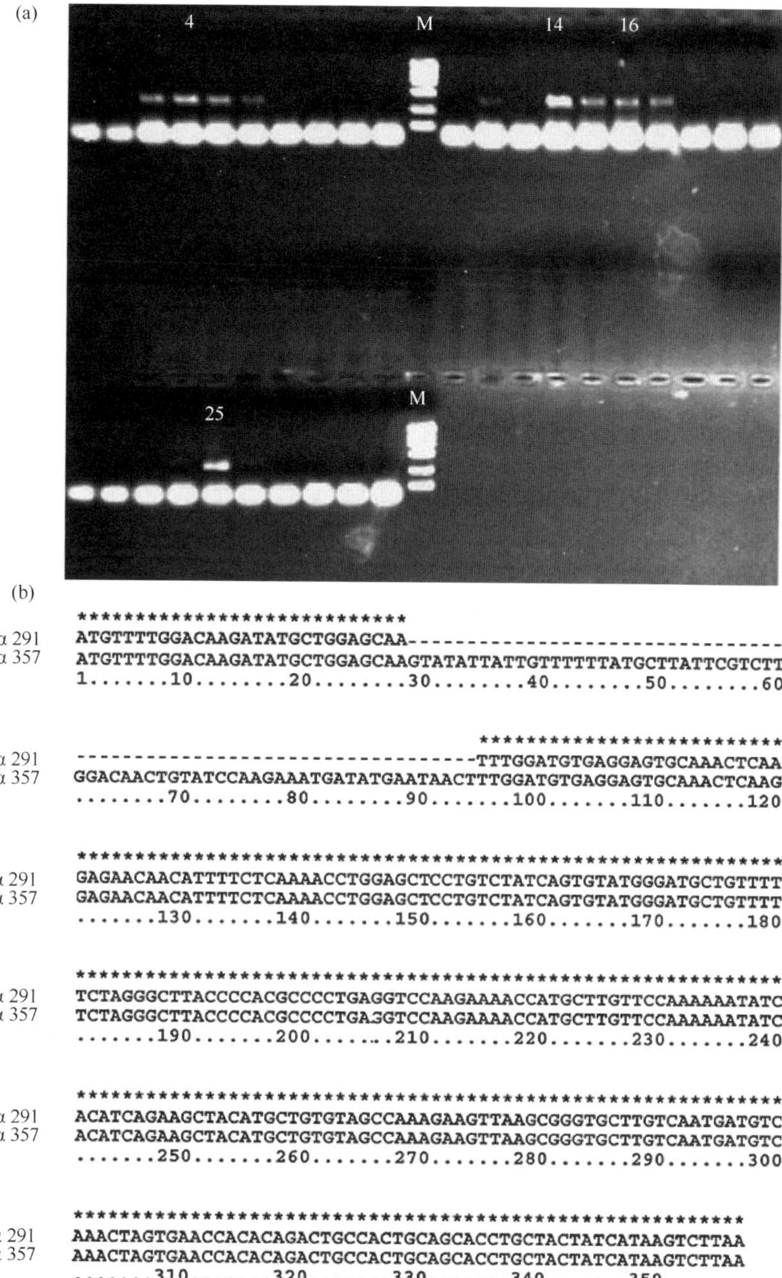

Fig. 2 A novel truncated GtH-α transcript, named *GtH-α291*, cloned from the common carp pituitary by RT-PCR. (a) The PCR products were separated with 1.5% agarose gel electrophoresis. Four clones, numbers 4, 14, 16 and 25 were cloned into pMD-18T vector and were sequenced. M, the DNA marker, whose size is separately 1450, 1000, 750, 500, 250 and 100 bp from the top to the bottom. (b) The sequence comparison between the number 25 clone (*GtH-α291*) and the published GtH α (*GtH-α357*). Clone number 16 was totally the same as the published sequence *GtH-α357* (NCBI number: X56497), while *GtH-α291* is 291 nt in size and 66 nt less than *GtH-α357*. *GtH-α291* should be a novel alternative splicing transcript of *GtH-α357* in terms of the sequence alignment analysis

of *GtH-α357*, while the 3′ ten nucleotides of P1′ were same with the 95 to 104 nt of *GtH-α357*, so it just spanned the 'missing segment' [Fig. 3(a)]. By using the *pT-α357* and *pT-α291* as templates, primer P1′ was confirmed very specific to *GtH-α291* because only the *pT-α291* but not *pT-α357* can be amplified if P1′ and P2 were used as primers.

In the pituitary, two completely different bands were obtained using P1′ and P2, P1 and P2 according to the same PCR programme and was confirmed by sequencing. Furthermore, when a mixture of P1 and P1′ in different proportions was used as the forward primers, results of RT-PCR was as following: when the ratio of P1′ and P1 is <50, only *GtH-α357* was detected; only when the ratio reached ⩾50 could *GtH-α291* be detected and the proportion began to become greater [Fig. 3(b)].

A recent study reported the novel expression of GtH subunit genes in ovary cells of the gilthead sea bream *Sparus aurata* L., and GtH subunits are expressed much lower in ovaries than in pituitaries (Wong & Zohar, 2004). In the RT-PCR results, both *GtH-α357* and *GtH-α291* was detected in ovary cells and the expression amount was less than in the pituitary [Fig. 3(c)]. By the combinations of P1 and P1′ in different ratios, only when the ratio of P1 and P1′ was 1∶1, could *GtH-α291* and *GtH-α357* coexist. Only when the proportion was smaller, was *GtH-α291* transcript detected [Fig. 3(c)].

GtH-α291 subunit lack of signal peptide sequence and failure to secrete out of cells

By deducing the amino acid sequence encoded by *GtH-α291*, the putative protein was predicted to contain 96 amino acid residues and was named GtH-α291. After the sequence comparison with GtH-α357, GtH-α291 lacked 22 amino acid residues beginning from Ser10 to Asn31 and it had a mutation from Phe32 into Ile [Fig. 4(a)].

SignalP server [www.cbs.dtu.dk\services\SignalP\] provides useful software for predicting the signal peptide information of proteins (Bendtsen *et al.*, 2004). Both of GtH-α357 and GtH-α291 subunits were analysed with this software. The amino acid residues between 20 and 30 of GtH-α357 are supposed to be core components of signal peptides which are involved in the secretion of GtH α subunit [Fig. 4(b)]. For the total 96 amino acid residues of GtH-α291, there are no exact amino acid residues which could work as a signal peptide [Fig. 4(c)].

By transfecting HA-tagged construct *pHM6-α291* and *pHM6-α357* into HeLa cells, the distribution patterns of GtH-α291 and GtH-α357 in the cell were revealed. As a positive control, GtH-α357 was detected both in cell lysate and medium, which indicated GtH-α357 subunit can be secreted out of cells, whereas GtH-α291 was detected only in cell lysate but not in the medium [Fig. 5].

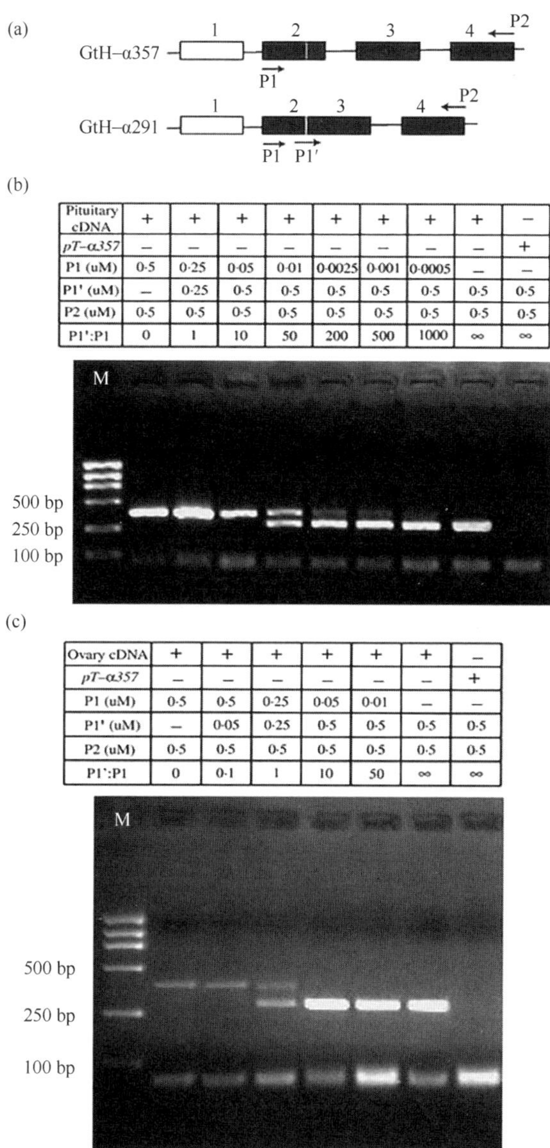

Fig. 3 The presence of *GtH-α291* transcript in the pituitary and ovary. (a) Primer P1′ was designed to amplify *GtH-α291*. The 5′ ten nucleotides of P1′ were from the end of *GtH-α291* exon 2 (19...28 nt of *GtH-α357*), while the 3′ ten nucleotides of P1′ were from the beginning of *GtH-α291* exon 2 (95...104 nt of *GtH-α357*). (b) Pituitary total RNA was extracted and the first cDNA strand was amplified with random primer. A mixture of P1 and P1′ in different proportions as well as P2 was used in the PCR reaction for 25 cycles with pituitary cDNA as the template. The data of schematics correspond to the RT-PCR band. The result indicated that the quantity of *GtH-α357* in the pituitary was much more than that of *GtH-α291*. (c) Ovary total RNA was extracted and the first cDNA strand was amplified with random primer. Both *GtH-α357* and *GtH-α291* were detected in the ovary with 35 PCR cycles. *GtH-α291* and *GtH-α357* could coexist when the ratio of P1 and P1′ was 1∶1. The data of schematics correspond to the RT-PCR band. M, the DNA marker

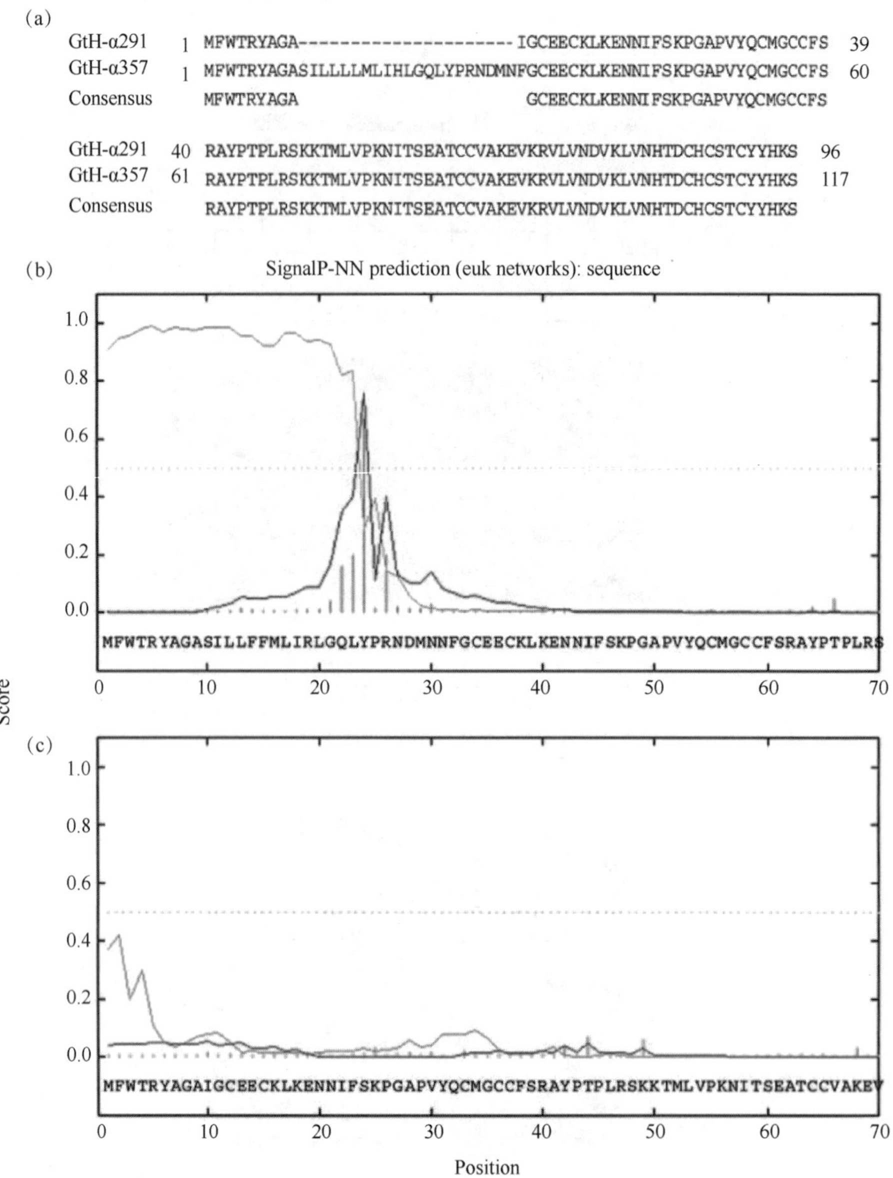

Fig. 4 (a) The amino acid sequence comparison between GtH-α357 and GtH-α291. Compared with the GtH-α357, GtH-α291 lacked 22 amino acid residues from Ser10 to Asn31 of GtH-α357 and it had a mutation from Phe32 into Ile. (b), (c) The SignalP-NN results of predicting signal peptide information by checking GtH-α357 and GtH-α291 at SignalP server [www.cbs.dtu.dk/services/SignalP/]. C (——), S (——) and Y (——) scores indicate cleavage sites, 'signal peptide-ness' and combined cleavage site predictions, respectively. (b) aa10 to aa30 of GtH-α357 might mostly serve as the signal peptide and the aa24 was the highest point and might most probably be the cleavage site. (c) GtH-α291 subunit had no signal sequences and therefore it might not be secreted out of the pituitary cells. So the 'missing segment' of GtH-α291, the 22 amino acids, is very likely to function as the signal peptide of GtH α subunit

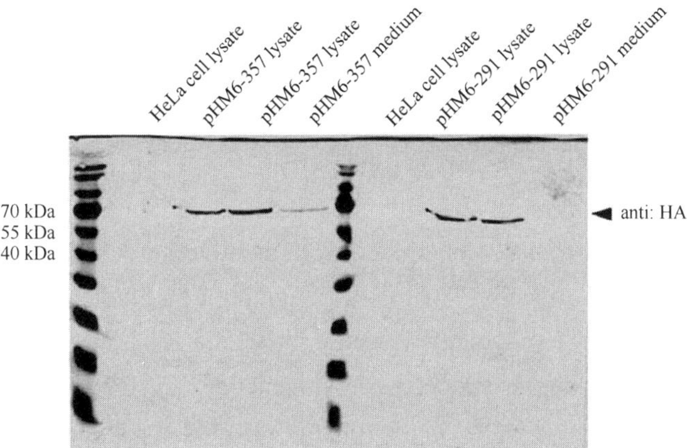

Fig. 5 HA-tagged constructs *pHM6-α291* and *pHM6-α357* were expressed in HeLa cells to test secretion activities of GtH-α291 and GtH-α357. Western blot analysis was performed using monoclonal anti-HA. HeLa cell lysate was negative control, while GtH-α357 was detected both in cell lysate and the condensed medium, which indicated GtH-α357 could be secreted out of cells. GtH-α291 subunit, however, was detected only in cell lysate but not in the condensed medium, which indicated it could not be secreted out of HeLa cells

GtH-α291 subunit can bind to both Β subunits *in vitro* as GtH-α357 does

To investigate whether there is difference between GtH-α357 and GtH-α291 subunit in their interaction with β subunits, Co-IP assay was performed. The blots indicated that pHM6-α291/pCMV-FSHβ, pHM6-α291/pCMV-LHβ, pHM6-α357/pCMV-FSHβ and pHM6-α357/pCMV-LHβ could be immunoprecipitated by HA-protein G beads (Fig. 6). These findings suggested that

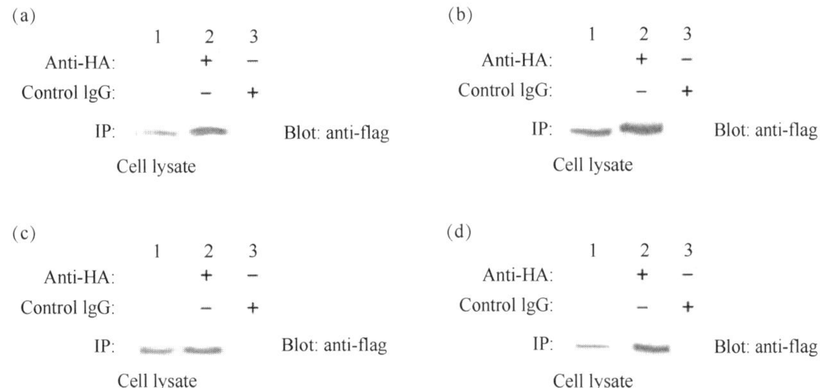

Fig. 6 Co-immunoprecipitation (Co-IP) assay with (a) α291-FSHβ, (b) α291-LHβ, (c) α357-FSHβ and (d) α357-LHβ. The coupling antibody was a monoclonal anti-HA (Santa Cruz Technology), and the negative comparison was control IgG without anti-HA. (a) Monoclonal anti-Flag antibody was used in the Co-IP western blot to detect the precipitated compound. Line 1, the cell lysate which was used as the positive control sample; line 2, the sample co-immunoprecipitated with the anti-HA monoclonal antibody (mAb)-protein G linked beads; line 3, the negative sample co-immunoprecipitated with the control IgG. The blots demonstrated that GtH-α291 could interact with both FSH-β and LH-β, as does GtH-α357

GtH-α291 interacts with both FSH-β and LH-β. GtH-α357 also binds to both β subunits, indicating that GtH-α291 can interact with either β subunit to form FSH and LH respectively.

Discussion

In the present study, *GtH-α291* is derived from an alternative splice variant of *GtH-α357* both in the pituitary and ovary cells. The missing segment was predicted as a signal peptide at the SignalP server. GtH-α291 subunit expressed in the HeLa cells failed to secrete out of cells, which confirmed the alternative spliced segment encodes a signal peptide exactly. Co-IP results demonstrated that the missing signal peptide of GtH α subunit does not influence its interaction with both GtH β subunits.

The novel *GtH-α291* transcript was cloned from pituitaries although it was much less than *GtH-α357*. Although the expression of GtH subunit genes in testes has been reported, FSH-β and common α in mouse testes and LH-β and common α in rat testes, the deduced peptides in rat testicular LH-β cDNAs were either truncated or initiated differently from pituitary LH-β (Markkula *et al.*, 1995; Zhang *et al.*, 1995). So GtH subunit genes still showed restricted tissue expression patterns; it has been generally accepted that the GtH subunits are exclusively synthesized in the anterior pituitary (Pierce & Parsons, 1981). Recently, the novel expression of *FSH-β*, *LH-β* and the common α subunits were detected in the gilthead sea bream ovary, but the quantity in the ovary was less than that in the pituitary (Wong & Zohar, 2004). In this study, both *GtH-α291* transcript and *GtH-α357* transcript were less in the ovary than in the pituitary, but it is surprising that in the ovary *GtH-α291* is expressed nearly as much as *GtH-α357*. This conservation of this alternative splicing in teleosts remains to be further investigated.

The mRNA alternative splicing of signal peptide was discovered recently to be a new way for regulating the diversity of protein function. A new PTH/PTHrP receptor gene was found which lacks the signal peptide sequence (Joun *et al.*, 1997). In a human cell, the presence of a new Tf transcript conducting to the elimination of the signal peptide sequence was characterized (Duchange *et al.*, 2002). With the signal peptide sequence missing, GtH-α291 subunit could not be secreted out of HeLa cells, but it still could bind with both GtH β subunits. This opens the possibility that a complex of GtH-α291 with the β subunits cannot be secreted. Moreover, the biological activity of heterodimer containing different α subunits could differ. GtH-α291 might interact with both kinds of β subunits to form α/β complexes which cannot be secreted out of the pituitary cells and play some physiological functions. Just like the reported α2 by Huang *et al.* (1991), the α1 subunit exhibits similar potency to the native α subunit purified from the pituitary, whereas the α2 subunit can associate with the β subunit, but only to form an inactive GtH. The competition of the α2 subunit with the α1 subunit for association with the β subunit decreases the GtH activity of the α/β complex. These results indicate that the difference in the four amino acids in the mature α2 protein affects the biological activity thus necessitating further study of the structure-function relationship of the GtH α subunit (Huang *et al.*, 1991).

The common α subunit is expressed in the primary and secondary growth oocytes, which may play a role in the intra-ovarian communication between oocytes and ovarian follicle cells (Matzuk *et al.*, 2002). The discovery of the novel *GtH-α291* expression in the ovary also makes the communication a complex network, where *GtH-α291* might function during GtH complex secretion from oocytes to ovarian follicle cells. The function of GtH α subunit is to direct the GtH α/β complex to appropriate cellular locations and secretion (Williams *et al.*, 2000), so the novel spliced α subunit might form GtH α/β complex in both pituitary cells and oocytes preventing them secreting out of cells.

In conclusion, this study described a novel alternative splicing transcript of the GtH α subunit in the common carp. The regulation mechanism and the conservation remain to be investigated.

Acknowledgements The authors thank X. Tong for the plasmids of pHM6 and pCMV-2b, T. T. Wong for the valuable suggestion of the experiment, Y. Liang and J. P. Yan for HeLa cell culture. This study was supported by the National Natural Science Foundation of China (Grant No. 30430540), the National Basic Research and Development Programme (Grant No. 2001CB109006) and the National '863' Programme (Grant No. 2006AA10Z141).

References

Bendtsen, J. D., Nielsen, H., von Heijne, G. & Brunak, S. (2004). Improved prediction of signal peptides: SignalP 3.0. Journal of Molecular Biology 340, 783-795.

Black, D. L. (2003). Mechanisms of alternative pre-messenger RNA splicing. Annual Review of Biochemistry 72, 291-336.

Burnside, J., Buckland, P. R. & Chin, W. W. (1988). Isolation and characterization of the gene encoding the alpha-subunit of the rat pituitary glycoprotein hormones. Gene 70, 67-74.

Chang, Y. S., Huang, C. J., Huang, F. L. & Lo, T. B. (1988). Primary structures of carp gonadotropin subunits deduced from cDNA nucleotide sequences. International Journal of Peptide and Protein Research 32, 556-564.

Duchange, N., Saleh, M. C., de Arriba Zerpa, G., Pidoux, J., Guillou, F., Zakin, M. M. & Baron, B. (2002). Alternative splicing in the brain of mice and rats generates transferrin transcripts lacking, as in humans, the signal peptide sequence. Neurochemistry Research 27, 1459-1463.

Fiddes, J. C. & Goodman, H. M. (1981). The gene encoding the common alpha subunit of the four human glycoprotein hormones. Journal of Molecular and Applied Genetics 1, 3-18.

Gen, K., Maruyama, O., Kato, T., Tomizawa, K., Wakabayashi, K. & Kato, Y (1993). Molecular cloning of cDNAs encoding two types of gonadotrophin alpha subunit from the masu salmon, *Oncorhynchus masou*: construction of specific oligonucleotides for the alpha 1 and alpha 2 subunits. Journal of Molecular Endocrinology 11, 265-273.

Godine, J. E., Chin, W. W. & Habener, J. F. (1982). Alpha subunit of rat pituitary glycoprotein hormones. Primary structure of the precursor determined from the nucleotide sequence of cloned cDNAs. Journal of Biological Chemistry 257, 8368-8371.

Goodwin, R. G., Moncman, C. L., Rottman, F. M. & Nilson, J. H. (1983). Characterization and nucleotide sequence of the gene for the common alpha subunit of the bovine pituitary glycoprotein hormones. Nucleic Acids Research 11, 6873-6882.

Horowitz, D. S. & Krainer, A. R. (1994). Mechanisms for selecting 5′ splice sites in mammalian pre-mRNA splicing. Trends in Genetics 10, 100-106.

Hsu, S. Y., Nakabayashi, K. & Bhalla, A. (2002). Evolution of glycoprotein hormone subunit genes in bilateral metazoa: identification of two novel human glycoprotein hormone subunit family genes, GPA2 and GPB5. Molecular Endocrinology 16, 1538-1551.

Huang, C. J., Huang, F. L., Chang, G. D., Chang, Y. S., Lo, C. H., Fraser, M. J. & Lo, T. B. (1991). Expression of two forms of carp gonadotropin a subunit in insect cells by recombinant baculovirus. Proceedings of National Academy of Sciences of the United States of America 88, 7486-7490.

Huang, C. J., Huang, F. L., Wang, Y. C., Chang, Y. S. & Lo, T. B. (1992). Organization and nucleotide sequence of carp gonadotropin alpha subunit genes. Biochimica et Biophysica Acta 6, 239-242.

Itoh, H., Suzuki, K. & Kawauchi, H. (1990). The complete amino acid sequences of alpha subunits of chum salmon gonadotropins. General and Comparative Endocrinology 78, 56-65.

Joun, H., Lanske, B., Karperien, M., Qian, F., Defize, L. & Abou-Samra, A. (1997). Tissue-specific transcription start sites and alternative splicing of the parathyroid hormone (PTH)/PTH-related peptide (PTHrP) receptor gene: a new PTH/PTHrP receptor splice variant that lacks the signal peptide. Endocrinology 138, 1742-1749.

Kamei, H., Ohira, T., Yoshiura, Y., Uchida, N., Nagasawa, H. & Aida, K. (2003). Expression of a biologically active recombinant follicle stimulating hormone of Japanese eel Anguilla japonica using methylotropic yeast, Pichia pastoris. General and Comparative Endocrinology 134, 244-254.

Kobayashi, M., Kato, Y., Yoshiura, Y. & Aida, K. (1997). Molecular cloning of cDNA encoding two types of pituitary gonadotropin alpha subunit from the goldfish, *Carassius auratus*. General and Comparative Endocrinology 105, 372-378.

Krovel, A. V. & Olsen, L. C. (2004). Sexual dimorphic expression pattern of a splice variant of zebrafish vasa during gonadal development. Developmental Biology 271, 190-197.

Lopez, A. J. (1998). Alternative splicing of pre-mRNA: developmental consequences and mechanisms of regulation. Annual Review of Genetics 32, 279-305.

Markkula, M., Hamalainen, T., Loune, E. & Huhtaniemi, I. (1995). The folliclestimulating hormone (FSH) β- and common α-subunits are expressed in mouse testis, as determined in wild-type mice and those transgenic for the FSH β-subunit/herpes simplex virus thymidine kinase fusion gene. Endocrinology 136, 4769-4775.

Matzuk, M. M., Burns, K. H., Viveiros, M. M. & Eppig, J. J. (2002). Intercellular communication in the mammalian ovary: oocytes carry the conversation. Science 296, 2178-2180.

Pierce, J. G. & Parsons, T. F. (1981). Glycoprotein hormones: structure and function. Annual Review of Biochemistry 50, 465-495.

Son, G. H., Jung, H., Seong, J. Y., Choe, Y., Geum, D. & Kim, K. (2003). Excision of the first intron from the gonadotropin-releasing hormone (GnRH) transcript serves as a key regulatory step for GnRH biosynthesis. Journal of Biological Chemistry 278, 18037-18044.

Song, G. J., Park, Y. S., Lee, Y. S., Lee, C. C. & Kang, I. S. (2002). Alternatively spliced variants of the follicle-stimulating hormone receptor gene in the testis of infertile men. Fertility and Sterility 77, 499-504.

Suzuki, K., Kawauchi, H. & Nagahama, Y. (1988). Isolation and characterization of subunits from two distinct salmon gonadotropins. General and Comparative Endocrinology 71, 302-306.

Swanson, P., Suzuki, K., Kawauchi, H. & Dickhoff, W. W. (1991). An isolation and characterization of two coho salmon gonadotropins, GTH I and GTH II. Biology of Reproduction 44, 29-38.

Van Der Kraak, G., Chang, J. P. & Janz, D. M. (1998). Reproduction. In The Physiology of Fishes (Evans, D. H., ed.), pp. 465-488. New York, NY: CRC Press.

Williams, E. J., Pal, C. & Hurst, L. D. (2000). The molecular evolution of signal peptides. Gene 253, 313-322.

Wong, T. T. & Zohar, Y. (2004). Novel expression of gonadotropin subunit genes in oocytes of the gilthead seabream

(*Sparus aurata*). Endocrinology 145, 5210-5220.

Zhang, F. P., Markkula, M., Toppari, J. & Huhtaniemi, I. (1995). Novel expression of luteinizing hormone subunit genes in the rat testis. Endocrinology 136, 2904-2912.

Zhen, S. J., Dunn, I. C., Wray, S. S., Liu, Y., Chappelli, P. E., Levinei, J. E. & Radovick, S. (1997). An alternative gonadotropin-releasing hormone (GnRH) RNA splicing product found in cultured GnRH neurons and mouse hypothalamus. Journal of Biological Chemistry 272, 12620-12625.

鲤促性腺激素α亚基新剪切异构体的鉴定和特征分析

王　蕴　胡　炜　刘维一　汪亚平　朱作言

淡水生态与生物技术国家重点实验室，中国科学院水生生物研究所，武汉 430072

摘　要　本研究通过RT-PCR鉴定出一个促性腺激素α亚基的可变剪切转录本*GtH-α291*，长度为291个碱基。该转录本不仅存在于鲤的脑垂体中，而且存在于卵巢中。SignalP软件对GtH-α291氨基酸序列的预测分析表明，“缺失片段”可能为一段信号肽序列。随后的分泌性实验数据显示，HeLa细胞能将GtH-α357分泌至胞外，而不能分泌GtH-α291，实验结果验证了上述推测。免疫共沉淀分析发现，GtH-α291和GtH-α357一样，与FSH-β和LH-β均可发生相互作用。本研究是涉及GtH α的可变剪切转录本的首次报道，下一步的研究有必要阐明GtH-α291在性腺发育和性成熟中特异的调控作用。

Antisense for Gonadotropin-Releasing Hormone Reduces Gonadotropin Synthesis and Gonadal Development in Transgenic Common Carp (*Cyprinus carpio*)

Wei Hu[1] Shuangfei Li[1] Bin Tang[1] Yaping Wang[1] Haoran Lin[2] Xiaochun Liu[2] Jun Zou[3] Zuoyan Zhu[1]

1 State Key Laboratory of Freshwater Ecology and Biotechnology, Institute of hydrobiology, Chinese Academy of Sciences, Wuhan 430072, China

2 College of Life Sciences, Zhongshan University, Guangzhou 510275, China

3 School of Biological Sciences, University of Aberdeen, Aberdeen AB24 2TZ, UK

Abstract Generating transgenic fish with desirable traits (e.g., rapid growth, larger size, etc.) for commercial use has been hampered by concerns for biosafety and competition if these fish are released into the environment. These obstacles may be overcome by producing transgenic fish that are sterile, possibly by inhibiting hormones related to reproduction. In vertebrates, synthesis and release of gonadotropin (GtH) and other reproductive hormones is mediated by gonadotropin-releasing hormone (GnRH). Recently two cDNA sequences encoding salmon-type GnRH (sGnRH) decapeptides were cloned from common carp (*Cyprinus carpio*). This study analyzed the expression of these two genes using real-time polymerase chain reaction (RT-PCR) in different tissues carp at varying developmental stages. Transcripts of both genes were detected in ovary and testis in mature and regressed, but not in juvenile carp. To evaluate the effects of sGnRH inhibition, the recombinant gene CAsGnRHpc-antisense, expressing antisense sGnRH RNA driven by a carp beta-actin promoter, was constructed. Blocking sGnRH expression using antisense sGnRH significantly decreased GtH in the blood of male transgenic carp. Furthermore, some antisense transgenic fish had no gonadal development and were completely sterile. These data demonstrate that sGnRH is important for GtH synthesis and development of reproductive organs in carp. Also, the antisense sGnRH strategy may prove effective in generating sterile transgenic fish, eliminating environmental concerns these fish may raise.

Keywords common carp (*Cyprinus carpio*); gonadotropin-releasing hormone (GnRH); transgene

1. Introduction

The commercial use of genetically altered fish with greater advantageous characteristics has been impeded by potential biosafety hazards. The first transgenic fish were produced in the 1980s (Maclean and Talwar, 1984; Zhu *et al*., 1985), and since then modified fish have

文章发表于 *Aquaculture*, 2007, 271(1): 498-506

been produced with genetically desirable traits such as rapid growth, high feed conversion, disease resistance and cold tolerance (Zhang *et al*., 1990; Wang *et al*., 1995; Devlin *et al*., 2001; Rahman *et al*., 2001; Nam *et al*., 2001, 2002; Zhong *et al*., 2002; Jhingan *et al*., 2003; Mao *et al*., 2004), all of which would prove beneficial commercially. However, these transgenic fish have not been implemented because if they were released into natural ecosystems, interbreeding could occur resulting in competition with native fish. Interbreeding could lead to genetic introgression of transgenes into the wild gene pool that may have potentially negative consequences on genetic diversity. Along these lines, the larger size of transgenic fish could increase their competitive ability and viability in the natural environment, so they may become predominant members of the aquatic community leading to an undesirable ecological impact on biodiversity (Maclean and Laight, 2000). The potential genetic and ecological risks could be minimized by producing sterile transgenic fish.

One strategy for rendering fish sterile is inhibiting gonadotropin-releasing hormone (GnRH; Maclean *et al*., 2003; Hu *et al*., 2006; Devlin *et al*., 2006). GnRH is a member of the conserved neuro-decapeptide family that links the brain and reproductive system. It plays an important role in controlling and maintaining gonadal development and reproductive function by stimulating the synthesis and release of gonadotropin hormones (GtH) from the pituitary (Somoza *et al*., 2002; Fernald and White, 1999; Dubois *et al*., 2002). GtH then acts directly on the gonads to regulate reproductive hormone levels. Two or three unique GnRH variants have been identified in the central nervous system of some teleost fish (Somoza *et al*., 2002; Dubois *et al*., 2002; Nabissi *et al*., 2000; Amano *et al*., 2002). The Cyprinid fish family, to which the common carp belongs, has two forms of GnRH molecules: salmon GnRH (sGnRH) and chicken typeII GnRH (cGnRH-II; Penlington *et al*., 1997; Lin and Peter, 1996; Powell *et al*., 1996; Steven *et al*., 2003; Li *et al*., 2004a, 2004b), which show regional differences in brain expression (e.g., the anterior parts of the cyprinid brain express more sGnRH than cGnRH-II; Yu *et al*., 1998). In addition to the brain, sGnRH present peripherally in terminal nerves is believed to be a neuron regulator of the reproductive system (Fernald and White, 1999).

Inhibiting or deleting GnRH causes loss of reproductive function and/or abnormal gonadal development in mammals. For example, deleting the mammalian GnRH gene in mice decreased GtH synthesis and disrupted development of reproductive organs (Mason *et al*., 1986a). Reproductive function was restored to this hypogonadal mouse line after gene therapy or supplementing with exogenous androgen (Mason *et al*., 1986b; Singh *et al*., 1995). Furthermore, blocking GnRH with immunization using GnRH antibodies suppressed both steroidogenesis and spermatogenesis in a number of mammalian species including mouse, dog, sheep, cat and horse (Kam *et al*., 2002).

Although inhibiting GnRH in mammals has successfully caused sterility, results have proved less effective in fish. While transgenic rainbow trout (*Oncorhynchus mykiss*) expressed sGnRH-antisense RNA, the induced ablation of endogenous sGnRH mRNA did not induce sterility (Uzbekova *et al.*, 2000). Similar work was performed in tilapia (*Oreochromis niloticus*) with more promising results, i.e., some fish expressing antisense sGnRH were sterile (Maclean *et al.*, 2003); however, more work is needed to determine whether using antisense sGnRH is an effective way to induce sterility in transgenic fish. To this end, our study applied the antisense approach to inhibit sGnRH gene expression in common carp (*Cyprinus carpio* L.) by gene transfer into fertilized eggs. The inhibitory effect of antisense sGnRH on expression of sGnRH was evaluated by RT-PCR, and its impact on GtH synthesis and gonadal development was investigated by RIA and histological assay.

2. Materials and Methods

2.1 Isolation of total RNA

Common carp were sampled from the Guanqiao Experimental Station of the Institute of Hydrobiology, Chinese Academy of Sciences. Juvenile, mature and regressed fish of each sex were used to investigate sGnRH gene expression using real-time polymerase chain reaction (RT-PCR). Three fish of each sex/stage were used, with the exception of only two mature females, of which the different developmental stage was identified according to the age and gonad development. Fish were transferred to the laboratory 2 days before the experiments and maintained at room temperature. Blood was taken from each fish and then the brain, pituitary and gonads were dissected immediately on ice. Samples (about 30 mg each) from five brain regions, pituitary and gonad were separately transferred to 1.5 ml tubes containing 175 μl lysis buffer (SV total RNA isolation systems, Promega) and immediately ground with a cone-shaped pestle. Total RNA was extracted according to the manufacturer's instructions (Promega). Quality and concentrations of total RNA were determined by agarose gel electrophoresis and optical density measurement at 260 and 280 nm.

2.2 Expression study of two sGnRH mRNAs by RT-PCR

Three specific primers were designed: sGSP1 (5′ATCTTGAGCAAAGACAGC-3′) in the conserved region (exon 1) of the two sGnRH genes, and sGSP2 (5′-TTGTTCCATTGGA-GAGTCTG-3′) and sGSP3 (5′-TCGGTGAAAGCCGCTCCATT-3′) in the variable region (exon 3) (Li *et al.*, 2004b). The specificity of primers for RT-PCR analysis to detect carp sGnRH gene 1 (sGnRH1) and gene 2 (sGnRH2) was confirmed by PCR amplification in TGradient PCR (Whatman, Biometra). The β-actin gene was used as an internal control. This allowed an equal amount of templates to be used for comparing expression levels of the two sGnRH genes. The sequences of β-actin primers were ACT-F, 5′-CACTGTGCCCATCTAC-GAG-3′ and ACT-R, 5′-CTGCATCCTGTCAGCAATGC-3′.

RT-PCR was performed according to the protocol of the mRNA selective PCR kit (Takara,

Japan). Briefly, 3 μg total RNA was used for the first strain cDNA synthesis in a volume of 50 μL and 10 μL of the cDNA sample used for the PCR reaction. The PCR program was as follows: 1 cycles of 90℃/1 min, 35 cycles of 90℃/20 s, 55℃/30 s, 72℃/50 s; 1 cycle of 72℃/5 min. PCR products were separated on a 1.3% agarose gel and visualized by staining with ethidium bromide.

2.3 Construction of GnRH antisense plasmid

The construct, pCAsGnRHpc-antisense, expressing antisense RNA against carp sGnRH genes contains (1) a carp β-actin gene promoter; (2) a 328-bp antisense DNA fragment including a 303-bp fragment targeting the sGnRH mRNA and a 25-bp fragment introduced for PCR detection of antisense sGnRH expression using the P1 primer; and (3) the 3′ flanking sequence of grass carp growth hormone gene (Fig. 1). For the exact sequence of the antisense fragment, refer to GenBank No. AF521130 and AY189960. The recombinant plasmid was linearized by restriction enzyme digestion with *Eco*RI *and Hin*dIII. The 3.5-kb fragment without pUC18 vector sequences was recovered from the agarose gel and dissolved in ST solution (88 mmol/L NaCl, 10 mmol/L Tris-HCl, pH 7.5) at the concentration of 50 ng/μL for gene transfer. By microinjection (Zhu *et al*., 1989), the linearized CAsGnRHpc-antisense fragment was injected into freshly fertilized carp eggs using a microscope. Fish derived from eggs injected with antisense sGnRH construct were raised in the secured pounds in Wuhan Duofu Scientific & Technological Farm Co., Ltd. After 8 months, fish were screened for the presence of the transgene by PCR with primers P1 (5′-CCATGGCGTATCGATGTCGAC-3′) and P2 (5′-CATGGCTTTGCCAGCATTGG-3′). The position of primers P1 and P2 is shown in Fig. 1, where P1 was in the introduced 25-bp sequence and P2 at the end of the inserted antisense GnRH fragment. Three hundred forty-two putative transgenic fish were assayed, of which 102 transgenic fish with antisense construct were obtained. Fish retaining the transgene were named AS(+).

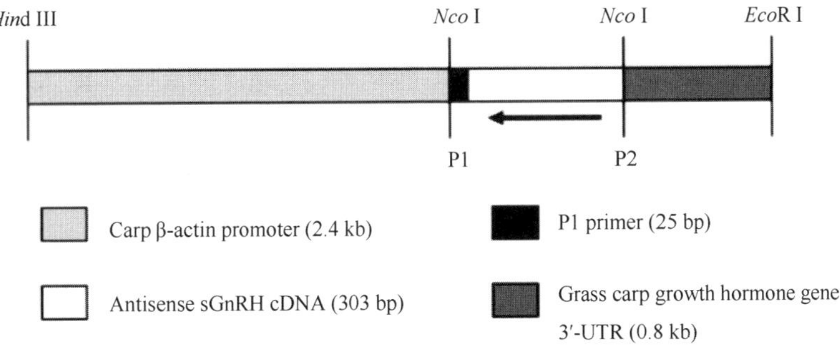

Fig. 1 The linear map of CAsGnRHpc-antisense recombinant gene. P1 and P2 were specific primers for the assay of transgenic carp and the expression analysis of antisense CAsGnRHpc recombinant gene

2.4 RT-PCR analysis of expression of the antisense sGnRH and the two endogenous sGnRH genes in AS(+) transgenic carp

Primers P1 and P2 were used to examine expression of the antisense sGnRH and two endogenous sGnRH (sGnRH1 and sGnRH2) genes in different tissues of 10 one-year-old AS(+) transgenic carp (4 females, 4 males and 2 fish without gonads) and 15 nontransgenic controls. Isolation of total RNA and the RT-PCR protocols are described above. To avoid contamination of genomic DNA, total RNA was treated with RNase-free DNase I (Promega) for 20 min at room temperature.

2.5 Gonadosomatic index (GSI) measurement and gonad histology

Gonads of the 14-month-old AS(+) transgenic and control fish were dissected, weighed and fixed in Bouin's solution for histological examination. Gonads of 14-month-old AS(+) transgenic (n=23) and control (n=24) fish were dissected and weighed for the calculation of gonadosomatic index (GSI). Of these, the gonads of 14 transgenic fish (5 female, 5 male and 4 undeveloped individuals) and 10 control fish (5 female and 5 male) were fixed in Bouin's solution for histological examination.

2.6 Radioimmunoassay

Sperm can be extracted using light abdominal pressure from 1-year-old, mature male common carp that were normally developed and bred in a local environment. Females at 1 year of age are still immature. The gender of one-year-old AS(+) transgenic fish was therefore determined by pressing lightly on the abdomen. The AS(+) transgenic fish without sperm were marked using a passive inductive transponder tag (PIT). Blood was extracted from the caudal vessel of the AS(+) transgenic fish without sperm (n=65) and controls (n=34) and stored at 4℃ for 3-4 h. Serum was separated by centrifugation and frozen at –70℃ until tested. The GtH content in 50 μl plasma was measured by specific radioimmunoassays (RIA) (Peter *et al.*, 1984; Wang *et al.*, 1997) using a standard carp GtH-II protein and a rabbit-anti-carp-GtH-II antibody (RacGtH-II). Goat-anti-γ-globulin serum (GAR) (Arnel, USA) was used as the secondary antibody.

2.7 Statistical analysis

Data were expressed as means±SEM. Variation in GSI and body weight among AS(+) transgenic and control groups was analyzed using a Student's t-test. Comparison of mean GtH concentration in plasma of AS(+) transgenic and control groups was performed using one-way analysis of variance and Duncan's multiple range test. Differences were considered significant at P<0.05.

3. Results

3.1 Expression of two sGnRH genes in brain, pituitary and gonads of carp at varying developmental stages

To detect the expression patterns of sGnRH1 and sGnRH2, two sets of primers were designed and used for RT-PCR analysis. Specific amplification of sGnRH1 and sGnRH2 transcripts was confirmed by TGradient PCR (Li *et al.*, 2004b). Fig. 2 shows the expression of sGnRH1 and sGnRH2 in male and female fish at different developmental stages. In male brains, juvenile fish expressed both sGnRH1 and sGnRH2 in telencephalon and hypothalamus, while only sGnRH1 was expressed in cerebellum. Other brain areas, including midbrain and medulla-spinal, as well as pituitary and testes, did not express either gene. In mature and regressed males, both sGnRH1 and sGnRH2 were expressed in testes and most brain tissues, including telencephalon, hypothalamus, midbrain and cerebellum, but not in medulla-spinal. In pituitary, only expression of sGnRH1 was detected. Thus, for males, differential patterns of sGnRH1 and sGnRH2 expression were observed in cerebellum of juveniles, and pituitary of mature and regressed fish. In female fish, expression patterns of both GnRH isotypes were similar at all developmental stages. In juveniles, both genes were expressed in telencephalon and hypothalamus but not in any other tissue studied. In mature and regressed females, all tissues excluding medulla-spinal, expressed both sGnRH1 and sGnRH2.

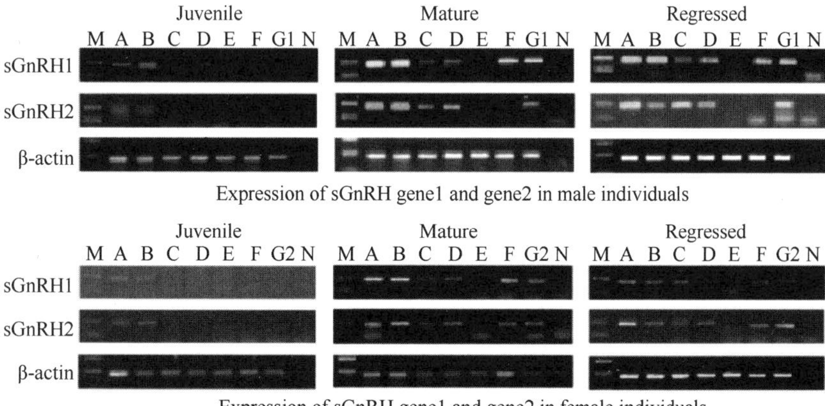

Fig. 2 Expression of sGnRH gene1 and gene2 in brain regions, pituitary and gonad of juvenile, mature and regressed carp. A, telencephalon; B, hypothalamus; C, midbrain; D, cerebellum; E, medulla-spinal; F, pituitary; G1, testis; G2, ovary; N, negative control without template. The length of PCR products for sGnRH cDNA1 and cDNA2 and for beta-actin cDNA was 222 bp, 222 bp and 460 bp, respectively. The loading volume was 8 μL for sGnRH genes amplification products. The first band of molecular weight marker (M) was 500 bp, followed by 250 bp and 100 bp

3.2 Effect of the sGnRH antisense on gene expression in transgenic fish

The sGnRH antisense construct was detected by PCR in fin tissues of 102 (30%) of 342 microinjected fish. RT-PCR was performed to examine the presence of the antisense sGnRH transcript in varying tissues of AS(+) transgenic fish with and without gonads. Antisense sGnRH expression varied among individuals, for example, among gonadally intact AS(+) male fish, one individual had limited sGnRH antisense RNA expression in brain, pituitary, heart and intestine, but whereas another showed expression in a wider range of tissues including brain, pituitary, heart, testes, liver and kidney. In one AS(+) transgenic female, sGnRH antisense RNA was mainly expressed in the brain, pituitary and heart, and also weakly expressed in the intestine, whereas in another it was expressed in the brain, pituitary, heart, ovary, liver, kidney and intestine. In AS(+) transgenic fish without gonad, sGnRH antisense RNA was detectable in brain, pituitary, muscle, heart and intestine, but not in liver and kidney in both individuals (Fig. 3).

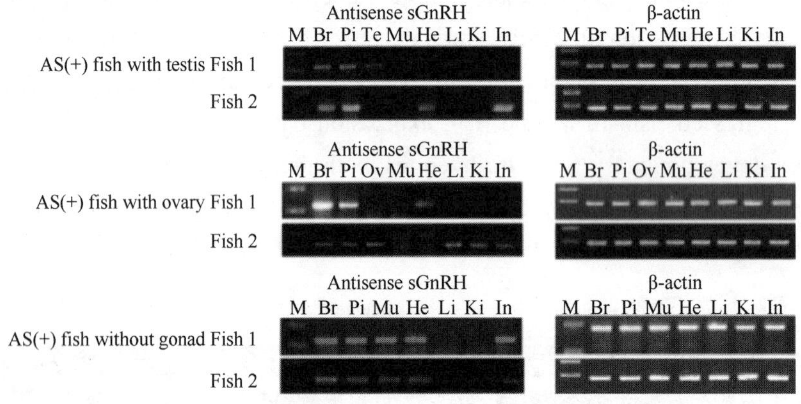

Fig. 3 Expression of CAsGnRHpc-antisense in different tissues of one-year-old AS(+) carp. The expression patterns of CAsGnRHpc-antisense gene were assayed in different tissues of two males, two females AS(+) carp and two AS(+) carp without gonads, respectively. Br, brain; Pi, pituitary; Ov, ovary; Te, testis; Mu, muscle; He, heart; Li, liver; Ki, kidney; In, intestine. The first band of molecular weight marker (M) was 500 bp, and the second was 250 bp

The expression patterns of two endogenous sGnRH genes were similar in all assayed tissues of AS(+) and control carp (Fig. 4), but the expression levels in brain, pituitary and gonad in AS(+) group were much lower than that of controls.

3.3 Plasma levels of GtH

The mean plasma GtH levels were about 2 × lower in the AS(+)fish without sperm (5.63± 0.44 ng/ml; range 1.19-17.23 ng/ml) than controls (12.41±1.44 ng/ml; range 2.97-44.92 ng/ml; P<0.05), as assessed by RIA.

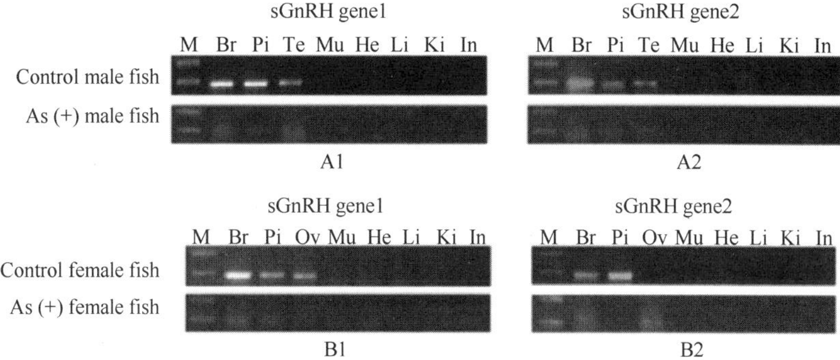

Fig. 4 Expression of two endogenous sGnRH genes in different tissues of one-year-old AS(+) and control carp. The first band of molecular weight marker (M) was 500 bp, and the second was 250 bp

3.4 Gonad development

All control fish had normal gonadal development, however 5 of 16 AS(+) transgenic carp did not have gonads, and some AS(+) fish was only unilateral or undeveloped gonads. Ovarian development did not differ between AS(+) and control groups, as evidenced by histological assay; ovaries consisted mainly of primary oocytes that were round or polygonal and had a yolk nucleus. Some primary oocytes had formed yolk globules (Fig. 5A, B). Similar to controls, male AS(+) fish contained many seminiferous tubules with sperm (Fig. 5C, D); however, some presented abnormal gonadal development in that the midsection of the testes stagnated at phase III and consisted mainly of primary spermatocytes (Fig. 5E). As shown in Table 1, comparing GSI of the AS(+) transgenic group and controls indicated no significant differences in females, but in males the GSI of the AS(+) transgenic group was 36% lower than that of controls.

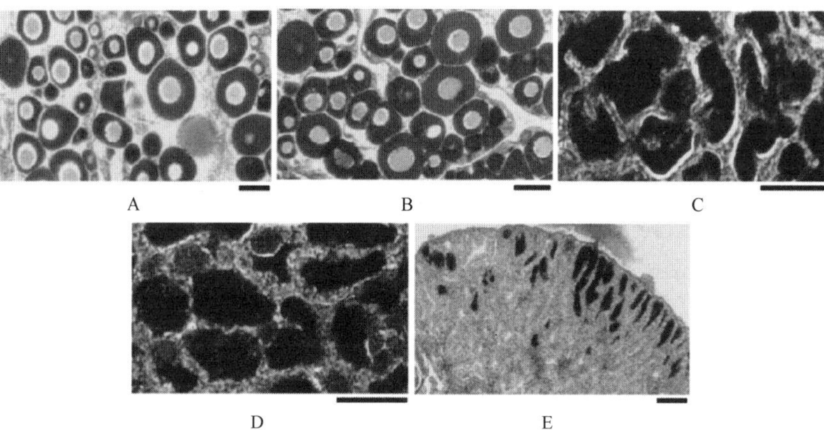

Fig. 5 Gonad structure of 14-month-old AS(+) and control fish. (A) normally developed ovary of control carp; (B) normally developed ovary of AS(+) carp; (C) normally developed testis of control carp; (D) normally developed testis of AS(+) carp; (E) abnormally developed testis of AS(+) carp. Black bar in bottom right corner of each micrograph represented 100 μm

Table 1 Gonadosomatic index (GSI) of AS(+) and control groups in 14-month-old fish

	Female		Male	
	Samples	Mean GSI (%)	Samples	Mean GSI (%)
AS(+) group	n=8	1.27±0.11	n=15	1.03±0.10
Control group	n=12	1.43±0.12	n=12	1.61±0.22
t-test	$P>0.05$		$P<0.05$	

4. Discussion

This is the first study to demonstrate sGnRH synthesis in gonads, brain and pituitary of the common carp. Furthermore, it has shown that the expression of two sGnRH genes differed according to tissue type and developmental stage in these fish. Both sGnRH genes are produced in brain and reproductive organs of mature and regressed fish, while only sGnRH1 was expressed in pituitary in male fish. In marked contrast, juvenile carp did not express either sGnRH gene in pituitary or reproductive organs. These results are consistent with previous studies showing that both sGnRH genes were expressed largely in forebrain (telencephalon and hypothalamus) of carp at all developmental stages (Steven *et al.*, 2003; Suetake *et al.*, 2000; Ferriere *et al.*, 2001).

This study implicates sGnRH in normal gonadal development. Most studies have focused on the function of the other form of GnRH carp gene, cGnRH-II (Li *et al.*, 2004a), and comparisons of its expression to sGnRH (Li *et al.*, 2004b) suggested that sGnRH was more important in regulating gonad development and GtH release. The present results support this statement in that sGnRH was found in the reproductive organs of mature but not juvenile fish. Furthermore, among the transgenic fish with sGnRH antisense, some AS(+) fish without sperm had no or abnormal gonadal development and lower GtH levels suggesting that inhibiting sGnRH produces completely, or partially, sterile transgenic fish. Thus, these data confirm previous studies showing that sGnRH is required for development of male gonads (Amano *et al.*, 1991; Ashihara *et al.*, 1995). No morphological or histological differences in ovarian development were observed among AS(+) females and controls, which suggests that further studies need to be performed to understand the role of sGnRH in female gonadal development.

Considering the potential risk transgenic fish may pose to the environment, the antisense GnRH strategy developed here is of particular interest regarding biological containment. A previous study used the antisense transgenic technique to obtain tilapia that were sterile or had very low fertility, but it only described the recombinant construct and it provide little other technical information (Maclean *et al.*, 2003). Similarly, another study used a recombinant vector which contained antisense sGnRH cDNA for Atlantic salmon (*Salmo salar*) and was driven by the sGnRH Pab promoter (Uzbekova *et al.*, 2000). Subsequently, the same group tried to inhibit endogenous sGnRH expression in transgenic rainbow trout (*Oncorhynchus mykiss*), but both of them they failed to induce sterility.

While the abovementioned studies failed or were inadequate, the success in present study could be due mainly to the presence of a strong ubiquitous expression promoter, selection of an effective long target sequence and sufficient basic research concerning the sGnRH gene and its function in carp. We developed a recombinant vector containing antisense DNA complementary to common carp sGnRH driven by the common carp beta actin promoter. The latter promoter was chosen as the regulatory component in the study because we reasoned that a highly appropriate regulatory sequence from the same genome may be necessary to effectively inhibit the specific targeting gene. Analyzing the presence of sGnRH antisense transcripts revealed that they were present in some, but not all, carp tissues. This was not surprising, as previous studies have shown that most transgenic fish are mosaics (Nam *et al.*, 2001; Gibbs and Schmale, 2000; Hamada *et al.*, 1998). In fish with detectable sGnRH antisense, levels of endogenous sGnRH transcripts were significantly reduced resulting in deficient GtH release. Synthesis of GtH was not entirely abolished in AS(+) fish probably because these fish were mosaics and not all AS(+)fish presented GnRH antisense in every tissue that synthesized GnRH.

Additionally, GnRH antisense did not entirely inhibit sGnRH transcription or affect cGnRH-II, which may also be involved in GtH production. The effects of sGnRH antisense on female fish were much less severe than in males. The fertility of some AS(+) female fish was delayed 1 year or 2 years but was not completely abolished. Based on the results of the current study, new designs are being developed to more completely inhibit carp fertility by targeting both sGnRH and cGnRH-II expression simultaneously using antisense transgenic fragments for both genes.

While the present study has shown that sterile transgenic fish could be generated by sGnRH antisense in carp, fertility recommencement is still being investigated. Generally, the antisense strategy for producing sterile transgenic fish for environmental control consists of two parts. First, transgenic fish are rendered sterile by suppressing expression of genes critical for gonad development and sexual maturity (e.g., GnRH). Second, the sterility is reversed by administering exogenous hormone to produce brood stock. This latter part of the strategy can be difficult since most transgenics are mosaics, thus resulting in founder transgenic fish that are not fully sterile or show individual variation in fertility. However, this difficulty can be solved if the transgenic fish line is produced by reversing infertility using hormone compensation method or other methods in founder sterile transgenic fish. In these lines, the transgenic fish are fertile physiologically but sterile genetically, so their progenies would be fully sterile.

5. Conclusion

sGnRH was expressed not only in the brain and pituitary, but also in the gonads of common carp. Inhibiting sGnRH expression using sGnRH antisense significantly reduced GtH release and subsequently diminished or severely hampered gonadal development. This demonstrates that sGnRH peptides are fundamentally important in regulating gonadal development via regulation of GtH synthesis and release. Importantly, sterile transgenic carp can be successfully generated using sGnRH antisense and thus could potentially provide a novel method for the biological containment of transgenic carp.

Acknowledgements This work was supported by the National Natural Science key Foundation of China (Grant No. 30430540, No. 30130050), the National Basic Research and Development Program (Grant No. 2001CB109006) and the '863' High Technology Project of China (Grant No. 2004AA213121). The authors thank Prof. Richard E. Peter for kindly providing the cGtH-II and RacGtH-II antibody, Dr. Shaojun Liu and Dr. Jiaofeng Peng for their technical assistance in histological study.

References

Amano, M., Oka, Y., Aida, K., Okumoto, N., Kawashima, S., Hasegawa, Y., 1991. Immunocyto-chemical demonstration of salmon GnRH and chicken GnRH-II in the brain of masu salmon, *Oncorhynchus masou*. Journal of Comparative Neurology 314, 587-597.

Amano, M., Oka, Y., Yamanome, T., Okuzawa, K., Yamamori, K., 2002. Three GnRH systems in the brain and pituitary of a pleuronectiform fish, the barfin flounder *Verasper moseri*. Cell Tissue Research 309, 323-329.

Ashihara, M., Suzuki, M., Kubokawa, K., Yoshiura, Y., Kobayashi, M., Urano, A., Aida, K., 1995. Two differing precursor genes for the salmon-type gonadotropin-releasing hormone exist in salmonids. Journal of Molecular Endocrinology 15, 1-9.

Devlin, R.H., Biagi, C.A., Yesaki, T.Y., Smailus, D.E., Byatt, J.C., 2001. Growth of domesticated transgenic fish. Nature 409, 781-782.

Devlin, R.H, Sundstrom, L.F., Muir, W.M., 2006. Interface of biotechnology and ecology for environmental risk assessments of transgenic fish. Trends in Biotechnology 24, 89-97.

Dubois, E.A., Zandbergen, M.A., Peute, J., Goos, H.J., 2002. Evolutionary development of three gonadotropin-releasing hormone (GnRH) systems in vertebrates. Brain Research Bulletin 57, 413-418.

Fernald, R.D., White, R.B., 1999. Gonadotropin-releasing hormone genes: phylogeny, structure and functions. Frontiers in Neuroendocrinology 20, 224-240.

Ferriere, F., Uzbekova, S., Breton, B., Jego, P., Bailhache, T., 2001. Two different messenger RNAs for salmon gonadotropin-releasing hormone are expressed in rainbow trout (*Oncorhynchus mykiss*) brain. General and Comparative Endocrinology 124, 321-332.

Gibbs, P.D., Schmale, M.C., 2000. GFP as a genetic marker scorable throughout the life cycle of transgenic zebra fish. Marine Biotechnology (NY) 2, 107-125.

Hamada, K., Tamaki, K., Sasado, T., Watai, Y., Kani, S., Wakamatsu, Y., Ozato, K., Kinoshita, M., Kohno, R., Takagi, S.,

Kimura, M., 1998. Usefulness of the medaka beta-actin promoter investigated using a mutant GFP reporter gene in transgenic medaka (*Oryzias latipes*). Molecular Marine Biology and Biotechnology 7, 173-180.

Hu, W., Wang, Y.P., Zhu, Z.Y., 2006. Perspective of fish gonad manipulation for biotechnical applications. Chinese Science Bulletin 51, 181-187.

Jhingan, E., Devlin, R.H., Iwama, G.K., 2003. Disease resistance, stress response and effects of triploidy in growth hormone transgenic coho salmon. Journal of Fish Biology 63, 806-823.

Kam, K.Y., Park, Y.B., Cheon, M.S., Kang, S.S., Kim, K., Ryu, K., 2002. Influence of GnRH agonist and neural antagonists on stressblockade of LH and prolactin surges induced by 17beta-estradiol in ovariectomized rats. Yonsei Medical Journal 43, 482-490.

Lin, X.W., Peter, R.E., 1996. Expression of salmon gonadotropin-releasing hormone (GnRH) and chicken GnRH-II precursor messenger ribonucleic acids in the brain and ovary of goldfish. General and Comparative Endocrinology 101, 282-296.

Li, S.F., Hu, W., Wang, Y.P., Zhu, Z.Y., 2004a. Cloning and expression analysis in mature individuals of two chicken type-II GnRH (cGnRH-II) genes in common carp (*Cyprinus carpio*). Science in China Series C 47, 349-358.

Li, S.F., Hu, W., Wang, Y.P., Zhu, Z.Y., 2004b. Cloning and expression analysis in mature individuals of salmon gonadotropin-releasing hormone (sGnRH) gene in common carp. Acta Genetica Sinica 31, 1072-1081.

Maclean, N., Laight, R.J., 2000. Transgenic fish: an evaluation of benefits and risks. Fish and Fisheries 1, 146-172.

Maclean, N., Talwar, S., 1984. Injection of cloned genes into rainbow trout eggs. Journal of Embryology and Experimental Morphology 82, 187.

Maclean, N., Hwang, G., Molina, A., Ashton, T., Muller, M., Aziz Rahman, M., Iyengar, A., 2003. Reversibly-Sterile Fish Via Transgenesis. Information Systems for Biotechnology News Report, December, 3-5.

Mao, W.F., Wang, Y.P., Wang, W.b., Wu, B., Feng, J.X., Zhu, Z.Y., 2004. Enhanced resistance to Aeromonas hydrophila infection and enhanced phagocytic activities in human lactoferrin-transgenic grass carp (*Ctenopharyngodon idellus*). Aquaculture 242, 93-103.

Mason, A.J., Hayflick, J.S., Zoeller, R.T., Young, W.S., Phillips, H.S., Nikolics, K., Seeburg, P.H., 1986a. A deletion truncating the gonadotropin-releasing hormone gene is responsible for hypogonadism in the hyp mouse. Science 234, 1366-1371.

Mason, A.J., Pitts, S.L., Nikolics, K., Szonyi, E., Wilcox, J.N., Seeburg, P.H., Stewart, T.A., 1986b. The hypogonadal mouse: reproductive functions restored by gene therapy. Science 234, 1372-1378.

Nabissi, M., Soverchia, L., Polzonetti-Magni, A.M., Habibi, H.R., 2000. Differential splicing of three gonadotropin-releasing hormone transcripts in the ovary of seabream (*Sparus aurata*). Biology of Reproduction 62, 1329-1334.

Nam, Y.K., Noh, J.K., Cho, Y.S., Cho, H.J., Cho, H.J., Kim, C.G., Kim, D.S., 2001. Dramatically accelerated growth and extraordinary gigantism of transgenic mud loach, *Misgurnus mizolepis*. Transgenic Research 10, 353-362.

Nam, Y.K., Cho, Y.S., Cho, H.J., Kim, D.S., 2002. Accelerated growth performance and stable germ-line transmission in androgenetically derived homozygous transgenic mud loach, *Misgurnus mizolepis*. Aquaculture 209, 257-270.

Penlington, M.C., Williams, M.A., Sumpter, J.P., Rand-Weaver, M., Hoole, D., Arme, C., 1997. Isolation and characterisation of mRNA encoding the salmon- and chicken-II type gonadotropin-releasing hormones in teleost fish *Rutilus rutilus* (Cyprinidae). Journal of Molecular Endocrinology 19, 337-346.

Peter, R.E., Nahorniak, C.S., Chang, J.P., Crim, L., 1984. Gonadotropin release from the pars distalis of goldfish, *Carassius auratus*, transplanted beside the brain or into the brain ventricles. Aquaculture 83, 193-199.

Powell, J.F., Krueckl, S.L., Collins, P.M., Sherwood, N.M., 1996. Molecular forms of GnRH in three model fishes: rockfish, medaka and zebrafish. Journal of Endocrinology 150, 17-23.

Rahman, M.A., Ronyai, A., Engidaw, B., Jauncey, Z., Hwang, K., Smith, G-L., Roderick, A., Penman, E., Varadi, D., Maclean, L., 2001. Growth and nutritional trials on transgenic Nile tilapia containing an exogenous fish growth hormone gene. Journal of Fish Biology 59, 62-78.

Singh, J., O'Neill, C., Handelsman, D.J., 1995. Induction of spermatogenesis by androgens in gonadotropin-deficient (*hyp*) mice. Endocrinology 136, 5311-5321.

Somoza, G.M., Miranda, L.A., Strobl-Mazzulla, P., Guilgur, L.G., 2002. Gonadotropin-releasing hormone (GnRH): from fish to mammalian brains. Cellular and Molecular Neurobiology 22, 589-609.

Steven, C., Lehnen, N., Kight, K., Ijiri, S., Klenke, U., Harris, W.A., Zohar, Y., 2003. Molecular characterization of the GnRH system in zebrafish (*Danio rerio*): cloning of chicken GnRH-II, adult brain expression patterns and pituitary content of salmon GnRH and chicken GnRH-II. General and Comparative Endocrinology 133, 27-37.

Suetake, H., Yoshiura, Y., Kikuchi, K., Gen, K., Ashihara, M., Kobayashi, M., Aida, K., 2000. Two salmon gonadotropin-releasing hormone genes and their differential expression in the goldfish *Carassius auratus*. Fisheries Science 66, 49-57.

Uzbekova, S., Chyb, J., Ferriere, F., Bailhache, T., Prunet, P., Alestrom, P., Breton, B., 2000. Transgenic rainbow trout expressed sGnRH-antisense RNA under the control of sGnRH promoter of Atlantic salmon. Journal of Molecular Endocrinology 25, 337-350.

Wang, D.S., Lin, H.R., Goos, H.J., 1997. Seasonal changes of the pituitary and serum basal gonadotropin (GtH) levels of the bagrid catfish *Mysius macropterus*. Transactions of the Chinese Ichthyological Society (No.6) Beijing. Science Press, pp. 22-27.

Wang, R., Zhang, P., Gong, Z., Hew, C.L., 1995. Expression of the antifreeze protein gene in transgenic goldfish (*Carassius auratus*) and its implication in cold adaptation. Molecular Marine Biology and Biotechnology 4, 20-26.

Yu, K.L., He, M.L., Chik, C.C., Lin, X.W., Chang, J.P., Peter, R.E., 1998. mRNA expression of gonadotropin-releasing hormones (GnRHs) and GnRH receptor in goldfish. General and Comparative Endocrinology 112, 303-311.

Zhang, P., Joyce, C., Gozalez-Villasenor, L.I., Lin, C.M., Dunham, R.A., Chen, T.T., Powers, D.A., 1990. Gene transfer, expression and inheritance of pRSV-rainbow trout-GH cDNA in the common carp *Cyprinus carpio* (Linnaeus). Molecular Reproduction and Development 25, 3-13.

Zhong, J.Y., Wang, Y.P., Zhu, Z.Y., 2002. Introduction of the human lactoferrin gene into grass carp (*Ctenopharyngodon idellus*) to increase resistance against GCH virus. Aquaculture 214, 93-101.

Zhu, Z.Y., Xu, K.S., Xie, Y.F., Li, G.H., He, L., 1989. A model of transgenic fish. Scientia Sinica. Series B. Chemical, Biological, Agricultural, Medical and Earth Sciences 2, 147-155.

Zhu, Z., Li, G., He, L., Chen, S., 1985. Novel gene transfer into the fertilized eggs of goldfish (*Carassius auratus* L. 1758). Z angew Ichthyol 1, 31-34.

促性腺激素释放激素的反义 RNA 降低受体鲤该激素的合成和性腺发育

胡 炜[1] 黎双飞[1] 汤 斌[1] 汪亚平[1] 林浩然[2]
刘晓春[2] 邹 钧[3] 朱作言[1]

1中国科学院水生生物研究所，武汉 430072，中国
2中山大学生命科学学院，广州 501275，中国
3 School of Biological Sciences, University of Aberdeen, Aberdeen AB24 2TZ, UK

摘 要 转基因鱼具有优良的经济性状，但是人们担心转基因鱼释放到自然环境中，可能会影响水生态系统，这也阻碍了转基因鱼的产业化。通过抑制生殖相关的激素进而获得不育的转基因鱼可以消除人们的担忧。在脊椎动物中，促性腺激素(GtH)和其他生殖相关激素的合成和释放受促性腺释放激素(GnRH)调控。最近，在鲤中克隆得到两种 GnRH(sGnRH)十肽。本研究通过 RT-PCR 分析了 GnRH 基因在鲤不同发育时期的不同组织中的表达模式。在性成熟鱼的卵巢和精巢中均检测到这两种基因的转录，但在幼年鲤中未检测到其表达。为了研究抑制 sGnRH 表达对鲤生殖发育的影响，本文构建了反义重组基因 CAsGnRHpc-antisense，用鲤 β-肌动蛋白基因启动子驱动 sGnRH 的反义 RNA 表达。反义 sGnRH 显著抑制了内源 GnRH 的表达，雄性转基因鲤血液中的 GtH 浓度显著下降，部分反义转基因鱼的性腺完全败育。实验结果表明 sGnRH 对鲤 GtH 的分泌和性腺发育具有重要作用。同时，应用反义 sGnRH 基因策略可以研制出不育转基因鱼，有望解决人们担忧的转基因鱼生态安全问题。

Production, Characterization, and Applications of Mouse Monoclonal Antibodies Against Gonadotropin, Somatolactin, and Prolactin from Common Carp (*Cyprinus carpio*)

Zhuwei Xu[1] Qi Li[1] Yun Wang[2] Chaojun Song[1] Tao Zhang[3] Lihua Chen[1] Jianguo Ji[4] Angang Yang[1] Zuoyan Zhu[2] Wei Hu[2] Boquan Jin[1]

1 Department of Immunology, The Fourth Military Medical University, Xi'an 710032

2 State Key Laboratory of Freshwater Ecology and Biotechnology, Institute of Hydrobiology, Chinese Academy of Sciences, Wuhan 430072

3 Department of Neurosurgery, Tangdu Hospital, The Fourth Military Medical University, Xi'an 710038

4 School of Life Science, Peking University, Beijing 100049

Abstract The gonadotropin α subunit (cGTHα), gonadotropinIIβ subunit (cGTHIIβ), somatolactin (cSL), and prolactin (cPRL) were isolated from the pituitaries of common carps, purified by traditional chromatographic analysis, identified by mass-chromatographic analysis, and used as immunogens in the B-lymphocyte hybridoma technique. Totally, 7, 11, 17, and 8 hybridoma cell lines were established, which were able to stably secrete monoclonal antibodies (mAbs) against cGTHα, cGTHIIβ, cSL, and cPRL, and designated as FMU-cGTHα1-7, FMU-cGTHIIβ1-11, FMU-cSL1-17, and FMU-cPRL1-8, respectively. The isotype, titer, and specificity were identified by enzyme-linked immunosorbent assay (ELISA), Western blot, and immunohistochemical staining, respectively, and application of these mAbs in the aforementioned tests has been proved. Furthermore, sensitive sandwich-ELISA systems for quantitative detection of the hormones mentioned above were also developed.

Keywords common carp; ELISA; gonadotropin; monoclonal antibody; prolactin; somatolactin

1. Introduction

As one of the oldest domesticated species of fish for food, the common carp (*Cyprinus carpio*) has been extensively studied during the past decades. Considering that the endocrinal system has a great impact on the growth and development of fish, the structure and function of the carp pituitary gland is one of the most intensively investigated subjects. Common-carp gonadotropin (cGTH), somatolactin (cSL), and prolactin (cPRL)—three hormones secreted by the pituitary gland—play crucial roles in stimulating and controlling the growth, metabolism,

文章发表于 *General & Comparative Endocrinology*, 2010, 167(3): 373-378 development, and maturation of the genitals and mammary glands. GTHs are glycoprotein hormones composed of a common alpha subunit and a hormone-specific β subunit, which confers its biological specificity. The hormonal activity is expressed only after a noncovalent association between these two subunits (Pierce and Parsons, 1981). PRL and SL are protein hormones of the growth hormone (GH) family, the latter considered to have resulted from gene duplication 400 million years ago (Miller and Eberhardt, 1983). Moreover, SL is the newest member of the GH/PRL/SL family of hormones (Ono *et al.*, 1990), but it has only been found in fish. In teleosts, expression of PRL and SL has been shown in many tissues, such as brain, gill, liver, ovary, and testis, apart from the pituitary gland (Santos *et al.*, 1999; Yang *et al.*, 1997; Zhang *et al.*, 2004). However, the full spectrum of their functions in teleosts is not completely understood (Bole-Feysot *et al.*, 1998). In this context, the application of cGTH, cSL, and cPRL in aquaculture by transgenesis has been receiving more attention recently (Hu *et al.*, 2007; Uzbekova *et al.*, 2003).

However, the production of antibodies against cGTH, cSL, and cPRL for studies of these hormones is lagging behind their expression studies. Although polyclonal antibody (pAb) against cGTH and cPRL was prepared years ago (Goos and van Oordt, 1975; Miyajima *et al.*, 1988), cross-reaction of the pAbs has always been a problem (Dubourg *et al.*, 1985; Naito *et al.*, 1983). To date, neither anti-cGTH/anti-cPRL monoclonal antibodies (mAbs) nor anti-cSL pAb or mAb has been reported, which causes the corresponding enzyme-linked immunosorbent assay (ELISA) systems to remain unavailable.

In the current study, 7, 11, 17, and 8 hybridoma cell lines, able to stably secrete mAbs against the gonadotropin α subunit (cGTHα), gonadotropinIIβ subunit (cGTHIIβ), cSL, and cPRL of the common carp, were established. Among these, some mAbs worked well in Western blots and/or immunohistochemical staining, and some were used to develop the corresponding sandwich-ELISA systems.

2. Materials and Methods

2.1 Animals

Common carps and BALB/c mice used in this study were obtained from the Laboratory Animal Center, Institute of Hydrobiology, Chinese Academy of Sciences, and Laboratory Animal Center of the Fourth Military Medical University, respectively. All animal experiments were carried out with the approval of the Committee of Animal Care and Use for Research and Education at the Institute of Hydrobiology, Chinese Academy of Sciences (Wuhan, PR China) or the Fourth Military Medical University (Xi'an, PR China). All efforts were made to minimize the number of animals used and their suffering.

2.2 Preparation, purification, and characterization of the native cGTHα, cGTHIIβ, cSL, and cPRL

Around 7000 matured common carps were killed and their pituitaries were obtained, frozen immediately in liquid nitrogen, and finally transferred to –70℃ for storage until use. Totally, about 10 g of pituitaries were used in the experiment to extract total proteins. After purification (refer Supplementary Material), target proteins were collected, characterized by SDS-PAGE and mass-spectrum sequencing, and freeze-dried for further use.

2.3 Production of hybridomas

Female BALB/c mice (8 weeks old) were immunized with 20 μg of immunogens (cGTHα, cGTHIIβ, cSL, or cPRL) in Freund's complete adjuvant (Sigma-Aldrich, USA) by subcutaneous (s.c.) injection. Subsequent immunizations were carried out twice with 20 μg of immunogens in Freund's incomplete adjuvant (Sigma-Aldrich) by s.c. injection and immunogens without adjuvant by intraperitoneal (i.p.) injection at 3-week intervals. Mice were bled from the caudal vein 10 days after the third immunization, and the antiserum titers were determined by indirect ELISA. Immunized mice were boosted with 20 μg of immunogens by i.p. injection. Three days later, splenocytes from the immunized mice and SP2/0 myeloma cells, cultured in RPMI 1640 (HyClone, USA) containing 20% fetal calf serum (GIBCO, USA), were fused in the presence of polyethylene glycol (MW4000, Merck, Germany). Positive hybrids were screened by indirect ELISA and subcloned four times by the limiting-dilution technique. mAbs were produced either from the supernatants of the hybridoma culture or from the ascites of BALB/c mice in which hybridomas had been injected by the i.p. route. Ig isotypes were identified using an Isotype kit (Sigma-Aldrich). Titers of the ascites were determined by indirect ELISA.

2.4 Western blot

The common carp were deeply anesthetized with 0.03% tricaine methane sulphonate (Crescent Research Chemicals, USA) buffered with 0.06% sodium bicarbonate (pH 7.2) and killed by decapitation. The pituitary glands were dissected, homogenized in an ice-cold lysis buffer containing 10 mM Tris-HCl (pH 7.4), 140 mM NaCl, 0.5 mM $CaCl_2$, 0.5 mM $MgCl_2$, 0.02% NaN_3, and a protease-inhibitor tablet (1 tablet/10 ml), and centrifuged at 13,000 rpm for 15 min. The proteins (50 μg/lane) in the supernatants or the purified antigens (cGTHα, cGTHIIβ, cSL, or cPRL) were denatured, subjected to SDS-PAGE, and then transferred onto polyvinylidene fluoride (PVDF) membranes (Bio-Rad). The membranes were blocked with PBS containing 5% skim milk at room temperature (RT) for 1 h with shaking, and then incubated with the corresponding mAbs (supernatants from the hybridoma cultures) or normal murine IgG (negative control) at 4℃ overnight. After washing with PBS containing 0.05% Tween 20

(PBS-T), membranes were incubated with 1 ∶ 1000 diluted horseradish peroxidase (HRP)-labeled goat anti-mouse Ig at RT for 1 h. An enhanced chemiluminescence (ECL, Roche, USA) kit was used to detect the bands.

2.5 Immunohistochemical staining

The common carp were anesthetized and killed by decapitation as mentioned above. The pituitary glands were dissected, postfixed in 4% paraformaldehyde in 0.1 M phosphate buffer (PB, pH 7.4), and further immersed in 0.1 M PB containing 30% sucrose (pH 7.4). Sagittal sections (15 μm) were cut by a freezing microtome (CM1850; Leica, Germany), thaw-mounted onto gelatinized glass slides, airdried for 20 min, and stored in 0.01 M PBS at 4℃. All sections were first treated with 0.3% H_2O_2 in Tris buffer for 10 min to inactivate endogenous peroxidase activity and incubated with the corresponding mAbs at 4℃ overnight. Then, the sections were incubated with 2 μg/ml peroxidase-conjugated goat anti-mouse secondary antibody (Pierce, USA) at RT for 4 h. Subsequently, the sections were washed 3 times and made to react for 15-20 min with 0.02% (*w*/*v*) 3,3′-diaminobenzidine tetrahydrochloride and 0.002% (*v*/*v*) H_2O_2 in 0.05 M Tris-HCl buffer (pH 7.6). The slides were then rinsed, dehydrated in ethanol, cleared in xylene, and coverslipped with a polystyrene-based medium, namely, DPX (Ajax Chemicals, Australia). Control experiments in which the primary mAbs were omitted or replaced with normal IgGs were also carried out to verify the specificity of the primary mAbs. The control experiments resulted in negligible background staining (data not shown).

2.6 Establishment of the sandwich ELISA

All the mAbs were purified from mouse ascites fluids using a Q fast-flow anion-exchange column (Amersham, UK) and conjugated with HRP (Sigma-Aldrich) by routine methods (Xu *et al.*, 2008). Subsequently, the mAb matched-pair array was carried out. In brief, purified anti-cGTHα, anti-cGTHIIβ, anti-cSL, or anti-cPRL mAbs (5 μg/mL) in coating buffer (0.05 M carbonate/bicarbonate buffer, pH 9.5) were coated at 100 μL/well on a 96-well ELISA plate (Nunc, Denmark) and incubated overnight at 4℃. After washing 3 times with PBS-T, the wells were blocked with 0.1% bovine serum albumin-PBS at RT for 1 h. After washing 3 times, 100 μL/well of the standards (cGTHα, cGTHIIβ, cSL, or cPRL) were serially diluted, added to the wells, and incubated at 37℃ for 1 h. After extensive washing, the wells were further incubated with 100 μL/well of anti-cGTHα, anti-cGTHIIβ, anti-cSL, or anti-cPRL mAbs conjugated with HRP at 37℃ for 1 h. After washing, 100 μL of 2,2′-azobis-3-ethylbenzthiazoline-6-sulfonic (ABTS, Vector) was added to each well and incubated at RT for 5-10 min. Absorbance at 410 nm (OD_{410}) was determined using a microplate reader (Bio-Rad, USA). The mAb pair with the most satisfactory result for each hormone was selected as the coating antibody and HRP- antibody to establish the corresponding double-antibody sandwich ELISA, and the standard curves were plotted using Microsoft Excel 2003.

3. Results

3.1 Purification and characterization of the native cGTHα, cGTHIIβ, cSL, and cPRL

All the hormones considered—cGTHα, cGTHIIβ, and cSL, are glycoproteins, which could be isolated from the bound proteins after affinity chromatography on ConA sephorose columns. Due to the different isoelectric points of cGTHα and the other two hormones, an ion-exchange chromatography was carried out. As an unbound protein, cGTHα was washed out (Fig. 1A, lane 1). On the basis of the different hydrophobicities, cGTHIIβ was separated from cSL in a subsequent hydrophobic-interaction chromatography. cGTHIIβ (Fig. 1A, lane 2) was first eluted out and cSL was the second (Fig. 1A, lane 3). Two bands (26 and 29 kDa) were observed in lane 3 on the SDS-PAGE gel, corresponding to cSL1 and cSL2, respectively. cPRL, a non-glycoprotein, was present in the unbound protein fraction during the ConA-affinity purification. In the next step, a Q-Sepharose chromatography was carried out to separate the proteins; the first fraction of elution (Fig. 1, lane 4) was the cPRL. All the purified proteins were confirmed by their mass spectra (Fig. 1B). Finally, 16.5 mg cGTHα, 15.5 mg cGTHIIβ, 7.1 mg cSL, and 7.7 mg cPRL were obtained.

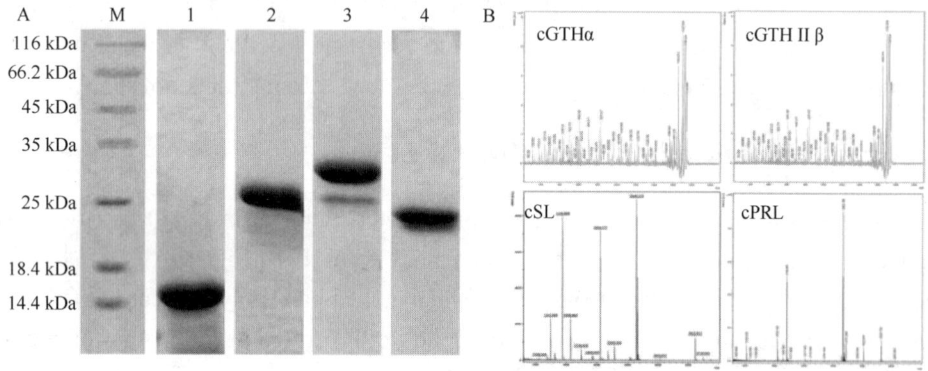

Fig. 1 Purification and characterization of the native cGTHα, cGTHIIβ, cSL, and cPRL. (A) 15% SDS-PAGE of the purified cGTHα (lane 1), cGTHIIβ (lane 2), cSL (lane 3), and cPRL (lane 4), M: marker; (B) mass-spectrum sequencing of cGTHα, cGTHIIβ, cSL, and cPRL

3.2 Preparation and characterization of the mAbs against cGTHα, cGTHIIβ, cSL, and cPRL

Totally, 7, 11, 17, and 8 hybridoma cell lines were established, which were able to stably secrete mAbs against cGTHα, cGTHIIβ, cSL, and cPRL, and these were designated as FMU-cGTHα1-7, FMU-cGTHIIβ1-11, FMU-cSL1-17, and FMU-cPRL1-8, respectively. The isotypes and ascites titers, in addition to the Western blot and immunohistochemistry results, are presented in Table 1. Furthermore, some of the mAbs could react with the freshly lysed

Table 1 Characterization of mAbs against cGTHα, cGTHIIβ, cSL, and cPRL

Clone	Isotype	Titers of ascites	Western blot[a]	Immunohistochemistry[a]
FMU-cGTHα1	IgG1,κ	10^{-6}	–	–
FMU-cGTHα2	IgG1,κ	10^{-6}	+	–
FMU-cGTHα3	IgG1,κ	10^{-6}	–	+
FMU-cGTHα4	IgG1,κ	10^{-6}	–	–
FMU-cGTHα5	IgG1,κ	10^{-7}	–	–
FMU-cGTHα6	IgG1,κ	10^{-7}	+	–
FMU-cGTHα7	IgG1,κ	10^{-7}	–	–
FMU-cGTHIIβ1	IgG1,κ	10^{-6}	+	–
FMU-cGTHIIβ2	IgM,κ	10^{-4}	–	–
FMU-cGTHIIβ3	IgG1,κ	10^{-6}	+	+
FMU-cGTHIIβ4	IgG1,κ	10^{-5}	–	–
FMU-cGTHIIβ5	IgG1,κ	10^{-6}	–	–
FMU-cGTHIIβ6	IgG1,κ	10^{-6}	–	–
FMU-cGTHIIβ7	IgG1,κ	10^{-2}	–	+
FMU-cGTHIIβ8	IgG1,κ	10^{-3}	–	+
FMU-cGTHIIβ9	IgG1,κ	10^{-6}	–	+
FMU-cGTHIIβ10	IgG1,κ	10^{-6}	–	–
FMU-cGTHIIβ11	IgG1,κ	10^{-7}	+	–
FMU-cSL1	IgG1,κ	10^{-5}	–	+
FMU-cSL2	IgG1,κ	10^{-5}	–	+
FMU-cSL3	IgG1,κ	10^{-4}	–	–
FMU-cSL4	IgG1,κ	10^{-5}	–	–
FMU-cSL5	IgG1,κ	10^{-5}	–	+
FMU-cSL6	IgG1,κ	10^{-5}	–	+
FMU-cSL7	IgG2a,κ	10^{-5}	+	–
FMU-cSL8	IgG1,κ	10^{-6}	+	–
FMU-cSL9	IgG1,κ	10^{-3}	–	–
FMU-cSL10	IgG1,κ	10^{-6}	–	+
FMU-cSL11	IgG1,κ	10^{-6}	–	+
FMU-cSL12	IgG2b,κ	10^{-6}	+	–
FMU-cSL13	IgG2b,κ	10^{-6}	+	–
FMU-cSL14	IgG2b,κ	10^{-6}	+	–
FMU-cSL15	IgG1,κ	10^{-5}	–	–
FMU-cSL16	IgG1,κ	10^{-5}	–	–
FMU-cSL17	IgG1,κ	10^{-3}	–	–
FMU-cPRL1	IgG1,κ	10^{-6}	–	+
FMU-cPRL2	IgG1,κ	10^{-7}	+	–
FMU-cPRL3	IgG1,κ	10^{-5}	–	+
FMU-cPRL4	IgG1,κ	10^{-6}	+	–
FMU-cPRL5	IgG1,κ	10^{-6}	+	–
FMU-cPRL6	IgG1,κ	10^{-4}	–	–
FMU-cPRL7	IgG1,κ	10^{-7}	–	–
FMU-cPRL8	IgG1,κ	10^{-7}	–	–

a +: positive result; –: negatine result

homogenates obtained from the carp pituitary gland, as shown in Fig. 2. FMU-cGTHα2, FMU-cGTHIIβ1, and FMU-cPRL2 could recognize bands of 16, 26, and 24 kDa (Fig. 2, lanes 2, 3, and 5), corresponding to the molecular weights of cGTHα, cGTHIIβ, and cPRL, respectively, whereas FMU-cSL7 recognized both cSL1 (26 kDa) and cSL2 (29 kDa). There was no band when the homogenates of carp pituitary gland were probed with normal murine IgG (Fig. 2, lane 1), and no cross-reaction was observed with unrelated proteins (data not shown), which proved their high specificity.

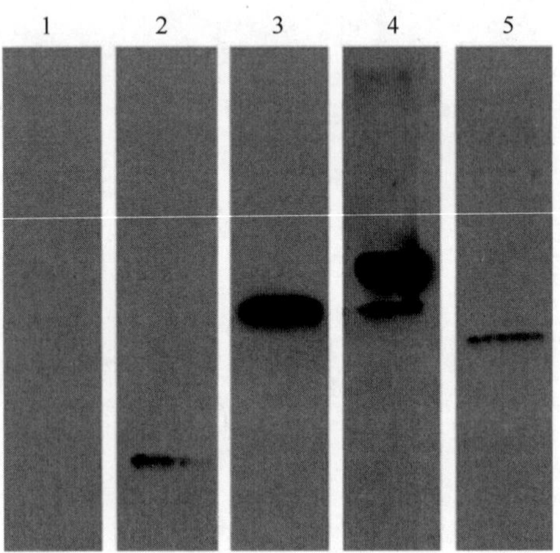

Fig. 2 Characterization of the reactivity of mAbs with homogenates from the carp pituitary gland by Western blot. The homogenates were probed with normal murine IgG (lane 1), FMU-cGTHα2 (lane 2), FMU-cGTHIIβ1 (lane 3), FMU-cSL7 (lane 4), and FMU-cPRL2 (lane 5)

3.3 Immunohistochemistry

The carp pituitary gland is composed of the neurohypophysis and the adenohypophysis, which is further subdivided into three parts that include the pars intermedia (PI), the proximal pars distalis (PPD), and the rostral pars distalis (RPD) based on their different cytological structures and hormone distributions (Follenius *et al.*, 1978). To investigate the immunohistochemical reactivity of the mAbs, sagittal sections of the pituitary gland of the common carp were prepared and probed with mAbs against cGTHα, cGTHIIβ, cSL, and cPRL; the immunostaining results showed that these four hormones were regionally expressed in a complementary manner (Fig. 3). cGTHα- and cGTHIIβ-immunoreactive cells were predominantly distributed throughout the PPD. Higher magnification images showed that these large polygonal GTH-immunoreactive cells were separated into many small clusters. cSL-immunoreactive cells were mainly localized in the PI, whereas cPRL-immunoreactive cells were mainly distributed in the RPD. Higher magnification images showed that the SL- and PRL-immunoreactive cells were small, being diverse in shape.

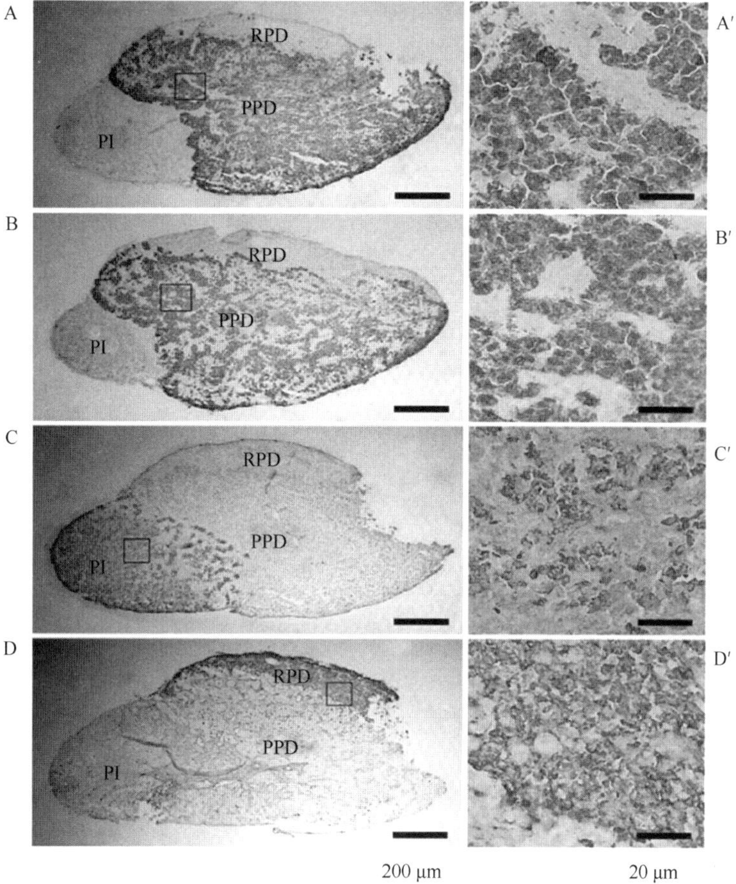

Fig. 3 Immunohistochemistry of sagittal sections of the pituitary gland of common carp. The sections were stained with FMU-cGTHα3 (A), FMU-cGTHIIβ9 (B), FMU-cSL10 (C), and FMU-cPRL1 (D). (A′-D′) Higher magnification images of the boxed areas in (A-D), respectively. Abbreviations: PI, pars intermedia; PPD, proximal pars distalis; RPD, rostral pars distalis. Scale bars = 200 μm (A-D) and 20 μm (A′-D′)

3.4 Establishment of the sandwich ELISA and standard curves

After carrying out the mAb matched-pairs array, although an ELISA test for detection of cPRL has not been successfully established due to the lack of a satisfactory mAb pair, three pairs of mAbs (FMU-cGTHα2 and FMU-cGTHα1; FMU-cGTHIIβ11 and FMU-cGTHIIβ6; FMU-cSL4 and FMU-cSL14) against cGTHα, cGTHIIβ, and cSL, respectively, were applied to establish the corresponding sandwich ELISAs. FMU-cGTHα2, FMU-cGTHIIβ11, and FMU-cSL4, which were immobilized on a 96-well plate, were used as the coating mAbs, whereas FMU-cGTHα1, FMU-cGTHIIβ6, and FMU-cSL14, which were conjugated with HRP, were used as the detecting mAbs. In Fig. 4, three typical standard curves obtained with cGTHα, cGTHIIβ, and cSL are shown, with the detection limits being 200, 50, and 100 ng/L, respectively. No cross-reaction was observed among these three proteins or with unrelated proteins (data not shown).

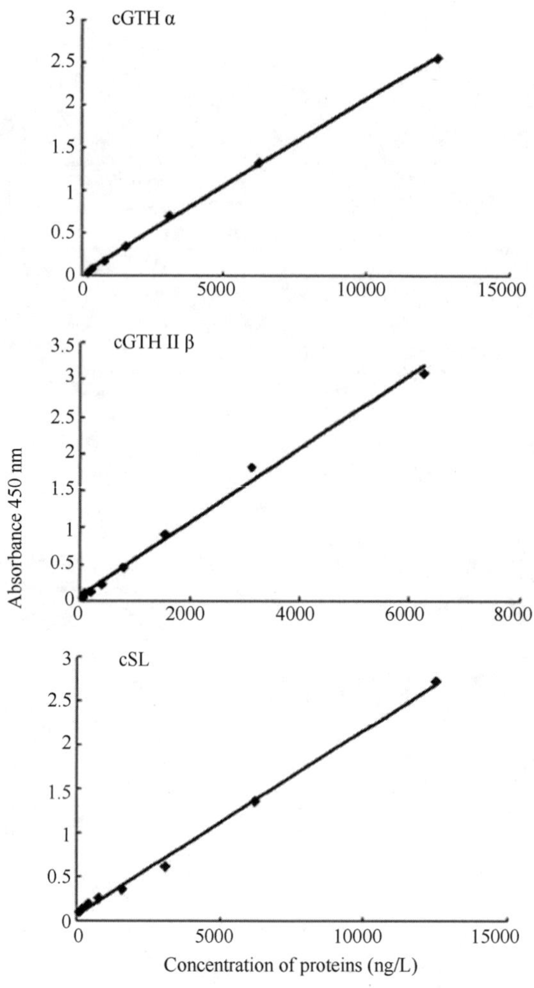

Fig. 4 Establishment of the sandwich ELISA. Typical standard curves of cGTHα, cGTHIIβ, and cSL obtained from the sandwich ELISA are shown

4. Discussion

Although there have been several reports regarding preparation and characterization of pAbs against the hormones mentioned above, the current authors are the first to prepare mAbs against cGTHα, cGTHIIβ, cSL, and cPRL. mAbs possess several advantages over their polyclonal counterparts, including high homogeneity, ease of characterization, and no batch-to-batch or lot-to-lot variability. Moreover, the greatest benefits are as follows: (1) mAbs react with a specific epitope, whereas pAbs interact with multiple epitopes on the antigen: this contrasting feature results in the higher specificity of mAbs; (2) in terms of production, the propagation of mAbs can be carried out in a culture medium or by transplantation of the hybridoma into the peritoneal cavity of syngeneic mice, from where the mAbs can be harvested through the ascites fluid. Thus, large and, theoretically at least, unlimited quantities of mAbs can be produced; (3) it is impossible to use pAb from one animal species alone for the sandwich-ELISA setup.

Although less effort is required to obtain immunogens by gene recombination, native proteins with natural conformation are much more suitable to produce mAbs with high quality. Thus, native antigens purified from the carp pituitary gland, rather than recombinant proteins, were used to immunize the mice. As a result, a series of satisfactory mAbs were obtained, among which some could be used in Western blots, whereas some worked well in immunohistochemistry. Attributed to the native antigens, it is not surprising that the mAbs could detect the hormones in the freshly lysed homogenates obtained from carp pituitary gland. In addition, no cross-reactivity was observed among these mAbs to the four different antigens when the whole homogenates were loaded onto an SDS-PAGE gel and transferred to PVDF membranes, which additionally showed the high purity of the antigens. Interestingly, using the Western blot technique, FMU-cSL7 could detect both cSL1 (26 kDa) and cSL2 (29 kDa) in the homogenates (Fig. 2). It is very probable that the epitope recognized by FMU-cSL7 is shared by both cSL1 and cSL2; so did FMU-cSL14. However, FMU-cSL8 only recognized the band of cSL2 (29 kDa), whereas both FMU-cSL12 and FMU-cSL13 only recognized the band of cSL1(26 kDa) (data not shown).

According to the mAbs obtained, sandwich-ELISA systems for the quantitative detection of cGTHα, cGTHIIβ, and cSL were successfully established. Levels of proteins as low as 200 ng/L of cGTHα, 50 ng/L for cGTHIIβ, and 100 ng/L for cSL are detected by the proposed system; this methodology is thus sensitive enough to detect the hormonal changes in common carp and probably other fishes. In the past, GTH changes were detected only by the sensitive radioimmunoassay method using the antiserum of the GTH beta subunit (Copeland and Thomas, 1992; Gur *et al.*, 2000). However, the ELISA established here is much more convenient than the radioimmunoassay system. It also provides the scientific community valuable tools to conduct many future researches, which could not be carried out before. For example, the physiological changes of cGTHα, cGTHIIβ, and cSL throughout a day, a month, a season, or a year can be studied now, in addition to the hormone changes throughout the complete developmental process of a fish. Furthermore, the system can also be largely applied in the field of transgenic fish. Various transgenic fish with different transgenes have been produced since the first transgenic human-GH fish was created, and it is inevitable that more transgenic fish with other different transgenes will be produced in future (Hu *et al.*, 2007). The potential ecological risks posed by the transgenic fish are always a big concern for a healthy aquaculture and closely related to the reproduction characteristics of transgenics (Hu *et al.*, 2007). Due to the status of these hormones in aquaculture, the proposed system will help in solving the question about how the transgene might influence the endogenous hormone changes in the pituitary, which will benefit the entire research community. For example, many labs in the world now try to place transgenic GH fish in the market (common carp, salmon, and tilapia) (Eppler *et al*., 2007; Hu *et al*., 2007; Raven *et al*., 2008). Previously, the current authors have prepared mAbs against cGH and established the corresponding ELISA system, which has been useful in the detection of cGH level in transgenic GH fish (Wu *et al*., 2008). In this study, the system can help to quantify the amounts of cGTHα, cGTHIIβ, and cSL, thus evaluating their biosafety. Moreover, the ELISA systems might have good potentials for the quantification of cGTHα, cGTHIIβ, and cSL expressed in different expression systems such as mammalian cell culture, insect cell culture, and silkworm larvae cultivation. Due to the lack of a satisfactory mAb pair, ELISA for the detection of cPRL has not been developed similar to that for the other three. The main reason is probably the shortage of positive hybridoma clones after screening and subcloning. The authors plan to reimmunize the mice with cPRL to produce more number of positive hybridoma clones, which would enhance the possibility of obtaining an excellent mAb pair.

In conclusion, the murine mAbs generated here were highly specific and had various applications, including the techniques of ELISA, Western blot, and immunohischemistry. The usefulness of the mAbs in detection of native hormones in the pituitary gland homogenates by Western blot and the quantification of cGTHα, cGTHIIβ, and cSL by sandwich ELISA have been established. The generation of mAbs in this study may help to accelerate the functional research and applications of cGTHα, cGTHIIβ, cSL, and cPRL.

Acknowledgements We express our appreciation to Dr. Bingen Ru and his laboratory staff for their help during the protein purification. This study was supported by the Development Plan of the State Key Fundamental Research of China (Grant No. 2007CB109206), Chinese National Programs for High Technology Research and Development (Grant No. 2006AA10Z141), the National Natural Science Foundation of China (Grant No. 30430540), and the Opening Project of State Key Laboratory of Freshwater Ecology and Biotechnology (Grant No. 2008FB011).

References

Bole-Feysot, C., Goffin, V., Edery, M., Binart, N., Kelly, P.A., 1998. Prolactin (PRL) and its receptor: actions, signal transduction pathways and phenotypes observed in PRL receptor knockout mice. Endocr. Rev. 19, 225-268.

Copeland, P.A., Thomas, P., 1992. Isolation of maturational gonadotropin subunits from spotted seatrout (*Cynoscion nebulosus*) and development of a betasubunit-directed radioimmunoassay for gonadotropin measurement in sciaenid fishes. Gen. Comp. Endocrinol. 88, 100-110.

Dubourg, P., Burzawa-Gerard, E., Chambolle, P., Kah, O., 1985. Light and electron microscopic identification of gonadotrophic cells in the pituitary gland of the goldfish by means of immunocytochemistry. Gen. Comp. Endocrinol. 59, 472-481.

Eppler, E., Caelers, A., Shved, N., Hwang, G., Rahman, A.M., Maclean, N., Zapf, J., Reinecke, M., 2007. Insulin-like growth factor I (IGF-I) in a growth-enhanced transgenic (GH-overexpressing) bony fish, the tilapia (*Oreochromis niloticus*): indication for a higher impact of autocrine/paracrine than of endocrine IGF-I. Transgenic Res. 16, 479-489.

Follenius, E., Doerr-Schott, J., Dubois, M.P., 1978. Immunocytology of pituitary cells from teleost fishes. Int. Rev. Cytol. 54, 193-223.

Goos, H.J., van Oordt, P.G., 1975. Proceedings: cross-reaction of rabbit anti-carp gonadotrophin globulin with gonadotrophic hormone of some teleost fish as tested by immunofluorescence. J. Endocrinol. 64, 45P.

Gur, G., Melamed, P., Gissis, A., Yaron, Z., 2000. Changes along the pituitary-gonadal axis during maturation of the black carp, Mylopharyngodon piceus. J. Exp. Zool. 286, 405-413.

Hu, W., Wang, Y., Zhu, Z., 2007. Progress in the evaluation of transgenic fish for possible ecological risk and its containment strategies. Sci. China C Life Sci. 50, 573-579.

Miller, W.L., Eberhardt, N.L., 1983. Structure and evolution of the growth hormone gene family. Endocr. Rev. 4, 97-130.

Miyajima, K., Yasuda, A., Swanson, P., Kawauchi, H., Cook, H., Kaneko, T., Peter, R.E., Suzuki, R., Hasegawa, S., Hirano, T., 1988. Isolation and characterization of carp prolactin. Gen. Comp. Endocrinol. 70, 407-417.

Naito, N., Takahashi, A., Nakai, Y., Kawauchi, H., Hirano, T., 1983. Immunocytochemical identification of the prolactin-secreting cells in the teleost pituitary with an antiserum to chum salmon prolactin. Gen. Comp. Endocrinol. 50, 282-291.

Ono, M., Takayama, Y., Rand-Weaver, M., Sakata, S., Yasunaga, T., Noso, T., Kawauchi, H., 1990. cDNA cloning of somatolactin, a pituitary protein related to growth hormone and prolactin. Proc. Natl. Acad. Sci. USA 87, 4330-4334.

Pierce, J.G., Parsons, T.F., 1981. Glycoprotein hormones: structure and function. Annu. Rev. Biochem. 50, 465-495.

Raven, P.A., Uh, M., Sakhrani, D., Beckman, B.R., Cooper, K., Pinter, J., Leder, E.H., Silverstein, J., Devlin, R.H., 2008. Endocrine effects of growth hormone overexpression in transgenic coho salmon. Gen. Comp. Endocrinol. 159, 26-37.

Santos, C.R., Brinca, L., Ingleton, P.M., Power, D.M., 1999. Cloning, expression, and tissue localisation of prolactin in adult sea bream (*Sparus aurata*). Gen. Comp.Endocrinol. 114, 57-66.

Uzbekova, S., Amoros, C., Cauty, C., Mambrini, M., Perrot, E., Hew, C.L., Chourrout, D., Prunet, P., 2003. Analysis of cell-specificity and variegation of transgene expression driven by salmon prolactin promoter in stable lines of transgenic rainbow trout. Transgenic Res. 12, 213-227.

Wu, G., Chen, L., Zhong, S., Li, Q., Song, C., Jin, B., Zhu, Z., 2008. Enzyme-linked immunosorbent assay of changes in serum levels of growth hormone (cGH) in common carps (*Cyprinus carpio*). Sci. China C Life Sci. 51, 157-163.

Xu, Z.W., Zhang, T., Song, C.J., Li, Q., Zhuang, R., Yang, K., Yang, A.G., Jin, B.Q., 2008. Application of sandwich ELISA for detecting tag fusion proteins in high throughput. Appl. Microbiol. Biotechnol. 81, 183-189.

Yang, B.Y., Arab, M., Chen, T.T., 1997. Cloning and characterization of rainbow trout (*Oncorhynchus mykiss*) somatolactin cDNA and its expression in pituitary and nonpituitary tissues. Gen. Comp. Endocrinol. 106, 271-280.

Zhang, W., Tian, J., Zhang, L., Zhang, Y., Li, X., Lin, H., 2004. cDNA sequence and spatio-temporal expression of prolactin in the orange-spotted grouper, *Epinephelus coioides*. Gen. Comp. Endocrinol. 136, 134-142.

鲤促性腺激素、生长促乳激素和催乳素的鼠源单克隆抗体制备、鉴定及应用

Zhuwei Xu[1] Qi Li[1] Yun Wang[2] Chaojun Song[1]
Tao Zhang[3] Lihua Chen[1] Jianguo Ji[4] Angang Yang[1]
朱作言[2] 胡 炜[2] Boquan Jin[1]

1第四军医大学，免疫学教研室，西安 710032
2中国科学院水生生物研究所，淡水鱼类生态与生物技术研究中心，武汉 430072
3第四军医大学，唐都医院神经外科，西安 710038
4北京大学，生命科学学院，北京 100049

摘 要 通过传统色谱分析方法，纯化获得了鲤促性腺激素 α 亚基(cGTHα)、促性腺激素Ⅱ型 β 亚基(cGTHIIβ)和促乳素(cSL)及催乳素(cPRL)蛋白。经质谱鉴定后，用作 B 淋巴细胞杂交瘤技术的抗原，进而获得能稳定分泌单克隆抗体的杂交瘤细胞系，分别命名为 FMU-cGTHα1-7、FMU-cGTHIIβ1-11、FMU-cSL1-17和 FMU-cPRL1-8。通过酶联免疫吸附试验(ELISA)，Western blot 和免疫组化染色方法鉴定了抗体对应的亚型、效价和特异性，并进一步证实了单克隆抗体的有效性，从而建立了灵敏的、定量检测上述激素的 ELISA 方法。

Defining Global Gene Expression Changes of the Hypothalamic-Pituitary-Gonadal Axis in Female sGnRH-Antisense Transgenic Common Carp (*Cyprinus carpio*)

Jing Xu[1,2] Wei Huang[1] Chengrong Zhong[1] Daji Luo[1]
Shuangfei Li[1] Zuoyan Zhu[1] Wei Hu[1]

1 State Key Laboratory of Freshwater Ecology and Biotechnology, Institute of Hydrobiology, Chinese Academy of Sciences, Wuhan

2 Graduate School of the Chinese Academy of Sciences, Beijing

Background The hypothalamic-pituitary-gonadal (HPG) axis is critical in the development and regulation of reproduction in fish. The inhibition of neuropeptide gonadotropin-releasing hormone (GnRH) expression may diminish or severely hamper gonadal development due to it being the key regulator of the axis, and then provide a model for the comprehensive study of the expression patterns of genes with respect to the fish reproductive system.

Methodology/Principal Findings In a previous study we injected 342 fertilized eggs from the common carp (*Cyprinus carpio*) with a gene construct that expressed antisense sGnRH. Four years later, we found a total of 38 transgenic fish with abnormal or missing gonads. From this group we selected the 12 sterile females with abnormal ovaries in which we combined suppression subtractive hybridization (SSH) and cDNA microarray analysis to define changes in gene expression of the HPG axis in the present study. As a result, nine, 28, and 212 genes were separately identified as being differentially expressed in hypothalamus, pituitary, and ovary, of which 87 genes were novel. The number of down- and up-regulated genes was five and four (hypothalamus), 16 and 12 (pituitary), 119 and 93 (ovary), respectively. Functional analyses showed that these genes involved in several biological processes, such as biosynthesis, organogenesis, metabolism pathways, immune systems, transport links, and apoptosis. Within these categories, significant genes for neuropeptides, gonadotropins, metabolic, oogenesis and inflammatory factors were identified.

Conclusions/Significance This study indicated the progressive scaling-up effect of hypothalamic sGnRH antisense on the pituitary and ovary receptors of female carp and provided comprehensive data with respect to global changes in gene expression throughout the HPG signaling pathway, contributing towards improving our understanding of the molecular mechanisms and regulative pathways in the reproductive system of teleost fish.

文章发表于 *PloS One*, 2011, 6(6): e21057

Introduction

The hypothalamic-pituitary-gonadal (HPG) axis is critical in the development and regulation of the reproductive, endocrine, and immune systems, in fish and other vertebrates [1]. The neuropeptide gonadotropin-releasing hormone (GnRH), which is secreted from the hypothalamus, is often described as the key regulator of this significant axis pathway [2]. Earlier observations have established that multiple forms of GnRH have the capacity to activate anterior pituitary receptors to stimulate the expression and release of reproductive hormones into the blood, including luteinizing hormone (LH), follicle-stimulating hormone (FSH) and growth hormone (GH). As a result, oogenesis and/or spermatogenesis processes are initiated, as well as sex steroid and inhibin production [2-6]. Recent studies in zerbafish (*Danio rerio*), medaka (*Oryzias latipes*) and goldfish (*Carassius auratus*) reported separately that some neuropeptides and neurotransmitters could interact with GnRH in the control of pituitary hormone release [7-9]. These results lead us to hypothesize that there may be a number of factors involved in the HPG signaling pathway that reflect the complexity of the teleost reproductive system.

In the common carp (*Cyprinus carpio*), which belongs to the family of Cyprinidae, two GnRH variants have been isolated and shown to be the same as GnRH-II (chicken GnRH-II, cGnRH-II) and GnRH-III (salmon GnRH, sGnRH), respectively. As in other teleosts, the two forms in common carp have different tissue distributions, expression levels, and regulation modes [10,11]. As in plants and mammals, the antisense transgenic technique has been applied in fish to determine the effectiveness of antisense and whether the inhibition of GnRHs is an effective tool to induce sterility [12-14]. Uzbekova *et al.* inhibited endogenous sGnRH expression by a recombinant vector containing antisense sGnRH cDNA in Atlantic salmon (*Salmo salar*) [12]. Maclean *et al.* obtained tilapia that were sterile or had very low fertility using the antisense transgenic technique [13]. Hu *et al.* [14] found that when sGnRH expression was inhibited by sGnRH antisense in common carp, the plasma LH (previously called gonadotropin-II or GtH-II) level in the males was significantly reduced, subsequently diminishing or severely limiting gonad development. As a result, this study generated some transgenic carp with abnormal gonad tissues. In the present study, we have begun to examine whether these fish might be a model to study changes in gene expression throughout the HPG axis with respect to fish reproduction.

The molecular techniques of suppressive subtractive hybridization (SSH) and microarray, have been preferentially used in recent studies as a high throughout screening approach for differentially expressed genes in different developmental or tissue samples. In the 1990s, an SSH library was constructed by Blázquez to identify candidate genes involved in the control of reproduction in the pituitary [15]. Later, SSH and microarray were combined and considered to have an advantage in generating an equalized representation of differentially expressed

ESTs, irrespective of the relatively disproportionate concentrations of the transcripts, hence guaranteeing the identification of the differentially expressed genes. For example, Villaret *et al.* [16] reported that 13 independent genes were significantly overexpressed in human tumor tissues in comparison with normal tissue when using subtractive and microarray technology. Similarly, 240 genes were identified as important in the development and maturation of the rainbow trout (*Oncorhynchus mykiss*) ovary, when using subtractive ovary and testis libraries and microarray analyses [17]. Furthermore, Vallée *et al.* [18] have reported that combined SSH and microarray techniques have even been utilized in three different vertebrate species (bovine, mouse, and frog), to identify the novel oocyte-specific genes.

In the present study, we used SSH combined with cDNA microarray analysis to screen for the changes in gene expression of the HPG axis between the female sGnRH-antisense transgenic and control common carp. Furthermore, we analyzed the biological processes associated with the activity of these reproductive system genes. In the case of some differentially expressed genes, real-time PCR was used to validate their differences. We hypothesize that differential expression patterns may account for the complex regulatory mechanisms of the reproductive system in teleost fish.

Results

The developmental status and histological characterization of the transgenic carp ovary

The transgenic common carp were generated by the injection of sGnRH antisense RNA and evaluated by tail clip PCR. Carp retaining the transgene were named AS(+) and totaled 102. Their gonadal development was checked in the next four breeding seasons. As shown in Table S1, 64 of the 102 carp had normal gonads with sperm or eggs. The other 38 carp had abnormal gonads, of which 14 were male carp, 12 were female carp, and 12 had no gonad tissues.

Twelve four-year-old female AS(+) carp with abnormal ovaries were selected and dissected. The sGnRH antisense RNA expression was confirmed in genomic DNA and hypothalamic cDNA (Shown in Figure S1). In comparison with the complete ovary tissues of normally developed control carp at the same age, the ovaries of AS(+) carp were found to be unilateral and undeveloped (Shown in Figure S2). Observation of histological sections under the microscope showed that AS(+) ovaries were pink, containing some atretic oocytes and primary oocytes, that were polygonal round or elliptic with a concentrated nucleus (Figure 1A). By contrast, control ovaries of the normal carp were creamy yellow in color, and contained a large number of oocytes at different developmental stages (Figure 1B). These results implied that the ovary of sGnRH-antisense transgenic carp AS (+) did not completely develop, being restricted in the early stages of development.

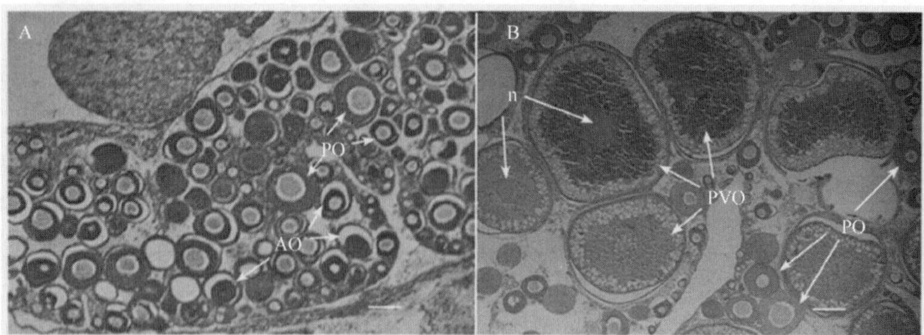

Figure 1 Histological sections of four-year-old AS(+) and control fish. (A) abnormally developed ovary section of AS(+) carp. (B) normally developed ovary section of control carp. The scale bar = 100 μm; n, nuclear; PO, primary oocyte; PVO, previtellogenic oocyte; AO, atretic oocyte

Suppressive subtractive hybridization and microarray hybridization

To identify the genes involved in HPG axis pathway differentially expressed between the transgenic and the control carp, six sublibraries of the hypothalami, pituitaries, and ovaries were created using SSH as a technique. From the six sublibraries, 15,998 single-insert clones were selected for construction of microarray cDNA chips. The hybridizations were then applied with cDNA probes labeled with fluorochromes. An example of one microarray hybridization is shown in Figure S3.

The duplicate spot intensities of the 15,998 cDNA spots for transgenic and control carp are presented as scatter plot charts, derived from Imagine software. In the scatter plot charts, the X coordinate value is the gene expression level (intensity value) in the test with the transgenic-Cy3 cDNA probes, and the Y coordinate value is the other test with the control-Cy5 cDNA probes. The green and red dots represent the clones up-regulated in AS(+) and control carp, respectively. The black dots indicate the clones with few changes between them (Figure 2). A total of 1,064 differentially expressed clones (FDR <0.01 and fold change ≥2) were obtained from the AS(+) using signal treatment. These include 23 clones from the hypothalamus, 198 from the pituitary, and 843 from the ovary. Additionally, the number of down-regulated and up-regulated clones in the three tissues was 16 and seven (hypothalamus), 171 and 27 (pituitary), 684 and 159 (ovary), respectively (data not shown).

Data analysis

The 1,064 differentially expressed clones obtained from the microarry hybridization were sequenced and analyzed in the National Center for Biotechnology Information (NCBI) for homology. After gene duplicate fragment checking, 249 valid ESTs were obtained, of which 162 ESTs had significant homology with known accessions GenBank database (*E* values ≤1e−5). This included 6 ESTs from the hypothalamus, 17 from the pituitary and 139

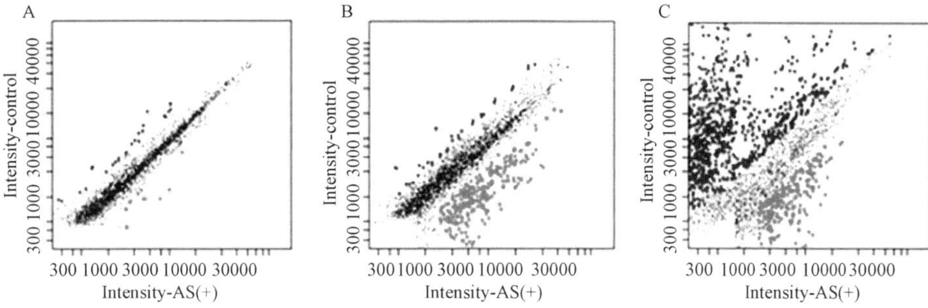

Figure 2 Scatterplot charts of the duplicate spot intensities obtained from cDNA microarray analyses. Cy3 and Cy5 dyes were used to label cDNA probes of the hypothalamus (A), pituitary (B), and ovary (C) prepared from AS(+) and control carp. The X coordinate value is the gene expression level (intensity value) in the test with the transgenic-Cy3 cDNA probes. The Y coordinate value is the other test with the control-Cy5 cDNA probes. Each dot represents a duplicate clone, and the green and red dots represent the clones up-regulated in AS(+) and control carp, respectively. The black dots indicate the clones with few changes between them

from the ovary. The remaining ESTs, about 87, were determined to be new, including 3 from the hypothalamus, 11 from the pituitary and 73 from the ovary. In additional, the confirmed amount of down-regulated and up-regulated ESTs of the three tissues were 5 and 4 for the hypothalamus, 16 and 12 for the pituitary, and 119 and 93 for the ovary, respectively. Details of each of the EST datasets are presented in Tables S2-S4. All of these ESTs have been submitted to NCBI and the GenBank Accession Numbers were JG017286-JG017531, JG390471-JG390473.

Of the ESTs explored for potential significant biological functions, 6 were expressed in the hypothalamus, and were found to be linked to the proteins arginine methyltransferase 1, hemoglobin alpha-globin, metallothionein II, Pro-melanin concentrating hormone, cytochrome oxidase subunit 1, and putative NADH dehydrogenase 5, respectively (Table 1). Of the pituitary level ESTs, 17 were primarily associated with hormones, metabolism and enzymes, such as gonadotropin subunits (gonadotropin common α subunit, FSHβ), growth hormone, cytochromec oxidase, and NADH dehydrogenase (Table 2). At the gonad level, the main molecular functions of 139 known ESTs were involved in a range of physiological and regulatory biological processes. These included enzymes, inflammation and immune factors, oogenesis and ovulation factors, cellular component organization and assembly molecules, apoptosis molecules, and other functional molecules, such as cathepsin L, zona pellucida glycoproteins, tissue inhibitor of metalloproteinase 4, ovulatory protein-2 precursor and C-type lectin (Table 3).

Comprehensive analysis of the differentially expressed genes

The functional category of the identified genes was determined based on sequence homologies and Gene Ontology (GO) enrichment analyses. The functional classifications showed that 10.4% were ribosomal/mitochondrial protein genes, 9.6% were related to protein,

Table 1 List of differentially expressed genes (FDR <0.01 and fold change ≥2) in AS(+) hypothalamus relative to control hypothalamus

Gategory and gene identity (Blast X)	Homolog species	GenBank Accession No.	*E*-value	Microarray fold change AS(+)/control
Hemoglobin subunit alpha	*Cyprinus carpio*	P02016	3.00E–75	4.14
Arginine methyltransferase	*Danio rerio*	BAE97650	2.00E–89	2.27
Metallothionein II	*Cyprinus carpio*	AF249875	4.00E–81	0.5
MCH 1 precursor	*Oncorhynchus kisutch*	P56943	3.00E–21	0.49
Differentially expressed in malignant melanoma	*Homo sapiens*	AJ293391	2.00E–16	0.46
Cytochromec oxidase subunit I	*Cyprinus carpio*	BAE97651	2.00E–47	0.39
NADH dehydrogenase	*Cyprinus carpio*	AP009047	3.00E–57	0.3

Table 2 List of differentially expressed genes (FDR <0.01 and fold change ≥2) in AS(+) pituitary relative to control pituitary

Gategory and gene identity (Blast X)	Homolog species	GenBank Accession No.	*E*-value	Microarray fold change AS(+)/control
Hormone				
Gonadotropin beta subunit 1	*Cyprinus carpio*	X59888	E–141	0.47
Glycoprotein hormones alpha chain	*Cyprinus carpio*	P01221	7.00E–67	0.45
Growth hormone	*Carassius auratus*	ABY71031	1.00E–18	0.4
Secretogranin III	*Danio rerio*	NP_957051	6.00E–32	0.4
Brain aromatase	*Carassius auratus*	BAA23757	2.00E–82	0.48
Metabolism				
Cytochrome c oxidase subunit I	*Cyprinus carpio*	BAE97651	2.00E–47	0.46
Cytochrome c oxidase subunit II	*Cyprinus carpio*	ABX72174	2.00E–43	0.42
Cytochrome c oxidase subunit III	*Carassius auratus*	AAP38173	1.00E–68	0.43
NADH dehydrogenase	*Cyprinus carpio*	AP009047	3.00E–57	0.21
Hemoglobin subunit alpha	*Cyprinus carpio*	P02016	3.00E–75	4.42
Others				
Arginine methyltransferase	*Danio rerio*	NM_200650	5.00E–35	2.37
DEAD (Asp-Glu-Ala-Asp) box polypeptide	*Danio rerio*	NP_998142	3.00E–72	4.23
Differentially expressed in malignant melanoma	*Homo sapiens*	AJ293391	2.00E–16	0.41
Pkm2 protein	*Danio rerio*	AAH67143	9.00E–30	2.13
Similar to melanoma inhibitory activity protein	*Danio rerio*	XP-001336607	4.00E–26	2.04

nucleic acid, or lipid metabolism, energy metabolism, or transport and/or signal transduction, 4.0% were related to inflammation and immune factors, 4.4% were related to oogenesis and ovulation processes, 3.6% were related to hormones, and 12% were related to other functions, such as binding proteins, ion transportation, cell signal transduction, and apoptosis. The function of 21.1% of genes was unknown, and 34.9% were termed novel ESTs as they bore no similarity to the known accessions of the GenBank database (Figure 3). Statistical analysis showed that 140 of these identified genes were down-regulated, while 109 were up-regulated.

Table 3 List of differentially expressed genes (FDR <0.01 and fold change ⩾2) in AS(+) ovary relative to control ovary

Gategory and gene identity (Blast X)	Homolog species	GenBank Accession No.	*E*-value	Microarray fold change AS(+)/control
Hormone				
Gonadotropin beta subunit 1	*Cyprinus carpio*	O13050	3.00E–63	0.26
Growth hormone	*Cyprinus carpio*	X51969	0	0.2
Isotocin precursor	*Cyprinus carpio*	AF322651	1.00E–18	3.94
MCH 1 precursor	*Oncorhynchus kisutch*	P56943	3.00E–21	0.32
Metabolism				
ATP synthase F0 subunit 6	*Ctenopharyngodon idella*	YP_001654971	1.00E–48	2.35
Cytochrome c oxidase subunit I	*Labeo batesii*	YP_913434	e–123	0.27
Cytochrome c oxidase subunit III	*Hemibartus barbus*	YP_913282	2.00E–64	2.51
Cytochrome c oxidase subunit II	*Cyprinus carpio*	ABX72174	2.00E–43	3.06
Geminin	*Danio rerio*	NM_200086	9.00E–177	0.01
NADH dehydrogenase subunit 1	*Barbus barbus*	YP_913406	3.00E–57	2.16
NADH dehydrogenase subunit 2	*Cyprinus carpio*	BAE97650	9.00E–58	2.57
Superoxide dismutase	*Danio rerio*	XP_698633	2.00E–30	0.04
Inflammation and immune				
Basigin	*Danio rerio*	AAH56721	2.00E–63	2.42
C-type-lectin	*Cyprinus carpio*	BAA95671	6.00E–80	0.01
Danio rerio high-mobility group box 1	*Danio rerio*	BC067193	e–105	2.7
Metallothionein-I	*Cyprinus carpio*	O13269	2.00E–10	0.21
Metallothionein-II	*Cyprinus carpio*	AF249875	e–126	0.04
Pentraxin	*Cyprinus carpio*	BAB69039	8.00E–55	0.02
Oogenesis and ovulation processes				
Cathepsin L	*Lates calcarifer*	ABV59078	5.00E–37	0.03
Tissue inhibitor of metalloproteinases 4	*Takifugu rubripes*	AAO17737	7.00E–22	0.03
ZP2	*Cyprinus carpio*	CAA96572	e–141	0.01
ZP3	*Cyprinus carpio*	CAA88735	e–142	0.07

Four of these genes were exhibited in all three libraries (i.e. hypothalamus, pituitary, and ovary), comprising cytochrome c oxidase subunit I, NADH dehydrogenase subunit I, hemoglobin subunit alpha, and an unknown factor homologous to a *Homo sapiens* mRNA differentially expressed in malignant melanoma (Figure 4). Furthermore, the cytochrome c oxidase subunit I and the unknown melanoma factor were found to be down-regulated while hemoglobin subunit alpha was up-regulated in all three tissue libraries. NADH dehydrogenase subunit I was down-regulated only in the hypothalamus and pituitary libraries (Tables 1-3).

Meanwhile, 11 genes were identified in the pituitary and ovary libraries, namely flavoprotein (*fp*), FSHβ, pro-opiomelanocortin, GH, DEAD (Asp-Glu-Ala-Asp) box

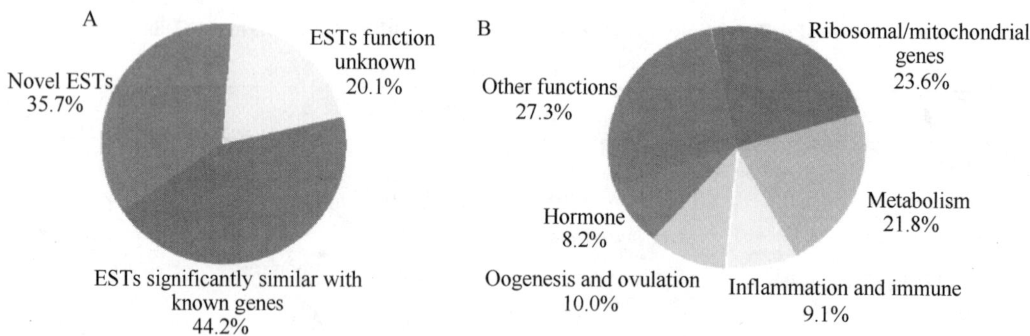

Figure 3 Pie diagram of the differentially expressed genes between AS(+) and control carp. The functional category was based on sequence homologies and Gene Ontology (GO) enrichment analyses. The percentages are shown under their categories. (A) Percentage of expressed sequence tags (ESTs) based on novel ESTs, ESTs significantly similar with known genes (*E* values ≤1e−5) and ESTs function unknown. (B) Function known ESTs were classified into ribosomal/mitochondrial genes, metabolism, inflammation and immune, oogenesis and ovulation, hormone, other molecular functions

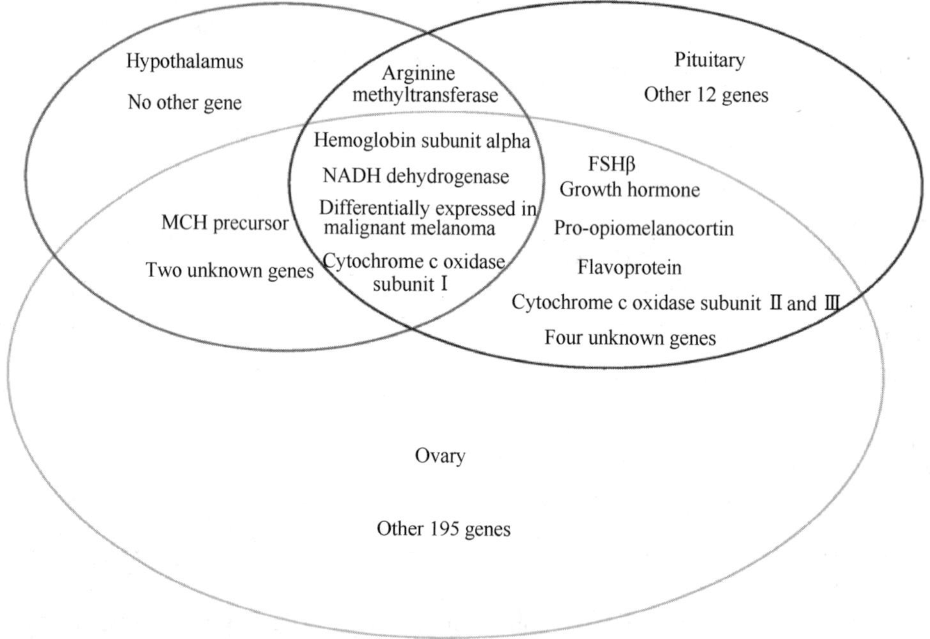

Figure 4 Schematic diagram comparing the differentially expressed genes of the hypothalamus, pituitary, and ovary. Cytochrome c oxidase subunit I, NADH dehydrogenase subunit I, hemoglobin subunit alpha, and an unknown factor homologous to a *Homo sapiens* mRNA differentially expressed in malignant melanoma were exhibited in all three libraries, flavoprotein, follicle stimulating hormone beta-subunit (FSHβ), pro-opiomelanocortin, growth hormone (GH), DEAD (Asp-Glu-Ala-Asp) box polypeptide, cytochrome c oxidase subunit II, and III, and four unknown factors were identified in the pituitary and ovary libraries. Melanin-concentrating hormone precursor (MCH) gene was exhibited in the hypothalamus and ovary libraries

polypeptide, cytochrome c oxidase subunit II, and III, and four unknown factors (Figure 4). Of these, *fp* was up-regulated in two libraries, while GH and FSHβ were down-regulated. Additionally, the MCH precursor gene (*mch*) was strongly down-regulated in the hypothalamus and ovary of AS(+) carp (Tables 1-3).

Validation of differentially expressed genes by real-time RT-PCR

To confirm the results obtained using combined SSH and microarray analyses, 25 candidates were selected from the six categories of GO enrichment analyses and amplified by real-time RT-PCR from three additional AS(+) and control carp samples. Among these genes, sGnRH, GH, gonadotropin beta subunit 1, and secretogranin III were associated with hormone functions. NADH dehydrogenase, superoxide dismutase, geminin, hemoglobin alpha, and a ribosomal gene were metabolic related factors. ZP2 and ZP3 were involved in oogenesis and ovulation directly. C-type lectin, pentraxin, basigin, and high-mobility group box 1 were inflammatory factors. MCH 1 precursor, melanoma inhibitory activity protein, S100 calcium binding protein A1, and GABA neurotransmitter transporter 1 were neuropeptides or factors associated with signal transduction. Others were arginine methyltransferase, pyruvate kinase M2, DEAD box polypeptide, B-cell translocation gene 4, and cystatin precursor. *β-actin* gene was used as control.

In some cases, a much higher-fold change was obtained from real-time RT-PCR than that from microarray results (Figure 5). This might be due to a low concentration of cDNA resulting in a relatively low dynamic range in microarrays [19,20]. The overall results of real-time RT-PCR agreed with the microarray data (Figure 5). In addition, the level of endogenous sGnRH expression was also detected using real-time RT-PCR of the hypothalamus of AS(+) and control carp. The result showed about 1.5-fold reduction from the former to the latter. These findings support the credibility of the SSH and microarray analysis results.

Plasma levels of LHβ

Microarray analysis revealed the steady-state expression of various important neuropeptides, hormones, metabolic, oogenesis, and inflammatory factors. Especially in the pituitary, gonadotropin common α subunits and FSHβ subunits were both down-regulated and repeated many times in the SSH library, but the expression of LHβ was not affected. To confirm this result, plasma levels of LHβ were checked by sandwich Enzyme-linked Immunosorbent Assay (ELISA) [21]. The mean plasma LHβ levels of AS(+) and the control group were 0.19±0.025 ng/ml and 0.37±0.048 ng/ml, $P>0.05$. This result showed that LHβ level of the two groups were not significantly different (Table 4). Therefore, these results support the patterns found in our analysis of SSH and microarray data.

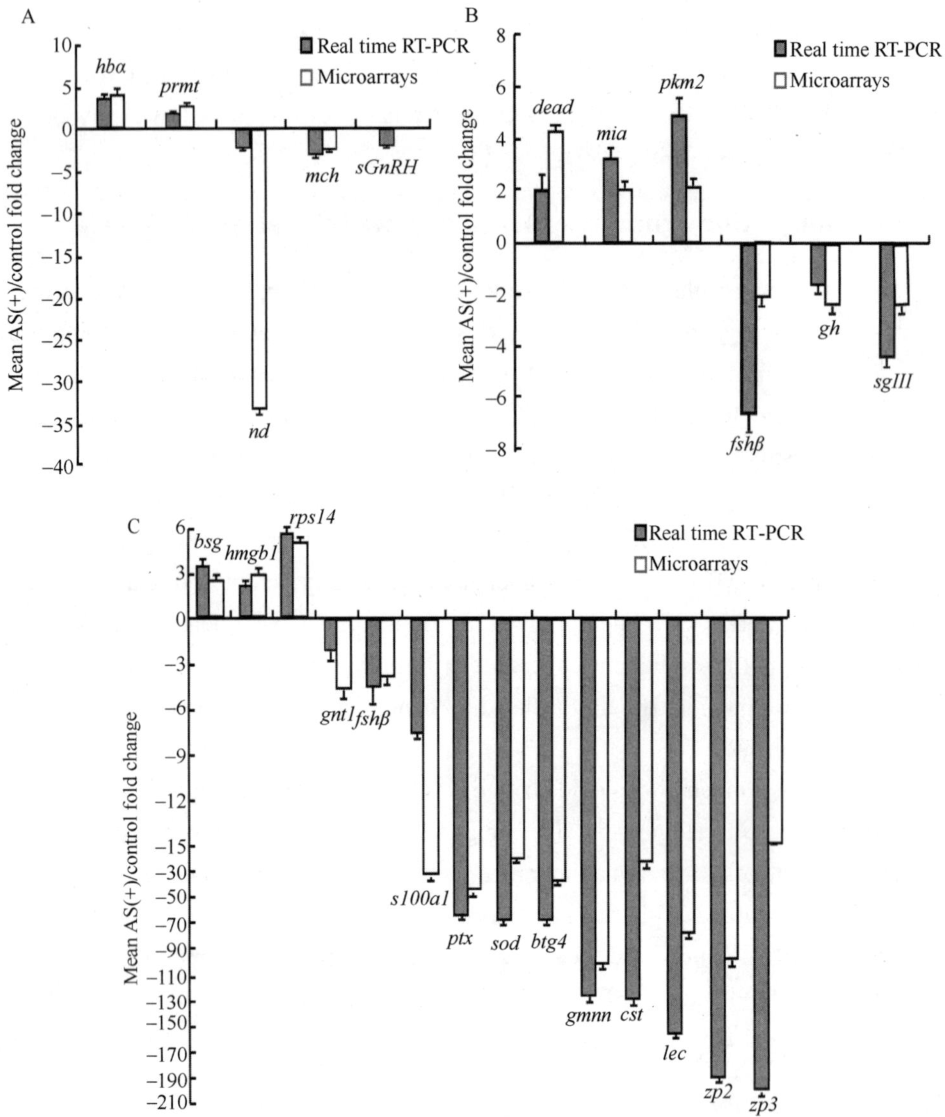

Figure 5 Validation of the differentially expressed genes using real-time quantitative RT-PCR. (A) Hemoglobin subunit alpha (*hbα*), NADH dehydrogenase (*nd*), arginine methyltransferase (*prmt*), melanin-concentrating hormone precursor (*mch*), salmon gonadotropin-releasing hormone (*sGnRH*) in hypothalamus. (B) Growth hormone (*gh*), DEAD box polypeptide (*dead*), similar to melanoma inhibitory activity protein (*mia*), follicle stimulating hormone beta-subunit (*fshβ*), pyruvate kinase M2 (*pkm2*), secretogranin III (*sgIII*) in pituitary. (C) Cystatin precursor (*cst*), C-type lectin (*lec*), B-cell translocation gene 4 (*btg4*), GABA neurotransmitter transporter 1 (*gnt1*), high-mobility group box 1 (*hmgb1*), basigin (*bsg*), ZP2 (*zp2*), follicle stimulating hormone beta-subunit (*fshβ*), 40S ribosomal protein S14 (*rps14*), S100 calcium binding protein A1 (*s100a1*), ZP3 (*zp3*), superoxide dismutase (*sod*), geminin (*gmnn*), pentraxin (*ptx*) in ovary. *β-actin* was amplified with the target gene as a positive control in each well

Table 4 Plasma LHβ level of AS(+) and control groups in 4-year-old female fish

	Samples	Mean LHβ level(ng/ml)
AS(+) group	n=6	0.19±0.025
Control group	n=6	0.37±0.048
t-test		P>0.05

Discussion

In teleosts, abnormal migration or mistargeting of GnRH neurons during embryogenesis has been shown by transgenic techniques in Atlantic salmon (*Salmo salar*), rainbow trout, and tilapia (*Oreochromis niloticus*), and the results showed that hypogonadism and infertility could occur [12,13,22]. Our laboratory has also generated transgenic common carp expressing sGnRH-antisense RNA, using an improved recombinant vector. As reported earlier, unilateral or undeveloped testes were found in some 14-month-old AS(+) male carp, while no morphological or histological differences were observed in the AS(+) and control females at the same age [14]. However, in the present study, unilateral and undeveloped ovaries were dissected from the remaining 4-year-old AS(+) female carp. One reason for the difference in females was that under natural conditions female carp took two to three years to mature sexually, while male carp could become sexually mature in one year. The gonads of 14-month-old females were incompletely developed for both the AS(+) and the control, and clearly morphological and histological changes could not be observed. Another reason may be that only 5 females were dissected randomly in the previous report and number was too small for us to obtain the carp with undeveloped ovaries.

Since the AS(+) carp with abnormal ovaries selected for SSH and microarray were founders, the antisense sGnRH was expressed as a mosaic. To eliminate the influence of mosaics, tail clip PCR and RT-PCR were performed to verify the functional integration of antisense sGnRH in AS(+) carp used in the present study. Tail clip PCR revealed that the construct was stably integrated in the genome. RT-PCR revealed the antisense sGnRH was expressed in the hypothalamus (Figure S1) and endogenous sGnRH was down-regulated (Figure 5A). The reduced sGnRH in the hypothalamus may influence gene expressions in the pituitary and ovary through the HPG axis. Gene expression surveys of ovary or testis have been carried out in recent years in a number of vertebrates, including coho salmon (*Oncorhynchus kisutch*), European sea bass (*Dicentrarchus labrax*), flatfish (*Heterosomata*), trout, zebrafish and kuruma shrimp (*Marsupenaeus japonicus*) [23-28]. Hypothalamic transcriptome analyses were applied in goldfish to study the neuroendocrine regulation. Changes of mRNA levels following various treatments were relatively modest in hypothalamus compared to other tissues [9,29,30], since disregulation of hypothalamus eventually leads to complete failure of control mechanisms. In the present study, AS(+) female carp were used as specific variations to check gene expression changes in the tissues of hypothalamus, pituitary and ovary. As a result, differentially expressed candidate genes were isolated specifically correlating to the hypothalamus, pituitary, and ovary subtraction libraries (Tables 1-3). The number of differentially expressed genes was smallest in the hypothalamus while was the largest number was in the ovary, indicating the progressive scaling-up effect of hypothalamic sGnRH antisense. The effect was consistent with an existing theory whereby sGnRH acts as the start signal molecule of the HPG axis pathway [31]. Of interest, most genes identified in the library were neuropeptides, enzymes, growth hormones, and immune molecules, in addition to a major cluster of ribosomal/mitochondrial genes.

In the present study, the neuropeptide melanin-concentrating hormone precursor (*mch*) is down-regulated in both hypothalamus and ovary libraries of AS(+). The protein MCH has been recorded in the brains of organisms ranging from mammals to fish, and functions in several physiological processes, such as food intake, osmoregulation and lactation [32]. With respect to reproduction, studies of female rats have shown that MCH is capable of stimulating the luteinizing hormone-releasing hormone (LHRH) which also known as GnRH from the median eminence and gonadotropin release from the pituitary [33]. In teleosts, Amiya *et al.* [34] have provided anatomical evidence in barfin flounder (*Verasper moseri*) showing that GnRH immunoreactive fibers were in close contact with MCH cell bodies in the hypothalamus. More recently, *in vitro* experiments on goldfish pituitary cells by Tanaka *et al.* [35] found that graded doses of MCH could stimulate the secretion of LH while suppressing somatolactin (SL) release via the MCH receptor. Hence our results, together with the published literature, suggest that hypothalamic MCH may play a role as a neuromodulator involved in controlling the release of gonadotropins associated with sGnRH.

FSH and LH, formerly termed GtH I and GtH II, are the two principal gonadotropins in vertebrates, and consist of a common α-subunit and the corresponding hormone-specific β-subunit FSHβ or LHβ [36]. Previous studies have indicated that these subunits play important, but different, roles in the reproductive system. It is believed that FSHβ controls growth stages, including puberty and gametogenesis. Its expression is at high levels during early vitellogenic stages. LHβ controls the maturation phases, including gonadal maturation and spawning. Its expression is lower during the vitellogenic stages and increases toward spawning [37,38,39]. In the present study, AS(+) and the control carp were sampled in January when normal gonads were at the previtellogenic or vitellogenic stages. Compared with the control group, common α and FSHβ were observed to be strongly down-regulated in the AS(+) pituitary (Table 2) while LHβ was not detected in the libraries and its plasma level in AS(+) indicated no significant difference (Table 4). The different expression patterns of FSHβ and LHβ observed in our study are consistent with their functional differences in controlling fish reproduction.

Meanwhile, growth hormone (GH), also known as somatotropin in fish, was found to be down-regulated in both pituitary and ovary libraries of the AS(+) carp. In recent years, it has been established that the pituitary growth hormone is involved in the reproductive system of a number of teleost species, in addition to the obligatory role in body growth. The secretion of GH is regulated by hypothalamic neuroendocrine factors, and is capable of stimulating ovarian steroidogenesis [2,40,41]. Therefore, a reduction in the expression of GH is not unexpected in transgenic carp, due to the degradation of sGnRH and the incomplete status of ovary development.

Alterations in steroid levels may serve as a stimulus signaling the alteration of teleosts' metabolic processes, because these fluctuations could be translated by neuroendocrine mechanisms into signals that, in turn, alter the pattern of food intake and subsequent energy balance [42,43]. Our study supported this viewpoint, with a large number of metabolic-related factors being detected, including cytochrome c oxidase subunits I, II, III, NADH dehydrogenase subunits, and ATP synthetase varying in AS(+) tissues, as well as a large number of ribosomal/mitochondrial genes (22.0%). Cytochrome c oxidase is a large transmembrane protein complex, containing three common subunits, I, II and III. As a key enzyme in aerobic metabolism, cytochrome c oxidase catalyzes the transfer of electrons from reduced cytochrome c to molecular oxygen [44]. NADH dehydrogenase, also called complex I, is an enzyme located in the inner mitochondrial membrane that catalyzes the transfer of electrons from NADH to coenzyme Q [45]. The two enzymes are considered to be vital to ATP synthesis in organisms. Recently, Tingaud-Sequeira *et al.* [25] and Lake *et al.* [46] independently reported that cytochromes and NADH were up-regulated during the vitellogenesis of flatfish and in the dietary phosphorus of the rainbow trout. These findings suggest the putative mitochondrial function of high energy production. In the present study, the cytochrome c oxidase subunit I was detected as down-regulated in all three tissue libraries. On the other hand, the NADH dehydrogenase subunit I was down-regulated only in the hypothalamus and pituitary libraries (Tables 1-3). These results may be related to the reduced energy expenditure requirement of the dysfunctional ovary of AS(+) carp. However, the NADH dehydrogenase subunit I and ATP synthase, a constituent of the respiratory chain responsible for ATP synthesis, were detected as up-regulated in the ovary library (Table 3). These adaptations may comprise a vital compensatory response to the abnormal conditions of the transgenic ovary.

An abnormal ovary is the greatest variation in the biology of AS(+) females as opposed to the control carp, hence ovarian transcripts involved in the development procession are likely to be affected. Indeed, a set of genes were confirmed as down-regulated in the ovary libraries based on the combined SSH and microarray analyses. These included cathepsin L (*catl*), zona pellucida glycoprotein 2 (*zp2*), zona pellucida glycoprotein 3 (*zp3*), tissue inhibitor of metalloproteinase 4 (TIMP4), and ovulatory protein-2 precursor (Table 3). Cathepsin L is a member of the papain-like family of cysteine proteinases. Robker *et al.* [47] has reported in mouse that Cathepsin L could be induced by FSH in the granulosa cells of growing follicles, while high levels of *catl* mRNA could be also induced by luteinizing hormone in a progesterone receptor-dependent fashion in pre-ovulatory follicles. In zebrafish, *catl* was also reported to be associated with ovarian follicle degeneration [48]. The zona pellucida (ZP) is an extracellular glycoprotein matrix which surrounds all mammalian oocytes. Recent teleost researches have shown the synthesis of ZP transcripts and proteins involved in yolk incorporation (vitellogenesis) and processing of primary growth [23,27]. Tissue inhibitors of metalloproteinases (TIMPs), which comprise a family of four proteases TIMP1, TIMP2, TIMP3 and TIMP4, are inhibitors of the matrix metalloproteinases (MMPs), a group of peptidases that are involved in the degradation of the extracellular matrix. Studies in mammals have indicated that MMPs/TIMPs might regulate normal follicular development and atresia to achieve the appropriate number of ovulatory follicles [49]. As to the ovulatory protein-2 precursor, previous studies in trout have described its dramatic up-regulation at the time of ovulation, and suggested a role in protecting ovulated eggs from bacterial infection, in addition to the concurrent function of ovulation [50]. Therefore, according to the published literature, the down-regulation of these genes in this study indicate that they are involved in the incomplete development process of the AS(+) ovary, related to the ovary adhesion tissues, vitellogenesis, oocyte maturation and ovulation processes, respectively.

In teleosts, several elegant studies of flatfish, rainbow trout, and salmon have indicated that inflammatory factors may be involved in the process of ovary development. These factors include serine protease 23 (*sp23*), disintegrin and metalloproteinase domain-containing protein 22 (*adam22*), Chemokine (C-X-C motif) ligand 14 (*cxcl14*), angiotensin I converting enzyme 2 (*ace2*), and leukocyte cell-derived chemotaxin 2 (*lect2*) [25,27,51]. The present study has provided further evidence validating the involvement of inflammatory factors in the process of ovary development, whereby several immune genes (C-type lectin, pentraxin, and metallothionein-I, -II) were detected to be strongly down-regulated in the AS(+) ovary (Table 3). It has been reported that the C-type lectins represent a large family, containing a common carbohydrate recognition domain that interacts with glycoproteins in a Ca^{2+}-dependent manner [52,53]. This family has been found throughout the animal kingdom, and is involved in many immune-system functions, such as innate immunity, or the inflammation and immunity response to tumor and virally infected cells [54]. The pentraxins are another type of immune gene, and are a family of proteins characterized by a pentagonal discoid arrangement of five non-covalently bound subunits similar to that of legume lectins [53]. Pentraxins have been found to be important for innate defense, particularly the acute phase response, in both mammals and fish [54,55]. The metallothioneins (MTs) have been described in a wide range of taxonomic groups, and are involved in heavy metal detoxification and homeostasis [56]. In our study these immune factors were obtained from the undeveloped ovary of the AS(+) carp, which based on the existing literature suggests that these genes may be involved in the process of teleost ovary development.

In contrast to the genes mentioned above, basigin was detected to be up-regulated in the AS(+) ovary. This gene is a member of the immunoglobulin superfamily that is also known as CD147, BSG, and EMMPRIN. The fundamental role of this gene is in intercellular recognition, including a range of immunologic phenomena, differentiation, and development. It has also been suggested that basigin may regulate mouse spermatogenesis, follicular development, and oocyte maturation [57]. In teleosts, basigin has been expressed in follicle/interstitial cells that are associated with previtellogenic growth in coho salmon [51]. Hence, our findings support those of the previous studies, whereby the increased expression of this gene may contribute to oocyte development in the abnormal ovary.

The data relating to the differentially expressed genes identified in this study with respect to the HPG signaling pathway are far more extensive than those mentioned above. Of note, the presented genes that have yet to be named with unknown functions, or span multiple functional categories have not been considered in this discussion, including tissue remodeling, ion transportation, apoptosis, cell-cycle progression, and growth. These have not been considered in this discussion and are a valuable resource for further investigation.

In conclusion, this study provided novel information on transgenic common carp with abnormal ovaries as a result of sGnRH-antisense RNA expression. This study also provided comprehensive data with respect to changes in gene expression throughout the HPG signaling pathway, as a result of employing combined SSH and microarrays. Moreover, significant gene families that were represented in the SSH libraries were highlighted, including neuropeptides, gonadotropins, metabolic, oogenesis and inflammatory factors. Finally, a list of 87 candidate novel genes was successfully generated, requiring further analyses to validate their uniqueness and roles. Therefore, this study not only indicated the progressive scaling-up effect of hypothalamic sGnRH antisense acting on the pituitary and ovary, but also provided new insights with respect to the understanding of the molecular mechanisms and regulative pathways in the reproductive system of teleosts.

Materials and Methods

Ethics statement

The animals are provided with the best possible care and treatment and are under the care of a specialized technician. Also, all animals are cared for and handled with respect. All procedures were conducted in accordance with the Guiding Principles for the Care and Use of Laboratory Animals and were approved by Institute of Hydrobiology, Chinese Academy of Sciences (Approval ID: keshuizhuan08529).

Screening of female transgenic common carp

The transgenic common carp with an antisense sGnRH construct were as described in our previous report [15]. In brief, the antisense sGnRH construct which contained a carp *β-actin* gene promoter, a 328-bp antisense DNA fragment and the 3′ flanking sequence of grass carp growth hormone gene was injected into freshly fertilized eggs from wild parent carp in April 2003. The same age non-transgenic siblings of the transgenic carp were raised as controls.

Transgenic and non-transgenic carp were raised communally in secure tanks in Wuhan Duofu Scientific & Technological Farm Co., Ltd. After eight months, the transgenic carp were screened for the presence of the transgene using PCR. A total of 342 putative transgenic carp were assayed, of which 102 were obtained with the antisense construct. Carp retaining the transgene were named AS(+).

In every breeding season (usually April) of the next four years, the AS(+) carp were pressed lightly on the abdomen and sperm could be extracted from the fertile males. Eggs could be extracted from those with normal ovaries with a tool made specially for female carp (copper, 10 cm long, with a 1.5 cm long and 0.3 cm wide groove in the anterior part). The remaining carp without sperm and eggs were tagged by coded wire tags (Northwest Marine Technology, Inc.) and the statistical data was put on record. In Jan 2008, the tagged AS(+) carp which could not produce mature gametes were dissected. Twelve AS(+) carp with abnormal ovaries and 10 four-year-old non-transgenic female siblings were selected for our analysis. Their blood was collected from the caudal vasculature and the tissues of hypothalamus, pituitary and gonads were dissected and frozen in liquid nitrogen immediately before storing at −80℃. At the same time, parts of the gonads were fixed in Bouin's solution for histological examination. Fixed gonads were dehydrated and embedded in paraffin, and 5 μm sections were cut and stained with Regaud Haematoxylin-Orange G-Aniline blue.

Tail clip PCR and RT-PCR were used to evaluate the expression of sGnRH antisense RNA in the genome and hypothalamus of 12 female AS(+) carp. Genomic DNA was extracted from the caudal fin using DNeasy Blood & Tissue Kit (Qiagen), and total RNA was extracted from the hypothalamus using Trizol (Invitrogen) reagent according to the manufacturer's instructions. Antisense sGnRH were amplified by the primers P1 (5′-CCATGG-CGTATCG-ATGTCGAC-3′) and P2 (5′-CATGGCTTTGCCAGCATTGG-3′). The PCR program was as follows: 1 cycle of 90℃ for 1 min, 35 cycles of 90℃ for 20 s, 55℃ for 30 s, 72℃ for 50 s; 1 cycle of 72℃ for 5 min. *β-actin* was amplified as a positive control and the sequences of *β-actin* primers were ACT-F, 5-CACTGTGCCCATCTACGAG-3 and ACT-R, 5-CTGCATCC-TGTCAGCAATGC-3.

Total RNA extraction and cDNA synthesis

Tissue samples of the hypothalamus, pituitary, and ovary were collected from three AS(+) and three control carp, respectively. Total RNAs were extracted using the TRIzol reagent (Invitrogen) and purified using PolyA Tract mRNA Isolation System (Promega) following the recommended guidelines. RNA quality and purity was measured using a spectrophotometer (Eppendorf Biometer) and electrophoresis on 1% agarose gels. RNA from the hypothalamus of three AS(+) carp were pooled together and RNA from three control carp were similarly pooled together, as were RNA from the pituitary and ovary. SMART cDNA was synthesized from 50 ng of total RNA using the SMART cDNA Library Construction Kit following a commercial protocol (Clontech) as described previously[58]. Subsequently, double-strand cDNAs were produced through PCR amplification and purified by phenol:chloroform:isoamyl alcohol (25 ∶ 24 ∶ 1) extraction.

Suppressive subtractive hybridization

To identify the genes involved in the HPG axis pathway that were differentially expressed between transgenic and control carp, we used SSH as a technique to create six sublibraries of the three levels of hierarchy from hypothalamus to pituitary, and to ovary, respectively. The purified double-stranded cDNA obtained from AS(+) was used as the "tester", while the cDNA obtained from control carp served as the "driver" (forward SSH library). Conversely, cDNAs from AS(+) and control carp were also used as driver and tester samples, respectively (reverse SSH library). Construction of the forward and reverse libraries was performed using a PCR-select cDNA subtraction kit (Clontech) according to the SSH procedure. Briefly, cDNA from each of the tester and driver populations were digested with *Rsa* I to produce shorter blunt-ended fragments. The digested tester cDNAs were then subdivided into two populations, each of which was ligated with a different adaptor, provided in the cDNA subtraction kit. PCR was performed to evaluate the efficiency of ligation using primers specific to *β-actin* and to the adaptor sequences. Following ligation, two rounds of hybridization and PCR amplification were performed. In the first hybridization step, the driver was added in excess to each tester, denatured, and allowed to anneal. In the second hybridization step, the two products from the first hybridization were mixed together, in addition to freshly denatured cDNA driver. Subsequently, the populations of normalized and subtracted single-strand target cDNA samples annealed with each other, forming double-stranded hybrids with different adaptor sequences at their 5′ ends. Finally, the subtracted molecules were specifically amplified using adaptor-specific primer pairs of 'nested PCR'.

Construction of subtracted cDNA libraries

PCR-amplified cDNAs produced by SSH were ligated into the pMD18-T plasmid vector (TaKaRa) and transformed into competent *Escherichia coli* (strain DH5α) by electroporation (Pulse Controller, BioRad, USA). The transformed bacteria were plated onto solid Luria Bertani medium containing ampicillin, X-gal and IPTG, and incubated overnight at 37℃. Of note, pMD18-T plasmid contains LacZ reporter which allows blue-white screening. About 20,000 recombinant white clones were randomly selected and amplified in a 100 μl PCR system using nested primer 1 and 2R (Clontech) for positive detection. Aliquots (1 μl) of the PCR products were analyzed in 1% agarose gel to verify the quality and quantity. Then, the single-insert clones with amplified fragments 200-1,000 bp were selected for the construction of cDNA chips.

Microarray construction

Numbers of cDNA microarray chips containing cDNA spots representing 15,998 SSH clones, of which 4,996 clones were from the hypothalamus libraries, 4,992 from the pituitary libraries, and 6,000 from the ovary counterparts, were constructed. Briefly, the PCR products of single-insert clones were purified by the chilled ethanol precipitation method and redissolved in 15 μl of 50% dimethyl sulphoxide (DMSO), and finally spotted onto amino-silaned glass slides with a SmartArrayer™ microarrayer (CapitalBio Corp., Beijing, China). Each clone was printed in duplicate. The slides were baked for 1 h at 80℃ and then stored at room temperature until use.

On each microarray chip, eight sequences derived from intergenic regions in yeast genome, showing no significant homology to common carp in GenBank, were spotted as external controls. And two housekeeping genes of common carp (*β-actin* and Glyceraldehye-3-Phosphate Dehydrogenase [GAPDH]) were used as internal controls. Additionally, 50% DMSO was used as a negative contol for subtracting the background, and Hex was positive control for nucleic acid fixation.

Preparation of fluorescent probes and hybridization of microarrays

Total RNA samples of the hypothalamus, pituitary and ovary, obtained from three additional AS(+) and control carp, were extracted using the standard TRIzol RNA isolation protocol and purified as mentioned above. Briefly, 5 μg of each isolated RNA was respectively reverse transcribed with an oligo (dT) 8-12 (Promega), Cy3/Cy5 CTP, and Superscript II reverse transcriptase (Invitrogen Life Technologies, Shanghai, P.R. China). In order to prepare the cDNA probes, Cy3 and Cy5 dyes (Amersham, Piscataway, NJ) were used to label the cDNA isolated from AS(+) and control tissue samples. All probes were purified and later hybridized with the spotted array at 42℃ for 16 h. The slides were then washed once with 2× SSC-0.1% SDS at 42℃ for 10 min, and four times with 0.1× SSC at room temperature for 1 min. Finally, the slides were washed with distilled water, ethanol, and then dried. Each hybridization step was performed twice for replicate dye-swaps.

Microarray data analysis

Arrays were scanned with a confocal laser scanner, LuxScan™ 10K (CapitalBio Corp.), and the resulting images were analyzed with SpotData Pro 2.0 software (CapitalBio Corp.). Spots with fewer than 50% of the signal pixels exceeding the local background value for both channels (Cy3 and Cy5) plus two standard deviations of the local background were removed. This step further ensured that spots with characteristic doughnut shapes, often encountered on microarrays, would not be part of the subsequent analysis. A spatial and intensity-dependent (LOWESS) normalization method was employed to normalize the ratio values [59]. Normalized ratio data were then log transformed. cDNA spots with less than four out of total six data points in each replicated hybridization were removed. Differentially expressed genes were identified using *t* test and multiple test corrections were performed using False Discovery Rate (FDR). Genes with FDR <0.01 and a fold change greater than or equal to two were identified as differentially expressed genes.

DNA sequence analysis and gene ontology analysis

The identified differentially expressed clones were sequenced using the M13± primer pairs (Invitrogen Life Technologies). Sequences were analyzed with the Basic Local Alignment Search Tool (BLAST) in NCBI for homology. cDNAs with *E* values ≤1e−5 were designated as having significant homology, and the higher score affirmed the corresponding gene. Functional categories of the identified genes were assigned based on the Gene Ontology annotations (http://www.geneontology.org/).

Validation of differentially expressed genes by real-time RT-PCR

The differentially expressed genes were further validated by real-time quantitative RT-PCR that were run on an ABI 7000 fluorescent sequence detection system (Perkin-Elmer, Foster City, CA), with SYBR green-based detection (ABI) using gene-specific primer pairs. In practice, 25 genes were chosen for the real-time RT-PCR analysis, five of which were tested in hypothalamus, six in pituitary, 14 in ovary. *β-actin* was used as the control housekeeping gene since it has been found not to vary across organs in carp [60] and is not affected by GnRH [61,62]. The primers of these genes are listed in Table S5. Total RNA came from three additional AS(+) and control carp specimens. Reactions were performed using the following conditions: an initial incubation at 95℃ for 5 min, followed by 40 cycles at 95℃ for 15 s, 55℃ for 15 s and 72℃ for 45 s. Output data generated by the instrument onboard software were transferred to a custom-designed Microsoft (Redmond, WA) Excel spreadsheet for analysis. The differential mRNA expression of each sample was calculated as previously described by the comparative Ct method with the formula $2^{(-\text{Delta Delta C(T)})}$ method [63,64]. The experiments were conducted independently for each of the hypothalamus, pituitary and ovary from the AS(+) carp with unilateral gonad and their control siblings. Each reaction was performed in triplicate, with the means being evaluated using the Student *t*-test (P<0.01).

Sandwich ELISA to detect plasma LHβ levels

To detect levels of LHβ, blood samples from an additional six AS(+) and six control carp were collected from the caudal vasculature, kept on ice, and later centrifuged to obtain plasma samples which were then examined by the sandwich ELISA system developed in our previous report [21]. In brief, Anti-LHβ mAb (5 μg/mL) in coating buffer (0.05 M carbonate/bicarbonate buffer, pH 9.5) was coated at 100 μL/well on a 96-well ELISA plate (Nunc, Denmark), incubated overnight at 4℃ and washed 3 times with PBS containing Tween-20 (PBS/Tween). Plates were then blocked with a 0.1% BSA-PBS. Standard LHβ serial dilutions with PBS containing BSA and Tween-20 or plasma samples of carp were added to the wells and incubated. After extensive washing with PBS/Tween, wells were incubated with 100 μL/well of anti-LHβ mAbs conjugated with HRP. Wells were washed once again for 3 times and 100 μL of 2,2′-azobis-3-ethylbenzthiazoline-6-sulfonic(ABTS, Vector) was added to each well. Plates were read at 410 nm on a microplate reader. Comparison of the mean LHβ concentration in the plasma of AS(+) and control groups was performed using the Student's *t*-test.

References

1. Sower SA, Freamat M, Kavanaugh SI (2009) The origins of the vertebrate hypothalamic-pituitary-gonadal (HPG) and hypothalamic-pituitary-thyroid (HPT) endocrine systems: new insights from lampreys. Gen Comp Endocrinol 161: 20-29.
2. Zohar Y, Muñoz-Cueto JA, Elizur A, Kah O (2010) Neuroendocrinology of reproduction in teleost fish. Gen Comp Endocrinol 165: 438-455.
3. Yaron Z, Gur G, Melamed P, Rosenfeld H, Elizur A, *et al.* (2003) Regulation of fish gonadotropins. Int Rev Cytol 225: 131-185.
4. Wong TT, Gothilf Y, Zmora N, Kight KE, Meiri I, *et al.* (2004) Developmental expression of three forms of gonadotropin-releasing hormone and ontogeny of the hypothalamic-pituitary-gonadal axis in gilthead seabream (*Sparus aurata*). Biol Reprod 71: 1026-1035.
5. Vickers ED, Laberge F, Adams BA, Hara TJ, Sherwood NM (2004) Cloning and localization of three forms of gonadotropin-releasing hormone, including the novel whitefish form, in a salmonid, *Coregonus clupeaformis*. Biol Reprod 70: 1136-1146.
6. Guilgur LG, Moncaut NP, Canario AV, Somoza GM (2006) Evolution of GnRH ligands and receptors in gnathostomata. Comp Biochem Physiol A: Mol Integr Physiol 144: 272-283.
7. Kitahashi T, Ogawa S, Parhar IS (2009) Cloning and expression of kiss2 in the zebrafish and medaka. Endocrinology 150: 821-831.
8. Popesku JT, Martyniuk CJ, Mennigen J, Xiong H, Zhang D (2008) The goldfish (*Carassius auratus*) as a model for neuroendocrine signaling. Mol Cell Endocrinol 293: 43-56.
9. Popesku JT, Martyniuk CJ, Denslow ND, Trudeau VL (2010) Rapid dopaminergic modulation of the fish hypothalamic transcriptome and proteome. PLoS One 20: e12338.
10. Li S, Hu W, Wang Y, Zhu Z (2004) Cloning and expression analysis in mature individuals of two chicken type-II GnRH (cGnRH-II) genes in common carp (*Cyprinus carpio*). Sci China C Life Sci 47: 349-358.
11. Li SF, Hu W, Wang YP, Sun YH, Chen SP, *et al.* (2004) Cloning and expression analysis in mature individuals of salmon gonadotropin-releasing hormone (sGnRH) gene in common carp. Acta Genetica Sinic. 31: 1072-1081. (In Chinese with English abstract).
12. Uzbekova S, Chyb J, Ferriere F, Bailhache T, Prunet P, *et al.* (2000) Transgenic rainbow trout expressed sGnRH-antisense RNA under the control of sGnRH promoter of Atlantic salmon. J Mol Endocrino 25: 337-350.
13. Maclean N, Hwang G, Molina A, Ashton T, Muller M, *et al.* (2003) Reversibly-sterile fish via transgenesis. Information Systems for Biotechnology News Report 12: 3-5.
14. Hu W, Li S, Tang B, Wang Y, Lin H, *et al.* (2007) Antisense for gonadotropin-releasing hormone reduces gonadotropin synthesis and gonadal development in transgenic common carp (*Cyprinus carpio*). Aquaculture 271: 498-506.

15. Blázquez M, Bosma PT, Chang JP, Docherty K, Trudeau VL (1998) Gamma-aminobutyric acid up-regulates the expression of a novel secretogranin-II messenger ribonucleic acid in the goldfish pituitary. Endocrinology 139: 4870-4880.
16. Villaret DB, Wang T, Dillon D, Xu J, Sivam D, *et al.* (2000) Identification of genes overexpressed in head and neck squamous cell carcinoma using a combination of complementary DNA subtraction and microarray analysis. Laryngoscope 110: 374-381.
17. Von Schalburg KR, Rise ML, Brown GD, Davidson WS, Koop BF (2005) A comprehensive survey of the genes involved in maturation and development of the rainbow trout ovary. Biol Reprod 72: 687-699.
18. Vallée M, Gravel C, Palin MF, Reghenas H, Stothard P, *et al.* (2005) Identification of novel and known oocyte-specific genes using complementary DNA subtraction and microarray analysis in three different species. Biol Reprod 73: 63-71.
19. Ozturk ZN, Talame V, Deyholos M, Michalowski CB, Galbaith DW, *et al.* (2002) Monitoring largescale changes in transcript abundance in drought and salt-stressed barley. Plant Mol Biol 48: 551-573.
20. Wu Z, Soliman KM, Bolton JJ, Saha S, Jenkins JN (2008) Identification of differentially expressed genes associated with cotton fiber development in a chromosomal substitution line (CS-B22sh). Funct Integr Genomics 8: 165-174.
21. Xu Z, Li Q, Wang Y, Song C, Zhang T, *et al.* (2010) Production, characterization, and applications of mouse monoclonal antibodies against gonadotropin, somatolactin, and prolactin from common carp (*Cyprinus carpio*). General and Comparative Endocrinology 167: 373-378.
22. Maclean N, Rahman MA, Sohm F, Hwang G, Iyengar A, *et al.* (2002) Transgenic tilapia and the tilapia genome. Gene 295: 265-277.
23. Luckenbach JA, Iliev DB, Goetz FW, Swanson P (2008) Identification of differentially expressed ovarian genes during primary and early secondary oocyte growth in coho salmon, *Oncorhynchus kisutch*. Reprod Biol Endocrinol 6: e2.
24. Darias MJ, Zambonino-Infante JL, Hugot K, Cahu CL, Mazurais D (2008) Gene expression patterns during the larval development of European sea bass (*dicentrarchus labrax*) by microarray analysis. Mar Biotechnol 10: 416-428.
25. Tingaud-Sequeira A, Chauvigné F, Lozano J, Agulleiro MJ, Asensio E, *et al.* (2009) New insights into molecular pathways associated with flatfish ovarian development and atresia revealed by transcriptional analysis. BMC Genomics 10: e434.
26. Bobe J, Nguyen T, Fostier A (2009) Ovarian function of the trout preovulatory ovary: new insights from recent gene expression studies. Comp Biochem Physiol A Mol Integr Physiol 153: 63-68.
27. Clelland E, Peng C (2009) Endocrine/paracrine control of zebrafish ovarian development. Mol Cell Endocrinol 312: 42-52.
28. Callaghan TR, Degnan BM, Sellars MJ (2010) Expression of Sex and Reproduction-Related Genes in *Marsupenaeus japonicus*. Marine Biotechnology 12: 664-677.
29. Martyniuk H, Xiong K, Crump S, Chiu R, Sardana A, *et al.* (2006) Gene expression profiling in the neuroendocrine brain of male goldfish (*Carassius auratus*) exposed to 17alpha-ethinylestradiol. Physiol Genomics 27: 328-336.
30. Marlatt VL, Martyniuk CJ, Zhang D, Xiong H, Watt J, *et al.* (2008) Auto-regulation of estrogen receptor subtypes and gene expression profiling of 17beta-estradiol action in the neuroendocrine axis of male goldfish. Mol Cell Endocrinol 283: 38-48.
31. Levavi-Sivan B, Bogerd J, Mañanós EL, Gómez A, Lareyre JJ (2010) Perspectives on fish gonadotropins and their receptors. Gen Comp Endocrinol 165: 412-437.
32. Amano M, Takahashi A (2009) Melanin-concentrating hormone: A neuropeptide hormone affecting the relationship between photic environment and fish with special reference to background color and food intake regulation. Peptides 30: 1979-1984.
33. Chiocchio SR, Gallardo MG, Louzan P, Gutnisky V, Tramezzani JH (2001) Melanin-concentrating hormone stimulates the release of luteinizing hormone-releasing hormone and gonadotropins in the female rat acting at both

median eminence and pituitary levels. Biol Reprod 64: 1466-1472.

34. Amiya N, Amano M, Yamanome T, Yamamori K, Takahashi A (2008) Effects of background color on GnRH and MCH levels in the barfin flounder brain. Gen Comp Endocrinol 155: 88-93.

35. Tanaka M, Azuma M, Nejigaki Y, Saito Y, Mizusawa K, *et al.* (2009) Melanin-concentrating hormone reduces somatolactin release from cultured goldfish pituitary cells. J Endocrinol 203: 389-398.

36. Pierce JG, Parsons TF (1981) Glycoprotein hormones: structure and function. Annu Rev Biochem 50: 465-495.

37. Wong AC, Van Eenennaam AL (2004) Gonadotropin hormone and receptor sequences from model teleost species. Zebrafish 1: 203-221.

38. Munakata A, Kobayashi M (2010) Endocrine control of sexual behavior in teleost fish. Gen Comp Endocrinol 165: 456-468.

39. Hellqvist A, Schmitz M, Mayer I, Borg B (2006) Seasonal changes in expression of LH-beta and FSH-beta in male and female three-spined stickleback, *Gasterosteus aculeatus*. Gen Comp Endocrinol 145: 263-269.

40. Degani G, Tzchori I, Yom-Din S, Goldberg D, Jackson K (2003) Growth differences and growth hormone expression in male and female European eels. Gen Comp Endocrinol 134: 88-93.

41. Campbell B, Dickey J, Beckman B, Young G, Pierce A, *et al.* (2006) Previtellogenic oocyte growth in salmon: relationships among body growth, plasma insulin-like growth factor-1, estradiol-17beta, follicle-stimulating hormone and expression of ovarian genes for insulin-like growth factors, steroidogenic-acute regulatory protein and receptors for gonadotropins, growth hormone, and somatolactin. Biol Reprod 75: 34-44.

42. Douglas AJ, Johnstone LE, Leng G (2007) Neuroendocrine mechanisms of change in food intake during pregnancy: a potential role for brain oxytocin. Physiol Behav 91: 352-365.

43. Barb CR, Hausman GJ, Lents CA (2008) Energy metabolism and leptin: effects on neuroendocrine regulation of reproduction in the gilt and sow. Reprod Domest Anim 43: 324-330.

44. Michel H (1999) Cytochrome c oxidase: catalytic cycle and mechanisms of proton pumping-a discussion. Biochemistry 38: 15129-15140.

45. Brandt U (2006) Energy converting NADH: quinone oxidoreductase (complex I). Annu Rev Biochem 75: 69-92.

46. Lake J, Gravel C, Koko GK, Robert C, Vandenberg GW (2010) Combining suppressive subtractive hybridization and cDNA microarrays to identify dietary phosphorus-responsive genes of the rainbow trout (*Oncorhynchus mykiss*) kidney. Comp Biochem Physiol Part D Genomics Proteomics 5: 24-35.

47. Robker RL, Russell DL, Espey LL, Lydon JP, O'Malley BW, *et al.* (2000) Progesterone-regulated genes in the ovulation process: ADAMTS-1 and cathepsin L proteases. Proc Natl Acad Sci U S A 97: 4689-4694.

48. Carnevali O, Cionna C, Tosti L, Lubzens E, Maradonna F (2006) Role of cathepsins in ovarian follicle growth and maturation. Gen Comp Endocrinol 146: 195-203.

49. Liu DL, Zhu C (2002) Regulation of ovarian function by the matrix metalloproteinase system. Chin Sci Bull 47: 1145-1149.

50. Coffman MA, Pinter JH, Goetz FW (2000) Trout ovulatory proteins: site of synthesis, regulation, and possible biological function. Biol Reprod 62: 928-938.

51. Adam-Luckenbach J, W-Goetz F, Swanson P (2008) Expressed sequence tags (ESTs) of follicle/interstitial cell enriched ovarian tissue from previtellogenic coho salmon. Cybium 32: 142-144.

52. Dong CH, Yang ST, Yang ZA, Zhang L, Gui JF (2004) A C-type lectin associated and translocated with cortical granules during oocyte maturation and egg fertilization in fish. Dev Biol 265: 341-354.

53. Fujiki K, Bayne CJ, Shin DH, Nakao M, Yano T (2001) Molecular cloning of carp (*Cyprinus carpio*) C-type lectin and pentraxin by use of suppression subtractive hybridisation. Fish Shellfish Immunol 11: 275-279.

54. Poisa-Beiro L, Dios S, Ahmed H, Vasta GR, Martínez-López A, *et al.* (2009) Nodavirus infection of sea bass (*Dicentrarchus labrax*) induces up-regulation of galectin-1 expression with potential anti-inflammatory activity. J Immunol 183: 6600-6611.

55. Mantovani A, Garlanda C, Doni A, Bottazzi B (2008) Pentraxins in innate immunity: from C-reactive protein to the long pentraxin PTX3. J Clin Immunol 28: 1-13.

56. Praveen B, Vincent S, Murty US, Krishna AR, Jamil K (2005) A rapid identification system for metallothionein proteins using expert system. Bioinformation 1: 14-15.

57. Ding NZ, He CQ, Yang ZM (2002) Quantification of basigin mRNA in mouse oocytes and preimplantation embryos by competitive RT-PCR. Zygote 10: 239-243.

58. Huang W, Zhou L, Li Z, Gui JF (2009) Expression pattern, cellular localization and promoter activity analysis of ovarian aromatase (Cyp19a1a) in protogynous hermaphrodite red-spotted grouper. Mol Cell Endocrinol 307: 224-236.

59. Yang YH, Dudoit S, Luu P, Lin DM, Peng V, *et al.* (2002) Normalization for cDNA microarray data: a robust composite method addressing single and multiple slide systematic variation. Nucleic Acids Res 30: e15.

60. Walsh JG, Barreda DR, Belosevic M (2007) Cloning and expression analysis of goldfish (*Carassius auratus* L.) prominin. Fish Shellfish Immunol 22: 308-317.

61. Gill JC, Wang O, Kakar S, Martinelli E, Carroll RS, *et al.* (2010) Reproductive hormone-dependent and -independent contributions to developmental changes in kisspeptin in GnRH-deficient hypogonadal mice. PLoS One 30: e11911.

62. Lim S, Pnueli L, Tan JH, Naor Z, Rajagopal G, *et al.* (2009) Negative feedback governs gonadotrope frequency-decoding of gonadotropin releasing hormone pulse-frequency. PLoS One 29: e7244.

63. Luo D, Hu W, Chen S, Xiao Y, Sun Y, *et al.* (2009) Identification of differentially expressed genes between cloned and zygote-developing zebrafish (*Danio rerio*) embryos at the dome stage using suppression subtractive hybridization. Biol Reprod 80: 674-684.

64. Livak KJ, Schmittgen TD (2001) Analysis of relative gene expression data using real-time quantitative PCR and the $2^{(-\text{Delta Delta C(T)})}$ method. Methods 25: 402-408.

转反义 sGnRH 基因雌性鲤的下丘脑-垂体-性腺轴全基因表达差异分析

徐　婧[1,2]　黄　伟[1]　钟成容[1]　罗大极[1]　黎双飞[1]　朱作言[1]　胡　炜[1]

1 中国科学院水生生物研究所，淡水生态与生物技术国家重点实验室，武汉
2 中国科学院大学，北京

摘　要　下丘脑-垂体-性腺(HPG)轴在鱼类发育及繁殖调控中具有重要作用。促性腺激素释放激素(GnRH)神经肽是 HPG 轴中的重要调节因子，抑制 GnRH 的表达可能减弱或严重抑制性腺发育，为系统研究鱼类生殖系统相关基因表达谱的变化提供了一个良好模型。

在以前的研究中，对324枚鲤受精卵注射了表达反义 sGnRH 的构建体。四年后，发现有38条转基因鲤的性腺败育。本研究选择其中12尾卵巢败育的雌性鲤作为实验材料，通过抑制性差减杂交和 cDNA 芯片分析相结合的方法，分析了 HPG 轴相关基因的表达变化。结果显示，下丘脑、垂体和卵巢分别有9、28和212个差异表达的基因，其中包括87个新基因。下丘脑中下调的基因有5个，上调的基因有4个；垂体中下调的基因有16个，上调的基因有12个；卵巢中下调的基因有119个，上调的基因有93个。功能分析结果表明这些基因主要参与以下几个生物过程，如生物合成、器官发生、新陈代谢通路、免疫系统、转运连接和凋亡。此外还发现一些与神经肽、促性腺激素、新陈代谢、卵子发生和炎症因子等相关的差异表达基因。

本研究揭示了下丘脑反义 sGnRH 基因对雌性鲤生殖的影响在垂体和卵巢水平具有渐进的级联放大效应，获得了系统的 HPG 信号通路基因表达变化的数据，加深我们对硬骨鱼类生殖系统中的分子机制和调控通路的理解。

Rapid Growth and Sterility of Growth Hormone Gene Transgenic Triploid Carp

YU Fan[1] XIAO Jun[1] LIANG XiangYang[1] LIU ShaoJun[1]
ZHOU GongJian[1] LUO KaiKun[1] LIU Yun[1] HU Wei[2]
WANG YaPing[2] ZHU ZuoYan[2]

1 Key Laboratory of Protein Chemistry and Fish Developmental Biology of Education Ministry of China, College of Life Sciences, Hunan Normal University, Changsha 410081
2 State Key Laboratory of Freshwater Ecology and Biotechnology, Institute of Hydrobiology, Chinese Academy of Sciences, Wuhan 430072

Abstract Triploid carp (100%) with 150 ($3n$=150) chromosomes were obtained by crossing the females of improved tetraploid hybrids (♀, $4n$=200) of red crucian carp (♀)×common carp (♂) with the males of diploid yellow river carp (♂, $2n$=100). The crosses yielded transgenic triploid carp (positive triploid fish, 44.2% of the progeny) and non-transgenic triploid carp (negative triploid fish). Histological examination of the gonads of 24-month-old positive triploid fish suggested they were sterile and the fish were not able to produce mature gametes during the breeding season. Morphologically, both the positive and negative triploid fish were similar. They had a spindle-shaped, laterally compressed, steel grey body with two pairs of barbells. Most of the quantifiable traits of the triploid carp were intermediate between those of the two parents. The positive and negative triploid fish were raised in the same pond for 2 years. The mean body weight of the positive triploid fish was 2.3 times higher than the negative triploid fish. The weight of the largest positive triploid fish was 2.91 times higher than that of the largest negative triploid fish. Thus, we produced fast-growing transgenic triploid carp that have a reduced ecological risk because of their inability to mate and produce progeny.

Keywords transgene; carp; diploid; triploid; improved tetraploid hybrids of red crucian carp×common carp; sterility; ecological safety

There have been substantial advancements in the development of transgenic commercial fish [1] in America, Canada and China. For example, researchers have produced transgenic carp that have an additional “all-fish” growth hormone gene [2], transgenic Atlantic Salmon (*Salmo salar*, and transgenic coho salmon (*Oncorhynchus kisutch*) [3,4]. Despite these advances, transgenic fish have never been successfully commercialized. The primary bottleneck for the use of transgenic fish is the concern about potential ecological safety. If transgenic fish were able to hybridize with a related species in the wild, it is possible that the transferred genes would pollute the wild gene pool and destroy species variety [5]. The only way to address this issue is to control

文章发表于 *Chinese Science Bulletin*, 2011, 56(16): 1679-1684

the fertility of transgenic fish. Thus, there has been an increase in research interest into the mechanism by which the fertility of transgenic fish can be controlled. There is no doubt that the production of sterile triploid transgenic fish will address concerns about ecological risk.

The College of Life Sciences at Hunan Normal University and cooperating institutions have successfully established a population of bisexual fertile tetraploid red crucian carp hybrids (4*n*=200) [6-8]. This population has been maintained for 18 generations (F_3-F_{18}). In recent years, a population of improved tetraploid red crucian carp hybrids [9,10] was obtained using gynogenetic methods. Both the tetraploid red crucian carp hybrids and the improved tetraploid red crucian carp hybrids were able to reproduce and both possessed stable genetic characteristics. Hybridization between the tetraploids and diploid fish species produced all-triploid progeny with three forms of abortive gonads, including sterile ovarian ovary-like, sterile testis-like, and sterile fat-like forms [11]. The development of tetraploid red crucian carp hybrid and improved tetraploid red crucian carp hybrid populations has ensured the availability of sufficient parents for producing fertile triploid fish on a large scale. Thus, crossing tetraploid red crucian carp hybrids and diploid transgenic carp should yield 100% sterile triploid progeny. If true, this would alleviate public concerns about the potential ecological safety of transgenic fish commercialization. We crossed improved tetraploid red crucian carp hybrids with diploid transgenic yellow river carp that possessed an additional grass carp growth hormone gene (*GCGH*). We obtained transgenic triploid carp (positive fish) and non-transgenic triploid carp (negative fish). We then evaluated the growth and reproductive development of these progeny. Our results offer insight into the expectation for successfully commercializing *GH* gene transgenic carp.

1. Materials and Methods

1.1 Experimental materials

We obtained transgenic diploid yellow river carp (2*n*=100) from the State Key Laboratory of Freshwater Ecology and Biotechnology, Institute of Hydrobiology, Chinese Academy of Sciences. These fish carried the grass carp growth hormone gene (abbreviated *CAgcGH*) and were the heterozygotes of the 3rd generation. We also obtained improved tetraploid red crucian carp hybrids (4*n*=200) from the National Resources Protection Base of Tetraploid Fish.

1.2 Experimental methods

We selected sexually mature male transgenic diploid yellow river carp (2*n*=100) and female improved tetraploid red crucian carp hybrids during the breeding season of 2008. These adults were artificially induced and the gametes were collected and fertilized. The fertilized eggs were incubated in a glass container with water at (20±2)℃ until hatching. We obtained a total of 10000 triploid fry. On April 20, 2008, the fry were transferred to a 28-m^2 cement pond and reared until they reached 3 cm in length. Seven days later, we randomly collected 500 juvenile fish of equal body size and transferred them into a 1200-m^2 experimental earthen pond for further rearing. The progeny were fed with commercial carp feed. A random sample of the progeny was collected each month to measure growth rates and check for the presence of CAgcGH.

1.3 Total DNA extraction

In April 2009, we collected 15 of the 12-month-old hybrid progeny from the experimental pool. These fish were euthanized and blood was collected from the caudal vein of each individual. We used acid citrate glucose solution (abbreviated ACD) to prevent coagulation of the blood samples. We isolated total DNA from blood cells using a UNIQ210 DNA Extraction Kit (Sangon, Shanghai, China), following the manufacturers protocol. The total DNA samples were stored at –20℃ after checking for quality and purity.

1.4 CAgcGH detection

The transgenic triploid fish (positive fish) and the non-transgenic triploid fish (negative fish) were identified by specific PCR amplification of a 328-bp fragment of the grass carp growth hormone gene from genomic DNA samples that were obtained. The primer sequences were as follows: P(+): 5′-CATTTACAGTTCAGCCATGGCTAGA-3′, P(–): 5′-AGCACCACCGACAA-CAGCACTAATG-3′.

The PCR cycling conditions were: 94℃ for 5 min; 35 cycles of 94℃ for 30 s, 60℃ for 30 s, and 72℃ for 30 s; and a final extension for 10 min at 72℃. We examined an additional 253 triploid hybrid progeny in November 2009.

1.5 Measurement of total genome DNA content

We measured the DNA content of red crucian carp, transgenic triploid carp, and non-transgenic triploid carp. We collected red blood cells from the caudal vein of the common carp for use as a control. The DNA content was measured using a flow cytometer (Partec GmbH).

1.6 Observation of chromosomes

To determine ploidy, we performed chromosome counts using the kidney tissue from transgenic triploid carp. After culture for 1-3 d at 18-22℃, the fish were injected 1-3 times with concanavalin (2-8 μg/g body weight). The interval between injections was 12-24 h. Each fish was also injected with colchicine (2-4 μg/g body weight) 2-6 h prior to dissecting. The kidney tissue was then removed and ground in 0.8% NaCl, immersed in a hypotonic solution of 0.075 mol L^{-1} KCl at 20℃ for 40-60 min, then fixed in 3∶1 methanol-acetic acid for three changes. The cells were plated on cold and wet slides and the chromosome spreads were observed and counted by Light Microscopy.

1.7 Gonad tissue section

We randomly sampled 20 24-month-old transgenic triploid carp and non-transgenic triploid carp from the population of hybrid progeny. In addition, we sampled 20 24-month-old diploid carp. The gonad tissue was removed from each individual and subjected to a series of steps. First, the samples were fixed in Bouin's solution, then dehydrated in alcohol, embedded in paraffin, sectioned (6-8 μm), and stained with haematoxylin and eosin. The gonads were observed by light microscopy and some were photographed with DM6000.M. We identified the developmental stage of each sample according to the standard for Cyprinid fish in [12].

2. Results

2.1 Morphological characters

There were few obvious differences in morphology between the transgenic triploid carp (positive fish) and the non-transgenic triploid carp (negative fish) (Figure 1). Both types of triploid carp possessed a spindle-shaped body with a high dorsum and rounded abdomen. Their body color was similar, being grey-back on the dorsum and silver grey on the abdomen. The mouth was terminal with two pairs of barbels. The triploid carp possessed small, upwardly inclined eyes and triangle-shaped short gill rakers. Their bodies covered by round scales had a slightly curved complete lateral line. Compared with the crucian carp, the triploid carp had two pairs of barbels, a lower back, and the fringe of their fin rays were light red.

Figure 1 The appearance of the transgenic triploid carp and non-transgenic triploid carp. The transgenic triploid carp (left) and the non-transgenic triploid carp (right) in the aquarium. Scale bar=5 cm

2.2 Transgenic triploid carp detection

The PCR products were separated by electrophoresis on a 1.2% agarose gel (Figure 2). A positive fragment of 328 bp was amplified from the genomic DNA sample of the transgenic triploid carp, but not from the non-transgenic triploid carp. The 328-bp fragment was purified, cloned and sequenced. The partial sequence matched that of the grass carp growth hormone gene.

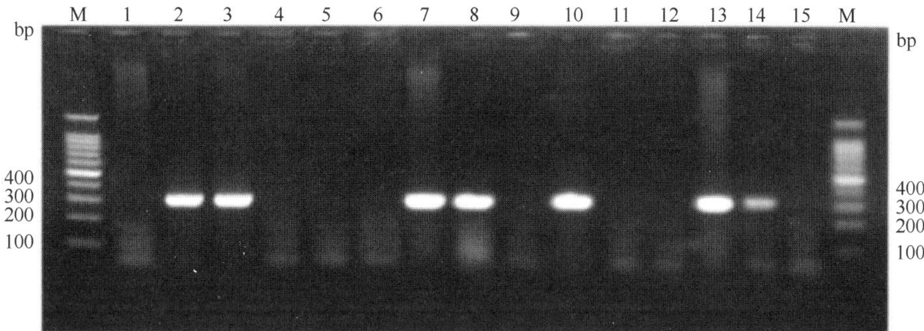

Figure 2 The results of PCR amplification using the total DNA of 15 triploid carp as the template. The positive fragments were obtained in samples 2, 3, 7, 8, 10, 13 and 14. Partial sequence of grass carp growth hormone gene is shown as follows: CATTTACAGTTCAGCCATGGCTAGAGGTATGGATATGATGCTTTCTGTTGAATGTT-ATTTGTGGATTTAAGACTTTTTGACAACTCCTTATTTCGTTTAGATGTTTTTTGTGCTTTCCTCTTAC-TTGTAAAACTAATTTCGTAGTTGTAAAAGAGTTTATTGGAGCATGGAAAACAAAACAGATGTATTT-GATCTCCAAAATCTGTTTTGATTAATCCAAAATCTTGATTTCATTGGTTATTTTATTAGGAACGATG-TTGTAGATGGTAAACAAACCTCTTTCTTTTCTTCTCCTAGCATTAGTGCTGTTGTCGGTGGTGCT

2.3 DNA content of transgenic triploid carp

The DNA content of the 3 groups is summarized in Table 1.

Table 1 DNA content of diploid and triploid carp[a)]

Number	Fish sample	DNA content	Ratio	Remarks
A	Red crucian carp	102.55		
B	Transgenic triploid carp	147.3	B/A=1.44*	P>0.5
C	Non-transgenic triploid carp	149.7	C/A=1.46*	P>0.5

a) *, compared to a ratio of 1/1.5, the observed ratio was not significantly different (P>0.05). The triploidy of the transgenic triploid carp was confirmed by their DNA content

2.4 Chromosome number of transgenic triploid carp

The chromosome counts were performed on kidney tissue from transgenic triploid carp. The chromosome number ranged from 145 to 150. The spreads with 150 chromosomes accounted for 88% of the 500 spreads that were counted. Thus, the triploid carp chromosome number was $3n$=150 (Figure 3).

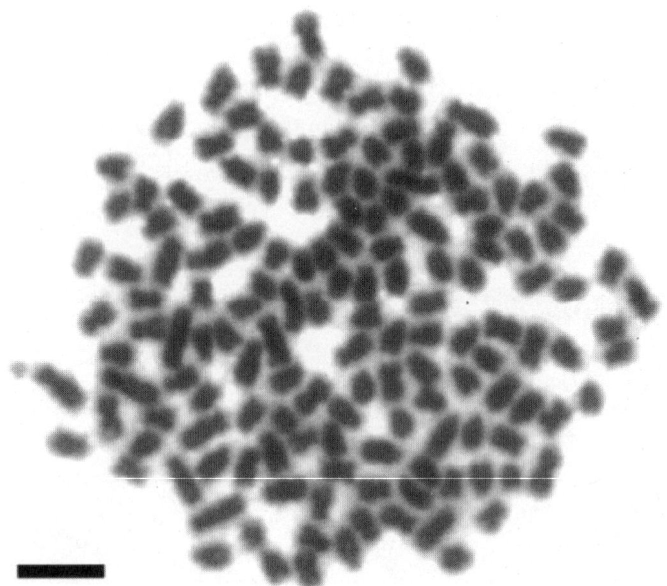

Figure 3 Chromosomes of the transgenic triploid carp. Chromosome spread of a 12-month-old transgenic triploid carp. The chromosome number was $3n$=150. Scale bar=4 μm

2.5 Microstructure observation on the gonads of transgenic triploid carp

We checked several transgenic and non-transgenic triploid carp for signs of reproductive maturation at 12- and 24-months of age during the breeding seasons in April 2009 and 2010. We were unable to express semen or eggs from any fish. Conversely, both the red crucian carp and the improved tetraploid red crucian carp hybrid produced semen and eggs during the breeding season.

There were no obvious differences in the testes or ovary of the transgenic and non-transgenic forms (Figure 4). The testes from 24-month-old transgenic and non-transgenic triploid carp were classified as sterile because of the lack of mature spermatozoa. However, we did observe some understained spermatogonium-like cells and some deeply stained spermatids. The ovarian tissue of 24-month-old transgenic and non-transgenic triploid carp was classified as being in stage I and was full of oocytes suggesting the females were also sterile. In contrast, we observed a large amount of mature spermatozoa in the testis of 24-month-old diploid carp. Similarly, the ovarian tissue of the diploid carp was steel grey and densely populated with vessels. We classified these as stage IV ovaries.

2.6 Growth rate

On 20 November, 253 fry of triploid carp were netted and tested for the presence of the *CAgcGH* gene using PCR. Of these, 112 were identified as transgenic triploid carp (44.2% of the random population) and had an average body weight of 1263 g. The remaining 141

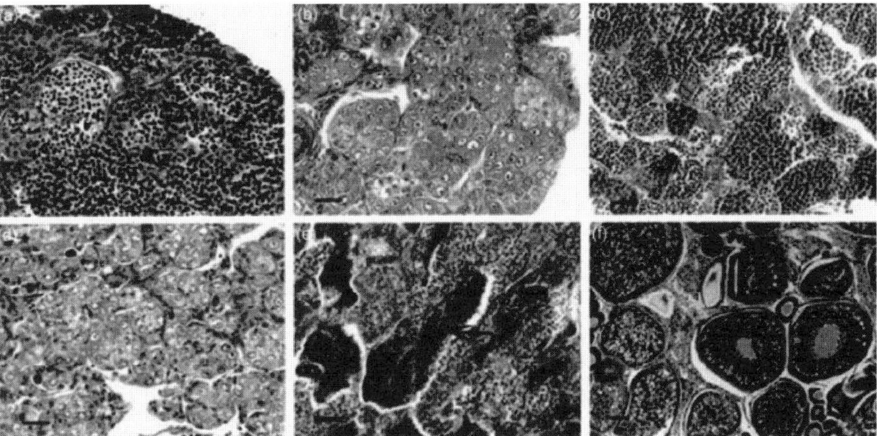

Figure 4 Triploid and diploid carp gonadal microstructure. (a) The testis of a 24-month-old transgenic triploid carp revealing the lack of mature spermatozoa in the lobules. Scale bar=10 μm. (b) The ovary of a 24-month-old transgenic triploid carp in which the oocytes were all at stage II or earlier. Scale bar=10 μm. (c) The testis of a 24-month-old non-transgenic triploid carp revealing the lack of mature spermatozoa in the lobules. Scale bar=10 μm. (d) The ovary of a 24-month-old non-transgenic triploid carp in which the oocytes were all at stage II or earlier. Scale bar=10 μm. (e) The testis of a 24-month-old diploid carp. The arrow points to the large amount of mature spermatozoa in the lobules. Scale bar=10 μm. (f) The stage IV ovary of a 24-month-old diploid carp in which the cytoplasm was filled with yolk. Scale bar=50 μm

individuals were non-transgenic triploid carp (55.8% of the random population) that had an average body weight of 547 g. The average body weight of the transgenic triploid carp was 2.3 times that of the non-transgenic triploid carp. The largest transgenic triploid carp individual was 2080 g whereas the largest non-transgenic triploid individual was 713 g.

3. Discussion

Our study is the first to document the production of transgenic triploid carp by crossing 2N and 4N parents. The offspring of these crosses were 100% triploid based on DNA content analysis and chromosome observations. It is possible to produce large numbers of triploid carp using this technique, thus satisfying the demands of commercial production. Previously, transgenic triploid fish have been artificially produced by physical or chemical induction. These techniques have several limitations, including an unstable introducing rate and variable induction of triploidy. Thus, it is not possible to control ecological risk or guarantee the supply of individuals for commercial aquaculture using these earlier techniques [5].

A number of researchers have documented the sterility of triploid fish such as triploid carp [13], triploid catfish [14], and triploid crucian carp [11]. The triploid crucian carp were obtained by distant crossing using maternal Japanese crucian carp and paternal tetraploid crucian carp. The triploid crucian carp were allotriploid, having inherited the germ plasm from the common carp, red crucian carp, and Japanese crucian carp. Triploid crucian carp have been cultured for ~20 years with no evidence that they are able to reproduce. Typically, fertile triploid fish have been the product of gynogenesis (e.g., triploid silver carp and Japanese Guandong crucian carp (*Carassius auratus langsdorfii*)) [15], not of the product of distant crossing.

Recently, we completed a comparative study of the tetraploid hybrids of red crucian carp, triploid crucian carp, and diploid red crucian carp. Based on our observations, we advance a theory to explain the mechanism of gonad abortion in triploid fish. We observed a disordered chromosome complement in the germ cells of triploid crucian carp during initial meiosis, forming 50 bivalents and 50 univalents. As a result, meiosis failed and gonadal development was stopped. During initial meiosis in the germ cells of red crucian carp and tetraploid hybrids of red crucian carp×common carp, 100 bivalents and 50 bivalents, respectively, were formed by the normal chromosome complement [16]. Based on a comparison of pituicyte ultrastructure among fish of different ploidies during the breeding season, the number of gonodotropin (GTH) secretory cells was highest in tetraploid crucian carp, followed by diploid red crucian carp, and lowest in triploid crucian carp. Similarly, in the GTH cells of the adenohypophysis, the density of secretory granules and secretory globules was highest in tetraploid crucian carp and diploid red crucian carp, and was significantly lower in triploid crucian carp [17]. Gonadotropin-releasing hormone (GnRH), gonadotropin, and the gonadotropin receptor (GTHR) is an important part of the hypothalamus-hypophysis-gonad axis. The three genes (*Gnrh2*, *Gth* and *Gthr*) play a role in the regulation of gonadal development. The expression of Gthr is known to be lower in the gonad of triploid crucian carp than in tetraploid crucian carp and diploid red crucian carp. Thus, the interaction between gonad GTHR and GTH release from the hypophysis is weakened, resulting in insufficient production of steroid hormones in the gonad of triploid crucian carp [18]. Accordingly, the triploid crucian carp that were obtained by crossing parents of different ploidies were rendered sterile [11,16,17] at 3 levels, chromosomal, hypophyseal, and gonadal. In this study, the transgenic triploid carp were also the product of distant-crossing hybrids that were obtained in a similar way. We did not observe mature spermatozoa, or oocytes that were beyond stage II, in any individuals during the breeding season. Similarly, we did not observe the production of sperm or eggs in any individual suggesting that all fish were sterile. In comparison, the diploid carp had large quantities of mature spermatozoa and stage IV oocytes that were full of yolk. All of the diploid carp were able to produce mature sperm or eggs.

Our results suggest that transgenic triploid carp have higher growth rates when compared to the control group under the same feeding conditions. The average body weight of the positive fish was 2.3 times of that of negative fish. The largest body weight of positive individuals was 2.91 times that of negative individuals. The increase in growth rate can be attributed to the presence of the transformed grass carp growth hormone gene which notably improved the feed conversion rate of the receptor fish. The feed conversion rates of F_2 transgenic red common carp and transgenic Atlantic salmon were 10% higher than that of the control [19]. Similarly, the conversion rates of transgenic Nile tilapia and transgenic Yellow River carp were 20% to 35% higher than that of the controls [20-22]. The feed conversion rate of 1-year-old transgenic coho salmon was 3.76 times of that of 1-year-old controls, and was 1.72 times of that of 2-year-old controls of the same body weight [3]. The feed conversion rates of transgenic mud loach (*Misgurnus mizolepis*) and transgenic tilapia (*Oreochromis* sp.) were also improved by 1 and 1.9 times, respectively [23,24].

Most of the measurable ratios characterizing the transgenic triploid carp were intermediate between those of the paternal diploid carp and maternal tetraploid crucian carp. The morphology of the transgenic crucian carp appeared normal and there were no obvious differences in the morphology of the transgenic triploid carp and the non-transgenic triploid carp. We did not observe malformed fish in the triploid population.

Our study was based on established populations of improved tetraploid red crucian carp hybrids and transgenic diploid Yellow River carp, the crossing of which produced 100% triploid offspring. The paternal transgenic diploid Yellow River carp were heterozygotes. Both transgenic and non-transgenic triploid hybrids were obtained from the crosses, providing a comparative system to study the biological characteristics of both kinds of fish. We crossed homozygote transgenic diploid carp with the improved tetraploid red crucian carp hybrid during the 2010 breeding season. In theory, we expect the progeny to be 100% transgenic triploid carp.

In conclusion, transgenic triploid carp have maintained the sterile characteristic of triploid hybrid fish but have a faster growth rate and a higher feed conversion rate. The use of sterile transgenic triploid carp poses significantly less risk to the environment but offers a number of benefits to commercial growers.

Acknowledgements This work was supported by the National Key Basic Research Program of China (2007CB109206) and the National Special Fund for Research in Public Welfare Sector (20090304608).

References

1 Maclean N, Laight R J. Transgenic fish: An evaluation of benefits and risks. Fish Fisher, 2000, 1: 146-172

2 Wang Y P, Hu W, Wu G, *et al*. Genetic analysis of "all-fish" growth hormone gene transferred carp (*Cyprinus carpio* L.) and its F1 generation. Chinese Sci Bull, 2001, 46: 1174-178

3 Devlin R H, Biagi C A, Yesaki T Y. Growth, viability and genetic characteristics of GH transgenic coho salmon strains. Aquaculture, 2004, 236: 607-632

4 Fletcher G L, Shears M A, Yaskowiak E S, *et al*. Gene transfer: Potential to enhance the genome of Atlantic salmon for aquaculture. Aus J Exp Agricul, 2004, 44: 1095-1100

5 Hu W, Wang Y P, Zhu Z Y. Progress in the evaluation of transgenic fish for possible ecological risk and its containment strategies. Sci China Ser C-Life Sci, 2007, 50: 573-579

6 Liu S J, Liu Y, Zhou G J, *et al*. The formation of tetraploid stocks of red crucian carp×common carp hybrids as an effect of interspecific hybridization. Aquaculture, 2001, 192: 171-186

7 Sun Y D, Liu S J, Zhang C, *et al*. The chromosome number and gonadal structure of F9-F11 tetraploid crucian-carp.

Acta Genet Sin, 2003, 30: 414-418
8 Liu S J. Distant hybridization leads to different ploidy fishes. Sci China Life Sci, 2010, 53: 416-425
9 Wang J, Qin Q B, Chen S, *et al*. Formation and biological characterization of three new types of improved crucian carp. Sci China Ser C-Life Sci, 2008, 51: 544-551
10 Liu S J, Sun Y D, Zhang C, *et al*. Production of gynogenetic progeny from tetraploid hybrids red crucian carp×common carp. Aquaculture, 2004, 236: 193-200
11 Liu S J, Hu F, Zhou G J, *et al*. Gonadal structure of triploid crucian carp produced by crossing tetraploid hybrids of *Carassium auratus* red var. (♀)×*Cyprinus carpio* (♂) with Japanese crucian carp (*Carassius auratus cavieri* t. et s). Acta Hydrobiol Sin, 2000, 24: 301-306
12 Liu Y. Breeding Physiology of Chinese Cultivated Fish. Beijing: Agricultural Publishing House, 1993, 22-30
13 Gervai J, Peter S, Nagy A, *et al*. Induction triploidy incarp, *Cyprinus carpio* L. J Fish Biol, 1980, 17: 667-671
14 Yin H B, Sun Z W, Pan Z W, *et al*. Study on gonadal development of triploid catfish (*Silurus asotus* L.). Oceanol Limnol Sin, 2000, 31: 123-129
15 Wu Q J, Gui J F. Fish Genetic Breeding Engineering. Shanghai: Shanghai Sicentific and Technology Press, 1999, 40-41
16 Zhang C, He X X, Liu S J, *et al*. The chromosome pairing in meiosis I in tetraploid hybrids and all triploid Crucian Carp. Acta Zool Sin, 2005, 51: 89-94
17 Long Y, Liu S J, Huang W R, *et al*. Comparative studies on histological and ultra-structure of the pituitary of different ploidy level fishes. Sci China Ser C-Life Sci, 2006, 49: 446-453
18 Long Y, Liu S J, Zhong H, *et al*. The distinct expression of *Gnrh*, *Gthβ* and *Gthr* gene in sterile triploids and fertile tetraploids. Cell Tissue Res, 2009, 338: 151-159
19 Cook J T, McNiven M A, Richardson G F, *et al*. Growth rate, body composition and feed digestibility/conversion of growth enhanced Atlantic salmon (*Salmo salar*). Aquaculture, 2000, 188: 15-32
20 Rahman M A, Ronyai A, Engidaw B Z, *et al*. Growth and nutritional trials of transgenic *Nile tilapia* containing an exogenous fish growth hormone gene. J Fish Biol, 2001, 59: 72-78
21 Fu C, Li D, Hu W, *et al*. Growth and energy budget of F2 "all-fish" growth hormone transgenic common carp. J Fish Biol, 2007, 70: 347-361
22 Kobayashi S, Alimuddin Morita T, Miwa M, *et al*. Transgenic *Nile tilapia* (*Oreochromis niloticus*) over-expressing growth hormone show reduced ammonia excretion. Aquaculture, 2007, 270: 427-435
23 Nam Y K, Noh J K, Cho Y S, *et al*. Dramatically accelerated growth and extraordinary gigantism of transgenic mud loach (*Misgurnus mizolepis*). Transgenic Res, 2001, 10: 353-362
24 Martinez R, Juncal J, Zaldivar C, *et al*. Growth efficiency in transgenic tilapia (*Oreochromis* sp.) carrying a single copy of an homologous cDNA growth hormone. Biochem Biophys Res Commun, 2000, 267: 466-472

转生长激素基因三倍体鲤的快速生长和不育性

于　凡[1]　肖　俊[1]　梁向阳[1]　刘少军[1]　周工建[1]
罗凯坤[1]　刘　筠[1]　胡　炜[2]　汪亚平[2]　朱作言[2]

1 湖南师范大学生命科学学院，教育部蛋白质化学及鱼类发育生物学重点实验室，长沙　410081
2 中国科学院水生生物研究所，淡水生态与生物技术国家重点实验室，武汉　430072

摘　要　将转草鱼生长激素基因黄河鲤(♂)与改良四倍体鲫鲤(♀)杂交，获得100%三倍体子代。通过转植基因 PCR 检测，三倍体子代中包括转基因个体(简称阳性鱼)和非转基因个体(简称阴性鱼)两部分，其中阳性鱼占总检测鱼数的44.2%；染色体检测表明，子代转基因鲤的染色体数目为$3n$=150；组织切片结果表明性腺不能发育成熟；在同龄鲤的繁殖季节(24月龄)，转基因三倍体鲤仍不能产生成熟配子。阳性和阴性三倍体鲤在形态上无明显差异，它们的体侧扁，近似纺锤形，体色青灰，有两对口须，大部分可测量性状和可数性状都介于父母本之间，表现出杂种特征。两年同塘养殖结果表明，阳性鱼的平均体重为阴性鱼的2.3倍；其中最大阳性鱼个体的体重为最大阴性鱼的 2.91倍。因此，本研究不仅证实了转基因三倍体鲤的生长快速特性，而且第一次用实验消除了转基因鲤养殖应用上的生态安全之虑。

Progress in Studies of Fish Reproductive Development Regulation

CHEN Ji HU Wei ZHU ZuoYan

State Key Laboratory of Freshwater Ecology and Biotechnology, Institute of Hydrobiology, Chinese Academy of Science, Wuhan, 430072

Abstract Mechanisms of the animal reproductive development are an important research field in the life sciences. The study of the reproductive development and regulatory mechanisms in fishes is important for elucidating the mechanisms of animal reproduction. This paper summarizes recent advances in the mechanisms of fish sex determination and differentiation, of fish gonad development and maturation, and of fish germ cell development, as well as the according regulating strategies. Fishes comprise an evolutionary stage that links invertebrates and higher vertebrates. They include diversiform species, and almost all vertebrate types of reproduction have been found in fishes. All these will lead to important advances in the regulatory mechanisms of animal reproduction by using fishes as model organisms. It will also enable novel fish breeding techniques when new controllable on-off strategies of reproduction and/or sex in fishes have been developed.

Keywords fish; reproduction; regulation

As low vertebrates, fishes are on an evolutionary stage linking invertebrates and higher vertebrates. There are as many as 28,000 known fish species, which show large diversity. A range of sexual characteristics and reproductive strategies has been described for fishes, and almost all reproductive types of vertebrates have been found in fishes. The study of fish reproductive development regulation is important for elucidating mechanisms of animal reproductive development. Molecular breeding will reveal finer varieties when genes and regulatory networks associated with economically important traits (such as fast growth, disease resistance, cold tolerance and hypoxia tolerance) in fishes are discovered[1]. Reproduction also provides a precondition for culturing exclusive fish breeds and expanding fish populations. Breeding techniques such as sexual control and fertility control have played important roles in developing new fish varieties. This paper will review advances of fish reproductive development regulation, with the aim of providing hints for mechanisms of fish reproductive development, as well as providing theoretical guidance for techniques of reproductive regulation and developing high-yielding, efficient, eco-friendly breeds.

文章发表于 *Chinese Science Bulletin*, 2013, 58(1): 7-16

1. Sexual Control of Fishes

1.1 Factors influencing sex determination

The sex differentiation pattern and determination mechanism of fishes are very diverse. Various sexual phenotypes are all present in fishes, such as hermaphroditism, unisexuality, gonochorism and dual reproduction mode including parthenogenesis and sexual reproduction[2]. For those fishes in which sex chromosome systems exist, there are single chromosome systems, such as XX/XY, ZZ/ZW, XX/XO and ZZ/ZO, and multiple chromosome systems, such as XX/XY_1Y_2, ZZ/ZW_1W_2, $X_1X_2X_1X_2/X_1X_2Y$, and $X_1X_2X_1X_2/X_1X_2\ X_1$. Like other animals, fish sex differentiation is controlled by their genetic system. However, fishes are oviparous, and their embryonic development proceeds in exposure to temperature-variable water. Thus, their external physical environment inevitably affects sex differentiation of fishes.

1.1.1 Genetic sex determination

In vertebrates, it appears that genes involved in sex determination and differentiation have the tendency to move onto the heteromorphic sex chromosomes, which may be a common feature of these genes[3,4]. Thus, when it comes to the study of fish sex determination mechanisms, a strategy of searching sex chromosomes for sex determination genes was always preferred. Based on the statistical data given by Arkhipchuk, among more than 1700 fish species that had been cytogenetically characterized until 1995, approximately 176 species had been found to have cytogenetically distinct sex chromosomes[5], which present important material for exploring mechanisms of fish sex determination. The guppy (*Poecilia reticulata*) has cytogenetically differentiated sex chromosomes. In the early pachytene stage, a segment on the Y chromosome is unpaired with the X chromosome. Comparative genomic hybridization (CGH) differentiated a large block of predominantly male-specific repetitive DNA in the region[6]. Artieri *et al.*[7] identified chromosome 2 of the Atlantic salmon (*Salmo salar*) as the sex chromosome by using probes designed from the flanking regions of sex-linked microsatellite markers. The region containing the sex-determining locus appeared to locate between the Ssa202DU marker and a region of heterochromatin on the long arm. They then used an *in silico* approach to draw a physical map of chromosome 2 with a view to sequencing and identifying the sex-determining gene. The X chromosome of the spiny eel (*Mastacembelus aculeatus*) was microdissected by Zhao *et al*[8]. They constructed an X chromosome library and isolated a repetitive sequence. The repetitive sequence was found by fluorescence *in situ* hybridization (FISH) to distribute on regions of the X chromosome different from the Y chromosome. Despite the studies described above, it is difficult to isolate sex-determining genes on heteromorphic sex chromosomes due to their heterochromatin containing large regions of repetitive sequences. How to locate sex-determining loci on sex chromosomes of fishes and how to discriminate sex-determining associated genes from large numbers of repetitive sequences is crucial for studying mechanisms of sex determination by using fish sex chromosomes.

Comparing sex determination-associated genes of other species with homologs of fishes is another strategy of searching for sex-determining genes of fishes. However, homologs of *wt1*, *sf-1*, *amh* and *dax1*, which play roles in mammalian sex determination, were found to only have indirect regulatory effects on sex differentiating signal pathways in fishes[9-12].

The *ff1d* gene of zebrafish (*Danio rerio*), a homolog of the mammalian *sf-1* gene, is expressed in the hypothalamus and gonads, especially in testicular Sertoli cells and Leydig cells, and can regulate the expression of *amh*[10]. In zebrafish testes, *amh* is expressed in presumptive Sertoli cells, which may be regulated by SF-1, GATA and WT1, and is related to the proliferation and differentiation of spermatogonia[13]. The *cyp19* gene encodes aromatase, the enzyme that converts androgens to estrogens. Fish *cyp19a1* participates in ovary development, and *cyp19a2* is indirectly involved in sexual differentiation via the hypothalamus-pituitary-gonad axis (HPG axis)[14]. The expression patterns of *cyp19a1a* and *cyp19a1b* of 3-month-old and 10-month-old common carp (*Cyprinus carpio*) were investigated in our laboratory. Important roles of *sf-1* and *cyp19a1b* in testis development and of *cyp19a1a* in ovary development were suggested[15]. Guiguen *et al.* [16] hypothesized that upregulation of *cyp19a1a* is required for triggering and maintaining ovarian differentiation, and that *cyp19a1a* downregulation was necessary for inducing a testicular differentiation pathway. Furthermore, SF-1, FOXL2 and AMH are believed to regulate expression of *cyp19a1* [17], which again suggests the indirect effect of those factors in fish sex determination.

Members of the SOX gene family are highly conserved in vertebrates and regulate extensive development pathways. However, the homologs of SOX in fishes are not crucial for sex determination. Sexually dimorphic expression modes of *sox9* have been reported for some fish species[18]. Although it plays a role in testis organogenesis, no evidence was found which could demonstrate that *sox9* directs sex determination.

Many members of the DMRT gene family are surmised to be involved in sex determination. The *dmrt1* gene is thought to be a primary sex-determining gene in vertebrates. In fishes, its expression was only found in gonads. Furthermore, it showed higher expression in testes than in ovaries for some species, and it might be related to the formation and maintenance of the testes. The ontogenetic expression pattern of *dmrt1* in *Gobiocypris rarus* was analyzed using semi-quantitative PCR in our laboratory, and the results showed a higher expression in testes than in ovaries since five days after hatching[19].

The DMY gene of medaka (*Oryzias latipes*), a functional copy of *dmrt1* on the Y chromosome, is a sex-determining gene with the solidest evidences to be responsible for formation and differentiation of the testes[20,21]. However, phylogenetic analysis indicated that the duplication of *dmrt1* occurred after medaka evolved from other Oryziatidae species, which means that DMY does not exist in other fish species.

Recently, a duplicated gene of *amh* on Y chromosome of *Odontesthes hatcheri*, named *amhy*, was isolated by Hattori *et al*[22]. The results from spatio-temporal expression analysis and gene knockdown suggested that *amhy* may play a critical role in the sex determination in this species. Compared with *sry*, *dmrt1* and *DMY*, which encode transcription factors to determine sex in some species, *amhy* may be a unique case of a hormone-related gene mastering sex determination in vertebrates.

Li *et al.*[23] identified sex chromosomes of *Salmo trutta*, *S. salar* and *Oncorhynchus mykiss* by FISH. The sex chromosomes showed different origins and the sex-determining genes might differ in these three salmonid species. Vandeputte *et al.*[24] proposed a polygenic hypothesis for sex determination in the European sea bass (*Dicentrarchus labrax*). Bradley *et al.*[25]reported a genome-wide linkage study of sex determination in zebrafish using a SNP geneticmap. They confirmed two loci on chromosomes 5 and 16, respectively, containing sex-determining related genes, which turned out to be *dmrt1* and *cyp21a2*. However, mutations of the two loci only accounted for about 16% of variance of the trait. Liew *et al.*[26] found that zebrafish sex was mainly determined genetically using repeated single pair crossings. Furthermore, using a FluoMEP assay or array comparative genomic hybridization, they failed to find universal sex-linked differences between the male and female genomes. They therefore proposed that zebrafish have a polygenic sex determination system.

Based on the studies described above, it appears that most fish species do not possess functionally specialized sex-determining genes. It is believed that fish sex may be determined by polygenic systems, with are influenced by the environment and other factors. This could explain why varieties of molecular mechanisms of sex determination are found in fishes.

1.1.2 Environmental sex determination

The sex differentiation of fishes is influenced by environmental factors, such as temperature, exogenous hormones and behavior control[27].

Among the environmental factors, temperature has a stronger influence on sex differentiation. *Menidia menidia* was the first reported fish species in which both temperature and genes played a role in sex differentiation during the mid-larval stage[28]. Further studies demonstrated that both genetic and environmental factors have effects on sex differentiation in fishes. In some extreme cases, temperature factor can even rise above genetic determination to induce sex change. At high temperatures (32-34℃), the level of cortisol in the entire body of medaka is elevated. This change inhibits female-type proliferation of germ cells as well as expression of follicle-stimulating hormone (FSH) receptor mRNA, which gives rise to masculinization of XX medaka[29]. It appears that the response to temperature depends on the fish species. In general, for most temperature-sensitive species, the ratio of males to females is increased at higher temperatures[30]. However, species such as *Gambusia affinis* and *Coregonus hoyi* are entirely temperature-independent[31]. Yamaguchi *et al.*[32] studied molecular mechanisms of temperature-dependent sex determination in the Japanese flounder, *Paralichthys olivaceus*.

They found that high temperature treatment inhibited expression of *foxl2* and FSH receptor, and surmised that the transcription of *cyp19a1* was suppressed indirectly by high temperature, which led to the masculinization of *P. olivaceus*.

The view that fish sex differentiation can be induced by exogenous sex steroids is widely accepted. In most cases, short time treatment with sex steroids at early stages of sex determination (embryonic development stage) can permanently change and maintain fish phenotypic sex. Sometimes, persistent administration of sex steroids could result in hypoplastic gonads or sterility[33]. In another case, male chinook salmon, *Oncorhynchus tshawytscha*, could be induced to become females with the androgen 17-MT (17α-methyltestosterone) [34]. This phenomenon has been interpreted as the result of the estrogen receptor recognizing aromatized 17-MT.

With regards to use of the term "estrogens" it should be pointed out that it commonly refers to the hormones possessed only by females, a view challenged by recent studies. Ito *et al.*[35] found two kinds of estrogen receptors in Sertoli cells and even in haploid germ cells of the Japanese common goby, *Acanthogobius flavimanus*, suggesting that estrogens might play a role in regulating male germ cells development. Chaves-Pozo *et al.*[36] treated gilthead seabream, *Sparus aurata* L., with high doses of 17β-estradiol (E2). They found that E2 could inhibit the proliferation of spermatogonia, induce the apoptosis of primary spermatogonia and accelerate spermatogenesis. Another study by Miura *et al.*[37] indicated the role of a kind of progestin (17-α-20-β-dihydroxy-4-pregnen-3-one, DHP) in initiating meiosis of eel spermatogenetic cells. These findings strongly suggest a role of estrogens in fish testes development and maturation. On the other hand, a report by Liu *et al.*[38], in which androgen receptors were found in ovary tissue of *Spinibarbus denticulatus*, suggested a role of androgens in ovary development. Taken together, it appears that a collaboration of estrogens and androgens regulate fish reproductive development, and that a balanced coordination of two classes of sex steroids determine sex differentiation.

In some fish species, social factors also have an effect on sex determination. In one group some individuals can influence other individuals by their behavior. The population usually consists of a group of smaller individuals of a common sex and fewer larger dominant individuals of the opposite sex. When the dominant individual leaves or cannot control the other individuals, the second dominant individual will change its sex and replace the former[39].

1.2 Sex reversal in fishes

Some fish species reverse their sex naturally at some developmental stage. Fishes such as *Epinephelus* sp., *Pagrus major* and *Monopterus albus*, have to undergo sex reversal during their ontogenesis. The molecular mechanism of sex reversal is a classical and interesting research topic in fish reproductive development biology. Bhandari *et al.*[40] showed that the body size of hermaphrodites correlated with age and sex, which led them to always remain in the sex with higher fertility. The *dmrt1* and *SOX* gene families might play important roles in sex change from female to male. The cDNA libraries of hypothalamus, pituitary and gonad tissue in several developmental stages of male orange-spotted groupers (*Epinephelus coioides*) were constructed by Zhou *et al*[41]. They identified more than 10 genes involved in sex reversal of the grouper, including *dmrt1* involved in spermatogonia differentiating into spermatocytes, and *sox3* involved in oogenesis regulation. The mRNA expressions of *dmrt1* and *foxl2* during female to male change of honeycomb groupers (*Epinephelus merra*) were analyzed by Alam *et al*[42]. The results suggested that the down-regulation of *foxl2* accelerated oocyte degeneration, while the up-regulation of *dmrt1* led to the differentiation of germ cells into spermatogonia and promoted sex reversal. Huang *et al.*[43] found multiple alternative splicing of *dmrt1* in gonad transformation of rice field eel, *M. albus*, two transcripts of which showed increased expression during female to male change. The mRNA expression pattern of *sox9* during artificial sex reversal of orange-spotted groupers was analyzed by our group, and we hypothesized that *sox9* was an important factor to initiate and maintain masculinization[44]. In conclusion, fishes that naturally reverse their sex present a unique experimental opportunity for studying mechanisms of animal sex determination and differentiation.

Sex steroids are involved in the development and maintenance of gonads. The levels of varieties of sex steroids fluctuate during sex reversal. Kokokiris *et al.*[45] found during the sex change from female to male of protogynous Mediterranean red porgy (*Pagrus pagrus*) that androgen (11-ketotestosterone and testosterone, 11-KT and T) levels in serum rose and estrogen (estrone and 17β-estradiol, E1 and E2) levels dropped. On the other hand, when protandrous black porgy (*Acanthopagrus schlegeli*) underwent sex change from male to female, the E2 levels increased[46]. It is obvious that sex steroids have an important role in fish sex reversal. The effect of sex steroids on sex differentiation of fishes was further confirmed by experiments of immersing or feeding fish with sex steroids. Li *et al.*[47] implanted MT, MDHT (non-aromatizable 17α-methydihydrotestosterone) and MT+AI (aromatase inhibitor) into the body of red-spotted groupers (*Epinephelus akaara*), respectively. One month later, they obtained groupers with intersex gonads containing aretic oocytes and spermatogenic cells in several developmental stages, whereas male *Halichoeres trimaculatus* turned into females after 12 weeks of feeding with E2[48]. Bhandari *et al.*[49] treated female honeycomb groupers (*Epinephelus merra*) with 11-KT by intraperitoneal injection, and the treated groupers were induced into 100% males after 75 days. Aromatase is encoded by *cyp19a1a*. Treating red-spotted groupers with aromatase inhibitors also resulted in female-to-male sex change[50]. Guiguen *et al.* [16] proposed that *cyp19a1a* and endogenous estrogens were crucial for natural sex reversal. Their reduced expression initiated female-to-male change and their elevation triggered male-to-female change.

Groupers and rice field eels are both economically important hermaphrodites, which produce eggs at first. In production, the late occurrence of natural sex reversal of grouper causes lack of male parents, whereas the early occurrence of sex change in rice field eel leads to smaller female parents with fewer numbers of eggs. In order to promote aquaculture of these two fish species, it will be important to establish a regulatory technology to control endogenous sex steroids levels, by which sex reversal of groupers can be brought forward or sex reversal of rice field eels can be postponed.

1.3 Sex-controlled breeding

Many fishes show sexual dimorphism in growth rate or age of sexual maturation. Before sexual maturation, the energy fishes take in is mainly used for body growth. In some species such as common carp, females show a faster growth rate, due to ovary maturation occurring later than testis maturation. In other species, such as *Pelteobagrus fulvjdraco* and *Tilapia nilotica*, males show a faster growth rate. For fishes with a high reproductive capacity, such as tilapia, raising both sexes in one pool will produce plenty of offspring, which wastes resources. For these reasons, it is important for fish culture to produce all-males or all-females by using sex-controlling technologies. All-female carp and all-male *P. fulvidraco* have been produced by using sex steroids and inducing gynogenesis[51,52].

Estrogens play an important role in fish sex differentiation, which are regarded as a natural inducer of ovary differentiation[11,53]. The amount of estrogens correlates closely to the expression levels of aromatases. Therefore, some researchers tried to control the level of estrogens by regulating aromatase expression to impact on sex differentiation to control sex.

Wang *et al.*[54] produced transgenic tilapia in which *dmrt1* was over-expressed to inhibit transcription of *cyp19a1a*. The E2 level in serum of transgenic fish was lower, the ovarian cavity development was retarded, follicles degenerated to varying degrees, and even about 5% of all individuals developed into males. FOXL2 promotes synthesis of estrogens and development of ovary by upregulating *cyp19a1a* expression in female fish. A dominant negative mutant of *foxl2* was transferred into female tilapia to interfere with the endogenous FOXL2 functions[17]. The expression of *cyp19a1a* was suppressed, as well as reduction of serum estrogens level, and transgenic fishes resulted in masculinization. Rodriguez-Mari *et al.*[55] inserted *Tol2* into the coding region of *Fancl* in zebrafish to inactivate it. They observed apoptosis of germ cells during sex differentiation, thus compromising oocyte survival, and lower expression of *cyp19a1a*. The gonads became testes and the mutant fishes developed into males. Taken together, the goal of controlling sex can be achieved by regulating the expression of factors in signal pathways of sex differentiation, which is hoped to develop into a new technology.

Other investigations showed that knockdown of *dnd*, a gene involved in development and migration of primary germ cells (PGC), caused apoptosis of PGCs in zebrafish and medaka. The germ cell-ablated fish developed into males and had the ability of inducing females to lay eggs[56,57].

2. Fish Reproductive Control

There is a close relationship between growth and reproduction in fishes[58]. Rendering fishes sterile by manipulating reproduction will reduce the energy consumption distributed to reproduction, thus the fishes will grow faster. On the other hand, the ecological risks have become a bottleneck restriction to introducing foreign species and industrialization of transgenic fishes. Strategies of creating sterile fish would avoid the ecological risk entirely[59,60]. Reproductive control is important for a sustainable development of fish varieties and for updating and breeding studies of transgenic fishes.

2.1 Development of fish gonads

2.1.1 Development of fish germ cells

The reproductive development of fishes includes generation, migration, proliferation, and differentiation of PGCs and production of gametes[61]. *Vasa* was the first verified gene specifically expressed in PGCs of fishes[62]. By marking germ cells with fluorescence using germ cell special promoters or 3′-UTR[63], the formation process of gonads could be tracked down. Combined with gene knock-down/-out, it is useful for exploring new regulatory factors of reproduction.

Whole mount *in situ* hybridization (WISH) of zebrafish embryos with probe designed from *vasa* showed four signals located on the marginal positions of the first cleavage plane at 2- or 4-cell stage[64]. During the cleavage stage until the 1K-cell stage, the number of *vasa*-positive cells remains four, with fixed locations in the embryo. The expression patterns of *vasa* in rare minnow (*Gobiocypris rarus*) were proved similar to those in zebrafish by our group[65]. In medaka, however, the *vasa* RNA was observed widely distributed in early embryos. It did not show a special distribution until the late gastrula stage with 10-25 *vasa*-positive cells concentrating on both sides of the posterior shield[66]. Two different expression patterns of *vasa* suggested a variety of occurrence patterns of germ cells.

In zebrafish, PGCs start to migrate since dome stage, accompanied by reduced synthesis of E-cadherin. Some genes are transcribed in PGCs, and the gene *dnd*, the product of which can combine with RNA, is essential for the transcription[67]. PGCs become polarized and the receptor CXCR4 on them can recognize the chemokine SDF1a, then PGCs are attracted to move towards SDF1a locations. The somatic cells in the path of migration and embryonic primordia express SDF1 one by one to introduce PGC to the destination[68]. At the same time, the orderly expression of CXCR7 in somatic cells, another receptor of SDF1a, can eliminate redundant SDF1 to guide the migration[69].

The migration of medaka PGCs is guided by SDF1a and SDF1b[70], similar to zebrafish. Kunwar *et al.*[71] suggested a potentially universal rule that the migration direction of PGCs is decided by the balance of attractive and repellant cues, which is mediated by G protein-coupled receptors. Once PGCs arrive at their destination, some of them proliferate to maintain the number of stem germ cells, while others differentiate into spermatocytes or oocytes, then enter into meiosis, and finally produce gametes.

Until now, few factors associated with fish germ cells differentiation are known. Herpin *et al.*[72] observed that DMY could arrest germ cells of male medaka in the G2 phase. *In situ* hybridization in medaka showed that *mis* mRNA was expressed in the somatic cells surrounding germ cells during sex differentiation, and that MIS protein was thought to promote germ cells differentiation by acting on the somatic cells[73]. In zebrafish, PIWI could combine with germ cell specific piRNA, which is essential for germ cells maintenance and differentiation[74].

Factors involved in germ cell maturation include pituitary hormones, sex steroids and growth factors. Pituitary hormones include FSH (follicle stimulating hormone), LH (luteinizing hormone), GH (growth hormone) and so on. Sex steroids mainly include progestin, androgens and estrogens. Growth factors include activin, GSDF (gonad soma derived factor), AMH (anti-Mullerian hormone), IGF-I (insulin-like growth factor I), etc. For detailed reviews, please see Schulz *et al.*[75] and Lubzens *et al.*[76].

2.1.2 Endocrine regulation of fish gonad development

The development and maturation of fish gonads are controlled by genetic and environmental factors, and regulated by a molecular network of signals. As in other vertebrates, the HPG axis of the neuroendocrine system plays a central regulatory role on the reproductive development of fishes. Gonadotropin-releasing hormone (GnRH), lying upstream of the HPG axis, is a key regulatory factor. Once hypothalamus-secreted GnRH arrives at the pituitary gland, the cAMP signal pathway in gonadotrophs will be activated, transcription, synthesis and secreting of FSH and LH are all up-regulated. Via blood circulation, FSH and LH enter the gonads to regulate gonad development and to promote sexual maturation by stimulating production of sex steroids. On the other hand, sex steroids can inhibit synthesis of GnRH in the hypothalamus through a negative feedback pathway.

In recent years, the discoveries of kisspeptin, which can positively regulate GnRH, and GnIH, which can negatively regulate GnRH, have been praised as two remarkable breakthroughs in vertebrate neuroendocrinology[77]. They have become attractive topics in fish reproduction and physiological research.

Kisspeptin is a neuropeptide encoded by the *kiss* gene, and with GPR54 as the receptor. The first research about *kiss* function was carried out in hypogonadism[78]. The patients had delayed pubertal maturation and reproductive function due to a mutation in the gene encoding GPR54. Subsequent studies in mice, rats and monkeys further confirmed the functions of kisspeptin and GPR54 in reproductive development. Currently, the roles kisspeptin is thought to play include: sexual differentiation of the brain, the initiation of puberty, secretion and release of GnRH, transmission of feedback signals of sex steroids and control of reproductive function by a photoperiodic factor[79].

Currently, functional research of Kisspeptin and GPR54 in fish reproduction is developing rapidly. Parhar *et al.*[80] found mRNA of *GPR54* in GnRH neurons of tilapia, which suggested a direct effect of kisspeptin on GnRH neurons. Kitahashi *et al.*[81] cloned and expressed the *kiss2* gene of zebrafish and medaka. Via intraperitoneal administration and real-time PCR, they observed a more important role of *kiss2* than *kiss1* in regulating zebrafish sex steroid synthesis. Li *et al.*[82] observed elevated levels of LH in goldfish serum after intraperitoneal administration of mature kisspeptin peptides. However, Pasquier *et al.*[83] confirmed an inhibitory effect of kisspeptin on LH levels of the eel, *Anguilla anguilla*. It is worth noting that two or three types of GnRH usually coexist in one fish species, and that two types of kisspeptin and GPR54, respectively, are also found. Thus, the relationship between the kisspeptin/GPR54 signal system and GnRH, as well as regulatory mechanisms on reproduction, might differ between fishes and mammals.

GnIH (gonadotropin-inhibitory hormone) was first found in birds by Japanese scientists in 2000. Subsequent studies showed that GnIH might regulate reproduction of birds and mammals on every level of the HPG axis. It can inhibit synthesis and release of LH and FSH, induce apoptosis of testis cells, and might have a regulatory effect on the production of sex steroids as well as on the differentiation and maturation of germ cells by the way of autocrine or paracrine mechanisms[84]. The functional study of GnIH in fishes has just started, and there are still many inconsistent findings. For instance, Zhang *et al.*[85] detected lower levels of LH in the serum of mature female goldfish after intraperitoneal administration of zebrafish GnIH peptides, which demonstrated that GnIH could negatively regulate release of LH in goldfish. On the contrary, the results from intraperitoneal administration of GnIH to goldfish at developmental stages from early to late gonad recrudescence showed that GnIH increased pituitary LH-β and FSH-β mRNA levels as well as reduced serum LH and pituitary GnIH-R mRNA levels[86]. It has been suggested that GnIH has a seasonal effect on LH and FSH. This could change our understanding of the regulation of fish reproductive axis. It would also help us to study the function of GnIH in fish reproduction and explore new factors negatively regulating GnIH.

There are also types of molecules that can indirectly regulate fish reproductive development by affecting the HPG axis. For example, in many teleosts the release of GtH can be inhibited by dopamine (DA) [87], or be accelerated by γ-aminobutyric acid (GABA) [88], and secretion of GtH can be stimulated by neuropeptide Y (NPY) [85]. In our group, suppression subtractive hybridization (SSH) and cDNA microarray were applied to sterile females of the common carp with abnormal gonads and to wild type common carp[89]. We found more than

200 genes expressed differently at the hypothalamus, pituitary or gonad level, which included 87 unknown genes, whereas the other genes were associated with growth, organogenesis, energy metabolization, immune response, signal transduction and cell apoptosis. The results showed a very complex molecular network regulating fish reproductive development.

2.2 Strategies of fish reproductive control

Since GnRH is known as a key factor in HPG axis, many trials have been undertaken to control fish reproduction by suppressing GnRH expression[90,91]. In our group, endogenous sGnRH expression was suppressed by using the anti-sense transgenic method, which led to a genesis of gonads[89-92]. The result demonstrated that fish reproduction could be controlled by suppressing critical signal molecules associated with reproductive development.

The development of gonads can be seen as a process of generation, migration, proliferation, differentiation of PGCs as well as a production of gametes. Apoptosis of germ cells can be induced by interfering with the expression of genes associated with germ cell development. In zebrafish, knockdown of *dnd* using morpholino caused PGCs losing their migratory ability and resulted in apoptosis[93]. Suppression of *nanos1* inhibited PGCs from migrating properly and reduced their number[94]. PGCs would migrate abnormally and induce apoptosis when *Stau* was knocked down[95]. PGCs of a *zili*-deficient mutant also induced apoptosis before reaching meiosis[74].

Another strategy targeting germ cells to control reproduction of fishes is to kill germ cells directly and specifically. Slanchev and Stebler.[56] expressed kid, a bacterial toxin, preferentially in PGCs and simultaneously expressed the antidote kis uniformly in zebrafish. They obtained normally developing embryos but abated PGCs. Hsu *et al.*[96] and Hu *et al.*[97] expressed the *nitroreductase* gene specifically in zebrafish gonads using *zp*, *asp*, *odf* or *sam* promoter. When transgenic zebrafish were immersed in metronidazole, the nitroreductase expressed in germ cells turned metronidazole into cytotoxins and the germ cells were killed. It should be noted that the technologies listed above involved cell apoptosis or cytotoxicity. The question remains how we can abate germ cells specifically and efficiently without affecting growth of somatic cells or expressing cytotoxic proteins. For reasons of food security, this is a problem that needs to be resolved before this strategy can be applied to fish breeding.

3. Cell Engineering in Fish Reproductive Technology

Cell engineering in fish reproductive technology includes the transplantation of germ cells and the induction of stem cells.

The transplantation technique of cell nuclei has been established in zebrafish and medaka[98-100]. Transplantation of sperm nuclei in medaka was established by Liu *et al.*[101], who immersed medaka sperms in a solution containing RFP-expressing plasmids and then transferred the sperm nuclei into unfertilized eggs. In the nuclear transplants, 50% expressed RFP and about 5% embryos developed into mature fish. This kind of germ cell nuclear transplant could be modified to be an efficient transgenic method. However, the survival rate of nuclei transplants is low, due to incomplete reprogramming of donor nuclei. Thus, this transplantation technique of cell nuclei needs to be improved. In our laboratory, we found that dome and shield stages were key stages in reprogramming zebrafish nuclei-transplanted cells. Proper expression of *mycb* and *klf* also determined whether transplants would develop successfully or not[102,103].

The process of "borrow belly raw son" in fishes means transplanting PGCs of one fish species into the abdomen of another. The transplanted donor PGCs can develop independently or participate in gonad development, which can produce mature gametes of the donor fish. To find an efficient transgenic method, the technique of "borrow belly of fish" was established by Takeuch *et al.*[104], and has been constantly modified and improved since then. Saito *et al.*[105] killed germ cells of zebrafish at the blastula stage by first using *dnd* morpholino and then transplanted a single PGC from a pearl danio (*Danio albolineatus*) into the zebrafish embryo. The transplanted cells migrated to gonad anlage, formed one side gonad and could produce functional gametes. Thereafter, Saito *et al.*[106] compared the efficiency of PGC transfer between the methods of transplanting a single PGC and transplanting blastomeres. They considered the mechanism of PGC migration to be highly conserved in fishes, and found that transplanting a single PGC was more efficient than transplanting blastomeres. Nobrega *et al.*[107] transplanted spermatogonial stem cells of zebrafish into male and female recipients. From the results of donor-derived spermatogenesis and oogenesis, they demonstrated that germ cell differentiation could be influenced by the environment. Successful transplantation of PGCs from *Odontesthes bonariensis* into the gonads of sexually mature *O. hatcheri* was conducted by Majhi *et al.*[108] After six months, the transplanted cells resumed spermatogenesis, which means that germ cell transplantation could shorten the reproduction time of donor fishes. Reports that the "borrow belly of fish" technique had been carried out successfully between salmon and rainbow trout as well as two tilapia species[109-110], which are economically important fish, are very exciting. Such research indicates that "borrow belly of fish" between different fish species has become increasingly common. It has become valuable for shortening breeding cycles of economically important fish species and for the conservation of endangered fish species.

The technique of *in vitro* stem cell culture facilitates animal gene manipulation at the level of individuals. Few fish stem cell lines have been obtained so far. An embryonic stem cell line from medaka has been established, which is stable and can differentiate into a variety of cell types to form chimeras of medaka embryos[111,112]. Chen *et al.*[113] reported several embryonic stem cell lines from sea fishes with chimerical competence. An embryonic stem cell line from zebrafish could be maintained *in vitro* for a time, but then lost its pluripotency[114]. A striking report was published by Yi *et al.*[115], who established a haploid embryonic stem cell line which maintained characteristics of stem cells and could form embryonic chimeras with diploid cells. Its nuclei could even be transferred into normal eggs to produce fertile adult fishes. This study used a technique of semi-cloning of animals, demonstrating that haploid embryonic stem cell nuclei could initiate fertilization and produce offspring. This is regarded as a major breakthrough in reproductive biology.

Regarding germ line stem cells, attempts were concentrated on isolating and culturing spermatogonial cells. The first established fish spermatogonial cell line was derived from medaka, and could be induced to differentiate into functional sperms cells [116]. Sakai[117] described a method of isolating and culturing spermatogonial cells from zebrafish. Later, Fan *et al.*[118] isolated RFP-labeled zebrafish PGCs by flow cytometry and cultured them *in vitro* for more than four months. Shikina and Yoshizaki.[119] used a similar approach to isolate spermatogonial cells from rainbow trout. Panda *et al.*[120] isolated and enriched spermatogonial cells from *Labeo rohita* and propagated them *in vitro*. No cases of culturing fish oogonia have been reported except a recent observation of germ line stem cells in medaka ovaries[121]. Techniques of culturing fish germ line stem cells *in vitro* and inducing them to differentiate in the correct orientation will provide new inspiration for fish breeding methods.

4. Conclusions and Perspectives

There are about 28,000 known fish species, which is more than all other vertebrate species. Fishes are on an evolutionary stage linking invertebrates and higher vertebrates. They also have varieties of sexual characteristics and reproductive strategies that are representative for all vertebrate species. It is expected that these features will lead to important advances in understanding the regulatory mechanisms of animal reproduction by using fishes as models.

An increasing number of studies suggest a complex network that regulates fish reproductive development. Genome sequencing of model fish species such as zebrafish and medaka as well as economically important species such as tilapia and grass carp will provide important information for isolating functional genes, exploring signal pathways and interpreting molecular networks that control or regulate fish reproduction.

Fish reproduction is a precondition for culturing specific fish breeds and increasing their populations. Many fish species with good economical traits are difficult to handle with artificial reproductive manipulation to become refined cultured strains. Establishing controllable on-off techniques of reproduction and sex would open new possibilities for fish breeding.

Manipulating fish reproduction is regulating the generation, maintenance and differentiation of germ cells, as well as the interaction between germ cells and somatic cells. Establishing controllable on-off techniques of reproduction and sex depends on various factors, including a thorough theoretical understanding of reproductive regulatory networks, of mechanisms of sex determination and differentiation, and of mechanisms of germ cell growth and development. Further required are improved techniques of manipulating fish genomes, regulating target gene expression consistently, and identifying genotypes and phenotypes accurately and efficiently.

Acknowledgements This work was supported by National Natural Science Foundation of China (30930069), the Knowledge Innovation Project of the Chinese Academy of Sciences (KSCX2-EW-N-004), and the National High Technology Research and Development Program of China (2011AA100404).

References

1. Gui JF, Zhu ZY. Molecular basis and genetic improvement of economically important traits in aquaculture animals. Chin Sci Bull, 2012, 57: 1751-1760
2. Gui JF, Zhou L. Genetic basis and breeding application of clonal diversity and dual reproduction modes in polyploidy *Carassius auratus gibelio*. Sci China Life Sci, 2010, 53(4): 409-415
3. van Doom GS, Kirkpatrick M. Turnover of sex chromosomes induced by sexual conflict. Nature, 2007, 449: 909-912
4. Ross JA, Urton JR, Boland J. Turnover of sex chromosomes in the stickleback fishes (gasterosteidae). PLoS Genet, 2009, 2: e1000391
5. Arkhipchuk VV. Role of chromosomal and genome mutations in the evolution of bony fishes. Hydrobiol J, 1995, 31: 55-65
6. Traut W, Winking H. Meiotic chromosomes and stages of sex chromosome evolution in fish: Zebrafish, platyfish and guppy. Chromosome Res, 2001, 9: 659-672
7. Artieri CG, Mitchell LA, Ng SH, *et al*. Identification of the sex-determining locus of Atlantic salmon (*Salmo salar*) on chromosome 2. Cytogenet Genome Res, 2006, 112: 152-159
8. Zhao G, Yu QX, Zhang WW, *et al*. The 5S rDNA related repetitive sequences in the sex chromosomes of the spiny eel (*Mastacembelus aculeatus*). Cytogenet Genome Res, 2008, 121: 143-148
9. Lee BY, Kocher TD. Exclusion of Wilms tumour (WT1b) and ovarian cytochrome P450 aromatase (CYP19A1) as candidates for sex determination genes in Nile tilapia (*Oreochromis niloticus*). Anim Genet, 2007, 38: 85-86
10. von Hofsten J, Larsson A, Olsson PE. Novel steroidogenic factor-1 homolog (*ff1d*) is coexpressed with anti-Mullerian hormone (*AMH*) in zebrafish. Dev Dyn, 2005, 233: 595-604
11. von Hofsten J, Olsson PE. Zebrafish sex determination and differentiation: involvement of *FTZ-F1* genes. Reprod Biol Endocrinol, 2005, 3: 63
12. Nakamoto M, Wang DS, Suzuki A, *et al*. Dax1 suppresses P450arom expression in medaka ovarian follicles. Mol Reprod Dev, 2007, 74: 1239-1246
13. Rogriguez-Mari A, Yan YL, Bremiller RA, *et al*. Characterization and expression pattern of zebrafish *Anti-Mullerian hormone* (*Amh*) relative to *sox9a*, *sox9b*, and *cyp19a1a*, during gonad development. Gene Expr Patterns, 2005, 5: 655-667
14. Sawyer SJ, Gerstner KA, Callard GV. Real-time PCR analysis of cytochrome P450 aromatase expression in zebrafish: gene specific tissue distribution, sex differences, developmental programming, and estrogen regulation. Gen Comp Endocrinol, 2006, 147: 108-117
15. Tang B, Hu W, Hao J, *et al*. Developmental expression of steroidogenic factor-1, cyp19a1a and cyp19a1b from common carp (*Cyprinus carpio*). Gen Comp Endocrinol, 2010, 167: 408-416

16. Guiguen Y, Fostier A, Piferrer F, *et al.* Ovarian aromatase and estrogens: a pivotal role for gonadal sex differentiation and sex change in fish. Gen Comp Endocrinol, 2010, 165: 352-366

17. Wang DS, Kobayashi T, Zhou LY. Foxl2 up-regulates aromatase gene transcription in a female-specific manner by binding to the promoter as well as interacting with ad4 binding protein/steroidogenic factor 1. Mol Endocrinol, 2007, 21(3): 712-725

18. Vizziano D, Randuineau G, Baron D, *et al.* Characterization of early molecular sex differentiation in rainbow trout, *Oncorhynchus mykiss*. Dev Dyn, 2007, 236: 2198-2206

19. Cao MX, Duan JD, Cheng NN, *et al.* Sexually dimorphic and ontogenetic expression of *dmrt1*, *cyp19a1a* and *cyp19a1b* in *Gobiocypris rarus*. Comp Biochem Physiol A Mol Integr Physiol, 2012, 162: 303-309

20. Matsuda M, Nagahama Y, Shinomiya A, *et al.* *DMY* is a Y-specific DM-domain gene required for male development in the medaka fish. Nature, 2002, 417: 559-563

21. Nanda I, Kondo M, Hornung U, *et al.* A duplicated copy of *DMRT1* in the sex-determining region of the Y chromosome of the medaka, *Oryzias latipes*. Proc Natl Acad Sci USA, 2002, 99: 11778-11783

22. Hattori RS, Murai Y, Oura M, *et al.* A Y-linked anti-Mullerian hormone duplication takes over a critical role in sex determination. Proc Natl Acad Sci USA, 2012, 109: 2955-2959

23. Li J, Phillips RB, Harwood AS, *et al.* Identification of the sex chromosomes of brown trout (*Salmo trutta*) and their comparison with the corresponding chromosomes in Atlantic salmon (*Salmo salar*) and rainbow trout (*Oncorhynchus mykiss*). Cytogenet Genome Res, 2011, 133: 25-33

24. Vandeputte M, Dupont-Nivet M, Chavanne H, *et al.* A polygenic hypothesis for sex determination in the European sea bass *Dicentrarchus labrax*. Genetics, 2007, 176: 1049-1057

25. Bradley KM, Breyer JP, Melville DB, *et al.* An SNP-based linkage map for zebrafish reveals sex determination loci. G3 (Bethesda), 2011, 1(1): 3-9

26. Liew WC, Bartfai R, Lim Z, *et al.* Polygenic Sex Determination System in Zebrafish. PLoS One, 2012, 7: e34397.

27. Devlin RH, Nagahama Y. Sex determination and sex differentiation in fish: an overview of genetic, physiological, and environmental influences. Aquaculture, 2002, 208: 191-364

28. Conover DO, Kynard BE. Environmental sex determination: interaction of temperature and genotype in a fish. Science, 1981, 213: 577-579

29. Hayashi Y, Kobira H, Yamaguchi T, *et al.* High temperature causes masculinization of genetically female medaka by elevation of cortisol. Mol Reprod Dev, 2010, 77(8): 679-686

30. Roemer U, Beisenherz W. Environmental determination of sex in Apistogramma (Cichlidae) and two other freshwater fishes (Teleostei). J Fish Biol, 1996, 48: 714-725

31. Conover DO, Daemond SB. Absence of temperature dependent sex determination in northern populations of two cyprinodontid fishes. Can J Zool, 1991, 69: 530-533

32. Yamaguchi T, Yamaguchi S, Hirai T, *et al.* Follicle-stimulating hormone signaling and Foxl2 are involved in transcriptional regulation of aromatase gene during gonadal sex differentiation in Japanese flounder *Paralichthys olivaceus*. Biochem Biophys Res Commun, 2007, 359: 935-940

33. Sehgal GK, Saxena PK. Effect of estrone on sex composition, growth and flesh composition in common carp, *Cyprinus carpio* communis (Linn.). J Aquacult Trop, 1997, 12: 289-295

34. Piferrer F, Baker IJ, Donaldson EM. Effects of natural, synthetic, aromatizable, and nonaromatizable androgens in inducing male sex differentiation in genotypic female chinook salmon (*Oncorhynchus tshawytscha*). Gen Comp Endocrinol, 1993, 91(1): 59-65

35. Ito K, Mochida K, Fujii K. Molecular cloning of two estrogen receptors expressed in the testis of the Japanese common goby, *Acanthogobius flavimanus*. Zool Sci, 2007, 24: 986-996

36. Chaves-Pozo E, Liarte S, Vargas-Chacoff L, *et al.* 17Beta-estradiol triggers postspawning in spermatogenically active gilthead seabream (*Sparus aurata* L.) males. Biol Reprod, 2007, 76: 142-148

37. Miura T, Higuchi M, Ozaki Y, *et al.* Progestin is an essential factor for the initiation of the meiosis in spermatogenetic cells of the eel. Proc Natl Acad Sci USA, 2006, 103: 7333-7338

38. Liu X, Su H, Zhu P, *et al*. Molecular cloning, characterization and expression pattern of androgen receptor in *Spinibarbus denticulatus*. Gen Comp Endocrinol, 2009, 160: 93-101
39. Quinitio GF, Caberoy NB, Reyes Jr DM. Induction of sex change in female *Epinephelus coioides* by social control. Isr J Aquacult, 1997, 49: 77-83
40. Bhandari RK, Komuro H, Nakamura S, *et al*. Gonadal restructuring and correlative steroid hormone profiles during natural sex change in protogynous honeycomb grouper (*Epinephelus merra*). Zool Sci, 2003, 20: 1399-1404
41. Zhou L, Yao B, Xia W, *et al*. EST-based identification of genes expressed in the hypothalamus of male orange-spotted grouper (*Epinephelus coioides*). Aquaculture, 2006, 256: 129-139
42. Alam MA, Kobayashi Y, Horiguchi R, *et al*. Molecular cloning and quantitative expression of sexually dimorphic markers Dmrt1 and Foxl2 during female-to-male sex change in *Epinephelus merra*. Gen Comp Endocrinol, 2008, 157: 75-85
43. Huang X, Guo Y, Shui Y, *et al*. Multiple alternative splicing and differential expression of *dmrt1* during gonad transformation of the rice field eel. Biol Reprod, 2005, 73: 1017-1024
44. Luo YS, Hu W, Liu XC, *et al*. Molecular cloning and mRNA expression pattern of *Sox9* during sex reversal in orange-spotted grouper (*Epinephelus coioides*). Aquaculture, 2010, 306: 322-328
45. Kokokiris L, Fostier A, Athanassopoulou F, *et al*. Gonadal changes and blood sex steroids levels during natural sex inversion in the protogynous Mediterranean red porgy, *Pagrus pagrus* (Teleostei: Sparidae). Gen Comp Endocrinol, 2006, 149: 42-48
46. Chang CF, Lee MF, Chen GL. Estradiol-17β associated with the sex reversal in protandrous black porgy, *Acanthopagrus schlegeli*. J Exp Zool, 1994, 268: 53-58
47. Li GL, Liu XC, Lin HR. Effects of aromatizable and nonaromatizable androgens on the sex inversion of red-spotted grouper (*Epinephelus akaara*). Fish Physiol Biochem, 2006, 32: 25-33
48. Kojima Y, Bhandari RK, Kobayashi Y, *et al*. Sex change of adult initial-phase male wrasse, *Halichoeres trimaculatus* by estradiol-17 beta treatment. Gen Comp Endocrinol, 2008, 156: 628-632
49. Bhandari RK, Alam MA, Soyano K, *et al*. Induction of female-to-male sex change in the honeycomb grouper (*Epinephelus merra*) by 11-ketotestosterone treatments. Zool Sci, 2006, 23: 65-69
50. Li GL, Liu XC, Lin HR. Seasonal changes of serum sex steroids concentration and aromatase activity of gonad and brain in red-spotted grouper (*Epinephelus akaara*). Anim Reprod Sci, 2007, 99: 156-166
51. Wu CJ, Chen RD, Ye YZ, *et al*. Production of all-female carp and its applications in fish cultivation. Aquaculture, 1990, 85: 327
52. Liu HQ, Cui SQ, Hou CC, *et al*. YY supermale generated gynogenetically from XY female in *Pelteobagrus fulvidraco* (Richardson) (in Chinese). Acta Hydrobiol Sin, 2007, 31: 718-725
53. Liu Z, Wu F, Jiao B, *et al*. Molecular cloning of doublesex and mab-3-related transcription factor 1, forkhead transcription factor gene 2, and two types of cytochrome P450 aromatase in Southern catfish and their possible roles in sex differentiation. J Endocrinol, 2007, 194: 223-241
54. Wang DS, Zhou LY, Kobayashi T, *et al*. Doublesex- and Mab-3-related transcription factor-1 repression of aromatase transcription, a possible mechanism favoring the male pathway in tilapia. Endocrinology, 2010, 151: 1331-1340
55. Rodriguez-Mari A, Canestro C, Bremiller RA, *et al*. Sex reversal in zebrafish *fancl* mutants is caused by Tp53-mediated germ cell apoptosis. PLoS Genet, 2010, 6: e1001034
56. Slanchev K, Stebler J, de la Cueva-Mendez G, *et al*. Development without germ cells: the role of the germ line in zebrafish sex differentiation. Proc Natl Acad Sci USA, 2005, 102: 4074-4079
57. Kurokawa H, Saito D, Nakamura S, *et al*. Germ cells are essential for sexual dimorphism in the medaka gonad. Proc Natl Acad Sci USA, 2007, 104: 16958-16963
58. Li WS, Lin HR. The endocrine regulation network of growth hormone synthesis and secretion in fish: Emphasis on the signal integration in somatotropes. Sci China Life Sci, 2010, 53: 462-470
59. Hu W, Wang YP, Zhu ZY. Progress in the evaluation of transgenic fish for possible ecological risk and its

containment strategies. Sci China Life Sci, 2007, 50: 573-579

60. Hu W, Zhu ZY. Integration mechanisms of transgenes and population fitness of GH transgenic fish. Sci China Life Sci, 2010, 53: 401-408

61. Xu HY, Li MY, Gui JF, *et al*. Fish germ cells. Sci China Life Sci, 2010, 53: 435-446

62. Olsen LC, Aasland R, Fjose A. A *vasa*-like gene in zebrafish identifies putative primordial germ cells. Mech Dev, 1997, 66: 95-105

63. Kawakami Y, Saito T, Fujimoto T, *et al*. Visualization and motility of primordial germ cells using green fluorescent protein fused to 3′UTR of common carp *nanos*-related gene. Aquaculture, 2011, 317: 245-250

64. Raz E, Hopkins N. Primordial germ-cell development in zebrafish. Results Probl Cell Differ, 2002, 40: 166-179

65. Cao MX, Yang YH, Xu HY, *et al*. Germ cell specific expression of Vasa in rare minnow, *Gobiocypris rarus*. Comp Biochem Physiol A Mol Integr Physiol,, 2012, 162: 163-170

66. Shinomiya A, Tanaka M, Kobayashi T, *et al*. The *vasa*-like gene, *olvas*, identifies the migration path of primordial germ cells during embryonic body formation stage in the medaka, *Oryzias latipes*. Dev Growth Differ, 2000, 42: 317-326

67. Blaser H, Eisenbeiss S, Neumann M, *et al*. Transition from non-motile behaviour to directed migration during early PGC development in zebrafish. J Cell Sci, 2005, 118: 4027-4038

68. Raz E, Reichman-Fried M. Attraction rules: germ cell migration in zebrafish. Curr Opin Genet Dev, 2006, 16: 355-359

69. Boldajipour B, Mahabaleshwar H, Kardash E, *et al*. Control of chemokine-guided cell migration by ligand sequestration. Cell, 2008, 132: 463-473

70. Herpin A, Fischer P, Liedtke D, *et al*. Sequential SDF1a and b-induced mobility guides medaka PGC migration. Dev Biol, 2008, 320: 319-327

71. Kunwar PS, Siekhaus DE, Lehmann R. *In vivo* migration: a germ cell perspective. Annu Rev Cell Dev Biol, 2006, 22: 237-265

72. Herpin A, Schindler D, Kraiss A, *et al*. Inhibition of primordial germ cell proliferation by the medaka male determining gene Dmrt1bY. BMC Dev Biol, 2007, 7: 99

73. Shiraishi E, Yoshinaga N, Miura T, *et al*. Mullerian inhibiting substance is required for germ cell proliferation during early gonadal differentiation in medaka (*Oryzias latipes*). Endocrinology, 2008, 149: 1813-1819

74. Houwing S, Kamminga LM, Berezikov E, *et al*. A role for Piwi and piRNAs in germ cell maintenance and transposon silencing in Zebrafish. Cell, 2007, 129: 69-82

75. Schulz RW, de França LR, Lareyre JJ, *et al*. Spermatogenesis in fish. Gen Comp Endocrinol, 2010, 165: 390-411

76. Lubzens E, Young G, Bobe J, *et al*. Oogenesis in teleosts: How fish eggs are formed. Gen Comp Endocrinol, 2010, 165: 367-389

77. Zohar Y, Munoz-Cueto JA, Elizur A, *et al*. Neuroendocrinology of reproduction in teleost fish. Gen Comp Endocrinol, 2010, 165: 438-455

78. de Roux N, Genin E, Carel JC, *et al*. Hypogonadotropic hypogonadism due to loss of function of the KiSS1-derived peptide receptor GPR54. Proc Natl Acad Sci USA, 2003, 100: 10972-10976

79. Tena-Sempere M, Felip A, Gomez A, *et al*. Comparative insights of the kisspeptin/kisspeptin receptor system: Lessons from non-mammalian vertebrates. Gen Comp Endocrinol, 2012, 175: 234-243

80. Parhar IS, Ogawa S, Sakuma Y. Laser-captured single digoxigenin-labeled neurons of gonadotropin-releasing hormone types reveal a novel G protein-coupled receptor (Gpr54) during maturation in cichlid fish. Endocrinology, 2004, 145: 3613-3618

81. Kitahashi T, Ogawa S, Parhar I. Cloning and expression of *kiss2* in the zebrafish and medaka. Endocrinology, 2009, 150: 821-831

82. Li S, Zhang Y, Liu Y, *et al*. Structural and functional multiplicity of the kisspeptin/GPR54 system in goldfish (*Carassius auratus*). J Endocrinol, 2009, 201: 407-418

83. Pasquier J, Lafont AG, Leprince J, *et al*. First evidence for a direct inhibitory effect of kisspeptins on LH expression

in the eel, *Anguilla anguilla*. Gen Comp Endocrinol, 2011, 173: 216-225

84. Tsutsui K, Bentley GE, Ubuka T, *et al*. The general and comparative biology of gonadotropin- inhibitory hormone (GnIH). Gen Comp Endocrinol, 2007, 153: 365-370

85. Zhang Y, Li S, Liu Y, *et al*. Structural diversity of the gnih/gnih receptor system in teleost: its involvement in early development and the negative control of LH release. Peptides, 2010, 31: 1034-1043

86. Moussavi M, Wlasichuk M, Chang JP, *et al*. Seasonal effect of GnIH on gonadotrope functions in the pituitary of goldfish. Mol Cell Endocrinol, 2012, 350: 53-60

87. Trudeau VL. Neuroendocrine regulation of gonadotrophin II release and gonadal growth in the goldfish, *Carassius auratus*. Rev Reprod, 1997, 2: 55-68

88. Popesku JT, Martyniuk CJ, Mennigen J, *et al*. The goldfish (*Carassius auratus*) as a model for neuroendocrine signaling. Mol Cell Endocrinol, 2008, 293: 43-56

89. Xu J, Huang W, Zhong CR, *et al*. Defining global gene expression changes of the hypothalamic-pituitary-gonadal axis in female sGnRH-antisense transgenic common carp (*Cyprinus carpio*). PLoS One, 2011, 6: e21057

90. Uzbekova S, Chyb J, Ferriere F, *et al*. Transgenic rainbow trout expressed sGnRH-antisense RNA under the control of sGnRH promoter of Atlantic salmon. J Mol Endocrinol, 2000, 25: 337-350

91. Maclean N, Hwang G, Molina A, *et al*. Reversibly-sterile fish via transgenesis. ISB News Report, 2003, 1-3

92. Hu W, Li SF, Tang B, *et al*. Antisense for gonadotropin-releasing hormone reduces gonadotropin synthesis and gonadal development in transgenic common carp (*Cyprinus carpio*). Aquaculture, 2007, 271: 498-506

93. Weidinger G., Stebler J, Slanchev K, *et al*. *dead end*, a novel vertebrate germ plasm component, is required for zebrafish primordial germ cell migration and survival. Curr Biol, 2003, 13: 1429-1434

94. Koprunner M, Thisse C, Thisse B, *et al*. A zebrafish *nanos*-related gene is essential for the development of primordial germ cells. Genes Dev, 2001, 15: 2877-2885

95. Ramasamy S, Wang H, Quach HN, *et al*. Zebrafish Staufen1 and Staufen2 are required for the survival and migration of primordial germ cells. Dev Biol, 2006, 292: 393-406

96. Hsu CC, Hou MF, Hong JR, *et al*. Inducible male infertility by targeted cell ablation in zebrafish testis. Mar Biotechnol, 2010, 12: 466-478

97. Hu SY, Lin PY, Liao CH, *et al*. Nitroreductase-mediated gonadal dysgenesis for infertility control of genetically modified zebrafish. Mar Biotechnol, 2010, 12: 569-578

98. Wakamatsu Y, Ju B, Pristyaznhyuk I, *et al*. Fertile and diploid nuclear transplants derived from embryonic cells of a small laboratory fish, medaka (*Oryzias latipes*). Proc Natl Acad Sci USA, 2001, 98: 1071-1076

99. Hu W, Wang YP, Chen SP, *et al*. Nuclear transplantation in different strains of zebrafish. Chin Sci Bull, 2002, 47: 1277-1280

100. Lee KY, Huang H, Ju B, *et al*. Cloned zebrafish by nuclear transfer from long-term-cultured cells. Nat Biotechnol, 2002, 20: 795-799

101. Liu TM, Liu L, Wei QW, *et al*. Sperm nuclear transfer and transgenic production in the fish medaka. Int J Biol Sci, 2011, 7: 469-475

102. Luo DJ, Hu W, Chen SP, *et al*. Identification of differentially expressed genes between cloned and zygote-developing zebrafish (*Danio rerio*) embryos at the dome stage using suppression subtractive hybridization. Biol Reprod, 2009, 80: 674-684

103. Luo DJ, Hu W, Chen SP, *et al*. Critical developmental stages for the efficiency of somatic cell nuclear transfer in zebrafish. Int J Biol Sci, 2011, 7: 476-486

104. Takeuchi Y, Yoshizaki G, Takeuchi T. Generation of live fry from intraperitoneally transplanted primordial germ cells in rainbow trout. Biol Reprod, 2003, 69: 1142-1149

105. Saito T, Goto-Kazeto R, Arai K, *et al*. Xenogenesis in teleost fish through generation of germ-line chimeras by single primordial germ cell transplantation. Biol Reprod, 2008, 78: 159-166

106. Saito T, Goto-Kazeto R, Fujimoto T, *et al*. Inter-species transplantation and migration of primordial germ cells in cyprinid fish. Int J Dev Biol, 2010, 54: 1481-1486

107. Nobrega RH, Greebe CD, van de Kant H, *et al*. Spermatogonial stem cell niche and spermatogonial stem cell transplantation in zebrafish. PLoS One, 2010, 5: e12808

108. Majhi SK, Hattori RS, Yokota M, *et al*. Germ cell transplantation using sexually competent fish: an approach for rapid propagation of endangered and valuable germlines. PLoS One, 2009, 4: e6132

109. Okutsu T, Shikina S, Kanno M, *et al*. Production of trout offspring from triploid salmon parents. Science, 2007, 317: 1517

110. Lacerda SM, Batlouni SR, Costa GM. A new and fast technique to generate offspring after germ cells transplantation in adult fish: the Nile tilapia (*Oreochromis niloticus*) model. PLoS One, 2010, 5: e10740

111. Hong YH, Winkler C, Schartl M. Pluripotency and differentiation of embryonic stem cell lines from the medakafish (*Oryzias latipes*). Mech Dev, 1996, 60: 33-44

112. Yi MS, Hong N, Li ZD, *et al*. Medaka fish stem cells and their applications. Sci China Life Sci, 2010, 53: 426-434

113. Chen SL, Sha ZX, Ye HQ, *et al*. Pluripotency and chimera competence of an embryonic stem cell line from the sea perch (*Lateolabrax japonicus*). Mar Biotechnol, 2007, 9: 82-91

114. Fan LC, Collodi P. Zebrafish embryonic stem cells. Methods Enzymol, 2006, 418: 64-77

115. Yi MS, Hong N, Hong YH. Generation of medaka fish haploid embryonic stem cells. Science, 2009, 326: 430-433

116. Hong YH, Liu TM, Zhao HB, *et al*. Establishment of a normal medakafish spermatogonial cell line capable of sperm production *in vitro*. Proc Natl Acad Sci USA, 2004, 101: 8011-8016

117. Sakai N. *In vitro* male germ cell cultures of zebrafish. Methods, 2006, 39: 239-245

118. Fan LC, Moon J, Wong TT, *et al*. Zebrafish primordial germ cell cultures derived from *vasa*: *RFP* transgenic embryos. Stem Cells Dev, 2008, 17: 585-597

119. Shikina S, Yoshizaki G. Improved *in vitro* culture conditions to enhance the survival, mitotic activity, and transplantability of rainbow trout type A spermatogonia. Biol Reprod, 2010, 83: 268-276

120. Panda RP, Barman HK, Mohapatra C. Isolation of enriched carp spermatogonial stem cells from *Labeo rohita* testis for *in vitro* propagation. Theriogenology, 2011, 76: 241-251

121. Nakamura S, Kobayashi K, Nishimura T, *et al*. Identification of germline stem cells in the ovary of the teleost medaka. Science, 2010, 328: 1561-1563

鱼类生殖发育调控研究进展

陈 戟 胡 炜 朱作言

中国科学院水生生物研究所，淡水生态与生物技术国家重点实验室，武汉 430072

摘 要 动物生殖发育机制一直是生命科学的重要研究课题之一。鱼类的生殖发育及其调控机制研究不仅具有重要的理论意义，而且在指导鱼类品种改良上具有广泛的应用前景。本文评述了鱼类性别决定和分化、生殖细胞和性腺发育，以及生殖发育调控策略等方面的研究进展。由于鱼类在动物演化中承上启下的地位、丰富的物种多样性和几乎具有所有脊椎动物生殖策略等特点，以鱼类为模型可望取得动物生殖调控机制研究的重要突破，并由此建立鱼类生殖开关和性别控制开关等生殖操作新技术，为鱼类养殖品种培育技术打开一扇新的大门。

Effects of Growth Hormone Over-Expression on Reproduction in the Common Carp *Cyprinus carpio* L.

Mengxi Cao[1,2] Ji Chen[1] Wei Peng[1,2] Yaping Wang[1] Lanjie Liao[1]
Yongming Li[1] Vance L. Trudeau[3] Zuoyan Zhu[1] Wei Hu[1]

1 State Key Laboratory of Freshwater Ecology and Biotechnology, Institute of Hydrobiology, Chinese Academy of Sciences, Wuhan 430072, China
2 University of the Chinese Academy of Sciences, Beijing 100049, China
3 Department of Biology, Centre for Advanced Research in Environmental Genomics, University of Ottawa, Ottawa, Canada, K1N 6N5

Abstract To study the complex interaction between growth and reproduction we have established lines of transgenic common carp (*Cyprinus carpio*) carrying a grass carp (*Ctenopharyngodon idellus*) growth hormone (GH) transgene. The GH-transgenic fish showed delayed gonadal development compared with non-transgenic common carp. To gain a better understanding of the phenomenon, we studied body growth, gonad development, changes of reproduction related genes and hormones of GH-transgenic common carp for 2 years. Over-expression of GH elevated peripheral *gh* transcription, serum GH levels, and inhibited endogenous GH expression in the pituitary. Hormone analyses indicated that GH-transgenic common carp had reduced pituitary and serum level of luteinizing hormone (LH). Among the tested genes, pituitary *lhβ* was inhibited in GH-transgenic fish. Further analyses *in vitro* showed that GH inhibited *lhβ* expression. Localization of *ghr* with LH indicates the possibility of direct regulation of GH on gonadotrophs. We also found that GH-transgenic common carp had reduced pituitary sensitivity to stimulation by co-treatments with a salmon gonadotropin-releasing hormone (GnRH) agonist and a dopamine antagonist. Together these results suggest that the main cause of delayed reproductive development in GH transgenic common carp is reduced LH production and release.

Keywords growth hormone; reproduction; interaction; luteinizing hormone; GH-transgenic common carp

1. Introduction

It is well known that growth and reproduction are closely related in vertebrates. There is also considerable 'cross-talk' between the neuroendocrine axes controlling growth and reproduction (Hull and Harvey, 2002; Klausen *et al.*, 2002; Le Gac *et al.*, 1993; Trudeau *et al.*, 1992). Growth hormone (GH) is synthesized and secreted mainly in the pituitary gland. In

文章发表于 *General and Comparative Endocrinology*, 2014, 195(1): 47-57 addition to the pituitary gland, GH is also detected in the gonads in many vertebrates (Harvey, 2010). Besides its role in growth promotion, GH acts directly on gonadal tissues to stimulate spermatogenesis and ovarian hormone synthesis (Miura *et al*., 2011; Van Der Kraak *et al*., 1990), or indirectly affects gonad development by stimulating the expression of IGF-1 (Berishvili *et al*., 2006). As GH participates in hormone synthesis, ovulation, growth and renewal of follicles, oocyte maturation, spermatogenesis, sperm motility, and other aspects of reproductive development, it may be considered a co-gonadotropin (Hull and Harvey, 2002). In the teleost pars distalis of the pituitary, somatotrophs and gonadotrophs are adjacent to each other, enabling paracrine communication (Wong *et al*., 2006). *In vitro* studies of grass carp (*Ctenopharyngodon idellus*) pituitary cells revealed the paracrine regulation of luteinizing hormone (LH) secretion by GH (Zhou *et al*., 2004). Unlike GH release in mammals, which is mainly regulated by the secretion of GH releasing hormone (GHRH) and somatostatin (SRIF) secreted by the hypothalamus, GH release in fish is also modulated by other neuroendocrine factors including gonadotropin-releasing hormone (GnRH), estradiol (E2), and testosterone (T) (Klausen *et al*., 2002; Lin *et al*., 1993; Peng and Peter, 1997; Trudeau *et al*., 1992; Wong *et al*., 2006; Xu *et al*., 2011). All these findings suggest a close connection between the neuroendocrine regulation of growth and reproduction (Trudeau, 1997). Growth hormone is therefore an important signal transducer connecting the growth and reproductive axes.

To date, five stable lines of GH-transgenic fish have been generated, including Atlantic salmon (*Salmo salar*) (Fletcher *et al*., 2004; Rokkones *et al*., 1989), coho salmon (*Oncorhynchus kisutch*) (Devlin *et al*., 2004), tilapia (*oreochromis niloticus*) (Rahman *et al*., 1998), mud loach (*Misgurnus mizolepis*) (Nam *et al*., 2002) and common carp (*Cyprinus carpio*) (Wang *et al*., 2001; Zhong *et al*., 2012). However, because of the potential ecological risk of transgenic fish, none of these fish have yet been produced for the intended goal of commercial use and human consumption (Devlin *et al*., 2006; Hu *et al*., 2006, 2007; Hu and Zhu, 2010). The over-expression of GH causes numerous biological effects such as changes in growth rates, feeding behavior, swimming speed, energy metabolism, osmoregulation, and hypoxia tolerance (Almeida *et al*., 2013; Caelers *et al*., 2005; Duan *et al*., 2011; Guan *et al*., 2008, 2011; Li *et al*., 2007; Lohmus *et al*., 2008; Mori *et al*., 2007; Rahman *et al*., 2001; Raven *et al*., 2008; Wang *et al*., 2001; Zhong *et al*., 2013, 2012). Growth hormone over-expression could also affect reproduction. Some of the male GH-transgenic tilapia have lower levels of sperm production while the female GH-transgenic tilapia have significantly lower gonadal-somatic index (GSI) (Rahman *et al*., 1998, 2001). GH-transgenic coho salmon reach sexual maturation one year earlier than cultured non-transgenic fish, although some transgenic coho salmon display less courtship and reduced spawning capacity than their non-transgenic counterparts (Bessey *et al*., 2004; Fitzpatrick *et al*., 2011). Male GH-transgenic Atlantic salmon show poor nest fidelity, quivering frequency, and spawn participation, as well as overall fertilization success compared with their wild counterparts (Moreau *et al*., 2011). In our laboratory, the "all-fish"

GH-transgenic common carp also have enhanced growth rate but with delayed gonadal development compared with their non-transgenic counterparts. Rahman *et al.* (2001) suggested that reduced reproductive performance in transgenic tilapia may be caused by increased energy allocation to somatic growth rather than gonad development. Reproductive performance is a key fitness parameter that is used to assess the potential ecological risks of GH-transgenic fish. However, the endocrine mechanisms causing of reproductive abnormalities in GH-transgenic fish are poorly understood.

In this study, we investigated the cause of reduced reproductive performance using the GH-transgenic common carp as our model system. Research on the reproductive performance will not only provide data for evaluation of the ecological risks of GH-transgenic fish but also enable further study of the complex interactions between the growth and reproductive axes.

2. Materials and Methods

2.1 Experimental fish

The GH-transgenic common carp used in the study was line #TG2, carrying the grass carp growth hormone gene (Zhong *et al.*, 2012). After fertilization in April 2009, GH-transgenic and non-transgenic common carp were reared under the same conditions at Guanqiao Experimental Depot, Wuhan, China. The #TG2 and non-transgenic lines were derived from the same non-transgenic mother. The transgenic genotypes were confirmed by PCR as described by Guan *et al.* (2008).

Fish were sampled once every month until the age of one year. The first sampling period was in May when the fish were one month old. Two-year-old fish were sampled in the March to April, which is the breeding season for common carp in Wuhan, China. Two samples were obtained in April (before and after ovulation/spermiation). At each time point, 15 individuals each were randomly selected from the transgenic and non-transgenic groups and the body weights were measured (Table S1). All procedures were conducted in accordance with the Guiding Principles for the Care and Use of Laboratory Animals and were approved by Institute of Hydrobiology, Chinese Academy of Sciences.

2.2 Gonadal-somatic index and gonadal histology

Body weights (W_b) and gonad weights (W_g) of the sampled common carp were obtained, and the GSI was calculated using the following formula: $I_G = 100 \times W_g / W_b$. The gonads were dissected, immersed in Bouin's fixative, dehydrated, embedded in paraffin, and cut into 8-μm sections with a microtome. The slides were stained with hematoxylin and eosin (H&E) and observed under a microscope (Zeiss, Axiovert 200) and recorded with a digital camera. The stage classification of gonad development was according to the method of Rodney *et al.* (2009).

2.3 Quantitative real-time PCR

At each time point, samples (including males and females, without the consideration of sex) from five organs, i.e. hypothalamus, pituitary, muscle, gonad and liver, were dissected out and immediately frozen in cryo tubes in liquid nitrogen and stored at –70℃. Total RNA was isolated with Trizol reagent (Invitrogen, Carlsbad, CA, USA), according to the manufacturer's instructions. Genomic DNA was eliminated from the samples using RNase-free DNase according to the manufacturer's description (Promega, Madison, WI, USA). The quality of the RNA was assessed with the NanoDrop 8000 UV-Vis Spectrophotometer (Thermo, Wilmington, DE, USA). The 260/280 nm absorbance ratio of 1.8-2.0 indicates a pure RNA sample. Integrity of the RNA was evaluated by native agarose gel electrophoresis with sharp, clear 28s and 18s rRNA bands and 2∶1 ratio (28S:18S) as a good indication. First-strand cDNA synthesis was carried out using 1μg of total RNA in a 20-μl reaction mixture using ReverTra Ace Reverse Transcriptase (Toyobo, Osaka, Japan) with 20 pmol oligo$(dT)_{20}$ primers (Toyobo) at 42℃ for 90 min.

Real-time PCR was performed using 1μl of 10-fold diluted cDNA in 20μl total volume. The relative gene expression levels were quantified using SYBR Green Realtime PCR Master Mix (Toyobo) on an ABI Prism 7000 Sequence Detection System (Applied Biosystems, Foster City, CA, USA), using the primers listed in Table S2. We chose the primers that have 60℃ annealing temperature, an $R^2 > 0.99$, and amplification efficiency between 95% and 105%. Amplicons from each primer set were sequenced to confirm specificity. PCR was achieved with a 2 min activation and denaturation step at 95℃, followed 40 cycles of 15s at 95℃, 15s at 60℃ and 40s at 72℃. We selected *β-actin* as a reference gene using the *geNorm* (Vandesompele *et al.*, 2002) VBA applet for Microsoft Excel (Fig. S1). The target gene expression was then calibrated/ normalized against *β-actin* by using the $2^{-\Delta Ct}$ calculation (Livak and Schmittgen, 2001): $\Delta Ct = Ct_{target\ gene} - Ct_{\beta\text{-}actin}$.

2.4 PCR analyses of gh mRNA levels

To evaluate *gh* expression in different tissues of GH-transgenic and non-transgenic common carp, six transgenic (three males and three females) and six non-transgenic (three males and three females) common carp were sampled at 12 months of age. cDNAs of each tissue were prepared as Section 2.3 described. Primers were designed to differentiation between endogenous common carp *gh*, exogenous grass carp *gh* and total *gh* expression in GH-transgenic and non-transgenic common carp. Total *gh* expression was checked by RT-PCR in different tissues of 12 month old GH-transgenic and non-transgenic common carp using ghF/ghR following 27 cycles of 15s at 95℃, 15s at 60℃ and 20s at 72℃. Exogenous and endogenous *gh* expression in the pituitary of GH-transgenic was determined by real-time

PCR using the primer sets g-ghF/g-ghR and c-ghF/c-ghR, respectively.

2.5 ELISA and Western blot analyses of tissue GH and LH protein levels

To evaluate GH and LH levels in GH-transgenic and non-transgenic common carp, three transgenic and non-transgenic female and male common carp were sampled at 12 months of age. Proteins from the hypothalamus, pituitary, liver, gonad, muscle, serum samples were extracted using a protein extraction kit (Sangon, Shanghai, China). Protein concentration was determined and normalized to 50 μg/μl using the bicinchoninic acid (BCA) method before ELISA and Western blot analysis. The ELISA system and Western blot method had been developed and validated in our previous studies (Wu *et al.*, 2008; Xu *et al.*, 2011).

2.6 Serum levels of hormones

Blood was collected at each time point from the caudal vasculature of common carp (n = 7-15) from GH-transgenic and non-transgenic common carp. Serum samples were obtained by centrifugation at 3000 g for 15 min. The GH and LH ELISA system, with the assay range of 1.56 – 50 ng/mL, were developed and validated in our previous study (Wu *et al.*, 2008; Xu *et al.*, 2011).

2.7 *In situ* hybridization and immunofluorescence

Probes for *in situ* hybridization were made following the protocol of Thisse and Thisse's (2008). A T7 RNA polymerase promoter was included in the appropriate primers (reverse primer for antisense probes, forward primer for sense probes; Table S3) used to amplify the probe template from cDNA. Antisense and sense single-stranded mRNA probes for *ghr* were obtained with DIG RNA labeling MIX (Roche Diagnostics, Indianapolis, IN, USA) by transcription with T7 polymerase (Promega, Madison, WI, USA).

Immunofluorescence was performed to compare the distribution of LHβ subunit in GH-transgenic and non-transgenic common carp. We prepared 8 μm-sections of the pituitaries of GH-transgenic and non-transgenic common carp (5 month old) as previously described (Cao *et al.*, 2012), with the exception that the secondary antibody was replaced with fluorescein isothiocyanate-conjugated anti-mouse IgG (Santa Cruz, CA, USA). Non-transgenic and GH-transgenic pituitaries were prepared on one slide to avoid different staining degree or other experimental errors. The LHβ monoclonal antibody used in the experiment was FMU- cGTHIIβ9 produced in mouse against common carp LHβ and had been validated for use in common carp as previously described (Xu *et al.*, 2010). The Image Pro Plus software (version 6.0) was used for image analysis. The integrated optical density (IOD) of the fluorescence for LHβ immunoreactivity was measured for each plane in at least three individuals of each group.

Colocalization of *ghr* and LH was performed using Dig-labeled *ghr* probe and LHβ monoclonal antibody. Sections were derived from non-transgenic pituitaries of 12 month old fish and were operated according to procedures validated by Servili *et al.* (2011) with slight modifications. Briefly, sections were washed with phosphate-buffered saline (PBS), treated with 20 μg/mL Proteinase K (Roche Diagnostics, Indianapolis, IN, USA) for 10 min at room temperature, and then washed with PBS twice for 10 min each time. Sections were then fixed with 4% PFA for 20 min, washed with PBST (containing 1% Triton) for 10 min, three times. Hybridization was performed at 65℃ overnight in a water bath using 50 μl hybridization buffer (2×SSC; 50% deionized formamide; 50 mg/mL of yeast tRNA, pH 8.0) containing the DIG-labeled *ghr* probe (3 μg/mL). Sense RNA probe was used as negative control. After hybridization, slides were washed in 2×SSC at 65℃, followed by two washes in 2×SSC /50% formamide for 15 min at 65℃. After two washes in PBST for 10 min at room temperature, slides were incubated with Anti-Digoxigenin-Fluorescein, Fab fragments from sheep (Roche Diagnostics, Indianapolis, IN, USA) antibody (30 μg/mL) and LHβ monoclonal antibody (1∶1000) at 37℃ for 1 h. Preabsorption (Pabs) of the primary antiserum with common carp LHβ (2 μM; overnight at 4℃) was taken as negative control which completely blocked the immunoreaction. Sections were washed three times in PBST and incubated with Fluorescein anti-sheep IgG (Vector Laboratories, Inc., Burlingame, CA, USA) and Dylight 549 goat anti-mouse IgG (1∶200) (EarthOx, LLC, San Francisco, CA, USA) for 1 h at 37℃. After washing in PBST, slides were coverslipped with Vectashield mounting medium with DAPI (Vector Laboratories, Inc., Burlingame, CA, USA). The slides were observed with LSM 710 (Karl Zeiss), and images were processed with the ZEN 2009 Light Edition software.

2.8 Primary cell culture and *in vitro* incubation

The pituitary glands were removed from GH-transgenic and non-transgenic common carp and placed in PBS. After washing three times, the tissue was cut into small pieces with scissors and incubated with collagenase (type III, 1 mg/mL, Roche Diagnostics, Indianapolis, IN, USA) for 30 min in a shaker water bath at 28℃. The collagenase was then removed by centrifuge at 100 *g* (Centrifuge 5424, Eppendorf, Hamburg, Germany) for 5 min. The tissue fragments were then suspended with a culture medium (Goor *et al.*, 1994) supplemented with fetal bovine serum (10%, Gibco, Carlsbad, CA, USA) and common carp serum (1%, prepared by centrifuging the common carp blood at 800 *g* and filtered), and were then seeded at a 24-well plates (Corning, Amsterdam, Netherland) coated with poly-L-Lysine. The cells were incubated at 28℃ with 5% CO_2. The culture medium was changed every two days. Two types of cells were cultured, namely, transgenic pituitary (TP), non-transgenic pituitary (NP) cells. After five days of culture, culture medium was replaced with serum-free medium and drug treatment was initiated. Genes (*gh*, *gthα*, *fshβ*, and *lhβ*) specifically related to the functional pituitary were checked by real-time PCR before treatment.

To determine the effects of GH on pituitary *gthα*, *fshβ*, and *lhβ* gene expression, NP cells were treated with 20 ng/mL ncGH (native common carp growth hormone, dissolved in PBS) that was purified previously in our laboratory (Wu *et al.*, 2008) and PBS only was added to the serum-free medium of control groups. The dosage of ncGH administration was chosen according to the serum GH content (~15 ng/mL) in GH-transgenic common carp. Cells were harvested at 2, 4, and 6 h after GH administration. Levels of *gthα*, *fshβ*, and *lhβ* were quantified by real-time PCR. Data presented in this study were the pooled results of at least three independent experiments and were normalized to the 2 h control sample as normalized fold expression.

To evaluate the response of pituitary in GH-transgenic common carp and non-transgenic common carp, TP and NP cells were treated with a commercial OVUPIN (Ningbo Sansheng Pharmaceutical Co., Ltd, Shanghai, China) which containss-GnRHa (Pro-His-Trp-Ser-Tyr-D-Arg^{6}-Trp-Leu-Pro-amide) 0.2 and 100 mg domperidone per vial. The dopamine antagonist was used to reduce dopaminergic inhibition and to allow maximal s-GnRHa-stimulated LH release (Trudeau, 1997; Popesku *et al.*, 2008). Levels of *gthα*, *fshβ*, and *lhβ* were taken as indicator for pituitary responsiveness to GnRH. The optimal dosage of s-GnRHa/domperidone was chosen in preliminary tests with the one (20 ng/mL s-GnRHa and 10 μg/mL domperidone) stimulating expression of the *gth* subunit mRNAs. For the control, PBS only was added to the serum-free medium. Cells were harvested at 2 and 4 h after incubation. Target gene expression was later quantified by real-time PCR. Data presented in this study were the pooled results of at least three independent experiments and were normalized to 2 and 4 h control sample as normalized fold expression at each time point.

2.9 Data analysis

Real-time PCR data were acquired and analyzed by ABI Prism 7000 SDS 1.1 and Microsoft Excel software. Data were considered reliable when the standard deviation of three replicated reactions was less than 0.5 Ct. Effects of GH on target gene expression was analyzed by two-way ANOVA. If there was a significant group*time interaction effect, the simple main effects of group (NT and T) was examined at different time. If there was no interaction effect, results from NT and T groups at each time point were compared by independent sample *t*-test. Data of the GH incubation experiments were analyzed by ANOVA followed by Student-Newman-Keuls method. Tamhane's T2 method was used whenever the variances were unequal. Differences were considered significant at $P < 0.05$. Data of the s-GnRH/domperidone incubation experiments were analyzed by independent sample *t*-test. Differences were considered significant at $P < 0.05$. All of the tests were carried out with SPSS 16.0 (SPSS Inc. Chicago, IL, USA).

3. Results

3.1 Gonadal development of GH-transgenic common carp

During the 2 year tested in our study, GH-transgenic common carp showed increased growth (Table S1) compared with the non-transgenic common carp, but the GSI was significantly lower than non-transgenic common carp at 12, 23, and 24 months pre-reproduction (Fig. 1a). Especially at 24 months pre-reproduction, the GSI of non-transgenic common carp reached 21%, but that in GH-transgenic common carp was only 3.3%. Analysis of the gonadal sections (Fig. 1b) revealed that most of (5 of 11) the non-transgenic common carp developed perinucleoar oocytes at 3 months while 9 of 11 of the GH-transgenic common carp were still in the oogonia/gonoocyte phase. At 4-5 months of age, all of the non-transgenic female common carp developed perinucleolar oocytes, while there were still gonoocytes (2 of 4 at 4 m; 2 of 7 at 5 m) in GH-transgenic carp, and the perinucleolar oocytes in GH-transgenic female common carp were visually smaller than those in the non-transgenic female common carp. The GH-transgenic male common carp at 4-5 months old were at the spermatocyte stage (4 of 5 at 4 m; 6 of 9 at 5 m) while the non-transgenic male common carp had already developed spermatids and showed obvious seminiferous lobules (4 of 5 at 4 m; 6 of 7 at 5 m). At 7, 9, and 11 months, while most of the non-transgenic female common carp showed cortical alveoli in the ooplasm, and some developed early vitellogenic oocytes, GH-transgenic female common carp of the same age developed perinucleolar oocytes and oogonia. The development status of the testes at 7, 9, and 11 months showed only slight variation between two groups, with the exception of a few transgenic individuals lagged slightly. At two years of age, when the ovaries of all non-transgenic female common carp developed mature/spawning oocytes, those of the GH-transgenic common carp developed non-uniformly and showed both perinucleolar and cortical alveolar oocytes in the same section. In light of the above results, we assessed the expression of genes related to growth and reproduction over a 2-year period.

3.2 GH and *ghr* expression in GH-transgenic common carp

In non-transgenic common carp, *gh* (using primer set ghF/ghR, detecting total gh level) mRNA was only detected in pituitary while in GH-transgenic common carp it was also found in muscle, hypothalamus, liver and gonad (Fig. 2a). Western blot analysis showed a decrease of pituitary GH content in GH-transgenic common carp compared with non-transgenic common carp (Fig. 2b). No GH was detected in other tissues using Western blot method. Considering the relatively low sensitivity of Western blot, we chose ELISA to measure the GH content in other tissues (Fig. 2c). Growth hormone content in GH-transgenic pituitary was significantly lower than non-transgenic pituitary ($P < 0.05$). In serum, gonad, muscle and hypothalamus, GH levels were higher in GH-transgenic common carp than in non-transgenic common carp.

Total *gh* expression in the pituitary gland was significantly lower in the transgenic common carp ($P < 0.05$) than in the non-transgenic common carp at all the growth stages (Fig. 3a). We also specifically quantified endogenous versus exogenous GH mRNA levels. In non-transgenic carp

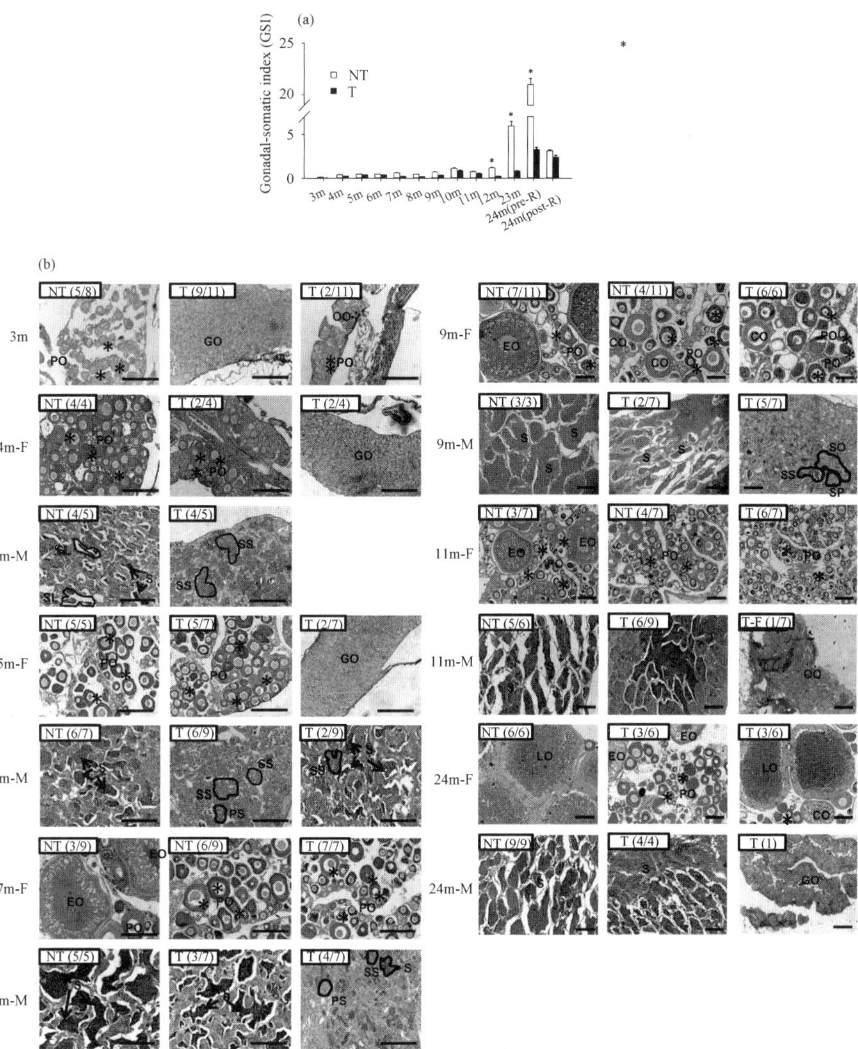

Fig. 1 Comparison of gonadal development between non-transgenic (NT) and GH-transgenic (T) common carp at different growth stages. (a) Comparison of the gonadalsomatic index (GSI) between non-transgenic and GH-transgenic common carp at different growth stages. Samples were collected from 1 month (1 m) to 24 months (24 m) of the age. Two samples were obtained at 24 months old, before (24 m pre-R) and after (24 m post-R) ovulation/spermiation. It is May when fish at their 1 month old, June at 2 months old, and so on. Values are represented as means ± S.E.M. (n = 10-15) at each sampling time, and analyzed by two-way ANOVA. Asterisks indicate statistically significant differences compared with those in NT and T at $P < 0.05$. (b) Comparison of gonad histology between non-transgenic and GH-transgenic common carp at different growth stages. The sections from one to three for male (M) and female (F) carps shown in the figure were taken from the NT and T groups in 8-16 sections. The ratio in each section represents the numbers of NT or T per total testing carps from each gender. "4m-F NT (4/5)" indicate that 5 female carps were tested in total, 4 of which were developed into the status showed in the figure. CO, cortical alveolar oocytes; EO, early vitellogenic oocytes; GO, gonocyte; LO, late vitellogenic oocytes; OO, oogonia; PO, perinucleolar oocyte; PS, primary spermatocytes; SS, secondary spermatocytes; S, sperms; SL, seminiferous lobules; SO, spermatogonia. ∗ represents perinucleolar oocyte, ●represents oogonia. Scale bar, 200 μm

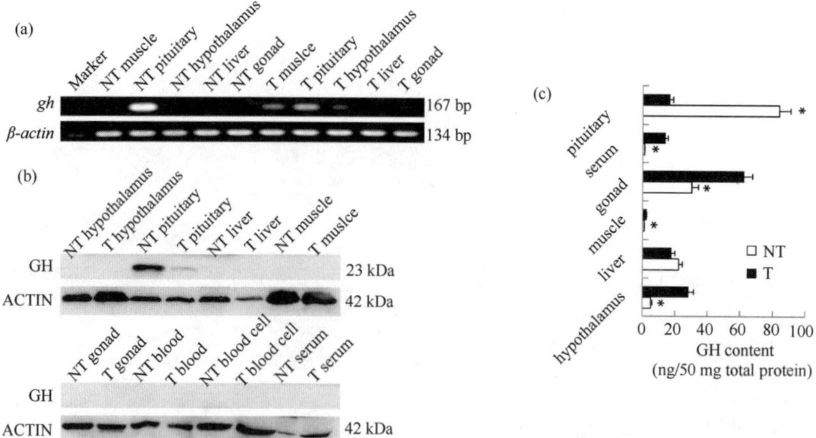

Fig. 2 Growth hormone distribution in 12 month old non-transgenic (NT) and GHtransgenic (T) common carp. Each tissue of NT and T common carp was a mixture of 3 males and 3 females. (a) RT-PCR results of total *gh* in adult tissues of NT and T common carp. (b) Western blotting results of GH in adult tissues of NT and T common carp. We chose ACTIN as an internal standard. (c) ELISA results of GH contents in adult tissues of NT and T common carp. The values represent GH content per 50 mg total protein for each tissue. Note that values for pituitary were not corrected for dilution (1 ∶ 10000), so that they could be plotted on the graph with other tissues. Actual values are 10^4 times higher. The results obtained were analyzed by independent sample *t*-test. Asterisks indicate statistically significant differences at $P < 0.05$ between NT and T groups in particular tissue

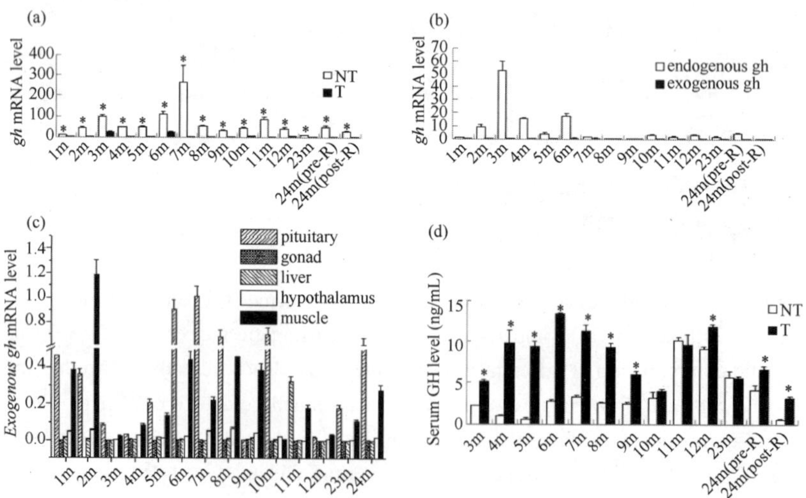

Fig. 3 Growth hormone distribution in non-transgenic (NT) and GH-transgenic (T) common carp at different growth stages. (a) Total *gh* expression in the pituitary of NT and T common carp; (b) Exogenous (grass carp *gh*) and endogenous (common carp *gh*) *gh* expression in the pituitary of T common carp; (c) Exogenous *gh* in different tissues of T common carp; (d) ELISA results of serum GH in NT and T common carp. Values of *gh* mRNA (mean ± S.E.M., n = 5-10) were determined by real-time PCR and expressed as the relative mRNA level normalized to *β-actin*. Values of GH level (mean ± S.E.M., n = 10-15) were determined by ELISA. The results obtained were analyzed by two-way ANOVA. Asterisks indicate statistically significant differences at $P< 0.05$ at a particular time point

increased sharply at 3 months, and then declined gradually to basal levels at 7 months of age. In contrast exogenous *gh* was consistently lowly expressed throughout the experimental period and was significantly lower than endogenous *gh* in GH-transgenic pituitary ($P < 0.05$) except for 8 and 9 months of age, and in the 24 months post-reproduction sample (Fig. 3b) when there were no detectable differences. The ranking of exogenous *gh* expression in extra-pituitary sites was pituitary > muscle > hypothalamus > gonad > liver (Fig. 3c). Serum GH levels were higher in the transgenic common carp than in the non-transgenic common carp ($P < 0.05$), except at 10, 11, and 23 months (Fig. 3d). The expression of *ghr* in GH-transgenic common carp was not significantly ($P > 0.05$) changed compared to pituitary, hypothalamus, muscle, gonad, and liver of non-transgenic fish (data not shown).

3.3 Changes of *kiss1*, *kiss2*, *gnrh3* levels in the hypothalamus of GH-transgenic common carp

In the hypothalamus of GH-transgenic common carp, levels of *kiss1*, *kiss2*, and *gnrh3* were variable at each time point compared with that of non-transgenic common carp. Expression of these genes were affected by both GH-transgenesis (*kiss1*, $P = 0.002$; *kiss2*, $P = 0.0001$; *gnrh3*, $P = 0.02$) and time (*kiss1*, $P = 0.0001$; *kiss2*, $P = 0.0001$; *gnrh3*, $P = 0.003$), but there were interactions between group and time (*kiss1*, $P = 0.003$; *kiss2*, $P = 0.00001$; *gnrh3*, $P = 0.0001$). The simple main effects of group (NT and T) were then examined at different time. Expression of *kiss1* was significantly higher in 1- and 6-month-old transgenic common carp compared with the non-transgenic common carp ($P < 0.05$) (Fig. 4a). The expression of *kiss2* increased significantly ($P< 0.05$) in 2-, 3-, 4-, 5-, and 6-month-old transgenic common carp compared with the non-transgenic variants (Fig. 4b). Both *gnrh2* (also called chicken GnRH-II) and *gnrh3* (also called salmon GnRH) exist in common carp, however, we found that only *gnrh3* could be reliably detected in the hypothalamus. We therefore report only results for *gnrh3*. Levels of *gnrh3* elevated ($P< 0.05$) in 6- and 12-month-old GH-transgenic common carp, but decreased in 8- and 10-month-old GH-transgenic common carp compared with non-transgenic common carp (Fig. 4c). The expression of *gpr54b* was not affected in GH-transgenic common carp ($P = 0.3$; data not shown).

3.4 Effects of GH on pituitary transcript levels and LH levels

The expression of *gthα* was not significantly changed in GH-transgenic common carp ($P = 0.48$). The expression of *fshβ* and *lhβ* was affected by GH transgenesis (*fshβ*, $P = 0.02$; *lhβ*, $P = 0.0001$) and time (*fshβ*, $P = 0.00001$; *lhβ*, $P = 0.0001$). There were interactions between group and time (*fshβ*, $P = 0.000001$; *lhβ*, $P = 0.00001$). The simple main effects of group (NT and T) were then examined at different time point. Compared with non-transgenic common carp, *fshβ* decreased ($P< 0.05$) in the GH-transgenic common carp at 3 months ($P = 0.00002$) and

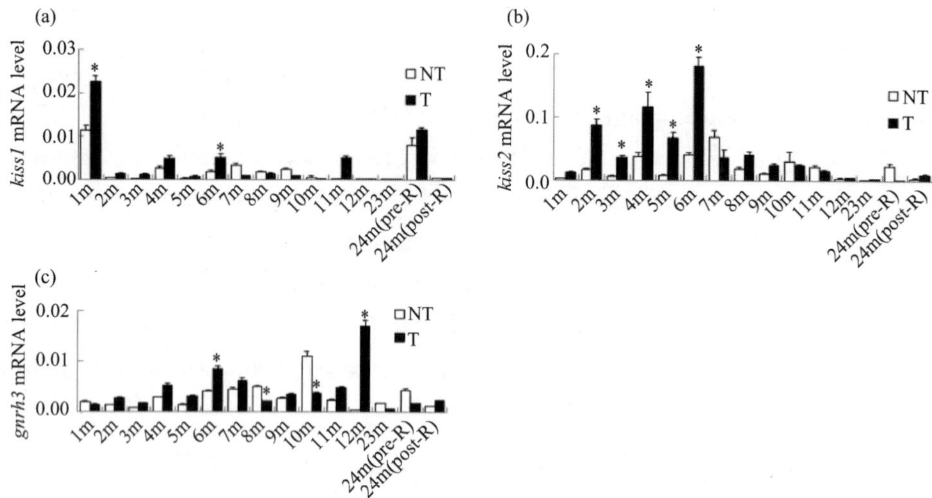

Fig. 4 Levels of *kiss1*, *kiss2* and *gnrh3* mRNA in the hypothalamus of non-transgenic (NT) and GH-transgenic (T) common carp at different growth stages. Values of target gene mRNA (mean ± S.E.M., n = 5-10) were determined by real-time PCR and expressed as the relative mRNA level normalized to *β-actin*. The results obtained were analyzed by two-way ANOVA. Asterisks indicate statistically significant differences at $P < 0.05$ at a particular time point

increased at 6 months ($P = 0.049$) of age (Fig. 5a). The expression of *lhβ* peaked at 4 and 7 months and before reproduction in non-transgenic common carp; however, no obvious peaks were seen in the GH-transgenic common carp (Fig. 5b). Except for an increase at 23 months of age, *lhβ* transcripts were significantly lower in the GH-transgenic common carp at 3-, 4-, 7-, and 24-month before reproduction and 24-month after reproduction ($P < 0.05$). Serum LH levels were significantly lower ($P < 0.05$) in GH-transgenic common carp compared to non-transgenic common carp except for 5 months (Fig. 5c). Western blot analysis showed a decrease of pituitary LH content in GH-transgenic compared with non-transgenic common carp (Fig. 5d). Immunofluorescence revealed that GH-transgenic common carp had fewer LH positive cells compared with non-transgenic common carp (Fig. 6). Two forms of GnRH receptor, *gnrhra* ($P = 0.07$) and *gnrhrb* ($P = 0.08$) were not found to be affected in the pituitary of GH-transgenic common carp (data not shown).

Non-transgenic pituitary cells incubated with GH showed decrease of *gthα* (Fig. 7a) and *fshβ* (Fig. 7b) expression at all the time points. Levels of *lhβ* expression decreased at 4-and 6-h after GH incubation (Fig. 7c). The expression of *ghr* was significantly inhibited after GH incubation at 2- and 4-h after GH incubation ($P < 0.05$) (Fig. 7d).

3.5 Responsiveness of pituitary to s-GnRHa/domperidone stimulation in GH-transgenic common carp

Pituitary cells were incubated with s-GnRHa/domperidone to test the differential responsiveness

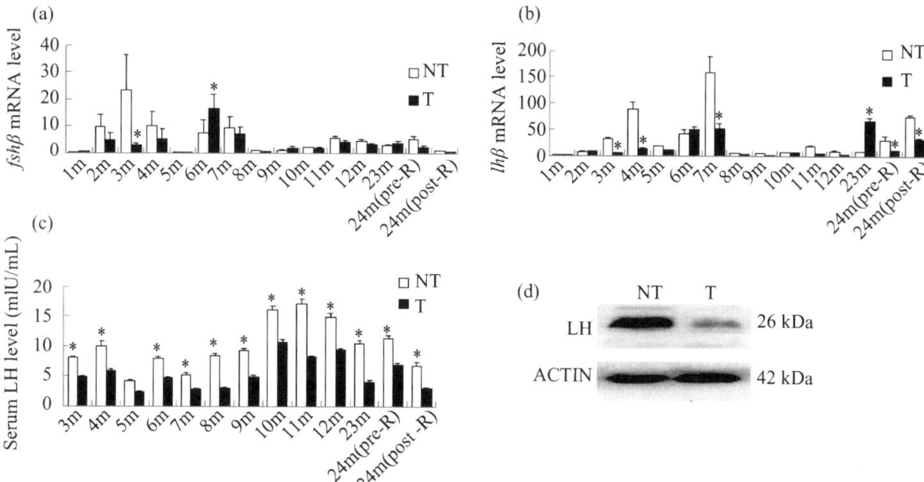

Fig. 5 Pituitary *fshβ*, *lhβ* mRNA and serum LH content in non-transgenic (NT) and GH-transgenic (T) common carp. (a) *fshβ*, and (b) *lhβ* expression at different growth stages of common carp. Values of target gene mRNA (mean ± S.E.M., n = 5-10) were determined by real-time PCR and expressed as the relative mRNA level normalized to *β-actin*. The results obtained were analyzed by two-way ANOVA. Asterisks indicate statistically significant differences at $P < 0.05$; (c) Serum LH in T and NT common carp at different growth stages; and (d) Western blotting results of LH protein in the pituitary of 12 month old NT and T common carp. The protein used in western blotting was a mixture of 3 male and 3 female pituitaries.

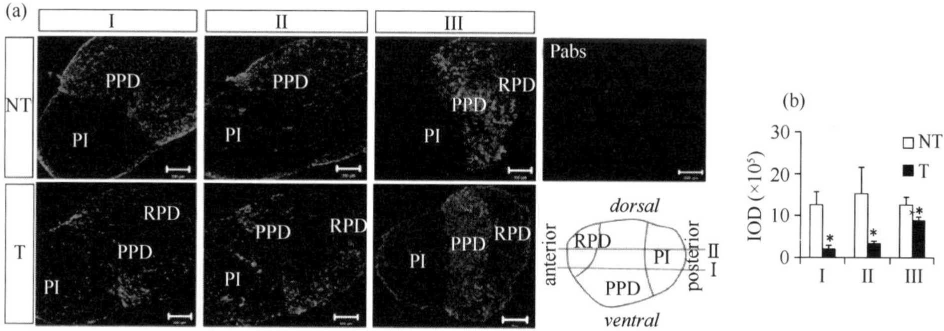

Fig. 6 Distribution of LH in the pituitary of non-transgenic (NT) and GH-transgenic (T) common carp. (a) Immunofluorescence of LH (green) in NT and T common carp. Panels I and II are horizontal sections of the pituitary. Panel III shows sagittal sections of the pituitary. (b) The integrated optical density (IOD) of the fluorescence for LHβ immunoreactivity in NT and T common carp. The micrographs were taken under the identical set of conditions for all groups. Asterisks indicate statistically significant differences at $P < 0.05$. n = 3 for all groups. Pabs: preabsorption of the primary antiserum with carp LH (2 μM; overnight at 4℃). Scale bar, 200 μm

of gonadotrophs from both non-transgenic and GH-transgenic common carp. As the expression of target genes at each time point were normalized to 2 and 4 h control sample, we chose independent sample t-test to do the statistical analysis. In non-transgenic pituitary cells, *gthα*, *fshβ*, and *lhβ* expression decreased at 2 h after incubation, while increasing at 4 h after incubation ($P < 0.05$).

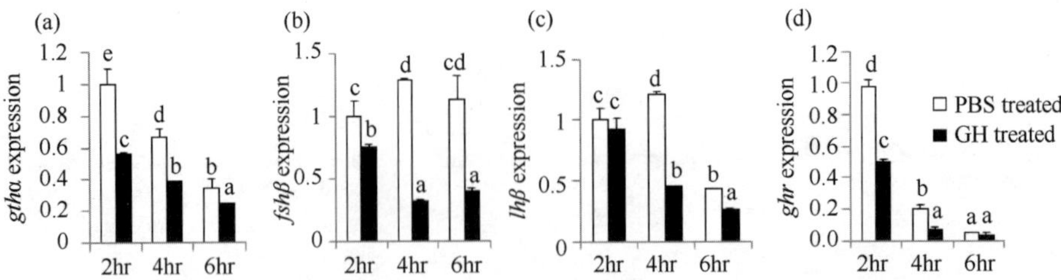

Fig. 7 Effects of GH on *gth* expression in non-transgenic (NT) pituitary cell *in vitro*. Pituitary cells were exposed to GH for 2, 4, and 6 h. The expression of the target genes (mean ± S.E.M.) were determined by real-time PCR and were expressed as fold change normalized to the 2 h controls. The results obtained were analyzed by ANOVA followed by Student-Newman-Keuls method. Symbols with the same indicate groups that are not significantly different

In GH-transgenic pituitary cells, *gthα* and *fshβ* expression were inhibited at 2 h after incubation ($P < 0.05$). The transcription of *gthα* and *lhβ* was increased after incubation for 4 h ($P < 0.05$), which was less than that in non-transgenic pituitary cells (Fig. 8). At 4 h after incubation, the non-transgenic pituitary cells showed 40-, 86-, and 28-fold elevations in the *gthα*, *fshβ*, and *lhβ* levels, respectively, while the corresponding levels in the GH-transgenic pituitary cells showed 2-, 1.6-, and 1.8-fold increases (Table. S4).

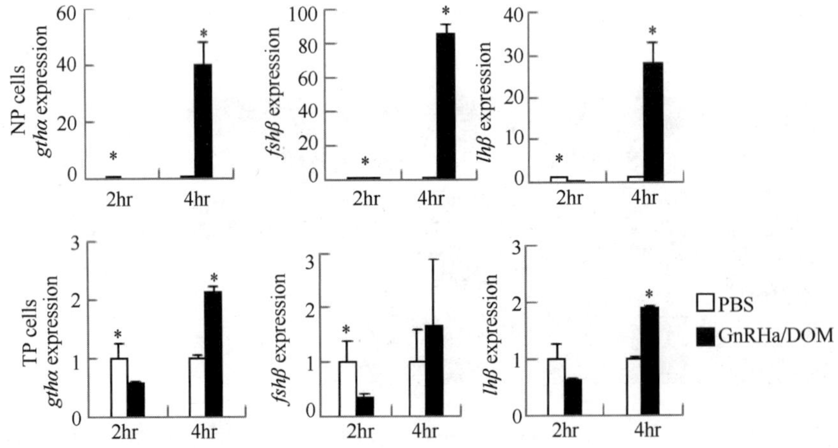

Fig. 8 Alterations in *gthα*, *fshβ*, and *lhβ* mRNA levels in non-tansgenic (NT) and GH-tansgenic (T) pituitary cells after incubation with s-GnRHa/domperidone. Gene expression in pituitary cells of NT and T common carp were analyzed after s-GnRHa/domperidone treatment. PBS represented PBS treated control. GnRHa/DOM represented s-GnRH-domperidone incubation. The expressions of the target genes (mean ± S.E.M.) were determined by real-time PCR and are expressed as fold change normalized to 2 and 4 h control sample. Results were analyzed by independent sample *t*-test. Asterisks indicate statistically significant differences at $P< 0.05$

3.6 Expression of ghr on LH positive cells in the pituitary

The common carp pituitary includes the rostral pars distalis (RPD), proximal pars distalis

(PPD) and neurointernediate lobe (NIL). Luteinizing hormone is located in PPD (Fig. 9). The signals of *ghr* were abundant in RPD, and some interspersed in the inner space of PPD (Fig. 9). Double staining revealed colocalization of the *ghr* and LH immunoreactivity in the pituitary of 12-month-old common carp.

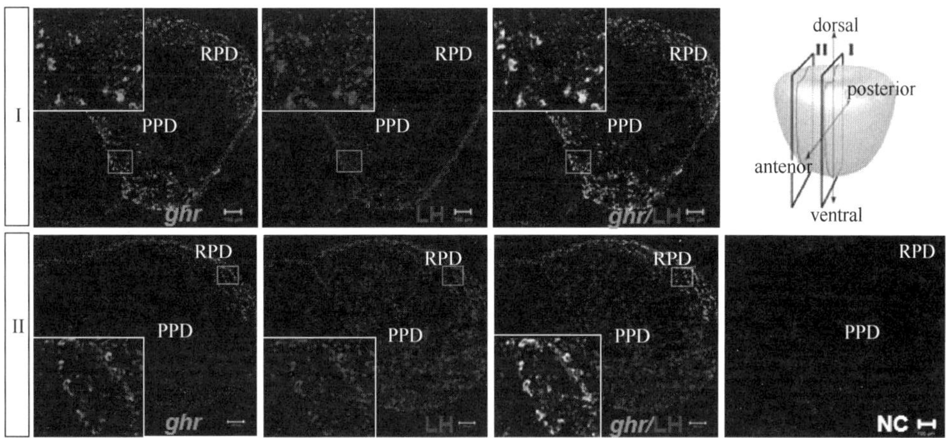

Fig. 9 Colocalization of *ghr* and LH in non-transgenic common carp pituitary. *In situ* hybridization of *ghr* (green) and immunofluorescence of LH (red) in 12 month old common carp pituitary. The insets reflect the dotted areas shown. The diagram (Kitahashi *et al.*, 2007) showing the sagittal sectioning plane of the pituitary. Yellow-orange color indicates colocalization. NC: Negative control using *ghr* sense probe and pre-absorbed primary antiserum with carp LH (2 μM; overnight at 4℃). Scale bar, 100μm

4. Discusion

4.1 GH-transgenesis delays the reproductive development of common carp

In this study by tracing the GSI and gonad histology we observed that reproductive development was slower in the GH-transgenic common carp than in the non-transgenic common carp. The fertility of human GH-transgenic male mice was significantly lower than that of normal mice, and the female transgenic mice exhibited a prolonged vaginal cycle, and mated but failed to become pregnant or showed pseudopregnancy (Bartke *et al.*, 1992, 1994). In fish, exogenous GH was found to influence the reproductive performance in some GH-transgenic species (Bessey *et al.*, 2004; Fitzpatrick *et al.*, 2011; Moreau *et al.*, 2011; Rahman *et al.*, 1998, 2001). Rahman *et al* (2001) proposed that limitations in energy availability for gonadal development may be the mechanism for reproductive changes in GH-transgenic tilapia. Together, these results suggest that over-expressed GH influences energy distribution in transgenic common carp, and finally resulted in delayed reproductive development.

4.2 GH expression patterns are altered in GH-transgenic common carp

In the present study, expression of grass carp GH was driven by the common carp β-actin promoter, which we refer to as the "all-fish" construct. The β-actin promoter drives exogenous *gh* expression in all the tissues of the transgenic common carp. Content of *gh* was highest in the pituitary, followed by muscle, hypothalamus, gonad, and liver. We hypothesized that extra-pituitary tissues, especially muscle, contributed the majority of the constant high level of serum GH. Although it was hypothesized that all cells in the pituitary should express GH in the GH-transgenic common carp, the total *gh* level in GH-transgenic common carp pituitary was much lower than total *gh* in non-transgenic common carp pituitary, which was 40-fold to 7415-fold less (Table S5). The GH protein content in pituitary was also lower in the transgenic common carp. In GH-transgenic tilapia (Caelers *et al.*, 2005) and GH-transgenic coho salmon (Mori and Devlin, 1999), there was a decrease of pituitary *gh* mRNA in transgenic fish. However, in contrast Raven *et al.* (2008) reported that small-sized (body weight of ~55g) GH-transgenic coho salmon had high and similar pituitary *gh* mRNA compared to non-transgenic fish. In our GH-transgenic common carp, pituitary GH may have been inhibited by negative feed-back effects of the high serum GH levels throughout most of the experimental period.

As GH exerts its effects by binding to specific receptors on target cells, we assessed the expression of *ghr* in different tissues of GH-transgenic common carp. We found no significant differences in *ghr* expression between two groups. However, the expression of *ghr* in GH-treated pituitaries *in vitro* decreased at 2 and 4 h after GH incubation, then no changes at 6 h after GH incubation. One possible explanation for these differences could be that there may be compensatory or feedback controls on *ghr in vivo* that are missing *in vitro*. There are reports on GH effects on GHR expression showing that there exists autoregulation of GH on GHR in the target cell with both down-regulatory effects (Hull and Harvey, 1998; Maiter *et al.*, 1988) and up-regulatory effects (Hull and Harvey, 1996) in some vertebrate species. In fish, GH effects on GHR expression are still unclear. In GH-transgenic coho salmon, *ghr* in pituitary is down-regulated, but no differences in muscle and liver (Raven *et al.*, 2012). In the present study, *ghr* was not changed in both pituitary and other tissues, perhaps indicative of species difference in teleost. Further study on the issue is warranted.

4.3 GH inhibited LH expression in GH-transgenic common carp

Development and maturation of fish gonads are controlled by LH and FSH secreted from the pituitary. In the present study, pituitary *lhβ* was down-regulated in GH-transgenic common carp at 3, 4, 7, and 24 months, and serum LH levels were decreased all through the year. Our *in vitro* results revealed that GH decreased the expression of *gthα*, *fshβ* and *lhβ* subunits. *In situ* hybridization combined with immunocytochemistry further supports the hypothesis of direct regulation of LH by GH in common carp because *ghr* was localized to LH-positive cells in the pituitary. In rats, LH expression was reduced after treatment with bovine GH via osmotic pumps (Chandrashekar and Bartke, 1998). There are data showing that LH secretion are decreased by *in vitro* treatments of grass carp pituitary cells in culture, indicating a paracrine signaling from the somatotroph to the gonadotroph (Zhou *et al.*, 2004). We speculated that the constant high levels of serum GH may directly inhibit LH expression through the GH receptor in common carp.

In fish, it is well known that FSH is expressed in early vitellogenesis, and is considered to control early gametogenesis and puberty onset while LH is mainly responsible for the final maturation of gonads, inducing ovulation and spermiation (Clelland and Peng, 2009; Schulz *et al.*, 2010). However, in the present study, *lhβ* expression peaked at 4 months, 7 months, and before reproduction in non-transgenic common carp, which was in accordance with the expression pattern of *fshβ*. Han found that *fshβ* and *lhβ* were both detectable in juvenile Japanese eels, and *lhβ* expression was higher than *fshβ* in the sub-adult stage when the ovary is in the growth stage (Han *et al.*, 2003). These results indicated the potential role of LH in early gonadal development in common carp. In GH-transgenic common carp, no peaks of *lhβ* exist at 3, 4, 7 months as non-transgenic animals did. Taken these results together, we propose that decreased LH is the reason for reduced reproductive performance in GH-transgenic common carp.

4.4 Reduced sensitivity of GH-transgenic pituitaries to stimulation *in vitro*

In the hypothalamus, several neurohormones such as GnRH, dopamine, gamma-aminobutyric acid (GABA) play important roles in the regulation of LH releasing (Trudeau, 1997; Zohar *et al.*, 2010). The decapeptide GnRH stimulates production and release of gonadotropins from the pituitary. Contrary to GnRH, dopamine is the principal inhibitory neurohormone controlling LH release (Popesku *et al.*, 2008; Trudeau, 1997; Trudeau *et al.*, 2000; Yaron, 1995). In the present study, to reduce the inhibitory effects of DA on LH release, we used the combination of s-GnRHa plus domperidone (DOM, antagonist of dopamine type 2 receptors) to maximally stimulate LH release. Despite maximal stimulation, the LH response of GH-transgenic pituitary was lower than non-transgenic pituitary cells.

The response pattern of gonadotrophs to GnRH stimulation is different among species depending on different experiment situations. In Sockeye Salmon, GnRH stimulated *lhβ* expression while had no effects on *fshβ* (Kitahashi *et al.*, 1998). In goldfish, GnRH stimulated both *fshβ* and *lhβ* gene expression *in vivo* and *in vitro* (no matter pituitary fragments or dispersed cells) (Klausen *et al.*, 2001). However in catfish, GnRH inhibited *lhβ* expression at 2h, 4h after GnRH injection and stimulated *lhβ* expression at 8h after GnRH injection *in vivo* while only inhibited *lhβ* expression *in vitro* (Rebers *et al.*, 2002). *In vivo* treatment of salmon with GnRH using fish from different developmental stages results in different *lhβ* expression pattern (Dickey and Swanson, 2000; Kitahashi *et al.*, 1998). These differing results indicate that gonadotroph responsiveness to GnRH is dependent on experiment conditions (*in vivo* or *in vitro*) and developmental stages. In our experiment, the transcription of *gthα*, *fshβ*, and *lhβ* (*lhβ* in TP cells decreased but not statistically significant) was decreased slightly at 2 h after incubation in NP and TP cells. Besides the species specificity, our use of passaged cells may have also influenced the responses observed.

4.5 Changes in hypothalamic gene expression do not explain reduced gonadal development in GH-transgenic common carp

Growth hormone overexpression could potentially lead to changes in multiple hypothalamic neuropeptides to cause changes in gonadotropin synthesis and release, and consequently disrupt gonadal development in the transgenic animals. We assessed the expression of *kiss1*, *kiss2*, and *gnrh3* in the hypothalamus of non-transgenic and GH-transgenic fish. Overall developmental patterns in expression in non-transgenic and GH-transgenic common carp do not parallel changes in gonadotropin gene expression in the pituitary or serum LH concentrations. While the robust stimulatory effects of GnRH on LH production are well established in teleost as in other vertebrates, the role of the kisspeptins is poorly understood. There is evidence for stimulatory effects of kisspeptin on LH transcription and secretion (Felip *et al.*, 2009; Kitahashi *et al.*, 2009; Li *et al.*, 2009; Zmora *et al.*, 2012). However, kisspeptin is an inhibitor of LH expression in eel (*Anguilla anguilla*) (Pasquier *et al.*, 2011). In our study, *kiss2* levels were higher in GH-transgenic common carp compared to non-transgenic animals during the early gonad development, so we cannot yet exclude the possibility of *kiss2* regulation on LH expression. It is also known that the GH regulatory network is under complex feedback control (Wong *et al.*, 2006), and perhaps overexpression of GH upsets some aspects of physiological control. Further research is required to clarify the regulation of kiss2 on LH production in common carp.

Taken together, we suggest that lower LH in GH-transgenic common carp may be due to the long-term exposure to high serum GH *in vivo*. It is likely that exogenous GH in the GH-transgenic common carp is acting via GH receptors on LH cells to reduce LH production.

5. Conclusions

We demonstrate that pituitary gonadotroph function is suppressed in GH-transgenic common carp. We propose that reduced reproductive performance of GH-transgenic common carp may be partially due to directly action of GH on the pituitary. Using GH-transgenic common carp as a GH over-expression model, we observed elevated serum GH in association with direct inhibition of pituitary *lhβ* expression and LH secretion. We hypothesize that this is an effect mediated by GH receptors in gonadotrophs. In fish, there is increasing evidence of GH interactions with the reproductive axes, including the modulation of GH by steroids, the direct or indirect action of GH on gonadal development and steroidogenesis, and the inhibitory effect of GH on LH secretion (Le Gac *et al.*, 1993; Trudeau *et al.*, 1997; Zhou *et al.*, 2004). Our results further support the central importance of GH for the coordinated regulation of growth and reproduction.

Acknowledgements This work was supported financially by the "863" High Technology Project (Grant No. 2011AA100404) National Natural Science Foundation (Grant No. 30930069), the Development Plan of the State Key Fundamental Research of China (Grant No. 2010CB126302), and the Key Research Program of the Chinese Academy of Sciences (Grant No. KSCX2-EW-N-004), Funding from the University of Ottawa International Research Acceleration Program (to VLT) is acknowledged with appreciation.

References

Almeida, D., Martinez Gaspar Martins, C., Azevedo Figueiredo, M., Lanes, C., Bianchini, A., Marins, L., 2013. Growth hormone transgenesis affects osmoregulation and energy metabolism in zebrafish (*Danio rerio*). Transgenic Res. 22, 75-88.

Bartke, A., Naar, E.M., Johnson, L., May, M.R., Cecim, M., Yun, J.S., Wagner, T.E., 1992. Effects of expression of human or bovine growth hormone genes on sperm production and male reproductive performance in four lines of transgenic mice. J. Reprod. Fertil. 95, 109-118.

Bartke, A., Cecim, M., Tang, K., Steger, R.W., Chandrashekar, V., Turyn, D., 1994. Neuroendocrine and reproductive consequences of overexpression of growth hormone in transgenic mice.Proc. Soc. Exp. Biol. Med. 206, 345-359.

Berishvili, G., D'Cotta, H., Baroiller, J.-F., Segner, H., Reinecke, M., 2006. Differential expression of IGF-I mRNA and peptide in the male and female gonad during early development of a bony fish, the tilapia *Oreochromis niloticus*. Gen. Comp. Endocrinol. 146, 204-210.

Bessey, C., Devlin, R.H., Liley, N.R., Biagi, C.A., 2004. Reproductive performance of growth-enhanced transgenic coho salmon. Trans. Am. Fish. Soc.133, 1205-1220.

Caelers, A., Maclean, N., Hwang, G., Eppler, E., Reinecke, M., 2005. Expression of endogenous and exogenous growth hormone (GH) messenger mRNA in a GH-transgenic tilapia (*Oreochromis niloticus*). Transgenic Res. 14, 95-104.

Cao, M., Yang, Y., Xu, H., Duan, J., Cheng, N., Wang, J., Hu, W., Zhao, H., 2012. Germ cell specific expression of vasa in rare minnow, *Gobiocypris rarus*. Comp. Biochem. Physiol., A: Mol. Integr. Physiol. 162, 163-170.

Chandrashekar, V., Bartke, A., 1998. The role of growth hormone in the control of gonadotropin secretion in adult male rats. Endocrinology. 139, 1067-1074.

Clelland, E., Peng, C., 2009. Endocrine/paracrine control of zebrafish ovarian development. Mol. Cell. Endocrinol. 312, 42-52.

Devlin, R.H., Biagi, C.A., Yesaki, T.Y., 2004. Growth, viability and genetic characteristics of GH transgenic coho salmon strains. Aquaculture. 236, 607-632.

Devlin, R.H., Sundström, L.F., Muir, W.M., 2006. Interface of biotechnology and ecology for environmental risk assessments of transgenic fish. Trends Biotechnol. 24, 89-97.

Dickey, J.T., Swanson, P., 2000. Effects of salmon gonadotropin-releasing hormone on follicle stimulating hormone secretion and subunit gene expression in coho salmon (*Oncorhynchus kisutch*). Gen Comp Endocrinol. 118, 436-449.

Duan, M., Zhang, T., Hu, W., Li, Z., Sundström, L.F., Zhu, T., Zhong, C., Zhu, Z., 2011. Behavioral alterations in GH transgenic common carp may explain enhanced competitive feeding ability. Aquaculture. 317, 175-181.

Felip, A., Zanuy, S., Pineda, R., Pinilla, L., Carrillo, M., Tena-Sempere, M., Gómez, A., 2009. Evidence for two distinct

KiSS genes in non-placental vertebrates that encode kisspeptins with different gonadotropin-releasing activities in fish and mammals. Mol. Cell. Endocrinol. 312, 61-71.

Fitzpatrick, J.L., Akbarashandiz, H., Sakhrani, D., Biagi, C.A., Pitcher, T.E., Devlin, R.H., 2011. Cultured growth hormone transgenic salmon are reproductively out-competed by wild-reared salmon in semi-natural mating arenas. Aquaculture. 312, 185-191.

Fletcher, G.L., Shears, M.A., Yaskowiak, E.S., King, M.J., Goddard, S.V., 2004. Gene transfer: potential to enhance the genome of Atlantic salmon for aquaculture. Aust J Exp Agric. 44, 1095-1100.

Goor, F., Goldberg, J.I., Wong, A.O.L., Jobin, R.M., Chang, J.P., 1994. Morphological identification of live gonadotropin, growth-hormone, and prolactin cells in goldfish (*Carassius auratus*) pituitary-cell cultures. Cell Tissue Res. 276, 253-261.

Guan, B., Hu, W., Zhang, T., Wang, Y., Zhu, Z., 2008. Metabolism traits of ‘all-fish’ growth hormone transgenic common carp (*Cyprinus carpio* L.). Aquaculture. 284, 217-223.

Guan, B., Ma, H., Wang, Y., Hu, Y., Lin, Z., Zhu, Z., Hu, W., 2011. Vitreoscilla hemoglobin (VHb) overexpression increases hypoxia tolerance in zebrafish (*Danio rerio*). Mar. Biotechnol. 13, 336-344.

Han, Y.-S., Liao, I.C., Huang, Y.-S., Tzeng, W.-N., Yu, J.Y.-L., 2003. Profiles of PGH-α, GTH I-β, and GTH II-β mRNA transcript levels at different ovarian stages in the wild female Japanese eel *Anguilla japonica*. Gen. Comp. Endocrinol. 133, 8-16.

Harvey, S., 2010. Extrapituitary growth hormone. Endocrine. 38, 335-359.

Hu, W., Wang, Y., Zhu, Z., 2006. A perspective on fish gonad manipulation for biotechnical applications. Chin. Sci. Bull. 51, 1-6.

Hu, W., Wang, Y., Zhu, Z., 2007. Progress in the evaluation of transgenic fish for possible ecological risk and its containment strategies. Sci. China, Ser C: Life Sci. 50, 573-579.

Hu, W., Zhu, Z., 2010. Integration mechanisms of transgenes and population fitness of GH transgenic fish. Sci. China, Ser C: Life Sci. 53, 401-408.

Hull. K.L., Harvey S., 1996. Autoregulation of growth hormone receptor and growth hormone binding protein transcripts in central and peripheral tissues of the rat. Growth Horm. IGF Res. 8, 167-173.

Hull, K.L., Harvey, S., 1998. Autoregulation of central and periph eral growth hormone receptor mRNA in domestic fowl. J. Endocrinol. 156, 323-329.

Johnson, R., Wolf, J., Braunbeck, T., 2009. OECD guidance dcument for the dagnosis of edocrine-related histopathology of fish gonads. http://www.oecd.org/chemicalsafety/testing/46801244.pdf. (Draft of January T, 2009).

Hull, K.L., Harvey, S., 2002. GH as a co-gonadotropin: the relevance of correlative changes in GH secretion and reproductive state. J Endocrinol. 172, 1-19.

Kitahashi, T., Alok, D., Ando, H., Kaeriyama, M., Zohar, Y., Ueda, H., Urano, A., 1998. GnRH analog Stimulates gonadotropin II gene expression in maturing sockeye salmon. Zool. Sci. 15, 761-765.

Kitahashi, T., Ogawa, S., Soga, T., Sakuma, Y., Parhar, I., 2007. Sexual maturation modulates expression of nuclear receptor types in laser-captured single cells of the cichlid (*Oreochromis niloticus*) pituitary. Endocrinology. 148, 5822-5830.

Kitahashi, T., Ogawa, S., Parhar, I.S., 2009. Cloning and expression of kiss2 in the zebrafish and medaka. Endocrinology. 150, 821-831.

Klausen, C., Chang, J.P., Habibi, H.R., 2001. The effect of gonadotropin-releasing hormone on growth hormone and gonadotropin subunit gene expression in the pituitary of goldfish, *Carassius auratus*. Comp. Biochem. Physiol. B, Biochem. Mol. Biol. 129, 511-516.

Klausen, C., Chang, J.P., Habibi, H.R., 2002. Time- and dose-related effects of gonadotropin-releasing hormone on growth hormone and gonadotropin subunit gene expression in the goldfish pituitary. Can. J. Physiol. Pharmacol. 80, 915-924.

Le Gac, F., Blaise, O., Fostier, A., Le Bail, P.-Y., Loir, M., Mourot, B., Weil, C., 1993. Growth hormone (GH) and reproduction: a review. Fish Physiol. Biochem. 11, 219-232.

Li, D., Fu, C., Hu, W., Zhong, S., Wang, Y., Zhu, Z., 2007. Rapid growth cost in "all-fish" growth hormone gene transgenic carp: Reduced critical swimming speed. Chin. Sci. Bull. 52, 1501-1506.

Li, S., Zhang, Y., Liu, Y., Huang, X., Huang, W., Lu, D., Zhu, P., Shi, Y., Cheng, C.H.K., Liu, X., Lin, H., 2009. Structural and functional multiplicity of the kisspeptin/GPR54 system in goldfish (*Carassius auratus*). J Endocrinol. 201, 407-418.

Lin, X., Lin, H., Peter, R.E., 1993. Growth hormone and gonadotropin secretion in the common carp (*Cyprinus carpio* L.): *in vitro* interactions of gonadotropin-releasing hormone, somatostatin, and the dopamine agonist apomorphine. Gen. Comp. Endocrinol. 89, 62-71.

Livak, K.J., Schmittgen, T.D., 2001. Analysis of relative gene expression data using real-time quantitative PCR and the $2^{-\Delta\Delta CT}$ method. Methods. 25, 402-408.

Lohmus, M., Raven, P., Sundstrom, L., Devlin, R., 2008. Disruption of seasonality in growth hormone-transgenic coho salmon (*Oncorhynchus kisutch*) and the role of cholecystokinin in seasonal feeding behavior. Horm Behav. 54, 506-513.

Maiter, D., Underwood, L.E., Maes, M., Ketelslegers, J.M., 1988. Acute down-regulation of the somatogenic receptors in rat liver by a single injection of growth hormone. Endocrinology. 122, 1291-1296.

Miura, C., Shimizu, Y., Uehara, M., Ozaki, Y., Young, G., Miura, T., 2011. Gh is produced by the testis of Japanese eel and stimulates proliferation of spermatogonia. Reproduction. 142, 869-877.

Moreau, D.T.R., Conway, C., Fleming, I.A., 2011. Reproductive performance of alternative male phenotypes of growth hormone transgenic Atlantic salmon (*Salmo salar*). Evol Appl. 4, 736-748.

Mori, T., Devlin, R.H., 1999. Transgene and host growth hormone gene expression in pituitary and nonpituitary tissues of normal and growth hormone transgenic salmon. Mol. Cell. Endocrinol.149, 129-139.

Mori, T., Hiraka, I., Kurata, Y., Kawachi, H., Mano, N., Devlin, R.H., Nagoya, H., Araki, K., 2007. Changes in hepatic gene expression related to innate immunity, growth and iron metabolism in GH-transgenic amago salmon (*Oncorhynchus masou*) by cDNA subtraction and microarray analysis, and serum lysozyme activity. Gen. Comp. Endocrinol. 151, 42-54.

Nam, Y.K., Cho, Y.S., Cho, H.J., Kim, D.S., 2002. Accelerated growth performance and stable germ-line transmission in androgenetically derived homozygous transgenic mud loach, *Misgurnus mizolepis*. Aquaculture. 209, 257-270.

Peng, C., Peter, R.E., 1997. Neuropeptide regulation of growth hormone secretion and growth in fish. Zool. Stud. 36, 79-89.

Pasquier, J., Lafont, A.G., Leprince, J., Vaudry, H., Rousseau, K., Dufour, S., 2011. First evidence for a direct inhibitory effect of kisspeptins on LH expression in the eel, *Anguilla anguilla*. Gen. Comp. Endocrinol. 173, 216-225.

Popesku, J.T., Martyniuk, C.J., Mennigen, J., Xiong, H., Zhang, D., Xia, X., Cossins, A.R., Trudeau, V.L., 2008. The goldfish (*Carassius auratus*) as a model for neuroendocrine signaling. Mol. Cell. Endocrinol. 293, 43-56.

Rahman, M.A., Mak, R., Ayad, H., Smith, A., Maclean, N., 1998. Expression of a novel piscine growth hormone gene results in growth enhancement in transgenic tilapia (*Oreochromis niloticus*). Transgenic Res. 7, 357-369.

Rahman, M.A., Ronyai, A., Engidaw, B.Z., Jauncey, K., Hwang, G.L., Smith, A., Roderick, E., Penman, D., Varadi, L., Maclean, N., 2001. Growth and nutritional trials on transgenic Nile tilapia containing an exogenous fish growth hormone gene. J. Fish Biol. 59, 62-78.

Raven, P.A., Uh, M., Sakhrani, D., Beckman, B.R., Cooper, K., Pinter, J., Leder, E.H., Silverstein, J., Devlin, R.H., 2008. Endocrine effects of growth hormone overexpression in transgenic coho salmon. Gen. Comp. Endocrinol. 159, 26-37.

Raven, P.A., Sakhrani, D., Beckman, B., Neregård, L., Sundström, L.F., Björnsson, B.T., Dvlin, R.H., 2012. Growth and endocrine effects of recombinant bovine growth hormone treatment in non-transgenic and growth hormone transgenic coho salmon. Gen. Comp. Endocrinol. 177, 143-152.

Rebers, F.E., Hassing, G.A., van Dijk, W., van Straaten, E., Goos, H.J., Schulz, R.W., 2002. Gonadotropin-releasing hormone does not directly stimulate luteinizing hormone biosynthesis in male African catfish. Biol Reprod. 66, 1604-1611.

Rokkones, E., Alestrom, P., Skjervold, H., Gautvik, K.M., 1989. Microinjection and expression of a mouse metallothionein human growth hormone fusion gene in fertilized salmonid eggs. J. Comp. Physiol. B, Biochem. Syst. Environ. Physiol. 158, 751-758.

Schulz, R.W., de França, L.R., Lareyre, J.-J., LeGac, F., Chiarini-Garcia, H., Nobrega, R.H., Miura, T., 2010. Spermatogenesis in fish. Gen. Comp. Endocrinol. 165, 390-411.

Servili, A., Le Page, Y., Leprince, J., Caraty, A., Escobar, S., Parhar, I.S., Seong, J.Y., Vaudry, H., Kah, O., 2011. Organization of two independent kisspeptin systems derived from evolutionary-ancient kiss genes in the brain of zebrafish. Endocrinology. 152, 1527-1540.

Thisse, C., Thisse, B., 2008. High-resolution *in situ* hybridization to whole-mount zebrafish embryos. Nat. Protoc. 3, 59-69.

Trudeau, V.L., Somoza, G.M., Nahorniak, C.S., Peter, R.E., 1992. Interactions of estradiol with gonadotropin-releasing hormone and thyrotropin-releasing hormone in the control of growth hormone secretion in the goldfish. Neuroendocrinology. 56, 483-490.

Trudeau, V.L., 1997. Neuroendocrine regulation of gonadotrophin II release and gonadal growth in the goldfish, *Carassius auratus*. Rev. Reprod. 2, 55-68.

Trudeau, V.L., Spanswick, D., Fraser, E.J., Larivière, K., Crump, D., Chiu, S., MacMillan, M., Schulz, R.W., 2000. The role of amino acid neurotransmitters in the regulation of pituitary gonadotropin release in fish. Biochem. Cell Biol. 78, 241-259.

Van Der Kraak, G., Rosenblum, P.M., Peter, R.E., 1990. Growth homone-dependent potentiation of gonadotropin-stimulated steroid production by ovarian follicles of the goldfish. Gen. Comp. Endocrinol. 79, 233-239.

Vandesompele, J., De Preter, K., Pattyn, F., Poppe, B., Van Roy, N., De Paepe, A., Speleman, F., 2002. Accurate normalization of real-time quantitative RT-PCR data by geometric averaging of multiple internal control genes. Genome Biol. 3, research 0034.1-research 0034.11.

Wang, Y., Hu, W., Wu, G., Sun, Y., Chen, S., Zhang, F., Zhu, Z., Feng, J., Zhang, X., 2001. Genetic analysis of "all-fish" growth hormone gene transferred carp (*Cyprinus carpio*) and its F1 generation. Chin. Sci. Bull. 46, a1-a4.

Wong, A.O.L., Zhou, H., Jiang, Y., Ko, W.K.W., 2006. Feedback regulation of growth hormone synthesis and secretion in fish and the emerging concept of intrapituitary feedback loop. Comp. Biochem. Physiol., Part A: Mol. Integr. Physiol. 144, 284-305.

Wu, G., Chen, L., Zhong, S., Li, Q., Song, C., Jin, B., Zhu, Z., 2008. Enzyme-linked immunosorbent assay of changes in serum levels of growth hormone (cGH) in common carps (*Cyprinus carpio*). Sci China, Ser C: Life Sci. 51, 157-163.

Xu, J., Huang, W., Zhong, C., Luo, D., Li, S., Zhu, Z., Hu, W., 2011. Defining global gene expression changes of the hypothalamic-pituitary-gonadal axis in female sGnRH-antisense transgenic common carp (*Cyprinus carpio*). PLoS ONE. 6, e21057.

Xu, Z., Li, Q., Wang, Y., Song, C., Zhang, T., Chen, L., Ji, J., Yang, A., Zhu, Z., Hu, W., Jin, B., 2010. Production, characterization, and applications of mouse monoclonal antibodies against gonadotropin, somatolactin, and prolactin from common carp (*Cyprinus carpio*). Gen. Comp. Endocrinol. 167, 373-378.

Yaron, Z., 1995. Endocrine control of gametogenesis and spawning induction in the carp. Aquaculture. 129, 49-73.

Zhong, C., Song, Y., Wang, Y., Zhang, T., Duan, M., Li, Y., Liao, L., Zhu, Z., Hu, W., 2013. Increased food intake in growth hormone-transgenic common carp (*Cyprinus carpio* L.) may be mediated by upregulating Agouti-related protein (AgRP). Gen. Comp. Endocrinol. 192. 81-88.

Zhong, C., Song, Y., Wang, Y., Li, Y., Liao, L., Xie, S., Zhu, Z., Hu, W., 2012. Growth hormone transgene effects on growth performance are inconsistent among offspring derived from different homozygous transgenic common carp (*Cyprinus carpio* L.). Aquaculture. 356-57, 404-411.

Zhou, H., Wang, X., Ko, W.K., Wong, A.O., 2004. Evidence for a novel intrapituitary autocrine/paracrine feedback loop regulating growth hormone synthesis and secretion in grass carp pituitary cells by functional interactions between gonadotrophs and somatotrophs. Endocrinology. 145, 5548-5559.

Zmora, N., Stubblefield, J., Zulperi, Z., Biran, J., Levavi-Sivan, B., Muñoz-Cueto, J.-A., Zohar, Y., 2012. Differential and gonad stage-dependent roles of kisspeptin1 and kisspeptin2 in reproduction in the modern teleosts, morone species.

Biol. Reprod. 86, 1-12.
Zohar, Y., Muñoz-Cueto, J.A., Elizur, A., Kah, O., 2010. Neuroendocrinology of reproduction in teleost fish. Gen. Comp. Endocrinol. 165, 438-455.

生长激素过度表达对鲤繁殖的影响

曹梦西[1,2] 陈　戟[1] 彭　伟[1,2] 汪亚平[1] 廖兰杰[1]
李勇明[1] Vance L. Trudeau[3] 朱作言[1] 胡　炜[1]

1中国科学院水生生物研究所，武汉 430072，中国
2中国科学院大学，北京 100049，中国
3 Department of Biology, Centre for Advanced Research in Environmental Genomics, University of Ottawa, Ottawa, Canada, K1N 6N5

摘　要　本文以转草鱼生长激素基因鲤家系为模型，研究鱼类生长与生殖之间的调控关系。与对照鲤相比，转基因鲤的性腺发育滞后。为了更好地解释这一现象，对转基因鲤和对照鱼的生长、性腺发育、生殖发育相关基因及激素的变化进行了为期两年的追踪检测。研究结果表明，外源草鱼生长激素基因过表达导致转基因鲤外周组织 *gh* 基因的转录水平及血清 GH 水平升高，而垂体内源 GH 表达受到抑制。转基因鲤血清及垂体中促黄体激素(LH)表达水平显著低于对照鱼。采用鲤 GH 蛋白体外孵育垂体细胞，*lhβ* 的转录受到显著抑制。进一步研究发现，在鲤垂体的 LH 分泌细胞上存在 GH 受体的表达，提示 GH 可能直接通过其受体 GHR 调控促性腺激素分泌细胞的活性。与此同时，研究发现转基因鲤垂体对于促性腺激素释放激素/多巴胺抑制剂复合物的反应敏感性显著低于对照鱼垂体。综合以上研究结果，转基因鲤性腺发育滞后的主要原因是垂体 LH 表达水平降低。

Endocrinology: Advances Through Omics and Related Technologies

Natàlia Garcia-Reyero[1] Angèle Tingaud-Sequeira[2] Mengxi Cao[3,4]
Zuoyan Zhu[3] Edward J. Perkins[5] Wei Hu[3]

1 Institute for Genomics Biocomputing and Biotechnology, Mississippi State University, Starkville, MS 39759, USA
2 Laboratoire MRMG, Maladies Rares: Génétique et Métabolisme, Université de Bordeaux, 33405 Talence Cedex, France
3 State Key Laboratory of Freshwater Ecology and Biotechnology, Institute of Hydrobiology, Chinese Academy of Sciences, Wuhan 430072, China
4 University of the Chinese Academy of Sciences, Beijing 100049, China
5 US Army Engineer Research and Development Center, Vicksburg, MS 39180, USA

Abstract The rapid development of new omics technologies to measure changes at genetic, transcriptomic, proteomic, and metabolomics levels together with the evolution of methods to analyze and integrate the data at a systems level are revolutionizing the study of biological processes. Here we discuss how new approaches using omics technologies have expanded our knowledge especially in nontraditional models. Our increasing knowledge of these interactions and evolutionary pathway conservation facilitates the use of nontraditional species, both invertebrate and vertebrate, as new model species for biological and endocrinology research. The increasing availability of technology to create organisms overexpressing key genes in endocrine function allows manipulation of complex regulatory networks such as growth hormone (GH) in transgenic fish where dysregulation of GH production to produce larger fish has also permitted exploration of the role that GH plays in testis development, suggesting that it does so through interactions with insulin-like growth factors. The availability of omics tools to monitor changes at nearly any level in any organism, manipulate gene expression and behavior, and integrate data across biological levels, provides novel opportunities to explore endocrine function across many species and understand the complex roles that key genes play in different aspects of the endocrine function.
Keywords omics; reproduction; transgenic organisms; integrative systems biology

1. Omics and Vertebrate Non-model Species

Omic technologies measure changes in genomes (genomics, epigenomics), global gene

---

文章发表于 *General and Comparative Endocrinology*, 2014, 203: 262-273

Abbreviations: DEGs, differentially expressed genes; DHP, 17α,20β-dihydroxy-4-pregnane-3-one; ERα, estrogen receptor α; FSH, follicle-stimulating hormone; GVBD, germinal vesicle breakdown; GH, growth hormone; GHR, growth hormone receptor; HPG, hypothalamus-pituitary-gonadal; IGF-1, insulin-like growth factor-1; LH, luteinizing hormone; Lmo4, Lim-domain only 4; MIH, maturation-inducing hormone; PBPK, physiologically based computational model; Q-PCR, quantitative polymerase chain reaction; SAGE, Serial Analysis of Gene expression; StAR, Steroidogenic Acute Regulatory protein; SF1, steroidogenic factor 1 expression (transcriptomics), global protein levels (proteomics), and global biochemical molecules involved in metabolism (metabolomics). These measurements have evolved from a focus on a single gene, protein or metabolite endpoint 30 years ago to a focus on rapidly measuring changes in thousands to millions of endpoints. Much of this development was initially driven by the human genome project, which began in 1991 (Collins and Galas, 1993).

An essential step for the development of omics applications for endocrine research was refining their use in model species used in understanding both the highly conserved and species-specific aspects of the endocrine system. A very well characterized model, the zebrafish (*Danio rerio*), has almost 1,800,000 expressed sequence tags deposited in Genbank (http://www.ncbi.nlm.nih.gov/nucest) and >20,000 publications in PubMed (http://www.ncbi.nlm.nih.gov/pubmed). Its genome-sequencing project started in 2001 at the Wellcome Trust Institute and a recent, high-quality assembly of its genome, was able to match 70% of its genes to human orthologs, further supporting its already recognized value as a model species for biomedical research (Howe *et al.*, 2013; Lieschke and Currie, 2007).

The classical cDNA microarray technologies, while simple to construct and permitting easy gene annotation, often presented problems of specificity such as an inability to distinguish close paralogs. Due to deep genomic understanding of this species, oligo-based microarrays were developed (often ≈60-mers mainly targeting the untranslated region sequences of transcripts) with commercial arrays available for more than 10 years. A less widely used transcriptomics approach, Serial Analysis of Gene expression (SAGE), is based on cloning 3′-end specific tags of 14-16 nucleotides that are often in the 3′-UTR and then correlates the number of tags to the degree of gene expression. This tool is dependent upon the extent to which the animal model has been sequenced since each single tag must be identified by matching to a corresponding gene. A recent study (Xu *et al.*, 2013) mapped more than 12 million tags and identified nearly 1500 disregulated genes in zebrafish liver in response to arsenic exposure using this approach. However, despite the deep genomic characterization of these model species, interest is growing for "alternative model species" in the fields ranging from evolution and adaptation to extreme environments to toxicology and endocrine research (Perkins *et al.*, 2013). Due to the enormous improvement of the sequencing yield with next-generation sequencing technologies, the ability to produce data for genomic and transcriptomic analysis is becoming increasingly affordable (Davey *et al.*, 2011). As a result, omics techniques can now be applied to "unusual" species to rapidly generate information that can be used to understand novel biological characteristics.

The extent of animal diversity is very large, with nearly 25,000 teleost species, and more than 1,000,000 arthropod species. Evolutionary concerns have driven this extensive interest for the genomic characterization of new species. In May 2010, the release 54 at http://www.ensembl.org/index.html involved 49 species with fully sequenced genome including 5 fish. In September 2013, Release 73 included 63 species with fully sequenced genome, 20 were ongoing and among them 10 were fish. Commercial interests (e.g. aquaculture, oyster-farming) have also driven the development of genome sequencing programs of species of economic interest. Indeed, several species have ongoing genome projects such as the European sea bass (Kuhl *et al.*, 2010), the Atlantic salmon (Ng *et al.*, 2005); the rainbow trout (Palti *et al.*, 2009), the Atlantic cod (Star *et al.*, 2011), eel (Henkel *et al.*, 2012), and the pacific oyster (Zhang *et al.*, 2012).

Often non-model species have poorly characterized, or a complete lack of, genomic/transcriptomic information. As a first step, researchers can use omics to produce large amounts of information that drastically increase their knowledge of a particular species. Non-model species can then be compared to other better-characterized genomes and/or transcriptomes to aid in their characterization. Comparative genomics was first applied between human and chimpanzee 10 years ago (Fujiyama *et al.*, 2002). This approach was based on synteny group search, a tool for evolutionary understanding and genome rearrangement identification such as duplication or reorientation events between species. By comparison and identification of similarity in genes and genomics, one can then infer functional relationships to aid in determining whether or not endocrine pathways are conserved across species (Burgess-Herbert and Euling, 2011). However, despite this advancement, not all genes, proteins, or metabolites can be identified and are therefore not interpreted.

A review of the early use of microarrays for aquatic species can be found in (Denslow *et al.*, 2007). The quality and quantity of sequence data generated by rapidly evolving sequencing techniques revolutionized research with non-model species, allowing researchers to construct high-quality microarrays for relevant species instead of focusing research efforts in model species. Such is the case of the development of a largemouth bass microarray (Garcia-Reyero *et al.*, 2008), used to study estrogenic response in males; or a goldfish cDNA array (Martyniuk *et al.*, 2006), used to study estrogenic effects on neuroendocrine function. The development of microarrays for non-model species though often faces the problem of poorly of annotated probes. For example, 37% of unique transcripts had no annotation in a 19,048 sequence database from European sea bass (Ferraresso *et al.*, 2010). Similarly, a microarray recently built to study embryogenesis of killifish, *Fundulus heteroclitus*, under air-exposed conditions versus water-exposed condition had 65% percent of probes missing annotations (Tingaud-Sequeira *et al.*, 2013). Therefore, investing in further sequencing for better coverage and annotation seems crucial when working with non-model species.

1.1 Use of omics technologies in endocrine research

The hypothalamus-pituitary-gonadal (HPG) axis is a clear example of a system highly conserved across vertebrates, as evidenced by sequence similarities among the estrogen receptor α (ERα) ligand-binding domain from several species (Fig. 1; Norris, 2006). Conservation of these molecules/receptors/signaling pathways involved in reproduction should allow non-mammalian vertebrate models to be used to explore and understand the function of the HPG axis and reproduction both in non-mammalian and mammalian species (Perkins *et al.*, 2013). It is a fact that there are still many unknowns in the processes involved in reproduction. For instance, the molecular pathways involved in oogenesis to produce competent female gametes are still incomplete. Omics technologies are being widely used to provide an overview of the processes underlying the reproductive success through gamete production

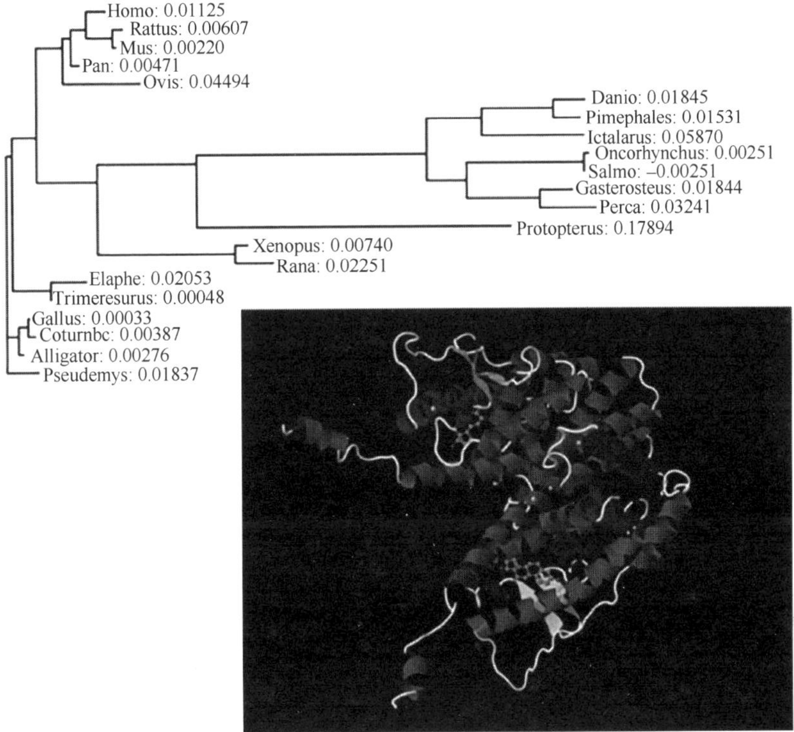

Fig. 1 Phylogenetic analysis of ERα ligand binding (LBD) domain. Sequences for different ERα LGB were obtained from NCBI databases (http://www.ncbi.nlm.nih.gov) and their amino acid sequences were aligned using Clustal Omega (http://www.ebi.ac.uk/Tools/msa/clustalo/). Structure shown using Cn3D macromolcular structure viewer (Wang *et al.*, 2000). Sequences are: *Homo sapiens* (NM_000125.3); *Rattus norvegicus* (NM_012689.1); *Mus musculus* (NM_007956.4); *Pan troglodytes* (XM_003311548.1); *Ovis aries* (AY033393.1); *Danio rerio* (NM_152959); *Pimephales promelas* (AY727528.1); *Ictalarus punctatus* (NM_001200074.1); *Oncorhynchus mykiss* (NM_001124349.1); *Salmo salar* (NM_001123592.1); *Gasterosteus aculeatus* (AB330740.1); *Perca flavescens* (DQ984124.1); *Protopterus annectens* (AB435636.1); *Xenopus laevis* (NM_001089617.1); *Rana rugosa* (AB491673.1); *Elaphe quadrivirgata* (AB548295.1); *Trimeresurus flavoviridis* (AB548294.1); *Gallus gallus* (HQ340611.1); *Coturnix japonica* (AF442965.1); *Alligator mississippiensis* (AB115909.1); *Pseudemys nelsoni* (AB301060.1). The numbers associated with genes are the genetic distance from a common ancestor scores. Genetic distances were initially calculated as percent identity scores and were converted to distances by dividing by 100 and subtracting from 1.0 to give the number of differences per site

quality. The mRNA content of an oocyte is shared between the need for oogenesis to be completed and the need for successful oocyte-embryo transition until zygotic genome activation, the maternal inheritance. Comparative transcriptomic analysis suggests that this maternally inherited pool of mRNA is evolutionary conserved whereas those mRNAs necessary for gamete production are divergent between species (Evsikov *et al.*, 2006; Sylvestre *et al.*, 2013).

Although many endocrinology aspects are often conserved across species, several specific aspects exist. In teleosts, reproductive strategies can be extremely diverse. For example, eggs can be benthic or pelagic; spawning can occur in salt or fresh water; fish ovaries can be synchronized or asynchronized; or embryos can develop immersed or aerially exposed. Some species still reproduce poorly in captivity, requiring the capture of young animals in the wild to maintain populations in captivity. Strong efforts are therefore being made to improve the quality of reproduction for species of economic and conservation interest. Transcriptomic patterns are now available to use as reference and track molecular marker of gametogenesis. Zebrafish was among the first species used for that purpose. One comparative transcriptomic approach of ovary versus testis revealed more than 3500 genes differentially expressed between the two organs with nearly 50 presenting a large difference in expression, suggesting gene silencing in one of the two sexes (Santos *et al.*, 2007a). Also, while individual variability between mature males does not appear to influence testis transcriptomic pattern, ovaries have been found to have significant variation between individuals (Santos *et al.*, 2007b). The zebrafish ovary is asynchronous, therefore the relative ratio of oogenesis stages across whole ovary is not stable and impacts transcriptomic patterns.

Omics approaches have also been used to understand processes involved in and affecting the endocrine system. For instance, Martyniuk *et al.* (2007) used a transcriptomics approach to identify novel genomic responses after exposure to 17α-ethinylestradiol (a potent estrogenic compound) in the zebrafish liver and telencephalon, demonstrating that multi-tissue gene profiling is needed to improve understanding of the effects of human pharmaceuticals on aquatic organisms (Martyniuk *et al.*, 2007). Garcia-Reyero *et al.* (2009a) used transcriptomics approaches to analyze gene expression profiles in male fathead minnow (*Pimephales promelas*) exposed to mixtures of estrogens and antiestrogens. Their results suggested that response to estrogens occurs via multiple mechanisms, including binding to soluble estrogen receptors and membrane estrogen receptors (Garcia-Reyero *et al.*, 2009a). Similarly, a transcriptional analysis of fathead minnow ovaries after exposure to a model androgen and antiandrogen identified transcriptional regulators as potential early molecular switches to control phenotypic changes in ovary in response to androgen or antiandrogen exposure (Garcia-Reyero *et al.*, 2009c). Reading *et al.* (2013) used proteomics to examine the ovary in striped bass and found that the proteasome-ribosome supercomplex known as translasome may play an important role of during the ovarian cycle. Flatfish are also of great interest in aquaculture and an oligo-based microarray for the senegalese sole based on multi-tissue library sequencing was developed (Cerdà *et al.*, 2008). Tingaud-Sequeira *et al.* (2009) applied this array to study ovarian function in sole and identified several new hypotheses involving chemoattractant, angiogenic and antiapoptotic pathways in follicular cells at atresia in addition to the potential atretic molecular markers *apoc1*, *lect2* and *thbs*. This newly developed Senegalese sole microarray was also used during spermatogenesis to identify over 400 differentially regulated genes through the annual cycle (Forné *et al.*, 2011; Marín-Juez *et al.*, 2011). Proteomic data obtained from the same samples revealed 49 differentially expressed proteins and had a weak similarity with transcriptomics since only 7 regulated proteins were correlated with regulated gene expression (Forné *et al.*, 2011). Forné *et al.* (2011) compared the testis proteomic pattern of the F0 generation (caught in the wild) versus the F1 generation (grown up in captivity). They found clear differences in the abundance of 3 proteins demonstrated to be involved in sperm motility, suggesting that lower reproduction in captivity of this species could partially be due to low quality sperm. Egg quality has also been suggested to be linked to maternal inheritance of transcripts (Aegerter *et al.*, 2005 and Mommens *et al.*, 2010). For instance, Lanes *et al.* (2013) performed a large-scale analysis of the transcriptome at early embryonic stage in Atlantic cod comparing fertilized eggs from wild and farmed animals finding significant differences in fructose metabolism, fatty acid metabolism, glycerophospholipid metabolism, and oxidative phosphorylation.

Gracey *et al.* (2004) used an integrative, multi-tissue analysis of the transcriptome of a poikilothermic vertebrate, the common carp *Cyprinus carpio*, identifying a range of candidate genes endowing thermotolerance and revealing a previously unknown scale and complexity of responses at the level of cellular and tissue function (Gracey *et al.*, 2004). Chojnowski *et al.* (2007) analyzed he patterns of vertebrate isochore evolution by comparing expressed mammalian, avian, and crocodilian genes. This analysis allowed them to reject the models that explained the evolution of GC content using changes in body temperature associated with the transition from poikilothermy to homeothermy (Chojnowski *et al.*, 2007). Hirakawa *et al.* (2013) used gene expression analysis to elucidate the molecular basis of testis-ova induction in *Silurana tropicalis* exposed to an estrogenic compound (17α-ethinylestradiol). They found that genes including genes involved in egg envelope composition, 42S particle genes, and regulation of female germ cells were associated with the testis-ova and sex-reversal situation (Hirakawa *et al.*, 2013).

Tuohimaa *et al.* (2013) used gene expression profiles in human and mouse primary cells to better understand the differential actions of vitamin D3 metabolites. Their results showed that there are three distinct vitamin D3 hormones with clearly different biological activities, presenting a new conceptual insight into the vitamin D3 endocrine system (Tuohimaa *et al.*, 2013). Another interesting study (Fullston *et al.*, 2013) investigated the effects of diet-induced paternal obesity, in the absence of diabetes, on the metabolic health of two resultant generations and the molecular profiles of the testes and sperm in mice. Their results showed that diet-induced paternal obesity modulated sperm microRNA content and germ cell methylation status, which are potential signals that initiate the transmission of obesity and impaired metabolic health to future generations, thus implicating paternal obesity in the transgenerational amplification of obesity.

Often the complexity of omics data can be overwhelming, and many approaches have been and are still being developed in order to understand and integrate the data from several biological levels of organization.

2. Approaches to Integrate Omics and Other Data

Systems biology is the study of an organism as interacting networks with the goal of understanding and predicting properties (Weston and Hood, 2004). It incorporates a top-down approach complementary to the traditional, reductionist bottom-up approach with the objective of defining the genetic, protein, and biochemical reactions as integrated and interacting networks, and to characterize the flow of information that links these elements, and their networks, to an emergent biological process (Hood and Perlmutter, 2004). A tandem "top-down and bottom-up" approach would be crucial to precisely manipulate a biological circuit and predict its outcomes (Garcia-Reyero and Perkins, 2011; Koide *et al.*, 2009). Systems analysis can be applied to many different levels of organization, including molecules, cells, organs, individuals, populations, or even ecosystems. One area of systems biology, network biology, has been given most of the attention. One objective of network biology is to reconstruct (or reverse-engineer) accurate models of gene regulation (gene regulatory networks) and to interrogate them to elucidate their physiological mechanism (reviewed in Lefebvre *et al.* (2012)). Regulation of molecules and processes can be extremely context specific, challenging attempts to represent them as universal chains of events, such as canonical pathway representations. There can be rather striking changes in the molecular interactions due to cell-specific expression of different protein isoforms leading to very different outcomes under similar perturbations (Bandyopadhyay *et al.*, 2010 and Mani *et al.*, 2008). Therefore, the idea of pathways is useful as a conceptual tool but could be potentially misleading. Consequently, as network analysis tools become more mature, there might be a gradual migration away from pathway biology into network biology (Lefebvre *et al.*, 2012).

Integrative systems biology approaches are used to elucidate the structure and the function of molecular interactions within a system, including the generation of complex networks and the discovery of novel interactions within those networks (Fig. 2). Recently, new efforts are being made in order to integrate these networks with each other and together with different molecular profiles to identify sets of molecules and interactions within a biological function (Mitra *et al.*, 2013). Network approaches provide powerful tools to examine the underlying biology driving disease and chemical toxicity to the endocrine system through the integration of different levels of organization (Stevens *et al.*, 2013). These approaches (reviewed in Mitra *et al.* (2013)) can be used to aggregate a wide range of measurements including known interactions, omics data, chemistry data, and histochemistry enabling both unsupervised and supervised explorations of complex data sets. While transcriptional data sets are routinely used to examine pathway networks using expert-curated databases, a more complete understanding of chemical effects on endocrine systems both in model and non-model systems can be gained using *de novo* networks or a combination of curated knowledge with new data. For example, Williams *et al.* (2011) were able to examine the effect of environmental pollutants on European flounder fish health through the integration of microarray, microsatellite, metabolomic, protein, and histological data, thereby linking exposures in the environment to disease outcomes. Linkage of hormones levels to gene expression levels via networks can also be used to discover mechanisms of toxicity, as demonstrated in analysis of a large fathead minnow microarray data set generated from

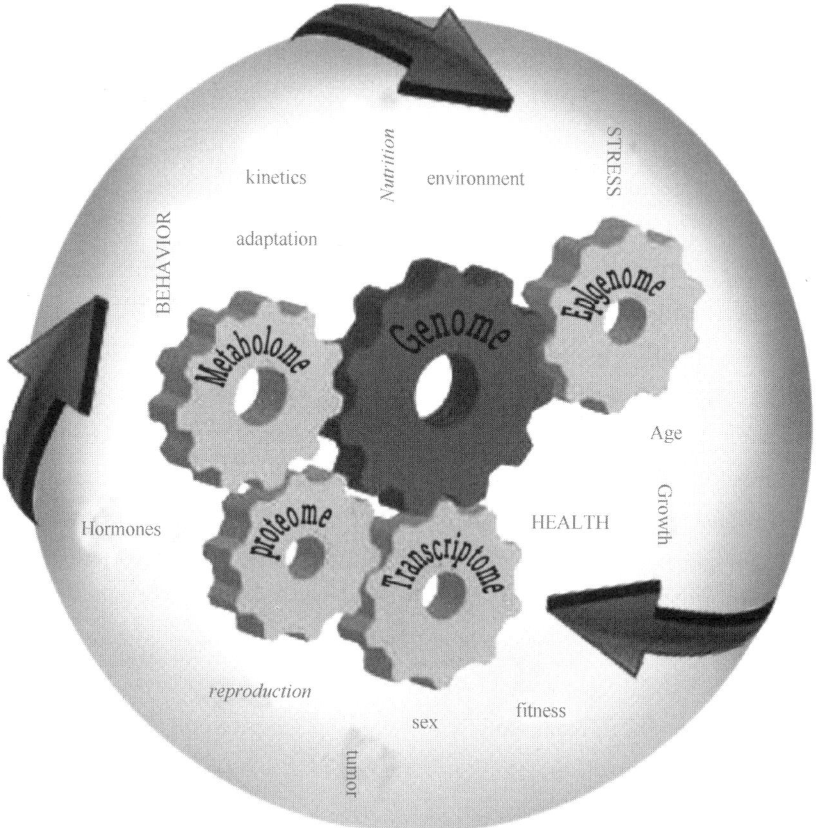

Fig. 2 The paradigm of integrated systems biology approaches. The figure represents the fact that the different parts of a system and different biological levels of organization act as a whole integrated system, influencing each other and being influenced by many internal and external factors

ovaries of fish exposed to different endocrine active chemicals (Perkins *et al.*, 2011). There are many examples of the use of network analysis in endocrinology. For instance, transcriptional network analyses from fathead minnow ovaries exposed to progesterone linked oocyte maturation to RNA post-transcriptional modification, and showed a transcriptional down-regulation of all ribosomal proteins present in the inferred network (Garcia-Reyero *et al.*, 2013). Another study (Nilsson *et al.*, 2013) used a systems approach to identify gene networks related to ovarian follicle assembly in rat, identifying among others highly interconnected genes previously linked to primary ovarian insufficiency and polycystic ovarian disease syndrome. These findings could help elucidate the molecular control of primordial follicle assembly, as well as provide potential targets for ovarian disease treatment. In order to elucidate the first steps of sexual specification, a recent study (Munger *et al.*, 2013) used transcriptomic profiling and probabilistic models to predict novel regulatory genes involved in the process in mice. They showed that Lmo4 (Lim-domain only 4) is a novel regulator of sex determination upstream of SF1 (steroidogenic factor 1). Popesku *et al.* (2012) performed a meta-type analysis using microarray datasets to characterize gene expression changes that underlie dopaminergic signaling in the goldfish brain. Specifically, they performed sub-network enrichment analysis (SNEA), an approach that builds subnetworks by mapping experimental gene expression data onto known biomolecular interactions, to identify common gene regulators and binding proteins associated with genes mediated by dopamine. Using this approach, they identified gene targets related to three major categories including cell signaling, immune response and cell proliferation and growth (Popesku *et al.*, 2012). Another study used omics technologies to examine the dopaminergic modulation of the goldfish hypothalamic transcriptome and proteome. They showed that dopamine rapidly regulates multiple hypothalamic pathways and processes known to be involved in pathologies of the central nervous system (Popesku *et al.*, 2010). A recent study used a mutual information-based network inference method, Context Likelihood of Relatedness (CLR; Faith *et al.*, 2007), to understand changes and infer connections at the transcript and metabolite level in the ovary of fathead minnow exposed to an aromatase inhibitor, fadrozole. This integrated approach revealed changes in gene expression consistent with increased testosterone due to aromatase inhibition. Furthermore, metabolites such as glycogen and taurine were strongly correlated with increased testosterone levels (Garcia-Reyero *et al.*, 2014).

Researchers have been using network analysis to identify conserved genetic modules and conserved patterns of protein interactions (Sharan *et al.*, 2005). Therefore, network analysis can also be useful to assess conservation of effects or processes on different species. For instance, (Garcia-Reyero *et al.*, 2011) used network inference and a meta-analysis with genomic data to compare the effects of a neurotoxicant throughout phylogenetically remote organisms. Their results suggested a common and conserved mode of action of the chemical in species as diverse and distant as rat, quail, fathead minnow, earthworm, and coral.

While network inference can provide valuable information about the regulation of genes in cells, the fact that the number of potentially inferred interactions exceeds the number of independent measurements can make network inference an underdetermined problem (De Smet and Marchal, 2010). Many tools have been developed to deal with this underdetermination. An extensive review of these tools and which of these tools are more appropriate for each specific research question can be found in De Smet and Marchal (2010). Marbach *et al.* (2012) characterized the performance, data requirements and inherent bases of different inference approaches. They observed that no single inference method performed optimally across all data sets, but integration of predictions from multiple inference methods showed robust and high performance across all data sets (Marbach *et al.*, 2012).

Another challenge when dealing with integrative approaches of different omics data, such as transcriptomics and proteomics, is the often-found lack of correlation between them (Piras *et al.*, 2012; Taniguchi *et al.*, 2010). For example, Martyniuk *et al.* (2009) explored the proteomic profiles of androgen receptor signaling in the liver of fathead minnows. Their results showed that, in general, protein and mRNA changes did not correlate (Martyniuk *et al.*, 2009). De Wit *et al.* (2010) had similar results using an integrated approach to characterize the effects of an estrogenic compound on the transcriptome and proteome of zebrafish liver, finding limited concordance between transcriptomics and proteomics datasets. Vogel and Marcotte, 2012 reviewed extensively the potential causes and solutions. Some of the discrepancies can be explained by temporal and other differences in mRNA and protein synthesis and degradation. For example, in mammalian cells, mRNA is produced at a much lower rate than proteins, i.e., 2 copies of mRNA per hour versus dozens of copies of the protein per hour (Vogel and Marcotte, 2012). Furthermore, mRNAs are less stable than proteins, with an average half-life of mammalian mRNA of 2.6-7 h versus 46 h for proteins (Eden *et al.*, 2011, Schwanhäusser *et al.*, 2011, Sharova *et al.*, 2009; Yen *et al.*, 2008). Moreover, protein rates or abundance may vary greatly. For example, RNAs and proteins from metabolic genes tend to be very stable and have high protein per mRNA ratios, while proteins that are involved in chromatin organization and transcriptional regulation tend to be rapidly degraded (Schwanhäusser *et al.*, 2011; Vogel *et al.*, 2010; Vogel and Marcotte, 2012). Therefore, a deep understanding of the basic properties of the system and the correlations between different types of data will be crucial for a successful development of truly integrative analyses and models. Despite these challenges, metabolite, hormone and proteins can be integrated with gene expression as demonstrated using mutual information network inference algorithms (Perkins *et al.*, 2011; Garcia-Reyero *et al.*, 2014).

Another useful approach to the challenge of interpreting proteomic, genomic, and/or other types of data is mathematical modeling, which provides an organizational platform to consolidate protein dynamics with transcriptional regulation. Magill *et al.* (2013) developed a mathematical model of the pulse-coded hormone signal responses in pituitary gonadotroph cells using *in vitro* data from a murine-derived cell line to identify key elements involved in the signaling that regulated transcription of LH and FSH. Simulations of the model identified a preference for FSH synthesis at low pulse frequencies (Magill *et al.*, 2013). Mathematical modeling of the endocrine system processes is useful not only to understand the processes involved but also to predict the effects of stress and perturbations. For example, Shoemaker *et al.* (2010) developed a mathematical model of HPG axis in female fathead minnow to evaluate effects of aromatase inhibition on transcriptional regulation of steroid production. Using simulation studies and sensitivity analysis, they were able to evaluate effects of aromatase inhibition on estradiol production and feedback control in the HPG axis. These analyses found that transport of cholesterol into mitochondria by the Steroidogenic Acute Regulatory protein (StAR) and conversion of cholesterol into pregnenolone by the cholesterol-side chain cleavage enzyme P450scc are likely key regulatory points for maintaining production of estradiol in response to global cues coming from the hypothalamus and pituitary (Shoemaker *et al.*, 2010), supporting experimental data that these proteins, especially P450scc, are rate limiting steps in cholesterol transport into the mitochondria and estradiol synthesis (see Miller and Bose (2011) for review). Watanabe *et al.* (2009) developed a physiologically based computational model (PBPK) of the HPG axis in order to understand how estrogen exposure affected reproductive endpoints in the male fathead minnow. Following the same process, Li *et al.* (2011a) developed a PBPK model of the HPG axis in female fathead minnows, in order to understand how estrogens and androgens affect plasma concentration of steroid hormones and vitellogenin (Li *et al.*, 2011b).

The advantage of systems approaches and mathematical modeling of the endocrine system is that it allows integration of temporal changes in gene regulation, protein production and biochemical fluxes to provide a more realistic interpretation of effects. A truly integrated systems approach would take advantage of all the previously described methods, following an iterative cycle of model development and experimental validation. Experimental results can be used to modify the model and develop new hypothesis to test until the computational system closely simulates the biological system (Garcia-Reyero and Perkins, 2011; Ghosh *et al.*, 2011). Each stage of the cycle has different needs and requires specifically tailored software tools that are extensively reviewed in Gosh (Ghosh *et al.*, 2011).

3. Use of Invertebrates As Alternative Model Species

Integrative systems approaches are being increasingly combined with the use of non-mammalian species, particularly invertebrate models, due to increased public concerns on the use of vertebrates for research. The use of invertebrate models for biomedical research, testing, and education is thoroughly reviewed by Wilson-Sanders (2011). Here we will focus on some of those species and their potential use to understand and characterize processes related to the endocrine system (Table 1). Two very well established invertebrate models are the fruit fly (*Drosophila melanogaster*) and the round worm (*Caenorhabditis elegans*). We may know more about Drosophila than any other animal, partly because of its genome sequencing and chromosome mapping together with a vast array of genetic and molecular tools that make *Drosophila* a very powerful model for the study of human genetics, disease, behavior or development (Bruneaux *et al*., 2013; Wilson-Sanders, 2011). The many available mutant strains have proved extremely useful in elucidating pathways, including embryonic development, and served as models for numerous human diseases, emphasizing the fact that the development of genetic/genomic and molecular tools play a crucial role in our capability to successfully use invertebrate species as models.

Table 1 Invertebrate species used as alternative for models or disorders in vertebrates

Model/disorder	Species used	References
Aggressive behavior	*Homarus americanus, Astacus astacus, Drosophila*	Huber *et al.* (1997), Kravitz and Huber (2003) and Nilsen *et al.* (2004)
Aging	*Bivalves, Drosophila, C. elegans, Macrostonum lignano*	Abele *et al.* (2009), Ballard (2005), Berryman *et al.* (2008), Brys *et al.* (2007) and Mouton *et al.* (2009)
Behavior	*Drosophila*	Jasinska and Freimer (2009) and Vosshall (2007)
Calcium signaling	*Asterina pectinifera*	Santella *et al.* (2008)
Cancer	*Drosophila*	Read (2011)
Development	*Dissostera Carolina, Ciona intestinalis,* sea urchin, *Drosophila*	Holland and Gibson-Brown (2003), Isbister and O'Connor (2000), Mallo and Alonso (2013), Peter and Davidson (2011) and Peter *et al.* (2012)
Embryotoxicity	Sea urchin	Ben-Tabou de-Leon and Davidson (2009)
Endocrine disruption	*Drosophila, Daphnia, Potamopgyrus antipodarum*	Avanesian *et al.* (2009), Duft *et al.* (2007), Gupta *et al.* (2007) and Tatarazako and Oda (2007)
Endocrine function and metabolism	*Drosophila, C. elegans, Bombix mori, Ciona intestinalis, Daphnia magna*	Dow (2007), LeBlanc and McLachlan (2000), Sherwood *et al.* (2006) and Yoshida *et al.* (1998)
Formation of the nervous system	*Drosophila*	Quan and Hassan (2005)
Gene regulation	*C. elegans, Drosophila*	Ercan and Lieb (2009) and Mendjan and Akhtar (2007)
Gene regulatory networks	*Echinoidea*	Peter and Davidson (2009)
Hypoxia	*Drosophila, Daphnia magna, C. elegans*	Gorr *et al.* (2006) and Romero *et al.* (2007)
Learning and memory	*Octopus, Aplysis californica, Hermissenda crassicornis, Apis mellifera*	Avarguès-Weber and Giurfa (2013), Bédécarrats *et al.* (2013), Crow and Xue-Bian (2011), De Lisa *et al.* (2012) and Sperduti *et al.* (2012)
MicroRNA function in embryogenesis	*C. elegans*	Wienholds and Plasterk (2005)
Muscle development	*Drosophila*	Maqbool and Jagla (2007)
Neurodegenerative diseases	*C. elegans, Drosophila*	Konsolaki (2013), Gilbert (2008) and Johnson *et al.* (2010)
Neurological diseases	*Drosophila*	Inlow and Restifo (2004)
Regulatory switches	*Drosophila*	Borok *et al.* (2010)
Sexual differentiation	*C. elegans, Drosophila*	Franco and Yao (2012)
Tumor suppression	*Drosophila*	Vaccari and Bilder (2009)

Drosophila is a well-established model for many human pathologies such as cancer, development, or the nervous system, including neurodegenerative diseases (Elmer and Meyer,

2011; Gilbert, 2008; Wilson-Sanders, 2011; Zhang *et al.*, 2013). For example, of the 282 genes associated with human mental retardation, 87% have one or more fly homologs and 76% have at least one fruit fly functional ortholog, suggesting that *Drosophila* could be an excellent model towards the discovery of new drugs for treating mental retardation (Bruneaux *et al.*, 2013; Inlow and Restifo, 2004; Sarropoulou and Fernandes, 2011; Wilson-Sanders, 2011). Some studies clearly show the use of *Drosophila* to study complex behaviors such as courtship, learning, alcoholism, and aggression (Ferraresso *et al.*, 2010; Gilbert, 2008), and even the signaling pathways regulating sleep patterns in a part of the fly brain that is developmentally and functionally analogous to the hypothalamus in vertebrates (Foltenyi *et al.*, 2007).

The worm *C. elegans* also has many advantages over vertebrate models. These worms are very prolific reproducers, have short generation times, grow easily in the laboratory, are inexpensive to care for, are anatomically simple and have a fully mapped nervous system. They also have virtually the same number of genes as humans and share many similarities at the genetic and molecular level (Garcia *et al.*, 2012; Wilson-Sanders, 2011). *C. elegans* are already being used by the US Environmental Protection Agency ToxCast program (http://www.epa.gov/ncct/toxcast/) to predict hazards and characterize toxicity pathways, and are used for *in vivo* testing of new drugs and chemicals, including high-throughput screening technologies. *C. elegans* also serves as a model for many diseases, such as Parkinson's, Alzheimer's, and Huntington's diseases, diabetes, cancer, or immune disorders and has been used to study the genetics and behavioral mechanisms of alcohol, nicotine and cocaine addictions (Kocabas *et al.*, 2006; Wilson-Sanders, 2011).

Cladocerans are increasingly gaining acceptance as alternative models, particularly since the water flea *Daphnia* was recently added by the National Institutes of Health (NIH) to its list of model organisms for biomedical research. While *Daphnia* has long been used in biomedical research to study cellular immunity, phenotypic plasticity, or genetic heritability among others (Ebert, 2011; Kocabas *et al.*, 2006), the recent advance on genetic tools coupled with the release of the *Daphnia pulex* genome (Colbourne *et al.*, 2011), the first crustacean genome ever sequenced, rapidly promoted the use of *Daphnia* as a model species. Transcriptomic profiling of *Daphnia magna* was rapidly embraced as a tool for ecotoxicogenomics research (Garcia-Reyero *et al.*, 2012, 2009b; Watanabe *et al.*, 2008, 2007). *Ceriodaphnia dubia* has been proposed as a cost-effective alternative to Daphnia due to their shorter life cycle, while providing similar outcomes in toxicological studies (Constantine and Huggett, 2010). This organism currently lacks genomic/genetic and molecular tools necessary for advanced research. A recent study by (Stanley *et al.*, 2013) used an integrative approach that included transcriptomics, lipidomics, and apical endpoints to link effects on lipid metabolism to hormetic effects (increased growth and reproduction) on *D. magna*. Similar findings and effects have been seen in other species including fathead minnows, quail and rat, supporting the suitability of Daphnia as an alternative model species (Deng *et al.*, 2011; Rawat *et al.*, 2010; Wintz *et al.*, 2006). Further supporting the Daphnia model, researchers from Japan (Kato *et al.*, 2011) used Daphnia to reveal how environmental sex determination is implemented by selective expression of a fundamental genetic component that is functionally conserved in animals using genetic sex determination, inferring that there is an ancient link between genetic and environmental sex determination.

4. Use of Transgenics in Endocrinology Research

Thanks to the advance of omics technologies, the use of transgenic organisms is becoming increasingly popular in endocrinology studies. For example, the use of several strains of genetically engineered mouse models has provided new insights into the functions of kisspeptin signaling in the hypothalamus (reviewed in Dungan Lemko and Elias (2012)). Kang *et al.* (2014) reviewed the strategies for genetic overexpression and knockout of specific genes in adipose tissues in order to better understand the role of adipose tissue in metabolic homeostasis in the context of different types of adipocytes. Another review (Piret and Thakker, 2011) described mouse models that have been generated for the study of human hereditary and metabolic disorders. McGonnell (2006) explored the use of genetically modified zebrafish as a novel and emerging model to study the endocrine system, often replacing the most commonly used mouse model. As these models have been extensively reviewed (Dungan Lemko and Elias, 2012; Kang *et al.*, 2014; McGonnell, 2006; Piret and Thakker, 2011), here we will focus on a fish transgenic model, the growth hormone transgenic fish, as an example of the advances in the field and the potential applications of transgenic organisms.

4.1 GH-transgenic fish

Growth hormone (GH) is an important hormone known to regulate body growth and metabolism. In GH-transgenic fish, the serum GH level ranges from three times (Rahman and Maclean, 1999) to as much as 40 times higher than non-transgenic fish, which leads to the fast growing of body weight (Devlin *et al.*, 2004). Nevertheless, GH-transgenic fish also show reduced reproductive performance besides fast growth rate. The physiological relevance, as well as the molecular mechanisms underlying these phenomena, however, is still poorly understood. Recently, studies on GH-transgenic common carp have revealed that the presence of a GH interaction with the hypothalamus and pituitary glands can regulate the HPG axis through an autocrine/paracrine mechanism. These findings may provide new insights into the functional interactions between the gonadotropic and somatotropic axes in fish models. Fig. 3 shows the sequence similarities for Growth Hormone receptor (GHR, Fig 3a) and Growth Hormone (GH, Fig. 3b). The understanding of similarities or differences among species can

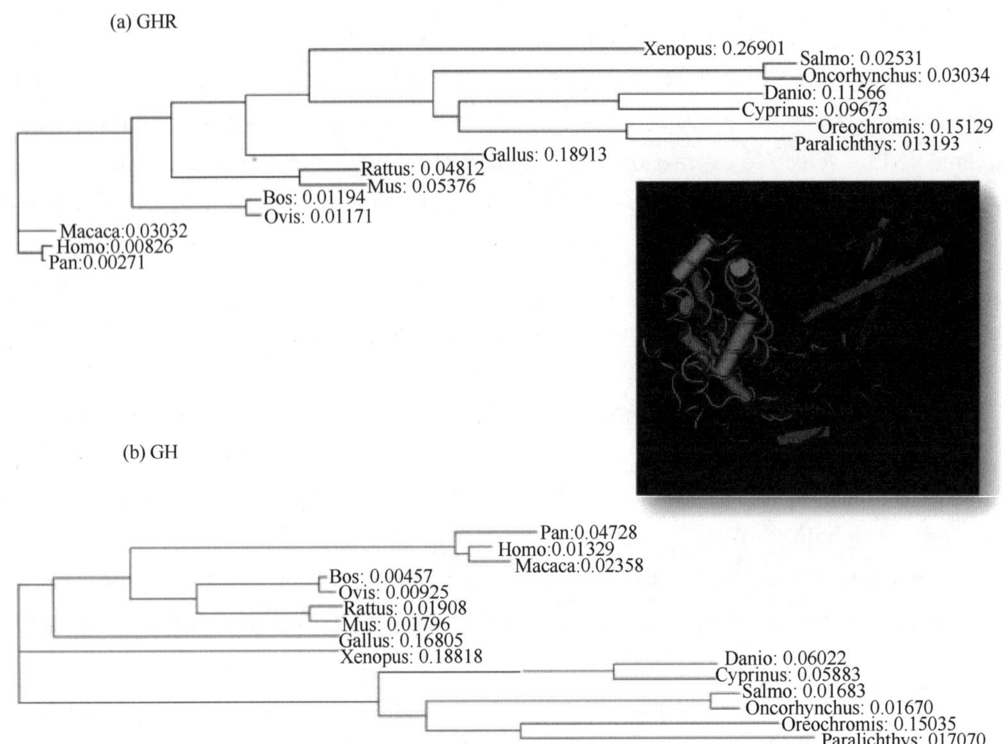

Fig. 3 Phylogenetic analysis of (a) growth hormone receptor (GHR) and (b) growth hormone (GH). Sequences were obtained from NCBI databases (http:// www.ncbi.nlm.nih.gov) and their amino acid sequences were aligned using Clustal Omega (http://www.ebi.ac.uk/Tools/msa/clustalo/). Structure shown using Cn3D macromolecular structure viewer (Wang *et al.*, 2000). Sequences are: *Homo sapiens* (GR: AAA98618.1; GHR: NP_000154.1); *Pan troglodytes* (GR: AAL72285.1; GHR: XP_003950612.1); *Macaca mulatta* (GR: AAA18842.1; GHR: NP_001036132.1); *Rattus norvegicus* (GR: ACX46986.1; GHR: NP_058790.1); *Mus musculus* (GR: CAA86658.1; GHR: EDL03404.1); *Bos taurus* (GR: AAA30544.1; GHR: NP_788781.1); *Ovis aries* (GR: AAB24467.2; GHR: NP_001009323.1); *Gallus gallus* (GR: AEZ51853.1; GHR: NP_001001293.1); *Paralychthys olivaceus* (GR: AAA49443.1; GHR: BAC76398.1); *Oreochromis niloticus* (GR: AAA49626.1; GHR: NP_001266530.1); *Cyprinus carpio* (GR: CAA36228.1; GHR: AAU43899.1); *Danio rerio* (GR: CAI79040.1; GHR: NP_001077047.1); *Oncorhynchus mykiss* (GR: AAA49554.1; GHR: NP_001118007.1); *Salmo salar* (GR: AAA49558.1; GHR: NP_001117048.1); *Xenopus laevis* (GR: AAF05773.1; GHR: AAF05775.1). The numbers associated with genes are the genetic distance from a common ancestor scores. Genetic distances were initially calculated as percent identity scores and were converted to distances by dividing by 100 and subtracting from 1.0 to give the number of differences per site

allow for a better understanding of the functions of GH in different species. Here, we review the reproductive performance of GH-transgenic fish and discuss the possible regulatory mechanism of GH on the HPG axis.

4.1.1 Interaction of GH and reproduction

It is well known that growth and reproduction are closely related. Fast growing GH-transgenic mice always show reduced reproductive performance (Bartke *et al.*, 1988). They also have enhanced growth, reduced life-span, as well as a number of endocrine and reproductive abnormalities (Bartke *et al.*, 1992). Research in mice indicates that GH might, directly or indirectly, act on the neuroendocrine system to influence gonadotropin and prolactin expression, which would affect reproduction as a result.

Le Gac *et al.* (2013) summarized the role of GH in puberty onset, gametogenesis, and fertility in mammals and fish (Le Gac *et al.*, 1993). In fish, GH actions on reproductive function are not clear. GH could be detected in many other tissues besides pituitary, which hints at the possible role of GH in other physical processes. Particularly, GH is expressed in the gonad, indicating GH might affect gonadal development. Here, we summarize the possible mechanisms of GH action on the HPG axis, on the basis of the results obtained from GH-transgenic common carp and others.

4.1.1.1 Hypothalamus

In GH-transgenic common carp, in addition to the reduced LH expression, hypothalamus *gnrh3* and *kiss2* mRNA also increased at 2-6 months during early gonadal development (Cao *et al.*, 2014). Interestingly, *gnrh3* and *kiss2* genes are known to promote gonadal development, which contradicts the delayed reproductive development observed in GH-transgenic carp. However, it is reminiscent of the potential interactions between GH and the kiss/GnRH system. Results from morpholino experiments suggested that GH might directly induce gnrh3 and *kiss2* gene expression at the hypothalamus level (Cao *et al.*, 2014).

4.1.1.2 Pituitary

The pituitary is the regulatory center of reproduction and it produces and secretes GH, FSH, and LH. Unlike mammals, fish gonadotrophs always exhibit a patchy distribution with somatotrophs, which provides the possibility of local interactions in between gonadotrophs and somatotrophs (Wong *et al.*, 2006). *In vitro* experiments using grass carp pituitaries confirmed the paracrine actions of LH on GH mRNA expression and GH action on LH expression (Zhou *et al.*, 2005; Zhou *et al.*, 2004). In GH-transgenic common carp, serum LH levels were decreased throughout the reproductive cycle. The inhibitory effects of GH on LH mRNA expression were further confirmed by *in vitro* experiments (Cao *et al.*, 2014). GHR has also been found in pituitary LH positive cells, indicating that GH might regulate LH expression through GHR in LH secreting cells. These results clearly show the paracrine effects of GH acting on GHR to inhibit LH transcription and secretion at the pituitary level.

4.1.1.3 Gonad

There are many reports of GH action on gonadal development (Berishvili *et al.*, 2006; Fukada *et al.*, 2003; Kajimura *et al.*, 2004), which suggest that there might be an independent GH system in the gonad. Gonadal GH might also exert its effects on reproduction through IGF-1, insulin-like growth factor-1, (Duan, 1997; Melamed *et al.*, 1999). In common carp, IGF-1 increased maturation-inducing hormone (MIH) and 17α,20β-dihydroxy-4-pregnane-3-one (DHP) which then induced oocytes to undergoing germinal vesicle breakdown (GVBD) *in vitro* (Mukherjee *et al.*, 2006). In rainbow trout (*Oncorhynchus mykiss*), testicular GH directly stimulated IGF-1 expression (Perrot and Funkenstein, 1999), and in turn stimulated DNA synthesis and cell proliferation (Loir, 1999). Recently, a newly discovered IGF, named IGF-3, was reported to be specifically expressed in the gonad (Wang *et al.*, 2008). It has been found to significantly enhance oocyte maturation in time-, dose- and stage-dependent manners (Li *et al.*, 2011a), and induce spermatogenesis through stimulating type A spermatogonial proliferation (graduation thesis). These results support the hypothesis that GH regulates gonadal development through IGFs inside the gonad.

The data presented in this section are all focused on genes already known to be the principle signaling factors in HPG axes. However, it is known that reproduction is under control of a much more complex system involving genes related with growth, metabolism, and immune system. Omics approaches discussed in previous sections facilitate the investigation of the functional complexity of transcriptomes and provide a powerful approach to comprehensively study the relationship between growth and reproduction. They will also provide new insights into genes involved in the regulation of reproduction. For example, Miao and collaborators (Miao and Luo, 2013) characterized the DEGs in Small-tail Han sheep and the Surabaya fur sheep in a genome-wide level using RNA-sequencing technology. They found multiple genes that were potentially critical for regulating sheep fecundity and prolificacy (Miao and Luo, 2013). In fish, Devlin *et al.* (2009) used microarrays to identify the growth traits of domesticated and GH-transgenic coho salmon.

GH-transgenic fish provide a special model for the study of interactions between growth and reproduction. Recent research indicates that GH interacts with the HPG axis at the hypothalamus, pituitary, and gonadal levels. Besides effects on the HPG axis, GH exerts its impact by affecting metabolism and energy allocation (Bartke *et al.*, 1999; Figueiredo *et al.*, 2013). Because the neuroendocrine system is a complex system that might involve many different gene regulatory networks, analysis of large amounts of omics data will be needed to infer those networks and connections so that we can better understand the involvement of GH on reproduction.

5. Conclusions

As omics technologies and integrated approaches continue to evolve, researchers will find new ways to overcome challenges, allowing for a better understanding of the comparative endocrinology field. Issues such as species sensitivity or specificity and reproductive strategies among others can greatly benefit from the use of transgenic organisms for research. In order to successfully apply these novel integrative approaches though, several issues must be taken into account. For instance, no single method or algorithm seems to perform optimally for every data set and research question. Therefore, a proper understanding of the different methods and the ability to choose the best approach will be crucial to get the most and more accurate information from the data. Also, an in depth understanding of the system and of the correlations between different types of omics data is crucial for the development of a truly integrative analysis. We believe that these new and evolving approaches are indeed very exciting and will lead to novel insights in endocrine function, which might not be approachable by traditional methods.

Acknowledgements This work was funded by the US Army Environmental Quality Research Program (including BAA 11-4838). Permission for publishing this information has been granted by the Chief of Engineers. This work was also supported financially by the "863" High Technology Project (Grant No. 2011AA100404), National Natural Science Foundation (Grant No. 30930069), the Development Plan of the State Key Fundamental Research of China (Grant No. 2010CB126302), and the Key Research Program of the Chinese Academy of Sciences (Grant No. KSCX2-EW-N-004). In addition, this work was supported by the Conseil Régional d'Aquitaine, France (Grant No. 20088130103l/TOD project) and the Agence Nationale de la Recherche (ANR Grant No. 2010BLAN 1126 01/LIGENAX project).

References

Abele, D., Brey, T., Philipp, E., 2009. Bivalve models of aging and the determination of molluscan lifespans. Exp. Gerontol. 44, 307-315.

Aegerter, S., Jalabert, B., Bobe, J., 2005. Large scale real-time PCR analysis of mRNA abundance in rainbow trout eggs in relationship with egg quality and postovulatory ageing. Mol. Reprod. Dev. 72, 377-385.

Avanesian, A., Semnani, S., Jafari, M., 2009. Can *Drosophila melanogaster* represent a model system for the detection of reproductive adverse drug reactions? Drug Discov. Today 14, 761-766.

Avarguès-Weber, A., Giurfa, M., 2013. Conceptual learning by miniature brains. Proc. Biol. Sci. 280, 20131907.

Ballard, J.W.O., 2005. *Drosophila simulans* as a novel model for studying mitochondrial metabolism and aging. Exp. Gerontol. 40, 763-773.

Bandyopadhyay, S., Mehta, M., Kuo, D., Sung, M.K., Chuang, R., Jaehnig, E.J., Bodenmiller, B., Licon, K., Copeland, W., Shales, M., Fiedler, D., Dutkowski, J., Guenole, A., van Attikum, H., Shokat, K.M., Kolodner, R.D., Huh, W.K., Aebersold, R., Keogh, M.C., Krogan, N.J., Ideker, T., 2010. Rewiring of genetic networks in response to DNA damage. Science 330, 1385-1389.

Bartke, A., Chandrashekar, V., Turyn, D., Steger, R.W., Debeljuk, L., Winters, T.A., Mattison, J.A., Danilovich, N.A., Croson, W., Wernsing, D.R., Kopchick, J.J., 1999. Effects of growth hormone overexpression and growth hormone resistance on neuroendocrine and reproductive functions in transgenic and knock-out mice. Proc. Soc. Exp. Biol. Med. 222, 113-123.

Bartke, A., Naar, E.M., Johnson, L., May, M.R., 1992. Effects of expression of human or bovine growth hormone genes on sperm production and male reproductive performance in four lines of transgenic mice. J. Reprod. Fertil. 95, 109-118.

Bartke, A., Steger, R.W., Hodges, S.L., Parkening, T.A., Collins, T.J., Yun, J.S., Wagner, T.E., 1988. Infertility in transgenic female mice with human growth hormone expression: evidence for luteal failure. J. Exp. Zool. 248,

121-124.

Ben-Tabou de-Leon, S., Davidson, E., 2009. Experimentally based sea urchin gene regulatory network and the causal explanation of developmental phenomenology. WIREs Syst. Biol. Med. 1, 13.

Berishvili, G., D'Cotta, H., Baroiller, J.-F., Segner, H., Reinecke, M., 2006. Differential expression of IGF-I mRNA and peptide in the male and female gonad during early development of a bony fish, the tilapia *Oreochromis niloticus*. Gen. Comp. Endocrinol. 146, 204-210.

Berryman, D.E., Christiansen, J.S., Johannsson, G., Thorner, M.O., Kopchick, J.J., 2008. Role of the GH/IGF-1 axis in lifespan and healthspan: lessons from animal models. Growth Horm. IGF Res. 18, 455-471.

Bédécarrats, A., Cornet, C., Simmers, J., Nargeot, R., 2013. Implication of dopaminergic modulation in operant reward learning and the induction of compulsive-like feeding behavior in Aplysia. Learn. Mem. 20, 318-327.

Borok, M.J., Tran, D.A., Ho, M.C.W., Drewell, R.A., 2010. Dissecting the regulatory switches of development: lessons from enhancer evolution in Drosophila. Development 137, 5-13.

Bruneaux, M., Johnston, S.E., Herczeg, G., Merilä, J., Primmer, C.R., Vasemägi, A., 2013. Molecular evolutionary and population genomic analysis of the ninespined stickleback using a modified restriction-site-associated DNA tag approach. Mol. Ecol. 22, 565-582.

Brys, K., Vanfleteren, J.R., Braeckman, B.P., 2007. Testing the rate-of-living/oxidative damage theory of aging in the nematode model *Caenorhabditis elegans*. Exp. Gerontol. 42, 845-851.

Burgess-Herbert, S.L., Euling, S.Y., 2011. Use of comparative genomics approaches to characterize interspecies differences in response to environmental chemicals: challenges, opportunities, and research needs. Toxicol.Appl. Pharm. 271, 372-385.

Cao, M., Chen, J., Peng, W., Wang, Y., Liao, L., Li, L., Trudeau, V.L., Zhu, Z., Hu, W., 2014. Effects of growth hormone over-expression on reproduction in the common carp *Cyprinus carpio* L. Gen. Comp. Endocrol 195, 47-57.

Cerdà, J., Mercadé, J., Lozano, J.-J., Manchado, M., Tingaud-Sequeira, A., Astola, A., Infante, C., Halm, S., Viñas, J., Castellana, B., Asensio, E., Cañavate, P., MartínezRodríguez, G., Piferrer, F., Planas, J.V., Prat, F., Yúfera, M., Durany, O., Subirada, F., Rosell, E., Maes, T., 2008. Genomic resources for a commercial flatfish, the Senegalese sole (*Solea senegalensis*): EST sequencing, oligo microarray design, and development of the Soleamold bioinformatic platform. BMC Genomics 9, 508.

Chojnowski, J.L., Franklin, J., Katsu, Y., Iguchi, T., Guillette, L.J., Kimball, R.T., Braun, E.L., 2007. Patterns of vertebrate isochore evolution revealed by comparison of expressed mammalian, avian, and crocodilian genes. J. Mol. Evol. 65, 259-266.

Colbourne, J.K., Pfrender, M.E., Gilbert, D., Thomas, W.K., Tucker, A., Oakley, T.H., Tokishita, S., Aerts, A., Arnold, G.J., Basu, M.K., Bauer, D.J., Caceres, C.E., Carmel, L., Casola, C., Choi, J.-H., Detter, J.C., Dong, Q., Dusheyko, S., Eads, B.D., Frohlich, T., Geiler-Samerotte, K.A., Gerlach, D., Hatcher, P., Jogdeo, S., Krijgsveld, J., Kriventseva, E.V., Kultz, D., Laforsch, C., Lindquist, E., Lopez, J., Manak, J.R., Muller, J., Pangilinan, J., Patwardhan, R.P., Pitluck, S., Pritham, E.J., Rechtsteiner, A., Rho, M., Rogozin, I.B., Sakarya, O., Salamov, A., Schaack, S., Shapiro, H., Shiga, Y., Skalitzky, C., Smith, Z., Souvorov, A., Sung, W., Tang, Z., Tsuchiya, D., Tu, H., Vos, H., Wang, M., Wolf, Y.I., Yamagata, H., Yamada, T., Ye, Y., Shaw, J.R., Andrews, J., Crease, T.J., Tang, H., Lucas, S.M., Robertson, H.M., Bork, P., Koonin, E.V., Zdobnov, E.M., Grigoriev, I.V., Lynch, M., Boore, J.L., 2011. The ecoresponsive genome of *Daphnia pulex*. Science 331, 555-561.

Cóllins, F., Galas, D., 1993. A new five-year plan for the United States: human genome program. Science 262, 43-46.

Constantine, L.A., Huggett, D.B., 2010. A comparison of the chronic effects of human pharmaceuticals on two cladocerans, *Daphnia magna* and *Ceriodaphnia dubia*. Chemosphere 80, 1069-1074.

Crow, T., Xue-Bian, J.J., 2011. Proteomic analysis of short-and intermediate-term memory in Hermissenda. Neuroscience 192, 102-111.

Davey, J.W., Hohenlohe, P.A., Etter, P.D., Boone, J.Q., Catchen, J.M., Blaxter, M.L., 2011. Genome-wide genetic marker discovery and genotyping using nextgeneration sequencing. Nat. Rev. Genet. 12, 499-510.

De Lisa, E., Paolucci, M., Di Cosmo, A., 2012. Conservative nature of oestradiol signalling pathways in the brain lobes of octopus vulgaris involved in reproduction, learning and motor coordination. J. Neuroendocrinol. 24, 275-284.

De Smet, R., Marchal, K., 2010. Advantages and limitations of current network inference methods. Nat. Rev. Microbiol., 1-13.

Deng, Y., Meyer, S.A., Guan, X., Escalon, B.L., Ai, J., Wilbanks, M.S., Welti, R., GarciaReyero, N., Perkins, E.J., 2011. Analysis of common and specific mechanisms of liver function affected by nitrotoluene compounds. PLoS One 6, e14662.

Denslow, N.D., Garcia-Reyero, N., Barber, D.S., 2007. Fish ''n'' chips: the use of microarrays for aquatic toxicology. Mol. Biosyst. 3, 172-177.

Devlin, R.H., Biagi, C.A., Yesaki, T.Y., 2004. Growth, viability and genetic characteristics of GH transgenic coho salmon strains. Aquaculture 236, 607-632.

Devlin, R.H., Sakhrani, D., Tymchuk, W.E., Rise, M.L., Goh, B., 2009. Domestication and growth hormone transgenesis cause similar changes in gene expression in coho salmon (*Oncorhynchus kisutch*). Proc. Natl. Acad. Sci. U.S.A. 106, 3047-3052.

De Wit, M., Keil, D., van der Ven, K., Vandamme, S., Witters, E., De Coen, W., 2010. An integrated transcriptomic and proteomic approach characterizing estrogenic and metabolic effects of 17 alpha-ethinylestradiol in zebrafish (*Danio rerio*). Gen. Comp. Endocr. 167, 190-201.

Dow, J.A.T., 2007. Model organisms and molecular genetics for endocrinology. Gen. Comp. Endocrinol. 153, 3-12.

Duan, C., 1997. The insulin-like growth factor system and its biological actions in fish. Am. Zool. 37, 491-503.

Duft, M., Schmitt, C., Bachmann, J., Brandelik, C., Schulte-Oehlmann, U., Oehlmann, J., 2007. Prosobranch snails as test organisms for the assessment of endocrine active chemicals—an overview and a guideline proposal for a reproduction test with the freshwater mudsnail *Potamopyrgus antipodarum*. Ecotoxicology 16, 169-182.

Dungan Lemko, H.M., Elias, C.F., 2012. Kiss of the mutant mouse: how genetically altered mice advanced our understanding of kisspeptin's role in reproductive physiology. Endocrinology 153, 5119-5129.

Ebert, D., 2011. A genome for the environment. Science 331, 539-540.

Eden, E., Geva-Zatorsky, N., Issaeva, I., Cohen, A., Dekel, E., Danon, T., Cohen, L., Mayo, A., Alon, U., 2011. Proteome half-life dynamics in living human cells. Science 331, 764-768.

Elmer, K.R., Meyer, A., 2011. Adaptation in the age of ecological genomics: insights from parallelism and convergence. Trends Ecol. Evol. (Amst.) 26, 298-306.

Ercan, S., Lieb, J.D., 2009. *C. elegans* dosage compensation: a window into mechanisms of domain-scale gene regulation. Chromosome Res. 17, 215-227.

Evsikov, A.V., Graber, J.H., Brockman, J.M., Hampl, A., Holbrook, A.E., Singh, P., Eppig, J.J., Solter, D., Knowles, B.B., 2006. Cracking the egg: molecular dynamics and evolutionary aspects of the transition from the fully grown oocyte to embryo. Genes Dev. 20, 2713-2727.

Faith, J.J., Hayete, B., Thaden, J.T., Mogno, I., Wierzbowski, J., Cottarel, G., Kasif, S., Collins, J.J., Gardner, T.S., 2007. Large-scale mapping and validation of Escherichia coli transcriptional regulation from a compendium of expression profiles. PLoS Biol. 5, e8.

Ferraresso, S., Milan, M., Pellizzari, C., Vitulo, N., Reinhardt, R., Canario, A.V.M., Patarnello, T., Bargelloni, L., 2010. Development of an oligo DNA microarray for the European sea bass and its application to expression profiling of jaw deformity. BMC Genomics 11, 354.

Figueiredo, M.A., Fernandes, R.V., Studzinski, A.L., Rosa, C.E., Corcini, C.D., Junior, A.S.V., Marins, L.F., 2013. GH overexpression decreases spermatic parameters and reproductive success in two-years-old transgenic zebrafish males. Anim. Reprod. Sci. 139, 162-167.

Foltenyi, K., Greenspan, R.J., Newport, J.W., 2007. Activation of EGFR and ERK by rhomboid signaling regulates the consolidation and maintenance of sleep in Drosophila. Nat. Neurosci. 10, 1160-1167.

Forné, I., Castellana, B., Marín-Juez, R., Cerdà, J., Abián, J., Planas, J.V., 2011. Transcriptional and proteomic profiling of flatfish (*Solea senegalensis*) spermatogenesis. Proteomics 11, 2195-2211.

Franco, H.L., Yao, H.H., 2012. Sex and hedgehog: roles of genes in the hedgehog signaling pathway in mammalian sexual differentiation. Chromosome Res. 20, 247-258.

Fujiyama, A., Watanabe, H., Toyoda, A., Taylor, T.D., Itoh, T., Tsai, S.-F., Park, H.-S., Yaspo, M.-L., Lehrach, H., Chen, Z., Fu, G., Saitou, N., Osoegawa, K., de Jong, P.J., Suto, Hattori, M., Sakaki, Y., 2002. Construction and analysis of a human chimpanzee comparative clone map. Science 295, 131-134.

Fukada, H., Dickey, J.T., Pierce, A.L., Hodges, N., Hara, A., Swanson, P., Dickhoff, W.W., 2003. Gene expression levels of growth hormone receptor and insulin-like growth factor-I in gonads of maturing coho salmon (*Oncorhynchus kisutch*). Fish Physiol. Biochem. 28, 335-336.

Fullston, T., Ohlsson Teague, E.M.C., Palmer, N.O., DeBlasio, M.J., Mitchell, M., Corbett, M., Print, C.G., Owens, J.A., Lane, M., 2013. Paternal obesity initiates metabolic disturbances in two generations of mice with incomplete penetrance to the F2 generation and alters the transcriptional profile of testis and sperm microRNA content. FASEB J. 27, 4226-4243.

Garcia, T.I., Shen, Y., Crawford, D., Oleksiak, M.F., Whitehead, A., Walter, R.B., 2012. RNA-Seq reveals complex genetic response to Deepwater Horizon oil release in *Fundulus grandis*. BMC Genomics 13, 474.

Garcia-Reyero, N., Ekman, D.R., Habib, T., Villeneuve, D.L., Collette, T.W., Bencic, D., Ankley, G.T., Perkins, E.J., 2014. Integrated approach to explore the mechanisms of aromatase inhibition and recovery in fathead minnows (*Pimephales promelas*). Gen. Comp. Endocrinol. 203, 182-191.

Garcia-Reyero, N., Escalon, B.L., Loh, P.R., Laird, J.G., Kennedy, A.J., Berger, B., Perkins, E.J., 2012. Assessment of chemical mixtures and groundwater effects on Daphnia magna transcriptomics. Environ. Sci. Technol. 46, 42-50.

Garcia-Reyero, N., Griffitt, R.J., Liu, L., Kroll, K.J., Farmerie, W.G., Barber, D.S., Denslow, N.D., 2008.Construction of a robust microarray from a non-model species (*largemouth bass*) using pyrosequencing technology. J. Fish Biol. 72, 2354-2376.

Garcia-Reyero, N., Habib, T., Pirooznia, M., Gust, K.A., Gong, P., Warner, C., Wilbanks, M., Perkins, E.J., 2011. Conserved toxic responses across divergent phylogenetic lineages: a meta-analysis of the neurotoxic effects of RDX among multiple species using toxicogenomics. Ecotoxicology 20, 580-594.

Garcia-Reyero, N., Kroll, K.J., Liu, L., Orlando, E.F., Watanabe, K.H., Sepulveda, M.S., Villeneuve, D.L., Perkins, E.J., Ankley, G.T., Denslow, N.D., 2009a. Gene expression responses in male fathead minnows exposed to binary mixtures of an estrogen and antiestrogen. BMC Genomics 10, 308.

Garcia-Reyero, N., Martyniuk, C.J., Kroll, K.J., Escalon, B.L., Spade, D.J., Denslow, N.D., 2013. Transcriptional signature of progesterone in the fathead minnow ovary (*Pimephales promelas*). Gen. Comp. Endocrinol. 192, 159-169.

Garcia-Reyero, N., Perkins, E.J., 2011. Systems biology: leading the revolution in ecotoxicology. Environ. Toxicol. Chem. 30, 265-273.

Garcia-Reyero, N., Poynton, H.C., Kennedy, A.J., Guan, X., Escalon, B.L., Chang, B., Varshavsky, J., Loguinov, A.V., Vulpe, C.D., Perkins, E.J., 2009b. Biomarker discovery and transcriptomic responses in *Daphnia magna* exposed to munitions constituents. Environ. Sci. Technol. 43, 4188-4193.

Garcia-Reyero, N., Villeneuve, D.L., Kroll, K.J., Liu, L., Orlando, E.F., Watanabe, K.H., Sepulveda, M.S., Ankley, G.T., Denslow, N.D., 2009c. Expression signatures for a model androgen and antiandrogen in the fathead minnow (*Pimephales promelas*) ovary. Environ. Sci. Technol. 43, 2614-2619.

Ghosh, S., Matsuoka, Y., Asai, Y., Hsin, K.-Y., Kitano, H., 2011. Software for systems biology: from tools to integrated platforms. Nat. Publishing Group 12, 821-832.

Gilbert, L.I., 2008. Drosophila is an inclusive model for human diseases, growth and development. Mol. Cell. Endocrinol. 293, 25-31.

Gorr, T.A., Gassmann, M., Wappner, P., 2006. Sensing and responding to hypoxia via HIF in model invertebrates. J. Insect Physiol. 52, 349-364.

Gracey, A.Y., Fraser, E.J., Li, W., Fang, Y., Taylor, R.R., Rogers, J., Brass, A., Cossins, A.R., 2004. Coping with cold: an integrative, multitissue analysis of the transcriptome of a poikilothermic vertebrate. Proc. Natl. Acad. Sci. U.S.A. 101, 16970-16975.

Gupta, S.C., Siddique, H.R., Mathur, N., Mishra, R.K., Mitra, K., Saxena, D.K., Chowdhuri, D.K., 2007. Adverse effect of organophosphate compounds, dichlorvos and chlorpyrifos in the reproductive tissues of transgenic *Drosophila melanogaster*: 70 kDa heat shock protein as a marker of cellular damage. Toxicology 238, 1-14.

Henkel, C.V., Dirks, R.P., de Wijze, D.L., Minegishi, Y., Aoyama, J., Jansen, H.J., Turner, B., Knudsen, B., Bundgaard, M., Hvam, K.L., Boetzer, M., Pirovano, W., Weltzien, F.-A., Dufour, S., Tsukamoto, K., Spaink, H.P., van den Thillart, G.E.E.J.M., 2012. First draft genome sequence of the Japanese eel, *Anguilla japonica*. Gene 511, 195-201.

Hirakawa, I., Miyagawa, S., Mitsui, N., Miyahara, M., Onishi, Y., Kagami, Y., Kusano, T., Takeuchi, T., Ohta, Y., Iguchi, T., 2013. Developmental disorders and altered gene expression in the tropical clawed frog (*Silurana tropicalis*) exposed to 17α-ethinylestradiol. J. Appl. Toxicol. 33, 1001-1010.

Holland, L.Z., Gibson-Brown, J.J., 2003. The Ciona intestinalis genome: when the constraints are off. Bioessays 25, 529-532.

Hood, L., Perlmutter, R.M., 2004. The impact of systems approaches on biological problems in drug discovery. Nat. Biotechnol. 22, 1215-1217.

Howe, K., Clark, M.D., Torroja, C.F., Torrance, J., Berthelot, C., Muffato, M., Collins, J.E., Humphray, S., McLaren, K., Matthews, L., McLaren, S., Sealy, I., Caccamo, M., Churcher, C., Scott, C., Barrett, J.C., Koch, R., Rauch, G.-J., White, S., Chow, W., Kilian, B., Quintais, L.T., Guerra-Assunção, J.A., Zhou, Y., Gu, Y., Yen, J., Vogel, J.H., Eyre, T., Redmond, S., Banerjee, R., Chi, J., Fu, B., Langley, E., Maguire, S.F., Laird, G.K., Lloyd, D., Kenyon, E., Donaldson, S., Sehra, H., Almeida-King, J., Loveland, J., Trevanion, S., Jones, M., Quail, M., Willey, D., Hunt, A., Burton, J., Sims, S., McLay, K., Plumb, B., Davis, J., Clee, C., Oliver, K., Clark, R., Riddle, C., Eliott, D., Threadgold, G., Harden, G., Ware, D., Mortimer, B., Kerry, G., Heath, P., Phillimore, B., Tracey, A., Corby, N., Dunn, M., Johnson, C., Wood, J., Clark, S., Pelan, S., Griffiths, G., Smith, M., Glithero, R., Howden, P., Barker, N., Stevens, C., Harley, J., Holt, K., Panagiotidis, G., Lovell, J., Beasley, H., Henderson, C., Gordon, D., Auger, K., Wright, D., Collins, J., Raisen, C., Dyer, L., Leung, K., Robertson, L., Ambridge, K., Leongamornlert, D., McGuire, S., Gilderthorp, R., Griffiths, C., Manthravadi, D., Nichol, S., Barker, G., Whitehead, S., Kay, M., Brown, J., Murnane, C., Gray, E., Humphries, M., Sycamore, N., Barker, D., Saunders, D., Wallis, J., Babbage, A., Hammond, S., Mashreghi-Mohammadi, M., Barr, L., Martin, S., Wray, P., Ellington, A., Matthews, N., Ellwood, M., Woodmansey, R., Clark, G., Cooper, J., Tromans, A., Grafham, D., Skuce, C., Pandian, R., Andrews, R., Harrison, E., Kimberley, A., Garnett, J., Fosker, N., Hall, R., Garner, P., Kelly, D., Bird, C., Palmer, S., Gehring, I., Berger, A., Dooley, C.M., Ersan-Ürün, Z., Eser, C., Geiger, H., Geisler, M., Karotki, L., Kirn, A., Konantz, J., Konantz, M., Oberländer, M., Rudolph-Geiger, S., Teucke, M., Osoegawa, K., Zhu, B., Rapp, A., Widaa, S., Langford, C., Yang, F., Carter, N.P., Harrow, J., Ning, Z., Herrero, J., Searle, S.M.J., Enright, A., Geisler, R., Plasterk, R.H.A., Lee, C., Westerfield, M., de Jong, P.J., Zon, L.I., Postlethwait, J.H., Nüsslein-Volhard, C., Hubbard, T.J.P., Roest Crollius, H., Rogers, J., Stemple, D.L., 2013. The zebrafish reference genome sequence and its relationship to the human genome. Nature 496, 498-503.

Huber, R., Smith, K., Delago, A., Isaksson, K., Kravitz, E.A., 1997. Serotonin and aggressive motivation in crustaceans: altering the decision to retreat. Proc. Natl. Acad. Sci. U.S.A. 94, 5939-5942.

Inlow, J.K., Restifo, L.L., 2004. Molecular and comparative genetics of mental retardation. Genetics 166, 835-881.

Isbister, C.M., O'Connor, T.P., 2000. Mechanisms of growth cone guidance and motility in the developing grasshopper embryo. J. Neurobiol. 44, 271-280.

Jasinska, A.J., Freimer, N.B., 2009. The complex genetic basis of simple behavior. J. Biol. 8, 71.

Johnson, J.R., Jenn, R.C., Barclay, J.W., Burgoyne, R.D., Morgan, A., 2010. *Caenorhabditis elegans*: a useful tool to decipher neurodegenerative pathways. Biochem. Soc. Trans. 38, 559-563.

Kajimura, S., Kawaguchi, N., Kaneko, T., Kawazoe, I., Hirano, T., Visitacion, N., Grau, E.G., Aida, K., 2004. Identification of the growth hormone receptor in an advanced teleost, the tilapia (*Oreochromis mossambicus*) with special reference to its distinct expression pattern in the ovary. J. Endocrinol. 181, 65-76.

Kang, S., Kong, X., Rosen, E.D., 2014. Adipocyte-specific transgenic and knockout models.Methods Enzymol. 537, 1-16.

Kato, Y., Kobayashi, K., Watanabe, H., Iguchi, T., 2011. Environmental sex determination in the branchiopod crustacean *Daphnia magna*: deep conservation of a Doublesex gene in the sex-determining pathway. PLoS Genet. 7, e1001345.

Kocabas, A.M., Crosby, J., Ross, P.J., Otu, H.H., Beyhan, Z., Can, H., Tam, W.-L., Rosa, G.J.M., Halgren, R.G., Lim, B., Fernandez, E., Cibelli, J.B., 2006. The transcriptome of human oocytes. Proc. Natl. Acad. Sci. U.S.A. 103, 14027-14032.

Koide, T., Pang, W.L., Baliga, N.S., 2009. The role of predictive modelling in rationally re-engineering biological systems. Nat Rev Microbiol. 7, 297-305.

Konsolaki, M., 2013. Fruitful research: drug target discovery for neurodegenerative diseases in Drosophila. Expert Opin. Drug Discov. 8, 1503-1513.

Kravitz, E.A., Huber, R., 2003. Aggression in invertebrates. Curr. Opin. Neurobiol. 13, 736-743.

Kuhl, H., Beck, A., Wozniak, G., Canario, A.V.M., Volckaert, F.A.M., Reinhardt, R., 2010. The European sea bass *Dicentrarchus labrax* genome puzzle: comparative BAC-mapping and low coverage shotgun sequencing. BMC Genomics 11, 68.

Lanes, C.F.C., Bizuayehu, T.T., de Oliveira Fernandes, J.M., Kiron, V., Babiak, I., 2013. Transcriptome of Atlantic Cod (*Gadus morhua* L.) early embryos from farmed and wild broodstocks. Mar. Biotechnol. 15, 677-694.

Le Gac, F., Blaise, O., Fostier, A., Le Bail, P.-Y., Loir, M., Mourot, B., Weil, C., 1993. Growth hormone (GH) and reproduction: a review. Fish Physiol. Biochem. 11, 219-232.

LeBlanc, G.A., McLachlan, J.B., 2000. Changes in the metabolic elimination profile of testosterone following exposure of the crustacean *Daphnia magna* to tributyltin. Ecotoxicol. Environ. Saf. 45, 296-303.

Lefebvre, C., Rieckhof, G., Califano, A., 2012. Reverse-engineering human regulatory networks. Wiley Interdiscip. Rev. Syst. Biol. Med. 4, 311-325.

Li, J., Liu, Z., Wang, D., Cheng, C.H.K., 2011a. Insulin-like growth factor 3 is involved in oocyte maturation in zebrafish. Biol. Reprod. 84, 476-486.

Li, Z., Kroll, K.J., Jensen, K.M., Villeneuve, D.L., Ankley, G.T., Brian, J.V., Sepulveda, M.S., Orlando, E.F., Lazorchak, J.M., Kostich, M., Armstrong, B., Denslow, N.D., Watanabe, K.H., 2011b. A computational model of the hypothalamic-pituitary-gonadal axis in female fathead minnows (*Pimephales promelas*) exposed to 17α-ethynylestradiol and 17β-trenbolone. BMC Syst. Biol. 5, 63.

Lieschke, G.J., Currie, P.D., 2007. Animal models of human disease: zebrafish swim into view. Nat. Rev. Genet. 8, 353-367.

Loir, M., 1999. Spermatogonia of rainbow trout: II. *in vitro* study of the influence of pituitary hormones, growth factors and steroids on mitotic activity. Mol. Reprod. Dev. 53, 434-442.

Magill, J.C., Ciccone, N.A., Kaiser, U.B., 2013. A mathematical model of pulse-coded hormone signal responses in pituitary gonadotroph cells. Math. Biosci. 246, 38-46.

Mallo, M., Alonso, C.R., 2013. The regulation of Hox gene expression during animal development. Development 140, 3951-3963.

Mani, K.M., Lefebvre, C., Wang, K., Lim, W.K., Basso, K., Dalla-Favera, R., Califano, A., 2008. A systems biology approach to prediction of oncogenes and molecular perturbation targets in B-cell lymphomas. Mol. Syst. Biol. 4, 169.

Maqbool, T., Jagla, K., 2007. Genetic control of muscle development: learning from Drosophila. J. Muscle Res. Cell Motil. 28, 397-407.

Marbach, D., Costello, J.C., Küffner, R., Vega, N.M., Prill, R.J., Camacho, D.M., Allison, K.R., Aderhold, A., Allison, K.R., Bonneau, R., Camacho, D.M., Chen, Y., Collins, J.J., Cordero, F., Costello, J.C., Crane, M., Dondelinger, F., Drton, M., Esposito, R., Foygel, R., de la Fuente, A., Gertheiss, J., Geurts, P., Greenfield, A., Grzegorczyk, M., Haury, A.-C., Holmes, B., Hothorn, T., Husmeier, D., Huynh-Thu, V.A., Irrthum, A., Kellis, M., Karlebach, G., Küffner, R., Lèbre, S., De Leo, V., Madar, A., Mani, S., Marbach, D., Mordelet, F., Ostrer, H., Ouyang, Z., Pandya, R., Petri, T., Pinna, A., Poultney, C.S., Prill, R.J., Rezny, S., Ruskin, H.J., Saeys, Y., Shamir, R., Sîrbu, A., Song, M., Soranzo, N., Statnikov, A., Stolovitzky, G., Vega, N., VeraLicona, P., Vert, J.-P., Visconti, A., Wang, H., Wehenkel, L., Windhager, L., Zhang, Y., Zimmer, R., Kellis, M., Collins, J.J., Stolovitzky, G., 2012. Wisdom of crowds for robust gene network inference. Nat. Methods 9, 796-804.

Marín-Juez, R., Castellana, B., Manchado, M., Planas, J.V., 2011. Molecular identification of genes involved in testicular steroid synthesis and characterization of the response to gonadotropic stimulation in the Senegalese sole (*Solea senegalensis*) testis. Gen. Comp. Endocrinol. 172, 130-139.

Martyniuk, C.J., Alvarez, S., McClung, S., Villeneuve, D.L., Ankley, G.T., Denslow, N.D., 2009. Quantitative proteomic profiles of androgen receptor signaling in the liver of fathead minnows (*Pimephales promelas*). J. Proteome Res. 8, 2186-2200.

Martyniuk, C.J., Gerrie, E.R., Popesku, J.T., Ekker, M., Trudeau, V.L., 2007. Microarray analysis in the zebrafish (*Danio rerio*) liver and telencephalon after exposure to low concentration of 17alpha-ethinylestradiol. Aquat. Toxicol. 84, 38-49.

Martyniuk, C.J., Xiong, H., Crump, K., Chiu, S., Sardana, R., Nadler, A., Gerrie, E.R., Xia, X., Trudeau, V.L., 2006. Gene expression profiling in the neuroendocrine brain of male goldfish (*Carassius auratus*) exposed to 17-ethinylestradiol. Physiol. Genomics 27, 328-336.

McGonnell, I.M., 2006. Fishing for gene function-endocrine modelling in the zebrafish. J. Endocrinol. 189, 425-439.

Melamed, P., Gur, G., Rosenfeld, H., Elizur, A., Yaron, Z., 1999. Possible interactions between gonadotrophs and somatotrophs in the pituitary of tilapia: apparent roles for insulin-like growth factor I and estradiol. Endocrinology 140, 1183-1191.

Mendjan, S., Akhtar, A., 2007. The right dose for every sex. Chromosoma 116, 95-106.

Miao, X., Luo, Q., 2013. Genome-wide transcriptome analysis between small-tail Han sheep and the Surabaya fur sheep using high-throughput RNA sequencing. Reproduction 145, 587-596.

Miller, W.L., Bose, H.S., 2011. Early steps in steroidogenesis: intracellular cholesterol trafficking. J. Lipid Res. 52, 2111-2135.

Mitra, K., Carvunis, A.-R., Ramesh, S.K., Ideker, T., 2013. Integrative approaches for finding modular structure in biological networks. Nat. Rev. Genet. 14, 719-732.

Mommens, M., Fernandes, J.M., Bizuayehu, T.T., Bolla, S.L., Johnston, I.A., Babiak, I., 2010. Maternal gene expression in *Atlantic halibut* (*Hippoglossus hippoglossus* L.) and its relation to egg quality. BMC Res. Notes 3, 138.

Mouton, S., Willems, M., Braeckman, B.P., Egger, B., Ladurner, P., Schärer, L., Borgonie, G., 2009. The free-living flatworm *Macrostomum lignano*: a new model organism for ageing research. Exp. Gerontol. 44, 243-249.

Mukherjee, D., Mukherjee, D., Sen, U., Paul, S., Bhattacharyya, S.P., 2006. *In vitro* effects of insulin-like growth factors and insulin on oocyte maturation and maturation-inducing steroid production in ovarian follicles of common carp, *Cyprinus carpio*. Comp. Biochem. Physiol. A 144, 63-77.

Munger, S.C., Natarajan, A., Looger, L.L., Ohler, U., Capel, B., 2013. Fine time course expression analysis identifies cascades of activation and repression and maps a putative regulator of mammalian sex determination. PLoS Genet. 9, e1003630.

Ng, S.H., Artieri, C.G., Bosdet, I.E., Chiu, R., Danzmann, R.G., Davidson, W.S., Ferguson, M.M., Fjell, C.D., Hoyheim, B., Jones, S.J., de Jong, P.J., Koop, B.F., Krzywinski, M.I., Lubieniecki, K., Marra, M.A., Mitchell, L.A., Mathewson, C., Osoegawa, K., Parisotto, S.E., Phillips, R.B., Rise, M.L., von Schalburg, K.R., Schein, J.E., Shin, H., Siddiqui, A., Thorsen, J., Wye, N., Yang, G., Zhu, B., 2005. A physical map of the genome of Atlantic salmon, *Salmo salar*. Genomics. 86, 396-404.

Nilsen, S.P., Chan, Y.-B., Huber, R., Kravitz, E.A., 2004. Gender-selective patterns of aggressive behavior in *Drosophila melanogaster*. Proc. Natl. Acad. Sci. U.S.A. 101, 12342-12347.

Nilsson, E., Zhang, Bin., Skinner, M.K., 2013. Gene bionetworks that regulate ovarian primordial follicle assembly.BMC Genomics 14 (1-1).

Norris, D.O., 2006. Vertebrate Endocrinology.Elsevier academic press.

Palti, Y., Luo, M.-C., Hu, Y., Genet, C., You, F.M., Vallejo, R.L., Thorgaard, G.H., Wheeler, P.A., Rexroad, C.E., 2009.A first generation BAC-based physical map of the rainbow trout genome. BMC Genomics 10, 462.

Perkins, E.J., Ankley, G.T., Crofton, K.M., Garcia-Reyero, N., Lalone, C.A., Johnson, M.S., Tietge, J.E., Villeneuve, D.L., 2013. Current perspectives on the use of alternative species in human health and ecological hazard assessments. Environ. Health Perspect. 121, 1002-1010.

Perkins, E.J., Chipman, J.K., Edwards, S., Habib, T., Falciani, F., Taylor, R., Van Aggelen, G., Vulpe, C., Antczak, P.,

Loguinov, A., 2011. Reverse engineering adverse outcome pathways. Environ. Toxicol. Chem. 30, 22-38.

Perrot, Funkenstein, B., 1999. Cellular distribution of insulin-like growth factor II (IGF-II) mRNA and hormonal regulation of IGF-I and IGF-II mRNA expression in rainbow trout testis (*Oncorhynchus mykiss*). Fish Physiol. Biochem. 20, 219-229.

Peter, I.S., Davidson, E.H., 2009. Modularity and design principles in the sea urchin embryo gene regulatory network. FEBS Lett. 583, 3948-3958.

Peter, I.S., Davidson, E.H., 2011. A gene regulatory network controlling the embryonic specification of endoderm. Nature 474, 635-639.

Peter, I.S., Faure, E., Davidson, E.H., 2012. Predictive computation of genomic logic processing functions in embryonic development. Proc. Natl. Acad. Sci. U.S.A. 109, 16434-16442.

Piret, S.E., Thakker, R.V., 2011. Mouse models for inherited endocrine and metabolic disorders. J. Endocrinol. 211, 211-230.

Piras, V., Tomita, M., Selvarajoo, K., 2012. Is central dogma a global property of cellular information flow? Front Physiol. 3, 439.

Popesku, J.T., Martyniuk, C.J., Denslow, N.D., Trudeau, V.L., 2010. Rapid dopaminergic modulation of the fish hypothalamic transcriptome and proteome. PLoS One 5, e12338.

Popesku, J.T., Martyniuk, C.J., Trudeau, V.L., 2012. Meta-type analysis of dopaminergic effects on gene expression in the neuroendocrine brain of female goldfish. Front. Endocrinol. (Lausanne) 3, 130.

Quan, X.-J., Hassan, B.A., 2005. From skin to nerve: flies, vertebrates and the first helix. Cell. Mol. Life Sci. 62, 2036-2049.

Rahman, M.A., Maclean, N., 1999. Growth performance of transgenic tilapia containing an exogenous piscine growth hormone gene. Aquaculture 173, 333-346.

Rawat, A., Gust, K.A., Deng, Y., Garcia-Reyero, N., Quinn, M.J., Johnson, M.S., Indest, K.J., Elasri, M.O., Perkins, E.J., 2010. From raw materials to validated system: the construction of a genomic library and microarray to interpret systemic perturbations in Northern bobwhite. Physiol. Genomics 42, 219-235.

Reading, B.J., Williams, V.N., Chapman, R.W., Williams, T.I., Sullivan, C.V., 2013. Dynamics of the striped bass (*Morone saxatilis*) ovary proteome reveal a complex network of the translasome. J. Proteome Res.

Read, R., 2011. *Drosophila melanogaster* as a model system for human brain cancers. Glia 59, 13.

Romero, N.M., Dekanty, A., Wappner, P., 2007. Cellular and developmental adaptations to hypoxia: a Drosophila perspective. Methods Enzymol. 435, 123-144.

Santella, L., Puppo, A., Chun, J.T., 2008. The role of the actin cytoskeleton in calcium signaling in starfish oocytes. Int. J. Dev. Biol. 52, 571-584.

Santos, E.M., Paull, G.C., Van Look, K.J.W., Workman, V.L., Holt, W.V., van Aerle, R., Kille, P., Tyler, C.R., 2007a. Gonadal transcriptome responses and physiological consequences of exposure to oestrogen in breeding zebrafish (*Danio rerio*). Aquat. Toxicol. 83, 134-142.

Santos, E.M., Workman, V.L., Paull, G.C., Filby, A.L., Van Look, K.J.W., Kille, P., Tyler, C.R., 2007b. Molecular basis of sex and reproductive status in breeding zebrafish. Physiol. Genomics 30, 111-122.

Sarropoulou, E., Fernandes, J.M.O., 2011. Comparative genomics in teleost species: knowledge transfer by linking the genomes of model and non-model fish species. Comp. Biochem. Physiol. D 6, 92-102.

Schwanhäusser, B., Busse, D., Li, N., Dittmar, G., Schuchhardt, J., Wolf, J., Chen, W., Selbach, M., 2011. Global quantification of mammalian gene expression control. Nature. 473, 337-342.

Sharan, R., Ideker, T., Kelley, B., Shamir, R., Karp, R.M., 2005. Identification of protein complexes by comparative analysis of yeast and bacterial protein interaction data. J. Comput. Biol. 12, 835-846.

Sharova, L.V., Sharov, A.A., Nedorezov, T., Piao, Y., Shaik, N., Ko, M.S.H., 2009. Database for mRNA half-life of 19 977 genes obtained by DNA microarray analysis of pluripotent and differentiating mouse embryonic stem cells. DNA Res. 16, 45-58.

Sherwood, N.M., Tello, J.A., Roch, G.J., 2006. Neuroendocrinology of protochordates: insights from Ciona genomics. Comp. Biochem. Physiol. A 144, 254-271.

Shoemaker, J.E., Gayen, K., Garcia-Reyero, N., Perkins, E.J., Villeneuve, D.L., Liu, L., Doyle, F.J., 2010. Fathead minnow steroidogenesis: in silico analyses reveals tradeoffs between nominal target efficacy and robustness to cross-talk.

BMC Syst. Biol. 4, 89.

Sperduti, A., Crivellaro, F., Rossi, P.F., Bondioli, L., 2012. ''Do Octopuses have a brain?'' knowledge, perceptions and attitudes towards neuroscience at school. PLoS One 7, e47943.

Stanley, J.K., Perkins, E.J., Habib, T., Sims, J.G., Chappell, P., Escalon, B.L., Wilbanks, M., Garcia-Reyero, N., 2013. The good, the bad, and the toxic: approaching hormesis in *Daphnia magna* exposed to an energetic compound. Environ. Sci. Technol. 47, 9424-9433.

Star,B., Nederbragt, A.J., Jentoft, S., Grimholt, U., Malmstrøm, M., Gregers, T.F., Rounge, T.B., Paulsen, J., Solbakken, M.H., Sharma, A., Wetten, O.F., Lanzén, A., Winer, R., Knight, J., Vogel, J.-H., Aken, B., Andersen, O., Lagesen, K., Tooming-Klunderud, A., Edvardsen, R.B., Tina, K.G., Espelund, M., Nepal, C., Previti, C., Karlsen, B.O., Moum, T., Skage, M., Berg, P.R., Gjøen, T., Kuhl, H., Thorsen, J., Malde, K., Reinhardt, R., Du, L., Johansen, S.D., Searle, S., Lien, S., Nilsen, F., Jonassen, I., Omholt, S.W., Stenseth, N.C., Jakobsen, K.S., 2011. The genome sequence of Atlantic cod reveals a unique immune system. Nature 477, 207-210.

Stevens, A., De Leonibus, C., Hanson, D., Dowsey, A.W., Whatmore, A., Meyer, S., Donn, R.P., Chatelain, P., Banerjee, I., Cosgrove, K.E., Clayton, P.E., Dunne, M.J., 2013. Network analysis: a new approach to study endocrine disorders. J. Mol. Endocrinol. 52, R79-R93.

Sylvestre, E.-L., Robert, C., Pennetier, S., Labrecque, R., Gilbert, I., Dufort, I., Léveillé, M.-C., Sirard, M.-A., 2013. Evolutionary conservation of the oocyte transcriptome among vertebrates and its implications for understanding human reproductive function. Mol. Hum. Reprod. 19, 369-379.

Taniguchi, Y., Choi, P.J., Li, G.-W., Chen, H., Babu, M., Hearn, J., Emili, A., Xie, X.S., 2010. Quantifying *E. coli* proteome and transcriptome with single-molecule sensitivity in single cells. Science 329, 533-538.

Tatarazako, N., Oda, S., 2007. The water flea *Daphnia magna* (Crustacea, Cladocera) as a test species for screening and evaluation of chemicals with endocrine disrupting effects on crustaceans. Ecotoxicology 16, 197-203.

Tingaud-Sequeira, A., Chauvigné, F., Lozano, J., Agulleiro, M.J., Asensio, E., Cerdà, J., 2009. New insights into molecular pathways associated with flatfish ovarian development and atresia revealed by transcriptional analysis. BMC Genomics 10, 434.

Tingaud-Sequeira, A., Lozano, J.-J., Zapater, C., Otero, D., Kube, M., Reinhardt, R., Cerdà, J., 2013. A rapid transcriptome response is associated with desiccation resistance in aerially-exposed killifish embryos. PLoS One 8, e64410.

Tuohimaa, P., Wang, J.-H., Khan, S., Kuuslahti, M., Qian, K., Manninen, T., Auvinen, P., Vihinen, M., Lou, Y.-R., 2013. Gene expression profiles in human and mouse primary cells provide new insights into the differential actions of vitamin D3 metabolites. PLoS One 8, e75338.

Vaccari, T., Bilder, D., 2009. At the crossroads of polarity, proliferation and apoptosis: the use of Drosophila to unravel the multifaceted role of endocytosis in tumor suppression. Mol. Oncol. 3, 354-365.

Vogel, C., Abreu Rde, S., Ko, D., Le, S.Y., Shapiro, B.A., Burns, S.C., Sandhu, D., Boutz, D.R., Marcotte, E.M., Penalva, L.O., 2010. Sequence signatures and mRNA concentration can explain two-thirds of protein abundance variation in a human cell line. Mol. Syst. Biol. 6, 400.

Vogel, C., Marcotte, E.M., 2012. Insights into the regulation of protein abundance from proteomic and transcriptomic analyses. Nat. Rev. Genet. 13, 227-232.

Vosshall, L.B., 2007. Into the mind of a fly. Nature 450, 193-197.

Wang, Y., Geer, L.Y., Chappey, C., Kans, J.A., Bryant, S.H., 2000. Cn3D: sequence and structure views for Entrez. Trends Biochem Sci. 25, 300-302.

Wang, D.-S., Jiao, B., Hu, C., Huang, X., Liu, Z., Cheng, C.H.K., 2008. Discovery of a gonad-specific IGF subtype in teleost. Biochem. Biophys. Res. Commun. 367, 336-341.

Watanabe, H., Kobayashi, K., Kato, Y., Oda, S., Abe, R., Tatarazako, N., Iguchi, T., 2008. Transcriptome profiling in crustaceans as a tool for ecotoxicogenomics: *Daphnia magna* DNA microarray. Cell Biol. Toxicol. 24, 641-647.

Watanabe, K.H., Li, Z., Kroll, K.J., Villeneuve, D.L., Garcia-Reyero, N., Orlando, E.F., Sepúlveda, M.S., Collette, T.W., Ekman, D.R., Ankley, G.T., Denslow, N.D., 2009. A computational model of the hypothalamic-pituitary-gonadal axis in male fathead minnows exposed to 17alpha-ethinylestradiol and 17beta-estradiol. Toxicol Sci. 109, 180-192.

Watanabe, H., Takahashi, E., Nakamura, Y., Oda, S., Tatarazako, N., Iguchi, T., 2007. Development of a *Daphnia magna* DNA microarray for evaluating the toxicity of environmental chemicals. Environ. Toxicol. Chem. 26, 669-676.

Weston, A.D., Hood, L., 2004. Systems biology, proteomics, and the future of health care: toward predictive, preventative, and personalized medicine. J. Proteome Res. 3, 179-196.

Wienholds, E., Plasterk, R.H.A., 2005. MicroRNA function in animal development. FEBS Lett. 579, 5911-5922.

Williams, T.D., Turan, N., Diab, A.M., Wu, H., Mackenzie, C., Bartie, K.L., Hrydziuszko, O., Lyons, B.P., Stentiford, G.D., Herbert, J.M., Abraham, J.K., Katsiadaki, I., Leaver, M.J., Taggart, J.B., George, S.G., Viant, M.R., Chipman, K.J., Falciani, F., 2011. Towards a system level understanding of non-model organisms sampled from the environment: a network biology approach. PLoS Comput. Biol. 7, e1002126.

Wilson-Sanders, S.E., 2011. Invertebrate models for biomedical research, testing, and education. ILAR J. 52, 126-152.

Wintz, H., Yoo, L.J., Loguinov, A., Wu, Y.-Y., Steevens, J.A., Holland, R.D., Beger, R.D., Perkins, E.J., Hughes, O., Vulpe, C.D., 2006. Gene expression profiles in fathead minnow exposed to 2,4-DNT: correlation with toxicity in mammals. Toxicol. Sci. 94, 71-82.

Wong, A.O.L., Zhou, H., Jiang, Y., Ko, W.K.W., 2006. Feedback regulation of growth hormone synthesis and secretion in fish and the emerging concept of intrapituitary feedback loop. Comp. Biochem. Physiol. A 144, 284-305.

Xu, H., Lam, S.H., Shen, Y., Gong, Z., 2013. Genome-wide identification of molecular pathways and biomarkers in response to arsenic exposure in zebrafish liver. PLoS One 8, e68737.

Yen, H.-C.S., Xu, Q., Chou, D.M., Zhao, Z., Elledge, S.J., 2008. Global protein stability profiling in mammalian cells. Science 322, 918-923.

Yoshida, H., Kong, Y.Y., Yoshida, R., Elia, A.J., Hakem, A., Hakem, R., Penninger, J.M., Mak, T.W., 1998. Apafl is required for mitochondrial pathways of apoptosis and brain development. Cell 94, 739-750.

Zhang, G., Fang, X., Guo, X., Li, L., Luo, R., Xu, F., Yang, P., Zhang, L., Wang, X., Qi, H., Xiong, Z., Que, H., Xie, Y., Holland, P.W.H., Paps, J., Zhu, Y., Wu, F., Chen, Y., Wang, J., Peng, C., Meng, J., Yang, L., Liu, J., Wen, B., Zhang, N., Huang, Z., Zhu, Q., Feng, Y., Mount, A., Hedgecock, D., Xu, Z., Liu, Y., Domazet-Lošo, T., Du, Y., Sun, X., Zhang, S., Liu, B., Cheng, P., Jiang, X., Li, J., Fan, D., Wang, W., Fu, W., Wang, T., Wang, B., Zhang, J., Peng, Z., Li, Y., Li, N., Wang, J., Chen, M., He, Y., Tan, F., Song, X., Zheng, Q., Huang, R., Yang, H., Du, X., Chen, L., Yang, M., Gaffney, P.M., Wang, S., Luo, L., She, Z., Ming, Y., Huang, W., Zhang, S., Huang, B., Zhang, Y., Qu, T., Ni, P., Miao, G., Wang, J., Wang, Q., Steinberg, C.E.W., Wang, H., Li, N., Qian, L., Zhang, G., Li, Y., Yang, H., Liu, X., Wang, J., Yin, Y., Wang, J., 2012. The oyster genome reveals stress adaptation and complexity of shell formation. Nature, 1-6.

Zhang, Y., Liu, S., Lu, J., Jiang, Y., Gao, X., Ninwichian, P., Li, C., Waldbieser, G., Liu, Z., 2013. Comparative genomic analysis of catfish linkage group 8 reveals two homologous chromosomes in zebrafish and other teleosts with extensive interchromosomal rearrangements. BMC Genomics 14, 387.

Zhou, H., Jiang, Y., Ko, W.K.W., Li, W., Wong, A.O.L., 2005. Paracrine regulation of growth hormone gene expression by gonadotrophin release in grass carp pituitary cells: functional implications, molecular mechanisms and signal transduction. J. Mol. Endocrinol. 34, 415-432.

Zhou, H., Wang, X., Ko, W.K.W., Wong, A.O.L., 2004. Evidence for a novel intrapituitary autocrine/paracrine feedback loop regulating growth hormone synthesis and secretion in grass carp pituitary cells by functional interactions between gonadotrophs and somatotrophs. Endocrinology 145, 5548-5559.

内分泌：通过组学及其相关技术研究进展

Natàlia Garcia-Reyero[1]　Angèle Tingaud-Sequeira[2]　曹梦西[3,4]
朱作言[3]　Edward J. Perkins[5]　胡　炜[3]

1 Institute for Genomics Biocomputing and Biotechnology, Mississippi State University, Starkville, MS 39759, USA
2 Laboratoire MRMG, Maladies Rares: Génétique et Métabolisme, Université de Bordeaux, 33405 Talence Cedex, France
3 中国科学院水生生物研究所，武汉 430072，中国
4 中国科学院大学，北京 100049，中国
5 US Army Engineer Research and Development Center, Vicksburg, MS 39180, USA

摘　要　转录组、蛋白质组、代谢组技术的快速发展，以及对各类组学数据的整合分析方法的发展，为生物学研究带来了革新。本文着重讨论组学技术在科学研究，尤其是在非模式生物的研究中发挥的重要作用。由于基因相互作用的方式及各个信号通路在演化中相对保守，因此可以利用非模式生物（包括脊椎动物和无脊椎动物）来研究生物的内分泌学过程。采用过表达技术，我们可以获得内分泌系统中关键基因过表达的模型来研究复杂的内分泌调控网络，如具有快速生长优势的转生长激素基因鱼，同时可以作为研究生长激素在精巢发育中功能的模型，结果表明生长激素可能通过胰岛素样生长因子来影响精巢发育。组学研究方法可以检测任何物种在不同情况下的基因、蛋白、代谢产物的表达模式，而且可以通过整合不同生物水平的数据，为研究不同物种内分泌功能及关键基因在内分泌不同方面的复杂功能提供新方法。

A Controllable On-Off Strategy for the Reproductive Containment of Fish

Yunsheng Zhang[1,2] Ji Chen[1] Xiaojuan Cui[1,2] Daji Luo[1,3]
Hui Xia[1,2] Jun Dai[1] Zuoyan Zhu[1] Wei Hu[1]

1 State Key Laboratory of Freshwater Ecology and Biotechnology, Institute of Hydrobiology, Chinese Academy of Sciences, No. 7 Donghu South Road, Wuhan 430072
2 University of Chinese Academy of Sciences, Beijing 100049
3 Department of Genetics, School of Basic Medical Sciences, Wuhan University, No. 185 Donghu East Road, Wuhan 430071

Abstract A major impediment to the commercialization and cultivation of transgenic fish is the potential ecological risks they pose to natural environments: a problem that could be solved by the production of sterile transgenic fish. Here, we have developed an on-off reproductive containment strategy for fish that renders the offspring sterile but leaves their parents fertile. TG1 (Tol2-CMV-GFP-pA-CMV-*gal4*-pA-Tol2) and TG2 (Tol2-CMV-RFP-pA-5×UAS-as/*dnd*-pA-Tol2) zebrafish lines were established using a GAL4/UAS system. While the parental lines remained fertile, in the hybrid offspring, GAL4 induced 5×UAS to drive the transcription of antisense *dnd*, which significantly down-regulated endogenous *dnd* expression. This disrupted the migration of primordial germ cells (PGCs), led to their apoptosis, and resulted in few or no PGCs migrating to the genital ridge. This process induced sterility or reduced fertility in adult fish. This on-off strategy is a potentially effective means of generating sterile fish for commercialization while retaining fertility in brood stocks, and offers a novel method to mitigate the ecological risks of fish introductions.

Fish are the last wild food available to humans[1]. However, wild fish stocks are under increasing pressure from overfishing as the global demand for fish increases alongside rapid population growth. Overfishing compromises the long-term sustainability of fisheries resources and results in biodiversity loss, potentially leading to ecosystem collapse[2,3]. Aquaculture is, therefore, considered to be the only long-term sustainable solution to supply our growing demand for fish[4]. The introduction of exotic fish species to the environment is an effective method of sourcing fish with valuable traits for aquaculture, but there is an ecological risk associated with it: the intoduction of the exotic species may be economically viable but can also cause substantial economic and ecological damage should the species become invasive. Transgenic technologies offer another means of producing new fish varieties that exhibit physiologically and commercially desirable traits for aquaculture, and are an important factor in the sustainable development of future aquaculture industries. Since the first rapid-growth

文章发表于 *Scientific Reports*, 2015, 5(5): 73-82

transgenic fish in the world was developed by our group 30 years ago[5], many fish breeds with commercially desirable traits (e.g. rapid growth, cold tolerance, enhanced disease resistance) have been generated using transgenic technologies[6,7]. However, to date, no transgenic fish has been approved for release into a natural environment, or for commercial cultivation as food. The "AquaBounty™" all-fish growth-hormone transgenic salmon (*Oncorhynchus tshawytscha*) is perhaps the most notable example of current attempts to commercialize the cultivation of transgenic fish. This variety was declared safe for consumption by the US Food and Drug Administration in 2010, and is nearing approval for commercialization[8]. The primary impediment to the commercialization of transgenic fish is the concern over their potential ecological risk to natural ecosystems. As a result of their superior viability and competitive ability, the inadvertent release or escape of transgenic fish into natural environments can alter natural community structure. Additionally, transgenic fish may also interbreed with native fish populations resulting in gene introgression to the wild.

A key means of eliminating the potential ecological risk posed by invasive and transgenic fish is fertility control to make individuals infertile[9-15]. At present, however, there are few cost-effective means of controlling reproduction in fish. Generating fish that are triploid rather than diploid is currently the most common strategy used to develop sterile fish. However, in commercially important species, the 'triploidization' rate (achieved via physical or chemical methods) rarely approaches 100%, and varies greatly among species, treatment methods and egg quality[16]. Although up to 99.8% of the all-female transgenic salmon produced by AquaBounty are triploid and sterile, some researchers are still concerned about the non-zero possibility that fertile escapees will be produced[17]. Moreover, in some fish species triploidization is known to produce defects such as impaired growth and reduced disease resistance[16]. For example, compared with diploids, the growth rate of triploid tilalia (*Oreochromis mossambicus*) was slow[18], and triploid rainbow trout and Atlantic salmon were more susceptible to disease[19,20]. The death rate of triploid Atlantic salmon was higher than that of the diploids in adverse conditions such as high temperature and hypoxia[21]. Therefore, developing new strategies for the reproductive containment of fish is the focus of extensive research with promising outcomes both for enabling the commercialization of transgenic fish and for protecting aquatic ecosystems.

Recently, the P0 generation of sterile transgenic common carp (*Cyprinus carpio* L.) and tilapia (*Oreochromis* sp.) were produced, both of which achieve sterility via the expression of antisense gonadotropin-releasing hormone (*GnRH*) RNA to inhibit the expression of endogenous *GnRH*[22-24]. Unfortunately, the sterile trait in these transgenic P0 founders is not heritable. The development of transgenic fish varieties in which sterility can be maintained as a heritable trait is critical for large-scale aquaculture. Wong and Collodi (2013) recently described a method to induce sterilization of zebrafish with heat shock treatment[25]. In this method the overexpression of the Stromal-derived factor 1a gene (*sdf1a*) driven by the heat-shock protein-70 promoter could impair the migration of PGCs during embryogenesis, such that the embryos developed into sterile adults[25]. However, heat shock treatment is not not convenient and practical for large-scale fish production in aquaculture. Hu *et al.*, (2010) established an inducible platform to control the reproduction of transgenic ornamental zebrafish using the *Escherichia coli* nitroreductase (NTR)/metronidazole (Mtz) system[26]. It should be noted that this platform is prodrug-dependent. In their system, an NTR-mediated germ cell ablation occurred only if the transgenic fish were raised in an Mtz solution for at least two weeks[26]. Mtz is an aquatic pollutant and the induced cytotoxic effects are prohibitive in the cultivation of transgenic fish as food. Therefore, a novel strategy to control reproduction in transgenic fish is urgently needed.

Primordial germ cells (PGCs) are the precursors of germ cells and appear early in an organism's development. PGCs migrate to the genital ridge where they differentiate into either sperm or eggs[27]. The *dead end* (*dnd*) gene is essential for the normal migration and survival of PGCs. When *dnd* is knocked down using *dnd* antisense morpholino oligonucleotide (MO), the migration of PGCs is disrupted and apoptosis of PGCs occurs. Zebrafish embryos devoid of *dnd* develop into sterile male adults[28-30]. Therefore, one means of controlling fish reproduction is to inhibit gamete development by disrupting the migration of PGCs via the down-regulation of *dnd* expression.

As a model organism, the zebrafish (*Danio rerio*) has been widely used in transgenic fish research[26, 31-35]. The GAL4/UAS system, in which the yeast transcriptional activator GAL4 can activate transcription of effector transgenes under the control of GAL4-responsive upstream activator sequences (UAS), has been adapted for the precise control of gene expression patterns *in vivo* in zebrafish[36]. In this study, we develop a novel on-off strategy to control fish reproduction, which renders the offspring sterile but ensures that the parental generation remains fertile. We generated two transgenic lines using the GAL4/UAS inducible system; one line expressed the transcriptional activator GAL4, and the other line expressed antisense *dnd* RNA under the control of the 5×UAS effector. Both transgenic lines were fertile. However, in their hybrid offspring, GAL4 induced the 5×UAS-driven antisense *dnd* transcription to knock-down the expression of endogenous *dnd*. This disrupted the migration of PGCs, brought about their apoptosis, and led to the sterility or reduced reproductive capacity of adults. Importantly, this novel method ensures the maintenance of transgenic lines that can serve as brood stocks alongside the production of sterile hybrid offspring that evade the problems of ecological risks that plague the use of transgenic fish.

Results

Endogenous *dnd* expression was inhibited during early embryogenesis in the TG3 transgenic line

A schematic representation of the on-off strategy for controlling reproduction in transgenic fish using the GAL4/UAS inducible system and the *dnd* antisense knock-down technique is shown in Figure 1. The TG1 line expressed the transcriptional activator GAL4, driven by the CMV promoter (Fig. 1a). The TG2 line expressed antisense *dnd* RNA under the control of 5×UAS (Fig. 1b). The expression level of the UAS-regulated gene was related to the number of UAS repeat units[36]. We evaluated the effect of a range of UAS repeat units (from 1 to 11) on *egfp* transgene activation induced by GAL4, and found the expression level of 5×UAS: *egfp* to be strongest (Supplementary Fig. 1). Based on this result, the TG2 line was established with a transplant of 5×UAS: antisense *dnd* to get the strongest expression of antisense *dnd* in the hybrids of TG1 and TG2 lines (which are hereafter referred to as the TG3 line; Fig. 1c).

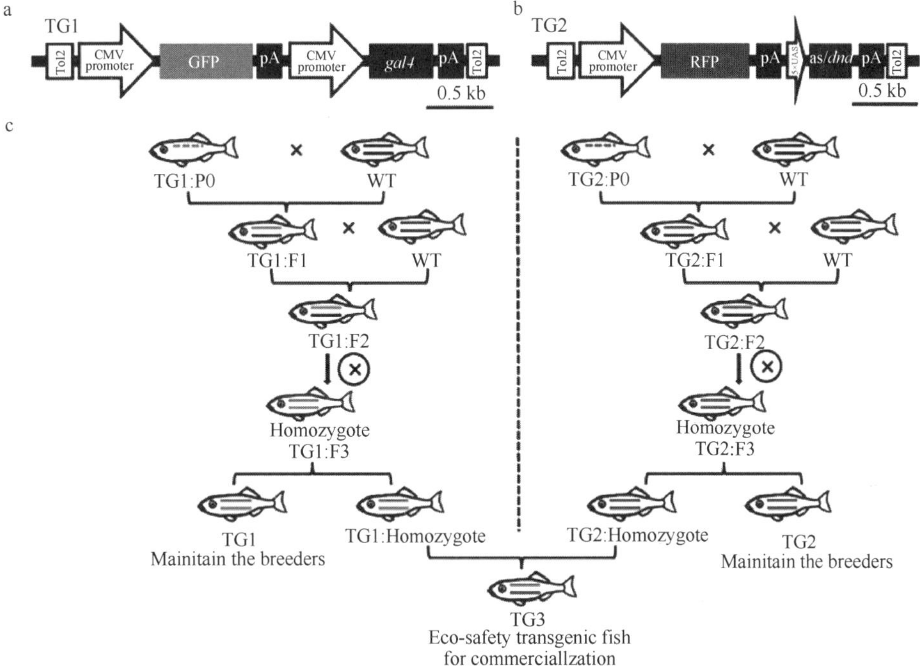

Figure 1 Schematic representation of the on-off strategy to control transgenic fish reproduction. Schematic of expression gene cassette construct for generating (a) the TG1 line and (b) the TG2 line. (c) Flow chart showing how the parental generation maintains their fertility while their hybrid offspring become sterile. This figure was created using Adobe Photoshop CS3 Extended (Adobe Systems, USA) by Yunsheng Zhang

During the development of wild-type (WT) embryos, the expression level of endogenous *dnd*, *nanos1* and *tdrd7* mRNA was initially high then decreased rapidly (Fig. 2a-c). There were no significant differences in the endogenous *dnd* mRNA level between the TG3 and the WT lines at the stages of 1-cell, 1k-cell and oblong (Fig. 2a). Meanwhile, *dnd* mRNA level in the TG3 line was markedly lower than that of the WT line at the stages of 50%-epiboly and 3-somite ($p<0.01$; Fig. 2a). The *dnd* gene encodes an RNA binding protein that can bind to the

3′ untranslated regions of *nanos1* and *tdrd7* mRNAs to protect them from miR-430-mediated repression[37,38]. Hence, the expression level of *nanos1* and *tdrd7* mRNA can mirror the level of the DND protein. Our study showed that the patterns of *nanos1* and *tdrd7* expression change were similar to those of *dnd* during early development, both in the TG3 line and in the WT embryos (Fig. 2b, c). The *nanos1* and *tdrd7* mRNA levels were normal at the 1-cell, 1k-cell and the oblong stage in the TG3 line, but were significantly lower than those of the WT line at the 50%-epiboly stage (P<0.05) and at the 3-somite stage (P<0.01; Fig. 2b,c). These results suggest that the endogenous *dnd* expression could be significantly repressed by GAL4/UAS induced antisense *dnd* in the TG3 line during early embryogenesis.

PGC migration was impaired and the number of PGCs was reduced in TG3 embryos

The *dnd* gene is essential for the migration and survival of PGCs in zebrafish[28]. We labeled PGCs with antisense *vasa* probes to detect the migration and number of PGCs during the embryogenesis. At 4.3 hours post-fertilization (hpf), when compared with the normal distribution of PGCs seen in the majority of WT embryos 87% (141/166), about 72% (94/130) of TG3 embryos showed unusual distribution of some PGCs, with most being located far from the yolk syncytial layer (YSL; Fig. 2d). At 20 hpf and 24 hpf, the number of PGCs which had migrated near the genital ridge varied among TG3 embryos. Moreover, when compared to the WT embryos the number of PGCs in TG3 embryos had visibly reduced in 79% (109/138) of the TG3 embryos at 20 hpf and 67% (127/155) of the TG3 embryos at 24 hpf. Furthermore, there were almost no PGCs remaining in about 14.8% (23/155) of the TG3 embryos at 24 hpf (Fig. 2e).

In order to label the PGCs using green fluorescent protein (GFP) in living embryos, TG4-construction was undertaken by replacing the GFP element of TG1-construction using a red fluorescent protein (RFP) element (Fig. 3a). Then a TG5 line was derived by crossing the TG2 and TG4 lines. To label the PGCs, the 5′end-capped mRNA encoding GFP along with *nanos1* 3′UTR was then injected into the TG5 line and the WT embryos at the 1-cell stage. At 10 hpf under a fluorescence stereoscope, the PGCs could be observed to converge into two clusters that were distributed symmetrically on both sides of the dorso axis in 83% (332/400) of the WT embryos and 32% (128/395) of the TG5 embryos, and were distributed far from the dorso axis in the remaining embryos. At 13 hpf, the PGCs lay close to the dorso axis and strongly arrayed in 86% (344/400) of the WT embryos and 36% (132/369) of the TG5 embryos, but lay far from the dorso axis in the remaining embryos (Fig. 3b). Meanwhile, at 20 hpf and 24 hpf there was a variable number of PGCs among the TG5 embryos, but a constant number of PGCs among the WT embryos. The number of PGCs visibly decreased in 80% (279/347) of the TG5 embryos at 20 hpf and 62% (169/273) of the TG5 embryos at 24 hpf

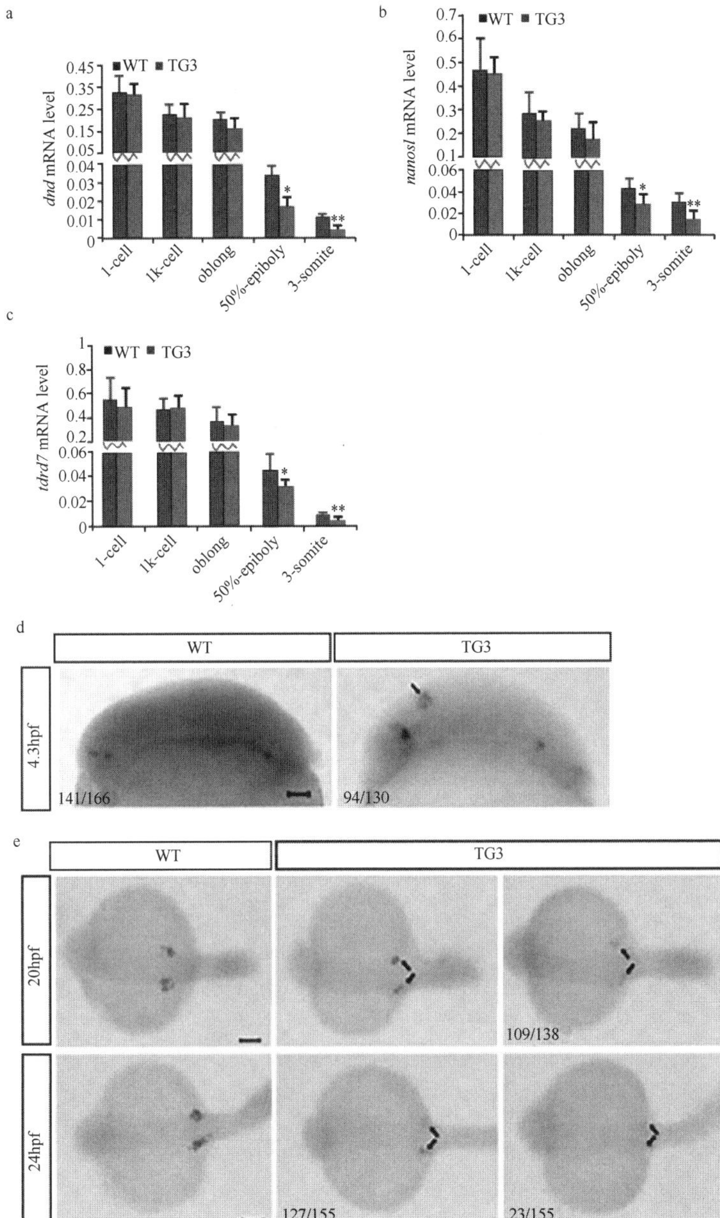

Figure 2 The down-regulation of *dnd* results in the disrupted migration and reduced number of PGCs in the TG3 embryos. The mRNA level of (a) *dnd*, (b) *nanos1*, (c) *tdrd7* in the TG3 and the WT embryos at the 1-cell, 1k-cell, oblong, 50%-epiboly, and 3-somite embryonic stages, validated by quantitative Q-PCR. The $2^{-\Delta Ct}$ values were plotted as the relative mRNA level normalized to *odc1*. Values are represented as the mean ± SEM. of five repeats. **P*<0.05, ***P*<0.01. (d) *Vasa*-positive PGCs were found close to the yolk syncytial layer (YSL) in the WT embryos. At 4.3 hours post-fertilization (hpf) some PGCs (indicated by arrows) were far from the YSL in the TG3 embryos. (e) The number of *vasa*-positive PGCs decreased in the TG3 embryos compared with the WT embryos at 20 hpf and 24 hpf in the dorsal view. (Scale bars: 100 μm)

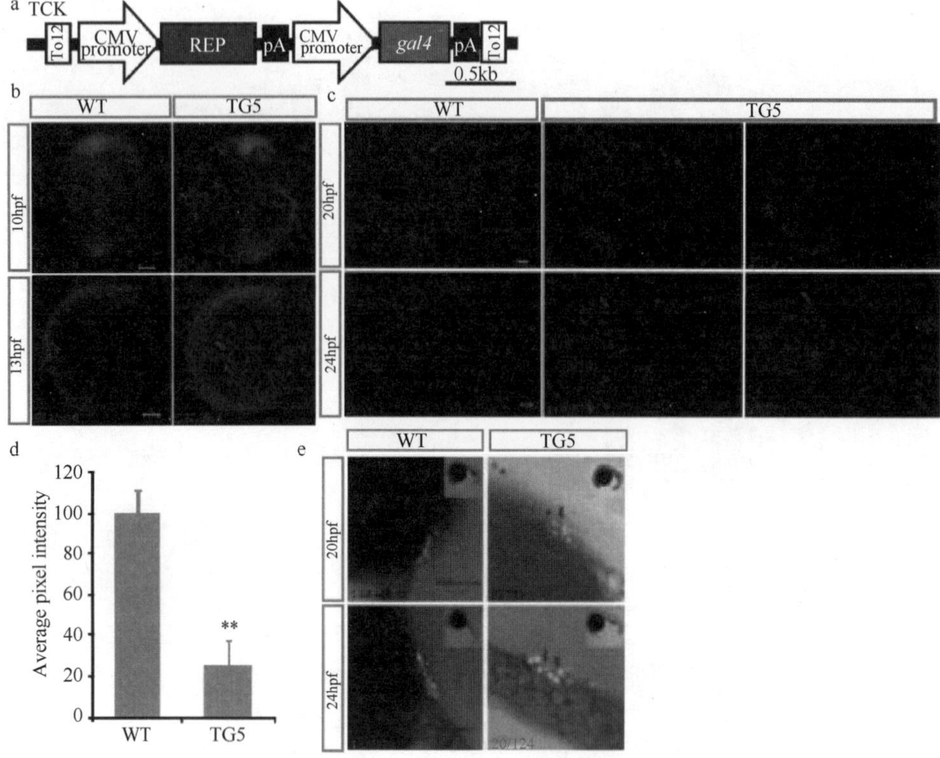

Figure 3 Validation of the migration, number and apoptosis of PGCs in the TG5 and the WT embryos. (a) Schematic of the expression gene cassette of the plasmids used to establish the TG4 line. (b) A part of PGCs labeled by GFP (red arrows) migrated abnormally or lagged behind in the TG5 embryos at 10 hpf and 13 hpf, compared with the WT embryos. (c) The number of PGCs decreased in the TG5 embryos at 20hpf and 24hpf compared with the WT embryos. The red arrow indicates the location of PGCs. (d) WT embryos contained approximately four times the number of PGCs as the TG5 embryos (detected by GFP fluorescence intensity of PGCs at 24 hpf (n=10). (e) Apoptosis of some PGCs was detected (red arrows) within some of the TG5 embryos. (Scale bars: 100 μm)

(Fig. 3b). No PGCs were detected in 17% (47/273) of the TG5 embryos at 24 hpf (Fig. 3c). We randomly selected 10 embryos with PGCs labeled by GFP from the TG5 and the WT lines at 24 hpf, and observed that the GFP fluorescence intensity of the PGCs within the WT embryos was four times that of the PGCs within the TG5 embryos (Fig. 3d).

To confirm whether the PGCs underwent apoptosis, 5′ end-capped mRNA encoding GFP followed by *nanos1* 3′ UTR was injected into the TG5 and the WT embryos at the 1-cell stage. The embryos were then analyzed for terminal deoxynucleotidyl transferase-mediated dUTP-biotinnick end labeling assay (TUNEL assay). The results showed that part of the PGCs underwent apoptosis in 13.4% (19/142) of the TG5 embryos at 20 hpf and 16% (20/124) of the TG5 embryos at 24 hpf, while apoptosis of PGCs occurred in a further 2% (3/143 and 3/126) of the WT embryos at 20 hpf and 24 hpf (Fig. 3e).

TG3 adult fish were sterile or exhibited limited reproductive capacity

The majority of juvenile zebrafish reach sexually mature within three to four months[39]. In this study, we evaluated reproductive capacity at 4.5 months of age as at this age all the WT fish were able to be naturally inseminated. In this experiment, the fertilization rate of WT fish was 89.27%. While all the TG3 males chased the WT females, 30.8% (53/172) failed to fertilize the eggs. About a third (31%; 15/48) of the TG3 females did not spawn after stimulus by the WT males. For convenience, fish that did not reproduce are termed TG3-1, while those that did reproduce are termed TG3-2. Although TG3-2 individuals could reproduce, the fertilization rate of TG3-2 males was significantly lower than that of the WT males (Table 1). Likewise, the relative fecundity of TG3-2 females was significantly lower than that of the WT females (Table 1).

Table 1 Body weight, gonad somatic index (GSI), fertilization and relative fecundity of WT, TG3-1, and TG3-2 fish. **P<0.01 vs. WT; ## P<0.01 vs. TG3-1

	Weight (g)	GSI	Fertilization rate (%)	Relative fecundity
WT (♂, n=10)	0.2886±0.067	1.13±0.13	89.27±5.00	—
TG3-1 (♂, n=10)	0.2681±0.03	0.41±0.04**	0	—
TG3-2 (♂, n=10)	0.2729±0.09	0.65±0.17**##	42.37±19.97**	—
WT (♀, n=10)	0.3674±0.07	10.04±2.74	89.27±5.00	2519±225
TG3-1 (♀, n=10)	0.3661±0.065	5.43±0.43**	No eggs	—
TG3-2 (♀, n=10)	0.3734±0.07	7.38±1.16**##	80.01±12.34	1035±441**

There was no significant difference between the body weight of the TG3-1 and TG3-2 individuals, but the gonad somatic indices (GSI) of TG3-1 and TG3-2 were significantly lower compared to the WT individuals (Table 1). Gonadal histological slices were stained by hematoxylin-eosin (HE). The result showed that in the testes of WT fish the lobular cavities were filled with mature sperm that were aligned in a tight and orderly manner. In the testes of the TG3-1 males, lobular cavities were fewer in number and smaller than in the WT fish, and contained very little or no mature sperm. In TG3-2 individuals, only a portion of the lobular cavities lacked mature sperm.

In the ovaries of WT females, most oocytes were at stage III. However, in the ovaries of TG3-1 females, the majority of oocytes were at stage I or II, with only a few oocytes having developed to stage III. In ovaries of TG3-2 females, meanwhile, most oocytes were at stage III, but were loosely aligned and fewer in number (Fig. 4a). The results of the TUNEL analysis revealed that apoptosis occurred more frequently in the gonads of the TG3-1 and TG3-2 fish than in the WT fish, regardless of sex. The cells undergoing apoptosis were not evenly distributed, and the oocytes undergoing apoptosis were always at stage I or stage II (Fig. 4b).

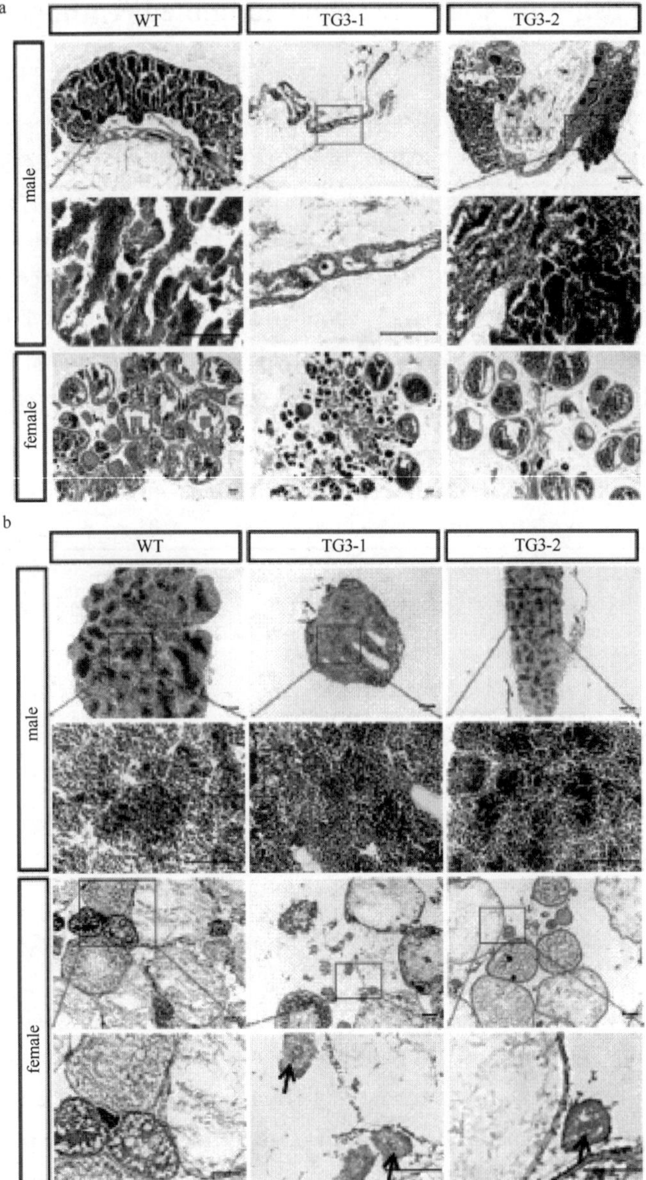

Figure 4 Histology and apoptosis analysis of the gonads of the TG3 adults. (a) Histology of the testes and ovaries from WT, TG3-1 and TG3-2 individuals by HE staining at 4.5 months of age. The lobular cavities in gonad of TG3-1 and TG3-2 males were observed to be in a single-row, were smaller than those of the WT males and contained none or very few mature sperm. When compare to the WT females, most oocytes were at stage I and widely dispersed throughout the gonads of the TG3-1 female fish, and at stage III and loosely arrayed in the gonads of the TG3-2 females. (b) Results of TUNEL staining of the testes and ovaries from the WT, TG3-1 and TG3-2 individuals. Multiple clusters of apoptotic spermatocytes were present in the testes of TG3-1 and TG3-2 males and some apoptotic oocytes were also present in the ovaries of TG3-1 and TG3-2 females. The red dotted line and black arrows indicate germ cells that were undergoing apoptosis. (Scale bars: 100 μm)

The hybrid offspring exhibited methylation of the UAS and the mosaic expression of the UAS-regulated gene

The reproductive capacity of TG3 adults varied from complete sterility (TG3-1) to partial reproductive capacity (TG3-2). To understand the cause of this variation, we established the TG6 transgenic line, in which 5×UAS directly regulated the expression of *egfp* (Fig. 5a). Mosaic expression of *egfp* was observed in the embryos of the hybrid offspring of the TG4 and the TG6 lines, and varied among individuals (Fig. 5b). Mosaic expression of *egfp* was also observed in the testes and the ovaries of the hybrids, again with variation among individuals (Fig. 5c, d). We detected little *dnd* mRNA level in the gonads of TG3 adults (TG3-1 P<0.01; TG3-2 P<0.05), and significantly lower *dnd* mRNA level in the gonads of TG3-1 individuals than in TG3-2 individuals (P<0.05; Fig. 5e, f). The *gal4* mRNA level did not differ significantly between the TG3-1 and the TG3-2 adults (Fig. 5g, h). For the UAS each repeat (CGGAGTACTGTCCTCCGAG) contained two CpG sites, which were the targets of methylation. Hence, we examined the methylation status of 5×UAS and the minimal promoter E1b (5×UAS-E1b). The results showed that the 5×UAS-E1b sequence was susceptible to methylation in 77.5% of the TG3-1 males, 81.2% of the TG3-1 females, 88.3% of the TG3-2 male, and 90.8% of the TG3-2 females (Fig. 5i).

Discussion

The potential ecological risks posed by transgenic fish are a key impediment to their commercial use and development. Controlling the fertility of transgenic fish offers a solution to this problem. However, to date, a method of transferring the 'infertility' trait of the transgenic fish to its offspring while maintaining a self-propagating genetic line has not been established. The galactose regulated upstream promoter element (GAL4) is a transcriptional activator in yeast that contains not only the DNA-binding domain but also the transcriptional activation domain. The upstream active sequence (UAS) is a type of regulatory sequence like a eukaryotic enhancer that occurs in yeast. GAL4 can specifically recognize UAS and drive the transcription of genes following UAS: a system that has strong specificity and easily controlled[36,40]. In our study, TG1 and TG2 transgenic lines were generated using this GAL4/UAS system with *dnd* as the target gene. We have shown that the expression of *dnd* mRNA was significantly suppressed, resulting in disrupted migration of the PGCs and their apoptosis in the hybrid embryos. This process led either to sterility (TG3-1) or to poor reproductive ability (TG3-2) in adults. Our novel strategy (which ensures that the parental transgenic generation remains fertile while their hybrid offspring become sterile) can effectively maintain the valuable traits of transgenic fish while avoiding issues of ecological risk via the production of sterile offspring.

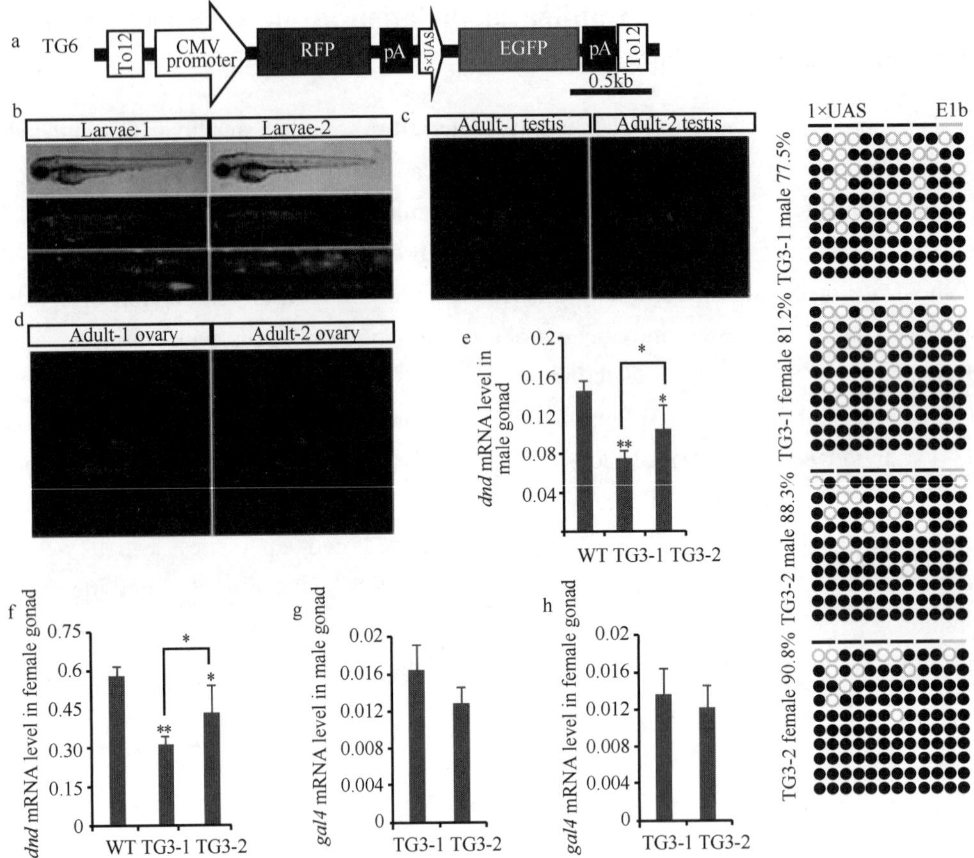

Figure 5 Mosaic expression of *egfp* regulated by 5×UAS and methylation analysis of 5×UAS-E1b by bisulfite sequencing. (a) Schematic of the expression gene cassette of the plasmids used to establish the TG6 line. (b) Variable mosaic expression of the 5×UAS-regulated *egfp* gene was detected in the offspring of the TG4 and TG6 lines (at 60 hpf). The variable mosaic expression of the 5×UAS-regulated *egfp* gene was detected in (c) the testes and (d) the ovaries of the offspring of the TG4 and the TG6 lines (at 3 months of age). (Scale bars: 100μm). The expression of *dnd* in (e) the male gonads and (f) female gonads of the WT, TG3-1 and TG3-2 lines was validated by quantitative PCR (Q-PCR). The expression of *gal4* in (g) the male gonads and (h) the female gonads of TG3-1 and TG3-2 individuals was validated by Q-PCR. The $2^{-\Delta Ct}$ values were plotted as the relative mRNA level normalized to *odc1*. Values are represented as the mean ± SEM of four repeats. **P*<0.05, ** *P*<0.01. (i) DNA from the gonads of the TG3-1 and TG3-2 adults (at 4.5 months of age) was subjected to bisulfite sequencing. CpG methylation patterns in the 5×UAS-E1b are indicated on the horizontal axis. The first ten CpG sites resided within the 5×UAS, the last two CpGs sites resided within the E1b promoter. Ten clones were tested and the results are shown on the vertical axis. Black circles represent methylated sites

The fact that both sterile individuals (TG3-1) and individuals with poor reproductive ability (TG3-2) appeared in the TG3 line is likely associated with the methylation of UAS sequence. Genes of interest driven by 5×UAS in the Gal4-UAS system showned strong expression in stable transgenic zebrafish lines[41,42]. However, repetitive sequences attract methylation. Li and colleagues reported the methylation of 5×UAS in a human cell culture system [43]. Moreover, Engineer and colleagues found transcriptional silencing of genes driven by 5×UAS in transgenic Arabidopsis, and they suggested that the silencing is due to methylation, possibly at the level of 5×UAS[44]. In zebrafish, transcriptional repression was correlated with the CpG methylation of 14×UAS[45], and single insertions containing 4×UAS were found to be far less susceptible to mythylation than insertions containing 14×UAS[46]. In our study, we detected the methylation of 5×UAS in TG3 transgenic zebrafish according to a reported method[47]. Methylation of the 5×UAS-E1b sequence appeared in the gonads of the TG3 line, and there was significantly less methylation of the 5×UAS-E1b sequence in the TG3-1 individuals than in the TG3-2 individuals. Accordingly, the level of endogenous *dnd* mRNA was significantly lower in the gonads of the TG3-1 line compared with the TG3-2 line. Moreover, the *gal4* gene driven by the CMV promoter was found to be effectively expressed in the gonads of TG3 transgenic fish and the level of *gal4* mRNA did not differ between the TG3-1 and TG3-2 lines; further, GFP or RFP under the control of the CMV promoter was strongly expressed in the next generation (F5) of these transgenic lines (see Supplementary Fig. 2a). These findings indicate that the CMV promoter is active in transgenic zebrafish. The mosaic expression of the *egfp* gene regulated by the 5×UAS occurred not only in the larvae but also in the gonads of adults and was variable among individuals. The mosaic expression of UAS-regulated genes has also been reported by other researchers, and similarly attributed to the methylation of the UAS[45,46]. The mosaic expression of the UAS-regulated gene is even known to occur in stable transgenic lines[48]. Different tissues and different cells from the same tissue have different methylation levels[45], and the methylation level of the promoter is also known to vary during an organism's development[49]. In mammals, DNA methylation levels differ among individuals and even in different cells[50,51]. In humans, meanwhile, almost every sperm cell has its own unique pattern of methylation[52]. We surmised that the patterns of methylation in zebrafish might be similar to mammals. Germ cells from different individuals or from the same individual might exhibit different levels of UAS methylation leading to different levels of mosaic expression of antisense *dnd* RNA. These differences could generate variation in the numbers of germ cells undergoing apoptosis in the gonads of the TG3 individuals, resulting in TG3 embryos that develop into adults with different reproductive capacities. To our knowledge, this is the first time methylation of 5xUAS has been identified in zebrafish, so these findings may prove useful for research on the regulation of gene expression with the Gal4/UAS system in zebrafish.

PGCs in zebrafish are located at the blastoderm layer and close to the yolk syncytial layer (YSL) and start to migrate at the dome embryonic stage. By the shield stage, PGCs have migrated close to the dorso axis at which point they begin to cluster. By the 8-somite stage, the PGCs have gathered into two clusters distributed symmetrically on both sides of the dorso axis[53,54.] In our study, some PGCs remained far from the YSL in the majority of the TG3 embryos at 4.3 hpf, which suggested that *dnd* expression had been inhibited by the antisense *dnd* RNA. Continuous transcription of the antisense *dnd* disrupted the migration of the majority of PGCs at 10 hpf and 13 hpf. We then observed that the PGCs underwent apoptosis at 20 hpf and 24 hpf, resulting in few or no PGCs migrating to the genital ridge. Similar results have been reported to occur when *dnd*-MO was injected into zebrafish embryos, resulting in a reduction or even disappearance of PGCs in the embryo stage and males that were sterile or exhibited very low reproductive capacity[28-30]. In general, fish fry containing no or few PGCs develop into males. For example, when the number of germ cells was reduced in *ziwi* mutants, the fish developed into sterile males[55]. In addition, transplantation of single PGC into embryos with no PGCs created male adults with a unilateral gonad[56]. However, in our study, it was interesting to note that the TG3 embryos developed into sterile and low reproductive adults. In zebrafish, *dnd* is a maternal gene and its expression can also be detected in the gonads of adult fish[28](see Fig. 2a, Fig. 5e, f). In MO-injected embryos, this inhibitory effect is strongest upon injection with MO and grows gradually weaker due to the degradation of MO[57], *dnd* mRNA level decreases rapidly after fertilization (see Fig. 2a). Thus almost no PGCs migrate to the genital ridge and all the fry develop into males. However, in our study, transcription of the gene (regulated by UAS) became gradually stronger when induced by GAL4 (see Supplementary Fig. 2b), which could explain the lower levels of *dnd* mRNA at the 50%-epiboly and 3-somite stages, but not at the 1-cell, 1k-cell and oblong stages. Additionally, the mosaic expression of the UAS-regulated gene appeared in our research, another factor that could result in the successful migration of some PGCs to the genital ridge. Indeed, the presence of some PGCs is essential for female differentiation in zebrafish[58], and explains why some females appeared in the TG3 line. Our results also suggest that *dnd* might play an important role in the survival of germ cell in adult zebrafish.

The mechanism by which the number of germ cells affects sex differentiation remains unknown in zebrafish. Based on the histological features of their gonads, zebrafish are thought to be a bony fish that exhibits a protogynous hermaphrodite at the larval stage[59]. 'Juvenile ovaries' are formed in all fries at 10-14 days post-fertilization (dpf), and oocyte apoptosis, which marks the transformation of 'juvenile ovaries' into testes occurs in some individuals at 21 dpf[60]. In *ziwi* mutant zebrafish, the number of germ cells has been reported to decrease rapidly or disappear before 21 dpf, such that all fry develop into sterile males[55]. In zebrafish, the mutation of *fancl* causes apoptosis of the oocytes that are undergoing meiosis and leads to female-to-male sex reversal[61]. Wong *et al.*, (2011) transplanted ovarian germ cells (N = 12 ± 4.7) obtained from three-month-old adults into two-week-old sterile larva and reported that the larva developed into both males and females, of which a few males had low fertility levels while all the females were sterile[62]. In our study, some sterile and low reproductive female adults also appeared in the TG3 line. According to the above lines of evidence, we hypothesized that when the reduction of germ cells occurs before sex differentiation, more individuals differentiate into males. When the reduction of germ cells occurs after sex differentiation, then if the germ cells are absent female-to-male sex reversal transpires. However, the presence of just a few germ cells in the ovaries was evidently enough to maintain the female characteristics and not trigger female-to-male sex reversal. The reasons for this might be that the process of de-differentiation (and re-differentiation) in specialized cells of the gonad tissues (such as sertoli cells in testis and granulosa cells in ovarian) is difficult.

Our research has shown that antisense *dnd* RNA regulated by 5×UAS can inhibit the expression of endogenous *dnd*. We observed that this process was affected by the methylation of UAS and resulted in partial sterility of the adults. However, we expect that the use of gene editing tools such as TALEN or CRISPR/Cas9 to refine the strategy implemented in this study (namely, to induce sterility in the offspring of two viable transgenic parental lines) will ensure that all hybrid offspring are sterile. This is because these tools enforce complete knockout of the *dnd* gene, allowing for the establishment of the reproductive switch in fish.

The novel reproductive containment strategy we have developed in this study will be of great use in controlling the reproduction capacity of introduced fish. The introduction of exotic species is a traditional way of obtaining fish with valuable traits for aquaculture, but this process can cause extensive economic and environmental harm. For example, it is estimated that up to 5.4 billion dollars are lost annually in USA due to the negative impacts of exotic fish[63]. Another notable example is the extent of environmental damage caused by the introduction of the Asian carp in the USA. Recently, Thresher *et al.*, (2014) reported a successful means of controlling invasive fish by reducing the effective population size of females[64]. The controllable on-off strategy for the reproductive containment of fish that we have developed provides a novel means of controlling the ecological and economic risks inherent to fish introductions.

Methods

Fish

AB strain zebrafish were maintained and raised in recirculation systems at about 28.5℃ with 14 hours light and 10 hours dark each day. All animal experiments were conducted in accordance with the Guiding Principles for the Care and Use of Laboratory Animals and were approved by the Institute of Hydrobiology, Chinese Academy of Sciences (Approval ID: keshuizhuan08529).

Generation of the transgenic zebrafish lines

The GAL4 and 5×UAS sequence were amplified by PCR from the TK5×C vector[36]. The

GAL4 sequence was cloned into lab stocks of the pSK-GFP (Tol2-CMV-GFP-pA-CMV-MCS-pA-Tol2) and the pSK-RFP (Tol2-CMV-RFP-pA-CMV-MCS-pA-Tol2) vectors to get TG1-construction (Tol2-CMV-GFP-pA-CMV-*gal4*-pA-Tol2; Fig. 1a) and TG4-construction (Tol2-CMV-RFP-pA-CMV-*gal4*-pA-Tol2; Fig. 3a). The 5×UAS sequence was cloned into the pSK-RFP vector for the construction of the pSK-RFP-5×UAS (Tol2-CMV-RFP-pA-5×UAS-MCS-pA-Tol2) vector. The primers used were GAL4-s (5′-CCCAAGCTTGCCGCCAC-CATGAAACTGCTC-3′) and GAL4-a (5′-CCCCCCGGGGGAGGATGTCCAGGTCGTAGTC-3′); 5×UAS-s (5′-ACGCGTCGACGTCGCGGAGTACTGTCCTCCGAG-3′) and 5×UAS-E1b-a (5′-CCCAAGCTTGGGAAAGTGAGGCTGAGACGCGATGCTCGGAGGACAGTAC-TCCG-3′).

The antisense *dnd* sequence (NCBI NM_212795.1, –74bp~257bp) was amplified by PCR from the complementary DNA (cDNA) obtained from the 1-cell stage of the embryos of the WT fish. This was then cloned into the pSK-RFP-5×UAS vector for TG2-construction (Tol2-CMV-RFP-pA-5×UAS-as/*dnd*-pA-Tol2; Fig. 1b). The primers used were: Anti-dnd-s (5′-CCGCTCGAGCGGAATGACCTTTTCTTGACTTTTCC-3′) and Anti-dnd-a (5′-CCGG-AATTCCGGGAGGCGAAACTCGTAAATGG-3′). The *egfp* gene (from pEGFP-C1) was cloned into the pSK-RFP-5×UAS vector for TG6-construction (Tol2-CMV-RFP-pA-5×UAS-EGFP-pA-Tol2; Fig. 5a). mRNA encoding Tol2 transposase (100 ng/μL) and Tol2 based constructs (TG1, TG2, TG4 and TG6; 50 ng/μL) were co-injected into zebrafish embryos at the 1-cell stage. The positive embryos expressed fluorescence and could be screened with a fluoroscope (Olympus MVX10).

Quantitative PCR and *in situ* hybridization

Total RNA was extracted from the embryos (each sample contained 20-30 embryos) of the TG3 and the WT zebrafish at the 1-cell, 1k-cell, oblong, 50%-epiboly and 3-somite stages, using TRIzol reagent (Invitrogen). Total RNA was also extracted from the gonads of adult TG3 and WT zebrafish. After a 30 minute treatment with RNase-free DNase, the total RNA was reverse-transcribed to cDNA using Rever Tra Ace M-MLV (TOYOBO) with random primers. The cDNA samples were used for quantitative PCR (Q-PCR) analysis using 2×SYBR green real-time PCR mix (TOYOBO). The primers used for Q-PCR were: 5′-GTCAAC-AGACTCGGCTCTCC-3′/5′-GCACAAGGTTTGGATCACCT-3′ for *dnd*, and 5′-TCCAAG-GAAAAGCCGAAAT-3′/5′-GGACGAATAAGCCAGTCAAAA-3′ for *gal4*. The primers for Q-PCR analysis of *nanos1*, *tdrd7* and the reference ornithine decarboxxylase 1 (*odc1*) gene were those described by Kedde *et al.*, 2007[37]. PCR cycling conditions were as follows: 2 min at 95℃, followed by 40 cycles of 15s at 94℃, 15s at 60℃, and 40s at 72℃. The $2^{-\Delta Ct}$ values were plotted as the relative mRNA level normalized to *odc1*[65]. To label the PGCs, we used *vasa* (nucleotides 779-1284, GenBank accession number AB005147) as the marker gene[66]. *In situ* hybridization was performed as described in Thisse Lab[67]. Images were observed under a fluoroscope (Olympus MVX10) and captured using a digital camera (Nikon, MS-SMC) controlled by the ACT-2U software (Nikon).

Fluorescence microscopy and imaging of live PGCs

About 200 pg of RNA, consisting of a 5′end-capped mRNA encoding GFP and a *nanos1* 3′UTR, was injected into the 1-cell stage embryos of the WT and the TG5 lines to visualize the PGCs. Images were observed under a fluoroscope (Olympus MVX10) and captured using a digital camera (Nikon, MS-SMC) controlled by ACT-2U software (Nikon). The average pixel intensity of the PGCs was measured using Image-pro plus 6.0 software (Media Cybernetics, Inc.), and subtracted from the background signal. The injected embryos were collected and fixed in 4% paraformaldehyde (PFA) overnight at 4℃, then dehydrated in 30% sucrose solution for 48 hours. A TUNEL assay of the PGCs was performed using the *In Situ* Cell Death Detection Kit (Roche), according to the manufacturer's instructions. Images were captured under confocal microscope LSM 710 (Karl Zeiss).

Histological analysis and fertility assessment

Gonadal tissues were dissected from the TG3 and WT fish and fixed overnight in 4% PFA at 4℃ before being embedded in paraffin. The samples were sectioned at a width of 7μm and stained with hematoxylin-eosin (HE) using standard protocols. Developmental stages of oocytes were identified as described by Hu *et al.*, 2010[26]. Fertility was assessed by pairing each TG3 male fish with a WT female fish (and vice versa) to mate naturally. If the WT females did not spawn or if their eggs were not fertilized after mating with the TG3 males the experiment was repeated 5 days later. TG3 males were considered sterile after three such failed attempts. Similarly, TG3 transgenic females were also considered sterile if they failed to spawn after three replicate mating experiments.

The fertilization rate and relative fecundity of TG3-2 adults were validated over a 15 day period. For each individual we calculated the gonad somatic index (GSI; gonad weight/body weight × 100), the fertilization rate (fertilized eggs/total eggs × 100%), and the relative fecundity [total eggs/body weight (g)].

TUNEL cell death assay of the gonadal tissues

Gonad tissues were dissected from 4.5 month-old adult TG3 fish and WT fish and embedded in Optimal Cutting Temperature (O.C.T. SAKURA). The samples were sectioned at a width of 7 μm using a freezing microtome. The TUNEL cell death assay was performed using the *in Situ* Cell Death Detection Kit (Roche) according to the manufacturer's instructions. Images were observed under a fluoroscope (Olympus MVX10), and captured using a digital camera (Nikon, MS-SMC) controlled by ACT-2U software (Nikon) after hematoxylin re-dying.

DNA bisulfite sequencing

Gonad tissues were dissected from 4.5 month-old TG3 adults (TG3-1 males and females, TG3-2 males and females). Each sample contained six gonads of TG3 fish. The mixed genomic DNA was extracted using a genomic DNA kit (TIANGEN). The 5×UAS-E1b sequence was amplified using the primers 5′-TTTTTTTTATTGTATTTTAGTTGTGGT-3′ and 5′-TAAAATTTTCATCAATCAAATCC-3′. The PCR products were cloned into a pMD-18T vector (TAKARA) and ten clones from each sample were sequenced. The percentage of methylation of all sites (CpG n=12) was counted in each sample.

Statistical analyses and image acquisition

T-tests were used to assess the significance of differences between the TG3 line and WT line in the following metrics: quantitative PCR data, body weight, GSI, fertilization rate and fecundity; t-tests were also used to assess the significance of differences between the TG5 line and WT line in fluorescence intensity and between TG3-1 and TG3-2 in quantitative PCR data, body weight, GSI and bisulfate. Images weve taken and observation of the embryos and slices was performed using a fluoroscope (Olympus MVX10) and a confocal microscope LSM 710 (Karl Zeiss). The brightness and contrast of the images were processed using Adobe Photoshop (Adobe Systems, USA).

Acknowledgements We greatly appreciate Mrs. Ming Li for raising the zebrafish. This work was supported financially by the “863” High Technology Project [grant number 2011AA100404], the National Natural Science Foundation [grant number 31325026] and the Key Research Program of the Chinese Academy of Sciences [grant numbers XDA08010106, KSCX2-EW-N-004, 2011FBZ19].

References

1. Holmlund, C. M. & Hammer, M. Ecosystem services generated by fish populations. Ecol. Econ. 29, 253-268 (1999).
2. Neubauer, P., Jensen, O. P., Hutchings, J. A. & Baum, J. K. Resilience and recovery of overexploited marine populations. Science 340, 347-349 (2013).
3. Worm, B.*et al.* Impacts of biodiversity loss on ocean ecosystem services. Science 314, 787-90 (2006).
4. Cressey, D. Future fish. Nature 458, 398-400 (2009).
5. Zhu, Z., Li, G., He, L. & Chen, S. Novel gene transfer into the fertilized eggs of gold fish (*Carassius auratus* L. 1758). J. Appl. Lchthyol. 1, 31-34 (1985).
6. Maclean, N. & Laight, R. J. Transgenic fish: An evaluation of benefits and risks. Fish. Fish. 1, 146-172 (2000).
7. Devlin, R. H., Biagi, C. A., Yesaki, T. Y., Smailus, D. E. & Byatt, J. C. Growth of domesticated transgenic fish. Nature 409, 781-782 (2001).

8. Ledford, H. Transgenic salmon nears approva. Nature 497, 17-18 (2013).
9. Check, E. Environmental impact tops list of fears about transgenic animals. Nature 418, 805-805 (2002).
10. Stokstad, E. Engineered Fish: Friend or Foe of the Environment? Science 297, 1797-1799 (2002).
11. Devlin, R. H., Sundstrom, L. F. & Muir, W. M. Interface of biotechnology and ecology for environmental risk assessments of transgenic fish. Trends. Biotechnol. 24, 89-97 (2006).
12. Hu, W., Wang, Y. & Zhu, Z. Progress in the evaluation of transgenic fish for possible ecological risk and its containment strategies. Sci. China. C. Life. Sci. 50, 573-579 (2007).
13. Hu, W. & Zhu, Z. Integration mechanisms of transgenes and population fitness of GH transgenic fish. Sci. China. Life.Sci. 53, 401-8 (2010).
14. Marris, E. Transgenic fish go large. Nature 467, 259 (2010).
15. Van Eenennaam, A. L. & Muir, W. M. Transgenic salmon: a final leap to the grocery shelf? Natuer. 29, 706-710 (2011).
16. Wong, A. C. & Van Eenennaam, A. L. Transgenic approaches for the reproductive containment of genetically engineered fish. Aquaculture 275, 1-12 (2008).
17. Fox, J. L. Transgenic salmon inches toward finish line. Nat. Biotechnol. 28, 1141-2 (2010).
18. Penman, D. J., Skibinski, D. O. F. & Beardmore, J. A. Survival, growth rate and maturity in triploid tilapia. In Proc. World Symp. on Selection, Hybridization and Genetic Engeneering in Aquaculture2, 277-288 (1987).
19. Ojolick, E. J., Cusack, R., Benfey, T. J. & Kerr, S. R. Survival and growth of allfemale diploid and triploid rainbow trout (*Oncorhynchus mykiss*) reared at chronic high temperature. Aquaculture 131, 177-187 (1995).
20. Ozerov, M. Y. *et al.* High *Gyrodactylus salaris* infection rate in triploid Atlantic salmon *Salmo salar*. Dis. Aquat. Organ. 91, 129 (2010).
21. Fraser, T. W., Fjelldal, P. G., Hansen, T. & Mayer, I. Welfare considerations of triploid fish. Rev. Fish. Sci. 20, 192-211 (2012).
22. Maclean, N. *et al.* Transgenic tilapia and the tilapia genome. Gene 295, 265-277 (2002).
23. Hu, W. *et al.* Antisense for gonadotropin-releasing hormone reduces gonadotropin synthesis and gonadal development in transgenic common carp (*Cyprinus carpio*). Aquaculture 271, 498-506 (2007).
24. Xu, J. *et al.* Defining Global Gene Expression Changes of the HypothalamicPituitary-Gonadal Axis in Female sGnRH-Antisense Transgenic Common Carp (*Cyprinus corpio*). Plos one 6, e21057 (2011).
25. Wong, T. T. & Collodi, P. Inducible Sterilization of Zebrafish by Disruption of Primordial Germ Cell Migration. Plos one 8, e68455 (2013)
26. Hu, S. Y. *et al.* Nitroreductase-mediated gonadal dysgenesis for infertility control of genetically modified zebrafish. Mar. Biotechnol. 12, 569-78 (2010).
27. Doitsidou, M. *et al.* Guidance of Primordial Germ Cell Migration by the Chemokine SDF-1. Cell. 111, 647-659 (2002).
28. Weidinger, G. *et al.* dead end, a Novel Vertebrate Germ Plasm Component, Is Required for Zebrafish Primordial Germ Cell Migration and Survival. Curr. Biol. 13, 1429-1434 (2003).
29. Slanchev, K., Stebler, J., de la Cueva-Mendez, G. & Raz, E. Development without germ cells: the role of the germ line in zebrafish sex differentiation. Proc. Nati. Acad. Sci. USA 102, 4074-9 (2005).
30. Liu, W. & Collodi, P. Zebrafish dead end possesses ATPase activity that is required for primordial germ cell development. J. FASEB 24, 2641-50 (2010).
31. Li, Y. *et al.* Progranulin regulates zebrafish muscle growth and regeneration through maintaining the pool of myogenic progenitor cells. Sci. Rep. 3, 1176; DOI:10.1038/srep01176 (2013).
32. Guan, B. *et al.* Vitreoscilla hemoglobin (VHb) overexpression increases hypoxia tolerance in zebrafish (*Danio rerio*). Mar. Biotechnol. 13, 336-44 (2011).
33. Haas, P. & Gilmour, D. Chemokine signaling mediates self-organizing tissue migration in the zebrafish lateral line. Dev. Cell.10, 673-80 (2006).
34. Gong, Z. *et al.* Green fluorescent protein expression in germ-line transmitted transgenic zebrafish under a stratified

epithelial promoter from keratin 8. Dev. Dyn. 223, 204-15 (2002).

35. Hsu, C. C., Hou, M. F., Hong, J. R., Wu, J. L. & Her, G. M. Inducible male infertility by targeted cell ablation in zebrafish testis. Mar. Biotechnol. 12, 466-78 (2010).

36. Distel, M., Wullimann, M. F. & Koster, R. W. Optimized Gal4 genetics for permanent gene expression mapping in zebrafish. Proc. Nati. Acad. Sci. USA 106, 13365-13370 (2009).

37. Kedde, M. *et al.* RNA-binding protein Dnd1 inhibits microRNA access to target mRNA. Cell131, 1273-86 (2007).

38. Mishima, Y. *et al.* Differential regulation of germline mRNAs in soma and germ cells by zebrafish miR-430. Curr. Biol. 16, 2135-2142 (2006).

39. Patton, E. E. & Zon, L. I. The art and design of genetic screens: zebrafish. Nat. Rev. Genet. 2, 956-966 (2001).

40. Osterwalder, T., Yoon, K. S., White, B. H. & Keshishian, H. A conditional tissuespecific transgene expression system using inducible GAL4. Proc. Natl. Acad. Sci. USA 98, 12596-601 (2001).

41. Asakawa, K. *et al.* Genetic dissection of neural circuits by Tol2 transposonmediated Gal4 gene and enhancer trapping in zebrafish. Proc. Nati. Acad. Sci. USA 105, 1255-1260 (2008).

42. Choe, S. K., Nakamura, M., Ladam, F., Etheridge, L. & Sagerström, C. G. A Gal4/UAS system for conditional transgene expression in rhombomere 4 of the zebrafish hindbrain. Dev. Dynam. 241, 1125-1132 (2012).

43. Li, F. *et al.* Chimeric DNA methyltransferases target DNA methylation to specific DNA sequences and repress expression of target genes. Nucleic. Acids. Res. 35, 100-112 (2007).

44. Engineer, C. B., Fitzsimmons, K. C., Schmuke, J. J., Dotson, S. B. & Kranz, R. G. Development and evaluation of a Gal4-mediated LUC/GFP/GUS enhancer trap system in Arabidopsis. BMC. Plant. Biol. 5, 9 (2005).

45. Goll, M. G., Anderson, R., Stainier, D. Y., Spradling, A. C. & Halpern, M. E. Transcriptional silencing and reactivation in transgenic zebrafish. Genetics 182, 747-55 (2009).

46. Akitake, C. M., Macurak, M., Halpern, M. E. & Goll, M. G. Transgenerational analysis of transcriptional silencing in zebrafish. Dev. Bio. 352, 191-201 (2011).

47. Deng, J. *et al.* Targeted bisulfite sequencing reveals changes in DNA methylation associated with nuclear reprogramming. Nat. Biotechnol. 27, 353-360 (2009).

48. Sagasti, A., Guido, M. R., Raible, D. W. & Schier, A. F. Repulsive interactions shape the morphologies and functional arrangement of zebrafish peripheral sensory arbors. Curr. Biol. 15, 804-841 (2005).

49. Fang, X., Corrales, J., Thornton, C., Scheffler, B. E. & Willett, K. L. Global and gene specific DNA methylation changes during zebrafish development. Comp. Biochem. Physiol. B Biochem. Mol. Biol. 166, 99-108 (2013).

50. Sandovici, I. *et al.* Interindividual variability and parent of origin DNA methylation differences at specific human Alu elements. Hum. Mol. Genet. 14, 2135-43 (2005).

51. Lister, R. *et al.* Human DNA methylomes at base resolution show widespread epigenomic differences. Nature 462, 315-22 (2009).

52. Flanagan, J. M. *et al.* Intra- and interindividual epigenetic variation in human germ cells. Am. J. Hum. Genet. 79, 67-84 (2006).

53. Weidinger, G., Wolke, U., Koprunner, M., Klinger, M. & Raz, E. Identification of tissues and patterning events required for distinct steps in early migration of zebrafish primordial germ cells. Development 126, 5295-5307 (1999).

54. Weidinger, G. *et al.* Regulation of zebrafish primordial germ cell migration by attraction towards an intermediate target. Development 129, 25-26 (2002).

55. Houwing, S. *et al.* A role for Piwi and piRNAs in germ cell maintenance and transposon silencing in Zebrafish. Cell 129, 69-82 (2007).

56. Saito, T., Goto-Kazeto, R., Arai, K. & Yamaha, E. Xenogenesis in teleost fish through generation of germ-line chimeras by single primordial germ cell transplantation. Biol. Reprod. 78, 159-66 (2008).

57. Nasevicius, A. & Ekker, S. C. Effective targeted gene ‘knockdown’ in zebrafish. Nat. Genet.26, 216-220 (2000).

58. Siegfried, K. R. & Nusslein-Volhard, C. Germ line control of female sex determination in zebrafish. Dev. Biol. 324, 277-87 (2008).

59. Maack, G. & Segner, H. Morphological development of the gonads in zebrafish. J. FISH. BIOL. 62, 895-906 (2003).
60. Uchida, D., Yamashita, M., Kitano, T. & Iguchi, T. Oocyte apoptosis during the transition from ovary like tissue to testes during sex differentiation of juvenile zebrafish. J. EXP. BIOL. 205, 711-718 (2002).
61. Rodriguez-Mari, A. *et al.* Sex reversal in zebrafish fancl mutants is caused by Tp53-mediated germ cell apoptosis. Plos. genet. 6, e1001034 (2010).
62. Wong, T. T., Saito, T., Crodian, J. & Collodi, P. Zebrafish germline chimeras produced by transplantation of ovarian germ cells into sterile host larvae. Biol. Reprod. 84, 1190-7 (2011).
63. Pimentel, D., Zuniga, R. & Morrison, D. Update on the environmental and economic costs associated with alien-invasive species in the United States. Ecol. Econ. 52, 273-288 (2005).
64. Thresher, R. *et al.* Sex-ratio-biasing constructs for the control of invasive lower vertebrates. Nat. Biotechnol. 32, 424-7 (2014).
65. Livak, K. J. & Schmittgen, T. D. Analysis of Relative Gene Expression Data Using Real-Time Quantitative PCR and the $2^{-\Delta\Delta CT}$ Method. methods 25, 402-408 (2001).
66. Yoon, C., Kawakami, K. & Hopkins, N. Zebrafish vasa homologue RNA is localized to the cleavage planes of 2- and 4-cell-stage embryos and is expressed in the primordial germ cells. Development 124, 3157-3165 (1997).
67. Thisse, C. & Thisse, B. High-resolution *in situ* hybridization to whole-mount zebrafish embryos. Nat. protoc. 3, 59-69 (2007).

鱼类生殖控制的可控开关策略

张运生[1,2] 陈 戟[1] 崔小娟[1,2] 罗大极[1,3] 夏 慧[1,2]
戴 军[1] 朱作言[1] 胡 炜[1]

1 中国科学院水生生物研究所，淡水生态与生物技术国家重点实验室，武汉430072
2 中国科学院大学，北京 100049
3 武汉大学，基础医学院，遗传学系，武汉 430071

摘 要 潜在的生态安全问题是制约转基因鱼产业化的瓶颈之一，控制转基因鱼的育性可以从根本上解决其潜在的生态安全问题。在此，我们提出亲本可育而杂交子代诱导不育的控制鱼类育性的新策略。我们利用 GAL4/UAS 诱导系统，分别建立了含有 GAL4基因和5×UAS-as/dnd 基因两种转基因斑马鱼家系 Tg1(Tol2-CMV-GFP-pA-CMV-*gal4*-pA-Tol2)和 Tg2 (Tol2-CMV-RFP-pA-5×UAS- as/*dnd*-pA-Tol2)。这两种转基因鱼家系都可以繁殖产生后代。但是，在这两种转基因鱼的杂交子代中，GAL4诱导 UAS 驱动 antisense-dnd 基因转录，显著抑制了内源 *dnd* mRNA 水平，引起杂交胚胎发育早期 PGCs 迁移紊乱并发生凋亡，迁移至生殖脊的 PGCs 数量显著减少，甚至没有 PGCs 迁移至生殖脊，最终导致成鱼的生殖败育或生殖能力显著降低。我们的研究结果提示，转基因鱼的生殖开关策略既可以保存转基因鱼可育亲本的优良性状，又可以生产大量不育的、具有生态安全性的转基因鱼，用于产业化养殖。

The Distribution of Kisspeptin (Kiss)1- and Kiss2-Postive Neurones and Their Connections with Gonadotrophin-Releasing Hormone-3 Neurones in the Zebrafish Brain

Yanlong Song[1,2] Xiaohai Duan[1,2] Ji Chen[1] Wei Huang[1]
Zuoyan Zhu[1] Wei Hu[1]

1 State Key Laboratory of Freshwater Ecology and Biotechnology, Institute of Hydrobiology, Chinese Academy of Sciences, Wuhan 430072
2 University of Chinese Academy of Sciences, Beijing 100039

Abstract Kisspeptin is a neuroendocrine hormone with a critical role in the activation of gonadotrophin-releasing hormone (GnRH) neurones, which is vital for the onset of puberty in mammals. However, the functions of kisspeptin neurones in non-mammalian vertebrates are not well understood. We have used transgenics to labell kisspeptin neurones (Kiss1 and Kiss2) with mCherry in zebrafish (*Danio rerio*). In *kiss1:mCherry* transgenic zebrafish, Kiss1 cells were located in the dorsomedial and ventromedial habenula, with their nerve fibres contributing to the fasciculus retroflexus and projecting to the ventral parts of the interpeduncular and raphe nuclei. In *kiss2: mCherry* zebrafish, Kiss2 cells were primarily located in the dorsal zone of the periventricular hypothalamus and, to a lesser extent, in the periventricular nucleus of the posterior tuberculum and the preoptic area. Kiss2 fibres formed a wide network projecting into the telencephalon, the mesencephalon, the hypothalamus and the pituitary. To study the relationship of kisspeptin neurones and GnRH3 neurones, these fish were crossed with *gnrh3:EGFP* zebrafish to obtain *kiss1:mCherry/gnrh3:EGFP* and *kiss2:mCherry/gnrh3:EGFP* double transgenic zebrafish. The GnRH3 fibres ascending to the habenula were closely associated with Kiss1 fibres projecting from the ventral habenula. On the other hand, GnRH3 fibres and Kiss2 fibres were adjacent but scarcely in contact with each other in the telencephalon and the hypothalamus. The Kiss2 and GnRH3 fibres in the ventral hypothalamus projected into the pituitary via the pituitary stalk. In the pituitary, Kiss2 fibres were directly in contact with GnRH3 fibres in the pars distalis. These results reveal the pattern of kisspeptin neurones and their connections with GnRH3 neurones in the brain, suggesting distinct mechanisms for Kiss1 and Kiss2 in regulating reproductive events in zebrafish.

Keywords Kiss1; Kiss2; GnRH3; transgenic; zebrafish

Kisspeptin is a neuroendocrine hormone encoded by the gene *kiss1* that was originally identified as a metastatic suppressor in human malignant melanoma cells [(1)]. Subsequent research found that *kiss1* encodes several short peptides named kisspeptins (i.e. kisspeptin-54, -14, -13, -10),

文章发表于 *Journal of Neuroendocrinology*, 2015, 27(3): 198-211

which share a common RFamide C-terminus, and are the natural ligands of the G protein-coupled receptor GPR54 (also known as Kiss1r) [(2-4)]. In 2003, the role of kisspeptin in reproduction was further elucidated when *GPR54* gene mutations in humans and mice were found to cause hypogonadotrophic hyp-ogonadism [(5-7)], with a similar phenotype observed in *kiss1* knockout mice [(8)].

In mammals, the critical role of kisspeptin in maintaining normal reproductive function is now widely recognised. Mechanistically, there is a large body of evidence suggesting that kisspeptins activate gonadotrophin-releasing hormone (GnRH) neurones to regulate gonadotrophin secretion, thereby exerting their effect on the reproductive axis [(9-13)]. In non-mammalian vertebrates, such as teleosts, in addition to *kiss1*, there exists a novel *kiss2* gene [(14-17)]. Three *gnrh* genes are recognised by phylogenetic analysis (*gnrh1*, *gnrh2* and *gnrh3*) [(18)]. In teleosts, such as zebrafish (*Danio rerio*) and common carp (*Cyprinus carpio* L.), GnRH1 is missing and, in addition to its nonhypophysiotrophic neuromodulatory function, GnRH3-positive neurones are also regarded as playing hypophysiotrophic roles for their expression in the preoptic area and the caudal hypothalamus [(19-22)].

Genomic organisational structure analysis of the *kiss1* and *kiss2* genes suggests that they originated from an ancient kiss gene, which was duplicated during the first round of whole-genome duplication approximately 250 million years ago, and that *kiss2* was lost in placental mammals during vertebrate evolution [(14,23,24)]. The two *kiss* genes are preserved in teleosts, and their pattern of cellular and axonal fibre distribution is known to exist in different regions of the brain [(25)]. In zebrafish, *kiss1* mRNA is expressed in the ventromedial habenula exclusively, whereas *kiss2* mRNA is expressed in the posterior tuberal nucleus and the dorsal zone of the periventricular hypothalamus, as detected by *in situ* hybridisation [(15,26)]. Recently, Servili *et al.* [(26)] generated antibodies to zebrafish Kiss1 and Kiss2, showing that Kiss1 localised to neurones in the ventral habenula and projected to the interpeduncular nucleus and the median raphe, and that Kiss2 localised to the dorsal and ventral hypothalamus, with fibres widely distributed in the forebrain and midbrain [(26)]. Whether these kisspeptin systems are regulative centres of GnRH is unknown in non-mammalian vertebrates.

To date, our understanding of kisspeptin function in the teleost is relatively limited, and an improved understanding of the relationship between kisspeptin and GnRH neurones would enhance our comprehension of their regulative function in fish reproduction. As such, the present study aimed to characterise the distribution of Kiss1- and Kiss2-positive neurones and their contacts with GnRH3 neurones in the zebrafish brain. To achieve this goal, we generated *kiss1:mCherry* and *kiss2:mCherry* transgenic zebrafish in which Kiss1- and Kiss2-expressing neurones and fibres were labelled with red fluorescent protein. These fish were also crossed with *gnrh3: EGFP* zebrafish to obtain *kiss1: mCherry/gnrh3: EGFP* and *kiss2:mCherry/gnrh3: EGFP* double transgenic zebrafish, in which GnRH3-expressing neurones were labelled with enhanced green fluorescent protein. Imaging of these transgenic fish allowed us to characterise the distribution of Kiss1- and Kiss2-neurones in the zebrafish brain, as well as determine their connectivity with GnRH3-positive neurones.

Materials and Methods

Animals

All zebrafish were of the AB strain and were kept in recirculating systems under an alternating 12∶12 h light/dark cycle at 28-29℃. Fish water was changed once a day. For whole-mount *in situ* hybridisation, post-gastrulation embryos were raised in water with 0.3 mM 1-phenyl-2-thiourea to prevent pigmentation. Mature zebrafish and larvae were anaesthetised with 100 mM tricaine methanesulfonate (MS222; Sigma-Aldrich, St Louis, MO, USA) prior to dissection or microscopy. All procedures were approved by the Institutional Animal Care and Use Committee of the Institute of Hydrobiology.

Generation of promoter-reporter constructs

A 2.4-kb sequence upstream of the zebrafish *kiss1* coding region was cloned and isolated to construct the *kiss1* promoter-reporter construct. This fragment corresponded to bases 2462541-2460062 (GenBank: NW_003334952.1), and contained 2.3 kb of the 5′ upstream flanking sequence and part of exon 1 of *kiss1*. The 2.1 kb kiss2 promoter-report con-struct corresponded to bases 583137-585286 (GenBank: NW_001878907.3) and contained 1.5 kb of the 5′ upstream flanking sequence, exon-1, intron-1 and part of exon-2 of *kiss2*. These two fragments were cloned into pSK MCS mCherry Tol2 vectors (kindly provided by Professor Yonghua Sun, Institute of Hydrobiology, Chinese Academy of Sciences, Wuhan, China) (Fig. 1).

Microinjection and production of transgenic zebrafish lines

To produce transgenic zebrafish, the constructs were linearised and diluted in nuclease-free water to a final concentration of 100 μg/ml. Before microinjection, Tol2 transposase mRNA was added to a final concentration of 200 μg/ml and phenol red (0.1%) was used for visualisation during microinjection. Approximately 5 nl of solution was injected into the animal pole of zebrafish embryos at the one cell stage. The P_0 fish injected with *kiss1: mCherry* or *kiss2:mCherry* constructs were bred to maturity and crossed with wild-type zebrafish. The F_1 fish with mCherry signals were picked out and crossed with wild-type zebrafish to obtain F_2 zebrafish. The F_3 lines were the cross between male and female F2 fish. The F2 and F3 zebrafish were used for all experiments. The lines of zebrafish generated in this experiment were referred to as *kiss1:mCherry* and *kiss2:mCherry* transgenics. The *gnrh3: EGFP* transgenic zebrafish were developed as reported by Abraham *et al.* [27]. Enhanced green fluorescent protein was specifically expressed in GnRH3 neurones and fibres in *gnrh3: EGFP*

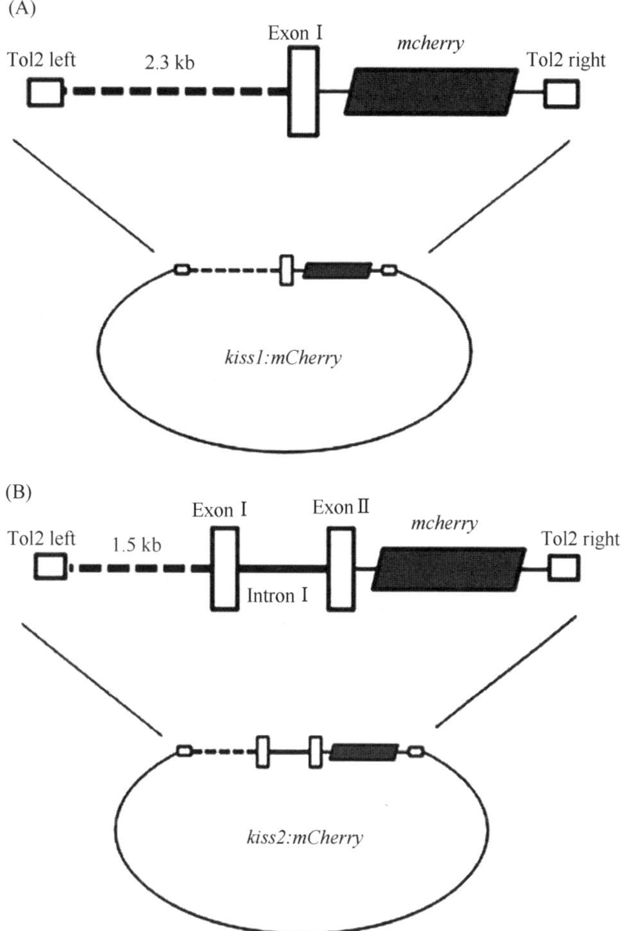

Fig. 1 Schematic of *kiss1:mCherry* and *kiss2:mCherry* transgenic vectors. Schematics of the *kiss1:mCherry* (A) and *kiss2:mCherry* (B) transgenic vectors are shown. The *kiss1:mCherry* vector contains a 2.4-kb sequence upstream of the kisspeptin1 coding region including 2.3 kb of 5′ flanking sequence and part of exon I (A). The *kiss2:mCherry* vector contains a 2.1-kb sequence upstream of the kisspeptin2 coding region including 1.5 kb of 5′ flanking sequence, exon I, intron I, and part of exon II (B)

zebrafish (see Supporting information, Fig. S1). Double transgenic lines of zebrafish were created with the genotypes *kiss1: mCherry/gnrh3:EGFP* and *kiss2:mCherry/gnrh3: EGFP*.

Preparation of frozen brain sections

Six sexually mature zebrafish of both sexes were anaesthetised and their brains were carefully taken out and rinsed in phosphate-buffered saline (PBS). The brains were then fixed in buffered 4% paraformaldehyde (PFA; Sigma-Aldrich) in PBS overnight, and dehydrated in 30% sucrose at 4℃. Next, the whole brains were embedded in Tissue Freezing Medium (Leica Biosystems, Heidelberger, Germany) and frozen by floating in an ethanol, dry ice slurry. Frozen sections of the embedded brains were cut using a cryostat (Leica CM1950) at –7℃. Transverse 20-lm tissue sections were collected and then mounted onto poly-L-lysine-coated slides. The slides were then air-dried for 1 h, rinsed in PBS, covered with Vectashield mounting medium containing 4′,6-diamidino-2-phenylindole (DAPI; Vector Laboratories, Burlingame, CA, USA) and coverslipped.

In situ hybridisation

Fragments of zebrafish kiss cDNA (*kiss1*: 88-532 GenBank: AB245404.1; *kiss2*: 94-513 GenBank: NM_001142585.1) were cloned and inserted into pCS2+ vectors to generate probes for *in situ* hybridisation. The constructs were linearised with *Hin*dIII as templates to synthesise digoxigenin (DIG) labelled anti-sense probes using T7 RNA polymerase, or linearised with *Xba*I to synthesise sense probes using SP6 RNA polymerase (DIG RNA Labelling Mix: Roche Diagnostics, Mannheim, Germany; T7 & SP6 RNA polymerase: Promega Biotech, Madison, WI, USA).

The protocol for whole-mount *in situ* hybridisation was performed as described previously (28). Briefly, 20 zebrafish larvae, aged 1 week old, were anaesthetised and fixed in buffered 4% PFA in PBS. Next, hybridisation with DIG-labelled riboprobes (500 μg/ml) was performed in hybridisation buffer [50% formamide, 5 × saline sodium citrate SSC, 50 μg/ml heparin, 500 μg/ml tRNA, 0.1% tween 20, 9 mM citric acid, pH 6.0] at 70℃, then incubated with anti- DIG antibodies conjugated to alkaline phosphatase (dilution 1︰5000) at 4℃, and finally stained with nitroblue phosphate and 5-bromo-4-chloro-3-indolil phosphate.

The protocol for *in situ* hybridisation in brain sections is described below. Serial transverse cryostat sections were washed in 0.1 M PBS and postfixed for 20 min in buffered 4% PFA. After washing, sections were treated with proteinase K for five min at room temperature [10 μg/ml in 50 mM Tris-HCl, pH 8.0, 5 mM ethylenediaminetetraacetic acid (EDTA)], rinsed, and fixed in buffered 4% PFA. Sections were rinsed twice in 2×SSC. Prehybridisation was performed at 65 ℃ in hybridisation buffer (50% formamide; 5×SSC; 5 mM EDTA; 0.1% tween 20: 0.1% Chaps; 50 μg/ml heparin; 1 mg/ml yeast RNA) for 5 h. Hybridisation was performed at 65 ℃ in a humidified chamber using 100 μl hybridisation buffer containing the DIG-labelled probe (1 μg/ml) overnight. After hybridisation, slides were rinsed in 50% formamide/ 2×SSC for 30 min, followed by one rinse in 2×SSC, and two rinses in 0.2×SSC at 65℃ for 15 min. Final rinses were made in PBS at room temperature and sections were processed for immunodetection. The sections were incubated in blocking buffer (2 mg/ml bovine serum albumin, 2% sheep serum and 0.2% Triton X-100 in PBS) for 5 h, then incubated overnight at 4℃ in anti-DIG-Fluorescein (Roche Diagnostics) diluted to 1︰100 in blocking buffer. Sections were washed five times for 5 min with 0.1% Tween 20 in PBS and covered with Vectashield mounting medium containing DAPI.

Fluorescence microscopy

Images of larval zebrafish and whole brains were captured using a Nikon digital sight DS-5Mc camera (Nikon Corporation, Tokyo, Japan) attached to an Olympus fluorescence macro-microscope MVX10 (Olympus Corp., Tokyo, Japan). Confocal microscopy of brain sections was conducted using a Zeiss LSM710 NLO microscope system (Carl Zeiss MicroImaging, Oberkochen, Germany).

Results

Distribution of Kiss1-positive cells and fibres in *kiss1:mCherry* zebrafish

To detect mCherry expression, we observed the transgenic zebrafish larva under a fluorescence macro-microscope. The red fluorescence was too weak to detect in the early larvae, so we used 7 days post-fertilisation (dpf) larvae in the present study. In *kiss1: mCherry* zebrafish larvae, the red fluorescent labelled Kiss1 neurones emerged symmetrically in the commissural nucleus of the telencephalon and optic tectum (Fig. 2A). Similarly, *in situ*

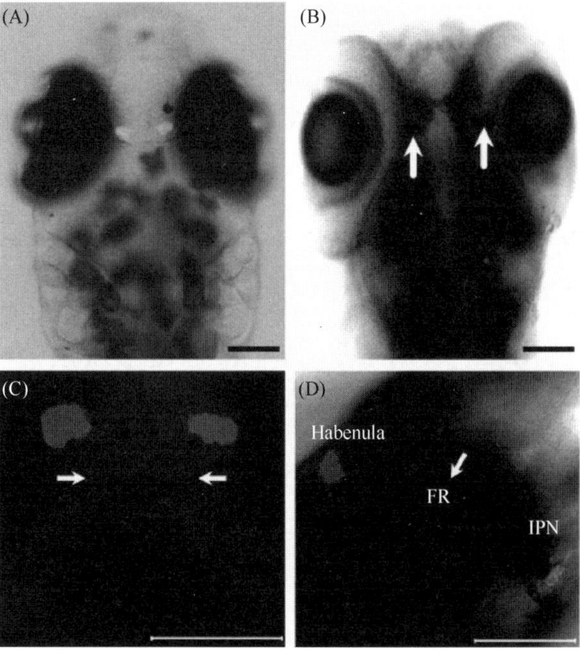

Fig. 2 Kiss1 expression in the brain of larval zebrafish. A representative image of red fluorescent Kiss1 neurones in the habenula in 7-dpf zebrafish is shown (A). Similarly, *in situ* hybridisation shows that *kiss1* mRNA is expressed in the same regions as the red fluorescent signal in the habenula in 7-dpf larvae (B). Kiss1- positive fibres emanate from the habenula (C) and project along the fasciculus retroflexus (FR) into the interpeduncular nucleus (IPN; D). Scale bars = 100 μm

hybridisation clearly showed *kiss1* mRNA expressing cells in the same regions (Fig. 2B). Anatomically, these areas correspond to the right and left habenula nuclei in zebrafish. In the 7-dpf larvae, two bundles of Kiss1-positive fibre were observed emanating from the Kiss1 perikarya and projecting into the interpeduncular nucleus (IPN) (Fig. 2C,D). A strong Kiss1-positive sig-nal was also observed in the ventral part of the IPN (Fig. 2D).

Imaging of the outer surface of the whole brain of mature *kiss1: mCherry* zebrafish revealed two strong red fluorescent signals in the habenula and one in the rear of the hypothalamus (Fig. 3A, B). To confirm that mCherry was expressed specifically in Kiss1 cells, we conducted *in situ* hybridisation in *kiss1: mCherry* zebrafish brain sections. The kiss1 mRNA signals and mCherry signals colocalised in the ventral habenula (Fig. 3C-E), convincingly showing that *kiss1: mCherry* zebrafish specifically drive mCherry expression in Kiss1 cells in the habenula. Imaging of transverse brain sections at the same level revealed that red fluorescent labelled Kiss1 cells were present in the medial habenula and the ventral habenula but not the dorsal habenula (Fig. 3F-H). The red fluorescent signal filled all the dense, small cells in the ventromedial habenula. At the upper extreme of the ventrolateral habenula, red fluorescent nerve fibres projected out from both sides of the ventral habenula (Fig. 3G, H). These axons made up the fasciculus retroflexus (FR), with red fluorescence detectable along its entire extent (Fig. 3I, J). At the caudal region of the FR bundles, these fibres bilaterally entered into the ventral tegmentum in the IPN (Fig. 3K, L), and a mass of red fluorescent fibres densely innervate the whole ventral extent of the IPN (Fig. 3M). By contrast, there was no obvious sig-nal detected in the dorsal part of the IPN. A relatively small number of red fluorescent fibres were also observed projecting to the raphe nucleus (Fig. 3N). These results revealed that Kiss1 cells are specifically located in the ventromedial habenula and project to the IPN and median raphe via the FR.

Connectivity between Kiss1 neurones and GnRH3 neurones

To explore the potential relationship between Kiss1-positive neurones and GnRH3-positive neurones in the zebrafish brain, we analysed *kiss1: mCherry/gnrh3:EGFP* double transgenic zebrafish. GnRH3 neurones originated in the olfactory bulb, and then migrated backward through the subpallium, ventral telencephalon, anterior commissure and optic commissure, and extended to the hypothalamus, as reported by Abraham *et al.* [27]. We observed many GnRH3-positive fibres going up in the epithalamus and two bundles of GnRH3-positive fibres migrating upward along the outside of the habenula, crossing the midline at the top of the habenular commissure (Fig. 4A,B). In the posterior habenula, multiple GnRH3-positive fibres extended to both sides of the ventral habenula and the GnRH3-positive fibres significantly arose around the Kiss1-positive fibres projecting from the ventral habenula. The GnRH3-positive fibres were closely associated with the Kiss1-positive fibres in the FR. Indeed, some

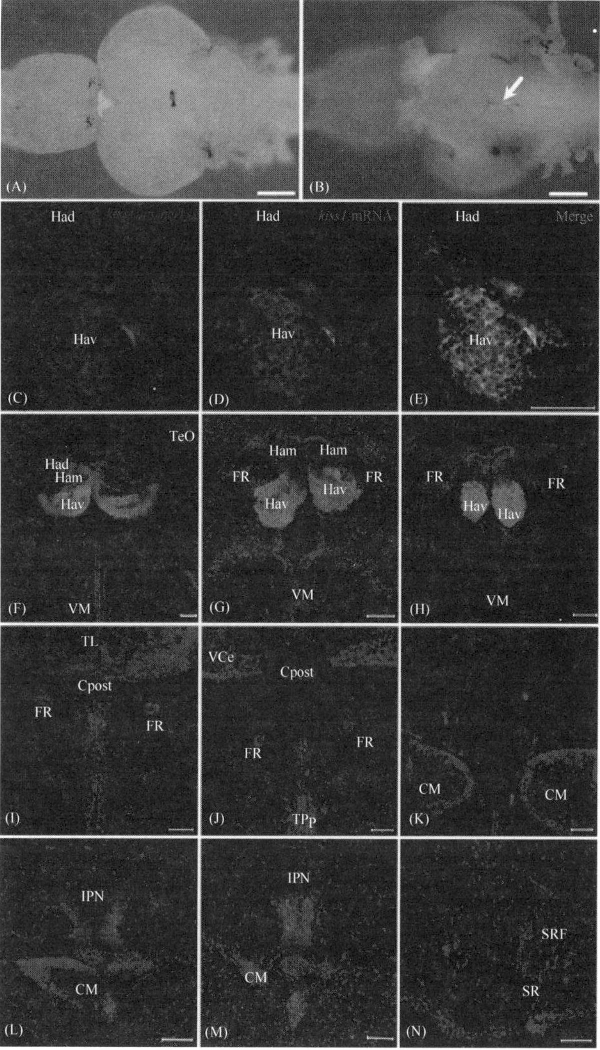

Fig. 3 Localisation of Kiss1 expression in the adult kiss1:mCherry zebrafish brain. In the adult zebrafish brain, the dorsal aspect shows the red fluorescent labelled Kiss1 neurones located in the habenula (A), and the ventral view shows red fluorescence in the caudal part of the hypothalamus (B). Representative images of Kiss1-positive neurones and fibres are shown, and cell nuclei are stained blue with DAPI. *Kiss1* mRNA-expressing cells (green) are in accordance with red fluorescent labelled cells in the ventral habenula (Hav), as indicated by *in situ* hybridisation (C-E). Kiss1-expressing cells are strictly located in the ventromedial habenula (F-H). The Kiss1 fibres emanating from the ventral habenula (H) go through the fasciculus retroflexus (FR) down to the interpeduncular nucleus (IPN; I-K). These fibres enter the IPN through its ventrolateral margin (K, L), and the Kiss1 fibres suffuse the entire ventral IPN (M). There are also scattered Kiss1 fibres projecting into the superior raphe nucleus (SR) and the superior reticular formation (SRF; N). CM, corpus mamillare; Cpost, posterior commissure; Had, dorsal habenula; Ham, medial habenula; TeO, tectum opticum; TL, torus longitudinalis; TPp, periventricular nucleus of posterior tuberculum; VCe, valvula cerebelli; VM, ventromedial thalamic nucleus. (A, B) Scale bars = 1 mm; (C-N) Scale bars = 50 μm

GnRH3-positive fibres directly contacted Kiss1-positive fibres in these regions (Fig. 4C, D). Furthermore, Kiss1-positive fibres were occasionally contacted by GnRH3- positive fibres along the FR entire extent (Fig. 4E, F). The connectivity between Kiss1 neurones and GnRH3 neurones was not observed in the olfactory bulb, preoptic area or other regions in the brain.

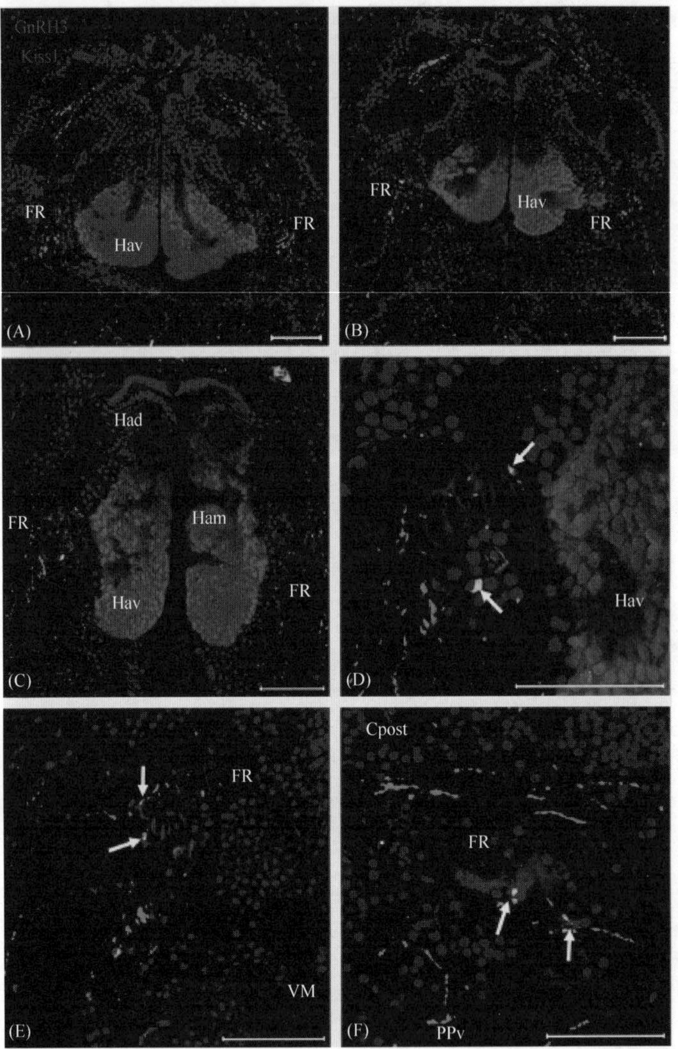

Fig. 4 Relationship between Kiss1 and gonadotrophin-releasing hormone (GnRH)-3 expression in the adult *kiss1:mCherry/gnrh3:EGFP* zebrafish brain. Representative images of Kiss1 (red), GnRH3 (green) and 4′,6-diamidino-2-phenylindole (DAPI) (blue) staining are shown. GnRH3 fibres extend into the anterior parts of habenula and distribute around the outer flank (A, B). In posterior parts of the habenula, dense GnRH3 fibres are present around the fasciculus retroflexus (FR) and some are in contact with Kiss1 fibres from the ventral habenula (C, D). The GnRH3 fibre network in the mesencephalon also occasionally contacts the FR (arrow; E and F). Scale bars = 50 μm. Had, dorsal habenula; Ham, medial habenula; PPv, vental part of the pretectal diencephalic cluster; VM, ventromedial thalamic nucleus

Distribution of Kiss2-positive cells and fibres in *kiss2:mCherry* zebrafish

In 7-dpf *kiss2:mCherry* zebrafish larvae, several symmetrical red fluorescent signals appeared in the ventral parts of the mesencephalon (Fig. 5A,B). This staining pattern was consistent with that obtained by *in situ* hybridisation of *kiss2* mRNA (Fig. 5C,D). In the larvae brain, Kiss2 axons consisted of projections extending bilaterally from Kiss2-positive perikarya clusters located in the hypothalamus (Fig. 5E). The Kiss2-positive fibres were detected projecting in the mesencephalon and some forward to the thalamus. Some Kiss2-positive cell clusters were observed in the posterior tuberculum (Fig. 5F). The Kiss2-positive fibres passed through the optic commissure and extended into the telencephalon (Fig. 5G). The Kiss2-positive neurones clusters were also observed in the preoptic region and their fibres extended forward to the anterior part of telencephalon (Fig. 5H).

Imaging of the outer surface of the brain of mature *kiss2:mCherry* zebrafish revealed two symmetrical and strong red fluorescent spots in the mediobasal hypothalamus (Fig. 6A). The results of *in situ* hybridisation in *kiss2:mCherry* zebrafish brain sections revealed that mCherry labelled cells also expressed the *kiss2* mRNA in the mediobasal hypothalamus (Fig. 6B-D). Imaging of transverse brain sections detailed the innervations of Kiss2 neurones and fibres in the brain. Some Kiss2-positive neurones and intense fibres were also detected in the preoptic area (Fig. 6E). In the diencephalon, a small number of Kiss2-positive cells were found in the periventricular nucleus of the posterior tuberculum (TPp) and the Kiss2- positive fibres projected out from both sides of these cells (Fig. 6F). The densest Kiss2-positive innervations were found surrounding the superior extent of the lateral recess (LR) in the mediobasal hypothalamus (Fig. 6G, H). The Kiss2-positive fibres in the anterior parts of the LR spread down along the lateral wings of the ventral hypothalamus to the bottom of the hypothalamus (Fig. 6G), and the fibres in the posterior part of the LR projected to the caudal zone of the periventricular hypothalamus (Hc) (Fig. 6I, J). Some Kiss2-positive cells and fibres were also detected in the caudal hypothalamus (Fig. 6I, J). Within the brain, the Kiss2-positive fibres formed a wide network of projections in the telencephalon and mesencephalon. Sporadic Kiss2-positive fibres even reached to the posterior parts of the olfactory bulbs (Fig. 6K). Signifi-cant numbers of Kiss2-positive fibres were detected in the subpallium, the ventral telencephalon and the optic commissure (Fig. 6L, M). The Kiss2-positive fibres were detected extending upward into the thalamus and some fibres reached the outside of the habenula (Fig. 6N). More posteriorly, Kiss2-positive projections were found in the anterior parts of the tectum opticum (Fig. 6O). Two conspicuous bundles of Kiss2-positive fibres were found leaving the TPp (Fig. 6P). Furthermore, Kiss2-positive fibres projected into many regions of the mesencephalon (Fig. 6Q), the torus semicircularis (Fig. 6R), the lateral longitudinal fascicle and the nucleus of the medial longitudinal fascicle (Fig. 6S). No significant red fluorescent labelled cells or fibres could be detected in other

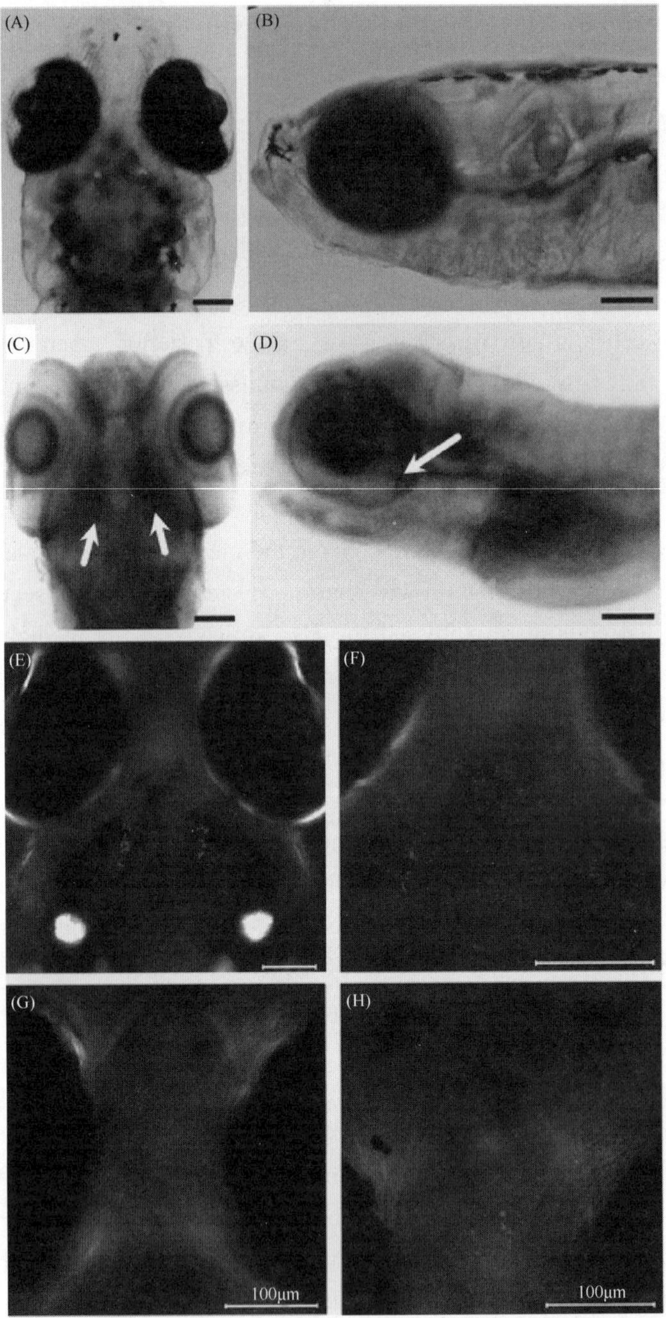

Fig. 5 Kiss2 expression in the brain of larval zebrafish. Representative images of red fluorescent labelled Kiss2 neurones in the hypothalamus are shown in dorsal (A) and lateral views (B) of 7-dpf larval zebrafish. Similarly, *in situ* hybridisation shows *kiss2* mRNA is expressed in the hypothalamus (C: dorsal view; D: lateral view). Kiss2-positive neurones and projections were also observed via confocal microscopy in the larval brain. Most Kiss2-positive neurones are located in the mesencephalon (E), with few in the posterior tuberculum (F) and the preoptic area (H). The Kiss2-positive fibres project widely in the mesencephalon and forward into the telencephalon (F-H). Scale bars = 100 μm

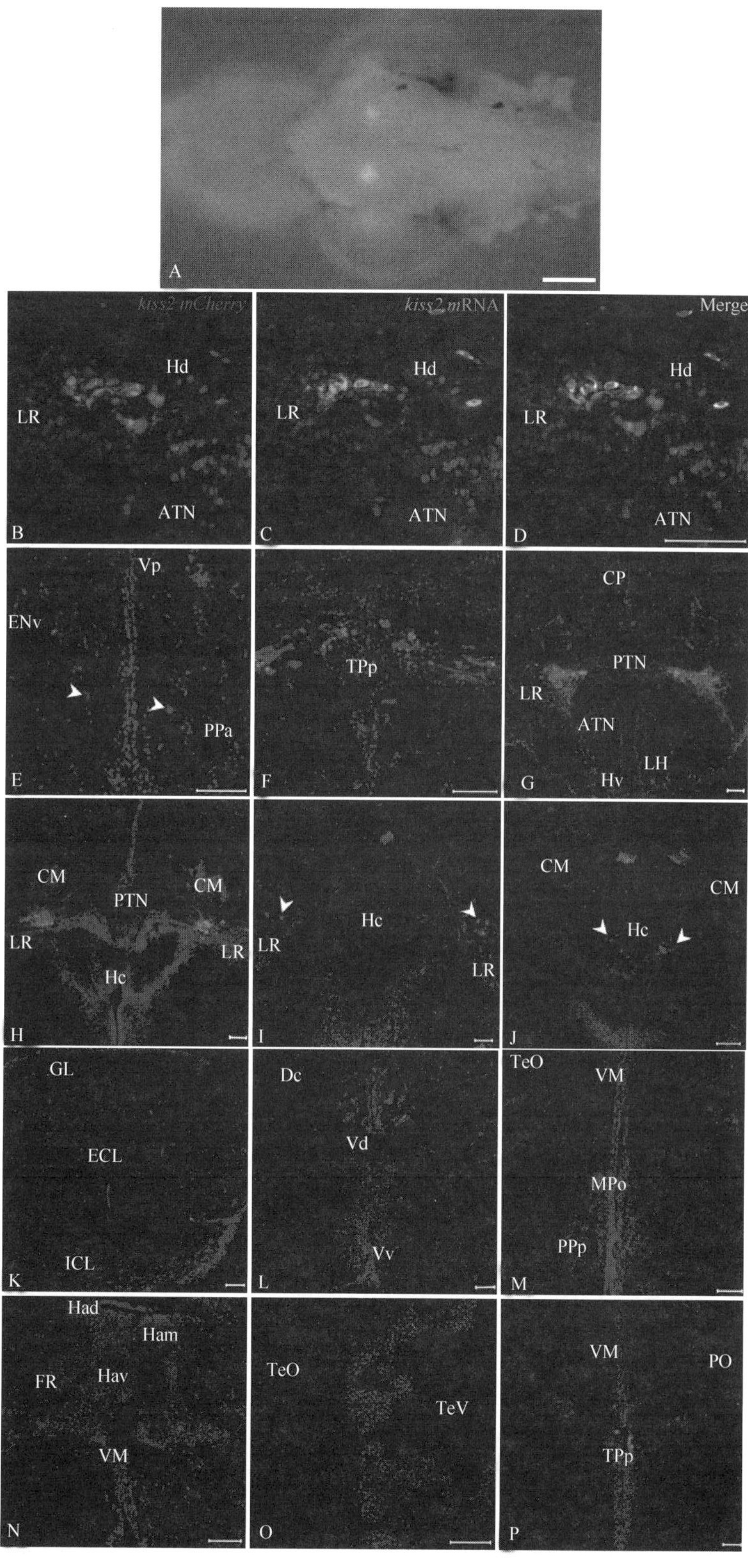
A
kiss2 mCherry
kiss2 mRNA
Merge
Hd
LR
ATN
B
C
D
Vp
ENv
PPa
E
TPp
F
CP
PTN
LR
ATN
LH
Hv
G
CM
PTN
CM
LR
LR
Hc
H
Hc
LR
LR
I
CM
CM
Hc
J
GL
ECL
ICL
K
Dc
Vd
Vv
L
TeO
VM
MPo
PPp
M
Had
Ham
FR
Hav
VM
N
TeO
TeV
O
VM
PO
TPp
P

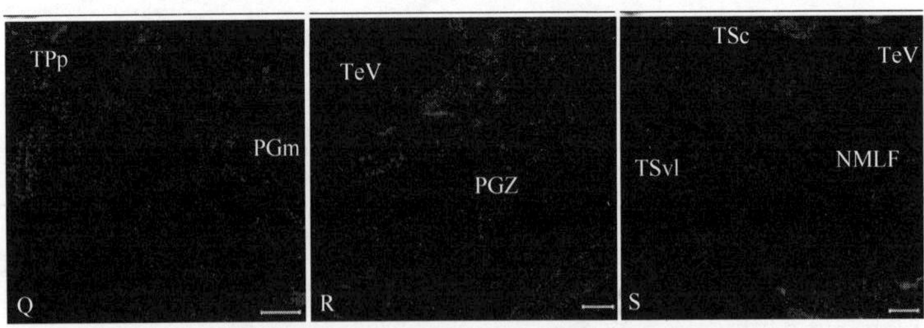

Fig. 6 Localisation of Kiss2 expression in the adult *kiss2:mCherry* zebrafish brain. In the adult brain, a ventral view of the brain shows red fluorescent labelled Kiss2 neurones (red) in both sides of the hypothalamus (A). Red fluorescent labelled Kiss2 signals (red) and kiss2 mRNA signals (green) are colocalised in the same cells in the hypothalamus as indicated by *in situ* hybridisation (B-D). Representative images of Kiss2 and 4′,6-diamidino-2-phenylindole (DAPI) (blue) expression are shown. Some Kiss2-positive cells were observed in the anterior parvocellular preoptic nucleus (PPa, arrowheads; E) and the periventricular nucleus of the posterior tuberculum (TPp; F). The prominent Kiss2-positive cells located dorsal to the lateral recess (LR) send fibres around the anterior tuberal nucleus (ATN) toward the ventromedial hypothalamus (Hv; G). At the posterior parts of the LR, the Kiss2 fibres project caudally (H). Some Kiss2 cells and fibres were also detected in the caudal hypothalamus (I, J). Within the telencephalon, the Kiss2 projections spread into the posterior parts of the olfactory bulbs (K), the ventral telencephalon (Vv; L) and the optic commissure (M). A wide distribution of Kiss2 fibres is seen in the thalamus and many of these fibres extend through the outside of the ventral habenula (N) to the anterior parts of the tectum opticum (TeO; O). The Kiss2-positive cells in the TPp send out fibres bilaterally (P). Furthermore, the Kiss2 fibres are detected in the preglomerular nucleus (Q), the tectum opticum (R) and the torus semicircularis (TS; S). Dc, central zone of dorsal telencephalic area; ENv, entopendoncular nucleus; external layer (ECL), glomerular layer (GL), internal cellular layer (ICL) of olfactory bulb; LH, lateral hypothalamic nucleus; MPo, magnocellular preoptic nucleus; NMLF, nucleus of MLF; PGm, medial preglomercular nucleus; PGZ, periventricular gray zone of the TeO; PO, posterior pretectal nucleus; PPp, parvocellular preoptic nucleus; PTN, posterior tuberal nucleus; TeV, tectal ventricle; central nucleus (TSc), ventrolateral nucleus (TSvl) of torus semicircularis; dorsal nucleus (Vd), postcommissural nucleus (Vp) of ventral telencephalic area. (A) Scale bars = 1 mm; (B-S) Scale bars = 50 μm

regions of the brain. The Kiss2 distribution patterns showed no significant difference between the sexes in the zebrafish brain.

Connectivity between Kiss2- and GnRH3-positive neurones

To explore the potential relationship between Kiss2 and GnRH3 neurones, we analysed the connection of Kiss2- and GnRH3-positive neurones in the *kiss2: mCherry/gnrh3: EGFP* double transgenic zebrafish. In the brain of mature transgenic zebrafish, GnRH3-positive perikarya clusters in the olfactory bulbs bilaterally sent out nerve tracts and these neurones migrated dorsocaudally along the ventral telencephalon toward the hypothalamus, forming a crossing at the anterior commissure. The nerve bundles consisted of GnRH3-postive axons at the bottom of the hypothalamus. Two Kiss2-positive perikarya clusters also located symmetrically in the hypothalamus in these double transgenic zebrafish (Fig. 7A). A more detailed relationship between Kiss2- and GnRH3-positive fibres was revealed via the imaging of transverse brain sections. GnRH3-positive neurones were vertically located in the both sides of the ventral telencephalon. Although the Kiss2-positive fibres were found in the posterior parts of the olfactory bulbs and the anterior parts of telencephalon, they rarely contact with GnRH3 fibres directly in these regions. Kiss2-positive fibres projected into the middle portions, whereas GnRH3-positive neurones were located in both sides of the ventral telencephalon (Fig. 7B). Intense Kiss2- and GnRH3-positive fibres were detected in the lateral preoptic area. Although few contacts were detected, these fibres were in close association with each other in these areas (Fig. 7C, D). In the ventral thalamus and anterior hypothalamus, the scattered Kiss2- and GnRH3-positive fibres were also closely distributed (Fig. 7E-G). In the hypothalamus, GnRH3-positive fibres migrated down along the outer margin of the lateral wings of the ventral hypothalamus. The Kiss2-positive projections from the perikarya surrounding the lateral recess extended along the same course (Fig. 7H-J). Both Kiss2- and GnRH3-positive fibres reached the ventral zone of the periventricular hypothalamus near the pituitary stalk. Furthermore, some Kiss2-positive cells were observed in the mediobasal hypothalamus just above the pituitary stalk (Fig. 7J). In the pituitary, the GnRH3-positive fibres were detected distributing widely in multiple regions. By contrast, the Kiss2-positive fibres were detected almost exclusively in the pars distalis. The Kiss2-positive fibres in the pituitary were directly contacted by GnRH3-positive fibres (Fig. 7K-M).

Discussion

The present study is the first to describe transgenic zebrafish in which Kiss1- and Kiss2-expressing neurones are labelled with fluorescent proteins. This unique approach allowes us to clearly observe the distribution of Kiss1- and Kiss2-expressing neurones and fibres within the zebrafish brain, as well as their interactions with GnRH3-expressing neurones.

The distribution of Kiss1-expressing neurones and fibres in the zebrafish brain

The *in situ* hybridisation data showed that kiss1 mRNA is expressed exclusively in the ventral habenula. Similarly, mCherry labelled Kiss1 neurones were also located in these areas in *kiss1: mCherry* zebrafish. Previously, preprokiss1-immunoreactive neurones were also

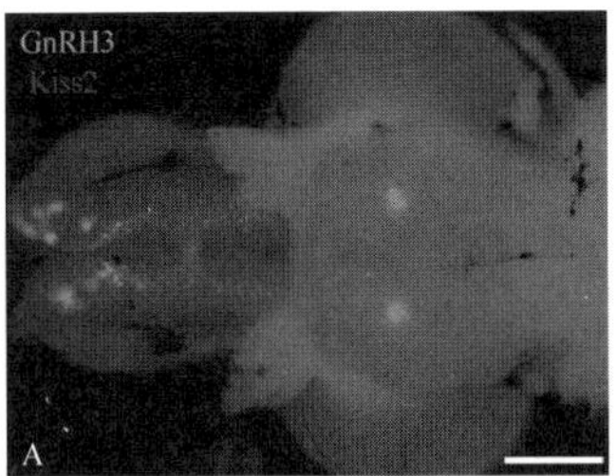

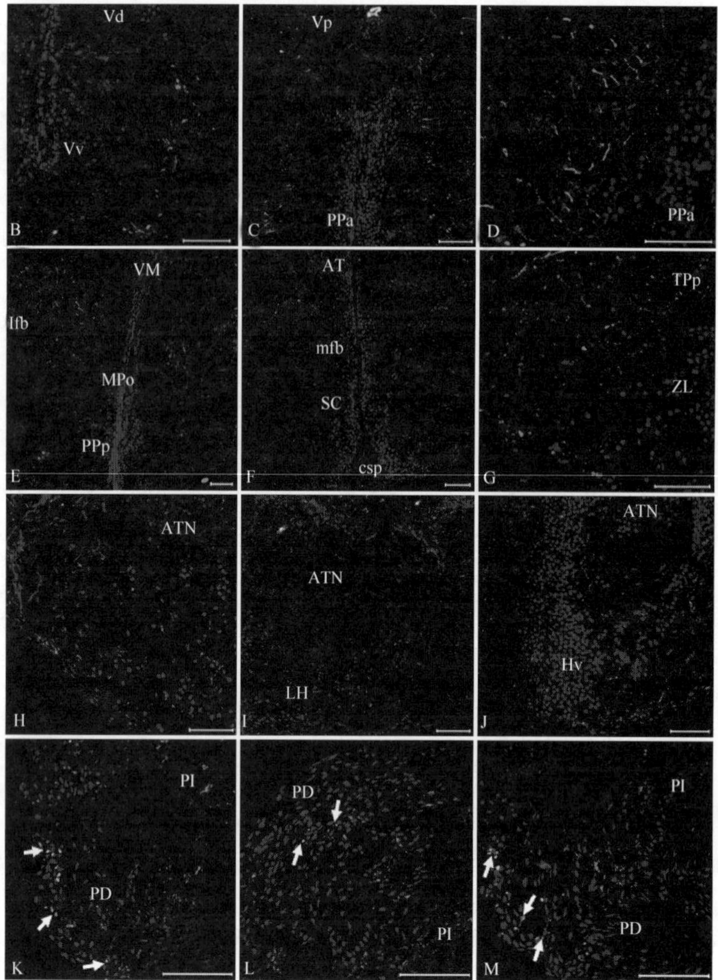

Fig. 7 Relationship between Kiss2 and gonadotrophin-releasing hormone (GnRH)-3 expression in the adult *kiss2:mCherry/gnrh3:EGFP* zebrafish brain. Representative images of Kiss2 (red), GnRH3 (green) and 4′,6-diamidino-2-phenylindole (DAPI) (blue) staining are shown. The transgenic zebrafish shows that GnRH3 perikarya originate in the olfactory bulbs and migrate caudally along the ventral telencephalon to the hypothalamus (A). In the ventral telencephalon, the Kiss2-positive fibres are located in the middle region and the GnRH3 are distributed to the sides (B). The Kiss2 and GnRH3 fibres distribute densely in the preoptic region (C, D) and the optic commissure (E), whereas less Kiss2 and GnRH3 fibres are detected in the anterior hypothalamus (F, G). Kiss2 projections accompanying GnRH3 fibres migrate along the lateral wings of the hypothalamus to the bottom of ventral hypothalamus (H-J). Although few contacts are detected, the Kiss2 fibres and GnRH3 fibres are closely associated in the preoptic region and hypothalamus. In the pituitary, GnRH3 fibres are detected in the pars intermedia (PI) and the pars distalis (PD). The Kiss2 fibres are also present in the PD and are almost contacted by GnRH3 fibres in these areas (arrows; K-M). AT, anterior thalamic nucleus; csp, commissure of superficial pretectum; lfb, lateral forebrain bundle; mfb, medial forebrain bundle; mlf, medial longitudinal fascicle; MPo, magnocellular preoptic nucleus; PPp, parvocellular preoptic nucleus, posterior part; SC, suprachiasmatic nucleus; ZL, zona limitans. (A) Scale bars = 1 mm; (B-M) Scale bars = 50 μm

specifically observed in the ventral habenula [26]. We observed that Kiss1 cells sent out two strands of nerve fibres that projected caudally to the IPN and raphe nucleus through the FR, which is also in agreement with previous immunohistochemical data [26]. Taken together, we conclude that the *kiss1:mCherry* zebrafish we created specifically drives mCherry expression in Kiss1-expressing neurones and fibres.

In zebrafish, the habenula is divided into the dorsal habenula (dHb) and ventral habenula (vHb), based on differences in the cytoarchitechture, which are homologous to the mammalian medial habenula and lateral habenula, respectively [29,30]. The dHb nuclei can be further subdivided into two asymmetric subnuclei: the left-side dHb projects to the dorsal region of the IPN and the right-side dHb projects to the ventral region of the IPN [31,32]. By contrast, the symmetric subnuclei in the vHb were shown to project to the ventral IPN and the median raphe [30]. In the present study, Kiss1 neurones were specifically located in the ventral habenula and established projections to the ventral IPN-median raphe as expected. Functionally, the dHb-IPN pathway plays a critical role in controlling fear and anxiety responses in zebrafish [33,34]. Interestingly, recent studies in zebrafish have demonstrated the possible role of habenular Kiss1 neurones in modulating the serotonergic raphe nuclei [35]. Furthermore, Kiss1 neurones were shown to play important roles in modulating fear response in zebrafish [36]. These studies reveal multiple functions of the habenular Kiss1 system beyond the control of reproduction; however, the complete mechanisms and pathways regarding how the Kiss1 system functions remain unclear. In this regard, the *kiss1: mCherry* zebrafish described in the present study provide a powerful tool for future studies.

The distribution of Kiss2-expressing neurones and fibres in the zebrafish brain

Kiss2 was expressed in the ventral mesencephalon of larval zebrafish (7 dpf), in the approximate location of the primordial hypothalamus. In the mature zebrafish, Kiss2-positive neurones were found predominantly in the dorsal and ventral hypothalamus, with their fibres projecting forward to the posterior parts of the olfactory bulbs, telencephalon, preoptic area, thalamus, optic tectum and a wide area of the mesencephalon. Our *in situ* hybridisation data reveal that *kiss2* mRNA is expressed in the same location as the mCherry-labelled neurones in the hypothalamus. Previously, *kiss2* mRNA was detected in the preoptic area, the posterior tuberal nucleus, the periventricular nucleus of the posterior tuberculum and the periventricular hypothalamus in zebrafish [15,26]. Similarly, *kiss2* mRNA expressing cells in the hypothalamus have also been noted in other teleosts such as medaka [15,37], goldfish [38], European sea bass [39], red seabream [40] and tilapia [41]. These expression patterns of *kiss2* mRNA further confirm the Kiss2-positive cells detected in the *kiss2: mCherry* transgenic zebrafish. The expression of Kiss2, as revealed by immunohisto-chemistry in zebrafish and

European sea bass, shows that most Kiss2-immunoreactive cells were detected surrounding the lateral recess in the hypothalamus and a small number were detected in the periventricular nucleus of the posterior tuberculum and the preoptic regions. The Kiss2-immunoreactive fibres were widely detected in the subpallium, preoptic region, thalamus, optic tectum, mesencephalon, and the ventral and caudal hypothalamus [(26,42)]. The close agreement between the published data and our results suggest that the *kiss2:mCherry* zebrafish we created specifically drives mCherry expression in Kiss2-expressing neurons.

Connectivity between Kiss1- and Kiss2-expressing neurones and GnRH3-expressing neurones in the zebrafish brain

The kisspeptin-GPR54 system is vital for the onset of puberty because it is essential for the activation of GnRH neurones in mammals. In teleosts, the distinct distribution of Kiss1- and Kiss2-expressing neurones suggested that these factors may carry out different functions in reproductive regulation. In mature zebrafish, kisspeptin2 administration increased FSH-β and LH-β mRNA expression; however, there was no obvious difference in FSH and LH mRNA level in zebrafish injected with kisspeptin1 [(15)]. In sea bass, it was similarly shown that the Kiss2 decapeptide was signifi-cantly more potent in promoting LH and FSH secretion than the Kiss1 decapeptide [(14)]. These studies suggest that Kiss2-expressing neurones might play more important role in regulating gonadotrophic hormone synthesis and secretion.

In zebrafish, Kiss1 neurones were located in the ventral habenula and the fibres projected via the FR to the IPN and raphe nuclei. In vertebrates, the habenula plays a pivotal role in adaptive behaviours by regulating serotonergic and dopaminergic activities [(43,44)]. In zebrafish, the important role of the habenular Kiss1 system in regulating the serotonergic system and, furthermore, modulating the fear response to maintain emotional aspects of reproductive behaviour in zebrafish has been reported previously [(35,36)]. Thus, in mammals and teleosts, the evidence suggests that the kisspeptin system plays an important role in integrating various environmental factors, such as photoperiods, steroid hormones, metabolic signals and stress with reproductive behaviour [(45-50)].

In teleosts, the nonhypophysiotrophic GnRH systems, such as the terminal GnRH3 nerve, also have been linked with the control of reproductive behaviour [(51)]. For example, lesions of GnRH terminal nerve cells affected the reproductive behaviour of male Dwarf Gouramis [(52)]; in male cichlids, immunoneutralisation of GnRH3 significantly suppressed nest-building ability and aggressive behavior [(53)]. Recently, a study in medaka showed that the terminal-nerve GnRH3 neurones function as a gate for activating mating preferences [(54)]. These results showed that extrahypothalamic GnRH3 is a potent neuromodulator of reproductive behaviour in teleosts. In *kiss1:mCherry/gnrh3:EGFP* transgenic zebrafish, the GnRH3 fibres extended close to the Kiss1 fibres projecting from the ventral habenula, and some Kiss1 fibres were directly contacted by GnRH3 fibres. Based on their intimate positional relationship, we believe the Kiss1 neurones may cooperate with GnRH3 neurones in the regulation of reproductive behaviour. Indeed recent electrophysiology studies demonstrated the potential action of Kiss1on GnRH3 neurones in zebrafish and medaka (55,56). However, this functional relationship remains to be established and we consider that the *kiss1: mCherry* and *kiss2: mCherry* zebrafish described in the present study provide a powerful tool for future studies.

In the teleost brain, the preoptic area and the ventromedial hypothalamus are important locations for hypophysiotrophic neurones. Indeed, GnRH3-expressing neurones distributed in these regions are responsible for regulating gonadotrophic hormone synthesis and release in zebrafish (21,57). Our observations in double transgenic *kiss2: mCherry/gnrh3: EGFP* zebrafish show that Kiss2 fibres accompany GnRH3 fibres throughout the preoptic area and hypothalamus. Specifically, both Kiss2 and GnRH3 fibres extended to the mediobasal hypothalamus, just above the pituitary stalk. In the pituitary, prominent Kiss2 fibres were observed in the pars distalis and these fibres were almost contacted by GnRH3 fibres. Therefore, it appears that Kiss2 neurones in the hypothalamus project to the pituitary via the pituitary stalk just as GnRH3 neurones do. These results confirm the stimulatory functions of kisspeptin2 on LH and FSH in zebrafish and sea bass (14,15). The close relationship between Kiss2 and GnRH3 neurones indicates that Kiss2 could be a regulatory factor of gonadotrophic functions in the pituitary. Of course, this functional relationship between Kiss2 and GnRH3 should be confirmed using electrophysiology. Similar results were obtained when the experiment was operated with mammals. The kisspeptin axons were intimately associated with GnRH neurones in the median eminence and kisspeptin-10 significantly stimulated GnRH release in monkeys (58,59). The kisspeptin-10 can also act directly at the GnRH nerve terminals and stimulate GnRH release in the mediobasal hypothalamus of the mouse (60). Furthermore, the wide distribution of Kiss2 fibres in the telencephalon and mesencephalon suggests that Kiss2 might have multiple functions in teleosts (25,26). This is supported by the widespread sites of *kiss-r2* mRNAs expression in zebrafish brain (26). The Kiss-R1 and Kiss-R2 showed characteristic distribution in isotocin and vasotocin neurones in medaka brain (61). The *kiss-r2* mRNAs are detectable in many chemically identified neurones, including those neurones expressing nitric oxide synthase, neuropeptide Y, tyrosine hydroxylase and somatostatin in the brain of sea bass (42). Taken together, we conclude that the Kiss2 system may be involved in the regulation of various physiological processes.

Summary

In the last decade, there has been considerable focus upon Kiss1 function in mammals; however, studies on kisspeptins in non-mammalian vertebrates are lacking. This is particularly true in fish, where two types of kisspeptin (Kiss1 and Kiss2) are expressed. In conclusion, the transgenic zebrafish that express mCherry in Kiss1- and Kiss2-expressing neurones characterised in the present study show clear distribution patterns of kisspeptin neurones and their relationship with GnRH3 neurones in zebrafish. We consider these transgenic zebrafish will provide useful models for studying the function of these factors in reproductive regulation, as well as other unknown functions.

Acknowledgements We thank Mr Wei Liu for guidance and assistance with the confocal microscopy. This work was supported financially by the National Natural Science Foundation (Grant No. 31325026), the '863' High Technology Project (Grant No. 2011AA100404) and the Key Research Program of the Chinese Academy of Sciences (Grant No. KSCX2-EW-N-004, 2011FBZ21). Wei Hu, Zuoyan Zhu and Yanlong Song conceived and designed the research. Yanlong Song developed methods and performed experiments, and also prepared Figs 1-7. Xiaohai Duan, Ji Chen and Wei Huang contributed to the preparation of transgenic zebrafish lines. Yonglong Song, Wei Hu analysed the data and wrote the manuscript. All authors reviewed the original data and agreed with the final version of the manuscript submitted for publication. The authors declare that they have no conflicts of interest.

References

1 Lee JH, Miele ME, Hicks DJ, Phillips KK, Trent JM, Weissman BE, Welch DR. *KiSS-1*, a novel human malignant melanoma metastasis-suppressor gene. J Natl Cancer Inst 1996; 88: 1731-1737.

2 Kotani M, Detheux M, Vandenbbogaerde A, Communi D, Vanderwinden JM, Le Poul E, Brezillon S, Tyldesley R, Suarez-Huerta N, Vandeput F, Blanpain C, Schiffmann SN, Vassart G, Parmentier M. The metastasis suppressor gene *KiSS-1* encodes kisspeptins, the natural ligands of the orphan G protein-coupled receptor GPR54. J Biol Chem 2001; 276: 34631-34636.

3 Muir AI, Chamberlain L, Elshourbagy NA, Michalovich D, Moore DJ, Calamari A, Szekeres PG, Sarau HM, Chambers JK, Murdock P, Steplewski K, Shabon U, Miller JE, Middleton SE, Darker JG, Larminie CGC, Wilson S, Bergsma DJ, Emson P, Faull R, Philpott KL, Harrison DC. AXOR12, a novel human G protein-coupled receptor, activated by the peptide KiSS-1. J Biol Chem 2001; 276: 28969-28975.

4 Ohtaki T, Shintani Y, Honda S, Matsumoto H, Hori A, Kanehashi K, Terao Y, Kumano S, Takatsu Y, Masuda Y, Ishibashi Y, Watanabe T, Asada M, Yamada T, Suenaga M, Kitada C, Usuki S, Kurokawa T, Onda H, Nishimura O, Fujino M. Metastasis suppressor gene *KiSS-1* encodes peptide ligand of a G-protein-coupled receptor. Nature 2001; 411: 613-617.

5 de Roux N, Genin E, Carel J-C, Matsuda F, Chaussain J-L, Milgrom E. Hypogonadotropic hypogonadism due to loss of function of the KiSS1-derived peptide receptor GPR54. Proc Natl Acad Sci USA 2003; 100: 10972-10976.

6 Funes S, Hedrick JA, Vassileva G, Markowitz L, Abbondanzo S, Golovko A, Yang SJ, Monsma FJ, Gustafson EL. The KiSS-1 receptor GPR54 is essential for the development of the murine reproductive system. Biochem Biophys Res Commun 2003; 312: 1357-1363.

7 Seminara SB, Messager S, Chatzidaki EE, Thresher RR, Acierno JS, Shagoury JK, Bo-Abbas Y, Kuohung W, Schwinof KM, Hendrick AG, Zahn D, Dixon J, Kaiser UB, Slaugenhaupt SA, Gusella JF, O'Rahilly S, Carlton MBL, Crowley WF, Aparicio SAJR, Colledge WH. The GPR54 gene as a regulator of puberty. N Engl J Med 2003; 349: 1614-1627.

8 de Tassigny XDA, Fagg LA, Dixon JP, Day K, Leitch HG, Hendrick AG, Zahn D, Franceschini I, Caraty A, Carlton

MB. Hypogonadotropic hypog-onadism in mice lacking a functional *Kiss1* gene. Proc Natl Acad Sci USA 2007; 104: 10714-10719.

9 Gottsch M, Cunningham M, Smith J, Popa S, Acohido B, Crowley W, Seminara S, Clifton D, Steiner R. A role for kisspeptins in the regulation of gonadotropin secretion in the mouse. Endocrinology 2004; 145: 4073-4077.

10 Irwig MS, Fraley GS, Smith JT, Acohido BV, Popa SM, Cunningham MJ, Gottsch ML, Clifton DK, Steiner RA. Kisspeptin activation of gonadotropin releasing hormone neurons and regulation of *Kiss-1* mRNA in the male rat. Neuroendocrinology 2005; 80: 264-272.

11 Messager S, Chatzidaki EE, Ma D, Hendrick AG, Zahn D, Dixon J, Thresher RR, Malinge I, Lomet D, Carlton MB. Kisspeptin directly stimulates gonadotropin-releasing hormone release via G protein-coupled receptor 54. Proc Natl Acad Sci USA 2005; 102: 1761-1766.

12 Clarkson J, Han S-K, Liu X, Lee K, Herbison AE. Neurobiological mechanisms underlying kisspeptin activation of gonadotropin-releasing hormone (GnRH) neurons at puberty. Mol Cell Endocrinol 2010; 324: 45-50.

13 Gopurappilly R, Ogawa S, Parhar IS. Functional significance of GnRH and kisspeptin, and their cognate receptors in teleost reproduction. Front Endocrinol 2013; 4: doi: 10.3389/fendo.2013.00024.

14 Felip A, Zanuy S, Pineda R, Pinilla L, Carrillo M, Tena-Sempere M, Gómez A. Evidence for two distinct *KiSS* genes in non-placental vertebrates that encode Kisspeptins with different gonadotropin-releasing activities in fish and mammals. Mol Cell Endocrinol 2009; 312: 61-71.

15 Kitahashi T, Ogawa S, Parhar IS. Cloning and expression of *kiss2* in the zebrafish and medaka. Endocrinology 2009; 150: 821-831.

16 Li S, Zhang Y, Liu Y, Huang X, Huang W, Lu D, Zhu P, Shi Y, Cheng CH, Liu X. Structural and functional multiplicity of the kisspeptin/GPR54 system in goldfish (*Carassius auratus*). J Endocrinol 2009; 201: 407-418.

17 Mechaly AS, Viñas J, Piferrer F. The kisspeptin system genes in teleost fish, their structure and regulation, with particular attention to the situation in Pleuronectiformes. Gen Comp Endocrinol 2013; 188: 258-268.

18 Fernald RD, White RB. Gonadotropin-releasing hormone genes: phylogeny, structure, and functions. Front Neuroendocrinol 1999; 20: 224-240.

19 Steven C, Lehnen N, Kight K, Ijiri S, Klenke U, Harris WA, Zohar Y. Molecular characterization of the GnRH system in zebrafish (*Danio rerio*): cloning of chicken GnRH-II, adult brain expression patterns and pituitary content of salmon GnRH and chicken GnRH-II. Gen Comp Endocrinol 2003; 133: 27-37.

20 Hu W, Li SF, Tang B, Wang YP, Lin HR, Liu XC, Zou J, Zhu ZY. Antisense for gonadotropin-releasing hormone reduces gonadotropin synthesis and gonadal development in transgenic common carp (*Cyprinus carpio*). Aquaculture 2007; 271: 498-506.

21 Abraham E, Palevitch O, Gothilf Y, Zohar Y. Targeted gonadotropin-releasing hormone-3 neuron ablation in zebrafish: effects on neurogenesis, neuronal migration, and reproduction. Endocrinology 2010; 151: 332-340.

22 Xu J, Huang W, Zhong CR, Luo DJ, Li SF, Zhu ZY, Hu W. Defining global gene expression changes of the hypothalamic-pituitary-gonadal axis in female sGnRH-antisense transgenic common carp (*Cyprinus carpio*). PLoS ONE 2011; 6: e21057.

23 Lee YR, Tsunekawa K, Moon MJ, Um HN, Hwang J-I, Osugi T, Otaki N, Sunakawa Y, Kim K, Vaudry H. Molecular evolution of multiple forms of kisspeptins and GPR54 receptors in vertebrates. Endocrinology 2009; 150: 2837-2846.

24 Van de Peer Y, Maere S, Meyer A. The evolutionary significance of ancient genome duplications. Nat Rev Genet 2009; 10: 725-732.

25 Ogawa S, Parhar IS. Anatomy of the kisspeptin systems in teleosts. Gen Comp Endocrinol 2013; 181: 169-174.

26 Servili A, Le Page Y, Leprince J, Caraty A, Escobar S, Parhar IS, Seong JY, Vaudry H, Kah O. Organization of two independent kisspeptin systems derived from evolutionary-ancient *kiss* genes in the brain of zebrafish. Endocrinology 2011; 152: 1527-1540.

27 Abraham E, Palevitch O, Ijiri S, Du S, Gothilf Y, Zohar Y. Early development of forebrain gonadotrophin-releasing hormone (GnRH) neurones and the role of GnRH as an autocrine migration factor. J Neuroendocrinol 2008; 20: 394-405.

28 Thisse C, Thisse B. High-resolution *in situ* hybridization to whole-mount zebrafish embryos. Nat Protoc 2008; 3: 59-69.

29 Hendricks M, Jesuthasan S. Asymmetric innervation of the habenula in zebrafish. J Comp Neurol 2007; 502: 611-619.

30 Amo R, Aizawa H, Takahoko M, Kobayashi M, Takahashi R, Aoki T, Okamoto H. Identification of the zebrafish ventral habenula as a homolog of the mammalian lateral habenula. J Neurosci 2010; 30: 1566-1574.

31 Aizawa H, Bianco IH, Hamaoka T, Miyashita T, Uemura O, Concha ML, Russell C, Wilson SW, Okamoto H. Laterotopic representation of left-right information onto the dorso-ventral axis of a zebrafish midbrain target nucleus. Curr Biol 2005; 15: 238-243.

32 Gamse JT, Kuan Y-S, Macurak M, Bröosamle C, Thisse B, Thisse C, Halpern ME. Directional asymmetry of the zebrafish epithalamus guides dorsoventral innervation of the midbrain target. Development 2005; 132: 4869-4881.

33 Agetsuma M, Aizawa H, Aoki T, Nakayama R, Takahoko M, Goto M, Sassa T, Amo R, Shiraki T, Kawakami K, Hosoya T, Higashijima S-I, Okamoto H. The habenula is crucial for experience-dependent modification of fear responses in zebrafish. Nat Neurosci 2010; 13: 1354-1356.

34 Lee A, Mathuru AS, Teh C, Kibat C, Korzh V, Penney TB, Jesuthasan S. The habenula prevents helpless behavior in larval zebrafish. Curr Biol 2010; 20: 2211-2216.

35 Ogawa S, Ng KW, Ramadasan PN, Nathan FM, Parhar IS. Habenular Kiss1 neurons modulate the serotonergic system in the brain of zebrafish. Endocrinology 2012; 153: 2398-2407.

36 Ogawa S, Nathan FM, Parhar IS. Habenular kisspeptin modulates fear in the zebrafish. Proc Natl Acad Sci USA 2014; 111: 3841-3846.

37 Mitani Y, Kanda S, Akazome Y, Zempo B, Oka Y. Hypothalamic Kiss1 but not Kiss2 neurons are involved in estrogen feedback in medaka (*Oryzias latipes*). Endocrinology 2010; 151: 1751-1759.

38 Kanda S, Karigo T, Oka Y. Steroid sensitive Kiss2 neurones in the goldfish: evolutionary insights into the duplicate kisspeptin gene-expressing neurones. J Neuroendocrinol 2012; 24: 897-906.

39 Escobar S, Felip A, Gueguen MM, Zanuy S, Carrillo M, Kah O, Servili A. Expression of kisspeptins in the brain and pituitary of the European sea bass (*Dicentrarchus labrax*). J Comp Neurol 2013; 521: 933-948.

40 Shimizu Y, Tomikawa J, Hirano K, Nanikawa Y, Akazome Y, Kanda S, Kazeto Y, Okuzawa K, Uenoyama Y, Ohkura S, Tsukamura H, Maeda K-I, Gen K, Oka Y, Yamamoto N. Central distribution of Kiss2 neurons and peri-pubertal changes in their expression in the brain of male and female red seabream Pagrus major. Gen Comp Endocrinol 2012; 175: 432-442.

41 Ogawa S, Ng KW, Xue X, Ramadasan PN, Sivalingam M, Li S, Levavi-Sivan B, Lin H, Liu X, Parhar I. Thyroid hormone upregulates hypothalamic *kiss2* gene in the male Nile tilapia, *Oreochromis niloticus*. Front Endocrinol 2013; 4: 184.

42 Escobar S, Servili A, Espigares F, Gueguen M-M, Brocal I, Felip A, Gömez A, Carrillo M, Zanuy S, Kah O. Expression of kisspeptins and Kiss receptors suggests a large range of functions for kisspeptin systems in the brain of the European sea bass. PLoS ONE 2013; 8: e70177.

43 Okamoto H, Agetsuma M, Aizawa H. Genetic dissection of the zebrafish habenula, a possible switching board for selection of behavioral strategy to cope with fear and anxiety. Dev Neurobiol 2012; 72: 386-394.

44 Aizawa H. Habenula and the asymmetric development of the vertebrate brain. Anat Sci Int 2013; 88: 1-9.

45 Greives TJ, Humber SA, Goldstein AN, Scotti MAL, Demas GE, Kriegsfeld LJ. Photoperiod and testosterone interact to drive seasonal changes in kisspeptin expression in Siberian hamsters (*Phodopus sungorus*). J Neuroendocrinol 2008; 20: 1339-1347.

46 Kanda S, Akazome Y, Matsunaga T, Yamamoto N, Yamada S, Tsukamura H, Maeda K-I, Oka Y. Identification of *KiSS-1* product kisspeptin and steroid-sensitive sexually dimorphic kisspeptin neurons in medaka (*Oryzias latipes*). Endocrinology 2008; 149: 2467-2476.

47 Gingerich S, Wang X, Lee PKP, Dhillon SS, Chalmers JA, Koletar MM, Belsham DD. The generation of an array of clonal, immortalized cell models from the rat hypothalamus: analysis of melatonin effects on kisspeptin and gonadotropin-inhibitory hormone neurons. Neuroscience 2009; 162: 1134-1140.

48 Kinsey-Jones JS, Li XF, Knox AMI, Wilkinson ES, Zhu XL, Chaudhary AA, Milligan SR, Lightman SL, O'Byrne KT. Down-regulation of hypothalamic kisspeptin and its receptor, Kiss1r, mRNA expression is associated with stress-induced suppression of luteinising hormone secretion in the female rat. J Neuroendocrinol 2009; 21: 20-29.

49 Chalivoix S, Bagnolini A, Caraty A, Cognie J, Malpaux B, Dufourny L. Effects of photoperiod on kisspeptin neuronal populations of the ewe diencephalon in connection with reproductive function. J Neuroendocrinol 2010; 22: 110-118.

50 Parhar I, Ogawa S, Kitahashi T. RFamide peptides as mediators in environmental control of GnRH neurons. Prog Neurobiol 2012; 98: 176-196.

51 Abe H, Oka Y. Mechanisms of neuromodulation by a nonhypophysiotropic GnRH system controlling motivation of reproductive behavior in the teleost brain. J Reprod Dev 2011; 57: 665-674.

52 Yamamoto N, Oka Y, Kawashima S. Lesions of gonadotropin-releasing hormone-immunoreactive terminal nerve cells: effects on the reproductive behavior of male dwarf gouramis. Neuroendocrinology 1997; 65: 403-412.

53 Ogawa S, Akiyama G, Kato S, Soga T, Sakuma Y, Parhar IS. Immunoneu-tralization of gonadotropin-releasing hormone type-III suppresses male reproductive behavior of cichlids. Neurosci Lett 2006; 403: 201-205.

54 Okuyama T, Yokoi S, Abe H, Isoe Y, Suehiro Y, Imada H, Tanaka M, Kawasaki T, Yuba S, Taniguchi Y, Kamei Y, Okubo K, Shimada A, Naruse K, Takeda H, Oka Y, Kubo T, Takeuchi H. A neural mechanism underlying mating preferences for familiar individuals in Medaka fish. Science 2014; 343: 91-94.

55 Zhao Y, Lin M-CA, Mock A, Yang M, Wayne NL. Kisspeptins modulate the biology of multiple populations of gonadotropin-releasing hormone neurons during embryogenesis and adulthood in zebrafish (*Danio rerio*). PLoS ONE 2014; 9: e104330.

56 Zhao Y, Wayne NL. Effects of kisspeptin1 on electrical activity of an extrahypothalamic population of gonadotropin-releasing hormone neu-rons in medaka (Oryzias latipes). PLoS ONE 2012; 7: e37909.

57 Zohar Y, Muñoz-Cueto JA, Elizur A, Kah O. Neuroendocrinology of reproduction in teleost fish. Gen Comp Endocrinol 2010; 165: 438-455.

58 Keen KL, Wegner FH, Bloom SR, Ghatei MA, Terasawa E. An increase in kisspeptin-54 release occurs with the pubertal increase in luteinizing hormone-releasing hormone-1 release in the stalk-median eminence of female rhesus monkeys in vivo. Endocrinology 2008; 149: 4151-4157.

59 Ramaswamy S, Guerriero KA, Gibbs RB, Plant TM. Structural interactions between kisspeptin and GnRH neurons in the mediobasal hypothalamus of the male rhesus monkey (*Macaca mulatta*) as revealed by double immunofluorescence and confocal microscopy. Endocrinology 2008; 149: 4387-4395.

60 d'Anglemont de Tassigny X, Fagg LA, Carlton MBL, Colledge WH. Kiss-peptin can stimulate gonadotropin-releasing hormone (GnRH) release by a direct action at GnRH nerve terminals. Endocrinology 2008; 149: 3926-3932.

61 Kanda S, Akazome Y, Mitani Y, Okubo K, Oka Y. Neuroanatomical evidence that kisspeptin directly regulates isotocin and vasotocin neurons. PLoS ONE 2013; 8: e62776.

斑马鱼 Kiss1和 Kiss2阳性神经元的分布及其与 GnRH3神经元的关联

宋焱龙[1,2] 段小海[1,2] 陈 戟[1] 黄 伟[1] 朱作言[1] 胡 炜[1]

1 中国科学院水生生物研究所，武汉 430072
2 中国科学院大学，北京 100049

摘 要 Kisspeptin 是脊椎动物调节青春期启动和生殖行为的一类重要神经内分泌激素，其在调控 GnRH 的合成和分泌中发挥着关键作用。斑马鱼具有两个 *Kisspeptin* 基因，*kiss1*和 *kiss2*基因，它们在生殖调控中的作用存在差异。为了在斑马鱼体内研究 *Kiss1*和 *Kiss2*神经元的分布及其与 GnRH3神经元的相互关系，我们克隆了 *Kiss1*和 *Kiss2*基因的启动子，将其分别与 *mCherry* 基因整合构建转基因载体，并成功制备了 *Kiss1:mCherry* 和 *Kiss2:mCherry* 转基因斑马鱼，以及 *Kiss1:mCherry/GnRH3:EGFP* 和 *Kiss2:mCherry/GnRH3:EGFP* 双重转基因斑马鱼。实验结果表明，Kiss1:mCherry 斑马鱼脑部红色荧光标记的 Kiss1神经细胞特异分布于 dorsomedial habenula 和 ventromedial habenula。该区域两侧散发出的 Kiss1神经纤维组成 fasciculus retroflexus(FR)神经束并向后延伸到 ventral part of interpeduncular nucleus(IPN)区。GnRH3神经元束围绕 habenula 分布，并有少量 GnRH3神经纤维深入 habenula 内部与 Kiss1神经细胞直接接触。尤其 ventral habenula 两侧伸发出的 kiss1 纤维与 GnRH3神经元接触紧密。Kiss2:mCherry 斑马鱼脑部红色荧光标记的 Kiss2神经细胞集中在 dorsal zone of periventricular hypothalamus，少量分布于 periventricular nucleus of poaterior tuberculum。Kiss2神经纤维形成一个广泛的网络，投射到 subpallium、preoptic area、thalamus、ventral and caudal hypothalamus 和 mesencephalon。Kiss2神经元和 GnRH3神经元从端脑到下丘脑一直伴随分布，尤其在 preoptic area 和 ventral and caudal hypothalamus 接触非常频繁。斑马鱼脑部 kiss1, kiss2和 GnRH3神经元的分布位置及接触方式，提示 kiss1调控 nonhypophysiotropic neuromodulatory GnRH3 neurons 功能，而 kiss2主要调控 hypophysiotropic GnRH3 neruons 功能。

Direct Production of XY^{DMY-} Sex Reversal Female Medaka (*Oryzias latipes*) by Embryo Microinjection of TALENs

Daji Luo[1,2] Yun Liu[3] Ji Chen[2] Xiaoqin Xia[2] Mengxi Cao[2] Bin Cheng[1] Xuejuan Wang[1] Wuming Gong[4] Chao Qiu[1] Yunsheng Zhang[2] Christopher Hon Ki Cheng[3] Zuoyan Zhu[2] Wei Hu[2]

1 Department of Genetics, School of Basic Medical Sciences, Wuhan University, Wuhan, Hubei, P. R. China, 430071

2 State Key Laboratory of Freshwater Ecology and Biotechnology, Institutes of Hydrobiology, Chinese Academy of Sciences, Wuhan, Hubei, P. R. China, 430072

3 School of Biomedical Sciences, The Chinese University of Hong Kong, Shatin, Hong Kong, P. R. China

4 Lillehei Heart Institute, University of Minnesota, Minneapolis, USA

Abstract Medaka is an ideal model for sex determination and sex reversal, such as XY phenotypically female patients in humans. Here, we assembled improved TALENs targeting the *DMY* gene and generated XY^{DMY-} mutants to investigate gonadal dysgenesis in medaka. *DMY*-TALENs resulted in indel mutations at the targeted loci (46.8%). *DMY-nanos3UTR*-TALENs induced mutations were passed through the germline to F_1 generation with efficiencies of up to 91.7%. XY^{DMY-} mutants developed into females, laid eggs, and stably passed the Y^{DMY-} chromosome to next generation. RNA-seq generated 157 million raw reads from WT male (WT_M_TE), WT female (WT_F_OV) and XY^{DMY-} female medaka (TA_F_OV) gonad libraries. Differential expression analysis identified 144 up- and 293 down-regulated genes in TA_F_OV compared with WT_F_OV, 387 up- and 338 down-regulated genes in TA_F_OV compared with WT_M_TE. According to genes annotation and functional prediction, such as *Wnt1* and *PRCK*, it revealed that incomplete ovarian functionand reduced fertility of XY^{DMY-}mutant is closely related to the wnt signaling pathway. Our results provided the transcriptional profiles of XY^{DMY-} mutants, revealed the mechanism between sex reversal and *DMY* in medaka, and suggested that XY^{DMY-} medaka was a novel mutant that is useful for investigating gonadal dysgenesis in phenotypic female patients with the 46, XY karyotype.

Keywords TALEN; *DMY*; sex reversal; medaka; male heterogametic sex determination system

Over two decades have passed since the male determining gene, Sry, was identified[1]. The SRY gene plays a pivotal role in sex determination: point mutations or deletions of the SRY gene are found in approximately 15% of XY females in humans[2,3]. XY female syndrome, phenotypic female patients with the XY karyotype, has been studied clinically, cytogenetically, hormonally, endoscopically and histologically[4-6]. However, the transcriptional and post-transcriptional regulation

文章发表于 *Scientific Reports*, 2015, 5: 14057

mechanisms of SRY-induced XY female syndrome remain largely unknown. A lack of animal models has meant that it is difficult to perform in-depth studies of the pathological changes and to determine the mechanisms underlying the XY female syndrome.

Sex-determining systems in fish are very diverse, which could be determined by heredity, environment, or both. Meanwhile, the pathway of sex determination can be manipulated by administering exogenous sex steroids[7-11]. In non-mammalian species, which also have a XX-XY sex-determination system, SRY is not present at all. Until *DMY* gene of medaka (*Oryzias latipes*) was identified in the teleost fish[12, 13], it is believed that the XX-XY sex-determination system was conserved in a wide range of animals, including *C. elegans*, *Drosophila*, fish, and mammals. Although much is known about the master male sex-determining (SD) gene in medaka[12,14-16], the precise mechanisms involved in primary sex determination and sex differentiation remain undefined. It is difficult to perform specific gene targeting in medaka, because of the lack of methodologies for homologous recombination and embryonic stem cell derivation, which has impeded its use in male heterogametic (XX-XY) sex determination system studies.

Modifications of genomes have laid the foundation of functional studies in modern biology and have led to significant discoveries[17]. Recently, Zinc finger nucleases (ZFNs)[18-20], transcription activator-like effector nucleases (TALENs)[21-24] and clusters of regularly interspaced short palindromic repeats (CRISPR)[25-27] have been shown to edit genomic DNA in a variety of cell types and different model organisms at stable efficiency and specificity. In our previous studies, we modified the Golden Gate method to disrupt the gene and edit the genome in zebrafish[28,29] and medaka[30]. Here, we directly producedTALEN-induced XY^{DMY-} females in medaka using TALENs. Phenotype of XY^{DMY-} mutant is very similar to human XY female syndrome[4-6] and SRY KO mouse[3-31], especially as several individuals in the population were fertile.

The specificity of targeted sites and the off-target phenomenon are core problems in gene knockout and gene editing research. To test for potential nonspecific mutations induced by TALENs, we also developed a simple and reliable off-target prediction program for confirming TALEN-induced mutation. The TALEN-induced XY^{DMY-} medaka developed into females and laid eggs. Although the sex-determining function of the DMY protein is already recognized; however, as a transcription factor, how does it regulate the downstream factor(s) to control testis differentiation, development and germ cell maturation? To better explain *DMY*'s regulatory functions as a transcription factor, we performed RNA-seq, a recently developed approach to transcriptome profiling based on deep-sequencing[32], and generated 157 million raw reads from WT male (WT_M_TE), WT female (WT_F_OV) and XY^{DMY-} female medaka (TA_F_OV) gonad libraries. These transcriptomic data will contribute to unravel the relationship and mechanism between sex reversal and the *DMY* gene. Our results suggest that the medaka XY^{DMY-} mutant is a novel mutant line that is useful for investigating XY to XX sex reversal and gonadal dysgenesis in phenotypically female patient with the 46, XY karyotype.

Results

***DMY*- and *DMY*- *nanos3UTR*-TALENs effectively induced *DMY* gene disruption in Medaka**

To improve the germline integration efficiency, we incorporated the zebrafish *nanos*-3′UTR into the TALEN construct (Figure 1A). Potential TALEN target sites were scanned and designed in the exon of the *DMY/DMRT1bY* gene (ENSORLT000000025382 and ENSORLT000000025383) (Figure 1B). We generated TALEN constructs using our previously published method[29]. The mixture containing a pair of TALEN mRNAs was microinjected into one-cell stage embryos of medaka (Figure 2A and File S1). 72 hours after microinjection, ten injected embryos were randomly pooled for extracting genomic DNA. As illustrated in Figure 2B, primers DMY F and DMY R bridge both the effector binding element (EBE) regions, while primers DMY F1 and DMY R link the spacer region and the downstream EBE. If primer DMY F and DMY R generated a 396 bp fragment, while primer DMY F1 and DMY R failed to generate the167 bp fragment, the result suggested that the targeted gene was disrupted by the TALENs (Figure 2C). Sequenced PCR positive clones had mutated sequences in the spacer (Figure 2D). Both *DMY*-TALENs and *DMY*-Nanos-3UTR TALENs were effective at disrupting the targeted genes in medaka embryos. Various concentrations (200, 400, 600 and 800 pg) of the TALENs mRNA were microinjected (Figure 2E, 2F, File S1 and S2). The targeting efficiency of the *DMY*-TALENs was good, with a higher TALEN-induced mutation ratio of 42.07% and lower levels of dead and deformed embryos when 600 pg mRNA was microinjected (Figure 2G and 2I). Similarly, the targeting efficiency of *DMY*-nanos-3UTR TALENs was about 41.96% (Figure 2H and 2I). Thus, 600 pg was determined as the appropriate concentration in medaka. These results indicated that the TALEN activity was dose-dependent, and high-dose microinjection of TALEN mRNA might cause nonspecific and toxic defects in medaka embryos (Figure 2E, 2F, 2G, 2H, 2I, File S1 and S2).

Successful germline transmission is essential to establish knockout lines. To evaluate the germline transmission efficiency of the TALEN-mediated gene disruption, ten embryos for *DMY* and *DMY*-nanos3UTR from each independent cross were individually collected at 3 days post fertilization (dpf), and genomic DNA was extracted from each cross to assess mutagenesis at the TALEN-targeted site (Figure 2C and 2D). 9.02% of *DMY*-TALEN-induced F1 embryos carried mutations; and 37.56% of *DMY*-nanos3UTR-TALEN-induced F1 embryos carried mutations in the *DMY* gene (Figure 2I and File S2). The higher proportion induced by *DMY*-nanos3UTR-TALEN indicated that a majority of gametes in the F0 medaka were mutant. These results indicated that there was no significant difference in the targeting efficiency of F0 somatic mutations at the targeted loci (Figure 2H and 2I); however, *DMY*-nanos3UTR-TALEN induced a higher portion of mutations in the germline than *DMY*-TALENs did.

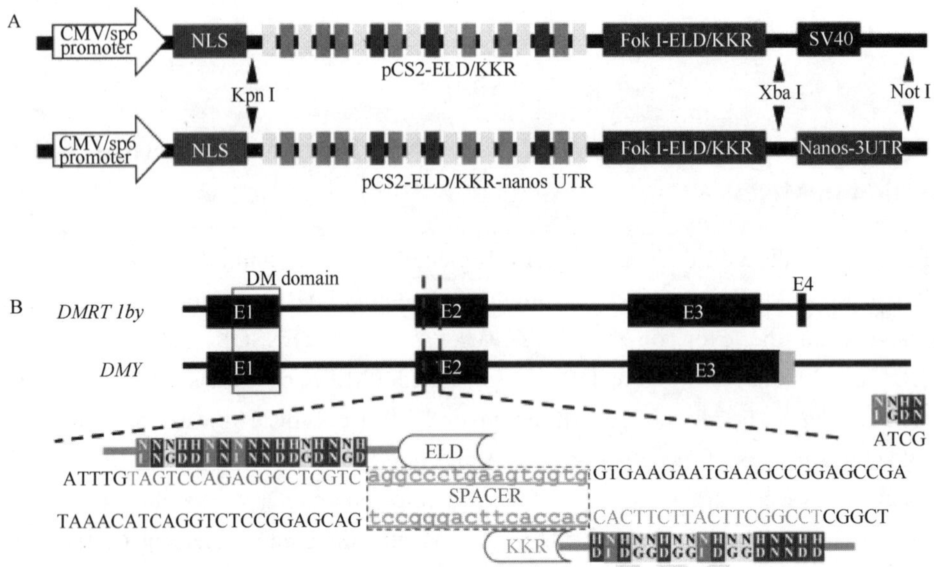

Figure 1 The design and assembly of *DMY*-TALEN sites. (A) Reconstructions of pCS2- *DMY*-TALENs-ELD/KKR-nanos UTR plasmids. (B) The location and sequence information of *DMY*-TALENs. Red uppercase letters indicate the DNA sequence of *DMY*-TALEN-L sites. Green uppercase letters indicate the complementary paired DNA sequence of *DMY*-TALEN-R sites. Lowercase letters indicate the spacer region of TALEN sites

A novel program to identify potential off-target sites of TALENs

To test for potential nonspecific mutations induced by TALENs, we designed a program to scan the medaka genomic sequence (http://www.ensembl.org/Oryzias_latipes) to identify potential off-target sites potentially targeted by *DMY* TALENs. Potential off-target sites of *DMY*-TALENs were searched using the program and 55 candidate sites were identified (File S3). When the spacers were less than 10 bp or more than 24 bp long, the scaffold of TALENs had lower disrupting activity[33,34]. Five of the 55 candidates had spacers less than 10 bp, and 43 of 55 candidates had spacers of more than 24 bp, indicating that it is unlikely that the TALENs could induce mutations at these sites. We analyzed one candidate site (Chr. 3: 36,197,357-36,197,403) that had 7-bp mismatches in the recognition sequences and a 12-bp spacer (File S4). PCR amplified the identified potential off-target regions using genomic DNA from TALEN-injected embryos as template; no mutations were found at these sites by DNA sequencing. This result suggested that the novel program could predict the potential off-target sites of TALENs; and that TALENs have high specificity for their target sequences.

Mating scheme of TALENs-induced *DMY*-mutants and mutant phenotypes

The mating scheme for the TALENs-induced *DMY*-mutant lines is shown in Figure 3A. The F0 generation was produced by microinjecting 600 pg *DMY*-nanos3UTR-TALENs into

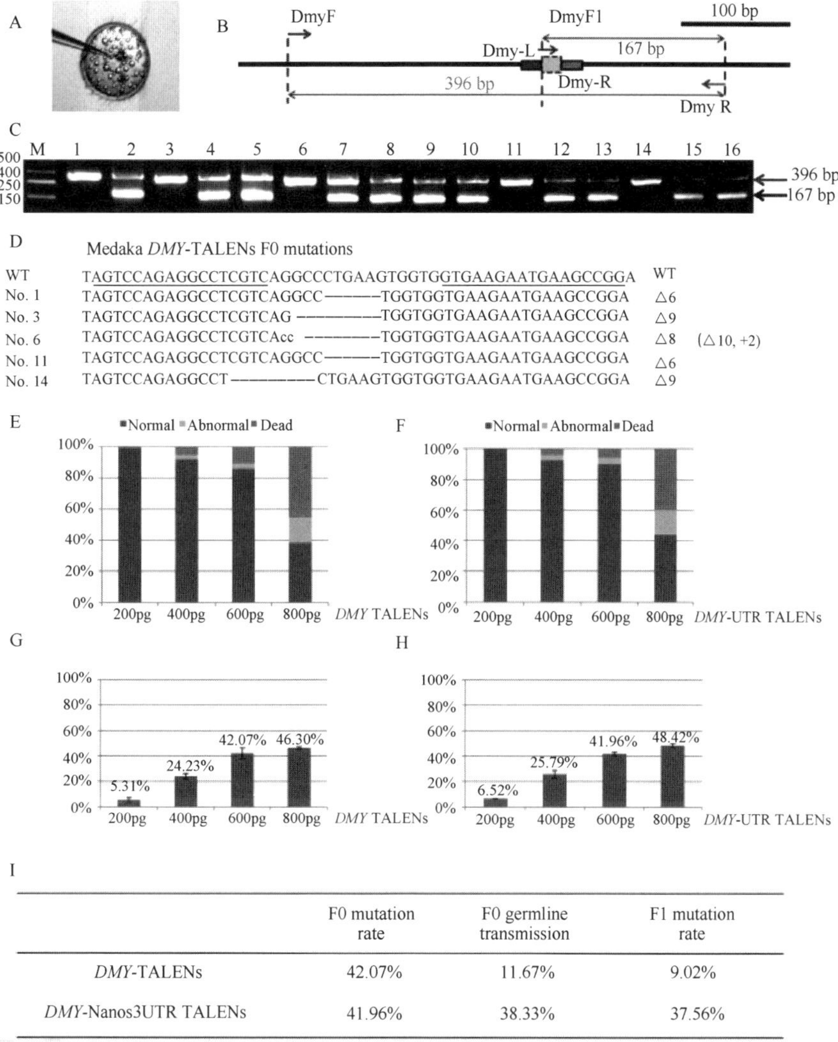

	F0 mutation rate	F0 germline transmission	F1 mutation rate
DMY-TALENs	42.07%	11.67%	9.02%
DMY-Nanos3UTR TALENs	41.96%	38.33%	37.56%

Figure 2 The dosage effects and efficiency evaluation of *DMY*-TALENs. (A) Microinjection of TALENs mRNA into medaka. (B) Detection of mutations in TALEN targeted medaka embryos. Primers DMY F, DMY F1 and DMY R were used to amply the *DMY* gene. Primers DMY F and DMY R bridge both EBE regions, while primers DMY F1 and DMY R link the spacer region and the downstream EBE. DMY F and DMY R generated a 396 bp PCR fragment. DMY F1 and DMY R generated a 167 bp PCR fragment. If primer DMY F and DMY R generated a PCR fragment, while primer DMY F1 and DMY R failed to do so, this suggested that the targeted gene is disrupted by the *DMY*-TALEN. (C) Electrophoretic detection of mutations in TALEN-injected medaka embryos. Line 1 to 16, TALENs injected embryos. 1, 3, 6, 11, 14 show mutated embryos. (D) Sequencing detection of mutations in TALEN-induced medaka embryos. -, deleted nucleotide; lowercase letter, added nucleotide. +, insertions; Δ, deletions. (E) Evaluation of embryonic toxicity of *DMY*-TALENs. (F) Evaluation of embryonic toxicity of *DMY*-nanos3UTR-TALENs. (G) The targeting efficiency statistics of *DMY*- TALENs. (H) The targeting efficiency statistics of *DMY*-nanos-3UTRTALENs. (I) Comparative analysis of mutation rate and germline transmission rate between *DMY*-TALENs and *DMY*-nanos3UTR-TALENs

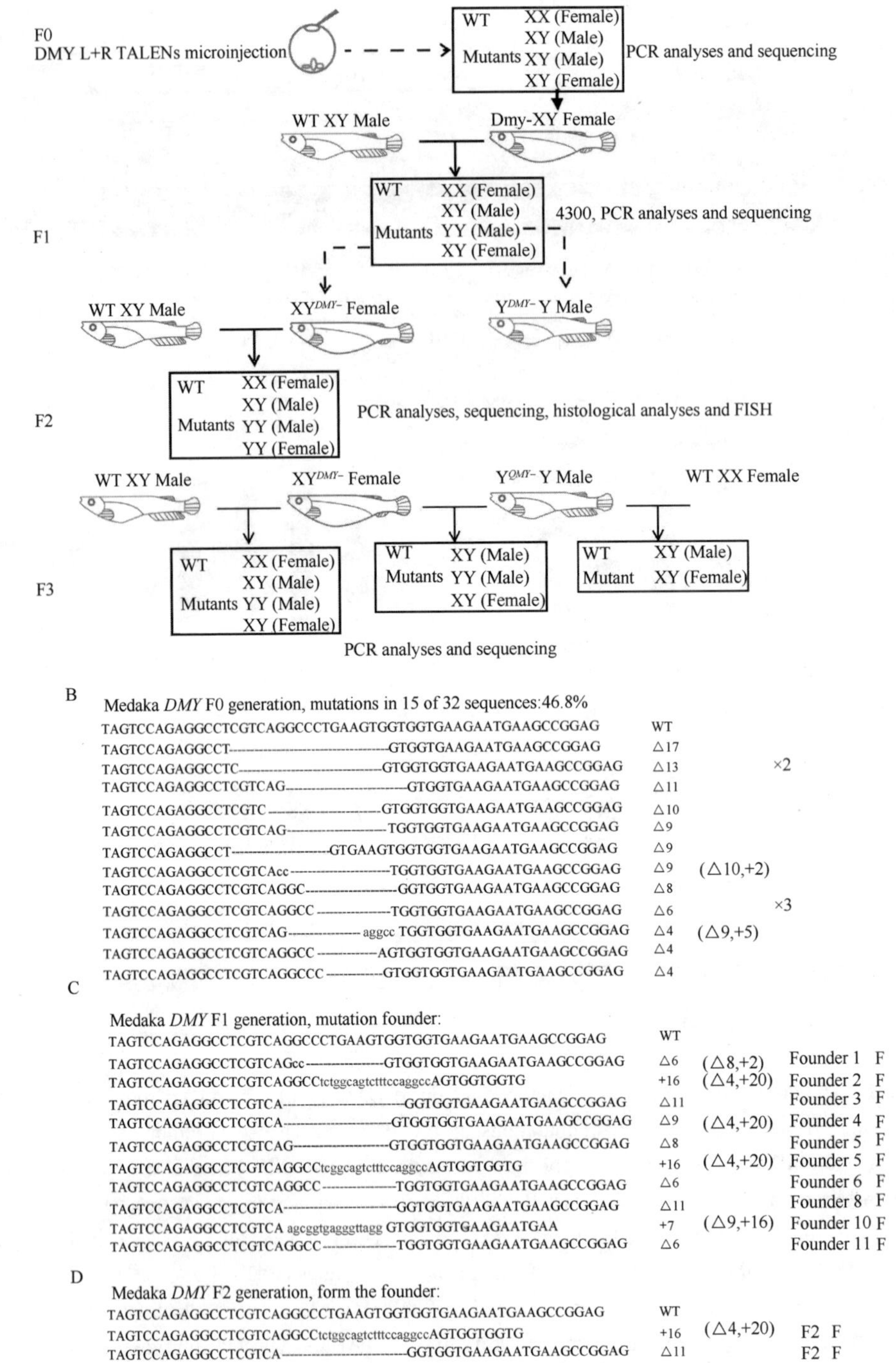

Figure 3 Flowchart and establishment of *DMY*-TALENs-induced mutant lines. (A) Flowchart of *DMY*-TALENs-induced mutant lines. (B) The genotypes of TALENs-induced mutations in the F0 generation. (C) The genotypes of TALENs induced F1 founders. (D) The genotypes of TALENs induced F2 mutant lines. Red lowercase letters indicate an additional nucleotide; "-" indicates a deleted nucleotide. +, insertions; Δ, deletions. F, female; M, male

one-cell stage embryos of medaka; the mutation rate was 46.8% (15/32) (Figure 3B). The *DMY* gene knockout medaka could develop into females. 11-bp deletions (named DMYΔ11) and 16-bp insertions (named DMY+16) were identified and chosen to establish mutant lines (Figure 3C and 3D). Notably, during the establishment of the mutant lines, two genotyping alleles of *DMY* gene fragments were identified in individual TALEN-induced F1 mutations using the Li-con 4300 system (Figure 4A). This result indicated that $Y^{DMY-}Y$ male mutants were present in the testcross F1 generation. The genetic males (XY) of the *DMY* gene mutants, DMYΔ11 and DMY+16 mutant types, developed into females in the F2 generation, which was identified using genomic PCR of the *DMY* gene (Figure 4B). To confirm un-expression and dysfunction of *DMY* gene in the XY^{DMY-} F2 generation, using RT-PCR, expression of *DMY* gene was not identified in the XY^{DMY-} female medaka (both DMYΔ11 and DMY+16) (Figure 4C). The first morphological sex difference manifested in the gonads is reflected in the number of germ cells[14, 35]. The number of germ cells in several DMY mutants identified from wild populations resembled that of the female[13, 16, 36]. To elucidate sex reversal during the development of XY^{DMY-} mutants, we evaluated the effect of DMY on germ cell number at 5 days after hatching (DAH) in the XY^{DMY-} F3 generation. The XY^{DMY-} mutant fry had more germ cells than that of the WT XY male at 5 DAH (Figure 4D). This implied sex reversal of XY^{DMY-} mutants took place in early developmental stages, and the increased number of germ cells in the XY^{DMY-} mutants may be due to the disruption of *DMY* gene expression.

According to the amino acid sequence of DMY, DMYΔ11 and DMY+16 are frame-shifted mutant alleles that would produce truncated DMY protein caused by a region of altered translation (Figure 4E). Thus, an error of the coding sequence has occurred in the mutated *DMY* gene that resulted in the loss function of the *DMY* gene. The XY^{DMY-} female or $Y^{DMY-}Y$ male (N=10) from F2 generation that had lost the *DMY* gene were crossed with the WT to obtain the testcross F3 generation. XX female, XY male, XY^{DMY-} mutant female and $Y^{DMY-}Y$ mutant male were identified in the XY female testcross F3 generation. XY males and XY^{DMY-} mutant females were identified in the $Y^{DMY-}Y$ male testcross F3 generation. This showed that the DMY gene in the Y chromosome of WT medaka rescued the female phenotype of *DMY* gene disruption in mutated XY female medaka. Unfortunately, an $Y^{DMY-}Y^{DMY-}$ mutant female with a genomic homozygous *DMY* gene mutation on the Y chromosome was never found.

Mature XY^{DMY-} mutants in the F3 generation were obtained for phenotype identification, histological analyses and fluorescence *in situ* hybridization (FISH) (Figure 5). There are significant differences between females and males in the size and shape of dorsal and anal fins[13, 14], which are main part of the secondary morphological sexual characteristics in medaka. The shape of both dorsal and anal fins of XY^{DMY-} sex reversal female were similar to that of the WT XX female, rather than to that of the WT XY male. The size of both dorsal and anal fins of XY^{DMY-} female was significant smaller than those of the WT XY males (Figure 5A). In addition, ovarian tissue was identified in XY^{DMY-} mutants (Figure 5B). To confirm the Y chromosome in the

TALENs-induced mutants, we analyzed the karyotypes of metaphase cells from the WT male, WT female and XY^{DMY-} female using FISH showing the male specific hybridization signal[12]. Compared with the two spots in females, three hybridization spots for the specific probe were identified in males (Figure 5C). The additional FISH signal in males is on the Y chromosome. XY^{DMY-} mutants did not express *DMY* gene, which were different from WT XY individuals (Figure 4C). The gonadosomatic index (GSI) and the maturation stages of oocytes are commonly used to evaluate the gonad and gonad development[37]. At 90 days after hatching, the GSI of XY^{DMY-} mutated female was 10.9 (N=10); that of the WT female and WT male

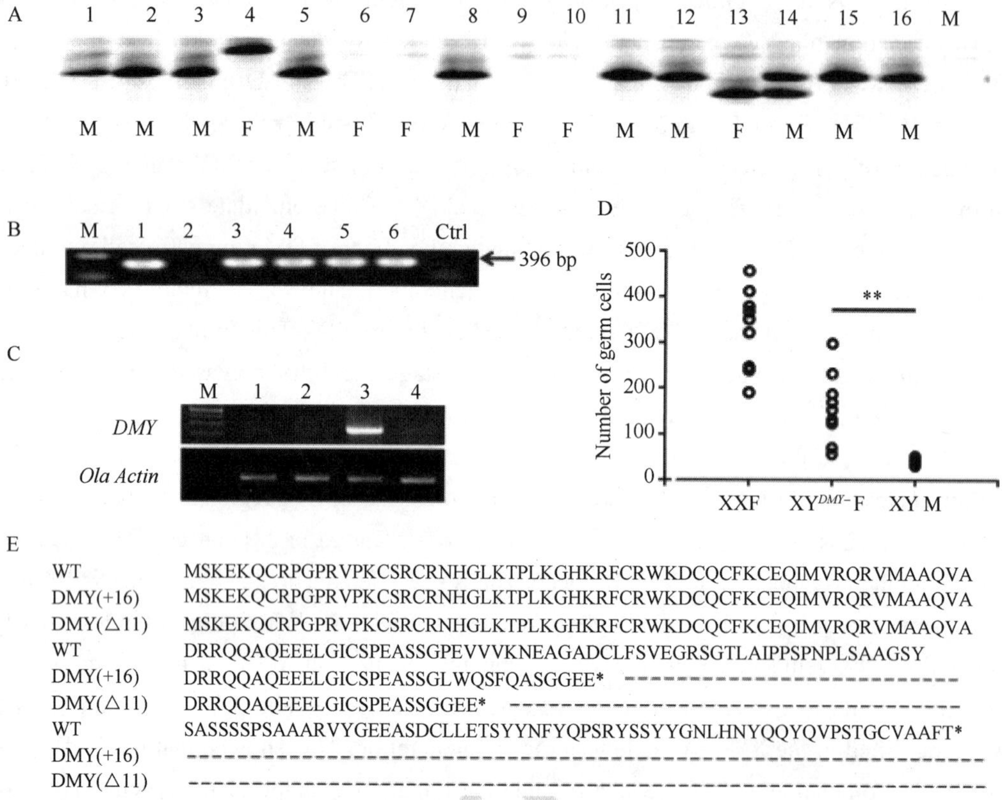

Figure 4 *DMY* genotyping, expression and CDS frameshift mutations of *DMY* mutants. (A) Genotyping of *DMY* gene fragments using Li-con 4300 system. 1 to 16, randomly selected individuals in F1 generation. F, female; M, male. (B) Genomic PCR confirmed the *DMY* gene. M: marker DL2000; 1: WT Male; 2: WT Female; 3: Founder 2; 4: the female mutant from the F2 generation of Founder 2; 5: Founder 3; 6: the female mutant from the F2 generation of Founder 3; Ctrl: no template control. (C) Reverse transcription-PCR with *DMY*-specific primers. *OLA Actin* expression was determined for calibration. M: 250 bp Marker; 1. XY^{DMY-} mutant (DMYΔ11); 2. XY^{DMY-} mutant (DMY+16); 3. WT XY male; 4. WT XX female. (D) Numbers of germ cells in mutants and WT medaka fry at 5DAH. XX F: WT XX females; XY^{DMY-} F: XY^{DMY-} matants; XY M: WT XY males. Open circles represent the number of germ cells in individuals (N= 9). $^{**}P< 0.001$. (E) The CDS frameshift mutations of TALEN-induced *DMY* gene. WT: CDS sequence of DMY

were 19.1 and 0.7, respectively (Figure 5D). During the generation of offspring from the XY^{DMY-} mutated female testcross, we found that it was difficult to obtain a sufficient number of offspring. Therefore, we performed the comparative analysis of the number of matured oocytes between the WT female and XY^{DMY-} mutants (N=10). During 10 days' embryos collecting, the XY^{DMY-} mutated female produced six eggs per day on average; whereas, the WT XX female produced 20 eggs per day (Figure 5E). The results from F2 phenotypes revealed that the genetic males (XY) of TALENs-induced *DMY* gene disruption mutants all developed into females and laid eggs (DMYΔ11 and DMY+16 mutants). Furthermore, histological analyses demonstrated that all XY mutants developed into females; however, there were significant differences in germ cells and gonad (Figure 5F). The ovary of XY^{DMY-} mutants (12/15) appeared to have

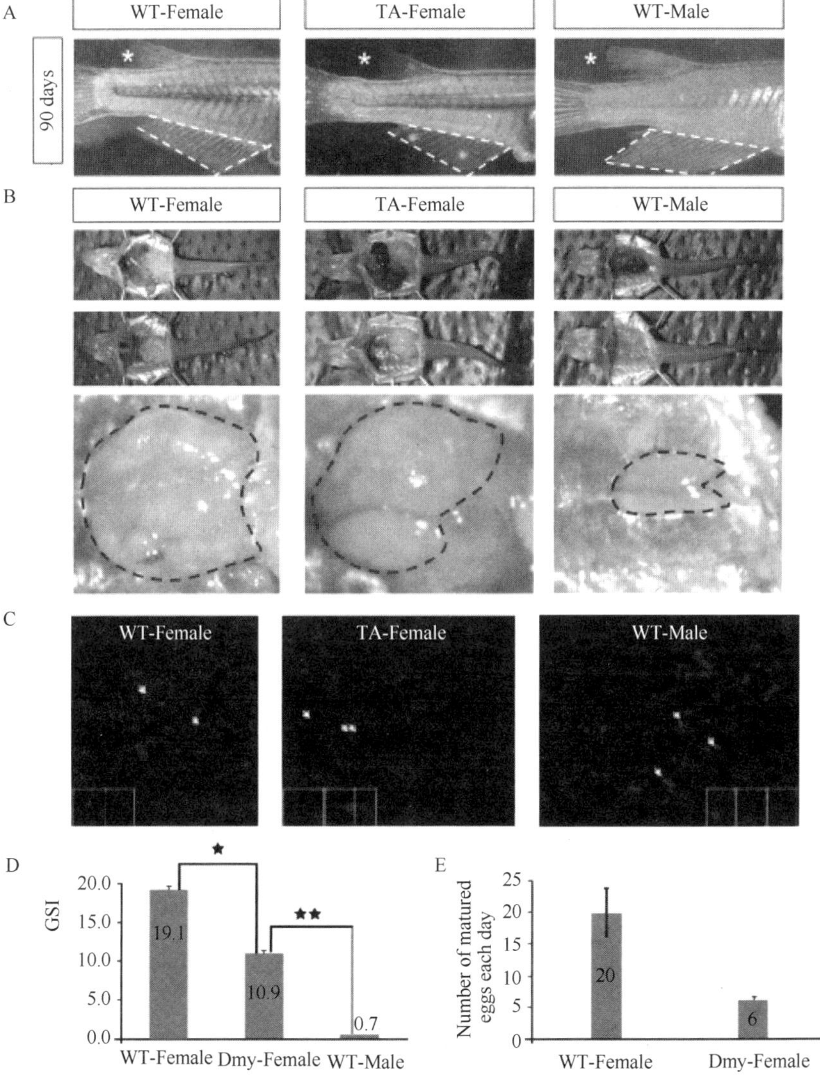

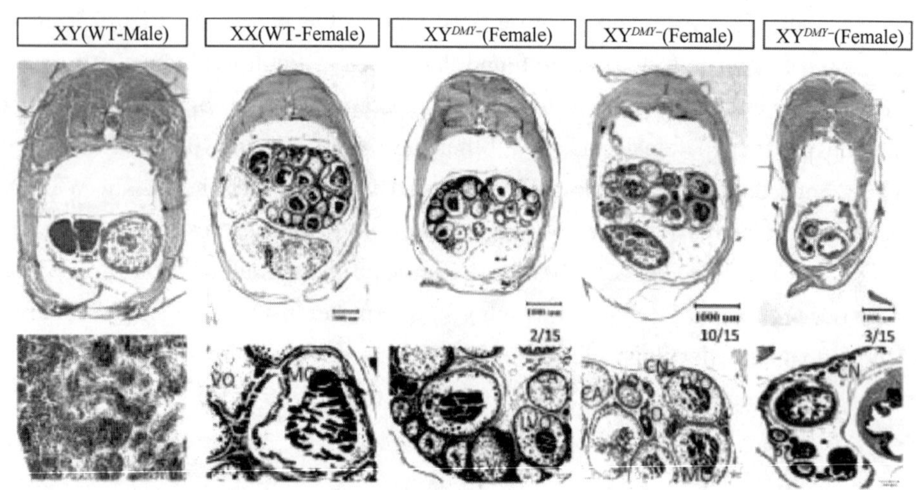

Figure 5 Phenotypic identification and analysis of XY^{DMY-} mutants. (A) Phenotypic diagnostics of thesecondary sexual characters in XY^{DMY-} mutants. White dashed area shows the anal fin of medaka. *shows the dosal fin of medaka. (B) The gonad of WT and TA mutants. The red dashed area shows the gonad tissue of medaka. (C) Fluorescence *in situ* hybridization of the karyotypes of metaphases cell from WT male, WT female and XY mutant female. The pink signal is the male specific hybridization signal. (D) The comparative analyses of gonadosomatic index (GSI) among WT-F, WT-M and XY mutant female (TA-F). *P< 0.5; **P< 0.001. (E) The comparative analyses of oocytes maturation between WT-F and TA-F. (F) Histological analyses of gonad tissue. CN, chromatin nucleolar oocytes; PO, perinucleolar oocytes; CA, cortical alveolar oocytes; EVO, early vitellogenic oocytes; VO, vitellogenic oocytes; LVO, late vitellogenic oocytes; MO, mature oocytes

fewer oocytes than that of the WT XX female (Figure 5F). According to the developmental stage of oocytes in medaka, there were equilibrium distributions in each stage of oocytes in the WT XX female ovary, containing cortical alveolar oocytes (CA), vitellogenic oocytes (VO) and mature oocytes (MO). Among 15 XY^{DMY-} mutants (15 mutants were identified in the same genotype, fed in the same tank and sampled at the same time), three did not form mature oocytes, with majority of chromatin nucleolar oocytes (CN), several perinucleolar oocytes (PO) and few CA oocytes in the ovary, and no further developmental stage oocytes (Figure 5F). Twelve of them successfully formed MOs, however, the size of ovary and total number of each developmental stage oocytes were significant smaller and fewer than that of WT XX female (Figure 5F). This can also explain it is difficult to collect embryos in XY^{DMY-} mutants, even there is sometimes no embryos could be collection (Figure 5E). The analysis of gonadal histology, GSI, and the statistics of mature eggs demonstrated significant differences in the gonadal development and maturity between WT females and XY^{DMY-} mutant females.

Transcriptomic analysis of TALENs-induced *DMY*– mutants and WT

To better explore the transcriptional regulation function of *DMY*, and identify *DMY*-related downstream factors that affect the generation, development and maturation of testis or ovary, we used RNA-seq to analyze the transcriptome of WT_M_TE, WT_F_OV and TA_F_OV after 90 days. For both the WT and *DMY*-mutants, ten samples of gonadal tissues were mixed for library construction. RNA-seq generated 157 million raw reads comprising 10333, 8746 and 8621 transcripts, respectively (Figure 6A). Using the Ensembl medaka genome database as the reference, 75.8% of the raw reads matched medaka genomic sequences (File S6). Twelve genes were selected randomly from hundreds of different expression transcripts between the gonad of WT and that of *DMY* mutants. The result of real-time quantitative PCR showed that the trends of these genes expression were the same as in the RNA-seq data (File S7). Thus, the RNA-seq information was accurate and reliable.

To isolate *DMY*-related gonad developmental-regulated genes, we conducted a comparative analysis of the three transcriptomic groups. Comparing WT_M_TE with TA_F_OV, there were 9407 differentially expressed transcripts (Data A, Dataset 1). The only difference between WT and TA medaka is the absence of the *DMY* gene, more or less, these differentially expressed genes were attributed to the loss of the *DMY* gene. There were 4629 upregulated genes and 4778 downregulated genes in WT-M-Te (Figure 6B and 6C). The functions of these genes were mainly related to cilium morphogenesis (191), cilium organization (152), cilium assembly (152), spermatid differentiation and development, wnt signal pathway (77) (Dataset 1). The master male SD marker, *DMY*[12, 13] (*Dmrt1Y*, Uingene20290/Unigene19692), did not expressed in TA_F_OV (Dataset 1), supporting previous observations that *DMY* gene was successfully disruptedin XY^{DMY-} mutants (Figure 4C). The male sex-differentiation or testes maintaining marker *Dmrt1a*[38] (Unigene42535/Unigene30831), *Gsdf*[39] (Unigene44032), *Sox9b*[40] (Unigene18419) were also significantly down-expressed in TA_F_OV (Dataset 1). Interestingly, the majority of genes in wnt signal pathway were differentially expressed between WT_M_TE and TA_F_OV. *Wnt1* (Unigene54610), *Wnt5a* (Unigene50823), *Wnt5b* (Unigene25302), *Wnt9* (Unigene32218), *Wnt11* (CL6243), *PRCK* (CL13707, PREDICTED: *Oryzias latipes* PRKC apoptosis WT1 regulator protein-like), *DKK* (Unigene58048) and *PRKCA* (Unigene17297) were up-expressed in TA_F_OV; *Wnt2* (Unigene 27916), *Wnt8a* (Unigene51222), *FZD6* (CL9539) and *FZD8* (Unigene49832) were down-expressed in TA_F_OV (Dataset 1). These genes also contribute significantly to the molecular supporting the sex-reversal female phenotype of XY^{DMY-} mutants (Figure 5A and 5B).

Theoretically, a number of differentially expressed genes between the testis (WT_M_TE) and ovary (WT_F_OV) (Data B) might confound Data A. More accurately, Data B must be excluded from Data A. The intersection of Data A and Data B identified the real differentially expressed genes between the WT XY male testis and XY^{DMY-} female ovary (Data C, Dataset 2), which were the sum of 309 transcripts, 276 transcripts, 144 transcripts, 293 transcripts, 62 transcripts and 78 transcripts, which are *DMY*-related or affected genes (Figure 6F). Relative to WT_F_OV, 144 ovary-specific transcripts were downregulated and 293 ovary-specific transcripts were upregulated in male-to-female reversed gonads (Figure 6F). 62 transcripts in

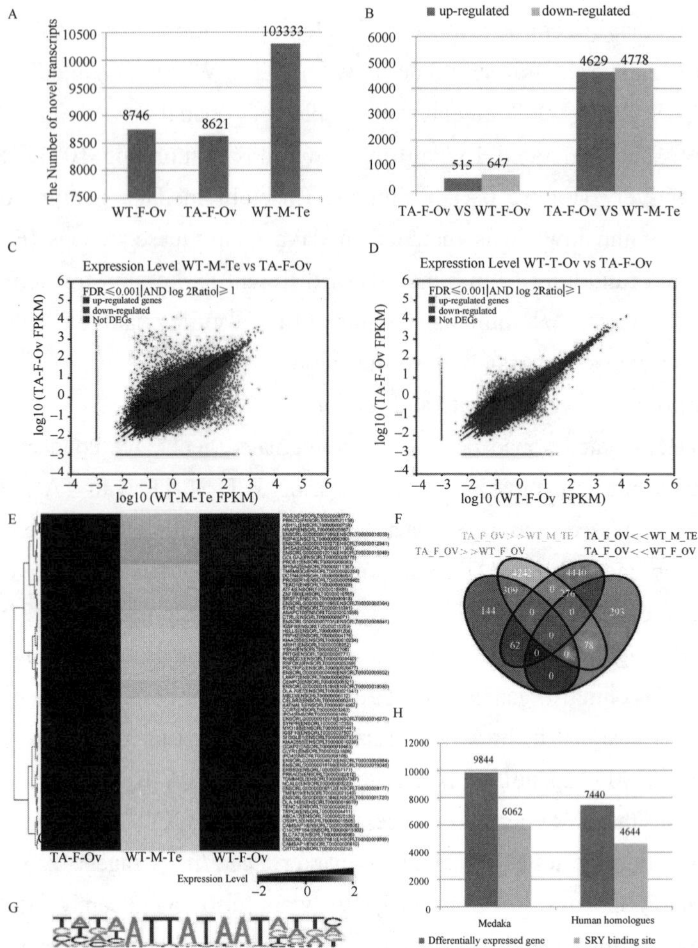

Figure 6 Bioinformatic analyses of RNA-seq data. (A) The number of novel transcripts in the RNA-seq data the WT female (WT_F_OV), the wild-type male (WT_M_Te) and XY^{DMY-} female medaka (TA_F_OV). (B) The differentially expressed transcripts between TA_F_OV and WT_F_OV/WT_M_Te. (C) Correlation of gene expression between WT_M_Te and TA_F_Ov. The up- and downregulated genes are shown in red and green, respectively. Non-differentially expressed genes are shown in blue. (D) Correlation of gene expression between WT_F_Ov and TA_F_Ov. The up- and downregulated genes are shown in red and green, respectively. Non-differentially expressed genes are shown in blue. (E) The cluster of testis-specific expressing transcripts. Cluster analyses of differentially expressed genes among WT_F_Ov, WT_M_Te and TA_F_Ov. The high- and low-expressed genes are shown form yellow to blue, corresponded to the expression level from negative 2 fold to positive 2 fold. (F) The differentially expressed transcripts among TA_F_Ov, WT_F_Ov and WT_M_Te. The up- and downregulated genes in WT_M_Te were shown in Fuchsia and yellow, respectively. The up- and downregulated genes in WT_F_Ov are shown in green and purple, respectively. TA_F_Ov ≫ WT_F_Ov means the genes were upregulated in TA_F_Ov. TA_F_Ov ≪ WT_F_Ov means the genes were downregulated in TA_F_Ov. TA_F_Ov ≫ WT_M_Te means genes were upregulated in TA_F_Ov. TA_F_Ov ≪ WT_M_Te means the genes were downregulated in TA_F_Ov. (G) SRY binding sites. (H) The SRY binding sites analyses of differentially expressed genes among TA_F_Ov, WT_F_Ov and WT_M_Te

XY^{DMY-} female ovary were downregulated relative to WT_M_TE and were upregulated relative to WT_F_OV. 78 transcripts in XY^{DMY-} female ovary were upregulated relative to WT_M_TE and were downregulated relative to WT_F_OV. Relative to WT_F_OV and WT_M_TE, 309 transcripts were upregulated in TA_F_OV, and 276 transcripts were downregulated in TA_F_OV (Figure 6F and Dataset 2). Seventy-three testis-specific transcripts were identified (Figure 6E and File S8). GO analysis and homologous annotation with human genes is a traditional way to further predict the functions of the genes in Data C (Dataset 2). According to the annotation of Blast2GO, genes of Data C were associated with the regulation of ubiquitination and fertilization. Moreover, a large number of genes, such as ENSORLT00000000529, have not been previously reported.

GO analysis is not a straightforward way to predict the relationships among genes. Sry and *DMY* are transcription factors; theoretically, their downstream genes should have the direct binding regions for Sry or *DMY*. The predicted binding site of human SRY is shown in Figure 6G. Human homologous genes of data C were scanned for SRY binding sites (Dataset 3). If SRY binding sites could be found in the human homologous genes, we could speculate that their medaka homologous genes may have the *DMY* binding sites or be a directly affected gene. There were 9844 unique genes in medaka that were analyzed: 7440 of which had homologous genes in the human genome. There were 4644 human homologous transcripts that may have more than one potential SRY binding site (Figure 6H and Dataset 3). A number of potential *DMY* regulated genes, such as SLC25A38, had a potential SRY binding site in the upstream region of its human homolog; and a novel transcript, ENSORLT00000000529, had six potential SRY binding sites in the upstream region of its human homolog. These genes are significant for investigating the transcriptional function of the *DMY* gene.

When we compared TA_F_OV with WT_F_OV, 1163 differentially expressed transcripts (Data D) were found in XY^{DMY-} female (Dataset 4). There were 515 upregulated genes and 647 downregulated genes in WT-F-Ov (Figure 6B and 6D). Using Blast2GO, the differentially expressed genes were blasted and annotated on biological process (BP), cell components (CC), and molecular function (MF). The genes of Data D are associated with the wnt receptor signaling pathway (Predicted: syntabbulin-like, axin-2-like), ovarian follicle development (Forkhead box O5, Predicted: beta-arrestin-1-like, adenomatous polyposis coli protein-like, ubiquitin-protein ligase E3A-like, bone morphogenetic protein receptor type-1B-like), and the follicle-stimulating hormone signaling pathway (luteinizing hormone receptor, lhr, Predicted: beta-arrestin-1-like) (Dataset 4). In terms of the number and repetitions of the genes, the genes of the wnt signaling pathway were the most significant proportion of Data D. *PRCKA* (CL423) and *DKK* (Unigene31972) were up-expressed in TA_F_OV; *Wnt1* (CL10671), *MAPK8* (CL4166), *FZD6* (CL9539) and *PRCK* (CL13707) were down-expressed in TA_F_OV. These genes, especially *Wnt1* and *PRCK* (CL13707), were down-expressed in

XY^{DMY-} mutants, compared with WT XX females (Dataset 4). Therefore, we reasonably believed that the wnt signaling pathway is one of the vital pathways on regulating the development and mature of oocytes in XY^{DMY-} mutants, why the ovarian function of XY^{DMY-} mutants are significant lower than that of WT females. Comprehensive Data A, B, C, and D, greatly promote research into the transcriptional function of the *DMY* gene in medaka, not only in sex reversal, but also in the normal development and maturation of gonads.

Discussion

Nanos3UTR- TALENs effectively generated a targeted gene mutant line in Medaka

Germline transmission of the knockout genotypes is critical to obtain homozygous gene knockout animals. Generally, a 46.8% mutation rate in the F0 is not ideal for gene disruption. Perhaps this low efficiency should be attributed to the *DMY* gene on the Y chromosome rather than our optimized TALENs. In general, there should be equal numbers of male and female embryos in a generation. This means that the targeting efficiency of *DMY*-TALENs could not beyond be greater than 50%. Indeed, in our results, the efficiency was never more than 50%, even if the concentration of TALENs was increased to 800pg (Figure 2I). To improve the germline integration efficiency, we incorporated the zebrafish *nanos*-3′UTR into the TALEN construct (Figure 1A), which was reported to protect mRNA from degradation in primordial germ cells and improve the germ cell targeting efficiency[28, 41, 42]. Compared with 9.02% *DMY*-TALEN-induced F1 embryos carrying mutations; there were 37.56% *DMY*-nanos3UTR-TALEN-induced F1 embryos that carried mutations in the *DMY* gene (Figure 2I and File S2). These results indicated that *DMY*-nanos3UTR-TALEN induced a high proportion, possible a majority, of mutant gametes in the F0 medaka. Thus, the incorporation of the *nanos*-3′UTR into the TALEN construct improved the germ cell targeting efficiency in medaka, permitting the generation of medaka knockout lines.

A novel program to identify potential off-target sites of TALENs

Specificity is essential to establish precisely targeted gene knockout lines. TALENs have become an accepted tool for targeted mutagenesis, but undesired off-targets, in addition to the targeted genomic region, remain an important issue[15,21,43-47]. Unfortunately, using e-PCR to perform BLAST searches, potential off-target sites were identified in several studies[46]. Using Primer3 and BLAST, potential off-target sites of the *DMY* gene in medaka were identified (File S5). According to this data, BLAST emphasized sequence similarity, but there were few base pair sites of TALENs that could match to potential off-target sites of the genomes. In fact, not only the similarity of sequence, but also that fact that the 0 position of EBEs must be T is very important to TALENs binding site[48]. Using the novel program, 55 candidates' off-target sites were identified in the medaka genome. We amplified the top predicted potential off-target locus by PCR, sequenced it and found no mutations (File S4). Using our data, our program was more efficient than ePCR to predict TALENs off-target sites. Recently, tools for predicting TALEN off-targets have been developed, such as idTALE (http://idtale.kaust.edu.sa), Paired Target Finder (https://tale-nt.cac.cornell.edu), and TALENoffer[49]. Compared with these, the advantage of our program is its simplicity and reliability. However, a more detailed investigation on the possible off-target effects of TALENs and more accurate program will be needed in the future.

Mutant phenotypes

Gonadal dysgenesis in 46, XY patients was first noted by Drash *et al.*[4]. Phenotypic female patient with an XY karyotype were initially reported by Kaplan[5]. Several phenotypic female patients with the XY karyotype were evaluated clinically, cytogenetically, hormonally, endoscopically and histologically[6]. The functional study of gonadal dysgenesis in phenotypic female patients with an XY karyotype has been hindered by a lack of animal models with specific mutations, except for the mouse sry mutant[3, 31]. Using TALENs, two different Y-linked genes were efficiently manipulated in mouse embryonic stem cells (mESCs)[31] and an *Sry* knockout mouse was generated[3]. The mutant mice are almost completely infertile, although the *Sry* knockout mouse is similar to humans in terms of their physiological phenotype. The phenotype of the *DMY* knockout medaka is also similar to that human XY female syndrome. In particular, a majority of fertile individuals were found in the population. XY^{DMY-} female medakas could help in studies of the mechanism of human XY female syndrome in genetics and reproductive biology. Benefiting from the number of embryos or offspring from the parents, fish are good for large sample analysis and for investigating individual differences of human XY female syndrome in populations.

DMY, a duplicated copy of *DMRT1*, is identified as the master male SD gene and shows all features of a SD gene in medaka[12]. In this study, we generated both insertions (+16) and deletions (−11) of the TALEN-induced *DMY* mutant line, which developed into females with the XY karyotype (Figure 5A, B, C and F). Sex reversal was also reported very early in medaka[50], which could be induced by steroid hormones or high temperature[51, 52]. To evaluate whether the sex reversal observed in XY^{DMY-} female medaka was caused by *DMY* gene mutation, the best method is rescue of the phenotypes of the XY^{DMY-} female medaka. In the former case, the expected mutant *DMY* lacks functionally important motifs of DMY; this mutated Y chromosome could pair with the WT X and Y chromosome to generate females and males, respectively. Crossing with the WT medaka (Figure 3), produced Y^{DMY-}Y male mutants in the next generation, which meant that dysfunction of the *DMY* gene was the unique difference between XY^{DMY-} female and $Y^{DMY-}Y$ male. This result confirmed that the sex reversal observed in XY^{DMY-} female medakas was caused by a *DMY* gene mutation.

Adults of two phenotypically female mutant lines were evaluated through mutation confirmation, phenotype diagnostic (Figure 5A) and histological analysis (Figure 5B and 5F). Using genomic PCR (Figure 4B) and FISH of *DMY* gene (Figure 5C) showing the specific hybridization signal of the Y chromosome, the two phenotypically female mutant lines were identified as genetic males with the XY karyotype. RT-PCR analysis (Figure 4C) and RNA-seq (Dataset 1) confirmed that XY^{DMY-} mutants did not express *DMY* gene, which were different from WT XY individuals. *Gsdf*, co-localized with *DMY* in the same somatic cells in the XY gonads, was expressed exclusively in primordial gonads of only the genetic males[39]. Both the increased number of germ cells (Figure 4D) and the significantly down-expression of *Gsdf* gene (Unigene44032) implied sex reversal of XY^{DMY-} mutants took place in early developmental stages, attributed to the disruption of *DMY* gene. As expected, the TALEN-induced *DMY* knockout medakas had female external and internal specificities. Unlike the *Sry* KO mouse, which did not produce any offspring[3], the majority of TALEN-induced *DMY* knockout medakas were fertile. Twelve of 15 XY^{DMY-} mutants developed functional ovaries; and the ovaries of three females showed incomplete development. There were significant differences among XY^{DMY-} mutants, WT female and WT male in GSI (Figure 5D) in terms of their GSI scores. Histological analysis showed that the ovary of XY^{DMY-} mutants displayed a reduced number of oocytes (Figure 5F). In three of the 15 mutants, no eggs could develop beyond CA oocytes. The other 12 could generate mature eggs, however, the mature oocytes number of XY^{DMY-} mutants are significantly fewer than that of WT XX females. Therefore, the ovary function of XY^{DMY-} mutantswas lower than that of WT females, not only in quality, but also in the quantity (Figure 5F). To investigate whether there is significant difference in the fertilization of eggs and between XY^{DMY-} female and WT female; test crosses were continuously and systemically recorded. During generation of offspring from the XY^{DMY-} female testcrosses, we found that it was always difficult to obtain a sufficient number of offspring. After 10 days of continuous monitoring, 6 eggs matured per day from XY^{DMY-} mutants; 20 eggs matured per day from WT XX females (Figure 5E). The results from F2 phenotypes revealed that a reduced number of mature oocytes were one of the most plausible explanations for the reduced fertility in the *DMY* KO medakas. Thus, the XY^{DMY-} mutants could develop functional ovaries; however, the development, maturation and fertilization capacity of their eggs were significantly lower than those of the WT female.

Transcriptomic analysis of TALENs induced *DMY*- mutants and the WT

Natural sex reversal has been reported in medaka[50], and mutation of the *DMY* gene was identified in several artificial mutants[14]. However, the molecular mechanism of sex reversal, inducing by loss of *DMY*, has not been resolved. RNA-Seq is a recently developed approach for transcriptome profiling that is based on deep-sequencing[32]. RNA-seq could quantify the changing expression levels of each transcript during development and under different conditions. XY^{DMY-} mutant female from the F3 generation were used for RNA-seq analysis, which minimized the off-target effects of *DMY*-TALENs. Using the RNA-seq approach, we obtained transcriptome information of XY^{DMY-} mutants, and through a comparative analysis of them, revealed the transcriptional function of the *DMY* gene in medaka.

Using RNA-seq, 157 million raw reads were generated from WT_M_TE, WT_F_OV and TA_F_OV libraries. However, only 75.8% of the processed reads were mapped to the reference genome of medaka in the Ensemble database (File S6). This indicated that the medaka genome information requires improvement, and transcriptomic sequencing could revise and promote the improvement of medaka genome. In addition, it implied that it would be better to use independent analysis with no reference genome in the analysis of medaka transcriptome, which we used in this study. The problem with the medaka genomic information had little bearing on the fact that medaka is a good model for human diseases. Although only 7481 of 9844 transcripts could match the medaka genome, 7440 transcripts had homologous genes in the human genome (Figure 6H). This suggested that medaka might be similar to humans in terms of sex determination and regulation; the medaka transcriptome data could play a role in the analysis of human sex reversal patients.

Medaka sex is primarily determined by the presence or absence of *DMY* gene[12, 13]. Several studies show that *Dmrt1*[16,38], *Gsdf*[39] and *Sox9b*[40] is essential to maintain testis differentiation or regulate testis development. Un-expression of *DMY* (Uingene20290) and lower expression of *Dmrt1a* (Unigene42535), *Gsdf* (Unigene44032) and *Sox9b* (Unigene18419) in RNA-seq analysis provided the molecular basis to the failure of male sex determination, male-to-female sex reversal, and the failure of testis differentiation or development in XY^{DMY-} mutants (Dataset 1). Among the differentially expressed genes in TA_F_OV compared with WT_F_OV and TA_F_OV compared with WT_M_TE (Figure 6F and Dataset 2), we found several potential factors may directly bind SRY and DMY, such as SLC25A38 and ENSORLT000 00000529. This validated the transcriptional function of the *DMY* gene. Theoretically, the *DMY* gene, a SD gene, might have a relationship with the sperm production and cilia assembly. However, from our comparative analysis, it was apparent that differences in the expressions of genes involved in these processes represented background differences between the WT ovary and WT testis tissue. This is an essential problem in the transcriptional function of *DMY*, but it beyond the scope of this study. Further, we may investigate developmental stages of gonad or different tissue of the HPG axis to explain the problem.

The main differentially expressed genes between WT_F_OV and TA_F_OV, were genes in the wnt receptor signaling pathway (Predicted: syntabbulin-like, axin-2-like) (Dataset 4). In mammals, beta-arrestin-1-like gene regulated IGF-1 affects human reproductive endocrinology. The medaka homolog of beta-arrestin-1-like gene was detected 10 times in the RNA-seq data. The high expression of the beta-arrestin-1-like gene could provide some clues to the mechanism of the degradation of the testis and the development of the ovary in XY *DMY* mutant females. Lhr and Forkhead box O5 factor were also differentially expressed between TA_F_OV and WT_F_OV. Lhris the luteinizing hormone receptor, and Forkhead box O5 factor plays an important role in metabolism and cell differentiation. The low expression of lhr and Forkhead box O5 factor might affect the incomplete ovary function of the XY *DMY* mutant female. The majority of genes in wnt signal pathway, such as *Wnt* genes, *PRCK* genes, *FZD* genes and *DKK* were differentially expressed between WT_M_TE and TA_F_OV (Dataset 1), it implied that the wnt signaling pathway is the main regulation pathway during male-to-female sex reversal process. In addition, the genes in wnt signaling pathway, especially *Wnt1* and *PRCK*, were up-expressed in XY^{DMY-} mutants compared with WT XY males, and down-expressed compared with WT XX females (Dataset 1 and 4). These results implied that the wnt signaling pathway also contributed primarily to the ovarian development, reduced fertility and ovarian maturation in the XY^{DMY-} mutants. Interestingly, *PRCKA* and *DKK* were up-expressed in TA_F_OV, whether compared with WT_M_TE (Dataset 1) or compared with WT_F_OV (Dataset 4). In summary, these results implied that the wnt signaling pathway is the root of sex reversal, the incomplete ovarian functionand reduced fertility in the XY^{DMY-} mutants.

Methods

Ethics statement on the use of animals

The research animals are provided with the best possible care and treatment and are under the care of a specialized technician. All procedures were approved by the Institute of Hydrobiology, Chinese Academy of Sciences, and were conducted in accordance with the Guiding Principles for the Care and Use of Laboratory Animals.

Medaka husbandry

The Orange strain of medaka was used in this study (Originally from National University of Singapore). Medakas were maintained in aquaria under an artificial 14-h light/10-h dark photoperiod at 26℃ in the Institute of Hydrobiology, Chinese Academy of Science[30].

Construction of TALENS

TALEN were assembled and transferred into vectors pCS2-KKR and pCS2-ELD[28]. To improve the germline integration efficiency, the zebrafish *nanos*-3′UTR was separately inserted into the 3′ end of the pCS2-ELD/KKR vector to replace the SV40 UTR using *NotI* and *XbaI*.

Manipulation of medaka embryos

The final TALEN plasmids were linearized using *Not*1, and the mMessage mMachine SP6 kit (Ambion) was used to synthesize mRNAs. TALENs mRNAs (half left and half right monomer mRNAs) were microinjected into medaka embryos at the one cell stage.

Detection of mutations in TALEN-targeted medaka embryos

72 hours after microinjection, TALEN targeted embryos were pooled for genomic DNA extraction (10 embryos for each pool). PCR was performed using primers DMY F and DMY R; PCR products were purified from the agarose gel using a gel extraction kit (QIAGEN). Amplicons harboring the targeted gene fragments were sub-cloned into pMD-18T using TA cloning (Takara), and single colonies were examined by PCR using primers DMY F, DMY F1 and DMY R. The PCR conditions were as follows: 5 min at 95℃; followed by 30 cycles of 15 sec at 95℃, 20 sec at 52℃, 30 sec at 72℃; and a final step of 5 min at 72℃. PCR products were electrophoresed in a 2% agarose gel and verified by DNA sequencing.

Off-target analysis

The criteria of the novel program for determining off-target sites were that the 0 position of the EBEs must be T, from 1 to 10 mismatch bases occur in the pairs of EBEs, and the spacer between the two putative EBE regions is less than 100 bp, because it has been suggested that longer spacers interfere with Fok I dimerization.

Founder screening

TALEN-injected medaka embryos were raised to sexual maturity. F0 *DMY* mutated females were crossed with wild-type males; and F1 embryos were collected at 72 hours post fertilization. Genomic DNA was extracted from ten randomly pooled embryos to assess mutagenesis at the TALEN-targeted site by PCR and sequencing. F1 embryos were individually collected at 7 days post fertilization (dpf), and genomic DNA was extracted from each individual embryo to assess mutagenesis at the TALEN-targeted site by PCR and sequencing.

Fluorescence *in situ* hybridization

FISH was performed using a *DMY* fragment labelled by PCR (Dmy Probe F: 5′-TGCCCAAGTGCTCCC-3′; Dmy Probe R: 5′-CCCCTTTTGTCTGTCCTCT-3′) with digoxigenin-11-dUTP (Boehringer). Standard nick translation using biotin-16-dUTP and digoxigenin-11-dUTP[12] separately labeled the probe. Before hybridizing with denatured medaka mitotic chromosomes, the probe was denatured and preannealed in the presence of excess genomic DNA. Rhodamine- conjugated avidin and antidigoxigenin (monoclonal)- conjugated fluorescein (Sigma) were used to detect the hybridization sites for the probe simultaneously. 4,6-diamidino-2-phenylindole (DAPI) was used to counterstain the chromosomes. EASY FISH 1.0 software (Applied Spectral Imaging, Mannheim, Germany) was used to capture and display digitized images of the rhodamine signals on DAPI-stained chromosomes.

Expression analysis

The TRIZOL reagent (Invitrogen) was used to extract total RNA from pooled organs of three adult medaka fish, according to the supplier's recommendation. After DNase treatment, reverse transcription was performed with 2 or 4 mg RNA by using Superscript II reverse transcriptase (Invitrogen) and random primers. cDNA from 10 ng (actin) to 200 ng (adult organs) of total RNA was used for PCR with gene-specific primers: *Ola Actin*, *Ola DMY* (Dmy F1 and PG17.89 primers[12]).

Histological analysis

For the histological analysis, identified mutants from the F2 generation of medaka were dissected into head and body segments (N=15). Each dissected head was used to determine the genetic sex. The body portions were fixed overnight in Bouin's fixative solution and then embedded in paraffin. Cross-sections were cut serially at 5μm thickness, and after hematoxylin and eosin (H&E) staining; the images were obtained under a microscope (Eppendorf).

Library construction and high-throughput sequencing

Medaka gonads (divided into WT-F, WT-M and TA-F, 10 samples pooled in each group, respectively) were collected for RNA extraction. TRIZOL (Invitrogen) performed the total RNA extraction, following the manufacturer's instructions. RNA library construction was then performed by BGI-Shenzhen, Shenzhen, China then constructed the RNA library. Before library construction, the integrity of RNA samples was confirmed using an Agilent 2100 Bioanalyzer; 10 μg of total RNA was used for isolation of mRNA with Sera-mag Magnetic Oligo (dT) beads from Illumina.

Quantitative real-time PCR

The TRIZOL reagent (Invitrogen) was used to isolate total RNA from embryos. The SYBR Green PCR Master Mix (Applied Biosystems) and the Real-Time PCR System (ABI 7900) measured gene transcription. The *actb1* gene was used as the endogenous control gene for normalizing expression of the target gene. Triplicate technical replicates were performed for duplicate cultures.

Functional annotation, classification, and enrichment analysis

Blast2GO software (http://www.blast2go.de) was used to perform BLAST searching, mapping, and annotation of proteins differentially expressed[53]. BINGO 2.44[54] plug-in in the Cytoscape platform[55] was used to perform functional enrichment analysis of differentially expressed proteins to determine the significantly enriched GO terms and relevant proteins. Enrichment analysis of GO term assignment was performed in reference to the entire annotated *P. tricornutum* proteome. The corrected P values were derived from a hypergeometric test followed by Benjamini and Hochberg false discovery rate correction. A corrected $P < 0.05$ was regarded as significant.

Acknowledgements We thank Ms. Li Ming (Institute of Hydrobiology, Chinese Academy of Sciences) for her kind help during the feeding of medaka and microinjection of medaka embryos. This work was supported by the National Science Foundation for Distinguished Young Scholars of China (31325026 to WH), the National High Technology Research and Development Program of China (2011AA100404 to WH), the National Science Foundation of China (31472263 to DL), and the Chinese Academy of Sciences (XDA08010106 and 2011FBZ19 to WH).

References

1. Sinclair AH, *et al.* A gene from the human sex-determining region encodes a protein with homology to a conserved DNA-binding motif. Nature 346, 240-244 (1990).
2. Hawkins JR, *et al.* Mutational analysis of SRY: nonsense and missense mutations in XY sex reversal. Hum Genet 88, 471-474 (1992).
3. Kato T, *et al.* Production of Sry knockout mouse using TALEN via oocyte injection. Sci Rep 3, 3136 (2013).
4. Drash A, Sherman F, Hartmann WH, Blizzard RM. A syndrome of pseudohermaphroditism, Wilms' tumor, hypertension, and degenerative renal disease. J Pediatr 76, 585-593 (1970).
5. Kaplan E. Gonadal dysgenesis in a phenotypic female with an XY chromosomal constitution. S Afr Med J 53, 552-553 (1978).
6. Portuondo JA, *et al.* Management of phenotypic female patients with an XY karyotype. J Reprod Med 31, 611-615 (1986).
7. Conover DO, Kynard BE. Environmental sex determination: interaction of temperature and genotype in a fish. Science 213, 577-579 (1981).
8. Bulmer M. Evolution: sex determination in fish. Nature 326, 440-441 (1987).
9. Conover DO, Heins SW. Adaptive variation in environmental and genetic sex determination in a fish. Nature 326, 496-498 (1987).
10. Loukovitis D, *et al.* Quantitative trait loci for body growth and sex determination in the hermaphrodite teleost fish *Sparus aurata* L. Anim Genet 43, 753-759 (2012).
11. Piferrer F, Ribas L, Diaz N. Genomic approaches to study genetic and environmental influences on fish sex determination and differentiation. Mar Biotechnol (NY) 14, 591-604 (2012).

12. Nanda I, *et al.* A duplicated copy of DMRT1 in the sex-determining region of the Y chromosome of the medaka, *Oryzias latipes*. Proc Natl Acad Sci U S A 99, 11778-11783 (2002).
13. Matsuda M, *et al.* DMY is a Y-specific DM-domain gene required for male development in the medaka fish. Nature 417, 559-563 (2002).
14. Matsuda M, *et al.* DMY gene induces male development in genetically female (XX) medaka fish. Proc Natl Acad Sci U S A 104, 3865-3870 (2007).
15. Otake H, *et al.* The medaka sex-determining gene DMY acquired a novel temporal expression pattern after duplication of DMRT1. Genesis 46, 719-723 (2008).
16. Masuyama H, *et al.* Dmrt1 mutation causes a male-to-female sex reversal after the sex determination by Dmy in the medaka. Chromosome Res 20, 163-176 (2012).
17. Esvelt KM, Wang HH. Genome-scale engineering for systems and synthetic biology. Mol Syst Biol 9, 641 (2013).
18. Bibikova M, Golic M, Golic KG, Carroll D. Targeted chromosomal cleavage and mutagenesis in Drosophila using zinc-finger nucleases. Genetics 161, 1169-1175 (2002).
19. Kim YG, Cha J, Chandrasegaran S. Hybrid restriction enzymes: zinc finger fusions to Fok I cleavage domain. Proc Natl Acad Sci U S A 93, 1156-1160 (1996).
20. Kim S, *et al.* Preassembled zinc-finger arrays for rapid construction of ZFNs. Nat Methods 8, 7 (2011).
21. Hockemeyer D, *et al.* Genetic engineering of human pluripotent cells using TALE nucleases. Nat Biotechnol 29, 731-734 (2011).
22. Huang P, *et al.* Heritable gene targeting in zebrafish using customized TALENs. Nat Biotechnol 29, 699-700 (2011).
23. Li T, *et al.* TAL nucleases (TALNs): hybrid proteins composed of TAL effectors and FokI DNA-cleavage domain. Nucleic Acids Res 39, 359-372 (2011).
24. Christian M, *et al.* Targeting DNA double-strand breaks with TAL effector nucleases. Genetics 186, 757-761 (2010).
25. Ishino S, *et al.* Nucleotide sequence of the meso-diaminopimelate D-dehydrogenase gene from Corynebacterium glutamicum. Nucleic Acids Res 15, 3917 (1987).
26. Hwang WY, *et al.* Efficient genome editing in zebrafish using a CRISPR-Cas system. Nat Biotechnol 31, 227-229 (2013).
27. Cong L, *et al.* Multiplex genome engineering using CRISPR/Cas systems. Science 339, 819-823 (2013).
28. Liu Y, *et al.* Inheritable and Precise Large Genomic Deletions of Non-Coding RNA Genes in Zebrafish Using TALENs. PLoS One 8, e76387 (2013).
29. Liu Y, *et al.* A highly effective TALEN-mediated approach for targeted gene disruption in *Xenopus tropicalis* and zebrafish. Methods 69, 58-66 (2014).
30. Qiu C, *et al.* Efficient knockout of transplanted green fluorescent protein gene in medaka using TALENs. Mar Biotechnol (NY) 16, 674-683 (2014).
31. Wang H, *et al.* TALEN-mediated editing of the mouse Y chromosome. Nat Biotechnol 31, 530-532 (2013).
32. Wang Z, Gerstein M, Snyder M. RNA-Seq: a revolutionary tool for transcriptomics. Nat Rev Genet 10, 57-63 (2009).
33. Ansai S, *et al.* Efficient targeted mutagenesis in medaka using custom-designed transcription activator-like effector nucleases. Genetics 193, 739-749 (2013).
34. Miller JC, *et al.* A TALE nuclease architecture for efficient genome editing. Nat Biotechnol 29, 143-148 (2011).
35. Kobayashi T, *et al.* Two DM domain genes, DMY and DMRT1, involved in testicular differentiation and development in the medaka, *Oryzias latipes*. Dev Dyn 231, 518-526 (2004).
36. Otake H, *et al.* Wild-derived XY sex-reversal mutants in the Medaka, *Oryzias latipes*. Genetics 173, 2083-2090 (2006).
37. Kagawa H, Young G, Nagahama Y. Relationship between seasonal plasma estradiol-17 beta and testosterone levels and *in vitro* production by ovarian follicles of amago salmon (*Oncorhynchus rhodurus*). Biol Reprod 29, 301-309 (1983).
38. Hornung U, Herpin A, Schartl M. Expression of the male determining gene dmrt1bY and its autosomal coorthologue

dmrt1a in medaka. Sex Dev 1, 197-206 (2007).

39. Shibata Y, *et al.* Expression of gonadal soma derived factor (GSDF) is spatially and temporally correlated with early testicular differentiation in medaka. Gene Expr Patterns 10, 283-289 (2010).
40. Nakamura S, *et al.* Sox9b/sox9a2-EGFP transgenic medaka reveals the morphological reorganization of the gonads and a common precursor of both the female and male supporting cells. Mol Reprod Dev 75, 472-476 (2008).
41. Koprunner M, Thisse C, Thisse B, Raz E. A zebrafish nanos-related gene is essential for the development of primordial germ cells. Genes Dev 15, 2877-2885 (2001).
42. Mishima Y, *et al.* Differential regulation of germline mRNAs in soma and germ cells by zebrafish miR-430. Curr Biol 16, 2135-2142 (2006).
43. Mussolino C, Cathomen T. On target? Tracing zinc-finger-nuclease specificity. Nat Methods 8, 725-726 (2011).
44. Mussolino C, *et al.* A novel TALE nuclease scaffold enables high genome editing activity in combination with low toxicity. Nucleic Acids Res 39, 9283-9293 (2011).
45. Tesson L, *et al.* Knockout rats generated by embryo microinjection of TALENs. Nat Biotechnol 29, 695-696 (2011).
46. Lei Y, *et al.* Efficient targeted gene disruption in Xenopus embryos using engineered transcription activator-like effector nucleases (TALENs). Proc Natl Acad Sci U S A 109, 17484-17489 (2012).
47. Osborn MJ, *et al.* TALEN-based gene correction for epidermolysis bullosa. Mol Ther 21, 1151-1159 (2013).
48. Boch J, *et al.* Breaking the code of DNA binding specificity of TAL-type III effectors. Science 326, 1509-1512 (2009).
49. Grau J, Boch J, Posch S. TALENoffer: genome-wide TALEN off-target prediction. Bioinformatics 29, 2931-2932 (2013).
50. Yamamoto T. Artificial induction of functional sex-reversal in genotypic females of the medaka (*Oryzias latipes*). J Exp Zool 137, 227-263 (1958).
51. Paul-Prasanth B, Shibata Y, Horiguchi R, Nagahama Y. Exposure to diethylstilbestrol during embryonic and larval stages of medaka fish (*Oryzias latipes*) leads to sex reversal in genetic males and reduced gonad weight in genetic females. Endocrinology 152, 707-717 (2011).
52. Hattori RS, *et al.* Temperature-dependent sex determination in Hd-rR medaka *Oryzias latipes*: gender sensitivity, thermal threshold, critical period, and DMRT1 expression profile. Sex Dev 1, 138-146 (2007).
53. Conesa A, *et al.* Blast2GO: a universal tool for annotation, visualization and analysis in functional genomics research. Bioinformatics 21, 3674-3676 (2005).
54. Maere S, Heymans K, Kuiper M. BiNGO: a Cytoscape plugin to assess overrepresentation of gene ontology categories in biological networks. Bioinformatics 21, 3448-3449 (2005).
55. Shannon P, *et al.* Cytoscape: a software environment for integrated models of biomolecular interaction networks. Genome Res 13, 2498-2504 (2003).

胚胎注射 TALENs 培育 XY^{DMY-} 性逆转的雌性青鳉

罗大极[1,2] 刘 云[3] 陈 戟[2] 夏晓勤[2] 曹梦西[2]
程 彬[1] 王雪娟[1] 龚午明[4] 邱 超[1] 张运生[2]
Christopher Hon Ki Cheng[3] 朱作言[2] 胡 炜[2]

1 武汉大学基础医学院遗传系，湖北武汉，中国，430071
2 中国科学院水生生物研究所，湖北武汉，中国，430072
3 香港中文大学生物医学院，香港沙田，中国
4 Lillehei Heart Institute, University of Minnesota, Minneapolis, USA

摘　要　青鳉是研究性别决定及性逆转相关疾病的理想模型，如人类 XY 女性患者。本研究中，我们运用改良的 TALENs 技术，构建了 XY^{DMY-} 突变个体用来研究青鳉性腺发育障碍。*DMY*-TALENs 在目标位点产生缺失或插入突变(46.8%)，91.7%的 *DMY-nanos3UTR*-TALENs 诱导突变可以通过生殖细胞系传代。XY^{DMY-} 突变的个体发育成雌性，能够正常产卵，且能稳定地将 Y^{DMY-} 染色体传给子代。对野生型雄性 (WT_M_TE)、野生型雌性 (WT_F_OV) 和 XY^{DMY-} 性反转雌性(TA_F_OV) 青鳉进行 RNA-seq，获得了157000000 raw reads。分析发现，与 WT_F_OV 相比，TA_F_OV 中有144个上调和293个下调表达的基因；而与 WT_M_TE 相比，TA_F_OV 中有387个上调和338个下调表达的基因。根据基因注解和功能预测，XY^{DMY-} 突变体的卵巢功能不全和生殖力下降与 wnt 信号通路，如 Wnt1和 PRCK 密切相关。我们的结果系统展示了 XY^{DMY-} 突变体的转录情况，揭示了青鳉性逆转与 *DMY* 之间的关系。XY^{DMY-} 突变体对于研究 XY 女性患者性腺发育障碍具有重要参考意义。

朱作言文集

（下卷）

汪亚平　胡　炜　孙永华　崔宗斌　主编

科 学 出 版 社

北　京

内 容 简 介

朱作言先生研究团队于20世纪80年代初期首创农业动物基因工程研究，在1985年发表了世界第一篇鱼类基因转移研究文章，享有世界声誉。本书从该团队发表的200多篇文章中，选取了100篇编辑成册，分上卷和下卷出版。本书的主线是基于转基因鱼理论模型的建立、转基因鲤研制及品系选育、转植基因功能，以及转基因鱼食用和生态安全研究。本书内容共包括7个部分。第一部分和第二部分建立了转基因鱼研究的完整理论模型，并以此为基础成功研制转“全鱼”生长激素基因黄河鲤家系，系统揭示了转植基因的整合、表达和遗传规律。第三部分从摄食、代谢、生长、免疫和游泳行为等方面，详细描述了转基因黄河鲤与野生黄河鲤的生物学特性。第四部分系统评估了转基因鱼在实验室及模拟自然环境条件下的生态学效应。第五部分系统研究了鱼类生长激素对性腺发育的调控作用、育性控制技术及研制不育转基因鱼的策略。第六部分包括草鱼全基因组草图绘制、草鱼病原病毒分子生物学和抗病草鱼基因工程育种的基础研究。第七部分主要以斑马鱼为模型，对鱼类基因工程的生物学原理、技术拓展和应用进行了全方位的探讨。为了便于广大读者阅读，英文文章后面均附有中文摘要。

本书内容是迄今为止关于转基因鱼，乃至整个转基因动物研究领域最系统、最全面和最深入的研究总结，可供鱼类学、鱼类生理学、鱼类内分泌学、农业动物基因工程育种等领域的科技工作者和师生们学习和参考。

图书在版编目(CIP)数据

朱作言文集：全2册/汪亚平等主编. —北京：科学出版社，2019. 3

ISBN 978-7-03-060461-3

Ⅰ. ①朱…　Ⅱ. ①汪…　Ⅲ. ①鱼类–基因工程–文集　Ⅳ. ①Q959. 403-53

中国版本图书馆CIP数据核字(2019)第014010号

责任编辑：王　静　罗　静　王　好 / 责任校对：郑金红
责任印制：肖　兴 / 封面设计：无极书装

科 学 出 版 社 出版
北京东黄城根北街16号
邮政编码：100717
http://www.sciencep.com

北京通州皇家印刷厂 印刷
科学出版社发行　各地新华书店经销
*
2019年3月第 一 版　开本：787×1092 1/16
2019年3月第一次印刷　印张：90 3/4
字数：2 110 000

定价：880.00元（全2册）

(如有印装质量问题，我社负责调换)

序　　言

《朱作言文集》收录了朱作言教授及其同事们在中国科学院水生生物研究所围绕鱼类基因转移所开展的研究工作，包括转基因鱼的研制、转基因鱼的基础生物学研究、转基因鱼的遗传和生态安全研究及对未来鱼类基因工程育种的展望等。

朱作言教授是全球鱼类转基因研究的开创者。20 世纪 80 年代中期，他带领团队率先成功研制快速生长的转基因鱼，他的这一研究开创了鱼类转基因育种的新领域。他建立了转基因鱼研究的完整理论模型，系统揭示了转基因的整合、表达和遗传规律。接下来，他带领团队克隆了草鱼的 β-肌动蛋白基因启动子和鲤生长激素基因，构建了全部由鲤科鱼类基因组元件构成的“全鱼”转基因载体，并成功研制出转“全鱼”生长激素基因黄河鲤家系，系统开展了外源转基因的生物学效应和转基因鲤的生态安全研究。朱作言团队独立完成了从转基因鱼的理论模型建立到生态安全评估的全链条研究。从这一角度来说，朱教授团队无疑是在全球范围内同领域中工作最为完整的一个研究团队。这些系统研究代表这一团队对转基因鱼生物学领域的巨大贡献。

朱作言教授还带领团队详细研究了转基因鱼的生态安全对策。他的团队一方面开发了具有普适效应的转基因鱼生殖控制策略，另一方面应用三倍体策略获得了完全不育的转基因黄河鲤。利用这些策略，将能够有效解决转基因鱼应用的遗传和生态安全问题，对于转基因鱼的商品化有着重要意义。

该文集共包含朱作言教授团队在过去 30 年内所产出的 100 篇同行评审论文，由 7 个非常具有逻辑性的部分组成，每一部分的主要内容简介如下。

第一部分：鱼类基因转移模型建立。朱作言团队在全球首次成功研制转基因鱼，并在多种重要经济鱼类中成功研制转生长激素基因鱼。他们延续了中国在鱼类细胞核移植方面的独创性工作，是利用这一工具开展早期发育过程中基因表达等基础生物学问题的世界权威。

第二部分：外源基因的整合与表达。接下来，他的团队通过一系列工作发展和优化了鱼类转基因技术。值得一提的是，这一团队深入研究了转基因在受体基因组中传代特征。他的团队永远保持技术先进性，将很多最为先进的基因操作技术引入到多种鱼类中，如转座子、Cre/loxP 重组系统、Gal4/UAS 转录激活系统、TALEN 和 CRISPR/Cas9 基因组编辑工具、RNA 干扰等(同时见第二、五、七部分的部分论文)。

第三部分和第四部分：外源基因表达的生理功能和转基因黄河鲤的生态安全研究。他的团队利用整体手段从摄食、代谢、生长到发育、免疫和游泳行为等方面详细描述了生长激素基因转基因黄河鲤。特别值得一提的是，他的团队花费了大量精力，详尽地评估了转基因鱼在不同条件下的生态学效应。这些研究为未来转基因鱼的产业化在食品安全和生态安全方面提供了大有裨益的信息。

第五部分：鱼类生殖调控与转基因鱼的育性控制研究。此外，他的团队还系统研究了生长激素对性腺发育的作用及开发不育转基因鱼的技术。这将有助于为转基因鱼的产业化提供有效的生态安全保障。

第六部分：抗病草鱼基因工程技术育种研究。他的团队进行了多种抗病的基因工程鲤和草鱼基础研究，这些工作为利用遗传改良手段开发抗病鱼新品系奠定了科学基础。他的团队完成了草鱼的全基因组草图绘制，这是对鱼类生物学领域的巨大贡献。这不仅为草鱼的遗传选育提供了基因组信息，而且将为鲤科鱼类的生物学研究提供全方位的帮助。

第七部分：鱼类基因转移育种技术展望。他们进一步利用斑马鱼模型开展了一些在水产育种中非常有价值的转基因研究，如激活生长激素信号通路、耐低氧、高 omega-3 多不饱和脂肪酸含量等。

总之，朱作言教授团队在多个方面都做出了重要的科学贡献，正如他们通过以上 7 个部分所展示的。他们的研究是将基础科研和水产应用科学相结合的典范——从技术研发到研制快速生长和抗病的转基因鱼，这是现代水产业亟须解决的两个重大问题。他们同时对鱼类生物学的很多基础生物学问题做出了多方面的贡献，如细胞核移植、草鱼全基因组测序、鱼类生物技术的研发和基因工程技术的优化等。该文集的出版，无疑是转基因鱼研究领域的一个重要里程碑。我们热切地期待着这本书的及时出版。

博士

新加坡国立大学荣誉教授

博士

新加坡国立大学教授

2018 年 12 月

Preface

The present monograph represents a major contribution to the transgenic fish study by Professor Zuoyan Zhu and his scientific colleagues at the Institute of Hydrobiology of the Chinese Academy of Sciences, Wuhan, China. The major topics include the development and generation of transgenic fish, the basic biological study of transgenic fish, the genetic and ecological studies of transgenic fish, and the perspective on the future development of fish genetic engineering.

Prof. Zhu is a pioneer in the field of the transgenic fish study. In the mid-1980s, his group succeeded in the generation of the first-case of fast-growing transgenic fish in the world, which initiated a new research field of transgenic fish biology. His team has established a complete research model of transgenic fish, including the mechanisms of transgene integration, expression, and inheritance. Subsequently, his team cloned grass carp *β-actin* promoter and common carp *growth hormone* (*gh*) gene, and generated an "all-fish" transgene construct which is composed of *Cyprinidae*-derived genomic elements. Based on the construct, his team successfully generated an "all-fish" *gh*-transgenic line of Yellow River common carp and conducted a series of studies concerning the biological effects of *gh*-transgene and the ecological evaluation of *gh*-transgenic Yellow River common carp. Zhu's team has accomplished the full-chain study from the theoretical model to the ecological evaluation of transgenic fish, and it should be the team that has done the complete study of transgenic fish worldwide. These studies are comprehensive and represent a major contribution to the field of transgenic fish biology by the Zhu team.

The Zhu team has thoroughly studied the biosafety containment of transgenic fish. On the one hand, his team developed some general approaches for reproductive control, and on the other hand, his team explored the triploidy to manipulate sterility in common transgenic. By using these strategies, it is possible to effectively solve the genetic and ecological problems in the application of transgenic fish, which is of great importance for the commercialization of these new species.

The monograph, compiling of 100 peer-reviewed papers by the Zhu team in the past 30 years, is well organized into 7 logical sections. The contents of each section are briefly highlighted below.

The 1st section: the establishment of a transgenic fish model. Zhu's group was the first one in the world to generate transgenic fish and has established *gh*-transgenic fish in many species, including aquaculture important species. They continued the strength of fish nuclear transplantation work initiated from China and are a world authority in this technology to study fundamental biological questions to investigate early gene expression in development.

The 2nd section: the transgene's integration and expression. His group has conducted a series of works to develop and optimize the transgenic technology. In particular, his groups carried out extensive studies to characterize the fate of transgenes in the host for multiple generations. His group always followed the most updated technological development in genetic engineering and has introduced many state-of-the-art gene modification technologies in various fish models, such as use of transposons, Cre/loxP recombination, Gal4/UAS binary system, genome editing by TALEN and CRISPR/Cas9, RNA interference, etc. (also Sections 2, 5, 7).

The 3rd and 4th sections: the physiological function of the expressed transgene and the ecological assessment of the transgenic Yellow River common carp. His group uses a holistic approach to characterize *gh*-transgenic fish from feeding, metabolism, growth to development, immunity, swimming behavior, etc. In particular, his group has conducted tremendous studies in the assessment of ecological consequence of *gh*-transgenic fish under various conditions. These studies have provided valuable information on food safety and ecological impact for future application of transgenic fish in aquaculture.

The 5th section: fish reproduction regulations and the control of transgenic fish fertility. Also, his group has conducted a series of experiments to investigate the hormonal effect on gonad development and the possibility for development of sterility in transgenic fish; this may help to develop an effective transgene terminator when transgenic fish is used in aquaculture.

The 6th section: the genetic engineering breeding of disease-resistant grass carp. His group also researched various genetically modified carp for disease resistance. These works provide the scientific basis for using the genetic modification approaches to develop disease-resistant fish strains in aquaculture. His group completed the draft genome sequence of the grass carp and this is an important contribution in fish biology, which is not only important genomic resources for selection of better carp strains in aquaculture, but also facilitates future biological studies in carp and other fish in all aspects.

The 7th section: the prospect of transgenic fish breeding. They also made a few interesting transgenic studies using the zebrafish model fish to test some potentially important features in aquacultures, such as the constitutively activated growth hormone pathways, hypoxia resistance and production of fish flesh containing high omega-3 fatty acids.

Overall, Zhu's group has made important scientific contributions in several areas as they classified in the content. Their research is an excellent combination of fundamental research and translational potential in aquaculture, as evident from the technology development to generate growth-accelerated and disease-resistant transgenic fish, the two most important aspects in modern aquaculture. They also made numerous contributions in fundamental fish biology, such as nuclear transplantation, carp genome sequencing, technical development and optimization of various genetic engineering tools for fish, etc. The monograph represents a major milestone in this field, and we look forward to the timely publication of this monograph.

PhD

Emeritus Professor, National University of Singapore

and

PhD

Professor, National University of Singapore

Dec. 2018

前　言

——从鱼类细胞核移植到基因转移

20 世纪五六十年代，科学家们发展了用于低等脊椎动物胚胎显微操作的细胞核移植技术，研究细胞核的全能性，以及胚胎发育中细胞核与细胞质的相互作用。其中，代表性科学家有美国的 R. Briggs 和 T. J. King，英国的 J. B. Gurdon (2012 年诺贝尔生理学或医学奖获得者)和我国的 T. C. Tung (童第周)。美英学者的实验对象主要是两栖类，而童第周先生的实验对象主要是鱼类，而且是在不同种的鱼类之间，探讨受体卵母细胞质与外来异种细胞核在促进胚胎发育中的作用。20 世纪 70 年代，我们在童第周先生指导下完成的重要研究成果之一，是发现了鱼类的卵细胞质可以在一定程度上配合外来的异种细胞核，共同促进胚胎早期发育；甚至在个别情况下，可以获得性成熟的个体，即“核质杂种鱼”，由此尝试开启新一代养殖鱼类育种的细胞核移植(克隆)技术。

也是在 20 世纪五六十年代，生物学研究出现了巨大的飞跃。以解析 DNA 双螺旋结构为先导的分子生物学兴起，引发了生命科学一日千里的蓬勃发展。新理论、新技术不断涌现，大大加深了人们对生命过程和生命本质的认识。半个多世纪以来，分子生物学研究的成就，就像 19 世纪达尔文物种演化论的创立一样，为生物学和现代生命科学竖起了光芒万丈的灯塔，为相关学科的深入发展指明了航向，也为以分子生物学为基础的现代生物技术及产业的发展夯实了宽广和坚实的基础。在这一大背景下，鱼类基因工程育种研究应运而生。

纵观人类文明和科学技术进步的历史，农业育种实践由引种驯化、品种选育、杂交育种、植物组织培养或动物人工授精和胚胎分割扩繁、染色体组的倍性操作，到细胞核移植(克隆)技术，从群体和个体的层面向细胞水平逐步深入。20 世纪 80 年代初期，以基因克隆为代表的 DNA 分子操作技术逐渐成熟，为品种改良从遗传物质的源头，即 DNA 和基因水平的精准操作提供了可能性。我们很幸运地在中国科学院水生生物研究所率先进行了第一次鱼类基因转移的成功尝试。中国科学院水生生物研究所的前身，是 20 世纪 30 年代初期建立的中央研究院所辖的自然历史博物馆，后来逐渐演替为中央研究院的动植物研究所、动物研究所，直至中国科学院成立后组建为水生生物研究所，并由上海迁至武汉。中国科学院水生生物研究所积淀了数十年的鱼类生物学研究成果，在养殖鱼类资源发掘、人工繁殖、品种驯化、杂交选育和细胞遗传育种方面有着独特的优势，鱼类基因工程育种由此发源乃属水到渠成。早在 20 世纪 70 年代中后期，我们在非常困难的研究条件下，利用显微操作技术，把内蒙古瓦氏雅罗鱼(*Leuciscus waleckii*，冷水性鱼类)的总 DNA 注射到广东鲮鱼(*Cirrhinus molitorella*，温水性鱼类)的受精卵，探索了从遗传物质载体——DNA 分子层面入手，提高鱼类耐寒性的可行性。此时，我虽然萌发了基因转移育种的概念，但苦于没有克隆的基因而无法付诸行动。1981~1983 年，我有机会去欧美学习，先到了英国南安普顿大学细胞遗传学家 Norman Maclean 博士实验室考察，向他介绍了鱼类细胞核移植实验的成果，就计划开展基因转移研究的设想做了广泛交流；接着去英国帝国癌症研究基金会(ICRF)转录实验室主任 Robert Kamen 博士的实验室学习分子克隆技术，他后来又推荐我到美国波士顿遗传研究所，与 Edward F. Fritsch (*Molecular Cloning* 作者之一)合作，从事人胰岛素和促红细胞生成素(erythropoietin，EPO)基因克隆研究，以全面掌握分子克隆技术。可是，当我于 1983 年按期回国的时候，还是没有获得所需的鱼类克隆基因，世界上也没有任何一个实验室完成了鱼类基因的克隆。可喜的是，我得到了美国国家卫生研究院 D. H. Hamer 博士赠予的人生长激素重组 DNA 载体。有了它，鱼类基因转移研究终于可以扬帆起航。从 1984 年起，我们把生长激素基因片段与载体质粒 DNA 分离，显微注射到鲫(*Carassius auratus*)等多种鱼类的受精卵，获得了具有明显快速生长效应的转基因鱼；通过随后几年的研究，建立了完整的转基因鱼研究模型，并得到了国际学术界的认可。后来，我带领研究团队克隆了鲤科(Cyprinidae)鱼类的 β-肌动蛋白基因(与 P. Hackett 教授合作)和生长激素基因(与 T. T. Chen 教授合作)，构建了“全鱼”基因表达载体，培育了生长快、饵料利用率高、生态和食用安全的转基因鲤，我们后来命名它为“冠鲤”，以彰显其卓越的养殖特性。本文集选编了我们研究团队的研究论文 100 篇，分上下两册，作为我们研究工作的一份总结。

这是在中国科学院水生生物研究所及淡水生态与生物技术国家重点实验室开创性的一项系统性研究工作，时任所长刘建康院士是启动这项研究最重要的支持者和推动者。早期有李国华、何玲和稍后加入的谢岳峰三位年轻助手跟我一起白手起家，因陋就简，克服重重困难，创建了第一个鱼类基因转移研究实验室，它也是我国最早的几个分子生物学实验室之一。在随后的 30 多年里，先后有几十位同事和同学们参与这些研究，并与本项研究共同成长。我衷心感谢汪亚平、胡炜、孙永华、崔宗斌 4 位研究员，他们不仅是我们研究团队的主力，而且共同主编了本文集。我衷心感谢张甫英、黄容等副研究员，许克圣、陈尚萍、廖兰杰等高级工程师，黎明、李勇明、祁德运、甄建设等实验助理的出色工作和辛勤付出。我衷心感谢河南省农业厅水产科学研究院张西瑞、冯建新和王宇锋高级工程师在提供黄河鲤底栖品种和养殖上的鼎力帮助。我衷心感谢武汉多福科技农庄股份有限公司和大湖水殖股份有限公司高瞻远瞩地按国家《农业转基因生物安全管理条例》与《农业转基因生物安全评价管理办法》的规定建立了合格的鱼类基因工程育种养殖试验基地，包括构建面积为 6.7 hm^2 的人工模拟试验湖泊；配合我们开展了转“全鱼 *gh*”基因黄河鲤的食品安全与生态安全评估试验，包括中间试验、环境释放试验和生产性试验。我还要衷心感谢中国科学院、科学技术部、国家自然科学基金及湖北省等不同渠道的研究经费资助。本文集出版得到了淡水生态与生物技术国家重点实验室的资助。

最后我想强调的是，从本文集所呈现的 30 余年的研究轨迹不难看出，转基因育种技术实质上是杂交育种的逻辑延伸和技术跃迁。转基因育种技术可理解为一个优良性状目的基因分子与一个生物个体整套基因组的杂交，是建立在现代分子生物学基础上的定向、精准、可预测、可追踪的分子杂交育种技术。由于种种原因，这一在我国开创并具有完全自主知识产权的鱼类基因工程育种新技术，在推广应用上却遇到了重重困难。“莫道浮云终蔽日，总有云开雾散时”，我坚信，科学精神终将驱散困扰公众的形形色色的迷雾。历史终将证明，转基因技术一定会在现代农业发展中发挥无可比拟和不可替代的重要作用。

中国科学院水生生物研究所　研究员
北京大学　博雅讲席教授
《中国科学》《科学通报》　总主编
中国科学院　发展中国家科学院　院士
2018 年 12 月

Introduction

——From nuclear transplantation to gene transfer of fish

The 1950s and 1960s, scientists developed a technique of nuclear transplantation for micro-manipulating lower vertebrate embryos to study the totipotency of nucleus and the interaction between nucleus and cytoplasm in early embryonic development. The representative scientists were R. Briggs and T. J. King of USA, J. B. Gurdon of UK (Nobel laureat in Physiology or Medicine in 2012) and T. C. Tung of China. The subjects of these experiments by American and British scholars are mainly amphibians. However, Professor T. C. Tung mainly utilized fish to explore the role of the enucleated oocyte with interspecies donor nucleus in promoting embryonic development. In the 1970s, one of the major achievements of our team under the guidance of Professor T. C. Tung was the discovery that fish egg cytoplasm with interspecies donor nucleus was able to go through the early development and, in some cases, the whole ontogenies, resulting in "nucleocytoplasmic hybrid fish". Thus, we attempted to develop a new generation of the technology of nuclear transplantation (cloning) for fish breeding in aquaculture.

During the same period, there was an enormous breakthrough in biological sciences. The rise of molecular biology evoked by the discovery of the structure of DNA double helix triggered the explosive development of life sciences. The emergence of new theories and technologies has greatly promoted people's understanding of the nature of life and the process of adapting to the surrounding environment. For more than half a century, the achievements in molecular biology, just like Darwin's theory of evolution in the 19th century, have shone as a beacon light for biology and modern life science, pointed out the direction of in-depth development of related disciplines, and laid a broad and solid foundation for modern biotechnology and industry. In this context, the genetic engineering of fish breeding came into being.

Throughout the history of human civilization and scientific and technological advances, the practice of agricultural breeding has evolved from introduction and domestication, variety breeding, crossbreeding, plant tissue culture, artificial animal fertilization and embryonic bisection for propagation, chromosomal ploidy operation to nuclear transplantation (cloning), gradually deepening from the level of population and individuals to the level of cells. In the early 1980s, techniques of DNA manipulation represented by gene cloning matured gradually, providing the possibility for variety improvement by way of accurately manipulating the genetic material itself. We were fortunate to have taken the lead in the first successful attempt of fish gene transfer at the Institute of Hydrobiology, Chinese Academy of Sciences. The predecessor of the Institute of Hydrobiology, Chinese Academy of Sciences was the Natural History Museum of Academia Sinica established in 1930. Later, it gradually became the Institute of Zoology and Botany in 1934 and the Institute of Zoology in 1944 of Academia Sinica, and was reorganized to form the Institute of Hydrobiology after the establishment of Chinese Academy of Sciences in 1950, and eventually moved from Shanghai to its current location in Wuhan in 1954. The Institute of Hydrobiology has accumulated decades of research achievements in fish biology and has unique advantages in the exploitation of cultured fish resources, artificial reproduction, variety domestication, crossbreeding and cytogenetic breeding, from which fish breeding by genetic engineering just happened that way. In the middle of the late 1970's, under very difficult research conditions, we used the micro-manipulation technique to explore the possibility of DNA transfer in fish: injecting the total DNA from *Leuciscus waleckii* (cold water fish) into the fertilized eggs of *Cirrhinus molitorella* (warm water fish) to improve cold tolerance of fish. At this time, the concept of transgenic breeding came into my mind, but I was unable to put it into practice because there was no cloned gene available for us. From 1981 to 1983, I had the opportunity to study in Europe and in USA. I first visited Dr. Norman Maclean, a professor of cytogenetics at the University of Southampton, UK and introduced him the findings of nuclear transplantation of fish. We made extensive exchanges on the idea of planning to carry out gene transfer in fish. I then studied molecular biology and gene cloning from Dr. Robert Kamen, director of Gene Transcription Laboratory of Imperial Cancer Research Fund laboratories in London. Dr. Robert Kamen then recommended me to work in the Genetics Institute of Boston, USA with Dr. Edward F. Fritsch (one of the authors of *Molecular Cloning*) in cloning human insulin and erythropoietin (EPO) genes to master the molecular cloning technology. However, when I returned to my homeland on schedule in 1983, I still did not obtain the necessary fish genes, and in fact no laboratory in the world had completed the cloning of any fish gene. Fortunately, I received a gift of recombinant human growth hormone DNA vector from Dr. D. H. Hamer of National Institutes of Health, USA. With it, fish gene transfer research finally started. From 1984 onwards, we isolated the growth hormone gene fragment from the vector plasmid DNA and injected it into the fertilized eggs of crucian carp and other kinds of fish to obtain transgenic fish with obvious rapid growth performance. We established a transgenic fish research model in the subsequent years, which was recognized by the international academic community. Later, we cloned β-actin gene (with Professor Perry Hackett) and growth hormone gene (with Professor T. T. Chen) of Cyprinidae, made a series of corresponding "all-fish" gene expression vector and generated the transgenic carp with rapid growth performance, high rate of feed utilization, ecological safety for cultivation and food safety for consumption. We now named it "crown carp" to highlight its outstanding characteristics for breeding and cultivation. This book has compiled 100 research papers from our research team, divided into two volumes, as a summary of our research work.

The systematic research has been mainly carried out in the Institute of Hydrobiology of the Chinese Academy of Sciences and the State Key Laboratory of Freshwater Ecology and Biotechnology. The then director, academician Dr. Jiankang Liu, was the most important supporter and promoter for the launch of the initiatives. In the early stage, Guohua Li, Ling He, and Yuefeng Xie, three young assistants who joined together with me, started from scratch and overcame countless difficulties to establish the first fish gene transfer research laboratory, which is also one of the earliest molecular biology laboratories in China. Over the next 30 years, dozens of colleagues and over ×× graduate students participated in these studies and grew up together with this program. I would like to thank Professors Yaping Wang, Wei Hu, Yonghua Sun and Zongbin Cui, who are not only the main force of our research team but also co-editors of this monograph. I sincerely thank Associate Professors Fuying Zhang, Rong Huang and others, the senior engineers Kesheng Xu, Shangping Chen, Lanjie Liao, and research assistants Ming Li, Yongming Li, Deyun Qi, Jianshe Zhen for their excellent work. I sincerely thank senior engineers Xirui Zhang, Jianxin Feng and Yufeng Wang of Henan Academy of Fishery Sciences for their great help in providing Yellow River common carp for stating varieties and assistance of fish breeding and cultivation. I sincerely thank Wuhan Duofu Technological Farm Co., Ltd. and the Dahu Aquaculture Co., Ltd. for their foresight in establishing a qualified experimental genetic engineering fish farm, including the construction of 6.7 hm^2 artificially simulated lake, in accordance with the provisions of the national regulations on safety of agricultural GMOs and the administrative measures on safety assessment of agricultural GMOs. In cooperation with them, we carried out the food safety and environmental assessment test of the "all-fish *gh*" transgenic Yellow River common carp, including restricted field testing, enlarged field testing, and productive testing. I would also like to thank the funding supports from the Chinese Academy of Sciences, the Ministry of Science and Technology, the National Natural Science Foundation of China and Hubei province. This book is published with the support of the State Key Laboratory of Freshwater Ecology and Biotechnology.

Last but not least, I would like to point out that transgenic breeding technology is essentially a logical extension and technological transition of the traditional crossbreeding technology, which is not difficult to be figured out from the 30-year's research presented in this book. Transgenic breeding technology can be understood as the hybridization between the target gene coding for the desired trait and a whole genome of an organism that needs to be improved. It is a type of molecular crossbreeding technology based on modern molecular biology, but more directional, accurate, predictable and traceable when compared with the traditional crossbreeding. For various reasons, this new technology, which was initiated in our country with all independent intellectual property rights, has encountered many difficulties in its popularization and application. "Do not say that the clouds always blot out the sun. There's always a time when the mist clears away. " I firmly believe that the spirit of science will eventually dispel all kinds of mist that troubles the public. History will eventually prove that transgenic technology will surely play an incomparable and irreplaceable role in the development of modern agriculture.

Professor of IHB, CAS
Peking University Boya Chair Professor
Editor General, *Science China* and *Science Bulletin*
Academician of CAS & TWOS
Dec. 2018

目　　录

上　卷

第一部分　鱼类基因转移模型建立

第二部分　外源基因的整合与表达

第三部分 外源基因表达的生理功能

第四部分　转基因黄河鲤的生态安全研究

第五部分　鱼类生殖调控与转基因鱼的育性控制研究

下 卷

第六部分 抗病草鱼基因工程技术育种研究

第七部分　鱼类基因转移育种技术展望

第六部分 抗病草鱼基因工程技术育种研究

Sequence of Genome Segments 1, 2, and 3 of the Grass Carp Reovirus (Genus *Aquareovirus*, Family *Reoviridae*)

Qin Fang[1] Houssam Attoui[2] Jean François Philippe Biagini[2] Zuoyan Zhu[1]
Philippe de Micco[2] Xavier de Lamballerie[3]

1 State Key Laboratory of Freshwater Ecology and Biotechnology, Hydrobiology Institute, Wuhan Institute of Virology, CAS, Wuchang, 430071, Wuhan, Hubei, China
2 Unité des Virus Emergents, Laboratoire de Virologie Moléculaire, EFS Alpes-Méditrranée
3 Laboratoire de Virologie Moléculaire, Tropicale et Transfusionnelle, Faculté de Médecinede Marseille, Université de la Méditerranée, 27 Boulevard Jean Moulin, 13005 Marseille cedex 5, France

Abstract The genome segments 1, 2, and 3 of the grass carp reovirus (GCRV), a tentative species assigned to genus *Aquabirnavirus*, family *Reoviridae*, were sequenced. The respective segments 1, 2, and 3 were 3949, 3877, and 3702 nucleotides long. Conserved motifs 5′(GUUAUUU) and 3′(UUCAUC) were found at the ends of each segment. Each segment contains a single ORF and the negative strand does not permit identification of consistent ORFs. Sequence analysis revealed that VP2 is the viral polymerase, while VP1 might represent the viral guanylyl/methyl transferase (involved in the capping process of RNA transcripts) and VP3 the NTPase/ helicase (involved in the transcription and capping of viral RNAs). The highest amino acid identities (26%-41%) were found with orthoreovirus proteins. Further genomic characterization should provide insight about the genetic relationships between GCRV, aquareoviruses, and orthoreoviruses. It should also permit to precise the taxonomic status of these different viruses.

Keywords *Aquabirnavirus*; GCRV; GCHV; grass carp; double-stranded RNA; polysegmented fish reoviruses

Introduction

Double-stranded (ds) RNA viruses that infect aquatic animals belong to two families: *Birnaviridae* (genus *Aquabirnavirus*, type species: infectious pancreatic necrosis virus), and *Reoviridae* (genus *Aquareovirus*, type species: Striped bass reovirus) [1,2]. The *Aquareoviruses* have genomes consisting of 11 segments of dsRNA contained in a core surrounded with a double layered icosahedral capsid with a T = 13 symmetry. These viruses physically resemble orthoreoviruses and infect a variety of aquatic animals, including finfish and Crustacea [2]. They

文章发表于 *Biochem Biophys Res Commun*, 2000, 274(3): 762-766

are classified into six genogroups (A to F) and the genus includes a number of tentative species [3]. Among unclassified species, figures the grass carp reovirus GCRV (or grass carp hemorrhage virus, GCHV) [2]. This virus was isolated from the grass carp (*Ctenopharyngodon idellus*) in the People's Republic of China [4,5] and provokes hemorrhage in about 85% of fingerling and yearling populations of the same fish [7]. It was assigned to the *Aquareovirus* genus based mainly on its natural hostand its genome composed of 11 dsRNA segments. Previous dsRNA hybridization analyses have ruled out the belonging of this virus to any of the established *Aquareovirus* genogroups and designated it as belonging to a new genogroup identified as G [3].

We have realized the characterization of the fulllength sequences of genome segments 1, 2, and 3 of the GCRV. These are the first reported genome sequences for this virus. This permitted the identification of the RNA-dependent RNA polymerase (RDRP) gene, and its comparison with RDRPs of other members of the family *Reoviridae*.

Materials and Methods

Cell culture and virus propagation

Ctenopharyngodon idellus kidney (CIK) cells were grown as monolayers at 28℃ in Eagle's minimum essential medium supplemented with 10% fetal bovine serum (FBS).

The virus propagation was realized as previously described [6]. Briefly, confluent monolayers of CIK were infected with a virus stock at 5 PFU/ml. The cells were overlaid with maintenance medium containing 2% FBS and incubated for 3 days at 28℃. Culture supernatant (500 ml) was recovered and clarified by centrifugation at 8000 *g* for 30 min. This was followed by pelleting the virus at 40,000 *g* for 3.5 h.

Sequence determination of segments 1, 2, and 3

Determination of partial sequences. Genomic RNA was extracted from virus using the proteinase K phenol-chloroform protocol as reported elsewhere [8]. The dsRNA segments were run for 1 h (7 V/cm) in a 15-cm-long 1% agarose gel-containing 0.5 μg/ml ethidium bromide in TAE buffer. The bands were visualized by UV transillumination and excised using a scalpel blade. The dsRNA was purified from agarose using the RNaid kit (Bio 101) as described by the manufacturer. Each segment (2 μg) was heat denatured at 70℃ for 15 min in 90% DMSO. The reverse transcription was performed at 42° for 1 h in a final volume of 20 μl containing 50 mM Tris-HCl (pH 8.3), 75 mM KCl, 3 mM $MgCl_2$, 10 mM DTT, 0.2 mM each dNTP, 0.2 μg of DMSO-denatured dsRNA (the final DMSO concentration in the reaction mixture was 4.5%), 40 U of RNase inhibitor (Gibco BRL), 0.1 μM hexanucleotide mixture, and 200 U of Superscript II reverse transcriptase (Gibco BRL). The resulting cDNA was treated with the Klenow fragment of DNA Polymerase I as described elsewhere [8], blunt end cloned into the *Eco*RV site of pZERO2.0 vector (Invitrogen) and recombinant vectors were transfected into competent TOP10 *E. coli* by electroporation. Colony screening permitted the identification of the longest cDNA insert from each segment which was sequenced using M13 universal primers, the D-Rhodamine DNA sequencing kit and an ABI Prism 377 sequence analyzer (Perkin-Elmer).

Determination of 5′ and 3′ end sequences. Virus dsRNA was extracted using a guanidinium isothiocyanate procedure, using the RNA Now kit (Biogentex) according to manufacturer's instructions. RNA segments 1, 2, and 3 were separated and extracted as described above. The single primer amplification technique was realized as described before [9]. Briefly, a 3′-amino blocked oligo deoxyribonucleotide (primer A: 5′-po_4-aggtctcgtagaccgtgcacc-nh2-3′) was ligated to both 3′-OH termini of the dsRNA, using 10 U of T4 RNA ligase (Boehringer Mannheim). The tailed dsRNA was recovered using the RNaid kit and denatured by heating at 99℃ for 1 min in presence of 15% dimethyl sulfoxide. cDNA copies of the genomic RNA were synthesized using a complementary primer (primer B: 5′-ggtgcacggtctacgagacct-3′) and 200 U of MMuLV Superscript reverse transcriptase (Gibco BRL).

Primers designated Pr5′ and Pr3′ were designed from the partial sequences obtained as described above. The remainder of each segment sequence was PCR amplified using primers B and Pr5′ or Pr3′ as necessary, according to the scheme shown in Fig. 1. PCR reactions were carried out using 2.5 U of Taq polymerase (Gibco BRL) and 0.5 μM of each primer. Thermal cycling parameters were as follows: one cycle of denaturation (90℃, 10 min) followed by 40 cycles of denaturation (94℃, 50 s), annealing (55℃, 50 s) and extension (72℃, 2 min). The cycling program ended by an extension step at 72℃ for 10 min.

The amplicons were analyzed by agarose gel electrophoresis, then ligated into the PGEM-T cloning vector. The recombinant vector was transfected into competent XL-blue *E. coli* and the insert sequence determined as described above.

Sequence analysis

The longest open reading frame (ORF) of each segment was determined. The putative protein sequences were compared to the sequences in the databases using the NCBI gapped BLAST 2.0 program. Sequence specific primers for the PCR amplification of the viral cDNA were designed with the help of the Oligo software (National Biosciences Inc.).

Results and Discussion

Segments 1, 2, and 3 of GCRV genome were characterized and their sequences deposited in the GenBank database under Accession Nos. AF260511, AF260512, and AF260513, respectively. The cloning of each segment was realized in three successive steps. In the first

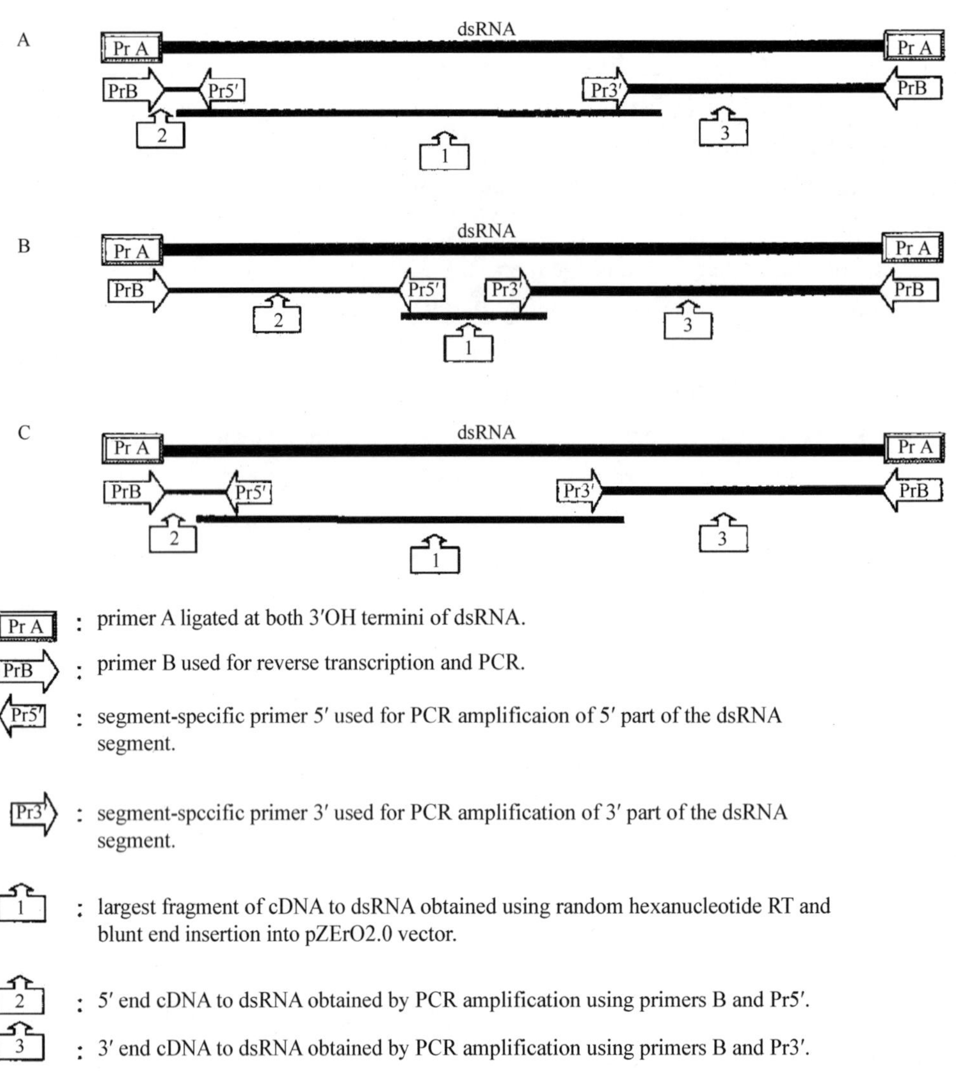

FIG. 1 Strategies for the cloning of segments 1 (A), 2 (B), and 3 (C) of GCRV

step, the cDNA resulting from reverse transcription of dsRNA using random hexanucleotides was cloned into the pZERO2.0 vector. The screening of the recombinant clones permitted to identify an insert of 2300 bp from segment 1, another insert of 800 bp from segment 2, and a third one of 2500 bp from segment 3. The sequence of each of these inserts was determined and used to design primers for the amplification of the 5′(second step) and 3′ (third step) ends of each segment. The sequences, locations and orientations of these primers are given in Table 1.

This three-step analysis permitted the full-length sequence determination of segments 1, 2, and 3 of GCRV. The length of the segments, those of the corresponding encoded putative proteins and the 5′ and 3′ non-coding regions are given in Table 2. Each segment was found to

Table 1 Primrs used for amplification of 5′and 3′ parts of segments 1, 2, and 3 of GCRV genome

Segment	Primer Pr5′			Primer Pr3′		
	Sequence	Orientation	Position	Sequence	Orientation	Position
1	AGGGCACGGTCATGAATG-AACTCC	Antisense	290-267	CCTCGTCCATTATGTT-CACTCGTG	Sense	2234-2257
2	CAGCGACGCGCTGGAACC-TCCTCGG	Antisense	1428-1404	TCCCGTTGCTCCCTAC-TCTAGACC	Sense	1926-1949
3	CGATTGCGGGTCAGAAAC-CACGG	Antisense	327-305	GTGTCCCACAGAGCTG-TCCGTCC	Sense	2284-2262

Table 2 Lengths of dsRNA segments 1, 2, and 3; encoded putative proteins; 5′ and 3′ non-coding regions of GCRV

	Segment length(dp)	Protein length (AA)	5′NCR		3′NCR	
			Length (bp)	Terminal sequences	Terminal sequences	Length (bp)
Segment 1	3949	1299	12	5′-GUUAUUUgcAuuAUG-----UUCAUC-3′		37
Segment 2	3877	1274	12	5′-GUUAUUUguAccAUG-----UUCAUC-3′		40
Segment 3	3702	1214	12	5′-GUUAUUUccAccAUG-----UUCAUC-3′		45
Consensus				5′-GUUAUUUsyAyyAUG-----UUCAUC-3′		

Note. Highly conserved sequences are in uppercase letters. Initiator AUG codons are shown in bold-faced letters. S represents C or G, and Y represents C or U

contain a single ORF and the analysis of the negative strand sequences did not permit to identify any consistent ORF.

As previously observed with other members of the family *Reoviridae* [2], segments 1, 2, and 3 of GCRV were found to have conserved terminal sequences. All positive strands of each dsRNA segment had the motif “5′-GUUAUUU-3′” in common at the 5′ end, and the motif “5′-UUCAUC-3′” in common at the 3′ end. Moreover, the first and last nucleotides of each segment are inverted complements. In previous studies of *Reoviridae* genomes, comparable conserved motifs have been reported [2]. They are supposed to act as sorting signals, bringing a single copy of each genome segment into the nascent viral capsid [10,11]. These GCRV specific motifs, together with a number of segment specific inverted terminal repeats that were found in the 5′ and 3′ regions of each segment, could interact by homologous base pairing thus holding the RNA transcripts in a circular form. Similar situations were previously reported in the case of other viruses with segmented genomes [10-14].

One interesting feature in the nucleic acid sequences was that the initiator AUG in the three segment was located at position 13 from the 5′end. Segments 1 and 2 AUGs obey to the Kozak’s rule [(–3)A/GNN-AUGG(+1)] of strong initiation codons, with an A at position –3. In the case of segment 3, position –3 corresponds to a C, and therefore partially conforms to Kozak’s rule.

The theoretical proteins deduced from segments 1, 2, and 3 of GCRV were compared to those of other *Reoviridae* members deposited in the databases. Significant amino acid identities (26-41%) were found with structural orthoreovirus proteins in the order given in

Table 3, with the highest identity values being with orthoreovirus polymerase sequences. This finding is in agreement with the physical resemblance in electron microscopy between orthoreoviruses and *aquareoviruses*. This finding can also permit to speculate on the physical position of these proteins in the viral structure with regard to the analysis of the orthoreovirus structure. Hence, VP1 of GCRV might represent the "turret" protein at the five-fold axis. It is also noticeable that this protein in orthoreoviruses carries the enzymatic function of guanylyl transferase and methyl transferase. It is involved in the capping process of the nascent RNA transcripts in the primary transcription cycle during infection. Further investigations are needed to confirm this function in the GCRV VP1.

Table 3 Comparison of GCRV VP1, VP2, and VP3 to orthoreovirus proteins

GCRV	REOV	% Protein identity	% Protein similarity	Putative function (similarity to orthoreoviruses)
Segment 1 (3949 bp) VP1 protein	Segment 2 (3916) λ2 protein	26	41	Guanylyl transferase,methyl transferase
Segment 2 (3877 bp) VP2 protein	Segment 1 (3854) λ3 protein	41	57	RNA-dependent RNA polymerase
Segment 3 (3702 bp) VP3 protein	Segment 3 (3896) λ1 protein	31	50	Binds dsRNA, NTPase, helicase

Note. GCRV, grass carp reovirus; REOV, orthoreovirus

The VP2 of GCRV contains the signature motifs for the *Reoviridae* RDRP: The motif SG is found at position 688 and the motif GDD at position 739. These features together with the high degree of protein identity (41%) with the RDRP of orthoreovirus, strongly support that VP2 of GCRV is the viral RDRP. This protein is most probably located in the virus core.

Based on comparison to the VP3 of orthoreoviruses, GCRV VP3 might represent the NTPase, helicase protein which is involved in the transcription and capping of viral RNA.

The comparison of GCRV sequences to homologous segments of aquabirnaviruses was hampered by the lack of genome sequence data in that genus. The only comparison that was possible is that of the G + C content of GCRV sequence with that of sequences currently available from aquabirnaviruses, namely segment 10 of Strip bass reovirus (*Aquareovirus* genogroup A, Accession No. U83396) and segment 10 of Coho Salmon reovirus (*Aquareovirus* genogroup B, Accession No. U90430). Similar values ranging between 52 and 55% were calculated from all of these sequences. The comparison of the G + C content of GCRV segments 1, 2, and 3 to that of homologous orthoreovirus genes was also realized. The average value of the G + C content of orthoreovirus genes 1, 2, and 3 was found to be 46%.

In conclusion, the significant similarities between GCRV and orthoreovirus proteins are most unusual concerning viruses belonging to separate genera. This situation has never been reported before in the family *Reoviridae*. In order to identify or rule out further homologies to orthoreoviruses and aquabirnaviruses, further genomic characterization is required. The remainder of the genome of GCRV (segments 4-11) and the genome of a representative member of each *Aquareovirus* genogroup should be sequenced. These further characterizations should provide insight about the genetic relationships between GCRV, aquareoviruses, and orthoreoviruses. It should also permit to precise the taxonomic status of these different viruses.

References

1. Kim, H. C., and Leong, J-O. (1999) Fish viruses. In Encyclopedia of Virology (Granoff, A., and Webster, R. G., Eds.), pp. 558-568, Academic Press, San Diego.
2. Mertens, P. P. C., Arella, M., Attoui, H., Belloncik, S., Bergoin, M., Boccardo, G., Booth, T. F., Chiu, W., Diprose, J. M., Duncan, R., Estes, M. K., Gorziglia, M., Gouet, P., Gould, A. R., Grimes, J. M., Hewat, E., Hill, C., Holmes, I. H., Hoshino, Y., Joklik, W. K., Knowles, N., López Ferber, M. L., Malby, R., Marzachi, C., McCrae, M. A., Milne, R. G., Nibert, M., Nunn, M., Omura, T., Prasad, B. V. V., Pritchard, I., Samal, S. K., Schoehn, G., Shikata, E., Stoltz, D. B., Stuart, D. I., Suzuki, N., Upadhyaya, N., Uyeda, I., Waterhouse, P., Williams, C. F., Winton, J. R., and Zhou, H. Z. (2000) *Reoviridae*. In Virus Taxonomy. Seventh Report of the International Committee for the Taxonomy of Viruses (Van Regenmortel, M. H. V., Fauquet, C. M., Bishop, D. H. L., Calisher, C. H., Carsten, E. B., Estes, M. K., Lemon, S. M., Maniloff, J., Mayo, M. A., McGeoch, D. J., Pringle, C. R., and Wickner, R. B., Eds.), Academic Press, San Diego.
3. Rangel, A. A. C., Rockemann, D. D., Hetrick, F. M., and Samal, S. K. (1999) Identification of grass carp hemorrhage virus as a new genogroup of *Aquareovirus*. J. Gen. Virol. 80, 2399-2402.
4. Chen, B. S., and Jiang, Y. (1984) Morphological and physicochemical characterization of the hemorrhagic virus of grass carp. Kexue Tonboga 29, 832-835.
5. Ke, L-H., Fang, Q., and Cai, Y-Q. (1990) Characteristics of a novel isolate of grass carp Hemorrhage Virus [J]. Acta Hydrobiol. Sini. 14, 153-159.
6. Fang, Q., Ke, L-H., and Cai, Y-Q. (1989) Growth characterization and high titre culture of GCHV. Virol Sini. 3, 315-319.
7. Jiang, Y., and Ahne, W. (1989) Some properties of the etiological agent of the hemorrhagic disease of grass carp and black carp. In Viruses of Lower Vertebrates (Ahne, W., and Kurstak, E., Eds.), pp. 227-239, Springer-Verlag, Berlin.
8. Sambrook, J., Fritsch, E. F., and Maniatis, T. (1989) Molecular Cloning, A Laboratory Manual, Cold Spring Harbor Laboratory Press, New York.
9. Attoui, H., Billoir, F., Biagini, P., De Micco, P., and de Lamballerie, X. (2000) Complete sequence determination and genetic analysis of Banna virus and Kadipiro virus: Proposal for assignment to a new genus (*Seadornavirus*) within the family *Reoviridae*. J. Gen. Virol. 81,1507-1515.
10. Anzola, J. V., Xu, Z., Asamizu, T., and Nuss, D. L. (1987) Segment-specific inverted repeats found adjacent to conserved terminal sequences in wound tumour virus genome and defective interfering RNAs. Proc. Natl. Acad. Sci. USA 84, 8301-8305.
11. Xu, Z., Anzola, J. V., Nalin, C. M., and Nuss, D. L. (1989) The 3′-terminal sequence of wound tumor virus transcripts can influence conformational and functional properties associated with the 5′ terminus. Virology170, 511-522.
12. Hsu, M. T., Parvin, J. D., Gupta, S., Krystal, M., and Palese, P. (1987) Genomic RNA of influenza viruses are held in a circular conformation in virions and in infected cells by a terminal panhandle. Proc. Natl. Acad. Sci. USA 84, 8140-8144.
13. Attoui, H., De Micco, P., and de Lamballerie, X. (1997) Complete nucleotide sequence of Colorado Tick Fever virus segments M6, S1, and S2. J. Gen. Virol. 78, 2895-2899.
14. Attoui, H., Charrel, R., Billoir, F., Cantaloube, J. F., De Micco, P., and de Lamballerie, X. (1998) Comparative

sequence analysis of American, European, and Asian isolates of viruses in the genus Coltivirus. J. Gen. Virol. 79, 2481-2489.

草鱼出血病病毒基因组第1、2和3节段的序列

方 勤[1] Houssam Attoui[2] Jean François Philippe Biagin[2]
朱作言[1] Philippe de Micco[2] Xavier de Lamballerie[3]

1中国科学院水生生物研究所，湖北武汉，中国 430072
2 Unité des Virus Emergents, Laboratoire de Virologie Moléculaire, EFS Alpes-Méditrranée;
3 Laboratoire de Virologie Moléculaire, Tropicale et Transfusionnelle, Faculté de Médecinede Marseille, Université de la Méditerranée, 27 Boulevard Jean Moulin, 13005 Marseille cedex 5, France

摘 要 草鱼呼肠孤病毒隶属呼肠孤病毒科水生呼肠孤病毒属。本研究对它的基因组片段1、2和3进行了测序分析。基因组片段1、2和3的长度分别为 3949 bp、3877 bp 和 3702 bp。每个片段的末端均具有 5′(GUUAUUU) 和 3′(UUCAUC) 保守序列，包括单一开放阅读框，并且不能从负链识别一致的开放阅读框。序列分析显示 VP2具有 RNA 聚合酶活性，VP1可能具有鸟苷酸/甲基转移酶活性(参与病毒RNAs的转录和加帽)。与正呼肠孤病毒蛋白相比，氨基酸相似性最高达到26%-41%。深入的基因组特征研究可以为草鱼呼肠孤病毒、水生呼肠孤病毒和正呼肠孤病毒的演化关系提供新视野，同时将为这些病毒提供更精确的分类信息。

Genome Segment S8 of Grass Carp Hemorrhage Virus Encodes A Virion Protein

Tao Qiu Jing Zhang Renhou Lu Zuoyan Zhu

State Key Laboratory of Freshwater Ecology and Biotechnology, Institute of Hydrobiology, Chinese Academy of Sciences, Wuhan

Abstract The complete nucleotide sequence of the genome segment S8 of grass carp hemorrhage virus (GCHV) was determined from cDNA corresponding to the viral genomic RNA. It is 1,287 nucleotides in length and contains a large open reading frame that could encode a protein of 409 amino acids with a predicted molecular mass of 44 kD. The S8 was expressed using the pET fusion protein vector and detected by Western blotting analysis using the chicken egg IgY against intact GCHV particles, indicating that S8 encodes a virion protein. Amino acid sequence comparisons revealed that the protein encoded by S8 is closely related to protein σ2 of mammalian reovirus, suggesting that the deduced protein of S8 is an inner capsid protein.

Keywords RNA virus; GCHV; segment S8; nucleotide sequence; virion protein; IgY

The members of the family Reoviridae are the segmented double-stranded RNA viruses. They are known in all major groups of organisms, from bacteria and fungi to animals and plants. Grass carp hemorrhage virus (GCHV) also called grass carp reovirus (GCRV), as a member of the genus *Aquareovirus*, is an important fish pathogen causing hemorrhagic disease [1]. GCHV genome consists of 11 double-stranded RNA (dsRNA) segments that are enclosed within a double-layered protein capsid [2]. The segments are transcribed by virus-associated RNA polymerase to form mRNAs [3] that also function as templates for a putative replicase in virus-infected cells [4]. GCHV was placed to the seventh genogroup of *Aquareovirus* because it could not hybridize to the type viruses of all the other six genogroups in reciprocal RNA-RNA dot blotting analysis [5]. Little information is available for the coding assignments for most genome segments except for the sequences of the three largest segments [4]. They are putative guanylyltransferase, viral polymerase and helicase genes. More recently, we have determined the segment S10, which was thought to encode an outer-capsid protein [6], whereas the structure and function of other segments remain unknown. In this report, the full-length cDNA of GCHV genome segment S8 were cloned and its characteristics of nucleotide and amino acid sequences are documented here.

The strain 873 of GCHV was adapted for growth in CIK cell [7]. Virus was purified from a continuous sucrose gradient (30-60% sucrose) and centrifuged at 100,000 *g* for 2 h.

文章发表于 *Intervirology*, 2001, 44(5): 317-320

Genomic dsRNA was extracted from the purified virus particles by proteinase K and phenol-chloroform extraction. The synthesis of the full-length cDNA of dsRNA was carried out with the method of Lambden *et al.* [8]. Amplified cDNA products were separated on agarose gel (fig. 1C) and the fragment approximately 1,300 bp, which corresponded in size to that calculated from the GCHV dsRNA molecular weight for S8 segment, was excised and purified by Glassmilk DNA purify kit (BioStar). The S8 cDNA was directly ligated into pGEM-T vector, and transformed into DH5α strain of *Escherichia coli* (Gibco BRL). The recombinant plasmid containing the full-length cDNA was identified according to the size of inserted segment by PCR and was purified using a plasmid DNA purification miniprep kit (Viogene). The nucleotide sequence of S8 was determined on an ABI 310 Genetic Analyzer (Perkin-Elmer). To verify the S8 cDNA, the genome dsRNAs were separated in a 1.0% agarose gel, transferred to a Nylon membrane (Hybond), and hybridized with the S8 cDNA probe labeled with digoxigenin-dUTP, followed by Northern blotting analysis using Dig High Primer Labeling and Detection Starter Kit (Boehringer-Mannheim) (fig. 1A, B).

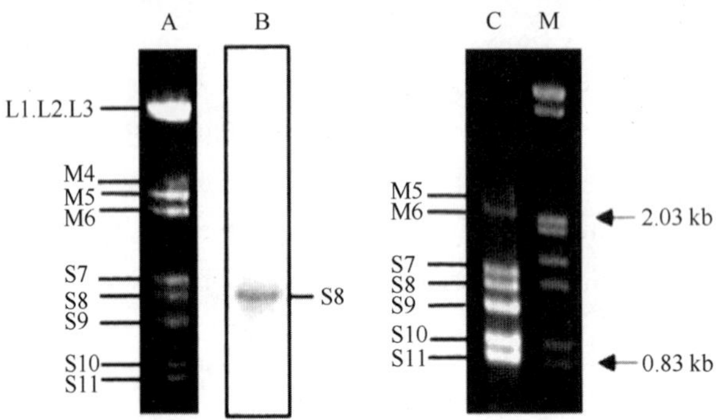

Fig. 1 Analysis of the GCHV S8. Lane A: viral genome dsRNA of GCHV analyzed by electrophoresis in a 1% agarose gel; lane B: result of Northern blotting with S8 cDNAs as probe; lane C: amplified products of cDNAs transcribed from viral RNA in a 1% agarose gel; lane M: lambda DNA/*Eco*R I+ *Hin*d III marker

The complete nucleotide and amino acid sequences of genome segment S8 of GCHV are shown in figure 2. The S8 sequence is 1,287 bp long and has a large open reading frame (ORF). The distribution of the four bases was found relatively rich in cytidine (21.3% A, 22.68% G, 23.66% U, 32.36% C). The ORF starts with the first AUG (at nucleotides 12-14) and terminates at nucleotides 1239-1241 with a UAA. The AUG appears to be a very favorable context for initiation of translation (AUC<u>AUG</u>G), according to the consensus sequence established by Kozak [9]. The terminal region of the GCHV S8 segment displays the nucleotide sequences 5′ GUUAUU and CAUC 3′, which are conserved in genome segments of GCHV [unpubl. data]. In addition, segment-specific inverted repeats, GUGAUGGCA at 9-17 and UGCCUCAC at 1270-1277, were identified adjacent to the terminal sequences.

The ORF of S8 encodes a 409 amino acid protein of calculated molecular mass of 44 kD.

To clarify the identifying of the protein, full-length segment of S8 was amplified with two primers (5′ AGGGATCCGCGGTAAATGTTATT 3′ and 5′ GGAAGCTTGGCGGTAAATG ATGA 3′) from the pGEM-T vector and inserted into the *Bam*H I and *Hin*d III sites of pET-28a vector and expressed in *Escherichia coli*. Protein expression was performed according to the pET system manual provided by Novagen Corp. The extracted *E. coli* proteins were resolved by SDS-PAGE, transferred to NC membrane (Gleman); Western blotting was performed with immunoglobulin (IgY) against intact virus particles that isolated from the yolks of eggs of GCHV-immunized hens according to the method of Horikoshi *et al.* [10]. Peroxidase-labeled rabbit anti-chicken IgY (Promage) was used for final detection. The results are shown in figure 3, in which an approximately 48-kD band was identified. It is very similar to the predicted molecular mass of the fusion protein because the sizes of the S8 product and foreign sequence in pET28a vector are 44 and 4.5 kD, respectively. The band, 48 kD, was also detected by Western blotting, and no obvious additional bands could be detected in the preparation, identifying the IgY that reacted specifically with the protein expressed by S8. It could be deduced that GCHV S8 encodes a virion protein.

Since prior work has revealed that the largest three segments and S10 segment are similar to the segments of mammalian reovirus (MRV), a member of another genus, *Orthoreovirus* in the family Reoviridae [4], we compared the deduced protein of GCHV S8 with the reported proteins of MRV in computer using CLUSTALW programs. The protein encoded by S8 showed 18.54% identity with that of the inner capsid protein σ2 of MRV [11] in a 426-amino-acid alignment length though no long identical predicted amino acid sequences were found between them.

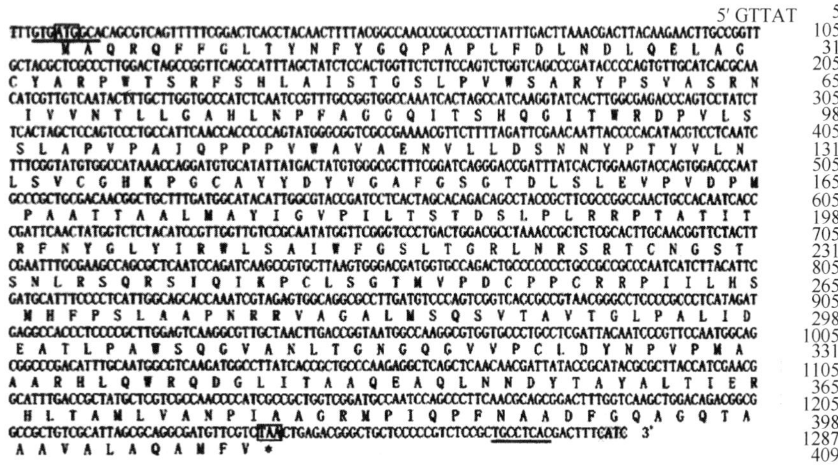

Fig. 2 Complete nucleotide sequence (presented in the cDNA form) of S8 RNA segment of GCHV. The conserved 5′- and 3′-terminal nucleotide sequences are indicated with a gray background and the inverted repeats are underlined. The initiation and termination sites are boxed

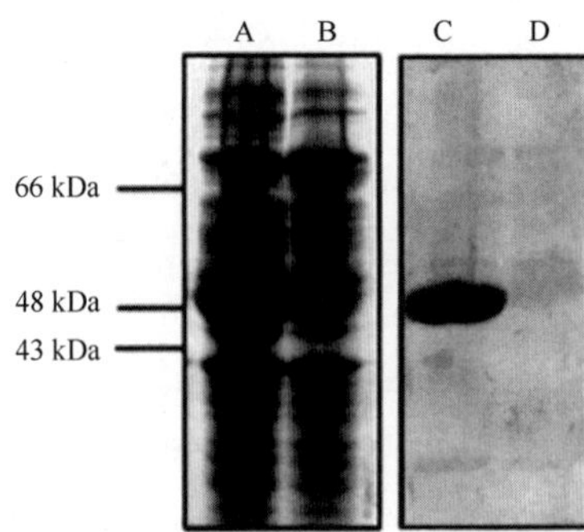

Fig. 3 Western blot analysis of expressed pET-S8 fusion protein. The extracted *E. coli* proteins were resolved by 10% SDS-PAGE and stained with Coomassie blue (lane A) or transferred to NC membrane and probed with IgY (lane C). An insert-minus plasmid control (lane B) was also probed with IgY (lane D)

More detailed comparisons revealed that it is more closely related to MRV σ2 protein. First, the secondary structure analysis of the amino acid sequence of the ORF [12] indicated that the protein included a carboxy-terminal region that is formed from α-helices and β-turns and a large amino-terminal region that is formed predominantly by β-strands and β-turns. A similar distribution of the secondary structure along the sequence was also observed in MRV σ2 [11], suggesting that the deduced proteins of GCHV S8 and MRV σ2 are structurally and functionally related. Second, the hydrophilic plots [13] showed very similar profiles among the deduced proteins of GCHV S8 and MRV σ2 (data not shown), suggesting that they may have an analogous molecular conformation. Third, the size of the predicted protein encoded by S8, 44 kD, is very similar to MRV σ2, 47 kD. These facts support the hypothesis that GCHV S8 may correspond to the MRV S2 gene, which encodes the inner capsid protein, MRV σ2 [11]. This implies that S8 encodes an analogous inner capsid protein of GCHV.

A search made against protein database using BLASTp algorithm [14] revealed that a region (amino acids 228-263) in deduced protein encoded by S8 has similarity with a region in the immunoglobulin gamma (amino acids 210-245) of cynomolgus monkey [15] with 38% identity. It does not appear to be an easily interpretable match. Whereas, interestingly, this short region is characterized by the largest hydrophilic and highest antigenic region and contains three cysteine residues (fig. 4). Moreover, it is identified as the most flexible portion, which corresponds to the sequences predicted to predominantly form β-turns which alternate with β-sheets. The region may be an important determinant for protein-protein interaction or for host immune response to this protein. Moreover, positive charges are concentrated in this region, implying that it also could be involved in protein RNA interaction. Some suggestions are given by prior work on MRV σ2 showing that it can interact with MRV λ1 protein to form the inner capsid [16] and has affinity for reovirus dsRNA [11]. However, extensive evidence has to be presented to strengthen the above views. Our study serves as a preliminary to future investigations to learn more about the structure and function of protein encoded by GCHV S8.

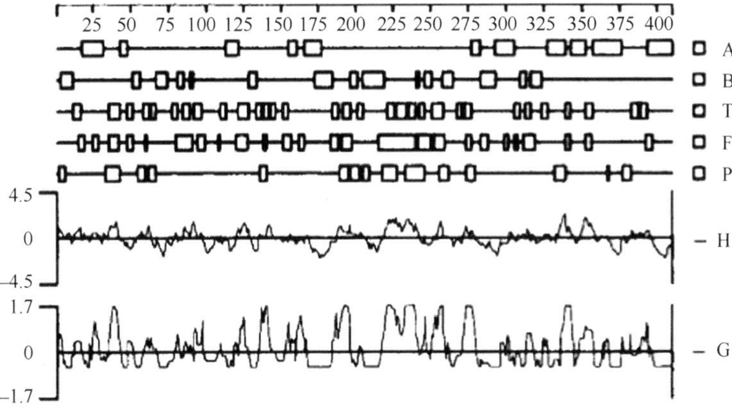

Fig. 4 Secondary structure, charge, hydrophilic plot, and antigenic index of the protein encoded by GCHV S8. A = Alpha regions; B = beta regions; T = turn regions; F = flexible regions; P = positive charge; H = hydrophilic plot; G = antigenic index

Acknowledgements This work was supported by grant No. ZYJ01-04 from the China-Israel Fund for Scientific and Strategic Research and Development, and by grant No. 990205-FB05 from FEBL, the State Key Laboratory of Freshwater Ecology and Biotechnology.

References

1. Chen Y, Jiang Y: Studies on the morphological and physico-chemical characterization of the hemorrhagic virus of grass carp. KeXue TongBao (foreign lang ed) 1983; 28: 1138-1140.
2. Ke L, Fang Q, Cai Y: Characteristics of a new isolation of hemorrhagic virus of grass carp (in Chinese with English abstract). Acta Hydrobiol Sin 1990; 14:153-159.
3. Zhang BY, Ke LH: Studies on *in vivo* translation of grass carp hemorrhage virus genome (in Chinese with English abstract). Virol Sin 1993; 8: 185-188.
4. Fang Q, Attoui H, Biagini JF, Zhu Z, de Micco P, de Lamballerie X: Sequence of genome segments 1, 2, and 3 of the grass carp reovirus (genus aquareovirus, family reoviridae). Biochem Biophys Res Commun 2000; 274: 762-766.
5. Rangel AA, Rockemann DD, Hetrick FM, Samal SK: Identification of grass carp haemorrhage virus as a new genogroup of aquareovirus. J Gen Virol 1999; 80: 2399-2402.
6. Qiu T, Lu RH, Zhang J, Zhu ZY: Complete nucleotide sequence of the S10 genome segment of grass carp reovirus (GCRV). Dis Aquat Org 2001; 44: 69-74.
7. Zuo W, Qian H, Xu Y, Du S, Yang X: A cell line derived from the kidney of grass carp (in Chinese with English abstract). J Fish China 1986; 10: 11-17.
8. Lambden PR, Cooke SJ, Caul EO, Clarke IN: Cloning of noncultivatable human rotavirus by single primer amplification. J Virol 1992; 66: 1817-1822.
9. Kozak M: Point mutations define a sequence flanking the AUG initiator codon that modulates translation by eukaryotic ribosomes. Cell 1986; 44: 283-292.
10. Horikoshi T, Hiraoka J, Saito M, Hamada S: IgG antibody from hen egg yolks. Purification by ethanol fractionation. J Food Sci 1993; 58: 739-742, 779.
11. Dermody TS, Schiff LA, Nibert ML, Coombs KM, Fields BN: The S1 gene nucleotide sequences of prototype strains of the three reovirus serotypes: Characterization of reovirus core protein σ2. J Virol 1991; 65: 5721-5731.
12. Chou PY, Fasman GD: Prediction of the secondary structure of proteins from their aminoacid sequence. Adv

Enzymol 1978; 47: 45-148.

13. Kyte J, Doolittle RF: A simple method for displaying the hydropathic character of a protein. J Mol Biol 1982; 157: 105-132.
14. Altschul SF, Madden TL, Schiffer AA, Zhang J, Zhang Z, Miller W, Lipman DJ: GappedBLAST and PSI-BLAST: A new generation ofprotein database search programs. Nucleic Acids Res 1997; 25: 338-340.
15. Lewis AP, Barber KA, Cooper HJ, Sims MJ, Worden J, Crowe JS: Cloning and sequence analysis of kappa and gamma cynomolgus monkey immunoglobulin cDNAs. Dev Comp Immunol 1993; 17: 549-560.
16. Matsuhisa T, Joklik WK: Temperature-sensitive mutants of reovirus. V. Studies on thenature of the temperature-sensitive lesion ofthe group C mutant ts447. Virology 1974; 60: 380-389.

草鱼出血病病毒基因组 S8节段编码一个病毒蛋白

邱　涛　Jing Zhang　陆仁厚　朱作言

中国科学院水生生物研究所，淡水生态与生物技术国家重点实验室，武汉

摘　要　本研究测定了草鱼出血病病毒 S8基因组节段的核苷酸序列。S8节段全长1287bp，包含一个编码409个氨基酸的开放阅读框，氨基酸分子量大小为44kD。通过 pET 融合蛋白载体，对 S8节段进行了原核表达，其能够被源于鸡的抗 GCHV IgY 抗体识别，表明 S8节段编码一个病毒蛋白。氨基酸序列比较分析显示，GCHV S8节段编码的蛋白与哺乳动物呼肠孤病毒编码的σ2蛋白较为接近，说明 S8节段编码的蛋白可能为内衣壳蛋白。

Molecular Characterization and Expression of the M6 Gene of Grass Carp Hemorrhage Virus (GCHV), An Aquareovirus

T. Qiu　R. H. Lu　J. Zhang　Z. Y. Zhu

State Key Laboratory of Freshwater Ecology and Biotechnology, Institute of Hydrobiology, Chinese Academy of Sciences,Wuhan

Summary The complete nucleotide sequence of M6 gene of grass carp hemorrhage virus (GCHV) was determined. It is 2039 nucleotides in length and contains a single large open reading frame that could encode a protein of 648 amino acids with predicted molecular mass of 68.7 kDa. Amino acid sequence comparison revealed that the protein encoded by GCHV M6 is closely related to the protein μl of mammalian reovirus. The M6 gene, encoding the major outer-capsid protein, was expressed using the pET fusion protein vector in *Escherichia coli* and detected by Western blotting using chicken anti-GCHV immunoglobulin (IgY). The result indicates that the protein encoded by M6 may share a putative Asn-42-Pro-43 proteolytic cleavage site with μl.

Aquareovirus is an important pathogen involved in diseases of aquatic animals. It is a genus in the *Reoviridae* family. The virions consist of a double-layered capsid containing a genome composed of 11 segments of double-stranded RNA. A large number of aquareoviruses have been isolated from different fish, shellfish and crustacean species. Not all these aquareoviruses have been associated with clinical disease. Many viruses were isolated from normal fish [17]. Grass carp hemorrhage virus (GCHV) [3], as a member of aquareovirus, is a pathogen causing hemorrhagic disease of grass carp, and was found to be the most pathogenic aquareovirus [16]. Thus it has obtained more attention. GCHV shares the common physicochemical properties and morphological characteristics with other aquareoviruses [6]. Its 11 segments genome was approximate 15.46×10^6 Dalton in total molecular mass and separated into three size classes: large (L1, L2, L3), medium (M4, M5, M6) and small (S7, S8, S9, S10, S11).

Unlike viruses of other genera in family *Reoviridae*, little molecular information is currently known about genome segments of members of *Aquareovirus*. Previous research has identified six different genogroups of aquareovirus using RNA-RNA blot hybridization [10,11,18]. Recently, GCHV was placed to the seventh genogroup [16] and its three largest genome segments and segment S10 were sequenced [4,15]. Therefore, it would be of interest to obtain the molecular information of the other segments. In this study, we determined the M6 gene of GCHV and performed immunoblotting

文章发表于 *Archives of Virology*, 2001, 146(7): 1391-1397

analysis on its encoding protein.

The strain 873 of GCHV was adapted for growth in CIK cell [22]. Virus was purified using a continuous sucrose gradient (30 to 60% sucrose in NTE buffer) and centrifuged at 100,000*g* for 2 hrs. Genomic dsRNA was extracted from the purified virus particles by proteinase K treatment and phenol-chloroform extraction. RNasefree DNase was used to eliminate the remaining cellular DNA. The synthesis of the full-length cDNA of dsRNA was carried out with the method of Lambden *et al.* [9]. Two primers used in the process were 5′-PO_4-ATTTACCGCCGAGCCTGACTT-NH_2-3′ and 5′-AAGTCAGGCTCGGCGGTAAAT- 3′. Amplified cDNA products were separated on agarose gel. The profile of fulllength cDNAs are found consistent with that of GCHV dsRNA (data not shown). The fragment approximately 2030 bp, which corresponded in position to the M6 segment and in size to that calculated from the GCHV dsRNA molecular weight for M6 segment [6], was excised and purified by Glassmilk DNA purification kit (DNAstar). The M6 cDNA was directly ligated into pGEM-T vector, and transformed into DH5α strain of *Escherichia coli* (GIBCO BRL). The recombinant plasmid containing the full-length cDNA was identified according to the size of inserted segment by PCR using two M13 primers on pGEM-T and was purified using a plasmid DNA purification miniprep kit (Viogene). The nucleotide sequence of M6 was sequenced on an ABI 377 (Perking Elmer).

The complete nucleotide sequence of genome segment M6 of GCHV is shown in Fig. 1. GCHV M6 is 2039 nucleotides in size with a single long open reading frame (ORF) starting with the first initiation codon at bases 31 to 33 and ending with a termination codon at bases 1977 to 1979. The ORF encodes 648 amino acids with a predicted molecular mass of 68.7 kDa. The distributions of the four bases of M6 were found relatively in favour of cytidine (21.28% A, 21.63% G, 23.74% T, 32.82% C). The translation from the ORF corresponds to translation of 95.5% of the M6 RNA. The first AUG appears to be not in a very favorable context for initiation of translation (ACAAUGU) according to the consensus sequence (A/G)NNAUGG (N=any nucleoside) established by Kozak [8]. The terminal ends of the GCHV M6 segment display the nucleotide sequences 5′ GUUAUU and CAUC 3′, which are conserved in genome segments of GCHV (unpublished data). In addition, segment-specific inverted repeats were identified adjacent to the terminal sequences. The 5′ end sequence, CGACACTTC at 8 to 16, is complementary to its 3′ end inverted repeat, GAAGTGTCG at 2021 to 2029. Terminal conserved sequence appears to be a basic property of each genome segment of most reoviruses while a domain of inverted repeat adjacent to it always plays an important role in distinguishing this genome segment from other segments in sorting functions [2].

A search made against the protein databases with BLAST [1] has revealed that the deduced protein of GCHV M6 share some homology with the major outer capsid protein μl of mammalian reovirus (MRV), a member of another genus *Orthoreovirusin* the family *Reoviridae*. The predicted amino acid sequence of GCHV M6 showed 24% identity in entire

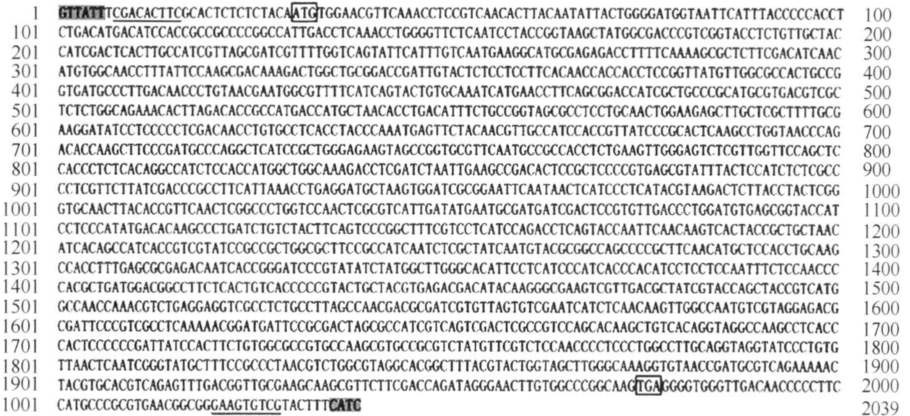

Fig. 1 Complete nucleotide sequence (presented in the cDNA form) of M6 RNA segment of GCHV. The conserved 5′- and 3′-terminal nucleotide sequences are indicated with a grey background and the inverted repeats are underlined. The initiation and termination sites are boxed

length overlap with MRV μl as shown in Fig. 2. As a reference, we have found other genome segments of GCHV excluding S11 and S7 also share homology with *orthoreovirus* (unpublished data). Previous report [14] indicated that it is possible that sequences sharing as little as 15% identity over their entire length are homologues. Taking this conclusion into consideration, GCHV M6 may correspond to MRV M2 gene, which encodes μl protein [12]. However, there was no long stretch of identical amino acid sequence found between the protein encoded by M6 and μl. The longest one is a 5 residue stretch at amino acids 14 to 18. The most comparable region was at the aminoterminal sequence of GCHV M6 (amino acids 1 to 61) and MRV μl (amino acids 1 to 63) with 46% identity and 67% similarity. In this region of MRV μl, a highly sensitive cleavage site has been demonstrated between asparagine (residue 42) and proline (residue 43) [19], at which μl can be degraded into two proteins, μlN and μlC, by proteolytic cleavage in infected cell. It is noteworthy that this site and several flanking residues are retained in the predicted amino acid sequence of M6 (Fig. 2), suggesting that protein encoded by M6 may be easy cleaved by proteinase.

In addition, the basic and acidic stretches in the protein encoded by M6 alternate along the sequence as do those in the MRV μl. There is a long acidic region (fragment from amino acids 1 to 533, charge −12.3 at pH 7.0) at the aminoterminal portion and a short basic region (fragment from amino acids 534 to 648, charge +9.07 at pH 7.0) at the carboxy-terminal portion of the protein. Similar stretches were also found in the μl. Thus, it can be predicted that the protein encoded by M6 and μl may be structurally and functionally related. Interestingly, at the junction of the basic and acidic stretches, μl can undergo another proteolytic cleavage *in vitro* [13], which is thought to be important for the virus to penetrate cell membranes. Whereas, no amino acid sequence similarity could be found in this junction between GCHV M6 and MRV μl.

```
                                               ▼
A:  1   MWNVQTSVNTYNITGDGNSFTPTSDMTSTAAPAIDLKPGVLNPTGKLWRPVG--TSVATIDSL------------------------AIVSDRFGQYSFVNEGMRETFSKALFDINM   91
        M N  + V T N+TGDGN F P+++ +STA P++ L PG+LNP G  W  +G  TSV +  +L                         A+V          V E    F+KA   +
B:  1   MGNASSIVQTINVTGDGNVFKPSAETSSTAVPSLSLSPGMLNPGGVPWIAIGDETSVTSPGALRRMTSKDIPETAIINTDNSSGAVPSESALVPYNDEPLVVVTEHAIANFTKAEMALEF  120

A:  92  WQPLFQATKTGCGPIVLSSFTTTTSGYVGATAGDALDNPVTNGVFISTVQIMNLQRTIAARMRDVALWQKHLDTAMTMLTPDISAGSASCNWKSLLAFAKDILPLDNLCLTYPNEFYNVA  211
        +       +         S   T    YVG +A  AL+N      I+  +       +I A ++ +  W+  L  A T+L  ++  G  SC  +S++     D LP D+L     YP E
B:  121 NREFLDKLRVLSVSPKYSDLLTYVDCYVGVSARQALNNFQKQVPVITPTRQTMYVDSIQAALKALEKWEIDLRVAQTLLPTNVPIGEVSCPMQSVVKLLDDQLPDDSLIRRYPKEAAVAL  240

A:  212 IHRYPALKPGNPDTKLPDAQAHPLGEVAGAFNAATSEVGSLVGSSSTLSQAISTMAGKDLDLIEADTPLPVSVFTPSLAPRSYRPAFIKPEDAKWIAEFNNSSLIRKTLTYSGATYTVQL  331
         R  ++   D           + VA +  A ++    L  S    QA+  +   + ++I +  P+P  VF      P  Y    +K ++A W+            +I KT+    G  + +Q+
B:  241 AKRNGGIQW--MDVSEGTVMNEAVNAVAASALALSASAPPLEEKSRLTEQAMDLVTAAEPEIIASLVPVPAPVFAIPPKPADYNVRTLKIDEATWLR------MIPKTM---GTPFQIQV  349

A:  332 GP-----------GPTRVIDMNAMIDSVLTLDVSGTILPYDTSPDLSTSVPAFVLIQTSVPIQQVTTAANITAITVVSAAGASAINLAINVRGQPRFNMLHLQATFERETITGIP---YI  437
                     G TRV++++ +        LD+ G      +TS D +      F++ Q+ +P +   T A+ I   TVV+       A + +          Q           L   +E E +
B:  350 TDNTGTNWHLNLRGGTRVVNLDQIAPMRFVLDLGGKSYK-ETSWDPNGKKVGFIVFQSKIPFELWTAASQIGQATVVNYVQLYAEDSSFTA--QSIIATTSLAYNYEPEQLNKTDPEMNY  466

A:  438 YGLGTFLIPSPTSSSNFSNPTLMDGLLTVTPVLLRETTYKGEVVDAIVPATVMANQTSEEVASALANDAIVLVSNHLNKLANVVGDAIPVASKTDDSA--------TSAIVSRLAVQHKL  549
        Y L TF+   +  + +N + P + D LLT++P+    E T KG VV  +VPA ++ + T E + ++L NDA     + + +K+A  +             K DD A          +  I  +LA+
B:  467 YLLATFIDSAAITPTNMTQPDVWDALLTMSPLSAGEVTVKGAVVSEVVPAELIGSYTPESLNASLPNDAARCMIDRASKIAEAI--------KIDDDAGPDEYSPNSVPIQGQLAISQLE  578

A:  550 SQVGQASPTPPDYPLLWRRAKRAASMFVSNPSLALQVGIPVLTQSGMLSALTSGVGTALRTGSLGKGVTDASEKLRARQSLTVAKQAFFDQIGNLWP                       646
        + G      P    +L + A RA   F+ +PS  +     PVL+        AL  GV T+LRT SL  GV   A  KL + +S+      Q F D++    +P
B:  579 TGYGVRIFNPKG--ILSKIASRAMQAFIGDPSTIITQAAPVLSDKNNWIALAQGVKTSLRTKSLSAGVKTAVSKLSSSESIQNWTQGFLDKVSTHFP                         673
```

Fig. 2 Comparison of the amino acid sequence of the deduced protein encoded by GCHV M6 (A) and MRV μl (reovirus serotype 1 [21]) (B). A dash indicates a gas inserted to obtain maximum alignment. The position of putative Asn-42-Pro-43 cleavage site is indicated with "▼"

The deduced protein encoded by GCHV M6 has a calculated molecular mass of 68.7 kDa. It is very similar to the size of the protein obtained from *in vitro* translation of M6 dsRNA estimated by SDS-PAGE [20], which is approximately 68.5 kDa (calculated according to the figure, no data detailed in the report). This 68.5 kDa protein was thought to be the protein VP67 [20], a major outer capsid protein of the virus described by Ke [7]. Thus it is supported that the protein encoded by M6, like μl, is a major outer capsid protein, consistent with the close relationship between the protein encoded by M6 and μl revealed above. In order to clarify the identity of protein encoded by M6, a 2011 bp fragment (nucleotides 29-2039, containing the fullength ORF) was amplified with two primers (5′ CTGCTAGCTCAATGTGGAACGTT 3′ and 5′ AAGAGCTCGGCGGTAAATGATGA 3′) from the pGEM-T vector and inserted into the NheI and SacI sites of pET-28a vector and expressed in *Escherichia coli*. Protein expression was performed according to the pET system manual provided by Novagen Corp. The extracted *E. coli* proteins were resolved by SDS-PAGE (Fig. 3A), transferred to NC membrane (Gleman), and probed with anti-GCHV immunoglobulin (IgY). that was isolated from the yolks of eggs of GCHV-immunized hen according to the method of Horikoshi[5]. Peroxidase-coupled Rabbit anti-chicken IgY (Promage) was used for final detection. The results were shown in Fig. 1, in which a 72 kDa band was identified. It was very similar to the predicted molecular mass of the pET-M6 fusion protein, of which the size is thought to be the sum of the M6 product and the foreign sequence in pET, which are 68.7 kDa and 2.7 kDa, respectively. The 72 kDa band was also detected by Western blotting (Fig. 3B), identifying the anti-GCHV IgY reacted specifically with the protein expressed by M6. Additional bands including a 64 kDa band and several smaller bands were also detected in the preparation. Excluding the smallest bands, most additional bands

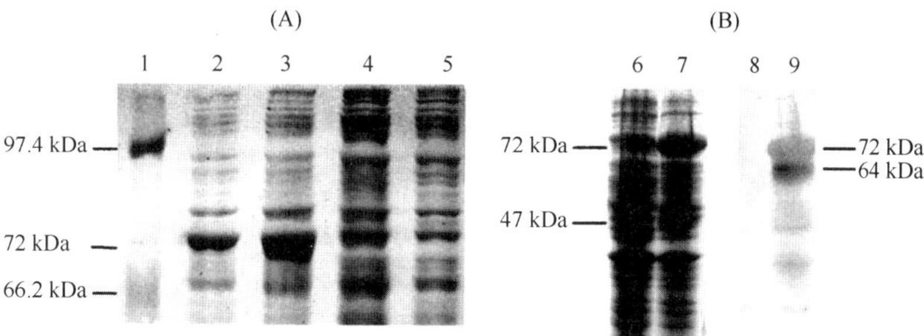

Fig. 3 Expression (7% SDS-PAGE (A)) and Western blotting analysis (10% SDS-PAGE(B)) of pET-M6 fusion protein produced by *E. coli*. 1 Protein marker; 2 induced for 1 h with IPTG; 3 induced for 3 h with IPTG; 4 incubated for 3 h without IPTG; 5 incubated 1 h without IPTG; 6 insert-minus plasmid control; 7 expression of pET-M6 fusion protein; 8 Western blotting of insert-minus plasmid control; 9 Western blotting of pET-M6 fusion protein

are probably degradation products from the 72 kDa protein and not due to non-specific trapping as they migrate differently to proteins of *E. coli*. Significantly, the 64 kDa band may be a degraded fragment of 72 kDa pET-M6 fusion protein that was cleaved at the putative Asn-42-Pro-43 cleavage site because calculated molecular mass of carboxy-terminal fragment is 64.3 kDa (amino acids 43 to 648). This is reasonable as the deduced protein of M6, like μl, possesses the Asn-42 and Pro-43 residues, suggesting that the cleavage site is shared by two proteins. Whereas, no extensive evidence could be presented to this point. Our study only serves as a preliminary for future investigations. To characterize the protein encoded by GCHV M6, further work including biochemical and cytological analyses would be required.

Acknowledgements This work was supported by Grant No. ZYJ01-04 from China-Israel fund for scientific and strategic research and development, and by Grant No. 990205-FB05 from FEBL, the state key laboratory of Freshwater Ecology and Biotechnology.

References

1. Altschul SF, Madden TL, Schiffer AA, Zhang J, Zhang Z, Miller W, Lipman DJ (1997) Gapped BLAST and PSI-BLAST: a new generation of protein database search programs. Nucleic Acids Res 25: 3389-340
2. Anzola JV, Xu Z, Asamizu T, Nuss DL (1987) Segment specific inverted repeats found adjacent to conserved terminal sequences in wound tumor virus genome and defective interfering RNAs. Proc Natl Acad Sci USA 84: 8301-8305
3. Chen Y, Jiang Y (1983) Studies on the morphologycal and physico-chemical characterization of the hemorrhagic virus of grass carp. KeXue TongBao (Foreign Lang Ed) 28: 1138-1140
4. Fang Q, Attoui H, Biagini JF, Zhu Z, de Micco P, de Lamballerie X (2000) Sequence of genome segments 1, 2, and 3 of the grass carp reovirus (Genus *aquareovirus*, family *reoviridae*). Biochem Biophys Res Commun 274: 762-766
5. Horikoshi T (1993) J Food Sci 58: 739-742, 779

6. Ke L, Fang Q, Cai Y (1990) Characteristics of a new isolation of hemorrhagic virus of grass carp. Acta Hydrobiol Sin 14: 153-159 (in Chinese with English abstract)
7. Ke L, Wang W, Fang Q, Cai Y (1992) Studies on the *in vitro* translation of grass carp hemorrhage virus and its proteins. Chin J Virol 8: 169-173 (in Chinese with English abstract)
8. Kozak M (1986) Point mutations define a sequence flanking the AUG initiator codon that modulates translation by eukaryotic ribosomes. Cell 44: 283-292
9. Lambden PR, Cooke SJ, Caul EO, Clarke IN (1992) Cloning of noncultivatable human rotavirus by single primer amplification. J Virol 66: 1817-1822
10. Lupiani B, Hetrick FM, Samal SK (1993) Genetic analysis of aquareovirus using RNA-RNA blot hybridization. Virology 197: 475-479
11. Lupiani B, Hetrick FM, Samal SK (1994) Identification of the angelfish Pomacanthus Semicirulatus aquareovirus as a member of aquareovirus genogroup. A using reciprocal RNA-RNA blot hybridization. J Fish Dis 17: 667-672
12. McCrea MA, Joklik WK (1978) The nature of the polypeptide encode by each of the ten double-stranded RNA segments of reovirus type 3. Virology 89: 578-593
13. Nibert ML, Fields BN (1992) A carboxy-terminal fragment of protein μl/μlC is present in infectious subvirion particles of mammalian reoviruses and is proposed to have a role in penetration. J Virol 66: 6408-6418
14. Pearson WR (1990) Rapid and sensitive sequence comparison with FASTP and FASTA. Methods Enzymol 183: 63-98
15. Qiu T, Lu RH, Zhang J, Zhu ZY (2001) Complete nucleotide sequence of the S10 genome segment of grass carp reovirus (GCRV). Dis Aquat Org 44: 69-74
16. Rangel AA, Rockemann DD, Hetrick FM, Samal SK (1999) Identification of grass carp haemorrhage virus as a new genogroup of aquareovirus. J Gen Virol 80: 2399-2402
17. Subramanian K, McPhillips TH, SAMAL SK (1994) Characterization of the polypeptides and determination of genome coding assignments of an aquareovirus. Virology 205: 75-81
18. Subramanian K, Hetrick FM, Samal SK (1997) Identification of a new genogroup of aquareovirus by RNA-RNA hybridization. J Gen Virol 78: 1385-1388
19. Tillotson L, Shatkin AJ (1992) Reovirus polypeptide σ3 and N-terminal myristoylation of polypeptide μl are required for site-specific cleavage to μlC in transfected cells. J Virol 66: 2180-2186
20. Wang W, Cai Y, Fang Q, Liu Y, Ke L (1994) Gene-protein-coding assignments of grass carp hemorrhage virus genome RNA species. Virologica Sinica 9: 356-361 (in Chinese with English abstract)
21. Wiener JR, Joklik WK (1988) Evolution of reovirus genes: A comparison of serotype 1, 2, and 3 M2 genome segments, which encode the major structural capsid protein mu-lC. Virology 163: 603-613
22. Zuo W, Qian H, Xu Y, Du S, Yang X (1986) A cell line derived from the kidney of grass carp. Fish China 10: 11-17 (in Chinese with English abstract)

草鱼出血病病毒 M6基因的分子特征和表达

邱 涛 陆仁厚 Jing Zhang 朱作言

中国科学院水生生物研究所，淡水生态与生物技术国家重点实验室，武汉

摘 要 本研究克隆了草鱼出血病病毒(GCHV)M6基因的全长核苷酸序列。该基因长2309 bp，开放阅读框编码648个氨基酸，分子量68.7 kD。氨基酸序列比对发现 GCHV M6蛋白与哺乳动物呼肠孤病毒的 μl 蛋白亲缘关系较近。M6为主要的外衣壳蛋白，本研究通过 pET 融合蛋白载体在大肠杆菌中进行了表达，并用鸡抗 GCHV 免疫球蛋白(IgY)进行了 Western 检测。推测 M6蛋白具有与 μl 蛋白相同的 Asn-42-Pro-43蛋白酶裂解位点。

Complete Nucleotide Sequence of the S10 Genome Segment of Grass Carp Reovirus (GCRV)

Tao Qiu　Ren-Hou Lu　Jing Zhang　Zuo-Yan Zhu

State Key Laboratory of Freshwater Ecology and Biotechnology, Institute of Hydrobiology, Chinese Academy of Sciences, Wuhan 430072

Abstract Hemorrhagic disease, caused by the grass carp reovirus (GCRV), is one of the major diseases of grass carp in China. Little is known about the structure and function of the gene segments of this reovirus. The S10 genome segment of GCRV was cloned and the complete nucleotide sequence is reported here. The S10 is 909 nucleotides long and contains a large open reading frame (ORF) encoding a protein of 276 amino acids with a deduced molecular weight of approximately 29.7 kDa. Comparisons of the deduced amino acid sequence of GCRV S10 with those of other reoviruses revealed no significant homologies. However, GCRV S10 shared a putative zinc-finger sequence and a similar distribution of hydrophilic motifs with the outer capsid proteins encoded by Coho salmon aquareovirus (SCSV) S10, striped bass reovirus (SBRV) S10, and mammalian reovirus (MRV) S4. It was predicted that this segment gene encodes an outer capsid protein.

Keywords grass carp reovirus (GCRV); S10 genome segment; nucleotide sequence

Grass carp reovirus (GCRV), an important fish pathogen involved in hemorrhagic disease (Chen &Jiang 1983), not only infects grass carp *Ctenopharyngodon idellus*, but also was found capable of infecting black carp *Mylopharyngodon piceus*, topmouth gudeon *Pseudorasbora parva* (Ding *et al.* 1991) and rare minnow *Gobiocypris rarus* (Wang *et al.* 1994), and historically has resulted in large losses in freshwater fish culture in China. The virions are resistant to chloroform and ether, non-sensitive to acid (pH 3) and alkaline (pH 10) treatment, and stable within a certain range of temperature. The virus belongs to the genus *Aquareovirus*, as a tentative member, and shares the physicochemical properties and morphological characteristics of the family Reoviridae. About 10 isolates reported to date (Li *et al.* 1999) had different electrophoretic patterns but the same number of genomic dsRNA. The GCRV possesses a double-stranded dsRNA genome consisting of 11 segments packaged into a nonenveloped icosahedral double capsid approximately 55 to 80 nm in diameter (Ke *et al.* 1990). The genome segments are approximately 25 000 nucleotide pairs in total size deduced by the dsRNA molecular weight and separated into 3 size classes: large (L1, L2, and L3); medium (M4, M5, and M6); and small (S7, S8, S9, S10, and S11) in order of mobility in polyacrylamide gels from the slowest to fastest, respectively.

Many reoviruses capable of infecting aquatic organisms have been identified (Plumb *et al.*

文章发表于 *Diseases of Aquatic Organisms*, 2001, 44(1): 69-74

1979, Hedrick *et al.* 1984, Ahne & Kölbl 1987, Winton *et al.* 1987, Varner & Lewis 1991, Neukirch *et al.* 1999), and using RNA-RNA blot hybridization, 7 different genogroups have been established among at least 45 aquareovirus isolates (Rangel *et al.* 1999). However, unlike those of mammalian and plant reoviruses in the family Reoviridae, which have been extensively studied, little information is available on the molecular characteristics of the genomes of members of the genus *Aquareovirus*. More recently, the sequences of the 3 largest segments of GCRV were determined (Fang *et al.* 2000); they are putative guanylyltransferase, viral polymerase and helicase genes. In addition, 2 sequences of other aquareoviruses have been reported. They are genome segment 10 of the Coho salmon aquareovirus (SCSV) and that of the striped bass reovirus (SBRV), which encode major outer capsid proteins (Lupiani *et al.* 1997a,b). Although these sequences provided some knowledge helpful for determining the diseases derived from aquatic viruses, they are still limited. In order to obtain further information on the GCRV genome and use available data on *Aquareovirus*, the full-length cDNA of GCHV genome segment S10 was cloned and the characteristics of this sequence are documented here.

Materials and Methods

Virus strain, purification and RNA extraction

The CIK cell strain (Zuo *et al.* 1986) was used to propagate GCRV-873 (Ke *et al.* 1990). Virus was purified from a continuous sucrose gradient (30 to 60% sucrose in NTE [NaCl-Tris-EDTA] buffer) and centrifuged at 100 000×*g* for 2 h. Genomic dsRNA was extracted from the purified virus particles by 1% SDS and 10 μg ml^{-1} Proteinase K at 37℃ for 3 h, as well as by phenol-chloroform extraction. The purified dsRNA was treated with RQ1 RNase-free DNase (Promage) at 37℃ for 1 h to remove the contamination of cell DNA, and displayed on 12.5% polyacrylamide gel (Fig. 1A).

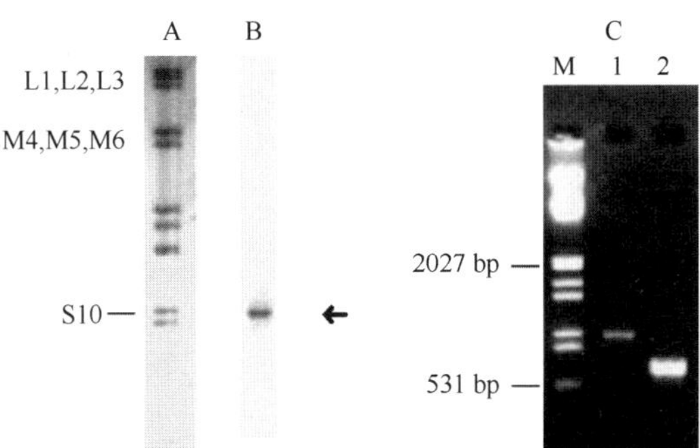

Fig. 1 Analysis of cloned S10 segment of GCRV. (A) PAGE profile of dsRNA genome of GCRV. (B) Result of Northern blotting with S10 cDNA as probe. The position of the S10 segment is indicated with←. (C) Results of RT-PCR amplification of dsRNA of GCRV. M: Lambda DNA/EcoR I+*Hind* III marker; Lane 1: PCR product amplified with Primer 2; Lane 2: RT-PCR product amplified with Primer 3 + Primer 4

Cloning of the S10 gene

The method described by Lambden *et al.* (1992) for amplification of the rotavirus dsRNA genome segments was used, with modification, to clone the GCRV S10 gene. The oligodeoxyribonucleotide primer 1 (5′-PO_4-ATTTACCGCCGAGCCTGACTT-NH_2-3′), which was ligated to both ends of the dsRNA genome segment, was chemically synthesized and modified (Sango) to prevent self-ligation and subsequent concatenation during the RNA ligase (Life Technologies) reaction, as described by Lambden *et al.* (1992). The primer 1-tailed dsRNA was denatured by heating to 94℃ for 5 min in 62% DMSO in the presence of primer 2 (5′-AAGTCAGGCTCGGCGGTAAAT-3′ complementary to primer 1) (Sango) and cooled rapidly on ice. The synthesis of cDNA was carried out in reverse transcriptase reaction mix (200 ng dsRNA, 50 mM Tris-HCL pH 8.3, 3 mM $MgCl_2$, 75 mM KCL, 1 mM DTT, 0.5 mM each deoxynucleoside of dNTP [Promega], 40 U RNAsin [SABC], and 200 U reverse transcriptase [Gibco BRL]). The mixture was incubated at 37℃ for 1 h and the reaction was stopped by addition of EDTA to finial 20 mM. The RNA was removed by NaOH and the cDNA was allowed to anneal. After extraction by phenol-chloroform and precipitation by ethanol, the annealed partial duplexes were filled in using DNA Polymerase I Large Fragment (Promage) and purified using a Glassmilk DNA purifying kit (BioStar). The amplification of cDNA was accomplished by PCR with primer 2 only using a PE9600, consisting of a denaturation step at 94℃ for 3 min followed by 30 cycles of 30 s at 94℃, 35 s at 60℃, and 2 min at 72℃. Amplified DNA products were separated on agarose gel and the 900 bp fragment, which corresponded in size to that calculated from the GCRV dsRNA molecular weight for S10, was excised and purified using the Glassmilk DNA purifying kit. The purified S10 was ligated directly into pGEM-T vector and transformed into DH5α strain of *Escherichia coli* (Gibco BRL).

S10 gene sequence determination and analysis

The recombinant plasmids containing the full-length cDNA of S10 were identified according to the size of inserted segment by PCR using 2 M13 primers on pGEM-T and were purified using a plasmid DNA purification miniprep kit (Viogene). Two clones that had the full-length cDNA of S10 were used for sequencing. The nucleotide sequence of S10 was determined by Taq dye primer cycle sequencing on an ABI 310 Genetic Analyzer (Perkin Elmer). The sequence of S10 was analysed using Lasergene sequence analysis package (DNAStar).

Northern blotting and RT-PCR

Assignment of the cDNA inserts to S10 was confirmed by alkaline blotting analyses as described by Li *et al.* (1987). In brief, the genome dsRNA was separated in 1.0% agarose gel in

Tris-acetate-EDTA buffer and transferred to a nylon memberane (Hybond) in 0.2 N NaOH for 1 h at room temperature. Transferred RNA was hybridized with a cDNA probe labeled with digoxigenin-dUTP, followed by an immunodetection using Dig High Primer Labeling and Detection Starter Kit (Boehringer Mannheim). Two primers for RT-PCR, Primer 3 (5′-CCCCGA-TCATCACCACGAT-3′) (from nucleotide 14 to 32) and Primer 4 (5′-CGCGTTCGCTGATG-TAAGG-3′) (from nucleotide 693 to 711), were synthesized according to the cDNA sequences of the cloned S10. The reverse transcription was carried out following the procedure described by Li *et al.* (1997).

Results and discussion

Comparing with the GCRV genome dsRNA in Fig. 1A, the clone containing the cDNA of S10 was confirmed by Northern blotting as shown in Fig. 1B, in which the cloned cDNA hybridized to genome segment S10 rather than the other genome segments. Fig. 1C indicates that an expected, approximately 700 bp DNA band was amplified in the result of RT-PCR using 2 primers within the nucleotide sequence of S10. No product was found in negative control of RT-PCR with genomic DNA of CIK cell as template (data not shown).

The complete nucleotide sequence and deduced amino acid sequence of the S10 genome segment of GCRV are shown in Fig. 2. The nucleotide sequence was obtained from 2 different recombinant plasmids and constitutes a consensus sequence. The S10 gene of GCRV is 909 nucleotides long and contains a large open reading frame (ORF) proceeded by a 30 bp untranslated region and followed by a 48 bp untranslated downstream sequence. The ORF starts with an ATG codon at nucleotides 31 to 33 and ends with a TGA at nucleotide 859 to 861. No additional ORF of significant length was detected in either the plus or minus strand RNA. Three AUG triplets, the first (ACG<u>ATG</u>C) located at 31, the second (CAC<u>ATG</u>A) at 43, and the third (GCT<u>ATG</u>G) at 67, were recognized in the same reading frame and near the 5′ end of the mRNA strand. Since in most cases the AUG nearest the 5′-terminal cap is exclusively used for initiation for protein synthesis in a ‘scanning model’ (Kozak 1980), the putative methionine start codon should be the first triplet rather than the others. Interestingly, the third triplet also appeared to be potentially suitable as a functional initiator because it was consistent with the strong initiation sequence of RNNATGG (R = purine nucleoside, N = any nucleoside) (Kozak 1981).

The GCRV S10 genome segment displays the terminal sequences 5′ GUUAUU and CAUC 3′, which were recognized to be conservative in the GCRV RNA segments (authors’ unpubl. data). Moreover, the 3′-terminal sequence CAUC 3′ was found to be the same as that of segment S10 of SBRV (Lupiani *et al.* 1997b), another member of *Aquareovirus*. In addition, a putative inverted repeat sequence was identified adjacent to the terminal sequence. The 5′ end sequence, GAGCCCCC at 10 to 17, and its 3′ end inverted

```
1      GTTATTTCTGAGCCCCCGATCATCACCACGATG CCA CTT CAC ATG ATT CCG CAA GTC GCC    60
1                                        M   P   L   H   M   I   P   Q   V   A    10

61     CAC GCT ATG GTG CGT GCA GCC GCT GCA GGA CGC CTT ACC TTA TAC ACA AGA ACT AGA ACT   120
11      H   A   M   V   R   A   A   A   A   G   R   L   T   L   Y   T   R   T   R   T    30

121    GAG ACC ACC AAC TTT GAT CAC GCT GAG TAC GTC ACC TGC GGG CGG TAC ACC ATC TGC GCC   180
31      E   T   T   N   F   D   H   A   E   Y   V   T   C   G   R   Y   T   I   C   A    50

181    TTC TGC CTT ACG ACT CTG GCT CCC CAC GCC AAC GTC AAG ACC ATT CAA GAC TCC CAC GCT   240
51      F   C   L   T   T   L   A   P   H   A   N   V   K   T   I   Q   D   S   H   A    70

241    TGT TCA CGT CAA CCA AAT GAA GCC ATT CGC TCA TTA GTC GAA GTG AGT GAC AAA GCG CAG   300
71      C   S   R   Q   P   N   E   A   I   R   S   L   V   E   V   S   D   K   A   Q    90

301    ACC GCC CTC GTC GGT AGC CGT ACT GTA GAC TAT CAC GAA TTG GAT GTG AAA GCT GGG TTC   360
91      T   A   L   V   G   S   R   T   V   D   Y   H   E   L   D   V   K   A   G   F   110

361    GTC GCC CCA ACT GCC GAT GAA ACA ATA GCC CCC TCT AAG GAT ATC GTC GAA CTT CCG TTT   420
111     V   A   P   T   A   D   E   T   I   A   P   S   K   D   I   V   E   L   P   F   130

421    CGC ACC TGT GAC TTG GAC GAT TCC TCT GCT ACC GCT TGC GTC CGA AAT CAC TGC CAG GCC   480
131     R   T   C   D   L   D   D   S   S   A   T   A   C   V   R   N   H   C   Q   A   150

481    GGT CAC GAC GGC GTT ATC CAC CTC CCG ATC CTT TCT GGA GAT TTC AAA TTG CCT AAC GAG   540
151     G   H   D   G   V   I   H   L   P   I   L   S   G   D   F   K   L   P   N   E   170

541    CAT CCC ACC AAA CCG TTG GAC GAT ACG CAT CCC CAC GAC AAG GTG CTG ACT CGC TGC CCC   600
171     H   P   T   K   P   L   D   D   T   H   P   H   D   K   V   L   T   R   C   P   190
601    AAG ACT GGT CTC CTC CTC GTC CAT GAC ACT CAC GCA CAC GCC ACC GCC GTA GTT GCC ACC   660
191     K   T   G   L   L   L   V   H   D   T   H   A   H   A   T   A   V   V   A   T   210

661    GCT GCT ACG AGA GCT ATC CTC ATG CAC GAC CTC CTT ACA TCA GCG AAC GCG GAT GAC GGC   720
211     A   A   T   R   A   I   L   M   H   D   L   L   T   S   A   N   A   D   D   G   230

721    CAT CAA GCA CGT TCC GCT TGC TGC GGT CCA GCG TTT AAC AAC CTG ACC TTC GCT TGC CAC   780
231     H   Q   A   R   S   A   C   C   G   P   A   F   N   N   L   T   F   A   C   H   250
                                                        *   *   *   *
781    TCC ACC TGT GCT TCA GAT ATG GCT CAC TTC GAC TGC GGC CAG ATC GTT GGA CTC GAC TTG   840
251     S   T   C   A   S   D   M   A   H   F   D   C   G   Q   I   V   G   L   D   L   270
                                                    #   #   #   #   #   #
841    CAT GTG GAG CCA TCC GAT TAAGCACCGGCCTACCCACCCGGTGCGGGTGTAGGGGTCTCTCGTTTCATC    909
271     H   V   E   P   S   D   .                                                     276
```

Fig. 2 Complete nucleotide sequence (presented in the cDNA form) and deduced amino acid sequence of S10 RNA segment of GCRV. The 5′- and 3′-terminal nucleotide sequences are **bold** and indicated with a grey background and the inverted repeats are **bold**. Three ATG triplets near the 5′ end of nucleotide sequence are underlined. In the amino acid sequence, the putative zincfinger motif (amino acids 49-71) is also indicated with a grey background. Possible N-glycoslation site (*) and N-myristoylation site (#) are indicated

repeat sequence, GGGGTCTC at 892 to 899, are almost completely complementary.

The conserved terminal sequences are broadly reported in members of the family Reoviridae. They may be important in sorting and packing functions of the virus (Anzola *et al.* 1987). Analysis of the terminal nucleotide sequences is of interest in relation to understanding mechanisms of transcription and replication. The same 3′ terminal sequence between GCRV S10 and SBRV S10 implied relatively closer relations between them. In addition, like other reoviruses, a domain of inverted repeat adjacent to the 5′ and 3′ terminus always plays an important role in distinguishing this genome segment from other segments (Anzola *et al.* 1987).

The ORF is considered to encode a protein of 276 amino acids with a deduced molecular weight of approximately 29.7 kDa. The possible modification sites, an N-glycosylation site present at 244 to 247 and an N-myristoylation site at 263 to 268, were indicated and near the

C terminus of the polypeptide (Fig. 2). Using the BLAST programs (Altschul *et al.* 1997), we were unable to find substantial similarity with either the nucleotide or the amino acid sequence of S10 and sequences of characterized animal reovirus and plantinfecting reovirus. However, some similarities in GCRV S10, SCSV S10 and SBRV S10 were observed. They not only have similar segment lengths, but the molecular weights of their deduced peptides are also analogous. Moreover, they are all segment 10 of the viruses in the same genus, i.e. *Aquareovirus*. We compared the deduced amino acid sequence of GCRV S10 with those of SCSV S10 and SBRV S10 using the Lasergene program (DNAStar). The predicted protein encoded by GCRV S10 showed a similarity index of 17.1 (Lipman & Pearson 1985) with SCSV S10 in a 173 amino acid overlap, and of 30 with SBRV S10 in only a 20 amino acid overlap.

Although it can be proposed that GCRV S10 may correspond to SCSV S10 to a low extent and may not be related to SBRV S10 with respect to amino acid sequences, the hydrophilic plots (Fig. 3) produced according to the method of Kyte & Doolittle (1982) significantly showed that the predicted proteins encoded by GCRV S10, SCSV S10 and SBRV S10 had very similar profiles, particularly in the regions of the N-terminal half. Thus, it can be predicted that their amino acid sequences may have an analogous molecular conformation. Furthermore, it is very likely that this analogous conformation leads to functional similarities among the deduced proteins of the S10 segments of these aquareoviruses.

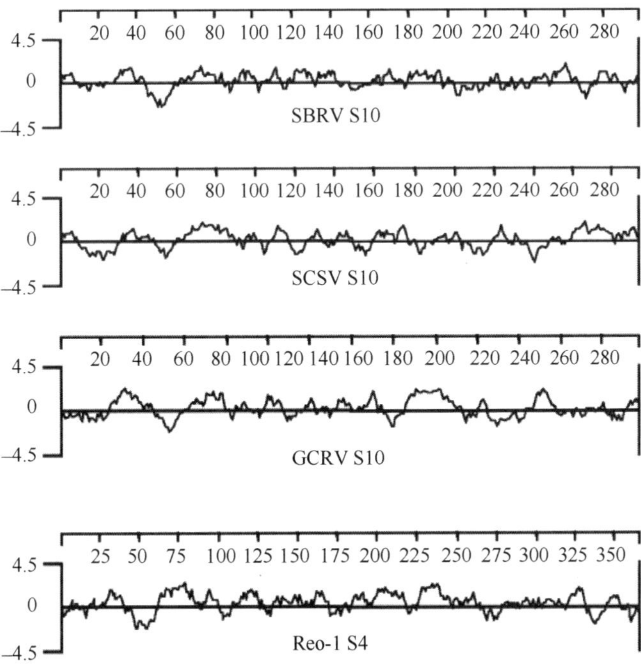

Fig. 3 Hydrophilic plots of deduced proteins encoded by SBRV S10, SCSV S10, GCRV S10 and Reo-1 S4

```
            45        55        65        75
SBRV S10   GAYVICACCFKVLLNWPGPPIHITHECHDSHGVCR
                *  *               * *

            45        55        65        75
SCSV S10   GQYQLCALCLKQVCSYHVTEPCYYPHECHHGQATR
                *  *                * *

            45        55        65        75
GCRV S10   RYTICAFCLTTLAPHANVKTIQDSHACSRQPNEAI
               *  *                * *

            45        55        65        75
Reo-1 S4   CGGAVVCMHCLGVVGSLQRKLKHLPHHRCNQQIRH
                 *  *               ** *
```

Fig. 4 Sequences of the putative zinc-binding sites in deduced amino acid sequences of GCRV S10, SBRV S10, SCSV S10, and Reo-1 S4. *Residues of the putative zinc fingers. Numbers refer to the amino acid positions

The deduced amino acid sequence of GCRV S10 possesses a $CX_2CX_{16}HX_1C$ sequence within the N terminus, from residues 49 to 71, which is analogous to the zinc-finger domain (CCHC) identified within the mammalian reovirus (MRV) σ3 protein encoded by genome S4 gene (Mabrouk & Lemay 1994). Comparable in position, the zinc-finger sequence $CX_2CX_{15}HHX_1C$ is from amino acids 51 to 73 within the MRV σ3 protein. Notably, we also recognized the zinc-finger domains within deduced amino acid sequences of SCSV S10 ($CX_2CX_{16}HX_1C$) and SBRV S10 ($CX_2CX_{15}HX_1C$) (Fig. 4), although they had not been indicated in the original reports (Lupiani *et al.* 1997a,b). Moreover, they are located at the same or a position very similar to that of GCRV S10, namely, at residues from 50 to 71 and from 49 to 71, respectively. Thus, the $CX_2CX_{15\text{-}16}HX_1C$ was found to be the special zinc- finger sequence shared by the deduced peptide of the S10 segments of these aquareoviruses and MRV σ3. Since the most probable functional motif was identified as being a form $CX_2CX_{15}HX_2C$ (Mabrouk & Lemay 1994), the function of the zinc-finger motifs indicated here requires further study. In previous reports, the zinc-finger domains may be required by some proteins to maintain proper conformation (McIntyre *et al.* 1993) and may also be involved in protein-protein interactions (Cunningham *et al.* 1991).

Since the comparable zincfinger domain exists in the MRV σ3 protein, the hydrophilic plot of the reovirus serotype-1 (Reo-1) S4 (Seliger *et al.* 1992), as an example of MRV, was given together with those of aquareoviruses. Surprisingly, many profiles were also similar to those of aquareoviruses, although MRV belongs to another genus, *Orthoreovirus*. In particular, the N-terminal 160 amino acids of Reo-1 S4 are very like those of GCRV S10, and the C-terminal is analogous to those of GCRV S10 and SBRV S10, suggesting that the deduced amino acid sequences of the aquareovirus S10 may share some functional similarity with MRV σ3 protein. In addition, it was notable that the terminal sequences of MRV S4, 5′ GCUAUU and CAUC 3′ are very similar to that of GCRV S10. From an evolutionary standpoint, it is likely that the GCRV S10 and the MRV S4 have evolved from a common ancestral precursor.

Since SCSV S10, SBRV S10, and Reo-1 S4 were previously revealed to encode major outer capsid proteins (Atwater *et al.* 1986, Lupiani *et al.* 1997a,b), and since similarities have been shown between the zincfinger sequences and hydrophilic plots of GCRV S10 and the above viral segments, it can be predicted that the deduced peptide of GCRV S10 is an outer capsid protein. In addition, the smallest segment of the GCRV genome, S11, has already been assigned to encode nonstructural proteins (authors' unpubl. data), suggesting that the deduced peptide encoded by GCRV S10 may be the smallest outer capsid protein.

We compared the molecular weight of the deduced peptide of GCRV S10 with that of the proteins isolated from GCRV virions as described by Ke *et al.* (1992). In that report, the smallest outer capsid protein of GCRV is 27 kDa, slightly smaller than the 29.7 kDa protein deduced. Considering previous studies on aquareoviruses (Winton *et al.* 1987, Hsu YL *et al.* 1989, Subramanian *et al.* 1994) which reported that the smallest structural proteins all were approximately 31 to 36 kDa, it is possible that a deviation from reading data on the SDS-PAGE existed in the Ke *et al.* report. Moreover, the same protein identified in another report (Wang *et al.* 1990) was 31 kDa, which approximates the size of deduced product of S10. Our study will serve as a preliminary for future investigations. In order to learn more about the structure and function of the protein encoded by GCRV S10 and its similarities with other proteins, additional work including biochemical assays and immunological analysis should be performed.

Acknowledgements This work was supported by Grant No. ZYJ01-04 from the China-Israel fund for scientific and strategic research and development, and by Grant No. 990205FB05 from FEBL, the state key laboratory of Freshwater Ecology and Biotechnology.

References

Ahne W, Kolbl O (1987) Occurrence of reovirus in European cyprinid fishes (*Tinca tinca* Lin., *Leucisus cephalus* Lin.) J Appl Ichthyol 3: 139-141

Altschul SF, Madden TL, Schiffer AA, Zhang J, Zhang Z, Miller W, Lipman DJ (1997) Gapped BLAST and PSIBLAST: a new generation of protein database search programs. Nucleic Acids Res 25: 3389-3402

Anzola JV, Xu Z, Asamizu T, Nuss DL (1987) Segment specific inverted repeats found adjacent to conserved terminal sequences in wound tumor virus genome and defective interfering RNAs. Proc Natl Acad Sci USA 84: 8301-8305

Atwater JA, Munemitsu SM, Samuel CE (1986) Biosynthesis of reovirus-specified polypeptides. Molecular complementary DNA cloning and nucleotide sequence of the reovirus serotype 1 Lang strain S4 messenger RNA which encodes the major capsid surface polypeptide σ3. Biochem Biophys Res Commun 136: 183-192

Chen Y, Jiang Y (1983) Studies on the morphological and physico-chemical characterization of the hemorrhagic virus of grass carp. KeXue TongBao (Foreign Lang Edn) 28: 1138-1140

Cunningham BC, Mulkerrin MG, Wells JA (1991) Dimerization of human growth hormone by zinc. Science 253: 545-548

Ding Q, Yu L, Wang X, Ke L (1991) Study on infecting other fishes with grass carp hemorrhagic virus. Chin J Virol 6(4): 371-373 (in Chinese with English abstract)

Fang Q, Attoui H, Biagini JF, Zhu Z, de Micco P, de Lamballerie X (2000) Sequence of genome segments 1, 2, and 3 of the grass carp reovirus (Genus *aquareovirus*, family *reoviridae*). Biochem Biophys Res Commun 274(3): 762-766

Hedrick RP, Rosemark R, Aronstein D, Winton JR, McDowell T, Amend DF (1984) Characteristics of a new reovirus from channel catfish (*Ictalurus punctatus*). J Gen Virol 65: 1527-1534

Hsu Y, Chen B, Wu J (1989) Characteristics of a new reo-like virus isolated from landlocked salmon *(Oncorhynchus Masou Brevoort*). Fish Pathol 24(1):37-45

Ke L, Fang Q, Cai Y (1990) Characteristics of a new isolation of hemorragic virus of grass carp. Acta Hydrobiol Sin 14(2): 153-159 (in Chinese with English abstract)

Ke L, Wang W, Fang Q, Cai Y (1992) Studies on the *in vitro* translation of grass carp hemorrhage virus and its proteins. Chin J Virol 8(2): 169-173 (in Chinese with English abstract)

Kozak M (1980) Evaluation of the ‘scanning model' for initiation of protein synthesis in eukaryotes. Cell 22: 459-467

Kozak M (1981) Possible role of flanking nucleosides in recognition of the AUG initiator codon by eukaryotic ribosomes. Nucleic Acids Res 6: 5233-5250

Kyte J, Doolittle RF (1982) A simple method for displaying the hydropathic character of a protein. J Mol Biol 157:105-132

Lambden PR, Cooke SJ, Caul EO, Clarke IN (1992) Cloning of noncultivatable human rotavirus by single primer amplification. J Virol 66: 1817-1822

Li JK, Parker B, Kowlik T (1987) Rapid alkaline blottransfer of viral dsRNAs. Anal Biochem 163: 210-218

Li J, Wang T, Yi Y, Liu H, Lu R, Chen H (1997) A detection method for grass carp hemorrhagic virus (GCHV) based on a reverse transcription-polymerase chain reaction. Dis Aquat Org 29: 7-12

Li J, Wang T, Lu R, Chen H (1999) Advance in research of hemorrhagic virus of grass carp. Oceanol Limnol Sin 30(4): 445-453 (in Chinese with English abstract)

Lipman DJ, Pearson WR (1985) Rapid and sensitive protein similarity searches. Science 227: 1435-1441

Lupiani B, Reddy SM, Samal SK (1997a) Sequence analysis of genome segment 10 encoding the major outer capsid protein (VP7) of genogroup B aquareovirus and its relationship with the VP7 protein of genogroup A aquareovirus. Arch Virol 142(12): 2547-2552

Lupiani B, Reddy SM, Subramanian K, Samal SK (1997b) Cloning, sequence analysis and expression of the major outer capsid protein gene of an aquareovirus. J Gen Virol 78(6): 1379-1383

Mabrouk T, Lemay G (1994) Mutations in a CCHC zinc-binding motif the reovirus σ3 protein decrease its intracellular stability. J Virol 68: 5287-5290

McIntyre MC, Frattini MG, Grossman SR, Laimins LA (1993) Human papillomavirus type 18 E7 protein requires intact Cys-X-X-Cys motif for zinc binding, dimerization, and transformation but not for Rb binding. J Virology 67: 3142-3150

Neukirch M, Haas L, Lehmann H, Messeling VV (1999) Preliminary characterization of a reovirus isolated from golden ide Leuciscus idus melanotus. Dis Aquat Org 35: 159-164

Plumb JA, Bowser PR, Grizzle JM, Mitchell AJ (1979) Fish viruses: a double-stranded RNA icosahedral virus from a North American cyrinid. J Fish Res Board Can 36: 1390-1394

Rangel AAC, Rockemann DD, Hetrick FM, Samal SK (1999) Identification of grass carp haemorrhage virus as a new genogroup of aquareovirus. J Gen Virol 80: 2399-2402

Seliger LS, Giantini M, Shatkin AJ (1992) Translation effects and sequence comparisons of the three serotypes of the reovirus S4 gene. Virology 187: 202-210

Subramanian K, McPhillips TH, Samal SK (1994) Characterization of the polypeptides and determination of genome coding assignments of an aquareovirus. Virology 205: 75-81

Varner PW, Lewis DH (1991) Characterazation of a virus associated with head and lateral line erosion syndrome in marine angel fish. J Aquat Anim Health 3: 198-205

Wang T, Chen H, Liu H, Yi Y, Guo W (1994) Preliminary studies on the susceptibility of *Gobiocypris rarus* to hemorrhagic virus of grass carp. Acta Hydrobiol Sin 18(2): 144-149 (in Chinese with English abstract)

Wang W, Chen Y, Ke L, Cai Y (1990) Fine structure, genome and polypeptides of a grass carp hemorrhage virus (GCHV) isolate from the South Lake in Wuhan Hemorrhage. Chin J Virol 6(1): 46-49 (in Chinese with English abstract)

Winton JR, Lannan CN, Fryer JL, Hedrick RP, Meyers TR, Plumb JA, Yamamoto T (1987) Morphological and biochemical properties of four members of a novel group of reovirus isolated from aquatic animals. J Gen Virol 68: 353-364

Zuo W, Qian H, Xu Y, Du S, Yang X (1986) A cell line derived from the kidney of grass carp. J Fish China 10(1): 11-17 (in Chinese with English abstract)

草鱼出血病病毒基因组 S10节段的完整序列

邱　涛　陆仁厚　Jing Zhang　朱作言

中国科学院水生生物研究所，淡水生态与生物技术国家重点实验室，武汉 430072

摘　要　草鱼呼肠孤病毒(GCRV)引起的出血病是草鱼养殖业主要病害之一。然而，有关草鱼呼肠孤病毒基因节段的结构和功能知之甚少。本研究中，我们克隆了 GCRV 的 S10基因组节段，全长为909个核苷酸，包含一个编码276个氨基酸的开放阅读框，氨基酸分子量约29.7 kD。将 GCRV S10节段编码的氨基酸序列与其他呼肠孤病毒的 S10节段进行比对，未发现明显的同源性。但是，GCRV S10编码的蛋白与银大麻哈鱼呼肠孤病毒 S10节段，鲈鱼呼肠孤病毒 S10节段，以及哺乳动物呼肠孤病毒 S4节段编码的外衣壳蛋白拥有同样的锌指序列和类似的亲水模体的分布模式。推测本研究克隆的 S10节段编码 GCRV 外衣壳蛋白。

Introduction of the Human Lactoferrin Gene into Grass Carp (*Ctenopharyngodon idellus*) to Increase Resistance Against GCH Virus

Jiayu Zhong Yaping Wang Zuoyan Zhu

State Key Laboratory of Freshwater Ecology and Biotechnology, Institute of Hydrobiology, Chinese Academy of Sciences, Wuhan 430072

Abstract Haemorrhage can be an epidemic and fatal condition in grass carp. It is known now that the Grass Carp Haemorrhage Virus (GCHV) triggers haemorrhage. Human lactoferrin (hLF) plays an important role in the non-specific immune system, making some organisms more resistant to some viruses. Sperm of grass carp was mixed with linearized pCAhLFc, which is a DNA construct containing an hLF cDNA and the promoter of common carp β-actin gene, and then electroporated. Then, mature eggs were fertilized *in vitro* with the treated sperm cells. The fry were sampled and analyzed by polymerase chain reaction (PCR). Results indicated that the foreign gene had been transferred successfully into the cells of some fry. Under optimal electroporation conditions, the efficiency of gene transfer was as high as 46.8%. About 35.7% of treated 5-month-old grass carp contained foreign genes. Most transgenic fry demonstrated significant delays in onset of symptoms of haemerrhage after injection of GCHV, suggesting a significant positive relationship between hLF cDNA and levels of disease resistance ($P< 0.01$). Results suggest that transgenic grass carp could be bred for increased resistance to haemorrhage.

Keywords grass carp (*Ctenopharyngodon idellus*); electroporation; sperm-mediated gene transfer; haemorrhage; human lactoferrin; artificial infection

1. Introduction

Farming of grass carp (*Ctenopharyngodon idellus*) is very important in Chinese freshwater aquaculture. Grass carp haemorrhage, which is caused by the Grass Carp Haemorrhage Virus (GCHV), is an epidemic and fatal disease. Grass carp farming has suffered badly as a result. No way has yet been found to overcome it.

Human lactoferrin (hLF) is an iron-binding protein which plays an important role in the non-specific immune system (Boxer *et al.*, 1981; Hashizume *et al.*, 1983; Gahr *et al.*, 1991; Panella *et al.*, 1991; Esaguy *et al.*, 1993; Kawasaki *et al.*, 2000). Not only does it inhibit the

文章发表于 *Aquaculture*, 2002, 214(1): 93-101

growth of many bacteria, it can even kill them (Kalmar and Arnold, 1988; Miehlke *et al.*, 1996; Groenink, 1999). It also can inhibit the infection of some organisms by some viruses (Chen *et al.*, 1987; Hasegawa *et al.*, 1994; Marchetti *et al.*, 1999). It has been used in gene transfer to produce virus-resistant transgenic tobacco (Zhang *et al.*, 1998a,b). Whether expression of an hLF transgene can make a vertebrate, such as fish, more resistant to a virus is an important basic and applied topic.

Among methods for gene transfer, microinjection is the most popular because it is visible, precise, and has a high success rate. However, among many technical problems associated with it are the need for experienced handlers and many subtle procedural steps. It is not viable for eggs of some species. Some fishes'eggs are too fragile, some are opaque and the pronuclei are invisible, and some have too hard a chorion to be injected. Grass carp is such an example. The chorion of grass carp eggs is too flexible to be physically removed, while naked eggs whose chorions were removed by trypsin digestion are too fragile. Most eggs treated with trypsin digestion and microinjection do not develop into a normal embryo (Z. Zhu, unpublished results). For these reasons, microinjection is not a suitable method for gene transfer in grass carp. In contrast, sperm-mediated gene transfer and electroporation are very simple. They have been successfully used for common carp, catfish, tilapia, salmon, goldfish, loach, zebrafish and other organisms (Khoo *et al.*, 1992; Hu *et al.*, 2000). However, the rate of successful gene transfer by these two methods is low and variable. Electroporated sperm can significantly enhance gene transfer (Muller *et al.*, 1992; Symonds *et al.*, 1994a,b). The success rate mainly depends on the structure of the membrane of sperm cells, and sperm cells of different species have different membranes. Success in electroporated sperm-mediated gene transfer in grass carp has not been reported. We set out to optimize the parameters of electroporation of grass carp sperm with the DNA construct pCAhLFc to establish a simpler, more efficient, and more stable method of gene transfer. We also examined the level of resistance to GCHV in transgenic fry.

2. Materials and Methods

2.1 Constuction of pCAhLFc

The plasmid skeleton and the common carp β-actin promoter, pCA, come from cleavage at the double *Nco*I sites within the lactoferrin "all-fish" gene construct, pCAgcGH (Zhu, 1992). The structural gene of human cDNA (hLFc) was isolated from a recombinant plasmid phLFc (Zhang *et al.*, 1998a,b). The 8.8 kb recombinant plasmid, pCAhLFc (Fig. 1), contains the hLFc under control of the promoter of the common carp β-actin gene. To linearize the DNA, pCAhLFc is digested with *Hin*dIII. The clone of phLFc was the kind gift from Prof. Mengmin Hong, at Shanghai Institute of Plant Physiology and Ecology, Chinese Academy of

Sciences, Shanghai, China.

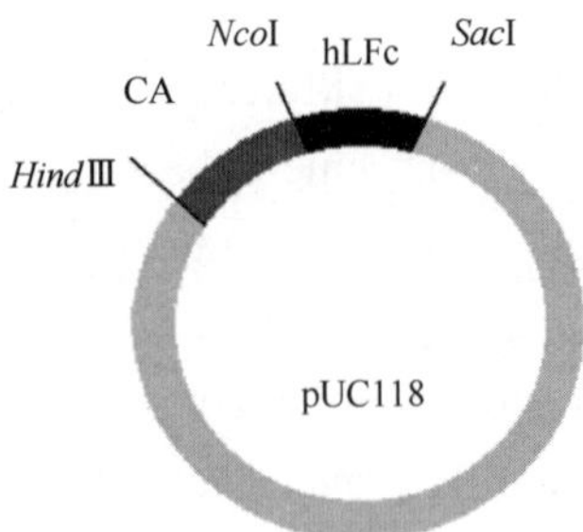

Fig. 1 Structure of the recombinant plasmid pCAhLFc used for gene transfer, containing the carp β-actin gene promoter (CA) and the lactoferrin structural gene cDNA (hLFc)

2.2 Electroporation, fertilization and incubation

Fresh sperm samples were obtained from Guanqiao fish farm, mixed with the same volume of sperm storage buffer (glucose 29 g/l, trisodium citrate dihydrate 10 g/l, sodium bicarbonate 2 g/l, potassium chloride 0.3 g/l, bovin serum albumin 30 g/l), and stored at 4℃ until electroporation within 3 h. Then, 100 μl batches of sperm were mixed with linearized pCAhLFc and electroporated using the Baekon 2000, Series I (Baekon, Saratoga, CA) electroporation device. Successful electroporation conditions are presented in Results below. The motility of sperm was checked with a microscope. Eggs were artificially fertilized with treated sperm and then incubated.

2.3 The polymerase chain reaction (PCR) with grass carp fry

Fry of 2-5 days age were sampled and put into PCR-clean microcentrifuge tubes, together with 50μl of freshly prepared homogenization buffer (Tris 0.05 M; NaCl 0.02 M; SDS 0.1%; proteinase K 1 mg/ml). Fry were digested at 55℃ for 1 h, then the tubes were gently shaken, and incubated for another 2-16 h. Tubes were spun intermittently and 150μl ultra-pure water was added and mixed gently. Tubes were boiled at 100℃ for 7 min, then allowed to cool to room temperature.

The 25μl PCR mixture contain 0.2 mM dNTPs, 1 μM primers, 0.5 U *Taq* DNA polymerase, 1.5 mM $MgSO_4$, 1μl homogenate, and 2.5 μl 10×buffer. The PCR reaction included 94℃ predenaturation for 4 min, and 30 cycles of amplification (denaturation at 94℃ for 1 min, annealing at 55℃ for 1 min, and extension at 72℃ for 1 min). The forward primer sequence was 5′-TGGCGTGATGAATGTCG-3′ which anneals on CA, and the reverse primer sequence was 5′-CTGTTTTCCGCAATGGC-3′ which anneals on hLFc. The expected PCR product length is 550 bp. The PCR products were separated by electrophoresis in a 0.8% agarose gel, stained with ethidium bromide, and analyzed under UV light.

2.4 Artificial infection with GCHV

Infected fish tissue was a kind gift from Mr. Zhang Xuewen and Xiao Tiaoyi of Hunan Agricultural University, Changsha, Hunan province, China. This was thoroughly mashed at 4℃, then a 10-fold volume of 0.65% NaCl solution was added. The suspension was twice frozen at –20℃ and melted at room temperature. It then was centrifuged at 8000×*g* for 30 min at 4℃. The supernatant was sieved with 0.22 μm mesh at 4℃, and the filtrate was stored at –80℃.

Grass carp of 3-4 months age were sampled randomly and fed at 28-30℃. The stored GCHV filtrate was melted and diluted with a nine-fold volume of 0.65% NaCl solution. Then, on two separate occasions, GCHV was injected into the abdomen of selected grass carp, at 0.1 ml for each 10 g of fish weight.

2.5 PCR analysis of artificially infected grass carp

Treated fishes were numbered according to the sequence of death within 360 h after injection of GCHV. The later death occurred, the higher the number assigned. Then, DNA was extracted from their tail fin and assayed using PCR. There were, respectively, 23 and 25 treated fish used in two artificial injection experiments. All treated fish were analyzed.

3. Results

3.1 Effect of electroporation on motility of sperm

By testing each parameter individually and then in combination, we found the following settings to give optimal electroporation results with the Baekon apparatus:amplitude, 10 kV; pulse frequency, 2^{10}; burst time, 0.4 s; cycle number, 12; pulse width, 160 μs; and distance of electrode from surface of buffer, 1 mm. Among these parameters, pulse frequency proved an important factor. Electroporation with an excessive pulse frequency, for example, pulse frequency 2^{11} or more, caused sparking at the surface of the mixture of sperm and foreign genes; thus, sperms were scorched. When electroporated with a suitable frequency, for example, a pulse frequency of $2^{9\text{-}10}$, no such problem was observed. While the precise percentage of motile sperm was not recorded, microscopic observation showed no observable effect of electroporation on the ability of the sperm to be activated.

3.2 Sperm-mediated gene transfer

Sperm were mixed with pCAhLFc and incubated for 10-30 min, then added to eggs for artificial fertilization. The efficiency of such gene transfer was between 2.2% and 4.3%. The rate of success of gene transfer among fry developing from eggs fertilized with electroporated-sperm was between 19.6% and 46.8%.

3.3 The level of disease resistance

Two trials of grass carp infection and death upon exposure to GCHV are given in Table 1. Table 2 shows the relation of PCR results to time of death for the two artificial infection experiments. The transgenic individuals lived longer before dying from haemorrhagic disease, and some transgenic individuals survived. The data of Table 1 were analyzed with non-parametric statistical testing (Wilcoxon). The results showed that the difference for GCHV resistance between transgenic individuals and non-transgenic was statistically significant (T=0, P< 0.01).

Table 1 Mortality of transgenic and non-transgenic grass carp following the artificial infection with grass carp haemorrhage virus

Time after injection (h)	Deaths in non-trangenic group		Deaths in trangenic group	
	Cumulative number	Cumulative percentage	Cumulative number	Cumulative percentage
Experiment 1				
78	0	0	0	0
84	0	0	0	0
94.5	2	14.28	1	11.11
130.5	4	28.57	1	11.11
154.5	12	85.71	2	22.22
166	12	85.71	2	22.22
177.5	13	92.86	3	33.33
192	13	92.86	5	55.56
198	14	100	5	55.56
206	14	100	6	66.67
288	14	100	6	66.67
Experiment 2				
56	0	0	0	0
72	0	0	0	0
88	0	0	0	0
96	3	20.00	0	0
128	5	33.33	0	0
152	5	33.33	0	0
168	5	33.33	1	10
172	7	46.66	1	10
176	15	100	3	30
184	15	100	3	30
192	15	100	3	30
200	15	100	4	40
208	15	100	5	50
216	15	100	6	60
272	15	100	6	60

Table 2 Presence/absence of trangene in each treated individual in first infected group

Experiment 1													
No.	1	2	3	4	5	6	7	8	9	10	11	12	
Transgene status	+	–	–	–	–	–	–	–	–	–	+	–	
No.	13	14	15	16	17	18	19	20	21[a]	22[a]	23[a]		
Transgene status	–	–	–	+	+	+	–	+	+	+	+		
Experiment 2													
No.	1	2	3	4	5	6	7	8	9	10	11	12	13
Transgene status	–	–	–	–	–	+	–	–	–	–	–	–	–
No.	14	15	16	17	18	19	20	21	22[a]	23[a]	24[a]	25[a]	
Transgene status	–	–	–	+	+	+	+	+	+	+	+	+	

No. –indicates order of mortality to disease challenge

a The survivors

4. Discussion

4.1 Electroporated sperm-mediated gene transfer is a suitable method in grass carp

The efficiency of gene transfer using electroporated sperm-mediated gene transfer ranged from 19.6% to 46.8%. Compared with microinjection, the classic method, it is much more simple; almost no special skill or experience is required to execute this method effectively. One can treat hundreds of thousand of eggs in 1 day and produce many transgenic fish. We also carried out microinjection on the same batch of grass carp eggs, although the result was unsatisfactory. Among the 48 embryos assayed, 46 contained pCAhLFc, so the success rate was as high as 95.8%. However, the survival ratio of these embryos was very low. About 2000 fertilized eggs were injected, but only 100 hatchlings were obtained. Only one of them proved able to feed normally. From this experience, we conclude that some treatments used during microinjection, for example, digestion by trypsin or puncture by glass needle can damage grass carp eggs, or that very high quality grass carp eggs are required for microinjection.

There have been many experiments on electroporation of fertilized eggs mixed with foreign genes. Xie *et al.* (1989) transferred gene in loach with an efficiency of 10%. The electroporation may hurt eggs and affect the survival and hatching rate. Inoue *et al.* (1990) demonstrated that 90% of medaka embryos survived the electric shock, but only 25% of the surviving embryos hatched, and only about 4% of the hatchlings were positively transgenic. Lu *et al.* (1992) reported that 70% of the surviving embryos hatched and 20% of the surviving fish showed integration of transgene. It is obvious that the volume of eggs is much larger than that of sperm; thus, to achieve the same final concentration of a foreign gene in the electroporated mixture, the eggs will require much more of the foreign gene than will do sperm. To get a large amount of transgenic fish, electroporation of eggs will waste much foreign gene. Also, eggs are not as strong as sperm to endure the shock of electroporation.

The efficiency of sperm-mediated gene transfer varies greatly. Li *et al.* (1996) showed the efficiency of this method was 33.3% in mirror carp and 37.0% in loach. Yu *et al.* (1994) showed that only 6.7% of treated goldfish embryos contained the transgene. And Kang *et al.* (1999) obtained a value of 8% in common carp. Our experiment with grass carp exhibited low efficiency, similar to results of the latter two reports.

The efficiency of electroporated sperm-mediated gene transfer is larger than sperm-mediated gene transfer or electroporation of fertilized eggs, without being more complex. Also, compared with other methods applied to gene transfer, this method is also simple and cheap. Other methods might include particle bombardment (Zelenin *et al.*, 1991), lipofection (Szelei *et al.*, 1994) and retroviral vector (Lin *et al.*, 1994). When a large number of transgenic fishes is needed, even if its efficiency will not be higher after further optimization, electroporated sperm-mediated gene transfer is one very good alternative candidate for species with eggs similar to those of grass carp. These eggs might be too fragile or have a chorion which is too hard for injection or whose survival rate at the larval stage is normally low. Hence, this method will be useful not only for grass carp, but also for many other species.

During the artificial infection, it was found that approximately 35% of 5-month old grass carp in the treated group contained the transgene. This suggests that the foreign gene transferred with electroporated sperm is more than transiently expressed in a host. Whether the transgene became integrated into the host genome can be determined only after Southern blot analysis.

4.2 This routine is viable to produce haemorrhage-resistant transgenic grass carp

The sequence of death of individuals in the treated group and PCR analysis of them showed that transgenic fish died later than non-transgenic fish. All surviving fish in the treated group were transgenic fish. However, some transgenic fish died earlier than nontransgenic individuals. Zhu *et al.* (1989) explained that there are three kinds of integration, toxic, silent, and efficient. Presumably, the transgenic fish whose transgene was not expressed showed no phenotypic difference from control. On the other hand, the intensity of amplification of the transgene is positively correlated with survival time (Fig. 2), which implied that the acquired character of transgenic fish is relevant to the copy number of transgene.

All these observations suggest that transgenic fish expressing pCAhLFc are more resistant to GCHV. We hope to extend this line of research to produce a transgenic line of GCHV-resistant grass carp.

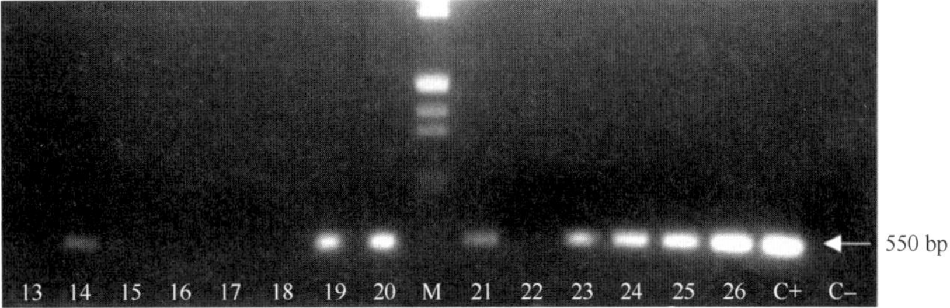

Fig. 2 PCR analysis of some samples from the putative transgenic group after the first artificial infection. Lane M: Marker (λ/*Eco*RI +*Hin*dIII); Lane C+: Positive control; Lane C–: Negative control; results indicate that nos.13, 15, 16, 17, 18 and 22 are transgene negative fish, and nos. 14, 19, 20, 21, 23, 24, 25 and 26 are transgene positive fish

Acknowledgements We thank Prof. Hong Mengmin for supplying the clone of phLFc. We also thank Mr. Zhen Jianshe and Qi Deyun for collaboration. The technical assistance by Mr. Zhang Xuewen, Mr. Xiao Tiaoyi and Mr. Liu Hanqin is appreciated.

References

Boxer, L.A., Oseas, R., Yang, H.H., Baehner, R.L., 1981. Lactoferrin: a promoter of polymorphonuclear leukocyte adhesiveness. Blood 57, 939-945.

Chen, L.T., Lu, L., Broxmeyer, H.E., 1987. Effects of purified iron-saturated human lactoferrin on spleen morphology in mice infected with Friend virus complex. Am.J. Pathol. 126, 285-292.

Esaguy, N., Freire, O., van Embden, J.D.A., Aguas, A.P., 1993. Lactoferrin triggers invitro proliferation of T cells of Lewis rats submitted to mycobacteria-induced adjuvant arthritis. Scand. J. Immunol. 38, 147-152.

Gahr, M., Speer, C.P., Damerau, B., Sawatzki, G., 1991. Influence of lactoferrin on the function of human polymorphonuclear leukocytes and monocytes. J. Leukoc. Biol. 49, 427-433.

Groenink, J., 1999. Cationic amphipathic peptides, derived from bovine and human lactoferrins, with antimicrobial activity against oral pathogens. FEMS Microbiol. Lett. 179, 217-222.

Hasegawa, K., Motsuchi, W., Tanaka, S., Dosako, S.I., 1994. Inhibition with lactoferrin of *in vitro* infection with human herpes virus. Jpn. J. Med. Sci. Biol. 47, 73-85.

Hashizume, S., Kuroda, K., Murakami, H., 1983. Identification of lactoferrin as an essential growth factor for human lymphocytic cell lines in serum-free medium. Biochim. Biophys. Acta 763, 377-382.

Hu, W., Yu, D., Wang, Y., Wu, K., Zhu, Z., 2000. Electroporation of sperm to introduce foreign DNA into the genome of *Pinctada maxima* (Jameson). Chin. J. Biotechnol. 16, 165-168.

Inoue, K., Yamashita, S., Hata, J.-I., Kabeno, S., Sada, S., Nagahisa, E., Fujita, T., 1990. Electroporation as a new technique for producing transgene fish. Cell Differ. Dev. 29, 123-128.

Kang, J.-H., Yoshizaki, G., Homma, O., 1999. Effect of an osmotic differential on the efficiency of gene transfer by eletroporation of fish spermatozoa. Aquaculture 173, 297-307.

Kalmar, J.R., Arnold, R.R., 1988. Killing of *Actinobacillus actinomycetemcomitans* by human lactoferrin. Infect. Immun. 56, 2552-2557.

Kawasaki, Y., Sato, K., Shinmoto, H., Dosako, S., 2000. Role of basic residues of human lactoferrin in the interaction with B lymphocytes. Biosci. Biotechnol. Biochem. 64, 314-318.

Khoo, H.W., Ang, L.H., Lim, H.B., 1992. Sperm cells as vectors for introducing foreign DNA into zebrafish. Aquaculture 107, 1-19.

Li, G., Cui, Z., Zhu, Z., Huang, S., 1996. Introduction of foreign gene carried by sperms. Acta Hydrobiol. Sin. 20, 242-247.

Lin, S., Gaiano, N., Gulp, P., Burns, J.C., Friedmann, T., Yee, J.-K., Hopkins, N., 1994. Integration and germline transmission of a pseudotyped retroviral vector in zebrafish. Science, 265-669.

Lu, J.-K., Chen, T.T., Chrisman, C.L., Andrisani, O.M., Dixon, J.E., 1992. Integration, expression, and germ-line transmission of foreign growth hormone genes in medaka (*Oryzias latipes*). Mol. Mar. Biol. Biotechnol. 1, 366-375.

Marchetti, M., Superti, F., Ammendolia, M.G., Rossi, P., Valenti, P., Seganti, L., 1999. Inhibition of poliovirus type 1 infection by iron-, manganese- and zinc-saturated lactoferrin. Med. Microbiol. Immunol. 187, 199-204.

Miehlke, S., Reddy, R., Osato, M.S., Ward, P.P., Conneely, O.M., Graham, D.Y., 1996. Direct activity of recombinant human lactoferrin against *Helicobacter pylori*. J. Clin. Microbiol. 34, 2593-2594.

Muller, F., Ivics, Z., Erdelyi, F., Papp, T., Varadi, L., Horvath, L., Maclean, N., 1992. Introducing foreign genes into fish eggs with electroporated sperm as a carrier. Mol. Mar. Biol. Biotechnol. 1, 276-281.

Panella, T.J., Liu, Y., Huang, A.T., Teng, C.T., 1991. Polymorphism and altered methylation of the lactoferrin gene in normal leukocytes, leukemic cells, and breast cancer. Cancer Res. 51, 3037-3043.

Symonds, J.E., Walker, S.P., Sin, F.Y.T., 1994a. Electroporation of salmon sperm with plasmid DNA: evidence of enhanced sperm/DNA association. Aquaculture 119, 313-327.

Symonds, J.E., Walker, S.P., Sin, F.Y.T., Sin, I., 1994b. Development of a mass gene transfer method in Chinook salmon: optimization of gene transfer by electroporated sperm. Mol. Mar. Bio. Biotechnol. 3, 104-111.

Szelei, J., Varadi, L., Muller, F., Erdely, F., Orban, L., Horvath, L., Dudo, E., 1994. Liposome-mediated genetransfer in fish embryos. Transgenic Res. 3, 116-119.

Xie, Y., Liu, D., Zou, J., Li, G., Zhu, Z., 1989. Gene transfer in the fertilized eggs of loach via electroporation. Acta Hydrobiol. Sin. 13, 387-389.

Yu, J., Yan, W., Zhang, Y., 1994. Sperm mediated gene transfer and method of detection of integrated gene by PCR. Acta Zool. Sin. 40, 96-99.

Zelenin, A.V., Alimov, A.A., Barmintez, V.A., Beniumov, A.O., Zelenina, I.A., Krasnov, A.M., Kolesnikov, V.A.,1991. The delivery of foreign genes into fertilized fish eggs using high-volecity microprojectiles. FEBS Lett. 287, 118-120.

Zhang, D., Jiang, Y., Wu, X., Hong, M., 1998a. Expression of humen lactoferrin cDNA in insect cells. Acta Biochim. Biophys. Sin. 30, 575-578.

Zhang, Z., Coyne, D.P., Vidaver, A.K., Mitra, A., 1998b. Expression of human lactoferrin cDNA confers resistance to Ralstonia solanacearum in transgenic tobacco plants. Phytopathology 88, 730-734.

Zhu, Z., 1992. Growth hormone gene and the transgenic fish. In: You, C.B., Chen, Z.Z. (Eds.), Agricultural Biotechnology. China Science and Technology Press, Beijing, pp. 106-116.

Zhu, Z., Xu, K., Xie, Y., Li, G., He, L., 1989. A model of transgenic fish. Sci. China, Ser. B2, 147-155.

转人乳铁蛋白基因草鱼增强抗草鱼出血病病毒感染能力

钟家玉　汪亚平　朱作言

中国科学院水生生物研究所，武汉 430072

摘　要　草鱼的出血病是具有传染性的，也是致命的。众所周知，草鱼出血病病毒(GCHV)能够引起出血症。人乳铁蛋白在非特异性免疫系统中发挥重要作用，使得某些生物能够抵抗病毒的侵染。我们将草鱼的精子与线性化的重组质粒 pCAhLFc(包括人乳铁蛋白基因的 cDNA 序列和鲤 β 肌动蛋白基因的启动子)进行混合，随后进行电穿孔操作将其转染到草鱼的精子内，然后将成熟的卵细胞与处理过的精子细胞在体外进行受精，并且通过聚合酶链式反应分析外源基因的整合情况。实验结果表明外源基因已经成功地整合到部分鱼苗细胞。最佳电穿孔条件下，基因整合效率高达46.8%。5月龄草鱼中约35.7%含有外源基因。大部分转基因鱼苗在注射草鱼出血病病毒后出现症状的时间延迟，提示人乳铁蛋白基因与抗病性状显著正相关 ($P<0.01$)。结果表明可以通过培育转基因草鱼来增强对出血症状的抗性。

Enhanced Resistance to *Aeromonas hydrophila* Infection and Enhanced Phagocytic Activities in Human Lactoferrin-Transgenic Grass Carp (*Ctenopharyngodon Idellus*)

Mao Weifeng　Wang Yaping　Wang Wenbo　Wu Bo
Feng Jianxin　Zhu Zuoyan

State Key Laboratory of Freshwater Ecology and Biotechnology, Institute of Hydrobiology, Chinese Academy of Sciences, 430072 Wuhan

Abstract Human lactoferrin (hLF) is an iron-binding protein with antimicrobial and immunomodulatory activities. *hLF* cDNA was transferred into grass carp via electroporated sperm. The production of transgenic fish was as high as 55% under the best parameters, 2^{11} pulses and 20-min incubation. The expression of the transgene was demonstrated by the detection of *hLF* mRNA by RT-PCR. We also investigated the response of G_0 transgenic grass carp to *Aeromonas hydrophila* infection. Serum lysozyme activities (P>0.05) and phagocytic activities of kidney cells (P<0.05) were measured in transgenic individuals. The transgenic fish not only cleared *A. Hydrophila* significantly faster than the control carp (P<0.05), but also showed enhanced phagocytic activities. The result shows that hLF has immunomodulatory activities in *hLF*-transgenic grass carp. The transgenic grass carp exhibited enhanced immunity to *A. hydrophila* infection. These results reveal that the mechanisms of disease resistance are different between *hLF*-transgenic plants and *hLF*-transgenic grass carp.
Keywords sperm-mediated gene transfer; lactoferrin; *Aeromonas hydrophila*; phagocytosis of kidney cells; grass carp; *Ctenopharyngodon idellus*

1. Introduction

Human lactoferrin (hLF) is an iron-binding protein mainly present in milk. It belongs to the iron transporter family and plays an important role in the human nonspecific immune system (Boxer *et al*., 1981; Hashizume *et al*., 1983; Gahr *et al*., 1991; Panella *et al*., 1991; Esaguy *et al*., 1993; Kawasaki *et al*., 2000). Several functions have been attributed to lactoferrin, including broad spectrum antimicrobial activities to both Gram-positive and Gram-negative bacteria (Kalmar and Arnold, 1988; Miehlke *et al*., 1996; Groenink, 1999; Levy, 1996; Nibbering *et al*., 2001; Arnold *et al*., 1977; Hoek *et al*., 1997; Ellison, 1994; Bellamy *et al*., 1992), antiviral activities (Chen *et al*., 1987; Hasegawa *et al*., 1994; Marchetti

文章发表于 *Aquaculture,* 2004, 242(1-4): 93-103

et al., 1999), and fungistatic activities (Levay and Viljoen, 1995; Lonnerdal and Iyer, 1995; Bellamy *et al.*, 1993; Kuipers *et al.*, 1999; Nikawa *et al.*, 1994; Soukka *et al.*,1992; Kirkpatrick *et al.*, 1971). Thus, this protein has been widely used to produce disease-resistant transgenic plants, such as tobacco and potato. (Mitra and Zhang, 1994; Zhang *et al.*, 1999). It also shows immunoregulatory activities in human, mammals, and fishes (Baveye *et al.*, 1999; Djeha and Brock, 1992; Richie *et al.*, 1987; Guillen *et al.*, 2002; Sakai *et al.*, 1993).

Aeromonas hydrophila is a Gram-negative bacterium usually found in aquatic habitats (Thune *et al.*, 1993). Motile *A. hydrophila* can cause disease in fishes, resulting in high mortality and losses (Angka, 1990). The bacteria can produce siderophores, including one recently characterized and named amonabactin (Barghouthi *et al.*, 1989; Telford and Raymond, 1998). Amonabactin can acquire iron from lactoferrin, and it was reported that the addition of lactoferrin can promote the growth of *A. hydrophila* (Stintzi and Raymond, 2000).

In general, two strategies are used to increase disease resistance in transgenic fishes. One is to boost the fish overall defense system or express broad spectrum antimicrobial or antibacterial substances, such as cecropin, interferon, cytokinin, or lysozyme (Dunham *et al.*, 2002). The other method is to enhance the resistance to certain pathogens in transgenic fishes. We have previously investigated the introduction of the lactoferrin gene into grass carp and observed the enhanced resistance against GCH virus in *hLF*-transgenic grass carp (Zhong *et al.*, 2002). Due to *hLF*'s several physiological functions, using this model, we investigated the resistance of *hLF*-transgenic grass carp to *A. hydrophila*. We also examined their serum lysozyme activities and phagocytic activities of their kidney cells to *A. hydrophila*, in order to understand the possible role of *hLF* as an immunomodulatory molecule in grass carp. The transcription of the Cec B promoter in channel catfish cells exhibited an inducible pattern and could be placed under the control of the immune system *in vivo* (Zhang *et al.*, 1998). The result suggests that the transgenesis might target improved disease resistance by stimulating grass carp's nonspecific immune system.

2. Materials and Methods

2.1 Fishes

The grass carp (*Ctenopharyngodon idellus*) used in these experiments were produced at the Institute of Aquaculture, Henan Province and hatched during the spring, 2001.

2.2 Construction of plasmid

The 8.8 kb recombinant plasmid, p*CAhLFc*, was constructed in our laboratory including the carp β-actin (*CA*) gene promoter and the *hLF* cDNA (Zhong *et al.*, 2002). The recombinant plasmid was digested by *Hin*dIII to linearize the DNA. The digested DNA fragments were purified using a glassmilk kit (MBI, Vilnius, Lithuania).

2.3 Sperm-mediated gene transfer

Fresh sperm samples were blended with the same volume of sperm storage buffer (glucose 29 g/l, trisodium citrate dihydrate 10 g/l, sodium bicarbonate 2 g/l, potassium chloride 0.3 g/l, bovine serum albumin 30 g/l); 100 μl batches of sperm were mixed with linearized p*CAhLFc* (100 ng/μl final concentration), incubated for 10-30 min at room temperature and electroporated using the Baekon 2000, Series^2 (Baekon, Saratoga, CA) electroporation device under conditions described in the results. Following electroporation, sperm motility following addition of water was checked with a microscope. Fresh eggs were artificially fertilized with treated sperm, and then incubated at room temperature for 10-30 min. The effects of different electroporation parameters on efficiency of fertilization and transgenesis were investigated.

2.4 Isolation of fry genomic DNA and detection of *hLF* DNA

Seven-day-old fry were sampled randomly. Individual fry was added to 300 μl extraction buffer (10 mmol/l Tris-HCl, pH 7.5, 10 mmol/l EDTA, 150 mmol/l 0.5% SDS, 0.3 mg/ml Proteinase K), homogenized, and added in a micro-centrifuge tube at 37℃ for 12 h. Fry genomic DNA was extracted by phenol/chloroform and purified by ethanol precipitation.

PCR was used to check for the presence of the introduced gene construct. The binding site for primer CL1 (5′-TGGCGTGATGAATGTCG-3′) is located at the *CA* promoter;primer CL2 (5′-CTGTTTTCCGCAATGGC-3′) anneals onto the *hLF* cDNA. The PCR mixture (25 μl in total) contained 100 ng template DNA, 0.2 mmol/l dNTP, 0.5 μmol/l, 1 U *Taq* DNA polymerase (MBI), and 1.5 mmol/l Mg^{2+}. The PCR protocol performed in a PE 9600 thermal cycler included 94℃ denaturation for 4 min, and 30 cycles of amplification (denaturation at 94℃ for 1 min, annealing at 55℃ for 0.5 min, and extension for 1 min at 72℃). The PCR products were separated using 0.8% agarose gel electrophoresis and visualized by a UVP GDS8000 system (UVP, CA, US). The expected PCR product length is 550 bp.

2.5 Isolation of total RNA from 5-month-old grass carp and RT-PCR

Total RNA was isolated from fins of transgenic carps with the SV Total RNA Isolation System (Promega, Cat Z3100). RT-PCR was carried out with the Access RT-PCR System (Promega, Cat A1250). The reaction (total 25 μl) contained 50 ng total RNA template, 10 mmol/l dNTP, 0.5 μmol/l Primer CL1 and CL2, 2.5 U AMV reverse transcriptase (Promega), 2.5 U Tfl DNA polymerase (Promega). The process of the reaction included reverse transcription at 48℃ for 45 min, reactivation of AMV reverse transcriptase at 94℃ for 2 min, 35 cycles of DNA synthesis (denaturing at 94℃ for 30 s, annealing at

55℃ for 30 s, extension at 68℃ for 1 min), and finally incubation at 48℃ for 7 min. The RTPCR products were separated using 0.8% agarose gel electrophoresis and visualized by a UVP GDS8000 system.

2.6 Artificial infection with *A. hydrophila*

Exposure to 10^7 clone-forming units (CFU)/ml of *A. hydrophila* yields mortality rate of control grass carp of approximately 50%; 0.2 ml suspensions of 10^7(CFU)/ml *A. hydrophila* were injected into control grass carp and transgenic grass carp identified by PCR. Their mortalities were recorded from 36 to 120 h.

2.7 Serum lysozyme assay

The turbidimetric assay for lysozyme (Parry *et al.*, 1965) was performed with 0.2 mg/ml *A. hydrophila* dissolved in sodium phosphate buffer (0.04 M, pH 5.75). Test serum (2 μl) was added to the phosphate buffer for a final volume of 3 ml. The reaction was carried out at room temperature, and the absorbance (530 nm) measured after 0.5 and 4.5 min. The unit of enzyme activity was defined as the amount of enzyme that caused a decrease in absorbance of 0.001 per min.

$$\text{U/ml} = \frac{OD1 - OD2}{4 \times 0.001 \times 2} \times 100$$

2.8 Phagocytic assay

Phagocytic activity was measured according to Rafiq *et al.* (2001). Macrophage cell suspensions were prepared according to the method of Secombes (1990), by teasing head kidney tissue through 100 μm nylon mesh into 1640 cell culture medium (Sigma). The head kidney tissue was layered for 1 h and centrifuged at 400×*g* for 25 min at 4℃. The macrophages, found at the interface of the concentrations, were collected and washed twice with 1640 medium and concentrated for 7 min at 1000×*g*. The cells were counted and their concentration adjusted to 5×10^6/ml with 1640 medium containing 1% penicillin/streptomycin. Macrophage monolayers were prepared by placing 200-μl aliquots of the suspension onto the circles of sterilized frosted glass slides and incubating at 22℃ for 3 h in a humid chamber. The slides were then washed three times with 1640 medium. The slides were placed in an incubator for 24 h at 22℃. The macrophage monolayers were washed twice with 1640 medium. *A. hydrophila* (Strain 12#), the bacterium used in the assay, was opsonized by formaldehyde adjusted to 5×10^7/ml. Phagocytosis of the bacteria was examined by placing 200-μl aliquots of opsonized bacterial suspension onto each macrophage monolayer and incubating for 1 h at 22℃. The macrophage monolayers were then washed five times with PBS (pH 7.2), fixed with formaldehyde, and stained with Giemsa. Finally, the slides were washed with PBS (pH 7.2), and allowed to air dry; 300 cells were counted microscopically and phagocytic activity was quantified by the following formula:

$$PA = \frac{\text{Number of phagocytizing cells}}{300} \times 100$$

2.9 Statistical analysis

Statistical significance was assessed using the Student's *t*-test.

3. Results

3.1 Sperm-mediated gene transfer

With the different combinations of electroporation parameters, the efficiency of gene transfer was between 33.3% and 55.0%, and the ratio of fertilization was between 21.43% and 86.0% (Table 1). The combination: amplitude 10 kV; pulse frequency, 2^{10}; burst 0.4 s; cycle number, 12; distance of electrode from surface of buffer, 1 mm; was found optimal. With this combination, the efficiency of gene transfer was 55.0% and the ratio of fertilization was 33.5%.

Table. 1 Percent fertifization and percent transgenic after eletroporation under various conditions

Group	Incubation time(min)	Amplitude (kV)	Pulse frequency	Burst time(s)	Cycle number	Pulse time	Distance of electrode from surface of buffer (mm)	Rates of fertilization (%)	Percent transgenic (%)
E3	30	10	2^9	0.4	12	fixed	1.5	86.0	40.00(6/15)
D4	20	10	2^{10}	0.4	12	fixed	1.5	73.8	33.30(5/15)
E4	30	10	2^{10}	0.4	12	fixed	1.5	74.1	40.00(6/15)
B5	0	10	2^{11}	0.4	12	fixed	1.5	64.0	42.90(6/14)
D5	20	10	2^{11}	0.4	12	fixed	1.5	33.5	55.0(11/20)
E5	30	10	2^{11}	0.4	12	fixed	1.5	21.43	54.45(12/22)

3.2 Detection of *hLF* DNA in transgenic fry and the efficiency of gene transfer

The number of fry in which *hLF* DNA was detected was recorded for every experimental group (Table 1). For example, 11 fry in group D5 exhibited the *hLF* DNA, all of which had undergone gene transfer with the optimal electroporation parameters (Fig. 1).

3.3 Detection of *hLF* mRNA in fins of transgenic grass carp

Total RNA was isolated from fins of transgenic grass carp. The results of RT-PCR showed *hLF* mRNA in fins of all detected transgenic grass carp (Fig. 2).

3.4 Infection with *A. hydrophila*

Six-month-old *hLF*-transgenic grass carp (20) and control non-transgenic grass carp (20) were injected with *A. hydrophila* solution. Death of individual control non-transgenic grass carp occurred between 36- and 120-h post-injection, while death of transgenic grass carp occurred between 36- and 60-h post-injection (Table 2). Cumulative death percentage in transgenic group is much lower than in non-transgenic group. The *hLF*-transgenic grass carp showed enhanced resistance to *A. hydrophila* relative to control non-transgenic grass carp ($P<0.05$).

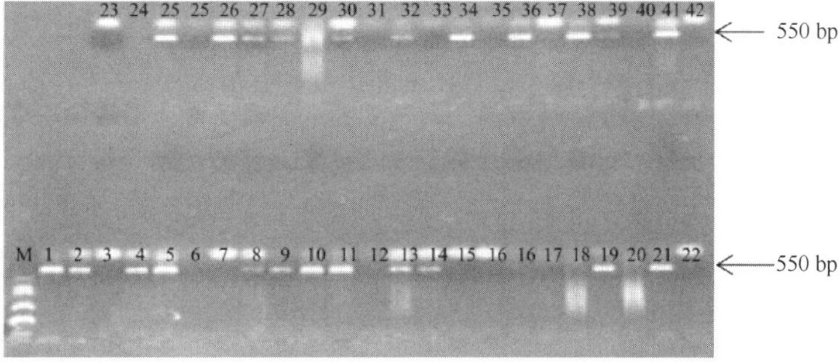

Fig. 1 Result of PCR for the positive individuals in the experimental group. M: DNA marker (λ/*Eco*RI+*Hin*d III); 1-21: tested individuals of the E5 group, the 550 bp band is the positive band; 23-41: tested individuals of the D5 group, the 550 bp band is the positive band; 22, 42: result of negative control (no sample)

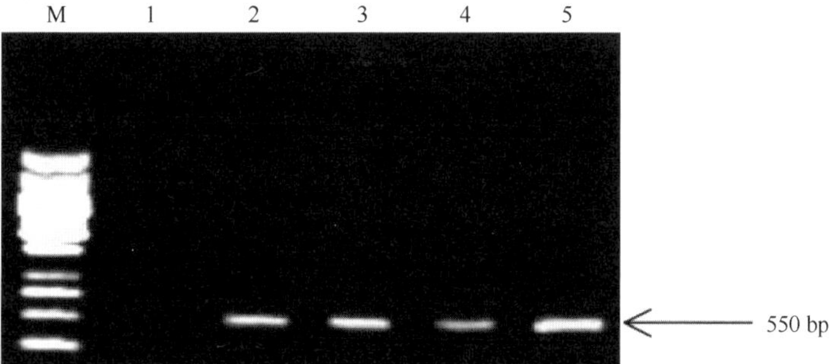

Fig. 2 Result of RT-PCR for human lactoferrin mRNA for 5-month-old transgenetic carp. M: DNA marker (1 kb ladder); 1: negative control (no sample); 2-5: RT-PCR of transgenetic carp; total RNA is from fins of transgenic grass carp

Table 2 Results of artificial infection with *A. hydrophila* in 1 week

Experimental group (number)	Deaths in non-transgenic group/total non-transgenic (% mortality)	Deaths in transgenic group/total transgenic (% mortality)
20	10/20(50%)	4/20(20%)

3.5 Serum lysozyme assay

Table 3 shows no evident difference between lysozyme activities of serum of *hLF*-transgenic and control non-transgenic grass carp (P>0.05).

Table 3 Serum lysozyme activities in control and transgenic carp

Cotrol carp		Transgenic carp	
No.	U/ml	No.	U/ml
1	125	1	125
2	500	2	600
3	125	3	250
4	125	4	125
5	250	5	125
6	250		
7	125		

3.6 Phagocytic assay

The average phagocytic activity of the kidney leucocytes from the *hLF*-transgenic grass carps was higher than that of the control non-transgenic fish (P<0.05). However, the phagocytic ctivity of four *hLF*-transgenic fish was similar to those of the control fish (Table 4).

Table 4 Phagocytic activities of kidney cells isolated from transgenic carp and control carp

	Phagocytic activities in transgenic grass carp (%)	Mean phagocytic activities in transgenic grass carp (%±S.D.)	Phagocytic activities in control grass carp (%)	Mean phagocytic activities in control grass carp (%±S.D.)
1	58.3	48.575±9.09	40.5	41.4±1.09
2	40.6		41.4	
3	46.7		39.7	
4	40.1		43.2	
5	42.3		41.5	
6	59.6		40.3	
7	59.9		41.7	
8	41.1		40.5	

4. Discussion

Transgenic technology has the potential to generate fast-growing (Zhu *et al.*, 1985), freeze-resistant (Fletcher *et al.*, 1988), and disease-resistant fish (Zhong *et al.*, 2002) for aquaculture. We previously demonstrated electroporated sperm-mediated gene transfer in grass carp (Zhong *et al.*, 2002). In the present study, the parameters of electroporation were investigated and adjusted. Using the optimal combination of parameters tested, the maximum efficiency of gene transfer was 55%, but the ratio of fertilization was only 33.5%. Because one mature female grass carp can produce millions of eggs at one time, it is advisable for us to sacrifice the ratio of fertilization for high efficiency of gene transfer. We also showed that 20-min incubation at room temperature before electroporation improves the efficiency of gene transfer.

Lactoferrin has been widely used in research on disease-resistance transgenic plants. Lactoferrin kills or inhibits the growth of bacteria and virus by its broad spectrum antimicrobial and antiviral activities in transgenic plants (Mitra and Zhang, 1994; Zhang *et al.*, 1999). We previously reported that *hLF*-transgenic grass carp shows increased resistance against GCH virus (Zhong *et al.*, 2002). In these previous studies, systemic modulation of immune response was not detected. We hypothesized that the broad spectrum antimicrobial and antiviral activities of *hLF* were responsible for the increased resistance against GCH virus. In this study, we showed that the *hLF*-transgenic grass carp showed enhanced ability to resist *A. hydrophila* compared with the control non-transgenic grass carp. We also investigated the mechanism of resistance to *A. hydrophila* infection in *hLF*-transgenic grass carp.

The antibacterial and antifungal activity of *hLF* is achieved by deprivation of iron from the pathogen's microenvironment and the cell walls of bacteria and fungi, which will cause membrane perturbation and leakage of intracellular components (Bellamy *et al.*, 1993; Levay and Viljoen, 1995; Lonnerdal and Iyer, 1995; Wakabayashi *et al.*, 1996). However, *A. hydrophila*'s siderophore, amonabactin, is able to remove iron from transferrin and lactoferrin (Massad *et al.*, 1991). Stintzi and Raymond (2000) showed that addition of lactoferrin promoted the growth of *A. hydrophila*. Our experimental results show that *hLF*-transgenic grass carp exhibited enhanced resistance to *A. hydrophila* infection. Since the hLF is not able to inhibit the growth of *A. hydrophilaby* itself, the fish defense system should be responsible for the higher resistance to *A. hydrophila* infection in *hLF*-transgenic grass carp. We examined two important factors of the nonspecific immune system in fish, lysozyme activity and phagocytic activity. The results of phagocytic assay show that the ability of head kidney macrophages to phagocytose *A. hydrophila* is higher in *hLF*-transgenic grass carp than in the control non-transgenic grass carp. No difference of lysozyme activities was found between *hLF*-transgenic grass carp and the control non-transgenic grass carp. These results suggest that expression of *hLF* in the transgenic grass carp stimulates enhanced phagocytic activity, which is responsible for the improved bacterial clearance. Our results support the working hypothesis that enhanced phagocytic activity in transgenic fish is stimulated by *hLF*. Further studies are needed to elucidate the antimicrobial and antiviral ability of *hLF* expressed in the *hLF* transgenic grass carp.

The phagocytosis activities in several *hLF*-transgenic grass carp were similar to those in non-transgenic grass carps. This might be explained by the fact that there are three kinds of integration in transgenic fish: toxic, silent, and efficient (Zhu *et al.*, 1986). Only efficient expression of the transgene can show its function.

Acknowledgements We thank Prof. Hong Mengmin for supplying the *hLF* gene. We also thank Miss Li Ming, Mr. Zen Jianshe and Qi Deyun for technical assistance. The valuable comments from Dr. Hu Wei, Dr. Sun Yonghua, Dr. Wu Gang are appreciated.

References

Angka, S.L., 1990. The pathology of the walking catfish *Clarias batrachus* (L.) infected intraperitoneally with *Aeromonas hydrophila*. Asian Fish. Sci. 3, 343-351.

Arnold, R.R., Cole, M.F., Mcghee, J.R., 1977. A bactericidal effect for human lactoferrin. Science 197, 263-265.

Barghouthi, S., Young, R., Olson, M.O., Arceneaux, J.E., Clem, L.W., Byers, B.R., 1989. Amonabactin, a novel tryptophan- or phenylalanine-containing phenolate siderophore in *Aeromonas hydrophila*. J. Bacteriol. 171,1811-1816.

Baveye, S., Elass, E., Mazurier, J., Spik, G., Legrand, D., 1999. Lactoferrin: a multifunctional glycoprotein involved in the modulation of the inflammatory process. Clin. Chem. Lab. Med. Mar. 37, 281-286.

Bellamy, W., Takase, M., Yamauchi, K., Wakabayashi, H., Kawase, K., Tomita, M., 1992. Identification of the bactericidal domain of lactoferrin. Biochim. Biophys. Acta 1121, 130-136.

Bellamy, W., Wakabayashi, H., Takase, M., Kawase, K., Shimamura, S., Tomita, M.,1993. Killing of *Candida albicans* by lactoferricin B, a potent antimicrobial peptide derived from the N-terminal region of bovine lactoferrin. Med. Microbiol. Immunol. 182, 97-105.

Chen, L.T., Lu, L., Broxmeyer, H.E., 1987. Effects of purified iron-saturated human lactoferrin on spleen morphology in mice infected with Friend virus complex. Am. J. Pathol. 126, 285-292.

Djeha, A., Brock, J.H., 1992. Effect of transferrin, lactoferrin and chelated iron on human T-lymphocytes. Br. J. Haematol. 80, 235-241.

Dunham, R.A., Warr, G.W., Nichols, A., Duncan, P.L., Argue, B., Middleton, D., Kucuktas, H., 2002. Enhanced bacterial disease resistance of transgenic channel catfish *Ictalurus punctatus* possessing cecropin genes. Mar. Biotechnol. 4, 338-344.

Esaguy, N., Freire, O., Van Embden, J.D.A., Aguas, A.P., 1993. Lactoferrin triggers *in vitro* proliferation of T cells of Lewis rats submitted to mycobacteria-induced adjuvant arthritis. Scand. J. Immunol. 38, 147-152.

Fletcher, G.L., Shears, M.A., King, M.J., Davies, P.L., Hew, C.L., 1988. Evidence for antifreeze protein gene transfer in Atlantic salmon. Can. J. Fish. Aquat. Sci. 45, 352-357.

Gahr, M., Speer, C.P., Damerau, B., Sawatzki, G., 1991. Influence of lactoferrin on the function of human polymorphonuclear leukocytes and monocytes. J. Leukoc. Biol. 49, 427-433.

Groenink, J., 1999. Cationic amphipathic peptides, derived from bovine and human lactoferrins, with antimicrobial activity against oral pathogens. FEMS Microbiol. Lett. 179, 217-222.

Guillen, C., McInnes, I.B., Vaughan, D.M., Kommajosyula, S., Van Berkel, P.H., Leung, B.P., Aguila, A., Brock, J.H., 2002. Enhanced *Th1* response to *Staphylococcus aureus* infection in human lactoferrin-transgenic mice. J. Immunol. 168, 3950-3957.

Hasegawa, K., Motsuchi, W., Tanaka, S., Dosako, S.I., 1994. Inhibition with lactoferrin of *in vitro* infection with human herpes virus. Jpn. J. Med. Sci. Biol. 47, 73- 85.

Hashizume, S., Kuroda, K., Murakami, H., 1983. Identification of lactoferrin as an essential growth factor for human lymphocytic cell lines in serum-free medium. Biochim. Biophys. Acta 763, 377-382.

Hoek, K.S., Milne, J.M., Grieve, P.A., Dionysius, D.A., Smith, R., 1997. Antibacterial activity of bovine lactoferrin-derived peptides. Antimicrob. Agents Chemother. 41, 54-59.

Kalmar, J.R., Arnold, R.R., 1988. Killing of *Actinobacillus actinomycetemcomitans* by human lactoferrin. Infect.Immun. 56, 2552-2557.

Kawasaki, Y., Sato, K., Shinomto, H., Dosako, S., 2000. Role of basic residues of human lactoferrin in the interaction with

B lymphocytes. Biosci. Biotechnol. Biochem. 64, 314-318.

Kirkpatrick, C.H., Green, I., Rich, R.R., Schade, A.L., 1971. Inhibition of growth of *Candida albicans* by ironunsaturated lactoferrin: relation to host-defense mechanisms in chronic mucocutaneous candidiasis. J. Infect. Dis. 124, 539-544.

Kuipers, M.E., de Vries-Hospers, H.G., Eikelboom, M.C., Meijer, D.K.F., Swart, P.J., 1999. Synergistic fungistatic effects of lactoferrin in combination with antifungal drugs against clinical *Candida isolates*. Antimicrob. Agents Chemother. 43, 2635-2641.

Levay, P.F., Viljoen, M., 1995. Lactoferrin: a general review. Haematologica 80, 252-267.

Levy, O., 1996. Antibiotic proteins of polymorphonuclear leukocytes. Eur. J. Haematol. 56, 263-277.

Lonnerdal, B., Iyer, S., 1995. Lactoferrin: molecular structure and biological function. Annu. Rev. Nutr. 15, 93-110.

Marchetti, M., Superti, F., Ammendolia, M.G., Rossi, P., Valenti, P., Seganti, L., 1999. Inhibition of poliovirus type 1 infection by iron-, manganese- and zinc-saturated lactoferrin. Med. Microbiol. Immunol.187, 199-204.

Massad, G., Arceneaux, J.E.L., Byers, B.R., 1991. Acquisition of iron from host sources by mesophilic Aeromonas species. J. Gen. Microbiol. 137, 237-241.

Miehlke, S., Reddy, R., Osato, M.S., Ward, P.P., Connelly, O.M., Graham, D.Y., 1996. Direct activity of recombinant human lactoferrin against *Helicobacter pylori*. J. Clin. Microbiol. 34, 2593-2594.

Mitra, A., Zhang, Z., 1994. Expression of a human lactoferrin cDNA in tobacco cells produces antibacterial protein(s). Plant Physiol. 106, 977-981.

Nibbering, P.H., Ravensbergen, E., Welling, M.M., van Berkel, L.A., van Berkel, P.H., Pauwels, E.K., Nuijens, J.H., 2001. Human lactoferrin and peptides derived from its N terminus are highly effective against infections with antibiotic-resistant bacteria. Infect. Immun. 69, 1469-1476.

Nikawa, H., Samaranayake, L.P., Tenovuo, J., Hamada, T., 1994. The effect of antifungal agents on the *in vitro* susceptibility of *Candida albicans* to apo-lactoferrin. Arch. Oral Biol. 39, 921-923.

Panella, T.J., Liu, Y., Huang, A.T., Teng, C.T., 1991. Polymorphism and altered methylation of the lactoferrin gene in normal leukocytes, leukemic cells, and breast cancer. Cancer Res. 51, 3037-3043.

Parry, R.M., Chandan, R.C., Shahani, K.M., 1965. A rapid and sensitive assay of muramidase. Proc. Soc. Exp. Biol. 118, 384-386.

Rafiq, M.I., Sarder, K.D., Thompson, D.J., Penman, B.J., McAndrew, 2001. Immune responses of Nile tilapia (*Oreochromis niloticus* L.) clones: I. Non-specific responses. Dev. Comp. Immunol. 25, 37-46.

Richie, E.R., Hilliard, J.K., Gilmore, R., Gillespie, D.J., 1987. Human milk-derived lactoferrin inhibits mitogen and alloantigen induced human lymphocyte proliferation. J. Reprod. Immunol. 12, 137-148.

Sakai, M., Otubo, T., Atsuta, S., Kobayashi, M., 1993. Enhancement of resistance to bacterial infection in rainbow trout, *Oncorhynchus mykiss* Walbaum by oral administration of bovine lactoferrin. J. Fish Dis. 16, 239-247.

Secombes, C.J., 1990. Isolation of salmonid macrophages and analysis of their killing activity. In: Stolen, J.S., Fletcher, T.C., Anderson, D.P., Robertson, B.S., van Muiswinkel, W.B. (Eds.), Techniques in Fish Immunology, vol. 1. SOS Publications, Fair Haven, NJ, pp. 137-154.

Soukka, T., Tenovuo, J., Lenander Lumikari, M., 1992. Fungicidal effect of human lactoferrin against *Candida albicans*. FEMS Microbiol. Lett. 69, 223-228.

Stintzi, A., Raymond, K.N., 2000. Amonabactin-mediated iron acquisition from transferrin and lactoferrin by *Aeromonas hydrophila*: direct measurement of individual microscopic rate constants. JBIC 5, 57-66.

Telford, J.R., Raymond, K.N., 1998. Coordination chemistry of the amonabactins, bis (catecholate) siderophores from *Aeromonas hydrophila*. Inorg. Chem. 37, 4578-4583.

Thune, R.L., Stanley, L.A., Cooper, R.K., 1993. Pathogenesis of Gram-negative bacterial infection in warm-water fish. In: Faisal, M., Hetrick, F.M. (Eds.), Annu. Rev. Fish Dis., pp. 37-68. Volume #.

Wakabayashi, H., Abe, S., Okutomi, T., Tansho, S., Kawase, K., Yamaguchi, H., 1996. Cooperative anti-*Candida* effects of lactoferrin or its peptides in combination with azole antifungal agents. Microbiol. Immunol. 40, 821-825.

Zhang, Q., Tiersch, T.R., Cooper, R.K., 1998. Inducible expression of green fluorescent protein within channel catfish cells by a cecropin gene promoter. Gene 216 (1), 207-213 (Aug 17).

Zhang, D., Zhang, J., Wang, Z., Hong, M., 1999. Expression of cloned human lacto ferrin cDNA in transgenic tobacco confers resistance to bacteria and viral disease. Acta Phytophysiologica Sin. 25, 234-243.

Zhong, J., Wang, Y., Zhu, Z., 2002. Introduction of the human lactoferrin gene into grass carp (*Ctenopharyngodon idellus*) to increase resistance against GCH virus. Aquaculture 214, 93-101.

Zhu, Z., Li, G., He, L., Chen, S., 1985. Novel gene transfer into the fertilized eggs of goldfish (*Carassius auratus L.*1758). Z. Angew. Ichthyol. 1, 31-34.

Zhu, Z., Xu, K., Li, G., Xie, Y., He, L., 1986. Biological effects of human growth hormone gene microinjected into the fertilized eggs of loach *Misgurnus anguillicaudatus* (Cantor). Kexue Tongbao 31, 988-990.

转人乳铁蛋白基因草鱼增强抗嗜水气单胞菌感染和吞噬细胞吞噬活性的能力

茅卫锋 汪亚平 Wang Wenbo 吴 波 Feng Jianxin 朱作言

中国科学院水生生物研究所，武汉 430072

摘 要 人乳铁蛋白是具有抗菌和免疫调节功能的铁结合蛋白。通过电脉冲精子介导法把人乳铁蛋白基因的 cDNA 序列转移到草鱼中。在 2^{11} 的脉冲频率和 20 分钟孵化期的最佳参数条件下，转基因鱼的发生率可以高达 55%。利用定量 PCR 对 *hLF* mRNA 进行检测可以检测转植基因的表达情况。本研究不仅分析了 G_0 转基因草鱼对嗜水气单胞菌感染的响应情况，还在转基因鱼个体水平检测了血清溶菌酶的活性($P>0.05$)和肾脏细胞的吞噬活性。转基因鱼不仅比对照鱼更快地清除了嗜水气单胞菌($P<0.05$)，而且显示出更强的吞噬功能。研究结果显示人乳铁蛋白在转基因草鱼中具有免疫调节活性。转基因草鱼对嗜水气单胞菌表现出免疫力增强的特性。这一结果提示人乳铁蛋白转基因植物和转基因草鱼具有不同的抗病机制。

Molecular Cloning, Characterization and Expression Analysis of the PKZ Gene in Rare Minnow *Gobiocypris rarus*

Jianguo Su[1,2] Zuoyan Zhu[1] Yaping Wang[1]

1 State Key Laboratory of Freshwater Ecology and Biotechnology, Institute of Hydrobiology, Chinese Academy of Sciences, Wuhan 430072

2 Department of Aquaculture, College of Animal Science and Technology, Northwest A&F University, Yangling 712100

Abstract Double-stranded RNA-activated protein kinase (PKR) plays an important role in interferon-induced antiviral responses, and is also involved in intracellular signaling pathways, including the apoptosis, proliferation, and transcription pathways. In the present study, a PKR-like gene was cloned and characterized from rare minnow *Gobiocypris rarus*. The full length of the rare minnow PKR-like (GrPKZ) cDNA is 1946 bp in length and encodes a polypeptide of 503 amino acids with an estimated molecular mass of 57,355 Da and a predicted isoelectric point of 5.83. Analysis of the deduced amino acid sequence indicated that the mature peptide contains two Zalpha domains and one S_TKc domain, and is most similar to the crucian carp (*Carassius auratus*) PKR-like amino acid sequence with an identity of 77%. Quantitative RT-PCR analysis showed that GrPKZ mRNA expression is at low levels in gill, heart, intestine, kidney, liver, muscle and spleen tissues in healthy animals and up-regulated by viruses and bacteria. After being infected by grass carp reovirus, GrPKZ expression was up-regulated from 24 h post-injection and lasted until the fish became moribund ($P<0.05$). Following infection with *Aeromonas hydrophila*, GrPKZ transcripts were induced at 24 h post-injection ($P<0.05$) and returned to control levels at 120 h post-injection. These data imply that GrPKZ is involved in antiviral defense and Toll-like receptor 4 signaling pathway in bacterial infection.

Keywords *Gobiocypris rarus*; Z-DNA binding protein kinase; antiviral gene; gene cloning; characterization

Introduction

The double-stranded RNA-activated protein kinase (PKR) is a critical mediator of antiproliferation and a key component of the innate antiviral response exerted by interferons. PKR integrates signals in response to Toll-like receptor (TLR) activation, growth factors,

文章发表于 *Fish & Shellfish Immunology*, 2008, 25(1-2): 106-113

differentiation, apoptosis, and diverse cellular stresses [1-5]. PKR phosphorylation of the α-subunit of eukaryotic initiation factor 2 (eIF2α) results in a blockade on translation initiation. This prevents viral replication and inhibits normal cell ribosome function, killing both the virus and the host cell if the response is active for a sufficient amount of time [6]. PKR cannot block the translation of some cellular and viral mRNAs bearing special features in their 5′ untranslated regions [7]. PKR also affects diverse transcriptional factors such as interferon regulatory factor 1, STATs, *p53*, activating transcription factor 3, and NF-B [8,9]. The extent and strength of the antiviral action of PKR are clearly understood by the findings that unrelated viral proteins of animal viruses have evolved diverse strategies to inhibit PKR action [7,10]. Knock-out mouse studies confirmed a role for PKR in antiviral innate immunity [11]. Recently, PKR has been implicated in TLR signal transduction in response to bacterial cell wall components [12].

The PKR gene was first identified from human in 1990 [13]. Now PKR sequence similarity genes have been discovered in many eukaryotes, such as mammals, birds amphibians, fishes, insects, and yeast (searched by blastp). Induction of the gene with sequence similarity in yeast is in response to starvation and activated by uncharged tRNAs and the Gcn1p-Gcn20p complex (accession No. AY293929; *E*-value = 6e–36). In insects, there are several similar sequences with *E*-value from 2e–34 to 3e–47. However, these sequence similarity genes in invertebrate may not have functional similarity, because of the lack of homologous interferon genes in several fully sequenced invertebrate genomes [14-16]. In fishes, PKR homologous genes are found in five species in GenBank, including crucian carp, zebrafish, Atlantic salmon, olive flounder and pufferfish (February, 2008).

Fish appear to have a PKR-like protein that has Z-DNA binding (Zalpha) domain(s) instead of dsRNA binding (DSRM) domain(s) in the regulatory domain, and has thus been designated Z-DNA binding protein kinase (PKZ) [17]. Zalpha domain(s) specifically bind dsDNA and dsRNA in the left-handed Z conformation, often with high affinity [17,18]. PKZ is able to phosphorylate eIF2α [19] and may play a role, like PKR, in host defense against virus infection [17,19].

Grass carp (*Ctenopharyngodon idellus*) is an important aquaculture species in China, but great economic loss is often caused by Grass Carp Reovirus (GCRV), a dsRNA virus. A better understanding of the immune defense mechanisms of grass carp may contribute to the development of management strategies for disease control and long term sustainability of grass carp farming. The rare minnow *Gobiocypris rarus*, which is a small cyprinid species, has been recognized as a useful model. It is very sensitive to GCRV [20]. Moreover, *Aeromonas hydrophila,* a Gramnegative bacterium of the family Aermonadaceae, is often found in association with hemorrhagic septicemia in cold-blooded animals including fish, reptiles and amphibians [21].

In this paper, we employed the rare minnow as a model fish to study the mechanism of

GCRV disease. A full-length cDNA sequence encoding a PKR-like protein (GrPKZ) was identified and characterized. We present GrPKZ mRNA expression profiles in different tissues and following GCRV and *A. hydrophila* infection.

Materials and Methods

Animals, immune challenge and sample collection

Rare minnows (2-3 g body weight) were obtained from a laboratory-breeding stock and acclimatized to new laboratory conditions for one week in a quarantine area. They were maintained in 25 l aerated aquaria at 28℃ and fed twice a day with commercial diet (feed composition: protein 32%, starch 63%, fat 3%, additive 2%). Before experiments, the fish were copparehtly healthy.

For challenge experiments with GCRV (991 strain) or *A. hydrophila* (C1 strain), fish were injected with 10 μl PBS per gram body weight, 10 μl GCRV suspended in PBS (2×10^8 PFU/ml) or 10 μl *A. hydrophila* resuspended in PBS (OD_{600} = 0.1) intraperitoneally. Non-injected animals were used as blank group. Three individuals were killed and tissues including gill, heart, intestine, kidney, liver, muscle, and spleen were collected at 0, 12, 24, 36, 48, 72, 96, 120, 144 and 168 h after injection. The samples were homogenized in TRIZOL® LS reagent (Invitrogen) and total RNA was extracted according to the manufacturer's instruction. Total RNA was incubated with RNase-free DN ase I (Roche) to remove contaminated genomic DNA before being reverse transcribed into cDNA using random hexamer primers with SuperScript™ III Reverse Transcriptase (Invitrogen).

Amplification of cDNA and nucleotide sequence analysis

To clone PKZ cDNA from rare minnow, degenerate primers were designed based on the multiple alignment of known fish PKR-like sequences including *Danio rerio* (Accession No. AJ852024), *Carassius auratus* (Accession No. AY293929), *Salmo salar* (Accession No. DQ182560), *Tetraodon nigroviridis* (Accession No. AJ544919). PCR was performed with primers F106a and R110a (Table 1) using the cDNA generated from liver. The PCR program was: 1 cycle of 94℃/4 min, 35 cycles of 94℃/30 s, 50℃/30 s, 72℃/45 s; 1 cycle of 72℃/5 min. The PCR product was ligated into the pMD18-T easy vector, transformed into the competent *E. coli* TOP10 cells, and plated on the LB-agar petri-dish. Positive clones containing inserts at the expected size were screened by colony PCR and plasmid prepared using an Axyprep™ plasmid miniprep kit (Axygen Biosciences). Three plasmid DNAs were sequenced by a commercial company (Shanghai Invitrogen Biotechnology Co., Ltd, China).

Rapid amplification of cDNA ends (RACE)

Using the BD SMART^{TM} RACE cDNA amplification kit (BD Biosciences Clontech), first strand cDNA synthesis and RACE were performed on liver-derived RNA. To obtain the 3′ unknown region, primer pairs, F141/adaptor primer UPM and F142/adaptor primer NUP (Table 1), were used for the primary PCR and the nested PCR respectively. The amplified PCR products were cloned and sequenced as described above. Similarly, the 5′ end of GrTLR3 was obtained by nested PCR using primer pairs R163/UPM and R164/NUP (Table 1). The full length cDNA sequences were confirmed by sequencing the PCR product amplified by primers F164a and R164b (Table 1) within the predicted 5′ and 3′ untranslated regions respectively.

Table 1 Primers used in the study

Primer	Sequence (5′→3′)
GrPKR	
F106a(forward)	TRYYTVTWYATYCAGATGGAG
R110a(reverse)	GACAGCATYTTMYYRATGARTTTAT
F141(forward)	CAAACAGAGAAACACAGAGGAAATC
F142(forward)	TAGAGAGGTCAAAGAAAAGAGGAACAC
R163(reverse)	ATATGTTATCAGGCTTCAGGTCTC
R164(reverse)	GTGGATGAGATTGTTTGAATGGATGT
F164a(forward)	TGCAAAATGTCTGACGAAACTGA
R164b(reverse)	ACCGAAATGAAACTTGTGAAATACT
F126(forward)	GGCGGCTCAGACCAATCA
R127(reverse)	CAAACCATACGAGTCCAAGGG
β-Actin	
F86(forward)	GATGATGAAATTGCCGCACTG
R87(reverse)	ACCAACCATGACACCCTGATGT
Universal adaptor primer	
UPM	Long:CTAATACGACTCACTATAGGGCAAGCAGTGGTATCAACGCAGAGT Short:CTAATACGACTCACTATAGGGC
NUP	AAGCAGTGGTATCAACGCAGAGT
5′-RACE adaptor	
OligodG	AAGCAGTGGTAACAACGCGAGTACGCGGG
3′-RACE primer	
3′-CDS	AAGCAGTGGTATCAACGCAGAGTAC$(T)_{30}$VN

Note: Y=C/T; M=A/C; R=A/G; W=A/T; V=A/G/C; N=A/G/C/T

Sequence analysis and phylogeny

Sequence homology was obtained using BLAST program (http://www.ncbi.nlm.nih.gov/blast). The deduced amino acid sequences were analyzed with the Expert Protein Analysis

System (http://www.expasy.org/) and the protein domain features were predicted by the Simple Modular Architecture Research Tool (SMART) (http://smart.embl-heidelberg.de/) [22]. Intra-domain features were predicted by a scan of the sequence against the PROSITE database (http://us.expasy.org/tools/scanprosite) [23]. Multiple sequence alignments were created using the ClustalW2 (http://www.ebi.ac.uk/Tools/clustalw2/index.html) and The sequence manipulation suite (http://www.bioinformatics.org/sms/) programs. Phylogenetic and molecular evolutionary analyses were conducted using MEGA version 3.1 [24] and optimized manually.

Quantification of gene expression

Expression of GrPKZ mRNA was assessed in different tissues and infection states using quantitative real-time RT-PCR (qRT-PCR) in an ABI Prism 7000 Sequence Detection System (Applied Biosystems). To check the invariant expressions of the putative house keeping gene β-actin under different situations, the expression of β-actin in 1μg of total RNAs was tested in healthy fish and fish infected by viruses and bacteria. β-Actin was utilized as an internal control for cDNA normalization. The gene specific primers for β-actin were forward primer F86 and reverse primer R87 (Table 1). GrPKZ gene expression was detected using forward primer F126 and reverse primer R127 (Table 1). The qRT-PCR mixture consisted of 1 μl of cDNA sample, 8 μl nuclease-free water, 10 μl of SYBR Green PCR master mix (Toyobo), and 0.5μl of gene specific primers (5 μM). The PCR cycling conditions were: 1 cycle of 95℃/2 min, 40 cycles of 95℃/25 s, 60℃/30 s, and 72℃/1 min, followed by dissociation curve analysis to verify the amplification of a single product. The threshold cycle (CT) value was determined using the manual setting on the ABI Sequence Detection System and exported into a Microsoft Excel Sheet for subsequent data analysis where the PCR efficiency and relative expression ratio of target gene in treated groups versus that in control were calculated by Pfaffl equation [25]. The expression data obtained from three independent biological replicates were subjected to one-way analysis of variance (one-way ANOVA) followed by an unpaired, two-tailed *t*-test. $P<0.05$ was considered statistically significant.

Nucleotide sequence deposition

The GrPKZ cDNA sequence was submitted to GenBank and has been designated accession number EF661570.

Results

Characterization of GrPKZ mRNA and the encoded protein

Initial amplification of GrPKZ gene sequences was conducted using liver cDNA as template and degenerate primers F106a and R110a. Of multiple amplified segments, a 506 bp product (trimmed vector sequence) was shown by blastx analysis to have significant homology (*E*-value = 4e–84) to the C-terminal region of *C. auratus* interferon-inducible and double-stranded-dependent eIF-2kinase gene (namely PKZ) (accession No. AAP49830). Based on the sequence of the amplified 506 bp segment, primers F141 and F142 were subsequently designed and used with adaptor primers UPM and NUP to amplify the downstream sequence. Alignment of sequences from all homologous clones yielded a 1034 bp consensus sequence. The clone included a poly(A) tail, suggesting that the expressed sequence tag (EST) represented the 3′ region of the PKZ gene.

The 5′ region of the gene was amplified by using 5′-RACE. A product of 1074 bp from nested PCR perfectly overlapped the initial 1034 bp consensus sequence. To confirm that the resulting 1946 bp consensus sequence represented a single mRNA expressed in the liver of adult *G. rarus*, primers were designed within the 5′ and 3′ UTRs of the sequence. Using these primers, a single 1740 bp product was amplified from liver cDNA. Sequence analysis of the product confirmed that the full length transcript in *G. rarus* comprises 1946 nucleotides including the 5′ untranslated region, coding sequence, 3′ untranslated region and poly(A) tail (Fig. 1). The longest open reading frame commences at nucleotide 7 and terminates at nucleotide 1518, encoding a putative 503 amino acid polypeptide. Its 3′ UTR contains a polyadenylation signal AATAAA at nucleotide positions 1752-1757 by polyadq software (http://rulai.cshl.edu/tools/polyadq/polyadq_form.html) [26], and two putative ATTTA instability sequences, a motif possibly involved in rapid message degradation [27]. Blastp analysis indicated that the translated sequence has significant homology to previously characterized PKR-like proteins of eukaryotic origin.

Analysis of the deduced amino acid sequence of GrPKZ

The 503 amino acid polypeptide encoded in the amplified gene has a calculated molecular weight of 57,355 Da and isoelectric point of 5.83. Blastp search showed that this gene was most similar to crucian carp PKR-like protein (*E*-value = 0.0), indicating that it might be a rare minnow homologue of PKR-like (named GrPKZ). GrPKZ has three main structural domains, two Zalpha domains in N-terminal regulatory region and one catalytic domain (S_TKc) in C-terminal (Fig. 1). We selected all the representative PKR homologous proteins from fish and one from human (minimum *E*-value) to compare their conserved domains and other sequence features (Fig. 2). All the sequences contain a conserved S_TKc domain. Fish PKZs contain two Zalpha domains and the protein sequences consist of 503-513 amino acids (aa). Fish PKRs hold three conserved double-stranded RNA binding motifs (DSRM) instead of Zalpha domains and are composed of 667-682 aa. Human PKR has two DSRM and 551 aa. Other parameters, namely molecular weight and isoelectric point, were also distinguishing. In zebrafish, both PKZ and PKR are identified. To study the molecular evolutionary aspects, all the representative vertebrate PKR homologous protein sequences, including all known fish PKR homologous protein sequences, human PKR standing for mammal and African clawed frog PKR representing amphibian, were used to construct a phylogenetic tree (Fig. 3). These homologue proteins can be divided into two groups, PKZ and PKR.

Expression analysis of GrPKZ mRNA in different tissues

RNA was extracted from gill, heart, intestine, kidney, liver, muscle and spleen at 36 h post-injection and the levels of GrPKZ transcripts were assessed using qRT-PCR. The expression of β-actin gene in healthy and infected animals was invariant (Fig. 4A-C).

```
   1       M  S  D  E  T  E  M  E  R  K  I  L  D  F  L  G  R  N
   1 TGCAAAATGTCTGACGAAACTGAAATGGAGAGGAAGATCTTAGACTTCTTGGGAAGAAAT
  19 G  K  S  K  A  L  I  I  S  K  E  F  G  L  N  R  S  T  V  N
  61 GGCAAAAGCAAAGCTTTGATCATTTCTAAAGAATTCGGACTGAACAGATCCACCGTGAAC
  39 R  H  L  Y  K  L  Q  K  S  K  Q  V  F  G  T  N  E  T  P  P
 121 AGACATTTGTATAAACTACAGAAATCAAAACAAGTGTTCGGAACTAATGAAACGCCTCCT
  59 V  W  D  L  M  E  K  R  N  E  I  K  P  E  E  R  S  P  T  T
 181 GTTTGGGATCTAATGGAGAAGAGGAATGAGATCAAACCAGAGGAAAGGTCCCCGACGACA
  79 R  D  T  C  E  E  K  R  V  R  D  L  L  R  S  G  G  L  K  A
 241 AGAGACACATGCGAAGAGAAACGTGTGAGGGATCTGCTGAGGTCAGGAGGTCTAAAAGCC
  99 H  Q  I  A  T  E  L  G  Q  P  T  K  P  I  K  K  Q  L  Y  S
 301 CATCAGATCGCCACAGAACTGGGACAGCCAACAAAACCCATAAAGAAACAGCTGTACAGT
 119 L  E  Q  E  G  K  V  Q  K  C  A  K  T  S  I  W  T  L  N  E
 361 CTGGAGCAGGAGGGAAAAGTGCAGAAATGTGCCAAAACCAGCATATGGACACTGAATGAA
 139 E  E  S  Y  E  E  S  D  H  R  L  G  S  N  S  G  L  S  Q  C
 421 GAGGAGAGCTATGAAGAGAGTGATCATAGACTGGGCTCAAATTCAGGGTTGTCCCAGTGT
 159 F  D  V  I  S  V  L  G  K  G  G  F  G  W  V  Y  K  V  K  H
 481 TTTGATGTGATCTCAGTGCTTGGTAAAGGAGGCTTCGGTTGGGTTTATAAAGTAAAGCAT
 179 K  F  D  G  K  I  Y  A  V  K  K  V  V  L  T  P  E  A  D  S
 541 AAATTTGATGGCAAGATCTACGCTGTGAAGAAAGTCGTCTTAACGCCGGAAGCTGATTCC
 199 E  V  K  A  L  A  R  L  D  H  P  H  I  V  R  Y  I  T  C  W
 601 GAGGTGAAGGCATTGGCCAGACTAGATCACCCACACATAGTGCGCTACATTACATGTTGG
 219 P  D  S  E  N  C  T  S  T  Q  E  R  N  K  V  S  K  T  S  G
 661 CCAGATTCTGAGAACTGCACATCAACGCAAGAAAGAAACAAAGTGTCCAAAACATCAGGC
 239 S  S  S  D  V  V  T  F  E  R  S  G  C  E  E  N  D  D  D  A
 721 TCTTCATCAGATGTGGTGACCTTTGAGAGATCCGGCTGTGAGGAGAATGATGATGACGCC
 259 D  D  D  D  D  E  D  D  D  V  S  D  V  T  S  G  M  A  S  L
 781 GATGATGATGATGATGAGGATGATGATGTCAGTGATGTCACATCAGGAATGGCAAGTCTG
 279 G  V  T  A  E  S  A  S  A  A  G  P  S  G  N  L  D  S  L  N
 841 GGTGTGACCGCAGAATCAGCATCTGCCGCTGGGCCCTCAGGAAACCTGGACTCTTTAAAT
 299 H  S  K  M  Y  L  L  I  Q  M  E  F  C  E  G  G  T  L  T  T
 901 CACAGCAAGATGTATTTGTTGATTCAGATGGAGTTCTGTGAGGGAGGAACACTGACTACG
 319 W  I  K  D  R  N  L  R  N  K  Q  R  N  T  E  E  I  H  K  V
 961 TGGATAAAGGACAGAAATCTGAGAAACAAACAGAGAAACACAGAGGAAATCCATAAAGTA
 339 F  H  E  I  I  I  G  V  E  Y  I  H  S  N  N  L  I  H  R  D
1021 TTTCATGAGATTATCATTGGAGTGGAATACATCCATTCAAACAATCTCATCCACCGAGAC
 359 L  K  P  D  N  I  L  F  G  T  D  G  K  V  K  I  G  D  F  G
1081 CTGAAGCCTGATAACATATTGTTTGGCACCGATGGCAAAGTGAAGATCGGAGACTTTGGG
 379 L  V  A  A  Q  T  N  H  S  G  D  P  I  E  R  S  K  K  R  G
1141 TTGGTGGCGGCTCAGACCAATCACAGCGGTGACCCTATAGAGAGGTCAAAGAAAAGAGGA
 399 T  L  Q  Y  M  S  P  E  Q  E  N  K  R  N  Y  N  E  K  T  D
1201 ACACTGCAGTATATGAGTCCTGAGCAGGAAAATAAGAGGAATTATAATGAAAAAACAGAC
 419 I  F  P  L  G  L  V  W  F  E  M  L  W  K  L  S  T  G  M  E
1261 ATTTTCCCCCTTGGACTCGTATGGTTTGAGATGCTCTGGAAATTGTCCACTGGTATGGAG
```

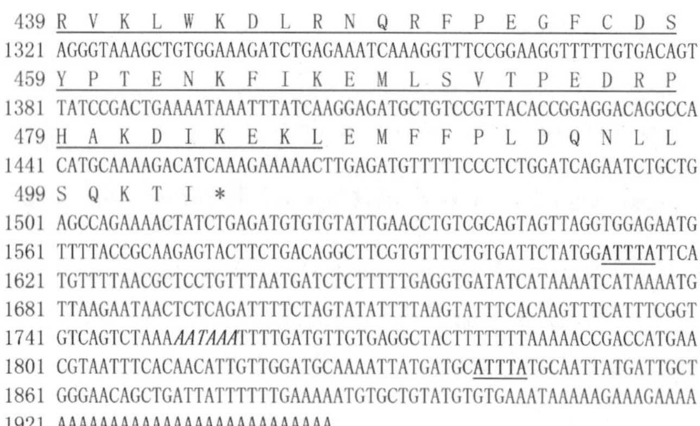

439 R V K L W K D L R N Q R F P E G F C D S
1321 AGGGTAAAGCTGTGGAAAGATCTGAGAAATCAAAGGTTTCCGGAAGGTTTTTGTGACAGT
459 Y P T E N K F I K E M L S V T P E D R P
1381 TATCCGACTGAAAATAAATTTATCAAGGAGATGCTGTCCGTTACACCGGAGGACAGGCCA
479 H A K D I K E K L E M F F P L D Q N L L
1441 CATGCAAAAGACATCAAAGAAAAACTTGAGATGTTTTTCCCTCTGGATCAGAATCTGCTG
499 S Q K T I *
1501 AGCCAGAAAACTATCTGAGATGTGTGTATTGAACCTGTCGCAGTAGTTAGGTGGAGAATG
1561 TTTTACCGCAAGAGTACTTCTGACAGGCTTCGTGTTTCTGTGATTCTATGGATTTATTCA
1621 TGTTTTAACGCTCCTGTTTAATGATCTCTTTTTGAGGTGATATCATAAAATCATAAAATG
1681 TTAAGAATAACTCTCAGATTTTCTAGTATATTTTAAGTATTTCACAAGTTTCATTTCGGT
1741 GTCAGTCTAAA***AATAAA***TTTTGATGTTGTGAGGCTACTTTTTTTAAAAACCGACCATGAA
1801 CGTAATTTCACAACATTGTTGGATGCAAAATTATGATGCATTTATGCAATTATGATTGCT
1861 GGGAACAGCTGATTATTTTTTGAAAAATGTGCTGTATGTGTGAAATAAAAAGAAAGAAAA
1921 AAAAAAAAAAAAAAAAAAAAAAAAAA

Figure 1 Nucleotide sequence and deduced amino acid sequence of GrPKZ cDNA. The start codon (ATG) is boxed and the stopcodon (TGA) is indicated with an asterisk. The polyadenylation signal sequence (AATAAA) is in italics and bold, the motifs associated with mRNA instability (ATTTA) are underlined and in bold. In the deduced amino acid sequence, both putative Zalpha domains (Z-DNA-binding domain) are shaded and S_TKc domain (Serine/Threonine protein kinases, catalytic domain) is underlined. S_TKc active-site signature sequence is marked by under dots

GrPKZ mRNA was detected in all fish tissues tested at low levels, and was significantly up-regulated by viral or bacterial infection ($P < 0.05$) (Fig. 4D). In the virally infected group, the PKZ expression in liver was most up-regulated 10.3-fold. In the bacterial infection group, the PKZ expression in spleen was most up-regulated 10.4-fold. For GCRV disease, liver is one of main affected organs, and GrPKZ was most up-regulated by viral infection in liver. Therefore, liver tissue was selected to investigate the temporary expression of GrPKZ gene after immune stimulation.

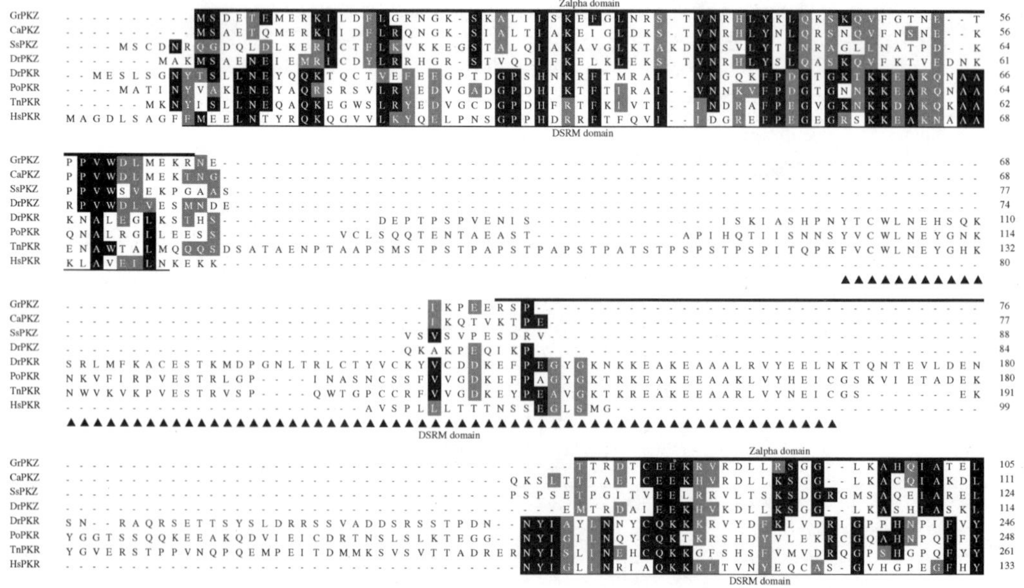

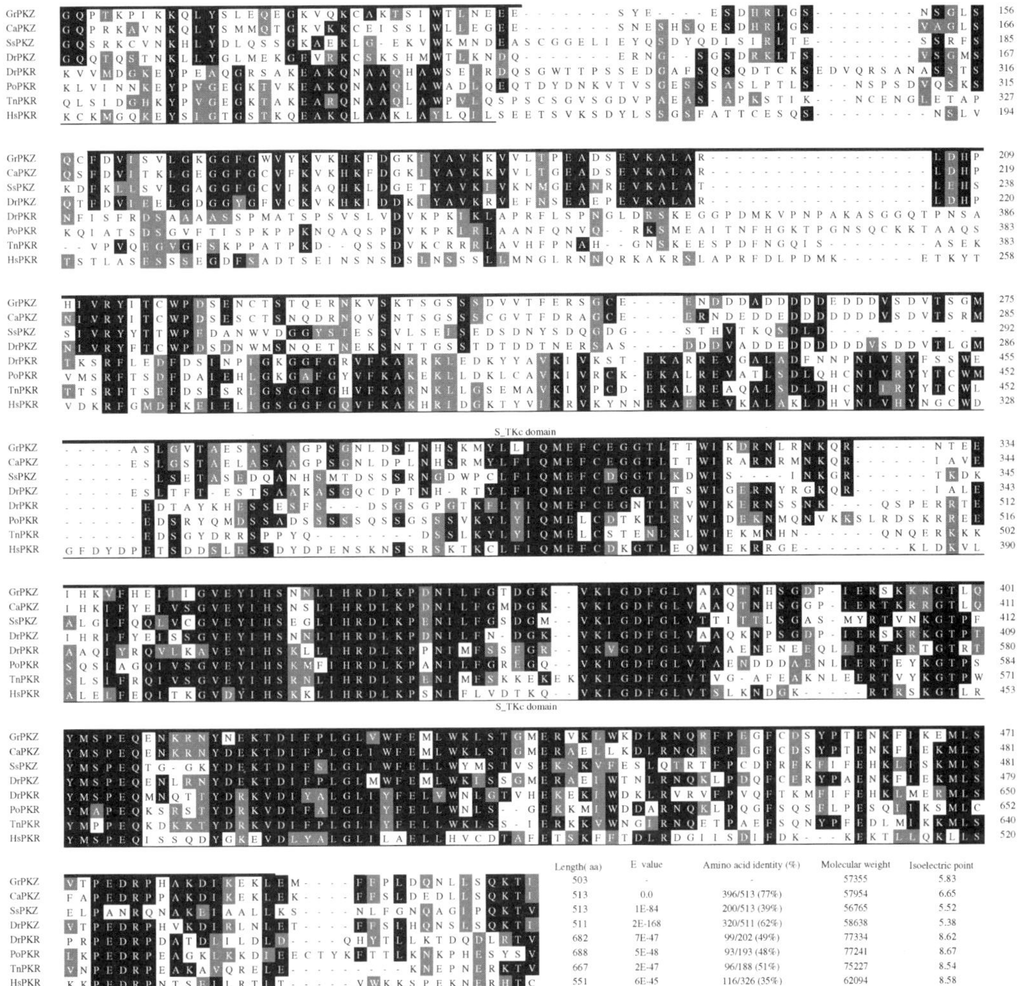

	Length(aa)	E value	Amino acid identity (%)	Molecular weight	Isoelectric point
GrPKZ	503	-	-	57355	5.83
CaPKZ	513	0.0	396/513 (77%)	57954	6.65
SsPKZ	513	1E-84	200/513 (39%)	56765	5.52
DrPKZ	511	2E-168	320/511 (62%)	58638	5.38
DrPKR	682	7E-47	99/202 (49%)	77334	8.62
PoPKR	688	5E-48	93/193 (48%)	77241	8.67
TnPKR	667	2E-47	96/188 (51%)	75227	8.54
HsPKR	551	6E-45	116/326 (35%)	62094	8.58

Figure 2 Multiple alignment of GrPKZ with all known PKR homologous proteins from fish and human PKR by ClustalW2 and the Sequence Manipulation Suite. Amino acid sequences are deduced from cDNA. Residues highlighted in black are identical and those in dark gray are similar. Percentage of sequences must agree for identity or similarity coloring to be added 50%. The right numbers represent the amino acid position in the corresponding species. *E*-value and amino acid identity were determined by blastp. Molecular weight and isoelectric point were calculated by the Expert Protein Analysis System software. Domains in PKZ were marked with an overbar with their names. Domains in PKR were underlined with corresponding names. The second DSRM in the fish PKR were marked with solid triangles (▲). The aligned sequences are as follows: GrPKZ, *G. rarus* PKZ (ABV26701); CaPKZ, *Carassius auratus* PKR-like protein (AAP49830); SsPKZ, *Salmo salar* dsRNA-activated Z-DNA binding protein kinase (ABA64562); DrPKZ, *Danio rerio* Z-DNA binding protein kinase (CAH68528); DrPKR, *D. rerio* double-stranded RNA-activated protein kinase (CAM07151); PoPKR, *Paralichthys olivaceus* dsRNA-dependent protein kinase (ABV21735); TnPKR, *Tetraodon nigroviridis* double-stranded RNA-activated protein kinase 1 (CAM07147); HsPKR, *Homo sapiens* eukaryotic translation initiation factor 2-alpha kinase 2 (NP_002750)

GrPKZ mRNA expression profile stimulated by viral or bacterial infection

To determine the effects of viral or bacterial infection on GrPKZ gene expression, three animals from each group were sacrificed at each time point (0, 12, 24, 36, 48, 72, 96, 120, 144, 168 h post-injection) and qRT-PCR assays were conducted to determine the levels of GrPKZ mRNAs in liver tissue. No significant differences were detected in control and blank groups at any of the time points ($P > 0.05$) (data not shown). In infected groups, the expressions were significantly up-regulated at 24 h post-injection, reached a peak at 36 h post-injection ($P < 0.05$), and then gradually decreased. In GCRV infection group, the GrPKZ mRNA expressions were significantly higher than those in the control group from 24 h until the fish became moribund ($P < 0.05$) (Fig. 5). In the bacterial infection group, the expression of GrPKZ recovered to the original level from 120 h post-injection ($P > 0.05$) (Fig. 5). The data clearly demonstrated that GrPKZ expression levels were stimulated in infected fish, for both viral and bacterial infection ($P < 0.05$). However, the levels and time-course of GrPKZ expression were somewhat different under the two types of infection.

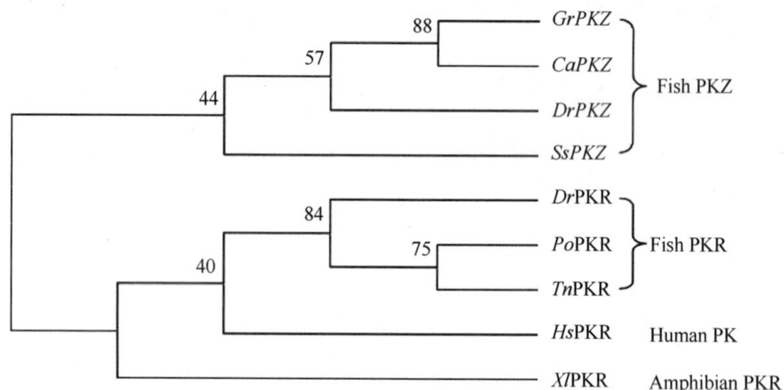

Figure 3 Phylogenetic relationships of PKZ and PKR. Phylogenetic tree was obtained from a ClustalW2 alignment and MEGA3.1 Neighbor-Joining of nine sequences. The bar indicated the distance. The accession number of XlPKR is AAR92030. Other abbreviations and accession numbers are the same as in Fig. 2

Discussion

Recently, cDNAs encoding PKZ have been isolated from crucian carp, zebrafish and Atlantic salmon [17,19,28]. Crucian carp PKZ is detected at a very low level of constitutive expression and can be induced by GCRV, poly(I:C) or interferon in CAB cells [28,29]. Zebrafish PKZ is transcribed constitutively at low levels and is highly induced after injection of poly(I:C). It also inhibits translation of proteins in transfected cells and plays a role in the host response to viruses [17]. Atlantic salmon PKZ expression is up-regulated by interferon in

Atlantic salmon TO cells and by poly(I:C) in head kidney. Recombinant PKZ is able to phosphorylate eIF2α *in vitro* and plays a role in host defense against virus infection [19]. These papers studied fish PKZ characterizations from different aspects.

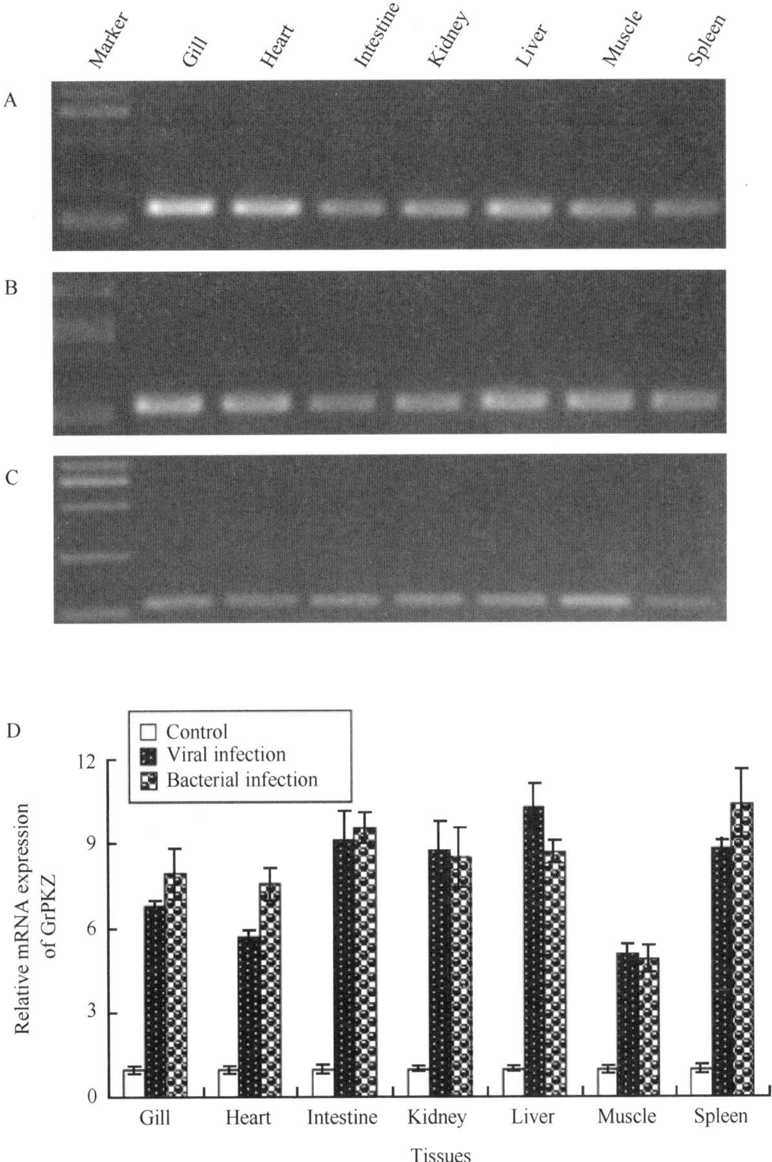

Figure 4 Tissue distribution of GrPKZ transcripts in rare minnow. A-C. The expressions of β-action in different tissues. PCR templates were from 1 μg of total RNA reverse transcription product, the cycles of PCR were 25. A. Healthy animal. B. Virus infected animals at 36 h post-injection. C. Bacteria infected animals at 36 h post-injection. D. Quantitative analysis of GrPKZ in different tissues at 36 h post-injection. Each bar represented the level of target gene mRNA relative to those in control group, expressed as the mean ±SE of triplicate from qRT-PCR assays for three different cDNA samples. Error bars indicate standard error

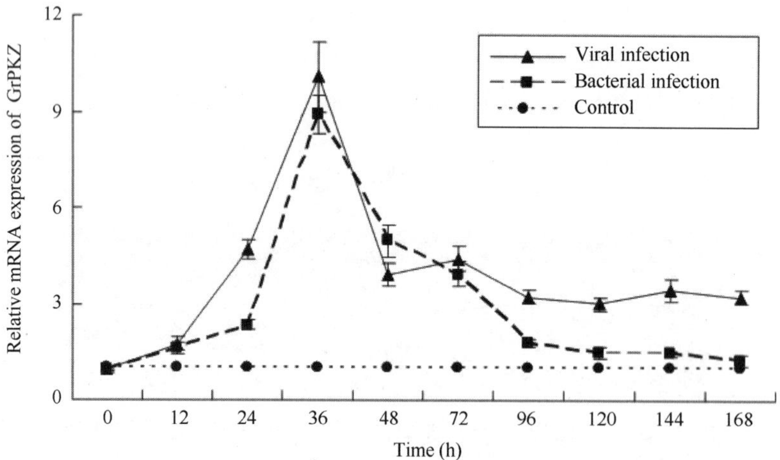

Figure 5 The expression profiles of GrPKZ mRNA in the liver by qRT-PCR. The controls were injected with PBS; the experimental fish were infected with GCRV or *A. hydrophila*. Error bars indicated standard error

In the present study, GrPKZ was remarkably similar to other fish PKZ (Fig. 2) with two Zalpha domains and an S_TKc domain. The protein sequence was also similar to fish and human PKR proteins with DSRM domains and an S_TKc domain (Fig. 2). Based on the phylogenetic analysis, PKZ is distictrt from PKR. Fish PKR firstly cluster with mammal and amphibian PKR, then cluster with fish PKZ (Fig. 3). The kinase domains of fish PKR genes are more closely related to those of fish PKZ than to the PKR kinase domains of other vertebrate species [30]. The similarity in structure inferred the analogy in functions. Both PKZ and PKR are present in zebrafish, and both of them can phosphorylate eIF2α in yeast [30]. In other fishes, either PKZ or PKR is reported. Why fish have both PKR and PKZ is not yet clear, but this may reflect early evolutionary solutions to produce an eIF2α-kinase sensitive to viral infection or to detect different pathogen associated molecule patterns.

GrPKZ expression was constitutively at low levels in the tissues tested, and induced after infection by viruses or bacteria. The liver is one of the main target organs for GCRV, which prompted us to study GrPKZ expression profiles in liver infected by viruses or bacteria. After infection with GCRV, GrPKZ expression was statistically up-regulated at 24 h post-injection ($P<0.05$), and lasted until the fish became moribund. The data imply that GrPKZ is induced by viruses and may participate in inhibition of viral protein translation. In mammals, LPS induces PKR expression by the TLR4 signaling pathway [9,12,31,32]. LPS is the main component of Gram-negative bacterial cell walls. After infection with *A. hydrophila*, GrPKZ transcription was induced at 24 h post-injection ($P<0.05$), and returned to control levels by 120 h post-injection. This feature was first reported in fish PKZ and is similar to mammalian PKR.

PKR's role as a critical antiviral effector protein leading to inhibition of translation and hence shut-down of infected cells make it a target for viral evasion strategies. The reovirus σ3 protein sequesters dsRNA and thus prevents it from activating PKR or other dsRNA receptors [33]. GCRV is a reovirus and whether the σ3 relatedprotein in GCRV has a similar function needs further research.

Acknowledgements The authors thank Wei Hu, Ming Li, Feng Xiong, Jun Dai, Bing Tang, Shangping Chen and other laboratory members for technical assistance and helpful discussion. This work was supported by grants from the National Natural Science Foundation of China (30740009, 30540084), from 973 National Basic Research Program of China (2006CB102100), from the Chinese Academy of Sciences (KSCX2-YW-N-021), from Northwest A & F University in China (01140309, 05ZR096 and 01140508), from China Postdoctoral Science Foundation (20070410298) and from the Institute of Hydrobiology, CAS (2007FB09).

References

[1] Mundschau LJ, Faller DV. Platelet-derived growth factor signal transduction through the interferon-inducible kinase PKR. J Biol Chem 1995; 270: 3100-6.

[2] Kronfeld-Kinar Y, Vilchik S, Hyman T, Leibkowicz F, Salzberg S. Involvement of PKR in the regulation of myogenesis. Cell Growth Differ 1999; 10: 201-12.

[3] Patel CV, Handy I, Goldsmith T, Patel RC. PACT, a stress-modulated cellular activator of interferon-induced double-stranded rna-activated protein kinase, PKR. J Biol Chem 2000; 275: 37993-8.

[4] Hardenberg G, Planelles L, Schwarte CM, van Bostelen L, Le Huong T, Hahne M, *et al.* Specific TLR ligands regulate APRIL secretion by dendritic cells in a PKR-dependent manner. Eur J Immunol 2007; 37: 2900-11.

[5] Zhang P, Samuel CE. Protein kinase PKR plays a stimulus- and virus-dependent role in apoptotic death and virus multiplication in human cells. J Virol 2007; 81: 8192-200.

[6] Wu S, Kumar KU, Kaufman RJ. Identification and requirement of three ribosome binding domains in dsRNA-dependent protein kinase (PKR). Biochemistry 1998; 37: 13816-26.

[7] García MA, Gil J, Ventoso I, Guerra S, Domingo E, Rivas C, *et al.* Impact of protein kinase PKR in cell biology: from antiviral to antiproliferative action. Microbiol Mol Biol Rev 2006; 70: 1032-60.

[8] Wong AH, Tam NW, Yang YL, Cuddihy AR, Li S, Kirchhoff S, *et al.* Physical association between STAT1 and the interferon-inducible protein kinase PKR and implications for interferon and double-stranded RNA signaling pathways. EMBO J 1997; 16: 1291-304.

[9] Williams BRG. Signal integration via PKR. Sci STKE 2001; 89: re2.

[10] Unterholzner L, Bowie AG. The interplay between viruses and innate immune signaling: recent insights and therapeutic opportunities. Biochem Pharmacol 2008; 75: 589-602.

[11] Balachandran S, Roberts PC, Brown LE, Truong H, Pattnaik AK, Archer DR, *et al.* Essential role for the dsRNA dependent protein kinase PKR in innate immunity to viral infection. Immunity 2000; 13: 129-41.

[12] Cabanski M, Steinmüller M, Marsh LM, Surdziel E, Seeger W, Lohmeyer J. PKR regulates TLR2/TLR4-dependent signaling in murine alveolar macrophages. Am J Respir Cell Mol Biol 2008; 38: 26-31.

[13] Meurs E, Chong K, Galabru J, Thomas NSB, Kerr IM, Williams BRG, *et al.* Molecular and characterization of the human double-stranded RNA-activated protein kinase induced by interferon. Cell 1990; 62: 379-90.

[14] *Caenorhabditis elegans* Sequencing Consortium. Genome sequence of the nematode *Caenorhabditis elegans*: a platform for investigating biology. Science 1998; 282: 2012-8.

[15] Adams MD, Celniker SE, Holt RA, Evans CA, Gocayne JD, Amanatides PG, *et al.* The genome sequence of

Drosophila melanogaster. Science 2000; 287: 2185-95.

[16] Holt RA, Subramanian GM, Halpern A, Sutton GG, Charlab R, Nusskern DR, *et al.* The genome sequence of the malaria mosquito *Anopheles gambiae*. Science 2002; 298: 129-49.

[17] Rothenburg S, Deigendesch N, Dittmar K, Koch-Nolte F, Haag F, Lowenhaupt K, *et al.* A PKR-like eukaryotic initiation factor 2α kinase from zebrafish contains Z-DNA binding domains instead of dsRNA binding domains. Proc. Natl Acad Sci USA 2005; 102: 1602-7.

[18] Rich A, Zhang S. Timeline: Z-DNA: the long road to biological function. Nat Rev Genet 2003; 4: 566-72.

[19] Bergan V, Jagus R, Lauksund S, Kileng Ø, Robertsen B. The Atlantic salmon Z-DNA binding protein kinase phosphorylates translation initiation factor 2 alpha and constitutes a unique orthologue to the mammalian dsRNA-activated protein kinase R. FEBS J 2008; 275: 184-97.

[20] Wang T, Liu P, Chen H. Preliminary study on the susceptibility of *Gobiocypris rarus* to hemorrhagic virus of grass carp (GCHV). Act Hydrobiol Sin 1994; 18: 144-9[in Chinese, English abstract].

[21] Austin B, Austin DA. Bacterial fish pathogens: disease in farmed and wild fish. New York. New York, Halsted Press: 1987.

[22] Letunic I, Copley RR, Pils B, Pinkert S, Schultz J, Bork P. SMART 5: domains in the context of genomes and networks. Nucleic Acids Res 2006; 34: D257-60.

[23] Hulo N, Bairoch A, Bulliard V, Cerutti L, Cuche B, De Castro E, *et al.* The 20 years of PROSITE. Nucleic Acids Res 2008; 36: D245-9.

[24] Kumar S, Tamura K, Nei M. MEGA3: Integrated software for molecular evolutionary genetics analysis and sequence alignment. Brief Bioinformatics 2004; 5: 150-63.

[25] Pfaffl MW. A new mathematical model for relative quantification in real-time RT-PCR. Nucleic Acids Res 2001; 29: e45.

[26] Tabaska JE, Zhang MQ. Detection of polyadenylation signals in human DNA sequences. Gene 1999; 231: 77-86.

[27] Shaw G, Kamen RA. Conserved AU sequence from the 3′ untranslated region of GM-CSF mRNA mediates selective mRNA degradation. Cell 1986; 46: 659-67.

[28] Hu C, Zhang Y, Huang G, Zhang Q, Gui J. Molecular cloning and characterization of a fish PKR-like gene from cultured CAB cells induced by UV-inactivated virus. Fish Shellfish Immunol 2004; 17: 353-66.

[29] Zhang YB, Jiang J, Chen YD, Zhu R, Shi Y, Zhang QY, *et al.* The innate immune response to grass carp hemorrhagic virus (GCHV) in cultured *Carassius auratus* blastulae (CAB) cells. Dev Comp Immunol 2007; 31: 232-43.

[30] Rothenburg S, Deigendesch N, Dey M, Dever TE, Tazi L. Double-stranded RNA-activated protein kinase PKR of fishes and amphibians: varying number of double-stranded RNA binding domains and lineage specific duplications. BMC Biology 2008; 6: 12 doi: 10.1186/1741-7007-6-12 [online].

[31] Gusella GL, Musso T, Rottschafer SE, Pulkki K, Varesio L. Potential requirement of a functional double-stranded RNA-dependent protein kinase (PKR) for the tumoricidal activation of macrophages by lipopolysaccharide or IFN-alpha beta, but not IFN-gamma. J Immunol 1995; 154: 345-54.

[32] Horng T, Barton GM, Medzhitov R. TIRAP: an adapter molecule in the Toll signaling pathway. Nat Immunol 2001; 2: 835-41.

[33] Langland JO, Cameron JM, Heck MC, Jancovich JK, Jacobs BL. Inhibition of PKR by RNA and DNA viruses. Virus Res 2006; 119: 100-10.

稀有鮈鲫 PKZ 基因的克隆、特征及表达分析

苏建国[1,2] 朱作言[1] 汪亚平[1]

1 中国科学院水生生物研究所，武汉 430072
2 西北农林科技大学，杨凌 712100

摘 要 双链 RNA 激活蛋白激酶(PKR)在干扰素介导的抗病毒反应中发挥着重要作用，同时也参与细胞内信号通路，包括细胞凋亡、增殖和转录途径。本研究中，一个 PKR 样基因在稀有鮈鲫中被克隆、鉴定。稀有鮈鲫 PKR 样基因(GrPKZ)cDNA 全长 1946 bp，编码 503 个氨基酸，分子量为 57.3 kDa，预测等电点为 5.83。氨基酸序列分析显示，成熟的多肽链包含两个 Zalpha 结构域和一个 S_TKc 结构域，这和鲫的 PKR 样基因氨基酸序列有 77%的相似度。实时定量 PCR 显示 GrPKZ 的 mRNA 表达量在健康鱼体内的鳃、心脏、肠、肾脏、肝脏、肌肉和脾脏组织中较低，在病毒和细菌感染的鱼体内上调表达。草鱼呼肠孤病毒感染后 GrPKZ 在 24 h 后上调表达直至鱼死亡，嗜水气单孢菌感染后，24 h 后诱导 GrPKZ 转录，120 h 后恢复正常水平。这些数据表明 GrPKZ 参与了稀有鮈鲫抗病毒应答反应和细菌感染的 Toll 样受体 4 的信号通路途径。

Toll-Like Receptor 3 Regulates Mx Expression in Rare Minnow *Gobiocypris rarus* After Viral Infection

Jianguo Su[1,2] Zuoyan Zhu[1] Yaping Wang[1] Jun Zou[3] Wei Hu[1]

1 State Key Laboratory of Freshwater Ecology and Biotechnology, Institute of Hydrobiology, Chinese Academy of Sciences, Wuhan 430072, China

2 Department of Aquaculture, College of Animal Science and Technology, Northwest A&F University, Yangling 712100, China

3 School of Biological Sciences, University of Aberdeen, Aberdeen AB24 2TZ, UK

Abstract Toll-like receptor 3 (TLR3) plays a key role in activating immune responses during viral infection. To study the genes involved in the regulatory function of TLR3 in the rare minnow *Gobiocypris rarus* after viral infection, a full-length cDNA of TLR3 (GrTLR3) with a splice variant (GrTLR3s) was identified by homologous cloning and RACE techniques. The antiviral effector molecule Mx gene was cloned and partially sequenced. The mRNA expression levels of GrTLR3, GrTLR3s and Mx were studied in different tissues before and after virus infection by real-time quantitative RT-PCR. The transcripts of all three genes in liver were significantly increased following GCRV infection ($P<0.05$). The mRNA levels in liver were upregulated at 24 h post-injection for GrTLR3 and GrTLR3s, and at 12 h for Mx. The upregulated expression levels were several folds for GrTLR3s, tens of folds for GrTLR3 and hundreds of folds for Mx. By semi quantitative RT-PCR, GrTLR3 and Mx expressed at all the developmental stages, whereas GrTLR3s could only be detected at later developmental stages. Using RNAi and transgenic techniques, GrTLR3 mediated Mx expression but GrTLR3s did not. The time-dependent upregulation of receptor and effector, and the Mx over-expression dependent on TLR3 indicated that GrTLR3 regulated Mx expression in viral infection through a configuration change in rare minnow, and its splice variant didnhot contribute to the process.

Keywords *Gobiocypris rarus*; TLR3; Mx; transgenic fish; RNAi

Introduction

Toll-like receptors (TLRs) are at the front line in the fight against invading microorganisms (Mak and Yeh 2002). They are evolutionarily conserved components and act as sentinels of the innate immune system, sensing a variety of ligands from lipopolysaccharide to flagellin to dsRNA by leucine-rich-repeat motifs (LRRs) in their extracellular domain (West *et al.* 2006; Xu

文章发表于 *Immunogenetics*, 2008, 60(3-4): 195-205

et al. 2000; Ishii *et al.* 2007). TLRs are type I transmembrane pattern recognition receptors (PRRs) that detect invading pathogens by binding conserved, microbially derived molecules (pathogen-associated molecular patterns, PAMPs) that are not found in higher eukaryotes, and that signal and induce expression of multiple host defense genes including proinflammatory cytokines and chemokines (Rudd *et al.* 2005). TLRs are the focus of considerable attention as potential regulators and controllers of the immune response through their ability to recognize PAMPs, and are arousing interest as potential targets for the development of new therapies for multiple diseases (Zuany-Amorim *et al.* 2002).

The *Drosophila* genome encodes nine Toll-related receptors, most of which appear to carry out developmental rather than immune functions. One exception may be Toll-9, which shares structural and functional similarities with mammalian TLRs. Identification of the stimuli to which it responds, and characterization of its function in *Drosophila*, could shed light on the ancestral function of TLRs. (Bilak *et al.* 2003). The ten human TLRs, which are essentially conserved in the mouse genome, appear to have evolved specifically to recognize conserved PAMPs (Aderem and Ulevitch 2000; Medzhitov 2001).

Toll-like receptor 3 (TLR3), a member of a Toll-like receptor family, is an indispensable recognition receptor for host defense against viral infection (Nishiya *et al.* 2005) and the first identified antiviral TLR (Schroder and Bowie 2005). TLR3, together with TLR7 and TLR8, constitutes a powerful system to detect RNA viruses. TLR3 has been shown to bind viral dsRNA whereas TLR7 and TLR8 are receptors for viral ssRNA (Tissari *et al.* 2005). dsRNAs derived from virus replication, associated with necrotic cells, generated by *in vitro* transcription, or mimicked by polyinosinic-polycytidylic acid (poly I/C) also induce immune activation via TLR3 (Kariko *et al.* 2004).

In fish, the zebrafish genome contains at least 17 expressed TLR genes (Meijer *et al.* 2004). A prediction of the TLR family in the genome of the pufferfish *Fugu rubripes* was reported, indicating that this teleost contains counterparts of most human TLRs, with the exception of the LPS-specific TLR4 gene (Oshiumi *et al.* 2003a). There is just one TLR3 orthologue in zebrafish and pufferfish genomes. Up to now, there have been six TLR3 sequences submitted to GenBank (August, 2007) in fishes, including zebrafish, pufferfish, goldfish, rainbow trout, channel catfish, and olive flounder.

Mx protein is an antiviral effector molecule and is induced by type I interferons (IFNα and β) in mice and human. It inhibits the replication of orthomyxoviruses and certain other ssRNA viruses (Haller *et al.* 1998). As in mammals, type I IFN is thought to be the primary inducer of Mx protein in fish (Nygaard *et al.* 2000). Piscine Mx is induced by viral infection and ploy I/C, and serves as a marker for *in vivo* production of IFN and virus infection (Jensen *et al.* 2002). Fish Mx genes have been cloned from many species, such as zebrafish, rainbow trout, Atlantic salmon, Atlantic halibut, grass carp, and goldfish.

Grass carp (*Ctenopharyngodon idellus*) is a crucial aquaculture species in China, but tremendous economical loss is often caused by Grass Carp Reovirus (GCRV), a dsRNA virus. Understanding the immune defense mechanisms of grass carp may contribute to the development of management strategies for disease control and long-term sustainability of grass carp farming. However, the bigger size and longer reproductive cycle make it difficult to study the antiviral breeding. The rare minnow *Gobiocypris rarus*, which is a small cyprinid species, has been recognized as a useful model. It's very sensitive to GCRV (Wang *et al.* 1994).

In the current study, we employed the rare minnow as a model fish to study the mechanism of GCRV disease. A full-length cDNA of the recognition receptor TLR3 was isolated and a splice variant (TLR3s) was unexpectedly identified. To investigate the functions of TLR3 and TLR3s, a partial sequence of the antiviral effector molecule Mx gene was identified. The TLR3, TLR3s and Mx mRNA expression profiles were examined. To investigate the regulatory function of TLR3(s), we hypo-expressed TLR3 and TLR3s by RNAi, over-expressed them by transgene, and studied the relationship between TLR3(s) and Mx.

Materials and Methods

Animals, immune challenge, and sample collection

Young adult rare minnows were selected from a laboratory breeding stock and acclimated for 1 week at the Institute of Hydrobiology, Chinese Academy of Sciences, Wuhan, China. Fish were kept in three aerated aquaria at 28℃. For challenge experiments with GCRV, the control group was injected with 10 μl PBS per gram body weight and the experimental group was injected with GCRV (2×10^8 PFU/ml in PBS) under the ventral fin. The blank group was not injected. Three individuals were randomly killed and tissues including gill, heart, intestine, kidney, liver, muscle, and spleen were collected at 0, 6, 12, 24, 36, 48, 72, 96, 120, and 144 h (moribund fish) post-injection respectively. For the mRNA expression studies in different developmental stages, embryos at the stage of zygote, dome, shield, 1-somite, prim-5, protruding mouth and adult fish were also taken. The tissue and embryo samples were immediately homogenized in TRIZOL® LS reagent (Invitrogen) and total RNAs were extracted according to the manufacturer's instruction. The isolated total RNAs were incubated with RNase-free DNase I (Roche) to remove contaminated genomic DNA. Reverse transcription was performed with SuperScript™ III Reverse Transcriptase (Invitrogen) and random hexamers.

Cloning of TLR3 cDNA

Degenerate primers for cloning TLR3 cDNA from rare minnow were designed based on the multiple alignment of known fish TLR3 sequences including *Danio rerio* (accession No.

NM_001013269), *Carassius auratus* (accession No. DQ291158), *Ictalurus punctatus* (accession No. AY741552), *Oncorhynchus mykiss* (accession No. AY883999), *Fugu rubripes* (accession No. AC156436). First round of PCR was performed with SCTF1 and SCTR4 primers followed by nested PCR using primers SCTF2 and SCTR3 (Table 1). The PCR product was processed as previously reported (Su *et al.* 2007). A 257-bp sequence was obtained and named as fragment 1. By means of a BLASTX search in the GenBank database, fragment 1 was seen to be significantly homologous to known TLR3 sequences.

The 5′ and 3′ unknown sequences of TLR3 were obtained using a BD SMART™ RACE cDNA Amplification Kit (BD Clontech) according to the manufacturer's instruction. Primers SGTF5 and adaptor primer UPM (Table 1) were used for the first round of PCR, and primers SGTF6 and adaptor primer NUP (Table 1) were used for the nested PCR to amplify the 3′ region. After gel electrophoresis, a strong band and one weak band were visualized and cloned as described above. Similarly, the 5′ end of TLR3 was obtained by nested PCR using primer pairs SGTR8/UPM and SGTR9/NUP (Table 1). A single band was generated and sequenced. The full-length cDNA sequences were confirmed by sequencing the PCR-transformed product amplified by primers SGTF68a and NUP (Table 1), and the template was the same as the 3′ RACE template. The positive colonies were screened with SGTF15 and SGTR16 (Table 1) as primers for TLR3 (477 bp) and for its splice variant (726 bp). The conventional sequence and its alternative splicing variant sequence were named GrTLR3 and GrTLR3s respectively.

Identification of a rare minnow Mx gene

To test the interferon pathway in GCRV infection, a 989-nucleotide partial sequence of the interferon indicator gene Mx was identified by homologous clone. The degenerate primers were designed according to the published Mx sequences in fishes. The forward SMFa and reverse SMRb primer sequences are listed in Table 1.

Table 1 Oligonucleotide primers used in the experiments

Primer name	Sequence (5′→3′)
GrTLR3	
SCTF1(forward)	CTCAGYAAYAAYAAYATYGCMAACAT
SCTF2(forward)	TKCAGCACAAYAAYTTDGC
SCTR3(reverse)	AYAGRATGCTCTCRCASGTGCA
SCTR4(reverse)	TTCCAGWAGAAYYGRATYCTCCA
SCTF5(forward)	CCTCAAGGATGCCACAAAACTCAC
SCTF6(forward)	TGATGCTTTGCGTGGCTTCTCTGA
SCTR8(reverse)	CAGCTGATCCAAGAGATTACCACGG
SCTR9(reverse)	GAGAAGCCACGCAAAGCATCAAGT
SCTF41(forward)	CACTTCACCAGGTAATGGAGGAC
SCTR16(reverse)	TGTTGGTAGAGGCTAATGCGGA

Continued

Primer name	Sequence (5′→3′)
GrTLR3	
SCTF15(forward)	TCGGGTCGATGAAGGATGAAAGT
SCTF68a(forward)	GCTGAACGTGGGAAATTAGATCG
SCTF68(forward)	ATCGGAATTCGCTGAACGTGGGAAATTAGATCG
SCTR80a(revere)	ATCGGGGCCCAAGCAGTGGTATCAACGCAGAGT
GrTLR3s	
SCTF17(forward)	GCAAACCACTGACCAGTTTTGG
SCTR17a(reverse)	GCATGATGTGCTTTGAATCTGC
Mx	
SMFa(forward)	TGACACGCTGTCCTCTGGTA
SMRb(reverse)	GKTTTCCTCCGTCTTWAWGG
SMF52a(forward)	CAACCGGACAAAGGACTCTGGG
SMR53a(reverse)	GATGTCTTGCTGGCCTCTGCAC
β-actin	
SAF40(forward)	ATCTGGCATCACACCTTCTACAAC
SAR39a(reverse)	GGTACGACCAGAGGCATACAGG
SAR39b(reverse)	TAACCCTCGTAGATGGGCACAGT
CMV promoter	
SCMVF70(forward)	ATCGAAGCTTTAGTTATTAATAGTAATCAATTACG
SCMVR71(reverse)	ATCGGAATTCCGATCTGACGGTTCACTA
SV40 poly A	
SVF81	ATCGGGGCCCAGCGGCCGCGACTCTAGATCAT
SVR82	ATCGCTCGAGGCAGTGAAAAAAATGCTTTATTTGTG
Universal adaptor primer	
UPM	Long:CTAATACGACTCACTATAGGGCAAGCAGTGGTATCAACGCAGAGT Short:CTAATACGACTCACTATAGGGC
NUP	AAGCAGTGGTATCAACGCAGAGT
3-′RACE primer	
3′-CDS	5′-AAGCAGTGGTATCAACGCAGAGTAC$(T)_{30}$ VN-3′

Y=C/T; M=A/C; K=G/T; D=A/T/G; R=A/G; S=G/C; W=A/T; V=A/G/C; N=A/G/C/T.

Sequence analyses

Sequence homology was analysed with the BLAST program (http://www.ncbi.nlm.nih.gov/blast); the deduced amino acid sequences were analyzed with the Expert Protein Analysis System (http://www.expasy.org/); and the protein domain features were predicted by the Simple Modular Architecture Research Tool (SMART; http://smart.embl-heidelberg.de/; Letunic *et al.* 2006). The three-dimensional (3D) structure was predicted by the SWISS MODEL (http://swissmodel.expasy.org/; Schwede *et al.* 2003) and visualized by the DeepView program. The Toll/interleukin-1 receptor (TIR) domain sequences of rare minnow, human, and zebrafish TLR3 were aligned with the ClustalW Multiple Alignment Program (http://www.ebi.ac.uk/clustalW/) and Multiple Alignment Show Program (http://www. biosoft. net/sms/ in dex.html) with manual optimization.

Expression analyses

The mRNA expressions of GrTLR3, GrTLR3s and Mx in different tissues in healthy rare minnows and 24 h postinjection, and their transcripts in liver after GCRV infection were determined by real-time quantitative RT-PCR (qRT-PCR). The housekeeping gene β-actin was utilized as an internal control for cDNA normalization. The forward GrTLR3 primer was SGTF41 and the reverse primer was SGTR16. The forward primer for GrTLR3s was SGTF17 and the reverse was SGTR17a. Mx was used to confirm immune system stimulation by virus infection (Mx primers: forward SMF52a and reverse SMR53a). β-actin primer sequences were forward SAF40 and reverse SAR39a. Primer sequences are listed in Table 1. In this study, the quantity of GrTLR3 is the total quantity of GrTLR3 and GrTLR3s. The qRT-PCR assay was carried out with an ABI Prism 7000 Sequence Detection System (Applied Biosystems). Each reaction consisted of 1 μl of cDNA sample, 8 μl nuclease-free water, 10 μl of SYBR Green PCR master mix (Toyobo), and 0.5 μl of each primer (5 μM). The PCR cycling conditions were 95℃ for 3 min, 40 cycles consisting of 95℃ for 15 s, 60℃ for 15 s, and 72℃ for 45 s, followed by dissociation curve analysis to verify the amplification of a single product. Reactions were run in triplicate. The threshold cycle (CT) values were determined with the setting on the ABI Sequence Detection System and were exported into a Microsoft Excel Sheet for subsequent data analyses.

The differences in the CT values of GrTLR3, GrTLR3s and Mx genes from the corresponding internal control β-actin gene, ΔCT (CT_{gene}–CT_{actin}), were calculated. The relative expression levels of target genes to β-actin were described using the equation $2^{-\Delta CT}$, and the value stood for a 1/n-fold difference relative to β-actin gene. Target gene expressions relative to β-actin were multiplied by 1000 to simplify the presentation of the data. The data obtained from qRT-PCR analyses were subjected to one-way analysis of variance (one-way ANOVA) followed by an unpaired, two-tailed *t*-test. $P < 0.05$ was considered statistical significance.

The expressions of GrTLR3, GrTLR3s, and Mx genes at different developmental stages was checked by semiquantitative RT-PCR. To distinguish the different fragments easily in the gel photos, we changed some primers for β-actin (SAR39b instead of SAR39a) and GrTLR3s (SGTR16 instead of SGTR17a; Table 1). Other primers were the same as qRT-PCR primers.

Microinjecting siRNAs

In our previous work, we found that si60 (GGGCCUAACCAGCAUUAAA) targeting

1447-1465 nt of GrTLR3 mRNA could inhibit more than 80% of the mRNA expression level of GrTLR3 at 48 h post-microinjection (unpublished). The siRNA sequence was identical in GrTLR3s. After the zygotes were dechorionated by 0.25% trypsinase, chemically modified si60 was introduced into zygotes via microinjection. The control was injected with scrambled si60 (UGAAAUCACUAACCACGGG), and normal embryos (without injection) were used as blank.

Constructing expression vectors

The cDNA synthesis was from liver total RNA at 24 h postinjection using reverse transcription with 3′-CDS primer (Table 1). The full-length cDNAs of TLR3 (s) were amplified using LA Taq™ DNA polymerase (TaKaRa) by primers SGTF68 with *Eco*RI site and SGTR80a with *Apa*I site (Table 1). After purification, the fragments were ligated into pMD-18 T vector and transformed into Top10 competent cells. The positive clones were screened by PCR using SGTF15 and SGTR16 primers for GrTLR3 (477 bp) and GrTLR3s (726 bp; Table 1). Three positive colonies from each construct were sequenced to validate the sequences without mutation.

The plasmid vector pCMV-EGFP (BD Clontech) containing an EGFP reporter gene directed by the CMV promoter was served as framework plasmid. The CMV promoter region was amplified by primers SCMVF70 with *Hin*dIII site and SCMVR71 with *Eco*RI site. The SV40 polyA fragment was produced by primers SVF81 with *Apa*I site and SVR82 with *Xho*I site. The primer sequences are listed in Table 1. The enzyme-digested CMV promoter, GrTLR3 (s) and SV40 ployA fragments were cloned into pCMV-EGFP vector between the restriction enzyme sites of *Hin*dIII and *Xho*I. The recombinant plasmid pCMV-EGFP-TLR3 (s) had two CMV promoters driving EGFP and GrTLR3 (s) respectively. It also contains SV40 polyA for EGFP and GrTLR3 (s) respectively. pCMV-EGFP and pCMV-EGFP-TLR3 (s) were microinjected zygotes as above.

mRNA expression profiles in RNAi and transgenic fish

The microinjection embryos were cultured in sterile 0.3 × Danieau's solution at 23°C. At prim-5 stage, poly I/C was added to a final concentration of 10 μg/ml and continued incubation for 2 h. The transgenic embryos with green fluorescence were selected under fluorescent microscope at 0, 6, 12, 24, and 48 h after poly I/C stimulation. RNAi embryos were collected at 0 and 12 h after stimulation. At least 30 embryos per sample were pooled and total RNA was extracted for qRT-PCR assays.

Nucleotide sequence accession numbers

The sequences of GrTLR3, GrTLR3s, and Mx mRNA were deposited in GenBank under accession numbers DQ885908, DQ885909, and EF095273, respectively.

Results

Cloning and analyzing of GrTLR3, GrTLR3s genes

The PCR product amplified by the degenerate primers was 257 bp. One fragment of 1849 bp using 5′ RACE and two fragments with sizes of 1422 bp and 1671 bp using 3′ RACE resulted respectively. The fragments of 3249 bp and 3498 bp were obtained by overlapping the sequences, and the sequences of both fragments were verified by PCR and sequencing. The fragment of 3249 bp represented the complete cDNA sequence of the GrTLR3 gene. The 3498 bp fragment was a splice variant of the GrTLR3 gene conforming to the principle of GT-AG consensus for intron splicing in eukaryotes (Shapiro and Senapathy 1987, see supplementary material 1).

The GrTLR3 sequence contains a 5′ untranslated region (UTR) of 85 bp with two termination codons, followed by an open reading frame (ORF) of 2718 bp and a 3′ UTR of 446 bp including a poly (A) tail. An instability motif (ATTTA) was observed in the 5′ UTR and 3′ UTR, respectively (see supplementary material 2). The GrTLR3s sequence consisted of a 5′ UTR, the same as GrTLR3, a 2493 bp ORF, and a long 3′ UTR of 920 bp holding four instability motifs and a poly (A) tail.

The GrTLR3 protein sequence, deduced from the ORF of 2718 bp, consisted of 905 amino acids (aa) with an estimated molecular mass (MM) of 102.74 kDa and a predicted isoelectric point (PI) of 8.39. It contained a putative signal peptide (1-23 aa), 14 LRR motifs (53-659 aa), an alternative LRR C-terminal motif (647-699 aa), a transmembrane motif (708-730 aa), and a TIR motif (761-900 aa). Sequence homology analysis revealed that the GrTLR3 protein had a significant similarity to multiple TLR3 molecules in high vertebrates including bony fishes, bird, mammals and so on (see supplementary material 3). The GrTLR3s protein was encoded by an alternatively spliced transcript, comprising 830 aa with a calculated MM of 93.87 kDa and a theoretical PI of 7.58. The GrTLR3s had a TIR domain of 70 aa, shorter than that of GrTLR3.

Homology structural modeling was performed to analyze the structures and functions further (see supplementary material 4). A large, horseshoe-shaped assembly consisting of 14 LRRs that adopted a right-handed solenoid structure was similar to the structure of the N-terminal region of *H. sapiens* TLR3. The C-terminal part of GrTLR3 with the TIR domain lay in cytoplasm. The GrTLR3 TIR domain contained a central four-stranded parallel β-sheet that was surrounded by a total of six helices, which was similar to the C-terminal structure of *H. sapiens* TLR3. The predicted secondary structure of the TIR domain of GrTLR3s lacked two β-sheets and two helices. The alignment of TIR domains of GrTLR3, zebrafish TLR3, and human TLR3 is shown in Fig. 1. Asimilar result was obtained for the conserved LRR domains (data not shown).

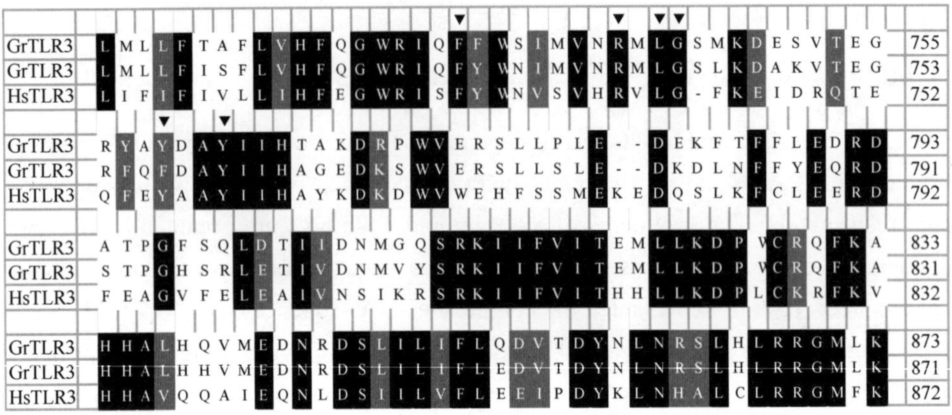

GrTLR3	L M L L F T A F L V H F Q G W R I Q F F W S I M V N R M L G S M K D E S V T E G	755
GrTLR3	L M L L F I S F L V H F Q G W R I Q F Y W N I M V N R M L G S L K D A K V T E G	753
HsTLR3	L I F I F I V L L I H F E G W R I S F Y W N V S V H R V L G - F K E I D R Q T E	752
GrTLR3	R Y A Y D A Y I I H T A K D R P W V E R S L L P L E - - D E K F T F F L E D R D	793
GrTLR3	R F Q F D A Y I I H A G E D K S W V E R S L L S L E - - D K D L N F F Y E Q R D	791
HsTLR3	Q F E Y A A Y I I H A Y K D K D W V W E H F S S M E K E D Q S L K F C L E E R D	792
GrTLR3	A T P G F S Q L D T I I D N M G Q S R K I I F V I T E M L L K D P W C R Q F K A	833
GrTLR3	S T P G H S R L E T I V D N M V Y S R K I I F V I T E M L L K D P W C R Q F K A	831
HsTLR3	F E A G V F E L E A I V N S I K R S R K I I F V I T H H L L K D P L C K R F K V	832
GrTLR3	H H A L H Q V M E D N R D S L I L I F L Q D V T D Y N L N R S L H L R R G M L K	873
GrTLR3	H H A L H H V M E D N R D S L I L I F L E D V T D Y N L N R S L H L R R G M L K	871
HsTLR3	H H A V Q Q A I E Q N L D S I I L V F L E E I P D Y K L N H A L C L R R G M F K	872

Fig. 1 Alignment of TIR domain sequences of TLR3 in *G. rarus*, *D. rerio*, and *H. sapiens*. Amino acid sequences were deduced from cDNA. Residues highlighted in black were identical and those in dark gray were similar. Numbers on the right represent the amino acid position in the corresponding species. The conserved functional sites are marked with *solid triangles*

Tissue distribution of GrTLR3, GrTLR3s, and Mx mRNA expression

qRT-PCR was used to quantify the expressions of GrTLR3, GrTLR3s, and Mx in different tissues of healthy and challenged rare minnows. GrTLR3, GrTLR3s and Mx gene mRNA expression was detected in a wide variety of tissues of adult rare minnows. All gene expression showed no significant difference between control and blank groups (P>0.05; data not shown).

The GrTLR3 gene expressions levels in control animals were: intestine > muscle > kidney > spleen > gill > heart >liver; the expressions levels in GCRV-infected animals were: intestine > liver > muscle > kidney > spleen > gill > heart (Fig. 2a). In contrast to control, the expressions increased statistically significantly in intestine and gill (P<0.05) and in liver (P<0.01). Expressions in other tissues were also upregulated but not statistically significant.

The GrTLR3s gene expressions in control animals were: intestine > gill > heart > muscle > spleen > kidney >liver; the expressions in GCRV-infected animals were: liver > intestine > gill > muscle > heart > kidney > spleen (Fig. 2b). After being infected by GCRV, the expressions of GrTLR3 decreased significantly in spleen (P<0.01) and increased significantly in liver (P<0.01). No significant differences were observed in other tissues (P>0.05).

The mRNA transcripts of Mx in control animals were: intestine > heart > spleen > kidney > gill > muscle >liver; the expression levels of Mx in GCRV-infected animals were: intestine > liver > kidney > gill > heart > spleen > muscle (Fig. 2c). In the infected fish, the expression increased significantly in liver, kidney, and gill (P<0.05). Expression in other tissues varied but not significant (P>0.05).

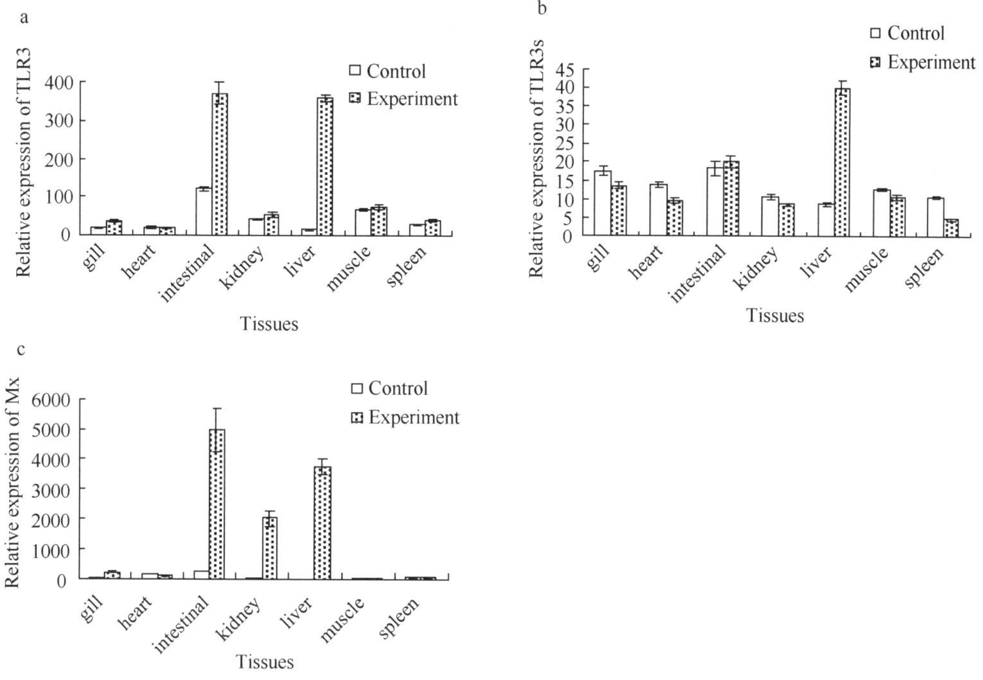

Fig. 2 Tissue distribution of GrTLR, GrTLR3s, and Mx transcripts in rare minnow. Each bar represents the level of target gene mRNA relative to β-actin mRNA, expressed as the mean±SE of triplicate from qRT-PCR assays ($2^{-\Delta CT}\times 1000$) for three different cDNA samples. Error bars indicated standard error. (a) relative expression level of the TLR3 gene; (b)relative expression level of the TLR3s gene; (c) relative expression level of the Mx gene

All together, only in liver tissue, did expression of the three target genes increase significantly (P<0.05). Liver tissue was used to investigate the temporary expression of the three genes after GCRV infection.

GrTLR3, GrTLR3s, and Mx transcriptional regulation in liver tissue after inoculation of GCRV

The gene expressions of liver tissue at 0, 6, 12, 24, 36, 48, 72, 96, 120, and 144 h post-injection was measured for GrTLR3, GrTLR3s and Mx, respectively. No significant differences were detected in control and blank groups among all the time points (P>0.05; data not shown). The peak of GrTLR3 mRNA was observed at 24 h post-injection (22.24-fold increase relatively to control). From 24 h to moribund fish, the GrTLR3 mRNA expressions was significantly higher than that in the control group (P<0.05; Fig. 3a). GrTLR3s expression increased significantly at 24 h post-injection (3.98-folds relative to control) (P<0.05; Fig. 3b), and this trend was maintained until the animals became moribund (P<0.05).

Mx was a low-abundance gene in healthy fish. GCRV induced a rapid and significant increase at 12 h post-injection ($P<0.05$), reached the maximum level at 36 h post-injection and continues throughout the course of infection. Compared with the control, the expressions were approximately 47-folds at 12 h post-injection and 570-to 934-folds from 24 h to 144 h (Fig. 3c).

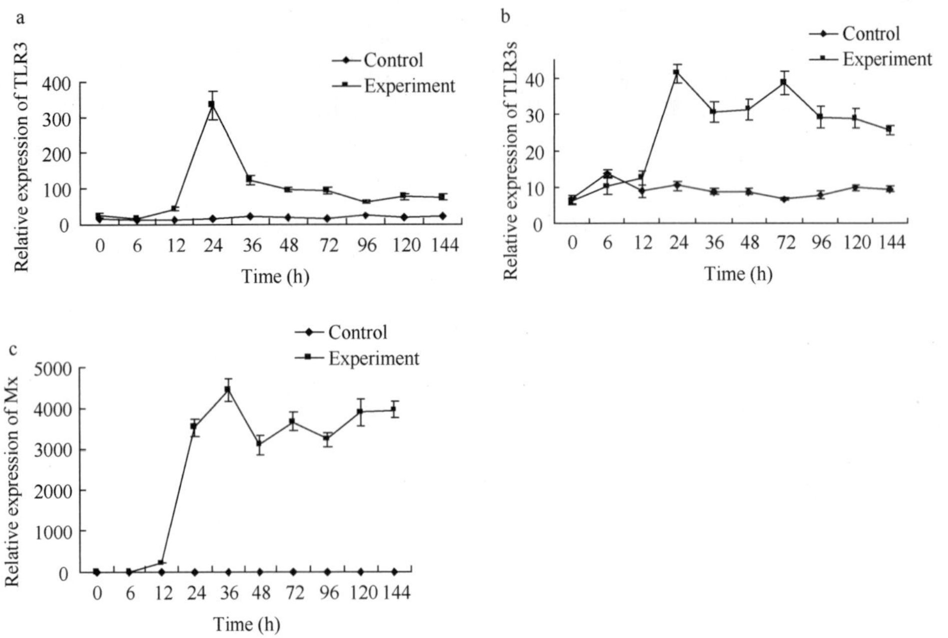

Fig. 3 Expression profiles of GrTLR3, GrTLR3s, and Mx mRNAs in liver by qRT-PCR. Controls were injected with PBS; the experiments suffered from GCRV infection. Error bars indicate standard error. (a) relative expression level of the TLR3 gene; (b) relative expression level of the TLR3s gene; (c) relative expression level of the Mx gene

GrTLR3, GrTLR3s, and Mx mRNA expression profiles at different developmental stages

Semi-quantitative RT-PCR was employed to determine the gene expression of three genes at different developmental stages (Fig. 4). GrTLR3 and Mx transcripts were detected from zygote to adult stages, whereas GrTLR3s expression was undetectable at early developmental stages. The zygote mRNA mainly came from maternal animals. The GrTLR3 and Mx expression levels were high at the zygote stage. GrTLR3 expression decreased at the dome stage and then increased gradually. At the prim-5 stage, the GrTLR3 transcript reached adult level. Mx expression decreased at the shield stage, and then climbed up again and reached adult level at the 1-somite stage. As both GrTLR3 and Mx expression levels approached those of an adult at the prim-5 stage, this stage was chosen to study the expression of GrTLR3 and Mx in transgenic fish.

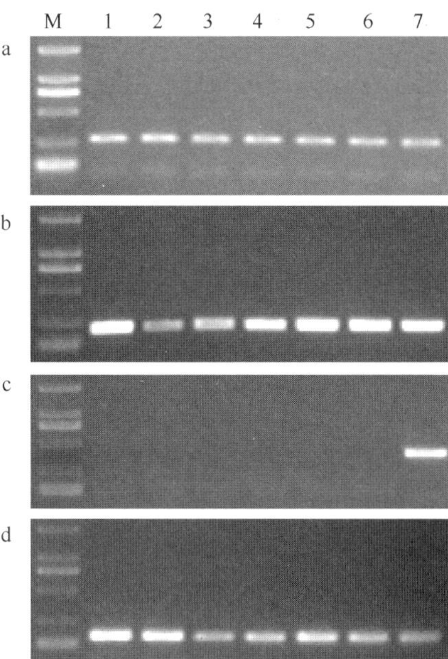

Fig. 4 Expression of TLR3, TLR3s, and Mx mRNAs at the different developmental stages by semi-quantitative RT-PCR. (a), (b), (c), and d showed the expression profiles of β-actin, TLR3, TLR3s, and Mx, respectively. We amplified 22 cycles for β-actin, and 32 cycles for other genes. M, 1-7 stands for DNA marker (DL2000), zygote, dome, shield, 1-somite, prim-5, protruding mouth, and adult stages, respectively

GrTLR3, GrTLR3s, and Mx mRNA expression study in RNAi and transgenic fish

In the RNAi embryos, TLR3 expression decreased by 76% relative to the control at 0 h post-stimulation, and decreased 85% relative to the control at 12 h post-stimulation. At 12 h post-stimulation, the expression of GrTLR3 increased 4.46-fold in the control group and 2.88 fold in experimental group (Fig. 5a). There was no significant difference for the Mx gene between the control and the experimental groups at 0 h post-stimulation. However, Mx expressions increased 8.08-fold in the control at 12 h, and 2.05-fold in the experimental group (Fig. 5b). GrTLR3s was undetectable in all the groups. There was no significant difference between control and blank groups for GrTLR3 and Mx (P>0.05) (data not shown).

In GrTLR3 transgenic fish, the GrTLR3s mRNA could not be detected by RT-PCR in all the groups and at all time points. GrTLR3 expression in control groups was considerably significantly lower than that in experimental groups (P<0.01). After stimulation by poly I/C, GrTLR3 expression in control groups increased significantly at 12 h (P< 0.05), while in experiment groups, it rose significantly at 6 h (P<0.05). GrTLR3 mRNA transcripts decreased to normal levels in both groups at 48 h (Fig. 6a). Before stimulation, Mx expression in the control and the experimental group were not significantly different (P>0.05). After challenge, Mx expression increased significantly in both groups (P<0.05), with a significantly lower level in the control group than that in the experimental group, from 6 to 24 h ($P<0.01$). Mx expression was recovered in both groups at 48 h (Fig. 6b).

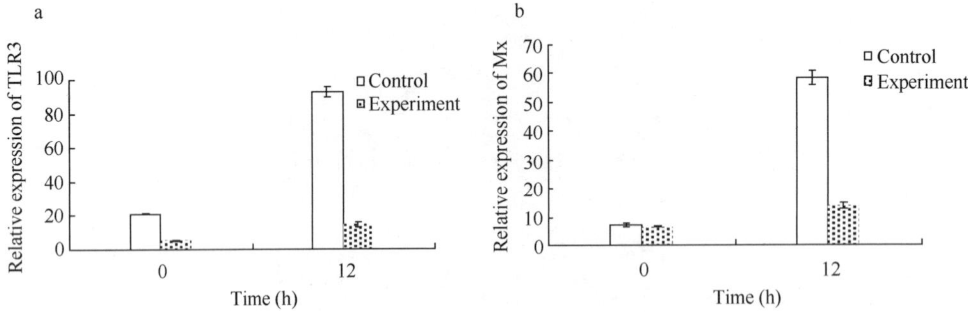

Fig. 5 Gene expression features in GrTLR3 knock-down embryos by qRT-PCR. All embryos were treated by poly I/C at the prim-5 stage. The experimental group was microinjected with si60, and the control group was microinjected with scrambled si60. Error bars indicated standard error. (a) relative expression level of the GrTLR3 gene. (b) relative expression level of the Mx gene

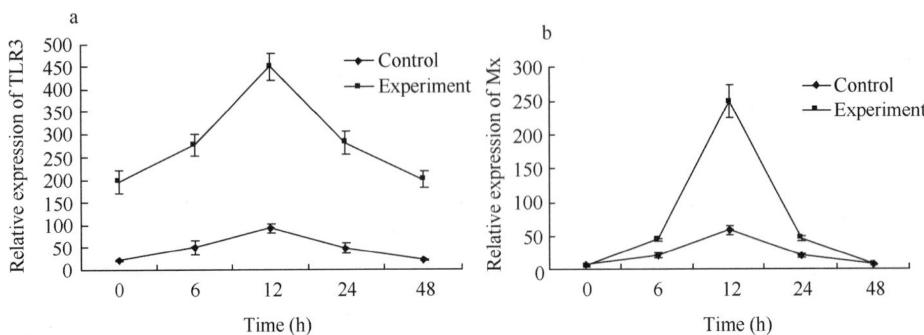

Fig. 6 Temporal expression of GrTLR3 and Mx mRNAs in transgenic fish of GrTLR3 by qRT-PCR. The embryos were stimulated by poly I/C at the prim-5 stage. The control group was injected with pCMV-GFP expression vector; the experimental group was injected with pCMV-GFP-TLR3 expression vector. Error bars indicate standard error. (a) relative expression level of the GrTLR3 gene. (b)relative expression level of the Mx gene

In GrTLR3s transgenic fish, GrTLR3s expression was detected easily in the experimental group. After stimulation by poly I/C, TLR3s expressions increased significantly at 12 h ($P<0.05$) compared to 0 h and recovered at 48 h (data not shown). TLR3 and Mx expressions increased significantly in the control and experimental group at 12 h ($P<0.05$) and recovered at 48 h post-stimulation. There was no significant difference between experimental and control groups at the same time points (data not shown).

Discussion

In the present study, a splice variant of GrTLR3, shortened at part of the TIR domain in the C-terminal, is first reported. The previously reported splice variants in human and chimpanzee were shortened at part of the LRR domain (Yang *et al.* 2004; XM_001165491). This newly identified splice variant in *Gobiocypris rarus* alerts us to the fact that further study is needed to dotermine its biological function.

Unstable mRNAs often contain AU-rich elements in their 3′ UTRs, and they are especially true for inflammatory-related mRNAs (Kracht and Saklatvala 2002; Jensen and Whitehead 2004). One instability motifs (ATTTA) was identified in the 3′ UTR of GrTLR3 and four in the 3′ UTR of GrTLR3s. Three were identified in human and none in the zebrafish. It usually takes three of these motifs to make mRNAs unstable. Our data suggests that GrTLR3 and the zebrafish TLR3 mRNAs might be stable, and GrTLR3s and the human TLR3 mRNAs might be unstable.

Analyzed with the SMART program, GrTLR3 was seen to contains 14 LRRs which is similar to the observations in zebrafish (15) and in human (18), although 23 LRRs were reported in human TLR3 by Bell *et al.* (2003). The discrepancy might be due to the different programs used to identify the repeats.

Eleven potential N-glycosylation sites were identified in the extracellular region of GrTLR3. One face of the horseshoe-shaped structure was completely devoid of any glycosylation (NetNGlyc 1.0 Server; Blom *et al.* 2004). When the surface electrostatic potential of GrTLR3 was calculated with Deepview (Schwede *et al.* 2003), the concave surface of the horseshoe-shaped structure and the glycosylation-free face were largely positively charged, which was consistent with its predicted interaction with dsRNA. Thus, the glycosylation-free face and the concave surface of the horseshoe-shaped structure could correspond to potential RNA binding sites (Bell *et al.* 2003; Choe *et al.* 2005).

The TIR domain is known to play a pivotal role in both receptor signaling and intracellular localization (Figs. 1; Funami *et al.* 2004; Sarkar *et al.* 2003), and the TIR domain might be considered as a cassette (Xu *et al.* 2000). Most of the conserved residues were located in the hydrophobic core. Signal transduction through TIR domain required receptor oligomerization, which was induced either by ligand binding or by over-expression (Xu *et al.* 2000). The TIR domain of GrTLR3s was short of two β-sheets and two helices, which caused the hydrophobic core to be exposed outside. Funami *et al.* (2004) identified three amino acid residues essential for ligand-inducing NF-κB and IFN-β promoter activation (Phe^{732}, Leu^{742}, and Gly^{743}) in human TLR3, all of which were present in the GrTLR3 sequence (Phe^{734}, Leu^{744}, and Gly^{745}). Sarkar *et al.* (2003) identified specific tyrosine residues in the TIR domain as being essential for the intracellular signaling in human TLR3. The most critical residue was Tyr^{759}, which was present in the rare minnow sequence (Tyr^{762}). In addition to Tyr^{759}, either tyrosine residues Tyr^{733} or Tyr^{858} was needed for maximal activity of the 561-gene promoter in human. A corresponding human Tyr^{858} site was conserved in GrTLR3 (Tyr^{859}), another site was variant in rare minnow. Additional tyrosine residues (Tyr^{756} and Tyr^{764}) could also complement Tyr^{759} function, but they were less effective. Of these, only Tyr^{756} was conserved in the rare minnow sequence (Tyr^{759}). Funami *et al.* (2004) identified two amino acids important for the intracellular localization in human TLR3, Arg^{740} and Val^{741}. The arginine residue was conserved in the rare minnow sequence (Arg^{742}). Further analysis is necessary to explore the physiological significance of the missing *Val* residue in rare minnow. GrTLR3s had all the conserved sites of GrTLR3 except for Tyr^{859} in GrTLR3.

In promoting cross-priming of viral antigens, TLR3 provided new insights into the mechanisms that allowed TLR signaling to bridge innate and adaptive immune responses (Salio and Cerundolo 2005). Sequence similarities and structural conservation of full-length GrTLR3 implied that TLR3 in rare minnow is involved in innate and adaptive antiviral immunity.

The tissue expression profiles of GrTLR3 and GrTLR3s were similar to that of zebrafish TLR3 (Jault *et al.* 2004; Phelan *et al.* 2005) with a wide range of tissues (Fig. 2a). On the other hand, pufferfish TLR3 expressions was limited to the liver and digestive organs (Oshiumi *et al.* 2003b). GrTLR3 expression in liver and intestine was much higher than that in other tissues after stimulation (Fig. 2a). Differences in TLR3 expression patterns have also been reported between mouse and human, with expression being restricted to dendritic cells in human (Muzio *et al.* 2000; Kadowaki *et al.* 2001), while highly expressed in murine macrophages (Alexopoulou *et al.* 2001; Heinz *et al.* 2003). A comparison of TLR expression should be interpreted with caution since such discrepancies might result not only from species variation but also from differences in immunological status, developmental stages, and genetic background (Renshaw *et al.* 2002; Dhara *et al.* 2007). The different expression profiles of TLR3 might result in the distinct susceptibility of viral diseases (Baoprasertkul *et al.* 2006).

Nonspecific activation of innate immunity can drastically enhance susceptibility to immune destruction of a solid organ. Additional innate immune signals orchestrated by TLR3 provoke liver damage (Bertolino and Holz 2007). The rare minnow liver turned gray on day 4 post-injection (not shown), which might have been caused by GCRV via TLR3 activation. GCRV significantly induced Mx expression in liver, kidney and gill in rare minnow ($P<0.05$), which was in accordance with studies of Mx expression in liver and head kidney of Atlantic salmon treated with poly I/C (Jensen *et al.* 2002). The gill is a potential port of entry for infectious agents, hance inducible expression of Mx in gill might represent a critical protection mechanism against virus invasion.

In mammals, TLR3 is strongly associated with the antiviral response via dsRNA intermediate during viral replication (Alexopoulou *et al.* 2001; Matsumoto *et al.* 2004; Oshiumi *et al.* 2003b), although a recent study by Edelmann *et al.* (2004) questioned the universal antiviral nature of TLR3. While this work was in progress, a paper appeared describing the signal pathway from TLR3 to IFN and Mx induction in zebrafish (Sullivan *et al.* 2007). In our study, GrTLR3, GrTLR3s, and Mx gene expression levels in the experimental group were not significantly different from those in control group during the early stage of infection ($P>0.05$). As time progressed and the virus multiplied, significant upregulation was observed from 12 h or 24 h post-injection. Furthermore, the GrTLR3s expression increased several folds, GrTLR3 expression elevated tens of folds, and Mx expression increased hundreds of folds. From receptor to effector, it's a strong immune cascade reaction. Based on the time sequence, the effector Mx and the receptor GrTLR3 were upregulated almost synchronously, which implied that the configuration change of the receptor guickly activated the downstream signal pathway or that Mx did not depend on TLR3 (s). All these gene expressions in blank and control fish might be due to the fact that the fish were inevitabiy exposed to viral particles presenting in the water.

The GrTLR3 and Mx expressions at zygote, and rapid recovery after depleting maternal GrTLR3 and Mx mRNAs, facilitated the protection of embryo development against virus invasion. The GrTLR3s could not be detected at early stages, which might indicate that the GrTLR3s was not an essential element in the early development stage.

The Mx expression profile in GrTLR3 transgenic fish indicated that GrTLR3 and Mx have no interaction in healthy fish. However, increased GrTLR3 could activate more Mx expression after stimulation by dsRNA. We speculated that TLR3 changed its configuration by binding to dsRNA and triggered a signaling pathway, which induced Mx gene expression. When the stimulation (dsRNA) disappeared, the configuration of GrTLR3 recovered gradually and the Mx transcription returned to a normal level. In the RNAi experiments, the results also supported Mx upregulation, depended on the GrTLR3 configuration change after binding dsRNA. This process was rapid and efficient. GrTLR3s expression in GrTLR3s transgenic fish was up-regulated after being stimulated by dsRNA, while TLR3 and Mx expressions were hardly affected. This result indicated that GrTLR3s was not involved in the process of activatingan interferon pathway.

Taken together, the sequence and structure conservation of GrTLR3 and GrTLR3 mRNA upregulation and antiviral effector gene Mx over-expression after dsRNA virus exposures indicated that GrTLR3 played a pivotal role in the antiviral immunity in rare minnow. The studies using RNAi and transgenic fish indicated that GrTLR3 regulated Mx expression, whereas GrTLR3s did not participate in the interferon pathway. The biological functions of GrTLR3s need be investigated further. Further immune precaution through enhancing or inhibiting TLR3 (s) expression might be apotential research direction and strategy for suppressing viral diseases.

Acknowledgements The authors would like to thank Ms. Ming Li and Mr. Renxiang Yan for fish administration and microinjection, and Ms. Shangping Chen and other laboratory staff for technical advice and helpful discussions. We are grateful to Dr, Yonghong Wang from CSIRO for critically reading the manuscript. This study was supported by the National

Natural Science Foundation of China (30428024, 30540084 and 30740009), 973 National Basic Research Program of China (2006CB102100 and 2004CB117406), Chinese Academy of Sciences project (KSCX2-YW-N-021), research foundation (08080262, 08080245 and 01140508) from Northwest A&F University in China, China Postdoctoral Science Foundation (20070410298), and the grant from the State Key Laboratory of Freshwater Ecology and Biotechnology (2007FB09).

References

Aderem A, Ulevitch RJ (2000) Toll-like receptors in the induction of the innate immune response. Nature 406: 782-787

Alexopoulou L, Holt AC, Medzhitov R, Flavell RA (2001) Recognition of double-stranded RNA and activation of NF-κB by Toll-like receptor 3. Nature 413: 732-738

Baoprasertkul P, Peatman E, Somridhivej B, Liu Z (2006) Toll-like receptor 3 and TICAM genes in catfish: species-specific expression profiles following infection with *Edwardsiella ictaluri*. Immunogenetics 58: 817-830

Bell JK, Mullen GE, Leifer CA, Mazzoni A, Davies DR, Segal DM (2003) Leucine-rich repeats and pathogen recognition in Toll-like receptors. Trends Immunol 24: 528-533

Bertolino P, Holz LE (2007) Toll-like receptor-3 and the regulation of intrahepatic immunity: implications for interferon-alpha therapy. Hepatology 45: 250-251

Bilak H, Tauszig-Delamasure S, Imler JL (2003) Toll and Toll-like receptors in *Drosophila*. Biochem Soc Trans 31: 648-651

Blom N, Sicheritz-Ponten T, Gupta R, Gammeltoft S, Brunak S (2004) Prediction of post-translational glycosylation and phosphorylation of proteins from the amino acid sequence. Proteomics 4: 1633-1649

Choe J, Kelker MS, Wilson IA (2005) Crystal structure of human Toll-like receptor 3 (TLR3) ectodomain. Science 309: 581-585

Dhara A, Saini M, Das DK, Swarup D, Sharma B, Kumar S, Gupta PK (2007) Molecular characterization of coding sequences and analysis of Toll-like receptor 3 mRNA expression in water buffalo (*Bubalus bubalis*) and nilgai (*Boselaphus tragocamelus*). Immunogenetics 59: 69-76

Edelmann KH, Richardson-Burns S, Alexopoulou L, Tyler KL, Flavell RA, Oldstone MB (2004) Does Toll-like receptor 3 play a biological role in virus infections? Virology 322: 231-238

Funami K, Matsumoto M, Oshiumi H, Akazawa T, Yamamoto A, Seya T (2004) The cytoplasmic 'linker region' in Toll-like receptor 3 controls receptor localization and signaling. Int Immunol 16: 143-154

Haller O, Frese M, Kochs G (1998) Mx proteins: mediators of innate resistance of RNA viruses. Rev Sci Tech 17: 220-230

Heinz S, Haehnel V, Karaghiosoff M, Schwarzfischer L, ller MM, Krause SW, Rehli M (2003) Species-specific regulation of Toll-like receptor 3 genes in men and mice. J Biol Chem 278: 21502-21509

Ishii A, Kawasaki M, Matsumoto M, Tochinai S, Seya T (2007) Phylogenetic and expression analysis of amphibian *Xenopus* Toll-like receptors. Immunogenetics 59: 281-293

Jault C, Pichon L, Chluba J (2004) Toll-like receptor gene family and TIR-domain adapters in *Danio rerio*. Mol Immunol 40: 759-771

Jensen LE, Whitehead AS (2004) The 3′ untranslated region of the membrane-bound IL-1R accessory protein mRNA confers tissue-specific destabilization. J Immunol 173: 6248-6258

Jensen I, Albuquerque A, Sommer A, Robertsen B (2002) Effect of poly I: C on the expression of Mx proteins and resistance against infection by infectious salmon anaemia virus in Atlantic salmon. Fish Shellfish Immunol 13: 311-326

Kadowaki N, Ho S, Antonenko S, Malefyt RW, Kastelein RA, Bazan F, Liu YJ (2001) Subsets of human dendritic cell precursors express different Toll-like receptors and respond to different microbial antigens. J Exp Med 194: 863-870

Kariko K, Ni H, Capodici J, Lamphier M, Weissman D (2004) mRNA is an endogenous ligand for Toll-like receptor 3. J Biol Chem 279: 12542-12550

Kracht M, Saklatvala J (2002) Transcriptional and post-transcriptional control of gene expression in inflammation. Cytokine 20: 91-106

Letunic I, Copley RR, Pils B, Pinkert S, Schultz J, Bork P (2006) SMART 5: domains in the context of genomes and networks. Nucl Acids Res 34: D257-D260

Mak TW, Yeh W (2002) Immunology: A block at the toll gate. Nature 418: 835-836

Matsumoto M, Funami K, Oshiumi H, Seya T (2004) Toll-like receptor 3: a link between toll-like receptor, interferon and viruses. Microbiol Immunol 48: 147-154

Medzhitov R (2001) Toll-like receptors and innate immunity. Nat Rev Immunol 1: 135-145

Meijer AH, Krens SFG, Rodriguez IAM, He S, Bitter W, Snaar-Jagalska BE, Spaink HP (2004) Expression analysis of the Toll like receptor and TIR domain adaptor families of zebrafish. Mol Immunol 40: 773-783

Muzio M, Bosisio D, Polentarutti N, D'amico G, Stoppacciaro A, Mancinelli R, van't VC, Penton-Rol G, Ruco LP, Allavena P, Mantovani A (2000) Differential expression and regulation of Toll-like receptors (TLR) in human leukocytes: selective expression of TLR3 in dendritic cells. J Immunol 164: 5998-6004

Nishiya T, Kajita E, Miwa S, DeFranco AL (2005) TLR3 and TLR7 are targeted to the same intracellular compartments by distinct regulatory elements. J Biol Chem 280: 37107-37117

Nygaard R, Husgård S, Sommer AI, Leong JAC, Robertsen B (2000) Induction of Mx-protein by interferon and double-stranded RNA in salmonid cells. Fish Shellfish Immunol 10: 435-450

Oshiumi H, Tsujita T, Shida K, Matsumoto M, Ikeo K, Seya T (2003a) Prediction of the prototype of the human Toll-like receptor gene family from the pufferfish, *Fugu rubripes*, genome. Immunogenetics 54: 791-800

Oshiumi H, Matsumoto M, Funami K, Akazawa T, Seya T (2003b) TICAM-1, an adaptor molecule that participates in Toll-like receptor 3 mediated interferon-beta induction. Nat Immunol 4: 161-167

Phelan PE, Mellon MT, Kim CH (2005) Functional characterization of full-length TLR3, IRAK-4, and TRAF6 in zebrafish (*Danio rerio*). Mol Immunol 42: 1057-1071

Renshaw M, Rockwell J, Engleman C, Gewirtz A, Katz J, Sambhara S (2002) Cutting edge: impaired Toll-like receptor expression and function in aging. J Immunol 169: 4697-4701

Rodriguez MF, Wiens GD, Purcell MK, Palti Y (2005) Characterization of Toll-like receptor 3 gene in rainbow trout (*Oncorhynchus mykiss*). Immunogenetics 57: 510-519

Rudd BD, Burstein E, Duckett CS, Li X, Lukacs NW (2005) Differential role for TLR3 in respiratory syncytial virus-induced chemokine expression. J Virol 79: 3350-3357

Salio M, Cerundolo V (2005) Viral immunity: cross-priming with the help of TLR3. Curr Biol 15: R336-R339

Sarkar SN, Smith HL, Rowe TM, Sen GC (2003) Double-stranded RNA signaling by Toll-like receptor 3 requires specific tyrosine residues in its cytoplasmic domain. J Biol Chem 278: 4393-4396

Schroder M, Bowie AG (2005) TLR3 in antiviral immunity: key player or bystander. Trends Immunol 26: 462-468

Schwede T, Kopp J, Guex N, Peitsch MC (2003) SWISS-MODEL: an automated protein homology-modeling server. Nucl Acids Res 31: 3381-3385

Shapiro MB, Senapathy P (1987) RNA splice junctions of different classes of eukaryotes: sequence statistics and functional implications in gene expression. Nucl Acids Res 15: 7155-7174

Su J, Ni D, Song L, Zhao J, Qiu L (2007) Molecular cloning and characterization of a short type peptidoglycan recognition protein (CfPGRP-S1) cDNA from Zhikong scallop *Chlamys farreri*. Fish Shellfish Immunol 23: 646-656

Sullivan C, Postlethwait JH, Lage CR, Millard PJ, Kim CH (2007) Evidence for evolving Toll-IL-1 receptor-containing adaptor molecule function in vertebrates. J Immunol 178: 4517-4527

Tissari J, Sirén J, Meri S, Julkunen I, Matikainen S (2005) IFN- enhances TLR3-mediated antiviral cytokine expression in human endothelial and epithelial cells by up-regulating TLR3 expression. J Immunol 174: 4289-4294

Wang T, Liu P, Chen H (1994) Preliminary study on the susceptibility of *Gobiocypris rarus* to hemorrhagic virus of grass carp (GCHV). Act Hydrobiol Sin 18: 144-149 (In Chinese, English abstract).

West AP, Koblansky AA, Ghosh S (2006) Recognition and signaling by Toll-like receptors. Ann Rev Cell Dev Biol 22: 409-437

Xu Y, Tao X, Shen B, Horng T, Medzhitov R, Manley JL, Tong L (2000) Structural basis for signal transduction by the Toll/interleukin-1 receptor domains. Nature 408: 111-115

Yang E, Shin JS, Kim H, Park HW, Kim MH, Kim SJ, Choi IH (2004) Cloning of TLR3 isoform. Yonsei Med J 45: 359-361

Zuany-Amorim C, Hastewell J, Walker C (2002) Toll-like receptors as potential therapeutic targets for multiple diseases. Nat Rev Drug Discov 1: 797-807

病毒感染后稀有鮈鲫 TLR3 可调控 Mx 的表达

苏建国[1,2]　朱作言[1]　汪亚平[1]　邹　钧[3]　胡　炜[1]

1. 中国科学院水生生物研究所，武汉，430072，中国

2. 西北农林科技大学，杨凌，712100，中国

3. University of Aberdeen, School of Biologycal Sciences, Abderdeen, AB242TZ, UK

摘　要　Toll 样受体 3(TLR3)在病毒感染时激活免疫反应的过程中起重要作用。为了研究病毒感染后稀有鮈鲫 TLR3 参与免疫调节的功能，通过同源克隆及 RACE 法获得 GrTLR3 及 GrTLR3s 的 cDNA 全长。克隆并部分测序了抗病毒因子 Mx 基因。使用定量 RT-PCR 的方法研究 GrTLR3、GrTLRs 及 Mx 基因 mRNA 在不同组织中病毒感染前后的表达水平。在肝脏中，三个基因的转录在草鱼呼肠孤病毒感染后显著上调($P<0.05$)。GrTLR3 和 GrTLR3s 的 mRNA 水平在感染后 24h 的肝脏中上调，Mx 则在感染 12h 后上调。上调水平在 GrTLR3s、GrTLR3 和 Mx 中分别表现为几倍、十几倍及上千倍。半定量结果显示，GrTLR3 和 Mx 在发育的所有阶段均表达，但 GrTLR3 只在发育晚期表达。通过使用 RNAi 及转基因技术，我们发现 GrTLR3 调节 Mx 的表达，但 GrTLR3s 不能。受体和因子的时序上调及 Mx 依赖 TLR3 的过表达表明，病毒感染后 GrTLR3 在稀有鮈鲫中调节 Mx 的表达是通过构象变化实现的，并且剪接突变体在这个过程中并不发挥作用。

Enhanced Grass Carp Reovirus Resistance of Mx-Transgenic Rare Minnow (*Gobiocypris rarus*)

Jianguo Su[1,2] Chunrong Yang[1] Zuoyan Zhu[2] Yaping Wang[2] Songhun Jang[2] Lanjie Liao[2]

1 Northwest A & F University, Shaanxi Key Laboratory of Molecular Biology for Agriculture, Yangling 712100

2 State Key Laboratory of Freshwater Ecology and Biotechnology, Institute of Hydrobiology, Chinese Academy of Sciences, Wuhan 430072

Abstract In the interferon-induced antiviral mechanisms, the Mx pathway is one of the most powerful. Mx proteins have direct antiviral activity and inhibit a wide range of viruses by blocking an early stage of the viral genome replication cycle. However, antiviral activity of piscine Mx remains unclear *in vivo*. In the present study, an Mx-like gene was cloned, characterized and gene-transferred in rare minnow *Gobiocypris rarus*, and its antiviral activity was confirmed *in vivo*. The full length of the rare minnow Mx-like cDNA is 2241 bp in length and encodes a polypeptide of 625 amino acids with an estimated molecular mass of 70.928 kDa and a predicted isoelectric point of 7.33. Analysis of the deduced amino acid sequence indicated that the mature peptide contains an amino-terminal tripartite GTP-binding motif, a dynamin family signature sequence, a GTPase effector domain and two carboxy-terminal leucine zipper motifs, and is the most similar to the crucian carp (*Carassius auratus*) Mx3 sequence with an identity of 89%. Both P0 and F1 generations of Mx-transgenic rare minnow demonstrated very significantly high survival rate to GCRV infection ($P<0.01$). The mRNA expression of Mx gene was consistent with survival rate in F1 generation. The virus yield was also concurrent with survival time using electron microscope technology. Rare minnow has Mx gene(s) of its own but introducing more Mx gene improves their resistance to GCRV. Mx-transgenic rare minnow might contribute to control the GCRV diseases.

Keywords rare minnow (*Gobiocypris rarus*); Mx gene; grass carp reovirus; transgenic fish; antiviral activity; gene cloning

1. Introduction

Type I interferons (IFNs) represent a crucial component of the innate immune response to viruses. An important downstream effector of IFN is the Mx (myxovirus resistance) gene, which is activated solely through this pathway [1]. Mx proteins form a distinct subclass of the

文章发表于 *Fish & Shellfish Immunology*, 2009, 26(6): 828-835

dynamin superfamily of large guanosine triphosphatases (GTPases) known to be involved in intracellular vesicle trafficking and organelle homeostasis. A unique property of some Mx GTPases is their antiviral activity against a wide range of viruses by blocking an early stage in their life cycle, soon after entry into host cells and before genome amplification. In general, Mx GTPases appear to detect viral infection by sensing nucleocapsid-like structures. As a consequence, these viral components are trapped and sorted to locations where they become unavailable for the generation of new virus particles [2-4].

Early work disclosed that mouse Mx1 was responsible for the resistance phenotype and that the Mx1 protein had intrinsic antiviral activity [5,6]. Antivirally active Mx proteins are found in most vertebrate species, including birds, fish and man. However, not all Mx proteins appear to possess antiviral activity. Human cells express two Mx GTPases which both reside in the cytoplasm and are referred to as MxA and MxB. Human MxB has no antiviral activity and may serve other functions. Human MxA, however, has a wide antiviral spectrum against different types of viruses [7]. Furthermore, Mx genes are polymorphic in most species. The resulting amino acid differences account for variations in the antiviral activities of the allelic gene products [8-12]. In addition, the antiviral activity of Mx proteins is virus specific, for example human MxA and rat Mx1 protect against influenza virus and vesicular stomatitis virus (VSV) while rat Mx2 protects against VSV but not influenza [6,13].

The available sequence alignment indicated it is considerable conservation among Mx proteins in different species. Vertebrate Mx proteins can be grouped into five subgroups according to their sequence similarities. These comprise the fish, avian and rodent subgroups as well as an MxA-like and MxB-like subgroup [4].

Recently, an antiviral role has also been described for piscine Mx proteins at cell level. Atlantic salmon Mx1 inhibits an influenza-like virus in fish, called infectious salmon anaemia virus (ISAV) and also infectious pancreatic necrosis virus (IPNV), an aquatic birnavirus [14,15]. Rainbow trout Mx can be induced by chum salmon reovirus (CSR), Poly I:C, or viral haemorrhagic septicaemia virus (VHSV) in trout monocyte/macrophage and fibroblast cell lines [16,17]. All the three Mx isoforms from grouper have the efficiency of reducing the titer of virus 10- to 100-fold in GB3 cells [18,19]. Barramundi Mx is able to inhibit the proliferation of fish nodavirus and birnavirus [20]. Sole aquabirnavirus is inhibited in cells expressing recombinant Senegalese sole Mx, gilt-head seabream Mx or turbot Mx [21-23]. However, the antiviral activity of Mx protein lacks direct evidence at individual level in fish.

Grass carp (*Ctenopharyngodon idella*) is a crucial fish species in Chinese aquaculture. In fish farming, grass carp reovirus (GCRV) infection is a constant threat therefore, breeding an antiviral strain and understanding the defense mechanisms are fairly important to decrease economic losses. However, the bigger size and longer reproductive cycle make it difficulty to study the antiviral breeding. The rare minnow *Gobiocypris rarus*, which is a small cyprinid species, has been recognized as a useful model creature. It's very susceptible to GCRV [24].

In the current work, we cloned and characterised the rare minnow *Gobiocypris rarus* Mx (GrMx) cDNA and its antiviral activity to GCRV was assessed by transgenic rare minnow over-expressing GrMx. The antiviral activity against this pathogen was demonstrated by reducing mortality and virus yield. These findings contribute to our understanding of antiviral mechanisms *in vivo* of piscine Mx gene and open the possibility of using this protein as a tool for fighting viral infections in aquaculture.

2. Materials and Methods

2.1 Cloning of a rare minnow Mx cDNA fragment

Adult rare minnow *Gobiocypris rarus* of approximately 2 g was injected with 20 μl GCRV suspended in PBS (2×10^8 PFU/ml) intraperitoneally and was maintained in an aerated aquarium at 28°C. Three individuals were killed and liver tissue was collected at 24 h post-injection. The samples were homogenized in TRIZOL® LS reagent (Invitrogen) and total RNA was extracted according to the manufacturer's instruction. Total RNA was stored at –80℃.

The degenerate primers for cloning Mx cDNA from rare minnow were designed, based on the published sequences of the Mx gene from grass carp *Ctenopharyngon idellus* (GenBank accession number, AY395698), crucian carp *Carassius auratus* (GenBank accession number, AY303812), and *Danio rerio* (accession No., NM_182942). Primer sequences were SMFc and SMRb (Table 1). Reverse transcription was conducted using random hexamer primers with SuperScript™ III Reverse Transcriptase (Invitrogen). PCR was carried out in an ABI 9700 Thermal Cycler Instrument in a 25 μl reaction volume containing 1 μl of cDNA, 2.5 μl of 10× buffer, 2.5 μl of Mg^{2+} (25mM), 1 μl of dNTPs (10mM), 1 μl of Taq polymerase (1 U/μl), 1 μl of each primer (10 mM), and 15 μL of nuclease-free water. The thermal cycling profile was a denaturing step of 94℃ for 5min followed by 35 cycles of 94℃ 30s, 60℃ 30s and 72℃ 1min, 5min at 72℃ for the final extension, and holding at 20℃. The PCR product was separated on a 1% agarose gel, purified using the Axygen gel purification kit (Axygen) and ligated into pMD-18 T vector (TaKaRa). The ligation product was transformed into competent Top10 cells (Invitrogen). After screening by PCR, three positive colonies were sequenced with an ABI 3730 DNA sequencer (Applied Biosystems). A 989 bp sequence was obtained. By BLASTX searching in the GenBank database, the sequence was significantly homologous to known fish Mx sequences.

2.2 Rapid amplification of cDNA ends (RACE)

Using the BD SMART™ RACE cDNA amplification kit (BD Biosciences Clontech), first strand cDNA synthesis and RACE were performed on liver-derived RNA. To obtain the 3′ unknown region, primer pairs, SMF52a/adaptor primer UPM and SMF58a/adaptor primer NUP

(Table 1), were used for the primary PCR and the nested PCR respectively. The amplified PCR product was cloned and sequenced as described above. Similarly, the 5′ end of GrMx was obtained by nested PCR using primer pairs SMR53a/UPM and SMR57a/NUP (Table 1). The full length cDNA sequences were confirmed by sequencing the PCR product amplified by primers SMFh and SMRi (Table 1) within the predicted 5′ and 3′ untranslated regions respectively.

Table 1 Oligonucleotide primers used in these experiments

Primer name	Sequence (5′→3′)	Amplicon length (nt) and primer information
Mx		
SMFc (forward)	TGACACGCTGTCCTCTKGTA	989
SMRb (reverse)	GKTTTCCTCCGTCTTWAWGG	Gene cloning
SMF52a (forward)	CAACCGGACAAAGGACTCTGGG	3′RACE
SMF58a (forward)	GAAGGGCTACATGATCGTGAAGTG	
SMR53a (reverse)	GATGTCTTGCTGGCCTCTGCAC	5′RACE
SMR57a (reverse)	CTGACCACTATATCCTCCAT	
SMFh (forward)	GACATCACAGAAGGCTGGTGAACTA	2050
SMRi (reverse)	AACAAAGGAACAGCCAGGAAAGAT	Confirming sequence
SMF79a (forward)	ATCGGTCGACGACATCACAGAAGGCTGGTGAACTA	2284
SMR80a (reverse)	ATCGGGGCCCAAGCAGTGGTATCAACGCAGAGT	Inserting GrMx
SMF100 (forward)	CACGCTGTCCTCTGGTATTGA	101
SMR101 (reverse)	CAGTTTCTTTGTTTGGCTCTGATAT	qRT-PCR
β-actin		
SGAF86 (forward)	GATGATGAAATTGCCGCACTG	135
SGAR87 (reverse)	ACCAACCATGACACCCTGATGT	qRT-PCR
CMV promoter		
SCF93a (forward)	CTCGAGTAGTTATTAATAGTAATCAATTACG	630
SCR94a (reverse)	AAGCTTATGGGCCCGTCGACCGATCTGACGGTTCACTA	Promoter
SV40 transcriptional termination site		
SVF81a (forward)	GGGCCCAGCGGCCGCGACTCTAGATCAT	224
SVR95a (reverse)	AAGCTTGCAGTGAAAAAAATGCTTTATTTGTG	Transcriptiona l termination
Universal adaptor primer		
UPM	Long: CTAATACGACTCACTATAGGGCAAGCAGTGGTATCAACGCAGAGT Short: CTAATACGACTCACTATAGGGC	RACE
NUP	AAGCAGTGGTATCAACGCAGAGT	
5′-RACE adaptor		
OligodG	AAGCAGTGGTAACAACGCAGAGTACGCGGG	5′RACE
3′-RACE primer		
3′-CDS	AAGCAGTGGTATCAACGCAGAGTAC(T)$_{30}$VN	3′RACE

Note: K=G/T; W=A/T; V=A/G/C; N=A/G/C/T

2.3 Sequence analysis

Sequence homology was obtained using BLAST program (http://www.ncbi.nlm.nih.gov/blast). The deduced amino acid sequences were analyzed with the Expert Protein Analysis System (http://www.expasy.org/) and the protein domain features were predicted by Simple Modular Architecture Research Tool (SMART) (http://smart.embl-heidelberg.de/) [25]. Intra-domain features were predicted by a scan of the sequence against the PROSITE database (http://us.expasy.org/tools/scanprosite) [26]. The subcellular localization of this protein was predicted using LOCtree software (http://cubic.bioc.columbia.edu/services/loctree/) [27].

2.4 Plasmid construction

The pCMV-EGFP plasmid (Clontech, USA) is used as the plasmid skeleton. The second CMV promoter was obtained and introduced restriction sites in both upstream and downstream of the sequence by PCR, using the pCMV-EGFP as template. A PCR sense primer was SCF93a with *Xho* I site, and an antisense primer was SCR94a with *Hin*d III, *Apa* I, and *Sal* I sites (Table 1). The PCR product of 630 bp was purified with gel extraction kit (Axygen), ligated into pMD-18T vector (TAKARA), transformed into TOP10 competent cells. Three positive colonies were selected and sequenced for verification of the insert without mutation. The plasmid with correct insert was extracted and digested with *Xho* I and *Hin*d III, meanwhile, the pCMV-EGFP plasmid was also digested with the same enzymes. The target fragments were purified, ligated with T4 ligase, and named as pCMV-EGFP-CMV.

The PCR to get the SV40 transcriptional termination site was set up with pCMV-EGFP plasmid as template. The forward primer was SVF81a with *Apa* I site, and the reverse primer was SVR95a with *Hin*d III site (Table 1). The amplicon was carried out and confirmed as above. The plasmid with correct insert was digested with *Apa* I and *Hin*d III, and cloned into the pCMV-EGFP-CMV plasmid. It was designated with pCMV-EGFP-CMV-SV40.

GrMx cDNA sequence was amplified by sense primer SMF79a with *Sal* I site and antisense primer SMR80a with *Apa* I site, using the reverse transcription product for 3′ RACE as a template. The amplicon was dealt with and verified as above. The plasmid with target sequence was digested with *Sal* I and *Apa* I, and inserted into pCMV-EGFP-CMV-SV40 plasmid. The recombinant plasmid was called after pCMV-EGFP-CMV-Mx-SV40 (Fig. 1).

2.5 Microinjection and culture of rare minnow embryos

The 7818 bp recombinant plasmid, pCMV-EGFP-CMV-Mx-SV40, was digested by *Cla* I to linearize the DNA. The linearized DNA fragment was purified using gel extraction kit (Axygen). After artificial insemination, about 1nl of linearized recombinant plasmid (100 pg/nl) was microinjected into rare minnow embryo at the 1-2 cell stage under a dissecting microscope with a micromanipulater [28-30]. The control group was injected with linearized pCMV-EGFP plasmid. We injected approximate 600 embryos for every group. The injected and uninjected embryos were subsequently incubated in sterile 0.3 × Danieau's solution (19.3 nM NaCl, 0.23 mM KCl, 0.13 mM $MgSO_4 \cdot 7H_2O$, 0.2 mM $Ca(NO_3)_2$, 1.67 mM HEPES pH 7.2) at 25℃. We changed the solution and removed dead embryos twice a day. The non-fluorescent injected embryos were removed under an Olympus SZ×12 fluorescent microscope at 48 h post-transgene. After hatching, we reared the fry in aerated tap water.

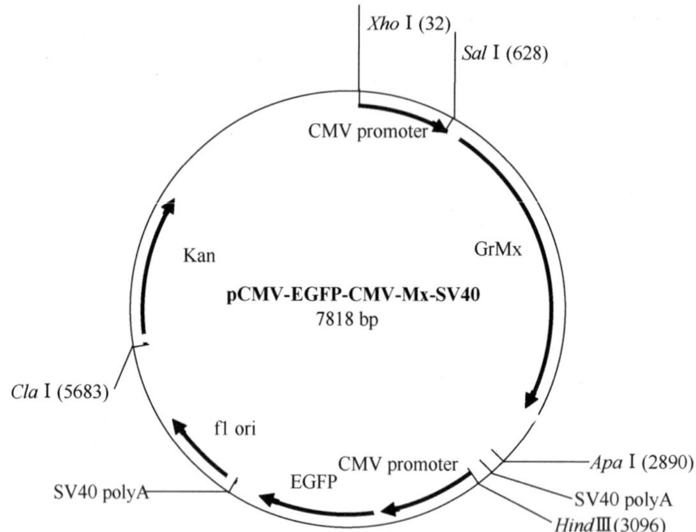

Fig. 1 The construct of rare minnow Mx expression vector. The transgene plasmid contained CMV promoter, GrMx sequence, SV40 transcription termination sequence, restriction sites and the skeleton component of original pCMV-EGFP plasmid

2.6 Artificial infection with GCRV and data analysis

Sixty five-month-old rare minnows were randomly sampled from each group, including Mx-transgenic group, control group (EGFP-transgenic group) and blank group (without any treatment). All the fish were injected with 10 μl GCRV suspended in PBS (2×10^8 PFU/ml) per gram body weight under the ventral fin and were kept in three aerated aquaria at 28℃. In order to test the average mortality rate among transgenic fish during experiment period, sixty Mx-transgenic animals were not infected and just kept in parallel to infected fish. Survival rate was defined with survival animals divided by total animals in one group (60 animals). Data were compared using CHITEST. The differences were considered statistically significant at $P < 0.05$.

2.7 Reproduction of positive transgenic fish

The embryos (P0 generation) injected with pCMV-EGFP-CMV-Mx-SV40 or pCMV-EGFP were raised to maturation and mated with wild type rare minnow. If there were some embryos showing green fluorescence under an Olympus SZ×12 fluorescent microscope, these embryos (F1 generation) were considered as pCMV-EGFP-CMV-Mx-SV40 or pCMV-EGFP transgenic embryos and sorted out for further research. The confirmation of GrMx transcription in F1 generation rare minnow was conducted by real time fluorescent quantitative reverse transcription polymerase chain reaction (qRT-PCR). The intestine tissue was homogenized in TRIZOL® LS reagent (Invitrogen) and total RNA was extracted according to the manufacturer's instruction. Total RNA was incubated with RNase-free DNase I (Roche) to remove contaminated genomic DNA before being reverse transcribed into cDNA using random hexamer primers with SuperScript™ III Reverse Transcriptase (Invitrogen). The Mx mRNA expressions were examined with primers SMF100 and SMR101 (Table 1). β-actin was utilized as an internal control for cDNA normalization. The gene specific primers for β-actin were forward primer SGAF86 and reverse primer SGAR87 (Table 1). The qRT-PCR methodology was as previously report [31].

2.8 Electron microscopy

Three individuals injected GCRV for each group were killed and the intestines were rapidly collected at 144 h post-injection (moribund fish for control group) and cut into 1-2 mm pieces. The samples were quickly and separately fixed in 2.5% glutaraldehyde solution. After fixation for 24 h, they were post-fixed in aqueous 1% osmium tetroxide (OsO_4), dehydrated, embedded and examined by routine ultramicroscopy under JEM-1230 transmission electron microscope.

2.9 Nucleotide sequence deposition

The GrMx cDNA sequence was submitted to GenBank and has been designated to accession number EF095273.

3. Results

3.1 Isolation and characterization of GrMx

The 989 bp fragment initially amplified with degenerate primers was highly homologous to known teleost Mx sequences. The 3′ and 5′ ends of the Mx gene were successfully obtained using RACE technique. These fragments were aligned to give the full length cDNA sequence of 2241 bp with a putative ATTAAA polyadenylation signal (Fig. 2). It contained an open reading frame of 1878 bp coding a protein of 625 amino acids with a predicted molecular weight of 70.928 kDa and a predicted isoelectric point of 7.33. Based on this sequence, primers to amplify the full coding sequence were designed, and RT-PCR was performed with liver tissue sample infected by GCRV. A unique band was obtained, indicating that there is no

```
   1                                      M  F  S  S  S  S  T  K
   1 GACATCACAGAAGGCTGGTGAACTACAGAGTTAAGACATGTTTTCAAGCAGTTCTACGAA
   9 G  K  S  S  G  L  N  Q  H  Y  E  E  K  V  R  P  C  I  D  L
  61 GGGAAAGAGTAGTGGTCTGAACCAGCACTATGAAGAGAAAGTGCGCCCATGCATTGATCT
  29 V  D  S  L  R  S  L  G  V  E  K  D  L  N  L  P  A  I  A  V
 121 AGTGGACTCTCTCAGGTCATTGGGTGTTGAAAAGGACCTGAACCTGCCAGCTATTGCTGT
  49 I  G  D  Q  S  S  G  K  S  S  V  L  E  A  L  S  G  V  A  L
 181 CATAGGTGACCAGAGCTCAGGAAAGAGTTCTGTGTTGGAAGCCCTGTCTGGAGTGGCGCT
  69 P  R  G  T  G  I  V  T  R  C  P  L  V  L  K  L  K  K  I  S
 241 GCCTAGGGGAACAGGTATTGTGACACGCTGTCCTCTGGTATTGAAACTGAAGAAAATTTC
  89 K  D  K  N  W  H  Q  W  H  G  L  L  S  Y  Q  S  Q  T  K  K
 301 AAAGGACAAAAATTGGCATCAGTGGCATGGATTGCTGTCATATCAGAGCCAAACAAAGAA
 109 L  K  D  P  S  E  I  E  D  A  V  L  N  A  Q  T  V  L  A  G
 361 ACTGAAAGACCCATCAGAAATAGAAGATGCTGTCTTAAATGCTCAGACAGTATTGGCTGG
 129 K  G  E  G  I  S  H  E  M  I  T  L  E  I  Q  S  S  D  V  P
 421 AAAGGGAGAAGGGATCAGTCATGAAATGATCACTCTGGAGATCCAGTCCAGTGATGTCCC
 149 D  L  T  L  I  D  L  P  G  I  A  R  V  A  T  G  N  Q  P  K
 481 TGACCTCACTCTCATTGATCTGCCAGGCATTGCTAGAGTTGCCACTGGCAACCAGCCAAA
 169 D  I  E  K  Q  I  K  D  L  I  E  K  F  I  K  R  Q  E  T  I
 541 AGACATCGAGAAACAAATAAAAGATCTAATTGAAAAGTTCATTAAAAGACAAGAAACCAT
 189 S  L  V  V  V  P  A  N  I  D  I  A  T  T  E  A  L  Q  M  A
 601 CAGCTTGGTTGTGGTGCCTGCAAACATTGACATCGCCACCACTGAGGCACTGCAGATGGC
 209 S  K  V  D  S  T  G  Q  R  T  L  G  I  L  T  K  P  D  L  V
 661 ATCCAAAGTAGATTCAACCGGACAAAGGACTCTGGGTATTCTGACTAAACCAGATTTAGT
 229 D  K  G  M  E  D  I  V  V  R  T  V  N  N  Q  V  I  Q  L  K
 721 GGACAAAGGCATGGAGGATATAGTGGTCAGAACAGTCAATAATCAAGTGATACAACTGAA
 249 K  G  Y  M  I  V  K  C  R  G  Q  Q  D  I  N  E  K  L  D  L
 781 GAAGGGCTACATGATCGTGAAGTGCAGAGGCCAGCAAGACATCAATGAGAAGCTTGATCT
 269 V  K  A  L  E  K  E  R  H  F  F  D  E  H  S  H  F  R  S  L
 841 GGTCAAAGCGTTGGAAAAAGAAAGACATTTTTTTGACGAACATTCTCATTTCAGGTCTCT
 289 L  E  E  G  K  A  T  I  P  L  L  A  E  R  L  T  K  E  L  V
 901 TCTTGAAGAAGGAAAAGCTACAATACCCCTTCTTGCAGAAAGACTCACAAAAGAATTGGT
 309 E  H  I  T  K  T  L  P  Q  L  Q  K  Q  L  E  M  K  L  E  K
 961 CGAACACATTACTAAAACACTACCACAGTTGCAGAAACAACTTGAGATGAAATTAGAGAA
 329 T  T  E  D  L  R  A  L  G  D  G  V  P  T  D  E  Q  E  K  I
1021 GACGACTGAGGATCTTAGAGCACTGGGAGATGGAGTTCCTACTGATGAACAAGAGAAGAT
 349 N  F  F  I  T  K  I  R  Q  F  N  D  A  I  E  G  V  K  R  A
1081 CAATTTTTTTATCACGAAAATTCGCCAGTTCAATGATGCCATTGAAGGAGTAAAGAGGGC
 369 E  E  D  L  K  N  S  D  K  R  V  F  T  K  I  R  E  E  F  G
1141 AGAAGAAGATCTAAAAAACTCAGACAAAAGGGTCTTTACCAAAATCAGGGAGGAATTTGG
```

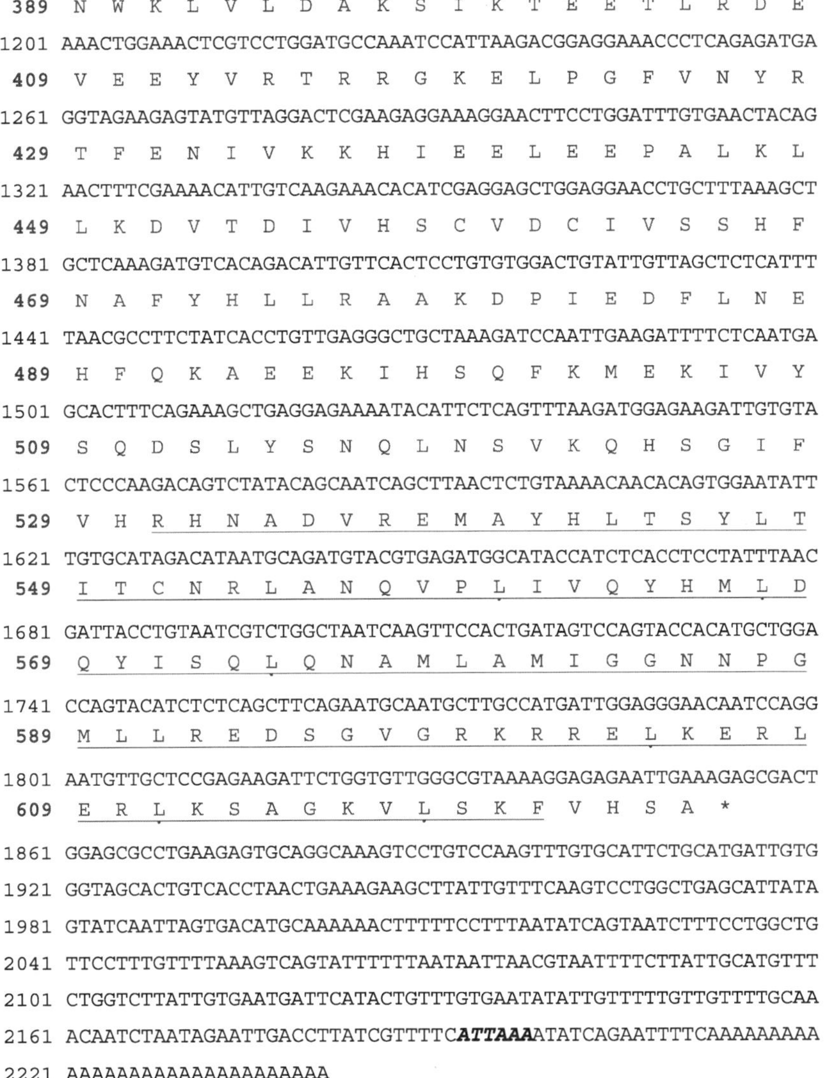

Fig. 2 Nucleotide sequence and translated amino acid sequence of the rare minnow Mx gene. The start codon (ATG) was boxed and the stop codon (TGA) was indicated with an asterisk. The polyadenylation signal sequence (ATTAAA) was in italics and bold. The positions of deduced amino acids were indicated in bold numbers. In the deduced amino acid sequence, the dynamin domain (DYNc) (*E*value = 7.92e-128) appeared with light grey background. The tripartite GTP-binding motifs were marked with double underlines and the dynamin family signature consensus sequence which was located at amino acid positions 68-77 was marked with a wavy underline. The GTPase effector domain (GED) (*E*value = 1.24e-21) was underlined. The leucine residues of the leucine zipper repeats were shown with underdots at the carboxy-terminal region

alternative splicing. The amplified product was cloned and sequenced in order to verify the Mx sequence obtained through overlapping fragments. GrMx protein exhibited its highest level of identity with crucian carp *Carassius auratus* Mx3 (AAP68827) (Identities = 89%, *E*-value = 0), followed by the grass carp *Ctenopharyngodon idella* Mx (Q6TKS7) (Identities =

88%, E value = 0). GrMx showed motifs characteristic of Mx proteins. The amino terminus contained a tripartite GTP binding motif (GDQSSGKS/DLPG/TKPD) and a dynamin family signature (LPRGTGIVTR). The carboxyl terminus contained two sets of putative leucine zipper motifs defined as in previous report [32]. No nuclear localization signal was found in the sequence.

3.2 Antiviral features of P0 generation in Mx-transgenic rare minnow

Five-month-old Mx-transgenic rare minnow, control group (EGFP-transgenic rare minnow) and blank group (without any treatment) were injected with GCRV. Death in control group and blank group occurred on d 5 (20 and 18 animals), d 6 (34 and 38 animals) and d 7 (6 and 4 animals) post-injection, while death of Mx-transgenic rare minnow occurred on d 5 (2 animals), d 6 (28 animals), d 7 (26 animals) and d 8 (4 animals) post-injection. The animals in control group and blank group died off on d 7 post-injection, and Mx-transgenic animals died off on d 8 post-injection. The survival rate was very significantly high in Mx-transgenic rare minnow in d 5, d 6 and d 7 ($P < 0.01$). There were no significant differences between control group and blank group ($P > 0.05$) (Fig. 3a). All the rare minnow appeared very obvious symptoms in control and blank groups, one-third Mx-transgenic rare minnow didn't show apparent symptoms, especially in dead fish latterly. The transgenic fish that were not infected and just kept in parallel to infected fish did not die at all.

3.3 Positive screening and survival rate of F1 generation in transgenic rare minnow

Sixty male P0 generation Mx-transgenic rare minnow artificially inseminated wild female rare minnow, respectively. Because P0 generation was chimera, we totally got about 360 embryos with fluorescence from 3 to 60 mates. In control group, we obtained 418 embryos with fluoresce. Five-month-old F1 generation Mx-transgenic rare minnow and control group were injected with GCRV. The control group died out in seven days as above (60 animals). The Mx-transgenic group started death from d 5 to d 8, and there were not dead from d 9 to d 14. The survival rate was 33.33% (20 animals), which was very significant difference from the control ($P < 0.01$) (Fig. 3b).

3.4 Mx mRNA expression pattern in F1 generation

We randomly dissected 60 five-month-old rare minnows, and collected the intestine tissues to check Mx expression in every group (20 animals per group). The expression levels in control and blank were similar. The Mx expression in 50% of F1 Mx-transgenic fish was six-fold higher than average expression in control group, and that in 30% of F1 Mx-transgenic fish was ten-fold higher than average expression in control group (Fig. 4).

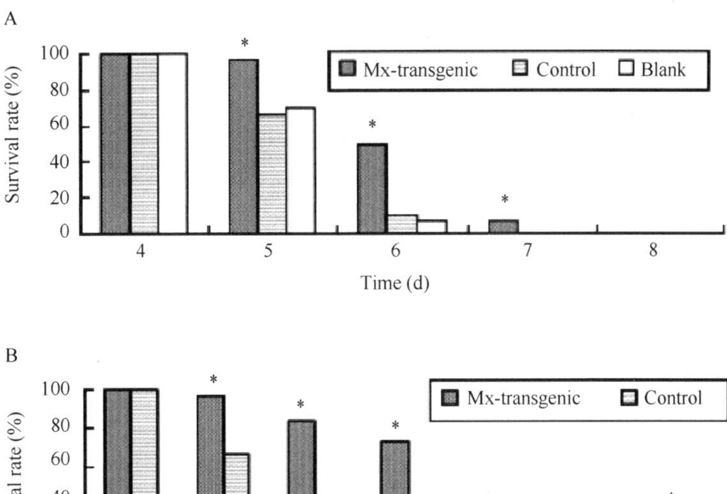

Fig. 3 Survival rate of rare minnow challenged with the GCRV. Rare minnows at an average weight of 1.5 g were five-months-old. They were challenged by injection with 10 μl GCRV suspended in PBS (2×10^{8} PFU ml^{-1}) per gram body weight under the ventral fin. Sixty fish were infected in each group. Data were analyzed using CHITEST and very significant differences were considered statistically significant at $P<0.01$. Asterisk indicated the very significant differences between control group and corresponding group. (A) P0 generation samples; (B) F1 generation rare minnows

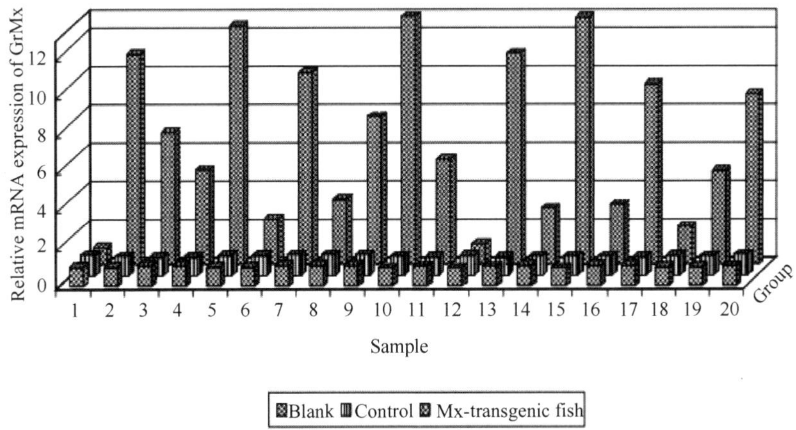

Fig. 4 Quantitative analysis of GrMx mRNA expression in intestine tissue. Blank group was not transgenic fish. Control group was EGFP-transgenic fish in F1 generation. Experimental group was Mx-transgenic fish in F1 generation. Each bar represented the expression level of Mx mRNA in sample relative to average expression level of Mx mRNA in blank group

3.5 The electron micrograph of the rare minnow's intestine injected with GCRV

The survival rare minnows have not any obvious symptoms. Virogenesis was studied in intestine of fish injected with GCRV experimentally. Subviral or viral particles could be easily observed in the intestine of moribund rare minnow with obvious symptoms, including transgenic group, control group, and blank group. We tried several times to find out very few subviral or viral particles in the transgenic moribund fish without evident symptoms. We could not find any subviral or viral particles in the survival rare minnows (Fig. 5).

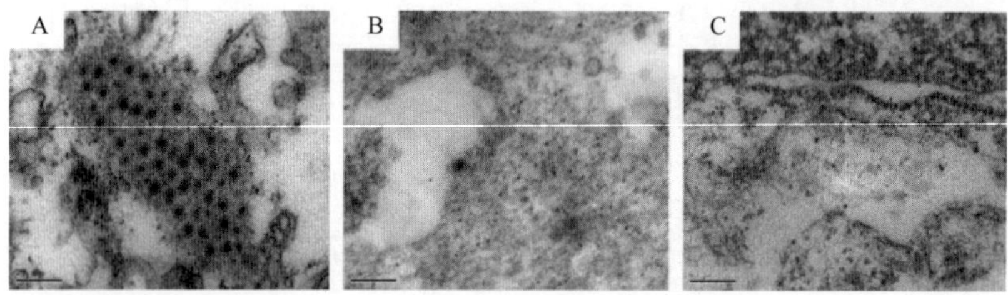

Fig. 5 The electron micrograph of the infected rare minnow mid-intestine showed virions at 144 h post-injection. (A) Moribund rare minnow with obvious symptoms in blank group; (B) Transgenic moribund fish without evident symptoms; (C) Survival fish

4. Discussion

GCRV is a most economically important viral pathogen for grass carp, causing mortalities mainly in juveniles. The pathogenesis and immune response of GCRV disease are not well known so far. Innate immunity is the first line of defense and therefore has a relevant role after infection.

In the current study, a cDNA from rare minnow, *Gobiocypris rarus*, homolog of Mx has been identified, characterized, and its antiviral function was verified *in vivo*. The translational start sequence of GrMx, GACATGT, does not conform to the consensus start sequence, ACCATGG, as determined by Kozak [33]. However, it abide by the reduced consensus start sequence of Cavener [34] (A/G)NCATG. The GrMx has 625 amino acids composed of five main functional domains: an amino-terminal tripartite GTP-binding motif GDQSSGKS, DLPG, TKPD, a dynamin family signature sequence LPRGTGIVTR, a GTPase effector domain (GED) and two carboxy-terminal leucine zipper motifs. The dynamin family signature sequence has been implicated in cellular transport processes [35]. Mutations in the GTP binding domain were found to markedly reduce antiviral activity [36]. The leucine zipper is thought to be responsible for oligomerisation, which is also essential for antiviral activity [37]. Without putative bipartite nuclear localization signal, the GrMx proteins located in the cytoplasm, which makes it to inhibit multiplication of GCRV easily. Highest amino acid identity was observed with the sequences from the crucian carp Mx3 (89%) and grass carp Mx (88%). The antiviral activity of Mx has been studies in some fish species *in vitro* [18,22,38], yet there is no clear evidence that piscine Mx protein is involved in the inhibition of virus *in vivo*.

Transgenic introduction of mouse Mx1 or human MxA is sufficient to turn susceptible mice into resistant animals [6,39]. They have Mx genes of their own but introducing more potent Mx genes might improve their antiviral resistance. Transgenic technology also has the potential to generate fast-growing [30], freeze-resistant [40], and disease-resistant fish [41] for aquaculture. To check the antiviral activity of the GrMx *in vivo*, we constructed the expression vector for transgenic fish and examined the survival rate after artificial infection. Although all the rare minnows died in 8 d post-injection in P0 generation, to some extent, the Mx-transgenic rare minnows exhibited some resistance to GCRV based on the statistic analysis (Fig. 3a). All survival fish were Mx-transgenic in the F1 generation. However, some transgenic fish died earlier than non-transgenic individuals. There are three kinds of integration, toxic, silent, and efficient in transgenic fish [42]. Presumably, the transgenic fish whose transgene was not expressed showed no phenotypic difference from the control. On the other hand, the Mx over-expression was coincident with survival rate, which implied that the antiviral character of transgenic fish, to some extent, related to the expression of Mx gene. Up to now, detection of GCRV in rare minnow is difficulty by RT-PCR. We reflected the quantification of viral particles with electron microscope technology.

All these observations suggested that transgenic fish over-expressing Mx were more resistant to GCRV. Further study will be needed to elucidate antiviral mechanism of this Mx protein in rare minnow in more details, which would be helpful to obtain more insight into the fish innate immune system and to develop new tools for long-lasting protection against virus infection.

Acknowledgements We thank Wei Hu, Ming Li, Lin Yang, Feng Xiong, Jun Dai, Bing Tang and other laboratory members for technical assistance and helpful discussion. This work was supported by grants from National Natural Science Foundation of China (30740009, 30871917), from the project-sponsored by SRF for ROCS, SEM (14110104), from Northwest A & F University in China (08080113,01140309 and 01140508), from China Postdoctoral Science Foundation (20070410298) and from Institute of Hydrobiology, CAS (2007FB09).

References

[1] Altmann SM, Mellon MT, Johnson MC, Paw BH, Trede NS, Zon LI, et al. Cloning and characterization of an Mx gene and its corresponding promoter from the zebrafish, *Danio rerio*. Dev Comp Immunol 2004; 28: 295-306.

[2] Haller O, Staeheli P, Kochs G. Interferon-induced Mx proteins in antiviral host defense. Biochimie 2007; 89: 812-8.

[3] Haller O, Kochs G, Weber F. Interferon, Mx, and viral countermeasures. Cytokine Growth Factor Rev 2007; 18: 425-33.

[4] Haller O, Stertz S, Kochs G. The Mx GTPase family of interferon-induced antiviral proteins. Microbes Infect 2007; 9: 1636-43.

[5] Staeheli P, Haller O, Boll W, Lindenmann J, Weissmann C. Mx protein: constitutive expression in 3T3 cells transformed with cloned Mx cDNA confers selective resistance to influenza virus. Cell 1986; 44: 147-58.

[6] Arnheiter H, Skuntz S, Noteborn M, Chang S, Meier E. Transgenic mice with intracellular immunity to influenza virus. Cell 1990; 62: 51-61.

[7] King MC, Raposo G, Lemmon MA. Inhibition of nuclear import and cell-type progression by mutated forms of the dynamin-like GTPase MxB. Proc Natl Acad Sci USA 2004; 101: 8957-62.

[8] Haller O, Acklin M, Staeheli P. Influenza virus resistance of wild mice: wild-type and mutant Mx alleles occur at comparable frequencies. J Interferon Cytokine Res 1987; 7: 647-56.

[9] Jin HK, Takada A, Kon Y, Haller O, Watanabe T. Identification of the murine Mx2 gene: interferon-induced expression of the Mx2 protein from the feral mouse gene confers resistance to vesicular stomatitis virus. J Virol 1999; 73: 4925-30.

[10] Nakajima E, Morozumi T, Tsukamoto K, Watanabe T, Plastow G, Mitsuhashi T. A naturally occurring variant of porcine mx1 associated with increased susceptibility to influenza virus *in vitro*. Biochem Genet 2007; 45: 11-24.

[11] Palm M, Leroy M, Thomas A, Linden A, Desmecht D. Differential anti-influenza activity among allelic variants at the *Sus scrofa* Mx1 locus. J Interferon Cytokine Res 2007; 27: 147-55.

[12] Seyama T, Ko JH, Ohe M, Sasaoka N, Okada A, Gomi H, et al. Population research of genetic polymorphism at amino acid position 631 in chicken Mx protein with differential antiviral activity. Biochem Genet 2006; 44: 437-48.

[13] Pavlovic J, Zurcher T, Haller O, Staeheli P. Resistance to influenza virus and vesicular stomatitis virus conferred by expression of human MxA protein. J Virol 1990; 64: 3370-5.

[14] Larsen R, Røkenes TP, Robertsen B. Inhibition of infectious pancreatic necrosis virus replication by Atlantic salmon Mx1 protein. J Virol 2004; 78: 7938-44.

[15] Lockhart K, McBeath AJA, Collet B, Snow M, Ellis AE. Expression of Mx mRNA following infection with IPNV is greater in IPN-susceptible Atlantic salmon post-smolts than in IPN-resistant Atlantic salmon parr. Fish shellfish immunol 2007; 22: 151-6.

[16] DeWitte-Orr SJ, Leong JC, Bols NC. Induction of antiviral genes, Mx and vig-1, by dsRNA and Chum salmon reovirus in rainbow trout monocyte/macrophage and fibroblast cell lines. Fish Shellfish Immunol 2007; 23: 670-82.

[17] Tafalla C, Chico V, Perez L, Coll JM, Estepa A. *In vitro* and *in vivo* differential expression of rainbow trout (*Oncorhynchus mykiss*) Mx isoforms in response to viral haemorrhagic septicaemia virus (VHSV) G gene, poly I:C and VHSV. Fish shellfish immunol 2007; 23: 210-21.

[18] Lin C, Christopher John JA, Lin C, Chang CY. Inhibition of nervous necrosis virus propagation by fish Mx proteins. Biochem Biophys Res Commun 2006; 351: 534-9.

[19] Chen Y, Su Y, Lin JH, Yang H, Chen T. Cloning of an orange-spotted grouper (*Epinephelus coioides*) Mx cDNA and characterisation of its expression in response to nodavirus. Fish Shellfish Immunol 2006; 20: 58-71.

[20] Wu YC, Chi SC. Cloning and analysis of antiviral activity of a barramundi (*Lates calcarifer*) Mx gene. Fish Shellfish Immunol 2007; 23: 97-108.

[21] Fernandez-Trujillo MA, Porta J, Borrego JJ, Alonso MC, Alvarez MC, Bejar J. Cloning and expression analysis of Mx cDNA from Senegalese sole (*Solea senegalensis*). Fish Shellfish Immunol 2006; 21: 577-82.

[22] Fernandez-Trujillo MA, Garcia-Rosado E, Alonso MC, Borrego JJ, Alvarez MC, Bejar J. *In vitro* inhibition of sole aquabirnavirus by Senegalese sole Mx. Fish Shellfish Immunol 2008; 24: 187-93.

[23] Garcia-Rosado E, Alonso MC, Bejar J, Manchado M, Cano I, Borrego JJ. Expression analysis of Mx protein and evaluation of its antiviral activity against sole aquabirnavirus in SAF-1 and TV-1 cell lines. Vet Immunol Immunopathol 2008; 121: 123-9.

[24] Wang T, Liu P, Chen H. Preliminary study on the susceptibility of *Gobiocypris rarus* to hemorrhagic virus of grass

carp (GCHV). Act Hydrobiol Sin 1994; 18: 144-9 (In Chinese, English abstract).

[25] Letunic I, Copley RR, Pils B, Pinkert S, Schultz J, Bork P. SMART 5: domains in the context of genomes and networks. Nucleic Acids Res 2006; 34: D257-60.

[26] Hulo N, Bairoch A, Bulliard V, Cerutti L, Cuche B, De Castro E, et al. The 20 years of PROSITE. Nucleic Acids Res 2008; 36: D245-9.

[27] Nair R, Rost B. Mimicking cellular sorting improves prediction of subcellular localization. J Mol Biol 2005; 348: 85-100.

[28] Chen TT, Powers DA. Transgenic fish. Trends Biotechnol 1990; 8: 209-15.

[29] Du SJ, Gong ZY, Fletcher GL, Shears MA, King MJ, Idler DR, et al. Growth enhancement in transgenic Atlantic salmon by the use of an "all fish" chimeric growth hormone gene construct. Biotechnology (N Y) 1992; 10: 176-81.

[30] Zhu Z, Li G, He L, Chen S. Novel gene transfer into the fertilized eggs of goldfish (*Carassius auratus* L. 1758). Z Angew Ichthyol 1985; 1: 31-4.

[31] Su J, Zhu Z, Wang Y. Molecular cloning, characterization and expression analysis of the PKZ gene in rare minnow *Gobiocypris rarus*. Fish shellfish immunol 2008; 25: 106-13.

[32] Landschultz WH, Johnson PF, McKnight SL. The leucine zipper: a hypothetical structure common to a new class of DNA binding proteins. Science, 1988; 240: 1759-64.

[33] Kozak M. Point mutations define a sequence flanking the AUG initiator codon that modulates translation by eukaryotic ribosomes. Cell 1986; 44: 283-92.

[34] Cavener DR, Ray SC. Eukaryotic start and stop translation sites. Nucleic Acids Res 1991; 19: 3185-92.

[35] Chieux V, Hober D, Chehadeh W, Wattre P. Alpha interferon, antiviral proteins and their value in clinical medicine. Annales de biologie clinique 1999; 57: 659-66.

[36] Melen K, Julkunen I. Mutational analysis of murine Mx1 protein: GTP binding core domain is essential for anti-influenza A activity. Virology 1994; 205: 269-79.

[37] Ponten A, Sick C, Weeber M, Haller O, Kochs G. Dominant-negative mutants of human MxA protein: domains in the carboxyl-terminal moiety are important for oligomerization and antiviral activity. J Virol 1997; 71: 2591-9.

[38] Caipang CMA, Hirono I, Aoki T. *In vitro* inhibition of fish rhabdoviruses by Japanese flounder, *Paralichthys olivaceus* Mx. Virology 2003; 317: 373-82.

[39] Pavlovic J, Arzet HA, Hefti HP, Frese M, Rost D, Ernst B, et al. Enhanced virus resistance of transgenic mice expressing the human MxA protein. J Virol 1995; 69: 4506-10.

[40] Fletcher GL, Shears MA, King MJ, Davies PL, Hew CL. Evidence for antifreeze protein gene transfer in Atlantic salmon. Can J Fish Aquat Sci 1988; 45: 352-7.

[41] Zhong J, Wang Y, Zhu Z. Introduction of the human lactoferrin gene into grass carp (*Ctenopharyngodon idellus*) to increase resistance against GCH virus. Aquaculture 2002; 214: 93-101.

[42] Zhu Z, Xu K, Xie Y, Li G, He L. A model of transgenic fish. Sci China Ser B 1989; 2: 147-55.

转 Mx 基因稀有鮈鲫具有增强的抗草鱼出血病病毒的能力

苏建国[1,2] 杨春荣[1] 朱作言[2] 汪亚平[2] 张成勋[2] 廖兰杰[2]

1 西北农林科技大学，陕西省农业分子生物学重点实验室，杨凌 712100
2 中国科学院水生生物研究所，武汉 430072

摘 要 在干扰素诱导的抗病毒机制中，Mx 通路是最有效的机制之一。Mx 蛋白具有直接抗病毒的活性，在病毒感染的早期阶段阻断病毒基因组的复制。然而，目前对鱼的 Mx 蛋白在体内的抗病毒活性还不清楚。在本研究中，我们克隆得到了一个 Mx 基因并且描述了其特征，同时把它导入到稀有鮈鲫中，验证了其在体内的抗病毒活性。稀有鮈鲫中 Mx 基因的全长 cDNA 为 2241 bp，该基因可编码一相对分子质量为 70.928 kDa、等电点为 7.33、含 625 个氨基酸的多肽片段。对氨基酸序列进行分析，发现成熟的 Mx 多肽在氨基酸末端有一个三联体 GTP 结合区域，一个发动蛋白家族的典型结构特征序列，一个 GTP 酶的效应结构域以及两个 C 端亮氨酸拉链基序，其与鲫的 Mx3 序列相似度最高，达到 89%。以草鱼呼肠孤病毒(GCRV)感染后，P0 和 F1 转 Mx 基因稀有鮈鲫均表现出非常显著的高成活率($P < 0.01$)。在 F1 中，Mx 基因的 mRNA 表达情况与 F1 的生存率保持一致。利用电镜技术观察到病毒产量也与存活期相一致。稀有鮈鲫自身拥有 Mx 基因，但是表达更多的 Mx 基因可以提高对草鱼呼肠孤病毒的抗性。转 Mx 基因稀有鮈鲫可能有助于控制 GCRV 疾病。

Isolation and Characterization of Argonaute 2: A Key Gene of the RNA Interference Pathway in the Rare Minnow, *Gobiocypris rarus*

Jianguo Su[1,2] Zuoyan Zhu[2] Yaping Wang[2] Songhun Jang[2]

1 College of Animal Science and Technology, Northwest A&F University, Shaanxi Key Laboratory of Molecular Biology for Agriculture, Yangling 712100

2 State Key Laboratory of Freshwater Ecology and Biotechnology, Institute of Hydrobiology, Chinese Academy of Sciences, Wuhan 430072

Abstract Argonaute 2 gene plays a pivotal role in RNAi in many species. Herein is the first report of the cloning and characterization of Agronaute 2 gene in fish. The full-length cDNA of *Gobiocypris rarus* Argonaute 2 (GrAgo2) consisted of 3073 nucleotides encoding 869 amino acid residues with a calculated molecular weight of 98.499 kDa and an estimated isoelectric point of 9.18. Analysis of the deduced amino acid sequence showed the presence of two signature domains, PAZ and PiWi. RT-PCR analysis indicated that GrAgo2 mRNA expression could be detected in widespread tissues. After infection with grass carp reovirus, GrAgo2 expression was up-regulated from 12 h post-injection ($p < 0.05$) and returned to control levels at 48 h post-injection ($p > 0.05$). These data imply that GrAgo2 is involved in antiviral defense in rare minnow.

Keywords rare minnow (*Gobiocypris rarus*); argonaute 2; RNA interference; gene cloning; mRNA expression

1. Introduction

Sequence-specific gene silencing triggered by double-stranded RNA (dsRNA) is a fundamental gene regulatory mechanism present in almost all eukaryotes. dsRNA may emerge during viral infection and replication or after transposition of mobile genetic elements. dsRNA is processed by Drosha and/or Dicer into small dsRNA molecules of specific length and structure and can enter into various gene-silencing pathways that are collectively referred to as RNA interference (RNAi) [1].

Argonaute is the key component and at the heart of the effector complexes in RNAi [2,3]. The number of Argonaute orthologs identified in different organisms ranges from 1 in *Schizosaccharomyces pombe* [4] to 27 in *Caencrhabditis elegans* [5,6]. Ten members

文章发表于 *Fish & Shellfish Immunology*, 2009, 26(1): 164-170

have been identified in *Arabidopsis thaliana* [7], five members in *Drosophila melanogaster* [8], and eight members in human [9]. Argonaute proteins contain a PAZ motif in the middle and Piwi motif in the C-terminal region [10,11]. Based on the sequence alignments, the Argonaute family was classified into two subfamilies referred to as the Ago and Piwi subfamilies [11]. Ago subfamily is ubiquitously expressed and participates in siRNAs (small interfering RNAs) and miRNAs (microRNAs) pathways, whereas Piwi subfamily is expressed specifically in the germline and early development of the embryo [12]. Human Argonaute proteins include the PiWi subfamily (*Piwil1/HIWI, Piwil2/HILI, Piwil3,* and *Piwil4/HIWI2*) and the eIF2C/AGO subfamily (*EIF2C1/hAGO1, EIF2C2/hAGO2, EIF2C3/hAGO3,* and *EIF2C4/hAGO4*) [9]. Human Argonaute 2 (Ago2) mediates RNA cleavage targeted by miRNAs and siRNAs [2]. The effector complex activity requires Ago2 but not Ago1, Ago3, or Ago4 [1]. In *D. melanogaster*, Ago2 but not Ago1 is required for dsRNA- and siRNA-triggered RNAi [13]. In *Aedes aegypti*, Ago2 is associated with RNAi [14].

Many Ago genes have been identified in eukaryotes. However, Ago2 gene has not previously been cloned in fish, except for a predicted Ago2 gene in *Danio rerio* during preparation of this manuscript.

Grass carp (*Ctenopharyngodon idellus*) is an important aquaculture species in China, but Grass Carp Reovirus (GCRV) seriously threatens its production. Better understanding of the immune defense mechanisms may contribute to the development of management strategies for disease control and long term sustainability of grass carp farming. The rare minnow *Gobiocypris rarus*, which is a small cyprinid species, has been recognized as a useful model. It is very sensitive to GCRV [15].

RNAi is a powerful tool to silence gene expression, which has been used as an antivirus agent to inhibit specific virus replication. Here, we employed the rare minnow as a model fish to explore whether RNAi plays a role in GCRV disease. A full-length cDNA sequence encoding an Argonaute 2 protein (GrAgo2) was identified and characterized. We present GrAgo2 mRNA expression profiles in different tissues and following GCRV infection.

2. Materials and Methods

2.1 Experimental animals and sample preparation

Rare minnow (*G. rarus*) from a laboratory-breeding stock was used in the experiments. Rare minnow (2-3 g) was acclimatized to laboratory conditions for one week in a quarantine area. They were kept in 25 L aerated aquaria with running freshwater at 28℃ and fed twice a day with commercial diet (feed composition: protein 32%, starch 63%, fat 3%, additive 2%)

at 3% body weight per day.

For the viral challenge, 10 μl of GCRV (991 strain) per gram body weight, suspended in PBS (2×10^8 PFU/ml), were injected intraperitoneally. The control animals were injected with PBS. The blank group was not injected. Three individuals from each group were killed and tissues including gill, heart, intestine, kidney, liver, muscle, and spleen were collected at 0, 12, 24, 36, 48, 72, 96, 120, 144 and 168 h after injection. The samples were homogenized in TRIZOL® LS reagent (Invitrogen) and total RNA was isolated according to the manufacturer's instructions. Total RNA was incubated with RNase-free DNase I (Roche) to remove any contaminating genomic DNA before being reverse transcribed into cDNA using random hexamer primers with SuperScript™ III Reverse Transcriptase (Invitrogen).

2.2 Cloning of *G. rarus* Ago2 (GrAgo2) cDNA

To identify Ago2 cDNA from rare minnow, degenerate primers were designed based on the multiple alignment of some Ago2 sequences including *Macaca mulatta* (accession No. XM_001100725), *Homo sapiens* (accession No. NM_012199), *Gallus gallus* (accession No. NM_001030900), *Rattus norvegicus* (accession No. XM_001058231), *Canis familiaris* (accession No. XM_532563), *Bos taurus* (accession No. XM_601262) and *Pan troglodytes* (accession No. XM_001167312). PCR was performed with primers SAF102a and SAR105a (Table 1) using the cDNA generated from liver. The PCR programme was: 1 cycle of 94℃/4 min, 35 cycles of 94℃/30 s, 60℃/30 s, 72℃/1 min 45 s; 1 cycle of 72℃/5 min. The PCR product was ligated into pMD18-T easy vector, transformed into competent *Escherichia coli* TOP10 cells, and plated on a LB-agar petridish. Positive clones containing inserts at the expected size were screened by colony PCR and plasmid prepared using an Axyprep™ plasmid miniprep kit (Axygen Biosciences). Three plasmid DNAs were sequenced by a commercial company (Shanghai Invitrogen Biotechnology Co., Ltd, China).

2.3 RACE PCR

Rapid amplification of cDNA ends (RACE) was performed using the BD SMART™ RACE cDNA amplification kit (BD Biosciences Clontech). The first strand cDNA synthesis and RACE were performed on liver-derived RNA. To obtain the 3′ unknown region, primer pairs, SAF128/adaptor primer UPM and SAF129/adaptor primer NUP (Table 1), were used for the primary PCR and the nested PCR respectively. The amplified PCR products were cloned and sequenced as described above. Similarly, the 5′ end of GrAgo2 was obtained by nested PCR using primer pairs SAR165/UPM and SAR166/NUP (Table 1). The full-length cDNA sequences were confirmed by sequencing the PCR product amplified by primers SAFu and SARd (Table 1) within the predicted 5′ and 3′ untranslated regions respectively.

Table 1 Primers used in the study

Primer name	Sequence (5′→3′)	Amplicon length (nt) and primer information
Ago2		
SAF102a(forward)	TGGAARATGATGCTSAAYATTGATGT	1742 Gene cloning
SAR105a(reverse)	ACGTADGTGTGRCAVAGCTGGTA	
SAF128(forward)	GTAGTAGTCCAGAAGAGACACCACAC	3′RACE
SAF129(forward)	GAATGAACGGGTTGGAAAGAGTG	
SAR165(reverse)	CACTTCTCATCAGTTTGCTGATCTC	5′RACE
SAR166(reverse)	ATGGCTTGCTGGTCTCCTAGTCA	
SAFu (forward)	TGGCATCAGAAGAGCAAAAC	2758 Confirming sequence
SARd (reverse)	AAATGAGGCAAACACATAAAAA	
SAF118 (forward)	GTGCCAACTTCAACACCGAT	144 qRT-PCR
SAR119 (reverse)	CCTGGACTGGGGTTGCTAT	
β-actin		
SGAF86 (forward)	GATGATGAAATTGCCGCACTG	135 qRT-PCR
SGAR87 (reverse)	ACCAACCATGACACCCTGATGT	
Universal adaptor primer		
UPM	Long: CTAATACGACTCACTATAGGGCAAGCAGTGGTATCAACGCAGAGT Short: CTAATACGACTCACTATAGGGC	RACE
NUP	AAGCAGTGGTATCAACGCAGAGT	
5′-RACE adaptor		
OligodG	AAGCAGTGGTAACAACGCAGAGTACGCGGG	5′RACE
3′-RACE primer		
3′-CDS	AAGCAGTGGTATCAACGCAGAGTAC$(T)_{30}$VN	3′RACE

Note: R=A/G; S=C/G; Y=C/T; D=A/G/T; V=A/G/C; N=A/G/C/T

2.4 Sequence analysis and alignment

Sequence homology was obtained using BLAST program (http://www.ncbi.nlm.nih.gov/blast). The deduced amino acid sequences were analyzed with the Expert Protein Analysis System (http://www.expasy.org/) and the protein domain features were predicted by Simple Modular Architecture Research Tool (SMART) (http://smart.embl-heidelberg.de/) [16]. Intra-domain features were predicted by a scan of the sequence against the PROSITE database (http://us.expasy.org/tools/scanprosite) [17]. Multiple sequence alignments were created using the ClustalW2 (http://www.ebi.ac.uk/Tools/clustalw2/index.html) and the sequence manipulation suite (http://www.bioinformatics.org/sms/) programs. Phylogenetic and molecular evolutionary analyses were conducted using MEGA version 4.0 [18] and optimized manually.

2.5 Quantification of gene expression

The real-time fluorescence quantitative reverse transcription polymerase chain reaction method (qRT-PCR) was used to quantify the GrAgo2 gene expression in different tissues and infection states in an ABI Prism 7000 Sequence Detection System (Applied Biosystems). The invariant expressions of the putative house keeping gene β-actin under different situations were reported previously [19]. β-actin was utilized as an internal control for cDNA normalization. The gene specific primers for β-actin were forward primer SGAF86 and reverse primer SGAR87 (Table 1). GrAgo transcripts were detected using forward primer SAF118 and reverse primer SAR119 (Table 1). The qRT-PCR mixture consisted of 1 μl of cDNA sample, 8 μl nuclease-free water, 10 μl of SYBR Green PCR master mix (Toyobo), and 0.5μl of gene specific primers (5 μM). The PCR cycling conditions were: 1 cycle of 95℃/2 min, 40 cycles of 95℃/25 s, 60℃/30 s, 72℃/1 min, followed by dissociation curve analysis to verify the amplification of a single product. The threshold cycle (CT) value was determined using the manual setting on the ABI Sequence Detection System and exported into a Microsoft Excel Sheet for subsequent data analysis where the PCR efficiency and relative expression ratio of target gene in treated groups versus that in control were calculated by Pfaffl equation [20]. The expression data obtained from three independent biological replicates were subjected to one-way analysis of variance (one-way ANOVA) followed by an unpaired, two-tailed *t*-test. $p<0.05$ was considered statistically significant.

2.6 Nucleotide sequence deposition

The GrAgo2 mRNA sequence was submitted to GenBank and has been designated accession number EF636801.

3. Results

3.1 Cloning and analysis of GrAgo2 full-length cDNA

The first fragment of GrAgo2 cDNA was amplified using degenerate primers SAF102a and SAR105a. Of multiple amplified segments, a 1742 bp fragment was shown by Blastx analysis to have significant homology (*E*-value = 0.0) to the middle region of *Bos taurus* eukaryotic translation initiation factor 2C 2 gene (namely Ago2) (accession No. AAI51492). Based on the sequence of the amplified 1742 bp segment, primers SAF128 and SAF129 were subsequently designed and used with adaptor primers UPM and NUP to amplify the downstream sequence. Alignment of sequences from all homologous clones yielded a 2084 bp consensus sequence. The clone included a poly(A) tail, suggesting that the expressed sequence tag (EST) represented the 3′ region of the Ago2 gene.

The 5′ region of the gene was amplified by using 5′-RACE. A product of 1232 bp from nested PCR perfectly overlapped the initial 2084 bp consensus sequence. To confirm that the resulting 3073 bp consensus sequence represented a single mRNA expressed in the liver of adult *G. rarus*, primers were designed within the 5′ and 3′ UTRs of the sequence. Using these primers, a single 2758 bp product was amplified from liver cDNA. Sequence analysis of the product confirmed that the full-length transcript in *G. rarus* comprises 3073 nucleotides including the 5′ untranslated region, coding sequence, 3′ untranslated region and poly(A) tail (Fig. 1). The

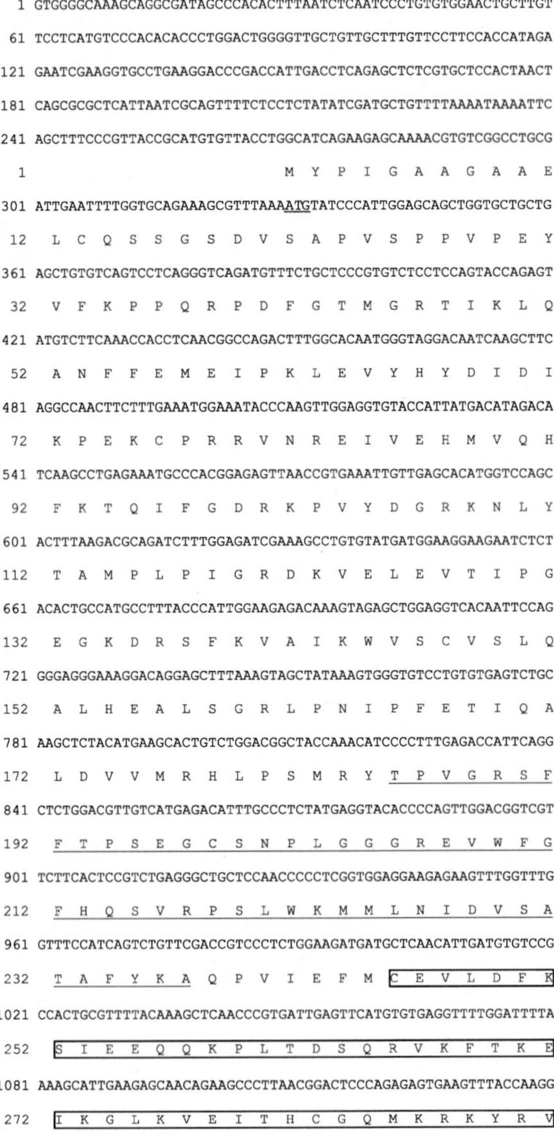

```
1141 AGATCAAAGGTCTGAAGGTTGAAATAACTCACTGTGGACAGATGAAGAGGAAATACAGGG
292   C  N  V  T  R  R  P  A  S  H  Q  T  F  P  L  Q  Q  E  N  G
1201 TGTGCAATGTGACTAGGAGACCAGCAAGCCATCAAACGTTTCCCTTGCAACAAGAGAATG
312   Q  T  I  E  C  T  V  A  Q  Y  F  K  D  K  Y  K  L  V  L  R
1261 GTCAGACCATTGAATGCACTGTCGCACAGTACTTCAAGGATAAGTACAAACTGGTGCTGC
332   Y  P  H  L  P  C  L  Q  V  G  Q  E  Q  R  H  T  Y  L  P  L
1321 GATATCCACATCTCCCATGTTTACAAGTTGGTCAGGAGCAGAGACACACCTACCTTCCTC
352   E  V  C  N  I  V  A  G  Q  R  C  I  K  K  L  T  D  N  Q  T
1381 TAGAGGTCTGTAACATAGTAGCCGGACAGAGATGCATCAAGAAACTGACAGACAATCAGA
372   S  T  M  I  R  A  T  A  R  S  A  P  D  R  Q  D  E  I  S  K
1441 CCTCCACTATGATACGTGCCACAGCCAGGTCTGCGCCTGACCGCCAGGATGAGATCAGCA
392   L  M  R  S  A  N  F  N  T  D  P  Y  V  R  E  F  G  V  M  V
1501 AACTGATGAGAAGTGCCAACTTCAACACCGATCCTTATGTGCGTGAGTTTGGAGTCATGG
412   R  D  E  M  T  E  V  N  G  R  V  L  Q  A  P  S  I  L  Y  G
1561 TAAGAGACGAGATGACTGAGGTCAATGGTCGGGTCCTTCAGGCACCTTCGATTCTCTATG
432   G  R  N  K  A  I  A  T  P  V  Q  G  V  W  D  M  R  N  K  Q
1621 GTGGAAGGAACAAAGCAATAGCAACCCCAGTCCAGGGTGTGTGGGACATGAGGAACAAGC
452   F  H  T  G  I  E  I  K  V  W  A  I  A  C  F  A  P  Q  R  Q
1681 AGTTCCACACAGGGATTGAGATTAAAGTGTGGGCTATCGCCTGCTTTGCCCCACAGAGGC
472   C  T  E  L  L  L  K  A  F  T  D  Q  L  R  K  I  S  R  D  A
1741 AGTGTACAGAACTCCTTCTGAAGGCGTTCACTGACCAGCTTCGTAAAATTTCACGTGATG
492   G  M  P  I  Q  G  Q  P  C  F  C  K  Y  A  Q  G  A  D  S  V
1801 CTGGGATGCCCATTCAGGGCCAGCCGTGTTTTTGCAAATATGCCCAAGGAGCAGACAGTG
512   E  P  M  F  K  H  L  K  Y  T  Y  Q  G  L  Q  L  V  V  V  I
1861 TGGAGCCCATGTTCAAACACCTCAAGTACACTTACCAAGGCTTACAGTTGGTGGTGGTTA
532   L  P  G  K  T  P  V  Y  A  E  V  K  R  V  G  D  T  V  L  G
1921 TCCTACCCGGGAAGACCCCTGTTTATGCTGAGGTGAAGCGTGTTGGAGACACTGTTCTTG
552   M  A  T  Q  C  V  Q  V  K  N  V  Q  K  T  T  P  Q  T  L  S
1981 GCATGGCCACGCAGTGTGTCCAGGTGAAGAATGTACAGAAGACCACACCTCAGACCCTTT
572   N  L  C  L  K  I  N  V  K  L  G  G  V  N  N  I  L  L  P  Q
2041 CCAACCTCTGCCTCAAGATTAATGTCAAACTGGGTGGAGTCAACAACATTCTTCTTCCAC
592   G  R  P  L  V  F  Q  Q  P  V  I  F  L  G  A  D  V  T  H  P
2101 AGGGCAGGCCCTTGGTGTTTCAGCAACCAGTCATCTTTCTAGGTGCAGATGTGACTCATC
612   P  A  G  D  G  K  K  P  S  I  A  A  V  V  G  S  M  D  A  H
2161 CACCAGCTGGAGATGGCAAGAAACCCTCAATTGCTGCTGTTGTTGGTAGTATGGACGCCC
632   P  S  R  Y  C  A  T  V  R  V  Q  Q  H  R  Q  D  I  I  Q  D
2221 ACCCAAGCCGATACTGTGCCACAGTGCGGGTGCAACAGCACCGTCAGGACATCATTCAGG
652   L  A  T  M  V  R  E  L  L  I  Q  F  Y  K  S  T  R  F  K  P
2281 ATCTGGCCACCATGGTGAGAGAGCTGCTCATCCAGTTCTACAAATCCACACGCTTTAAAC
672   T  R  I  I  Y  Y  R  D  G  I  S  E  G  Q  F  N  Q  V  L  Q
2341 CAACGCGCATAATCTACTACAGAGATGGCATCTCGGAGGGCCAGTTCAACCAGGTTCTAC
692   H  E  L  L  A  I  R  E  A  C  I  K  L  E  K  D  Y  Q  P  G
2401 AGCATGAACTGCTGGCTATCCGTGAGGCCTGCATTAAACTGGAGAAAGATTATCAACCAG
712   I  T  F  V  V  V  Q  K  R  H  H  T  R  L  F  C  M  D  R  N
2461 GAATTACCTTTGTAGTAGTCCAGAAGAGACACCACACCAGGCTCTTCTGCATGGACAGGA
732   E  R  V  G  K  S  G  N  I  P  A  G  T  T  V  D  T  K  I  T
```

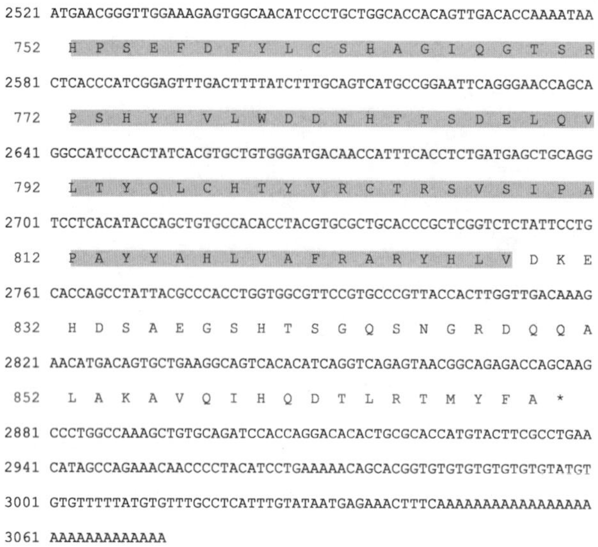

Fig. 1 Nucleotide sequence and deduced amino acid sequence of GrAgo2 cDNA. The start codon (ATG) is marked by wavy underline and the stop codon (TGA) is indicated with an asterisk. In the deduced amino acid sequence, DUF1785 domain is underlined, PAZ domain is marked by box, and Piwi domain is shaded

longest open reading frame commences at nucleotide 330 and terminates at nucleotide 2939, encoding a putative 869 amino acid polypeptide. Blastp analysis indicated that the translated sequence has significant homology to previously characterized Argonaute proteins.

3.2 Amino acid sequence analysis and phylogenetic relationships

The 869 amino acid polypeptide encoded in the amplified gene has a calculated molecular weight of 98,499 Da and isoelectric point of 9.18. Blastp search showed that this gene was most similar to *Mus musculus* eukaryotic translation initiation factor 2C, 2 (*E*-value = 0.0), indicating that it might be a rare minnow paralog of Ago2 (named GrAgo2). GrAgo2 has three main structural domains, one DUF1785 domain and one PAZ domain in the N-terminal region and one Piwi domain in the C-terminal (Fig. 1). We selected some representative Argonaute family member sequences, including from human, fish, tunicate, purple sea urchin, fruit fly, nematode, and thale cress, to compare the sequence similarities (Table 2). To compare the conserved domains and other sequence features, we chose and aligned some representative Ago2 sequences in vertebrates (Fig. 2). To study the molecular evolution, all the fish and human Argonaute family homologous protein sequences were used to construct a phylogenetic tree (Fig. 3). These homologue proteins can be divided into two subfamilies: Ago subfamily and Piwi subfamily. Every subfamily contains different groups. All the Ago2 proteins clustered together.

Table 2 Comparison of Argonaute amino acid sequences from different species

	1	2	3	4	5	6	7	8	9	10	11	12	13	14	15	16	17	18	19	20	21	22	23	24	25
1. GrAgo2																									
2. DrAgo1	79																								
3. DrAgo2	98	79																							
4. DrAgo3	79	83	79																						
5. DrAgo4	79	82	79	81																					
6. DrPiwi1	18	20	19	19	18																				
7. DrPiwi2	17	16	17	18	16	38																			
8. Tnupp	96	80	95	79	79	18	17																		
9. OmPiwi2	17	18	18	17	17	39	72	14																	
10. HsAgo1	80	95	81	83	83	19	18	81	16																
11. HsAgo2	91	80	91	78	78	19	17	91	16	82															
12. HsAgo3	79	83	80	95	81	19	18	80	17	84	79														
13. HsAgo4	78	82	78	81	97	17	17	78	17	83	78	81													
14. HsPiwi1	17	18	17	18	16	46	34	17	36	19	17	18	17												
15. HsPiwi2	18	18	19	19	19	39	55	18	56	17	17	19	19	34											
16. HsPiwi3	15	17	16	16	17	46	32	15	32	16	18	16	17	40	31										
17. HsPiwi4	17	17	17	18	16	46	34	17	36	19	17	18	17	99	34	40									
18. OdAgo1	30	30	30	29	30	16	18	30	17	29	30	30	29	17	16	16	17								
19. OdAgo2	72	66	71	70	69	18	16	71	14	71	72	70	70	17	17	18	17	30							
20. SpAgo1	67	65	67	69	68	18	17	67	17	68	69	69	68	18	16	17	18	30	64						
21. DmAgo1	73	68	72	72	71	17	16	73	16	74	74	72	71	18	17	18	18	31	65	66					
22. DmAgo2	28	30	28	28	29	16	17	28	16	29	29	28	29	15	18	17	15	24	29	27	28				
23. DmAgo3	15	16	16	15	17	34	35	15	34	18	18	15	17	31	32	31	31	15	14	15	15	15			
24. DmPiwi	18	20	19	18	17	33	31	18	29	18	19	18	17	31	30	32	31	15	17	18	17	11	29		
25. CeAgo2	65	62	64	64	64	17	17	65	13	65	65	64	64	16	17	17	16	30	65	61	67	27	13	18	
26. AtAgo2	29	29	30	29	27	16	18	29	16	29	28	27	27	13	16	15	14	22	25	27	27	25	14	16	29

Note: The species and accession numbers of the above protein sequences are as follows: GrAgo2, *Gobiocypris rarus* Argonaute 2 (ABV22635); DrAgo1, *Danio rerio* PREDICTED similar to Eif2c1 protein (XP_699384); DrAgo2, *Danio rerio* PREDICTED similar to Argonaute 2 (XP_699226); DrAgo3, *Danio rerio* PREDICTED similar to eukaryotic translation initiation factor 2C, 3 (XP_696563);DrAgo4, *Danio rerio* PREDICTED similar to eukaryotic translation initiation factor 2C, 4 (XP_691861); DrPiwi1, *Danio rerio* Piwi-like 1 (NP_899181); DrPiwi2, *Danio rerio* Piwi-like 2 (NP_001073668); Tnupp; *Tetraodon nigroviridis* unnamed protein product (CAG11109); OmPiwi2, *Oncorhynchus mykiss* ′Piwil2 protein (NP_001117714); HsAgo1, *Homo sapiens* eukaryotic translation initiation factor 2C, 1 (AAH63275); HsAgo2, *Homo sapiens* eukaryotic translation initiation factor 2C, 2 (NP_036286); HsAgo3, *Homo sapiens* eukaryotic translation initiation factor 2C, 3 isoform a (NP_079128); HsAgo4, *Homo sapiens* eukaryotic translation initiation factor 2C, 4 (NP_060099); HsPiwi1, *Homo sapiens* Piwi-like 1 (BAC81341); HsPiwi2, *Homo sapiens* Piwi-like 2 (NP_060538); HsPiwi3, *Homo sapiens* Piwi-like protein 3 (Q7Z3Z3); HsPiwi4, *Homo sapiens* Piwi-like 4 (NP_689644); OdAgo1, *Oikopleura dioica* Argonaute 1 (CAP07636); OdAgo2, *Oikopleura dioica* Argonaute 2 (CAP07637); SpAgo1, *Strongylocentrotus purpuratus* Argonaute 1 (ACE63524); DmAgo1, *Drosophila melanogaster* Argonaute I CG6671-PA, isoform a (NP_725341); DmAgo2, *Drosophila melanogaster* Argonaute 2 (ABB54719); DmAgo3, *Drosophila melanogaster* Argonaute 3 (ABO27430); DmPiwi, *Drosophila melanogaster* Piwi (AAD08705); CeAgo2, *Caenorhabditis elegans* Argonaute (plant)-like gene protein 2, isoform b (AAO38604); AtAgo2, *Arabidopsis thaliana* Argonaute 2 (NP_174413)

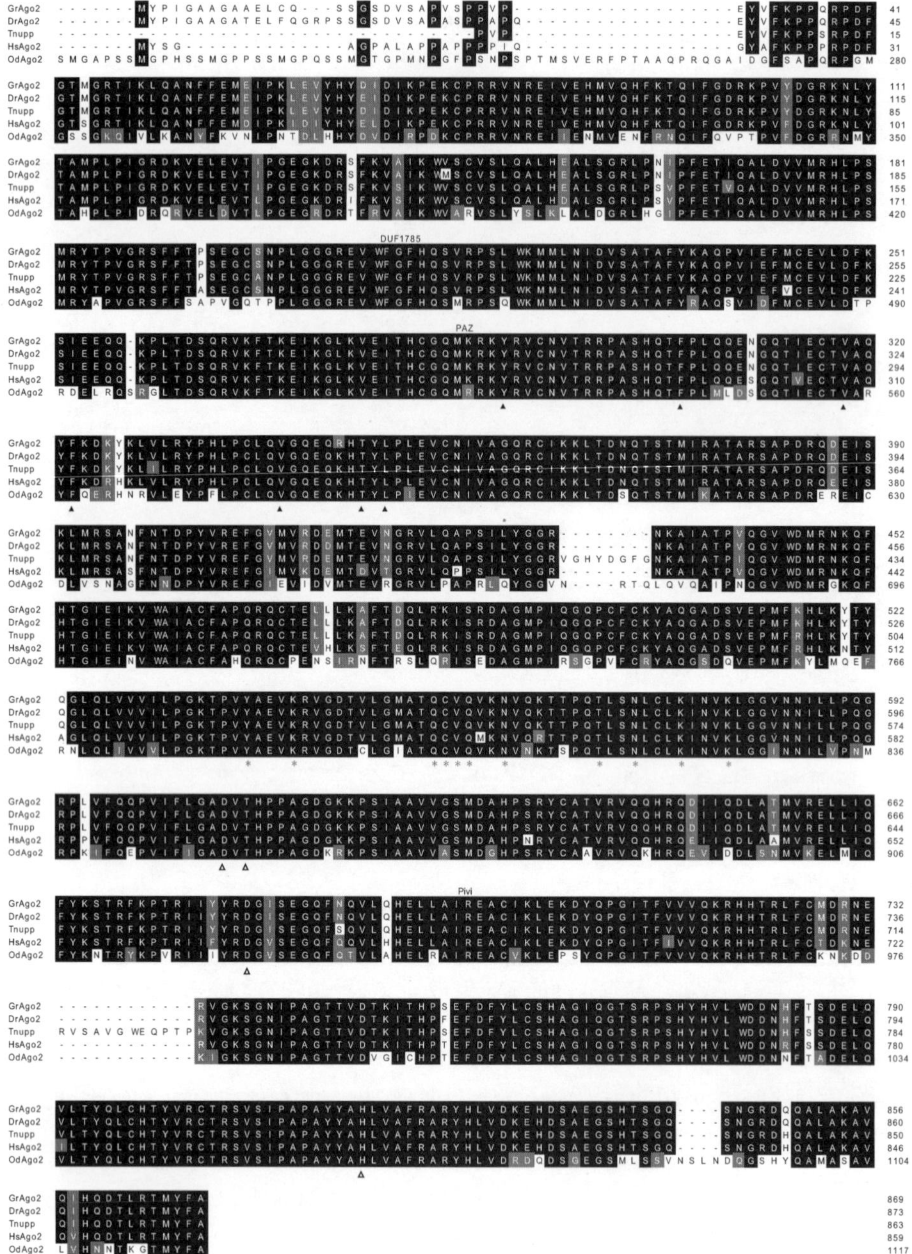

Fig. 2 Multiple alignment of GrAgo2 with some other representative Ago2 homologous proteins by ClustalW2 and the Sequence Manipulation Suite. Amino acid sequences are deduced from cDNA. Residues highlighted in black are identical and those in dark gray are similar. Percentage of sequences must agree for identity or similarity coloring to be added 80%. The right numbers represent the amino acid position in the corresponding species. Domains in Ago2 are toplined with their names. The nucleic acid-binding interface site is marked with solid triangles (▲). The 5′ RNA guide strand anchoring site is indicated with asterisks (*). The RNAase H active site is shown with hollow triangles (△). The accession numbers of aligned sequences are as in Table 2

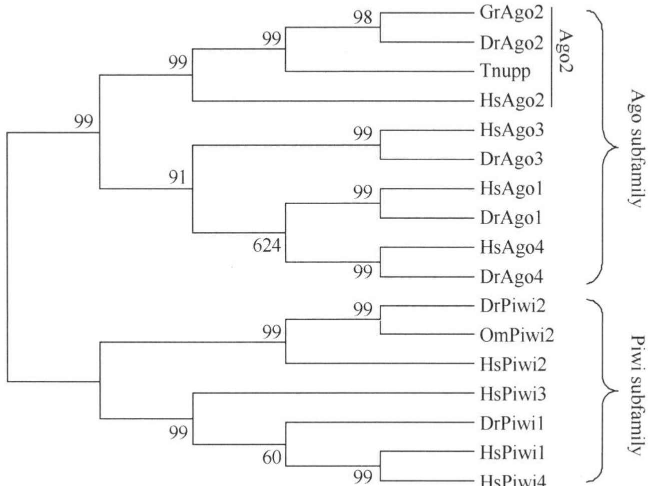

Fig. 3 Phylogenetic relationships of Argonaute family in fish and human. Phylogenetic tree was obtained from a ClustalW2 alignment and MEGA4.0 Neighbor-Joining of 17 sequences. The bar indicates the distance. The abbreviations and accession numbers are the same as in Table 2

3.3 GrAgo2 expression in different tissues

Total RNA was extracted from gill, heart, intestine, kidney, liver, muscle and spleen and the levels of GrAgo2 transcripts were examined using qRT-PCR. GrAgo2 expression was detected in all fish tissues tested at low levels, and was significantly up-regulated by GCRV ($p < 0.05$) (Fig. 4). In the virally infected group, the Ago2 expression in liver was most

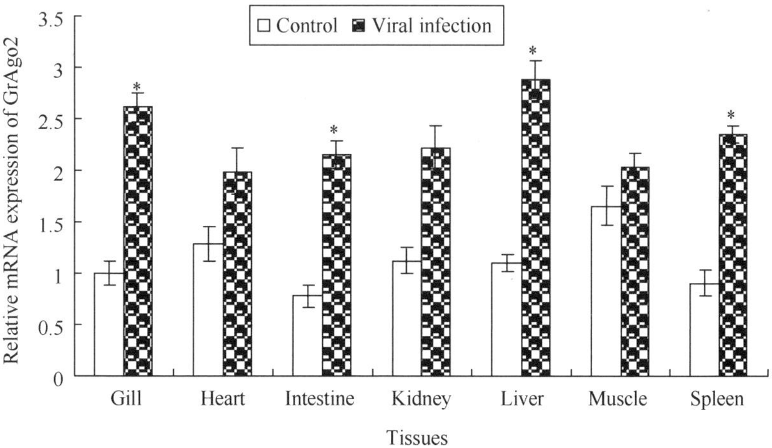

Fig. 4 Quantitative analysis of tissue distribution of GrAgo2 transcripts in rare minnow. Healthy group is uninjected and infected group is sampled at 24 h post-injection. Each bar represented the level of target gene mRNA relative to those in gills from healthy animals, expressed as the mean ± SE of triplicate qRT-PCR assays for three different cDNA samples. Asterisk (*) marks the significant difference between experimental group and control group ($p < 0.05$). Error bars indicate standard error

up-regulated 2.88-fold. For GCRV disease, liver is one of the main affected organs, and GrAgo2 was most up-regulated by viral infection in liver. Therefore, liver tissue was selected to investigate the temporary expression of GrAgo2 gene after immune stimulation.

3.4 GrAgo2 mRNA expression profile infected by virus

To determine the effects of viral infection on GrAgo2 gene expression, three animals from each group were dissected at each time point (0, 12, 24, 36, 48, 72, 96, 120, 144, 168 h post-injection) and qRT-PCR was employed to determine the levels of GrAgo2 mRNAs in liver tissue. No significant differences were detected in control and blank groups at any of the time points ($p > 0.05$) (data not shown). In infected groups, the expressions were significantly up-regulated at 12 h post-injection, reached a peak at 24 h post-injection ($p < 0.05$), and then recovered to the original level at 48 h post-injection ($p > 0.05$) (Fig. 5). The data obviously showed that GrAgo2 expression levels were stimulated in viral infection ($p < 0.05$).

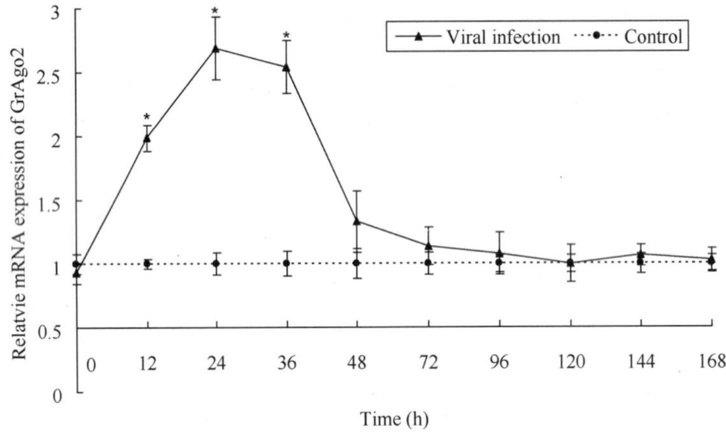

Fig. 5 The expression profiles of GrAgo2 mRNA in liver by qRT-PCR. The controls were injected with PBS; the experimental fishes were infected with GCRV. Asterisk (*) marks the significant difference between experimental group and control group ($p < 0.05$). Error bars indicate standard error

4. Discussion

The Argonaute family of proteins play a vital role in some biological processes especially RNAi. Ago2 as a central catalytic component in RNAi is essential for antiviral defense in adult *Drosophila melanogaster.* Ago2-defective flies are hypersensitive to infection with a major fruit fly pathogen, *Drosophila* C virus (DCV), Flock House virus (FHV), and with Cricket Paralysis virus (CrPV). Increased mortality in *ago2* mutant flies was associated with a dramatic increase in viral RNA accumulation and virus titers [21,22]. In fishes, Argonaute family homologous sequences were only found from three species in GenBank searched by Blastp. One is an unnamed protein product from *Tetraodon nigroviridis* (CAG11109). Another is predicted similar to Argonaute family from *Danio rerio* (XP_699384, XP_699226, XP_696563, XP_691861, NP_899181, NP_001073668) during preparation of this paper. The third is Piwi2 protein product from *Oncorhynchus mykiss* (NP_001117714). However, there is no report of Ago2 published for fish.

In the present study, we cloned and characterized the first Ago2 gene in fish. GrAgo2 was remarkably similar to other Ago subfamily proteins in vertebrates (Table 2), and firstly clustered with Ago2 clade (Fig. 3). The deduced protein has 92% sequence homology with mouse Ago2 protein (EDL29413) and 72% sequence similarity to the tunicate *Oikopleura dioica* Ago2 protein (CAP07637). GrAgo2 protein contains one DUF1785 domain, one PAZ domain and one Piwi domain. DUF1785 region is found in Argonaute proteins. The PAZ domain has been named after the proteins PiWi, Argonaut, and Zwille. PAZ is found in two families of proteins that are essential components of RNA-mediated gene-silencing pathways, including Argonaute and Dicer families. PAZ functions as a nucleic acid binding domain, with a strong preference for single-stranded nucleic acids (RNA or DNA) or RNA duplexes with single-stranded 3′ overhangs. It has been suggested that the PAZ domain provides a unique mode for the recognition of the two 3′-terminal nucleotides in single-stranded nucleic acids and buries the 3′ OH group, and that it might recognize characteristic 3′ overhangs in siRNAs within RISC and other complexes. The Piwi domain is the C-terminal portion of Argonaute and consists of two subdomains, one of which provides the 5′ anchoring of the guide RNA and the other is the catalytic site for slicing exerted by RNAase H activity [23]. All the aligned Ago2 contained all the conserved active sites (Fig. 2).

GrAgo2 expression was constitutively expressed at low levels in the tissues tested, and was induced after infection by virus. Liver is one of the main target organs for GCRV, which prompted us to study GrAgo2 expression profiles in liver infected by viruses. After infection with GCRV, GrAgo2 expression was statistically up-regulated at 12 h post-injection ($p<0.05$), and recovered to control levels at 48 h post-injection ($p>0.05$). The data suggest that GrAgo2 may be a key component in RNAi induced by viruses and imply that RNAi pathway may exert antiviral function in GCRV disease. However, its effect might be restricted by inclusion bodies of GCRV. The inclusion bodies of GCRV form within 24 h post-injection (data not published). The virus inclusion bodies are the sites for virus synthesis and assembly [24,25]. After virus inclusion bodies are formed, the interaction between virus and RNAi components is prevented, thus, the antiviral role of RNAi can be suppressed (data not published). If we could inhibit or block the formation of virus inclusion bodies, GCRV disease might be decreased or eliminated by RNAi pathway or other antiviral systems, which needs further research.

Acknowledgements The authors thank Wei Hu, Ming Li, Feng Xiong, Jun Dai, Bing Tang,

Shangping Chen and other laboratory members for technical assistance and helpful discussion. This work was supported by grants from National Natural Science Foundation of China (30740009, 30871917 and 30540084), from 973 National Basic Research Program of China (2006CB102100), from Chinese Academy of Sciences (KSCX2-YW-N-021), from Northwest A&F University in China (01140309, 05ZR096 and 01140508), from China Postdoctoral Science Foundation (20070410298) and from Institute of Hydrobiology, CAS (2007FB09).

References

[1] Meister G, Landthaler M, Patkaniowska A, Dorsett Y, Teng G, Tuschl T. Human Argonaute2 mediates RNA cleavage targeted by miRNAs and siRNAs. Mol Cell 2004; 15: 185-97.

[2] Ikeda K, Satoh M, Pauley KM, Fritzler MJ, Reeves WH, Chan EK. Detection of the argonaute protein Ago2 and microRNAs in the RNA induced silencing complex (RISC) using a monoclonal antibody. J Immunol Methods 2006; 317: 38-44.

[3] Faehnle CR, Joshua-Tor L. Argonautes confront new small RNAs. Curr Opin Chem Biol 2007; 11: 569-77.

[4] Verdel A, Jia S, Gerber S, Sugiyama T, Gygi S, Grewal S, Moazed D. RNAi-Mediated targeting of heterochromatin by the RITS Complex. Science, 2004; 303: 672-6.

[5] Yigit E, Batista PJ, Bei Y, Pang KM, Chen CC, Tolia NH, Joshua-Tor L, Mitani S, Simard MJ, Mello CC. Analysis of the *C. elegans* Argonaute family reveals that distinct Argonautes act sequentially during RNAi. Cell 2006; 127: 747-57.

[6] Grishok A, Pasquinelli AE, Conte D, Li N, Parrish S, Ha I, et al. Genes and mechanisms related to RNA interference regulate expression of the small temporal RNAs that control *C. elegans* developmental timing. Cell 2001; 106: 23-34.

[7] Morel J, Godon C, Mourrain P, Béclin C, Boutet S, Feuerbach F, et al. Fertile hypomorphic argonaute (ago1) mutants impaired in post-transcriptional gene silencing and virus resistance. Plant cell 2002; 14: 629-39.

[8] Williams RW, Rubin GM. ARGONAUTE1 is required for efficient RNA interference in *Drosophila* embryos. PNAS, 2002; 99: 6889-94.

[9] Sasaki T, Shiohama A, Minoshima S, Shimizu N. Identification of eight members of the Argonaute family in the human genome. Genomics 2003; 82: 323-30.

[10] Cerutti L, Mian N, Bateman A. Domains in gene silencing and cell differentiation proteins: the novel PAZ domain and redefinition of the Piwi domain. Trends Biochem Sci 2000; 25: 481-2.

[11] Carmell MA, Xuan Z, Zhang MQ, Hannon GJ. The Argonaute family: tentacles that reach into RNAi, developmental control, stem cell maintenance, and tumorigenesis. Genes Dev 2002; 16: 2733-42.

[12] Zhou X, Liao Z, Jia Q, Cheng L, Li F. Identification and characterization of Piwi subfamily in insects. Biochem Biophys Res Commun 2007; 362: 126-31.

[13] Okamura K, Ishizuka A, Siomi H, Siomi MC. Distinct roles for Argonaute proteins in small RNA-directed RNA cleavage pathways. Genes Dev 2004; 18: 1655-66.

[14] Adelman ZN, Anderson MA, Morazzani EM, Myles KM. A transgenic sensor strain for monitoring the RNAi pathway in the yellow fever mosquito, *Aedes aegypti*. Insect Biochem Mol Biol 2008; 38: 705-13.

[15] Su J, Zhu Z, Wang Y, Zou J, Hu W. Toll-like receptor 3 regulates Mx expression in rare minnow *Gobiocypris rarus* after viral infection. Immunogenetics 2008; 60: 195-205.

[16] Letunic I, Copley RR, Pils B, Pinkert S, Schultz J, Bork P. SMART 5: domains in the context of genomes and networks. Nucleic Acids Res 2006; 34: D257-60.

[17] Hulo N, Bairoch A, Bulliard V, Cerutti L, Cuche B, De Castro E, et al. The 20 years of PROSITE. Nucleic Acids Res 2008; 36: D245-9.

[18] Tamura K, Dudley J, Nei M, Kumar S. MEGA4: molecular evolutionary genetics analysis (MEGA) software version

4.0. Mol Biol Evol 2007; 24: 1596-9.

[19] Su J, Zhu Z, Wang Y. Molecular cloning, characterization and expression analysis of the PKZ gene in rare minnow *Gobiocypris rarus*. Fish Shellfish Immunol 2008; 25: 106-13.

[20] Pfaffl MW. A new mathematical model for relative quantification in real-time RT-PCR. Nucleic Acids Res 2001; 29: e45.

[21] Gitlin L. Andino R. Nucleic acid-based immune system: the antiviral potential of mammalian RNA silencing. J Virol 2003; 77: 7159-65.

[22] van Rij RP, Saleh M, Berry B, Foo C, Houk A, Antoniewski C, et al. The RNA silencing endonuclease Argonaute 2 mediates specific antiviral immunity in *Drosophila melanogaster*. Genes Dev 2006; 20: 2985-95.

[23] Marchler-Bauer A, Anderson JB, Derbyshire MK, DeWeese-Scott C, Gonzales NR, Gwadz M, et al. CDD: a conserved domain database for interactive domain family analysis. Nucleic Acids Res 2007; 35: D237-40.

[24] Brookes SM, Hyatt AD, Eaton BT. Characterization of virus inclusion bodies in bluetongue virus-infected cells. J Gen Virol 1993; 74: 525-30.

[25] Rivas C, Noya M, Dopazo CP. Replication and morphogenesis of the turbot aquareovirus (TRV) in cell culture. Aquaculture 1998; 160: 47-62.

稀有鮈鲫中 RNA 干扰途径关键基因 Argonaute 2 的分离和特征分析

苏建国[1,2]　朱作言[2]　汪亚平[2]　张成勋[2]

1 西北农林科技大学，杨凌 712100

2 中国科学院水生生物研究所，武汉 430072

摘　要　Argonature 2 (GrAgo2)在许多物种的 RNA 干扰途径中都起着重要作用。本文首次在鱼类中克隆和鉴定了 Argonature 2 基因。在稀有鮈鲫中，Argonature 2 基因 cDNA 全长 3073 个核苷酸，编码 869 个氨基酸，分子量为 98.499kDa，等电点为 9.18。经氨基酸序列分析共有两个结构域 Paz 和 Piwi。实时定量 RT-PCR 表明 GrAgo2 mRNA 的表达可以在大部分组织中检测出来，草鱼呼肠孤病毒感染 12 h 后 GrAgo2 会上调表达，48 h 后恢复正常水平。这些数据表明 GrAgo2 参与稀有鮈鲫的抗病毒防御机制。

Transcriptome Analysis of Head Kidney in Grass Carp and Discovery of Immune-Related Genes

Jin Chen[1,2] Cai Li[1,2] Rong Huang[1] Fukuan Du[1,2] Lanjie Liao[1]
Zuoyan Zhu[1] Yaping Wang[1]

1 State Key Laboratory of Freshwater Ecology and Biotechnology, Institute of Hydrobiology, Chinese Academy of Sciences, Wuhan 430072
2 Graduate School of Chinese Academy of Sciences, Beijing 100039

Background Grass carp (*Ctenopharyngodon idella*) is one of the most economically important freshwater fish, but its production is often affected by diseases that cause serious economic losses. To date, no good breeding varieties have been obtained using the oriented cultivation technique. The ability to identify disease resistance genes in grass carp is important to cultivate disease-resistant varieties of grass carp.

Results In this study, we constructed a non-normalized cDNA library of head kidney in grass carp, and, after clustering and assembly, we obtained 3,027 high-quality unigenes. Solexa sequencing was used to generate sequence tags from the transcriptomes of the head kidney in grass carp before and after grass carp reovirus (GCRV) infection. After processing, we obtained 22,144 tags that were differentially expressed by more than 2-fold between the uninfected and infected groups. 679 of the differentially expressed tags (3.1%) mapped to 483 of the unigenes (16.0%). The up-regulated and down-regulated unigenes were annotated using gene ontology terms; 16 were annotated as immune-related and 42 were of unknown function having no matches to any of the sequences in the databases that were used in the similarity searches. Semi-quantitative RT-PCR revealed four unknown unigenes that showed significant responses to the viral infection. Based on domain structure predictions, one of these sequences was found to encode a protein that contained two transmembrane domains and, therefore, may be a transmembrane protein. Here, we proposed that this novel unigene may encode avirus receptor or a protein that mediates the immune signalling pathway at the cell surface.

Conclusion This study enriches the molecular basis data of grass carp and further confirms that, based on fish tissue-specific EST databases, transcriptome analysis is an effective route to discover novel functional genes.

Keywords grass carp; head kidney; cDNA; EST; immune-related gene

Background

Grass carp (*Ctenopharyngodon idella*) is one of the most important freshwater fish, with

文章发表于 *Bmc Veterinary Research*, 2012, 8(1): 108

fast growth, low cost of breeding, and delicious meat. It is widely distributed in China's major river systems. Grass carp is a farmed species that is easily affected by diseases induced by viruses and bacteria; this can cause tremendous economic losses. To date, no excellent breeding varieties have been obtained by the oriented cultivation technique. Because of the long breeding cycle (4-5years), a hybrid breeding strategy is not feasible. Further, because of the lack of understanding of the genetic background of grass carp, no molecular breeding technology has been applied. The discovery of economically important trait-related genes and their functional study may help to establish a molecular breeding technology system in the fish.

ESTs (expressed sequence tags) are partial cDNA sequences obtained after sequencing the ends of random cDNA clones. ESTs were first used in 1991 as an effective new method to discover human genes. Using EST sequences, unknown genomes could be explored at a relatively low cost [1]. With the development of DNA sequencing technology, the cost of sequencing is becoming lower, and the application of large-scale EST sequencing combined with bioinformatics tools for analyzing data is being widely used in different species to find novel genes, for genome annotation, for the identification of gene structure and expression, and in the development of type I molecular markers [2]. In fish, large scale EST sequencing was used in channel catfish (*Ictalurus punctatus*) [3], common carp (*Cyprinus carpio*) [4], and zebrafish (*Danio rerio*) [5].

In recent years, high-throughput data analysis methods have gradually improved and the genomes of many kinds of fishes have been studied. The fishes that have been studied include zebrafish [6] and fugu [7], as model organisms, and the commercial fishes such as Atlantic salmon [8], sea bass [9,10], rainbow trout [11], Atlantic halibut [12], bluefin tuna [13], turbot [14,15], and Senegal sole fish [16]. In contrast, the molecular biology of grass carp is relatively unknown; currently, there are only 6,915 grass carp ESTs in NCBI's dbEST database. Most functional genomic research on economically important fish is focused mainly on the development of molecular markers, genetic map construction and gene interval mapping, and other basic data accumulation. Research into gene function and its application to breeding is still in the initial stages.

Head kidney is an important immune organ in teleost fish; its role is equivalent to mammalian bone marrow [17]. Head kidney contains a large number of T and B lymphocytes, macrophages and granulocytes that are the basis upon which specific and non-specific immunity is acquired.

In this study, we constructed a non-normalized cDNA library for the head kidney of grass carp and obtained 3,027 unigenes including 221 genes of unknown function. We compared the head kidney expression profiles of grass carp infected with grass carp reovirus (GCRV) with normal controls and obtained 22,144 differential expressed tags. Based on a comparison of the differential expressed tags and potential genes with unknown function in the cDNA library, and by identifying gene expression response to GCRV and predicting protein structure, we discovered a novel immune-related gene. This study provides a method for the discovery of novel genes, and reveals the function and the network regulation mechanism of immune-related genes. The results provide a theoretical foundation for molecular design breeding in grass carp.

Methods

RNA extraction and construction of the cDNA library

Total RNA was extracted from the head kidney of healthy adult grass carp using Trizol reagent (Invitrogen, Carlsbad, CA, USA). The mRNA was isolated using the Oligotex mRNA Kit (QIAGEN, Hilden, Germany). Full length cDNA was synthesized by the Creator^{TM} SMART^{TM} cDNA Library Construction Kit (Clontech, CA, USA) following the method described previously [18]. cDNA segments longer than 1 kb were isolated by electrophoresis, then ligated into pDNR-LIB vector (Clontech) and used to transform competent *E. coli* DH5α cells. After growing the colony for 12 hours on an LB plate containing chloramphenicol, the cDNA library was constructed by selecting mono-clones from the 96-well plate. Ethical approval for the work was obtained from Expert Committee of Biomedical Ethics, Institute of Hydrobiology of the Chinese Academy of Sciences. The Reference number obtained was Y12202-1-303.

DNA sequencing and processing of the EST sequences

10,464 clones were randomly selected from 109 96-well plates. After extracting the recombinant plasmids, 5′ terminal sequencing was performed using the T7 universal primer (T7: 5′-TAATACGACTCACTATAGGG-3′; Tm=53.2℃).

An optimal peak chart was obtained by processing the raw sequence data with basecalling. Next, FASTA format sequences (raw ESTs) were obtained by processing the optimal peak chart using the Phrap program [19] with the Q20 standard. We used crossmatch (Smith and Green, unpublished observations) to remove the pDNR-LIB vector sequences and after excluding EST sequences that were less than 100 bp long, we obtained a cleaned EST data set. Clustering of the cleaned ESTs was performed using UI cluster [20]. The UIcluster sequences were assembled using the Phrap program to build a unigene data set for the ESTs from the head kidney of grass carp.

BLAST searches, GO functional classification and KEGG pathway analysis

We used the NCBI BLAST server [21] to identify sequences that were similar to the sequences in the NCBI nucleotide sequence database (Nt), the protein sequence database (Nr) [22] and the Swissprot database [23] using BLASTN, and BLASTX [24]. Using the EST sequence with the highest homology as a guide, we set the threshold E-value to E<1e–6.

We used the BLASTX search results from the Swissprot database and the Blast2GO tool [25] to assign GO functional classification to the unigene sequences. Blast2GO parameters were set as follows: E-Value-Hit-Filter<1e–6; annotation cutOff=55; other parameters remained at the default values.

KAAS [26] was used to assign the unigene ESTs to pathways based on KEGG Orthology (KO) [27]. Unigenes were mapped to the corresponding KEGG pathways using the comparison method of bi-directional best hit.

GCRV infection of grass carp and preparation of RNA sample

The GCRV-873 strain was provided by the Gaobo biotechnology company (Wuhan, China). One-year-old grass carp with an average weight of 180-210 g were intraperitoneally injected with 150-200 μL GCRV, a dosage of approximately 10^6 $TCID_{50}$ kg^{-1} body weight. The injected grass carp were raised in clean tanks at 28℃. Three infected grass carp with typical hemorrhage symptoms (infected group, n=3) and three uninfected grass carp (healthy control group, n=3) were selected at 5d after infection for further study. Total RNA was extracted from the head kidney of both groups using Trizol reagent. cDNA was obtained after reverse transcription and used for Solexa sequencing.

Three-month-old grass carp with an average weight of 30-60 g were intraperitoneally injected with 50-80 μL GCRV, a dosage of approximately 10^6 $TCID_{50}$ kg^{-1} body weight; fish in the control group were injected with same amount of saline. The grass carp were raised in clean tanks at 28℃. At 1 d, 2 d, 3 d, 4 d, 5 d after infection ten GCRV-infected carp were selected for further study (n=10). Ten uninfected fish were selected from the control group at 0 d (n=10). The whole fish was immediately used for RNA isolation. cDNA was obtained after reverse transcription and used for the detection of gene expression.

Solexa sequencing and expression profile analysis

The *Nla*III and *Mme*I digestion method [28] was used to build a 21-bp cDNA tag library of the two groups (one-year-old), the control group and the GCRV-infected group. The tags in the two libraries end with different Illumina adapter sequences. The raw sequencing read length was 35 bp. The Solexa sequencing was performed by the Beijing Genomics Institute (BGI, Shenzhen, China).

The raw sequence data was processed through basecalling, the adapter and low quality sequences were removed, and cleaned 21-bp tags were obtained. We converted the cleaned tag number into the standard (relative) number of transcripts per million (TPM), and calculated the logarithm of TPM for each of the cleaned tags from the control and GCRV-infected groups. We used a dual limit of P <0.01 and FPR (false positive rate) <0.01, to find cleaned tags with log2Ratio ⩾ 1 or log2Ratio ⩽ –1 [29]. The selected tags have differential expression levels of more than 2-fold in both groups. We then compared the differential expressed tags with the unigenes from the cDNA library using SeqMap [30]; mismatch was set to 0, and sense and antisense strands were considered in the mapping.

Semi-quantitative RT-PCR and RACE cloning

Total RNA was used to synthesize the first strand cDNA. Upstream and downstream primers (Table 1) were designed based on the unigene sequences. β-actin (primers, β-actin-F and β-actin-R) was used as the internal reference. PCR and electrophoresis was used to detect the change of expression level.

3′ and 5′ RACE was performed using the BD SMART RACE cDNA Amplification Kit (Clontech) according to the manufacturer's instructions. Upstream and downstream primers used in the 3′ and 5′ RACE were designed based on the EST sequences (Table 1). Full length cDNA sequences of each gene were assembled using the 3′ and 5′ terminal sequences.

Table 1 Primers used for semi-quantitative RT-PCR and RACE

Primer	Sequence (5′ to 3′)	Application
291-F1	ATGTGGGTGATAGTTGGTTTACAAT	Expression study
291-R1	GTAATTTCAGAAGCACAGTTGAGAG	Expression study
357-F1	CTATCGCATGATTGCCTACTCAGACT	Expression study
357-R1	ACAACATTTTCCATCTCAATCTCAG	Expression study
788-F1	GGTCTTAACGGAGAGAAGTGCGA	Expression study
788-R1	GACTCTTCCGGCACGTAACT	Expression study
153-F1	CCAGCATCACAGTGTTCAGGCAG	Expression study
153-R1	AGTGTGTAGTTGTGTTCACCCTCC	Expression study
β-actin-F	CAGATCATGTTTGAGACC	Expression study
β-actin-R	ATTGCCAATGGTGATGAC	Expression study
291-F2	CTCTCAACTGTGCTTCTGAAATTAC	3′RACE PCR
291-R2	ATTGTAAACCAACTATCACCCACAT	5′RACE PCR
357-F2	GGTATGATTATGACTAAAGCAGGAC	3′RACE PCR
357-R2	GTCCTGCTTTAGTCATAATCATACC	5′RACE PCR
788-F2	AGTTACGTGCCGGAAGAGTC	3′RACE PCR
788-R2	TCGCACTTCTCTCCGTTAAGAC	5′RACE PCR
153-F2	GGAGGGTGAACACAACTACACACT	3′RACE PCR
153-R2	CTGCCTGAACACTGTGATGCTGG	5′RACE PCR

Results

Head kidney cDNA library of grass carp

The storage capacity of the original library was 6×10^5, in the form of the *E. coli* DH5α cells that were stored on the 532 96-well plates in a total of 51,072 clones. One hundred randomly selected clones were used for further study. The PCR test results showed that the size of inserts was between 1-3 kilobases, the library reorganization was 97.85% and the no-load rate was 2.15%.

EST sequence analysis

10,464 EST clones were sequenced, and 10,282 FASTA sequences (raw ESTs) with an average read length of 470 bp were obtained. After removing the vector and sequences less than 100 bp long, 7,918 cleaned ESTs (accession No. JK847435-JK855352) were obtained. After clustering and assembly, we obtained 3,027 unigene EST sequences, 802 (26.5%) of which were contigs and 2,225 (73.5%) of which were singletons; the library redundancy was 61.78%. Most genes in the library exhibited low-level expression, only a small number of genes exhibited high-abundance expression. The number of low expression unigenes, the singletons, was 2,225 (73.5%); the number of medium expression unigenes, those containing 2-5 ESTs was 641 (21.2%); and the number of high expression unigenes, those that contained six or more ESTs, was 161 (5.3%). Only 23 unigenes contained more than 20 ESTs. The average length of the unigenes was 431 bp and 77.33% of the unigenes were 300-500 bp long (Figure 1).

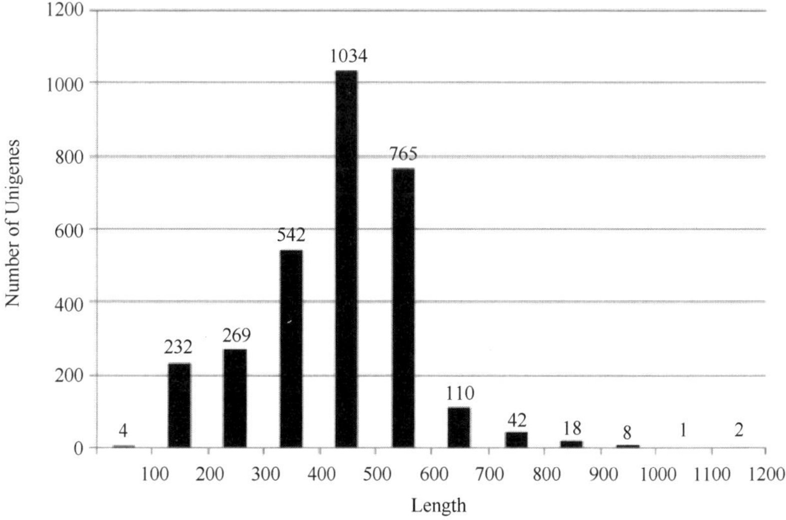

Figure 1 Length distribution of the assembled EST unigenes. The abscissa indicates the length of the unigenes, the ordinate indicates the number of unigenes

BLAST searches and GO functional classification

The 3,027 unigenes were used as queries in BLAST searches of the NCBI nucleotide and protein sequence databases and the Swissprot database. 2,713 unigenes (89.6%) matched sequences in the nucleotide sequence database, 2,162 unigenes (71.4%) matched sequences in protein sequence database and 1,845 unigenes (61.0%) matched sequences in the Swissprot database. In all, 2,806 unigenes (92.7%) matched sequences in at least one of the three databases; the remaining 221 unigenes (7.3%) were not found (E-value <1e–6) in any of the three databases and may be novel gene sequences.

Using the gene ontology (GO) classification, we successfully assigned functional annotations to 1,323 of the unigene sequences. In the GO biological process ontology, three terms accounted for the larges tproportion of unigenes, they were cellular process, metabolic process and biological regulation; in the GO molecular function ontology, the three most commonly occurring terms were binding, catalytic activity and structural molecule activity; and in the GO cellular component ontology, cell, cell part and organelle were the terms that occurred most frequently (Table 2). Of the 1,323 GO-annotated unigenes, 53 were immune system process-related genes (Table 3), 4 were response to virus, and 9 were response to bacterium process-related genes (Tables 4 and 5). Some unigenes were assigned multiple functions. Not all of the unigenes could be mapped to the lower level GO terms.

Table 2 GO functional classification of the unigene data set

	Go term	Number of unigenes	%
Biological process	cellular process	899	68.0
	metabolic process	672	50.8
	biolohical regulation	379	28.6
	regulation of biological process	356	26.9
	localization	250	18.9
	establishment of localization	232	17.5
	develomental process	135	10.2
	response to stimulus	126	9.5
	multicellular organismal process	100	7.6
	positive regulation of biological process	80	6.0
	anatomical structure formation	71	5.4
	negative regulation of biological process	66	5.0
	immune system process	53	4.0
	multi-organism process	22	1.7
	growth	18	1.4
	biological adhesion	14	1.1
	locomotion	14	1.1
	reproduction	14	1.1
	reproductive process	14	1.1
	viral reproduction	4	0.3

Continued

	Go term	Number of unigenes	%
Cellular component	cell part	1084	81.9
	cell	1084	81.9
	organnelle	733	55.4
	macromolecular complex	402	30.4
	organelle part	384	29.0
	membrane-enclosed lumen	140	10.6
	envelope	80	6.0
	extracellular region	32	2.4
	extracellular region part	12	0.9
	synapse	6	0.5
	synapse part	4	0.3
Molecular function	binding	770	58.2
	catalytic activity	440	33.3
	structural molecule activity	77	5.8
	transporter activity	70	5.3
	transcription regulator activity	46	3.5
	molecular transducer activity	46	3.5
	enzyme regulator activity	30	2.3
	translation regulator activity	29	2.2
	electron carrier activity	11	0.8
	antioxidant cativity	5	0.4

Table 3 Unigenes annotated with the GO term immune system process

Sequence name	Sequence description	Hit AC	Clustered EST
Cluster1088	fucolectin	Q7SIC1	1
Cluster1225	endoplasmic reticulum aminopeptidase 1	Q9NZ08	1
Cluster1249	transcription factor sp2	Q02086	1
Cluster1357	complement c3	P98093	1
Cluster1410	b-cell lymphoma 6 protein homolog	P41183	1
Cluster1474	matrix metalloproteinase-9	P14780	1
Cluster1562	serine threonine-protein phosphatase subunit	P30153	1
Cluster1638	inosine-5-monophosphate dehydrogenase 2	Q3SWY3	1
Cluster1667	chemokine-like factor	Q9UBR5	1
Cluster1692	60 kda heat shock mitochondrial	Q5ZL72	1
Cluster1821	transcription elongation factor	Q4KLL0	1
Cluster1865	serine threonine-protein kinase tbk1	Q9WUN2	1
Cluster1872	dedicator of cytokinesis protein 2	Q92608	1
Cluster1891	complement c3	P98093	1
Cluster1908	interferon regulatory factor 4	Q64287	1

Continued

Sequence name	Sequence description	Hit AC	Clustered EST
Cluster2109	sh2 domain-containing protein 1a	B2RZ59	1
Cluster2173	bisphosphate phosphodiesterase gamma-2	Q8CIH5	1
Cluster2189	ig heavy chain v-iii region cam	P01768	2
Cluster2214	complement c3	P12387	2
Cluster2253	calreticulin	P18418	2
Cluster2255	ap-2 complex subunit sigma-1	P62744	2
Cluster2335	myosin-if	O00160	2
Cluster2337	adenylate kinase mitochondrial	Q1L8L9	2
Cluster2342	40s ribosomal protein s14	P62263	2
Cluster2345	apoptotic chromatin condensation inducer	Q9UKV3	2
Cluster244	MHC I-related gene protein	Q95460	1
Cluster2440	ubiquitin thioesterase otub1	Q96FW1	2
Cluster2466	nf-kappa-b inhibitor alpha	P25963	2
Cluster2474	toll-interacting protein	A2RUW1	2
Cluster2602	moesin	P26038	3
Cluster2612	beta-2-microglobulin	Q04475	3
Cluster265	myosin-9	P14105	1
Cluster2659	proteasome maturation protein	Q3SZV5	3
Cluster2663	apoptotic chromatin condensation inducer	Q9UKV3	3
Cluster2706	cd81 antigen	P35762	3
Cluster2717	complement -binding mitochondrial	Q07021	4
Cluster2828	integrin alpha-l	P24063	6
Cluster2869	moesin	P26038	8
Cluster2872	beta-2-microglobulin	O42197	8
Cluster2877	c-x-c chemokine receptor type 4	P61072	8
Cluster2908	fucolectin	Q7SIC1	12
Cluster311	proteasome subunit beta type-9	Q8UW64	1
Cluster33	inosine-5 -monophosphate dehydrogenase 1	P20839	1
Cluster490	paired box protein pax-5	Q02548	1
Cluster493	nucleosome assembly protein 1-like 1-a	Q4U0Y4	1
Cluster588	cysteine-rich protein 2	Q9DCT8	1
Cluster634	interleukin enhancer-binding factor 2 homolog	Q6NZ06	1
Cluster668	zinc finger e-box-binding homeobox 1	P36197	1
Cluster780	kinase catalytic subunit delta isoform	O35904	1
Cluster789	cd81 antigen	P35762	1
Cluster812	interferon regulatory factor 1	P15314	1
Cluster937	high mobility group protein b3	Q32L31	1
Cluster999	aminoacyl trna synthetase protein	Q12904	1

Table 4 Unigenes annotated with the GO term response to virus

Sequence name	Sequence description	Hit AC	Clustered EST
Cluster 2255	ap-2 complex subunit sigma-1	P62744	2
Cluster 2287	interferon-induced gtp-binding ptotein	Q8JH68	2
Cluster 2379	40s riboomal protein s15a	P62244	2
Cluster 2877	c-x-c chemokine receptor type 4	P61072	8

Table 5 Unigenes annotated with the GO term response to bacterium

Sequence name	Sequence description	Hit AC	Clustered EST
Cluster 12	histone h2a	P02264	1
Cluster 1225	endoplasmic reticulum aminopeptidase 1	Q9NZ08	1
Cluster 1269	lysozyme c	P85045	1
Cluster 1910	akirin-2	Q25C79	1
Cluster 2173	phosphatidylinositol phosphodiesterase gamma-2	Q8CIH5	1
Cluster 2335	myosin-if	O00160	2
Cluster 2543	histone h1	P06350	2
Cluster 2861	histone h1	P06350	7
Cluster 566	histone h1	P06350	1

KEGG pathway analysis

A total of 989 of the 3,027 were assigned a KEGG ontology (KO) annotation; they were mapped to 201 KEGG pathways. Three most frequently occurring KEGG pathways were ribosome, oxidative phosphorylation, and proteasome. 68 unigenes mapped to immune-related pathways including leukocyte transendothelial migration, antigen processing and presentation, chemokine signalling pathway, and T cell receptor signalling pathway (Table 6). We found that 28 unigenes from head kidney in grass carp have been reported to be involved in the following pathways, Toll-like receptor signalling pathway, RIG-I-like receptor signalling pathway and the NOD-like receptor signalling pathway (Table 7).

Expression profiling analysis

By Solexa sequencing, we obtained 7,696,804 and 6,136,889 raw tags from the transcriptomes of head kidney tissue from grass carp before and after GCRV infection, respectively. After removing low quality sequences, adapter sequences and single copy sequence the cleaned tag numbers were 7,188,005 and 5,724,526, respectively. The final numbers of non-redundant distinct tags were 152,826 and 105,653 before and after GCRV infection, respectively. All tags were submitted to SRA at NCBI under the accession no. SRA052520.2. Of the distinct tags, 22,144 were differentially expressed by more than 2-fold between the GCRV-infected and uninfected groups.

Table 6 The most represented KEGG pathways in the unigene data set

Pathway	Mapping ganes	Categories
Ribosome	60	Genetic Information Processing
Oxidative phosphorylation	53	Metabolism
Proteasome	32	Genetic Information Processing
Spliceosome	31	Genetic Information Processing
Lysosome	28	Cellular Processes
Purine matabolism	25	Metabolism
Endocytosis	24	Cellular Processes
Regulation of actin cytoskeleton	24	Cellular Processes
Cell cycle	19	Cellular Processes
Leukocyte transendothelial migration	18	Organismal Systems
Pyrimidine metabolism	17	Metabolism
MAPK signalling pathway	17	Environmental Information Processing
Antigen processing and presentation	17	Organismal Systems
Chemokine signalling pathway	17	Organismal Systems
Tight junction	16	Cellular Processes
T cell receptor signalling pathway	16	Organismal Systems

Table 7 Mapping genes in fish primary non-specific immune pathways

Pathway	Mapping genes	Containing ESTs
Toll-like receptor signalling pathway	8	16
RIG-like receptor signalling pathway	11	20
NOD-like receptor signalling pathway	9	17

These 22,144 differentially expressed tags mapped to 3,027 unigenes using SeqMap [30]. Of the differentially expressed tags, 679 (3.1%) mapped to 483 differentially expressed unigenes (16.0%); 145 of the unigenes were up-regulated genes, 307 were down-regulated genes. The remaining 31 unigenes mapped to tags that exhibited both up and down regulation, and so these unigenes were not included in the statistics. The up- and down-regulated genes were mainly annotated with the GO terms, genetic information processing, metabolism, and cellular processes and 16 unigenes were annotated with the GO term immune-related (Table 8). We found 54 tags that mapped onto 42 of the 221 unknown unigenes. These are potentially infection related novel genes; 15 of them were up-regulated between the GCRV-infected and uninfected groups, and 27 were down-regulated genes (Table 9).

Table 8 Differentially expressed unigenes annotated as immune-related

Sequence name	Description	log2Ratio (VP/CP)	Up -Down
cichka_Cluster2189.seq.Contig 1	ig heavy chain v-iii region cam	9.552669098	Up
cichka_Cluster2214.seq.Contig 1	Complement c3	–1.234417227	Down
cichka_Cluster2335.seq.Contig 1	Mtosin-if	–1.616395009	Down
cichka_Cluster2337.seq.Contig 1	adenylate kinase mitochondrial	–1.622261042	Down
cichka_Cluster2612.seq.Contig 1	beta-2-microglobulin	14.96510786	Up
cichka_Cluster2717.seq.Contig 1	complement-binding mitochondrial	2.831849484	Up
cichka_Cluster2828.seq.Contig 1	integrin algha-l	–3.476196501	Down
cichka_Cluster2872.seq.Contig 1	beta-2-microglobulin	–2.257387843	Down
cichka_Cluster2877.seq.Contig 1	c-x-c chemoline receptor type 4	–2.941536738	Down
cichka_Cluster2908.seq.Contig 1	fucolectin	–3.57091306	Down
cichka_Cluster2379.seq.Contig 1	40s ribosonmal protein s15a	–2.133495724	Down
cichka_Cluster1269	lysozyme c	–5.60930435	Down
cichka_Cluster634	interleukin enhancer-binding factor 2 homolog	–8.383704292	Down
cichka_Cluster812	interferon regulatory factor 1	2.652601218	Up
cichka_Cluster1474	matrix metalloproteinase-9	–5.851050959	Down
cichka_Cluster1667	chemokine-like factor	–8.189824559	Down

Table 9 Potentially novel differentially expresed unigenes

Sequence name	log2Ratio (VP/CP)	Up -Down
cichka_Cluster1	–1.30897451703681	Down
cichka_Cluster1004	–8.18982455888002	Down
cichka_Cluster1074	–4.68008319087111	Down
cichka_Cluster1080	1.5772610962369	Up
cichka_Cluster1095	1.70760741456741	Up
cichka_Cluster1139	–3.01282922395069	Down
cichka_Cluster1321	–1.17687776208408	Down
cichka_Cluster1418	2.57740490960702	Up
cichka_Cluster144	–2.37056287013824	Down
cichka_Cluster1502	–2.69938241135805	Down
cichka_Cluster153	9.00842862207058	Up
cichka_Cluster155	–4.35320513151951	Down
cichka_Cluster1567	–3.23219204494701	Down
cichka_Cluster1689	–1.24599865006401	Down
cichka_Cluster18	–8.55458885167764	Down
cichka_Cluster1830	5.67155018571725	Up
cichka_Cluster1847	–1.8154025874359	Down
cichka_Cluster19	–704998458870832	Down
cichka_Cluster1931	1.442222322860508	Up

Continued

Sequence name	log2Ratio (VP/CP)	Up -Down
cichka_Cluster2016	2.84923580318831	Up
cichka_Cluster2063	−3.29278174922784	Down
cichka_Cluster219	8.44708322620965	Up
cichka_Cluster2432.seq.Contig 1	−1.12271915825313	Down
cichka_Cluster2506.seq.Contig 1	−2.26096007759593	Down
cichka_Cluster2646.seq.Contig 1	−8.18982455888736	Down
cichka_Cluster2651.seq.Contig 1	−8.32192809488736	Down
cichka_Cluster2765.seq.Contig 1	−8.38370429247405	Down
cichka_Cluster291	3.07771266869725	Up
cichka_Cluster2966.seq.Contig 1	−5.46317402032312	Down
cichka_Cluster317	−2.29418310440446	Down
cichka_Cluster357	3.48519281620541	Up
cichka_Cluster468	−1.41853954357293	Down
cichka_Cluster482	1.43096228428556	Up
cichka_Cluster559	7.82654848729092	Up
cichka_Cluster613	−1.71345884128158	Down
cichka_Cluster619	−3.62148817674627	Down
cichka_Cluster625	1.46068016483455	Up
cichka_Cluster751	1.83711846346595	Up
cichka_Cluster788	1.13154390971446	Up
cichka_Cluster790	−5.47619650111671	Down
cichka_Cluster837	−2.25442127552909	Down
cichka_Cluster891	−1.33599920243744	Down

Cloning and expression regulation analysis of the novel genes

Using semi-quantitative RT-PCR, we examined the gene expression changes of the 42 potentially novel unigenes that were detected in the head kidney after viral infection. By comparing the 1, 2, 3, 4, and 5 day post-infection samples and the samples from the control group, we found four unigenes that showed a significant response to the viral infection: cichka_Cluster153 and cichka_Cluster291 were up-regulated in days 1 and 2 post-infection after which their expressions returned to the starting level; cichka_Cluster357 and cichka_Cluster788 were up-regulated in days 1 and 2 post-infection, and the increased expression levels were maintained till day 5 (Figure 2).

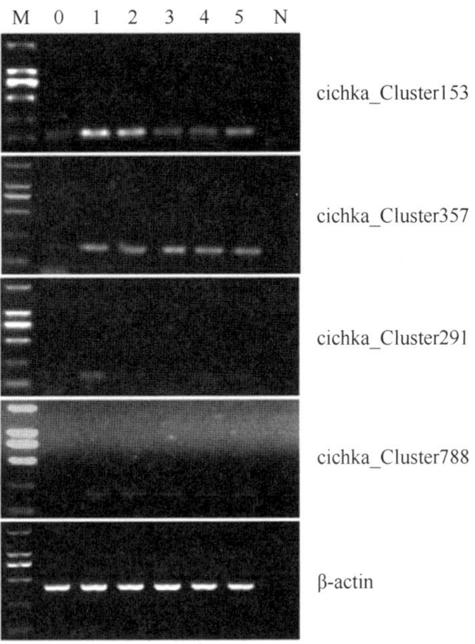

Figure 2 RT-PCR verification of the novel infection-related genes. M, maker; 0, non-infected tissue; 1, 1 day after infection; 2, 2 days after infection; 3, 3 days after infection; 4, 4 days after infection; 5, 5 days after infection; N, the negative control

The full-length cDNA sequences of these four unigenes were 2,057 bp (cichka_Cluster291, JQ412736), 2,288 bp (cichka_Cluster357, JQ412737), 1,044 bp (cichka_Cluster788, JQ412738) and 1,387bp (cichka_Cluster153, JQ412739) encoding polypeptides of 586, 322, 142 and 155 amino acids, respectively. BLAST searches revealed that cichka_Cluster291 can encode a protein that is similar to the vertebrate endonuclease domain containing protein, cichka_Cluster357 can encode a protein that is similar to the vertebrate ankyrin repeat domain 10 protein, cichka_Cluster788 can encode a protein that is similar to the CST complex subunit TEN1; for the cichka_Cluster153 encoded protein, no similar sequences were found in the databases that we searched, suggesting that cichka_Cluster153 may represent a novel gene in grass carp. We used the SMART server[31] to predict the domain structure of the 42 novel unigenes and found that 83.02% of them contained the endonuclease domain 1 that is found in proteins that are involved in the apoptosis pathway, and 35.22% contained the ankyrin repeat domain that is present in proteins that are involved in pathways that include the B cell receptor signalling pathway, the T cell receptor signalling pathway, and the apoptosis pathway. The cichka_Cluster788 unigene contained no obvious structural domains; the cichka_Cluster153 encoded protein contained two transmembrane domains and may be a transmembrane protein.

Discussion

Currently, there are about 6,915 sequences of grass carp in the public databases. This situation does not reflect the extremely important breeding position of grass carp. In this study, we built a head kidney non-normalized cDNA library of healthy grass carp and obtained 3,027 unigene EST sequences. This library greatly enriches the available genomic data for grass carp and lays an important foundation for the discovery of novel genes and for their functional investigation.

GO analysis revealed that the annotated unigenes were mainly related to genes involved in basic biological processes such as cellular process (25.5%), metabolic process (19.1%) and biological regulation (10.8%). This functional distribution is similar to the EST distributions reported earlier in the head kidney of zebrafish [32] and sea bass[10].

Of the unigenes that were similar to immune-related genes, 66 unigenes were annotated as associated with the immune process; 53 were related to the immune system process, 4 were annotated as response to virus, and 9 were related to response to bacteria. Among the 989 unigenes that were assigned KO annotations, 68 were mapped to immune-related pathways that included leukocyte transendothelial migration, antigen processing and presentation, chemokine signalling pathway and T cell receptor signalling pathway. By examining the literature, we found that 28 of the unigenes in grass carp head kidney were related to fish genes that were reported to be involved in the Toll-like receptor signalling pathway, the RIG-I-like receptor signalling pathway and the NOD-like receptor signalling pathway. Clearly, head kidney tissue plays an important role in immune processes in fish. EST databases of head kidney tissue are likely to become important resources in which immune-related genes can be identified.

In the 3,027 unigene library of head kidney in grass carp, 7.3% (221) failed to match any of the sequences in the three public databases that were searched. Of the 10 unigenes that were the most highly expressed in grass carp head kidney, 9 were unknown sequences (Table10). This could be partly because sequence data for fish is still very scarce, and partly because fish head kidney tissue may contain tissue-specific or species-specific genes. EST databases can be important resources for identifying unknown genes in fish [33-35]. In recent years, the fish transcriptome has been used to study the regulation of gene expression. Pardo *et al* (2008) conducted a comparative study of turbot expression profiles in the main immune tissue before and after pathogen infection to find genes that were related to immune response and disease resistance [36]. Chini *et al* (2008) carried out a comparative study of reproductive development-related tissues in bluefin tuna using transcriptome research methods to explore the molecular mechanism of gonadal development and maturity split [13]. Indeed, comparative transcriptome analysis can be used, not only to investigate the mechanisms of expression and regulation of known genes, but also as an effective means to find important and novel function-related genes.

Table 10　Ten most highly expressed unigenes in the head kidney of healthy grass carp

Sequence name	ORF length	Clustered ESTs	Description
Cluster2971	159	1114	Unknown
Cluster2970	267	251	hybrid granulin
Cluster2969	132	166	hypothetical 18K protein
Cluster2968	282	109	Unknown
Cluster2967	273	123	Unknown
Cluster2966	267	78	hypothetical protein
Cluster2965	132	85	Unknown
Cluster2964	0	83	Unknown
Cluster2963	279	63	Unknown
Cluster2962	108	55	Unknown

Conclusion

We carried out a comparative analysis to find differences in the Solexa expression profiles of head kidney in grass carp before and after infection, and identified 42 unigenes of unknown function that showed differential expression in response to the pathogen. After RT-PCR validation of the cDNA and gene structure analysis, we found a potentially novel immune-related gene. Based on its response to viral infection and the prediction that it might encode a membrane protein, we speculate that this novel gene may encode a virus receptor or a protein that mediates the immune signalling pathway at the cell surface. We intend to further investigate the function of this gene in a future study. Our findings confirm that fish tissue-specific EST databases combined with comparative transcriptome analysis are effective tools that can direct the discovery of novel functional genes.

Authors' contributions　CJ carried out the experiments and drafted the manuscript. LC and DFK conducted the database searches and bioinformatics analysis. HR and ZZY participated in the study design and in the manuscript preparation. LLJ was involved in the experiments. WYP was overall responsible for the project and finalized the manuscript. All authors read and approved the final manuscript.

Acknowledgements　The research was financially supported by the Innovation Project of the Chinese Academy of Sciences (KSCX2-EW-N-004-3), the National Key Basic Research Program (2009CB118701), and the Autonomous Project of State Key Laboratory of Freshwater Ecology and Biotechnology (2011FBZ18).

References

1. Adams MD, Kelley JM, Gocayne JD, Dubnick M, Polymeropoulos MH, Xiao H, Merril CR, Wu A, Olde B, Moreno RF, Kerlavage AR, Richard Mccombie W, Craig Venter J: Complementary DNA sequencing: expressed sequence tags and human genome project. *Science* 1991, 252: 1651-1656.
2. Nagaraj SH, Gasser RB, Ranganathan S: A hitchhiker's guide to expressed sequence tag (EST) analysis. *Brief Bioinform* 2007, 8: 6-21.
3. Karsi A, Li P, Dunham R, Liu ZJ: Transcriptional activities in the pituitaries of channel catfish before and after induced ovulation by injection of carp pituitary extract as revealed by expressed sequence tag analysis. *J Mol Endocrinol* 1998, 21: 121-129.
4. Savan R, Sakai M: Analysis of expressed sequence tags (EST) obtained from common carp, *Cyprinus carpio* L., head kidney cells after stimulation by two mitogens, lipopolysaccharide and concanavalin-A. *Comp Biochem Physiol B Biochem Mol Biol* 2002, 131: 71-82.
5. Zeng S, Gong Z: Expressed sequence tag analysis of expression profiles of zebrafish testis and ovary. *Gene* 2002, 294: 45-53.
6. Lo J, Lee S, Xu M, Liu F, Ruan H, Eun A, He Y, Ma W, Wang W, Wen Z, Peng J: 15, 000 Unique Zebrafish EST Clusters and Their Future Use in Microarray for Profiling Gene Expression Patterns During Embryogenesis. *Genome Res* 2003, 13: 455-466.
7. Clark MC, Edwards YJK, Peterson D, Clifton SW, Thompson AJ, Sasaki M, Suzuki Y, Kikuchi K, Watabe S, Kawakami K, Sugano S, Elgar G, Johnson SL: Fugu ESTs: New Resources for Transcription Analysis and Genome Annotation. *Genome Res* 2003, 13: 2747-2753.
8. Adzhubei AA, Vlasova AV, Hagen-Larsen H, Ruden TA, Laerdahl JK, Høyheim B: Annotated Expressed Sequence Tags (ESTs) from pre-smolt Atlantic salmon (*Salmo salar*) in a searchable data resource. *BMC Genomics* 2007, 8: 209.
9. Chini V, Rimoldi S, Terova G, Saroglia M, Rossi F, Bernardini G, Gornati R: EST-based identification of genes expressed in the liver of adult seabass (*Dicentrarchus labrax*, L.). *Gene* 2006, 376: 102-106.
10. Sarropoulou E, Sepulcre P, Poisa-Beiro L, Mulero V, Meseguer J, Figueras A, Novoa B, Terzoglou V, Reinhardt R, Magoulas A, Kotoulas G: Profiling of infection specific mRNA transcripts of the European seabass Dicentrarchus labrax. *BMC Genomics* 2009, 10: 157.
11. Govoroun M, Gac FL, Guiguen Y: Generation of a large scale repertoire of Expressed Sequence Tags (ESTs) from normalised rainbow trout cDNA libraries. *BMC Genomics* 2006, 7: 196.
12. Douglas SE, Knickle LC, Kimball J, Reith ME: Comprehensive EST analysis of Atlantic halibut (*Hippoglossus hippoglossus*), a commercially relevant aquaculture species. *BMC Genomics* 2007, 8: 144.
13. Chini V, Cattaneo AG, Rossi F, Bernardini G, Terova G, Saroglia M, Gornati R: Genes expressed in blue fin tuna (*Thunnus thynnus*) liver and gonads. *Gene* 2008, 410: 207-213.
14. Pardo BG, Fernández C, Millán A, Bouza C, Vázquez-López A, Vera M, Alvarez-Dios JA, Calaza M, Gómez-Tato A, Vázquez M, Cabaleiro S, Magariños B, Lemos ML, Leiro JM, Martínez P: Expressed sequence tags (ESTs) from immune tissues of turbot (*Scophthalmus maximus*) challenged with pathogens. *BMC Vet Res* 2008, 4: 7.
15. Park KC, Osborne JA, Tsoi SCM, Brown LL, Johnson SC: Expressed sequence tags analysis of Atlantic halibut (*Hippoglossus hippoglossus*) liver, kidney and spleen tissues following vaccination against Vibrio anguillarum and Aeromonas salmonicida. *Fish Shellfish Immunol* 2005, 18: 393-415.
16. Cerdà J, Mercadé J, Lozano JJ, Manchado M, Tingaud-Sequeira A, Astola A, Infante C, Halm S, Viñas J, Castellana B, Asensio E, Cañavate P, Martínez-Rodríguez G, Piferrer F, Planas JV, Prat F, Yúfera M, Durany O, Subirada F, Rosell E, Maes T: Genomic resources for a commercial flatfish, the Senegalese sole (*Solea senegalensis*): EST sequencing, oligo microarray design, and development of the Soleamold bioinformatic platform. *BMC Genomics* 2008, 9: 508.

17. Press CMcL, Evensen Ø: The morphology of the immune system in teleost fishes. *Fish Shellfish Immunol* 1999, 9: 309-318.
18. Zhu YY, Machleder EM, Chenchik A, Li R, Siebert PD: Reverse transcriptase template switching: a SMART approach for full-length cDNA library construction. *Biotechniques* 2001, 30: 892-897.
19. Ewing B, Hillier L, Wendl MC, Green P: Base-calling of automated sequencer traces using phred. I. Accuracy assessment. *Genome Res* 1998, 8: 175-185.
20. Trivedi N, Bishof J, Davis S, Pedretti K, Scheetz TE, Braun TA, Roberts CA, Robinson NL, Sheffield VC, Bento Soares M, Casavant TL: Parallel creation of non-redundant gene indices from partial mRNA transcripts. *Future Gen Comp Sys* 2002, 18: 863-870.
21. *BLAST*: http: //blast.ncbi.nlm.nih.gov/Blast.cgi.
22. Benson DA, Boguski MS, Lipman DJ, Ostell J, Ouellette BF, Rapp BA, Wheeler DL: GenBank. *Nucleic Acids Res* 1999, 27: 12-17.
23. Boeckmann B, Bairoch A, Apweiler R, Blatter M, Estreicher A, Gasteiger E, Martin MJ, Michoud K, O'Donovan C, Phan I, Pilbout S, Schneider M: The SWISS-PROT protein knowledgebase and its supplement TrEMBL in 2003. *Nucleic Acids Res* 2003, 31: 365-370.
24. Altschul SF, Gish W, Miller W, Myers EW, Lipman DJ: Basic Local Alignment Search Tool. *J Mol Biol* 1990, 215: 403-410.
25. Conesa A, Goetz S, Garcia JM, Terol J, Talon M, Robles M: Blast2GO: a universal tool for annotation, visualization and analysis in functional genomics research. *Bioinformatics* 2005, 21: 3674-3676.
26. Moriya Y, Itoh M, Okuda S, Yoshizawa A, Kanehisa M: KAAS: an automatic genome annotation and pathway reconstruction server. *Nucleic Acids Res* 2007, 35: W182-W185.
27. Kanehisa M, Goto S: KEGG: Kyoto encyclopedia of genes and genomes. *Nucleic Acids Res* 2000, 28: 27-30.
28. Sorana Morrissy A, Morin RD, Delaney A, Zeng T, McDonald H, Jones S, Zhao Y, Hirst M, Marra MA: Next-generation tag sequencing for cancer gene expression profiling. *Genome Res* 2009, 19: 1825-1835.
29. Audic S, Claverie JM: The significance of digital gene expression profiles. *Genome Res* 1997, 7: 986-995.
30. Jiang H, Wong WH: SeqMap: mapping massive amount of oligonucleotides to the genome. *Bioinformatics* 2008, 24: 2395-2396.
31. *SMART*: http: //smart.embl-heidelberg.de/.
32. Song HD, Sun XJ, Deng M, Zhang GW, Zhou Y, Wu XY, Sheng Y, Chen Y, Ruan Z, Jiang CL, Fan HY, Zon LI, Kanki JP, Liu TX, Look AT, Chen Z: Hematopoietic gene expression profile in zebrafish kidney marrow. *Proc Natl Acad Sci USA* 2004, 101: 16240-16245.
33. Yang AF, Zhou ZC, He CB, Hu JJ, Chen Z, Gao XG, Dong Y, Jiang B, Liu WD, Guan XY, Wang XY: Analysis of expressed sequence tags from body wall, intestine and respiratory tree of sea cucumber (*Apostichopus japonicus*). *Aquaculture* 2009, 296: 193-199.
34. Chen SL, Xu MY, Hu SN, Li L: Analysis of immune-relevant genes expressed in red sea bream (*Chrysophrys major*) spleen. *Aquaculture* 2004, 240: 115-130.
35. Cao D, Kocabas A, Ju Z, Karsi A, Li P, Patterson A, Liu Z: Transcriptome of channel catfish (*Ictalurus punctatus*): initial analysis of genes and expression profiles of the head kidney. *Anim Genet* 2001, 32: 169-188.
36. Pardo GB, Fernández C, Millán A, Bouza C, Vázquez-López A, Vera M, Alvarez-Dios AJ, Calaza M, Gómez-Tato A, Vázquez M, Cabaleiro S, Magariños B, Lemos LM, Leiro MJ, Martínez P: Expressed sequence tags (ESTs) from immune tissues of turbot (*Scophthalmus maximus*) challenged with pathogens. *BMC Vet Res* 2008, 4: 37.

草鱼头肾组织的转录组分析及免疫相关基因的发掘

Jin Chen[1,2] 李 偲[1,2] 黄 容[1] 杜富宽[1,2]
廖兰杰[1] 朱作言[1] 汪亚平[1]

1 中国科学院水生生物研究所，武汉 430072
2 中国科学院大学，北京 100039

摘 要 草鱼是中国最重要的淡水经济鱼类之一，但其养殖极易受病害影响，经济损失非常严重，草鱼抗病性相关基因的发掘对草鱼抗病品种的培育具有重要意义。本研究中，我们构建了草鱼头肾组织的非均一化 cDNA 文库，测序得到 3027 条高质量 unigene；同时，进行了草鱼头肾组织在健康及草鱼呼肠孤病毒感染状态下的表达谱比较分析，获得差异表达 tag 22,144 条。基于差异表达 tag 与草鱼头肾 EST 数据库的比对结果，经过基因应答表达检测和基因蛋白结构分析，发现了一个免疫相关新基因。本研究在丰富草鱼分子生物学基础资料的同时，进一步证实，基于鱼类特异组织 EST 数据库，结合比较转录组分析是发掘新功能基因的有效途径。

Cloning and Characterization of the Grass Carp (*Ctenopharyngodon idella*) Toll-Like Receptor 22 Gene, A Fish-Specific Gene

Jianjian Lv[1, 2] Rong Huang[1] Huaying Li[1] Daji Luo[1] Lanjie Liao[1] Zuoyan Zhu[1] Yaping Wang[1]

1 State Key Laboratory of Freshwater Ecology and Biotechnology, Institute of Hydrobiology, Chinese Academy of Sciences, No 7 Donghu South Road, Wuhan 430072

2 Graduate School of the Chinese Academy of Sciences, Beijing 100049

Abstract Toll-like receptor 22 (TLR22) is a fish-specific TLR which recognizes double-strand (ds) RNA and participates in the innate immune response through the Toll-IL-1R homology domain-containing adaptor protein 1 (TICAM-1). To further investigate how the innate immune system of teleosts responds to viral infections, we cloned the full-length cDNA sequence of grass carp (*Ctenopharyngodon idella*) TLR22 (CiTLR22). The complete cDNA sequence of CiTLR22 was 3647 bp and encodes a polypeptide of 954 amino acids. Analysis of the deduced amino acid sequence indicated that CiTLR22 has typical structural features of proteins belonging to the TLR family. These included 17 LRR domains (residues 88-634) and one C-terminal LRR domain (LRR-CT, residues 694-745) in the extracellular region, and a TIR domain (residues 801-944) in the cytoplasmic region. Comparison with homologous proteins showed that the deduced CiTLR22 has the highest sequence identity to common carp TLR22 (82.9%). Genomic DNA of CiTLR22 was obtained by long-distance (Ld) PCR and structure analysis revealed that the CiTLR22 gene is encoded by uninterrupted exons. Reverse transcriptase-PCR (RT-PCR) revealed that CiTLR22 is a non-maternal gene. It is prominently expressed in immune relevant tissues such as spleen and head kidney. Quantitative RT-PCR analysis showed that CiTLR22 transcripts were upregulated significantly in immune relevant tissues and blood following grass carp reovirus (GCRV) infection. In the whole genomic sequence, nine single nucleotide polymorphisms (SNPs) were detected. Seven of them were sited in the coding region, and the other two located in the 5′ and 3′ untranslated region (UTR) respectively. None of the SNPs was associated with the resistance of grass carp to GCRV. These results suggested a role for CiTLR22 in mediating immune protection against viral infection in grass carp.

Keywords grass carp (*Ctenopharyngodon idella*) TLR22; grass carp reovirus; mRNA expression; single nucleotide polymorphisms

文章发表于 *Fish & Shellfish Immunology*, 2012, 32(6): 1022-1031

1. Introduction

The innate immune system is an efficient first line defense against invading microbial pathogens. These pathogens are recognized by conserved microbial features, termed pathogen-associated molecular patterns (PAMPs), which are detected by pattern recognition receptors (PRRs) [1]. Toll-like receptors (TLRs) are one class of PRRs, which play an essential role in the innate immune system by initiating and directing immune responses to pathogens and by controlling activation of adaptive immune responses [2]. The original Toll receptor was detected in *Drosophila melanogaster* as an essential factor to establish the dorsoventral polarity during early embryogenesis [3]. Later it was shown that the insect Toll receptor participated in antifungal immunity [4] and subsequently, mammalian TLRs were identified [5,6]. At present, 13 TLRs (TLR1-TLR13) have been identified in mammals. Interestingly, to date in teleost fish species 15 TLRs have been identified [7-11]. Teleost fish possess mammalian TLR orthologs (TLR1, TLR2, TLR3, TLR5, TLR7,TLR8 and TLR9) as well as fish-specific novel TLRs, which have not been identified in mammals, such as TLR14, TLR21, TLR22 and the soluble TLR5 [7,12-14].

The TLR22, fish-specific gene, which is not present in mammals and birds, was first identified in goldfish (*Carassius auratus*) [15]. Subsequently, it was identified in fugu (*Takifugu rubripes*) [16], Japanese flounder (*Paralichthys olivaceus*) [12], rainbow trout (*Oncorhynchus mykiss*) [14], zebrafish (*Danio rerio*) [9] and large yellow croaker (*Pseudosciaena crocea*) [17]. It was postulated that these TLRs were lost in the mammalian lineage ever since the separation from the fish lineage, approximately 400 million years ago [18]. This hypothesis was augmented by the identification of TLR22 in the genome sequence of *D. rerio* [19,20]. Concomitant with the identification of TLR22, their function was being elucidated. Previous research has shown that TLR22 gene expression can be modulated by various PAMPS. In goldfish, TLR22 gene expression can be significant upregulated by LPS, heat-killed *Aeromonas salmonicida*, and live *Mycobacterium chelonei* in cultured macrophages [15] In Japanese flounder, the expression of TLR22 gene was shown to be induced after stimulation with peptidoglycan and poly (I:C) [12]. In rainbow trout, formalin-inactivated *A. salmonicida* pathogens induce TLR22 expression up to eight-fold in vitro in peripheral blood lymphocytes and in tissues from spleen and head kidney [14]. In large yellow croaker, upon stimulation with poly (I:C), the TLR22 expression was obviously upregulated in head kidney and spleen tissues, and also in primary head kidney cells [17].

Recent research has shown that in fugu TLR22 is a homolog of TLR3 which preferentially recognizes relatively long-sized double stranded RNA (dsRNA) on the cell surface and induces IFN by recruiting Toll-IL-1R homology domain-containing adaptor protein 1 (TICAM-1) [16]. When fish cells expressing fgTLR22 are exposed to dsRNA or aquatic dsRNA viruses, cells induce IFN responses to acquire resistance to virus infection. TLR22 may be a functional substitute of human cell surface TLR3 and serve as a surveillant for infection with dsRNA virus to alert the immune system for antiviral protection in fish [16,21]. These studies have contributed to deepen our understanding of the function of TLR22.

Grass carp (*Ctenopharyngodon idella*) is an important aquaculture species in China, but tremendous economical loss is often caused by infections with grass carp reovirus (GCRV), a dsRNA virus [22,23]. Preventing this infectious disease is still difficult to date. An involvement of grass carp TLR22 (CiTLR22L) in immune protection against GCRV has recently been described by Janguo Su *et al.* [24]. In this study, we report the molecular cloning, sequence analysis and characterization of another grass carp TLR22 (CiTLR22) which different from CiTLR22L. To further understand the function, we compared similarities and differences of expression pattern of CiTLR3 and CiTLR22 in early embryonic development, in immune relevant tissues of grass carp, and followed their expression pattern subsequent to infection with GCRV. The results implied that CiTLR22 participates in anti-GCRV immune processes, and future work will aim to elucidate antiviral immune mechanisms in grass carp.

2. Materials and Methods

2.1 Animals and GCRV challenge experiments

Four month old grass carp, averaging 10 cm in body length, were collected from fish farms in Wuhan, China. Prior to performing infection experiments, the animals were kept in 3000 L aerated aquaria at 28-30℃ and were fed twice daily using a commercial diet. Animals were acclimatized to laboratory conditions for one week preceding the experiments.

The GCRV challenge experiments were carried out as described by Su *et al.* slight modifications [22,25]. Briefly, fish were injected intraperitoneally with 1.8×10^7 PFU/g body weight of GCRV strain 991 or 50 ul PBS per gram body weight as control, and non-injected animals were used as blank group. Three individuals per group were sampled, and trunk kidney, head kidney, intestine, and spleen tissue were collected on days 0, 1, 2, 3, 4, and 5 post-injection. Samples were homogenized in TRIZOL LS reagent (Invitrogen, Carlsbad, CA, USA) and kept frozen at −80℃ until RNA extraction.

Grass carp died in the first 3 day post-injection and show obvious symptoms of hemorrhage disease were classified as susceptible individuals whereas the animals survived after 15 days challenge experiments were considered as resistant group. Fin clips were collected and kept in absolute ethanol. Genomic DNA was isolated following the standard phenol-chloroform method slight modifications [26]. Genomic DNA was dissolved in double-distilled water, the quality of DNA was checked on 1% agarose gel and the concentration was adjusted to 50 ng/μl according to the OD 260 measured with a GeneQuant spectrophotometer (Amersham Biosciences Inc.).

2.2 Cloning of full-length cDNA of CiTLR22

Degenerate primers were designed based on multiple alignments of TLR22 sequences from *D. rerio* (accession No. AAI63527.1), *T. rubripes* (accession No. NP_001106664.1), and *C. auratus* (accession No. AAO19474.1). PCR was performed using degenerate primers tlr22f1 and tlr22r1 (Supplementary table) and cDNA generated from the head kidney of grass carps. The PCR product was ligated into a pMD18-T easy vector. Competent *Escherichia coli* DH5α cells were transformed and plated on LB-agar. Positive colonies containing an insert of the expected size were screened by colony PCR. Three colonies were sequenced by a commercial company (BGI Wuhan Corporation, China).

To obtain the full-length cDNA sequence, the BD SMART™ RACE cDNA amplification kit (BD Biosciences Clontech, Santa Clara, CA, USA) was used. The 3′ region was amplified using primer pairs TLR22-3-GSP1/adaptor primer UPM and TLR22-3-GSP2/adaptor primer NUP (Supplementary table) for primary and nested PCR reaction, respectively. The PCR products were cloned and sequenced as described above. Similarly, the 5′ part of CiTLR22 was obtained by nested PCR using primer pairs TLR22-5-GSP1/UPM and TLR22-5-GSP2/NUP (Supplementary table). The full-length cDNA sequence was confirmed by sequencing PCR products amplified with three pairs of primers (T22qF5 and 22-5-1, Tlr22f and Tlr22r, STLR22F and T22qR) which covered 3384 bp sequence (8-3391 bp) of TLR22 (Supplementary table).

2.3 Cloning of genomic DNA of CiTLR22

Genomic DNA of grass carp was prepared from muscle using to long distance (Ld) PCR [27]. Three pairs of primers (T22qF5 and 22-5-1, Tlr22f and Tlr22r, STLR22F and T22qR) were designed based on the CiTLR22 cDNA sequence. Fifty ng of genomic DNA were used as template for PCR with an initial denaturation step for 5 min at 95℃, followed by 40 cycles of 95℃ for 30s, 58℃ for 30s and 72℃ for 3 min. The PCR products were analyzed by 1% agarose electro-phoresis and cloned into a pMD18-T vector. At least three clones were sequenced using M13 primers for sequencing.

2.4 Sequence analysis

Homologous sequences were obtained using the BLAST program (http://www.ncbi.nlm.nih.gov/blast). Deduced amino acid sequences were analyzed with the Expert Protein Analysis System (http://www.expasy.org/) and the features of protein domains were predicted by Simple Modular Architecture Research Tool (SMART) (http://smart.embl-heidelberg.de/) [28]. Multiple sequences were aligned using CLUSTAL W (http://align.genome.jp/).

2.5 Expression profiles of CiTLR22 and CiTLR3 in tissue and during embryonic development

Various tissues including gills, liver, spleen, intestine, trunk kidney, head kidney, muscle and brain were collected from at least three healthy and three GCRV infected animals. Total RNA was isolated according to the manufacturer's instructions of TRIZOL LS reagent (Invitrogen, Carlsbad, CA, USA). Total RNA was incubated with RNase-free DNase I (Roche, Mannheim, Germany) to remove any contaminating genomic DNA before reverse transcription into cDNA using oligo-DT (18) primers and Super-Script^{TM} III Reverse Transcriptase (Invitrogen). To amplify CiTLR22 and CiTLR3, PCR was performed with gene specific primer sets: STLR22F1 and STLR22R1 and TF192 (forward) and TR193 (reverse), respectively [22] (Supplementary table). For PCR amplification, an initial denaturation step of 3 min at 94℃ was used, followed by 25 cycles of denaturation at 94℃ for 30 s, annealing at 59℃ for 30s, and an extension at 72℃ for 30s. As a positive control for RT-PCR and to determine the concentration of template, the β-actin gene was amplified using the primer set of SGAF86 (forward) and SGAR87a (reverse) (Supplementary table).

200±20 embryos of each group were sample at 0, 8, 16, 24, 32, 40, 48, 56, and 64 h post-fertilization, the expression profile of CiTLR22 and CiTLR3 mRNA in embryos was assessed using RT-PCR as described above and real-time quantitative reverse transcription PCR (qRT-PCR) as described in the following section. The relative expression ratio of the target gene versus the β-actin gene were calculated using the ΔCT method.

2.6 mRNA expression profiles following infection GCRV

Total RNA was extracted from trunk kidney, head kidney, intestine and spleen tissue sampled on days 0, 1, 2, 3, 4 and 5 post-stimulation with GCRV. Levels of CiTLR22 and CiTLR3 mRNA were assessed using qRT-PCR in an ABI Prism 7000 Sequence Detection System (Applied Biosystems). The house keeping gene β-actin was utilized as an internal control for cDNA normalization. The gene specific primers for β-actin were forward primer SGAF86 and reverse primer SGAR87a (Supplementary table). CiTLR22 and CiTLR3 gene expression was detected using primer STLR22F1 and STLR22R1, TF192 and TR193, respectively (Supplementary table). The qRT-PCR mixture consisted of 1 μl of cDNA sample, 8.2 μl of nuclease-free water, 10 μl of SYBR Green PCR master mix (Toyobo, Osaka, Japan), and 0.4 μl of gene specific primers (10 μM). The PCR cycling conditions were: 1 cycle of 94℃ for 2 min, 40 cycles of 94℃ for 15s, 60℃ for 15s, 72℃ for 45s, followed by dissociation curve analysis to verify the amplification of a single product. The threshold cycle (CT) value was determined using the manual setting on the ABI Sequence Detection System and exported into a Microsoft Excel Sheet for subsequent data analysis where the PCR efficiency and relative expression ratio of the target gene in treated groups versus that in controls were calculated using the $2^{-\Delta\Delta CT}$ method. The expression data obtained from three independent biological replicates were subjected to an unpaired, two-tailed *t*-test. $P < 0.05$ was considered statistically significant.

2.7 Identification of single nucleotide polymorphism loci in CiTLR22

The whole sequence of CiTLR22 gene was scanned with three pairs of gene specific primers (Supplementary table), which were designed based on the CiTLR22 genomic sequence. PCR reaction was performed in 50 ul reaction volume containing 50 ng of DNA template. Annealing temperatures were calculated by primer premier 5.0 software. The PCR products, from 15 susceptible individuals and 15 resistant individuals, were purified and sequenced with the automated sequencer ABI3730 (Applied Bio-system). Analyze the sequencing result, the polymorphic loci were detected from the sequence alignments of different individuals with Vector NTI Suite 11.0 (Invitrogen). SPSS13.0 was employed to analyze the genotype frequencies and their associations with susceptibility/resistance to GCRV. χ^2 test was utilized for the significance tests. *P* value less than 0.05 was considered as statistically significant.

3. Results

3.1 CiTLR22 cDNA and gene structure

CiTLR22 cDNA (HQ676542) was found to be 3647 bp long including a 5′-untranslated region (UTR) of 120 nt, a coding region of 2862 nt and a 3′-UTR of 665 nt. The 3′-UTR had an ATTTA motif which mediates mRNA degradation. The polyadenylation signal (AATAAA) was located 92 bp upstream of the polyA tail. The 2862 nt open reading frame (ORF) encodes a protein with 954 amino acid residues which exhibits a typical TLR domain architecture. This includes 17 LRR domains (residues 88-634) and one C-terminal LRR domain (LRR-CT, residues 694-745) at the extracellular region, and a TIR domain (residues 801-944) in the cytoplasmic region (Fig. 1). The domain structures of CcTLR22 (*Cyprinus carpio*; ADR66025.1), OmTLR22 (*O. mykiss*; NP_001117884.1), TrTLR22 (*T. rubripes*; NP_001106664.1), DrTLR22 (*D. rerio*; AAI63527.1), CaTLR (*C. auratus*; AAO19474.1), PoTLR22 (*P. olivaceus*; BAD01045.1) and LcTLR22 (*Larimichthys crocea*; GU324977) were predicted using the web-based software program SMART (Fig. 2). Multiple sequence alignments showed that the deduced CiTLR22 protein had 82.9%, 46.7%, 40.8%, 76.5%, 80.8%, 42.3% and 41.5% identity to sequences CcTLR22, OmTLR22, TrTLR22, DrTLR22, GaTLR, PoTLR22 and LcTLR22, respectively. The region of highest amino acid identity of the different TLR22 genes was evidently within the TIR domain (Table 1 and Fig. 3). Genomic DNA of CiTLR22 was obtained by LdPCR, the electrophoresis and sequencing results showed that the gene was encoded by uninterrupted exons. Structure analysis revealed that the genomic organization of CiTLR22 was similar to that of DrTLR22 and OmTLR22 (Fig. 4).

GAAATTCAAAGCCAAACACTGCTGGGTGAAGGGCACAGAATAAAAACATTTCAAAGACAAAAAAATTTAAAAATAAATAAATAAAAAATC 90

AAGGCAAATCCAATAGATCTTCAGTTTGACATGAAAAAAGAATCAAGAAAAAGCATGGGAAAACTACAACAAATCACATTATATATCGTT 180

M K K E S R K S M G K L Q Q I T L Y I V 20

CTCTGTGGTTTCATTTCCGCATGTAGTGCATTTTCCTTAAAGAACTGCACAATCAGTACTCCCCTTAGAGATATCCAACCGAAAGTGCTT 270

L C G F I S A C S A F S L K N C T I S T P L R D I Q P K V L 50

TGCTACAACATGGGCTTTTTTAGGATTCCGTGGTGGATACCCAGAAATACACGGATTTTAGATATTTCCTTCAATGATTTTGCACAGATC 360

C Y N M G F F R I P W W I P R N T R I L D I S F N D F A Q I 80

CAGATTGGAGACTTTAGGCATTTGTCAAACTTACAGGACTTGAACATATCCAACAACAGGATCTCACAAATTCAAGAGGGTGCACTTGAC 450

Q I G D F R H L S N L Q D L N I S N N R I S Q I Q E G A L D 110

GATCTTTCCAACTTGACCTATCTCAATCTGGCCAGCAACAGACTGAAAGCGGTCTCAAGCGGGATGCTACACGGCCTAAGCAACCTGCTG 540

D L S N L T Y L N L A S N R L K A V S S G M L H G L S N L L 140

GTGCTACGTCTGGATGAAAATAATATCAAAGACATTGAAGAGTCAGCTTTCAGCACGCTTCAGAATTTAAAGGTGTTAAACCTAACAAAA 630

V L R L D E N N I K D I E E S A F S T L Q N L K V L N L T K 170

AATCACCTCCATTACATAGACAAAGTGAAGCCGGTTCTTGCATCACCACTTTTGGAGGAACTCTACATCGGAAGCAACAATTTTGATGTT 720

N H L H Y I D K V K P V L A S P L L E E L Y I G S N N F D V 200

TTCAACTCATATGAGATGTCAACAAAGCCCTTGTCATTAAAAAAGCTTGATTTTTCCAACAACCCTTTGGCAACATTTCAGCTCACTGAC 810

F N S Y E M S T K P L S L K K L D F S N N P L A T F Q L T D 230

AACATATTTCCATCCCTCAATCACCTTGATTTATCTTATTGTGGTCAAAATGGAAGCATGACGTGGAATGTTACAGAAAAGACATACTTC 900

N I F P S L N H L D L S Y C G Q N G S M T W N V T E K T Y F 260

TCCTCAGTGCAAACATTATACTTCATGGATGTTAATATGTCACCCCAGAATGTTGCTAATGTGCTTCTGAGCTTCAAGAATTCACTGAAT 990

S S V Q T L Y F M D V N M S P Q N V A N V L L S F K N S L N 290

AAAATCAGGTTTAATGGAAACGTTGAGCTTAATAAGACCAACCTTCTGCTGAGCGCATGTTCTCCAATGCTACGAGTTGTACGGCTAAAT 1080

K I R F N G N V E L N K T N L L L S A C S P M L R V V R L N 320

GCTAACAAAATAAAACACCTCACCGACAACATGTTTGATCCCTGTTCTGACCTGACGGAATTAGATTTAGGAGATAATGAAATATCCAAG 1170

A N K I K H L T D N M F D P C S D L T E L D L G D N E I S K 350

TTATCACCGAGTATGTTCAGAGGTTTCACTCAACTAAAAAAACTGCTTCTGCAAATCAATAAACTGACTCAAATAACAAACTCCTTTCAA 1260

L S P S M F R G F T Q L K K L L L Q I N K L T Q I T N S F Q 380

ATTCTTACCACGCTTGAGTTCATAGATCTCAGTAGAAACAGCATCAACAAGCTCACCTGTAATGACTTTGCCAATTTAACACAGGTGAAA 1350

I L T T L E F I D L S R N S I N K L T C N D F A N L T Q V K 410

ACTCTGTACTTGTATGGTAATAAAATCTCTCTCATTAGATCCTGTTTGTTTAAGGACCTCAAGAGTCTTGAAGTCCTAAAGCTTGGAACT 1440

T L Y L Y G N K I S L I R S C L F K D L K S L E V L K L G T 440

AATGATCTGTTAAGGATCGACGATGCTTTCAGCAATGGGCCACATTCTCTGAAGGATCTGCAAATTAATTTTAATAAGCTAAGCAAAATA 1530

N D L L R I D D A F S N G P H S L K D L Q I N F N K L S K I 470

GAAAAATACACTTTTAGAAATTTGTCACAGTTAAACAGTTTGACCTTAAATGACAATCAGATTTCAGAGATCGAAGCCCAAGCTTTCGAA 1620

E K Y T F R N L S Q L N S L T L N D N Q I S E I E A Q A F E 500

GGACTGAAGAATCTGACTTCTCTGTTTCTGTCATCAAACAAAATAACAGCTAAAACATTAACCAGACACCCGAATGTGTTCTCAGGCATG 1710

G L K N L T S L F L S S N K I T A K T L T R H P N V F S G M 530

CCCAACCTCCAAAATCTGGATTTGTACGCCAACAGCATTTCATTTGCAGATAATAAATTGAAACACCCTCCTTTTAAGGATCTGAAGCAA 1800

P N L Q N L D L Y A N S I S F A D N K L K H P P F K D L K Q 560

CTCAGGGTATTGACCCTTCACAGTCAACGCCGTGGAATCAACAAAATACCCTCAAATCTGCTTCAAGGTTTATCCTCCATGGAAATGTTT 1890

L R V L T L H S Q R R G I N K I P S N L L Q G L S S M E M F 590

TATGTTGGAAACACAAATCTTGGTCATCTAAATCCTGATACATTCAAGTTCAGCCCCCAGCTCTGGTTCTTGGATCTCTCCAAGAACGCA 1980

Y V G N T N L G H L N P D T F K F S P Q L W F L D L S K N A 620

CTGTCTGAAGACAACTCGATTCCGGCTGAGCTCTTCCACCCCATTTCGAGGCTAACCAAACTGATCCTTTCAAGAACGCAGCTTCGCTCG 2070

L S E D N S I P A E L F H P I S R L T K L I L S R T Q L R S 650

CTGAATTTCCTGTTGAATGCAAATCTCTCCAGACTCTCGACCTTAAGAGCCCCGGGCAATGAGATTGACACGATCAACAAAACTCTGATT 2160

L N F L L N A N L S R L S T L R A P G N E I D T I N K T L I 680

CAGTCACTGCCTCGACTAGAAGTCCTCGACTTGCAGAGAAACACCTTCACTTGCGACTGTCACAACGAGTTCTTCATTGAATGGGCCATG 2250

Q S L P R L E V L D L Q R N T F T C D C H N E F F I E W A M 710

AAAACTAATTCTACTCAGGTGTTTTATTTTAACAGGTACACGTGTAGCTATCCTCGTTCCTTACGAGGCATGAGCTTGACAGCTTTCAAC 2340

K T N S T Q V F Y F N R Y T C S Y P R S L R G M S L T A F N 740

ATCGAATCCTGCACCTTGAACATTGACTTCATCTGCTTTCTTTGCAGCAGTATTGTGGTCACTCTCACCCTCCTCTTGTCGTTTGTCTGG 2430

I E S C T L N I D F I C F L C S S I V V T L T L L L S F V W 770

CATTTTCTGCGCTATCAGGTGATTTATGCATACTACCTCTTCTTAGCCTTCCTTTACGACAACAAGAAGAAGCAGACCGTTTCAACGATC 2520

H F L R Y Q V I Y A Y Y L F L A F L Y D N K K K Q T V S T I 800

CGGTACGACACCTTCATTTCCTACAACACCGAAGACGAGCCTTGGGTAATGGAGGAACTCGTTCCCAAATTAGAAGGAGAACAGGGCTGG 2610

R Y D T F I S Y N T E D E P W V M E E L V P K L E G E Q G W 830

AAATTGTGCCTGCACCATAGAGACTTTGTACCAGGAAGGCCAATAATAGATAACATTATTGATGGCATATACAGCAGCCGCAAAACCATC 2700

K L C L H H R D F V P G R P I I D N I I D G I Y S S R K T I 860

TGCCTGATCACTAGGAACTACCTGAAGAGCAACTGGTGTTCAAGTGAGGTTCAGGTAGCGAGCTATAGGCTCTTTGATGAGCAGAAGGAT 2790

C L I T R N Y L K S N W C S S E V Q V A S Y R L F D E Q K D 890

GTGCTGATCCTGGTGTTCCTGGAGGATATTCCTGCACATCAGCTCTCTCCGCATCACAGGATGCGGAAGCTGGTGAAGAAACGAACTTAC 2880

V L I L V F L E D I P A H Q L S P H H R M R K L V K K R T Y 920

CTCCGCTGGCCCAAACCTGGAGAAGATACTAAGATCTTCTGGCAAAAGCTAAAAATGGCTTTAGAGACCAAAGAGGGTCATAACCCAGAG 2970

L R W P K P G E D T K I F W Q K L K M A L E T K E G H N P E 950

AGTGCAATTCTGTGAGGGTATTTTAACAATATTGTACACTATCACCTTTGAGAAAACTATGGGGGGTCATATTTAAATTTTGTTGTAATT 3060

S A I L * 954

CCTTTAAGTGCAAATGCATCCATTATATGTAAAGTAGACTTCTAGTAGATTATTATGCATTTTATTCTGTTGTAGTACTCAAGACCGGTC 3150

TTGGTCTTAAAGGGTTAGTTCACCCAAAAATGAAAATTATGTCATTAATTACTCACCCTCATGTCGTTCTACACCCGTGGGACCTTCGTT 3240

CGTCTTCGGAACACAAGTTGAGATGTTTTTGGTGAAATCCGATGGCTCAGTGAGGCCGCTATTGCTAGTAAGATAATTGGCACTTTCGGA 3330

TGCCCAGAAAGGTACTGGGAGCATGTTTAGGACAGTCCGTGTGACTATATTGGTTCAACCTTGATATTTTAGAGCGACGAGAGTATGTTT 3420

TCTGTGCCGAGGGAGACAGAATGATGACTTTTGAACAAT**ATTTA**GTGATGGCCAATTCTTCTGCTTTTGTTTTGAATCAGCACCAAAGTC 3510

ACATGATTTC**AATAAA**CGAGGCTTCGTTATGTTATAAGTGTTTCAAAATTTCAATAGTTCAAGTGACTTTGGCTTTTTGATACGCGATCC 3600

GAACCACTGATTCAAAGCAAAAAAAAAAAAAAAAAAAAAAAAAAAAAA 3647

Fig. 1 Nucleotide sequence and deduced amino acid sequence of CiTLR22 cDNA. The ''stop codons'' in front of the start codon was marked by a dotted underline. The start codon (ATG) was boxed and the stop codon (TAA) was marked with an asterisk. The polyadenylation signal motif (AATAAA) is given in bold. The motif associated with mRNA instability (ATTTA) is given in bold. In the deduced amino acid sequence, the signal peptide is shaded (1-30aa). The LRR and C-terminal LRR motifs are shown as underscore (88-111aa, 112-135aa, 136-159aa, 160-181aa, 312-334aa, 335-356aa, 359-381aa, 382-404aa, 406-427aa, 430-452aa, 454-480aa, 478-501aa, 502-538aa, 530-552aa, 558-583aa, 584-607aa, 609-634aa, 694-745aa).The transmembrane domain is shown as dotted underscore (750-772aa). The TIR domain is indicated by a wavy underline (801-944aa)

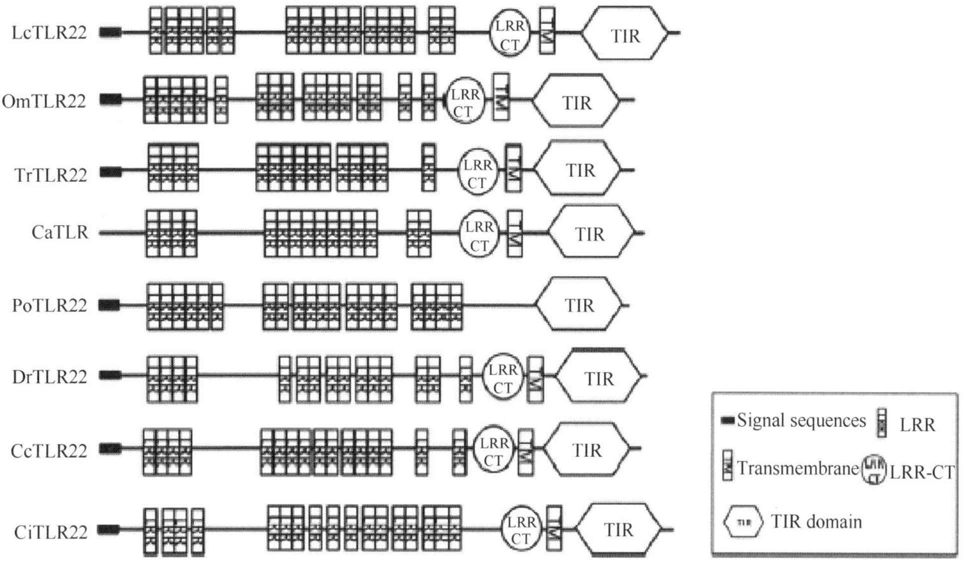

Fig. 2 Structural features of TLR22 genes. Schematic representation of TLR22 domains predicted by SMART. The GenBank accession numbers used here are as follows: *C. idella* (CiTLR22, ADX97523.1), *C. carpio* (CcTLR22, ADR66025.1), *T. rubripes* (TrTLR22, NP_001106664.1), *C. auratus* (CaTLR, AAO19474.1), *P. olivaceus* (PoTLR22, BAD01045.1), *Larimichthys crocea* (LcTLR22, ADK77870.1), *O. mykiss* (OmTLR22, NP_001117884.1), *D. rerio* (DrTLR22, AAI63527.1)

Table 1 Amino acid sequences identities (%) between CiTLR22 and other TLR22s

CiTLR22	CcTLR22	CaTLR22	DrTLR22	OmTLR22	PoTLR22	LcTLR22	TrTLR22
(658aa)LRR	81.9	79.2	76.1	43.0	38.0	37.1	37.4
(144aa)TIR	89.6	89.6	90.3	75.0	69.4	70.1	65.3
(954aa)LRR&TIR	82.9	80.8	76.5	46.7	41.9	41.5	40.7

3.2 Expression profiles of CiTLR22 in tissue and during embryonic development

RT-PCR was performed to analyze the expression levels of CiTLR22 and CiTLR3 in various tissues. In healthy fish, CiTLR22 was mainly expressed in the head kidney, followed by relatively high expression levels in spleen, trunk kidney and gills. Lower expression levels were observed in intestine, weak expression in brain and heart, and no expression in liver and muscle. In comparison, CiTLR3 was mainly expressed in the trunk kidney and head kidney, followed by gills and brain. Expression was weaker in liver, spleen, intestine and heart, and no expression was observed in muscle.

In infection fish, CiTLR22 and CiTLR3 were ubiquitously expressed in all grass carp tissues examined, and the CiTLR22 and CiTLR3 transcripts were generally upregulated in the various tissues, as shown in Fig. 5.

To investigate CiTLR22 and CiTLR3 expression pattern in different phases of embryonic development, 200±20 embryos were sample at 0, 8, 16, 24, 32, 40, 48, 56, and 64 h post-fertilization. The expression pattern of CiTLR22 and CiTLR3 in embryos was assessed using qRT-PCR. CiTLR22 mRNA was not detected in the first 24 h of embryonic development, but a significant increase was noted subsequently. CiTLR3′s expression was detected in all developmental stages, and there were no significant difference between any of the phases (Fig. 6).

3.3 CiTLR22 mRNA expression profiles stimulated by GCRV infection

To determine the effects of viral infection on CiTLR22 and CiTLR3 gene expression, three animals from each group were sacrificed on day 0, 1, 2, 3, 4 and 5 post-injection of GCRV. qRT-PCR was conducted to determine the levels of mRNA in trunk kidney, head kidney, blood, and spleen tissue. The results show that GCRV induced CiTLR22 and CiTLR3 expression in these

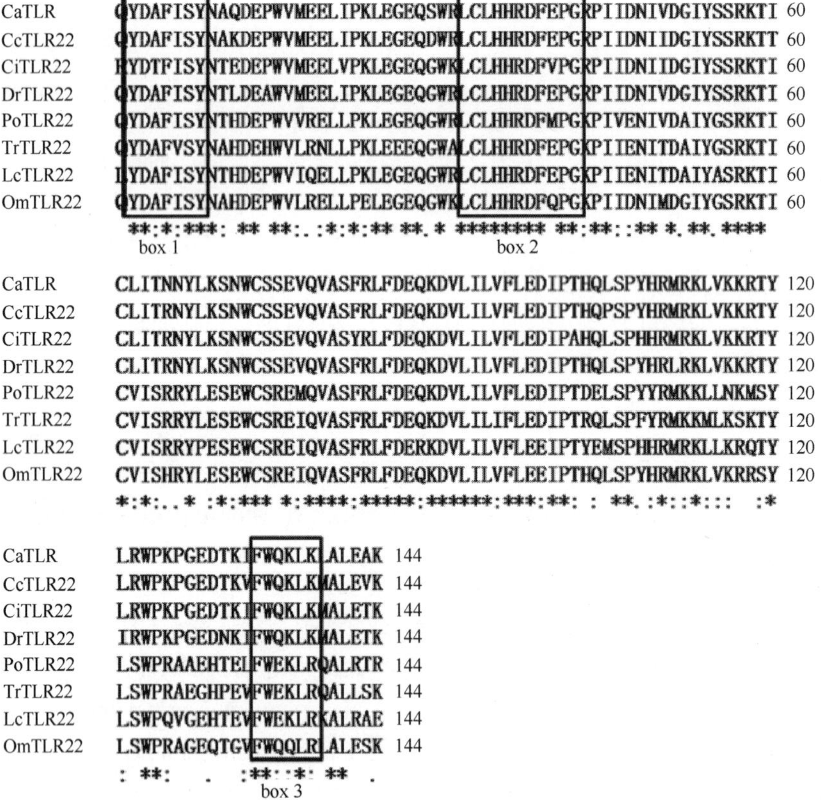

Fig. 3 Alignment of grass carp TLR22 TIR domain with that of other fish species TLR22. The three active motifs are boxed: box 1 (YDAFISY), box 2 (LC-RD-PG) and box 3 (a conserved W surrounded by basic residues). The GenBank accession numbers used here are as follows: *C. idella* (CiTLR22, ADX97523.1), *C. carpio* (CcTLR22, ADR66025.1), *T. rubripes* (TrTLR22, NP_001106664.1), *C. auratus* (CaTLR, AAO19474.1), *P. olivaceus* (PoTLR22, BAD01045.1), *Larimichthys crocea* (LcTLR22, ADK77870.1), *O. mykiss* (OmTLR22, NP_001117884.1), *D. rerio* (DrTLR22, AAI63527.1)

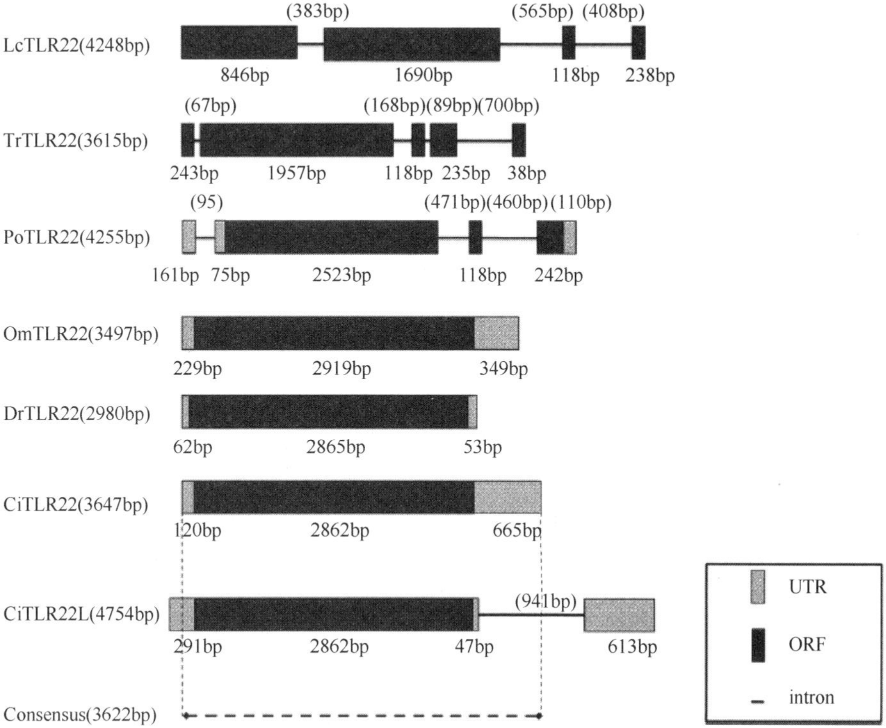

Fig. 4 Comparison of gene structures of grass carp TLR22 gene and previously reported TLR22 genes. Exons are represented by boxes, introns are represented by lines. The Consensus region of CiTLR22 and CiTLR22L are represented by dotted line. Numbers indicate the length of exons and introns in bp. The DNA sequences used in this study were obtained from GenBank: *C. idella* (CiTLR22, ADX97523.1, CiTLR22L, ACT78472), *T. rubripes* (TrTLR22, NP_001106664.1), *P. olivaceus* (PoTLR22, BAD01045.1), *Larimichthys crocea* (LcTLR22, ADK77870.1), *O. mykiss* (OmTLR22, NP_001117884.1), *D. rerio* (DrTLR22, AAI63527.1)

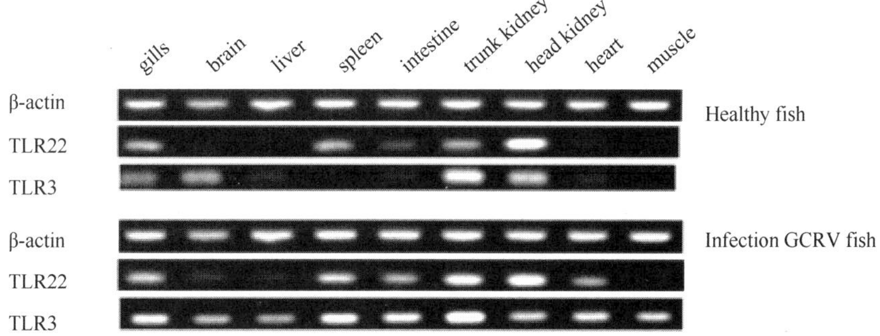

Fig. 5 Tissue-specific expression of CiTLR22 mRNA. Total RNAs were extracted in various tissues from three healthy and three infection fish. cDNAs were equally mixed from three healthy fish and three infection fish respectively in corresponding tissues. RT-PCR was employed to detect the transcription levels of CiTLR22 in 9 tissues. As an internal reference for RT-PCR, β-actin was amplified to normalize each template. The examined 9 tissues were marked above each lane

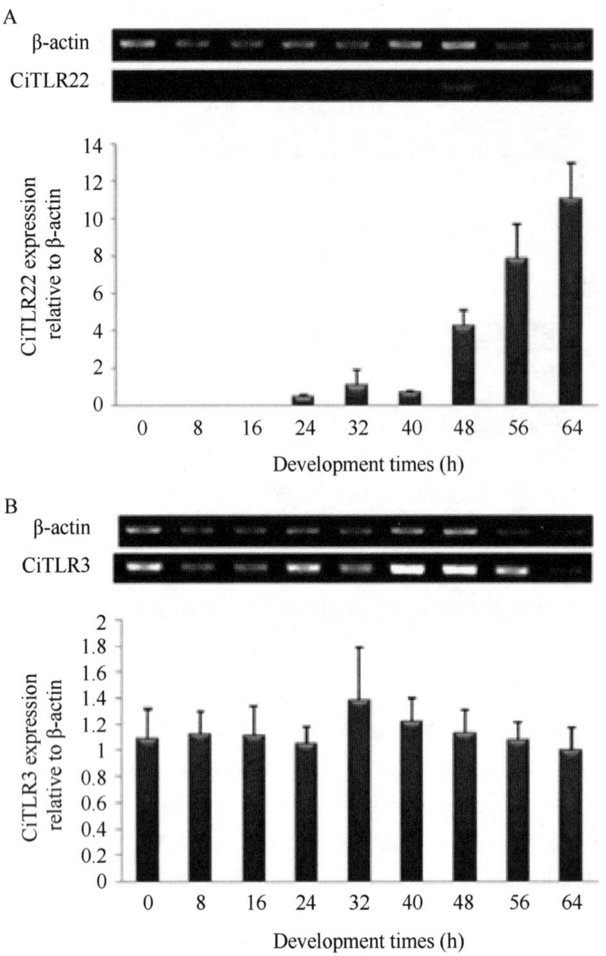

Fig. 6 Expression analysis of TLRs during embryonic development. About 200 embryos each group were sample at 0, 12, 24, 48 and 72 h post-fertilization. qRT-PCR was employed to detect the transcription levels. β-actin was employed as an internal control. A. Expression analysis of CiTLR22 during embryonic development. B. Expression analysis of CiTLR3 during embryonic development

tissues. While CiTLR22 transcripts reached peak levels in head kidney and spleen on day three ($P < 0.05$), in blood on day two and in trunk kidney on day four post-injection, CiTLR3 transcripts reached peak levels in all tissues on day three. The highest up-regulation of expression of CiTLR22 and CiTLR3 was 4.8-fold in spleen and 8.8-fold in blood, respectively (Fig. 7).

3.4 Single nucleotide polymorphism and the association between the polymorphisms and the susceptibility/resistance to GCRV

Nine SNPs (–8 A/T, 203 A/T, 526 A/T, 863 C/T, 1248 C/T, 1491 A/C, 1923 G/T, 2180 A/T, 2409 C/T, 2941 C/T) were discovered. Seven of them located in the coding sequence, 3 were synonymous and 4 were non-synonymous. one lies in the 5′UTR, one lies in 3′UTR. All the correlations between resistance/susceptibility to GCRV and the genotype were analyzed with SPSS13.0 and χ^2 test. The statistical results were shown in Table 2, which indicated that no SNP was significantly associated with the resistance of grass carp to GCRV.

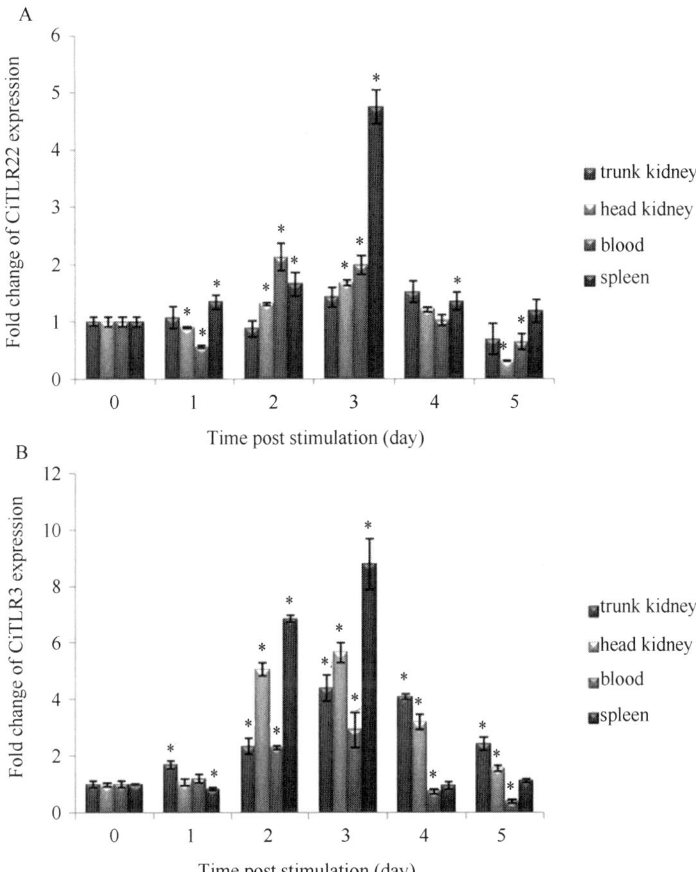

Fig. 7 Expression analysis of TLRs in different tissues (trunk kidney, head kidney, blood and spleen) of grass carp following infection with GCRV. The controls were injected with PBS; the experimental animals were infected with GCRV. β-actin was employed as an internal control. Asterisk (*) marks the significant difference between experimental group and control group ($P < 0.05$). Error bar indicate standard error. A. CiTLR22 expression in different tissues of grass carp following infection with GCRV. B. CiTLR3 expression in different tissues of grass carp following infection with GCRV

4. Discussion

In the present study we report the cloning of the full-length TLR22 cDNA of grass carp, the characterization of its structure and expression. Analysis of the sequence revealed a large ORF translating into a protein of 954 amino acid residues which possessed typical structural features of the TLR family. The domain structure of CiTLR22 was similar to CcTLR22 (*C. carpio*,

ADR66025.1), DrTLR22 (*D. rerio,* AAI63527.1) and CaTLR (*C. auratus*, AAO19474.1), but the number of LRR varies slightly between different fish species. There are 18, 17, 15, 15, 20, 15 and 16 LRRs in LcTLR22 [17], OmTLR22 [14], TrTLR22 [16], CaTLR22 [15], PoTLR22 [12], DrTLR22 [29] and CcTLR22 [30], respectively. The LRR domain of TLR has been implicated in ligand binding [31,32], and it is an open question whether the variation in the LRR domain changes the ligand affinity of fish TLR22.

Multiple alignments of the different TLR22 revealed that the highest level of identity is evidently within the TIR domain, likely because the cytoplasmic TIR domain evolves more slowly than the LRR domain-containing extracellular regions [33]. The TIR domain, which is functionally important for TLR signal transduction, contains three active motifs: box 1 (FDAFISY), box 2 (LC-RD-PG), and box 3 (a conserved W surrounded by basic residues) [34]. It has been suggested that box 1 and box 2 mediate binding to receptor molecules of signaling transduction pathways, while box 3 has been shown to be primarily involved in directing localization of receptors [34]. These active motifs were basically conserved in CiTLR22 (box 1, $Y^{802}DAFISY^{808}$; box 2, $L^{832}C\text{-}RD\text{-}PG^{842}$ and box 3, W^{934} surrounded by basic residue K940), and TLR22 of other fish species. However, it is noteworthy that there are two sites which were not completely conserved in CiTLR22. The A in box1 in TLR22 genes of other fish species is the site of an A→T change in CiTLR22, and the E in box 2 in TLR22 genes of cyprinid fish is the site of an E→V change in CiTLR22 (Fig. 3). Of the cyprinid fish species, *D. rerio, C. auratus* and *C. carpio*, show resistance to GCRV infection, while *C. idella* is susceptible [35]. This suggests that the mutation of single residue in the TIR domain may affect resistance to GCRV infection.

Genomic DNA of CiTLR22 was obtained by LdPCR suggesting that the coding region of CiTLR22 represents an uninterrupted exon. A similar result was found for DrTLR22 [29] and OmTLR22 [14], but it was different for TrTLR22 [16], PoTLR22 [12], and LcTLR22 [17], which had four, three and three introns, respectively (Fig. 4). These observations demonstrated that the genomic organization of TLR22 is not highly conserved among teleost fish in general, but it was highly conserved in the freshwater species *D. rerio*, *C. idella* and *O. mykiss*. In other words, when comparing freshwater (*D. rerio*, *C. idella* and *O. mykiss*) and seawater teleost species (*T. rubripes*, *P. olivaceus* and *P. crocea*) the latter tended to have more introns in TLR22 gene. Similar results were found for TLR2 [36], TLR5 [37] and TLR9 [38] in fish species and this warrants further investigation. Diversity in TLR gene organization was also reported for TLR1, TLR2, TLR3 and TLR5 within and between teleosts as well as in mammals [22,36,37,39], but it is as yet unclear whether this diversity causes functional differences.

While during the course of our study, su *et al*. (2011) reported another TLR22 gene sequence of grass carp containing an intron, full-length was 4754 bp [24], we named it CiTLR22L. The amino acid sequence of CiTLR22L was highly consistent with CiTLR22

(99.5%) reported in this study. However, the sequence differences and genomic organization provided pretty convincing evidence that those were two different genes in grass carp. May be the two grass carp TLR22 genes had different expression abundance in different grass carp populations living in different environment at different developmental stages, which resulted in two different genes were cloned in the two studies, respectively. The detection of two TLR22 genes in grass carp is not entirely unexpected, similar phenomenon was also found in rainbow trout [14]. Further analysis found the sequence of CiTLR22L gene intron and 3′UTR region of CiTLR22 gene is highly consistent. Therefore, we speculated that CiTLR22 gene of grass carp may be having different splicing types. In order to verify this hypothesis, we designed primers in the sequence of CiTLR22 ORF and 3′UTR respectively, genomic DNA and cDNA from different individuals and tissues (spleen, head kidney, liver and intestinal) were used as template for PCR. The Electrophoresis showed that all samples amplified the same target band (date was not shown), which proved the CiTLR22 gene in this study was not existence different splicing types between the ORF and 3′UTR region.

Recent research revealed that fish species possess a dual dsRNA recognition system. Both, TLR3 and TLR22 serve as a dsRNA recognition receptor and recruit a common adaptor, TICAM-1. But, as shown by the phylogenetic analysis, TLR22 does not belong to the TLR3 family [16,21,40]. Why are there two TLRs that respond to dsRNA in teleosts? We speculate that there are some differences in their function. Indeed, a previous study suggested that fgTLR3 and fgTLR22 may recognize different sizes of dsRNA with TLR22 preferentially recognizing long RNA molecules. An additional difference was that fgTLR22 localized to the surface of the cell membrane while fgTLR3 was localized inside the cell [23] being consistent with differences in function of TLR3 and TLR22. In the study presented here, we further investigated the function of these TLRs by comparing similarities and differences between CiTLR22 and CiTLR3 in tissue-specific expression profiles, following infection with GCRV and during early embryonic development.

The tissue-specific expression profiles indicated that both, CiTLR22 and CiTLR3, are highly expressed in immune relevant tissue such as spleen and head kidney. This result was partially consistent with gene expression patterns in other teleost fish species. In puffer fish, rainbow trout, Japanese flounder and large yellow crocker, the TLR22 gene was most prominently expressed in one or both of these tissues [8,12,14,17]. In contrast, the TLR3 gene was mainly expressed in liver, intestine, and trunk kidney in common carp [35], and in catfish the highest expression level was in liver and muscle [41]. These findings support the notion that tissue-specific expression pattern of TLR22 are more conserved in different teleost species compared to TLR3 which might also indicate that the function of TLR22 is more conserved than the function of TLR3. Nevertheless, CiTLR22 and CiTLR3 are mainly expressed in immune relevant tissues suggesting their functional relevance to counteract infections. This assumption is strongly supported by the fact that the expression of both genes is stimulated by the fish pathogenic dsRNA virus GCRV. Following infection with GCRV, the increase of CiTLR22 and CiTLR3 mRNA levels was highly significant in selected tissues (trunk kidney, head kidney, blood and spleen). In addition, we found that the extent of induction was significantly higher for CiTLR3 than that of CiTLR22, with the average expression increasing by 5.4-fold in CiTLR3 and 2.5-fold in CiTLR22. These results indicated that CiTLR3 was more sensitive than CiTLR22 to GCRV infection.

Table 2 The SNP distributions of CiTLR22 gene in susceptible and resistant groups

Locus	Located	Genotype	Synonymous(Y/N)	Susceptible N(%)	Resistant N(%)	χ^2(P)
-8A/T	5′UTR	AA AT	—	13(86.7) 2(13.3)	13(86.7) 2(13.3)	0.00(1.000)
203A/T	ORF	GG AG	N(Arg-Glu)	13(86.7) 2(13.3)	15(100) 0(0)	2.14(0.143)
526A/T	ORF	AA AT	N(Ile-Leu)	15(100) 0(0)	14(93.3) 1(6.7)	1.03(0.309)
863C/T	ORF	CC CT	N(Ser-Leu)	13(86.7) 2(13.3)	13(86.7) 2(13.3)	0.00(1.000)
1248C/T	ORF	TT CT	Y	13(86.7) 2(13.3)	12(80) 3(20)	0.24(0.624)
1491A/C	ORF	AA AC	N(Gln-His)	15(100) 0(0)	13(86.7) 2(13.3)	2.14(0.143)
1923G/T	ORF	GG GT	Y	13(86.7) 2(13.3)	14(93.3) 1(6.7)	0.37(0.543)
2409C/T	ORF	CC CT	Y	12(80) 3(20)	13(86.7) 2(13.3)	0.24(0.624)
2841C/T	3′UTR	CC CT	—	12(80) 3(20)	13(86.7) 2(13.3)	0.24(0.624)

The expression pattern of TLR during embryonic development revealed that CiTLR22 was a non-maternal gene, as it was not recognized in embryonic development until 24 h post-fertilization. Subsequently, the relative expression of CiTLR22 increased significantly with embryonic development. In contrast, CiTLR3 appears to be a maternal gene, and there were no significant difference between the different phases. These disparities in gene expression patterns of CiTLR22 and CiTLR3 are interesting and deserve further characterization.

Nine polymorphic loci (SNPs) were found in CiTLR22. Plus five SNPs detected in CiTLR22L gene [24], a total of 12 SNPs of grass carp TLR22 gene were found, of which only two SNPs were detected simultaneously. Association analysis between polymorphism and susceptibility resistance to GCRV showed that one SNP(417G/T) was significantly associated with the resistance of grass carp to GCRV in previous study [24], however, in our study, no SNP associated with the resistance of grass carp to GCRV, which showed that SNP polymorphisms had significant difference in different populations. In addition, sample size was small in this study and need further validation in large sample.

In conclusion, we have cloned entire TLR22 coding region from grass carp, and found that its expression was significantly induced in immune relevant tissues following GCRV stimulation. Our data imply that CiTLR22 may play a pivotal role in the innate antiviral immune response in grass carp. Meanwhile, the distribution of polymorphisms in CiTLR22 and their association with the resistance of grass carp to GCRV were investigated. However, further studies are required to clarify the mechanism(s) that regulate CiTLR22 expression.

Acknowledgements The authors thank Wei Hu, Yonghua Sun, Ming Li, Feng Dong, Shangping Chen and other laboratory members for technical assistance and helpful discussion. This work was supported by grants from National Natural Science Foundation of China (Grant no. 2009CB118701 and 2011AA100403).

Appendix A. Supplementary material Supplementary data related to this article can be found, in the online version at doi:10.1016/j.fsi.2012.02.024.

References

[1] Subramaniam S, Stansberg C, Cunningham C. The interleukin 1 receptor family. Developmental & Comparative Immunology 2004; 28: 415-28.

[2] Schnare M, Barton GM, Holt AC, Takeda K, Akira S, Medzhitov R. Toll-like receptors control activation of adaptive immune responses. Nature Immunology 2001; 2: 947-50.

[3] Belvin MP, Anderson KV. A conserved signaling pathway: the *Drosophila* Toll-dorsal pathway. Annual Review of Cell and Developmental Biology 1996; 12: 393-416.

[4] Hoffmann JA, Reichhart JM. *Drosophila* innate immunity: an evolutionary perspective. Nature Immunology 2002; 3: 121-6.

[5] Akira S. Mammalian Toll-like receptors. Current Opinion in Immunology 2003; 15: 5-11.

[6] Medzhitov R, Preston-Hurlburt P, Janeway CA. A human homologue of the *Drosophila* Toll protein signals activation of adaptive immunity. Nature 1997; 388: 394-7.

[7] Hwang SD, Kondo H, Hirono I, Aoki T. Molecular cloning and characterization of Toll-like receptor 14 in Japanese flounder, *Paralichthys olivaceus*. Fish & Shellfish Immunology 2011; 30: 425-9.

[8] Oshiumi H, Tsujita T, Shida K, Matsumoto M, Ikeo K, Seya T. Prediction of the prototype of the human Toll-like receptor gene family from the pufferfish, *Fugu rubripes*, genome. Immunogenetics 2003; 54: 791-800.

[9] Jault C, Pichon L, Chluba J. Toll-like receptor gene family and TIR-domain adapters in *Danio rerio*. Molecular Immunology 2004; 40: 759-71.

[10] Rebl A, Goldammer T, Seyfert HM. Toll-like receptor signaling in bony fish. Veterinary Immunology and Immunopathology 2010; 134: 139-50.

[11] Takano T, Hwang SD, Kondo H, Hirono I, Aoki T, Sano M. Evidence of molecular Toll-like receptor mechanisms in teleosts. Fish Pathology 2010; 45: 1-16.

[12] Hirono I, Takami M, Miyata M, Miyazaki T, Han HJ, Takano T, *et al*. Characterization of gene structure and expression of two Toll-like receptors from Japanese flounder, *Paralichthys olivaceus*. Immunogenetics 2004; 56: 38-46.

[13] Tsukada H, Fukui A, Tsujita T, Matsumoto M, Iida T, Seya T. Fish soluble Toll-like receptor 5 (TLR5S) is an acute-phase protein with integral flagellin-recognition activity. International Journal of Molecular Medicine 2005; 15: 519-25.

[14] Rebl A, Siegl E, Köllner B, Fischer U, Seyfert H- M. Characterization of twin Toll-like receptors from rainbow trout (*Oncorhynchus mykiss*): evolutionary relationship and induced expression by Aeromonas salmonicida salmonicida. Developmental & Comparative Immunology 2007; 31: 499-510.

[15] Stafford JL, Ellestad KK, Magor KE, Belosevic M, Magor BG. A Toll-like receptor (TLR) gene that is up-regulated

in activated goldfish macrophages. Developmental & Comparative Immunology 2003; 27: 685-98.

[16] Matsuo A, Oshiumi H, Tsujita T, Mitani H, Kasai H, Yoshimizu M, *et al*. Teleost TLR22 recognizes RNA duplex to induce IFN and protect cells from birnaviruses. Journal of Immunology 2008; 181: 3474-85.

[17] Xiao XQ, Qin QW, Chen XH. Molecular characterization of a Toll-like receptor 22 homologue in large yellow croaker (*Pseudosciaena crocea*) and promoter activity analysis of its 5′-flanking sequence. Fish & Shellfish Immunology 2011; 30: 224-33.

[18] Litman GW. Sharks and the origins of vertebrate immunity. Scientific American 1996; 275: 67-71.

[19] Tanekhy M, Kono T, Sakai M. Cloning, characterization, and expression analysis of Toll-like receptor-7 cDNA from common carp, *Cyprinus carpio* L. Comparative Biochemistry and Physiology D-Genomics & Proteomics 2010; 5: 245-55.

[20] Meijer AH, Gabby Krens S, Medina Rodriguez IA, He S, Bitter W, Ewa SJ. Expression analysis of the Toll-like receptor and TIR domain adaptor families of zebrafish. Molecular Immunology 2004; 40: 773-83.

[21] Seya T, Matsumoto M, Ebihara T, Oshiumi H. Functional evolution of the TICAM-1 pathway for extrinsic RNA sensing. Immunological Reviews 2009; 227: 44-53.

[22] Su JG, Jang SH, Yang CR, Wang YP, Zhu ZY. Genomic organization and expression analysis of Toll-like receptor 3 in grass carp (*Ctenopharyngodon idella*). Fish & Shellfish Immunology 2009; 27: 433-9.

[23] Attoui H, Fang Q, Jaafar FM, Cantaloube JF, Biagini P, de Micco P, *et al*. Common evolutionary origin of aquareoviruses and orthoreoviruses revealed by genome characterization of golden shiner reovirus, grass carp reovirus, striped bass reovirus and golden ide reovirus (genus Aquareovirus, family reoviridae). Journal of General Virology 2002; 83: 1941.

[24] Su J, Heng J, Huang T, Peng L, Yang C, Li Q. Identification, mRNA expression and genomic structure of TLR22 and its association with GCRV susceptibility/resistance in grass carp (*Ctenopharyngodon idella*). Developmental & Comparative Immunology; 2011.

[25] Heng JF, Su JG, Huang T, Dong J, Chen LJ. The polymorphism and haplotype of TLR3 gene in grass carp (*Ctenopharyngodon idella*) and their associations with susceptibility/resistance to grass carp reovirus. Fish & Shellfish Immunology 2011; 30: 45-50.

[26] Sambrook J, Russell DW. Molecular cloning: a laboratory manual. CSHL Press; 2001.

[27] Sambrook J, Fritsch E, Maniatis T. Molecular cloning: a laboratory manual. 2nd ed. New York: Cold Spring Harbor Laboratory; 1989. 18: 58.

[28] Letunic I, Copley RR, Pils B, Pinkert S, Schultz J, Bork P. SMART 5: domains in the context of genomes and networks. Nucleic Acids Research 2006; 34: D257.

[29] Strausberg RL, Feingold EA, Grouse LH, Derge JG, Klausner RD, Collins FS, *et al*. Generation and initial analysis of more than 15,000 full-length human and mouse cDNA sequences. Proceedings of the National Academy of Sciences U S A 2002; 99: 16899.

[30] Kongchum P, Palti Y, Hallerman EM, Hulata G, David L. SNP discovery and development of genetic markers for mapping innate immune response genes in common carp (*Cyprinus carpio*). Fish & Shellfish Immunology 2010; 29: 356-61.

[31] Akira S, Takeda K. Toll-like receptor signalling. Nature Reviews Immunology 2004; 4: 499-511.

[32] O'Neill LAJ, Bowie AG. The family of five: TIR-domain-containing adaptors in Toll-like receptor signalling. Nature Reviews Immunology 2007; 7: 353-64.

[33] Johnson GB, Brunn GJ, Tang AH, Platt JL. Evolutionary clues to the functions of the Toll-like family as surveillance receptors. Trends in Immunology 2003; 24: 19-24.

[34] Slack JL, Schooley K, Bonnert TP, Mitcham JL, Qwarnstrom EE, Sims JE, *et al*. Identification of two major sites in the type I interleukin-1 receptor cytoplasmic region responsible for coupling to pro-inflammatory signaling pathways. Journal of Biological Chemistry 2000; 275: 4670.

[35] Yang C, Su J. Molecular identification and expression analysis of Toll-like receptor 3 in common carp *Cyprinus carpio*. Journal of Fish Biology 2010; 76: 1926-39.

[36] Baoprasertkul P, Peatman E, Abernathy J, Liu ZJ. Structural characterisation and expression analysis of Toll-like receptor 2 gene from catfish. Fish & Shellfish Immunology 2007; 22: 418-26.

[37] Hwang SD, Asahi T, Kondo H, Hirono I, Aoki T. Molecular cloning and expression study on Toll-like receptor 5 paralogs in Japanese flounder, *Paralichthys olivaceus*. Fish & Shellfish Immunology 2010; 29: 630-8.

[38] Takano T, Kondo H, Hirono I, Endo M, Saito-Taki T, Aoki T. Molecular cloning and characterization of Toll-like receptor 9 in Japanese flounder, *Paralichthys olivaceus*. Molecular Immunology 2007; 44: 1845-53.

[39] Palti Y, Rodriguez MF, Gahr SA, Purcell MK, Rexroad CE, Wiens GD. Identification, characterization and genetic mapping of TLR1 loci in rainbow trout (*Oncorhynchus mykiss*). Fish & Shellfish Immunology 2010; 28: 918-26.

[40] Oshiumi H, Matsuo A, Matsumoto M, Seya T. Pan-vertebrate Toll-like receptors during evolution. Current Genomics 2008; 9: 488-93.

[41] Baoprasertkul P, Peatman E, Somridhivej B, Liu ZJ. Toll-like receptor 3 and TICAM genes in catfish: species-specific expression profiles following infection with *Edwardsiella ictaluri*. Immunogenetics 2006; 58: 817-30.

草鱼 tlr22 基因的克隆及特征分析

吕建建[1,2] 黄 容[1] 李华英[1] 罗大极[1] 廖兰杰[1]
朱作言[1] 汪亚平[1]

1 中国科学院水生生物研究所，武汉 430072
2 中国科学院大学，北京 100049

摘 要 Toll样受体22(TLR22)是鱼类特有的toll家族受体，能够通过TICAM-1受体蛋白识别双链 RNA 并参与先天性免疫反应。为了进一步了解硬骨鱼类免疫系统对病毒感染的反应，我们克隆了草鱼 TLR22(CiTLR22)的 cDNA 序列。CiTLR22 的 cDNA 序列全长 3647bp，编码 954 个氨基酸。对 CiTLR22 编码的氨基酸序列进行分析，显示其有 TLR 家族典型的结构特征，包括 17 个 LRR 结构域(88-634)，一个 C 端胞外区的 LRR 结构域(LRR-CT，694-745)和一个细胞质内的 TIR 结构域(801-944)。将 CiTLR22 与其同源蛋白比对，结果显示 CiTLR22 与鲤的 TLR22 有最高的同源性(82.9%)。通过长链 PCR 技术，克隆了 CiTLR22 的基因组序列，对其结构进行分析，发现其只有一个外显子。RT-PCR 显示 CiTLR22 是一个非母源表达的基因，并且显著表达于一些免疫组织如脾脏和头肾中。荧光定量 PCR 表明，在草鱼呼肠孤病毒(GCRV)刺激下，CiTLR22 的表达水平在免疫相关组织和血液中显著上调表达。在 CiTLR22 的全基因组序列中，发现了 9 个单核苷酸多态性位点，其中 7 个位于编码区，2 个位于 5′和 3′端的非翻译区域(UTR)，但这 9 个 SNP 位点均与草鱼对 GCRV 的抗性无关。以上结果说明 CiTLR22 在介导草鱼应对病毒感染的免疫保护中发挥了作用。

Identification, Characterization and the Interaction of Tollip and IRAK-1 in Grass Carp (*Ctenopharyngodon idellus*)

Rong Huang[1] Jianjian Lv[1,2] Daji Luo[1] Lanjie Liao[1]
Zuoyan Zhu[1] Yaping Wang[1]

1 State Key Laboratory of Freshwater Ecology and Biotechnology, Institute of Hydrobiology, Chinese Academy of Sciences, Wuhan 430072
2 Graduate School of Chinese Academy of Sciences, Beijing 100039

Abstract Tollip and IRAK-1 are key components of the TLR/IL-1R signaling pathway in mammals, which play crucial roles as mediators of the TLR/IL-1R signal transduction pathways. Although several TLRs have been found in fish, molecular associations, protein-protein interactions or the role of the TLR signaling pathway in infection-induced immunity in fish has received little attention. In this study, Tollip and IRAK-1 sequences of grass carp were isolated from a head kidney cDNA library. Full length transcripts and sequences of promoter regions were obtained by 3′ and 5′ RACE and genome walking, respectively. Reporter gene-promoter constructs and real-time RT-PCR analysis was used to determine grass carp Tollip and IRAK-1 transcription pattern in tissues. Recombinant proteins were used for antibodies production. Phylogenetically, the grass carp loci clustered with previously reported Tollip and IRAK-1 genes, respectively, and their sequences shared the highest identity with the genes of zebrafish (*Danio rerio*). The promoter region of grass carp Tollip and IRAK-1 proved to be active. After viral infection transcript levels of both loci were upregulated in most immune-related tissues in a time-dependent manner. Using antibodies produced in this study, immunofluorescence analysis indicated that Tollip and IRAK-1were uniformly distributed and co-localized in the cytoplasm of CIK cells. After viral infection, however, Tollip and IRAK-1 both trended toward the cell membrane. Our results demonstrate the existence of Tollip and IRAK-1 proteins in teleost species, and suggest that Tollip-IRAK-1 complexes are being recruited to receptor complexes after stimulation with virus. These results provide novel insights into the role of the TLR signaling pathway in teleosts, especially the action of teleost Tollip and IRAK-1 and the interaction of these molecules as part of this pathway.

Keywords tollip; IRAK-1; molecular cloning; viral infection; grass carp

文章发表于 *Fish & Shellfish Immunology*, 2012, 33(3): 459-467

1. Introduction

Innate immune responses from *Drosophila* to *Man* utilize remarkably conserved molecular components, many of which contain homologous protein-protein interaction domains [1,2] such as the Toll-IL-1 receptor (TIR) domain and the death domain (DD). Toll-like receptors (TLRs), interleukin-1 receptor (IL-1R), and interleukin-18 receptor (IL-18R) are single-transmembrane proteins with conserved cytoplasmic TIR domains, which facilitate the activation of a similar high molecular mass complex when stimulated with their cognate ligands [3]. Toll-interacting protein (Tollip) and IL-1 receptor-associated kinase 1 (IRAK-1) may be incorporated into this high molecular mass complex [4]. It has been shown that in untreated cells, Tollip forms a complex with IRAK-1. Recruitment of Tollip-IRAK-1 complexes to activated receptor complex occurs through association of Tollip with cytoplasmic TIR domain. Co-recruited MyD88 then triggers IRAK-1 autophosphorylation, which in turn leads to the dissociation of IRAK-1 from Tollip (and TLRs, IL-1Rs, IL-18Rs) [4], and its association with TNF receptor-associated factor 6 (TRAF-6) to signal activation of either NF-κB or mitogen-activated protein kinases [5].

The Tollip gene was first described in the mouse, human and *Caenorhabditis elegans* [5], while the IRAK-1 gene was first found in humans [6]. To date, Tollip and IRAK-1 protein sequences have been isolated from mammals, birds, nematodes and fish. In mammals, Tollip has been shown to be involved in two main functions. The first is that Tollip was found to play a role in the activated IL-1 receptor complex, suggesting that it was an integral part of the IL-1RI signaling cascade [5]. Later, Tollip was shown to associate with TLR2 and TLR4 receptors as well, which triggers suppression of TLR-mediated cellular responses [4]. In 2004, the role of Tollip in negative regulation of the IL-1β and TNF-α signaling pathway was proposed [7]. The second function concerns the role of Tollip in protein sorting through interaction with target of Myb1 (Tom1), ubiquitin and clathrin [8,9]. An endosomal function of the protein was first suggested by Katoh *et al.* [10,11]. In addition, Ciarrocchi and D'Angelo [12] have shown the involvement of Tollip in the sumoylation process, and in the control of both nuclear and cytoplasmic protein trafficking.

IRAKs play crucial roles as mediators in the TLR/IL-1R signal transduction pathways [13]. IRAKs constitute a protein family with four members, IRAK-1, IRAK-2, IRAK-3/M, and IRAK-4. Lin *et al.* (2010) reported the crystal structure of the MyD88/IRAK-4/IRAK-2 death domain (DD) complex, which surprisingly reveals a left-handed helical oligomer that consists of 6 MyD88, 4 IRAK-4 and 4 IRAK-2 DDs. Assembly of this helical signalling tower is hierarchical [14]. IRAK-1 has been identified as a key component of the IL-1R signaling pathway in mammals [6]. Upon ligand binding to TLRs or IL-1R, IRAK-1 is recruited to the receptor complex to activate the downstream signaling pathway leading to proinflammatory responses [15]. Furthermore, IRAK-1 is involved in activation and nuclear translocation of signal transducer and activator of transcription 3 (STAT3), STAT1 and interferon regulation factor 7 (IRF7), as well as in downstream gene expression [13,16,17]. Phosphorylated IRAK-1 also undergoes ubiquitin-mediated degradation or sumoylation resulting in nuclear translocation and transcriptional activation of inflammatory target genes [18]. In 2000, Burns *et al.* [5] provided evidence that Tollip forms a complex with IRAK-1 in resting cells using yeast two-hybrid interaction assays and by coimmunoprecipitation from 293T cells.

Based on molecular structure and phylogenetic analysis, Rebl *et al.* identified two closely related Tollip-encoding genes in Atlantic salmon (*Salmo salar*) and the respective orthologous mRNA molecules in rainbow trout (*Oncorhynchus mykiss*) [19]. Subsequently, an IRAK-1 cDNA sequence named ScIRAK-1 was also identified in mandarin fish [20,21]. Although several TLRs [22-28] have been found in fish, the role of the TLR signaling pathway, the molecular associations and protein-protein interactions in infection-induced immunity in fish has received little attention.

In the study reported here, we therefore characterized Tollip and IRAK-1 genes of grass carp and compared their amino sequence with other known Tollip and IRAK-1 proteins. We analyzed their expression at the transcript and protein level in virus infected grass carp and in control and viral infected CIK (*Ctenopharyngodon idellus* kidney) cells. The aim of our study was to provide further insight into the genetic basis and protein-protein interactions participating in the TLR signaling pathway in response to viral infection in teleost fish. The investigation of molecular associations and interactions in this study provide novel insights into the role of the TLR signaling pathway, especially with regards to teleost host responses to infection.

2. Materials and Methods

2.1 Cell culture and animals

CIK cells were grown in Eagle's minimum essential medium (MEM) supplemented with 10% fetal bovine serum (MEM-10) at 28℃ as described previously [29]. Grass carp were supplied by the Guanqiao farm in Wuhan, China.

2.2 Virus and Viral infection of grass carp

The Grass carp reovirus (GCRV) 873 strain was generously provided by Prof. Q. Fang of the Wuhan Institute of Virology, Chinese Academy of Science [30]. The virus titers, given as tissue culture infective dose ($TCID_{50}$/0.1 ml), were calculated by the method of Reed and Muench [31]. Nine-month-old grass carp with an average weight of 126 g were raised in clean tanks at 28°C. Sixty grass carp were randomly divided into two groups, one control group and a GCRV-infected group. In the GCRV-infection group, each fish was intraperitoneally injected with 150-200 μl GCRV for a dose of approximately 10^7 $TCID_{50}$ kg^{-1} body weight. Fish in the control group were injected with the same amount of saline. The various tissue samples of control (1 d after injection, control grass carp, n = 3) and virus-infected grass carp (at 2 d, 3 d, 4 d, 5 d, 7 d post-infection, infected grass carp, each time point n = 3) were immediately removed and frozen in liquid nitrogen, or stored at −80℃ until RNA isolation.

2.3 Screening of cDNA libraries, 3′ and 5′ RACE

The full cDNA sequence of Tollip was isolated from a head kidney cDNA library of grass carp. Through PCR amplification using primers shown in Table 1 (IRAF-F1/IRAK-R1) designed from conserve regions of known IRAK-1 genes, one cDNA fragment of 1051 bp was found to be homologous to known IRAK-1 genes. The cDNA template was transcribed using AMV reverse transcriptase (TaKaRa, Japan) and an oligo dT adapter (Table 1). The forward primer, IRAF (Table 1) was designed according to the partial cDNA sequence of IRAK-1. After reverse transcription, 3′ RACE PCR was performed with primers IRAF and 3′ adapter (Table 1). 5′ RACE was performed using the BD SMART RACE cDNA Amplification Kit (Clontech, USA) according to the manufacturer's instruction. The primer for 5′ CDS, BD SMART oligo, UPM and a Gene-Specific Primer IRAR (Table 1) were used to obtain the 5′ sequence of IRAK-1.

Table 1 Primers used for cloning and expression studies

Primer	Sequence (5′-3′)	Application
IRAF-F1	ATCAGATCAGACTGAACTACGCCTT	
IRAK-R1	CTCTAGCATGACCACTCCAAAACT	
IRAF	ATCGCCCTCTCTTGGTTACAGCG	3′ RACE
IRAR	TGAGGATGTCTGTGCGTCTGGG	5′ RACE
GT-R1	CTTACCTGTCCCCGTTGCGTGCT	Genomic walking (first round PCR)
GT-R2	GTTGCGTGCTAATTGTTGTTGCC	Genomic walking (second round PCR)
GI-R1	TAATGGCGAAGAATGCCTGGCTC	Genomic walking (first round PCR)
GI-R2	TCATAAGCCACTGCTACACCGC	Genomic walking (second round PCR)
ETo-F1	GGTCGCCTCAGCATCACTGTA	RT-PCR primer used in expression study
ETo-R1	CATAGACGGCATACCCAAGTCG	RT-PCR primer used in expression study
EIR-F1	ATACAGGCATCCCAACATAATGG	RT-PCR primer used in expression study
EIR-R1	ACAGCGTAGTCGGTCTTCTAAAG	RT-PCR primer used in expression study
β-actin-F	GGATGATGAAATTGCCGCACTGG	RT-PCR control used in expression study
β-actin-R	ACCGACCATGACGCCCTGATGT	RT-PCR control used in expression study
ETo-F2	AGGATCCGATTAGCACGCAACGAGGAC	Expression in *E. coli*

Continued

Primer	Sequence (5′-3′)	Application
ETo-R2	TAAGCTTCTCTGCCATCTGTAGCAAGG	Expression in *E. coli*
EIR-F2	CATGAATTCCAAGCAGATTGGAGAAGGAGGT	Expression in *E. coli*
EIR-R2	CATCTCGAGGCCAGAGGTCTTCAACCCTG	Expression in *E. coli*
5′ CDS	$(T)_{25}$ VN-3′ N = A, C, G, or T; V = A, G, or C	5′ RACE PCR
BD SMART oligo	AAGCAGTGGTATCAACGCAGAGTACGCGGG	5′ RACE PCR
5′ UPM	CTAATACGACTCACTATAGGGCAAGCAGTG-GTATCAACGCAGAGT	5′ RACE PCR
	CTAATACGACTCACTATAGGGC	
3′ Adapter	GGCCACGCGTCGACTAGTAC	3′ RACE PCR adaptor
Oligo dT adapter	GGCCACGCGTCGACTAGTACT$_{17}$	First strand cDNA synthesis
AP1	GTAATACGACTCACTATAGGGC	Genomic walking adaptor primer 1
AP2	ACTATAGGGCACGCGTGGT	Genomic walking adaptor primer 2

2.4 Cloning and detection of the promoter region

The promoter regions of Tollip and IRAK-1 were obtained using a gene walking approach by constructing genomic libraries with a Genome Walker^{TM} Universal Kit (Clontech, USA). Two pairs of primers (GT-R1 and GT-R2, GI-R1 and GI-R2) (Table 1) were designed from the 5′ end of grass carp Tollip and IRAK-1 cDNAs, priming upstream amplification through two rounds of PCR with the adaptor primers AP1 and AP2 (Table 1). The secondary PCR was carried out with 1 μl of the first round PCR mixture. Then, the promoter regions of Tollip and IRAK-1 were cloned into a eukaryotic expression vector which containing a GFP or RFP reporter, respectively. The constructed plasmids were microinjected to zebrafish embryos; 12 h post-fertilization, fluorescent microscopy was used to examine for GFP and RFP reporter expression.

2.5 Sequence analysis of DNAs

PCR products were separated by agarose gel electrophoresis and purified using the Gel Extraction kit (OMEGA, USA). The purified products were ligated into a pMD18-T vector (TaKaRa, Japan) which was used to transform competent *E. coli* DH5α cells. Colonies were screened by PCR. Recombinant plasmids were sequenced by the dideoxy chain termination method with M13 universal primers.

Sequences were analyzed for similarity with other known sequences using BLAST and multiple sequence alignments were generated using ClustalW. Conserved domains of grass carp Tollip and IRAK-1 were predicted by the SMART Server software [32]. The protein family signature was identified using the software InterPro [33]. Sequences were further analysed using the program Match^{TM} [34] and the software ProtParam (http://web.expasy.org/protparam/). The phylogenetic tree was constructed based on the full length amino acid sequences of known Tollips and IRAK-1s using the neighbor-joining algorithm in MEGA version 3.1 [35].

2.6 Quantification of Tollip and IRAK-1 mRNAs in GCRV-infected and control grass carp

Expression of Tollip and IRAK-1 genes were determined by real-time RT-PCR. Briefly, 1 μg total RNA was isolated from various tissues of grass carp and used to synthesize the first strand cDNAs using an oligo dT primer. The first strand cDNAs were used as template for PCR amplification using the primers ETo-F1/ETo-R1, EIR-F1/EIR-R1 (Table 1), which specifically amplify a part of the Tollip and IRAK-1 cDNAs, respectively. Serial 2-fold dilutions of the first strand cDNA, ranging from 2^5 down to 2^0 cDNA copies, were used to produce a standard curve for each PCR run. Reactions without template were used as blanks. The real-time RT-PCR was conducted on a Chromo 4 Real-Time Detection System (MJ Research, USA). The expected amplified fragment sizes of Tollip and IRAK-1were 103 bp and 112 bp, respectively. Primers β-actin-F and β-actin-R (Table 1) for β-actin cDNA amplification were used as the internal reference; the expected amplified fragment was 136 bp. Each tissue sample was examined in triplicates.

2.7 Expression of rTollip and rIRAK-1 in *E. coli* and preparation of antibody

The cDNA fragment coding for a partial Tollip protein (5-274 residues) was amplified using the primers ETo-F2 and ETo-R2 (Table 1) and digested with *Bam*HI and *Hin*dIII. The digested fragment was inserted in frame to the *Bam*HI and *Hin*dIII double-digested pET-21b expression vector. The cDNA fragment coding for a partial IRAK-1 protein (194-502 residues) was amplified using the primers EIR-F2 and EIR-R2 (Table 1), and digested with *Eco*RI and *Xho*I. The digested fragment was subcloned into the pET-21b expression vector. After the recombinant constructs were confirmed by DNA sequencing, they were transformed into *E. coli* BL21 (DE3), respectively. Protein expression was induced with IPTG (final concentration 1 mM). His-tagged rTollip and rIRAK-1 proteins were purified by Ni^{2+}-NTA agarose (Novagen, USA). Purified rTollip and rIRAK-1 proteins were used to immunize one rabbit and five mice, respectively.

2.8 Western blot analysis of endogenous grass carp Tollip and IRAK-1

The endogenous grass carp Tollip and IRAK-1 were detected by Western blotting analysis. Head kidney of infected grass carp were lysed in protein extraction buffer (Boster,

China). After centrifugation at 12000 *g* for 10 min, 20μg supernatant proteins per lane were separated by 10% SDS-PAGE and subsequently transferred to NC membrane (Millipore, USA). The blotting membrane was blocked with 5% dry milk in Tris buffered saline (TBS) buffer for 1 h, followed by an incubation with TBS containing 1.0% milk and antiserum (diluted 1∶2000) from rabbit, mouse, or negative control serum. As negative control, the Tollip and IRAK-1 antiserum were replaced by the antiserum pre-adsorbed with pure fusion protein for 16 h at 4℃. Membranes were washed three times for about 30 min followed by an incubation for 1 h with 1∶2000 diluted alkaline phosphatase conjugated Goat anti-rabbit IgG and Goat anti-mouse IgG (SABC, China), respectively. Membranes were washed three times (approximately 10 min each) in TBS-Tween buffer, and protein was detected using BCIP/NBT (SABC, China) as substrate.

2.9 Immunofluorescence and confocal microscopy

CIK cell infection with GCRV was performed as described previously [36,37]. Non-infected CIK cells were used as a control in the co-localization assay. CIK cells were grown on glass slides and 24 h post-infection, slides were washed in PBS, fixed in 4% paraformaldehyde and permeabilized with 0.2% Triton X-100 for 15 min. Cells were incubated in 5% goat serum in PBS at 4℃ for 12 h and then incubated overnight with rabbit anti-Tollip antibody and mouse anti-IRAK-1 antibody (diluted 1∶1000). Cells were washed three times with PBS and incubated with FITC-conjugated goat anti-rabbit IgG and cy3-conjugated goat anti-mouse IgG (Pierce, UAS). Following two washes in PBST (0.1% Triton X-100 in PBS) and three washes in PBS, cells were analyzed using a Revolution XD confocal microscope (Andor, USA).

3. Results

3.1 Sequence analysis

Characteristics of the genes and their deduced amino acid sequences were given in Table 2. For the Tollip gene, a possible signal sequence for polyadenylation, ATTAAA, was located 16 bp upstream of the poly (A) tail. The highest amino acid identity (97%) was shared with zebrafish Tollip protein and identities of 85%, 83%, 82% and 38% were determined for mouse (*Mus musculus*), chicken (*Gallus gallus*), human (*Homo sapiens*), and *C. elegans* Tollips. Amino acid sequence analysis indicated the existence of an N-terminal Tom1 binding domain (TBD) (amino acids 1-54), a conserved 2 (C2) domain (amino acids 55-154) and a consensus CUE domain (amino acids 237-267) (Fig. 1 and supplementary data Fig. 1). In general, the Tollip amino acid sequence was more conserved in the central region and the C-terminus than the N-terminus. In Tollip, the content of the amino acids Val (9.1%) and Gln

(9.1%) were high while that for Cys (1.1%), Trp (1.1%) and His (0.7%) was low.

A number of transcription regulatory elements were predicted in the 2.3 kb promoter region of the grass carp Tollip gene, including potential recognition sequences for transcription factors including CR1-BP1/c-Jun, Pax-6, Elf-1, Cart-1, Oct-1, FOXD3, CREB. Expression of the GFP reporter was observed 12-h post-fertilization by fluorescent microscope, and was not tissue specific even when observed at following times (data not shown) (cDNA sequence of grass carp Tollip GenBank accession number: JQ239167; promoter sequence of grass carp Tollip GenBank accession number: JQ239169).

In IRAK-1 (Table 2), a possible signal sequence for polyadenylation, AATAAA, was located 15 bp upstream of the poly (A) tail. The highest amino acid sequence similarity (76%) was found with zebrafish IRAK-1 like protein and identities of 47%, 46% and 39% were determined for mouse, western clawed frog (*Xenopus* (*Silurana*) *tropicalis*), and human IRAK-1s. Amino acid sequence analysis indicated the existence of an N-terminal death domain (DD) (amino acids 1-95), a central kinase domain (amino acids 177-505), a C-terminal C1 (amino acids 506-614) and a C2 domain (amino acids 615-701) (Fig. 1 and supplementary data Fig. 2). In IRAK-1 the amino acid sequence was more conserved in the central kinase region than at the N-terminus or C-terminus. The Ser (12.4%) and Leu (9.1%) contents were high while that of Met (2.3%) and Trp (1.0%) was low.

Table 2 Characteristics of the grass carp Tollip and IRAK-1

Feature	Tollip	IRAK-1
Open reading frame	1325 bp	3976 bp
Number of amino acids	276 aa	701 aa
Start of 5′ UTR upstream of ATG	93 bp	397 bp
Length of 3′ UTR	401 bp	1473 bp
Poly(A) signal, upstream of poly(A)	16 bp	15 bp
Sequence identity to zebrafish protein	97%	76%
MW of polypeptide	30,460 Da	78,304 Da
Isoelectric point	4.93	6.09
Negatively charged amino acids (Asp, Glu)	28	86
Positively charged amino acids (Arg, Lys)	21	76

A number of transcription regulatory elements are predicted in the 3.3 kb promoter regions of the grass carp IRAK-1 gene, including Nkx2-5, Elk-1, myogenin/NF-1, E2, USF, CDP, Clox, COMP1, Evi-1, HNF-3beta, FOXD3, HNF-1, VBP, HLF, CRE-BP1, CRE-BP1/c-Jun. Expression of the RFP reporter was observed 12 h post-fertilization by fluorescent microscopy, and was not tissue specific even when observed at following times (data not shown) (cDNA sequence of grass carp IRAK-1 GenBank accession number: JQ239168; promoter sequence of grass carp IRAK-1 GenBank accession number: JQ239170).

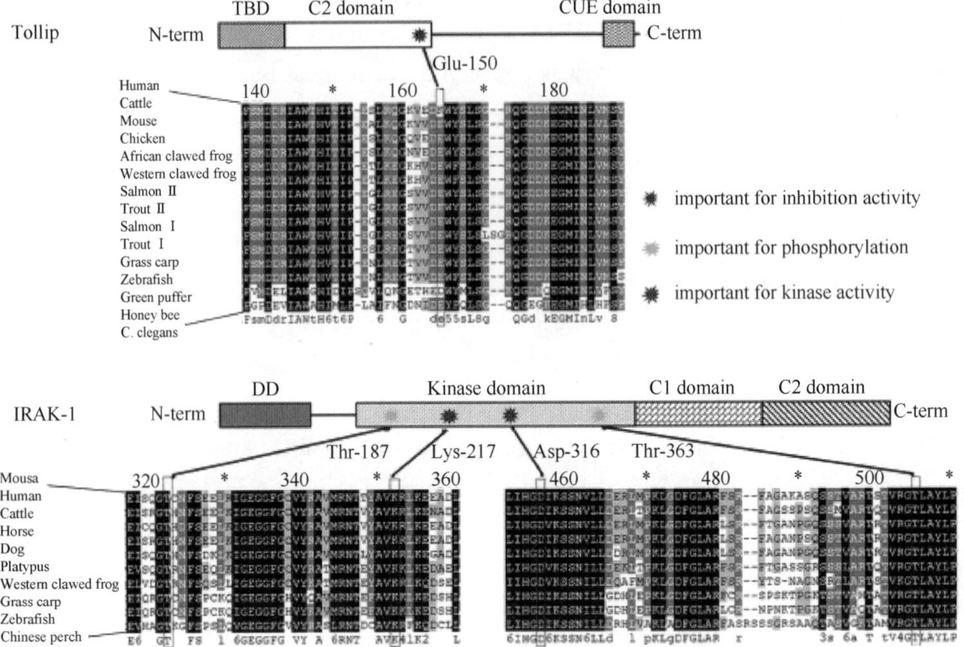

Fig. 1 Schematic representation and multiple alignments of the partial deduced amino acid sequences of grass carp Tollip and IRAK-1. GenBank accession numbers for the sequences used are listed in legend of Fig. 2. Black shaded sequences indicate positions where all the sequences share the same amino acid residue, gray shaded sequence indicates conserved amino acid substitutions; light gray shaded sequences indicate semi-conserved amino acid substitutions, and dashes indicate gaps. Several important domains are shown in boxes. The key residues in the structure are highlighted by the star symbol

3.2 Phylogenetic analysis

To further analyze the relationship of grass carp Tollip and IRAK-1 with the respective proteins of other species, two phylogenetic trees were constructed using the neighbor-joining method. In salmonid fish, Tollip proteins have been phylogenetically divided into Tollip Is and Tollip IIs. Although in the phylogenetic analysis grass carp Tollip clustered with previously reported teleost Tollip, it did not clearly fall into the Tollip I or II subgroup but clustered closely to zebrafish Tollip protein. The phylogenetic tree of IRAK separates the polypeptides into four clades, comprising IRAK-1s, IRAK-2s, IRAK-3s and IRAK-4s. Grass carp IRAK-1 fell into the cluster with previously reported teleost IRAK-1 genes. It showed the highest identity with zebrafish IRAK-1 and clustered accordingly in the phylogeny (Fig. 2). This suggested that the grass carp Tollip and IRAK-1 that were cloned in the present study were related to Tollip and IRAK-1 of other species.

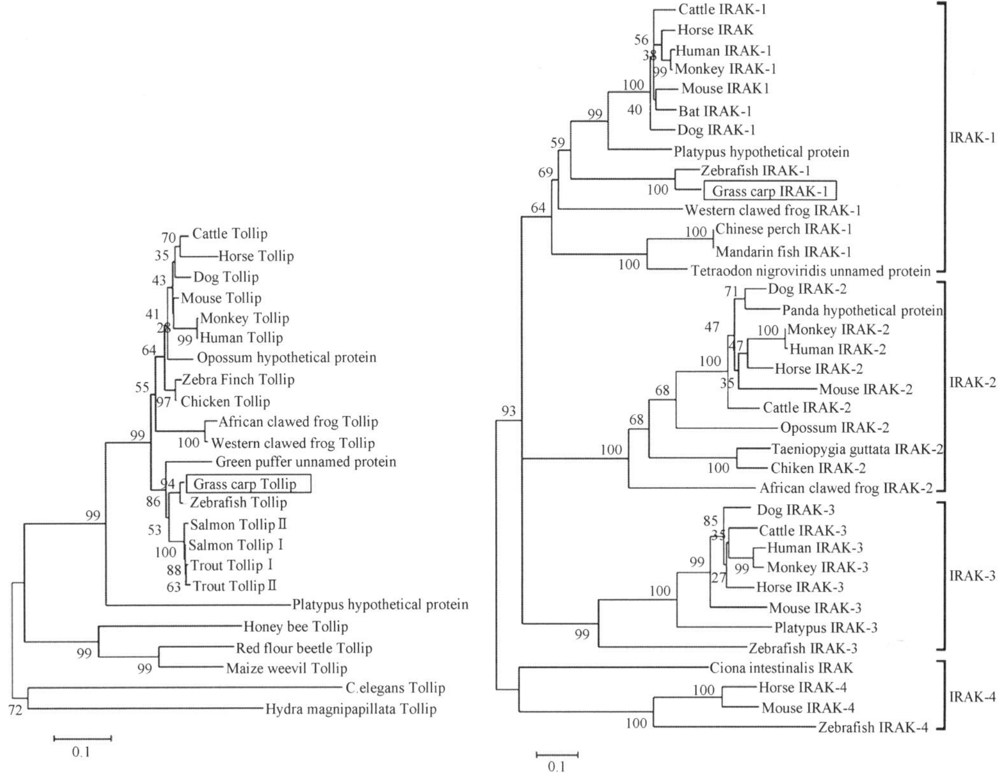

Fig. 2 Phylogenetic tree produced using the neighbor-joining algorithm. The GenBank accession numbers used are as follows: Zebrafish Tollip, NP_996994; Salmon Tollip I, CAM88654; Salmon Tollip, CAM88656; Trout Tollip I, NP_001117892; Dog Tollip, XP_540778; Monkey Tollip, XP_001090075; Trout Tollip II, Q4LBC7; Human Tollip, BAD92778; Mouse Tollip, NP_076253; Zebra Finch Tollip, XP_002199221; Chicken Tollip, NP_001006471; Cattle Tollip, Q2LGB5; Green puffer unnamed protein, CAF93239; Opossum hypothetical protein, XP_001362652; Horse Tollip, XP_001488144; African clawed frog Tollip, NP_001089737; Western clawed frog Tollip, NP_001006923; Red flour beetle Tollip, XP_975168; Maize weevil Tollip, ABZ80666; Honey bee Tollip, XP_624417; Platypus hypothetical protein, XP_ 001509306; *Hydra magnipapillata* Tollip, XP_002167406; *C. elegans* Tollip, NP_492757. Zebrafish IRAK-1, XP_697688; Chinese perch IRAK-1, ACN64942; Cattle IRAK-1, NP_001035645; Mouse IRAK-1, EDL29847; Dog IRAK-1, XP_549367; Human IRAK-1, AAC41949; Horse IRAK, XP_001491966; Platypus hypothetical protein, XP_001508862; Bat IRAK-1, ACC68898; Monkey IRAK-1, ACF60537; Western clawed frog IRAK-1, NP_001006713; *Tetraodon nigroviridis* unnamed protein product, CAF93411; Dog IRAK-2, XP_541772, *Taeniopygia guttata* IRAK-2, XP_002187461; Chicken IRAK-2, NP_001025776; Zebrafish IRAK-3, XP_001921584; Panda hypothetical protein, EFB29729; *Ciona intestinalis* IRAK, XP_002128882; Mouse IRAK-3, AAM83393; Cattle IRAK-2, NP_001069164; Monkey IRAK-2, XP_001090790; Human IRAK-2, NP_001561; Dog IRAK-3, XP_538271; Horse IRAK-2, XP_001491238; Mouse IRAK-2, AAO24761; Human IRAK-3, Q9Y616; Opossum IRAK-2, XP-001375625; Monkey,IRAK-3, XP_001117080; Horse IRAK-3, XP_001491198; Platypus IRAK-3, XP_001510859; Cattle IRAK-3, XP_587469; African clawed frog IRAK-2, NP_001079489; Horse IRAK-4, XP_001488489; Mouse IRAK-4, NP_084202; Mandarin fish IRAK-1, ACN64942; Zebrafish IRAK-4, AAH45381

3.3 Tissue specific expression pattern of grass carp Tollip and IRAK-1 mRNA

The tissue specific expression of grass carp Tollip and IRAK-1 mRNA was examined using real-time RT-PCR on ten tissues from healthy carp (Fig. 3). Reactions were normalized using β-actin. Transcripts of both genes were detectable in all tissues examined, but were predominantly detected in liver and peripheral blood lymphocytes (PBLs), followed by brain, muscle, gills, spleen, intestine, kidney and skin. The lowest expression levels were detected in head kidney. No Tollip and IRAK-1 transcripts were detected in negative controls.

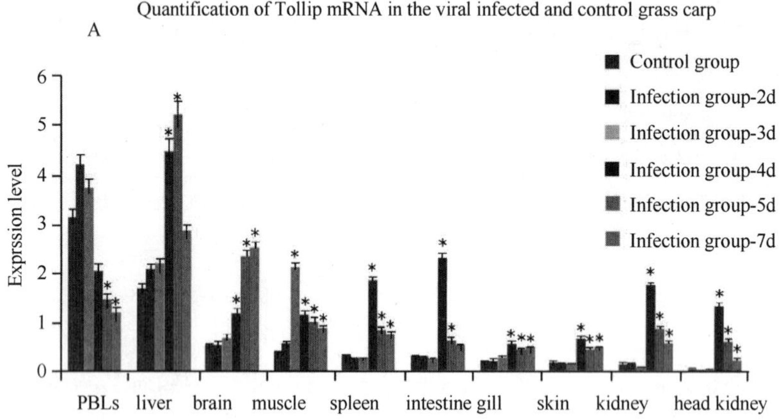

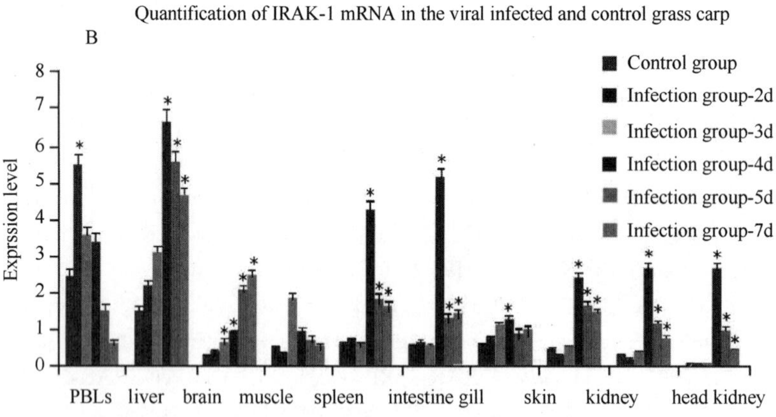

Fig. 3 Analysis of tissue expression of Tollip and IRAK-1 genes in control and GCVR infected groups by real-time RT-PCR. The mRNA level of Tollip and IRAK-1 are expressed as a ratio to β-actin mRNA levels. 2d, 3d, 4d, 5d, 7d indicate the time post-infection. $^*P < 0.05$, $n = 3$

3.4 Tollip and IRAK-1 mRNA is upregulated in the virus infected grass carp

Some symptoms were observed in virus infected grass carps, such as hyperanemia in muscle and caudal/ventral fin, and a marked desquamating of scales. The expression of β-actin in all tissues is shown in supplementary data Fig. 3. The expression of grass carp Tollip and IRAK-1 in various tissues at different time points after GCRV infection is shown in Fig. 3. After infection, Tollip and IRAK-1 mRNA levels were elevated in gills on day 2 with the highest expression occurring on day 4 (2.5-fold and 2.1-fold, respectively; $P<0.05$). In spleen, intestine, kidney, head kidney and skin, the levels of Tollip and IRAK-1 mRNAs remained nearly the same before day 3 but were significantly upregulated on day 4 (5.8-fold and 8.0-fold, 9.1-fold and 26.4-fold, 3.2-fold and 6.1-fold, 8.6-fold and 8.4-fold, 25.6-fold and 4.8-fold, respectively; $P<0.05$). The highest expression of grass carp Tollip and IRAK-1 in brain occurred simultaneously on day 6 (4.7-fold and 7.5-fold, respectively; $P<0.05$). In PBLs, Tollip and IRAK-1 mRNAs were induced on day 2 (1.3-fold and 2.2-fold, respectively) followed by a slow decrease. In muscle, grass carp Tollip and IRAK-1were significantly up regulated on day 3 (5.6-fold and 4.7-fold, respectively; $P<0.05$). It is worth to note that in the liver Tollip mRNA was induced on day 1, and was significantly upregulated on day 5 (3.1-fold; $P<0.05$), while IRAK-1 mRNA was induced on day 1, and was significantly upregulated on day 4 (4.3-fold; $P<0.05$).

3.5 Expression in *E. coli* and detection of endogenous proteins

The grass carp rTollip and rIRAK-1 were expressed in inclusion bodies of *E. coli*. The highest expression levels were detected about 2 h and 4 h post-IPTG induction, as shown in Fig. 4A. The purified rTollip and rIRAK-1 were shown in Fig. 4B.

The endogenous grass carp Tollip and IRAK-1 were detected by Western blot analysis using antibodies raised against the recombinant proteins. The 30 kDa and 78 kDa protein polypeptide bands, respectively, were recognized in head kidney of grass carp (Fig. 4C). In negative controls, the polypeptide band was hardly detectable.

3.6 Viral infection changes the localization of Tollip and IRAK-1

Antibodies raised against grass carp rTollip and rIRAK-1 enabled us to trace the localization and distribution of endogenous Tollip and IRAK-1 in CIK cells and to investigate whether this would change in response to viral-infection. In non-infected control CIK cells, Tollip and IRAK-1were uniformly distributed in the cytoplasm and co-localized. However, after infection with GCRV virus, both molecules moved toward the cell membrane, with the relocation of IRAK-1 being more remarkable than that of Tollip. However, colocalization was also observed near the cell membrane in virus infected cells (Fig. 5).

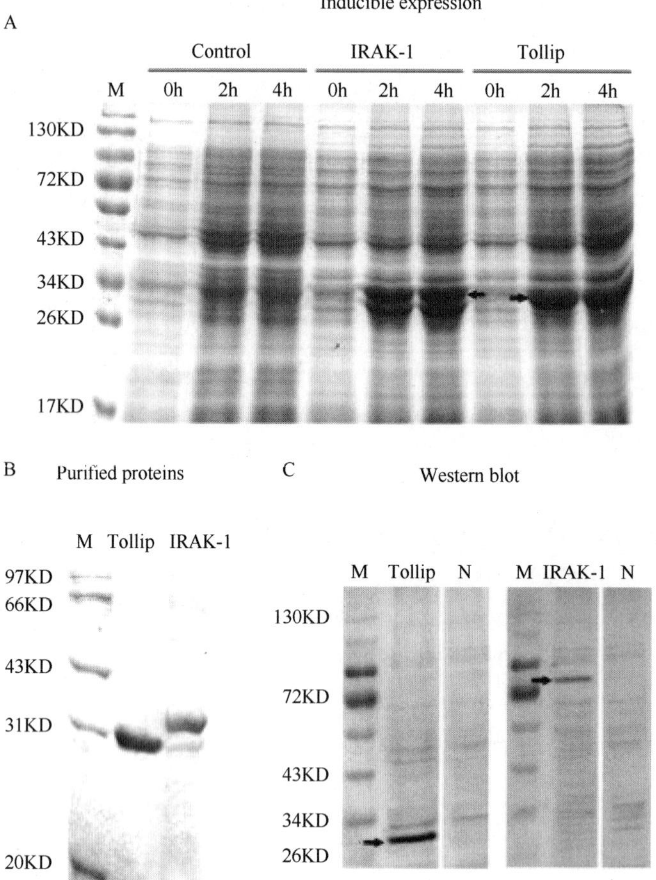

Fig. 4 (A) Expression of rTollip and rIRAK-1 in *E. coli* at different times after induction by IPTG. 0 h, 2 h, 4 h indicate the post-induction time. Control indicates the use of empty vector with no Tollip and IRAK-1 sequences. M = protein standard. (B) The purified rTollip and rIRAK-1. (C) Detection of endogenous Tollip and IRAK-1 by Western blot analysis. N = Head kidney lysate treated with antiserum pre-adsorbed with pure fusion protein for 16 h at 4℃. IRAK-1 and Tollip = Head kidney lysate treated with anti-IRAK-1 and anti-Tollip antiserum, respectively

4. Discussion

Several findings show that the components of the TLR and class II cytokine signaling systems known from mammals are also found in teleosts [38]. It was shown that salmonid Tollip and MyD88 factors can functionally replace their mammalian orthologues [39]. But how these molecules function in the TLR signaling transduction pathway in fish is not clear. In the present study, we described the molecular characterization of Tollip and IRAK-1 in grass carp, their expression profiles induced by GCRV, and their distribution in CIK cells. Most importantly, we confirmed the existence of Tollip and IRAK-1 proteins in grass carp and demonstrated their interaction by co-localizing them in cytoplasm and at the cell membrane.

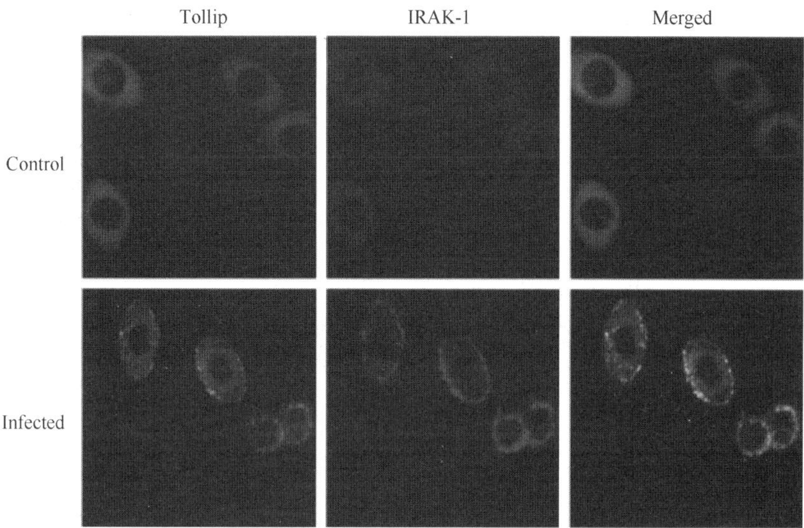

Fig. 5 Co-localization of Tollip and IRAK-1. CIK cells were grown on glass slides for 24 h, fixed in 4% paraformaldehyde, and immunofluorescence staining was performed using rabbit anti-Tollip antibody and mouse anti-IRAK-1 antibody. Samples were further incubated with FITC-conjugated goat anti-rabbit IgG (green) and cy3-conjugated goat anti-mouse IgG (red). Co-localization of Tollip (green) and IRAK-1 (red) appears as yellow. Scale bar = 5 μm. (For interpretation of the references to colour in this figure legend, the reader is referred to the web version of this article.)

Sequence analysis showed that the deduced polypeptide of Tollip had highly conserved structural characteristics, it shared the highest percentage similarity to Tollip protein of zebrafish. In addition, grass carp Tollip although being phylogenetically closely related to salmon Tollip I and II, trout Tollip I and II, did not clearly fall into either Tollip I or II subgroup. Our result also suggests that either Tollip in Cyprinidae represents a common ancestral gene which during evolution separated into Tollip I and II among the family Salmonidae [19], or if salmonid Tollips were more ancestral gene which may have been lost due to functional redundancy in Cyprinidae. However, these options remain to be elucidated in further studies. Grass carp IRAK-1 was phylogenetically closely related to members of the IRAK-1 subfamily. Therefore, the sequence obtained from grass carp appears to be an IRAK-1 homologue, and was named grass carp IRAK-1. Our results support the notion that teleost IRAK separated into IRAK-1, -3 and -4 before the separation of ray-finned fish. Whether this suggests that teleost IRAK-1, -3 and -4 are likely to be ancestral needs further investigation.

Similar to mammalian Tollip, grass carp Tollip contained an N-terminal TBD, a central C2 domain, and a C-terminal CUE domain. According to previous study, murine Tollip-Tom1 complex functioned as a factor that linked polyubiquitinated proteins to clathrin [8]. The CUE domain of murine Tollip inhibited IRAK signaling by targeting it for ubiquitination [4,40]. This domain is also important for the interaction of Tollip with Tom1 and the ubiquitination pathway [8]. So we speculate that the TBD and CUE domain of grass carp Tollip may also have similar function in ubiquitination of IRAK or other target proteins [4,19,41]. Overexpression of a human Tollip mutant lacking C2 domain failed to evoke an inhibitory effect on LPS-induced NF-Kb activation [19,42]. It was also shown for the human C2 domain that a point mutation of the lysine residue at position 150 (K150) to glutamic acid abrogated the Tollip function to inhibit LPS-induced NF-κB activation [19,42]. While mouse, fish and aves Tollip factors all feature a glutamic acid residue at this position, in insect Tollip, an aspartic acid was found (Fig. 1). It remains to be seen if this peculiar feature in the human C2 domain indicates a principal functional difference between human Tollip factors and that of other species (including fish).

Grass carp IRAK-1 similarly contained an N-terminal DD, which allows it to bind other DD-containing proteins and to oligomerise, a feature associated with the activation of IRAK-1 [43-45]. Although all IRAKs contain a kinase-like domain, only IRAK-1 and IRAK-4 exhibit kinase activity with two critical residue [13]. A point mutation at these residues (K239S or D340N) in human IRAK-1 completely abolished IRAK-1 kinase activity [46,47]. This two residues (Lys-217 and Asp-316) were also found in grass carp IRAK-1 kinase domain. In addition, grass carp IRAK-1 contained a serine/threonine kinase catalytic domain. There were two conserved residues (Thr-187 and Thr-363) at the catalytic site which is required for the phosphorylation of IRAK-1 by IRAK-4 and thus the formation of hyper-phosphorylated IRAK-1 via auto phosphorylation [6,48]. Grass carp IRAK-1 also contained a C-terminal C1 and a C2 domain. In mammals, the C-terminal domain of IRAK-1 could mediate the interaction of IRAK-1 with the downstream signaling molecule TRAF-6 [49]. Nguyen *et al.*, (2009) found that the C-terminal domain of mammalian IRAK-1 appears to play a role in negatively regulating the activation state of IRAK-1 [50]. Therefore, we speculate that grass carp IRAK-1 could regulate itself, be phosphorylated during TLR activation and may possess kinase activity.

In this study, we raised two polyclonal antibodies against grass carp rTollip and rIRAK-1. The data of Western blot analyses and immunofluorescence assays confirmed that IRAK-1 and Tollip do exist in fish at the protein level for the first time, although the mRNA of Tollip and IRAK-1 have been identified in Atlantic salmon, rainbow trout and mandarin fish [19,20]. Additionally, the co-localization of IRAK-1 and Tollip in CIK cells suggests that IRAK-1 can interact with Tollip, which was shown in mammalian cells but has not been demonstrated for fish species. Tollip is present in a membrane-proximal signaling complex with IRAK [5]. We also observed that under viral-infection condition, IRAK-1 moved from the cytoplasm to the membrane of CIK cells, and co-localization of IRAK and Tollip was also observed near the cell membrane. It has been shown in mammals that Tollip-IRAK-1 complexes can be recruited to the activated receptor complex upon stimulation by IL-1β [5]. We speculate that the association between Tollip and IRAK-1 exists in fish, and that their function in TLR/IL-1R signaling pathways is also conserved.

Except for cells of the immune system, TLRs are present in epithelial cells of the skin, respiratory tract, intestinal and genital-urinary tracts, all of which form the first protective barrier to invading pathogens [51]. Mammalian Tollip and IRAK-1 were widely expressed in the tissues tested [5,52]. In our study we were able to demonstrate by real-time PCR that grass carp Tollip and IRAK-1 were also expressed in all the tissues we examined. Both molecules were predominantly detected in PBLs and liver, while the lowest expression level for both was detected in head kidney. After viral infection, Tollip and IRAK-1 expression was upregulated in all tissues. This indicated that the expression profile of IRAK-1 and Tollip is similar. The high expression level of grass carp Tollip and IRAK-1 gene in PBLs may be related to the presence of lymphocytes, granulocytes, and monocytes in PBLs which is in agreement with reports by Zhang *et al.* in mandarin fish [20]. Liver is one of the main affected organs in GCRV disease [53], and the main source of acute-phase proteins [19]. In other system it has been shown that the TLR superfamilies are an important class of at least ten cell surface receptors which provide critical molecular linkage between the innate and adaptive immune system in the liver [54]. Among the immune-related tissues, the upregulated expression of grass carp Tollip and IRAK-1 in liver is the most significant, probably because of a great increase of activated components in TLR signaling pathway triggered by GCRV-infection, suggesting that TLR/IL-1R signaling is also important for immune responses in the liver of teleost. Rebl *et al.* (2008) observed that Tollip expression was upregulated in the liver of the viral infected trout [19]. Regarding to the negatively regulatory function of mammalian Tollip [4], they interpreted trout Tollip may served as an indicator for the activation of mechanisms counteracting the infection-related proinflammatory activation of the innate immune cascade [19]. In the present study, the upregulation of grass carp Tollip may control the activation of mechanisms to limit the production of proinflammatory mediators and to protect against immunopathological effects, this assumption has been proposed by Rebl *et al.* in 2010 [25]. The IL-1R mediated pathway has been clearly described in the peripheral immune system, but scattered information is available concerning the molecular composition and distribution of its members in neuronal cells [55]. However, the upregulated expression of Tollip and IRAK-1were detected in brain, we suggest that the teleost Tollip and IRAK-1 loci or genes down stream of the IL-1R pathway may be expressed in neuronal cell populations. In addition, it is noteworthy that in the present study Tollip and IRAK-1 were significantly upregulated in intestine, muscle and skin suggesting that in these sites of potential virus entry, Tollip and IRAK-1 may play an important role in epithelium-mediated cytolysis in viral clearance.

In conclusion, the results obtained in the present study broaden our knowledge on the role of genes and protein-protein interactions in the TLR signaling pathway in teleost responses to infection. These may provide new molecular markers for the presence of epithelial and neuronal mediated immune response, which may assist in analyzing the molecular basis of the relationship between immune, epithelial and neuronal cells and anti-viral defenses of fish.

Acknowledgements This work was supported by the National Natural Science Foundation of China (No. 31130055) and the National Basic Research Program of China (No. 2009CB118701).

Appendix A. Supplementary material Supplementary material associated with this article can be found, in the online version, at doi:10.1016/j.fsi.2012.05.025.

References

[1] Hoffmann JA, Kafatos FC, Janeway CA, Ezekowitz RA. Phylogenetic perspectives in innate immunity. Science1999; 284: 1313-8.

[2] Wilson I, Vogel J, Somerville S. Signalling pathways: a common theme in plants and animals? Curr Biol 1997; 7: R175-8.

[3] O'Neill LA, Greene C. Signal transduction pathways activated by the IL-1 receptor family: ancient signaling machinery in mammals, insects, and plants. J Leukoc Biol 1998; 63: 650-7.

[4] Zhang G, Ghosh S. Negative regulation of toll-like receptor-mediated signaling by Tollip. J Biol Chem 2002; 277: 7059-65.

[5] Burns K, Clatworthy J, Martin L, Martinon F, Plumpton C, Maschera B, *et al.* Tollip, a new component of the IL-1RI pathway, links IRAK to the IL-1 receptor. Nat Cell Biol 2000; 2: 346-51.

[6] Cao Z, Henzel WJ, Gao X. IRAK: a kinase associated with the interleukin-1 receptor. Science 1996; 271: 1128-31.

[7] Yamakami M, Yokosawa H. Toml (target of Myb 1) is a novel negative regulator of interleukin-1- and tumor necrosis factor-induced signaling pathways. Biol Pharm Bull 2004; 27: 564-6.

[8] Yamakami M, Yoshimori T, Yokosawa H. Toml, a VHS domaincontaining protein, interacts with Tollip, ubiquitin, and clathrin. J Biol Chem 2003; 278: 52865-72.

[9] Brissoni B, Agostini L, Kropf M, Martinon F, Swoboda V, Lippens S, *et al.* Intracellular trafficking of interleukin-1 receptor I requires Tollip. Curr Biol 2006; 16: 2265-70.

[10] Katoh Y, Shiba Y, Mitsuhashi H, Yanagida Y, Takatsu H, Nakayama K. Tollip and Toml form a complex and recruit ubiquitin-conjugated proteins onto early endosomes. J Biol Chem 2004; 279: 24435-43.

[11] Katoh Y, Imakagura H, Futatsumori M, Nakayama K. Recruitment of clathrin onto endosomes by the Toml-Tollip complex. Biochem Biophys Res Commun 2006; 341: 143-9.

[12] Ciarrocchi A, D'Angelo R, Cordiglieri C, Rispoli A, Santi S, Riccio M, *et al.* Tollip is a mediator of protein sumoylation. Plos One 2009; 4(2): e4404.

[13] Gottipati S, Rao NL, Fung-Leung WP. IRAK1: a critical signaling mediator of innate immunity. Cell Signal 2008; 20: 269-76.

[14] Lin SC, Lo YC, Wu H. Helical assembly in the MyD88-IRAK4-IRAK2 complex in TLR/IL-1R signaling. Nature 2010; 465: 885-91.

[15] Akira S, Takeda K. Toll-like receptor signalling, Nat Rev Immunol 2004; 4: 499-511.

[16] Huang Y, Li T, Sane DC, Li L. IRAK1 serves as a novel regulator essential for lipopolysaccharide-induced

interleukin-10 gene expression. J Biol Chem 2004; 279: 51697-703.

[17] Nguyen H, Chatterjee-Kishore M, Jiang Z, Qing Y, Ramana CV, Bayes J, *et al.* IRAK-dependent phosphorylation of Stat1 on serine 727 in response to interleukin-1 and effects on gene expression. J Interferon Cytokine Res 2003; 23: 183-92.

[18] Su J, Richter K, Zhang C, Gu Q, Li L. Differential regulation of interleukin-1 receptor associated kinase 1 (IRAK1) splice variants, Mol Immunol 2007; 44: 900-5.

[19] Rebl A, Høyheim B, Fischer U, Köllner B, Siegl E, Seyfert H. Tollip, a negative regulator of TLR-signalling, is encoded by twin genes in salmonid fish. Fish Shellfish Immunol 2008; 25: 153-62.

[20] Zhang C, Yin Z, He W, Chen W, Luo Y, Lu Q, *et al.* Cloning of IRAK1 and its upregulation in symptomatic mandarin fish infected with ISKNV. Biochem Biophys Res Commun 2009; 383: 298-302.

[21] Leong JS, Jantzen SG, von Schalburg KR, Cooper GA, Messmer AM, Liao NY, *et al. Salmo salar* and *Esox lucius* full-length cDNA sequences reveal changes in evolutionary pressures on a post-tetraploidization genome. BMC Genomics 2010; 11: 279.

[22] Hwang SD, Kondo H, Hirono I, Aoki T. Molecular cloning and characterization of Toll-like receptor 14 in Japanese flounder, *Paralichthys olivaceus*. Fish Shellfish Immunol 2011; 30: 425-9.

[23] Oshiumi H, Tsujita T, Shida K, Matsumoto M, Ikeo K, Seya T. Prediction of the prototype of the human Toll-like receptor gene family from the pufferfish, *Fugu rubripes*, genome. Immunogenetics 2003; 54: 791-800.

[24] Xiao XQ, Qin QW, Chen XH. Molecular characterization of a Toll-like receptor 22 homologue in large yellow croaker (*Pseudosciaena crocea*) and promoter activity analysis of its 5 '-flanking sequence. Fish Shellfish Immunol 2011; 30: 224-33.

[25] Rebl A, Goldammer T, Seyfert HM. Toll-like receptor signaling in bony fish. Vet Immunol Immunopathology 2010; 134: 139-50.

[26] Takano T, Hwang SD, Kondo H, Hirono I, Aoki T, Sano M. Evidence of Molecular toll-like receptor mechanisms in teleosts. Fish Pathol 2010; 45: 1-16.

[27] Baoprasertkul P, Peatman E, Abernathy J, Liu ZJ. Structural characterisation and expression analysis of toll-like receptor 2 gene from catfish. Fish hellfish Immunol. 2007; 22: 418-26.

[28] Hwang SD, Asahi T, Kondo H, Hirono I, Aoki T. Molecular cloning and expression study on Toll-like receptor 5 paralogs in Japanese flounder, *Paralichthys olivaceus*. Fish Shellfish Immunol. 2010; 29: 630-8.

[29] Fang Q, Ke LH, Cai YQ. Growth characterization and high titre culture of GCHV. Virol Sin 1989; 4: 315-9.

[30] Fang Q, Attoui H, Biagini JFP, Zhu ZY, de Micco P, de Lamballerie X. Sequence of genome segments 1-3 of the grass carp reovirus (genus *Aquareovirus*, family *Reoviridae*). Biochem Biophys Res Commun 2000; 274: 762-6.

[31] Reed RJ, Muench H. A simple method of estimating fifty percent endpoints. Am J Hyg 1938; 27: 493-503.

[32] Schultz, Milpetz F, Bork P, Ponting CP. SMART, a simple modular architecture research tool: Identification of signaling domains. Proc Natl Acad Sci USA 1998; 95: 5857-64.

[33] Apweiler R, Attwood TK, Bairoch A, Bateman A, Birney E, Biswas M, *et al.* The InterPro database, an integrated documentation resource for protein familes, domains and functional sites. Nucleic Acids Res 2001; 29: 37-40.

[34] Matys V, Fricke E, Geffers R, Gossling E, Haubrock M, Hehl R, *et al.* TRANSFAC: transcriptional regulation, from patterns to profiles. Nucleic Acids Res 2003; 31: 374-8.

[35] Kumar S, Tamura K, Jakobsen IB, Nei M. MEGA2: Molecular evolutionary genetics analysis software. Bioinformatics 2001; 17: 1244-5.

[36] Yang J, Guo S, Pan F, Geng H, Gong Y, Lou D, *et al.* Prokaryotic expression and polyclonal antibody preparation of a novel Rab-like protein mRabL5. Protein Expr Purif 2007; 53: 1-8.

[37] Cohen BJ, Audet S, Andrews N, Beeler J. Plaque reduction neutralization test for measles antibodies: description of a standardised laboratory method for use in immunogenicity studies of aerosol vaccination. Vaccine 2007; 26: 59-66.

[38] Stein C, Caccamo M, Laird G, Leptin M. Conservation and divergence of gene families encoding components of innate immune response systems in zebrafish, Genome Biol 2007; 8: R251.

[39] Rebl A, Rebl H, Liu S, Goldammer T, Seyfert HM. Salmonid Tollip and MyD88 factors can functionally replace

their mammalian orthologues in TLR-mediated trout SAA promoter activation. Dev Comp Immunol 2011; 35: 81-7.

[40] Wells CA, Chalk AM, Forrest A, Taylor D, Waddell N, Schroder K, *et al.* Alternate transcription of the Toll-like receptor signaling cascade. Genome Biol 2006; 7: R10.

[41] Shih SC, Prag G, Francis SA, Sutanto MA, Hurley JH, Hicke L. A ubiquitin-binding motif required for intramolecular monoubiquitylation, the CUE domain. EMBO J 2003; 22: 1273-81.

[42] Li T, Hu J, Li L. Characterization of Tollip protein upon lipopolysaccharide challenge. Mol Immunol 2004; 41: 85-92.

[43] Dong W, Liu Y, Peng J, Chen L, Zou T, Xiao H, *et al.* The IRAK-1-BCL10-MALT1-TRAF6-TAK1 cascade mediates signaling to NF-κB from toll-like receptor 4. J Biol Chem 2006; 281: 26029-40.

[44] Neumann D, Kollewe C, Resch K, Martin MU. The death domain of IRAK-1: an oligomerization domain mediating interactions with MyD88, Tollip, IRAK-1, and IRAK-4. Biochem Biophys Res Commun 2007; 354: 1089-94.

[45] Ross K, Yang L, Dower S, Volpe F, Guesdon F.Identification of threonine 66 as a functionally critical residue of the Interleukin-1 receptor-associated kinase. J Biol Chem 2002; 277: 37414-21.

[46] Knop J, Martin MU. Effects of IL-1 receptor-associated kinase (IRAK) expression on IL-1 signaling are independent of its kinase activity. FEBS Lett 1999; 448(1): 81-5.

[47] Maschera B, Ray K, Burns K, Volpe F. Overexpression of an enzymically inactive interleukin-1-receptor-associated kinase activates nuclear factor-kappaB. Biochem J 1999; 339: 227-31.

[48] Cheng H, Addona T, Keshishian H, Dahlstrand E, Lu C, Dorsch M, *et al.* Regulation of IRAK-4 kinase activity via autophosphorylation within its activation loop. Biochem Biophys Res Commun 2007; 352: 609-16.

[49] Janssens S, Beyaert R. Functional Diversity and regulation of different Interleukin-1 receptor-associated kinase (IRAK) family members. Mol Cell 2003; 11: 293-302.

[50] Nguyen T, De Nardo D, Masendycz P, Hamilton JA, Scholz GM. Regulation of IRAK-1 activation by its C-terminal domain. Cell Signal 2009; 21: 719-26.

[51] Sandor F, Buc M. Toll-like receptors. II. Distribution and pathways involved in TLR signalling. Folia Biol (Praha) 2005; 51(6): 188-97.

[52] Rao N, Nguyen S, Ngo K, Fung-Leung WP. A novel splice Variant of Interleukin-1 receptor (IL-1R)-Associated kinase 1 plays a negative regulatory role in Toll/IL-1R-induced inflammatory signaling. Molcell Biol 2005; 25: 6521-32.

[53] Su J, Yang C, Xiong F, Wang Y, Zhu Z. Toll-like receptor 4 signaling pathway can be triggered by grass carp reovirus and aeromonas hydrophila infection in rare minnow *Gobiocypris rarus*. Fish Shellfish Immunol 2009; 27: 33-9.

[54] Dufour JF, Clavien PA. Signaling pathways in liver diseases. 2th ed. Berlin Heidelberg: Springer; 2010.

[55] Gardoni F, Boraso M, Zianni E, Corsini E, Galli CL, Cattabeni F, *et al.* Distribution of interleukin-1 receptor complex at the synaptic membrane driven by interleukin-1b and NMDA stimulation. J Neuroinflammation 2011; 8: 14-9.

草鱼 Tollip 和 IRAK-1 的鉴定及相互作用检测

黄 容[1] 吕建建[1,2] 罗大极[1] 廖兰杰[1] 朱作言[1] 汪亚平[1]

1 中国科学院水生生物研究所，武汉 430072

2 中国科学院大学，北京 100039

摘 要 Tollip 和 IRAK-1 是哺乳动物 TLR / IL-1R 信号通路的关键组成部分，在 TLR / IL-1R 信号通路中起调节作用。目前已发现了几个鱼类 TLR，但是其分子结构、蛋白质相互作用，以及 TLR 信号途径在鱼类感染免疫中的作用报道较少。本研究通过 3′和 5′RACE 及基因组步移技术从草鱼头肾组织中分离了 Tollip 和 IRAK-1 cDNA 序列，基因组序列和启动子区序列；采用定量 RT-PCR 分析测定草鱼组织中 Tollip 和 IRAK-1 的实时转录模式；原核表达 Tollip 和 IRAK-1 重组蛋白用于制备多抗。在系统发育上，草鱼 Tollip 和 IRAK-1 与以前报道的其他物种 Tollip 和 IRAK-1 聚为一簇，且与斑马鱼的相似性最高。病毒感染后，草鱼 Tollip 和 IRAK-1 的转录水平在大多数免疫相关组织中上调。免疫荧光分析表明，Tollip 和 IRAK-1 均匀分布并共定位于 CIK 细胞的细胞质中，病毒感染后，Tollip 和 IRAK-1 有朝向细胞膜迁移的趋势。研究结果证明 Tollip 和 IRAK-1 蛋白存在于硬骨鱼类物种，并表明在受到病毒刺激后 Tollip-IRAK-1 复合物被招募至受体复合物附近。本研究提供了一些关于 TLR 信号传导途径在硬骨鱼类中功能的新见解。

Isolation and Analysis of A Novel Grass Carp Toll-Like Receptor 4 (*tlr4*) Gene Cluster Involved in the Response to Grass Carp Reovirus

Rong Huang[1] Feng Dong[1,2] Songhun Jang[1,2] Lanjie Liao[1]
Zuoyan Zhu[1] Yaping Wang[1]

1 State Key Laboratory of Freshwater Ecology and Biotechnology, Institute of Hydrobiology, Chinese Academy of Sciences, Wuhan 430072

2 Graduate School of Chinese Academy of Sciences, Beijing 100039

Abstract The mammalian response to lipopolysaccharide (LPS) is mainly mediated by Toll-Like Receptor 4 (TLR4). Fish and mammalian TLR4 vary; fish TLR4 ligands are unknown. Isolation of fish *tlr4* genes is difficult due to their complex genomic structure. Three bacterial artificial chromosome (BAC) clones containing grass carp *tlr4* were obtained. Four *tlr4* genes, with a varied genomic structure and different protein domains were subsequently isolated by constructing a subcloned library and rapid amplification of cDNA ends (RACE). The four *tlr4* genes were expressed during development from 12 h post-fertilization, in all healthy adult fish tissues tested, and significantly increased in grass carp reovirus (GCRV)-infected liver and muscle, suggesting the *tlr4* genes play a role in GCRV infection. This study effectively separated each gene in the *tlr4* gene cluster, implies that grass carp TLR4 proteins have different ligand recognition specificities to mammalian TLRs, and provides information on the functional evolution of TLRs.

Keywords grass carp; BAC library; *tlr4* gene cluster; grass carp reovirus

1. Introduction

Grass carp (*Ctenopharyngodon idellus*) is one of the most economically important fish in China. Grass carp reovirus (GCRV) causes hemorrhages and severely affects the development of grass carp aquaculture. Therefore, the identification of GCRV resistance genes has important implications for understanding the immune system of fish, antiviral drug development and molecular breeding research.

The Toll-Like Receptor (TLR) family is an important class of pattern recognition receptors (PRRs), which specifically recognize a series of highly conserved pathogenic microorganism structures, termed pathogen-associated molecular patterns (PAMPs) (Janeway,

文章发表于 *Developmental & Comparative Immunology*, 2012, 38(2): 383-388

1989; Skjæveland *et al.*, 2009; Palti *et al.*, 2010). PAMPs play an important role in the generation of innate immunity and inheritance of acquired immunity (Su *et al.*, 2012; Akira *et al.*, 2001). TLRs are typical type I transmembrane proteins, and contain three major domains: a tandem repeat leucine-rich repeat (LRR) motif which identifies PAMPs, a transmembrane region and an intracellular Toll/IL-1 receptor (TIR) domain which transmits signals (Yamamoto and Takeda, 2010).

Mammalian TLR4 recognizes lipopolysaccharide (LPS) from bacterial cell walls (Medzhitov *et al.*, 1999; Poltorak *et al.*, 1998) and certain viral proteins (Kurt-Jones *et al.*, 2000; Haynes *et al.*, 2001; Rassa *et al.*, 2002). To date, fish *tlr4* genes have only been identified in cyprinids (Meijer *et al.*, 2004; Jault *et al.*, 2004; Su *et al.*, 2009). After searching the Zebrafish Information Network (ZFIN), we surprisingly found that zebrafish *tlr4* exists in the form of a cluster containing three *tlr4* tandem repeat genes which are closely arranged in the same direction within the chromosome (Supplementary data, Fig. 1C). In practice, tandem homologous genes with a high similarity are difficult to isolate and clone. Full-length of fish *tlr4* gene cluster have not yet been identified in other species, with the exception of zebrafish with deep sequencing. Comprehensive sequence and phylogenetic analyses have revealed that the zebrafish *tlr4* genes are paralogous, rather than orthologous, to human *tlr4* (Sullivan *et al.*, 2009). A fusion protein containing the zebrafish TLR4 extracellular region and mouse TLR4 transmembrane and intracellular regions cannot identify LPS (Sullivan *et al.*, 2009) and it has been suggested that zebrafish TLR4 recognizes unknown PAMPs (Sullivan *et al.*, 2009; Sepulcre *et al.*, 2009). Although most fish lost *tlr4* during evolution (Oshiumi *et al.*, 2003; Baoprasertkul *et al.*, 2007), the structure and function study of the small number of fish which retained *tlr4* may provide evolutionary clues. Research of grass carp *tlr4* and its function may potentially provide useful information on the evolution of the fish TLR family and its role in the innate immune system.

In this study, we isolated and cloned four *tlr4* genes from a gene cluster by screening a grass carp BAC library. We compared the genomic structure and protein domains of the four grass carp *tlr4* genes with other known TLRs. We also analyzed expression of the four *tlr4* genes during early embryonic development, in healthy grass carp tissues and in the liver and muscle of GRCV-infected grass carp. This study implies that grass carp TLR4s have different ligand recognition specificities during immune defense than mammalian TLRs, and also provides new evidence on the functional evolution of the TLR family.

2. Materials and Methods

2.1 Experimental grass carp, BAC library and the nomenclature

Five-month-old grass carp (average weight 28 g) were supplied by Guanqiao Experimental

Fish Breeding Base, Institute of Hydrobiology, Chinese Academy of Sciences, China. The fish were allowed to acclimatize in clean 300 L tanks at approximately 28℃ for one week, before being used in experiments. The grass carp *Hin*d III BAC library was constructed previously in our laboratory (Jang *et al*., 2010). To avoid confusion, the nomenclature of *tlr4* genes in all species was that the lowercase italic letters indicated the gene; the uppercase letters indicated the protein encoded. The names of the *tlr4* gene in zebrafish (*Drtlr4ba* and *Drtlr4bb*) were presented according to the nomenclature rule proposed by Sullivan *et al*. (2009) and Palti (2011).

2.2 Probe preparation and BAC library screening

Based on existing sequences, including zebrafish *tlr4ba* (EU551724), zebrafish *tlr4bb* (AY388400), human *tlr4* (AAF07823) and mouse *tlr4* (NM_021297), we designed degenerate primers (tlr4F1 and tlr4R1) (Supplementary data, Table 1), which amplified a partial grass carp *tlr4* cDNA fragment, and the products were ligated into the pMD18-T vector and sequenced. The internal primers 38F and 39R (Supplementary data, Table 1) were designed for re-amplification, and this PCR product was labeled with α-^{32}P-dCTP and used a probe for in situ hybridization and DNA fingerprinting. The BAC library was copied to a membrane and screened by colony in situ hybridization as previously described (Yang *et al*., 2010), Briefly, two copies of each clone were copied from the wells of a 384 well plate onto a 12 × 8 cm nylon membrane using the Genet ix QPix system (BioRobotics, Cambridge, UK), cultured at 37℃ for 16-20 h, the bacteria were lysed in situ, bacterial DNA was fixed to the membrane by UV cross-linking, hybridized with the probe and exposed to X-ray to develop the colorimetric reaction. A *Not* I digested positive BAC clone plasmid, and the insert size of the positive clones was detected by pulsed-field gel electrophoresis (PFGE).

2.3 Positive BAC clone identification by DNA fingerprinting, subcloned library construction and screening

The QIAGEN Plasmid Maxi Kit (Qiagen, Hilden, Germany) was used to prepare plasmid DNA from the positive BAC clones. Briefly, the plasmid DNA was completely digested with *Hin*d III, *Eco*R I, *Bam*H I and *Pst* I, separated on 0.8% agarose gels, transferred to a nylon membrane, hybridized with the probe and exposed to X-ray to develop the colorimetric reaction.

Based on the size of the digested fragments of the positive clones, we selected *Pst* I to construct a subcloned library. Both the pUC19 plasmid and the BAC clone (H041G22) plasmid DNA were digested with *Pst* I, plated onto LB plates containing IPTG and X-gal after ligation and transformation, then 384 white clones were selected to build the subcloned library. The positive subclones were re-screened by colony in situ hybridization as described in

Section 2.2. Positive subclone plasmid DNA was prepared, digested with *Pst* I, and fragments of different sizes were selected for sequencing using the primer-walking method until the entire sequence was obtained.

2.4 Full-length sequencing of the grass carp *tlr4* gene cluster

Specific reverse and forward primers (Supplementary data, Table 1) for the 5′ and 3′ ends were designed for each of the subclones (Supplementary data, Fig. 1A). Any of the forward primers could be matched to any of the reverse primers, using the positive BAC clones as a template. PCR product size analysis and sequencing were used to determine the order of the genes in the gene cluster, and obtain the sequences between the subcloned sequences.

A specific reverse primer (specific 5d-3) (Supplementary data, Table 1) was designed for the 5′ end of the gene cluster, and a forward primer (3-24F3) (Supplementary data, Table 1) was designed based on the conserved regions of the tandem gene to amplify the 5′ end sequence of the gene cluster, using the positive BAC clones as template (Supplementary data, Fig. 1B).

2.5 Full-length cDNA cloning and sequencing of the four grass carp *tlr4* genes in the gene cluster

Alignment of the four *tlr4* gene sequences from the gene cluster was performed using ClustalW, and the tlr4f and tlr4r primer pair (Supplementary data, Table 1) was designed within the conserved region. A mixture of cDNA from several grass carp organs in 2.7 was used as a template for PCR, and the products were ligated into the pMD18-T vector and sequenced.

Four primers were designed within the obtained cDNA region, two forward primers (3drace-F1/3drace-F2) (Supplementary data, Table 1) for 3′ rapid amplification of cDNA ends (RACE) and two reverse primers (5drace-R5/5drace-R6) (Supplementary data, Table 1) for 5′ RACE. RACE was performed using the BD SMART RACE cDNA Amplification Kit (Clontech, CA, USA) according to the manufacturer's instructions. The 5′ and 3′ RACE PCR products were ligated into pMD18-T, sequenced and spliced.

2.6 Sequence analysis

Similarity with other known sequences was analyzed using BLAST (http://www.ncbi.nlm.nih.gov/blast) and multiple sequence alignments were generated using ClustalW (http://www.ebi.ac.uk/Tools/clustalw2/index.html). The conserved protein domains were predicted using the Simple Modular Architecture Research Tool (http://www.smart.embl-heidelberg.de/) (Letunic *et al.*, 2006). The phylogenetic tree was constructed using the neighbor-joining algorithm of MEGA version 4.0 (Tamura *et al.*, 2007).

2.7 Expression analysis

Unfertilized eggs, embryos 6, 12, 24, 30, 36, 48, 54 and 60 h post fertilization (hpf) and various tissue samples from three healthy grass carp were collected, immediately placed in TRIzol (Invitrogen, Carlsbad, CA, USA), total RNA was extracted and cDNA was transcribed using Oligo(dT)$_{18}$. Semi-quantitative RT-PCR was used to measure the expression levels of *Citlr4.1*, *Citlr4.2*, *Citlr4.3* and *Citlr4.4* using the primers 3-24QF2/3-24QR2, E8F2/E8R2, 3draceF3/E14R1 and E7F3/E7R3, respectively (Supplementary data, Table 1). The primers β-actinF and β-actinR (Supplementary data, Table 1) were used to amplify β-actin as an internal reference gene.

2.8 Expression analysis in GCRV-infected grass carp

One hundred carp were randomly divided into two groups, control and GCRV-infected. The grass carp in the GCRV-infected group were subjected to osmotic treatment in 6% NaCl solution for 1 min, then placed in 10% virus suspension (strain GCRV 873) for 30 min. The fish in the control group were treated with an UV-inactivated virus suspension using the same method. Both groups were raised in 300 L tanks at 28℃ in water containing 100 units/ml medical-grade penicillin and streptomycin, with full aeration. The liver and muscle tissues of control (1 d after infection, control grass carp, n = 3) and virus-infected grass carp (at every daypost-infection until peak of death, infected grass carp, each time point n = 3) were immediately removed and frozen in liquid nitrogen, the expression detection of each gene was performed by the method descried in section 2.7.

3. Results

3.1 Screening of the grass carp BAC library

Three positive clones (H041G22, H047B1, and H047C1) were identified from the 52,216 BAC clones screened. PFGE demonstrated that the insert sizes of the three positive BAC clones were approximately 140 Kb, 250 Kb and 210 Kb, respectively (Supplementary data, Fig. 2A).

3.2 Identification of positive BAC clones by DNA fingerprinting and screening of a subcloned library

*Hin*d III, *Eco*R I, *Bam*H I and *Pst* I digestion of plasmid DNA from the three positive BAC clones and the DNA fingerprinting results are shown in Supplementary data, Fig. 2B. Based on the size of the digested fragments of the positive clones, we used *Pst* I to construct a subclone library from clone H041G22.

In total, 26 positive subclones were screened from the subcloned library by colony in situ hybridization. After *Pst* I digestion and agarose gel electrophoresis, four subclone plasmids with different insert sizes were obtained (Supplementary data, Fig. 2C), and the subclones were picked for sequencing.

3.3 Full-length sequences and features of the grass carp *tlr4* gene cluster

The order of the four positive subclones (subH3, subO20, subE22, subG23) in BAC clone H041G22 is shown in Supplementary data, Fig. 1B, and they were named *Citlr4.1*, *Citlr4.2*, *Citlr4.3* and *Citlr4.4*. By sequencing, we obtain the sequences between the subcloned sequences and the 5′ end sequence of the gene cluster. Then we obtained the full-length sequence of the gene cluster by splicing. The full-length genomic sequences of these four *tlr4* genes are shown in Supplementary data, Fig. 1B (GenBank accession numbers: JF965429, JF965431, JF965433, JF965435).

Analysis of the cDNA and genomic sequences revealed the structural features of each gene in the grass carp *tlr4* gene cluster (Fig. 1). *Citlr4.1* contains two exons, *Citlr4.2* contains one exon, *Citlr4.3* and *Citlr4.4* both contains three exons. *Citlr4.4* has a high similarity to *Drtlr4bb*, *Hmtlr4* and *Mmtlr4*, with a coding region of approximately 90 bp in exon 1, 167 bp in exon 2 and approximately 2,200 bp in exon 3 (Fig. 1).

3.4 Sequence and phylogenetic analysis of the four *tlr4* genes

The full-length cDNA sequences of *Citlr4.1*, *Citlr4.2*, *Citlr4.3* and *Citlr4.4* were obtained by RACE. The full length cDNAs are 2,312 bp, 2,939 bp, 2,641 bp and 3,343bp and encode 585, 683, 818 and 820 amino acids, respectively.

Structural analysis suggested that CiTLR4.3 and CiTLR4.4 are typical TLR family proteins containing a signal peptide, leucine-rich domain, transmembrane domain and TIR domain. CiTLR4.1 only contains the leucine-rich domain, and may be a soluble protein. CiTLR4.1 and CiTLR4.2 contain no signal peptide, similarly to zebrafish DrTLR4ba and LOC795671 (Fig. 2).

A phylogenetic tree was constructed based on the amino acid sequence of the leucine-rich domain (Supplementary data, Fig. 3). The grass carp TLR4s initially clustered with the zebrafish TLR4s, then with mammalian and bird TLR4s.

3.5 Expression of the four *tlr4* genes during early development and in adult tissues

Expression of the four grass carp *tlr4* genes was first detected at 12 hpf. During development, the expression of *Citlr4.2*, *Citlr4.3* and *Citlr4.4* 3 gradually increased, reaching the highest value at 48 hpf, and then remained stable. In contrast, *Citlr4.1* reached a high expression level at 12 hpf, and then remained stable (Fig. 3A). Expression of *Citlr4.1*, *Citlr4.2*, *Citlr4.3* and *Citlr4.4* was detected in all of the adult tissues tested, with the highest levels observed in the gill and spleen and the lowest levels in the liver and muscle (Fig. 3B).

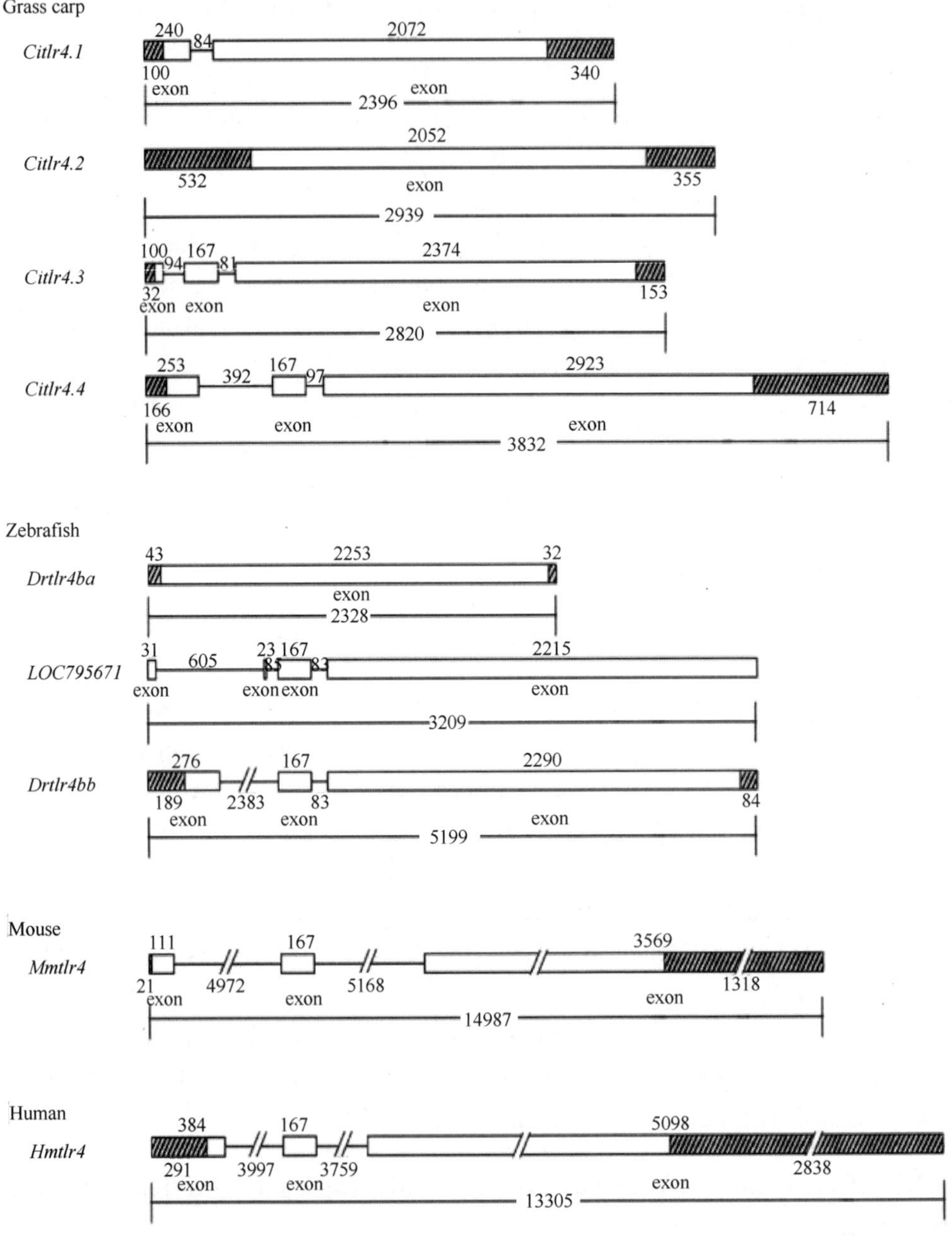

Fig. 1 Schematic representation of the *tlr4* genes in fish and mammals. Exons are indicated by boxes; coding regions are indicated by white boxes

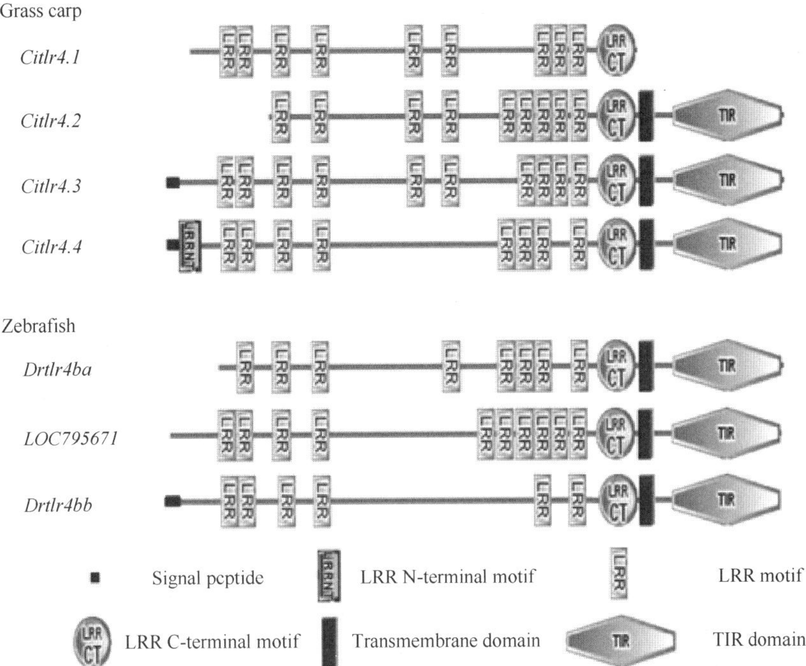

Fig. 2 Schematic representation of the organization of the predicted protein domains in the grass carp and zebrafish TLR4s

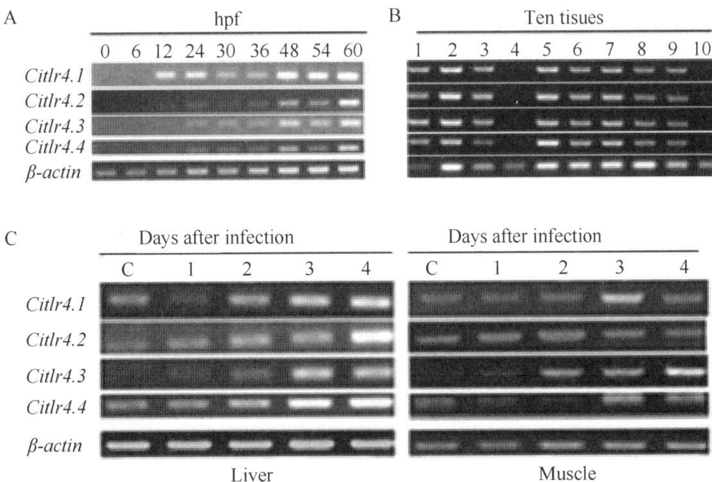

Fig. 3 RT-PCR analysis of the expression profile of the four grass carp *tlr4* genes. (A) Expression profile of the four *tlr4* genes during early grass carp embryonic development. The lane numbers indicate the number of hours post fertilization (hpf). (B) Expression profile of the four *tlr4* genes in different adult grass carp tissues. Lane 1, blood; lane 2, gill; lane 3, brain; lane 4, liver; lane 5, spleen; lane 6, intestine; lane 7, kidney, lane 8, head kidney; lane 9, heart; lane 10, muscle. (C) RT-PCR analysis of the temporal expression of the four *tlr4* genes after infection of grass carp with GCRV. “C” indicates the expression in control grass carp, “1”, “2”, “3”, “4” indicate the expression in grass carp at 1 d, 2 d, 3 d and 4 d after infection

3.6 Expression of the four *tlr4* genes in liver and muscle after infection with GCRV

Expression of *Citlr4.1*, *Citlr4.2*, *Citlr4.3* and *Citlr4.4* were upregulated in liver and muscle tissue after infection with GCRV. In the liver, *Citlr4.1* mRNA was downregulated at 1 d after infection, then increased with the highest expression level observed 4 d after infection. *Citlr4.2* mRNA was gradually upregulated from 1 d after infection, reaching the highest level 4 d after infection. In muscle, *Citlr4.1* mRNA was significantly upregulated 3 d after infection, and then decreased to normal levels 4 d after infection. *Citlr4.2* mRNA was upregulated by 2 d after infection then decreased to normal levels. The expression of *Citlr4.3* and *Citlr4.4* in infected liver and muscle were similar; both were continually induced between 1 and 3 d after infection, with the highest expression levels observed 4 d after infection (Fig. 3C).

4. Discussion

Some researchers have speculated that most fish lost *tlr4* genes during evolution (Oshiumi *et al.*, 2003; Baoprasertkul *et al.*, 2007), and to date, *tlr4* has only been found in cyprinids (Meijer *et al.*, 2004; Jault *et al.*, 2004; Su *et al.*, 2009). The zebrafish *tlr4* cluster contains three genes, while the grass carp *tlr4* cluster contains four genes, which suggests that fish *tlr4* genes have actively evolved. Phylogenetic analysis showed that CiTLR4.4 initially clustered with DrTLR4bb. LOC795671 and DrTLR4ba form a cluster subsequently, and then clustered with CiTLR4.1, CiTLR4.2 and CiTLR4.3 (Supplementary data, Fig. 3). Therefore, we speculate that a one to one relationship does not exist between grass carp and zebrafish TLR4s with the exception of CiTLR4.4 and DrTLR4bb. By comparing the flanking sequences of the grass carp and zebrafish *tlr4* genes, only the 3′ regions of *Citlr4.4* and *Drtlr4bb* exhibit a high homology, and we could not determine the relationship between the other grass carp and zebrafish *tlr4* genes. Structural analysis of grass carp, zebrafish and mammalian *tlr4* genes indicated that the structure of *Citlr4.4*, *Drtlr4bb*, *Hmtlr4* and *Mmtlr4* are very consistent, with a similar length of exon coding regions, which may represent the original characteristics of the ancestral *tlr4*. This study suggests that the grass carp and zebrafish *tlr4* genes have a common ancestral gene, which is likely to paralogous, not orthologous, to mammalian *tlr4*, in agreement with the *tlr4* gene evolutionary model proposed by Sullivan *et al.* (2009). The appearance of other *tlr4* genes in grass carp and zebrafish, apart from *Citlr4.4* and *Drtlr4bb*, necessitates further in-depth study of the structure, function and evolution of these genes.

Typical TLRs contain a signal peptide, which enables cell surface or intracellular membrane localization (Yamamoto and Takeda, 2010). CiTLR4.3 and CiTLR4.4 are typical TLR family proteins containing a signal peptide, leucine-rich domain, transmembrane domain and TIR domain. In contrast, grass carp CiTLR4.1, CiTLR4.2, zebrafish DrTLR4ba and

LOC795671 are predicted to have no signal peptide (Fig. 2), and therefore may exist as cytoplasmic proteins. In addition to TLRs, NOD-like receptors (NLRs) and RIG-I-like receptors (RLRs) are also major cytoplasmic pattern recognition receptors (Creagh and O'Neill, 2006). As fish have a complex and diverse evolutionary history and environment (Palti, 2011), we suspect that some fish TLR4 proteins may also function as cytoplasmic pattern recognition receptors. The spatial orientation of proteins is closely related to their function (Zou *et al.*, 2007), therefore, immunofluorescence or immuno-colloidal gold labeling could be used to determine the subcellular localization of fish TLR4 proteins, in order to more comprehensively understand their function.

Citlr4.2, *Citlr4.3*, *Citlr4.4* were expressed at low levels at 12 hpf and then expression gradually increased during embryonic development; however, the expression of *Citlr4.1* was highest at 12 hpf. In general, the expression pattern of *Citlr4.1* is consistent with the expression patterns of *Drtlr4ba* and *Drtlr4bb* during zebrafish embryonic development (Jault *et al.*, 2004). The end of the gastrula stage and beginning of neurula stage occurs at about 12 hpf in grass carp, suggesting that the *tlr4* genes play an important role during this developmental stage.

The four *tlr4* genes were widely expressed in various grass carp tissues, with basically similar expression patterns. Zebrafish *Drtlr4ba* is expressed in all tissues, while *Drtlr4bb* is only expressed in the skin and heart (Jault *et al.*, 2004). These different expression patterns suggest that the regulation or function of the zebrafish and grass carp *tlr4* genes vary. After GCRV-infection, the expression levels of the four *tlr4* genes was significantly upregulated in liver and muscle, implying that the grass carp TLR4s are related to GCRV infection. Sullivan *et al.* (2009) concluded that TLR4 does not recognize LPS in fish, and proposed that fish TLR4 may respond to unidentified bacterial, fungal, protistan, helminthic, and/or viral pathogen-associated molecular patterns (Sullivan *et al.*, 2009). It is also possible that TLR4a and TLR4b contribute to the ligand-recognizing diversity of other zebrafish TLR proteins, by functioning as co-receptors (Sullivan *et al.*, 2009). The results of this study indicate that grass carp TLR4s are associated with GCRV infection; but also provided valid evidence for the speculation of Sullivan *et al.* (2009).

Currently, only bird and mammalian TLR4s have been investigated in detail. Both identify LPS; however, the response of bird TLR4 to LPS is much weaker than mammalian TLR4 (Keestra and van Putten, 2008). As paralogs of mammalian TLR4, fish TLR4 does not have the ability to identify LPS (Sepulcre *et al.*, 2009); however, we have demonstrated that that grass carp TLR4s are associated with GCRV infection. This implies that the function of TLR4 varies significantly in different species. Further study of the structure and function of *tlr4* in amphibians, reptiles and other evolutionally ancient fish will provide more information on the evolution and function of the *tlr4* gene.

Acknowledgements This work was supported by the National Basic Research Program of China (No. 2009CB118701) and the Direction Program of Chinese Academy of Sciences (KSCX2-EW-N-004-3).

Appendix A. Supplementary data Supplementary data associated with this article can be found,in the online version,at http://dx.doi.org/10.1016/j.dci.2012.06.002.

References

Akira, S., Takeda, K., Kaisho, T., 2001. Toll-like receptors: critical proteins linking innate and acquired immunity. Nat. Immunol. 2, 675-680.

Baoprasertkul, P., Xu, P., Peatman, E., Kucuktas, H., Liu, Z., 2007. Divergent toll-like receptors in catfish (*Ictalurus punctatus*): TLR5S, TLR20, TLR21. Fish Shellfish Immunol. 23, 1218-1230.

Creagh, E.M., O'Neill, L.A., 2006. TLRs, NLRs and RLRs: a trinity of pathogen sensors that co-operatein innate immunity. Trend[s] in Immunology 27, 352-357.

Haynes, L.M, Moore, D.D, Kurt-Jones, E.A., Finberg, R.W., Anderson, L.J., Tripp, R.A., 2001. Involvement of toll-like receptor 4 in innate immunity to respiratory syncytial virus. J. Virol. 75, 10730-10737.

Janeway, Jr., C.A., 1989. Approaching the asymptote? Evolution and revolution in immunology. Cold Spring Harb: Symposia on Quantitative Biology 54, 1-13.

Jang, S.H., Liu, H., Su, J.G., Dong, F., Xiong, F., Liao, L., Wang, Y., Zhu, Z., 2010. Construction and characterization of two bacterial artificial chromosome libraries of grass carp. Mar. Biotechnol. 12, 261-266.

Jault, C., Pichon, L., Chluba, J., 2004. Toll-like receptor gene family and TIR-domain adapters in *Danio rerio*. Mol. Immunol. 40, 759-771.

Keestra, A.M., van Putten, J.P., 2008. Unique properties of the chicken TLR4/MD-2 complex: selective lipopolysaccharide activation of the MyD88-dependent pathway. The Journal of Immunology 181, 4354-4362.

Kurt-Jones, E.A, Popova, L., Kwinn, L., Haynes, L.M., Jones, L.P., Tripp, R.A., Walsh, E.E., Freeman, M.W., Golenbock, D.T., Anderson, L.J., Finberg, R.W., 2000. Pattern recognition receptors TLR4 and CD14 mediate response to respiratory syncytial virus. Nat. Immunol. 1, 398-401.

Letunic, I., Copley, R.R., Pils, B., Pinkert, S., Schultz, J., Bork, P., 2006. SMART 5: domains in the context of genomes and networks. Nucleic. Acids. Res. 34, D257-D260.

Medzhitov, R., Preston-Hurlburt, P., Janeway, C.A., 1999. A human homologue of the *Drosophila* Toll protein signals activation of adaptive immunity. Nature 388, 394-397.

Meijer, A.H., Gabby Krens, S.F., Medina Rodriguez, I.A., He, S., Bitter, W,, Snaar-Jagalska, B.E., Spaink, H.P., 2004. Expression analysis of the Toll-like receptor and TIR domain adaptor families of zebrafish. Mol. Immunol. 40, 773-783.

Oshiumi, H., Tsujita, T., Shida, K., Matsumoto, M., Ikeo, K., Seya, T., 2003. Prediction of the prototype of the human Toll-like receptor gene family from the pufferfish, *Fugu rubripes*, genome. Immunogenetics 54, 791-800.

Palti, Y., Gahr, S.A., Purcell, M.K., Hadidi, S., Rexroad Iii, C.E., Wiens, G.D., 2010. Identification, characterization and genetic mapping of TLR7, TLR8a1 and TLR8a2 genes in rainbow trout (*Oncorhynchus mykiss*). Dev. Comp. Immunol. 34, 219-233.

Palti, Y., 2011. Toll-like receptors in bony fish: From genomics to function. Dev. Comp. Immunol. 35, 1263-1272.

Poltorak, A., He, X., Smirnova I, Liu, M.Y., Van Huffel, C., Du, X., Birdwell, D., Alejos, E., Silva, M., Galanos, C., Freudenberg, M., Ricciardi-Castagnoli, P., Layton, B., Beutler, B., 1998. Defective LPS signaling in C3H/HeJ and C57BL/10ScCr mice: mutations in *Tlr4* gene. Sci. J. 282, 2085-2088.

Rassa, J.C., Meyers, J.L., Zhang, Y.M., Kudaravalli, R., Ross, S.R., 2002. Murine retroviruses activate B cells via interaction with toll-like receptor 4. Proc. Natl. Acad. Sci. USA 99, 2281-2286.

Sepulcre, M.P., Perez, F.A., Munoz, A.L., Roca, F.J., Meseguer, J., Cayuela, M.L., Mulero, V., 2009. Evolution of lipopolysaccharide (LPS) Recognition and Signaling: fish TLR4 does not recognize LPS and negatively regulates NF-κB activation. J. Immunol. 182, 1836-1845.

Skjæveland, I., Iliev, D.B., Strandskog, G., Jørgensen, J.B., 2009. Identification and characterization of TLR8 and MyD88 homologs in Atlantic salmon (*Salmo salar*). Dev. Comp. Immunol. 33, 1011-1017.

Su, J.G., Zhu, Z.Y., Wang, Y.P., 2009. cDNA cloning and characterization of Toll-like receptor 3 in bluntnose black bream, *Megalobrama Amblycephala*. Acta Hydrobiol Sin. 33, 986-997.

Su, J. G., Heng, J. F., Huang, T., Peng, L. M., Yang, C. R., Li, Q. M., 2012. Identification, mRNA expression and genomic structure of TLR22 and its association with GCRV susceptibility/resistance in grass carp (*Ctenopharyngodon idella*). Dev. Comp. Immunol. 36, 450-462.

Sullivan, C., Charette, J., Catchen, J., Lage, C.R., Giasson, G., Postlethwait, J.H., Millard, P.J., Kim, C.H., 2009. The gene history of zebrafish tlr4a and tlr4b Is predictive of their divergent functions. J. Immunol. 183, 5896-5908.

Tamura, K., Dudley, J., Nei, M., Kumar, S., 2007. MEGA4: molecular evolutionary genetics analysis (MEGA) software version 4.0. Mol. Biol. Evol. 24, 1596-1599.

Yamamoto, M., Takeda, K., 2010. Current views of toll-like receptor signaling pathways. Gastroenterol. Res. Pract. 3,8 (Article ID 240365).

Yang, C.R., Su, J.G., Zhang, R.F., 2010. Screening and expression analyses of TLR3 from grass carp bacterial artificial chromosome (BAC) library. J. Northwest Agric. Univ. 38, 83-87.

Zou, L.Y., Wang, Z.Z., Huang, J.M., 2007. Prediction of subcellular localization of eukaryotic proteins using position-specific profiles and neural network with weighted inputs. J. Genetics 34, 1080-1087.

一个新的草鱼 *tlr4* 基因簇的分离及对草鱼出血病病毒的应答

黄　容[1]　董　锋[1,2]　张成勋[1,2]　廖兰杰[1]　朱作言[1]　汪亚平[1]

1 中国科学院水生生物研究所，武汉 430072

2 中国科学院大学，北京 100039

摘　要　哺乳动物对脂多糖(LPS)的响应主要由 Toll 样受体 4(TLR4)介导。鱼类和哺乳动物的 TLR4 存在较大差异，目前鱼类 TLR4 的配体仍未知。鱼类 *tlr4* 基因具有复杂的基因组结构，克隆与分离比较困难。本研究在获得三种含有草鱼 *tlr4* 基因的细菌人工染色体后，通过构建亚克隆文库和 cDNA 末端快速扩增分离得到了 4 个 *tlr4* 基因，它们具有不同的基因组结构和不同的蛋白质结构域。受精后 12 小时 4 个 *tlr4* 基因开始在发育过程中表达。在健康的成年鱼组织中均可检测到，且在被草鱼呼肠孤病毒(GCRV)感染后的肝脏和肌肉中表达量显著增加，表明 *tlr4* 基因在 GCRV 感染过程中起到了作用。本研究有效地从 *tlr4* 基因簇中分离出每个基因，且通过感染实验表明草鱼 TLR4 具有不同于哺乳动物的配体。

Cloning and Preliminary Functional Studies of the *JAM-A* Gene in Grass Carp (*Ctenopharyngodon idellus*)

Fukuan Du[1,2] Jianguo Su[1] Rong Huang[1] Lanjie Liao[1]
Zuoyan Zhu[1] Yaping Wang[1]

1 State Key Laboratory of Freshwater Ecology and Biotechnology, Institute of Hydrobiology, Chinese Academy of Sciences, No. 7 Donghu South Road, Wuhan 430072
2 University of Chinese Academy of Sciences, Beijing 100039
3 College of Animal Science and Technology, Northwest A&F University, Yangling 712100

Abstract Grass carp (*Ctenopharyngodon idellus*) is a very important aquaculture species in China and other South-East Asian countries; however, disease outbreaks in this species are frequent, resulting in huge economic losses. Grass carp hemorrhage caused by grass carp reovirus (GCRV) is one of the most serious diseases. Junction adhesion molecule A (*JAM-A*) is the mammalian receptor for reovirus, and has been well studied. However, the *JAM-A* gene in grass carp has not been studied so far. In this study, we cloned and elucidated the structure of the *JAM-A* gene in grass carp (*GcJAM-A*) and then studied its functions during grass carp hemorrhage. *GcJAM-A* is composed of 10 exons and 9 introns, and its full-length cDNA is 1833 bp long, with an 888 bp open reading frame (ORF) that encodes a 295 amino acid protein. The *GcJAM-A* protein is predicted to contain a typical transmembrane domain. Maternal expression pattern of *GcJAM-A* is observed during early embryogenesis, while zygote expression occurs at 8 h after hatching. *GcJAM-A* is expressed strongly in the gill, liver, intestine and kidney, while it is expressed poorly in the blood, brain, spleen and head kidney. Moreover, lower expression is observed in the gill, liver, intestine, brain, spleen and kidney of 30-month-old individuals, compared with 6-month-old. In a *GcJAM-A*-knockdown cell line (CIK) infected with GCRV, the expression of genes involved in the interferon and apoptosis pathways was significantly inhibited. These results suggest that *GcJAM-A* could be a receptor for GCRV. We have therefore managed to characterize the *GcJAM-A* gene and provide evidence for its role as a receptor for GCRV.

Keywords *Ctenopharyngodon idellus*; *JAM-A*; grass carp hemorrhage

1. Introduction

Grass carp (*Ctenopharyngodon idellus*) is a very important aquaculture species that accounts for 20% of the freshwater aquaculture production in China. However, disease

文章发表于 *Fish & Shellfish Immunology*, 2013, 34(6): 1476-1484

outbreaks in this species are frequent, resulting in huge economic losses, which severely restrict the development of grass carp farming. Grass carp hemorrhage caused by grass carp reovirus (GCRV) is one of the most serious diseases. Hence, there is a need to study the pathogenesis of GCRV infection and look for preventative measures to promote sustainable culture of grass carp.

GCRV is a non-enveloped icosahedral virus that belongs to the family Reoviridae, genus *Aquareovirus*, and it has a dsRNA genome [1]. Reoviruses infect most mammalian species [1], and the receptor for reovirus in humans and other mammals is junction adhesion molecule A (*JAM-A*) [2,3], a member of the immunoglobulin superfamily. *JAM-A* regulates the formation of intercellular tight junctions [4] and is involved in platelet activation, leukocyte transmigration and angiogenesis [5,6]. Reovirus recognizes structural features that are present in *JAM-A*, but not in *JAM-B* or *JAM-C* [7]. The crystal structure of the extracellular region of *JAM-A* reveals the presence of two concatenated Ig-type domains (D1 and D2) with a pronounced bend at the domain interface. Two *JAM-A* molecules form a dimer that is stabilized by extensive ionic and hydrophobic contacts between the D1 domains. Binding and infection experiments using chimeric and domain-deletion mutant receptor molecules indicate that the amino-terminal D1 domain of *JAM-A* is required for reovirus attachment, infection and replication in mouse [8-10].

Mammalian reovirus and GCRV belong to different genera [11], however, they share some structural similarities. Although the functions of mammalian *JAM-A* have been studied in viral infection, no functional studies have been carried out on fish *JAM-A*.

This study is therefore the first to clone the *JAM-A* gene in grass carp and study its functions during grass carp hemorrhage. Our results suggest that *GcJAM-A*, like mammalian *JAM-A*, could also be a receptor for GCRV.

2. Materials and Methods

2.1 Experimental animals

Grass carp (average weight, 9.6 g) were adapted to the conditions in a 55.5 × 45.0 × 34.8 cm^3 aquarium with a water temperature of 28.5±0.5℃, pH 7.0, and dissolved oxygen concentration of 5.5±0.3 mg O_2/L dechlorinated and aerated water. The grass carp were fed twice daily, at 9:00 AM and 5:00 PM, during the experimental period. After an acclimatization period of 3 days, the fish were challenged with GCRV.

2.2 Cloning and sequencing of grass carp *JAM-A* (*GcJAM-A*)

The total RNA of grass carp liver was purified using TRIzol reagent (Invitrogen, Carlsbad, CA, USA), and treated with RNase-free DNase (Promega, Madison, WI, USA).

First-strand cDNA was synthesized from RNA using moloney murine leukemia (M-MLV) reverse transcriptase (Toyobo, Osaka, Japan). Sequence alignment of *JAM-A* nucleotide sequences from a variety of species was performed with the ClustalX 1.83 multiple-alignment software. The primers were designed based on the conserved nucleotides of the sequences of seven species reported before: *Homo sapiens* (NP_058642.1) *Pan troglodytes* (XP_001172741.1), *Canis familiaris* (XP_536132.2), *Bos taurus* (NP_776520.1), *Mus musculus* (NP_766235.1), *Rattus norvegicus* (NP_446248.1), *Danio rerio* (NP_001004667.1) (Table 1). A 750 bp fragment of the *JAM-A* gene was amplified from grass carp liver cDNA using the primer pair 25/34 (Table 1). The resulting fragments were separated on 1.0% agarose gel and purified using the Axygen DNA gel extraction kit (Axygen, Union City, CA, USA). The purified fragments were cloned into the pMD-18T vector (Takara, Dalian, China) by the TA cloning strategy and sequenced (BGI, Shenzhen, China). The 750 bp fragment was confirmed to be a homologous sequence of the *JAM-A* gene from *D. rerio* by BlastX. The 5′ and 3′ ends of the *JAM-A* cDNA were obtained using the rapid amplification of cDNA ends (RACE) approach (Zhuan Dao, Wuhan, China).

Table 1 Sequences of the primers used in this study

Primer	Sequence	Amplicon length (bp)	Use
25	5′-ACACCCAGAGTAGAATGGAAGTT-3′		
34	5′-ACACCACAAAAGATGATTTCTGTCTGAA-3′	750	*GcJAM-A* cloning
79	5′-TTTCGTCTTTGTGTGCCTCTCTATCTC-3′		
75	5′-AAATGTGGGGAACTTGCTGGGAT-3′	1627	*GcJAM-A* cloning
74	5′-ATCCCAGCAAGTTCCCCACATTT-3′		
71	5′-CCCTTTTTAGTGGCAAACCAGAGAG-3′	2834	*GcJAM-A* cloning
70	5′-ATTGCTCCTCTTTGCTCTCTGGTTT-3′		
68	5′-CATTCATAGTACCTTCTTTTCCCCAGA-3′	1170	*GcJAM-A* cloning
93S	5′-TGTGGCTTTGCTGGCAGTAG-3′		
93A	5′-TGGCTTGCTTTCTGCTATTTTT-3′	91	qRT-PCR for *JAM-A*
120S	5′-GGATGATGAAATTGCCGCACTGG-3′		
120A	5′-ACCGACCATGACGCCCTGATGT-3′	136	qRT-PCR for β-actin
107S	5′-GGGGTACCATGTTGACTTTCGT CTTTGTGTGCC-3′		
107A	5′-CCCAAGCTTCCAAAGCAGGCTACA CCACAAAAG-3′	898	Vector constructing
108S	5′-CGGGGTACCTCTGTGGATAACCG TATTACCGCC-3′		
108A	5′-CGGGATCCATGTTGACTTTCGT CTTTGTGTGCC-3′	1555	Vector constructing
112	5′-GATCTTAAAATAAAGCAATAGCATCACAAATTTC ACAAATAAAGCATTTTTTTCACTGCA-3′		
113	5′-AGCTTGCAGTGAAAAAAATGCTTTATTTGTGAAATT-3′		
RF230	5′-ACTACACTGAACACCTGCGGAA-3′		
RR231	5′-GCATCTTTAGTGCGGGCG-3′	106	qRT-PCR for RIG-1

Continued

Primer	Sequence	Amplicon length (bp)	Use
132S	5′-CCAGCATCAAACTTCACTACCG-3′		
132A	5′-TTACTGAGCGTGTGTCTCCAAA-3′	97	qRT-PCR for JAK1
138S	5′-ATCTGCGATGGGCTTTTGTC-3′		
138A	5′-GGCTTGGTCTCGTTTGGTTTC-3′	87	qRT-PCR for caspase 8
140S	5′-CAACCGAAAAGGCACTGAGAA-3′		
140A	5′-CGAGAGGACACAGCAGACAAAA-3′	170	qRT-PCR for caspase 9
142S	5′-AGTGAGGAAGATGCGGCTATTT-3′		
142A	5′TGTTGAGGGCACAGCGAAG-3′	116	qRT-PCR for IRF9
145S	5′-GACACATACAGTAGGATATTCACTCGC-3′		
145A	5′-TTGCCTGGGAAGTAGTTTTCTTG-3′	122	qRT-PCR for IFNγ
147S	5′-TGCCTGGATTGAGAAAAGAAACT-3′		
147A	5′-TTCCATCACTGCCGAACATTAT-3′	167	qRT-PCR for PKR
148S	5′-GCAGGGGACAAAAAGAGATTATAGA-3′		
148A	5′-AGCCAACTTAGGAATAGTAGCAAAAC-3′	134	Mxa
158S	5′-TGAGCAAAGACCCCAACGAG-3′		
158A	5′-TCGACTGCAGAATTCGAAGCT-3′	139	EGFP

2.3 *GcJAM-A* genomic DNA sequence amplification and organization analysis

Grass carp genomic DNA was extracted from the tail fin using phenol-chloroform. The primers 79/75, 74/71 and 70/68 (Table 1) were designed to amplify the *GcJAM-A* genomic DNA sequence according to the full-length cDNA of *GcJAM-A*. PCR amplification was carried out using LA *Taq*™ Polymerase (Takara, China) under the following cycling parameters: 94℃ for 5 min; followed by 36 cycles of 30s at 94℃, 40 s at 62℃, and 7 min at 72℃; and a final extension step for 10 min at 72℃. The resulting fragments were purified and sequenced, and the genomic DNA sequence obtained was analyzed to determine the exons and introns based on the cDNA sequence of *GcJAM-A*.

2.4 Analysis of nucleotide and amino acid sequences

The nucleotide and predicted amino acid sequences of *GcJAM-A* were analyzed using DNA figures software (http://www.bio-soft. net/sms/index.html). The similarity of *JAM-A* from carp with *JAM-A* from other organisms was analyzed using the BLASTP search program of NCBI (http://www.ncbi.nlm.nih.gov/blast). The domain structures were predicted using the SMART program (http://smart. embl-heidelberg.de/). The alignment was compared with that of multiple amino acid sequences of *JAM-A* reported, using ClustalX 1.83 (http://www.

ebi.ac.uk/ clustalW/) and the GeneDoc software. The phylogenetic tree was constructed using the MEGA 3.1 software (http://megasoftware.net).

2.5 RT-qPCR analysis of *GcJAM-A* mRNA expression profiles

For tissue distribution analysis of *GcJAM-A*, total RNA was extracted from the gill, liver, spleen, kidney, head kidney, brain, intestine and blood of 6-month-old and 30-month-old healthy grass carp using TRIzol Reagent (Invitrogen, USA). Each of the samples was obtained from three individuals to eliminate inter-individual differences.

Total RNA was also extracted from eggs at different time points (0, 8, 16, 24, 32, 40, 48, 56, 64 and 72h) after fertilization. Unfertilized eggs were obtained at time point 0 h, while the eggs obtained at the other time points were after fertilization. Each of the samples contained about 200 eggs.

The first-strand cDNA was synthesized with the ReverTra Ace® qPCR RT kit (Toyobo, Japan), and RT-qPCR was employed to detect the *GcJAM-A* expression profiles using β-actin as a reference gene. The RT-qPCR primers 93S/93A for *GcJAM-A* and 120S/120A for β-actin (Table 1) shared similar Tm values and were designed to amplify 91 bp and 136 bp fragments, respectively. RT-qPCR was performed on the ABI 7000 real-time PCR system (ABI, USA) using 2×SYBR green real-time PCR mix (Toyobo, Japan). PCR amplification was performed in triplicate, using the following cycling parameters: 94℃ for 2 min; followed by 40 cycles of 15 s at 94℃, 15 s at 59℃, and 45 s at 72℃. The expression of target genes was calculated as relative folds with the $2^{-\Delta\Delta CT}$ method.

2.6 RT-qPCR analysis of *GcJAM-A* mRNA expression profiles in 6-month-old and 30-month-old grass carp challenged with GCRV

A GCRV suspension was obtained from infected fish homogenate in our lab. GCRV-infected fish homogenates were centrifuged at 4000 *g* for 30 min at 4℃, and the supernatant was used for the challenge experiments. One hundred and twenty fish were divided into four groups of 40 fish each. Group I contained 6-month-old carp that were intraperitoneally (*i.p.*) injected with 20 μl/g body weight of the GCRV suspension. Group II contained 6-month-old carp that were injected *i.p.* with 20 μl/g body weight of the 0.7% NaCl as the control treatment. Group III contained 30-month-old carp that were injected *i.p.* with 20 μl/g body weight of the GCRV suspension, and Group IV contained 30-month-old carp that were injected *i.p.* with 0.7% NaCl per gram body weight.

Three individuals were sampled from each group, kidney and intestine tissue was collected at days 1, 2, 3, 4, 5, 6 and 7 after injection. The samples were homogenized using TRIzol reagent (Invitrogen), and total RNA was isolated according to the manufacturer's instructions. Total RNA was incubated with RNase-free DNase I (Promega) to remove contaminating genomic DNA. First-strand cDNA was synthesized with the ReverTra Ace® qPCR RT kit (Toyobo), and RT-qPCR was employed to detect *GcJAM-A* expression using β-actin as a reference gene. The process as described before.

2.7 *GcJAM-A*-knockdown and expression analysis of the interferon system and apoptosis pathway genes in CIK cells

The pCMV-anti-*JAM-A* vector was constructed by inserting the antisense sequence of *GcJAM-A* (amplified by the primers 107S/ 107A; Table 1) into PEGFP-N1 (Clontech, USA), this plasmid was used as the PCR template for amplifying a CMV-anti-*JAM-A* sequence using the primers 108S/108A (Table 1). P-anti-JAM-A was finally constructed by inserting the CMV-anti-*JAM-A* sequence into pCMV-EGFP-polyA, which was constructed by inserting the polyA sequence (Table 1) into PEGFP-C1 (Clontech).

To obtain cells that stably express antisense *GcJAM-A*, CIK cells were transfected with p-anti-JAM-A or pMV-eGFP (Clontech, USA) and then selected with 800 μg/ml G418 (Gibco, USA). In detail, CIK cells cultured in 24-well plates for 16-18 h were transiently transfected with either 2 μg p-anti-JAM-A or pCMV-eGFP, using Lipofectamine™ 2000 Reagent (Invitrogen) according to the manufacturer's instructions. At 48 h post-transfection, G418 was added to the medium at a final concentration of 800 g/ml. After five generations of selective culturing, the cells were observed under a fluorescence microscope (Leica, Germany); the number of transfected cells accounted for more than 50% of the total number of cells. The cells that were stably transfected with p-anti-JAM-A and pCMVeGFP were termed CIK-W5 and CIK-control cells, respectively.

CIK-W5 and CIK-control cells were cultured in separate 96-well plates for 16-18 h at 28℃. Both groups of cells were challenged with five GCRV concentrations, 3.75×10^4, 7.5×10^4, 1.5×10^5, 3.0×10^5, and 6.0×10^5 PFU/ml. The virus titer was determined by the end-point dilution method [12]. After 72 h, the cells were stained with crystal violet and images were taken.

CIK-W5 and CIK-control cells were cultured in 24-well plates for 16-18 h at 28℃. GCRV at a concentration of 1.5×10^5 PFU/ml was selected based on the results of the staining to challenge the cells. The virus-infected cells were collected at the indicated time points (0, 24, 48, 72 and 96 h post-infection), and each sample was harvested in triplicate. Meanwhile, aliquots of the same cell samples were used for RNA extraction and cDNA synthesis. The RT-qPCR primers used were 93S/93A for Gc*JAM-A* and 120S/120A for β-actin. The primers RF230/RR231, 132S/132A, 142S/142A, 138S/138A, 140S/140A, 147S/147A, 145S/145A, 148S/148A and 158S/158A were used to amplify 106 bp, 97 bp, 116 bp, 87 bp, 170 bp, 167 bp, 122 bp, 134 bp and 139 bp fragments of the genes *RIG-1, JAK1, IRF9*, caspase 8, caspase 9, PKR, IFNγ, *Mxa* and EGFP respectively (Table 1). β-Actin was used as the reference gene for all amplifications. Primers for *RIG-1* have been reported before [13], and other sequences of homologous genes (in supplement) for zebrafish were searched in the grass carp genomic bank using a BLAST search.

2.8 Statistical analysis

All mRNA levels were expressed as mean ± SD and subjected to the Student's *t*-test. Differences were considered significant at $p \leqslant 0.05$.

3. Results

3.1 Cloning and sequence characterization of the Gc*JAM-A* gene

A 750 bp cDNA fragment was amplified from the grass carp liver cDNA. BlastX analysis of the fragment showed that it shares high similarity to other reported *JAM-A* genes. Based on this conserved sequence, the full-length cDNA of grass carp *JAM-A* was obtained by RACE. The full-length cDNA of *GcJAM-A* was 1833 bp long and contained an 888 bp ORF that encoded a protein of 295 amino acids (Fig. 1). The 5′ and 3′ untranslated regions were 147 bp and 798 bp, respectively. Six instability motifs (ATTTA) have been found which are associated with mRNA instability and one polyadenylation signal (AATAAA) was present 12 nucleotides upstream of the poly(A) tail. The full-length genomic DNA sequence of *GcJAM-A* is 7116 bp long, and the schematic diagram for the *GcJAM-A* gene is shown in Fig. 2. The *GcJAM-A* gene is composed of 10 exons and 9 introns, and all exon/intron junctions conform to the splicing consensus sequence (GT-donor/AG-acceptor) rule.

3.2 Homology analysis of *GcJAM-A*

The deduced amino acid sequence of *GcJAM-A* showed 43%, 43%, 43%, 44%, 44%, 45% and 76% identity with the *JAM-A* homologs from *H. sapiens*, *P. troglodytes*, *C. familiaris*, *B. taurus*, *R. norvegicus*, *M. musculus* and *D. rerio*, respectively. In addition, the *GcJAM-A* protein sequence shares three conserved domains with other reported *JAM-A*s, including an 18 residue N-terminal signal peptide, an IG domain (95 residues), and an Igc2 domain (75 residues) followed by one transmembrane domain (22 residues) and a 43-residue intracellular domain (Fig. 3).

The phylogenetic tree of vertebrate *JAMs* consists of three major branches: *JAM-A, JAM-B* and *JAM-C*. *JAM-A* from *H. sapiens, P. troglodytes, C. familiaris, B. taurus, R. norvegicus* and *M. musculus* were clustered into the mammalians subgroup; *JAM-A* from *D. rerio* and *C. idellus* were clustered into the fish subgroup. In the tree, the sequences we had cloned were closer to *JAM-A* from *D. rerio*, which suggested that it was indeed *GcJAM-A*, not *B* or *C* (Fig. 4).

```
1    tttatttccgggtgttcggcgggtgtcgctttttgattttggctcagttttgtgggaaca
61   taacagataaacaactgcaactaccataggtctaatataaaccagacatcattttaaaa
1                                M   L   T   F   V   F   V   C   L   S   I
121  gagaaacatttactctcgagagtggatATGTTGACTTTCGTCTTTGTGTGCCTCTCTATC
12   S   V   T   G   L   H   A   S   F   T   V   S   I   S   N   P   L   I   T   V
181  TCAGTCACAGGCCTACATGCATCTTTTACGGTTAGCATTAGTAATCCCCTAATAACAGTG
32   K   E   N   E   G   V   D   L   K   C   S   Y   T   A   D   F   G   A   T   P
241  AAAGAGAATGAGGGAGTTGACTTGAAATGTTCCTACACTGCTGACTTTGGAGCAACACCC
52   R   V   E   W   K   F   K   D   L   K   G   Y   Q   F   F   I   Y   F   D   S
301  AGAGTCGAATGGAAGTTCAAGGATCTCAAGGGATATCAGTTTTTCATCTACTTTGATAGC
72   R   L   T   A   E   Y   E   G   R   I   T   V   Y   N   G   G   L   R   F   D
361  AGGCTAACTGCTGAATATGAAGGCCGTATCACTGTGTATAATGGAGGTCTGAGATTTGAC
92   K   V   T   R   A   D   T   G   D   Y   D   C   E   V   S   G   N   G   G   Y
421  AAGGTGACGAGAGCAGACACTGGAGACTATGACTGTGAGGTGTCTGGAAATGGTGGTTAT
112  D   E   K   T   I   K   L   M   V   L   V   P   P   S   K   P   V   S   R   I
481  GATGAGAAGACCATCAAACTCATGGTCCTTGTGCCTCCTTCCAAGCCCGTATCCAGAATT
132  P   S   S   V   T   T   G   S   N   V   L   L   T   C   F   D   N   V   G   S
541  CCTTCTTCAGTCACAACAGGCAGTAACGTCCTTCTGACCTGCTTCGACAATGTTGGATCT
152  P   P   P   T   Y   K   W   Y   K   D   N   T   P   L   P   D   D   P   S   K
601  CCCCCACCCACCTATAAGTGGTACAAAGACAACACACCTCTCCCTGATGATCCCAGCAAG
172  F   P   T   F   K   N   L   T   Y   K   M   N   V   F   N   G   N   L   E   F
661  TTCCCCACATTTAAAAACCTCACCTACAAGATGAATGTTTTCAATGGAAACCTGGAGTTC
192  P   S   V   S   K   L   D   I   G   S   Y   F   C   E   A   N   N   G   E   G
721  CCGAGTGTGTCTAAGCTGGATATCGGTTCATATTTCTGTGAGGCCAATAATGGAGAAGGT
212  A   P   Q   R   S   D   A   V   Q   M   D   V   R   D   L   N   V   G   G   I
718  GCCCCTCAGCGTAGTGACGCAGTGCAGATGGATGTCCGTGATCTAAATGTTGGTGGCATT
232  V   A   G   V   I   V   A   L   L   A   V   G   L   L   L   F   A   L   W   F
841  GTTGCGGGAGTTATTGTGGCTTTGCTGGCAGTAGGATTGCTCCTCTTTGCTCTCTGGTTT
252  A   T   K   K   G   Y   M   P   K   I   A   E   S   K   P   K   T   Q   A   V
901  GCCACTAAAAAGGGATATATGCCAAAAATAGCAGAAAGCAAGCCAAAAACACAAGCTGTT
272  Y   T   Q   P   R   A   D   E   G   V   D   G   D   G   E   F   K   Q   K   S
961  TATACACAGCCTAGAGCAGATGAAGGGGTTGATGGGGATGGGGAATTCAAACAGAAATCA
292  S   F   V   V   *
1021 TCTTTTGTGGTGTAGcctgctttggatgaaatatcaagatttgttgcctagatcaatttt
1081 tcctttttctatttaagttcatacaaatcaggatcagtttgtatgtgaagagtgtgaca
1141 tcaaatgttcctccctgccctgcctaccacaccacttgaccatagaaatgtccaaaagct
1201 ttaaaccatcgtgtgtgtgtgtgtgtgtgtgtgtgtgtgtgtgcgtgtgtgtgttagctt
1261 aaacttacagatttagttttaggaatctaaatcctctctcccatttgcagaaatgaattg
1321 aaatgtaatgtatttaagtacaaataatctgtgagtggaaaacagacttaaagggtttat
1381 tatcgatattttattaatataatttgctcaaattatattaatttgtttagttatttatgg
1441 attgaggttaattctgtgtgactgcacaatgggatgaaagcttctggggaaaagaaggta
1501 ctatgaatgttccccattcaggaaataatccccccttcctactgatataaacatttgctgt
1681 aatttaaaaatcatgctttattgacctctctctgaattgttcaactggaaactaattacg
1741 aaacagtttttcttgtttgatttgaagttttaatgtcagacttgatcctgttcttccaat
1801 aaattaaacttactaaaaaaaaaaaaaaaaaaa
```

Fig. 1 Nucleotide and deduced amino acid sequences of *GcJAM-A*. The full-length cDNA of *GcJAM-A* was 1833 bp long and contained an 888 bp ORF that encoded a protein comprising 295 amino acids. The 5′ and 3′ untranslated regions were 147 bp and 798 bp long, respectively. Capital letters above the nucleotide sequences indicate the corresponding amino acid sequence. The start codon (ATG), stop codon (TGA) are boxed, the motif associated with mRNA instability (ATTTA) are in bold and poly(A) signal sequence (AATAA) is in bold and under line

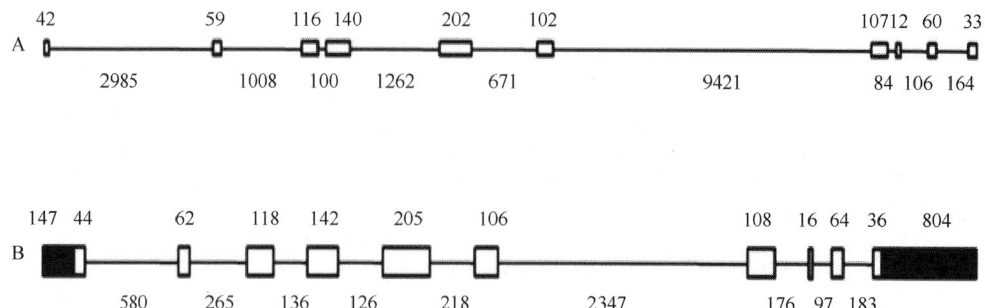

Fig. 2 A. Schematic of zebrafish *JAM-A*. B. Schematic of *GcJAM-A*. Exons are represented by white bars, and introns by lines. The 5′UTR and 3′UTR are indicated by dark bars. The number of nucleotides in each exon or intron are shown above or below the corresponding element. Both genes contain 10 exons and 9 introns

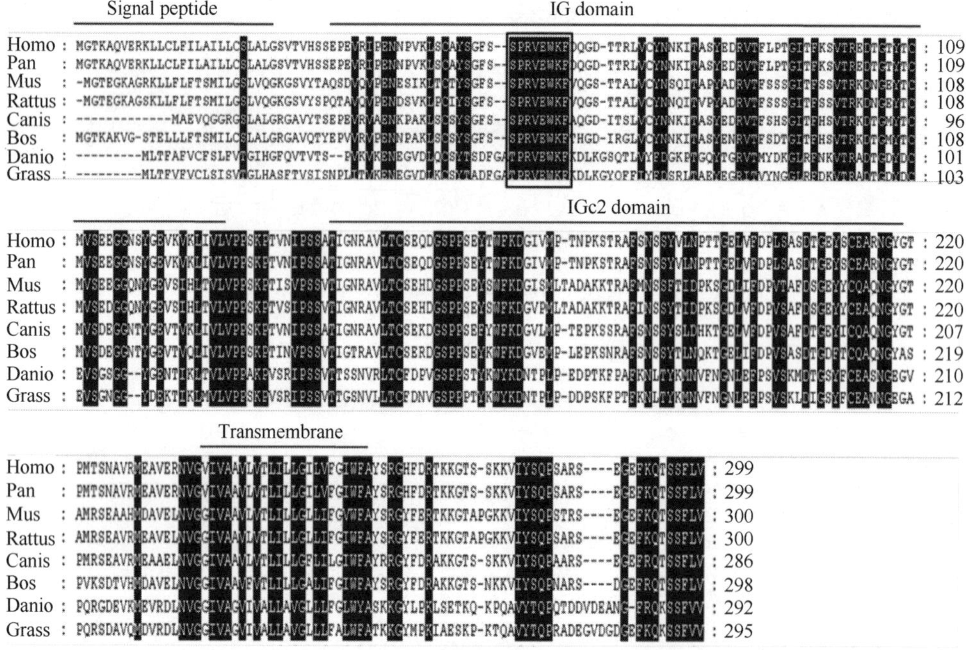

Fig. 3 Alignment of the deduced amino acid sequence of *GcJAM-A* and other reported JAM-A sequences. GcJAM-A is a membrane protein, comprising an extracellular signal peptide, immunoglobulin G and Igc2 domain, followed by a transmembrane region and the intracellular region. Completely conserved residues across all species are aligned and shaded in black. The conserved R(V,I,L)E motif in the IG domain (D1 domain) is boxed

3.3 Expression profiles of *GcJAM-A*

The expression profiles of *GcJAM-A* transcripts were assessed in different tissues by RT-qPCR using β-actin as a reference gene. In healthy fish, *GcJAM-A* was strongly expressed in the intestines and gill; comparatively highly in the liver and kidney; and at low levels in the blood, brain, spleen and head kidney. Moreover, lower expression was observed in the gills, liver, intestine, brain, spleen, and kidney of 30-month-old individuals, compared with 6-month-old individuals ($p < 0.05$; Fig. 5).

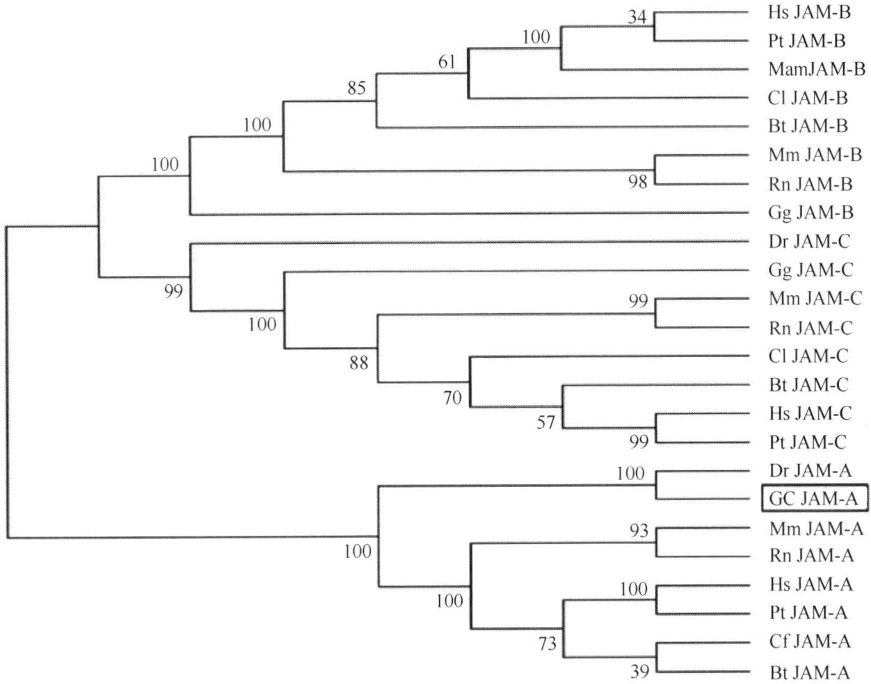

Fig. 4 Phylogenetic analysis of *GcJAM-A* and other reported *JAM-A*, B and C sequences in vertebrates. The phylogenetic tree was constructed by the neighbor-joining method with MEGA 3.1. The GenBank accession numbers of the analyzed sequences are as follows: HsJAM-B, *Homo sapiens* JAM-B, NP_067042.1; PtJAM-B, *Pan troglodytes* JAM-B, XP_001157948.1; MamJAM-B, *Macaca mulatta* JAM-B, XP_002803217.1; ClJAM-B, *Canis lupus familiaris* JAM-B, XP_535568.1; BtJAM-B, *Bos taurus* JAM-B, NP_001077205.1; MmJAM-B, *Mus musculus* JAM-B, NP_076333.3; RnJAM-B, *Rattus norvegicus* JAM-B, NP_001029176.1; GgJAM-B, *Gallus gallus* JAM-B, NP_001006257.1; DrJAM-C, *Danio rerio* JAM-C, XP_001338123.1; Gg JAM-C, *Gallus gallus* JAM-C, XP_417876.3; MmJAM-C, *Mus musculus* JAM-C, NP_075766.1; RnJAM-C, *Rattus norvegicus* JAM-C, NP_001004269.1; ClJAM-C, *Canis lupus familiaris* JAM-C, XP_546389.3; BtJAM-C, *Bos taurus* JAM-C, NP_001098834.1; HsJAM-C, *Homo sapiens* JAM-C, NP_116190.3; PtJAM-C, *Pan troglodytes* JAM-C, XP_529458.3; DrJAM-A, *Danio rerio* JAM-A, NP_001076451.1; MmJAM-A, *Mus musculus* JAM-A, NP_766235.1; RnJAM-A, *Rattus norvegicus* JAM-A, NP_446248.1; HsJAM-A, *Homo sapiens* JAM-A, NP_058642.1; PtJAM-A, *Pan troglodytes* JAM-A, XP_001172741.1; CfJAM-A, *Canis familiaris* JAM-A, XP_536132.2; BtJAM-A, *Bos taurus* JAM-A, NP_776520.1

To determine whether *GcJAM-A* was involved in immune responses to GCRV, different challenges and the corresponding responses were determined in different age groups. RT-qPCR was performed to examine the expression profile of *GcJAM-A* transcripts in fish intestines and kidney after the GCRV challenge. The GCRV-infected fish showed more severe disease symptoms (such as lower feed intake and intestinal hemorrhage) 7 days after infection. No clinical signs of disease were detected in the control fish.

The RT-qPCR results revealed 1.1- to 23.6-fold decreased expression of *GcJAM-A* in the intestines of GCRV-stimulated 6-month-old grass carp (Fig. 6). In detail, after the GCRV challenge, the level of *GcJAM-A* transcripts in the intestine decreased by 1.6-fold on post-infection day 1 and then showed a slight increase on day 2 compared to day 1. A similar fluctuating pattern was seen on days 3-7. This fluctuating expression pattern was also observed in the intestines of 30-month-old individuals infected with GCRV, with a 1.7- to 8.7-fold decrease in expression on days 1-7. Compared to the 6-month-old carp, significantly lower expression was observed on days 0, 2, and 4 in the 30-month-old individuals ($p < 0.05$).

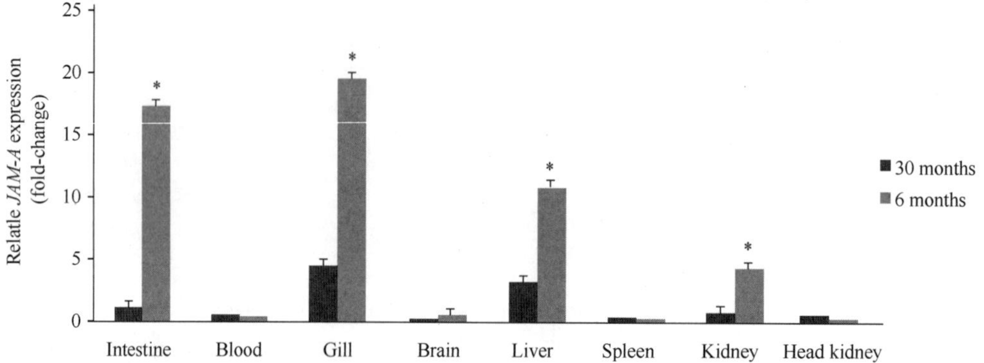

Fig. 5 RT-qPCR analysis of the expression profiles of *GcJAM-A* in different grass carp tissues. Data were expressed as the ratio of *GcJAM-A* mRNA expression in the tissue to its expression in the intestine for 30-month-old fish (mean±SD). Significant differences in *GCJAM-A* expression between the samples from the 6-month-old and 30-month-old fish are indicated with an asterisk (*), $p < 0.05$

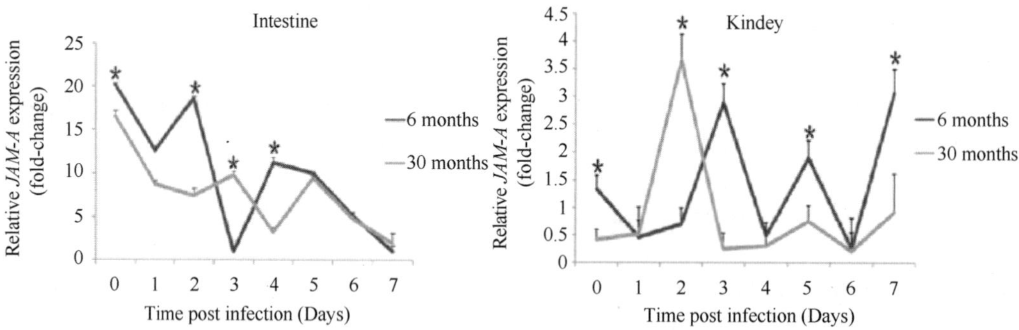

Fig. 6 Temporal expression analysis of *GcJAM-A* mRNA in the intestines and kidney of 6-month-old and 30-month-old grass carp at day 0, day 1, day 2, day 3, day 4, day 5, day 6 and day 7 after the GCRV challenge. Vertical bars represent the mean±SD, and significant differences are indicated with an asterisk (*), $p < 0.05$

A 2.3- to 6.5-fold increase in the expression of *GcJAM-A* was observed in the kidneys of GCRV-stimulated 6-month-old grass carp (Fig. 6), and a 2.0- to 9.3-fold increase was observed in the 30-month-old grass carp. As observed in the intestine, fluctuating expression patterns were also observed in the kidney. On days 0, 3, 5 and 7, significantly lower *GcJAM-A* expression was observed in the 30-month-old grass carp, compared with the 6-month-old grass carp ($p < 0.05$).

The maternal expression pattern was observed during early embryogenesis, while zygote *GcJAM-A* expression occurred at 8h after hatching (Fig. 7).

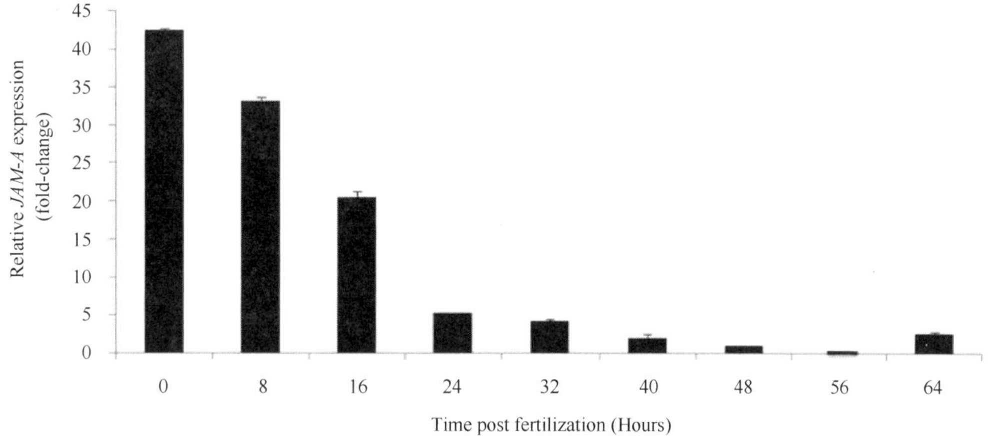

Fig. 7 Relative expression of *GcJAM-A* in embryos. Fold changes up to post-fertilization 40 h are shown (mean ± SD). The sample at 0 h is the unfertilized egg

3.4 Effect of *GcJAM-A*-knockdown in CIK cells

CIK cell lines were transfected with p-anti-*JAM-A* (CIK-W5) and pCMV-EGFP (CIK-control), and the CIK-W5 cell lines that expressed anti-JAM-A to silence *GcJAM-A* were selected and established using G418 resistance as a marker. A pCMV-EGFP-transfected cell line, the CIK-control cell line, was also established as a control (Fig. 8A).

Various titers of GCRV were used: apoptosis was limited in the CIK-W5 cell lines compared to the CIK-control cell line at titers of 3.75×10^4 PFU/ml, 7.5×10^4 PFU/ml and 1.5×10^5 PFU/ml (Fig. 8B).

To verify the results, *GcJAM-A* expression levels were detected at 0 and 96 h after GCRV infection by RT-qPCR, and they were normalized to the unchallenged CIK-control levels. The *GcJAM-A* transcript levels in CIK-control cells were more than 1.5 times higher than those in CIK-W5 cells at 0 h, and more than 98.9 times higher at 72h. *GcJAM-A* expression was therefore significantly inhibited in the CIK-W5 cells ($p < 0.05$; Fig. 8C).

3.5 Response of the interferon system and apoptosis pathway genes to GCRV in CIK cells

The expression of genes related to the interferon and apoptosis pathways was detected by RT-qPCR. The expression of these genes in CIK-control cells appeared to up- or downregulated to varying degrees. Compared to CIK-W5 cells, in CIK-control cells, the expression of *GcJAM-A* was significantly higher at 0 h, 72h and 96 h after the GCRV challenge (Fig. 9): it was 98 times and 18 times higher at 72 and 96 h respectively. The expression of *Mxa* and *IFNγ* in CIK-control cells showed an increase and reached a 35-fold and 120-fold peak at 72 h, compared to CIK-W5 cells. The expression of *RIG-1*, caspase 8, *IRF9*, and *PKR* were also significantly higher in CIK-control cells than CIK-W5 cells at 0, 72 and 96 h.

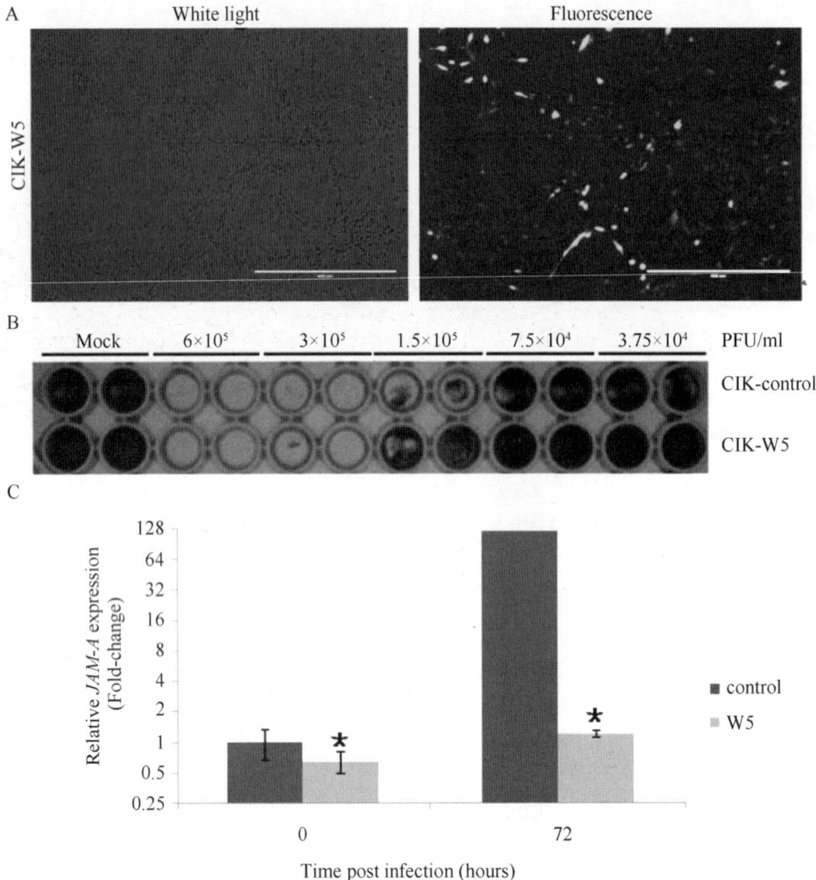

Fig. 8 A. Images of p-anti-JAM-A-transfected CIK cells in green fluorescence and white light; the transfection efficiency was more than 50%. B. CIK-w5 and CIK-control cells were challenged with different concentrations of GCRV; the apoptosis rates at 72 h are shown. C. Relative *GcJAM-A* expression (mean ± SD) in CIK-w5 and CIK-control cells at different time points after GCRV challenge according to real-time RT-PCR. *GcJAM-A* expression levels were significantly different in W5 and control cells ($*p < 0.05$). (For interpretation of the references to color in this figure legend, the reader is referred to the web version of this article.)

4. Discussion

JAM-A is a broadly expressed immunoglobulin superfamily protein that forms stable homodimers and regulates tight-junction permeability and lymphocyte trafficking [14,15]. The D1 domain (Ig domain) of human *JAM-A* is required for reovirus attachment, infection and replication. We found a high level of similarity in the D1 domain between *GcJAM-A* and mammalian *JAM-A* (Fig. 3). In agreement with this, it has been reported that the R(V,I,L)E motif in the D1 domain, which is essential for the formation of *JAM-A* dimers in solution [8], is highly conserved between fish and mammals. This conserved, critical motif possibly has a conserved function.

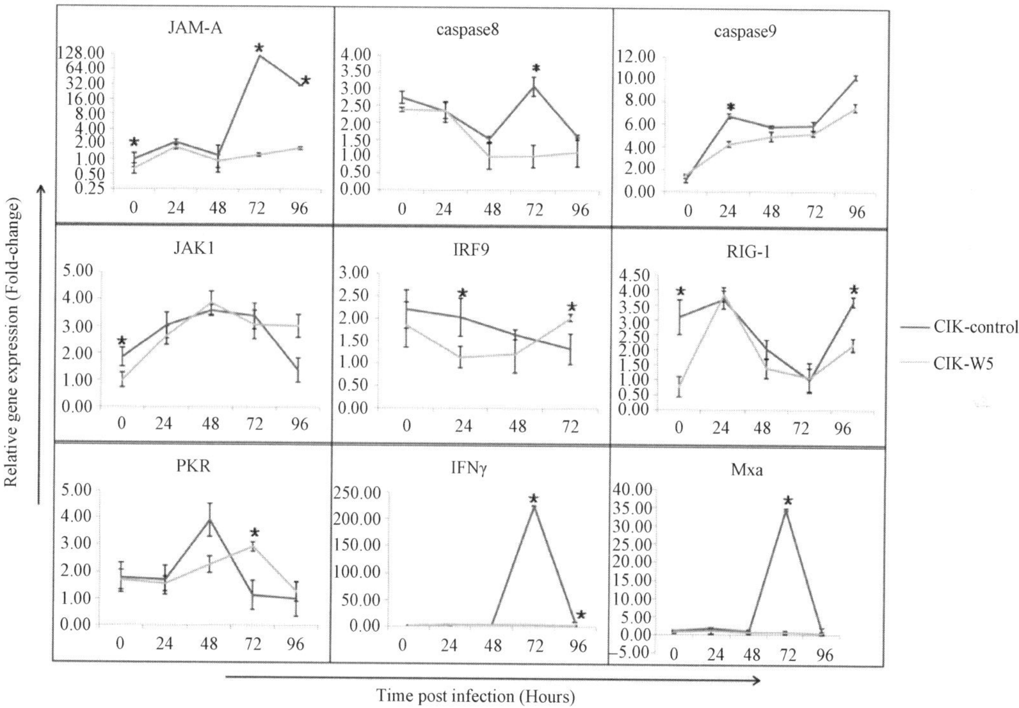

Fig. 9 Relative expression of *GcJAM-A*, and IFN and apoptosis pathway genes in CIK-control and CIK-w5 cells after the GCRV challenge. The vertical axis shows the relative gene expression levels (means ± SD), and the horizontal axis shows the time after inoculation (h)

The intron phases of the *JAM-A* gene in zebrafish are indeed quite different from those in grass carp, which is a very interesting phenomenon. Some research has been carried out to resolve this issue. According to our GCRV challenge experiment, the mortality rate in zebrafish was 40-45% after being challenged with GCRV compared to 60-90% in grass carp. The *JAM-A* cDNA of grass carp and zebrafish share up to 90% similarities. The question remains if there is any relationship between the introns and the mortality due to GCRV. Introns are known to regulate gene expression. Gene methylation is an important way to regulate gene expression, and methylation in nine introns and the first two exons of *JAM-A* have been detected, but no result has been found. However, resistance to GCRV is not affected by methylation of introns. More research is required in this field.

Similar to our results for *GcJAM-A* expression distribution in organs (Fig. 5), murine *JAM-A* has also been reported to be highly expressed in the liver, intestines and kidney [16].

Newborn mice are susceptible to reovirus, while adult mice are resistant to it [17,18]. These age-dependent features were also observed in grass carp: the mortality rate was as high as 60-90% in one-year-old fish, while it was almost 0% in the adult groups (2 years old or more; date not shown). Tardieu *et al.* speculated that the age-dependent resistance to reovirus appears to be related to an intrinsic resistance to viral replication by cells, but had no relation with modification or loss of viral receptors [18]. In contrast, Barton *et al.* were of the opinion that mouse *JAM-A* is highly expressed at birth, after which its expression declines, and this is responsible for the higher resistance in adults [2,6]. However, *JAM-A* expression during development has not been clearly examined in the murine model. In grass carp, lower *JAM-A* expression was observed in the gill, liver, intestine, brain, spleen, and kidney of 30-month-old individuals compared with the 6-month-old individuals (Fig. 5). The results of our experiments support Barton's speculation. Possibly, the expression of *JAM-A* declines in adults, which inhibits the entry of the virus into cells.

Thomas *et al.* reported that *JAM-A*-encoding mRNA is expressed at the two-cell stage and the protein is detectable from the eight-cell stage in the mouse [19]; however, Aurrand *et al.* found no expression of *JAMA* mRNA until the 17-day embryo stage in mice [16]. The one thing that both papers agree on is that murine *JAM-A* shows no maternal expression pattern, but is expressed during early embryogenesis. However, our results showed the maternal expression pattern of *GcJAM-A* during early embryogenesis and *zygote GCJAM-A* expression at 8 h after hatching (Fig. 7). This difference could be related to the mode of fertilization, *in vivo* or *in vitro*. *JAM-A* is present in elongated spermatids and in the plasma membrane of the head and flagellum of sperm [20], and plays an important role in cell adhesion [21]. Since *in vitro* fertilization occurs in fish, the high expression of *JAM-A* in fish eggs may be more conducive to egg and sperm cell adhesion and subsequent sperm penetration into the egg membrane. With development of the embryo, the expression of *GcJAM-A* declines sharply, and its expression is lowest in the fertilized egg membrane, which is conducive for hatching. *GcJAM-A* expression then increases again at the zygote stage, when it may be required for regulating tight-junction permeability and other physiological processes [4,22].

Genes that play a role in the interferon system and apoptosis pathway are activated during reovirus invasion in mammals [23-28]. Our results also show that these pathways are activated by GCRV in grass carp as evident from the responses of these genes to GCRV in CIK-control cells (Fig. 9). The inhibition of the interferon system and apoptosis pathway in the *GcJAM-A*-knockdown CIK cells and apoptosis was limited after GCRV challenge, which indicated that *GcJAM-A* could be involved in the entry of GCRV into cells.

Mx and IFNγ regulate the expression of antiviral genes, and are therefore involved in antiviral activities [24,25,27,29,30]. The expression of these genes in control cells was significantly higher than in the *GcJAM-A*-knockdown cells at 72 h (Fig. 9); however, the control cells showed a higher rate of apoptosis (Fig. 8B); the reason for this is not clear. IFNγ may play a role in apoptosis via activation of the TNF-related apoptosis-inducing ligand (TRAIL)-mediated apoptosis pathway [31-35]. Although there is high expression of Mx which is involved in antiviral activities in control cells, strong expression of caspase 8 is induced by IFNγ (72 h); this causes cell death through TRAIL-mediated apoptosis. Lower expression of IFNγ induces lower expression of caspase 8 (Fig. 9) and therefore lower rates of apoptosis in the *GcJAM-A*-knockdown cells (Fig. 9B). These results indicate that apoptosis induced by IFNγ may be dose-dependent. Low to intermediate levels of IFNγ play a role in activating antiviral mechanisms [36,37], while high levels induce apoptosis through the TRAIL apoptosis pathway; both mechanisms have an antiviral effect. When cells are infected by a small amount of virus, low-dose IFNγ is produced in response; this induces the expression of PKR, which inhibits viral replication. When a large amount of virus is present within cells, an excess of IFNγ is produced, which results in cell death via TRAIL-mediated apoptosis. As a result, viral replication, assembly and proliferation cannot be completed. However, this antiviral strategy is rather risky. An excessive amount of virus in the cells induces high rates of cell apoptosis, which causes organ malfunction and even death.

In summary, *GcJAM-A* was found to have similar expression patterns and similar functions to mammalian *JAM-A*, and the results of this study indicate that *GcJAM-A* could be a receptor for GCRV.

Acknowledgements This work was supported by the National Natural Science Foundation of China (No. 31101922) and the '863' High Technology Project (No. 2011AA100403).

References

[1] Qiya Z, Hongmei R, Zhenqiu L, Jing Z, Jianfang G. Detection of grass carp hemorrhage virus (GCHV) from Vietnam and comparison with GCHV strain from China. High Technol Lett 2003; 2: 001.

[2] Barton ES, Forrest JC, Connolly JL, Chappell JD, Liu Y, Schnell FJ, *et al.* Junction adhesion molecule is a receptor for reovirus. Cell 2001; 104: 441-51.

[3] Forrest JC, Dermody TS. Reovirus receptors and pathogenesis. J Virol 2003; 77: 9109-15.

[4] Liu Y, Nusrat A, Schnell FJ, Reaves TA, Walsh S, Pochet M, *et al.* Human junction adhesion molecule regulates tight junction resealing in epithelia. J Cell Sci 2000; 113: 2363-74.

[5] Del Maschio A, De Luigi A, Martin-Padura I, Brockhaus M, Bartfai T, Fruscella P, *et al.* Leukocyte recruitment in the cerebrospinal fluid of mice with experimental meningitis is inhibited by an antibody to junctional adhesion molecule (JAM). J Exp Med 1999; 190: 1351-6.

[6] Martin-Padura I, Lostaglio S, Schneemann M, Williams L, Romano M, Fruscella P, *et al.* Junctional adhesion molecule, a novel member of the immunoglobulin superfamily that distributes at intercellular junctions and modulates monocyte transmigration. J Cell Biol 1998; 142: 117-27.

[7] Prota AE, Campbell JA, Schelling P, Forrest JC, Watson MJ, Peters TR, *et al.* Crystal structure of human junctional adhesion molecule 1: implications for reovirus binding. Proc Natl Acad Sci U S A 2003; 100: 5366-71.

[8] Kostrewa D, Brockhaus M, D'Arcy A, Dale GE, Nelboeck P, Schmid G, *et al.* Xray structure of junctional adhesion molecule: structural basis for homophilic adhesion via a novel dimerization motif. EMBO J 2001; 20: 4391-8.

[9] Chappell JD, Prota AE, Dermody TS, Stehle T. Crystal structure of reovirus attachment protein sigma1 reveals

evolutionary relationship to adenovirus fiber. EMBO J 2002; 21: 1-11.

[10] Forrest JC, Campbell JA, Schelling P, Stehle T, Dermody TS. Structuree-function analysis of reovirus binding to junctional adhesion molecule 1. Implications for the mechanism of reovirus attachment. J Biol Chem 2003; 278: 48434-44.

[11] Mertens P. The dsRNA viruses. Virus Res 2004; 101: 3-13.

[12] Su J, Zhu Z, Wang Y, Zou J, Wang N, Jang S. Grass carp reovirus activates RNAi pathway in rare minnow, *Gobiocypris rarus*. Aquaculture (Amsterdam, Netherlands) 2009; 289: 1-5.

[13] Yang C, Su J, Huang T, Zhang R, Peng L. Identification of a retinoic acidinducible gene I from grass carp (*Ctenopharyngodon idella*) and expression analysis *in vivo* and *in vitro*. Fish Shellfish Immunol. 30: 936-943.

[14] Yeung D, Manias JL, Stewart DJ, Nag S. Decreased junctional adhesion molecule-A expression during blood-brain barrier breakdown. Acta Neuropathol 2008; 115: 635-42.

[15] Ueki T, Iwasawa K, Ishikawa H, Sawa Y. Expression of junctional adhesion molecules on the human lymphatic endothelium. Microvasc Res 2008; 75: 269-78.

[16] Aurrand-Lions M, Johnson-Leger C, Wong C, Du Pasquier L, Imhof BA. Heterogeneity of endothelial junctions is reflected by differential expression and specific subcellular localization of the three JAM family members. Blood 2001; 98: 3699-707.

[17] KL T. Mammalian reoviruses. In: Fields BNKD, Howley PM, editors. Fields Virology. Philadelphia: LippincottRaven Publisher; 2001. p. 1729-47.

[18] Tardieu M, Powers ML, Weiner HL. Age dependent susceptibility to Reovirus type 3 encephalitis: role of viral and host factors. Ann Neurol 1983; 13: 602-7.

[19] Thomas FC, Sheth B, Eckert JJ, Bazzoni G, Dejana E, Fleming TP. Contribution of JAM-1 to epithelial differentiation and tight-junction biogenesis in the mouse preimplantation embryo. J Cell Sci 2004; 117: 5599-608.

[20] Shao M, Ghosh A, Cooke VG, Naik UP, Martin-DeLeon PA. JAM-A is present in mammalian spermatozoa where it is essential for normal motility. Dev Biol 2008; 313: 246-55.

[21] Babinska A, Kedees MH, Athar H, Ahmed T, Batuman O, Ehrlich YH, *et al.* F11-receptor (F11R/JAM) mediates platelet adhesion to endothelial cells: role in inflammatory thrombosis. Thromb Haemost 2002; 88: 843-50.

[22] Liang TW, DeMarco RA, Mrsny RJ, Gurney A, Gray A, Hooley J, *et al.* Characterization of huJAM: evidence for involvement in cell-cell contact and tight junction regulation. Am J Physiol Cell Physiol 2000; 279: C1733-43.

[23] Clarke P, Meintzer SM, Gibson S, Widmann C, Garrington TP, Johnson GL, *et al.* Reovirus-induced apoptosis is mediated by TRAIL. J Virol 2000; 74: 8135-9.

[24] Samuel CE. Antiviral actions of interferons. Clin Microbiol Rev 2001; 14: 778-809 [table of contents].

[25] Zhang YB, Jiang J, Chen YD, Zhu R, Shi Y, Zhang QY, *et al.* The innate immune response to grass carp hemorrhagic virus (GCHV) in cultured *Carassius auratus* blastulae (CAB) cells. Dev Comp Immunol 2007; 31: 232-43.

[26] Smith PL, Lombardi G, Foster GR. Type I interferons and the innate immune response-more than just antiviral cytokines. Mol Immunol 2005; 42: 869-77.

[27] Malmgaard L. Induction and regulation of IFNs during viral infections. J Interferon Cytokine Res 2004; 24: 439-54.

[28] DJ K. Reovirus-induced apoptosis requires both death receptor- and mitochondrial-mediated caspase-dependent pathways of cell death. Cell Death Differ 2002; 9: 926-33.

[29] Nygaard R, Husgard S, Sommer A-I, Leong J-AC, Robertsen B. Induction of Mx protein by interferon and double-stranded RNA in salmonid cells. Fish Shellfish Immunol 2000; 10: 435-50.

[30] Saint-Jean SR, Perez-Prieto SI. Effects of salmonid fish viruses on Mx gene expression and resistance to single or dual viral infections. Fish Shellfish Immunol 2007; 23: 390-400.

[31] Lacour S, Hammann A, Wotawa A, Corcos L, Solary E, Dimanche-Boitrel MT. Anticancer agents sensitize tumor cells to tumor necrosis factor-related apoptosis-inducing ligand-mediated caspase-8 activation and apoptosis. Cancer Res 2001; 61: 1645-51.

[32] Johnsen JI, Pettersen I, Ponthan F, Sveinbjornsson B, Flaegstad T, Kogner P. Synergistic induction of apoptosis in

neuroblastoma cells using a combination of cytostatic drugs with interferon-gamma and TRAIL. Int J Oncol 2004; 25: 1849-57.

[33] Ruiz de Almodovar C, Lopez-Rivas A, Ruiz-Ruiz C. Interferon-gamma and TRAIL in human breast tumor cells. Vitam Horm 2004; 67: 291-318.

[34] Chou AH, Tsai HF, Lin LL, Hsieh SL, Hsu PI, Hsu PN. Enhanced proliferation and increased IFN-gamma production in T cells by signal transduced through TNF related apoptosis-inducing ligand. J Immunol 2001; 167: 1347-52.

[35] Shin EC, Ahn JM, Kim CH, Choi Y, Ahn YS, Kim H, *et al*. IFN-gamma induces cell death in human hepatoma cells through a TRAIL/death receptor-mediated apoptotic pathway. Int J Cancer 2001; 93: 262-8.

[36] Kuhen KL, Samuel CE. Isolation of the interferon-inducible RNA-dependent protein kinase Pkr promoter and identification of a novel DNA element within the 5′-flanking region of human and mouse Pkr genes. Virology 1997; 227: 119-30.

[37] Kuhen KL, Samuel CE. Mechanism of interferon action: functional characterization of positive and negative regulatory domains that modulate transcriptional activation of the human RNA-dependent protein kinase Pkr promoter. Virology 1999; 254: 182-95.

草鱼 *JAM-A* 基因的克隆及其功能的初步研究

杜富宽[1,2] 苏建国[3] 黄 容[1] 廖兰杰[1] 朱作言[1] 汪亚平[1]

1 中国科学院水生生物研究所，武汉 430072

2 中国科学院大学，北京 100049

3 西北农林科技大学，动物科技学院，杨凌 712100

摘 要 在中国和东南亚国家，草鱼是重要的水产养殖品种。然而，草鱼养殖过程中疾病频繁暴发，导致巨大的经济损失。草鱼呼肠孤病毒(GCRV)引起的草鱼出血病是最严重的疾病之一。连接黏附分子 A(JAM-A)是哺乳动物呼肠孤病毒的受体，在哺乳动物研究较多，尚未在草鱼报道。本研究中，我们克隆了草鱼 *JAM-A* 基因(*GcJAM-A*)并研究了其在草鱼出血病发病过程中的功能。*GcJAM-A* 基因由 10 个外显子和 9 个内含子组成，cDNA 全长 1833 bp，包含 888 bp 开放阅读框，编码 295 个氨基酸，预测 *GcJAM-A* 蛋白质包含一个典型的跨膜域。观察到 *GcJAM-A* 在早期的胚胎发生呈母系表达模式，而受精卵的表达发生在孵化后 8 h。*GcJAM-A* 在鳃、肝、肠和肾脏中表达强烈，而在血液、脑、脾脏和头肾中表达较低。此外，与 6 月龄草鱼相比，在 30 月龄中观察到其在鳃、肝、肠、脑、脾脏和肾脏低表达。敲除 *GcJAM-A* 的细胞系(CIK)感染 GCRV 后，参与干扰素和细胞凋亡通路的基因表达明显受到抑制。实验结果表明 *GcJAM-A* 可能是 GCRV 的受体。

RNA-Seq Profiles from Grass Carp Tissues After Reovirus (GCRV) Infection Based on Singular and Modular Enrichment Analyses

Mijuan Shi[1,2] Rong Huang[1] Fukuan Du[1] Yongyan Pei[1]
Lanjie Liao[1,2] Zuoyan Zhu[1] Yaping Wang[1]

1 State Key Laboratory of Freshwater Ecology and Biotechnology, Institute of Hydrobiology, Chinese Academy of Sciences, Wuhan 430072
2 University of Chinese Academy of Sciences, Beijing 100039

Abstract Hemorrhagic disease of the grass carp, *Ctenopharyngodon idella*, is a fatal disease in fingerlings and yearlings caused by a reovirus, GCRV. RNA-seq data from four diseased grass carp tissues (gill, intestine, liver and spleen) were obtained at 2 h before and six times after (2 h, 24 h, 48 h, 72 h, 96 h and 120 h) GCRV challenge. A total of 7.25 ± 0.18 million (M) clean reads and 3.53 ± 0.37 M unique reads were obtained per RNA-seq analysis. Compared with controls, there were 9060 unique differentially expressed genes (DEGs) in the four tissues at the six time points post-GCRV challenge. Hierarchical clustering analysis of the DEGs showed that the data from the six time points fell into three branches: 2 h, 24 h/48 h, and 72 h/96 h/120 h. Singular (SEA) and modular enrichment analyses of DEGs per RNA-seq dataset were performed based on gene ontology. The results showed that immune responses occurred in all four tissues, indicating that GCRV probably does not target any tissues specifically. Moreover, during the course of disease, disturbances were observed in lipid and carbohydrate metabolism in each of the organs. SEA of DEGs based on the Kyoto Encyclopedia of Genes and Genomes database was also performed, and this indicated that the complement system and cellular immunity played an important role during the course of hemorrhagic disease. The qPCR of pooled samples of duplicate challenge experiment were used to confirm our RNA-seq approach.

Keywords *Ctenopharyngodon idella*; hemorrhagic disease; innate immunity; complement; cellular immunity; metabolic disturbances

1. Introduction

Grass carp (*Ctenopharyngodon idella*) is an important economic fish species in China. Fingerlings and yearlings are predisposed to a hemorrhagic disease caused by the grass carp reovirus (GCRV), a double-stranded RNA virus belonging to the genus *Aquareovirus*

文章发表于 *Molecular Immunology*, 2014, 61(1): 44-53

(Subramanian *et al.*, 1997; Qiu *et al.*, 2001). Although GCRV is the leading cause of death in grass carp (Wang *et al.*, 2012), the focuses of GCRV research at present are on genome sequencing, antibody production and more sensitive methods to detect virus particles. The common symptoms of hemorrhagic disease are hemorrhages in muscles and gills and enteritis, and this disease is divided into three categories based on the different symptoms. Most studies of this disease have examined the functions of several immune associated genes, especially pattern recognition receptors (Chen *et al.*, 2012; Su *et al.*, 2012; Yang *et al.*, 2012a,b). However, systematic understanding is needed for the changes that occur at the molecular level in different tissues of grass carp during hemorrhagic disease.

Innate immunity in fish is investigated mainly because of its importance in the early stages of embryogenesis and of the fundamental defense mechanisms against pathogens (Uribe *et al.*, 2011; Magnadottir, 2006; Rombout *et al.*, 2005). In fact, fish occupy an apparent crossroads between the innate immune response and the evolution of the adaptive immune response (Tort *et al.*, 2003). The immune system of teleosts is very similar to higher vertebrates, not at the level of organs but at the molecular level (Tort *et al.*, 2003; Holland and Lambris, 2002). Some molecules related to adaptive immunity, such as IgM (Acton *et al.*, 1971), IgD (Wilson *et al.*, 1997) and CD8 (Fischer *et al.*, 2006), were identified in fish decades ago.

The complement system is an ancient mechanism of defense (Zhu *et al.*, 2005) and it is an important component of both the innate and adaptive immune systems. The complement system of teleost fish consists of three pathways, making it similar to of the system found in higher vertebrates (Boshra *et al.*, 2004; Nonaka and Smith, 2000). Many complement proteins have been isolated from various teleost species (Holland and Lambris, 2002; Dodds and Petry, 1993; Endo *et al.*, 1998; Yeo *et al.*, 1997), and it has been suggested the complement system of fish may have wider recognition functions than in mammals (Sunyer and Tort, 1995) due to the larger diversity of complement proteins (Nakao *et al.*, 2002).

Except for B cells, lymphocytes are the main players in cellular immunity, and these can be divided roughly into natural killer (NK) cells and T cells. NK cells, belonging to innate immune system, can kill target cells independent of antibodies. Activated cytotoxic T lymphocytes can induce apoptosis of target cells to prevent the production of new infectious particles (Fischer *et al.*, 2013), a process that relies on antigen-presentation. T cell activation is an important component of adaptive immunity. The major histocompatibility complex (MHC), including class I and class II MHCs, are the key molecules that present peptides to T cells (Klein, 1986). mRNA expression data from some fish species have shown that the presence of cytotoxic T lymphocytes in fish function similarly to corresponding cells in higher vertebrates (Fischer *et al.*, 2013).

Some virus infections are associated with metabolic disturbances. For instance, the human immunodeficiency virus-I can cause disorders of carbohydrate and lipid metabolism in patients (Grunfeld *et al.*, 1992). However, metabolic disturbance in grass carp infected by GCRV has not been investigated previously.

The gills and intestines are the organs that show bleeding symptoms during GCRV infection. In addition, intestine and liver tissues are susceptible to GCRV (Su *et al.*, 2008), while the spleen is an important immune-organ in this fish species (Fänge and Nilsson, 1985; Press and Evensen, 1999). Therefore, in this present study, RNA-seq sequencing was performed on these four tissues in grass carp before GCRV infection and at six time points post-challenge. The results showed that there were changes of gene expression in these four tissues during the course of GCRV infection, thus providing a basis for the systematic analysis of the response of grass carp to GCRV infection.

2. Materials and Methods

2.1 Animals

Full-sib 188-day grass carp weighing ~30 g were chosen for this study. All fish were divided into two groups: (1) 15 grass carp were used for the negative control; (2) the GCRV-infected group contained 135 grass carp; culture water and air temperatures were kept at 28℃.

One hundred full-sib 6-month grass carp were chosen for the duplicate experiment. Ten individuals were used for negative control with the rest of grass carp belonging to the infected group. Throughout the two experiments, all fish were fed once a day.

2.2 Preparation of viral inoculum and infection

Grass carps infected with GCRV and showing typical symptoms of disease (e.g., muscle bleeding) were selected and the internal organs were discarded. A volume of 0.7% saline three times the mass of the remaining fish was added and a tissue homogenate was prepared by mixing on a shaker at 28℃, 140 rpm for 30 min to release the virus particles. Then the homogenate was centrifuged at 4000×*g* for 30 min. The supernatant, considered to be the GCRV stock, was stored at –70℃.

The virus stock was recovered at 28℃ for 2 h and diluted 10-fold to give a working solution. Then, two antibiotics (streptomycin and penicillin) were added to a final concentration of 100 U/ml. All three groups of grass carps were immersed in 8% NaCl for 2 min, and then groups 2 and 3 were bathed with the virus working solution (containing antibiotics) for 30 min. Finally, all three groups were moved into aerated water at 28℃ containing 100 U/ml of streptomycin and penicillin.

The gills, intestines, livers and spleens were harvested from the 15 grass carps in group 1, and each tissue was pooled and cryopreserved at –80℃. These samples were used for the control profiles. These four tissues were harvested from 15 grass carps in group 2 at 2 h after GCRV infection. Thereafter, every 24 h until 120 h, the four tissues were taken from 15 individuals in group 2. Thus, including the controls, seven time points were examined in total. In the duplicate experiments, four tissues of 10 individuals were obtained and pooled respectively at the same time points as previously.

2.3 RNA extraction and construction of RNA-seq libraries

After grinding each sample in liquid nitrogen, 1 ml of Trizol reagent was added per 100 mg of tissue. All samples were stored at −80℃ until RNA extraction was performed following a standard protocol. mRNA was enriched from the total RNA using oligo(dT) magnetic beads. After addition of fragmentation buffer to generate short mRNA fragments (each of approximately 200 bp) as templates, the first strand cDNA was synthesized using random hexamer primers. Buffer, dNTPs, RNase H and DNA polymerase I were added to synthesize the second strand. The double stranded cDNA was purified with the QiaQuick PCR extraction kit and washed with EB buffer for end repair and single nucleotide adenine addition. Finally, sequencing adaptors were ligated onto the fragments. The desired fragments were purified by agarose gel electrophoresis and enriched by PCR amplification. The library products obtained were submitted for sequencing analysis on an Illumina HiSeq^{TM} 2000 (Wang *et al*., 2009). The accession number of the raw data is SRA099702.

2.4 Data processing and statistical analysis

After eliminating non-useful data by fast QC, the remaining clean tags were mapped to the reference genome (Bioproject: PRJNA39737; unpublished data) and the reference gene set using SOAPaligner/soap2 (Li *et al*., 2009). The number of tags that mapped to genes was recorded. Expression levels of genes were determined according to the reads per kb per million reads method (Mortazavi *et al*., 2008). Sequences were Blast searched, and mapped genes with an e-value $\leqslant 10^{-6}$ were annotated.

2.5 Identification of differentially expressed genes (DEGs) and cluster analysis

Gene expression profiles at the six time points post-infection were compared to the profile at 0 h for each of the corresponding tissues. Significant DEGs were defined and identified according to the method of Audic and Claverie (1997). The criteria for the data in this present study were a false discovery rate (FDR) value of $\leqslant 0.001$ and an absolute log2 ratio value of $\geqslant 1$, where the ratio is the ratio of gene expression at a non-zero time point profile relative to the corresponding control. MeV software (Saeed *et al*., 2003) was used for hierarchical clustering using Euclidean distance as the metric.

2.6 Enrichment of gene ontology (GO) terms and pathways

Traditional singular enrichment analysis (SEA) was used for the enrichment analysis of GO terms and pathways. The enrichment *P*-value calculation was performed with the Fisher's exact test. TopGO (Alexa *et al.*, 2006; Alexa and Rahnenfuhrer, 2010) was used for modular enrichment analysis (MEA) and drawing term-to-term relationships.

2.7 Validation of expression profiles using qPCR

Based on the RNA-seq data, four genes which significantly differentially expressed at not less than two time points in one tissue were selected for the qPCR. The length of PCR products was about 75-150 bp and T_m-55℃ (Primer list in Table S4).

3. Results

3.1 Preliminary statistical analysis of RNA-seq data

The expression profiles of gill, intestine, liver and spleen tissues from grass carp were obtained at 2h before infection with GCRV and at six time points after challenge. Analysis of these 28 RNA-seq datasets gave 203 million (M) reads in total. After qualification of the raw data, we obtained 7.25±0.18 M clean reads and 3.53±0.37 M unique reads per RNA-seq that aligned uniquely to the reference genes (Table S1). The number of mapped genes for every RNA-seq dataset reached saturation levels and all the expression profiles were qualified by the essential parameters defined in the Section 2. The sum of mapped genes from the seven time points differed across the four tissues examined (Table 1). Significant differences in the total number of mapped genes between any two tissues were tested by one-way analysis of variance. The liver showed significantly lower numbers of mapped genes compared with the

Table 1 The number of mapping genes in all profiles

Timepoint	Gill (×1000)	Intestine (×1000)	Liver (×1000)	Spleen (×1000)
Control	20.34	20.14	16.99	19.59
Day0	20.00	20.05	16.55	19.41
Day1	19.93	20.08	17.22	19.57
Day2	19.81	20.05	17.02	19.37
Day3	19.49	19.93	16.97	19.02
Day4	19.77	19.76	16.84	18.85
Day5	19.86	19.77	17.04	18.76
Ave	19.89	19.97	16.95	19.22
St.	0.26	0.15	0.21	0.34
Total	22.77	22.68	20.45	22.27

other three tissues ($P < 0.01$), which is a phenomenon similar to observe for *Pagothenia borchgrevinki* (Bilyk and Cheng, 2013). Furthermore, the spleen differed significantly from the other tissues ($P < 0.05$); however, there was no significant difference between the gill and intestine tissues.

3.2 Selection and initial analysis of significant DEGs

The expression level of genes at every time point after virus challenge was compared with the control (i.e., 2 h before infection) for each tissue type. DEGs were defined and identified based on digital gene expression profiles as described previously (Audic and Claverie, 1997). As a result, we obtained 9060 unique DEGs after removing redundant DEGs. The number of DEGs in each tissue showed temporal variations. During 72-120 h after infection, the number of DEGs sharply increased in all organs except the liver, while DEGs in the liver increased sharply at day 1 (Fig. 1). In the gills, the number of DEGs remained

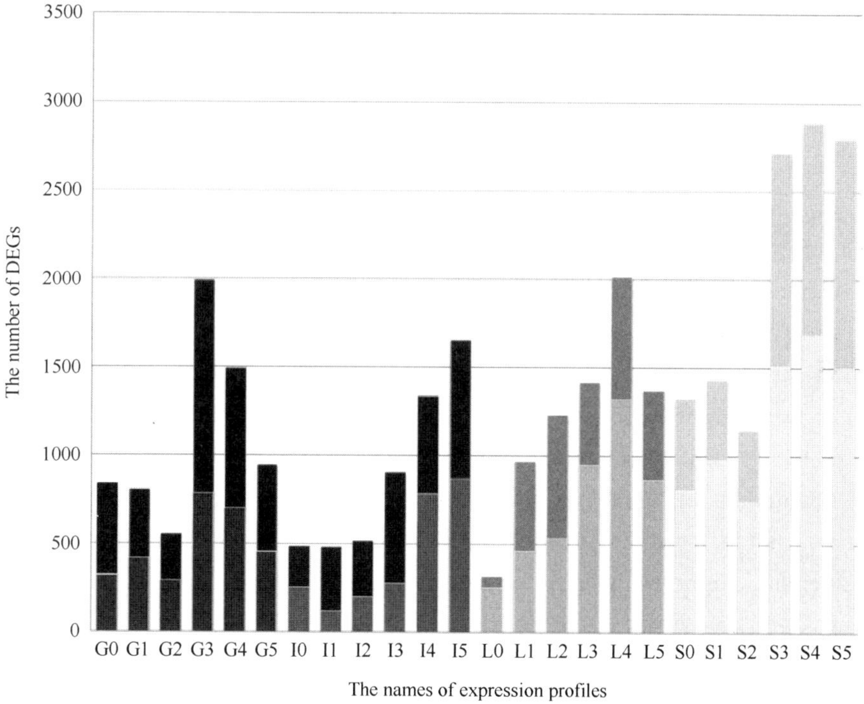

Fig. 1 Number of DEGs in GCRV-challenge profiles. The name of the abscissa consists of a letter and a numeral. The letter corresponds to the tissue (G, gills; I, intestine; L, liver; S, spleen), while the numeral indicates the time point (values between 0 and 5 correspond to 2 h, 24 h, 48 h, 72 h, 96 h and 120 h, respectively). This naming convention is the same as found in all tables and figures. Different shades of color mark different tissues and red bars indicate the number of significantly up-regulated DEGs while green bars indicates significantly down-regulated DEGs. (For interpretation of the references to color in this figure legend, the reader is referred to the web version of the article.)

relatively stable from 2 to 48 h and then increased sharply at 72 h, before decreasing gradually until levels were similar at 120h as at 2 h. In the intestines, the number of DEGs was stable and similar to the gills during the first three time points but then increased sharply and continued to increase between 72 and 120 h. In the liver, as mentioned above, the number of DEGs increased sharply at 24 h and these continued increasing until the peak value was observed at 96 h. In the spleen, the number of DEGs was relatively stable between 2 and 48 h, but then increased sharply before stabilizing again.

The ratio of DEGs is equal to the number of DEGs divided by the number of mapped genes under the same background. The ordination of the mean number of mapped genes at the six time points in each tissue was: intestine (19.97 k) > gill (19.89 k) > spleen (19.22 k) > liver (16.95 k). Interestingly, the order of the mean ratio of DEGs at all of the time points in the four tissues was considerably different: spleen (10.75%) > liver (7.18%) > gill (5.59%) > intestine (4.51%).

3.3 Hierarchical clustering of the different tissues

We performed hierarchical cluster analysis of the log2 ratios of the 9060 significant DEGs in each tissue at each time point using the MeV software (Saeed *et al.*, 2003) with Euclidean distance as the metric. The 24 processed profiles were divided into four classes corresponding to the four tissues. According to the cluster results, the profiles of the gills and intestines were most similar amongst the four tissues. The profile of the spleen was closer to the gill and intestine profiles than the liver (Fig. 2).

The profiles of the four tissues at every time point were mostly approximate except for slight differences in the liver at 2 h and intestine at 48h. For the liver, the six time points divided into two categories: 2 h/24 /48 h belonged to a cluster while 72 h/96 h/120 h profiles fell into another branch; however, for the other organs a branch fork also fell between the 2 h and 24 h time points. Excluding the 0 h control data, there were two branches in the trees of the other time points for all the organs except the intestine: 24 h/48 h and 72 h/96 h/120 h. In the intestine, the 72 h profile was closer to the 48 h profile rather than the 96 h profile. After integrating all of the cluster analyses, the time points after challenge broadly divided into three categories: 2 h, 24 h/48 h, and 72 h/96 h/120 h.

3.4 Analysis of tissue specificity of gene expression

Among the 24,072 mapped genes (Table S2), some were tissue-specific but most were non-tissue specific. The significant DEGs were divided into 15 categories according to their distribution patterns in the four tissues (based on the Venn diagram in Fig. S1). Then the chi-square test was applied to the numbers of significant and non-significant DEGs in each category (Table 2). All of the *P*-values were far less than 0.001, which indicated that regardless of the category, significant DEGs were not identified due to random sampling. Nearly 97% of DEGs were expressed in all four tissues and only 3% of DEGs showed some kind of tissue specificity. This indicates that GCRV likely causes a multi-organ disease rather than a tissue-specific one, although obvious symptoms are only reported in the gills and intestines (Su *et al.*, 2011, 2009; Zhang *et al.*, 2010).

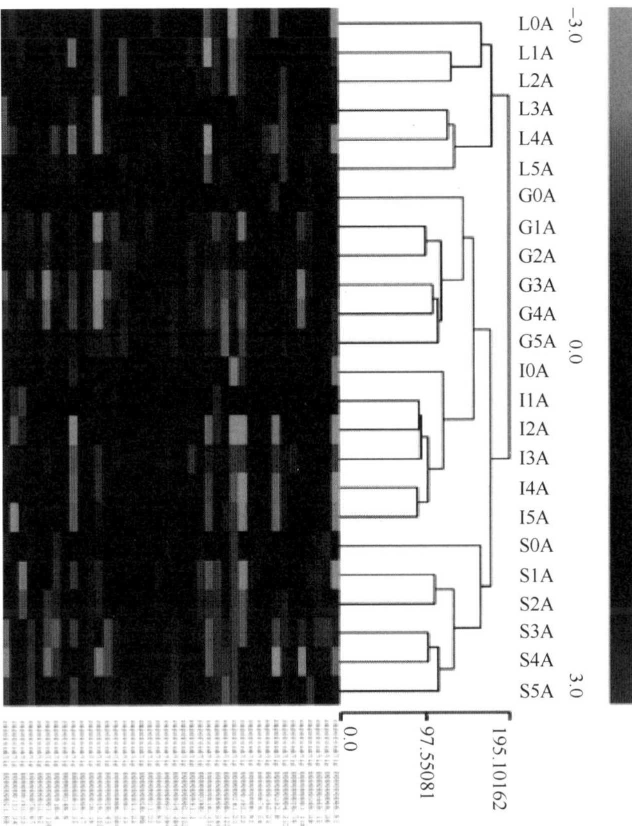

Fig. 2 Hierarchical cluster analysis of significant DEG expression profiles from different tissues at the six time points

3.5 SEA of GO terms

After obtaining the GO functional annotations of the significant DEGs, a classic SEA of GO process ontology was performed for every RNA-seq dataset to determine the significantly enriched GO terms with the Fisher's exact test (P <0.05, FDR adjustment). Statistics were obtained on the terms from component, function and process ontology (Table S3). In total, 56 unique component terms, 57 function terms and 138 process terms were enriched. At the level of component ontology, 38 enriched terms appeared in the liver and this was the greatest number for a tissue, while the minimum number was just one and this was in the spleen samples. At the level of functional ontology, the most enriched terms were recognized in the spleen while the least enrichment was in the gills.

Table 2 Chi-square tests of significantly and non-significantly expressed genes in different tissues

Tissue	Expr.		Sig.*		Non-sig.	*P*-value
	Num.	Perc.	Num.	Perc.		
G	481	2.00%	8	0.09%	473	2.12E−78
I	336	1.40%	5	0.06%	331	6.96E−86
L	72	0.30%	0	0.00%	72	3.13E−96
S	257	1.07%	2	0.02%	255	6.41E−90
G&I	542	2.25%	39	0.43%	503	1.11E−78
G&L	56	0.23%	3	0.03%	53	1.21E−101
G&S	309	1.28%	7	0.08%	302	1.29E−87
I&L	85	0.35%	0	0.00%	85	1.49E−99
I&S	252	1.05%	13	0.14%	239	1.55E−91
L&S	71	0.29%	1	0.01%	70	1.67E−100
G&I&L	230	0.96%	28	0.31%	202	1.24E−94
G&I&S	1452	6.03%	134	1.48%	1318	4.16E−45
I&L&S	227	0.94%	29	0.32%	198	6.26E−95
G&L&S	144	0.60%	5	0.06%	139	9.71E−97
G&I&L&S	19,558	81.25%	8786	96.98%	10,772	0.00E+00
Total	24,072	100%	9060	100.00%	15,012	

Tissue: G, gills; I, intestines; L, liver; S, spleen. Data represent DEGs found in all tissues listed. For example, gene data for "G&L" means that the genes were expressed both in gills and liver, excluding genes only expressed in one of the two tissues. Expr: the number of genes expressed in the category. Sig*: number of significant DEGs in the category. Non-sig.: number of non-significant DEGs in the category

Further analysis of 138 unique process terms was performed (Table 3). According to the curves fitted to the enriched GO term numbers for each tissue, the infection appeared at 48h except in the liver (Fig. 3). These results are consistent with the initial analysis of the significant DEGs and hierarchical clusters.

Table 3 Some of the enriched metabolism-related pathways*

	Gills	Intestine	Liver	Spleen	
Metabolic pathways	386	404	537	613	Metabolism
Glycolysis/Gluconeogenesis	29	41	38	48	
Cysteine and methionine metabolism	21	19	20	28	
Steroid biosynthesis	14	15	23	20	
Steroid hormone biosynthesis	24	24	28	35	
PPAR signaling pathway	42	55	56	62	Signaling pathway related to metabolism
Adipocytokine signaling pathway	32	49	48	46	
Insulin signaling pathway	63	55	70	95	

*The color represents the *P*-value range: green, greater than 5E−2; light pink, 5E−2-1E−3; pink, 1E−3-1E−5; red, 1E−5-1E−10; dark red, less than 1E−10

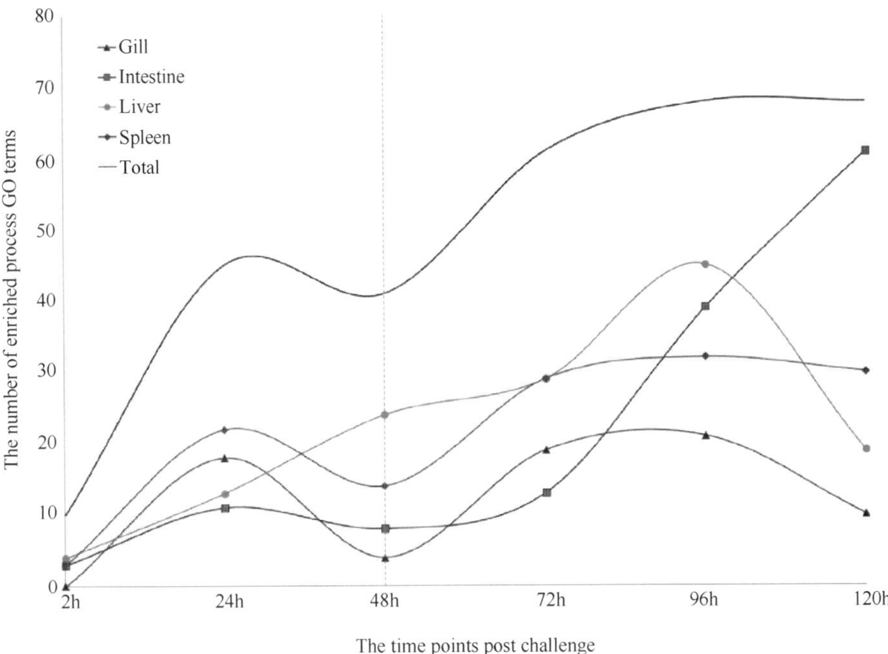

Fig. 3 The number of enriched process GO terms in different tissues at the six time points

The 138 GO terms were divided into four categories: responses to stimuli and the immune system; transport and location; the cell cycle; and metabolism. A total of 83 enriched terms were related to metabolism, and this included the synthesis and metabolism of carbohydrates, lipids, amino acids and ubiquitin. DNA replication, mismatch repairing and the synthesis of macromolecules were all attributed to the cell cycle and this group consisted of 21 GO terms. Finally, there were 18 GO terms related to responses to stimuli and the immune system (Fig. 4).

3.6 MEA of interested GO terms

The network of 138 unique GO terms was huge and it was difficult to build the network for all of these and, as described previously, the hemorrhagic disease of grass carp likely affects all tissues. Thus, the 18 GO terms that were enriched in all four tissues (ignoring time points) were collected for MEA (Fig. 5). This set of GO terms was divided into four categories: response to biotic stimuli (class a); small molecule metabolism (class b); lipid and steroid metabolism (class c); and carbohydrate metabolism (class d). From the only infection-related class (class a), all four tissues responsed to the GCRV invasion. Similar to the SEA data, the metabolism-related terms were still the majority of all-tissue-enriched terms (Hotamisligil, 2006). GO terms from classes b, c and d were closely associated with metabolism, which suggests that during the course of hemorrhagic disease, infected fish suffer metabolic disturbances at least in the gills, intestines, liver and spleen.

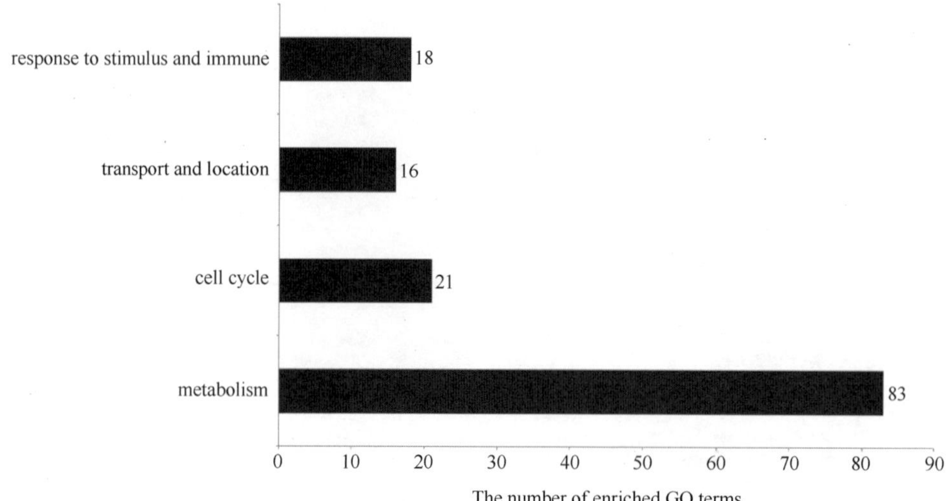

Fig. 4 Classification of all enriched process GO terms

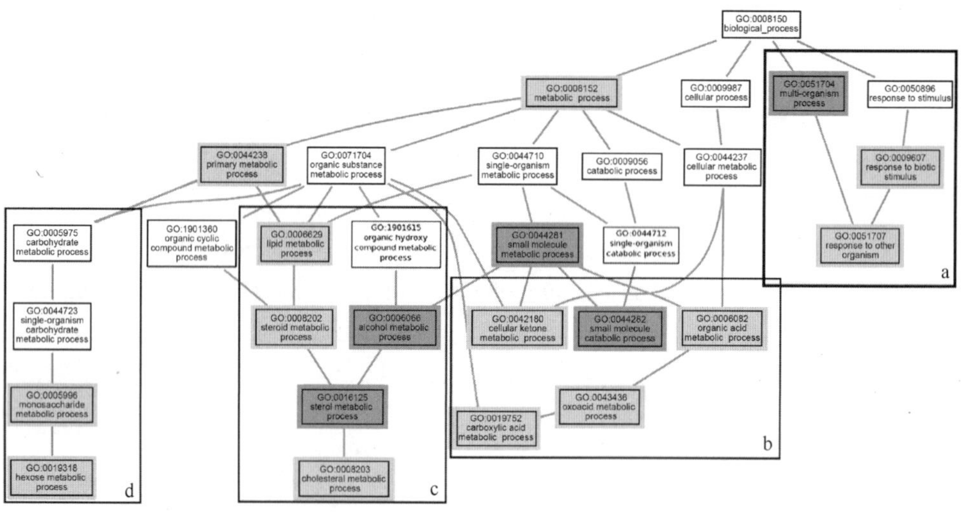

Fig. 5 Graphical view of enriched function GO terms in four grass carp tissues during the course of GCRV infection. The red box marks the category for responses to stimuli, while black boxes mark the other three categories: small molecule metabolism; lipid and steroid metabolism; and carbohydrate metabolism. The color represents the P-value range: blue, 1E−2-5E−2; pink, 1E−4-1E−2; yellow, 1E−4-1E−6. (For interpretation of the references to color in this figure legend, the reader is referred to the web version of the article.)

3.7 SEA of pathways

As the DEGs in the same pathway have their own time sequence and so SEA of a single time point may not work well. Thus, the DEGs of all the time points for each tissue were amalgamated and the unique ones were saved for SEA of pathways in order to obtain the pathways that were enriched during the whole course of infection. In total, 94 pathways were enriched in the four tissues (Fig. 6), and 58.51% of these were associated with metabolism.

This is in accordance with the GO analysis results. In the category of pathogen and disease, there were six pathways related to pathogen invasion. The measles virus pathway was the only one that was enriched in every tissue with a very highly significant *P*-value (*P*-value < E−10).

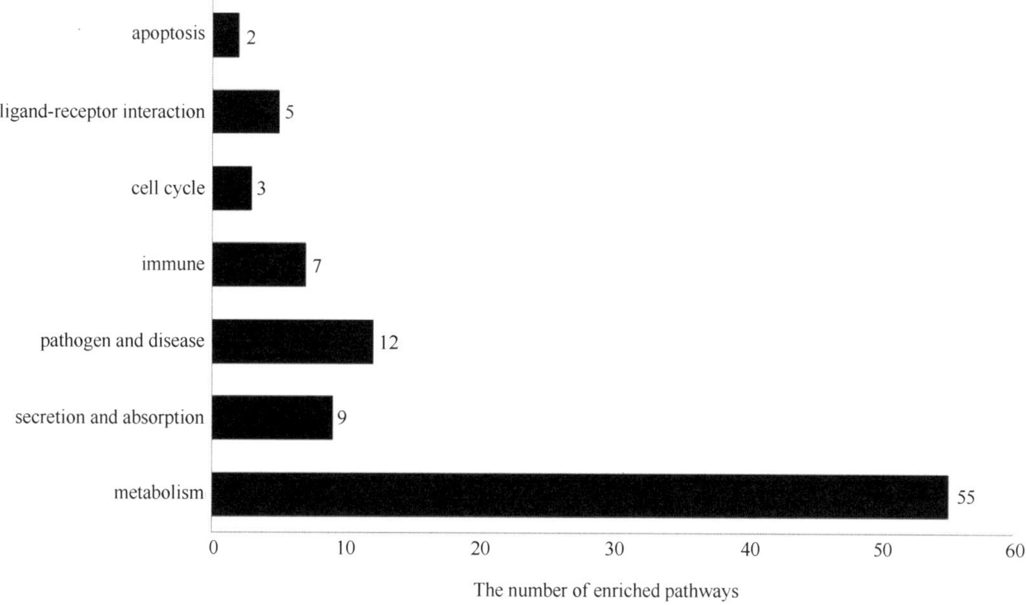

Fig. 6 Classification of all enriched pathways

Among the 55 enriched metabolism-related pathways, five were enriched in all four tissues and these were involved in glycolysis, cysteine metabolism and steroid/steroid hormone biosynthesis. Only three signaling pathways were enriched and these were involved in peroxisome proliferator-activated receptor (PPAR), adipocytokine and insulin, which are all related to lipid/glucose metabolism (Table 3).

In the category of immunity, three pathways associated with immunity, proteasome system and three organelles (peroxisome, lysosome and phagosome) were obtained (Table 4). There was no pathway enriched in all of the tissues. The enrichment of the complement system and lysosome appeared in three tissue types. As an immune organ of teleosts, 75% of immunity-related pathways were significantly enriched in the spleen, while the intestines showed lowest enrichment of these pathways. Humoral immune pathways were not enriched perhaps due to the sampling times that occurred between 2 and 120 h after virus challenge.

3.8 qPCR of multi-time-points DEGs

The pooled samples from biological repeated challenge experiments were used for qPCR validation. For each organ, we selected one differentially expressed gene which were agouti-related protein (AGRP) in gill, claudin b in spleen, influenza virus NS1A-binding

Table 4 Enriched immune-related and apoptosis pathways*

	Gills	Intestine	Liver	Spleen	
Complement and coagulation cascades	61	63	46	91	Immune response
Antigen processing and presentation	51	28	53	54	
Natural killer cell mediated cytotoxicity	41	38	42	83	
Proteasome	7	16	19	25	Vesicles
Lysosome	59	39	85	84	
Peroxisome	18	33	42	43	
Phagosome	79	77	85	127	

* The color represents the *P*-value range: green, greater than 5E−2; light pink, 5E−2-1E−3; pink, 1E−3-1E−5; red, 1E−5-1E−10; dark red, less than 1E−10

protein homolog B (NS1BB) in intestine and claudin 5 in liver. The qPCR results of seven time points of the four DEGs in related organs were obtained. As Fig. 7 shows that the expression trends of the four genes obtained by qPCR and RNA-seq were basically consistent. Obviously, the expressed trend of NS1BB in intestine seems to be almost consistent between aPCR and RNA-seq results. Although the result reveals one different data point (72h) of claudin b in spleen between two experiments, the rest data from qPCR and RNA-seq appears congruent. Similarly, each of both AGRP in gill and claudin 5 in liver shows the same expression trend in the duplicated experiment except a slight time axis misalignment. In consideration of the facts that the course of disease in two repeated experiments maybe different, the sight misalignment of the time axis can be ignored. Consequently, we consider the results of qPCR confirm the validity of the analyze of the RNA-seq.

4. Discussion

4.1 Metabolic disturbances during hemorrhagic disease of grass carp

GCRV can infect nearly all organs of the grass carp and cause hemorrhagic disease that likely affects many organs in the fish. The function of the liver could be disturbed during the course of disease. In our data, the mean ratio of DEGs at the six time points in liver was similar to that of the spleen, but this is inconsistent with the total mapped genes where the liver had the lowest value of these. The results indicate that some biological reactions may have become active during the period between 2 and 24 h after virus infection and this lasted for several days in the liver, which is an organ important for lipid and glucose metabolism. Metabolic disturbance, especially of lipid and glucose metabolism, appeared not only in the liver but also in the three other tissues. This indicates that disturbance of metabolism may be one reason underlying the death of infected fish.

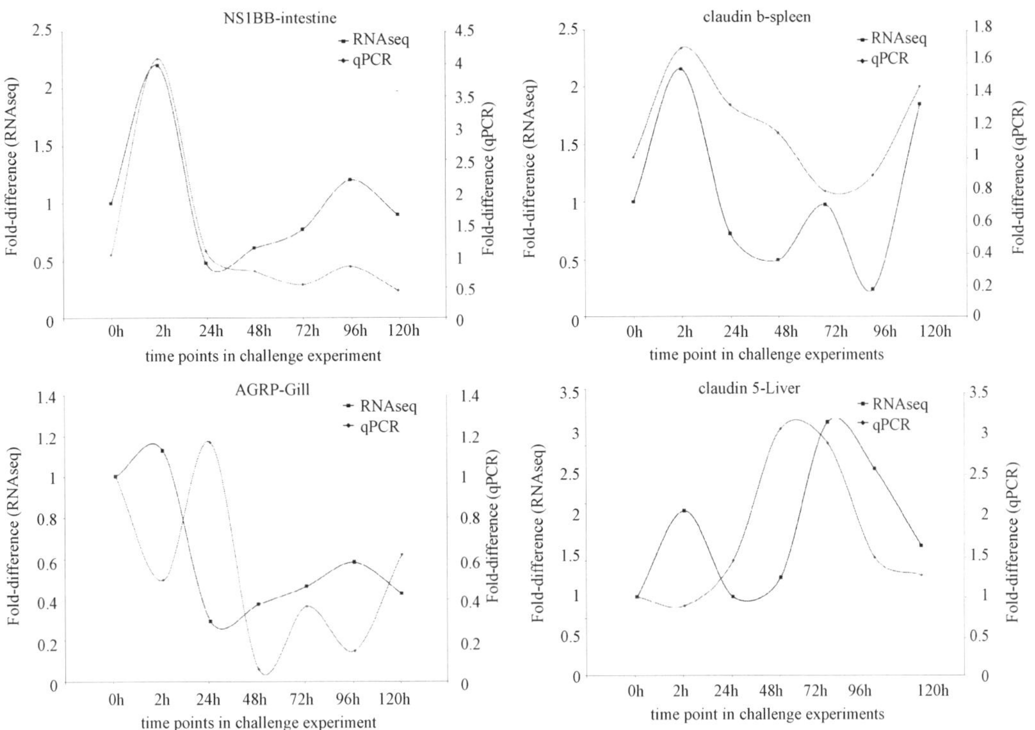

Fig. 7 The fitted curves of fold-difference of four selected genes using RNA-seq and qPCR results respectively

In our data, metabolic pathways that were enriched in all tissues at a very high level of significance (P-value < 10E−10) were involved in almost all metabolism-related pathways. The enrichment of these pathways may confirm that metabolic disturbances occur during the course of hemorrhagic disease. Glycolysis is one of the important parts of glucose metabolism, and this not only generates a small amount of energy for the body but it also provides a number of important precursor metabolites for other metabolic pathways. Gluconeogenesis is a synthesis pathway that creates glucose from non-carbohydrate precursors (Michal, 1999; Invalid, 2014). This pathway is related to demand for large amounts of energy that cannot be satisfied by normal glucose metabolic pathways. Methionine is one of the essential amino acids that can not be synthesized in animals. Some studies have shown that the level of methionine can affect the weight of the liver and intestines and the length of the latter during the middle growth stage of the grass carp. The appropriate level of methionine can improve growth performance and digestion and absorption capacity (Bingrong *et al.*, 2012). Steroid hormones are fat soluble hormones that play roles in maintaining homeostasis (Funder *et al.*, 1997), regulating glucose metabolism (Gupta and Lalchhandama, 2002), immunomodulation (Casto *et al.*, 2001) and ontogenetics (Holmes and Shalet, 1996).

PPARs are nuclear hormone receptors activated by fatty acids and their derivatives, and there are three subtypes in vertebrates: α, β and γ. These are all related to lipid metabolism but have different functions. PPARα plays a role in the clearance of circulating or cellular lipids and PPARβ/δ is involved in lipid oxidation and cell proliferation. Meanwhile, PPARγ is involved in gluconeogenesis and regulation of blood glucose concentration (Takahashi *et al.*, 2005; Desvergne and Wahli, 1999; Savage, 2005). In our data, the PPAR signaling pathway was enriched in three organs, except the gills which are known to lack lipids. PPARγ was the only PPAR not defined as a DEG. According to previous research, the expression of PPARγ is tissue-specific and is expressed only at a low level. Even though it appeared to be differentially expressed in some of our RNA-seq data, the low read counts from this gene would cause a large FDR value (>0.01), meaning that it would then be excluded from DEGs. However, based on the significant enrichment of the gluconeogenesis pathway, it is difficult to judge whether PPARγ plays an important role in the hemorrhagic disease.

The adipocytokine pathway is activated by the recognition of adiponectin, leptin or TNFα. Leptin is primarily an important regulator of energy intake and metabolic rate (Frederich *et al.*, 1995). When leptin activates the adipocytokine pathway, several neuropeptides are released that bring about an anorectic effect. The known neuropeptides are proopiomelanocortin (α-MSH), neuropeptide Y (NPY) and AGRP which also exist in teleosts (Munzberg and Myers, 2005; Gonzalez-Nunez *et al.*, 2003; Breton *et al.*, 1991; Cerdá-Reverter and Peter, 2003). In our profile data, NPY was only sequenced in the intestine and it showed a significant downregulation of expression, while AGRP was only detected in the gills and it was also down-regulated. The similar expression patterns of these two neuropeptides may be related to the increase in food intake and tissue-specific expression of these two genes may suggest that different neuropeptides are needed for different organs in the adipocytokine pathway (Servin *et al.*, 1989). The downstream molecules in the adiponectin-induced pathway are GLUT1/4 (an MFS transporter, member 1/4) and carnitine Opalmitoyltransferase 1. The former is related to gluconeogenesis while the latter is related to lipid catabolism. Both of these enzymes were up-regulated in all four tissues during the course of disease perhaps to increase the uptake of glucose. TNFα can interfere with the early stages of insulin signaling to accelerate lipid catabolism (Gual *et al.*, 2005).

4.2 The enriched virus and immune pathways

Measles is a highly contagious viral disease that remains the leading vaccine-preventable cause of child mortality (Bryce *et al.*, 2005; Moss *et al.*, 2004). Infection induces profound immunosuppression that can lead to serious secondary infections (Vidalain *et al.*, 2000). Measles infection is also associated with a vigorous antibody response without the production of significant measles virus-specific delayed type hypersensitivity through the inhibition of mitogen-induced T cell proliferation (Yanagi *et al.*, 1992; Griffin and Ward, 1993). The measles pathway was the only one enriched in all tissues, and measles and GCRV both cause serious illness in non-adult individuals. Therefore, it is suspected that there are similarities between the diseases caused by GCRV and measles.

4.2.1 Complement and coagulation cascades

The complement system is a key component of innate immunity in teleosts and its activation leads to opsonization of pathogens (Sakai, 1984; Lammens *et al.*, 2000), recruitment of inflammatory and immunecompetent cells (Watts *et al.*, 2001), and the direct killing of pathogens (Harris *et al.*, 1998). The majority of complements are secreted by the liver in mammals, but it is different in fish as multiple organs are related to complement secretion. The complement cascade consists of three pathways: classical (Holland and Lambris, 2002), lectin and alternative pathways (Sakai, 1992). In our data, the enrichment of these three pathways was tissue-specific. In the gills and spleen, all three pathways were enriched and there was differential expression of downstream membrane receptors (C3AR1, C5R1, CR2 and CR1) that are related to inflammation and humoral immune responses (Prodinger, 1999; Holland and Lambris, 2004). However, the alternative pathway was enriched only in the intestine while the classical pathway was only enriched in the liver, and the C3AR1, C5R1, CR2 and CR1 receptors were not expressed differentially in these two tissues. Liver and intestine tissues are most susceptible to GCRV and this may be due to the reduced activation of complement pathways in these two tissues.

The kallikrein-kinin system is an endogenous metabolic pathway that has been detected in several teleosts (Lipke and Olson, 1990; Masini *et al.*, 1997), and it consists of blood proteins that play roles in inflammation (Haussmann and Figueroa, 2011) and blood pressure control. The downstream molecules of this pathway are plasminogen (PLG), plasminogen activator (PLAUR/uPAR) and bradykinin receptor (BDKR). Activated PLG enzymatically degrades fibrin, leading to hemolysis. PLAUR is the key receptor for coagulation and fibrinolysis and the generation of fibrinolytic activity is accompanied by a high concentration of uPAR in leukemia patients (Lopez-Pedrera *et al.*, 1997). In our profiles, PLAUR was differential expressed in all four tissues while PLG was a DEG in gill, intestine and spleen tissues. BDKR was not a DEG.

The blood coagulation pathway controls the coagulation and fibrinolysis of blood through a series of sequentially activated serine kinases. The downstream effector molecule is coagulation factor II (F2), which can interact with the coagulation factor II receptor (F2R) to activate coagulation. The F2-receptor is a G protein-coupled receptor located on the cell membrane and it plays a role in signal transmission during hemostasis. In our data, this gene was significantly up-regulated in gill and intestine tissues, which are the organs that show hemorrhagic symptoms, while it was not a DEG in the liver and spleen. That the kallikrein-kinin system and coagulation cascade were enriched may be related to the bleeding symptoms observed in grass carp because angiectasis and weakening of blood coagulation both occur in infected fish.

4.2.2 Antigen presentation and NK cells

The course of grass carp hemorrhage disease is usually 1 week but it can vary from 4 to 15 days. Thus, innate immunity is the major focus in this disease because humoral immunity will not function until at least 1 week after virus infection. From the pathways enriched in this present study, cellular immunity belonging to acquired immunity also plays a role in the disease. There are two ways of presenting an antigen to the two kinds of T lymphocytes. Cytotoxic T lymphocytes (CTLs) can kill host cells infected with pathogens by binding to the T cell receptor while CD8 binds to MHC I and this induces apoptosis of the target cell. CD4 (+) T helper cells can bind MHC II on the professional antigen-presenting cells to cause immunity responses. In our data, MHC I and MHC II were both DEGs in all tissues, but MHC I was up-regulated in the tissues while MHC II was down-regulated. The measles virus can cause prolonged immune suppression through inhibition of T helper (Th) cell proliferation (Dubois *et al.*, 2001), which is also extremely significantly enriched. Therefore, we speculate the possible existence of some mechanism that inhibits Th cells in the hemorrhagic disease caused by GCRV.

NK cells are lymphocytes of the innate immune system that are involved in early defenses against foreign surfaces. They are activated through a variety of pathways. Activated NK cells, like CTLs, can release toxic particles to target cells for apoptosis. This pathway was only enriched in the spleen.

Acknowledgements This work was supported by the National Basic Research Program of China (No. 2009CB118701), the National Natural Science Foundation of China (No. 31101922) and the '863' High Technology Project of China (No. 2011AA100403).

Appendix A. Supplementary data Supplementary material related to this article can be found, in the online version, at http://dx.doi.org/10.1016/j.molimm. 2014.05.004.

References

Acton, R.T., *et al.*, 1971. Tetrameric immune macroglobulins in three orders of bony fishes. Proc. Natl. Acad. Sci. U.S.A. 68 (1), 107-111.

Alexa, A., Rahnenfuhrer, J., 2010. TopGO: Enrichment Analysis for Gene Ontology. Rpackage Version 2.12.0.

Alexa, A., Rahnenfuhrer, J., Lengauer, T., 2006. Improved scoring of functional groupsfrom gene expression data by decorrelating GO graph structure. Bioinformatics 22 (13), 1600-1607.

Audic, S., Claverie, J.M., 1997. The significance of digital gene expression profiles. Genome Res. 7 (10), 986-995.

Bilyk, K.T., Cheng, C.H.C., 2013. Model of gene expression in extreme cold – reference transcriptome for the high-Antarctic cryopelagic notothenioid fish *Pagothenia borchgrevinki*. BMC Genomics 14, 634-650.

Bingrong, T., *et al.*, 2012. Methionine requirement of Grass Carp (*Ctenopharyngodon idella*) during middle growth stage.

Chin. J. Anim. Nutr. 11, 2263-2271.

Boshra, H., Gelman, A.E., Sunyer, J.O., 2004. Structural and functional characterization of complement C4 and C1s-like molecules in teleost fish: insights into the evolution of classical and alternative pathways. J. Immunol. 173 (1), 349-359.

Breton, B., *et al.*, 1991. Neuropeptide Y stimulates *in vivo* gonadotropin secretion inteleost fish. Gen. Comp. Endocrinol. 84 (2), 277-283.

Bryce, J., *et al.*, 2005. WHO estimates of the causes of death in children. Lancet 365(9465), 1147-1152.

Casto, J.M., Nolan, V., Ketterson, E.D., 2001. Steroid hormones and immune function: experimental studies in wild and captive dark-eyed juncos (*Junco hyemalis*). Am.Nat. 157 (4), 408-420.

Cerdá-Reverter, J.M., Peter, R.E., 2003. Endogenous melanocortin antagonist in fish: structure, brain mapping, and regulation by fasting of the goldfish agouti-related protein gene. Endocrinology 144 (10), 4552-4561.

Chen, L.J., *et al.*, 2012. Functional characterizations of RIG-I to GCRV and viral/bacterial PAMPs in grass carp *Ctenopharyngodon idella*. Plos One 7 (7), e42182.

Desvergne, B., Wahli, W., 1999. Peroxisome proliferator-activated receptors: nuclearcontrol of metabolism. Endocr. Rev. 20 (5), 649-688.

Dodds, A., Petry, F., 1993. The Phylogeny and Evolution of the First Component of Complement C1, 93. Behring Institute, Mitteilungen, pp. 87.

Dubois, B., *et al.*, 2001. Measles virus exploits dendritic cells to suppress CD4$^+$ T-cell proliferation via expression of surface viral glycoproteins independently of T-cell transinfection. Cell Immunol. 214 (2), 173-183.

Endo, Y., *et al.*, 1998. Two lineages of mannose-binding lectin-associated serineprotease (MASP) in vertebrates. J. Immunol. 161 (9), 4924-4930.

Fänge, R., Nilsson, S., 1985. The fish spleen: structure and function. Experientia 41(2), 152-158.

FastQC http://www.bioinformatics.babraha m.ac.uk/projects/fastqc/

Fischer, U., Koppang, E.O., Nakanishi, T., 2013. Teleost T and NK cell immunity. Fish Shellfish Immunol. 35 (2), 197-206.

Fischer, U., *et al.*, 2006. Cytotoxic activities of fish leucocytes. Fish Shellfish Immunol. 20 (2), 209-226.

Frederich, R.C., *et al.*, 1995. Leptin levels reflect body lipid content in mice: evidencefor diet-induced resistance to leptin action. Nat. Med. 1 (12), 1311-1314.

Funder, J.W., *et al.*, 1997. Mineralocorticoid receptors, salt, and hypertension. Recent Progress in Hormone Research, Proceedings of the 1996 Conference, 52., pp.247-262.

Gonzalez-Nunez, V., Gonzalez-Sarmiento, R., Rodriguez, R.E., 2003. Identification of two proopiomelanocortin genes in zebrafish (*Danio rerio*). Mol. Brain Res. 120(1), 1-8.

Griffin, D.E., Ward, B.J., 1993. Differential Cd4 T-cell activation in measles. J. Infect. Dis. 168 (2), 275-281.

Grunfeld, C., *et al.*, 1992. Lipids, lipoproteins, triglyceride clearance, and cytokines in human-immunodeficiency-virus infection and the acquired-immuno deficiency-syndrome. J. Clin. Endocrinol. Metab. 74 (5), 1045-1052.

Gupta, B.B.P., Lalchhandama, K., 2002. Molecular mechanisms of glucocorticoid action. Curr. Sci. 83 (9), 1103-1111.

Gual, P., Le Marchand-Brustel, Y., Tanti, J.F., 2005. Positive and negative regulation of insulin signaling through IRS-1 phosphorylation. Biochimie 87 (1), 99-109.

Haussmann, D., Figueroa, J., 2011. Glandular kallikrein in the innate immune system of Atlantic salmon (*Salmo salar*). Vet. Immunol. Immunopathol. 139 (2), 119-127.

Harris, P., Soleng, A., Bakke, T., 1998. Killing of *Gyrodactylus salaris* (Platyhelminthes, Monogenea) mediated by host complement. Parasitology 117 (2), 137-143.

Holland, M.C.H., Lambris, J.D., 2002. The complement system in teleosts. Fish Shell-fish Immunol. 12 (5), 399-420.

Holland, M.C.H., Lambris, J.D., 2004. A functional C5a anaphylatoxin receptor in a teleost species. J. Immunol. 172 (1), 349-355.

Holmes, S.J., Shalet, S.M., 1996. Role of growth hormone and sex steroids in achieving and maintaining normal bone mass. Horm. Res. 45 (1-2), 86-93.

Hotamisligil, G.S., 2006. Inflammation and metabolic disorders. Nature 444 (7121), 860-867.

Klein, J., 1986. Natural History of the Major Histocompatibility Complex, XV. Wiley, New York, pp. 775.

Lammens, M., Decostere, A., Haesebrouck, F., 2000. Effect of *Flavobacterium psy-chrophilum* strains and their metabolites on the oxidative activity of rainbow trout *Oncorhynchus mykiss* phagocytes. Dis. Aquat. Organ. 41(3), 173-179.

Lipke, D.W., Olson, K.R., 1990. Enzymes of the kallikrein-kinin system in rainbow trout. Am. J. Physiol. Regul. Integr. Comp. Physiol. 258 (2), R501-R506.

Li, R., *et al.*, 2009. SOAP2: an improved ultrafast tool for short read alignment. Bioin-formatics 25 (15), 1966-1967.

Lopez-Pedrera, C., *et al.*, 1997. Tissue factor (TF) and urokinase plasminogen activator receptor (uPAR) and bleeding complications in leukemic patients. Thromb. Haemost. 77 (1), 62-70.

Magnadottir, B., 2006. Innate immunity of fish (overview). Fish Shellfish Immunol. 20 (2), 137-151.

Masini, M., Sturla, M., Uva, B., 1997. Key enzymes of the kallikrein-kinin system in Antarctic teleosts. Polar Biol. 17 (4), 358-362.

Michal, G., 1999. Biochemical Pathways: An Atlas of Biochemistry and Molecular Biology, xi. Wiley, Spektrum, New York, Heidelberg, pp. 277.

Mortazavi, A., *et al.*, 2008. Mapping and quantifying mammalian transcriptomes by RNA-Seq. Nat. Methods 5 (7), 621-628.

Moss, W.J., Ota, M.O., Griffin, D.E., 2004. Measles: immune suppression and immuneresponses. Int. J. Biochem. Cell Biol. 36 (8), 1380-1385.

Munzberg, H., Myers, M.G., 2005. Molecular and anatomical determinants of centralleptin resistance. Nat. Neurosci. 8 (5), 566-570.

Nakao, M., *et al.*, 2002. Diversity of complement factor B/C2 in the common carp (*Cyprinus carpio*): three isotypes of B/C2-A expressed in different tissues. Dev. Comp. Immunol. 26 (6), 533-541.

Nonaka, M., Smith, S.L., 2000. Complement system of bony and cartilaginous fish. Fish Shellfish Immunol. 10 (3), 215-228.

Press, C.M., Evensen, Ø., 1999. The morphology of the immune system in teleost fishes. Fish Shellfish Immunol. 9 (4), 309-318.

Prodinger, W.M., 1999. Complement receptor type two (CR2, CR21). Immunol. Res. 20 (2), 187-194.

Qiu, T., *et al.*, 2001. Complete nucleotide sequence of the S10 genome segment of grass carp reovirus (GCRV). Dis. Aquat. Organ. 44 (1), 69-74.

Rombout, J.H.W.M., *et al.*, 2005. Phylogeny and ontogeny of fish leucocytes. FishShellfish Immunol. 19 (5), 441-455.

Saeed, A.I., *et al.*, 2003. TM4: a free, open-source system for microarray data man agement and analysis. Biotechniques 34 (2), 374-378.

Sakai, D., 1984. Opsonization by fish antibody and complement in the immune phagocytosis by peritoneal exudate cells isolated from salmonid fishes. J. Fish Dis. 7 (1), 29-38.

Sakai, D., 1992. Repertoire of complement in immunological defense mechanisms of fish. Ann. Rev. Fish Dis. 2, 223-247.

Savage, D.B., 2005. PPAR gamma as a metabolic regulator: insights from genomics and pharmacology. Expert Rev. Mol. Med. 7 (1), 1-16.

Servin, A.L., *et al.*, 1989. Peptide-YY and neuropeptide-Y inhibit vasoactive intestinal peptide-stimulated adenosine 3′, 5′-monophosphate production in rat smallintestine: structural requirements of peptides for interacting with peptide-YY-preferring receptors. Endocrinology 124 (2), 692-700.

Sunyer, J.O., Tort, L., 1995. Natural hemolytic and bactericidal activities of sea breamSparus-Aurata serum are effected by the alternative complement pathway. Vet. Immunol. Immunopathol. 45 (3-4), 333-345.

Su, J.G., *et al.*, 2008. Toll-like receptor 3 regulates Mx expression in rare minnow *Gobiocypris rarus* after viral infection. Immunogenetics 60 (3-4), 195-205.

Su, J.G., *et al.*, 2009. Grass carp reovirus activates RNAi pathway in rare minnow, *Gobiocypris rarus*. Aquaculture 289 (1-2), 1-5.

Su, J.G., *et al.*, 2011. Evaluation of internal control genes for qRT-PCR normalizationin tissues and cell culture for antiviral studies of grass carp (*Ctenopharyngodon idella*). Fish Shellfish Immunol. 30 (3), 830-835.

Su, J.G., *et al.*, 2012. Identification, mRNA expression and genomic structure of TLR22 and its association with GCRV susceptibility/resistance in grass carp (*Ctenopharyngodon idella*). Dev. Comp. Immunol. 36 (2), 450-462.

Subramanian, K., Hetrick, F.M., Samal, S.K., 1997. Identification of a new genogroup of aquareovirus by RNA-RNA hybridization. J. Gen. Virol. 78, 1385-1388.

Takahashi, N., *et al.*, 2005. [The structures and functions of peroxisome proliferator-activated receptors (PPARs)]. Nihon Rinsho 63 (4), 557-564.

Tort, L., Balasch, J., Mackenzie, S., 2003. Fish immune system. A crossroads betweeninnate and adaptive responses. Inmunología 22 (3), 277-286.

Uribe, C., *et al.*, 2011. Innate and adaptive immunity in teleost fish: a review. Agric. J. 56 (10), 486-503.

Vidalain, P.O., *et al.*, 2000. Measles virus induces functional TRAIL production byhuman dendritic cells. J. Virol. 74 (1), 556-559.

Wang, Z., Gerstein, M., Snyder, M., 2009. RNA-Seq: a revolutionary tool for transcrip-tomics. Nat. Rev. Genet. 10 (1), 57-63.

Wang, Q., *et al.*, 2012. Complete genome sequence of a reovirus isolated fromgrass carp, indicating different genotypes of GCRV in China. J. Virol. 86 (22), 12466.

Watts, M., Munday, B.L., Burke, C.M., 2001. Immune responses of teleost fish. Aust. Vet. J. 79 (8), 570-574.

Wilson, M., *et al.*, 1997. A novel chimeric Ig heavy chain from a teleost fish shares similarities to IgD. Proc. Natl. Acad. Sci. U.S.A. 94 (9), 4593-4597.

Yanagi, Y., Cubitt, B.A., Oldstone, M., 1992. Measles virus inhibits mitogen-induced T cell proliferation but does not directly perturb the T cell activation processinside the cell. Virology 187 (1), 280-289.

Yang, C.R., *et al.*, 2012a. Identification and expression profiles of ADAR1 gene, respon-sible for RNA editing, in responses to dsRNA and GCRV challenge in grass carp (*Ctenopharyngodon idella*). Fish Shellfish Immunol. 33 (4), 1042-1049.

Yang, C., *et al.*, 2012b. Identification and expression profiles of grass carp *Ctenopharyngodon idella* TLR7 in responses to double-stranded RNA and virusinfection. J. Fish Biol. 80 (7), 2605-2622.

Yeo, G.S.H., *et al.*, 1997. Cloning and sequencing of complement component C9(1) and its linkage to DOC-2 in the pufferfish *Fugu rubripes*. Gene 200 (1-2), 203-211.

Zhang, L.L., *et al.*, 2010. An improved RT-PCR assay for rapid and sensitive detection of grass carp reovirus. J. Virol. Methods 169 (1), 28-33.

Zhu, Y., *et al.*, 2005. The ancient origin of the complement system. EMBO J. 24 (2), 382-394.

草鱼出血病病毒感染后草鱼组织的转录组测序分析

石米娟[1,2]　黄　容[1]　杜富宽[1]　裴永艳[1]　廖兰杰[1,2]
朱作言[1]　汪亚平[1]

1 中国科学院水生生物研究所，武汉 430072
2 中国科学院大学，北京 100039

摘　要　草鱼出血病是水生呼肠孤病毒引起的严重病害，主要感染幼鱼和一龄鱼。本研究取草鱼攻毒前 2 h 和攻毒后 2、24、48、72、96 和 120 h 的鳃、肠、肝、脾四个组织样品进行转录组测序，每个转录组测序获得 7.25 ± 0.18 M 过滤 reads 和 3.53 ± 0.37 M 单一 reads，通过与对照组比较，四个组织攻毒后 6 个时间点共有 9060 个差异表达基因。这 6 个时间点差异表达基因数据分层聚类分析分为三支：2 h，24 h/48 h，72 h/96 h/120 h。通过基因本体论对差异表达基因每个转录组数据集进行单一和模块化富集分析，结果显示在四个组织中都可以发生免疫应答反应，表明草鱼呼肠孤病毒没有专一性的攻击某个组织。另外，在发病的过程中，每个组织出现脂肪和二氧化碳代谢受干扰现象。差异表达基因基于京都基因与基因组百科全书数据库单一性分析结果表明，在草鱼出血病发病过程中，补体系统和细胞免疫发挥重要作用。对攻毒后混合样品进行荧光定量 PCR 验证了转录组测序数据的可靠性。

Isolation and Expression of Grass Carp Toll-Like Receptor 5a (CiTLR5a) and 5b (CiTLR5b) Gene Involved in the Response to Flagellin Stimulation and Grass Carp Reovirus Infection

Yao Jiang[1,3] Libo He[1] Changsong Ju[1,2] Yongyan Pei[1,3] Myonghuan Ji[2]
Yongming Li[1] Lanjie Liao[1] Songhun Jang[1,2] Zuoyan Zhu[1] Yaping Wang[1]

1 State Key Laboratory of Freshwater Ecology and Biotechnology, Institute of Hydrobiology, Chinese Academy of Sciences, Wuhan 430072, China

2 Department of Zoology, College of Life Sciences, Kim II Song University, Pyongyang, Democratic People's Republic of Korea

3 University of Chinese Academy of Sciences, Beijing 100049, China

Abstract Toll-like receptor 5 (TLR5), a member of Toll-like receptors (TLRs) family and is responsible for the bacterial flagellin recognition in vertebrates, play an important role in innate immunity. In the study, two TLR5 genes of grass carp (*Ctenopharyngodon idellus*), named CiTLR5a and CiTLR5b, were cloned and analyzed. Both CiTLR5a and CiTLR5b are typical TLR proteins, including LRR motif, transmembrane region and TIR domain. The full-length cDNA of CiTLR5a is 3054 bp long, with a 2646 bp open reading frame (ORF), 78 bp 5′ untranslated regions (UTR), and 330 bp 3′ UTR. The full-length cDNA of CiTLR5b is 3326 bp, with a 2627 bp ORF, 95 bp 5′ UTR, and 594 bp 3′ UTR. Phylogenetic analysis showed that CiTLR5a and CiTLR5b were closed to the TLR5 of *Cirrhinus mrigala*, *Cyprinus carpio*, and *Danio rerio.* Subcellular localization indicated that CiTLR5a and CiTLR5b shared similar localization pattern and may locate in the plasma membrane of transfected cells. Real-time quantitative PCR revealed CiTLR5a and CiTLR5b were constitutively expressed in all examined tissues, whereas the highest expressed tissue differed. Following exposure to flagellin and GCRV, CiTLR5a and CiTLR5b were up-regulated significantly. Moreover, the downstream genes of TLR5 signal pathway such as MyD88, NF-κB, IRF7, IL-1β, and TNF-α also up-regulated significantly, whereas the IκB gene was down-regulated, suggesting that CiTLR5a and CiTLR5b involved in response to flagellin stimulation and GCRV infection. The results obtained in the study would provide a new insight for further understand the function of TLR5 in teleost fish.

Keywords grass carp; toll-like receptor 5; flagellin; GCRV

文章发表于 *Fish & Shellfish Immunology*, 2015, 44(1): 88-99

1. Introduction

The grass carp (*Ctenopharyngodon idellus*) has been an important aquaculture species in China for over 60 years, accounting for more than 18% of total freshwater aquaculture production. The production of grass carp reached 478.2 million tons in 2012, making it the most highly consumed freshwater fish worldwide [1]. However, disease outbreaks are frequently and resulted in huge economic losses to the aquaculture of grass carp. The grass carp haemorrhagic disease, caused by the grass carp reovirus (GCRV), is one of the most serious diseases [2]. In the GCRV infected fish, gills and intestines are the organs that show haemorrhagic symptoms [3]. The liver, spleen, kidney, intestine, and muscle are susceptible to GCRV and had a higher number of viral RNA copies [4]. Nevertheless, no effective drugs or vaccines against GCRV were developed until now. Therefore, identification of grass carp genes that involved in innate immunity is important for antiviral drug development and fish breeding programs.

Toll-like receptors (TLRs) family is an important class of pattern recognition receptors (PRRs), which specifically recognize a series of highly conserved pathogenic microorganism structures, termed pathogen-associated molecular patterns (PAMPs) [5]. TLRs family is the best understand receptors and play an important role in the innate immunity of vertebrate and invertebrate [6]. TLRs are typical type I transmembrane proteins that included three major domains: tandem repeat leucine-rich repeat (LRR) motifs that identifies PAMPs, a transmembrane region, and an intracellular Toll/IL-1 receptor domain (TIR) domain that responsible for signal transmission [7]. When the corresponding ligands were recognized by TLRs, the subsequent signaling pathway was activated [8]. The signaling pathway of TLRs was divided into two types: myeloid differentiation primary response protein 88 (MyD88)-dependent and MyD88-independent signaling pathway [9]. The MyD88-dependent signaling pathway induced inflammatory-cytokine production via NF-κB activation, whereas the MyD88-independent signaling pathway leaded the production of interferon inducible gene via interferon regulatory factor 3 (IRF3) [9].

Toll-like receptor 5 (TLR5), a member of TLRs family and is responsible for the bacterial flagellin recognition in vertebrates, played an important role in the host defense against bacterial pathogens [10]. After the flagellin was recognized by TLR5, NF-κB was activated through the MyD88-dependent signaling pathway, and some inflammatory-cytokines such as IL-1 and TNF-α were induced [11]. In mammalians, TLR5 was expressed in the surface of many kinds of cells, such as intestinal epithelial cells, dendritic cells, monocytes cells, and splenic macrophages and so on [12, 13]. In teleost fish, two types of TLR5 were existed: the membrane form of TLR5 (TLR5M) and the soluble form of TLR5 (TLR5S). Until now, TLR5S was cloned from rainbow trout (*Onchorhynchus mikiss*) [14], atlantic salmon (*Salmo salar*) [15], catfish (*Ictalurus punctatus*) [16], Japanese flounder (*Paralichthys olivaceus*) [17], and gilthead seabream (*Sparus aurata*) [18]. However, in zebrafish (*Danio rerio*), common carp (*Cyprinus carpio*), and Indian major carp (*Cirrhinus mrigala*), only TLR5M were found [19, 20]. The results obtained in the above studies showed that TLR5 of teleost fish could also recognize flagellin of bacterial and activate the immunity response. However, it is remain unknown that whether other ligands could be recognized by TLR5 and the particular role of TLR5 in teleost fish was still unclear.

In the study, two TLR5 genes of grass carp (*Ctenopharyngodon idellus*), named CiTLR5a and CiTLR5b, were cloned and analyzed. The gene structure, tissue distribution, localization pattern, and the response to flagellin stimulation and GCRV infection were investigated. Moreover, the responses of the TLR5 downstream genes to flagellin stimulation and GCRV infection were also analyzed. Our study would provide a new insight for further understand the function of TLR5 in teleost fish.

2. Materials and Methods

2.1 Ethical procedures

Animal experimental procedures were conducted under the institutional guidelines of Hubei Province. The protocol was approved by the committee of institute of hydrobiology, Chinese Academy of Sciences (CAS). All surgery was performed under eugenol anesthesia, and all efforts were made to minimize suffering.

2.2 Experiment fish, sample collection, and virus exposure

Healthy grass carp at three months old were used in the study. The grass carp, weighing about 2-4 g with an average length of 5 cm, were obtained from the Guanqiao Experimental Station, institute of hydrobiology, CAS and acclimatized in aerated freshwater at 28℃ for one week before processing. The fish were fed with commercial feed twice a day, and the water was exchanged daily. After no abnormal symptom was observed, the grass carp was subjected to further study.

The virus exposure was conducted by feeding as follows. Dead fish with apparent symptoms of GCRV infection were collected and pestled together with an equal volume of 0.75% saline water, and then mixed with an equal amount of commercial feed. The resulting feed mixture was used as the source virus. The experimental group of fish was fed with the feed mixture on the first day, then with commercial feed on the other days. The temperature was maintained at 26-28℃ throughout the experiment.

Five uninfected fish were selected and samples from the gill, spleen, liver, intestine, kidney, head kidney, heart, muscle, skin and brain were prepared. RNA from these tissues was prepared to analyze the tissue distribution of CiTLR5a and CiTLR5b. In addition, kidney, gill, head, spleen and intestine samples were isolated from five infected fish at 0-6 days post infection. RNA from these tissues was prepared to analyze the response of CiTLR5a and CiTLR5b and their downstream genes to GCRV infection.

2.3 Cloning the full-length cDNA of CiTLR5a and CiTLR5b

Total RNA was extracted from the tissues of healthy grass carp using Trizol reagent (Invitrogen, USA). The first-strand cDNA synthesis was carried out using DNase I (Promega, USA)-treated total RNA as a template and oligo (dT)-adaptor primer as the control for the reverse transcriptase (TOYOBO, Japan). Specific primers (Table 1) were designed according to the cDNA sequences of zebrafish TLR5a and TLR5b and the deduced cDNA sequences of grass carp TLR5a and TLR5b. PCR amplification was conducted using the above cDNA as a template and the following program: 94℃ for 2 min, 31 cycles of 94℃ for 30 s, annealing at 60℃ for 30 s, and 72℃ for 1 min, followed by a final extension at 72℃ for 10 min. The PCR products were gel purified, cloned into pMD18-T vector (Takara, Japan), transformed into DH5α competent cells. The positive clones were picked and sequenced in both directions. The coding sequences of CiTLR5a and CiTLR5b were obtained by sequence assembly and the 5′- and 3′-untranslated region (UTR) sequences were then obtained using a SMARTer™ RACE cDNA Amplification Kit (Invitrogen, USA) and a 5′ RACE System for Rapid Amplification of cDNA Ends kit (Clontech, USA).

2.4 Homology and phylogenetic analysis

The amino acid sequences of CiTLR5a and CiTLR5b were translated using the National Center for Biotechnology Information online software. The conserved protein domains were predicated by Simple Modular Architecture Research Tool (SMART). The alignment search tool Blastn (National Center for Biotechnology Information) was used to search for homologous sequences of CiTLR5a and CiTLR5b in GenBank. ClustalX2.0 was used to perform multiple sequence alignments of CiTLR5a and CiTLR5b among different species. DNASTAR was used to calculate the percentage homology of different sequences. GeneDoc software was used to analyze the alignment results. Phylogenetic tree was constructed using the neighbor-joining method with Mega5.1 software, and analysis reliability was assessed by 1000 bootstrap replicates.

2.5 Cells, plasmids construction, and transfection

C. idellus kidney (CIK) cells used in the study were maintained in TC199 medium (Hyclone, USA) supplemented with 10% fetal bovine serum (FBS) at 25℃.

To analyze the subcellular location of CiTLR5a and CiTLR5b protein, DNA fragments that contained complete open reading frame (ORF) of CiTLR5a and CiTLR5b were amplified from grass carp cDNA that obtained above. Each fragment was cut with corresponding restriction enzymes and inserted into pEGFP-N3 (Clontech, USA) vector that has been treated with the same enzymes. The obtained plasmids were named as pEGFP-5a and pEGFP-5b. The resulted plasmids were confirmed by restriction enzymes and DNA sequencing. The primers for the plasmids construction were listed in Table 1.

Table 1 Primers used in the study (enzyme cleavage site is underlined)

Primers	Sequences (5′ to 3′)	Usage
T5a-F	TACCCCAACTGACTCTGAGAAG	Tlr5a cDNA cloning
T5a-R	TCTACAGGGATCTCACAATGGC	
T5aI-F	CTTGCTTTGCTCCCAGGTTCT	Tlr5a 3′-race
T5aI-R	GTCCAGGAATATGAGTACGTGG	Tlr5a 5′ -race
T5b-F	GGCTGAAGGATTATTGAAGCGT	Tlr5b cDNA cloning
T5b-R	CAGCCCTATTGCTACATTTCC	
T5bI-F	CCCGCTGCCAGTGGATAAAGAC	Tlr5b 3′-race
T5bI-R	CGAGACCGTGGACTACTTCTG	Tlr5b 5′-race
pTLR5a-F	GGAGAATTCAATGGCAACAATACAC	pEGFP-5a
pTLR5a-R	TTAGGATCCCACTGCAGTGTCTG	
pTLR5b-F	TTCGAATTCAATGGGATTTACATTTAT	pEGFP-5b
pTLR5b-R	TCAGGATCCTACTGATGTGTTTGCA	
Qt5a-F	TGACGCAGCAAATGTTCAAGC	qPCR of Tlr5a
Qt5a-R	GAGAACCTGGGAGCAAAGCAA	
Qt5b-F	CATTATCTCATGTTTCCTATG	qPCR of Tlr5b
Qt5b-R	GTTGAAGATGTTAAGATTTTGC	
Qβ-actin-F	AGCCATCCTTCTTGGGTATG	qPCR of β-actin
Qβ-actin-R	GGTGGGGCGATGATCTTGAT	
QMyD-F	TGGAGGACTGTCGCCGAAATG	qPCR of MyD88
QMyD-R	TGTGGCCTCTGGACGAGTTTC	
QIκB-F	GGCAGATGTAAACGCAAAG	qPCR of IκB
QIκB-R	GCCGAAGGTCAGGTGGTA	
QNF-κBF	GGCAGATGTAAACGCAAAG	qPCR of NF-κB
QNF-κBR	GCCGAAGGTCAGGTGGTA	
QIRF7-F	GGTGGAAAGTGGGCGGTAT	qPCR of IRF7
QIRF7-R	TCGTTAGGGTGCTCGTTGA	
QTNF-α-F	CATCCATTTAACAGGTGCATAC	qPCR of TNF-α
QTNF-α-R	GCAGCAGATGTGGAAAGAGAC	
QIL-1β-F	GATTCGAAAGTTCGATTCAATCT	qPCR of IL-1β
QIL-1β-R	TTCAGTGACCTCCTTCAAGAC	

Transfection was performed as descried previously [21]. Briefly, CIK cells that grown on coverslips in 6-well plates were transfected with plasmid pEGFP-5a or pEGFP-5b by Lipofectamine 3000 reagent (Invitrogen, USA) according to the manufacturer's instructions. After 24h post transfection, cells were fixed with 4% paraformaldehyde, permeabilized with 0.2% Triton X-100, and stained by Hoechst 33342. Finally, the cells were mounted with 50% glycerol and observed by fluorescence microscope (Leica, Germany).

2.6 Expression level of CiTLR5a and CiTLR5b in different tissues

Total RNA was extracted from nine tissues (gill, spleen, liver, intestine, kidney, head kidney, heart, muscle, skin, and brain) of five healthy grass carp, and reverse transcribed to obtain cDNA. cDNA from the same tissues were mixed and served as the template for Real-time quantitative PCR (qPCR), which was used to examine CiTLR5a and CiTLR5b expression level in the different tissues. The qPCR reaction mixture was as follows: 0.6 μl each of sense and reverse primers, 1 μl template, 10 μl 2×SYBR mix (TOYOBO, Japan), and 7.8 μl ddH_2O. Three replicates were conducted for each sample and β-actin was used as an internal control. The program for qPCR was as follows: 95℃ for 10 s, 40 cycles of 95℃ for 5 s, and 60℃ for 20 s. All data are given as mean ± standard deviation of three replicates. The qPCR primers for examination of CiTLR5a and CiTLR5b expression level and the primers for β-actin are listed in Table 1.

2.7 The response of TLR5s and TLR5 downstream genes after flagellin stimulation

CIK cells seeded in 6-well plates were stimulated with flagellin from *Salmonella typhimurium* (sigma, USA) at a concentration of 0.5 μg/ml. Cells were harvested at 0, 12, 24, 36, and 48 h post stimulation. Total RNA was extracted from the harvested cells and reverse transcribed to cDNA. The cDNA were used as a template for qPCR to determine the expression level of CiTLR5a and CiTLR5b.

The signal transmission of TLR5 via MyD88-dependent signaling pathway and then activate the downstream genes. In order to understand the response of TLR5 downstream genes after flagellin stimulation, the expression level of MyD88, IκB, NF-κB, IRF7, IL-1β, and TNF-α before and after flagellin stimulation was also determined by qPCR. The program and reaction mixture for qPCR were the same as above. The primers for the qPCR were listed in Table 1. Data are also given as mean ± standard deviation of three replicates. The significant difference between the control (0h) and treated groups at each time point (12, 24, 36, and 48 h post stimulation) was determined by one-way ANOVA. Significance was set at $p < 0.05$.

2.8 The response of CiTLR5 and TLR5 downstream genes after exposure to GCRV

Total RNA was extracted from four tissues (gill, liver, spleen, and intestine) of five grass carp at 0-6 days post GCRV infection, and then reverse transcribed to obtain cDNA. cDNA from the same tissues were mixed and served as the template for qPCR, which was used to examine the expression level of CiTLR5a, CiTLR5b, and TLR5 downstream genes (MyD88, IκB, NF-κB, IRF7, IL-1β, and TNF-α) in the different tissues post GCRV infection. The program, reaction mixture, and primers for qPCR were the same as above. Data are also as mean ± standard deviation of three replicates. The significant difference between the control (0 day) and treated groups at each time point (1, 2, 3, 4, 5 and 6 days post infection) was determined by one-way ANOVA. Significance was set at $p < 0.05$.

3. Result

3.1 Cloning and characterization of grass carp TLR5s

The full-length cDNA of two grass carp TLR5 genes, named CiTLR5a and CiTLR5b, was obtained by PCR and RACE. The full-length cDNA of CiTLR5a is 3054 bp long, with a 2646 bp ORF, 78 bp 5′ UTR, and 330 bp 3′ UTR (Genbank accession number: KF736231) (Fig.1a). The full-length cDNA of CiTLR5b is 3326 bp, with a 2627 bp ORF, 95 bp 5′ UTR, and 594 bp 3′ UTR (Genbank accession number: KF736232) (Fig.1b). The genome DNA sequences of CiTLR5a and CiTLR5b were obtained and compared with the cDNA sequences, the result showed that only one exon was found in CiTLR5a and CiTLR5b (Fig. 2a). Moreover, the genome DNA sequences of CiTLR5a and CiTLR5b were compared with grass carp reference genome (Bioproject: PRJNA39737, unpublished data), the result showed that CiTLR5a and CiTLR5b located in the same contig and at a distance of 6944 bp (Fig. 2a).

The amino acid sequences of CiTLR5a and CiTLR5b were analyzed and the results revealed that both of them are membrane form of TLR5. CiTLR5a contained 12 LRR motifs, a LRR-CT motif, a low complexity region, a transmembrane region, and a TIR domain. CiTLR5b included 12 LRR motifs, a LRR-CT motif, a transmembrane region, and a TIR domain. The schematic diagrams of CiTLR5a and CiTLR5b were shown in Fig. 2b.

3.2 Homology and phylogenetic analysis of TLR5s

To determine the identities between grass carp TLR5s and other TLR5 genes, the TLR5 homologous from teleost fish (including *Cyprinus carpio*, *Danio rerio*, *Cirrhinus mrigala*, *Ictalurus punctatus*, *Oncorhynchus mykiss*, *Salmo salar*) and so on), aves (including *Columba livia*, *Numida meleagris*, and *Anser anser*), repitilia (*Chelonia mydas*),

```
                                                                         M  A  T  I
   1 aaaaatctttgcgtaaatgataacattttcatttgtgggtaaactgttttaacctttgtattctacagggatctcacaATGGCAACAATA
   5 H  T  L  S  L  I  L  L  G  L  C  V  S  T  Q  I  V  K  C  T  S  V  C  S  V  G  A  S  V  A
  91 CACACATTATCTCTGATCCTCCTTGGATTATGCGTCAGCACTCAAATTGTGAAATGCACCTCAGTGTGTTCGGTTGGTGCTTCTGTCGCC
  35 L  C  I  D  K  G  L  Q  D  V  P  E  L  P  P  Y  V  N  K  V  D  L  S  K  N  N  I  A  E  L
 181 CTCTGCATCGATAAAGGTCTTCAAGATGTGCCAGAGCTTCCTCCATATGTAAATAAAGTGGATTTGAGTAAGAACAATATTGCTGAACTC
  65 N  E  T  S  F  S  H  L  E  G  L  Q  V  L  I  L  M  H  Q  T  T  R  L  V  I  R  N  N  T  F
 271 AATGAAACATCCTTTTCTCATCTGGAAGGTCTACAGGTCCTTATACTCATGCACCAAACAACAAGACTTGTGATCAGAAACAACACATTC
  95 R  R  L  S  N  L  T  S  L  Q  L  D  Y  N  N  F  L  R  M  D  T  G  A  F  N  G  L  S  N  L
 361 AGAAGACTCTCTAACTTAACATCACTTCAGCTTGACTACAACAACTTCCTGCGAATGGATACAGGAGCGTTTAACGGATTATCCAACCTT
 125 K  N  L  T  L  T  Q  C  S  L  E  D  T  I  L  S  G  D  F  L  K  P  L  V  S  L  E  M  L  V
 451 AAAAATCTCACTCTTACTCAGTGCAGTTTAGAGGATACCATTTTGTCTGGTGACTTCCTCAAACCTCTGGTGTCTCTTGAGATGCTTGTC
 155 L  R  E  N  N  I  K  R  I  Q  P  A  L  L  F  L  N  M  R  R  F  H  V  L  D  L  S  R  N  K
 541 TTACGTGAAAACAACATTAAAAGAATCCAGCCAGCATTGCTCTTTTTAAATATGAGGAGATTCCATGTGCTCGATCTCTCTCGCAACAAA
 185 V  K  S  I  C  E  E  D  L  L  S  F  Q  G  K  H  F  T  L  L  K  L  S  S  V  T  L  Q  D  M
 631 GTGAAGAGCATCTGTGAAGAAGACCTCCTCAGCTTTCAGGGTAAACATTTCACGCTTCTGAAGCTGTCCTCAGTGACACTGCAAGACATG
 215 N  E  Y  W  L  G  W  E  K  C  G  N  P  F  K  N  M  S  V  S  V  L  D  L  S  G  N  G  F  N
 721 AATGAGTACTGGTTAGGATGGGAAAAGTGTGGAAACCCATTTAAGAACATGTCCGTAAGTGTATTGGACTTATCTGGAAATGGCTTTAAT
 245 D  N  N  A  K  L  F  F  D  A  I  T  G  T  K  I  Q  S  L  I  L  S  N  S  H  S  M  G  S  S
 811 GATAACAATGCAAAGCTTTTCTTTGATGCAATCACTGGTACCAAAATACAAAGTCTCATTCTCAGTAACAGTCACAGCATGGGCAGTTCT
 275 S  G  N  N  S  K  D  P  N  K  F  T  F  K  G  L  E  A  S  G  I  K  I  F  D  L  S  N  S  S
 901 TCTGGTAATAATTCAAAAGATCCAAACAAATTCACATTTAAGGGTCTTGAGGCGAGTGGTATTAAGATTTTCGATCTCTCCAATTCAAGC
 305 I  F  A  L  S  Y  S  V  F  S  C  L  S  D  L  E  Q  I  T  L  A  E  S  R  I  N  K  I  E  K
 991 ATTTTTGCTCTGTCATATTCAGTATTTAGTTGTTTGTCAGATCTAGAACAAATTACATTAGCAGAAAGTCGGATCAACAAGATTGAAAAA
 335 S  A  F  L  G  M  A  N  L  L  K  L  N  L  S  K  N  F  L  G  N  I  D  S  N  T  F  Q  N  L
1081 AGTGCATTTTTGGGTATGGCAAATTTGCTGAAGCTAAACCTGTCCAAAAACTTCCTTGGTAATATTGATTCTAATACATTTCAGAATCTA
 365 E  K  L  E  V  L  D  L  S  Y  N  H  I  W  M  L  G  Y  E  S  F  R  G  L  P  N  L  L  N  L
1171 GAGAAGCTTGAGGTGCTTGATTTGTCTTATAATCATATATGGATGCTTGGATATGAATCATTTCGAGGGCTTCCAAATCTACTCAACCTA
 395 N  L  T  G  N  A  L  K  H  L  H  A  F  A  T  L  P  R  L  E  K  L  Y  L  G  D  N  K  I  L
1261 AATTTAACAGGAAATGCTCTCAAACATTTACATGCATTCGCAACCTTACCAAGACTGGAGAAGCTCTACTTGGGTGACAACAAAATTTTA
 425 S  V  F  Y  L  I  K  I  S  K  Y  L  T  T  L  Y  L  E  H  N  I  L  S  S  L  S  D  L  F  T
1351 TCTGTCTTTTATTTAATCAAAATTTCCAAATATCTTACAACCCTTTACCTGGAACATAACATACTTTCTTCCTTGTCAGATCTCTTCACA
 455 I  L  E  E  F  P  Q  I  E  E  I  V  F  R  G  N  E  L  L  Y  C  P  N  E  R  H  K  V  L  S
1441 ATACTAGAGGAATTTCCTCAAATTGAGGAAATTGTTTTTCGAGGTAATGAGCTTCTTTATTGCCCTAATGAAAGACACAAAGTGCTTTCA
 485 Q  K  I  Q  I  L  D  L  A  F  A  G  L  E  V  I  W  S  E  G  K  C  L  N  V  F  N  N  L  H
1531 CAAAAAATACAAATCCTTGATCTTGCATTTGCAGGTTTGGAAGTTATCTGGTCAGAAGGAAAATGTTTAAACGTGTTTAACAATCTTCAC
 515 Q  L  K  Q  L  S  L  S  H  N  L  L  Q  S  L  P  K  D  I  F  K  D  L  T  S  L  Y  F  L  D
1621 CAGTTAAAACAGCTTTCTCTGAGTCACAACTTGCTACAGTCTCTTCCCAAAGACATTTTTAAAGACCTTACCTCTTTGTACTTTTTGGAT
 545 L  S  F  N  S  L  K  Y  L  P  N  G  I  F  P  E  S  L  Q  I  L  N  L  E  Y  N  S  I  Y  S
1711 TTGTCCTTCAACTCTTTGAAGTACCTTCCAAACGGTATATTCCCTGAAAGTCTTCAAATTCTTAATCTTGAATATAATTCTATTTATTCA
 575 V  D  P  N  L  F  S  T  L  S  Y  L  S  L  I  K  N  D  F  R  C  D  C  K  L  R  D  F  Q  T
1801 GTAGATCCAAATCTCTTTAGCACCCTCAGCTACCTCAGCCTGATAAAAAACGATTTCCGTTGTGATTGCAAATTAAGGGATTTCCAAACT
 605 W  L  N  Q  T  N  V  I  I  S  H  S  I  E  D  V  I  C  A  S  P  E  D  Q  Y  M  V  P  V  V
1891 TGGCTAAACCAAACCAATGTAATCATTTCTCACTCCATTGAGGATGTGATATGTGCCAGTCCTGAGGATCAGTACATGGTTCCGGTTGTG
 635 R  S  S  I  Q  C  E  D  E  E  D  E  R  N  A  E  K  L  R  L  V  L  F  I  F  C  T  A  L  I
1981 AGATCCAGCATACAATGTGAGGATGAAGAGGACGAGAGAAACGCTGAAAAACTGAGGCTTGTGCTTTTTATTTTCTGTACCGCACTTATC
 665 T  L  L  T  A  S  A  I  I  Y  V  R  R  R  G  Y  I  F  K  L  Y  K  K  L  I  G  K  L  V  D
2071 ACGTTACTCACTGCTAGCGCCATCATTTATGTCCGTCGACGTGGCTACATCTTCAAGCTTTACAAAAAACTCATTGGCAAACTTGTGGAT
 695 G  K  R  E  E  P  D  P  D  Q  F  L  Y  D  V  F  L  C  F  S  S  N  D  I  K  W  V  E  R  A
2161 GGAAAGCGAGAGGAGCCTGATCCTGATCAATTCTTGTATGATGTGTTTCTTTGTTTTAGTTCCAATGATATTAAGTGGGTAGAAAGAGCA
 725 L  L  N  R  L  D  S  Q  F  S  E  Q  N  T  L  R  C  C  F  E  E  R  D  F  I  P  G  E  D  N
2251 CTGCTGAACAGGCTAGACTCTCAGTTCTCAGAGCAGAACACACTCCGCTGCTGCTTTGAGGAGCGAGACTTCATACCCGGGGAGGACAAT
 755 L  T  N  M  R  N  A  I  Q  N  S  H  K  T  L  C  V  V  S  E  H  F  L  K  D  G  W  C  L  E
2341 CTTACCAACATGCGAAATGCTATCCAGAATAGTCATAAAACCCTTTGTGTGGTGTCTGAGCATTTCCTGAAGGACGGCTGGTGCCTAGAA
 785 T  F  T  L  A  Q  C  R  M  L  V  E  L  K  D  I  L  V  V  L  V  V  G  N  I  P  Q  Y  R  L
2431 ACCTTCACCCTGGCACAGTGCAGGATGCTAGTGGAGCTAAAGGACATTCTGGTGGTGCTGGTTGTAGGGAACATACCGCAGTACAGGCTA
 815 L  K  Y  E  Q  L  R  S  Y  I  E  N  R  R  Y  L  L  W  P  D  D  S  Q  D  L  E  W  F  Y  D
2521 CTGAAGTACGAACAACTGAGATCCTACATTGAGAACAGAAGATACCTTCTGTGGCCCGATGACAGCCAGGACTTGGAGTGGTTTTATGAC
 845 Q  L  L  H  K  I  R  K  N  T  K  V  K  Q  T  N  T  K  V  N  E  G  E  K  N  L  D  A  A  N
2611 CAACTTCTGCATAAAATTAGAAAAAACACCAAGGTTAAACAAACAAATACTAAAGTCAATGAGGGTGAAAAAAATCTTGACGCAGCAAAT
 875 V  Q  A  D  T  A  V  *
2701 GTTCAAGCAGACACTGCAGTGTAActataaaggactttgtataattccttgctttgctcccaggttctcatccgtcctactttgttgaaa
2791 ttctctccattggacaatgacatcaacaaatattaagcatataaagctggaaaccatactgtattgtatagttataaatgtttattttgt
2881 attatccataatgtgtatgtacttctcagagtcagttggggtagttaaatgtcagtggaacggtgacatttttatctattatgtaaata
2971 ctccatttatatcacctttttcatcatagttttccaggatttggcaaataaagtcttagaataaaaaaaaaaaaaaaaaaaa
```

a

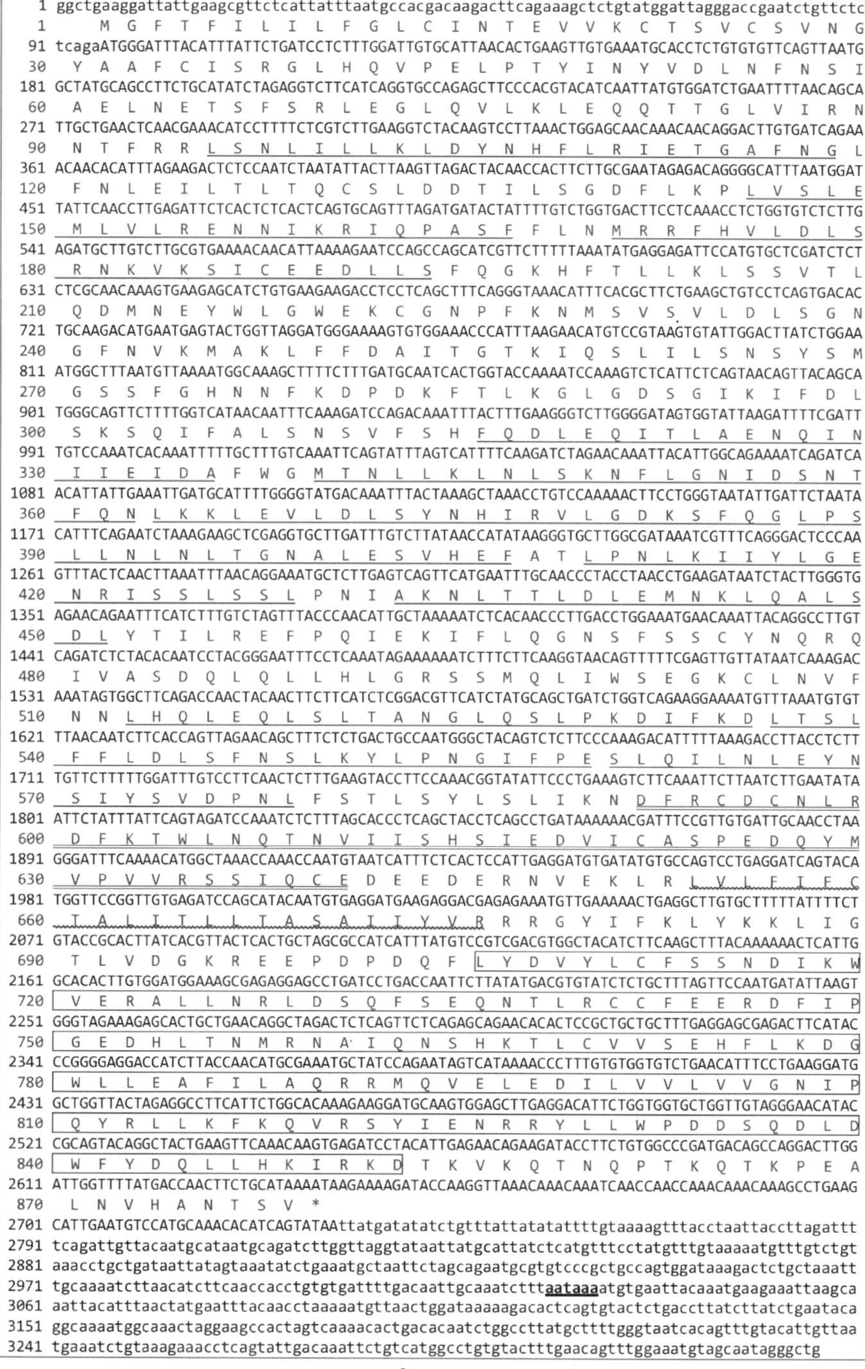

Fig. 1 Full length cDNA and deduced amino acid sequences of TLR5a (a) and TLR5b (b). The underlined amino acids indicated the LRR motifs, the double underlined amino acids are LRR C-terminal motif, the amino acids with wave line are transmembrane region, and the boxed amino acids are TIR domain, respectively. The underlined and bold bases are poly (A) signal sequence

a

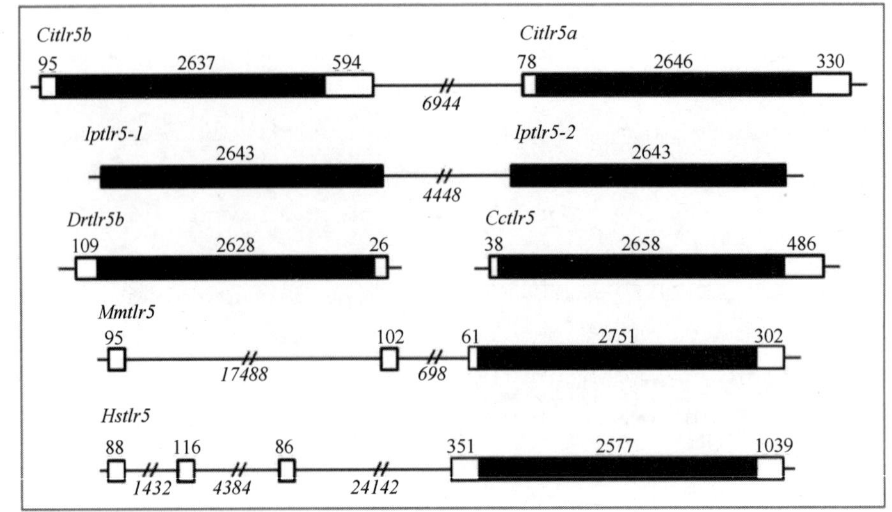

b

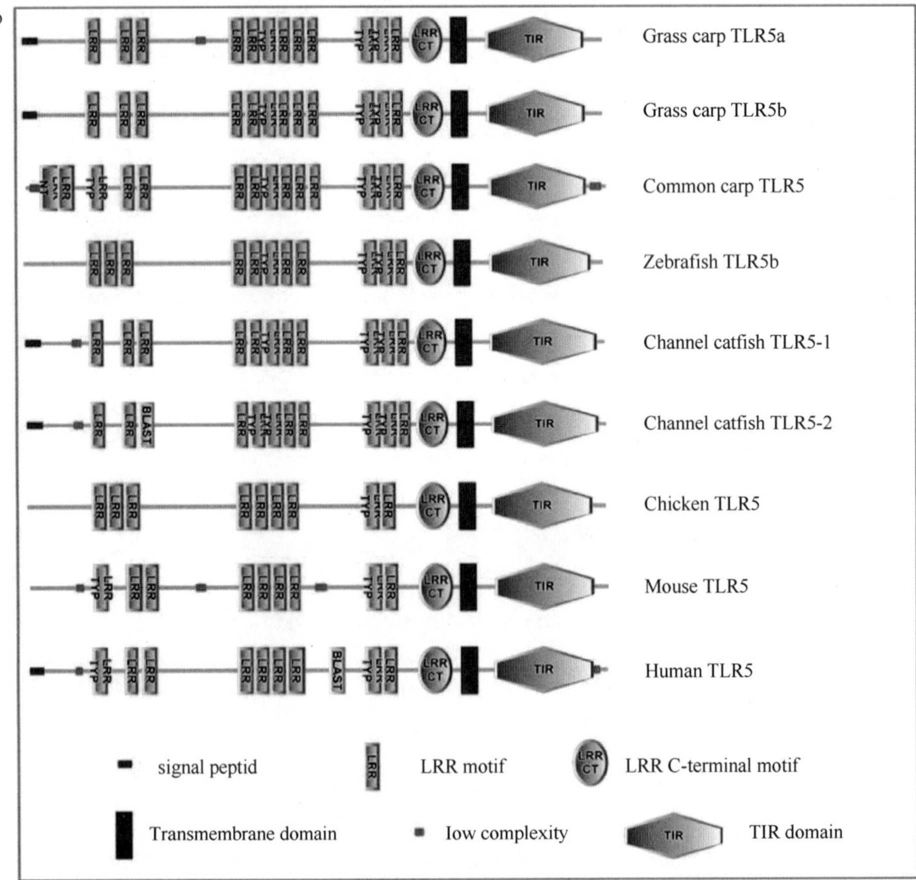

Fig. 2 Genome structure and schematic diagram of TLR5a and TLR5b. (a) Genomic structure of TLR5a and TLR5b. (b) Schematic diagram of TLR5a and TLR5b

amphibia (*Xenopus_laevis*), and mammalia (including *Sus scrofa*, *Homo sapiens*, *Mus musculus*, and *Canis lupus familiaris*) were compared with grass carp TLR5s and the results were shown in Table 2. CiTLR5a and CiTLR5b showed the highest identities with the homologous from teleost fish (identity>41.6%). The identities between grass carp TLR5s and the *Chelonia mydas* TLR5 were 40.4% and 39.7%. CiTLR5a and CiTLR5b showed ~38% identities with TLR5s from aves. In addition, the identities between grass carp TLR5s and mammalia TLR5s ranged from 35.2% to 38.4%.

Table 2 The identity between grass carp TLR5s and their homologues

Homologues	Identity (%)	
	TLR5a	TLR5b
Ctenopharhyngodon idella-5a	100.0	81.7
Ctenopharhyngodon idella-5b	81.7	100.0
Cyprinus carpio AGH15501.1	80.0	72.3
Danio rerio precursor NP-00112	72.4	81
Cirrhinus mrigala AEQ92867.1	72.5	75.1
Ictalurus punctatus 1 AEI59668	54.5	55.2
Ictalurus punctatus 2 AEI59669	54.2	55.2
Oncorhynchus mykiss NP-001118216.1	49.0	50.1
Salmo salar AEE38253.1	48.0	49.1
Paralichthys olivaceus AEN71825.1	42.1	43.6
Plecoglossus altivelis altivel BAI68383.1	43.0	43.6
Takifugu rubripes AAW69374.1	41.8	42.4
Oreochromis niloticus AFP44844	41.6	41.7
Chelonia mydas EMP25733.1	40.4	39.7
Columba livia EMC90359.1	38.1	38.3
Numida meleagris AEK75350.1	38.0	38.8
Anser anser AFO83527.1	38.4	38.1
Perdix perdix AFQ40032.1	38.6	39
Phasianus colchicus AEK75349.1	38.2	39.1
Gallus gallus ACR26269.1	38.1	38.5
Gallus lafayetii ACR26270.1	38.2	38.6
Anas platyrhynchos AFJ04295.1	37.7	38.3
Meleagris gallopavo ADX33343.1	37.6	38
Tadorna tadorna AGR50898.1	37.7	38.3
Xenopus laevis NP-001088449.1	37.1	37.3
Sus scrofa AGT79978.1	36.8	38.4
Homo sapiens AB060695.1	35.2	36.6
Mus musculus NP-058624.2	36.6	37.7
Canis lupus familiaris ACG60715.1	35.8	36.5

Phylogenetic tree was constructed to determine the evolutionary status of grass carp TLR5s. As shown in Fig. 3, the phylogenetic tree could be divided into two branches: TLR5 genes from the teleost fish fell into one branch; TLR5 genes from mammalia, repitilia, aves, and amphibia fell into another branch. CiTLR5a and CiTLR5b were close to TLR5 from *Cirrhinus mrigala*, *Cyprinus carpio*, and *Danio rerio* in the phylogenetic tree (Fig. 3).

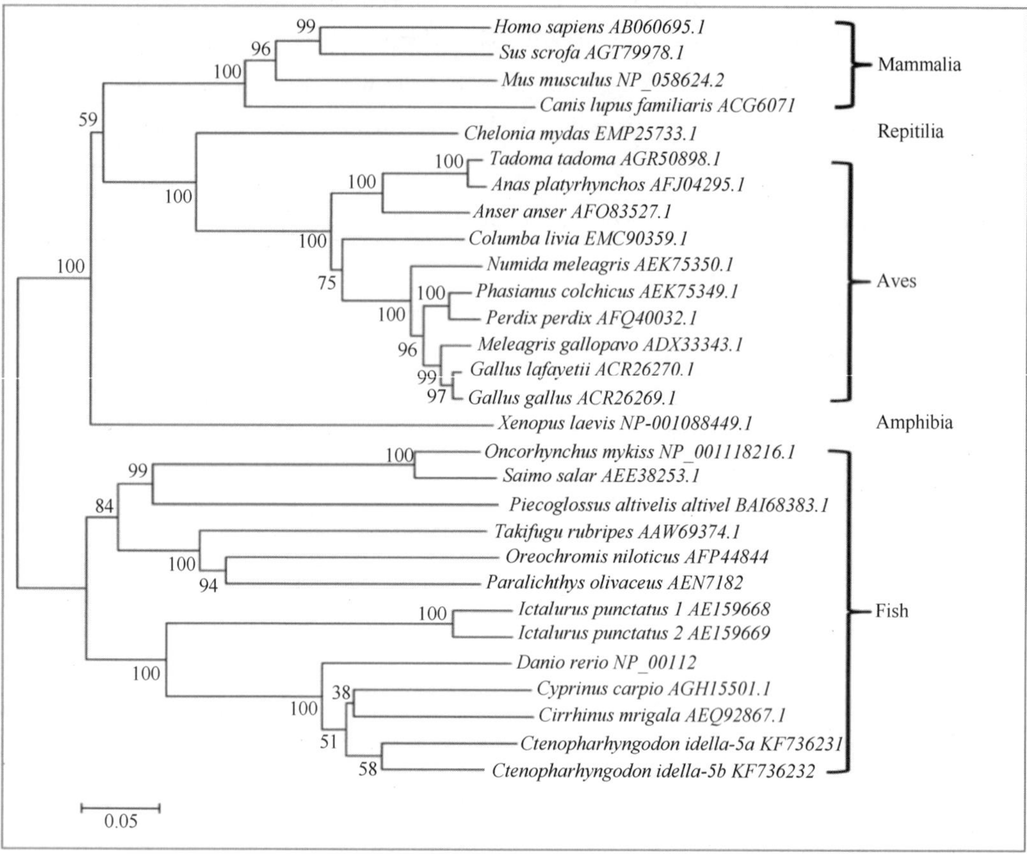

Fig. 3 Phylogenetic analysis of grass carp TLR5a and TLR5b. Phylogenetic tree was constructed by using the neighbor-joining method with Mega 5.1. The Genbank accession numbers of TLR5 homologues are given after the species names in the tree

3.3 Subcellular localization of TLR5s

To investigate the subcellular localization of TLR5 proteins and the difference between CiTLR5a and CiTLR5b, CIK cells were transfected with plasmid pEGFP-5a or pEGFP-5b and subjected to fluorescence observation at 24h post transfection. The empty plasmid pEGFP-N3 was transfected at the same time as a negative control. As shown in Fig.4, green fluorescence of 5a-EGFP distributed in the cytoplasm of transfected cells and presented as filiform or reticulation, suggesting that 5a-EGFP may locate in the plasma membrane of CIK cells. 5b-EGFP showed similar localization pattern to 5a-EGFP. As a control, the naked EGFP distributed uniformity in the whole cells (Fig. 4).

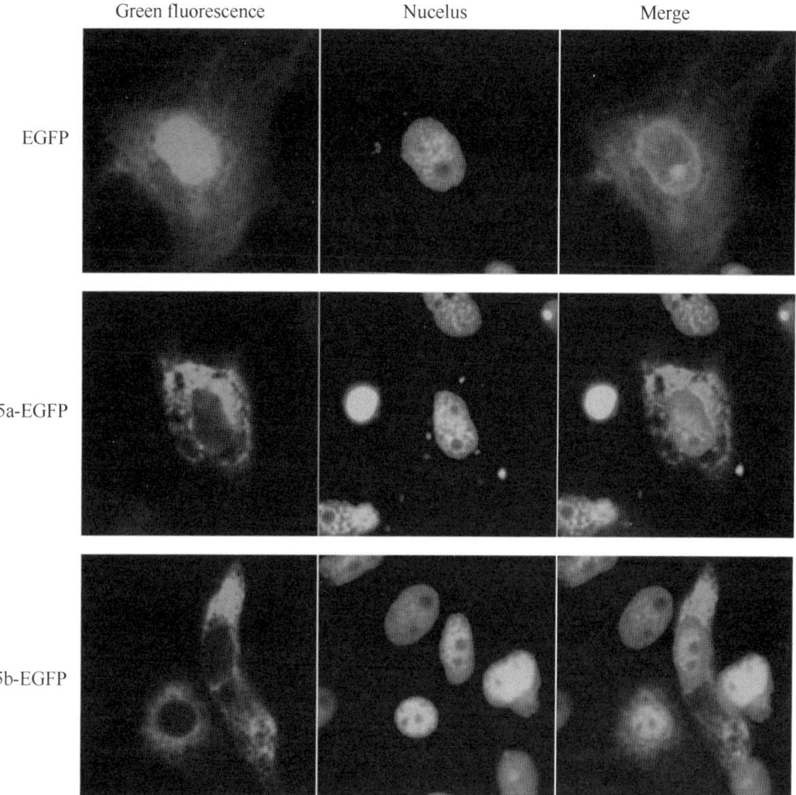

Fig. 4 Subcellular localization of CiTLR5a and CiTLR5b in CIK cells. Cells were transfected with plasmids and subjected to fluorescence observation at 24h post transfection. Green fluorescence shows the distribution of EGFP or EGFP–tagged proteins, blue fluorescence shows the nucleus that stained by Hoechst 33342.Magnification ×100 (oil-immersion objective). (For interpretation of the references to color in this figure legend, the reader is referred to the web version of this article.)

3.4 Expression of TLR5s in different tissues

qPCR was carried out to study the tissue distribution of grass carp TLR5s in different tissues. As shown in Fig. 4, CiTLR5a and CiTLR5b were constitutively expressed in all the nine examined tissues. CiTLR5a was highly expressed in head kidney and kidney, but was lowly expressed in liver, brain, intestines, and spleen. Expression of CiTLR5b was higher in gill, heart, head kidney, kidney, and muscle, and was also lower in liver, brain, intestines, and spleen (Fig. 5).

3.5 Expression of TLR5s and TLR5 downstream genes after flagellin stimulation

In order to understand the response of CiTLR5a, CiTLR5b, and their downstream genes after flagellin stimulation, total RNA from the cells that stimulated by flagellin and harvested at different time points was extracted for qPCR analysis. As shown in Fig.6, compared with the control (0h), expression level of CiTLR5a and CiTLR5b was significantly up-regulated following flagellin stimulation (p<0.05). The response pattern of CiTLR5a and CiTLR5b was similar, but the absolute mRNA expression level and durations differed ($p<0.05$). Moreover, the response of TLR5 downstream genes such as MyD88, IκB, NF-κB, IRF7, IL-1β, and TNF-α was also investigated. Expression level of MyD88, NF-κB, IRF7, IL-1β, and TNF-α was also significantly up-regulated after exposure to flagellin ($p<0.05$). However, the expression level of IκB, an inhibitor of NF-κB, was down-regulated ($p<0.05$) (Fig. 6).

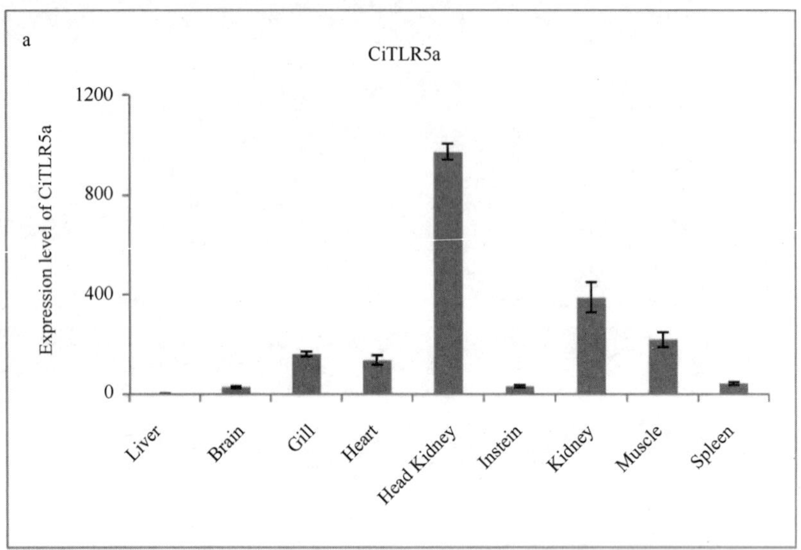

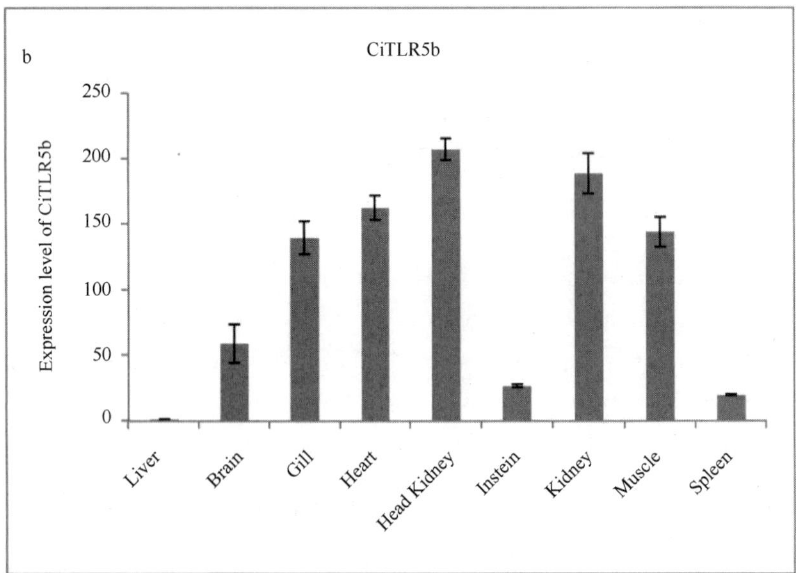

Fig. 5 Expression level of CiTLR5a (a) and CiTLR5b (b) in different tissues of grass carp. RNA was isolated from liver, brain, gill, heart, head kidney, intestine, kidney, muscle, and spleen and was reverse-transcribed into cDNA for qPCR analysis. The relative expression is the ratio of gene expression in different tissues relative to that in the liver. The β-actin gene was used as the internal control

3.6 Expression of TLR5s and TLR5 downstream genes in response to GCRV infection

To reveal the response of CiTLR5a and CiTLR5b after exposure to GCRV, total RNA was extracted from four tissues (gill, liver, spleen, and intestine) of five grass carp at 0-6 days post GCRV infection and reverse transcribed for qPCR analysis. As shown in Fig.7, compared with the control (0 day), the expression of CiTLR5a and CiTLR5b was significantly altered in gill, liver, spleen, and intestine kidney following GCRV infection. The overall expression patterns were similar for CiTLR5a and CiTLR5b genes, but the absolute mRNA expression level and durations differed. The expression level of CiTLR5a and CiTLR5b was significantly up-regulated in the liver at first days post infection (p<0.05). At the 3-5 days post infection, expression level of CiTLR5a and CiTLR5b was evidently increased in intestine (p<0.05). However, the expression level of CiTLR5b was decreased in some tissues after GCRV infection (p<0.05) (Fig. 7).

In addition, the expression level of MyD88, IκB, NF-κB, IRF7, IL-1β, and TNF-α before and after GCRV infection was also determined by qPCR. As shown in Fig.5, expression of MyD88, NF-κB, and IRF7 was up-regulated in different tissues after exposure to GCRV (p<0.05), whereas expression level of IκB was decreased (p<0.05). Moreover, expression of IL-1β and TNF-α was signifantly up-regulated in all the four tissues at the first day after GCRV infection (p<0.01), suggesting a strong immune response was induced (Fig. 7).

4. Discussion

There are two types of TLR5 existed in teleost fish: the membrane form of TLR5 (TLR5M) and the soluble form of TLR5 (TLR5S). The TLR5M of teleost fish was orthologous to TLR5 of mammalians, which contained LRR motifs, transmembrane region, and TIR domain. However, only LRR motifs were found in TLR5S, which was only presented in some teleost fish [22]. In the study, two TLR5 genes were cloned from grass carp. Both of them included LRR domain, transmembrane region, and TIR domain, so they are membrane form of TLR5. By homology search in then reference genome of grass carp (data not shown), no soluble form of TLR5 was detected. Previous study showed that only membrane form of TLR5 was found in zebrafish (*Danio rerio*), common carp (*Cyprinus carpio*), and Indian major carp (*Cirrhinus mrigala*) [19,20]. Grass carp, zebrafish, common carp, and Indian major carp belong to the *cyprinidae* family, and then the soluble form of TLR5 may not be existed in grass carp. However, further research is needed for confirmation.

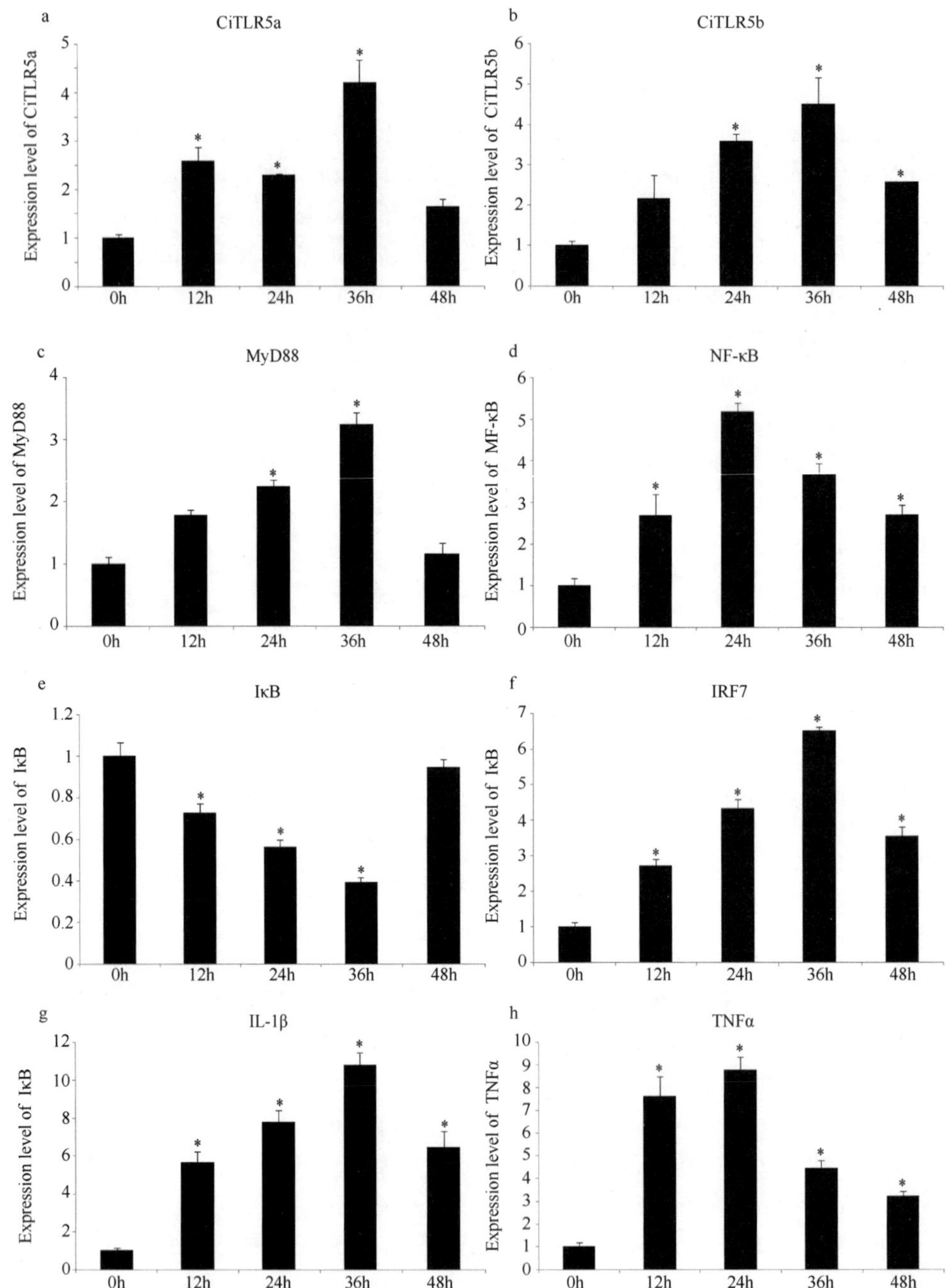

Fig. 6 Expression level of CiTLR5a, CiTLR5b, and TLR5 downstream genes after flagellin stimulation. RNA was isolated from cells after flagellin stimulation and harvested at 0, 12, 24, 36, and 48 h and was reverse-transcribed into cDNA. The expression level of CiTLR5a (a), CiTLR5b (b), MyD88 (c), NF-κB (d), IκB (e), IRF7 (f), IL-1β (g), and TNF-α (h) was determined by qPCR. The relative expression is the ratio of gene expression after exposure to flagellin (12, 24, 36, and 48 h) to that in the control group (0 h), normalizedto the β-actin gene. Significant difference (p<0.05) between the control and treated group was indicated with asterisks (*)

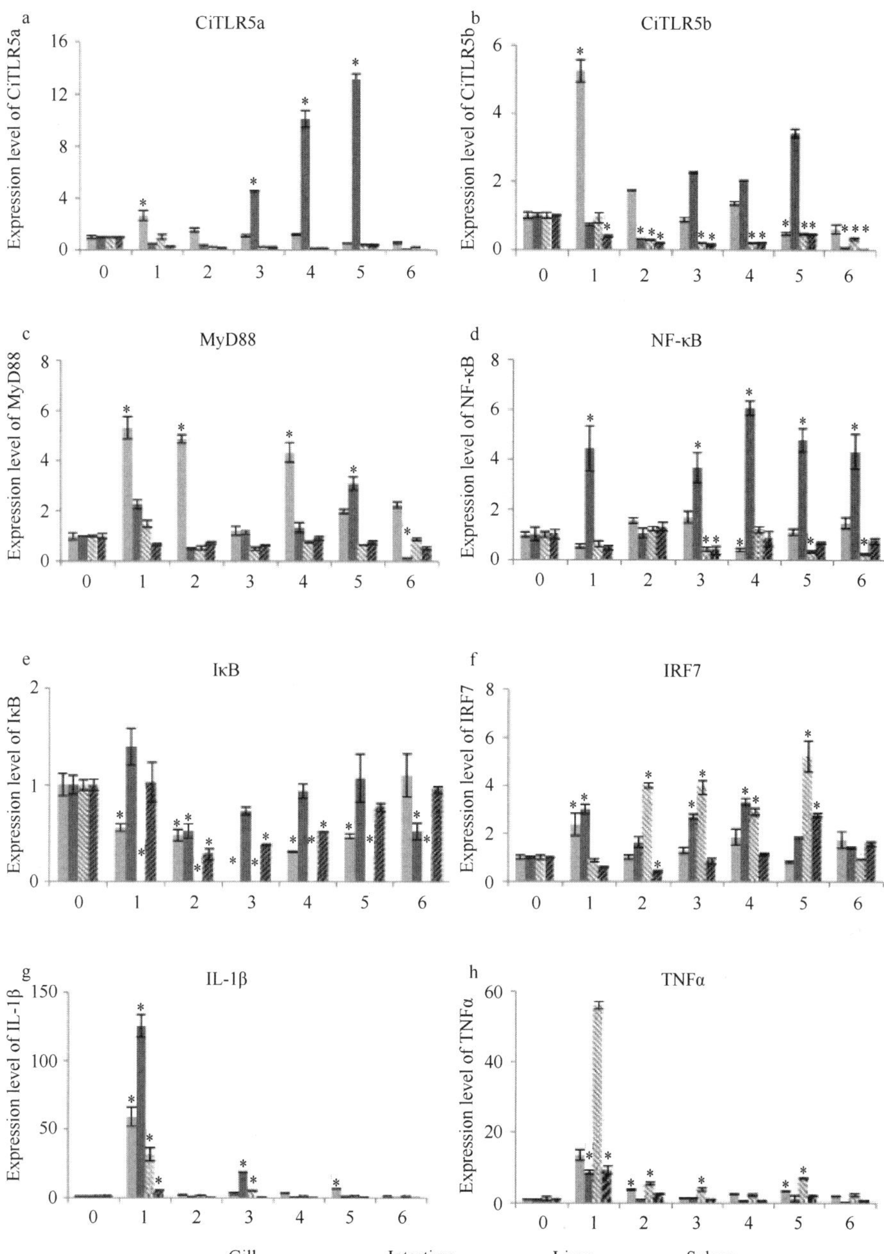

Fig.7 Expression level of CiTLR5a, CiTLR5b, and TLR5 downstream genes in response to GCRV infection. RNA was isolated from gill, liver, spleen, and intestine after exposure to GCRV at 0, 1, 2, 3, 4, 5, and 6 days and was reverse-transcribed into cDNA. The expression level of CiTLR5a (a), CiTLR5b (b), MyD88 (c), NF-κB (d), IκB (e), IRF7 (f), IL-1β (g), and TNF-α (h) was determined by qPCR. The relative expression is the ratio of gene expression after exposure to GCRV (1, 2, 3, 4, 5, and 6 days) to that in the control group (0 day) at same tissue, normalized to the β-actin gene. Significant difference (p<0.05) between the control and treated group was indicated with asterisks (*)

Both TLR5M and TLR5S were existed in rainbow trout (*Onchorhynchus mikiss*). The study of TLR5 in rainbow trout revealed that flagellin could be recognized by TLR5M, and then NF-κB was activated to induce immune responsive genes and TLR5S. Furthermore, the induced TLR5S could recognize flagellin in fluid phase and bind to TLR5M to amplify the signaling cascades [14]. Compared with the terrestrial animals, the survival environment of fish is more complex and diversity [22]. The two forms of TLR5 may be a beneficial strategy for teleost fish to adapt the complex environment. Although no TLR5S was found in grass carp and zebrafish, two copies of TLR5M were existed. The two copies of TLR5 genes in grass carp and zebrafish may be an altered strategy for them to adapt the complex environment. CiTLR5a and CiTLR5b had similar structure and localization pattern. Moreover, both CiTLR5a and CiTLR5b could response to flagellin stimulation and GCRV infection quickly and induced downstream immune response. These results suggested that CiTLR5a and CiTLR5b may have similar function.

In mammalians, TLR5 could recognize flagellin and via the MyD88-dependent signaling pathway to induce immune response [23]. In order to investigate whether TLR5s of grass carp had similar functions in signal transmission and ligand recognition, CIK cells were stimulated by flagellin from *Salmonella typhimurium* and the response of TLR5s and their downstream genes were determined by qPCR. Expression level of CiTLR5a, CiTLR5b, and their downstream genes such as MyD88, NF-κB, IRF7, IL-1β, and TNF-α were up-regulated, whereas the expression level IκB gene was down-regulated. These results suggested that TLR5s of grass carp may have similar functions to TLR5 of mammalians in signal transmission and ligand recognition.

Interestingly, expression level of CiTLR5a and CiTLR5b was up-regulated after GCRV infection. In addition, expression level of TLR5 downstream genes such as MyD88, NF-κB, IRF7, IL-1β, and TNF-α also up-regulated. Especially the IL-1β and TNF-α, which were up-regulated significantly in the first day after GCRV infection. GCRV belongs to the *Reoviridae* family, which contains double-strands RNA and two capsid layers [24]. In TLR, the recognition of double-strands RNA was mediated by TLR3, which activated the NF-κB and other downstream genes via the MyD88-independent signaling pathway [25, 26]. In our study, the up-regulation of TLR5s and their downstream genes may be due to the signaling feedback system. Similar phenomenon was also observed in TLR4, TLR7, TLR8, and TLR22 of grass carp, which were also up-regulated after GCRV infection [27-30].

In conclusion, two TLR5 genes were cloned from grass carp. Both of them are membrane form of TLR5 and shared similar structure and localization pattern. CiTLR5a and CiTLR5b were constitutively expressed in all examined tissues, but the highest expressed tissue was different. After exposure to flagellin and GCRV, CiTLR5a, CiTLR5b and their downstream genes such as MyD88, NF-κB, IRF7, IL-1β, and TNF-α were up-regulated in different tissues. Our study would provide a new insight into the understanding of TLR5 in teleost fish.

Acknowledgements This work was funded by the National Natural Science Foundation of China (No. 31130055), National High Technology Research and Development Program (No. 2011AA100403), and Direction Program of Chinese Academy of Sciences (No. KSCX2-EW-N-004-3).

References

[1] FAO. Fishery and aquaculture statistics yearbook. Rome: Food and Agriculture Oranization of the United Nations; 2013.

[2] Tian Y, Ye X, Zhang L, Deng G, Bai Y. Development of a novel candidate subunit vaccine against grass carp reovirus Guangdong strain (GCRV-GD108). Fish Shellfish Immunol 2013, 35; 351-356.

[3] Shi MJ, Huang R, Du FK, Pei YY, Liao LJ, Zhu ZY, et al. RNA-seq profiles from grass carp tissues after reovirus (GCRV) infection based on singular and modular enrichment analyses. Mol Immunol 2014, 61(1): 44-53.

[4] Liang HR, Li YG, Zeng WW, Wang YY, Wang Q, Wu SQ. Pathogenicity and tissue distribution of grass carp reovirus after intraperitoneal administration. Virol J 2014, 11: 178.

[5] Palti Y. Toll-like receptors in bony fish: from genomics to function. Dev Comp Immunol 2011, 35(12): 1263-1272.

[6] Purcell MK, Smith KD, Hood L, Winton JR, Roach JC. Conservation of toll-Like receptor signaling pathways in teleost fish. Comp Biochem Physiol Part D Genomics Proteomics 2006, 1(1): 77-88.

[7] Yamamoto M, Takeda K. Current views of toll-like receptor signaling pathways Gastroenterol Res Pract 2010, 2010: 240365.

[8] Medzhitov R. Toll-like receptors and innate immunity. Nat Rev Immunol 2001, 1: 135-145.

[9] Akira S, Takeda K. Toll-like receptor signalling. Nat Rev Immunol 2004, 4: 499-511.

[10] Sebastiani G, Leveque G, Larivière L, Laroche L, Skamene E, Gros P, et al. Cloning and characterization of the murine toll-like receptor 5 (Tlr5) gene: sequence and mRNA expression studies in Salmonella-susceptible MOLF/Ei mice. Genomics 2000, 64(3): 230-240.

[11] Didierlaurent A, Ferrero I, Otten LA, Dubois B, Reinhardt M, Carlsen H, et al. Flagellin promotes myeloid differentiation factor 88-dependent development of Th2-typeresponse. J Immunol 2004, 172 (11): 6922-6930.

[12] Means TK, Hayashi F, Smith KD, Aderem A, Luster AD. The Toll-like receptor 5 stimulus bacterial flagellin induces maturation and chemokine production in human dendritic cells. J Immunol 2003, 170(10): 5165-5175.

[13] Vijay-Kumar M, Aitken JD, Gewirtz AT. Toll like receptor-5: protecting the gut from enteric microbes. Semin Immunopathol 2008, 30(1): 11-21.

[14] Tsujita T, Tsukada H, Nakao M, Oshiumi H, Matsumoto M, Seya T. Sensing bacterial flagellin by membrane and soluble orthologs of Toll-like receptor 5 in rainbowtrout (*Onchorhynchus mikiss*). J Biol Chem 2004, 279(47): 48588-48597.

[15] Tsoi S, Park KC, Kay HH, O'Brien TJ, Podor E, Sun G, et al. Identification of a transcript encoding a soluble form of toll-like receptor 5 (TLR5) in Atlantic salmon during *Aeromonas salmonicida* infection. Vet Immunol Immunopathol 2006, 109(1-2): 183-187.

[16] Baoprasertkul P, Xu P, Peatman E, Kucuktas H, Liu Z. Divergent Toll-like receptors in catfish (*Ictalurus punctatus*): TLR5S, TLR20, TLR21. Fish Shellfish Immunol 2007, 23(6): 1218-1230.

[17] Hwang SD, Asahi T, Kondo H, Hirono I, Aoki T. Molecular cloning and expression study on Toll-like receptor 5 paralogs in Japanese flounder, Paralichthys olivaceus. Fish Shellfish Immunol 2010, 29(4): 630-638.

[18] Muñoz I, Sepulcre MP, Meseguer J, Mulero V. Molecular cloning, phylogenetic analysis and functional characterization of soluble Toll-like receptor 5 in gilthead seabream, *Sparus aurata*. Fish Shellfish Immunol 2013, 35(1): 36-45.

[19] Basu M, Swain B, Maiti NK, Routray P, Samanta M. Inductive expression of toll-like receptor 5 (TLR5) and

associated downstream signaling molecules following ligand exposure and bacterial infection in the Indian major carp, mrigal (*Cirrhinus mrigala*). Fish Shellfish Immunol 2012, 32(1): 121-131.

[20] Duan D, Sun Z, Jia S, Chen Y, Feng X, Lu Q. Characterization and expression analysis of common carp *Cyprinus carpio* TLR5M. DNA Cell Biol 2013, 32(10): 611-620.

[21] He LB, Ke F, Wang J, Gao XC, Zhang QY. *Rana grylio* virus (RGV) envelope protein 2L: subcellular localization and essential roles in virus infectivity revealed by conditional lethal mutant. J Gen Virol 2014, 95(3): 679-690.

[22] Rebl A, Goldammer T, Seyfert HM. Toll-like receptor signaling in bony fish. Vet Immunol Immunopathol 2010, 134(3-4): 139-150.

[23] Barton GM, Medzhitov R. Toll-like receptor signaling pathways. Science 2003, 300(5625): 1524-1525.

[24] Attoui H, Mertens PPC, Becnel J, Belaganahalli S, Bergoin M, Brussaard CP, et al. Family *reoviridae*. In: King, AMQ, Adams MJ, Carstens EB, Lefkowitz EJ. editors, Virus taxonomy: ninth report of the international committee on taxonomy of viruses. San Diego, CA, Elsevier; 538-637.

[25] Vercammen E, Staal J, Beyaert R. Sensing of viral infection and activation of innate immunity by toll-like receptor 3. Clin Microbiol Rev 2008, 21(1): 13-25.

[26] Xia J, Winkelmann ER, Gorder SR, Mason PW, Milligan GN. TLR3- and MyD88-dependent signaling differentially influences the development of West Nile virus-specific B cell responses in mice following immunization with RepliVAX WN, a single-cycle flavivirus vaccine candidate. J Virol 2013, 87(22): 12090-12101.

[27] Chen X, Wang Q, Yang C, Rao Y, Li Q, Wan Q, et al. Identification, expression profiling of a grass carp TLR8 and its inhibition leading to the resistance to reovirus in CIK cells. Dev Comp Immunol 2013, 41(1): 82-93.

[28] Huang R, Dong F, Jang S, Liao L, Zhu Z, Wang Y. Isolation and analysis of a novel grass carp toll-like receptor 4 (tlr4) gene cluster involved in the response to grass carp reovirus. Dev Comp Immunol 2012, 38(2): 383-388.

[29] Lv J, Huang R, Li H, Luo D, Liao L, Zhu Z, et al. Cloning and characterization of the grass carp (*Ctenopharyngodon idella*) Toll-like receptor 22 gene, a fish-specific gene. Fish Shellfish Immunol 2012;32(6):1022-1031.

[30] Yang C, Su J, Zhang R, Peng L, Li Q. Identification and expression profiles of grass carp *Ctenopharyngodon idell*a tlr7 in responses to double-stranded RNA and virus infection. J Fish Biol 2012, 80(7): 2605-2622.

草鱼 CiTLR5a 和 CiTLR5b 基因的分离及对鞭毛和草鱼出血病病毒的应答

江　遥[1,3]　何利波[1]　朱昌成[1,2]　裴永艳[1,3]　Myonghuan Ji[2]
李勇明[1]　廖兰杰[1]　张成勋[1,2]　朱作言[1]　汪亚平[1]

1 中国科学院水生生物研究所，武汉 430072，中国
2 Department of Zoology, College of Life Sciences, Kim Il Song University, Pyongyang, Democratic People's Republic of Korea
3 中国科学院大学，北京 100049，中国

摘　要　Toll 样受体 5(TLR5)是 Toll 样受体家族的成员，在脊椎动物中负责细菌鞭毛蛋白的识别，在先天免疫中起着重要的作用。本研究克隆并分析了草鱼的两种 TLR5 基因，CiTLR5a 和 CiTLR5b。CiTLR5a 和 CiTLR5b 都是典型的 Toll 样受体蛋白质，包含富含亮氨酸重复序列基序，跨膜区和 Toll 受体同源结构域。CiTLR5a cDNA 全长 3054 bp，包含 2646 bp 开放阅读框(ORF)，78 bp 5′ 非翻译区(UTR)和 330 bp 3′ UTR。CiTLR5b 的 cDNA 全长为 332 6bp，包含 2627 bp ORF，95 bp 5′ UTR，594 bp 3′ UTR。系统发育树分析显示 CiTLR5a 和 CiTLR5b 与鲮、鲤和斑马鱼的 TLR5 相近。亚细胞定位表明 CiTLR5a 和 CiTLR5b 共享类似定位模式和可能存在于转染细胞的质膜区域。实时定量 PCR 显示 CiTLR5a 和 CiTLR5b 在所有检测的组织中持续表达，然而在各组织中表达量不同。鞭毛蛋白和草鱼呼肠孤病毒(GCRV)感染后，CiTLR5a 和 CiTLR5b 显著上调。此外，TLR5 下游信号通路的基因，如 MyD88、NF-κB、IRF7、IL-1β 和 TNF-α 也显著上调，而 IκB 基因下调，表明 CiTLR5a 和 CiTLR5b 参与应对鞭毛蛋白刺激和 GCRV 感染。本研究获得的结果将为进一步理解 TLR5 在硬骨鱼中发挥的作用提供新的见解。

Characterizations of Four Toll-Like Receptor 4s in Grass Carp *Ctenopharyngodon idellus* and Their Response to Grass Carp Reovirus Infection and Lipopolysaccharide Stimulation

Y Y Pei[1, 2] R Huang[1] Y M Li[1] L J Liao[1] Z Y Zhu[1] Y P Wang[1]

1 State Key Laboratory of Freshwater Ecology and Biotechnology, Institute of Hydrobiology, Chinese Academy of Sciences, Wuhan 430072

2 Graduated University of Chinese Academy of Sciences, Beijing 100049

Abstract In this study, the subcellular localization, tissue distribution, and responses to grass carp reovirus (GCRV) infection and lipopolysaccharide (LPS) stimulation of four grass carp *Ctenopharyngodon idellus* toll-like receptor 4 (*tlr4*) genes were investigated. All four genes were constitutively expressed in all tissues studied, but the subcellular localization and tissue exhibiting the highest expression differed for each protein. Following GCRV infection, all the four *tlr4s* were upregulated in all tissues examined, and stimulation of *C. idellus* kidney (CIK) cells with LPS resulted in downregulation of all four *tlr4s*. These results provide a foundation for further investigation of *tlr4* genes in bony fishes

Keywords gene expression; grass carp reovirus; subcellular localization; tissue distribution; virus infection

Introduction

The grass carp *Ctenopharyngodon idellus* (Valenciennes 1844) has been an important aquaculture species in China for over 60 years, and accounting for > 18% of total freshwater aquaculture production. *Ctenopharyngodon idellus* production reached 478.2×10^6 t in 2012, making it the most highly consumed freshwater fish worldwide (FAO, 2013). Disease outbreaks, however, are frequency and result in huge economic losses. Grass carp haemorrhagic disease, caused by the grass carp reovirus (GCRV), is one of the most damaging diseases. Therefore, identification of *C. idellus* genes involved in disease resistance and immunity is important for antiviral drug development and fish breeding programmes.

Tlr4 was the first member of the toll-like receptor (TLR) family to be identified and characterized, and this protein plays an important role in innate immunity (Medzhitov*et al.*, 1997; Takeda & Akira, 2003). Tlr4 comprises three major domains: a tandem leucine-rich repeat (LRR) motif, a transmembrane region (TM) and an intracellular Toll/IL-1 receptor (TIR)

文章发表于 *Journal of Fish Biology*, 2015, 86(3): 1098-1108

domain (Yamamoto & Takeda, 2010). In mammals, lipopolysaccharide (LPS) derived from bacterial cell walls is recognized by Tlr4 (Poltorak *et al.*, 1998; Chow *et al.*, 1999; Hoshino *et al.*, 1999). Lower vertebrates such as fishes and amphibians, however, are resistant to the toxic effects of LPS (Berczi *et al.*, 1996). Moreover, viral proteins are also recognized by the mammalian Tlr4 protein, and recognition of respiratory syncytial virus (RSV) by mammalian Tlr4 leads to activation of the cellular immune response (Kurt-Jones *et al.*, 2000; Haynes *et al.*, 2001). Similarly, recognition of mouse mammary tumour virus (MMTV) by Tlr4 also activates innate immunity (Rassa *et al.*, 2002). In rare minnow, *Gobiocypris rarus* Ye & Fu 1983, *tlr4b* was up-regulated significantly following GCRV infection (Su *et al.*, 2009).

Mammals appear to contain a single copy of *tlr4* (Temperley *et al.*, 2008), whereas no *tlr4* gene was identified in the genome of three-spined stickleback *Gasterosteus aculeatus* L. 1758, green-spotted pufferish *Tetraodon nigroviridis* Marion de Procé 1822, or Japanese pufferish *Takifugu rubripes* (Temminck & Schlegel 1850) (Oshiumi *et al.*, 2003; Baoprasertkul *et al.*, 2007). In contrast, three *tlr4* tandem repeat-containing genes are encoded in a cluster in the zebrafish *Danio rerio* (Hamilton 1822) genome (Huang *et al.*, 2012). *Ctenopharyngodon idellus* has a cluster of four *tlr4* genes (*tlr4.1*, *tlr4.2*, *tlr4.3*, *and tlr4.4*). Tlr4.2, Tlr4.3 and Tlr4.4 exhibit the typical Tlr family architecture of a signal peptide, an LRR motif, a TM domain, and a TIR domain. In contrast, Tlr4.1 only contains an LRR motif, the other 207 bp being absent (Huang *et al.*, 2012). All the four *C. idellus tlr4* genes have been cloned, but their functions unknown and structure-function relationships are yet to be investigated.

In this study, the subcellular localization and tissue expression of four *C. idellus tlr4s* were investigated, and their behavior in response to GCRV infection and LPS stimulation was characterized.

Materials and Methods

Fish and cell lines

Three-month-old *C.idellus* (average mass 8 g) were supplied by the Guan qiao Experimental Fish Breeding Base, Institute of Hydrobiology, Chinese Academy of Sciences, China. *Ctenopharyngodon idella* kidney (CIK) cells were purchased from the Cell Preservation Center of Wuhan University and grown in M199 medium (HyClone; https://promo.gelifesciences.com/gl/hyclone/) containing 10% fetal bovine serum (HyClone) at 28℃, 5% CO_2.

Plasmid construction and cell transfection

Because of the numerous restriction enzyme sites in the *tlr4* sequences, the sequence and ligation-independent cloning (SLIC) method described previously by Li & Elledge (2007) was used for plasmids construction. Open reading frames (ORF) for the four *C. idellus tlr4s* were cloned into the plasmid expressing green flullorescent protein (pGFP)-N3 expression vector (Clontech; www.clontech.com) and confirmed by DNA sequencing. Cells were seeded onto cover slips in six-well plates and grown for 24 h. Transfection was performed using Lipofectamine 2000 (Invitrogen; www.lifetechnologies.com) according to the manufacturer's instructions. At 48 h post-transfection, the medium was removed and cells were fixed for 15 min in 4% formaldehyde solution. Cells were washed three times in phosphate-buffered saline (PBS) for 3-5min, labelled with 0.2 mg ml^{-1} 4′,6-diamidino-2-phenylindole dihydrochloride (DAPI) (Beyotime; www.bio-equip.com) in PBS for 1 min, and washed a further three times in PBS. Imaging was performed with a laser-scanning confocal microscope (Leica-NOL-LSM 710; www.leica-microsystems.com).

Expression, tissue distribution and response of *tlr4* genes to GCRV infection

A viral challenge experiment was performed using the feeding method. Briefly, dead fish exhibiting obvious hemorrhagic symptoms were homogenized to provide a source of GCRV. Normal feed was mixed with an equivalent amount of virus-infected material and fed to fish twice per day for 3 days, after which normal feed alone was provided. The temperature was maintained at 26-28℃ throughout the experiment. Three uninfected fish were selected and samples from gill, spleen, liver, intestine, kidney, head kidney, heart, muscle, skin and brain were prepared. In addition, kidney, gill, head, spleen and intestine samples were isolated from three infected fish at 2, 4, 6, and 8 days post infection. Samples were homogenized using TRIzol reagent (Invitrogen), and total RNA was isolated according to the manufacturer's instructions. Total RNA was incubated with RNase-free DNase I (Promega;www.promega.com) to remove genomic DNA, and reverse transcribed into cDNA using random hexamer primers with the ReverTra Ace kit (Toyobo; www.toyobo-global.com). Real-Time quantitative PCR (qPCR) was then employed to measure expression of the four *tlr4* genes using gene-specific qPCR primers (Table 1). β-Actin was used as an internal control. The relative expression level of each gene was calculated using the $2^{-\Delta\Delta Ct}$ method (Livak & Schmittgen, 2001). Data are expressed as mean ± *S.D.* A *t*-test was used to determine differences in expression, with a *P*-value <0.05 indicating statistical significance.

Expression of *tlr4* genes in CIK cells following LPS stimulation

CIK cells were transferred into a 60 mm cell culture dish containing 5 mL M199 media (Hyclone) and 10% fetal bovine serum (Hyclone) and cultured for 24 h at 28℃, 5% CO_2. Cells were incubated with (Sigma; www.sigmaaldrich.com) and collected 0, 6, 12, 24, 36, 48, and 72 h after incubation at 28℃, 5% CO_2. RNA was extracted and subjected to qPCR as described above.

Table I DNA sequences of PCR primers used in the study

Primers	Sequences(5′- 3′)	Use
Tlr4.1f1	CTCGAGCTCAAGCTTATGGGAAGAAACCTCAGC	Tlr4.1Vector constructing
Tlr4.1r1	CAATGTGCTTGATGTGGCATCTTGACAGATCAAGAACTCG	
Tlr4.1f2	CGAGTTCTTGATCTGTCAAGATGCCACATCAAGCACATTG	
Tlr4.1r2	GGTGGCGATGGATCCTGACCCAAAGTCTGCTGC	
Tlr4.2f	CTCGAGCTCAAGCTTATGAATAACCTAACCAAGC	Tlr4.2Vector constructing
Tlr4.2r	GGTGGCGATGGATCCTTGCTTTGTGGCAATAATAG	
Tlr4.3f1	CTCGAGCTCAAGCTTATGAGTTTCTTCACTCTAAGTG	Tlr4.3 Vector constructing
Tlr4.3r1	CAAAGTAGTCAAATTCTTCACATTGTAGAAAGCATC	
Tlr4.3r2	GATGCTTTCTACAATGTGAAGAATTTGACTACTTTG	
Tlr4.3r2	GGTGGCGATGGATCCTTGCTTTGTGGAAATAATAGC	
Tlr4.4f1	CTCGAGCTCAAGCTTATGATCATGTCATATTGGG	Tlr4.4 Vector constructing
Tlr4.4r1	TTTTCAATTTGCTTGATGTGGCATCTTGTGAGATCCAAAACTTGTAA	
Tlr4.4f2	TTACAAGTTTTGGATCTCACAAGATGCCACATCAAGCAAATTGAAAA	
Tlr4.4r2	GGTGGCGATGGATCCTTGCTTTATAATAGCTCTC	
Tlr4.1dF	GGGTTCTTTGCCAGGTAAACAACTG	Tlr4.1 qPCR
Tlr4.1dR	TGCACAGCATCGTTTTAATT	
Tlr4.2dF	ATTTGGAAAGTTCAGGGGA	Tlr4.2 qPCR
Tlr4.2dR	ACCATCCTTCACAACCACAA	
Tlr4.3dF	ATAAGATTTTCCCACCAGATTCAG	Tlr4.3 qPCR
Tlr4.3dR	TGTCACATACGGAATACCACG	
Tlr4.4dF	GTTACAGAATGCAATGGAAAAGCAGTG	Tlr4.4 qPCR
Tlr4.4dR	GGCAAAAAGGAACACGTTCCATTCTC	
β-actinF	GGATGATGAAATTGCCGCACTGG	β-actin qPCR
β-actinR	ACCGACCATGACGCCCTGATGT	

Results

Cellular localization of *C. idellus* Tlr4

Confocal microscopy was used to investigate the expression of enhanced green fluorescent protein (EGFP)-tagged Tlr4 proteins in CIK cells. Tlr4.1 was primarily present in the cytoplasm and nucleus (Fig. 1a), whereas Tlr4.2 was only observed in the cytoplasm (Fig. 1b). Tlr4.3 and Tlr4.4 were also present in the cytoplasm (Fig. 1c, d); however, they were localized in dense regions or spots, suggesting that they may be located at the plasma membrane and endosomes. As expected, expression of the negative control plasmid pEGFP-N3 resulted in a uniform green fluorescence throughout the cell (Fig. 1e).

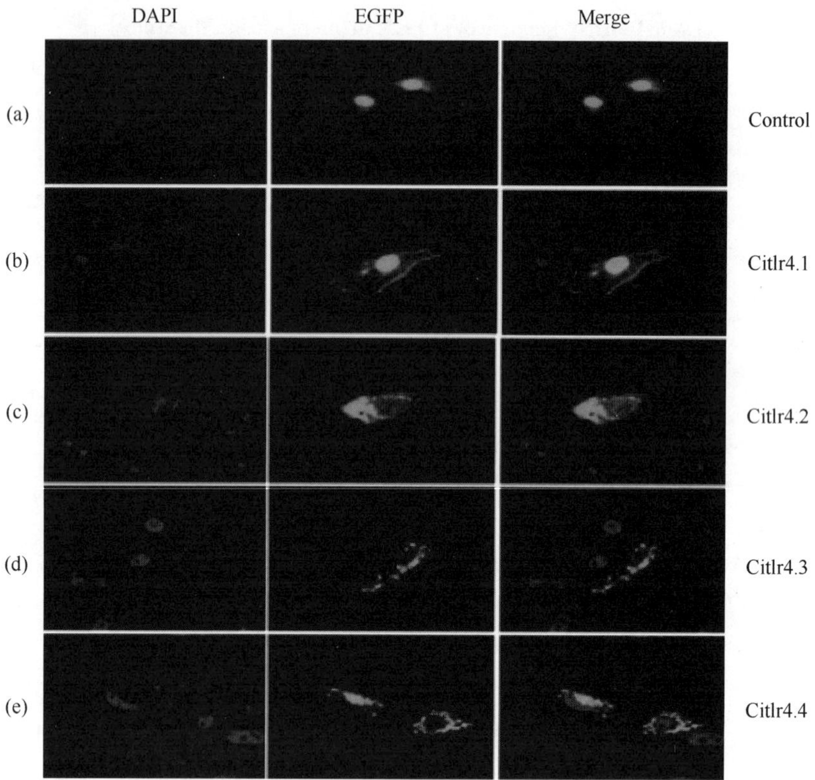

Fig. 1 Subcellular localization of *Ctenopharyngodon idellus* (Ci) tlr4s in *C. idellus* kidney (CIK) cells. After 48 h of transfection, cells were fixed in 4% formaldehyde solution for 15min, and nuclei were stained using 4′, 6-diamidino-2-phenylindole dihydrochloride (DAPI). Green and blue indicate localization of the green fluorescent protein (GFP) tag or GFP-tagged Citlr4s in the nucleus, respectively. (a) Control plasmid enhanced green flueorescent protein (pEGFP)-N3, (b) cellular location of CiTLR4.1, (c) cellular location of CiTLR4.2, (d) cellular location of CiTLR4.3 and (e) cellular location of CiTLR4.4

Expression of *tlr4* genes in different tissues

Real-time quantitative qPCR was used to investigate the tissue distribution of *tlr4* mRNA transcripts in samples from *C. idellus* gill, spleen, liver, intestine, kidney, head kidney, heart, muscle, skin, and brain. In healthy fish, *tlr4* genes were constitutively expressed in all tissues examined. Expression of *tlr4.1* was high in heart, brain and liver, and lower in kidney and intestine (Fig. 2a). *tlr4.2* mRNA was high in skin, liver and intestine, and low in head kidney and heart (Fig. 2b). Expression of *tlr4.3* was high in gill, skin and muscle, and low in head kidney and kidney (Fig. 2c). *tlr4.4* was most highly expressed in heart, skin and liver, and expression levels were lower in head kidney and kidney (Fig. 2d).

Expression of *tlr4s* in response to GCRV infection

In order to determine the expression patterns of *tlr4s* in response to GCRV infection *in vivo*, mRNA levels were measured in five immune organs using RT-qPCR. Severe disease symptoms were detected in GCRV-infected fish, including intestinal and muscular hemorrhage, at 8 days after infection. Meanwhile, disease-free control fish showed no disease symptoms. The expression of all four *tlr4s* was significantly altered in gill, liver, intestine, spleen and kidney following GCRV infection (Fig. 3). The overall expression patterns were similar for all the four *tlr4* genes, but absolute mRNA levels and expression duration differed. In the gill, up-regulation of *tlr4.2* was most significant, which peaked 2 days after infection. In liver, *tlr4.1* and *tlr4.2* were the most upregulated, again peaking at two days postinfection. In spleen, *tlr4.4* upregulation was most prominent, peaking 6 days after infection. *Tlr4.4* was also significantly upregulated in kidney, peaking 4 days after infection. In the intestines, *tlr4.1* and *tlr4.4* were the most upregulated, peaking 2 and 8 days after GCRV infection, respectively.

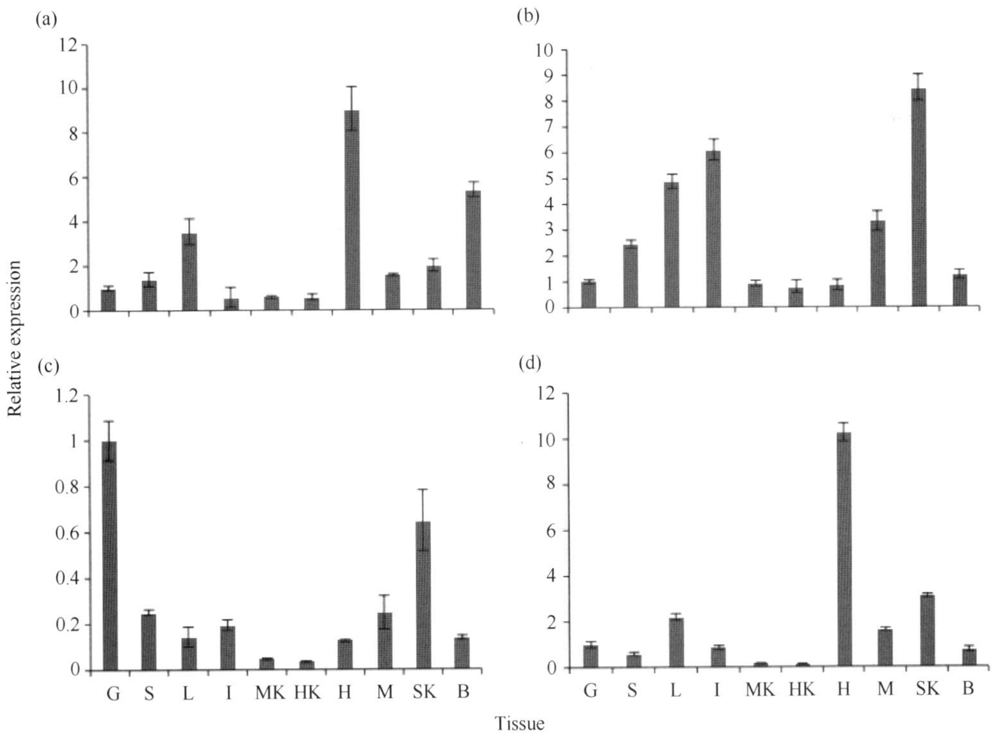

Fig. 2 Tissue distribution of (a) *Ctenopharyngodon idellus* (Ci) *tlr4.1*,(b) *Citlr4.2*,(c) *Citlr4.3* and (d) *Citlr4.4*. The *x*-axis indicates different tissues of *Ctenopharyngodon idellus*. RNA was isolated from gill (G), spleen (S), liver (L), intestine (I), kidney (MK), head kidney (HK), muscle (M), heart (H), skin (SK) and brain (B). The *y*-axis indicates the mean±S.E. fold change of expression

Expression of *C. idellus tlr4s* in response to LPS stimulation

Transcription of *C. idellus tlr4s* in CIK cells challenged by LPS was measured using qPCR. Following stimulation, cells were collected at 0, 6, 12, 24, 36, 48, and 72 h. Total RNA was extracted and reverse transcribed. All the four genes were generally downregulated at all time points, although *tlr4.1* and *tlr4.2* were upregulated slightly at 72 h (Fig. 4).

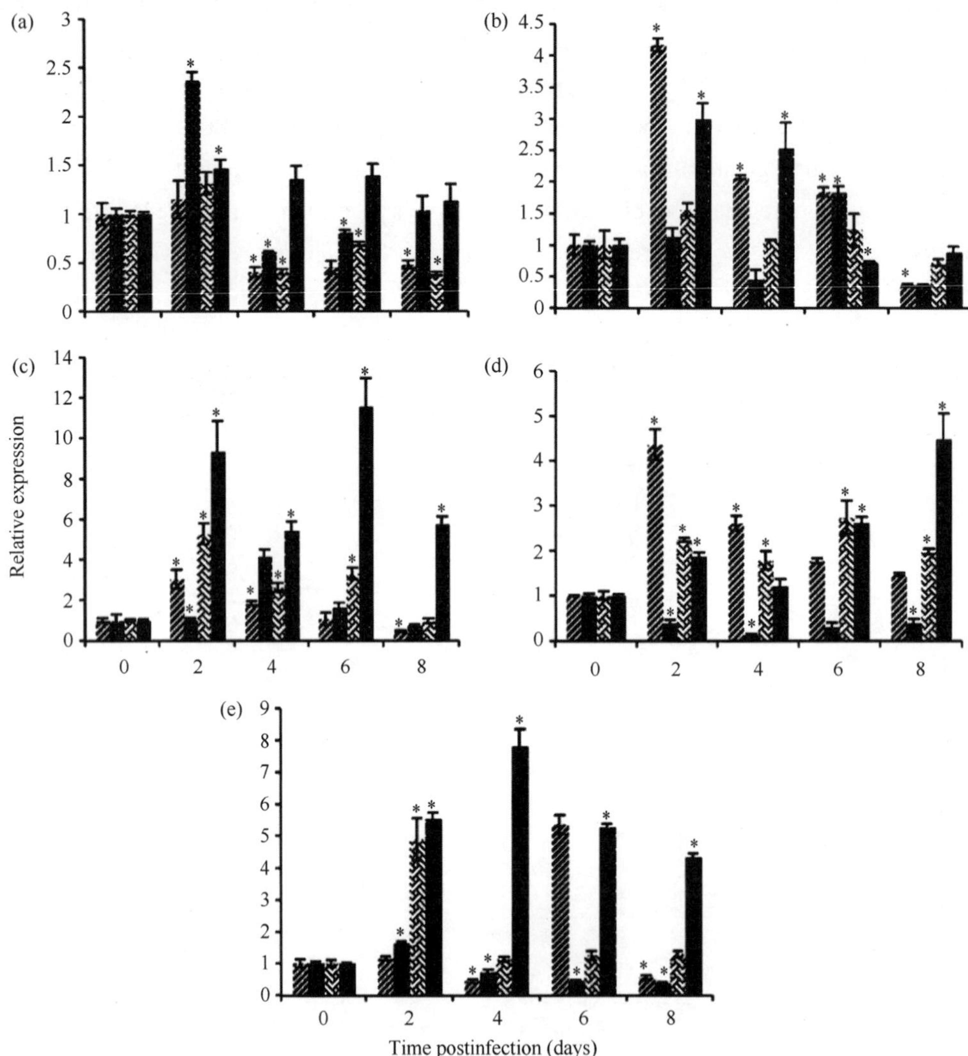

Fig. 3 Expression (mean±S.E.) of *Ctenopharyngodon idellus* (Ci) *tlr4s* genes (*Citlr4.1*; *Citlr4.2*; *Citlr4.3*; *Citlr4.4*) following grass carp reovirus (GCRV) infection in (a) gill, (b) liver, (c) spleen, (d) intestine and (e) kidney. *, Signiicant differences from the control (P<0.05)

Discussion

In mammals, the distribution of TLRs is coordinated with the ligands recognized by the receptor proteins, as would be expected. Based on the type of ligand bound, Tlrs are classified into two subclasses. One subclass is localized at the cell membrane and includes Tlr1, Tlr2, Tlr4, Tlr5 and Tlr6. Members of this subclass are associated with recognizing pathogen-associated molecular patterns (PAMP) such as peptidoglycan, LPS and flagella presented on the surface of pathogens. The other subclass is found in the membranes of intracellular compartments, and includes Tlr3, Tlr7, Tlr8 and Tlr9. These Tlrs recognize PAMPs such as nucleic acids, CpG polyunmethylated deoxyoligonucleotides and double-stranded RNAs (Janssens & Beyaert, 2003; Roach *et al.*, 2005; Akira *et al.*, 2006; Hughes & Piontkivska, 2008; Guan *et al.*, 2010; Zhang Y., *et al.*, 2013a).

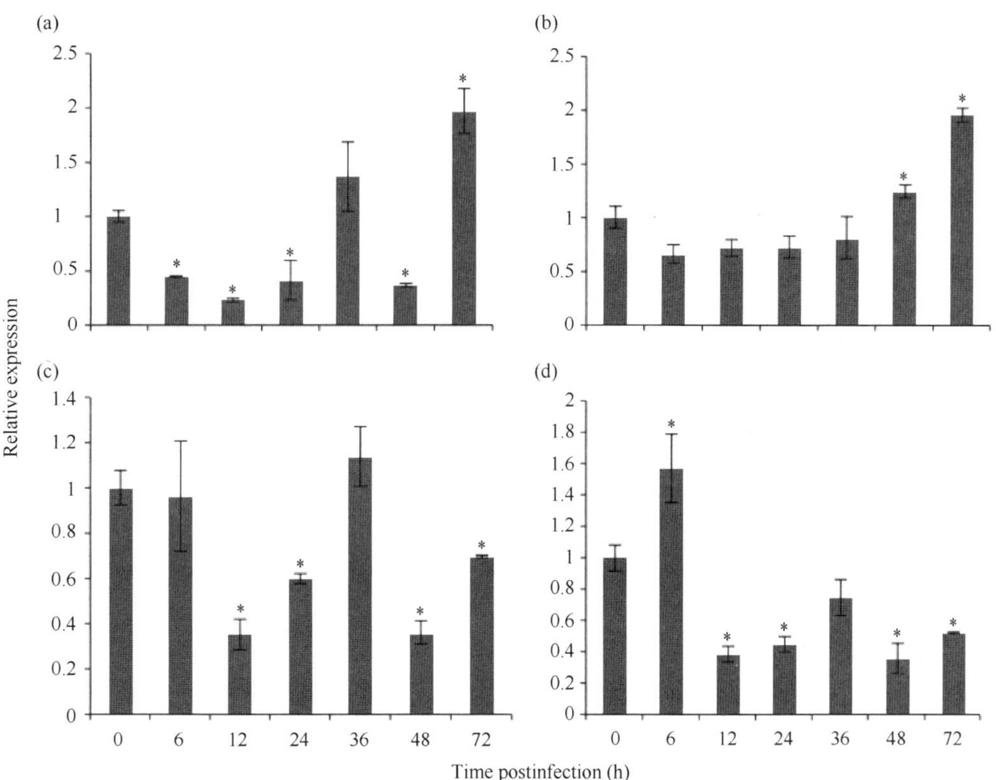

Fig. 4 Expression of (a) *Ctenopharyngodon idellus* (Ci) *tlr4.1*, (b) *Citlr4.2*, (c) *Citlr4.3* and (d) *Citlr4.4* following stimulation with lipopolysaccharide (LPS) in *Ctenopharyngodon idellus* kidney (CIK) cells. The y-axis indicates the mean±s.e. fold change of expression. The x-axis indicates the time after LPS stimulation. CIK cells were collected 0, 6, 12, 24, 36, 48 and 72 h after stimulation. *, Signiicant differences from the control (P<0.05)

In this study, four *C. idellus* Tlr4s exhibited different localization patterns in CIK cells. Tlr4.1 was present in the cytoplasm and nucleus, and lacked both a signal peptide and transmembrane domain. Tlr4.2 was primarily present in the cytoplasm and also lacked a signal peptide. Tlr4.3 and Tlr4.4 were found in the cytoplasm in dense zones or spots, and were probably localized at the plasma membrane and endosomes. In addition, both Tlr4.3

and Tlr4.4 were predicted to include a signal peptide and transmembrane domain, both of which appear to influence the location of Tlr4s in *C. idellus*. Import into the nucleus is usually dependent on the presence of a nuclear localization signal (NLS) (Dingwall *et al.*, 1982; Gasiorowski & Dean, 2003). some proteins however are imported into the nucleus by directly binding to the nuclear envelope (Fagotto *et al.*, 1998). In addition, many proteins may include an NLS that has not yet been identified (Zheng *et al.*, 2007). Tlr4.1 appears to lack an NLS, but may be imported into the nucleus in this way. Tlr4.3 and Tlr4.4 were present as dense spots, suggesting that they may be located at the plasma membrane and endosomes. Tlr4.2 was only present in the cytoplasm, possibly because of the lack of a signal peptide. Differences in localization may also reflect differences in structure. The localization of Tlr4.1 and Tlr4.2 suggests that these proteins may function as cytoplasmic pattern recognition receptors. Indeed, mammalian Tlr3 and Tlr7-9 are endosomal and are involved in recognition of virus particles (Takeda & Akira, 2005). The similar localization patterns of Tlr4.3 and Tlr4.4 with Tlr3 and Tlr7-9 suggests that they may recognize the same ligands.

All four *C. idellus tlr4s* were expressed in all 10 of the tissues examined, consistent with previous findings (Huang *et al.*, 2012). Similar results have been reported for *tlr4* genes in other fish; *D. rerio tlr4ba* was found to be expressed in all examined tissues (Jault *et al.*, 2004). Similarly, *tlr4s* are expressed constitutively in most tissues in birds and mammals, including liver, kidney, spleen, thymus, and lung, where the constitutive distribution is believed to contribute to a variety of immune responses (Zhao *et al.*, 2013). The ubiquitous expression of *tlr4*s indicates that they may play an important function. *tlr4.1* and *tlr4.2* expression was most abundant in heart, *tlr4.3*was highest in skin, and *tlr4.4* was most prevalent in the gill, suggesting different roles for different *tlr4s* in the tissues and organs of *C. idellus*.

Following GCRV infection, the virus was detected in gill, liver, intestines, spleen and kidney (Wang *et al.*, 1993; Li *et al.*, 1999), as was observed for all the four *tlr4s* in this study. Similarly, mammalian *tlr4* was upregulated after RSV infection (Kurt-Jones *et al.*, 2000; Haynes *et al.*, 2001), domestic *tlr4* was upregulated after a bacterial or viral challenge (Zhao *et al.*, 2013), and *tlr4b* was also be upregulated following GCRV infection in *G. rarus* (Su *et al.*, 2009). Taken together, these results strongly indicate that *tlr4* plays an important role in response to, virus infection. The expression of the different *tlr4s* in response to GCRV infection was different in gill, liver, intestine, spleen and kidney (Fig. 3). Although the absolute expression levels differed in different tissues, a similar overall pattern was present for all four genes, and resembled that of *tlr4b* in *G. rarus* (Su *et al.*, 2009). These results suggested that despite the likely functional differences, the four *tlr4s* may cooperate in the defense against viral infection.

Following stimulation by LPS, all the four genes were generally downregulated at all time points. In birds and mammals, *tlr4* was up-regulated by LPS (Lien *et al.*, 2000; Leveque *et al.*, 2003; Keestra *et al.*, 2008; Zhao *et al.*, 2013). MD-2 and CD14 were shown to be required for recognition of LPS by Tlr4 (Lu, *et al.*, 2008), but neither of these genes have been identified in fish (Rebl, *et al.*, 2010). This may explain why upregulation of *C. idellus tlr4s* was minimal following LPS stimulation. Moreover, amino-acid sequence alignment showed that the main differences between *C. idellus* and mammalian TLR4 are found in the N-terminal region, which includes residues Glu24 and Pro34 that are critical for interaction with MD-2 in mammalian Tlr4 (Nishitani *et al.*, 2005, 2006). this region was not present in Tlr4.1 or Tlr4.2, while Tlr4.3 and Tlr4.4 do contain this region, but only share a sequence similarity of 29% with the human protein This may also explain why *tlr4s* were not significantly upregulated following LPS stimulation. Interestingly, in *T. nigroviridis*, Tlr4 is replaced by scavenger receptors that recognize LPS and negatively regulate NF-κB activity (Meng *et al.*, 2012). Whether this phenomenon exists in other fishes is currently unknown. Following LPS stimulation, all the four *tlr4* genes were down regulated, as was observed previously in channel catish *Ictalurus punctatus* (Rainesque 1818) (Peatman *et al.*,2007; Ligas, 2008), infected with *Edwardsiella ictaluri* (Zhang, J., *et al.*, 2013). The response to LPS therefore differs among fishes, mammalsand birds.

In summary, subcellular localization and tissue expression of *C. idellus tlr4* genes were investigated, and the responses to GCRV infection and LPS stimulation were characterized. All the four *tlr4* genes were widely expressed in all tissues examined, but absolute expression levels and timing differed. Following GCRV infection, all the four *tlr4* genes were upregulated, but expression levels and timing again differed. Following LPS stimulation, all the four *tlr4* genes were downregulated slightly. The differences among the four *tlr4* genes suggest independent functions, while the similarities indicate that they may cooperate in virus recognition and defense. The response to LPS appears to proceed *via* a different pathway from the mammalian receptors.

Acknowledgements This work was funded by the National High Technology Research and Development Program (No. 2011AA100403), the National Natural Science Foundation of China (No. 31130055), and the Direction Program of Chinese Academy of Sciences (No. KSCX2-EW-N-004-3).

References

Akira, S., Uematsu, S., & Takeuchi, O. (2006). Pathogen recognition and innate immunity. *Cell* 124, 783-801.

Baoprasertkul, P., Xu, P., Peatman, E., Kucuktas, H. & Liu, Z. (2007). Divergent Toll-like receptors in catfish (*Ictalurus punctatus*): TLR5S, TLR20, TLR21. *Fish & Shellfish Immunol* 23, 1218-1230.

Berczi, I., Bertok, L. & Bereznai, T. (1966). Comparative studies on the toxicity of *Escherichia coli* lipopolysaccharide endotoxin in various animal species. *Canadian Journal of Microbiology* 12, 1070-1071.

Chow, J. C., Young, D. W. & Golenbock, D. T. (1999). Toll-like Receptor-4 Mediates lipopolysaccharide-induced Signal transduction. *Journal of Biological Chemistry* 274, 10689-10692.

Dingwall, C., Sharnick, S. V. & Laskey, R. A. (1982). A poly peptide domain that specifies migration of nucleoplasmin into the nucleus. *Cell* 30, 449-458.

Fagotto, F., Gluck, U. & Gumbiner, B. M. (1998). Nuclear localization signal-independent and importin/karyopherin-independent nuclear import of beta-catenin. *Current Biology* 8, 181-190.

FAO. (2013). Food and Agriculture Oranization of the United Nations. Fishery and Aquaculture Statistics Yearbook. Rome.

Gasiorowski, J. Z. & Dean, D. A. (2003). Mechanisms of nuclear transport and interventions. *Advanced Drug Delivery Reviews* 55, 703-716.

Guan, Y., Ranoa, D. R., Jiang, S., Mutha, S. K., Li, X., Baudry, J. & Tapping, R. I. (2010). Human TLRs10 and 1 share common mechanisms of innate immune sensing but not signaling. *The Journal of Immunology* 184, 5094-5103.

Haynes, L. M., Moore, D. D., Kurt-Jones, E. A., Finberg, R. W., Anderson, L. J. & Tripp, R. A. (2001). Involvement of toll-like receptor 4 in innate immunity to respiratory syncytial virus. *Journal of Virology* 75, 10730-10737.

Hoshino, K., Takeuchi, O., Kawai, T., Sanjo, H., Ogawa, T., Takeda, Y., Takeda, K. & Akira, S. (1999). Cutting edge: toll-like receptor4 (TLR4)-dficient mice are hyporesponsive to lipopolysaccharide: evidence for TLR4 as the LpsGeneProduct. *Journal of Immunology* 162, 3749-3752.

Huang, R., Dong, F., Jang, S. H., Liao, L. J., Zhu, Z. Y. & Wang, Y. P. (2012). Isolation and analysis of a novel grass carp toll-like receptor 4 (TLR4) gene cluster involved in the response to grass carp reovirus. *Developmental & Comparative Immunology* 38, 383-388.

Hughes, A. L. & Piontkivska, H. (2008). Functional diversification of the toll-like receptor gene family. *Immunogenetics* 60, 249-256.

Janssens, S. & Beyaert, R. (2003). Role of Toll-like receptors in pathogen recognition. *Clinical Microbiology Reviews* 16, 637-646.

Jault, C., Pichon, L. & Chluba, J. (2004). Toll-like receptor gene family and TIR-domain adapters in *Danio rerio*. *Molecular Immunology* 40, 759-771.

Keestra, A. M. & van Putten, J. P. (2008). Unique properties of the chicken TLR4/MD-2 complex: selective lipopolysaccharide activation of the MyD88-dependent pathway. *Journal of Immunology* 181, 4354-4362.

Kurt-Jones, E. A., Popova, L., Kwinn, L., Haynes, L. M., Jones, L. P., Tripp, R. A., Walsh, E. E., Freeman, M. W., Golenbock, D. T., Anderson, L. J. & Finberg, R. W. (2000). Pattern recognition receptors TLR4 and CD14 mediate response to respiratory syncytial virus. *Nature Immunology* 1, 398-401.

Leveque, G., Forgetta, V., Morrol, L. S., Smith, A. L., Bumstead, N., Barrow, P., Loredo-Osti, J. C., Morgan, K. & Malo, D. (2003). Allelic variation in TLR4 is linked to susceptibility to Salmonella entericaserovar *Typhimurium* infection in chickens. *Infection and Immunity* 71, 1116-1124.

Li, M. Z. & Elledge, S. J. (2007). Harnessing homologous recombination *in vitro* to generatere combinant DNA via SLIC. *Nature Methods* 4, 251-256.

Li, J., Wang, T. H. & Lu, R. H. (1999). Advances in research of hemorrhagic virus of grass carp. *Oceanologia et Limnologcian Sinica* 30, 445-453.

Lien, E., Means, T. K., Heine, H., Yoshimura, A., Kusumoto, S., Fukase, K., Fenton, M. J., Oikawa, M., Qureshi, N., Monks, B., Finberg, R. W., Ingalls, R. R. & Golenbock, D. T. (2000). Toll-like receptor 4 imparts ligand-specific recognition of bacterial lipopolysaccharide. *The Journal of Clinical Investigation* 105, 497-504.

Ligas, A. (2008). First record of the channel catfish, *Ictalurus punctatus* (Rafinesque, 1818), incentral Italian waters. *Journal of Applied Ichthyology* 24,632-634.

Livak, K. J. & Schmittgen, T. D. (2001). Analysis of relative gene expression data using real-time quantitative PCR and the $2^{-\Delta\Delta Ct}$ method. *Methods* 25, 402-408.

Lu, Y. C., Ye, W. C. & Ohashi, P. S. (2008). LPS/TLR4 signal transduction pathway. *Cytokine* 42, 145-151.

Medzhitov, R., Preston-Hurlburt, P. & Janeway, C. A. Jr. (1997). A human homologue of the *Drosophila* toll protein signals activation of adaptive immunity. *Nature* 388, 394-397.

Meng, Z., Zhang, X. Y., Guo, J., Xiang, L. X. & Shao, J. Z. (2012). Scavenger receptor in fish is a lipopolysaccharide recognition molecule involved in negative regulation of NF-κB activation by competing with TNF receptor-associated factor 2 recruitment into the TNF-α signaling pathway. *Journal of Immunology*189, 4024-4039.

Nishitani, C., Mitsuzawa, M., Hyakushima, N., Sano, H., Matsushima, N. & Kuroki, Y. (2005). The Toll-like receptor 4 region Glu24-Pro34 is critical for interaction with MD-2. *Biochemical and Biophysical Research Communications*. 328, 586-590.

Nishitani, C., Mitsuzawa, M., Sano, H., Shimizu, T., Matsushima, N. & Kuroki, Y. (2006). Toll-like receptor 4 region Glu24-Lys47 is a site for MD-2 binding: importance of CYS29 and CYS40. *Journal of Biological Chemistry* 281, 38322-38329.

Oshiumi, H., Tsujita, T., Shida, K., Matsumoto, M., Ikeo, K. & Seya, T. (2003). Prediction of the prototype of the human toll-like receptor gene family from the pufferfish, *Fugurubripes*, genome. *Immunogenetics* 54, 791-800.

Peatman, E., Baoprasertkul, P., Terhune, J., Xu, P., Nandi, S., Kucuktas, H., Li, P., Wang, S., Somridhivej, B., Dunham, R. & Liu, Z. (2007). Expression analysis of the acute phase response in channel catfish (*Ictalurus punctatus*) after infection with a Gram-negative bacterium. *Developmental &Comparative Immunology* 31, 1183-1196.

Poltorak, A., He, X., Smirnova, I., Liu, M. Y., Van, Huffel. C., Du, X., Birdwell, D., Alejos, E., Silva, M., Galanos, C., Freudenberg, M., Ricciardi-Castagnoli, P., Layton, B. & Beutler, B. (1998). Defective LPS Signaling in C3H/HeJ and C57BL/10ScCr Mice: mutations in Tlr4 gene. *Science* 282, 2085-2088.

Rassa, J. C., Meyers, J. L., Zhang, Y., Kudaravalli, R. & Ross, S. R. (2002). Murine retroviruses activate B cells via interaction with toll-like receptor 4. Proceedings of the National Academy of Sciences of the Lnited States of America qq, 2281-2286.

Rebl, A. Goldammer, T. & Seyfert, H. M. (2010). Toll-like receptor signaling in bony fish. *Veterinary Immunology and Immunopathology*134, 139-150.

Roach, J. C., Glusman, G., Rowen, L., Kaur, A., Purcell, M. K., Smith, K. D., Hood, L. E. & Aderem, A. (2005). The evolution of vertebrate toll-like receptors. *Proceedings of the National Academy of Sciences of the United States of America* 102, 9577-9582.

Su, J. G., Yang, C. R., Xiong, F., Wang, Y. P. & Zhu, Z. Y. (2009). Toll-like receptor 4 signaling pathway can be triggered by grass carp reovirus and *Aeromonashydrophila* infection in rare minnow *Gobiocypris rarus*. *Fish & Shellfish Immunology* 27, 33-39.

Takeda, K. & Akira, S. (2003). Toll receptors and pathogen resistance. *Cellular Microbiology* 5, 143-153.

Takeda, K. & Akira, S. (2005). Toll-like receptors in innate immunity. *International Immunology*17, 1-14.

Temperley, N. D., Berlin, S., Paton, I. R., Griffin, D. K. & Burt, D. W. (2008). Evolution of the chicken Toll-like receptor gene family: a story of gene gain and gene loss. *BMC Genomics* 9, 62.

Wang, T. H., Chen, H. X., Liu, P., Liu, H., Guo, W. & Yi, Y. (1993). Observations on the ultra-thin sections of the main organs and tissues of hemorrhagic *gobiocyp risrarus* artificially infectied by grass carp hemorrhagic virus (GCHV). *Acta Hydrobiologica Sinica* 17, 343-346.

Yamamoto, M. & Takeda, K. (2010). Current Views of Toll-Like Receptor Signaling Pathways. *Gastroenterology Research and Practice*, 2010 240365.

Zhang, Y., He, X., Yu, F., Xiang, Z., Li, J., Thorpe, K. L.& Yu, Z. (2013a). Characteristic and functional analysis of toll-like receptors (TLRs) in the lophotrocozoan, *Crassostreagigas*, reveals ancient origin of TLR-mediated innate immunity. *PLoS One* 8, e76464.

Zhang, J., Liu, S., Rajendran, K. V., Sun. L., Zhang, Y., Sun, F., Kucuktas, H., Liu, H. & Liu, Z J. (2013). Pathogen recognition receptors in channel catfish: III Phylogeny and expression analysis of Toll-like receptors. *Developmental & Comparative Immunology* 40, 185-194.

Zhao, W., Huang, Z., Chen, Y., Zhang, Y., Rong, G., Mu, C., Xu, Q. & Chen, G. (2013). Molecular cloning and functional analysis of the duck TLR4 gene. *International Journal of Molecular Sciences* 14, 18615-18628.

Zheng, X., Zhu, J., Kapoor, A. & Zhu, J. K. (2007). Role of arabidopsis AGO6 in siRNA accumulation, DNA methylation and transcriptional gene silencing. *The EMBO Journal* 26, 1691-1701.

草鱼四个 *tlr4* 基因的特征分析及对草鱼出血病病毒和脂多糖的应答

裴永艳[1,2] 黄 容[1] 李勇明[1] 廖兰杰[1] 朱作言[1] 汪亚平[1]

1 中国科学院水生生物研究所，淡水生态与生物技术国家重点实验室，武汉 430072
2 中国科学院大学，北京 100049

摘　要 本研究对草鱼 *tlr4* 基因簇中四个 *tlr4* 基因的亚细胞定位、组织分布、以及草鱼呼肠孤病毒(GCRV)和脂多糖诱导后的响应差异进行了比较分析。四个 *tlr4* 基因在被检测的十个组织中均有表达，然而它们的亚细胞定位和最高表达的组织却不尽相同。GCRV 感染后，四个 *tlr4* 基因在被检测的组织中表达量均呈现上升趋势。脂多糖刺激草鱼肾细胞后，这四个基因均下调表达。研究结果为进一步了解硬骨鱼类 *tlr4* 基因的功能提供了基础。

The Draft Genome of the Grass Carp (*Ctenopharyngodon idellus*) Provides Insights into Its Evolution and Vegetarian Adaptation

Yaping Wang[1] Ying Lu[2] Yong Zhang[3,4] Zemin Ning[5] Yan Li[2] Qiang Zhao[2] Hengyun Lu[2] Rong Huang[1] Xiaoqin Xia[1] Qi Feng[2] Xufang Liang[6,7] Kunyan Liu[2] Lei Zhang[2] Tingting Lu[2] Tao Huang[2] Danlin Fan[2] Qijun Weng[2] Chuanrang Zhu[2] Yiqi Lu[2] Wenjun Li[2] Ziruo Wen[2] Congcong Zhou[2] Qilin Tian[2] Xiaojun Kang[1,8] Mijuan Shi[1] Wanting Zhang[1] Songhun Jang[1,9] Fukuan Du[1] Shan He[6,7] Lanjie Liao[1] Yongming Li[1] Bin Gui[1] Huihui He[1] Zhen Ning[1] Cheng Yang[1,8] Libo He[1] Lifei Luo[1] Rui Yang[10] Qiong Luo[10] Xiaochun Liu[3,4] Shuisheng Li[3,4] Wen Huang[3,4] Ling Xiao[3,4] Haoran Lin[3,4] Bin Han[2] Zuoyan Zhu[1]

1 State Key Laboratory of Freshwater Ecology and Biotechnology, Institute of Hydrobiology, Chinese Academy of Sciences, Wuhan, China 430072

2 National Center for Gene Research, Shanghai Institute of Plant Physiology and Ecology, Shanghai Institutes for Biological Sciences, Chinese Academy of Sciences, Shanghai, China 200233

3 State Key Laboratory of Biocontrol, School of Life Sciences, Sun Yat-Sen University, Guangzhou, China 510527

4 Guangdong Province Key Laboratory for Aquatic Economic Animals, School of Life Sciences, Sun Yat-Sen University, Guangzhou, China 510527

5 Wellcome Trust Sanger Institute, Wellcome Trust Genome Campus, Hinxton, UK

6 Key Laboratory of Freshwater Animal Breeding of the Ministry of Agriculture, College of Fisheries, Huazhong Agricultural University, Wuhan, China 430070

7 Freshwater Aquaculture Collaborative Innovation Center of Hubei Province, College of Fisheries, Huazhong Agricultural University, Wuhan, China 430074

8 School of Computer Science, China University of Geosciences, Wuhan, China 430074

9 College of Life Science, Kim Illinois Sung University, Pyongyang, North Korea

10 College of Plant Protection, Yunnan Agricultural University, Kunming, China 650201

Abstract The grass carp is an important farmed fish, accounting for ~16% of global freshwater aquaculture, and has a vegetarian diet. Here we report a 0.9-Gb draft genome of a gynogenetic female adult and a 1.07-Gb genome of a wild male adult. Genome annotation identified 27,263 protein-coding gene models in the female genome. A total of 114 scaffolds consisting of 573 Mb are anchored on 24 linkage groups. Divergence between grass carp and zebrafish is estimated to have occurred 49-54 million years ago. We identify a chromosome fusion in grass carp relative to zebrafish and report frequent crossovers between the grass carp X and Y chromosomes. We find that transcriptional activation of the mevalonate pathway and steroid biosynthesis in liver is associated with the grass carp's adaptation from a carnivorous

文章发表于 *Nature Genetics*, 2015, 47(6): 625-631

to an herbivorous diet. We believe that the grass carp genome could serve as an initial platform for breeding better-quality fish using a genomic approach.

Introduction

Constituting a member of the Cyprinidae family and the only species of the genus *Ctenopharyngodon*, the grass carp *Ctenopharyngodon idellus* is one of the most important aquaculture species, having great commercial value and a worldwide distribution [1] (Fig. 1a). Global production of cultured or farmed grass carp is approximately 4.6 million tons per year, accounting for 15.6% of global freshwater aquaculture production in 2011 [2]. The completion of the zebrafish (*Danio rerio*) genome sequence [3] has accelerated studies on the genomes of other members of the Cyprinidae family. In grass carp, progress included the construction of a genetic linkage map[4] and the identification of 3,027 UniGene entries [5] and 6,269 ESTs [6]. Such studies have enriched genome research on grass carp. Recent work has focused on genes involved in the immune system [7, 8] control of food intake [9, 10] and nutrition and growth [11, 12]. However, the lack of a complete genome sequence has made it difficult to conduct an in-depth investigation of grass carp biology and breeding for better-quality fish. As a first step toward this goal, we report a draft genome sequence and transcriptomic analysis of grass carp, adding this important species to the other sequenced teleosts: cod [13], fugu [14], medaka [15], tetraodon [16], stickleback [17] and zebrafish [3]. Taken together, this information provides genomic insights into the evolutionary history of the grass carp and its unique adaptation to a vegetarian diet.

Results

Genome assembly

The grass carp genome is composed of 24 pairs of chromosomes [18, 19] ($2n$ = 48). Adopting a whole-genome shotgun sequencing strategy, we generated approximately 132 Gb of Illumina sequence reads on genomic DNA isolated from the blood of a gynogenetic female adult grass carp and 136 Gb of reads from a wild, water-captured male adult (Supplementary Table 1 and Supplementary Note). We constructed the final assemblies of the female (0.90-Gb) and male (1.07-Gb) genomes using the modified *de novo* Phusion-meta assembly pipeline, as previously described [20] (Supplementary Fig. 1 and Supplementary Table 2). The draft genome of the female was fully annotated to mine genomic information (Fig. 1) and was applied to the anchoring of scaffolds on the genetic linkage map, whereas the male genome was used to detect sequence variation between the male and female genomes.

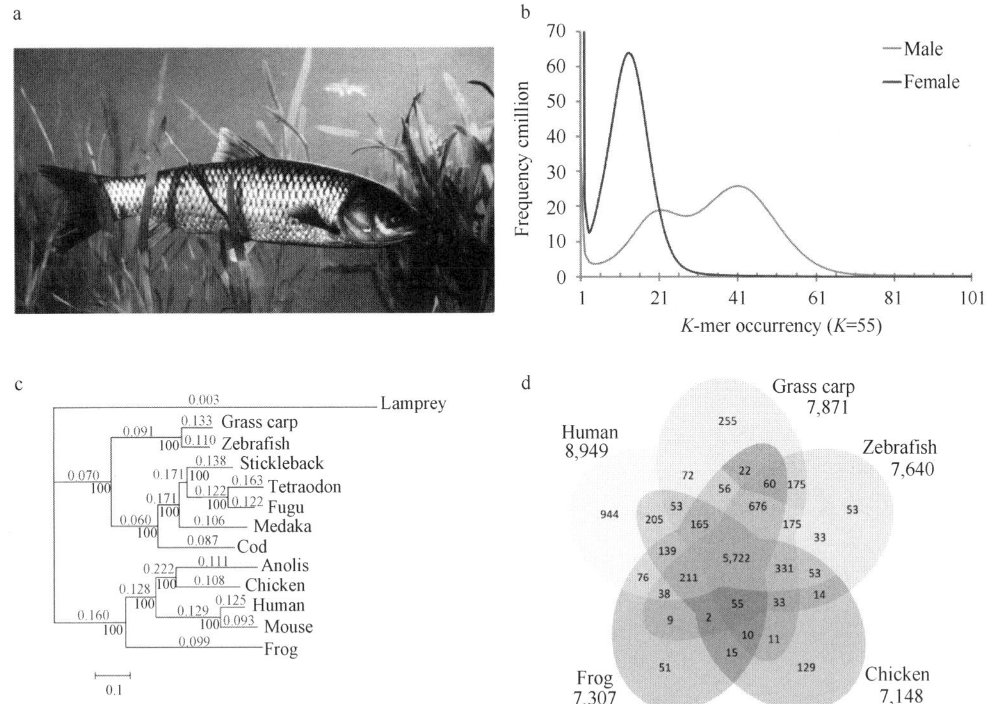

Figure 1 Assemblies and evolution of the grass carp genome. (a) Image of a grass carp adult. (b) Distribution of 55-mer frequency. The distribution of *K*-mer frequency in the reads was derived from libraries of short insert size (350-400 bp). Values for *K*-mers are plotted against the frequency (*y* axis) of their occurrence (*x* axis). The leftmost truncated peak at low occurrence (1-2) was mainly due to random base errors in the raw sequencing reads. (c) Reconstructed phylogeny of 13 vertebrate genomes. The *dN/dS* ratio of each branch is shown in blue. The numbers in black correspond to values of bootstrap support. The lamprey is used as an outgroup. Branch length is measured in expected substitutions per site. (d) Venn diagram of gene clusters for five selected vertebrate genomes. Each number represents the number of orthologous gene families shared by the indicated genomes

The female genome assembly had scaffolds with an N50 length greater than 6.4 Mb, and 90% of the assemblies were composed of 301 scaffolds, which were all greater than 179 kb in length (Table 1). Estimation of genome size by distribution of *K*-mer frequency showed that the female genome was about 891 Mb, close to the size of the assemblies (Fig. 1b and Supplementary Note). We assessed the accuracy of the genome assembly by alignment of the scaffolds to 3,027 published UniGene entires [5] and 11 BACs [21] (Supplementary Fig. 2 and 3, and Supplementary Table 3), which indicated that the coverage by the initial contigs and scaffolds was approximately 95% and 97%, respectively. Sequence errors were predominantly from insertions or deletions introduced by short-read assembly (Supplementary Table 3a).

We identified a total of 644,817 heterozygous SNPs and 66,101 short indels (10 nucleotides in length or less) in the female genome. In the male genome, we identified

1,465,819 SNPs and 166,867 short indels. The estimated overall heterozygous rates were approximately 0.9 and 2.5 polymorphisms per kilobase in the female and male genomes, respectively (Supplementary Table 4). Clearly, the wild male genome had a much higher heterozygosity rate, which caused a bimodality in the distribution of the *K*-mer frequency (Fig. 1b) and a shorter length for the assembled scaffolds.

Table 1 Overview of assembly and annotation for the grass carp genome

Female	
Total length	900,506,596 bp
Length of unclosed gaps	35,069,100 bp
N50 length (initial contigs)	40,781 bp
N50 length (scaffolds)	6,456,983 bp
N90 length (scaffolds)	179,941 bp
Quantity of scaffolds (>N90 length)	301
Largest scaffold	19,571,558 bp
GC content	37.42%
Quantity of predicted protein-coding genes	27,263
Quantity of predicted non-coding RNA genes	1,579
Content of transposable elements	38.06%
Length of scaffolds anchored on linkage groups	573,471,712bp (64%)
Quantity of scaffolds anchored on linkage groups	114
Male	
Total length	1,076,149,922bp
N50 length (initial contigs)	18,252 bp
N50 length (scaffolds)	2,279,965bp
N90 length (scaffolds)	3,052bp
Quantity of scaffolds (>N90 length)	6,950
Largest scaffold	16,339,329bp

Genome annotation

We annotated a total of 27,263 protein-coding genes in the female genome. The evidence used in gene prediction included 27 Gb of RNA sequencing (RNA-seq) data from 6 tissues (embryo, liver, spleen, brain, kidney and head kidney), over 3,000 known UniGene entries and homologous gene information from zebrafish (Ensembl release 67; Supplementary Fig. 4, Supplementary Table 5 and Supplementary Data Set). We predicted 1,538 tRNA, 24 rRNA, 207 small nucleolar RNA (snoRNA), 136 small nuclear RNA (snRNA) and 444 microRNA genes in our annotation of noncoding RNA genes (Supplementary Tables 6 and 7) and 467,783 simple-sequence repeats (Supplementary Tables 8). *De novo* repeat annotation indicated an overall repeat content of 38%, in comparison to that of 43% in BACs (Supplementary Table 9).

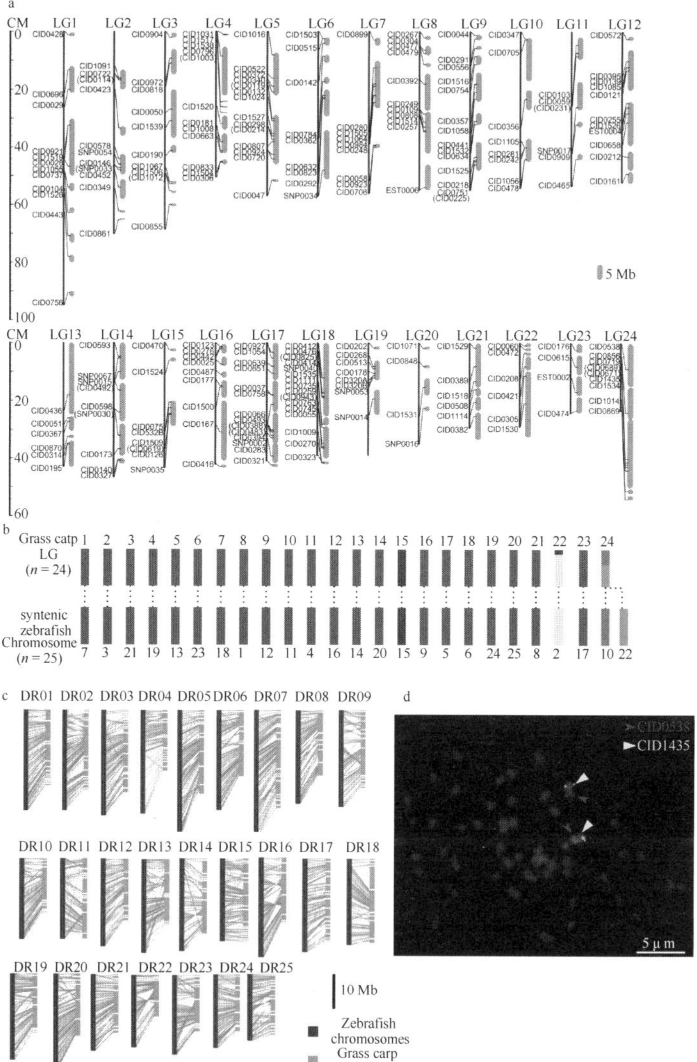

Figure 2 Female scaffolds anchored on the genetic map. (a) The scaffolds were anchored on a published consensus linkage map [4]. The blue lines indicate the length of each linkage group (LG) to which the markers are mapped. Map distances between markers are depicted on a Kosambi cM scale. Orange bars represent the anchored scaffolds. The black lines linking markers and scaffolds show the locations of the markers on the scaffolds. The length of each scaffold is shown relative to a 5-Mb scale bar. (b) Syntenic relationship between the zebrafish chromosomes and the grass carp linkage groups. Linkage group 22 is aligned to zebrafish chromosomes 2 and 15, and linkage group 24 is aligned to zebrafish chromosomes 10 and 22. (c) Gene collinearity between zebrafish and grass carp. The zebrafish chromosomes are represented by blue blocks (for example, DR01). The grass carp scaffolds (length > 50 kb) are represented by orange blocks. Aligned genes are connected by green lines. The lengths of the chromosomes and scaffolds are shown relative to a 10-Mb scale bar. (d) FISH study of linkage group 24. The yellow marker CID1435 is located on the region aligned to zebrafish chromosome 10, and the red marker CID0538 is aligned to zebrafish chromosome 22. Scale bar, 5 μm

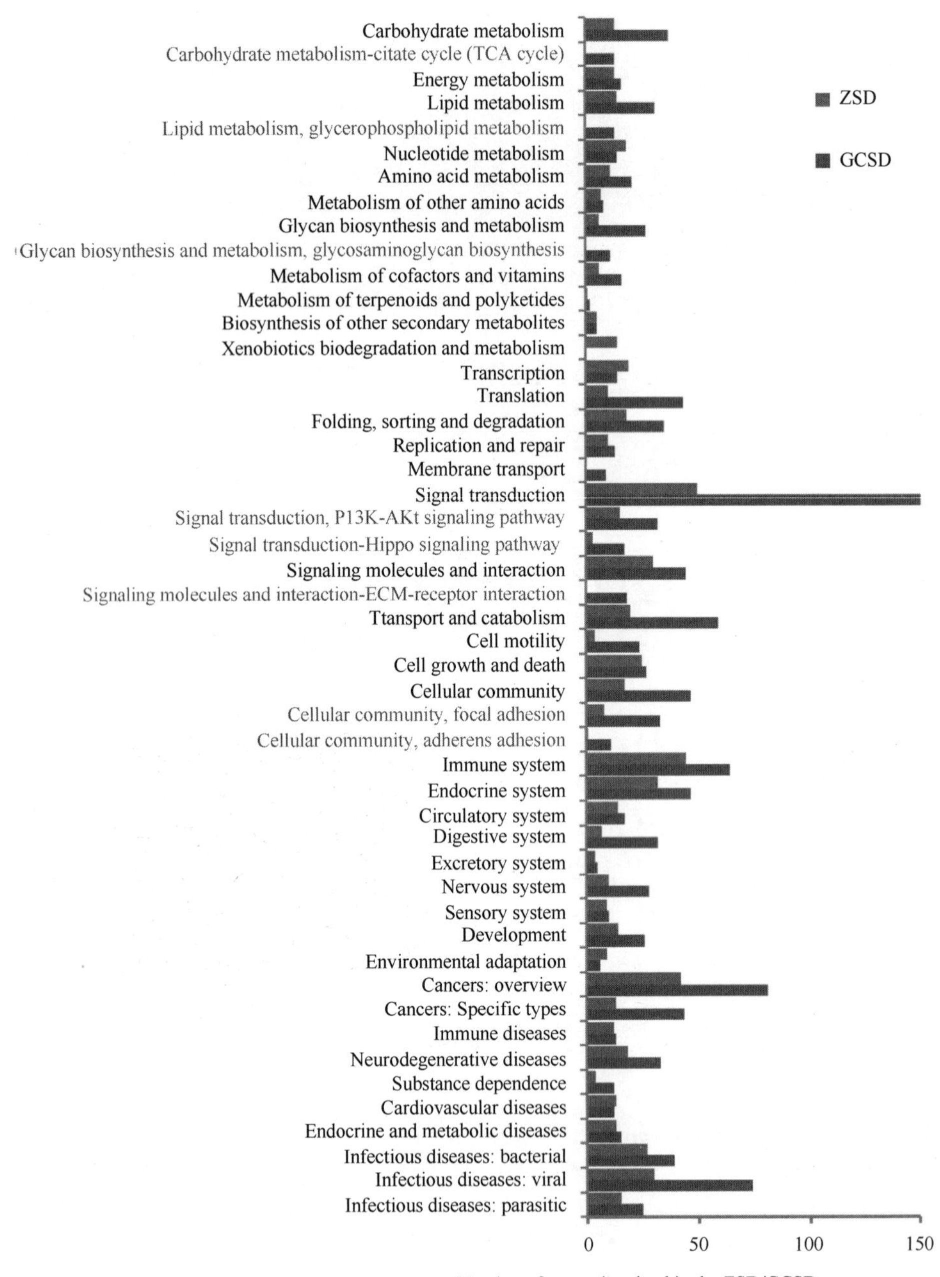

Figure 3 Distribution of GCSD- and ZSD-related genes by pathway. All of the genes were involved in the GCSD or ZSD. Pathways were determined by searching the KEGG pathway database. The x axis indicates the numbers of genes involved in each pathway. Pathways with a label in blue highlight metabolic processes that are potentially important to development or diet adaptation

This proportion is less than the 52.2% repeat content observed in zebrafish[3]. This difference might be due to the exclusion of repetitive sequences located in unclosed gaps and on small fragments (<200 bp) of the grass carp assemblies. The majority of transposable elements found in the grass carp genome were type II DNA transposable elements, covering over 20% of the genome, similar to in the zebrafish genome [3].

Using the published genetic linkage map of grass carp [4], we anchored 114 scaffolds on the 24 linkage groups (Fig. 2a, Supplementary Table 10 and Supplementary Note), covering 573 Mb (64%) of the female assembly with 17,456 (64%) annotated genes localized. Gene synteny over the anchored scaffolds showed that most of the grass carp linkage groups had extensive collinearity with corresponding zebrafish chromosomes (Fig. 2b). Alignment of genes showed high synteny for grass carp and zebrafish, and up to 24,018 grass carp genes (88% of the 27,263 total genes) were located on syntenic blocks (Fig. 2c). Although this result was similar to a previous report [4], we found two cross-chromosome arrangements for linkage groups 22 and 24. It is noteworthy that linkage group 24 aligned to zebrafish chromosomes 22 and 10 but not to any other grass carp linkage group. FISH analysis of grass carp chromosomes demonstrated that two grass carp markers aligning to zebrafish chromosomes 10 and 22 were indeed located on the single linkage group 24 (Fig. 2d), explaining why the chromosome number is 25 in zebrafish but 24 in grass carp.

Evolutionary anulysis

To examine grass carp evolution, we clustered the grass carp gene models with the genes from 12 other vertebrate genomes and used 202 single-copy genes with one-to-one correspondence in the different genomes to reconstruct a phylogenetic tree (Fig. 1c and Supplementary Note). As a species of the Cyprinidae family, grass carp had the closest relationship to zebrafish. According to the TimeTree [22] database, the estimated divergence time between zebrafish and grass carp was around 49-54 million years ago (Supplementary Table 11). Most of the selected teleostei genomes showed similar selection pressures, according to calculated dN/dS values (the ratio of the rate of nonsynonymous substitution to the rate of synonymous substitution).

We determined gene families using the TreeFam database [23] (Supplementary Note). We performed a five-way comparison among the gene families of a representative mammal (human), bird (chicken), amphibian (frog) and two fishes (zebrafish and grass carp) to quantify the shared or species-specific families present in each genome (Fig. 1d). Zebrafish and grass carp shared 7,227 families, more than the 5,772 families shared by all 5 vertebrate species. Of the 10,184 families identified, 7,171 (70%) carried the same number of gene members in grass carp and zebrafish. Specific comparison of the Hox [24,25], Sox [26] and Toll-like receptor [27] gene clusters among human, medaka, zebrafish and grass carp indicated that the zebrafish and grass carp genomes carried an identical copy number for most subfamilies (Supplementary Figs. 5-7 and Supplementary Table 12). We determined the number of human, zebrafish and grass carp gene members in each family (Supplementary Table 13). The 1,047 families in the class having many grass carp members relative to one human member were composed of 2,658 grass carp genes and 1,047 human genes, with an average ratio of 2.53 grass carp genes to one human gene. Interestingly, the 832 families in the class having many zebrafish members relative to one human member consisted of 2,077 zebrafish genes and 832 human genes, with nearly the same average ratio of 2.50 zebrafish genes to one human gene. It was suggested that the grass carp genome underwent a whole-genome duplication similar to zebrafish after the teleost radiation [28].

We estimated the expansion and contraction of gene families to examine their evolutionary history in comparison to the zebrafish, stickleback, tetraodon, fugu, medaka and cod genomes (Supplementary Fig. 8). The significantly expanded families in grass carp included many immune-associated functional domains ($P < 0.001$; Supplementary Table 14), consistent with the adaption of grass carp to variable environments. Among the 10,184 gene families generated, 2,346 included teleost-specific duplications in zebrafish or grass carp as determined by comparison of the number of gene copies within each family. Of the gene families involved in the teleost-specific duplications, 695 and 295 showed evidence of undergoing a grass carp-specific duplication (GCSD) or a zebrafish-specific duplication (ZSD), respectively (Supplementary Fig. 9 and Supplementary Note), with additional gene duplications found in grass carp and zebrafish. The 695 grass carp families contained 2,561 genes, whereas the 295 zebrafish families consisted of 1,029 genes. We annotated all of these genes using the KEGG [29] (Kyoto Encyclopedia of Genes and Genomes) pathway database. Functional analyses of these pathways indicated that genes involved in the ZSD were mainly composed of immune-related genes. Comparably, the grass carp genes involved in the GCSD were not only associated with immune-related genes but also with development-related genes (Fig. 3 and Supplementary Table 15) and were involved in cell proliferation and differentiation (for example, the focal adhesion pathway and the extracellular matrix (ECM)-receptor interaction pathway [30,31]), nutritional homeostasis (for example, the protein digestion and absorption pathway [32,33]) and organ size control (for example, the Hippo signaling pathway [34,35]). Comparison of genes involved in the overview maps of metabolism (reference map ko01100 of the KEGG database) also indicated that grass carp genes involved in the GCSD clustered in carbohydrate metabolism and nucleotide metabolism (Supplementary Fig. 10a), whereas the zebrafish genes only clustered in nucleotide metabolism (Supplementary Fig. 10b). These results indicate that the GCSD was important for adaptation to a vegetarian diet and for some developmental characteristics of grass carp.

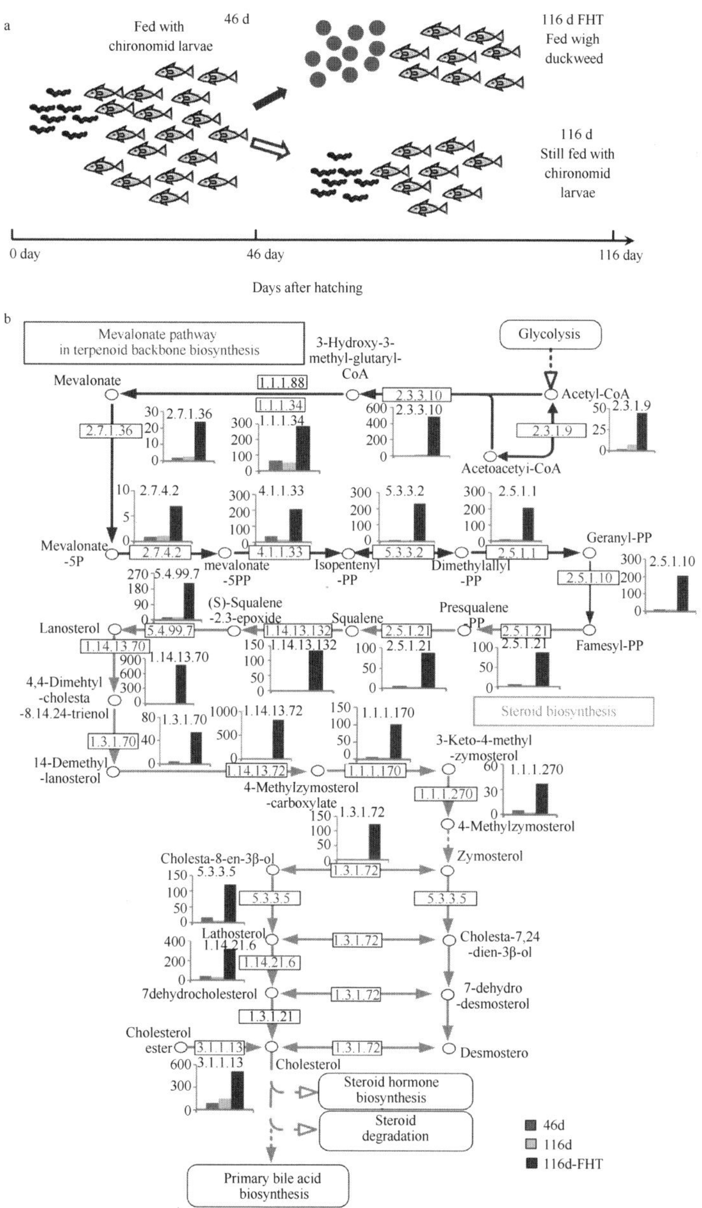

Figure 4 Characterization of gene expression during the FHT period. (a) Design of the FHT experiments. At 46 d after hatching, fish not undergoing FHT (fed with chironomid larvae) were collected as sample "46 d". During the period from 46 to 116 d after hatching, the fish were divided into two groups-one fed with duckweed and the other still fed with chironomid larvae-which were collected at 116 d after hatching as sample "116 d-FHT" and sample "116 d," respectively. (b) Activation of the mevalonate pathway and steroid biosynthesis. The blue and orange arrows indicate the reaction steps in different pathways. The number in each rectangle shows the EC code of the enzyme catalyzing that transfer. Red codes indicate enzyme genes with significantly increased expression after the FHT (q<0.001). Gene expression as measured by quantification of transcription levels (reads per kilobase of exon model per million mapped reads, RPKM [45]) is shown in the histograms. Compound names are shown beside the corresponding circles

A potential sex-determination mechanism

By comparison of the assemblies for the male and female grass carp, we identified 206 contigs with a total length of 2.38 Mb that were carried by the male adult but not by the gynogenetic female (Supplementary Fig. 11 and Supplementary Tables 16 and 17). We confirmed each contig by PCR-based sequencing. We also performed PCR amplification of these regions in an extended group of 24 male and 24 female individuals, identifying frequent chromosome crossovers between the X and Y chromosomes in grass carp (Supplementary Fig. 12). Sex in grass carp may be determined not by an entire chromosome but by a few critical genes. Noticeably, we identified a male genome-specific probe that mapped to one of the contigs (probe 184 in Supplementary Fig. 12). We did not find a sequence alignment of this region to any other vertebrate genome, suggesting its unique origin in grass carp.

Gene modeling showed that the male-specific contigs mainly contained genes with domains related to the immunoglobulin V-set, ABC transporter, proteasome subunit and NACHT domains (Supplementary Table 18). Alignment of these predicted genes to the female gene model set showed that 40 genes had homologs in the female genome, of which 22 clustered on linkage group 24 (Supplementary Fig. 13). Gene collinearity and FISH analysis demonstrated that zebrafish chromosomes 10 and 22 fused to form a single chromosome-linkage group 24-in grass carp, which had the largest physical size (Fig. 2d) but the smallest genetic distance, at 23.4 cM, of all of the grass carp linkage groups. Zebrafish has an all-autosome karyotype of 25 chromosomes [36], whereas grass carp has differential X and Y chromosomes and a karyotype of 24 chromosomes [18,19]. Interestingly, linkage group 24 carried most of the predicted genes located in the regions with variable sequence for the male and female genomes, indicating a potential connection between linkage group 24 and a sex-determination chromosome. The identity of the sex-determination chromosome(s) in grass carp is still unclear, but comparison of the male and female genomes may provide further insights into grass carp sex determination, as well as linkage group 24.

Transcriptome analyses of food habit transition

Grass carp are typically herbivorous, a characteristic that has contributed to making them a popular breeding species. How grass carp effectively absorb nutrients from plants to support their rapid growth is an unanswered research question [37]. Grass carp complete the transition from a carnivorous to an herbivorous diet when they become 3 to 5.5 cm in body length, around 1.5 months after hatching. We analyzed RNA-seq data derived from gut, liver and brain to characterize variations in gene expression before and after the change (Fig. 4a and Online Methods). Genes with differential expression were significantly enriched in pathways associated with circadian rhythm in gut and with steroid biosynthesis, terpenoid backbone biosynthesis and glycerophospholipid metabolism pathways in liver (DAVID Bioinformatics

Resources [38]; $P< 0.05$; Supplementary Table 19).

After the change in diet, it is essential for grass carp to maintain a continuous feeding rhythm so that they can obtain sufficient nutrients from their food. The analyses of liver RNA-seq data identified significant activation of the steroid biosynthesis pathway downstream of the mevalonate pathway in terpenoid backbone biosynthesis, as reflected by an average expression level 32-fold higher than before the change in diet (q value [39] < 0.001; Fig. 4b). In comparison, control individuals not undergoing a change in diet had an average 0.94-fold increase in the expression levels of these genes (Supplementary Table 19b,c). We confirmed the differential expression levels of most of these genes by quantitative RT-PCR (qRT-PCR; Supplementary Fig. 14), finding an average 58-fold increase in expression after the change in diet. The mevalonate pathway is employed to transfer the products of glycolysis to precursors of terpenoids (such as farnesyl-PP), which can potentially be metabolized by downstream steroid biosynthesis pathways and used for different processes, such as the biosynthesis of hormones and vitamin D [40,41]. During the diet change experiment, we observed that grass carp fed with duckweed had considerably greater growth than those fed with chironomid larvae (fish before the transition and fish not undergoing a transition) in terms of body length, gut length, body weight and the rate of gut length/body length (data not shown). Metabolic adaption of these pathways apparently supports the effective use of plant-derived nutrients in grass carp.

In addition, a previous report indicated that grass carp fed with a plant diet spend a longer time on feeding[37]. We also observed that the grass carp that received a plant diet fed almost continuously throughout any given 24-h period. Analyses of transcriptome data showed that genes involved in the circadian rhythm pathway were activated after the food habit transition (FHT), which included the *clock-bmal1* heterodimer and the *ror* and *clock* genes (Supplementary Fig. 15a). The related genes showed high similarity to those of zebrafish (such as the *rorca* [42] and *clocka* [43] genes), with the exception of some promoter regions of genes carrying different elements (Supplementary Fig. 15b). Although the relationship between the genes involved in circadian rhythm and continuous feeding is unclear, this finding might suggest that feeding frequency should be examined during the diet change of grass carp.

We investigated the non-grass carp reads for potential gut microbiota by removing all gut RNA-seq reads with alignment to the grass carp assemblies (Supplementary Tables 20 and 21, and Supplementary Note). This analysis did not identify any genes encoding predicted cellulose-digesting enzymes in the gut, suggesting that the grass carp intestine may not digest and absorb cellulose, which reinforces the view that cellulase may be developed as an aquatic additive to promote the growth of grass carp [44]. However, sequencing of gut biota will likely prove to be more powerful in understanding how the grass carp digests plant materials.

Online Methods

DNA library preparation and sequencing

Genomic DNA was isolated from the blood cells of a gynogenetic female adult grasscarp and a wild-captured male adult grass carp, using the DNeasy Blood and Tissue kit (Qiagen). An amplification-free approach [46] was applied to prepare sequencing libraries with short inserts of 350-450 bp for paired-end reads, following the manufacturer's protocol for Illumina. To construct the libraries with insert sizes of 1, 3, 5 and 8-10 kb for mate-paired reads (Supplementary Fig. 16), the protocols of the Mate-Pair Library v2 Sample Preparation Guide (Illumina) and the Paired-End Library Preparation Method Manual (Roche) were combined. Raw data were generated by an Illumina Hiseq 2000 sequencer and an Illumina Genome Analyszer IIx sequencer.

Sequence assembly

A de novo assembly pipeline was developed to assemble the short reads (Supplementary Fig. 1), which was a modified version of the Phusion-meta method as previously reported [20]. Briefly, when the paired-end reads were screened to remove low-quality reads containing ten or more unique *K*-mers, they were clustered into thousands of groups by Phusion2 with *K*-mer set at 51 bp. The reads grouped into each cluster were assembled in parallel into contigs by ABYSS [47], Fermi [48] and SOAPdenovo [49]. All these contigs were merged to form initial draft contigs. Consensus contigs were obtained by aligning all the reads back to the draft contigs using the assembly management tool GAP5 [50]. The mate-paired reads were then hierarchically and iteratively assembled into contigs to build preliminary scaffolds with SOAPdenovo (*K*-mer set at 61 bp). Final scaffolding was conducted with Spinner.

Prediction of protein-coding genes

The construction of gene models was based on expression sequences (UniGene entries and RNA-seq data) and homologs of zebrafish genes (Ensembl release 67). We built a seven-step pipeline to facilitate the gene model set. (1) A total of 3,027 published UniGene sequences were mapped to the repeat-masked scaffolds by an mRNA/EST genome mapping program in GMAP [51] with the parameters set to '-n1-f2-B1-A-t4'. (2) Protein sequences form zebrafish were aligned to the grass carp genome using TBLASTN with an *E*-value cutoff of 1×10^{-10}. (3) The Illumina RNA-seq sequences from six organs or tissues (embryo, liver, spleen, brain, kidney and head kidney) were mapped to the repeat-masked scaffolds using the SMALT aligner with the parameters set to a minimum Smith-Waterman (-m) value of 35, a maximum insert size (-i) of 5,000 and a minimum insert size (-j) of 20. Only uniquely matched reads were selected to assist in gene prediction. (4) On the basis of the alignment results from steps 1-3, the target gene fragments found by extending the aligned regions by 3,000 bp on both ends were selected to build preliminary gene models with FgeneSH++ (Softberry) using parameters trained on fish genomes. If an overlap was found for adjacent regions, the fragments were selected in order of priority for alignment from UniGene sequences to zebrafish homologs to transcriptome sequences. (5) To check the accuracy of the preliminary gene models, the models were aligned to the zebrafish gene model set with BLASTP with an E-value cutoff of 1×10^{-20}. When a grass carp gene mapped to two or more adjacent zebrafish genes, this chimeric gene was split according to information from the zebrafish homologs. When two or more adjacent grass carp genes were aligned to a single zebrafish gene, they were combined by Genewise [52] using the corresponding zebrafish gene as the reference sequence. The longest translation product was chosen to represent each gene. (6) The coding sequence for each gene model was aligned to the Repbase library and a grass carp repeat library created with RepeatModeler, using BLASTN results with an E-value cutoff of 1×10^{-5}, to remove transposon genes. (7) Genes with a coding sequence of less than 60 amino acids were removed. The genes comprising three or fewer exons and not supported by expression sequences or zebrafish homologs were also removed. In this manner, the grass carp gene models were generated.

Transcriptome sequencing

Six tissues (embryo, liver, spleen, brain, kidney and head kidney) were collected from a male adult to perform RNA-seq, with the resulting data used for gene modeling. All these samples, as well as those from the FHT experiments, were subjected to RNA isolation using SV TRIzol Reagent (Invitrogen). Poly(A)$^+$ mRNA was purified using the DynaBeads mRNA Purification kit (Life Technologies). Paired-end cDNA libraries were constructed using the RNA Seq NGS Library Preparation Kit for Whole-Transcriptome Discovery (Gnomegen). The resulting paired-end cDNA libraries were sequenced using the Illumina HiSeq 2000 system.

Identification of male contigs located on the potential sex-determination chromosome

Male contigs (length $\geqslant$ 500 bp) were aligned to the female contigs (length $\geqslant$ 500 bp) using the MUMmer [53] aligner with default parameters. Coverage of each male contig by the female ones was then estimated using an identity threshold of $\geqslant$95% and requiring that any unaligned gap within the aligned region be $\leqslant$10 bp in length. Male contigs with coverage less than 0.05 were selected to be masked by RepeatMasker against the grass carp repetitive sequence library. The masked contigs were used to design the primers for the following PCR analysis. Using male and female genomic DNA as template, the primer pairs primed PCR amplification by Easy Taq DNA Polymerase (Transgen Biotech) to verify each of the contigs. Primer sequences are listed in Supplementary Table 16. Male contigs with amplification from male DNA but no amplification from female DNA were confirmed by sequencing of the PCR products. In total, 206 contigs were identified that potentially carried sequence variation between the male and female genomes, and these contigs were annotated to predict protein-coding genes using FgeneSH++.

Sample preparation and identification of differentially expressed genes in the FHT experiments

Grass carp larvae were raised in tanks at 25 ± 2℃ and fed with chironomid larvae (*Chironomus tentans*). At 46 d after hatching (body weight = 0.39 ± 0.05 g, body length = 28.05 ± 0.99 mm), 30 fish were randomly selected for sample collection before the FHT (designated as "46 d"). The rest of the fish were then randomly divided into 2 groups (n = 1,000 for each group) and fed with either chironomid larvae as fish undergoing no transition (designated as "116 d") or duckweed *Lemna minor* (designated as "116 d-FHT") as fish undergoing a transition to herbivory. Fish had free access to food 24 h a day and fed for 70 d. At 116 d after hatching (body weight = 2.97–7.34 g, body length = 53.96–72.78 mm), 30 fish were randomly selected from each group for sample collection. The collection time for all samples was at 9 a.m. Total RNA was isolated from brain, liver and gut tissues using SV TRIzol Reagent (Invitrogen) according to the manufacturer's protocol. Equal amounts of total RNA from nine fish in each group were pooled for each tissue and used to construct the libraries for transcriptome analysis.

The resulting RNA-seq reads were aligned to the gene model set with the SMALT aligner using the parameters of a minimum Smith-Waterman (-m) value of 35, a maximum insert size (-i) of 5,000 and a minimum insert size (-j) of 20. Numbers of uniquely mapped reads (mapping score ⩾ 50) were converted to quantified transcript levels by RPKM. The R package DEGseq [54] was used to digitally measure differential expression at the annotated loci. When a gene's expression increased or decreased in 116 d-FHT liver by more than two fold (q value [39] < 0.001) compared to that in the 46 d sample (before FHT), Δ1 was defined as log2(RPKM(116 d-FHT)/RPKM (46 d)). When a gene's expression increased or decreased in 116 d liver (not FHT) by more than two fold compared to that in the 46 d sample (q value < 0.001), Δ2 was defined as log2(RPKM(116 d)/RPKM(46 d)). If the value of Δ1 - Δ2 was less than −1 or greater than 1, the corresponding gene was considered to be differentially expressed by the FHT and non-FHT samples. All identified genes were functionally annotated by DAVID to group them into significantly clustered pathways ($P < 0.05$). Ethical approval for this work was obtained from the Expert Committee of Biomedical Ethics, Institute of

Hydrobiology of the Chinese Academy of Sciences (Y11201-1-301).

Discussion

We have generated two draft grass carp genome assemblies, a male one and a fully annotated female assembly. Comparison of a large number of gene models and synteny analysis demonstrated that zebrafish and grass carp share a similar genomic evolutionary history. However, a chromosome fusion resulting in linkage group 24 and the occurrence of a GCSD may be responsible for the substantial differences in development (for example, in body size) and other characteristics (for example, sex determination) between grass carp and zebrafish. Characterization of diet change-associated transcriptomes has provided new genomic information on the metabolic adaption of grass carp to the shift to a vegetarian diet during its life history. These genome sequences also bring grass carp breeding into a genomic phase.

Methods

Methods and any associated references are available in the online version of the paper.

URLs

Kyoto Encyclopedia of Genes and Genomes (KEGG), http://www.genome.jp/kegg/; KEGG Automatic Annotation Server (KAAS), http://www.genome.jp/tools/kaas/; DAVID, http://david.abcc.ncifcrf.gov/summary.jsp; SMALT, http://www.sanger.ac.uk/resources/software/smalt/; SOAPdenovo, http://soap.genomics.org.cn/; RepeatModeler, http://www.repeatmasker.org/RepeatModeler.html; RepeatMasker, http://www.repeatmasker.org/; Repbase, http://www.girinst.org/repbase/; TimeTree, http://www.timetree.org/; TreeFam, http://treefam.genomics.org.cn/; fermi, https://github.com/lh3/fermi; Spinner, ftp://ftp.sanger.ac.uk/pub/users/zn1/spinner/; Ensembl, ftp://ftp.ensembl.org/pub/; PHYLIP, http://evolution.genetics.washington.edu/phylip.html.

Accession codes

All of the Illumina short-read sequencing data for this project have been deposited at the DNA Data Bank of Japan (DDBJ), the European Molecular Biology Laboratory (EMBL) and GenBank under accession PRJEB5920, which includes whole-genome shotgun sequence (ERS428355 for the female and ERS428356 for the male; Sequence Read Archive (SRA)) and RNA-seq data (ERS430059, SRA). All of the grass carp data have been released at the official National Center for Gene Research website (http://www.ncgr.ac.cn/grasscarp/). The current version of the data set is the first version (CI01).

Acknowledgements We thank S. Liu (Hunan Normal University) for providing a gynogenetic female individual. We thank Y. Xue and J. Torrance for their help in English-language editing of the manuscript. This work is supported by the National High-Technology Research and Development Program (863 Program, 2011AA100403), the National Natural Science Foundation of China (31130055), the Strategic Pilot Science and Technology Projects (A) Category, Chinese Academy of Science (XDA08030203), the Guangdong Provincial Science and Technology Program (2012B090500008) and the State Key Laboratory of Freshwater Ecology and Biotechnology (2011FBZ18).

References

1. Chilton, E.W. & Muoneke, M.I. Biology and management of grass carp (*Ctenopharyngodon idella, Cyprinidae*) for vegetation control: a North American perspective. *Rev. Fish Biol. Fish*. 2, 283-320 (1992).
2. Statistics and Information Service, Fisheries and Aquaculture Department, Food and Agriculture Organization of the United Nations. *FAO Yearbook of Fishery and Aquaculture Statistics 2011* (Food and Agriculture Organization of the United Nations, 2013).
3. Howe, K. *et al.* The zebrafish reference genome sequence and its relationship to the human genome. *Nature* 496, 498-503 (2013).
4. Xia, J.H. *et al.* A consensus linkage map of the grass carp (*Ctenopharyngodon idella*) based on microsatellites and SNPs. *BMC Genomics* 11, 135 (2010).
5. Chen, J. *et al.* Transcriptome analysis of head kidney in grass carp and discovery of immune-related genes. *BMC Vet. Res.*8, 108 (2012).
6. Xu, B. *et al.* Generation and analysis of ESTs from the grass carp, *Ctenopharyngodon idellus*. *Anim. Biotechnol.* 21, 217-225 (2010).
7. Guo, T. *et al.* Cloning, molecular characterization, and expression analysis of the signal transducer and activator of transcription 3 (*STAT*$_3$) gene from grass carp (*Ctenopharyngodon idellus*). *Fish Shellfish Immunol.* 35, 1624-1634 (2013).
8. Wang, T.T. *et al.* Molecular characterization, expression analysis, and biological effects of interleukin-8 in grass carp *Ctenopharyngodon idellus*. *Fish Shellfish Immunol.* 35, 1421-1432 (2013).
9. Chen, Y., Pandit, N.P., Fu, J., Li, D. & Li, J. Identification, characterization and feeding response of peptide YYb (PYYb) gene in grass carp (*Ctenopharyngodon idellus*). *Fish Physiol. Biochem.* 40, 45-55 (2014).
10. Feng, K., Zhang, G.R., Wei, K.J. & Xiong, B.X. Molecular cloning, tissue distribution, and ontogenetic expression of ghrelin and regulation of expression by fasting and refeeding in the grass carp (*Ctenopharyngodon idellus*). *J. Exp. Zool. A Ecol. Genet. Physiol.* 319, 202-212 (2013).
11. Yu, E.M. *et al.* Molecular cloning of type I collagen cDNA and nutritional regulation of type I collagen mRNA expression in grass carp. *J. Anim. Physiol. Anim. Nutr.* (*Berl.*) 98, 755-765 (2014).
12. Zhong, S.S., Jiang, X.Y., Sun, C.F. & Zou, S.M. Identification of a second follistatin gene in grass carp (*Ctenopharyngodon idellus*) and its regulatory function in myogenesis during embryogenesis. *Gen. Comp. Endocrinol.*185, 19-27 (2013).
13. Star, B. *et al.* The genome sequence of Atlantic cod reveals a unique immune system. *Nature* 477, 207-210 (2011).
14. Aparicio, S. *et al.* Whole-genome shotgun assembly and analysis of the genome of *Fugu rubripes*. *Science* 297, 1301-1310 (2002).
15. Kasahara, M. *et al.* The medaka draft genome and insights into vertebrate genome evolution. *Nature* 447, 714-719 (2007).
16. Jaillon, O. *et al.* Genome duplication in the teleost fish *Tetraodon nigroviridis* reveals the early vertebrate

proto-karyotype. *Nature* 431, 946-957 (2004).

17. Jones, F.C. *et al.* The genomic basis of adaptive evolution in threespine sticklebacks. *Nature* 484, 55-61 (2012).
18. Nocusa, S. A comparative study of the chromosomes in the fishes with particular consideration on taxonomy and evolution. *Mem. Hyogo Univ. Agric.* 3, 1-62 (1960).
19. Stanley, J.G. Female homogamety in grass carp (*Ctenopharyngodon idella*) determined by gynogenesis. *J. Fish. Res. Board Can.* 33, 1372-1374 (1976).
20. Peng, Z. *et al.* The draft genome of the fast-growing non-timber forest species moso bamboo (*Phyllostachys heterocycla*). *Nat. Genet.* 45, 456-461 (2013).
21. Jang, S. *et al.* Construction and characterization of two bacterial artificial chromosome libraries of grass carp. *Mar. Biotechnol. (NY)*12, 261-266 (2010).
22. Kumar, S. & Hedges, S.B. TimeTree2: species divergence times on the iPhone. *Bioinformatics* 27, 2023-2024 (2011).
23. Li, H. *et al.* TreeFam: a curated database of phylogenetic trees of animal gene families. *Nucleic Acids Res.* 34, D572-D580 (2006).
24. Crow, K.D., Stadler, P.F., Lynch, V.J., Amemiya, C. & Wagner, G.P. The "fish-specific" Hox cluster duplication is coincident with the origin of teleosts. *Mol. Biol. Evol.* 23, 121-136 (2006).
25. Kurosawa, G. *et al.* Organization and structure of *hox* gene loci in medaka genome and comparison with those of pufferfish and zebrafish genomes. *Gene* 370, 75-82 (2006).
26. Wegner, M. All purpose Sox: the many roles of Sox proteins in gene expression. *Int. J. Biochem. Cell Biol.* 42, 381-390 (2010).
27. Akira, S., Yamamoto, M. & Takeda, K. Role of adapters in Toll-like receptor signalling. *Biochem. Soc. Trans.* 31, 637-642 (2003).
28. Braasch, I. & Postlethwait, J.H. in. *Polyploidy and Genome Evolution* (eds. Soltis, P.S. and Soltis, D.E.) 341-383 (Springer, 2012).
29. Kanehisa, M. *et al.* Data, information, knowledge and principle: back to metabolism in KEGG. *Nucleic Acids Res.* 42, D199-D205 (2014).
30. Petit, V. & Thiery, J.P. Focal adhesions: structure and dynamics. *Biol. Cell* 92, 477-494 (2000).
31. Danen, E.H. & Yamada, K.M. Fibronectin, integrins, and growth control. *J. Cell. Physiol.* 189, 1-13 (2001).
32. Palacín, M. *et al.* The genetics of heteromeric amino acid transporters. *Physiology* (*Bethesda*) 20, 112-124 (2005).
33. Goodman, B.E. Insights into digestion and absorption of major nutrients in humans. *Adv. Physiol. Educ.* 34, 44-53 (2010).
34. Zhao, B., Tumaneng, K. & Guan, K.L. The Hippo pathway in organ size control, tissue regeneration and stem cell self-renewal. *Nat. Cell Biol.*13, 877-883 (2011).
35. Zhao, B., Lei, Q.Y. & Guan, K.L. The Hippo-YAP pathway: new connections between regulation of organ size and cancer. *Curr. Opin. Cell Biol.* 20, 638-646 (2008).
36. Traut, W. & Winking, H. Meiotic chromosomes and stages of sex chromosome evolution in fish: zebrafish, platyfish and guppy. *Chromosome Res.* 9, 659-672 (2001).
37. Cui, Y., Chen, S., Wang, S. & Liu, X. Laboratory observations on the circadian feeding patterns in the grass carp (*Ctenopharyngodon idella* Val.) fed three different diets. *Aquaculture* 113, 57-64 (1993).
38. Huang, W., Sherman, B.T. & Lempicki, R.A. Systematic and integrative analysis of large gene lists using DAVID Bioinformatics Resources. *Nat. Protoc.*4, 44-57 (2009).
39. Benjamini, Y. & Hochberg, Y. Controlling the false discovery rate: a practical and powerful approach to multiple testing. *J. R. Stat. Soc., B* 57, 289-300 (1995).
40. Dempsey, M.E. Regulation of steroid biosynthesis. *Annu. Rev. Biochem.* 43, 967-990 (1974).
41. Lange, B.M., Rujan, T., Martin, W. & Croteau, R. Isoprenoid biosynthesis: the evolution of two ancient and distinct pathways across genomes. *Proc. Natl. Acad. Sci. USA* 97, 13172-13177 (2000).
42. Flores, M.V. *et al.* The zebrafish retinoid-related orphan receptor (*ror*) gene family. *Gene Expr. Patterns* 7, 535-543 (2007).

43. Tan, Y., DeBruyne, J., Cahill, G.M. & Wells, D.E. Identification of a mutation in the *Clock1* gene affecting zebrafish circadian rhythms. *J. Neurogenet.* 22, 149-166 (2008).

44. Zhou, Y. *et al.* Enhancement of growth and intestinal flora in grass carp: the effect of exogenous cellulase. *Aquaculture* 416-417, 1-7 (2013).

45. Mortazavi, A., Williams, B.A., McCue, K., Schaeffer, L. & Wold, B. Mapping and quantifying mammalian transcriptomes by RNA-seq. *Nat. Methods* 5, 621-628 (2008).

46. Kozarewa, I. *et al.* Amplification-free Illumina sequencing-library preparation facilitates improved mapping and assembly of (G+C)-biased genomes. *Nat. Methods* 6, 291-295 (2009).

47. Simpson, J.T. *et al.* ABySS: a parallel assembler for short read sequence data. *Genome Res.* 19, 1117-1123 (2009).

48. Li, H. Exploring single-sample SNP and INDEL calling with whole-genome *de novo* assembly. *Bioinformatics* 28, 1838-1844 (2012).

49. Li, R. *et al.* SOAP: short oligonucleotide alignment program. *Bioinformatics* 24, 713-714 (2008).

50. Bonfield, J.K. & Whitwham, A. Gap5-editing the billion fragment sequence assembly. *Bioinformatics* 26, 1699-1703 (2010).

51. Wu, T.D. & Watanabe, C.K. GMAP: a genomic mapping and alignment program for mRNA and EST sequences. *Bioinformatics* 21, 1859-1875 (2005).

52. Birney, E., Clamp, M. & Durbin, R. GeneWise and Genomewise. *Genome Res.* 14, 988-995 (2004).

53. Kurtz, S. *et al.* Versatile and open software for comparing large genomes. *Genome Biol.* 5, R12 (2004).

54. Wang, L., Feng, Z., Wang, X., Wang, X. & Zhang, X. DEGseq: an R package for identifying differentially expressed genes from RNA-seq data. *Bioinformatics* 26, 136-138 (2010).

草鱼基因组框架图谱绘制及其草食性适应机制研究

汪亚平[1] 陆 颖[2] 张 勇[3,4] Zemin Ning[5] 李 艳[2]
赵 强[2] 陆恒运[2] 黄 容[1] 夏晓勤[1] 冯 旗[2] 梁旭方[6,7]
刘坤艳[2] 张 垒[2] 陆婷婷[2] 黄 涛[2] 范丹林[2] 翁崎峻[2]
祝传让[2] 陆怡琪[2] 李文俊[2] 文子若[2] Congcong Zhou[2] 田琪琳[2] 康晓军[1,8]
石米娟[1] 张婉婷[1] 张成勋[1,9] 杜富宽[1] 何 珊[5] 廖兰杰[1] 李勇明[1]
桂 彬[1] 何慧慧[1] 宁 真[1] 杨 诚[1,6] 何利波[1] 罗丽飞[1] Rui Yang[10]
Qiong Luo[10] 张晓川[3,4] 李水生[3,4] 黄 文[3,4] 肖 玲[3,4]
林浩然[3,4] 韩 斌[2] 朱作言[1]

1 中国科学院水生生物研究所，武汉，中国 430072
2 中国科学院国家基因研究中心，上海，中国 200233
3 中山大学生命科学学院生物防治国家重点实验室，广州，中国 510527
4 中山大学生命科学学院水生经济动物广东省重点实验室，广州，中国 510527
5 Wellcome Trust Sanger Institute, Wellcome Trust Genome Campus, Hinxton, UK
6 华中农业大学水产学院农业部淡水动物育种重点实验室，武汉，中国 430072
7 华中农业大学水产学院湖北省淡水养殖协同创新中心，武汉，中国 430072
8 中国地质大学计算机学院，武汉，中国 430074
9 College of Life Science, Kim Illinois Sung University, Pyongyang, North Korea
10 云南农业大学植物保护学院，昆明，中国 650201

摘 要 草鱼是重要的淡水养殖鱼类，产量约占全球淡水养殖总量的 16%，因其典型的草食性特征而得名。本研究采用鸟枪法测序策略，分别对一尾雌核发育雌性和一尾野生雄性草鱼进行了全基因组测序，获得雌性(0.9 GB)和雄性(1.07 GB)基因组组装序列，并在雌性基因组中注释了 27,263 个蛋白编码基因，完成了其中 17,456 个基因在染色体上的定位。与现有 12 种脊椎动物基因组的比较研究发现，草鱼与斑马鱼两者的分化时间距今 49 百万-54 百万年。草鱼的第 24 号染色体对应于斑马鱼的第 10 和第 22 号染色体，提示草鱼基因组在演化过程中发生了一次染色体融合。比较转录组分析发现，草鱼在草食性转化过程中，肠道中昼夜节律相关基因的表达模式发生了重设，肝脏中甲羟戊酸通路和类固醇生物合成通路被激活。草鱼可能通过持续高强度的食物摄入，获取足够多的可利用营养以维持其快速生长。草鱼全基因组序列的解析，将从基因组层面为培育遗传改良的养殖品种提供重要支撑。

第七部分
鱼类基因转移育种技术展望

Embryonic and Genetic Manipulation in Fish

ZUO YAN ZHU   YONG HUA SUN

State Key Laboratory of Freshwater Ecology and Biotechnology, Institute of Hydrobiology, Chinese Academy of Sciences, Luojiashan, Wuhan 430072

Abstract Fishes, the biggest and most diverse community in vertebrates are good experimental models for studies of cell and developmental biology by many favorable characteristics. Nuclear transplantation in fish has been thoroughly studied in China since 1960s. Fish nuclei of embryonic cells from different genera were transplanted into enucleated eggs generating nucleo-cytoplasmic hybrids of adults. Most importantly, nuclei of cultured goldfish kidney cells had been reprogrammed in enucleated eggs to support embryogenesis and ontogenesis of a fertile fish. This was the first case of cloned fish with somatic cells. Based on the technique of microinjection, recombinant MThGH gene has been transferred into fish eggs and the first batch of transgenic fish were produced in 1984. The behavior of foreign gene was characterized and the onset of the foreign gene replication occurred between the blastula to gastrula stages and random integration mainly occurred at later stages of embryogenesis. This eventually led to the transgenic mosaicism. The MThGH-transferred common carp enhanced growth rate by 2-4 times in the founder juveniles and doubled the body weight in the adults. The transgenic common carp were more efficient in utilizing dietary protein than the controls. An "all-fish" gene construct CAgcGH has been made by splicing the common carp β-actin gene (CA) promoter onto the grass carp growth hormone gene (gcGH) coding sequence. The CAgcGH-transferred Yellow River Carp have also shown significantly fast-growth trait. Combination of techniques of fish cell culture, gene transformation with cultured cells and nuclear transplantation should be able to generate homogeneous strain of valuable transgenic fish to fulfil human requirement in 21st century.

Keywords fish; nuclear transplantation; transgenic fish; gene targeting

Introduction

Lives on earth first appeared over 3 billion years ago. As a result of evolution, vertebrates are the utmost advanced life form distributing from seabed to high-mountain. Fishes, at the lower stage of evolution but being the biggest community in vertebrates, include about 21,700 to 28,000 species that take over almost half of the total number of vertebrates[1]. Fishes are also models for experimental study as the eggs are large, fertilization and

---

文章发表于《细胞研究》(英文版), 2000, 10(1): 17-27

development are externally and the embryos are transparent. In addition, mono-sexual breeding and crossbreeding between many distantly related species could be done in fishes.

Fish culture is one of the earliest activities of human civilization. Time dating back to 2500 years ago when China was at the Spring and Autumn Period, an ancient Chinese named Fan Li created a great literature "Handbook of Fish Culture" in which domestication and cultivation of common carp in ponds were described in detail. This is the first monograph of fish culture in the world. Since then, Chinese farmers enjoyed a long tradition of selective domestication resulting in very valuable "four farming species" of fish for pond culture. These are black carp (*Mylopharyngodon poceus*), grass carp (*Ctnopharyngodon idellus*), silver carp (*Hypophthalmichihys molivrix*) and big-head carp (*Aristichthys nobilis*). Based on the feeding habits of these fishes, Chinese farmers also invented "multi-culture method" of stocking the four farming species together in the same water body. This is in fact a sustainable ecological system by that farmers are able to gain maximum output with minimum cost. These domesticated species and the multi-culture method have been favorably adopted by other nations abroad.

In addition to the traditional experiences of fish farming, scientists in China initiated some fundamental research in fish cell and developmental biology and molecular genetics leading to a break-through in fish biotechnology. The major achievements are pointed as follows. In 1960's, Professor Tung T. C. first introduced the art of nuclear transplantation with fish to study the interaction between nucleus and cytoplasm[2]. Tung and his colleagues succeeded in generation of "nuclear-cytoplasmic hybridized fish" between different species (reviewed in [3]). In early 1980's, the first somatic cell cloned fish was derived from kidney cells by nuclear transplantation[4]. In 1984, the first batch of transgenic fish was also generated[5]. The development and combination of technologies of nuclear transplantation and gene transfer in fish predicate that a new era in fish breeding is coming; fish directional breeding will revolutionize the traditional fish farming industry.

Nuclear Transplantation in Fish

The concept of nuclear transplantation dates back to 1938 when a German biologist, Hans Spermann[6], proposed an experiment that would evaluate the relative importance of the nucleus and cytoplasm in controlling early developmental events. In the early 1950's, the technology of nuclear transplantation was first demonstrated with frogs by Briggs and King[7] and amphibians were becoming most commonly used for studying nuclear transplantation (reviewed in [8]). These studies revealed that the nuclei undergo restriction in developmental capacity as cells became differentiated. Although several experiments suggested that differentiated somatic cells still had developmental totipotency[9-11], the cell differentiation didn't necessitate any irreversible changes in nuclear genetic material. Some subsequent experiments with differentiated somatic cells from Xenopus[12,13] and Ranna pipens[14-17] could hardly supported the previous cases because that nuclei from these cells only supported the ontogenesis to feeding-stage of tadpole but not to an adulthood. Thus, whether ultimately differentiated cells have the capacity to support the reconstructed embryos to complete their full ontogenesis still remains unanswered. Illmensee and Hoppe in 1981 first succeeded in nuclear transplantation in mammals[18]. They reported that an inner cell mass nucleus could support early development of an enucleated egg in mouse. Three years latter, McGrath and Solter[19] reported nuclear transplantation in mice by using micrioinjection coupled with cell fusion technique. Since then, embryonic nuclear transplantation in sheep, calve, rabbit, porcine and goat were successively done[20-24]. In 1997, the birth of cloned lamb "Dolly" derived from an adult mammary gland cell marked the beginning of a "golden-age" in animal cloning[25]. Recently, success has also been announced in cloning mice, goat, and cattle derived from differentiated cells: fetal fibroblasts[26], muscle cells[27], cumulus cells[28,29] or oviductal cells[29]. Successes on somatic cloning conclusively revealed that differentiated adult cells still remain totipotent and maintain the whole genome to support normal development to term.

In amphibian and mammals, nuclear transplantation has been successfully done within the same species. In other words, the donor of nucleus and the host of enucleated eggs must come from the same species. On the other hand, "Tung's fish" brought an absolutely new story. Tung's embryonic nuclear transplantation in fish involves three steps in general[2,3]. (1) Preparation of donor cells: the blastoderm was separated from the yolk with a fine glass needle, and carefully placed in Holtfreter's dissociation solution for further separating into individual cells. (2) Preparation of unfertilized host eggs: the mature eggs stripped from the female were dichorionized with a pair of forceps, and enucleated by inserting a sharp glass needle into the egg cytoplasm just underneath the site of the second polar body. (3) Nuclear transplantation: after donor cell was sucked into the micropipette by a slight negative pressure and microinjected into the enucleated eggs, the reconstructed embryos were put into Holtfreter's solution for further development. Great successes of fish nuclear transplantation were achieved between different genus (common carp (*Cyprinus Carpio* Linnaeus) and crucian carp (*Carassius aurantus*)[30, 31] and different subfamilies (grass carp (*Ctenopharyngodon idellus*) and blunt-snout bream (*Megalobrama amblycephals*))[32,33]. The nuclear-cytoplasmic hybrid fish revealed that while most phenotypic characteristics were controlled by the nucleus, a few were controlled by the cytoplasm or by both (reviewed in [3]).

In 1984, the first somatic cell cloned fish derived from short-term cultured kidney cells of triploid crucian carp was produced[4]. In this experiment two rounds of nuclear transfer were carried out. In the first round, the nucleus ($3n$) was transferred into the enucleated eggs ($2n$) of crucian carp. There were 41% of the injected eggs developed into blastulae but no future development occurred. Nuclei from the blastulae were taken for second round of transfer into another batch of enucleated eggs ($2n$). In the second round, 8 gastrula generated and one of them developed into a fertile female fish (1.2%) with normal morphological features of crucian carp. The chromosome number was triploid ($3n$=150). This was the first case of cloned animal with somatic cells in fishes. It suggested that some nuclei of somatic cells, following two rounds of nuclear transplantation, could be reprogrammed to totipotent status as zygotic nucleus does.

Nuclear transplantation in fishes, as an approach for studying the relative roles of nucleus and cytoplasm in controlling the characteristics, had also been considered as a helpful method in obtaining new farming strains of fishes. Yan[3] indicated that the nuclear-cytoplasmic hybrid was 22% higher in growth rate, 3.8% higher in protein content and 5.58% lower in lipid content than that of the control, respectively. The art of nuclear transplantation in fish, however, not only requires very skilful manipulators but also bring about very low efficiency in producing the nuclear-cytoplasmic hybrids. The application of nuclear transfer fish in fishery is, thus, under restriction to a certain extent.

Gene Transfer in Fish

People never give up their effort on pursuing of breeding new strains of farming species with high quality. Meanwhile, researchers are seeking for some genetic materials other than the whole cell nucleus for transfer. Around the late 1970's, researchers in China introduced the genomic DNA of common carp into the fertilized eggs of Mud carp (*Cirrhina molitorella*), a tropical species. About 8% of the "total DNA-transferred" founder Mud carp showed increase of cold-resistance [Zuoyan Zhu, *et al.* unpublished data].

In early 1980's, with the advancement of techniques in molecular cloning and embryonic micro-manipulation, recombinant genes are able to be constructed and transferred into the host animals. The transgenic "super mouse" was the most stimulating report in transgenic studies[34].

The first batch of faster growing transgenic fish was generated by introducing a recombinant human growth hormone (hGH) gene capped with a mouse metallothionein-1 (MT) gene promoter into goldfish in 1984[5]. In this study, fertilized eggs of goldfish were obtained by artificial spawning and insemination and the chorion of eggs were removed by digestion in 0.25% trypsin solution. About 1-2 nl of DNA solution containing about 10^5-10^6 copies of the MThGH gene was delivered with a micromanipulator into the germinal disc just underneath the second polar body. All the manipulated eggs were carefully put into Holtfreter's solution for further development. By Southern hybridization, more than 50% of the founders are transgenics. Since then, dozens of laboratories all over the world began to show great zeal for the study of transgenic fish[35], and gene transfer into fish embryos were performed in several species, such as rainbow trout. (*Salmo irdeus* Gibbonsi), Atlantic salmon (*Salmo salar* Linnaeus), tilapia (*Oreochromis nilotica*), medaka (*Oryzias latipes*), common carp, zebrafish (*Danio rerio*), loach (*Misgurnus anguillicaudatus*), catfish (*Parasilurus asoltus* Linnaeus), etc[36]. Some other techniques, e.g. electroporation[37] and sperm-mediating[38,39], were also successfully employed in producing of transgenic fish. In addition to gene transfer with GH gene, other types of genes were also employed in gene transfer in fish. For example, GFP (green fluorescent protein) gene was transferred into fertilized eggs of zebrafish as a reporter[40] and AFP (antifreeze proteins) gene was introduced into Atlantic salmon to gain freeze-resistant salmon[41]. However, gh is so far the most commonly used and most thoroughly investigated type of transgenes.

It was in a transgenic fish model that the behavior of a foreign gene, this time the MThGH gene, in embryogenesis was intensively studied by Zhu *et al*[42]. Southern blotting revealed that the behaviors of MThGH gene in host fish were a dynamic process, including replication, degradation, concatemer formation, and integration during embryogenesis. The replication began at very early stage of cleavage and suddenly took place at late-blastula to early-neurula stages. After neurula stage, most of the foreign gene were migrating with the host chromosomal DNA as revealed on the agarose gel by electrophoresis. It was suggested that foreign gene at these stages was in form of either large concatemers situated outside the chromosomes or integrated into the host genome. If the integration occurred at germ line cells of the host fish, the transgene could be inherited to their offspring. Northern hybridization showed that the transcripts of hGH gene could only be found at post-late-gastrula stage, which was consistent with the timing of the differentiation of fish embryonic cells. As a result of the expression of MThGH gene, transgenic fish showed significantly faster growing trait. On the other hand, a certain proportion of the founders did not show growth enhancement and a few even grew slower than the control. This unexpected observation was reasonable when the multi-sited integration and transgenic mosaicism were taken into account. In the case of the generation of transgenic mammals, foreign gene had been microinjected into the zygotic pronucleus, which resulted in the integration of foreign gene occurring mostly before the first cleavage[43]. It was believed that the status of transgene in founder mammals was in a homozygous form, which could be transmitted to next generation in a Mendelian manner. However, pronucleus of fish eggs is not visible and the foreign gene could only be microinjected into their cytoplasm. The foreign gene integration in fish was found to be spanned over a long time course from gastrula to late developmental stages resulting in multi-sited integration and consequent transgenic mosaicism. Three categories of transgene integration could be deduced: functional integration, silent integration, and toxic integration. It is only in the functional integration that the transgene was integrated into host genomic sites suitable for expression and therefore showing "fast-growth" trait. Thus, the founder generation (P_0) of transgenic fish was far from a genetically homogeneous strain.

The specific growth rate (SGR) of P_0, F_1, and F_2 MThGH-transgenic fish, as well as of F_4 generation was significantly higher than that of the controls[42,44-47]. Bioenergetic analysis on MThGH-transgenic fish compared with controls was thoroughly worked out. When feeding with fresh tubificid worms, the energy budget of both transgenic and control fishes can be expressed by the following equations[46]:

MThGH-transgenic

F_2 fish $100C = 8.9F+0.63U+49.03R+41.44G$

Control fish $100C = 7.37F+1.14U+53.36R+38.13G$

In which, C is the total energy from food, F is the energy lost in faeces, U is the energy lost in nitrogenous exaction, R is the energy channeled to metabolism and G is the energy channelled to growth.

Compared with the controls, transgenic fish had a significantly higher proportion of food energy channelled to G and a significantly lower proportion of that channelled to R and U. The transgenic fish saved 6.62% of the total energy from eaten food for growth improvement. That phenomenon was named as “fast-growing and less-eating” effect. Growth and feed utilization by MThGH-transgenic F4 fish feeding with diets containing different protein levels had also been carried out[47]. Protein and energy intakes were significantly higher in the transgenics than in the controls fed with 20% protein diet, and recovered energy, as a proportion of protein intake, was also significantly higher in the transgenics than in the controls fed with 40% protein diet. It was thus concluded that at a lower dietary protein level, transgenics achieved higher growth rates mainly by increasing food intake; but at a higher dietary protein level, transgenics achieved higher growth rates mainly through higher energy conversion efficiency. That is to say, transgenics are more efficient in utilizing dietary protein than the controls, which leads to transgenics getting a significantly higher specific growth rate than the controls. For the body composition, it was revealed that the transgenic fish had body contents of dry matter 1.6%, and protein 2.2% to 4.3% higher than those of the controls, but contents of lipid 3.9% to 13.1% lower than that of the controls[48]. The apparent digestibility of amino acids tends to be higher in the transgenics than in the controls, especially in fish fed diets with lower protein levels. While taking a look at the proportion of amino acids in transgenics and controls, there was no difference whatever betwen 17 amino acids. Thus, transgenic fishes would have much more nutritious value than control fishes. It is reasonable that fishes with "fast growing and less eating" as well as "high protein content and low lipid content" traits will fulfil human increasing requirement on protein source from fishes.

Nevertheless, both the mouse metallothionein-1 (MT-1) gene promoter and the hGH structural gene are not suitable for the purpose of producing farming species of transgenic fish. It was urged to construct “all-fish” gene for transfer[42]. Researchers have cloned both common carp *β*-actin gene (CA)[49], and grass carp growth hormone gene (gcGH)[50], and subsenquetly made a new construct of pCAgcGH, an "all-fish" genomic construct with a powerful promoter of β-actin gene from common carp and the whole transcription unit of GH gene from grass carp. In the spring of 1997, this construct has been microinjected into the fertilized eggs of Yellow River Carp (*Cyprinus carpio* L.), and a batch of CAgcGH-transgenic was produced. As these fishes grew up to 5-month-old, body weight of the heaviest transgenic individual was 2.75 kg, while that for the controls was 1.1 kg. About 10% of the transgenics were over 2.0 kg, while the controls were about 0.7 kg on average. It is more exciting that the heaviest body weight of 17-month-old transgenics, 7.65 kg, was two fold and more than that of the control siblings [unpublished data]. Some experiments revealed further that the gene constructs with genetic elements derived from fish expressed more efficiently in fish cells than that from mammalian sp., e.g. the transcriptional activity of mouse MT promoter was only 1/2 of that of carp MT promoter in CAT (chloramphenicol acetyl-transferase) gene-transferred fish cells[51]. Till now, more than ten “all-fish” recombinant genes have been constructed all over the world[35]. The most dramatic growth acceleration came from Delvin *et al*., in which their pOnMTGH1-ransgenic salmon were more than 11-fold heavier than the controls on average[52].

It is very sensitive to talk about the biosafety of transgenic fish. There are food, genetic and ecological safeties, each of which should be concerned seriously. At present, the widely accepted principle on safety evaluation of foods produced by modern biotechnology was the "substantial equivalence principle" delivered by European OECD (Organization for Economic Cooperation and Development) in 1993[53]. According to this principle, "all-fish" transgenic carp is included in "level I", the safest level[54]. Studies from Cui's investigation revealed that the transgene could only flow among individuals within a species but not between species by natural reproduction[55]. Fish crossbreeding has been widely used in aquaculture, two sets of whole genome of different species mixing with each other, i.e. 10^5 genes of one species crossing with 10^5 genes of another species. On the other hand, the "all-fish" gene transfer can also be regarded as a crossbreeding, but 10^5 genes from one species cross with 1 gene of another species instead. It can be simply figured out that the hetrozygosity of genome in the CAgcGH-transgenic common carp is about 10^5 times less than that in the hybrids between common carp and grass carp. In other words, the risk of stocking "all-fish" gene transgenics is considerably on a lower level in comparison with stocking the hybrid fish. Nevertheless, people should take a cautious attitude towards the application of transgenics, since transgenic fish is still a "newborn" in comparison with naturally existing species that have undergone a long evolutionary and selective course. As polyploid-breeding in fish has been very popular during the past[56], this technique can also be employed in the breeding of transgenic fish. By crossing tetraploid individual with haploid transgenics, the infertile triploid strain of transgenic fish could be generated. It is reasonable to consider that stocking infertile strain of transgenic fish will lessen their impact on water ecosystem to the least degree.

The Prospect of Transgenic Fish

Just as what has been discussed at the preceding part of this article, the transgenic fish generated at present is far from a genetic homogenous strain. Gene transfer in fish has not succeeded in the site-specific integration, controllable expression and stable transmission of the transgene. One of the most efficient ways to solve this problem is to use gene targeting technique in fish gene transfer. Gene targeting, homologous recombination between DNA sequence residing in the chromosome and newly introduced cloned DNA sequence, allows the transfer of any modified gene into the host genome of living cells. Since embryonic cell culture[4] and embryonic cell nuclear transplantation[2,3] techniques have been developed in fishes, gene targeting is hopeful to be carried out to gain embryonic cell lines carrying artificially modified and site-specific integrated gene. By nuclear transplantation with gene-targeted embryonic cells, the genetic homogenous strain of transgenic fish can be generated. Additionally, to establish stem cell like lines in some model fish species, such as zebrafish and medakafish, were successfully reported[57-59]. If researchers can make great progress on the study of stem cell in farming fishes, it will be more convenient to generate homogenous strain of transgenic fish for aquaculture.

References

[1] Nelson, JS. Fishes of the world. 2nd edition, A Wiley-interscience Pub: USA 1984.

[2] Tung TC, Wu SC, Tung YYF, Yan SY, Tu M, Lu TY. Nuclear transplantation in fish. *Science bulletin, Academia Sinica (in Chinese)* 1963; 60-1.

[3] Yan SY. Cloning in fish-Nuclear-cytoplasmic Hybrids. Educational and Cultural Press: Hong Kong 1998.

[4] Chen H, Yi Y, Chen M, Yang X. Studies on the developmental potentiality of cultured cell nuclei of fish. Acta Hydrobiol Sin (in Chinese) 1986; 10(1): 1-7.

[5] Zhu Z, Li G, He L, Chen S. Novel gene transfer into the fertilized eggs of goldfish (*Carassius auratus* L. 1758). Z Angew Ichthyol 1985; 1: 31-4.

[6] Spemann H. Embryonic development and induction. New Haven, CT: Yale University Press; 1938.

[7] Briggs R, King TJ. Transplantation of living nuclei from blastula cells into enucleated frog's eggs. Proc Natl Acad Sci USA 1952; 38: 455-63.

[8] DiBerardino MA. Genomic potential of differentiated cells. New York: Columbia University Press; 1997.

[9] Gurdon JB. Adults frogs derived from the nuclei of single somatic cells. Dev Biol 1962; 4: 256-73.

[10] Gurdon JB. The developmental capacity of nuclei taken from intestinal epithelium cells of feeding tadpoles. J Embryol exp Morph 1962; 10: 622-40.

[11] Gurdon JB, Uehlinger V. "Fertile" intestine nuclei. Nature 1966; 210: 1240-1.

[12] Gurdon JB, Laskey RA, Reeves OR. The development capacity of nuclei transplanted from keratinized skin cells of adults frogs. J Embryol exp Morph 1975; 34: 93-112.

[13] Wabl MR, Brun RB, DuPasquier L. Lymphocytes of the toad Xenopus laevis have the gene set for promoting tadpole development. Science 1975; 190: 1310-2.

[14] Di Berardino MA, Hoffner NJ. Gene reactivation in erythrocytes. Nuclear transplantation in oocytes and eggs of Rana. Science 1983; 219: 862-4.

[15] Orr Hoffner N, Di Berardino MA, Mchinnell RG. The genome of frog erythrocytes displays centuplicate replications. Proc Natl Acad Sci USA1986; 83: 1369-73.

[16] Di Berardino MA, Orr Hoffner N, Mckinnell RG. Feeding tadpoles cloned from Rana erythrocyte nuclei. Proc Natl Acad Sci USA 1986; 83: 8231-4.

[17] Di Berardino MA, Orr Hoffner N. Genomic potential of erythroid and leukocytic cells of Rana pipiens analyzed by nuclear transfer into diplotene and maturing oocytes. Differentiation 1992; 50: 1-13.

[18] Illmensee K, Hoppe PC. Nuclear transplantation in Musmusculus: developmental potential of nuclei from preimplantation embryos. Cell 1981; 23: 9-18.

[19] McGrath J, Solter D. Nuclear transplantation in the mouse embryo by microsurgery and cell fusion. Science 1983; 220: 1300-2.

[20] Willadsen SM. Nuclear transplantation in sheep embryos. Nature 1986;320: 63-5.

[21] Prather RS, BarnesFL, Sims MM, *et al.* Nuclear transplantation in the bovine embryo: assessment of donor nuclei and recipient oocyte. Biol Reprod 1987; 37: 859-66.

[22] Stice SL, Robl JM. Nuclear reprogramming in nuclear transplant rabbit embryos. Biol Reprod 1988; 39: 657-64.

[23] Prather RS, Sims MM, First NL. Nuclear transplantation in early pig embryos. Biol Reprod 1989; 41: 123-32.

[24] Zhang Y, Wang JC Qian JF. Nuclear transplantation in goats. Theriogenology1991; 35: 289.

[25] Wilmut I, Schnieke AE, McWhir J, Kind AJ, Campbell KHS. Viable offspring derived from fetal and adult mammalian cells. Nature 1997; 385: 810-3.

[26] Cibelli JB, Stice SL, Golueke PJ, *et al.* Cloned transgenic calves produced from nonquisent fetal fibroblast. Science 1998; 280: 1256-8.

[27] Butler D French clone provides support for Dolly. Nature 1998; 392: 113.

[28] Wakayama T, Perry AC, Zuccottti M, Johnson KR, Yanagimachi R. Full-term development of mice from enuleated

oocytes injected with cumulus cell nuclei. Nature 1998; 394(6691): 396-74.

[29] Kato Y, Tani T, Sotomaru Y, *et al*. Eight calves cloned from somatic cells of a single adult. Science 1998; 282: 2095-8.

[30] Tung TC. Nuclear transplantation in teleosts. I. Hybrid fish from the nucleus of carp and the cytoplasm of crucian. Scientia Sinica (in Chinese) 1980;23(4): 517-23.

[31] Yan SY, Lu DY, Zhu ZY, *et al*. Nuclear transplantation in teleosts. II. Hybrid fish from the nucleus of crucian and the cytoplasm of carp. Scientia Sinica (B) (in Chinese) 1984; 27(10): 1029-34.

[32] Tung TC, Tung YFY, Lu TY, Tung SM, Tu M. Transplantation of nuclei between two subfamilies of teleosts (Goldfish-domesticated *Carassius auratus* and Chinese bitterling, *Rhodeus sinensis*) Acta Zool Sin (in Chinese) 1973; 19(3): 201-12.

[33] Yan SY, Lu DY, Du M, *et al*. Nuclear transplantation in teleosts. Nuclear transplantation between different subfamilies-hybrid fish from the nucleus of grass carp (*Ctenopharyngoden idellus*) and the cytoplasm of blunt-snout bream (*Megalobrama amblycaephala*). Chin Biotech (in Chinese) 1985; 1(4): 15-26.

[34] Palmiter RD, Brinster RL, Hammer RE, Trumbauer ME, Rosenfeld MG, Bimberg NC, Evans RM. Dramatic growth of mice that develop from eggs microinjected with metallothionein-growth hormone fusion gene. Nature 1982; 300: 611-5.

[35] Cui Z, Zhu Z. Several interesting questions about breeding transgenic fish. J Biotech (in Chinese) 1998; (5): 1-10.

[36] Houdebine LM, Chourrout D. Transgenesis in fish. Experentia 1991; 47(9): 891-7.

[37] Xie Y, Liu D, Zou J, Li G, Zhu Z. Gene transfer via electroporation in fish. Aquaculture 1993; 111: 207-13.

[38] Khoo HW, Ang LH, Lim HB, *et al*. Sperm cells as vectors for introducing foreign DNA into zebrafish. Aquaculture, 1992; 107: 1-19.

[39] Li G, Cui Z, Zhu Z, Huang S. Introduction of foreign gene carried by sperms. Acta Hydrobiol Sin (in Chinese) 1996; 20(3): 242-7.

[40] Amsterdam A, Lin S, Hopkins N. The Aequorea victoria green fluorescent protein can be used as a reporter in live zebrafish embryos. Dev Biol 1995; 171(1): 123-9.

[41] Hew CL, Davies PL, Fletcher G. Antifreeze protein gene transfer in Atlantic salmon. Mol Mar Biol Biotechnol 1992; 1: 309-17.

[42] Zhu Z, Xu K, Xie Y, Li G, He L. A model of transgenic fish. Scientia Sinica (B) (in Chinese) 1989; 2: 147-55.

[43] Hew CL. Transgenic fish: present status and future directions. Fish Physiology and Biochemistry 1989; 7: 409-13.

[44] Xu K, Wei Y, Guo L, Zhu Z. The effects of growth enhancement of human growth hormone gene transfer and human growth administration on crucian carp (*Carassius auratus gibelio*, Bloch). Acta Hydrobiol Sinica 1991; 15: 103-9.

[45] Wei Y, Xie Y, Xu K, *et al*. Heredity of human growth hormone gene in transgenic carp (*Cyprinus carpio* L) Chinese J Biotech (in Chinese) 1992; 8: 140-4.

[46] Cui Z, Zhu Z, Cui Y, Li G, Xu K. Food consumption and energy budget in MThGH-transgenic F2 red carp (*Cyprinus carpio* L. red var.). Chinese Science Bulletin (in English) 1996; 41: 591-6.

[47] Fu C, Cui Y, Hung SSO, Zhu Z. Growth and feed utilization by F4 human growth hormone transgenic carp fed diets with different protein levels. J Fish Biol 1998; 53: 115-29.

[48] Fu C. Growth and feed utilization by F4 human growth hormone transgenic red carp, *Cyprinus Carpio* L: effects of dietary protein level. Thesis for MSc (Supervised by Zuoyan Zhu) submitted to the Institute of Hydrobiology, Chinese Academy of Sciences 1998.

[49] Li Z, Zhu Z, Roberg K, *et al*. Isolation and characterization of β-actin gene of carp (*Cyprinus carpio*). DNA Sequence 1990; 1: 125-36.

[50] Zhu Z, He L, Chen TT. Primary-structural and evolutionary analyses of growth-hormone gene from grass carp (*Ctenopharyngodon idellus*). Eur J Biochem 1992; 207: 643-8.

[51] Li H, Liu DM, Ju DH, *et al*. Studies on the function of the promoter of carp (*Cyprinus carpio*) metallothionein gene. Acta Zool Sin (in Chinese) 1997;43(2): 197-202.

[52] Devlin RH, Yesakl TY, Blagl CA, Donaldson EM, Swanson P, Chan WK. Extraordinary salmon growth. Nature 1994; 371: 209-10.

[53] OECD (Organization for Economic Cooperation and Development), Safety evaluation of foods produced by modern biotechnology: concepts and principles. OECD Paris 1993.

[54] Zhu Z, Zeng Z. Open a door for transgenic fish to market. J Biotech (in Chinese) (in press).

[55] Cui Z. Biosafety assessment of GH-transgenic common carp (*Cyprinus carpio* L.). Thesis for Ph. D. (Supervised by Zuoyan Zhu) submitted to the Institute of Hydrobiology, Chinese Academy of Sciences 1998.

[56] Zhang J, Sun X. The survey and prospect of fish genetics and breeding research. In: The Selected Paper of Breeding in Jian Carp (*Cyprinus Carpio* var. Jian) Written by Zhang J. Sun X. *et al*. Science Press Beijing 1994: pp1-10.

[57] Wakamatsu Y, Ozato K, Sasado T. Establishment of a pluripotent cell line derived from a medaka (*Oryzias latipes*) blastula embryo. Mol Mar Biol Biotechnol 1994; 3(4): 185-91.

[58] Sun L, Bradford CS, Ghosh C, Cotllodi P, Barnes DW. ES-like cell cultures derived from early zebrafish embryos. Mol Mar Biol Dev 1995; 4: 193-9.

[59] Hong Y, Schartl M. Establishment and growth responses of early medakafish (*Oryzias latipes*) embryonic cells in feeder layer-free cultures. Mol Mar Biol Biotechnol 1996; 5(2): 93-104.

鱼类胚胎及遗传操作

朱作言　孙永华

中国科学院水生生物研究所，淡水生态及生物技术国家重点实验室，武汉 430072

摘　要　作为脊椎动物中最具多样性的一个生物类群，鱼类由于拥有许多优良性状，它们已成为研究细胞及发育生物学的良好实验模型。自 1960 年以来，鱼类细胞核移植相关研究已在中国获得了较为广泛的研究。不同种属间的鱼类胚胎细胞核被移植到去核的卵子中产生了核质异种的成鱼。更重要的是，体外培养的金鱼肾细胞核在去核的卵子中进行了重编程且能够支撑胚胎形成和个体发育，最终得到了可育的成鱼。这是以体细胞得到的首条克隆鱼。1984 年，通过显微注射技术将重组 MThGH 基因转移到鱼类受精卵中，获得了首批转基因鱼。外源基因的行为特征表现为在囊胚到原肠胚期间复制，而在胚胎发育后期阶段则主要表现为外源基因的随机整合，这最终会导致转基因嵌合体的形成。MThGH 转基因鲤在幼年期的生长速率可增加 2-4 倍，而成年后体重则会加倍。与对照组相比，转基因鲤的食物蛋白利用效率更高。我们将鲤的 β-肌动蛋白基因(CA)启动子与草鱼的生长激素基因(gcGH)编码序列进行拼接，从而得到了一种“全鱼”基因载体 CAgcGH。CAgcGH 的转基因黄河鲤同样表现出显著的快速生长特性。在 21 世纪，结合鱼类细胞培养、培养细胞的基因改造以及核移植技术应该可以得到具有经济价值的转基因鱼纯系，进而满足人类对鱼类蛋白更高的需求。

Nuclear Transplantation with Early-Embryonic Cells of Transgenic Fish

CHEN Shangping ZHAO Haobin SUN Yonghua
ZHANG Fuying ZHU Zuoyan

State Key Laboratory of Freshwater Ecology and Biotechnology, Institute of Hydrobiology, Chinese Academy of Sciences, Wuhan 430072

Abstract Like other transgenic animals, transgenic fishes produced by microinjection are transgenic mosaics. In order to produce homogenous transgenic fish, the transgenic blastula or gastrula cells were dissociated from *Carassius auratus*, Pengze var. and *Cyprinus carpio*, Huanghe var., and the nuclei were transferred into the mature eggs of the same species via microinjection or electro-fusion. Five nuclear-transferred *Carassius auratus*, Pengze var. and one *Cyprinus carpio*, Huanghe var. were obtained and the existence of the transgene was detected. The possibility of generating homogenous strain of transgenic fish by nuclear transplantation with transgenic early-embryonic cells is discussed.

Keywords fish; gene transfer; nuclear transplantation

Since the first transgenic fish was born in China [1], human growth hormone (hGH) gene has been successfully introduced into fertilized eggs of loach, common carp and crucian carp, and the integration expression, biological function and inheritance of transgenes in these transgenic fishes have been extensively investigated [2,3]. The experimental results have revealed that transferred foreign gene can start transcription in embryonic cells whether integration occurs or not, and the integrated transgene can be inherited to its sub-generation through germ-line transmission. It has also been found that growth hormone gene transgenic fishes have the traits of fast-growing and less-eating, and they contain high protein and low lipid contents [4,5]. Therefore, the transgenic fish will meet the requirement of increased animal protein supply. While some problems have to be solved in producing transgenic fish, such as to achieve homozygous integration and to control expression and to stable inheritance of transgene. Using of gene transfer combined with nuclear transplantation will solve these problems, which will lead to a breakthrough in the traditional fish breeding [6]. Recently, green fluorescent protein (GFP) gene was used as a reporter in the study on the integration of foreign gene in host fish, the result revealed that the integration began from blastula stage [7]. Here we report the experimental results of the nuclear transplantation with transgenic early-embryonic cells in *Carassius auratus*, Pengze var. and *Cyprinus carpio*, Huanghe var.

文章发表于《自然科学进展：国际材料》(英文), 2000, 10(12): 925-930

1. Materials and Methods

1.1 Fishes

Carassius auratus, Pengze var. were supplied by Saisi Freshwater Farming Center of Wuhan, and Yellow River carp (*Cyprinus carpio,* Huanghe var.) by Institute of Aquaculture, Henan Province. All experimental fishes were spawned by injection of common carp pituitary homogenates.

1.2 Genes and gene transfer

The genes for transfer are "all-fish" gene constructs: pCAgcGH, pCAgcGHc and pCAsIGFc. They are recombinant constructs of grass carp growth hormone gene (gcGH), grass carp growth hormone gene cDNA (gcGHc), and salmon insulin-like growth factor cDNA (sIGFc) (Unpublished data). All are driven by the β-actin gene promoter of common carp (fig. 1). Gene transfer was performed by microinjection [1].

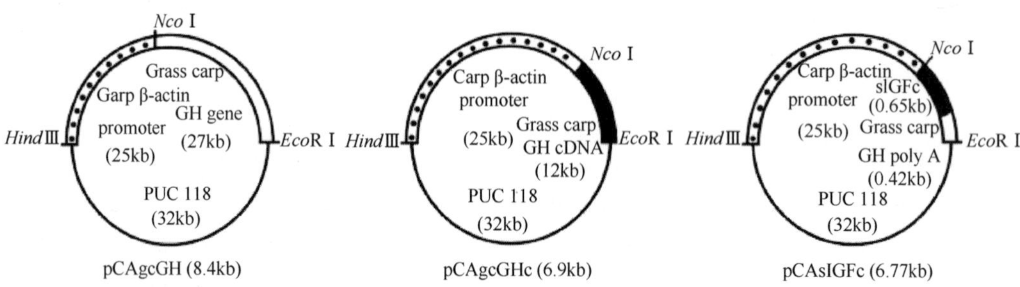

Fig. 1 Recombinant plasmids

1.3 Nuclear transplantation

Both methods of microinjection and electro-fusion were utilized in this study. The microinjection was carried out according to Yu *et al.* [8]. Unfertilized mature eggs served as nuclear recipients; early-embryonic cells (blastula or gastrula cells) were dissociated and used as nuclear donors for transplanting into the animal pole of the recipient eggs. The reconstructed eggs were put into Holtfreter's solution for development at the first 24h, then transferred into "aged" boiled water for examination. The numbers of reconstructed embryos at each developmental stage were counted. The electro-fusion was carried out according to Yi *et al.* [9] with a few modifications. The unfertilized eggs were placed in a fusion-chamber that was filled with Holtferter's solution, and the separated embryonic cells were added. They were then subjected to a DC pulse with electric field intensity of 1 kV/cm. The electro-fused embryos were kept in Holtferter's solution for development.

1.4 DNA preparation

The embryonic DNA was prepared as follows: 100 μL DNA extraction buffer [3] was added into the Eppendorf tube containing the embryos, and incubated at 50℃ for 3h. Then the DNA was extracted by phenol/chloroform and purified by ethanol precipitation. Fry's DNA was prepared with routine method [1].

1.5 Detection of foreign gene

PCR was used to examine the existence of foreign gene. Primer P1 (5′-TGGCGTGATGAATGTCG-3′) is located at the β-actin promoter sequence of common carp; primer P2 (5′-GCAGTCATACGTGTT-3′) at the coding sequence of grass carp GH, and primer P3 (5′-AGGAACTACCGAATGTAG-3′) at the cDNA sequence of salmon IGF. The sizes of PCR products amplified by P1 and P2 are 0.41 kb for pCAgcGHc and 0.68 kb for pCAgcGH. The PCR product amplified by P1 and P3 is 0.8 kb for pCAsIGFc. The PCR reaction (25 μL in total) contained each of the primers 25 pmol, each of dNTP 250 pmol and 1 U Taq DNA polymerase. The reaction process included a pre-denaturing at 94℃ for 2 min, 30 cycles of denaturing at 94℃ for 30 s, annealing at 60℃ for 30 s and extension at 72℃ for 1min, and an extension at 72℃ for 5 min. The PCR products were separated on a 0.8% agarose gel by electrophoresis and visualized by a UVP GDS8000 system.

2. Results and Analysis

2.1 Existence of foreign gene in host embryos

After the "all-fish" genes were transferred into the fertilized eggs of *Carassius auratus*, Pengze var., 30 embryos at each developmental stage were sampled for DNA preparation. The PCR results (table 1 and fig. 2) showed that 100% of embryos at myotome formation stage carried foreign gene, confirming that the foreign gene was successfully transferred into the host fertilized eggs by microinjection. Along with the process of embryogenesis, the foreign gene became less detectable, but it was still found in 70% of the frys.

Table 1 Detection of the foreign gene in the process of embryogenesis and the fry of *Carassius auratus*, Pengze var.

	Developmental stages					
	Blastula	Gastrula	Myotome formation	Muscular reaction	Heart beat	Fry[a)]
Detected(No.)	30	30	30	30	30	30
pCAsIGFc(%)	100	100	90	80	80	70
pCAgcGH(%)	100	90	90	80	70	70
pCAgcGHc(%)	100	100	100	90	80	70

a) Six-day-old

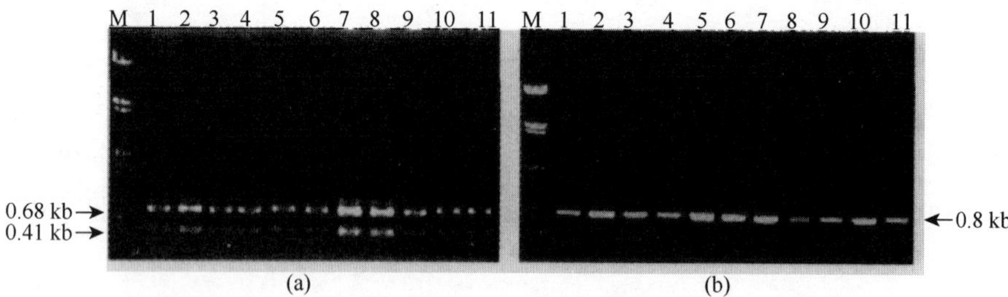

Fig. 2 PCR amplification of pCAgcGH, pCAgcGHc(a) and pCAsIGFc(b) in blastulae of *Carassius auratus*, Pengze var. M, λDNA (*Eco*RI/*Hin*dIII); 1-11, samples

2.2 Development of nuclear transplants

The development of nuclear transplants is shown in table 2. The eggs from two species served as nuclear recipients, and the embryonic cells of transgenic blastula or gastrula served as nuclear donors. Among 908 nuclear-transferred eggs, 12.8% developed to blastulas, 2.3% to gastrulas, 2.0% to neural plate stage, 0.7% to myotome formation stage, and 0.4% to frys. Three nuclear-transferred *Carassius auratus*, Pengze var. developed to frys, among which one died at its age of one month. And one nuclear-transferred Yellow River carp developed to a fry, but it died soon after its birth. In contrast, for 570 unfertilized eggs, which were needle-punctured as the controls, no one was found to develop to blastula stage. In addition, as the nuclear donors, there was no difference between blastula cells and gastrula cells.

Table 2 Development of nuclear transplants by microinjection

Nuclear donors		Number of transplant	Developmental stages				
			Blastula/%	Gastrula/%	Neural plate/%	Myotome formation/%	Fry/%
Carassius auratus. Pengze var.	blastula	330	39(11.8)	14(4.2)	10(3.0)	3(0.9)	2(0.6)
	gastrula	412	49(11.9)	2(0.5)	2(0.5)	1(0.3)	1(0.3)
	total	742	88(11.9)	16(2.2)	12(1.6)	4(0.5)	3(0.4)
	control	570	0				
Cyprinus carpio. Huanghe var.	blastula	77	20(26.0)	4(5.2)	1(1.3)	1(1.3)	0
	gastrula	89	8(9.0)	1(1.1)	1(1.1)	1(1.1)	1(1.1)
	total	166	28(16.9)	5(3.0)	2(1.2)	2(1.2)	1(0.6)

Numbers in parentheses are developmental rates at such stage (number of nuclear-transferred embryos at such developmental stage/number of transplant)

A total of 1335 eggs in 26 groups were electro-fused with embryonic cells with various parameters for nuclear transplantation. Only two nuclear-transferred *Carassius auratus*, Pengze var. were gained (table 3) due to the difficulty of defining the optimal parameters of electro-fusion.

Table 3 Nuclear transplantation of *Carassius auratus*, Pengze var. by electro-fusion (pulse strength = 1 kV/cm)

Treatment	Pulse/μs				Control
	20	40	80	160	
No. treated	44	43	42	41	63
Blastula	0	1	0	1	0
Fry	0	1	0	1	0
Survival rate of fry/%	0	2.3	0	2.4	0

2.3 Existence of foreign gene in nuclear transplants

To confirm the transmission of foreign gene from gene-transferred embryos to nuclear-transferred embryos, the nuclear transplants supported by blastula nuclei of *Carassius auratus*, Pengze var. were taken for examining the existence of foreign genes with PCR. The results showed the pCAsIGFc gene could be detected in 40% of nuclear-transferred blastulas, and for pCAgcGH and pCAgcGHc genes, the detection rates were 80% and 70%, respectively (figs. 3(a) and (b)). The results indicate that the foreign genes carried by blastula cells can be transmitted to the reconstructed embryos through nuclear transplantation.

PCR was also used to examine the existence of foreign gene in nuclear-transferred fishes. Among 6 fishes tested, only one was pCAgcGHc positive. The detection rate of foreign gene in nuclear-transferred fish was much lower (16.7%) than that in nuclear-transferred blastulas (fig. 3(c)). It implies that most foreign genes in donor cells derived from blastulas were not integrated into the genome, and they were subsequently lost or degraded during the cell division; or that foreign genes did not distribute evenly in blastula cells, resulting in only partial blastula cells containing foreign genes.

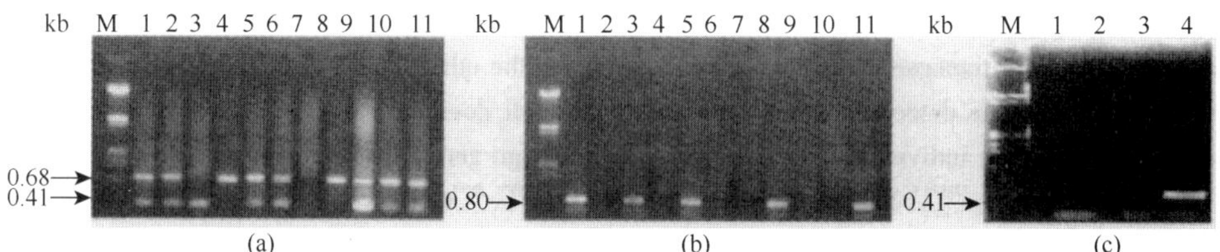

Fig. 3 PCR amplification of foreign genes in nuclear transplants (a) pCAgcGH and pCAgcGHcin nuclear-transferred blastulae of *Carassius auratus*, Pengze var.; (b) pCAsIGFc in nuclear-transferred blastulae of *Carassius auratus*, Pengze var.; M, λDNA (*Eco*RI/*Hin*dIII); 1, positive control; 2-11, samples. (c) pCAgcGH and pCAgcGHc in nuclear-transferred fish of *Carassius auratus*, Pengze var., aged 2-month. M, λDNA (*Eco*RI/*Hin*dIII); 1-3, samples

3. Discussion

Our results showed that the existence of foreign gene in embryos at pre-myotome stage was 100%. Nevertheless, the nuclear-transferred fishes supported by blastula or gastrula nuclei were not all transgenics. Here two questions need to be answered: one is whether the nuclear-transferred fishes come out from the genetic materials of donor nuclei; and the other is what is the frequency of the existence/integration of foreign genes in the nuclear-transferred fishes.

Many studies revealed that donor cells did participate in the development of nuclear transplants reconstructed from the non-enucleated eggs. When non-enucleated eggs were used as nuclear recipients in nuclear transplantation in insects, nuclear-transferred insects with normal development were gained and the female pronucleus did not fuse with the transferred nucleus [10,11]. In amphibians, non-enucleated eggs also served as nuclear recipients, and nucleatransplants were shown to be triploid, indicating that female pronucleus participate in the development of reconstructed embryos together with the donor nucleus [10,12]. Yu *et al.* [9] performed inter-species nuclear transplantation with non-enucleated eggs in fish, and found that the karyotypes and the electrophoretic patterns of serum proteins of nuclear-transferred fishes were the same as those of the donors. They considered that the haploid pronucleus had been excluded by the donor nucleus when non-enucleated eggs were used as nuclear recipients. Qi and Xu's study [13] on nuclear transplantation between *Aristichthys nobilis* and *Mamblycepjhala* also supported Yu *et al.*'s conclusion. Niwa *et al.* [14] transplanted blastula nuclei into non-enucleated eggs of medakafish, and analyzed the isozyme patterns and kayotypes of nuclear-transferred fishes, they found that the nuclear-thansferred fishes were thiploid. It is obvious that the nuclear-transfenred individuals are either triploid or diploid, the genetic materials carried by the nuclear donors can be preserved and the development of reconstructed embryos can be sustained. In our study, 570 needle-punctured unfertilized eggs served as controls and none of them could develop to the blastula stage (table 2), suggesting strongly that the development of nuclear transplants resulted from the reprogramming of the donor nuclei.

Although the stud with GFP as a reporter gene indicates that the integration of transgene takes place at blastula stage in the host fish [7], the embryonic cells integrated with foreign gene are merely a very small part of the blastula cells and most of the foreign gene integrations take place at the late stages of embryogenesis [3]. Therefore, it is difficult to gain homogeneous transgenic fish by nuclear transplantation with transgenic early-embryonic cells. On the other hand, as a whole body, early embryos are foreign gene detectable, but each embryonic cell does not carry with foreign gene. Moreover, considering the individual cell that carries with foreign gene, the copies of foreign gene in such a cell are much fewer than that in a gene transferred eggs (about 10^6 copies). Therefore, by nuclear transplantation with such embryonic cells carrying with a few copies of foreign gene situated outside the chromosomes, transgenic nuclear-transferred fish can hardly be generated. It is likely that homogeneous transgenic fish could be produced through nuclear transplantation with transgenic late-embryonic cells or transgenic somatic cells. Additionally, since embryonic stem cell-like cell lines have been successfully established in some model fishes [15-17], combination of the techniques of gene transformation with embryonic stem cells and nuclear transplantation may offer a practical approach to produce homogeneous strain of transgenic fish [18].

References

1 Zhu, Z., Li, G., He, I., *et al.*, Novel gene transfer into the fertilized eggs of goldfish (*Carassius auratus* L. 1758), *Z. Angew, Icluhyo.l*, 1985, 1: 31.

2 Zhu, Z., Xu, K., Li, G., *et al.*, Biological effects of human growth hormone gene microinjected into the fertilized eggs of loach *Misgurnus anguillicaudatus* (Cantor), *Chinese Science Bulletin*, 1985, 31(14): 988.

3 Zhu, Z., Xu, K., Xie, Y., *et al.*, A model of transgenic fish, *Science in China* (in Chinese) (B), 1989, 2: 147.

4 Cui, Z., Zhu, Z., Cui, Y., *et al.*, Food consumption and energy budget in MThGH- transgenic F2 red carp(*Cyprinus carpio* L. red var.), *Chinese Science Bulletin,* 1996, 41(16): 591.

5 Fu, C., Cui, Y., Hung, S. S. O., *et al.*, Growth and feed utilization by F4 human growth hormone transgenic carp fed diets with different protein levels, *J. Fish Biol.*, 1998, 53: 115.

6 Zhu, Z., Generation of fast growing transgenic fish: methods and mechanisms, *In Transgenic Fish,* Singapore: World Scientific Publishing, 1992, 92.

7 Zhao, H., Chen, S., Sun, Y., *et al.*, Time course of foreign gene integration and expression in fish, *Chinese Science Bulletin*, 2000, 45(8): 734.

8 Yu, I., Yang, Y., Liu, L., *et al.*, Studies on the nuclear transplantation in fish by using unenucleated eggs as recipient, *Fresh Water* (in Chinese), 1989, 3: 3.

9 Yi, Y., Liu, P., Liu, H., *et al.*, Electric fusion between blastula cells and unfertilized eggs in fish, *Acta Hydrobiol. Sinica* (in Chinese), 1988, 12(2): 189.

10 Gurdon, J. B., Nuclear transplantation in eggs and oocytes, *J. Cell Sci. Suppl.* , 1986, 4: 287.

11 Schubiger, M., Schneiderman, H. A., Nuclear transplantation in *Drosophila melanogaster,* Nuture, 1971, 230: 185.

12 Kroll, K. L., Gerhart, J. C., Thansgenic X. laevis embryos from eggs transplanted with nuclei of transfected cultured cells, *Science*, 1994, 266: 650.

13 Qi, F. Y., Xu, G. Z., Genetic character and individual growth of the trans-nucleus fish of bighead and the blunt-snout bream, *Acta Zool. Sinica* (in Chinese), 1997, 43(2): 211.

14 Niwa, K., Ladygina, T., Kinoshita, M., *et al.*, Transplantation of blastula nuclei to non-enncleated eggs in the madaka, *Oryzias latipes, Dev. Growth Differ.*, 1999, 41(2): 163.

15 Wakamatsu, Y., Ozato, K., Sasado, T., *et al.*, Establishment of a pluripotent cell line derived from a medaka (*Oryzias latipes*) blastula embryo, *Mol. Mar. Biol. Biotechnol.*, 1994, 3(4): 185

16 Sun, L., Bradford, C. S., Ghosh, C., *et al.*, ES-like cell cultures derived from early zebrafish embryos, *Mol. Mar. Biol. Dev.*, 1995, 4: 193.

17 Hong, Y., Schartl, M., Establishment and growth responses of early medaka fish (*Oryzias latipes*) embryonic cells in feeder layer-free cultures, *Mol. Mar. Biol. Biotechnol.*, 1996, 5(2): 93.

18 Zhu, Z., Sun, Y., Embryonic and genetic manipulation in fish, *Cell Research*, 2000, 10(1): 17.

转基因鱼早期胚胎细胞的核移植

陈尚萍　赵浩斌　孙永华　张甫英　朱作言

中国科学院水生生物研究所，淡水生态与生物技术国家重点实验室，武汉 430072

摘　要　像许多转基因高等生物一样，用显微注射方法研制的转基因鱼均为转植基因嵌合体，培育纯合转基因品系已成为研究者们向往和追求的目标。以澎泽鲫(*Carassius uratus* var. *pengze*)和黄河鲤(*Cyprinus carpio* var. *huanghe*)为实验材料，以转基因囊胚细胞核或原肠胚细胞核为供体，进行了细胞核移植。检测了转植基因在所获得的移核胚胎和5尾澎泽鲫移核鱼及1尾黄河鲤移核鱼中的存在状况，分析了用转基因早期胚胎细胞核移植方法研制纯合转基因鱼的可能性。

The Onset of Foreign Gene Transcription in Nuclear-Transferred Embryos of Fish

SUN Yonghua CHEN Shangping WANG Yaping ZHU Zuoyan

1 State Key Laboratory of Freshwater Ecology and Biotechnology, Institute of Hydrobiology, Wuhan 430072

Abstract The transcriptional onset of *hGH*-transgene in fish was studied in the following three cases: the first is in *MThGH*-transgenic F_4 common carp (*Cyprinus carpio*) embryos, the second is in nuclear-transferred embryos supported by the transgenic F_4 embryonic nuclei, and the third is in nuclear-transferred embryos supported by the transgenic F_4 tail-fin nuclei. RT-PCR results show that the *hGH*-transgene initiates its transcriptional activity from early-gastrula stage, the early blastula stage and even 16-cell stage in the first, second and third cases, respectively. It looks like that fish egg cytoplasm could just offer a very restricted reprogramming on transcriptional activity of specific gene in differentiated cell nuclei by nuclear transplantation.

Keywords transgenic fish; nuclear transplantation; serial nuclear transplantation; transcription; reverse transcriptional PCR; reprogramming

Nuclear transplantation plays an important role in studies on developmental totipotency of cells, the interaction of cytoplasm and nucleus during embryogenesis and ontogenesis, and the molecular mechanism of nuclear reprogramming. To compare the patterns of gene transcription in normal embryos with that in nuclear-transferred embryos could provide evidence for what extent the differentiated cells could be reprogrammed to in host eggs. Gurdon *et al.*[1] studied in *Xenopus* that the expression of *α-actin* gene initiated from late-gastrula stage in normal embryo. When genetically marked nuclei of larval muscle cells were transplanted into wild-type enucleated eggs, the *α-actin* gene became transcriptionally inactive immediately. It was reactivated when the reconstructed embryos reached late-gastrula stage. In mammals, similar experiments have been done to study the changes of transcriptional activity of donor nuclei. Kanka *et al.*[2] found that the transcriptional activity of donor nucleus derived from 32-cell stage morula substantially decreased after 4.5 h and was completely inhibited at last 15 h after the nuclear transfer. The transcription resumed thereafter in 2-cell stage embryos and rapidly increased at 16-cell stage, reaching the transcriptional level typically as that of the 32-cell stage nuclei used for the transfer. In bovine, Lavoir *et al.*[3] transferred nuclei of 16- to 32-cell stages *in vitro*-fertilized embryos into

文章发表于《中国科学：生命科学》, 2000, 43(6): 597-605

enucleated oocytes and they found that the hnRNA production in nuclear transfer embryos was much higher than that in control *in vitro*-fertilized embryos at 1 to 4-cell stages. They concluded that it was because the reprogramming of the transferred nuclei was absent or incomplete that the patterns of gene transcription in nuclear-transferred embryos did not behave like those in *in vitro*-fertilized embryos.

Zhu *et al.*[4] initiated transgenic studies in fish and generated the "model of transgenic fish"[5] subsequently. In this model they pointed out that the transcripts of foreign gene could only be found after late-gratrula stage of embryogenesis. Recently, nuclear transplantation was performed to study the timing of foreign gene integration, and it was found that the integration started from blastula stage[6]. This finding gave new evidence to the "model of transgenic fish". Zou *et al.*[7] considered that the transgenic fish embryos could be used as a new system to study the gene expression and regulation during embryogenesis. By giving a comparison on the transcriptional behavior of the *hGH*-transgene in transgenic carp F_4 generation against that in nuclear-transferred embryos, this article intends to discuss whether differentiated cell nuclei are able to be reprogrammed through nuclear transplantation and to what extent the donor nuclei can be reprogrammed.

1. Materials and Methods

1.1 Fishes

MThGH-transgenic red common carp F_3 generation and Yellow River carp (*Cyprinus carpio* Huanghe Var.) were from Guanqiao Fish Farm, Institute of Hydrobiology, Chinese Academy of Sciences. Sea carp (*Cyprinus acutidorsalis*) were supplied by Prof. Chen Xianglin, Department of Biology, Huanan Normal University. All experimental fishes were spawned by injection with common carp pituitary homogenates at spring time.

1.2 Treatment with eggs, embryos and cells

The F_3 generation of *MThGH*-transgenic common carp was used to produce the transgenic F_4 embryos with the technique of artificial fertilization. The fertilized eggs were dichorionized with 0.25% trypsine digestion and put into Holtfreter's solution for further development. When the fertilized eggs developed to blastula or gastrula, blastoderms were separated from the yolk with a fine glass needle and carefully placed in Holtfreter's dissociation solution for further separating into individual cells that were used as donor cells in embryonic nuclear transplantation. The transgenic F_4 cultured tail-fin cells were isolated from the flasks by digestion with 0.25% trypsine and collected as donor cells for somatic nuclear transplantation. As the tail-fin nuclear-transferred embryos developed into blastulas, their blastoderms were cut with fine glass needle and placed into Holtfreter's dissociation solution for further separating into individual cells that were used as donor cells for "serial nuclear transplantation". Being dichorionized with 0.25% trypsine digestion, mature eggs of Yellow River carp and Sea carp were used as the nuclear recipients. Nuclear transplantation was carried out according to Yu *et al.*[8] with a few improvements. The nuclear-transferred embryos developed in Holtfreter's solution.

1.3 RNA samples

The transgenic F_4 embryos and the embryonic nuclear-transferred embryos were taken as samples from early-blastula stages. The nuclear-transferred embryos supported by the transgenic F_4 tail-fin nuclei and the following serial nuclear-transferred embryos were taken as samples from 16-cell stages. All the samples were frozen in liquid nitrogen and then stocked at –70℃. The total RNAs of the samples were isolated with the SV Total RNA Isolation System (Promega, Cat#Z3100).

1.4 RT-PCR

RT-PCR was carried out with the Access RT-PCR System (Promega, Cat# A1250). Two primers (P1 and P2) for detection of *hGH* transcripts locate at the first exon and the second exon of *hGH* gene respectively. The upstream primer sequence is 5′-GGTAAGCGCC-CCTAAAATCC-3′, and the downstream primer sequence is 5′-TTGAAGATCTGCC-CAGTCCG-3′. The expected RT-PCR product length is 512 bp. The mixture of RT-PCR system (50 μL in total) contained template RNA 100 ng, dNTPs 40 μmol, primers 100 pmol, AMV reverse transcriptase 5 U, and Tfl DNA polymerase 5 U. The reaction mixture was then covered with 40 μL nuclease-free mineral oil. The process of the reaction included reverse transcription at 48℃ for 45 min, reactivation of AMV reverse transcriptase at 94℃ for 2 min, 40 cycles of DNA synthesis (denaturing at 94℃ for 30 s, annealing at 60℃ for 1 min and extension at 68℃ for 2 min), and finally incubation at 48℃ for 7 min. The RT-PCR products were separated on 0.8% agarose gel electrophoresis, stained with EB and analyzed under UVP GDS8000 system.

1.5 Probe labeling and Southern blotting

Recombinant plasmid pMThGH was digested with *Eco*R I, separated on 0.8% agarose gel electrophoresis. A 3.5 kb fragment containing the *MT*-1 promoter sequence and *hGH* structure sequence was purified from the agarose gel with Advantage^{TM} PCR-Pure Kit (CloneTech Co. Ld.). The probe was labeled with DIG DNA Labeling and Detection Kit (Boehringer Mannheim). The RT-PCR products of each sample were separated on 0.8% agarose gel electrophoresis and transferred onto nylon membrane with Bio-Rad 785 Vacuum blotter. Southern blotting and detection were carried out according to the user's manual (DIG DNA Labeling and Detection Kit, Boehringer Mannheim).

2. Results

2.1 Development of nuclear transplants supported by the transgenic F_4 embryonic nuclei

In April and May, the male and female individuals of transgenic carp F_3 generation that had been examined to be *MThGH*-positive were induced for spawning and artificial fertilization was carried out. The transgenic F_4 embryos were put into 20℃ Holtfreter's solution for further development. At blastula and gastrula stages, embryonic cells were separated as nuclear donors and transferred into mature unfertilized eggs of Yellow River Carp. Among 132 nuclear-transferred embryos supported by the blastula nuclei, 96 embryos (72.7%) developed to blastula stage,16 (12.1%) to gastrula stage, 2 (1.5%) to tail bud stage, and 1 (0.75%) to muscular reaction stage. Among 221 nuclear-transferred embryos supported by the gastrula nuclei, 140 embryos (63.3%) developed to blastula stage, 29 (13.1%) to gastrula and 1 (0.45%) to tail bud stage (table 1).

Table 1 Development of nuclear transplants supported by transgenic F_4 embryonic nuclei

Nuclear donors	Total transfer	Blastula(%)	Gastrula(%)	Tail bud(%)	Muscular reaction(%)	Fry(%)
F_4 blastula	132	96(72.7)	16(12.1)	2(1.5)	1(0.75)	0(0.0)
F_4 gastrula	221	140(63.3)	29(13.1)	1(0.45)	0(0.0)	0(0.0)

The figures shown here do not include the part sampled for RT-PCR; percentage was figured out as follows: number of nuclear transplants at such developmental stage / number of total transfer

2.2 Development of nuclear transplants supported by the transgenic F_4 tail-fin nuclei

Around the middle May, the transgenic F_4 tail-fin nuclei were transferred into mature unfertilized eggs of Sea carp, and the reconstructed eggs were put into 23℃ Holtfreter's solution for further development. Among 194 reconstructed eggs, 84 embryos (43.2%) developed to blastula stage, 12 (6.2%) to early-gastrula stage, 3 (1.5%) to mid-gastrula stage, and 1 (0.5%) to late gastrula stage. The blastula cells of the reconstructed embryos supported by the tail-fin nuclei were separated as nuclear donors for serial nuclear transplantation. Among 51 serial nuclear transferred embryos, 28 embryos (54.9%) developed to blastula stage, 4 (7.8%) to gastrula stage, and 1 (1.9%) to late-grastrula stage (table 2).

Table 2 Development of nuclear transplants supported by transgenic F_4 tail-fin nuclei

Nuclear donors	Total transfer	Blastula(%)	Early-grastrula (%)	Mid-grastrula (%)	Late-grastrula (%)	Tail bud (%)
Culatured tail-fin cells	194	84(43.2)	12(6.2)	3(1.5)	1(0. 5)	0(0.0)
Tail-fin nuclear-transferred blastula	51	28(54.9)	4(7.8)	2(3.9)	1(1.9)	0(0.0)

The figures shown here do not include the part sampled for RT-PCR; percentage was figured out as follows: number of nuclear transplants at such developmental stage / number of total transfer

2.3 *hGH* transcription in transgenic F_4 embryos

From early-blastula stage of the transgenic F_4 development, 30 embryos at each developmental stage were taken for preparation of total RNA. The RT-PCR results showed that the *hGH* transcription was initiated since the early-gastrula stage (fig. 1).

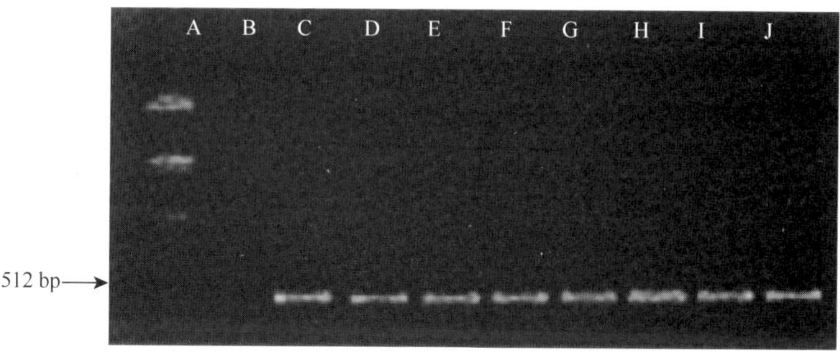

Fig. 1 *hGH*RT-PCR results of transgenic F_4 embryos. A, Marker (λDNA/*Eco*R I+*Hin*d III); B-J, RT-PCR results of transgenic F_4 embryos at blastula, early-gastrula, mid-gastrula, late-gastrula, neural plate, myotomic, muscular reaction, blood circulation and fry stages, respectively

2.4 *hGH* transcription in reconstructed embryos supported by the transgenic F_4 embryonic nuclei

The total RNA of embryos was prepared from the early-blastula to myotomic stages. For sampling, 5 embryos of the test group plus 25 non-transgenic embryos were taken for each stage from early-blastula to late gastrula stages. And 1 embryo of the test group plus 29 non-transgenic embryos at each stage was taken from neural to myotomic stages. RT-PCR showed that the *hGH* transcripts could be found in embryonic nuclear-transferred embryos from early-blastula stage, lasting to all developmental stages being tested (fig. 2).

2.5 *hGH* transcription in reconstructed embryos supported by the transgenic F_4 tail-fin nuclei

Since reconstructed embryos supported by the transgenic F_4 tail-fin nuclei developed to 16-cell stage, 1 of the tested embryos plus 19 non-transgenic embryos at each developmental stage was taken for preparation of total RNA. RT-PCR showed that *hGH* transcripts could be found in somatic nuclear-transferred embryos from 16-cell stage, lasting to all developmental stages being tested (fig. 3).

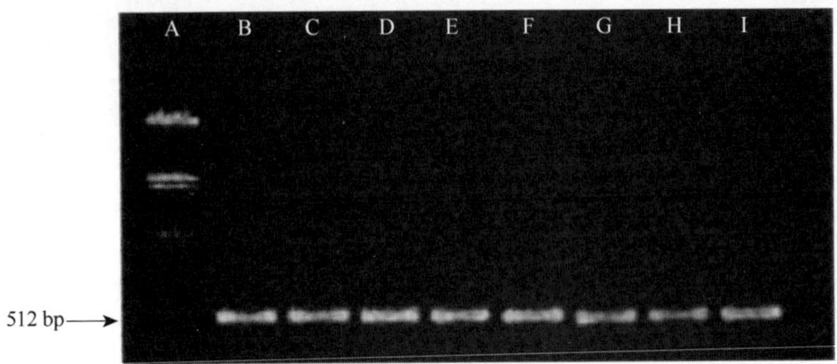

Fig. 2 *hGH* RT-PCR results of embryonic nuclear-transferred embryos. A, Marker (λDNA/*Eco*R I+*Hin*d III); B-I, RT-PCR results of embryonic nuclear-transferred embryos at early-blastula, mid-blastula, late-blastula, early-gastrula, mid-gastrula, late-gastrula, neuralplate and myotomic stages, respectively

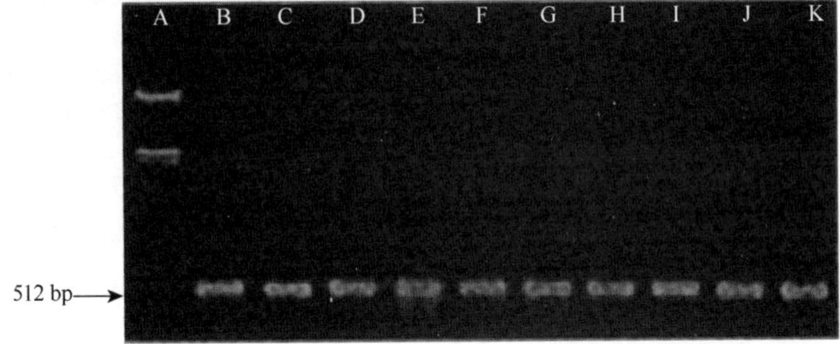

Fig. 3 *hGH* RT-PCR results of somatic nuclear-transferred embryos. A, Marker (λ DNA/*Eco*R I+*Hin*d III); B-K, RT-PCR results of somatic nuclear-transferred embryos at 16-cell, 32-cell, 64-cell, 128-cell, 512-cell, blastula, early-gastrula, mid-gastrula and late-gastrula stages, respectively

2.6 *hGH* transcription in serial nuclear-transferred embryos

Since the "serial nuclear-transferred" embryos developed to 16-cell stage, 1 embryo plus 19 non-transgenic embryos was taken for preparation of total RNA. The RT-PCR showed that the *hGH* transcripts could be found in serial nuclear-transferred embryos from 16-cell stage, lasting to all developmental stages being tested (fig. 4). In the sample of 16-cell stage, one more band with lower molecular weight was revealed on the electrophoresis gel. Southern blotting approved that this band could not hybridize against the DIG-labeled probes and it therefore came from an unspecific amplification.

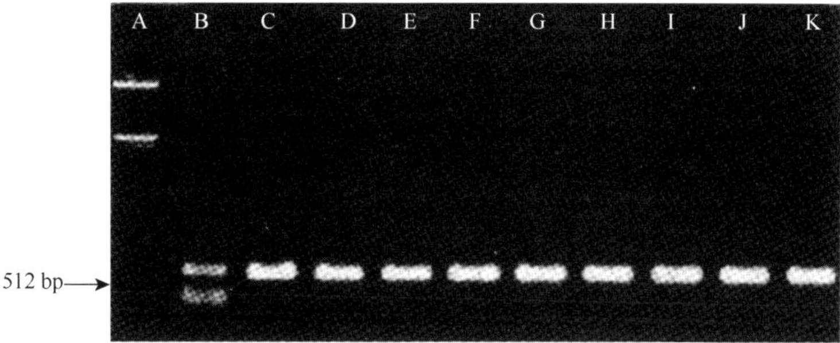

Fig. 4 *hGH* RT-PCR results of serial nuclear-transferred embryos. A, Marker (λDNA/*Eco*R I+*Hin*d III); B-K, RT-PCR results of serial nuclear-transferred embryos at 16-cell, 32-cell, 64-cell, 128-cell, 512-cell, blastula, early-gastrula, mid-gastrula and late-gastrula stages, respectively

2.7 Southern blotting of RT-PCR samples

Southern blotting results showed that all the representative bands of RT-PCR on electrophoresis gel were distinguishably hybridized against the DIG-labeled probes, which locate in the structural region of *hGH* gene in pMThGH (fig. 5). Hereby, all the specified bands should result from the reverse-transcription of the faithful transcripts of *hGH*.

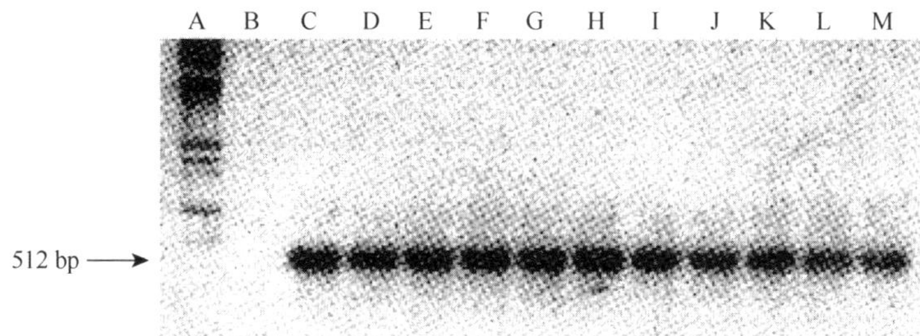

Fig. 5 Southern blotting results of *hGH* RT-PCR samples. A, Marker (λ DNA/*Hin*d III+*Eco*R I); B-D, hybridization against RT-PCR samples of transgenic F4 blastula, gastrula and fry, respectively; E-G, hybridization against RT-PCR samples of embryonic nuclear-transferred blastula, gastrula and myotome, respectively; H-J, hybridization against RT-PCR samples of somatic nuclear-transferred 16-cell, blastula and gastrula, respectively; K-M, hybridization against RT-PCR samples of serial nuclear-transferred 16-cell, blastula and gastrula, respectively

3. Discussion

In the study of "model of transgenic fish", it was indicated by Northern hybridization that the transcripts of *hGH* could only be found after post-gastrula stage in the transgenic embryos[5]. The present study found by RT-PCR that the *hGH* transcription was initiated from early-gastrula stage, which was consistent with Zou *et al.* [7] with the transgenic crucian carp. This may be due to the sensitivity of examination approaches, since RT-PCR can detect a few copies of *hGH* transcripts, but Northern blotting was incapable of doing that. In fish embryogenesis, some endogenous transcripts, e.g. hnRNA and tRNA, start to transcribe since blastula stage[9]. Going with the transgene's transmission from generation to generation, the integration and inheritance of the transgene tend to be more and more stable[10]. This phenomenon can be named "the endogenous domestication of foreign gene". According to this principle, we did examine the transcripts of integrated transgene in transgenic F_4 common carp. It is clear that the foreign gene is able to pass to the following generations by germ-line transmission and maintains the transcription properly, leading to the faithfully spliced transcripts of *hGH*. Meanwhile, the results revealed that mouse *MT*-1 promoter turned on transcriptional activity from early-gastrula stage in common carp embryos, for *hGH*-transgene was situated under the regulation of *MT*-1 promoter sequence.

Since the transcriptional activity of *hGH* was already turned on in the transgenic F_4 gastrula, what kind of influence of egg cytoplasm would give the gastrula nuclei when they had been transferred into host eggs? This study indicated that *hGH* gene at least started its transcriptional activity from early-blastula stage in the embryonic nuclear-transferred embryos. The gene transcriptional activity of the donor nuclei did not revert to completely suppressive status as being in the zygote. Ten times of cell division (at blastula stage) is enough for the *hGH* activity reaching the status of zygotic genome activation (ZGA, at gastrula stage)[11]. On the other hand, when using transgenic common carp blastula cells as nuclear donors, the transcripts of *hGH* in nuclear-transferred embryos were detected at the early-blastula stage. Tung *et al.*[12] considered that embryonic cells before blastula stage did not get any irreversible differentiation and had developmental totipotency. Nevertheless, the genomic DNA in blastula cells appears not to act as the same as the zygotic DNA does. When the blastula cells were transferred into mature eggs, they did not need the proper times of cell division to reach the status of genomic activation. Meanwhile, since embryonic nuclear-transferred embryos reconstructed by blastula cells, which did not turn on the *hGH* transcription, also initiated transcription from blastula stage, it was impossible that *hGH* transcripts in the nuclear-transferred embryos came from the transmission of very few *hGH* transcripts of donor cells through cell division.

When the transgenic common carp tail-fin cells were used as nuclear donors, the *hGH* transcripts could even be found in 16-cell stage of the reconstructed embryos. This means that 4 times of cell division are enough to turn on the genomic DNA transcription in the somatic nuclear-transferred embryo. The egg cytoplasm is not able to revert the differentiated somatic nucleus to the status equivalent to the zygotic one immediately. It is partially the reason that nuclear-transferred embryos have an incomplete developmental capacity since the activated gene transcription cannot be completely suppressed in the cytoplasm of eggs. Additionally, it is considerably giving an answer to why somatic nuclear transplantation gets such a low efficiency in developmental ratio. Serial nuclear transplantation in *Xenopus* could improve the developmental capacity of the reconstructed embryos supported by differentiated cell nuclei[13, 14]. In mammals, however, Stice and Keefer[15] found that developmental ratio varied along with different rounds of nuclear transplantation when serial nuclear transplantation was performed in bovine. In the case of tail-fin cells used for serial nuclear transplantation, the *hGH* transcription was still initiated from 16-cell stage. The egg cytoplasm seems not to be able to revert the one round of serial transferred nuclei to the status of zygotic counterparts. In other words, only one round of serial nuclear transfer could not significantly improve the developmental totipotency of fish somatic nuclei in the nuclear-transferred embryos.

Complete reprogramming indicates that the donor nucleus in host egg cytoplasm can be forced to "re-programme" to status equivalent to zygotic counterpart, which entails "expression of the correct gene at the right times as it occurs during normal embryogenesis"[16]. All sampled nuclear-transferred embryos in this study were different from normal embryos in the patterns of *hGH* transcription. It can be deduced that only few donor nuclei can be completely reprogrammed in fish nuclear transplantation. Maybe in a few nuclear-transferred embryos, donor nuclei could and did take complete reprogramming, in which *hGH* transcribed as the same as it did in the control embryos. Fortunately, these reconstructed embryos could complete their ontogenesis through the complete reprogramming of donor nuclei. Of course, whether in mammals[17] and amphibians[18], or in fishes[19], this kind of ratio was terribly low. The previous studies in mammals indicated that most donor nuclei could be remodelled and transcriptionally suppressed to a certain degree under the circumstance of egg cytoplasm, and commonly turned on their transcriptional activity at the moment similar to the timing of ZGA[2,16,20,21]. However, the result in fish nuclear transplantation is quite different. In most cases, the transcriptional activity in reconstructed fish embryos was turned on much earlier than the timing of ZGA. A reasonable interpretation may be as follows: cell division in fish early development goes too fast to have enough space-time for remodelling the donor nuclei and reprogramming the donor genome. On the other hand, as soon as fish eggs encounter with water, they are activated immediately and coupled with the subsequent decline of maturation promoting factor (MPF). The MPF is considered as a key factor for the reprogramming of donor nucleus in the nuclear transplants[22].

To examine the onset of *hGH* transcription, samples were taken from early-blastula stage in nuclear-transferred embryos supported by transgenic F_4 embryonic nuclei, and from 16-cell stage in nuclear-transferred embryos supported by transgenic F_4 tail-fin nuclei. It was unexpected that all the sampled embryos had turned on the transcriptional activity of *hGH*. Obviously, to make sure whether nuclear-transferred embryos turn on their transcriptional activity from even earlier developmental stages, or just from the stages when we took samples, further investigations are needed. Nevertheless, it is clear that the patterns for *hGH* transcription in the nuclear-transferred embryos were different from those in the control embryos. In conclusion, this article first discusses the reprogramming of donor nucleus in fish nuclear transplantation on the level of gene transcription, and generates the interpretation that fish egg cytoplasm could just offer a very restricted reprogramming on specific gene in differentiated cell nucleus.

Acknowledgements We thank Prof. Chen Xianglin for providing partial experimental fishes, and we thank Hu Wei, Zeng Zhiqiang, Wang Wei, Zhao Haobin, Huang Mei and Li Ming for their help and advice. This work was supported by the National Natural Science Foundation of China (Grant No. 39730290).

References

[1] Gurdon, J. B., Brennan, S., Fairman, S. *et al.*, Transcription of muscle-specific actin genes in early *Xenopus* development: nuclear transplantation and cell dissociation, Cell, 1984, 38: 691.

[2] Kanka, J., Hozak, P., Heyman, Y. *et al.*, Transcriptional activity and nucleolar ultrastructure of embryonic rabbit nuclei after transplantation to enucleated oocytes, Mol. Reprod. Dev., 1996, 43: 135.

[3] Lavoir, M. C., Kelk, D., Rumph, N. *et al.*, Transcription and translation in bovine nuclear transfer embryos, Biol. Reprod.,1997, 57: 204.

[4] Zhu, Z., Li, G., He, L. *et al.*, Novel gene transferred into the fertilized eggs of gold fish (*Carassius auratus* L. 1758), J. Applied Ichthyology, 1985, 1: 31.

[5] Zhu, Z., Xu, K., Xie, Y. *et al.*, A model of transgenic fish, Science in China (in Chinese), Ser. B, 1989, (2): 147.

[6] Zhao, H., Chen, S., Sun, Y. *et al.*, The time course of foreign gene integration and expression in transgenic fish embryos, Chinese Science Bulletin, 1999, 44(22): 2414.

[7] Zou, J., Xie, Y., Liu, D. *et al.*, Foreign gene expression during embryogenesis of crucian carp, Acta Hydrobiol. Sinica (in Chinese), 1991, 15(4): 372.

[8] Yu, L., Yang, Y., Liu, L. *et al.*, Studies on the nuclear transplantation in fish by using unenucleated eggs as recipitant, Fresh Water (in Chinese), 1989, (3): 3.

[9] Davidson, E. H., Gene activity in early development, 3rd ed., New York: Academic Press, Inc., 1986, 181-185.

[10] Cui, Z., Zhu, Z., Several interesting questions about breeding transgenic fish, Biotech. Information (in Chinese), 1998, (5): 1.

[11] Telford, N. A., Watson, A. J., Schultz, G. A., Transition from maternal to embryonic control in early mammalian development: a comparison of several species, Mol. Reprod. Dev., 1990, 26: 90.

[12] Tung, T. C., Wu, S. C., Tung, Y. Y. F. *et al.*, Nuclear transplantation in fish, Kexue Tongbao (Chinese Science Bulletin) (in Chinese), 1963, 13(7): 60.

[13] Gurdon, J. B., Woodland, H. R., The transplantation of nuclei from single cultured cells into enucleated frog's eggs, J. Embryol. Exp. Morphol., 1970, 24: 227.

[14] Gurdon, J. B., Laskey, R. A., Reeves, O. R., The development capacity of nuclei transplanted from keratinized skin cells of adult frogs, J. Embryol. Exp. Morphol., 1975, 34: 93.

[15] Stice, S. L., Keefer, C. L., Multiple generational bovine embryo cloning, Biol. Reprod., 1993, 48: 715.

[16] Kanka, J., Smith, S. D., Soloy, E. *et al.*, Nucleolar ultrastructure in bovine nuclear transfer embryos, Mol. Reprod. Dev.,1999, 52: 253.

[17] Wolf, E., Zakhartchenoko, V., Berm, G., Nuclear transfer in mammals: Recent developments and future perspectives, J. Biotech., 1998, 65: 99.

[18] DiBerardino, M. A., Genomic Potential of Differential Cells, USA: Columbia University Press, 1997.

[19] Yan, S. Y., Cloning in Fish-nuclear-cytoplasmic Hybrids, Hongkong: Educational and Cultural Press, 1998.

[20] Kanka, J., Fulka, J. Jr., Fulka, J. *et al*., Nuclear transplantation in bovine embryos: fine structure and autoradiographic studies, Mol. Reprod. Dev., 1991, 29: 110.

[21] Sun, F. Z., Moor, R. M., Nuclear-cytoplasmic interactions during ovine oocyte meiotic maturation, Development, 1991,111: 171.

[22] Fulka, J. Jr., First, N. L., Moor, R. M., Nuclear transplantation in mammals: remodelling of transplanted nuclei under the influence of maturation promoting factor, Bioessays, 1996, 18: 835.

鱼类核移植胚胎中外源基因转录的启动

孙永华　陈尚萍　汪亚平　朱作言

中国科学院水生生物研究所，淡水生态及生物技术国家重点实验室，武汉 430072

摘　要　本研究比较了 *hGH* 转植基因在转 *MThGH* 基因鱼、胚胎细胞核移植后代，以及尾鳍培养细胞核移植后代中的转录时序差异。RT-PCR 实验结果表明，*hGH* 基因在转基因鱼 F4 胚胎中从原肠早期开始转录； F4 胚胎细胞核移植的后代在囊胚早期已能检测到 *hGH* 基因转录本; F4 尾鳍培养细胞核移植的后代自 16 胞期就出现 *hGH* 基因的转录。上述结果表明，鲤卵细胞质对分化细胞核特定基因的再程序化能力是有限的，进而推论鱼类细胞核移植试验中仅有少部分的供体核能够发生完全再程序化。

Nuclear Transplantation of Somatic Cells of Transgenic Red Carp (*Cyprinus carpio haematopterus*)

Zhao Haobin Zhu Zuoyan

The National Key Laboratory of Freshwater Ecology and Biotechnology, Institute of Hydrobiology, Chinese Academy of Sciences, Wuhan 430072

Abstract Nuclear transplantation was performed using kidney cells, tail fin cells and cultured tail fin cells of 18th passage of the F_4 *hGH* gene transferred red carp (*Cyprinus carpio haematopterus*) as donors and the eggs of loach (*Misgurnus anguillicaudatus*) of Huanghe carp (*Cyprinus carpio haematopterus*) as recipients. Using loach unfertilized eggs as recipients 0.33% of reconstructed embryos developed in to neural plate from kidney cells, and 0.1% of reconstructed embryos developed in to muscular reaction stage from tail fin cells of F_4 *hGH* gene transferred red carp. A total of 0.4% reconstructed embryos from cultured tail fin cells of 18th passage and unfertilized eggs of Huanghe carp developed in to myotome stage. All the nuclear transfer embryos were positive for *hGH* gene by PCR analysis. The results suggest that exogenous *hGH* gene is stable in the F_4 *hGH* gene transferred red carp, *hGH* gene exists in all the cells of the F_4 transgenic red carp. Somatic cell nuclear transfer is an effective pathway for producing homogenous transgenic fish, but with low efficiency. The reasons of low efficiency of somatic nuclear transplantation may be incompatibility of nuclear and ooplasm, incongruity of cell cycle, aberrance of chromosomes, etc.

Key words fish; carp; loach; somatic cell; nuclear transfer; transgene; homogenous

Zhu *et al* firstly introduced human growth hormone (*hGH*) gene driven by mouse metallothionein promoter into fertilized eggs of gold fish (*Carassius auratus*), common carp (*Cyprinus carpio haematopterus*), etc, and obtained fast growing transgenic fish. They studied the integration, expression, heredity and growth promoting mechanism of novel gene, and established a model of transgenic fish[1,2]. The behavior of novel gene is sophisticated after microinjection, only a little part can get into host genome finally. The integrated novel genes can insert into the host genome randomly or by recombination, and can exist as monomer of polymer. All founder transgenic animals are mosaic for novel gene. Ordinarily, the heritance of foreign gene is not as Mendel's law, and the transgenic fish with stable inheritance can be established at F_2 generation[3]. However, Wei *et al* showed that inheritance of novel gene in F_1 transgenic fish was not in Mendel's model, the F_2 transgenic fish was also mosaic, and different tissues had different copies of novel gene[4]. Cui *et al* detected the *MThGH* gene in

文章发表于 *Acta Genetica Sinica*, 2002, 29(5): 406-412

transgenic red carp population from P_0 to F_4 by PCR (Polymerase Chain Reaction), and they found that the positive ratios for *hGH* were increased with generation. The positive ratios in P_0, F_1, F_2, F_3 and F_4 were 58.3%, 70.8%, 77.3%, 90% and 94.3%, respectively[5]. These studies suggested that it was difficult to get homogenous transgenic fish with steadily integrated novel gene by traditional breeding.

Tung *et al* got the first nuclear transfer fish in 1963[6], and Chen *et al* obtained the first somatic nuclear transfer goldfish (*Carassius auratus*) from kidney cell in 1986[7]. Nuclear transfer may be the appropriate method to establish homogenous transgenic fish, and to make transgenic fish clone.

We did nuclear transfer to study the stability of novel gene in F_4 *MThGH* gene transferred fish and to explore the possibility of cloning transgenic fish.

1. Materials and Methods

1.1 Fish

F_4 *MThGH* gene transferred red carp (*Cyprinus carpio haematopterus* Tem.et Schl.) was natural breeding from F_3 red carp which was positive for *MThGH* gene.

Loach (*Misgurnus anguillicaudatus* Cantor) was purchased from market at Dadongmen in Wuhan, Huanghe carp (*Cyprinus carpio*, Huanghe var,) was provided by Henan Institute of Fisheries. These fishes were induced for reproduction by injection of a homogenate of carp pituitary gland.

1.2 Preparation of somatic cells

All cells for nuclear transfer were prepared freshly and used once.

The cells including kidney cells, tail fin cells, and cultured tail fin cells were from F_4 transgenic fish that were positive for *MThGH* detected by PCR.

The kidney from newly killed fish was dissected and digested by 0.25% trypsin, of pipetted in 0.7% physiological salt water, and then the suspension was centrifuged at 1000rpm for 5 min, and the sediment of kidney cells was resuspended in phosphate balanced solution (PBS) or Holtfreter's solution.

The tail fin of the fish was dissected and digested by 0.25% trypsin and centrifuged, the upper part of sediment, tail fin cell layer, was collected and resuspended in Holtfreter's solution for nuclear transfer.

The cultured tail fin cells were cultured from *hGH* positive fish in M199 medium containing 25mmol/L HEPES plus 20% fetal bovine serum and antibiotics at 28℃, and the cells were passed by one to three in culture. The confluent cells at 18th passage were dissolved by 0.25% trypsin, washed, and preserved in PBS for nuclear transfer.

1.3 Nuclear transfer

The mature oocytes were squeezed out from loach or Huanghe carp, and the chorions of eggs were removed by trypsinization, the eggs used for nuclear transfer were freshly prepared and used in 30min. The cells, kidney cells or fin cells, were sucked into a fine glass pipette to rupture the cell membrane, and then injected into animal pole of oocytes. After nuclear transfer, the eggs were cultured in Holtfreter's soulution.

1.4 Detection of *hGH* gene by PCR

The reaction volume of PCR was 25 μl containing template DNA 0.25 μg, dNTP 50 μmol/L for each, two primers 20pmol for each, and 1～2 U Taq enzyme. The PCR reaction was proceeded in 94℃ 30s, 60℃ 30s, and 72℃ 1min for 30 cycles. The PCR products were identified by electrophoresis on 0.8% agarose and UVP GDS8000 gel image system. The primer 1 is 5′-AAGCGTCACCACGACT-3′ at MT promoter, and the primer 2 is 5′-AAAAGCCAGGAGCAG-3′ at the *hGH* sequence, the length of PCR product between these two primers is 450 bp.

1.5 Preparation of chromosome

The method was adopted from Yu *et al*[8].

2. Results and Analysis

2.1 Positive ratio of F_4 transgenic fish

We did nuclear transfer with F_4 transgenic red carp for 19 times, and all donors were detected by PCR. The results showed all F_4 transgenic fish were positive for *hGH* (Fig. 1,A), the positive ratio was 100%. Different tissues of same fish were also detected for 5 fish, all detected tissues were positive for *hGH* too (Fig.1,B). These results suggested that novel gene was steady in F_4 transgenic red carp after selection and directional breeding for four generations.

2.2 Development of nuclear transfer embryos

The results were shown in table 1 and Fig. 2. The reconstructed embryos from kidney cell nuclei or tail fin cell nuclei of red carp and loach eggs could develop forward, two embryos developed to neural plate stage from kidney cells, and one embryo developed to muscular reaction stage from tail fin cell. However, no control egg developed. We randomly selected some nuclear transfer embryos, and detected *hGH* gene in these embryos by PCR. All embryos detected were positive for *hGH* (See Fig. 3). These results suggest that heterogonous nuclear transfer embryo from different species can develop, but the developmental ability is limited. The heterogonous nucleus can dedifferentiate and be reprogrammed in oocyte plasma.

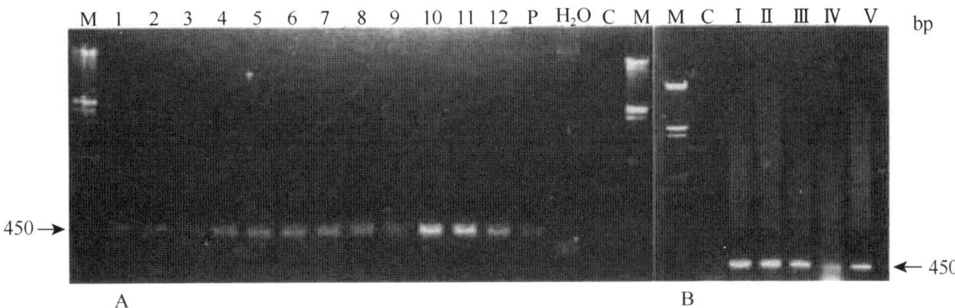

Fig. 1 PCR results of F_4 generation of *MThGH* gene transferred red carp A: the result for different individuals; B: the result for different tissues of one fish; M: λDNA/*Eco*RI+ *Hin*dIII; P: *pMhGH* plasmid; C: negative control. The numbers presented different individual.I: the tail fin; II: the muscle; III: the kidney; IV: the brain; V: the liver

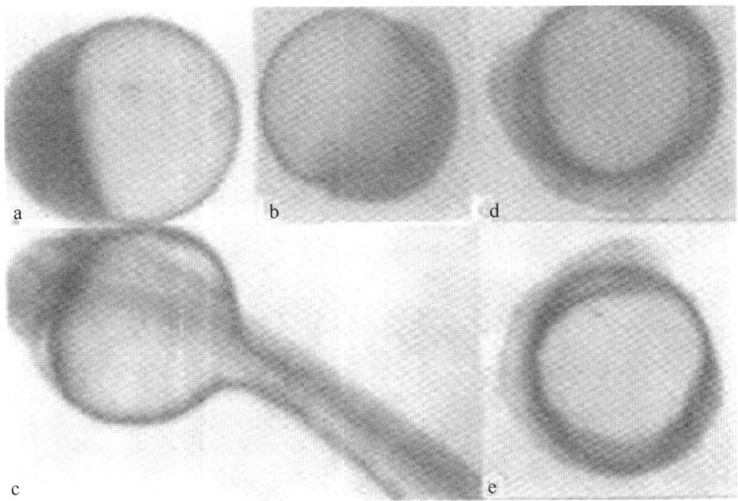

Fig. 2 The nuclear transfer embryos derived from somatic cells of F_4 generation of *MThGH* transgenic red carp.The photos a-c were nuclear transfer embryos from tail fin cell using loach oocytes as recipients. a: a blastula; b: a gastrula; c: a embryo at muscular reaction stage. The photos d and e were nuclear transfer embryos at neural plate stage derived from the cultured F_4 red carp tail fin cells using Huanghe carp oocytes as recipients

More embryos developed to blastula and gastrula stage from eggs of Huanghe carp than from loach eggs with tail fin cells of F_4 transgenic red carp as nuclear donors. A total of 50.53% and 24.5% embryos developed in to blastula stage, 5.69% and 1.3% embryos developed into gastrula stage from eggs of Huanghe carp and loach, respectively. The difference of development rate may be the cause of incompatibility of recipient ooplasm and donor nucleus. Moreover, the variation of karyotype might limit the development of nuclear transfer embryo, the reconstructed embryos from cultured red carp cell nuclei and eggs of Huanghe carp only developed in to myotome stage.

Table 1 The results of development of nuclear transfer embryos derived from somatic cells of the F_4 *hGH* gene transferred red carp

Donors	Recipients	Total No.	Developmental stage				
			Blastula (%)	Gastrula (%)	Neural plate (%)	Myotome formation(%)	Muscular effect(%)
Kidney	Loach eggs	611	76(12.4)	2(0.33)	2(0.33)	0	0
Tai fin	Loach eggs	1119	274(24.5)	12(1.1)	3(0.3)	2(0.2)	1(0.1)
Control		591	0				
Cultured tail fin cells	Eggs of carp	756	382(50.53)	43(5.69)	4(0.53)	3(0.4)	0

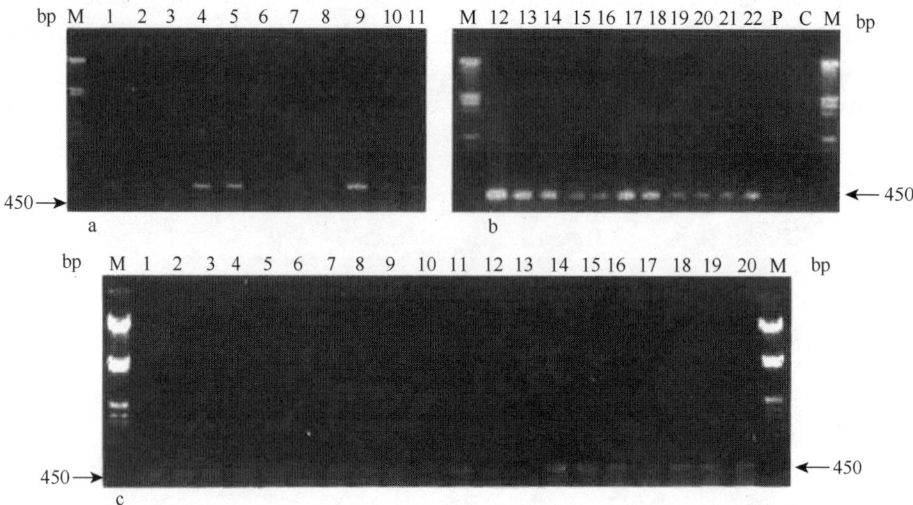

Fig. 3 The PCR results of *hGH* gene in nuclear transfer embryos. Panel a and b showed the PCR results of *hGH* gene in nuclear transfer embryos derived from somatic cells of the F_4 transgenic red carp with loach oocytes as recipients. 1-11: blastulas; 12-17: gastrulas; 18-20: embryos at neural plate stage; 21: an embryo at myotome stage; 22: an embryo at muscular reaction stage; M: λDNA/*Eco*RI/*Hin*dIII; P: plasmid; C: control. Panel c showed the PCR results of *hGH* gene in nuclear transfer embryos from the cultured tail fin cells of transgenic red carp with oocytes of Huanghe carp as recipients. 1-12: blastulas, 13-17: gastrulas, 18: an embryo at neural plate stage, 19-20: embryos at myotome stage

We prepared the chromosome sample of cultured cells of passage 18, and selected 32 clear mitosis phases to count chromosome numbers. The mean number of chromosome was 98.6±17.1, it deviated the standard number, 100, of carp chromosome. It suggested the cells were aneuploid in culture for a long time.

2.3 Detection of *hGH* in nuclear transfer embryos by PCR

We randomly selected 100 blastulas, 10 gastrulas, some embryos at stage of myotome and muscular reaction from somatic cells of F_4 transgenic red carp and loach eggs, all embryos were positive for *hGH* gene detected by PCR. We also randomly selected 30 blastulas, 5 gastrulas, some embryos at stage of neural plate and myotome from tail fin cells of F_4 transgenic red carp and eggs of Huanghe carp, 100% embryos were positive for *hGH* gene. The results were showed in Fig.3. It referred to that donor cell nuclei took part in the embryonic development, *hGH* gene steadily existed in all cells of F_4 transgenic red carp, *hGH* gene could be inherited to nuclear transferred embryos.

As reconstructed embryos from tail fin cells of red carp and loach eggs developing to late gastrula, some embryos were selected and the karyotype of embryonic cells was analyzed. The majority of chromosome numbers were centralized in 100. The mean number was 94±35. This was not different with standard chromosome number of carp ($P > 0.05$).

3. Discussion

3.1 Stability of novel gene in F_4 generation

Is the novel gene stable in F_4 transgenic red carp after selection for four generation? Cui analyzed the positive ratio of P_0, F_1, F_2, F_3, and F_4 transgenic red carp by PCR, he found that the positive ratio of foreign gene was increased with generation increase and trended towards stable[5]. We randomly selected 19 F_4 transgenic red carp, and analyzed *hGH* gene by PCR. Consequently, *hGH* gene exists in every F_4 transgenic red carp and every tissue of five carp was analyzed. All nuclear transfer embryos from cells of F_4 transgenic red carp contained the novel gene too. These results suggest that the novel gene has been stable and evenly distributed into all cells of transgenic fish after selective breeding for four generations. However, it should be further studied whether the copy numbers and integration sites of novel gene were the same in one transgenic fish.

3.2 The origin of genetic matter of nuclear transfer embryo

Yu *et al* made nuclear transfer fish from blastula cells and eggs whose nuclei were not removed, they found that the karyotype and hemoglobin pattern of nuclear transfer fish were as same as the donor fish. They arrived a conclusion that the haploid nucleus of egg was rejected in development of nuclear transfer embryo[9]. Gasaryan *et al* did nuclear transfer in *Misgurnus fossilis* L using eggs that nuclei were not removed as recipients. The majority of nuclear transfer embryos they made were diploid, very few embryos kept the egg nucleus and were triploid. They inferred that egg nucleus was excluded in development of reconstructed embryos[10]. Niwa *et al* did nuclear transfer of blastula of medaka (*Oryzias latipes*) using eggs that nuclei were not removed as recipients. Isozyme and karyotype analysis showed that nuclear transfer fish were triploid, so they deduced that haploid nucleus of egg fused with diploid donor nucleus. However, they once obtained one diploid male nuclear transfer fish whose appearance was the same as donor fish[11]. These examples showed that transplanted nucleus could completely remain in reconstructed embryo.

In our research, because the two PCR primers were at *MT* promoter sequence and *hGH* construct sequence respectively, a 450 bp product could only amplified from transgenic embryos or fish, and the endogenous *GH* gene could not be amplified by PCR. There was not one control egg developed, but nuclear transfer embryos could develop in to myotome and muscular reaction stage, and all nuclear transfer embryos were positive for *hGH.* It was illustrated that developmental embryos keeping the donor nuclei, because only donor nucleus held *hGH.* The chromosome numbers of nuclear transfer embryos from cells of red carp and loach eggs were concentrated in 100, a few were far from100. There fore the majority of nuclear transfer embryos are diploid. It is deduced that the genetic matter of nuclear transfer embryo mainly consists of donor genetic matter, but it is not excluded that few embryos contain egg nuclear matter.

3.3 Development of hybrid embryo with different species

Some researchers in China have got nucleusooplasm hybrid fish by transferring cell nuclei of blastula of one species into eggs of another species, e.g. nuclear transfer fish from carp and goldfish, the success rate was 0.9%-3.2%[12,13]. Lin *et al* transferred kidney cell nuclei of goldfish, *Cirrhinus molitorella* and *Tilapia nilotica* into enucleated mature carp eggs, and obtained larva with blood circulation, larva with heart beating, and embryos with muscular reaction, respectively, by double nuclear transfer[14]. We transferred somatic nuclei of red carp into loach eggs, also got embryos at neural plate and muscular reaction stage. However, the developmental ratios of blastula and gastrula were higher in nuclear transfer embryos from red carp somatic cells and eggs of Huanghe carp than in nuclear transfer embryos from somatic cell of red carp and loach eggs. It is suggested that mature ooplasm can reprogram exogenous nucleus no matter their kindred and make the nucleus into embryogenesis. However, the relationship of donor nucleus and recipient egg affects the development of reconstructed embryo, the nuclear transfer embryo develop well if nucleus has near relation with egg, vice versa.

3.4 Causes may affect somatic nuclear transfer efficiency

Except incompatibility of donor nucleus and recipient egg, the reasons of low success rate of somatic nuclear transfer of fish may be as follows: (1) Egg plasma do not reprogram donor nucleus enough. In mammal, long time exposure of somatic nucleus to plasma of

M II oocyte can strengthen reaction of reprogramming factors and nucleus, and can improve development of nuclear transfer embryo. Nuclear transfer embryos of mouse developed well from cumulus cells exposed to M II oocytes for 1-6 h, and some developed in to term[15]. However, fish egg can be activated quickly in water, and fish embryos develop quickly too, the first division of zygote may be in 30min[16]. So, there are few chances and little time for reprogramming factors acting on exogenous nucleus to reprogramit enough, and it results in poor development of nucleartransfer. (2) Donor nuclear chromosome is abnormal. Gurdon realized that abnormal development of nuclear transfer embryo resulted from chromosome variation, and this variation occurred in first division because the difference of dividing speed of donor cell and recipient cell[17]. Comparing with embryonic cells, fish somatic cells cleave very slowly. Embryonic cells divide one time every 30min, but somatic cells cleave one time more than 10 h. This large difference may be one important cause for low efficiency of somatic nuclear transferof fish. In addition, cultured tail fin cells of red carp at passage 18 were an aneuploid cell population; very few cells were normal diploid, so the majority of nuclear transfer embryos from culture cells developed difficultly. (3) Donor cells are asynchronous. In mammal, G_1 and G_0 cell are optimum for nuclear transfer. Dolly was the example from G_0 cell of mammary gland synchronized by serum starvation[18]. In fish, there is not a report about effects of cell cycle on development of nuclear transfer embryos, there it is difficult to determine the optimum cell cyele stage of fish cells for nuclear transfer, and it should be further studied in future.

Acknowledgements We would like to express appre our colleagues, Chen Shangping, Huang Mei, Sun Yonghua, Wang Yaping, for their support and contribution to this paper.

References

[1] Zhu Z, Li G, He L, Chen S. Novel gene transfer into the fertilized eggs of goldfish (*Carassius auratus* L.1758). Z Angew Ichthyol, 1985, 1: 31-34.

[2] Zhu Z, Xu K, Xie Y *et al.* A model of transgenic fish. Science in China Ser B, 1989, (2): 147-155.

[3] Zhao H, Chen S, Zhu Z. Advances in transgenic fish research. J. Agri. Biotech., 1999, 7(3): 301-306.

[4] Wei Y, Xie Y, Xu K *et al.* Heredity of human growth hormone gene in ransgenic carp (*Cyprinus carpio* L). Chinese J. Biotech., 1992, 8(2): 140-144.

[5] Cui Z, Zhu Z. Several problems of transgenic fish breeding. Biotech. Bulletin, 1998, (5): 1-10.

[6] Tung TC, Wu S C, Tung YF *et al.* Nuclear transplantation in fish. Kexue Tongbao, 1963, (7): 60-61.

[7] Chen H, Yi Y, Chen M *et al.* Studies on the developmental potentiality of cultured cell nuclei of fish. Acta Hydrobiol Sinica, 1986, 10(1): 1-6.

[8] Yu X, Zhou T, Li Y *et al.* Chromosome of freshwater fish in China (in Chinese). Science Press, Beijing: 1989.

[9] Yu L, Yang Y, Liu L *et al.* Research on nuclear transplantation in fish using eggs whose nuclei were not removed as recipients. Freshwater fisheries, 1989, (3): 3-7.

[10] Gasaryan K G. N M Hung, A A Neyfakh, V V Ivanenkov. Nuclear transplantation in teleost *Misgurnus fossilis* L. Nature, 1979, 280: 585-587.

[11] Niwa K, T Ladygina, M Kinoshita *et al*. Transplantation of blastula nuclei to non- enucleated eggs in the medaka, *Oryzias latipes*. Dev Growth Differ, 1999, 41(2): 163-172.

[12] Research group of Cytogenetics, Institute of Zoology, CAS, Research group of Somatic cell genetics, Institute of Hydrobiology, CAS, Research group of Nuclear transplantation, Changjiang Fisheries Research Institute. Nuclear transplantation in teleosts — hybrid fish from the nucleus of carp and cytoplasm of crucian. Scientia Sinica, 1980, 23(4): 377-380.

[13] Yan S, Lu D, Du M. Nuclear transplantation in teleosts — hybrid fish from the nucleus of crucian and the cytoplasm of carp. Scientia Sinica (B) (in Chinese), 1984, 27(8): 729-732.

[14] Lin L, Xia S, Zhu X. Studies on nuclear transplantation of somatic cells in teleost. Zoological Res, 1996, 17(3): 337-340.

[15] Wakayama T, A C EPerry, M Zuccotti *et al*. Full- term development of mice from enucleated oocytes injected with cumulus cell nuclei. Nature, 1998, 394: 369-374.

[16] Tchou- su P, C Chao- his, W Yu- lan. Etude cytologique de la fecondation sur l'oeuf des poisons osseux: *Carassius auratus* et *Megalobrama terminalis*. Acta Biologiae Experimentalis simica, 1960, 7(1): 29-46.

[17] Gurdòn J B. Nuclear transplantation in eggs and oocytes. J. Cell Sci. Suppl., 1986, 4: 287-318.

[18] Wilmut I, A E Schnieke, J McWhir *et al*. Viable offspring derived from fetal and adult mammalian cells. Nature, 1997, 385: 810-813.

转基因红鲤体细胞的核移植

赵浩斌　朱作言

中国科学院水生生物研究所，淡水生态与生物技术国家重点实验室，武汉 430072

摘　要　以 F_4代转 *hGH* 基因红鲤体细胞(肾脏和尾鳍) 及培养18 代的 F_4代转 *hGH* 基因红鲤尾鳍细胞为核供体，泥鳅或黄河鲤成熟卵为受体，进行了核移植，以探讨外源 F_4代转基因鱼体外源基因的分布与存在形式、稳定性和克隆转基因鱼的可能性。F_4 代红鲤肾脏细胞核与泥鳅卵配合的核移植胚胎有 12.14%发育到囊胚，0.33%发育到神经胚；F_4代尾鳍细胞核移入泥鳅卵后的重组胚发育到囊胚、神经胚、肌节期和肌肉效应期的胚胎分别为 24.5%、0.3%、0.2%和 0.1%；对照卵无发育。F_4代红鲤尾鳍培养细胞与黄河鲤卵子配合的重组胚胎有 50.53%发育到囊胚，5.69%发育到原肠胚，0.53%发育到神经胚，0.4%发育到肌节期。说明由于同种细胞核与卵细胞的相容性高于异种核卵的相容性，早期发育率高；而由于培养细胞的异倍化，后期的发育率降低。用 PCR 技术对供体鱼不同个体及同一个体不同组织外源基因检测，结果 100%个体为阳性鱼，而且不同组织的阳性率也是 100%，说明外源基因均匀分布在不同组织中。无论 F_4代转基因鱼的肾脏细胞、尾鳍细胞还是培养的尾鳍细胞作核移植供体，核移植胚胎中 *hGH* 基因的检出率为 100%。说明 F_4 代转基因红鲤个体不同细胞都存在 *hGH* 基因，而且经长期培养不会丢失。表明 F_4代转基因红鲤中的外源 *hGH* 基因已基本稳定，体细胞核移植可以作为获得同质化转基因鱼的有效手段，但核移植效率还很低。另外还讨论了核质的相容性、细胞周期的协调、染色体的变异等因素对核移植的影响。

Nuclear Transplantation in Different Strains of Zebrafish

HU Wei WANG Yaping CHEN Shangping ZHU Zuoyan

State Key Laboratory of Freshwater Ecology and Biotechnology, Institute of Hydrobiology, Chinese Academy of Sciences, Wuhan 430072

Abstract Single later blastula nuclei from AB strain of zebrafish (*Danio rerio*) were transplanted into enucleated unfertilized eggs of Long fin strain. Of 1119 cloning embryos, 14 reconstructed embryos developed into fry. DNA fingerprinting systems of the cloned fish were similar to those of the nuclear donor fish, but were distinctly different from those of the nuclear recipient fish. It confirmed that the genetic material originated from nuclear donor cell other than from nuclear recipient egg. The research suggested that the basic technique for nuclear transplantation performed with different strains of zebrafish has made a breakthrough. It should be helpful for the study of some important developmental problems such as gene function, the regulation of gene expression during animal development, the developmental potential of a nucleus and the interactions between the donor nucleus and the recipient cytoplasm, etc.

Keywords Zebrafish; different strains; nuclear transplantation; DNA fingerprint

A fundamental question in cell and developmental biology concerns the function and interactions between nuclei and cytoplasm in the process of development, differentiation and heredity. Nuclear transplantation has been recognized as the most powerful experimental approach to answer this question. Since Tung *et al.* [1] initiated and established the technique of nuclear transplantation in fish in 1963, successful production of nuclear transplant fish by transplanting the nuclei of one species into enucleated eggs of different species, varieties, genera and subfamilies has been achieved. These nuclear transplants were even developed into adult fishes that produced viable off-spring [2-6]. In 1986, by means of the serial nuclear transplantation technique, Chen *et al.*[7] obtained a sexually mature adult nuclear transplant fish from short-term cultured kidney cell nucleus of an adult crucian carp (*Carassius auratus* L.). The results confirmed that the specialized somatic cell nuclei of adult fish still retained their developmental totipotency [7]. Indubitably, nuclear transplantation has made great success when used with fish. Nevertheless, these studies were mainly performed with loach (*Paramisgurnus dabryaus*) and/or some farmed cyprinid fish. As is well known, it takes these fish 2 years to get mature while they spawn only once a time per year in general. On the other hand, the fish used for nuclear transplantation is not the model organism. They might be purchased from markets, collected from the fields or provided by merchant farms. They are in

---

文章发表于《科学通报》(英文版), 2002, 47(15): 1277-1280

a heterozygous population. Consequently, it is very inconvenient for further research used with them [8]. In the basic developmental biology, the use of fishes such as medaka (*Oryzias iatipes*) and most notably zebrafish (*Danio rerio*) as model animals is becoming so popular now. It is very urgent to establish the technique of nuclear transplantation with model fishes [9]. In 1999, Niwa *et al.* [10] established the basic technique for nuclear transplantation in medaka by transplanting nuclei from embryonic cells into nonenucleated eggs. Then, by using blastula embryonic cell nuclei from transgenic fish carrying the green fluorescent protein (*gfp*) gene as donors, fertile and diploid nuclear transplants were generated when the unfertilized eggs of medaka enucleated by X-ray irradiation were used as recipient [11]. The zebrafish has emerged as a new model organism for vertebrate developmental biology [12,13]. Meanwhile, the introduction of nuclear transplantation to zebrafish has become hotspot and urgent in development. However, it is very difficult to make nuclear transplantation with zebrafish and there is not convincing research available yet. Under the circumstance, the present study was conducted to establish the technique for nuclear transplantation in zebrafish using different strains and obtain nuclear transplants. DNA fingerprinting systems showed that the cloned fish were derived solely from the transplanted nuclei. The technique of nuclear transplantation, beyond all doubt, provides the powerful support for researching in development with zebrafish. Whereas ES-like cell cultures had been derived from early zebrafish embryos [14], it will be also beneficial to developing a gene targeting technique in zebrafish.

1. Materials and Methods

(i) Zebrafish strains and collection of eggs

AB strain (a gift of Prof. Meng at Tsinghua University) and Long fin strain (supplied by our laboratory) of zebrafish were used as the donor and recipient for nuclear transplantation, respectively. Long fin strain has dominant phenotypic characteristics. Its fins are much longer than those of AB strain (fig. 1(a) and (b)). According to the general methods for zebrafish care, the fish were maintained in 18L aquaria under a cycle of 14 h light and 10 h dark at 28.5℃. On the day before nuclear transplantation, the males and females of AB strain were transferred into the same tank at a ratio of 1 male to 1 female. At the beginning of the next light cycle, the embryos were collected by siphoning the bottom of the tank. At the beginning of the experiments, the males were transferred into the tank with females of Long fin strain at a ratio of 2 males to 1 female. Females were separated as soon as mature eggs were ovulated from the female into the water. The abdomens of the females were extruded to artificially collected unfertilized eggs. Then the unfertilized eggs were treated with some approaches such as hypotonic solution, trypsin digestion, etc. Finally, chorions of eggs were removed with sharp forceps and a sharp needle.

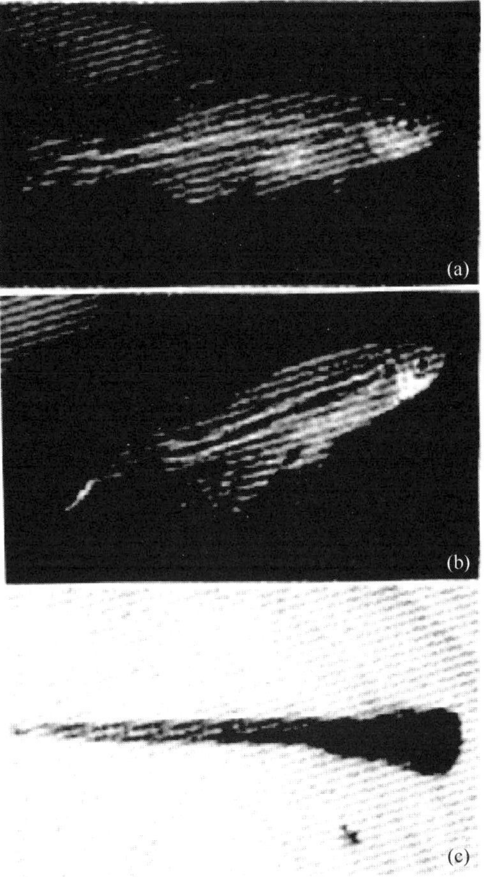

Fig. 1 Two different stains of zebrafish and cloned fish at the age of 37d. (a) Nuclear donor, AB strain; (b) nuclear recipient, Long fin strain; (c) cloned fish

(ii) Nuclear transplantation

The cells providing the donor nuclei were taken from the late blastula of AB strain. Separation of the blastoderm from the yolk was performed with a fine glass needle and a hair loop. The isolated blasoderm was carefully placed in Holtfreter's dissociation solution (Ca^{2+}-free Holtfreter's solution with 0.15 mmol/L EDTA) for further separating into individual cells, and then transferred into Holtfreter's standard solution (0.35% NaCl, 0.01% $CaCl_2$, 0.005% KCl (*w/v*), 50 iu/mL streptomycin, and 100 iu/mL ampicillin). The enucleated unfertilized snaked eggs of Long fin strain were used as recipients. Inserting a sharp needle into the egg cytoplasm just underneath the polar body carefully removed the egg nucleus. Nuclear transplantation was carried out under stereoscopic microscope at 30×magnification. Both donor cells and recipient eggs are transferred into Holtfreter's solution in a 1.2% agar layered Petri dish. During the operation, the single donor cell was slightly sucked into the micropipette by a slight negative pressure, and then it was microinjected into the cytoplasm of the recipient eggs at the animal pole. The reconstructed embryos were allowed to develop in Holtfreter's solution up to early neurula stage and then transferred into "aged" boiling water (boiled and then left at room temperature for more than 2 d).

(iii) DNA extraction and random amplified polymorphic DNA (RAPD) analysis of cloned fish

Total genomic DNA of mature AB and long fin strains was isolated from tail fin by standard phenol-chloroform extraction. The cloned fish genomic DNA was prepared using the same method. RAPD was performed in a volume 20 HL containing approximately 60 ng template DNA, 0.2 Hmol/L primer, 2 U Taq polymerase, 0.1 mmol/L of each dNTP, and 2×Taq polymerase buffer (BioAsia, China). RAPD primers were obtained from Operon Technologies (BioAsia, China). PCR amplifications were performed in a Perkin-Elmer DNA GeneAmp PCR System 9600, with an initial denaturation step of 94℃ for 3 min; followed by 40 cycles of 45 s at 94℃, 30 s at 36℃, and 1 min at 72℃; and a final extension step at 72℃ for 10 min. Amplification products were separated by electrophoresis on 2% agarose gels. Sizes in kilobase (kb) pairs were inferred by comparisons with a λDNA digested with *Eco*RIand *Hin*dIII markers (SABC). Gels were stained by ethidium bromide and recorded using a computer and a white/ultraviolet transilluminator (Ultra-Violet Products).

2. Results

(i) Development of embryos reconstructed with two different zebrafish strains

The nuclei of late blastula cells derived from AB strain were transplanted into enucleated eggs of Long fin strain. Of the total of 1119 operated eggs, 51.9% reached the blastula stage. However, the development of most of them was blocked. Only 15.7% surmounted the blastula stage and reached the gastrula stage. Hereafter, 3.8% developed to the segmentation stage, 1.6% to the heart beating stage. 1.25% were hatched, among which 3 died after 5d living for enlargement of pericardial chamber, 10 died after 15-30 d living. At the age of 33 d, the feeding of the last transplant was significantly reduced. It swam slowly and its body got gradually slimming. 4 d later, its pictures were taken (fig. 1(c)).

(ii) DNA fingerprint analysis of the cloned fish

At first, 10 adult fish of the AB strain and Long fin strain were screened using 40 primers (S1-S20, S121-S140, BioAsia, China) of 10 nucleotides in length, respectively. Of these primers, 2 primers (S129: CCAAGCTTCC; S138: TTCCCGGGTT) generated strain-specific band patterns. So they were selected to carry out the DNA fingerprint analysis to determine the origin of the donor cell used in nuclear transfer. As shown in fig. 2, DNA fingerprinting systems of cloned fish were similar to those of the nuclear donor fish, distinctly different from those of the nuclear recipient fish. The results confirmed that the genetic material of the cloned fish was originated from nuclear donor cell.

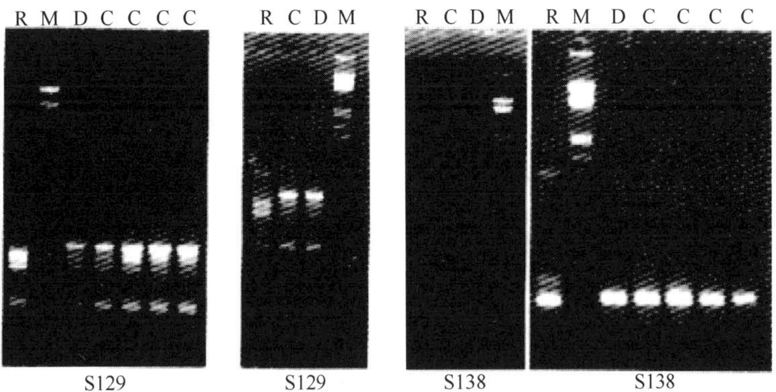

Fig. 2 DNA fingerprint analysis of cloned fish. D, Nuclear donor, AB stain; C, cloned zebrafish; R, nuclear recipient, Long fin stain; M, marker, λDNA/*Hin*dIII+*Eco*RI

3. Discussion

The zebrafish has emerged as a popular model organism for studies of vertebrate development. As an excellent model, it possesses many characteristics including short generation time, easy availability of eggs and embryos, external fertilization and development, transparent embryos during most of embryogenesis. These advantages make zebrafish be suited for manipulations involving DNA transfer, cell labeling, and transplantation. Also, methods have been established for performing large-scale mutagenesis screens in zebrafish to identify developmentally important genes, and the zebrafish genome plan has also made great progress. In mice, the use of pluripotent embryonic embryonic stem (ES) cell cultures for the production of knockout mutants has provided a powerful approach to the study of gene function during embryogenesis. However, the gene targeting technique in zebrafish has not been developed yet. Therefore, development of nuclear transplantation for gene targeting would be especially of benefit to analyzing functions of cloned genes at the individual, also to study the developmental potential of a nucleus and the interactions between the donor nucleus and the recipient cytoplasm.

Unfortunately, it is very difficult to make nuclear transplantation with zebrafish. There is very high pressure in the unfertilized eggs. The egg is so fragile that nuclear transplantation is usually abortive for breakdown of the egg. To solve the problem, we treated the eggs with hypotonic solution, etc. Subsequently, the enucleated unfertilized snaked eggs adaptive to nuclear transplantation have been obtained. Then, nuclear transplantation in different strains of zebrafish was made in which single later blastula nuclei were used as donor cells.14 reconstructed embryos developed into fry, of which 11 lived for about 2-5 weeks.

In general, phenotype identification, karyotype analysis and allozyme detection were usually used to analyze the results of nuclear transplantation [15]. However, phenotype of fish is affected by the surroundings. On the other hand, intuitionistic phenotype identification could not detect the change arisen from gene rearrangement, gene translocation and gene losing during the develop-mental process of fish. There are so many small chromosomes in fish, which makes it very difficult to analyze the karyotype of fish. Allozyme could only detect limited gene loci and could not detect the difference arisen from degeneracy and variance of the 3rd base of genetic codes. RAPD is an easy, rapid and sensitive PCR-based technique that has been widely used for polymorphism analyses of intraspecific and inbred strains genetic structures [16,17]. We have therefore carried out a DNA fingerprint analysis by RAPD to determine the origin of the donor cell used in nuclear transfer. RAPD band patterns generated from the cloned fish were all the same as those of the donor strain, which were distinctly different from those of the recipient strain. This gave solid evidence that the cloned fish were resulted from the reprogramming of donor cells. *Gobiocyopris rarus* is a different species from zebrafish. According to the same techniques, single gastrula nuclei from *G. rarus* were transplanted into enucleated unfertilized eggs of zebrafish. Cloned fish determined by DNA fingerprint were also achieved (unpublished data in our laboratory). Therefore, it is considered that the first step to establish the techniques for nuclear transplant in zebrafish was achieved. It is noticeable that the normally developed larval fish would live on *Artemia salina* after they begin to feed 2 d later. However, all of the 11 cloned fish could only feed yolk. It was very difficult for them to feed baby brine shrimp, which implied that the cloned fish change its feeding behaviour. Moreover, they died off in 2-5 weeks. It was worthwhile to make further research whether the cloned fish died from innutrition or other unknown reasons.

There is only one report about the nuclear transplantation in zebrafish till now. In 2000, Li *et al.* [18] transplanted nuclei of zebrafish from blastulae into the non-enucleated unfertilized eggs of the same strain. 10 larval fish were generated while one of them lived for 26 d. The chromosome number analysis showed that the majority of the recipient embryos were diploid. The histological examination also revealed that the original female pronuclei in the recipient eggs became degenerated at early cleavage. However, it has been controversial whether the non-enucleated unfertilized eggs were suitable for recipient in nuclear transplantation or not. There were contrary results. Yu *et al.* [19] performed inter-species nuclear transplantation in fish with non-enucleated eggs. The karyotype of transplants was diploid and its eletrophoretic patterns of serum proteins were the same as those of the donors. So they considered that the haploid pronucleus had been excluded by the donor nucleus when non-enucleated eggs were used as nuclear recipients. Nevertheless, in amphibians, the nuclear transplants were shown to be triploid if non-enucleated eggs were served as nuclear recipient. The results indicated that female pronucleus participate in the development of reconstructed embryos together with the donor nucleus [20,21]. In 1999, Niwa *et al.* [10] transplanted blastula nuclei to non-enucleated eggs in the medaka and generated cloned fish. The allozyme analysis of phosphoglucomutase, measurements of relative DNA content by microfluorometry and chromosome counts consistently indicated that the nuclear transplants were triploids that originated from both the diploid donor nuclei and the haploid recipient pronuclei. Also, several gynogenesis zebrafish, loach and golden fish had been induced when non-enucleated eggs were used as nuclear recipients in inter-species nuclear transplantation in our laboratory (unpublished data). On the other hand, in nuclear transplantation using non-enucleated eggs in loach, a small number of haploids, triploids and tetraploids nuclear transplant embryos were even produced in addition to many diploids [22]. Therefore, if the nuclei without any marker to be distinguished were transplanted into non-enucleated eggs of the same species fish, the results would be very complicated. It might be relative to some reasons, such as materials in experiments, cell cycle of donor nuclei and nucleocytoplasmic interaction between donor cells and recipient eggs. It was the cause that the technique of nuclear transplantation in medaka was developed from non-enucleated eggs to enucleated eggs as recipient. Under the circumstance, it is an acceptable principle that enucleated eggs should be used as recipient in nuclear transplantation.

Acknowledgements We thank Prof. Meng Anming for providing zebrafish AB strain and Dr. Wang, J. W., He, S. P., Sun, Y. H., Wu, G. and Li, M. for their help and advice. This work was supported by the National Natural Science Foundation of China (Grant No. 30000090), the Chinese Academy of Sciences (Grant No. KSCX2-SW-303) and by "973" Project of the Ministry of Science and Technology (Grant No. G2000016109).

References

1. Tung, T. C., Wu, S. C., Ye, Y. F. *et al.*, Nuclear transplantation in fishes, *Chinese Science Bulletin*, 1963, 7: 60.
2. Study Group of Cell Genetics, Institute of Zoology, Chinese Academy of Sciences, Nuclear transplantation in teleosts. I. Hybrid fish from the nucleus of carp and the cytoplasm of crucian, *Science in China* (in Chinese), Ser. B, 1980, 4: 377.
3. Yan, S. Y., Lu, D. Y., Du, M. *et al.*, Nuclear transplantation in teleosts. II. Hybrid fish from the nucleus of crucian and the cytoplasm of carp, *Science in China, Ser. B*, 1984, 8: 729.
4. Tung, T. C., Ye, Y. F., Lu, D. Y. *et al.*, Transplantation of nuclei between two subfamilies of teleosts, *Acta Zool. Sin.*, 1973, 19(3): 201.
5. Yan, S. Y., Lu, D. Y., Du, M. *et al.*, Nuclear transplantation in teleosts, IVA. Nuclear transplantation between different subfamilies-hybrid fish from the nucleus of grass carp (*Ctenopharyngoden idellas*) and the cytoplasm of blunt-snout bream (*Megalobrama amblycephala*), *Chinese J. Biotechnology*, 1985,1(4): 15.

6. Qi, F. Y., Xu, G. Z., Genetic character and individual growth of the transnucleus fish of the bighead and the bluntsnout bream, *Acta Zool. Sinica*, 1997, 43(2): 211.
7. Chen, H. X., Yi, Y. L., Chen, M. R. *et al.*, Studies on the developmental potentiality of cultured cell nuclei of fish, *Acta Hydrobiologica Sinica*, 1986, 10(1): 1.
8. Yan, S. Y., A historical review and some comments on the nuclear transplantation in fish, *Chinese J. Biotechnology*, 2000, 16(5): 541.
9. Okada, T. S., Introduction remarks, in Cloning in Fish-Nucleocy toplasmic Hybrids (ed. Yan, S. Y.), *Hong Kong: IUBS Educational and Cultural Press Ltd, 1998.*
10. Niwa, K., Ladygina, T., Kinoshita, M. *et al.*, Transplantation of blastula nuclei to non-enucleated eggs in the Medaka, *Oryzias Iatipes*, *Develop. Growth Differ.*, 1999, 41: 163.
11. Wakamatsu, Y., Ju, B. S., Pristyaznhyuk, I. *et al.*, Fertile and diploid nuclear transplants derived from embryonic cells of a small laboratory fish, medaka (*Oryzias Iatipes*), *PNAS*, 2001, 98(3): 1071.
12. Kahn, P., Zebrafish hit the big time, *Science*, 1994, 264: 904.
13. Concordet, J. P., Ingham, P. Catch of the decade, *Nature*, 1994, 369: 19.
14. Sun, L., Bradford, C. S., Ghosh, C. *et al.*, ES-like cell cultures derived from early zebrafish embryos, *Mol. Mar. Biol. Dev.*, 1995, 4: 193.
15. Yan, S. Y., Cloning in Fish: Nucleo-cytoplasmic Hybrids, *Hong Kong: IUBS Educational and Cultural Press Ltd,* 1998.
16. Williams, J. G. K., Kubelik, A. E., Licak, K. J. *et al.*, DNA polymorphism amplicied by arbitrary primers are useful as genetic markers, *Nuc. Acids Res.*, 1990, 18: 6531.
17. Caccone, A., Allegrucci, G., Fortunato, C. *et al.*, Genetic differentiation within the European sea bass (*D. Labrax*) as revealed by RAPD-PCR assays, *J. Hered.*, 1997, 88: 316.
18. Li, L., Zhang, S. C., Wang, R. *et al.*, Cloning of zebrafish: development of non-enucleated unfertilized eggs provided with nuclei from blastula cells, *High Technology Letters*, 2000, 10(115): 24.
19. Yu, L., Yang, Y., Liu, L. *et al.*, Studies on the nuclear transplantation in fish by using unenucleated eggs as recipient, *Fresh Water* (in Chinese), 1989, 3: 3.
20. Gurdon, J. B., Nuclear transplantation in eggs and oocytes, *J. Cell Sci. Suppl.*, 1986, 4: 287.
21. Kroll, K. I., Gerhart, J. C., Transgenic X. *Laevis* embryos from eggs transplanted with nuclei of transfectd cultured cells, *Science*, 1994, 266: 650.
22. Gasaryan, K. G., Hung, N. M., Neyfakh, A. A. *et al.*, Nuclear transplantation in teleost Misgurnus fossilis L., *Nature*, 1979, 280: 585.

斑马鱼不同品系的细胞核移植

胡　炜　汪亚平　陈尚萍　朱作言

中国科学院水生生物研究所，淡水生态与生物技术国家重点实验室，武汉 430072

摘　要　以斑马鱼AB品系囊胚晚期胚胎细胞为细胞核供体，以斑马鱼长鳍品系未受精的去核卵为受体，进行不同品系间斑马鱼的细胞核移植。采用显微注射法，操作胚胎1119枚，获得克隆鱼14尾。RAPD分析显示，克隆鱼与细胞核供体鱼的DNA扩增带纹一致而与受体鱼不同，表明克隆鱼的遗传物质来源于供体细胞核。模式动物斑马鱼的细胞核移植技术的建立，有望在研究动物发育过程的基因功能、细胞核的发育潜能及核质关系等重要问题上发挥作用。

Cytoplasmic Impact on Cross-Genus Cloned Fish Derived from Transgenic Common Carp (*Cyprinus carpio*) Nuclei and Goldfish (*Carassius auratus*) Enucleated Eggs

Yong-Hua Sun Shang-Ping Chen Ya-Ping Wang Wei Hu Zuo-Yan Zhu

State Key Laboratory of Freshwater Ecology and Biotechnology, Institute of Hydrobiology, Chinese Academy of Sciences, Wuhan 430072

ABSTRACT In previous studies of nuclear transplantation, most cloned animals were obtained by intraspecies nuclear transfer and are phenotypically identical to their nuclear donors; furthermore, there was no further report on successful fish cloning since the report of cloned zebrafish. Here we report the production of seven cross-genus cloned fish by transferring nuclei from transgenic common carp into enucleated eggs of goldfish. Nuclear genomes of the cloned fish were exclusively derived from the nuclear donor species, common carp, whereas the mitochondrial DNA from the donor carp gradually disappeared during the development of nuclear transfer (NT) embryos. The somite development process and somite number of nuclear transplants were consistent with the recipient species, goldfish, rather than the nuclear donor species, common carp. This resulted in a long-lasting effect on the vertebral numbers of the cloned fish, which belonged to the range of goldfish. These demonstrate that fish egg cytoplasm not only can support the development driven by transplanted nuclei from a distantly related species at the genus scale but also can modulate development of the nuclear transplants.

Keywords cloned fish; common carp; cross-genus; cytoplasmic impact; developmental biology; early development; embryo; goldfish

Introduction

In general, nuclear transplantation is a technique of recombination of an enucleated egg with a novel diploid nucleus. The recombined egg may have the potential to go through embryo genesis and even develop into an adult. The nuclear donor and the recipient egg can come from either the same species, i.e., intraspecies nuclear transplantation, or different species, i.e., cross-species nuclear transplantation. Intraspecies nuclear transplantation offers a powerful approach to study the totipotency or pluripotency of differentiated nuclei. In contrast, by taking the advantage of the developmental differences between two species, cross-species

文章发表于 *Biology of Reproduction*, 2005, 72(3): 510-515

nuclear transplantation provides an access to probe into the interaction between nucleus and cytoplasm involved in development. The art of nuclear transplantation was first demonstrated in frogs [1], and great achievements were subsequently gained with amphibians. However, all of them were successfully accomplished as intraspecies nuclear transplantation (reviewed by Gurdon [2]). In mammals, despite the great success of cloning of numerous species [3-7], only a few cases of cross-species nuclear transplantation between very closely related species have been successful [8-10]. Consequently, cross-species nuclear transplantation in fish is a means to explore the contributions of nucleus and egg cytoplasm to vertebrate development (reviewed by Zhu and Sun [11]). In our previous study, led by Tung [12], we found that the vertebral numbers of some nuclear transfer (NT) fish were consistent with those of the egg-providing species, but no conclusive evidence was provided and the results have been challenged by the scientific community [2, 13]. There were suggestions, with no supporting evidence, that the cross-species NT fish were nucleo-cytoplasmic hybrid fish [11,14]. Recently, as one of the important model animals for developmental studies, the zebrafish was successfully cloned from cultured cells [15]. Since then, however, there has been no successful report on fish cloning.

Just as in other vertebrates, fish bone skeletal system is made up of repeating patterns, among which the most obvious are the vertebrae. Vertebral patterning is a result of somite patterning during embryogenesis, and the vertebral number varies a lot among different fishes but is relatively stable within a given species [16]. For example, the vertebral number of common carp is 33-36 and that of goldfish is 26-28 [17]. Thus, the vertebral number has been considered an important element in taxonomic study. Nevertheless, we still lack a comprehensive understanding of the molecular mechanism of action that controls the somite number and vertebral number. Nuclear transplantation between two species with different vertebral numbers may provide novel insights into the mechanisms underlying this process.

In the present study, we conducted experiments of cross-genus nuclear transplantation between two fish species with different vertebral numbers, *Cyprinus* and *Carassius* [17,18]. We tested whether the enucleated eggs of goldfish could adapt and reprogram common carp nuclei to direct the embryogenesis and ontogenesis of resulted nuclear transplants. We extensively analyzed the origination of the nuclear and mitochondrial genomes in the NT animals.

Materials and Methods

Experimental fishes

The founder of transgenic red common carp, with a long body in shape and two barbels on each side of the mouth, was produced by introducing the recombinant construct *MThGH*, composed of the *human growth hormone* gene (*hGH*) driven by a mouse metallothionein-1 gene promoter (*MT*), into fertilized eggs via microinjection [19]. The subsequent F1-F3 generations were produced as previously reported [20]. The red-dragoneye strain of goldfish, which has protruding eyes, triangle tail, and spherical body in shape, was used in this experiment. Use of these animals for experimental purposes was approved by the Scientific Committee at the Institute of Hydrobiology, Chinese Academy of Sciences.

Preparation of donor nuclei

Fertilized eggs from transgenic F3 red common carp were cultured in Holtfreter solution (0.35% NaCl, 0.01% $CaCl_2$, 0.005% KCl [*w/v*], 50 IU/ml streptomycin, and 100 IU/ml ampicillin) up to the blastula stage. The blastoderms were cut from the yolk with a fine glass needle and placed into Holtfreter dissociation solution (Ca^{2+}-free Holtfreter solution with 0.15 mM EDTA). After 2 min, cells of the blastoderms were dissociated and would be used as donors for nuclear transplantation.

Preparation of recipient enucleated eggs

Goldfish were artificially induced to spawn for egg collection. The unfertilized eggs were placed into a trypsin solution of 0.25% (*w/v*; Sigma, St. Louis, MO) for 3 min. The softened chorion was subsequently removed by microsurgery. The second polar body of the dechorionated egg was visible under a 40×stereomicroscope. The egg nucleus underneath the second polar body was removed by picking with a sharp glass needle. Enucleated eggs were held in an agar plate filled with Holtfreter solution for further manipulation. Successful enucleation was proved in parallel experiments with Hoechst 33342 (Sigma) staining and examination under an ultraviolet light using Olympus SZX-12 microscope (Japan).

Nuclear transfer

Using a micromanipulator designed in our lab, a single donor cell was ruptured by gently sucking into a micropipette and then was microinjected into the animal pole of an enucleated egg. The manipulated eggs were carefully placed into Holfreter solution for development at 19℃, and subsequent development was followed by microscopy. Deformed embryos were removed periodically. The fry that appeared to be developing normally were raised with great care in a glass tank and then in a small pond. For controls, sexual hybrid fish were produced by artificial mating of red common carp males with goldfish females.

Total DNA extraction

Total DNAs from embryos and fry was prepared as follows: each sample was centrifuged onto the bottom of a microfuge tube and digested with 1 μg each of proteinase K (Sigma) and RNase A (Sigma) in 100 μl of DNA extraction buffer (10 mM Tris·Cl, pH 8.0, 300 mM NaCl, 10 mM EDTA, 2.0% [*w/v*] SDS). DNA was recovered by serial extractions with phenol and chloroform, precipitated with ethanol, and finally dissolved in 20 μl of TE buffer (10 mM Tris·Cl, pH 8.0, 1 mM EDTA). Total DNAs of the cloned fish were extracted from the tailfin as described [19].

Polymerase chain reaction analysis

Two primers (forward: 5′-GGTAAGCGCCCCTAAAATCC-3′ and reverse: 5′-TTGAAGATCTGCCCAGTCCG-3′) for detection of *hGH*-transgene are both located in the *hGH* coding sequence. The expected amplification size of polymerase chain reaction (PCR) product was 712 base pairs (bp). PCR was performed by using 0.5 U of Taq DNA polymerase (BioAsia, Shanghai, China), 10 pmol of each primer, and 50 ng of total fish DNA (for embryonic and fry samples, 5 μl DNA solution instead) as template in a volume of 25 μl. The reaction process was 94℃ for 4 min, 30 cycles of 94℃, 30 sec; 58℃, 30 sec; and 72℃, 1 min. All PCR products were separated by electrophoresis on 0.8% (*w/v*) agarose gels and visualized using a UVP GDS8000 system.

Random amplification of polymorphic DNA analysis

We used a comparative random amplification of polymorphic DNA (RAPD) analysis to distinguish genomic DNAs of common carp and gold-fish. Specific primers that could give unique patterns to common carp or goldfish were selected from 20 random primers (S121-S140; Sangon, Shanghai, China). RAPD reaction mixture contained 1 U of Taq DNA polymerase (BioAsia), 10 pmol of oligonucleotide primer, and 60 ng of total DNA as template in a volume of 20 μl. The process of reaction included 94℃ for 4 min; 40 cycles of 94℃, 45 sec; 36℃, 1 min; and 72℃, 1 min. RAPD products were separated by electrophoresis on 1.5% (*w/v*) agarose gels for visualization.

Amplification of mtDNA

The full sequences of mtDNAs for both common carp and goldfish were downloaded from GenBank (Accession no. NC-001606 and NC-002079, respectively). Based on the DNA alignment with DNATools soft-ware (5.1 version, S.W. Rusmussen), PCR primers for distinguishing mtDNA from common carp and goldfish were designed. For common carp, the forward primer was 5′-GGAGGTAGCACTCCC-3′ (5′-3′ position: 1-15) and the reverse primer was 5′-GGGGTTTGTCGCGCA-3′ (5′-3′ position: 688-702).

Both primers are located in the D-loop region of common carp mtDNA and the expected length of the PCR product was 702 bp. For goldfish, the forward primer was 5′-CCTGGCTGCCGGTAT-3′ (5′-3′ position: 7002-7106) and the reverse primer was 5′-CGTGGTATTCCTGCT-3′ (5′-3′ position: 7698-7712). Both primers are located in the goldfish *cytochrome c oxidase* subunit I gene and the expected length of PCR product was 711 bp. The PCR amplifications were conducted under the following parameters: 0.5 U of Taq DNA polymerase (BioAsia), 10 pmol of each primer, and 100 ng of total DNA (for embryonic and fry samples, 5 μl of DNA solution instead) in a total volume of 25 μl; 94℃, 4 min, 30 cycles of 94℃, 30 sec; 50℃, 30 sec; and 72℃, 1 min.

Developmental observation and phenotypic analysis

The whole processes of embryonic development of common carp, goldfish, the hybrid fish, and the nuclear transplants were examined under an Olympus SZX-12 microscope. The serial timings of somite development were recorded. The adults of common carp, goldfish, the hybrid fish, and the cloned fish were sampled for phenotypic analysis. These fishes were immobilized in 80 ppm MS-222 (Sigma) and x-ray photographed using Super Soft X-ray Inspection System (Model CMB-2; Softex Co., Tokyo, Japan). X-ray films were developed as described in the user's manual and scanned using a Microtek ScanMaker 4800i (Shanghai, China). The vertebral numbers were counted from the scanned pictures.

Results

Generation of cross-genus cloned fish

Larval red common carp did not show any pigmentation (Fig. 1A), while larval goldfish did (Fig. 1B). Thus, pigmentation could serve as a marker to determine whether the red common carp nuclei contribute to the development of NT larval fish. In total, five batches of successful nuclear transplantations were conducted, in which 52.9% of 501 transplanted eggs developed to the blastula stage. However, 62.6% of these blastulae failed to develop further to gastrulation, which seems to be a critical stage in the development of NT embryos, just as the midblastula transition is crucial to normal development of fish [21]. In some cases, the NT embryos excluded a small proportion of recipient yolk during gastrulation (Fig. 1, C and D), but this occurrence did not interrupt the subsequent development. A total of 99 blastulae (19.8% of transplanted eggs) developed to the gastrula stage, of which 12 (2.4% of transplanted eggs) were hatched. All the hatched nuclear transplants did not show any pigmentation (Fig. 1E), indicating that the red common carp nuclei contributed to the development of NT fry. Among the hatched fry, three failed to reach the blood-circulation stage (Fig. 1F) and two failed to feed, while the remaining seven (1.4% of transplanted eggs) reached adulthood. When these cloned fish were put into the same tank with red common carp and goldfish, it was difficult to find any difference between the cloned fish and the red common carp that provided nuclei. The exterior phenotypic characteristics of red common carp, such as two pairs of barbells, long body shape, normal tail, and normal eyes were present in the cloned fish, but there was almost no visible contribution of distinctive goldfish characteristics, such as spherical body shape, triangle tail, and dragon eyes (Fig. 1G). None of the 2-yr-old cloned fish could produce sperm or mature eggs.

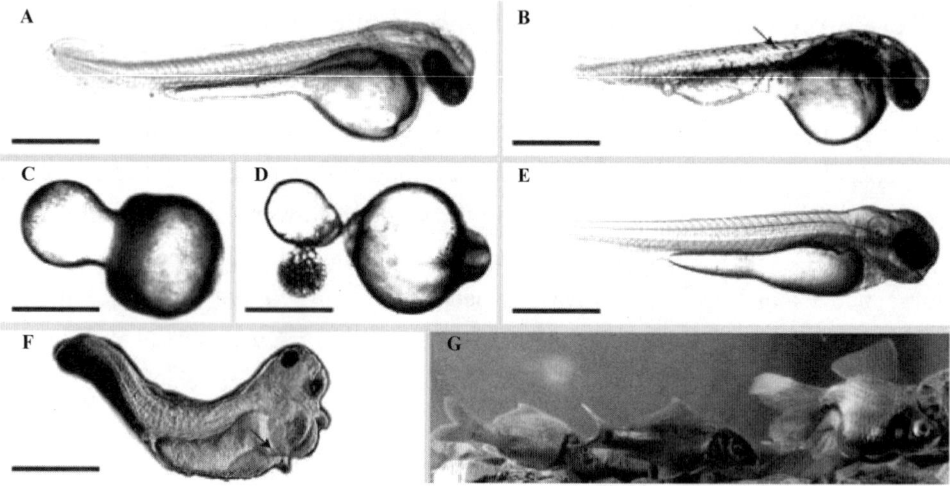

Fig. 1 Cross-genus cloned fish derived from red common carp nucleus and goldfish enucleated egg. **A**) Red common carp larval fish showing no pigmentation; scale bar=1 mm. **B**) Goldfish larval fish showing pigmentation: the arrow indicates the pigmentation; bar=1 mm. **C, D**) NT embryos excluding portions of the recipient yolks; bar=1 mm. **E**) NT larval fish developed just like a normal red common carp, showing no pigmentation; bar=1 mm. **F**) NT larval fish could not process blood circulation properly. The arrow indicates the thrombus in the abnormal NT embryo; bar=1 mm. **G**) A cloned fish (left), a donor cell providing red common carp (middle), and a recipient egg providing goldfish (the right)

Nuclear DNA genotypes of nuclear transplants

PCR amplification showed that all of 50 randomly sampled NT embryos had the characteristic *hGH*-transgene band (data now shown). The transgene was also detected in all of the cloned fish, in line with transgenic common carp that provided the donor nuclei (Fig. 2A). In addition, a comparative RAPD assay was developed to distinguish common carp, goldfish, and hybrid fish. Among 20 oligo-nucleotide primers, four of them (S121, S123, S128, S136) could produce different and distinguishable patterns for common carp and goldfish to identify the origins of the nuclear genomes of the cloned fish. However, the RAPD pattern of the hybrid fish did not often present both bands of common carp and goldfish, which may be due to the recombination between the genomes of common carp and goldfish in the hybrid genome. We found that the RAPD pattern resulting from each primer of the cloned fish DNA was identical to those from common carp and distinctly different from those of goldfish and the hybrid fish (Fig. 2B). Both results of transgene-based PCR amplification and comparative RAPD analysis proved that nuclear DNA (nDNA) of the cloned fish was exclusively derived from common carp, the source of the transplanted nuclei.

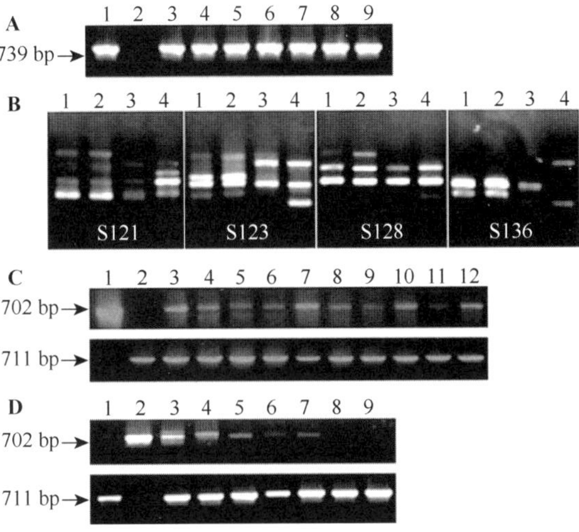

Fig. 2 Identification of the nuclear and mitochondrial genotypes of the nuclear transplants. **A**) PCR detection of pMThGH-transgene among the cloned fish. Lane 1: positive control from DNA from transgenic common carp; lane 2: negative control from DNA from recipient goldfish; lanes 3-9: amplified DNA from different cloned fish. 712 bp refers to theamplification band of pMThGH. **B**) Comparative RAPD analysis of the cloned fish. S121, S123, S128, and S136 refer to different random primers used for RAPD analysis. The primer sequences (from 5′ to 3′) are ACGGATCCTG, CCTGATCACC, GGGATATCGG, and GGAGTACTGG, respectively. Lane 1: common carp; lane 2, cloned fish; lane 3: hybrid fish; lane 4: goldfish. **C**) PCR amplification of mtDNA from cloned blastulae. Lane 1: common carp; lane 2: goldfish; lanes 3-12: different cloned blastulae. 702 bp is the amplification band from common carp mtDNA specific primers, and 711 bp is the amplification band from goldfish mtDNA specific primers. **D**) PCR amplification of mtDNA in cloned embryos at different developmental stages. Lane 1: goldfish; lane 2: common carp; lanes 3-9: cloned embryos at blastula, gastrula, somite, muscular-reaction, blood-circulation, larval and adult stages, respectively. 702 bp is the amplification band from common carp mtDNA-specific primers, and 711 bp is the amplification band from goldfish mtDNA-specific primers

The mtDNA genotypes of nuclear transplants

The mtDNA genotypes of the cloned embryos at different stages and the cloned fish were analyzed by PCR with two sets of species-specific primers. The result showed that each NT blastula contained a mixture of two types of mtDNA genome, one from goldfish and another from common carp. The amplified yields with carp-specific primers were tiny and varied among the embryos, while the amplified yields with goldfish-specific primers were abundant and uniform among the embryos (Fig. 2C). These data indicate that the NT blastulae were all mtDNA heterogeneous, containing abundant recipient-type mtDNA while harboring a relatively small amount of donor-type mtDNA. This may be due to variation in the mtDNA copy numbers that accompanied transplanted nuclei. In contrast, in late-stage NT embryos after blood circulation, only goldfish-derived mtDNA genotype could be detected (Fig. 2D). This suggests that the contaminating mtDNA from the nuclear donor cells was eliminated during development in the nuclear transplants. Similar results were found during the development of the hybrid fish. Common carp-derived mtDNA could be detected before the blastula stage but not in gastrula and the following developmental stages of the hybrid fish (data not shown).

Table 1 Developmental timing (hour : min) of nuclear transplants (at 19℃)[a]

	Two-cell	Blastula	Midgastrula	Somitogenesis initiation	13-somite	Somitogenesis complete
Common carp	1:15	7:20	11:30	22:00	29:00	45:50 (34-36)
Goldfish	1:20	8:00	13:30	24:00	33:00	45:00 (28-30)
NT[b]	1:12	8:00	13:30	24:00	32:40	45:00 (29)
Hybrid[c]	1:18	7:55	13:15	23:50	32:00	45:00 (28-30)

a Each group except the NT group includes samples of 20 embryos. Numbers in parentheses show the somite numbers of the embryos after they have finished somitogenesis

b NT indicates the nuclear transplants derived from common carp nuclei and goldfish enucleated eggs

c Hybrid indicates the hybrid embryos derived from crosses of male common carp with female goldfish

Somite development and vertebral number of nuclear transplants

From the view of somite development, the embryonic development rate of nuclear transplants was a little slower than that of the nuclear-donor species, common carp, but was similar to that of the recipient species, goldfish (Table 1). The somite number in the only NT embryo that developed to larval stage was 29, which is within the range of goldfish. Similar results were found among the hybrid embryos, which had somite numbers that ranged from 28 to 30. X-ray photographs showed that the vertebral number of six cloned fish was of the enucleated egg providing goldfish type, ranging from 26 to 28. Of all the seven cloned fish, the vertebral number was 26 for one fish, 27 for two fish, 28 for three fish, and 31 for one fish. For one of the survival hybrid fish, the vertebral number was 28. In contrast, the vertebral number of nuclear-donor common carp was 33-36 (Fig. 3). These data suggest that the goldfish egg cytoplasm plays an important role in regulating the somite development and vertebral number in the nuclear transplants.

Discussion

Though common carp and goldfish can be artificially hybridized, they hardly mate in nature, the survival ratio of the offspring is very low, and the hybrid offspring are generally sterile [22]. In fact, common carp and goldfish belong to different genera, the *Cyprinus* and *Carasius* [17,18]. Hybrid offspring can be obtained from distantly related species of fishes through artificial fertilization. The present study demonstrates that nuclei from embryonic fish cells can support the embryogenesis and ontogenesis following transplantation into a recipient egg cytoplasm of a distantly related species at the genus scale. This has never been done in other vertebrates. In amphibians, nuclear transplantationcould be done only between the same species (reviewed by Gurdon [2]). In mammals, cross-species nuclear transplantation has succeeded between two closely related species [8-10], but most experiments do not yield cloned animals [23,24]. In contrast, fish cross-species nuclear transplantation has been reported among several species [11,14]. In our present study, when *MThGH*-transgene served as a genetic marker and comparative RAPD assay between common carp and goldfish was conducted, the origins of nuclei of the cross-genus cloned fish were easy to clarify. The results verified that the donor nuclear genome contributed to the nuclear genome of the cloned fish instead of the recipient egg. As in our previous studies, most of the donor nuclei could not attain complete reprogramming [25] and only a few NT embryos developed to term. Nevertheless, as a result of the reprogramming and adaptation of the donor nuclei of common carp in the recipient eggs of goldfish, about 1% (7/501) NT embryos developed to adult stage and most of the apparent characteristics of the resulting fish resembled those of the nuclear donor, common carp. The morphological data are solid evidence that the common carp nuclei directed the development of the cross-genus cloned fish. In ongoing complementary studies, we have also generated cross-genus cloned fish derived from goldfish nuclei and common carp enucleated eggs. It seems that the recombination of nucleus and egg cytoplasm from different species of fish is more feasible than those of other vertebrates, perhaps because fish were the first vertebrates to evolve.

In mammalian cloning, studies on the mtDNA genotypes of cloned animals are quite controversial. Although the first somatic cloned mammal [26] and the cross-species somatic cloned mammals [8,9] showed to be mtDNA homoplasmy that contained just recipient cytoplasm-derived mtDNA, some cloned mammals were found to be mtDNA heteroplasmy—they contained mtDNA representative of both the donor cells as well as the recipient eggs [27,28]. The present study demonstrates that the goldfish-derived mtDNA can exist in cloned embryos until the blood-circulation stage, after which it faded away. In other words, the mtDNA heteroplasmy in cloned embryos converted to mtDNA

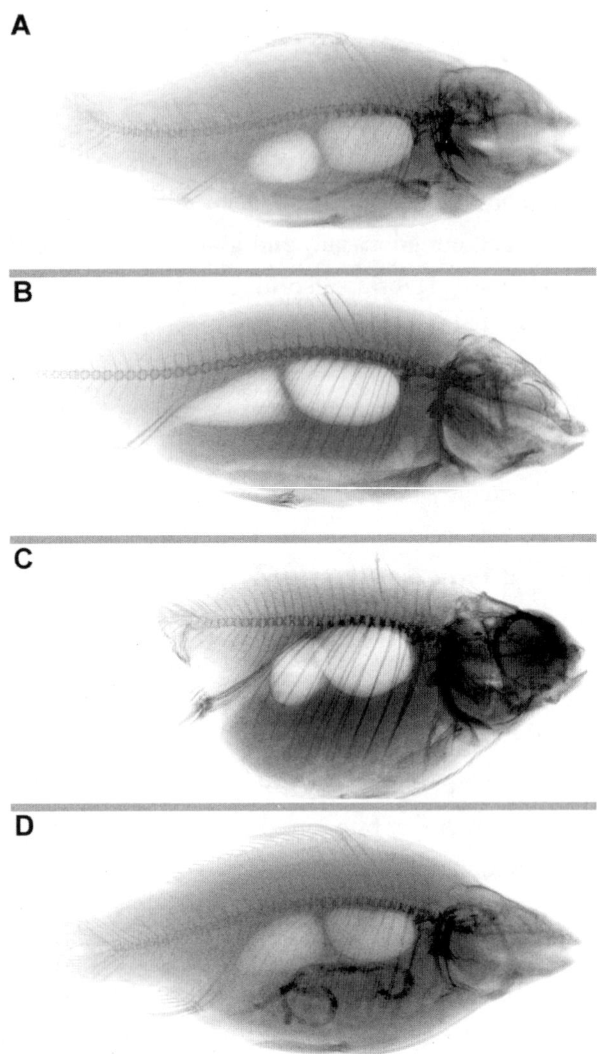

Fig. 3 X-ray photographs of the cloned fish (**A**), common carp (**B**), goldfish (**C**), and the hybrid fish. The vertebral numbers of the cloned fish, common carp, goldfish, and the hybrid fish (**D**) are 27, 33, 26, 28, respectively

homoplasmy over the course of development. From this point of view, the cross-genus cloned fish could be properly referred to as a nucleo-cytoplasmic hybrid fish that contains a combination of common carp-derived nuclear genome and goldfish-derived mitochondrial genome. However, the mechanism underlying the absence of donor cell-derived mtDNA in most cloned animals was unclear. In the present study, the mtDNAs from the nuclear donor cells were eliminated during the development of the cloned fish, just mimicking the destiny of the sperm-derived mtDNA in the sexual hybrid fish. In addition, the cross-genus cloned fish is as healthy as their peers of nuclear donors. This indicated that goldfish-derived mtDNA could not only work together with common carp nDNA but also be responsible for all the deserved pathways, though mitochondria control many fundamental metabolic pathways [29], just as nonhuman primate-derived mtDNA was able to survive in mtDNA-less human cells [30].

In previous studies of animal cloning, as expected, most cloned animals were identical to their nuclear donor species in phenotype [3-8]. Likewise, in the present study, most development characteristics of the cloned fish were the same as those of nuclear donor common carp. But, strikingly, analysis of somite development and vertebral number led to an unexpected result: vertebral development resembled that of the cytoplasmic recipient. In ongoing studies of reciprocal NT, the cloned fish need to grow bigger for analysis of vertebral numbers. Until now, the mechanism of vertebrate somitogenesis has been a mystery, although several theoretical models have been proposed [31]. The segmentation clock is the essence of many recently proposed models, which could explain most aspects of somitogenesis, such as the variation of somite numbers [32,33]. However, little is known about the clock, especially the mechanism that drives it [33]. According to our data, the clock is likely regulated by cytoplasmic factors in the egg cytoplasm. This resulted in vertebral numbers in most cloned fish that were the same as those of egg-donating goldfish and different from that of nuclear-donor common carp, based on the concept that the vertebrae can be aligned with somite in vertebrates [34]. In previous studies, most embryonic induction factors are expressed in the maternal body and accumulated in mature oocytes [35]. Our data suggest that the somite number or segmentation clock of fish is determined in early embryogenesis under the regulation of egg cytoplasmic components during the formation of presomitic mesoderm [36]. During the somite stages, somitogenesis-related genes are expressed cyclically, resulting in the formation of somites [33]. Meanwhile, as a result of hierarchical activation of cascades of nuclear genes, cell fates and most of the developmental characteristics of fish are controlled by the temporal and spatial expression of nuclear genes. Therefore, the cross-genus cloned fish showed to be almost the same as donor common carp in exterior appearance.

In addition, in the studies of fish gene transfer, foreign gene integration occurs gradually and randomly in host genomes, which results in many problems, such as multi-site integration, positional effects, and transgenic mosaicism [11,19,37]. Here, the first case of successful cloning with nuclei from valuable, fast-growing *GH*-transgenic fish has been demonstrated. This is a crucial step for further cloning with *in vitro* genetically modified, cultured cells–a step that is considered to be an efficient method to solve these problems [11, 38]. On the other hand, some species of fish are near the edge of extinction due to environmental pollution and overfishing [39]. The present study provides a promising way to preserve endangered fish species through cross-species cloning by transplanting the nuclei of the endangered species into the enucleated eggs of another well-populated species.

Overall, the present study reveals that goldfish enucleated eggs could not only support the development of the cross-genus nuclear transplants receiving common carp nuclei but also have an evident impact on certain developmental characteristics, especially the somite development and vertebral number of the nuclear transplants.

Acknowledgements We thank Ms. Ming Li for technical help in generation of the cloned fish and three anonymous reviewers for valuable comments.

REFERENCES

1. Briggs R, King TT. Transplantation of living nuclei from blastula cells into enucleated frog's eggs. Proc Natl Acad Sci U S A 1952; 38: 455-463.
2. Gurdon JB. Nuclear transplantation in eggs and oocytes. J Cell Sci Suppl 1986; 4: 287-318.
3. Wilmut I, Schnieke AE, McWhir J, Kind AJ, Campbell KH. Viable offspring derived from fetal and adult mammalian cells. Nature 1997; 385: 810-813.
4. Wakayama T, Perry AC, Zuccotti M, Johnson KR, Yanagimachi R. Full-term development of mice from enucleated oocytes injected with cumulus cell nuclei. Nature 1998; 394: 369-374.
5. Kato Y, Tani T, Sotomaru Y, Kurokawa K, Kato JY, Doguchi H, Yasue H, Tsunoda Y. Eight calves cloned from somatic cells of a single adult. Science 1998; 282: 2095-2098.
6. Baguisi A, Behboodi E, Melican DT, Pollock JS, Destrempes MM, Cammuso C, Williams JL, Nims SD, Porter CA, Midura P, Palacios MJ, Ayres SL, Denniston RS, Hayes ML, Ziomek CA, Meade HM, Godke RA, Gavin WG, Overstrom EW, Echelard Y. Production of goats by somatic cell nuclear transfer. Nat Biotechnol 1999; 17: 456-461.
7. Polejaeva IA, Chen SH, Vaught TD, Page RL, Mullins J, Ball S, Dai Y, Boone J, Walker S, Ayares DL, Colman A, Campbell KH. Cloned pigs produced by nuclear transfer from adult somatic cells. Nature 2000; 407: 86-90.
8. Loi P, Ptak G, Barboni B, Fulka J Jr, Cappai P, Clinton M. Genetic rescue of an endangered mammal by cross-species nuclear transfer using post-mortem somatic cells. Nat Biotechnol 2001; 19: 962-964.
9. Lanza RP, Cibelli JB, Diaz F, Moraes CT, Farin PW, Farin CE, Hammer CJ, West M, Damiani P. Cloning of an endangered species (*Bos gaurus*) using interspecies nuclear transfer. Cloning 2000; 2: 79-90.
10. Woods GL, White KL, Vanderwall DK, Li GP, Aston KI, Bunch TD, Meerdo LN, Pate BJ. A mule cloned from fetal cells by nuclear transfer. Science 2003; 301: 1063.
11. Zhu ZY, Sun YH. Embryonic and genetic manipulation in fish. Cell Res 2000; 10: 17-27.
12. Tung TC. Nuclear transplantation in teleosts. I. Hybrid fish from the nucleus of carp and the cytoplasm of crucian. Scientia (Peking) 1980; 14: 1244-1245.
13. Wakamatsu Y, Ju B, Pristyaznhyuk I, Niwa K, Ladygina T, Kinoshita M, Araki K, Ozato K. Fertile and diploid nuclear transplants derived from embryonic cells of a small laboratory fish, medaka (*Oryzias latipes*). Proc Natl Acad Sci U S A 2001; 98: 1071-1076.
14. Yan SY. Cloning in Fish—Nucleocytoplasmic Hybrids. Hong Kong: Educational and Cultural Press; 1998.
15. Lee KY, Huang H, Ju B, Yang Z, Lin S. Cloned zebrafish by nuclear transfer from long-term-cultured cells. Nat Biotechnol 2002; 20: 795-799.
16. Richardson MK, Allen SP, Wright GM, Raynaud A, Hanken J. Somite number and vertebrate evolution. Development 1998; 125: 151-160.
17. Wu HW. The Cyprinid Fishes of China. Shanghai: Scientific Technical Press; 1982: 412-433.
18. Nelson JS. Fishes of the World, 2nd ed. New York: Wiley-Interscience; 1984: 125.

19. Zhu ZY, Xu KS, Xie YF, Li GH, He L. A model of transgenic fish. Sci China (Ser B) 1989; 19: 147-155.

20. Fu C, Cui Y, Hung SSO, Zhu Z. Whole-body amino acid pattern of F_4 human growth hormone gene-transgenic red common carp (*Cyprinus carpio*) fed diets with different protein levels. Aquaculture 2000; 189: 287-292.

21. Kane DA, Kimmel CB. The zebrafish midblastula transition. Development 1993; 119: 447-456.

22. Zhang J, Sun X. The Selected Paper of Breeding in Jian Carp (*Cyprinus carpio* var. Jian). Beijing: Science Press; 1994: 54-61.

23. Waksmundzka M. Development of rat x mouse hybrid embryos produced by microsurgery. J Exp Zool 1994; 269: 551-559.

24. Dominko T, Mitalipova M, Haley B, Beyhan Z, Memili E, McKusick B, First NL. Bovine oocyte cytoplasm supports development of embryos produced by nuclear transfer of somatic cell nuclei from various mammalian species. Biol Reprod 1999; 60: 1496-1502.

25. Sun YH, Chen SP, Wang YP, Zhu ZY. The onset of foreign gene transcription in fish nuclear transferred embryos. Sci China (Ser C) 2000; 43: 597-605.

26. Evans MJ, Gurer C, Loike JD, Wilmut I, Schnieke AE, Schon EA. Mitochondrial DNA genotypes in nuclear transfer-derived cloned sheep. Nat Genet 1999; 23: 90-93.

27. Steinborn R, Schinogl P, Wells DN, Bergthaler A, Muller M, BremG. Coexistence of *Bos taurus* and *B. indicus* mitochondrial DNAs in nuclear transfer-derived somatic cattle clones. Genetics 2002; 162: 823-829.

28. Hiendleder S, Zakhartchenko V, Wenigerkind H, Reichenbach HD, Bruggerhoff K, Prelle K, Brem G, Stojkovic M, Wolf E. Heteroplasmy in bovine fetuses produced by intra- and inter-subspecific somatic cell nuclear transfer: neutral segregation of nuclear donor mitochondrial DNA in various tissues and evidence for recipient cow mitochondria in fetal blood. Biol Reprod 2003; 68: 159-166.

29. Enriquez JA, Fernandez-Silva P, Montoya J. Autonomous regulation in mammalian mitochondrial DNA transcription. Biol Chem 1999; 380: 737-747.

30. Kenyon L, Moraes CT. Expanding the functional human mitochondrial DNA database by the establishment of primate xenomitochondrial cybrids. Proc Natl Acad Sci U S A 1997; 94: 9131-9135.

31. Schnell S, Maini PK. Clock and induction model for somitogenesis. Dev Dyn 2000; 217: 415-420.

32. Saga Y, Takeda H. The making of the somite: molecular events in vertebrate segmentation. Nat Rev Genet 2001; 2: 835-845.

33. Stern CD, Vasiliauskas D. Clocked gene expression in somite formation. Bioessays 1998; 20: 528-531.

34. Morin-Kensicki EM, Melancon E, Eisen JS. Segmental relationship between somites and vertebral column in zebrafish. Development 2002; 129: 3851-3860.

35. Lemaire P, Gurdon JB. Vertebrate embryonic inductions. Bioessays 1994; 16: 617-620.

36. Pourquie O. Vertebrate somitogenesis. Annu Rev Cell Dev Biol 2001; 17: 311-350.

37. Wu B, Sun YH, Wang YP, Wang YW, Zhu ZY. Sequences of transgene insertion sites in transgenic F4 common carp. Transgenic Res 2004; 13: 95-96.

38. Melamed P, Gong Z, Fletcher G, Hew CL. The potential impact of modern biotechnology on fish aquaculture. Aquaculture 2002; 204: 255-269.

39. Casey JM, Myers RA. Near extinction of a large, widely distributed fish. Science 1998; 281: 690-692.

细胞质对跨属克隆鱼的影响

孙永华　陈尚萍　汪亚平　胡　炜　朱作言

中国科学院水生生物研究所，淡水生态与生物技术国家重点实验室，武汉 430072

摘　要　在之前的核移植研究中，已报道的多数为通过种内核移植而得到的克隆动物，其表型与核供体的表型非常一致。自克隆斑马鱼的报道之后，没有见到更多关于克隆鱼的研究报道。本研究中我们将转基因鲤的细胞核移植到去核的金鱼卵中，成功获得了 7 例克隆鱼。克隆鱼的核基因组完全来自于其核供体，即转基因鲤，然而来自于核供体的线粒体 DNA 则在发育过程中逐渐消失了。核移植胚的体节发育过程和体节数目则与受体金鱼的一致，且对克隆鱼的后期发育产生了持久的影响。这些结果表明，鱼类的卵细胞质不仅能支持来自于属间的亲缘关系较远的物种的细胞核的发育，还能调节核移植胚的发育。

Identification of Differentially Expressed Genes from the Cross-Subfamily Cloned Embryos Derived from Zebrafish Nuclei and Rare Minnow Enucleated Eggs

D. S. Pei[1,2] Y. H. Sun[1] S. P. Chen[1] Y. P. Wang[1] W. Hu[1] Z. Y. Zhu[1]

1 State Key Laboratory of Freshwater Ecology and Biotechnology, Institute of Hydrobiology, Chinese Academy of Sciences, Wuhan 430072

2 Group of Environmental Genomics, Institute of Hydrobiology, Chinese Academy of Sciences, Wuhan 430072

Abstract Cross-species nuclear transfer (NT) has been used to retain the genetic viability of a species near extinction. However, unlike intra-species NT, most embryos produced by cross-species NT were unable to develop to later stages due to incompatible nucleocytoplasmic interactions between the donor nuclei and the recipient cytoplasm from different species. To study the early nucleocytoplasmic interaction in cross-species NT, two laboratory fish species (zebrafish and rare minnow) from different subfamilies were used to generate cross-subfamily NT embryos in the present study. Suppression subtractive hybridization (SSH) was performed to screen out differentially expressed genes from the forward and reverse subtractive cDNA libraries. After dot blot and real-time PCR analysis, 80 of 500 randomly selective sequences were proven to be differentially expressed in the cloned embryos. Among them, 45 sequences shared high homology with 28 zebrafish known genes, and 35 sequences were corresponding to 22 novel expressed sequence tags (ESTs). Based on functional clustering and literature mining analysis, up- and down-regulated genes in the cross-subfamily cloned embryos were mostly relevant to transcription and translation initiation, cell cycle regulation, protein binding, etc. To our knowledge, this is the first report on the determination of genes involved in the early development of cross-species NT embryos of fish.

Keywords nuclear transfer; nuclear reprogramming; nucleo-cytoplasmic interaction; suppression subtractive hybridization; cross-subfamily cloned embryos

1. Introduction

Nuclear transfer (NT) between two fish species, i.e., cross-species cloning, provides a useful tool to study nucleocytoplasmic interaction [1,2]. A pioneering study on fish NT was carried out by Tung *et al.* [3] and extensive studies on fish cross-species NT were mainly conducted in *Cyprinid* [4,5]. In addition, NT in loach, medaka and zebrafish was also reported [6-8]. Recently, we generated cross-genus cloned fish derived from transgenic common carp nuclei and goldfish

文章发表于 *Theriogenology*, 2007, 68(9): 1282-1291

enucleated eggs and reported that the somitogenesis and vertebral number of the cloned fish were consistent to the egg-providing species, goldfish (*Carassius auratus*), instead of the donor cell species, common carp (*Cyprinus carpio*) [2]. Common carp and goldfish belong to different genera, the *Cyprinus* and the *Carasius*. We inferred that fish egg cytoplasm not only supported the development driven by transplanted nuclei from a distantly related species, it also affected developmental morphology of the NT individuals. Therefore, it is of great importance to compare gene expression patterns between the cross-species cloned fish and the donor nucleus providing fish to investigate differentially expressed genes.

NT between two laboratory fish species, rare minnow (*Gobiocypris rarus*) and zebrafish (*Danio rerio*), provides an ideal model to conduct the study of cross-species NT. Rare minnow and zebrafish belong to different subfamilies, the *Gobioninae* and the *Danioninae* [9-10]. Zebrafish is an important model for developmental and genetic studies, due to its short maturation cycle, high reproductive capacity, and transparent eggs, etc. [11-12]. Rare minnow, a special local species in China, not only shares the aforementioned advantages with zebrafish, but also has many unique traits for laboratory study, such as typical eurytherm and high adaptation [13], and sensitivity to toxins and viruses [14-15]. Such advantages enable rare minnow to be an excellent experimental fish [16].

In previous studies, cross-species NT, even between two closely related species, usually failed to produce viable offspring [17-18]. That was not unexpected, considering the incompatible interaction between the transferred nucleus and the recipient egg cytoplasm from different species, in addition to incomplete reprogramming of the transferred nucleus. To further explore essential mechanisms underlying cross-species cloning, we used two laboratory fish species (zebrafish and rare minnow) and used suppression subtractive hybridization (SSH) to study the transcriptional profiles of the cross-subfamily cloned embryos at the sphere stage.

2. Materials and Methods

2.1 Preparation of the cross-subfamily cloned embryos and control embryos

The cross-subfamily cloned embryos were generated by NT, as previously described [2], with the blastula nuclei of zebrafish and the enucleated eggs of rare minnow. Roscovitine (Rosco) was used to synchronize the cell cycle of donor blastula cells, as described by Gibbons *et al.* [19]. Three batches of the cloned embryos were used to monitor embryonic development. Both batches of the control embryos (zebrafish and rare minnow) were produced by IVF. All embryos were incubated in Holtfreter's solution at 28℃ and collected at the sphere stage [20]. The manipulations in the experiment adhered to Guidelines for Animal Use in Biomedical Research Laboratories (ILAR, 1996).

2.2 Evaluation of the success of NT with sequence characterized amplified region (SCAR) markers

Total DNA from a single embryo of the non-cloned embryos (zebrafish and rare minnow) and a cloned embryo at the sphere stage were prepared as described [21]. First, a RAPD (random amplification of polymorphic DNA) random primer S8 (Sangon, Shanghai, China; Table 1) was used to get comparative RAPD patterns of zebrafish and rare minnow, then we subcloned two unique fragments that resulted from the RAPD amplification into pGEM-T vector (Promega, Madison, WI, USA) for sequencing. Subsequently, the RAPD marker was converted successfully into SCAR markers, in terms of the sequences of two unique fragments (AY753569, AY753568) according to Paran and Michelmor [22]. The SCAR1 and SCAR2 (Table 1) successfully distinguished the genomic DNA between zebrafish and rare minnow, and gave unique patterns (548 and 500 bp amplicons, respectively). Herein, the successfully cloned embryos could be only amplified with SCAR1 primer, but not SCAR2 primers. The SCAR PCR was performed with 0.5 U of Taq DNA polymerase (BioAsia, Shanghai, China),10 pmol of SCAR1 or SCAR2 primers, and 50 ng of template DNA in a volume of 25 μL. The reaction process was 94℃ for 5 min, 30 cycles of 94℃, 30 s; 61℃, 30s; and 72℃, 60 s. All PCR products were separated by electrophoresis on 1.0% (*w*/*v*) agarose gel.

Table 1 Primers used to generate corss-subfamily NT embryors

Name	Sequence	Amplicon size (bp)	Usage
NP1	5-TCGAGCGGCCGCCCGGGCAGGT-3		SSH libraries construction, reference [24]
NP2R	5-AGCGTGGTCGCGGCCGAGGT-3		
S8	5-GTCCACACGG-3		RAPD patterns analysis
SCAR1	5-GTCCACACGGCAGGGAATAA-3 5-ACCTTGCCCCGTGTGGAC-3	548	SCAR cloning
SCAR2	5-TGACGATAGCCAGCGAGACC-3 5-TGTGTGATTGGACCCCGTCG-3	500	SCAR cloning
GAPDH	5-GTGTAGGCGTGGACTGTGGT-3 5-TGGGAGTCAACCAGGACAAATA-3	121	Expression analysis (E), Reference [23]
B11	5-GTAAGTCACAGCAGGTCACAGGA-3 5-GTGCTGGAAAACCGAAAAATG-3	110	E
B30	5-CGTGAGATTGCCTTTGATGCT-3 5-TTCAACATTCATCACTTCAGACCA-3	246	E
B75	5-AAGAAGTGCGTTTGGTGCCT-3 5-CAGGCGTCAGTGCCAAATC-3	163	E
B7	5-TCTTTTGAACGAGAAGCCACC-3 5-GCCGACCTGATAGAGGAATGG-3	100	E
B96	5-CCAGGAGTCCCACAAACAGAG-3 5-GCTGGTTGGAGGTGGAAGTAA-3	177	E
B70	5-GTTCGCTATGACCGAGAGATGA-3 5-AAGACAAAGAAAGCAGATGGTGG-3	117	E
B49	5-CACGCTGTTGGGGATTCTGT-3 5-TCACCGAAGAACCAGCCATT-3	130	E
B73	5-AAGGGTTCAGCGAGTCCGA-3 5-TTGAAGAGGCTGCTGAGGATG-3	100	E
K15	5-TCTTCGCTTGCTGAACCATCT-3 5-GGGTGGGAGGCACAGTAACA-3	218	E

Continued

Name	Sequence	Amplicon size (bp)	Usage
Kg3	5-CAGCCAAATGGACCAGCAA-3 5-GGAATGAAGGGATTGAGCACC-3	146	E
K68	5-GACAGGCGTTAGTTGGCTTTG-3 5-CATTCCCGCTGCTCCGT-3	95	E
Kg1	5-CAATACGACCGAAAGGGCAC-3 5-CGGGCAACACGGCATAGT-3	260	E
Kg4	5-TGTCCACCAGCCGAGTCC-3 5-AGTCAACACCTCCGAGAACTAACG-3	86	E
Kg6	5-CTTCACTCTGTCTTTGGCTTCG-3 5-CTACGGACTGGTGAACAAGCAA-3	87	E
Bg2	5-GAAACAGCGAAGGTGAAGTATGC-3 5-CGTGGAGGAAGGAAGGGATT-3	119	E
Bg4	5-AATCCGCCGCTGAAAGTCT-3 5-TGGGTGGCTCTGTCTCAAGTT-3	125	E
K41	5-TTATTGTATGAGCCCACCAC-3 5-CGAAATACCTTTACGCCAGTT-3	107	E
B59	5-CTTCTCCACCTACCCTCCTCT-3 5-CCAGCACCACCGATTTTCT-3	105	E
B72	5- GCACATTCATCGTTTCTCCC-3 5- TCATCTGGATTCTCAAAGGTCA-3	218	E
B94	5- TTCAGTTTCTGGGACTTTGTGC-3 5- ATGAGGGCGATGGGGATG-3	98	E
B125	5- CAGTAACTGCAAAAGACA-3 5- GAGGCTAAGTAAATGAAGAC-3	75	E
B35	5- TGGTGTAACAGCAGGGTA-3 5- GTTAGCAGAGGCATTTGT-3	68	E
K18	5- CTACCAGAACCAGCTTAAA-3 5- CTTCCTCGATCACTTCAA-3	107	E
K20	5- CAAAGTCCTTGAACTGCGATTA-3 5- TTTTCTCCAAGCCTTTTGTTAG-3	143	E
K32	5- GGACGGGATGCTGTATTAGTG-3 5- GGCGGACGTGGTTTGTTT-3	229	E
Kg2	5- GCCAAACGATGCCACGAT-3 5- GGGGCTTCATTCCATCAACA-3	164	E
K34	5-GTTGAGCCGAGTCTGATTCTT-3 5-TCCTCCACTTATGCCACGA-3	78	E
Kg5	5-TGGCGGAGATGCTGAAAG-3 5-GCAAAGTGAAGTCCAGGGTG-3	201	E

2.3 Construction of SSH cDNA libraries

Total RNA of two kinds of embryos (50 of the cloned embryos and 50 of zebrafish embryos) was extracted using SV^{TM} total RNA kit (Promega). The quality of sample total RNA was vital to the successful construction of the SSH cDNA libraries. Thus, the quantification and integrity of total RNA were strictly examined by spectrophotometer and 1.0% agarose electrophoresis. Poly(A) + RNA was isolated with the PolyATtract mRNA Isolation System III (Promega) and the purity of poly(A) RNA was verified by spectrophotometer. Complementary DNA (cDNA) was synthesized from poly(A) RNA using the SMART cDNA synthesis kit (Clontech, Mountain View, CA, USA). Double strand cDNA of the cloned embryos and zebrafish embryos were used as the driver and tester, respectively. The SSH was carried out using a PCR-Select™ cDNA subtraction kit (Clontech), according to the manufacturer's protocol. After secondary hybridization, subtraction efficiency was examined by performing 15, 19, 23, and 27 rounds of PCR using GAPDH primers (Table 1) in the subtracted samples and unsubtracted samples [23]. All PCR reactions were carried out using the Advantage Polymerase Mix (Clontech) on the GeneAmp PCR System-9700 (Applied Biosystems, Foster City, CA, USA). After the secondary PCR, the products were cloned into pGEM-T Vector (Promega) to generate the SSH cDNA library.

2.4 Screening of differentially expressed sequences by dot blot analysis

Bacterial colony PCR was performed on the GeneAmp PCR System-9700 with 1 μL of bacteria culture in a total volume of 25 μL (400 nM each of NP1 and NP2R primers [24], 0.2 mM of each dNTP, 0.5 μL Taq polymerase Mix, 1×PCR buffer). Then, the forward and reverse subtracted cDNA were digested with *R*saI to remove the SSH adapters. The adapter-free cDNA was labeled by dig high prime labeling and detection starter kit (Roche Molecular Biochemicals, Mannheim, Germany) by a random priming method. An aliquot (1 μL) of each positive PCR product (10 ng) was dropped on to a nylon membrane (Amersham, Braunschweig, Germany) in duplicate, and fixed by irradiation under a UV transilluminator (Vilber Lourmat, Marne-La-Valle′e, France) for 8 min. Detailed operation for hybridization was carried out according to the manufacturer's protocol.

2.5 Sequencing analysis of differentially expressed sequences

Following dot blot hybridization, the differentially expressed sequences were sequenced using SP6 and T7 primers by Sagon Corporation (Sangon). Homology searches against GenBank databases were done using the BLAST server (http://www.ncbi.nlm.nih.gov/BLAST/) at the National Center for Biotechnology Information (NCBI).

2.6 Confirmation of differentially expressed genes by real-time PCR

Total RNA of both zebrafish embryos and the cloned embryos was isolated and normalized on the basis of GAPDH expression. The relative quantification with real-time RT-PCR was done as described by the manufacturer (Applied Biosystems) with slight adaptation. In brief, the samples were placed in 96-well plates and amplified in ABI PRISM 7000 sequence detection system (Applied Biosystems). Each PCR proceeded in 30 μL of SYBR Green PCR buffer (Applied Biosystems) containing 400 nM each of forward primers and reverse primers, 1 U AmpliTaq Gold DNA polymerase, 2.5 mM dNTPs, 0.5 U AmpErase UNG, 3 mM $MgCl_2$, and 2 μL of the diluted cDNA sample. Amplification conditions were 2 min at 50℃, 10 min at 95℃, 40 cycles of 30 s at 95℃, and 60 s at 60℃. The expression level was calculated using

$2^{[-\text{delta delta C(T)}]}$ method with correction for different amplification efficiencies [25]; GAPDH was used as endogenous reference in all quantitative real-time PCR as described [23]. Moreover, every reaction was done in triplicate and the means of three independent experiments between the cloned embryos and the control embryos were evaluated by a Student's *t*-test ($P < 0.01$).

2.7 Functional clustering and analysis of differentially expressed genes

The differentially expressed genes proven by expression analysis were nominated according to the definition of ZFIN (http://zfin.org) and classified according to the definition of Gene Ontology (http://www.geneontology.org/) at the aspects of biological function, molecular function and cellular component. All abstracts harboring the gene symbols were downloaded in text format from Pubmed on NCBI. Data mining was performed using the programs of WordStat 4.0 (Provalis Research, Montreal, QC, Canada), SimStat 4.0 (Provalis Research), Cluster and Tree View [26] as described [27-29]. In brief, after all the downloaded texts were transformed to the special format by WordStat 4.0 software, a statistical analysis was performed to get the output of the frequency percentum of each keyword targeting each bait gene by SimStat 4.0 software. Clusting analysis was processed on the frequency values by Cluster software. The relationship among the target genes and the keywords was shaped by Tree View software.

3. Results

3.1 Development of the cloned embryos and the efficiency of cross-subfamily NT

Regarding development of cloned embryos, the percentage of viable embryos gradually decreased from 4-cell stage to 50% epiboly. A dramatic decrease occurred between sphere stage and 50% epiboly; only ~3% of the cloned embryos could reach the 50% epiboly, although ~40% developed to the sphere stage (Table 2). Total DNA from three kinds of single embryos (zebrafish, rare minnow, and the cloned embryo) at sphere stage was used as templates for PCR reaction with SCAR1 or SCAR2 primers, respectively. The SCAR1 primers gave a unique amplicon of 548 bp in zebrafish, whereas SCAR2 primers gave an amplicon of 500 bp in rare minnow. In 10 samples that were randomly selected from the cloned embryos, as expected, a 548 bp fragment was amplified with SCAR1 primers, but a 500 bp fragment did not appear with SCAR2 primers (Fig. 1), indicating the success of cross-subfamily NT.

3.2 Construction of SSH cDNA libraries and sequence analysis

The band intensity of 28S rRNA was approximately twice that of 18S rRNA, indicating that total RNA was of high quality (Fig. 2A). Since 27 cycles of amplification of the subtracted

Table 2 Development Proportions of the cross-subfamily cloned embryos embryos in three transplant manipulations

Number of embryos	4-Cell stage	64-Cell stage	High stage	Sphere stage	50%-Epiboly
80	70([a] 87.5%)	60 (75%)	50 (62.5%)	31 (39%)	1 (1.25%)
100	80 (80%)	67 (67%)	55 (55%)	40(40%)	3(3%)
120	98 (82%)	81 (68%)	70 (59%)	50 (42%)	4 (3.2%)

a. Numbers in parentheses represent the percentage of the total number of transplants that produced developing embryos

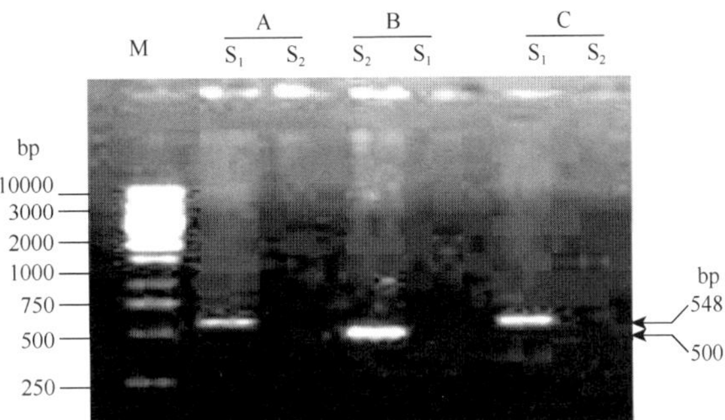

Fig. 1 Evaluation of the efficiency of cross-subfamily NT with SCAR markers. The PCR was performed with genomic DNA from single embryos of three kinds (zebrafish, rare minnow, and cloned embryo) at sphere stage. SCAR1 were special primers for zebrafish, with SCAR2 for rare minnow, respectively. (A) Zebrafish genomic DNA was amplified using SCAR1 and SCAR2, respectively; (B) rare minnow genomic DNA was amplified using SCAR2 and SCAR1, respectively; (C) the genomic DNA of cloned embryo was amplified using SCAR1 and SCAR2, respectively. S1: SCAR1; S2: SCAR2; M: 1 kb DNA Marker (Fermentas Inc., Glen Burnie, MD, USA)

samples produced a similar intensity of band as 19 cycles of the unsubtracted samples, the subtraction efficiency of SSH efficiency was at least 2^7 folds (Fig. 2B). For further analysis of the subtracted genes, 500 clones were randomly selected and bacterial colony PCR were performed (Fig. 3A). The amplified cDNA was dotted onto nylon membranes and hybridized separately with digoxigenin-labeled forward or reverse subtracted cDNA. Of 500 clones, 80 gave different signal intensities when probed with the forward and reverse subtracted cDNA probes (Fig. 3B). All 80 clones were sequenced and subjected to BLAST analysis against the GenBank databases. Of the 80 differentially expressed sequences, 45 corresponded to 28 different genes, which shared 73-100% homology to known zebrafish genes, e.g. *ef1a*, *dazl*, *cth1* (full names of the differentially expressed genes are shown in Tables 3 and 4), etc. The other 35 sequences corresponded to 22 novel ESTs. These 50 sequences were all submitted to the GenBank database (Table 7).

3.3 Real-time RT-PCR confirmation

To confirm the accuracy of dot blot screening, the above-mentioned 28 genes were assayed in the cloned embryos by real-time RT-PCR. Thirteen genes were upregulated (Table 3) and 15 genes were down-regulated in the cloned embryos (Table 4). Interestingly, *mt-co1*, a

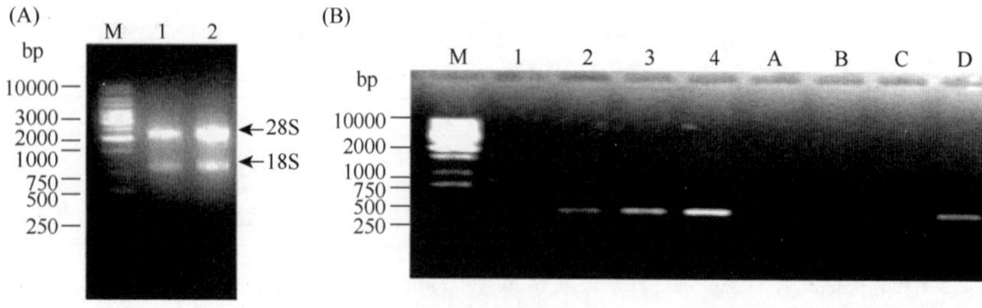

Fig. 2 The quality and efficiency of suppression subtractive hybridization. (A) The integrity of total RNA was detected by agarose gelelectrophoresis. M: 1 kb DNA Ladder Marker; 1: total RNA of cloned embryos; 2: total RNA of zebrafish embryos; 28S and 18S rRNA were shown with arrows. (B) Reduction of GAPDH abundance was tested in the subtractive cDNA library. PCR analysis was performed on unsubtracted (Lanes 1-4) and subtracted (Lanes A-D) samples. M: 1 kb DNA Marker; Lane 1 and Lane A (15 cycles); Lane 2 and Lane B (19 cycles); Lane 3 and Lane C (23 cycles); Lane 4 and Lane D (27 cycles)

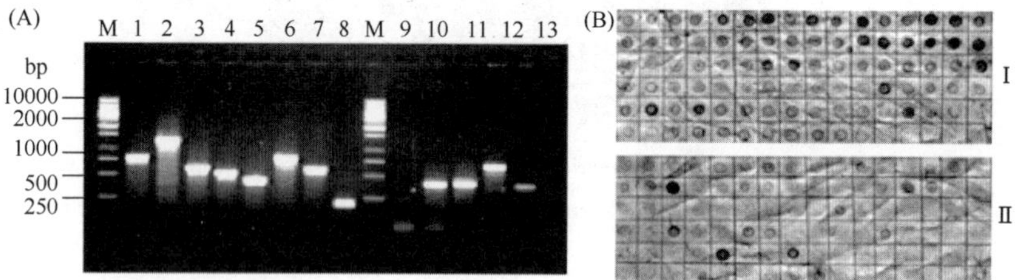

Fig. 3 Screening of the subtracted library. (A) The subtracted library was screened by PCR. M was 1 kb DNA Ladder Marker; 1-13 were PCR results by NP1/NP2R primers. (B) The subtracted library was screened by dot blot. (I) The PCR fragments of zebrafish embryos were hybridized by forward-subtracted cDNA probes; (II) the PCR fragments of the cloned embryos were hybridized by forward-subtracted cDNA probes

Table 3 Up-regulated genes differentially expressed in the cloned embryos (and supporting statistics)

Clones/accession No.	[a] Gene symbol	[a] Gene name	Identities	Blast database No.	Up folds
K15/ CV576861	*cobra1*	*Cofactor of BRCA1*	268/307 (87%)	XM_702544	15
B35/ CV576854	*zgc:73179*	*zgc:73179*	157/158 (99%)	BC059529	20
Kg4/ CV576871	*psmb7*	*Proteasome subunit, beta type, 7*	422/468 (90%)	AF155581	200
Kg3/ CV576872	*gcdh*	*Glutaryl-coenzyme A dehydrogenase*	422/481 (87%)	BC050168	10
Kg6/ CV576869	*rtn4a*	*Reticulon 4a*	145/164 (88%)	AY164754	47.6
kg1/ CV576873	*zgc:56547*	*zgc:56547*	158/159 (99%)	BC059529	300
K41/ DY544186	*mt-co1*	*Cytochrome c oxidase I, mitochondrial*	263/302 (87%)	AY116187	19
K18/DY544178	*ran*	*ras-related nuclear protein*	129/138 (93%)	BC050517	10.5
K20/DY544180	*cth1*	*cysteine-three-histidine*	113/129 (87%)	AL928799	16
K32/ DY544181	*tfa*	*Transferrin-a*	139/158 (87%)	XM_687953	21
K34/DY544182	*pus7*	*Pseudouridylate synthase 7 homolog*	115/123 (93%)	BC065861	12
Kg2/CV576850	*zgc:112166*	*zgc:112166*	159/175 (90%)	BC093237	8.9
Kg5/ CV576870	*ssra13*	*Similar to stimulated by retinoic acid 13*	141/162 (87%)	XM_682342	14.6

a. Gene symbol and gene name were nominated according to ZFIN (http://zfin.org) standard

Table 4 Down-regulated genes differentially expressed in the cloned embryos (and supporting statistics)

Clones/ Accession No. [a]	Gene symbol	[a] Gene name	Identities	Blast database No.	Down folds
B75/CV576863	*snrpb2*	*Small nuclear ribonucleoprotein polypeptide B2*	57/63 (90%)	XM419331	6.2
B30/CV576855	*gmnn*		330/339 (97%)	BC055552	89.9
B96/CV576853	*sepw1*	*Selenoprotein W, 1*	394/397 (99%)	AY216582	54.6
B11/CV576856	*dazl*	*daz-like gene*	232/317 (73%)	BC076423	80
B70/CV576876	*hmgb1*	*High-mobility group box 1*	185/224 (82%)	NM204902	12
B7/CV576862	*rbmx*	*RNA binding motif protein, X-linked*	285/285 (100%)	AJ717349	43.6
B49/CV576865	*zgc:110251*	*zgc:110251*	174/193 (90%)	BX000999	52.7
B73/CV576864	*bzw1l*	*Basic leucine zipper and W2 domains 1*	424/427 (99%)	NM213491	169
Bg2/CV576866	*rcc1*	*Regulator of chromosome condensation 1*	346/347 (99%)	BC066721	8.2
Kbg4/CV576874	*eif4a*	*Eukaryotic translation initiation factor 4-alpha*	337/338 (99%)	BC042330	55.6
B94/ DY544174	*zp2*	*Zona pellucida glycoprotein 2*	679/684 (99%)	BC097083	14
B125/ DY544176	*smc1a*	*Structural maintenance of chromosomes 1A*	146/146 (100%)	AY648730	26
B72/ DY544171	*cirbp*	*Cold inducible RNA binding protein*	348/349 (99%)	BC095030	12.5
B59/DY544172	*ef1a*	*Elongation factor 1-alpha*	694/699 (99%)	BC064291	28
K68/CV576858	*eif2s2*	*Eukaryotic translation initiation factor 2, subunit 2 beta*	244/268 (91%)	NM_212675.	50

a. Gene symbol and gene name were nominated according to ZFIN (http://zfin.org) standard

mitochondrial gene, was 19-fold up-regulated in the cloned embryos. The changes of transcription levels of *psmb7* and *zgc*:56547 exceeded 200-fold.

3.4 Functional clustering analysis and data mining

Twenty-eight genes proved by expression analysis were classified in terms of Gene Ontology (Tables 5 and 6). Interestingly, in 28 genes, the cellular component of eight genes was the nucleus, the molecular function of four genes was protein binding, and the biological function of eight genes was related to the regulation of translation and/or transcription. Genes without specific data available were denoted as ND.

For literature mining, there were few articles related to six genes (*zgc:73179*, *zgc:56547*, *zgc:112166*, *zgc:110251*, *ssra13*, and *bzw1l*). However, a total of 11416 abstracts (in text format) relevant to the other 22 differentially expressed genes were downloaded. Literature mining analysis showed that the keywords and genes were clustered in terms of function (Fig. 4). Therein, *dazl*, *zp2*, *gmnn*, and *cobra1* were highlighted on protein and binding keywords, in agreement with their molecular functions in Tables 5 and 6. Interestingly, five genes (*eif4a*, *cirbp*, *cth1*, *dazl*, and *rbmx*) link protein binding and RNA binding together. Many genes, such as *eif4a*, *eif2s2*, and *ef1a* were highly associated with the keyword translation, whereas *ran* and *rcc1* clustered together in the keywords of GTP, exchange, protein and binding (suggesting potential interaction between these two genes) (Fig. 4).

Table 5 Classification of up-regulated genes in cloned embryos based on gene ontology

Genes	Molecular Function	Biological Process	Cellular component
cobra1	Protein binding	Negative regulation of transcription	Nucleus
zgc:73179	Calcium channel activity	Calcium ion transport	ND[a]
psmb7	Endopeptidase activity	Ubiquitin-dependent protein catabolic process	cytosol
gcdh	FAD binding	Electron transport	ND[a]
rtn4a	ND[a]	ND[a]	Endoplasmic reticulum
zgc:56547	Transcription regulator activity	Regulation of transcription	Nucleus
mt-co1	Copper ion binding	Response to cadmium ion	Integral to membrane
ran	GTP binding	Intracellular protein transport	Exosome
cth1	Nucleic acid binding zinc ion binding	Cellulariron ion homeostasis	Nucleus
tfa	Ferric iron binding transferase activity	Iron ion homeostasis and transport	Extracellular region
pus7	Pseudouridylate synthase activity	tRNA processing	Nucleus
zgc:112166	Hydrolase activity	Protein amino acid dephosphorylation	ND[a]
loc559042	Signal transduction	Similar to stimulated by retinoic acid 13	ND[a]

a. No biological data available

Table 6 Classification of down-regulated genes in cloned embryos based on gene ontology

Genes	Molecular function	Biological process	Cellular component
snrpb2	Nucleic acid binding	mRNA processing	Nucleus
gmnn	Protein binding	Negative regulation of DNA replication and regulation of progression through cell cycle	Nucleus
sepw1	selenium binding, oxidoreductase activity	Cell redox homeostasis	Cytoplasm
dazl	RNA binding nucleic acid binding	Cell differentiation,oogenesis,spermatogenesis	Cytoplasm
hmgb1	DNA binding	Regulation of transcription, DNA-dependent	Chromatin
rbmx	Nucleic acid binding	ND[a]	Nucleus
zgc:110251	Iron ion binding, lipoxygenase activity	Electron transport	ND[a]
bzw1l	ND[a]	Regulation of transcription Transcription termination	ND[a]
rcc1	Chromatin binding and Ran GTPase binding	Regulation of G1/S transition of mitotic cell cycle	Nuclear chromatin
eif4a	ATP-dependent helicase activity	Translational initiation	Nucleus and cytosol
zp2	Coreceptor activity and protein binding	Binding of sperm to zona pellucida	Extracellular space
smc1a	ATP binding,ATPase activity,protein binding	Chromosome organization and biogenesis	Chromosome
cirbp	Nucleic acid binding	Response to cold	ND[a]
ef1a	GTP binding	Translational elongation	Cytoplasm
eif2s2	Translation initiation factor activity	Translational initiation	Ribosome

a. No biological data available

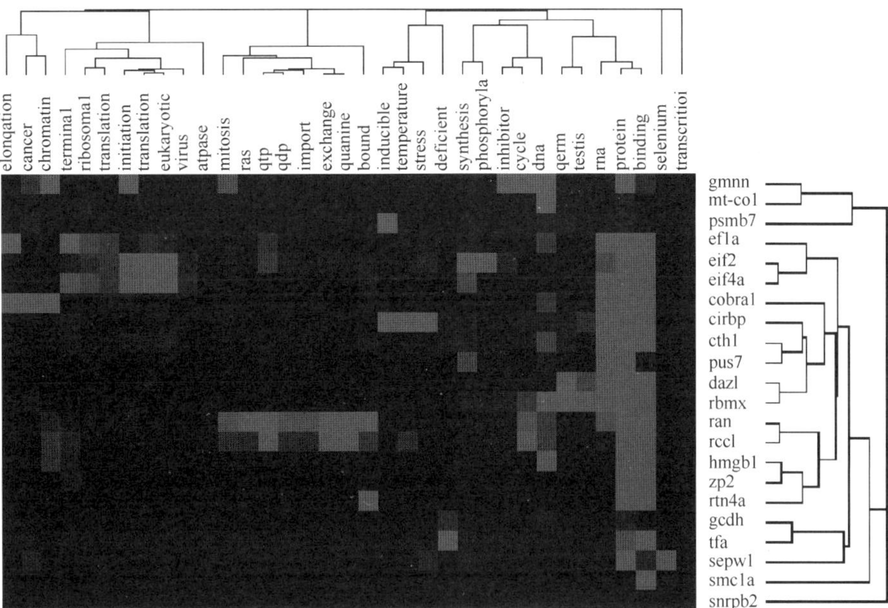

Fig. 4 The gene expression profiling using literature mining. The tree on the right represented the cluster of genes, whereas the tree on the top represented the cluster of keywords. The correlation grade between keywords and genes was reflected by the brightness; the brighter the pane, the higher the frequency of keywords

4. Discussion

Suppression subtractive hybridization (SSH) has been widely used in the screening of differentially expressed genes in developmental studies [30-31]. In the present study, this technique was used to study the differential transcript profiles in the fish cross-subfamily cloned embryos by comparison with normal embryos. The detection of the differentially expressed genes will assist to elucidate molecular mechanisms associated with cross-species NT. Obviously, those identified genes might be related to the interaction between donor nucleus and recipient egg cytoplasm and/or the reprogramming of donor nucleus. Regarding the limitations of SSH technique, we did not compare three batches of samples (cross-species cloned embryos, intra-species cloned embryos, and non-cloned embryos) in a single system; this could have enabled us to distinguish genes involved in nucleo-cytoplasmic interaction and nuclear reprogramming. For the sampling of embryos, we chose embryos at the sphere stage for SSH analysis, since previous studies have shown that many essential genes for embryonic patterning and organizer formation were expressed just following or before the sphere stage in zebrafish [32-33], and more importantly, only a few of the cloned embryos of sphere stage could develop to 50% epiboly, indicating that the failure of proper embryonic patterning of the cloned embryos happened during this stage.

Among the identified genes, three translation factor genes (*ef1a*, *eif4a*, and *eif2s2*) were down-regulated in the cloned embryos. It was reported that *ef1a*, *eif4a*, and *eif2s2* played a key regulatory role in translation initiation [34-36]. Down-regulation of these three genes in the present study indicated that translation failure occurred in cloned embryos. In addition, some genes related to chromatin organization and cell cycle regulation, *rcc1* and *smc1a* were down-regulated in the cloned embryos, in agreement with a previous study in somatic cell NT embryos [37-39]. Some other genes related to the regulation of cell cycle were also differentially expressed in the cloned embryos, e.g. *gmnn*; it was noteworthy that *gmnn* deficiency caused a Chk1-dependent G2 arrest immediately after the midblastula transition in Xenopus [40]. Furthermore, it was recently determined (using the SSH method) that *gmnn* was also differentially expressed at the early implantation stage in the rhesus monkey [31]. Therefore, *gmnn* may be important in the development of cloned embryos. In addition, we found that *mt-co1*, a mitochondrial gene, was up-regulated in the cloned embryos [41], suggesting that the aberrant regulation of mitochondrial genes was also involved in the incomplete reprogramming and/or incompatible nucleocytoplasmic interaction in cross-species cloned embryos. Interestingly, a novel gene, K31, was identified as an up-regulated gene in the cloned embryos using RACE method. Over-expression of K31 gene can cause epithelioma papulosum cyprinid (EPC) cells to die in the cell culture, consistent with the inefficient reprogramming events that occurred in cloned embryos (unpublished data).

In summary, we utilized two laboratory fish species, zebrafish and rare minnow, to establish a study model of cross-species NT; using an SSH approach, we identified 50 genes which were differentially expressed in the cloned embryos. Therein, down-regulation of translation factors (*ef1a*, *eif4a*, and *eif2s2*) indicated that translation failure could occur in cloned embryos. Some genes related to chromatin organization, cell cycle regulation and mitochondria function might be involved in the incomplete reprogramming. Further, partial novel genes such as K31 may participate in incompatible nucleo-cytoplasmic interaction in the cloned embryos. We are in the process of conducting a functional study of these genes to provide more molecular evidence for this event.

Table 7 The ESTs from the present study that were submitted to Genbank

User_id	Genbank no.	User_id	Genbank no.
EST00001	CV576850	EST00026	CV576875
EST00002	CV576851	EST00027	CV576876
EST00003	CV576852	EST00028	CV576877
EST00004	CV576853	EST00029	CV576878
EST00005	CV576854	EST00030	CV576879
EST00006	CV576855	EST00031	CV576880
EST00007	CV576856	EST00032	CV576881
EST00008	CV576857	EST00033	DY544171
EST00009	CV576858	EST00034	DY544172

Continued

User_id	Genbank no.	User_id	Genbank no.
EST00010	CV576859	EST00035	DY544173
EST00011	CV576860	EST00036	DY544174
EST00012	CV576861	EST00037	DY544175
EST00013	CV576862	EST00038	DY544176
EST00014	CV576863	EST00039	DY544177
EST00015	CV576864	EST00040	DY544178
EST00016	CV576865	EST00041	DY544179
EST00017	CV576866	EST00042	DY544180
EST00018	CV576867	EST00043	DY544181
EST00019	CV576868	EST00044	DY544182
EST00020	CV576869	EST00045	DY544183
EST00021	CV576870	EST00046	DY544184
EST00022	CV576871	EST00047	DY544185
EST00023	CV576872	EST00048	DY544186
EST00024	CV576873	EST00049	DY544188
EST00025	CV576874	EST00050	DY544187

Acknowledgements This work was supported by the State Key Fundamental Research of China (Grant No. 2004CB117406) and the National Natural Science Foundation of China (Grant Nos. 90208024 & 30623001). In addition, the authors are grateful to Ming Li for supplying experimental materials.

References

[1] Zhu ZY, Sun YH. Embryonic and genetic manipulation in fish. Cell Res 2000; 10: 17-27.

[2] Sun YH, Chen SP, Wang YP, Hu W, Zhu ZY. Cytoplasmic impact on cross-genus cloned fish derived from transgenic common carp (*Cyprinus carpio*) nuclei and goldfish (*Carassius auratus*) enucleated eggs. Biol Reprod 2005; 72: 510-5.

[3] Tung TC, Wu SC, Tung YYF, Yan SY, Tu M, Lu TY. Nuclear transplantation in fish. Science Bulletin Academia Sinica (in Chinese) 1963; 7: 60-1.

[4] Yan SY. The nucleo-cytoplasmic interaction as revealed by nuclear transplantation in fish. In: Malacinski GM, editor. Cytoplasmic organization systems: a primer in developmental biology. New York: McGraw-Hill; 1989. p. 61-81.

[5] Yan SY. Cloning in fish: nucleocytoplasmic hybrids, Rev. ed., H.K.: Educational and Cultural Press; 1998.

[6] Gasaryan KG, Hung NM, Neyfakh AA, Ivanenkov VV. Nuclear transplantation in teleost *Misgurnus fossilis* L. Nature 1979; 280: 585-7.

[7] Niwa K, Ladygina T, Kinoshita M, Ozato K, Wakamatsu Y. Transplantation of blastula nuclei to non-enucleated eggs in the medaka, *Oryzias latipes*. Dev Growth Differ 1999; 41: 163-72.

[8] Lee KY, Huang H, Ju B, Yang Z, Lin S. Cloned zebrafish by NT from long-term-cultured cells. Nat Biotechnol 2002; 20: 795-9.

[9] He S, Liu H, Chen Y, Kuwahara M, Nakajima T, Zhong Y. Molecular phylogenetic relationships of Eastern Asian *Cyprinidae* (*pisces*: *cypriniformes*) inferred from cytochrome *b* sequences. Sci China C Life Sci 2004;47:130-8.

[10] Pei DS, Sun YH, Chen SP, Wang YP, Zhu ZY. Cloning and characterization of cytochrome *c* oxidase subunit I

(COXI) in *Gobiocypris rarus*. DNA Seq 2007; 18: 1-8.

[11] Grunwald DJ, Eisen JS. Headwaters of the zebrafish—emergence of a new model vertebrate. Nat Rev Genet 2002; 3: 717-24.

[12] Key B, Devine CA. Zebrafish as an experimental model: strategies for developmental and molecular neurobiology studies. Methods Cell Sci 2003; 25: 1-6.

[13] Wang J. Acute effects of high concentration of dissolved free carbon dioxide and low dissolved oxygen on rare minnow. Acta Hydrobiol Sinica (in Chinese) 1995; 19: 84-8.

[14] Qun-Fang Z, Gui-Bin J, Ji-Yan L. Effects of sublethal levels of tributyltin chloride in a new toxicity test organism: the Chinese rare minnow (*Gobiocypris rarus*). Arch Environ Contam Toxicol 2002; 42: 332-7.

[15] Zhong J, Wang Y, Zhu Z. Introduction of the human lactoferrin gene into grass carp (*Ctenopharyngodon idellus*) to increase resistance against GCH virus. Aquaculture 2002; 214: 93-101.

[16] Wang J, Cao W. *Gobiocypris rarus* and fishes as laboratory animals. Trans Chin Ichthyol Soc 1997; 6: 144-52.

[17] Yan SY, Tu M, Yang HY, Mao ZG, Zhao ZY, Fu LJ, *et al*. Developmental incompatibility between cell nucleus and cyto- plasm as revealed by nuclear transplantation experiments in teleost of different families and orders. Int J Dev Biol 1990; 34: 255-66.

[18] Loi P, Ptak G, Barboni B, Fulka Jr J, Cappai P, Clinton M. Genetic rescue of an endangered mammal by cross-species nuclear transfer using post-mortem somatic cells. Nat Biotechnol 2001; 19: 962-4.

[19] Gibbons J, Arat S, Rzucidlo J, Miyoshi K, Waltenburg R, Respess D, *et al*. Enhanced survivability of cloned calves derived from roscovitine-treated adult somatic cells. Biol Reprod 2002; 66: 895-900.

[20] Kimmel CB, Ballard WW, Kimmel SR, Ullmann B, Schilling TF. Stages of embryonic development of the zebrafish. Dev Dyn 1995; 203: 253-310.

[21] Blin N, Stafford DW. A general method for isolation of high molecular weight DNA from eukaryotes. Nucleic Acids Res 1976; 3: 2303-8.

[22] Paran I, Michelmore RW. Development of reliable PCR-based markers linked to downy mildew resistance genes in lettuce. TAG Theor Appl Genet 1993; 85: 985-93.

[23] Pei DS, Sun YH, Chen SP, Wang YP, Hu W, Zhu ZY. Zebrafish GAPDH can be used as a reference gene for expression analysis in cross-subfamily cloned embryos. Anal Biochem 2007; 363: 291-3.

[24] Rebrikov DV, Britanova OV, Gurskaya NG, Lukyanov KA, Tarabykin VS, Lukyanov SA. Mirror orientation selection (MOS): a method for eliminating false positive clones from libraries generated by suppression subtractive hybridization. Nucleic Acids Res 2000; 28: E90.

[25] Livak KJ, Schmittgen TD. Analysis of relative gene expression data using real-time quantitative PCR and the $2^{(-\text{Delta Delta C(T)})}$ method. Methods 2001; 25: 402-8.

[26] Eisen MB, Spellman PT, Brown PO, Botstein D. Cluster analysis and display of genome-wide expression patterns. Proc Natl Acad Sci USA 1998; 95: 14863-8.

[27] Schulze A, Downward J. Analysis of gene expression by microarrays: cell biologist's gold mine or minefield? J Cell Sci 2000; 113 (Pt 23): 4151-6.

[28] Masys DR, Welsh JB, Lynn Fink J, Gribskov M, Klacansky I, Corbeil J. Use of keyword hierarchies to interpret gene expression patterns. Bioinformatics 2001; 17: 319-26.

[29] Semeiks JR, Grate LR, Mian IS. Text-based analysis of genes, proteins, aging, and cancer. Mech Ageing Dev 2005; 126: 193-208.

[30] Diatchenko L, Lau YC, Campbell AP, Chenchik A, Mooadam F, Huang B, *et al*. Suppression subtractive hybridization: a method for generating differentially regulated or tissue-specific cDNA probes. Proc Natl Acad Sci USA 1996; 93: 6025-30.

[31] Sun XY, Li FX, Li J, Tan YF, Piao YS, Tang S, *et al*. Determination of genes involved in the early process of embryonic implantation in rhesus monkey (*Macaca mulatta*) by suppression subtractive hybridization. Biol Reprod 2004; 70: 1365-73.

[32] Kane DA, Kimmel CB. The zebrafish midblastula transition. Development 1993; 119: 447-56.

[33] Schulte-Merker S, Hammerschmidt M, Beuchle D, Cho KW, De Robertis EM, Nusslein-Volhard C. Expression of zebrafish goosecoid and no tail gene products in wild-type and mutant no tail embryos. Development 1994; 120: 843-52.

[34] Negrutskii BS, El'skaya AV. Eukaryotic translation elongation factor 1 alpha: structure, expression, functions, and possible role in aminoacyl-tRNA channeling. Prog Nucleic Acid Res Mol Biol 1998; 60: 47-78.

[35] Svitkin YV, Pause A, Haghighat A, Pyronnet S, Witherell G, Belsham GJ, *et al*. The requirement for eukaryotic initiation factor 4A (eIF4A) in translation is in direct proportion to the degree of mRNA 5′ secondary structure. RNA 2001; 7: 382-94.

[36] Proud CG. Regulation of eukaryotic initiation factor eIF2B. Prog Mol Subcell Biol 2001; 26: 95-114.

[37] Moreira PN, Robl JM, Collas P. Architectural defects in pronuclei of mouse nuclear transplant embryos. J Cell Sci 2003; 116: 3713-20.

[38] Sullivan EJ, Kasinathan S, Kasinathan P, Robl JM, Collas P. Cloned calves from chromatin remodeled *in vitro*. Biol Reprod 2004; 70: 146-53.

[39] Nolen LD, Gao S, Han Z, Mann MR, Gie Chung Y, Otte AP, *et al*. X chromosome reactivation and regulation in cloned embryos. Dev Biol 2005; 279: 525-40.

[40] McGarry TJ. Geminin deficiency causes a Chk1-dependent G2 arrest in Xenopus. Mol Biol Cell 2002; 13: 3662-71.

[41] Pei DS, Sun YH, Chen SP, Wang YL, Zhu ZY. Dramatic overexpression of cytochrome *C* oxidase subunit I gene (COXI) in fish cross-subfamily cloned embryos. Dev Growth Differ; in press.

斑马鱼与稀有鮈鲫跨亚科克隆胚胎中差异表达基因的鉴定

裴得胜[1,2]　孙永华[1]　陈尚萍[1]　汪亚平[1]　胡　炜[1]　朱作言[1]

1 中国科学院水生生物研究所，淡水生态与生物技术国家重点实验室，武汉 430072

2 中国科学院水生生物研究所，环境基因组学科组，武汉 430072

摘　要　种间细胞核移植被认为是维持濒临灭绝物种遗传存在的一种方式。然而，不像种内克隆，来自不同物种的供体细胞核和受体细胞核存在互相作用和不相容性，种间的核移植相对种内核移植困难很多，大多数情况下胚胎无法发育到后期。为研究在早期跨亚科克隆胚胎的核质互作机理，我们采用属于不同亚科的斑马鱼和稀有鮈鲫进行跨亚科克隆研究。本实验使用抑制性消减杂交的方法，从差减 cDNA 文库中筛选得到差异表达的基因。采用斑点杂交和实时荧光定量 PCR 的方法，我们从 500 个随机选取的序列中筛选出了 80 个在克隆胚中差异表达的序列。其中，45 个差异基因片段序列与斑马鱼中 28 个已知基因高度同源，35 条序列对应于 22 个全新的基因表达标签。通过基因功能分析和文献发掘，我们发现在跨亚科克隆胚胎中上调或下调的基因，大多数都与转录、翻译起始、细胞周期调控、以及蛋白质结合活性相关。迄今为止，这是首次报道有关鉴定跨亚科克隆胚胎早期发育过程中的相关基因研究。

Identification and Characterization of A Novel Gene Differentially Expressed in Zebrafish Cross-Subfamily Cloned Embryos

De-Sheng Pei[1,2] Yong-Hua Sun[1] Chun-Hong Chen[1,3] Shang-Ping Chen[1]
Ya-Ping Wang[1] Wei Hu[1] Zuo-Yan Zhu[1]

1 State Key Laboratory of Freshwater Ecology and Biotechnology, Institute of Hydrobiology, Chinese Academy of Sciences, Wuhan 430072

2 Group of Environmental Genomics, Institute of Hydrobiology, Chinese Academy of Sciences, Wuhan 430072

3 College of Life Science, Wuhan University, Wuhan 430072

Background Cross-species nuclear transfer has been shown to be a potent approach to retain the genetic viability of a certain species near extinction. However, most embryos produced by cross-species nuclear transfer were compromised because that they were unable to develop to later stages. Gene expression analysis of cross-species cloned embryos will yield new insights into the regulatory mechanisms involved in cross-species nuclear transfer and embryonic development.

Results A novel gene, K31, was identified as an up-regulated gene in fish cross-subfamily cloned embryos using SSH approach and RACE method. K31 complete cDNA sequence is 1106 base pairs (bp) in length, with a 342 bp open reading frame (ORF) encoding a putative protein of 113 amino acids (aa). Comparative analysis revealed no homologous known gene in zebrafish and other species database. K31 protein contains a putative transmembrane helix and five putative phosphorylation sites but without a signal peptide. Expression pattern analysis by real time RT-PCR and whole-mount in situ hybridization (WISH) shows that it has the characteristics of constitutively expressed gene. Sub-cellular localization assay shows that K31 protein can not penetrate the nuclei. Interestingly, over-expression of K31 gene can cause lethality in the epithelioma papulosum cyprinid (EPC) cells in cell culture, which gave hint to the inefficient reprogramming events occurred in cloned embryos.

Conclusion Taken together, our findings indicated that K31 gene is a novel gene differentially expressed in fish cross-subfamily cloned embryos and over-expression of K31 gene can cause lethality of cultured fish cells. To our knowledge, this is the first report on the determination of novel genes involved in nucleo-cytoplasmic interaction of fish cross-subfamily cloned embryos.

Background

Nuclear reprogramming is used to describe that the transferred nucleus from partially or

---

文章发表于 *Bmc Developmental Biology*, 2008, 8(1): 1-10

Abbreviations: ORF, open reading frame; SSH, suppression subtractive hybridization; WISH, whole-mount *in situ* hybridization; RACE, rapid amplification cDNA ends; SMART, switch mechanism at the 5′ end of RNA templates fully differentiated cell has the potential to direct the reconstructed embryo to develop like a normal embryo [1]. Although successful production of animal clones from somatic cells has been achieved in various species, many problems in offspring could not be hurdled due to incomplete nuclear reprogramming[2]. Cross-species nuclear transfer involves transferring cell nuclei of one species into enucleated oocytes of another species, which has been shown to be a potent approach to retain the genetic viability of a certain species near extinction[3]. However, most embryos produced by cross-species nuclear transfer were compromised because they were unable to develop to later developmental stages. To study inefficient reprogramming of the donor nuclei in the recipient cytoplasm from another species, nuclear transfer (NT) between two fish species was used as a model in the present study. A pioneering study on fish NT was carried out by Tung *et al* [4] and extensive studies on fish cross-species NT were mainly conducted in *Cyprinid* [5]. Recently, cross-genus cloned fish derived from transgenic common carp nuclei and goldfish enucleated eggs were generated and the somitogenesis and vertebral number of the cloned fish were consistent to the egg providing species, goldfish (*Carassius auratus*), instead of the donor cell species, common carp (*Cyprinus carpio*) [6]. Gene expression analysis of cross-species cloned embryos will shed light on the regulatory mechanisms involved in cross-species nuclear transfer and embryonic development.

Nuclear transfer between two laboratory fish species, rare minnow (*Gobiocypris rarus*) and zebrafish (*Danio rerio*), provides an ideal model for the study of cross-species nuclear transfer. Rare minnow and zebrafish belong to different subfamily-the *Gobioninae* and the *Danioninae* [7-8]. Zebrafish is a notable model for developmental and genetic studies for its short sex-maturity cycle, high reproductive capacity, and transparent eggs, *etc* [9-10]. Rare minnow, a special local species in China, not only shares aforementioned advantages with zebrafish, but also has many unique traits for laboratory study such as typical eurytherm and high adaptation [11], and sensitivity to toxicity and virus [12,13]. Such advantages enable rare minnow to be an excellent type of experimental fish [14].

In the present study, we performed cross–subfamily nuclear transfer between zebrafish and rare minnow and obtained nuclear transfer embryos derived from zebrafish nuclei and rare minnow enucleated eggs. Using a suppression subtractive hybridization (SSH) approach, we found a novel gene–K31 over-expressed in cloned embryos, potentially participating in the improper reprogramming of transferred nuclei.

Results

Identification of K31 as an up-regulated gene

To better understand the molecular events in cloned embryos, we performed nuclear transfer between two laboratory fish, zebrafish and rare minnow. As reported in our previous study, most of the cloned embryos were arrested at between sphere and 50%-epiboly stages [15]. By using a SSH approach, we have totally screened out 50 differentially expressed genes in the cloned embryos at sphere stage. Among them, about 10% are related to redox function, such as selenoprotein W1, 5-lipoxygenase and glutaryl-coenzyme dehydrogenase *etc*; about 6% are responsible for cell growth and division, including geminin, daz-like gene and cofactor of BRCA2 *etc*. Interestingly, a novel gene, K31, was found to be up-regulated in the cloned embryos at sphere stage. Real-time RT-PCR analysis showed that the mRNA abundance of K31 gene in the cloned embryos was about 15-fold than that in normally fertilized zebrafish embryos (Fig. 1), which was agreement with the dot blotting assay.

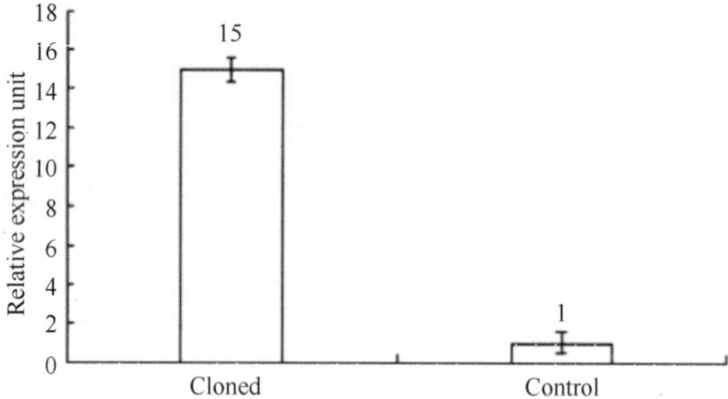

Figure 1 Real time PCR analysis of K31 gene. The expression of K31 gene in cloned embryos is 15 times fold than in zebrafish embryos; GAPDH was used as an endogenous reference

Cloning and characterization of K31 gene

Full-length cDNA of K31 gene was obtained from a SMART cDNA library. It is 1106 bp in length with an open reading frame (ORF) of 342 bp encoding a putative protein of 113 aa, a 5′ untranslated region (UTR) of 307 bp, and a 3′ UTR of 457 bp. It contains an mRNA instable motif (ATTTA) and a poly(A) signal (AATAAA) followed by a poly(A) tail (Fig. 2).

Homology search of public database found three novel zebrafish protein sequences (CAK10721, CAK04277 and XP_696518) to be closer to K31 but no homologous known gene in other species (Fig. 3A). Secondary structure prediction by DNAstar software shows that K31 protein contains more α-helices and β-folds but less turns structure and hydrophilicity regions, indicating its lipophilic trait. Predicated by SignalP 3.0 software, there is no signal peptide in K31 protein. Interestingly, the transmembrane helices analysis by TMHMM2.0 software shows that K31 protein contains a TMhelix of 23 amino acids (from aa32 to aa54) (Fig. 3B). Moreover, phosphorylation sites analysis by NetPhos 2.0 Server shows that there is a phosphorylation site (Tyr_9) in the N-terminal and four phosphorylation sites (Ser_{82}, Ser_{83}, Ser_{93} and Tyr_{97}) in the C-terminal but no phosphorylation sites in the transmembrane region. The data above indicates that K31 protein may participate in signal transduction.

```
GGACATACTGTCCTGTCTGCGCATGTGCAGACACATACCGCCATTTTTTTCTGAAGGCGCCAAG 64
GAGTATTTTGAAGCACTCTTTTGAATGCTGCATCCAGCTCTGCATTTTCGGTTATTCTCATATCTGATTGTGTTGGTGTGA 145
TGTGACCTAATTCTTAGGGGAGCTCATATTGCTTTGGCTAGAAGTGACATCTATCTTAAAAGGGCAGCAGTCATCAAGGCA 226
CTTTGCATCTACCTCGGTGAGGATGATGGACATCTCATTCAGGAGTATAAGGACATTGAAGGGGATGACATCCTGAGAGAC 307
 M  Q  K  S  T  M  G  I  Y  V  I  N  K  E  G  E  E  I  G  A  H  D  D  I  G  I  Y 27
ATGCAGAAATCCACCATGGGCATATATGTCATCAACAAGGAGGGTGAAGAAATTGGAGCTCATGATGACATTGGCATTTAT 388
 V  E  G  V  I  I  L  D  N  I  G  S  V  A  Q  A  C  A  M  M  L  G  V  I  Y  V  L 54
GTTGAAGGAGTAATCATTCTGGACAACATTGGATCTGTAGCTCAGGCATGTGCAATGATGCTTGGAGTCATCTATGTACTG 469
 N  M  A  Y  P  K  K  L  K  Y  F  Y  E  F  I  Q  K  V  L  L  K  M  D  G  E  R  L 81
AACATGGCTTACCCCAAAAAGCTGAAATACTTCTATGAATTCATTCAAAAAGTGCTCTTGAAAATGGATGGGGAGAGGCTT 550
 S  S  K  V  L  G  L  K  N  L  N  S  M  L  D  Y  D  F  L  S  C  Y  Y  E  P  I  V 108
TCCTCCAAAGTCCTTGGACTAAAGAACCTAAATTCAATGCTGGACTATGATTTCCTCAGCTGCTATTATGAACCCATTGTT 631
 I  V  F  T  E  *                                                                113
ATTGTGTTTACGGAGTGA                                                                649
TAAAGCATATTTAGTATGATGTGAAATGCAGCTGTAGAAGATGTTCCAAAGTGAGTTATGGACCTTAGTAATGTGTGTGGC 730
TATATGTAACAACAACAGTGGTTAACATTTAAATCTGTACTAGGTATGTTTTGGAGTATTGACTCAATATAAGTGTTTGCA 811
GTTTAATGCAATCTCACACAAGTTCATCAGTGTTTGCTGTCTTTCAGCTATTCTGGATCAATTGTTTTTGTCACGAATGGC 892
ACAATGTGAAATGAATGAAAATTAATGAAATTAATGTATTTTATGTTAATGGATGATACTAGGGCATTTTGTTCTAAGAAA 973
GCAATTTGAGAAATGGTGACTACTGAGTTACATTACTGGCACTCTTGTTTGGGAGAGAAAAAATCTATCTCTTGGATTAGA 1054
AAAATAAAGTTACAGTTGAAAATCAAAAAAAAAAAAAAAAAAAAAAAAAAA                               1106
```

Figure 2 Nucleotide sequence and deduced amino acid sequence of K31 gene. The accession no. was AY885255 at Genbank. K31 cDNA was 1106 bp. Its ORF has 342 nucleotides, which codes 113 aa. The nucleotides (lower row) and deduced amino acids (upper row) are numbered at the right side of the sequences, respectively. Poly (A) signal (in the 3′ UTR) is underlined. The start codon (ATG) is in bold and the stop codon (TAA) is indicated by an asterisk. An unstable motif (ATTTA) is doubly underlined

Expression profile of K31 during development

To figure out the expression pattern of K31 gene, real-time RT PCR was performed. As shown in Fig. 4A, K31 transcripts were found in all the checked developmental stages and it was expressed maternally in the oocytes, indicating that K31 gene presents a characteristic of constitutive expression. WISH was performed to obtain the temporalspatial expression pattern of K31 gene and the results showed that K31 expression was detected in all stages, which was agreement with real-time RT-PCR results. In the ovary samples, the signals were obviously detected in the oocytes of different stages, clearly demonstrating the maternal expression of K31 (Fig. 4B). The maternally transmitted mRNA could be detected at 8-cell stage after fertilization (Fig. 4C). After mid-blastula transition of zygotic genome activation, K31 was ubiquitously expressed in the whole embryos at blastula (Fig. 4D), gastrula (Fig. 4E), and mid-somitogeneis stages (Fig. 4F). At 1 dpf (day post-fertilization) and 2 dpf of development, the signals were detected in the whole embryo with more intensive labeling at the anterior tissues (Fig. 4G, H), including an obvious labeling of the hatching gland (Fig. 4I) and the pectoral fin territory (Fig. 4J).

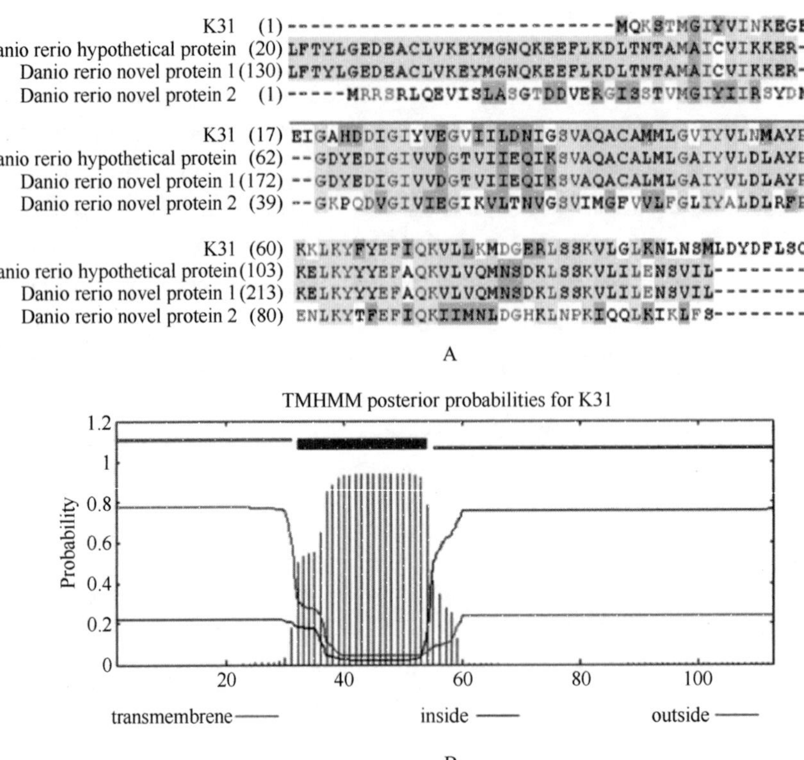

Figure 3 Bio-information analysis of K31 amino acids. A. Homologous alignment is analysed by BLAST program in NCBI. Therein, *Danio rerio* hypothetical protein (XP_696518), novel protein 1 (CAK10721) and novel protein 2 (CAK04277) is 57%, 57% and 39% identity with K31, respectively. B. The transmembrane helices are analysed by TMHMM2.0 software. A transmembrane helices lies in the region from aa 32 to aa 54

Sub-cellular localization and ectopic over-expression of K31 gene

To study the sub-cellular localization of K31 gene, we constructed pK31-EGFP plasmid with K31 coding sequence fused to EGFP gene. The pK31-EGFP and pEGFP-N3 (as negative control) were transfected in EPC cells, respectively. At 24 h after transfection, Hochest 33342 was used to stain cell nuclei and sub-cellular localization of K31 was judged by the co-localization of GFP protein. The results showed that K31-EGFP fusion protein located in cytoplasm, rather than in nucleus (Fig. 5).

To study the biological effects after ectopic over-expression of K31 gene, we visualized the EPC cells nuclei using Hochest 33342 at 24 h, 28 h and 32 h after transfection. We found the cells were dead after 32 h, which was validated by trypan blue staining. Following comparative analysis in white field and the overlapping of the stained nuclei with green fluorescence which indicates the overexpression of K31-GFP fusion protein, we conclude that ectopic over-expression of K31 in cytoplasm causes the lethality of cultured fish cells (Fig. 6).

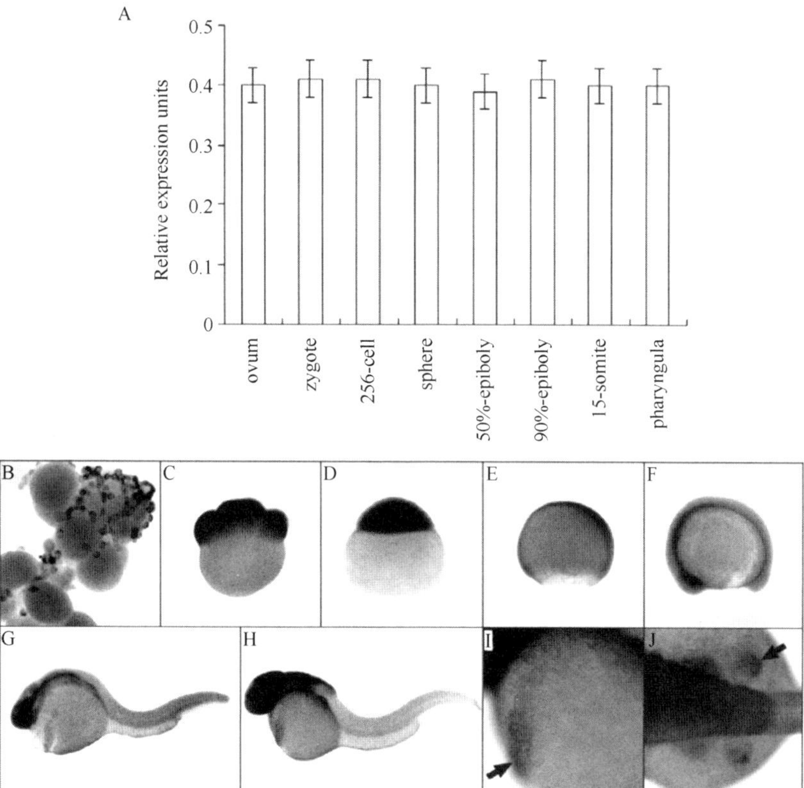

Figure 4 Analysis of the expression level of K31 gene at different developmental stages in zebrafish. (A) Real time RT-PCR analysis of K31 transcription during development. K31 transcripts are expressed maternally in the ovum. From zygote to pharyngula period, K31 has a characteristic of constitutive expression gene. GAPDH was used as endogenous reference. (B-J) Expression of K31 transcripts as detected by WISH during embryogenesis in zebrafish. (B) ovary, (C) 8-cell stage (1.25 hpf, hours post-fertilization), (D) sphere stage (4 hpf), (E) 75% epiboly (8 hpf), (F) 10-somite stage (13 hpf), (G) 1 dpf (day post-fertilization), (F) 2 dpf, (I) 2 dpf embryo with arrow indicating the hatching gland, (J) 2 dpf embryo with arrow indicating the pectoral fin. Embryos in C-I are lateral views, C-E with the animal pole to the top, and the dorsal to the right, F-I with dorsal to the top and anterior to the left; embryo in J is dorsal view, with anterior to the left

Discussion

Nuclear transfer in fish has been successfully manipulated for over 40 years [16]. Cross-genus cloned fish derived from common carp nuclei and goldfish enucleated eggs was given birth as reported by Sun *et al* [6]. In our recent study, we not only demonstrated the success of cloned embryos derived from zebrafish nuclei and rare minnow enucleated eggs, but also the success of cloned embryos derived from rare minnow nuclei and zebrafish enucleated eggs by SCAR approaches [17]. Samples used in the present study were strictly chosen at sphere stage of embryogenesis as described by Kimmel *et al* [18] to ensure the

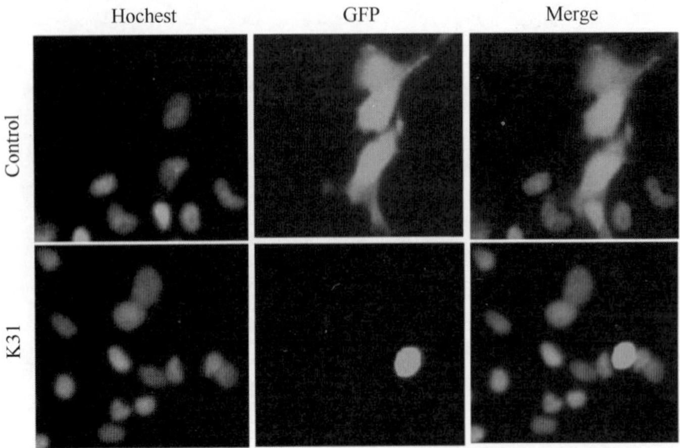

Figure 5 Sub-cellular localization of pEGFP-K31 expressed in EPC cells. The sub-cellular localization of control (pEGFP-N3) and pEGFP-K31 expressed GFP signals in EPC cells was in the upper and lower rows, respectively. Therein, blue signals represented the cell nuclei stained by Hochest 33342; Green signals represented the expression of pEGFP-N3 and pEGFP-K31 fluorescence proteins in EPC cells, respectively; Merge represented overlapping the images of pEGFP-N3 or pEGFP-K31 fluorescent protein with the images of cell nuclei stained by Hochest 33342. All three panels had the same view field at 24 h after transfection

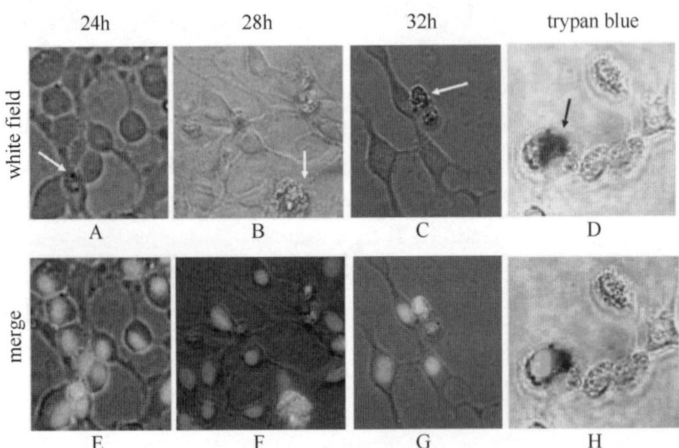

Figure 6 Ectopic over-expression of K31 protein at different times after transfection. (A-C) the shapes of EPC cells taken in white field at 24 h, 28 h and 32 h after transfection by pEGFP-K31, respectively. The white arrows indicate the supposed dying cells. (E-G) the overlaps of the EGFP fluorescence and the cell nuclei stained by Hochest 33342 for A-C, respectively. (D, H) confirmation of the supposed dying cells using trypan blue dye; D is the shapes of EPC cells taken in white field at 32 h after transfection and stained by trypan blue; The black arrow indicate the dead cell. H is the overlaps of the EGFP fluorescence and the dead cell stained by trypan blue. Green, blue and dark blue signals represented GFP fluorescence, Hochest 33342 stained nuclei and trypan blue stained dead cell, respectively

accuracy. In single cloned embryo, SCAR PCR was performed with primers distinguished rare minnow from zebrafish, revealing the success of nuclear transfer [17]. Consequently, the novel gene, K31, is certainly resulted from the nucleo-cytoplasmic interaction between zebrafish and rare minnow, excluding any artificial interfere.

Previous studies demonstrated that the majority of embryos produced by nuclear transfer were compromised because they were unable to develop past the early development stages[19]. A common hypothesis is that inefficient reprogramming of the donor nucleus results in inappropriate expression of genes required for embryonic development. Gene expression analysis of individual embryos will undoubtedly yield new insights into the regulatory mechanisms involved in nuclear transfer inducing reprogramming and embryonic development. In the present study, a novel gene, K31 was found to be overexpressed in cross-subfamily cloned embryos by SSH approach. To our knowledge, this is the first report of gene expression analysis in cross-subfamily cloned embryos. Interestingly, K31 transcripts were found in all embryonic development stages of zebrafish and K31 was also expressed maternally in ovum, but its ectopic over-expression caused the EPC cells to die during the cell culture. Result of sub-cellular localization indicates that K31 protein can not penetrate the nuclei, and it may be participate in signal transduction by its five putative phosphorylation sites. Previous studies indicated that the reprogramming efficiency of cloned embryos was influenced by many factors, such as epigenetic changes like DNA methylation and XCI patterns [20-21], failure to suppress previously active gene transcription as well as failure to activate previously inactive genes [22-24], failure of chromatin remodeling [21,25-27] and mitochondria effects of proper reprogramming [28], *etc*. Yet, we can conclude that nucleo-cytoplasmic interaction in such cross-subfamily cloned embryos caused the over-expression of K31 gene. Further reliable studies in a comprehensive way should provide solid evidences to unveil whether K31 gene can affect the nuclear reprogramming in cloned embryos.

Conclusion

Cross-species nuclear transfer can be used to maintain limited populations of highly endangered species, especially when the oocytes of these species are difficult to obtain. However, most embryos produced by cross-species nuclear transfer were compromised because they were unable to develop to later developmental stages. Therefore, gene expression analysis of cross-species cloned embryos is necessary. Here we used two laboratory fish species, rare minnow and zebrafish as cross-subfamily nuclear transfer model and report K31 gene as an up-regulated gene in fish cross-subfamily cloned embryos. Importantly, ectopic over-expression of K31 gene can cause lethality of EPC cells in the cell culture, which gave hint that why most of the cloned embryos were developmentally arrested in between the stages of sphere and 50% epiboly.

Methods

Preparation of cross-subfamily cloned embryos and noncloned embryos

The cross-subfamily cloned embryos were generated by nuclear transfer as described by Sun *et al* [6], with nuclei derived from zebrafish at blastula stage and the enucleated unfertilized eggs of rare minnow. Meanwhile, batches of non-cloned zebrafish embryos were produced by *in vitro* fertilization. All embryos were incubated in Holtfreter's solution at 28℃, and collected at sphere stage [18]. The manipulations in the experiment adhered to Guidelines for Animal Use in Biomedical Research Laboratories (ILAR, 1996).

SMART cDNA synthesis and construction of SSH cDNA libraries

Total RNA was extracted from cloned and zebrafish embryos by SV™ total RNA kit (Promega, WI, USA). Poly(A)+ RNA was purified with Poly(A)Tract mRNA Isolation system (Promega WI, USA) and then used to synthesize SMART cDNA according to the instructions of BD SMART cDNA Library Construction Kit (Clontech, CA, USA). The forward subtracted cDNA library was obtained using tester dscDNAs from cloned embryos and driver dscDNA from zebrafish embryos. At the same time, the reverse subtracted cDNA library was obtained using tester dscDNAs from zebrafish embryos and driver dscDNA from cloned embryos. To evaluate the efficiency of the cDNA subtraction, reverse transcriptase PCR was performed with *GAPDH* (glyceraldehyde-3-phosphate dehydrogenase) primers in forward subtracted and unsubtracted cDNA. After the secondary PCR, the PCR products generated by SSH were cloned into the pGEM-T Easy vector (Promega, WI, U.S.A). PCR and dot blots were applied to screen improper reprogramming genes from the subtracted cDNA library as described by Sung *et al* [29].

RACE-PCR and real-time PCR analysis

RACE-PCR was used to clone the full-length cDNA of K31 gene. Using SMART cDNA as templates, the combination of universal primer SMART F and K31 R, universal primer SMART R and K31 F was used for 5′ and 3′ RACE PCR, respectively (Table 1). The generated PCR products were sequenced, and the full-length cDNA of K31 was composed of 5′ RACE sequence and 3′ RACE sequence.

The relative quantification with real-time RT-PCR was done as described by the manufacturer (Applied Biosystems, USA) with slight modification. In brief, the samples were placed in 96 well plates and amplified in an automated fluorometer (ABI PRISM 7000 Sequence Detection System, Applied Biosystems). Each PCR proceeded in 30 μl SYBR Green PCR buffer (Applied Biosystems) containing 400 nM K31F and K31R primers, 1 U

Table 1 Primers used in the present study

Names	Sequences
SMART F	5-CAACGCAGAGTACGCGGG-3
SMART R	TCAACGCAGAGTACT(16)
K31 F	5-CTTGAAAATGGATGGGGAGA-3
K31 R	5-ACAATGGGTTCATAATAGCAGC-3
K31DW F	5-AACTGCAGATGCAGAAATCCACCATGGG-3
K31DW R	5-CGGGATCCCTCCGTAAACACAATAACAATGG-3
GAPDH F	5-GTGTAGGCGTGGACTGTGGT-3
GAPDH R	5-TGGGAGTCAACCAGGACAAATA-3

AmpliTaq Gold DNA polymerase, 2.5 mM dNTPs, 0.5 U AmpErase UNG, 3 mM $MgCl_2$ and 50 ng cloned embryos or non-cloned embryos cDNA template. Amplification conditions were 2 min at 50℃, 10 min at 95℃, 40 cycles of 30 s at 95℃ and 60 s at 60℃. All samples were analyzed in triplet and the results were expressed as relative fold of the expression of the GAPDH gene with $2^{[-\mathrm{deltadeltaCT}]}$ method with correction for different amplification efficiencies (the amplification efficiency of cloned embryos or zebrafish embryos is 0.995 or 0.990, respectively) [30].

Data mining and bio-information analyses

Homology search of K31 gene was performed on the sequences listed in EMBL/GenBank/ DDBJ databases using PHI- and PSI-BLAST, EST-BLAST and Protein-protein BLAST (blastp) at the web site of the National Center of Biotechnology Information (NCBI) [31]. Secondary structure analysis of K31 protein sequence was performed by DNAstar software (Lasergene, Madison, Wis.). Transmembrane helices, phosphorylation sites and signal peptide were predicted by TMHMM2.0 software [32], NetPhos 2.0 software [33] and SignalP 3.0 software [34], respectively.

Analysis of expression pattern by real time RT-PCR and whole-mount in situ hybridization (WISH)

Zebrafish embryos were obtained by *in vitro* fertilization and raised in Holtfreter's solution (0.35% NaCl, 0.01% KCl, and 0.01% $CaCl_2$) at 28℃. Embryos were staged according to Kimmel *et al* [18]. Total RNA of 8 samples (ovum, zygote, 256-cell stage, sphere stage, 50%-epiboly stage, 90%-epiboly stage, 15-somite stage, pharyngula period) was separately isolated using SV™ total RNA kit (Promega, WI, USA). Then, 3 μg total RNA was reverse transcribed (RT) for each sample and 2 μl of the RT product was amplified to quantify

K31 transcripts by real-time RT-PCR. The manipulation of real-time RT-PCR was all the same as described above (to see materials and methods 2.3).

For WISH, embryos were fixed in MEMPFA (100 mM Mops (Sigma), pH 7.4; 2 mM EGTA (Sigma); 1 mM $MgSO_4$ (Merck); 4% (*w/v*) paraformaldehyde (Sigma) at different developmental stages. For generation of full length K31 antisense probes, pBluescripts KS II-K31 plasmids were linearized and used as templates for synthesis of DIG-labeled antisense RNA (Roche). RNA probe was purified using RNeasy columns (QIAGEN). In situ hybridization essentially following the protocol described by Thisse *et al* [35]. WISH was performed with a probe concentration of 100 ng/mL at 65℃. As a control, WISH with similarly produced sense probes was performed. Images of zebrafish embryos were recorded using an Olympus SZX12 microscope and a digital camera.

Cell culture and sub-cellular localization and ectopic overexpression of K31

EPC cells from carp (*Cyprinus carpio*) were cultured in medium 199 supplemented with 10% fetal calf serum (FCS) and antibiotics (100 U/ml penicillin and 100 μg/ml streptomycin). Cultures were maintained at 28℃ in an atmosphere of 5% CO_2 in air.

For fluorescence microscopy, the coding region of K31 gene was amplified using K31DW primers (Table 1) and cloned into pEGFP-N3 (BD Biosciences, PaloAlto, CA, USA) using *Pst*I and *Bam*HI sites. After sequencing validation, the p K31-EGFP construct was transfected into EPC cells using Lipofectamine 2000 reagent (Invitrogen). After 24, 28 and 32 h culture, cells were removed by trypsin/ EDTA and analysed for sub-cellular localization and ectopic over-expression with 5 mg/L Hochest 33342 (Calbiochem) to stain the nuclei of cells as described [36] and 0.4% trypan blue dye (sigma) to stain the dead cells as described [37], respectively.

Acknowledgements This work was supported by the State Key Fundamental Research of China (Grant No 2004CB117406) and the National Natural Science Foundation of China (Grant No. 30771100 and 30700607). In addition, the authors are grateful to Ming Li for supplying experimental materials and Qiya Zhang for providing EPC cells.

References

1. Gurdon JB, Byrne JA, Simonsson S: Nuclear reprogramming and stem cell creation. Proc Natl Acad Sci U S A 2003, 100 Suppl 1:11819-11822.
2. Smith SL, Everts RE, Tian XC, Du F, Sung LY, Rodriguez-Zas SL, Jeong BS, Renard JP, Lewin HA, Yang X: Global gene expression profiles reveal significant nuclear reprogramming by the blastocyst stage after cloning. Proc Natl Acad Sci U S A 2005, 102(49):17582-17587.
3. Loi P, Ptak G, Barboni B, Fulka J Jr., Cappai P, Clinton M: Genetic rescue of an endangered mammal by cross-species nuclear transfer using post-mortem somatic cells. Nat Biotechnol 2001, 19(10):962-964.

4. Tung TC, Wu SC, Tung YYF, Yan SY, Tu M, Lu TY: Nuclear transplantation in fish. Science Bulletin, Academia Sinica (in Chinese) 1963, 7:60-61.
5. Yan SY: The Nucleo-cytoplasmic Interaction as Revealed by Nuclear Transplantation in Fish. In Cytoplasmic Organization Systems: A Primer in Developmental Biology Edited by: Malacinski GM. New York, McGraw-Hill; 1989:61-81.
6. Sun YH, Chen SP, Wang YP, Hu W, Zhu ZY: Cytoplasmic impact on cross-genus cloned fish derived from transgenic common carp (*Cyprinus carpio*) nuclei and goldfish (*Carassius auratus*) enucleated eggs. Biol Reprod 2005, 72(3):510-515.
7. He S, Liu H, Chen Y, Kuwahara M, Nakajima T, Zhong Y: Molecular phylogenetic relationships of Eastern Asian Cyprinidae (pisces: cypriniformes) inferred from cytochrome b sequences. Sci China C Life Sci 2004, 47(2):130-138.
8. Pei DS, Sun YH, Chen SP, Wang YP, Zhu ZY: Cloning and characterization of cytochrome c oxidase subunit I (COXI) in Gobiocypris rarus. DNA Seq 2007, 18(1):1-8.
9. Grunwald DJ, Eisen JS: Headwaters of the zebrafish–emergence of a new model vertebrate. Nat Rev Genet 2002, 3(9):717-724.
10. Key B, DevineCA: Zebrafish as an experimental model: strategies for developmental and molecular neurobiology studies. Methods Cell Sci 2003, 25(1-2):1-6.
11. Wang J: Acute effects of high concentration of dissolved free carbon dioxide and low dissolved oxygen on rare minnow. Acta Hydrobiologica Sinica (in Chinese) 1995, 19:84-88.
12. Qun-Fang Z, Gui-Bin J, Ji-Yan L: Effects of sublethal levels of tributyltin chloride in a new toxicity test organism: the Chinese rare minnow (*Gobiocypris rarus*). Arch Environ Contam Toxicol 2002, 42(3):332-337.
13. Zhong J, Wang Y, Zhu Z: Introduction of the human lactoferrin gene into grass carp (*Ctenopharyngodon idellus*) to increase resistance against GCH virus. Aquaculture 2002, 214:93-101.
14. Wang J, Cao W: *Gobiocypris rarus* and fishes as laboratory animals. Trans Chin Ichthyol Soc 1997, 6:144-152.
15. Pei DS, Sun YH, Chen SP, Wang YP, Hu W, Zhu ZY: Identification of differentially expressed genes from the cross-subfamily cloned embryos derived from zebrafish nuclei and rare minnow enucleated eggs. Theriogenology 2007, 8(9):1282-1291.
16. Zhu ZY, Sun YH: Embryonic and genetic manipulation in fish. Cell Res 2000, 10(1):17-27.
17. Hu W, Pei DS, Dai J, Chen SP, Sun YH, Wang YP, Zhu ZY: Identification and application of SCAR markers in detecting crossspecies cloned embryos between zebrafish and rare minnow. In Chinese High Technology Letters Volume 16. Issue 9 United States; 2006:959-963.
18. Kimmel CB, Ballard WW, Kimmel SR, Ullmann B, Schilling TF: Stages of embryonic development of the zebrafish. Dev Dyn 1995, 203(3):253-310.
19. Hill JR, Winger QA, Long CR, Looney CR, Thompson JA, Westhusin ME: Development rates of male bovine nuclear transfer embryos derived from adult and fetal cells. Biol Reprod 2000, 62(5):1135-1140.
20. Santos F, Zakhartchenko V, Stojkovic M, Peters A, Jenuwein T, Wolf E, Reik W, Dean W: Epigenetic marking correlates with developmental potential in cloned bovine preimplantation embryos. Curr Biol 2003, 13(13):1116-1121.
21. Nolen LD, Gao S, Han Z, Mann MR, Gie Chung Y, Otte AP, Bartolomei MS, Latham KE: X chromosome reactivation and regulation in cloned embryos. Dev Biol 2005, 279(2):525-540.
22. Sun Y, Chen S, Wang Y, Zhu Z: The onset of foreign gene transcription in nuclear-transferred embryos of fish. Sci China C Life Sci 2000, 43:597-605.
23. Boiani M, Eckardt S, Scholer HR, McLaughlin KJ: Oct4 distribution and level in mouse clones: consequences for pluripotency. Genes Dev 2002, 16(10):1209-1219.
24. Bortvin A, Eggan K, Skaletsky H, Akutsu H, Berry DL, Yanagimachi R, Page DC, Jaenisch R: Incomplete reactivation of Oct4-related genes in mouse embryos cloned from somatic nuclei. Development 2003, 130(8):1673-1680.
25. Moreira PN, Robl JM, Collas P: Architectural defects in pronuclei of mouse nuclear transplant embryos. J Cell Sci 2003, 116(Pt 18):3713-3720.

26. Sullivan EJ, Kasinathan S, Kasinathan P, Robl JM, Collas P: Cloned calves from chromatin remodeled *in vitro*. Biol Reprod 2004, 70(1):146-153.

27. Zhang LS, Zhang KY, Yao LJ, Liu SZ, Yang CX, Zhong ZS, Zheng YL, Sun QY, Chen DY: Somatic nucleus remodelling in immature and mature Rassir oocyte cytoplasm. Zygote 2004, 12(2):179-184.

28. Hiendleder S, Zakhartchenko V, Wenigerkind H, Reichenbach HD, Bruggerhoff K, Prelle K, Brem G, Stojkovic M, Wolf E: Heteroplasmy in bovine fetuses produced by intra- and inter-subspecific somatic cell nuclear transfer: neutral segregation of nuclear donor mitochondrial DNA in various tissues and evidence for recipient cow mitochondria in fetal blood. Biol Reprod 2003, 68(1):159-166.

29. Sung YK, Moon C, Yoo JY, Moon C, Pearse D, Pevsner J, Ronnett GV: Plunc, a member of the secretory gland protein family, is upregulated in nasal respiratory epithelium after olfactory bulbectomy. J Biol Chem 2002, 277(15):12762-12769.

30. Livak KJ, Schmittgen TD: Analysis of relative gene expression data using real-time quantitative PCR and the $2^{(-\mathrm{Delta\ Delta\ C(T)})}$ Method. Methods 2001, 25(4):402-408.

31. Altschul SF, Madden TL, Schaffer AA, Zhang J, Zhang Z, Miller W, Lipman DJ: Gapped BLAST and PSI-BLAST: a new generation of protein database search programs. Nucleic Acids Res 1997, 25(17):3389-3402.

32. Sonnhammer EL, von Heijne G, Krogh A: A hidden Markov model for predicting transmembrane helices in protein sequences. Proc Int Conf Intell Syst Mol Biol 1998, 6:175-182.

33. Blom N, Gammeltoft S, Brunak S: Sequence and structure-based prediction of eukaryotic protein phosphorylation sites. J Mol Biol 1999, 294(5):1351-1362.

34. Bendtsen JD, Nielsen H, von Heijne G, Brunak S: Improved prediction of signal peptides: SignalP 3.0. J Mol Biol 2004, 340(4):783-795.

35. Thisse C, Thisse B, Schilling TF, Postlethwait JH: Structure of the zebrafish snail1 gene and its expression in wild-type, spadetail and no tail mutant embryos. Development 1993, 119(4):1203-1215.

36. Song Z, Wu M: Identification of a novel nucleolar localization signal and a degradation signal in Survivin-deltaEx3: a potential link between nucleolus and protein degradation. Oncogene 2005, 24(16):2723-2734.

37. Dohn M, Jiang J, Chen X: Receptor tyrosine kinase EphA2 is regulated by p53-family proteins and induces apoptosis. Oncogene 2001, 20(45):6503-6515.

斑马鱼跨亚科间克隆胚胎中一个新基因的鉴定及特征分析

裴得胜[1,2] 孙永华[1] 陈春红[1,3] 陈尚萍[1] 汪亚平[1] 胡 炜[1] 朱作言[1]

1 中国科学院水生生物研究所，淡水生态与生物技术国家重点实验室，武汉 430072
2 中国科学院水生生物研究所，环境基因组学科组，武汉 430072
3 武汉大学，生命科学学院，武汉 430072

摘 要 人们期望跨种核移植是保持频临灭绝物种遗传生存能力的有效方法。然而，大多数跨种核移植产生的胚胎不能发育到后期。跨亚科克隆胚胎的基因表达分析有助于剖析跨种核移植胚胎发育的调控机制。通过 SSH 和 RACE 技术发现一个新基因 K31，其在斑马鱼跨亚科克隆胚胎中表达上调。K31 完整的 cDNA 序列长 1106 bp, 342 bp 的 ORF 区编码一个 113 个氨基酸的蛋白质。比较分析显示在斑马鱼和其他物种数据库中 K31 没有已知的同源基因。K31 蛋白包含一个跨膜螺旋和 5 个磷酸化位点，但是不含有信号肽。通过荧光定量 PCR 和整体原位杂交表达模式分析，K31 具有母源性持续表达特性。亚细胞定位实验表明 K31 蛋白仅在细胞核中表达。有趣的是，K31 基因的过表达可以导致 EPC 细胞出现凋亡，这为克隆胚胎中低效率重编程的原因提供了解释原因。

Identification of Differential Transcript Profiles Between Mutual Crossbred Embryos of Zebrafish (*Danio rerio*) and Chinese Rare Minnow (*Gobiocypris rarus*) by cDNA-AFLP

J. Liu[1] Y. H. Sun[1] Y. W. Wang[1,2] N. Wang[1] D. S. Pei[1]
Y. P. Wang[1] W. Hu[1] Z. Y. Zhu[1]

1 State Key Laboratory of Freshwater Ecology and Biotechnology, Institute of Hydrobiology, Chinese Academy of Sciences, Luojiashan, Wuhan 430072
2 College of Life Science, Wuhan University, Wuhan 430072

Abstract The crosstalk between naive nucleus and maternal factors deposited in egg cytoplasm before zygotic genome activation is crucial for early development. In this study, we utilized two laboratory fishes, zebrafish (*Danio rerio*) and Chinese rare minnow (*Gobiocypris rarus*), to obtain mutual crossbred embryos and examine the interaction between nucleus and egg cytoplasm from different species. Although these two types of crossbred embryos originated from common nuclei, various developmental capacities were gained due to different origins of the egg cytoplasm. Using cDNA amplified fragment length polymorphism (cDNA-AFLP), we compared transcript profiles between the mutual crossbred embryos at two developmental stages (50%- and 90%-epiboly). Three thousand cDNA fragments were generated in four cDNA pools with 64 primer combinations. All differentially displayed transcript- derived fragments (TDFs) were screened by dot blot hybridization, and the selected sequences were further analyzed by semi-quantitative RT-PCR and quantitative real-time RT-PCR. Compared with ZR embryos, 12 genes were up-regulated and 12 were down-regulated in RZ embryos. The gene fragments were sequenced and subjected to BLASTN analysis. The sequences encoded various proteins which functioned at various levels of proliferation, growth, and development. One gene (ZR6), dramatically down-regulated in RZ embryos, was chosen for loss-of-function study; the knockdown of ZR6 gave rise to the phenotype resembling that of RZ embryos.

Keywords zebrafish; chinese rare minnow; crossbred embryo; cDNA-AFLP; differential expression

1. Introduction

In previous studies of animal cloning, the majority of intra-species nuclear transfers (NTs)

文章发表于 *Theriogenology*, 2008, 70(9): 1525-1535

yielded cloned individuals identical to their nuclear donor species either on the basis of phenotype or genotype [1-5]. However, this was not the case forcross-species NT, wherein enucleated eggs cytoplasm had an obvious impact on the mitochondrial genetic materials and the development of NT embryos [6-7].

Our previous study revealed that the somitogenesis and vertebral number of the cross-genus cloned fish resembled those of the cytoplasmic recipient species, goldfish, instead of the donor nuclear species, common carp [6]. Therefore, the crosstalk between the donor nucleus and the recipient egg cytoplasm from different species could modify the early development of NT embryo, possibly by affecting gene expression of the transplanted nucleus. Studies on inter-species NT between bovine and pig likewise demonstrated that the embryonic development was driven by the host oocyte, up to the stage when zygotic genome activation should occur [8]. Therefore, it is necessary to develop an appropriate research system to study the impact on the transplanted nucleus induced by egg cytoplasm from another species. As described in our previous study, we utilized cross-species NT to address this question [9]. However, there were two scientific issues, i.e., nuclear reprogramming of the donor nucleus and interaction between the nucleus and cytoplasm from different species. This presented a difficulty in distinctly clarifying genes that were solely involved in the nucleo-cytoplasmic interaction, prompting us to seek a novel system.

Unlike higher vertebrates, fishes could be easily used to produce crossbred embryos between two different species. In the present study, we utilized two laboratory fish species, zebrafish and Chinese rare minnow, to obtain two types of mutual crossbred embryos, zebrafish ♀× Chinese rare minnow ♂ (ZR) and Chinese rare minnow ♀ × zebrafish ♂ (RZ). As one of the model animals, zebrafish (*Danio rerio*) has an important role in the basic research of genetics and developmental biology [10]. The Chinese rare minnow (*Gobiocypris rarus*), meanwhile, is a small cyprinid fish which has proved to be an ideal species for laboratory study [9]. Given that the two types of mutual crossbred embryos have the same composition of nucleus but different egg cytoplasm, they could provide us with novel materials to study cross-species nucleo-cytoplasmic interaction. More importantly, when compared with cross-species NT, since there are no issues related to nuclear reprogramming in this event, mutual crossbred embryos are more appropriate for the study of nucleo-cytoplasmic interaction.

The differential expression analysis approach of cDNA amplified fragment length polymorphism (cDNA-AFLP) is based on the selective PCR-amplification of adapter-ligated restriction fragments derived from cDNA [11]. It is proven to be a powerful transcript profiling approach for genome-wide analysis without the need for prior sequence knowledge [12]. In this study, we applied the cDNA-AFLP approach to identify differential transcript profiles between ZR and RZ embryos.

2. Materials and Methods

2.1 Experimental fish and embryos

Zebrafish and Chinese rare minnow were raised in our laboratory. Crossbred embryos ZR and RZ were obtained by artificial fertilization and cultured to a specific stage for RNA and DNA isolation. Embryos of the zebrafish, the Chinese rare minnow, and the crossbred varieties were incubated at 28, 25, and 26℃, respectively. To compare the developmental timing of these embryos, they were incubated at 26℃ and staged according to the literature descriptions [13].

2.2 Extraction of total RNA and DNA

Total RNA was extracted using the SV Total RNA Isolation System (Promega, Madison, WI, USA) from the crossbred embryos at 50%- and 90%-epiboly stages, and zebrafish or Chinese rare minnow embryos at the 50%-epiboly stage. The integrity and concentration of the total RNA were verified by electrophoresis and quantified with a spectrophotometer.

For DNA isolation, larval fish were lysed with 300 μL extraction buffer (10 mmol/L Tris-Cl, pH 8.0, 0.1 mol/L EDTA, pH 8.0, 0.5% SDS, 10 mg/mL Proteinase K), homogenized, and incubated at 55℃ for 3 h. Total DNA was extracted by phenol/chloroform, purified by ethanol precipitation, and dissolved in TE buffer (10 mM Tris-Cl, pH 7.5, 1 mM EDTA). The DNA concentration was estimated through agarose gel electrophoresis and stored at –20℃.

2.3 Molecular confirmation of the crossbred embryos

Before cDNA-AFLP was performed, total DNA from the crossbred embryos was PCR-amplified with a species-specific sequence characterized amplified region (SCAR) primers of zebrafish and Chinese rare minnow [14] (Table 1). The SCAR PCR with appropriate primers was expected to yield a 500 bp fragment in the Chinese rare minnow and a 248 bp band in the zebrafish. The PCR reactions included 100 ng DNA, 0.2 mM dNTP, 1×LA buffer, 0.5 U LA Taq (Takara, Dalian, China), and 10 mM of each primer. The PCR parameters were as follows: a pre-denaturation of 94℃ for 2 min, 30 cycles of amplification (94℃ for 30 s, 56℃ for 30 s, and 72℃ for 30 s). The PCR products were separated on 1.5% agarose gel by electrophoresis.

2.4 cDNA-AFLP

Reverse-transcription was performed with 2 μg total RNA using reverse transcriptase (Takara). First-strand and second-strand cDNA syntheses were carried out according to standard protocol with 2-base anchored oligo (dT) primers. Double-stranded cDNA was employed in the

Table 1 Primers used in SCAR, cDNA-AFLP, ZR6 cloning, and expression analysis

Primers	Sequence information	Usage
SCAR A	5-TGACGATAGCCAGCGAGACC-3 5-TGTGTGATTGGACCCCGTCG-3	Identification of crossbred embryos
SCAR B	5-TTTCGGCATTCAACCTTATTC-3 5-AAGACCGCATCCATTCAGC-3	Identification of crossbred embryos
*Eco*RI adapters	5-GACTGCGTACCAATTC-3	cDNA-AFLP analysis
*Mse*I adapters	5-GATGAGTCCTGAGTAA-3	
E-pre	5-GACTGCGTACCAATTC-3	cDNA-AFLP analysis
M-pre	5-GATGAGTCCTGAGTAA-3	
E-AN	5-GACTGCGTACCAATTCAN-3	cDNA-AFLP analysis
E-GN	5-GACTGCGTACCAATTCGN-3	
M-CN	5-GATGAGTCCTGAGTAACN-3	
M-GN	5-GATGAGTCCTGAGTAAGN-3	
ZR6u	5-ACRGGCCTGATGGAGGAG-3	Cloning of ZR6 coding region
ZR6d	5-CAGTYTTGTTAGACTATGGCG-3	
ZR1	5-GACGATAAAATCATAAGCGC-3 5-TTGCCCGTATGGACAATGT-3	Expression analysis
ZR3	5-ATGTAGCTCTGACCTGAATA-3 5-GCAGAATGTCAGATTTGG-3	Expression analysis
ZR4	5-ACCAATTCGATCTCAGTTGA-3 5-CCTGAAGCCAGTCTTCTTT-3	Expression analysis
ZR5	5-AGGCGGCGTTCACTTACC-3 5-CCGGCAGCAGTTTGTGATTA-3	Expression analysis
ZR6	5-TGTCCACAAAAATGATGCT-3 5-GCCTCAATATCACGGGCT-3	Expression analysis
ZR7	5-AGCACTGCTTTGTGGGT-3 5-GCAGCATCAGCATCATC-3	Expression analysis
ZR8	5-TGCCAATGGTTTGTCCAAC-3 5-CTGACAATAATATCTTACTTCCC-3	Expression analysis
ZR9	5-TCCTGCAGCAGTTTGTGATT-3 5-AGGAGGCGGCGTTCGC-3	Expression analysis
ZR10	5-TGAGGTCGAGAGCAAGG-3 5-CCACAGGCAAGATTTTACT-3	Expression analysis
ZR11	5-AGGTGATTCATAGTGTTCTC-3 5-ATGTGGGTGACCGCGT-3	Expression analysis
ZR12	5-TGATGCTGTTTTCTCCAA-3 5-TCATATATGAGGAACTGAA-3	Expression analysis
RZ1	5-TCAGTAAACACGTAGCCTAG-3 5-GGGTTCGTGGACAGGA-3	Expression analysis
RZ2	5-GCGCTGGTTCTAATGGAC-3 5-GGACCTCTGAATCAGTCG-3	Expression analysis
RZ4	5-GACACTCCCCAGATGATATG-3 5-ATGACGCCCAACCCTGG-3	Expression analysis

Continued

Primers	Sequence information	Usage
RZ6	5-TCTCATGGGACAGGTCTA-3 5-GCAGGAATCTTCCATTCT-3	Expression analysis
RZ7	5-TGATATCTGTGTTGCAGG-3 5-AAGGTTTCTGCCCTGAAA-3	Expression analysis
RZ11	5-CGAGGGATTCGTGGACG-3 5-CAGTAAGTACAAGAAACAG-3	Expression analysis
RZ12	5-TGAGTAATACGATACCAAAGTG-3 5-TAATACCAGACCCTATTGAGAA-3	Expression analysis
β-actin	5-TCACCACCACAGCCGAAAG-3 5-GGTCAGCAATGCCAGGGTA-3	Expression analysis

AFLP analysis following the previous study [11]. The cDNA was digested with restriction endonuclease *Eco*RI and *Mse*I (MBI, Vilnius, Lithuania) for 6 h. After ligation of adapters, pre-amplification of purified cDNA templates was performed for 30 cycles (94℃, 30 s; 45℃, 20 s; and 72℃, 60 s) with primers corresponding to the *Eco*RI and *Mse*I adapters, E-pre and M-pre (Table 1) in four cDNA pools. These cDNA pools came from ZR and RZ embryos at 50%- and 90%-epiboly stages, respectively.

After pre-amplification, the products were diluted 10 times in TE buffer, and 2 μL of each diluted product was subsequently utilized for selective amplification. A total of 64 primer combinations composed of eight pair primers were utilized to generate selective PCR amplification products. The PCR was performed in a 25 μL volume with 1×Taq buffer, 1.5 mM $MgCl_2$, 0.2 mM of each dNTP, 30 ng of each primer (Table 1), and 0.5 units of Taq DNA polymerase. Selective amplification was carried out according to the following parameters: 2 min at 94℃; 35 cycles of 30 s at 94℃, 30 s at 65℃ touchdown by 0.7℃ per cycle, and 1 min at 72℃; and 5 min at 72℃. The products were separated on 6% polyacrylamide gel containing urea (8.0 M) at 60 W. The cDNA bands were stained with silver nitrate following the protocol prescribed in the DNA Sequencing System kit (Promega). The desired bands were cut out from the gel and each gel slice was eluted with 50 μL ddH_2O at 37℃ for 30 min. The eluted cDNA was re-amplified with the same primers under the same conditions as the cDNA-AFLP selective amplifications, except that the PCR condition consisted of 30 cycles of 1 min at 94℃, 1 min at 60℃, and 1 min at 72℃. The amplified fragments were cloned into PMD18-T vector (Promega).

2.5 Dot blot, sequencing, and real-time RT-PCR analysis

Probe labeling and dot blot were performed using a DIG High Prime DNA Labeling and Detection Starter Kit I (Roche, Mannheim, Germany) according to the manufacturer's protocol. The cloned cDNA-AFLP products were sequenced and subjected to BLASTN analysis against a public database (http://www.ncbi. nlm.nih.gov/blast/).

Reverse-transcription PCR (RT-PCR) was utilized to verify the differential transcription of the isolated fragments in four cDNA pools. The primers which had perfect matches according to the sequences information pre-derived from zebrafish and Chinese rare minnow were utilized for RT-PCR analysis (Table 1). Meanwhile, β-actin was used in parallel as an internal reference for RT-PCR reactions, which could allow us to normalize the mRNA amount in each sample. The RT-PCR was performed using the RNA PCR Kit (AMV) Ver. 2.1 (Takara), following these parameters: for first-strand cDNA synthesis, 30℃ for 10 min, 42℃ for 30 min with random primers; and for PCR, 94℃ for 3 min and 35 cycles of 94℃ for 30 s, 56℃ for 40 s, and 72℃ for 1 min. The amplified fragments were separated by 1.5% (*w*/*v*) agarose gel electrophoresis.

Real-time PCR was used to analyze the expression changing levels of differentially transcribed genes in 90%-epiboly ZR and RZ embryos. The expression level was calculated using $2^{[-\text{delta delta C (T)}]}$ method [15]. For the –delta delta C (T) calculation to be valid, we assessed the absolute slope value of the log cDNA dilution versus delta C (T). The primer efficiencies of target and reference genes could be regarded as equal if the absolute slopes are <0.1. The real-time PCR reaction contained 1 μL of first-strand cDNA of 90%-epiboly stage ZR or RZ embryos, 1 μL SYBR Green (ABI, Foster City, CA, USA), 0.2 mM dNTP, 1×LA buffer, 0.5 U LA Taq (Takara), and 10 mM of each primer. PCR condition was 94℃ for 3 min, and 40 cycles of 94℃ for 30 s, 56℃ for 40 s, and 72℃ for 1 min. β-actin was utilized as an endogenous reference in all quantitative real-time PCR. Each gene was analyzed in three parallel samples and the mean values and standard deviations were calculated.

2.6 Cloning of ZR6 cDNA and its morpholino (MO) knockdown in zebrafish embryos

To clone the ZR6 coding region, the primers (Table 1) were designed according to the BLASTN hit sequence (BC093436). RT-PCR was performed using the RNA PCR Kit (AMV) Ver. 2.1 (Takara). PCR was carried out with 30 cycles of 94℃ for 30 s, 58℃ for 30 s, and 72℃ for 1 min. Purified PCR products were cloned into PMD18-T vector (Promega) by TA cloning and sequenced.

The deduced protein sequence of ZR6 was subjected to BLASTP analysis in the NCBI webserver. ZR6 cDNA was sub-cloned into pCS2+ vector for *in vitro* synthesis of mRNA. ZR6 mRNA was synthesized with *Not*I-linearized ZR6-pCS2+ by using the Ambion SP6 mRNA MESSAGE MACHINE kit (Ambion, Austin, TX, USA).

MO oligonucleotides of ZR6 were ordered from GeneTools (LLC, Philomath, OR, USA) according to translation start site sequence. Antisense ZR6 morpholino (5′-TGATTTTCTCATG TGTCTCCTCCAT-3′) and a control morpholino with four mispaired bases (5′-TGATgTTCT-

CcTGTGTaTCCTCtAT-3′) were used. MO and mRNA were injected into one-cell stage ZR embryos with an Eppendorf pressure injector (Model 5246; Eppendorf, Hamburg, Germany).

2.7 Analysis of blood vessel development in RZ embryos

Development of the RZ embryos' blood vessel system was analyzed by RNA whole-mount *in situ* hybridization with blood specific probe *draculin* (*drl*) at the developmental stage of 1-day post-fertilization (dpf). A digoxin-labeled antisense RNA probe of *drl* was synthesized from linearized plasmid with T3 RNA polymerase (Promega). *In situ* hybridization was carried out as described [16].

3. Results

3.1 Early development and molecular characterization of zebrafish, Chinese rare minnow, and crossbred embryos

The fertilization rates of zebrafish, Chinese rare minnow, and ZR embryo were 100%, whereas the rate of RZ was only ~10%. Microscopic observation revealed that Chinese rare minnow embryos developed more slowly than zebrafish embryos. Under 26℃ water temperature, zebrafish embryos developed to 50%- and 90%-epiboly within 6.5 and 9.5 h, respectively, whereas Chinese rare minnow embryos required 8 and 11 h. Interestingly, the ZR embryos displayed the same developmental pace as zebrafish embryos, whereas the RZ exhibited the same pace as the Chinese rare minnow (Fig. 1).

With SCAR PCR, the two types of crossbred embryos gave amplification of both 500 and 248 bp bands, in comparison to the presence of a 248 bp band in zebrafish and a 500 bp fragment in Chinese rare minnow. Therefore, although the mutual crossbred embryos had different origins of egg cytoplasm, they carried the same markers of nuclear genome—a combination of markers from zebrafish and Chinese rare minnow. However, the developmental capacities of both types of crossbred embryos were fairly different. The ZR, zebrafish, and Chinese rare minnow developed normally until the hatching stage, whereas the majority of RZ embryos died between the 50%- and 90%-epiboly stages, and the surviving RZ embryos manifested abnormal morphology after the tail bud stage. After 36 h, the RZ embryos were unable to display further development (Fig. 1L).

3.2 Isolation, sequencing, and identification of differentially expressed genes

A total of 83 transcript-derived fragments (TDFs), ranging from 70 to 400 bp in length, were differentially displayed by selective amplification using 64 primer combinations from four cDNA pools. Based on dot blot results, 24 TDFs manifested pronounced differences of

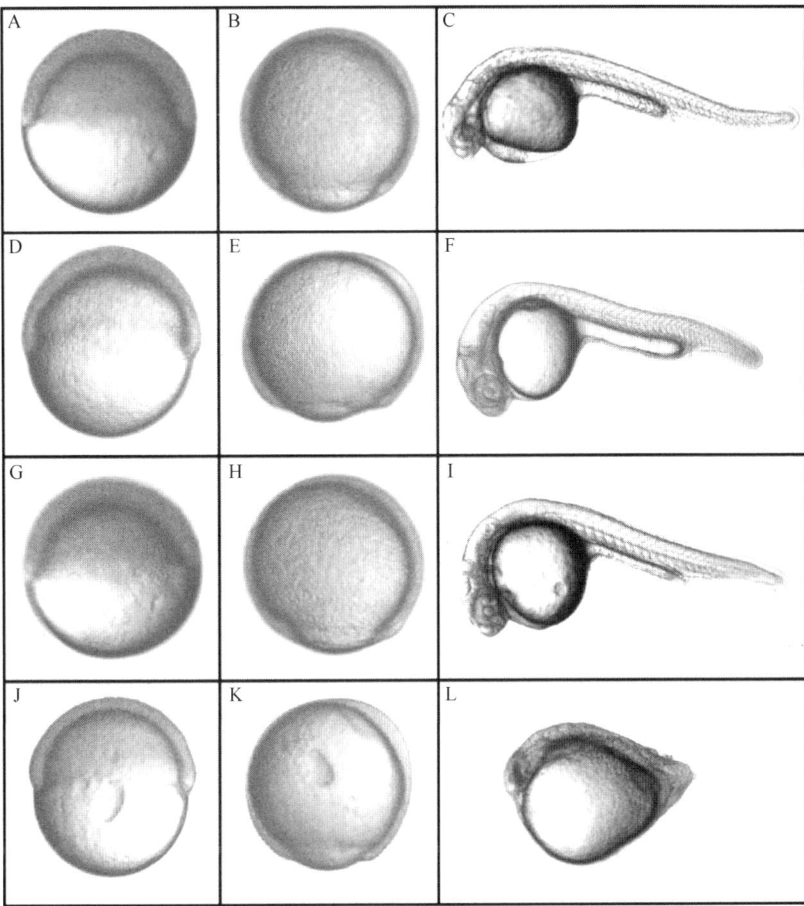

Fig. 1 Development of zebrafish embryos, Chinese rare minnow embryos, and crossbred embryos. A-C: zebrafish embryos; D-F: Chinese rare minnow embryos; G-I: ZR embryos; J-L: RZ embryos. Embryos shown in panels A, D, G, and J are at 50%-epiboly; B, E, H, and K are at 90% epiboly embryos; C, F, I, and L are 1-day embryos. Note that all the embryos developed normally before 90%-epiboly and RZ embryos showed abnormal morphology after 1 day of development (L)

staining intensity against cDNA pools between the mutual crossbred embryos. Sequencing and database mining showed that the 24 TDFs corresponded to 24 unigenes, of which 12 were up-regulated and 12 downregulated in RZ embryos (Tables 2 and 3). Of the 24 differentially expressed genes, the deduced amino acid sequences of 10 genes displayed high homology to proteins with known functions previously described in zebrafish. They shared significant similarities with a broad range of gene families encoding mitochondrial proteins, transferases, structural proteins, regulatory factors, and transport and protective proteins (Tables 2 and 3).

To confirm the up-regulation and down-regulation of specific genes in RZ embryos, 18 primer pairs that had no binding site polymorphism in the two species were designed and used to perform RT-PCR analysis. All the genes nominated as ZRx and RZx were down-regulated and up-regulated in RZ embryos, respectively (Fig. 2). Interestingly, genes such as ZR3, ZR5,

ZR7, ZR10, RZ2, RZ7, and RZ10 did not display any amplification in 50%-epiboly embryos of zebrafish and Chinese rare minnow, but manifested minimal expression in either ZR or RZ embryos within the same stage.

We selectively analyzed some genes by quantitative real-time RT-PCR in 90%-epiboly embryos because more remarkable changes of expression levels occurred at 90%-epiboly from the RT-PCR results. As illustrated in Fig. 3A, ZR1 (Class III histocompatibility antigen RD), ZR3 (glycosyltransferase), and ZR4 (16S ribosomal RNA gene), ZR6, ZR8, ZR9, ZR10, and ZR12 were down-regulated approximately 3-, 10-, 5-, 202-, 18-, 39-, 14-, and 25-fold, respectively, in RZ embryos at 90%-epiboly (Fig. 3A). Furthermore, RZ1, RZ2 (soluble neuropilin 2b1), RZ4, RZ6 (glycylpeptide *N*-tetradecanoyltransferase), and RZ10 (nucleosome assembly protein 1-like 4) were up-regulated approximately 10-, 9-, 5-, 16-, and 4-fold, respectively, in RZ embryos at 90%-epiboly (Fig. 3B). Overall, the quantitative real-time RT-PCR results were consistent with the data from dot blot analysis and RT-PCR analysis.

Table 2 Down-regulated TDFs in RZ embryos compared with ZR embryos

Fragments	Blast results	Length	Identities (%)	Gene types
ZR1	AY391454	73	83	*Danio rerio* class III histocompatibility antigen RD
ZR2	BX571712	205	80	*Danio rerio* similar to RNA pseudouridylate synthase domain
ZR3	XM_689125	209	91	*Danio rerio* similar to glycosyltransferase 25 domain
ZR4	AF036006	143	98	*Danio rerio* 16S ribosomal RNA gene
ZR5	XM_682775	205	99	*Danio rerio* similar to mitochondrial protein of bilateral origin
ZR6	BC093436	139	100	*Danio rerio* hypothetical protein LOC553509
ZR7	XM_698670	164	98	*Danio rerio* hypothetical protein LOC556595 (LOC556595)
ZR8	None	148		ND[a]
ZR9	XM_695567	141	100	*Danio rerio* similar to acid alpha glucosidase (LOC571921)
ZR10	None	185		ND[a]
ZR11	None	166		ND[a]
ZR12	None	90		ND[a]

a No data available

Table 3 Up-regulated TDFs in RZ embryos compared with ZR embryos

Fragments	Blast results	Length	Identities (%)	Gene types
RZ1	None	124		ND[a]
RZ2	AY437401	356	97	*Danio rerio* soluble neuropilin 2b1 (nrp2b)
RZ3	XM_690843	148	94	*Danio rerio* similar to methylenetetrahydrofolate reductase
RZ4	None	152		ND[a]
RZ5	None	253		ND[a]

Continued

Fragments	Blast results	Length	Identities (%)	Gene types
RZ6	XM_677981	181/193	93	*Danio rerio* similar to glycylpeptide *N*-tetradecanoyltransferase 2
RZ7	BX323464	180	98	*Danio rerio* DNA sequence
RZ8	AY654014	105	97	*Penaeus monodon* sex and growth traits AFLP marker sequence
RZ9	XM_704066	152	100	*Danio rerio* hypothetical protein
RZ10	NM_00108934	178	79	*Danio rerio* nucleosome assembly protein 1-like 4 (nap114)
RZ11	CR9258107	234	96	Zebrafish DNA sequence from clone DKEY-250M6
RZ12	BC124467	117	94	*Danio rerio* zgc:153818, mRNA

a No data available

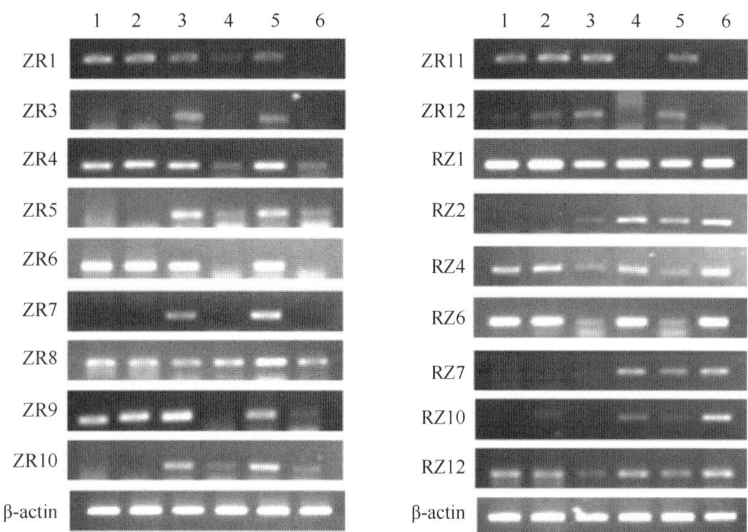

Fig. 2 RT-PCR analysis of differentially expressed genes at different developmental stages and different embryos. Clone names are listed on the left side of each gel. Lane 1, 50%-epiboly zebrafish; lane 2, 50%-epiboly Chinese rare minnow; lane 3, 50%-epiboly ZR; lane 4, 50%-epiboly RZ; lane 5, 90%-epiboly ZR; lane 6, 90%-epiboly RZ

3.3 Characterization and knockdown of ZR6 gene

ZR6 was characterized as a novel cDNA of 774 bp coding region (accession number: EF371470), encoding a putative protein composition of 257 amino acids. BLASTP analysis through the NCBI server showed that ZR6 contained two types of conserved domains. One was COG5189, SFP1, a putative transcriptional repressor regulating G2/M transition, and the other was COG5048, a zinc finger DNA-binding domain.

After injecting ZR6 MO at a concentration of 0.5 mM, ZR embryos could not continue developing after 24 h because of the various defects in development. In contrast, embryos

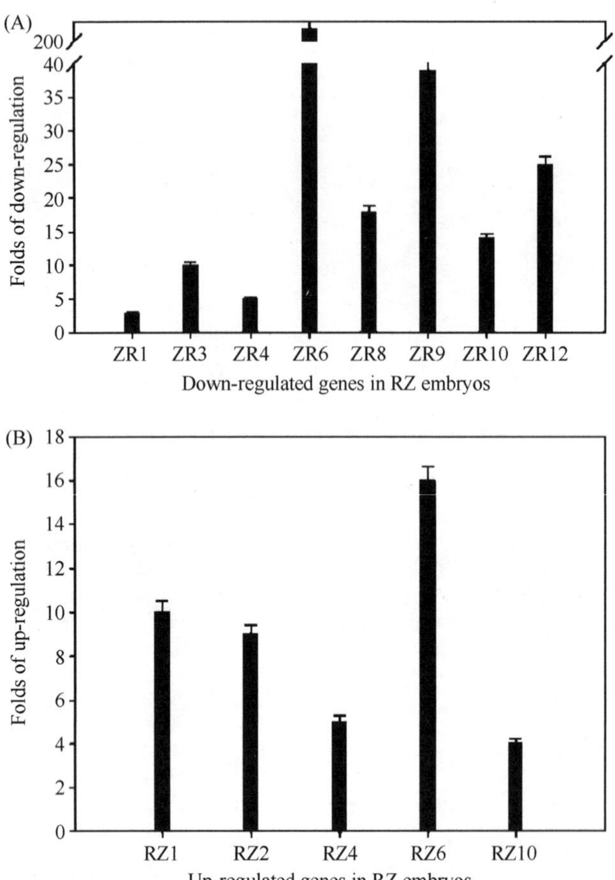

Fig. 3 Real-time PCR analysis of some differentially expressed genes at 90%-epiboly stage in crossbred embryos. (A) Quantitative real-time RT-PCR of ZR genes (down-regulated genes in RZ embryos). The *Y*-axis value represents the down-regulation fold of a specific gene in RZ embryos when compared with the expression level of the same gene in ZR embryos. (B) Quantitative real-time RT-PCR of RZ genes (upregulated genes in RZ embryos). The *Y*-axis value represents the up-regulation fold of a specific gene in RZ embryos when compared with the expression level of the same gene in ZR embryos. The error bars show the mean ± S.D.

injected with misMO at the same concentration developed normally. More importantly, the phenotype induced by ZR6 MO injection was, to a certain extent, similar to that of RZ embryos by showing a reduced head size, short tail, and disorganized somites. To validate if the phenotype resulting from ZR6 MO injection was morpholino specific, we co-injected *in vitro* synthesized ZR6 mRNA and observed an effective rescue effect (data not shown).

3.4 Abnormal development of blood system in RZ embryos

We analyzed the vascular vessel development of RZ embryos at 1 dpf because one of the up-regulated genes, RZ2, was found to be *nrp2b*, a gene related to vascular development [17].

The *drl*-labeling vascular system of RZ embryos was disorganized, and the *drl* staining was much stronger in RZ embryos than in ZR or normal embryos (Fig. 4).

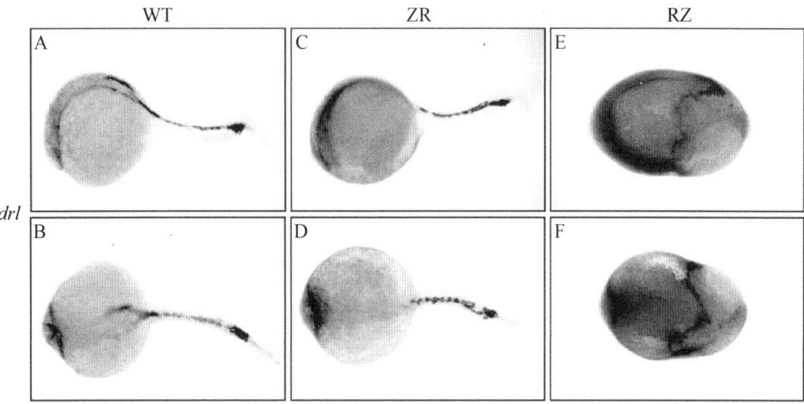

Fig. 4 The development of blood system was disrupted in RZ embryos. All embryos were fixed at 1 dpf and checked by *drl* antisense RNA probe. A and B: wild type zebrafish embryo; C and D: ZR embryo; E and F: RZ embryo. All the embryos are oriented with anterior to the left, the embryos in A, C, and E are lateral view, and the embryos in B, D, and F are dorsal view. Note that the *drl* staining in the embryos in E and F are disrupted when compared with embryos in A, B, C, and D

4. Discussion

In the past few decades, substantial efforts have been undertaken to understand the interaction between nucleus and cytoplasm in fish, with the majority of studies performed using NT embryos [6,18-21]. In the present study, a relatively innovative system anchored on crossbred embryos of zebrafish and Chinese rare minnow was utilized to address this question, since RZ embryos could be safely considered as the cross-species NT embryos derived from ZR nuclei and Chinese rare minnow egg cytoplasm. We inferred that the early development of fish embryos was largely affected by the maternal factors deposited in eggs, since the developmental pace of ZR embryos was similar to that of zebrafish embryos, and RZ embryos similar to Chinese rare minnow embryos.

With this system, we discovered that soluble neuropilin 2b1 (*nrp2b*) (RZ2) was up-regulated 9-fold in RZ embryos. Previous studies have shown that *nrp2b* was involved in VEGF-mediated vessel development as a receptor [22], and its knockdown in zebrafish caused vessel malformations and a pericardial defect [17]. Likewise, RZ embryos showed disruption of the blood vessel system as revealed by blood specific staining, which might have resulted from the over-expression of *nrp2b*.

In addition, when a novel zinc finger protein gene, ZR6 (down-regulated gene in RZ embryos), was knocked down by antisense morpholino injection in ZR embryos, the resulting phenotype was analogous with RZ embryos. This suggested that the down-regulation of ZR6 in RZ embryos should be partially responsible for the developmental malformation of RZ embryos. Based on the analysis of conserved domains in ZR6, ZR6 is likely a novel zinc finger transcriptional repressor involved in cell cycle regulation. Further study is certainly needed to characterize the molecular function of ZR6 gene in the developmental regulation of zebrafish embryo.

In RZ embryos, there were expression changes of two mitochondrial genes (ZR4, 16S ribosomal RNA gene; ZR5, mitochondrial protein of bilateral origin). Previous studies have shown that, mitochondria have a complex symbiotic relationship with their eukaryotic host cell, and changes in mitochondrial function could lead to a variety of nuclear changes, such as lowering of DNA repair that leads to increased mutation rates [23], changes in gene expression, simple loss of energy sources required for normal chromosome transactions or chromosome segregation, and even the occurrence of apoptosis [24], which possibly result from increases in reactive oxygen species [25]. This suggested that the expression alteration of mitochondrial genes in RZ embryos might lead to the developmental arrest and even the apoptosis.

In cross-species NT embryos, whether heteroplasmic or not, incompatibility between the nuclear and cytoplasmic counterpart may be partially responsible for the developmental arrest [26], and the interaction of mitochondria from one species with the nucleus of an unrelated species may result in serious metabolic disruption [26-28]. In accordance with these, we identified five differentially expressed genes (ZR2, ZR3, ZR9, RZ3, and RZ6) related to metabolism. For example, RNA pseudouridylate synthase (ZR2) belongs to a gene family involved in modifying bases in cytoplasmic and mitochondrial tRNAs, U2 snRNA, and rRNAs from cytoplasmic ribosomes [29]. Glycosyl-transferase (ZR3) is an enzyme that catalyzes the transfer of glycosyl residues to an acceptor, both during degradation and during biosynthesis of polysaccharides, glycoproteins, and glycolipids [30]. Acid alpha glucosidase (ZR9) is an active enzyme in lysosomes that breaks down glycogen into glucose, to provide an important energy source for cells [31]. Methylenetetrahydrofolate reductase (Mthfr) (RZ3) is a crucial enzyme that regulates the metabolism of homocysteine and methionine by catalyzing the reduction of methylenetetrahydrofolate to methyltetrahydrofolate, the methyl donor for methionine synthesis from homocysteine [32]. Glycylpeptide *N*-tetradecanoyltransferase 2 (RZ6), which belongs to the *N*-myristoyltransferase (NMT) family, catalyzes the reaction of *N*-terminal myristoylation of many proteins that have diverse biological functions such as signal transduction, cellular transformation, and oncogenesis [33]. Here, the up- or down-regulation of these genes would strongly affect the metabolic behavior of the cells of RZ embryos.

Overall, we identified 24 differentially expressed genes between mutual crossbred embryos of zebrafish and Chinese rare minnow. Expression changes of these genes were likely induced by the improper nucleo–cytoplasmic interaction and might lead to various developmental malformations of RZ embryos. Thus, this gave a possible explanation of why we could not obtain cross-species cloned fish after years of nuclear transferring of zebrafish nuclei to Chinese rare minnow eggs. Our results, combined with those of other studies, should contribute to an understanding of the molecular mechanism underlying the interaction between nucleus and cytoplasm from different species, as well as the difficulties to obtain cross-species cloned animals.

Acknowledgements We thank Ming Li for providing us assistance with the fish. We are also grateful to Perry Hackett for his valuable comments. This study was supported by the State Key Fundamental Research of China (Grant No. 2004CB117406) and the National Natural Science Foundation of China (Grant No. 30771100).

References

[1] Wilmut I, Schnieke AE, McWhir J, Kind AJ, Campbell KH. Viable offspring derived from fetal and adult mammalian cells. Nature 1997; 385: 810-3.

[2] Wakayama T, Perry AC, Zuccotti M, Johnson KR, Yanagimachi R. Full-term development of mice from enucleated oocytes injected with cumulus cell nuclei. Nature 1998; 394: 369-74.

[3] Kato Y, Tani T, Sotomaru Y, Kurokawa K, Kato J, Doguchi H, *et al.* Eight calves cloned from somatic cells of a single adult. Science 1998; 282: 2095-8.

[4] Baguisi A, Behboodi E, Melican DT, Pollock JS, Destrempes MM, Cammuso C, *et al.* Production of goats by somatic cell nuclear transfer. Nat Biotechnol 1999; 17: 456-61.

[5] Loi P, Ptak G, Barboni B, Fulka Jr J, Cappai P, Clinton M. Genetic rescue of an endangered mammal by cross-species nuclear transfer using post-mortem somatic cells. Nat Biotechnol 2001; 19: 962-4.

[6] Sun YH, Chen SP, Wang YP, Hu W, Zhu ZY. Cytoplasmic impact on cross-genus cloned fish derived from transgenic common carp (*Cyprinus carpio*) nuclei and goldfish (*Carassius auratus*) enucleated eggs. Biol Reprod 2005; 72: 510-5.

[7] Mastromonaco GF, Perrault SD, Betts DH, King WA. Role of chromosome stability and telomere length in the production of viable cell lines for somatic cell nuclear transfer. BMC Dev Biol 2006; 6: 41.

[8] Lagutina I, Brunetti D, Lazzari G, Galli C. Preliminary data on pig–bovine interspecies nuclear transfer embryo development. Reprod Fertil Dev 2005; 18: 134-5.

[9] Pei DS, Sun YH, Chen SP, Wang YP, Hu W, Zhu ZY. Identification of differentially expressed genes from the cross-subfamily cloned embryos derived from zebrafish nuclei and rare minnow enucleated eggs. Theriogenology 2007; 68: 1282-91.

[10] Schier AF, Talbot WS. Molecular genetics of axis formation in zebrafish. Annu Rev Genet 2005; 39: 561-613.

[11] Bachem CW, van der Hoeven RS, de Bruijn SM, Vreugdenhil D, Zabeau M, Visser RG. Visualization of differential gene expression using a novel method of RNA fingerprinting based on AFLP: analysis of gene expression during potato tuber development. Plant J 1996; 9: 745-53.

[12] Vuylsteke M, Peleman JD, van Eijk MJ. AFLP-based transcript profiling (cDNA-AFLP) for genome-wide expression analysis. Nat Protoc 2007; 2: 1399-413.

[13] Westerfield M. The zebrafish book. Eugene, OR: University of Oregon Press; 1995.

[14] Hu W, Pei DS, Dai J, Chen SP, Sun YH, Wang YP, *et al.* Identification and application of SCAR markers in detecting cross-species cloned embryos between zebrafish and rare minnow. High Tech Lett 2006; 16: 959-63.

[15] Livak KJ, Schmittgen TD. Analysis of relative gene expression data using real-time quantitative PCR and the $2^{(-\text{delta delta C(T)})}$

method. Methods 2001; 25: 402-8.

[16] Thisse B, Heyer V, Lux A, Alunni V, Degrave A, Seiliez I, *et al.* Spatial and temporal expression of the zebrafish genome by large-scale in situ hybridization screening. Methods Cell Biol 2004; 77: 505-19.

[17] Martyn U, Schulte-Merker S. Zebrafish neuropilins are differentially expressed and interact with vascular endothelial growth factor during embryonic vascular development. Dev Dyn 2004; 231: 33-42.

[18] Zhu ZY, Sun YH. Embryonic and genetic manipulation in fish. Cell Res 2000; 10: 17-27.

[19] Lee KY, Huang H, Ju B, Yang Z, Lin S. Cloned zebrafish by nuclear transfer from long-term cultured cells. Nat Biotechnol 2002; 20: 795-9.

[20] Ju B, Pristyazhnyuk I, Ladygina T, Kinoshita M, Ozato K, Wakamatsu Y. Development and gene expression of nuclear transplants generated by transplantation of cultured cell nuclei into non-enucleated eggs in the medaka *Oryzias latipes*. Dev Growth Differ 2003; 45: 167-74.

[21] Bureau WS, Bordignon V, Leveillee C, Smith LC, King WA. Assessment of chromosomal abnormalities in bovine nuclear transfer embryos and in their donor cells. Cloning Stem Cells 2003; 5: 123-32.

[22] Rossignol M, Gagnon ML, Klagsbrun M. Genomic organization of human neuropilin-1 and neuropilin-2 genes: identification and distribution of splice variants and soluble isoforms. Genomics 2000; 70: 211-22.

[23] Flury F, von Borstel RC, Williamson DH. Mutator activity of petite strains of *Saccharomyces cerevisiae*. Genetics 1976; 83: 645-53.

[24] Brenner C, Kroemer G, Apoptosis. Mitochondria—the death signal integrators. Science 2000; 289: 1150-1.

[25] Karthikeyan G, Resnick MA. Impact of mitochondria on nuclear genome stability. DNA Repair (Amst) 2005; 4: 141-8.

[26] Dey R, Barrientos A, Moraes CT. Functional constraints of nuclear–mitochondrial DNA interactions in xenomitochondrial rodent cell lines. J Biol Chem 2000; 275: 31520-7.

[27] Kenyon L, Moraes CT. Expanding the functional human mitochondrial DNA database by the establishment of primate xenomitochondrial cybrids. Proc Natl Acad Sci USA 1997; 94: 9131-5.

[28] Barrientos A, Kenyon L, Moraes CT. Human xenomitochondrial cybrids. Cellular models of mitochondrial complex I deficiency. J Biol Chem 1998; 273: 14210-7.

[29] Ansmant I, Massenet S, Grosjean H, Motorin Y, Branlant C. Identification of the *Saccharomyces cerevisiae* RNA:pseudouridine synthase responsible for formation of psi(2819) in 21S mitochondrial ribosomal RNA. Nucleic Acids Res 2000; 28: 1941-6.

[30] Shibayama K, Ohsuka S, Tanaka T, Arakawa Y, Ohta M. Conserved structural regions involved in the catalytic mechanism of *Escherichia coli* K-12 WaaO (RfaI). J Bacteriol 1998; 180: 5313-8.

[31] Fukuda T, Roberts A, Plotz PH, Raben N. Acid alpha-glucosidase deficiency (Pompe disease). Curr Neurol Neurosci Rep 2007; 7: 71-7.

[32] Goyette P, Sumner JS, Milos R, Duncan AM, Rosenblatt DS, Matthews RG, *et al.* Human methylenetetrahydrofolate reductase: isolation of cDNA, mapping and mutation identification. Nat Genet 1994; 7: 195-200.

[33] Rajala RV, Datla RS, Moyana TN, Kakkar R, Carlsen SA, Sharma RK. *N*-Myristoyltransferase. Mol Cell Biochem 2000; 204: 135-55.

斑马鱼与稀有鮈鲫杂交胚胎中的差异表达基因

刘 静[1] 孙永华[1] 王燕舞[1,2] 王 娜[1] 裴得胜[1]
汪亚平[1] 胡 炜[1] 朱作言[1]

1 中国科学院水生生物研究所，淡水生态与生物技术国家重点实验室，武汉 430072
2 武汉大学，生命科学学院，武汉 430072

摘　要　在合子基因组激活之前，细胞核和卵细胞质中母源因子的通讯和互作对早期发育至关重要。本研究利用斑马鱼和稀有鮈鲫的正反交胚胎，检测两个物种中细胞核和卵细胞质之间的相互作用。虽然这两种杂交胚胎具有相同的细胞核，但是由于不同卵细胞质的影响而获得不同的发育能力。采用 cDNA-AFLP 技术，我们比较了正反杂交胚胎在不同发育时期(50%外包期和 90%外包期)的转录组差异。利用 4 种 cDNA 文库和 64 个引物组合产生 3000 个 cDNA 片段，差异显示的 TDFs 用斑点印迹和半定量及定量 RT-PCR 作进一步分析。与 ZR 胚胎相比，在 RZ 胚胎中有 12 个基因上调表达和 12 个基因下调表达。将这些片段进行测序和 BLASTN 分析，结果发现这些序列编码与增殖、生长、发育相关的多种蛋白。我们选择在 RZ 胚胎中显著下调的基因 ZR6 进行基因敲除研究，结果显示 ZR6 的基因敲除胚出现了类似于 RZ 胚胎的表型。

Identification of A Novel Gene K23 Over-Expressed in Fish Cross-Subfamily Cloned Embryos

De-Sheng Pei Yong-Hua Sun Zuo-Yan Zhu

State Key Laboratory of Freshwater Ecology and Biotechnology, Institute of Hydrobiology, Chinese Academy of Sciences, Wuhan 430072

Abstract A novel gene–K23, differentially expressed in cross-subfamily cloned embryos, was isolated by RACE-PCR technique. It had 2,580 base pairs (bp) in length, with a 1,425 bp open reading frame (ORF) encoding a putative protein of 474 amino acids (aa). Bioinformatic analysis indicated that K23 had 22 phosphorylation sites, but it had no signal peptides. Developmental expression analysis in zebrafish showed that K23 transcripts were maternally expressed in ovum and the amount of K23 transcripts increased gradually from zygote to pharyngula period. Subcellular localization analysis revealed that K23 protein was homogeneously distributed both in nuclei and cytoplasm. Taken together, our findings indicate that K23 gene is a novel gene differentially expressed in fish crosssubfamily cloned embryos.

Keywords nuclear reprogramming; nuclear transfer; rare minnow (*Gobiocypris rarus*); zebrafish (*Danio rerio*)

Introduction

To retain the genetic viability of a certain species near extinction, cross-species nuclear transfer (NT) becomes a potent approach. Utilization of oocytes for recipient cytoplasts from other species that are accessible and abundant is an exciting possibility for endangered species with limited availability of oocytes [1]. However, most embryos produced by cross-species NT are unable to develop to later stages because reproductive NT required many basic biological necessities, such as the compatibility of nuclear-mitochondrial interaction and the reversal of the differentiated state of the transferred nucleus [2-3]. Efficient nuclear reprogramming needs appropriate gene expression during embryonic development. Thus, differential gene analysis in cross-species cloned embryos can give hint for unveiling the regulatory mechanisms of the feasibility of rescuing genetically endangered animal species.

In the previous study, we have chosen two laboratory fish species, rare minnow (*Gobiocypris rarus*) and zebrafish (*Danio rerio*), as a model to study the cross-subfamily NT [4-5]. By using a suppression subtractive hybridization (SSH) approach, we have totally screened out 50

文章发表于 *Molecular Biology Reports*, 2009, 36(6): 1375-1380

differentially expressed genes in the cloned embryos at sphere stage [6]. After dot blotting confirmation, real-time RT-PCR analysis showed that a novel gene—K23 was over-expressed in the cloned embryos more 15-folds than that in normally fertilized zebrafish embryos [6]. In this paper, we report the identification of K23 gene and investigate its expression profiles.

Materials and Methods

RACE-PCR amplification

The full-length cDNA of K23 gene was isolated by RACE-PCR technique. Using the SMART cDNA as template [6], the combination of universal primer SMART F and K23 R or universal primer SMART R and K23 F was used for 5′ or 3′ RACE PCR, respectively (Table 1). The generated PCR products were sequenced, and the full-length cDNA of K23 was composed by both the coding sequence (CDS) and the 5′ and 3′ UTRs.

Table 1 Primers used in the present study

Names	Sequences	Length (bp)
SMART F	5-CAACGCAGAGTACGCGGG-3	18
SMART R	5-TCAACGCAGAGTACT(16)-3	30
K23 F	5-AAAACAGTGGGTGATGGGTAGA-3	22
K23 R	5- CGGTGCCAAAAGGGACAA-3	18
K23DW F	5-AACTGCAGATGACGCATGGAACTTACAAC-3	29
K23DW R	5-CGGGATCCCTTCAGTCTTTGTAAGGCGG-3	28
GAPDH F	5-GTGTAGGCGTGGACTGTGGT-3	20
GAPDH R	5-TGGGAGTCAACCAGGACAAATA-3	22

Bio-information analyses

Homology search for K23 gene was performed on the sequences listed in EMBL/GenBank/DDBJ databases using EST-BLAST and Protein–protein BLAST (blastp) at the web site of the National Center of Biotechnology Information (NCBI) (http://www.ncbi. nlm.nih. gov/blast/Blast.cgi). Secondary structure analysis of K23 protein sequence was performed by DNAstar software (Lasergene, Madison, Wis.). Phosphorylation sites and signal peptide were predicted by NetPhos 2.0 software (http://www.cbs.dtu.dk/ services/NetPhos/) and SignalP 3.0 software (http://www.cbs.dtu.dk/services/SignalP/), respectively.

Semi-quantitative RT-PCR assay

Zebrafish embryos were generated by *in vitro* fertilization and raised as previously described [6]. Embryos were staged according to Kimmel *et al.* [7]. Total RNA of eight samples (ovum, zygote, 256-cell stage, sphere stage, 50%-epiboly stage, 90%-epiboly stage, 15-somite stage, pharyngula period) was separately isolated using SV^{TM} total RNA kit (Promega). Then, 3 μg total RNA was reverse transcribed (RT) for each sample and 2 μl of the RT product was amplified to analyze the expression patterns of K23 by RT-PCR assay. Primers used for amplifying the cDNA fragments of K23 were K23 F and K23 R (Table 1). As an internal standard, GAPDH F and GAPDH R primers were used to amplify the constitutively expressed gene-GAPDH (Table 1). In order to ensure the reactions remained in log-linear range, 20 PCR cycles were used.

Whole-mount in situ hybridization (WISH) analysis

For WISH, embryos were fixed in MEMPFA (100 mM Mops (Sigma), pH 7.4; 2 mM EGTA (Sigma); 1 mM $MgSO_4$ (Merck); 4% (*w*/*v*) paraformaldehyde (Sigma)) at different developmental stages. The methods for generation of full-length K23 antisense probes and WISH analysis followed the protocol described by Pei *et al.* [8]. Images of zebrafish embryos were recorded using an Olympus SZX12 microscope and a digital camera.

Cell culture and sub-cellular localization of K23

Epithelioma papulosum cyprinid (EPC) cells from carp (*Cyprinus carpio*) were cultured as described by Zhou *et al.* [9]. For fluorescence microscopy, the coding region of K23 gene was amplified using K23DW primers (Table 1) and cloned into pEGFP-N3 (BD Biosciences) using *Pst*I and *Bam*HI sites. After sequencing validation, the pK23-EGFP and pEGFP-N3 (as negative control) were transfected into EPC cells using Lipofectamine 2000 reagent (Invitrogen), respectively. At 24 h after transfection, cells were removed by trypsin/EDTA and cell nuclei were stained with 5 mg/l Hochest 33342 (Calbiochem). Sub-cellular localization of K23 was judged by the co-localization of GFP protein.

Results

Cloning and characterization of K23 gene

Full-length cDNA of K23 gene was obtained from a SMART cDNA library. It is 2,580 bp in length with an open reading frame (ORF) of 1,425 bp encoding a putative protein of 474 aa, a 5′ untranslated region (UTR) of 309 bp, and a 3′ UTR of 846 bp. It contains an mRNA instable motif (ATTTA) and a poly(A) signal (AATAAA) followed by a poly(A) tail (Fig. 1).

Homology search of public database found no homologous known gene in other species. The secondary structure analysis by DNAstar software shows that K23 protein has much more β-folds and antigenic regions than α-helices and hydrophilicity regions (Fig. 2a). Predicted by SignalP 3.0 software, K23 protein has no signal peptide. Interestingly, phosphorylation sites analysis by NetPhos 2.0 Server shows that there are 22 phosphorylation sites (12 Ser, 6 Thr and 4 Tyr) (Fig. 2b).

```
GTTACCAGGAATACACAATCCCCATTATTGTATGACTAATCATATTTAGATCTGCATTAATAACAT 66
ATGGGATTCCTGAACAAACACAGTGGTTATTATATACCGATAACTTATTGTGTAGGTGAATGTTCAAATAACATAGACAAT 147
TTCACCAATCCACAAAATTGTACATAGACATCTAAAAATTTTGAATTGACCACCTGTGGATACACCAAATCCAAAGAATGT 228
CCAAAATTACACCCAATTCCTCCTGGTGGACTGATTAGAGGGAACCTTAACAAGACTCTATTACATGATGTAAATTTGGTG 309
M T H G T Y N I S N I F S A Y T K H C N A N D E S Y A 27
ATGACGCATGGAACTTACAACATTTCCAATATTTTTTCTGCTTATACAAAACATTGTAATGCCAACGATGAATCATATGCA 390
W L Q S R I N D L I S D Y K D R L A V Q I P S A R S K 54
TGGCTACAGTCACGAATCAATGATTTAATTTCTGATTACAAAGACAGACTTGCTGTCCAAATTCCATCTGCTCGTTCAAAA 471
R D L L G N I A G L F G S I N S A A N T Y Q I T K Q S 81
CGAGACCTACTTGGTAACATAGCAGGCTTATTCGGATCAATTAATTCAGCAGCTAACACCTACCAAATAACCAAACAATCA 552
Q F S T W L M D Q V A T G F Q H I T N S N D N L I K A 108
CAATTCTCAACATGGTTGATGGACCAGGTGGCCACTGGGTTTCAACACATCACCAATAGCAATGATAACCTAATCAAAGCG 633
V R S E A Q A L L T I S H T L F N Q T R T I E R D L A 135
GTTAGATCCGAAGCGCAGGCGCTTCTAACGATCAGCCATACATTGTTCAATCAGACCCGTACCATCGAACGTGACTTAGCC 714
C R S Y T Q D L F T A T R Q E I F D L R L H K T P R H 162
TGTAGATCATATACACAAGATTTGTTCACTGCCACAAGACAGGAAATTTTTGACCTTCGTCTTCATAAAACTCCAAGACAT 795
V L N D L I E I L D L H R W F Y A E K M K N V R Y S E 189
GTTTTAAACGATTTGATTGAGATTTTAGACCTACATAGATGGTTCTATGCGGAGAAAATGAAAAATGTAAGATATTCAGAA 876
L L S T I M M Y T G K E C T G C I G F F A T F' P L I H 216
CTGTTATCTACAATTATGATGTATACTGGAAAAGAATGTACTGGTTGCATTGGGTTTTTTGCCACCTTTCCTTTGATCCAC 957
P D Q V Y P N S T T I R S I G M V V K D Q V I K W D H 243
CCAGACCAAGTTTATCCAAATTCCACTACTATCCGTTCCATCGGCATGGTAGTTAAAGATCAGGTAATTAAATGGGACCAC 1038
L T G Y M T L K G T E T L F T S R T C C Q E T H N Y V 270
CTCACTGGTTATATGACTTTGAAGGGTACTGAGACCTTATTTACCAGTCGCACTTGTTGTCAAGAAACTCATAATTATGTT 1119
V C T C N T L Q P F S P N D N K L V N V Q S L H G H S 297
GTTTGTACATGTAACACATTACAACCTTTCTCTCCCAATGATAACAAACTTGTGAATGTCCAATCATTGCATGGCCATTCG 1200
N A V Q V S H T Q W C I I S E M N S F T Y G G M T C P 324
AATGCTGTTCAAGTGTCTCATACACAGTGGTGCATCATCAGTGAGATGAATTCTTTCACATATGGAGGAATGACCTGTCCT 1281
A N Y S F C L E V T E D F S M G Q I D I L G R M P Q D 351
GCCAATTACTCCTTTTGCTTGGAAGTGACAGAGGACTTTTCTATGGGTCAGATTGACATCCTTGGAAGGATGCCGCAGGAC 1362
A E V S P W W Y D T F Y E H G T Q A L V E T M D L V Q 378
GCAGAGGTCTCTCCATGGTGGTATGACACCTTCTATGAACACGGAACACAAGCACTGGTCGAGACGATGGACTTGGTACAG 1443
K V I L Q T E Y H L S Q A Q V E T N L A R K T A Q I L 405
AAAGTCATCCTTCAAACAGAGTACCACCTTTCTCAAGCGCAAGTGGAGACAAACTTGGCGCGAAAGACTGCACAGATTCTG 1524
T S S S I R S A Q T A Y S W W D W I L R G C A V G S A 432
ACTAGCTCATCTATTCGATCGGCACAGACTGCATACTCTTGGTGGGACTGGATACTTCGAGGATGTGCTGTGGGAAGTGCC 1605
L I F F L T V F Q C C Y F R H L I R S V K S S T N A I 459
CTCATCTTCTTCCTCACAGTATTCCAATGTTGTTACTTCAGACATCTCATCAGATCTGTGAAATCATCCACCAACGCTATC 1686
L A L S P L Q L P A L Q R L K * 474
TTGGCCCTCAGTCCCCTGCAACTACCCGCCTTACAAAGACTGAAGTGA 1734
CACCAGCGAAACCCTTGTGAATGGCTCACGGTGCCAAAAGGGACAAATCTGTAAGGAAGTTGCCAGGGCGTCTGGCATGAG 1815
AAACGATGACTTCCTGGGAACAAAGGACTTTTTCTAAGGATCAACATTTGAACACTCTTCTCAATACATAACAGCCACACC 1896
CAAAACACACCATACTGAGACACACACATCTCACACTCACAATCTACCCATCACCCACTGTTTTGCCCATTTGCTGCCACA 1977
CTGAAAACCCATTGTGAACCTACGCAATGCTAAATATGTACAGTGTGTAACCATTCATGTTTAGTATCATTTGAAACGTGT 2058
CAGTGCGTCTAGTAAAATGGTCTCATTCTCCAAACCAGAATACAAACCTCTAAGAAGTTCTATAAGTTGAGAAAATATTGT 2139
GTGACGTTAAGGAGGTGATCCCCCACAGATCTCTAATTCAGGATACCTCCTGTATGTAAATACGTATTCTCCTTCAAAAAT 2220
GGTGAGAACCCATTCAAACACAAATTCTGATTGCCCATTTGTGACCCCCTGAATAAGGGGCCAAAACCTAATTATAATCAC 2301
AGTAAACCACCATTTGGTATCTATTGAATTATCTGGGTATTTAAAGGAAAGTGACACTAAACTCGGGGAGTCTGTATCTAG 2382
ATATCCCACACTGTTTGTGTGTAGTCTTTCACGACCAGGTAAAGATTTCTGGAATCCTTGATTCTGGAATTATTGTTTCTG 2463
TATAGGATTGATTGATGCTTTTTATCCTTTAAATTGAACTTTGATCGATGTTTCTTACCTGGTTAATAAATCGTTATGTTT 2544
AAACCTAAAAAAAAAAAAAAAAAAAAAAAAAAAAA 2580
```

Fig. 1 Nucleotide sequence and deduced amino acid sequence of K23 gene (Accession No. AY882988). The nucleotides (lower row) and deduced amino acids (upper row) are numbered at the right side of the sequences. The start codon (ATG) is in bold and the stop codon (TGA) is indicated by an asterisk. An unstable motif (ATTTA) and Poly (A) signal (AATAAA) are underlined

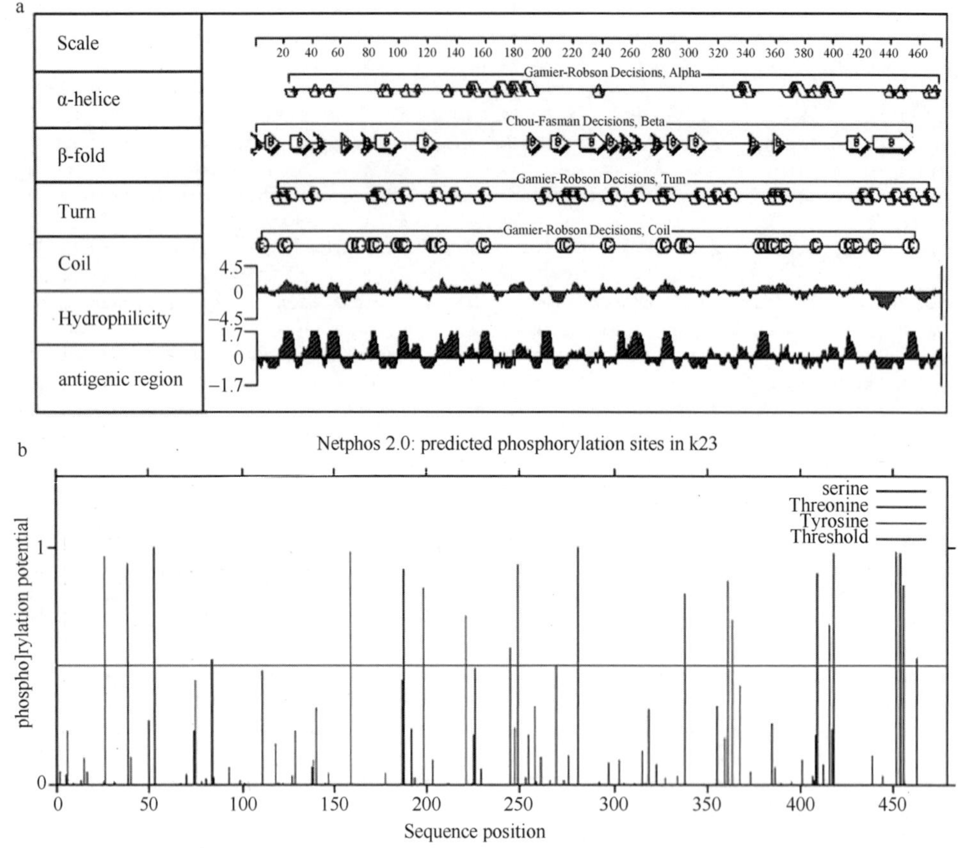

Fig. 2 Bioinformatic analysis of K23 amino acids. (a) Secondary structure, hydrophilicity and antigenic region are analysed by DNAstar program. (b) Phosphorylation sites are analysed by NetPhos 2.0 software

Expression profile of K23 during development

To figure out the expression pattern of K23 gene during embryonic development, semi-quantitative RT-PCR was performed. As shown in Fig. 3a, K23 transcripts were found in all the checked developmental stages and it was expressed maternally in the oocytes. Interestingly, K23 transcripts increased gradually from zygote to pharyngula period (24 hpf) and reached a maximum level at pharyngula period (24 hpf) during embryogenesis in zebrafish. WISH was performed to obtain the temporalspatial expression pattern of K23 gene and the results showed that K23 expression was detected in all stages (Fig. 3b), which was agreement with semi-quantitative RT-PCR results.

Sub-cellular localization of K23 gene

To study the sub-cellular localization of K23 gene, we constructed pK23-EGFP plasmid with K23 coding sequence fused to EGFP gene. As shown in Fig. 3c, blue signals represented

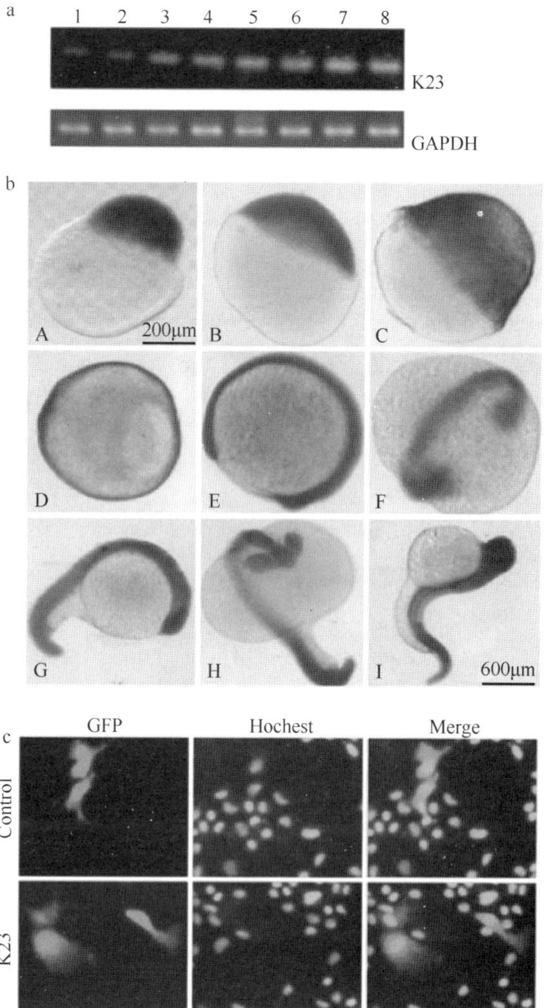

Fig. 3 Expression pattern analysis and sub-cellular localization of K23 gene. (a) Reverse transcriptase-PCR analysis of K23 transcription during development. 1-ovum; 2-zygote; 3-256-cell stage (2.5 hpf); 4-sphere stage (3.8 hpf); 5-50%-epiboly stage (5.25 hpf); 6-90%-epiboly stage (9 hpf); 7-15-somite stage (16.5 hpf); 8-pharyngula period (24 hpf). GAPDH was used as endogenous reference. (b) Expression of K23 transcripts as detected by WISH during embryogenesis in zebrafish. (A) high stage (3.3hpf, hours post fertilization), (B) dome stage (4.3hpf), (C) 50%-epiboly stage (5.25hpf), (D) bud stage (10hpf), (E) 4-somite stage (11.3hpf), (F) 8-somite stage (13hpf), (G) 17-somite stage (17.5hpf), (H) 25-somite stage (17.5hpf), (I) 2 dpf (day post-fertilization). Embryos in (A)-(G) and (I) are lateral views, (A)-(C) with the animal pole to the top; embryo in (H) is dorsal view, with anterior to the top. Bar 200 μm is for (A)-(H); bar 600 μm is for (I). (c) Sub-cellular localization of pEGFP-K23 expressed in EPC cells. The sub-cellular localization of control (pEGFP-N3) and pEGFP-K23 expressed GFP signals in EPC cells was in the upper and lower rows, respectively. All three panels had the same view field at 24h after transfection

the cell nuclei stained by Hochest 33342, while green signals represented the expression of pEGFP-N3 or pEGFP-K23 fluorescence proteins in EPC cells, respectively. Overlapping of the images of pEGFP-N3 or pEGFP-K23 fluorescent protein with the images of cell nuclei stained by Hochest 33342, the results showed that K23-EGFP fusion protein located both in cytoplasm and nucleus.

Discussion

Cross-species NT in fish has been successfully manipulated for over 40 years [10]. Recently, cross-genus cloned fish derived from common carp nuclei and goldfish enucleated eggs was given birth [11]. The result showed that the somitogenesis progress and vertebral numbers of the cloned fish were consistent to the egg-providing species, but not to the donor cell species. It suggested that the maternal factors deposited in matured eggs contributed largely to early pattern formation and had a long-lasting effect on the development of vertebrates. Interestingly, another study showed that the maternal nucleolus was essential for early embryonic development and nucleoplasmic components were essential for the successful development of zygote-somatic cell NT embryos [12]. In the previous study, K23 transcripts were expressed maternally in ovum and over-expressed (about 15-fold) in cross-subfamily cloned embryos. Sub-cellular localization of K23 protein indicated that K23 protein could penetrate the nuclei, and might be participate in signal transduction by its twenty-two putative phosphorylation sites. Moreover, major breakthroughs were reported whereby expression of four transcription factors *Oct4/Sox2/c-myc/Klf4* or *OCT4/SOX2/NANOG/LIN28* could reprogram mouse or human somatic cells into induced pluripotent stem (iPS) cells [13-18]. Those transcription factors induced the genome of differentiated cells to be reprogrammed into an embryonic state. Therefore, whether K23 may function as transcription factor deserves us for further study.

To date, the molecular mechanisms of nuclear reprogramming remain elusive. Gene expression analysis of individual embryos will undoubtedly yield new insights into the regulatory mechanisms involved in cross-species NT in a direct way [8]. In our previous study, we have totally screened out 50 differentially expressed genes in the cloned embryos at sphere stage by using a SSH approach [6]. Among them, about 10% are related to redox function, such as selenoprotein W1, 5-lipoxygenase and glutaryl-coenzyme dehydrogenase etc; about 6% are responsible for cell growth and division, including geminin, daz-like gene and cofactor of BRCA2, etc. In addition, nuclear-mitochondrial compatibility played an important role for the success of cross-species NT, which was agreement with recent review [2]. Other studies indicated that epigenetic modification such as DNA methylation and histone deacetylation might contribute to cross-species nuclear reprogramming [19-21]. Remodeling of donor cell centrosomes and the centrosome-associated cytoskeleton was also found to be crucially important for nuclear cloning [22]. Currently nuclear reprogramming in cross-species cell constructs was frequently associated with inappropriate gene expression during embryonic development [23-24]. Although we found no solid evidences that over- expression of K23 gene obviously affected the nuclear reprogramming in such cross-subfamily cloned embryos, further studies on other differentially expressed genes will give more data in a comprehensive way.

Acknowledgements This work was supported by the National Natural Science Foundation of China (Grant No. 30700607 and 30771100) and the State Key Fundamental Research of China (Grant No. 2004CB117406). In addition, the authors are grateful to Ming Li for supplying experimental materials and Qiya Zhang for providing EPC cells.

References

1. Loi P, Ptak G, Barboni B, Fulka J Jr, Cappai P, Clinton M (2001) Genetic rescue of an endangered mammal by cross-species nuclear transfer using post-mortem somatic cells. Nat Biotechnol 19:962-964. doi:10.1038/nbt1001-962
2. Tecirlioglu RT, Guo J, Trounson AO (2006) Interspecies somatic cell nuclear transfer and preliminary data for horse-cow/mouse iSCNT. Stem Cell Rev 2:277-287. doi:10.1007/BF02698054
3. Wrenzycki C, Wells D, Herrmann D, Miller A, Oliver J, Tervit R *et al* (2001) Nuclear transfer protocol affects messenger RNA expression patterns in cloned bovine blastocysts. Biol Reprod 65:309-317. doi:10.1095/biolreprod65.1.309
4. Pei DS, Sun YH, Chen SP, Wang YP, Zhu ZY (2007) Cloning and characterization of cytochrome c oxidase subunit I (COXI) in *Gobiocypris rarus*. DNA Seq 18:1-8. doi:10.1080/10425170600699752
5. Pei DS, Sun YH, Chen SP, Wang YP, Hu W, Zhu ZY (2007) Zebrafish GAPDH can be used as a reference gene for expression analysis in cross-subfamily cloned embryos. Anal Biochem 363:291-293. doi:10.1016/j.ab.2006.12.005
6. Pei DS, Sun YH, Chen SP, Wang YP, Hu W, Zhu ZY (2007) Identification of differentially expressed genes from the cross-subfamily cloned embryos derived from zebrafish nuclei and rare minnow enucleated eggs. Theriogenology 68:1282-1291. doi:10.1016/j.theriogenology.2007.08.027
7. Kimmel CB, Ballard WW, Kimmel SR, Ullmann B, Schilling TF (1995) Stages of embryonic development of the zebrafish. Dev Dyn 203:253-310
8. Pei DS, Sun YH, Chen CH, Chen SP, Wang YP, Hu W *et al* (2008) Identification and characterization of a novel gene differentially expressed in zebrafish cross-subfamily cloned embryos. BMC Dev Biol 8:29. doi:10.1186/1471-213X-8-29
9. Zhou GZ, Li ZQ, Zhang QY (2006) Characterization and application of monoclonal antibodies against turbot (*Scophthalmus maximus*). Rhabdovirus.Viral Immunol 19:637-645. doi:10.1089/vi2006.19.637
10. Zhu ZY, Sun YH (2000) Embryonic and genetic manipulation in fish. Cell Res 10:17-27. doi:10.1038/sj.cr.7290032
11. Sun YH, Chen SP, Wang YP, Hu W, Zhu ZY (2005) Cytoplasmic impact on cross-genus cloned fish derived from transgenic common carp (*Cyprinus carpio*) nuclei and goldfish (*Carassius auratus*) enucleated eggs. Biol Reprod 72:510-515. doi:10.1095/biolreprod.104.031302
12. Ogushi S, Palmieri C, Fulka H, Saitou M, Miyano T, Fulka J Jr (2008) The maternal nucleolus is essential for early embryonic development in mammals. Science 319:613-616. doi:10.1126/ science.1151276
13. Wernig M, Meissner A, Foreman R, Brambrink T, Ku M, Hochedlinger K *et al* (2007) *In vitro* reprogramming of fibro-blasts into a pluripotent ES-cell-like state. Nature 448:318-324. doi:10.1038/nature05944
14. Takahashi K, Okita K, Nakagawa M, Yamanaka S (2007) Induction of pluripotent stem cells from fibroblast cultures. Nat Protocols 2:3081-3089. doi:10.1038/nprot.2007.418
15. Okita K, Ichisaka T, Yamanaka S (2007) Generation of germline-competent induced pluripotent stem cells. Nature 448:313-317. doi:10.1038/nature05934
16. Nakagawa M, Koyanagi M, Tanabe K, Takahashi K, Ichisaka T, Aoi T *et al* (2008) Generation of induced pluripotent stem cells without Myc from mouse and human fibroblasts. Nat Biotechnol 26:101-106. doi:10.1038/nbt1374
17. Lowry WE, Richter L, Yachechko R, Pyle AD, Tchieu J, Sridharan R *et al* (2008) Generation of human induced pluripotent stem cells from dermal fibroblasts. Proc Natl Acad Sci USA 105:2883-2888. doi:10.1073/

pnas.0711983105

18. Aoi T, Yae K, Nakagawa M, Ichisaka T, Okita K, Takahashi K *et al* (2008) Generation of pluripotent stem cells from adult mouse liver and stomach cells. doi:10.1126/science.1154884
19. Enright BP, Sung LY, Chang CC, Yang X, Tian XC (2005) Methylation and acetylation characteristics of cloned bovine embryos from donor cells treated with 5-aza-2′-deoxycytidine. Biol Reprod 72:944-948. doi:10.1095/biolreprod.104.033225
20. Han YM, Kang YK, Koo DB, Lee KK (2003) Nuclear reprogramming of cloned embryos produced *in vitro*. Theriogenology 59:33-44. doi:10.1016/S0093-691X(02)01271-2
21. Santos F, Zakhartchenko V, Stojkovic M, Peters A, Jenuwein T, Wolf E *et al* (2003) Epigenetic marking correlates with developmental potential in cloned bovine preimplantation embryos. Curr Biol 13:1116-1121. doi:10.1016/S0960-9822(03)00419-6
22. Zhong Z, Spate L, Hao Y, Li R, Lai L, Katayama M *et al* (2007) Remodeling of centrosomes in intraspecies and interspecies nuclear transfer porcine embryos. Cell Cycle 6:1510-1520
23. Mann MR, Chung YG, Nolen LD, Verona RI, Latham KE, Bartolomei MS (2003) Disruption of imprinted gene methylation and expression in cloned preimplantation stage mouse embryos. Biol Reprod 69:902-914. doi:10.1095/biolreprod. 103.017293
24. Smith SL, Everts RE, Tian XC, Du F, Sung LY, Rodriguez-Zas SL *et al* (2005) Global gene expression profiles reveal significant nuclear reprogramming by the blastocyst stage after cloning. Proc Natl Acad Sci USA 102:17582-17587. doi:10.1073/pnas.0508952102

一个新基因 K23 的鉴定及在跨亚科间克隆胚胎中的过量表达

裴得胜[1,2]　孙永华[1]　朱作言[1]

1 中国科学院水生生物研究所，淡水生态与生物技术国家重点实验室，武汉 430072

2 中国科学院水生生物研究所，环境基因组学科组，武汉 430072

摘　要　通过 RACE-PCR 技术在亚科克隆胚胎中获得一个差异表达的新基因 K23。该基因全长 2580 bp，包含一个 1425 bp 的开放阅读框，编码一个推测的由 474 个氨基酸组成的蛋白质。生物信息学分析显示，K23 基因带有 22 个磷酸化位点，但不带有信号肽。在斑马鱼胚胎发育中的表达分析显示，K23 基因在卵子中呈现母源性表达，其转录产物从受精卵到咽胚期逐渐增强。亚细胞定位分析表明 K23 蛋白均匀分布于细胞核和细胞质中。本研究表明 K23 基因是在鱼类亚科间克隆胚胎中差异性表达的新基因。

Identification of Differentially Expressed Genes Between Cloned and Zygote-Developing Zebrafish (*Danio rerio*) Embryos at the Dome Stage Using Suppression Subtractive Hybridization

Daji Luo[1,2] Wei Hu[2] Shangping Chen[2] Yi Xiao[1]
Yonghua Sun[2] Zuoyan Zhu[1,2]

1 College of Life Sciences, Wuhan University, Wuhan
2 State Key Laboratory of Freshwater Ecology and Biotechnology, Institute of Hydrobiology, Chinese Academy of Sciences, Wuhan

Abstract Comparative analyses of differentially expressed genes between somatic cell nuclear transfer (SCNT) embryos and zygotedeveloping (ZD) embryos are important for understanding the molecular mechanism underlying the reprogramming processes. Herein, we used the suppression subtractive hybridization approach and from more than 2900 clones identified 96 differentially expressed genes between the SCNT and ZD embryos at the dome stage in zebrafish. We report the first database of differentially expressed genes in zebrafish SCNT embryos. Collectively, our findings demonstrate that zebrafish SCNT embryos undergo significant reprogramming processes during the dome stage. However, most differentially expressed genes are down-regulated in SCNT embryos, indicating failure of reprogramming. Based on Ensembl description and Gene Ontology Consortium annotation, the problems of reprogramming at the dome stage may occur during nuclear remodeling, translation initiation, and regulation of the cell cycle. The importance of regulation from recipient oocytes in cloning should not be underestimated in zebrafish.

Keywords embryo; differentially expressed genes; reprogramming; somatic cell nuclear transfer; suppression subtractive hybridization

Introduction

Nuclear transfer has been studied for more than 50 years since it was first demonstrated in frogs [1], but the use of differentiated somatic cells has not yet produced individuals that can survive beyond the tadpole stage. Fully differentiated cells can undergo reprogramming after fusion with a matured oocyte through a process commonly known as somatic cell nuclear transfer (SCNT). The competence of blastocysts produced by SCNT was first demonstrated by the production of living animals [2]; since then, several successful animal cloning

文章发表于 *Biology of Reproduction*, 2009, 80(4): 674-684

experiments using SCNT have been documented [3-6]. Despite success in the cloning of many species of vertebrates, animal cloning remains an inefficient process, with a preponderance of reconstructed mammalian embryos failing at the early to midgestational stages of development [7]. The major problem is the reprogramming of a somatic nucleus after fusion with an oocyte [8], and the reprogramming mechanism is poorly understood. Therefore, a better understanding of the molecular mechanisms underlying the reprogramming process is a central scientific issue and will enhance the efficiency of cloning and the ability of cloned animals to survive development.

The zebrafish is an important model for research in vertebrate developmental biology and genetics because of its ease of use in forward genetics and embryonic manipulations. The transparent embryos are well suited to manipulation such as cell labeling, microinjection, and transplantation. Because most zebrafish genes have homologs in humans and other vertebrates, the expression patterns of zebrafish genes can be very instructive in mammals. Nuclear transfer in fish has been studied since the 1960s [4,9-11]. Recently, Sun *et al.* [12] transplanted nuclei from a transgenic common carp into enucleated goldfish eggs. Despite success in fish interspecies and crossgenus cloning, the process remains as inefficient in fish as it is in mammals, with most cloned embryos failing at the midblastula to early gastrula stages of development [12]. Zebrafish SCNT research began in the late 1990s; Lee *et al.* [3] transplanted long-term cultured cells into enucleated eggs and first obtained feeding zebrafish. Our laboratory established zebrafish nuclear transfer technology in 2002 [13]. The research herein—a comparative analysis of differentially expressed genes between SCNT and zygote-developing (ZD) embryos at the dome stage—is the first reported *in vivo* study (to our knowledge) of the gene expression pattern of the reprogramming process in zebrafish.

Gene expression studies in SCNT embryos have been conducted primarily with cattle and mice. Large-scale analysis of gene expression in preimplantation embryos and in cloned animals was reported first in mouse models [14]. To our knowledge, no investigation has included systematic analysis of the gene expression pattern of the reprogramming process using SCNT embryos at the stage when abortion is most prominent *in vivo*. Therefore, systematic analysis of differentially expressed genes in SCNT embryos at the dome stage may yield new insights into the process of reprogramming. Suppression subtractive hybridization (SSH) [15,16] is an efficient and widely used PCR-based method to obtain subtracted libraries and identify differentially expressed genes under two biological conditions. Modified SSH methods had been widely used in functional genomic studies [17-19]. Suppression subtractive hybridization includes a normalization step, which makes this approach preferable for cloning lowabundance transcripts [20]. Therefore, SSH is particularly well suited to identifying differentially expressed genes at the early developmental stage when only a tiny amount of RNA is available from SCNT embryos.

Using the SSH approach, we performed a large-scale search for differentially expressed genes between SCNT embryos and ZD embryos at the dome stage. We then validated the differentially expressed genes using dot blot assays of more than 2900 clones from the SSH libraries, followed by DNA sequencing and clustering analyses. Based on Ensembl description [21] and Gene Ontology Consortium (GO) annotation [22-23], we achieved an all-around analysis and identified some of the key genes responsible for and/or indicative of successful reprogramming. The available data suggest that most differentially expressed genes are downregulated in SCNT embryos at the dome stage, indicating failure of reprogramming. Reprogramming problems at the dome stage may occur during nuclear remodeling, translation initiation, and regulation of the cell cycle. In a crosscomparison with the existing database for mice and cattle [24-25], we found that problems of reprogramming from translation initiation-related genes are shared among these SCNT animals. Furthermore, we found that *mycb* and *klf4* were differentially expressed in zebrafish SCNT embryos and that the expression trends *in vivo* are similar to the successful reprogramming of differentiated somatic cells into a pluripotent state *in vitro* [26-27]. These findings suggest that the balance between *mycb* and *klf4* expression may be important for the reprogramming process in zebrafish SCNT embryos.

Materials and Methods

Ethics statement on the use of animals

The research animals are provided with the best possible care and treatment and are under the care of a specialized technician. All procedures were approved by the Institute of Hydrobiology, Chinese Academy of Sciences, and were conducted in accord with the Guiding Principles for the Care and Use of Laboratory Animals.

Zebrafish strain and maintenance

We used the AB/Tübingen zebrafish (*Danio rerio*) for these experiments. Zebrafish were raised and maintained under standard laboratory conditions, and embryos were staged by morphological features [28].

Media for nuclear transfer

Eggs were held in Hanks saline solution (0.137 M NaCl, 5.4 mM KCl, 0.025 mM Na_2HPO_4, 0.44 mM KH_2PO_4, 1.3 mM $CaCl_2$, 1.0 mM $MgSO_4$, 4.2 mM $NaHCO_3$) supplemented with 1.5% bovine serum albumin (*w/v*; Sigma, St. Louis, MO). This working medium was kept at 4℃ until nuclear transfer. Before the nuclear transfer, streptomycin (100 IU/ml) and ampicillin (100 IU/ml) were added to the working medium and mixed briefly.

Preparation of donor cells

On the day before nuclear transfer, primary cells were collected from kidney tissues of adult male zebrafish. Briefly, kidney tissues were placed into a 0.25% trypsin solution (*w*/*v*; Sigma) for 15 min at 20-25℃, dissociated in Holtfreter dissociation solution (Ca^{2+}-free Holtfreter solution containing 0.15 mM edetic acid [EDTA]), collected by centrifugation, and washed several times using Holtfreter solution (0.35% NaCl, 0.01% $CaCl_2$, 0.005% KCl [*w*/*v*], 100 IU/ml of streptomycin, and 100 IU/ml of ampicillin). The dissociated cells were maintained at 4℃ in JM199 medium until nuclear transfer. Normally, the dissociated cells were used for nuclear transfer within 60 min.

Preparation of recipient eggs

For egg collection, zebrafish were artificially induced to spawn. The quality of eggs has an important role in SCNT. High-quality eggs are slightly granular and yellowish in color, whereas immature eggs are whitish or withered, and the best eggs appear intact and smooth on the yolk surface. Unfertilized embryos were placed into a trypsin solution of 0.25% (*w*/*v*; Sigma) for 3 min, and the softened chorion was subsequently removed by microsurgery. Once activated, the egg cytoplasm coalesces, moves toward the animal pole, and forms the blastodisc. The blastodisc of the zebrafish requires approximately 12 min to form at 25℃ and becomes a full-sized one-cell egg after 40 min. Therefore, these eggs can act as recipients for up to 40 min following activation at 28℃.

Somatic cell nuclear transfer

To remove the egg pronuclei, we placed recipient eggs in an agar plate filled with Hanks saline solution. The second polar body of the dechorionated egg was visible under a 403 stereomicroscope. The egg nucleus underneath the second polar body was removed by aspiration with a fine glass needle. Enucleated eggs were maintained in a 1.5% agar (*w*/*v*; Sigma) plate filled with Hanks saline solution for further manipulation.

All nuclear transfers were conducted using either a microinjection system (model 5171/5246; Eppendorf, Hamburg, Germany) with a Nikon (Melville, NY) TE300 microscope or a Narishige system (NT-188NE; Leeds Precision Instruments, Minneapolis, MN) with an Axiovert 200 microscope (Carl Zeiss, Thornburg, NY). Donor cells were ruptured by aspiration into the transfer needle, which had an inner diameter smaller than the cell (approximately 12 μm). Next, they were transplanted into the cytoplasm of the enucleated eggs at the animal pole. Each of the nuclear transplants was transferred into the agar plate filled with Holtfreter solution. Nuclear transplants were cultured in Holtfreter solution at 28℃ before collection. The embryos were collected at the dome stage and were shield-stage staged by morphological features; SCNT embryos were usually cultured 10 h after nuclear transfer.

Embryo collection and tRNA extraction

For SSH analysis, embryos were collected at the dome stage from SCNT embryos and from ZD embryos. Total RNA was extracted from batches of embryos (n = 100) using the SV tRNA isolation system kit (Promega, Madison, WI). We analyzed the integrity of the RNA by examining its electrophoretic mobility on 1.5% agarose gels in 1×Tris-acetate-EDTA buffer. The UV absorbance of the RNA was measured (using an Eppendorf biometer) at 260 nm (A260) and 280 nm (A280), and RNA purity was determined using the A260 ∶ A280 ratio.

For real-time quantitative RT-PCR analysis, three independent groups of new embryos (n = 30) were collected at the dome stage from SCNT embryos and from ZD embryos (SCNT1, ZD1, SCNT2, ZD2, SCNT3, and ZD3). Total RNA was extracted from batches of embryos (n = 30) using the SV tRNA isolation system kit. Either RNA integrity or RNA purity criteria were met.

Construction of SSH cDNA libraries

The tester to driver hybridization steps in the SSH procedure require 100 ng of tester and driver cDNA; however, a preimplantation early developmentalstage embryo contains only a few picograms of mRNA. Furthermore, SCNT embryos are difficult to achieve, which prohibits larger-scale acquisition of mRNA. For this reason, we used the SMART PCR cDNA synthesis kit (Clontech, Palo Alto, CA), starting with tRNA as the template. The optimum number of PCR cycles using the Perkin-Elmer (Norwalk, CT) GeneAmp PCR system 9600 was used as suggested in the SMART PCR cDNA synthesis kit protocol.

Suppression subtractive hybridization libraries were generated using the reagents and protocols included with the Clontech PCR-Select cDNA subtraction kit. In one SSH library (referred to herein as ZB-CB), the RNA from ZD embryos (referred to herein as ZB) was used as the tester, and the RNA from SCNT embryos (referred to herein as CB) was used as the driver. In another SSH library (referred to herein as CB-ZB), CB was used as the tester, and ZB was used as the driver. The PCR analysis of the SSH products showed that the level of the housekeeping gene *gapdh* decreased significantly in the ZB-CB and CB-ZB cDNA libraries relative to unsubtracted cDNA (data not shown), suggesting that the subtraction procedure was very effective. As the final step, the subtracted DNAs were ligated into the pMD18-T vector (Takara, Gennevilliers, France), and the plasmid was used to transform *Escherichia coli* DH5α by electroporation (Pulse Controller; BioRad, Hercules, CA).

Preparation of templates and probes

We plated cDNA libraries onto solid Luria Bertani medium containing ampicillin. Clones were randomly selected and amplified in a 25-μl PCR system using nested PCR primer 1 and nested PCR primer 2R (Clontech). We selected single-insert clones as templates for the dot blot assays.

After the second hybridization (performed following the Clontech PCRSelect cDNA subtraction kit protocol), we used cDNAs from two libraries as templates to prepare the digoxigenin (DIG) probe. One microgram of cDNA was denatured by heating in a boiling water bath for 10 min, followed by quick chilling on ice. Next, 4 μl of thoroughly mixed DIG-high prime were added to the denatured DNA, followed by mixing and brief centrifugation. Following overnight incubation at 37℃, the reaction was stopped by adding 2 μl of 0.2 M EDTA (all reagents were supplied by Roche Applied Science, Penzberg, Germany). The efficiency of DIG-labeled DNA was determined using the direct detection method (DIG-High Prime DNA Labeling and Detection Starter Kit I; Roche Applied Science). When the expected labeling efficiency was achieved, we used the labeled probe at the recommended concentration in the hybridization reaction.

Dot blot assays of differentially expressed genes in tester and driver materials

For dot blot assays, all reagents were supplied by Roche Applied Science. The PCR products were spotted (1 μl per dot) onto the same position on each of two positive nylon membranes (Millipore Corporation, Bedford, MA) and denatured by 0.6 M NaOH in situ. Spotted membranes were presoaked for 2-5 min at room temperature in 2× saline-sodium citrate solution and were baked at 80℃ for 2 h. Dried filters were prehybridized with preheated DIG Easy Hyb solution at 42℃ in the hybridization incubator (model 40; Robbins Scientific, Sunnyvale, CA), followed by hybridization with the probe (about 25 ng/ml) at 42℃ overnight. After stringent washes, the immunological detection protocol was conducted following the manufacturer's instructions. For color detection using nitroblue tetrazolium/ 5-bromo-4-chloro-3-indolyl phosphate, the reaction was stopped using Tris-EDTA buffer (pH, 8.0) when the desired spot or band intensity was achieved.

Sequencing and clustering analysis

Each candidate DNA fragment was sequenced in a single pass using the M13+ primer site, and reverse sequencing was conducted when the forward sequences failed or the fractions were larger than 800 bp using the M13-primer (Applied Invitrogen, Shanghai, China). Vector DNA sequences were removed automatically using computer program routines. Sequence analysis and clustering were performed as the data set.

Validation of dot blot assays by real-time quantitative RT-PCR analysis

A set of 16 genes was chosen to validate the dot blot assays using real-time quantitative RT-PCR analysis. Real-time quantitative RT-PCR was performed using an Applied Biosystems (Foster City, CA) 7000 real-time PCR system. Table 1 lists the primers used in this analysis.

Complementary DNA samples and a pair of primers were diluted in bidistilled H_2O and plated in triplicate in adjacent wells. β-actin (*actb*) was amplified together with the target gene as an endogenous control in each well with a VIC-labeled probe (Applied Biosystems) to normalize expression levels among samples. Reactions were performed using the following conditions: an initial incubation at 95℃ for 10 min, followed by 40 cycles at 95℃ for 10 sec and 60℃ for 1 min. Output data generated by the instrument onboard software were transferred to a customdesigned Microsoft (Redmond, WA) Excel spreadsheet for analysis. The differential mRNA expression of each candidate gene was calculated by the comparative Ct method using the formula $2^{(-\text{Delta Delta C(T)})}$ method [29]. Moreover, every reaction was performed in triplicate, and the means of three independent experiments between SCNT embryos and ZD embryos were evaluated using Student *t*-test ($P < 0.01$).

Table 1 Primers used to detect 16 chosen genes in real-time RT-PCR

Detector	Primer name	Primer sequence (5′-3′)	Amplicon size (bp)
B20(*napa*)	*napa+*	CAAATCCTCGCAGTCGT	94
	napa–	TCGCAGCCCTTCCATAC	
B27(*ddx41*)	*ddx41+*	GTGCCGTATGTTCCTGTC	87
	ddx41–	TTCTCCACCACTGTCCTTC	
B44(*ccna 1*)	*cca 1+*	GAACCAACGCACCAGG	98
	cca 1–	GAAGGCAGCAGGAATGT	
B56(*abcf2*)	*abcf2+*	AGCCTACCAATCACCTCG	82
	abcf2–	ACCAGCATCATTCCACCC	
B76(*cdc20*)	*cdc20+*	GGTCATTCAGCAAGGGTG	98
	cdc20–	GTGTCCGCCGAAGGTA	
B88(*arpc3*)	*arpc3+*	AAATCCGCCAGGAGACC	96
	arpc3–	ATCCACATCCCGCACA	
B92(*klf4*)	*klf4+*	GTTGGGAAGGTTGTGG	96
	klf4–	ATCTGAGCGGGAGAAA	
B94(*mycb*)	*mycb+*	TGCGATGATGCGGACTA	89
	mycb–	TCAGCGTGCAAAGACG	
B101(*psmd13*)	*psmd13+*	GCAAGTATTACCGCATCAT	100
	psmd13–	CTTCTGGCAAGTCTTTAGC	
B103(*ccne*)	*ccne+*	GCTGGGAAAGGTTCACTC	94
	ccne–	GCTTGGTGGTGGCGTA	
B104(*mtch2*)	*mtch2+*	GTTGGACTCCTAACCCTTCT	92
	mtch2–	GCTGATGCTGGAAACTGA	
B112(*rpp21*)	*rpp21+*	GATTCGCAATGAAGTGAT	98
	rpp21–	GGGAAGCCATAAAGAGTT	

Continued

Detector	Primer name	Primer sequence (5′-3′)	Amplicon size (bp)
CB7(*tp53*)	*tp53+*	TGTGGCTGAAGTGGTC	91
	tp53–	TTTGCTCGCTGATTGC	
CB111(*eif4a1a*)	*eif4a 1a+*	GGTGGTTGAAGGCATTAG	100
	eif4a 1a–	GTGAGGTAGGTTACAGGAGC	
CB120(*nat13*)	*nat13+*	GCGGGAAACTGCTCGTGT	100
	nat13–	CGTGCTTTTGGAGGTGGC	
CB123(*bzw1*)	*bzw1+*	GTATCTGCCGCCTTCG	98
	bzw1–	TCCATTAGCCTGTTGTCC	
actb	*bactin+*	GATGATGAAATTGCCGCACTG	134
	bactin-	ACCAACCATGACACCCTGATGT	

Results

Development of SCNT embryos derived from kidney cells

Following the transfer of nuclei from differentiated cells to enucleated eggs, whether in lower vertebrates or in mammals, only a few of the nuclear transplants are able to develop into adult animals. The SCNT experiments in fish are limited by the inability to directly label the SCNT embryos. To address this issue, we created a negative control (Table 2) by microinjecting a small amount of Hanks saline solution, rather than the kidney cell, into the enucleated egg; 4 h later, none of the Hanks saline solution-injected embryos survived. Tables 2 and 3 summarize that the embryos we collected at the dome stage following injection with the kidney cell were a result of SCNT. When kidney cells as donor cells were injected to the enucleated eggs, the SCNT embryos were sampled at the dome stage and the shield stage to identify some key factors underlying the reprogramming processes, so no successful SCNT offspring were produced. Although 72.1% (2640/3660) of the transplanted eggs cleaved, 16.8% (444/2640) of these transplants developed to the dome stage, and 80.8% (359/444) of these blastulae failed to undergo gastrulation (Table 3); thus, most of the SCNT embryos we collected at the dome stage were a failure of reprogramming. In the investigation of nuclear transplants, there were differences in the speed of development between zebrafish SCNT embryos and ZD embryos. The development of SCNT embryos lagged behind that of ZD embryos at the high stage. Significant differences were found at the dome stage and shield stage (Luo *et al.*, unpublished results).

Table 2 The efficiency of the enucleated procedure

Donor cells (buffer) [a]	No. of enucleated eggs	No. of embryos at 1 h	No. of embryos at 2 h	No. of embryos at 4 h
Control experiment 1	20	0	0	0
Control experiment 2	20	0	0	0

a Zebrafish eggs were enucleated and used as the recipient cells

Using the SSH approach to identify differentially expressed genes in SCNT embryos

To conduct the comparative transcriptomic study of genes present at low levels in abnormally developed SCNT embryos, we followed the approach shown in Figure 1. We used a PCR based SSH approach to comparatively analyze the whole genome of embryos at the dome stage and to enrich for differentially expressed cDNA clones between SCNT embryos and ZD embryos at the dome stage. The SSH was conducted in a forward and reverse manner: Total RNA prepared from ZD embryos and SCNT embryos (Fig. 2A) was used as the tester and driver, respectively, to yield ZD subtracted amplicons, and tRNA prepared from SCNT embryos and ZD embryos was used as the tester and driver, respectively, to yield SCNT subtracted amplicons. The optimum number of PCR cycles was 17 cycles for ZD embryos and 20 cycles for SCNT embryos. We plated cDNA libraries onto Luria Bertani medium containing ampicillin. More than 3000 clones were randomly selected and amplified using nested PCR primer 1 and nested PCR primer 2R (Fig. 2B). The PCR amplification showed that 97% of the clones contained the insert fragments and that 93% of the clones contained the single-insert fragment. We selected 2900 single-insert clones as templates for the dot blot assays.

Figure 3 shows representative results obtained from dot blot assays. Genes expressed at the dome stage of ZD embryos were used as the standard, so that up- or down-regulated genes were identified relative to the ZD embryos. Both up-regulated and down-regulated genes were detected in zebrafish SCNT embryos (Fig. 3). The differentially expressed genes in the ZB-CB and CB-ZB libraries were identified with direct vision based on a qualitative judgment. A more sensitive detection of these differentially expressed genes was obtained using realtime quantitative RT-PCR analysis. We repeated 100 clones randomly selected from 2900 clones using dot blot assays; no significant differences were observed (data not shown).

Identification of differentially expressed genes between SCNT embryos and ZD embryos at the dome stage

Using dot blot assays of 2900 random clones, we identified 242 gene fragments representing significant differences between SCNT embryos and ZD embryos at the dome stage, and we selected these fragments for sequencing. After sequencing these clones and

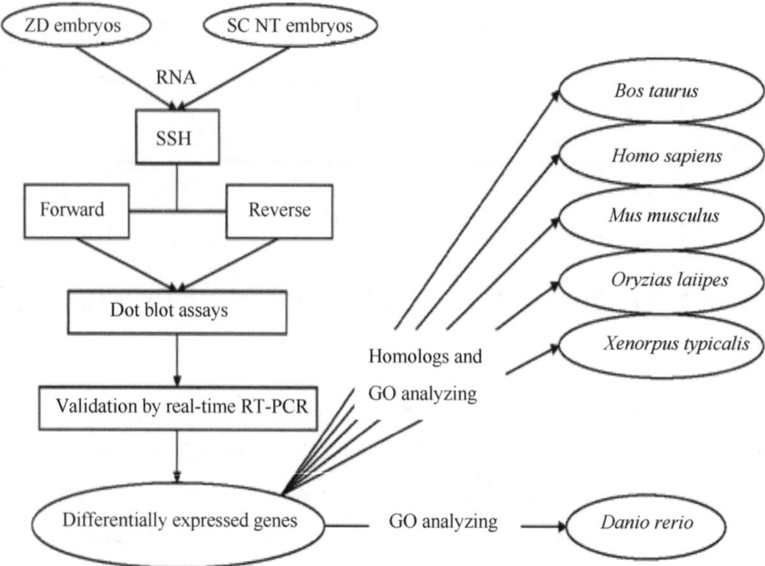

Fig. 1 Experimental flowchart. Total RNA was prepared from ZD embryos and SCNT embryos at the dome stage for subjection to PCR-based SSH. The resulting subtracted cDNA libraries were labaled with DIG for dot bolt assays. The SSH was performed in both the forward (ZB as tester) and reverse (CB as tester) manners to enrich up-regulated (ZB-CB amplicon) and down-regulated (CB-ZB amplicon) transcriptomes, respectively, in the reprogramming process. The two subtracted amplicons were selected as targets for dot blot assays. The results were futher confirmed by real-time quantitative RT-PCR analysis using three independent SCNT embryos and ZD embryos. Differentially expressed genes were characterized using GO analyses and homologs with *Bos taurus*, *Homo sapiens*, *Mus musculus*, *Oryzias latipes*, and *Xenopus typicalis*

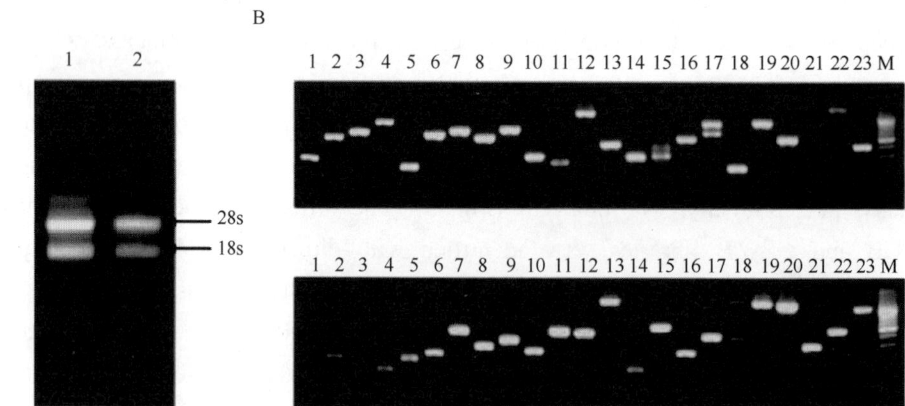

Fig. 2 The quality of tRNA and the subtracted library. A) The integrity of tRNA was detected by agarose gel electrophoresis: 1 shows tRNA of fertilized control embryos, and 2 shows tRNA of SCNT embryos; 28s and 18s rRNA are marked with arrows. A260:A280=1.97:1.94. B) The subtracted library was screened by PCR. M is the DL2000 DNA ladder marker; 1-23 are PCR results by nested PCR primers 1 and 2R. The PCR fragments were 200-2000 bp, and the single-insert PCR fragments were selected as templates for dot blot assays

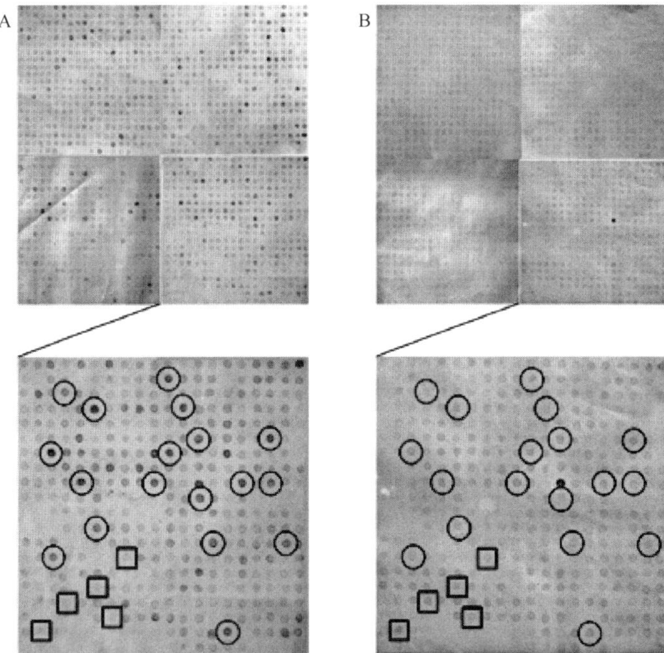

Fig. 3 Representative results of SSH/dot blot assays. DNA was prepared from ZD embryos and SCNT embryos as drivers of SSH libraries, and DNA samples were dotted on the same location in two membranes: A) Representative results using ZB-DIG-labeled probes. B) Representative results using CB-DIG-labeled probes. Squares indicate clones that are up-regulated in SCNT embryos; circles,clones that are down-regulated in SCNT embryos

clustering them according to sequence similarity (BLAST cutoff, 1e–50), we found 96 fragments that were differentially expressed. Of these 96 fragments, 25 were unknown zebrafish expressed sequence tags, and the remaining 71 had high homology to defined or predicted zebrafish genes. In dot blot assays, 17 fragments were up-regulated in zebrafish SCNT embryos, and 79 fragments were down-regulated.

For each cluster, the longest representative clone was chosen from results of BLAST against the National Institutes of Health National Center for Biotechnology Information RefSeq database for zebrafish. All hits with an E value less than 1e–5 were displayed. When we analyzed these differentially expressed fragments in terms of gene ontology using GeneInfoViz constructing [21], we found that these fragments at the dome stage in zebrafish SCNT embryos were involved mainly in physiological processes and regulation of biological process (Fig. 4A, a). The differentially expressed genes were concentrated in pathways, including regulation of DNA-dependent transcription, the ubiquitin cycle, protein modification, protein folding, regulation of the cell cycle, chromosome organization and biogenesis, cytokinesis, the cell cycle, transport, and nucleosome assembly (Fig. 4B, a). The main molecular function of these differentially expressed genes was related to binding and transporter activity (Fig. 4A, b); the top 10 overrepresented GO molecular functions associated with these differentially expressed genes in zebrafish SCNT embryos were nucleic acid binding, protein domain-specific binding, zinc ion binding, DNA binding, ligase activity, ubiquitin-conjugating enzyme activity, ATP binding, monooxygenase activity, translation initiation factor activity, and isomerase activity.

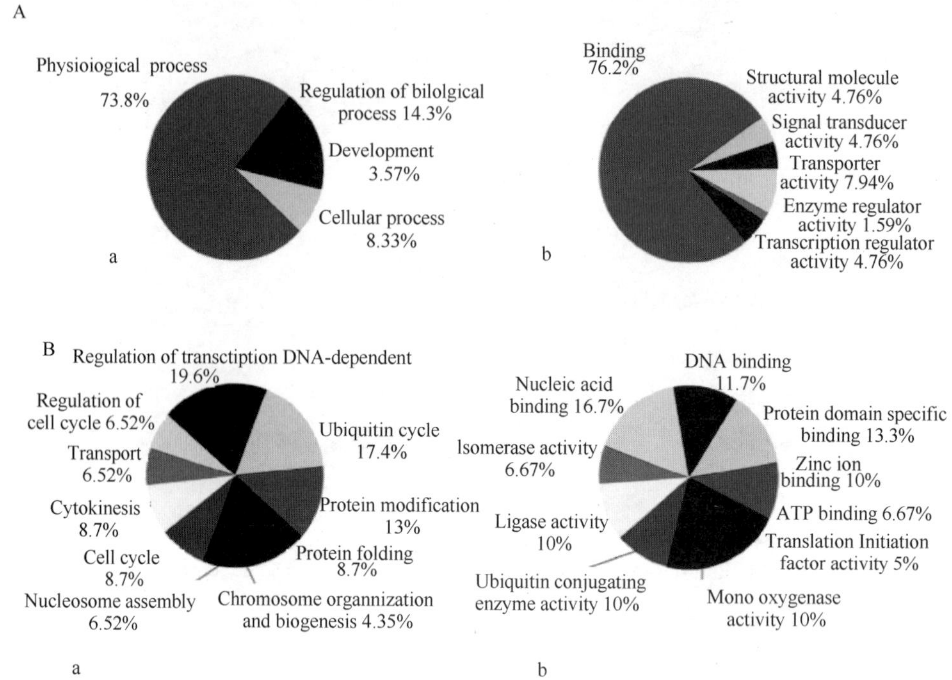

Fig. 4 Gene ontology analysis of differentially expressed genes showing functional categories over represented among differentially expressed genes in zebra fish SCNT embryos. Gene ontology was detected for biological processes and molecular funcations using GenelnfoViz for construcying and viualizing gene relation networks. A) Distribution of differentially expressed genes (biological process [a] and molecular function [b]). B) Top 10 functional categories overrepresented among differentially expressed genes (biological process [a] and molecular function [b])

Table 3 Nuclear transplants generated using kidney cells

Donor cells (kidney cells)	No. of eggs transplanted	No. of embryos cleaved	No. of embryos developed to 256-cell stage	No. of embryos developed to blastulae at dome stage (partial)[a]	No. of embryos developed to gastrula at shield stage (partial)[a]
Experiment 1	1600	997	787	149(30)	24(0)
Experiment 2	2060	1643	1228	295(49)	61(0)

a Partial: In several SCNT embryos, some of the animal pole cells did not participate in the development of SCNT embryos

Validation of dot blot assays using real-time quantitative RT-PCR analysis

We selected 16 genes (Table 1) for real-time quantitative RT-PCR analysis to validate the results from the dot blot assays. We used the *actb* gene as the endogenous control, and the analysis was performed using the comparative Ct method with the formula $2^{(-\text{Delta Delta C(T)})}$ method as the calibrator. Table 1 gives the primers used for the real-time quantitative RT-PCR. Using Statistica 6.0 (Statsoft, Krakow, Poland), three independent experiments between SCNT embryos and ZD embryos were evaluated by means of statistical methods of the nonparametric correlation coefficients (Spearman r); the results showed significant correlation among three independent experiments ($P<0.01$). Sixteen genes showed real-time quantitative RT-PCR results that were consistent with those from the dot blot assays (Fig. 5). These results provide strong support for the differential gene expression data obtained using dot blot assays, and significant differences in dot blot assays show more than 3-fold change compared with the normal expression level.

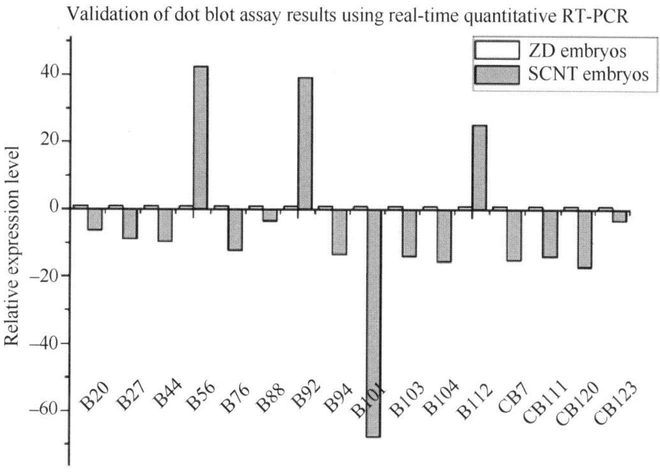

Fig. 5 Validation of dot blot assay results using real-time quantitative RT-PCR. Sixteen genes were selected for real-time quantitative RT-PCR analysis. All genes showed the same pattern of expression as that on the dot blot assays

Discussion

Gene expression profiles of SCNT embryos at the dome stage

In our study, we used the SSH approach to study the differential transcript profiles in zebrafish SCNT embryos relative to ZD embryos at the dome stage. The identification of differentially expressed genes may provide new insights into the reprogramming process of donor nuclei in SCNT embryos.

Nuclear reprogramming is tightly linked to chromatin structure, and chromatin remodeling is extensively involved in epigenetic reprogramming in mammalian cells [30]. Chromatin remodeling is not an isolated event: it is highly coordinated by many binding processes and forms large multiprotein complexes [31-34]. Using the SSH approach, we first identified a number of differentially expressed genes related to binding processes in zebrafish

SCNT embryos; among them, nine genes were implicated in protein binding (*nat13*, *ywhab*, *ywhabl*, *ywhae*, *ywhag*, *ywhaq*, *ywhaz*, *ywhai*, and *ywhah*), 17 in nucleic acid binding (*ddx4l*, *eif4a1a*, *hnrpab*, *klf2a*, *klf2b*, *klf4*, *klfd*, *nono*, *atf3*, *cnr1*, *zgc:100951*, *zgc:66483*, *mycb*, *h3f3d*, *h3f3c*, *zgc:66241*, and *eif4e3*), one in lipid binding (*rgl1*), eight in ion binding (*klf2a*, *klf2b*, *klf4*, *klfd*, *zgc:100951*, *zgc:66483*, *rcc1*, and *zgc:92066*), and five in ATP or guanosine triphosphate binding (*abcf2*, *ddx41*, *eif4a1a*, *gtf2f2*, and *zgc:76988*). In addition, some factors were identified as related to isomerase activity, ligase activity, ubiquitin-conjugating enzyme activity, helicase activity, oxidoreductase activity, ATPase activity, and ATP-dependent helicase activity, all of which are indispensable in the binding process. Therefore, our findings suggest that nuclear remodeling occurs at the early developmental stage in SCNT embryos, which is consistent with previous studies [35-36] proposing that nuclear reprogramming requires prior remodeling of nuclear structures. Notably, some genes related to chromatin organization, including *rcc1*, were down-regulated in the cloned embryos, which is in agreement with the results of other studies [37-38]. Most of these genes were downregulated in SCNT embryos, which indicates a failure of nuclear remodeling in SCNT embryos. Thus, it is highly possible that factors related to nuclear remodeling have important roles in the reprogramming of zebrafish SCNT embryos.

Among the genes identified in our study, several involved in the process of translation, including *bzw1* and *eif4a1a*, were down-regulated in SCNT embryos. Other researchers have reported that *bzw1* and *eif4a* have key regulatory roles in translation initiation [39-40]. Down-regulation of these genes in the present study indicates that the failure in translation may contribute to the reprogramming problems that occur in SCNT embryos. In a cross-comparison with the existing database of mice and cattle [24-25], translation initiation-related genes such as *eif4a* were shared. It is possible that the failure in translation may represent a root cause of reprogramming failure in SCNT embryos across species.

Moderate abnormality of nuclear transplants has been well documented in studies [41-42] of nuclear transfer, and it is thought to be caused by asynchrony between the cell cycles of the recipient egg and donor nucleus. In lower vertebrates, this asynchrony is conspicuous [43]. To confirm the importance of cell cycle synchrony on the reprogramming process in SCNT embryos, we chose kidney cells without culturing them as donor cells, and then we identified many genes related to regulation of the cell cycle. In our study, *ccna1*, *ccnb1*, *ccne*, and *cdc20* were down-regulated in SCNT embryos. A previous study [44] demonstrated that oscillation of cyclins A and B caused the loss of the gap phase and the alternation of the M phase and S phase. Furthermore, in mammalian cells the G1-S transition requires the activation of cyclins D and E, which form a complex with cdk4/6 and cdk2, respectively [45]. *Xenopus* blastomeres undergoing cleavage exhibit a distinctive cell division cycle that is characterized by an extended S phase accompanied by very short or absent G1 and G2 phases, and a similar dynamic process was observed in mammalian embryos [46]. Within this context, it is conceivable that undergoing a distinctive cell division cycle such as that which occurs in *Xenopus* may also be important to zebrafish SCNT embryos. Therefore, the factors that regulate the cell cycle may have important roles in the reprogramming of SCNT embryos. Our data provide molecular evidence suggesting that a number of failures in SCNT embryos may contribute to asynchrony between the cell cycles of the recipient egg and donor nucleus. Synchronization between the cell cycles of the recipient egg and donor nucleus could improve efficiency in cloning.

In our study, *mycb* and *tp53* (B94 and CB7, respectively) were identified and were down-regulated in SCNT embryos. In addition, *klf4* and *rpp21* (B92 and B112, respectively) were identified and up-regulated in SCNT embryos. In mammals, the helix-loop-helix/leucine zipper transcription factor MYC is associated with a number of cellular functions, including cell growth, differentiation, and proliferation, and *Myc* has been shown to alter cellular responses to oncogenes in a culture system [47]. MYC has been proposed as a major downstream target for two pathways that support the maintenance of pluripotency. The Krüppel-type zinc finger transcription factor KLF4, like MYC, is a downstream target of activated STAT3 in leukemia inhibitory factor-induced embryonic stem (ES) cells. KLF4 overexpression leads to the inhibition of differentiation in ES cells [48]. KLF4 has been shown to repress TP53 directly [49], and TP53 protein has been shown to suppress NANOG during ES cell-differentiation [50]. Takahashi and Yamanaka [26] found that induced pluripotent stem cells showed levels of TP53 protein that were lower than those in mouse embryonic fibroblasts. Thus, KLF4 may contribute to the activation of NANOG and other ES cell-specific genes through TP53 repression. Alternatively, KLF4 might function as an inhibitor of MYC-induced apoptosis through the repression of TP53 [51]. In contrast, KLF4 activates $CDKN1A^{CIP1}$, thereby suppressing cell proliferation [52]. This antiproliferative function of KLF4 may be inhibited by MYC, which suppresses the expression of $CDKN1A^{CIP1}$ [53]. Compared with fertilized control embryos, KLF4 was overexpressed in mouse SCNT embryos during the first two cell cycles of development [54]. This suggests that KLF4 may be important for the reprogramming process *in vivo*. As discussed previously, the balance between MYC and KLF4 may be important for the generation of induced pluripotent stem cells in mammals. Surprisingly, four factors (B92, B94, B112, and CB7) in zebrafish SCNT embryos were differentially expressed in the same manner as those in the reprogrammed mammalian cells *in vitro*. To date, there is no evidence that these factors perform similar functions in zebrafish, but the differentially expressed genes may be related to the reprogramming process of donor nuclei in the SCNT embryos. Therefore, the balance between *mycb* and *klf4* may be important for the reprogramming process in zebrafish SCNT embryos. In addition, five recent studies [26, 27, 55-57] show that transfection of mammalian somatic cells with special factors was sufficient to induce the cells to show stem cell characteristics *in vitro*. However, there is no evidence (to our knowledge) that these specific factors have a role in the reprogramming of SCNT embryos *in vivo*. Based on our data from zebrafish SCNT embryos, we provide *in vivo* evidence that *mycb* and *klf4* were retrovirally transduced into donor cells. However, the embryos in our study failed to complete the reprogramming process. This implies that, although B92 and B94 (*klf4* and *mycb*, respectively) were regulated in SCNT embryos, their expression was not sufficient to induce a complete reprogramming process.

Traditionally, it was believed that the genome from the donor cell has an exclusive role in the regulation of SCNT embryos. However, recent evidence from crossgenus-cloned fish has shown that the somite development process and somite number of nuclear transplants were consistent with the recipient species (goldfish) rather than the nuclear donor species (common carp) [12]. We speculated that there might be a subset of genes from recipient oocytes that could have a role in reprogramming. To test our hypothesis, we performed a temporal expression analysis of 16 differentially expressed genes, 12 of which could be identified in unfertilized embryos (data not shown). This result provides some molecular evidence that the fish egg cytoplasm not only supports the development driven by transplanted nuclei but also modulates development of the nuclear transplants.

Successful cloning in zebrafish [3], as well as in other species, indicates that the oocytes possess all of the components required for the establishment of a pluripotent phenotype. According to our data, 82.3% (79/96) of the identified genes were down-regulated in SCNT embryos, and this may shed light on the low survival rate of zebrafish clones. With this result in mind, we speculate that coinjection with specific factors may be the most efficient approach for increasing the viability of zebrafish clones.

A new model of reprogramming research using zebrafish SCNT embryos

Identification of the differentially expressed genes between SCNT embryos and ZD embryos is important for the comprehensive study of reprogramming. Although growing numbers of studies [14,24,25,54,58,59] have conducted large-scale analyses of gene expression in preimplantation embryos and in cloned mammals in mouse and bovine models, the mechanism and process of reprogramming in SCNT embryos remain largely unknown. The differences in gene expression reported in these studies may be due to not only the stages of development analyzed or the different species used but also the stochastic nature of nuclear reprogramming after nuclear transfer [59]. Large-scale analyses of gene expression in cloned mammals usually analyzed single nuclear transfer embryos and were restricted in ovulation quantity in mammals. Fish are generally fecund and can produce hundreds of eggs on a periodic basis. Zebrafish and medaka especially have a good research base on SCNT [3,4,13,60,61], which facilitated our analyzing 3660 zebrafish SCNT embryos and sampling 444 SCNT embryos at the dome stage (Table 3). These SCNT embryos covered various elements of the stochastic nature of nuclear reprogramming after nuclear transfer. From the perspective of statistical analysis, the findings from a large number of SCNT embryos have good reproducibility. Many key factors of our findings in zebrafish SCNT embryos were found in mice and cattle [24,25,54]; compared with fertilized control embryos, *klf4* is overexpressed not only in zebrafish SCNT embryos but also in mouse SCNT embryos [54]. Using zebrafish SCNT embryos as a model would increase our understanding of more comprehensive reprogramming processes. In addition, the analysis of early developmental embryos after nuclear transfer is very important to elucidate the reprogramming processes; the natural fertilization and development *in vitro* in fish facilitate observation and sampling of the development of SCNT embryos after nuclear transfer. Initial reprogramming research has been reported in medaka, including chromosomal abnormalities after nuclear transfer [60]. Therefore, fish offer special advantages as a model of reprogramming research.

Regarding the development of zebrafish SCNT embryos, two dramatic decreases in the number of surviving embryos occurred in the dome and shield stages: only 10%-15% of the SCNT embryos reached the dome stage, and 1%-3% reached the shield stage, whereas 65%-80% entered the cleavage stage (Table 3). Recent observations show a similar decrease at midblastulation to early gastrulation with SCNT embryos from amphibians [62,63] and cross-species nuclear transfer fish [12,64]. Obviously, two dramatic decreases point to incomplete reprogramming processes after nuclear transfer. Therefore, we chose embryos at the dome and shield stages for our SSH analysis to reveal the molecular mechanisms underlying reprogramming processes. It is widely accepted that reprogramming can be divided into two major events that occur just after SCNT: 1) the reversal to pluripotency and 2) the establishment of new differentiation programs[8]. In our study, we used the SSH approach to study the differential transcript profiles of zebrafish SCNT embryos relative to those of ZD embryos at the dome stage. Identification of differentially expressed genes between SCNT embryos and ZD embryos (Table 4) provides evidence for a failure to reverse to pluripotency at the dome stage. Based on the available data, we speculate that the two major reprogramming events are associated with the dome stage and the shield stage. Furthermore, abortion before the dome stage is due to failed reversal to pluripotency, and abortion between the dome stage and the shield stage is due to a failure to establish new differentiation programs. Future studies will focus on the analysis of differentially expressed genes at the shield stage (Luo *et al.*, unpublished data). Taken together, our findings provide molecular evidence to explain two dramatic decreases in the number of surviving embryos during the development of SCNT embryos, and our results will enable the use of zebrafish SCNT embryos as a new model of reprogramming research.

Using the SSH approach, the results herein provide a systemic database of gene expression profiles of zebrafish SCNT embryos compared with ZD embryos at the dome stage. Striking differences in gene expression profiles were identified between SCNT embryos and ZD embryos. Our data suggest that the one broadly defined major reprogramming event—reversal to pluripotency—occurred at the dome stage in zebrafish SCNT embryos. Based on

Table 4 Ninety-six genes differentially expressed between SCNT and ZD embryos at dome stage

User_Id	dbEST_Id	Sequence identifier	Gene description (gene symbol)
Seventy-nine genes down-regulated in SCNT embryos[a]			
B4	FD487114	NM_001004555.1	*Danio rerio* reticulon 4a *(rtn4a),* mRNA
B5	FD487115	NM_200294.1	*Danio rerio* UDP-N-acteylglucosamine pyrophosphorylase 1, like 1 (*uaptlt*), mRNA
B6	FD487116	XM_696928.1	PREDICTED:*Danio rerio* hypothetical protein LOC555084 (LOC555084), mRNA
B7	FD487117	XM_704560.1	PREDICTED:*Danio rerio* similar to AKT1 substrate 1 (proline-rich),transcript variant 2 (LOC566559), mRNA
B8	FD487118	XM_688491.1	PREDICTED:*Danio rerio* similar to host cell factor C1 (LOC565206)， mRNA
B13	FD487123	XM_683028.1	PREDICTED:*Danio rerio* similar to chromosome adhesion protein SMC1-like (LOC559665),mRNA
B17	FD487127	NM_001004548.1	*Danio rerio* T-cell activation GTPase activeating protein (*tagap*), mRNA
B18	FD487128	NM_212670.1	*Danio rerio* zgc:*77560 (zgc:77560)*, mRNA
B20	FD487130	NM_199766.1	*Danio rerio* N-ethylmaleimide senstitive fusion protein attachment protein alpha (*napa*), mRNA
B27	FD487137	XM_695585.1	PREDICTED: *Danio rerio* similar to rhamnose-binding lectin OLL (LOC571939), mRNA
B30	FD487140	XM_701747.1	PREDICTED: *Danio rerio* similar to restricted expression proliferation associated protein-100, transcript variant 3 (LOC557321), mRNA
B44	FD487154	NM_212818.1	*Danio rerio* cyclin A 1 (*ccna1*), mRNA
B48	FD487158	XM_698991.1	PREDICTED: *Danio rerio* hypothetical protein LOC561719,transcript variant 1 (LOC561719), mRNA
B50	FD487160	NM_214817.1	*Danio rerio* zgc:85812 (*zgc:85812*), mRNA
B54	FD487164	NM_001017593.1	*Danio rerio* zgc:110304 (*zgc:110304*), mRNA
B59	FD487169	XM_687848.1	PREDICTED: *Danio rerio* similar to LOC431817 protein (LOC564521), mRNA
B61	FD487171	NM_001002127.1	*Danio rerio* general transcription factor IIF, polypeptide 2 (*gtf2f2*), mRNA
B66	FD487176	NM_201513.1	*Danio rerio* tyrosine 3-monooxygenase/tryptophan 5-monooxygenase activation protein, theta polypeptide (*ywhaq*), mRNA
B67	FD487177	XM_692196.1	PREDICTED: *Danio rerio* similar to zinc finger protein 91 (HPF7, HTF10) (LOC568840), mRNA
B74	FD487184	XM_683556.1	PREDICTED: *Danio rerio* similar to ubiquitin -conjugating enzyme E2D4 (putative), transcript variant 1 (LOC573457),mRNA
B76	FD487186	XM_697535.1	PREDICTED:*Danio rerio* hypothetical protein LOC554402 (LOC554402), mRNA
B80	FD487190	XM_680521.1	PREDICTED:*Danio rerio* similar to Zona pellucida sperm-binding protein 3 precursor (Zona pellucida glycoprotein ZP3) (Zona pellucida protein C) (Sperm receptor) (ZP3A/ZP3B), transcript variant 1 (LOC563179),mRNA

Continued

User_Id	dbEST_Id	Sequence identifier	Gene description (gene symbol)
Seventy-nine genes down-regulated in SCNT embryos[a]			
B84	FD487194	XM_678002.1	PREDICTED:*Danio rerio* similar to Mknkl-prov protein, transcript variant 1 (LOC555974), mRNA
B87	FD487197	NM_001013300.1	*Danio rerio* zgc:112980 (*zgc:112980*), mRHA
B88	FD487198	NM_001002114.1	*Danio rerio* actin related protein 2/3 complex, subunit 3 (*arpc3*), mRHA
B91	FD487201	XM_696483.1	PREDICTED: *Danio rerio* hypothetical protein LOC554649 (LOC554649), mRHA
B94	FD487204	NM_200172.1	*Danio rerio* myelocytomatosis oncogene b (*mycb*), mRHA
B96	FD487206	NM_001002378.1	*Danio rerio* zgc: 92006 (*zgc: 92066*), mRHA
B100	FD487210	NM_212996.1	*Danio rerio* H3 histone, family 3A (*h3f3a*), mRHA
B101	FD487211	XM_200948.1	*Danio rerio* proteasome (prosome, macropain) 26S subunit non-ATPase,13 (*psmd13*),mRNA
B102	FD487212	XM_678430.1	PREDICTED:*Danio rerio* similar to ADP-ribosylation factor-like 6 interacting protein 2, transcript variant 1 (LOC555608), mRNA
B103	FD487213	XM_702578.1	PREDICTED:*Danio rerio* similar to G/2mitotic-specific cyclin A, transcript variant 2 (LOC561391), mRNA
B104	FD487214	XM_131382.1	*Danio rerio* mitochondrial carrier homolog 2 (*mtch2*), mRNA
B111	FD487221	XM_679269.1	PREDICTED: *Danio rerio* similar to c-src tyrosine kinase (LOC556454),mRNA
B112	FD487222	NM_001003530.1	*Danio rerio* ribonuclease P 21 subunit (*rpp21*), mRNA
B113	FD487223	NM_001002124.1	*Danio rerio* PSMC3 interacting protein (*psmc3ip*), mRNA
B114	FD487224	NM_213178.1	*Danio rerio* regulator of chromosome condensation 1 (*rcc1*), mRNA
B115	FD487225	NM_001004589.1	*Danio rerio* eukaryotic translation initiation factor 4E family member 3 (*eif4e3*) ,mRNA
B116	FD487226	XM_695667.1	PREDICTED: *Danio rerio* similar to expressed sequence AV312086 (LOC572014), mRNA
CB2	FD487229	NM_001003991.1	*Danio rerio* Yip1 domain family, member 1 (*yipf1*), mRNA
CB4	FD487231	NM_213527.1	*Danio rerio* hippocampus abundant transcript 1 (*hiat1*), mRNA
CB7	FD487234	XM_680315.1	PREDICTED: *Danio rerio* similar to novel apoptosis-stimulating protein of p53 (*tp53*), mRNA
CB10	FD487237	NM_201514.1	*Danio rerio* zgc:55813 (*zgc:55813*), mRNA
CB12	FD487239	XM_697083.1	PREDICTED: *Danio rerio* hypothetical protein LOC554918 (LOC554918),mRNA
CB16	FD487243	XM_684580.1	PREDICTED:*Danio rerio* similar to KH-type splicing regulatory protein (FUSE binding protein 2) (LOC573065), mRNA

Continued

User_Id	dbEST_Id	Sequence identifier	Gene description (gene symbol)
Seventy-nine genes down-regulated in SCNT embryos[a]			
CB25	FD487252	NM_214805.1	*Danio rerio* pKU-alpha protein kinase (TLK2),mRNA
CB51	FD487276	XM_694025.1	PREDICTED:*Danio rerio* similar to bromodomain-containing protein 4 isoform long (LOC570531), mRNA
CB65	FD487290	NM_212587.2	*Danio rerio* heterogeneous nuclear ribonucleoprotein A/B (*hnrnpab*), mRNA
CB73	FD487298	NM_201468.1	*Danio rerio* SET translocation (myeloid leukemia-associated) B (*setb*), mRNA
CB75	FD487300	NM_212758.1	*Danio rerio* peptidylprolyl isomerase A (cyclophilin A) (*ppia*), mRNA
CB76	FD487301	NM_001017797.1	*Danio rerio* zgc112425 (*zgc112425*), mRNA
CB81	FD487306	NM_200866.1	*Danio rerio* catenin, beta like 1 (*ctnnbl1*), mRNA
CB100	FD487325	XM_697434.1	PREDICTED: *Danio rerio* hypothetical protein LOC554490 (LOC55490), mRNA
CB101	FD487326	XM_681649.1	PREDICTED: *Danio rerio* hypothetical protein LOC558435 (LOC558435), mRNA
CB106	FD487331	NM_199863.1	*Danio rerio* ubiquitin-conjugating enzyme E2G2 (*ube2g2*), mRNA
CB107	FD487332	XM_686123.1	PREDICTED: *Danio rerio* hypothetical protein LOC553758 (LOC553758), mRNA
CB111	FD487336	NM_198366.1	*Danio rerio* eukaryotic translation initiation factor 4A, isoform 1A (*eif4a1a*),mRNA
CB112	FD487337	XM_680455.1	PREDICTED: *Danio rerio* similar to UNR protein (N-ras upstream gene protein), transcript variant 1 (LOC558349), mRNA
CB114	FD487339	XM_682985.1	PREDICTED: *Danio rerio* hypothetical protein LOC553496, transcript variant 1 (LOC553496), mRNA
CB115	FD487340	XM_684187.1	PREDICTED: *Danio rerio* similar to Poly [ADP-ribose] polymerase-1 (PARP-1) (ADPRT) (NAD (+) ADP-ribosyltransferase-1) (Poly[ADP-ribose] synthetase-1) (LOC560788), mRNA
CB118	FD487343	NM_200675.2	*Danio rerio* RER1 retention in endoplasmic reticulum 1 homolog (S. cerevisiae) (*rer1*), mRNA
CB120	FD487345	NM_001003623.1	*Danio rerio* N-acetylatranferase 13 (*nat13*), mRNA
CB122	FD487347	NM_205717.2	*Danio rerio* solute carrier family 31 (copper transporters), member 1 (*slc31a1*), mRNA
CB123	FD487348	NM_199708.1	*Danio rerio* basic leucine zipper and W2 domains 1 (*bzw1*),mRNA
CB124	FD487349	NM_201469.1	*Danio rerio* FK506 binding protein 4 (*fkbp4*), mRNA
CB125	FD487350	XM_696969.1	PREDICTED: *Danio rerio* hypothetical protein LOC554501 (LOC554501), mRNA
CB127	FD487352	NM_199719.1	*Danio rerio* zgc:55512 *(zgc:55512),* mRNA

Continued

User_Id	dbEST_Id	Sequence identifier	Gene description (gene symbol)
Seventeen genes up-regulated in SCNT embryos[b]			
B3	FD487113	NM_200014.1	*Danio rerio* zgc:56134 (*zgc:56134*), mRNA
B29	FD487139	NM_200507.2	*Danio rerio* family with sequence similarity 46 , member C (*fam46c*), mRNA
B56	FD487166	NM_201315.1	*Danio rerio* ATP-binding cassette, sub-family F (GCN20),member 2 (*abcf2*),mRNA
B92	FD487202	NM_131723.1	*Danio rerio* Kruppel-like factor 4 (*klf4*), mRNA

a Twelve of seventy-nine genes down-regulated in SCNT embryos, which have no similarity sequences using BLAST the NCBI Refseq database of zebrafish, were not shown here

b Thirteen of seventeen genes up-regulated in SCNT embryos, which have no similarity sequences using BLAST the NCBI Refseq database of zebrafish, were not shown here

GO analyses, the factors related to nuclear remodeling, translation initiation, and regulation of the cell cycle may represent the most important roles in the reprogramming of SCNT embryos. In addition, we propose that *klf4* and *mycb*, whose homologs in mice provide a powerful combination for more efficient reprogramming of somatic cells, are of significant importance in the reprogramming process in zebrafish SCNT embryos [8]. Although the donor cell genome seems capable of exclusively regulating the process of nuclear reprogramming after SCNT, the importance of regulation from recipient oocytes in cloning experiments should not be underestimated in zebrafish.

Using zebrafish SCNT embryos, we provide initial molecular evidence that abortion of SCNT embryos at the dome stage is due to failed major reprogramming processes after nuclear transfer. In a future article, we will describe the gene expression profiles of SCNT embryos compared with ZD embryos at the shield stage in zebrafish. Collectively, the results of these studies should provide new insights that will enhance our understanding of the reprogramming mechanism after nuclear transfer.

Acknowledgements We greatly appreciate Ming Li for supplying experimental materials and Wuming Gong (Department of Genetics, Cell Biology and Development, University of Minnesota) for his assistance regarding bioinformatics questions.

References

1. Briggs R, King TJ. Transplantation of living nuclei from blastula cells into enucleated frogs' eggs. Proc Natl Acad Sci U S A 1952; 38: 455-463.
2. Wilmut I, Schnieke AE, McWhir J, Kind AJ, Campbell KH. Viable offspring derived from fetal and adult mammalian cells. Nature 1997; 385: 810-813.

3. Lee KY, Huang H, Ju B, Yang Z, Lin S. Cloned zebrafish by nuclear transfer from long-term-cultured cells. Nat Biotechnol 2002; 20: 795-799.
4. Wakamatsu Y, Ju B, Pristyaznhyuk I, Niwa K, Ladygina T, Kinoshita M, Araki K, Ozato K. Fertile and diploid nuclear transplants derived from embryonic cells of a small laboratory fish, medaka (*Oryzias latipes*). Proc Natl Acad Sci U S A 2001; 98: 1071-1076.
5. Wakayama T, Perry AC, Zuccotti M, Johnson KR, Yanagimachi R. Fullterm development of mice from enucleated oocytes injected with cumulus cell nuclei. Nature 1998; 394: 369-374.
6. Tian XC, Xu J, Yang X. Normal telomere lengths found in cloned cattle. Nat Genet 2000; 26: 272-273.
7. Rideout WM III, Eggan K, Jaenisch R. Nuclear cloning and epigenetic reprogramming of the genome. Science 2001; 293: 1093-1098.
8. Alberio R, Campbell KH, Johnson AD. Reprogramming somatic cells into stem cells. Reproduction 2006; 132: 709-720.
9. Gasaryan KG, Hung NM, Neyfakh AA, Ivanenkov VV. Nuclear transplantation in teleost *Misgurnus fossilis* L. Nature 1979; 280: 585-587.
10. Yan SY, Lu DY, Du M, Li GS, Lin LT, Jin GQ, Wang H, Yang YQ, Xia DQ, Liu AZ, Zhu ZY, Yi YL, Chen HX. Nuclear transplantation in teleosts: hybrid fish from the nucleus of crucian and the cytoplasm of carp. Sci Sin [B] 1984; 27: 1029-1034.
11. Zhu ZY, Sun YH. Embryonic and genetic manipulation in fish. Cell Res 2000; 10:17-27.
12. Sun YH, Chen SP, Wang YP, Hu W, Zhu ZY. Cytoplasmic impact on cross-genus cloned fish derived from transgenic common carp (*Cyprinus carpio*) nuclei and goldfish (*Carassius auratus*) enucleated eggs. Biol Reprod 2005; 72: 510-515.
13. Hu W, Wang YP, Chen SP, Zhu ZY. Nuclear transplantation in different strains of zebrafish. Chin Sci Bull 2002; 47: 1277-1280.
14. Humpherys D, Eggan K, Akutsu H, Friedman A, Hochedlinger K, Yanagimachi R, Lander ES, Golub TR, Jaenisch R. Abnormal gene expression in cloned mice derived from embryonic stem cell and cumulus cell nuclei. Proc Natl Acad Sci U S A 2002; 99: 12889-12894.
15. Diatchenko L, Lukyanov S, Lau YF, Siebert PD. Suppression subtractive hybridization: a versatile method for identifying differentially expressed genes. Methods Enzymol 1999; 303: 349-380.
16. Diatchenko L, Lau YF, Campbell AP, Chenchik A, Moqadam F, Huang B, Lukyanov S, Lukyanov K, Gurskaya N, Sverdlov ED, Siebert PD. Suppression subtractive hybridization: a method for generating differentially regulated or tissue-specific cDNA probes and libraries. Proc Natl Acad Sci U S A 1996; 93: 6025-6030.
17. Hepworth PJ, Leatherbarrow H, Hart CA, Winstanley C. Use of suppression subtractive hybridisation to extend our knowledge of genome diversity in *Campylobacter jejuni*. BMC Genomics 2007; 8: e110.
18. Altincicek B, Vilcinskas A. Analysis of the immune-inducible transcriptome from microbial stress resistant, rat-tailed maggots of the drone fly *Eristalis tenax*. BMC Genomics 2007; 8: e326.
19. Herrero S, Gechev T, Bakker PL, Moar WJ, de Maagd RA. *Bacillus thuringiensis* Cry1Ca-resistant *Spodoptera exigua* lacks expression of one of four Aminopeptidase N genes. BMC Genomics 2005; 6: e96.
20. Ji W, Wright MB, Cai L, Flament A, Lindpaintner K. Efficacy of SSH PCR in isolating differentially expressed genes. BMC Genomics 2002; 3: e12.
21. Kasprzyk A, Keefe D, Smedley D, London D, Spooner W, Melsopp C, Hammond M, Rocca-Serra P, Cox T, Birney E. EnsMart: a generic system for fast and flexible access to biological data. Genome Res 2004; 14: 160-169.
22. Ashburner M, Ball CA, Blake JA, Botstein D, Butler H, Cherry JM, Davis AP, Dolinski K, Dwight SS, Eppig JT, Harris MA, Hill DP, *et al.* Gene Ontology Consortium. Gene ontology: tool for the unification of biology. Nat Genet 2000; 25: 25-29.
23. Zhou M, Cui Y. GeneInfoViz: constructing and visualizing gene relation networks. In Silico Biol 2004; 4: 323-333.
24. Smith SL, Everts RE, Tian XC, Du F, Sung LY, Rodriguez-Zas SL, Jeong BS, Renard JP, Lewin HA, Yang X. Global gene expression profiles reveal significant nuclear reprogramming by the blastocyst stage after cloning. Proc

Natl Acad Sci U S A 2005; 102: 17582-17587.

25. Beyhan Z, Ross PJ, Iager AE, Kocabas AM, Cunniff K, Rosa GJ, Cibelli JB. Transcriptional reprogramming of somatic cell nuclei during preimplantation development of cloned bovine embryos. Dev Biol 2007; 305: 637-649.

26. Takahashi K, Yamanaka S. Induction of pluripotent stem cells from mouse embryonic and adult fibroblast cultures by defined factors. Cell 2006; 126: 663-676.

27. Takahashi K, Tanabe K, Ohnuki M, Narita M, Ichisaka T, Tomoda K, Yamanaka S. Induction of pluripotent stem cells from adult human fibroblasts by defined factors. Cell 2007; 131: 861-872.

28. Kimmel CB, Ballard WW, Kimmel SR, Ullmann B, Schilling TF. Stages of embryonic development of the zebrafish. Dev Dyn 1995; 203: 253-310.

29. Livak KJ, Schmittgen TD. Analysis of relative gene expression data using real-time quantitative PCR and the $2^{(-\mathrm{Delta\ Delta\ C(T)})}$ method. Methods 2001; 25: 402-408.

30. Oliveri RS, Kalisz M, Schjerling CK, Andersen CY, Borup R, Byskov AG. Evaluation in mammalian oocytes of gene transcripts linked to epigenetic reprogramming. Reproduction 2007; 134: 549-558.

31. Nan X, Ng HH, Johnson CA, Laherty CD, Turner BM, Eisenman RN, Bird A. Transcriptional repression by the methyl-CpG-binding protein MeCP2 involves a histone deacetylase complex. Nature 1998; 393: 386-389.

32. Wade PA, Gegonne A, Jones PL, Ballestar E, Aubry F, Wolffe AP. Mi-2 complex couples DNA methylation to chromatin remodelling and histone deacetylation. Nat Genet 1999; 23: 62-66.

33. Robertson KD, Ait-Si-Ali S, Yokochi T, Wade PA, Jones PL, Wolffe AP. DNMT1 forms a complex with Rb, E2F1 and HDAC1 and represses transcription from E2F-responsive promoters. Nat Genet 2000; 25: 338-342.

34. Rountree MR, Bachman KE, Baylin SB. DNMT1 binds HDAC2 and a new co-repressor, DMAP1, to form a complex at replication foci. Nat Genet 2000; 25: 269-277.

35. Alberio R, Johnson AD, Stick R, Campbell KH. Differential nuclear remodeling of mammalian somatic cells by *Xenopus laevis* oocyte and egg cytoplasm. Exp Cell Res 2005; 307: 131-141.

36. Kikyo N, Wade PA, Guschin D, Ge H, Wolffe AP. Active remodeling of somatic nuclei in egg cytoplasm by the nucleosomal ATPase ISWI. Science 2000; 289: 2360-2362.

37. Moreira PN, Robl JM, Collas P. Architectural defects in pronuclei of mouse nuclear transplant embryos. J Cell Sci 2003; 116: 3713-3720.

38. Sullivan EJ, Kasinathan S, Kasinathan P, Robl JM, Collas P. Cloned calves from chromatin remodeled *in vitro*. Biol Reprod 2004; 70: 146-153.

39. Svitkin YV, Pause A, Haghighat A, Pyronnet S, Witherell G, Belsham GJ, Sonenberg N. The requirement for eukaryotic initiation factor 4A (elF4A) in translation is in direct proportion to the degree of mRNA 5′ secondary structure. RNA 2001; 7: 382-394.

40. Lister JA, Robertson CP, Lepage T, Johnson SL, Raible DW. Nacre Encodes a zebrafish microphthalmia-related protein that regulates neuralcrest-derived pigment cell fate. Development 1999; 126: 3757-3767.

41. Kurosaka S, Imai H. Cell-cycle coordination and nuclear reprogramming in nuclear transfer embryos [in Japanese]. Tanpakushitsu Kakusan Koso 2002; 47: 1804-1809.

42. Wells DN, Laible G, Tucker FC, Miller AL, Oliver JE, Xiang T, Forsyth JT, Berg MC, Cockrem K, L'Huillier PJ, Tervit HR, Oback B. Coordination between donor cell type and cell cycle stage improves nuclear cloning efficiency in cattle. Theriogenology 2003; 59: 45-59.

43. Di Berardino D, Ramunno L, Jovino V, Pacelli C, Lioi MB, Scarfi MR, Burguete I. Spontaneous rate of sister chromatid exchanges (SCEs) in mitotic chromosomes of sheep (*Ovis aries* L.) and comparison with cattle (*Bos taurus* L.), goat (*Capra hircus* L.) and river buffalo (*Bubalus bubalis* L.). Hereditas 1997; 127: 231-238.

44. Hartley RS, Rempel RE, Maller JL. *In vivo* regulation of the early embryonic cell cycle in *Xenopus*. Dev Biol 1996; 173: 408-419.

45. Sherr CJ. G1 phase progression: cycling on cue. Cell 1994; 79: 551-555.

46. Campbell KH, Loi P, Cappai P, Wilmut I. Improved development to blastocyst of ovine nuclear transfer embryos reconstructed during the presumptive S-phase of enucleated activated oocytes. Biol Reprod 1994; 50: 1385-1393.

47. Rawson C, Shirahata S, Collodi P, Natsuno T, Barnes D. Oncogene transformation frequency of nonsenescent SFME cells is increased by cmyc. Oncogene 1991; 6: 487-489.

48. Li Y, McClintick J, Zhong L, Edenberg HJ, Yoder MC, Chan RJ. Murine embryonic stem cell differentiation is promoted by SOCS-3 and inhibited by the zinc finger transcription factor Klf4. Blood 2005; 105: 635-637.

49. Rowland BD, Bernards R, Peeper DS. The KLF4 tumour suppressor is a transcriptional repressor of p53 that acts as a context-dependent oncogene. Nat Cell Biol 2005; 7: 1074-1082.

50. Lin T, Chao C, Saito S, Mazur SJ, Murphy ME, Appella E, Xu Y. p53 Induces differentiation of mouse embryonic stem cells by suppressing Nanog expression. Nat Cell Biol 2005; 7: 165-171.

51. Zindy F, Eischen CM, Randle DH, Kamijo T, Cleveland JL, Sherr CJ, Roussel MF. Myc signaling via the ARF tumor suppressor regulates p53-dependent apoptosis and immortalization. Genes Dev 1998; 12: 2424-2433.

52. Zhang W, Geiman DE, Shields JM, Dang DT, Mahatan CS, Kaestner KH, Biggs JR, Kraft AS, Yang VW. The gut-enriched Krüppel-like factor (Krüppel-like factor 4) mediates the transactivating effect of p53 on the p21WAF1/Cip1 promoter. J Biol Chem 2000; 275: 18391-18398.

53. Seoane J, Le HV, Massague J. Myc suppression of the p21(Cip1) Cdk inhibitor influences the outcome of the p53 response to DNA damage. Nature 2002; 419: 729-734.

54. Vassena R, Han Z, Gao S, Baldwin DA, Schultz RM, Latham KE. Tough beginnings: alterations in the transcriptome of cloned embryos during the first two cell cycles. Dev Biol 2007; 304: 75-89.

55. Takahashi K, Okita K, Nakagawa M, Yamanaka S. Induction of pluripotent stem cells from fibroblast cultures. Nat Protoc 2007; 2: 3081-3089.

56. Hanna J, Wernig M, Markoulaki S, Sun CW, Meissner A, Cassady JP, Beard C, Brambrink T, Wu LC, Townes TM, Jaenisch R. Treatment of sickle cell anemia mouse model with iPS cells generated from autologous skin. Science 2007; 318: 1920-1923.

57. Yu J, Vodyanik MA, Smuga-Otto K, Antosiewicz-Bourget J, Frane JL, Tian S, Nie J, Jonsdottir GA, Ruotti V, Stewart R, Slukvin II, Thomson JA. Induced pluripotent stem cell lines derived from human somatic cells. Science 2007; 318: 1917-1920.

58. Pfister-Genskow M, Myers C, Childs LA, Lacson JC, Patterson T, Betthauser JM, Goueleke PJ, Koppang RW, Lange G, Fisher P, Watt SR, Forsberg EJ, *et al.* Identification of differentially expressed genes in individual bovine preimplantation embryos produced by nuclear transfer: improper reprogramming of genes required for development. Biol Reprod 2005; 72: 546-555.

59. Somers J, Smith C, Donnison M, Wells DN, Henderson H, McLeay L, Pfeffer PL. Gene expression profiling of individual bovine nuclear transfer blastocysts. Reproduction 2006; 131: 1073-1084.

60. Kaftanovskaya E, Motosugi N, Kinoshita M, Ozato K, Wakamatsu Y. Ploidy mosaicism in well-developed nuclear transplants produced by transfer of adult somatic cell nuclei to nonenucleated eggs of medaka (*Oryzias latipes*). Dev Growth Differ 2007; 49: 691-698.

61. Bubenshchikova E, Kaftanovskaya E, Motosugi N, Fujimoto T, Arai K, Kinoshita M, Hashimoto H, Ozato K, Wakamatsu Y. Diploidized eggs reprogram adult somatic cell nuclei to pluripotency in nuclear transfer in medaka fish (*Oryzias latipes*). Dev Growth Differ 2007; 49: 699-709.

62. Gurdon JB, Byrne JA, Simonsson S. Nuclear reprogramming by *Xenopus oocytes*. Novartis Found Symp 2005; 265:129-141, 204-111.

63. Gurdon JB, Byrne JA, Simonsson S. Nuclear reprogramming and stem cell creation. Proc Natl Acad Sci U S A 2003; 100(suppl 1): 11819-11822.

64. Pei DS, Sun YH, Chen SP, Wang YP, Hu W, Zhu ZY. Identification of differentially expressed genes from the cross-subfamily cloned embryos derived from zebrafish nuclei and rare minnow enucleated eggs. Theriogenology 2007; 68: 1282-1291.

斑马鱼体细胞克隆胚胎和受精胚胎在囊胚期的差异表达基因

罗大极[1,2] 胡 炜[2] 陈尚萍[2] 萧 逸[1] 孙永华[2] 朱作言[1,2]

1 武汉大学，生命科学技术学院，武汉 430072

2 中国科学院水生生物研究所，武汉 430072

摘 要 分析体细胞核移植(SCNT)胚胎和受精卵发育(ZD)胚胎之间的差异性表达基因，对于理解再程序化的分子机制具有重要作用。本文利用抑制性削减杂交技术，从囊胚期 SCNT 和 ZD 的 2900 多个克隆中分离出了 96 个差异表达的基因，由此构建了第一个斑马鱼体细胞核移植胚胎的差异表达基因数据库。已有的数据表明，斑马鱼的 SCNT 胚胎在囊胚期经历了明显的再程序化过程，但大部分差异表达基因在此时期胚胎中下调表达，提示再程序化的失败。基于 Ensembl 的描述和 Gene Ontology Consortium 的注解，推测斑马鱼囊胚期胚胎再程序化的失败可能与细胞核重塑、翻译起始和细胞周期调控有关，受体卵母细胞的调控作用在斑马鱼克隆的重要性不容小觑。

Critical Developmental Stages for the Efficiency of Somatic Cell Nuclear Transfer in Zebrafish

Da-Ji Luo[1,2] Wei Hu[1] Shang-Ping Chen[1] Zuo-Yan Zhu[1]

1. State Key Laboratory of Freshwater Ecology and Biotechnology, Institute of Hydrobiology, Chinese Academy of Sciences, Wuhan
2. School of Basic Medical Science, Wuhan University, Wuhan

Abstract Somatic cell nuclear transfer (SCNT) has been performed extensively in fish since the 1960s with a generally low efficiency of approximately 1%. Little is known about somatic nuclear reprogramming in fish. Here, we utilized the zebrafish as a model to study reprogramming events of nuclei from tail, liver and kidney cells by SCNT. We produced a total of 4,796 reconstituted embryos and obtained a high survival rate of 58.9-67.4% initially at the 8-cell stage. The survival rate exhibited two steps of dramatic decrease, leading to 8.7-13.9% at the dome stage and to 1.5-2.96% by the shield stage. Concurrently, we observed that SCNT embryos displayed apparently delayed development also at the two stages, namely the dome stage (1:30 ± 0:40) and the shield stage (2:50 ± 0:50), indicating that the dome and shield stage are critical for the SCNT efficiency. Interestingly, we also revealed that an apparent alteration in *klf4* and *mycb* expression occurred at the dome stage in SCNT embryos from all the three donor cell sources. Taken together, these results suggest that the dome stage is critical for the SCNT efficiency, and that alternated gene expression appears to be common to SCNT embryos independently of the donor cell types, suggesting that balanced *mycb* and *klf4* expression at this stage is important for proper reprogramming of somatic nuclei in zebrafish SCNT embryos. Although the significant alteration in *klf4* and *mycb* expression was not identified at the shield stage between ZD and SCNT embryos, the importance of reprogramming processes at the shield stage should not be underestimated in zebrafish SCNT embryos.

Keywords SCNT; reprogramming; dome stage; shield stage; *klf4* and *mycb*

Introduction

Somatic cell nuclear transfer (SCNT) has widely been done mostly in aquaculture fish species since the 1960s [1-5]. Recently, SCNT has successfully been applied to laboratory fish models, leading to the production of cloned zebrafish and medaka [6-11]. A generally low efficiency of approximately 1% has been documented for SCNT in fish and other species [12-13]. However, this can be compensated since many eggs are readily available in fish [14]. The ability and convenience

文章发表于 *International Journal of Biological Sciences*, 2011, 7(4): 476-486 for dynamic observation would bring even greater utility to fish in SCNT research, especially the zebrafish, which due to its transparent embryos and suitability for forward-genetics studies is already used extensively as a model vertebrate organism for studying development and disease.

In 2002, Lee *et al.* obtained first feeding SCNT zebrafish by transplanting long-term cultured cells into enucleated eggs [6], and a similar SCNT procedure was independently established also in our laboratory [11]. Recently, Siripattarapravat *et al.* developed a method using laser-ablated metaphase II eggs as recipients and an egg activation protocol after nuclear reconstruction in zebrafish, producing only SCNT embryos but not adults [15]. Although apparently healthy SCNT embryos were reported [6,10-11,15-16], a major obstacle of zebrafish SCNT experiments is the paucity of knowledge about somatic nuclear reprogramming in reconstituted eggs. The SCNT efficiency may vary considerably with the source and types of somatic donor cells, because they may have different capacities to be reprogrammed by ooplasma. It is therefore intriguing to identify factors that regulate the potential of somatic donor cells for reprogramming. One way is to analyze the gene expression profile [17]. Reprogramming-related alterations of gene expression have been documented in mammalian SCNT embryos [18-20]. Previously, we have observed altered gene expression also in fish SCNT [10,21], which is most evident at the dome stage [10], when the majority of SCNT embryos derived from kidney cells exhibited incomplete reprogramming processes.

Developmental retardation of various mammalian SCNT embryos during the pre-implantation stages is also a well-documented phenomenon [20]. Concerns also existed as to whether a similar form of retardation occurs in zebrafish. Based on the data of kidney cell derived SCNT embryos, we speculated that these SCNT embryos would finally fail to develop into adult animals, and that the dome stage is a significant developmental stage in these zebrafish SCNT embryos [10]. Although these previous studies have provided a solid basis for understanding the reprogramming process in zebrafish SCNT embryos, it leaves the unresolved issue of how the dome stage retardation and incomplete reprogramming progresses could commonly occur in zebrafish SCNT embryos derived from other cells. It was suggested that the type of donor cells could affect the development of nuclear transferred embryos or the somatic nuclear reprogramming process in the oocyte [12,22]; thus, this issue should be verified in zebrafish SCNT embryos derived from different cells.

Previously, we have reported that *mycb* and *klf4* have altered expression in the kidney cell-derived zebrafish SCNT embryos [10], which is similar to reprogramming of differentiated cells into a pluripotent state *in vitro* [23-24]. These observations suggest that a balance between *mycb* and *klf4* expression may be important for the reprogramming process in zebrafish SCNT embryos. We have also revealed that an apparent difference in *klf4* and *mycb* expression occurs at the dome stage [10]. Since the dome and shield stage are critical for the SCNT efficiency, proper *mycb* and *klf4* expression at the dome stage may be important for reprogramming. In this study, we made use of three different sources of donor cells from tail, kidney and liver to produce zebrafish SCNT embryos. We analyzed their development and gene expression profile of SCNT embryos at critical stages. We found that SCNT embryos derived from different donor cell sources were similar in development and gene expression.

Results

Early development of kidney cell derived SCNT embryos in zebrafish

In the present study, eight stages of early development were chosen for monitoring SCNT embryogenesis. These are 2-, 8-, 256-cell stages, high, dome, 30% epiboly, shield and 75% epiboly stages. SCNT embryos were staged on the basis of morphological features, by comparison to the developmental stages of the ZD embryos [25].

Observations of the early development of SCNT embryos obtained by injection of enucleated eggs with nuclei from donor kidney cells (KC SCNT embryos) as recorded in Table 1 showed that in the three batches 69.0% (1044/1511) of the transplanted eggs cleaved. The overwhelming majority, 94.1% (983/1044) of these cleaved embryos easily developed to the 8-cell stage. The great majority, 79.6% (783/983) of 8-cell stage SCNT embryos developed to the 256-cell stage. There were 46.94% to 50.44% of the SCNT embryos that developed to the high stage, although only 13.3% (201/1511) of these transplants developed to the dome stage. The developmental block apparently occurred between the high and dome stage, which seemed to be a critical stage in the development of SCNT embryos [21,26-27]. Interestingly, 88.57-97.47% of these dome stage embryos easily developed to the 30% epiboly stage. Unfortunately, just 2.60-2.96% of the transplanted eggs could complete blastulae to undergo gastrulation, and 14.28-18.75% of these shield stage embryos could further develop to 75% epiboly. In this study, we focused on the early development of SCNT, so we did not further investigate the SCNT embryos. However, it is worth mentioning that SCNT embryos beyond gastrulation usually develop into adult individuals [27]. Our findings indicated that dome stage and shield stage are the barrier stages that results in low SCNT efficiency [6,10,16,21,27-30].

Table 1 Nuclear transplants generated using kindney cells

No. of Egg operated	No. of 2-cell stage (%)	No. of 8-cell stage (%)	No. of 256-cell stage (%)	No. of High stage (%)	No. of Dome stage (%)	No. of 30% Epiboly (%)	No. of Shied stage (%)	No. of 75% Epiboly (%)
405	271	257	211	197	55	51	12	2
(Exp.1)	(66.91)	(63.45)	(52.10)	(48.64)	(13.58)	(12.59)	(2.96)	(0.49)
539	375	344	273	253	70	62	14	2
(Exp.2)	(69.57)	(63.82)	(50.65)	(46.94)	(12.99)	(11.50)	(2.60)	(0.37)
567	398	382	299	286	79	77	16	3
(Exp.3)	(70.19)	(67.37)	(52.73)	(50.44)	(13.93)	(13.58)	(2.82)	(0.53)

Early development of liver cell derived SCNT embryos in zebrafish

Table 2 summarizes the early development of liver cell derived SCNT embryos (LC SCNT embryos), where 67.1% (818/1219) of the transplanted eggs cleaved after nuclear transfer; and 90.6% (741/818) of these embryos easily developed to the 8-cell stage. The cleavage rate of LC SCNT embryos was very similar to that of the KC SCNT embryos. However, 84.40% (628/744) of the 8-cell stage embryos developed to the 256-cell stage, which was significantly higher than the 79.6% rate of the KC SCNT embryos. The 8-cell stage and 256-cell stage are subdivided stages of the cleavage period and blastulae period, respectively. The different developmental rates demonstrated that KC and LC SCNT embryos undergo different reprogramming processes between the cleavage and blastulae periods.

Interestingly, there were 47.5-49.1% SCNT embryos that developed to the high stage, and only 11.15%-11.89% of these transplants developed to the dome stage. Therefore, there was no significant difference between the KC and LC SCNT embryos until the blastulae stage, as 96.5% (140/145) of these dome-stage embryos easily developed to the 30% epiboly stage. Notably, only 2.06-2.50% of the transplanted eggs could complete blastulae to undergo gastrulation, and 7.69-14.2% of them developed to 75% epiboly. Compared with that of the KC SCNT embryos, more LC SCNT embryos underwent incomplete reprogramming from blastulae to the gastrula periods.

Table 2 Nuclear transplants generated using liver cells

No. of Egg operated	No. of 2-cell stage (%)	No. of 8-cell stage (%)	No. of 256-cell stage (%)	No. of High stage (%)	No. of Dome stage (%)	No. of 30% Epiboly (%)	No. of Shield stage (%)	No. of 75% Epiboly (%)
387	266	228	201	190	49	46	8	1
(Exp.1)	(68.73)	(58.91)	(51.94)	(49.09)	(12.66)	(11.89)	(2.06)	(0.25)
520	342	317	263	247	58	58	13	1
(Exp.2)	(65.77)	(60.96)	(50.57)	(47.50)	(11.15)	(11.15)	(2.50)	(0.19)
312	210	196	164	151	38	36	7	1
(Exp.3)	(67.31)	(62.82)	(52.56)	(48.40)	(12.18)	(11.54)	(2.24)	(0.14)

Early development of tail cell derived SCNT embryos in zebrafish

The early development of tail cell derived SCNT embryos (TC SCNT embryos) are summarized in Table 3, where 67.2% (1389/2066) of the transplanted eggs cleaved after nuclear transfer; and 89.8% (1246/1389) of these embryos, easily developed to the 8-cell stage. Of these 8-cell stage SCNT embryos, 76.2% (950/1246) developed to the 256-cell stage. There were 39.7-44.8% SCNT embryos that developed to the high stage; however, only 8.9% (183/2066) of these transplants developed to the dome stage. From cleavage to the blastulae period, the developmental rate of TC SCNT embryos was the lowest among the three types of SCNT embryos at every subdivided stage. Interestingly, 96.1-98.4% of these dome stage embryos also easily developed to the 30% epiboly stage. Unfortunately, only 1.5-1.8% of the transplanted eggs could complete blastulae to undergo gastrulation, resulting in the lowest efficiency recorded between the blastulae and gastrula stages. However, 9.1-14.2% of them could develop to 75% epiboly in the gastrula period, indicating that the block between the blastulae and gastrula periods had more of an effect on the development of SCNT embryos than that in the gastrula period.

Main features in early development of zebrafish SCNT embryos

A staging series is a tool that provides accuracy in developmental studies [25]. According to the staging series of zebrafish, we recorded the early development of kidney, liver and tail cell derived SCNT embryos. Fig. 1A shows normal morphology in the blastulae period of zebrafish SCNT embryos. During development, several SCNT embryos displayed abnormal morphology in the blastulae period (Fig. 1 B-D). In following the abnormal SCNT embryos, we observed that they could not complete the normal progression through the blastulae period, and remained at any given subdivision of the blastulae period without any morphological changes for several hours.

Although there was no significant morphological difference among different types of donor cells in the gastrula period under the microscope, the morphogenetic cell movements of involution, convergence, and extension occurred, producing the primary germ layers and the embryonic axis [25]. Several SCNT embryos also displayed abnormal morphology in the gastrula period (Fig. 2 A-D). The significant abnormal morphology suggested that a number of developmental pathways could have progressed in an aberrant manner, and there were some alterations in the temporal and spatial dynamics of gene expression [31]. Therefore, some of these SCNT embryos underwent incomplete reprogramming processes that prevented them from developing into adults.

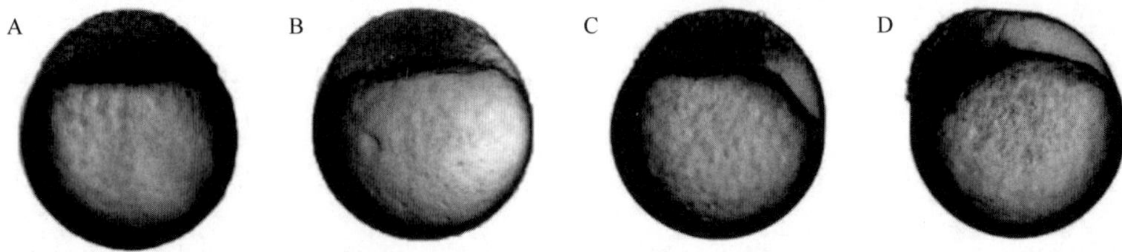

Figure 1 The morphology of zebrafish SCNT embryos during the blastulae period. (A) The normal morphology of zebrafish SCNT embryos in the blastulae period; (B-D) The abnormal morphology of zebrafish SCNT embryos in blastulae period; part of the animal pole cells of these SCNT embryos did not participate in the development of SCNT embryos

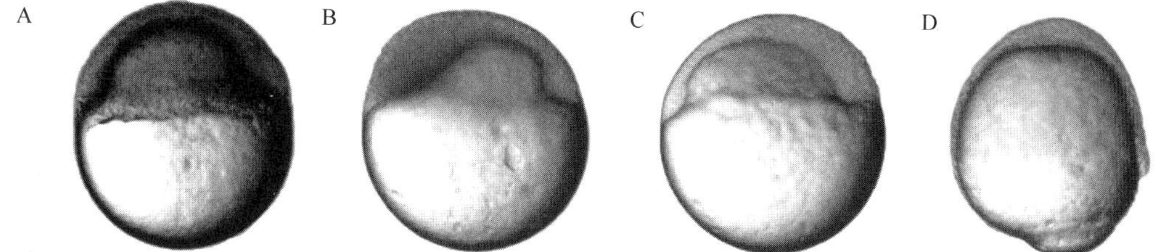

Figure 2 The abnormal morphology of zebrafish SCNT embryos during the gastrula period. (A-C) The abnormal morphology of zebrafish SCNT embryos at the 40% to 50% epiboly stage; (D) The abnormal morphology of zebrafish SCNT embryos at the 80% epiboly stage

We previously mentioned that there have been differences observed in the speed of development between zebrafish SCNT embryos and ZD embryos; however, there are no detailed and accurate experimental data on these differences. In the present study, a systematic analysis was performed on the developmental timing of the SCNT embryos derived from the different cell types (Table 4). During the cleavage period, the developmental timing of SCNT embryos (0:50 ± 0:05) was a slightly slower than that of the ZD embryos (0:45), while the timing of SCNT embryos had no significant difference among these SCNT embryos derived from the different cell types. From the 256-cell stage, the timing of the TC SCNT embryos was different from that of the KC and LC SCNT embryos. At the dome stage, differences in developmental timing appeared between the KC and LC SCNT embryos, and several SCNT embryos were blocked in this stage. As development proceeds, developmental delay became more obvious in SCNT embryos, with the majority of affected SCNT embryos being at the dome stage (1:30 ± 0:40), shield stage (2:50 ± 0:50) and the 75% epiboly stage (3:30 ± 1:00) Combining results in Tables 1, 2 and 3, the developmental speed of SCNT embryos sharply decreased at these stages, implying that the developmental delay in zebrafish SCNT embryos occurred primarily during early stages of development, and that these stages play an important role in the development of SCNT embryos.

Table 3 Nuclear transplants generated using tail cells

No. of Egg operated	No. of 2-cell stage (%)	No. of 8-cell stage (%)	No. of 256-cell stage (%)	No. of High stage (%)	No. of Dome stage (%)	No. of 30% Epiboly (%)	No. of Shield stage (%)	No. of 75% Epiboly (%)
587	391	348	261	233	51	49	9	1
(Exp.1)	(66.61)	(59.28)	(44.46)	(39.69)	(8.69)	(8.35)	(1.53)	(0.17)
767	512	465	346	317	72	70	14	2
(Exp.2)	(66.75)	(60.63)	(45.11)	(41.33)	(9.39)	(9.13)	(1.83)	(0.26)
712	486	433	343	319	63	62	11	1
(Exp.3)	(68.26)	(60.81)	(48.17)	(44.80)	(8.85)	(8.71)	(1.54)	(0.14)

Table 4 Developmental timing (hour:min) of nuclear transplants[a]

	2-cell stage	8-cell stage	256-cell stage	High stage	Dome stage	30% Epiboly	Shield stage	75% Epiboly
ZD embryos	0:45	1:15	2:30	3:20	4:20	4:40	6:00	8:00
KC SCNT embryos[b]	0:50±0:05	1:25±0:05	2:40±0:05	3:50±0:10	5:30±0:20;∞	5:55±0;30;∞	8:30±0:50;∞	10:45±1:00;∞
LC SCNT embryos[c]	0:50±0:05	1:25±0:05	2:40±0:05	3:50±0:10	5:30±0:25;∞	5:55±0;30;∞	8:35±0:50;∞	11:00±1:00;∞
TC SCNT embryos[d]	0:50±0:05	1:25±0:05	2:50±0:10	4:00±0:20	5:50±0:40;∞	6:30±0;40;∞	8:35±0:50;∞	11:30±1:00;∞

a. Each stage except the shield stage and 75% epiboly includes samples beyond 30 embryos, the timing is the average timing of this stage adjusted by the floating range;

b. KC SCNT embryos: kidney cell derived SCNT embryos;

c. LC SCNT embryos: liver cell derived SCNT embryos;

d. TC SCNT embryos: tail cell derived SCNT embryos;

∞. embryos developed in several hours without any morphological changes until death

Molecular features of early development in zebrafish SCNT embryos

We have previously showed that *mycb* (B94) is down-regulated but *klf4* (B92) is up-regulated in KC SCNT embryos, leading to a notion that KLF4 and MYCB proteins are of significant importance in the reprogramming process in zebrafish SCNT embryos [10]. In the present study, we examined whether gene *klf4* and *mycb* were similarly expressed in LC and TC SCNT embryos as in KC SCNT embryos.

We performed a semi-quantitative RT-PCR analysis of *klf4* and *mycb* expression in ZD and SCNT embryos at the cleavage, dome and shield stages (Fig. 3). At the cleavage stage, there was no difference between ZD and SCNT embryos, as they all exhibited a comparably low level of expression for both *klf4* and *mycb* (Fig. 3A). At the dome stage, there was a significant difference between ZD and SCNT embryos, because the klf4 expression was dramatically higher in SCNT embryos than ZD embryos, whereas no significant difference was detected among the KC, LC and TC SCNT embryos (Fig. 3B). When development proceeded to the shield stage, the difference disappeared between ZD and SCNT embryos on the expression of these genes (Fig. 3C). These results indicate that there is a transient upregulation of *klf4* expression at the dome stage in SCNT embryos from donor nuclei of the three different cell types used.

To validate the semi-quantitative RT-PCR data, we performed the real-time RT-PCR analysis by using the comparative Ct method with the formula $2^{-\Delta\Delta CT}$ with the *actb* gene as an endogenous control. Using Statistica 6.0 (Statsoft, Krakow, Poland), three independent batches of SCNT embryos and ZD embryos at the dome stage were evaluated and statistically compared by calculating the nonparametric correlation coefficients (Spearman r). The results showed significant correlation among the three independent experiments ($P < 0.01$). Compared to ZD embryos, SCNT embryos exhibited significant upregulation of klf4 expressions (Fig. 4A), but significant

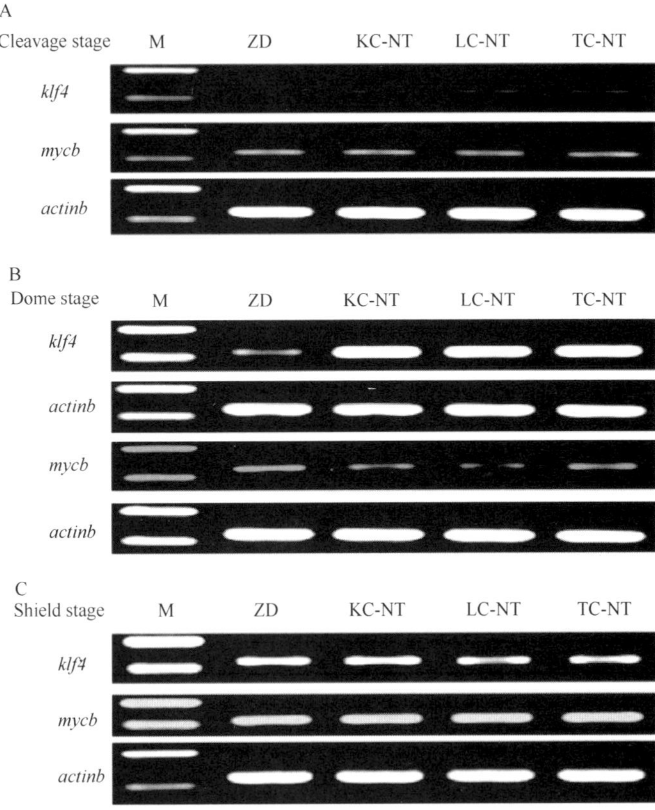

Figure 3 Gene expression between SCNT embryos and ZD embryos by semi-quantitative RT-PCR analysis. RNA expression was identified using RT-PCR during three developmental stages of zebrafish embryos among ZD embryos, KC SCNT embryos, LC SCNT embryos and TC SCNT embryos. (A) The relative expression of klf4 and mycb gene at the cleavage stage; (B) The relative expression of klf4 and mycb gene at the dome stage; (C) The relative expression of klf4 and mycb gene at the shield stage

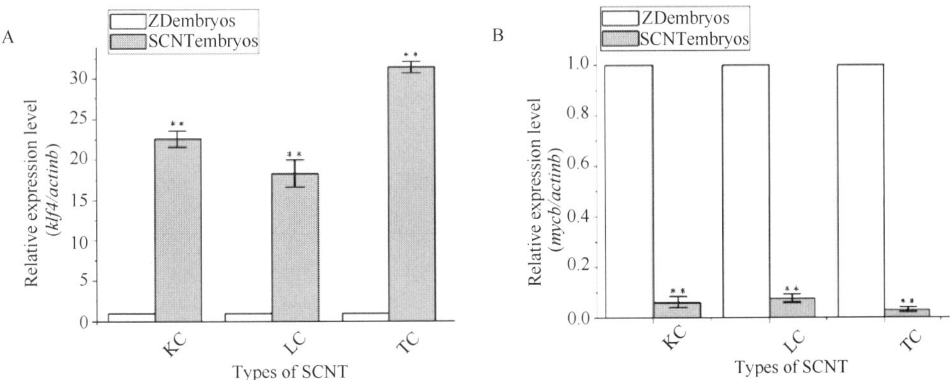

Figure 4 Real-time RT-PCR analyses of klf4 and mycb gene expression. RNA expression was identified using real-time RT-PCR among ZD embryos, KC SCNT embryos, LC SCNT embryos and TC SCNT embryos at the dome stage. (A) Gene *klf4*. (B) Gene *mycb*. Data are means ± s.d (bars above columns) of three samples; *, $0.01 \leqslant p \leqslant 0.05$; **, $p \leqslant 0.01$

downregulation of *mycb* expression (Fig. 4B). Again, these changes were found to be specific to SCNT embryos, because no significant differences in the expression of both genes were found among LC, TC and KC SCNT embryos that had received donor nuclei of different cell types.

Discussion

In the present study, we investigated the early developmental fate of different donor cells following their transfer into recipient enucleated eggs in zebrafish SCNT embryos. It is known that failures in the development are usually the main causes of incomplete reprogramming processes [10,28]. However, few if any studies have focused on the early developmental events or significant reprogramming stage of zebrafish SCNT embryos after nuclear transfer. Here, we provided the first detailed analysis of the early developmental characteristics of SCNT embryos in zebrafish.

Early development of zebrafish SCNT embryos

Following the transfer of nuclei from differentiated cells to enucleated eggs in zebrafish, only a few of the nuclear transplants are able to develop into adult animals. Although apparently healthy zebrafish SCNT embryos were reported [6,10-11, 15-16], the early development of zebrafish SCNT embryos has remained poorly characterized up until now. Different ZD embryos, even together within a single clutch, develop at slightly different rates [25], so it would not be surprising that SCNT embryos would develop differently as well. Therefore, identifying significant developmental stages for SCNT embryos would be a valuable tool to provide accurate timing in reprogramming studies after nuclear transfer.

Although we demonstrated that zebrafish SCNT embryos undergo significant reprogramming processes during the dome stage after comparative analyses of differentially expressed genes between SCNT embryos and ZD embryos, the evidence that the dome stage is a significant developmental period of zebrafish SCNT embryos remains to be determined. First, there are only five stages to be chosen, which is too broad to provide this evidence. Second, the type of donor cells could affect the development of nuclear transferred embryos [12,22] and leaves doubt of how the dome stage retardation could occur in zebrafish SCNT embryos derived from other cells. To clarify these issues and to achieve a more comprehensive overview of differentially expressed genes between KC SCNT and ZD embryos, we chose kidney, liver and tail cells to transplant. According to the developmental rate after NT (Table 1, 2, 3), a total of 4,796 reconstructed embryos were produced. Remarkably, there were little differences between the developmental rates of the KC, LC and TC SCNT embryos at the early developmental stages. However, the survival rate, which was initially high (58.9-67.4%) at the 8-cell stage, sharply decreased to 8.7-13.9% at the dome stage and to 1.50-2.96% by the shield stage, regardless of the origin of the donor cells, indicating that there are some important biological or regulatory processes which occurred at these stages. Because the essential difference between the SCNT and ZD embryos is the reprogramming processes after nuclear transfer, the dome stage and shield stage may be the critical periods during which a blockade in development could cause the low nuclear transfer efficiency.

Main features of early development in zebrafish SCNT embryos

During development, several SCNT embryos display abnormal morphology in the blastulae period (Fig. 1B-D) and gastrula period (Fig. 2A-D). Although there was no significant morphological characteristic of the cells in the early developmental period, the spatial placement of the cell in the embryo could affect the accuracy of the temporal and spatial gene expression. The significant abnormal morphology indicated that these SCNT embryos had undergone the incomplete reprogramming processes and could not develop into adulthood.

Even if zebrafish SCNT embryos were strictly incubated in standardized conditions as ZD embryos [10,25], there would be significant differences in the rate of development between these two type of embryos. We had previously mentioned this developmental lag phenomenon between KC SCNT embryos and ZD embryos [10], but a systematic analysis of the developmental timing was only first demonstrated in the present study (Table 4). During the cleavage period, the developmental timing of SCNT embryos (0:50 ± 0:05) was slightly slower than that of ZD embryos (0:45). As time progressed, the developmental lag phenomenon was more and more obvious. Interestingly, from the dome stage, a significant difference of developmental timing appeared between the SCNT embryos and ZD embryos. The large developmental lag occurred at the dome stage (1:30 ± 0:40), shield stage (2:50 ± 0:50) and 75% epiboly (3:30 ± 1:00) stage. Combining results in Tables 1, 2 and 3, the developmental rate of the SCNT embryos showed sharp decreases at these stages. Regardless of the developmental rate and timing in zebrafish SCNT embryos, there were significant events which apparently occurred at the dome and shield stages. The essential difference between the SCNT and ZD embryos is the reprogramming processes after nuclear transfer. It is widely accepted that reprogramming can be divided into two major events that occur just after SCNT: 1) the reversal to pluripotency and 2) the establishment of new differentiation programs [28]. In addition, we have previously provided gene expression evidence of the failure of embryos to reverse pluripotency at the dome stage [10]. Accordingly, we speculate the processes required to reverse pluripotency is critical for the long developmental lag at the dome stage. Thus, the dome stage may play an important role on the development of SCNT embryos after nuclear transfer.

Molecular characterization of early development in zebrafish SCNT embryos

We have previously shown that KC SCNT embryos displayed normal morphology but underwent incomplete reprogramming processes. After comparative analyses of differentially expressed genes between SCNT embryos and ZD embryos, *mycb* and *klf4* were identified [10]. In mammals, the balance between MYC and KLF4 may be important for the generation of induced pluripotent stem cells [23-24]. Surprisingly, *mycb* and *klf4* in zebrafish KC SCNT embryos were differentially expressed in the same manner as those in the reprogrammed mammalian cells *in vitro*. To date, there is no evidence that *mycb* and *klf4* perform reprogramming functions in zebrafish. However, it is possible that *mycb* and *klf4* may participate in the reprogramming process of the donor nuclei, as it constitutes the essential difference between SCNT and ZD embryos. Wang *et al* identified seven medaka pluripotency genes (containing *klf4*), and speculated the homologs/paralogs of *klf4* gene is perhaps also the pluripotency gene in other lower vertebrates [32]. Furthermore, *klf4* may participate in the reversal of the donor cell to pluripotency in the recipient cell after NT, and the balance between the effects of *mycb* and *klf4* may be important for the reprogramming process in zebrafish SCNT embryos. In the present study, we examined whether the *klf4* and *mycb* genes were correctly expressed as the reprogramming related marker in zebrafish LC and TC SCNT embryos. The semi-quantitative RT-PCR analysis in the expression of *klf4* and *mycb* was performed among the ZD, KC-SCNT, LC-SCNT, TC-SCNT embryos (Fig. 3). Interestingly, there are significant differences between ZD and SCNT embryos just at the dome stage (Fig. 3B). And the expressions of *klf4* and *mycb* in LC and TC SCNT embryos were similar to that in KC SCNT embryos at the dome stage, which are confirmed by the quantitative real-time RT-PCR analysis (Fig. 4). These results provided molecular evidence that the dome stage is a significant developmental stage for SCNT embryos. Thus, *klf4* and *mycb* could be used to evaluate the reprogramming process at the dome stage in zebrafish SCNT embryos.

In summary, we utilized the zebrafish as a model to study reprogramming events of nuclei from tail, liver and kidney cells. The initially high survival rate of zebrafish SCNT embryos at the 8-cell stage (58.9-67.4%) sharply decreased thereafter (8.7-13.9% at the dome stage and 1.50-2.96% by the shield stage). The developmental lag phenomenon was first reported in zebrafish (Table 4), and a large developmental lag was observed at the dome stage (1:30 ± 0:40), shield stage (2:50 ± 0:50) and 75% epiboly (3:30 ± 1:00) stage. These indicating that the dome and shield stage are critical for the SCNT efficiency. However, there was an apparent difference in the *klf4* and *mycb* expression profiles between SCNT embryos and ZD embryos just at the dome stage, suggesting that the *klf4* and *mycb* genes could be the reprogramming related marker in zebrafish SCNT embryos at this stage, and this period is the significant reprogramming stage after NT that results in low nuclear transfer efficiency in zebrafish. It is worth mentioning that it could not be underestimated that the shield stage is a critical stage for the efficiency of SCNT in zebrafish.

Materials and Methods

Zebrafish strain and maintenance

The AB/Tubingen zebrafish (*Danio rerio*) was used for these experiments. Zebrafish were raised and maintained under standard laboratory conditions, and embryos were staged by morphological features [25].

The research animals were provided with the best possible care and treatment and are under the care of a specialized technician. All procedures were approved by the Institute of Hydrobiology, Chinese Academy of Sciences, and were conducted in accordance with the Guiding Principles for the Care and Use of Laboratory Animals.

Media for nuclear transfer

Eggs were maintained in Hank's saline solution (0.137 M NaCl, 5.4 mM KCl, 0.025 mM Na_2HPO_4, 0.44 mM KH_2PO_4, 1.3 mM $CaCl_2$, 1.0 mM $MgSO_4$, 4.2 mM $NaHCO_3$) supplemented with 1.5% BSA (*w/v*; St. Louis, MO, USA). This working medium was kept at 4℃ until nuclear transfer. Prior to nuclear transfer, streptomycin (100 U/ml) and ampicillin (100 U/ml) were added to the working medium and mixed briefly.

Preparation of donor cells

On the day prior to nuclear transfer, primary cells were collected from the tail, liver and kidney tissues of adult male zebrafish (AB strain). Briefly, these tissues were placed into a 0.25% trypsin solution (*w/v*; Sigma, St. Louis, MO, USA) for 15 min at 20-25℃, dissociated in Holtfreter dissociation solution (Ca^{2+}-free Holtfreter solution containing 0.15 mM EDTA), collected by centrifugation, and washed several times using Holtfreter solution (0.35% NaCl, 0.01% $CaCl_2$, 0.005% KCl [*w/v*], 100 IU/ml streptomycin, and 100 IU/ml ampicillin). The dissociated cells were maintained at 4℃ in JM199 medium until nuclear transfer. Normally, the dissociated cells were used for nuclear transfer within 60 min.

Preparation of recipient eggs

For egg collection, zebrafish were artificially induced to spawn. The quality of eggs plays an important role in SCNT. High quality eggs are slightly granular and yellowish in color, whereas immature eggs are whitish or withered, and the best eggs appear intact and smooth on the yolk surface. Unfertilized embryos were placed into a trypsin solution of 0.25% (*w/v*; Sigma) for 3 min, and the softened chorion was subsequently removed by microsurgery. Once activated, the egg cytoplasm coalesced, moved toward the animal pole, and formed the blastodisc. The blastodisc of the zebrafish required approximately 12 min to form at 25℃ and became a full-sized one-cell egg after 40 min. Therefore, these eggs could act as recipients for up to 40 min following activation at 28℃.

Somatic cell nuclear transfer

To remove the egg pronucleus, we placed recipient eggs in an agar plate filled with Hank's saline solution. The second polar body of the dechorionated egg was visible under a 403 stereomicroscope. The egg nucleus underneath the second polar body was removed by aspiration with a fine glass needle. Enucleated eggs were maintained in a 1.5% agar (*w/v*; Sigma) plate filled with Hank's saline solution. Nuclear transfer was conducted using either an Eppendorf microinjection system (Model 5171/5246, Hamburg, Germany) with a Nikon TE300 microscope (Nikon, Melville, NY, USA) or a Narishige system (NT-188NE, Leeds Precision Instruments, Minneapolis, MN, USA) with an Axiovert 200 microscope (Carl Zeiss). Donor cells were ruptured by aspiration into the transfer needle, which had an approximately 12-μm inner diameter smaller than the cell, and were transplanted into the cytoplasm of the enucleated eggs at the animal pole. Nuclear transplants were transferred into an agar plate filled with Holtfreter's solution. SCNT was performed three times for each batch of donor cells and recipient eggs. Nuclear transplants were cultured in Holtfreter's solution at 28℃ prior to collection.

The SCNT experiments in fish are limited by the inability to directly label the SCNT embryos. To address this issue, we created a negative control as in our previous study [10]. Four hours later, none of the Hanks saline solution-injected embryos survived, demonstrating successful SCNT.

Developmental observation and collection of zebrafish SCNT embryos

The whole processes of embryonic development of zebrafish ZD embryos and zebrafish SCNT embryos cultured in Holtfreter's solution at 28℃ were examined under an Olympus SZX-12 microscope. The series of stages for development of the zebrafish embryo were determined as previously described [25]. The serial timings of the early development of embryos were recorded and statistically analyzed. For real-time quantitative RT-PCR analysis, embryos were collected at the dome stage from SCNT embryos derived from tail, liver and kidney cells.

Total RNA extraction and cDNA synthesis

Total RNA was extracted from batches of embryos (n = 100) using the SV Total RNA Isolation System Kit (Promega, CA, USA). We analyzed the integrity of the RNA integrity by its electrophoretic mobility on 1.5% agarose gels in 1× TAE buffer. The UV absorbance of the RNA was also measured at 260 nm (A260) and 280 nm (A280), and the RNA purity was determined using the ratio of A260:A280 (Eppendorf Biometer, Hanburg, Germany).

Gene expression by semi-quantitative RT-PCR and quantitative real-time RT-PCR analysis

Two genes (*klf4* and *mycb*, *klf4* forward: 5′-GTT GGG AAG GTT GTG G-3′, *klf4* reverse: 5′-ATC TGA GCG GGA GAA A-3′; *mycb* forward: 5′-TGC GAT GAT GCG GAC TA-3′, *mycb* reverse: 5′-TCA GCG TGC AAA GAC G-3′) were analyzed in the samples by semi-quantitative RT-PCR which was performed using an Applied Biosystems 9700 (Applied biosystems, Foster City, CA, USA). β-actin was amplified as an endogenous control (*actb* forward: 5′-GAT GAT GAA ATT GCC GCA CTG-3′, *actb* reverse: 5′-ACC AAC CAT GAC ACC CTG ATG T-3′). Reactions were performed using the following conditions: an initial incubation at 94℃ for 5 min, followed by 30-35 cycles (30 cycles for all genes at the cleavage and shield stage, 28 cycles for *klf4* and *actb* gene at the dome stage, 35 cycles for *mycb* and *actb* gene at the dome stage) at 94℃ for 10 sec, 50-60℃ for 30 sec (*klf4* at 50℃, *mycb* at 55℃ and *actb* at 60℃) and 72℃ for 30 sec, followed by holding at 72℃ for 7 min and ending at 20℃ forever.

Two genes (*klf4* and *mycb*) were chosen to be analyzed in the samples by quantitative real-time RT-PCR which was performed using an Applied Biosystems 7000 Real-Time PCR System (Applied biosystems, Foster City, CA, USA) [8]. cDNA samples and a pair of primers were diluted in ddH_2O and plated in triplicate in adjacent wells. Three wells without any templates were also included on each plate as negative controls. β-actin was amplified together with the target gene as an endogenous control in each well with a VIC-labeled probe to normalize expression levels among samples. Reactions were performed using the following conditions: an initial incubation at 95℃ for 10 min, followed by 40 cycles at 95℃ for 10 sec and 60℃ for 1 min. Output data generated by the instrument on board software was transferred to a custom designed Microsoft Excel spreadsheet for analysis. The differential mRNA expression of each candidate gene was calculated by the comparative Ct method using the formula $2^{-\Delta\Delta CT}$ method [33].

Ethics Committee Approval The research animals are provided with the best possible care and treatment and are under the care of a specialized technician. All procedures were approved by the Institute of Hydrobiology, Chinese Academy of Sciences, and were conducted in accord with the Guiding Principles for the Care and Use of Laboratory Animals.

Acknowledgements We thank Ms. Ming Li for technical assistance in nuclear transfer and Ms. Chao Qiu for valuable suggestions. This work was supported by National Natural Science Foundation of China (Grant No. 30900853), the Specialized Research Fund for the Doctoral Program of Higher Education of China (Grant No. 20090141120015) and China Postdoctoral Science Foundation funded project (Grant No. 201003505).

References

1. Gasaryan KG, Hung NM, Neyfakh AA, *et al*. Nuclear transplantation in teleost *Misgurnus fossilis* L. Nature. 1979; 280: 585-587.
2. Zahnd JP and Porte A. Morphologic signs of nuclear material transfer in the cytoplasm of the ovocytes of certain species of fish. C R Acad Sci Hebd Seances Acad Sci D. 1966; 262: 1977-1978.
3. Yan SY, Lu DY, Du M, *et al*. Nuclear transplantation in teleosts. Hybrid fish from the nucleus of crucian and the cytoplasm of carp. Sci Sin B. 1984; 27(10): 1029-1034.
4. Chen H, Yi Y, Chen M, *et al*. Studies on the developmental potentiality of cultured cell nuclei of fish. Int J Biol Sci. 2010; 6: 192-198.
5. Deng C and Liu H. An unknown piece of early work of nuclear reprogramming in fish eggs. Int J Biol Sci. 2010; 6: 190-191.
6. Lee KY, Huang H, Ju B, *et al*. Cloned zebrafish by nuclear transfer from long-term-cultured cells. Nat Biotechnol. 2002; 20: 795-799.
7. Wakamatsu Y, Ju B, Pristyaznhyuk I, *et al*. Fertile and diploid nuclear transplants derived from embryonic cells of a small laboratory fish, medaka (*Oryzias latipes*). Proc Natl Acad Sci U S A. 2001; 98: 1071-1076.
8. Murphey RD and Zon LI. Attack of the fish clones. Nat Biotechnol. 2002; 20: 785-786.
9. Yi M, Hong N and Hong Y. Generation of medaka fish haploid embryonic stem cells. Science. 2009; 326: 430-433.
10. Luo D, Hu W, Chen S, *et al*. Identification of differentially expressed genes between cloned and zygote-developing zebrafish (*Danio rerio*) embryos at the dome stage using suppression subtractive hybridization. Biol Reprod. 2009; 80: 674-684.
11. Hu W, Wang YP, Chen SP, Zhu ZY. Nuclear transplantation in different strains of zebrafish. Chin Sci Bull. 2002; 47: 1277-1280.
12. Wakayama T and Yanagimachi R. Cloning of male mice from adult tail-tip cells. Nat Genet. 1999; 22: 127-128.
13. Yan SY, Tu M, Yang HY, *et al*. Developmental incompatibility between cell nucleus and cytoplasm as revealed by nuclear transplantation experiments in teleost of different families and orders. Int J Dev Biol. 1990; 34: 255-266.
14. Manabu Hattori HH, Ekaterina Bubenshchikova, Yuko Wakamatsu. Nuclear Transfer of Embryonic Cell Nuclei to Non-enucleated and Activated Eggs in Zebrafish, *Danio rerio*. Int J Biol Sci. 2011.
15. Siripattarapravat K, Pinmee B, Venta PJ, *et al*. Somatic cell nuclear transfer in zebrafish. Nat Methods. 2009; 6: 733-735.
16. Ju B, Huang H, Lee KY, *et al*. Cloning zebrafish by nuclear transfer. Methods Cell Biol. 2004;77:403-411.
17. Zhou W, Sadeghieh S, Abruzzese R, *et al*. Transcript levels of several epigenome regulatory genes in bovine somatic donor cells are not correlated with their cloning efficiency. Cloning Stem Cells. 2009; 11: 397-405.
18. Vassena R, Han Z, Gao S, *et al*. Tough beginnings: alterations in the transcriptome of cloned embryos during the first two cell cycles. Developmental biology. 2007; 304: 75-89.
19. Humpherys D, Eggan K, Akutsu H, *et al*. Abnormal gene expression in cloned mice derived from embryonic stem cell and cumulus cell nuclei. Proc Natl Acad Sci U S A. 2002; 99: 12889-12894.
20. Zeng F, Baldwin DA and Schultz RM. Transcript profiling during preimplantation mouse development. Developmental biology. 2004; 272: 483-496.
21. Pei DS, Sun YH, Chen SP, *et al*. Identification of differentially expressed genes from the cross-subfamily cloned embryos derived from zebrafish nuclei and rare minnow enucleated eggs. Theriogenology. 2007; 68: 1282-1291.
22. Eggan K, Akutsu H, Loring J, *et al*. Hybrid vigor, fetal over-growth, and viability of mice derived by nuclear cloning and tetraploid embryo complementation. Proc Natl Acad Sci U S A. 2001; 98: 6209-6214.
23. Takahashi K and Yamanaka S. Induction of pluripotent stem cells from mouse embryonic and adult fibroblast cultures by defined factors. Cell. 2006; 126: 663-676.
24. Takahashi K, Tanabe K, Ohnuki M, *et al*. Induction of pluripotent stem cells from adult human fibroblasts by defined factors. Cell. 2007; 131: 861-872.

25. Kimmel CB, Ballard WW, Kimmel SR, *et al*. Stages of embryonic development of the zebrafish. Dev Dyn. 1995; 203: 253-310.

26. Kane DA and Kimmel CB. The zebrafish midblastula transition. Development (Cambridge, England). 1993; 119: 447-456.

27. Sun YH, Chen SP, Wang YP, *et al*. Cytoplasmic impact on cross-genus cloned fish derived from transgenic common carp (*Cyprinus carpio*) nuclei and goldfish (*Carassius auratus*) enucleated eggs. Biol Reprod. 2005; 72: 510-515.

28. Alberio R, Campbell KH and Johnson AD. Reprogramming somatic cells into stem cells. Reproduction. 2006; 132: 709-720.

29. Ju B, Pristyazhnyuk I, Ladygina T, *et al*. Development and gene expression of nuclear transplants generated by transplantation of cultured cell nuclei into non-enucleated eggs in the medaka *Oryzias latipes*. Dev Growth Differ. 2003; 45: 167-174.

30. Zhao HB and Zhu ZY. Nuclear transplantation of somatic cells of transgenic red carp (*Cyprinus carpio* haematopterus). Yi Chuan Xue Bao. 2002; 29: 406-412.

31. Yin C, Ciruna B and Solnica-Krezel L. Convergence and extension movements during vertebrate gastrulation. Curr Top Dev Biol. 2009; 89: 163-192.

32. Wang D, Dwarakanath MA, Wang T, *et al*. Identification of Pluripotency Genes in the Fish Medaka. Int J Biol Sci. 2010.

33. Livak KJ and Schmittgen TD. Analysis of relative gene expression data using real-time quantitative PCR and the $2^{(-\text{Delta Delta C(T)})}$. Method Methods. 2001; 25: 402-408.

影响斑马鱼体细胞核移植效率的关键发育阶段

罗大极[1,2]　胡　炜[1]　陈尚萍[1]　朱作言[1]

1 中国科学院水生生物研究所，武汉

2 武汉大学，基础医学院，武汉

摘　要　体细胞核移植(SCNT)技术自 1960 年代出现以来，已经广泛应用于鱼类相关研究，但体细胞核移植的成功效率仅约 1%，对鱼类体细胞核的再程序化机制仍知之甚少。本研究以斑马鱼为模型，以尾鳍、肝脏和肾脏细胞为供体核，进行体细胞核移植实验，以探究体细胞核的再程序化机制。核移植实验获得了 4796 个重构胚，这些胚胎在 8-细胞期的成活率为 58.9-67.4%，在随后的胚胎早期发育过程中，重构胚的存活率出现两次锐减，第一次发生于 dome stage，存活率降至 8.7-13.9%，第二次发生于 shield stage，存活率降至 1.5-2.96%。与此同时，我们观察到 SCNT 胚胎在这两个时期均表现出明显的发育滞后，dome stage(1:30 ± 0:40)和 shield stage(2:50 ± 0:50) 是 SCNT 胚胎早期发育的关键时期，直接影响重构胚的存活率。有趣的是，我们还观察到不论是哪一种体细胞作为供体，在 dome stage 胚胎中，*klf4* 和 *mycb* 表达都发生了明显变化。上述实验结果表明，dome stage 是影响 SCNT 效率的关键时期，重构胚中 *klf4* 和 *mycb* 的平衡表达对斑马鱼体细胞克隆胚的再程序化至关重要。虽然 shield stage 时期受精胚与重构胚间 *klf4* 和 *mycb* 的表达没有显著改变，但斑马鱼体细胞克隆胚在 shield stage 时期再程序化的重要性不可忽视。

Cross-Species Cloning: Influence of Cytoplasmic Factors on Development

Yong-Hua Sun　Zuo-Yan Zhu

State Key Laboratory of Freshwater Ecology and Biotechnology, Institute of Hydrobiology, Chinese Academy of Sciences, Wuhan 430072

Abstract It is widely accepted that the crosstalk between naive nucleus and maternal factors deposited in the egg cytoplasm before zygotic genome activation is crucial for early development. This crosstalk may also exert some influence on later development. It is interesting to clarify the relative roles of the zygotic genome and the cytoplasmic factors in development. Cross-species nuclear transfer (NT) between two distantly related species provides a unique system to study the relative role and crosstalk between egg cytoplasm and zygotic nucleus in development. In this review, we will summarize the recent progress of cross-species NT, with emphasis on the cross-species NT in fish and the influence of cytoplasmic factors on development. Finally, we conclude that the developmental process and its evolution should be interpreted in a systemic way, rather than in a way that solely focuses on the role of the nuclear genome.

Introduction: environmental factors in genetics and development

It is widely accepted that the crosstalk between naive nucleus and maternal factors deposited in the egg cytoplasm of vertebrates before zygotic genome activation is crucial for early development (Dosch *et al.* 2004). This crosstalk may also exert some influence on later developmental characteristics. Therefore, it is interesting to clarify the relative roles of zygotic genome and cytoplasmic factors in development. The direct evidence of maternal control on development comes from the screening of maternal-effect mutants in zebrafish (Dosch *et al.* 2004; Wagner *et al.* 2004). Cross-species nuclear transfer (NT) between two distantly related species, which have distinct appearances or phenotypes, provides a unique system to study the relative role and crosstalk between the egg cytoplasm and the zygotic nucleus in development (Fig. 1; Pei *et al.* 2007). In this review, we will summarize the recent progress of cross-species NT, with emphasis on the influence of cytoplasmic factors on development. We conclude that the developmental process and its evolution should be interpreted in a systemic way, for example, to consider the genome and the environment at different levels, rather than in a way that solely focuses on the role of the nuclear genome.

文章发表于 *The Journal of Physiology*, 2014, 592(11): 2375-2379

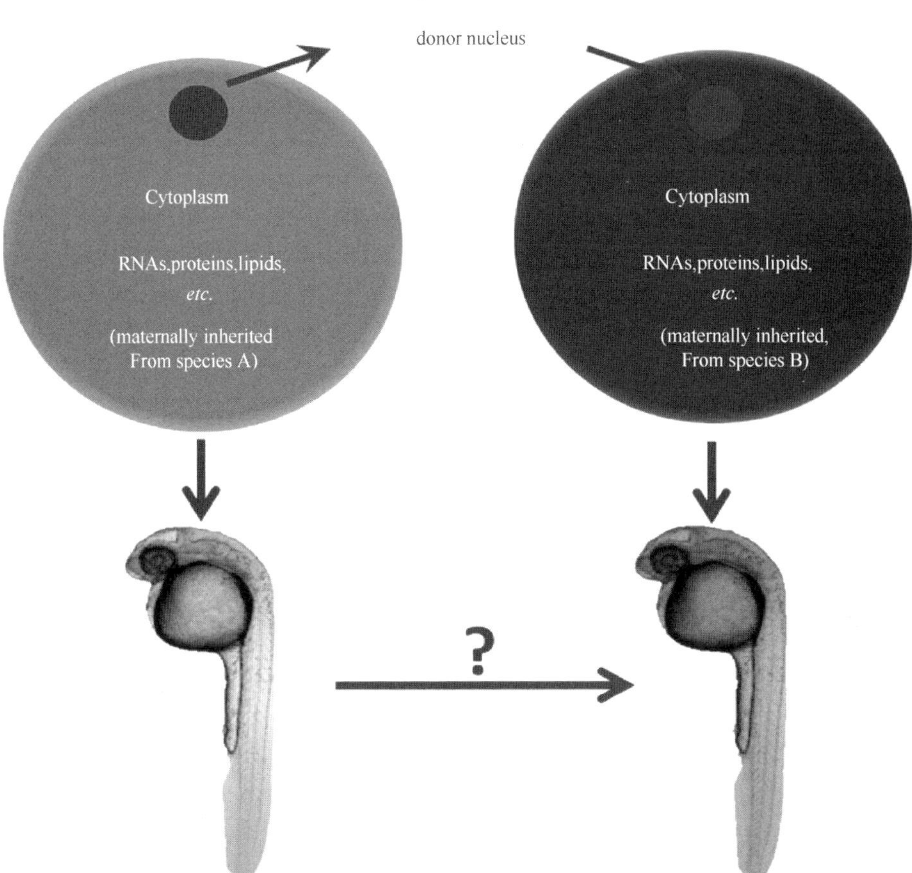

Figure 1 Diagram of cross-species nuclear transfer (NT) Left: development of a regularly fertilized embryo of species A; right: development of a cross-species NT embryo with a nucleus from species A combined with an enucleated egg from species B. The egg cytoplasm of species A and species B contains different types of maternal factors, such as RNAs, proteins and lipids, etc. After nuclear transfer, the reconstructed embryo that contains the nucleus of species A and the cytoplasm of species B may develop into an animal that does not fully resemble species A

Cross-species NT in amphibians and mammals

A species, as a basic unit in biological taxonomy, is considered as a group of organisms that can breed naturally and produce fertile offspring. Different species usually have different genomic materials and distinct developmental shapes. NT is defined as transferring the nuclei of donor cells into enucleated oocytes or eggs to generate reconstructed embryos, which may have the ability to develop to term. If NT is done within one species, i.e. donor cells and oocytes (eggs) come from the same species, it is called inter-species NT. Inter-species NT has been used to study developmental plasticity and nuclear reprogramming of the donor nucleus and to generate reprogrammed stem cells from differentiated cells (Gurdon & Wilmut, 2011).

However, if the oocytes and donor cells come from two different species, the NT will be defined as cross-species NT. Cross-species NT was first described in amphibians, within the genera of *Rana* or *Xenopus* (Moore, 1960; Gurdon, 1962). In those studies, all the NT embryos showed early developmental arrest, probably due to the incomplete reprogramming of the donor nuclei and/or incompatibility between the nuclei and the egg cytoplasm that contains mitochondria from different species. In mammals, although some reprogramming events such as sperm demethylation occur in cross-species NTs (Beaujean *et al.* 2004), the NT embryos usually die at the stage when zygotic transcription starts, suggesting that the egg cytoplasmic environment is crucial for the proper development of transferred nuclei. Nevertheless, cross-species NT has succeeded in cloning some endangered mammals, such as the gaur (Lanza *et al.* 2000), the mouflon (Loi *et al.* 2001), the African wild cat (Gomez *et al.* 2004), the sand cat (Gomez *et al.* 2008) and the coyote (Hwang *et al.* 2012), by using the oocytes from closely related species. It is commonly reported that the cloned animals are identical to their nuclear donors in genotypes and phenotypes, indicating the significant dominance of the nuclear genome in phenotypic determination.

Cross-species NT in fishes–effect of cytoplasmic factors on development

In fish, a type of relatively primitive vertebrate, cross-species NT could be achieved in quite a few genetically distant species. The art of fish NT was first demonstrated with goldfish and bitterling fish by Tung *et al.* (1963). Later, cross-species NT was conducted between two different genera, such as a combination of common carp (*Cyprinus carpio*, genus *Cyprinus*) nuclei with crucian carp (*Carassius auratus*, genus *Carassius*) egg cytoplasm (Tung & Tung, 1980), as well as crucian carp nuclei with common carp egg cytoplasm (Yan *et al.* 1984), in order to obtain nucleo-cytoplasmic hybrid fish with improved economical traits. In those studies led by Tung, it was found that the vertebral numbers of some NT fish were consistent with those of the egg-providing species, but, unfortunately, no conclusive evidence was provided, and the results have been challenged by the scientific community to a certain extent (Gurdon, 1986; Wakamatsu *et al.* 2001). Cross-species fish NT was even conducted between members of two different families, such as the goldfish (*Carassius auratus*, family *Cyprinidae*, order *Cypriniformes*) and the loach (*Paramisgurnus dabryanus*, family *Cobitidae*, order *Cypriniformes*), and between two orders, such as the tilapia (*Oreochromis nilotica*, order *Perciformes*) and the goldfish, as well as the tilapia and the loach (Yan *et al.* 1990, 1991). However, there were only suggestions, with no confirming evidence, that the cross-species NT fish actually were nucleo-cytoplasmic hybrids.

In recent years, with the development of transgenic fish (Zhu & Sun, 2000), we were able to generate cross-genus cloned fish by transferring the nuclei of transgenic common carps into the enucleated eggs of goldfish (*Carassius auratus*) (Sun *et al.* 2005). By analysing transgene and comparative DNA fingerprint markers, we proved that the nuclear genomes of the cloned fish were exclusively derived from the nuclear donor species, the transgenic common carp, instead of the egg-providing species, the goldfish, whereas the mitochondrial DNA from the donor carp gradually disappeared during the development of NT embryos, and only the mitochondrial DNA from recipient goldfish existed in the NT adults. Therefore, the cross-genus cloned fish is really a type of nucleo-cytoplasmic hybrid, with a nuclear genome from the transgenic donor and egg cytoplasm from the recipient species. All the NT fish were identical with the nucleus-providing common carp regarding exterior phenotypic characteristics, such as long body shape, two pairs of barbels, a normal tail and normal eyes. By contrast, there was almost no visible contribution of distinctive goldfish characteristics, such as spherical body shape, triangular tail and 'dragon' eyes (Fig. 2A). Strikingly, somite development and somite number of nuclear transplants were consistent with the recipient species, the goldfish, rather than the nuclear donor species, the common carp. This resulted in a long-lasting effect on the vertebral numbers of the cloned fish, as vertebrae develop from the embryonic somites. The vertebral numbers of the cloned fish belonged within the range of goldfish, which has 28-30 vertebrae, and were distinctly different from those of the common carp, with vertebral numbers of 32-36 (Fig. 2B). This demonstrates that fish egg cytoplasm can not only support the development

Figure 2 Influence of cytoplasmic factors on development of the cloned fish. A, exterior phenotype of the egg-providing goldfish (left), the common carp (middle) and the cloned fish (right). Note that the NT fish was identical to the nucleus-providing common carp regarding exterior phenotypic characteristics, such as long body shape, two pairs of barbels, normal tail and normal eyes, but there was almost no visible contribution of distinctive goldfish characteristics, such as spherical body shape, triangular tail and dragon eyes, B, X-ray analysis of the egg-providing goldfish (left), the common carp (middle) and the cloned fish (right). Note that the vertebral numbers of the cloned fish belonged to the range of goldfish, which has 28-30 vertebrae distinctly different from those of the common carp, with vertebral numbers of 32-36

driven by transplanted nuclei from a distantly related species at the genus scale, but can also significantly modulate development of the nuclear transplants.

In order to study the cross-species NT in more detail, we established two research models. First, for the study of the early nucleo-cytoplasmic interaction in cross-species NT, we utilized two laboratory fish species from different subfamilies, the zebrafish (*Danio rerio*) and the rare minnow (*Gobiocypris rarus*), in order to generate cross-subfamily NT embryos (Pei *et al*. 2007). We further used suppression subtractive hybridization (SSH) to screen out differentially expressed genes from the forward and reverse subtracted cDNA libraries. After dot blot and real-time PCR analysis, 80 of 500 randomly selected sequences were proven to show transcriptional differences in the cloned embryos. Among them, 45 sequences shared high homology with 28 known zebrafish genes, and 35 sequences were corresponding to 22 novel expressed sequence tags (ESTs). Based on the analysis of gene ontology and literature mining, up- and down-regulated genes in the cross-subfamily cloned embryos were shown to be relevant to transcription and translation initiation, cell cycle regulation, protein binding, etc. Therefore, we concluded that the fish egg cytoplasm can not only support the division and development of nuclei from distantly related species, but can also exert a certain impact on the genetic modulation and cellular homeostasis of the transferred nuclei.

Second, zebrafish and Chinese rare minnow were also utilized to produce mutual crossbred embryos in order to examine the impact of the cytoplasm from different species on a common type of nucleus. Although these two types of crossbred embryos originated from common nuclei of the same genetic materials, diverse developmental capacities were gained due to different cytoplasmic environments from different species (Liu *et al*.2008). Using the cDNA amplified fragment length polymorphism (cDNA-AFLP) approach, we compared transcript profiles between the mutual crossbred embryos at two developmental stages (50%- and 90%-epiboly). Three thousand cDNA fragments were generated in four cDNA pools with 64 primer combinations. Compared with ZR (zebrafish ♀ × Chinese rare minnow ♂) embryos, 12 genes were up-regulated and 12 were down-regulated in RZ (Chinese rare minnow ♀ × zebrafish ♂) embryos. The sequences encoded variant proteins which function at different levels of proliferation, growth and development. This strongly suggests that different egg cytoplasms exert completely different impacts on a common nucleus.

Conclusions

Overall, the recent studies of cross-species NT fish experimentally revealed that the nucleus placed in the circumstance of different egg cytoplasm can be strongly influenced by the cytoplasmic factors, at the levels of both genetic regulation and phenotypic determination.

On the other hand, introgressive hybridization in animals, especially in fishes, is one of the driving forces of evolution (Smith, 1992; Dowling & DeMarais, 1993), and this type of hybridization certainly includes the crosstalk between nucleus and cytoplasm from different species. Therefore, any type of developmental process and its evolution should be interpreted in a systemic way rather than in a way that solely focuses on the role of the nuclear genome.

Funding This work was supported by the China 973 Project (Grant Nos. 2010CB126306 and 2012CB944504) and the National Science Fund for Excellent Young Scholars of the Natural National Science Foundation of China (NSFC) (Grant No. 31222052) to Y.H.S.

References

Beaujean N, Taylor JE, McGarry M, Gardner JO, Wilmut I, Loi P, Ptak G, Galli C, Lazzari G, Bird A, Young LE & Meehan RR (2004). The effect of interspecific oocytes on demethylation of sperm DNA. Proc Natl Acad Sci U S A 101, 7636-7640.

Crabbe JC, Wahlsten D & Dudek BC (1999). Genetics of mouse behavior: interactions with laboratory environment. Science 284, 1670-1672.

Dosch R, Wagner DS, Mintzer KA, Runke G, Wiemelt AP & Mullins MC (2004). Maternal control of vertebrate development before the midblastula transition: Mutants from the zebrafish I. Dev Cell 6, 771-780.

Dowling TE & DeMarais BD (1993). Evolutionary significance of introgressive hybridization in cyprinid fishes. Nature 362, 444-446.

Gó mez MC, Pope CE, Giraldo A, Lyons LA, Harris RF, King AL, Cole A, Godke RA & Dresser BL (2004). Birth of African Wildcat cloned kittens born from domestic cats. Cloning Stem Cells 6, 247-258.

Gó mez MC, Pope CE, Kutner RH, Ricks DM, Lyons LA, Ruhe M, Dumas C, Lyons J, Ló pez M & Dresser BL (2008). Nuclear transfer of sand cat cells into enucleated domestic cat oocytes is affected by cryopreservation of donor cells. Cloning Stem Cells 10, 469-484.

Gurdon JB (1962). The transplantation of nuclei between two species of Xenopus. Dev Biol 5, 68-83.

Gurdon JB (1986). Nuclear transplantation in eggs and oocytes. J Cell Sci Suppl 4, 287-318.

Gurdon JB & Wilmut I (2011). Nuclear transfer to eggs and oocytes. Cold Spring Harb Perspect Biol 3 a002659.

Hwang I, Jeong YW, Kim JJ, Lee HJ, Kang M, Park KB, Park JH, Kim YW, Kim WT & Shin T (2012). Successful cloning of coyotes through interspecies somatic cell nuclear transfer using domestic dog oocytes. Reprod Fertil Dev 25, 1142-1148.

Lanza RP, Cibelli JB, Diaz F, Moraes CT, Farin PW, Farin CE, Hammer CJ, West MD& Damiani P (2000). Cloning of an endangered species (*Bos gaurus*) using interspecies nuclear transfer. Cloning 2, 79-90.

Liu J, Sun YH, Wang YW, Wang N, Pei DS, Wang YP, Hu W & Zhu ZY (2008). Identification of differential transcript profiles between mutual crossbred embryos of zebrafish (*Danio rerio*) and Chinese rare minnow (*Gobiocypris rarus*) by cDNA-AFLP. Theriogenology 70, 1525-1535.

Loi P, Ptak G, Barboni B, Fulka J Jr, Cappai P & Clinton M (2001). Genetic rescue of an endangered mammal by cross-species nuclear transfer using post-mortemsomatic cells. *Nat Biotechnol* 19, 962-964.

Miller W, Schuster SC, Welch AJ, Ratan A, Bedoya-Reina OC, Zhao F, Kim HL, Burhans RC, Drautz DI & Wittekindt NE (2012). Polar and brown bear genomes reveal ancient admixture and demographic footprints of past climate change. Proc Natl Acad Sci U S A 109, E2382-E2390.

Moore JA (1960). Serial back-transfers of nuclei in experiments involving two species of frogs. Dev Biol 2, 535-550.

Pei DS, Sun YH, Chen SP, Wang YP, Hu W & Zhu ZY (2007). Identification of differentially expressed genes from the cross-subfamily cloned embryos derived from zebrafish nuclei and rare minnow enucleated eggs. Theriogenology 68, 1282-1291.

Roche HM, Phillips C & Gibney MJ (2005). The metabolic syndrome: the crossroads of diet and genetics. Proc Nutr Soc

64, 371-377.

Schlessinger J (2000). Cell signaling by receptor tyrosine kinases. Cell 103, 211-225.

Slack JMW (2012). Essential Developmental Biology, 3rd edn. Wiley-Blackwell.

Smith GR (1992). Introgression in Fishes: Significance for Paleontology, Cladistics, and Evolutionary Rates. Syst Biol 41, 41-57.

Sun YH, Chen SP, Wang YP, Hu W & Zhu ZY (2005). Cytoplasmic impact on cross-genus cloned fish derived from transgenic common carp (*Cyprinus carpio*) nuclei and goldfish (*Carassius auratus*) enucleated eggs. Biol Reprod 72, 510-515.

Tung T & Tung Y (1980). Nuclear transplantation in teleosts. I. Hybrid fish from the nucleus of carp and the cytoplasm of crucian. Scientia (Peking) 14, 1244-1245.

Tung T, Wu S, Tung Y, Yan S, Tu M & Lu T (1963). Nuclear transplantation in fish. Chin Sci Bull, Academia Sinica 7, 60-61. (In Chinese.)

Wagner DS, Dosch R, Mintzer KA, Wiemelt AP & Mullins MC (2004). Maternal control of development at the midblastula transition and beyond: mutants from the zebrafish II. Dev Cell 6, 781-790.

Wakamatsu Y, Ju B, Pristyaznhyuk I, Niwa K, Ladygina T, Kinoshita M, Araki K & Ozato K (2001). Fertile and diploid nuclear transplants derived from embryonic cells of a small laboratory fish, medaka (*Oryzias latipes*). Proc Natl Acad Sci USA 98, 1071-1076.

Wei CY, Wang HP, Zhu ZY & Sun YH (2014). Transcriptional factors Smad1 and Smad9 act redundantly to mediate zebrafish ventral specification downstream of Smad5. J Biol Chem 289, 6604-6618.

Yan S, Lu D, Du M, Li G, Lin L, Jin G, Wang H, Yang Y, Xia D& Liu A (1984). Nuclear transplantation in teleosts. Hybrid fish from the nucleus of crucian and the cytoplasm of carp. Sci Sin B 27, 1029.

Yan SY, Mao ZR, Yang HY, Tu MA, Li SH, Huang GP, Li GS, Guo L, Jin GQ, He RF, *et al.* (1991). Further investigation on nuclear transplantation in different orders of teleost: the combination of the nucleus of Tilapia (*Oreochromis nilotica*) and the cytoplasm of Loach (*Paramisgurnus dabryanus*). Int J Dev Biol 35, 429-435.

Yan SY, Tu M, Yang HY, Mao ZG, Zhao ZY, Fu LJ, Li GS, Huang GP, Li SH, Jin GQ, *et al.* (1990). Developmental incompatibility between cell nucleus and cytoplasm as revealed by nuclear transplantation experiments in teleost of different families and orders. Int J Dev Biol 34, 255-266.

Zhu ZY & Sun YH (2000). Embryonic and genetic manipulation in fish. Cell Res 10, 17-27.

跨物种克隆：细胞质因子对发育的影响

孙永华　朱作言

中国科学院水生生物研究所，淡水生态与生物技术国家重点实验室，武汉 430072

摘　要　众所周知，在合子基因激活之前，细胞核和卵细胞质之间的物质交流和相互作用，对早期胚胎发育至关重要，这种相互作用对后期的发育也会产生一定的影响。理清合子基因组和卵细胞质因子之间的互作关系非常重要，异种间的细胞核移植技术为研究这一机制提供了一个独特的系统。在这篇综述中，我们总结了近年来异种核移植的进展，特别是在鱼类中的研究进展，以及细胞质因子对移植胚发育的影响。最后，我们得出的结论是：解读任何发育和演变过程应该从一种系统角度来进行，而不仅仅只关注于核基因组的功能和作用。

Efficient RNA Interference in Zebrafish Embryos Using siRNA Synthesized with SP6 RNA Polymerase

Wei-Yi Liu[1,2] Yan Wang[1] Yong-Hua Sun[1] Yun Wang[1,2]
Ya-Ping Wang[1] Shang-Ping Chen[1] Zuo-Yan Zhu[1]

1 State Key Laboratory of Freshwater Ecology and Biotechnology, Institute of Hydrobiology, Chinese Academy of Sciences, Wuhan 430072
2 Graduate School of the Chinese Academy of Sciences, Beijing 100039

Abstract Double-stranded RNA (dsRNA) has been shown to be a useful tool for silencing genes in zebrafish (*Danio rerio*), while the blocking specificity of dsRNA is still of major concern for application. It was reported that siRNA (small interfering RNA) prepared by endoribonuclease digestion (esiRNA) could efficiently silence endogenous gene expression in mammalian embryos. To test whether esiRNA could work in zebrafish, we utilized *Escherichia coli* RNaseIII to digest dsRNA of zebrafish no tail (*ntl*), a mesoderm determinant in zebrafish and found that esi-ntl could lead to developmental defects, however, the effective dose was so close to the toxic dose that esi-ntl often led to non-specific developmental defects. Consequently, we utilized SP6 RNA polymerase to produce si-ntl, siRNA designed against *ntl*, by *in vitro* transcription. By injecting *in vitro* synthesized si-ntl into zebrafish zygotes, we obtained specific phenocopies of reported mutants of *ntl*. We achieved up to a 59% *no tail* phenotype when the injection concentration was as high as 4 μg/μL. Quantitative reverse transcription-polymerase chain reaction (RT-PCR) and whole-mount *in situ* hybridization analysis showed that si-ntl could largely and specifically reduce mRNA levels of the *ntl* gene. As a result, our data indicate that esiRNA is unable to cause specific developmental defects in zebrafish, while siRNA should be an alternative for downregulation of specific gene expression in zebrafish in cases where RNAi techniques are applied to zebrafish reverse genetics.

Keywords esiRNA; *no tail*; RNAi; siRNA; zebrafish

Introduction

Double-stranded RNA interference (RNAi) has become a powerful genetic tool for selectively silencing gene expression in many eukaryotes (Fire *et al*. 1998; Tuschl *et al*. 1999). In the RNAi reaction, the cellular RNaseIII enzyme Dicer cleaves the double-stranded RNA (dsRNA) into 21- to 25-nt RNA called siRNA (small interfering RNA) (Elbashir S. *et al*. 2001;

文章发表于 *Development Growth & Differentiation*, 2005, 47(5): 323-331

Hannon 2002). With its cognate mRNA, siRNA pairs lead to degradation of target mRNA and amplification of gene-specific silencing signals (Zamore *et al.* 2000; Hammond *et al.* 2001). In contrast to the non-specific effect of long dsRNA, siRNA can mediate selective gene silencing in mammals (Bernstein *et al.* 2001; Elbashir S.M. *et al.* 2001). Chemically synthesized siRNA of 21 bp with a-3′ TT overhang and siRNA transcribed by T7 RNA polymerase *in vitro* also selectively silences expression of genes that are homologous to the siRNA sequence (Billy *et al.* 2001). However, inhibition of gene expression by synthetic siRNA is limited because of the position effect of siRNA and off-target gene regulation patterns (Jackson *et al.* 2004; Luo & Chang 2004). Therefore, each mRNA must be screened for an efficient siRNA, a laborious and costly process.

Processing of long dsRNA can generate a great variety of siRNA by endoribonuclease digestion to produce what are called esiRNA (endoribonuclease digestion). The ensemble of esiRNA from the cleavage of a long dsRNA is capable of interacting with multiple sites on target mRNA and thereby increasing the interfering chance that at least one siRNA will be against an efficient target. In *Drosophila* cells and mammalian embryos, esiRNA could silence endogenous gene expression efficiently and specifically (Calegari *et al.* 2002; Kawasaki *et al.* 2004).

To understand gene function in a physiological context, methods are required to study complex systems such as the whole animals. In zebrafish, gene function can be studied by the antisense morpholino technique (Nasevicius & Ekker 2000; Bauer *et al.* 2001). However, the generation of gene-knockdown in zebrafish by morpholinos is useful only for embryonic functions, is quite costly and is non-heritable. Consequently, new knockdown methods by RNAi have been explored in zebrafish to elucidate gene function (Li *et al.* 2000; Boonanuntanasarn *et al.* 2003). Unfortunately, injection of dsRNA has been reported to trigger non-specific interference and toxicity to embryos, suggesting that zebrafish cells may treat the dsRNA as a warning sign of viral infection just as it does in mammalian cells (Oates *et al.* 2000; Zhao *et al.* 2001). Recently, the *dmd* gene of zebrafish was temporarily suppressed following the injection of chemically synthesized siRNA, which indicated the possibility of siRNA-based gene silencing in zebrafish (Dodd *et al.* 2004). However, the question of whether the siRNA may really be suitable for studying novel gene function in zebrafish still remains to be resolved until siRNA-mediated gene knockdown can mimic the specific phenotypes of known mutants without an accompanying ‘background noise’ of other phenotypic effects. Additionally, to produce siRNA with lab routine methods such as *in vitro* transcription will facilitate the application of siRNA approach in zebrafish.

Here, we present the potential of esiRNA and siRNA synthesized with SP6 RNA polymerase to specifically block the expression of *no tail* (*ntl*) to produce a phenotype similar to the known no tail mutant (Schulte-Merker *et al.* 1992).

Materials and Methods

Preparation of esiRNA

Plasmids of pZ-ntl, pZanti-ntl, pZ-GFP and pZanti-GFP were constructed as templates to transcribe *in vitro* the sense and antisense *ntl* mRNA and enhanced green fluorescent protein (eGFP) mRNA. Plasmid pBSK-ntl (gift from S. Schulte-Merker), was digested with *Xho*I and *Nco*I. The smaller fragments (about 1 kb from 3′-end of *ntl*) were inserted into the pZeroAmp (GenBank accession number AY569776) *Xho*I and *Nco*I sites to form pZ-ntl, and inserted into *Sal*I and *Nco*I sites to form pZanti-ntl, respectively. The eGFP fragments of Plasmid pCMVeGFP digested with *Bam*HI and *Xba*I were inserted into the pZeroAmp *Xba*I and *Bam*HI sites to form pZ-GFP. The eGFP fragments of Plasmid pCMVeGFP digested with *Bam*HI and *Xho*I were inserted into the pZeroAmp *Sal*I and *Bam*HI sites to form pZanti-GFP. The sense and antisense mRNA fragments of *ntl and egfp* were transcribed with SP6 RNA polymerase, purified and hybridized to generate ds-ntl and ds-GFP. We mixed 10 μg of ds-ntl or ds-egfp with 2.6 U of shortcut RNaseIII (New England Biolabs, Beverly, MA, USA) in 20 μL reaction buffer (20 mM $MnCl_2$. 1 mM dithiothreitol (DTT), 50 mM Tris-HCl (pH 7.5)). The mixture was incubated at 37℃ for 30 min. Then, 20 μL of the reaction mixture was fractionated by electrophoresis on a non-denaturing 12% polyacrylamide gel electrophoresis (PAGE) gel. The esiRNA band was excised with a clean scalpel and transferred to approximately 350 μL of an RNase free solution (0.5 M ammonium acetate, 1 mM ethylenediamine tetraacetic acid (EDTA), 0.2% sodium dodecylsulfate (SDS)). To elute the nucleic acids, the gel slice was incubated in the above solution at 37℃ overnight. The gel slice was removed from the elution solution and the esiRNA was precipitated with 0.1 volume of 5 M ammonium acetate and 3 volume of 100% ethanol. After incubation of the precipitation reaction at −20℃ for 15 min, the esiRNA were precipitated by centrifuging at 14 000 *g* for 15 min. After centrifuging, the ethanol was aspirated and the pellet was washed with 70% ethanol. The recovered 21-23 bp esi-RNA were dissolved in nuclease free water and stored at −70℃ until they were needed.

Design of siRNA

The siRNA included in the analysis were determined using Ambion online siRNA design tool (www.ambion.com/techlib/misc/siRNA_design.html, Ambion, Austin, TX, USA). We preferably chose the sites with a-3′ UU terminal end overhangs, and GC% between 40% and 60%. Water solubility was maintained by using less then 36% guanine content and no runs of more than triplets of GCG (Gene Tools LLC, Philomath, OR, USA). Also we used the selected sequences to search GenBank to confirm their specificity. Following the above criteria, three specific siRNA target sites for *ntl* were determined.

Generation of siRNA by *in vitro* transcription

According to the designed siRNA targeting sites, three pairs of DNA oligonucleotides were designed respectively to produce siRNA (Table 1). Two 29 mer DNA oligonucleotides with 21 nt of si-ntl template and 8 nt complementary to the SP6 promoter primer were synthesized and desalted (Sangon, Shanghai, China). In separate reactions, the two template oligonucleotides were hybridized to a SP6 promoter primer (28 nt) respectively. The 3′ ends of the hybridized DNA oligonucleotides were extended by the Klenow fragment of DNA polymerase (Takara, Otsu, Japan) to create double-stranded si-ntl transcription templates. The sense and antisense single-strand si-ntl were transcribed with SP6 RNA polymerase (Ambion) and the resulting RNA transcripts were hybridized at 37℃ for 5 h to create double strand si-ntl. The si-ntl consists of 5′-terminal single-stranded leader sequences, a 19 bp target specific dsRNA, and 3′ terminal UU. The leader sequences were removed by digesting the double strand si-ntl with RNaseI (New England Biolabs) in a 20 μL reaction buffer (10 mM $MgCl_2$, 100 mM NaCl, 1 mM DTT, 50 mM Tris-HCl (pH 7.9)). The DNA templates were removed at the same time by DNaseI (Sangon) digestion. The resulting si-ntl were recovered by electrophoresis on 12% PAGE, purified by ethanol precipitation (the same purification process as esiRNA), and measured by PAGE electrophoresis in comparison with standard pBR322/*Msp*I (SABC, Luoyang, China) marker. Si-ntl at different concentrations were prepared for microinjection.

Table 1 List of DNA molecules synthesized

Primers	Sequences of DNA molecules	Target position in *ntl* cDNA
T13A	AATGCAATGTACTCGGTCCTG	232-252
T13S	AACAGGACCGAGTACATTGCA	
T29A	AACGGAGGAGGACAGATTATG	436-456
T29S	AACATAATCATGTCCTCCTCCG	
T36A	AATGAAGAGATTACCGCTCTG	577- 597
T36S	AACAGAGCGGTAATCTCTTCA	

According to the Ambion design tool, a total 62 siRNA target sites were provided. We selected the 13th, 29th and 36th siRNA target sites to test the experiments. (A) antisense strand, (S) sense strand of *ntl* cDNA

Microinjection of zebrafish embryos

RNA was introduced into zebrafish zygotes by pressure injection under a normal microscope. The injection concentration of esi-ntl and si-ntl was noted for selection of the most effective dose for downregulation of the target gene expression. The injection volume was 1 nL per embryo. The zebrafish embryos were cultured according to Westerfield (1989).

Semi-quantitative PCR (SqPCR)

SqPCR was used to evaluate the most efficient si-ntl and as-ntl which specifically knockdown *ntl* in zebrafish. Chemical synthesized si-GFP (targeted mRNA sequence GCAAGCUGACCCUGAAGUUCA,122 nt to 140 nt of *egfp* mRNA; gift from Y. Chen) was injected as the control group. Forty zebrafish embryos injected with 450 ng/μL si-ntl, as-ntl, si-GFP or uninjected embryos were collected at 8 hours postfertilization (hpf). Total RNA was extracted and purified by SV Total RNA Isolation System (Promega, Madison, WI, USA). The reverse transcription was carried out at 30℃ for 5 min, 42℃ for 60 min with 50 ng of total RNA, 5 pmol of oligo (dT), 2 U RNasin, 200 μM dNTP, 20 U reverse transcriptase XL (Takara), 75 mM KCl, 3 mM $MgCl_2$, 20 mM DTT in 50 mM Tris-HCl, pH 8.3, at a final volume of 10 μL. The reaction was stopped by heating at 80℃ for 5 min. The PCR was performed in 25 μL containing 1 μL cDNA products from the reverse transcription, 200 μM dNTP, 10 pmol primers, and 1 unit of *Taq* polymerase (Fermentas, Vilnius, Lithuania), 50 mM KCl, 1.5 mM $MgCl_2$, and 0.001% gelatin in 10 mM Tris-HCl, pH 8.3. A GenAmp PCR System 9700 (Perkin Elmer, Pomona, CA, USA) was used with the following program: a predenaturation at 94℃ for 3 min, 30 cycles of amplification (94℃ for 20 s, 56℃ for 30 s, 72℃ for 30 s) and a final extension at 72℃ for 7 min. The reverse transcription–polymerase chain reaction (RT-PCR) products were separated using 1.5% agarose gel electrophoresis. The marker was 250 bp marker (Sangon) and the DNA bands were scanned in GeneGenius (Syngene, Frederick, MD, USA). The PCR primer sequences of *ntl* were according to Li *et al.* (2000). For *ntl* (GenBank accession number NM_131162), the forward primer was 5′-TTGGAACAACTTGAGGGTGA-3′ and the reverse primer was 5′-CGGTCACTTTTCAAAGCGTAT-3′; for *β-actin* (GenBank accession number AF025305), the forward primer was 5′-TCACCACCACAGCCGAAAG-3′ and the reverse primer was 5′-GGTCAGCAATGCCAGGGT A-3′.

Real-time quantitative PCR (RT-qPCR)

RT-qPCR was used to quantify the related mRNA level of *ntl* at different zebrafish developmental stages following the induction dose of 4 μg/μL si-ntl13. Zebrafish embryos were collected at 50% epiboly stage (5.25 hpf), 90% epiboly stage (9 hpf), 8-somite stage (13 hpf) and 24-somite stage (22 hpf), respectively. Protocols for total RNA extraction and reverse transcription were the same as the above SqPCR. RT-qPCR were performed using an Applied Biosystems (Foster City, CA, USA) model 7000HT platform and SYBR Green dye chemistries (Applied Biosystems). All reactions (25 μL) were performed in triplicate. A standard PCR mix using LA Taq DNA polymerase (Takara) supplemented with 1 μL/25 μL reaction of SYBR Green dye. Relative expression of *ntl* was calculated using 2^ (-delta delta C(T)) method with correction for different amplification efficiencies (Livak & Schmittgen 2001).

Whole-mount *in situ* hybridizations

Sense and antisense RNA probes of ntl cDNA were synthesized *in vitro* in the presence of digoxin (DIG) RNA labeling mix (Roche, Basel, Switzerland). The efficiency of the probes was determined and the sense RNA probe was used as a negative control. The embryos were fixed in 4% paraformaldehyde and used for *in situ* hybridization essentially following the protocol described by Thisse *et al.* (1993).

Light microscopy

Images of zebrafish embryos were recorded using an Olympus SZX12 microscope and a digital camera.

Results

esiRNA leads to non-specific developmental interference of zebrafish embryos

Following the digestion of ds-ntl by RNaseIII, 21-23 bp fragments of esi-ntl were selectively obtained and injected with a concentration ranging from 20 ng/μL to 200 ng/μL. The total *no tail* phenotype could easily be observed when the total RNA concentration was 50 ng/μL or higher. The ratio of no tail phenotype increased from 7% (13/187) to 18% (8/46) as the esi-ntl concentration increased from 20 ng/μL to 100 ng/μL. At the same time, we took plasmid pZ-ntl and esi-GFP as the control group. Both kinds of treatments could not lead to the *no tail* phenotype.

However, the *no tail* phenotype was overwhelmed by the non-specific developmental defects. Zebrafish eggs injected with more than 20 ng/μL of esi-ntl developed normally up to approximately the 30% epiboly stage, but thereafter we observed non-specific interference in the phenotype. For those embryos that survived to the somite stage, esi-ntl caused pleiotropic morphological defects including anteriorization, malformed eyes, anterior/posterior truncations, delayed development and misshapen notochord (Fig. 1A). Just like ds-ntl, which can cause nonspecific interference in zebrafish embryos (Oates *et al.* 2000), esi-ntl had similar toxicity to zebrafish embryos that led to the death of most embryos. Regardless of the injection dose, toxicity of esi-ntl to embryos was observed. Doses between 20 ng/μL and 200ng/μL resulted in survival rates that decreased in proportion to the increase of esi-ntl (Fig. 1B). 20 ng/μL of esi-ntl led to about 40% embryo death by 36 hpf, 50 ng/μL of esi-ntl led to about 70% embryo death, and 200 ng/μL of esi-ntl killed nearly all embryos. Approximately 40% of the total injected embryos died before the early somite stage following the injection of more than 50 ng/μL esi-ntl. The plasmidinjection groups of embryos developed normally, while the esi-GFP led to similar toxicities as esi-ntl to zebrafish embryo (Fig. 1C).

siRNA could knockdown gene expression specifically and cause specific developmental defects in zebrafish

High yields of short sense RNA (ss-ntl) and antisense RNA (as-ntl) of si-ntl were

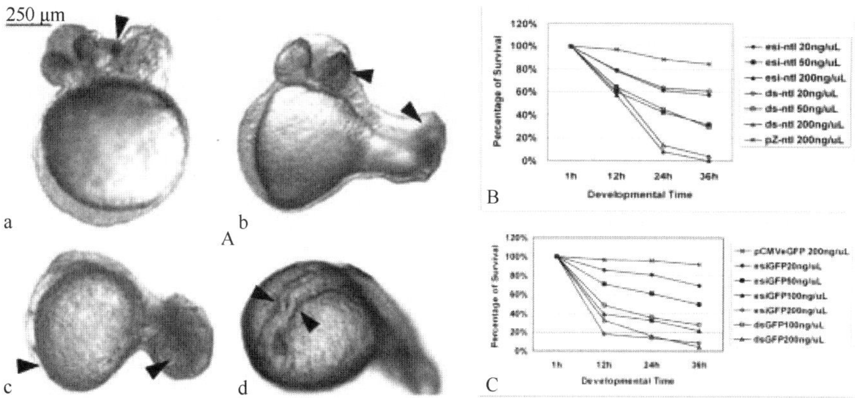

Fig. 1 esi-ntl led to different kinds of abnormal development and toxicities of zebrafish embryos. (A) 50 ng/μL esi-ntl led to the death of most embryos, with those surviving also showing pleiotropic morphological defects, the figure showing typical kinds of abnormal development: (a) an anterioried embryo; (b) malformed eyes, posterior truncation; (c) lack of head and posterior truncation; (d) misshapen notochord. Arrows show the malformed positions. (B) Different doses of esi-ntl and ds-ntl were injected into zygotes of zebrafish, separately, and the survival percentages of embryos at different stages of the total number. 200 ng/μL pZ-ntl per embryo served as a control. As was shown, the toxicities between ds-ntl and esi-ntl of the same injection concentration are surprisingly similar. (C) At the same time, we compared the toxicities of esi-green fluorescent protein (GFP) and ds-GFP, the same results can be concluded

obtained by *in vitro* transcription using SP6 RNA polymerase. Corresponding ss-ntl and as-ntl were hybridized to generate highly purified and concentrated double-stranded short interference *ntl* RNA (si-ntl), in which si-ntl13 targets against 232-252 nt of *ntl* mRNA, si-ntl29 against 436-456 nt and si-ntl36 against 577-597 nt. The presence of asRNA and si-ntl were assayed and confirmed by gel electrophoresis.

Because the *no tail* phenotype can be easily observed if the *ntl* gene is knocked out or sufficiently knocked down in zebrafish (Halpern *et al.*1993; Griffin & Kimelman 2003), we chose it as the target gene again to study the possibility of utilizing siRNA to inhibit expression of endogenous genes. A partial *no tail* or total *no tail* phenotype was generated for the zebrafish groups subjected to si-ntl microinjection (Fig. 2). The si-ntl-injected embryos showed the lack of posterior tissue and posterior body shortening, which are similar to the known phenotypes of *ntl* mutants (Amacher *et al.* 2002) (Fig. 2G). When the concentration of the si-ntl13 was up to 4 μg/μL, we observed 23 total *no tail* embryos (6.4%) and 190 partial *no tail* embryos (52.6%) among 361 viable embryos at 72 hpf. Meanwhile, SqPCR showed that 450 ng/μL si-ntl13 could block 76% mRNA level of *ntl* of injected embryos on average at mid-gastrula stage (Fig. 3). Considering there should be some embryos in which the mRNA level of *ntl* was not largely affected (see next paragraph), we propose that the blocking efficiency must be higher than 76% in those effectively knockdown embryos. To analyze whether siRNA influence expression of other nonrelated genes, RT-PCR was used to test the levels of *β-actin* mRNA and as a result there was no difference between the si-ntl injected group and each other group (Fig. 3).

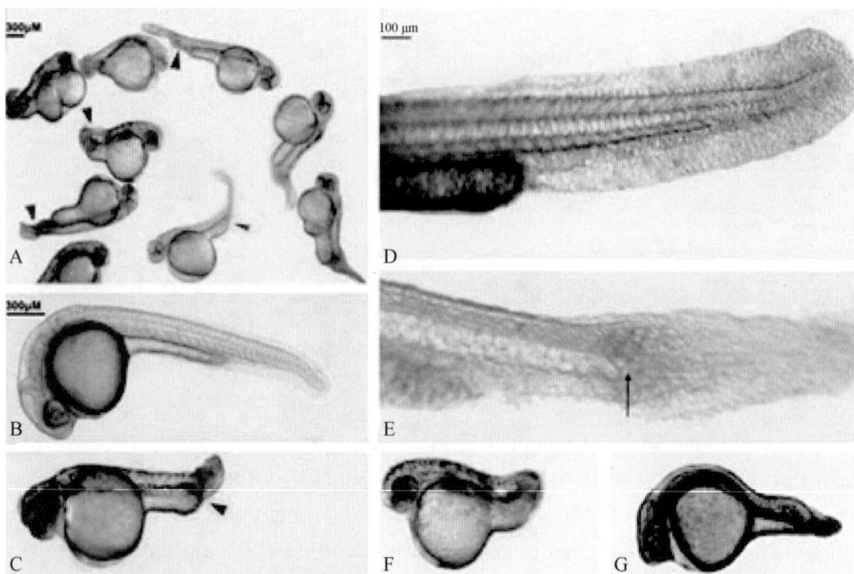

Fig. 2 Induction of no tail phenocopies by injection of si-ntl RNA. (A) Embryos were injected with 450 ng/μL si-ntl13 and analyzed at 72 hours postfertilization (hpf). (B) si-GFP injected embryo. (D) Tail of si-GFP injected embryo. (C, E, F) si-ntl13 injected embryos with *no tail* phenotypes to different extent: (C) deficiency of posterior tail, partially *no tail* phenotype; (E) malformed posterior tails morphologically associated with the *no tail* phenotype; (F) total no tail phenotype. (G) a typical phenotype of reported *ntl* mutant (Amacher *et al.* 2002). Arrows showed the initial sites of posterior tail lack. (B) bar, 300 μm (for B, C, F); (D) bar, 100 μm (for D, E)

In situ hybridization results revealed that *ntl* mRNA message was totally undetectable in 61% (11/18) of the si-ntl injected gastrulas, while the knockdown effects were attenuated along the development of embryos (Fig. 4). At the early somite stage, *ntl* mRNA message was weakly detectable in 76% (16/21) of the injected embryos. However, at the late somite stage, the *ntl* mRNA message was recovered and strong at the most-posterior part of all embryos, although *ntl* mRNA message along the notochord was largely decreased in 83% (45/54) of the injected embryos. RT-qPCR analysis showed that at 50% epiboly and 90% epiboly stage, 4 μg/μL si-ntl13 could block 83% and 74% *ntl* mRNA, while only 33% and 12% *ntl* mRNA could be blocked at the 8-somite stage and 24-somite stage, respectively (Fig. 5). Therefore, we suggest that si-ntl were degenerated along the development of embryos, and the stages between blastula and gastrula are the best time for si-ntl to knockdown *ntl* expression.

In an attempt to determine whether short antisense RNA (as-ntl) could yield a similar effect, we injected the as-ntl13, as-ntl29 and as-ntl36 which target the corresponding 232-252nt, 436 - 456nt and 577-597nt of *ntl* mRNA respectively. The as-ntl did not suppress *ntl* expression as assayed by phenotype observation and RT-PCR (Fig. 3).

Statistical analysis of the RT-PCR results and the phenotype changes indicated that the effects of knocking down the expression of *ntl* gene varied among different si-ntl sites (Table 2,

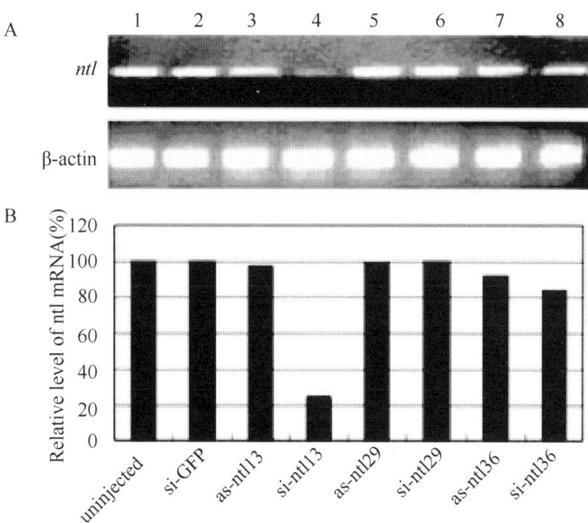

Fig. 3 Reverse transcriptase–polymerase chain reaction (RT-PCR) analysis of ntl mRNA in zebrafish embryos injected with 450 ng/μL si-ntl or as-ntl RNA against different targets. Total RNA of 8 hpf embryos was isolated and analyzed by RT-PCR. *β-actin* was used as the endogenous control. Si-GFP injected embryos and uninjected embryos were used as the control group. The injection concentration is 450 ng/μL. (A) Agarose gel electrophoresis of RT-PCR of *ntl* mRNA and *β-actin* mRNA. Numbers 1-8 represent different RNA treatment groups. (1) uninjected embryos; (2) si-GFP injected embryos; (3) as-ntl13; (4) si-ntl13; (5) as-ntl29; (6) si-ntl29; (7) as-ntl36; (8) si-ntl36. (B) The agarose gel was scanned by GeneGenius and analyzed by genesnap software (Syngene), and the intensity of each DNA band was calculated based on a 250 bp marker. *ntl* mRNA level for the uninjected embryos was defined as 100%. The relative level of *ntl* mRNA of each experimental group was calculated by comparing the intensity of the group with the uninjected group

Fig. 3). Si-ntl13 led to 15% *ntl* phenotype if the injected concentration was up to 450 ng/μL, while at the same injection concentration si-ntl29 and si-ntl36 lead to 4% and 8% *ntl* phenotype, respectively (Table 2). As the concentration of si-ntl13 rises, the percentage of embryos with the *ntl* phenotype also rises. This was not the case with si-ntl29 and si-ntl36. Therefore, to choose an

Table 2 si-ntl injections and the resulting phenotype changes

Injected material	Egg injected	Egg viable at 72 hpf	Normal	Phenotype Abnormal	No tail
si-GFP(450 ng/μL)	242	196	195	1	0
si-GFP(4 μg/μL)	324	274	267	7	0
si-ntl13(450ng/μL)	487	374	314	5	55 (15%)
si-ntl13 (4 μg/μL)	479	361	137	11	213 (59%)
si-ntl29(450ng/μL)	288	210	198	3	9 (4%)
si-ntl36(450ng/μL)	132	101	90	3	8 (8%)

Percentages were calculated by dividing the number of the no tail phenotype by the number of those surviving. Abnormal embryos non-related to the no tail phenotype could be observed in each group including the control one, but the ratios of the abnormal to the total that survived were no more than 3% and therefore negligible. All experiments were repeated at least three times, and the percentage shown is a mean. hpf, hours postfertilization

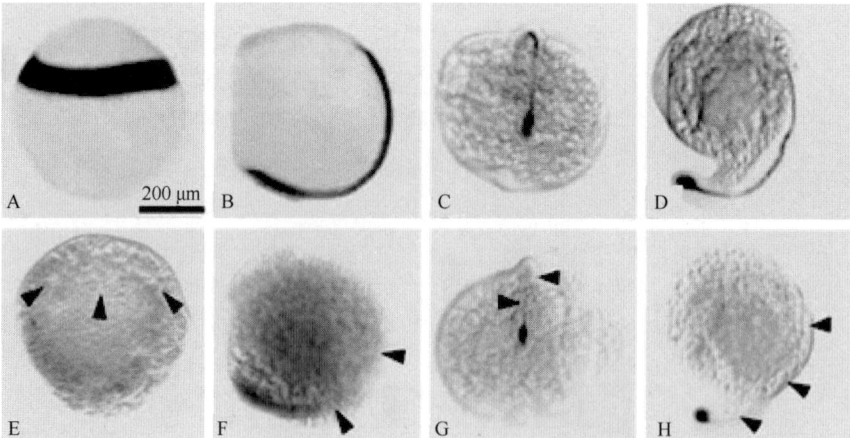

Fig. 4 Whole-mount *in situ* hybridization of *ntl* mRNA revealed that *ntl* mRNA were decreased in zebrafish embryos following the injection of si-ntl13 RNA. (A, B, C, D) si-GFP-injected embryos; (E, F, G, H) si-ntl13-injected embryos. The injection concentration is 4 μg/μL. (A, E) *ntl* expression at blastula embryos, 4.7 hpf) in si-GFP injected and si-ntl13 injected embryos. *ntl* mRNA showed an undetectable message in 61% (11/18) embryos injected of si-ntl13. (B, F) *ntl* mRNA expression at late early somite stage (1-2 somite, 10 hpf) in si-GFP-injected and si-ntl13-injected embryos. *ntl* mRNA showed a weak message in 76% (16/21) embryos injected with si-ntl13. (C, D, G, H) *ntl* mRNA expression at late somite stage (16 somite stage, 16 hpf) in si-GFP injected and si-ntl13-injected embryos. *ntl* mRNA showed relatively a weak message along the notochord axis in 83% (45/54) of the embryos injected with si-ntl13, while all the control embryos showed a strong hybridization message. The arrows show the areas where the *ntl* signal is normally expected. Embryos (A, B, D, E, F, H) are lateral views, with the anterior to the top, and the dorsal to the right. Embryos (C, G) are dorsal views, with anterior to the top. Bar, 200 μm

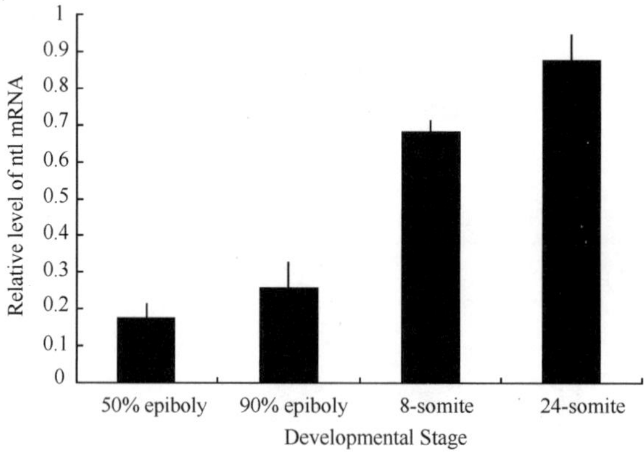

Fig. 5 RT-qPCR analysis of *ntl* mRNA in zebrafish embryos injected with 4 μg/μL si-ntl13 at different developmental stage. We utilize the methods of $2^{-\delta\delta CT}$ to evaluate the relative level of *ntl* mRNA at different developmental stages. The abscissa represents different developmental stages of 4 μg/μL si-ntl13-treated embryo group. Embryos of four developmental stages were analyzed. The ordinate represents the relative *ntl* mRNA level. For each developmental stage, the *ntl* mRNA level for the control was defined as 1.0. Control groups were the uninjected embryos and the si-GFP-injected embryos. *β-actin* was used as a reference gene. Each value is the average of two independent assays. Data show mean values ± SD (n = 3)

effective siRNA site appeared to be the most important step to significantly knockdown gene expression in order to observe obvious phenotype changes of zebrafish embryos for studying the function of a specific gene.

Discussion

As an excellent model system of developmental biology, the demand for gain- and loss-of-function studies in zebrafish is of great interest. Previous reverse genetics studies in zebrafish were mainly based on knockdown technology like ribozyme and antisense morpholino (Hudziak *et al.* 1996; Xie *et al.*1997; Nasevicius & Ekker 2000). The ribozyme mediated gene knockdown strategy is based on a T7 expression system and could identify gene function in zebrafish, however, the system was too complicated to apply in a larger field. Up to now, antisense morpholino technology has been the most successful knockdown strategy for loss-of-function studies in zebrafish. Although the success rate of morpholino has been up to 90% for reported genes (Heasman 2002), the application of the technique is limited to the first few days of embryo development. Unlike morpholino-mediated gene-knockdown, siRNA-mediated expre-ssion vector could inhibit gene expression in a stable and inheritable way if siRNA prove to suppress expression of target genes.

Consequently, we utilized RNaseIII *in vitro* to digest dsRNA into siRNA of 21-23 bp, which was reported to significantly suppress exogenous and endogenous gene expression in mammalian cell (Yang *et al.* 2002). However, our results indicate that this approach is not efficient in zebrafish. We found that esiRNA leads to non-specific defects and toxicity to zebrafish embryos. Because the esiRNA-mediated knockdown strategy is a homology-dependent gene-silencing response, a possible reason is that esiRNA containing those siRNA sequences that are homologous with other endogenous genes, leading to their down-regulation synchronously. Therefore, great care should be taken to make target esiRNA non-homologous with genes necessary for normal development, which might alleviate the non-specific side-effects of esiRNA treatment in zebrafish. Alternatively, a siRNA pool, formed by mixing different sites of siRNA together to knockdown specific gene expression, perhaps appears to be an efficient knockdown strategy.

EsiRNA failed to knockdown the *ntl* gene specifically, so we tried to test whether siRNA could work better than esiRNA in zebrafish. The resulting zebrafish phenotype of si-ntl treatment was a phenocopy of the known mutants of the *ntl* gene. The target mRNA (*ntl* mRNA) was degraded while that of a non-related gene (*β-actin*) was unchanged. Our *in situ* hybridization and RT-qPCR results showed that siRNA in zebrafish embryos acted mainly during gastrulation and the early somite stage. After this stage, siRNA may be diluted or degenerated. So it would be limited for siRNA-mediated RNAi by just directly injecting small RNA molecules. A vector stably expressing siRNA would be more effective in the downregulation of specific gene expression, and would be more useful in the study of gene function during zebrafish development. Although all the candidate si-ntl sites (si-ntl13 at 78-84aa, si-ntl29 at 146-152aa, si-ntl36 at 193-209aa) lie in the T-box DNA-binding domain (DBD, amino acids 1-229) that the T-box family of transcriptional regulators share (Goering *et al.* 2003), we found no related phenotypes observed such as *spt* mutant spade tail (Marlow *et al.* 2004). However, of the total 361 surviving embryos at 72 hpf, 11 abnormal embryos non-related to the no tail phenotype also could be observed in the 4 μg/μL siRNA treatment group, but the ratio was no more than 3% and therefore is negligible. Additionally, we found that asRNA in zebrafish could not inhibit gene expression (though, of course, compared to dsRNA, single strand RNA seemed much more unstable and to be easily decomposed *in vitro* according to our observation and it is still the case under physiological conditions), while similar behaviors of single-strand and double-strand siRNA were observed for mammalian cells (Martinez *et al.* 2002; Holen *et al.* 2003).

As was reported, different siRNA target sites lead to different inhibitory effects (Amarzguioui & Prydz 2004; Luo & Chang 2004). We also observed the phenomenon in our experiment. We synthesized three pairs of si-ntl against different targeting positions, but only one of them was effective. To screen an efficient siRNA target site, more than useful siRNA have to be synthesized, which can be costly and time-consuming. In order to avoid this problem, we should make further studies to have a comprehensive knowledge on the molecular mechanism of siRNA and the efficient target sites.

In conclusion, RNAi techniques do hold tremendous promise for unleashing the dormant potential of sequenced genomes. Here, we demonstrate an alternative siRNA knockdown method based on *in vitro* transcription, which is convenient and accessible to most molecular biology labs and will likely provide us a promising way to study gene functions at a large scale in zebrafish.

Acknowledgements We express our appreciation to Perry Hackett for his valuable comments, Cunming Duan and Shuangwei Li for their revision of the manuscript, and Ming Li for technical assistance. This study was supported by the State Key Fundamental Research of China (Grant no. 2004CB117406 and G2000016109), the National Natural Science Foundation of China (Grant no. 90208024), and the Frontier Science Projects Program from the Institute of Hydrobiology, Chinese Academy of Sciences (Grant no. 220311).

References

Amacher, S. L., Draper, B. W., Summers, B. R. & Kimmel, C. B. 2002. The zebrafish T-box genes no tail and spade tail are required for development of trunk and tail mesoderm and medial floor plate. Development 129, 3311-3323.

Amarzguioui, M. & Prydz, H. 2004. An algorithm for selection of functional siRNA sequence. Biochem. Biophys. Res. Commun. 316, 1050-1058.

Bauer, H., Lele, Z., Rauch, G. J., Geisler, R. & Hammerschmidt, M. 2001. The type I serine/threonine kinase receptor Alk8/Lost-a-fin is required for Bmp2b/7 signal transduction during dorsoventral patterning of the zebrafish embryo. Development 128, 849-858.

Bernstein, E., Denli, A. M. & Hannon, G. J. 2001. The rest is silence. RNA. 7, 1509-1521.

Billy, E., Brondani, V., Zhang, H., MuÈller, U. & Filipowicz, W. 2001. Specific interference with gene expression induced by long, double-stranded RNA in mouse embryonal teratocarcinoma cell lines. Proc. Natl Acad. Sci. USA 98,14428-14433.

Boonanuntanasarn, S., Yoshizaki, G. & Takeuchi, T. 2003. Specific gene silencing using small interfering RNAs in fish embryos. Biochem. Biophys. Res. Commun. 310, 1089-1095.

Calegari, F., Haubensak, W., Yang, D., Huttner, W. B. & Buchholz, F. 2002. Tissue-specific RNA interference in postimplantation mouse embryos with endoribonuclease-prepared short interfering RNA. Proc. Natl Acad. Sci. USA 99, 14236-14240.

Dodd, A., Chambers, S. P. & Love, D. R. 2004. Short interfering RNA-mediated gene targeting in the zebrafish. FEBS Letters 561, 89-93.

Elbashir, S., Harborth, M., Lendeckel, J. W., Yalcin, A., Weber, K. & Tuschl, T. 2001. Duplexes of 21-nucleotide RNAs mediate RNA interference in mammalian cell culture. Nature 411, 494-498.

Elbashir, S. M., Lendeckel, W. & Tuschl, T. 2001. RNA interference is mediated by 21- and 22-nucleotide RNAs. Genes Dev. 15, 188-200.

Fire, A., Xu, S., Montgomery, M. K., Kostas, S. A., Driver, S. E. & Mello, C. C. 1998. Potent and specific genetic interference by double stranded RNA in Caenorhabditis elegans. Nature 39, 1806-1811.

Goering, L. M., Hoshijima, K., Hug, B., Bisgrove, B., Kispert, A. & Grunwald, D. J. 2003. An interacting network of T-box genes directs gene expression and fate in the zebrafish mesoderm. Proc. Natl Acad. Sci. USA 100, 9410-9415.

Griffin, K. J. P. & Kimelman, D. 2003. Interplay between FGF, one-eyed pinhead, and Ntl transcription factors during zebrafish posterior development. Dev. Biol. 264, 456-466.

Halpern, M. E., Ho, R. K., Walker, C. & Kimmel, C. B. 1993. Induction of muscle pioneers and floor plate is distinguished by the zebrafish no tail mutation. Cell 75, 99-111.

Hammond, S. M., Caudy, A. A. & Hannon, G. J. 2001. Posttranscriptional gene silencing by double-stranded RNA. Nature Rev. Genet. 2, 110-119.

Hannon, G. J. 2002. RNA interference. Nature 418, 244-251.

Heasman, J. 2002. Morpholino oligos: making sense of antisense? Dev. Biol. 243, 209-214.

Holen, T., Amarzguioui, M., Babaie, E. & Prydz, H. 2003. Similar behaviour of single-strand and double-strand siRNAs suggests they act through a common RNAi pathway. Nucleic Acids Res. 31, 2401-2407.

Hudziak, R. M., Barofsky, E., Barofsky, D. F., Weller, D. L., Huang, S. B. & Weller, D.D. 1996. Resistance of morpholino phos-phorodiamidate oligomers to enzymatic degradation. Antisense Nucleic Acid Drug Dev. 6, 267-272.

Jackson, A. L., Bartz, S. R., Schelter, J. *et al.* 2004. Expression profiling reveals off-target gene regulation by RNAi. Nature Biotechnol. 21, 635-637.

Kawasaki, H., Suyama, E., Iyol, M. & Taira, K. 2004. siRNAs generated by recombinant human Dicer induce specific and significant but target site-independent gene silencing in human cells. Nucleic Acids Res. 31, 981-987.

Li, Y. X., Farrell, M. J., Liu, R., Mohanty, N. & Kirby, L. 2000. Double-stranded RNA injection produces null phenotypes in zebrafish. Dev. Biol. 217, 394-405.

Livak, K. J. & Schmittgen, T. D. 2001. Analysis of relative gene expression data using real-time quantitative PCR and the $2^{[-\mathrm{delta\ delta\ C(T)}]}$ Method. Methods 25, 402-408.

Luo, K. Q. & Chang, D. C. 2004. The gene-silencing efficiency of siRNA is strongly dependent on the local structure of mRNA at the targeted region. Biochem. Biophys. Res. Commun. 318, 303-310.

Marlow, F., Gonzalez, E. M., Yin, C. Y., Rojo, C. & Solnica-Krezel, L. 2004. No tail co-operates with non-canonical Wnt signaling to regulate posterior body morphogenesis in zebrafish. Development 131, 203-226.

Martinez, J., Patkaniowska, A., Urlaub, H., Luhrmann, R. & Tuschl, T. 2002. Single-stranded antisense siRNAs guide target RNA cleavage in RNAi. Cell 110, 563-574.

Nasevicius, A. & Ekker, S. C. 2000. Effective targeted gene 'knockdown' in zebrafish. Nat. Genet. 26, 216-220.

Oates, A. C., Bruce, A. E. E. & Ho, R. K. 2000. Too much interference: injection of double-stranded RNA has nonspecific effects in the zebrafish embryo. Dev. Biol. 224, 20-28.

Schulte-Merker, S., Ho, R. K., Herrmann, B. G. & Nusslein-Volhard, C. 1992. The protein product of the zebrafish homologue of the mouse T gene is expressed in nuclei of the germ ring and the notochord of the early embryo. Development 116, 1021-1032.

Thisse, C., Thisse, B., Schilling, T. F. & Postlethwait, J. H. 1993. Structure of the zebrafish *snail1* gene and its expression in wild type, *spadetail* and *no tail* mutant embryos. Development 119, 1203-1215.

Tuschl, T., Zamore, P. D., Lehmann, R., Bartel, D. P. & Sharp, P.A. 1999. Targeted mRNA degradation by double-stranded RNA *in vitro*. Genes Dev. 13, 3191-3197.

Westerfield, M. 1989. The Zebrafish Book: A Guide for the Laboratory Use of Zebrafish (*Brachydanio Rerio*). University of Oregon Press, Eugene.

Xie, Y. F., Chen, X. Z. & Wagner, T. E. 1997. A ribozyme-mediated gene knockdown strategy for the identification of gene function in zebrafish. Proc. Natl Acad. Sci. USA 94, 13777-13781.

Yang, D., Buchholz, F., Huang, Z. *et al.* 2002. Short RNA duplexes produced by hydrolysis with *Escherichia coli* RNase III mediate effective RNA interference in mammalian cells. Proc. Natl Acad. Sci. USA 99, 9942-9947.

Zamore, P., Tuschl, T., Sharp, P. & Bartel, D. 2000. RNAi: double-stranded RNA directs the ATP-dependent cleavage of mRNA at 21- to 23-nucleotide intervals. Cell. 101, 25-33.

Zhao, Z., Cao, Y., Li, M. & Meng, A. M. 2001. Double-Stranded RNA injection produces nonspecific defects in zebrafish. Dev. Biol. 229, 215-223.

斑马鱼 SP6 RNA 聚合酶介导 siRNA 合成可引起有效的 RNA 干扰作用

刘维一[1,2]　王　艳[1]　孙永华[1]　王　蕴[1,2]
汪亚平[1]　陈尚萍[1]　朱作言[1]

1 中国科学院水生生物研究所，淡水生态及生物技术国家重点实验室，武汉 430072
2 中国科学院大学，北京 100039

摘　要　双链 RNA 是一种有效的用于沉默基因表达的工具，然而其特异性仍是阻碍其应用的主要因素。据报道，通过核糖核酸内切酶消化的 siRNA(esiRNA) 能够高效地沉默哺乳动物胚胎中内源基因的表达。为了研究 esiRNA 是否能够在斑马鱼中起作用，我们使用大肠杆菌 RNaseIII消化斑马鱼中胚层决定因子 *ntl* 基因的 dsRNA。实验结果发现 esi-ntl 可以导致发育缺陷，但由于有效剂量非常接近毒性剂量，esi-ntl 也导致了非特异性的发育缺陷。为此，我们使用 SP6 RNA 聚合酶，体外转录合成 si-ntl。通过向斑马鱼受精卵中注射体外合成的 si-ntl，我们获得了与报道一致的 *ntl* 突变的特定表型，当注射浓度高达 4 μg/μl 时，无尾突变表型比率高达 59%。定量 RT-PCR 和胚胎整体原位杂交分析均显示，si-ntl 能够显著且特异性地减少 *ntl* 基因的 mRNA 水平。我们的数据表明，esiRNA 在斑马鱼中无法引起特异性发育缺陷，而在将 RNA 干扰技术用于斑马鱼反向遗传学研究时，siRNA 是下调斑马鱼特定基因表达的一种有效方法。

Cloning, Characterization and Promoter Analysis of Common Carp *Hairy/Enhancer-of-Split*-Related Gene, *her6*

JING LIU[1,2] YONG-HUA SUN[1] NA WANG[1,2]
YA-PING WANG[1] ZUO-YAN ZHU[1]

1 State Key Laboratory of Freshwater Ecology and Biotechnology, Institute of Hydrobiology, Chinese Academy of Sciences, Wuhan 430072
2 Graduate School of the Chinese Academy of Sciences, Beijing 100039

Abstrat Some members of *hairy/Enhancer-of-split*-related gene (HES) family have important effects on axial mesoderm segmentation and the establishment and maintenance of the somite fringe. In fishes, the *her6* gene, a member of the HES family, is the homologue of *hes1* in mammals and chicken. In this study, the *her6* gene and its full-length cDNA from the common carp (*Cyprinus carpio*) were isolated and characterized. The genomic sequence of common carp *her6* is approximately 1.7kb, with four exons and three introns, and the full-length cDNA of 1314 bp encodes a putative polypeptide of 271 amino acids. To analyse the promoter sequence of common carp *her6*, sequences of various lengths upstream from the transcription initiation site of *her6* were fused to enhanced green fluorescent protein gene (*eGFP*) and introduced into zebrafish embryos by microinjection to generate transgenic embryos. Our results show that the upstream sequence of 500 bp can direct highly effcient and tissue-specific expression of *eGFP* in zebrafish embryos, whereas a fragment of 200 bp containing the TATA box and a partial suppressor of hairless paired site sequence (SPS) is not sufficient to drive *eGFP* expression in zebrafish embryos.
Keywords *her6*; common carp; promoter; genomic walking; expression pattern; development

Introduction

With the identification and characterization of segmentation genes in *Drosophila melanogaster* (Lewis 1978), many researchers have studied the mechanisms underlying vertebrate segmentation, especially somitogenesis (Nusslein-Volhard and Wieschaus 1980; Tam 1981; Tautz and Sommer 1995; Kimmel 1996; De Robertis 1997; Leimeister *et al.* 2000; Vasiliauskas and Stern 2000; Davis *et al.* 2001; Elmasri *et al.* 2004; Oates *et al.* 2005). *hairy/ Enhancer-of-split*-related gene (HES) family, which encode E-type basic helix-loop-helix

文章发表于 *Journal of Genetics*, 2006, 85(3): 171-178

(bHLH) transcription factors (Ledent and Vervoort 2001), and are required for early embryonic segmentation and determination of several cell types (Pasini *et al.* 2001), play an important role in somitogenesis. HES family proteins are defined by three characters: (i) a proline in the sixth residue within the basic domain allowing the factors to bind preferentially to 'N-boxes' in their target genes (Van Doren *et al.* 1994), (ii) an Orange domain mediating the specificity of their biological action *in vivo* (Dawson *et al.* 1995), and (iii) a four-amino acid motif, WRPW, located at the general corepressor protein, that has repressor activity (Jimenez *et al.* 1997). Since the *hairy* gene was first discovered in *Drosophila* (Lewis 1978), many genes of the *hairy* family have been isolated from different vertebrates including mouse (Sasai *et al.* 1992), chicken (Palmeirim *et al.* 1997; Jouve *et al.* 2000) and zebrafish (Gajewski and Voolstra 2002). The *hairy* genes were named after *hes* in mammals and *her* in fish, with a number that most probably reflects the temporal order of cloning (Minguillon *et al.* 2003). Most *hairy* genes are cyclically expressed in the presomitic mesoderm (PSM) during somitogenesis, such as *mhes1*, *mhes7*, *c-hairy1*, *c-hairy2*, *her1*, *her7* (Palmeirim *et al.* 1997; Takke and Campos-Ortega 1999; Jouve *et al.* 2000; Leimeister *et al.* 2000; Bessho *et al.* 2001a,b, 2003; Saga and Takeda 2001; Henry *et al.* 2002; Oates and Ho 2002). In zebrafish, *her6* is a homologue of the mammalian *hes1* gene, but its expression pattern is different from that of mammalian *hes1*. Expression of mouse *hes1* is cyclically regulated by Notch signalling (Jarriault *et al.* 1995; Ohtsuka *et al.* 1999) in developing nephrons (Honjo 1996; Gaiano and Fishell 2002; Selkoe and Kopan 2003; Piscione *et al.* 2004). *hes1* is also expressed at an early stage prior to Notch activity (Kageyama *et al.* 2005), indicating that Notch is not the sole regulator of *hes1* expression. Zebrafish *her6* is expressed in the half-posterior of every somite (PSM) and is dynamically expressed in the romdomeres in the hindbrain (Pasini *et al.* 2001). Feedback regulation of *notch1a* in the anterior PSM is required for maintaining the synchronization of cyclic gene expression among adjacent cells within the PSM but this regulation of *her6* is not present in somites (Pasini *et al.* 2004).

In a recent study, we demonstrated that the somitogenesis process and vertebral numbers of cross-genus cloned fish derived from transgenic common carp (*Cyprinus carpio*) nuclei and goldfish (*Carassius auratus*) enucleated eggs were both consistent with those of goldfish but different from those of common carp (Sun *et al.* 2005). This suggests that somite development might be affected by cytoplasmic factors from the maternal parent. Thus, it is important to study somitogenesis in common carp which will likely provide novel insights into somite development. In the study reported here, the full-length cDNA and genomic sequences of common carp *her6* were characterized and the upstream regulatory sequences amplified and linked to a reporter gene *eGFP*. To analyse the promoter activity of common carp *her6*, two expression constructs, pcher6eGFPa and pcher6eGFPb, were injected into zebrafish embryos. By identifying key sequences that can direct tissue-specific *GFP* expression that mimics normal patterns of *her6* expression, we have extended our insights into the roles of *hairy* genes in fish development.

Materials and Methods

Embryos

Embryos from both common carp (*Cyprinus carpio*) and zebrafish (*Danio rerio*) were obtained by artificial fertilization. Zebrafish embryos were incubated at 28℃ and the developmental stages were determined according to the descriptions in Westerfield (1995).

Isolation of common carp total RNA and DNA

Total RNA of common carp embryos at the 20-somite stage was extracted and purified using the SV Total RNA Isolation System (Promega, Madison, USA). Common carp tail fins were broken into pieces and added to 300-μl extraction buffer (10 mmol/l TrisCl pH 8.0, 0.1 mol/l EDTA pH 8.0, 0.5% SDS, 10 mg/ml proteinase K), homogenized, and incubated at 37℃ for 12 h. Genomic DNA was extracted by phenol/chloroform, precipitated with ethanol, and dissolved in TE. The concentration of total DNA was estimated using 0.8% agarose gel electrophoresis and the DNA samples were stored at −20℃.

Cloning of full-length *her6* cDNA, genomic sequence, and upstream regulatory sequence

To clone the full-length *her6* cDNA, degenerate primers her6A and her6B (table 1) were designed according to the conserved region of several *hairy* gene sequences. RT-PCR was carried out with the RNA PCR Kit (AMV) Ver.2.1 (Takara, Otsu, Japan) with 30 cycles of amplification (94℃ 30 s, 57℃ 40 s, 72℃ 1 min). The primers for 5′ RACE and 3′ RACE were designed with the known coding sequence (table 1). The first strand cDNA in 5′ RACE was synthesized in the presence of her65A, and the cDNA was purified using a glassmilk kit (MBI, Vilnius, Lithuania). Thirty cycles of amplification (94℃ 30 s, 55℃ 40 s, 72℃ 1 min) were carried out with oligdT and her65A for the first PCR. The second PCR (94℃ 30 s, 58℃ 40 s, 72℃ 1 min) with 50× diluted first PCR solution was carried out with her63B and her65B. The first strand cDNA in 3′ RACE was synthesized in presence of oligo(dT). Thirty cycles of amplification (94℃ 30 s, 57℃ 40 s, 72℃ 1 min) was performed with her63A and her63B. All products were cloned into pMD18-T vector (Takara) by TA cloning and sequenced.

For cloning of *her6* genomic sequence, the primers her6a and her6b (table 1) were designed from the known *her6* cDNA sequence. The underlined portions in table 1 indicate additional endonuclease sites (*Hin*dIII and *Eco*RI). The PCR reaction had 100 ng of total DNA, 0.2 mM of dNTP, 10 mM of primers and 0.5 U of LA *Taq* (Takara), and PCR was carried out in 30 cycles of amplification (94℃ 30 s, 60℃ 40s, 72℃ 2 min). The PCR product was cloned into pMD18-T vector and sequenced.

Table 1 Primers for common carp *her6* cDNA sequence, genomic sequence and upstream regulatory sequence

DNA region	Primer sequence
Partial coding region	her6A ATGCCWGCYGATATMATGGA
	her6B GCRAAGGCMSCGTTGGGRAT
5′ race	her65A CCAGCTCTGTATTTCCCAAGA
	her65B TCAGAGCATCCAAGATTAGCG
3′ race	her63A CRGCMACCGACGGACAGT
	her63B CTGATCTAGAGGTACCGGATCC
	oligdT CTGATCTAGAGGTACCGGATCCTTTTTTTTTTTTTTT
Genomic sequence	her6a ATCTGCAGGACTGTGACATCATTGCC
	her6b ATCTGCAGTCTTTGGCATCACAACGT
Upstream sequence	walking1 TCAGGTGTAGTGTTCATGCT
	walking2 TCCATGATATCAGCTGGCAT
For constructs	CHPA ATGCAAGCTTGATCTATTGTGCATTGAAT
	CHPB ATGCAAGCTTGAAGTTTCACACGAGCCGTT
	CHPC ATGCGAATTCCTTCTGTGGGAAGTATCCTT

For cloning of the 5′-flanking region of *her6*, genomic walking was carried out with LA PCR *in vitro* cloning kit using the primers walking1 and walking2 (Takara) (table 1). The genomic DNA was digested with *Sau3*AI (Fermentas, Vilnius, Lithuania), and the PCR product was cloned into pMD18-T vector and sequenced.

Fusion of the presumptive promoters of *her6* to eGFP reporter constructs

From sequence information for the obtained genome-walking products, primers CHPA and CHPB (table 1) were designed to amplify the 500-bp presumptive promoter of *her6*, and the upstream primer CHPC (table 1) was used to amplify the 200-bp presumptive promoter. In these primers also, the underlined portions in table 1 indicate additional endonuclease sites (*Hin*dIII and *Eco*RI). The amplified fragments were digested with *Hin*dIII and *Eco*RI, and ligated into pEGFP1 (ClonTech, Basingstoke, UK) with T4 ligase (Fermentas). Competent cells of *E. coli* Top10 (Invitrogen, Carlsbad, USA) were transformed with the ligation products. Positive clones were screened by PCR and the fragments sequenced. The resulting plasmids were extracted from the positive clones, linearized using *Hin*dIII, purified using a glassmilk kit (MBI), and resuspended in ST buffer (5 mM Tris, 0.5 mM EDTA, 0.1 M KCl) at a final concentration of 100 ng/μl for microinjection.

Microinjection and fluorescence observation

Linearized recombinant plasmids were microinjected into the animal poles of fertilized zebrafish eggs. Till the embryos developed to early-somite stage, presence of GFP in transgenic zebrafish was directly observed under a 480-nm fluorescence microscope (SZX-12, Olympus, Tokyo, Japan) and images were captured with a digital camera.

Results

Structural analysis of common carp *her6* gene

The genomic sequence of common carp *her6* gene is 1718 bp and has four exons and three introns (figure 1; table 2). The full length of *her6* cDNA is 1314 bp, which comprises the coding region of 813 bp encoding a putative polypeptide of 271 amino acids (figure 1), the 5′UTR of 190 bp, and the 3′UTR of 311 bp. The upstream regulatory sequence of *her6* is about 400 bp (figure 2).

Table 2 Intron-exon organization of the *her6* gene

Exon no.	Exon size (bp)	3′ end of the exon	5′ end of the intron	Intron size (bp)	3′end of the intron	5′ end of the next exon
1.	198	gcacagaaag	gtttgtattga	125	gcattttag	tcttcgaaac
2.	96	gaaaaaagat	gtaagtagcc	112	ttcacacag	agctccaggc
3.	88	caaatgaccg	gtaagtcgcc	167	ctcccaacag	ctgccctgaa
4.	838	taaataaattacatcaaaatgagaaaaaaaaaaaaa				

Sequence alignment of some homologous proteins

The predicted amino acid sequence of the *her6* product is homologous to WRPW-bHLH proteins of zebrafish, and other vertebrate homologues belonging to the HES family proteins. The three conservative domains, bHLH, the Orange domain and WRPW, are all present in common carp *her6* product (figure 3).

Sequence analysis of her6 promoter

Using genome walking, we isolated the *her6* upstream regulatory sequence (figure 2). The transcriptional initiation site was determined by comparison of the genomic and cDNA sequences. A TATA box (TATAAA) was identified at 27-33 bp upstream from the transcription initiation site; a CCAAT box is located at 160-163 bp upstream from the initiation site and 126 bp upstream from the TATA box; and a CATTGG box is located at 188-193 bp upstream from the transcription initiation site and 154 bp upstream from the TATA box. Between the TATA box and the CCAAT box, there is a potential suppressor of hairless paired site (SPS). By digesting the PCR products from the promoter sequence with *Hin*dIII and *Eco*RI, we were able to construct two expression constructs, pcher6eGFPa and pcher6eGFPb. The two constructs, of 4709 bp and 4409 bp, respectively, include 500 kb and 200 kb of *her6* upstream sequences fused to the *eGFP* reporter gene (figure 4).

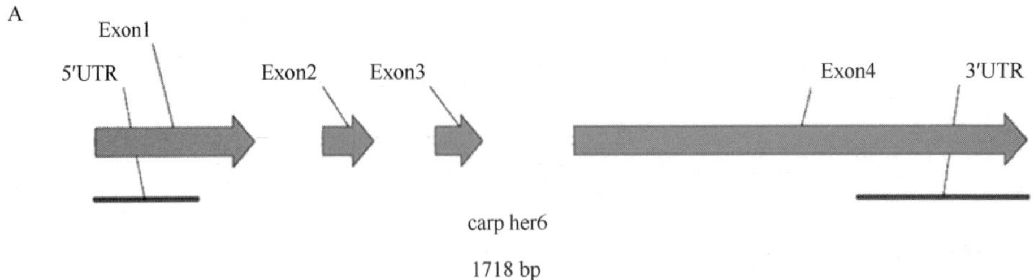

B

AGTTGCTACGACCGCTGGCAAACGGTGAAATCAGAGACTGTGACATCATTGCCGCACCAGTTGAA
CTCGGGACACTTCGTGCGGATATCCATTCATATCTGGAACTGTATCTGCCTACAGCGTTCACTCT
AGTAGAAGATACCGGAGGATTTCTTTCTAAACAAATCTCAAAGGATACTTCCCAGAGAAGATGCC
TGCCGATATCATGGAAAAAAACTCATCTTCTCCGGTTGCCGCGACTCCGGCGAGCATGAAcacta
cacctgataaacccaaaacggcttctgagcacagaaaggtttgtattgattctgacctgtagcac
atttattttcgaacagcgcgcatatgcgctataagatgcattgagaaatagttcccgaagtagtt
ttgacttatttatttttttcttttgcattttagTCTTCGAAACCTATTATGGAGAAAAGAAGAAG
AGCGAGAATCAACGAAAGCTTGGGTCAGCTGAAAACGCTAATCTTGGATGCTCTGAAAAAAGATg
taagtagcctactggatgtctgagtgattcttacaattacatctggattgaatactgatgaatct
catcacatagtcgcctatatctaaatcaatccctattttcacacagAGCTCCAGGCATTCTAAAC
TTGAGAAAGCGGACATCCTGGAGATGACAGTGAAACATCTTAGAAATATGCAGCGGGTACAAATG
ACCGgtaagtcgccagtccaatgaacaatcagacaataaatagctttatgagaacttttcctcta
ctgaattcaactagattgaacgcagttttagaaatgactcagaacatattaatgttcatctacag
cgaatgatctgaccgatggtcttacctcattctcccaacagCTGCCCTGAACACGGATCCCACAG
TTCTTGGGAAATACAGAGCTGGATTCAGTGAATGCATGAACGAGGTGACCCGGTTCCTGTCCACC
TGTGAAGGGGTTAACACCGAGGTCAGGACCCGGCTGCTGGGTCACTTAGCCAGCTGCATGACACA
GATCAACGCCATGAATTATCCAACACAGCACCAGATACCTGCCGGGCCTCCTCATCCATCCTTCA
GTCAGCCAATGGTTCAGATCCCCAGCGCCAATCAGCAAGCCAACGTTGTGCCTCTTAGCGGAGTC
CCCTGCAAAAGCGGATCTTCCTCCAACTTGACTTCTGACGCAACAAAAGTATACGGAGGCTTCCA
GCTCGTGCCGGCAACCGACGGACAGTTCGCCTTTTTGATTCCTAACGCTGCCTTTGCTCCAAACG
GTCCCGTTATTCCAGTGTATGCCAACAATTCCAGCACACCGGTGCCGGTGGCCGTGTCTCCGGGA
GCACCGTCCGTCACGTCAGATTCCGTTTGGCGGCCTTGGTAAACTATGAAAAAAAGAAAAAAAAA
CTCTTTAAAGGCAAATGACACTTGTTTTCTTTGTTAAGCGTTACTTTTTGTTCTTTTTGTATAAA
AAATGTGTTTTAAGAGATGATGCACTATATTTGTATAGATCAAAAGGGAGAGGAGTTCATATTGA
AATAAGTTTTGTATTGTTAAATTCCGTTCAATGCATTTTTTTATATATATTTGCGGTATCTTTTC
CACGTTGTGATGCCAAAGATATGTGAATGCGCTTTCAAGTTTCTTCTTTTTGGAAGATAAATAAA
TTACATCAAAATGAGAAAAAAAAAAAAAA

Figure 1 (A) Genomic organization of the common carp *her6* gene. (B) Common carp *her6* genomic sequence. Uppercase letters indicate the four exons and lowercase letters indicate the three introns. The coding regions are indicated by panes

AAACTGAATGATTGAAGGCAGTCCCGGTTGTGCCCTAAACCCGGCCGCG

CGCCATTGGCCACCCGCTCAAACGCTAAACAACCAATGGAGGGGCGCCG

ACCACGAGCCATCCTCCCCTCGGGCCGCGTGTCCCTCACGCTGATTGGT

SPS sequence

AGAAAGTTACTGTGGGAAAGAAAGTTTGGGAAGTTTCACACGAGCCGTT

CGCGTGCAGTCGGAGATATAAATAAGACCAACACGACGCTGAGGCAGGC

AGTTGCTACGACCGCTGGCAAACGGTGAAATCAGAGACTGTGACATCAT

TGCCGCACCAGTTGAACTCGGGACACTTCGTGCGGATATCCATTCATAT

CTGGAACTGTATCTGCCTACAGCGTTCACTCTAGTAGAAGATACCGGAG

GATTTCTTTCTAAACAAATCTCAAAGGATACTTCCCAGAGAAG

Figure 2 Nucleotide sequence of the *her6* promoter region. The transcriptional initiation site, designated as +1, is indicated by an arrow. The CATTGG-box, CCAAT-box, SPS sequence and TATA-box are indicated by panes

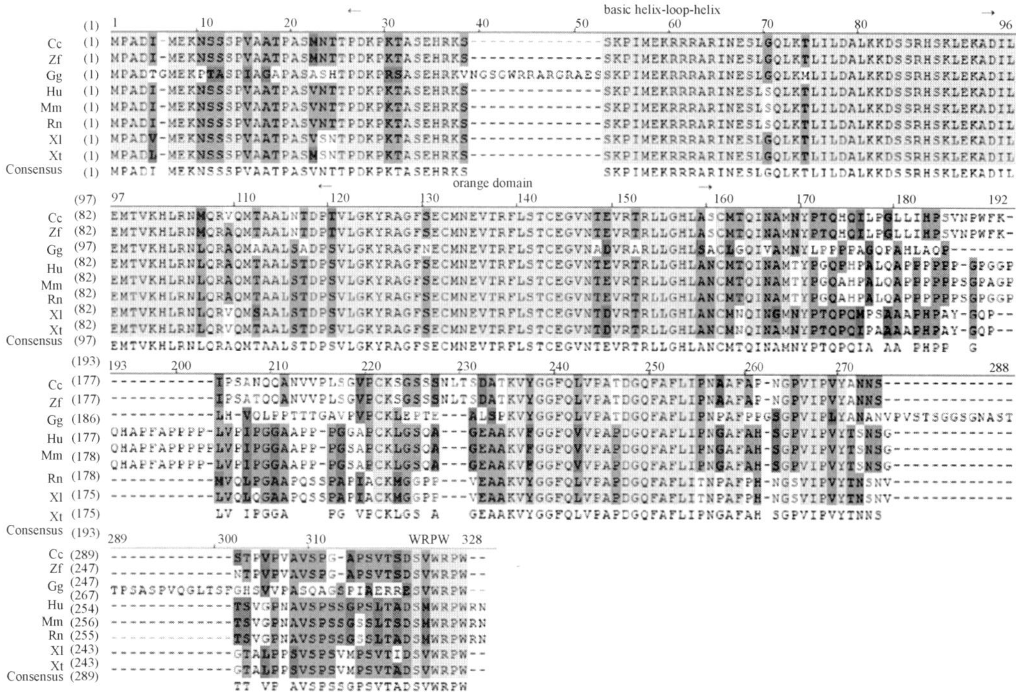

Figure 3 Sequence alignment of the different c-hairy type proteins. Common carp *her6* product was aligned with HES family proteins from *Danio rerio* and other vertebrates. Common carp *her6* product shows high homology with zebrafish *her6* product and the mammalian HES homologies. The three conservative domains, the basic helix-loop-helix domain, the Orange domain and the WRPW motif, are shown. The abbreviations Cc, Zf, Gg, Hu, Mm, Rn, Xl, Xt refer to common carp, zebrafish, *Gallus gallus*, *human*, *Mus musculus*, *Rattus norvegicus*, *Xenopus laevis* and *Xenopus tropicalis*, respectively

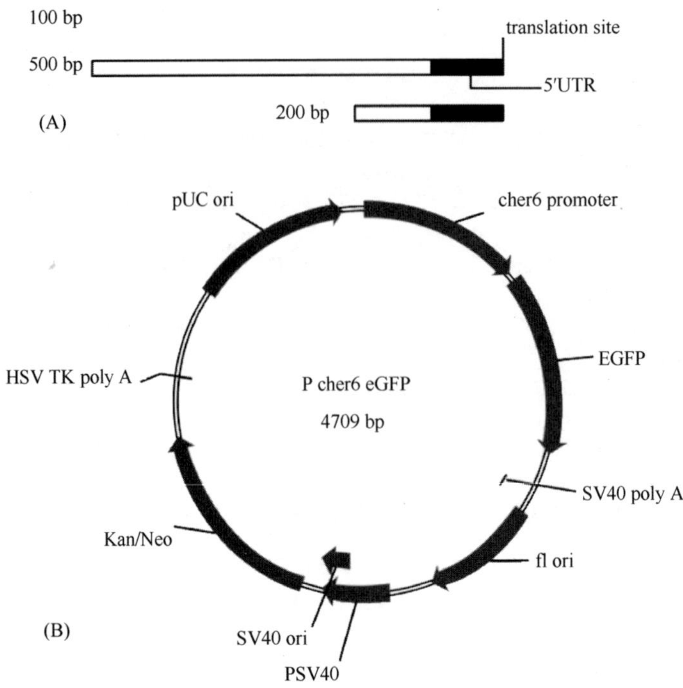

Figure 4 Maps of two different-length fragments of the 5′ sequence of (A) common carp *her6* gene, and (B) the expression vector pcher6eGFP

GFP expression in P_0 generation transgenic zebrafish embryos

We tracked the patterns of expression from pcher6eGFP in transgenic zebrafish embryos. A typical strong expression was first observed in blastula (figure 5, A) but there was no obvious tissue-specific expression from blastula to the tailbud stage (figure 5, B). From the tailbud stage on, a clear segmental pattern of expression started to appear and specific expression was observed from the later stages. GFP was observed in head including eyes and brain (figure 5, C), caudal half of somites (figure 5, D&E), notochord (figure 5, F&I), PSM, and even the area around Kupffer's vesicle (figure 5, G&H) and muscle (figure 5, J&K). The pcher6eGFPb transgenic embryos showed practically no green fluorescence (figure 5, L).

Discussion

Studies over the past twenty years have provided major insights into molecular mechanisms responsible for somitogenesis. The *hairy* family of genes encodes one of the most important groups of transcriptional factors involved in somitogenesis. These factors are expressed in specific tissues and/or at specific developmental stages (Takke and Campos-Ortega 1999; Vasiliauskas and Stern 2000; Bessho *et al.* 2001 a,b; Pasini *et al.* 2001; Minguillon *et al.* 2003; Elmasri *et al.* 2004; Kageyama *et al.* 2005; Oates *et al.* 2005). In the present study, we have described cloning and

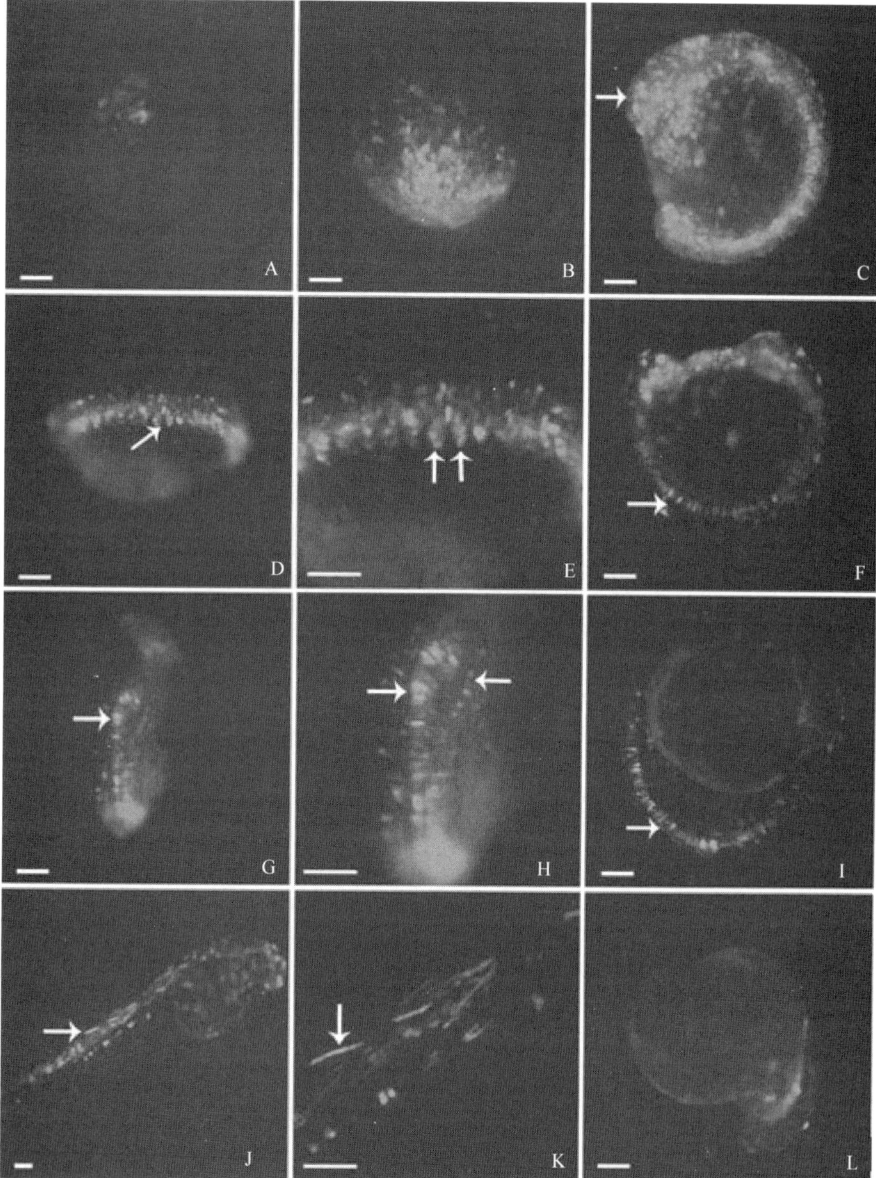

Figure 5 Transient *GFP* expression from pcher6eGFPa and pcher6eGFPb constructs in zebrafish embryos: (A) 30%-epiboly stage; (B) 50%-epiboly stage; (C-I,L) somitogenesis stages; (J,K) the prim-12 stage. The two reporter constructs were microinjected into zebrafish embryos at 1-cell stage. Zebrafish embryos with GFP fluorescence at several stages were selected for image analysis. From the pcher6eGFPa transgenic embryos, we concluded that *her6* promoter began to have transcriptional activity from blastula (A). B is in gastrula. Although P_0 generation of transgenic zebrafish are transgenic chimaeras, we are still able to observe the tissue-specific expression of *GFP* from the later stages. GFP is distributed in head (C), somites (D, E), notochord (F, I), PSM (G, H) and muscle (J, K). The pcher6eGFPb transgenic embryos had nearly no green fluorescence (L). Scale bars represent 300 μm

characterization of common carp *her6* gene and its transcriptional regulatory sequences. The *her6* gene has three introns and four exons. The protein sequence is highly homologous to zebrafish *her6* product, and mouse and human *hes1* products. The 500 bp of upstream regulatory sequence of common carp *her6* gene was assayed for activity in zebrafish embryos where it directed highly efficient and tissuespecific *GFP* expression. This region contains several crucial transcription-factor-binding sites, including CATTGG, CCAAT, SPS and TATA motifs. In contrast, the first 200 bp of upstream sequence, which only has a partial SPS sequence and the TATA box, could only direct faint and nonspecific *GFP* expression. This indicates that the entire SPS sequence, CATTGG and CCAAT boxes in the 5′ regulatory region of *her6* are important for tissue-specific and stage-specific features of normal transcription of *her6*. Because GFP protein could be observed in 30%-epiboly embryos, *her6* might be transcribed much earlier in normal development. From the bud stage, a majority of embryos had the green fluorescence in their heads, somites and notochords, which is consistent with the sites previously described by *in situ* hybridization in zebrafish (Pasini *et al*. 2001).

In previous studies, some ubiquitous promoters (such as *Xenopus* eIF1, β-actin) and tissue-specific promoters (such as GATA-2 promoter and SHH promoter) were used to drive *GFP* expression in zebrafish (Amsterdam *et al*. 1995; Higashijima 1997; Hamada *et al*. 1998; Gibbs and Schmale 2000; Kinoshita *et al*. 2000; Meng and Lin 2000; Teng *et al*. 2004). In this study, we isolated the 500 bp of 5′ flanking region of common carp *her6* and found that it was capable of driving *GFP* expression in a tissue-specific manner in zebrafish embryos. Once a stable transgenic line is obtained, it will be helpful in studying transcriptional regulation of *her6* and fish somitogenesis. Since our previous studies revealed that the somitogenesis process and vertebral numbers of the cloned fish derived from common carp nuclei and goldfish enucleated eggs were consistent with those of goldfish instead of those of common carp, we can probe into the molecular basis of this finding by using the pcher6eGFP transgenic fish. If nuclei of cher6eGFP-transgenic common carp are transferred into enucleated eggs of goldfish, we expect that *GFP* expression patterns in the cloned fish will be consistent with those of the transgenic goldfish instead of common carp.

Acknowledgements This study was supported by the State Key Fundamental Research of China (Grant No. G2004CB117406) and the National Natural Science Foundation of China (Grant No. 90208024 and 30123004) and the Chinese Academy of Sciences (Grant No. KSCX2-SW-303).

References

Amsterdam A., Lin S. and Hopkins N. 1995. The *Aequorea victoria* green fluorescent protein can be used as a reporter in live zebrafish embryos. *Dev. Biol*. 171, 123-129.

Bessho Y., Miyoshi G., Sakata R. and Kageyama R. 2001a. *Hes7*: a bHLH-type repressor gene regulated by *Notch* and expressed in the presomitic mesoderm. *Genes Cells* 6, 175-185.

Bessho Y., Sakata R., Komatsu S., Shiota K., Yamada S. and Kageyama R. 2001b. Dynamic expression and essential functions of *Hes7* in somite segmentation. *Genes Dev.* 15, 2642-2647.

Bessho Y., Hirata H., Masamizu Y. and Kageyama R. 2003. Periodic repression by the bHLH factor Hes7 is an essential mechanism for the somite segmentation clock. *Genes Dev.* 17, 1451-1456.

Davis R. L., Turner D. L., Evans L. M. and Kirschner M. W. 2001. Molecular targets of vertebrate segmentation: two mechanisms control segmental expression of *Xenopus hairy2* during somite formation. *Dev. Cell* 1, 553-565.

Dawson S. R., Turner D. L., Weintraub H. and Parkhurst S. M. 1995. Specificity for the hairy/enhancer of split basic helix-loop-helix (bHLH) proteins maps outside the bHLH domain and suggests two separable modes of transcriptional repression. *Mol. Cell. Biol.* 15, 6923-6931.

De Robertis E. M. 1997. Evolutionary biology: the ancestry of segmentation. *Nature* 387, 25-26.

Elmasri H., Liedtke D., Lucking G., Volff J. N., Gessler M. and Winkler C. 2004. *her7* and *hey1*, but not *lunatic fringe* show dynamic expression during somitogenesis in medaka (*Oryzias latipes*). *Gene Expr Patterns* 4, 553-559.

Gaiano N. and Fishell G. 2002. The role of Notch in promoting glial and neural stem cell fates. *Annu. Rev. Neurosci.* 25, 471-490.

Gajewski M. and Voolstra C. 2002. Comparative analysis of somitogenesis related genes of the *hairy/Enhancer of split* class in Fugu and zebrafish. *BMC Genomics* 3, 21.

Gibbs P. D. and Schmale M. C. 2000. GFP as a genetic marker scorable throughout the life cycle of transgenic zebrafish. *Mar. Biotechnol.* 2, 107-125.

Hamada K., Tamaki K., Sasado T., Watai Y., Kani S., Wakamatsu Y. *et al.* 1998. Usefulness of the medaka beta-actin promoter investigated using a mutant GFP reporter gene in transgenic medaka (*Oryzias latipes*). *Mol. Mar. Biol. Biotechnol.* 7, 173-180.

Henry C. A., Urban M. K., Dill K. K., Merlie J. P., Page M. F., Kimmel C. B. and Amacher S. L. 2002. Two linked *hairy/Enhancer of split* related zebrafish genes, *her1 and her7*, function together to refine alternating somite boundaries. *Development* 129, 3693-3704.

Higashijima S., Okamoto H., Ueno N., Hotta Y. and Eguchi G. 1997. High-frequency generation of transgenic zebrafish which reliably express GFP in whole muscles or the whole body by using promoters of zebrafish origin. *Dev. Biol.* 192, 289-299.

Honjo T. 1996. The shortest path from the surface to the nucleus: RBP-J kappa/Su(H) transcription factor. *Genes Cells* 1, 1-9.

Jarriault S., Brou C., Logeat F., Schroeter E. H., Kopan R. and Israel A. 1995. Signalling downstream of activated mammalian Notch. *Nature* 377, 355-358.

Jimenez G., Paroush Z. and Ish-Horowicz D. 1997. Groucho acts as a corepressor for a subset of negative regulators, including Hairy and Engrailed. *Genes Dev.* 11, 3072-3082.

Jouve C., Palmeirim I., Henrique D., Beckers J., Gossier A., Ish-Horowicz D. and Pourquie O. 2000. Notch signaling is required for cyclic expression of the *hairy-like* gene *HES1* in the presomitic mesoderm. *Development* 127, 1421-1429.

Kageyama R., Ohtsuka T., Hatakeyama J. and Ohsawa R. 2005. Roles of bHLH genes in neural stem cell differentiation. *Exp. Cell Res.* 306, 343-348.

Kimmel C. B. 1996. Was Urbilateria segmented? *Trends Genet.* 12, 329-331.

Kinoshita M., Kani S., Ozato K. and Wakamatsu Y. 2000. Activity of the medaka translation elongation factor 1alpha-A promoter examined using the GFP gene as a reporter. *Dev. Growth Differ.* 42, 469-478.

Ledent V. and Vervoort M. 2001. The basic helix-loop-helix protein family: comparative genomics and phylogenetic analysis. *Genome Res.* 11, 754-770.

Leimeister C., Dale K., Fischer A., Klamt B., Hrabe de Angelis M., Radtke F. *et al.* 2000. Oscillating expression of *c-Hey2* in the presomitic mesoderm suggests that the segmentation clock may use combinatorial signaling through multiple interacting bHLH factors. *Dev. Biol.* 227, 91-103.

Lewis E. B. 1978. A gene complex controlling segmentation in Drosophila. *Nature* 276, 565-570.

Meng A. and Lin S. 2000. Generation of germ-line transgenic zebrafish expressing GFP in a tissue-specific manner by using GATA-2 regulatory sequences. *Chin. Sci. Bull.* 45, 31-34.

Minguillon C., Jimenez-Delgado S., Panopoulou G. and Garcia-Fernandez J. 2003. The amphioxus Hairy family: differential fate after duplication. *Development* 130, 5903-5914.

Nusslein-Volhard C. and Wieschaus E. 1980. Mutations affecting segment number and polarity in *Drosophila*. *Nature* 287, 795-801.

Oates A. C. and Ho R. K. 2002. *Hairy/E(spl)-related* (*Her*) genes are central components of the segmentation oscillator and display redundancy with the Delta/Notch signalling pathway in the formation of anterior segmental boundaries in the zebrafish. *Development* 129, 2929-2946.

Oates A. C., Mueller C. and Ho R. K. 2005. Cooperative function of deltaC and her7 in anterior segment formation. *Dev. Biol.* 280, 133-149.

Ohtsuka T., Ishibashi M., Gradwohl G., Nakanishi S., Guillemot F. and Kageyama R. 1999. Hes1 and Hes5 as Notch effectors in mammalian neuronal differentiation. *EMBO J.* 18, 2196-2207.

PalmeirimI., Henrique D., Ish-Horowicz D. and Pourquie O. 1997. Avian *hairy* gene expression identifies a molecular clock linked to vertebrate segmentation and somitogenesis. *Cell* 91, 639-648.

Pasini A., Henrique D. and Wilkinson D. G. 2001. The zebrafish *Hairy/Enhancer-of-split*-related gene *her6* is segmentally expressed during the early development of hindbrain and somites. *Mech. Dev.* 100, 317-321.

Pasini A., Jiang Y. J. and Wilkinson D. G. 2004. Two zebrafish Notch-dependent *hairy/Enhancer-of-split*-related genes, *her6* and *her4*, are required to maintain the coordination of cyclic gene expression in the presomitic mesoderm. *Development* 131, 1529-1541.

Piscione T. D., Wu M. Y. and Quaggin S. E. 2004. Expression of Hairy/Enhancer of split genes, Hes1 and Hes5, during murine nephron morphogenesis. *Gene Expr. Patterns* 4, 707-711.

Saga Y. and Takeda H. 2001. The making of the somite: molecular events in vertebrate segmentation. *Nat. Rev. Genet.* 2, 835-845.

Sasai Y., Kageyama R., Tagawa Y., Shigemoto R. and Nakanishi S. 1992. Two mammalian helix-loop-helix factors structurally related to *Drosophila* hairy and Enhancer of split. *Genes Dev.* 6, 2620-2634.

Selkoe D. and Kopan R. 2003. Notch and presenilin: regulated intramembrane proteolysis links development and degeneration. *Annu. Rev. Neurosci.* 26, 565-597.

Sun Y. H., Chen S. P., Wang Y. P., Hu W. and Zhu Z. Y. 2005. Cytoplasmic impact on cross-genus cloned fish derived from transgenic common carp (*Cyprinus carpio*) nuclei and goldfish (*Carassius auratus*) enucleated eggs. *Biol. Reprod.* 72, 510-515.

Takke C. and Campos-Ortega J. A. 1999. *her1*, a zebrafish pair-rule like gene, acts downstream of notch signalling to control somite development. *Development* 126, 3005-3014.

Tam P. P. 1981. The control of somitogenesis in mouse embryos. *J. Embryol. Exp. Morphol.* 65, 103-128.

Tautz D. and Sommer R. J. 1995. Evolution of segmentation genes in insects. *Trends Genet.* 11, 23-27.

Teng P., Cao Y., Wang W. X., Zhu S. X., Meng A. M. and Zhang J. P. 2004. The shh promoter of zebrafish directs the expression of GFP in notochord. *Acta Genet. Sin.* 31, 39-42.

Van Doren M., Bailey A. M., Esnayra J., Ede K. and Posakony J. W. 1994. Negative regulation of proneural gene activity: hairy is a direct transcriptional repressor of achaete. *Genes Dev.* 8, 2729-2742.

Vasiliauskas D. and Stern C. D. 2000. Expression of mouse HES-6, a new member of the Hairy/Enhancer of split family of bHLH transcription factors. *Mech. Dev.* 98, 133-137.

Westerfield M. 1995. *The zebrafish book*. University of Oregon Press, Eugene.

鲤分裂增强相关基因 *her6* 的克隆、特征分析和启动子分析

刘　静[1,2]　孙永华[1]　王　娜[1,2]　汪亚平[1]　朱作言[1]

1 中国科学院水生生物研究所，淡水生态与生物技术国家重点实验室，武汉 430072
2 中国科学院大学，北京 100049

摘　要　很多 *hairy/Enhancer-of-split* (HES)相关基因在体轴中胚层的分节以及体节边缘的建立和维持方面有着重要作用。在鱼类中，*her6* 是 HES 家族中的一员，与人类和鸡的 *hes1* 基因的同源性很高。本研究对鲤 *her6* 基因进行了克隆和鉴定。鲤 *her6* 基因的基因组全长大约 1.7 kb，包含 4 个外显子和 3 个内含子，其 cDNA 序列 1314 bp，编码 271 个氨基酸的多肽。为了研究鲤 *her6* 启动子序列的功能，我们将包含转录起始位点上游的不同长度的序列与绿色荧光蛋白基因串联重组，通过显微注射方式导入到斑马鱼胚胎基因组中。实验结果显示，上游 500 bp 的序列可以直接驱动绿色荧光蛋白在斑马鱼胚胎的特定组织中高效表达，而上游 200 bp 序列(包含 TATA 重复和 SPS 位点)则无法有效促使绿色荧光蛋白在斑马鱼胚胎中表达。

Knock Down of *gfp* and *No Tail* Expression in Zebrafish Embryo by *In Vivo*-Transcribed Short Hairpin RNA with T7 Plasmid System

Na Wang[1,2] Yong-Hua Sun[1] Jing Liu[1,2] Gang Wu[1,2]
Jian-Guo Su[1] Ya-Ping Wang[1] Zuo-Yan Zhu[1]

1 State Key Laboratory of Freshwater Ecology and Biotechnology, Institute of Hydrobiology, Chinese Academy of Sciences, Wuhan 430072
2 Graduate School of the Chinese Academy of Sciences, Beijing 100039

Abstract A short-hairpin RNA (shRNA) expression system, based on T7 RNA polymerase (T7RP) directed transcription machinery, has been developed and used to generate a knock down effect in zebrafish embryos by targeting *green fluorescent protein (gfp)* and *no tail (ntl)* mRNA. The vector pCMVT7R harboring T7RP driven by CMV promoter was introduced into zebrafish embryos and the germline transmitted transgenic individuals were screened out for subsequent RNAi application. The shRNA transcription vectors pT7shRNA were constructed and validated by *in vivo* transcription assay. When pT7shGFP vector was injected into the transgenic embryos stably expressing T7RP, *gfp* relative expression level showed a decrease of 68% by analysis of fluorescence real time RT-PCR. As a control, injection of chemical synthesized siRNA resulted in expression level of 40% lower than the control when the injection dose was as high as 2 μg/μl. More importantly, injection of pT7shNTL vector in zebrafish embryos expressing T7RP led to partial absence of endogenous *ntl* transcripts in 30% of the injected embryos when detected by whole mount *in situ* hybridization. Herein, the T7 transcription system could be used to drive the expression of shRNA in zebrafish embryos and result in gene knock down effect, suggesting a potential role for its application in RNAi studies in zebrafish embryos.
Keywords short hairpin RNA; T7 RNA polymerase; T7 promoter; zebrafish; fluorescence real time RT-PCR; whole mount *in situ* hybridization

Introduction

RNA interference (RNAi) is the phenomena of gene silencing in which double-stranded RNA triggers the degradation of a homologous mRNA. This post-transcriptional gene silencing phenomena was first met with Guo and Kemphues in 1995, but they did not give a reasonable explanation [1]. Until 1998, Fire and his colleagues validated that the phenomena of

文章发表于 *Journal of Biomedical Science*, 2007, 14(6): 767-776

gene expression inhibition also by sense RNA was induced by contaminative double-stranded RNA generated during *in vitro* transcription [2]. This finding led to a series of RNAi-mediated gene silencing studies in many organisms such as worm [3,4], fly [5,6], mammalian [7-9] and plant [10] as well. Now RNAi has become a powerful tool to study gene function and explore the gene expression regulation mechanism, as well as provide a new approach for gene therapy.

The phage T7 transcription system is specific in that T7 promoter is recognized only by the T7 RNA polymerase existing in phage [11]. Davanloo *et al.* isolated the T7 RNA polymerase gene from T7 phage and analyzed its transcription activity [12]. T7 RNA polymerase has the identification stringency for its own promoter and specifically transcribes DNA linked to such a promoter. Based on this specificity, the T7 RNA polymerase/T7 promoter system has been successfully applied in the study for the expression and regulation of the foreign genes [13-15].

In recent years, many studies have shown that siRNAs transcribed *in vitro* by T7 system have a preferable effect in cells or organisms [16-19] and thus become an economic and effective method for production of siRNA instead of chemically synthesis. Nevertheless, *in vitro* synthesized siRNA could only give a temporary effect, and the T7 based siRNA transcription cassettes need to be introduced into the cell and organism in order to achieve a long-lasting effect. Recently, various investigators made attempts in this aspect and succeeded in *in vitro* cultured cells [20-22].

In one of the important animal models, zebrafish, just a few studies have been focused on the application of RNAi and controversy opinions appear. Some researchers held the opinion that dsRNA or siRNA could silence target gene specifically in zebrafish [23-27], whereas some other researchers found dsRNA may trigger non-specific effects in zebrafish embryo [28,29]. Previously, we found that *in vitro* synthesized siRNA could knock down the expression level of *ntl* and phenocopy the *ntl* mutant. Nevertheless, *in vitro* synthesized siRNA cannot be viable beyond 10-somite stage and is not able to function as an inheritable fashion [24]. Here we conducted the RNAi study in zebrafish embryos by constructing a shRNA expression system including the T7RP expression vector and the T7shRNA vectors, which target the foreign *green fluorescent protein* (*gfp*) transgene and endogenous *no tail* (*ntl*) gene, respectively. The results revealed that T7 system could effectively drive the transcription of shRNA in zebrafish embryo, and subsequently knock down the expression level of specific genes.

Materials and Methods

Zebrafish embryos and microinjection

Embryos from zebrafish AB line were obtained by artificial fertilization. Zebrafish embryos were incubated at 28.5℃ and the developmental stages were determined according to the standard descriptions [30]. The fertilized zebrafish eggs were injected with DNA or RNA in dose of 1-2 nl/egg. The injection concentration of pT7shGFP, pT7shNTL and pT7Bmp2b was 200 ng/μl. The chemical synthesized siRNA was introduced into the fertilized eggs with the concentration of 1 μg/μl and 2 μg/μl. The injected embryos were incubated at 28.5℃ in 0.3×Danieau's solution (10.17% NaCl, 0.156% KCl, 0.297% $MgSO_4$, 0.441% Ca $(NO_3)_2$, 3.57% HEPES [*w*/*v*], 50 IU/ml streptomycin, and 100 IU/ml ampicillin).

pCMVT7R and pT7shRNA constructs

The T7RP expression vector pCMVT7R (Figure 1A) was reconstructed from pZ NRS and pZ CEG. In this vector, *T7RP* is under the control of CMV promoter. Sense and antisense oligonucleotides of GFP and NTL, SIGT_SH, SIGB_SH, SINT_SH, SINB_SH (Table 1) used for shRNA construction were synthesized by Sangon Co. and their sequences were designed according to the published sequences [23,31]. About 1 nmol of sense and antisense oligonucleotides (GFP or NTL) were mixed, heated to 95℃ for 10 min, and then allowed to cool down to form double-stranded DNA. The primers T7P+S and T7T-H (shown in Table 1) were designed to conduct inverse PCR to amplify the vector pZ T7T which contains a T7 promoter and a T7 terminator as well as the multiple cloning sites between the two introduced restriction endonuclease sites *Sal*I and *Hin*dIII. The cycling parameters used for PCR were as follows: a pre-denaturation of 94℃ 5 min, 30 cycles of 94℃ 30 s, 68℃ 3 min 50 s and a final extension of 72℃ 5 min. The Taq polymerase used in PCR amplification was highfidelity KOD Plus (Toyobo). The PCR products were digested by *Sal*I and *Hin*dIII, purified using the Axyprep™ DNA Gel Extraction Kit (Axygen) and ligated with GFP or NTL double-stranded oligonucleotides. Positive clones were identified by PCR amplification and verified by sequencing analysis. The detailed construction process was shown in Figure 1B.

To construct the vector pT7Bmp2b (Figure 1C), the primers T3 and SP6 were used to amplify the CDS of zebrafish *Bmp2b* from pCS2-Bmp2b vector. The PCR product was ligated with reverse PCR product amplified from pT7eGFP with primers T7+B and T7-N (Table 1) at the endonuclease sites of *Bam*HI and *Not*I.

Transgenic lines stably expressing T7 RNA polymerase

The embryos injected with pCMVT7R were raised to maturation and mated with wild type zebrafish. For every mating pair, at least 100 embryos were examined for GFP reporter expression. If there were some embryos showing GFP expression, these embryos were considered as pCMVT7R-transgenic embryos and sorted out for further breeding. The F1 adult zebrafish was mated with wild type to produce the F2 embryos. The F2 embryos expressing GFP were raised to adult and mated with wild type to produce F3 embryos, which could express T7 RNA polymerase stably.

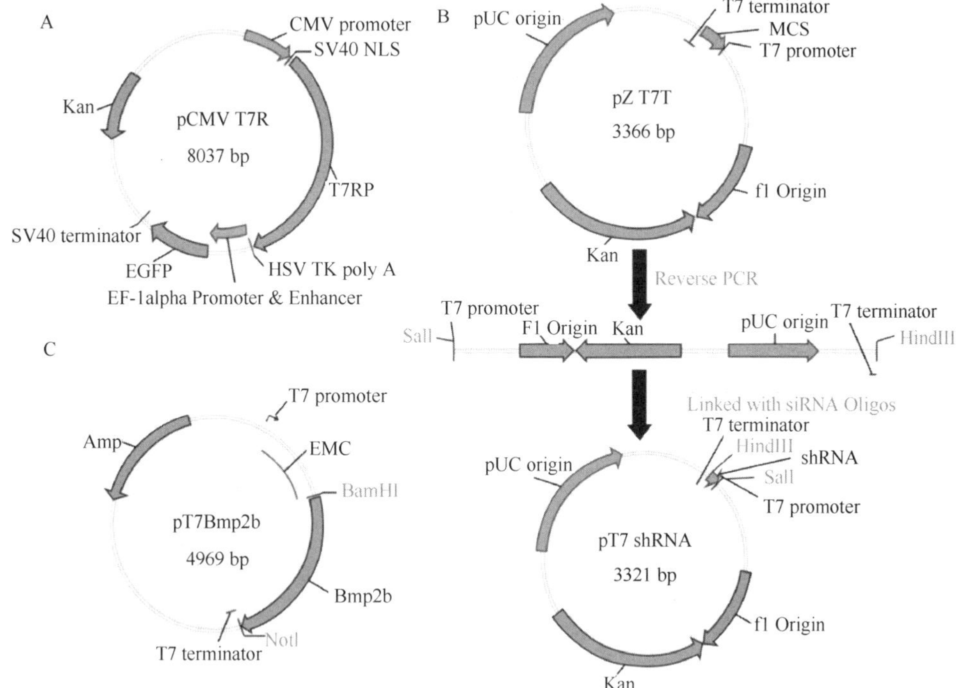

Figure 1 The structure of pCMVT7R, pCMVT7R shNTL and pT7Bmp2b. (A) Shows the plasmid pCMVT7R. T7 RNA polymerase is driven by CMV promoter and the expression of EGFP is driven by EF-1alpha promoter and enhancer as a reporter gene for selection of germline inherited individuals. (B) Shows the plasmid pT7Bmp2b. Bmp2b CDS was driven by T7 promoter. (C) Shows schematic diagram of pT7shRNA vector construction. Firstly, primers T7P+S, T7T-H were designed for inverse PCR to amplify the vector pZ T7T to introduce the restriction site *Sal*I and *Hin*dIII. GFP and NTL oligos were linked with PCR products cleaved by *Sal*I and *Hin*dIII to form the new plasmid pT7shRNA (shGFP/NTL)

Table 1 Primer used in the present study

primers	Sequences(5′-3′)
SINT_SH	TCGACTGCAATGTACTCGGTCCTGTTCAAGAGACAGGACCGATTACATTGCATTA
SINB_SH	AGCTTAATGCAATGTACTCGGTCCTGTCTCTTGAACAGGACCGAGTACATTGCAG
SIGT_SH	TCGACGCAAGCTGACCCTGAAGTTCTTCAAGAGAGAACTTCAGGGTCAGCTTGCTTA
SIGB_SH	AGCTTAAGCAAGCTGACCCTGAAGTTCTCTCTTGAAGAACTTCAGGGTCAGCTTGCG
T7P+S	CAGGTCGACCCCTATAGTGAGTCGTATTACAATTCACTG
T7T–H	CAGAAGCTTCTAGCATAACCCCTTGGGGCCTCTAAA
T7+B	ACGGGATCCTATTATCGTGTTTTTCAAAGG
T7–N	ACGGCGGCCGCCTAGCATAACCCCTTGGGGCC
GFPA	TCCAGGAGCGCACCATCTT
GFPB	TGCTCAGGTAGTGGTTGTCGG
ActinA	CACTGTGCCCATCTACGAG

Continued

primers	Sequences(5′-3′)
ActinB	ATTGCCAATGGTGATGAC
T7R+	CTGGTGAGGTTGCGGATAA
T7R–	TACCGAAGGAGTCGTGAAT
7G1	GTCGACGCAAGCTGACC
7N1	GTCGACTGCAATGTACT
7GN2	CCCCTCAAGACCCGTTT

The confirmation of T7RP transcription in F3 embryos was conducted by RT-PCR with primers T7R+ and T7R– (seen in Table 1). The further confirmation of T7RP function in zebrafish embryos was approved by injection of pT7Bmp2b into the embryos of F3 generation. If the T7 system could function in zebrafish, the injection of pT7Bmp2b would mimic the phenotype of over-expression of Bmp2b in zebrafish embryos, which would result in ventralized phenotype [32].

In vivo transcription assay of pT7shRNA vectors

Every 50 embryos injected with pT7shGFP or pT7shNTL were colleted at 16-cell, dome, 30%-epiboly, 50%-epiboly, 75%-epiboly, bud and 1d stage, respectively. Total RNA was isolated with SV Total RNA Isolation Kit (Promega) and 1 μg RNA was reverse transcribed to cDNA with random 9mers primer. PCR was conducted with the primers G1, N1 and GN2 which were designed at transcription start sites and terminator respectively for detection of shGFP and shNTL transcripts. PCR conditions were as follows: 94℃ for 5 min, 30 cycles of 94℃ 30 s, 56℃ 30 s, 72℃ 30 s, and 72℃ for 5 min. The PCR products were analyzed with 1.5% agarose gel and the predicted length of PCR products was 104 bp.

Fluorescence real time RT-PCR

About 50 embryos were collected to isolate total RNA with SV Total RNA Isolation Kit (Promega). Then 1 μg total RNA was reverse transcribed to cDNA with Random 9 mers primer. The cDNA could be used for the template to detect the expression level of *gfp* mRNA with *β-actin* as the internal control. The 20 μl PCR reaction system consisted of 1 μl cDNA, 0.8 μM primers, and 10 μl SYBR Green Realtime PCR Master Mix (Toyobo). PCR was conducted as follows: 94℃ for 5 min, 40 cycles of 94℃ 30 s, 58℃ 40 s, 72℃ 40 s. Every sample was repeated for three times and the results were analyzed with $2^{-\Delta\Delta Ct}$ method [33].

Whole mount in situ hybridization

The vector pBKS-ntl (a kind gift of SchulteMerker S) was linearized with *Xho*I and transcribed to antisense probe with T7 RNA polymerase in the presence of Dig RNA Labeling Mix (Roche). The embryos injected with 200 ng/μl of pT7shNTL or pCMVT7R shNTL were fixed with 4% polyformaldehyde and whole mount *in situ* hybridization was carried out as described [34]. Images were captured with a digital camera connected to stereo microscope (Olympus SZX12, Japan).

Results

Construction of pCMVT7R and pT7shRNA vectors

In the plasmid pCMVT7R (shown in Figure 1A), T7 RNA polymerase was driven by CMV promoter, and it was linked with EGFP reporter driven by EF-1alpha promoter and enhancer.

The shRNA expression plasmids based on T7 promoter were constructed as materials and methods. The plasmids pT7shRNA were successfully constructed (Figure 1B) after double-stranded oligonucleotides of GFP and NTL were ligated with the reverse PCR products from the vector pZ T7T.

Screening of transgenic line expressing T7 RNA polymerase

When the pCMVT7R transgenic zebrafish embryos were raised to maturation, we screened out the P0, F1 and F2 generations by the facilitation of GFP expression. As shown, the GFP proteins of P0 generation were mosaically expressed (Figure 2C, P0), while the GFP proteins of F1 and F2 generation were uniformly expressed (Figure 2C, F1/F2). The transcription of T7RP RNA in GFP-positive embryos has been validated by RT-PCR analysis with primers T7R+ and T7R-in transgenic F3 embryos expressing GFP (Figure 2A, lane 1). As a control, the T7RP transcripts were not detectable from F3 embryos that did not express GFP (Figure 2A, lane 2), supporting the evidence that T7RP was transcribed in the F3 generation of pCMVT7R transgenic zebrafish showing GFP expression.

In order to validate the biological function of T7RP expressed in transgenic pCMVT7R zebrafish F3 embryos, the vector pT7Bmp2b was injected into the F3 embryos, with the injection into WT embryos as the negative control. The results showed that the phenotypes of Bmp2b over-expression were observed in F3 embryos (Figure 3A), with the most severe phenotype of ventralization, Class III, of 52% (Figure 3B). In contrast, there were only 7% of the embryos showing the severely ventralized phenotype in the control embryos. This confirmed that the T7RP existed in F3 generation embryos could specifically recognize the T7 promoter in the vector pT7Bmp2b and the T7 transcription system was functionally validated in zebrafish embryos.

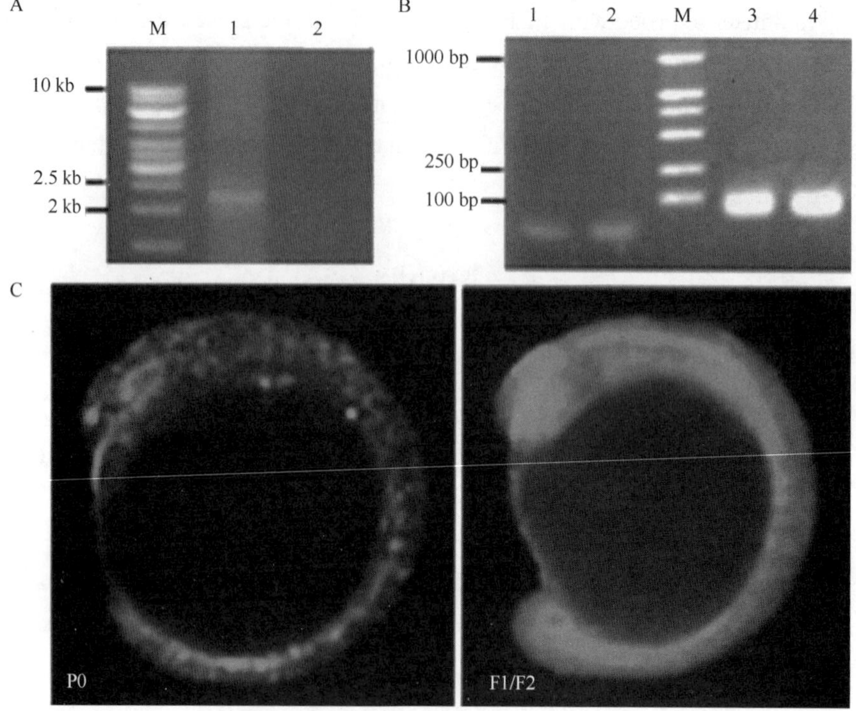

Figure 2 The confirmation of the vector pCMVT7R and pT7shRNA. (A) RT-PCR detection of T7 RNA polymerase in the pCMVT7R transgenic zebrafish F3 embryos. Lane 1 shows that GFP positive embryos could express T7 RNA polymerase mRNA, while lane 2 shows that GFP negative embryos could not express T7R. (B) The shRNA transcripts detection by RT-PCR are shown in lane 3 and 4, resembling shGFP and shNTL, with lane 1 and 2 resembling the negative control. (C) GFP expression pattern in pCMVT7R transgenic zebrafish. The left embryo shows the P0 embryo of mid-somite stage expressing GFP mosaically in the embryo. The right one shows the F1, F2 embryo of mid-somite stage expressing GFP uniformly in the whole embryo

Detection of GFP and NTL shRNA in transgenic embryos

Since T7RP mRNA in transgenic pCMVT7R zebrafish was detected and its protein was approved to be functional, the transgenic pCMVT7R F3 generation could be used as the material for microinjection for shRNA vectors. In order to analyze whether the pCMVT7R transgenic embryos could direct the transcription of shRNA in zebrafish embryos, we conducted *in vivo* transcription analysis for the pT7shGFP and pT7shNTL. The F3 generation embryos injected with the vectors of pT7shRNA were collected at seven developmental stages as materials and methods. The total RNAs of these embryos were isolated and conducted RT-PCR with primers G1, GN2 and N1, GN2, respectively. The results revealed that the PCR products in correct length were amplified from the GFP positive embryos injected with pT7shGFP and pT7shNTL (Figure 2B, lane 3 and 4) but not the negative embryos (Figure 2B, lane 1 and 2). These revealed that the functional plasmid pT7shGFP and pT7shNTL have been constructed and the pCMVT7R transgenic embryos could be applied successfully for *in vivo* transcription of shRNA. Since the transcription of shRNA was continuously detectable from 16-cell stage to 1d stage, we only showed the expression of shRNA at one stage to resemble the production of shRNA.

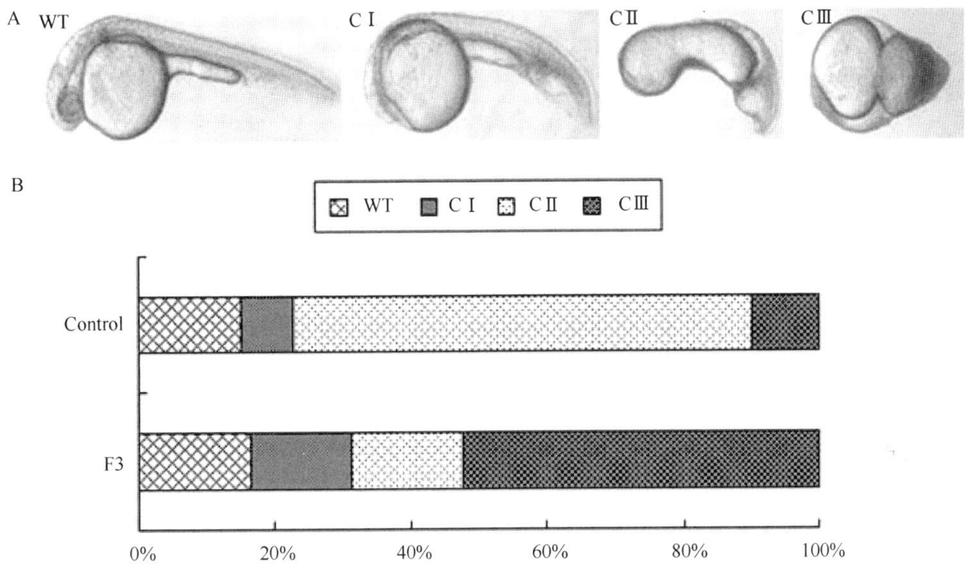

Figure 3 The validation of the function of T7RP *in vivo*. (A) Shows the wild type embryos in 1d stage and the three kinds of ventralized phenotypes injected with pT7Bmp2b, with CI, CII, CIII representing phenotypes from weakly to severely ventralization. (B) Shows the percentages of three kinds of phenotypes appeared in F3 embryos and WT embryos injected with pT7Bmp2b

Inhibition of GFP expression by shGFP transcribed by T7 system

After validation by sequencing and *in vivo* transcription, the shRNA vectors were microinjected into the F3 embryos of pCMVT7R transgenic zebrafish and the knock down efficiency of shRNA vectors was analyzed.

With regard to the embryos injected with pT7shGFP, the expression of GFP were firstly examined with fluorescent microscope and the result showed that the percentage of fluorescent embryos did not change significantly at 24 hpf, while the fluorescence intensity was remarkably reduced compared with the control embryos in the same time of exposure (Figure 4A). The further analysis of fluorescence real time RT-PCR revealed that the *gfp* mRNA of embryos injected with pT7shGFP was decreased 68% when compared with the control embryos (Figure 4B).

In order to compare the inhibition efficiency between shRNA transcribed by T7 system and chemically synthesized siRNA, siGFP was microinjected into WT embryos at 1 μg/μl and 2 μg/μl concentrations. The plasmid pCMVeGFP was taken as a control at 200 ng/μl.

After injection with G (pCMVeGFP 200 ng/μl), G1 (siGFP 1 μg/μl, pCMVeGFP 200 ng/μl) and G2 (siGFP 2 μg/μl, pCMVeGFP 200 ng/μl), *gfp* mRNA relative expression levels were analyzed with fluorescence real time RT-PCR. The results were shown in Figure 4C, that the decrease of expression level of G1 and G2 were 36% and 51%, respectively. In the F3 generation embryos, siGFP at 2 μg/μl was also injected and real time RT-PCR analysis showed that the relative expression of *gfp* mRNA was 60% (Figure 4B) when the control expression was set as 100%. The results suggested that the inhibition efficiency of injected siRNA was lower than the injection of pT7shGFP into F3 generation of the transgenic pCMVT7R embryos.

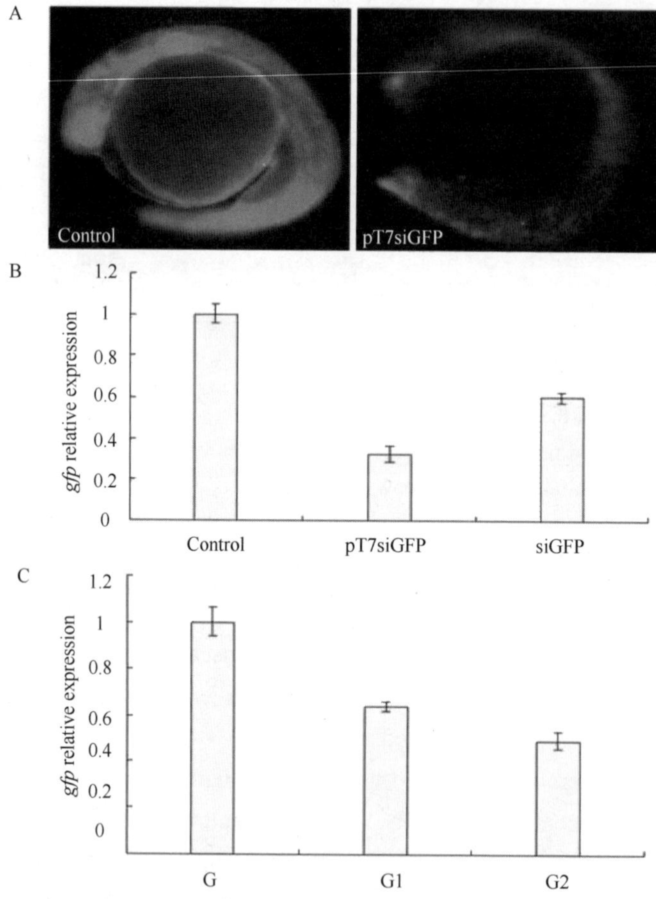

Figure 4 The inhibition effect of pT7shGFP and chemical synthesized siGFP. (A) The F3 embryo microinjected with pT7shGFP shows weaker fluorescence than the control of F3 embryo at the same exposure time. (B) The fluorescence real time RT-PCR reveals that the relative *gfp* mRNA levels of the F3 embryos injected with pT7shGFP and siGFP (200 ng/μl) are 0.32 and 0.60, respectively while the control is set as 1. (C) The real time RT-PCR analysis of *gfp* mRNA expression levels of G (pCMVeGFP 200 ng/μl), G1 (siGFP 1 μg/μl + pCMVeGFP 200 ng/μl and G2 (siGFP 2 μg/μl + pCMVeGFP 200 ng/μl) are 1, 0.64, and 0.49. The error bars show the mean±S.D. (standard deviation)

Knock down of ntl expression by shNTL transcribed by T7 system

After the inhibition effect of pT7shRNA vector on foreign gene *gfp* was studied, the inhibition effect of pT7shRNA on endogenous gene *ntl* was further analyzed. *Ntl* is the zebrafish homologue of the mouse *T* (*Brachuyury*) gene [35] and the *ntl* mutant embryos lack differentiated notochord and the most posterior 11-13 of their normal 30 somites [36]. The pT7shNTL vector was injected into the transgenic F3 embryos expressing functional T7RP at a concentration of 200 ng/μl. At shield stage, these embryos were fixed and whole mount *in situ* hybridization assay revealed that 30% (36/120) embryos showed partial absence of *ntl* transcripts in germ ring (Figure 5C, D), while the *ntl* signal was detected in the whole germ ring in 100% of wild type embryos (Figure 5A, B). At 25-somite stage, about 14% (11/77) of the injected embryos showed various extents of the phenotype of *ntl* mutant, such as lacking of the most posterior part of the notochord and somites (Figure 5F, G). Given the injected embryos were transgenic mosaics and the introduced pT7shNTL could not induce a complete inhibition of *ntl* transcription, it is reasonable that only a small proportion (less than 30%) of the injected embryos showed partial *ntl* phenotype. The wild type phenotype and typical phenotype of reported *ntl* mutant were shown in Figure 5E and Figure 5H, respectively [37]. Both the *in situ* hybridization results and morphological observation suggested that the pT7shNTL could transcribe shNTL in the zebrafish transgenic embryos stably expressing T7RP, and as a result partially knock down the expression of endogenous gene *ntl*.

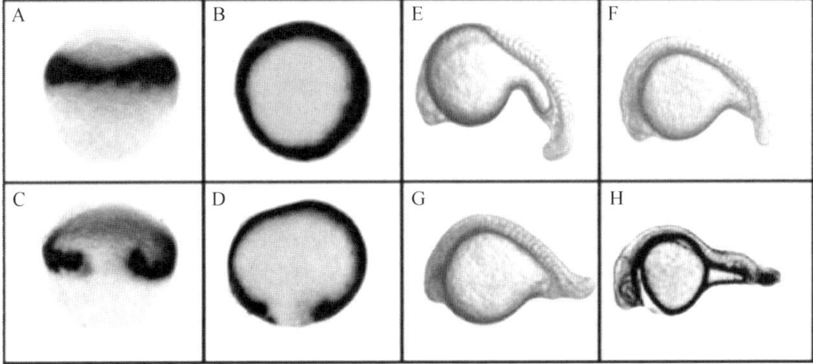

Figure 5 The inhibition effect on endogenous gene *ntl*. (A-D) Show the *ntl* expression pattern in pCMVT7R transgenic F3 embryos injected with pT7shNTL. In the control embryos, *ntl* transcripts are detected in the whole germ ring of shield stage (A, B). While in the F3 embryos injected with pT7shNTL *ntl* transcripts show partial absent in the germ ring (C, D). A and C are lateral view. B and D are animal view. E-H show the phenotype appeared at 25-somite stage. The similar *ntl* phenotype was also observed in 14% (11/77) embryos (F-G). E shows a wild type embryo and H presents the *ntl* mutant

Discussion

In the present work, we try to discover the possibility of T7 system in RNAi study of zebrafish embryos. Based on the specific recognition of T7RP to T7 promoter, firstly, the transgenic zebrafish line stably expressing T7RP was established; secondly, shRNA vectors which targeted foreign *gfp* gene and endogenous *ntl* gene were constructed; and finally the shRNA constructs were injected into the F3 embryos of the pCMVT7R transgenic line. Our results revealed that this T7 system based siRNA approach could function in zebrafish embryos.

As to the T7 system, although some investigators found the T7 promoter could be recognized by RNA polymerase II in cells [38,39], the specific recognition and transcription of T7 RNA polymerase for T7 promoter has been successfully applied to shRNA synthesis [20-22]. In addition, the gene expression application of T7 system in zebrafish embryos is also approved by Verri *et al.* [40]. In our present study, the validation of T7RP recognition on T7 promoter was validated by injection of pT7Bmp2b into the pCMVT7R transgenic F3 embryos. Although the over-expression phenotype was observed in WT embryos, the percentage of the most severe phenotype CIII was significantly lower than that in pCMVT7R transgenic F3 embryos. This suggested that the T7RP in F3 transgenic embryos could recognize the T7 promoter in the construct pT7Bmp2b and then transcribe an elevated level of *Bmp2b* mRNA, resulting in more severely ventralized phenotype. As to the phenomena of ventralized phenotype observed in WT embryos, we doubted that it was caused by the random insertion of *Bmp2b* to the downstream of any promoters or the basal expression of T7 promoter in zebrafish cells, since even an extreme low dose (4 pg per embryo) of injection of *Bmp2b* could result in remarkably ventralized phenotype [41]. On the other hand, RT-PCR analysis confirmed that shRNA could be *in vivo* transcribed in pCMVT7R transgenic zebrafish embryos stably expressing T7RP.

To study the application of *in vivo* transcription directed siRNA by T7 system, both foreign *gfp* gene and endogenous *ntl* gene were chosen as the target genes. In the embryos injected with pT7shGFP, both the results of GFP fluorescence observation and fluorescence real time RT-PCR approved that GFP expression was partially inhibited. In addition, this inhibition efficiency was higher than that of the chemically synthesized siGFP either in F3 embryos or in WT embryos. As to the effect on the endogenous *ntl* gene, we chose the most effective siRNA target sequence of *ntl* as previously reported [23], and placed it between the T7 promoter and T7 terminator. In transgenic zebrafish stably expressing T7RP, shNTL RNA was transcribed as detection by RT-PCR analysis. We found that there were 30% embryos (36/120) showing partial absence of *ntl* signal in the germ ring at shield stage. As a result, at 25-somite stage, various extents of *ntl* phenotype were observed in 14% (11/77) of the injected embryos. Obviously, the percentage of embryos showing *ntl* phenotype was much lower than that showing absence of *ntl* expression. Since the *ntl* expressing cells will give rise to the formation of posterior mesoderm and notochord, our results suggest that partial inhibition of *ntl* expression at gastrula stage may not give any obvious developmental defects at later stages. Nevertheless, both results of *in situ* hybridization and morphological observation demonstrated that the endogenous *ntl* was knocked down in some degree. But the efficiency was lower than the one caused by siRNA synthesized with SP6 RNA polymerase [23]. The bacteriophage SP6 RNA polymerase also has the specific recognition to its promoter SP6 promoter and has been used to produce synthetic RNA [42]. However, the SP6 RNA polymerase has diverged from other polymerases such as T7, T3 and so on [43]. We proposed that the efficiency difference may be partly owing to the different structure of RNA transcribed with different RNA polymerases. A more obvious and reasonable interpretation is, that the shRNA transcribed by shRNA vectors in the present study was mosaically distributed in the embryo because of the transgenic mosaicism, thus it could only induce a partial inhibition effect.

As discussed above, the inhibition effect of T7 system to *gfp* and *ntl* was just the results of temporary transcription by direct microinjection. If *red fluorescence protein (rfp)* marker was linked with the plasmid pT7shRNA and the resulted plasmid was injected into zebrafish embryos. Then, we may screen out germline transmitted pT7shRNA transgenic zebrafish and a more effective RNAi effect could be achieved if this transgenic line was mated with pCMVT7R transgenics. Furthermore, if tissue-specific promoter is used to drive the expression of T7RP, the tissue-specific gene inhibition could be obtained, which would be more helpful in the study of gene function and gene therapy.

Acknowledgements This work was supported by the State Key Fundamental Research of China (Grant No. 2004CB117406) and the National Natural Science Foundation of China (Grant No. 90208024).

References

1. Guo S. and Kemphues K.J., Par-1, a gene required for establishing polarity in *C. elegans* embryos, encodes a putative Ser/Thr kinase that is asymmetrically distributed. Cell 81: 611-620, 1995.
2. Fire A., Xu S., Montgomery M.K., Kostas S.A., Driver S.E. and Mello C.C., Potent and specific genetic interference by double-stranded RNA in *Caenorhabditis elegans*. Nature 391: 806-811, 1998.
3. Tabara H., Grishok A. and Mello C.C., RNAi in *C. elegans*: soaking in the genome sequence. Science 282: 430-431, 1998.
4. Tabara H., Sarkissian M., Kelly W.G., Fleenor J., Grishok A., Timmons L., Fire A. and Mello C.C., The rde-1 gene, RNA interference, and transposon silencing in *C. elegans*. Cell 99: 123-132, 1999.
5. Caplen N.J., Fleenor J., Fire A. and Morgan R.A., dsRNA-mediated gene silencing in cultured Drosophila cells: a tissue culture model for the analysis of RNA interference. Gene 252: 95-105, 2000.
6. Kennerdell J.R. and Carthew R.W., Heritable gene silencing in Drosophila using double-stranded RNA. Nat. Biotechnol. 18: 896-898, 2000.
7. Yang S., Tutton S., Pierce E. and Yoon K., Specific doublestranded RNA interference in undifferentiated mouse embryonic stem cells. Mol. Cell. Biol. 21: 7807-7816, 2001.
8. Yu J. and McMahon A.P., Reproducible and inducible knockdown of gene expression in mice. Genesis 44: 252-261, 2006.
9. Berezhna S.Y., Supekova L., Supek F., Schultz P.G. and Deniz A.A., siRNA in human cells selectively localizes to target RNA sites. Proc. Natl. Acad. Sci. USA 103: 7682-7687, 2006.
10. Chuang C.F. and Meyerowitz E.M., Specific and heritable genetic interference by double-stranded RNA in *Arabidopsis thaliana*. Proc. Natl. Acad. Sci. USA 97: 4985-4990, 2000.

11. Moffatt B.A., Dunn J.J. and Studier F.W., Nucleotide sequence of the gene for bacteriophage T7 RNA polymerase. J. Mol. Biol. 173: 265-269, 1984.

12. Davanloo P., Rosenberg A.H., Dunn J.J. and Studier F.W., Cloning and expression of the gene for bacteriophage T7 RNA polymerase. Proc. Natl. Acad. Sci. USA 81: 2035-2039, 1984.

13. Davison J., Chevalier N. and Brunel F., Bacteriophage T7 RNA polymerase-controlled specific gene expression in Pseudomonas. Gene 83: 371-375, 1989.

14. Pattnaik A.K. and Wertz G.W., Replication and amplification of defective interfering particle RNAs of vesicular stomatitis virus in cells expressing viral proteins from vectors containing cloned cDNAs. J. Virol. 64: 2948-2957, 1990.

15. McBride K.E., Schaaf D.J., Daley M. and Stalker D.M., Controlled expression of plastid transgenes in plants based on a nuclear DNA-encoded and plastid-targeted T7 RNA polymerase. Proc. Natl. Acad. Sci. USA 91: 7301-7305, 1994.

16. Yu J.Y., DeRuiter S.L. and Turner D.L., RNA interference by expression of short-interfering RNAs and hairpin RNAs in mammalian cells. Proc. Natl. Acad. Sci. USA99: 6047-6052, 2002.

17. Sioud M. and Leirdal M., Potential design rules and enzymatic synthesis of siRNAs. Methods Mol. Biol. 252: 457-469, 2004.

18. Li H., Fu X., Chen Y., Hong Y., Tan Y., Cao H., Wu M. and Wang H., Use of adenovirus-delivered siRNA to target oncoprotein p28GANK in hepatocellular carcinoma. Gastroenterology 128: 2029-2041, 2005.

19. Hamazaki H., Ujino S., Miyano-Kurosaki N., Shimotohno K. and Takaku H., Inhibition of hepatitis C virus RNA replication by short hairpin RNA synthesized by T7 RNA polymerase in hepatitis C virus subgenomic replicons. Biochem. Biophys. Res. Commun. 343: 988-994, 2006.

20. Holle L., Hicks L., Song W., Holle E., Wagner T. and Yu X., Bcl-2 targeting siRNA expressed by a T7 vector system inhibits human tumor cell growth *in vitro*. Int. J. Oncol. 24: 615-621, 2004.

21. Prabhu R., Vittal P., Yin Q., Flemington E., Garry R., Robichaux W.H. and Dash S., Small interfering RNA effectively inhibits protein expression and negative strand RNA synthesis from a full-length hepatitis C virus clone. J. Med. Virol. 76: 511-519, 2005.

22. Hamdorf M., Muckenfuss H., Tschulena U., Pleschka S., Sanzenbacher R., Cichutek K. and Flory E., An inducible T7 RNA polymerase-dependent plasmid system. Mol. Biotechnol. 33: 13-21, 2006.

23. Liu W.Y., Wang Y., Sun Y.H., Wang Y., Wang Y.P., Chen S.P. and Zhu Z.Y., Efficient RNA interference in zebrafish embryos using siRNA synthesized with SP6 RNA polymerase. Dev. Growth. Differ. 47: 323-331, 2005.

24. Acosta J., Carpio Y., BorrotoI., Gonzalez O. and Estrada M.P., Myostatin gene silenced by RNAi show a zebrafish giant phenotype. J. Biotechnol. 119: 324-331, 2005.

25. Wargelius A., Ellingsen S. and Fjose A., Double-stranded RNA induces specific developmental defects in zebrafish embryos. Biochem. Biophys. Res. Commun. 263: 156-161, 1999.

26. Boonanuntanasarn S., Yoshizaki G. and Takeuchi T., Specific gene silencing using small interfering RNAs in fish embryos. Biochem. Biophys. Res. Commun. 310: 1089-1095, 2003.

27. Dodd A., Chambers S.P. and Love D.R., Short interfering RNA-mediated gene targeting in the zebrafish. FEBS Lett. 561: 89-93, 2004.

28. Zhao Z., Cao Y., Li M. and Meng A., Double-stranded RNA injection produces nonspecific defects in zebrafish. Dev. Biol. 229: 215-223, 2001.

29. Mangos S., Vanderbeld B., Krawetz R., Sudol K. and Kelly G.M., Ran binding protein RanBP1 in zebrafish embryonic development. Mol. Reprod. Dev. 59: 235-248, 2001.

30. Westerfield M., The Zebrafish Book. University of Oregon Press, Eugene, OR, USA, 1995.

31. Tiscornia G., Singer O., Ikawa M. and Verma I.M., A general method for gene knockdown in mice by using lentiviral vectors expressing small interfering RNA. Proc. Natl. Acad. Sci. USA 100: 1844-1848, 2003.

32. Nguyen V.H., Schmid B., Trout J., Connors S.A., Ekker M. and Mullins M.C., Ventral and lateral regions of the zebrafish gastrula, including the neural crest progenitors, are established by a bmp2b/swirl pathway of genes. Dev. Biol. 199: 93-110, 1998.

33. Livak K.J. and Schmittgen T.D., Analysis of relative gene expression data using real-time quantitative PCR and the $2^{(-\text{Delta Delta C(T)})}$ method. Methods 25: 402-408, 2001.

34. Thisse C., Thisse B., Schilling T.F. and Postlethwait J.H., Structure of the zebrafish snail1 gene and its expression in wild-type, spadetail and no tail mutant embryos. Development 119: 1203-1215, 1993.
35. Schulte-Merker S., van Eeden F.J., HalpernM.E., Kimmel C.B. and Nusslein-Volhard C., No tail (ntl) is the zebrafish homologue of the mouse T (Brachyury) gene. Development 120: 1009-1015, 1994.
36. Halpern M.E., Ho R.K., Walker C. and Kimmel C.B., Induction of muscle pioneers and floor plate is distinguished by the zebrafish no tail mutation. Cell 75: 99-111, 1993.
37. Amacher S.L., Draper B.W., Summers B.R. and Kimmel C.B., The zebrafish T-box genes no tail and spadetail are required for development of trunk and tail mesoderm and medial floor plate. Development 129: 3311-3323, 2002.
38. Sandig V., Lieber A., Bahring S. and Strauss M., A phage T7 class-III promoter functions as a polymerase II promoter in mammalian cells. Gene 131: 255-259, 1993.
39. Lieber A., Sandig V. and Strauss M., A mutant T7 phage promoter is specifically transcribed by T7-RNA polymerase in mammalian cells. Eur. J. Biochem. 217: 387-394, 1993.
40. Verri T., Argenton F., Tomanin R., Scarpa M., Storelli C., Costa R., Colombo L. and Bortolussi M., The bacteriophage T7 binary system activates transient transgene expression in zebrafish (*Danio rerio*) embryos. Biochem. Biophys. Res. Commun. 237: 492-495, 1997.
41. Schmid B., Furthauer M., Connors S.A., Trout J., Thisse B., Thisse C. and Mullins M.C., Equivalent genetic roles for bmp7/snailhouse and bmp2b/swirl in dorsoventral pattern formation. Development. 127: 957-967, 2000.
42. Krieg P.A. and Melton D.A, Functional messenger RNAs are produced by SP6 *in vitro* transcription of cloned cDNAs. Nucleic Acids Res. 12: 7057-7070, 1984.
43. Chen Z.H. and Schneider T.D., Information theory based T7-like promoter models: classification of bacteriophages and differential evolution of promoters and their polymerases. Nucleic Acids Res. 33: 6172-6187, 2005.

T7 启动子驱动的体内转录 shRNA 可介导 *gfp* 和 *ntl* 的敲降

王　娜[1,2]　孙永华[1]　刘　静[1,2]　吴　刚[1,2]　苏建国[1]　汪亚平[1]　朱作言[1]

1 中国科学院水生生物研究所，武汉 430072
2 中国科学院大学，北京 100049

摘　要　本研究将一种依赖 T7 RNA 聚合酶的 shRNA 表达系统，应用于斑马鱼胚胎中敲降外源绿色荧光蛋白 *gfp* 和内源 *ntl* 基因。首先，将 CMV 启动子与 T7 聚合酶的重组载体 pCMVT7R 注射到斑马鱼胚胎中，筛选稳定表达的转基因个体用于下一步的 RNA 干扰实验。然后，构建 shRNA 表达载体 pT7shRNA，并通过体内实验验证了其转录活性。当 pT7shGFP 载体注射到表达 T7 RNA 聚合酶的转基因胚胎中时，实时荧光定量 RT-PCR 的检测结果显示 *gfp* 的表达量下调了 68%；注射化学合成的小干扰 RNA(2 μg/μl)时(阳性对照)，*gfp* 表达量也下降了 40%。更为重要的是，原位杂交的结果显示：在表达 T7 RNA 聚合酶的斑马鱼胚胎中，注射 pT7shNTL 载体会导致 30%的内源性 *ntl* 基因表达缺失。由此可见，T7 转录表达系统可以用于斑马鱼胚胎中表达 shRNA，并且可以发挥敲降基因的功能，这也提示该系统在斑马鱼胚胎中研究 RNA 干扰具有应用前景。

Hybrid Cytomegalovirus-U6 Promoter-Based Plasmid Vectors Improve Efficiency of RNA Interference in Zebrafish

Jianguo Su[1,2] Zuoyan Zhu[1] Feng Xiong[1] Yaping Wang[1]

1 State Key Laboratory of Freshwater Ecology and Biotechnology, Institute of Hydrobiology, Chinese Academy of Sciences, Wuhan 430072

2 Department of Aquaculture, College of Animal Science and Technology, Northwest A&F University, Yangling 712100

Abstract Short hairpin RNA (shRNA) directed by RNA polymerase III (Pol III) or Pol II promoter was shown to be capable of silencing gene expression, which should permit analyses of gene functions or as a potential therapeutic tool. However, the inhibitory effect of shRNA remains problematic in fish. We demonstrated that silencing efficiency by shRNA produced from the hybrid construct composed of the CMV enhancer or entire CMV promoter placed immediately upstream of a U6 promoter. When tested the exogenous gene, silencing of an enhanced green fluorescent protein (EGFP) target gene was 89.18 ± 5.06% for CMVE-U6 promoter group and 88.26 ± 6.46% for CMV-U6 promoter group. To test the hybrid promoters driving shRNA efficiency against an endogenous gene, we used shRNA against no tail (NTL) gene. When vectorized in the zebrafish, the hybrid constructs strongly repressed NTL gene expression. The NTL phenotype occupied 52.09 ± 3.0% and 51.56 ± 3.68% for CMVE-U6 promoter and CMV-U6 promoter groups, respectively. The NTL gene expression reduced 82.17 ± 2.96% for CMVE-U6 promoter group and 83.06 ± 2.38% for CMV-U6 promoter group. We concluded that the CMV enhancer or entire CMV promoter locating upstream of the U6-promoter could significantly improve inhibitory effect induced by the shRNA for both exogenous and endogenous genes compared with the CMV promoter or U6 promoter alone. In contrast, the two hybrid promoter constructs had similar effects on driving shRNA.

Keywords hybrid promoter; CMV promoter; U6 promoter; RNAi; zebrafish

Introduction

In past years, the most exciting development in gene regulation was the discovery of RNA interference (RNAi), through which short interfering RNAs (siRNAs) mediate selective gene inactivation by mRNA destruction (Song *et al.*, 2004). This highly efficient and specific

文章发表于 *Marine Biotechnology*, 2008, 10(5): 511-517

technology opens up a broad spectrum of potential implications for both therapeutics and experimental research (Dorn *et al.*, 2004). First described in plants (Napoli *et al.*, 1990), it has since been applied to a wide variety of invertebrate and vertebrate models (Hutvagner and Zamore, 2002). In mammalian cells, the introduction of long dsRNAs into mammalian cells activates protein kinase PKR and RNase L, leading to an interferon response and hence to the non-specific extinction of genes resulting in cell death. This non-specific effect of long dsRNAs into mammalian cells can be bypassed by using small RNA duplexes of 19-21 nt, which are sufficient to trigger specific RNAi in mammalian cells without activating the interferon response (Elbashir *et al.*, 2001).

Short hairpin RNAs (shRNAs) transcribed *in vivo* under the control of RNA polymerase III (Pol III) promoters can trigger degradation of corresponding mRNAs similar to siRNAs and inhibit specific gene expression (Brummelkamp *et al.*, 2002; Jacque *et al.*, 2002). In addition to synthesizing mRNA, RNA polymerase II (Pol II) is responsible for synthesis of many noncoding RNAs, including small nuclear and nucleolar RNAs (Song *et al.*, 2004). Recently, siRNA transcripts expressed from a RNA polymerase II promoter, cytomegavirus (CMV) promoter, have been shown to be capable of reducing gene expression in mammalian cells and zebrafish (Xia *et al.*, 2002; Su *et al.*, 2007).

We designed shRNA-expressing construct controlled by a Pol III U6 promoter (Wang, 2007) and shRNA-expressing construct controlled by CMV promoter (Su *et al.*, 2007) to inhibit an exogenous transgenic EGFP and an endogenous no tail (NTL) expressions in zebrafish. While testing the efficacy of these constructs, we found they selectively inhibited the expression of target gene but did not affect others. However, the efficacy of RNAi produced by these constructs was modest, which might affect the ultimate application. One way to overcome this problem was to increase the dose of the shRNA by enhancing the promoter activity. Some snRNAs are synthesized by Pol II whereas others are synthesized by Pol III, and they share similar enhancer elements (Bark *et al.*, 1987; Carbon *et al.*, 1987; Das *et al.*, 1988; Kunkel and Pederson, 1988; Mattaj *et al.*, 1988; Lobo and Hernandez, 1989). Hence, a Pol II enhancer might be able to enhance the Pol III transcription. One previous study demonstrated that the enhancer from the cytomegalovirus immediate-early promoter, when placed near the Pol III U6 promoter in a plasmid vector, could enhance the activity of the U6 promoter, increase shRNA expression, and strengthen the gene-silencing effect in human cells (Xia *et al.*, 2003).

The efficiency of shRNA-carrying plasmids in terms of target gene inhibition varies with the composition of promoter elements, the promoter position relative to the transcriptional start site, the location of a promoter within a given vector, and probably the type of tissues or cells tested (Ilves *et al.*, 1996; Arendt *et al.*, 2003; Boden *et al.*, 2003; Koper-Emde *et al.*, 2004). Each tissue and condition (*in vivo* versus *in vitro)* requires optimization for the use of shRNAs-triggered RNAi, too (Hassani *et al.*, 2007). Therefore, the choice of promoter is vital in achieving effective gene silencing by intracellular expression of shRNA in various organisms.

In the present study, we constructed a hybrid promoter by placing a 465-bp fragment of the CMV enhancer and a hybrid promoter by appending the CMV enhancer and early promoter 5′ to the U6 promoter and explored whether these modified promoters could improve gene silencing effects in zebrafish.

Materials and Methods

Plasmid constructions

pCMV-EGFP plasmid (referred in the following as pG), carrying the CMV promoter and enhanced green fluorescent protein (EGFP) gene as reporter mark, was from Clontech (USA) as backbone. To ensure the correct sequence of the inserts, the clones selected were sequenced after inserting each following fragment.

To construct the hybrid promoters, including hybrid CMV enhancer U6 (CMVE-U6) promoter and hybrid CMV enhancer and early promoter U6 (CMV-U6) promoter, a CMV enhancer element (1-465 nt of human cytomegalovirus immediate early promoter) and a CMV promoter (1-589 nt of human cytomegalovirus immediate early promoter) were PCR-amplified from pG with forward and reverse primers. For cloning purposes, the forward sequence SCF128 (Table 1) was flanked with *Xho*I site at the 5′ end, and the reverse sequence SCR94c (for CMV enhancer) or SCR94d (for CMV promoter) (Table 1) was flanked with *Hin*dIII, *Sal*I, and *Kpn*I sites at the 5′ end in order. PCR-amplified products were purified and digested with *Xhol*I and *Hin*dIII sites, and then subcloned into pG at the *Xhol*I and *Hin*dIII sites to form pCMVE-CMV-EGFP or pCMV-CMV-EGFP.

Table 1 Oligonucleotides used in this study

Name	Sequence (5′-3′)	Sequence information
SCF128	AATCGCTCGAGTAGTTATTAATAGTAATCAATTACG	CMV promoter/enhancer primer
SCR94c	AATCGAAGCTTGTCGACGGTACCCAAAACAAACTCCCATT	CMV enhancer primer
SCR94d	AATCGAAGCTTGTCGACGGTACCGATCTGACGGTTCACTA	CMV promoter primer
SUF130	AATCGGGTACCTCTTTAGCCTCCGAGAG	U6 promoter primer
SUR131	AATCGGTCGACGAACTAGGAGCCTGGAG	U6 promoter primer
SG134a	TCGACGCAAGCTGACCCTGAAGTTCTTCAAGAGAGAACTTCAGGGTC-AGCTTGCTTTTTA	ShGFP
SG135a	AGCTTAAAAAGCAAGCTGACCCTGAAGTTCTCTCTTGAAGAACTTCAG-GGTCAGCTTGCG	ShGFP
SG134b	TCGACAGCATCTAAGGCGACCTCGTTTCAAGAGAGGTCAGTAGGTCAA-CCTCCATTTTTA	ShScrambled
SG135b	AGCTTAAAAAACCTCCAACTGGATGACTGGAGAGAACTTTGCTCCAGC-GGAATCTACGAG	ShScrambled

Continued

Name	Sequence (5′-3′)	Sequence information
SN120b	TCGACTGCAATGTACTCGGTCCTGTTCAAGAGACAGGACCGAGTACAT-TGCATTTTTA	ShNTL
SN121b	AGCTTAAAAAATGTACTCGGTCCTGTCTCTTGAACAGGACCGAGTACA-TTGCAG	ShNTL
SBAF86	GATGATGAAATTGCCGCACTG	β-actin Q-PCR primer
SBAR87	ACCAACCATGACACCCTGATGT	β-actin Q-PCR primer
SGF114	CAAGCAGAAGAACGGCATCA	GFP Q-PCR primer
SGR115	AGGTAGTGGTTGTCGGGCA	GFP Q-PCR primer
SNF122	CAGCACTGACAACCAGCAATC	NTL Q-PCR primer
SNR123	GAACCCGAGGAGTGAACAGG	NTL Q-PCR primer

The Zebrafish U6 promoter was obtained from zebrafish genomic DNA with sense primer SUF130 carrying *Kpn*I site and antisense primer SUR131 carrying *Sal*I site (Table 1). The amplicon was purified, digested, and inserted in the pCMVE-CMV-EGFP and pCMV-CMV- EGFP at 3′ region of the inserted CMV enhancer or CMV promoter between the *Kpn*I and *Sal*I sites. The vectors were named pCMVE-U6-CMV-EGFP and pCMV-U6-CMV-EGFP, respectively.

Two complementary oligonucleotides SG134a and SG135a (Table 1) directed against the EGFP-coding region with cohesive *Sal*I and *Hin*dIII sites were chemically synthesized according to the sequences previously described (Su *et al.*, 2007), denatured, annealed, and cloned into pCMVE-U6-CMV-EGFP and pCMV-U6-CMV-EGFP. The two plasmid vectors were named pCMVE-U6-siEGFP-CMV-EGFP (pEUsiG) and pCMV-U6-siEGFP-CMV-EGFP (pCUsiG), respectively. pCMVE-U6-siScrambled-CMV-EGFP (pEUsiS) and pCMV-U6-siScrambled-CMV-EGFP (pCUsiS) with a scrambled fragment (SG134b and SG135b) (Table 1) of siEGFP were used as the control. To silence the endogenous NTL gene expression, two cohesive complementary oligonucleotides SN120b and SN121b (Table 1) were synthesized and cloned into plasmid vectors as described above. These vectors were named pCMVE-

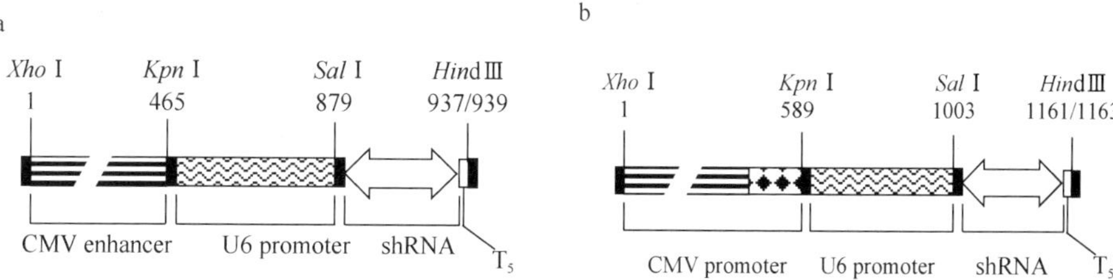

Fig. 1 Scheme of the different promoter constructions driving shRNA. The two promoter constructs tested for driving-shEGFP, -shScrambled or -shNTL mediated inhibition in zebrafish were shown. a. Hybrid CMV enhancer and U6 promoter; b. Hybrid entire CMV promoter upstream of U6 promoter. In each construction, the shRNA was followed by a TTTTT sequence required for the U6-transcription arrest

U6-siNTL-CMV-EGFP (pEUsiN) and pCMV-U6-siNTL-CMV-EGFP (pCUsiN), respectively (see scheme, Fig. 1). All the oligonucleotides for shRNA transcription were followed by 5 thymidines (Ts) as transcriptional termination signal.

Gene transfer *in vivo*

The protocol of gene transfer was as described previzously (Su *et al.*, 2007). Briefly, Individual 1 to 2-cell embryos were microinjected under a dissecting microscope with pulled microcapillary pipettes to deliver approxionately 1nl of shRNA expression vector solution (100 ng/μl). We injected at least 450 embryos and cultured in three dishes for every group. Totally we used approxionately 4,000 embryos. The injected and uninjected embryos were subsequently incubated in sterile 0.3 × Danieau's solution at 28℃. We refreshed the solution and disposed of dead embryos twice per day.

Phenotypic observations

Phenotypes were evaluated at 48 hpf (hours post fertilization). Images of zebrafish embryos were recorded using an Olympus SZX12 fluorescent microscope and a digital camera (Liu *et al.*, 2007). Features analyzed for NTL: disrupted notochord, abnormal somites and reduced tail (Halpern *et al.*, 1993). EGFP expression was monitored using fluoresce microscopy. For phenotype count and fluorescent observation, we employed all the embryos.

Q-PCR assays

Target mRNA levels in embryos were quantitatively analyzed using real time fluorescent quantitative reverse transcription polymerase chain reaction (Q-PCR). Zebrafish embryos injected and uninjected were collected at 48 and 120 hpf, and total RNA was extracted from pools of 30-40 embryos per sample with triplicate repeat using Trizol (Invitrogen). The RNA samples were further treated with RNase-free DNase I (Roche) to remove contaminated genomic DNA, followed by phenol/chloroform extraction and ethanol precipitation. 1 μg of total RNA from each sample was used for the first-strand cDNA, using a Superscript III first-strand synthesis system (Invitrogen) according to the manufacturer's protocol. For the Q-PCR detection of EGFP or NTL mRNA, cDNA samples were diluted 1∶5.

Q-PCRs were performed in an ABI Prism 7000 Sequence Detection System (Applied Biosystems) using the following thermal cycling profile: 95℃ 3 min, followed by 40 cycles of amplification (95℃ 15 s, 58℃ 15 s, 72℃ 45 s), followed by dissociation curve analysis to validate the amplification of a single product. Each reaction contained 4 μl of cDNA sample, 10 μl of SYBR Green PCR master mix (Toyobo), 1μl each of forward and reverse primers (2.5 μM), as well as 4 μl of nuclease-free water to make up a 20-μl reaction volume. All reactions were performed in triplicate. All primers were designed to conform to a universal cycling program by the Primer Express 2.0 software (Applied Biosystems). The endogenous housekeeping gene encoding β-actin was used as an internal standard to normalize differences in template amounts. The oligonucleotides used for PCR amplification of β-actin were as follows: forward primer SBAF86 and reverse primer SBAR87 (Table 1). The PCR product size was 135 bp. For the Q-PCR detection of EGFP mRNA, the primers used were SGF114 and SGR115 (Table 1). The PCR product of the EGFP gene was 138 bp. NTL primer sequences were sense SNF122 and antisense SNR123 (Table 1). The NTL amplicon was 126 bp.

Relative expression was calculated using a modified comparative cycle threshold (CT) method, where CT was defined as the cycle numbers at which fluorescence reached a set threshold value. The differences in the CT value of the target genes from the corresponding internal control β-actin gene, ΔCT ($CT_{gene}- CT_{actin}$), were calculated. The changes in ΔCT of the experiment group to the control group, ΔΔCT ($\Delta CT_{sample}- \Delta CT_{control}$), were computed. The relative expression level of the experiment group to the control group was described by using the equation $2^{-\Delta\Delta CT}$, and the value was determined for a $1/n$-fold difference relative to the control. The expression of experiment group relative to the control group was multiplied by 100% to simplify the presentation of the data.

Statistical analysis

The data obtained from Q-PCR analysis were shown by means ± SE. Statistical analysis was done using unpaired Student's *t*-test for comparison between the control and treated groups. Error bars indicated standard errors. $P < 0.05$ was considered statistically significant, and $P < 0.01$ was very significant difference.

Results

Design of different promoter constructs driving the transcription of shRNA

Two different hybrid promoter constructs were used for the production of shRNAs, including shEGFP, shNTL and shSrambled control (Fig. 1). The first was a CMVE-U6 hybrid promoter transcribing shRNA. In this construction, the type III RNA polymerase synthesized the shRNA, and the transcription was enhanced by the CMV enhancer. Second, in order to explore whether the CMV enhancer and early promoter could improve the efficacy of the U6-mediated production of shRNAs in fish, we constructed a CMV-U6 hybrid construct driving the production of the shRNA. In the constructions, the full enhancer and promoter sequences of CMV have been integrated upstream of the U6 promoter. The transcription of shRNA was terminated by 5 Ts of transcriptional termination sequence at the 3′ end of the shRNA.

pEUsiG and pCUsiG trigger the exogenous EGFP gene knockdown

We tested and compared the efficiency for pEUsiG and pCUsiG to trigger the inhibition of a tandem target gene by RNAi in zebrafish. As described in methods, two EGFP and two control shRNA expression vectors described in Fig. 1, and pG control plasmid, were microinjected in zebrafish embryos. Under a fluorescence microscope, the fluorescence was very weak in the experimental groups (Fig. 2f). As shown in Fig. 3, quantification of EGFP expressions revealed strong inhibition efficiency for both vectors at 48 hpf. The 89.18 ± 5.06% inhibition of the target gene was achieved with pEUsiG, and the 88.26 ± 6.46% reduction of the EGFP was obtained with pCUsiG, which were significantly stronger than the controls ($P < 0.01$). There was no significant difference in gene silence between pEUsiG and pCUsiG ($P > 0.05$). The control pEUsiS and pCUsiS did not provide inhibition of EGFP.

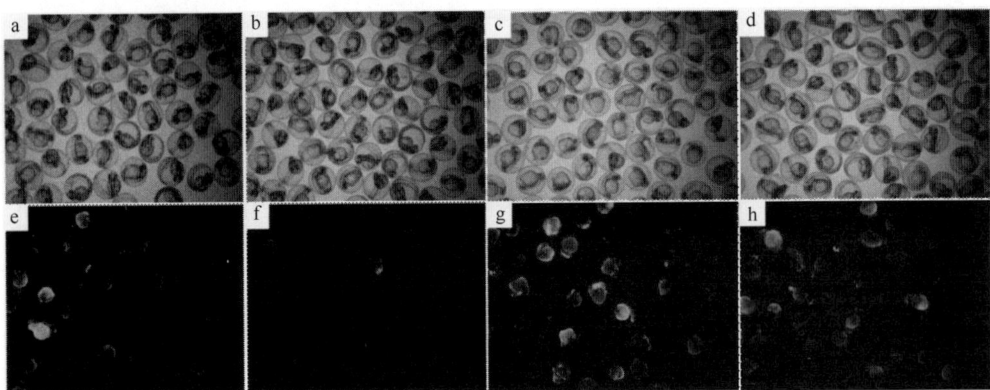

Fig. 2 The effect of hybrid CMV-U6 promoter constructs injection on the expression of EGFP and NTL in zebrafish embryos at 48 hpf. Upper row (a-d) showed the same field of embryos by light microscopy as seen under fluorescence in lower row (e-h). Column 1 was injected with EGFP expression vector (pG). Column 2 was injected with shGFP construct (pCUsiG). Column 3 was injected with shNTL construct (pCUsiN). Column 4 was injected with shScrambled construct (pCUsiS). The hybrid CMVE-U6 promoter constructs had the similar results. shScrambled construct and shNTL construct didn't interfere with EGFP expression. The embryos injected with shGFP or shScrambled construct hardly showed the no tail phenotype

pEUsiN and pCUsiN inhibit the endogenous NTL gene expression

After having demonstrated that the presence of a CMV enhancer/promoter upstream of a U6 promoter strongly increased the inhibition efficiency of shRNAs, we further exploited this technology to interfere in the endogenous NTL gene expression in fish. We designed shNTL expression vectors directed against NTL, and checked their efficiency and their specificity in zebrafish. The embryos of no tail phenotype were counted under a dissection microscope, and occupied 52.09 ± 3.06% at 48 hpf in pEUsiN group, 51.56 ± 3.68% in pCUsiN group, and 1.12 ± 0.15% in other groups ($P<0.05$) (Fig. 2c and 4a). Using Q-PCR, we observed a significant down-regulation of NTL under the control of the hybrid promoter constructs. The inhibitory efficiency was 82.17 ± 2.96% for pEUsiN and 83.06 ± 2.38% for pCUsiN compared with the uninjected group (P<0.01). The control pEUsiS and pCUsiS also were tested in the same situation, where no inhibition was visible (Fig. 4b).

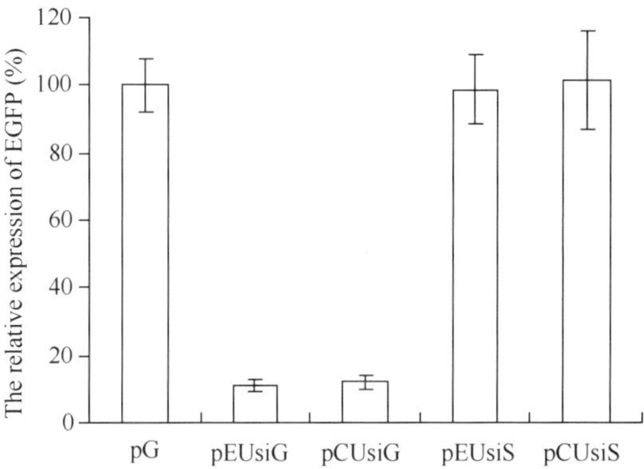

Fig. 3 pEUsiG and pCUsiG showed similar inhibitory efficiency for exogenous gene in zebrafish. Plasmids containing shEGFP or shScrambled were microinjected into zebrafish as described in methods, along with the reporter gene EGFP. Only pEUsiG and pCUsiG lead to an inhibition. Data showed mean values ± SE (n=3). Error bars indicate standard error

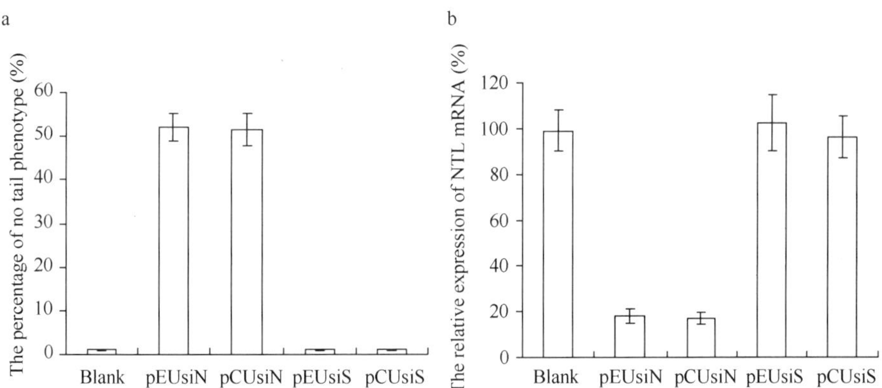

Fig. 4 pEUsiN and pCUsiN demonstrated analogous silencing activity for endogenous gene in zebrafish. a. pEUsiN and pCUsiN were microinjected as described in the methods section, along with the irrelevant pEUsiS and pCUsiS as controls. The no tail phenotype was very high in pEUsiN and pCUsiN groups. b. zebrafish was microinjected by the different plasmid constructions described in (a), only pEUsiN and pCUsiN provided significant inhibition of the endogenous NTL expression. Means ± SE are shown. Error bars indicated standard error

Discussion

Some scientists tried to exploit RNAi technology in fish since it was discovered; however, until now there have still some obstacles to be overcome, such as nonspecificity and low efficacy. Some findings suggested dsRNA-mediated interference could inhibit the gene expression in dose-dependent mode (Wargelius *et al.*, 1999; Acosta *et al.*, 2005), others reported nonspecific effect (Oates *et al.*, 2000; Li *et al.*, 2000; Zhao *et al.*, 2001; Mangos *et al.*, 2001). These views might be resulted from different gene sequences studied. siRNA-based gene silencing was effective, but siRNA duplexes were unsteady *in vivo* (Boonanuntanasarn *et al.*, 2003; Dodd *et al.*, 2004; Liu *et al.*, 2005).

The majority of the mammalian RNAi systems are driven by Pol III promoters that express ubiquitously (Song *et al.*, 2004). However, under certain conditions, the activity provided by an unmodified Pol III promoter is not sufficient in providing satisfactory gene-silencing outcomes (Paul *et al.*, 2002; Boden *et al.*, 2003; Xia *et al.*, 2003). In fish, the shRNA under the control of Pol III promoters (H1 and U6 promoters) (Wang, 2007) also inhibited the exogenous or endogenous gene modestly. To improve potency, we designed two hybrid promoters by fusing the CMV enhancer or entire CMV promoter 5′ to the U6 promoter and observed improved inhibition by siRNA of target gene expression. The U6 promoter belongs to the Pol III promoter and is located only in the 5′-flanking region of the gene transcribed. This gene-external feature resembles that of Pol II promoters. Whereas the Pol III U6 promoter directs the synthesis of the U6 small nuclear RNA (snRNA), Pol II promoters are used for transcription of the U1 to U5 snRNAs. Strikingly, not only is the spatial organization of the U6 promoter elements similar to that of the U2 elements, they also share sequence homologies in the enhancer (Hernandez, 2001). These findings from the snRNA promoters have led to the concept that the Pol II and III transcription machineries share common factors.

In the current study, we modified the U6 promoter by placing the enhancer or full promoter from the CMV immediate-early promoter immediately upstream of a U6 promoter. The hybrid promoters driving the synthesis of a hairpin sequence provided a significant improvement in terms of inhibitory effect for both exogenous and endogenous genes. The 89.18 ± 5.06% and 88.26 ± 6.46% reduction of the exogenous EGFP gene with hybrid promoters were stronger than the 70% and 32% knockdown obtained with pCMV-shEGFP (Su *et al.*, 2007) and pU6-shEGFP (Wang, 2007), respectively. shNTL produced by the modified U6 promoter inhibited endogenous NTL gene expression up to 82.17 ± 2.96% and 83.06 ± 2.38%, which was much higher than the 54.52 ± 2.05% suppression with pCMV-shEGFP (Su *et al.*, 2007) and approximately 16% reduction with U6 promoter alone (Wang, 2007). Both the hybrid promoter configurations directed synthesis of scrambled shRNA did not show any inhibitory activity towards the target genes. The findings from the current study supported the notion that a CMV enhancer, even thorough CMV promoter, was able to enhance the transcriptional activity of a U6 promoter in fish.

Attaching the CMV enhancer or entire promoter at the upstream of U6 promoter appeared highly effective gene-specific knockdown, and the reduction activities of target gene expressions were similar in zebrafish (Fig. 3 and 4). The underlying mechanism is as yet unknown. It is well recognized that enhancer and promoter recognition by transcription factors, repressors, and auxiliary proteins is a complex process, involving both primary and secondary sequence characteristics of a gene expression regulatory element. The number, diversity, orientation, and placement of transcription factor-binding sites within an enhancer and/or a promoter are critical parameters that define gene expression levels.

Conclusion

We have constructed two hybrid promoters carrying the U6 promoter fused to the CMV enhancer or complete CMV promoter and demonstrated the improved silencing effects in fish. In addition to functioning as a powerful instrument for biological studies, this modified promoter might be useful where limited choices of shRNA sequences preclude the selection of a highly efficient RNAi target region and might also serve as a potentially useful therapeutic means to silence disease-related genes and correct disease phenotypes.

Acknowledgements The authors thank Ming Li for microinjection. We also appreciate Wei Hu, Jun Dai, Na Wang, Shangping Chen, and other laboratory members for technical assistance and helpful discussion. This work was supported by grants from National Natural Science Foundation of China (30740009, 30540084 and 30428024), from 973 National Basic Research Program of China (2006CB102100), from Chinese Academy of Sciences (KSCX2-YW-N-021), from Northwest A & F University in China (08080262, 08080245 and 01140508), from China Postdoctoral Science Foundation (20070410298) and from Institute of Hydrobiology, CAS (2007FB09).

References

Acosta J, Carpio Y, Borroto I, Gonzalez O, Estrada MP (2005) Myostatin gene silenced by RNAi show a zebrafish giant phenotype. J Biotechnol 119: 324-331

Arendt CW, Tang G, Zilberstein A (2003) Vector systems for the delivery of small interfering RNAs: Managing the RISC. Chembiochemistry 4: 1129-1136

Bark C, Weller P, Zabielski J, Janson L, Pettersson U (1987) A distant enhancer element is required for polymerase III transcription of a U6 RNA gene. Nature 328: 356-359

Boden D, Pusch O, Lee F, Tucker L, Shank PR, Ramratnam B (2003) Promoter choice affects the potency of HIV-1 specific RNA interference. Nucleic Acids Res 31: 5033-5038

Boonanuntanasarn S, Yoshizaki G, Takeuchi T (2003) Specific gene silencing using small interfering RNAs in fish embryos. Biochem Biophys Res Commun 310: 1089-1095

Brummelkamp TR, Bernards R, Agami R (2002) A system for stable expression of short interfering RNAs in mammalian cells. Science 296: 550-553

Carbon P, Murgo S, Ebel JP, Krol A, Tebb G, Mattaj LW (1987) A common octamer motif binding protein is involved in the transcription of U6 snRNA by RNA polymerase III and U2 snRNA by RNA polymerase II. Cell 51: 71-79

Das G, Henning D, Wright D, Reddy R (1988) Upstream regulatory elements are necessary and sufficient for transcription of a U6 RNA gene by RNA polymerase III. EMBO J 7: 503-512

Dodd A, Chambers SP, Love DR (2004) Short interfering RNA-mediated gene targeting in the zebrafish. FEBS Lett 561: 89-93

Dorn G., Patel S, Wotherspoon G, Hemmings-Mieszczak M, Barclay J, Natt FJ, Martin P, Bevan S, Fox A (2004) siRNA relieves chronic neuropathic pain. Nucleic Acids Res 32: e49

Elbashir SM, Harborth J, Lendeckel W, Yalcin A, Weber K, Tuschl T (2001) Duplexes of 21-nucleotide RNAs mediate RNA interference in cultured mammalian cells. Nature 411: 494-498

Halpern ME, Ho RK, Walker C, Kimmel CB (1993) Induction of muscle pioneers and floor plate is distinguished by the zebrafish no tail mutation. Cell 75: 99-111

Hassani Z, Francois J, Alfama G, Dubois GM, Paris M, Giovannangeli C, Demeneix BA (2007) A hybrid CMV-H1 construct improves efficiency of PEI-delivered shRNA in the mouse brain. Nucleic Acids Res 35: e65

Hernandez N (2001) Small nuclear RNA genes: a model system to study fundamental mechanisms of transcription. J Biol Chem 276: 26733-26736

Hutvagner G, Zamore PD (2002) RNAi: nature abhors a double-strand. Curr Opin Genet Dev 12: 225-232

Ilves H, Barske C, Junker U, Bohnlein E, Veres G (1996) Retroviral vectors designed for targeted expression of RNA polymerase III-driven transcripts: a comparative study. Gene 171: 203-208

Jacque JM, Triques K, Stevenson M (2002) Modulation of HIV-1 replication by RNA interference. Nature 418: 435-438

Koper-Emde D, Herrmann L, Sandrock B, BeNecke BJ (2004) RNA interference by small hairpin RNAs synthesized under control of the human 7S K RNA promoter. Biol Chem 385: 791-794

Kunkel GR, Pederson T (1988) Upstream elements required for efficient transcription of a human U6 RNA gene resemble those of U1 and U2 genes even though a different polymerase is used. Genes Dev 2: 196-204

Li YX, Farrell MJ, Liu R, Mohanty N, Kirby ML (2000) Double-stranded RNA injection produces null phenotypes in zebrafish. Dev Biol 217: 394-405

Liu WY, Wang Y, Sun YH, Wang Y, Wang YP, Chen SP, Zhu ZY (2005) Efficient RNA interference in zebrafish embryos using siRNA synthesized with SP6 RNA polymerase. Dev Growth Differ 47: 323-331

Liu WY, Wang Y, Qin Y, Wang YP, Zhu ZY (2007) Site-directed gene integration in transgenic zebrafish mediated by cre recombinase using a combination of mutant lox sites. Mar Biotechnol 9: 420-428

Lobo SM, Hernandez N (1989) A 7 bp mutation converts a human RNA polymerase II snRNA promoter into an RNA polymerase III promoter. Cell 58: 55-67

Mangos S, Vanderbeld B, Krawetz R, Sudol K, Kelly GM (2001) Ran binding protein RanBP1 in zebrafish embryonic development. Mol Reprod Dev 59: 235-248

Mattaj IW, Dathan NA, Parry HD, Carbon P, Krol A (1988) Changing the RNA polymerase specificity of U snRNA gene promoters. Cell 55: 435-442

Napoli C, Lemieux C, Jorgensen R (1990) Introduction of a chimeric chalcone synthase gene into petunia results in reversible co-suppression of homologous genes in trans. Plant Cell 2: 279-289

Oates AC, Bruce AE, Ho RK (2000) Too much interference: injection of double-stranded RNA has nonspecific effects in the zebrafish embryo. Dev Biol 224: 20-28

Paul CP, Good PD, Winer I, Engelke DR (2002) Effective expression of small interfering RNA in human cells. Nat Biotechnol 20: 505-508

Song J, Pang S, Lua Y, Chiu R (2004) Poly(U) and polyadenylation termination signals are interchangeable for terminating the expression of shRNA from a pol II promoter. Biochem Biophys Res Commun 323: 573-578

Su J, Zhu Z, Wang Y, Xiong F, Zou J (2007) The cytomegalovirus promoter-driven short hairpin RNA constructs mediate effective RNA interference in zebrafish *in vivo*. Mar Biotechnol DOI 10.1007/s10126-007-9059-4

Wang N (2007) Study of short interfering RNA (siRNA) produced in transcription system of zebrafish embryo. Dissertation of doctor of science, Chinese Academy of Sciences

Wargelius A, Ellingsen S, Fjose A (1999) Double-stranded RNA induces specific developmental defects in zebrafish embryos. Biochem Biophys Res Commun 263: 156-161

Xia H, Mao Q, Paulson HL, Davidson BL (2002) siRNA-mediated gene silencing *in vitro* and *in vivo*. Nat Biotechnol 20: 1006-1010

Xia XG, Zhou H, Ding H, Affar EB, Shi Y, Xu Z (2003) An enhanced U6 promoter for synthesis of short hairpin RNA. Nucleic Acids Res 31: e100

Zhao Z, Cao Y, Li M, Meng A (2001) Double-stranded RNA injection produces nonspecific defects in zebrafish. Dev Biol 229: 215-223

CMV-U6 复合启动子可增强斑马鱼 RNA 干扰的效率

苏建国[1,2] 朱作言[1] 熊 凤[1] 汪亚平[1]

1 中国科学院水生生物研究所，武汉 430072

2 西北农林科技大学，杨凌 712100

摘 要 研究发现由 RNA 聚合酶 III(Pol III)或 Pol II 启动子驱动的短发夹结构 RNA(shRNA)具有沉默基因表达的作用，因此可以用于基因功能研究和作为潜在的治疗工具。然而 shRNA 在鱼类中抑制基因表达的作用尚不明确。我们构建了复合启动子，即在 U6 启动子上游插入 CMV 增强子(CMVE-U6)或完整的 CMV 启动子(CMV-U6)，用于驱动 shRNA 的表达，并检测了其沉默效率。CMVE-U6 构建体对外源绿色荧光蛋白(EGFP)目标基因的沉默效率为 89.18±5.06%，CMV-U6 构建体的沉默效率为 88.26±6.46%。同时也将复合启动子驱动 shRNA 的载体注射到斑马鱼胚胎中，检测了其对内源 NTL 基因的沉默效率。CMVE-U6 和 CMV-U6 的 NTL 表型率分别为 52.09±3.06%和 51.56±3.68%，而 NTL 基因的表达量分别下降 82.17±2.96%和 83.06±2.38%。这些结果表明，与单独的 CMV 或 U6 启动子相比，CMV 增强子或 CMV 启动子与 U6 启动子结合使用可以显著增强抑制效应，而 CMV 增强子和完整的 CMV 启动子在驱动 shRNA 的表达中作用相近。

The Cytomegalovirus Promoter-Driven Short Hairpin RNA Constructs Mediate Effective RNA Interference in Zebrafish *In Vivo*

Jianguo Su[1,2] Zuoyan Zhu[1] Yaping Wang[1]
Feng Xiong[1] Jun Zou[3]

1 State Key Laboratory of Freshwater Ecology and Biotechnology, Institute of Hydrobiology, Chinese Academy of Sciences, Wuhan 430072, China
2 Department of Aquaculture, College of Animal Science and Technology, Northwest A&F University, Yangling 712100, China
3 School of Biological Sciences, University of Aberdeen, Aberdeen AB24 2TZ, UK

Abstract The ability to utilize the RNA interference (RNAi) machinery for silencing target-gene expression has created a lot of excitement in the research community. In the present study, we used a cytomegalovirus (CMV) promoter-driven DNA template approach to induce short hairpin RNA (shRNA) triggered RNAi to block exogenous Enhanced Green Fluorescent Protein (EGFP) and endogenous No Tail (NTL) gene expressions. We constructed three plasmids, pCMV-EGFP-CMV-shGFP-SV40, pCMV-EGFP-CMV-shNTL-SV40, and pCMV-EGFP-CMV-shScrambled-SV40, each containing a CMV promoter driving an EGFP reporter cDNA and DNA coding for one shRNA under the control of another CMV promoter. The three shRNA-generating plasmids and pCMV-EGFP control plasmid were introduced into zebrafish embryos by microinjection. Samples were collected at 48 h after injection. Results were evaluated by phenotype observation and real-time fluorescent quantitative reverse-transcription polymerase chain reaction (Q-PCR). The shGFP-generating plasmid significantly inhibited the EGFP expression viewed under fluorescent microscope and reduced by 70.05±1.26% of exogenous EGFP gene mRNA levels compared with controls by Q-PCR. The shRNA targeting endogenous NTL gene resulted in obvious NTL phenotype of 30±4% and decreased the level of their corresponding mRNAs up to 54.52±2.05% compared with nontargeting control shRNA. These data proved the feasibility of the CMV promoter-driven shRNA expression technique to be used to inhibit exogenous and endogenous gene expressions in zebrafish *in vivo*.

Keywords CMV promoter; EGFP; no tail; RNAi; shRNA; zebrafish

Introduction

RNA interference (RNAi), a conserved antiviral immunity of plants and animals (Ding *et*

文章发表于 *Marine Biotechnology*, 2008, 10(3): 262-269

al. 2004), has been developed into an effective method of sequence-specific gene knockdown for analyzing gene functions in plants, invertebrates, and mammalian cells (Napoli *et al.* 1990; Fire *et al.* 1998; Elbashir *et al.* 2001). RNAi also holds great promise as a powerful therapeutic tool (Banan and Puri 2004; Kondraganti *et al.* 2006). The conserved RNAi pathway involves the processing of double-stranded RNA (dsRNA) duplexes into 21-23 nucleotide (nt) molecules known as small interfering RNAs (siRNA) to initiate gene knockdown (Hannon 2002). The long dsRNA in lower eukaryotes, especially in the model organism *Caenorhabditis elegans*, has been used to determine gene functions (Ashrafi *et al.* 2003). However, long dsRNA in mammalian systems induces an antiviral defence mechanism initiated by interferon (IFN), leading to nonspecific translational shutdown and apoptosis (Gil and Esteban 2000).

Direct transfection of either chemically synthesized or in vitro transcribed siRNAs of approximately 21 nt in length does not activate the IFN response but can induce reliable and efficient transient knockdown of target genes in mammalian cells (Tuschl 2002; Dykxhoorn *et al.* 2003). Compared with siRNA, the DNA-based vectors for expression of short hairpin RNA (shRNA) offer additional advantages in silencing longevity, cost, and delivery options (McIntyre and Fanning 2006). As a consequence, the development of shRNA molecules that are processed within the cell to produce active siRNA molecules has progressed rapidly (Brummelkamp *et al.* 2002). Such DNA expression constructs have achieved highly gene knock-down efficiency without induction of the IFN response.

Expressed shRNA is transcribed in cells from a DNA template as a single-stranded RNA molecule. Complementary regions spaced by a small loop cause the transcript to fold back on itself, forming a short hairpin in a manner analogous to natural micro-RNA. Recognizing and processing by the RNAi machinery convert the shRNA into the corresponding siRNA. shRNA expression vectors have been engineered by using both viral and plasmid systems. These vectors often utilize promoters from a small class of the RNA polymerase III-type (pol. III) promoters (Schramm and Hernandez 2002) to drive the expression of shRNA. Promoters of this type are preferred because they naturally direct the synthesis of small, highly abundant noncoding RNA transcripts, with defined termination sequences consisting of 4 to 5 thymidines (Ts) and have no requirement for downstream termination elements (Geiduschek and Kassavetis 2001).

For siRNA expression, the RNA polymerase II-type (pol. II) CMV promoter (human cytomegalovirus immediate-early promoter) has several advantages over pol. III promoters, such as U6 or H1. First, pol. II will tolerate strings of 4 or more Ts within the siRNA sequence, unlike pol. III, which will terminate transcription after incorporation of a stretch of Ts. Second, the CMV promoter does not interfere with other transcription events (such as expression of the antibiotic resistance gene), making it easier to perform long-term gene silencing studies (Xia *et al.* 2002; Jonathan *et al.* 2007).

The zebrafish is a simple vertebrate that has many attributes that make it ideal for the study of the immune system (Yoder *et al.* 2002). Long dsRNA causes nonspecific regression (Zhao *et al.* 2001), U6 promoter-driven shRNA is not high efficient (Xie *et al.* 2005), and siRNA last short time *in vivo*, which hinder the application of RNAi technique in fish.

In the current study, we explored the CMV promoter to drive high-level expression of shRNA molecules in zebra-fish embryos. We constructed the plasmid vectors targeting the exogenous expressing enhanced green fluorescent protein (EGFP) gene and the endogenous no tail (NTL) gene. The inhibition assays used phenotype observation and real-time fluorescent quantitative reverse-transcription polymerase chain reaction (Q-PCR).

Materials and Methods

Introducing CMV promoter sequence

All the oligonucleotides in this study were synthesized on an applied biosystem model 380B automated DNA synthesizer by Shanghai Sangon Biological Engineering Technology & Services Co., Ltd. (Table 1; Figure 1a-c). All the following inserts, including the CMV promoter and SV40 transcriptional termination sequence produced by PCRs and shRNAs chemically synthesized, were sequenced on a Model 3730 DNA Sequence System (Shanghai Invitrogen Co., Ltd., China) to guarantee the correct sequences.

The CMV promoter was obtained and introduced restriction sites both upstream and downstream of the sequence by PCR, using the pCMV-EGFP plasmid (Clontech, USA) as template. A PCR sense primer was SCF93a with *Xho* I site, and an antisense primer was SCR94 with *Hin*d III, *Apa* I, and *Sal* I sites (Table 1). The PCR product of 630 bp was purified with gel extraction kit (Axygen), ligated into pMD18-T vector (TAKARA), transformed into TOP10 competent cells. Three positive colonies were selected and sequenced for verification of the insert without mutation. The plasmid with correct insert was extracted and digested with *Xho* I and *Hin*d III; meanwhile, the pCMV-EGFP plasmid was digested with the same enzymes. The target fragments were purified, ligated with T4 ligase, and named as pCMV-EGFP-CMV.

Inserting SV40 transcriptional termination sequence

The PCR to get the SV40 transcriptional termination site was set up with pCMV-EGFP plasmid as template. The forward primer was SVF81a with *Apa* I site, and the reverse primer was SVR95a with *Hin*d III site (Table 1). The amplicon was performed and confirmed as above. The plasmid was digested with *Apa* I and *Hin*d III, and cloned into the pCMV-EGFP-CMV plasmid. The target fragments were purified, ligated, and named as pCMV-EGFP-CMV-SV40.

Table 1 Oligonucleotides used in this study

Name	Sequence
SCF93a	CTCGAGTAGTTATTAATAGTAATCAATTACG
SCR94a	AAGCTTATGGGCCCGTCGACCGATCTGACGGTTCACTA
SVF81a	GGGCCCAGCGGCCGCGACTCTAGATCAT
SVR95a	AAGCTTGCAGTGAAAAAAATGCTTTATTTGTG
SBAF86	GATGATGAAATTGCCGCACTG
SBAR87	ACCAACCATGACACCCTGATGT
SGF114	CAAGCAGAAGAACGGCATCA
SGR115	AGGTAGTGGTTGTCGGGCA
SNF122	CAGCACTGACAACCAGCAATC
SNR123	GAACCCGAGGAGTGAACAGG

a

		Sal I	Sense strand	Loop	Reverse complement	*Apa* I		
SGFP97	5′ -	TCGAC	GCAAGCTGACCCTGAAGTTC	TTCAAGAGA	GAACTTCAGGGTCAGCTTGC	GGGCC	–3′	Upper oligos
SGFP98	3′ -	C	GCAAGCTGACCCTGAAGTTC	TTCAAGAGA	GAACTTCAGGGTCAGCTTGC	G	–5′	Lower oligos

b

		Sal I	Sense strand	Loop	Reverse complement	*Apa* I		
SGFP97a	5′ -	TCGAC	AGCATCTAAGGCGACCTCGT	TTCAAGAGA	GGTCAGTAGGTCAACCTCCA	GGGCC	–3′	Upper oligos
SGFP98a	3′ -	C	AGCATCTAAGGCGACCTCGT	TTCAAGAGA	GGTCAGTAGGTCAACCTCCA	G	–5′	Lower oligos

c

		Sal I	Sense strand	Loop	Reverse complement	*Apa* I		
SNTL120	5′ -	TCGAC	TGCAATGTACTCGGTCCTG	TTCAAGAGA	CAGGACCGAGTACATTGCA	GGGCC	–3′	Upper oligos
SNTL121	3′ -	C	TGCAATGTACTCGGTCCTG	TTCAAGAGA	CAGGACCGAGTACATTGCA	G	–5′	Lower oligos

d

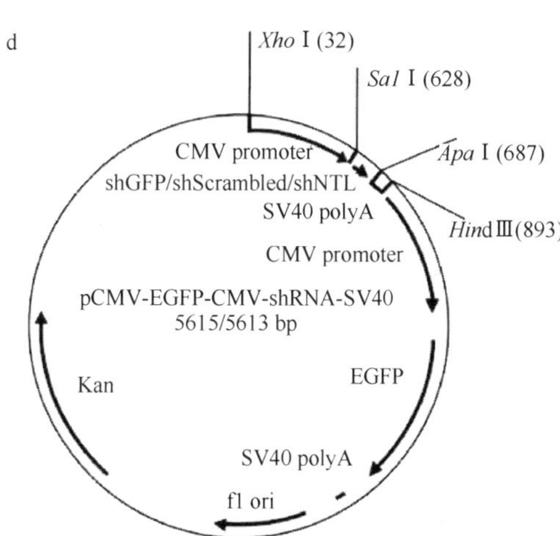

Fig. 1 The constructs of shRNA expression vectors. a. Illustration of shRNA expression plasmid, containing CMV promoter, SV40 transcription termination sequence, shRNA template, restriction sites, and the skeleton component of original pCMV-EGFP plasmid. b-d. The transcription template sequence of shGFP, shScrambled, shNTL unit, respectively

shRNA design and expression vectors construct

The EGFP and NTL siRNA sequences used for construction of shRNA vectors had previously been reported to be effective in silencing gene expression (Tiscornia *et al.* 2003;

Liu *et al.* 2005). shRNA oligonucleotides were designed as a synthetic duplex with overhanging ends identical to those created by restriction enzyme digestion (upper oligo: *Sal* I at the 5′ and *Apa* I at the 3′). The shRNA coding region contained a sense strand of 20 (EGFP), 19 (NTL), or 20 (scrambled) nucleotide sequences followed by a short spacer (TTCAAGAGA), and the reverse complement sequence of the sense strand (Fig. 1a-c). Forward and reverse oligos for EGFP, Scrambled and NTL were SGFP97, SGFP98, SGFP97a, SGFP98a, SNTL120, and SNTL121, respectively (Fig. 1a-c). The nonspecific control was shuttling DNA sequence of siGFP and analyzed by BLASTN. Each oligo was suspended in water (50 mM) and 5 μl from each (SGFP97 and SGFP98 for EGFP) was put together, the mix was heated to 95℃ for 5 min, and then slowly equilibrated to room temperature. The duplex was inserted to the *Sal* I and *Apa* I sites of pCMV-EGFP-CMV-SV40. Similarly, the scrambled and NTL shRNA vectors were produced, respectively. The shRNA vectors for EGFP, scrambled, and NTL were confirmed by sequencing and called after pCMV-EGFP-CMV-shGFP-SV40, pCMV-EGFP-CMV-shScrambled-SV40, and pCMV-EGFP-CMV-shNTL-SV40 (Fig. 1d), respectively.

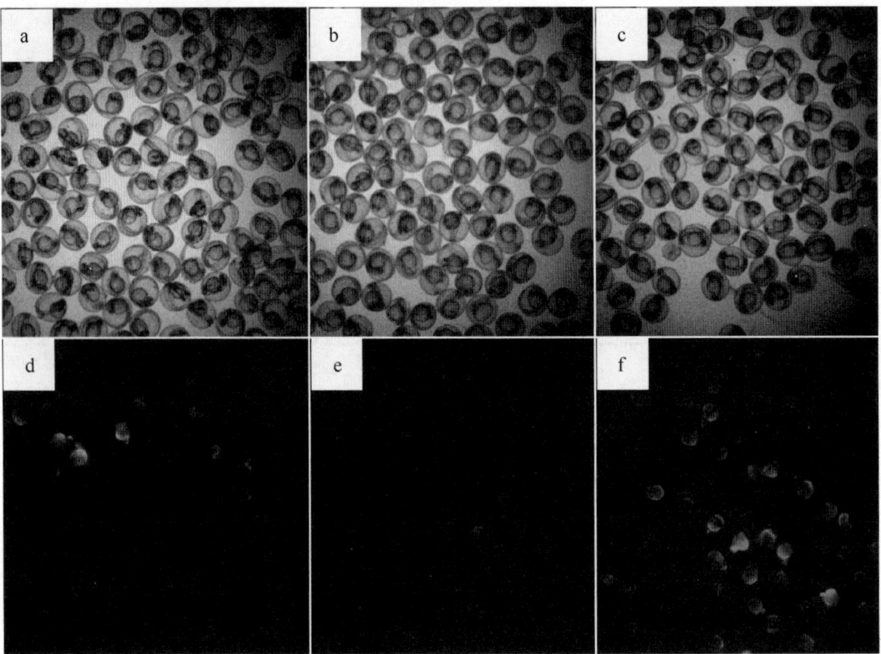

Fig. 2 The effect of shGFP and shScrambled constructs injection on the expression of EGFP in zebrafish embryos at 48 hpf. Upper row (a-c) showed the same field of embryos by light microscopy as seen under fluorescence in lower row (d-f). Column 1 was injected with EGFP expression vector (pCMV-EGFP). Column 2 was injected with shGFP construct (pCMV-EGFP-CMV-shGFP-SV40). Column 3 was injected with shScrambled construct (pCMV-EGFP-CMV-shScrambled-SV40). shScrambled construct did not interfere with EGFP expression and these embryos hardly showed the no tail phenotype

Microinjection of zebrafish embryos

After artificial insemination, approximately 1 nl of shRNA expression vector solution (100 pg/nl) was microinjected into the embryo at the 1- to 2-cell stage under a dissecting microscope using a pulled microcapillary pipette as previously report (Wu *et al.* 2006; Liu *et al.* 2006). We injected at least 450 embryos and cultured in three dishes for every group. Totally we used approximately 3,000 embryos. The injected and uninjected embryos were subsequently incubated in sterile 0.3 × Danieau's solution (19.3 nM NaCl, 0.23 mM KCl, 0.13 mM $MgSO_4 \cdot 7H_2O$, 0.2 mM $Ca(NO_3)_2$, 1.67 mM HEPES pH 7.2) at 28℃. We changed the solution and removed dead embryos twice per day.

Phenotypic analyses

Phenotypes were evaluated at 48 hpf (hours post fertilization). Images of zebrafish embryos were recorded by using an Olympus SZX12 fluorescent microscope and a digital camera (Liu *et al.* 2007). Features analyzed for NTL were: disrupted notochord, abnormal somites, and reduced tail (Halpern *et al.* 1993). EGFP expression was monitored by using fluoresce microscopy. For phenotype count and fluorescent observation, we used all the embryos.

Q-PCR assays

Target mRNA levels in embryos were quantitatively analyzed by using Q-PCR. Zebrafish embryos injected with pCMV-EGFP-CMV-shGFP-SV40, pCMV-EGFP-CMV-shNTL-SV40, pCMV-EGFP-CMV-shScrambled-SV40, and pCMV-EGFP and uninjected controls were collected at 48 hpf, and total RNA was extracted from pools of 30 to 40 embryos per sample with triplicate repeat using Trizol (Invitrogen). The RNA samples were further treated with RNase-free DNase I (Roche) to remove contaminated genomic DNA, followed by phenol/chloroform extraction and ethanol precipitation. A total of 1 μg RNA from each extraction was reverse transcribed with Superscript III reverse transcriptase (Invitrogen) and random primers in a 20-μl reaction volume according to the manufacturer's instructions. cDNA samples were diluted 1∶5 before use in Q-PCR assays.

Q-PCRs were performed in an ABI Prism 7000 Sequence Detection System (Applied Biosystems) using the following thermal cycling profile: 95℃ 3 min, followed by 40 cycles of amplification (95℃ 15 s, 58℃ 15 s, 72℃ 45 s), followed by dissociation curve analysis to validate the amplification of a single product. Each reaction consisted of 4 μl of cDNA sample, 4 μl of nuclease-free water, 10 μl of SYBR Green PCR master mix (Toyobo), and 1 μl of each primer set (2.5 μM). All reactions were performed in triplicate. All primers were designed to conform to a universal cycling program by the Primer Express 2.0 software (Applied Biosystems). The house-keeping gene β-actin was used as an internal standard to normalize differences in template amounts. Primers for β-actin were upstream primer

SBAF86 and downstream primer SBAR87 (Table 1). The PCR product size was 135 bp. The forward primer sequence for EGFP was SGF114 and the reverse primer was SGR115 (Table 1). The PCR product of the EGFP gene was 138 bp. NTL primer sequences were forward SNF122 and reverse SNR123 (Table 1). The NTL amplicon was 126 bp.

Relative expression was calculated by using a modified comparative cycle threshold (CT) method, in which CT was defined as the cycle numbers at which fluorescence reached a set threshold value. The differences in the CT value of the target genes from the corresponding internal control β-actin gene, ΔCT ($CT_{gene}-CT_{actin}$), were calculated. The changes in ΔCT of the experiment group to the control group, ΔΔCT ($\Delta CT_{sample}-\Delta CT_{control}$), were computed. The relative expression level of the experiment group to the control group was described using the equation $2^{-\Delta\Delta CT}$, and the value standed for a 1/*n*-fold difference relative to the control. The expression of experiment group relative to the control group was multiplied by 100% to simplify the presentation of the data.

Statistical analysis

The data obtained from Q-PCR analysis were subjected to an unpaired, two-tailed, Student's *t* test. Error bars indicated standard errors. $P<0.05$ was considered statistically significant, and $P<0.01$ was a very significant difference.

Results

Construction of shRNA expression vectors

Using the pCMV-EGFP as template, PCR based cloning strategies were performed to generate another set of CMV promoter, SV40 transcription termination sequence, and suitable restriction sites. Insert the DNA template for expressing shRNA between CMV promoter and SV40 terminator to produce shRNA expression vectors targeting EGFP and NTL. Similarly, the nonspecific control shRNA vector (pCMV-EGFP-CMV-shScrambled-SV40) also was generated (Fig. 1d). All final shRNA expression constructs consisted of a CMV promoter, shRNA sense sequence, loop sequence, shRNA antisense sequence, SV40 termination sequence, and the pCMV-EGFP vector skeleton. The transcripted shRNA composed of two complementary 19- or 20-nucleotide sequence motifs in an inverted orientation, separated by a 9-bp spacer to form a hairpin dsRNA.

Activity of CMV promoter measured by knockdown of exogenous EGFP expression

To analyze the function of CMV promoter, the level of EGFP expression in zebrafish embryos injected pCMV-EGFP-CMV-shGFP-SV40 was directly compared with that injected pCMV-EGFP-CMV-shScrambled-SV40 and pCMV-GFP. Knockdown of EGFP in embryos was visualized by fluorescence microscopy (Fig. 2). Results showed that the fluorescence in the embryos injected pCMV-EGFP-CMV-shGFP-SV40 exhibited reduction compared with that injected pCMV-EGFP-CMV-shScrambled-SV40 and pCMV-GFP.

Q-PCR was used to quantify the EGFP mRNA transcript. The relative EGFP mRNA expression level reduced 70.05± 1.76% (P<0.01) in the experiment group. The mRNA transcripts were not significant differences between pCMV-EGFP-CMV-shScrambled-SV40 group and pCMV-GFP group (P>0.05; Fig. 3).

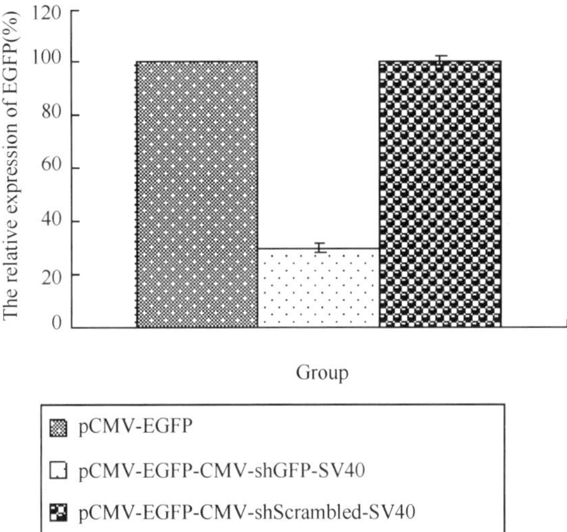

Fig. 3 Q-PCR analysis of exogenous EGFP mRNA expression levels in zebrafish embryos injected with 100 ng/μl plasmids. Data showed mean values±SE (n=3). Error bars indicated standard error

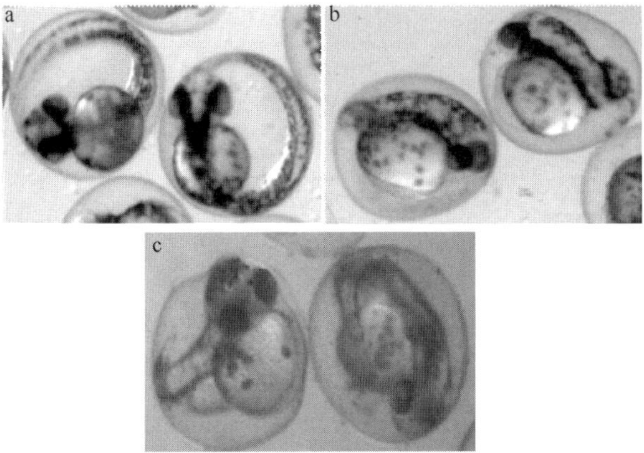

Fig. 4 The phenotypes of normal and no tail embryos. a. Wild-type embryos. b. Typical NTL mutant embryos by injection of pCMV-EGFP-CMV-shGFP-SV40. c. Intermediate phenotypes of NTL

Function of CMV promoter measured by knockdown of endogenous NTL expression

To further validate the function of the CMV promoter for RNAi in fish, shRNA expression vector targeting the endogenous NTL gene was produced. The embryos of no tail phenotype (Fig. 4) were counted under a dissection microscope, and occupied 30±4% at 48 hpf in pCMV-EGFP-CMV-shNTL-SV40 group, and 1±0.11% in other groups ($P<0.05$). The relative mRNA expression levels were determined by Q-PCR in the embryos injected pCMV-EGFP-CMV-shNTL-SV40 and compared with those injected the irrelevant control plasmid pCMV-EGFP-CMV-shScrambled-SV40 and uninjected groups. The NTL mRNA expression decreased 54.52±2.05% by comparison with the control ($P<0.01$; Fig. 5).

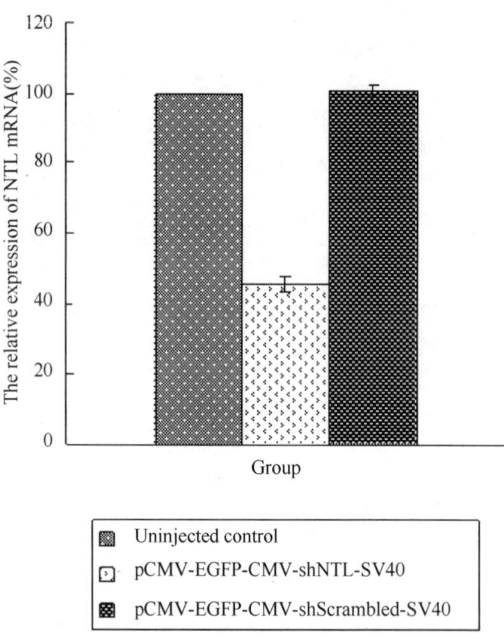

Fig. 5 Q-PCR analysis of endogenous NTL mRNA transcription level in zebrafish embryos injected with 100 ng/μl plasmids. Data showed mean values ± SE (n=3). Error bars indicated standard error

Discussion

RNAi, the promising therapeutic tool, has achieved remarkable progress on human diseases, especially in viral infections, cancers, and inherited genetic disorders (Wang *et al.* 2006). RNAi in mammalian cells is achieved through introduction or expression of 21-23 bp siRNAs (Banan and Puri 2004). Permanent gene suppression can be achieved by siRNAs as stem-loop precursors transcribed from pol. III promoter, such as H1 and U6 based on vector. This approach, however, has a major drawback: inhibition cannot allow for time- or tissue-specific manner because the pol. III promoter is ubiquitously expressed (Yuan *et al.* 2006). To overcome these limitations, pol. II promoter can be employed (Shinagawa and Ishii 2003); however, RNAi technique is still in the exploring stage in fishes.

Some findings suggested that dsRNA-mediated interference could be used to analyze the functional roles of genes in zebrafish, and the interference of gene function showed a strong dependence on the amount of dsRNA (Wargelius *et al.* 1999; Acosta *et al.* 2005). Other reports indicated that dsRNA had a nonspecific effect at the posttranscriptional level, and it seemed that RNAi was not a viable technique for studying gene function in zebrafish embryos (Oates *et al.* 2000; Li *et al.* 2000; Zhao *et al.* 2001; Mangos *et al.* 2001). Maybe some siRNAs, generated by RNAi mechanism from the dsRNAs of various genes or different region of the sequences, do not have other corresponding homologous genes to be inactivated, whereas others do.

siRNA-based gene silencing was effective in rainbow trout embryos (Boonanuntanasarn *et al.* 2003), zebrafish embryos (Dodd *et al.* 2004; Liu *et al.* 2005), and fathead minnow cell (Xie *et al.* 2005). siRNAs effectively inhibited tiger frog virus replication in fish cells (Xie *et al.* 2005). However, siRNA duplexes were unstable *in vivo*, therefore any effects of injected siRNA were short-term and dose-dependent. For example, NTL mRNA in zebrafish embryos injected with 4 μg/μl siNTL was reduced by 83% at 50% epiboly stage (5.25 hpf) and by 12% at 24-somite stage (22 hpf). No tail phenotype was 15% in injected 450 ng/μl siNTL group and 59% in 4 μg/μl group (Liu *et al.* 2005).

Although pol. III U6 promoter worked in fish cell, the silencing effect caused by siRNAs was approximately 8-fold that of shRNAs driven by U6 promoter (Xie *et al.* 2005). This was probably because the transfection efficiency for siRNAs was generally much greater than that of large plasmid vectors in fish cell.

In our previous study, T7 transcription system was used to drive the expression of shRNA in zebrafish embryos (Wang *et al.* 2007). When pT7shGFP vector was injected into the transgenic embryos stably expressing T7 RNA polymerase (T7RP), EGFP relative expression level showed a decrease of 68% by analysis of Q-PCR. Injection of pT7shNTL vector in zebrafish embryos expressing T7RP led to partial absence of endogenous ntl transcripts in 30% of the injected embryos when detected by whole mount in situ hybridization. Because the activity of T7 promoter depends on the T7RP, we first obtained the transgenic fish stably expressing T7RP driven by CMV promoter, then studied that T7 promoter drove shRNA expression. This procedure was long.

In this work, we designed a strategy that allowed synthesis of shRNA by pol. II promoter CMV to efficiently knock down expression of both exogenous and endogenous genes in zebrafish *in vivo*. The off-targeting of the shRNA construct driven by CMV promoter did not significantly affect the expression of the target gene.

Our expression vector (derived from pCMV-EGFP, Clontech) contained a CMV promoter-driven EGFP expression open reading frame and transcription termination sequence of SV40, which ensured the equal transgenic efficiencies of shRNA and EGFP, and was convenient to supervise the transgenic embryos under a fluorescent micro-scope. We introduced another set of CMV promoter and transcription termination sequence—between them the shRNA expression duplex was inserted. Each duplex contained a *Sal* I site, an shRNA encoding region (sense stem, loop sequence, and anti-sense stem), and an *Apa* I site.

As assayed by both fluorescence observing and Q-PCR, the protein and mRNA products of exogenous gene EGFP were effectively inhibited (up to 70.05±1.76%); phenotype count and Q-PCR results respectively demonstrated that endogenous NTL protein and mRNA was suppressed without global down-regulation of protein synthesis (up to 54.52±2.05%). When siNTL was injected with 450 ng/μl, the no tail phenotype occupied 15% (Liu *et al*. 2005). In present study, the shNTL was injected with 100 ng/μl. The plasmid was 5613 bp, so the efficient siNTL concentration was approximately 0.677 ng/μl (((19×2)×100 ng/μl)/5613), which was 1/665 (0.677/450) of the reported siNTL concentration, but the no tail phenotype was 30±4%. Compared with our previous work (Wang *et al.* 2007), the efficiencies were similar, but the process in present study was simple. These results appeared that CMV-driven shRNA vector was preferable to siRNA and T7 transcription system in inhibition assay in fish *in vivo*.

Normally the CMV promoter is long lasting, because the construction can integrate the genome, continuously transcript RNA, and inherit next generation. Cecropin B, an insect antimicrobial peptide, was expressed in medaka, driven by a CMV promoter. The F2 transgenic fish had acquired elevated resistance to bacterial infection (Sarmasik *et al*. 2002). In present study, we combined the activity of CMV promoter with shRNA sequence. Actually, we also tested the expressions at day 5 after injection. The inhibition efficiency reduced a little. To compare with other results mentioned, we just showed the results at 48 h after injection.

Considered together, the CMV promoter, the pol. II promoter, could be used in the construction of plasmid based shRNA expression vector. The CMV promoter-driven shRNA constructs could lead to effectively and specifically inhibition of exogenous and endogenous gene expressions in zebrafish *in vivo*. These vectors efficiently induced RNAi in zebrafish through production of shRNA molecules targeted at the exogenous expressing reporter gene EGFP, and the endogenous zebrafish NTL gene. That the use of this promoter sequence and the shRNA vector cloning strategy described will be advantageous in RNAi functional genomic experiments and an important step in the development of novel RNAi technology for immune therapeutics in the transgenic delivery of shRNA molecules in fishes. This method may provide a novel perspective for the application of RNAi technology in suppressing gene expression in fish.

Acknowledgements The authors thank Miss Yuejiao Lu for critically reading the manuscript. The technical assistance provided by Ming Li, Na Wang, Jun Dai, Shangping

Chen, and other laboratory members was greatly appreciated. This work was supported by (30428024 and 30540084) from National Natural Science Foundation of China, (KSCX2-YW-N-021) from Chinese Academy of Sciences, (08080262, 08080245, and 01140508) from Northwest A & F University in China, (20070410298) from China Postdoctoral Science Foundation and (2007FB09) from Institute of Hydrobiology, CAS.

References

Acosta J, Carpio Y, Borroto I, Gonzalez O, Estrada MP (2005) Myostatin gene silenced by RNAi show a zebrafish giant phenotype. J Biotechnol 119: 324-331

Ashrafi K, Chang FY, Watts JL, Fraser AG, Kamath RS, Ahringer J, Ruvkun G (2003) Genome-wide RNAi analysis of Caenorhabditis elegans fat regulatory genes. Nature 421: 268-272

Banan M, Puri N (2004) The ins and outs of RNAi in mammalian cells. Curr Pharm Biotechnol 5: 441-450

Boonanuntanasarn S, Yoshizaki G, Takeuchi T (2003) Specific gene silencing using small interfering RNAs in fish embryos. Biochem Biophys Res Commun 310: 1089-1095

Brummelkamp TR, Bernards R, Agami R (2002) A system for stable expression of short interfering RNAs in mammalian cells. Science 296: 550-553

Ding S, Li H, Lu R, Li F, Li W (2004) RNA silencing: a conserved antiviral immunity of plants and animals. Virus Res 102: 109-115

Dodd A, Chambers SP, Love DR (2004) Short interfering RNA-mediated gene targeting in the zebrafish. FEBS Lett 561: 89-93

Dykxhoorn DM, Novina CD, Sharp PA (2003) Killing the messenger: short RNAs that silence gene expression. Nat Rev Mol Cell Biol 4: 457-467

Elbashir SM, Harborth J, Lendeckel W, Yalcin A, Weber K, Tuschl T (2001) Duplexes of 21-nucleotide RNAs mediate RNA interference in cultured mammalian cells. Nature 411: 494-498

Fire A, Xu S, Montgomery MK, Kostas SA, Driver SE, Mello CC (1998) Potent and specific genetic interference by double-stranded RNA in *Caenorhabditis elegans*. Nature 391:806-811

Geiduschek EP, Kassavetis GA (2001) The RNA polymerase III Transcription Apparatus. J Mol Biol 310: 1-26

Gil J, Esteban M (2000) Induction of apoptosis by the dsRNA-dependent protein kinase (PKR): mechanism of action. Apoptosis 5:107-114

Halpern ME, Ho RK, Walker C, Kimmel CB (1993) Induction of muscle pioneers and floor plate is distinguished by the zebrafish no tail mutation. Cell 75: 99-111

Hannon GJ (2002) RNA interference. Nature 418:244-251

Jonathan EP, Foster JS, Kestler D, Alan S, Wall JS (2007) Inhibition of Bence-Jones protein synthesis by RNA interference. J Immunol 178: S88-S89

Kondraganti S, Gondi CS, McCutcheon I, Dinh DH, Gujrati M, Rao JS, Olivero WC (2006) RNAi-mediated downregulation of urokinase plasminogen activator and its receptor in human meningioma cells inhibits tumor invasion and growth. Int J Oncol 28: 1353-1360

Liu J, Sun Y, Wang N, Wang Y, Zhu Z (2006) Upstream regulatory region of zebrafish lunatic fringe: isolation and promoter analysis. Mar Biotechnol 8: 357-365

Li YX, Farrell MJ, Liu R, Mohanty N, Kirby ML (2000) Double-stranded RNA injection produces null phenotypes in zebrafish. Dev Biol 217: 394-405

Liu WY, Wang Y, Qin Y, Wang YP, Zhu ZY (2007) Site-directed gene integration in transgenic zebrafish mediated by cre recombinase using a combination of mutant lox sites. Mar Biotechnol 9: 418-420

Liu WY, Wang Y, Sun YH, Wang Y, Wang YP, Chen SP, Zhu ZY (2005) Efficient RNA interference in zebrafish embryos using siRNA synthesized with SP6 RNA polymerase. Dev Growth Differ 47: 323-331

Mangos S, Vanderbeld B, Krawetz R, Sudol K, Kelly GM (2001) Ran binding protein RanBP1 in zebrafish embryonic development. Mol Reprod Dev 59: 235-248

McIntyre GJ, Fanning GC (2006) Design and cloning strategies for constructing shRNA expression vectors. BMC Biotechnol 6: 1-8

Napoli C, Lemieux C, Jorgensen R (1990) Introduction of a Chimeric Chalcone Synthase Gene into Petunia Results in Reversible Co-Suppression of Homologous Genes *in trans*. Plant Cell 2: 279-289

Oates AC, Bruce AE, Ho RK (2000) Too much interference: injection of double-stranded RNA has nonspecific effects in the zebrafish embryo. Dev Biol 224: 20-28

Sarmasik A, Warr G, Chen T (2002) Production of transgenic medaka with increased resistance to bacterial pathogens. Mar Biotechnol 4: 310-322

Schramm L, Hernandez N (2002) Recruitment of RNA polymerase III to its target promoters. Genes Dev 16: 2593-2620

Shinagawa T, Ishii S (2003) Generation of Ski-knockdown mice by expressing a long double-strand RNA from an RNA polymerase II promoter. Genes Dev 17: 1340-1345

Tiscornia G, Singer O, Ikawa M, Verma IM (2003) A general method for gene knockdown in mice by using lentiviral vectors expressing small interfering RNA. Proc Natl Acad Sci USA 100: 1844-1848

Tuschl T (2002) Expanding small RNA interference. Nat Biotechnol 20: 446-448

Wang N, Sun Y, Liu J, Wu G, Su J, Wang Y, Zhu Z (2007) Knock down of gfp and no tail expression in zebrafish embryo by *in vivo*-transcribed short hairpin RNA with T7 plasmid system. J Biomed Sci 14: 767-776

Wang L, Wu G, Yu L, Yuan J, Fang F, Zhai Z, Wang F, Wang H (2006) Inhibition of CD147 expression reduces tumor cell invasion in human prostate cancer cell line via RNA interference. Cancer Biol Ther 5: 608-614

Wargelius A, Ellingsen S, Fjose A (1999) Double-stranded RNA induces specific developmental defects in zebrafish embryos. Biochem Biophys Res Commun 263: 156-161

Wu Y, Zhang G, Xiong Q, Luo F, Cui C, Hu W, Yu Y, Su J, Xu A, Zhu Z (2006) Integration of double-fluorescence expression vectors into zebrafish genome for the selection of site-directed knockout/knockin. Mar Biotechnol 8: 304-311

Xia H, Mao Q, Paulson H, Davidson B (2002) siRNA-mediated gene silencing in vitro and in vivo. Nat Biotechnol 20: 1006-1010

Xie J, Lu L, Deng M, Weng S, Zhu J, Wu Y, Gan L, Chan SM, He J (2005) Inhibition of reporter gene and Iridovirus-tiger frog vinus in fish cell by RNA interference. Virology 338: 43-52

Yoder JA, Nielsen ME, Amemiya CT, Litman GW (2002) Zebrafish as an immunological model system. Microbes Infect 4: 1469-1478

Yuan J, Wang X, Zhang Y, Hu X, Deng X, Fei J, LiN (2006) shRNA transcribed by RNA Pol II promoter induce RNA interference in mammalian cell. Mol Biol Rep 33: 43-49

Zhao Z, Cao Y, Li M, Meng A (2001) Double-stranded RNA injection produces nonspecific defects in zebrafish. Dev Biol 229: 215-223

CMV 启动子驱动的 shRNA 构建体在斑马鱼中介导高效的 RNA 干扰

苏建国[1,2] 朱作言[1] 汪亚平[1] 熊 凤[1] 邹 钧[3]

1 中国科学院水生生物研究所，淡水生态与生物技术国家重点实验室，武汉 430072，中国
2 西北农林科技大学，杨凌，712100，中国
3 School of Biological Sciences, University of Aberdeen, Aberdeen AB24 2TZ, UK

摘 要 RNA 干扰沉默目标基因的表达在生命科学研究中取得了重要进展。本研究应用 CMV 启动子驱动 DNA 模板的方法，诱导短发夹结构 RNA(shRNA)介导的 RNA 干扰，进而抑制外源增强型绿色荧光蛋白(EGFP)和内源 NTL 基因的表达。我们构建了 pCMV-EGFP-CMV-shGFP-SV40，pCMV-EGFP-CMV-shNTL-SV40 和 pCMV-EGFP-CMV-shScrambled-SV40 等 3 个质粒，它们都包含一个 CMV 启动子驱动的 EGFP 报告基因和由另一个 CMV 启动子驱动的 shRNA。将这 3 个 shRNA 编码质粒和 pCMV-EGFP 对照质粒显微注射到斑马鱼胚胎中，注射后 48h 收集样品，通过表型观察和定量 PCR 方法检测 shRNA 的作用。荧光显微镜观察发现，注射的 shGFP 编码质粒显著抑制了 EGFP 的表达；定量 PCR 检测结果表明，与对照样品相比外源 EGFP 基因的 mRNA 水平下降了 70.05 ± 1.26%。与对照 shRNA 相比，靶向 NTL 基因的 shRNA 导致 30 ± 4%的胚胎产生明显的 NTL 表型，其 mRNA 水平下降 54.52 ± 2.05%。这些结果表明，CMV 启动子驱动的 shRNA 技术可用于在斑马鱼中抑制外源和内源基因的表达。

Molecular Cloning of Growth Hormone Receptor (GHR) from Common Carp (*Cyprinus carpio* L.) and Identification of Its Two Forms of mRNA Transcripts

SUN Xiaofeng[1,2] GUO Qionglin[1] HU Wei[1]
WANG Yaping[1] ZHU Zuoyan[1]

1 State Key Laboratory of Freshwater Ecology and Biotechnology, Institute of Hydrobiology, Chinese Academy of Sciences, Wuhan 430072
2 Graduate School of Chinese Academy of Sciences, Beijing 100039

Abstract The cDNA of growth hormone receptor (GHR) was cloned from the liver of 2-year common carp (*Cyprinus carpio* L.) by reverse transcription-polymerase chain react ion (RT-PCR) and rapid amplification of cDNA end (RACE). Its open reading frame (ORF) of 1806 nucleotides is translated into a putative peptide of 602 amino acids, including an extracellular ligand-binding domain of 244amino acids (aa), a single transmembrane domain of 24 aa and an int racellular signal-transduction domain of 334 aa. Sequence analysis indicated that common carp GHR is highly homologous to goldfish (*Carassius auratus*) GHR at both gene and protein levels. Using a pair of genespecific primers, a GHR fragment was amplified from the cDNA of 2-year common carp, a 224 bp product was identified in liver and a 321 bp product in other tissues. The sequencing of the products and the partial genomic DNA indicated that the difference in product size was the result of a 97 bp intron that alternatively spliced. In addition, the 321 bp fragment could be amplified from all the tissues of 4-month common carp including liver, demonstrating the occurrence of the alternative splicing of this intron during the development of common carp. Moreover, a semi-quantitative RT-PCR was performed to analyze the expression level of GHR in tissues of 2-year common carp and 4-month common carp. The result revealed that in the tissues of gill, thymus and brain, the expression level of GHR in 2-year common carp was significantly lower than that of 4-month common carp.

Keywords growth hormone receptor (GHR); molecular cloning; transcript; expression analysis; common carp

Growth hormone receptor (GHR) belongs to the hematopoietic receptor superfamily[1]. The action of growth hormone (GH) in regulating growth[2], reproduction[3] and immunity[4] has been elucidated. The binding of GH to the GHR on target tissues triggers a cascade of tyrosine and protein phosphorylation events, which culminates in the biological action of GH[5-6]. Up to date GHR cDNAs have been cloned from many species[7-9], including various

文章发表于《自然科学进展：国际材料》(英文), 2006, 16(11): 1156-1163

kinds of mammalian animals; avian of chicken and domestic pigeon; reptilian of soft-shelled turtle (*Pelodiscus sinensis japonicus*) and amphibian of African clawed frog (*Xenopus laevis*). In the year of 2001, Lee *et al*. cloned the full length cDNA of GHR from goldfish (*Carassius auratus*)[10]. This is the first report about the cloning of GHR in teleost fish. Subsequently Tse-successfully cloned the GHR cDNA from black seabream (*Acanthopagrus schlegelii*), and found two kinds of GHR cDNAs in all the tissues, which resulted from a 93 bp intron alternatively spliced[11]. The diversity of GHR has been proved in mammals. For example, 8 GHR mRNAs (V1-V8) have been identified in human[12], 10 different GHR cDNAs were cloned from a cow endometrium cDNA library, and the expression of the 5′UTR of GHR in ratliver has shown sexual dimorphism. Besides black seabream, two GHR cDNAs have been successfully cloned from rainbow trout (*Oncorhynchus mykiss*) recently[13].

In this study, we cloned GHR from common carp (*Cyprinus carpio* L.) and identified its two forms of transcripts, which will provide more valuable information on the gene structure, function and expression of freshwater fish.

1. Material and Methods

1.1 Fish and sampling

The 2-year and 4-month common carp (*Cyprinus carpio* L.) were cultured in the Guanqiao Experimental Statition in Wuhan. When the fishes were transferred to the laboratory, the liver, spleen, kidney, headkidney, thymus, gill and brain were carefully removed and immediately stored in the liqulid nitrogen.

1.2 RT-PCR and RACE

About 50mg of each tissue was used for extraction of total RNAs, which was performed with Trizol (Invitrogen, Japan) following the user’s manual.

A primer GP1 (5′-CCATGGGTGGAGTTCATC-3′) was designed according to GHR Box2 region which is highly conservative. The oligo(dT) adaptor primer (AP) used for the first strain cDNA synthesis was 5′-GTTTTCCCAGTCACGAC(T) n-3′, so the specific adaptor primer AP was 5′-GTTTTCCCAGTCACGAC-3′. The first strain cDNA synthesis and the PCR amplification of 3′RACE were performed with TaKaRa RNA PCR kit (AMV)Ver 3.0, referring to the protocols. The template was from the liver of controlled carp. Then PCR was performed with the primer of GP1 and AP, under the condition of 94℃ denaturation for 5 min, running 30 cycles of 94℃ 30 sec; 60℃ 30 sec; 72℃ 1.5 min, and 72℃ elongation for 5 min.

To obtain the GHR cDNA with a complete coding region, the sense primer 5′UTR (5′-GAAACGATGTTCGGGTGATT-3′) was designed according to the 5′UTR of goldfish and grass carp (*Ctenopharyngoden idella*). At the same time the reverse gene specific primer GP2

(5′-CTCTGCAGGGTCATCAAGGT-3′) was designed according to the partial cDNA sequence of common carp obtained by 3′RACE. Two primers were used to amplify the coding region and part of non-coding region at 5′end.

1.3 Cloning and sequencing

PCR products were separated by agarose gel electrophoresis, and the amplified products were purified from the gels using the Glass Milk Extraction Kit (Fermentas). The extracted products were ligated into PMD18-T vector (TaKaRa) and used to transform competent *E. coli* DH5α cells. Positive colonies were screened by the method of PCR. The recombinant plasmids were sequenced by the dideoxy chain termination method with M13 universal primers. The data were automatically collected on the ABI PRISM3730 Genetic Analyzer.

1.4 Tissue expression of GHR

Of the 1 μg total RNA isolated from liver, spleen, kidney, headkidney, thymus, gill and brain of 2-year common carp was used to synthesize the first strain cDNA. The first strain cDNA was used as PCR amplification template with the primers of Gf: 5′-GTGCGTGAGAACATAACC-3′ and Gr: 5′-CAGTGGGAGTTGTTTCTG-3′ which could specifically amplify a part of GHR cDNA. The negative control contained no template. The expected amplified fragment size was 224 bp. Amplification of β-actin was as the internal reference in PCR. The primers for β-actin cDNA amplification were actinF 5′-CAGATCATGTTTGAGACC-3′ and actinR 5′-ATTGCCAATG-GTGATGAC-3′ which covered an intron in genome, and the expected amplified fragment from cDNA was 460 bp. The reaction was performed with an initial denaturation of 5 min at 94℃, followed by 30 cycles of 45 sec at 94℃; 30 sec at 60℃; 45 sec at 72℃. The final step was 10 min at 72℃. The PCR products were fractionated on a 1. 5% agarose gel.

The same procedures were performed using the total RNAs isolated from liver, spleen, kidney, headkidney, thymus, gill and brain of 4-month common carp.

1.5 Genomic DNA amplification

Genomic DNA was prepared from the liver of 2-year common carp[14]. The PCR was performed to trap the intron with the template of the genomic DNA and a pair of primers of Gf and Gr, under the condition of 94℃ denaturation for 10 min, running 30 cycles of 94℃ 45 sec; 60℃ 30 sec; 72℃ 1.5 min, and 72℃ elongation for 10 min.

1.6 Computer-aided sequence analysis of cloned DNA

The sequences were analyzed for similarity with other known sequences by BLAST program. The signal peptide prediction was performed by SignalP program. The protein family signature was identified by InterPro[15] program. The phylogenetic tree was constructed based on the full length amino acid sequences of partial known GHRs using neighbor-joining algorithm within MEGA version 3.0[16].

2. Results

2.1 Cloning of GHR cDNA from the liver of 2-year common carp

A partial sequence of 1102 bp was obtained by 3′RACE, and a product with the size of 1196 bp obtained using the primers of 5′UTR and GP2. The BLAST searches on the NCBI database indicated that the two sequences resemble to GHR cDNA of other species (http://www. ncbi. nlm. nih. gov/BLAST/). Putting the two fragments together and omitting the identical nucleotide sequence of the overlapping, the GHR cDNA including a complete coding region of 2252bp was obtained (GenBank accession number: AY741100) (Fig. 1) .

2.2 Expression of two transcripts in 2-year and 4-month common carp revealed an alternative splicing

RT-PCR was performed using the primers of Gf and Gr, and the result showed that the amplification product in the liver of 2-year common carp was 224bp in size, which differs from the transcripts in other tissues where a 321 bp fragment was identified. In all the tissues of 4-month common carp including liver, only the 321 bp fragment was identified (Fig. 2), which implies that a 97 bp sequence was deleted in the transcript of liver. The sequencing result confirmed this implication. However, the expression level in gill, thymus and brain of 2-year common carp was significantly lower when compared with that in 4-month common carp (Figs. 2 and 3).

Through the genomic DNA analysis by PCR and sequencing, we obtained the same result as indicated above. From the result, we considered that the alternative splicing of the intron might be correlated with the development of the common carp.

2.3 Homology and phylogenetic analysis

Amino acids sequence of GHR in common carp shared 92.2% identity to GHR of goldfish, 90% to grass carp, 36.4% to human, and 36.2% to rat. Morever, the extracellular domain is more conserved than the intracellular domain (Table 1).

A phylogenetic tree constructed based on the full length amino acid sequences of common carp and other known GHR showed the relationship of common carp with other species (Fig. 4).

1 GAAACGATGTT

12 CGGGTGATTTTTGAGGTTGATCTGACCACGTTTTTGCATCGTTTAAAAGGGGGAGAATAC

72 CGAGAGACCCACAACACGCAAGTCTGTTGATCCGGATGAACGAACTGTTGAGAAAGTAAA

132 AACTCGCAACAGATTTTTCTCGCGGACAACTTCTCTGGAGCTGAGGAGACAGCAGAAGCT

192 ATGGCTTACTCTCTCTCGCTCGGTCTGCTCTACCTGGGCTTGCTGTGTGGAAACGGACTG

M A Y S L S L G L L Y L G L L C G N G L

252 GTGTCTGCAAGATCCGAGCTGTTCACTCCAGATCCAAGCAGAGGACCTCATTTTACAGGC

V S A R S E L F T P D P S R G P H F T G

312 TGCCGCTCCAGAGAGCAGGAGACCTTCCGTTGCTGGTGGAGCGCTGGGATCTTCCAGAAC

C R S R E Q E T F R C W W S A G I F Q N

372 CTCACCGAGCCTGGAGCTCTCAGGGTCTTCTACCAGACAAAAAATTTCCTCTCTGAGTGG

L T E P G A L R V F Y Q T K N F L S E W

432 CAGGAGTGTCCAGACTACACACGTACTGTGAAAAATGAGTGCTACTTCAACAAAACCTTC

Q E C P D Y T R T V K N E C Y F N K T F

492 ACACAGATCTGGACCTCGTACTGCATTCAGCTGCGCTCAGTGCGTGAGAACATAACCTAT

T Q I W T S Y C I Q L R S V R E N I T Y

552 GACGAGGCCTGCTTTACAGTAGAGAACATAGTGCATCCTGACCCACCAATTGGGCTGAAC

D E A C F T V E N I V H P D P P I G L N

612 TGGACTCTATTAAATGTGAGTCGCTCGGGGTTGCACTTTGACGTCCTTGTGCGCTGGGCT

W T L L N V S R S G L H F D V L V R W A

672 CCCCCTCCGTCAGCAGATGTGCAGATGGGCTGGATGAGCCTGGTGTACCAGGTTCAGTAC

P P P S A D V Q M G W M S L V Y Q V Q Y

732 CGGGTCAGAAACAACTCCCACTGGGAAATGCTGGACCTGGAGAGTGGCACACAGCAGTCC

R V R N N S H W E M L D L E S G T Q Q S

792 ATCTACGGTTTACATACTGACAAAGAGTATGAAGTCCGGGTGCGCTGCAAGATGTCAGCC

I Y G L H T D K E Y E V R V R C K M S A

852 TTTGACAACTTTGGCGAATTCAGTGACAGCATCATTGTGCATGTGGCACAGATACCAAGC

F D N F G E F S D S I I V H V A Q I P S

912 AAAGAATCAACGTTCCCGACGACGTTGGTGTTGATTTTTGGAGTGATTGGAGTGGTGATT

K E S T F P T T L V L I F G V I G V V I

972 CTTCTTGTCCTCCTCATCTTCTCTCAACAACAGAGGTTGATGGTAATATTTTTACCACCT

L L V L L I F S Q Q Q R L M V I F L P P

1032 ATTCCTGCACCTAAAATAAAAGGCATCGACCCAGAGCTGCTGAAGAATGGAAAGCTTGAC

I P A P K I K G I D P E L L K N G K L D

1092 CAGCTCAATTCTTTGCTGAGCAGTCAGGATATGTACAAGCCGGACTTCTACCATGAGGAT

Q L N S L L S S Q D M Y K P D F Y H E D

1152 CCATGGGTGGAGTTCATCCAGCTGGACCTTGATGACCCTGCAGAGAAGAATGAGAGTTCT

P W V E F I Q L D L D D P A E K N E S S

1212 GATACACAACATCTGCTGGGCTTGTCTCGCTCAGGCTCTTCTCACTTCCTTAATTTCAAA

D T Q H L L G L S R S G S S H F L N F K

```
1272  AGTGACAACGATTCGGGTCGTGCTAGCTGCTACGACCCAGAAATCCCAAATCCCAAGGAC
       S  D  N  D  S  G  R  A  S  C  Y  D  P  E  I  P  N  P  K  D
1332  TTGGCTTCTTTTCTGCCTGGCCATTCAGGACGAGGAGATAACCACCCTCTGGTTTCCAGA
       L  A  S  F  L  P  G  H  S  G  R  G  D  N  H  P  L  V  S  R
1392  AGCAGCTCATCCATCCCTGATCTTGGTTTCCAGCAGACATCAGAAGTGGAGGAGACTCCC
       S  S  S  S  I  P  D  L  G  F  Q  Q  T  S  E  V  E  E  T  P
1452  ATTCAAACGCAACCAGCTGTGCCCAGCTGGGTTAACATGGACTTTTATGCCCAAGTAAGT
       I  Q  T  Q  P  A  V  P  S  W  V  N  M  D  F  Y  A  Q  V  S
1512  GATTTCACACCAGCAGGAGGTGTCGTGCTTTCACCTGGACAACTGAACAGCTCTCCAGTG
       D  F  T  P  A  G  G  V  V  L  S  P  G  Q  L  N  S  S  P  V
1572  AAAAAGAAGGGAGAAGGGAATGAGAAGAAGATACAATTCCAGTTGCTTTCCGATGGAGCC
       K  K  K  G  E  G  N  E  K  K  I  Q  F  Q  L  L  S  D  G  A
1632  TACACCTCAGAGAACACAGCCAGGCTGCTTTCTGCCGATGTGCCACCCAGCCCTGGTCCT
       Y  T  S  E  N  T  A  R  L  L  S  A  D  V  P  P  S  P  G  P
1692  GAGCAGGGGTACCAAGCATTCCCAACCCAAGCCGTTGAGGGGAACCTCTGGAATGGTGAG
       E  Q  G  Y  Q  A  F  P  T  Q  A  V  E  G  N  L  W  N  G  E
1752  TACCTGGTGTCCGCCAATGATTCCCAGACGCCGTGCCTGGTTCCTGAAGCTCCTCCAGCC
       Y  L  V  S  A  N  D  S  Q  T  P  C  L  V  P  E  A  P  P  A
1812  CCCATACTGCCACCAGTATCAGACTATACTGTAGTGCAGGAAGTGGATGCCCAGCACAGC
       P  I  L  P  P  V  S  D  Y  T  V  V  Q  E  V  D  A  Q  H  S
1872  CTCCTCCTGAATCCTCCTTCCTCACAGCCTGCGATATGCCCTCACAGCCCAAACAAACAT
       L  L  L  N  P  P  S  S  Q  P  A  I  C  P  H  S  P  N  K  H
1932  CTCCCTGTAATCCCAACCATGCCCATGGGGTACCTCACCCCAGACCTTCTGGGAAACCTG
       L  P  V  I  P  T  M  P  M  G  Y  L  T  P  D  L  L  G  N  L
1992  AACCCATGAAGGGACTAAAAAGCATAAAGTTCATGGTCTTGTTGTACATTTTCACTGCTG
       N  P  *
2052  GAAAGTTGCATGAATGGCGTGACACGTGACGGAAATTCAGCACATGAATTTTCTCTGCTG
2112  TACGCAACAAAATGGAAGAAGCTGAAAACGTTTCTTTTCTGTCAGCTGTTGCAGTAGTGA
2172  TACGCTAGCTCACCGAGAGATTTTAAATGTGCTTTCGATTTACAAACAGAAACAGCTGGT
2232  TACATGCAAAAAAAAAAAAAA
```

Fig. 1 Nucleotide and deduced amino acid sequences of common carp GHR. Signal peptide predicted by SignalP is marked by a black line. Conserved cysteine residues in the extracellular domain are marked by pink letters. Potential *N*-glycosylation sites are marked by blue letters. Transmembrane domain is shadowed. Box1 and Box2 regions are double underlined

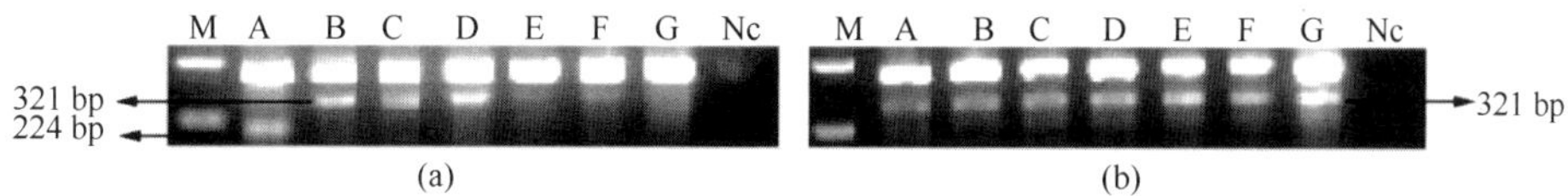

Fig. 2 Expression of two GHR genes in different tissues. (a) Expression of GHR in 2 year common carp. (b) Expression of GHR in 4-month common carp. M, molecular weight marker (DL2000); A, liver; B, spleen; C, kidney; D, headkidney; E, thymus; F, gill; G, brain; Nc, negative control without the template

3. Discussion

The GHR cDNA cloned from the liver of 2-year common carp was translated into a

4-mon:gtgcgtgagaacataacctatgacgaggcctgctttacagtggagaacataggtgagttaatagctgtgaatctg ctgaatgaagg

2-year:gtgcgtgagaacataacctatgacgaggcctgctttacagtggagaacata---

4-mon: tttacacagttacagaatgatctgtgttggaggtctgacttcaaatgttttgtgactttccagtgcatcctgacccaccaattgggctga

2-year: --gtgcatcctgacccaccaattgggctga

4-mon: actggactctattaaatgtgagtcgctcggggttgcactttgacgtccttgtgcgctgggctccccctccgtcagcagatgtgcaga

2-year: actggactctattaaatgtgagtcgctcggggttgcactttgacgtccttgtgcgctgggctccccctccgtcagcagatgtgcaga

4-mon: tgggctggatgagcctggtgtaccaggt tcagtaccgggtcagaaacaactcccactg

2-year: tgggctggatg agcctggtgtaccaggttcagtaccgggtcagaaacaactcccactg

Fig. 3 Comparison of the amplified products from the liver of 2-year and 4-month common carp using the primers of Gf and Gr. '-'means deleted nucleotides

transmembrane glycoprotein of 602 aa (Fig. 1), including an extra-cellular ligand-binding domain of 244 amino acids (aa), a single transmembrane domain of 24 aa and an intracellular signal-transduction domain of 334 aa. The characteristic landmark of GHR is the YGEFS motif found within common carp GHR in the position of aa224-228. In common carp, the first aa of this motif is a phenylalanine instead of tryosine as goldfish GHR[10]. So the motif turns to a FGEFS motif accordingly. The result of CLUSTAL showed that in mammalian GHRs the aa in the situation is tyrosine without exception. However, in avian, reptilian, amphibian and all the teleost the aa turns to phenylalanine. Though the significance of a phenylanine instead of a tyrosine in the motif is unknown yet, it is considered a conservative change for both aa residues containing an aromatic side chain[8].

In the extracellular domain of common carp GHR, the 6 conserved cysteine residues are believed to play significant roles. These cysteine residues are probably engaged in forming disulfide bonds between C41 and C51, between C83 and C94, and between C108 and C126 according to their homologous positions when compared with human GHR[17]. The aa216 in the extracellular domain is an unpaired cysteine found in all GHRs identified so far. Interestingly, in mammalian, avian and reptilian GHRs the unpaired cysteine occurs after the FGEFS motif in a position proximal to the transmembrane domain. In amphibian and all the teleost GHRs, the unpaired cysteine is located upstream the FGEFS motif in a position about 30 aa away from the transmembrane domain. It appears, therefore, that the FGEFS motif and the occurrence of the unpaired cysteine upstream the FGEFS motif are the characteristic of lower vertebrate GHRs.

There are 6 potential *N*-glycosylation sites in the extracellular domain of common carp GHR. The first one is located at aa60. A homologous site for this is found in avian, reptilian, amphibian and all the teleost, but not in mammalian. The second one is located at aa97. A homologous site for this is found in mammalian, avian, reptilian and some of the teleost. The third one is located at aa117. A homologous site for this is found in goldfish, grass carp and *Silurus meridionalis* only. The fourth one is located at aa140. A homologous site for this is

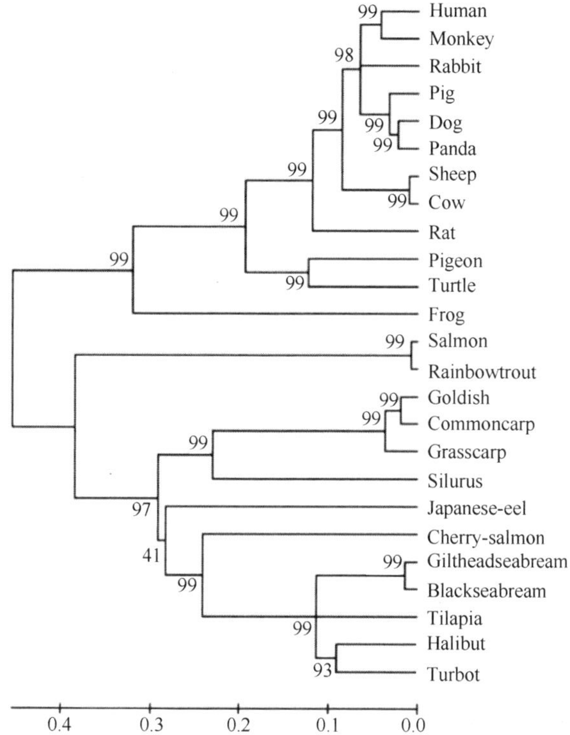

Fig. 4 Phylogenetic tree showing the relationship of common carp with other species analyzed by GHR amino acids comparison. The sequences were aligned by CLUSTAL Wprogram and the phylogenetic tree was constructed by neighbor-joining methods using MEGA version 3. 0. The species included: human (Human, AAA52555), monkey (Monkey, AAK62288), rabbit (Rabbit, 1401239A), pig (Pig, AAZ15732), dog (Dog, NP_001003123), giant panda (Panda, AAK72050), sheep (Sheep, NP_001009323), cow (Cow, O46600), Norway rat (Rat, NP_058790), domestic pigeon (Pigeon, BAC43750), soft-shelled turtle (Turtle, AAG43525), African clawed frog (Frog, AAF05775), coho salmon (Salmon, AAK95625), rainbow trout (Rainbow trout, AAW27914), goldfish (Goldfish, AAK60495), common carp (Commoncarp, AY741100), grass carp (Grasscarp, AAP37033), Silurus meridionalis (Silurus, AAP97011), Japaneseeel (Japanese-eel, BAD20706), cherry salmon (Cherry-salmon, BAD51998), gilthead seabream (Giltheadseabream, AAM 00431), black seabream (Blackseabream, AAN77286), Mozambique tilapia (Tilapia, BAD83668), bast ard halibut (Halibut , BAC76398), turbot (Turbot , AAK72952)

found in all species except for cherry salmon. The fifth one is located at aa145. A homologous site for this is found in all species GHRs except for Japanese eel. The sixth one is located at aa184. A homologous site for this is found in all species. There are two additional potential *N*-glycosylation sites in common carp GHR compared with mammalian GHR. The physiological significance of these additional potential *N*-glycosylation sites remains to be investigated, particularly in view of the possible involvement of glycosylation in ligand binding[18,19].

Two highly conserved regions named Box1 and Box2 are found within the intracellular domain. Box1 is a site for JAK2 binding[20]. The PPVPVP sequence conserved in mammalian, avian, reptilian and amphibian is changed to PPIPAP in common carp, and changed to PPVPAP in some other teleost. However, this is a relatively conservative change as valine, isoleucine and alanine all carry non polar aliphatic side chains. The Box2 region is believed to be involved in the proliferative response of the receptor[21]. In fact, the sequence around the critical phenylalanine residue (WVEFI) is identical in all GHRs except that in Mozambique tilapia (*Oreochronus mossambicus*) GHR. The conserved Box1 and Box2 regions in fish GHRs suggest a post-receptor signaling mechanism in fish akin to mammalian GHRs.

Out of the 8 conserved intracellular tyrosine residues found in most GHRs, seven are found in common carp, namely Y312, Y317, Y371, Y436, Y481, Y549 and Y591. Conservation of these tyrosine residues varies to some extent among species. Y436 and Y591 are found in all GHRs, indicating their essential role in mediating some common biological functions across species. Y312 is found in all species except for guinea pig, rainbow trout and coho salmon. Y317 is found in all species except for human and monkey. Y371 is less conservative and changed into cysteine in avian, reptilian and amphibian. Y481 is found only in all fish species. Y549 is found in all species except for cherry salmon.

The phylogenetic tree (Fig. 4) based on GHR amino acid sequences puts together all fish tested, and displays the same clustering as the present hierarchy of vertebrate species. The evolution of mammalian, avian, reptilian, amphibian and fishes accords with the current evolution law from lower to higher. In view of the phylogenetic diversity of fish, the elucidation of the GHR sequences of other fish species would provide a molecular means for fish classification and taxonomy.

As can be seen in Table 1, the extracellular domain, which defines the ligand binding site, is more conserved than the intracellular domain which defines the signaling events. The

Table 1 Amino acid identities between common carp GHR and GHRs of other species (%)

Species	Whole receptor	Extracellular domain	Intracellular domain
Common carp	100	100	100
Grass carp	90.4	92.3	89.0
Goldfish	92.2	95.9	89.2
Cherry salmon	51.1	59.7	44.1
Gilthead seabream	45.9	49.6	43.5
African clawed frog	35.1	40.2	32.5
Soft-shelled turtle	36.4	41.9	32.6
Domestic pigeon	36.3	41.2	32.3
Human	36.4	40.4	33.2
Norway rat	36.2	38.6	33.0
Sheep	35.8	37.9	33.2
Monkey	34.7	40.4	30.2
Pig	36.9	40.4	32.9
Dog	37.1	42.1	32.1

conservation of extracellular domain partly explains why fish GHRs can recognize GHs of other species[10,11].

Because of intron alternatively splicing, sexual dimorphism and so on, GHR has variants. This is because GHR accommodates the diversity of GH. Two different amplified fragments were found in common carp. The sequencing results of the fragments and partial genomic DNA indicated that the two fragments are generated by alternative splicing of the GHR gene in which an intron of 97 bp is either removed or retained during mRNA processing. There is a stop codon in the intron included in the larger one of PCR amplified fragments, making the translation terminate and forming the truncated GHR. The truncated GHR corresponds to the partial sequence of extra-cellular domain of common GHR (Fig. 3), while the smaller one of PCR amplication fragments, because of the intron spliced, corresponds to the whole GHR. Besides GHR, another protein termed growth-hormone binding protein (GHBP), which can form complexes with circulating GH, was also found in serum and cytoplasm of most species. GHBP is either identical, or highly homologous, to the extracellular domain of the GHR. In mouse and rat two different mRNAs encoding GHR and GHBP respectively were found. They both formed through alternative splicing of the primary GHR transcript[22]. It is believed that GHBP can enhance the growth-promoting effects of GH *in vivo*, probably by increasing the half-life of GH in the circulation. Though the truncated GHRs were found in many higher animals[22] and turbot (*Scophthalmus maximus*)[23-25], most of the truncated GHRs including a section of hydrophobic amino acids are membrane-anchored proteins, except the GHBP. So they are still a membrane-anchored protein. A truncated GHR was also found in monkey[22]: the absence of a hydrophobic transmembrane domain suggests that the protein would not function as membrane-bound receptor, while the absence of the amino acids critical for GH binding suggests that it would not function as circulating GHBP. The exact biological role of the truncated protein is unclear. A 0.7 kb mRNA transcript was also found in chicken[22]. Only the first 95 amino acids of the 221 amino acids hormone-binding domain of the receptor are encoded. These 95 amino acids are considered important for GH binding by mutagenesis and crystallography studies[26]. In the 3′ UTR of the 0.7 kb transcript, there is a domain of AAUAAA and a GU-rich region. The 0.7 kb transcript should be produced by RNA cleavage between the domain of AAUAAA and a GU-rich region.

Expression analysis revealed that in the tissues of gill, thymus and brain, the expression level of GHR in 2-year common carp was significantly lower than that of 4-month common carp, which corresponds to the intensity of thymus's function in different development stage. Thymus is developing in 4-month common carp, and the immune function of thymus in this stage is increasing. However, thymus is in a degenerated stage when the common carp grows to an age of 2-year. Thymus is one of the most important immune organs. But the expression of GHR in fish thymus has not been reported. The expression experiment provides us a direct proof that GHR does exist in thymus. RT-PCR performed with the primers of Gf and Gr showed that the amplified products from all the tissues of 4-month common carp are in the same size of 321 bp. But the amplified products from 2-year common carp showed different size, a 224 bp product from liver and a 321 bp product from other tissues. Though the exact mechanism of the phenomenon is unclear, we consider that it is associated with development: the membrane-anchored GHR occurs in liver due to the intron alternatively spliced mechanism when common carp develops in some degree, or more possibly the membrane-anchored GHR exists from the very beginning, but in the early stage of development this form of GHR is not dominant so it cannot be detected by the common PCR method. This is the first report about the variety of GHR in freshwater fish. The biological significance and action mechanism of this truncated GHR need further investigation.

References

1. Moutoussamy S., Kelly P A. and Finidori J. Growth hormone receptor and cytokine receptor family signaling. Eur. J. Biochem., 1998, 255: 1-11.
2. Cavari B., Funkenstein B., Chen T. T. *et al.* Effect of growth hormone on the growth rate of the gilthead seabream (*Sparus aurata*), and use of different constructs for the production of transgenic fish. Aquaculture, 1993, 111(1-4): 189-197.
3. Trudeau V. L. Neuroendocrine regulation of gonadotrophin II release and gonadal growth in the goldfish *carassius auratus*. Reviews of Reproduction, 1997, 2: 55-68.
4. Yada T., Nagae M., Moriyama S. *et al.* Effects of prolactin and growth hormone on plasma immunoglobulin M levels of hypophysectomized rainbow trout, *Oncorhynchus mykiss*. General and Comparative Endocrinology, 1999, 115(1): 46-52.
5. Argetsinger L. S. and Carter Su C. Mechanism of signaling by growth hormone receptor. Physiological Reviews, 1996, 76: 1089-1107.
6. Zhu T., Goh E. L. K., Graichen R. *et al.* Signal transduction via the growth hormone receptor. Cellular Signalling, 2001, 13 (9): 599-616.
7. Leung D. W., Spencer S. A., Cachianes G. *et al.* Growth hormone receptor and serum binding protein: purification, cloning and expression. Nature, 1987, 330: 537-543.
8. Zhang X. N., Lu X. B., Jing N. H. *et al.* cDNA cloning and functional expression of growth hormone receptor from soft shelled turtle (*Pelodiscus sinensis japonicus*). General and Comparative Endocrinology, 2000, 119(3): 265-275.
9. Huang H. C. and Brown D. D. Overexpression of *Xenopus laevis* growth hormone stimulates growth of tadpoles and frogs. Proc.Nat. Aca. Sci. USA, 2000, 97(1): 190-194.
10. Lee L. T. O., Nong G., Chan Y. H. *et al.* Molecular cloning of a teleost growth hormone receptor and its functional interaction with human growth hormone. Gene, 2001, 270(1-2): 121-129.
11. Tse D. L. Y., T se M. C. L., Chan C. B. *et al.* Seabream growth hormone receptor: molecular cloning and functional studies of the full length cDNA, and tissue expression of two alternatively spliced forms. Biochimica et Biophysica Acta (BBA) Gene Structure and Expression, 2003, 1625(1): 64-76.
12. Zou L., Burmeister L. A., Sperling M. A. Isolation of a liver specific promoter for human growth hormone receptor gene. Endocrinology, 1997, 138: 1771-1774.
13. Very N. M., Kittilson J. D., Norbeck L. A. *et al.* Isolation, characterization, and distribution of two cDNAs encoding for growth hormone receptor in rainbow trout (*Oncorhy nchus mykiss*). Comparative Biochemistry and Physiology Part B: Biochemistry and Molecular Biology, 2005, 140(4): 615-628.
14. Lu S. D. Current Protocols for Molecular Biology (in Chinese). 2nd ed., Beijing: Peking Union Medical College Press, 1999,

102-104.

15. Apweiler R., Attwood T. K., Bairoch A. *et al.* The InterPro database, an integrated documentation resource for protein familes, domains and functional sites. Nucleic Acids Research, 2001, 29(1): 37-40.
16. Kumar S., Tamura K., Jakobsen I. B. *et al.* MEGA2: Molecular evolutionary genetics analysis software. Bioinformatics, 2001, 17(12): 1244-1245.
17. Fuh G., Mulkerrin M. G., Bass S. *et al.* The human growth receptor. Secretion from *Escherichia coli* and disulfide bonding pattern of the extracellular binding domain. Journal of Biological Chemistry, 1990, 265(6): 3111-3115.
18. Szecowka J., T ai L. R. and Goodman H. M. Effects of tunicamycin on growth hormone binding in rat adipocytes. Endocrinology, 1990, 126(4): 1834-1841.
19. Harding P. A., Wang X. Z., Kelder B. *et al. In vitro* mutagenesis of growth hormone receptor Asn-linked glycosylation sites. Molecular and Cellular Endocrinology, 1994, 106(1-2):171-180.
20. Frank S. J., Gilliland G., Kraft A. S. *et al.* Interaction of the growth hormone receptor cytoplamsmic domain with the JAK2 tyrosine kinase. Endocrinology, 1994, 135: 2228-2239.
21. Ihle J. N., Witthuhn B. A., Quelle F. W. *et al.* Signaling through the hematopoietic cytokine receptors. Annual Review of Immunology, 1995, 13: 369-398.
22. Edens A. and Talamantes F. Alternative processing of growth hormone receptor transcripts. Endocrine Review, 1998, 19 (5) :559-582.
23. Calduch Giner J. A., Duval H., Chesnel F. *et al.* Fish growth hormone receptor: molecular characterization of two membrane anchored forms. Endocrinology, 2001, 142: 3269-3273.
24. Clevenger C. V. and Kline J. B. Prolactin receptor signal transduction. Lupus, 2001, 10(10): 706-718.
25. Sweeney G. Leptin signalling. Cell Signal, 2002, 14(8): 655-663.
26. De Vos A. M., Ultsch M. and Kossiakoff A. A. Human growth hormone and extracellular domain of its receptor: crystal structure of the complex. Science, 1992, 255(5042): 306-312.

鲤生长激素受体GHR的克隆及两个mRNA转录本的鉴定

孙晓凤[1,2] 郭琼林[1] 胡 炜[1] 汪亚平[1] 朱作言[1]

1 中国科学院水生生物研究所，武汉 430072
2 中国科学院大学，北京 100049

摘 要 本研究通过 RT-PCR 和 RACE 获得 2 龄鲤肝脏中生长激素受体(GHR)的 cDNA 序列，其开放阅读框(ORF)由 1806 个碱基组成，编码 602 个氨基酸的多肽，包括 244 个氨基酸的胞外配体结合区域，24 个氨基酸的单一跨膜结构域和 344 个氨基酸的胞内单一信号转导结构域。测序结果显示，鲤 GHR 与金鱼 GHR 在基因和蛋白质水平均具有高度同源性。运用一对特异引物，从 2 龄鲤的 cDNA 中扩增得到一段 GHR 序列，在肝脏中的扩增片段长度为 224 bp，在其他组织中扩增产物为 321 bp。扩增产物序列和基因组序列比较发现，造成两种产物的原因是由于一段 97 bp 内含子发生了选择性剪切。321 bp 的片段可以从 4 月龄鲤鱼的所有组织，包括肝脏中扩增出来，表明在鲤鱼发育期间发生了该段内含子的选择性剪切。通过半定量 TR-PCR 分析了 2 龄和 4 月龄鲤组织中 GHR 的表达水平，检测结果显示，在鳃、胸腺和脑中，2 龄鲤的 GHR 表达水平明显低于 4 月龄鲤。

Comparative Expression Analysis of GHR Signaling Related Factors in Zebrafish (*Danio rerio*) and An *In Vivo* Model to Study GHR Signaling

Abu Shufian Ishtiaq Ahmed   YU Li-Qun   ZHU Zuo-Yan   SUN Yong-Hua

State Key Laboratory of Freshwater Ecology and Biotechnology, Institute of Hydrobiology, Chinese Academy of Sciences, Wuhan 430072

Abstract Growth hormone receptor (GHR) signaling pathway plays an important role in postnatal growth of animals. Although significant progress has been made in elucidating the signaling pathways activated by GHR in recent years, a comparative expression analysis of all the GHR signaling related genes and evaluation of GH-signal activation (GHSA) in an *in vivo* model still remain elusive. The zebrafish (*Danio rerio*) is an excellent model organism to study both developmental and physiological processes. In the present study, we comparatively analyzed the expression of GHR signal related genes, such as *gh*, *ghra*, *ghrb*, *jak2a*, *jak2b*, *stat5.1*, *stat5.2*, *igf1*, *c-fos*, *socs1* and *socs2* in adult tissues, and during embryonic development by means of RT-PCR. Maternal expression was found in most of them and their zygotic expression mostly started from early-somite stage, far before the differentiation of pituitary somatotrophs and before the establishment of a functional circulatory system. In adults, the tissue distribution abundance of *socs* mRNA was negatively correlated with the GH-signal targets, *igf1* and *c-fos*. By analyzing the transcription level of *c-fos* and *igf1* and the promoter activity of *spi2.1*, in zebrafish embryos injected with different constructs, i.e., the expression constructs of *gh* and *ghr*, we established an *in vivo* model to evaluate GH-signal activation (GHSA) in early development of zebrafish embryos. Overexpression of *gh* or *ghr* alone in zebrafish embryos could significantly stimulate GHSA at 1 day post-fertilization (dpf) and 3 dpf, strongly indicating the existence of functional GH and GHR proteins in zebrafish embryos from 1 dpf, far prior to the formation of functional pituitary gland. A synergetic overexpression of *gh* and *ghr* could amplify the GHSA effects of *gh* or *ghr* overexpression.

Keywords GH; GHR; GHR signaling; zebrafish; transcriptional analysis

Growth hormone (GH) is a member of a large class of evolutionarily related protein hormones that includes the prolactins (PRL) and placental lactogens (PL)[1]. Growth hormone (GH), secreted by the anterior pituitary into the circulation, binds to membrane receptors that is growth hormone receptors (GHRs) in target tissues which activates the tyrosine kinase Janus

文章发表于 *Acta Hydrobiologica Sinica*, 2011, 35(5):727-738

kinase 2 (JAK2), thus initiating a multitude of signaling cascades that result in a variety of biological responses including cellular proliferation, differentiation and migration, prevention of apoptosis, cytoskeletal reorganization and regulation of metabolic pathways[2]. GHR is a member of the class I cytokine receptor family[3] with the typical features of (i) an extracellular domain (ECD) with 3 pairs of conserved cysteine residues, a (Y/F)GEFS signature motif and a fibronectin type III domain, followed by a single transmembrane domain (TMD) and an intracellular domain (ICD) with the Proline-rich Box 1 and Box 2 sequences and multiple tyrosine residues[1]. In mammals, GHR is functionally coupled with the JAK2/STAT5, MAPK, PI3-K, and PKC/Ca^{2+} signaling cascades[2].

In the GHR signaling pathway, one GH molecule binds sequentially to two GHR molecules[4] and causes the intracellular domains of the GHR to undergo relative rotation[5]. As it is thought that the cytoplasmic domain of each GHR molecule bind a single JAK2 molecule, this rotation is postulated to bring two JAK2 molecules into sufficient proximity to allow each JAK2 molecule to phosphorylate the kinase domain on the other JAK2 molecule, thereby activating JAK2[5]. STATs 5a, and 5b bind to particular phospho-tyrosines in the cytoplasmic domain of the GHR following their phosphorylation by JAK2, are themselves tyrosine phosphorylated by JAK2, dimerize, and translocate to the nucleus to activate *igf1* transcription in particular. JAK2 also directly activates STAT1 and 3 by tyrosine phosphorylation, and again, these STATs translocate to the nucleus, bind to particular STAT responsive elements and together with other transfactors, activate the transcription of target genes such *c-fos*[6,7]. However, the suppressors of cytokine signaling (SOCS) proteins play an important role in regulating GH actions[8]. The SOCS protein family comprises of at least eight proteins namely, CIS and SOCS-1 to SOCS-7. Different members of the SOCS family utilize different mechanisms to regulate GHR signaling[9].

In the present study, by utilizing zebrafish as an animal model, we studied the expression pattern of *growth hormone* (*gh*), *growth hormone receptor* (*ghr*) and other molecules related to GH signaling pathway and also their target genes namely, *igf1* and *c-fos* in different tissues of adults as well as in a series of embryonic stages by RT-PCR method. We also developed an *in vivo* model system to study GH-signal activation (GHSA) by analyzing the transcriptional level of GHR signaling target genes and the promoter activity of the GH responsive *spi2.1* promoter after overexpressing *gh* and *ghr* in zebrafish embryos.

1. Materials and Methods

1.1 Experimental fish and embryo injection

Zebrafish (*Danio rerio*) were cultured in the fish culture laboratory, Institute of Hydrobiology, according to the Zebrafish Book[10]. Embryos were obtained from artificial fertilization. The fertilized zygotes were microinjected with indicated reagents with a pressure microinjector at

1-cell stage. The injected larvae were transferred to small glass aquaria, kept for 1-3 days and preserved in the TRIzol® reagent (Invitrogen, USA) in –70℃ refrigerator pending for total RNA isolation.

1.2 Computer-aided sequence analysis of zebrafish GHRs

Sequences were analyzed for similarity with other known sequences by protein BLAST program and the multiple sequence alignments were generated using CLUSTALW program. The phylogenetic tree was constructed based on the full-length amino acid sequences of GHR using neighbor-joining algorithm within MEGA version 3.1[11].

1.3 Constructs design and mRNA synthesis

The 3 dpf (day-post-fertilization) zebrafish was used to amplify the full length cDNA of *GH* and *GHRa* via high-fidelity PCR by using Kod-plus enzyme (Toyobo, Japan), dNTPs, Kod-plus buffer, $MgSO_4$ and appropriate primers (Tab. 1). The primers were designed according to zebrafish *GH* and *GHRa* cDNA sequence available in GenBank (Tab. 1). For overexpression of *GH* and *GHR*, the PCR products were cloned into pCS2+ expression vector by double digestion with restriction enzymes (NEB, USA) followed by ligation in double digested pCS2+ expression vector by T4 DNA ligase (Fermentas, Canada) (Fig. 1A and B). All the expression constructs were confirmed by double digestion, PCR identification and sequencing analysis. For mRNA synthesis, *GH*_pCS2+ and *GHRa*_pCS2+ constructs were linearized with *Kpn*I and capped mRNA was synthesized by SP6 RNA polymerase mMessage mMachine kit (Ambion, USA). All the DNA constructs have the same constitutive CMV promoter.

Tab. 1 Primers used in the study

Gene (Purpose) (GenBank Accession No)	Primer name	Sequence (5′—3′)	Amplicon size(bp)
GH (RT-PCR) (NM_00102492)	GH_P1	GTTGGTGGTGGTTAGTTTGCT	331
	GH_P2	CAGGCTGTTTGAGATAGTGGAG	
GH (full length cloning)	GH_P3	GCAGGATCCATGGCTAGAGCATTGGTGCTG	651
	GH_P4	AGGCTCGAGCTACAGGGTACAGTTGGAATC	
GHRa (RT-PCR) (NM_001083578)	GHRa_P1	ACCGATAAAGAGTATGAAGTGCG	541
	GHRa_P2	TGTGAAGAAGGGAAGCCAAG	
GHRa (full length cloning)	GHRa_P3	GCAGGATCCATGGCCCACTCGCTCTCTCTC	1731
	GHRa_P4	AGGCTCGAGTTATCTGGGTTGCGCAGATAAG	
GHRb (RT-PCR) (NM_001111081)	GHRb_P1	TACAACTCCCAGAGGAATGAA	496
	GHRb_P2	CAGCAGTAAAGACCAGCACA	
jak2a (RT-PCR) (NM_191093)	jak2a_P1	TGTTTGGGTTTGGCGGTTCT	491
	jak2a_P2	AGTGTCTGACTGTCCTGTCGGTTT	
jak2b (RT-PCR) (NM_191087)	jak2b_P1	CGGTATCAGTCAATATCCTGTG	314
	jak2b_P2	AATCTCCCATAAAGTTGTCCC	

Continued

Gene (Purpose) (GenBank Accession No)	Primer name	Sequence (5′-3′)	Amplicon size(bp)
stat5.1 (RT-PCR) (NM_194387)	Stat5.1_P1	CATCCTTTACACGGAGCAGA	492
	Stat5.1_P2	AGTATGTCCAGTCCTCCCTCA	
stat5.2 (RT-PCR) (NM_001003984)	Stat5.2_P1	AGACCTGGCTGACACGAGAA	427
	Stat5.2_P2	AACCGAACTGTTGCTGAAAA	
socs1 (RT-PCR) (NM_001003467)	Socs1_P1	CAGAGCGACGTTTTCTTTACACT	291
	Socs1_P2	GAGGAAGTCTTTGAGGATTGGTT	
socs2 (RT-PCR) (NM_001114550.1)	Socs2_P1	AAAACACTGGCTGGTATTGGG	181
	Socs2_P2	AACTTGCCGTCCTTGTATTCG	
c-fos (RT-PCR) (BC163170)	c-fos_P1	GTGGGAGCAGGAACTGAGGG	618
	c-fos_P2	GGTTCTTGTTTACTGGAGGGATA	
c-fos (Real-time)	c-fos_P3	TCGCTGATTCGCAGACACGC	199
	c-fos_P4	ATCGCTCTACATCCATCTCACAGTCC	
igf1 (RT-PCR) (NM_131825)	igf1_P1	GTCTAGCGGTCATTTCTTCCA	320
	igf1_P2	CAGGCGCACAATACATCTCG	
igf1 (Real-time)	igf1_P3	TTCAGCAAACCGACAGGATA	124
	igf1_P4	TCTTCACAGGCGCACAATAC	
β-actin (RT-PCR) (AF057040)	BA_P1	ATCTGGCATCACACCTTCTACAAC	254
	BA_P2	TAACCCTCATAGATGGGCACGGT	
β-actin (Real-time)	BA_P3	TCACCACCACAGCCGAAAG	98
	BA_P4	AGAGGCAGCGGTTCCCAT	

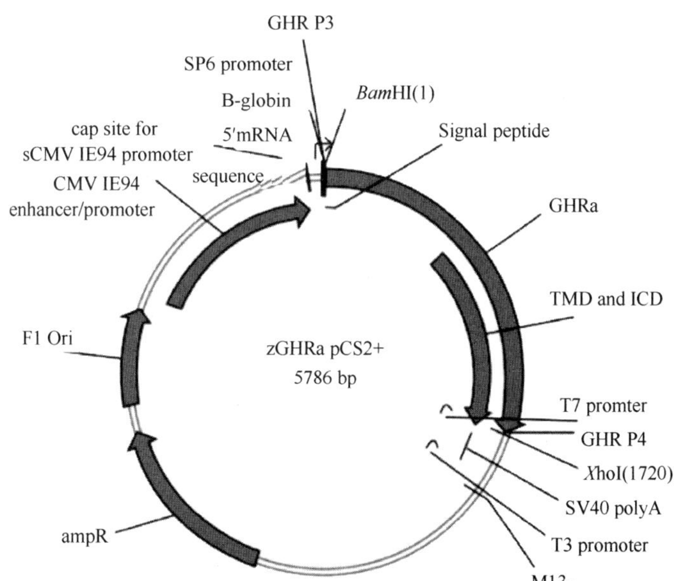

(b)

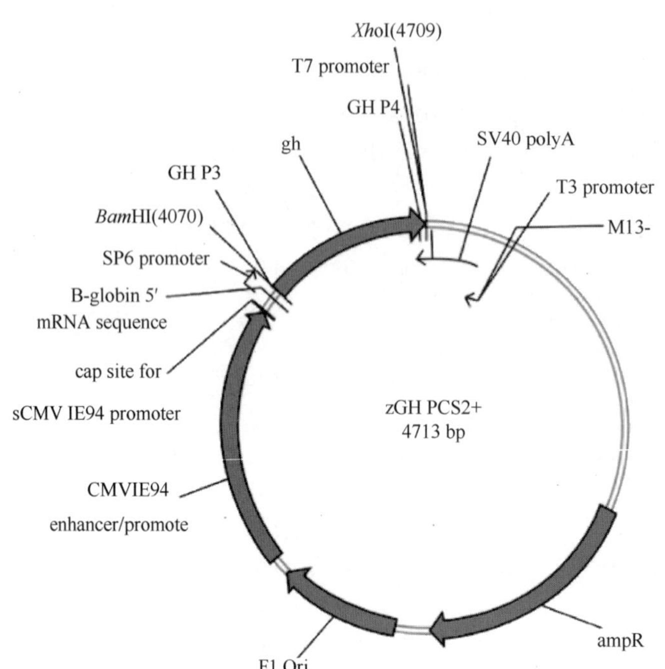

Fig. 1 Map of plasmid zGHRa_PCS2+ (A) and zGH_PCS2+ (B)

1.4 RT-PCR (reverse-transcription PCR) analysis of GHR signaling related genes

Total RNA of 50 zebrafish embryos from different developmental stages such as 2-cell, high, 80%-epiboly, early somites, 1-day, 2-day and 3-day of wildtype zebrafish were extracted using TRIzol® reagent (Invitrogen, USA) according to the manufacturer's instructions. All RNA samples were treated with RNase-Free DNase (Promega, USA), to remove genomic DNA, according to manufacturer's protocol. In the same way, total RNA was extracted from different organs/tissues such as, brain, heart, liver, kidney, gonad, muscle and fin of the wild type adult zebrafish. cDNA was synthesized by reverse transcription (RT) from 1 μg of total RNA using the ReverTra Ace enzyme (TOYOBO, Japan), dNTPs and Oligo (dT) 20 RT primer (TOYOBO, Japan). The RT reaction was performed at 42ºC for 60min followed by 95ºC for 5min. PCR was performed to study the expression of different genes, such as *gh*, *ghra*, *ghrb*, *jak2a*, *jak2b*, *stat5.1*, *stat5.2*, *socs1*, *socs2*, *c-fos* and *igf1* in different samples. GenBank accession numbers and primer sequences were described in Tab. 1. Each pair of primer sequences locate in different exons of the respective gene, to avoid the amplification of genomic DNA potentially contaminated in the samples. PCR was carried out in a 25 μL reaction volume containing 2.5 μL of 10× PCR buffer, 0.2 μmol/L of each primer, 0.2 mmol/L of each dNTP, 0.75 mmol/L of $MgCl_2$, 0.5 unit of *Taq* DNA polymerase (Fermentas, Canada) and 50 ng of cDNA solution. The PCR reaction consists of a denaturation at 94℃ for 5min,

35 cycles (30 cycles for *jak2a*, *jak2b*, *socs1* and *socs2*) of 30s at 94℃, 30s at 56℃ and 45s at 72℃, with a final elongation step of 10min at 72℃. *β-actin* primer pairs (Tab. 1) were used to normalize the cDNA concentration of all samples. The products were run on 1% agarose gels stained with ethidium bromide (0.5 μg/mL), and amplified bands were visualized by ultraviolet transillumination and semi-quantified by Glyko Bandscan software (Novato, CA). A molecular weight marker DNA (DL2000; 100 ng/μL conc.) was run on left side of the gel.

1.5 Real-time quantitative PCR analysis for the stability of *β-actin* transcript and target genes

Fertilized zebrafish zygotes were microinjected with the constructs of zebrafish *GHRa* (*zGHRa*) and *GH* (*zGH*) at a concentration of 50 ng/μL for each. mRNA of *zGH* and *zGHRa* were also microinjected at a concentration of 800 ng/μL. Each embryo was injected with 1 nL of DNA or RNA sample and each sample was injected with 300 embryos. Microinjection was performed with the aid of an Eppendorf 5242 (Germany) microinjector. The microinjected eggs were allowed to develop in Danieau's buffer (58 mmol/L NaCl, 0.7 mmol/L KCl, 0.4 mmol/L $MgSO_4$, 0.6 mmol/L $Ca(NO_3)_2$, 5 mmol/L Hepes, pH 7.6) and cultured in incubator at 28.5℃ until 3 dpf. cDNA samples from the manipulated zebrafish embryos at 2-cell stage and 3 dpf or 1 dpf were analyzed with real-time quantitative PCR (Q-PCR) for *igf*1 and *c-fos* expression level. Real-time PCR was performed on an ABI PRISM® 7000 Sequence Detector (Applied Biosystems, Inc. USA) according to the manufacturer's instructions. Reactions were performed in a 20 μL volume with 150 ng of cDNA, 0.2 μmol/L primers and 10 μL of SYBR® Green Real-time PCR Master Mix (TOYOBO, Japan). The primer pairs were shown in Tab. 1. All reactions were run using the following conditions: 1min denaturation step followed by 40 cycles with a 95℃ denaturation for 15s, 55℃ annealing for 15s, and 72℃ extension for 45s. Detection of the fluorescent product was carried out at the end of the 72℃ extension period. The data (Ct value), obtained from the software (7000 system SDS software), was transferred to Windows Excel (Microsoft co.) sheet and fold-change was calculated using $2^{-\Delta\Delta Ct}$method, with *β-actin* as the calibrator. To validate the stability of *β-actin* for real-time PCR normalization, several pairs of primers specific for *PSB*7, *eTIF-2B*, *eEF-1A* and *GAPDH* were designed as described in the previous study and used for GeNorm analysis[12]. The gene name, accession number, primer sequences, and amplicon length are all given in Tab. 2. Three samples were run for each analysis and all real-time PCR reactions were run in triplicates. The significance of the mean differences between various experimental groups was determined by one way ANOVA followed by Duncan's multiple range test analyses. A P value < 0.05 was considered statistically significant.

1.6 *In vivo* luciferase assay of *spi2.1* promoter activity

Each zebrafish zygote was co-injected with *spi2.1-luc*[13,14] of 50 pg and a constitutively expressed *TK-Renilla* luciferase construct (Promega) of 0.5 pg and the manipulated embryos were subsequently injected with indicated DNA samples. Embryos only co-injected with 50 pg *spi2.1-luc* and 0.5 pg TK-Renilla construct (Promega) were considered as control. Embryos were allowed to develop until 3 dpf, then sets of 20 embryos were lysed in passive lysis buffer (Dual-Luciferase Reporter Assay System, Promega) and the luciferase activity were measured with a Berthold luminometer and relative luciferase activity was calculated as described[15]. Each sample was analyzed in triplicate, and mean value and standard deviation were calculated. The significance of the mean differences between various experimental groups was determined by one way ANOVA followed by Duncan's multiple range test analyses. A *P* value < 0.05 was considered statistically significant.

2. Results

2.1 Phylogenetic and sequence analysis of zebrafish GHR

Phylogenetic analysis of GHRs from various vertebrates indicated that the two types of GHRs of zebrafish that was GHRa and GHRb could be clustered into two clades (Fig. 2). The GHR1 clade encompassed the sequences previously reported in a number of fish species including zebrafish GHRa (GenBank accession number ACF60805), the goldfish GHR (AAK60495), common carp GHR (AAU43899), grass carp GHR (AY283778), Atlantic halibut (AAZ14785), rohu GHR (AAU93895), gilthead seabream GHRI (AAR01947), Nile tilapia GHRI (AAY86769), Mozambique tilapia GHR (BAD 83668), and Japanese eel (GHR1: BAD20706; GHR2: BAD20707). The GHRs within this clade were structurally more homologous to GHRs found in the non-teleost vertebrates. The GHR2 clade, which was unique to teleosts, encompassed the zebrafish GHRb (NP_001104551), channel catfish GHR (AAZ80471), Nile tilapia GHRII (AAY86770), gilthead seabream GHRII (AAU00110), Japanese medaka GHR (NP_001116377), coho salmon GHR (isoform 1: AAK95624; isoform 2: AAK95625), rainbow trout GHR (isoform 1: NP_001118007; isoform 2: NP_001118203), and cherry salmon GHR (BAB64911).

Tab. 2 Primer used for the stability analysis of *β-actin*

Gene (GenBank Accession No.)	Primer name	Sequence (5′—3′)	Amplicon size (bp)
PSB7	PSB7_P1	TGTCCACCAGCCGAGTCC	86
(CV576871; AF155581	PSB7_P2	AGTCAACACCTCCGAGAACTAACG	
eTIF-2B	eTIF-2B_P1	GACAGGCGTTAGTTGGCTTTG	95
(CV576858; NM_212675)	eTIF-2B_P2	CATTCCCGCTGCTCCGT	
eEF-1A	eEF-1A_P1	CTTCTCCACCTACCCTCCTCT	105
(DY544172; BC064291)	eEF-1A_P2	CCAGCACCACCGATTTTCT	
GAPDH	GAPDH_P1	GTGTAGGCGTGGACTGTGGT	121
(AY818346)	GAPDH_P2	TGGGAGTCAACCAGGACAAATA	

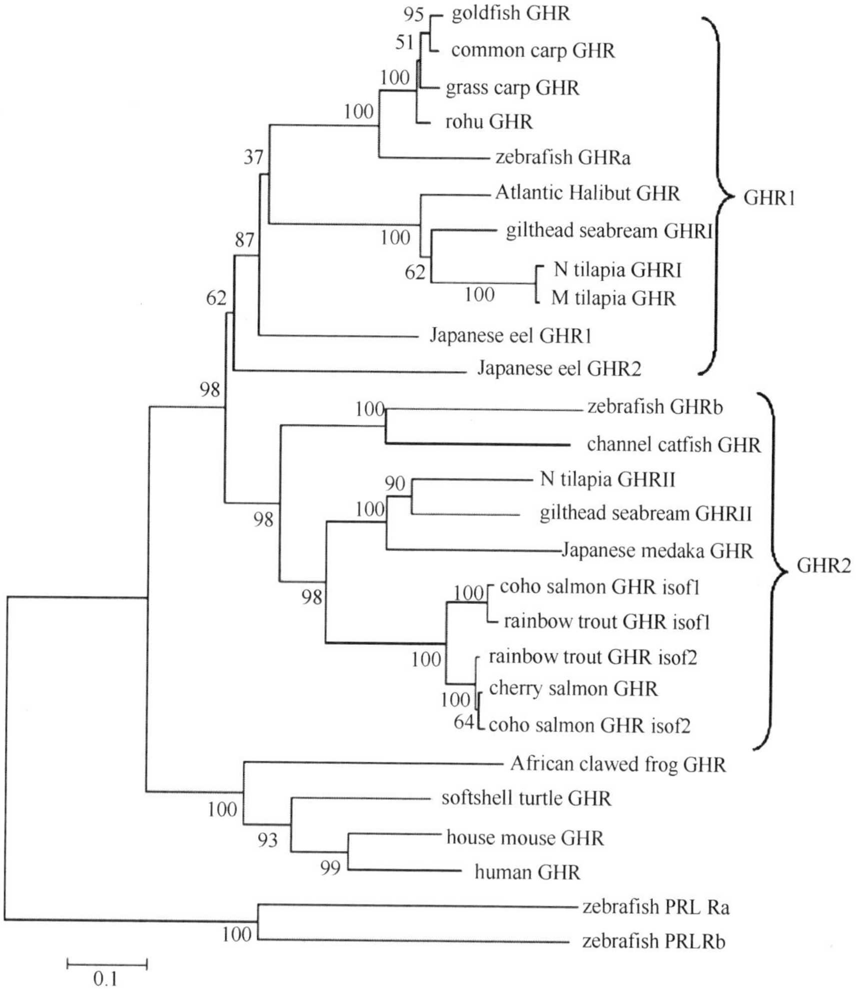

Fig. 2 Phylogenetic tree produced using the neighbor-joining algorithm The two fish GHR clades are named as GHR1 and GHR2. The GHR sequences used in the present study are: zebrafish (*Danio rerio*) GHRa, ACF60805; zebrafish GHRb, NP_001104551; goldfish (*Carassius auratus*), AAK60495; grass carp (*Ctenopharyngodon idella*), AY283778; common carp (*Cypinus carpio*), AAU43899; Atlantic halibut (*Hippoglossus hippoglossus*), AAZ 14785; rohu (*Labeo rohita*), AAU93895; gilthead seabream (*Sparus aurata*) GHRI, AAR01947; gilthead seabream GHRII, AAU00110; Nile tilapia (*Oreochromis niloticus*) GHRI, AAY86769; Nile tilapia GHRII, AAY86770; Mozambique tilapia (*Oreochromis mossambicus*), BAD83668; Japanese eel (*Anguilla japonica*) GHR1, BAD20706; Japanese eel GHR2, BAD20707; channel catfish (*Ictalurus punctatus*), AAZ80471; Japanese medaka (*Oryzias latipes*), NP_001116377; coho salmon (*Oncorhynchus kisutch*) isoform 1, AAK95624; coho salmon isoform 2, AAK95625; rainbow trout (*Oncorhynchus mykiss*) isoform 1, NP_001118007; rainbow trout isoform 2, NP_001118203; cherry salmon (*Oncorhynchus masou*), BAB64911; African clawed frog (*Xenopus laevis*), NP_001081978; softshell turtle (*Pelodiscus sinensis japonicus*), AAG43525; house mouse (*Mus musculus*), AAH75720 and human (*Homo sapiens*), AAA52555

Structural homology among all the so far known members of the fish GHR1 clade was shown in Fig. 3. By using the protein BLAST program on the NCBI database, we found that the deduced amino acid sequence of zebrafish GHRa shared 80% identity with grass carp (*Ctenopharyngodon idella*), 79%, 79% and 77% identity with carp GHR (*Cyprinus carpio*), rohu (*Labeo rohita*), and goldfish (*Carassius auratus*), respectively. In the same way, zebrafish GHRb shared 52% identity with channel catfish GHR (*Ictalurus punctatus*). Alignment of deduced amino acid sequences of the fish GHR1 clade showed the existence of six consensus cysteine residues and three conserved intracellular tyrosine residues. Their extracellular domains all contained the conserved GHR ligand-binding motif (F/A)GEFS (Fig. 3) and they also shared intracellular conserved region of Box 1.

2.2 Spatio-temporal expression of GHR signaling components

RNA from various tissues was extracted and RT-PCR was used to analyze the transcription of positive molecules in GHR signaling pathway like *gh*, *ghra*, *ghrb*, *jak2a*, *jak2b*, *stat5.1*, *stat5.2* (Fig. 4 and Fig. 6) and negative molecules like *socs*1 and *socs*2 (Fig. 5 and Fig. 7). We also checked the target genes of GHR signaling pathway like *c-fos* and *igf*1 (Fig. 5 and Fig. 7). These all represent the ligand, receptor, transcription factor, negative regulator and targets in the GHR signaling (Fig. 6 and 7).

In adult fish, *gh* was expressed in brain, liver, ovary and muscle with the highest expression level in ovary by analysis with Glyko Bandscan software (Fig. 4). *ghra* transcript was found in all tissues examined except for heart and *ghrb* transcript was found in all the tissues tested with the light expression in fin tissues (Fig. 4). The highest levels of *ghra* expression were observed in liver, followed by brain and ovary whereas the highest levels of *ghrb* expression were detected in ovary followed by liver. *jak2a* and *jak2b* transcript also found in all the tissues tested but highest expression *jak2a* and *jak2b* found in the kidney and ovary respectively (Fig. 4). Likely, presence of *stat5.1* and *stat5.2* transcripts were found in all the tested tissues with the highest expression of *stat5.1* found in liver and *stat5.2* found in ovary and fin, followed by brain (Fig. 4). The presence of *socs1* and *socs2* found with low levels of expression in all the tissues tested except that *socs*1 showed the highest expression in ovary and *socs2* showed highest expression in kidney (Fig. 5). In case of *c-fos*, the highest expression level was observed in ovary followed by brain, and in all other tissues except the kidney (Fig. 5). *igf1* transcripts were detected in all tissues except kidney with the highest expression level in liver (Fig. 5), indicating that *socs2* might be the main negative regulator for *igf1*, since *socs2* is highly expressed in kidney.

During different stages of embryonic development, *gh* transcript was detected in 2-cell stage embryos but not at later stages, reappearing from early somites, and increasing with development (Fig. 6). This indicated that *gh* was maternally expressed and its zygotic expression started from early somites. *Ghra* and *ghrb* were also maternally expressed although the expression level

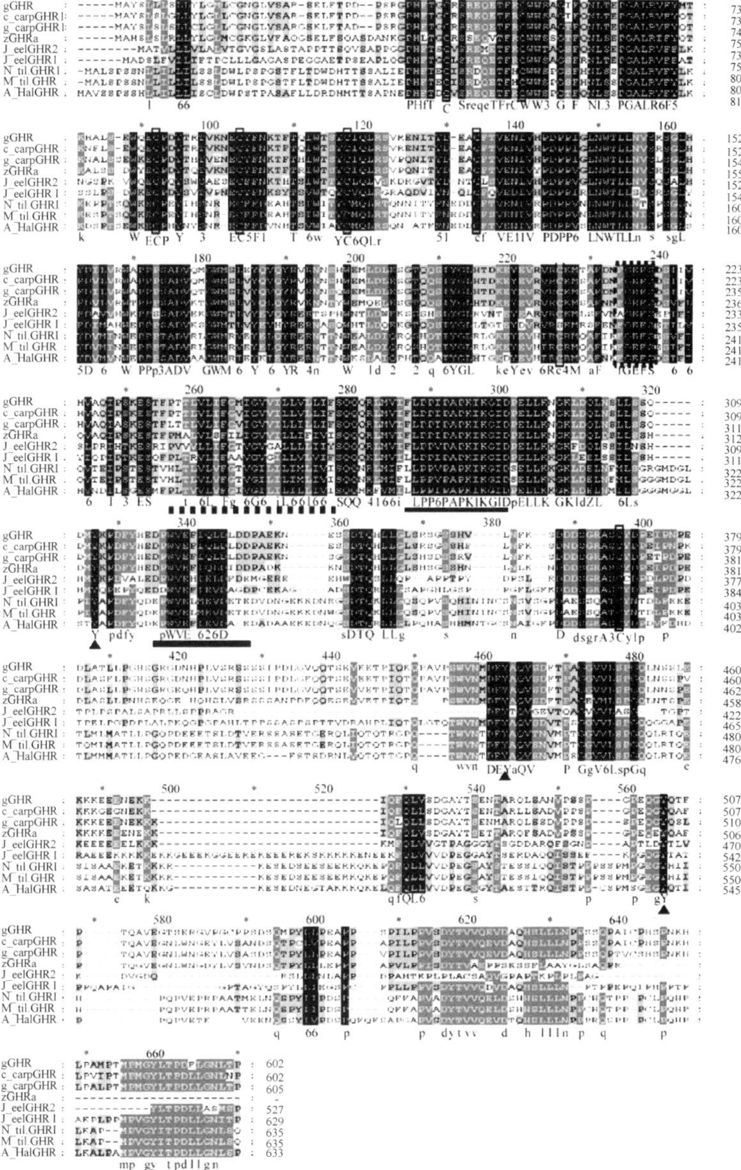

Fig. 3 Multiple alignment of all the fish GHR sequences in the clade GHR1 known to date Black shaded sequence indicates positions where all the sequences share the same amino acid residue, gray shaded sequence indicates conserved amino acid substitutions; Light gray shaded sequence indicates semi-conserved amino acid substitutions, and dashes indicate gaps. The conserved cysteine residues are boxed by an open rectangle. The (F/A) GEFS motif is boxed by an open dotted rectangle. The transmembrane domain is underlined with one dotted line. The Box 1 and Box 2 regions are underlined with one line and two lines respectively. The conserved tyrosine resides in the intracellular domain are highlighted by dark triangles underneath the sequences. Gaps were introduced for maximal similarity. The fish species included in this comparison are: zebrafish (zGHRa), goldfish (gGHR), grass carp (g_carp_GHR), common carp (c_carp_GHR), Japanese eel GHR1 (J_eelGHR1), Japanese eel GHR2 (J_eelGHR2), Nile tilapia (N_til.GHR), Mozambique tilapia (M_til.GHR), and Atlantic halibut (A_HalGHR)

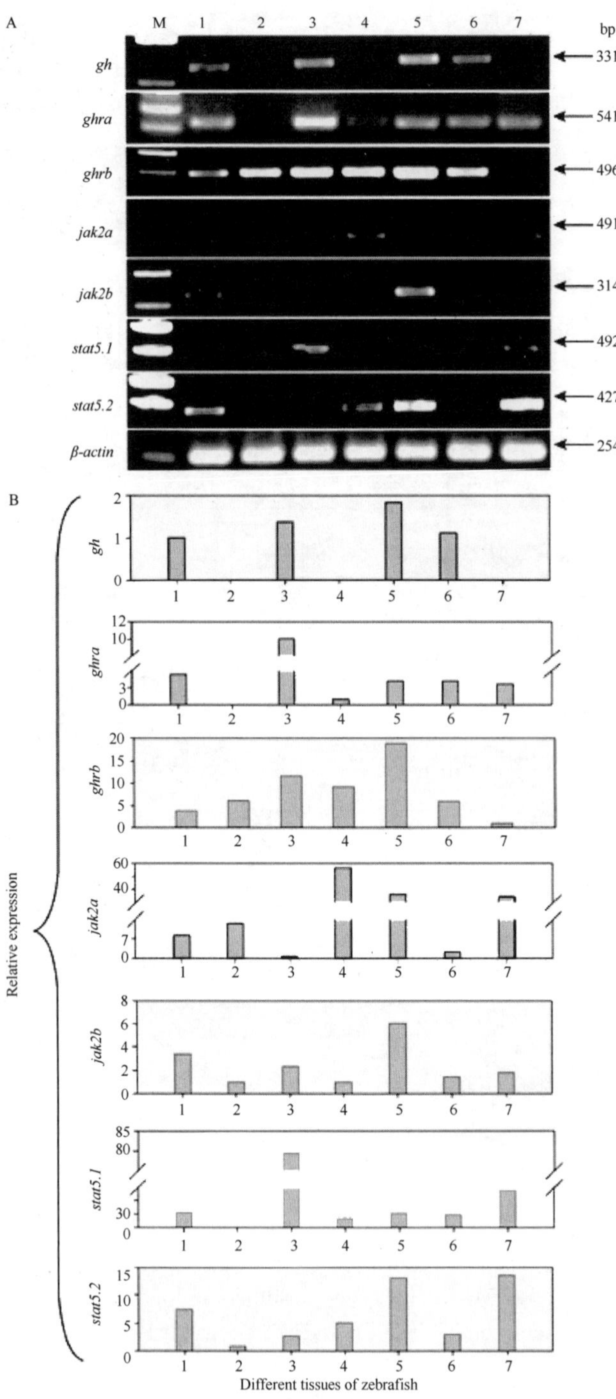

Fig. 4 The transcription analysis of *gh, ghra, ghrb, jak2a, jak2b, stat5.1* and *stat5.2* in different tissues of adult zebrafish A. Gel electrophoregram of the transcription of different genes. B. Graph showing the relative expression of different genes. 1, brain; 2, heart; 3, liver; 4, kidney; 5, ovary; 6, muscle; 7, fin; and M, DNA Ladder (DL 2000). *β-actin* was used as the reference gene

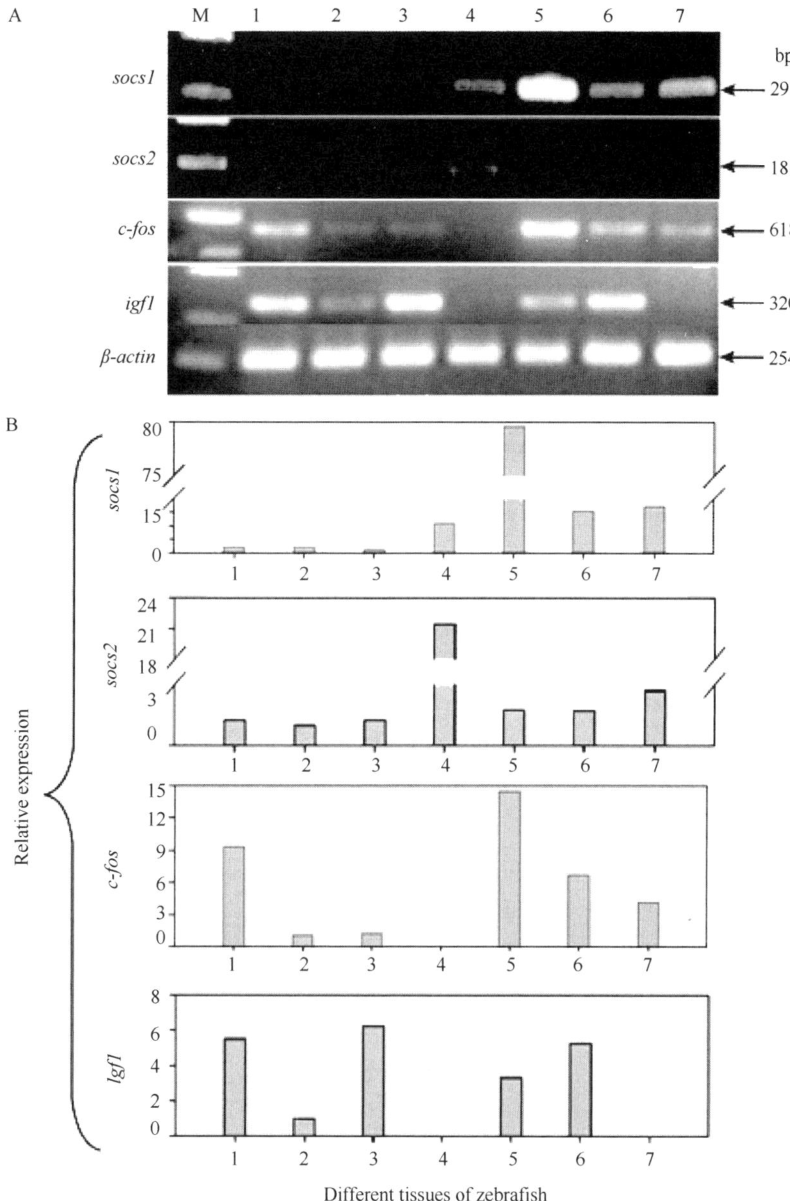

Fig. 5 The transcription analysis of negative regulator *socs1*, *socs2* and target genes, *c-fos* and *igf1* of GHR signaling pathway in different tissues of adult zebrafish A. Gel electrophoregram of the transcription of *socs1*, *socs2*, *c-fos and igf1* genes. B. Graph showing the relative expression of *socs1, socs2, c-fos* and *igf1* genes. 1, brain; 2, heart; 3, liver; 4, kidney; 5, ovary; 6, muscle; 7, fin; and M, DNA Ladder (DL 2000). *β-actin* was used as thereference gene

might be not as high as that of *gh* (Fig. 6). The embryonic expression of *ghra* started from the early somite stage, and in contrasted to *gh*, the expression of *ghra* decreased from the early somite stage to 3 dpf. Interestingly, *ghrb* started its embryonic expression from high stage and its expression increased with time up to 2 dpf and after that the expression decreased rapidly, as we detected the light expression of *ghrb* transcript in 3 dpf (Fig. 6). In case of *jak2a,* the zygotic expression started from the 80%-epiboly whereas the presence of *jak2b* was observed in all the stages starting from 2-cell stage (Fig. 6). More or less similar type of expression observed in between *stat5.1* and *stat5.2* (Fig. 6), both of them were maternally expressed and their zygotic expression started from the high stage in *stat5.1* and for *stat5.2*. *socs1* was found to be maternally expressed and its transcript was detected in all through its development up to 3 dpf with a little higher expression level in 80%-epiboly stage (Fig. 7). Whereas, *socs2* started its expression from 80%-epiboly stage and slightly increased with the development (Fig. 7). Although *c-fos* transcript was detected in 2-cell embryos, from the early somite stage, the expression of *c-fos* increased gradually with the time, mimicking the *GH* expression profile (Fig. 7). *igf1* was minimally expressed maternally, with expression from 1 dpf, and increasing there after (Fig. 7).

2.3 Expression stability of *β-actin* gene

The melting-curve analysis of real-time PCRs after 40 cycles of amplification and agarose gel electrophoresis showed that all the five primer pairs amplified single PCR product of desired size in manipulated embryos and wildtype zebrafish embryos. No primer dimers or nonspecific amplification products could be observed (data not shown). Analysis of the gene expression stability over the manipulated embryo groups vs. wildtype zebrafish embryo groups was done by using the geNorm software. The gene with the lowest M value was considered the most stable gene and the gene with the highest M value was excluded. The ranking of the five control genes according to their M value was equivalent in the three assays. ACTB (*β-actin*) and GAPDH were the two most stable genes, but eTIF-2B was the least stable gene. The ranking of the examined genes was ACTB (GAPDH)> eEF-1A> PSB7> eTIF-2B (Fig. 8A). Thus, in our study we used *β-actin* as reference genes in order to quantify the relative expression levels of *igf*1 and *c-fos*.

2.4 Target genes analysis in GH and GHR overexpressed embryos

In order to quantify GHR signaling in transgenic fish, we measured expression of the GH target genes. For, *igf*1 and *c-fos*, we used 3 dpf embryos and 1 dpf embryos respectively in order to check the status of GHSA, normalizing values to 2-cell embryo levels, since these should not be altered by the transgene. First, we injected *GH* expression construct into zebrafish embryos and found that *igf1* expression was 3.1-fold in the *GH*-injected embryos in comparison to wildtype embryos (Fig. 8B), suggesting significant GHSA in the injected embryos. Second, we overexpressed zebrafish *GHR* and found 3.5-fold of *igf1* expression, demonstrating that overexpression of *GHR* could also activate GHR signaling in the presence of endogenous GH. Third, we used a combined overexpression of *GH* and *GHR*, and this overexpression showed a 4.6-fold of *igf1* expression, higher than these constructs separately. Similar results also found from the *GH*, *GHRa* and *GH+GHRa* mRNA

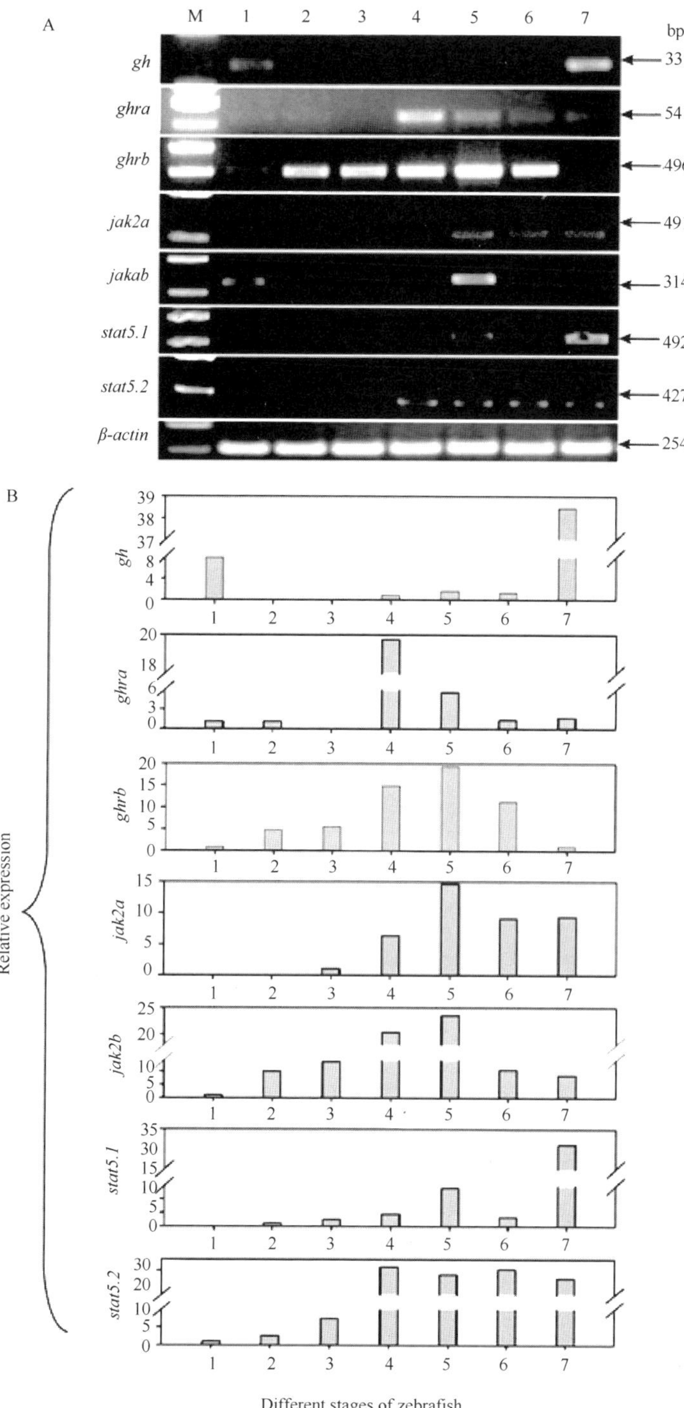

Fig. 6 The transcription analysis of *gh, ghra, ghrb, jak2a, jak2b, stat5.1 and stat5.2* in different developmental stages of zebrafish A. Gel electrophoregram of the transcription of different genes. B. Graph showing the relative expression of different genes. 1, 2-cell; 2, high; 3, 80%-epiboly; 4: early somite, 5, 1 dpf; 6, 2 dpf; 7, 3 dpf; and M, DNA ladder (DL 2000). *β-actin* was used as the reference gene

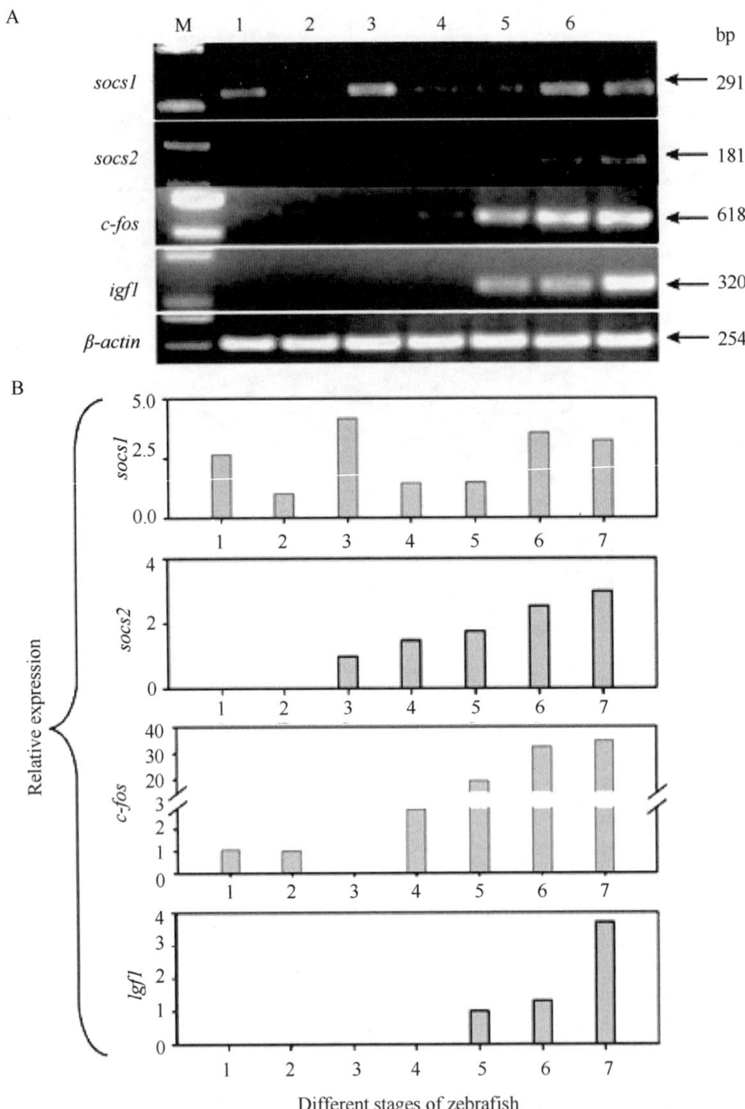

Fig. 7 The transcription analysis of negative regulator *socs1*, *socs2* and target genes, *c-fos* and *igf1* of GHR signaling pathway in different developmental stages of zebrafish A. Gel electrophoregram of the transcription of *socs1, socs2, c-fos* and *igf1* genes. B. Graph showing the relative expression of *socs1 socs2*, *c-fos* and *igf1* genes. 1, 2-cell; 2, high; 3, 80%-epiboly; 4: early somite, 5, 1 dpf; 6, 2 dpf; 7, 3 dpf; and M, DNA ladder (DL 2000). *β-actin* was used as the reference gene

injected 1 dpf zebrafish (Fig. 8C). For *c-fos*, *GH* overexpressed 1 dpf zebrafish showed 1.7-fold higher than wildtype embryos (Fig. 8C). Similar to *igf1*, overexpressed zebrafish *GHR* found higher *c-fos* expression (1.72-fold) than overexpressed *GH* (1.7-fold). However, in the combined overexpression of *GH* and *GHRa*, showed a 2-fold of *c-fos* expression, higher than these constructs separately and it showed similarity in *igf1* expression with the combined overexpressed of *GH* and *GHRa*.

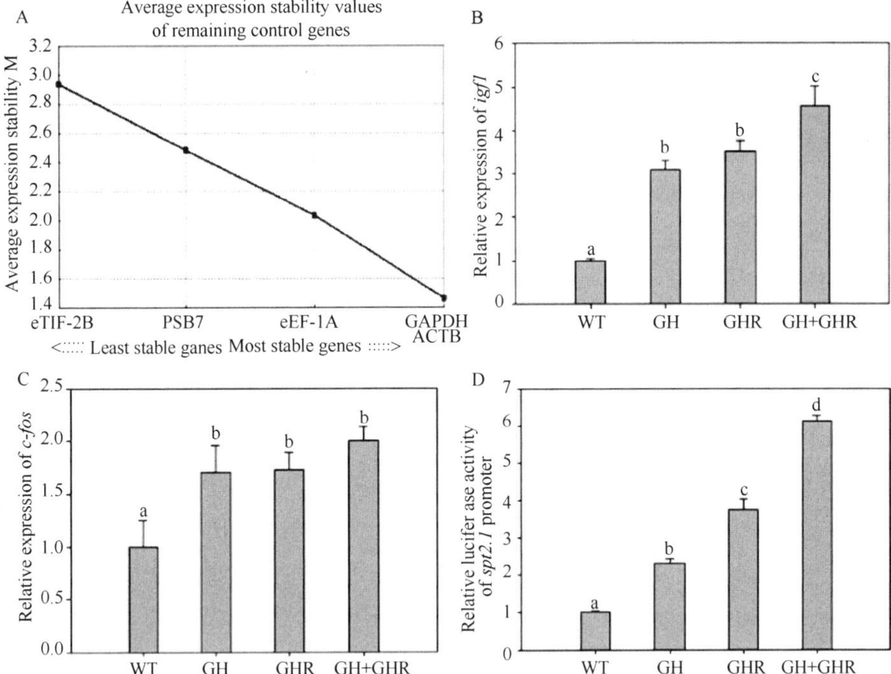

Fig. 8 Relative expression analysis of the target genes and promoter *spi2.1* after overexpression of GH and GHR A. Gene expression stability of the five candidate reference genes analyzed by the geNorm program. Average expression stability values (M) of the control genes, plotted from least stable (left) to most stable (right). B. Relative expression analysis of *igf1* in zebrafish embryos at 3 dpf by real-time PCR analysis. C. Relative expression of *c-fos* in zebrafish embryos at 1 dpf by real-time PCR analysis. D. Relative promoter activities of *spi2.1* in zebrafish embryos at 3 dpf by luciferase assay. Each bar presents the mean value and the corresponding standard deviation from triplicate analysis. A statistically significant difference ($P< 0.05$) is indicated by a different letter above the bar, as determined by one-way ANOVA followed by Duncan's test

2.5 *Spi 2.1* promoter assay in GH and GHR overexpressed embryos

We conducted luciferase assay by utilizing *spi2.1-luc* construct[13-14], to evaluate the overall GHSA level in zebrafish embryos. As shown in figure 8D, *GH* overexpression induced 2.3-fold GHSA in comparison with wildtype embryos, z*GHR* overexpression induced 3.7-fold GHSA and the synergetic overexpression of GH and GHR could even result in a 6.1-fold GHSA. These clearly demonstrated that the *in vivo* analysis of GHR signaling activity was validated by *spi2.1* luciferase assay. Thus all the three experiments, i.e., transcription analysis of *c-fos* and *igf1*, and promoter assay of *spi2.1*, showed that higher GHR signaling activation could be achieved by synergetic expression of GH and GHR.

3. Discussion

In the present work, we obtained some molecular structure information of zebrafish *ghra* and *ghrb* and analyzed the expression profiles of different molecules involved GHR signaling pathway. We also established an *in vivo* model, i.e., zebrafish embryos, by anlyzing on their target genes in order to study the overexpression effect of GHR signaling related factors. Phylogenetic analysis of GHRs from various vertebrates indicated that the fish GHRs could be clustered into two clades and zebrafish GHRa and GHRb belonged to the clade GHR1 and GHR2 respectively. The deduced amino acid sequence of zebrafish GHRa shared the highest percentage identity to grass carp GHR (80%) which belonged to the same class of cyprinidae. Similar to other fish GHR like goldfish GHR, common carp GHR, Atlantic halibut, rohu GHR, gilthead seabream GHRI, Nile tilapia GHRI, Mozambique tilapia GHR, and Japanese eel GHR1 & GHR2 contained two consensus (F/A) GEFS motif. They shared intracellular conserved proline-rich region of Box 1, the site of Janus kinase 2 (JAK2) binding and was essential for the signaling functions of GHR[16].

It was previously demonstrated that GH and GHR were present from a very early embryonic stage in mice and common carp[17,18], and it was suggested the potential contributions of GH to prenatal development in mammals[19]. In this study, the expression of *gh*, *ghra*, *ghrb*, *jak2a*, *jak2b*, *stat5.1*, *stat5.2*, *socs1*, *socs2*, *c-fos* and *igf1* genes were extensively analyzed during 7 stages of embryonic development in zebrafish. Transcription of *gh, ghra*, *ghrb*, *jak2a*, *jak2b*, *stat5.2*, *socs1* and *c-fos* was observed at the 2-cell stage, indicating maternal expression of these genes, and detection of their expression in the ovary further suggested that they started to be accumulated in the oocytes during maturation. For most of them, the zygotic expression was detected from the gastrulation, i.e., 80%-epiboly stage and some even earlier than that. Most remarkable thing was that all these expression started far prior to the formation of functional pituitary gland of zebrafish. Expression of *GH* increased with expression of *igf1* at 3 dpf. The maternal expression and early expression of *GH* and *igf1* had also been reported by several investigators[6,20]. These observations suggested that GH/GHR pathway also played a role in early development of fish, and in our present study, overexpression of GHR or GH alone led to the increase of GHSA in 1 dpf and 3 dpf embryos, strongly indicating the existence of functional GH and GHR expression in zebrafish embryos from 1 dpf.

It was found that *GHRa* was predominantly expressed in liver. The expression of the GHRs in liver was reported in all species of fish and mammals analyzed to date, since liver was the main organ contributing to IGF-I circulating levels in response to GH[21]. Furthermore, *GH*, *GHRa* and *GHRb* expression were found in ovary, which may indicate their role in fish reproduction. Although, there was considerable evidence that the GH/IGF-I axis played an important role in fish reproduction, it was not clear if it is *via* endocrine or paracrine/autocrine mechanisms. In teleost, specific localization of *ghr* transcripts was observed in the cytoplasm and nucleus of the immature oocytes, and in the granulosa and theca cells surrounding vitellogenic oocytes of tilapia[22,23]. Although *GHRa* and *GHRb* have been detected in almost all tissues in zebrafish like other teleost species, we could not discard species-specific differences, since zebrafish has a determinate growth[24]. While the immunomodulatory effect of GH was reported in fish[25], the mechanisms of GH interaction remain elusive. Surprisingly, although *GHRa* and *GHRb* were transcribed in the immune relevant tissues kidney of zebrafish, we did not detect transcription of *igf*1 or *c-fos* in the kidney, which suggested that the zebrafish kidney may lack some intracellular transducer for the GH/GHR pathway.

Of the seven STATs described for mammals, the one involved specifically with GHR signaling was STAT5b which was functionally equivalent to STAT5.1 in zebrafish (*Danio rerio*)[26]. We studied the expression of *jak2a, jak2b, stat5.1* and *stat5.2* genes in different tissues and different developmental stages. Although, it was found that they were transcribed in all the tested tissues, we did not detect transcription of *igf1* or *c-fos* in the kidney. This might be due to the highest expression of *socs2* transcription observed in the kidney (Fig. 5). It was also found that the lowest transcription level of *socs1* and *socs2* in liver while the highest transcription level of *igf1* was observed in this organ. Different members of the SOCS family utilized different mechanisms to regulate GHR signaling[9] and SOCS-2[27,28] was found to be one of the major regulators for GH actions among other SOCS protein family members. These findings may give another reason of no transcription activity of *c-fos* and *igf1* in kidney.

In recent studies, the geNorm Visual Basic application for Microsoft Excel was well performed to search the most stable reference genes[29]. It presented an adequate way of working around the problem of accurate normalization of gene expression levels. In this study, the analysis of geNorm Visual Basic application indicated that ACTB (*β-actin*) and GAPDH were the two most stable genes. Similar result also found in our previous study in identifying the reference gene[12]. Therefore, in our study *β-actin* gene was used as reference gene.

In most of previous studies, the transcriptional activity of GHR signaling in fish has been measured in cultured cells or tissues[13,30]. Here we have utilized zebrafish embryos as a host system to analyze the GHR signaling target gene *igf1*, *c-fos* and the promoter activity of *spi2.1*, after injection of different combination of transgene constructs, such as *GH*, *GHR*, and *GH+GHR* into zygotes. This has allowed us to compare the signaling activity of GH/GHR pathway upon treatment with different GH pathway-related components *in vivo*. It was found that significant GHSA in the *GH*- and *GHR*- overexpressed zebrafish embryos. These results validate the efficacy of this *in vivo* system for studying GHR signaling.

Acknowledgements We are grateful to Prof. Cheng CHK for providing the *spi2.1-luc* construct.

References

[1] Kopchick J J, Andry J M. Growth hormone (GH), GH receptor, and signal transduction [J]. *Molecular Genetics and Metabolism*, 2000, 71(1-2): 293-314

[2] Lanning N J, Carter-Su C. Recent advances in growth hormone signaling [J]. *Reviews in Endocrine & Metabolic*

Disorders, 2006, 7(4): 225-235

[3] Moutoussamy S, Kelly P A, Finidori J. Growth-hormone-receptor and cytokine-receptor-family signaling [J]. *European Journal of Biochemistry*, 1998, 255(1): 1-11

[4] Wells J A. Binding in the growth hormone receptor complex [J]. *Proceedings of the National Academy of Sciences of the United States of America*, 1996, 93(1): 1-6

[5] Brown R, Adams J, Pelekanos R, *et al.* Model for growth hormone receptor activation based on subunit rotation within a receptor dimer [J]. *Nature structural & molecular biology*, 2005, 12(9): 814-821

[6] Cesena T I, Cui T X, Piwien-Pilipuk G, *et al.* Multiple mechanisms of growth hormone-regulated gene transcription [J]. *Molecular Genetics and Metabolism*, 2007, 90(2): 126-133

[7] Ihle J N, Gilliland D G. Jak2: normal function and role in hematopoietic disorders [J]. *Current Opinion in Genetics & Development*, 2007, 17(1): 8-14

[8] Endo T A, Masuhara M, Yokouchi M, *et al.* A new protein containing an SH2 domain that inhibits JAK kinases [J]. *Nature*, 1997, 387(6636): 921-924

[9] Krebs D L, Hilton D J. SOCS proteins: negative regulators of cytokine signaling [J].*Stem Cells*, 2001, 19(5): 378-387

[10] Westerfield M. The zebrafish book. A guide for the laboratory use of zebrafish (*Danio rerio*) [M]. 4th ed. Eugene: University of Oregon Press. 2000

[11] Kumar S, Tamura K, Jakobsen I B, *et al.* MEGA2: molecular evolutionary genetics analysis software [J]. *Bioinformatics*, 2001, 17(12): 1244-1245

[12] Pei D S, Sun Y H, Chen S P, *et al.* Zebrafish GAPDH can be used as a reference gene for expression analysis in cross-subfamily cloned embryos [J]. *Analytical Biochemistry*, 2007, 363(2): 291-293

[13] Jiao B, Huang X, Chan C B, *et al.* The co-existence of two growth hormone receptors in teleost fish and their differential signal transduction, tissue distribution and hormonal regulation of expression in seabream [J]. *Journal of Molecular Endocrinology*, 2006, 36(1): 23-40

[14] Chan Y H, Cheng C H, Chan K M. Study of goldfish (*Carassius auratus)* growth hormone structure-function relationship by domain swapping [J]. *Comparative Biochemistry and Physiology - Part B: Biochemistry & Molecular Biology*, 2007, 146(3): 384-394

[15] Liu Y, Bourgeois C, Pang S, *et al.* The Germ Cell Nuclear Proteins hnRNP G-T and RBMY Activate a Testis-Specific Exon [J]. *Plos Genetics*, 2009, 5(11): e1000707

[16] VanderKuur J A, Wang X, Zhang L, *et al.* Growth hormone-dependent phosphorylation of tyrosine 333 and/or 338 of the growth hormone receptor [J]. *The Journal of Biological Chemistry*, 1995, 270(37): 21738-21744

[17] Pantaleon M, Whiteside E J, Harvey M B, *et al.* Functional growth hormone (GH) receptors and GH are expressed by preimplantation mouse embryos: a role for GH in early embryogenesis [J]. *Proceedings of the National Academy of Sciences of the United States of America*, 1997, 94(10): 5125-5130

[18] Lu Y J, Hu W, Zhu Z Y. Gene expression profiles of growth and reproduction related genes during the early development of common carp (*Cyprinus carpio* L.) [J]. *Acta Hydrobiologica Sinica*, 2009, 33(6): 1126-1131

[19] Waters M J, Kaye P L. The role of growth hormone in fetal development [J]. *Growth Hormone & IGF Research*, 2002, 12(3): 137-146

[20] Li M, Greenaway J, Raine J, *et al.* Growth hormone and insulin-like growth factor gene expression prior to the development of the pituitary gland in rainbow trout (*Oncorhynchus mykiss*) embryos reared at two temperatures [J]. *Comparative Biochemistry and Physiology Part A: Molecular & Integrative Physiology*, 2006, 143(4): 514-522

[21] Di Prinzio C M, Botta P E, Barriga E H, *et al.* Growth hormone receptors in zebrafish (*Danio rerio*): Adult and embryonic expression patterns [J].*Gene Expression Patterns*, 2010,10(4-5): 214-225

[22] Ma X, Liu X, Zhang Y, *et al.* Two growth hormone receptors in Nile tilapia (*Oreochromis niloticus*): molecular characterization, tissue distribution and expression profiles in the gonad during the reproductive cycle [J]. *Comparative Biochemistry and Physiology - Part B: Biochemistry & Molecular Biology*, 2007, 147(2): 325-339

[23] Li M, Raine J C, Leatherland J F. Expression profiles of growth-related genes during the very early development of rainbow trout embryos reared at two incubation temperatures [J]. *General and Comparative Endocrinology*, 2007,

153(1-3): 302-310

[24] Biga P R, Goetz F W. Zebrafish and giant danio as models for muscle growth: determinate vs. indeterminate growth as determined by morphometric analysis [J]. *American Journal of Physiology - Regulatory, Integrative and Comparative Physiology*, 2006, 291(5): R1327-1337

[25] Yada T. Growth hormone and fish immune system [J]. *General and Comparative Endocrinology*, 2007, 152(2-3): 353-358

[26] Lewis R S, Ward A C. Conservation, duplication and divergence of the zebrafish stat5 genes [J]. *Gene*, 2004, 338(1): 65-74

[27] Metcalf D, Greenhalgh C J, Viney E, *et al.* Gigantism in mice lacking suppressor of cytokine signalling-2 [J]. *Nature*, 2000, 405(6790): 1069-1073

[28] Greenhalgh C J, Metcalf D, Thaus A L, *et al.* Biological evidence that SOCS-2 can act either as an enhancer or suppressor of growth hormone signaling [J]. *The Journal of Biological Chemistry*, 2002, 277(43): 40181-40184

[29] Vandesompele J, De Preter K, Pattyn F, *et al.* Accurate normalization of real-time quantitative RT-PCR data by geometric averaging of multiple internal control genes [J]. *Genome Biology*, 2002, 3(7): RESEARCH0034

[30] Björnsson B, Johansson V, Benedet S, *et al.* Growth hormone endocrinology of salmonids: regulatory mechanisms and mode of action [J].*Fish Physiology and Biochemistry*, 2002, 27(3): 227-242

斑马鱼 GHR 信号通路相关因子的表达分析及生长激素受体信号的在体研究模型

Abu Shufian Ishtiaq Ahmed　于力群　朱作言　孙永华

中国科学院水生生物研究所, 淡水生态与生物技术国家重点实验室, 武汉 430072

摘　要　GHR 信号通路在动物出生后的生长中扮演着重要角色。本研究以斑马鱼为模型，对其成体组织、胚胎及幼体的 GHR 信号相关基因进行了比较分析，这些基因包括 *gh*、*ghra*、*ghrb*、*jak2a*、*jak2b*、*stat5.1*、*stat5.2*、*igf1*、*c-fos*、*socs1* 和 *socs2*。值得关注的是，上述大部分基因都存在母源性表达，且它们的合子表达均起始于体节早期之前，这说明在脑垂体和循环系统建立之前，GH 及 GH 信号相关因子就已经在早期胚胎中发挥作用。因此，GH 可能是控制胚胎发育的一系列自分泌/旁分泌生长因子中的一员。同时，我们发现成体组织的 *socs* 表达水平与 GH 信号靶基因 *igf1* 和 *c-fos* 的表达呈某种程度的负相关。利用实时定量 PCR 技术和荧光素酶分析技术，通过注射 GH 和 GHR 表达载体，分析了它们促进 GH 信号靶基因 *c-fos* 和 *igf*1 转录活性及 GH 应激启动子 *spi2.1* 活性的能力。由此，我们利用斑马鱼胚胎建立了一个在体研究模型，用以评估发育过程中的 GH 信号激活(GHSA)。实验证实，在受精后 1 天(dpf)和 3dpf 斑马鱼胚胎中，单独过表达 *gh* 或 *ghr* 均可以显著刺激 GHSA，这表明在 1dpf 的斑马鱼胚胎中即存在功能性 GH 和 GHR 蛋白表达，而这一时期是在功能性垂体形成之前，而且，*gh* 及 *ghr* 的协同过表达则可以显著放大 *gh* 或 *ghr* 单独过表达的 GHSA 效果。

Activation of GH Signaling and GH-Independent Stimulation of Growth in Zebrafish by Introduction of A Constitutively Activated GHR Construct

A. S. Ishtiaq Ahmed Feng Xiong Shao-Chen Pang Mu-Dan He
Michael J. Waters Zuo-Yan Zhu Yong-Hua Sun

State Key Laboratory of Freshwater Ecology and Biotechnology, Institute of Hydrobiology, Chinese Academy of Sciences, 430072 Wuhan, Hubei, China
Physiology & Pharmacology Department and Centre for Molecular & Cellular Biology, University of Queensland, St. Lucia, QLD 4072, Australia
Graduate School of the Chinese Academy of Sciences, 100039 Beijing, China

Abstract *Growth hormone* (*GH*) gene transfer can markedly increase growth in transgenic fish. In the present study we have developed a transcriptional assay to evaluate GH-signal activation (GHSA) in zebrafish embryos. By analyzing the transcription of *c-fos* and *igf1*, and the promoter activity of *spi2.1*, in zebrafish embryos injected with different constructs, we found that overexpression of either *GH* or growth hormone receptor (*GHR*) resulted in GHSA, while a synergetic overexpression of *GH* and *GHR* gave greater activation. Conversely, overexpression of a C-terminal truncated dominant-negative GHR (ΔC-*GHR*) efficiently blocked GHSA epistatic to *GH* overexpression, demonstrating the requirement for a full GHR homodimer in signaling. In view of the importance of signal-competent GHR dimerization by extracellular GH, we introduced into zebrafish embryos a constitutively activated *GHR* (CA-*GHR*) construct, which protein products constitutively dimerize the GHR productively by Jun-zippers to activate downstream signaling *in vitro*. Importantly, overexpression of CA-*GHR* led to markedly higher level of GHSA than the synergetic overexpression of *GH* and *GHR*. CA-*GHR* transgenic zebrafish were then studied in a growth trial. The transgenic zebrafish showed higher growth rate than the control fish, which was not achievable by *GH* transgenesis in these zebrafish. Our study demonstrates GH-independent growth by CA-*GHR in vivo* which bypasses normal IGF-1 feedback control of GH secretion. This provides a novel means of producing growth enhanced transgenic animals based on molecular protein design.

Keywords GH; GHR; constitutively active GHR; protein design; zebrafish.

Introduction

The potential economic benefits of transgenic technology to aquaculture are enormous,

文章发表于 *Transgenic Research*, 2011, 20(3): 557-567

following the introduction of novel desirable traits to farmed fishes (Chen *et al.* 1996). With the expansion of the global population and overfishing, advanced aquaculture techniques are needed to meet the increasing demand for fish protein. Transgenic technology offers the opportunity to improve both the quantity and quality of conventional fish strains currently exploited in aquaculture (Fu *et al.* 2005). It is also an important methodology for studying the function of genes and genomes in model animals (Kikuta and Kawakami 2009). Successful production of transgenic fish was first demonstrated in goldfish (Zhu *et al.* 1985) and 3 years later in zebrafish (Stuart *et al.* 1988). More than 30 fish species, including many of the major aquaculture species like carp, tilapia, catfish and salmonids, have been genetically engineered with most efforts targeted to enhancing growth and feed conversion efficiency through the transfer of *growth hormone* (*GH*) gene constructs (Zhu and Sun 2000; Wu *et al.* 2003; Devlin *et al.* 2006). *GH* transgenesis has shown to result in different effects of growth enhancement in host fish depending on different genetic backgrounds (Devlin *et al.* 2001, 2009). For instance, one report showed that *GH* transgenesis is effective to growth only in hemizygous but not in homozygous individuals of those transgenic zebrafish (Studzinski *et al.* 2009) and we have never obtained fast-growing transgenic zebrafish in our laboratory by *GH* transgenesis.

It is well known that GH is the major regulator of postnatal growth and metabolism (Lichanska and Waters 2008) via GH receptor (GHR) signaling pathways (Rowland *et al.* 2005). The GHR is a type I cytokine receptor consisting of extracellular, transmembrane and intracellular domains. GH activates the GHR by realigning two identical receptor subunits in a constitutive dimer through binding with their extracellular domains, leading to the activation of JAK2 kinase associated with the intracellular domain of GHR (Herrington and Carter-Su 2001; Waters *et al.* 2006). STATs 5a, and 5b bind to particular phosphortyrosines in the cytoplasmic domain of the GHR following their phosphorylation by JAK2, are themselves tyrosine phosphorylated by JAK2, dimerize, and translocate to the nucleus to activate *igf1* transcription in particular. JAK2 also directly activates STAT1 and 3 by tyrosine phosphorylation, and again, these STATs translocate to the nucleus, bind to particular STAT responsive elements and together with other transfactors, activate the transcription of target genes such *c-fos* (Cesena *et al.* 2007; Ihle and Gillil and 2007). Intriguingly, in our previous study, GH-independent activation of GHR can be achieved in cell culture by fusing the transmembrane and intracellular domains of GHR to the leucine zippers to achieve an active dimer orientation. This stabilizes the receptor dimer in a conformation that holds the box 1 sequences in proximity, facilitating signaling (Behncken *et al.* 2000). Here we have utilized the leucine zipper constructs to demonstrate for the first time enhanced growth by a molecularly designed constitutively activated *GHR* (CA-*GHR*) in zebrafish, which we have found is normally resistant to *GH* transgene-mediated growth.

In previous studies, the expression profiles of *GH* and *GHR* have been analyzed either during development or in different tissues of zebrafish (Zhu *et al.* 2007; Di Prinzio *et al.*

2010). However, a comparative expression analysis of *GH*, *GHR* and the signaling targets, *c-fos* and *igf1*, in zebrafish, has not been described. In our present study, we utilized zebrafish to study the comparative expression of GH signaling factors during embryonic development and in different tissues of adults. By setting up an *in vivo* model to evaluate the GH-signal activation (GHSA) levels in zebrafish embryos, furthermore, we were able to perform functional analysis of GH signaling during zebrafish early development by overexpression of *GH* or *GHR*.

Materials and Methods

Fish maintenance and embryo injection

Fish were cultured in the fish culture facility, Institute of Hydrobiology, according to the *Zebrafish Book* (Westerfield 2000). Embryos were obtained from artificial fertilization and microinjected with indicated reagents with a pressure microinjector as described (Liu *et al.* 2005). mRNA was injected in the yolk below the first developing cell and DNA was injected inside the cytoplasm at 1-cell stage.

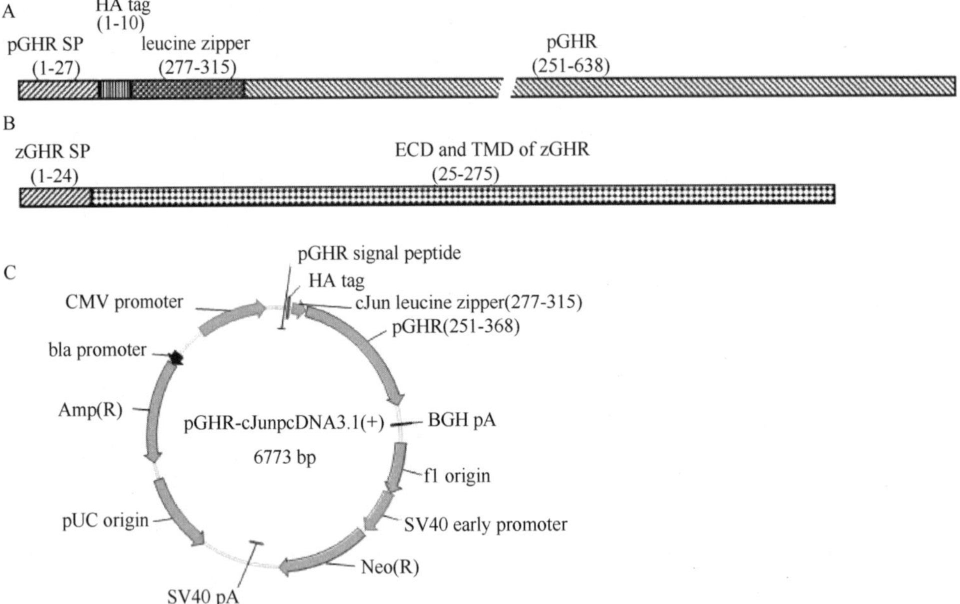

Fig. 1 Schematic description of leucine zipper transgene chimera *Jun-GHR* (a) and C-terminal truncated zebrafish *GHR* (b) and the plasmid pGHR-cJun-pcDNA3.1(+) (c). Numbers in parentheses refer to the amino acid sequence from which the respective cDNA segments were taken. SP signal peptide, *pGHR* porcine GHR, *zGHR* zebrafish GHR, *ECD* extracellular domain, *TMD* transmembrane domain, *BGH pA* bovine GH polyA

Constructs design and mRNA synthesis

The CA-*GHR* construct was described previously (Behncken *et al.* 2000). As shown in

Fig. 1a, the transgene codes for a porcine GHR signal peptide (amino acids 1-27), the mouse *c-jun* leucine zipper (amino acids 277-315) fused upstream of the porcine GHR transmembrane and cytoplasmic domains (amino acids 251-638), cloned into the pcDNA3.1(+) vector. The transgene expression cassette is driven by a CMV promoter and utilizes a bovine GH polyA (bGH pA) (Fig. 1c) and the construct pGHR-cJun-pcDNA3.1 is hereafter termed as *Jun-GHR*. cDNA of 3 dpf (day-post-fertilization) zebrafish was used to amplify the full length cDNA and the sequence coding the extracellular and transmembrane domains (dominant negative C-terminal truncated protein, ΔC-*GHR*, Fig. 1b) of *GHRa* via high-fidelity PCR by using Kod-plus enzyme (Toyobo, Japan) and appropriate primers (Table 1). The primers were designed according to zebrafish *GHRa* cDNA sequence available in GenBank (NM_001083578). Common carp *GH* (c*GH*) cDNA was amplified from cDNA pool of common carp pituitary with primers cGH_P1 and cGH_P2 (Table 1). For overexpression of c*GH* and z*GHR*, the PCR products were cloned into pCS2+ Expression vector after double digestion of the PCR products. For mRNA synthesis, ΔC-*GHR*-pCS2+ construct was linearized with *Kpn*I and capped mRNA was synthesized using the Message Machine kit (Ambion Inc., USA) as previously described (Chen *et al.* 2009). All the DNA constructs have the same constitutive CMV promoter.

Table 1 Primers used in the study

Primer name	Sequence (5′-3′)	Gene (purpose)	Expected size (bp)
GH_P1	GTTGGTGGTGGTTAGTTTGCT	*GH* (RT-PCR)	331
GH_P2	CAGGCTGTTTGAGATAGTGGAG		
GHRa_P1	ACCGATAAAGAGTATGAAGTGCG	*GHR* (RT-PCR)	541
GHRa_P2	TGTGAAGAAGGGAAGCCAAG		
c-fos_P1	GTGGGAGCAGGAACTGAGGG	*c-fos* (RT-PCR)	618
c-fos_P2	GGTTCTTGTTTACTGGAGGGATA		
igf1_P1	GTCTAGCGGTCATTTCTTCCA	*igf1* (RT-PCR)	320
igf1_P2	CAGGCGCACAATACATCTCG		
BA_P1	ATCTGGCATCACACCTTCTACAAC	*β-actin* (RT-PCR)	254
BA_P2	TAACCCTCATAGATGGGCACGGT		
Jun_GHR_P1	TGCTCAGGGAACAGGTGG	Transgene detection	715
Jun_GHR_P2	TGGGCAGTTTGATGAGTTGA		
GHR_P3	GCAGGATCCATGGCCCACTCGCTCTCTCTC	*GHR* (full length cloning)	1731
GHR_P4	AGGCTCGAGTTATCTGGGTTGCGCAGATAAG		
GHR_P5	AGGCTCGAGTCACCTCTGCTGCTGGGAGATGAC	*GHR* (C-terminal truncated cloning)	846

Continued

Primer name	Sequence (5′-3′)	Gene (purpose)	Expected size (bp)
cGH_P1	TTCGAATTCTGAGCGAAATGGCTAGAGTA	*carp GH* (cDNA cloning)	924
cGH_P2	AGTTCTAGATAAATTGCTTAGACACCACTGT		
c-fos_P3	TCGCTGATTCGCAGACACGC	*c-fos* (real-time PCR)	199
c-fos_P4	ATCGCTCTACATCCATCTCACAGTCC		
igf1-P3	TTCAGCAAACCGACAGGATA	*igf1* (real-time PCR)	124
igf1-P4	TCTTCACAGGCGCACAATAC		
BA_P3	TCACCACCACAGCCGAAAG	*β-actin* (real-time PCR)	98
BA_P4	AGAGGCAGCGGTTCCCAT		

RT-PCR (reverse-transcription PCR) analysis of GHR signaling related genes

Total RNA of 50 zebrafish embryos, pooled tissues from 3 wildtype individuals or tissues from one transgenic individual was extracted using TRIzol® reagent (Invitrogen, USA) according to the manufacturer's instructions. All RNA samples were treated with RNase-Free DNase (Promega, USA) according to manufacturer's protocol. cDNA was synthesized by reverse transcription (RT) from 1 μg of total RNA using the ReverTra Ace enzyme (TOYOBO, Japan), dNTPs and Oligo(dT)20 RT primer (TOYOBO, Japan). The RT reaction was performed at 42℃ for 60 min followed by 95℃ for 5 min. PCR was performed to study the expression of different genes, such as *GH*, *GHR*, *c-fos* and *igfl* in different samples. The primer sequences are shown in Table 1. Each pair of primer sequences locate in different exons of the respective gene, to avoid the amplification of genomic DNA potentially contaminated in the samples. PCR was carried out in a 25 μl reaction volume containing 2.5 μl of 10× PCR buffer, 0.2 μM of each primer, 0.2 mM of each dNTP, 0.75 mM of $MgCl_2$, 0.5 unit of *Taq* DNA polymerase (Fermentas, Canada) and 50 ng of cDNA solution. The PCR reaction consists of a denaturation at 94℃ for 5 min, 35 cycles of 30 s at 94℃, 30 s at 56℃ and 45 s at 72℃, with a final elongation step of 10 min at 72℃. *β-actin* primer pairs (Table l) were used to normalize the cDNA concentration of all samples. The products were run on l% agarose gels stained with ethidium bromide (0.5 μg/ml), and amplified bands Were visualized by ultraviolet transillumination and semiquantified by Glyko Bandscan software (Novato, CA).

Real-time quantitative PCR analysis of *c-fos* and *igf1*

Fertilized zebrafish zygotes were microinjected with the DNA constructs *Jun-GHR*, zebrafish *GHRa* (*zGHRa*), carp *GH* (c*GH*), c*GH* and z*GHRa* at a concentration of 50 ng/μl for each construct. To illustrate the importance of functional *GHR*, c*GH* expression construct of 50 ng/μl and ΔC-*GHR* mRNA of 800 ng/μl were co-injected into zebrafish embryos. Each embryo was injected with 1 nl of DNA or RNA sample and each sample was injected with 300 embryos. cDNA from the manipulated zebrafish embryos at 2-cell stage, 1 and 3 dpf and from different tissues of adult transgenic zebrafish was synthesized as described above and analyzed with real-time quantitative PCR for *c-fos* and *igf1* expression level. Real-time PCR was performed on an ABI PRISM® 7000 Sequence Detector (Applied Biosystems, Inc. USA) according to the manufacturer's instructions. Reactions were performed in a 20 μl volume with 150 ng of cDNA, 0.2 μM primers and 10 μl of SYBR® Green Realtime PCR Master Mix (TOYOBO, Japan). The primer pairs were shown in Table 1. All reactions were run using the following conditions: 1 min denaturation at 95℃ followed by 40 cycles of 95℃ denaturation for 15 s, 55℃ annealing for 15 s, and 72℃ extension for 45 s. Detection of the fluorescent product was carried out at the end of the 72℃ extension period. The data (Ct value), obtained from the software (7000 system SDS software), was transferred to Windows Excel (Microsoft co.) sheet and fold-change was calculated using $2^{-\Delta\Delta Ct}$ method, with *β-actin* as the calibrator. To validate the stability of *β-actin* for real-time PCR normalization, several pairs of primers specific for *PSB7*, *eTIF-2B*, *eEF-1A* and *GAPDH* were designed as described in our previous study and used for GeNorm analysis (Vandesompele *et al.* 2002; Pei *et al.* 2007). Three samples were run for each analysis and all real-time PCR reactions were run in triplicates. The significance of the mean differences between various experimental groups was determined by one way ANOVA followed by Duncan's multiple range test analyses. A P value < 0.05 was considered statistically significant.

In vivo luciferase assay of *spi2.1* promoter activity

Each zebrafish zygote was co-injected with *spi2.1-luc* (Jiao *et al.* 2006) of 50 pg and a constitutively expressed *TK-Renilla* luciferase construct (Promega) of 0.5 pg and the manipulated embryos were subsequently injected with indicated DNA or RNA samples. Embryos were allowed to develop until 3 dpf, then sets of 20 embryos were lysed in passive lysis buffer (Dual-Luciferase Reporter Assay System Promega) and the luciferase activity were measured with a Berthold luminometer and relative luciferase activity was calculated as described (Liu *et al.* 2009). Each sample was analyzed in triplicate, and mean value and standard deviation were calculated. The significance of the mean differences between various experimental groups was determined by one way ANOVA followed by Duncan's multiple range test analyses. A P value < 0.05 was considered statistically significant.

Generation, screening and growth trial of transgenic zebrafish

Fertilized zebrafish zygotes were microinjected with *Jun-GHR* construct of 50 ng/μl.

Transgenic embryos and fish were raised according to standard procedure (Westerfield 2000). Transgene positive P0 fish were screened out by PCR assay of the total DNA of caudal fin. DNA samples were extracted from caudal fin samples or embryo samples by means of treatment with DNA extraction buffer and phenol/chloroform purification, followed by ethanol precipitation as described (Sun *et al.* 2005). PCR reactions were performed in a reaction mix of 25 μl, containing 50 ng of total DNA, 0.5 unit of Taq DNA polymerase (Fermentas, Canada), 1×Taq buffer, 0.5 μl of 10 mM dNTPs (Generay Biotech, China), 1 μl of each 10 μM primer, and a suitable amount of sterile deionized water. The PCR protocol used for transgene detection was as follows: 5 min denaturation at 95℃, 30 cycles of 30 s at 95℃, 30 s at 55℃ and 45 s at 72℃ and a final elongation step for 7 min at 72℃. *Jun-GHR*_P1 and *Jun-GHR*_ P2 (Table 1) were used as PCR primers. All PCR products were separated by 1% agarose gel electrophoresis and visualized by ultraviolet transillumination.

Each matured P0 transgenic zebrafish was mated with wildtype zebrafish and a pool of 10 F1 embryos from each cross was checked by PCR detection of the transgene. Each batch of embryos showing transgene positive was considered as an individual F1 transgenic batch. At 1 month-post-fertilization (mpf), tail fin DNA was extracted to check the presence of transgene in each F1 fish. 100 transgenic individuals from one batch of F1 embryos and 100 non-transgenics were raised in aquaria at the same condition. The body length and weight were measured from 20 individuals of transgenic and non-transgenic zebrafish after random selection from 1 to 4 mpf. Student's *t*-test was applied for statistical comparison and the differences were considered significant at $P < 0.01$. All values in the text and figures refer to mean ± standard deviation (SD).

Results

Spatio-temporal expression of GH signaling components

RNA from various tissues was extracted and RT-PCR was used to analyze the transcription of *GH*, *GHR*, *c-fos* (Cesena *et al.* 2007) and *igf1* (Wood *et al.* 2005), which represent the ligand, receptor and targets in the signaling, in different tissues as well as in a series of developmental stages of wildtype zebrafish (Fig. 2a, b).

In adult fish (Fig. 2a), *GH* was expressed in brain, liver, ovary and muscle with the highest expression level in the brain that includes pituitary gland by analysis with Glyko Bandscan software. *GHR* transcript was found in all tissues examined except for heart. The highest levels of *GHR* expression were observed in liver, followed by brain and ovary. In the case of *c-fos*, the highest expression was observed in ovary followed by brain, and in all other tissues except the kidney. *igf1* transcripts were present in all tissues except kidney with the highest expression level in liver.

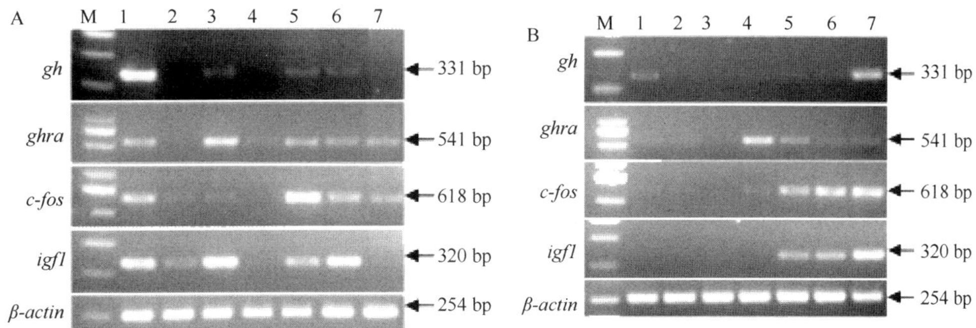

Fig. 2 The transcription analysis of *GH*, *GHR*, *c-fos* and *igf1* in wildtype zebrafish. a RT-PCR analysis in different tissues of adult zebrafish. 1 brain, 2 heart, 3 liver, 4 kidney, 5 ovary, 6 muscle, 7 fin, and M DNA Ladder (DL 2000). b RT-PCR analysis in different developmental stages of zebrafish embryo. 1 2-cell, 2 high, 3 80%-epiboly, 4 early somite, 5 1 dpf, 6 2 dpf, 7 3 dpf, and M DNA ladder (DL 2000). *β-actin* was used as the internal control

During different stages of embryonic development (Fig. 2b), *GH* transcript was detected in 2-cell stage embryos but not at later stages, reappearing from 1 dpf, and increasing with development. This indicates that *GH* is maternally expressed and its zygotic expression starts from 1 dpf. *GHR* is also maternally expressed although the expression level might be not as high as that of *GH*. The embryonic expression of *GHR* starts from the early somite stage, and in contrast to *GH*, the expression of *GHR* decreased from the early somite stage to 3 dpf. Although *c-fos* transcript was in 2-cell embryos, from the early somite stage expression of *c-fos* increased gradually with the time, mimicking the *GH* expression profile. *igf1* was minimally expressed maternally, with expression from 1 dpf, and increasing thereafter.

Activation of GHR signaling by combined overexpression of GH and GHR in zebrafish embryos

In order to quantify GH signaling in transgenic fish, we measured expression of the GH target genes *c-fos* and *igf1*. Since *c-fos* reached a rather high expression level at 1 dpf and *igf1* reached its highest expression level at 3 dpf, we used 1 dpf and 3 dpf embryos to check the expression levels of *c-fos* and *igf1*, respectively, by normalizing values to 2-cell embryo levels, since these should not be altered by the transgene. By GeNorm analysis, we found that *β-actin* is the most stable gene among all the candidate reference genes (data not shown) and it could be used to quantify the relative expression levels of *c-fos* and *igf1*. First, we injected c*GH* expression construct into zebrafish embryos and found that *c-fos* expression was 1.7-fold and *igf1* expression was 3.1-fold in the *GH*-injected embryos in comparison to wildtype embryos (Fig. 3a, b), suggesting significant GHSA in the injected embryos. Second, we overexpressed zebrafish *GHR* and found 1.7-fold of *c-fos* expression and 3.5-fold of *igf1* expression, demonstrating that overexpression of *GHR* could also activate GH signaling in the presence of endogenous GH. Third, we used a synergetic expression of *GH* and *GHR*, and this overexpression showed 2.0-fold of *c-fos* expression and 4.5-fold of *igf1* expression, higher than these constructs separately. In contrast, overexpression of ΔC-*GHR* efficiently attenuated the GHSA activity of *GH* overexpression, as shown by a lower expression level of *igf1* (1.5-fold) than the overexpression of *GH* alone (3.1-fold). These results clearly indicate the importance of GHR level in GH signaling, and by implication, fish growth, since Igf1 is one of the major effectors of growth performance (Eppler *et al.* 2010).

On the other hand, we conducted luciferase assay by utilizing *spi2.1-luc* construct (Jiao *et al.* 2006), to evaluate the overall GHSA level in zebrafish embryos. As shown in Fig. 3c, c*GH* overexpression induced 2.3-fold GHSA in comparison with wildtype embryos, z*GHR* overexpression induced 3.7-fold GHSA and the synergetic overexpression of GH and GHR could even result in a 6.1-fold GHSA. Thus the *in vivo* analysis of GHR signaling activity is validated by the luciferase assay of *spi2.1* promoter activity. Both expression analysis of *c-fos* and *igf1* and promoter activity analysis of *spi2.1* showed that higher GHR signaling activation could be achieved by synergetic expression of GH and GHR.

Elevated activation of GHR signaling by Jun-GHR overexpression in zebrafish embryos

To test whether our designed *Jun-GHR* constructs could activate GH signaling, we injected *Jun-GHR* construct into zebrafish embryos and checked the expression of *c-fos* in 1 dpf and *igf1* in 3 dpf embryos (Fig. 3a, b). The relative expression levels of *c-fos* and *igf1* were 3.9-fold and 10.8-fold, respectively, in *Jun-GHR* transgenic embryos, relative to the expression level of wildtype embryos. By contrast, the synergetic overexpression of *GH* and *GHR*, which showed the highest relative expression levels of *c-fos* and *igf1* in our earlier injection combinations, were only 2.0-fold and 4.5-fold of wildtype embryos. We also checked the promoter activity of *spi2.1* in *Jun-GHR* injected embryos. The relative luciferase activity of the *Jun-GHR* injected embryos was 26.8-fold, significantly higher than that of the cGH + zGHR injected embryos, which was only 6.1-fold. Herein the overexpression of *Jun-GHR* gave a higher efficiency of activation of signaling in zebrafish embryos when compared with the synergetic overexpression of *GH* and *GHR*. This result has been revealed by the stimulation of two GH target genes, *c-fos* and *igf1*, and the activation of GH signaling responsive *spi2.1* promoter. Thus it is expected that we would obtain transgenic zebrafish with accelerated growth performance using *Jun-GHR* construct as the transgene.

Activation of GH targets and accelerated growth of transgenic fish

As shown by RT-PCR analysis during development of transgenic embryos, the transcription of *Jun-GHR* starts from high stage, reached the highest level in 3 dpf embryos (Fig. 4a). By PCR

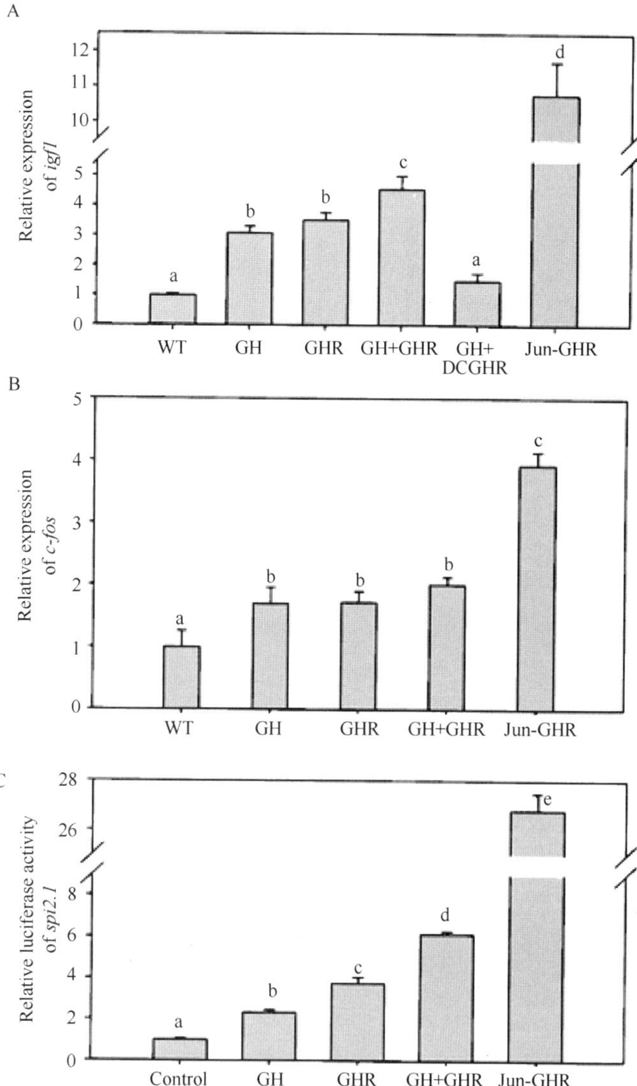

Fig. 3 Analysis of GH-signal activation levels in zebrafish embryos receiving different constructs. A Relative expression of *igf1* in zebrafish embryos at 3 dpf by real-time PCR analysis. B Relative expression of *c-fos* in zebrafish embryos at 1 dpf by real-time PCR analysis. C Relative promoter activities of *spi2.1* in zebrafish embryos at 3 dpf by luciferase assay. Control refers to wildtype embryos that were only injected with *spi2.1-luc* and *TK-Renilla* constructs. In a, b and c, each *bar* presents the mean value and the corresponding standard deviation from triplicate analysis. A statistically significant difference ($P < 0.05$) is indicated by a *different letter* above the bar, as determined by one-way ANOVA followed by Duncan's test

screening of transgenic P0 generation we obtained 16 *Jun-GHR* transgene positive founders. We further sampled one P0 *Jun-GHR* transgenic fish, and *Jun-GHR* transcript expression was found in almost all the tissues tested, although expression in the brain was weak (Fig. 4b). More interestingly, by both bandscan analysis of the agarose gel bands (Fig. 4b) and real-time PCR assay (Fig. 4c, d), we detected higher expression of GH target genes, *c-fos* and *igf1* in

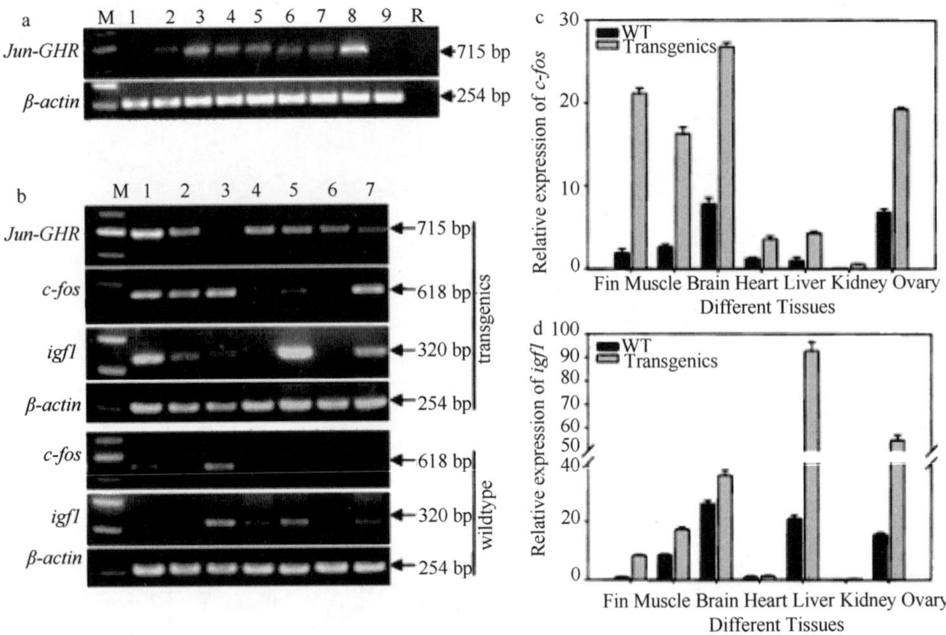

Fig. 4 The transcription of *Jun-GHR*, *c-fos* and *igf1* in *Jun-GHR* transgenic zebrafish. a *Jun-GHR* transcription in transgenic embryos at different developmental stages. *1* 2-cell, *2* high, *3* shield, *4* 80%-epiboly, *5* early somite, *6* 1 dpf, *7* 2 dpf, *8* 3 dpf, *9* wildtype embryos, *R* template of total RNA without reverse transcription. b The transcription of *Jun-GHR* transgene, *c-fos* and *igf1* in different tissues of transgenic zebrafish. *1* fin, *2*muscle, *3* brain, *4* heart, *5* liver, *6* kidney, *7* ovary, and *M* DNA Ladder (DL 2000). *β-actin* amplified from the same samples was used as the internal control. c The relative expression levels of *c-fos* in different tissues of wildtype and in transgenic zebrafish. d The relative expression levels of *igf1* in different tissues of wildtype and in transgenic zebrafish. In c and d, each *bar* presents the mean value and the corresponding standard deviation from triplicate analysis

different tissues of transgenic zebrafish. For instance, higher expression of *c-fos* was observed in all the tissues except kidney (Fig. 4b, c). Ectopic expression of *igf1* was even found in the fin tissue of transgenic fish, since there was nearly no expression of *igf1* in the fin tissue of wildtype fish (Fig. 4b, d). These results strongly indicate that *Jun-GHR* transgene is faithfully transcribed in transgenic fish, and as a result strongly activates GH/GHR target genes in various tissues.

All the transgene positive founders were crossed with wildtype zebrafish to produce the F1 generation. Among them, 2 crosses gave transgene positive F1 embryos. At 1 mpf, we screened out 100 transgene positive zebrafish from 1 cross and raised them in the aquaria,with 100 non-transgenic zebrafish in the same condition as control. From 2 mpf, the average body weight and body length (mean ± SD) of *Jun-GHR* transgenic F1 fish showed to be significantly higher ($P < 0.01$) than those of the non-transgenics (Fig. 5a, b). At 4 mpf, the transgenic fish weighed 0.56 ± 0.08 g at average, 86.1% higher than the average body weight of controls (0.30 ± 0.08 g) ($P < 0.01$), and the average body length of the transgenics was 3.61 ±

0.18 cm, 25.2% higher than that of the controls (2.88 ± 0.29 cm) ($P < 0.01$). As shown in Fig. 5c, the transgenic fish showed wider lateral bodies with wider and thicker dorsal bodies than the non-transgenic control fish as what we found in *GH*-transgenic common carp (Wang *et al.* 2001). These demonstrate that the expression of *Jun-GHR* transgene resulted in growth acceleration of transgenic zebrafish.

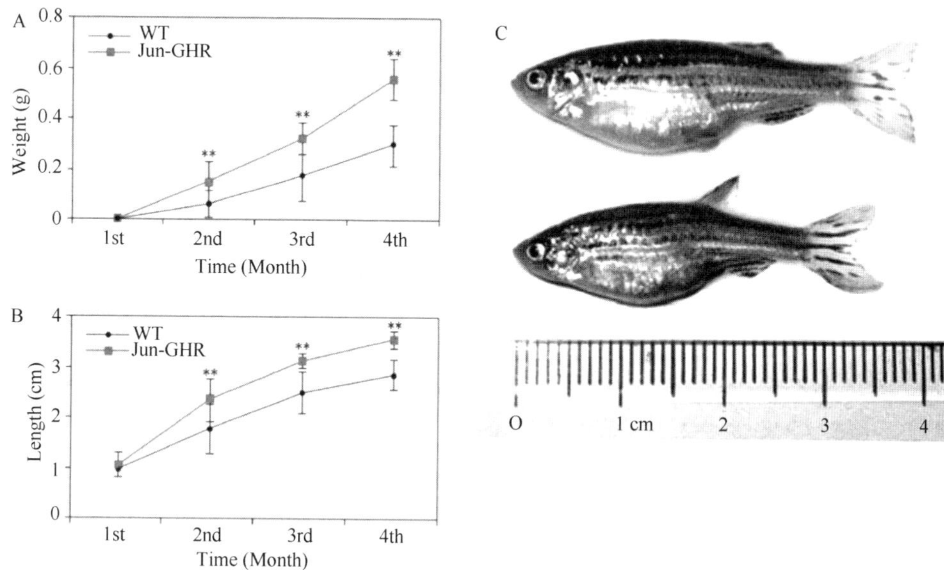

Fig. 5 Growth trial of F1 transgenic and non-transgenic zebrafish. a The increase of body weight along development. b The increase of body length along development. c The typical phenotype of *Jun-GHR* transgenic (*upper*) and non-transgenic (*lower*) zebrafish. Weight and length of transgenic and non-transgenic zebrafish were statistically compared by student's *t*-test (** $P< 0.01$)

Discussion

We have previously demonstrated that GH and GHR are present from a very early embryonic stage in mice (Pantaleon *et al.* 1997), and summarized potential contributions of GH to prenatal development in mammals (Waters and Kaye 2002). In this study, the expression of *GH, GHR*, *c-fos* and *igf1* genes were extensively and comparatively analyzed during 7 stages of embryonic development and in 7 different tissues of adult zebrafish. Transcription of *GH, GHR* and *c-fos* was observed at the 2-cell stage, indicating maternal expression of these genes, and detection of their expression in the ovary further suggests that they start to be accumulated in the oocytes during maturation. For most of them, the zygotic expression was detected from the end of gastrulation, i.e., early somite stage, far prior to the formation of functional pituitary gland. Expression of *GH* increased with expression of *igf1* at 3 dpf. The maternal expression and early expression of *GH* and *igf1* in fishes have also been reported by several investigators (Li *et al.* 2006, 2007). These observations suggest that GH/GHR pathway also plays a role in early development of fish, and in our present study, overexpression of GHR or GH alone led to increased GHSA in 1 and 3 dpf embryos, strongly indicating the existence of functional GH and GHR expression in zebrafish embryos from 1 dpf. Moreover, blocking GHSA in zebrafish during early development by overexpression a dominant-negative *GHR*, ΔC-*GHR*, resulted in embryos with early developmental defects (Ishtiaq Ahmed *et al.*, unpublished data). In previous studies, it has been found that GH can stimulate actin rearrangement (Goh *et al.* 1997), microtubule polymerization (Goh *et al.* 1998), and the assembly of multiprotein complex involved in cell adhesion and cell movement (Zhu *et al.* 1998a, b). Given that the fish embryos undergo dynamic cell movement in early development, e.g., gastrulation movements (Chen *et al.* 2009), it is possible that the GH/GHR pathway may participate in the regulation of cell movement during early development. It was also reported that GH/Stat5b can directly regulate the transcription of a Wnt signaling element (*frizzled4*) and suppressors of cytokine signaling (Vidal *et al.* 2007), thus it is also possible that GH/GHR pathway may crosstalk with the other pathways to regulate the early development of fish.

In most of previous studies, the transcriptional activity of GH signaling in fish has been measured in cultured cells or tissues (Björnsson *et al.* 2002; Jiao *et al.* 2006). Here we have utilized zebrafish embryos as a host system to analyze the GH signaling target gene *igf1* and the promoter activity of *spi2.1*, after injection of different combination of transgene constructs, such as *GH*, *GHR*, *GH* + *GHR*, and *GH* + ΔC-*GHR* into zygotes. This has allowed us to compare the signaling activity of GH/GHR pathway upon treatment with different GH pathway-related components *in vivo*. As compared with overexpression of *GH* or *GHR* alone, the combined overexpression of *GH* and *GHR* increased GHSA, and overexpression of ΔC-*GHR* which lacks the intracellular signal transducer domain efficiently attenuated the transcriptional activity. These results validate the efficacy of this *in vivo* system for studying GH/GHR signaling.

Because GH-induced activation of the GHR is the critical step in GH signaling, we utilized a constitutively activated form of *GHR*, *Jun-GHR* to ascertain the effect of continuous activation of GHR in the absence of the IGF-1 feedback on GH secretion. Although shown to be highly active in cell cultures (Behncken *et al.* 2000), this construct has not been studied *in vivo*. By using zebrafish as a host animal, we found that *Jun-GHR* transgenics could not only strongly activate the GH/GHR signaling but also stimulate the growth performance of the transgenic fish. Notably, we have never previously been able to stimulate growth of these zebrafish with *GH* transgenes alone. In terms of GH target gene activation, *Jun-GHR* increased the transcriptional activity of the GH/GHR pathway very significantly more than the overexpression of *GH*, *GHR*, and even *GH* + *GHR*. This demonstrates that CA-*GHR* transgenesis should be much more powerful than the common *GH* transgenesis in the aspect of signaling activation and growth stimulation. Intriguingly, *igf1* was even ectopically activated in the fin tissue of transgenic zebrafish, which tissue did not display detectable *igf1* transcription in wildtype zebrafish. Surprisingly, although *Jun-GHR* was transcribed in the kidney of transgenic fish, we could not detect visible transcription of *igf1* or *c-fos* in the kidney, which suggests that the zebrafish kidney may lack some intracellular transducer for the GH/GHR pathway. The broad distribution of *Jun-GHR* transcripts in various tissues and strong activation of GH/GHR target genes have presumably contributed to the allometric growth stimulation of transgenic zebrafish, which will be tremendously useful for increasing yields of transgenic farmed fish. Nevertheless, in future studies we should use "all-fish" CA-*GHR* constructs for production of transgenic commercial fishes, as CMV promoter was used in our present study.

In previous studies of transgenic animals, researchers usually utilized chimeric constructs. Although the promoters and the coding sequences are from different genes (Laible 2009), the final products are usually the wildtype forms of certain proteins. Here we propose that researchers may produce transgenic animals by introducing the conception of molecular designed protein, e.g., the *Jun-GHR* product does not exist in the nature but show higher efficiency than the natural proteins. As the comprehensive understanding of many pathways has been extensively advanced, it is certain that other transgenes could be usefully and efficiently modified based on protein design.

Acknowledgements The authors thank Prof. Christopher HK Cheng (the Chinese University of Hong Kong) and Prof. Nils Billestrup (University of Copenhagen) for sharing *spi2.1-luc* construct, and Prof. Cunming Duan (University of Michigan) and Prof. Yong Zhu (East Carolina University) for valuable suggestions. This work was supported by the China 973 Project (2010CB126306), the National Natural Science Foundation of China (30972248 & 30771100), China 863 project (2007AA10Z186), the Youth Sunshine Project of Wuhan City and the Youth Innovation Project of Institute of Hydrobiology, CAS to Y.H.S.

References

Behncken SN, Billestrup N, Brown R *et al* (2000) Growth hormone (GH)-independent dimerization of GH receptor by a leucine zipper results in constitutive activation. J Biol Chem 275(22): 17000-17007

Björnsson B, Johansson V, Benedet S *et al* (2002) Growth hormone endocrinology of salmonids: regulatory mechanisms and mode of action. Fish Physiol Biochem 27(3): 227-242

Cesena TI, Cui TX, Piwien-Pilipuk G *et al* (2007) Multiple mechanisms of growth hormone-regulated gene transcription. Mol Genet Metab 90(2): 126-133

Chen TT, Vrolijk NH, Lu JK *et al* (1996) Transgenic fish and its application in basic and applied research. Biotechnol Ann Rev 2: 205-236

Chen CH, Sun YH, Pei DS *et al* (2009) Comparative expression of zebrafish lats1 and lats2 and their implication in gastrulation movements. Dev Dyn 238(11): 2850-2859

Devlin RH, Biagi CA, Yesaki TY *et al* (2001) Growth of domesticated transgenic fish. Nature 409(6822): 781-782

Devlin RH, Sundstrom LF, Muir WM (2006) Interface of biotechnology and ecology for environmental risk assessments of transgenic fish. Trends Biotechnol 24(2): 89-97

Devlin RH, Sakhrani D, Tymchuk WE *et al* (2009) Domestication and growth hormone transgenesis cause similar changes in gene expression in coho salmon (*Oncorhynchus kisutch*). Proc Natl Acad Sci USA 106(9): 3047-3052

Di Prinzio CM, Botta PE, Barriga EH *et al* (2010) Growth hormone receptors in zebrafish (*Danio rerio*): adult and embryonic expression patterns. Gene Expr Patterns 10(4-5): 214-225

Eppler E, Berishvili G, Mazel P *et al* (2010) Distinct organ-specific up- and down-regulation of IGF-I and IGF-II mRNA in various organs of a GH-overexpressing transgenic Nile tilapia. Transgenic Res 19(2): 231-240

Fu C, Hu W, Wang Y *et al* (2005) Developments in transgenic fish in the People's Republic of China. Rev Sci Tech 24(1): 299-307

Goh EL, Pircher TJ, Wood TJ *et al* (1997) Growth hormone-induced reorganization of the actin cytoskeleton is not required for STAT5 (signal transducer and activator of transcription-5)-mediated transcription. Endocrinology 138(8): 3207-3215

Goh EL, Pircher TJ, LobiePE (1998) Growth hormone promotion of tubulin polymerization stabilizes the microtubule network and protects against colchicine-induced apoptosis. Endocrinology 139(10): 4364-4372

Herrington J, Carter-Su C (2001) Signaling pathways activated by the growth hormone receptor. Trends Endocrinol Metab 12(6): 252-257

Ihle JN, Gilliland DG (2007) Jak2: normal function and role in hematopoietic disorders. Curr Opin Genet Dev 17(1): 8-14

Jiao B, Huang X, Chan CB *et al* (2006) The co-existence of two growth hormone receptors in teleost fish and their differential signal transduction, tissue distribution and hormonal regulation of expression in seabream. J Mol Endocrinol 36(1): 23-40

Kikuta H, Kawakami K (2009) Transient and stable transgenesis using Tol2 transposon vectors. In: Graham JL, Andrew CO, Kawakami K (eds) Zebrafish, methods in Molecular Biology, vol 546. Humana Press, New York, pp 69-84

Laible G (2009) Enhancing livestock through genetic engineering—recent advances and future prospects. Comparative immunology. Microbiol Infect Dis 32(2): 123-137

Li M, Greenaway J, Raine J *et al* (2006) Growth hormone and insulin-like growth factor gene expression prior to the development of the pituitary gland in rainbow trout (*Oncorhynchus mykiss*) embryos reared at two temperatures. Comp Biochem Physiol A Mol Integr Physiol 143(4): 514-522

Li M, Raine JC, Leatherland JF (2007) Expression profiles of growth-related genes during the very early development of rainbow trout embryos reared at two incubation temperatures. Gen Comp Endocrinol 153(1-3): 302-310

Lichanska AM, Waters MJ (2008) How growth hormone controls growth, obesity and sexual dimorphism. Trends Genet 24(1): 41-47

Liu WY, Wang Y, Sun YH *et al* (2005) Efficient RNA interference in zebrafish embryos using siRNA synthesized with SP6 RNA polymerase. Dev Growth Differ 47(5): 323-331

Liu Y, Bourgeois C, Pang S *et al* (2009) The germ cell nuclear proteins hnRNP G-T and RBMY activate a testis-specific exon. Plos Genetics 5(11): e1000707

Pantaleon M, Whiteside EJ, Harvey MB *et al* (1997) Functional growth hormone (GH) receptors and GH are expressed by preimplantation mouse embryos: a role for GH in early embryogenesis? Proc Natl Acad Sci USA 94(10): 5125-5130

Pei DS, Sun YH, Chen SP *et al* (2007) Zebrafish GAPDH can be used as a reference gene for expression analysis in cross-subfamily cloned embryos. Anal Biochem 363(2): 291-293

Rowland JE, Lichanska AM, Kerr LM *et al* (2005) In vivo analysis of growth hormone receptor signaling domains and their associated transcripts. Mol Cell Biol 25(1): 66-77

Stuart GW, McMurray JV, Westerfield M (1988) Replication, integration and stable germ-line transmission of foreign sequences injected into early zebrafish embryos. Devel-opment 103(2):403-412

Studzinski AL, Almeida DV, Lanes CF *et al* (2009) SOCS1 and SOCS3 are the main negative modulators of the somatotrophic axis in liver of homozygous GH-transgenic zebrafish (Danio rerio). Gen Comp Endocrinol 161: 67-72

Sun YH, Chen SP, Wang YP *et al* (2005) Cytoplasmic impact on cross-genus cloned fish derived from transgenic com-mon carp (Cyprinus carpio) nuclei and goldfish (*Carassius auratus*) enucleated eggs. Biol Reprod 72(3): 510-515

Vandesompele J, De Preter K, Pattyn F *et al* (2002) Accurate normalization of real-time quantitative RT-PCR data by geometric averaging of multiple internal control genes. Genome Biol 3(7): RESEARCH0034

Vidal OM, Merino R, Rico-Bautista E *et al* (2007) In vivo transcript profiling and phylogenetic analysis identifies suppressor of cytokine signaling 2 as a direct signal transducer and activator of transcription 5b target in liver. Mol Endocrinol 21(1): 293-311

Wang Y, Hu W, Wu G *et al* (2001) Genetic analysis of all fish growth hormone gene transferred carp (*Cyprinus carpio* L.) and its F1 generation. Chinese Sci Bull 46(14): 1174-1177

Waters MJ, Kaye PL (2002) The role of growth hormone in fetal development. Growth Horm IGF Res 12(3):137-146

Waters MJ, Hoang HN, Fairlie DP *et al* (2006) New insights into growth hormone action. J Mol Endocrinol 36(1):1-7

Westerfield M (2000) The zebrafish book. A guide for the laboratory use of zebrafish (Danio rerio). Eugene, University of Oregon Press

Wood AW, Duan C, Bern HA (2005) Insulin-like growth factor signaling in fish. Int Rev Cytol 243: 215-285

Wu G, Sun Y, Zhu Z (2003) Growth hormone gene transfer in common carp. Aquat Living Resour 16: 416-420

Zhu ZY, Sun YH (2000) Embryonic and genetic manipulation in fish. Cell Res 10(1): 17-27

Zhu Z, Li G, He L *et al* (1985) Novel gene transfer into the fertilized eggs of gold fish (*Carassius auratus* L. 1758). J Appl Ichthyol 1(1): 31-34

Zhu T, Goh EL, LeRoith D *et al* (1998a) Growth hormone stimulates the formation of a multiprotein signaling complex involving p130(Cas) and CrkII. Resultant actiation of c-Jun N-terminal kinase/stress-activated protein kinase (JNK/SAPK). J Biol Chem 273(50): 33864-33875

Zhu T, Goh EL, Lobie PE (1998b) Growth hormone stimulates the tyrosine phosphorylation and association of p125 focal adhesion kinase (FAK) with JAK2. Fak is not required for stat-mediated transcription. J Biol Chem 273(17): 10682-10689

Zhu Y, Song D, Tran NT et al (2007) The effects of the members of growth hormone family knockdown in zebrafish devel-opment. Gen Comp Endocrinol 150(3): 395-404

组成型激活的GHR构建体介导斑马鱼GH信号通路的活化

A. S. Ishtiaq Ahmed 熊 凤 庞少臣 何牡丹
Michael J. Waters 朱作言 孙永华

中国科学院水生生物研究所，武汉 430072，中国
Physiology & Pharmacology Department and Centrefor Molecular & Cellular Biology, UniversityofQueensland, St. Lucia, QLD 4072, Australia
中国科学院大学，北京 100049，中国

摘 要 已有的研究证实，转植生长激素基因能够显著促进转基因鱼的生长。在本研究中，我们建立了一种通过转录活性来评估生长激素基因活性(GHSA)的方法。在斑马鱼胚胎中，我们导入由 *c-fos*、*igf-1* 和 *sp2.1* 启动子等元件构成的不同类型表达载体，实验发现，无论是过表达生长激素(GH)还是生长激素受体(GHR)都能激活生长激素信号通路的活性，而协同表达 GH 和 GHR 会更大程度地激活生长激素信号通路的活性。相反地，过表达 C 末端缺失的 GHR 则能有效阻断其与 GH 的结合，此结果表明，GHSA 依赖于 GHR 二聚体的形成。鉴于细胞膜外的 GH 对 GHR 二聚体形成的重要性，我们构建了一个在斑马鱼胚胎中能够持续性激活的 GHR 载体(CA-GHR)，该载体的表达蛋白产物通过 Jun 拉链结构使 GHR 持续以二聚体方式形成，从而激活 GH 信号通路下游基因的表达。更重要的是，过表达 CA-GHR 比协同表达 GH 和 GHR 更能显著提高 GH 信号通路的水平。转 CA-GHR 斑马鱼的生长测定结果显示，其生长指数高于对照鱼，而这种提高并不依赖于 GH 的高表达。我们的研究结果表明，这种不依赖于 GH 的 CA-GHR 促生长作用不受 IGF-1 的反馈调控，同时也提供了一种新型的通过分子设计构建促生长的转基因动物模型的方法。

Vitreoscilla Hemoglobin (VHb) Overexpression Increases Hypoxia Tolerance in Zebrafish (*Danio rerio*)

Bo Guan[1,2] Hong Ma[1] Yaping Wang[1] Yuanlei Hu[1]
Zhongping Lin[1] Zuoyan Zhu[1] Wei Hu[1]

1 State Key Laboratory of Freshwater Ecology and Biotechnology, Institute of Hydrobiology, Chinese Academy of Sciences, Wuhan 430072

2 Graduate School of the Chinese Academy of Sciences, Beijing 100049

Abstract Aquaculture farming may benefit from genetically engineering fish to tolerate environmental stress. Here, we used the vector pCVCG expressing the *Vitreoscilla* hemoglobin (*vhb)* gene driven by the common carp β-actin promoter to create stable transgenic zebrafish. The survival rate of the 7-day-old F_2 transgenic fish was significantly greater than that of the sibling controls under 2.5% O_2 (dissolved oxygen (DO) 0.91mg/l). Meanwhile, we investigated the relative expression levels of several marker genes (hypoxia-inducible factor alpha 1), heat shock cognate 70-kDa protein, erythropoietin, beta and alpha globin genes, lactate dehydrogenase, catalase, superoxide dismutase, and glutathione peroxidase) of transgenic fish and siblings after hypoxia exposure for 156 h. The expression profiles of the *vhb* transgenic zebrafish revealed that VHb could partially alleviate the hypoxia stress response to improve the survival rate of the fish. These results suggest that that *vhb* gene may be an efficient candidate for genetically modifying hypoxia tolerance in fish.

Keywords *Vitreoscilla* hemoglobin (VHb); hypoxia; zebrafish (*Danio rerio*); transgenic fish

Introduction

Dissolved oxygen is one of the most important environmental factors in aquaculture and fish ecology. Other environmental factors, such as temperature, circadian rhythm, seasonal change, and eutrophication can cause hypoxia in water bodies. Because hypoxia can cause mortality, reduce growth rates, induce endocrine disruption and alter distributions and behaviors of fish, it can lead to large reductions in the abundance, diversity, and harvest of various species within affected waters (Breitburg 2002; Wu 2002, 2003; Shang and Wu 2004; Thomas *et al*. 2007). Therefore, it becomes very important to find ways to cultivate new fish lines with hypoxia tolerance in aquaculture. Transgenic technology can be used to breed the new fish strains to enhance factors which would allow tolerance to extreme environmental

文章发表于 *Marine Biotechnology*, 2011, 13(2): 336-344

conditions (Zhu *et al.* 1989; Hew and Fletcher 2001). However, enhancing hypoxia tolerance in fish by transgenic technologies has not yet been described.

The hypoxia tolerance of fish may depend on the oxygen affinity, content and types of globin family (Terwilliger 1998; Skjæraasen *et al.* 2008). Four types of globins, differing in structure, tissue distribution and likely in function, have been discovered in man and other vertebrates: hemoglobin (Hb), myoglobin (Mb), neuroglobin and cytoglobin (Pesce *et al.* 2002). Myoglobin and hemoglobin are induced by hypoxia or ischemia (Hoppeler and Vogt 2001; Vogt *et al.* 2001; Nitta *et al.* 2003). In a hypoxia-tolerant fish model, the common carp (*Cyprinus carpio* L.), ectopic expression of the Mb gene was shown to be substantially enhanced in nonmuscle tissues such as the liver, gills, and brain (Fraser *et al.* 2006). Therefore, transfer of specific globin genes to hosts could improve the hypoxia tolerance in vertebrates (Yoshizaki *et al.* 1991; Khan *et al.* 2006).

Vitreoscilla stercoraria, a filamentous bacterium in the *Beggiatoa* family, is strictly aerobic, and to cope with the hypoxic conditions, this species synthesizes a soluble hemoglobin (VHb) protein which has two identical subunits each with a relative molecular mass of 15.8 kDa and two b hemes per molecule (Wakabayashi *et al.* 1986). In hypoxic conditions, VHb is proposed to enhance respiration and energy metabolism by promoting oxygen delivery with a high oxygen dissociation rate constant (Bülow *et al.* 1999; Frey and Kallio. 2003). VHb has also been shown to improve growth, protein secretion and metabolite productivity when the *vhb* gene is expressed in various organisms including microorganisms, plants and mammalian cells (Khosla *et al.* 1990; DeModena *et al.* 1993; Pendse and Bailey. 1994; Holmberg *et al.* 1997; Wilhelmson *et al.* 2006). In plants, such as cabbage (*Brassica oleracea* var. *Capitata*), alfalfa (*Medicago sativa* cv. Regen SY), and petunias, overexpressing VHb can enhance the growth and survival of the plant under hypoxia stress (Dordas *et al.* 2003; Mao *et al.* 2003; Li *et al.* 2005). In microbes, attempts have been made to overcome the problem of limiting oxygen by metabolically engineering bacteria to express the *vhb* transgene under hypoxic conditions (Patel *et al.* 2000; Nasr *et al.* 2001; So *et al.* 2004; Urgun-Demirtas *et al.* 2004). Recently, it was also suggested that VHb can improve tolerance to oxidative, nitrosative and submergence stresses in host cells, which is based on the integration of new enzymatic functions into the host metabolic network (Kaur *et al.* 2002; Mao *et al.* 2003; Frey *et al.* 2004; Li *et al.* 2005).

Zebrafish (*Danio rerio*) have become one of an ideal vertebrate model organisms used to study for transgenic technology, genetic development, normal body function, and disease (Alestrom *et al.* 2006; Kusik *et al.* 2008; Pan *et al.* 2008; Hsu *et al.* 2009; Zhan *et al.* 2010). In this study, we investigated the ability of VHb to improve the tolerance to hypoxia in the transgenic zebrafish model. Our findings could provide the basis for potential genetic modifications of farmed species to survive in oxygen reduced environments.

Materials and Methods

Construction of the pCVCG expression vector

The pCVCG vector was constructed to express the 441-bp *vhb* gene driven by a common carp β-actin promoter as well as the enhanced green fluorescent protein (eGFP) reporter under the control of a CMV promoter and enhancer (Fig. 1). The primers vhb1-F and vhb1-R (shown in Table 1) were designed to amplify the 441-bp *vhb* DNA fragment from a standard strain of *V. stercoraria* (ATCC, 15128. Mao *et al.*, 2003), according to a published sequence (Khosla and Bailey 1988). A β-actin gene promoter of common carp was cloned from the pCAgcGH vector (Guan *et al.* 2008).

Gene transfer

The pCVCG vector was dissolved in ST solution (88 mmol/L NaCl, 10 mmol/L Tris-HCl, pH 7.5) at a concentration of 85 ng/μl for gene transfer. Microinjections were carried out before the first cell division of freshly fertilized zebrafish eggs according to the method described by Zhu *et al.* (1989).

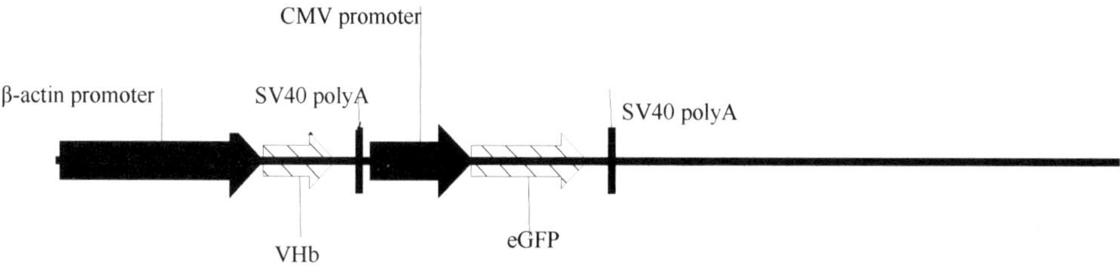

Fig. 1 Schematic of the pCAVG expression gene cassette

Transgenic lines stably expressing VHb

Twenty-four hours post-fertilization, about 3000 zebrafish embryos injected with the pCVCG vector were examined for green fluorescent protein (GFP) reporter expression. The GFP-positive embryos were considered pCVCG transgenic embryos (P_0) and isolated for further breeding. The P_0 transgenic adult zebrafish were mated with wildtype to produce the F_1 embryos, and the F_1 embryos expressing GFP were raised. We also applied the same method to produce the F_2 and F_3 pCVCG transgenic zebrafish. There are 8 stable transgenic lines in F_2 pCVCG transgenic zebrafish. And the No. 5 stable transgenic lines were used for hypoxia exposure test.

Total DNA were prepared from each fry as the PCR templates by the Chelex 100 method (Walsh *et al.* 1991). These fries were tested for the presence of the transgene by PCR with primers Pu (5′-ATCTGCCTGTAACCCATTCT-3′) and Pd (5′-AATACTTCTTTAATCGCACCC-3′), and the PCR conditions were as follows: 94℃ for 5 min, 35 cycles of 94℃ 30 s, 58℃ 30 s, 72℃ 30 s, and 72℃ for 5 min. The PCR products were analyzed for the presence of the specific 383 bp DNA band by electrophoresis on an agarose gel. The confirmation of *vhb* transcription in F_2 fries was conducted by reverse-transcription PCR (RT-PCR). Total RNA was extracted from 10 fries per sample using Trizol (Invitrogen). A total of 1 μg of RNA from each extraction was reverse-transcribed with ReverTra Ace (TOYOBO) and random primers in a 10 μl reaction volume according to the manufacturer's instructions. cDNA samples were diluted 1∶4 before use in the RT-PCR reaction performed with primers vhbu1 (5′-CGTTACCATTACCACGACTTT-3′) and vhbd1 (5′-GCATCGCCCAATACTTCTT-3′). The PCR conditions were as follows: 94℃ for 5 min, 35 cycles of 94℃ 30 s, 56℃ 30 s, 72℃ 30 s, and 72℃ for 5 min. The PCR products were analyzed for the presence of the specific 276 bp DNA band by electrophoresis on an agarose gel. The F_2 embryos expressing eGFP proteins and *vhb* transcripts were raised to the adult stage and mated with wild type to produce F_3 embryos, which could stably express the *vhb* gene.

Hypoxia exposures and real-time RT-PCR assays

The ability of the zebrafish embryos to survive in a hypoxia environment was recently documented, and the level of hypoxic stress that is tolerated by 24 hpf embryos for 24 h is 5% O_2 (Ton *et al.* 2003). For the purposes of our study, it was necessary to establish an appropriate level of oxygen tension (pO_2) low enough to induce hypoxia but sufficient to ensure the survival of some of the fries and the death of others. 24 hour before each hypoxic treatment, oxygen was set to the required concentration using nitrogen gas (99.999% purity) within the incubator (3131/Thermo; Forma Scientific, Inc., Marietta, OH). So the dissolved oxygen levels would be relatively stable. And we also measured the dissolved oxygen and temperature by an YSI550A Dissolved Oxygen Analyzer (YSI USA). Therefore, we chose a pO_2 of 2.5% and 97.5% N_2, levels which were achieved with many preliminary experiments for induction of hypoxic stress.

The No. 5 F_2 transgenic line stably expressing *vhb* was mated with wild-type animals to produce F_3. Twenty-four hours after fertilization the *vhb* transgenic fish and sibling control fish were divided according to detection of eGFP signals by fluorescence microscopy. About 50% of the F_3 offspring were wild type, and about 50% of the F_3 offspring were pCVCG transgenic fish. The two groups were separately raised and handled according to standard methods under the same condition. Experimental fish had respectively been bred three batches at different times.

For hypoxic exposure experiments, experimental fish at 7 days of age were simultaneously exposed to 2.5% O_2 and 97.5% N_2 using an incubator. And the dissolved oxygen was 0.91 mg/l, and the temperature was 28.5℃. The other 30-40 transgenic fish and 30-40 sibling control fish were raised in a normoxic environment (DO, 7.6 mg/l) at 28.5℃. These fish were starved during the hypoxia experiment. The number of dead fish were recorded every 12 h, and the dead fish were displaced. Three hypoxia exposure experiments were respectively done at different times. The first and second hypoxia exposure experiment lasted 144 h. The third hypoxia exposure experiment lasted 156 h.

When the third hypoxia exposure experiment was terminated, 10 fries per sample were triplicately collected in Trizol (Invitrogen) under normoxic environment and hypoxia environment. And total RNA was extracted from pools of 10 fries per sample using Trizol (Invitrogen). About 9 μg of total RNA from each extraction were then reverse-transcribed with ReverTra Ace (TOYOBO) and random primers in a 60 μl reaction volume according to the manufacturer's instructions. cDNA samples were diluted 1∶4 before use in fluorescent real-time RT-PCR reactions, and the cDNA was used as the template to detect the expression level of hypoxia related candidate genes as wells as that of β-actin as the internal control. A part of candidate genes and primers were referenced for gene expression profile of zebrafish exposed to hypoxiaduring development (Ton *et al.* 2003). These genes included hypoxia-inducible factor alpha 1 (HIF-1), heat shock cognate 70-kDa protein (HSP70), erythropoietin (EPO), beta and alpha globin genes, lactate dehydrogenase (LDH), cytochrome C oxidase subunit 1, calcium adenosine triphosphatase (ATPase), creatine kinase 3, skeletal muscle myosin light chain 3, catalase, superoxide dismutase 1, superoxide dismutase 2, and glutathione peroxidase.

Real-time RT-PCRs were performed in an ABI Prism 7000 Sequence Detection System (Applied Biosystems) using the following thermal cycling profile: 95℃ for 3 min, followed by 40 cycles of amplification (95℃ 15 s, 60℃ 15 s, 72℃ 45 s). Dissociation curve analysis was performed after the PCR run to validate that a single product was specifically amplified. Each reaction consisted of 2 μl of the cDNA sample, 7.2 μl of nuclease-free water, 10 μl of a 2 × SYBR Green PCR master mix (Toyobo), and 0.4 μl of each primer set (10 μM). All reactions were performed in triplicate. All primers were designed to conform to a universal cycling program by the Primer Express 2.0 software (Applied Biosystems). Relative expression was analyzed by the $2^{-\Delta\Delta CT}$ method (Livak and Schmittgen 2001).

Table 1　Reference genes and primer sequences

Gene name	Accession no.	Abbreviation	Primer sequences (5′ → 3′)
Beta actin	NM_131031	β-actin	(F) TTCTGGTCGTACTACTGGTATTGTG
			(R)ATCTTCATCAGGTAGTCTGTCAGGT
Hypoxia-inducible factor 1 alpha	NM_200233.1	HIF-1	(F) CCAGTGGAACCAGACATCAG
			(R) AGGAGGGTAAGGGTTGGAAT
Heat shock cognate 70-kDa protetin	NM_131397	HSP70	(F) CAACAACCTGCTGGGCAAA
			(R) GCGTCGATGTCGAAGGTCA
Erythropoietin, transcript variant 1	NM_001038009	EPO	(F) TCCAACCAGTAGTGGAGAAG
			(R) GTTGACATGGACCTGAAACA

Continued

Gene name	Accession no.	Abbreviation	Primer sequences (5′ → 3′)
Beta and alpha globin genes	U50382	Globin	(F) TGACCGTCTACCCTCAAACCA
			(R) GTCCTCCCACAAGGTCGTCT
Lactate dehydrogenase A4	NM_131246	LDHa	(F) CACAGTTGAAGATGGCCTCA
			(R) TGTGCGTCTTGAGAAACAGG
Cytochromc c oxidase subunit 1	AY996924	COXI	(R) ACTTAGCCAACCAGGAGCAC
			(R) GGGTGGAAGAAGTCAGAAGC
ATPase	NM_001077533	ATPase	(F) TGGTGATGGTGTCAATGATG
			(R) TAGCTCTGCCTTCCTCAACA
Creatine kinase	NM_130932	CKM	(F) AGGCAACTGACAAGCACAAG
			(R) ACTCTCCATCCAAGCTGCTC
Myosin, light polypentide 3, skeletal muscle	NM_131619	MYL3	(F) GTAAGAACCCCACCAACAAG
			(R) GGTTCGCTCATCTTCTCACC
Catalase	NM_130912	CAT	(F) ACCAACAACCCTCCAGACAG
			(R) GCCGTCATCGCATTTAACTT
Superoxide dismutase 1	NM_131294	SOD (Cu^{2+})	(F) GTCCGCACTTCAACCCTCA
			(R) CCTCATTGCCACCCTTCC
Superoxide dismutase 2	NM_199976	SOD (Mn^{2+})	(F) TTCAGGGCTCAGGCTGG
			(R) ATGGCTTTAACATAGTCCGGT
Glutathione peroxidase 1a	BC083461	GPX1a	(F) ACGACCCTGTGTCCCTTAT
			(R) CTTCTGCTGTACCTCTTGAATG

Statistical analyses

Statistical significance of the survival rates between sibling and *vhb* transgenic fish under hypoxia were determined by the Chi-square test. The data obtained from real-time RT-PCR analysis were subjected to Duncan's multiple-range tests. Statistical analyses of data were performed with STATISTICA 6.0, and illustrations were performed with SigmaPlot 10.0. $P < 0.05$ was considered statistically significant, and $P < 0.01$ represented a highly significant difference. Error bars indicate standard errors.

Results

Creation of transgenic zebrafish lines expressing VHb

Using the pCVCG plasmid expressing VHb and eGFP constructed as shown in Fig. 1, we generated and raised transgenic zebrafish embryos to sexual maturity. The F_1 and F_2 generations' offspring were then screened for GFP expression. As shown, the GFP proteins of F_1 or F_2 generation were uniformly expressed in the embryo (Fig. 2a). The integration of pCVCG in GFP-positive embryos was validated by PCR analysis with primers Pu and Pd in transgenic F_2 embryos expressing GFP (Fig. 2b). As a control, the integration of pCVCG was not detectable in F_2 embryos that did not express GFP (Fig. 2b). Similarly, the transcription of VHb in GFP-positive offspring was validated by RT-PCR analysis with primers vhbu1 and vhbd1 in transgenic F_2 embryos expressing GFP (Fig. 2c), and the transcription of VHb could not be detected from F_2 embryos that did not express GFP (Fig. 2c). Therefore, we demonstrated that VHb was transcribed in the F_2 generation of pCAVG transgenic zebrafish showing GFP expression, and confirmed that VHb was not transcribed in the F_2 generation siblings that did not express GFP.

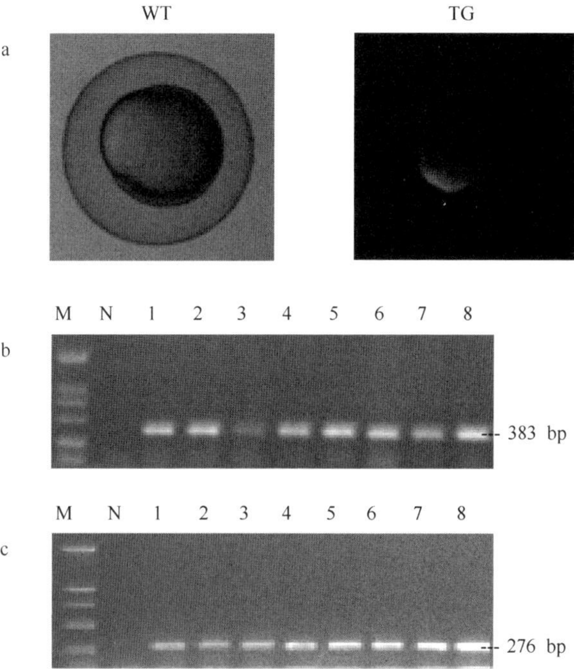

Fig. 2 Confirmation of the presence of the pCAVG vector and its gene expression. (a) WT (wild-type fish) and TG (pCAVG transgenic fish) at 75% epiboly stage (8 h). (b) Detection of *vhb* transgene by PCR in F_2 embryos. (c) Detection of *vhb* transcripts by RT-PCR. Lane *N* indicates negative amplification from offspring negative for GFP; lanes 1-8 show positive *vhb* amplification from individual offspring positive for GFP. The first lane (*M*) is the molecular weight marker (DNA Marker DL2000)

Hypoxia exposures and real-time RT-PCR assays

In the present study, the survival and tolerance between the 7-day-old F_3 transgenic fish and sibling control fish under 2.5% O_2 (DO, 0.91 mg/l) were compared. Although the survival rate of transgenic fish or sibling control fish varied greatly among different batches of hypoxia exposure experiment, the survival rate of the transgenic fish was significantly greater than those of the sibling control under hypoxia (Table 2).

Table 2 Survival rates of *vhb* transgenic and sibling zebrafish under hypoxia

Batch	Survival rate of the control[a]	Survival rate of transgenic fish[a]	Duration (h)	Chi-squared test
1	15.69% (16/102)	34.57% (28/81)	144	P<0.000000[b]
2	19.57% (18/92)	49.41% (42/85)	144	P<0.000000[b]
3	65% (65/100)	92% (92/100)	156	P<0.000000[b]

The survival rates of the two groups under normoxia were 100%

a Survival/total number

b Indicates a highly statistically significant difference (P<0.01)

We used real-time RT-PCR to investigate the changes in mRNA levels of 13 candidate genes affected under hypoxia conditions, as shown in Fig. 3. The relative HIF-1 mRNA levels of both groups were upregulated under hypoxia, but the relative expression levels of transgenic fish were significantly lower than that of sibling under hypoxia in comparison with normoxia (Fig. 3). The relative HSP 70 and EPO mRNA levels of sibling group were also upregulated under hypoxia in comparison with normoxia; meanwhile, these levels in the transgenic group did not significantly change, but they were lower than those of the sibling under hypoxia (Fig. 3). The relative beta and alpha globin mRNA levels of the sibling group were significantly upregulated under hypoxia in comparison with normoxia, and those of the transgenic group were significantly downregulated; however, the relative beta and alpha globin gene expression levels of the sibling group were significantly higher than those of the transgenic group under hypoxia (Fig. 3). The relative LDH, ATPase and creatine kinase 3 mRNA levels of both groups did not significantly change under hypoxia in comparison with normoxia (Fig. 3). The relative MYL3, SOD(Cu^{2+}), SOD(Mn^{2+}) and GPX1a mRNA levels of sibling and transgenic fish were significantly downregulated under hypoxia in comparison with normoxia, but the relative MYL3, SOD(Cu^{2+}), SOD(Mn^{2+}) and GPX1a gene expression levels of transgenic fish were significantly higher than those of sibling fish under hypoxia (Fig. 3). The relative chloramphenicol acetyl transferase (CAT) mRNA levels of transgenic fish were not significantly different under hypoxia in comparison with normoxia, and those of sibling fish were significantly downregulated under hypoxia in comparison with normoxia; meanwhile, the relative CAT gene expression levels of transgenic fish were significantly higher than those of sibling fish under hypoxia (Fig. 3).

Discussion

In this paper, we constructed a transgenic zebrafish line stably expressing VHb and found that these fish had significantly higher survival rates under a prolonged period of hypoxia treatment. The expression profiling of several differentially expressed genes in *vhb* transgenic zebrafish under hypoxia suggested that VHb could partially alleviate the hypoxia stress response to improve the survival rate of fish under partial oxygen deprivation. Therefore, we believe that over-expression of the *vhb* gene by transgenic technology can help to improve hypoxia tolerance in fish.

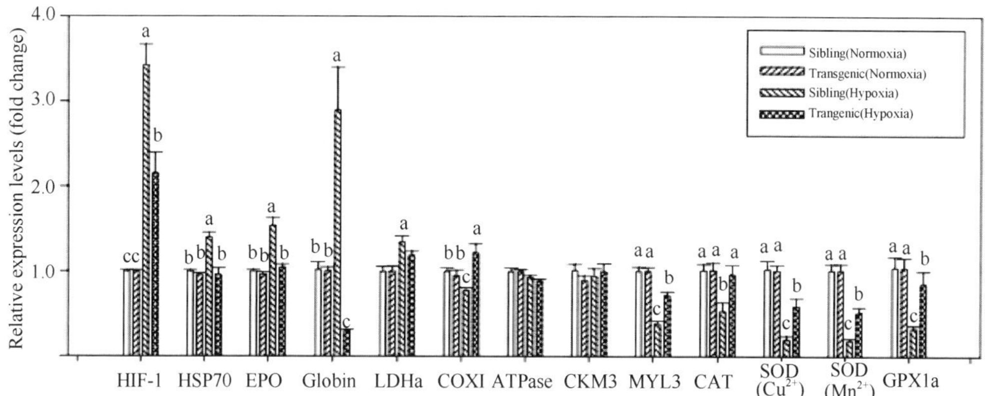

Fig. 3 Expression of selected genes by real-time RT-PCR. Abundance of mRNA levels in *vhb* transgenic fish was compared against sibling zebrafish under normoxia. Full names of the abbreviated genes are listed in Table 1. Duncan's multiple-range test was applied to data between two groups under hypoxia (2.5% oxygen), and a, b, c indicate statistically significant differences between two groups ($P<0.05$)

The survival time of zebrafish under hypoxia partly depends on the different hypoxia exposure method and the different developmental stages. Because early life stage (ELS) is one of the most sensitive life stages to toxicity during the life cycle, ELS tests with fish species are commonly used as routine toxicity tests in risk assessment (OECD 1998; Roex *et al.* 2002). *D. rerio* was recommended as a model fish species for short-term embryo-larval toxicity testing (OECD 1998), and there are many studies investigating the effects of hypoxia on zebrafish embryo or larvae (Padilla and Roth. 2001; Ton *et al.* 2003; Rombough and Drade. 2009; Steele *et al.* 2009; Van Rooijen *et al.* 2009; Yaqoob *et al.* 2009). We found that the 7-day-old F_3 *vhb* transgenic larvae may be suitable for study of the hypoxia tolerance traits. It was certain that the survival rate of the transgenic fish was significantly greater than those of the sibling control under hypoxia. But why did the survival rate of transgenic fish or sibling control fish vary greatly among different batches of hypoxia exposure experiment? Good-quality eggs have been defined as those exhibiting low mortalities at fertilization, eyeing, hatching and first feeding (Bromage *et al.*, 1992). The quality of an egg is determined by the intrinsic properties of the egg itself, by its genes, and by the maternal mRNA transcripts and nutrients contained within the yolk, all of which are provided by the mother (Brooks *et al.* 1997). The experimental conditions are consistent during the three hypoxia exposure experiments, except the female parent. So we speculate that the egg quality of the female parent could be associated with the variation of *vhb* transgenic fry survival rate under hypoxia.

However, the precise mechanisms underlying the enhanced hypoxia tolerance of the *vhb* transgenic zebrafish remains to be seen. There are numerous studies investigating mechanisms of VHb function. Frey and Kallio (2003) suggested that VHb are surprisingly versatile proteins serving several biological functions, including protection of the host cell from nitrosative and oxidative stress and delivering oxygen to respiring cells. Compared with other hemoglobins, VHb has an average oxygen association rate constant (k_{on} 78/μM s), while its oxygen dissociation rate constant (k_{off} 5000/s) reflects a moderate affinity for oxygen; however, its much larger k_{off} enables the hemoglobin to release the bound oxygen more readily than almost all other hemoglobins (Frey and Kallio 2003). Additionally, there is other evidence to show that VHb interacts with bacterial respiratory membranes and cytochrome bo proteoliposomes (Kim *et al.* 2000; Park *et al.* 2002). Accordingly, we presume that VHb can enable transgenic zebrafish to survive in oxygen-limited environments by increasing the intracellular effective dissolved oxygen concentration and facilitating oxygen diffusion.

In our study expression profiles of selective genes previously validated to be regulated under oxygen-limiting conditions in zebrafish (Ton *et al.* 2003) were examined between our *vhb* transgenic zebrafish and control fish under hypoxia. In mammals and fish, HIF-1 receives signals from the molecular oxygen sensor, and in turn, regulates the transcription of a number of hypoxia-inducible genes, including genes involved in erythropoiesis, angiogenesis, glycolysis and antigrowth mechanisms, which together constitute the core signaling pathways of the hypoxia response (Wu 2002; Lendahl *et al.* 2009). These molecular responses then cascade into a series of biochemical and physiological adjustments, enabling the animal to survive better under hypoxic conditions (Wu 2002). We found that the relative HIF-1 mRNA expression levels of transgenic fish were lower than that of wild-type siblings under hypoxia, which indicated that VHb could partially alleviate the hypoxia stress response in zebrafish by enhancing the efficiency or utility of the oxygen transportation. In mammals and fish, HIF-1 can promote EPO production and increase erythropoiesis under hypoxia (Wu 2002; Lai *et al.* 2006). Interestingly, we also observed that while hypoxia resulted in upregulation of genes encoding for EPO and hemoglobin in control zebrafish fry, the EPO mRNA expression levels did not significantly change in *vhb* transgenic fish during hypoxia and the hemoglobin mRNA expression level was significantly down-regulated. Accordingly, the relatively lower HIF-1 mRNA expression levels could not activate the EPO mRNA expression of *vhb* transgenic fish under hypoxia. HSP70 functions as a chaperone and is also known to protect cells against apoptosis (Hohfeld 1998). Hypoxic conditions represent high stress levels that can lead to the induction of genes involved in the cellular stress response such as heat shock proteins (Williams and Benjamin 2000; Ton *et al.* 2003; van der Meer *et al.* 2005). The relative HSP 70 mRNA levels of the sibling group were upregulated under hypoxia, and that of the transgenic group did not significantly change, which also indicated that *vhb* transgenes in the zebrafish could partially alleviate the hypoxia stress response.

One of the most important defense mechanisms against hypoxia in animals is the reduction of energy costs (Hochachka and Lutz 2001). We showed that the relative LDH, ATPase and creatine kinase 3 mRNA levels of both groups did not significantly change under hypoxia in comparison with normoxia. However, the expression profiles of these genes (LDH, ATPase, and creatine kinase 3) in *vhb* transgenic zebrafish did not truly reflect the changes in their complex energy metabolic pathways.

The expression profiles of antioxidative enzymes (CAT, SOD (Cu^{2+}), SOD (Mn^{2+}) and GPX1a) gene in *vhb* transgenic zebrafish were generally similar. The relative CAT, SOD(Cu^{2+}), SOD(Mn^{2+}) and GPX1a gene expression levels of *vhb* transgenic fish were significantly higher than those of sibling fish under hypoxia, and the relative SOD(Cu^{2+}), SOD(Mn^{2+}) and GPX1a mRNA levels of sibling and transgenic fish were significantly downregulated under hypoxia in comparison with those in the normoxia condition. Many studies have shown that tissues and cells in humans and rats respond to hypoxia by decreasing activities of antioxidative enzymes, as well as cellular total GSH content (Plateel *et al* 1995; Jackson *et al.* 1996; Jessup *et al.* 1999; Li and Jackson. 2002), but the mechanism by which antioxidant enzyme expression is regulated by hypoxia is not known with certainty. Some studies suggested that VHb could protect cells from nitrosative and oxidative stress both in bacteria and plants (Kaur *et al.* 2002; Wang *et al.* 2009). Therefore, VHb may function to scavenge oxygen radicals, reducing peroxidases under hypoxia (Bülow *et al.* 1999). We suggest that VHb can alleviate oxidative stress by the relatively higher mRNA expression levels of antioxidant enzymes in zebrafish, which could help to reduce damage by the reactive oxygen species to the tissue cells.

In conclusion, this is the first report on the introduction of the *vhb* transgene into zebrafish for improvement of hypoxia tolerance. This study demonstrated that VHb could partly alleviate the hypoxia stress response to improve the survival rate of the zebrafish under hypoxia, and that the *vhb* gene may be a good candidate for genetic engineering methods for efficiently modifying hypoxia tolerance in fish.

Acknowledgements This work was financially supported by "863" High-Technology Project (grant number 2007AA10Z186); the National Natural Science Foundation of China (grant number 30623001) and the Development Plan of the State Key Fundamental Research of China (grant number 2009CB118804).

References

Alestrom P, Holter JL, Nourizadeh-Lillabadi R (2006) Zebrafish in functional genomics and aquatic biomedicine. Trends Biotechnol 24: 15-21

Breitburg DL (2002) Effects of hypoxia, and the balance between hypoxia and enrichment, on coastal fishes and fisheries. Estuar Coast 25(4b): 767-781

Bromage N, Jones J, Randall C, Thrush M, Davies B, Springate J, Duston J, Barker G (1992) Broodstock management, fecundity, egg quality and the timing of egg production in the rainbow trout (*Oncorhynchus mykiss*). Aquaculture 100: 141-166

Brooks S, Tyler CR, Sumpter JP (1997) Egg quality in fish: what makes a good egg? Rev Fish Biol Fish 7: 387-416

Bülow L, Holmberg N, Lilius G, Bailey JE (1999) The metabolic effects of native and transgenic hemoglobins on plants. Trends Biotechnol 17: 21-24

DeModena JA, Gutierrez S, Velasco J, Fernandez FJ, Fachini RA, Galazzo JL, Hughes DE, Martin JF (1993) The production of cephalosporin C by *Acremonium chrysogenum* is improved by the intracellular expression of a bacterial hemoglobin. Nat Biotechnol 11: 926-929

Dordas C, Hasinoff BB, Igamberdiev AU, Manac'h N, Rivoal J, Hill RD (2003) Expression of a stress-induced hemoglobin affects NO levels produced by alfalfa root cultures under hypoxic stress. Plant J 35: 763-770

Fraser J, Vieira de Mello L, Ward D, Rees HH, Williams DR, Fang Y, Brass A, Gracey AY, Cossins AR (2006) Hypoxia-inducible myoglobin expression in nonmuscle tissues. Proc Natl Acad Sci U S A 103: 2977-2981

Frey AD, Kallio PT (2003) Bacterial hemoglobins and flavohemoglobins: versatile proteins and their impact on microbiology and biotechnology. FEMS Microbiol Rev 27(4): 525-545

Frey AD, Oberle BT, Farrés J, Kallio PT (2004) Expression of *Vitreoscilla* haemoglobin in tobacco cell cultures relieves nitrosative stress *in vivo* and protects from NO *in vitro*. Plant Biotechnol J 2: 221-231

Guan B, Hu W, Zhang T, Wang Y, Zhu Z (2008) Metabolism traits of 'all-fish'growth hormone transgenic common carp (*Cyprinus carpio* L.). Aquaculture 284: 217-223

Hew CL, Fletcher GL (2001) The role of aquatic biotechnology in aquaculture. Aquaculture 197: 191-204

Hochachka PW, Lutz PL (2001) Mechanism, origin, and evolution of anoxia tolerance in animals. Comp Biochem Physiol B Biochem Mol Biol 130: 435-459

Hohfeld J (1998) Regulation of the heat shock conjugate Hsc70 in the mammalian cell: The characterization of the anti-apoptotic protein BAG-1 provides novel insights. Biol Chem 3: 269-274

Holmberg N, Lilius G, Bailey JE, Bülow L (1997) Transgenic tobacco expressing *Vitreoscilla* hemoglobin exhibits enhanced growth and altered metabolite production. Nat Biotechnol 15: 244-247

Hoppeler H, Vogt M (2001) Muscle tissue adaptations to hypoxia. J Exp Biol 204: 3133-3139

Hsu CC, Hou MF, Hong JR, Wu JL, Her GM (2009) Inducible male infertility by targeted cell ablation in zebrafish testis. Mar Biotechnol (NY) doi: 10.1007/s10126-009-9248-4

Jackson RM, Parish G, Ho YS (1996) Effects of hypoxia on expression of superoxide dismutases in cultured ATII cells and lung fibroblasts. Am J Physiol Lung Cell Mol Physiol 271: L955-L962

Jessup JM, Battle P, Waller H, Edmiston KH, Stolz DB, Watkins SC, Locker J, Skena K (1999) Reactive nitrogen and oxygen radicals formed during hepatic ischemia-reperfusion kill weakly metastatic colorectal cancer cells. Cancer Res 59: 1825-1829.

Kaur R, Pathania R, Sharma V, Mande SC, Dikshit KL (2002) Chimeric *Vitreoscilla* hemoglobin (VHb) carrying a flavoreductase domain relieves nitrosative stress in *Escherichia coli*: new insight into the functional role of VHb. Appl Environ Microbiol 68: 152-160

Khan AA, Wang Y, Sun Y, Mao XO, Xie L, Miles E, Graboski J, Chen S, Ellerby LM, Jin K, Greenberg DA (2006) Neuroglobin-overexpressing transgenic mice are resistant to cerebral and myocardial ischemia. Proc Natl Acad Sci USA 103:17944-17948

Khosla C, Bailey JE (1988) The *Vitreoscilla* hemoglobin gene: molecular cloning, nucleotide sequence and genetic expression in *Escherichia coli*. Mol Gen Genet 214: 158-161

Khosla C, Curtis JE, DeModena J, Rinas U, Bailey JE (1990) Expression of intracellular hemoglobin improves protein synthesis in oxygen-limited *Escherichia coli*. Nat Biotechnol 8: 849-853

Kim K J, Chi PY, Hwang KW, Stark BC, Webster DA (2000) Study of cytochrome *bo* function in *Vitreoscilla* using a cyo^- knockout mutant. J Biochem 128: 49-55

Kusik B, Carvan Iii M, Udvadia A (2008) Detection of mercury in aquatic environments using EPRE reporter zebrafish. Mar Biotechnol (NY) 10: 750-757

Lai JCC, Kakuta I, Mok HOL, Rummer JL, Randall D (2006) Effects of moderate and substantial hypoxia on erythropoietin levels in rainbow trout kidney and spleen. J Exp Biol 209: 2734-2738.

Lendahl U, Lee KL, Yang H, Poellinger L (2009) Generating specificity and diversity in the transcriptional response to hypoxia. Nature Rev Genet 10: 821-832

Li C, Jackson RM (2002) Reactive species mechanisms of cellular hypoxia-reoxygenation injury. Am J Physiol Cell Physiol 282: C227-C241

Li X, Peng RH, Fan HQ, Xiong AS, Yao QH, Cheng ZM, Li Y (2005). *Vitreoscilla* hemoglobin overexpression increases submergence tolerance in cabbage. Plant Cell Rep23: 710-715

Livak KJ, Schmittgen TD (2001) Analysis of relative gene expression data using real-time quantitative PCR and the 2(-Delta Delta C (T)) method. Methods 25: 402-408

Mao ZC, Hu YL, Zhong J, Wang LX, Guo JY, Lin ZP (2003) Improvement of the hydroponic growth and waterlogging tolerance of petunias by the introduction of *vhb* gene. Acta Bot Sin 45: 205-210

Nasr MA, Hwang KW, Akbas M, Webster DA, Stark BC (2001) Effects of culture conditions on enhancement of 2,4-dinitrotoluene degradation by *Burkholderia* engineered with the *Vitreoscilla* hemoglobin gene. Biotechnol Prog 17: 359-61

Nitta T, Xundi X, Hatano E, Yamamoto N, Uehara T, Yoshida M, Harada N, Honda K, Tanaka A, Sosnowski D (2003) Myoglobin gene expression attenuates hepatic ischemia reperfusion injury. J Surg Res 110: 322-331

OECD (1998) Guideline for testing of chemicals: fish, short-term toxicity test on embryo and sac-fry stages. OECD 212, Paris

Padilla PA, Roth MB (2001) Oxygen deprivation causes suspended animation in the zebrafish embryo. Proc Natl Acad Sci USA 98: 7331-7335

Pan X, Zhan H, Gong Z (2008) Ornamental expression of red fluorescent protein in transgenic founders of white skirt tetra (*Gymnocorymbus ternetzi*). Mar Biotechnol (NY) 10: 497-501

Park KW, Kim KJ, Howard AJ, Stark BC, Webster DA (2002) *Vitreoscilla* hemoglobin binds to subunit I of cytochrome *bo* ubiquinol oxidases. J Biol Chem 277: 33334-33337

Patel SM, Stark BC, Hwang KW, Dikshit KL, Webster DA (2000) Cloning and expression of *Vitreoscilla* hemoglobin gene in *Burkholderia* sp. strain DNT for enhancement of 2,4-dinitrotoluene degradation. Biotechnol Prog 16: 26-30

Pendse GJ, Bailey JE (1994) Effect of *Vitreoscilla* hemoglobin expression on growth and specific tissue plasminogen activator productivity in recombinant Chinese hamster ovary cells. Biotechnol Bioeng 44: 1367-1370

Pesce A, Bolognesi M, Bocedi A, Ascenzi P, Dewilde S, Moens L, Hankeln T, Burmester T (2002) Neuroglobin and cytoglobin: Fresh blood for the vertebrate globin family. EMBO Rep 3(12): 1146-1151

Plateel M, Dehouck MP, Torpier G, Cecchelli R, Teissier E (1995) Hypoxia increases the susceptibility to oxidant stress and the permeability of the blood-brain barrier endothelial cell monolayer. J Neurochem 65: 2138-2145

Roex EW, de Vries E, van Gestel CA (2002) Sensitivity of the zebrafish (*Danio rerio*) early life stage test for compounds with different modes of action. Environ Pollut 120: 355-362

Rombough P, Drader H (2009) Hemoglobin enhances oxygen uptake in larval zebrafish (*Danio rerio*) but only under conditions of extreme hypoxia. J Exp Biol 212: 778-784

Shang E, Wu RSS (2004) Aquatic hypoxia is a teratogen and affects fish embryonic development. Environ Sci Technol 38: 4763-4767

Skjæraasen JE, Nilsen T, Meager JJ, Herbert NA, Moberg O, Tronci V, Johansen T, Salvanes AGV (2008) Hypoxic avoidance behaviour in cod (*Gadus morhua* L.): The effect of temperature and haemoglobin genotype. J Exp Mar Biol Ecol 358: 70-77

So J, Webster DA, Stark BC, Pagilla KR (2004) Enhancement of 2, 4-dinitrotoluene biodegradation by *Burkholderia* sp. in sand bioreactors using bacterial hemoglobin technology. Biodegradation 15: 161-171

Steele SL, Lo KH, Li VW, Cheng SH, Ekker M, Perry SF (2009) Loss of M2 muscarinic receptor function inhibits development of hypoxic bradycardia and alters cardiac β-adrenergic sensitivity in larval zebrafish (*Danio rerio*). Am J Physiol Regul Integr Comp Physiol 297: R412-R420

Terwilliger NB (1998) Functional adaptations of oxygen-transport proteins. J Exp Biol 201(8): 1085-1098

Thomas P, Rahman MS, Khan IA, Kummer J (2007) Widespread endocrine disruption and reproductive impairment in an estuarine fish population exposed to seasonal hypoxia. Proc R Soc B 274: 2693-2701

Ton C, Stamatiou D, Liew CC (2003) Gene expression profile of zebrafish exposed to hypoxia during development. Physiol Genomics13: 97-106

Urgun-Demirtas M, Pagilla KR, Stark BC (2004) Enhanced kinetics of genetically engineered *Burkholderia cepacia*: the role of *vhb* in the hypoxic metabolism of 2-CBA. Biotechnol Bioeng 87: 110-118

van der Meer DL, van den Thillart GE, Witte F, de Bakker MA, Besser J, Richardson MK, Spaink HP, Leito JT, Bagowski CP (2005) Gene expression profiling of the long-term adaptive response to hypoxia in the gills of adult zebrafish. Am J Physiol 289: R1512-R1519

Van Rooijen E, Voest EE, Logister I, Korving J, Schwerte T, Schulte-Merker S, Giles RH, Van Eeden FJ (2009) Zebrafish mutants in the von Hippel-Lindau tumor suppressor display a hypoxic response and recapitulate key aspects of *Chuvash polycythemia*. Blood 113: 6449-6460

Vogt M, Puntschart A, Geiser J, Zuleger C, Billeter R, Hoppeler H (2001) Molecular adaptations in human skeletal muscle to endurance training under simulated hypoxic conditions. J Appl Physiol 91: 173-182

Wakabayashi S, Matsubara H, Webster DA (1986) Primary sequence of a dimeric bacterial hemoglobin from *Vitreoscilla*. Nature 322: 481-483

Walsh PS, Metzger DA, Higuchi R (1991) Chelex 100 as a medium for simple extraction of DNA for PCR-based typing from forensic material. Biotechniques 10(4): 506-513

Wang ZN, Xiao Y, Chen WS, Tang KX, Zhang L (2009) Functional expression of *Vitreoscilla* hemoglobin (VHb) in *Arabidopsis* relieves submergence, nitrosative, photo-oxidative stress and enhances antioxidants metabolism. Plant Science 176: 66-77

Wilhelmson A, Hälkkinen ST, Kallio PT, Oksman-Caldentey KM, Nuutila AM (2006) Heterologous expression of *Vitreoscilla* hemoglobin (VHb) and cultivation conditions affect the alkaloid profile of *Hyoscyamus muticus* hairy roots. Biotechnol Prog 22: 350-358

Williams RS, Benjamin IJ (2000) Protective responses in the ischemic myocardium. J Clin Invest 106: 813-818

Wu RSS (2002) Hypoxia: from molecular responses to ecosystem responses. Mar Pollut Bull 45: 35-45

Wu RSS, Zhou BS, Randall DJ, Woo NYS, Lam PKS (2003) Aquatic hypoxia is an endocrine disruptor and impairs fish reproduction. Environ Sci Technol 37: 1137-1141

Yaqoob N, Holotta M, Prem C, Kopp R, Schwerte T (2009) Ontogenetic development of erythropoiesis can be studied non-invasively in GATA-1: DsRed transgenic zebrafish. Comp Biochem Physiol A Mol Integr Physiol 154: 270-278

Yoshizaki G, Oshiro T, Takashima F (1991) Introduction of carp α-globin gene into rainbow trout. Bull Jpn Soc Sci Fish 57(5): 819-824

Zhan H, Spitsbergen J, Qing W, Wu Y, Paul T, Casey J, Her G, Gong Z (2010). Transgenic expression of walleye dermal sarcoma virus rv-cyclin gene in zebrafish and its suppressive effect on liver tumor development after carcinogen treatment. Mar Biotechnol (NY) doi:10.1007/s10126-009-9251-9

Zhu Z, Xu K, Xie Y, Li G, He L (1989) A model of transgenic fish. Sci China, Ser B (in Chinese) 2: 147-155

VHb 的过表达增强斑马鱼的低氧耐受能力

管　波[1,2] Hong Ma[2] 汪亚平[1] 胡鸢雷[2]
林忠平[2] 朱作言[1] 胡　炜[1]

1 中国科学院水生生物研究所，淡水生态与生物技术国家重点实验室，武汉 430072
2 北京大学，生命科学院，北京 100049

摘　要　遗传改良技术可用于提高养殖鱼类对环境胁迫的耐受能力。我们将鲤β-肌动蛋白启动子驱动的透明颤菌血红蛋白基因(*vhb*) 的表达载体 pCVCG 转移到斑马鱼受精卵中，研制出转VHb基因的斑马鱼家系。在2.5% O_2 (0.91 mg/L)条件下，经过 156 h 的低氧胁迫后，7 日龄的 F_2 转 VHb 斑马鱼的生存率明显高于同胞对照鱼。同时，本文研究了低氧胁迫下相关标志基因的表达模式，包括低氧诱导因子(HIF)、热激蛋白 70 (HSP70)、促红细胞生成素(EPO)、α,β 球蛋白、乳酸脱氢酶(LDH)、过氧化氢酶(CAT)、超氧化剂物歧化酶(SOD)和谷胱甘肽过氧化物酶等基因，结果表明，转植 VHb 基因提高了氧和能量供给，缓解了低氧胁迫下转基因鱼的氧化应激水平，削弱了氧化损伤，从而增强了转基因鱼在低氧胁迫条件的生存力。因此，*vhb* 基因可以用于研制具有低氧耐受特性的鱼类新品种。

Double Transgenesis of Humanized *fat1* and *fat2* Genes Promotes Omega-3 Polyunsaturated Fatty Acids Synthesis in a Zebrafish Model

Shao-Chen Pang[1,3] Hou-Peng Wang[1] Kuo-Yu Li
Zuo-Yan Zhu[1] Jing X. Kang[2] Yong-Hua Sun[1]

1 State key Laboratory of Freshwater Ecology and Biotechnology, Institute of Hydrobiology, Chinese Academy of Sciences, Wuhan 430072, China

2 Laboratory for Lipid Medicine and Technology, Department of Medicine, Massachusetts General Hospital and Harvard Medical School, Boston, Massachusetts, USA

3 University of Chinese Academy of Sciences, Beijing 100049, China

Abstract: Omega-3 long-chainpolyunsaturated fatty acid (*n*-3 LC-PUFA), especially eicosapentaenoic acid (EPA) and docosahexaenoic acid (DHA), are essential nutrients for human health. However, vertebrates including human, have lost the abilities to synthesize EPA and DHA *de novo*, majorly due to the genetic absence of *delta-12 desaturase* and *omega-3 desaturase* genes. Fishes, especially those naturally growing marine fish, are major dietary source of EPA and DHA. Because of the severe decline of marine fishery and the decrease in *n*-3 LC-PUFA content of farmed fishes, it is highly necessary to develop alternative sources of *n*-3 LC-PUFA. In the present study, we utilized transgenic technology to generate *n*-3 LC-PUFA-rich fish by using zebrafish as an animal model. Firstly, *fat1* was proved to function efficiently in fish culture cells, which showed an effective conversion of *n*-6 PUFA to *n*-3 PUFA with the *n*-6/*n*-3 ratio that decreased from 7.7 to 1.1. Secondly, expression of *fat1* in transgenic zebrafish increased the 20:5*n*-3 and 22:6*n*-3 content to 1.8- and 2.4-fold respectively. Third, co- expression of *fat2*, a fish codon-optimized *delta-12 desaturase gene*, and *fat1* in fish culture cell significantly promoted *n*-3 PUFA synthesis with the decreased *n*-6/*n*-3 ratio from 7.7 to 0.7. Finally, co-expression of *fat1* and *fat2* in double transgenic zebrafish increased 20:5*n*-3 and 22:6*n*-3 contents to 1.7- and 2.8-fold respectively. Overall, we generated two types of transgenic zebrafish rich in endogenous *n*-3 LC-PUFA, *fat1* transgenic zebrafish and *fat1*/*fat2* double transgenic zebrafish. Our results demonstrate that application of transgenic technology of humanized *fat1* and *fat2* in farmed fishes can largely improve the *n*-3 LC-PUFA production.

Keywords: omega-3 fatty acids; omega-6 fatty acids; desaturase; transgenic fish; transcriptional regulation

---

文章发表于 *Marine Biotechnology*, 2014, 16(5): 580-593

Introduction

Omega-3 long chain-polyunsaturated fatty acids (*n*-3 LC-PUFA), including eicosapentaenoic acid (20:5*n*-3, EPA) and docosahexaenoic acid (22:6*n*-3, DHA), play important roles throughout life—from late gestation to adulthood. Substantial amounts of LC-PUFA, especially DHA, are critical for neurodevelopment, such as neurite outgrowth, brain and retina development (Marszalek and Lodish 2005). Increasing epidemiological evidence and clinical trials have indicated protective roles of EPA and DHA in the prevention of the numbers of disease including atherosclerosis, cardiovascular disease, stroke, neurological disorders, obesity, type-2 diabetes, cancers, inflammatory and autoimmune disease (Kris-Etherton *et al.* 2003; Whelan and Rust 2006; Fritsche 2006). Unfortunately, all vertebrates, including human beings, have lost the capacity to synthesize *n*-3 LC-PUFA *de novo* due to genetic absence of *delta-12 desaturase* and *omega-3 desaturase* (or *delta-15 desaturase*) (Nakamura and Nara 2004). As a result, human need to obtain these *n*-3 LC-PUFA from the daily diet.

Marine fish and seafood are the major dietary sources of EPA and DHA, which are stored in fish flesh through the food chain. According to the latest report of the FAO, there is a growing decline of wild marine fish fishery globally (FAO 2012). The ever-increasing expansion of the aquaculture industry, which is dependent on fish oil derived from marine reduction fisheries, is bringing more and more pressure on the natural marine resource. The increasing pressure makes the aquaculture industry to feed fish with vegetable oil as alternative oil, and, as a result, the replacement of fish oil with linoleic acid-rich (18:2*n*-6, LA) vegetable oil significantly decreases *n*-3 LC-PUFA content in the fish's flesh. Therefore, there is an obvious requirement of an alternative and sustainable source of *n*-3 LC-PUFA for human consumption. Application of transgenic technology to produce *n*-3 LC-PUFA-rich fish has been mainly reported in the zebrafish model. The main strategies of these studies are to elevate the activities of LC-PUFA biosynthesis pathway by overexpression of fish-derived desaturase and elongase involved LC-PUFA synthesis (Alimuddin *et al.* 2005, 2007, 2008; Kabeya *et al.* 2014). Actually, freshwater fish species, including zebrafish, possess fatty acid desaturase (Hastings *et al.* 2001) and elongase (Monroig *et al.* 2009) to produce *n*-6 polyunsaturated fatty acid (*n*-6 PUFA), such as arachidonic acid (20:4*n*-6, ARA) and docosapentaenoic acid (22:5*n*-6, DPA), from their 18:2*n*-6 fatty acid precursor, and to produce *n*-3 polyunsaturated fatty acid (*n*-3 PUFA), such as EPA and DHA, from their precursor-α-linolenic acid (18:3*n*-3, ALA). However, both the fish-derived desaturase and elongase act on the *n*-3 and *n*-6 fatty acid series (Monroig et al. 2011), and this generally does not substantially change the ratio of *n*-6 PUFA against *n*-3 PUFA, which is in fact the most important health issue for consideration (Wan *et al.* 2010; Deckelbaum 2010). On the other hand, *n*-6 PUFAs are relatively abundant in both the routine dietary and the flesh of freshwater fish (our data and reference (Özogul *et al.* 2007)). Therefore, the directive and rational strategy to elevate *n*-3 PUFA content is to convert *n*-6 PUFA to *n*-3 PUFA through omega-3 desaturase.

Quite a lot of lower life, such as plants, microorganisms and *Caenorhabditis elegans* are capable of converting *n*-6 to *n*-3 fatty acids by their endogenous omega-3 desaturase genes, which have been well characterized (Spychalla *et al.* 1997). The *C. elegans fat1* (*omega-3 desaturase*) gene was used to produce transgenic mice, which were able to convert *n*-6 PUFA to *n*-3 PUFA with high efficiency (Kang *et al.* 2004). This has provided a new paradigm to produce *n*-3 PUFA in animal meat. Until now, several *fat1* transgenic live stocks rich in *n*-3 LC-PUFA have been developed as potential suppliers of *n*-3 LC-PUFA (Lai *et al.* 2006; Zhang *et al.* 2012, 2013; Wu *et al.* 2012). All of these transgenic animals are generated by introducing humanized omega-3 desaturase gene (*fat1*) of *C. elegans*. Nevertheless, the *fat1*-transgenic animals still lack the ability of *de novo* synthesis of 18:2*n*-6 PUFA, the substrate for both the transgene-based Fat1 enzyme and certain endogenous desaturases and enlongases, which can produce 18:3*n*-3 PUFA and synthesize longer *n*-6 PUFA, such as ARA and DPA, respectively. Several plants and *C. elegans* have delta-12 desaturase (Fat2) activity which can efficiently convert 18:1 to 18:2*n*-6 fatty acids (Peyou-Ndi *et al.* 2000; Okuley *et al.* 1994). *fat2*-transgenesis has been done in several mammals, in which the expressed Fat2 protein could efficiently convert 18:1 to 18:2*n*-6 fatty acids (Chen *et al.* 2009; Saeki *et al.* 2004). Nevertheless, as the major source of omega-3 LC-PUFA, fish has never been succeeded in *fat1* or *fat2* transgenesis.

Zebrafish has shown to be an excellent animal model, not only to study vertebrate genetics and development (Wei *et al.* 2014; Zhu and Sun 2000), but also to investigate human disease and fish directional breeding (Ahmed *et al.* 2011; Xiong *et al.* 2013; Lieschke and Currie 2007). The primary aim of the present study is to use zebrafish to establish a *fat1*-transgenic fish model, with elevated level of *n*-3 LC-PUFA. Moreover, we are interested to determine the capacities for *n*-3 LC-PUFA biosynthesis when *fat2* and *fat1* were co-expressed in a vertebrate model, when compared to expression of *fat1* or *fat2* alone. Our study demonstrates that *fat1* and *fat2* double transgenesis can fulfill the gap of *de novo* LC-PUFA biosynthesis pathway in vertebrates and result in robust production of *n*-3 LC-PUFA in a freshwater fish, zebrafish. To our knowledge, this is the first successful report of *fat1* transgenic fish and the first report of *fat1* and *fat2* double transgenesis in a fish species.

Materials and Methods

Zebrafish

Zebrafish embryos of *AB* genetic background were obtained from the China Zebrafish

Resource Center (http://zfish.cn; Wuhan, China), raised, and staged as previously described (Kimmel *et al.* 1995). The experiments involving zebrafish were approved by the Institutional Animal Care and Use Committee of the Institute of Hydrobiology, Chinese Academy of Sciences.

Preparation of DNA constructs

The *fat1* gene was an artificial *omega-3 desaturase* gene previously described (Kang *et al.* 2004). A fish codon-optimized *delta-12 desaturas*e gene (*fat2*) was artificially synthesized according to the predicated Fat2 protein sequence of *C. elegans* (Peyou-Ndi *et al.* 2000) by GenScript Corporation (Shanghai, China). The humanized *fat2* sequence was deposited in Genbank at accession number KJ081691.

The *fat1* expression construct, pTol2-CMV:fat1-SV40 poly(A); CMV:EGFP-SV40 poly(A), was a Tol2 transposonbased, bipartite construct containing CMV promoter and SV40 poly(A)-regulated *fat1* expression cassette as well as a *cis*-linked CMV promoter and SV40 poly(A)-regulated EGFP expression cassette (Fig. 1a). Briefly, the *fat1* coding sequence was amplified from pCAGGS-*fat1* (Kang *et al.* 2004) with two primers (fat1Fr and fat1Re) and inserted into the construct pTol2-CMV-SV40 poly(A); CMV:EGFP-SV40 poly(A) at *Eco*RI and *Xba*I sites. The pTol2-CMV-SV40 poly(A); CMV:EGFP-SV40 poly(A) construct was generated by three steps. Firstly, to generate pTol2-SV40 construct, the vector backbone containing Tol2 transposon element and SV40 poly(A) signal were obtained by digestion of 4×KGFP vector from Distel *et al.* (2009) with *Xho*I and *Xba*I. A DNA fragment containing 4 restriction enzyme sites of *Sal*I, *Eco*RI, *Hin*dIII, and *Bam*HI was generated by self-annealing extension of two overlapping primers (MCS-F and MCS-R) with *Xho*I and *Xba*I sites at 5′ ends and inserted into 4×KGFP vector at the sites of *Xho*I and *Xba*I. Subsequently, to generate pTol2-CMV-SV40 construct, CMV promoter from pEGFP-N1 (Invitrogen) was amplified by primers of CMV-F and CMV-R containing *Xho*I and *Sal*I sites and inserted into the pTol2-SV40. Finally, the CMV:EGFP-SV40 reporter cassette was amplified by overlapping extension PCR (Heckman and Pease 2007) using four primers of SP1, SP2, SP3, and SP4, with the elements of CMV promoter and EGFP coding sequence from pEGFP-C1 (Invitrogen), and SV40 poly(A) signal fragment from pCS2+. Then the EGFP reporter cassette was inserted into pTol2-CMV-SV40 at *Bgl*II and *Xho*I sites to generate pTol2-CMV-SV40 poly(A); CMV:EGFP-SV40 poly(A).

The *fat2* expression construct, pTol2-CMV-fat2-SV40 poly(A); CMV:mCherry-SV40 poly(A), was likewise a Tol2 transposon-based, bipartite construct consisting of CMV promoter and SV40 poly(A) regulated *fat2* expression cassette as well as a *cis*-linked CMV:mCherry-SV40 poly(A) reporter (Fig. 3a). Briefly, the *fat2* coding sequence was subcloned into the construct pTol2-CMV-SV40 poly(A); CMV:mCherry-SV40 poly(A) at

*Hin*dIII and *Bam*HI sites. To generate pTol2-CMV-SV40 poly(A); CMV:mCherry-SV40 poly(A), the mCherry fragment was amplified from TK5xC of Distel *et al.* (2009) with two primers (Cherry-F and Cherry-R) and the EGFP fragment of pTol2-CMV-SV40 poly(A); CMV:EGFP-SV40 poly(A) was replaced by mCherry coding sequence at the sites of *Nhe*I and *Xho*I. The primers sequences for PCR were listed in Supplemental Table S1.

Cell culture and transfection

Epithelioma Papulosum *Cyprini* (EPC) was cultured in TC199 medium supplemented with 10% fetal bovine serum (FBS) at 28.5℃ as previously described (Pei *et al.* 2008). Fish cells were transfected with FuGENE HD (Roche) in a 10-mm culture dish. After 10 hours incubation, transfection complex medium were replaced with fresh medium supplemented with 10 uM 18∶1 fatty acids (Nu-Chek Prep). 48 hours after transfection, cells were collected for lipid extraction and analysis as previously described (Kang *et al.* 2001).

Generation of transgenic zebrafish

Approximately 1 nl of DNA and RNA mixtures containing 25ng/ul expression constructs and 25ng/ul Tol2 transposase mRNA was injected into fertilized embryos at one-cell stage as previous described (Kawakami *et al.* 2004; Xiong *et al.* 2013). About 24 h after microinjection, embryos showing strong expression of fluorescent proteins were raised to adulthood. The mature F0 transgenic fish was mated with wildtype zebrafish and their offspring were screened using a fluorescence microscope (Olympus MVX10). F1 embryos expressing fluorescence protein were raised to adulthood. For each expression construct, two transgenic lines from two different F0 individual founders were established. To generate *fat1* and *fat2* double transgenic zebrafish, F2 heterozygous transgenic zebrafish of *Tg(CMV:fat1, CMV:EGFP)* and *Tg(CMV:fat2, CMV:mCherry)* were crossed and the offspring with four genotypes, wildtype, *Tg(CMV:fat1, CMV:EGFP)*$^{+/-}$, *Tg(CMV:fat2, CMV:mCherry)*$^{+/-}$, *Tg(CMV: fat1, CMV:EGFP)*$^{+/-}$/*Tg(CMV:fat2, CMV:mCherry)*$^{+/-}$ were raised together.

Feeding of transgenic fish

Transgenic fish of three genotypes and wildtype fish, which could be identified by visible fluorescent protein expression were raised in the same aquaria with density of 5 pieces/L and fed with brine shrimp (Heading, Tianjin, China) 3 times a day for 3 months. The fatty acids composition of diets was extracted and analyzed.

Lipid extraction and analysis

Lipid of cells, fish diets, and dorsal muscle tissue were extracted with chloroform/ methanol (2∶1, *Vol*/*Vol*) containing 0.005% butylated hydroxytoluene (Sigma). Fatty acid methyl esters (FAMEs) were prepared by incubating the total lipid with 1 ml of methylation reagent containing 5% H_2SO_4 (vol / vol) and 95% methanol for 1 hour at 100℃ and extracting by hexane. Fatty acids methyl esters were separated and quantified by gas chromatography (TRACE GC, Thermo Scientific, Milan, Italy) equipped with a capillary column (60 m × 0.25 mm) (DB-23, J&W Scientific, USA), a flame ionization detector (FID) and a split/splitless injector. Nitrogen was used as carrier gas with an oven thermal gradient from an initial 50℃ to 170℃ at 40℃ / min and then to a final temperature of 210℃ at 18℃ / min and isothermal for 28 min.

RNA isolation and Real-Time PCR

Total RNA of cells or muscle tissue were extracted using TRIzol reagent (Invitrogen) and treated with DNase I (Promega). cDNA was synthesized by reverse transcription using ReverTra Ace enzyme (TOYOBO). The real-time PCR was carried with SYBR Mix (Bio-Rad) and performed by CFX Real Time PCR System (Bio-Rad). The $2^{-\Delta\Delta CT}$ method was used to analyze the expression levels of target genes with β-actin as the internal control (Livak and Schmittgen 2001). Each real-time PCR analysis was repeated in triplicates. The primers sequences for real-time PCR were listed in Supplemental Table S1.

Statistical analysis

Statistical analyses were conducted by SPSS statistics 14.0 version software. The data obtained from each experiment were analyzed by one-way ANOVA. $P<0.05$ was considered as statistically significant.

Results

Fat1 converted *n*-6 to *n*-3 in fish-cultured cells

To test whether omega-3 desaturase acts its function in fish cells, a *fat1* over-expression construct pTol2-CMV:SV40 poly(A); CMV:EGFP-SV40 poly(A), namely *CMV:fat1* (Fig. 1a), was generated and transfected to common carp Epithelioma Papulosum Cyprini (EPC) cells. The EGFP expression cassette was used to evaluate the efficiency of cells transfection. 48h after transfection, approximately 30%-40% cells transfected with *CMV:fat1* or control construct (*CMV:EGFP*) showed strong expression of green fluorescent protein (Fig. 1 b, c). As the *fat1* and *EGFP* coding sequences were both driven by the *CMV* promoter and the two expression cassettes were *cis*-linked, the high-level expression of EGFP should indicate the high efficiency of gene transfection and high expression level of the *fat1* gene. The expression of *fat1* mRNA was determined by RT-PCR analysis, and the *fat1* mRNA could be detected in all the cells transfected with *CMV:fat1* (Fig. 1d).

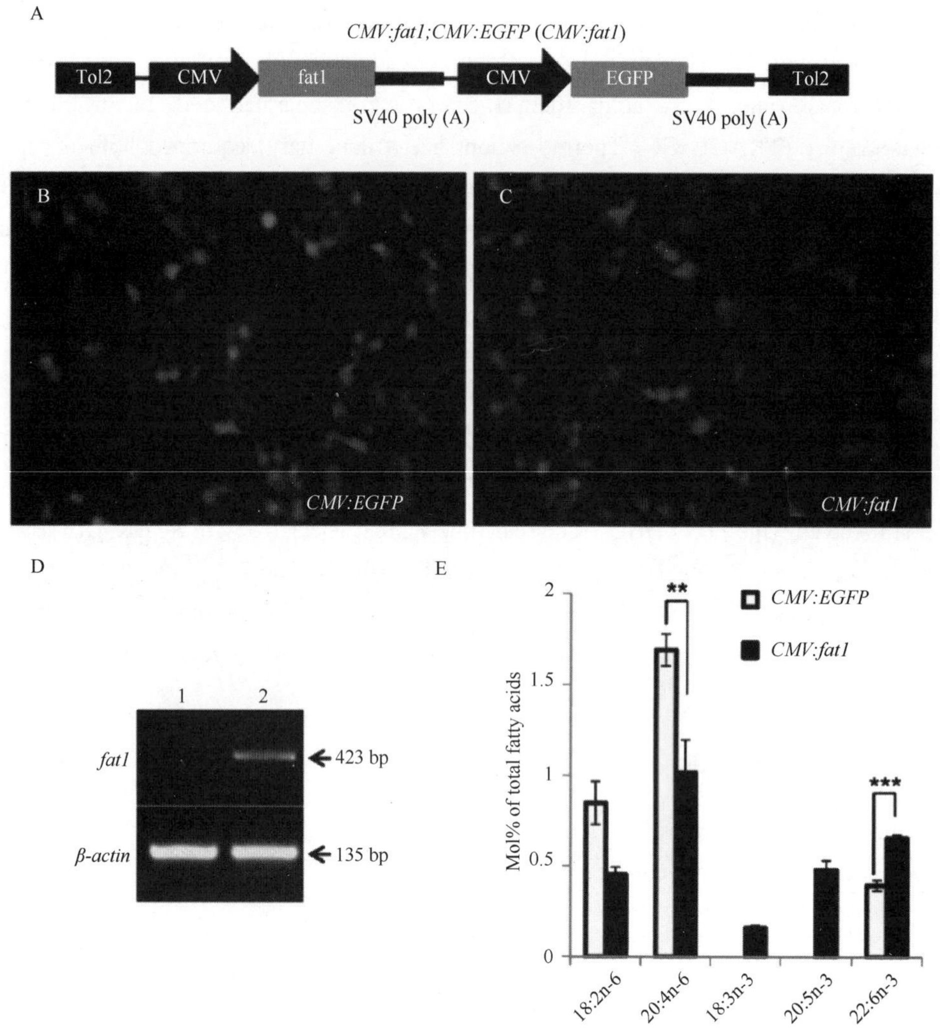

Fig. 1 Overexpression of *fat1* converted *n*-6 to *n*-3 in fish culture cells. **a** Schema of *fat1* expression construct for cell transfection and generation of *Tg(CMV:fat1)* transgenic zebrafish. **b** Expression of GFP in control cells transfected *CMV:EGFP* construct. **c** Expression of GFP in cells transfected with *CMV:fat1* construct. **d** Detection of *fat1* expression in *CMV:EGFP* and *CMV:fat1*-transfected cells by RT-PCR analysis. 1 *CMV:EGFP*-transfected cells, 2 *CMV:fat1*-transfected cells. **e** Fatty acids composition of *CMV:EGFP* and *CMV:fat1*-transfected cells. Results were presented as means±SD, *n*=3. The *asterisks* labeled *above the error bars* indicated significant differences (***P*<0.01, ****P*<0.005)

The fatty acid compositions of total cellular lipids were analyzed in different groups of cells (Fig. 1e) (Table 1). Although 18:3*n*-3 and 20:5*n*-3 was not detectable in the control cells that was transfected with *CMV:EGFP* construct , the amounts of 18:3*n*-3 and 20:5*n*-3 were increased to 0.2% and 0.5%, respectively, in the *CMV:fat1* transfected group. Moreover, the amount of 22:6*n*-3 was increased by 75.0% compared with that of *CMV:EGFP*-transfected cells (0.7% vs. 0.4%; $P<0.005$). On the other hand, the amount of 18:2*n*-6 was decreased by 37.5% (0.5% vs. 0.8%; $P>0.05$) though there was no statistical significance, and the amount of 20:4*n*-6 was decreased by 41.2% (1.0% vs. 1.7%; $P<0.01$) in the cells expressing *fat1* gene, when compared with those in *CMV:EGFP*-transfected cells (Fig. 1e). As a result, the ratio of *n*-6/*n*-3 was decreased to 1.1 in the *CMV:fat1*-transfected cells from 7.7 in the *CMV:EGFP*-transfected cells (Table 1). These results demonstrated that the *n*-6 fatty acids, including 18:2*n*-6 and 20:4*n*-6, could be efficiently converted to corresponding *n*-3 fatty acids, namely 18:3*n*-3 and 20:5*n*-3, in fish-cultured cells by humanized *fat1* gene.

Table 1 Fatty acids composition of total lipids extracted from transfected EPC fish cell supplemented with 18: 1 substrate

Mol% of Total fatty acids	WT	*CMV:fat1*	*CMV:fat2*	*CMV:fat1+CMV:fat2*
Delta-9 MUFA				
18:1	40.4 ± 2.1^{A}	39.0 ± 2.2^{A}	27.5 ± 3.1^{B}	30.6 ± 1.0^{B}
n-6 PUFA				
18:2*n*-6	0.8 ± 0.1^{A}	0.5 ± 0.0^{A}	7.1 ± 0.2^{B}	3.2 ± 0.2^{C}
18:3*n*-6	0	0	0	0
20:2*n*-6	0	0	0.3 ± 0.2	0
20:3*n*-6	0.5 ± 0.1^{A}	0	0.5 ± 0.1^{A}	0
20:4*n*-6	$1.7\pm0.1^{A,a}$	$1.0\pm0.2^{B,b}$	$1.5\pm0.1^{AB,a}$	$0.8\pm0.1^{B,b}$
Total	3.0 ± 0.2^{AC}	1.5 ± 0.2^{A}	9.3 ± 0.4^{B}	4.0 ± 0.1^{C}
n-3 PUFA				
18:3*n*-3	0	0.2 ± 0.0^{A}	0	3.8 ± 0.2^{B}
20:5*n*-3	0	$0.5\pm0.1^{AB,a}$	$0.1\pm0.0^{A,b}$	$0.6\pm0.0^{B,c}$
22:5*n*-3	0	0	0	0.1 ± 0.0
22:6*n*-3	0.4 ± 0.0^{A}	0.7 ± 0.0^{B}	0.6 ± 0.1^{B}	0.9 ± 0.0^{C}
Total	$0.4\pm0.0^{A,a}$	$1.3\pm0.0^{B,b}$	$0.7\pm0.1^{AB,a}$	5.5 ± 0.3^{Cc}
n-6/*n*-3 ratio	$7.7\pm0.4^{A,a}$	$1.1\pm0.2^{B,b}$	$14.3\pm1.1^{C,c}$	$0.7\pm0.0^{B,b}$

Lipids were analyzed in three independent samples. Results are presented as means± SD. The differences between values in the same row with different superscripts letter were statistically significant determined by one-way ANOVA. The superscripts labeled with capital letter represent the $P<0.005$, and the superscripts labeled with small letter represent the $P<0.05$

Fat1 converted *n*-6 to *n*-3 in transgenic zebrafish

Since the products of *fat1* gene could efficiently convert *n*-6 PUFA to *n*-3 PUFA in fish-cultured cells, we asked whether it was possible to generate *fat1* transgenic zebrafish and how the *fat1* transgene would act in the transgenic fish. To generate *fat1* transgenic zebrafish lines, we injected the Tol2 transposone based *CMV:fat1* expression construct into zebrafish embryos. The transgenic founders of F0 generation were screened by observing the expression of green fluorescence in their offspring embryos. As a result, two germline etransmitted transgenic founders were obtained and two independent transgenic lines were established. The F2 transgenic embryos showed ubiquitous expression of EGFP (Fig. 2b), indicating the high expression level of *fat1* and *EGFP* in the whole embryos, since both transgenes were *cis*-linked and driven by the same CMV promoter. To examine the expression of transgene in adult fish, total RNA was extracted from muscle tissue of transgenic fish, and RT-PCR analysis results revealed that *fat1* mRNA was strongly expressed in the 3 randomly sampled fluorescent individuals (Fig. 2c).

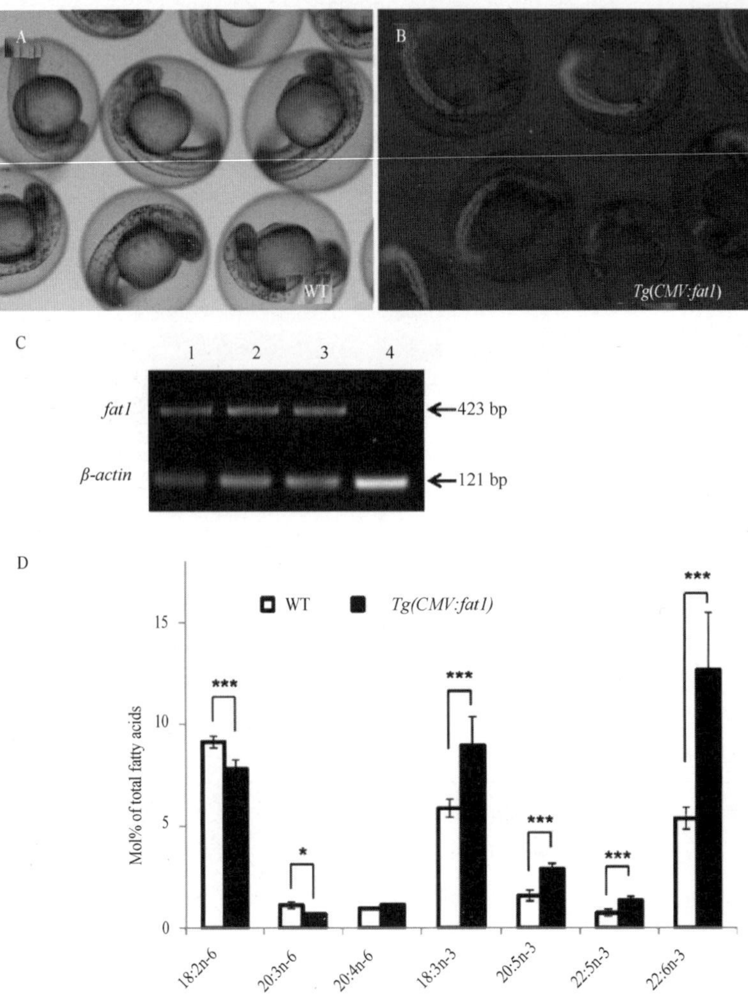

Fig. 2 Expression of *fat1* converted *n*-6 PUFA to *n*-3 PUFA in *Tg(CMV:fat1)* zebrafish. **a** Wildtype (WT) embryos at 36hpf. **b** Expression of GFP in *Tg(CMV:fat1)* zebrafish embryos at 36hpf. **c** RT-PCR analysis of fat1 expression in muscle tissue of WT and *Tg(CMV:fat1)* zebrafish. 1-3 Transgenic samples, 4 control sample. **d** Fatty acids composition of muscle tissue in WT and *Tg(CMV:fat1)* zebrafish. Results were presented as means ± SD, n=3. The *asterisks* labeled *above the error bars* indicated significant differences (*P<0.05, ***P<0.005)

Tg(CMV:fat1) fish and wildtype fish were raised in the same aquaria and fed with the same daily diets, which contain various levels of fatty acids (Table 2). Fatty acids composition analysis of the muscle tissues showed that the amount of 18:3*n*-3 was increased to 1.5-fold (9.0% vs. 5.9%; $P < 0.005$), the amount of 20:5*n*-3 was increased to 1.8-fold (2.9% vs. 1.6%; $P < 0.005$), and the amount of 22:6*n*-3 was increased to 2.4-fold (12.7% vs. 5.4%; $P < 0.005$) in *Tg(CMV:fat1)* zebrafish than those in the wildtype (Fig. 2d). The total amount of *n*-3 PUFA is increased from 13.6% in the wild type to 25.9% in *Tg(CMV:fat1)* fish, with an increase of 90.0%. Meanwhile, 18:2*n*-6 was decreased by 14.5% in the *Tg(CMV:fat1)* zebrafish, when compared with that in the wild type (7.8% vs. 9.1%; $P < 0.005$). As a result, the ratio of *n*-6 and *n*-3 (*n*-6/*n*-3) in the transgenic fish was 0.4, which was less than half of that in the wild type (WT ratio=0.9) (Table 3). These results demonstrate that *n*-6 PUFA could be efficiently converted to *n*-3 PUFA in *Tg(CMV:fat1)* zebrafish, even if the zebrafish is fed with daily diet containing elevated *n*-3 PUFA content (Table 2).

Table 2 Fatty acids composition of total lipids extracted from brine shrimp (zebrafish diet)

Mol% of total fatty acids	OA	LA	GLA	ALA	EDA	DGLA	ARA	EPA	DHA
Brine shrimp	25.6	8.64	1.0	36.0	0.3	0.1	0.2	0.2	0.0

OA: 18:1 oleic acid, *LA*:18:2*n*-6 linolenic acid, *GLA*: 18:3*n*-6 (γ-Linolenic acid), *ALA*: 18:3*n*-3α-Linolenic acid, *EDA*: 20:2*n*-6 eicosadienoic acid, *DGLA*: 20:3*n*-6 di-homo γ-linolenic acid, *ARA*: 20:4*n*-6 arachidonic acid, *EPA*:20:5*n*-3 eicosapentaenoic acid, *DHA*: 22:6*n*-3 docosahexaenoic acid

Co-expression of *fat1* and *fat2* promoted *n*-3 synthesis in fish-culture cells

In order to get a general idea of the fatty acids composition in freshwater fish, we analyzed the fatty acids composition in the flesh of several major freshwater fish species, such as grass carp (*Ctenopharyngodon idella*), blunt snout bream (*Megalobrama amblycephala*), common carp (*Cyprinus carpio* L.) and rare minnow (*Gobiocypris rarus*) (Supplement S2). It was found that oleic acids (18:1, OA), which could be *de novo* synthesized by fish body, were relatively abundant with unsaturated fatty acids in those fish species. To exploit plenty of 18:1 in fish body and to test whether it is possible to utilize endogenous 18:1 for *de novo* synthesis of 18:2*n*-6, we utilized *delta-12 desaturase* gene (*fat2*) for further cell transfection and transgenic studies. First, a fish codon-optimized *fat2* cDNA was synthesized according to the protein sequence of *C. elegans* delta-12 desaturase (Fat2) enzyme, and a *fat2* expression construct, pTol2-CMV-fat2-SV40 poly(A), CMV:mCherry-SV40 poly(A), namely *CMV:fat2* was generated (Fig. 3a). *CMV:fat2* was transfected into EPC fish cells supplemented with 18:1 fatty acid. The mCherry fluorescent protein was used as a reporter to evaluate the transfection efficiency. 24 hours after transfection, approximately 30% of the *CMV:fat2* transfected cells showed expression of red fluorescent protein, indicating high level of cell transfection efficiency and high level of *fat2* expression (Fig. 3b). The RT-PCR analysis showed that *fat2* mRNA was highly expressed in the transfected cells (Fig. 3e). The control and *CMV:fat2* transfected cells were subject to analysis of fatty acid composition (Table 1). In *fat2* transfected cells, the amount of 18:1 was significantly decreased from 40.4% (in the control group) to 27.5% (in the *CMV:fat2* transfected group) ($P < 0.005$), and the amount of 18:2*n*-6 was dramatically increased by almost 7.9-fold compared with the control (7.1% vs. 0.8%; $P < 0.005$) (Fig. 3f). As shown in table 1, 20:2*n*-6 was increased to 0.3% from zero level in the control cells, and the total amount of *n*-6 PUFA was increased to 3.1-fold in the *CMV:fat2* transfected cells compared with the control (9.3% vs. 3.0%; $P < 0.005$). Whereas,

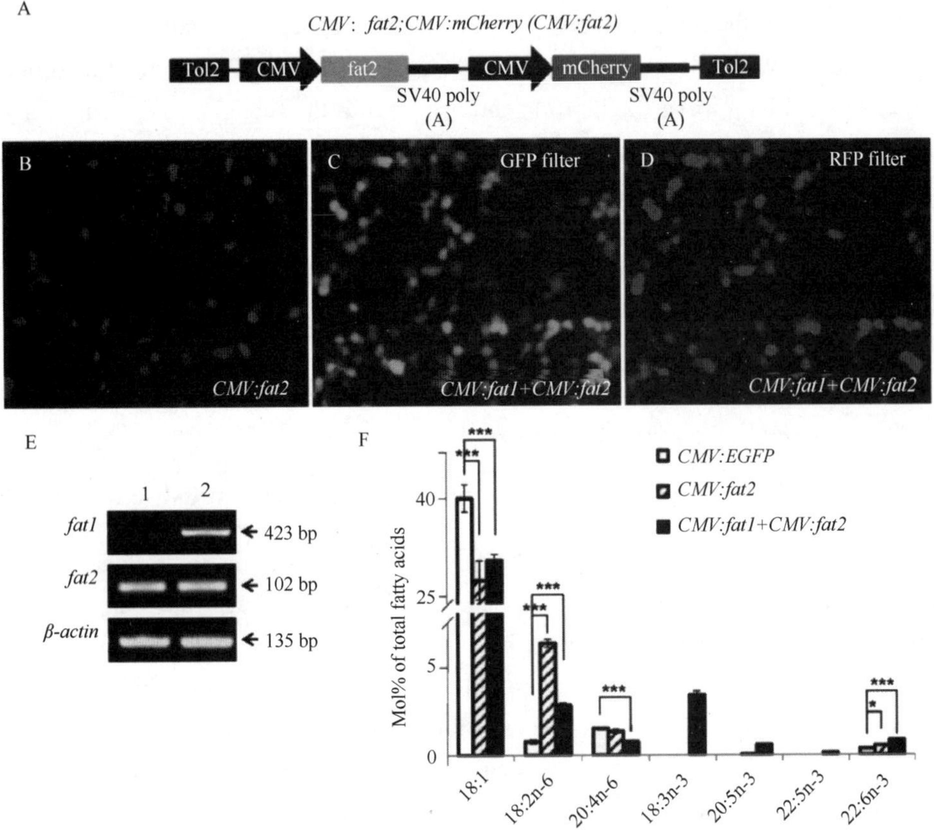

Fig. 3 Co-expression of *fat2* and *fat1* promoted *n*-3 PUFA synthesis in fish culture cells. **a** Schema of *fat2* expression construct for cell transfection and generation of *Tg(CMV:fat2)* transgenic zebrafish. **b** Expression of mCherry in cells transfected with *CMV:fat2* construct. **c** Expression of GFP in *CMV:fat1* cells co-transfected with *CMV:fat1* and *CMV:fat2*. **d** Expression of mCherry in cells co-transfected with and *CMV:fat2*. **e** Detection of *fat1* and *fat2* expression in cells co-transfected with *CMV:fat1* and *CMV:fat2* by RT-PCR analysis. **f** Fatty acids composition of cells transfected with *CMV:EGFP*, *CMV:fat2*, and co-transfected with *CMV:fat1* and *CMV:fat2*. Results were presented as means ± SD, *n*=3. The *asterisks* labeled *above the error bars* indicated significant differences (*$P<0.05$, ***$P<0.005$) determined by One-way ANOVA

the amount of 18:3*n*-3 was still undetectable in the *CMV:fat2* transfected cells. Totally, the ratio of *n*-6 and *n*-3 (*n*-6/*n*-3) was 14.3 in *CMV:fat2* transfected cells, which was nearly 2.0-fold of that in the control cells (*n*-6/*n*-3=7.7) (Table 1). Taken together, these results indicate that 18:1 was converted to 18:2*n*-6 effectively by Fat2 desaturase in fish cells, and the latter could be minimally utilized for synthesis of longer-chain PUFA through the endogenous desaturases and elongases.

Because *fat1* encodes a type of omega-3 desaturase and the increased amount of 18:2*n*-6 fatty acid could be utilized as the substrate for *fat1* transgene, the expression constructs *CMV:fat1* and *CMV:fat2* were co-transfected into fish culture cells. In the *CMV:fat1* and *CMV:fat2* co-transfected cells, green fluorescence and red fluorescence were both observed in a large proportion of cells (Fig 3c, d), indicating the co-expression of *EGFP* and *mCherry*, and *fat1* and *fat2* as well. RT-PCR analysis results showed that *fat1* and *fat2* mRNA was strongly expressed in the co-transfected cells (Fig. 3e). Analysis of the fatty acid composition showed that in *fat1* and *fat2* co-transfected cells, 18:1 was 24.2% decreased (30.6% vs. 40.4%; $P < 0.005$), and 18:2*n*-6 was 3.0-fold increased (3.2% vs. 0.8%; $P < 0.005$) when compared with the control cells (Fig. 3f). The content of 18:3*n*-3 in co-transfected cell was increased to 3.8%, from zero levels in the control group and the *CMV:fat2* transfected group. These results strongly indicate that 18:1 fatty acid was converted to 18:2*n*-6 by *fat2*-encoded delta-12 desaturase, and the 18:2*n*-6 fatty acid was subsequently converted to 18:3*n*-3 efficiently by *fat1*-encoded omega-3 desaturase. Meanwhile, the amount of 20:5*n*-3 fatty acid was increased to 0.6% of total fatty acids from zero level in the control group, and the amount of 22:6*n*-3 was 1.3-fold increased, as a result of the conversion of *n*-6 PUFA into *n*-3 PUFA by *fat1* desaturase (Fig. 3f). More importantly, the total amount of *n*-3 PUFA in the *CMV:fat1* and *CMV:fat2* co-transfected cells was increased to 14.0-fold compared with control cells and was increased to 3.3-fold compared with *CMV:fat1* transfected cells (Table 1). The ratio of *n*-6 and *n*-3 (*n*-6/*n*-3) in the *CMV:fat1* and *CMV:fat2* co-transfected cells were 0.7, which were significantly lower than that in the control (7.7), in the *fat2*-transfected cells (14.3) and in the *fat1*-transfected cells (1.1) (Table 1). These results demonstrate that co-expression of *fat1* and *fat2* could significantly promote *n*-3 PUFA synthesis in fish culture cells, especially the 18:3*n*-3, which could be utilized for synthesis of *n*-3 LC-PUFA.

Double transgenesis of *CMV:fat1* and *CMV:fat2* promoted *n*-3 synthesis in zebrafish

To test the effect of *f*at2 overexpression in the fish's body, the *CMV:fat2* DNA construct was injected into zebrafish embryos and *Tg(CMV:fat2)* transgenic zebrafish lines were generated. The red fluorescence was ubiquitously detected in the F2 *Tg(CMV:fat2)* embryos (Fig. 4a) and the *fat2* transcript was detected in the muscle tissue of *Tg(CMV:fat2)* adults (Fig. 4d). As

Table 3 Fatty acids composition of total lipids extracted from dorsal muscle tissues of wildtype and different transgenic zebrafish

Mol% of total fatty acids	WT	*CMV:fat1*	*CMV:fat2*	*CMV:fat1/CMV:fat2*
Delta-9 MUFA				
18:1	$28.2\pm1.2^{A,a}$	$23.5\pm0.9^{B,b}$	$25.3\pm2.1^{A,b}$	$17.6\pm0.5^{C,c}$
n-6 PUFA				
18:2*n*-6	9.1 ± 0.3^{A}	7.8 ± 0.4^{B}	11.9 ± 0.1^{C}	9.7 ± 0.2^{A}
18:3*n*-6	$0.4\pm0.0^{AB,ac}$	$0.3\pm0.1^{A,b}$	$0.5\pm0.0^{B,c}$	$0.3\pm0.1^{AB,ab}$
20:2*n*-6	$0.2\pm0.0^{A,a}$	$0.3\pm0.0^{BC,b}$	$0.3\pm0.0^{AB,a}$	$0.4\pm0.0^{C,b}$
20:3*n*-6	$1.1\pm0.2^{A,ac}$	$0.7\pm0.1^{B,b}$	$1.3\pm0.1^{A,a}$	$1.0\pm0.0^{AB,c}$
20:4*n*-6	$1.0\pm0.0^{A,a}$	$1.1\pm0.0^{A,a}$	$1.1\pm0.0^{A,a}$	$1.5\pm0.0^{A,b}$
Total	$11.8\pm0.6^{A,a}$	$10.2\pm0.2^{B,b}$	$15.1\pm0.4^{C,c}$	$12.9\pm0.2^{A,d}$
n-3 PUFA				
18:3*n*-3	$5.9\pm0.4^{A,a}$	$9.0\pm1.4^{A,b}$	$5.4\pm0.6^{A,a}$	$8.6\pm2.2^{A,b}$
20:5*n*-3	1.6 ± 0.3^{A}	2.9 ± 0.2^{B}	1.5 ± 0.1^{A}	2.7 ± 0.2^{B}
22:5*n*-3	0.8 ± 0.2^{A}	1.4 ± 0.2^{B}	0.7 ± 0.0^{A}	1.5 ± 0.0^{B}
22:6*n*-3	5.4 ± 0.5^{A}	12.7 ± 2.8^{B}	5.5 ± 0.7^{A}	14.9 ± 1.7^{B}
Total	13.6 ± 1.0^{A}	25.9 ± 2.0^{B}	13.5 ± 1.1^{A}	27.7 ± 1.1^{B}
n-6/*n*-3 ratio	$0.9\pm0.0^{A,a}$	$0.4\pm0.0^{B,b}$	$1.1\pm0.1^{C,c}$	$0.5\pm0.0^{B,d}$

Lipids were analyzed in three independent samples. Muscle tissue of three to five fishes were taken for every sample. Results were presented as means±SD. The differences between values in the same row with different superscripts letter were statistically significant determined by one-way ANOVA. The superscripts labeled with capital letter represent the $P<0.005$, and the superscripts labled with small letter represent the $P<0.05$

expected, fatty acids composition analysis of muscle tissue showed that the amount of 18:1 was decreased by 10.3% (25.3% vs. 28.2%; $P < 0.05$), and the amount of 18:2*n*-6 was increased by 30.8% in *Tg(CMV:fat2)* zebrafish (11.9% vs. 9.1%; $P < 0.005$) compared with wildtype, as a result of conversion of 18:1 into 18:2*n*-6 by Fat2 desaturase (Fig. 4f). The total amount of *n*-6 PUFA was increased by 28.0% (15.1% vs. 11.8%; $P < 0.005$), whereas the total amount of *n*-3 PUFA was not significantly changed (13.5% in *fat2* transgenesis and 13.6% in wildtype). Finally, the ratio of *n*-6/*n*-3 was 1.1 in the *Tg(CMV:fat2)* fish , which was increased by 22.2% compared with that in wild type (*n*-6/*n*-3=0.9). All these demonstrate that *CMV:fat2* transgene functions efficiently in the transgenic zebrafish model, which has elevated content of *n*-6 PUFA.

To generate *fat1* and *fat2* double transgenic zebrafish, *Tg(CMV:fat1)*$^{+/-}$ transgenic zebrafish was crossed with *Tg(CMV:fat2)*$^{+/-}$. The *fat1* and *fat2* double transgenic fish expressing both red and green fluorescent proteins (Fig. 4b, 4c) were raised with *fat1* or *fat2* single transgenic fish as well as wildtype fish in the same aquaria and fed with the same diet (Table 2). As the 4 types of transgenic fish showed expression of different types of

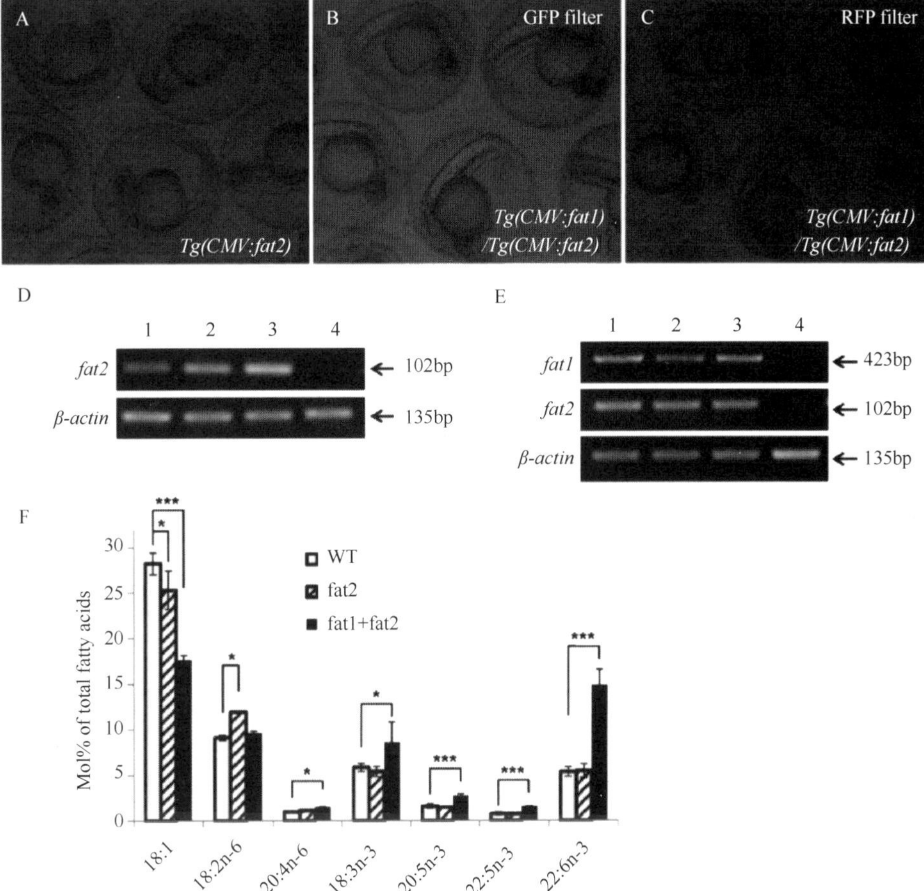

Fig. 4 Co-expression of *fat2* and *fat1* promoted *n*-3 PUFA synthesis in double transgenic zebrafish. **a** Expression of mCherry in *Tg(CMV:fat2)* zebrafish embryos at 36 hpf. **b-c** Co-expression of GFP **(b)** and mCherry **(c)** in *Tg(CMV:fat1)/Tg(CMV:fat2)* zebrafish embryos at 36 hpf. **d** Detection of *fat2* expression in muscle tissue of *Tg(CMV:fat2)* zebrafish by RT-PCR analysis. 1-3 Transgenic samples, 4 control sample. **e** RT-PCR analysis of *fat1* and *fat2* expression in muscle tissue of *Tg*(*CMV:fat1*)/*Tg*(*CMV:fat2*) zebrafish. 1-3 Transgenic samples, 4 control sample. **f** Fatty acids composition of muscle tissues in WT, *Tg(CMV:fat2)* zebrafish, and *Tg*(*CMV:fat1*)/*Tg*(*CMV:fat2*) zebrafish. Results were presented as means ± SD, *n*=3. The *asterisks* labeled above the error bars indicated significant differences (**P*<0.05, ****P*<0.005) determined by one-way ANOVA

fluorescent proteins, i.e., no fluorescence in wild type, EGFP in $Tg(CMV{:}fat1)^{+/-}$, mCherry in $Tg(CMV{:}fat2)^{+/-}$, and both EGFP and mCherry in $Tg(CMV{:}fat1)^{+/-}$;$Tg(CMV{:}fat2)^{+/-}$, they were easily distinguished by naked eye observation or fluorescent microscopy. The RT-PCR analysis results showed that the transgenes of *fat1* and *fat2* were co-expressed in the muscle tissue of transgenic zebrafish (Fig. 4e). In the *fat1* and *fat2* double transgenic zebrafish, the amount of 18:1 was decreased by 37.6%, when compared with that in wild type (17.6% vs. 28.2%; $P < 0.005$) (Fig. 4f). The amount of 18:2*n*-6 was almost the same as that in wild type

(9.7% vs. 9.1%; $P > 0.05$), however, it was less than that in *Tg*(*CMV:fat2*) zebrafish (9.7% vs. 11.9%; $P < 0.005$), which indicated that part of 18:2*n*-6 produced by Fat2 enzyme was converted to 18:3*n*-3 by Fatl desaturase in *Tg*(*CMV:fat1*)$^{+/-}$; *Tg*(*CMV:fat2*)$^{+/-}$ fish. The amount of 18:3*n*-3 was 45.8% increased (8.6% vs. 5.9%; $P < 0.05$), the amount of 20:5*n*-3 was increased to 1.7-fold (2.7% vs. 1.6%; $P < 0.005$), and the amount of 22:6*n*-3 was increased to 2.8-fold (14.9% vs. 5.4%; $P < 0.005$) in muscle tissue of double transgenic zebrafish, when compared with that in the wildtype. The amount of 22:6*n*-3 in double transgenesis (14.9) was 17.3% higher than that in *Tg*(*CMV:fat1*) zebrafish (12.7), although the difference was not statistically significant (Table 3). The total amount of *n*-6 PUFA was decrease by 14.6% compared with that in the *Tg*(*CMV:fat2*) (12.9% vs. 15.1%; $P < 0.005$), and total amount of *n*-3 PUFA in double transgenic fish was increased by 105.2%, 6.9% and 103.7% when compared to *Tg*(*CMV: fat2*), *Tg(CMV:fat1)* and wild type, respectively. In general, the ratio of *n*-6 and *n*-3 (*n*-6/*n*-3) in the *fat1* and *fat2* double transgenic zebrafish was 0.4, which was significantly lower than that in the wildtype zebrafish (*n*-6 / *n*-3=0.9) and that in the *Tg*(*CMV:fat2*) (*n*-6/*n*-3=1.1). Therefore, we concluded that the *n*-6 PUFA synthesized by *fat2*-encoded detla-12 desaturase could be efficiently converted to *n*-3 PUFA by *fat1*-encoded omega-3 desaturase in *fat1* and *fat2* double transgenic zebrafish.

Transcription activity of LC-PUFA synthesis related genes was altered in transgenic zebrafish

In order to analyze the effects of different fatty acids compositions on the transcriptional activities of PUFA synthesis related factors in different types of transgenic fish, we examined the mRNA expression profiles of key genes on catalyzing desaturation and elongation reaction of fatty acids in liver tissue of those fish by real-time PCR. For instance, stearoyl CoA desaturase (*scd*) introduces the first double bond at the delta-9 position of 18:0 (stearic acid) saturated carbon chains to produce 18:1 monounsaturated fatty acids, and the delta-5 and delta-6 fatty acid desaturase (*fadsd2*), elongase-2 (*elovl2*) and elongase-5 (*elovl5*) are involved in consecutive desaturation and elongation reaction that convert 18:2*n*-6 and 18:3*n*-3 to 20:4*n*-6 and 20:5*n*-3 (Monroig *et al*. 2011). As shown in Fig. 5a, the transcripts of all these desaturase and elongase genes were strongly up-regulated in the *Tg*(*CMV:fat1*), *Tg*(*CMV:fat2*), and *Tg(CMV:fat1)/Tg(CMV:fat2)* zebrafish. Especially, the *scd* transcripts were increased by 24-fold in *Tg(CMV:fat2)* zebrafish when compared with the wildtype, in which 18:1 was converted to 18:2*n*-6 fatty acids effectively. This indicates that the consumption of 18:1 fatty acids by Fat2 enzyme will have a positive effect on the transcriptional regulation of *scd*.

Furthermore, two major transcription factors controlling desaturase and elongase, peroxisome proliferator-activated receptor α (PPARα) and sterol regulatory element binding protein-1 (SREBPF-1) (Rosen *et al*., 2000), were analyzed by real-time PCR in different genotypes of

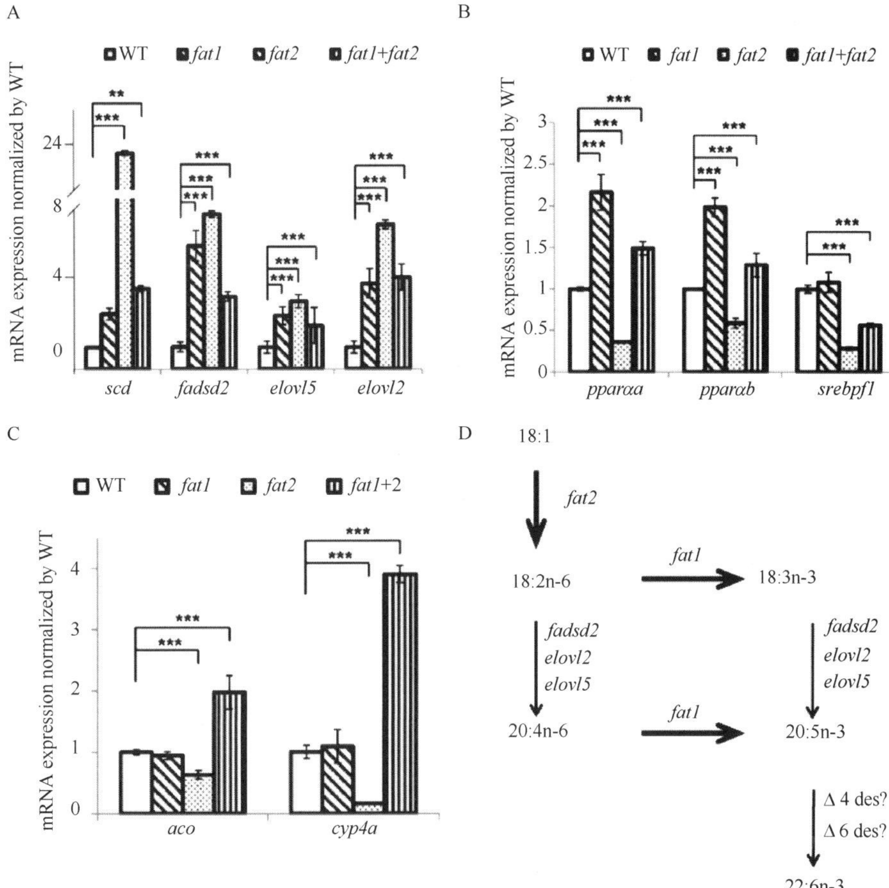

Fig. 5 Transcription activities of LC-PUFA synthesis related factors were altered in transgenic zebrafish. a-c The relative expression level of desaturase and elongase **(a)**, transcriptional regulatory factors involved in fatty acids metabolism **(b)**, and fatty acids oxidation **(c)** genes in liver tissue of wildtype, *Tg(CMV:fat1)*, *Tg(CMV:fat2)*, and *Tg(CMV:fat1)/Tg(CMV:fat2)* zebrafish were analyzed by Real-time PCR. Results were presented as means ± SD, n=3. The *asterisks* labeled above the error bars indicated significant differences (**P<0.01, ***P<0.005) determined by one-way ANOVA **(d)** The double transgenesis of *fat1* and *fat2* fullfills the gap of *de novo* PUFA biosynthesis pathway. In zebrafish body, the long chain n-6 and n-3 PUFA (20:4n-6 and 20:5n-3) are synthesized by fatty acids desaturase (*fadsd2*), fatty acids elongase (*elovl2*, *elovl5*) using the precursor of 18:2n-6 and 18:3n-3. The 22:6n-3 is potentially synthesized by Δ4 desaturase pathway or Δ6 desaturase pathway. The *bold arrows* and *texts* of the *fat1* and *fat2* indicate the enzyme activities that do not endogenously exist in fish body

zebrafish. Transcription of *pparαa* and *pparαb* was likewise significantly up-regulated in *Tg*(*CMV:fat1*) and *Tg*(*CMV:fat1*)/*Tg*(*CMV:fat2*), and was significantly downregulated in *Tg*(*CMV:fat2*) zebrafish. Transcription of *srebpf1* was significantly downregulated in *Tg*(*CMV:fat2*) and *Tg*(*CMV:fat1*)/ *Tg*(*CMV:fat2*), and was almost unchanged in *Tg*(*CMV:fat1*) zebrafish (Fig. 5b). In addition, PPARα and SREBPF-1 also promote a shift of fatty acids metabolism toward fatty acid oxidation and away from fatty acid desaturase, elongase and storage (Jump 2008). We therefore examined mRNA expression of two fatty acid oxidation genes, *aco* and *cyp4a* in the liver tissues of transgenic zebrafish. As shown in Fig. 5c, the transcription activities of *aco* and *cyp4a* were unchanged in *Tg(CMV:fat1)* zebrafish, decreased in *Tg(CMV:fat2)* zebrafish, and increased in *Tg(CMV:fat1)/Tg(CMV:fat2)* double transgenic zebrafish. All these indicate that the *CMV:fat1* and *CMV:fat2* transgenesis not only has a strong effect on the fatty acids composition of the transgenic fish, but also strongly impairs the transcription of a set of genes related to fatty acids synthesis and metabolism.

Discussion

Human and most animals lack omega-3 desaturase (delta-15 desaturase), which is required for *de novo* synthesis of ALA (18:3*n*-3), the essential precursor for synthesis of most *n*-3 LC-PUFA, such as EPA and DHA. Since the first report of *fat1*-transgenic mice model, which gained the ability to convert *n*-6 PUFA to *n*-3 PUFA (Kang *et al.* 2004), *n*-3 PUFA rich *fat1*-transgenic animals have been reported in many mammalian species, for example, pig (Lai *et al.* 2006; Zhang *et al.* 2012), cow (Wu *et al.* 2012) and sheep (Zhang *et al.* 2013). However, *fat1* transgenesis has not been succeeded in fish species, which is one of the major protein sources and *n*-3 PUFA for human being. On the other hand, *fat1* transgenic animals still need *n*-6 PUFA as the substrate for *n*-3 PUFA synthesis. For instance, LA (18:2*n*-6) is considered as an essential fatty acid for human and other vertebrates, since vertebrates are also absent of the activity of delta-12 desaturase to convert 18:1*n*-9 to 18:2*n*-6. Therefore, it is of great importance to establish a transgenic animal model which harbors the activity of both delta-12 desaturase and omega-3 desaturase and fulfills the gaps in the whole pathway of *de novo* synthesis of *n*-6 PUFA and *n*-3 PUFA (Fig. 5d). In the present study, we have not only established a *fat1* transgenic fish model rich in *n*-3 PUFA, but also realized an effective *fat1* and *fat2* double transgenesis in a fish model for the first time.

To reach our final goal, we have conducted a series of studies from *in vitro*-cultured cells to *in vivo* transgenic animals, from single gene transgenesis to double transgenesis. First, *fat1* is expressed in the culture cells of common carp, which belongs to the same family as zebrafish. Although 18:2*n*-6 and 20:4*n*-6 have not been supplemented in the culture medium, *fat1* converts cellular *n*-6 PUFA to *n*-3 PUFA effectively, which significantly increases the contents of 18:3*n*-3 in the control to 0.2% in the *CMV:fat1* transfected cells. Secondly, several *fat1* transgenic zebrafish lines have been established and the transgenic fish shows to be rich in *n*-3 PUFA, especially the 22:6*n*-3 was increased by 135.2% when compared with the controls. In previous study of transgenic zebrafish, however, 22:6*n*-3 was onlyincreased by 10% to 26.8% in salmon *delta-6 like desaturase* transgenic zebrafish (Alimuddin *et al.* 2007). These results demonstrate that the omega-3 desaturase from *C. elegans* is functionally active not only in fish cells but also in fish body, and more importanty, *fat1* transgenesis is an effective way to promote *n*-3 PUFA synthesis in fish model.

Second, we have proved that delta-12 desaturase acts efficiently in fish cultured cells and fish body as well. In fish cultured cells, transfection of *fat2* gene efficiently decreased the content of 18:1 from 40.4% to 27.5% and increased the content of 18:2*n*-6 from 0.8% to 7.1%. In the *Tg(CMV:fat2)* zebrafish, the amount of 18:2*n*-6 in muscle tissue increased by 30.8% when compared with wild-type fish. This result is even more obvious than the previous work in transgenic pig and mice. In transgenic pig, constitutive expression of spinach *delta-12 desaturase* gene led to the content of 18:2*n*-6 in white adipose tissue of transgenic pigs 20% higher than that of nontransgenic control (Saeki *et al.* 2004). In the study of cotton *delta-12 desaturase* transgenic mice, the content of 18:2*n*-6 in muscle tissue is increased by 19% than that of nontransgenic control (Chen *et al.* 2009). Different activities of Fat2 enzymes in different animal models may be due to the fatty acid metabolisms differences in different animals, or the different origins of *fat2* genes, i.e., *fat2* from *C. elegans* in our study and *fat2* from plants in previous studies.

Third, our study demonstrates that co-expression of delta-12 and omega-3 desaturases dramatically promotes 18:2*n*-6 and *n*-3 PUFA synthesis in fish culture cells. In previous study, co-expression of delta-12 and omega-3 desaturase of *C. elegans* in HC11 mouse mammary epithelial cells also promoted *n*-3 PUFA synthesis (Morimoto *et al.* 2005). Our study suggests that fish cells have a similar fatty acid biosynthesis pathway as mammalian cells. In a previous report, transgenic mice expressing *fat2* and *fat1* genes with a pIRES-FAT1-FAT2 expression vector was generated (Chen *et al.* 2013). It was shown that the content of *n*-3 PUFA were slightly increased in the double transgenic mice, however, the detail effects of *fat1* and *fat2* double transgenesis and single transgenesis were not analyzed in a comparable way. In our study, to obtain comprehensive effect of single transgene and cooperation of double transgenes, we generate *Tg(CMV:fat1)*, *Tg(CMV:fat2)* and *Tg(CMV:fat1)/Tg(CMV:fat2)* and analyze the fatty acids composition of every transgenic line. With the help of co-expressed fluorescent proteins, we were able to breed 4 types of transgenic fish in the same aquarium, which facilitates further classification and comprehensive analysis of 4 genotypes and their corresponding phenotypes. In *Tg(CMV:fat1)/Tg(CMV:fat2)* zebrafish, the total amount of *n*-3 PUFA was increased by 103.7% than that of wildtype fish, and the amount of 22:6*n*-3 was even increased by 176%. These results indicate that co-transgenesis of *fat1* and *fat2* is most promising technology to promote *n*-3 PUFA synthesis in fish.

The increased content of *n*-3 PUFA may have positive effect on the transcriptional regulation of certain desaturases and elongases. In a type of freshwater fish, murray cod, the fatty acid desaturase and elongase activity on *n*-3 PUFA are increased with the elevated dietary 18:3*n*-3/18:2*n*-6 ratios (Senadheera *et al.* 2011), as a result, the products of *n*-3 PUFA biosynthesis pathway including 18:4*n*-3, 20:3*n*-3, and 20:4*n*-3 are increased. In rainbow trout, the LC-PUFA biosynthetic pathway is substrate limited (Thanuthong *et al.* 2011). The level of

n-3 LC-PUFA in the muscle is increased with the increasing dietary supply of 18:3n-3. In our study, cooperation of Fat1 and Fat2 or Fat1 alone *in vivo* provided the zebrafish a high level of endogenous 18:3n-3, which made the 18:3n-3/18:2n-6 ratio increase and supplied more substrate to n-3 LC-PUFA biosynthesis pathway, therefore leading to high level transcription of various desaturases and elongases (Fig. 5a). In the previous study of *fat1*-transgenic mammal (Kang *et al.* 2004; Lai *et al.* 2006; Wu *et al.* 2012; Zhang *et al.* 2013), the 20:4n-6 was abundant in dietary and should be converted to 20:5n-3 by *fat1*-encoded omega-3 desaturase effectively. In our study, however, the 20:4n-6 content of wildtype zebrafish was much low, because little 20:4n-6 was provided by the routine diet of zebrafish. Although Fat1 desaturase converts 20:4n-6 to 20:5n-3 effectively in fish culture cells, there is no significant change of 20:4n-6 in the *Tg(CMV:fat1)/Tg(CMV:fat2)* zebrafish, when compared with that in the *Tg(CMV:fat1)* zebrafish. This suggests that the majority of the increased n-3 LC-PUFA in transgenic zebrafish is likely synthesized from 18:3n-3 PUFA rather than from both C_{18} and C_{20} n-6 PUFA as that in *fat-1* transgenic mammals.

Regulation of whole body fatty acids synthesis and metabolism is complex in vertebrates. Dietary fatty acids regulate hepatic gene expression in liver. The transcription of desaturase and elongase is increased in fish fed with vegetable oil rich in C_{18} PUFA compared to fish fed with fish oil (Zheng *et al.* 2004; Monroig *et al.* 2010; Morais *et al.* 2011; Ren *et al.* 2013), suggesting that C_{18} PUFA may exert a positive effect on the transcription of desaturases and elongases. In the present study, both the fish are fed with the same diet, transcription activity of both the desaturases and elongases is up-regulated in different types of transgenic zebrafish. This result indicates that the transgenesis of omega-3 desaturase and delta-12 desaturase not only modifies the fatty acids composition of zebrafish tissue, but also influences the fatty acids metabolic pathway. Moreover, the increasing C_{18} PUFA including 18:2n-6 or 18:3n-3 in the *Tg(CMV:fat2)*, *Tg(CMV:fat1)*, and *Tg(CMV:fat1)/Tg(CMV:fat2)* zebrafish may push the transcriptional activities of endogenous desaturase and elongase to be elevated. In *Tg(CMV:fat1)* and *Tg(CMV:fat1)/Tg(CMV:fat2)* zebrafish, the increased transcription activities of desaturase and elongase were consistent with the increased n-3 LC-PUFA product. The 22:6n-3 has shown to have different regulation effects on different transcriptional factors, for instance, it suppresses *srebp-1* expression in mammals, and it is shown to be a weak activator of PPARα (Jump 2008). In our study, likewise, *pparα* is unregulated and *srebp-1* is down-regulated in *Tg(CMV:fat1)/Tg(CMV:fat2)* zebrafish. Both of these two fatty acid feedback factors are able to regulate the expression of desaturase, elongase and oxidation genes. We may conclude that in the *fat1* and *fat2* double transgenic zebrafish, both of the desaturase, elongase and oxidation genes are upregulated through the regulation of SREBP-1 and PPARα.

The dietary 20:5n-3 and 22:6n-3 not only suppress LC-PUFA biosynthesis pathway (Buzzi *et al.* 1996), but also mask the n-3 LC-PUFA from the biosynthesized pathway. In the present study, the advantage of brine shrimp as zebrafish diet is deficient in 20:5n-3 and

22:6*n*-3 fatty acids (Table 1). However, the brine shrimp rich in 18:3*n*-3 is not an ideal diet for the double transgenic study, because 18:3*n*-3 also masks part of 18:3*n*-3 generated from 18:1 by Fat1 and Fat2 or 18:2*n*-6 by Fat1. In the fish culture cell supplemented with 18:1 substrate, cooperation of Fat1 and Fat2 synthesized remarkable amount of 18:3*n*-3, which is converted to LC-PUFA by desaturase and elongase *in vivo*. However, in the *fat1* and *fat2* double transgenic zebrafish, 18:3*n*-3 was not significantly increased when compared to *fat1* transgenics, which might be due to the mask of the abundant dietary 18:3*n*-3. Over all, fatty acid composition of diet needed to be taken into carefully consideration if future studies need to be conducted.

Fish is very valuable source of protein and essential micronutrients including vitamins, minerals and *n*-3 PUFA for balanced nutrition and good health. The declining wild fisheries worldwide and increasing demand of fish oil for aquaculture make the aquaculture industry to feed fish with vegetable oil as alternative oil. Replacement of fish oil with C_{18}-rich vegetable oil decreases *n*-3 LC-PUFA content in the flesh, and it reduces nutritional benefit for human health. In present study, we propose that while feeding 18:2*n*-6-rich vegetable oil, such as soybean oil and sunflower-seed oil (Miller *et al.* 2008), *fat1* transgenic fish should be able to convert the 18:2*n*-6 fatty acid to 18:3*n*-3, and result in higher content of *n*-3 LC-PUFA by endogenous desaturases and enlongases. If 18:1-rich vegetable oil, such as rapeseed oil and olive oil, was used in fish diet, *fat1* and *fat2* double transgenic fish would be able to convert 18:1 to 18:3*n*-3 fatty acids. In other words, two transgenic technologies are offered to improve the *n*-3 LC-PUFA content directing toward different types of vegetable oil.

The *fat1* transgenic mice rich in endogenous *n*-3 PUFA have been an excellent research model for studying omega-3 fatty acids and disease (Yamashita *et al.* 2013; Bousquet *et al.* 2011; Kang 2007). In the present study, we generated two types of transgenic zebrafish, *fat1* transgenic zebrafish and *fat1* and *fat2* double transgenic zebrafish, both of which were rich in endogenous *n*-3 PUFA. The optical transparency of zebrafish embryos and larvae enables *in vivo* monitoring of vascular pathological processes in live zebrafish subjected to atherosclerosis (Stoletov *et al.* 2009). Zebrafish has also shown to be ideal vertebrate model to study various human diseases (Lieschke and Currie 2007). Therefore, the *fat1* and *fat1*/*fat2* transgenic zebrafish rich in omeaga-3 PUFA may serve as an ideal research model of *n*-3 PUFA and human disease.

Acknowledgement The authors thank Hanhua Hu, Xian-Tao Fang, Chun Cai, Jun-Jie Zhang, Guang-Zhao Chen for their technical support throughout the research project. This work was supported by the China 863 High-Tech Program Grant 2011AA100404, the China 973 Basic Research Program Grants 2010CB126306 and 2012CB944504, the National Science Fund for Excellent Young Scholars of NSFC Grant 31222052 and FEBL Grant 2011FBZ23 to Y. H. S.

Reference

Ahmed AS, Xiong F, Pang SC, He MD, Waters MJ, Zhu ZY, Sun YH (2011) Activation of GH signaling and GH-independent stimulation of growth in zebrafish by introduction of a constitutively activated GHR construct. Transgenic Res 20 (3): 557-567.

Alimuddin, Kiron V, Satoh S, Takeuchi T, Yoshizaki G (2008) Cloning and over-expression of a masu salmon (*Oncorhynchus masou*) fatty acid elongase-like gene in zebrafish. Aquaculture 282 (1-4): 13-18.

Alimuddin, Yoshizaki G, Kiron V, Satoh S, Takeuchi T (2005) Enhancement of EPA and DHA biosynthesis by over-expression of masu salmon delta6-desaturase-like gene in zebrafish. Transgenic Res 14 (2): 159-165.

Alimuddin, Yoshizaki G, Kiron V, Satoh S, Takeuchi T (2007) Expression of masu salmon delta5-desaturase-like gene elevated EPA and DHA biosynthesis in zebrafish. Mar Biotechnol (NY) 9 (1): 92-100.

Bousquet M, Gue K, Emond V, Julien P, Kang JX, Cicchetti F, Calon F (2011) Transgenic conversion of omega-6 into omega-3 fatty acids in a mouse model of Parkinson's disease. Journal of Lipid Research 52 (2): 263-271.

Buzzi M, Henderson RJ, Sargent JR (1996) The desaturation and elongation of linolenic acid and eicosapentaenoic acid by hepatocytes and liver microsomes from rainbow trout (*Oncorhynchus mykiss*) fed diets containing fish oil or olive oil. Biochim Biophys Acta 1299 (2): 235-244.

Chen Q, Liu Q, Wu Z, Wang Z, Gou K (2009) Generation of fad2 transgenic mice that produce omega-6 fatty acids. Science in China Series C, Life sciences / Chinese Academy of Sciences 52 (11): 1048-1054.

Chen Y, Mei M, Zhang P, Ma K, Song G, Ma X, Zhao T, Tang B, Ouyang H, Li G, Li Z (2013) The generation of transgenic mice with fat1 and fad2 genes that have their own polyunsaturated fatty acid biosynthetic pathway. Cell Physiol Biochem 32 (3): 523-532.

Deckelbaum RJ (2010) *n*-6 and *n*-3 Fatty acids and atherosclerosis: ratios or amounts? Arterioscler Thromb Vasc Biol 30 (12): 2325-2326.

Distel M, Wullimann MF, Koster RW (2009) Optimized Gal4 genetics for permanent gene expression mapping in zebrafish. Proc Natl Acad Sci U S A 106 (32): 13365-13370.

FAO (2012) The State of world fisheries and aquaculture. FAO.

Fritsche K (2006) Fatty Acids as Modulators of the Immune Response. Annual Review of Nutrition 26 (1): 45-73.

Hastings N, Agaba M, Tocher DR, Leaver MJ, Dick JR, Sargent JR, Teale AJ (2001) A vertebrate fatty acid desaturase with Delta 5 and Delta 6 activities. Proc Natl Acad Sci U S A 98 (25): 14304-14309.

Heckman KL, Pease LR (2007) Gene splicing and mutagenesis by PCR-driven overlap extension. Nat Protoc 2 (4): 924-932.

Jump DB (2008) *n*-3 polyunsaturated fatty acid regulation of hepatic gene transcription. Current opinion in lipidology 19 (3): 242-247.

Kabeya N, Takeuchi Y, Yamamoto Y, Yazawa R, Haga Y, Satoh S, Yoshizaki G (2014) Modification of the *n*-3 HUFA biosynthetic pathway by transgenesis in a marine teleost, nibe croaker. Journal of Biotechnology 172: 46-54.

Kang JX (2007) Fat-1 transgenic mice: a new model for omega-3 research. Prostaglandins Leukot Essent Fatty Acids 77 (5-6): 263-267.

Kang JX, Wang J, Wu L, Kang ZB (2004) Transgenic mice: Fat-1 mice convert *n*-6 to *n*-3 fatty acids. Nature 427 (6974): 504-504.

Kang ZB, Ge Y, Chen Z, Cluette-Brown J, Laposata M, Leaf A, Kang JX (2001) Adenoviral gene transfer of *Caenorhabditis elegans* *n*-3 fatty acid desaturase optimizes fatty acid composition in mammalian cells. Proc Natl Acad Sci U S A 98 (7): 4050-4054.

Kawakami K, Takeda H, Kawakami N, Kobayashi M, Matsuda N, Mishina M (2004) A transposon-mediated gene trap approach identifies developmentally regulated genes in zebrafish. Dev Cell 7 (1): 133-144.

Kimmel CB, Ballard WW, Kimmel SR, Ullmann B, Schilling TF (1995) Stages of embryonic development of the zebrafish. Dev Dyn 203 (3): 253-310.

Kris-Etherton PM, Harris WS, Appel LJ, Committee ftN (2003) Fish Consumption, Fish Oil, Omega-3 Fatty Acids, and Cardiovascular Disease. Arteriosclerosis, Thrombosis, and Vascular Biology 23 (2): e20-e30.

Lai L, Kang JX, Li R, Wang J, Witt WT, Yong HY, Hao Y, Wax DM, Murphy CN, Rieke A, Samuel M, Linville ML, Korte SW, Evans RW, Starzl TE, Prather RS, Dai Y (2006) Generation of cloned transgenic pigs rich in omega-3 fatty acids. Nat Biotech 24 (4): 435-436.

Lieschke GJ, Currie PD (2007) Animal models of human disease: zebrafish swim into view. Nature Reviews Genetics 8 (5): 353-367.

Livak KJ, Schmittgen TD (2001) Analysis of Relative Gene Expression Data Using Real-Time Quantitative PCR and the $2^{-\Delta\Delta CT}$ Method. Methods 25 (4): 402-408.

Marszalek JR, Lodish HF (2005) Docosahexaenoic acid, fatty acid-interacting proteins, and neuronal function: breastmilk and fish are good for you. Annu Rev Cell Dev Biol 21: 633-657.

Miller MR, Nichols PD, Carter CG (2008) n-3 Oil sources for use in aquaculture—alternatives to the unsustainable harvest of wild fish. Nutrition research reviews 21 (2): 85-96.

Monroig Ó, Navarro JC, Tocher DR (2011) Long-Chain Polyunsaturated Fatty Acids in Fish: Recent Advances on Desaturases and Elongases Involved in Their Byosinthesis. Avances en Nutrición Acuícola 11: 257-283.

Monroig O, Rotllant J, Sanchez E, Cerda-Reverter JM, Tocher DR (2009) Expression of long-chain polyunsaturated fatty acid (LC-PUFA) biosynthesis genes during zebrafish *Danio rerio* early embryogenesis. Biochim Biophys Acta 1791 (11): 1093-1101.

Monroig Ó, Zheng X, Morais S, Leaver MJ, Taggart JB, Tocher DR (2010) Multiple genes for functional Δ6 fatty acyl desaturases (Fad) in Atlantic salmon (*Salmo salar* L.): Gene and cDNA characterization, functional expression, tissue distribution and nutritional regulation. Biochimica et Biophysica Acta (BBA) - Molecular and Cell Biology of Lipids 1801 (9): 1072-1081.

Morais S, Pratoomyot J, Taggart JB, Bron JE, Guy DR, Bell JG, Tocher DR (2011) Genotype-specific responses in Atlantic salmon (*Salmo salar*) subject to dietary fish oil replacement by vegetable oil: a liver transcriptomic analysis. BMC genomics 12: 255.

Morimoto KC, Van Eenennaam AL, DePeters EJ, Medrano JF (2005) Endogenous production of *n*-3 and *n*-6 fatty acids in mammalian cells. Journal of dairy science 88 (3): 1142-1146.

Nakamura MT, Nara TY (2004) Structure, function, and dietary regulation of delta6, delta5, and delta9 desaturases. Annu Rev Nutr 24: 345-376.

Okuley J, Lightner J, Feldmann K, Yadav N, Lark E, Browse J (1994) Arabidopsis FAD2 gene encodes the enzyme that is essential for polyunsaturated lipid synthesis. The Plant Cell Online 6 (1): 147-158.

Özogul Y, Özogul F, Alagoz S (2007) Fatty acid profiles and fat contents of commercially important seawater and freshwater fish species of Turkey: A comparative study. Food Chemistry 103 (1): 217-223.

Pei DS, Sun YH, Chen CH, Chen SP, Wang YP, Hu W, Zhu ZY (2008) Identification and characterization of a novel gene differentially expressed in zebrafish cross-subfamily cloned embryos. BMC Dev Biol 8: 29.

Peyou-Ndi MM, Watts JL, Browse J (2000) Identification and characterization of an animal delta(12) fatty acid desaturase gene by heterologous expression in *Saccharomyces cerevisiae*. Arch Biochem Biophys 376 (2): 399-408.

Ren H-t, Zhang G-q, Li J-l, Tang Y-k, Li H-x, Yu J-h, Xu P (2013) Two Δ6-desaturase-like genes in common carp (*Cyprinus carpio* var. Jian): Structure characterization, mRNA expression, temperature and nutritional regulation. Gene 525 (1): 11-17.

Rosen ED, Walkey CJ, Pujgserver P, Spiegelman BM (2000) Transcriptional regulation of adipogenesis. Cenes Dev 14(11): 1293-1307.

Saeki K, Matsumoto K, Kinoshita M, Suzuki I, Tasaka Y, Kano K, Taguchi Y, Mikami K, Hirabayashi M, Kashiwazaki N, Hosoi Y, Murata N, Iritani A (2004) Functional expression of a Delta12 fatty acid desaturase gene from spinach in transgenic pigs. Proc Natl Acad Sci U S A 101 (17): 6361-6366.

Senadheera SD, Turchini GM, Thanuthong T, Francis DS (2011) Effects of dietary alpha-linolenic acid (18: 3*n*-3)/linoleic acid (18: 2*n*-6) ratio on fatty acid metabolism in Murray cod (*Maccullochella peelii peelii*). Journal of agricultural and food chemistry 59 (3): 1020-1030.

Spychalla JP, Kinney AJ, Browse J (1997) Identification of an animal omega-3 fatty acid desaturase by heterologous expression in *Arabidopsis*. Proc Natl Acad Sci U S A 94 (4): 1142-1147.

Stoletov K, Fang L, Choi SH, Hartvigsen K, Hansen LF, Hall C, Pattison J, Juliano J, Miller ER, Almazan F, Crosier P, Witztum JL, Klemke RL, Miller YI (2009) Vascular lipid accumulation, lipoprotein oxidation, and macrophage lipid uptake in hypercholesterolemic zebrafish. Circulation research 104 (8): 952-960.

Thanuthong T, Francis DS, Senadheera SP, Jones PL, Turchini GM (2011) LC-PUFA biosynthesis in rainbow trout is substrate limited: use of the whole body fatty acid balance method and different 18: 3*n*-3/18: 2*n*-6 ratios. Lipids 46 (12): 1111-1127.

Wan JB, Huang LL, Rong R, Tan R, Wang J, Kang JX (2010) Endogenously decreasing tissue *n*-6/*n*-3 fatty acid ratio reduces atherosclerotic lesions in apolipoprotein E-deficient mice by inhibiting systemic and vascular inflammation. Arterioscler Thromb Vasc Biol 30 (12): 2487-2494.

Wei C-Y, Wang H-P, Zhu Z-Y, Sun Y-H (2014) Transcriptional factors Smad1 and Smad9 act redundantly to mediate zebrafish ventral specification downstream of Smad5. Journal of Biological Chemistry.

Whelan J, Rust C (2006) Innovative Dietary Sources of *n*-3 Fatty Acids. Annual Review of Nutrition 26 (1): 75-103.

Wu X, Ouyang H, Duan B, Pang D, Zhang L, Yuan T, Xue L, Ni D, Cheng L, Dong S, Wei Z, Li L, Yu M, Sun QY, Chen DY, Lai L, Dai Y, Li GP (2012) Production of cloned transgenic cow expressing omega-3 fatty acids. Transgenic Res 21 (3): 537-543.

Xiong F, Wei ZQ, Zhu ZY, Sun YH (2013) Targeted Expression in Zebrafish Primordial Germ Cells by Cre/loxP and Gal4/UAS Systems. Mar Biotechnol (NY) 15 (5): 526-539.

Yamashita A, Kawana K, Tomio K, Taguchi A, Isobe Y, Iwamoto R, Masuda K, Furuya H, Nagamatsu T, Nagasaka K, Arimoto T, Oda K, Wada-Hiraike O, Yamashita T, Taketani Y, Kang JX, Kozuma S, Arai H, Arita M, Osuga Y, Fujii T (2013) Increased tissue levels of omega-3 polyunsaturated fatty acids prevents pathological preterm birth. Sci Rep 3: 3113.

Zhang P, Liu P, Dou H, Chen L, Chen L, Lin L, Tan P, Vajta G, Gao J, Du Y, Ma RZ (2013) Handmade cloned transgenic sheep rich in omega-3 Fatty acids. PloS one 8 (2): e55941.

Zhang P, Zhang Y, Dou H, Yin J, Chen Y, Pang X, Vajta G, Bolund L, Du Y, Ma RZ (2012) Handmade cloned transgenic piglets expressing the nematode fat-1 gene. Cell Reprogram 14 (3): 258-266.

Zheng X, Tocher DR, Dickson CA, Bell JG, Teale AJ (2004) Effects of diets containing vegetable oil on expression of genes involved in highly unsaturated fatty acid biosynthesis in liver of Atlantic salmon (*Salmo salar*). Aquaculture 236 (1-4): 467-483.

Zhu ZY, Sun YH (2000) Embryonic and genetic manipulation in fish. Cell Research 10 (1): 17-27.

fat1 和 *fat2* 双转基因斑马鱼可以促进 Omega-3 多不饱和脂肪酸的合成

庞少臣[1,3]　王厚鹏[1]　朱作言[1]　康景轩[2]　孙永华[1]

1 中国科学院水生生物研究所，武汉 430072，中国
2 Laboratory for Lipid Medicine and Technology, Department of Medicine, Massachusetts General Hospital and Harvard Medical School, Boston, Massachusetts, USA
3 中国科学院大学，北京 100049，中国

摘　要　Omega-3 长链多不饱和脂肪酸(*n*-3 LC-PUFA)，尤其是二十碳五烯酸(EPA)和二十二碳六烯酸(DHA)，是维持人体健康的必要营养物质。但是，包括人类在内的脊柱动物，由于缺少 Δ-12 脱氢酶和 ω-3 脂肪酸去饱和酶基因，失去了体内合成 EPA 和 DHA 的能力。鱼类，尤其是自然生长的海洋鱼类，是人类获取 EPA 和 DHA 的主要来源。目前，海洋渔业资源逐渐枯竭。另一方面，鱼粉价格的上升，使得水产养殖中使用植物油脂替代传统生产中使用的鱼油，这导致水产养殖鱼类中 omega-3 多不饱和脂肪酸的含量下降。这些矛盾和问题迫使人们不断寻找和开发新的 omega-3 多不饱和脂肪酸的食物来源。在本研究中，我们以斑马鱼作为动物模型，人工培育出富含 *n*-3 LC-PUFA 的转基因鱼。首先，*fat1* 基因被证实可以在鱼类培养细胞中有效地发挥功能，表现为将 *n*-6 不饱和脂肪酸有效地向 *n*-3 多不饱和脂肪酸进行转化，使得 *n*-6/*n*-3 比例从 7.7 减少到 1.1。同时，我们构建 *fat1* 转基因斑马鱼，发现转基因鱼中 EPA 和 DHA 的含量分别增加 1.8 和 2.4 倍。接下来，我们在体外培养细胞中共表达 *fat1* 基因和 *fat2*(*fat2* 是密码子优化的 Δ-12 脱氢酶基因)基因，显著促进了鱼类培养细胞中 *n*-3 多不饱和脂肪酸的合成，使得 *n*-6/*n*-3 比例从 7.7 降到了到 0.7。最后，我们构建了 *fat1* 和 *fat2* 共表达的双转基因斑马鱼，其中 EPA 和 DHA 的含量分别增加和 1.7 和 2.8 倍。总之，我们构建了两种富含内源性 *n*-3 LC-PUFA 的转基因斑马鱼模型，*fat1* 转基因斑马鱼和 *fat1*/ *fat2* 双转基因斑马鱼。我们的研究结果表明，运用人工优化的 *fat1* 和 *fat2* 转基因技术将可以大大提高养殖鱼类 *n*-3 LC-PUFA 的含量。

Developments in Transgenic Fish in the People's Republic of China

C. Fu W. Hu Y. Wang Z. Zhu

State Key Laboratory of Freshwater Ecology and Biotechnology, Institute of Hydrobiology, Chinese Academy of Sciences, Wuhan 430072

Summary In the People's Republic of China, genetically modified (GM) fish are being developed primarily to produce desirable alterations to growth rates or feed-conversion efficiency. Up to the present, no transgenic fish have been commercially approved for human consumption. This review introduces advances in the People's Republic of China in transgenic fish studies, biosafety studies of fast-growth GM fish, and the regulation of GM fish.

Keywords common carp; growth hormone; People's Republic of China; transgenic fish

Introduction

With the expansion of the global population and overfishing, advanced aquaculture is needed to meet the increasing demand for high-quality fish protein. Genetically modified (GM) fish (transgenic fish) offer the opportunity of improving both the production and characteristics of conventional fish strains currently exploited in aquaculture. Since Zhu *et al.* (41) produced the first batch of fast-growth transgenic fish, many laboratories throughout the world have been successful in generating transgenic fish in a variety of species using different foreign gene constructs. At present, biotechnology has made considerable advances in producing transgenic animals, and fish may be considered the best candidate for the first marketable transgenic animal for human consumption (36). Before such a product can enter the marketplace, many aspects of the biosafety of transgenic fish have to be carefully evaluated (16, 18, 36). The Food and Drug Administration in the United States of America is reviewing the application of a fast-growth transgenic salmon (19). Because growth-enhanced transgenic fish are near the point of application to aquaculture, many countries are discussing or developing safety assessment strategies for fast-growth transgenic fish.

Recently, many reports have focused on the biosafety considerations of transgenic fish (5, 21, 22, 27, 28), which are also a major concern for the government, the public and scientists

文章发表于 *Revue Scientifique et Technique*, 2005, 24(1): 299-307 in the People's Republic of China. To overcome these concerns, it is important to develop a reliable and widely accepted method of assessing the potential for harm that might be caused by GM fish escaping into the wild. Recently, the National Natural Science Foundation of China and the Ministry of Science and Technology of the People's Republic of China have provided funding to Chinese scientists for the assessment of the biosafety of a fast-growth transgenic common carp that contain 'all-fish' growth hormone (GH) gene constructs. In this review, the authors will introduce advances in transgenic fish studies in the People's Republic of China, biosafety studies of fast-growth GM fish, and Chinese regulations regarding GM fish.

Transgenic Fish Studies in the People's Republic of China

Integration, expression and inheritance of transgenes

In the late 1980s, the integration and expression of foreign genes were thoroughly studied in a model system in which the common carp (*Cyprinus carpio*) was the host fish and a recombinant human growth hormone (hGH) gene was the transgene (43). Using *in situ* Dot blotting, Southern blotting, Northern blotting and radio-immunoprecipitation analysis experiments, the study found that the foreign gene underwent a dynamic process during embryogenesis. The *in situ* Dot blotting experiment showed that the replication of the foreign gene started as soon as it was introduced into the fertilised eggs, and the strongest signal of replication occurred from late blastula to early neurula. The Southern blotting experiment showed that the integration of the foreign gene was most likely to occur at the early blastula stage and last for a long time, resulting in transgenic mosaicism; in other words, the integrated foreign genes were distributed in different tissues and organs of the transgenic fish. Only those foreign genes integrated into the genome of germlines could be transmitted to the offspring via sexual reproduction. The transcription of the foreign gene could be observed at the late-gastrula stage by Northern hybridisation, and radio-immuno-precipitation analysis revealed that different individuals had different levels of foreign gene expression. Depending on the positional effects related to the expression and function of the transgene, foreign gene integration can be categorised in three different ways: functional integration, silent integration and toxic integration. Silent integration shows no visible effect and toxic integration blocks the normal development of the fish; only functional integration results in hGH expression that leads to growth enhancement (43).

Due to the mosaic integration of transgenes in the founder fish, the frequencies of transgene transmission to F_1 progeny are usually less than the frequencies in Mendelian ratios. When 'all-fish' GH-transgenic carp founders were crossed with non-transgenic controls, the transgenic ratios in the F_1 generation were from 72% to 88%, and genetic analysis showed that two or three chromosomes of each founder were integrated with transgenes (33). No matter how the integration of the transgene occurs or how many integration sites the transgenic founder owns, a transgenic line with stable germline transmission has to be established. Many studies have shown that F_1 transgenics produced from crosses between a wild type female and a transgenic male are in a heterozygous state, since they give birth to the next generation with a transgene positive ratio of about 50% (3, 9, 13, 25). In the authors' laboratory, it was also recently discovered that frequencies of transgene transmission to F_2 progeny from a fast-growth 'all-fish' GH F_1 transgenic germline were about 50% (unpublished findings). To mate a pair of F_2 transgenic siblings from the same parents has been shown to be an efficient way to establish a homozygous line of transgenic fish (3, 10). In another study of metallothionein-1 (MThGH)-transgenic common carp, transgenics were mated to each other and reproduced generation by generation, producing F_4 offspring. Most transgenes were reported to be steadily inherited over four generations of transmission, although a small proportion of the transgenes were rearranged, deleted and/or inserted with host sequences, and appeared to be highly polymorphic (38). It was also observed that a MThGH-transgene in the F_4 generation could still initiate transcription properly, and produced faithful transcript products (26). Therefore, after several generations of transmission, the transgene tends to be stably inherited and to function as an endogenous gene. This phenomenon could be called the 'endogenous domestication' of a novel gene.

Growth of enhanced transgenic fish

In the middle 1980s, a research group led by Dr Zuoyan Zhu successfully transferred a recombinant human growth gene under the control of mouse MThGH into the fertilised eggs of goldfish (41) and loach (42), which led to the birth of fast-growth transgenic fish.

For application purposes, a new 'all-fish' genomic construct (pCAgcGH) was made, which included the common carp β-actin gene, CA (15), and grass carp growth hormone gene (gcGH) (44). This 'all-fish' GH construct was microinjected into the fertilised eggs of yellow river carp, a local strain of common carp, and the growth performance was examined at different growth stages (33). The study found that the frequency distribution of body weight (BW) of nontransgenic fish (n = 359) was normal, while that of transgenics (n = 324) was not normal at the 120-day stage. The BW of the largest non-transgenics was 1,414 g, and that of the smallest was 264 g. The BW of the largest transgenics was 2,750 g, although that of the smallest was only 84 g. Among the transgenic individuals, 8.7% had a higher BW than the largest non-transgenic individual, and 6.4% of the transgenics weighed more than 2 kg, which is more than double the mean BW of non-transgenics. Fast-growth transgenic founders were crossed with non-transgenics. Results showed that the BW frequency distribution of F_1 individuals was normal at the 80-day stage, with a mean BW of 417.89 ± 79.72 g. Among all

F_1 individuals, 60% were above the average BW of the controls (260.4 ± 22.47 g). The study also found that fast-growth individuals of the P_0 transgenic fish had much thicker muscles on the back and an obvious hump behind the head (33). The study indicated that 'all-fish' GH-transgenic common carp could attain higher growth rates than the controls.

In addition, our laboratory successfully produced P_0 GH autotransgenic blunt-snout bream (*Misgurnus mizolepis*) with a construct of blunt-snout bream β-actin gene and its GH complementary deoxyribonucleic acid (cDNA) in 2003. Preliminary studies showed that some individual P_0 GH autotransgenic blunt-snout bream displayed fast-growth (12).

Bioenergetic analysis of growth-enhanced transgenic fish

Why and how can the growth rate of GH transgenic fish be dramatically improved? Is there any difference in body composition between transgenics and controls? Several studies have been carried out to analyse the bioenergetics of transgenic fish to try to answer these questions. There has been a thorough bioenergetic analysis of F_2 MThGH transgenic common carp as compared with controls (2). When fed with fresh tubificid species, the energy budget of transgenic and control common carp can be expressed by the following equations:

a) F_2 MThGH-transgenic common carp: $100C = 8.9F + 0.63U + 49.03R + 41.44G$

b) control carp: $100C = 7.37F + 1.14U + 53.36R + 38.13G$

where C is the total energy from food, F is the energy lost in faeces, U is the energy lost in nitrogen excretion, R is the energy channelled to metabolism, and G is the energy channelled to growth.

Compared with the controls, transgenic fish had a significantly higher proportion of food energy channelled to G and a significantly lower proportion channelled to R and U. The transgenic fish saved 6.62% of the total energy from food for growth improvement. This phenomenon was named the 'fast-growing and less-eating' effect.

A study on the growth and feed utilisation by F_4 MThGH transgenic common carp fed with diets containing 20%, 30% and 40% protein has also been carried out (6). The study showed that at each protein level the transgenics had higher specific growth rates than the controls. Feed intake was significantly higher in the transgenics than in the controls fed a low protein diet (20%), but feed intake did not significantly differ between transgenics and the controls when the diets contained levels of 30% or 40% protein. It was thus demonstrated that at a lower dietary protein level, transgenics achieved higher growth rates mainly by increasing feed intake; but at higher dietary protein levels, transgenics achieved higher growth rates mainly through higher energy conversion efficiency. The study also showed that the transgenics contained significantly higher amounts of dry matter and protein, but lower amounts of lipids than the controls at all dietary protein levels. The apparent digestibility of amino acids tended to be higher in the transgenics than in the controls, especially in fish that were fed diets with lower protein levels. When studying the whole-body amino acid pattern in transgenics and controls, Fu *et al.* (7) found no differences in 17 amino acids between the transgenics and controls.

Field trials of transgenic fish

Although great advances have been made in fast-growth GH transgenic fish in the People's Republic of China, no GH transgenic strains have yet been applied for commercial purposes. To determine whether transgenic fish have the potential for commercial exploitation, trials must be performed in conditions that are similar to those of commercial aquaculture. Recently, the authors' laboratory has completed a medium-scale trial of 'all-fish' (CAgcGH) transgenic common carp in Wuhan, a major city in the centre of the country, authorised by the Ministry of Agriculture of the People's Republic of China (40). New rearing ponds with an area of 2 hectares were built by professional aquaculturists, with banks high enough to withstand floods and enclosed with wire netting. All of the water channels into and out of the ponds were carefully designed to prevent the escape of transgenic fish, and strict measures were adopted to ensure no escapes occurred. Each pond was equipped with oxygen supplies and autofeeding machines, which greatly facilitated aquatic rearing. In the spring of 2000, two-year-old mature founders of 'allfish' GH transgenic common carp were examined with polymerase chain reaction (PCR) to detect those transgene carriers whose sexual gonads were transgene positive. Two hundred transgenic individuals that had shown significant growth enhancement were artificially spawned to produce F_1 transgenics. The F_1 transgenics were reared in the ponds with a total area of 1.67 hectares, and non-transgenics were reared as controls under the same conditions. From 16 June to 7 September 2000, 50 transgenics and 50 controls were randomly sampled for analysis at 20-day intervals. The results showed that on average the transgenics had growth rates of 80%, 55%, 77%, 60%, and 42% greater than the controls in the serial samplings. On 7 September, when the fish were 142 days old, most of the transgenics had reached market size, while the controls needed another year to reach the same size. Furthermore, the feeding coefficient (total food weight per unit of gained BW) of the transgenics was 1.10 and that of the controls was 1.35. The study indicated that GH transgenic common carp not only had faster growth rates but also were more efficient at feed utilisation than the controls in the pond culture. Relatively detailed descriptions of studies on fastgrowth GH transgenic common carp have also been recorded by Wu *et al.* (35).

Biosafety Studies of Genetically Modified Fish

Producing sterile transgenic fish

Sterile transgenic fish pose no risks of any genetic impacts on the local gene pools. In the

People's Republic of China, work on the production of sterile transgenic fish has focused on polyploidy manipulation and use of transgenesis.

Polyploidy manipulation techniques are easily applied to fish, thus offering fish biologists an approach for producing various useful reproductive characteristics for commercial aquaculture. In many fish species, triploid males and females generally fail to produce mature gonads and turn out to be sterile. In the People's Republic of China, sterile transgenic fish are produced by hybridising tetraploids and diploids. A study of interspecific hybridisation between red crucian carp (*Carassius auratus* red var.) and common carp found that the F_3-F_8 generations of the hybrids were allotetraploids, and capable of producing tetraploid offspring (14). 'All-fish' GH gene construct had been transferred into the fertilised eggs of the allotetraploids, and the transgenic tetraploids showed significantly better growth performance than non-transgenic tetraploids. Transgenic tetraploids could produce spermatozoa at the age of 240 days (37). Subsequently, transgenic triploids were successfully produced by crossing transgenic diploid common carp with tetraploids (40). Transgenic triploids were found to be sterile, and had higher growth rates than non-transgenics. Since transgenic tetraploids are able to produce successive generations of offspring and still maintain the tetraploidy, the technology is available to produce sterile transgenic triploids for aquaculture.

It is obvious that the triploid strategy is not suitable for all species of transgenic fish, and more feasible techniques need to be developed to ensure the environmental safety of transgenic fish. Fish gonadotropin-releasing hormones (GnRHs) are well known to be decapeptides that play critical roles in fish gonadal development and in regulation of the reproductive cycle (1, 31). Repressing the expression of GnRHs is likely to cause sterility in the fish. To this end, the most promising method at present appears to be through antisense technology, which has achieved success in transgenic plants (23, 32). The introduction of short DNA or ribonucleic acid (RNA) sequences corresponding to part of the coding sequence can block the expression of a particular gene, either by binding the double-stranded DNA to form a triplex to block transcription, or by binding the mRNA to block the processing of transportation. After introducing constructs containing salmon GnRH promoter fused to GnRH antisense cDNA into rainbow trout, Uzbekova *et al.* (29) found the expression of antisense GnRH RNA in the brain and a decrease in the production of endogenous GnRH mRNA in the brain and pituitaries. Unexpectedly, the levels of the gonadotropins, folliclestimulating hormone and luteinising hormone, were not affected in the antisense GnRH transgenics, and the fish reached maturity at the same time as non-transgenic individuals. On the other hand, when histone H3 promoter was used to drive antisense GnRH cDNA, some transgenic rainbow trout were sterile, although their fertility could be restored by hormonal treatment (30). In the People's Republic of China, sterile transgenic fish have been produced by antisense technology to repress the expression of GnRHs. GnRH cDNA derived from the common carp was isolated, and a construct with common carp β-actin promoter and antisense GnRH cDNA was subsequently generated. Introducing this antisense GnRH construct into fast-growing transgenic common carp may cause sterility. Furthermore, the fertility of antisense GnRH-transgenics with desirable performance could be restored by exogenous hormone administration, resulting in a physiologically reversible fertility strain of transgenic fish that could serve as brood stock for aquaculturists (11).

Food safety of transgenic fish

At present, the generally accepted principle for evaluating the safety of foods produced by modern biotechnology is the 'substantial equivalence principle', which was proposed by the Organization for Economic Cooperation and Development in 1993 (20). In 1995, a World Health Organization (WHO) consulting group that convened to provide practical guidance for the safety evaluation of plants derived from modern biotechnology used the substantial equivalence principle in this guidance (34). In 1996, the WHO and the Food and Agriculture Organization came together to provide practical and concrete recommendations for international guidelines for the safety assessment of foods derived from biotechnology, and suggested that the substantial equivalence principle be used as a general guidance for the safety evaluation of all genetically modified organisms (GMOs) (4).

In the People's Republic of China, the food safety of 'all-fish' GH-transgenic common carp was evaluated through studies of three groups of mice that were fed with 'all-fish' GH-transgenic common carp, non-transgenic common carp, and physiological saline respectively for six weeks. The experiments strictly followed the pathological rules for testing new medicines issued by the Ministry of Health of the People's Republic of China. The results indicated that the test mice showed no significant differences when compared with the two groups of control mice, including in terms of growth performance, biochemical analysis of blood, histochemical assay of twelve organs, or reproductive ability (39). Information for assessing the food safety of transgenic fish is still limited, and more studies may need to be undertaken.

Environmental risk assessments of fast-growth genetically modified fish

At present, Chinese scientists are developing environmental safety assessment strategies for fast-growth GM fish. Since the beginning of 2005 many studies have been taking place to compare the morphology, life history and behaviour of fast-growth GM fish with those of wild fish. Scientists hope to obtain complete information on these aspects of fast-growth transgenic fish in order to develop a generally accepted model for assessing the environmental risk of such fish.

In the authors' laboratory, some preliminary information has been obtained on the morphology and life history of fast-growth transgenic carp. In a study conducted with fast-growth 'all-fish' GH-transgenic common carp, the gonadosomatic index (ovary weight/BW), fertilisation rate (number of fertilised eggs/number of eggs) and hatchability (number of hatched fry/number of fertilised eggs) were compared with wild-type non- transgenics. The results showed that the quality of eggs of the transgenics was similar to that of the non-transgenics, while the gonadosomatic index of the transgenics was significantly lower than that of the non-transgenics. For both transgenics and non-transgenics, the fertilisation rate was higher than 80% and hatchability was over 60%, and there were no significantly differences in these parameters between the transgenics and non-transgenics (33). Fastgrowth 'all-fish' GH-transgenic common carp also promoted thymus development and thymocyte proliferation, and retarded thymus degeneration (8). Recently, the authors found that sexual maturity in nontransgenic carp male individuals occurred at the age of six months (body mass = 1,079 ± 46 g), while sexual maturity in the fast-growth 'all-fish' GH-transgenic common carp male individuals had still not occurred at seven months (body mass = 3,143 ± 405 g) (unpublished findings). The results indicated that fast-growth 'all-fish' GH-transgenic carp did not have a lower age of sexual maturity.

Regulation of Genetically Modified Fish

There is a comprehensive framework of committees and legislation to regulate and provide advice on GMOs in the People's Republic of China. Under existing legislation, all GMOs developed in the People's Republic of China, whatever their intended use, must be assessed by the committees. The regulations in place are outlined briefly here.

On 24 December 1993, the first 'Safety Administration Regulation on Genetic Engineering' was issued by the State Science and Technology Commission of the People's Republic of China. The aims of this regulation were as follows:

- to promote research and development of biotechnology in the People's Republic of China
- to tighten safety controls on genetic engineering work
- to protect the health of both the public as a whole andgenetic engineering workers
- to prevent environmental pollution
- to maintain ecological balance.

The regulation made specific stipulations about the management of technologies that utilise carrier systems to reorganise DNA and that import the DNA from other sources into organisms in physical and chemical ways.

On 10 July 1996, the second measure, the 'Safety Administration Implementation Regulation on Agricultural Biological Genetic Engineering', was issued by the Ministry of Agriculture of the People's Republic of China. This regulation provides a classification of safety considerations relating to different genetic engineering carriers, and prescribes relevant management measures. In particular, it designates the procedures and rules for the registration and safety assessment of agricultural biological genetic engineering. The regulation is effectively enforced in the management of agricultural biological genetic engineering across the country. The above-mentioned regulations are available at www.biosafety.gov.cn (English-language versions are also available: 17, 24).

On 9 May 2001, the Chinese State Council promulgated the 'Regulations on Safety of Agricultural Genetically Modified Organisms'. The purpose of these regulations was to strengthen the safety administration of agricultural GMOs, safeguard human health and the safety of animals, plants and microorganisms, protect the environment, and promote research on agricultural GMOs. The activities of research, testing, production, processing, marketing, import or export with respect to agricultural GMOs within the territories of the People's Republic of China must conform with these regulations (an English-language version is available: www.biosafety.gov.cn/image200105 18/3107.doc).

On 5 January 2002, the 'Implementation Regulations on the Labelling of Agricultural Genetically Modified Organisms' were issued by the Ministry of Agriculture. The purpose of these regulations was to strengthen the labelling administration of agricultural GMOs, standardize the marketing activities of agricultural GMOs, guide the production and consumption of agricultural GMOs, and protect consumers' rights to full access to information about the product. All listed agricultural GMOs for marketing should be labelled. Any agricultural GMO that bears no label or whose label is not in conformity with the requirements of the regulation shall be banned from import or marketing (an English-language version is available: www.biosafety.gov.cn/image20010518/3110.doc).

On 8 April 2002, the 'Health Administration Regulation on Transgenic Food' was issued by the Ministry of Health. The aims of this regulation were to strengthen supervision and management, and to protect consumers' rights to access information about GM food. This regulation licenses and monitors all GM food from the human safety point of view. It also designates the procedures for safety assessment of GM food (a Chinese-language version is available: www.moh.gov.cn/public/open.aspx?n_id=7504&seq=).

On 8 July 2002, the 'National Transgenic Biosafety Committee of the People's Republic of China' was founded. The committee is responsible for the biosafety assessment of GMOs and for biotechnology consultation and supervision.

All these regulations are used to regulate, license and monitor the use of GM fish in the People's Republic of China. At some point in the future, when sterilityreversible transgenic fish have been developed and environmental release testing has been successfully performed, a scientist or company wishing to market a GM fish for food will have to apply to the National Transgenic Biosafety Committee of the People's Republic of China.

Conclusion

In the People's Republic of China, GM fish are being developed primarily to produce desirable alterations to growth rates or feed-conversion efficiency. Up to the present, no transgenic animals, and certainly no transgenic fish, have been commercially produced or approved for human consumption. The Chinese government and scientists are very cautious in assessing the biosafety of GM fish. Further studies will be conducted to comprehensively assess the environmental impacts and food safety of fastgrowth GM carp. Those involved in the technology, whether in the development of legislation or in the application of scientific developments, need to engage in an open and frank debate with the public, and to recognise and address public concerns about these issues.

Acknowledgements The authors would like to thank anonymous reviewers for comments on the manuscript. This work was supported by the National Natural Science Foundation of China (30130050), the State '863' High-Tech Project (2004AA213120), and the Development Plan of the State Key Fundamental Research (2001CB109006) of the Ministry of Science and Technology, the People's Republic of China.

References

1. Alestrom P., Kisen G., Klungland H. & Andersen O. (1992). – Fish gonadotropin-releasing hormone gene and molecular approaches for control of sexual maturation: development of a transgenic fish model. *Molec. mar. Biol. Biotechnol.*, **1** (4-5), 376-379.
2. Cui Z., Zhu Z., Cui Y., Li G. & Xu K. (1996). – Food consumption and energy budget in MThGH transgenic F_2 red carp (*Cyprinus carpio* L. red var.). *Chin. Sci. Bull.*, **41**, 591-596.
3. Culp P., Nusslein-Volhard C. & Hopkins N. (1991). – Highfrequency germ-line transmission of plasmid DNA sequences injected into fertilized zebrafish eggs. *Proc. natl Acad. Sci. USA*, **88** (18), 7953-7957.
4. Food and Agriculture Organization of the United Nations (FAO) & World Health Organization (WHO) (1996). – Biotechnology and food safety. Report of 2nd joint FAO/WHO consultation. FAO/WHO, Rome.
5. Food and Agriculture Organization of the United Nations (FAO) & World Health Organization (WHO) (2003). – FAO/WHO expert consultation on the safety assessment of foods derived from genetically modified animals, including fish. FAO/WHO consultation. FAO/WHO, Rome, 41 pp.
6. Fu C., Cui Y., Hung S. & Zhu Z. (1998). – Growth and feed utilization by F_4 human growth hormone transgenic carp fed diets with different protein levels. *J. Fish Biol.*, **53**, 115-129.
7. Fu C., Cui Y., Hung S. & Zhu Z. (2000). – Whole-body amino acid pattern of F_4 human growth hormone gene-transgenic red common carp (*Cyprinus carpio*) fed diets with different protein levels. *Aquaculture*, **189** (3-4), 287-292.
8. Guo Q., Wang Y., Jia W. & Zhu Z. (2003). – Transgene for growth hormone in common carp (*Cyprinus carpio* L.) promotes thymus development. *Chin. Sci. Bull.*, **48**, 1764-1770.
9. Hew C., Davies P. & Fletcher G. (1992). – Antifreeze protein gene transfer in Atlantic salmon. *Molec. mar. Biol. Biotechnol.*, **1** (4-5), 309-317.
10. Hew C., Poon R., Xiong F., Gauthier S., Shears M., King M., Davies P. & Fletcher G. (1999). – Liver-specific and seasonal expression of transgenic Atlantic salmon harboring the winter flounder antifreeze protein gene. *Transgenic Res.*, **8** (6), 405-414.
11. Li S. (2004). – Isolation of gonadotropin-releasing hormone from common carp (*Cyprinus carpio* L.) and primary study of the controlled reversible sterile transgenic fish. PhD thesis. Institute of Hydrobiology, Chinese Academy of

Sciences, Wuhan.

12. Li X. (2003). – Growth-hormone autotransgenic studies in blunt-snout bream (*Megalobrama amblycephala*). PhD thesis. Institute of Hydrobiology, Chinese Academy of Sciences, Wuhan.

13. Lin S., Gaiano N., Culp P., Burns J., Friedmann T., Yee Y. & Hopkins N. (1994). – Integration and germ-line transmission of a pseudotyped retroviral vector in zebrafish. *Science*, **265** (5172), 666-669.

14. Liu S., Liu Y., Zhou G., Zhang X., Luo C., Feng H., He X., Zhu G. & Yang H. (2001). – The formation of tetraploid stocks of red crucian carp × common carp hybrids as an effect of interspecific hybridization. *Aquaculture*, **192** (2-4), 171-186.

15. Liu Z., Zhu Z., Roberg K., Faras A., Guise K., Kapuscinski A. & Hackett P. (1990). – Isolation and characterization of β-actin gene of carp (*Cyprinus carpio*). *DNA Sequence*, **1** (2), 125-136.

16. MacLean N. (2003). – Genetically modified fish and their effects on food quality and human health and nutrition. *Trends Food Sci. Technol.*, **14**, 242-252.

17. Ministry of Agriculture, People's Republic of China (1996). –Safety administration implementation regulation on agricultural biological genetic engineering. Website: binas.unido.org/binas/show.php?id=8&type=html&table=regulation_sources&dir=regulations (accessed on 30 May 2005).

18. Muir W. (2004). – The threats and benefits of GM fish. *EMBO Rep.*, **5** (7), 654-659.

19. Niiler E. (2000). – FDA, researchers consider first transgenic fish. *Nature Biotechnol.*, **18** (2), 143.

20. Organization for Economic Cooperation and Development (OECD) (1993). – Safety evaluation of foods produced by modern biotechnology: concepts and principles. OECD, Paris.

21. Pew Initiative on Food and Biotechnology (2003). – Future fish: issues in science and regulation of transgenic fish. Pew Initiative on Food and Biotechnology, Washington, DC. Website: pewagbiotech.org/research/fish/fish.pdf (accessed on 29 March 2005).

22. Royal Society of Canada (Expert Panel on the Future of Food Biotechnology) (2001). – Elements of precaution: recommendations for the regulation of food biotechnology in Canada. Royal Society of Canada, Ottawa. Website: www.rsc.ca/foodbiotechnology/GMreportEN.pdf (accessed on 29 March 2005).

23. Schmulling T., Rohrig H., Pilz S., Walden R. & Schell J. (1993). – Restoration of fertility by antisense RNA in genetically engineered male sterile tobacco plants. *Molec. gen. Genet.*, **237** (3), 385-394.

24. State Science and Technology Commission of the People's Republic of China (1993). – Safety administration regulation on genetic engineering. Website: binas.unido.org/binas/show.php? id=9&type=html&table=regulation_sources&dir=regulations (accessed on 30 May).

25. Stuart G., McMurray J. & Westerfield M. (1988). –Replication, integration and stable germ-line transmission of foreign sequences injected into early zebrafish embryos. *Development*, **103** (2), 403-412.

26. Sun Y., Chen S., Wang Y. & Zhu Z. (2000). – The onset of foreign gene transcription in nuclear transferred embryos of fish. *Science in China (Series C)*, **43**, 597-605.

27. United States Office of Science and Technology Policy (OSTP) (2001). – CEQ/OSTP assessment: case studies of environmental regulation for biotechnology. Case study No. 1. Growth-enhanced salmon. Website: www.ostp.gov/html/ceq_ostp_study2.pdf (accessed on 29 March 2005).

28. United States National Research Council (2002). – Animal biotechnology: science-based concerns. National Academy Press, Washington, DC. Website: www.nap.edu/books/ 0309084393/html (accessed on 29 March 2005).

29. Uzbekova S., Chyb J., Ferriere F., Bailhache T., Prunet P., Alestrom P. & Breton B. (2000). – Transgenic rainbow trout expressed sGnRH-antisense RNA under the control of sGnRH promoter of Atlantic salmon. *J. mol. Endocrinol.*, **25** (3), 337-350.

30. Uzbekova S., Hanley S., Cauty C., Smith T. & Breton B. (2001). – Functional study of salmon histone H3 promoter in transgenic rainbow trout: constitutive expression of GnRH antisense mRNA could lead to a reversible sterility. *In* Proc. 3rd International Union of Biological Sciences (IUBS) symposium on molecular aspect of fish genomes and development, 18-21 February, Singapore.

31. Uzbekova S., Lareyre J., Madigou T., Davail B., Jalabert B. & Breton B. (2002). – Expression of prepro-GnRH and

GnRH receptor messengers in rainbow trout ovary depends on the stage of ovarian follicular development. *Molec. Reprod. Dev.*, **62** (1), 47-56.

32. Van der Meer I., Stam M., Van Tunen A., Mol J. & Stuitje A. (1992). – Antisense inhibition of flavonoid biosynthesis in petunia anthers results in male sterility. *Plant Cell*, **4** (3), 253-262.

33. Wang Y., Hu W., Wu G., Sun Y., Chen S., Zhang F., Zhu Z., Feng J. & Zhang X. (2001). – Genetic analysis of 'all-fish' growth hormone gene transferred carp (*Cyprinus carpio* L.) and its F_1 generation. *Chin. Sci. Bull.*, **46**, 1174-1177.

34. World Health Organization (WHO) (1995). – Application of the safety evaluation of foods or food components from plants derived by modern biotechnology. Report of a WHO workshop. WHO, Geneva.

35. Wu G., Sun Y. & Zhu Z. (2003). – Growth hormone gene transfer in common carp. *Aquat. living Resour.*, **16**, 416-420.

36. Zbikowska H. (2003). – Fish can be first: advances in fish transgenesis for commercial applications. *Transgenic Res.*, **12** (4), 379-389.

37. Zeng Z., Hu W., Wang Y., Zhu Z., Zhou G., Liu S., Zhang X., Luo C. & Liu Y. (2000). – The genetic improvement of tetraploid fish by pCAgcGHc-transgenism. *High Technol. Lett.*, **10**, 6-12.

38. Zeng Z. & Zhu Z. (2001). – Transgenes in F_4 pMThGH transgenic common carp (*Cyprinus carpio* L.) are highly polymorphic. *Chin. Sci. Bull.*, **46**, 143-147.

39. Zhang F., Wang Y., Hu W., Cui Z., Zhu Z., Yang J. & Peng R. (2000). – Physiological and pathological analysis of mice fed with 'all-fish' gene transferred Yellow River carp. *High Technol. Lett.*, **10**, 17-19.

40. Zhu Z. (2000). – Collection of the technical materials of the national 863 high-tech project of China – 'Middle-scale trial of fast-growing transgenic common carp'. Institute of Hydrobiology, Chinese Academy of Sciences, Wuhan.

41. Zhu Z., Li G., He L. & Chen S. (1985). – Novel gene transfer into the fertilized eggs of goldfish (*Carassius auratus* L. 1758). *J. appl. Ichthyol.*, **1**, 31-34.

42. Zhu Z., Xu K., Li G., Xie Y. & He L. (1986). – Biological effects of human growth hormone gene microinjected into the fertilized eggs of loach, *Misgurus anguillicaudatus* (Cantor). *Chin. Sci. Bull.*, **31**, 988-990.

43. Zhu Z., Xu K., Xie Y., Li G. & He L. (1989). – A model of transgenic fish. *Scientia sinica B*, **2**, 147-155.

44. Zhu Z., He L. & Chen T.T. (1992). – Primary-structural and evolutionary analyses of growth-hormone gene from grass carp (*Ctenopharyngodon idellus*). *Eur. J. Biochem.*, **207**, 643-648.

中国转基因鱼研制的发展历程

傅萃长　胡　炜　汪亚平　朱作言

中国科学院水生生物研究所，淡水生态与生物技术国家重点实验室，武汉 430072

摘　要　中国的鱼类遗传改良主要集中在提高生长速率或饵料转化效率方面。目前，尚没有一例转基因鱼被批准上市。本文主要介绍了中国转基因鱼研究的进展、快速生长转基因鱼的生物安全研究及其管理。

Father of Biological Cloning in China

Zuoyan Zhu[1,2] Ming Li[3] Le Kang[3]

1 Institute of Hydrobiology, Chinese Academy of Sciences, Wuhan 430072
2 College of Life Science, Peking University, Beijing 100871
3 Beijing Institutes of Life Science, Chinese Academy of Sciences, Beijing 100101

Dizhou Tong (T. C. Tung) is a well-familiar name in China, as there is a story about him in one of the elementary school textbooks. The story describes how he came from a poor family and could not go to middle school until the age of 17, and how his resolution and diligence made him a successful biologist. But in the academic community, he is remembered and respected for another reason—the outstanding achievement of cross-species cloning in fish.

When talking about the issue of cloning, the first thing that come to the minds of most people is probably the sheep Dolly, the breakthrough in animal cloning by Ian Wilmut, Keith Campbell and the collaborators at the Roslin Institute, Scotland in 1997. However, Dolly is not the first cloned animal because the art of animal cloning in vertebrates has been developing for more than half a century. The first successfully cloned vertebrate is a northern leopard frog *Rana pipiens*, which was done by Robert Briggs and Thomas King through introducing nuclei from cells of the hemisphere blastula into enucleated eggs of the same species in 1952. Nevertheless, people wondered whether a fully differentiated somatic cell would contain the same genetic information as an embryonic cell did, i.e., whether cloning with somatic cells was possible. To answer this question, John Gurdon and his colleagues used the nucleus of an adult intestinal epithelial cell to clone South African frogs *Xynopus* in 1962, proving that even a well-specialized cell still maintains the full genetic information of developmental totipotency. On the other hand, the incompatibility of nuclei and enucleated eggs among different species had been puzzling experimental biologists.

As an experimental embryologist, during his decades of research experience, Dizhou Tong laid out his research in a wide spectrum including embryology, cell biology and genetics with different models such like sea urchins, amphioxux, and teleosts. No matter what subject he was studying and what techniques he was using, Dizhou Tong was very much curious about the compatibility of nucleus and cytoplasm in embryogenesis and development—did the nucleus alone have all the genetic materials necessary to support the development of an organism? What would happen if the egg nucleus was replaced by another one from different species? He thought of this issue over and over again, but the lack of techniques in those years

文章发表于 *Protein & Cell*, 2010, 1(9):793-794

did not allow him to test his curiosity.

When he read the paper about nuclear transplantation performed on amphibians by Briggs and King in 1952, Tong Dizhou realized that the nuclear micro-manipulation of eggs could be an appropriate way to approach his question. In the 1950s when all kinds of resources for research were extremely deficient in China, it was almost impossible to perform highly precise research like nuclear transplantation. After a few years of enduring efforts and countless failures, Tong Dizhou and his colleagues successfully made their own set of micro-manipulator. With this not-so-pretty but very handy equipment, they managed to transfer the nuclei from cells of a male Asian carp to enucleated eggs of a female Asian carp, and therefore, generated the first cloned fish in the world. The cloned fish were perfectly healthy—they developed to full term, swam around and finally produced offspring. This happened in 1963, thirty-four years before Dolly the sheep was born. Sadly, the significance of this study was not recognized by the international scientific community, as the result was only published in Chinese even without English abstract.

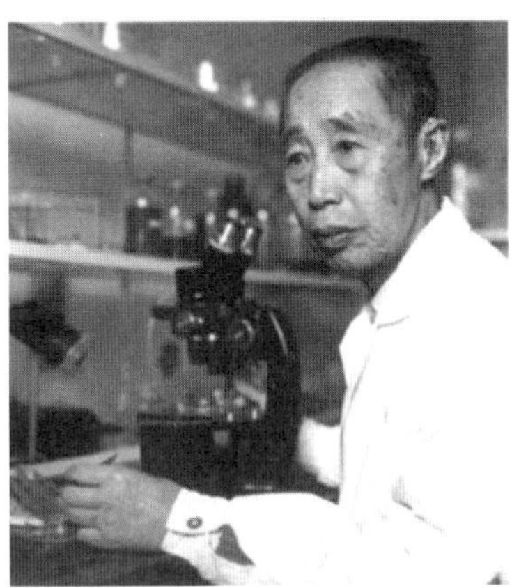

Dizhou Tong

Dizhou Tong and his wife, embryologist Yufen Ye

Dizhou Tong did not stop at the success of fish cloning. He moved onto the next page—cross-species cloning. Ten years later, in 1973, he and his coworkers isolated nuclei from embryonic cells of a common carp *Cyprinus carpio* and transferred it into enucleated eggs of crucian carp *Carassius auratus*. Some of the nuclear-transplants later fully developed into adulthood. It was the first case of cross-species animal cloning in the world! However, the study did not attract much attention as it should have been, due to the difficult circumstances in China in the 1970s. In fact, the result of this study did not get published until 1980, one year after Dizhou Tong passed away. Dizhou Tong was very happy to see the nucleo-cytoplasmic hybrid fish with some distinct characteristics from the nuclear donor species, and dreamed that the novel traits, such as faster growth rate and nutritional contents, would be beneficial to aquaculture in the future.

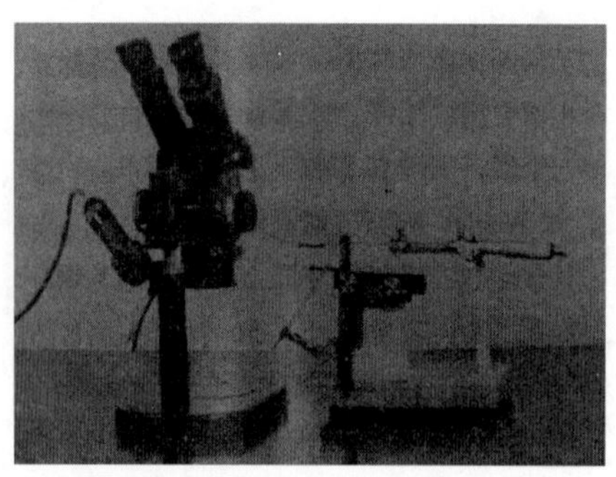

The micro-injection machine designed by Dizhou Tong and his colleagues (From *Dizhou Tong Collected*, Institute of Developmental Biology, Chinese Academy of Sciences, published by Academic Journal Press, Beijing, 1989)
(《童第周文集》, 中国科学院遗传与发育生物学研究所, 学术期刊出版社出版, 北京, 1989)

中国的生物克隆之父

朱作言[1,2] Ming Li[3] Le Kang[3]

1 中国科学院水生生物研究所，武汉 430072
2 北京大学生命科学学院，北京 100871
3 中国科学院北京生命科学研究院，北京 100101

在中国，童第周这个名字家喻户晓，因为在小学教科书上，讲述了一个关于他如何通过毅力和勤奋克服种种困难最终成为一个优秀生物学家的故事。但在学术界，他被记住并尊敬是因为另一个原因——在鱼类异种克隆中的杰出成就。

作为实验胚胎学家，在几十年的研究生涯中，童第周利用不同的模式生物，如海胆、文昌鱼和硬骨鱼等，将他的研究广泛扩展到了包括胚胎学、细胞生物学和遗传学等领域。无论他研究的是什么物种，运用的什么技术，童第周都对胚胎形成和发育过程中的细胞核与细胞质的兼容性十分好奇。

1963 年，在艰苦的实验环境下，童第周和他的团队利用简陋却方便的实验仪器，成功地将一尾雄性鲤鱼的细胞核转移到去核的雌性鲤鱼的卵细胞中，由此产生了世界首例克隆鱼。遗憾的是，这项重要的研究成果并没有被国际科学界所知，因为当时是中文发表，并且没有英文摘要。

1973 年，童第周和他的同事从鲤鱼胚胎细胞中分离出细胞核，然后移植到去核的鲫鱼卵细胞中，一些核移植个体成功发育成成鱼，这是世界首例属间核质杂交鱼。看到核质杂交鱼具有来自细胞核供体物种的明显特征，童第周十分兴奋，并希望一些新的特征，例如生长速率快，营养成分高，能够应用在未来水产行业，从而促进水产养殖业的发展。

Fish Genome Manipulation and Directional Breeding

YE Ding   ZHU ZuoYan   SUN YongHua

State Key Laboratory of Freshwater Ecology and Biotechnology, Institute of Hydrobiology, Chinese Academy of Sciences, Wuhan 430072

Abstract Aquaculture is one of the fastest developing agricultural industries worldwide. One of the most important factors for sustainable aquaculture is the development of high performing culture strains. Genome manipulation offers a powerful method to achieve rapid and directional breeding in fish. We review the history of fish breeding methods based on classical genome manipulation, including polyploidy breeding and nuclear transfer. Then, we discuss the advances and applications of fish directional breeding based on transgenic technology and recently developed genome editing technologies. These methods offer increased efficiency, precision and predictability in genetic improvement over traditional methods.

Keywords fish directional breeding; polyploidy breeding; nuclear transfer; transgenic fish; genome editing

Fish are one of the most important sources of protein for humans. As a result of declines in wild fisheries, aquaculture has become one of the fastest developing agricultural industries worldwide [1]. To increase the sustainability of aquaculture, culturists commonly conduct selective breeding to develop strains that perform well in captivity. Unfortunately, the sustainability of many sectors of the industry is negatively affected by inbreeding depression, disease outbreaks, under production, and low meat quality [2]. To address these issues, there is an urgent need for the development of high quality fish strains that have high growth rates, disease-resistance, and/or higher nutritional value.

Traditional crossbreeding methods such as intra-species crossbreeding [3] and inter-species hybridization [4] have been successfully used for several decades. However, crossbreeding requires multiple-generations of hybridization to introduce a desirable trait to a given strain. Additionally, the outcomes of these methods are unpredictable because the underlying mechanisms controlling desirable traits are unknown. Thus, there is a need to develop more efficient, precise and predictable techniques for producing production scale numbers of high-quality fish. We review the history of fish breeding methods based on classical genome manipulation approaches, including polyploidy breeding [5] and nuclear transfer [6]. Then, we discuss directional breeding of fish based on transgenic technology and recently developed genome editing technologies. Advances in

文章发表于《中国科学: 生命科学》, 2015, 58(2): 170-177

breeding methodology based on classical genome manipulation and recently developed genome editing methods will likely play a major role in the future of genetic breeding in fish.

1 Polyploidy Breeding

Polyploidy, a way to artificially duplicate the chromosome, is considered to be a classic approach to genome manipulation [5]. In some fish species, growth rates differ between females, males, or infertile individuals. To exploit this, culturists produce monosex or infertile populations to increase productivity. A number of methods have been used to achieve sex- and fertility-control during fish breeding. Of these, polyploidy breeding was one of the earliest and the most efficient. This method can also be used to rapidly obtain a homozygous population, thereby decreasing the generational length of the genetic breeding process.

The concept of polyploidy breeding was developed based on research into gynogenesis. In fish, the second meiotic division is completed shortly after ovulation or fertilization. This process can be inhibited by cold-shock treatment, resulting in duplication of chromosomes in the oocyte [7-8]. The combination of physical treatments such as heat-shock and hydrostatic pressure yields the best results [9-10]. Gynogenesis in fish is induced by thermal shock or hydrostatic pressure treatment after fertilization of the egg with inactivated sperm [11-12]. Conversely, androgenesis is induced by thermal shock or hydrostatic pressure treatment of an inactivated egg fertilized with normal sperm. Both gynogenesis and androgenesis have been used on a variety of fish species to produce double haploids (reviewed by [9]).

The production of double haploids can be used to maintain a monosex population. In species with XX-XY sex determination, offspring that are produced by gynogenesis are expected to be all female and the offspring derived by androgenesis have an XX or YY genome. These traits are reversed in species with ZW-ZZ sex determination. The use of gynogenesis and androgenesis for sex-control is only suitable for production of small populations because of the relatively high mortality rate, which is likely caused by irradiation damage, the side-effects from thermal/pressure shock, and inbreeding depression [9]. Notably however, polyploidy breeding methods can be combined with other classical methods, such as induced sex-reversal, to control the sex of offspring and decrease the duration of the breeding process. One successful example of this approach is the production of super, all-male yellow catfish (*Pelteobagrus fulvidraco*), in which males grows faster than females [13]. The YY-super male yellow catfish population is obtained from gynogenesis of induced physiological XY females, and the YY-strain is maintained by incrossing with induced sex-reversed YY females.

As with other animals, sterile fish typically have higher growth rates than fertile individuals. Thus, a number of researchers have evaluated methods to produce sterile triploid fish. The earliest studies used thermal shock treatment after fertilizing the eggs with normal sperm. However, this method has a relatively low success rate and is often accompanied by high mortality. A higher rate of triploidy can be achieved by mating tetraploid fish with diploid fish [14]. Because most fish species are diploid, the most critical step in this process is the production of fertile tetraploids. Researchers successfully obtained fertile allotetraploid fish by hybridizing crucian carp (*Carassius auratus*) with common carp (*Cyprinus carpio*) until the F3 generation [15-16]. The same group also successfully obtained another tetraploid fish by crossing crucian carp with blunt snout bream (*Megalobrama amblycephala*) [17-18]. Although polyploidy breeding has been used for a long time and has yielded several valuable strains, it is not widely applicable because only a few fish species can be used for inter-species crossing to generate tetraploid fish. As this technique is used primarily to breed growth-enhanced strains, other desirable non-growth characters are unlikely to be obtained through this method. Additionally, this method is not of use in species that do not exhibit growth differences between males, females, or infertile individuals.

2 Nuclear Transfer

Nuclear transfer (NT), the transfer of one nucleus into another enucleated egg resulting in a re-constructed egg, is a method for whole-genome manipulation. Intra-species NT refers to transfer of donor cells and oocytes (eggs) from the same species. Intra-species NT has been used to study developmental plasticity and nuclear reprogramming of a nucleus and to produce reprogrammed stem cells from differentiated nuclei [19]. Although animal cloning studies have successfully used stem cell nuclei for several decades (reviewed by [6]), this approach was not widely applied until the birth of the sheep, "dolly"—the first mammal cloned from a somatic nucleus [20]. In fish, the "art" of NT was first demonstrated by Tung *et al.* in goldfish (*Carassius auratus auratus*) and bitterling (*Rhodeus amarus*) [21]. The first animal successfully cloned from a short-term cultured somatic nucleus was born in 1984. This study was originally written in Chinese [22], and was not translated and republished in English until 2010 [23]. Intra-species NT has promised significance for generating genetically manipulated fish from genetically modified *in vitro* cultured cells. The first report of successful intra-species nuclear transfer was in zebrafish (*Danio rerio*) using long-term cultured cells [24]. Interestingly, semicloning technology was recently successfully applied by using medaka (*Oryzias latipes*) haploid stem cells to conduct NT and generate haploid cloned fish [25]. However, this technique currently has a low success rate and cloned offspring often exhibit a range of defects.

If the oocytes and donor cells are derived from two different species, the NT will be defined as cross-species NT. Cross-species NT results in the combination of the nuclear genome from one species and the cytoplasmic factors from another species, so offers a range of unique possibilities in breeding programs [26]. Cross-species NT was first described in amphibians, within the genera of *Rana* and *Xenopus* [27-28]. In those studies, all the NT embryos exhibited early developmental arrest, likely due to the incomplete reprogramming of the donor nuclei and/or incompatibility between the nuclei and the egg cytoplasm. In mammals, cross-species NT has been successfully applied to cloning of endangered mammals within a few closely related species (reviewed by Sun & Zhu [29]). The cloned animals are totally identical to their nuclear donors, highlighting the importance of the nuclear genome in phenotypic determination. In fish, however, cross-species NT between two distantly related species, which have distinct appearances or phenotypes, has resulted in some interesting and different outcomes.

In fishes, cross-species NT can be successfully conducted between distantly related species. For example, NT was conducted between two genera by combining common carp (*Cyprinus carpio*, genus *Cyprinus*) nuclei with crucian carp (*Carassius auratus*, genus *Carassius*) egg cytoplasm [26] or crucian carp nuclei with common carp egg cytoplasm [30]. In those studies, and in our recent study of cross-genus cloned common carp derived from transgenic common carp nuclei and goldfish enucleated eggs [31], the vertebral number of some cross-genus NT individuals was consistent with that of the egg-donor species, goldfish. This suggests that the fish egg cytoplasm can not only support development driven by the transplanted nuclei from a distantly related species at the genus scale, but can also significantly modulate development of the nuclear transplants. Notably, cross-species NT has also been conducted between members of two different families, such as the goldfish (*Carassius auratus*, family *Cyprinidae*, order *Cypriniformes*) and the loach (*Paramisgurnus dabryanus*, family *Cobitidae*, order *Cypriniformes*), and between two orders, such as the tilapia (*Oreochromis nilotica*, order *Perciformes*) and the goldfish, and the tilapia and the loach [32-33]. However, offspring of these cross-family or cross-order NT experiments were unable to develop to term. Therefore, current evidence suggests that cross-species NT can only be successfully applied to a few species that can be artificially hybridized. However, the success rate of cross-species NT is relatively low, so it is unlikely that cross-species NT can be used to introduce desirable traits from a distantly related species into a target species of commercial importance. Nevertheless, cross-species NT has significant potential for cloning or genetic breeding of endangered fish species.

Because of technical difficulties with the process of nuclear transfer, some studies have focused on alternative methods. Among these, cell fusion is a means of obtaining a large number of re-constructed eggs at a single time. This method has been successfully adapted to create hybrid fish with common carp nuclei and crucian carp egg cytoplasm [34]. The hybrid fish has similar morphology to the nuclear donor, common carp. Additionally, a hybrid fish was generated by fusing grass carp (*Ctenopharyngodon idellus*) hemorrhagic virus (FRV) resistant liver cells with unfertilized eggs [35]. Unfortunately, however, the resistant ability of this strain has not yet been reported. As with nuclear transfer, the success rate of cell fusion-based technologies is extremely low, which limits its application in breeding research.

3 Transgenic Breeding

Recently, there has been rapid progress in functional genomic studies in a range of organisms including plants, invertebrates, and vertebrates. As a result, researchers have deduced the function of several thousand genes and evaluated the degree of conservation among different species. Because transgenic methods can be easily applied to incorporate the function of a specific gene, this approach is potentially the most direct and rapid method of obtaining a stable and genetically inherited trait in fish.

A range of gene-delivery methods have been used to conduct transgenesis in fish, including electroporation [36-38], sperm-mediation [39-40], electroporated-sperm-mediation [41-42], retrovirus [43-45], and liposome-mediated methods [46]. However, microinjection is currently the most popular method for generation of transgenic fish [47].

The first transgenic fish was generated by overexpression of humanized growth hormone (hGH) gene driven by a mouse metallothionein-1 (MT) gene promoter in Chinese goldfish [47]. Since then, growth hormone (GH) transgenic fish have been created using a range of fish species, including loach [48], common carp [49-53], channel catfish (*Silurus asotus*) [37], Atlantic salmon (*Salmo salar*) [54], and tilapia [55-56]. In most cases, the *GH*-transgenic fish grow faster and have higher feed conversion efficiency than their non-transgenic siblings, demonstrating that higher GH levels induce fish growth.

In addition to transferring growth hormone genes to promote growth rate, a number of other genes have also been successfully transferred into fish. For example, the anti-freeze protein gene (AFP) was transferred to promote cold-tolerant traits [54-57] and the lysozyme gene was introduced into Atlantic salmon to confer disease resistance [58-59]. Similarly, the cecropin B gene from *hyalophora cecropia* was inserted into channel catfish genome to increase the survival rate [60]; the human lactoferrin (hLF) gene was transferred into grass carp to promote resistance to grass carp hemorrhagic virus (GCHV) [42]; the *vitreoscilla hemoglobin* (*vhb*) gene was transferred into zebrafish to increase hypoxia tolerance [61].

Because of concerns surrounding transgenic safety and bioethics, researchers have focused on using endogenous fish genes ("all fish" transgenesis) rather than exogenous genes such as human *GH*. For example, an "all-fish" transgenic construct was re-designed by using a common carp *β-actin* promoter to drive expression of the grass carp growth hormone gene (gcGH) [52-53]. Similarly, an AFP promoter from ocean pout (*Zoarces americanus*) was linked to a Chinook salmon (*Oncorhynchus keta*) GH cDNA clone [54]. In both studies, transgenic fish had significantly higher growth rates than non-transgenic controls.

In addition to transgenesis with natural gene, molecularly-designed genes have also been delivered into fish to test their viability. For example, we designed a constitutively activated growth hormone receptor (CA-GHR) gene and transferred it into zebrafish, in which two GHR molecules maintain dimerization by Jun-zippers and constitutively activate downstream signaling. The CA-GHR transgenic fish exhibit higher growth rates than *GH* transgenic fish [62]. Thus, there is considerable scope for future studies of transgenic fish to evaluate the utility of highly-activated transgenes using a "molecular design" approach.

Another application of transgenic technology was derived from synthetic biology approaches used to improve the nutritional value of fish. N3 and n6 polyunsaturated fatty acids such as Omega-3 and Omega-6 are beneficial to human health, and particularly important for brain and retina development [63]. Fish-derived desaturase and elongase were transferred to develop n3 polyunsaturated fatty acid (PUFA) and n6 PUFA rich fish strains [64-66]. However, the production of high levels of n3 or n6 PUFA is dependent on the level of n3 or n6 PUFA in the diet. Thus, feeding costs are increased because of the need to supplement diets with extra n3 or n6 PUFA. Recently, *de novo* LC-PUFA biosynthesis was induced in zebrafish by using fat1 and fat1/fat2 double transgenic fish, resulting in robust n-3 LC-PUFA production even when using low PUFA food [67]. Transgenic fish can also be used as a bioreactor. Several studies have used zebrafish eggs to produce recombinant proteins such as human coagulation factor VII [68], luteinizing hormone [69] and insulin-like growth factors [70].

To obtain multiple traits-related genetic improvement within one fish species, researchers could use of 2A peptides to combine those traits, instead of outcrossing between different transgenic strains. 2A peptides allow for more efficient expression of multiple genes, separated by a 2A sequence, within the same cell. This approach has been used in fish [71], mice [72], and pigs [73]. Introducing 2A peptides into a transgene construct allows for incorporation of multiple desirable characters and significantly shortens the breeding process by combining two or more phenotypes.

Genetically modified animals should be subject to ecological safety assessment because of the risk of escape [74]. Evaluation and analysis should be conducted on a case-by-case basis. In addition to an intensive evaluation of the impact of GM fish on other species [75], the production of sterile triploid transgenic fish can reduce such impacts [76-77]. At least two transgenic fish are close to being market-ready. *Growth hormone* transgenic salmon produced by the AquaBounty Technology Company have been submitted to the Food and Drug Administration (FDA) for approval (http://www.fda.gov/AnimalVeterinary/ DevelopmentApprovalProcess/ GeneticEngineering/GeneticallyEngineered-Animals/ucm280853.htm). Similarly, sterile triploid *GH*-transgenic carp are close to satisfying regulatory requirements [77].

4 Genome Editing

In recent years, there has been rapid development of targeted nuclease technologies, such as ZFNs (zinc finger nucleases) [78], TALEN (transcription activator-like effector nucleases) [79-80], and CRISPR (clustered regularly interspersed short palindromic repeats)/Cas9 [81-83]. These methods can be used for targeted knock out or targeted genome editing, and they have been applied to quite a few species including zebrafish. Compared to ZFNs and TALENs, the components of the CRISPR/Cas9 system are much simpler but the CRISPR/Cas9 system can achieve similar or even higher efficiencies. Furthermore, the time to prepare constructs for CRISPR/Cas9 is significantly shorter than for the other two. As a result, the CRISPR/Cas9 system has become widely adopted by researchers. The CRISPR/Cas9 system uses a short guide RNA, which contains an~20bp target sequence, to bind to its complementary DNA target and direct the Cas9 nuclease to the target site to make double-strand breaks (DSBs). The DSBs are typically repaired by either homology-directed repair (HDR), which results in precise genome editing if an exogenous DNA repair template exists, or non-homologous end-joining (NHEJ), which usually results in indel mutations [84]. To date, there have been limited reports documenting the successful application of this method for directional breeding purposes. For instance, the Celtic POLLED, a non-horned allele, was introduced into the genome of horned dairy cattle breeds, and the endogenous horned POLLED allele was removed by TALEN specific cleavage to obtain the non-horned trait [85]. Several genes were edited in pig, sheep and cattle, in order to obtain virus-resistant or growth-enhanced characteristics in those livestock species [86-87]. In fish, using zebrafish as model and using TALEN technology, we knocked out *socs*2 which belongs to the SOCS superfamily—the major negative regulators of the GH signaling pathway. The mutation of *socs*2 resulted in increased stimulation of the GH signaling pathway and the zebrafish mutant had higher growth rates during early larval stages [88].

Single stranded oligonucleotides are an effective repair template for HDR via a single strand annealing (SSA) mechanism. Co-injection of targeted nucleases and single stranded DNA was successfully used to introduce single nucleotide alterations in zebrafish and mice, an approach that is comparable to the occurrence of single nucleotide polymorphisms (SNP) in nature [79-89]. This method can also be used to introduce a small DNA fragment such as the HA (Hemagglutinin) tag or loxP sequence at specific sites [79-81-90]. However, off-target effects are common and are caused primarily by NHEJ repair. Instead of constructing DSBs with the general Cas9, a recently improved Cas9 can create single-strand breaks by point mutating the RuvC or HNH nuclease domains on the Cas9 [91-93]. The use of mutated $\mathrm{Cas9}^{\mathrm{HNH+/RuvC-}}$ along with pairs of guide RNAs allows efficient indel formation while reducing off-target effects and improves specificity by up to 1500 fold when compared to the general Cas9 [94-95].

TALEN mediated homologous recombination (HR) has been successfully applied in zebrafish [96]. CRISPR/Cas9 mediated HR has been successfully applied in *C. elegans* [97], *Drosophila*[98-100], and even human cells [101-102]. Although, targeted nucleases have largely improved the efficiency of HR, the rate of HR is less than satisfactory. To address this, studies have shown that inhibition of NHEJ by disruption of specific genes involved in the NHEJ process significantly increase the efficiency of HR [103-107]. A number of studies have shown that homology-directed DNA repair occurs primarily in somatic tissue, which makes the screening process longer and more costly. The ability to detect germ cell specific HDR in F0 and subsequent generations would save significant effort screening individuals. We generated a primordial germ cell (PGC) specific manipulation system based on UAS/Gal4 and Cre/LoxP in zebrafish, a tool that may prove useful to fill this gap [108].

The recently developed genome editing techniques allow researchers to modify multiple genes at precise sites with high efficiency and in a comparably short time [109]. These characteristics make the approach suitable for improving aquaculture strains. More importantly, the process is based on homology-directed DNA repair so does not bring in any foreign DNA elements, but instead modifies the endogenous DNA itself. Therefore, when combined with PGC specific manipulation and conventional genome manipulation techniques (such as polyploidy manipulation), this technique should make fish breeding (and other animals) more efficient, more precise and more predictable.

5 Conclusion

The development of scientific technology accelerates scientific research and turns “impossible” into “possible”. Although the commercialization of transgenic fish faces significant non-scientific concerns, there is currently no conclusive evidence of a safety problem associated with commercialized genetically modified organisms (GMO). Nevertheless, there is a need to conduct a careful and long-term evaluation before authorizing GM animals into the commercial market. In addition to transgenesis, recently developed genome editing techniques provide an enormously valuable tool for fish breeding. In the near future, the introduction of genome editing into conventional fish breeding will allow researchers to directly and precisely improve specific traits without affecting other traits. Because this approach no longer uses exogenous gene fragments, but instead modifies the genetic information itself in a minimal manner, it deserves to play a major role in the future of fish genetic breeding and the breeding of other animals.

Acknowledgements This work was supported by the National Basic Research Program of China (2010CB126306, 2012CB944504), the National Science Fund for Excellent Young Scholars of the National Natural Science Foundation of China (31222052), the Chinese Academy of Sciences Grant KSCX2-EW-N-004-4, and the State Key Laboratory of Freshwater Ecology and Biotechnology grant 2011FBZ23.

References

1 Cressey D. Aquaculture: Future fish. Nature, 2009, 458: 398-400

2 Gui JF, Zhu ZY. Molecular basis and genetic improvement of economically important traits in aquaculture animals. Chin Sci Bull, 2012, 57: 1751-1760

3 Bakos J, Gorda S. Genetic-improvement of common carp strains using intraspecific hybridization. Aquaculture, 1995, 129: 183-186

4 Bartley DM, Rana K, Immink AJ. The use of inter-specific hybrids in aquaculture and fisheries. Rev Fish Biol Fisher, 2000, 10: 325-337

5 Wu CJ, Ye YZ, Chen RD. Genome manipulation in carp (*Cyprinus carpio* L.). Aquaculture, 1986, 54: 57-61

6 Zhu ZY, Sun YH. Embryonic and genetic manipulation in fish. Cell Res, 2000, 10: 17-27

7 Nagy A, Rajki K, Horvath L, Csanyi V. Investigation on carp, *Cyprinus carpio* L. gynogenesis. J Fish Bio, 1978, 13: 215-224

8 Sajiro Makino YO. Formation of the diploid egg nucleus due to suppression of the second maturation division, induced by refrigeration of fertilized eggs of the carp, *cyprinus carpio*. Cytologia, 1943, 13: 55-60

9 Komen H, Thorgaard GH. Androgenesis, gynogenesis and the production of clones in fishes: a review. Aquaculture, 2007, 269: 150-173

10 Streisinger G, Walker C, Dower N, Knauber D, Singer F. Production of clones of homozygous diploid zebrafish (*brachy danio rerio*). Nature, 1981, 291: 293-296

11 Chen SL, Ji XS, Shao CW, Li WL, Yang JF, Liang Z, Liao XL, Xu GB, Xu Y, Song WT. Induction of mitogynogenetic diploids and identification of ww super-female using sex-specific ssr markers in half-smooth tongue sole (*Cynoglossus semilaevis*). Mar Biotechnol, 2012, 14: 120-128

12 Chen SL, Tian YS, Yang JF, Shao CW, Ji XS, Zhai JM, Liao XL, Zhuang ZM, Su PZ, Xu JY, Sha ZX, Wu PF, Wang N. Artificial gynogenesis and sex determination in half-smooth tongue sole (*Cynoglossus semilaevis*). Mar Biotechnol, 2009, 11: 243-251

13 Wang D, Mao HL, Chen HX, Liu HQ, Gui JF. Isolation of y- and x-linked scar markers in yellow catfish and application in the production of all-male populations. Anim Genet, 2009, 40: 978-981

14 Horvath L, Orban L. Genome and gene manipulation in the common carp. Aquaculture, 1995, 129: 157-181

15 Guo XH, Liu SJ, Liu Y. Evidence for recombination of mitochondrial DNA in triploid crucian carp. Genetics, 2006, 172: 1745-1749

16 Liu SJ, Liu Y, Zhou GJ, Zhang XJ, Luo C, Feng H, He XX, Zhu GH, Yang H. The formation of tetraploid stocks of red crucian carp×common carp hybrids as an effect of interspecific hybridization. Aquaculture, 2001, 192: 171-186

17 Liu SJ, Qin QB, Xiao J, Lu WT, Shen JM, Li W, Liu JF, Duan W, Zhang C, Tao M, Zhao RR, Yan JP, Liu Y. The formation of the polyploid hybrids from different subfamily fish crossings and its evolutionary significance. Genetics, 2007, 176: 1023-1034

18 Qin QB, He WG, Liu SJ, Wang J, Xiao J, Liu Y. Analysis of 5S rDNA organization and variation in polyploid hybrids from crosses of different fish subfamilies. J Exp Zool Part B, 2010, 314B: 403-411

19 Gurdon JB, Wilmut I. Nuclear transfer to eggs and oocytes. CSH Perspect in Biol, 2011, 3: a002659

20 Wilmut I, Schnieke AE, McWhir J, Kind AJ, Campbell KH. Viable offspring derived from fetal and adult mammalian cells. Nature, 1997, 385: 810-813

21 Tung TC, Wu SC, Tung YYF, Yan SY, Tu M, Lu TY. Nuclear transplantation in fish. Chinese Sci Bull, 1963, 7: 60-61

22 Chen HX, Yi YL, Chen MR, Yang X. Studies on the developmental potentiality of cultured cell nuclei of fish. Acta Hydrobiol Sin, 1986,10: 1-7

23 Chen H, Yi Y, Chen M, Yang X. Studies on the developmental potentiality of cultured cell nuclei of fish. Int J Biol Sci, 2010, 6: 192-198

24 Lee KY, Huang H, Ju B, Yang Z, Lin S. Cloned zebrafish by nuclear transfer from long-term-cultured cells. Nat Biotechnol, 2002, 20: 795-799

25 Yi M, Hong N, Hong Y. Generation of medaka fish haploid embryonic stem cells. Science, 2009, 326: 430-433

26 Tung TC, Tung YYF. Nuclear transplantation in teleosts. I. Hybrid fish from the nucleus of carp and the cytoplasm of crucian. Sci Sin,1980, 14: 1244-1245

27 Gurdon JB. The transplantation of nuclei between two species of xenopus. Dev Biol, 1962, 5: 68-83

28 Moore JA. Serial back-transfers of nuclei in experiments involving two species of frogs. Dev Biol, 1960, 2: 535-550

29 Sun YH, Zhu ZY. Cross-species cloning: Influence of cytoplasmic factors on development. J Physiol, 2014, 592: 2375-2379

30 Yan S, Lu D, Du M, Li G, Lin L, Jin G, Wang H, Yang Y, Xia D, Liu A. Nuclear transplantation in teleosts. Hybrid fish from the nucleus of crucian and the cytoplasm of carp. Sci Sin Series B, 1984, 27: 1029

31 Sun YH, Chen SP, Wang YP, Hu W, Zhu ZY. Cytoplasmic impact on cross-genus cloned fish derived from transgenic common carp (*Cyprinus carpio*) nuclei and goldfish (*Carassius auratus*) enucleated eggs. Biol Reprod, 2005, 72: 510-515

32 Yan SY, Mao ZR, Yang HY, Tu MA, Li SH, Huang GP, Li GS, Guo L, Jin GQ, He RF, *et al.* Further investigation on nuclear transplantation in different orders of teleost: the combination of the nucleus of tilapia (*Oreochromis nilotica*) and the cytoplasm of loach (*Paramisgurnus dabryanus*). Int J Dev Biol, 1991, 35: 429-435

33 Yan SY, Tu M, Yang HY, Mao ZG, Zhao ZY, Fu LJ, Li GS, Huang GP, Li SH, Jin GQ, *et al.* Developmental incompatibility between cell nucleus and cytoplasm as revealed by nuclear transplantation experiments in teleost of different families and orders. Int J Dev Biol, 1990, 34: 255-266

34 Yi YL, Liu PL, Liu HQ, Chen HX. Electric fusion between blastula cells and unfertilized eggs in fish. Acta Hydrobiol Sin, 1988, 12: 189-192

35 Yu LL, Zuo WG, Fang YL, Zheng WD. Cell-engineering grass carp produced by the combination of electric fusion and nuclear transplantation. J Fish China, 1996, 20: 314-318

36 Inoue K, Yamashita S, Hata J, Kabeno S, Asada S, Nagahisa E, Fujita T. Electroporation as a new technique for producing transgenic fish. Cell Diff Dev, 1990, 29: 123-128

37 Powers DA, Hereford L, Cole T, Chen TT, Lin CM, Kight K, Creech K, Dunham R. Electroporation: a method for transferring genes into the gametes of zebrafish (*Brachy danio rerio*), channel catfish (*Ictalurus punctatus*), and common carp (*Cyprinus carpio*). Mol Mar Biol Biotechnol, 1992, 1: 301-308

38 Xie YD, Liu J, Zou GL, Zhu Z. Gene transfer via electroporation in fish. Aquaculture, 1993, 111: 207-213

39 Khoo HW, Ang LH, Lim HB, Wong KY. Sperm cells as vectors for introducing foreign DNA into zebrafish. Aquaculture, 1992, 107: 1-19

40 Lavitrano M, Camaioni A, Fazio VM, Dolci S, Farace MG, Spadafora C. Sperm cells as vectors for introducing foreign DNA into eggs—genetic-transformation of mice. Cell, 1989, 57: 717-723

41 Tsai HJ. Electroporated sperm mediation of a gene transfer system for finfish and shellfish. Mol Reprod Dev, 2000, 56: 281-284

42 Zhong JY, Wang YP, Zhu ZY. Introduction of the human lactoferrin gene into grass carp (*Ctenopharyngodon idellus*) to increase resistance against GCH virus. Aquaculture, 2002, 214: 93-101

43 Lin S, Gaiano N, Culp P, Burns JC, Friedmann T, Yee JK, Hopkins N. Integration and germline transmission of a pseudotyped retroviral vector in zebrafish. Science, 1994, 265: 666-669

44 Linney E, Hardison NL, Lonze BE, Lyons S, DiNapoli L. Transgene expression in zebrafish: a comparison of retroviral-vector and DNA-injection approaches. Dev Biol, 1999, 213: 207-216

45 Lu JK, Burns JC, Chen TT. Pantropic retroviral vector integration, expression, and germline transmission in medaka (*Oryzias latipes*). Mol Mar Biol Biotechnol, 1997, 6: 289-295

46 Lu JK, Fu BH, Wu JL, Chen TT. Production of transgenic silver sea bream (*Sparus sarba*) by different gene transfer methods. Mar Biotechnol, 2002, 4: 328-337

47 Zhu ZY, Li GH, He L, Chen S. Novel gene transfer into the fertilized eggs of goldfish (*Carassius auratus* L. 1758). J Appl Ichthyol, 1985, 1: 31-34

48 Zhu Z, Xu K, Li G, Xie Y, He L. Biological effects of human growth hormone gene microinjected into the fertilized eggs of loach, *misgurus anguillicaudatus* (cantor). Chinese Sci Bull, 1986, 31: 988-990

49 Zhang PJ, Hayat M, Joyce C, Gonzalez-Villasenor LI, Lin CM, Dunham RA, Chen TT, Powers DA. Gene transfer, expression and inheritance of pRSV-rainbow trout-GH cDNA in the common carp, *cyprinus carpio* (Linnaeus). Mol Reprod Dev, 1990, 25: 3-13

50 Feng H, Fu YM, Luo J, Wu H, Liu Y, Liu S. Black carp growth hormone gene transgenic allotetraploid hybrids of *Carassius auratus* red var. (female symbol)×*Cyprinus carpio* (male symbol). Sci China Life Sci, 2011, 54: 822-827

51 Zhu Z, Xu K, Xie Y, Li G, He L. A model of transgenic fish. Sci Sin B, 1989, (2): 147-155

52 Wang Y, Hu W, Wu G, Sun Y, Chen S, Zhang F, Zhu Z, Feng J, Zhang X. Genetic analysis of "all-fish" growth hormone gene transferred carp (*Cyprinus carpio* L.) and its F1 generation. Chinese Sci Bull, 2001, 46: 1174-1177

53 Zhu Z, He L, Chen TT. Primary-structural and evolutionary analyses of the growth-hormone gene from grass carp (*Ctenopharyngodon idellus*). Eur J Biochem, 1992, 207: 643-648

54 Du SJ, Gong ZY, Fletcher GL, Shears MA, King MJ, Idler DR, Hew CL. Growth enhancement in transgenic atlantic salmon by the use of an "all fish" chimeric growth hormone gene construct. Biotechnology (NY), 1992, 10: 176-181

55 Martinez R, Estrada MP, Berlanga J, Guillen I, Hernandez O, Cabrera E, Pimentel R, Morales R, Herrera F, Morales A, Pina JC, Abad Z, Sanchez V, Melamed P, Lleonart R, de la Fuente J. Growth enhancement in transgenic tilapia by ectopic expression of tilapia growth hormone. Mol Mar Biol Biotechnol, 1996, 5: 62-70

56 Rahman MA, Mak R, Ayad H, Smith A, Maclean N. Expression of a novel piscine growth hormone gene results in growth enhancement in transgenic tilapia (*Oreochromis niloticus*). Transgenic Res, 1998, 7: 357-369

57 Wang R, Zhang P, Gong Z, Hew CL. Expression of the antifreeze protein gene in transgenic goldfish (*Carassius auratus*) and its implication in cold adaptation. Mol Mar Biol Biotechnol, 1995, 4: 20-26

58 Fletcher GL, Hobbs RS, Evans RP, Shears MA, Hahn AL, Hew CL. Lysozyme transgenic atlantic salmon (*Salmo salar* L.). Aquac Res, 2011, 42: 427-440

59 Hew CL, Fletcher GL, Davies PL. Transgenic salmon: tailoring the genome for food production. J Fish Biol, 1995, 47: 1-19

60 Dunham RA, Warr GW, Nichols A, Duncan PL, Argue B, Middleton D, Kucuktas H. Enhanced bacterial disease resistance of transgenic channel catfish *Ictalurus punctatus* possessing cecropin genes. Mar Biotechnol, 2002, 4: 338-344

61 Sun CF, Tao Y, Jiang XY, Zou SM. IGF binding protein 1 is correlated with hypoxia-induced growth reduce and developmental defects in grass carp (*Ctenopharyngodon idellus*) embryos. Gen Comp Endocrinol, 2011, 172: 409-415

62 Ishtiaq Ahmed AS, Xiong F, Pang SC, He MD, Waters MJ, Zhu ZY, Sun YH. Activation of gh signaling and gh-independent stimulation of growth in zebrafish by introduction of a constitutively activated ghr construct. Transgenic Res, 2011, 20: 557-567

63 Marszalek JR, Lodish HF. Docosahexaenoic acid, fatty acid-interacting proteins, and neuronal function: Breastmilk and fish are good for you. Annu Rev Cell Dev Biol, 2005, 21: 633-657

64 Alimuddin, Kiron V, Satoh S, Takeuchi T, Yoshizaki G. Cloning and over-expression of a masu salmon (*Oncorhynchus masou*) fatty acid elongase-like gene in zebrafish. Aquaculture, 2008, 282: 13-18

65 Alimuddin, Yoshizaki G, Kiron V, Satoh S, Takeuchi T. Expression of masu salmon delta5-desaturase-like gene elevated EPA and DHA biosynthesis in zebrafish. Mar Biotechnol, 2007, 9: 92-100

66 Alimuddin, Yoshizaki G, Kiron V, Satoh S, Takeuchi T. Enhancement of EPA and DHA biosynthesis by over-expression of masu salmon delta6-desaturase-like gene in zebrafish. Transgenic Res, 2005, 14: 159-165

67 Pang SC, Wang HP, Li KY, Zhu ZY, Kang JX, Sun YH. Double transgenesis of humanized fat1 and fat2 genes promotes omega-3 polyunsaturated fatty acids synthesis in a zebrafish model. Mar Biotechnol, 2014, 16: 580-593

68 Hwang GL, Muller F, Rahman MA, Williams DW, Murdock PJ, Pasi KJ, Goldspink G, Farahmand H, Maclean N. Fish as bioreactors: transgene expression of human coagulation factor VII in fish embryos. Mar Biotechnol, 2004, 6: 485-492

69 Morita T, Yoshizaki G, Kobayashi M, Watabe S, Takeuchi T. Fish eggs as bioreactors: the production of bioactive luteinizing hormone in transgenic trout embryos. Transgenic Res, 2004, 13: 551-557

70 Hu SY, Liao CH, Lin YP, Li YH, Gong HY, Lin GH, Kawakami K, Yang TH, Wu JL. Zebrafish eggs used as bioreactors for the production of bioactive tilapia insulin-like growth factors. Transgenic Res, 2011, 20: 73-83

71 Provost E, Rhee J, Leach SD. Viral 2A peptides allow expression of multiple proteins from a single ORF in transgenic zebrafish embryos. Genesis, 2007, 45: 625-629

72 Trichas G, Begbie J, Srinivas S. Use of the viral 2A peptide for bicistronic expression in transgenic mice. BMC Biol, 2008, 6: 40

73 Deng W, Yang D, Zhao B, Ouyang Z, Song J, Fan N, Liu Z, Zhao Y, Wu Q, Nashun B, Tang J, Wu Z, Gu W, Lai L.

Use of the 2A peptide for generation of multi-transgenic pigs through a single round of nuclear transfer. PLoS One, 2011, 6: e19986

74 Devlin RH, Sundstrom LF, Muir WM. Interface of biotechnology and ecology for environmental risk assessments of transgenic fish. Trends Biotechnol, 2006, 24: 89-97

75 Sundstrom LF, Vandersteen WE, Lohmus M, Devlin RH. Growth-enhanced coho salmon invading other salmon species populations: effects on early survival and growth. J Appl Ecol, 2014, 51: 82-89

76 Hu W, Wang Y, Zhu Z. Progress in the evaluation of transgenic fish for possible ecological risk and its containment strategies. Sci China Ser C-Life Sci, 2007, 50: 573-579

77 Yu F, Xiao J, Liang XY, Liu SJ, Zhou GJ, Luo KK, Liu Y, Hu W, Wang YP, Zhu ZY. Rapid growth and sterility of growth hormone gene transgenic triploid carp. Chin Sci Bull, 2011, 56: 1679-1684

78 McCammon JM, Amacher SL. Using zinc finger nucleases for efficient and heritable gene disruption in zebrafish. Methods Mol Biol, 2010, 649: 281-298

79 Bedell VM, Wang Y, Campbell JM, Poshusta TL, Starker CG, Krug RG 2nd, Tan W, Penheiter SG, Ma AC, Leung AY, Fahrenkrug SC, Carlson DF, Voytas DF, Clark KJ, Essner JJ, Ekker SC. *In vivo* genome editing using a high-efficiency talen system. Nature, 2012, 491: 114-118

80 Huang P, Xiao A, Zhou M, Zhu Z, Lin S, Zhang B. Heritable gene targeting in zebrafish using customized talens. Nat Biotechnol, 2011, 29: 699-700

81 Chang N, Sun C, Gao L, Zhu D, Xu X, Zhu X, Xiong JW, Xi JJ. Genome editing with RNA-guided Cas9 nuclease in zebrafish embryos. Cell Res, 2013, 23: 465-472

82 Hwang WY, Fu Y, Reyon D, Maeder ML, Tsai SQ, Sander JD, Peterson RT, Yeh JR, Joung JK. Efficient genome editing in zebrafish using a CRISPR-cas system. Nat Biotechnol, 2013, 31: 227-229

83 Zhang LL, Zhou Q. CRISPR/cas technology: a revolutionary approach for genome engineering. Sci China Life Sci, 2014, 57: 639-64

84 Hsu PD, Lander ES, Zhang F. Development and applications of CRISPR-cas9 for genome engineering. Cell, 2014, 157: 1262-1278

85 Tan W, Carlson DF, Lancto CA, Garbe JR, Webster DA, Hackett PB, Fahrenkrug SC. Efficient nonmeiotic allele introgression in livestock using custom endonucleases. Proc Natl Acad Sci USA, 2013, 110: 16526-16531

86 Lillico SG, Proudfoot C, Carlson DF, Stverakova D, Neil C, Blain C, King TJ, Ritchie WA, Tan W, Mileham AJ, McLaren DG, Fahrenkrug SC, Whitelaw CB. Live pigs produced from genome edited zygotes. Sci Rep, 2013, 3: 2847

87 Proudfoot C, Carlson DF, Huddart R, Long CR, Pryor JH, King TJ, Lillico SG, Mileham AJ, McLaren DG, Whitelaw CB, Fahrenkrug SC. Genome edited sheep and cattle. Transgenic Res, 2015, 24: 147-153

88 Wang HL, Zhu ZY, Sun YH. TALEN-mediated knock out of zebrafish SOCS2 and the growth performance of SOCS2 mutants. Acta Hydrobiol Sin, 2015, in press

89 Xiao A, Wang Z, Hu Y, Wu Y, Luo Z, Yang Z, Zu Y, Li W, Huang P, Tong X, Zhu Z, Lin S, Zhang B. Chromosomal deletions and inversions mediated by talens and CRISPR/cas in zebrafish. Nucleic Acids Res, 2013, 41: e141

90 Hruscha A, Krawitz P, Rechenberg A, Heinrich V, Hecht J, Haass C, Schmid B. Efficient CRISPR/cas9 genome editing with low off-target effects in zebrafish. Development, 2013, 140: 4982-4987

91 Gasiunas G, Barrangou R, Horvath P, Siksnys V. Cas9-crRNA ribonucleoprotein complex mediates specific DNA cleavage for adaptive immunity in bacteria. Proc Natl Acad Sci USA, 2012, 109: E2579-2586

92 Jinek M, Chylinski K, Fonfara I, Hauer M, Doudna JA, Charpentier E. A programmable dual-RNA-guided DNA endonuclease in adaptive bacterial immunity. Science, 2012, 337: 816-821

93 Sapranauskas R, Gasiunas G, Fremaux C, Barrangou R, Horvath P, Siksnys V. The streptococcus thermophilus CRISPR/cas system provides immunity in escherichia coli. Nucleic Acids Res, 2011, 39: 9275-9282

94 Mali P, Aach J, Stranges PB, Esvelt KM, Moosburner M, Kosuri S, Yang L, Church GM. Cas9 transcriptional activators for target specificity screening and paired nickases for cooperative genome engineering. Nat Biotechnol, 2013, 31: 833-838

95 Ran FA, Hsu PD, Lin CY, Gootenberg JS, Konermann S, Trevino AE, Scott DA, Inoue A, Matoba S, Zhang Y,

Zhang F. Double nicking by RNA-guided CRISPR cas9 for enhanced genome editing specificity. Cell, 2013, 154: 1380-1389

96 Zu Y, Tong X, Wang Z, Liu D, Pan R, Li Z, Hu Y, Luo Z, Huang P, Wu Q, Zhu Z, Zhang B, Lin S. Talen-mediated precise genome modification by homologous recombination in zebrafish. Nat Methods, 2013, 10: 329-331

97 Dickinson DJ, Ward JD, Reiner DJ, Goldstein B. Engineering the caenorhabditis elegans genome using Cas9-triggered homologous recombination. Nat Methods, 2013, 10: 1028-1034

98 Bassett AR, Tibbit C, Ponting CP, Liu JL. Mutagenesis and homologous recombination in drosophila cell lines using CRISPR/cas9. Biol Open, 2014, 3: 42-49

99 Bottcher R, Hollmann M, Merk K, Nitschko V, Obermaier C, Philippou-Massier J, Wieland I, Gaul U, Forstemann K. Efficient chromosomal gene modification with CRISPR/cas9 and PCR-based homologous recombination donors in cultured drosophila cells. Nucleic Acids Res, 2014, 42: e89

100 Yu Z, Chen H, Liu J, Zhang H, Yan Y, Zhu N, Guo Y, Yang B, Chang Y, Dai F, Liang X, Chen Y, Shen Y, Deng WM, Chen J, Zhang B, Li C, Jiao R. Various applications of talen- and CRISPR/cas9-mediated homologous recombination to modify the drosophila genome. Biol Open, 2014, 3: 271-280

101 Rong Z, Zhu S, Xu Y, Fu X. Homologous recombination in human embryonic stem cells using CRISPR/cas9 nickase and a long DNA donor template. Protein Cell, 2014, 5: 258-260

102 Xue H, Wu J, Li S, Rao MS, Liu Y. Genetic modification in human pluripotent stem cells by homologous recombination and CRISPR/cas9 system. Methods Mol Biol, 2014, 1114: 37-55

103 Kretzschmar A, Otto C, Holz M, Werner S, Hubner L, Barth G. Increased homologous integration frequency in *Yarrowia lipolytica* strains defective in non-homologous end-joining. Curr Genet, 2013, 59: 63-72

104 Ninomiya Y, Suzuki K, Ishii C, Inoue H. Highly efficient gene replacements in neurospora strains deficient for nonhomologous end-joining. Proc Natl Acad Sci USA, 2004, 101: 12248-12253

105 Nishizawa-Yokoi A, Nonaka S, Saika H, Kwon YI, Osakabe K, Toki S. Suppression of Ku70/80 or Lig4 leads to decreased stable transformation and enhanced homologous recombination in rice. New Phytol, 2012, 196: 1048-1059

106 Verbeke J, Beopoulos A, Nicaud JM. Efficient homologous recombination with short length flanking fragments in Ku70 deficient *Yarrowia lipolytica* strains. Biotechnol Lett, 2013, 35: 571-576

107 Wei ZQ, Xiong F, He MD, Wang HP, Zhu ZY, Sun YH. Suppression of ligase 4 or XRCC6 activities enhances the DNA homologous recombination efficiency in zebrafish primordial germ cells. Acta Hydrobiol Sin, 2015, in press

108 Xiong F, Wei ZQ, Zhu ZY, Sun YH. Targeted expression in zebrafish primordial germ cells by cre/loxP and Gal4/UAS systems. Mar Biotechnol, 2013, 15: 526-539

109 Wang H, Yang H, Shivalila CS, Dawlaty MM, Cheng AW, Zhang F, Jaenisch R. One-step generation of mice carrying mutations in multiple genes by CRISPR/cas-mediated genome engineering. Cell, 2013, 153: 910-918

鱼类基因操作和定向育种

叶 鼎 朱作言 孙永华

中国科学院水生生物研究所，淡水生态及生物技术国家重点实验室，
武汉 430072

摘 要 水产养殖已成为全球范围内发展最快的农业产业之一，可持续发展水产养殖的关键在于培育具有优良性状的养殖品种。基因组操作技术为快速、定向的鱼类遗传育种提供了一条重要的可行性途径。本文回顾了基于经典基因组操作技术的鱼类育种方法学历史，如多倍体育种及细胞核移植等。然后重点介绍并展望了基于转基因技术及新近发展的基因组编辑技术的鱼类定向育种方法。两种技术的发展和应用将会为未来的鱼类种业带来更加高效和更具预见性的育种新方法。

后　记

20 世纪 80 年代初期，当我国现代科学正从 10 年停歇后苏醒过来的时候，震撼世界学术界的“超级鼠”赫然问世，拉开了动物基因工程的序幕。与此同时，我们的恩师朱作言先生已在筹划养殖鱼类基因工程育种研究，并于 1985 年在世界上首次完成了农艺性状的基因转移研究，发表了第一篇转基因鱼的研究论文，建立了一个高效、简洁、完整的转基因鱼育种理论模型，由此开拓了鱼类基因工程育种新领域。1990 年，美国《纽约时报》在长篇述评中指出，中国研究组关于鱼类基因工程的研究成果领先美国 3 年。“世界首例转基因鱼的诞生”更是被作为近代中国科技领域的两大重要科学技术成就之一，载入了美国的世界科学年史（*Alexander Hellemans and Bryan Bunch*. 1988. The Timetable of Science: a Chronology of the Most Important People and Events in the History of Science. New York: Simon & Schuster Inc.: 600）。此后，全球鱼类转基因研究如雨后春笋般蓬勃发展。转基因鱼也不仅限于养殖鱼类品种的遗传改良，在基础生物学、生命医学等领域都扮演了重要角色。

2015 年 10 月，编者在武汉组织召开了“转基因鱼诞生 30 周年”暨鱼类遗传与发育学术研讨会。在会上，我们萌生了选编《朱作言文集》的想法。直至今日，这一想法终于实现。在这部论文集付梓之际，编者首先要对朱作言先生致以最高的敬意。我们分别在不同的时间点有幸加入到朱先生领导的充满温暖、富有远见、具有执着和卓越执行力的集体，不仅接触到鱼类基因工程前沿研究的脉搏，而且参与和见证了转基因鱼从实验室的研究逐渐走向实用化的艰难历程。这些年来，追随朱先生学习和工作，先生言传身教我辈以严谨、求实、创新、勤勉的治学精神，关爱和提携后进的高尚人格魅力，教诲我们做研究和做人的道理，我们有幸。今天，我们又有机会选编本文集，何其荣焉！

斗转星移，朱先生开创鱼类基因工程育种领域已越 30 多年。美国和加拿大的同仁研发的转基因大西洋鲑已于 2015 年在北美上市，但遗憾的是，朱先生领导培育出的冠鲤仍然静静地游淌在实验鱼池。诚然，我国社会上存在着源于不了解而对包括转基因鱼在内的转基因生物的误解与争议。近年来，朱作言先生指导我们发表了一系列有影响的科普论文，借此把深奥的科学道理用通俗易懂的语言准确地传播给公众，希望减轻或消除人们因不了解而担忧、排斥，甚至对转基因产品产生虚幻的恐惧。但是，因为篇幅所限，这些启迪人们思想、闪耀着精彩与睿智科普之光的论文没有收录于本文集中。

我们希望藉由本文集的出版，能展示诞生在中国，并且相关研究一直处于国际前沿、为世界同行所公认的鱼类基因工程育种领域的研究成果，并期盼着有朝一日我们培育的优质转基因鱼能够游上老百姓的餐桌。

我们要感谢先后参与这项工作的新老同事和同门兄弟姐妹：李国华、许克圣、何玲、魏彦章、张甫英、陈尚萍、谢岳峰、刘东、邹钧、崔宗斌、汪亚平、胡炜、孙永华、舒少武、廖兰杰、黎明、祁德运、李勇明、甄建设、傅萃长、王泽群、赵浩斌、曾志强、钟家玉、方勤、王伟、吴刚、吴波、黎双飞、刘军建、黄梅、茅卫锋、廖莎、刘航、汤斌、刘维一、刘静、王娜、潘锦波、钟山、李小明、陈芸、秦瑶、管波、戴军、李德亮、罗大极、王蕴、裴得胜、黄容、连灏、徐婧、苏建国、张成勋、王燕舞、陈春红、鲍海蓉、杜富宽、辜自荣、黄伟、罗青、罗余山、吕健健、杨琳、董锋、江遥、康晓军、李华英、石米娟、卢肖男、邱超、邱涛、卢月娇、裴永艳、钟成容、陈戟、曹梦西、宋焱龙、张运生、崔小娟、彭伟、段小海、陶彬彬、Abu Shufian Ishtiaq Ahmed、魏昌勇、熊凤、庞少臣、何牡丹、魏志强、杨明宇，以及张堂林、段明、龙勇、叶鼎、何利波、王厚鹏等，是你们的出色工作成就了本文集。特别感谢黄容副研究员在汪亚平研究员指导下完成论文编录、修改等方面所做的细致工作。

由于篇幅及编者水平有限，本文集难免有遗珠之憾。编者特别要向做出出色工作却未被选编入本文集的团队成员表示歉意。

汪亚平　胡　炜　孙永华　崔宗斌

2018 年 12 月